369 0160144

PRACTICAL FLOW CYTOMETRY

Fourth Edition

PRACTICAL FLOW CYTOMETRY

Fourth Edition

HOWARD M. SHAPIRO

A JOHN WILEY & SONS, INC., PUBLICATION

Published by John Wiley & Sons, Inc., Hoboken, New Jersey.

Library of Congress Cataloging-in-Publication Data:

Shapiro, Howard M. (Howard Maurice), 1941–
Practical flow cytometry / Howard M. Shapiro. — 4th ed.
Includes bibliographical references and index.
ISBN 0-471-41125-6 (alk. paper)
1. Flow cytometry.
[DNLM: 1. Flow Cytometry. QH 585.5.F56 S529p 2002] I. Title.
QH585.5.F56 S48 2002
571.6'0287—dc21 2002002969

Printed in the United States of America.

10 9

DEDICATION

To the memory of Bart de Grooth, who built, and inspired his students to build, small, simple, elegant cytometers; I wish we had played more guitar duets.

To the memory of Mack Fulwyler, who gave us the cell sorter and became a biologist in the bargain, always eager to learn more and to put the knowledge to good use.

To the memory of Janis Giorgi, a very classy lady who made great advances in the immunology of HIV infection, never forgetting she was working for the patients.

And for Jacob, Benjamin, and Anna – "לתקן עולם במלכות שדי"

CONTENTS

10.6 Applications of Flow Cytometry (continued)

TABLES AND FIGURES

TABLES

FIGURES

FIGURES (continued)

FIGURES (continued)

FIGURES (continued)

FIGURES (continued)

PREFACE TO THE FOURTH EDITION: WHY YOU SHOULD READ THIS BOOK – OR NOT

Cytometry, as you probably know if you are looking at this page, is a process for measuring the physical and chemical characteristics of biological cells. In **flow cytometry**, the measurements are made as cells flow through the instrument in a fluid stream. This is the Fourth Edition of what I am proud and honored to have had many colleagues call the "bible" of flow cytometry. Of course, the real Bible doesn't need new editions.

Cytometry's (and *Practical Flow Cytometry's*) Genesis

Cells were discovered in the 1600's by gentleman scientists who made their own microscopes, which, for the first time, allowed objects of such small dimensions to be seen. Within a few decades, microscopes became available to affluent amateurs, as well as to an emerging class of professional scientists. For the next two hundred years or so, visual observation was the only means of acquiring information about individual cells. The 1800's brought us the cell theory and the germ theory, which provided additional impetus for learning more about cells, and synthetic dyes and photography, which, with improvements in optics, facilitated microscopy. However, it was not until the 1930's that then primitive electronics first permitted objective, quantitative measurements of cellular characteristics to be made. Within another two decades, electron microscopes and television had become commonplace, the latter much more so than the former, and the first practical instruments for counting cells began to appear in research and clinical laboratories.

The 1960's saw the development of the first flow cytometers capable of making quantitative measurements of the physical and chemical properties of cells, and of the first **cell sorters**, which allowed individual cells with selected characteristics to be isolated for further study. By the 1970's, flow cytometers and cell sorters were commercially produced and in some demand, but relatively few institutions were able to afford them and relatively few people were able to operate and maintain the large, expensive, and user-unfriendly first-generation instruments. Others sought to add to the limited measurement capabilities of existing flow cytometers by building their own and/or by modifying commercial systems. The predecessor of this book, *Building and Using Flow Cytometers* (1983), and the first two editions of *Practical Flow Cytometry* (1985 and 1988) were written for this cadre of adventurers; the books included more or less complete instructions for building a flow cytometer with performance comparable to that of a then-current commercial instrument.

By 1993, when I began preparing the Third Edition (1995), it was clear that only a few dozen of the several thousand purchasers of the previous editions of the book had actually built flow cytometers from the included plans. Benchtop flow cytometers, equipped with personal computers or their equivalent for data acquisition and analysis, and operable by individuals unburdened by degrees in physics and/or engineering, had appeared in thousands of laboratories. It was decided, by mutual agreement of the author and publisher, that the emphasis should be shifted, expanding the explanations of the chemistry, physics, optics, electronics, statistics, computer science, and even a little of the biology that makes flow cytometers work – or not work, and surveying a wide range of applications. The do-it-yourself flow cytometer portion of the book disappeared.

The basic science behind cytometry hasn't changed much since the Third Edition appeared, but numerous new instruments, new reagents, new software, new applications, and new users have appeared on the scene, and one cannot meet the needs of the last of these without providing improved coverage of the first four.

Is this Book for You?

These days, much analytical flow cytometry and almost all cell sorting are done in core laboratories by instrument operators who, in theory, are trained not only to run the apparatus, but to provide the researchers and clinicians who need the cells and the information from the cells with as many helpful hints as might be needed about experimental design, choice of reagents, analysis of data, etc. The operators are, at least ideally, supervised by people in whose research and/or clinical work flow cytometry is critical, and, together, these individuals are responsible for many of the advances and refinements in flow cytometric technique. A lot of them have read previous editions of this book, and I would hope to keep my old readers and attract some new ones from this community.

But suppose you are a clinician, researcher, or student lucky enough to have a well-staffed and well-equipped cytometry facility to which to bring your problems. Do you really need to know the details when you can follow cookbook protocols with cookbook reagents and feed racks full of cells, or multiwell plates, into "black box" benchtop instruments that you can't adjust? Can't you just lean on the local experts, and/or get by on the page or so about flow cytometry and cell sorting that it is now *de rigeur* to include in textbooks of cellular and molecular biology and immunology?

You can rely on the experts, perhaps, but not the books. I have looked at around a dozen such texts, all produced within the past few years by groups of distinguished scientists, and have found exactly one in which the above mentioned one- or two-page description does not contain at least one glaring error. The winner here is the 5th Edition of *Immunobiology: The Immune System in Health and Disease*[2315], by Janeway et al (alas, the book survives its first author); the losers will remain anonymous and uncited.

There is no question that in cytometry, as in many other areas of what I like to call analytical biology, new instruments conceal most of the details of their operation, reagents come prepackaged, and procedural details may appear in package inserts, be accessible on line, or be passed down by oral tradition. New users presented with what seems to be the equivalent of a set of building blocks should remember that there is a difference between what a toddler can build with a set of blocks and what a mechanically inclined and informed teenager or adult can accomplish with the same set. Most of cytometry isn't cutting edge science, but the cytometry that has been used to do cutting edge science is often cutting edge cytometry, and if you expect to be involved in the design and interpretation of an experiment that involves a ten-color fluorescence measurement, you're about at the point where you're trying to build a scale model of the Empire State Building out of those blocks. If there isn't anybody in the group who is well versed in cytometry, there's a good chance that, if you don't get a good idea of what's in the black boxes and the reagent bottles, and how and how not to use them, the pile of blocks will fall down before the boss makes it to Stockholm.

If you really just want a quick introduction to the basics of flow cytometry, you can get it from less weighty volumes than this one; Alice Givan's *Flow Cytometry: First Principles*[2316] comes to mind. However, I have also taken a cut at a brief introduction to the basics, and incorporated it as the first chapter (*Overture*) of this edition of *Practical Flow Cytometry*. The rest of the book will keep a myriad of details you didn't think you needed to know handy for those moments of panic that will almost surely occur if you keep doing cytometry for any length of time.

What's in the Book, What's Not, and Why

The basic task of cytometry is to extract information about cell populations, and about differences in physical and chemical characteristics (**parameters**, in cytometry jargon) between cells in those populations. We typically do this by making measurements of optical properties of the cells, usually after applying one or more fluorescent reagents (to which we often refer as **probes**), translating information about cells' structure and function into pulses of light. The detectors in a cytometer produce electrical signals in response to those pulses of light; electronic hardware and computer software extract numbers from the electrical signals. In cell sorting, we have the option of converting the results of our number crunching back into electrical signals that will physically separate cells with preselected characteristics from the rest of the population.

Understanding how all this stuff works requires more background in physics and chemistry than most people who work in the biological sciences have learned, can remember, and/or have thought about. It also helps to know at least a little about math, statistics, computer science, and electronics. *Practical Flow Cytometry* goes over the background material in detail, aiming for intelligible explanations in colloquial English, with a minimum of math, and that as unintimidating as possible.

Arthur C. Clarke once said that any sufficiently advanced technology was indistinguishable from magic. Cytometry is an advanced technology, but I'll start out right now by giving you:

Shapiro's Zeroth Law of Flow Cytometry:

There is no Magic!

Absolutely everything in the boxes and bottles has to, and does, follow the laws of physics and chemistry; I do my best to show you how. If you come from the chemistry/physics/engineering side, and need an introduction to cell biology, immunology, etc., you'll have to look elsewhere, at least for the most part.

Chapter One (*Overture*) has been extensively rewritten to present an introduction to cytometry and flow cytometry that stresses the relation of cytometry to microscopy and the emergence of cytometry in the context of what people wanted and want to know about cells and what technologies were and are available to provide the information. Chapter Two introduces other sources and resources helpful for learning or learning more about cytometry. The detailed history of the field is covered in Chapter Three.

Chapter Four (*How Flow Cytometers Work*) discusses light and its interactions with matter; optical systems in general; and the light sources, illumination and collection optics, and detectors and electronics used in cytometry, including expanded coverage of static and scanning cytometry as well as the details of flow cytometry. The chapter includes are information on newer diode and solid-state lasers and detectors, and discussions of high-resolution digital signal processing (DSP), of hardware and software approaches to logarithmic transformation of signals and fluorescence compensation, and of measures of cytometer performance. These discussions continue in Chapter Five (*Data Analysis*), which confronts the problems of how to display and present data, and of how to evaluate data displayed and presented by others.

Chapter Six (*Flow Sorting*) adds additional detail about well-established fluidic and droplet cell sorting mechanisms, as well as an introduction to the microfluidic and pneumatic systems recently described for sorting large molecules and bacteria, on the one hand, and multicellular organisms, such as *C. elegans* and *Drosophila*, on the other.

Chapter Seven (*Parameters and Probes*) features an expanded discussion of the large and growing list of fluorescent labels and tandem labels now available, which have enabled cytometry's most sophisticated practitioners to measure fluorescence from as many as thirteen fluorescent antibodies or other probes bound to a single cell. New nucleic acid stains and probes and methods for functional parameters such as membrane potential and intracellular calcium are also covered, as are green fluorescent protein (GFP) and its variants and relatives.

In recognition of the fact that the vast majority of readers of this book, new and old, will be doing their flow cytometry using commercially produced flow cytometers rather than home- or laboratory-built instruments, I have, with enthusiastic and much appreciated help from the manufacturers, packed Chapter Eight with far more detail about available instruments than appeared in previous editions. Chapter Nine, which, in the First and Second Editions, included all the do-it-yourself stuff, now presents only a very brief discussion of the pros and cons of building your own instrument.

The discussion of applications of cytometry in Chapter Ten adds details on new uses in cell biology, clinical medicine, biotechnology and drug development, including multiplex analysis, kinetic analysis, sorting for gene expression, and approaches to process monitoring. Stem cells, hematopoietic and otherwise, and rare event detection are discussed, as is sperm sorting for sex selection. I have emphasized new and potential applications for cytometric analysis of bacteria, fungi, parasites, and viruses, reflecting my own focus on this area in recent work, and hoping to provide some guidance for others well acquainted with the organisms but relatively new to cytometric methods for their study.

Contact information for manufacturers and vendors of cytometers, accessories, reagents, etc., appears in Chapter Eleven. I know the biotech sector has had its ups and downs; however, the chapter numbering was pure coincidence.

Chapter Twelve is an Afterword, containing some late corrections, really new stuff, details on book production, and my thoughts on some of cytometry's unfinished business.

In compiling the Third Edition of *Practical Flow Cytometry*, I extracted and read through over 15,000 titles and abstracts relevant to the subject, representing all of the articles added to the MEDLINE database between 1988, when the Second Edition appeared, and mid-1994, when the Third Edition went to press. I had to access the database from a set of CD-ROM's; my 9,600 baud modem just wasn't up to the job. Now, I can use broadband connections to participate in near real time, via e-mail, in the design of experiments done, and the interpretation of data collected, halfway around the world. I could almost certainly have downloaded all of the flow cytometry references that were added to MEDLINE since mid-1994 in some reasonable time, but, since there appear to be about 50,000 of them, it would have been hard for me to look at them all, much less give them thumbs up or down. Since neither I nor the publisher was prepared to triple the size of the book, I had little choice but to be fairly selective in preparing this Fourth Edition, to which more than 1,200 new references have been added. Reference *n* in a previous edition is reference *n* in this one. Because older and newer references are mixed, reference numbers do not appear in numerical sequence in the text. I have noted a few duplicated references; there may be others.

In the era of CD's, DVD's, and the Web, it might be argued that the paper book is an archaic medium; some wanted to dispense with the hard copy and put the whole thing on line. I disagreed. The real Bible, or at least the first five books of it, has been dutifully and faithfully copied by scribes onto parchment scrolls, which are at least as archaic as 8-inch floppy discs, for thousands of years, and continues to present a picture of its time; there are numerous external sources, on paper and in electronic media, to provide translation and commentary, much of it aimed at bringing ancient messages up to date. And, moving from the sublime way over in the direction of the ridiculous, even a book like this one is useful in that it, too, presents a picture of its time; a compilation of the same information on a Web site, frequently updated, will lose historical perspective.

A compromise was obvious. It was agreed that this Fourth Edition would appear in book form, and be supplemented by a Web site, which would contain supplementary material, and allow information likely to change – the lists of suppliers of apparatus, reagents, and accessories, for example – to be kept current. Bottom line: you can still read *Practical Flow Cytometry* in the bathroom. You're on your own about Web access there.

The puns, bad jokes, and occasional poems and lyrics that readers of previous editions have come to expect are still here, starting below. I have tried to hit the high points – and some of the low ones; it is, unfortunately, sometimes the case that bad cytometry happens to good journals.

You can actually read this book from cover to cover, although I suspect not many people do. However, whether you are a flow novice or an old hand, you are likely to benefit from skimming the book from cover to cover. You may find things you wouldn't have found just using the Table of Contents and the Index; these are likely to be the most idiosyncratic parts of any book, and this book is more idiosyncratic than most.

ACKNOWLEDGEMENTS

Leslie Hochberg Shapiro continues to keep me in love and at large, with additional legal help from Peter. Jill and Mike offered moral support when they weren't busy taking care of patients and/or kids, and such kids (they have added "There's No Business Like Flow Business" [p. 566] to their repertoire of Broadway classics and Gilbert and Sullivan)! But let's face it, the rest of my family doesn't do flow cytometry, although the newest generation might, some day.

So how did I write this edition? I got by with a lot of help from my friends, who, like most people in cytometry, are almost always eager to help other people learn about what they're doing and how to do it. At meetings, and via e-mail, I asked a whole bunch of people to send me their good stuff and to put me in touch with other people who were doing interesting things in their facilities. I did the same thing, on a much more limited basis, when I wrote the Third Edition, but the response this time around was much more dramatic; this edition is really a cooperative venture.

(Music: the Toreador Song from Bizet's *Carmen)*

As I'd say, one May in Montpellier,
There's more to know of flow
Than I wrote years ago.
Of compensation,
Polarization,
GFP and DSP and bead assays galore.

Lists of probes now tax our frontal lobes,
And we've got laser beams
Beyond our wildest dreams.
With my book slated
To be updated,
I requested help in getting on, on with my chore!

Colleagues heard, and when they got the word,
They e-mailed files of cells
And piles of URL's,
Sent me plots of what they'd stained,
Told me what should be explained,
And all I've gained is here contained.

Edition Four is finally out the door,
Fully revised; now supersized.
Adding references and figures, I have stressed
How the field has progressed.

I've done as good a job as I could do,
All thanks to help from you:

Thanks to Luna Han, Kristen Hauser, and Andrew Prince at Wiley, who had the unenviable job of reminding me that the book was overdue, more overdue, and still more overdue. Well, it's here.

Phil Stein read the sections on electronics, computers, and statistics, and suggested the many corrections and clarifications I have made therein. Rob Webb did the same for the optics sections, helping me make the discussion more intelligible and providing several new and improved figures.

Information for the section on flow cytometer sensitivity was provided by Bob Hoffman and Jim Wood ("Q and B"), with an assist from Chantell Kuhlmann.

Several figures illustrating data display came from present and past members of Len Herzenberg's lab at Stanford. Mario Roederer provided the data for the front cover figure, which was generated by Jennifer Wilshire using Tree Star's FlowJo Software, written by Adam Treister and Mario. The "Late Breaking News" segment on BiExponential data transforms came from Dave Parks and Wayne Moore. Jennifer also helped out with some of the references. Thanks also to Len and Lee Herzenberg, Marty Bigos, and Dick Stovel.

I'd like to acknowledge the support and efforts of several groups of Cytomutt builders and/or users: Edgar Milford and Isabelle Wood (Brigham and Women's Hospital); Ira Farber, Dan O'Mara, Ruibao Ren, and Alex Stewart (Rosenstiel Basic Sciences Center, Brandeis University); Frank Mandy, Michèle Bergeron, and Zoran Solajic (Health Canada); Adolfas Gaigalas and LiLi Wang (National Institute of Standards and Technology [NIST]), abetted by Jerry Marti (FDA), Abe Schwartz, and Bob Vogt (CDC); and Ted Calvin, Van Chandler, and Wayne Roth (Luminex).

Thanks to George Janossy, David Barnett, Bruno Brando, Debbie Glencross, and Ilesh Jani for the AffordCD4 project, in which Frank Mandy and I also participate.

My lab doesn't run without the focused attention of Nancy Perlmutter, who has compiled the index for this edition and also did the first round of copyediting. Paul Savin helped get some of the material and references together in the early stages, and Dave Novo, now an alumnus, provided stimulating discussions.

I appreciate the grant support received from NIH through the National Center for Research Resources and the

National Institute of Allergy and Infectious Diseases during part of the time I was working on this edition. At one time or another, I have consulted and/or done research for almost all of the companies in the cytometry business, and I still have financial relationships with some of them.

Last, but not least, I'd like to thank the many people who made suggestions, dug out references, sent data and figures, and/or otherwise provided help and comfort during the multiyear gestation period: Lilia Alberghina, Nancy Allbritton, Mohamed Al-Rubeai, John Altman, Stefan Andreatta, Anders Arekrans, Bob Ashcroft, Ann Atzberger, Bob Auer, Bruce Bagwell, Keith Bahjat, Lesley Barber, Jeffery Barker, Marc Barnard, Phil Barren, David Basiji, Ken Bauer, Kevin Becker, Andrew Beernink, Kurt Benirschke, Gideon Berke, Diana Bianchi, Nigel Blackhall, Wade Bolton, Mike Borowitz, Byron Brehm-Stecher, Dennis Broud, Corinna Brussaard, Jenny Bryant, Christine Bunthof, Roger Burger, Leo Burke, Don Button, Cleo Cabuz, Alberto Cambrosio, Lisa Campbell, Nigel Carter, Kent Cavender-Bares, Eric Chase, Ho-Pou Chou, Dave Coder, Greg Colella, Jaume Comas, Bunny Cotleur, Harry Crissman, Nick Crosbie, Katie Crosby, John Daley, Sandor Damjanovich, Zbigniew Darzynkiewicz, Patrick Daugherty, Hazel Davey, Derek Davies, Bruce Davis, Ken Davis, Phil Dean, Sue DeMaggio, Tom Denny, Peter Dervan, Motti Deutsch, Jaroslav Doležel, David Dombkowski, Albert Donnenberg, Richard Doornbos, Norm Dovichi, Janet Dow, Jack Dunne, Gary Durack, Sunny Dzik, Marie Earley, Tricia Echeagaray, Kerry Emslie, Lorrie Epling, Mike Evans, Don Evenson, Shuli Eyal, Belinda Ferrari, Nigel Ferrey, Sheila Frankel, Bruce Freedman, Tom Frey, Anne Fu, Steve Gaffin, David Galbraith, Vanya Gant, Duane Garner, Becky Gelman, George Georgiou, Alice Givan, Alex Glazer, Roland Göhde, Wolfgang Göhde, Jr., Wolfgang Göhde, Sr., Philippe Goix, Peggy Goodell, Jan Gratama, Steve Graves, Jan Grawé, Jan Greve, Kate Griffiths, Emanuel Gustin, Rob Habbersett, Martin Hadam, Brian Hall, Maris Handley, Peter Hansen, Kristi Harkins, Dick Haugland, Bob Hawley, David Hedley, Don Herbert, Ray Hester, Ray Hicks, Elizabeth Hill, Jim Ho, Phil Hodgkin, Ron Hoebe, Tom Huard, Ruud Hulspas, Bill Hyun, Sujata Iyer, Jake Jacobberger, Charlotte Nexmann Jacobsen, Molly James, Janine Jason, Bob Johnson, Eric Johnson, Iain Johnson, Peter Tibor Jung, Jon Kagan, Mike Kagan, Lou Kamentsky, David Kaplan, Arieh Karger, Ken Kaufmann, Mike Keeney, Douglas Kell, Kathi Kellar, Oliver Kenyon, Young-Ran Kim, Louis King, Laurie Kittl, Adriaan Klinkenberg, Rich Konz, Stan Korsmeyer, Jennifer Kramer, Petra Krauledat, Michael Kuhn, Lily Lai, Alan Landay, Joanne Lannigan, Ray Lannigan, Gretchen Lawler, Jim Leary, Alain Le Hérissé, Bob Leif, Rodica Lenkei, Dorothy Lewis, Charles Lin, Swee Kim Lin, Daniela Livnat, Peter Lopez, Mark Lowdell, Ed Luther, George Malachowski, Valeri Maltsev, Eric Martz, Tom McCloskey, Phil McCoy, John McCullough, Perran McDaniel, David McFarland, Pat McGrath, John McInerney, Rita McManamon, Jim McSharry, Norbert Meidenbauer, Mike Melamed, Michael Melnick, Steve Mentzer, Steve Merlin, Sue Merrill, Béla Molnár, Dick Montali, Simon Monard, Jonni Moore, Jane Morris, Matt Morrow, Don Mosier, Kathy Muirhead, Rick Mumma, Mark Munson, Bob Murphy, Rick Myers, Thomas Nebe, Gerhard Nebe-von-Caron, Jan Nicholson, Garry Nolan, John Nolan, Robert Nordon, Randy Offord, Mo O'Gorman, Betsy Ohlsson-Wilhelm, Mark Olsen, Geoff Osborne, Volker Ost, Gunnar Ostgaard, Matt Ottenberg, Roy Overton, Erlina Pali, Glenn Paradis, Florentin Paris, Allen Parmelee, Omar Perez, Steve Perfetto, Jim Phillips, Gene Pizzo, Susan Plaeger, Rebecca Poon, Danilo Porro, Fred Preffer, Jeff Price, Calman Prussin, Steve Quake, Peter Rabinovitch, Andy Rawstron, Marcus Reckermann, Diether Recktenwald, Willem Rens, Bruce Rideout, Richard Riese, Art Roberts, Paul Robinson, Bill Rodriguez, Oystein Ronning, Tony Rossini, Dan Rosson, Eli Sahar, Misha Salganik, Yael Schiffenbauer, Ingrid Schmid, Jörn Schmitz, Dan Schrag, Tom Schulte, Luca Scorrano, Jeff Scott, Kirill Sergueev, Janine Sharpe, John Sharpe, Chris Sieracki, Karel Sigler, Liz Simons, Stephanie Sincock, Vicki Singer, Larry Sklar, Brad Smith, Dan Smith, Paul Smith, Randy Smith, Kit Snow, Lydia Sohn, Ulrik Sprogøe-Jakobsen, Friedrich Srienc, Edward Srour, Alan Stall, Harald Steen, Dana Stein, Henrik Stender, Maryalice Stetler-Stevenson, Carl Stewart, Sigi Stewart, Lisa Stoiano-Coico, Willem Stokdijk, Pete Stopa, Rob Sutherland, Akos Szilvasi, Janos Szöllösi, Shuichi Takayama, Karen Tamul, Attila Tarnok, Michael Taubert, Majid Tebianian, Bill Telford, Leon Terstappen, Wayne Thibaudeau, Rick Thomas, Arjan Tibbe, Frank Traganos, Barb Trask, Joe Trotter, Edward Tufte, Joerg Ueckert, Marc Unger, Lari Vähäsalo, Fred Valentine, Dirk van Bockstaele, Ger van den Engh, Lucia Vasconcellos, Duncan Veal, Marco Vecoli, Ben Verwer, Graham Vesey, Josep Vives-Rego, Sharon Vogt, Joe Voland, John Voorn, Alan Waggoner, Mette Walberg, Andy Watson, Jim Watson (the flow one), Jim Weaver (FDA), Jim Weaver (MIT), Andrew Wells, Leon Wheeless, Reed Wicander, Imogen Wilding, Jerry Wilson, Dane Wittrup, Gajus Worthington, Sonja Wulff, Gülderen Yanikkaya-Demirel, Hopi Yip, Ted Young, Stephen Yue, Dave Zelmanovic, Qing Zeng, Yu-Zhong Zhang, Bob Zucker, and Naomi Zurgil.

If your name isn't in this list and it should be, I'll buy you a beer or its equivalent the next time I see you. Of course, I make the decision about "should." Warn me in advance at <hms@shapirolab.com>.

Now to some serious stuff. The previous edition was dedicated to the memories of my father, Alfred Shapiro, "who goaded me and guided me in the study of a wide variety of subjects," my mother, Jennie Shapiro, "a supermom before it was fashionable, from whom I also learned a lot about science," and Jonas Gullberg, "who taught me a lot about microscopes and their users in too short a time."

I would have loved to keep that dedication for this edition, but I felt compelled to memorialize the recent premature loss of three friends and colleagues who contributed a huge amount to the field of flow cytometry.

I had hoped that Mack Fulwyler, one of the three fathers of cell sorting, would join the other two, Len Herzenberg and Lou Kamentsky, in contributing a Foreword to this book. (Len wrote one for the Third Edition and Lou wrote one for the First). However, by the time I got around to asking him, he was too sick to do it, and I decided not to ask anyone else. I think cell sorting is worth a Nobel Prize (see Chapter 12); I was hoping to see it split three ways.

Bart de Grooth and Janis Giorgi have been eulogized by their colleagues in the pages of *Cytometry*[3671-2], and Janis was profiled in *The New Yorker* in 1998[3673]. I wish I had had more time to hang out with all of them.

Thanks for listening.

HOWARD SHAPIRO

HMS 5.03

West Newton, Massachusetts
May 11, 2003

FOREWORD TO THE THIRD EDITION

This is a light-hearted and very useful book on a complex but very widely used technology. When we had in hand the first working model of a fluorescence- activated cell sorter in 1969, we expected that the major application would be the sorting of live fluorescently stained cells to obtain pure populations of cells that would then be further analyzed off-line. However, implicit in the technology was the on-line analytical capability, so nicely described in this book by Howard Shapiro.

The first two editions from Dr. Shapiro's prolific pen maintained the fiction that working scientists, especially biologists and medically oriented scientists, would build their own flow cytometers. In this Third Edition, Shapiro has bowed to practical reality, and does not predicate this excellent text on flow cytometry and sorting on the "Cytomutt" and improvements as he had in the First Edition, but continues the tradition he started in the Second Edition of presenting the principles of modern multiparameter analysis (importantly, explaining what a parameter is), so that working biomedical scientists can understand how to get the best machines for their money, how to evaluate capabilities of these machines, how and where errors can come in, how to use the instruments most effectively in their important biomedical experiments, and, finally, how some technologically-minded folk are trying to advance the art of flow cytometry and sorting.

This technology has spawned an estimated four hundred million dollars a year in sales of instruments and reagents in 1994. More than 900 participants, mostly machine operators, engineers, technological buffs, staff members of principal investigators, and a relative few of the principal investigators themselves, are expected to attend the International Society for Analytical Cytology's meeting in Lake Placid in the Fall of 1994. Also present will be many members of the large and small companies that hope to provide the material base for this field. Rubbing elbows at this meeting will be immunologists, both basic and clinical, oncologists and cell biologists, as well as molecular biologists, AIDS specialists (and activists), pharmacologists, and too many types of flow cytometrists to name in this Foreword. Nevertheless, all will find information that interests and helps them throughout this book.

Highly capable computers are vitally important components that must be included in modern cell analysis and sorting. Two-parameter analysis is the minimum that any flow cytometer offers. Three, four, and five fluorescence parameters are available from the major producers in 1994. Six, seven, and even ten such parameters are available on some experimental machines being put into practical this same year. Soon thereafter, there may be considerably more than ten measurement parameters on the more advanced instruments.

Consider the data taken at the Stanford Shared FACS Facility in mid-1994 as typical of a heavily used multi-user flow cytometry center. About 125 experiments are analyzed per week, averaging 30 samples per experiment, or 3,750 samples/week. This requires approximately 300 megabytes (30,000 cells, 6 measurements, and 9 bit resolution produce about 200 kilobytes/sample). Thus, we must store 15 gigabytes of new data per year. This creates a need for very extensive and sophisticated means of data management, retrieval, and analysis.

We soon will have all these many gigabytes of data available on-line, with access to investigator names, dates, experimental parameters, etc.; all of the ancillary information needed for analysis of the accumulated data from current as well as previous experiments will be easily accessible to the investigator.

In order for all this data to be meaningful, excellent standardization, compensation, and stability of measurement will have to be featured in the specifications of all serious machines. This is done now in the Shared FACS Facility at Stanford and should be done everywhere flow cytometry is used.

Shapiro covers many technical and scientific considerations in this excellent book and, as I said, treats them with light-hearted humor. Take, for example, "Flow's Golden Oldies" as a heading on page 63, or aphorisms like "Shapiro's First Law of Flow Cytometry: A 51 μm Particle CLOGS a 50 μm Orifice", on page 11.

I recommend a thorough reading for all who are using and plan to use flow cytometry in analysis and sorting of cells and other biological particles.

Leonard A. Herzenberg
Stanford University
July 29, 1994

PREFACE TO THE THIRD EDITION

אפתח בכנור חידתי

--Psalm XLIX

LARGO AL FACSTOTUM

Turn on the lasers, turn on the flow.
Turn on the lasers, turn on the flow.

My book will give you a broad overview of flow;
I'll tell you more than you think that you need to know.

When not at leisure, here's what I treasure:
It gives me pleasure quickly to measure
Cells as they go, cells as they go,

In single file in a rapidly flowing stream,
Through the intense focal spot of a laser beam.

They scatter light, and absorb, and fluoresce;
All this can be quantified with success.

Though forward scattering gives us a smattering
Of data related to particle size,
Change in refraction comes into action,
Decreasing signals, when a cell dies.

Light scattered wider gives us insider
Information about cells' detail,
Irregularity and granularity,
Which we can use and still stay out of jail.

But to learn most, we measure fluorescence,
Which is now flow cytometry's essence,
Much as tumescence is to male adolescence.

Each fluorescent label I list in my table,
As long as it's stable, dispels the fable;
Honestly tells what's in the cells.

DNA ploidy, cell cycle position,
Chromatin structure, base composition.

RNA content, protein as well,
With DNA, all in the same cell.

If your cytometer's clean, or is cleanable,
All these parameters now are amenable,
And there's one more, which I cannot ignore,
'Cause it's flow's major chore:

Antigens, antigens, antigens...

CD's, six score, a CD-ROM can't store,
And labels in cherry and orange and lemon and lime,
Each with a reason and a rhyme,
No need to see cells; our industry sells
Things which count T-cells, one at a time.

Cytometry is fun; It keeps me on the run.
Shapiro here, Shapiro there;
Two days a week, I'm in the air.
Now I count cells instead of sheep,
And I give lectures while I'm still asleep,
And so it goes.

Before I close, I'll switch to prose,
But I propose one thing more to disclose:
I can tell a cell grew, using BrdU.
And I know you can, too.

Up to this junction, I've said naught of function,
Of enzyme kinetics, or cell energetics,
Or how you can spy on the calcium ion,
pH and sulfhydryls inside of a cell,
Nor have I mentioned that polarization
Can help you detect lymphocyte activation,
Or that viability, permeability,
And surface charge can be measured as well.

Though our field's barely out of its teens,
We can now look at microbes and genes,
An end that justifies our means -
And our machines!

[Music: "Largo al Factotum" (Rossini: *The Barber of Seville*)]

This is the Third Edition of a book in which I have tried to include almost everything anyone might want or need to know about flow cytometry, with enough bad jokes interspersed to give the reader a chance to stay awake.

The previous editions were well received, but it occurred to me that I should have done some things differently. In the older books, I dragged the reader through decades of history before I explained the barest detail of the gadgetry being discussed. This time around, I've tried to explain what flow cytometry is and flow cytometers are first, then consider how they got to be that way, and get into the real details after that.

I've retained a lot of the practices I adopted in the previous editions. When I want a word or phrase to catch your attention, I've put it in **boldface**. I still emphasize the fact that flow cytometry rests on the same foundations as other techniques of analytical cytology; the optics and spectroscopy are the same, as are many of the parameters measured and the probes used for their measurement. If you want to do image analysis or confocal microscopy with DNA stains or calcium probes, you can learn about the probes here in about as much detail as you could get anywhere.

The First Edition had 623 references; I added 404 to the Second Edition, and I have added another 1,288 to this edition, selecting most of them from some 15,600 papers dealing with flow cytometry which were entered into the MEDLINE database between July 1987 and June 1994. I still had to leave out a lot of good stuff.

This edition omits the details of how to build flow cytometers; thousands of people have read the previous editions, in which this material did appear, but only a few dozen people have built "Cytomutts" following the designs in the books. If you're interested in building an instrument, help is still available; see Chapters 9 and 11.

I have put in more, and I hope, better illustrations. They include diagrams, photographs, and displays of flow cytometric data. In the last two editions, there were only a couple of figures contributed by other people. In this edition, I decided I couldn't get by without a lot of help from my friends, and called a lot of people, asking them to send me stuff representative of their areas of expertise. The response was enthusiastic, and I think the book is better for it.

I have continued to give priority to including references describing new techniques or refinements, whether in the area of instrumentation, sample preparation, cytochemistry, or data analysis, or pertaining to unusual applications. There are now a couple of dozen other books available which deal with the bread-and-butter applications and the technical details, and I don't see any point in duplicating their contents. What I have tried to do, instead, is to provide my readers with enough information to enable them to make informed decisions about choosing instruments, designing experiments, and believing what comes out in the literature.

The book has been called *Practical Flow Cytometry* since 1985, and, if you have any acquaintance with the field and the previous editions, you are probably aware that things which weren't practical nine and six years ago are practical now. In 1988, there weren't more than a few dozen papers dealing with three-color immunofluorescence; in 1994, four-color immunofluorescence is becoming commonplace. We have to think carefully in order to design experiments which don't involve hundreds of tubes; multiplex labeling, as described on pp. 293-5, may be a practical solution to some problems of this sort.

Quantitative immunofluorescence measurements (pp. 28-9 and 302-6), which really weren't practical in 1988, are now. This will make it easier to standardize measurements made in large numbers of laboratories, and to analyze cellular processes which are characterized by quantitative, rather than qualitative, changes in antigen expression.

New, simplified, powerful methods for analysis of cell proliferation have become available. The SBIP method for detection of DNA synthesis using bromodeoxyuridine as a tracer (p. 325) offers considerable advantages over existing cytochemical and immunochemical methods. Tracking dyes (pp. 312-3) now permit identification of successive generations of cells *in vivo* and *in vitro*, and can be tremendously helpful in clarifying the heterogeneity of cellular immune responses.

Although the technology has not yet reached the clinical microbiology laboratory, the utility of flow cytometry for analyses of bacteria (pp. 412-25) has been facilitated greatly by the increased fluorescence and scatter measurement sensitivity available in newer instruments, resulting in rapid growth of the literature in this area.

While I have been described as, and am, opinionated, I'm willing to change positions I've taken when presented with new evidence. I have done so, e.g., on the subject of mathematical models for DNA histogram analysis (p. 372). If you disagree with me on other points, let me know. I'd like the next edition to include doggerel, but not dogma.

ACKNOWLEDGEMENTS

Leslie Hochberg Shapiro has continued to keep me in love and at large. Jill made valiant attempts to organize my collection of reprints and set up the framework for the book's index; she also had the good taste and fortune to marry Mike Fischer, adding a third source of gross jokes to my input stream. Peter demonstrated conclusively that a pre-law student could assemble, disassemble, and reassemble a Cytomutt in a couple of afternoons.

I still appreciate the perspectives on the design, construction, and use of cytometers and software I have gotten from Dick Adams, Bob Auer, Bruce Bagwell, Leo Burke, Dave Coder, Annette Coleman, Wallace Coulter, Phil Dean, Bart de Grooth, Motti Deutsch, Fred Elliott, Mack Fulwyler, Wolfgang Göhde, Brian Hall, Mike Hercher, Dick Hiebert, Bob Hoffman, Lou Kamentsky, Joan McDowell, Lew Nowitz, Len Ornstein, Dave Parks, Dan Pinkel, Peter Rabinovitch, Diether Recktenwald, Tom Sharpless, Kit Snow, Harold Steen, Phil Stein, John Sullivan, Dick Sweet, Barb Trask, Joe Trotter, Ger van den Engh, Jim Watson, Rob Webb, and Leon Wheeless.

I'd like to thank John Brandes, Nancy Perlmutter, Chris Spychalski, and Dennis Way, who have maintained and improved the Cytomutt breed during the past few years. Nancy and Chris have been more than tolerant of the level of entropy associated with my rewriting job, and respectively took on the chores of indexer-in-chief and scanmeister, as well.

I am happy that Len Herzenberg undertook to write a Foreword in the spirit of the book. For data, pictures, preprints, reprints, stories, and advice, my thanks go to Yosh Agrawal, Judy Andrews, Jim Bacus, Bruce Bagwell, Andrew Beavis, Kevin Becker, Becton-Dickinson Immunocytometry Systems, Bentley Instruments, Marty Bigos, Eric Brown, Don Button, Alberto Cambrosio, Penny Chisholm, Dave Coder, Greg Colella, Coulter Corporation, Cytomation, Inc., John Daley, Zbigniew Darzynkiewicz, Lauren Ernst, Marcia Etheridge, Excitech, Ltd., Foss Electric, Don Frankel, Sheila Frankel, Tom Frey, Mitch Friedman, Duane Garner, Becky Gelman, Janis Giorgi, Alex Glazer, Chuck Goolsby, Martha Gray, Rob Habbersett, Brian Hall, Dick Haugland, David Hedley, Don Herbert, Iain Johnson, Larry Johnson, Norman Jones, Jon Kagan, Lou Kamentsky, Douglas Kell, Dick Keller, Keith Kelley, Pam Kidd, Marc Lalande, Alan Landay, Heidrun Lewalski, Frank Mandy, Eric Martz, Edgar Milford, Kathy Muirhead, Jan Nicholson, Stephen O'Brien, Betsy Ohlsson-Wilhelm, Rob Olsen, Oriel Corp., Ortho Diagnostic Systems, Helene Paxton, Kenneth Pennline, Ken Petersen, Peter Rabinovitch, Doug Redelman, Betsy Robertson, Ingrid Schmid, Angela Schultz, Abe Schwartz, Kit Snow, Carl Stewart, Leon Terstappen, Joe Trotter, Ger van den Engh, Bob Vogt, Alan Waggoner, Jim Weaver, John Williams, Isabelle Wood, Yasuhiro Yamamura, and Heddy Zola.

My staff and I and the Cytomutts have been fed in part by grants AI30853, ES05895, GM44421, HG00441, RR03015, and RR07751 from the National Institutes of Health, and also by the rewards for doing high-tech odd jobs for several industrial concerns which would prefer to remain unnamed.

Finally, my thanks go to Brian Crawford, of Wiley-Liss, who suggested I start this project, and especially to Susan King, to whom he turned over the unenviable job of waiting for me to cough up the finished product.

HOWARD SHAPIRO
HMS 8·94

West Newton, Massachusetts
August, 1994

PREFACE TO THE SECOND EDITION

This book can tell you almost everything you need to know about flow cytometry: what it is, how it works, what you can and can't do with it, and how to buy or build and use flow cytometers. A pretty tall order? Well, I've got a track record now. A lot of people bought the First Edition, and most of them seemed to like it - even the reviewers.

The things the reviewers liked least in the First Edition were the dot matrix print and the sloppy illustrations, so I spent the royalties from the First Edition on the laser printer and other desktop publishing hardware and software with which this Second Edition is being generated. I hope you appreciate the sacrifices I've made for you...

The changes in this Edition, however, are a lot more than cosmetic. A lot has happened in nearly all aspects of the field of flow cytometry. When the First Edition appeared in 1985, flow cytometry and cell sorting were already in demand by biomedical researchers, by workers in biotechnology, and even in clinical laboratories. More people wanted flow cytometers than could afford them; now, even with the advent of smaller, more user-friendly, and somewhat less expensive instruments from the surviving manufacturers, there are still a lot of people who can't afford flow cytometers.

Before my original book, *Building and Using Flow Cytometers*, and its successor, the First Edition of *Practical Flow Cytometry*, appeared, building flow cytometers was not considered feasible by most people because they thought it required sophisticated skills and extensive resources, and because the parts of many laboratory-built instruments were at least as costly as are some commercial instruments. A few dozen people have now built "Cytomutts" following the designs in the books, and it is now established that it isn't all that hard or that expensive to do. I've left the details on instrument construction in the book, with improvements added, but this Edition is written for the users, who outnumber the builders by about a hundred to one. Even if you have no intention of building a flow cytometer, read the chapter on the subject; you'll probably pick up a few tips which will help you keep your commercial machine running. If you are interested in building apparatus, this edition will tell you how to build a smaller, cheaper, easier-to-build, and better performing multiparameter instrument than was described in the last one.

I've tried to make the Second Edition a self-contained treatise on flow cytometry, eliminating the "required reading list" I had in the First Edition. You'll still get something out of reading some of the landmark papers in the field, but, if you can't lay hands on them, it won't stunt your growth. I've actually expanded the reference list for background subjects such as optics, computers, and spectroscopy, but I've also put in a lot of new material on the interactions of light and matter, on optical systems, and on data analysis. When I want a word or phrase to catch your attention, I've put it in **boldface**.

This time around, I emphasize the fact that flow cytometry rests on the same foundations as other techniques of analytical cytology, such as microspectrophotometry and image analysis. The spectroscopy is the same, the optics are the same, and many of the parameters measured and the probes used are the same. This book is thus, in many respects, as much a book on cytometry in general as a book on flow cytometry. In keeping with this orientation, I have included some material on alternative techniques to flow cytometry and situations in which these alternatives may be preferable to flow cytometry.

The discussion of applications has been expanded considerably, reflecting the rapid growth in the use of flow cytometry in routine research and clinical applications as well as the change in the orientation of the book. There are more references; it's been nearly impossible to keep track of all the papers which involve flow cytometry, and it would have been impossible to mention all 2,000 listings in the computer data base to which John Maples and Pat Reynolds kindly let me have access.

I have tried, and I mean tried hard, to include references which describe new techniques or refinements in technique, whether they be in the areas of instrumentation, sample preparation, cytochemistry, or data analysis. I have also tried to provide thorough coverage of some fields in which the application of flow cytometry is in itself relatively novel. I haven't tried to include a reference to every article in which flow cytometric immunofluorescence, or analysis of DNA content, has been mentioned. The purpose of this book is to enable the reader to evaluate the level of the flow cytometry in such papers and/or to reproduce experiments reported in the literature – not that experiments involving flow cytometry are going to be any more reproducible than others.

I obviously can't read all of the journals all of the time. Toward the end of the First Edition, I asked readers to please send me reprints. A dozen or so people, mostly in Europe, took me seriously enough to do this, for which I thank them. I doubt that I've referred to everything they sent me, but their courtesy did insure that I'd have the chance to see things I might otherwise not have seen. My address is still 283 Highland Avenue, West Newton, Massachusetts 02465; keep those reprints coming.

The commercial aspects of flow cytometry have changed a bit since the First Edition, with one major manufacturer out of the picture and several new models available from the survivors. There are also a few people, myself included, selling add-on hardware and software for the acquisition and analysis of flow cytometric data. The chapter on "Sources of Supply" still lists suppliers for a wide range of things you might need if you do flow cytometry.

Even with all of this lovely word processing and publishing hardware and software (the computer system and accessories which produced this Edition cost about seven times as much as what was used for the First Edition), writing this book hasn't been easy. Thus, I have tried to make this edition one which will wear better, in the sense that future developments will require that things be added rather than changed. That way, in another two or three years, it will be time for a supplementary volume rather than a rewrite, unless everybody's deserted flow cytometry for molecular biology by then.

The First Edition sold some 2,500 copies, which boggles my mind, but what boggles my mind even more is that the gross proceeds from sales would not have paid for one top-of-the-line commercial dual laser flow cytometer. The royalties wouldn't have bought a plasma tube, much less a big laser, but the laser printer is a much nicer toy, so I'm not complaining. I still think commercial flow cytometers are too expensive, and I've told the manufacturers how to make them less expensive; if they don't, I will.

ACKNOWLEDGEMENTS

I again first thank Leslie Hochberg Shapiro, with whom I am still in love and thanks to whom I am still at large. Jill and Peter, in addition to screening even grosser jokes, respectively helped with attempting to organize huge piles of paper and with construction of prototypes of the new electronic designs, for which I thank them.

I still appreciate the perspectives on the design, construction, and use of cytometers I have gotten over the years from Dick Adams, Bob Auer, Leo Burke, Annette Coleman, Wallace Coulter, Phil Dean, Bart de Grooth, Fred Elliott, Mack Fulwyler, Mike Hercher, Dick Hiebert, Bob Hoffman, Lou Kamentsky, Joan McDowell, Len Ornstein, Dave Parks, Dan Pinkel, Diether Recktenwald, Tom Sharpless, Phil Stein, John Sullivan, Dick Sweet, Barb Trask, Ger van den Engh, and Rob Webb.

Thanks to Ron Weinstein, Alan Landay, and John Coon for their interest and support, viewpoints on clinical applications, and forbearance in beta testing my software.

Paulette Cohen and Reed Elfenbein, of Alan R. Liss, Inc., had the unenviable task of noodging me as the book fell farther and farther behind schedule; I appreciate their not having me kidnapped and held hostage until I finished.

Finally, I'd like to acknowledge Geoff Caine, David Feinstein, Terry Fetterhoff, Sheila Frankel, Carol Hirschmann Levenson, Ed Luther, Bob McCarthy, Nancy Perlmutter, Siobhan Spillane, and Sandy Stephens, who have maintained and improved the Cytomutt breed during the past few years, and especially thank Ed and Siobhan for minding the store when things were even squirrelier than usual.

Howard Shapiro
West Newton, Massachusetts
March, 1988

FOREWORD TO THE FIRST EDITION

It is now just 21 years since I first saw clusters of dots on an oscilloscope screen which I thought represented the total nucleic acids and proteins of cells flowing single file through an ultraviolet light beam. As Howard Shapiro correctly points out, the early work, including my own, in what was then called cytology automation was based on the pioneering studies done during the 1930's and 1940's in Sweden by Caspersson and Thorell. I had the good fortune to have worked with Bo Thorell in both his lab in Stockholm and mine in New York. I believe much of the early development of the biophysics and metrology of analytical cytology was influenced by his pioneering work and his often understated advice. No discussion of the history of this field could be complete without acknowledging Bo Thorell's contributions and I would like to do that on this 21st anniversary. Howard's very complete treatise on how to succeed with flow cytometry without trying too hard also provides me with a vehicle to reflect on what we expected of this technology and what was achieved as it reached maturity.

I got into trying to quantify cell properties while attempting to build a device to automate cervical cancer cytology. Automating existing clinical laboratory tasks by microscopic image analysis or flow cytometry was the driving force for many of us during this first stage of development. I would categorize the next years as the childhood of flow cytometry, during which the technique was applied to a broader range of tasks in analytical cytology. Next, with adolescence, came a search for power - bigger lasers, higher resolution, sophisticated computation, multiple beams and measurement parameters. This got us to the present generation of expensive, complex, hard-to-use systems with which we must now contend. This book deals with simpler approaches to many of these goals, which I believe makes it timely as well as useful.

I would like to briefly discuss each of the phases of the early life of flow cytometry and describe what I expect during its adulthood. Cytology automation, beginning with the slide scanning studies of Mellors using fluorescent dyes, Tolles et al and Mendelsohn et al, using a variety of staining techniques, and Coulter, using electrical resistance measurements, had been concerned with automating two tasks, cancer detection and blood cell counting. Although I was involved in designing scanners and programming computers to read text when I was introduced to the problem of automating cancer detection by Mike Melamed, I decided that making a machine to mimic a microscopist was not appropriate to the state of the art in optical scanning and computer science at that time. I could neither get enough light through a microscope, nor process enough data fast enough, to examine the required large populations of cells in a reasonable time. Also, I believed that flow measurements inherently gave better representations of the biophysical characteristics of cells.

This became the basis for the second phase of development, analytical cytology by flow. Except for various instruments that simply counted total red or white blood cells, cytology automation did not succeed. We sought better ways to prepare cells and better markers differentiating cell types, but the markers never proved specific enough to yield acceptable false positive rates. Parameters such as DNA content, while of some clinical utility, did not fulfill their early promise as specific markers of cancer. Thus, although there are now flow cytometers in clinical laboratories, their applications remain limited by a lack of specific markers. It is my hope that the new techniques of biotechnology will provide new reagents which will bind to individual antigenic determinants with greater specificity, and other reagents which will hybridize to appropriate specific gene sequences, and that this will somehow break this bottleneck. I am also hopeful that progress will result from the application of newer biophysical techniques, many of which originated during the second phase of development.

I tend to look upon the late 1960's, which I define as the era of analytical cytology, as the golden age of flow cytometry. During this time, we began to understand the physical factors responsible for resolution and reproducibility of measurements, the effects of interactions of different cellular parameters upon signals, and the quantitative relationships between signal intensities and constituents or physical properties of cells. Fluorescence measurements replaced absorption measurements because of the linear relationship observed between fluorescence signal intensities and quantities of various constituents present in cells, and because artifacts due to light scattering could more readily be eliminated from fluorescence measurements using optical filters. This made it necessary to find new dyes and histochemical methods to replace such problematic techniques as the Feulgen reaction, and many dyes applicable to living and fixed cells came into use. Other properties of fluorescence were exploited to provide information relating to structure and function, in addition to quantity, of cellular constituents. During this time, the application of Coons' fluorescent antibody technique to immunologically mark cells containing specific protein variants probably did more to revolutionize flow cytometry than did any other development, and vaulted the field into adolescence.

During the 1970's, the do-it-yourself flow cytometry community sought after ultimate sensitivity and resolution as research goals. I believe that was desirable, but I also believe it was overdone. In addition, the emphasis shifted from analytical measurements to physical separation of cells by flow cytometers and the devices began to be called sorters. Having built and run sorters based on a variety of different principles, I learned of the problems of paying for them and keeping them running and wrote trying to make arguments for other, more efficient bulk techniques for isolating cell populations, the results of which could be monitored by flow cytometry. I believe many biologists who bought sorters have used them only on rare occasions after understanding the parameters and quirks which affected their performance. The desire for sorting capacity whether or not it was actually needed, the mistaken belief that bigger lasers automatically produced better results, and the irrational demand for instruments to do everything anyone else had done at the time they were ordered led to an inflation of flow cytometers' complexity and cost which outpaced even the high rate of inflation of health care costs.

I believe that there is much room for development of new methodologies to study the interactions of cells with light and, perhaps, with other energy fields; I believe some progress toward this may come through an understanding of principles described in this book. Technology has come a long way since 1963, when I discarded methods involving imaging and computer analysis because they couldn't be implemented with existing hardware. The equivalent hardware is now orders of magnitude faster and/or cheaper. It would be nice to develop a new generation of instruments which could use some of the principles of flow cytometry to answer questions in biology which cannot be addressed using flow cytometry. Among these are questions related to the kinetics of individual cell functions in heterogeneous populations and to cell-to-cell communication and control. I look forward to some readers going beyond the building and use of flow cytometers, to develop new methodologies which may solve a whole new range of biologic problems.

Louis A. Kamentsky
Cambridge, Massachusetts
April, 1984

PREFACE TO THE FIRST EDITION

This book can tell you almost everything you need to know about flow cytometry: what it is, how it works, what you can do with it, and how to buy or build and use flow cytometers. A pretty tall order? Maybe, but I can deliver. Bear with me.

If you're perusing this book in hope of learning something about flow cytometry, you may know that flow cytometry and cell sorting have become useful to biomedical researchers and to workers in industrial and clinical laboratories. You may also know that many people who would like to use these techniques cannot afford commercial apparatus at present prices.

Building flow cytometers was not considered a feasible alternative by most of these people because they thought it required sophisticated skills and extensive resources, and because the parts of many laboratory-built instruments are at least as costly as are some commercial instruments.

Some years ago, I set out to develop a multiple illumination beam, multiparameter flow cytometer with capabilities not then available on the market. I made an effort to make that system as simple as possible to build, maintain, and operate. I was surprised to find that a flow cytometer competitive with some of the most sophisticated instruments described at that time could be made from readily available, relatively inexpensive parts, using minimal electronic and machine shop facilities. I built one and taught a few others to build them. We named the instruments "Cytomutts".

Since we had already done the work, we followed standard procedure and applied for a grant to build and use a Cytomutt and compare its performance with that of some fancier apparatus. The reviewers decided that what we proposed to do couldn't be done, and didn't fund us. We then took a Cytomutt to a meeting and showed that it worked. Then we applied for a grant again, requesting money to make the results of our work available to interested parties. The reviewers, at least some of whom had seen the machine in operation, again decided it couldn't be done.

By that time, some other people were believers; I was spending more time than I wanted to building flow cytometers, teaching people how to build flow cytometers, and advising people lucky enough to get funded about which flow cytometers to buy. I had promised a few people some detailed technical information about the hardware and about flow cytometry in general, so I generated a privately printed volume called *Building and Using Flow Cytometers*. Knowing full well that that book would be handed to some electronikers who would be told to go build a flow cytometer without being told what a flow cytometer was, I included all the information about the history, operation, and uses of flow cytometers which I thought these uninitiated readers might need.

I also put in really detailed instructions for building flow cytometers. It seemed to me that the people I had taught had learned by hanging around my laboratory, watching me build things and asking questions. It thus seemed logical to make the "how-to" book out of my side of the dialogue, including answers to questions I had been asked. The book was, therefore, written and illustrated in a very informal style, but it contained as many of the relevant details as I could think of.

Quite a few people who saw *Building and Using Flow Cytometers* expressed interest in buying a similar book which would go into more detail about dyes, staining procedures, and clinical and research applications. They expressed even more interest in buying such a book for less than the price I was charging for *Building and Using*. It sounded good to me, and it sounded good to Paulette Cohen at Alan R. Liss, Inc., and so *Practical Flow Cytometry* was conceived. The book had a longer gestation period than I envisioned because

I wanted to begin at the beginning and work up to the state of the art for different groups of readers, all of whom would have to cross the boundaries of the disciplines in which they were originally trained.

I think that the best way to deal with such disparate constituencies is to cover the fundamentals in detail and tell people where they can learn more if and when they need to. Accordingly, after the "Overture," which introduces themes which will recur in the book's main body, there is a Chapter on "Prerequisites," covering what you may need to know to get started, and where to find the information. In that chapter, I list a few important papers which I think everybody doing flow cytometry ought to read.

The book continues with a history of the field; this includes a lot of personal perspective because I think it is informative to understand why different people did or didn't do things at various stages in the development of flow cytometry. I then go on to discuss how flow cytometers work, and the whys and hows of data analysis and cell sorting.

The Chapter on "Parameters and Probes" includes detailed discussions of the cellular characteristics measurable by flow cytometry, the dyes and other reagents used for such measurements, and the results obtained. It is followed by an overview of the applications of flow cytometry in various areas of biological and medical research and in clinical medicine.

The discussion of hardware begins by considering the merits of several commercial flow cytometers. Then we get down to the do-it-yourself manual; I say "we" because this is the part where you can, as it were, "sing along with Howard". Since *Building and Using Flow Cytometers* was published, a number of people have built working Cytomutts from the plans it contained. You will learn how to build everything from a simple, microscope-based "Cytopup" to multistation Cytomutts, complete with multiparameter analyzers and computer interfaces. These inexpensive and efficient analyzers and interfaces can be used with commercial flow cytometers as well as with Cytomutts.

At the back of the book, the Chapter on "Sources of Supply" tells you where to find not only parts for Cytomutts, but reagents, calibration particles, and accessories which may be useful for any flow cytometry laboratory. So, there you go. Soup to nuts. Take my book. Please.

ACKNOWLEDGEMENTS

Let me first thank Leslie Hochberg Shapiro, who has kept me in love and out of jails and mental institutions for all these years, and Jill and Peter, who screened all the gross jokes they heard while I was writing and only interrupted me with the really good ones.

My thanks go to Phil Stein and to Tom Sharpless for helpful discussions about electronics, with some strings attached.

Thanks also go to Mike Hercher and Rob Webb, who taught me things about optics I couldn't have learned at school.

I appreciate the perspectives on the design, construction, and use of flow cytometers I have gotten over the years from Wallace Coulter, Phil Dean, Fred Elliott, Mack Fulwyler, Lou Kamentsky, Joan McDowell, Dave Parks, Dan Pinkel, John Sullivan, and Dick Sweet.

Paulette Cohen and Sharon Freund of Alan R. Liss, Inc., helped in many ways, not least of which was cracking long whips between New York and Massachusetts when the schedule slipped. I was very pleased that Zbigniew Darzynkiewicz agreed to read the manuscript, and delighted that Lou Kamentsky accepted my plea for a Foreword.

Finally, I'd like to acknowledge the contributions of Cytomutt breeders and trainers Lisa Christenson, Jay Connor, David Feinstein, Sheila Frankel, Al Kirsch, Kathy Mead, Rob Olson, Nancy Perlmutter, Larry Scherr, and Bob Young, without whom this book could never have been written. Don't blame them, though.

Howard Shapiro
West Newton, Massachusetts
October, 1984

1. OVERTURE

This is a book about **cytometry**, in general, emphasizing **flow cytometry**, in particular. In it, I hope to tell you what cytometry is, how it works, why and how to use it, when you should favor one type of cytometry or another, and when cytometry won't solve your problem. This chapter, like the overture to an opera or a musical, presents important themes from the body of the work, but may also stand alone.

1.1 WHAT (AND WHAT GOOD) IS CYTOMETRY?

Cytometry is a process in which physical and/or chemical characteristics of **single cells**, or by extension, of other biological or nonbiological particles in roughly the same size range, are **measured**. In **flow cytometry**, the measurements are made as the cells or particles pass through the measuring apparatus, a **flow cytometer**, in a fluid stream. A **cell sorter**, or **flow sorter**, is a flow cytometer that uses electrical and/or mechanical means to divert and collect cells (or other small particles) with measured characteristics that fall within a user-selected range of values.

Neither the cells nor the apparatus are capable of putting the process of cytometry in motion; the required critical element for that is a human interested in obtaining information about a cell sample and, in the case of sorting, extracting cells of interest from the sample. At the most basic level, a cytometer might be considered to be a "black box" with cells as "inputs" and numbers as "outputs"; the outputs of a cell sorter would include both numbers and cells. However, while some modern cytometers (and some modern users) can obtain the desired results while running unattended in "black box" mode, it is fair to say that most of the applications, and all of the interesting applications, of cytometry call for some understanding and some intellectual effort on the part of the user.

Tasks and Techniques of Cytometry

From the time of van Leeuwenhoek and Hooke until the mid-20th century, determining:

1) whether cells were present in a specimen,
2) how many were there,
3) what kinds of cells were represented, and
4) what their functional characteristics might be

required that a human observer interpret a microscope image. The same tasks remain for modern cytometry.

Although electrical and acoustic properties of, and nuclear radiation emission from, single cells can be measured, it is fair to say that **optical** measurements are by far the most common in cytometry. A typical cytometer is thus a specialized **microscope**; the degree of physical resemblance is dictated by the requirements of the measurement(s) to be made, which in turn are dictated by what the user needs to know about the cell sample. In successful applications of cytometry, electro-optics, electronics, and computers are employed to improve on what could be obtained "by eye," although interpretation is required more often than not. The successful applications are many, increasing in number, and commonplace in locales as diverse as clinical laboratories and breweries.

Some Notable Applications

Cytometry is currently used to obtain the helper T lymphocyte counts needed to monitor the course and treatment of HIV infection, and to determine tumor cell DNA content and proliferative activity, which may aid in assessing prognosis and determining treatment for patients with breast cancer and other malignant diseases. The technology has also been used to crossmatch organs for transplantation, to isolate human chromosomes for the construction of genetic libraries, to separate X- and Y-chromosome bearing sperm for sex selection in animal breeding and *in vitro* fertilization in humans, to identify the elusive hematopoietic stem cell and an expanding family of other stem cell types, and to reveal several widely distributed but previously unknown genera of marine microorganisms.

Biological particles that have been subjected to cytometric analysis range, in order of decreasing size, from multicellular organisms (e.g., *Drosophila* embryos and adult *Caenorhabditis elegans* nematodes) through cell aggregates (e.g., pancreatic islets and tumor cell spheroids), eukaryotic cells, cellular organelles (e.g., mitochondria), bacteria, liposomes, individual virus particles and immune complexes, down to the level of single molecules of proteins, nucleic acids, and organic dyes. Cytometers can also be used for sensitive chemical analyses involving the binding of suitably labeled ligands to solid substrates or to particles such as polystyrene beads.

The first practical applications of flow cytometry, beginning in the 1940's, were to counting blood cells in liquid suspension, on the one hand, and bacteria and other small particles in aerosols, on the other, based on measurements of **light scattering** or **electrical impedance**; these signals were also used to provide estimates of cell size.

In the early 1960's, **light absorption** measurements were used for quantitative flow cytometric analyses of cellular nucleic acid and protein. Flow cytometers in modern clinical hematology laboratories perform counts of red cells (erythrocytes), white cells (leukocytes), and platelets (thrombocytes) in blood, as well as differential leukocyte counts, using combinations of electrical impedance, light scattering, and light absorption measurements.

However, many people who know the term "flow cytometer" tend to use it – incorrectly – to describe only instruments that measure **fluorescence** as well as light scattering. The first fluorescence flow cytometers were built in the late 1960's; although there are now well over 10,000 in use in clinical and research laboratories worldwide, they are still outnumbered by impedance and scattering-based hematology analyzers. So much for fluorescence chauvinism.

What is Measured: Parameters and Probes

The novice should not be intimidated by the jargon of cytometry; there are no native speakers, and he or she can soon enough become as fluent in it as the rest of us. The term **parameter** is, unfortunately, used in several different senses in our jargon. It can refer to a **physical or chemical characteristic of a cell** (e.g., cytoplasmic granularity or nuclear DNA content) that is measurable by cytometry; it can also describe a **physical property, measured by a sensor,** defined broadly (e.g., light scattering or fluorescence), or more narrowly (e.g., orthogonal light scattering or red fluorescence), or a **physical property of a cell-associated reagent** (e.g., propidium fluorescence). A fairly comprehensive list of measurable cellular parameters appears as Table 1-1 on the facing page.

I have characterized cellular parameters as **intrinsic** or **extrinsic**, depending upon whether they can or cannot be measured without the use of **reagents**, which are often referred to in cytometric jargon as **probes**. Some parameters can, at least in principle, be measured either with or without probes; cellular DNA content, for example, can be estimated from ultraviolet (UV) absorption at 260 nm in unstained cells, but it's much more practical to use a fluorescent dye probe such as propidium iodide. A deeper philosophical dilemma arises when considering fluorescence from *Aequorea* green fluorescent protein (GFP) or one of its genetically engineered offshoots, introduced by cloning into cells of other species to report gene expression; one could characterize this as intrinsic or extrinsic, but I lean toward the latter.

Parameters can also be defined as **structural** or **functional**, again with some ambiguity. For example, the glycoprotein efflux pump responsible for multidrug resistance in tumor cells can be detected, and the amount present in a cell quantified, using fluorescent antibodies, but such antibodies might also bind to an inactive mutant protein, and thus provide a measurement (in this case, inaccurate) based on structure. The function of the glycoprotein pump can be demonstrated by measurement of uptake or loss of fluorescent drugs or dyes by cells over periods of time.

In a **kinetic measurement** such as that just described, **time** itself can be used as a parameter. When such analyses are done by flow cytometry, the dynamic behavior of a cell population must be inferred from observations of different cells at different times, because conventional flow cytometers cannot make successive measurements of a single cell over time periods exceeding a few microseconds.

Both the novice and the expert in flow cytometry should be aware that almost every parameter that can be measured by flow cytometry can also be measured by alternative cytometric methods such as **microspectrophotometry, confocal microscopy, image analysis,** and **scanning cytometry.** These methods are often applicable where flow cytometric methods are not, e.g., for true kinetic analyses involving repeated examination of the same cell or cells over a period of time, or for *in situ* analyses of cells growing in aggregates attached to solid substrates. In general, the fluorescent probes used for flow cytometry can be used with alternative measurement techniques. However, most dyes and other reagents that are commonly employed in absorption microspectrophotometry are not readily usable in fluorescence flow cytometers.

1.2 BEGINNINGS: MICROSCOPY AND CYTOMETRY

It recently (i.e., since the last time I wrote an introduction to cytometry) occurred to me that the best way in

PARAMETER	MEASUREMENT METHOD AND PROBE IF USED
Intrinsic Structural Parameters (no probe)	
Cell Size	Electronic (DC) impedance, extinction, small angle light scattering; image analysis
Cell shape	Pulse shape analysis (flow); image analysis
Cytoplasmic granularity	Large angle light scattering, Electronic (AC) impedance
Birefringence (e.g., of blood eosinophil granules)	Polarized light scattering, absorption
Hemoglobin, photosynthetic pigments, porphyrins	Absorption, fluorescence, multiangle light scattering
Intrinsic Functional Parameter (no probe)	
Redox state	Fluorescence (endogenous pyridine and flavin nucleotides)
Extrinsic Structural Parameters (probe required)	
DNA content	Fluorescence (propidium, DAPI, Hoechst dyes)
DNA base ratio	Fluorescence (A-T and G-C preference dyes, e.g., Hoechst33258 and chromomycin A_3)
Nucleic acid sequence	Fluorescence (labeled oligonucleotides)
Chromatin structure	Fluorescence (fluorochromes after DNA denaturation)
RNA content (single and double-stranded)	Fluorescence (acridine orange, pyronin Y)
Total protein	Fluorescence (covalent- or ionic-bonded acid dyes)
Basic protein	Fluorescence (acid dyes at high pH)
Surface/Intracellular antigens	Fluorescence; scattering (labeled antibodies)
Surface sugars (lectin binding sites)	Fluorescence (labeled lectins)
Lipids	Fluorescence (Nile red)
Extrinsic Functional Parameters (probe required)	
Surface/intracellular receptors	Fluorescence (labeled ligands)
Surface charge	Fluorescence (labeled polyionic molecules)
Membrane integrity (not always a sign of "viability")	Fluorescence (propidium, fluorescein diacetate [FDA]); absorption or scattering (Trypan blue)
Membrane fusion/turnover	Fluorescence (labeled long chain fatty acid derivatives)
Membrane organization (phospholipids, etc.)	Fluorescence (annexin V, merocyanine 540)
Membrane fluidity or microviscosity	Fluorescence polarization (diphenylhexatriene)
Membrane permeability (dye/drug uptake/efflux)	Fluorescence (anthracyclines, rhodamine 123, cyanines)
Endocytosis	Fluorescence (labeled microbeads or bacteria)
Generation number	Fluorescence (lipophilic or covalent-bonded tracking dyes)
Cytoskeletal organization	Fluorescence (NBD-phallacidin)
Enzyme activity	Fluorescence; absorption (fluorogenic/chromogenic substrates)
Oxidative metabolism	Fluorescence (dichlorofluorescein)
Sulfhydryl groups/glutathione	Fluorescence (bimanes)
DNA synthesis	Fluorescence (anti-BrUdR antibodies, labeled nucleotides)
DNA degradation (as in apoptosis)	Fluorescence (labeled nucleotides)
"Structuredness of cytoplasmic matrix"	Fluorescence (fluorescein diacetate [FDA])
Cytoplasmic/mitochondrial membrane potential	Fluorescence (cyanines, rhodamine 123, oxonols)
"Membrane-bound" Ca^{++}	Fluorescence (chlortetracycline)
Cytoplasmic [Ca^{++}]	Fluorescence ratio (indo-1), fluorescence (fluo-3)
Intracellular pH	Fluorescence ratio (BCECF, SNARF-1)
Gene expression	Fluorescence (reporter proteins)

Table 1-1. Some parameters measurable by cytometry.

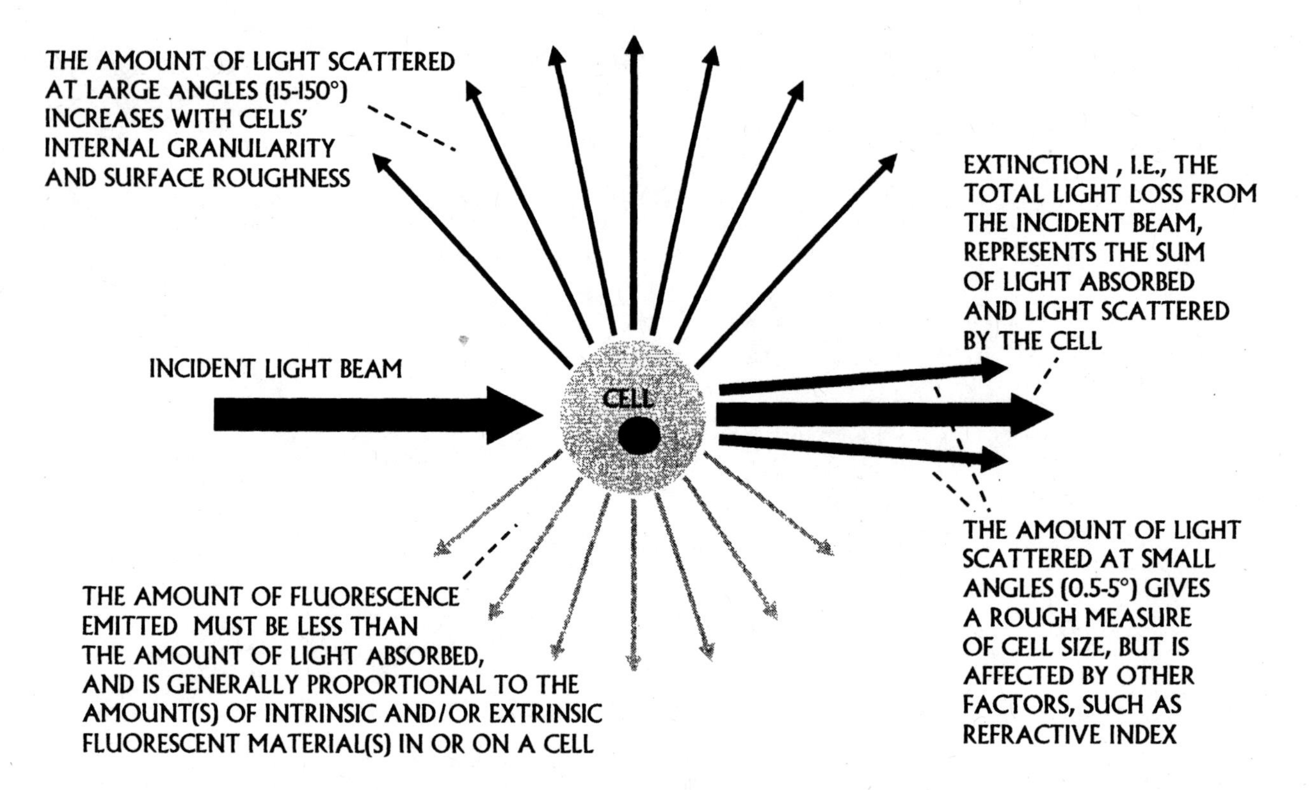

Figure 1-1. Interaction of light with a cell.

which to introduce the subject might be to consider how cytometry developed from microscopy, emphasizing both the similarities and the differences between the two, and stressing how the information gets from the cells to the user. That is what I will try to do in the remainder of this chapter. I hope this will be helpful for the uninitiated reader, but, also, that it will be equally thought-provoking, informative, and at least moderately amusing to those who have been over the terrain one or many times before.

The first order of business in both microscopy and cytometry is discriminating between the cells and whatever else is in the sample; the next is often discriminating among a number of different cell types that may be present. Optical microscopes first allowed cells to be discovered and described in the seventeenth century, and were refined in design in the eighteenth and early nineteenth, but the capacity of microscopy to discriminate among different cell types remained limited by the relative difficulty of obtaining **contrast** between cells and the background in microscope images.

A Little Light Music

While all the senses can provide us with pleasure and discomfort, it is predominantly vision that shapes our perception of the world around us, and, without light, our visual imagery is restricted to memories, dreams, and hallucinations. According to the Book of Genesis, the discrimination of light from darkness is the divine achievement of the first day of creation, and we humans, despite taming fire and inventing light bulbs and lasers, remain aware of and profoundly affected by the daily difference, not least during power outages.

What most of us know as **light** is defined by physicists as **electromagnetic radiation** with wavelengths ranging between about 400 and about 700 nanometers (nm). Other species can detect shorter and longer wavelengths, but most lack our ability to discriminate among wavelengths, i.e., color vision, and some of us have genetic deficiencies that restrict this capacity.

When we look at the macroscopic world, most of our retinal images are formed by light that we say is **reflected** from objects around us, and an early concept of light was that of **rays** traveling in straight lines, and reflecting from a surface at the same angle at which they strike it. If we look at an object under water and attempt to grab it, we find that it is not exactly where it appears to be; this is explained by the concept of **refraction**, according to which light passing from one material medium into another is bent at an angle depending on a macroscopic property of the medium known as the **refractive index**, and on the wavelength of the light. The "white" light emitted by the sun and by incandescent and fluorescent bulbs comprises a range of visible wavelengths; objects and materials that **absorb** some, but not all, wavelengths reflect others, and thus appear colored.

As we turn our attention to smaller and smaller objects, the concepts of reflection and refraction become less and less useful, and we instead make use of the concept of **light scattering**. Figure 1–1 describes the interaction of light with a cell in terms of **scattering**, **absorption**, and **fluorescence**. The last of these phenomena is not readily explicable in terms of either ray (geometrical) or wave optics, and can only be dealt with properly by the theory of **quantum electrodynamics**, which considers light as particles, or **photons**, which interact with **electrons** in atoms and molecules. The energy of a photon is inversely proportional to the corresponding wavelength; i.e., photons of short-wavelength, 400 nm violet light have a higher energy content than photons of long-wavelength, 700 nm red light.

Scattering, which explains both reflection and refraction, typically involves a brief interaction between a photon and an electron, in which the photon is annihilated, transferring its energy to the electron, which almost immediately releases all of the energy in the form of a new photon. Thus, light scattered by an object has the same (or almost exactly the same) wavelength, or color, as the incident light. However, the new photon does not necessarily travel in the same direction as the old one, so scattered light usually appears to be at an angle to the incident beam.

In empty space, there are, by definition, no atoms or molecules, and there are thus no electrons available to interact with photons. Although, according to quantum electrodynamics, a photon has a finite probability of going in any direction, when we actually calculate the probabilities that apply in the case of photons in empty space, we come up with what look like rays of light traveling in straight lines.

As a general rule, the density of atoms and molecules in atmospheric air is fairly low, meaning that there are few opportunities for light to be scattered as it appears to traverse distances of a few meters or tens of meters. However, we note the blue appearance of a cloudless sky, resulting from light scattering throughout the atmosphere; the color results from the fact that shorter wavelengths of light are more likely to be scattered than longer ones, with the intensity of scattering inversely proportional to the fourth power of the wavelength.

The well-known laws of reflection and refraction emerge from quantum electrodynamics applied to objects substantially bigger than the wavelength of light. Materials that appear transparent to the human eye, e.g., glass and water, still contain relatively high densities of atoms and molecules, and thus provide numerous opportunities for scattering.

Some light appears to be reflected at the interfaces between layers of different materials, with the angle of reflection equal to the angle of incidence. The total amount of light reflected is found to be a function of the thickness of the layers and the wavelength of the incident light; that is, layers of different thicknesses reflect different colors of light to different extents. This **interference** effect, explained by the theory of wave optics, accounts for the patterns of color seen in peacock feathers, butterfly wings, diffraction gratings in spectrophotometers, on credit cards, and in cheap jewelry, and in opals in somewhat more expensive jewelry. It is exploited in optical design, notably in the production of **interference filters** used to select ranges of wavelengths to be observed and/or detected in microscopes and other optical instruments. Quantum electrodynamics comes up with the same results for interference and reflection as wave optics, even while taking into account that the phenomena are due to scattering throughout objects, not just from front and back surfaces.

The apparent bending of light striking an interface between two materials is described in classical optics with the aid of invented quantities, called **refractive indices**, which are characteristic of the materials involved. Light appears to travel more slowly through a material of higher refractive index than through a material of lower index, and a "ray" appears to "bend" toward the normal (i.e., toward a line perpendicular to the interface) when passing from a lower-index medium to a higher one, and away from the normal when passing from a higher-index medium to a lower one. The apparent velocity of light in a material is less than in empty space; the higher the refractive index, the lower the apparent velocity. Light of a shorter wavelength is "bent" more than light of a longer one, allowing a transparent object with surfaces that are not parallel (i.e., a **prism**) to **disperse** light of different wavelengths in different directions.

Armed with ray optics and the classical law of refraction, we can calculate how an object with appropriately curved surfaces, i.e., a **lens**, will "bend" light originating from two points separated in space. If the surfaces are convex, divergent "rays" coming through the lens from two points a given distance apart on the "input" side can be made to converge at two points a greater distance apart on the "output" side; this provides us with a **magnified image**. A magnifying lens is, of course, the fundamental ingredient of a microscope.

Not surprisingly, everything useful that classical optics tells us about refraction can be obtained using quantum electrodynamics. Although actually doing this usually involves a great deal of advanced mathematics, Richard Feynman, who received his Nobel Prize for work in the field, wrote a small book called *QED*[641], in which he used simple diagrams and concepts to make the subject accessible to a lay audience (which, in this context, includes me). What I am writing here paraphrases the master.

The light scattering behavior of objects of dimensions near the wavelength of light is not predictable from ray optics. For spherical particles ranging in diameter from one or two wavelengths to a few tens of wavelengths, most of the light scattering occurs at small angles (0.5° to 5°) to the incident beam; the intensity of this "**small angle**," or "**forward**," light scattering is dependent on the refractive index difference between the particle and the medium, and on particle size. However, the relationship between particle size and small angle scattering intensity is not monotonic, meaning that, although a particle 10 μm in diameter will probably produce a bigger signal than one of the same composition 5 μm in diameter, a particle 5.5 μm in diameter might produce a smaller signal than one 5 μm in diameter. It is thus wise to avoid thinking of the small angle scatter signal as an accurate measure of cell size.

Smaller particles scatter proportionally more light at larger angles (15° to about 150°) to the incident beam; the amplitude of such signals, variously described as "**side**," "**orthogonal**," "**large angle**," "**wide angle**," or "**90°**" light scattering, is, all other things being equal, larger for cells with internal granular structure, such as blood granulocytes, than for cells without it, such as blood lymphocytes.

Ray optics and wave optics break down when we consider the process of **light absorption**. This comes down to photons and electrons, period. Quantum theory tells us that the electrons in a given atom or molecule can exist only in discrete energy states. The lowest of these is referred to as the **ground state**, and the absorption of a photon by an electron in the ground state raises it to a higher energy **excited state**. An electron in an excited state can absorb another photon, ending up in a still higher energy excited state.

Like scattering, and all other quantum phenomena, absorption is probabilistic. We cannot say that a particular electron will absorb a particular photon; the best we can do is calculate the probability that an electron in a particular energy state will absorb a photon of a particular energy, or wavelength. This probability increases as the difference in energy between the current energy state of the electron and the next higher energy state gets closer to the energy of the photon involved.

In many molecules, the energy difference between states is greater than the energy in a photon of visible light. Such molecules may exhibit substantial absorption of higher energy, shorter wavelength photons, e.g., those with wavelengths in the **ultraviolet (UV)** region between about 200 and 400 nm. Substances made up of such molecules appear transparent to the human eye; smearing them on exposed skin decreases the likelihood that ultraviolet photons will interact with electrons in DNA and other macromolecules of dermal cells, and reduces the likelihood of sunburn (yay!) and tanning (boo!). We're not sure yet about skin cancer.

For a molecule to absorb light in the visible region, the energy differences between electronic energy states have to be rather small. This condition is satisfied in some inorganic atoms and crystals, which have unpaired electrons in *d* and *f* orbitals, in metals, which have large numbers of "free" electrons with an almost continuous range of energy states, resulting in high absorption (and high reflectance) across a wide spectral range, and in organic molecules with large systems of conjugated π orbitals, including natural products such as porphyrins and bile pigments, and synthetic dyes such as those used to stain cells.

The interaction of light with matter must obey the law of conservation of energy; the amount of light transmitted should therefore be equal to the amount of incident light minus the amount scattered and the amount absorbed. But what happens to the absorbed light? One would not expect the electrons involved in absorption to remain in the excited state indefinitely, and, indeed, they do not. In some cases, all of the absorbed electronic energy is converted to vibrational or rotational energy, and lost as heat. In others, some energy is lost as heat, but the remainder is emitted in the form of photons of lower energy (and, therefore, longer wavelength) than those absorbed. Depending on the details of the electronic energy transitions involved, this emission can occur as **fluorescence** or as **phosphorescence**. Fluorescence emission usually occurs within a few tens of nanoseconds of absorption; phosphorescence is delayed, and may continue for seconds or longer. As is the case with absorption, fluorescence and phosphorescence are inexplicable by ray and wave optics; they can only be understood in terms of quantum mechanics.

Making Mountains out of Molehills: Microscopy

When we are not looking at luminous displays such as the one I face as I write this, most of our picture of the world around us comes from reflected light. Contrast between objects comes from differences in their reflectivities at the same and/or different wavelengths. When ambient light levels are high, we utilize our retinal cones, which give us color vision capable of prodigious feats of spectral discrimination (humans with normal vision can discriminate millions of colors), at the expense of relatively low sensitivity to incident light. The high light levels bleach the visual pigments in our more sensitive retinal rods; if the light level is decreased abruptly, it takes some time for the rod pigment to be replenished, after which we can detect small numbers of photons, sacrificing color vision in the process. Thus, while we can perceive large numbers of 450 nm photons, 550 nm photons, and 650 nm photons, respectively, as red, green, and blue light, using our cones, we cannot distinguish individual photons with different energy levels as different colors. Night vision equipment typically utilizes monochromatic green luminous displays because the rods are most sensitive to green light, but the cone system also exhibits maximum sensitivity in the green region, making the spot from a green laser pointer much more noticeable than that from a red one emitting the same amount of power.

While the spectral discrimination capabilities of the unaided human visual system are remarkable, its spatial discrimination power is somewhat limited. The largest biological cells, e.g., ova and large protists, are just barely visible, and neither the discovery of cells nor the appreciation of their central role in biology would have occurred had the light microscope not been invented and exploited.

When unstained, unpigmented cells are examined in a traditional **transmitted light**, or **bright field**, microscope, light absorption is negligible; contrast between cells and the background is due solely to scattering of light by cells and subcellular components, and the only information we can get about the cells is thus, in essence, contained in the scattered light. Some of this is scattered out of the field of view; we must therefore rely on slight differences in transmission between different regions of the image to detect and characterize cells. We are working against ourselves by presenting our eyes (or the detector(s) in a cytometer) with a large amount of light that has been transmitted by the specimen.

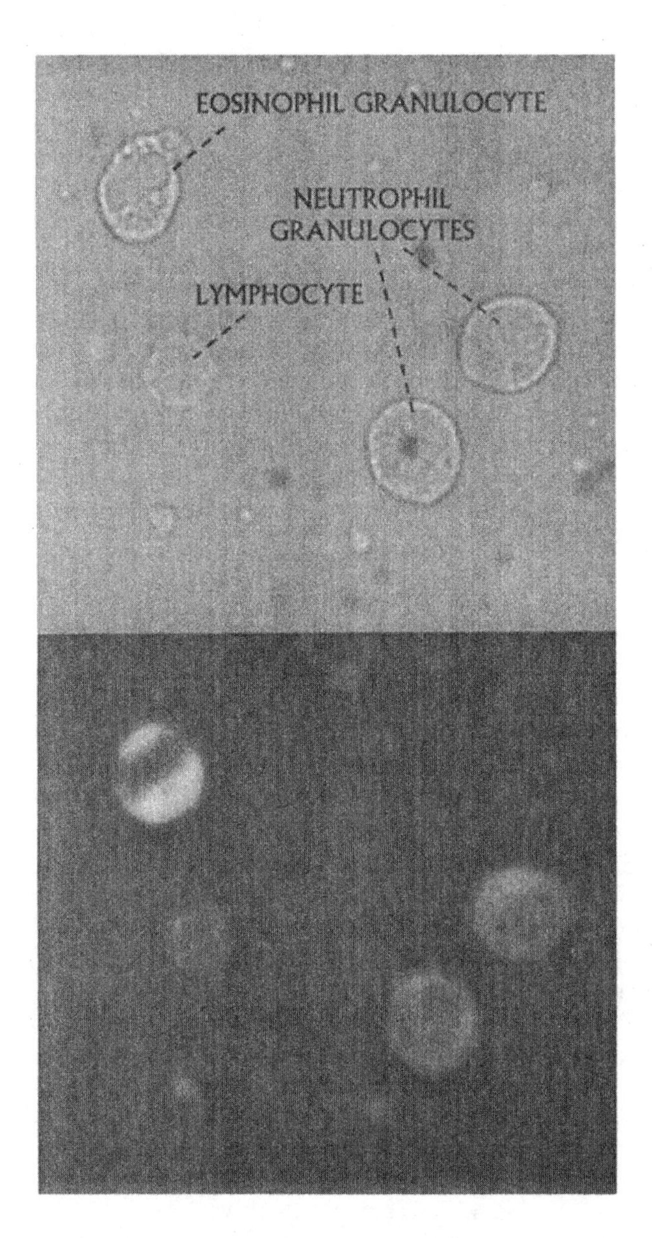

Figure 1-2. Transmitted light (bright field) (top panel) and dark field (bottom panel) images of an unstained suspension of human peripheral blood leukocytes. The objective magnification was 40 ×.

As it happens, the maximum spatial resolution of a microscope is achieved, i.e., the distance at which two separate objects can be distinguished as separate is minimized, when illuminating light reaches, and is collected from, the specimen at the largest possible angle. The **numerical aperture (N.A.)** of microscope condensers and objectives is a measure of the largest angle at which they can deliver or collect light. However, when the illumination and collection angles in a transmitted light microscope are large, much of the light scattered by objects in the specimen finds its way back into the microscope image, increasing resolution, but decreasing contrast. The top panel of Figure 1-2 shows a bright field microscope image of a suspension of human peripheral blood leukocytes; the condenser was stopped down to increase contrast between the cells and background. The cytoplasmic granules in the eosinophil and neutrophil granulocytes are not particularly well resolved, nor is it easy to distinguish the nuclei from the cytoplasm. Increasing the level and angle of illumination might, as just mentioned, increase resolution, but this would not be useful, as contrast would not be increased.

Modern microscopy exploits both differences in **phase** and **polarization** of **transmitted light** and the phenomenon of **interference** to produce increased contrast in bright field images. However, **staining**, which came into widespread use in the late 1800's, largely due to the emergence of synthetic organic dyes, was the first generally applicable practical bright field technique for producing contrast between cells and the medium, and between different components of cells in microscope images. Paul Ehrlich, known for his later researches on chemotherapy of infectious disease, stained blood cells with mixtures of acidic and basic dyes of different colors, and identified the three major classes of blood granulocytes, the basophils, eosinophils (which he termed acidophils), and neutrophils, based on the staining properties of their cytoplasmic granules.

Stained elements of cells are visually distinguishable because of their **absorption** of incident light, even when the refractive index of the medium is adjusted to be equal or nearly equal to that of the cell. The dyed areas transmit only those wavelengths they do not absorb, resulting in a difference in spectrum, or color, between them and undyed areas or areas that take up different dyes. Absorption by pigments within cells, such as the hemoglobin in erythrocytes, also makes the cells more distinguishable from the background.

Microscopy of opaque specimens, such as samples of minerals, obviously cannot use transmitted light bright field techniques. Instead, specimens are illuminated from above, and the image is formed by light reflected (i.e., scattered) from the specimen. In **incident light bright field microscopy**, illumination comes through the objective lens, using a partially silvered mirror, or **beam splitter**, to permit light to pass between source and specimen and between specimen and eyepiece at the same time. In **dark field microscopy**, illumination is delivered at an oblique angle to the axis of the objective by a separate set of optics. The bottom panel of Figure 1-2 is a dark field image of the same cells as are shown in the top panel. In the dark field microscope, none of the illuminating light can reach the objective unless it is scattered into its field of view by objects in the specimen. The illumination geometry used in this instance ensured that the only light contributing to the dark field image was light scattered at relatively large angles to the illuminating beam. It has already been noted that this is the light represented in the side scatter signal, and it can be seen that the lymphocyte, which would have the smallest side scatter signal, appears dimmer than the neutrophil granulocytes and the eosinophil, which would have higher side scatter signals. Although the cytoplasmic granules within the granulocytes are not well resolved, the intensity of light coming from the

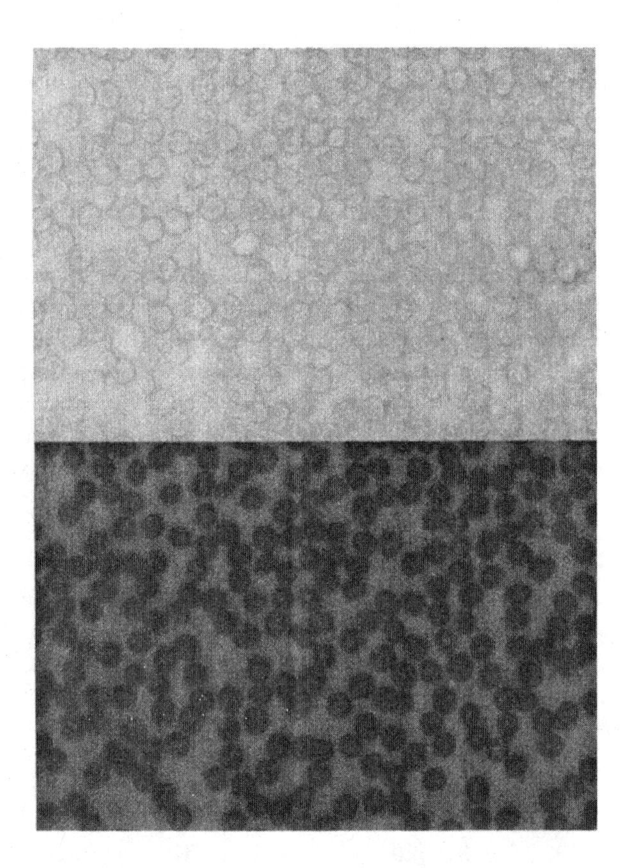

Figure 1-3. Transmitted light microscope images of an unstained smear of human peripheral blood. The picture in the top panel was taken with "white light" illumination; that in the bottom panel was taken with a violet (405 nm, 15 nm bandwidth) band pass optical filter, and demonstrates the strong absorption of intracellular hemoglobin in this wavelength region. Objective: 40 ×.

cytoplasm provides an indication of their presence; indeed, it is much easier to resolve nucleus from cytoplasm in the granulocytes in the dark field image than in the bright field image. Thus, we can surmise that it may be possible to get information about subcellular structures from a cytometer operating at an optical resolution that would be too low to allow them to be directly observed as discrete objects. In fact, using dark field microscopy, one can observe light scattered by, and fluorescence emitted from, particles well below the limit of resolution of an optimally aligned, high-quality optical microscope; the dark field "ultramicroscope" of the 1920's allowed researchers to see and count viruses, although it was obviously impossible to discern any structural detail.

Absorption measurements are bright field measurements, and they work best, especially for quantification, when the absorption signal is strong. The material being looked for should have a high likelihood of absorbing incident light, as indicated by a high **molar extinction coefficient**, and there should be a lot of it in the cell. Figure 1-3 shows the absorption of hemoglobin in the cytoplasm of unstained red blood cells. Note that the "white light" image in the top panel gives little hint of strong absorption, which is restricted to the violet region known as the Soret band; the "white" light used here, which came from a quartz-halogen lamp, contains very little violet, and the exposure time used for the picture in the bottom panel was about 100 times as long as that for the picture in the top panel.

Fluorescence microscopy is inherently a dark field technique; even in a "transmitted light" fluorescence microscope, **optical filters** are employed to restrict the spectrum of the illuminating beam to the shorter wavelengths used for fluorescence **excitation**, and also to allow only the longer-wavelength fluorescence **emission** from the specimen to reach the observer. As is the case in dark field microscopy, fluorescent cells (ideally) appear as bright objects against a dark background.

Most modern fluorescence microscopes employ the optical geometry shown in Figure 1-4. Excitation light is usually supplied by a mercury or xenon arc lamp or a quartz-halogen lamp, equipped with a lamp condenser that **collimates** the light, i.e., produces parallel "rays." These components are not shown, but would be to the left of the excitation filter in the figure. The excitation filter passes light at the excitation wavelength, and reflects or absorbs light at other wavelengths. The excitation light is then reflected by a **dichroic mirror**, familiarly known simply as a **dichroic**, which transmits light at the emission wavelength. The microscope objective is used for both illumination of the specimen and collection of fluorescence emission, which is transmitted through both the dichroic and the emission filter.

In any microscope, a real image of the specimen is formed by the objective lens; the eyepiece and the lens of the observer's eye then project an image of this image onto the retina of the observer. Light falling on sensitive cells in the retina produces electrical impulses that are transmitted along the optic nerves. What happens next is the province of neurology, psychology, and, possibly, psychiatry.

It has already been noted that humans are very good at color discrimination, and we also know that humans, with some training, can get pretty good at discriminating cells from other things. With more training, we can become proficient at telling at least some kinds of cells from others, usually on the basis of the size, shape, color, and texture of cells and their components in microscope images; it is not always easy to program computers to make the same distinctions on the same basis.

The human visual system can detect light intensities that vary over an intensity range of more than nine decades; in other words, the weakest light we can perceive is on the order of one-billionth the intensity of the strongest perceptible light. However, we can't cover the entire range at once; as previously mentioned, we need dark-adapted rods to see the least intense signals, and do so only with monochromatic vision. And we aren't very good at detecting small changes in light intensity. This has forced us to invent instruments to make precise light intensity measurements to meet the needs

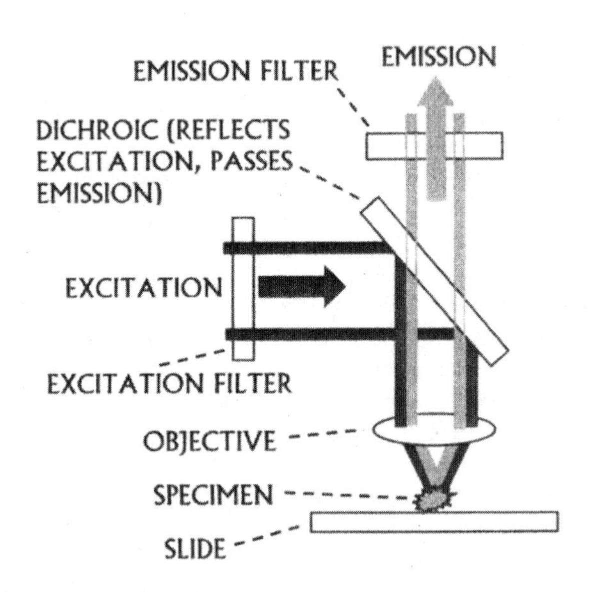

Figure I-4. Schematic of a fluorescence microscope.

of science, technology, medicine, and/or art (remember when the light meter was not built into the camera?). It was this process that eventually got us from microscopy to cytometry.

Why Cytometry? Motivation and Machinery

In the 1930's, by which time the conventional histologic staining techniques of light microscopy had already suggested that tumors might have abnormalities in DNA and RNA content, Torbjörn Caspersson[34], working at the Karolinska Institute in Stockholm, began to study cellular nucleic acids and their relation to cell growth and function. He developed a series of progressively more sophisticated **microspectrophotometers**, which could make fairly precise measurements of DNA and RNA content based on the strong intrinsic UV absorption of these substances near 260 nm, and also found that UV absorption near 280 nm, due to aromatic amino acids, could be used to estimate cellular protein content. When Caspersson began working, it had not yet been established that DNA was the genetic material; he helped move others toward that conclusion by establishing, through precise measurement, that the DNA content of chromosomes doubled during cellular reproduction[7].

A conventional optical microscope incorporates a **light source** and associated optics that are used to illuminate the specimen under observation, and an **objective** lens, which collects light transmitted through and/or scattered, reflected and/or emitted from the specimen. Some means are provided for moving the specimen and adjusting the optics so that the specimen is both properly illuminated and properly placed in the field of view of the objective. In a microscope, a mechanical stage is used to position the specimen and to bring the region of interest into focus.

A microspectrophotometer was first made by putting a small "pinhole" aperture, or **field stop**, in the image plane of a microscope, restricting the field of view to the area of a single cell, and placing a photodetector behind the field stop. The diameter of the field stop could be calculated as the product of the magnification of the objective lens and the diameter of the area from which measurements were to be taken. If a 40× objective lens were used, measuring the transmission through, or the absorption of, a cell 10 μm in diameter would require a 400 μm diameter field stop.

Using a substantially smaller field stop, it would be possible to measure the transmission through a correspondingly smaller area of the specimen; for example, a 40 μm field stop would permit measurement of a 1 μm diameter area of the specimen. By moving the specimen in the x and y directions (i.e., in the plane of the slide) in the raster pattern now so familiar to us from television and computer displays, and recording and adding the measurements appropriately, it was possible to measure the integrated absorption of a cell, and/or to make an image of the cell with each pixel corresponding in intensity to the transmission or absorption value. This was the first, and, at the time, the only feasible approach to **scanning cytometry**.

The use of stage motion for scanning made operation extremely slow; it could take many minutes to produce a high-resolution scanned image of a single cell, and there were no computers available to capture the data. Somewhat higher speed could be achieved by using moving mirrors, driven by galvanometers, for image scanning, and limiting the tasks of the motorized stage to bringing a new field of the specimen into view and into focus; this required some primitive electronic storage capability, and made measurements susceptible to errors due to uneven illumination across the field, although this could be compensated for.

Since the late 1940's and early 1950's had already given us Howdy Doody, Milton Berle, and the Ricardos, it might be expected that somewhere around that time, someone would have tried to automate the process of looking down the microscope and counting cells using video technology. In fact, image analyzing cytometers were developed; most of them were not based on video cameras, for a number of reasons, not the least of which was the variable light sensitivity of different regions of a camera tube, which would make quantitative measurements difficult. There was also the primitive state of the computers available; multimillion dollar mainframes had a processor speed measured in tens of kilohertz, if that, and memory of only a few thousand kilobytes, and this made it difficult to acquire, store, and process the large amount of data contained even in a digitized image of a single cell.

By the 1960's, a commercial version of Caspersson's microspectrophotometer had been produced by Zeiss, and several groups of investigators were using this instrument and a variety of laboratory-built scanning systems in attempts to automate analysis of the Papanicolaou smear for cervical cancer screening, on the one hand, and the differential white blood cell count, on the other[42,43,52,53,57-60]. It was felt that both of these tasks would require analysis of cell images with resolution of 1 μm or better, to derive measures of such characteristics as cell and nuclear size and shape, cytoplasmic texture or granularity, etc., which could then be used to develop the cell classification algorithms needed to do the job. Although it was widely recognized that practical instruments for clinical use would have to be substantially faster than what was then available, this was not of immediate concern in the early stages of algorithm development, and few people even bothered to calculate the order of magnitude of improvement that might be necessary.

Flow Cytometry and Sorting: Why and How

Somewhat simpler tasks of cell or particle identification, characterization, and counting than those involved in Papanicolaou smear analysis and differential white cell counting had attracted the attention of other groups of researchers at least since the 1930's. During World War II, the United States Army became interested in developing devices that could rapidly detect bacterial biowarfare agents in aerosols; this would require processing a relatively large volume of sample in substantially less time than would have been possible using even a low-resolution scanning system. The apparatus that was built in support of this project[29-31] achieved the necessary rapid specimen transport by injecting the air stream containing the sample into the center of a larger (**sheath**) stream of flowing air, confining the particles of interest to a small region in the center, or **core**, of the stream, which passed through the focal point of what was essentially a dark-field microscope. Particles passing through the system would scatter light into a collection lens, eventually producing electrical signals at the output of a photodetector. The instrument could detect at least some *Bacillus* spores, objects on the order of 0.5 μm in diameter, in specimens, and is generally recognized as having been the first flow cytometer used for observation of biological cells; similar apparatus had been used previously for studies of dust particles in air and of colloidal solutions.

By the late 1940's and early 1950's, the same principles, including the use of **sheath flow**, as just described, for keeping cells in the center of a larger flowing stream of fluid, were applied to the detection and counting of red blood cells in saline solutions[48]. This paved the way for automation of a diagnostic test notorious for its imprecision when performed by a human observer using a counting chamber, or **hemocytometer**, and a microscope.

Neither the bacterial counter nor the early red cell counters had any significant capacity either for discriminating different types of cells or for making quantitative measurements. Both types of instrument were measuring what we would now recognize as side scatter signals; although larger particles would, in general, produce larger signals than smaller ones composed of the same material, the correlations between sizes and signal amplitudes were not particularly strong. In the case of the bacterial counter, a substantial fraction of the spores of interest would not produce signals detectable above background; the blood cell counters had a similar lack of sensitivity to small signals, which was advantageous in that blood platelets, which are typically much smaller than red cells, would generally not be detected. White cells, which are larger than red cells, would be counted as red cells; however, since blood normally contains only about 1/1000 as many white cells as red cells, inclusion of white cells in the red cell count would not usually introduce any significant error.

An alternative flow-based method for cell counting was developed in the 1950's by Wallace Coulter[49]. Recognizing that cells, which are surrounded by a lipid membrane, are relatively poor conductors of electricity as compared to the saline solutions in which they are suspended, he devised an apparatus in which cells passed one by one through a small (< 100 μm) orifice between two chambers filled with saline. A constant electric current was maintained across the orifice; when a cell passed through, the **electrical impedance** (similar to **resistance**, which is the inverse of **conductance**) increased in proportion to the volume of the cell, causing a proportional increase in the measured voltage across the orifice. The Coulter counter was widely adopted in clinical laboratories for blood cell counting; it was soon established that it could provide more accurate measurements of cell size than had previously been available[50-1].

In the early 1960's, investigators working with Leitz[61] proposed development of a hematology counter in which a fluorescence measurement would be added to the light scattering measurement used in red cell counting. If a fluorescent dye such as acridine orange were added to the blood sample, white cells would be stained much more brightly than red cells; the white cell count could then be derived from the fluorescence signal, and the raw red cell count from the scatter signal, which included white cells, could, in theory, be corrected using the white cell count. It was also noted that acridine orange fluorescence could be used to discriminate mononuclear cells from granulocytes. However, it does not appear that the device, which would have represented a new level of sophistication in flow cytometry, was ever actually built.

A hardwired image analysis system developed in an attempt to automate reading of Papanicolaou smears had been tested in the late 1950's; although it was nowhere near accurate enough, let alone fast enough, for clinical use, it showed enough promise to encourage executives at the International Business Machines Corporation to look into producing an improved instrument.

Assuming this would be some kind of image analyzer, IBM gave technical responsibility for the program to Louis Kamentsky, who had recently developed a successful optical character reader. He did some calculations of what would be required in the way of light sources, scanning rates, and computer storage and processing speeds to solve the problem using image analysis, and concluded it couldn't be done that way.

Having learned from pathologists in New York that cell size and nucleic acid content should provide a good indicator of whether cervical cells were normal or abnormal, Kamentsky traveled to Caspersson's laboratory in Stockholm and learned the principles of microspectrophotometry. He then built a flow cytometer that used a transmission measurement at visible wavelengths to estimate cell size and a 260 nm UV absorption measurement to estimate nucleic acid content[1,65].

Subsequent versions of this instrument, which incorporated a dedicated computer system, could measure as many as four cellular parameters[78]. A brief trial on cervical cytology specimens indicated the system had some ability to discriminate normal from abnormal cells[77]; it could also produce distinguishable signals from different types of cells in blood samples stained with a combination of acidic and basic dyes, suggesting that flow cytometry might be usable for differential leukocyte counting.

Although impedance (Coulter) counters and optical flow cytometers could analyze hundreds of cells/second, providing a high enough data acquisition rate to be useful for clinical use, scanning cytometers offered a significant advantage. A scanning system with computer-controlled stage motion could be programmed to reposition a cell on a slide within the field of view of the objective, allowing the cell to be identified or otherwise characterized by visual observation; it was, initially, not possible to extract cells with known measured characteristics from a flow cytometer. Until this could be done, it would be difficult to verify any cell classification arrived at using a flow cytometer, especially where the diagnosis of cervical cancer or leukemia might be involved.

This problem was solved in the mid 1960's, when both Mack Fulwyler[67], working at the Los Alamos National Laboratory, and Kamentsky, at IBM[66], demonstrated **cell sorters** built as adjuncts to their flow cytometers. Kamentsky's system used a syringe pump to extract selected cells from its relatively slow-flowing sample stream. Fulwyler's was based on ink jet printer technology then recently developed by Richard Sweet[68] at Stanford; following passage through the cytometer's measurement system (originally a Coulter orifice), the saline sample stream was broken into droplets, and those droplets that contained cells with selected measurement values were electrically charged at the droplet breakoff point. The selected charged droplets were then deflected into a collection vessel by an electric field, while uncharged droplets went, as it were, down the drain.

Fluorescence and Flow: Love at First Light

Fluorescence measurement was introduced to flow cytometry in the late 1960's as a means of improving both quantitative and qualitative analyses. By that time, Van Dilla et al[79] at Los Alamos and Dittrich and Göhde[83] in Germany had built fluorescence flow cytometers to measure cellular DNA content, facilitating analysis of abnormalities in tumor cells and of cell cycle kinetics in both neoplastic and normal cells. Kamentsky had left IBM to found Bio/Physics Systems, which produced a fluorescence flow cytometer that was the first commercial product to incorporate an argon ion laser; Göhde's instrument, built around a fluorescence microscope with arc lamp illumination, was distributed commercially by Phywe.

Leonard Herzenberg and his colleagues[2], at Stanford, realizing that fluorescence flow cytometry and subsequent cell sorting could provide a useful and novel method for purifying living cells for further study, developed a series of instruments. Although their original apparatus[82], with arc lamp illumination, was not sufficiently sensitive to permit them to achieve their objective of sorting cells from the immune system, based on the presence and intensity of staining by fluorescently labeled antibodies, the second version[86], which used a water-cooled argon laser, was more than adequate. This was commercialized as the FACS in 1974 by a group at Becton-Dickinson (B-D), led by Bernard Shoor.

By 1979, B-D, Coulter, and Ortho (a division of Johnson & Johnson that bought Bio/Physics Systems) were producing flow cytometers that could measure small- and large-angle light scattering and fluorescence in at least two wavelength regions, analyzing several thousand cells per second, and with droplet deflection cell sorting capability. DNA content analysis was receiving considerable attention as a means of characterizing the aggressiveness of breast cancer and other malignancies, and monoclonal antibodies had begun to emerge as reagents for dissecting the stages of development of cells of the blood and immune system. Instruments with two lasers were used to detect staining of cells by different monoclonal antibodies conjugated with spectrally distinguishable dyes.

Image cytometers existed; they were much slower and even less user-friendly than the early flow cytometers, weren't easily adapted for immunofluorescence analysis, and couldn't sort. Meanwhile, the early publications and presentations based on flow cytometry and sorting created a large demand for cell sorters among immunologists and tumor biologists. By the early 1980's, when a mysterious new disease appeared, best characterized – using flow cytometry and monoclonal antibodies – by a precipitous drop in the numbers of circulating T-helper lymphocytes, clinicians, as well as researchers, had become anxious to obtain and use fluorescence flow cytometers – and, often, to avoid sorting!

In the decades since, confocal microscopes, scanning laser cytometers, and image analysis systems have come into use. They can do things flow cytometers cannot do; they typically have better spatial resolution and can be used to examine cells repeatedly over time, but they cannot analyze cells as rapidly, and there are many fewer of them than there are flow cytometers. They are also, unlike flow cytometers, not subject to:

Shapiro's First Law of Flow Cytometry:
A 51 µm Particle CLOGS a 51 µm Orifice!

That notwithstanding, in these first years of the 21st century, most cytometry is flow cytometry, and, for almost all applications except clinical hematology analysis, flow cytometry involves fluorescence measurement.

Fluorescence and flow are made for each other for several reasons, but primarily because fluorescence, at least from organic materials, is a somewhat ephemeral measurement. Recall that fluorescence occurs when a photon is absorbed by an atom or molecule, raising the energy level of an associated electron to an excited state, after which a small amount of the energy is lost as heat, and the remainder is emitted in the form of a longer wavelength photon, as fluorescence. However, there is a substantial chance that a photon at the excitation wavelength will not excite fluorescence but will, instead, **photobleach** a fluorescent molecule, producing a nonfluorescent product by breaking a chemical bond. In general, you can expect to get only a finite number of cycles of excitation and emission out of each fluorescent molecule (**fluorophore**) before photobleaching occurs.

If you look at a slide of cells stained with a fluorescent dye under a fluorescence microscope, you are likely to notice that, each time you move to a new field of view, the fluorescence from the cells in the new field is more intense than the fluorescence from the field that you had been looking at immediately before, which has undergone some photobleaching. This effect makes it difficult to get precise quantitative measurements of fluorescence intensity from cells in a static or scanning cytometer if you have to find the cells by visual observation before making the measurement, because the extent of photobleaching prior to the measurement will differ from cell to cell. In a flow cytometer, each cell is exposed to excitation light only for the brief period during which it passes through the illuminating beam, usually a few microseconds, and the flow velocity is typically nearly constant for all the cells examined. These uniform conditions of measurement make it relatively easy to attain high **precision**, meaning that one can expect nearly equal measurement values for cells containing equal amounts of fluorescent material; this is especially desirable for such applications as DNA content analysis of tumors, in which the abnormal cells' DNA content may differ by only a few percent from that of normal stromal cells.

A basis for the compatibility between fluorescence measurements and cytometry in general is found in the dark field nature of fluorescence measurements. It has already been noted that precise absorption measurements are best made when the concentration of the relevant absorbing material is relatively high. When one is trying to detect a small number of molecules of some substance in or on a cell, this condition is not always easy to satisfy. In the 1930's, unsuccessful attempts were made to detect antibody binding to cellular structures by bright field microscopy of the absorption of various organic dyes bound to antibodies. In 1941, Albert Coons, Hugh Creech, and Norman Jones successfully labeled cells with an antibody containing a fluorescent organic molecule[44], enabling structures binding the antibody to be visualized clearly against a dark background. In general, fluorescence measurements, when compared to absorption measurements, offer higher **sensitivity**, meaning that they can be used to detect smaller amounts or concentrations of a relevant analyte; this is of importance in attempting to detect many cellular antigens, and also in identifying genetic sequences and/or fluorescent protein products of transfected genes present in small copy numbers.

It is also usually easier to make simultaneous measurements of a number of different substances in cells, a process referred to as **multiparameter** cytometry, by fluorescence than by absorption, and the trend in recent years in both flow and static cytometry has been toward measurement of an increasingly large number of characteristics of each cell subjected to analysis, as can be appreciated from Table 1-1, way back on page 3.

Conflict: Resolution

When I first got into cytometry in the late 1960's, and for the next twenty years or so, there was a "farmer vs. rancher" feud going on between the people who did image analysis and the people who did flow, especially in the areas of development of differential white cell counters and Pap smear analyzers.

The first automated differential counters to hit the market were, in fact, image analyzers that scanned blood smears stained with the conventional Giemsa's or Wright's stains. Most of them are gone, now; modern hematology counters, which produce total red cell, reticulocyte (immature red cell), white cell, and platelet counts and red cell and platelet size (and, in at least one case, red cell hemoglobin) distributions, in addition to the differential white cell count, are typically flow based. Various instruments may measure electrical impedance (AC as well as DC), light absorption, scattering (polarized or depolarized), extinction, and/or fluorescence. None of them uses Giemsa's or Wright's stain.

Of course, with hundreds of monoclonal antibodies available that react with cells of the blood and immune system in various stages of development, we can use fluorescence flow cytometry to count and/or classify stem cells and other normal and abnormal cells in bone marrow, peripheral blood, and specimens from patients with leukemias and lymphomas, taking on tasks in hematology that few of the pioneers seriously believed could be approached using instruments. However, while the hematology counters run in a highly automated mode and produce numbers that can go directly into a hospital chart, most of the more sophisticated fluorescence-based analyses require considerable human intervention at stages ranging from the selection of a panel of antibodies to be used to the performance of the flow cytometric analysis and the interpretation of the results. This may facilitate reimbursement for the tests, but it leaves some of us unfulfilled, although perhaps better paid.

Cytometric apparatus that facilitated the performance and interpretation of the Papanicolaou (Pap) smear reached the market much later than did differential white cell counters. The first improvements were limited to automation of sample preparation and staining; there are now several image analysis based systems approved for clinical use in aiding screening (locating cells and displaying images of them to a human observer), and at least one approved for performing screening itself. All use the traditional Papanicolaou stain, a witches' brew of highly nonspecific acidic and basic dyes known since the 19th century and blended for its present purpose before the middle of the 20th.

Why the difference? What made the Pap smear survive the smear campaign and the Wright's stained blood smear go with the flow? The answer is simple. Both Pap smear analysis and blood smear analysis on slides depend heavily on morphologic information about the internal structure of cells. Criteria for cell identification in these tasks may include cell and nuclear size and shape, cytoplasmic granularity or texture, and, especially in dealing with abnormal blood cells, finer details such as whether nucleoli or intracellular inclusions are present.

Some of these characteristics, e.g., cytoplasmic granularity (which, as has already been noted, is a major contributor to a side scatter signal), can be determined using flow cytometers. While the fluorescent antibodies used for such tasks as leukemia and lymphoma classification using flow are highly specific (although not, in general, specific to a single cell type), most of the instruments that perform the differential leukocyte count do not need to use particularly specific reagents. In fact, it is possible, using only a combination of polarized and depolarized light scattering measurements, to do a differential white cell count with no reagents other than a diluent containing a lysing agent for red cells.

In the case of differential leukocyte counting, we have learned to substitute measurements that can be made of whole cells in flow, requiring only low-resolution optics, for those that would, if we were dealing with a stained smear, require that we make and analyze a somewhat higher-resolution image of each cell. Flow is faster, simpler, and cheaper, and, although morphologic hematologists still look at stained smears of blood and bone marrow from patients in whom abnormal cells have been found, we no longer need to look at a stained smear by eye or by machine to perform a routine white cell differential count. Although there may be combinations of low-resolution flow-based measurements that could provide a cervical cancer screening test comparable in performance to Pap smear analysis, none have yet been clinically validated; we therefore still rely on image analysis in approaches to automation of cervical cancer cytology and on visual observation where automation is not available.

Researchers face problems similar to those faced by clinicians. If you want to select and sort the 2,000 cells out of 10,000,000 cells in a transfected population that express the most green fluorescent protein (GFP), you will probably use a flow cytometer with high-speed sorting capability and settle for a low-resolution optical measurement that detects all of the GFP in or on the cell without regard to its precise location. If you have arranged for the GFP to be coexpressed with a particular structural protein involved, say, in the formation of the septum in dividing bacterial cells, you will very likely want to look at images of those cells at as high a **resolution** as you can achieve in order to get the information you need from the cells. There are, of course, tradeoffs.

It is January 1, 2002, as I write this, and therefore particularly appropriate to continue this New Year's resolution discussion. In the age of the personal computer and the digital camera and camcorder, there is little need to introduce the concepts of **digital images** and their component **pixels** (the term originally came from "picture elements"); most of us are exposed to at least 1024 × 768 almost 24/7. In this instance, the familiar 1024 × 768 figure describes the **pixel resolution** of an image acquisition or display device, with the image made up of 768 rows, each containing 1,024 pixels (or of 1,024 columns, each containing 768 pixels). However, the pixel resolution of the device doesn't, in itself, tell us anything about the **image resolution**, i.e., the area in the specimen represented by each pixel.

This depends to a great extent on what's in the image. In an image from the Hubble Space Telescope, each pixel could be light-years across; in an image from an atomic force microscope, each pixel might only be a few tenths of a nanometer (Ångstroms) across. But the image resolution also depends on the combination of hardware and software used to acquire and process the image. We are free to collect a transmitted light microscope image of a 10.24 μm by 7.68 μm rectangle (close quarters for a single lymphocyte) somewhere on a slide containing a stained smear of peripheral blood, using a digital camera chip with 1024 × 768 resolution, but we are not free to assume that each of the pixels in the image represents an area of approximately 0.01 by 0.01 μm. In this instance, the optics of the light microscope will limit our effective resolution to somewhere between 0.25 and 0.5 μm, and using a camera with a high pixel resolution won't help resolve smaller structures any more than would projecting the microscope image on the wall. Either strategy provides what microscopists have long known as **empty magnification**; the digital implementation, by allowing us to collect many more bits worth of information than we need or can use, slows down the rate at which we can process samples by a factor of at least several hundred, and is best avoided.

So what do we do when we really need high-resolution images? As it turns out, one of the physical factors that limits resolution in a conventional fluorescence microscope, in which the entire thickness of the specimen is illuminated, is fluorescence emission from out-of-focus regions of the specimen above and below the plane of what we are trying to look at. In a **confocal microscope**, the illumination and light collection optics are configured to minimize the contributions from out-of-focus regions; this provides a high-resolution image of a very thin slice of the specimen. Resolution is improved further in **multiphoton confocal microscopy**, in which fluorescence is excited by the nearly simultaneous absorption of two or more photons of lower energy than would normally be needed for excitation. The illumination in a multiphoton instrument comes from a tightly focused high-energy pulsed laser, and it is only in a very small region near the focal spot of the laser that the density of low energy photons is sufficient for multiphoton excitation to occur. This produces an extremely high-resolution image (pixel dimensions of less than 0.1 µm are fairly readily achieved), and also minimizes bleaching of fluorescent probe molecules and photodamage to cells.

As usual, we pay a price for the higher-resolution images. We are now looking at slices of the specimen so thin that we need to construct a three-dimensional image from serial slices of the specimen to fully visualize many cellular structures. Instead of two-dimensional pixels, we must now think in terms of three-dimensional **voxels**, or volume elements. Let's go back to the single lymphocyte which, on the blood smear discussed on the previous page, was confined in a two-dimensional, 10.24 µm by 7.68 µm, rectangular area. For three-dimensional imaging, we would prefer that the cell not be flattened out, especially if we want to look at it while it is alive, so we will assume it to be roughly spherical, and imprison it in a cube 10 µm on a side. If we used a multiphoton microscope with each voxel representing a cube 0.1 µm on a side, building a 3-D image of that single cell would require us to collect data from 100 × 100 × 100 voxels, or 10^6 voxels, and, even if it only took one microsecond to get data from each voxel, it would take a second just to collect the data.

This is a perfectly acceptable time frame for an investigator who needs information about subcellular structures; even with the computer time required for image processing, one can examine hundreds, if not thousands, of cells in a working day. However, even this is feasible only if the experimenter and/or the hardware and software in the instrument first scan the specimen at low resolution to find the cells of interest.

1.3 PROBLEM NUMBER ONE: FINDING THE CELL(S)

Continuing with the scenario just described, suppose we have cells at a concentration of 10^6/mL, dispersed on a slide in a layer 10 µm thick. A 1 × 1 cm area of the slide will contain 10,000 of the 10 µm cubicles in which we could cache a lymphocyte. Recalling that 10 µm is 1/1,000 cm, and that 1 cm^3 is 1 mL, we can calculate the aggregate volume of these 10,000 little boxes as 1/1,000 mL. If the cell concentration is 10^6/mL, we can only expect to find about 1,000 cells in 1/1,000 mL, and it would take us 16 minutes, 40 seconds to scan all of them at high resolution. However, if we adopted the brute force approach and did 3-D scans over the entire 1 × 1 cm area, instead of finding the locations of the cells and restricting the high-resolution scanning to those regions, we would waste 9,000 seconds, or 2 hours and 15 minutes, scanning unoccupied cubicles.

There's another problem; although we may arbitrarily divide the 1 × 1 cm × 10 µm volume into 10 µm cubicles, we have not created actual physical boundaries on the slide, and we can expect the cells to be randomly distributed over the surface, which means that parts of the same cell could lie in more than one cubicle. If we deal with a specimen thicker than 10 µm or so, the positional uncertainty extends to a third dimension, further compounding the problem of finding the cells, which gets even more difficult if we are trying to get high-resolution images of specific cell types in a tissue section, or in a small living organism such as a *Drosophila* embryo or a *C. elegans* worm.

When I first got into the cytometry game, in the late 1960's, my colleagues and I at the National Bureau of Standards and the National Institutes of Health built a state-of-the art computerized microscope, with stage position and focus, among other things, under computer control[73]. The instrument could be operated in an interactive mode, which allowed an experimenter to move the stage and focus the microscope using a small console that included a keypad and a relatively primitive joystick; the actual motion remained under computer control at all times. This made it possible to scan a slide visually, find cells of interest, store their locations in the computer, and have the instrument come back and do the high-resolution scans (resolution, in this instance, was better than 0.25 µm) needed for an experiment.

We didn't have a computer algorithm for finding cells automatically; since scanning the area immediately surrounding a cell took us not one second, but two minutes, there would have been little point to automating cell finding. The actual scanning time required to collect integrated absorption measurements of the DNA content of 100 cells, stained by the Feulgen method, was 3 hours, 20 minutes. We could find the cells that interested us by eye in a few minutes; scanning the slide looking for them might have taken days.

We were able to make life a little easier for ourselves by developing an algorithm to remove objects from the periphery of an image. A typical microscope field would contain a cell of interest, which we had positioned in the center of the field, surrounded by other cells, parts of cells, or dirt and/or other junk. Since the algorithm was relatively simple-minded, our visual selection process required us to exclude cells that touched or were overlapped by other cells. Figure 1-5, on the next page, shows the results of applying the algorithm.

The figure also shows how difficult it might be to develop algorithms to find cells. Even among the few cells present in the image shown, there are substantial differences in size and shape, and there are marked inhomogeneities in staining intensity within cells. Humans get very good very fast at finding cells and at discriminating cells from junk, even when cell size, shape, and texture vary. If staining (or whatever else produces contrast between the cell and the background) were relatively uniform, recognizing a cell by computer would be fairly easy; one would only have to find an appropriately sized area of the image in which all the pixel values were above a certain threshold level. This simple approach clearly won't work with cells such as those shown in the figure.

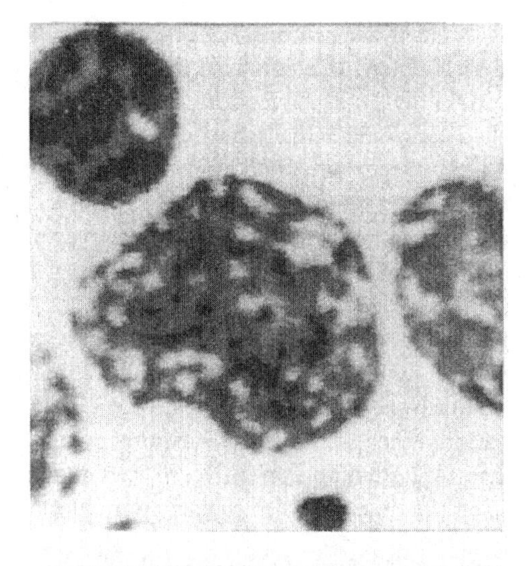

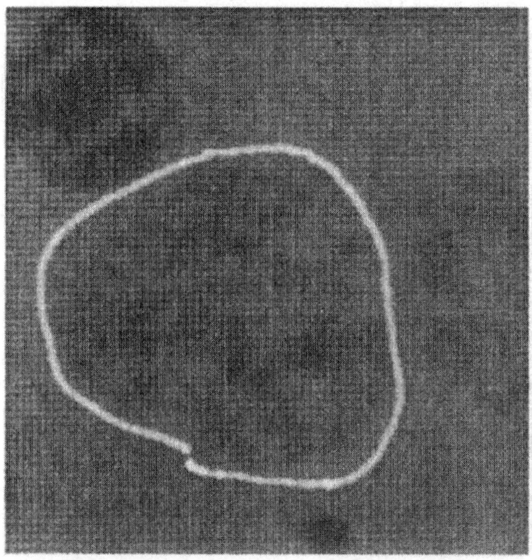

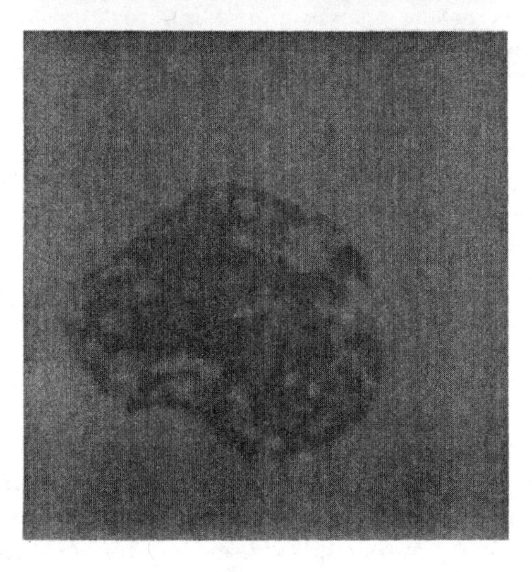

Figure 1-5. Top panel: scanned image of Feulgen-stained lymphoblastoid cells. In the middle panel, a boundary drawn around the cell of interest is shown; the bottom panel shows results of applying an algorithm to remove all objects except the cell of interest.

Before we developed the procedure for removing unwanted material from images, we had the option of picking out the cell of interest by drawing a boundary around it using a light pen, as shown in the middle panel of Figure 1-5. Many researchers working with cell images still find it convenient to locate cells and define boundaries for analysis in this fashion, and essentially the same procedure is used to draw the boundaries of regions of interest in two-parameter data displays from flow cytometers. This sidesteps the issue of automated cell finding (or of automated cluster finding, in the case of data displays). The boundary drawing is now commonly done using a personal computer and a mouse; in 1970, there were no mice, at least not the computer kind, and the interactive display and light pen we used cost tens of thousands of dollars, and had to be attached to the mainframe computer we needed to do the image processing. Very few laboratories could have afforded to duplicate our apparatus; today, you can introduce your children and grandchildren to the wonders of the microscopic world using a digital video microscope that costs less than $100 and attaches to your computer's USB port. But, although your computer is probably hundreds of times faster than the one we used and has thousands of times the storage capacity, which could allow it to be used to implement cell finding algorithms of which we could only dream, it still takes a long time to capture high-resolution cell images, and the detail in those images makes it more difficult for those algorithms to define the boundaries of a cell or a nucleus than it would be if the images used for cell finding were of lower resolution.

A cell 10 μm in diameter occupies thousands of contiguous pixels in a high-resolution image with 0.1×0.1 μm pixels, such as might be obtained from a multiphoton confocal microscope, but fewer than 100 contiguous pixels in a lower-resolution image with 1×1 μm pixels, such as might be obtained from a scanning laser cytometer. The high-resolution image may contain many pixels with intensity near that of the background (as is the case with the image shown in the top panel of Figure 1-5), making it necessary to do fairly convoluted analyses of each pixel in the context of its neighbor pixels to precisely define the area of a cell or an internal organelle. However, each of the 1×1 μm pixels of the lower-resolution image can be thought of as representing contributions from a hundred 0.1×0.1 μm areas of the cell, and, since it is unlikely that all of these are at background intensity, it is apt to be easier to define an area as composed of contiguous pixels above a certain intensity level if one uses larger pixels.

When one is working with isolated cells, it becomes attractive to attempt to confine them to defined areas of a slide rather than to have to scan the entire surface to find cells distributed at random. By the 1960's, it had occurred to more than one group of investigators that depositing cells in a thin line on a glass or plastic tape would allow an automated cytology instrument to restrict stage motion to one dimension instead of two, potentially speeding up processing. The concept is illustrated in Figure 1-6 (next page).

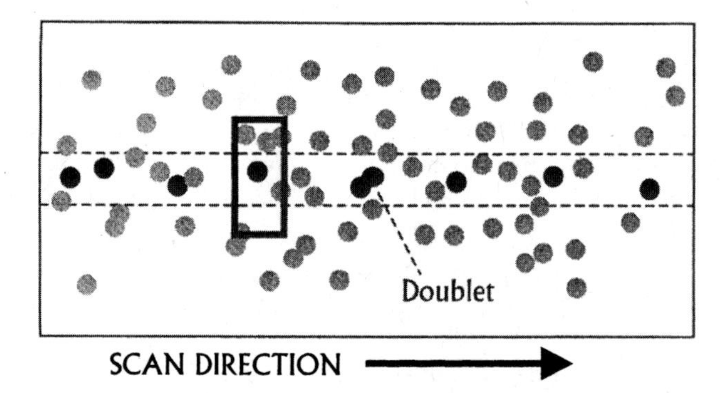

Figure 1-6. One-dimensional scanning of cells deposited in a narrow line (between dotted lines) on a slide or tape simplifies finding cells in a specimen. Black dots represent cells deposited in the line, gray dots represent cells deposited at random, and the small rectangle shows the field of view.

You can actually try this trick at home, if you happen to keep a microscope there, or in the lab, if you don't. Simulate the "cells" with dots in different colors made by a permanent marker with a fine or extra fine point; make dots in one color, corresponding to the black dots in the figure, along a straight edge placed parallel to the long edge of a slide, and make dots of another color (or enlist a [much] younger associate to do so), corresponding to the gray dots in the figure, all over the slide. Put the slide under the microscope, using a low- (10× or lower) power objective; place one of the "black" dots in the center of the field of view. Stop down the substage iris diaphragm until you get a field a few times the diameter of the "cell." Then move only the horizontal stage motion control. You should note that, although the "black" cells you encounter as you scan along the slide in one dimension remain entirely in the field of view (up to a point; if the line along which you scribed wasn't exactly parallel to the edge of the slide, there will be some drift), you will almost certainly find "gray" cells cut off at the edges of the field of view. Now, looking at Figure 1-7, we can consider what a photodetector "looking" at the field of view would "see" if the slide in Figure 1-6 were scanned. We can regard this signal as a series of images, each made up of a single pixel that is considerably larger than the cells of interest.

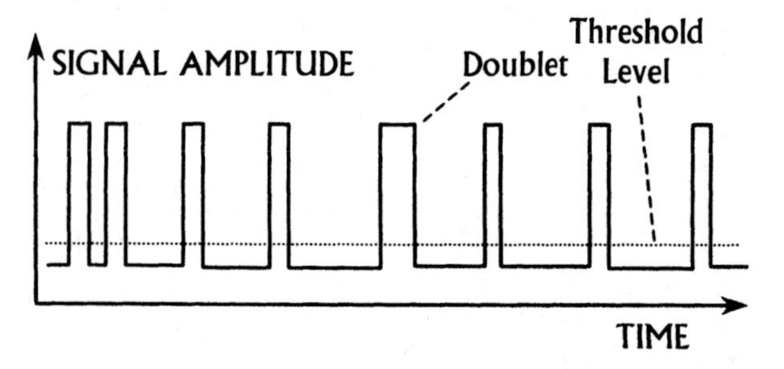

Figure 1-7. Idealized plot of signal amplitude vs. time representing a scan at constant speed along the cell deposition line of Figure 1-6; only signals corresponding to the "black" cells in that figure are shown.

For the moment, we can make believe the "gray" cells in Figure 1-6 aren't there; the simulated detector signal shown in Figure 1-7 only goes above threshold when it scans a "black" cell. This would actually happen if, for example, the slide were illuminated with blue light, and the "gray" and "black" cells were, respectively, unstained and stained with a green fluorescent dye. Figure 1-7 could then represent the electrical signal from a photodetector with a green filter in front of it.

When we look at the slide by eye, we don't scan very rapidly, and we almost never scan at uniform speed, so we don't instinctively relate what we see to the exact time at which we see it. When we scan with a cytometer, it is at least an advantage, and often an imperative, to scan at a constant speed, putting the times at which signals from objects appear at the detector output(s) in a fixed and precise relation to the positions of the objects in space.

In constructing Figure 1-7, the assumption was made that both the illumination intensity in the field of view and the scanning rate remained constant. If we look at the signal amplitude in the figure, it remains at a relatively low **baseline** level most of the time, and there are eight **pulses** during which the amplitude rises to a higher level and returns to the baseline value after a brief interval. If we glance up from Figure 1-7 to Figure 1-6, we notice that the positions of the pulses in time correspond to the positions of the black cells on the slide.

Flow Cytometry: Quick on the Trigger

The signal(s) used to detect cells' presence in the field of view (also called the **measurement point**, **region**, **station**, or **zone**, or the **analysis point**, **interrogation zone or point**, or **observation point**) of a cytometer is (are) called **trigger signal(s)**. The amplitude of a trigger signal must be substantially different in the cases in which a cell is and is not present at the observation point; in other words, it must be possible to define a **threshold level** above which the amplitude will invariably rise when a cell is present. If we pick a threshold level indicated by the dotted horizontal line in Figure 1-7, we see that the signal shown in the figure can serve as a trigger signal; its amplitude is well above the threshold level whenever a cell or cells are present in the field of view, and comfortably below that level when the field of view contains no cells.

Now, suppose that, instead of scanning cells deposited in a line on a slide or tape, we confine cells to the center of a flowing stream, and look at that through a microscope. We'll get rid of the gray cells this time, and only consider the black ones. And, if we want to draw a schematic picture of this, what we get is Figure 1-6, except that the gray cells aren't there, and the arrow indicates "Flow Direction" instead of "Scan Direction." Instead of defining the boundaries of the cell deposition area, the dotted lines define the diameter of the core stream containing the cells. We have sneakily built ourselves a **flow cytometer**.

Of course, if we were actually looking at the stream of cells in a flow cytometer, it would probably be flowing fast enough so that we couldn't distinguish the individual cells as they went by; remember that the visual system makes a "movie" out of images displayed at rates of 25-30/second (/s). Most photodetectors don't have this problem; they can respond to changes in light intensity that occur in nanoseconds (ns). So we could get a signal pretty much like the signal in Figure 1-7 out of a photodetector in a flow cytometer; the major difference would be in the time scale.

When scanning a slide by eye, we are apt to take at least 100 milliseconds (ms) to examine each cell; slide-scanning apparatus is substantially faster, producing pulse durations of hundreds of microseconds (μs) or less. Flow cytometers are faster still; most current commercial instruments produce pulses with durations in the range between 0.5 and 12 μs. Thus, the hardware and software responsible for detecting the presence of a cell need to do their job in a relatively short time, particularly in cell sorters, where the cell must be detected and analyzed, and the decision to sort it or not made and implemented, in the space of a few microseconds. If the signal in Figure 1-7 were coming from a detector in a flow cytometer, we could use it as a trigger signal.

Many of the signals of most interest to users of flow cytometers are of very low amplitude. Routine immunofluorescence measurements often require detection of only a few thousand fluorescently labeled antibody molecules bound to a cell surface. In such cases, the signal from the fluorescence detectors may be only slightly above background or baseline levels, and their use as trigger signals is likely to result in an unacceptably high level of **false triggering**, resulting in accumulation of spurious data values, due to the influences of stray light and electronic noise fluctuations. Even in cases when relatively weak fluorescence signals can be used as trigger signals to indicate the presence of stained cells, they will be of no help in detecting unstained cells. It has thus become customary to use a small-angle (forward) light scattering signal as the trigger signal when measuring immunofluorescence; all cells scatter light.

When none of the pulses from cells of interest are expected to be of high amplitude, requiring that a threshold level be set close to the baseline, discrimination of cells from background noise may be improved by using multiple triggers, requiring that two or more signals go above threshold at the same time to indicate a cell's presence. I almost always use forward light scattering and fluorescence as dual trigger signals when working with bacteria.

The Main Event

Looking back at Figures 1-6 and 1-7, though, we can see that there is another catch to triggering; it is not Catch-22, but Catch-2. Two of the black cells in Figure 1-6 are stuck together, and delineated as a "doublet" in that figure; the corresponding pulse, similarly delineated in Figure 1-7, is, though wider than the other pulses, still only a single pulse. Since cells going through a flow cytometer (or cells deposited on a slide) arrive (or appear) at more or less random intervals, there is always the chance that two or more cells will be close enough in space, and their corresponding output signals close enough in time, so that they produce only a single pulse at the detector output. Note that the cells do not have to be physically stuck together for such **coincidences** to occur, they must simply be close enough so that the detector signal does not fall below the threshold value between the time the first cell enters the measurement region and the time the second (or last, if there are more than two) cell leaves it.

When we get technical about what we are really measuring in a flow cytometer (and now is one of those times), rather than saying that a pulse above threshold level represents a **cell**, we say that it represents an **event**, which might correspond to the passage of one cell, or multiple cells, or one or more pieces of noncellular junk capable of generating an equivalent optical/electronic signal, through the system, or which might result from stray light and/or electronic noise or some other glitch in the apparatus.

The Pulse Quickens; The Plot Thickens

There are ways of identifying pulses that result from coincidences; the height, width, and/or area of such pulses is/are typically different from those resulting from the transit of single cells, and, with the aid of appropriate hardware and/or software, it is possible to identify coincidences and correct counts. And now is probably an opportune time for me to confess that the pulses of Figure 1-7 are highly idealized, in that all of the pulses from single cells look pretty much the same; that definitely isn't the way things really are.

In fact, **all** of the information about a cell that can be gotten from flow cytometers is contained in, and must be extracted from, the **height**, or **amplitude**, the **area**, or **integral**, and the **width** and **shape** of the pulses produced at the detector(s) as the cell passes through the measurement region(s). Generally speaking, there isn't much point to doing flow cytometry if you expect all of the cells you analyze to look alike; the usual purpose of an experiment is the characterization of **heterogeneity** within a cell population, and the rest of this book is intended to help you make sure that the differences in pulses you see from cell to cell represent biological differences you are looking for, rather than reflecting vagaries of apparatus, reagents, and technique.

And now, at last, we have gotten our fingers on the pulse of flow cytometry. For the fact is that, while the information in scanning and imaging cytometers ultimately makes its way into the processing electronics in the form of a series of pulses, often referred to as a **pulse train**, it is only in flow cytometers and in the lowest resolution scanning devices that all of the information a detector gets about a cell (or, more accurately, an event) is contained in a single pulse. This was recognized early on as an important and distinctive characteristic of flow cytometry; before the term "flow cytometry" itself was coined in the 1970's, many workers in the field referred to it as **pulse cytophotometry**.

1.4 FLOW CYTOMETRY: PROBLEMS, PARAMETERS, PROBES, AND PRINCIPLES

Since the 1970's, it has become possible for users blissfully unconcerned with the nuts and bolts (or the atoms and bits) of instrumentation to buy flow cytometers capable of extracting more and more pulses from an increasingly diverse variety of objects, ranging downward from eukaryotic cells and microorganisms to organelles and large molecules, and upward to pancreatic islets, *C. elegans*, *Drosophila* embryos, and multicellular plankton organisms.

From reading the manufacturers' brochures and visiting their Web sites, interested researchers and clinicians can learn that it is possible to analyze and sort over a hundred thousand cells per second, to identify rare cells that represent only one of every ten million cells in mixed populations, to simultaneously measure light scattering at two or three angles and fluorescence in twelve or more spectral regions, to measure fluorescence with a precision better than one percent, and to detect and quantify a few hundred molecules of fluorescent antibody bound to a cell surface. It is somewhat harder to discern that it may be difficult or impossible to accomplish two or three of these amazing feats at once. If you're contemplating pushing the envelope, you definitely need to look at the **problem(s)** you're trying to solve, the measurement **parameters** and **probes** with which you can extract the necessary information from the cells, and the **principles** that may allow you to get your answers – or prevent you from getting them. I will take this approach in considering how the technology has gotten to its present state, starting with relatively simple problems and the relatively simple systems for solving them.

Since flow cytometers are designed to analyze single cells in suspension, it is not surprising that their development and evolution have been directed in large part by workers in the fields of hematology and immunology, who deal primarily with cells that are either in suspension, as is the case in blood samples, or relatively easy to get into suspension, as is the case when it is necessary to examine cells from bone marrow or lymphoid tissues or tumors.

In addition to being conveniently packaged, cells from the blood and immune system provide us with a number of models for fundamental biological processes. With the analysis of the genome behind us, we still need the details of differentiation that allow politically sensitive fertilized ova to develop through the politically sensitive embryonic stem cell stage into multicellular organisms who, after some years, can be dropped from the welfare rolls with the blessings of the same legislators who so staunchly defended them at smaller cell numbers. Cells in the blood and immune system develop from a single class of stem cells, which were hypothesized about and sought for years, and were finally identified with the aid of flow cytometry, and we now traffic in blood stem cells for patients' benefit as well as studying the cells' development in the interest of science. Differentiation gone wrong, with the aid of somatic mutation, produces leukemias and lymphomas, and we use flow cytometry both to clarify the biology of neoplasia and to determine the prognosis and treatment in individual instances. The processes of clonal selection underlying both cellular and humoral immune responses provide a picture of evolution at work, as well as examples of a wide variety of mechanisms of inter- and intracellular signaling.

Counting Cells: Precision I (Mean, S.D., CV)

The simplest flow cytometers, and the first to be widely used, solved the **problem** of providing precise counts of the number of cells per unit volume of a sample, without explicitly characterizing the cells otherwise. Such instruments have only a single detector, and, because they measure an **intrinsic parameter**, typically **light scattering** or **extinction** or **electrical impedance**, do not require that the cells be treated with any reagent, or **probe**.

The **sample** used for cell counting may be taken directly from the **specimen** containing the cells, or may be an aliquot of that specimen **diluted** by a known amount, or **dilution factor**. If, for example, the specimen is diluted 1:20 to produce the sample, the dilution factor is 20.

The **principle** of operation of a cell counter is almost embarrassingly simple. An electronic counter is set to zero at the beginning of each run. Next, sample is passed through the system at a known, constant flow rate. As cells go through the measurement system, they produce pulses at the detector output; the count is increased by one whenever the output from the detector goes above the threshold level. Those cells that produce pulses with amplitudes above threshold are counted; those that do not are not. Any particle other than a cell that produces a signal above threshold is counted as a cell; any transient electrical disturbance or noise that causes the sensor output to go above threshold is also counted as a cell.

Although this sounds like a very simple-minded approach, it usually works, can be implemented using relatively primitive electronics, and can deal with thousands of cells per second. And, as will be amply illustrated later in this section, it is relatively easy to get from this point to a flow cytometer that makes one or several additional measurements of cells. The principal requirement is that, in addition to (or instead of) being used to increase the number in the counter, the trigger signal(s) initiate(s) the capture and recording of information about the height, area, and/or width of pulses from one or more detectors.

In the late 1950's and 1960's, the first optical and electronic (Coulter) cell counters reached the market. They were designed to count blood cells; I have already noted that red cell counts were done by setting a threshold high enough to prevent platelets from triggering, and that white cells were counted with red cells, but did not normally introduce significant inaccuracy into the red cell count because of their relatively low numbers. White cell counts were done on samples in which the red cells had been lysed by addition of a chemical such as saponin or one of a number of detergents to the diluent.

Before counters became available, people did cell counts by examining diluted blood (or another cell sample) in a **hemocytometer** under a microscope. A hemocytometer is a specially designed microscope slide with a ruled grid that defines square or rectangular areas, each fractions of a millimeter on a side, and with ridges on either side of the ruled area that insure that the thickness of the layer of diluted blood under the cover slip will be constant (usually 0.1 mm). For a white cell count, blood is typically diluted 1:20 with a solution that lyses red cells and stains white cells; the number of cells in four 1 × 1 mm squares is counted. The total volume of diluted blood counted is therefore 0.4 mm^3, or 0.4 μL. To obtain the count of white cells/mm^3 (the old-fashioned unit used when I was a medical student), one divides this number by 0.4 (the volume counted) and multiplies the result by 20 (the dilution factor). Because red cells are so much more numerous than white cells, blood is diluted 1:200 for red cell counts (without lysis, obviously), and a smaller area of the slide is used for counting.

Poisson Statistics and Precision in Counting

So what's wrong with hemocytometer counts, apart from the fact that they used to be done by slave labor (for which read medical students, or at least those of my generation)? The problem is with the **precision** of the counts. Precision, as was noted on p. 12, refers to the degree to which replicate measurements agree with one another. The precision of a measurement is often characterized by a statistic called the **coefficient of variation (CV)**, which, expressed as a percentage, is 100 times the **standard deviation (S.D.)** divided by the **mean** (and by **mean** I mean the **arithmetic mean**, or **average**, i.e., the sum of the individual measurements divided by the number of measurements). Well, you might say, "What mean and standard deviation? The count is only done once; how much time do you think those overworked medical students can spare?"

Enter another Student; not a 1960's medical student, this time, but a man of an earlier generation named William Sealy Gossett, who published his basic statistical works as "Student" because his employers at the Guinness Brewery worried that their competitors might improve their positions by using statistics if they discovered his identity. He showed in 1907[2317] that, if one actually counted n cells in a hemocytometer (that's before the division and multiplication steps), one should expect the standard deviation of the measurement to be the square root of n (I will use the notation $n^{1/2}$ rather than $\sqrt{n}$ for this quantity for typographic reasons), meaning that the coefficient of variation, in percent, would be $100/n^{1/2}$. We would now say that the statistics of counts conform to the **Poisson distribution**, which was described by Siméon Poisson in 1837[2318], but "Student" was apparently unaware of Poisson's work, and reached his conclusions independently. In fact, the Poisson distribution was only given that name seven years after Gossett's paper appeared[2319]. We will encounter the Poisson distribution in several other contexts related to cytometry, flow and otherwise.

Now, if we consider looking at a sample with a white cell count of 5,000/mm^3, which is in the normal range, the number of cells you would actually have counted in the hemocytometer to obtain that value would be 5,000 divided by the dilution factor (20) and multiplied by the volume (in mm^3) counted (0.4), which works out to 100 cells. The standard deviation would therefore be the square root of 100, or 10; the CV would be 10 percent. If you were dealing with an abnormally low white cell count, say one that you read as 1,250/mm^3, you would only have counted 25 cells; the standard deviation would be 5, and the CV would be 20 percent. And all of this assumes that the counting process is perfect; we know that it isn't, and we also know that other factors, such as dilution and pipetting errors, will further decrease precision. So the precision of a hemocytometer white cell count in the normal range is barely acceptable. Getting a CV of 1 percent, which is more than respectable, would require that you count 10,000 objects, which would be 100 hemocytometers' worth if you were dealing with our original white cell count of 5,000/mm^3. Nobody is going to sit there and do that by eye, but it's a piece of cake for an electronic or optical counter.

A typical hematology counter uses a **constant volume pump**, such as a syringe pump, to deliver sample at a constant flow rate. The flow rate is the volume of sample analyzed per unit time; dividing the number of cells counted per unit time by the flow rate gives the number of particles per unit volume. Blood specimens are usually diluted before being run in a counter, so the raw value must be multiplied by the dilution factor to get the particle count per unit volume of blood. For example, if the counter's sample flow rate is 1 μL/s, and a blood sample is diluted 1:20 (with a solution that lyses red cells) to count white cells, and running the counter for 40 seconds yields a raw count of 10,000 cells, the white cell count in the blood is:

$$\frac{10{,}000 \text{ (\# of cells counted in 40 s)} \times 20 \text{ (dilution)}}{40 \text{ (\# of } \mu\text{L counted in 40 s)}}$$

or 5,000/μL. Since the raw count is 10,000, the standard deviation is 100, and the CV is 1 percent.

Rare Event Analysis: The Fundamental Things Apply as Cells Go By

Many of the tasks in modern cytometry are examples of **rare event analysis**. Examples are looking for primitive stem cells, leukemic cells or cancer cells in blood or bone marrow, for fetal cells in maternal blood, or for transfected cells present at low frequency in a culture. In comparing different samples, it is frequently necessary to determine the statistical significance of small differences between large numbers. Some people seem to think that counting hundreds of thousands or millions of cells lets them beat the Poisson statistics; what's important, however, is the number of cells of interest you count, not the total. Suppose, for example, that you find your cells of interest present at a frequency of 0.04% positives in one sample of 200,000 cells and 0.15% in another. Simple arithmetic tells you that 0.01% of 200,000 is 20 cells, so the first sample has 80 cells of interest and the second has 300. The Poisson standard deviations for the numbers of cells of interest counted would be about 9 for the 80 cells in the first sample and about 18 for the 300 cells in the second. The two values are thus separated by several standard deviations, which is to say that there is a statistically significant difference between them. However, the statistics provide no information as to the source of the difference. If the cells came from the same pot, one would suspect instrumental factors related to data collection and/or analysis, unless there is reason to believe that a process such as differential settling of the rare cell type would change the composition of a sample aliquot with time. A mild degree of paranoia is probably an asset when dealing with rare event analysis.

Poisson statistics apply to counting anything, from cells to photons and photoelectrons, and even to votes. Digressing briefly from rare event analysis to not-so-current event analysis, if 3,000,000 votes are counted, one expects a Poisson standard deviation of 1,732 votes, or roughly 6 parts in 10,000, meaning that if the vote counting process is supposedly even less reliable or accurate than Poisson statistics would predict (Florida's was said to be 99.9% reliable, or accurate to 10 parts in 10,000), neither candidate had a strong claim to having won the state's Presidential vote.

We have a little more control over cell counting than over vote counting. If you count enough cells, you can accurately discriminate between, say, .01% and .02%. If you only count 10,000 cells total, you'd expect to find one cell (and a CV of 100%) in the sample with .01% and 2 cells (CV of 70.7%) in the sample with .02%; so 10,000 cells total is too small a sample to let you discriminate. If you count 1,000,000 cells total, you end up with 100 cells in the .01% sample (10% CV) and 200 cells (7.1% CV) in the .02% sample, and this difference will be statistically significant.

Count Constant Numbers for Constant Precision

The best way to do counts, although almost nobody does them this way, is to always count the same number of cells of interest, which gives you equal precision no matter what the value is. Normally, we do absolute counts by analyzing a fixed volume of blood (or other sample) and percentage counts by analyzing a fixed number of cells. The alternative is to decide on the level of precision you want – suppose it is 5%. Then you have to count 400 cells (the square root of 400 is 20, and 100/20 = 5). What you do is measure the volume of sample (in the case of absolute counts), or the total number of cells (in the case of percentage counts), which has to be analyzed to yield 400 of the cells of interest. If the cells of interest are at .01%, you'll have to count 4,000,000 cells total to find your 400 cells of interest; if they are at 1%, you'll only have to count 40,000 cells, but, instead of the .01% value being much less precise than the 1% value, both will have the same 5% precision. The down side of doing things this way is that it may require some reprogramming of the apparatus, and probably uses more reagent, but, if you want good numbers, there is simply no better way to get them.

Alternative Counting Aids: The Venerable Bead

As it happens, most fluorescence flow cytometers do not use constant volume pumps for sample delivery, nor do they provide an alternative means of measuring the sample volume flow rate with sufficient precision to allow calculation of cell counts per unit volume by the method described above. Carl Stewart, being a leukocyte biologist, must have felt deprived of one of the major tools of his trade when he arrived at Los Alamos National Laboratory many years ago and discovered that the very fancy fluorescence flow cytometers built there did not provide a cell count. He and John Steinkamp solved that problem by adding fluorescent beads at known concentrations to cell samples[1539]. If you have a bottle full of beads that contains a known number of beads per unit volume, adding a known volume of bead suspension (and it had better be well-mixed bead suspension) to a known volume of cell sample allows you to calculate the number of beads per unit volume in the sample. You can then run the sample for an arbitrary length of time, tallying the total numbers of beads and cells counted. The cell count per unit volume is then given by:

$$\frac{\text{\# of cells counted} \times \text{\# of beads per unit volume}}{\text{\# of beads counted}}$$

and the number of cells per unit volume in the original material from which the cells were taken can be obtained by multiplying by the dilution factor, as in previous examples.

There are a few caveats here. If the determination of the concentration of beads per unit volume is done by a relatively imprecise method (Stewart and Steinkamp used a hemocytometer), the precision of the cell count cannot be improved by counting large numbers of cells and beads. One must also take into account the frequency of clumps and coincidences among both cells and beads, which affect the **accuracy** of the count, i.e., the degree to which the measured value agrees with the "true" value. And, of course, the cytometer must be capable of accurately identifying and counting both cells and beads.

Addition of beads to the sample is now widely practiced in the context of counting CD4 antigen-bearing (**CD4-positive**, or **CD4+**) T lymphocytes in HIV-infected individuals. The identification of these cells is most often done by staining with fluorescently labeled monoclonal anti-CD4 antibody (and, usually, at least one other monoclonal antibody labeled with a different fluorescent label). Before counting beads became available, the standard procedure was the so-called "two-platform" method, in which a hematology counter with a constant volume sample feed is used to obtain both the total white cell count per unit volume of blood, and the differential white cell count, which includes the percentage of lymphocytes among the white cells; the number of lymphocytes per unit volume is then calculated. The fluorescence flow cytometer is used to define the lymphocyte population and the fraction of that population represented by CD4+ T-cells, allowing calculation of the number of these cells per unit volume. Using counting beads, the procedure can be done on a single platform, i.e., the fluorescence flow cytometer, and this appears to improve accuracy.

And Now to See with Eye Serene the Very Pulse of the Machine: Display, Digitization, and Distributions

In general, people who use flow cytometry want to know more about their samples than how many cells are contained in each milliliter, and that translates into getting more information about the signal pulses than whether their amplitudes exceed the threshold level. In a single-parameter electronic (Coulter) counter, the heights of pulses are proportional to the volumes of the cells passing through. However, whereas only relatively simple circuitry, triggered by the rise above threshold in the signal, is required to increment and store the cell count, more complex hardware and software are needed to capture and store measured values of the volumes of cells. Information about the measured particle may be extracted from the peak amplitude (height), the integral (area), the duration (width), and the shape of signal pulses.

The earliest electronic counters did not come equipped with the means to collect and display **distributions**, i.e., **histograms**, or **bar graphs**, of cell volumes; investigators interested in such information acquired it by feeding the pulse train from a counter into a gadget called a **pulse height analyzer**, a hardwired digital computer originally used by nuclear physicists to measure and discriminate among gamma ray energies.

The prerequisite to pulse height analysis, and to just about anything else that one might want to do in the way of data analysis in cytometry, is the conversion of information from an **analog** form, usually a **voltage** representing one of the pulse characteristics mentioned above, to a **digital** form, i.e., a **number**, using a device appropriately named an **analog-to-digital converter (ADC)**. Digital processing in the flow cytometers of the early 1970's was pretty much restricted to the use of pulse height analyzers, which had the disadvantage that their single ADCs (ADCs were expensive in those days) could only provide information on one measured quantity, or parameter, at a time. It was, however, possible to use live display and storage **oscilloscopes**, without benefit of digitization, to provide simple **dot plots** showing the interrelation of two parameters.

The pulses produced during a cell's passage through the measurement system typically last for only a few microseconds at most (making them veritable "phantoms of de light"), and, until recently, the only ADCs that could practically be used in flow cytometers required more time than this to digitize signals. As a result, it was necessary to use hybrid circuits, which combine analog and digital electronics, to store the appropriate analog values for long enough to permit analog-to-digital conversion. These **peak detector, integrator**, and **pulse width measurement** circuits must be **reset** as each particle passes through the illuminating beam, allowing new analog signal levels to be acquired; it is then necessary to **hold** their outputs at a constant level until digitization is complete. The "reset" and "hold" signals must be delivered to the analog storage circuits at the proper times by additional hybrid "**front end**" electronics, which compare one or more trigger signal levels with preset threshold values to determine when a cell is present.

Luckily, a flow cytometer is an example of what is known as a **low duty cycle** device. Even when a sample is being run, cells pass the sensors rather infrequently; what goes by the sensors, most of the time, is the water or saline suspending medium, meaning that a certain amount of **dead time**, during which the pulse measurement circuits are occupied with data from one cell and cannot respond to signals from a second, is tolerable. Because cells arrive at random times, rather than at fixed intervals, **coincidences**, when a second cell arrives before processing of signals from the first is complete, are inevitable. The probability of coincidences can be calculated from – guess what – the **Poisson distribution**, and, while they cannot be eliminated entirely, it is possible to reduce them to acceptable levels by limiting the number of cells analyzed per unit time in accordance with the instrument's dead time.

Once held signals have been digitized, further analysis is accomplished with a digital computer, which, in modern instruments, is typically either an Intel/Microsoft-based or Apple Macintosh personal computer. The necessary software is now available from both flow cytometer manufacturers and third parties, in some cases at no cost. In recent years, as inexpensive, fast, high-resolution ADCs have become available (due largely to the needs of the consumer electronics and telecommunications markets), **digital signal processing (DSP)** hardware and software have replaced analog and hybrid circuits for peak detection, integration, and pulse width measurement, and for some other common tasks in flow cytometry, such as **fluorescence compensation** and **logarithmic conversion** of data. There will be a great deal more said about this further on in the book; for now, however, we will go back to another old problem, its old and newer solutions, and their implications for science, medicine, and society.

DNA Content Analysis: Precision II (Variance)

Most users of flow cytometers and sorters have at least a passing acquaintance with measurements of the **DNA content** of cells and chromosomes, which can be done rapidly and precisely by flow cytometry using a variety of fluorescent stains.

As a rule, all normal diploid cells (nonreplicating or G_0 cells and those in the G_1 phase of the cell cycle) in the same eukaryotic organism should have the same DNA content; this quantity is usually expressed as 2C. DNA syn-

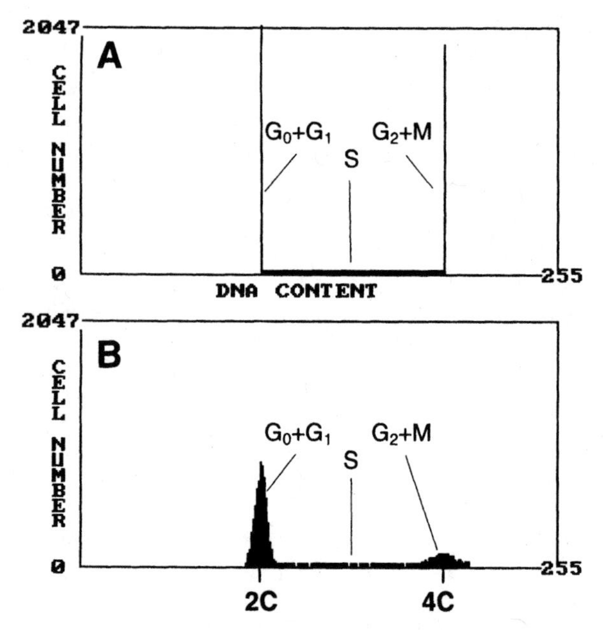

Figure 1-8. Ideal (A) and "real" (B) DNA content distributions, with the same ratios of $(G_0+G_1)/S/(G_2+M)$ cells represented in both.

thesis during the S phase of the cell cycle results in an increase in cellular DNA content, which reaches 4C at the end of S phase and remains at this value during the G_2 phase and during mitosis (M phase), at the completion of which the original cell has been replaced by two daughter cells, each of which has a DNA content of 2C. The haploid germ cells have a DNA content of approximately C; there are approximately equal populations of sperm with DNA content slightly greater than and slightly less than C due to differences in the DNA content of male and female sex chromosomes.

An idealized DNA content **distribution**, that is, a **bar graph** or **histogram** of values of DNA content that would be expected to be observed in a population of cells, some or all of which were progressing through the cell cycle, is shown in panel A of Figure 1-8, above. A "real" distribution, actually synthesized by a mathematical model, but more like those actually obtained from flow cytometry, appears in Panel B. Real (really real) DNA content distributions always exhibit some variance in the G_0/G_1 peak, which may be due to staining procedures, to instrumental errors, and/or to cell-to-cell differences in DNA content. The belief in the constancy of DNA content in diploid cells has been strengthened by the observation that the variances have diminished in magnitude with improvements in preparative and staining techniques and in instrumentation since the first DNA content distributions were published in the 1960's.

When I used the word "variance" in the paragraph above, I meant it, and you probably took it, to denote variability from measurement to measurement. However, the term also has a defined (and related) meaning in statistics; the **variance** of a set of measurements is the sum of squares of the differences between the individual measurements and the **arithmetic mean**, or **average**, divided by one less than the number of measurements. In fact, the statistical **variance** is the square of the **standard deviation**, or, to put it more accurately, the standard deviation is the square root of the variance, and is calculated from it instead of the other way round. For purposes of this discussion, and in most of the rest of the book, I will try to use "variance" to mean the statistical entity unless I tell you otherwise. I may slip; word processors have spelling checkers and grammar checkers, but not intention checkers.

The Normal Distribution: Does the Word "Gaussian" Ring a Bell?

Although the number of cells counted does have some effect on the observed variance of a set of measurements, we are not dealing with Poisson statistics here; the variance of a Poisson distribution is not independent of the mean, but is always equal to it. The peak representing the G_0 and G_1 phase cells of a real DNA content distribution is generally considered to be best approximated by what statisticians define as a **normal** or **Gaussian** distribution, sometimes popularly known as a **bell curve**. The normal distribution is **symmetric**; the arithmetic **mean**, the **median** (the value separating the upper and lower halves of the distribution), and the **mode** (the highest point, or most common value) coincide. The coefficient of variation (CV) (which, you may recall, is expressed in percentage terms as 100 times the S.D. divided by the mean) remains a valid measure of precision, but there is an obvious problem in calculating the CV for a G_0/G_1 peak in a DNA content distribution. The peak falls off as one would expect on the left (low) side, but, on the right (high) side, it merges into the part of the distribution made up of S phase cells, and there isn't a convenient way to decide where the G_0/G_1 cells leave off and the S cells begin.

Because the anatomy of the normal distribution is well known and predictable, we have a statistical trick available to us. The width of the distribution between the two points on the curve at half the maximum (modal, mean, median) value, often referred to as the **full width at half maximum (FWHM)**, is 2.36 standard deviations, and the width between the two points at 0.6 times the maximum value is very nearly two standard deviations.

Binned Data: Navigating the Channels

The process of analog-to-digital conversion that occurs in a pulse height analyzer or in a modern flow cytometer's computer-based data acquisition and analysis system puts data into **bins**, to which we frequently refer as **channels**. These **binned data** are used to compile distributions of measured values of cellular parameters. The distributions in Figure 1-8 are broken into 256 channels, which, by convention, are numbered from 0 to 255. That is the number of bins, or channels, into which an 8-bit ADC distributes its output; an ADC with *m* bits resolution will have 2^m possible outputs, which, by convention, would be described as channels 0 to 2^{m-1}. Although the outputs of ADCs are often the same unsigned binary numbers between 0 and 2^{m-1} that denote the channel numbers, ADCs with outputs in different binary formats are not uncommon. For our purposes, it is safest to think in terms of channel numbers, and leave the raw binary formats to the engineers and computer people.

Suppose that the maximum value, i.e., the largest number of cells, in the G_0/G_1 peak of such a distribution is 500 cells, occurring at channel 100, and that channels 97 and 103 each contain 300 cells (that is, 0.6 times the maximum number, 500). It is assumed here that each of the channels between 98 and102 contains 300 or more cells, and that each of the channels below 97 and above 103 contains fewer than 300 cells. The width of the distribution at 0.6 times the maximum value, representing two standard deviations, is then 7 channels, one standard deviation is 3.5 channels, and the CV, expressed as a percentage, is 100 × (3.5/100), or 3.5 percent. It is obviously easier for most people to calculate a CV in their heads using the width at 0.6 times maximum than it is using the width at half maximum, and a real piece of cake if you set the gain so that the maximum value ends up at channel 100, but we've all got calculators, anyway.

So what's the big deal about precision in DNA content measurement? To appreciate this, we go back to the 1960's again. The first cell counters had become available, and they were being used for counting and sizing blood cells. The 1960's also saw a great deal of progress in the field of tissue culture, resulting in substantial numbers of investigators having ready supplies of cells other than blood cells that were either in suspension or could conveniently be put into suspension. People became interested in the details of the cell cycle in cells derived from healthy tissues and from tumors, and in the effect of drugs on the cell cycle.

Once it became convenient to culture cells, it was possible to observe enough mitotic figures to establish that humans had 46 chromosomes, and not 48, as had once been believed, and to establish that cells from many tumors had more or fewer chromosomes, whereas cells from others had what appeared to be chromosomal deletions and translocations. This would mean that the amount of DNA in G_0/G_1 cells from a tumor could be different from the amount in G_0/G_1 cells from the normal stromal elements found in the tumor, potentially providing at least a means of identifying the tumor cells, and, possibly, an objective measurement with prognostic implications.

The catch here is that, as the difference you are trying to detect between two populations becomes smaller, you need better and better precision (lower CVs) in the measurement process. Generally speaking, two populations are resolvable if their means are a few standard deviations apart. If a tumor cell has one or two small chromosomes duplicated, adding, say, two percent to its G_0/G_1 DNA content, you would need a measurement process with a CV well under one percent to resolve separate G_0/G_1 peaks, although you might get a hint of the existence of two populations in a tumor specimen by observing broadening and/or **skewness (asymmetry)** in the peak of a distribution measured with a less precise process. A triploid tumor population, with 50% more DNA than was found in stromal cells, could, of course, be resolved using very imprecise measurements.

DNA Content: Problem, Parameter, Probes

So, the **problem** became one of measuring DNA content with reasonably high precision. It was then necessary to find a suitable measurement **parameter** to solve it. Although Caspersson, in his microspectrophotometers, and Kamentsky, in his early flow cytometers, had used absorption at 260 nm for nucleic acid content measurement, the absorption measurements were difficult to make (among other things, they required special, very expensive quartz optics, because the UV wavelength used is strongly absorbed by glass), and not precise enough to detect small differences.

In the 1920's, Feulgen[35] developed a staining method that coupled a dye to the backbone of the DNA molecule, allowing DNA content in cells to be quantified by measuring absorption of visible light, but some fundamental problems with absorption measurements still limited the precision of DNA analysis. However, in one of the first publications describing fluorescence flow cytometry, in 1969, Van Dilla et al, at Los Alamos National Laboratory, reported the use of a modified **Feulgen procedure**, with fluorescent stains (acriflavine and auramine O) and an argon laser source flow cytometer, to produce DNA content distributions with a coefficient of variation of 6% for the G_0/G_1 cell peak[79]. The Feulgen staining procedure was relatively technically intensive, due to its requirement for fairly elaborate chemical treatment of the cells, and the search for dyes that were easier to use began almost immediately. The first step in this direction was taken in 1969, when Dittrich and Göhde published a relatively sharp DNA content distribution obtained using their arc source flow cytometer to measure the fluorescence of fixed cells stained with **ethidium bromide**[83]. Thus, **fluorescence** became the **parameter** of choice for DNA content measurement.

Ethidium, which increases its fluorescence about thirtyfold when intercalated into double-stranded DNA or RNA, quickly replaced the fluorescent Feulgen stains as the **probe** of choice, and was then largely supplanted by a close chemical relative, **propidium**[217], which remains widely used as a DNA stain. Both dyes require that the cell be fixed, or that its membrane be permeabilized, in order to achieve good stoichiometric staining; they are frequently used to stain nuclei released from cells by treatment with one of a variety of **nonionic detergents**, such as Nonidet P-40 or Triton X-100[223]. Precise measurement of DNA in whole cells, and the best precision measurements in nuclei, require treatment of the sample with RNAse to remove any residual double-stranded RNA.

Once cell sorters became available, in the 1970's, it was realized that a dye that could enter living cells and stain DNA stoichiometrically would make it possible to sort cells in different phases of the cell cycle and analyze their subsequent biological behavior and/or their chemical composition. Several *bis*-benzimidazole compounds originally synthesized as antiparasitic drugs by Hoechst AG turned out to meet these requirements[238]; the one that has been most widely used, by far, is **Hoechst 33342**[239]. This dye, like the other Hoechst dyes, has two characteristics that limit its use in some situations. Ultraviolet light is required to excite its blue fluorescence, preventing its use in the majority of fluorescence flow cytometers, which are equipped only with a 488 nm (blue-green) argon ion laser as a light source. And, although Hoechst dye staining is highly specific for DNA, the dyes, which do not intercalate but instead bind to the minor groove of the macromolecule, are selective for sequences of three **adenine-thymine (A-T)** base pairs[268]. The latter characteristic is disadvantageous for such applications as DNA content determination in plants, which is widely used as an aid in classification of species, because the Hoechst dyes would yield different results for two species having the same amount of DNA but different **base compositions**, i.e., different ratios of A-T and **G-C (guanine-cytosine)** base pairs. However, the base specificity of the Hoechst dye is an advantage in other circumstances; the combination of the A-T-selective Hoechst 33258 and G-C selective, DNA-specific dyes such as **chromomycin A_3** and **mithramycin**[230], has been used to stain chromosomes from humans and other species, enabling chromosomes with similar total DNA content but different base composition to be distinguished and sorted separately[278]. **High-speed sorting** of dual-stained human chromosomes[904] provided a valuable set of DNA libraries in the early phases of the Human Genome Project, but I'm getting ahead of myself. We can't get into that until we take at least a first look at one- and two-parameter **data displays**.

One-Parameter Displays: Pulse Height Distributions

The cells represented in Figures 1-9 are from the CCRF-CEM T-lymphoblastoid line. They were incubated with **Hoechst 33342**, which, as has already been mentioned, stains DNA stoichiometrically (neglecting, for the moment, differences in base composition). The cells were also exposed to **fluorescein diacetate (FDA)**, a nonfluorescent ester of fluorescein, which should more properly be called diacetylfluorescein but which almost never is. Both compounds are taken up by living cells; once inside cells, FDA is hydrolyzed by nonspecific esterases to **fluorescein**, which exhibits intense green fluorescence when excited with blue or blue-green light, and which, because of its anionic character, is retained in cells for minutes to hours. The cells were measured in a flow cytometer with two separated laser beams; they were first illuminated by a UV laser beam, and the blue fluorescence of Hoechst 33342 (panel A of Figure 1-9) excited by this beam was used as the trigger signal. The cells then passed through the 488 nm beam of a second laser, which provided excitation for the fluorescein fluorescence signal (panel B of Figure 1-9). The histograms of the distributions were collected at different times during a single sample run, using a multichannel pulse height analyzer. The horizontal axis of each histogram indicates fluorescence intensity, on a 512-channel scale; the vertical axis of each histogram represents the number of cells with the corresponding fluorescence intensity. This, by the way, is not a historically informed modern performance on period instruments; the histograms are from around 1980, when one pulse height analyzer and a storage oscilloscope (see Figure 1-11) were all I had to work with for data capture and analysis.

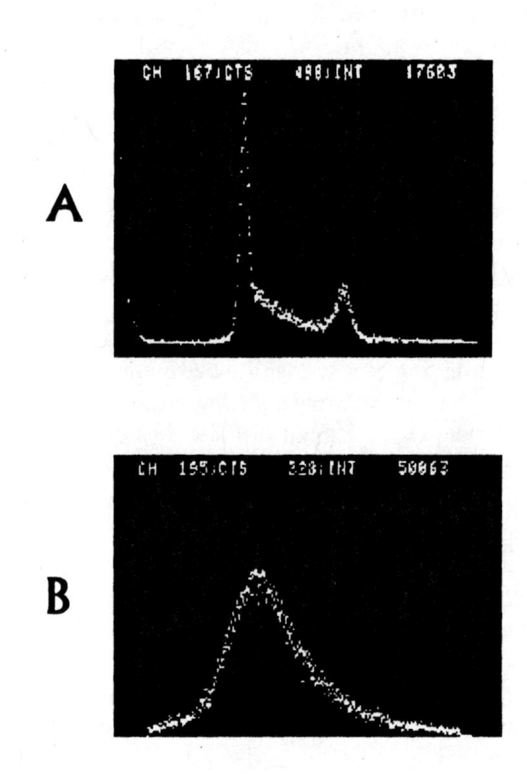

Figure 1-9. Two single parameter histogram displays from the oscilloscope screen of a multichannel pulse height analyzer. A: Fluorescence of the stoichiometric DNA stain Hoechst 33342. B: Fluorescence of intracellular fluorescein. Cells from the same sample are represented in the two histograms.

That said, the data are pretty respectable; their quality is determined primarily by the design and the state of alignment of the flow cytometer optics and fluidics. The CV of the G_0/G_1 peak of the histogram is about 3%, which is excellent for live cells stained with Hoechst 33342.

I have often described sharp peaks, such as the G_0/G_1 peak of a DNA content distribution, as being shaped like a **needle**. Such distributions are not common in flow cytometry data, unless one happens to be analyzing populations of objects that have been intentionally designed to be highly homogeneous, such as the fluorescent plastic microspheres used for instrument alignment and calibration. Although nuclei stained for DNA content, which exemplify one of the best of nature's own quality control processes, yield needles, the shapes of the distributions of most cellular parameters are closer to that of the fluorescein fluorescence distribution in panel B of Figure 1-9, which resembles a **haystack**, in which it will be unlikely to find a needle.

The pulse height analyzer used to accumulate and display the histograms shown in Figure 1-9 is a specialized computer system that also incorporates some of the features of a flow cytometer's front end electronics and a peak detector. It can accept as input signal a train of pulses ranging in height from 0 to 10 V, using the input signal or another pulse train as a trigger signal, with a threshold set by the operator. Once a signal above threshold is encountered, the peak height is captured by the peak detector, and the signal is digitized by an ADC that, in this instance, produces a **9-bit** output, i.e., a number between 0 and 511.

The pulse height analyzer stores its histograms in 512 memory locations. The **program**, or, more accurately, the procedure, or **algorithm**, for calculating a histogram is fairly simple. First, set the contents of all memory locations to zero. Then, every time a new numerical value emerges from the A-D converter, add one to the contents of the memory location corresponding to that numerical value. Stop when the total number of cells reaches a preset value.

This particular analyzer actually had several options on when to stop: at a preset value for the total number of cells, or for the number of cells in a single channel or memory location, or for the number of cells in a **region of interest**, a range of contiguous channels settable by the operator. It also had some refinements in its display; it would show the channel location of a cursor (CH) and the number of counts in that channel (CTS), as well as the total number of counts in the region of interest (INT). The histogram, sans numbers, could also be drawn on an X-Y plotter; several could be compared by eye in overlays using different color pens.

Pretty much the same algorithm is used for histogram computation today as was used in the analyzer. The difference is that in 1973, when the pulse height analyzer was built, a small startup company called Intel had just begun to ship samples of the first 4-bit microprocessor, and computer memory costs were on the order of 10 cents a byte. The smallest minicomputers available cost around $10,000. The pulse height analyzer didn't have a central processing unit, couldn't process alphanumeric data, couldn't calculate a sine or a logarithm; it used special-purpose hardware to implement the algorithm, and, even at that and even then, it sold for about $5,000. I'm not sure you can even buy a standalone pulse height analyzer today; instead, there are boards containing the necessary front end electronics that plug into standard personal computers. But, even if I could have afforded a second pulse height analyzer in 1980, it wouldn't have helped me do correlated analyses of two parameters.

Mathematical Analysis of DNA Histograms: If It's Worth Doing, It's Worth Doing Well

It was noted on p. 22 that, when one looks at a DNA content histogram, there isn't a convenient way to decide where the G_0/G_1 cells leave off and the S cells begin; there also isn't a convenient way to decide where the S cells leave off and the G_2/M cells begin, or identify debris and cell aggregates in a sample, and things get worse in tumor samples.

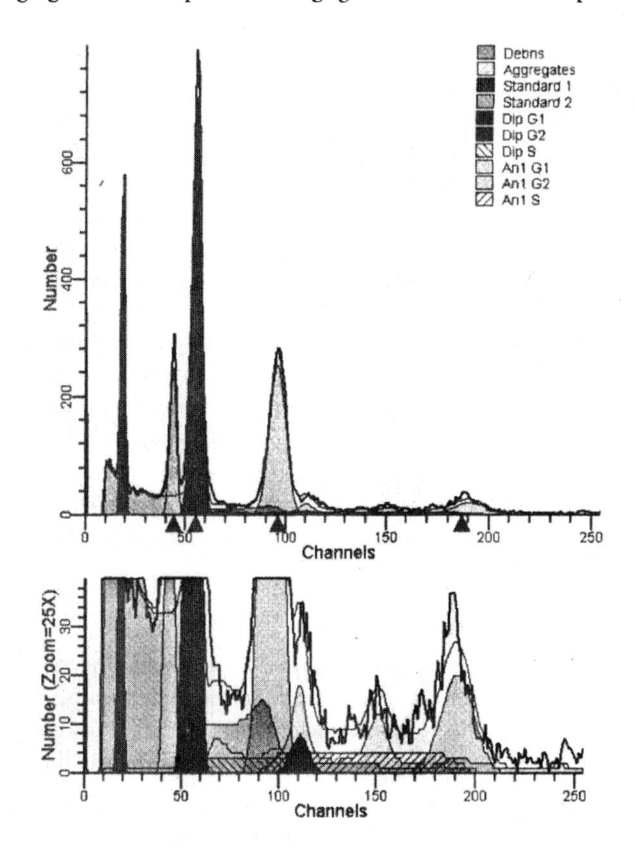

Figure 1-10. Use of a mathematical model to determine fractions of DNA-aneuploid breast cancer cells and normal stromal cells in different cell cycle phases in a sample from a tumor. Chicken and trout erythrocytes are added to the sample to provide standards with known DNA content. Contributed by Verity Software House.

Although tumor cells with abnormal numbers of chromosomes are correctly described as **aneuploid**, a tumor in which the neoplastic and stromal G_0/G_1 cells have different DNA contents is, by convention[741], referred to as **DNA aneuploid**. Mathematical models for DNA histogram analysis have been developed over the years, first, to estimate the fractions of cells in different cell cycle phases in an otherwise homogeneous population, and, later, to determine cell cycle distributions of both stromal cells and DNA aneuploid tumor cells. Further refinements allow for modeling of cellular debris and cell aggregates, enabling them to be largely excluded from analysis. An example of the application of one of the more sophisticated such models (ModFit LT™, from Verity Software House) appears in Figure 1-10.

The earliest publications on fluorescence flow cytometry[79,83] dealt with DNA analysis, and cancer researchers and clinicians began to use the technique almost immediately to attempt to establish the prognostic significance of both DNA aneuploidy and the fraction of cells in S phase in tumors. The development of a method for extracting nuclei from paraffin-embedded tissue for flow cytometric analysis of DNA content[610,891] allowed these issues to be approached by retrospective as well as by prospective studies. By the early 1990's, DNA analysis of breast cancer had come into reasonably widespread clinical use as a prognostic tool. However, in 1996, the American Society of Clinical Oncology recommended against the routine use of flow cytometry in breast cancer[2320], and the volume of specimens analyzed has declined substantially since then. Bagwell et al[2321] have recently demonstrated, based on reanalysis of data from several large studies of node-negative breast cancer, that, after application of a consistent method of analysis and adjustment of some previously used criteria, DNA ploidy and S phase fraction again become strong prognostic indicators. This is not the only publication that shows that how and how well a laboratory test is done can profoundly affect its clinical significance, and that message is important whether or not flow cytometric DNA analysis comes back into vogue.

Linear Thinking

Noncycling cells with known DNA content, such as chicken and trout erythrocytes, can be added to a sample to serve as standards, as was done in the sample shown in Figure 1-10. Such standards are useful in establishing the **linearity** of the instrument and data acquisition system. A system is said to be linear when a proportional change in its input changes its output by the same proportion. In a simple DNA histogram, if the system is linear, and the mean or mode of the G_0/G_1 peak, representing cells with 2C DNA content, is at channel n, the mean or mode of the G_2/M peak, representing cells with 4C DNA content, will be at channel $2n$, or, because of the inherent error of ADCs, within one channel of channel $2n$. In practice, somewhat larger degrees of nonlinearity can be tolerated and corrected for, provided the nonlinearities are stable over time.

Lineage Thinking: Sperm Sorting

Since X- and Y-chromosomes in most species do not contain the same amount of DNA, one would expect a highly precise fluorescence flow cytometer to be able to distinguish them. The necessary precision has been achieved in high-speed sorters by modifications to flow chamber geometry and light collection optics, and sperm vitally stained with Hoechst 33342 have been successfully sorted by sex chromosome type and used for artificial insemination and/or *in vitro* fertilization in animals and, more recently and with a great deal more attention from the media, in humans[2322-3]. Gender selection in humans using sorted sperm, while still under attack from some quarters, is now deemed preferable to other methods that involve determination of the sex of preimplantation embryos. Gender selection in animals using the same methodology appears not to have generated as much controversy as has introducing a foreign gene or two into tomatoes, and may yet become big business[2324].

Two-Parameter Displays: Dot Plots and Histograms

Histograms of the individual parameters do not provide any indication of correlations between Hoechst 33342 and fluorescein fluorescence values on a cell-by-cell basis. In modern flow cytometers, computer-based data acquisition and analysis systems make it trivial to capture, display, and analyze correlated multiparameter data from cells, but, until the 1980's, many instruments could only obtain correlated data on two parameters in the form of a display on an oscilloscope. Such a display was called a **cytogram** by Kamentsky and is now more commonly known as a **dot plot**. One showing both Hoechst 33342 and fluorescein fluorescence values for the cells from the same sample analyzed to produce Figure 1-9, appears in Figure 1-11, below.

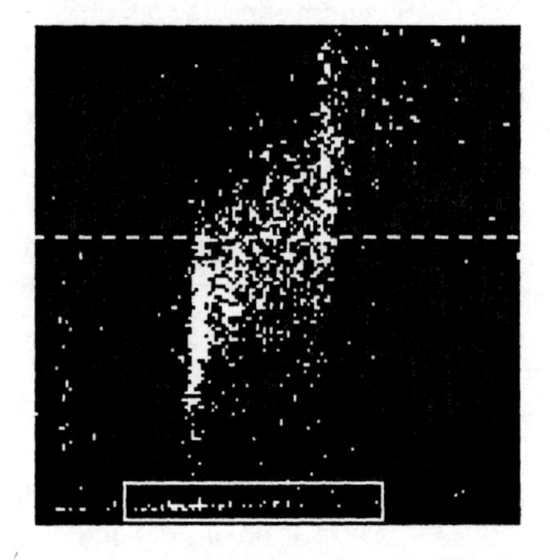

Figure 1-11. Dot plot (cytogram) of Hoechst 33342 fluorescence (x-axis) vs. fluorescein fluorescence (y-axis) for CCRF-CEM cells from the same sample shown in Figure 1-9. Cells in the box are dead; the dotted line is explained below.

Dot plots were the first, and remain the simplest, multiparameter displays in cytometry, and, as we shall presently see, tell us more than we could find out simply by looking at single-parameter histograms. In order to demonstrate this point, we should keep the histograms of Figure 1-9 in mind as we proceed.

In order to appreciate why two parameters are better than one, we need only look at the dot plot in Figure 1-11. One of the first things we notice is that cells with higher Hoechst dye fluorescence intensities, i.e., cells containing more DNA, show higher fluorescein fluorescence intensities. This shouldn't be surprising; if cells didn't get bigger during the process of reproduction, they'd eventually vanish, and it would seem logical that the amounts of FDA cells would take up, and the amounts of fluorescein they would produce and retain, would be at least roughly proportional to cell size. The horizontal dotted line across the dot plot defines two ranges of fluorescein fluorescence values that almost completely separate the diploid and tetraploid populations.

Even more significant, but less obvious to the untrained eye, are the cells represented in the box near the bottom of the cytogram. These exhibit Hoechst 33342 fluorescence, but not fluorescein fluorescence; they are **dead cells**, or would be so defined by the criteria of a **dye exclusion test**. Such tests actually detect a breach in the cell membrane, which allows dyes such as propidium iodide and Trypan blue, which normally do not enter intact cells, to get in. In this case, the hole in the membrane allows the fluorescein produced intracellularly to leak out very rapidly. As a result, the dead cells exhibit little or no fluorescein fluorescence; their Hoechst dye fluorescence intensities remain indicative of their DNA content.

Dot plots, then, could readily generate an appetite for multiparameter data analysis capability which, given the state of instrumentation and computers in the early days of flow cytometry, was not readily satisfied. A few people could afford what were called **two-parameter pulse height analyzers**. These devices could produce distributions tabulating the number of events (cells, in this case) corresponding to each possible pair of values for two variables. They were about ten times the price of single-parameter pulse height analyzers; they also didn't have great resolution, due to the high cost of memory. Even if the two variables were digitized to only 6 bits' precision, with each yielding a number between 0 and 63, storage of the two-parameter, or **bivariate**, distribution would require 64 × 64, or 4,096, memory locations. However, much of the information contained in bivariate distributions could be obtained, at much lower cost, by adding relatively simple **gating electronics** to the circuitry used to generate dot plots.

Multiparameter Analysis Without Computers: Gates Before Gates

Multiparameter analysis and **gating** may be the most important concepts in flow cytometry. Overall progress in the field was undoubtedly slowed during the 1970's and early 1980's because many of the people studying the really interesting biological problems didn't have either information about or access to the tools needed to implement even relatively simple multiparameter analysis and gating, let alone the sophisticated schemes that are now commonplace.

A dot plot, made using an oscilloscope, and demonstrating simple electronic gating, is shown in Figure 1-12. In order to understand how the gating works, we need first to consider how the dot plot is generated. An oscilloscope, like a television set, is built around a **cathode ray tube**. Electrons are accelerated toward a screen coated with a **phosphor** by the electric field generated by a high voltage applied between the cathode and the screen. The electrons are focused into a beam by a magnetic field. The trajectory of the beam, i.e., the horizontal and vertical locations at which it will hit the screen, is determined by voltages applied to pairs of **deflection plates** inside the tube. A **modulation voltage** may be applied to control how much of the beam reaches the screen. Electrons that do reach the screen are absorbed by the phosphor, which subsequently emits some of the absorbed energy as light, by the process of **phosphorescence**.

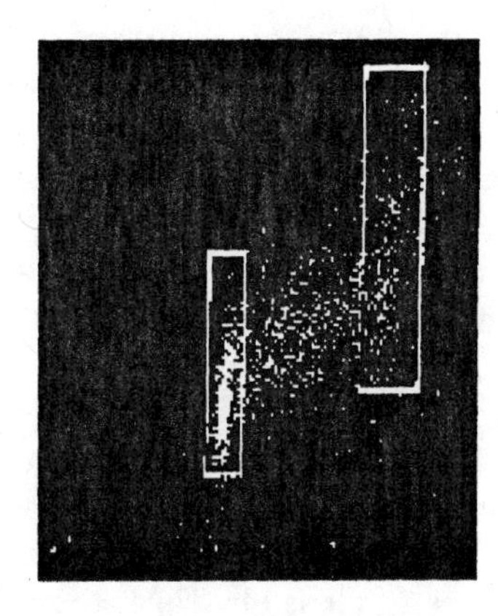

Figure 1-12. Gating regions for counting or sorting set electronically and drawn electronically on an oscilloscope display of a dot plot of DNA content (Hoechst 33342 fluorescence, shown on the x-axis) vs. RNA content (pyronin Y fluorescence, shown on the y-axis) in CCRF-CEM cells.

The dot plot above displays Hoechst 33342 fluorescence on the horizontal or x-axis, and the fluorescence of **pyronin Y**, which stains RNA, on the vertical or y-axis. To generate it, the output from the Hoechst dye fluorescence peak detector was connected to the horizontal deflection plate drive electronics, and the output from the pyronin fluorescence peak detector was connected to the vertical deflection plate drive electronics. The peak detector outputs are both **analog** signals; when applied to the deflection plates, they determine the x- and y- coordinates of the point at which the electron beam will hit the oscilloscope screen. Whether or not an intensified spot, representing the Hoechst 33342 and pyronin fluorescence values associated by the cell, is produced on the screen is determined by the oscilloscope's modulation voltage, which, in this instance, is controlled by what is called a **strobe** signal, generated by the same front end electronics that send the reset and hold signals to the peak detectors.

The strobe signal is a **digital** signal, or **logic pulse**, meaning that its output voltage values are in one of two narrow ranges, or **states**. In this case, voltages at or near about 5 volts (V) represent a "**(logical) 1**," or "**on**," or "**true**" output state, and voltages at or near 0 V, or **ground**, represent a "**(logical) 0**," or "**off**," or "**false**" output condition. The transitions between those two voltage ranges are made rapidly, which, in this instance, means within a small fraction of a microsecond; the interval required is known as the **rise time**.

Some systems use a **positive going** or **positive true** strobe signal, i.e., the strobe output is at ground when the strobe is "off" or "false" and at 5 V when the strobe is "on" or "true"; others use a **negative going** or **negative true** strobe signal, with the output at 5 V when the strobe is off and at ground when the strobe is on. The strobe signal described above is positive true.

The front end electronics are designed so that the strobe signal does not start until the analog signal value in the peak detectors, which can vary continuously between ground and 10 V, has stabilized, and the hold signal is applied to the peak detectors to keep this value from changing during the time the strobe is "on."

When a computer is used for data acquisition and analysis, the beginning, or **leading edge**, of the strobe signal is used to start analog-to-digital (A-D) conversions of the data in the peak detectors; when a dot plot is generated on an oscilloscope, no computer is used, and no digitization is done. The modulation electronics are set so that the beam will reach the screen when the strobe is on and not reach the screen when the strobe is off. Thus, every strobe signal received by the modulation electronics causes a dot to appear on the screen in a position corresponding to the values of the parameters represented on the x- and y-axes.

Early flow cytometers often used **analog storage oscilloscopes**, which incorporated special tubes, with long-persistence phosphors, and associated circuitry that could keep any region of the screen already intensified by the beam "on" until the screen was cleared, or erased, by the user. When an oscilloscope without such storage capacity was used, the dot plot could be recorded by taking a time exposure photograph of the screen.

A dot plot, whether it is recorded on an oscilloscope or using a digital computer (and today's oscilloscopes are increasingly likely to be special-purpose digital computers), does not contain as much information as a **bivariate distribution**. When you see a dot at a given position on the display, you know only that at least one cell in the sample had values of the two measured parameters corresponding to the position of the dot; and you can't get a better estimate of the actual number of cells that shared those values. That's where **gating** comes in. The strobe signal itself can be connected to a digital electronic **counter**, which will store a count and increase the count by one each time a strobe pulse is received. If the value in the counter is set to zero before analysis of a sample begins, the counter will maintain a tally of the total number of cells counted during the analysis.

Now, suppose we were interested in finding out how many of the cells in our dot plot had Hoechst dye fluorescence signals in the range between 3.5 and 4.75 V and pyronin fluorescence signals between 2.5 and 7 V. We could do this if we connected the relevant peak detector signals to an electronic circuit called a **window comparator**.

A **comparator** is a circuit element with two analog inputs, termed **positive** and **negative**, and a **digital**, or **logic level** (e.g., ground for "0" or "off"; 5V for "1" or "on") output. The digital output is on when the voltage at the positive input is higher than the voltage at the negative input, and off otherwise. Comparators are used in the analog front end circuitry of a flow cytometer to determine when the trigger signal (positive input) rises above the threshold level (negative input); one comparator is required for each trigger signal used.

A window comparator is made by connecting the logical outputs of four comparators together in a logical "AND" configuration. The inputs to the the individual comparators are appropriate combinations of the two input signals and two sets of upper and lower limits such that the combined output is "on" only when both signals fall within the limits. The limits would typically be set by turning the knobs of **variable resistors**, or **potentiometers**, which are best known in their roles as volume controls in relatively unsophisticated and older radios and television sets.

Gating is accomplished by connecting the digital output of the window comparator to one input, and the digital strobe signal to the other input, of a purely digital circuit called an **AND gate**. The output of an AND gate is on only when both inputs are on; in this case, the output of the AND gate will be a pulse train containing only the strobe pulses from those cells with parameter values falling between the set limits.

While one counter, working off the strobe signal, is counting all the cells in the sample, another counter, connected to the output of the AND gate, accumulates a count of the cells falling within the gating region. The output of the AND gate can also be used as an input to the electronics that control cell sorting, allowing the cells with values within the set limits to be physically separated from the rest of the sample.

By incorporating a few other bits of analog and digital circuitry into the window comparator modules of my earliest "Cytomutt" flow cytometers, I could, at the press of a button, draw the boundaries of rectangular gating regions on the oscilloscope, as is shown in Figure 1-12; this greatly facilitated setting the upper and lower boundaries of the gating regions. Early commercial instruments had similar features. Of course, they were still limited to rectangular gating regions, and there were clearly situations when one could not separate the cells one wanted to count or sort from the unwanted cells using rectangular gates.

It was possible, by adding still more analog electronics to generate sums and differences of signals from two parameters, and feeding the sums and differences, rather than the original signals, into a window comparator, to define a gating region that corresponded to a parallelogram or other quadrilateral, rather than a rectangle, in the two-dimensional measurement space. This feature was incorporated in the instruments Kamentsky built at Bio/Physics Systems in the early 1970's.

Kamentsky also described, but did not put into production, a clever alternative counting/sort control circuit made by placing opaque black tape over all of an oscilloscope screen except the area corresponding to the gating region, and mounting a photodetector in front of the screen. The gating region defined in this manner could be any arbitrary shape, or even a set of disconnected arbitrary shapes, limited in size and scope only by the user's dexterity with scissors or a knife blade and black tape. Every time a cell lying within the region was encountered, the uncovered portion of the screen was intensified, generating an output pulse at the photodetector that could be sent to a counter and/or used to initiate sorting. In the era of Bill Gates, we describe freeform gating regions of this type, implemented with mice and computers rather than blades and tape, as one type of **bit-mapped (or bitmap) gates**.

Well, most of the above is all ancient history, right? You must be wondering why I've devoted so much time to searching the souls of old machines when we do everything with computers now.

There are two reasons. The first is that the computers, in most cases, are doing the same things we did with hardwired electronics years ago, and if you understand how things worked then, you'll understand how they work now. The second is that there were, and still are, a few advantages to the old-fashioned electronics, especially for time-critical tasks.

I should mention that, then and now, nothing precluded or precludes us from defining a one-dimensional gating region, using either a simpler window comparator or a computer, and I did note that one-dimensional gating capability, allowing definition of a "region of interest," was typically built into pulse height analyzers. One-dimensional gating was widely used to control cell sorting in the earliest cell sorters, a logical choice when one considers that they typically measured only one relevant parameter.

Two-Parameter Histograms: Enter the Computer

As I wrote in 1994 for the 3rd Edition of this book, "Digital computers are extremely versatile. The same notebook computer on which I am writing this book with the aid of word processing software can be, and has been, used to acquire and analyze data from my flow cytometer. All I have to do is load and start a different program; I can even continue writing while I wait for the cytometer to get data from a new sample. Using additional telecommunications hardware and software, I can, and have, set gates on the cytometer, which is in Massachusetts, from a conference room in Maryland. However, while the computer's overall speed and its ability to switch rapidly between tasks make it appear as if it's doing many things at once, this is an illusion. About the only thing a computer can really do while it is running whatever program is occupying its attention is read or write data from or to a single source. Otherwise, digital computers do one thing at a time, even if they do that one thing really fast."

It's all still pretty much true. Of course, the notebook computer on which the 3rd Edition was written cost nearly $5,000, weighed about seven pounds, had a 50 MHz 80486 processor, at most a couple of MB of RAM, a 500 MB disk drive, and a 640 by 480 pixel screen, and the one I use now cost about $2,000, weighs three pounds, has a 750 MHz Pentium III processor, 256 MB of RAM, a 30 GB disk drive, and a 1024 by 768 pixel screen. The operating system and word processing software have also supposedly been improved. Last time around, my telecommunications were limited to what I could do over standard telephone lines using a 9,600 baud modem; now, I gripe when my cable modem or DSL connections slow to even fifty times that speed. So, what I or you can do with a single computer can be done faster than what could be done eight years ago. But there's more; cytometry today can take advantage of both digital signal processing and multiprocessor systems in ways that, while obvious, were simply infeasible then.

A window comparator implemented in electronics is really making four comparisons at the same time, and they are accomplished in well under a microsecond. If you build a sorter using two window comparators to control deflection into left and right droplet streams, the two comparators work simultaneously. If you want to use a digital computer for sort control in a brute force kind of way, the computer has to fetch the value of the x-axis parameter for the left gating region, check it against the lower and upper bounds for that region, fetch the value of the y-axis parameter, check it against both bounds, and repeat the same steps for the right gating region. Obviously, the computation for a particular gating region can be stopped as soon as a parameter value is found to be out of bounds, but, if you think about it, the full four comparisons for one gating region or another have to be done for any cell that falls in either region, and, until they get done, no signal can be sent to initiate droplet deflection.

In droplet cell sorting, a sort decision has to be made within the few dozen microseconds it takes at most for a cell to get from the observation point to the point at which droplets break off from the cell stream. Up to 10 µs may be required for the signals from the peak detectors (or integrators) to become stable. When hardwired electronics, e.g., window comparators, are used to control sorting, the sort decision signal is sent within a microsecond or so after this time. When a digital computer is used to control sorting, another time interval of at least a few µs is required for A-D conversion before the computer can process the data. And, although the computers have gotten faster, the emergence of high-speed sorting has made it necessary for them to respond within even shorter time intervals.

Until the late 1970's, even minicomputers weren't really fast enough to be competitive with hardwired electronics for sort control. Today's much faster personal computers can easily accomplish the computations required for the window comparison described above well within the time period in which a sort decision must be made. The same computers, however, might not be able to get through a more complex computation, which, say, involved calculating four logarithms and solving quadratic equations to determine whether a cell falls in an elliptical gating region, in time to issue a sort signal, largely because while modern computer hardware is extremely fast, often requiring less than 1 ns to execute a machine instruction, the real time response of the hardware is literally slowed to a crawl by the design of the graphical user interface (GUI) based operating systems (various versions of Microsoft's Windows™, Linux with GUI extensions, and Apple's Macintosh™ OS) now in most widespread use. The sorting problems are now solved by using some combination of external analog and digital electronics, frequently including one or more digital signal processors, or DSP chips, to implement time-critical processes such as sorting decisions, taking the load off the personal computer's central processing unit (CPU), leaving it free to do what it does best, namely, display the data informatively and attractively.

For plain old flow cytometric data analysis, in which there is no need to initiate action within a few microseconds after a cell actually goes through the beam, computers have always been better than hardwired electronics. That's why Kamentsky used one in his original instrument at IBM. Computers for the rest of us only came along as we could afford them. A few lucky souls, myself included, had computers on their cytometers in the mid-1970's; they were minicomputers, and they were expensive. Now, it's virtually impossible to buy a flow cytometer that doesn't have at least one computer external to the box; most have one or more inside, as well.

Figure 1-13 is a histogram, collected, displayed, and annotated using my own competent, if ancient, MS-DOS-based **4Cyte™** data acquisition software, showing 90° (side) scatter values from a human leukocyte population. The data are plotted on a linear scale. The sample was prepared by incubating whole blood with fluorescently labeled antibodies to the CD3, CD4, and CD8 antigens, and lysing the erythrocytes by addition of an ammonium chloride solution. The "Cytomutt" cytometer used 488 nm excitation from an air-cooled argon ion laser, and measured forward and side scatter and fluorescence in 30 nm bands centered at 520 nm (green; principally fluorescence from **anti-CD4** antibody labeled with **fluorescein**), 580 nm (yellow, principally fluorescence from **anti-CD8** antibody labeled with the **phycobiliprotein, phycoerythrin**), and 670 nm (red, principally fluorescence from **anti-CD3** antibody labeled with a **tandem conjugate** of **phycoerythrin** and the cyanine dye **Cy5**). The forward scatter signal was used as the trigger signal.

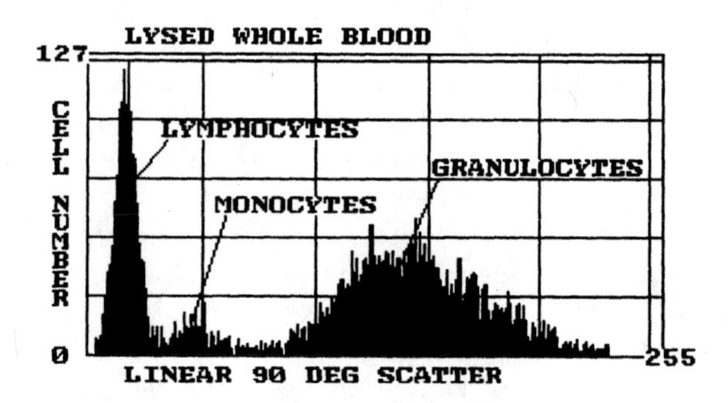

Figure 1-13. Histogram of 90° (side) scatter from leukocytes in lysed whole blood stained with fluorescent antibodies to lymphocyte antigens.

Modern Multiparameter Analysis: List Mode

The histograms and dot plots appearing in Figures 1-9, 1-11, and 1-12 are preserved for posterity only in the form of photographs. Figure 1-9A was photographed after the Hoechst dye fluorescence signal was connected to the pulse height analyzer and some 17,000 cells were run through the cytometer. The analyzer's memory was then cleared, the input was connected to the fluorescein fluorescence signal, and another 50,000 cells from the same sample were run through the instrument to generate the histogram of Figure 1-9B. The dot plots are taken from photos of the screen of an analog storage oscilloscope. I don't suppose the fact that we and a lot of other people stopped buying all of that Polaroid black-and-white film for our oscilloscope cameras loomed large in the company's eventually going bankrupt, but you never know. In the context in which we were using it, the film was a highly unsatisfactory archival medium.

The data represented in the histogram of Figure 1-13 were acquired in **list mode**, meaning that values of all parameters from all cells were stored in the computer's memory and, subsequently, on disk. List mode data acquisition doesn't preclude generating histograms, dot plots, or multivariate distributions while a sample is being run, and it does offer the user the considerable advantage of being able to reanalyze data well after they were acquired. The histogram in Figure 1-13 was generated months after the data were taken. Years ago, even after people had gotten used to having computers attached to their flow cytometers, they used to make a big fuss about acquiring data in list mode. There may have been some flimsy excuse for that attitude before mass storage media such as recordable CDs became available; today, there is simply no reason not to collect data from every run in list mode. Period. All currently available instruments have the necessary software for list mode data storage, and can write files compliant with one or another revision of the **Flow Cytometry Standard (FCS)** established by the Data File Standards Committee of the **International Society for Analytical Cytology (ISAC)**, an organization to which most serious flow cytometer users either belong or should. The standard makes it possible for analysis software from both cytometer manufacturers and third parties to read data from any conforming instrument.

As to the actual data in Figure 1-13, we notice that the histogram, like the DNA histograms in Figures 1-8B and 1-9A, is **multimodal**, meaning not that it has multiple identical maxima, but that it contains multiple peaks. Only one of these, that to the far left, would even be suspected of being a needle rather than a haystack. From the labels in Figure 1-13, it can be surmised that there is good reason to suspect that the peaks at increasingly higher values of 90° scatter represent lymphocytes, monocytes, and granulocytes; we can even go back to page 7 and look at Figure 1-2 to convince ourselves that this is the case. However, just as we can't readily separate the G_0/G_1 cells from the S cells, or the S cells from the G_2/M cells, by looking at a DNA histogram alone, we can't readily separate the lymphocytes from the monocytes and the monocytes from the granulocytes by looking only at the histogram of 90° scatter.

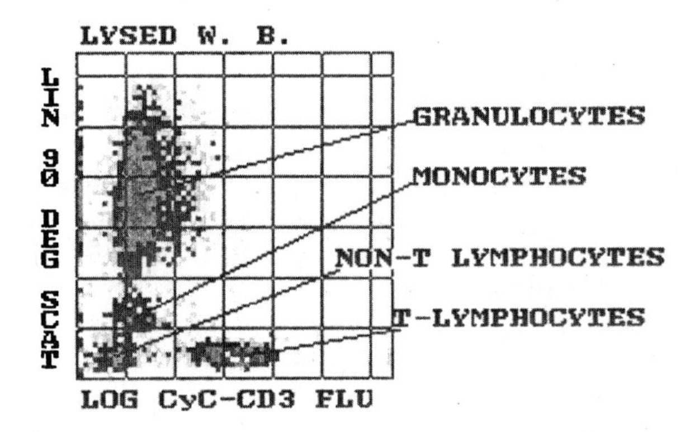

Figure 1-14. Bivariate distribution of anti-CD3 antibody fluorescence intensity vs. 90° (side) scatter for the same leukocyte population shown in Figure 1-13.

The picture gets a lot clearer when we look at the **bivariate distribution**, or **two-parameter histogram**, shown in Figure 1-14. The raw data in this distribution came from the same **list mode** file used to compute the histogram of 90° scatter shown in Figure 1-13; meaning that, thanks to the ready availability of computers and data storage media, we are able to look at the same cells from many different points of view. Figure 1-14 shows clearly identifiable **clusters** of cells; it provides a much clearer separation of lymphocytes, monocytes, and granulocytes than one could obtain using 90° scatter alone, and it also clearly separates the lymphocytes into those that bind the anti-CD3 antibody, i.e., the T cells, and those that do not, most accurately identified as "non-T" lymphocytes.

While a similar separation of cell clusters would be discernible on a dot plot, the bivariate distribution provides a more detailed picture of the relationship between two measured parameters, because the distribution provides an indication of the number of cells and/or the fraction of the cell population sharing the data values corresponding to each point in the two-dimensional measurement space, whereas the dot plot only indicates that one or more cells share the data values corresponding to a point in that space.

A bivariate distribution is computed by setting aside n^2 storage locations, where n is the number of bits of resolution desired for the data. Obviously, n cannot be greater than the number of bits of resolution available from the ADC; in practice, a lower value is typically used, for two reasons. First, the memory requirements are substantial. If each parameter has values ranging from 0 to 1,023, it is necessary to use 1,048,576 storage locations for a single distribution; this requires 2 megabytes if each location uses two bytes, or 16 bits, which would allow up to 65,535 cells or events to be tallied in any given location. If each location uses four bytes, or 32 bits, 4 megabytes of storage are required, but the maximum number of cells that can be tallied in a location is increased to over 4 billion.

While the issue of memory requirements for distribution storage would seem moot at a time when a computer can be equipped with a gigabyte of RAM for a couple of hundred dollars, a second consideration remains. When a two-parameter histogram is computed at high resolution, it is usually necessary to include a very large number of events in order to have more than a few events in each storage location; computing at a lower resolution may actually make it easier to appreciate the structure of the data from smaller cell samples.

For a relatively long time, it was common to compute two-parameter histograms with a resolution of 64 × 64; these require 4,096 storage locations per histogram, which was a manageable amount of memory even in the early days of personal computers. Now, resolutions of 128 × 128 (16,384 storage locations) and 256 × 256 (65,536 storage locations) are widely available. The distribution displayed in Figure 1-13 has 64 × 64 resolution; values on a 1,024-channel scale, such as would be produced by a 10-bit ADC, would be divided by 16 to produce the appropriate value on a 64-channel scale, while the 8-bit (256 channels) values yielded by the lower-resolution converters found in older instruments would be divided by 4.

The data presentation format used in the display of Figure 1-14 is that of a **gray scale density plot**; the different shades of gray in which different points are displayed denote different numbers of cells sharing the corresponding data values. There is an alternative display format for density plots in which different frequencies of occurrence are represented by different colors instead of different shades of gray; this type of plot is described as a **chromatic** or **color** density plot. One can think of the gray levels or different colors in density plots as analogous to the scales that indicate different altitudes on topographical maps. Unfortunately, although the altitude scale is displayed on almost every published topographical map, the analogous scale of cell numbers or frequencies rarely finds its way into print alongside cytometric density plots.

Since computers now used for flow cytometric data analysis have color displays and color printers, chromatic plots are more common than gray scale plots. However, although color pictures are eye-catching and useful for presentations and posters, they can cost you money when included in publications. Those of us who run on lower budgets can almost always use a well-chosen gray scale for published displays without losing information; those lucky enough to not have to think about the cost of color plates might want to choose color scales that will not become uninformative when viewed by readers with defects in color vision.

Figure 1-15 (next page) displays the two-parameter histogram data of Figure 1-14 in an **isometric plot**, or **three-dimensional projection**, also commonly called a **peak-and-valley plot**; Figure 1-16 (next page) shows the same histogram as a **contour plot**.

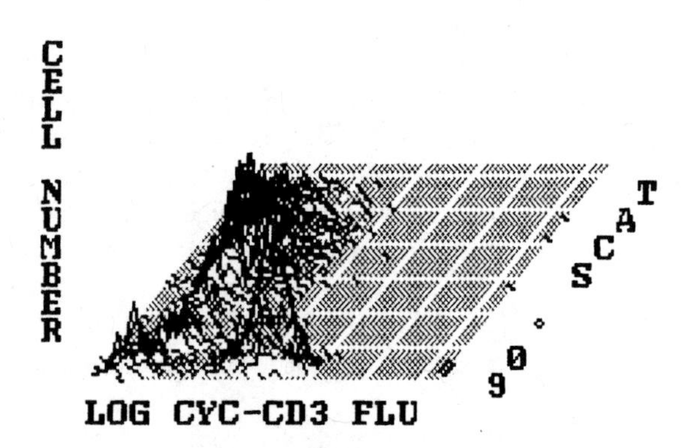

Figure 1-15. The two-parameter histogram of Figure 1-14 shown as an isometric or "peak-and-valley" plot.

In a peak-and-valley plot, a simulated "surface" is created; the apparent "height," or z-value, corresponding to any pair of x- and y-coordinates is made proportional to the frequency of occurrence of the corresponding paired data values in the sample. In a contour plot, a direct indication of frequency of occurrence is not given for each point in the x-y plane. Instead, a series of **contour lines**, or **isopleths**, are drawn, each of which connects points for which data values occur with equal frequency. A contour plot, like a density plot, resembles a topographic map; a peak-and valley plot is more like a relief map. The fact is, though, that there is really no more information in one type of bivariate histogram display than in another. Take this as a mantra.

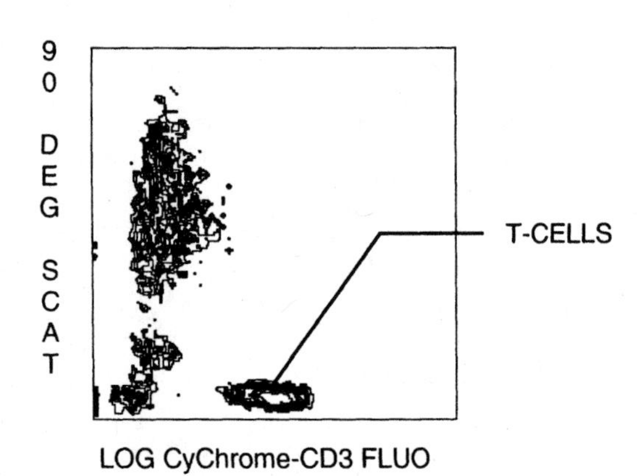

Figure 1-16. The two-parameter histogram of Figures 1-14 and 1-15 displayed as a contour plot.

Peak-and-valley plots seem to have largely fallen out of favor; none too soon, say I. One of their major disadvantages lies in the *a priori* unpredictability of where peaks and valleys will turn up. Big peaks in the simulated "front" block the view of smaller peaks in the simulated "back," unless you use the 3-D graphics capabilities of your computer display to tilt and rotate the display. I suspect it's the dedicated computer gamers who are saving peak-and-valley plots from extinction.

Contour plots require more computation than either chromatic or peak-and-valley plots. Although contour plots may appear to have higher resolution, this appearance is deceiving, resulting as it does from the necessity to **smooth** the data, i.e., average over neighboring points, in order to get the plot to look respectable. Contour plots also do not normally show single occurrences, although these can be superimposed as dots on a contour plot; some such adaptation is essential when dealing with rare events. I think people tend to use contour plots in publications because they look more detailed than dot plots and often reproduce better than gray scale density plots.

I have always favored density plots, using chromatic plots for primary computer output and presentations, and, as a rule, gray scale plots for publication. I routinely use a **binary logarithmic intensity scale**, with one color or gray level indicating single occurrences, the next 2-3 occurrences, and subsequent colors indicating 4-7, 8-15, 16-31, 32-63, 63-127, and more than 127 occurrences. This makes it very easy to spot cells that occur with frequencies of less than one in 10,000.

Although commercially available flow cytometers are now equipped to display sixteen or more parameters (which would typically include light scattering at two angles and fluorescence in twelve spectral regions, with the balance possibly made up of different characteristics of the same pulses, such as width or height and area, and/or of ratios of the heights or areas of two signals from the same cell), almost all analysis is done using two-dimensional histograms or dot plots of two parameters in various combinations.

Three-Dimensional Displays: Can We Look at Clouds from Both Sides? No.

Humans aren't very good at visualizing spaces of more than three dimensions, but you'd certainly expect that, with everybody doing five- and six-parameter measurements in flow cytometry, three-dimensional displays would be commonplace. Some software packages produce a "three-dimensional dot plot," which I have called a **cloud plot** (see Figure 5-11, p. 241). Cloud plots have the same disadvantage as peak-and-valley plots; when one cloud gets in front of another, you have to recompute and change the viewing angle to see what's where. A few people have gone so far as to generate stereo pair images to improve the three-dimensional quality of the displays; they may be the same folks who have kept peak-and valley plots alive.

Three-parameter histograms are problematic for several reasons. First, even a 64 × 64 × 64 3-parameter histogram requires 262,144 storage locations, although there are some tricks that can reduce the storage requirements. Once you do, though, there's still a problem with how to display the data. Isometric plots would require four dimensions, which is out, and contour and chromatic plots demand x-ray vision on the part of the observer. As a result, what people have generally done when they need to represent something analogous to a 3-parameter histogram is show 2-parameter histograms in "slices," with the resolution along the "sliced" axis often lower than that of the 2-parameter histograms, so that there might be four to sixteen 64 × 64 histograms rather than sixty-four. "Slicing" a histogram, if you stop to think about it, is exactly equivalent to defining a series of rectangular gating regions along the z-axis. And, speaking of gating regions, the slicing technique just mentioned is about the only practical way of setting gates in a three-dimensional space.

There's a lot of information to deal with when you're just looking at two-parameter displays; three-parameter displays could quickly get you to the point of information overload. If we are dealing with *n* parameters, the number of possible two-parameter displays, counting those showing the same two parameters with the x- and y-axis switched, and omitting those in which the same parameters are on both axes, is $n \times (n\text{-}1)$, while the number of possible three-parameter displays, counting those showing the same three parameters with axes switched and omitting those with the same parameter on two or three axes, is $n \times (n\text{-}1) \times (n\text{-}2)$. For the five-parameter data we have been looking at, we have 20 possible two-parameter displays and 60 possible three-parameter displays; for 16-parameter data, we would have 240 possible two-parameter displays, which is frightening enough, and a mind-boggling 3,360 three-parameter displays. It could take months to run an analysis on a single tube if we had to look at all of them. So, as usual, it is best to get our heads out of the clouds.

Identifying Cells in Heterogeneous Populations: Lift Up Your Heads, Oh Ye Gates!

Most of the interesting applications of flow cytometry involve identifying cells in heterogeneous populations; what varies from case to case is the basis of the heterogeneity. We have already noted several varieties of heterogeneity in our brief examination of DNA content analysis. Cells in a presumably pure, clonally derived, unsynchronized culture will contain different amounts of DNA because they are in different stages of the cell cycle. A DNA aneuploid tumor contains stromal and tumor cells with different G_0/G_1 DNA contents, and both stromal and tumor cells may be in different cell cycle phases. Sperm differ in DNA content depending on which sex chromosome is present. Heterogeneous populations of microorganisms such as are encountered in seawater contain many different genera and species, each with its characteristic genome size. In all of the above cases, it is possible to identify cell subpopulations based on differences in DNA content.

In the widely studied heterogeneous cell populations that comprise blood, the majority of cells are neither DNA aneuploid nor progressing through the cell cycle. Thus, when the **problem** is the identification of different cell types in blood, DNA content is generally not a **parameter** of choice. Figure 1-17, on the next page, illustrates the use of several better suited parameters and of multiple gating methods in one of the most common clinically relevant applications of flow cytometry, the identification of T lymphocytes bearing CD4 and CD8 antigens in human peripheral blood.

If we simply stained cells with a combination of differently colored acidic and basic dyes, as Paul Ehrlich, who developed the basic technique, did in the late 1800's, we would be able to use transmitted light microscopy (with relatively strongly absorbing dyes at high concentrations) or fluorescence microscopy (with fluorescent dyes, probably at lower concentrations) to do a classical differential white blood cell count. The presence or absence of cytoplasmic granules would let us distinguish the granulocytes from the mononuclear cells (monocytes and lymphocytes). The relative amount of staining of those granules by the acidic and the basic dye would allow us to identify eosinophilic (acidophilic to Ehrlich), basophilic, and neutrophilic granulocytes. The size of the cells, amount of cytoplasm, and nuclear shape would allow us to distinguish most of the monocytes from most of the lymphocytes. But that's about as far as we would get. A typical peripheral blood lymphocyte is a small, round cell with a relatively thin rim of cytoplasm surrounding a compact, round nucleus. The nucleus, like the nuclei of all cells, stains predominantly with the basic dye (one of the methylene azure dyes in a typical Giemsa or Wright's stain), which is attracted to the acidic phosphate groups of the nuclear DNA. The basic cytoplasmic proteins attract some of the acidic dye (eosin in the mixtures commonly used for staining blood), but RNA in the cytoplasm also attracts the basic dye. And the staining pattern of most peripheral blood lymphocytes is pretty much the same, whether they are B lymphocytes or T lymphocytes, and, if T lymphocytes, whether they bear the CD4 or the CD8 antigen (although both antigens are present on developing T lymphocytes in the thymus, almost all of the T lymphocytes present in the peripheral blood have lost one or the other).

The optical flow cytometers used for differential white cell counting in hematology laboratories, which typically measure forward and side scatter, can distinguish lymphocytes from monocytes and granulocytes using these measurements alone, but cannot thereby distinguish different types of lymphocytes. However we have already seen from Figures 1-13 through 1-16 that the combination of side scatter measurements and measurements of fluorescence of cell-bound antibodies allows us to distinguish T lymphocytes from other lymphocytes. It should therefore come as no surprise that the **probes**, or reagents, that allow us to define lymphocyte subpopulations, and most other subclasses of cells in the blood, bone marrow, and organs of the immune system, are **antibodies**, and that we detect antibodies bound to cells by the **fluorescence** of labels attached, usually covalently, to the antibody molecules. Flow cytometry greatly facilitated the development of monoclonal antibody reagents, and flow cytometry has since been indispensable for defining the specificities of these reagents and, thereby, allowing their routine use for cell classification in clinical and research laboratories.

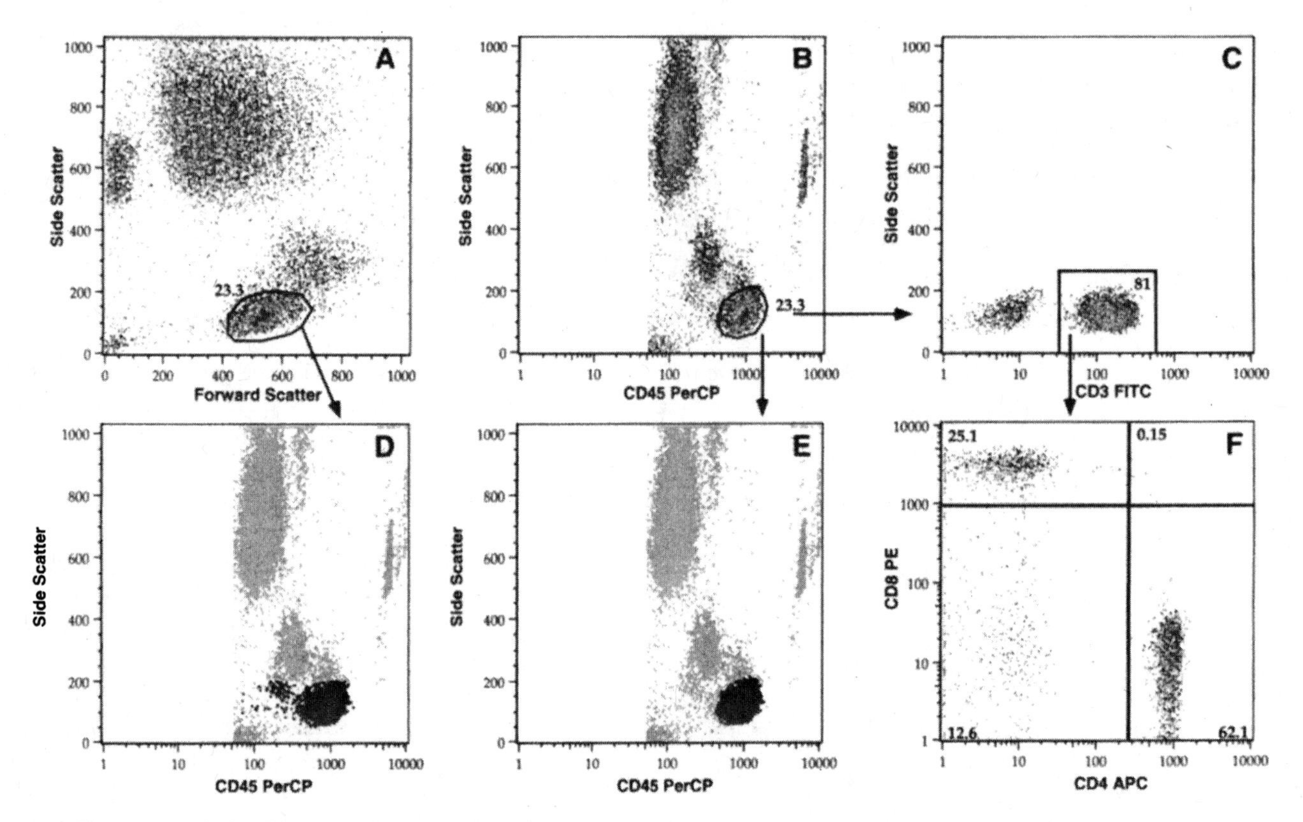

Figure 1-17. Identification of human peripheral blood T lymphocytes bearing CD4 and CD8 antigens. Data provided by Frank Mandy; analysis and displays done by Jennifer Wilshire using FlowJo software (Tree Star, Inc.).

The displays in Figure 1-17 show an older and a newer gating method for defining a lymphocyte population. The gates are drawn with the aid of a mouse or other pointing device. Flow cytometric software typically provides for several types of **bitmap gating**, which allows the user to define more or less arbitrarily shaped gating regions on a dot plot or two-parameter histogram. Almost all programs allow the user to draw **polygons** to define the boundaries of gating regions; most also allow definition of regions bounded by **rectangles**, **ellipses**, or **free-form curves**. While most cell sorters make use of no more than four gating regions at any given time, data analysis software typically provides for a larger number, to facilitate deriving counts of a reasonable number of cell subpopulations in heterogeneous samples such as are obtained from blood.

Cluster Headaches

The objective of gating is the isolation, in the measurement space, of a **cluster** of cells. The term **cluster** is used in flow cytometry (and in multivariate data analysis in general) to denote any relatively discernible, reasonably contiguous region of points in a bivariate display; that may sound imprecise, but there isn't any more precise definition. You're supposed to know a cluster when you see one.

In panel A of Figure 1-17, a polygonal gate is drawn around a cluster of cells with intermediate values of forward scatter and low values of side scatter; it was established by sorting experiments in the 1970's that most of the cells in such a cluster were lymphocytes[157], and lymphocyte gating was incorporated into analysis of lymphocyte subsets at a fairly early stage in the game[175-6]. However, there was some concern that cells other than lymphocytes might be found in the gate. If one were interested only in T lymphocytes, it would be possible, as Mandy et al demonstrated in the early 1990's[1027], to define a well isolated cluster of these cells on a display of anti-CD3 antibody fluorescence vs. side scatter (look back at Figures 1-13 to 1-16). This did not satisfy the HIV immunologists, who wanted to know not only the absolute number of CD4-bearing T cells per unit volume of blood, but also what percentage of total lymphocytes the CD4-bearing T cells represented. The current practice for defining a lymphocyte cluster uses a two-dimensional display of anti-CD45 antibody fluorescence vs. side scatter, as shown in panel B of Figure 1-17, taking advantage of the fact that lymphocytes have more CD45 antigen accessible on their surfaces than do monocytes and granulocytes[1251].

Painting and White- (or Gray-) Washing Gates

The gates in panels A and B of Figure 1-17 have been painstakingly drawn so that each includes 23.3 percent of the total number of events (where events include cells, doublets, debris, and the counting beads added to the sample). We have decided to accept the CD45/side scatter gate in panel B as the "true" lymphocyte gate; the question then comes up as to whether the forward scatter/side scatter gate in Panel A contains cells that would not fit into this gate.

In order to answer this question, we need a way of finding the cells in the forward scatter/side scatter gate on a display of CD45 vs. side scatter. Most modern data analysis programs incorporate the means to do this; the user can associate a different color with each gate set, thus allowing cells falling within that gate to be distinguished on plots of parameters other than those used to set the gates. Becton-Dickinson's "Paint-A-Gate" program was one of the first to provide this facility.

When you are working on a low budget, and restricted to monochrome displays, you can always emulate Whistler and use shades of gray instead of colors, as has been done in panels D and E of Figure 1-17. In this instance, the cells from the gates in panels A and B have, respectively, been shown in black in panels D and E; all of the other cells appear in light gray. The panel D and E displays also use a convenient feature found in the FlowJo program; the dots can be, and here are, made larger. This can be useful when one tries to show very small subpopulations in dot plots, and, indeed, we see that there are a few cells from the forward scatter/side scatter gate of panel A that show up in panel D outside the "true" lymphocyte gate as defined using CD45/side in panel B. Of course, the cells from the gate in panel B remain in the same positions in panel E; since we started out assuming that the gate in panel B was the true gate, you can't really call that a whitewash.

Moving right along, in this case to panel C, we will look only at the cells from the lymphocyte gate defined in panel B, on a plot of anti-CD3 antibody fluorescence vs. side scatter. We can now draw a rectangular gate around those that bear the CD3 antigen; these are the T cells.

The Quad Rant: Are You Positive? Negative!

In panel F, the T cells defined by the gate in panel C are shown on a plot of anti-CD4 antibody fluorescence vs. anti-CD8 antibody fluorescence. This plot is broken into **quadrants**, i.e., four rectangular gating regions that intersect at a single central point. The percentages of events that fall in each of the quadrants are indicated. There are clear clusters of events with high levels of CD8 and low levels of CD4 and of events with high levels of CD4 and low levels of CD8, a small but respectable number of events with low levels of both, and a few events with high levels of both. At first, it seems as if all's right with the world. But maybe there's a problem with our world view.

Dividing measurement spaces into quadrants is, in part, a throwback to the old days of flow cytometry without computers, when gates were implemented using hardware, and it was much easier to make them rectangular than it was to make them any other shape. Quadrants work best when the data fall neatly into rectangular regions, and when cells either have a lot of a particular antigen or other marker, making them **positive**, or very little or none, making them **negative**. The CD4-CD8 distribution of peripheral blood lymphocytes is about as good an example of this situation as can be found, but, even here, we see that, while the events divide clearly into positives and negatives on the CD4 axis, there are some events with intermediate levels of CD8.

If we were looking at cells from the thymus, quadrants wouldn't work well at all, because there are a lot of immature T cells in the thymus that have both CD4 and CD8, some of which are acquiring the antigens and some of which are losing them, and where one draws the quadrant boundaries is pretty much arbitrary. But problems with immunofluorescence data go beyond that, and beyond quadrants.

Deals with the Devil: Logarithmic Amplifiers and Fluorescence Compensation

The need and desire to measure immunofluorescence have motivated much of the development of modern flow cytometry. However, two problems associated with immunofluorescence measurement, and the less than satisfactory techniques applied to their solution, have been frustrating to beginners and experts alike.

The first problem is that of making and representing the results of measurements encompassing a large **dynamic range**. The first flow cytometers used to make immunofluorescence measurements weren't very sensitive. The green fluorescent dye **fluorescein** was used to label antibodies, and fluorescence was measured through color glass long pass filters, which, in addition to fluorescein fluorescence, let through **cellular autofluorescence**, probably due primarily to intracellular **flavins**. The filters themselves also emitted some fluorescence when struck by stray laser light. This made it impossible to detect fewer than several thousand antibody-bound fluorescein molecules on an unstained cell. However, since the maximum number of antibody-bound fluorescein molecules present on a cell might be a million or more, it was desirable, even before sensitivity was increased, to have some useful way of expressing results that varied over the three decade range between 1,000 and 1,000,000.

One obvious technique was to report and display results on a **logarithmic scale**. You can see examples of this in Figure 1-17, if you look at the numeric values and the positions of the tick marks on the axes of panels B, C, D, E, and F. Although the linear scales shown for forward and side scatter measurements (as in panel A) are accurate, the logarithmic scales may only be approximate. When analog data are digitized to relatively high resolution (16 to 20 bits), it is possible to convert signals accurately from a linear to a four decade (range 1 to 10,000) logarithmic scale and back using a digital computer; some modern cytometers employ this technique. However, in the 1970's and 1980's, the high-resolution ADCs needed to implement this procedure simply weren't available. The stopgap solution, which is still in use by some manufacturers, was to employ **logarithmic amplifiers**, commonly if not affectionately known as **log amps**.

A log amp is an analog electronic circuit that, in principle, puts out a voltage or current proportional to the voltage or current at its input. So far, so good. The bad news is that the proportionality constant may vary with time, temperature, input voltage, and, I suspect, the experimenter's astrological sign. A log amp isn't a log table, or even approximately like one. The worse news is that nobody much cared how bad log amps were until the late 1980's, when people got interested in trying to make quantitative measurements of immunofluorescence and got really screwed up trying to convert from logarithmic to linear scales and back. We can expect the trend toward digital processing will continue, allowing logarithmic amplifiers to be replaced or, alternatively, monitored and calibrated; either approach should result in increased accuracy of representation of measurements on logarithmic scales.

A different set of complications was introduced by the development of antibody labels that enabled flow cytometers with a single illuminating beam (488 nm) to be used to make simultaneous measurements of immunofluorescence from several cell-bound antibodies. The first of these labels was the yellow fluorescent **phycoerythrin (PE)**, a protein found in the photosynthetic apparatus of algae. By attaching dyes to this molecule, making what are called **tandem conjugates**, it is possible to obtain fluorescence emission at longer wavelengths. When the rhodamine dye **Texas red** is attached to phycoerythrin, the resulting conjugate emits in the orange spectral region (620 nm); phycoerythrin with the **indodicarbocyanine** dye **Cy5** attached emits in the red (670 nm). Tandem conjugates of phycoerythrin with the cyanine dyes **Cy5.5** and **Cy7** emit even farther in the red or near infrared, at 700 and 770 nm. Some flow cytometers now in commercial production can be used to make simultaneous measurements of cells labeled with fluorescein, phycoerythrin, and all of the tandem conjugates just mentioned; most allow fluorescence in at least three spectral regions to be measured. The raw measurements, however, will not leave us in a state of conjugate bliss; we still have to contend with the problem of **compensation** for **fluorescence emission spectral overlap** between the labels, which only gets worse as the number of labels excited at a single wavelength increases. Figures 1-18 and 1-19 (pages 37 and 38) should provide some understanding of the problem and its solution.

Most fluorescent materials emit over a fairly broad range of wavelengths. When we describe fluorescein as green fluorescent, what we really mean is that if you look at it under a fluorescence microscope, the fluorescence looks green, and that if you measure the spectrum in a spectrofluorometer, the emission maximum is in the green spectral region. When we try to measure fluorescein fluorescence in a flow cytometer, we typically use a detector fitted with a green filter that passes wavelengths between 515 and 545 nm. However, as can be seen in Figure 1-18, the emission spectrum of fluorescein doesn't start abruptly at 515 nm and stop abruptly at 545 nm; it extends out well beyond 600 nm, although the fluorescence at the longer wavelengths is considerably less intense. There's quite a bit of emission from fluorescein in the 560-590 nm spectral region that we call yellow, and in which the emission maximum of phycoerythrin lies. That means that if we were to stain cells or other particles with fluorescein and nothing else, and measure them in a flow cytometer with both green and yellow detectors, we'd pick up a strong signal in the green detector, and also detect some signal in the yellow detector.

The phycoerythrin emission spectrum also extends well beyond the 560-590 nm yellow wavelength range we use for measurements of phycoerythrin fluorescence. There is some emission below 560 nm, in the 515-545 nm green region, and considerably more above 580 nm. If we put cells stained with phycoerythrin and nothing else into the cytometer, we'd get the strongest signals from the yellow detector, and some signals from the green, orange, and possibly the red detectors as well. The same argument holds for the orange and red fluorescent tandem conjugates; each of these will definitely be detectable in the detector intended to measure the other, and signals from the orange conjugate will show up at the yellow detector as well, and possibly also in the green one.

If we put a cell sample stained with antibodies labeled with fluorescein, phycoerythrin, and the orange and red PE-Texas red and PE-Cy5 conjugates into the machine, the signal we get from the green detector is going to be comprised mostly of fluorescein fluorescence, with a smaller contribution from phycoerythrin fluorescence, possibly a wee bit from the orange conjugate, and some from cellular autofluorescence. The signal from the yellow detector will represent mostly phycoerythrin fluorescence, with substantial contributions from fluorescein and the orange conjugate, possibly some from the red conjugate, and some from autofluorescence. And so on for the signals from the orange (600-620 nm) and red (660-680 nm) detectors. Now, how much is "some," "a substantial contribution," or "a wee bit"?

That will depend on the gain settings used for the various detectors. Once these are set, it is fairly simple to quantify the degree of spectral overlap. For example, suppose we measure cells or beads stained only with fluorescein, and they produce signals with a mean intensity (peak height or area) of 5 V from the green detector and signals with a mean intensity of 1 V from the orange detector. If we were then to measure cells stained with fluorescein and phycoerythrin, and we wanted to remove the fluorescein contribution from the orange detector signal, we could subtract 1/5 of the green signal intensity. If a doubly stained cell yielded a signal of 1 V from the green detector, we'd subtract 0.2 V from the orange signal, no matter what the value of the orange signal was; if the cell yielded a 4 V signal from the green detector, we'd subtract 0.8 V from the orange signal, and so on. Well, actually, we'd also have to do the reciprocal calculations to figure out how much of the orange signal to subtract from the green signal to remove the contribution due to phycoerythrin. In principle, though, we could figure out the whole business using high school algebra, by solving **simultaneous linear equations**. Linear equations... aye, there's the rub.

All of the operations involved in fluorescence compensation must be performed on **linear** signals. You have to make the measurements on a linear scale to determine the fractions of fluorescence signal at each detector due to each fluores-

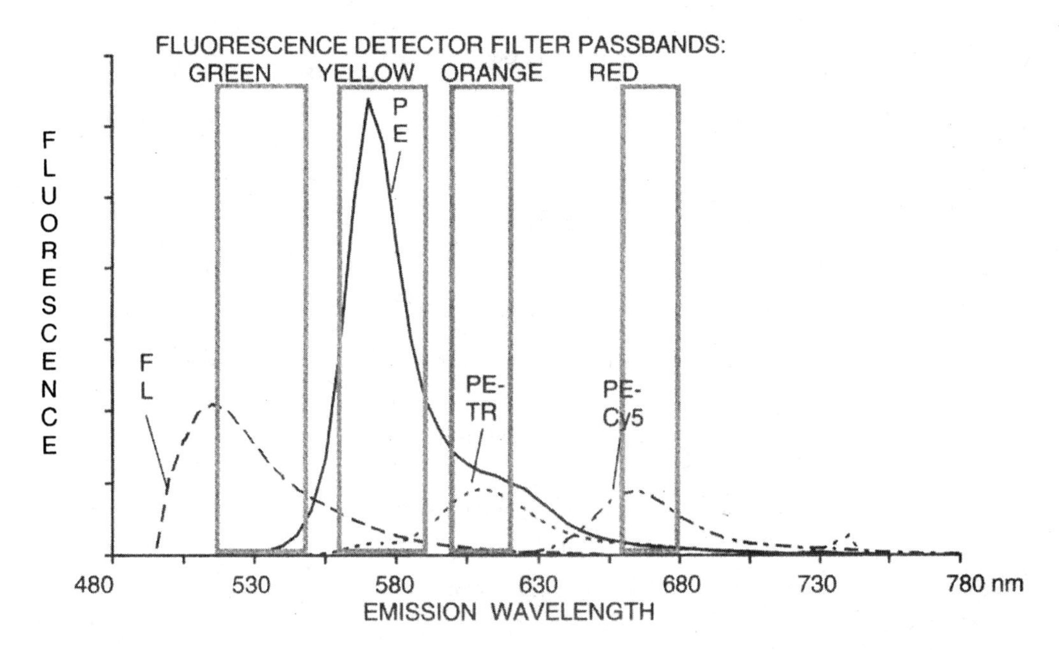

Figure 1-18. Why fluorescence compensation is necessary. The emission spectra shown are for equal concentrations (mg antibody/mL) of mouse anti-human IgG directly conjugated with fluorescein (FL), phycoerythrin (PE), and tandem conjugates of phycoerythrin with Texas red (PE-TR) and Cy5 (PE-Cy5), with excitation at 490 nm. Boxes demarcate the passbands of the green (530 nm), yellow (575 nm), orange (610 nm), and red (670 nm) filters used on the fluorescence detectors. Spectra are corrected for PMT responsivity differences at different wavelengths.

cent label, and the subtractions needed to make the necessary corrections also have to be done in the linear domain. But, as you remember, we usually tend to feed signals from immunofluorescence measurements through logarithmic amplifiers. How, then, do we introduce the fluorescence overlap compensation?

What happens in most older flow cytometers is that yet another analog circuit is built in between the preamplifier outputs and the log amp inputs. The circuit is something like an audio mixer, except that it subtracts signals instead of adding them; the operator adjusts one knob to determine the amount of green signal to subtract from yellow, another to determine the amount of yellow signal to subtract from green, etc. For two colors, this isn't all that hard to do. For three colors, you need six knobs, although you can get away with four if you ignore the green-orange and orange-green interactions. For four colors, you should have twelve knobs, though you might get away with eight. Each knob, of course, is attached to a **potentiometer**, or **variable resistor**, which, as was noted in the discussion of window comparators on p. 28, is basically a volume control. That starts to add up to a lot of electronic circuitry. Things may look neater if you let a computer control the compensation using **digital-to-analog (D-A) converters**, but you still end up with a lot of electronics at the input of your log amps.

Now, the whole reason we bother using log amps is to get a large dynamic range for our measurements. If we want a four decade dynamic range, with the top of the highest decade at 10 V, we end up with the top of the next highest at 1 V, the top of the next highest at 100 mV, and the lowest decade encompassing signals between 1 and 10 mV. If you want to process signals between 1 and 10 mV, you have to keep the noise level below 1 mV. I've measured noise in a number of older flow cytometers from a number of manufacturers, and I haven't run across one with noise below 1 mV at the preamplifier outputs. The more electronic components you stick in the circuit, the more opportunities there are to increase the noise level, and my considered opinion is that it is unlikely that a system that implements four-color compensation in electronics will be able to maintain the low noise level needed to insure a true four decade dynamic range.

Quite aside from all that, though, most people can't solve simultaneous linear equations in their heads, and those few who can probably can't manage to solve equations and twiddle knobs on compensation circuitry at the same time. You have a reasonable chance of getting two-color compensation close to right by eye; three-color compensation gets a little tougher, and you're kidding yourself if you think you can do four-color compensation correctly without solving equations. As far as I know, the manufacturers have capitulated completely on the subject of compensation for more than four colors; the knobs are gone.

There was really no choice. If you keep all the electronic measurements linear, using an A-D converter with 16 or more bits' precision, you can dispense with 1) all of the knobs and their associated electronics, 2) all of the log amps, and 3) the semiempirical process of knob twiddling for fluo-

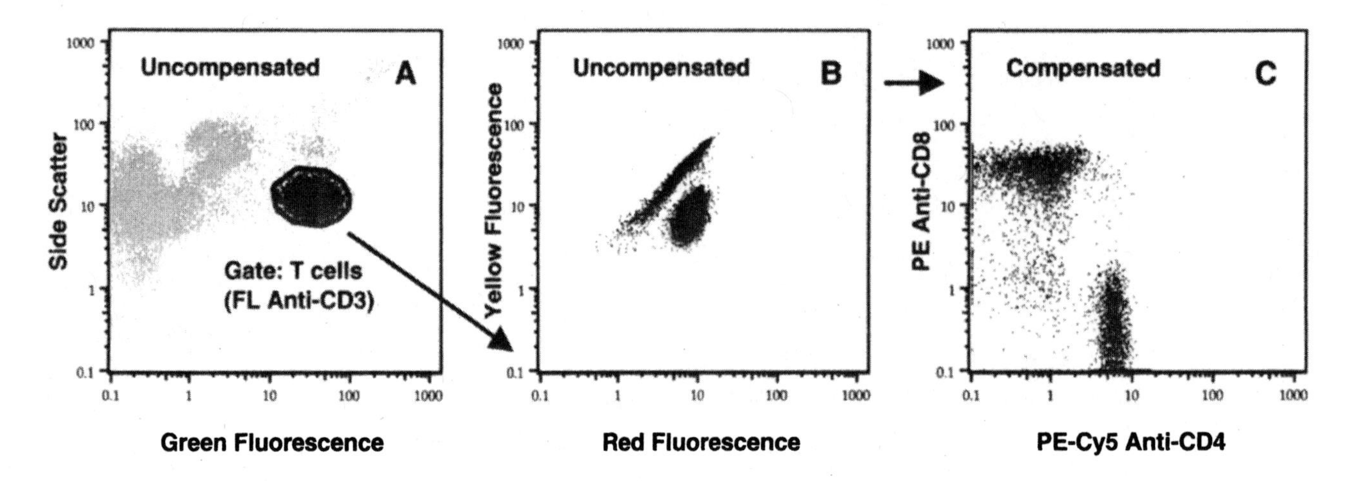

Figure 1-19. How compensation gets data to fit into quadrants.

rescence compensation. The simultaneous linear equations can be solved using digital computation, which can also do highly accurate conversions to a logarithmic scale. Once you have access to high-resolution digitized data, the logarithmic scale is only really needed for display, anyway; any statistical calculations that need to be done can be done on the linear data.

A significant advantage of high-resolution digitization of data is that you can go back to data that were not properly compensated when they were collected, transform them from a log scale to a linear one, if necessary, and redo the compensation. There are flow cytometry software packages that will let you play this game with log scale data that were digitized using 8- or 10-bit ADCs, but you end up with plots that have "holes" in the clusters due to the substantially larger, unavoidable digitization errors associated with lower resolution converters. The plots can be, and usually are, made more lovely to look at by **dithering**, adding random numbers to the data values. This technically degrades the quality of the data, but not by that much. I used to disapprove of it; I am now willing to accept it as yet another of the many deals with the devil that have to be made at the current state of the art. Within a few more years, almost all of the instruments in use will have higher resolution data analysis, and the plots of flow cytometric data will look pretty without benefit of dithering and diddling.

Evils of Axes: Truth in Labeling Cells and Plots

Mislabeling of axes, usually unintentional (I hope), is seen all too often in plots of flow cytometric data. Beginners and old-timers do it, and the mislabeling gets by journal reviewers, editors, and proofreaders. With the aid of Figure 1-19, which illustrates the effects of compensation, we can consider why mislabeling may occur and how to avoid it.

The data in Figure 1-19 were taken from a sample of whole blood stained with fluorescein anti-CD3 antibody, phycoerythrin anti-CD8 antibody, and phycoerythrin-Cy5 anti-CD4 antibody. Erythrocytes in the sample were lysed, and the sample was fixed with a low concentration of formaldehyde, before analysis. Panel A shows a dot plot of green fluorescence vs. side scatter, with both parameters displayed on a 4-decade logarithmic scale. A polygonal gate is drawn around a cluster I claim are T cells; the cells (events, if we want to be more precise) in this gate are plotted in black, while the remainder of the population is plotted in light gray.

If you look at Figure 1-14 (p. 31), you will notice that it is also a plot, in this case, a two-dimensional histogram, with anti-CD3 fluorescence on the x-axis and side scatter on the y-axis. In Figure 1-14, the y-axis is explicitly labeled as linear, and the x-axis as log, since one cannot tell whether the scale is log or linear simply by looking at the superimposed grid. The logarithmic scales on the axes of the panels in Figure 1-19 provide us with tick marks that would tell us that the scale was logarithmic even without the associated numbers, which are simply arbitrary indicators of intensity.

However, the x-axis of Figure 1-14 is labeled as "Log CyC-CD3 Flu," which means that this axis represents the intensity of fluorescence, on a logarithmic scale, of an anti-CD3 antibody, labeled in this case with PE-Cy5, with CyC being an abbreviation for one of the trademarked versions of this tandem conjugate label. The x-axis in panel A of Figure 1-19 is labeled "Green Fluorescence." What's the difference?

The difference is that fluorescence compensation has been applied to the data in Figure 1-14, but not to the data in panel A (or panel B) of Figure 1-19. So what is displayed on the x-axis in panel A is really green fluorescence, most of which is from the fluorescein label on the anti-CD3 antibody, but some of which is from the PE anti-CD8 and PE-Cy5 anti-CD4 antibodies. And some is probably from cellular autofluorescence, but we'll neglect that for the time being. We can get away with drawing a T cell gate using the uncompensated data because the fluorescein fluorescence pretty much dominates the uncompensated signal.

The situation is quite different when we look at panel B of Figure 1-19. The cells in this dot plot are only those with side scatter and fluorescence values falling within the T-cell gate shown in panel A. The axes of panel B are labeled as showing red and yellow fluorescence, both on logarithmic scales, and I labeled them that way because the data are not compensated. There are two major clusters of cells/events visible in panel B, but points in each display significant intensities of both red and yellow fluorescence.

Panel C of Figure 1-19 shows the same cells, i.e., those in the original T cell gate, after compensation has been applied. What compensation has done is solve three linear equations in three unknowns; this gives us the fluorescence intensities of the fluorescein anti-CD3, PE anti-CD8, and PE-Cy5 anti-CD4 antibodies, which can now be plotted as such, allowing the x- and y-axes of panel C to be labeled "PE-Cy5 anti-CD4" and "PE anti-CD8." The major clusters of cells, representing CD-4 bearing T lymphocytes (often described as CD3$^+$CD4$^+$ cells, where the superscript "+" denotes positive) and CD-8 bearing T lymphocytes (CD3$^+$CD8$^+$ cells), are clearly visible, and could be fit nicely into quadrants.

Now, it would probably be perfectly legitimate to label the x-axis of panel C as "PE-Cy5 CD4," or even just "CD4," and the y-axis as "PE CD8," or just "CD8." However, if you want to be picky, what you are looking at is antibody bound to the cells. There's little doubt that almost all of the anti-CD4 antibody bound to T cells is bound to CD4 antigen on the cell surfaces, or that almost all of the anti-CD8 antibody bound to T cells is bound to cell surface CD8 antigen. On the other hand, both Figure 1-14 and panel A of Figure 1-19 show apparent binding of anti-CD3 antibody to monocytes and granulocytes; this is almost certainly **nonspecific binding**, which can occur via a variety of different mechanisms, and if we haven't got "truth in labeling" for the cells, we won't have it for the axes.

Some labels for axes should get the axe right away. The first candidates on my hit list are "FL1," "FL2," "FL3," etc., which usually mean green (515-545 nm), yellow (564-606 nm), and red (635 to about 720 nm, by my guess, limited by the characteristics of the 650 nm long pass filter at the short end and by the fading response of the detector at the long end) fluorescence. These were the fluorescence measurement ranges in the Becton-Dickinson FACScan, the first really popular benchtop 3-color fluorescence flow cytometer. The fluorescence filters in this instrument could not be changed, so at least FL1, FL2, and FL3 always meant the same thing – to FACScan users. However, in the B-D FACSCalibur, which has replaced the FACScan, while FL1 and FL2 still represent the same wavelength ranges, FL3 is different for 3- and 4-color instrument setups (650 long pass for 3-color; 670 long pass for 4-color). I think it's perfectly appropriate to use, for example, "green fluorescence," "515-545 nm fluorescence" (indicating the approximate range), or "530 nm fluorescence" (indicating the center wavelength), or even "Green (530-545 nm) fluorescence)," but let's lose FL1, FL2, FL3, etc. If you're using a long pass filter, then say, for example, ">650 nm fluorescence." If the data come from a flow cytometer with multiple excitation beams, then you might want a label like "UV-excited blue fluorescence," or "355→450 nm fluorescence." And also remember that the fluorescence color designation or bandwidth range is only really appropriate if you're displaying or talking about uncompensated data; the whole point of compensation is to get you a new set of variables that represent the amounts of probes or labels in or on the cells, rather than the measurement ranges in the cytometer.

I've already been through the labels once, in the discussion in the previous column about whether to use the antigen name or the antibody name as an axis label. However, I will return to this area to skewer the next victim on my hit list, which is "FITC." Almost everybody uses this; I have done so myself, but I have seen the error of my ways. "FITC" is a perfectly valid abbreviation for fluorescein isothiocyanate, which is the most popular reactive fluorescein derivative used to attach a fluorescein label to antibodies and other probe molecules. Once the FITC reacts with the antibody, it isn't FITC anymore, and one typically dialyzes the fluorescent antibody conjugate, or runs it over a column, in order to remove free fluorescein (the FITC is pretty much all hydrolyzed by the time you finish, anyway). Oh, yes, FITC can also be applied directly to cells, to stain proteins; once again, what you end up with bound to the proteins is fluorescein, not FITC. It would seem simple enough to use "FL" as an abbreviation for fluorescein, the way we use "PE" for phycoerythrin. I guess the problem here is that nobody wants to describe a fluorescent antibody as, say, "FL anti-CD3," rather than "FITC anti-CD3," because that might get it confused with "FL1," "FL2," "FL3," etc. Well, after I take over the world, we won't have that problem.

Then there are the scatter signals. "Forward Scatter," "Small Angle Scatter," "FALS," and "FSC" are all acceptable as axis labels; however, unless you have calibrated your measurement channel and have derived a cell size measurement from forward scatter, "Cell Size" is really inappropriate. In the same vein, I'd use "Side Scatter," "Large Angle Scatter," "90° scatter," "RALS," or "SSC" without much hesitation, but avoid "Granularity." People knowledgeable about flow will know what you are measuring; if your audience is uninitiated, you should provide a brief explanation.

It's also about time that people stopped referring to data collected with flow cytometers as "FACS data" instead of "flow cytometry data." "FACS" is the abbreviation for "Fluorescence-Activated Cell Sorter (or Fluorescence-Activated Cell sorting"), originally used by Herzenberg et al, and has been a Becton-Dickinson trade name since B-D commercialized their instrument in the 1970's. All FACSes are flow cytometers, but not all flow cytometers are FACSes, and some FACSes, such as FACScans and FACSCounts, aren't even Fluorescence-Activated Cell Sorters.

And, finally, as long as I'm ticked off, I should remind you that the tick marks on the log scale will almost certainly not represent the real scale if the instrument uses log amps without compensating for their deviations from ideal response.

When Bad Flow Happens to Good Journals

Well, you might ask, does it really matter that much whether the axis labels are absolutely correct? Won't the more egregious mistakes be picked up before manuscripts get accepted and published? Unfortunately not; there has been a great deal of weeping and wailing in the cytometry community of late about this issue, because we see a lot of bad cytometry data presentation in a lot of the more prestigious general interest and cell biology journals, and even in some of the tonier titles in hematology and immunology.

To be sure, flow cytometry may not be the only technical area in which there are such problems. A typical paper with ten or more authors might include data from gel electrophoresis, gene array scanning, confocal microscopy, etc., as well as flow data. It will probably have been reviewed by no more than three people, and they can't know all of the methodology in detail. There may be gel curmudgeons and array curmudgeons out there grumbling at least as loudly as the flow curmudgeons and the confocal curmudgeons.

In preparing this edition of *Practical Flow Cytometry*, I asked several people to send me corrected versions of data displays that appeared in papers dealing with significant refinements in technology that were critical to the biological or medical applications discussed. The referees didn't pick up the original mistakes; neither did the authors, who were good sports about responding to my requests.

Most of the time, bad flow data presentations, or even minor errors in interpretation, don't invalidate the principal conclusion(s) of a paper. When they do, the obvious remedy is for the original authors to correct their errors, or for some other people to produce another paper using better technique to reach the right conclusion. But it's much better all around if the mistakes are corrected before the manuscripts get sent in.

Meanwhile, it is incumbent upon us all to maintain a certain level of vigilance, not only when preparing cytometric data for presentation and publication, but when looking at data that others have presented or published. If it's important to you to know the details of an experiment, either because you want to duplicate it and/or adopt the methodology or because its conclusions form part of the foundation for something you want to do, work through the details. These days, it's not that uncommon to find multiparameter flow data in a paper in which little details such as the source of the antibodies used, or even which antibodies had which labels, are omitted from the "Materials and Methods" section.

Now, in an ideal world, in which everything has been done correctly, it shouldn't matter that much; I've already come out in favor of simple axis labels such as "anti-CD4" or "CD4," and, assuming that the reagents and cell preparation, initial measurements, gating, and compensation were not flawed, it shouldn't matter which antibody or label was used in an experiment. But it does. If the details you need aren't in the published paper, contact the author. That's why the e-mail address, and the snail mail address, are there. There is also an increasing likelihood that there will be another option; the journal and/or the authors may maintain a web site from which you can get technical details that were omitted from the published work.

Sorting Sorting Out

Flow sorting extends gated analysis to isolate pure populations of viable cells with more homogeneous characteristics than could be obtained by any other means. If you can get the cells that interest you into a gate in your multiparameter measurement space, you can get them into a test tube, or into the wells of a multiwell plate. Flow sorting is especially useful in circumstances in which further characterization of the selected cells requires short- or long-term maintenance in culture or analytical procedures that cannot be accomplished by flow cytometry.

A flow cytometer is equipped for sorting by the addition of a mechanism for diverting cells from the sample stream and of electronics and/or computer hardware and software that can determine, within a few microseconds after a cell passes by the cytometer's sensors, whether the values of one or more measurement parameters fall within a range or ranges (called a **sort region**, or **sort gate**) preset by the experimenter, and generate a signal that activates the sorting mechanism. The selected cells can then be subjected to further biochemical analysis, observed in short- or long-term culture, or reintroduced into another biological system (as was mentioned on p. 26, a substantial number of animals and more than a few babies have been conceived from sorted sperm).

The range of particles that can be sorted has been extended substantially in recent years; laboratory-built[2325] and commercially available instruments are now in routine use for sorting *C. elegans* nematodes and *Drosophila* embryos, while laboratory-built microfluidic apparatus has been used to sort bacteria[2326] and could, in principle, sort DNA fragments, other macromolecules, or viruses[2327]. Sorting of beads, rather than cells, has also come into use for various applications of combinatorial chemistry; the work of Brenner et al[2328-9] on gene expression analysis presents a good example.

The first generation of practical sorters accomplished cell separation by breaking the sample stream up into **droplets**, applying an electric charge to the droplets containing the selected cells, and passing the stream through an electric field, which would divert the charged droplets into an appropriate collecting vessel. A few older, and some newer, instruments use mechanical actuators to collect cells from a continuous fluid stream; while such mechanical sorters operate at lower rates (hundreds versus thousands of cells/s) than droplet sorters, their closed fluidic systems are better adapted for work with potentially infectious or otherwise hazardous materials that might be dispersed in the aerosols inevitably generated by droplet sorters. Large-particle sorters are typically mechanical, but not all of them have closed fluidic systems.

In general, sorting larger objects limits you to lower sorting speeds. If you're sorting lymphocytes, or something smaller, in a droplet sorter, you can use a 50 µm orifice, and generate droplets at rates of 100,000 droplets/s. If you're sorting pancreatic islets, which may be a few hundred µm in diameter, you'll need a 400 µm orifice, and you probably won't be able to go much above 1 kHz for a droplet generation frequency. If you're sorting *Drosophila* embryos, using a mechanical sorter (they're probably a little too big for a droplet sorter), you can measure your sort rate in dozens per second, rather than thousands.

Since cells arrive at the observation point at random times, at least approximately following Poisson statistics, there is always some probability of coincidences, which, as was noted on pp. 17 and 20-21, can pose some problem in flow cytometric analysis. Coincidences pose a fairly obvious problem in sorting, as well; they can result in your getting cells you don't want in the same droplet/well/tube as cells you do want. If the sorter is operated in the so-called **coincidence abort** mode, in which a wanted cell accompanied by an unwanted cell is not sorted, the **purity** of sorted cells is maintained, but the **yield** is decreased, while if wanted cells coincident with unwanted ones are sorted, yield is maintained at the expense of purity. All other things being equal, working at higher cell analysis rates ultimately ends up increasing the likelihood of coincidences, but there may be times when the best strategy is to sort twice, first for enrichment of a rare subpopulation, and then to increase purity of the cells recovered during the first sort.

In many cases in which flow sorting comes to mind as an obvious way of answering questions about a cell subpopulation, multiparameter analysis may allow the desired information to be obtained expeditiously without physically isolating the cells. Since the 1990's, most flow cytometry is multiparameter flow cytometry, as should be obvious from the content of the past dozen or so pages. Things were different in the bad old days.

In the 1970's, a method that was likely to come to mind for determining the distribution of DNA content in a lymphocyte subpopulation defined by the presence of a particular cell surface antigen involved staining cells with the appropriate fluorescent antibody, and then flow sorting to isolate those cells bearing the surface antigen. The sorted cells would subsequently be stained with a DNA fluorochrome such as propidium iodide; the restained sorted cells could then be run through the flow cytometer once more to determine the DNA content distribution.

This procedure was actually followed when Ellis Reinherz and Stuart Schlossman wanted to know whether there was any difference in DNA synthetic patterns between CD4- (then T4-) and CD8- (then T8-) bearing T cells; cells were stained with fluorescein-labeled monoclonal antibodies, sorted on a Becton-Dickinson FACS fluorescence-activated cell sorter, then sent to my lab, stained with propidium iodide, and analyzed on my recently built flow cytometer, which, at that time, wasn't sensitive enough to measure immunofluorescence. The chart recorder attached to my "Cytomutt" duly produced histograms of DNA content for the CD4-positive and CD8-positive cells and the antigen-negative cells, which had also been sorted.

The technically demanding and tedious exercise just described, which required at least an hour's combined use of the two instruments, did get the desired results. However, it would have been much easier to stain the entire cell population with both the fluorescent antibody and the DNA fluorochrome, making correlated multiparameter measurements of antibody fluorescence and DNA fluorescence in each cell, and using gated analysis to compile the DNA content distributions of antibody-positive and antibody-negative cells, eliminating the sorting. There was even an instrument available to us that could have done the job.

To be fair, most immunologists, faced with the same problem today, would instinctively look toward multiparameter measurement for the solution. When some colleagues and I recently had occasion to revisit the issue of DNA content of peripheral blood CD4-positive and CD8-positive T cells in the context of HIV infection and response to multidrug therapy, it was reasonably simple to deal with cells simultaneously stained for CD3, CD4 or CD8, and DNA (and RNA) content[2330].

However, those of us who have been in the flow cytometry and sorting business for a long time are likely to experience a sense of *déjà vu* when the cell and molecular biologists and geneticists bring in samples to be sorted on the basis of expression of *Aequorea* green fluorescent protein (GFP) or, more likely, one of its variants. When I wrote the previous edition of this book, Martin Chalfie et al[1648] had just demonstrated the use of GFP as a reporter of gene expression; as far as he or I knew, nobody had yet done flow cytometry on cells transfected with GFP. Most cell sorting involved selection of cells bearing one or more surface antigens. Today, people who run sorting facilities tell me that a substantial amount of their time is now spent sorting samples for cells expressing GFP or its relatives. And they also mention that the people who bring in those samples often initially contemplate sorting the cells, staining them again to measure some other parameter, and reanalyzing them.

So, although multiparameter cytometry is now old hat for the immunologists, there are some other folks out there who haven't made it that far along the learning curve. I hope the above cautionary tale, and the lengthy discussion of multiparameter cytometry that has preceded it in this chapter, will help prevent unnecessary sorting. When in doubt, work with your sorter operator and facility manager.

The nuts and bolts details of sorting will be covered at length in Chapter 6; I'll devote the rest of this discussion to what is probably the most important step in designing a sorting experiment: doing the math. A lot of people think they know that state-of-the-art high-speed cell sorters can analyze at least 16 parameters and sort (into four streams) at rates of 100,000 cells/s. However, when I polled a select group of people who actually run state-of-the-art high-speed sorters in various labs at universities, medical facilities, and biotech and pharmaceutical companies in the Boston area, I found that nobody had done more than 8-parameter analysis, and that, while a few people had run 40,000 cells/s on occasion, 20,000 cells/s was a more typical analysis rate. Ger van den Engh, who has played and continues to play an important role in high-speed sorter development, recommends that experimenters assume analysis rates no higher than 10,000 cells/s when assessing the feasibility of proposed experiments.

Now, a lot of people want to use sorting to isolate cells that make up a very small fraction of the population being analyzed. Gross et al[2331] showed that it was possible to detect and sort cells from a human breast cancer line seeded into peripheral blood mononuclear cells at frequencies ranging from 1 cell in 10^5 to 1 cell in 10^7; they reported 40% yield and 22% purity for the sorts of cells at the lowest frequency. The raw numbers may be more impressive; a sample of 1.2×10^8 cells, which should have contained 12 cancer cells, was analyzed, giving rise to 23 sort decisions, of which 5 yielded cancer cells identifiable as such by microscopy. That sounds encouraging; even at 10,000 cells/s, it would only take about 3 hours to get 5 cells. Or about 6 hours to get 10 cells. And if you wanted to get 1,000 cells, you'd have to sort for about 25 days, 24/7.

You may have noticed that, when you're looking for cells present at low frequencies, while it is advantageous to be able to analyze at high speeds, there isn't much need for a high-speed sorting mechanism. In the above example, the sort frequency was 8/hr. There are a lot of people taking up time on very expensive, multiparameter high-speed sorters doing low frequency sorts based on one- or two-parameter measurements; sooner or later, somebody is going to make money selling simpler instruments for those jobs. Of course, if there is a method of enriching the population for the cells of interest before you start sorting – immunomagnetic separation, for example – you should take advantage of it.

A surprisingly large number of folks seem not to be doing the math before they write and submit grant applications involving sorting, which, for example, propose to isolate 10^6 cells initially present at a frequency of 1 cell/10^7. Even if you had a 100% yield, that would require analysis of 10^{13} cells *in toto*, and, even if you ran the high-speed sorter at 10^5 cells/s, it would take 10^8 seconds, or a little over three years, to do the sort. And, amazing though it seems, some of these cockamamie proposals actually get funded. A grant application is typically reviewed by a few more people than review a manuscript, but, if there are enough other high-tech gimmicks in the application, there may not be a reviewer who knows enough about sorting to ask the right questions. So, do the math. Whether as an applicant or as a reviewer, you could save the taxpayers some money.

Parameters and Probes II: What is Measured and Why

Most flow cytometers used for research, and the majority of such instruments used in clinical immunology applications, measure only three physical parameters, namely, forward (or small angle) and side (or large angle) light scattering and fluorescence, even if they measure 16 colors of fluorescence using excitation from four separate light sources. A few instruments can also measure light loss (extinction), or sense electronic impedance to measure cell volume. The remainder of the discussion of parameters and probes in this chapter will deal only with scatter and fluorescence measurements; Chapter 7 is more ecumenical and more comprehensive.

In the course of introducing cytometry in general and flow cytometry in particular, I have already covered DNA content determination using various fluorescent dyes and the identification of cells in mixed populations using fluorescently labeled antibodies. If you will flip back to Table 1-1 (p. 3), you will see that there are a great many parameters and probes about which I have, thus far, said nothing at all. However, DNA stains, on the one hand, and labeled antibodies, on the other, do represent two fundamentally different types of probes.

Probes versus Labels

The chemical properties of the DNA dyes themselves determine the nature and specificity of their interactions with the target molecule. The nature and specificity of interactions of labeled antibodies with their targets is, ideally, determined solely by the structure of their combining sites; labels are added to facilitate detection and quantification of the amount of bound antibody based on the amount of fluorescence measured from the label. Under various circumstances, the labels themselves may decrease the specificity of antibody binding; this is always at least slightly disadvantageous and may be intolerable. DNA dyes can fairly be classified as probes; molecules such as fluorescein more often serve as labels. But, as usual, there are gray areas.

Fluorescein diacetate (FDA), actually diacetylfluorescein, was discussed on pp. 24-27; this is an example of a **fluorogenic enzyme substrate**. The nonfluorescent, uncharged FDA diester freely crosses intact cell membranes; once inside cells, it is hydrolyzed by nonspecific esterases to produce the fluorescein anion, which is highly fluorescent and which leaves intact cells slowly. Since most cells contain nonspecific esterases, FDA is not terribly useful as an indicator of enzyme activity; other nonfluorescent fluorescein derivatives can be used as probes for the activity of more interesting enzymes, such as beta-galactosidase. Different derivatives of fluorescein and other dyes can be introduced into cells and cleaved by esterases to produce indicators of pH, oxidation-reduction (redox) state, and the concentration of sulfhydryl groups or of ions such as calcium and potassium. So the best I can do to clarify the status of fluorescein is to say that it is a label when it is used covalently bound to a relatively large molecule such as an antibody, oligonucleotide, or protein ligand for a cellular receptor, and a probe when introduced into cells in a slightly chemically modified, low molecular weight form. The detailed discussion of probes in Chapter 7 provides examples of when this distinction breaks down.

We will now embark on a quick tour of selected parameters and probes for their measurement. Details and spectra appear in Chapter 7. It is appropriate to mention that the single most useful reference on fluorescent probes is the *Handbook of Fluorescent Probes and Research Products*[2332], edited by Richard P. Haugland; this is the catalog of Molecular Probes, Inc. (Eugene, OR). The latest printed version is the 9th Edition, which appeared in 2002. A CD-ROM version is available as well, and all the information in the handbook, and more, with updates, can also be found at Molecular Probes' Web site (www.probes.com).

Living and Dyeing: Stains, Vital and Otherwise

Before getting down to specific (and not-so-specific) stains, it's probably a good idea to define some terms relevant to staining cells and what does or does not have to be done to the cells in order to get them to stain. A dye or other chemical that can cross the intact cytoplasmic membranes of cells is said to be **membrane-permeant**, or, more simply, **permeant**; a chemical that is excluded by intact cytoplasmic membranes is described as **membrane-impermeant**, or just **impermeant**. Because permeant dyes stain living cells, they (the dyes) are also described as **vital dyes**, or **vital stains**. You will occasionally find an opposite, incorrect definition of a vital stain as a stain that does not stain living cells; don't believe it. This seems to be one of the few urban legends of cytometry.

There are numerous transport proteins that concentrate certain chemicals in, or extrude other chemicals from, cells. Many commonly used dyes, including Hoechst 33342, serve as substrates for the glycoprotein pump associated with multidrug resistance in tumor cells, and may not readily stain cells in which this pump is active; the general lesson is that the action of transporters may make it appear that a permeant compound that is efficiently extruded is impermeant. Microorganisms may have a broader range of transporters than do mammalian cells, making it risky to assume that they will handle dyes in the same way.

Staining cells with impermeant dyes requires that the membrane be **permeabilized**. This can be accomplished in the context of **fixation** of the cells. "Fixation" originally described a process that made tissue tough enough to section for microscopy and prevented it from being autolyzed by internal hydrolytic enzymes and/or chewed up by contaminating microorganisms. Most fixatives act either by denaturing proteins (e.g., ethanol and methanol) or by cross-linking them (e.g., formaldehyde and glutaraldehyde); since this is likely to change the structure of cell-associated antigens, it is common practice to stain with fluorescent antibodies before fixing cells. In general, the fixation procedures used for flow cytometry are relatively mild; one principal objective is to kill HIV and other viruses that may be present in specimens, and another is to allow samples to be kept for several days before being analyzed. In recent years, the real pathologists have been using microwave radiation as a fixative or adjunct; I have not run across reports of its use for flow cytometry.

Permeabilization without fixation can be accomplished using agents such as the nonionic detergents Triton X-100 and Nonidet P-40; permeabilizing agents may also be added to a mixture of one or more fixatives to make cytoplasmic membranes permeable to fluorescent antibodies while retaining cellular constituents, allowing staining of intracellular antigens. Several proprietary mixtures, some of which include red cell lysing agents, are available from manufacturers and distributors of antibodies.

Most sorting is done with the intention of retrieving living cells, so fixation is not an option. However, there are procedures, such as lysolecithin treatment and electroporation, which can transiently permeabilize living cells, allowing otherwise impermeant reagents to enter while preserving viability of at least some of the cells in a sample. In this context, it is important to remember that a permeant "vital" stain may eventually damage or kill cells. It is always advisable to establish that measurement conditions do not themselves perturb what one is attempting to measure.

Nucleic Acid (DNA and RNA) Stains

Although a large number of fluorescent dyes can be used to stain DNA and/or RNA, relatively few of them are specific for DNA, and most of these are sensitive to base composition (A-T/G-C ratio). **DAPI** (4', 6-diamidino-2-phenylindole), **Hoechst 33258**, and **Hoechst 33342** increase fluorescence approximately 100 times when bound to A-T triplets in DNA. All these dyes are excited by UV light (325-395 nm), and emit in the blue spectral region with maxima between 450 and 500 nm.

Chromomycin A_3 and **mithramycin** exhibit increased fluorescence on binding to G-C pairs in DNA; they are excited by violet or blue-violet light (400-460 nm) and emit in the green between 525 and 550 nm. The combination of Hoechst 33258 and chromomycin A_3 has been used with dual excitation-beam flow cytometers to discriminate the majority of human chromosomes based on differences in DNA base composition, and to demonstrate differences in base composition among bacterial species. **7-amino-actinomycin D** (**7-AAD**[735]) also enhances fluorescence (maximum around 670 nm) on binding to G-C pairs in DNA; although it is best excited by green light (500-580 nm), it can be excited at 488 nm.

Dyes such as **ethidium bromide** (**EB**) and **propidium iodide** (**PI**), both excitable over a range from 325 to 568 nm and emitting near 610 nm, increase fluorescence on binding to double-stranded nucleic acid, whether DNA or RNA, and the latter property is shared by a large number of asymmetric cyanine nucleic acid stains (e.g., the **TO-PRO-** and **TOTO-** series (impermeant), **SYTO**-series (permeant), **Pico Green**, etc.) introduced by Molecular Probes. These dyes can be used to stain total nucleic acid in cells; specific staining of DNA requires RNAse treatment. Many of the cyanine nucleic acid dyes increase fluorescence several thousandfold; they have been used for detection of DNA fragments[1144,2327,2333-4] and viruses[2335-7].

Until recently, Hoechst 33342 was the only dye that could be used reliably to determine DNA content in living cells. However, in 1999 and 2000, Smith et al[2338-9] reported that **DRAQ5**, an anthraquinone dye with an excitation maximum around 650 nm and an emission maximum near 700 nm when bound to DNA, could also provide a reasonably good DNA content histogram. DRAQ5 can also be excited at 488 nm, albeit somewhat inefficiently.

DRAQ5 does not increase fluorescence significantly on binding to DNA; it stains nuclei because it is present in higher concentrations in association with nuclear DNA than elsewhere in the cell, and the quality of staining is thus relatively more dependent on relative concentrations of dye and cells than is the case for most other DNA dyes. **Acridine orange (AO)**, like DRAQ5, does not increase fluorescence on binding to either DNA or RNA, but stains by virtue of its concentration on the macromolecules.

Darzynkiewicz et al showed, beginning in the mid-1970's, that, after cell membrane permeabilization and acid treatment, AO could be used for stoichiometric staining of DNA and RNA in cells[262-3,525,1348-9]. On excitation with blue light (488 nm is eminently suitable), the DNA-bound monomer fluoresces green (about 520 nm); the RNA-bound dye forms red (>650 nm) fluorescent aggregates. The combination of DNA and RNA staining allows the cell cycle to be subdivided into stages that are not distinguishable on the basis of DNA content alone, permitting discrimination between G_0 and G_1 cells.

Relatively specific staining of double-stranded (predominantly ribosomal) RNA in cells can be achieved using a combination of **pyronin Y** (excitable at 488 nm with emission in the yellow around 575 nm), which stains RNA, with one of the Hoechst dyes, which binds to DNA and prevent DNA staining by pyronin Y. In a dual excitation-beam instrument (UV and 488 nm), DNA and RNA content in living cells can be estimated simultaneously from pyronin Y and Hoechst 33342 dye fluorescence, providing information that is substantially equivalent to what could be obtained using AO (Fig. 1-2, p. 27) without requiring that the cells be sacrificed[113]. Cells stained with this dye combination have been sorted with retention of viability[2340-2].

Toba et al[2343-5] found that DNA and RNA could be measured in permeabilized cells using the combination of 7-AAD and pyronin Y in a system with a single 488 nm excitation beam; Schmid et al modified the staining conditions and reported improved precision and reproducibility[2346].

Fluorescence and Fluorescent Labels

Because the fluorescent label on a probe is usually not intended to interact directly with the structure to which the probe binds, labels are developed and/or synthesized predominantly for their desirable spectral characteristics.

In order for an atom or molecule – or part of a molecule; the all-inclusive term would be **fluorophore** – to emit fluorescence, it must first absorb light at a wavelength shorter than or equal to the wavelength of the emitted light, raising an electron to an excited state. Absorption requires only about a femtosecond. In order to have a high likelihood of fluorescing, a material must have a high likelihood of absorbing the excitation light; the likelihood that a molecule will absorb is quantified as the **absorption cross-section** or the **molar extinction coefficient**.

Fluorescence results from the loss of at least some of the absorbed energy by light emission. The period between absorption and emission is known as the **fluorescence lifetime**; for organic compounds, this is typically a few nanoseconds. Some of the absorbed energy is almost always lost nonradiatively, i.e., unaccompanied by emission, by transitions from higher to lower vibrational energy levels of the electronic excited state. The fluorescence emission will then be less than the energy absorbed; in other words, emission will occur at a wavelength longer than the excitation wavelength. The difference between the absorption and emission maxima is known as the **Stokes shift**, honoring George Stokes, who first described fluorescence in the mid-1800's. Stokes shifts are typically only a few tens of nanometers.

Fluorescence is an intrinsically quantum mechanical process; the absorbed and emitted energy are in the form of photons. The **quantum yield** and **quantum efficiency** of fluorescence are, respectively, the number and percentage of photons emitted per photon absorbed; they typically increase with the cross section and extinction coefficient, but are also dependent on the relative likelihoods of the excited molecule losing energy via fluorescence emission and nonradiative mechanisms. The quantum yields of some dyes used in cytometry are quite high, above 0.5, but it is important to note that quantum yield, particularly for organic fluorophores, is affected by the chemical environment (i.e., the pH, solvent polarity, etc.) in which the molecule finds itself. If an excited molecule that might otherwise fluoresce instead loses energy nonradiatively, for example, by collision with solvent molecules, it is said to be **quenched**; once returned to the electronic ground state, it can be reexcited. However, there is usually a finite probability that light absorption will be followed by a change in molecular structure, making further cycles of fluorescence excitation and emission impossible; this is called **(photo)bleaching**.

In principle, increasing the illumination intensity can increase the intensity of light scattering signals without limit. However, this is not even theoretically possible for fluorescence signals, because, at some level of illumination, all the available molecules will be in excited states, leaving no more to be excited if illumination intensity is further increased. This condition of **photon saturation** is often reached in cytometers which use laser powers of 100 mW or more; bleaching, which may also make the dependence of emission intensity on excitation intensity less than linear, is noticeable at power levels of tens of milliwatts. Saturation and bleaching are discussed at length by van den Engh and Farmer[1130].

When an excited fluorophore is in close proximity (typically no more than a few nanometers) to another fluorophore, nonradiative energy transfer (**fluorescence resonance**

energy transfer, or **FRET**) from the excited (donor) molecule to the nearby acceptor molecule may occur, followed by fluorescence emission from the acceptor in its emission region. The probability of energy transfer increases with the degree of overlap between the absorption spectrum of the second fluorophore and the emission spectrum of the first. I have said "fluorophore" rather than "molecule" here because energy transfer can occur between different structures within the same molecule. An accessible review of FRET is provided by Szöllösi et al[2347].

In the intact photosynthetic apparatus of algae and cyanobacteria, absorbed blue-green and green light is utilized for photosynthesis by a series of intra- and intermolecular energy transfers via **phycobiliproteins** to **chlorophyll**, without subsequent emission. In 1982, Oi, Glazer, and Stryer[114] reported that extracted algal phycobiliproteins could be used as highly efficient fluorescent labels with large Stokes' shifts. As you might have noticed from the extensive previous discussion, it has become common practice to attempt to improve on nature by conjugating dyes to phycobiliproteins to add an additional phase of energy transfer and further shift the emission spectrum of the **tandem conjugates**. The first such tandem conjugate, described by Glazer and Stryer in 1983[306], was made by linking **phycoerythrin (PE)** to **allophycocyanin (APC)**, a phycobiliprotein which absorbs relatively efficiently, although not maximally, at phycoerythrin's yellow (575 nm) emission wavelength and which emits maximally in the red at 660 nm.

Until both flow cytometers and monoclonal antibodies became widely available in the early 1980's, the most widely used fluorescent label was **fluorescein**, usually conjugated to proteins as the **isothiocyanate (FITC)**; second labels were only infrequently needed. Fluorescein is nearly optimally excited at 488 nm, and emits in the green near 525 nm. While rhodamine dyes had been used for two-color immunofluorescence analysis by microscopy, they were not suitable for 488 nm excitation. A small number of studies were done with yellow-excited dyes, which needed a second excitation beam, making flow cytometers substantially more expensive. Phycoerythrin (PE), which emits in the yellow near 575 nm, is maximally excited by green light but absorbs reasonably well at 488 nm. Its extinction coefficient is high enough to make the fluorescence signal from PE-labeled antibody substantially higher than that from an equivalent amount of fluorescein-labeled antibody (Fig. 1-18, p. 37).

We have already encountered tandem conjugates of PE suitable for 488 nm excitation (PE-Texas red, emitting near 610 nm; PE-Cy5, near 670 nm; PE-Cy5.5, near 700 nm; PE-Cy-7, near 770 nm). Allophycocyanin absorbs maximally in the red near 650 nm, and is well excited by red diode (635-640 nm) and He-Ne (633 nm) lasers. Tandem conjugates of APC with Cy5.5 and Cy7 emit in the far red and near infrared, as do the PE conjugates with the same dyes.

A principal disadvantage of phycobiliproteins as fluorescent labels is their large size; with a molecular weight near 240,000, PE binding increases the molecular weight of an immunoglobulin G antibody by about 150 percent. This may not be an issue when labeled antibodies or lectins are used to stain cell surface structures, but becomes one when it is necessary to use labeled reagents to demonstrate intracellular constituents. A number of lower molecular weight labels have been developed for this purpose. The symmetric cyanines[1361-4] include Cy5, Cy5.5, and Cy7, and their shorter wavelength absorbing cousins, e.g., Cy3, which can be excited at 488 nm and emits in the same region as PE; we have already run across them as acceptors in tandem conjugates. Molecular Probes has recently developed the Alexa series of dyes[2348] (also see the Molecular Probes handbook/Web site[2322]); different members of this series are excitable at wavelengths ranging from the UV to the near infrared. Alexa dyes, used alone or as acceptors in tandem conjugates, are reported to have better fluorescence yields and photostability (resistance to bleaching) than more commonly used labels with similar spectral characteristics, and seem to be coming into wider use. Low, rather than high, molecular weight labels are almost always used on oligonucleotide probes, which allow demonstration and quantification of specific nucleic acid sequences in cells or on beads or solid substrates (e.g., in gene arrays).

As was mentioned previously, it is the probe, not the label, that confers specificity; dyes must be derivatized into forms that contain a functional group, such as an isothiocyanate or sulfonyl chloride, that will allow the **reactive dye** to bind covalently to the probe. FITC, applied to cells, will stain accessible proteins. Staining of intact cells will be limited to the cell surface; in fixed or permeabilized cells, both surface and intracellular proteins will be stained.

Binary Fishin': Tracking Dyes Through Generations

Otherwise nonspecific, but persistent fluorescent staining of cellular proteins or lipids has recently been put to good use in studying cell proliferation. Since cellular proteins and lipids are apportioned more or less equally to each daughter cell during cell division, analysis of the fluorescence of cells after staining with a so-called **tracking dye** should allow determination of how many cycles of division have occurred since its ancestor was stained. The dye first widely used for such studies was **PKH26**[1551-5], a yellow fluorescent cyanine dye with long alkyl side chains that incorporates itself tightly enough into lipid bilayers that it is not readily lost from cells. It was called a tracking dye because it could also be used to follow cells that had been removed from animals, labeled, and reinjected. Estimation of the numbers of cells in various daughter generations after PKH26 labeling requires application of a mathematical model[1555].

An alternative to PKH26, **carboxyfluorescein diacetate succinimidyl ester (CFSE)**[2349], is a nonfluorescent fluorescein ester that enters cells and is hydrolyzed to a reactive dye by nonspecific esters; the end result is that fluorescein molecules are bound covalently to intracellular protein. Distributions of CFSE fluorescence in proliferating populations usually show peaks indicating the positions of cells in different daughter generations; these can be analyzed with mathematical models, but it is also possible to combine sorting with CFSE labeling to isolate cells from different generations[2350], which cannot be done reliably when PKH26 is used as a tracking dye.

Membrane Perturbation: A Matter of Life and Death?

The integrity of the cytoplasmic membrane is essential to cell function. Although at least some cells can survive transient small breaches of the membrane, longer-term and/or larger defects may deprive the cell of materials it would normally accumulate, and may also expose it to toxins it would normally exclude. Thus, we tend to think that cells with a demonstrable loss of membrane integrity are dead.

Trypan blue has been the preferred probe for a **dye exclusion test for "viability,"** i.e., retention of membrane integrity, performed by visual inspection of cells under the microscope; the Bio/Physics Systems Cytograf, made in the early 1970's, measured extinction and scattering using a red He-Ne laser source, and could detect trypan blue uptake by cells. These days, people who want to do dye exclusion testing by flow cytometry typically use impermeant nucleic acid dyes such as **propidium iodide** or **7-aminoactinomycin D**, both excitable at 488 nm, and, emitting, respectively, at about 620 and about 670 nm, or the red-excited dye **TO-PRO-3**, emitting at about 670 nm. Cells that take up the dye and become fluorescent are considered to be nonviable.

Fluorescein is anionic, and, therefore, relatively impermeant; when produced intracellularly by hydrolysis of fluorescein diacetate (FDA), it leaves cells slowly, giving us a **dye retention test for "viability."** Cells with intact membranes accumulate and retain fluorescein after exposure to FDA and become (green) fluorescent; cells with membrane damage do not retain fluorescein and do not fluoresce. The fluorescein derivative **calcein**, produced in cells by esterase action after exposure to the **acetoxymethyl ester**, **calcein-AM**, is retained much more effectively than fluorescein and is now preferred for dye retention tests.

The problem with dye exclusion and retention tests is that, while the methodology works well for cells that are killed by freezing or heat or by interaction with cytotoxic T or NK cells, all of which inflict early and usually lethal damage on the cytoplasmic membrane, cells that are killed by other means, e.g., those rendered reproductively nonviable by such agents as ionizing radiation, may retain membrane integrity for days after exposure. Uptake of impermeant dyes is therefore a better indicator of nonviability than retention is of viability, but there are situations in which impermeant dyes can end up in viable cells[2351].

One can, of course, combine dyes, for example, propidium iodide and calcein-AM, which will result in cells with intact membranes exhibiting green cytoplasmic fluorescence while cells with damaged membranes show red nuclear fluorescence, but this does not solve the basic problem. And, in part thanks to cytometry, we can now distinguish one kind of death (**necrosis**) from another (**apoptosis**), making the issue of viability assays even more contentious. Darzynkiewicz et al have discussed the cytometry of cell necrobiology in detail[2352]. Disturbances in membrane organization in apoptosis, resulting in the exposure of phosphatidylserine, are usually looked for using fluorescently labeled **annexin V**[2353-4].

When viability is not an issue, measurements of fluorescence of cells over time after exposure to fluorescent dyes, drugs, or labeled drug analogs can be useful in detecting the presence of various transport proteins. Uptake or efflux kinetics in themselves can only suggest a mechanism; when the transporter or pump being investigated has been well characterized, establishing that known substrates and inhibitors affect fluorescence kinetics as predicted is critical for confirmation of the initial hypothesis.

Cytoplasmic/Mitochondrial Membrane Potential

Electrical potential differences are present across the cytoplasmic membranes of most living prokaryotic and eukaryotic cells, and also between the cytosol and the interior of organelles such as chloroplasts and mitochondria. Membrane potential ($\Delta\Psi$) is generated and maintained by transmembrane concentration gradients of ions such as sodium, potassium, chloride, and hydrogen.

Changes in cytoplasmic $\Delta\Psi$ play a role in transmembrane signaling in the course of surface receptor-mediated processes related to the development, function, and pathology of many cell types. Cytoplasmic $\Delta\Psi$ is reduced to zero when the membrane is ruptured by chemical or physical agents; mitochondrial $\Delta\Psi$ is reduced when energy metabolism is disrupted, notably in apoptosis. In bacteria, $\Delta\Psi$ reflects both the state of energy metabolism and the physical integrity of the cytoplasmic membrane.

Flow cytometry can be used to estimate membrane potential in eukaryotic cells, mitochondria *in situ*, isolated mitochondria, and bacteria[424,2355]. Older methods, using lipophilic cationic dyes such as the symmetric cyanines **dihexyloxacarbocyanine [DiOC$_6$(3)]** and **hexamethylindodicarbocyanine [DiIC$_1$(5)]** or **rhodamine 123**, or lipophilic anionic dyes such as **bis (1,3-dibutyl-barbituric acid) trimethine oxonol [DiBAC$_4$(3)]** (which is often, incorrectly, referred to as **bis-oxonol**), can detect relatively large changes in $\Delta\Psi$, and identify heterogeneity of response in subpopulations comprising substantial fractions of a cell population. All of the dyes just mentioned can be excited at 488 nm and emit green fluorescence, with the exception of DiIC$_1$(5), which is red-excited and emits near 670 nm. Newer techniques that use energy transfer and/or ratios of fluorescence emission at different wavelengths allow precise measurement of $\Delta\Psi$ to within 10 mV or less[2356-7].

Since, in most eukaryotic cells, $\Delta\Psi$ across mitochondrial membranes is larger than $\Delta\Psi$ across cytoplasmic membranes, exposure of cells to lipophilic cationic dyes results in higher concentrations of dye in the cells than in the suspending medium, and higher concentrations in mitochondria than in the cytosol. If cells are washed after being loaded with dye, staining of the cytosol may be minimized while mitochondrial staining persists. This is the basis for the use of $DiOC_6(3)$, $DiIC_1(5)$, rhodamine 123, and other cationic dyes to estimate mitochondrial ΔΨ; the procedure has become commonplace for studies of apoptosis, in which early increases in mitochondrial membrane permeability result in loss of ΔΨ. **JC-1**, a cyanine, exhibits green fluorescence in monomeric form and red fluorescence when aggregated at higher concentrations[1681-2], and has become popular for work on mitochondria in apoptosis.

Among other factors, action of efflux pumps, changes in membrane structure, and changes in protein or lipid concentration in the medium in which cells are suspended can produce changes in cellular fluorescence which may be interpreted erroneously as changes in ΔΨ. For example, it was observed in the 1980's that hematopoietic stem cells were not stained by rhodamine 123, and some people concluded that this reflected low mitochondrial ΔΨ; it was later found that the dye was being actively extruded by a glycoprotein pump. Getting good results from cytometric techniques for estimation and measurement of ΔΨ demands careful control of cell and reagent concentrations and incubation times and selection of appropriate controls.

Indicators of Cytoplasmic [Ca^{++}]: Advantages of Ratiometric Measurements

The importance of calcium fluxes in cell signaling was appreciated when flow cytometry was in a relatively early stage of development, but it was not until some years later that suitable probes became available[2358]. The first probes exhibited differences in the intensity of fluorescence in the presence of low and high intracellular [Ca^{++}], but did not change either their fluorescence excitation or emission spectral characteristics to a significant degree. Since the distribution of fluorescence intensity from cells loaded with the probes was typically quite broad (a problem also associated with membrane potential probes), it was possible to appreciate large changes in cytoplasmic [Ca^{++}] affecting all or most of the cells in a population, which would shift the entire distribution substantially, but not to detect even a large change in cytoplasmic [Ca^{++}] involving only a small subpopulation of cells. This came as a disappointment to immunobiologists who hoped to use flow cytometry to detect calcium responses associated with activation of lymphocytes by specific antigens.

Roger Tsien and his colleagues, who had developed some of the earlier calcium probes, came to the rescue in 1985 with **Indo-1**[858]. This, like other probes, is a selective calcium chelator, but does not significantly perturb cellular calcium metabolism. Its fluorescence is excited by UV light; wavelengths between 325 and about 365 nm, which pretty well covers the range of UV sources available for flow cytometry, are suitable. Indo-1's attraction, however, is due primarily to the fact that there are substantial differences in emission spectra between the free dye, which shows maximum emission at about 480 nm, and the calcium chelate, which emits maximally at about 405 nm. The **ratio** of emission intensities at 405 and 480 nm in cells loaded with Indo-1 [it is introduced as an acetoxymethyl (AM) ester] can, therefore, provide an indication of cytoplasmic [Ca^{++}]. The **ratiometric** measurement cancels out many extraneous factors, most notably including the effect of cell-to-cell variations in dye content, which plague older techniques for calcium measurement and for measurement of ΔΨ. Effects of uneven illumination and of light source noise also are eliminated by virtue of their equal influences on the numerator and denominator of the ratio. This advantage, it should be noted, is common to other ratiometric measurements (e.g., of ΔΨ and of intracellular pH) in which both parameters used in the ratio are measured at the same time in the same beam.

If aliquots of loaded cells are placed in solutions with various known Ca^{++} concentrations and treated with a **calcium ionophore** such as **A23187** or **ionomycin**, it is possible to calibrate the fluorescence ratio measurement to yield accurate molar values of cytoplasmic [Ca^{++}]. Indo-1 is widely used, at least by people with UV excitation sources in their flow cytometers[862, 1714-8].

Since there are probably more than 10,000 fluorescence flow cytometers out there that don't have UV sources, that's small comfort. Luckily, there are alternatives. In 1989, Tsien and his collaborators described a series of fluorescein- and rhodamine-based calcium indicators suitable for use with 488 nm excitation[1719]. The most widely used of these is **Fluo-3**, which has the spectral characteristics of fluorescein, but which is almost nonfluorescent unless bound to calcium. Unlike Indo-1, Fluo-3 does not exhibit a spectral shift with changes in calcium concentration. A Fluo-3 fluorescence distribution is a haystack; if you're stimulating a cell population, the haystack moves to the right when the cytoplasmic [Ca^{++}] goes up and back to the left when it goes back down. However, there is another dye, **Fura red**, also suitable for 488 nm excitation, which exhibits high fluorescence when free in solution (or cytosol) and low fluorescence when bound to calcium; a Fura red haystack moves in the opposite direction from a Fluo-3 haystack with changes in cytoplasmic [Ca^{++}]. More to the point, the ratio of fluo-3 to Fura red fluorescence provides a precise, calibratable indicator of cytoplasmic [Ca^{++}] that can be used in the majority of fluorescence flow cytometers[2358]. Both Fluo-3 and Fura red, like Indo-1, are loaded into cells as AM esters.

Finding Antigen-Specific Cells Using Tetramers

While ratiometric probes did improve the precision of intracellular calcium measurements, they did not get them quite to the point of being able to detect specific responses of very small numbers of lymphocytes to antigens. As it turned out, a more direct approach was to succeed. In 1996, Altman et al[2359] described identification of antigen-specific cytotoxic ($CD3^+CD8^+$) T cells using a fluorescently labeled complex containing four each of 1) a class I major histocompatibility complex (MHC) α chain, 2) β_2-microglobulin, and 3) an antigenic peptide. Since that gets to be a lot to write or say, the probes are now universally described as **tetramers**.

Antigen presentation to T cells requires binding of antigenic peptides associated with HLA proteins (class I proteins for cytotoxic [$CD3^+CD8^+$] T cells, class II proteins for helper [$CD3^+CD4^+$] T cells) on the antigen presenting cell to the T cell receptor; attempts to bind a labeled monomeric complex (1 each) of α chain, β_2-microglobulin, and peptide to cytotoxic T cells were unsuccessful because the binding affinity of the monomers was too low. Tetramers did the trick, and have come into wide use since they were originally described[2360-2]. We now have not only **class I tetramers**, reactive with cytotoxic T cells, but also **class II tetramers**, which contain MHC class II proteins, and react in an antigen-specific fashion with helper T cells[2363-4]. They're not available at the corner store quite yet, but rumor has it that it was tetramers and their possibilities that made one of the major instrument companies decide to stay in the fluorescence flow cytometry business.

Hip, Hip Arrays: Multiplexing on Slides and in Bead Suspensions

If you have been keeping up with biology at all over the past few years, it's unlikely that you have not run across gene array technology[2365-7], which allows the expression of hundreds or thousands of genes to be studied by, for example, hybridizing different colors of labeled cDNA derived from the same cells grown under different circumstances to a slide on which the requisite genetic sequences have been synthesized or deposited in small spots. The slides are then scanned, allowing differences in expression to be detected by color differences resulting from the presence of different amounts of the cDNAs on each spot. The array concept has taken off; we have gene arrays, protein arrays, cell arrays, and even tissue microarrays, which allow high-throughput molecular profiling of tumors[2368].

Multiplex analysis allows flow cytometry to accomplish some of the same tasks for which gene arrays are now used. It occurred to various people in the mid-1980's[1820-34] that various types of ligand binding assays could be done in a flow cytometer by using fluorescence measurements to quantify binding to appropriately coated beads. By using a different size and/or color bead for each of a number of assays, it would be possible to perform all of them at once on a single sample in a single tube[2369].

The latest incarnation of multiplex analysis uses a small, dedicated flow cytometer capable of identifying as many as 100 different colors of beads, and has been applied successfully to both protein[2370-1] and nucleic acid[2372-4] analysis. In a study monitoring multiple pathogenesis-related genes simultaneously in chemical-treated and control *Arabidopsis* samples, Yang et al[2375] reported that a multiplexed flow cytometric assay they developed yielded results comparable to those obtained from a slide-based gene array.

GFP and Its Relatives: Mild-Mannered Reporters

The 1994 report by Chalfie et al[1648] on the use of *Aequorea* green fluorescent protein (GFP) as a reporter of gene expression quickly spawned a growth industry. GFP mutants are now available with cyan, green, and yellow fluorescence and with excitation characteristics far better suited to flow cytometry (and imaging, confocal microscopy, etc.) than the wild type protein. Moreover, GFP variants have been engineered to behave as sensors of such functional parameters as intracellular (or intracompartmenal) pH, [Ca^{++}], etc., and, using energy transfer between molecules with different spectra, for quantitative measurements of protein-protein interactions[2376-8]. The mild-mannered reporter has shed glasses and business suit and emerged from the phone booth as Supermolecule. I have already mentioned that sorting for fluorescent protein expression now seems to account for a significant amount of flow facilities' time; this trend can be expected to continue.

Beyond Positive and Negative: Putting the -Metry in Cytometry

If you spend most of your flow cytometer time doing immunofluorescence analysis, you can pick up some bad habits. Given an instrument that often costs upwards of a hundred thousand dollars, is full of fancy electronics, has its own computer attached to it, and can probably detect a few hundred molecules of fluorescent dye in or on a cell, it does seem that we underutilize its capacities when we report the results of highly sensitive and precise fluorescence measurements as "positive" and "negative."

To be sure, sometimes "positive" and "negative" are good enough to get the job done. In the previous examples of counting various types of T lymphocytes in human peripheral blood, we defined the subpopulation of T cells by their scattering characteristics and by the presence of the CD3, CD4, or CD8, cell surface antigens, and, in general, the cells we're looking at either have a substantial amount of the antigen or have little or none. When we look at our "CD3-positive" cells, they either do or do not have substantial amounts of the CD4 and CD8 antigens. We don't need to be experienced in flow cytometry to know "positive" and "negative" when we see them in these contexts, and, using these concepts, we can obtain a satisfactory answer to the question, "What are the relative proportions of ($CD3^+CD8^+$) and ($CD3^+CD4^+$) T cells in this blood sample?"

However, if the question we are asking is, "What proportion of ($CD3^+CD4^+$) T cells are activated?," we may need to extend our conceptual framework somewhat, both in terms of biology and in terms of cytometry. "What is an activated lymphocyte?," "What is a cancer cell?," and "What is a dead cell?" are major quasitheological questions guaranteed to provoke debate among analytical cytologists for a while to come. But let's suppose we have decided to define activation in terms of expression of the CD25 antigen, which is the cell surface receptor for the cytokine interleukin-2. Well, then, we can just gate the T cells, further gate the CD4-positive cells, and then count the CD25-positive and negative cells, right? Unfortunately not. The number of CD25 molecules on an inducer T cell seems to range from hundreds or less to many thousands; the problem in defining "positive" and "negative" is that there is no clear breakpoint.

Well, then, perhaps we could say that a cell with more than 5,000, or 10,000, or some other seemingly arbitrary number of molecules of CD25 on its surface is activated. That might work, provided we had a way of determining the number of molecules from the immunofluorescence measurement. As it turns out, this can be done, but it isn't always as easy as it looks.

The hematology counters are ahead of the fluorescence flow cytometers in this department. They all report **red cell indices**, including erythrocytes' **mean corpuscular volume (MCV)** in femtoliters and **mean corpuscular hemoglobin (MCH)** in picograms. Every instrument in every lab everywhere uses the same units. Way back in 1977, I suggested that we should have "**white cell indices,**" which didn't go over resoundingly well in the Dark Ages of polyclonal antisera. The proposition has been better received of late, for several reasons. The need is more apparent, our apparatus and reagents are better, and there are people interested in developing and testing standardized materials that will make it possible for everyday users of flow cytometry to do quantitative immunofluorescence measurements. Figure 1-20 illustrates one technique, which uses beads with known numbers of antibody binding sites as standards.

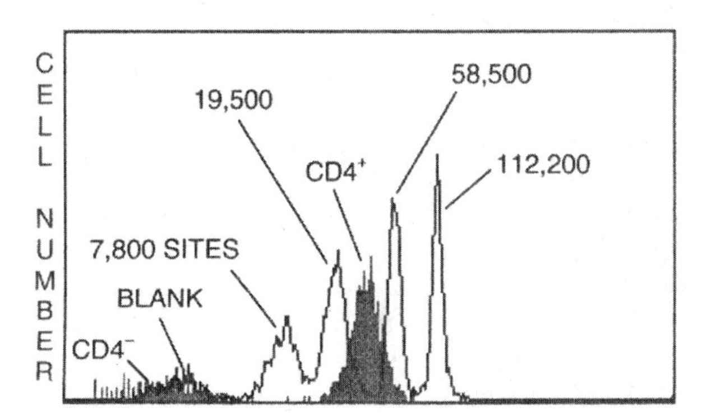

Figure 1-20. Fluorescence intensities of CD4-positive and negative cells (plotted as bars) compared with intensities of beads bearing known numbers of antibody binding sites, stained with the same fluorescein-anti-CD4 antibody as was used to stain the cells.

What Figure 1-20 shows is that most of the CD4-positive cells in the same lysed whole blood sample as is depicted in Figures 1-13 through 1-16 exhibit fluorescein fluorescence intensities consistent with there being somewhere between 19,500 and 58,500 antibodies bound to the cell surface. Is that a good number? Actually, it's probably a little low; people who've done the experiments carefully seem to come up with an average of about 50,000 molecules of CD4 per CD4-positive cell. I may have come up with the lower number because there wasn't enough antibody added to the blood sample to bind to all of the available CD4 molecules; I didn't **titrate** the antibody, i.e., determine whether adding more antibody would have increased the cells' fluorescence intensities. So, as I said, it isn't always as easy as it looks.

However, there has been a great deal of work done on improving quantitative fluorescence measurement since the last edition of this book was written; for now, it's probably enough to mention that an entire issue of the journal *Cytometry* was devoted to the topic in October, 1998[2379].

1.5 WHAT'S IN THE BOX: FLOW CYTOMETER ANATOMY, PHYSIOLOGY, AND PATHOLOGY

It may have occurred to you that I have spent a great deal of time dealing with history, data analysis, parameters, and probes without getting into the details of how a flow cytometer works. That fits in with my idea that what we should be concerned with, first and foremost, is what information we want to get out of the cells and what we have to do to the cells to get it. It is now fairly clear that, although we can derive some information about cell size and morphology from light scattering signals, getting the details about biochemistry and physiology will require treating the cells with one or more fluorescent probes. We are now ready to consider more of the details of how the fluorescence of those probes is measured.

Light Sources for Microscopy and Flow Cytometry

There are substantial differences in time scale between flow cytometry and microscopy. A human observer at a microscope moves different cells into and out of the field of view at a rate that is, under any circumstances, much slower than the rate at which cells are transported through the observation region (or, if you prefer, past the "interrogation point," which always seems to me to describe a "?") of a flow cytometer. The response time of the human observer is pretty long, i.e., hundredths of seconds, or tens of thousands of microseconds. That's why movies and television work; changing the picture a few dozen times a second produces the illusion of continuous motion. In flow cytometry, a cell passing through the apparatus is typically illuminated for somewhere between one and ten microseconds. This disparity in observation times means, among other things, that flow cytometers need more intense light sources than are commonly used in microscopes.

Both the **sensitivity** (i.e., how much light can be detected) and **precision** (i.e., how reproducibly this can be done) of light measurements are functions of the **amount of light**, i.e., the **number of photons**, reaching the detector. The human eye is an extremely sensitive photodetector; when properly dark-adapted, a person with good eyesight may well perceive single photons emitted from weakly fluorescent or luminescent objects. The quantum nature of light obviously does not allow for any improvement upon this level of sensitivity in the electro-optical photodetectors used in flow cytometers.

Therefore, to make a flow cytometer comparable in sensitivity to a human observer, we would expect to have to get approximately the same amount of light from the observation region of the flow cytometer in a few microseconds as is collected by the observer at the microscope in a few milliseconds. Since the amount of light collected is, in general, directly dependent on the intensity of illumination, a cytometer needs a light source approximately a thousand times as bright as would be needed in the microscope.

The term **brightness**, when used in a technical sense, denotes the amount of light emitted from or through a unit surface area or solid angle, rather than the total amount of light emitted from a source. By this criterion, the 800 μW **laser** in a supermarket bar code scanner is brighter than the sun, and practically any laser can potentially be used as a light source for flow cytometry. The requisite brightness is also found in some kinds of **arc lamps** (high-pressure mercury and xenon lamps, sometimes specified as "short arc" lamps).

The majority of fluorescence flow cytometers now in use are benchtop models with a single blue-green (488 nm) illuminating beam, derived from an air-cooled argon ion laser. If a benchtop apparatus has a second illuminating beam, it is usually red (nominally 635 nm), coming from a diode laser. Larger instruments, such as high-speed sorters, use water-cooled argon and krypton ion lasers, which can be tuned to produce emission at a variety of UV (350-364 nm) and visible wavelengths; some systems obtain UV emission at 325 nm from an air-cooled helium-cadmium laser. Typical laser powers range from 10 to 25 mW in benchtop cytometers and up to hundreds of milliwatts in larger systems.

Instrument Configurations: The Orthogonal Geometry

Flow cytometers using arc lamp sources have been and still may be built around upright or inverted microscopes, simply by placing the **flow cell** or **flow chamber** in which cells are observed where the slide would normally go. Most modern fluorescence flow cytometers, however, use laser sources, and employ a different optical geometry, which is shown schematically in the intimidating but informative Figure 1-21 (the uncaptioned color version of the figure on the back cover may be helpful). The cytometer shown in the figure is designed to measure light scattering at small and large angles and fluorescence in four spectral regions.

The figure is a top view. If you look carefully along the left side, about halfway up from the bottom, you'll see the cell, which is, or at least should be, the *raison d'être* for the instrument and for our mutual efforts. The **core** or **sample stream** of cells would pass through the system in a direction perpendicular to the plane of the drawing, and the axes of the sample stream, the focused laser beam used for illumination, and the lens used to collect orthogonal scatter signals are all at right angles to one another, which is why the cytometer is described as having an **orthogonal** geometry. For the time being, we won't go into the details of how the cell gets into the center of the rectangular quartz cuvette in which the measurements are made.

Laser Beam Geometry and Illumination Optics

The beam coming out of the laser is radially symmetric, but the intensity varies with distance from the axis of the beam. If you plotted intensity versus distance from the axis, you'd come up with the familiar bell-shaped **Gaussian** or **normal** distribution.

It helps our cause to illuminate the cell and as little of the region surrounding it as possible. Most cells that are subjected to flow cytometry are less than 20 μm in diameter, so it would be advantageous to focus the illuminating beam to a spot not much bigger than this. This could be done using a single convex spherical lens. However, problems arise due to the Gaussian intensity profile of the laser beam and to the vagaries of fluid flow.

In order to measure scatter and fluorescence signals from cells with a precision of a few percent, it is necessary that illumination be uniform within that same few percent over the entire width of the sample or **core** stream. As long as the sample is flowing, we know that cells will get through the plane, defined by the intersection of the axes of the illuminating beam and the collection lens, in which the observation point lies. However, while, under ideal conditions, we'd like to have the cells strung out along the axis of flow like beads on a string, in practice, there's apt to be some variation in lateral position of cells in the core stream. If the beam is focused to a very small spot, the variation in intensity of illumination reaching cells at different positions will be too high to permit precise measurements.

Calculations show that if the diameter of the focused beam is about 100 μm, there will be only about 2% variation in intensity over the width of a 20 μm sample stream. There are, however, good reasons not to use a 100 μm round spot. If cells travel through the apparatus at velocities in the range of 2-5 m/sec, it will take 20-50 μs for a cell to traverse a 100 μm beam. During this time, most of the beam will be illuminating things other than the cell, and any scatter and fluorescence signals from these things will increase background noise levels.

Since variations in intensity over the Gaussian profile of the laser beam along the axis of fluid flow aren't a problem, because each cell goes through the whole beam, it makes sense to use a relatively small focal spot dimension in the direction of the axis of flow. A spot size of 20 μm allows cells to traverse the beam in 4-10 μs, increasing illumination of the cells during their **dwell time** in the beam and decreasing background as well. If the spot is made smaller than a cell diameter, say 5 μm, cells of different sizes spend different lengths of time in the beam – everybody isn't famous for the same number of microseconds – and **pulse width** can be used to measure cell size.

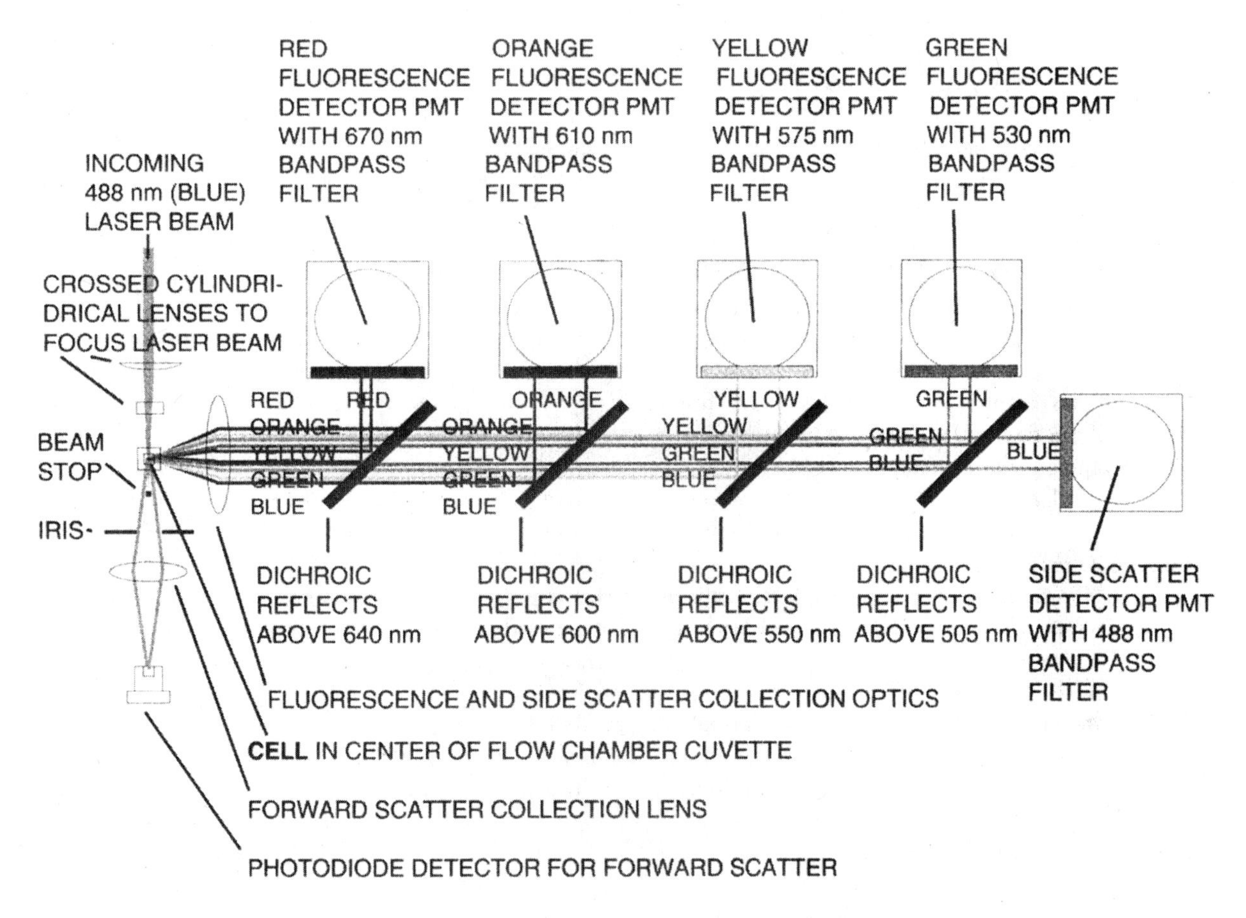

Figure 1-21. Schematic of the optical system of a fluorescence flow cytometer.

Using a really small spot, say 2 μm, you can extract a substantial amount of information about cell shape and structure by digitizing the signal at very high rates. Until recently, the processing electronics required for this technique, which is called **slit-scanning** flow cytometry, were too complex and expensive to be widely used, but the hardware and software are now more accessible should a compelling application come along. Current conventional instruments settle for **elliptical focal spots** 5-20 μm high and about 100 μm wide; these are obtained using **crossed cylindrical lenses** of different focal lengths, each of which focuses the beam in only one dimension. The crossed cylindrical lenses are shown at the left of Figure 1-21, above the cuvette. The lens closest to the cuvette is placed one focal length away from the sample stream, and focuses the beam in the dimension perpendicular to the plane, which is why you can't see the lens's curvature. The other lens, in this diagram, is placed so that its focal point is at the **beam stop**, which is a component of the forward scatter collection optics.

Flow Chamber and Forward Scatter Collection Optics

Earlier instruments examined cells in cylindrical quartz capillaries, or in a cylindrical **stream in air** following passage of fluid through a round orifice; the observation point in most cell sorters is still in a stream in air. However, in the benchtop instruments that are most widely used, observation is done in **flat-sided quartz cuvettes** with a square or rectangular cross section. The internal dimensions of the cuvettes are typically 100-200 by 200-400 μm; they are essentially small spectrophotometer cells and are, not surprisingly, produced for the flow cytometer manufacturers by the same companies that make spectrophotometer cells for other purposes. Cylindrical capillaries or streams in air themselves act like cylindrical lenses, and refract substantial amounts of light from the illuminating beam, which greatly increases the background noise level in scatter measurements and may also interfere with fluorescence measurements. Flat-sided cuvettes scatter relatively little of the incident light, minimizing such interferences.

The **beam stop** in the cytometer shown here is a vertical bar; we're looking at its cross section in the top view. What a beam stop needs to do is block the illuminating beam, once the beam has traversed the cuvette, so that as little of the beam as possible will reach the forward scatter detector and interfere with the measurement of light scattered by the cell at small angles to the beam. In an instrument in which observation is done in a round capillary or in a stream in air, the beam stop has to be horizontal, to block light refracted by the capillary or stream; the forward scattered light that is detected is light scattered "up and down," i.e., out of the plane of Figure 1-21. The laws of physics that govern focused laser beams end up dictating that we can collect light scattered at smaller angles using a flat-sided cuvette and a vertical beam stop than we can using a capillary or round stream and a horizontal beam stop.

The actual range of angles over which small-angle or forward scatter signals are collected varies considerably from instrument to instrument. The lower end of the range is set by the placement and dimensions of the beam stop; in many flow cytometers, the upper end of the range is adjustable by manipulating an **iris diaphragm**, shown below the beam stop at the left of the figure. The light that gets around the beam stop and through the diaphragm is converged by the **forward scatter collection lens**, which, in the apparatus shown in the figure, is bringing the light to a focus at the **forward scatter detector**.

The detector illustrated here is a **photodiode**, a silicon solid-state device that takes photons in and puts electrons out, usually at the rate of about 5 electrons out for every 10 photons in, giving it a quantum efficiency of 50 percent. The actual sensing area of the detector is in the neighborhood of 1 mm^2. When you make the same kind of silicon chip with a larger surface area, you can get some fairly serious electric currents out of the resulting **solar cell**. The photodiodes used as forward scatter detectors in most flow cytometers typically have output currents of a few microamperes, not because they're smaller than solar cells, but because there aren't enough photons, even in the relatively strong forward scatter signal, to produce higher currents. When you're trying to measure forward scatter signals from relatively small particles, e.g., bacteria, a photodiode may not be up to the job, and it may be better to use a more sensitive detector, such as a **photomultiplier tube (PMT)**. These are used for side scatter and fluorescence detection, but are larger, more complicated, and - probably most important from the commercial point of view - more expensive than photodiodes. In an ideal world, the flow cytometer manufacturers would offer a high-sensitivity PMT forward scatter detector option on all models; turn on the news if you still think ours is an ideal world.

Fluorescence and Side Scatter Optics

The really hairy part of Figure 1-21, and of the average flow cytometer, is the part that deals with the collection of fluorescence and side scatter signals and the diversion of light in different spectral regions to the appropriate photomultiplier tube detectors. The first task is to collect the light. I have shown a single, simple **collection lens** for fluorescence and side scatter, but the optics actually used are somewhat more complicated.

As was noted in Figure 1-1, light is scattered, and fluorescence emitted, in all directions, i.e., over a solid angle corresponding to the entire surface of a sphere. In principle, we'd like the lens to collect light over as large a solid angle as possible, so we can collect as much of the fluorescence as possible. One way to do this is to use a high-N.A. microscope lens to collect the light; this is done in many instruments, some of which even use a functional equivalent of oil immersion to get the highest possible N.A. Another is to place the collection lens at its focal distance from the sample stream. Various experimenters have used parabolic or ellipsoidal reflectors and high-N.A. fiber optics for light collection in attempts to increase the total amount of light collected.

As has already been suggested in the discussion of forward scatter detectors, ideal solutions are hard to come by. Every decision made in the design of a flow cytometer involves tradeoffs. In the case of light collection optics, the problem we run into is usually that, as we collect more light, we have less control over where we collect it from. What we really need to do is collect as much light from the immediate region of the cell, and as little from elsewhere, as possible, because any light we collect from elsewhere will only contribute to the background or noise. Thus, the all-important **signal-to-noise ratio** will decrease, even though the signal itself increases. Flow cytometer designs using ellipsoidal or parabolic reflectors or fiber optics for light collection have, so far, run into this problem.

The simple collection lens shown in the figure is illustrated as producing a **collimated beam** of light, i.e., one in which rays entering the lens at all angles come out parallel, with a so-called "focus at infinity". In most real flow cytometers, the light collected from the collection lens is either not collimated or is converged by a second lens, and then passes through a small aperture, or **field stop** (see p. 9), that lets most of the light collected from the region near the cell through and blocks most of the light collected from elsewhere. Some instruments incorporate an additional lens behind the field stop to recollimate the collected light, because there is some advantage in presenting a collimated beam to the **dichroics** and **optical filters** used direct light collected at different wavelengths to different detectors.

Optical Filters for Spectral Separation

The lens that collects the fluorescence emitted from, and the light scattered at large angles by, cells transmits light encompassing a range of wavelengths. Most of the light is scattered laser light, at 488 nm; much of the rest should be fluorescence from the cells, which will of necessity be at wavelengths above 488 nm. The choice of wavelength regions for fluorescence measurements is based on the fluorescence emission spectral characteristics of the available fluorescent probes or labels that can be excited at 488 nm.

The apparatus illustrated in Figure 1-21 is designed to detect fluorescence in four spectral regions, which we call **green (515-545 nm)**, **yellow (560-590 nm)**, **orange (600-620 nm)**, and **red (660-680 nm)**. It also detects scattered light at the excitation wavelength, 488 nm. Each of the detectors is a photomultiplier tube, and all of the detectors are fitted with **bandpass optical filters** that transmit light in the appropriate wavelength ranges.

There are basically two kinds of optical filters that can be used for wavelength selection; they are **color glass**, or **absorptive**, filters and **dielectric**, or **interference**, filters. Color glass filters are made of glass or plastic impregnated with **dyes** that absorb light in the unwanted wavelength regions and transmit most of the light in the desired regions. Dielectric filters are made by depositing thin layers of dielectric materials on a glass or quartz substrate; within some wavelength range, which is determined by the thickness of these layers, there will be **destructive interference**, resulting in light of these wavelengths being reflected from, rather than transmitted through, the filter.

Filters can be made with several kinds of transmission characteristics. There are **edge** filters, which may be either **long pass** or **short pass** types; long pass filters block shorter and transmit longer wavelengths and short pass filters block longer and transmit shorter wavelengths. Long pass and short pass filters are usually specified by the wavelength at which their transmission is either 50% of the incident light or 50% of their maximum transmission. There are **bandpass** filters, which block wavelengths above and below the desired region of transmission; they are specified by the wavelength of maximum transmission and by the **bandwidth**, which defines the range of transmission, usually expressed as the range between the points below and above the peak at which transmission is 50% of maximum. There are also **notch** filters, which are designed to exclude a narrow range of wavelengths.

Absorptive filters can be very effective at getting rid of light outside their desired **passbands**, i.e., those regions in which they transmit light (many transmit less than 0.01% outside the passband), and can also be made to have good (>90%) light transmission in the passband. However, the dyes incorporated into the filter to absorb the unwanted light may fluoresce; this phenomenon can (and did, in the earlier fluorescence flow cytometers) interfere with the detection of weak fluorescence signals from cells. As a result, most modern instruments now use interference filters, which reflect rather than absorb unwanted light.

Real interference filters used as long pass or bandpass filters frequently incorporate an absorptive layer behind the dielectric layers to get rid of the last little bit of unwanted light, because it's difficult to get rid of more than 99% of it by interference and reflection alone. Fluorescence in these filters is not a big problem because the interference layers get rid of most of the light that might excite fluorescence before it hits the absorptive layer – provided, that is, that you mount the filter **shiny side out**, that is, with the interference layers facing where the light's coming from and the colored absorptive side facing where it's going.

Dichroics, also called **dichroic mirrors** or **dichroic beamsplitters**, are interference filters, usually without an added absorptive layer. They can be made with either **long reflect** (i.e., short pass) or **short reflect** (i.e., long pass) characteristics, and both kinds are used in flow cytometers. As is the case with other types of interference filters, it's easier to make a filter that reflects 97% of unwanted light than it is to make one that transmits 90% of wanted light. When flow cytometers measured fluorescence in only two spectral regions, they only needed one dichroic (maybe two, if you count one to reflect blue (488 nm) light to the orthogonal scatter detector and keep it away from the fluorescence collection optics). When you start measuring fluorescence in three or four regions, it becomes advisable to do careful calculations to make sure you don't lose a lot of the light you want in the dichroics. The Devil, as we all know, is in the details, and more deals with the Devil are made in the details of dichroics and filters than in most other areas of flow cytometer design.

The layout shown in Figure 1-21 assumes that the strongest signal, or the one with the most light we can waste, is the blue orthogonal scatter signal, and that the green, yellow, orange, and red fluorescence signals are progressively weaker. Even if all of the dichroics transmit 90% of the incident blue light, only 65% of the light coming through the collection lens will reach the filter in front of the orthogonal scatter detector PMT. About 70% of the green fluorescence will make it to the filter in front of the green detector PMT, while 77% of the yellow, 86% of the orange, and 96% of the red fluorescence will get to the filters in front of the detectors for those spectral regions. We therefore lose the least light from the weakest signal.

There are other ways to improve light transmission; one is to ditch the in-line arrangement of PMTs shown in the figure, instead first splitting the red/orange and the blue/green/yellow regions, so that the green fluorescence signal passes through two dichroics and the others through only a single dichroic. Another, which I routinely use in the "Cytomutt" flow cytometers I build, is to place a second fluorescence collection lens at 180° from the first one, so that each lens collects light for at most three detectors.

The spacing between the dielectric layers of interference filters and dichroics determines the wavelengths at which interference will occur, and, therefore, the wavelengths that will be transmitted or reflected by these components. The distance between the layers changes with the angle at which light hits the filter (remember trigonometry?), and, as a result, the passband of the filter changes with the angle of incidence of the light. In theory, light should be collimated before it gets to the dichroics and filters; this is generally not done because the light coming from the collection lens is contained within a fairly small solid angle. Problems with dichroics and filters are more likely to result from using the wrong filters or from mounting filters incorrectly. Dielectric filters also degrade over time, as moisture gets in between the dielectric layers, but, when this occurs, the filters tend to look ugly enough so that you'd think about ordering new ones.

I hope, by now, to have conveyed the impression that dichroics and filters are among the most critical parts of a flow cytometer; not surprisingly, the right – or wrong – selection of dichroics and filters can also make a big difference when you're doing fluorescence microscopy, by eye or with image analyzers, etc. A few hundred dollars spent on good filters may dissuade you from smashing tens or hundreds of thousands of dollars worth of instrument to smithereens out of frustration.

Multistation Flow Cytometers

Before going on to a discussion of detectors and electronics, I will point out that, whereas most flow cytometers have a single excitation beam, and you can have any color you want as long as it's 488 nm, there are systems available that offer a wider choice of excitation wavelengths. Some of these can use two or more illumination beams, separated by a small distance in space. A good way to conceptualize such a **multistation flow cytometer** might be to imagine two or more copies of Figure 1-21 stacked one on top of another. Because the beams in a multistation instrument are separated by a short distance, it takes a short time for cells to travel from one beam to another, and the signals are therefore separated in time. Since the velocity of cells through the system is approximately constant, the time interval between signals from different beams is also approximately constant.

In flow cytometers that form an image of the sample stream, as most now do, it is customary to form separate images of the intersections of two or more beams with the sample stream, and divert light from each observation point to the appropriate detectors. In instruments in which no image is formed, and in which light from multiple observation points reaches all the detectors, a **time-gated amplifier** is used. This allows signals from the detectors that measure events at the downstream observation point to reach the signal processing electronics only at a set time interval after signals are detected at the upstream observation point.

Multistation instruments have also been built that incorporate electronic volume sensors as well as laser or arc lamp illumination; cell sorters are also multistation instruments, as are cell "zappers" or **photodamage cell sorters**. These use a high energy pulsed laser beam downstream from the measurement beam and switch the beam on to destroy cells with selected characteristics.

Flow cytometers with multiple illumination beams are used primarily for multiparameter measurements involving probes that cannot be excited at the same wavelength. For example, sorting human chromosomes stained with combinations of dyes that preferentially stain A-T and G-C rich regions of DNA requires separated ultraviolet (325-363 nm) and blue-violet (436-457 nm) illuminating beams. Other applications use ultraviolet and 488 nm beams and 488 and red (633 or 635 nm) beams; as many as five beams have been used in a single apparatus. The current trend is toward multiple illumination beams, even in benchtop instruments.

Photomultipliers and Detector Electronics

A **photomultiplier tube (PMT)**, like a photodiode, takes in photons and puts out electrons. However, whereas a plain photodiode never does much better than 7 electrons out for every 10 photons in, a PMT may get as many as a few hundred thousand electrons out for each photon that reaches its **photocathode**. PMTs, like cathode ray television tubes and the tubes favored by audiophiles and rock musicians who can't see the trees for DeForest, are among the last survivors of the vacuum tube era. They incorporate a photocathode, which is placed behind a glass or quartz window so light can reach it, a series of intermediate electrodes, or **dynodes**, and another electrode called the **anode**. A voltage is applied to each electrode; the photocathode is at the lowest voltage, with each dynode at a successively more positive voltage and the anode at the most positive voltage of them all – which is usually ground, because the photocathode is generally a few hundred to a couple of thousand volts negative.

Photons hitting the photocathode result in **photoelectrons** being emitted from the photocathode, and accelerated toward the first dynode by the electric field resulting from the difference in electric potential (voltage) between these electrodes. The electrons acquire energy during this trip, so, when they whack into the dynode, they dislodge more electrons from it, which are accelerated toward the next dynode, and so on. The bigger the difference in potential, i.e., applied voltage, between stages, the more energy is imparted to the electrons at each stage, and the more electrons are released from the receiving electrode. This gives the PMT a mechanism for **current gain** that is relatively noise-free. The PMTs used in most flow cytometers have current gains as high as 10^6. However, the **quantum efficiency** of PMT photocathodes is typically lower than that of photodiodes, with peak values of 25% (i.e. 25 electrons out for 100 photons in) in the blue spectral region, and, usually, much lower values in the red. Detector quantum efficiency is important because the sensitivity and precision with which fluorescence (or any other optical signal) can be measured ultimately depend on the number of electrons emitted from the detector photocathode.

Why is it that at detectors, we measure success one electron at a time? Because detection is subject to the same Poisson statistics we ran into on p. 19. When you count (or detect) n of anything, including photoelectrons, there is an associated standard deviation of $n^{1/2}$. When you detect 10,000 photoelectrons, the standard deviation is $10{,}000^{1/2}$, or 100, and the coefficient of variation (CV) is 100 × (100/10,000), or 1%. When you detect 10 photoelectrons, the standard deviation is $10^{1/2}$, or about 3.16, and the CV is 100 × (3.16/10), or 31.6%. I am talking about photoelectrons, rather than photons, here, because, while the detector, whether diode or PMT, "sees" photons, if you will, all the electronics lets us "see" is electrons.

If we had reliable low-noise amplifiers with gains of several million, we'd always be better off with the 50-70 electrons we could get out of the photodiode for every 100 photons hitting it than we would with the 8-25 electrons emitted from the PMT cathode under the same conditions; all the gain in the PMT doesn't get around the imprecision introduced by the lower number of electrons it starts with and, in fact, there is also a statistical aspect to the PMT's gain mechanism.

Unfortunately, the high-gain, low noise amplifiers we'd need to use photodiodes as sensitive fluorescence detectors don't exist. There are, however, solid-state devices called **avalanche photodiodes (APDs)**, which combine high quantum efficiency with a mechanism that can produce gains as high as a few thousand when a voltage is applied across the diode. While APDs are now used for both scatter and fluorescence detection in some commercial flow cytometers, they do not match the sensitivity of PMTs.

The photodetectors we have been talking about are sources of **electric current**. A **preamplifier**, which is the first stage in the **analog signal processing electronics**, converts the current output from its associated detector to a voltage. The preamplifier also accomplishes the important task of **DC baseline restoration**.

An ideal flow cytometer is something like an ideal dark field microscope; when there's no cell in the observation region, the detector shouldn't be collecting any light at all. In practice, there's always some small amount of light coming in. In the case of the scatter detectors, most of this light is stray scattered light from the illuminating beam; in the case of the fluorescence detectors, the light background may come from fluorescence excited in various optical elements such as the flow chamber, lenses, and filters, from fluorescence due to the presence of fluorescent materials in the medium in which cells are flowing, and from **Raman scattering**, which produces light at frequencies corresponding to the difference between the illumination frequency and the frequencies at which absorption changes molecular vibrational states. In flow cytometry, the major interference due to Raman scattering results from scattering by water; when 488 nm illumination is used, this scattering occurs at about 590 nm, and may interfere with detection of signals from probes labeled with phycoerythrin, which fluoresces near this wavelength.

The net result of the presence of all of the abovementioned stray light sources is that there are some photons reaching the detectors in a flow cytometer even when there isn't a cell at the observation point, producing some current at the detector outputs. There may also be some contribution from the so-called **dark current** of the detector, which results from the occasional electron breaking loose from the cathode due to thermal agitation. There are some situations in which performance of photodetectors is improved by refrigerating them to reduce dark current; flow cytometry in the contexts we're discussing isn't one of them. Even with the detectors in liquid nitrogen, we'd have to deal with the background light, which will contribute a signal with an average value above zero to whatever signal we collect from the cells.

The background signal can be considered as the sum of a constant **direct current (DC)** component and a variable **alternating current (AC)** component, representing the fluctuations due to photon statistics and to other sources of variation in the amount of stray light reaching the detector. One important source of such variation may be **light source noise**, i.e., fluctuations in the light output of the laser or lamp used for illumination; in some circumstances, particularly scatter measurements of small particles, source noise can be the major factor limiting sensitivity.

What we'd like to measure when a cell does pass by the observation station is the amount of light coming from the cell, not this amount plus the background light. We can do this, to a first approximation, by incorporating an electronic circuit that monitors the output of the detector and uses negative feedback to subtract the slowly varying component of the output from the input, thereby eliminating most of the DC background signal, and restoring the **baseline** value of the preamplifier output to ground.

In practice, baseline restorers will keep their voltage outputs within a few millivolts of ground when no cells are coming by. When a cell does arrive, it will scatter and probably emit small amounts of light, which will be collected and routed to the various detectors, producing transient increases, or **pulses**, in their output currents, which will result in voltage **pulses** at the preamplifier outputs. At this point, as was noted on p. 17, all of the information we wanted to get from the cell resides in the heights, areas, widths, and shapes of those pulses; we will ultimately convert these to digital values, in which form they can be dealt with by the computers that are almost universally used for data analysis in flow cytometry. However, before we get into the details of how pulse information is processed, we ought to consider the only element of Figure 1-21 that has been neglected to this point, namely, the cell flowing through the apparatus, and how it gets there.

Putting the Flow in Flow Cytometry

Figure 1-21 describes the cell as being in the center of the cuvette, and I have already talked about a **core** or sample stream of cells that is about 20 µm wide, while mentioning that the internal dimensions of the cuvette are on the order of 200 by 200 µm. The space between the core and the inner walls of the cuvette is occupied by another stream of flowing fluid, called the **sheath**. How the core and sheath get where they are can be appreciated from a look at Figure 1-22.

Fluid mechanics tells us that, if one smoothly flowing stream of fluid (i.e., the core stream) is injected into the center of another smoothly flowing stream of fluid (i.e., the sheath stream), the two streams will maintain their relative positions and not mix much, a condition called **laminar flow**. There are generally differences in fluid flow velocity from the inside to the outside of the combined stream, but the transitions are even. If the velocities of the two streams are initially the same, and the cross-sectional area of the vessel in which they are flowing is reduced, the cross-sectional areas of both streams will, obviously, be reduced, but they will maintain the same ratio of cross-sectional areas they had at the injection point. If the sheath stream is flowing faster than the core stream at the injection point, the sheath stream will impinge on the core stream, reducing its cross-sectional area. In the flow chamber of a flow cytometer, both mechanisms of constricting the diameter of the core stream may be operative.

The core stream, which contains the cell sample, is injected into the flowing water or saline sheath stream at the top of a conical tapered region that, in the flow chamber shown in the figure, is ground into the cuvette. The areas of both streams are reduced as they flow through the tapered region and enter the flat-sided region in which cells are observed. Core and sheath streams may be driven either by gas pressure (air or nitrogen), by vacuum, or by pumps; most instruments use air pressure. **Constant volume pumps**, e.g., **syringe pumps**, which, if properly designed, deliver a known volume of sample per unit time with minimum pulsation, provide finer control over the sample flow rate. Since knowing the sample flow rate makes it easy to derive counts of cells per unit volume, flow cytometric hematology analyzers incorporate constant volume pumps; why fluorescence flow cytometers, in some cases made by the same manufacturers, do not remains something of a mystery.

The overall velocity of flow through the chamber is generally determined by the pressure or pump setting used to drive the sheath. If the sheath flow rate is increased with no change in the core flow rate, the core diameter becomes smaller and the cells move faster; if the sheath flow rate is decreased under the same circumstances, the core diameter becomes larger and the cells move more slowly. In some circumstances, it is desirable to adjust sheath flow rates; if cells move more slowly, they spend more time in the illuminating beam, receive proportionally more illumination, and they therefore scatter and emit proportionally more light. If the amount of light being collected from cells is the limiting factor determining sensitivity, slowing the flow rate can improve sensitivity, allowing weaker signals to be measured.

This aside, it is generally preferable to be able to control the core diameter, and therefore the volume of sample and number of cells analyzed per unit time, without changing the velocity at which cells flow through the system. This is done by leaving the sheath flow rate constant and changing the driving pressure or pump speed for the core fluid. More drive for the core results in a larger core diameter; more cells can be analyzed in a given time, but precision is likely to be decreased because the illumination from a Gaussian beam is less uniform over a larger diameter core. Less drive for the core gives a smaller core diameter and a slower analysis rate, but precision is typically higher. When the cytometer is being used to measure DNA content, precision is important; when it is being used for immunofluorescence measurement, precision is usually of much less concern.

The use of **sheath flow** as just described has proven essential in making flow cytometry practical. Without sheath flow, the only way of confining 10 μm cells within a 20 μm diameter stream would be to observe them in a 20 μm diameter capillary or in a stream in air produced by ejecting the cells through a 20 μm diameter orifice. This would very quickly run afoul of Shapiro's First Law (p. 11). As a matter of fact, even with sheath flow, Shapiro's First Law frequently came into play when cell sorters were typically equipped with 50 μm orifices. That orifice size was fine for analyzing and sorting carefully prepared mouse lymphocytes, but people interested in analyzing things like disaggregated solid tumors might encounter mean intervals between clogs of two minutes or so. With the larger cross-sectional areas of the flow chambers now used in most flow cytometers, clogs are not nearly the problem they once were.

Clogs, however, are not the only things that can disturb the laminar flow pattern in the flow chamber. Air bubbles perturb flow, as do objects stuck inside the chamber but not large enough to completely obstruct it. In the first commercial cell sorters, the standard method for getting rid of air bubbles was to remove the chamber from its mount while the apparatus was running, and turn it upside down; the bubble would rise to the top and emerge from the nozzle along with a stream of sheath and sample fluid that would spray all over the lab. This technique became inappropriate with the emergence of AIDS in the 1980's. Now, even droplet sorters incorporate an air outlet (which I have referred to elsewhere as a "burp line") for getting rid of bubbles. In some flow cytometers with closed fluidic systems, the air bubble problem is minimized by having the sample flow in

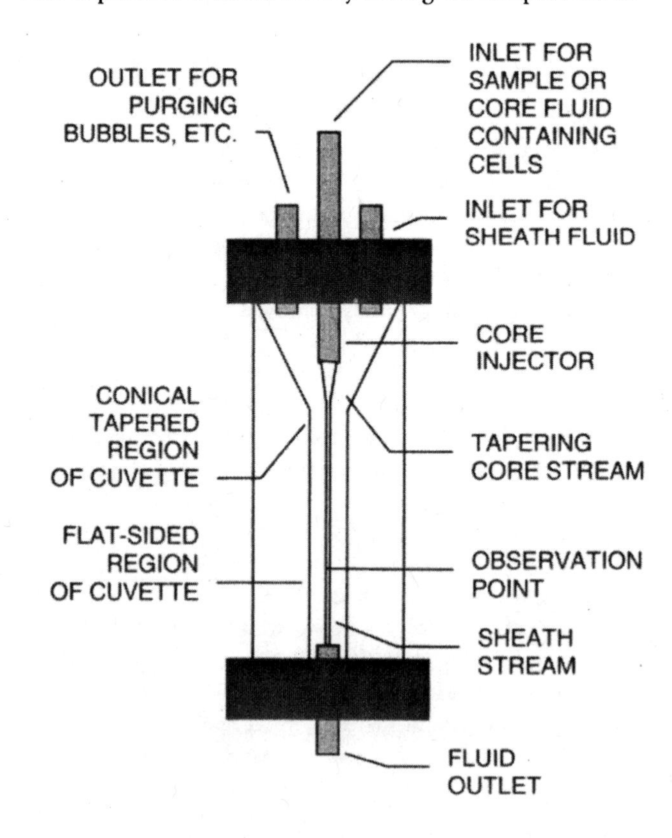

Figure 1-22. A typical flow chamber design.

at the bottom and out at the top, essentially turning Figure 1-22 upside down; bubbles are more or less naturally carried out of the flow chamber.

Disturbances in laminar flow, whether due to bubbles or junk, often result in the core stream deviating from its central position in the flow chamber and in differences in velocity between different cells at different points within the core. Turbulent fluid flow is now described mathematically using chaos theory; you can recognize turbulent flow in the flow chamber by the chaos in your data.

For the present, we will assume the flow is laminar, the optics are aligned, and the preamplifiers are putting out pulses with their baselines restored, and consider the next step along the way toward getting results you can put into prestigious journals and/or successful grant applications.

Signal Processing Electronics

We have already mentioned that a cell is going to pass through the focused illuminating beam in a flow cytometer in something under 10 μs, during which time the detectors will produce brief current pulses, which will be converted into voltage pulses by the preamplifiers. Using analog peak detectors, integrators, and/or pulse width measurement circuits, followed by analog-to-digital conversion, or, alternatively, rapid A-D conversion followed by digital pulse processing (p. 21), we will reduce pulse height, area, and width to numbers, at least some of which will, in turn, be proportional to the amounts of material in or on the cell that are scattering or emitting light. But which numbers?

First, let's tackle the case in which the focal spot, in its shorter dimension, along the axis of flow, is larger than the cell, meaning that there is some time during the cell's transit through the beam at which the whole cell is in the beam. Because the beam is Gaussian, the whole cell may not be uniformly illuminated at any given time, but intuition tells us that when the center of the cell goes through the center of the beam, we should be getting the most light to the cell and the most light out of it. The preamplifier output signal, after baseline restoration, is going to be roughly at ground before the cell starts on its way through the beam, and rise as the cell passes through, reaching its **peak value** or **height** when the center of the cell is in the center of the beam, and then decreasing as the cell makes its way out of the beam. Since the whole cell is in the beam when the pulse reaches its peak value, this value should be proportional to the total amount of scattering or fluorescent material in or on the cell.

Things get a little more complicated when the beam is the size of the cell, or smaller. In essence, different pieces of the cell are illuminated at different times as the cell travels through the beam. In order to come up with a value representing the signal for the whole cell, we have to take the **area**, or **integral**, rather than the height of the pulse. There are two ways to do this with analog electronics. One is to change the frequency response characteristics of the preamplifier, slowing it down so that it behaves as an integrator, in the sense that the height of the pulse coming out of the slowed-down preamplifier is proportional to the area or integral of the pulse that would come out of the original fast preamplifier. Putting the slowed pulse into a peak detector then gives us an output proportional to the area or integral we're trying to measure. Alternatively, we can keep the fast preamplifier, and feed its output into an analog **integrator** instead of a peak detector.

If we decide to do digital pulse processing, we have to digitize the pulse trains from the preamplifier outputs rapidly enough so that we have multiple samples or "slices" of each pulse. We can then add the values of a number of slices from the middle of the pulse to get an approximation of the area, or integral; eight slices will do, but sixteen are better. This works pretty well. However, if we're only taking eight or sixteen slices of a pulse, we may not get as accurate a peak value or a pulse width value as we could using analog electronics.

The peak value we get from digital processing is simply the largest of our eight or sixteen slices. These provide us with only a fairly crude connect-the-dots "cartoon" of the pulse, thus, while there is a substantial likelihood that the largest digitized slice is near the peak value, there is a relatively low probability that the digitization will occur exactly when the peak value is reached.

Similarly, if we estimate pulse width from the number of contiguous slices above a set threshold value, we will have a fairly coarse measurement; if the digitization rate gives us at most sixteen slices, our range of pulse widths runs from 1 to 16, with each increment representing at least a 6 percent change over the previous value. If we had fast enough analog-to-digital converters to be able to take a few hundred slices of each pulse, and fast enough DSP chips to process the data, we could get rid of analog peak detectors and pulse width measurement circuits, but we're not there yet. The digital integrals are already good enough to have been incorporated into commercial instruments.

Is It Bigger than a Breadbox?

I have been referring to benchtop flow cytometers and big sorters, but I haven't shown you any pictures. Now's the time to fix that.

Figure 1-23, on the next page, shows the Becton-Dickinson FACScan, the first really successful benchtop flow cytometer, introduced in the mid-1980's. It uses a single 488 nm illuminating beam from an air-cooled argon ion laser, and measures forward and side scatter and fluorescence at 530 and 585 and above 650 nm. The data analysis system is an Apple Macintosh personal computer, shown in front of the operator.

Figure 1-24 (courtesy of Cytomation) shows that company's MoFlo high-speed sorter. The optical components, including two water-cooled ion lasers and a large air-cooled helium-neon laser, are on an optical table in front of the operator. Most of the processing electronics are in the rack to the operator's left; the two monitors to her right display data from an Intel/Microsoft type personal computer.

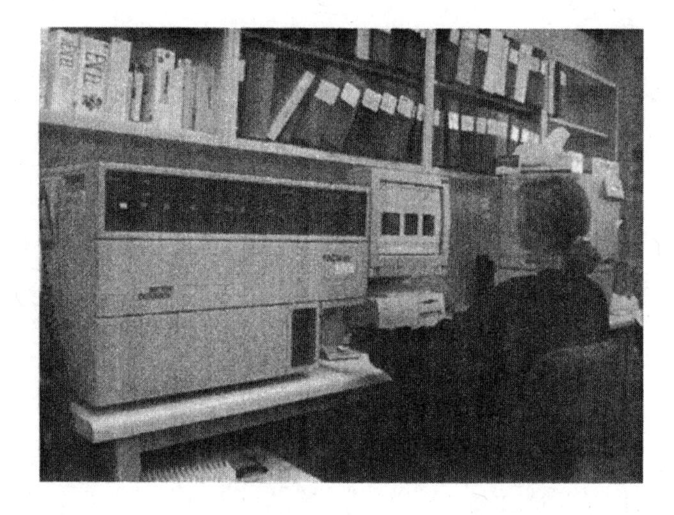

Figure 1-23. FACScan Analyzer (Becton-Dickinson)

Figure 1-24. MoFlo High-Speed Sorter (Cytomation)

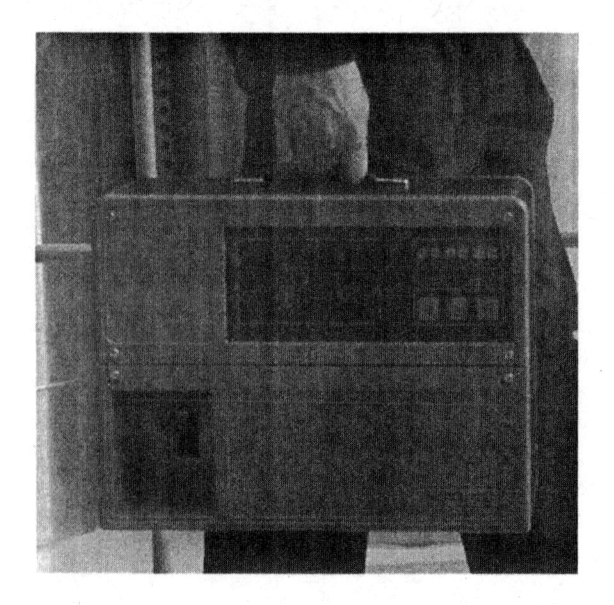

Figure 1-25. Microcyte Cytometer (Optoflow)

Neither the FACScan nor the MoFlo risks being mistaken for a breadbox. However, the Microcyte analyzer shown in Figure 1-25 (photo courtesy of Optoflow AS) comes close. It is a two-parameter instrument with a red diode laser source, and measures medium angle scatter using a photodiode and fluorescence using an avalanche photodiode. As you might guess, it can be run on batteries.

Flow Cytometer Pathology and Diagnostics

As the benchtop flow cytometer starts to look more and more like a "black box" (okay, a "beige box," "gray box," or whatever from some manufacturers), it becomes increasingly important for a user to know how to verify that the instrument is running properly. It is, of course, equally important to know when a big sorter is and is not running properly, but the larger instruments tend to make their operators aware of problems.

An instrument in proper alignment, running particles through an unobstructed flow system at a rate within the manufacturer's specifications, should get nearly identical measurements from nearly identical particles. There are now several companies producing nearly identical particles in the form of plastic microspheres, i.e., beads, impregnated with fluorescent dyes. If everything's right, one ought to be able to make scatter and fluorescence measurements of such particles with high precision, meaning coefficients of variation no higher than a few percent. The only biological objects that are likely to yield CVs in that range are noncycling cells, such as peripheral blood lymphocytes, stained with a fluorescent DNA stain; most people stick with beads.

An instrument in which optical alignment is adjustable by the operator will typically yield the lowest measurement CVs at the point at which signal amplitudes are maximized. However, optical misalignment is not the only potential cause of poor measurement precision. Fluctuations in the power output of the light source will decrease precision, as will the presence of cell aggregates, large pieces of debris, and/or gas bubbles in the flowing stream. These create turbulence, resulting in the measured particles being distributed over an excessively large portion of the stream and/or traveling at different velocities; under these conditions, nearly identical particles will obviously not produce nearly identical signals.

Sensitivity, which, in the context of flow cytometry, basically means the degree to which fluorescence distributions from dimly stained cells (or beads) can be discriminated from distributions from unstained (control) cells or (blank) beads will usually be degraded if precision falls substantially short of the mark. Loss of sensitivity may also be due to degradation and/or incorrect choice or installation of optical filters.

Precision of instruments should always be determined using beads carrying fairly large amounts of dye, to minimize the contribution to variance from photoelectron statistics. Determination of instrument sensitivity virtually demands that at least some of the test objects used produce low-intensity signals. Beads used for sensitivity testing typically come in sets containing an undyed or blank bead and beads loaded with four or more different levels of fluorescent dye.

Flow cytometer manufacturers and third parties also supply beads that can be used to optimize fluorescence compensation settings, and, as was previously noted in the discussion of quantitative fluorescence measurements, beads that allow the scale of the instrument to be calibrated in terms of numbers of molecules of a particular probe or label.

1.6 ALTERNATIVES TO FLOW CYTOMETRY; CYTOMETER ECOLOGY

In order to use flow cytometry to study characteristics of **intact cells** from solid tissues or tumors, or of cultured cells that grow attached to one another and/or to a solid substrate, various methods are used to prepare **single cell suspensions** from the starting material. Flow cytometry itself can provide a good indication of the efficacy of such preparative procedures. In a similar fashion, the technique can be very useful in **monitoring bulk methods** for purifying cell subpopulations, e.g., **sedimentation** and **centrifugation** techniques and **affinity-based separations.** If large cell yields are more important than high purity, bulk separation with flow cytometric monitoring may be preferable to sorting as a preparative method.

We have learned and can probably continue to learn a great deal by dissociating tissues and even organisms into suspensions of intact cells that can be characterized in flow cytometers, sorted, and subsequently studied in culture. However, the procedures used for cell dissociation, by nature, have to remove most of what holds the cells together. Since such **adhesion molecules** are probably as important as anything else for our understanding of cells' behavior, it is inevitable that there will come a point at which we won't be able to answer critical questions using cells stripped of these essential components.

It will make sense, at that point, to find instrumental alternatives to flow cytometry in a new generation of image analyzers and scanning cytometers, designed with an emphasis on preserving cell viability, which allow us to use the armamentarium of analytical techniques and reagents, in the development of which flow cytometry has played a major role, to study cells in organized groups.

We may also, of course, run up against the limits of flow cytometry simply by developing a desire to measure something repeatedly in one cell over an interval greater than a few hundred microseconds. This can be accomplished by combining static cytometry with **kinetic analysis techniques**, such as **flow injection analysis**, adapted from analytical chemistry.

I am reminded that one of the Mayo brothers said that a good surgeon had to know when to stop cutting and when not to cut; a good analytical cytologist will have to know when to put aside flow cytometry. Not now, though. Keep reading.

When we do consider the alternatives to flow cytometry, and even the availability of different types of flow cytometers, we run into something of an information gap. In the last edition of this book, I described flow cytometry as having been a growth industry since about 1985, based on census data compiled by Kit Snow of Beckman Coulter Corporation and shown in Figure 1-26.

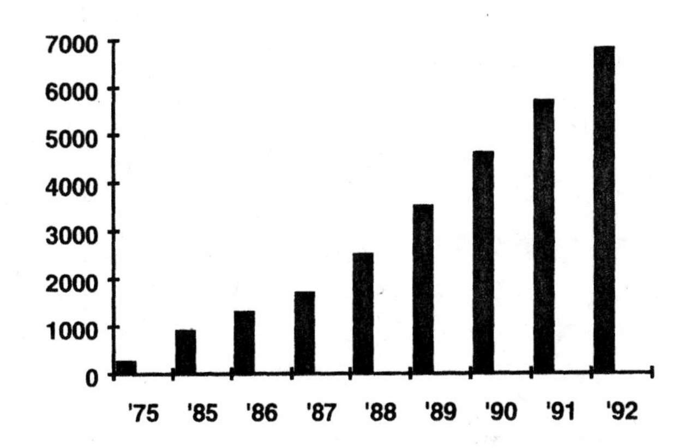

Figure 1-26. Estimated numbers of fluorescence flow cytometers in use worldwide, 1975-1992.

I tried to get updates on these numbers from various manufacturers, and even expanded my search to look for data about scanning laser cytometers, confocal microscopes, etc. Nobody's talking. The best I could do was come up with numbers that nobody would say were way too high or way too low. So here goes.

The great majority of fluorescence flow cytometers now in use are benchtop models similar to the one shown in Figure 1-23; they use low-power, air-cooled argon ion laser light sources operating at a fixed emission wavelength of 488 nm, and measure forward and orthogonal light scattering and fluorescence in three or four (green, yellow and/or orange, and red) spectral regions. Most of these systems have been designed for ease of use, with the needs of the clinical laboratory market foremost in mind. Newer instruments in the same class have added features such as a second (red) laser and closed fluidic sorting systems. The estimate is that there are somewhere between 12,000 and 20,000 such flow cytometers in use worldwide.

There are also probably around 2,000 larger, more elaborate fluorescence flow cytometers, which may use one or more air-cooled or water-cooled laser sources, can be equipped to measure eight or more parameters, and offer droplet sorting capability. These instruments are typically used in research laboratories rather than in clinical settings. Then, there are several hundred commercially produced fluorescence flow cytometers using arc lamp rather than laser sources, at least an equal number of instruments designed for multiplexed assays on beads, and one to two hundred laboratory-built flow cytometers.

The confocal microscopy folks seemed happy with the estimate that there are between 3,000 and 5,000 confocal systems worldwide; only two or three hundred of these are equipped for multiphoton excitation.

The area of relatively low-resolution scanning laser cytometry[2380] has gotten more active in recent years. The CompuCyte Laser Scanning Cytometer (LSC), developed by Lou Kamentsky[2047], is generating an increasing number of interesting publications[2381-2], about which I will say more later. I estimate that there are 100 to 250 LSCs now in circulation, and probably a similar number of volumetric capillary cytometers[1365], built by Biometric Imaging, now part of B-D. And there is at least one promising scanning system that hasn't yet made it into production but is worth watching[2383].

1.7 THE REST OF THE BOOK

In Chapter 2, I will point you toward some sources of information that may be of use to you in learning more of the details of cytometry, flow and otherwise, discussion of which began in this Chapter and will continue in Chapter 4. Chapter 2 will also provide brief descriptions of a bunch of books on cytometry and related topics that have appeared since the last edition of this tome.

I have devoted the intermediate Chapter 3 to the history of flow cytometry, because I think that an appreciation of how things came to be as they are is as important to further progress as is an understanding of the science and technology. Chapters 5 and 6, respectively, provide additional material on data analysis and flow sorting.

Parameters and the probes used for their measurement are discussed in Chapter 7, which also presents some basic applications of flow cytometry and of some alternative methods. Chapter 8 considers flow cytometers, software, and related accessories now available from commercial manufacturers, and criteria that may influence buying decisions.

Chapter 9 briefly discusses the option of building flow cytometers; although the details on the construction, care and feeding of "Cytomutts" featured in the earlier editions have been omitted, some material that may help users understand their apparatus better has been retained.

Current and proposed applications of, and alternatives to, flow cytometry in biomedical research and laboratory medicine are considered in Chapter 10. Chapter 11 lists "Sources of Supply," while Chapter 12 is an Afterword, containing afterthoughts, aftershocks, and late breaking news. That's all I wrote. Well, almost.

Lis(z)t Mode

When cells are in such altered states
You don't know where to set the gates,
It's best to minimize the risk
And store them all on your hard disk.
If there's a clog before you're done,
You'll save some data from a run,
And, thus, you may stay out of jams
You'd get in with live histograms.

List mode, just work in list mode;
When you consider all the options, it's the only thing to do.
This mode, and only this mode,
Lets you make sense of samples that, at first, leave you without a clue.

Once we're in list mode, anyway,
With prices as they are today,
It isn't putting on the Ritz
To digitize to sixteen [or more] bits.
It's clear that, once we've made this change,
We'll have enough dynamic range
To transform data digitally,
So log amps will be history.

List mode, we'll work in list mode,
And go from linear to log and back without the log amps' ills.
Once we've got list mode, our only pissed mode
May be when we try pinning down which agencies will pay the bills.

List mode can help us analyze
How many molecules of dyes
And antibodies will be found
On each cell type to which they're bound.
At long last, different labs can see
Results compared objectively,
Advancing science as a whole
And aiding quality control.

List mode, by using list mode,
We'll all get heightened sensitivities and much reduce the fears
And trepidation of calibration,
Although the folks who make the particles may have us by the spheres.

From East to West, from South to North,
We'll send our data back and forth,
Why, we'll soon have it in our reach
To run our samples from the beach.
But, unless they've been well prepared,
When they are run, we'll run them scared,
List mode or not, there's still no doubt
That garbage in gives garbage out.

List mode, we all need list mode,
Though there are ends for which list mode itself can never be the means.
Even with list mode, there won't exist code
That gets good data from bad samples and/or misaligned machines.

("List Mode" © Howard Shapiro; used by permission. The music is derived from Liszt's Hungarian Rhapsody No. 2.)

2. LEARNING FLOW CYTOMETRY

When you stop to think about it, a lot of different areas of technology are involved in flow cytometry, and you have to know something about several of those fields to be comfortable using a flow cytometer, let alone building one. The overall design of an experiment requires some appreciation of the biology of the cells to be studied, of the organic and physical chemistry of the dyes and reagents involved, of the biochemical and/or immunologic bases of the reaction between reagents and cells, and of the statistical methods needed to draw valid conclusions from the data obtained. Keeping an instrument in good working order requires some understanding of the fluid mechanics of the flow system, of the optics involved in illuminating the cells in the sample and collecting light emitted, scattered, or transmitted by those cells, and of the electronics used to detect, process, and analyze those optical signals. You may also have to learn more about lasers and computers than you ever wanted to know.

If this prospect is intimidating to you as a newcomer, take heart; everybody who has ever done flow cytometry has come to the field trained in one or two areas and has had to absorb a lot of practical information in several others. We have all learned to ask for help, and most of us have learned to give it as well. There is, however, a persistent communication problem in cytometry that arises from the interdisciplinary nature of the field. Although its practitioners, despite their diverse educational backgrounds, now appear to speak a common language, the same words may not necessarily have exactly the same meanings to any two people. If the specialists in the field don't really understand one another, it becomes very difficult indeed for newcomers to make any sense of much of the technical material discussed at meetings and in the literature.

It was once a fairly widespread practice in higher education to require all undergraduates to study a core curriculum in addition to taking courses in their major fields of study. This established a basis for communication between individuals who might follow very different career paths after leaving college. I used to view core curricula as coercive; I now believe that best way to get a group of people to understand the rationale for doing things a given way in a given field is to expose all of them to as much as possible of the background material that shaped the development of the field. In the first edition of this book, I carried the core curriculum analogy far enough to list several articles[1-8] about flow cytometry as "required reading." In the later editions, I have tried to put everything I think you need to start with, and then some, into the book, even though core curricula, in general have been creeping back into favor.

Learning from History: Take One

While the history of flow cytometry will be covered in some detail in the next chapter, I would still recommend that those of who you have the opportunity take a look at two of my older references, the introductory article by Herzenberg, Sweet, and Herzenberg[2] on "Fluorescence-activated Cell Sorting" (*Scientific American*, **234**, No. 3:108-117, March, 1976) and Kamentsky's extensive review[1] on "Cytology Automation" (*Advances in Biological and Medical Physics*, **14**:93-161, 1973). Both articles clearly demonstrate their authors' vision and foresight; the first anticipates the modern development and present value of flow sorting as a preparative tool, while the second, an enlightening discussion of multiparameter flow cytometry and computer-based data acquisition and analysis, readily dispels any illusions readers may have that these now-popular techniques are new. You might look at other older reviews[6,8] or earlier editions of this book (1985, 1988, and 1995) to get an idea of how things were midway between the 1970's and now – or you might not.

Who Should Read this Book?

It is certainly possible for one person to operate (or even assemble) a flow cytometer and design and do meaningful biological experiments with it, but most sophisticated flow cytometry is done by groups, usually including one or more biologists and one or more people with some experience in instrumentation who run, fix, and/or build the apparatus. Things go more smoothly when the cell biologists and immunologists pick up some electronics and the electronikers acquire some cell biology and immunology. If everybody involved in your project learns something about flow cytometry in general, even if he or she is only going to be concerned with one aspect of what you are doing, there should be a sound basis for communication, which experience suggests will be beneficial to your progress. This is true even now, when flow cytometry has gotten a lot easier to do, because it is now also easier to do it badly.

I have tried to make *Practical Flow Cytometry* accessible, unintimidating, and as enjoyable as possible to readers with diverse backgrounds and levels of expertise. You can learn the basics from this book, but there are also things in it that aren't covered in the dozens of other available sources on cytometry and flow cytometry. What's in the other books (and Web sites, and CDs) has shaped my thinking about what to put into the current edition of this one.

There are now several books devoted to flow cytometric methodology, in which the experts who have devised specific protocols provide details that I see little point in duplicating here. There are also several books on clinical applications of flow cytometry, which discuss the diagnostic and prognostic significance of flow cytometric tests as well as the procedural details; again, it does not seem useful to duplicate much of this material. What I have tried to do is point out controversial areas and issues, and otherwise facilitate readers' navigation through the literature.

In the remainder of this chapter, I will point you toward some books and other information sources and resources that may make cytometry and the scientific disciplines underlying it more understandable and/or more interesting. I have tried, where possible, to populate this optional "reading list" with material you can read with at least some enjoyment.

2.1 INFORMATION SOURCES AND RESOURCES

There are a lot of really good information sources on the World Wide Web – and a lot of really bad ones. Since, like most of us, I don't have a large plasma display on at least one wall of every room in the house, I find reading old-fashioned paper books much easier than surfing the Web from the bedroom or the bathroom. Also, while the beach is the only place for real surfing, as opposed to the Web variety, sand does much less damage to a book than to a computer keyboard. And they don't make you start up your book when you go through airport security, although I suppose that could change.

Books on Flow Cytometry in General

The first big book in this field was a thick yellow tome called *Flow Cytometry and Sorting*, edited by Melamed et al[9]. The first edition, which appeared in 1980, is now out of print; since there were only 1,000 copies printed, many of which seem to have been stolen from libraries, it's hard to find. The second edition[1028] appeared in 1990. As is true of most books that include contributions from multiple authors, the chapters in *Flow Cytometry and Sorting* contributed on time by the more conscientious authors were not, by the time that book eventually went to press, quite as up-to-date as those extracted from the procrastinators. At this late date, the whole compendium is showing its age, but it has been kept in print.

Some single author texts other than mine are worthy of note. The first (1992) edition of Alice Givan's *Flow Cytometry: First Principles*[1029] was a well-written, very readable, relatively brief introduction to the field; the second edition[2316], which has kept up the good work, appeared in 2001. Michael Ormerod produced an even briefer introduction as Volume 29 in the Royal Microscopical Society's Handbook series; the current (1999) version is Volume 44[2384]. Jim Watson (not the double helix one; cytometry has its own) published his *Introduction to Flow Cytometry*[1030], which, in fact, goes well beyond the introductory level, in 1991; the book is apparently now out of print. These books are written with enthusiasm and reflect their authors' philosophical outlooks about flow cytometry and science in general; I enjoyed reading them, and not just because we philosophers have to stick together.

Books on Flow Cytometric Methodology and Protocols

Several volumes in the *Methods in Cell Biology* series, prepared under the auspices of the American Society for Cell Biology, have been devoted to cytometry. The two most recent, (Volumes 63[2385] and 64[2386]) edited by Darzynkiewicz, Crissman, and Robinson, appeared in 2001; previous volumes date from 1994[2387-8] and 1990[1034]. All are or have been available in cloth or paperback; I have seen them (at least the earlier versions) on the shelves of many bookstores and of many labs, and use them frequently. They contain a lot of detailed methodological information as well as the necessary background.

The International Society for Analytical Cytology (ISAC)'s *Handbook of Flow Cytometry Methods*[1035], edited by Robinson and five associates, appeared in 1993; it has the feel and some of the look of a cookbook. The recipes are complete down to catalog numbers of reagents, instrument setups, and phone and fax numbers of the contributors, which suggests a high level of confidence in their methods.

It was the editors' intention to revise the *Handbook* frequently, possibly metamorphosing it from its present spiral binding into a looseleaf format. A slightly different pattern of metamorphosis was followed; what emerged was *Current Protocols in Cytometry*[2389], a continuing series available by subscription in looseleaf format and/or on CD-ROM. This is probably the most complete of the protocol books; sections are revised and corrected as needed, and updates appear several times a year.

In Living Color[2390], a book of protocols edited by Diamond and DeMaggio, though smaller than either *Current Protocols* or the two-volume *Methods in Cell Biology* offerings, has a certain amount of both physical and intellectual heft. Finally, there are a couple of less voluminous protocol books that manage to hit the high points, edited, respectively, by Ormerod[2391] and Jaroszeski and Heller[2392]. A few older protocol books[1031-3], including an older edition of Ormerod's, are somewhat dated by now.

Clinical Flow Cytometry Books

There are now a respectable number of books dedicated to clinical applications of flow cytometry; almost all of them deal exclusively with applications of fluorescence flow cytometers. If you're looking for a good book on hematology counters and their use, see Bessman's *Automated Blood Counts and Differentials: A Practical Guide*[985].

The senior clinical book, *Flow Cytometry in Clinical Diagnosis*[1036], edited by Keren, appeared in 1989, and was supplanted by *Flow Cytometry and Clinical Diagnosis*[1037], edited by Keren, Hanson, and Hurtubise, in early 1994. A 3rd edition, once again called *Flow Cytometry in Clinical Diagnosis*[2393], has just appeared.

Owens and Loken's *Flow Cytometry Principles for Clinical Laboratory Practice. Quality Assurance for Quantitative Immunophenotyping*[2394] (1995) focuses on a major clinical application, as does a newer volume, *Immunophenotyping*[2395], edited by Stewart and Nicholson (2000). The latter book is one of a series on "Cytometric Cellular Analysis" from Wiley-Liss; this also includes the titles *Phagocyte Function*[2396] and *Cellular Aspects of HIV Infection*[2397]. There is also a recent (2001) special issue of the periodical *Clinics in Laboratory Medicine* devoted to "New Applications of Flow Cytometry"[2398]; since it's hardbound, I'll call it a book.

Among older books, *Flow Cytometry: Clinical Applications*[1043], edited by Macey, appeared in 1994, and *Clinical Flow Cytometry. Principles and Application*[1038], edited by Bauer, Duque, and Shankey, and *Clinical Applications of Flow Cytometry*[1039], edited by Riley, Mahin, and Ross, both date from 1993. Other older books on clinical applications include *Diagnostic Flow Cytometry*[1040], edited by Coon and Weinstein (1991), *Flow Cytometry*[1041], edited by Vielh (1991), *Clinical Flow Cytometry*[1042], edited by Landay, Ault, Bauer, and Rabinovitch (1993), *Flow Cytometry in Hematology*[1044], edited by Laerum and Bjerknes (1992), and Sun's more narrowly focused *Color Atlas-Text of Flow Cytometric Analysis of Hematologic Neoplasms*[1045] (1993).

Other Flow Cytometry Books

The 1985 volume edited by Van Dilla et al, *Flow Cytometry: Instrumentation and Data Analysis*[624] is, as its title suggests, primarily oriented toward hardware; what was around then is discussed in great detail. However, despite the title, coverage of data analysis is somewhat sketchy. There is some discussion of hardware and software in *Current Protocols in Cytometry*[2389], in the most recent *Methods in Cell Biology* volumes on Cytometry[2385-6], and in *In Living Color*[2390], but, if the rest of the tome you're now reading doesn't do it for you, you might want to look at another book in the "Cytometric Cellular Analysis" series, *Emerging Tools for Single-Cell Analysis: Advances in Optical Measurement Technologies*[2399], edited by Durack and Robinson.

Some data analysis methods are covered in two general works, the two-volume set on *Flow Cytometry: Advanced Research and Clinical Applications*[1046], edited by Yen (1989), and *Flow Cytometry. New Developments*[1047], edited by Jacquemin-Sablon. However, I'd look first at Chapter 5 here and then at (our) Jim Watson's *Flow Cytometry Data Analysis. Basic Concepts and Statistics*[1185], a 1992 book still, as far as I know, in print.

In an attempt, futile thus far, to broaden my linguistic horizons, I have tried without success to get a copy of the first complete volume on flow cytometry written in French, *La Cytométrie en Flux pour l'Étude de la Cellule Normale ou Pathologique*[1048], edited by Métézeau, Ronot, Le Noan-Merdrignac, and Ratinaud (1988), which has been favorably reviewed[1049]. I don't know whether there have been later editions.

Three more specialized books deal with applications of flow cytometry to particles smaller than eukaryotic cells. *Flow Cytogenetics*[1050], edited by Gray (1989), discusses analysis and sorting of chromosomes, while uses in marine microbiology and in microbiology in general are covered in *Particle Analysis in Oceanography*[1051], edited by Demers (1991), and *Flow Cytometry in Microbiology*[1052], edited by Lloyd (1993). All of these are well past their prime, although the cytogenetic methodology has probably changed least.

Newer, and reflecting the growing importance of flow cytometry and sorting in biotechnology, is *Flow Cytometry Applications in Cell Culture* (1996)[2400], edited by Al-Rubeai and Emery.

There are now flow cytometry books available to suit every taste and almost every pocketbook; although I feel obligated to buy every one that comes out, I would hesitate to recommend this procedure to readers, especially after computing the cost per page for a few of these volumes. As the French don't say about sorting, ***chacun à son goutte***.

Flow's Golden Oldies

There are several older collections of articles on flow cytometry that may be helpful. The proceedings of the Engineering Foundation Conferences on Automated Cytology, published in the July 1974 (Vol. 22, No.7), January 1976 (Vol. 24, No. 1), July 1977 (Vol. 25, No. 7), and January 1979 (Vol. 27, No.1) issues of *The Journal of*

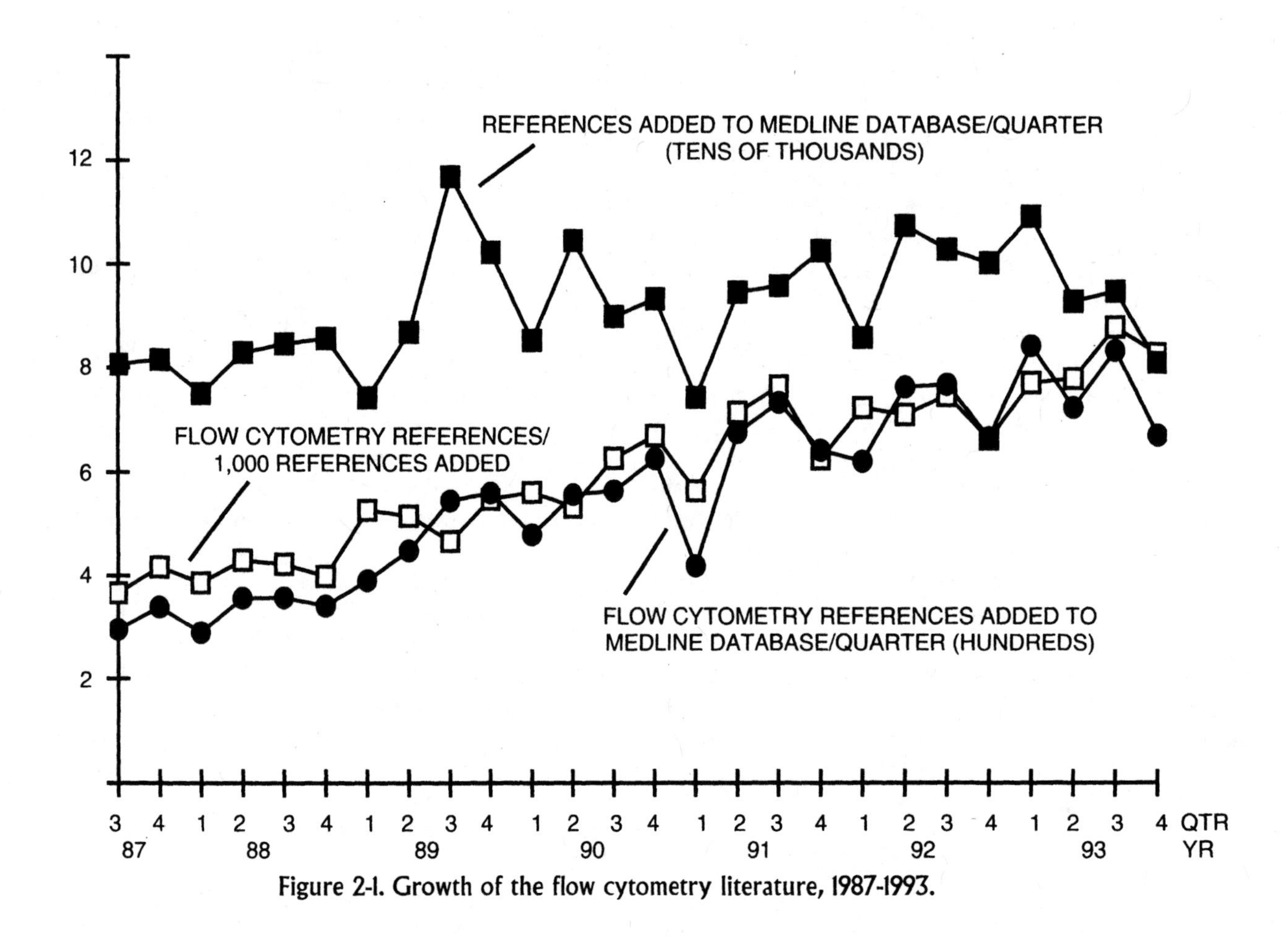

Figure 2-1. Growth of the flow cytometry literature, 1987-1993.

Histochemistry and Cytochemistry, include many of the landmark papers.

The proceedings of the first three European Pulse Cytophotometry Symposia were published by European Press Medikon, Ghent; I found it impossible to obtain copies when I tried to many years ago. The proceedings of Flow Cytometry IV are available in book form from Universitetsforlaget, Oslo, and also appeared as Supplement 274 to *Acta Pathologica et Microbiologica Scandinavica*, Section A, 1980.

2.2 THE READER'S GUIDE TO PERIODICAL LITERATURE

From the mid-1960's to the late 1970's, much of what was done, and most of what was novel, in flow cytometry was accomplished by people with a strong commitment to the technological side of the field. During the 1980's, the technology passed into the hands of users whose primary interests were and are in diverse biological fields. In cytometry, as in cytology, maturation was accompanied by differentiation. At the zygote stage, it was relatively easy for anybody who had a mind to keep track of the entire field. It is now much harder. We don't all go to the same meetings, and we don't all publish in the same journals. "Flow Cytometry" has been a subject heading in the *Index Medicus* for many years, but, for much of that time, only a small fraction of the papers published each month in which flow cytometry was used could be found under that heading. When I started doing the literature search needed for the 3rd edition of this book, I found things dramatically changed.

I searched the National Library of Medicine's MEDLINE database for articles entered between July 1987 and December 1993, in which "flow cytometry" appeared in the title, as a keyword, and/or in the body of the on-line abstract. Over 14,500 entries matching the search criterion were found among the total of 2,386,416 references added during that time period. I should probably thank everybody who didn't send me reprints.

Figure 2-1 gives a breakdown, by quarters, of the total number of references entered (in tens of thousands, indicated by closed squares), the number of flow cytometry references, i.e., those meeting the criteria mentioned above (in hundreds, indicated by closed circles), and the number of flow cytometry references per 1,000 references added (indicated by open squares). In mid-1987, approximately four of every 1,000 entries dealt with flow cytometry; by the end of 1993, almost eight of every 1,000 entries involved flow cytometry.

For those of you interested in the mechanics of this literature analysis, the relevant titles and abstracts were extracted using a CD-ROM version of MEDLINE and Knowledge Finder® software (Aries Systems Corporation, North Andover, MA). EndLink and EndNote® Plus software (Niles and Associates, Inc., Berkeley, CA) were used to compile and maintain a database of the flow cytometry articles. I really did read all of the 14,500 titles and all available on-line abstracts, and found more good stuff than could possibly be included in one book.

Even at that, I know I missed things, because I later ran across a number of papers published in the time interval covered by the search that weren't retrieved. In these cases, the words "flow cytometry" did not appear in the title or the abstract, and, although results of flow cytometric analyses were prominent in these publications, the people who compile MEDLINE didn't include "flow cytometry" as a keyword. Also, because I didn't have access to *Biological Abstracts* on CD-ROM, I couldn't begin to do the same kind of search for flow-related papers there that would not have made it into MEDLINE.

Note that I haven't mentioned on-line access. I work in a small, freestanding laboratory, and/or from home, and, until a few years ago, the only Internet access I had was via modem. When I was working on the 3rd edition, I had some very pricey 9,600 baud modems; I might have actually made the jump to 14,400 late in the process, but it's hard to remember. For the past few years, I have had a high-speed cable modem connection at the house, but was unable, at least until mid-2001, to get anything faster than dial-up access (usually 28,800 or slower) at the lab. There was no way I was going to be able to download 14,500 titles and abstracts over a phone line in the time I had available for the job. These days, I can not only use PubMed instead of my CD-ROM version of MEDLINE, I can also get access to a large number of journals and abstracting services from either the house or the lab. It's good to have bandwidth.

However, in terms of what I was prepared to do in terms of a literature search for this edition, bandwidth didn't help. In the last edition, I noted that "In the second edition, I singled out various journals for their relatively high content of articles dealing with flow cytometry; I won't even attempt that now. The technology seems to be everywhere." This time around, I tried the same trick I used the last time – pulling out everything in which there was any mention of "flow cytometry" anywhere at all – for a few selected months worth of MEDLINE, and concluded that I would have to look at 40,000 to 50,000 titles and abstracts if I wanted to cover the whole time interval between the beginning of 1994 and the end of 2001, and threw in the towel. Most of us have Internet access, and most of us who work in places where the budget can support a flow cytometer have reasonably speedy Internet access, meaning that you can search the literature about as effectively as I can. Since bad flow does happen to good journals, what we all need to know these days is not how to find the citations, but how to tell the good stuff from the bad stuff.

A good place to look for trustworthy articles dealing with flow cytometry (and other aspects of analytical cytology) is in *Cytometry*, which is published by Wiley-Liss for the International Society for Analytical Cytology (ISAC). If you have or intend to have more than a passing acquaintance with flow cytometry, you should join the Society, which will get you a subscription to the journal. Well, to most of it; several issues a year are set aside under the title *Clinical Cytometry* (called *Communications in Clinical Cytometry* until 2002), and, to get these, you either have to pay extra or join another organization, the Clinical Cytometry Society (CCS). The reviewers for *Cytometry* and *Clinical Cytometry*, though not perfect, are less likely to let in bad data or bad data presentations than are reviewers for many other journals.

Before 1980, when *Cytometry* started publication, the articles from the Automated Cytology conferences and a substantial number of other papers on flow cytometry were, as mentioned, published in the Histochemical Society's *The Journal of Histochemistry and Cytochemistry*, which still carries some work on the subject. During the 1970's, the International Academy of Cytology's *Acta Cytologica* carried articles on both analytical and clinical cytology; since 1979, the Academy has published the more specialized *Analytical and Quantitative Cytology* (now *Analytical and Quantitative Cytology and Histology*) in addition to *Acta*. Dyes and staining techniques useful for cytometry are frequently discussed in *Biotechnic & Histochemistry* (formerly *Stain Technology*), which is the official publication of the Biological Stain Commission. There is also *Analytical Cellular Pathology*, published by the European Society for Analytical Cellular Pathology (ESACP). Finally, the laboratory hematologists, including those hardy souls who do flow cytometry without benefit of fluorescence, have their own society, the International Society for Laboratory Hematology, with its own journal, *Laboratory Hematology*.

I previously noted the October 1, 1998 special issue of *Cytometry* on "Quantitative Fluorescence Cytometry: An Emerging Consensus"[2379]; other journals have also published special issues dedicated to or featuring cytometry. The most recent that come to mind are an issue of *The Journal of Immunological Methods* on "Flow Cytometry"[2400] (the regular issues of this journal also have a lot of good cytometry articles), an issue of *Methods*, a companion journal to *Methods in Cell Biology*, on "Flow Cytometry: Measuring Cell Populations and Studying Cell Physiology"[2401], and an issue of *Scientia Marina* on "Aquatic Flow Cytometry: Achievements and Prospects"[2402], all appearing in 2000. In the same year, there was also a special issue of *The Journal of Microbiological Methods* on "Microbial Analysis at the Single-Cell Level"[2403], with papers from a conference on that topic, which includes work in flow and image cytometry and other techniques; the complete text of all articles is available free of charge on the World Wide Web. Another issue, with papers from a June, 2002, conference on the same topic, is due out in 2003.

2.3 RESOURCES AND COURSES

There are flow cytometry prodigies who walk into a lab cold and are competent operators within a few days; they are probably about as rare as mathematical or musical prodigies. The rest of us (I am definitely not a flow cytometry prodigy) need hands-on help learning. The best way to find such help is to hook into the literal and physical networks of the flow cytometry community. The best place for a beginner to start is probably with the instrument manufacturers.

Flow Cytometer Manufacturers

The manufacturers of flow cytometers run training courses for their customers, and also maintain files of information on applications and techniques. It is entirely sporting to get such information from people who have sold or might sell you a flow cytometer; their names, addresses and URL's appear in the chapters on "Buying Flow Cytometers" and "Sources of Supply." Especially if you are new to flow cytometry, it certainly doesn't hurt to take the manufacturer's training course for your instrument.

The rub is that the manufacturers' courses aren't available for everybody who wants to learn flow cytometry. If you buy a new instrument, the price typically includes training for one or two people. An organization that already has an instrument can pay the manufacturer to get new people trained, but the manufacturers don't give training courses for instruments they no longer sell, and, if you don't happen to work for an organization with an instrument, you're unlikely to be able to take a manufacturer's course, even if you're willing to pay for it. College and university courses on flow cytometry for beginners are also scarce.

As a result, there are too many laboratories in which flow cytometers are run by people who were hired after the trained operator(s) left, and who had to pick up the basics by reading manuals and books and talking to better trained operators elsewhere in town. The good news is that there are more training opportunities for survivors of this hazing process, and for others who know at least a little about flow cytometry, than there are for novices.

The International Society for Analytical Cytology

Anybody with a serious interest in flow cytometry ought to join The International Society for Analytical Cytology (ISAC). ISAC publishes *Cytometry* and *Clinical Cytometry*, both now edited by Charles Goolsby, and also issues a Newsletter on-line; these publications include announcements of courses and meetings. As mentioned earlier, ISAC is also a moving force behind *Current Protocols in Cytometry*. Workshops on specific topics, sponsored by ISAC and by various instrument manufacturers, are held before and during meetings. ISAC's journals and Membership Directory are well worth the cost of dues; as a member, you'll also pay less to attend one of the ISAC Congresses held every two years, usually alternating between sites in the United States and Europe (more specialized meetings, memorializing Sam Latt, are held in alternate years). For further information, contact:

International Society for Analytical Cytology (ISAC)
60 Revere Drive, Suite 500
Northbrook, IL 60062-1577
USA
Phone: 847-205-4722
Fax: 847-480-9282
www.isac-net.org
E-mail: ISAC@isac-net.org

Cytometry (incorporating *Bioimaging*)
Editor-in-Chief:
Charles L. Goolsby, Ph.D.
c/o Patricia Sullivan
ISAC
(see mailing address above)
E-mail: cytometry@isac-net.org; cytometry@nwu.edu

ISAC is an international society; there are also continental, national, and regional organizations devoted to flow cytometry and other aspects of analytical cytology. Information about meetings and other activities of many of these groups finds its way into ISAC's publications.

The Clinical Cytometry Society

The Clinical Cytometry Society, which shares custody of *Clinical Cytometry* with ISAC, was organized at one of the annual conferences on Clinical Applications of Cytometry that started in Charleston, South Carolina in 1986. The current contact information for the society is:

Clinical Cytometry Society (CCS)
www.cytometry.org
P.O. Box 25456
Colorado Springs, CO 80936-5456
USA
Shipping Address:
5610 Towson View
Colorado Springs, CO 80918
USA
Phone: 719-590-1620
Fax: 719-590-1619
Business E-mail: admin@cytometry.org

Short courses and workshops focused on particular topics of clinical interest, and a longer course on clinical cytometry, are given before and during these meetings.

The National Flow Cytometry Resource

Since 1982, the United States Government, through the Department of Energy and the National Institutes of Health, has funded the National Flow Cytometry Resource (NFCR) at Los Alamos National Laboratory. Among other things, the NFCR makes several sophisticated, multibeam flow cytometers and related apparatus available to the research community for collaborative work and publishes the *Flow Systems Newsletter*, which includes abstracts of papers accepted for publication and announcements of various activities related to flow cytometry. Further information can be obtained from:

National Flow Cytometry Resource
Bioscience Division, M-888
Los Alamos National Laboratory
Los Alamos, NM 87545
Telephone (505) 667-1623
FAX (505) 665-3024
http://lsdiv.lanl.gov/NFCR/

"The Annual Courses" and Others

Annual courses on flow cytometry, one oriented toward research applications and the other toward clinical applications, have been offered since the late 1970's under the rotating sponsorship of a group of organizations including Dartmouth Medical School, the National Flow Cytometry Resource, Northwestern University Medical School, and Verity Software House. Paul Horan and Kathy Muirhead were prime movers in establishing these courses. and have continued to

Kathy and various colleagues now organize these weeklong, hands-on workshops under the aegis of Cytometry Educational Associates, Inc. (CEAI), a name reflecting both the inclusion of technologies other than flow cytometry in course curricula and the need to go through the legal system to set up a nonprofit organization, as opposed to an organization that doesn't profit.

Enrollment is usually limited to 40-80 people with at least some prior experience in flow cytometry; the courses include lab work using machines provided by the major manufacturers, and recent workshops have included image cytometry and/or hybrid instruments as well. The research course is now given in alternate years at Bowdoin College, in Brunswick, ME (next in 2004; see, and at Los Alamos (next in 2005). The clinical course alternates between Northwestern University Medical Center in Chicago (next in 2004) and Dartmouth-Hitchcock Medical Center in Hanover, NH (next in 2003).

Information on the courses is now available through the link to "Cytometry Courses" on Verity's Web site at www.vsh.com); announcements also appear in *Cytometry* and *Clinical Cytometry* and in various on-line resources, e.g., the Purdue Web site and Mailing List (see next page).

The Royal Microscopical Society (www.rms.org.uk) now has a Cytometry Section, and conducts courses on cytometry on at least an annual basis. Michael Ormerod, a principal in this enterprise, also independently offers courses on-line and *ad hoc*; he will come to you. See: <http://ourworld.compuserve.com/homepages/Michael_Ormerod/ormerod3.htm> .

FloCyte Associates (see Chapter 11 for contact information) is developing courses that will be offered yearly in four regions of the United States; they will presumably come to you if you're willing to pay the freight.

Various other organizations and institutions have offered and offer courses on various aspects of flow cytometry, which are generally announced in scientific periodicals; I particularly enjoyed lecturing in the Australasian Flow Cytometry Group Course in 1998.

Other Societies and Programs

American Society for Clinical Pathology (ASCP)
2100 West Harrison Street
Chicago IL 60612
USA
(312) 738-1336
info@ascp.org
ASCP certifies Medical Technologists and Medical Technicians, and offers a Qualification in Cytometry that requires between 6 and 18 months full-time acceptable experience in cytometry and satisfactory completion of an examination. Qualification in Cytometry is available to individuals without ASCP certification who have baccalaureate degrees from a regionally accredited college or university and 18 months of acceptable experience.

College of American Pathologists (CAP)
325 Waukegan Road
Northfield, IL 60093
USA
800-323-4040
847-832-7000 in Illinois
www.cap.org
CAP conducts proficiency studies and certifies clinical laboratories for both flow cytometry and quantitative image analysis. Since the CAP subcategories for flow cytometry include "FL1 – Lymphocyte Immunophenotyping, FL2 – DNA Content and Cell Cycle Analysis, FL3 – Leukemia/ Lymphoma, [and] FL4 – CD34+," those of you who have not been convinced not to use FL1, FL2, etc. as axis labels may want to reconsider.

European Society for Analytical Cellular Pathology (ESACP)
www.esacp.org
Secretariat:
Dr. Walter Giaretti
Laboratory of Biophysics and Cytometry
National Cancer Institute (IST)
Largo Rosanna Benzi, n. 10
16132 Genoa
Italy
Tel: +39/10/5600969, Fax: +39/10/5600711
E-Mail: walter.giaretti@istge.it
Membership (&ACP) 100 Euros/yr
Journal: *Analytical Cellular Pathology*
Editor-in-Chief: Prof.Albrecht Reith
Norwegian Radium Hospital & Institute of Cancer Research

Montebello
N-0310 Oslo 3
Norway
Tel: +47/22/934217, Fax: +47/22/730164
E-mail: albrecht.reith@labmed.uio.no
Publisher: IOS Press, Amsterdam

International Society for Laboratory Hematology
www.islh.org
Executive Office:
599 B Yonge Street, Suite 345
Toronto, Ontario M4Y 1Z4, Canada
Tel: (416) 586- 5120
Fax: (416) 586- 5125
e-mail: mail@islh.org
(Journal: *Laboratory Hematology*)

The Purdue Mailing List, Web Site, and CD-ROMs

Paul Robinson <jpr@flowcyt.cyto.purdue.edu>, Director of the Purdue University Cytometry Laboratories, has probably done as much to lead cytometry into the information age as anybody. The Purdue Cytometry Website (http://www.cyto.purdue.edu), which was up and running when the last edition of this book was being written, received 1,800,000 hits in 2000, and over 5,000,000 hits in 2001; it has links to a huge number of academic, commercial, governmental, and institutional sites related to cytometry.

Purdue also maintains a Cytometry Mailing List with several thousand subscribers that provides a forum in which cytometry people can pose questions, get answers, and/or just vent. It is maintained the old-fashioned way by the redoubtable Steve Kelley. To get added to the list, send e-mail to <subscribe@flowcyt.cyto.purdue.edu>; once you have subscribed, you will be given the e-mail address for the List (this helps keep trolls and spammers away).

Over the years, Paul et al have, after skillfully coaxing sponsorship out of various vendors and manufacturers, produced 6 CD-ROMs full of cytometry information contributed by various people in the field, and another CD-ROM on microscopy; the most recent cytometry CD-ROM appeared in May, 2002.

The most recent Robinson venture, Multimedia Knowledge, Inc., (www.ylearn.com) was originally developed to commercialize a CD-ROM and web-based teaching program for high school biology developed at Purdue with National Science Foundation funding. Beginning in 2002, on-line courses in cytometry will be added to the product line; for information, go to www.ylearn.com/elearn, or e-mail info@ylearn.com.

2.4 EXPLORING THE FOUNDATIONS

There's a lot of science behind flow cytometry, and there are a lot of books about the science, but there are relatively few books, particularly in the physical sciences, that offer much hospitality to readers from other fields. The authors of most of the books discussed in this section have at least made an effort at intelligibility to the general reader.

Optics and Microscopy

Optics may not cover a multitude of sins, but optics texts tend to cover a multitude of sines, integrals, and other mathematics, which may intimidate many people coming from the biological side. In previous editions, I recommended an optics text by Hecht and Zajac[10] for its clear illustrations and lucid prose; there's another very good technical optics book[625] by Meyer-Arendt, who happens to be a physics professor with a medical degree. A less intimidating, largely nonmathematical, and relatively entertaining treatment of optics can be found in a gorgeous book by Falk, Brill, and Stork called *Seeing the Light: Optics in Nature, Photography, Color Vision, and Holography*[626]. If you're really interested in photonics, there's a nice book by Saleh and Teich[1053], but it's definitely hard going. An easier-to-read, but thorough and informative, introduction to lasers can be found in the Second Edition of Hecht's *The Laser Guidebook*[1054] or his slightly later *Understanding Lasers: An Entry-Level Guide*[2405]. Harbison and Nahory's *Lasers: Harnessing the Atom's Light*[2406] is a 1997 book aimed at the interested layman, beautifully illustrated, and featuring a detailed discussion of recent developments in semiconductor lasers.

However, I would recommend that, before you try digging up any of the optics books, you take a look the Molecular Expressions Web site at www.microscopy.fsu.edu, which has an extensive tutorial on optics and microscopy, with a lot of interactive applets. This is a very well constructed site, because a lot of money was put into it. As I hear the story, Michael Davidson, of the National High Magnetic Field Laboratory (NHMFL) at Florida State University, took a lot of polarized light photomicrographs of crystals of various common materials – common bar cocktails, Ben & Jerry's ice cream, etc. – and licensed the pictures, which are very pretty and colorful, to commercial organizations. One of these is Stonehenge, Ltd., which produces silk neckties, scarves, and boxer shorts with the crystal patterns printed on them; some of the proceeds from their Molecular Expressions Cocktail Collection are even donated to Mothers Against Drunk Driving. I'm not sure whether those of us with ice cream abuse problems get discounts on the Ben and Jerry's line, but you get the idea. Anyway, there are apparently enough people willing to shell out a few dozen bucks a pop to wear crystals around their necks and elsewhere (one can readily imagine a conversation that starts out with "What are you drinking?" and progresses through "Want to see what the crystal structure looks like?") to have generated millions of dollars in royalties for FSU and the lab. Small potatoes, perhaps, compared to what the Seminoles usually bring in during football season, especially when they do make those field goals, but more than enough to produce a dynamite Web site.

A lot of practical information on the optical components (light sources, lenses, filters, etc.) used in flow cytometers can be found in catalogues from optical supply houses such as Edmund Scientific, Melles Griot, Newport Corporation, Optosigma, Oriel Corporation, and Thorlabs; see Chapter 11 for their contact information.

Everybody thinks it's a good idea to know how to use a microscope; almost nobody is taught how. If the Molecular Expressions Web site isn't your cup of tea, I can recommend some older and newer books. Virtually all of the basic theory can be acquired from a brief acquaintance with Spencer's slim *Fundamentals of Light Microscopy*[633]. For the practical details, it's hard to beat Smith's *Microscopy and Photomicrography. A Working Manual*[1055], which, with the aid of numerous photographs, tells you which knobs to turn how far to get optimal image quality from your microscope. Newer offerings include. Murphy's *Fundamentals of Light Microscopy and Electronic Imaging*[2407], Rost and Oldfield's *Photography with a Microscope*[2408], and Herman's *Fluorescence Microscopy*[2409]. With books like this around, plus the Web site, there's no excuse for not doing it right.

Electronics

When I played around with audio and ham radio equipment in the 1950's, I was totally mystified by vacuum tubes and discrete transistor circuitry, and I remain so to this day. Luckily, the phenomenal progress that has occurred since the advent of integrated circuit (IC) electronics has made it possible to build very sophisticated equipment using operational amplifiers, hybrid devices such as comparators, and small- to large-scale digital integrated circuits without understanding very much about transistors. Building the electronics for a multistation, multiparameter flow cytometer is no more difficult than building many of the gadgets described in various electronic, amateur radio and computer magazines; a hobbyist-level knowledge of electronics will equip you to take on this project, and you don't even need that much to appreciate most aspects of flow cytometer electronics.

One of the easiest and most enjoyable introductions to electronics is Hoenig's book[11], *How to Build and Use Electronic Devices without Frustration, Panic, Mountains of Money, or an Engineering Degree.* I haven't seen this around the bookstores lately. Another option, equally enjoyable, although somewhat more difficult because it covers virtually the whole field, is Horowitz and Hill's *The Art of Electronics*[12,1056]; this is unquestionably the best existing text. Anybody who already has some experience in electronics, or gets it from Horowitz and Hill, will also find pearls of wisdom and some good laughs in Pease's *Troubleshooting Analog Circuits*[1057] and in *Analog Circuit Design: Art, Science, and Personalities,* edited by Williams[1058].

Practical circuit details and a seat-of-the-pants introduction to various aspects of electronics suitable for anybody past junior high school age have been available from Radio Shack stores in Mims' *Getting Started in Electronics*[13] and *Engineer's Notebook II*[14]; various revisions of these may still be found in the chain's better-stocked stores.

Other practical and helpful information about analog and digital circuits appears in "cookbooks" by Jung[15] and Lancaster[16,17]. Armed with an introduction from some or all of the books just mentioned, you will find it possible to extract useful information from manufacturers' literature, which frequently omits important practical details because it is assumed that the reader will be sufficiently sophisticated to supply them. I keep most of my list of books handy in my electronics shop, and I heartily recommend this practice.

The problem with electronics these days is that integrated circuits are being built on a larger and larger scale, with most of them no longer available as chips with pins that plug into sockets. The development process instead requires that you use CAD software to design and test (by simulation) a circuit, and lay out a printed circuit board onto the surface of which the ICs are soldered directly. Much better for the real engineers; much worse for the hobbyists and others of us who have neither the in-house facilities to play this game nor the cash to pay outside contractors.

Computers: Hardware and Software

Some of my misadventures with computers, and more of other people's, were discussed in a book by Stein and Shapiro[18] (don't even bother looking for it), in which it was noted and lamented that the same misadventures have befallen people in the mainframe, mini, and microcomputer eras. I mention this because, having painfully entered binary programs into a vacuum tube mainframe from the console many years ago, and no less painfully entered binary programs into minis some years later, I was unenthusiastic about repeating the same unpleasant scenario with microprocessors, which seemed inevitable as I prepared to take the classical electronic engineering route to microprocessor system development.

Luckily, personal and home computers developed rapidly enough to provide cheap, user-friendly and otherwise convenient alternatives to microprocessor development for those of us who build and/or use flow cytometers and practically any other kind of instrumentation. Considering the amount of money now being spent on home and personal computers, there are pretty few readable, informative books on the topic, especially when it comes to hardware details. In previous editions, I recommended a few books[19-21,627-8,1059-64] that helped me and others to understand what it takes in the way of circuitry to connect things to microcomputer systems, and how operating systems and programming work. The specifics in these books may relate to obsolete hardware and software; the general principles are still valid.

Most bookstores are full of titles such as *Macs for Morons, DOS for Dummkopfs, Crays for Cretins*, and the like, which purportedly tell you how to use these computers and do not get down to how they work at the electronic level. I figure I could pick up some easy royalties by writing 500 pages or so of *The Power User's Complete Guide to Disk Formatting*, but that will have to wait until this book is done. I have bought a lot of computer books; my criterion for purchase is that the book helps me solve an existing problem with one of my computer systems. The trouble is that each book solves only one or two problems.

If you're interested in an accessible introduction to digital computers, starting with the basics of binary numbers and digital circuitry and moving right up to graphical user interfaces, read Charles Petzold's *Code*[2410]. Petzold is best know for his multiple editions of *Programming Windows*[2411], generally accepted as the definitive and the best-written book on writing programs for Microsoft's monopoly operating system; it's not an easy read, but *Code* is.

If you're looking for a book that will teach you how to program a computer, specifically an IBM-compatible or a Macintosh, you're really out of luck. Most books on programming are written by programmers. Although this should give the reader the benefit of the authors' expertise, programmers articulate enough to get a book past an editor are usually also smart enough to realize that, if more readers really learn how to program, it means less job security for programmers. I've been programming computers for over 40 years, and I can't understand the gobbledygook in most of the recent books about programming. There still are a few small books from which you can learn C and its extension, C++, which are the programming languages in favor today, which will get you up to speed if you're running the UNIX operating system on a minicomputer or DOS in an IBM-compatible. And Petzold's *Programming Windows* requires that you be fluent in C/C++, really meaning Microsoft's version, before you open the book.

What you really need to learn these days, however, is how to write programs that run with either Microsoft Windows or the Macintosh operating system. Many of the features of these graphical interfaces that make life easier for users make life much more complicated for programmers. There were never more than one or two books that attempted to teach a novice to program in C or C++ for the Macintosh or Windows, and, unfortunately, these books went out of date because the compiler developers brought out new versions of the C and C++ compilers with new bells and whistles that weren't explained in the books and aren't very clearly explained in the documentation which comes with the compilers. I can't recommend any of them.

The major drawback of many computer languages and applications is the tendency of their adherents to view them as religions rather than as examples of useful information technology. I personally use a computer language called Forth[22,629,1066-8], which was developed specifically to facilitate instrument control and data acquisition and analysis using small computers. It worked fine with DOS computers, and reasonably well with Macs, but, although there are good Windows versions available, they're hard to use, because it's just hard to write Windows programs, in any language.

The word among some programmers I trust is that Borland's Windows programming language tools, C++ Builder and Delphi (which uses the Pascal language), are substantially easier to work with than Microsoft's Visual C/C++ and Visual Basic. I have a lot of books about all of these; they are remarkably uninformative about some of the first things you'd think people would want to know about, such as how to get information in and out of the computer. But maybe I'm just a curmudgeon.

Digital Signal Processing

In theory, digital signal processing (DSP) is just another form of computation. In practice, it's a revolution and, arguably, even a religion to some. With the introduction of compact discs, consumer audio switched from analog processing and information storage to digital processing and storage; the newest generation of digital camcorders and high-definition television sets are bringing the digital revolution to consumer video. Data communication obviously requires digital processing, but so does most of voice communication in the age of the cellular phone and personal transceiver.

This has made the hardware and software necessary for digital processing affordable to manufacturers of scientific apparatus, such as cytometers, whose aggregate component needs don't even show up as a blip on the semiconductor manufacturers' radar screens. Those of us who use the instruments are beneficiaries, but we also have to evaluate instrument manufacturers' conflicting claims about when it is better to use digital processing and when we should stick with our old fashioned analog stuff. I'll try to sort that out at a basic level in subsequent chapters.

Books about DSP tend to be long on the math and short on plain language explanations of what's going on. A notable exception, which I wholeheartedly recommend, is Stephen Smith's *The Scientist and Engineer's Guide to Digital Signal Processing*[2412], which can be downloaded free from the author's Web site. If you are at all interested in learning more about the topic, this is the place to start.

Data Presentation and Display

A series of books by Edward Tufte, *The Visual Display of Quantitative Information*[1026,2418], *Envisioning Information*[1189], and *Visual Explanations*[2419], should be required reading for everybody with any need to present data. Tufte is a Professor Emeritus at Yale; he taught courses on statistical evidence, information design, and interface design. The books, printed by his own Graphics Press, are coffee table quality works of art, and, if the principles contained and expounded in them were more widely adhered to, viewing a poster session might be less like running the gauntlet. For about twice the price of the set of books, you can take Tufte's one-day course on Presenting Data and Information, which will get you the books plus the opportunity to watch and hear the man and see some of his collection of rare books. Information is available at www.edwardtufte.com.

Spectroscopy, Fluorescence and Dye Chemistry

A scholarly, entertaining, even, you should pardon the expression, absorbing treatment of the interactions of light with matter is given by Kurt Nassau in *The Physics and Chemistry of Color*[630]. A more formal coverage of this topic and its applications appears in Campbell and Dwek's *Biological Spectroscopy*[631]. Although there are still gaps in the literature when one looks for information about the biological applications of fluorescence and fluorescent dyes, considerably more is available now than was when the last edition of this book was written.

The first edition of an otherwise fairly comprehensive book on fluorescence spectroscopy by Lakowicz[23] contained relatively little information about biological applications; the second edition[2413] has improved somewhat in this respect. There is a smaller book on fluorescence spectroscopy by Sharma and Schulman[2414], and also Valeur's new, compact but comprehensive *Molecular Fluorescence*[2415].

The place to start in fluorescence microscopy is Rost's *Fluorescence Microscopy*[1070], a two-volume treatise with the second volume in gestation. This is an admirable work with lucid treatments of the physics and chemistry of fluorescence, technical details of fluorescence microscopy and microphotography, and such helpful goodies as a German-English vocabulary and suggestions on how to make text slides for talks that won't spoil your audience's dark adaptation. For those of you who take my advice and look at Rost's book, let me add that the cover of the 3rd edition of *Practical Flow Cytometry* was designed before I ever saw *Fluorescence Microscopy*.

Quantitative Fluorescence Microscopy[1071], also by Rost, continues in the style of his earlier work, introducing microspectrofluorometry, scanning and confocal scanning fluorescence microscopy, image analysis, and, in a chapter written by Tanke, flow cytometry. This is another good book to have. I have already mentioned a small, relatively new book by Herman[2409] that may be a good choice for beginners.

Books on biological stains by Gurr[24] and Lillie[25] and Horobin's newer, otherwise excellent *Understanding Histochemistry*[1069] contain few details about fluorescence or fluorescent dyes. The symposium volume on fluorescence applications edited by Taylor et al[632] , helpful in some areas, has largely been supplanted by newer works. These include *Cell Structure and Function by Microspectrofluorometry*[1073], edited by Kohen and Hirschberg, the two volumes edited by Wang and Taylor on *Fluorescence Microscopy of Living Cells in Culture*[1072], and *Fluorescent and Luminescent Probes for Biological Activity, a Practical Guide to Technology for Quantitative Real-Time Analysis*[1074,2416], edited by Mason.

Finally, one should neither underestimate nor overlook the magnificent *Handbook of Fluorescent Probes and Research Chemicals*[1075,2332], which, although nominally the catalog of Molecular Probes, Inc. (Eugene, OR; www.probes.com), contains more information on the structure, spectra, and use of fluorescent dyes (many thousands of references on several thousand compounds) than is available anywhere else.

A more general survey of methods in microscopy, which includes some discussion of histochemistry and analytical methodology, appears in *Light Microscopy in Biology*[1076], edited by Lacey; some chapters in this book are easier to read than others.

Cell and Molecular Biology and Immunology

It was pointed out to me that the first edition of this book provided no guidance for the hapless physicist trying to locate the nuclear membrane among the quarks and gluons. DeDuve's beautiful, two-volume *Guided Tour of the Living Cell*[635] is a good place to start. *Molecular Biology of the Cell*[636], by Alberts et al, is as helpful for those of us who learned biology more than a few years ago as for those who never learned it, and Darnell, Lodish, and Baltimore's *Molecular Cell Biology*[1077] covers similar territory and is even more lavishly illustrated. *Recombinant DNA*[1078], by Watson et al, provides a well-written and well-illustrated introduction to genes and their manipulation.

The clear winners among immunology books are the large volume *Immunology*[637], by Roitt, Brostoff, and Male (now in a new edition[1079]), and its pocket-sized sibling, Male's *Immunology: An Illustrated Outline*[638], both of which contain color illustrations which set a new standard for other texts to follow. Another useful and highly readable book, with a more philosophical bent, is Golub and Green's *Immunology: A Synthesis*[1080].

I haven't included the new editions of the cellular/ molecular biology and immunology books here. All of these seem to have a one or two page description of flow cytometry that gets something wrong at a basic level. As I mentioned in the Preface, the one book I found that got it right was *Immunobiology*, by Janeway et al[2315].

2.5 ALTERNATIVES TO FLOW CYTOMETRY

Flow cytometry has come into wide use at least in part because manufacturers presented ready-made solutions to instrumentation problems that few users would have tried to solve. The many tasks to which flow cytometers are unsuited, e.g., measurement of attached cells, repeated measurements of a single cell over time, and high-resolution localization of probes in or on cells, are readily approached using such devices as confocal microscopes and microscope-based imaging systems. Application of such apparatus was initially hindered because, while it was generally necessary for investigators to build their own systems from components, little guidance was available in print. Shinya Inoué produced a dramatic cure for this deficiency with a magnificent book called *Video Microscopy*[634], which covers foundations, history, practical details, results, and sources of supply. A second edition has been written by Inoué with Kenneth Spring[2417].

There is some discussion of instrumentation for fluorescence image analysis, confocal microscopy, and analysis of fluorescence recovery after photobleaching (FRAP) in references 1070-1072 and 1074. These techniques are also covered in several other books, including *Optical Methods in Cell Physiology*[1081], edited by Weer and Salzberg, Russ' *Computer-Assisted Microscopy: The Measurement and Analysis of Images*[1082], the first edition of the *Handbook of Biological Confocal Microscopy*[1083], edited by Pawley, and *New Techniques of Optical Microscopy and Microspectroscopy*[1084], edited by Cherry. All of these are well out of date at this point.

More recently, Wilkinson and Schut have edited a fairly comprehensive book on *Digital Image Analysis of Microbes*[2420], which has a lot of foundation material equally applicable to a wider range of biological specimens, and Wang and Herman have edited a volume on *Fluorescence Spectroscopy and Microscopy*[2421]. And the *Handbook of Biological Confocal Microscopy*, which remains a standard in its field, has emerged in a new edition[2422].

For now, however, we will move back toward classical microscopy, as we consider the history of flow cytometry.

3. HISTORY

The quotation attributed to George Santayana, "Those who cannot remember the past are condemned to repeat it," has already found its way into the literature of flow cytometry[26]. To judge from the amount of repetition of the past that has also found its way into the literature, people aren't reading as much as they are writing. I have always liked to pursue my fields of interest back to their original sources. For one thing, it does help you to avoid repeating other people's mistakes; for another, it improves your perspective and your personality to find out that the great idea which occurred to you last night occurred to Paul Ehrlich in the 1890's.

Since flow cytometry is a relatively new field with a relatively small number of hard-core practitioners, it is possible to gain some insight into why the technology has developed as it has from talking to the people who developed it and asking them why they did things as they did. I have now been collecting historical anecdotes in this fashion for more than a third of a century. I suspect that everyone writes history with what he or she calls a perspective and others describe as a bias. I will admit to a perspective.

Microscopy from Leeuwenhoek's time to the 1800's was as much the province of gentleman naturalists (with perhaps an occasional gentlewoman) as of physicians. Since then, most of the technical developments in microscopy, including flow cytometry and the rest of analytical cytology, have been motivated by both scientific and economic interests in improving medical diagnosis and treatment. Flow cytometry, in particular, was envisioned as an ideal method for counting and, eventually, for classifying blood cells, and also as a technique for making reliable distinctions between normal and malignant or premalignant cells in cytologic specimens. If you keep these sources of motivation in mind, you'll find it easier to understand why things have happened as they have.

This chapter is divided into sections called "Ancient History," which covers the period from Leeuwenhoek to the 1950's, "Classical History," which describes events of the 1950's and 1960's, and "Modern History," in which I consider what has happened in analytical cytology and flow cytometry from the time I started watching through the early 1990's. Events more current than that, "History in the Making," will be discussed in connection with the technical topics covered in later chapters.

3.1 ANCIENT HISTORY

Flow Cytometry: Conception and Birth

I am not the only revisionist author who has considered the history of flow cytometry; Derjaguin and Vlasenko[27], in discussing a flow system using light scatter measurements for counting and sizing hydrosols and aerosols, give one M. V. Lomonosov credit for describing what we in the West call the Tyndall effect as far back as 1742, and also for anticipating dark field microscopy and light scattering measurement of submicroscopic particles.

American historians of flow cytometry usually cite a 1934 paper in *Science* by Moldavan[28] as the first description of flow cytometry. This introduces the concept of counting cells, e.g., blood cells, flowing through a capillary tube, using a photoelectric sensor to make extinction measurements; the wording of the article strongly suggests that the author had never succeeded in getting the device working, at least at the time of publication.

Flow cytometry of biological specimens was actually accomplished in the 1940's; the cells analyzed were bacterial cells, and the suspending medium was air rather than water. It had been established by the 1920's that dark-field microscopes could be used to visualize objects, e.g., viruses, that were not resolvable by transmitted light microscopy.

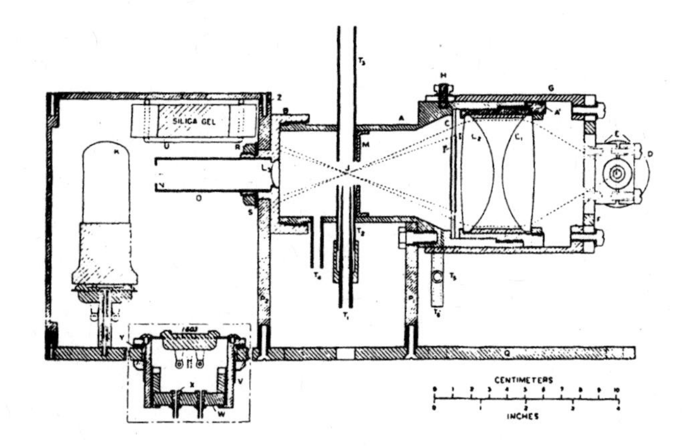

Figure 3-1. The first working flow cytometer. Reprinted with permission from F. T. Gucker, Jr., et al, *J.A.C.S.* 69:2422-31[29]. Copyright 1947 American Chemical Society.

In the 1920's and 1930's, colloid chemists and physical scientists built instruments incorporating such "ultra-microscopes" for analysis of flowing colloidal suspensions and for detection, counting, and sizing of particles in aerosols such as mine dusts.

A 1947 paper by Gucker et al[29] reported success in flow cytometric detection of bacteria in aerosols. The work, sponsored by the U.S. Army with the aim of rapid identification of airborne bacteria and spores used as biological warfare agents[30,31], was done during World War II at Camp (now Fort) Detrick and Harvard Medical School; the results could not be published until they were declassified after the war.

The original Gucker particle counter, shown in Figure 3-1, incorporated a sheath of filtered air to confine the air sample stream to the central portion of the flow chamber, in which it was subjected to dark-field illumination. The light source (far right), one of the most powerful then known, was a Ford headlight; a photomultiplier tube, then a newly developed device, was introduced as a detector, although the detector shown at the left of the figure is a thallium sulfide photocell. The observation point is at the intersection of the cones of light in the center of the figure. The instrument had about a 60 percent probability of detecting a particle 0.6 μm in diameter. Interestingly enough, history has been repeating itself in recent years, as the Army has regained interest in flow cytometric detection of airborne microbial pathogens[2423].

Until the 1950's, the electro-optical technology available for use by analytical cytologists was, as illustrated by the description above, rather primitive. Given this level of technology, it is somewhat surprising that so much was learned about the chemistry and physics of cells by that time.

Staining Before and After Paul Ehrlich

In preparing earlier editions of this book, I used Baker's *Principles of Biological Microtechnique*[32], which I would still recommend to the reader, as a primary source. That book was dedicated to the memory of Paul Ehrlich, who was a central figure in the field from his student days until his death. I have gained additional perspective from Clark and Kasten's revised third edition of Conn's *History of Staining*[1085], which I would likewise recommend.

From Leeuwenhoek's time until the mid-1800's, very little work was done on staining cells. Leeuwenhoek himself used saffron to improve contrast of muscle specimens; others focused primarily on uptake of naturally colored materials by living cells and tissues. Although it was possible to apply some of the color reactions being devised by analytical chemists to qualitative analysis of tissues and cells, the techniques and reagents used did not generally allow localization of chemical constituents of cells at the cellular and subcellular levels.

Rapid progress was made in this area from the 1850's on due to the availability of a large number of newly synthesized dyes, beginning with William H. Perkin's mauve in 1856, which represented the first technological fruits of the emerging science of organic chemistry. In a very real sense, synthetic dyes had the same status in the late 1800's that monoclonal antibodies and the products of genetic engineering have today. The textile industry represented a large market, enabling a synthetic chemical industry to develop with the production of dyes as its primary goal; as the chemical factories made new organic structures available, new applications could be found. Simon Garfield's recent popular book, *Mauve*[2424], provides an entertaining history of both the nascent dye and chemical industry and its spinoffs and progeny, including some accounts of Ehrlich's work.

Ehrlich studied the reactions of dyes with living tissues as well as with materials fixed by heat or chemical treatment. In studies of blood[33], he used mixtures of acidic and basic dyes to distinguish what have continued to be known as acidophilic, or eosinophilic, basophilic, and neutrophilic granular leukocytes. Principles he elucidated were applied by Malachowski and Romanowsky to develop mixtures of eosin and azure dyes which allowed visualization of malaria parasites in blood cells as well as identification of different types of leukocytes; the Giemsa, Leishman, MacNeal, and Wright stains for blood and bone marrow smears evolved from Romanowsky's.

Ehrlich also injected dyes into living animals, and studied the rate at which different cells and organs decolorized dyes by metabolic oxidation-reduction (redox) reactions. These studies anticipated the later development of tracer methods in which radioisotopes, rather than dyes, would be used, and provided a basis for the use of dyes as drugs, resulting in the first specific chemotherapy for syphilis. In the course of his work on immunology and chemotherapy, Ehrlich developed a concept of specific ligand-receptor interactions that anticipated much of what has been done in this area in more recent years.

Ehrlich employed the fluorescence of fluorescein, shortly after this dye was first synthesized in the 1880's, to study the dynamics of ocular fluids; it is sobering to contemplate what he might have accomplished had he had access to ultraviolet and fluorescence microscopes, which were invented shortly before his death, and which set the stage for the next great advances in analytical cytology.

Nostalgia now lets me recall my initial introduction to the world of Leeuwenhoek and Ehrlich, Paul de Kruif's *Microbe Hunters*[2425], which I read as a boy and still recommend, despite the fact that it reflects the prevailing prejudices of the era in which it was written (late 1920's). Paul de Kruif served as a technical adviser to Sinclair Lewis when the latter wrote *Arrowsmith*, which was also must reading for pre-meds, or at least those of my generation.

Most of the classical staining techniques for examination of blood cells, tissues, and bacteria had been developed by the beginning of this century. Since that time, the major thrust in histochemical technique has been toward procedures of increasing specificity. Of particular interest with regard to flow cytometry are staining methods for **nucleic acids**. To gain some perspective on the history of developments in this area, it should be remembered that, although the role of the nucleus in development and heredity had become apparent by the turn of the century, DNA was not conclusively identified as the genetic material until the mid 1940's. Until the 1920's, it was believed that DNA was present only in animals, while plants contained RNA.

In the early 1900's, Pappenheim and Unna adapted a combination of two basic dyes that had been used by Ehrlich, **methyl green** and **pyronin**, to produce green (methyl green) staining of nuclei and red (pyronin) staining of cytoplasm and nucleoli. Brachet[36] subsequently demonstrated, by comparison of ribonuclease-treated and untreated specimens, that pyronin, when combined with methyl green, bound to RNA, and also showed that RNA was present in the cytoplasm of animal as well as plant cells. Methyl green was shown to bind to polymerized DNA; we now know that the molecule binds to adenine-thymine pairs or triplets in a fashion similar to the UV-excited, blue fluorescent **Hoechst dyes 33258 and 33342**, both of which are used for DNA staining in flow cytometry.

In 1925, Robert Feulgen[35] developed a presumably stoichiometric procedure for staining DNA which involved derivatizing a dye, originally fuchsin, to a Schiff base, and reacting this with DNA from which the purine bases had been removed by mild acid hydrolysis. Feulgen was the first to demonstrate that DNA was present in both animal and plant cell nuclei. Refinements of Feulgen's procedure followed over the subsequent decades; a variant using fluorescent dyes such as **auramine O** and **acriflavine** was developed by Kasten[1085] and employed in some of the earliest flow cytometric fluorescence measurements[79].

Origins of Modern Microscopy

The optical "microscope" with which Leeuwenhoek visualized protozoa and bacteria was a **simple microscope**, essentially a very high power magnifying lens in a holder that allowed a specimen to be brought into the field of view by turning a screw. Leeuwenhoek was unusually successful in making observations at high magnification with his apparatus. It proved easier for others to make and use **compound microscopes**, in which an **objective lens** makes a modestly magnified image of the specimen; a magnified visual image of this image is then produced by a second lens, the **ocular**, or **eyepiece**.

The first compound microscope was built in 1590; the apparatus was refined over the next three centuries, with many of the features we now associate with modern microscopes being introduced by the Carl Zeiss works in Jena, Germany, during the late 1800's[1055]. Ernst Abbe, working with Zeiss, developed both the theory of microscopy and many refinements of optical design and technique, including apochromatic color-corrected lenses and oil immersion. The implementation of Abbe's designs was made possible by the chemist Otto Schott, who produced the special glasses needed to make the lenses and other optical components.

The resolution of a transmitted light microscope is a function of the illumination wavelength, and improves at shorter wavelengths. By the beginning of this century, microscopes employing ultraviolet light sources and quartz optics had been produced in an effort to resolve finer detail than could be observed with visible light. Transmitted light microscopy with ultraviolet light required that the image be photographed rather than observed directly, since ultraviolet light is invisible to the human eye. Fluorescence emission excited by ultraviolet light is, in general, visible, and was first observed in an ultraviolet microscope by August Köhler of Zeiss in 1904[1085]. By the start of World War I, several firms had refined ultraviolet microscopes into fluorescence microscopes.

Making Cytology Quantitative: Caspersson et al

Between the 1930's and the 1960's, the basis for much of modern analytical cytology was established by Torbjörn Caspersson and his colleagues in Stockholm, whose work was alluded to in Chapter 1. Caspersson's 1950 monograph, *Cell Growth and Cell Function*[34], describes detailed studies of nucleic acid and protein metabolism during normal and abnormal cell growth. These were done by highly precise microspectrophotometric measurement of the absorption of unstained cells in the ultraviolet and visible regions of the spectrum.

Caspersson's results, remarkable enough in themselves, are even more remarkable in that they were obtained using apparatus which seems strange and almost hopelessly primitive to those of us who have grown up with lasers and solid-state electronics. Cadmium spark sources were used for ultraviolet illumination; photocurrent measurements were done with string electrometers, unless the signal was strong enough to permit use of a vacuum-tube amplifier. Analytical cytology has obviously come a long way since the 1950's;

many of the advances in the field since then have been made by people who learned the basics in Stockholm.

It was possible by 1950 to determine the content of nucleic acids and protein in living cells by making measurements near 260 nm and 280 nm, although DNA and RNA could not be distinguished from one another in intact cells when this procedure was employed. Hemoglobin production in immature red blood cells was studied, by Thorell among others, by measurement of the strong absorption of heme porphyrins in the Soret band near 420 nm (see Figure 1-3, p. 8). The very nature of absorption measurements, however, restricted the range of application of this technique.

As was previously mentioned, the photodetector in a microspectrophotometer measures light transmitted through the specimen; such a measurement cannot always discriminate between light loss due to absorption and light loss due to scattering. Precise absorption measurements were shown by Caspersson et al to require optics of relatively high (>0.85) numerical aperture (N.A.), in order to collect as much of the scattered light as possible. It was also found desirable to match the refractive indices of the specimen, suspending or mounting medium, and immersion fluid used, to minimize scattering at the interfaces between them. In some cases, as when the cytoplasm of cells contained refractile granules, it remained impossible to measure absorption with the precision required for quantitative analyses of cellular constituents.

From the vantage point of a new century in which nucleic acid chemistry is as much a technology as a science, it is too easy to underestimate the significance of the work of Caspersson and other pioneers for the development of molecular biology and molecular genetics. The Feulgen staining procedure for DNA, described in the 1920's[35], was not universally accepted as quantitative; Brachet's studies of cellular growth and development, using methyl green and pyronin, respectively, to stain DNA and RNA[36], were also regarded with suspicion in some quarters. The ultraviolet absorption technique was less vulnerable to criticism, because it was based upon characteristics demonstrable in purified preparations of the macromolecules involved and because no reagent was used. Results obtained by all of these methods led to the same conclusions; i.e., that the content of both DNA and RNA was increased in actively growing cells. Caspersson and Schultz[37] showed in 1938 that the nucleic acid content of chromosomes doubled during the mitotic cycle, verifying that this chemical constituent exhibited the stoichiometry required of genetic material; it was not until 1944 that Avery et al[38] published the experiment usually regarded as establishing DNA as the carrier of genetic information.

Origins of Cancer Cytology: The Pap Smear

The clinical relevance of Caspersson's work was far from obvious in the late 1930's; even had it been obvious, it would have been almost impossible to implement UV microspectrophotometric measurements for routine cancer diagnosis at that time. The first practical procedure for the cytologic diagnosis of cancer instead made use of conventional transmitted light microscopy and an empirically derived mixture of acidic and basic dyes.

George Papanicolaou developed the first of several staining mixtures for use in studies of the primate estrous cycle, observing that staining characteristics of cells exfoliated from the female genital tract changed during the cycle. He later applied his procedures to material of human origin, and observed that exfoliated cells from patients with cervical dysplasias and cancer could be distinguished from normal cells.

A 1941 report by Papanicolaou and Traut[39] established the clinical relevance of nuclear chemistry and morphology for exfoliative cytologic diagnosis of cervical carcinoma. This provided a rationale for development of automated apparatus for clinical cytology that has persisted until the present. During the 1940's, it was necessary to train pathologists in the interpretation of smears stained according to Papanicolaou's procedure[40]; by the end of that decade, a number of investigators were devoting their energies to the development of new staining techniques which might better distinguish normal from malignant cells. By the early 1950's, some of these workers had turned their attention to possible applications of fluorescent dyes and fluorescence microscopy in cancer cytology.

At this time, it was not clear whether fluorescence measurements offered any significant advantage over absorption measurements for analytical cytology. The fluorescence microscope, developed in 1911, had been used until the 1940's largely for the same kinds of descriptive studies of which dyes stained which parts of which cells as had been done during the late 1800's based on transmitted light microscopy.

The development of fluorescence assay was given some impetus during World War II, when it was necessary to find new antimalarial drugs and new sources for older ones. Quinine, the natural product most widely used for malaria treatment, was found only in areas of Asia controlled by the Japanese, while the most commonly used synthetic substitute, the acridine derivative quinacrine (atebrine), was produced in Germany. A number of American medical scientists, many of whom would later form the core staff of the National Institutes of Health, conducted an extensive search for synthetic substitutes. Both quinine and quinacrine were highly fluorescent; this property could be exploited for quantitative analysis of these materials and of structural analogs that were screened for antimalarial activity. Improved spectrofluorometers developed for such analyses were also used to characterize fluorescent dyes[2426].

In 1950, Friedman described the use of a combination of acid fuchsin, acridine yellow, and berberine sulfate for uterine cancer detection by fluorescence microscopy[41]. He found that nuclei of malignant cells stained more intensely with berberine than did nuclei of normal cells. This stimulated Mellors and Silver[42] to develop a scanning microfluorometer capable of making quantitative measurements of berberine fluorescence; the instrument was then investigated for use in cancer cytodiagnosis by Mellors, Keane, and Papanicolaou[43].

The Fluorescent Antibody Method

Another extremely important application of fluorescence microscopy developed during the 1940's was the fluorescent antibody technique developed by Albert Coons, Hugh Creech, and Norman Jones[44]. Other workers[1087-8] had demonstrated that azo dye-conjugated antisera to bacteria retained their reactivity with the organisms and would agglutinate them to form faintly colored precipitates; however, the absorption of the dye-conjugated sera was not strong enough to permit visual detection of bacterial antigens in tissue preparations.

Coons surmised that it might be easier to detect small concentrations of antibody labeled with fluorescent material against a dark background using fluorescence microscopy. He consulted Louis Fieser of the chemistry department at Harvard for aid in preparing conjugates, and was told to "talk to two fellows in the basement who are already busy hooking fluorescent compounds to proteins"[1086,1089].

The two fellows were Hugh Creech, a cancer researcher interested in the biologic properties of conjugates of carcinogenic hydrocarbons and serum proteins, and Norman Jones, a spectroscopist who had brought new techniques of ultraviolet spectroscopy to bear on the analysis of polycyclic hydrocarbons (R. N. Jones, personal communication, 1993). Coons, Creech, and Jones labeled antipneumococcal antibodies with anthracene and could detect both isolated organisms[44] and, more importantly, antibody bound to antigen in tissue specimens[1090], by the UV-excited blue fluorescence of this label, as long as tissue autofluorescence was not excessive.

In 1950, Coons and Kaplan reported that fluorescein, conjugated as the isocyanate, gave better results than did anthracene, because the blue-excited yellow-green fluorescence of fluorescein was easier to discriminate from autofluorescence[45]. The requirement for the highly toxic gas phosgene in the isocyanate conjugation procedure delayed the widespread use of fluorescent antibody techniques until less hazardous alternative conjugation methods[46,47] were developed; from that point on, fluorescein became and has remained the most widely used immunofluorescent label.

Blood Cell Counting: Theory and Practice

Until the 1950's, there was no method not based on visual observation for counting erythrocytes (red cells), leukocytes (white cells), and thrombocytes (platelets) in blood. The apparatus employed for visual counting was the **hemocytometer** (see pp. 18-19). Erythrocytes, typically present in whole blood at concentrations around $5 \times 10^6/\mu L$, were counted at a 1:200 dilution in an isotonic saline solution. Leukocytes, at concentrations around $5 \times 10^3/\mu L$, were counted at a 1:10 dilution in a fluid containing a chemical agent to lyse the erythrocytes and a dye to color the leukocyte nuclei. Platelets, at concentrations near $2 \times 10^5/mm^3$, were counted at 1:100 dilution in a fluid that swelled the platelets and made the red cells appear translucent in a phase contrast microscope[1091]. The standard procedure was to mouth-pipette blood and diluent, something I did innumerable times as a medical student in the 1960's. Those days are gone forever.

Hemocytometer counts are subject to numerous sources of imprecision, due to errors in pipetting, dilution, and introduction of samples into the chamber, to imperfectly calibrated chambers, and last, but rarely least to the **Poisson statistical sampling error** associated with counting, which was discussed on p. 19. The expected standard deviation of a count of n items is $n^{1/2}$. It is generally impractical to do visual counts of more than 500 cells in a specimen; this would yield a standard deviation of 22 cells, and a coefficient of variation (CV) of 100 × 22/500, or 4.4%, in the absence of any other sources of error. The added effects of dilution errors, etc., raised CVs for erythrocyte counts to values near 10% under the best of circumstances; CVs were correspondingly higher for leukocyte counts, in which only 100-200 cells would be counted. The imprecision of erythrocyte counts, in particular, made accurate diagnosis of anemias difficult.

It had been observed that the size and color of blood cells varied in different types of anemia. The anemia of iron deficiency was characterized by smaller than normal, or **microcytic**, erythrocytes, which were also **hypochromic**, i.e., contained less hemoglobin than normal. In so-called **pernicious anemia**, now known to be due to vitamin B_{12} deficiency, the cells were larger than normal, or **macrocytic**, and **hyperchromic**, appearing to contain more than the normal amount of hemoglobin.

The hemoglobin content of blood could be estimated by colorimetry. The total mass of red cells, a function of both cell size and cell number, could be estimated by centrifuging whole blood and observing the **volume of packed red cells (VPRC)**, i.e., the fraction of the total volume occupied by cells. A calibrated tube in which such measurements were made was called a **hematocrit**; this term is now used more or less synonymously with VPRC.

Believing that such studies might shed some light on cell size variations in anemias (M. M. Wintrobe, personal communication), Wintrobe, during the 1920's, examined relationships between red cell numbers, size and hemoglobin content in diverse vertebrate species[1092]. He found that, although VPRC and hemoglobin were relatively constant, cell sizes and numbers showed considerable variation; animals with larger cells had lower cell counts and *vice versa*.

At that time, Wintrobe also[1093] defined three quantitative parameters called the **red cell indices**, to which I referred earlier (p. 49). These are the **mean cell (or corpuscular) volume (MCV)**, **mean cell (corpuscular) hemoglobin** [content] (MCH), and **mean corpuscular hemoglobin concentration (MCHC)**; values of these parameters, with MCV in femtoliters (10^{-15} L), MCH in pg, etc.; are reported by all modern laboratory hematology counters. In principle, once the red cell indices had been defined, hematologists could differentiate microcytic and macrocytic anemias from normal cells on the basis of measured values of MCV. In practice, this was not possible.

When the red cell indices were first defined, it was not possible to measure either the volume or hemoglobin content of individual cells with any precision. Instead, MCV was calculated by first obtaining the **hematocrit**, i.e., the fraction of blood volume occupied by red cells, and dividing it by the erythrocyte count. MCH was similarly calculated by measuring the hemoglobin content of the blood in bulk, in units such as g/dL, and dividing it by the erythrocyte count. However, the imprecision of the erythrocyte count was high enough to prevent clear distinctions being made between microcytic and normal, macrocytic and normal, etc. There was thus a perceived need for instruments which could improve the precision of erythrocyte counts, even if the improvement came solely from counting more cells than could be conveniently counted visually.

The imperfections of other cell counting procedures in hematology were also recognized. **Differential leukocyte counts**, i.e., enumeration of the percentages of various cell types present in blood (or bone marrow), were done by counting 100-200 cells on a thin smear stained with an eosin-azure dye mixture such as Giemsa's or Wright's stain. This resulted in imprecision due to sampling statistics, especially in counts of relatively rare cells such as eosinophil and basophil granulocytes, which typically account, respectively, for 2-5% and less than 1% of a total white cell population.

Sampling statistics were an even greater concern in the case of the blood **reticulocyte count**. Reticulocytes are erythrocytes newly released from the bone marrow into the blood. Before entering the circulation, they extrude their nuclei; however, they still retain remnants of the ribosomal RNA and protein used for synthetic purposes during their development in the marrow. The RNA is degraded in the course of a day or two; the average lifespan of an erythrocyte in circulation is about 120 days. This means that, under normal circumstances, about 1% of the erythrocytes in peripheral blood are reticulocytes. When red cell production is increased, as when the marrow compensates for cell loss due to hemolysis or bleeding, the percentage of reticulocytes is higher; when red cell production is decreased, as in vitamin B_{12} deficiency, the percentage of reticulocytes may approach zero.

Reticulocytes were shown to be identifiable by the formation of a netlike (reticular) intracellular precipitate of ribonucleoprotein and dye following brief incubation with **new methylene blue, brilliant cresyl blue**, or other dyes of similar structure. Their percentage was typically estimated by counting 1,000 red cells and noting the number of reticulocytes encountered. However, even when 1,000 red cells are counted, the number of reticulocytes counted in a normal is likely to be around 10, giving a standard deviation of 3.2, or a sample CV of 32%. Things get worse as the reticulocyte percentage decreases.

Both the differential leukocyte count and the reticulocyte count require somewhat more sophistication on the part of the observer, in terms of being able to discriminate among different cell types, than does either simple erythrocyte or leukocyte counting with a hemocytometer. While it was not clear in the early 1950's that computers might be able to perform the cell identification tasks needed to automate differentials or reticulocyte counts, this idea's time would come during the next decade.

Video and Electron Microscopy

The 1940's saw increasing exploitation of two related technologies developed in the preceding decades, both of which were to have a great impact on analytical cytology. The first was electron microscopy; the second was television. Both benefited from advances in electronics made during World War II. The electron microscope, in the late 1940's, occupied the ecological niche that a multilaser cell sorter might have occupied in the early 1980's; it was a coveted prize for a research laboratory whether or not it was really necessary for the laboratory's research. Price precluded introduction of electron microscopes into the clinical laboratory. Television was different; people had television sets in their homes and began to attach them to telescopes and microscopes as well. It did not seem illogical to develop a blood cell counter for clinical laboratory use in which cells in a hemocytometer were counted by an image analyzer.

Optical Cell Counters and the Coulter Orifice

It was no less logical to develop flow systems for blood cell counting. The sheath flow principle used in the Gucker aerosol counter was adopted by Crosland-Taylor[48] for a blood cell counter in which cells were detected by light scattering with dark-field illumination. During the late 1940's and early 1950's, several industrial organizations in England, Germany, and the United States developed or attempted to develop similar apparatus.

One American electrical engineer pursuing this goal (W. Coulter, personal communication) encountered some problems with optics and explored another means of cell detection, based upon the fact that the electrical conductivity of cells is lower than that of saline solutions. This phenomenon had been exploited since the 1890's in procedures for estimating the hematocrit from the conductivity of whole blood. Wallace Coulter reasoned that blood cells, suspended in a saline solution and passing one at a time through a small orifice, would be detectable by the change in electrical conductance or impedance of the orifice produced as the nonconducting cells passed through, displacing the conducting saline.

I am told that the first Coulter orifice was made in the cellophane wrapper from a cigarette package. The Coulter counter[49] proved accurate for counting[50] and sizing[51] blood cells and, as I have mentioned previously, apparatus based on this principle is now used worldwide in clinical and research laboratories.

By the mid-1950's, which I regard as the beginning of the "Classical Period" of flow cytometric history, much of both the methodology and the motivation of the field as we know it today already existed.

3.2 CLASSICAL HISTORY

Analytical Cytology in the 1950's

It was during the 1950's that analytical cytology acquired its name, coined by Francis O. Schmitt of M.I.T.; the first and second editions of a book entitled *Analytical Cytology*, edited by Mellors, appeared in 1955 and 1959[52]. The book included chapters by Mellors on "Fluorescent-antibody Method," by Novikoff on "The Intracellular Localization of Chemical Constituents," by Barer on "Phase, Interference, and Polarizing Microscopy," and by Pollister and Ornstein on "The Photometric Chemical Analysis of Cells," in addition to material on autoradiography and on electron and X-ray microscopy. The chapter by Pollister and Ornstein on the theory and practice of absorption measurements is well worth reading even today.

Another volume that provides a picture of the state of the art of analytical cytology in the 1950's contains the proceedings of a New York Academy of Sciences conference on *Cancer Cytology and Cytochemistry*[53]. At this 1955 meeting, several presentations dealt with instrumentation applied to the problem of discriminating malignant from benign cells in cytology specimens. It had become apparent that malignant cells were likely to contain more nucleic acid than normal cells. Mellors, having evaluated UV absorption, interference microscopy for nuclear dry mass measurement, and berberine fluorescence as an indicator of nucleic acid content, proposed construction of an automatic scanning instrument for screening cytological smears.

The Cytoanalyzer

Tolles and Bostrom, at Airborne Instruments Laboratory, described the "Cytoanalyzer" built for this purpose. A series of apertures in a disc that rotated in the image plane of a microscope system were used to produce a raster scan of a specimen with approximately 5 µm resolution. A hardwired analyzer extracted nuclear size and density information; cells were then classified as normal or malignant using these parameters. The Cytoanalyzer was, to make a long story short, right more of the time than it was wrong, but its false positive and false negative rates were too high for it to be suitable for clinical use. The results were encouraging enough for the American Cancer Society and the National Cancer Institute to continue funding research on cytology automation.

A different approach to high-resolution imaging was taken by Kopac, who equipped his microscope with a vidicon television camera. A single raster line from the television scan could be displayed on an oscilloscope screen, providing a density curve of absorption in a selected portion of the specimen. Differences in illumination intensity across the field of observation and differences in sensitivity in different portions of the camera tube limited the accuracy and precision of absorption measurements made with the television-based system; its obvious advantage over electromechanical scanning was its higher speed.

Acridine Orange as an RNA Stain: Round One

One cytologic development of the mid-1950's which was to have a great influence on the subsequent development of analytical cytology and flow cytometry was the demonstration by von Bertalanffy and Bickis[54] that the **metachromatic fluorescence** of **acridine orange** could be used to identify and quantitate **RNA content** in tissues. Armstrong, working independently, reported similar results a few months later[55]; by that time, von Bertalanffy et al had reported that acridine orange staining allowed good visual discrimination between normal and malignant cells in exfoliated smears[56].

At the state of the art as of the mid-1950's, any of several staining procedures and scanning methods could probably have supplied adequate input data to computer programs for cell classification. At that time, however, the few computers in existence were largely inaccessible to cytologists and there were no classification programs. Between the mid-1950's and the mid-1960's, progress in cytology automation was evident more in the automation than in the cytology.

How I Got Into this Mess

I started to get involved in analytical cytology as a spectator around this time. My mother, who was originally trained as a microbiologist, had been operating an electron microscope and had gone back to graduate school; her thesis work involved histochemical staining procedures. I was in high school, where I edited an underground newspaper and wrote songs about scientific topics. Although I expected to study medicine, I was also interested in mathematics and in building audio and amateur radio equipment. At that time, power transistors didn't exist; one could only use vacuum tubes. It was best if the tubes were selected for characteristics like low noise. I found the electron microscope in my mother's lab fascinating for several reasons, not the least of which was that it was manufactured by RCA and that it and its spare parts kit contained several tube types highly prized by builders of audio and radio equipment. I would often spend afternoons hanging around the lab, helping out with staining and darkroom work, after which new tubes would mysteriously appear in various apparatus that my friends and I built. In this way, I managed to learn a fair amount of biology while supporting my electronics habit. When I heard about Kopac's television microscope, it occurred to me that this line of research could be a great way for a biomedical scientist to keep supplied with up-to-date electronic components. Little did I know.

Most of my partners in crime were interested in physics or chemistry; several of them accumulated broken pinball machines in their basements in order to build computers. The digital computers of the 1940's were primarily electromechanical, built of switches and relays; a pinball machine was a good source of such components. The most advanced computers of the 1950's were electronic; they used vacuum tubes, cost millions of dollars, and occupied entire rooms at the few institutions lucky enough to have them. In the company of friends who lusted after such machines, I developed a desire to work with computers long before I could think of anything useful to do with them. My father, a practicing physician with a broad interest in science, encouraged my interests in mathematics, physical science, and computers; he was sure they would be of great use to me in my medical work.

The Rise of Computers

During the late 1950's and early 1960's, computers were acquired by more and more institutions, and people working in a variety of fields began to explore what computers could do to help them. To do this, they had to learn how to use computers; this process generally did not occur in a vacuum but required some interaction with people who already knew how. In this area, as in others, one's world view is apt to be derived from one's teachers'. I became interested in mathematical modeling of metabolizing systems; I learned about computers from people who had worked in mathematical economics and statistics. The emphasis in their work, and mine, was on multivariate analytical methods that could never have been put to practical use without computers.

Since, in those bygone days, there were few computers around and few people interested in computer applications in biology and medicine, it was possible to keep abreast of what everybody else was doing, if you had a mind to. There were only one or two meetings each year on the general topic, and they included the entire range of subject material. Mathematical models, computer diagnosis, computer analysis of electrocardiograms and electroencephalograms, and computer image processing, as applied to hematology, pathology, and radiology, were all discussed in front of the same audience. It was thus readily apparent to an interested observer that the successful application of computers in diagnosis in different fields of medicine would be based on overcoming a central problem common to all of those fields, i.e., the necessity to depend upon the diagnostic expertise of a trained observer in order to decide whether the computer's diagnosis was "correct."

Computers in Diagnosis: A Central Problem

This problem arises in any situation in which absolute objective criteria for classification do not exist. Where such criteria exist, it is easy to establish a diagnosis and to reconcile the findings of an instrument system and a human observer. To establish, for example, that a patient has sickle cell disease, one can perform a hemoglobin electrophoresis which will demonstrate the abnormal hemoglobin if it is present. One cannot diagnose mumps with anything approaching this degree of accuracy. In the days before mumps vaccine became available, about 95 percent of the population showed delayed hypersensitivity to mumps antigen, indicating previous infection with mumps virus. Only about 20 percent of the population would report having had symptomatic mumps. Mumps as a disease was originally defined by its symptoms; it is clear that an individual can be infected with mumps virus without exhibiting those symptoms. It is also known that someone previously infected with and immunized by mumps virus can lose immune reactivity to mumps virus antigen as a result of some disturbance of immune function. If mumps were redefined to mean infection with mumps virus, one could still not be sure that a member of the small fraction of the population which does not exhibit delayed hypersensitivity to mumps antigen had not previously been infected with the virus.

The notion of diagnosis, in the sense in which they perceived physicians as making diagnoses, was attractive to many of the people who developed the "systems approach" to engineering and management. Indeed, it is possible, using a binary decision tree, to arrive at a diagnosis of what went wrong with your car or television set, or with the space shuttle, or perhaps with the XYZ Widget Company. Faced with more complex problems, both the systems thinkers and the computer-oriented physicians were quick to adopt statistical methods for their solution.

Diagnosis and Classification: Statistical Methods

The general approach to computer diagnosis was similar to that used for such tasks as optical character recognition and the classification of animals and plants. Attributes of the populations of interest were selected which could be reproducibly measured; a formal statistical analysis was then carried out to define a **discriminant function**, i.e., some algebraic combination of the measured variables which assumed different values when applied to individuals from different classes. Despite this similarity in methodology, the three classification tasks just mentioned are fundamentally different in nature.

Optical character recognition, i.e., automated interpretation of the elements (not the content!) of printed text, required identification of features which, in combination, could be used to tell one letter or number from others. In actual practice, the subject material for analysis would be restricted to one or a few type fonts and sizes and to legible material. Under these circumstances, characters could be identified by an observer with almost absolute certainty, and one could readily assess the performance of a computer program for character recognition.

In the application of computers to classification of animals and plants, a field formally known as **numerical taxonomy**, the individual objects under study can not be precisely classified; the numerical analysis is oriented toward defining distances between objects in the feature space used. If two individuals differ slightly in characteristic A and greatly in characteristic B, they will appear to be more closely related if characteristic A is given more weight and less closely related if characteristic B is given more weight. Most controversies in numerical taxonomy arise because different people assign different importance to different characteristics. In some cases, it is clear that one characteristic, DNA sequence homology, for example, is more relevant to the analysis than another, e.g., hair color. When the DNA of every extant organism has been sequenced, there may be no controversies left among numerical taxonomists, assuming there are numerical taxonomists left by that time. Until then, this field will serve as my example of one in which the proper procedure is to let the data do the classifying for you.

The classification problems involved in the application of computers to medical diagnosis were often treated, particularly during the early days, as analogous to character recognition. It was assumed, particularly by people not intimately familiar with clinical medicine, that there was some physician who could say with certainty that a cell was or was not malignant or that a cardiogram was or was not normal or that a patient did or did not have heart disease. It was obvious, even at that time, that in the most difficult cases, the "definitive" diagnosis was established by fiat of the most senior of the physicians involved. Since it was clear that these experts arrived at less difficult diagnoses by application of objective criteria, and that they could more or less successfully define those criteria for the benefit of the students and house staff under their tutelage, there was a general tendency to give clinicians undue deference and the benefit of the doubt in the more difficult cases. The alternative was to assume that the experts were not only fallible, but also occasionally arbitrary.

In order to create a computerized diagnostic system for clinical use, whether it was designed to interpret electrocardiograms, Pap smears, blood smears, or chest radiograms, it would be necessary to demonstrate agreement between the instrument system and the human interpreter. The system was not likely to be accepted if the medical experts in its field of application were not convinced that it worked. If the computer and the experts agreed in all but the most difficult cases, the computer system might be regarded as suitable for routine use; naturally, the human experts would have to be called in as consultants for the remaining problem situations. The instrument developers could then take objective criteria as far as they could without any risk of confrontation with a professional community that might influence not only eventual acceptance of an instrument in the marketplace, but the initial grant funds for its development as well.

To suggest that an element of arbitrariness was involved in difficult cases in which the computer, using the clinicians' supposedly objective criteria, could not match their diagnoses, would have been politically and economically inexpedient, to say the least. In the 1960's, there was an additional sound reason to avoid this issue; there simply weren't computer-based systems that could do as well as a not-too-well-trained human interpreter, either for automated cytology or for any of the other tasks to which computer technology was being applied. To my mind, the best indication of the progress which has been made since that time is the present willingness of clinicians in many areas of medicine to rely on automated and semiautomated systems for a great deal of diagnostic information. It is now possible to use the computer to do the "numerical taxonomy" tasks in medicine as well as the "character recognition."

Cytology Automation in the 1960's

Most of the effort expended on automated cytology during the 1960's was, not surprisingly, directed toward the development of instruments that posed no threat to expert or inexpert physicians. The partial success of the Cytoanalyzer provided motivation and funding for attempts to produce a system that could match the performance of a cytotechnologist in screening cervical cytology specimens; support was also given for studies aimed at automating the differential leukocyte count, another laboratory test performed, not always adequately, by medical technologists rather than by physicians.

First Steps toward Automated Differentials

Marylou Ingram, then at the University of Rochester, began studies on automated analysis of leukocyte images in collaboration with scientists at the Perkin-Elmer Corporation in the early 1960's. The initial motivation for this work was the finding that exposure to radiation resulted in the appearance of increased numbers of binucleate lymphocytes in peripheral blood; the frequency of these cells was quite low (less than 1/10,000 leukocytes) in exposed and unexposed populations, and it would therefore be necessary to count hundreds of thousands of cells to derive reliable information as to whether the frequency of binucleate lymphocytes was abnormally high. This project thus represents an early example of what we now call **rare event analysis**.

The scanning apparatus used in these studies was largely conceived by Kendall Preston, Jr., who had previously been associated with Airborne Instruments Laboratory, where the Cytoanalyzer was built. Vidicon-based and, later, vibrating-mirror scanners were used to produce digitized images of leukocytes conventionally stained with eosin-methylene azure dye combinations; several illumination wavelengths were used to allow color information to be collected[57,58].

A second effort at automated differential leukocyte analysis was also an outgrowth of the Cytoanalyzer work. The CYDAC scanner, built by Airborne Instruments Laboratory, was used by Mortimer Mendelsohn, Brian Mayall, and Judith Prewitt at the University of Pennsylvania to produce high-resolution digitized images of leukocytes. The CYDAC operated only at a single fixed wavelength, and cells were stained with a combination of gallocyanin chrome alum and naphthol yellow S, rather than with a conventional eosin-azure stain[59,60].

Pattern Recognition Tasks in Cell Identification

The problem of cell identification by **image analysis** incorporates two separate **pattern recognition** tasks. The first of these is **feature extraction**, i.e., processing of the digitized cell image to extract a set of parameters or **descriptors**. These may be features that correspond to known cytologic parameters, e.g., the size of the nucleus and cell or cytoplasm, the degree of cytoplasmic basophilia, or the shape of the nucleus. They may also be features derived from the image that indirectly provide data corresponding to what would be described by a human observer. For example, in eosinophil granulocytes, the cytoplasm contains numerous refractile granules; the refractive index differences between these and the cytosol manifest themselves as differences in optical density within the cytoplasm in the scanned image. If one calculates the average difference in **optical density** between each point (i.e., **pixel**, or **picture element**) of the image and the points or pixels adjacent to it, this will provide an indicator of **texture** which can be used to aid differentiation of eosinophils from other cell types.

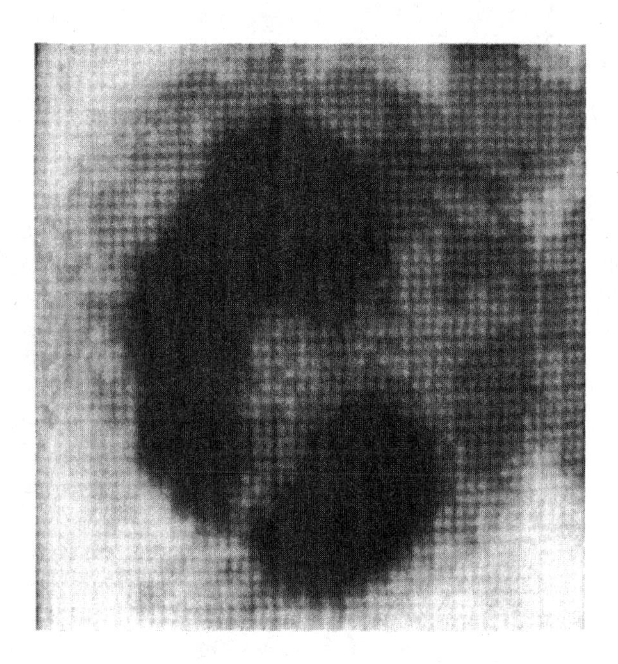

Figure 3-2. Digitized image of a neutrophil polymorphonuclear leukocyte stained with an eosin-azure dye mixture (courtesy of J. Bacus).

Figure 3-2 shows a leukocyte image digitized at the resolution used in commercial image analyzing differential leukocyte counters such as the Corning LARC™, which was introduced in 1969. This instrument was developed by James Bacus and his colleagues, then at Rush-Presbyterian-St. Luke's Medical Center in Chicago. Two lobes of the nucleus are visible as the darkest areas; cytoplasmic texture is evident from differences in intensity of different areas. Red cells adjacent to the leukocyte are seen at the right and upper right.

The feature extraction tasks that must be accomplished to obtain descriptors from a digitized image such as that shown in the figure require fairly complex methodology. Even a simplistic definition of a nuclear lobe, for example, must specify a content of a certain minimum number of contiguous pixels of a certain minimum optical density. Determining where the leukocyte ends and the adjacent red cells begin, a necessary step in defining the cell size, is also not a simple task. This process of feature extraction, however, is peculiar to image analysis.

The second pattern recognition task in cell identification, that of **cell classification**, is accomplished by statistical analysis of numeric data derived from the feature extraction procedure, and, as implemented by the developers of automated differential counters, used the same multivariate statistical procedures which others had applied to tasks such as optical character recognition and differentiation between normal and abnormal electrocardiograms. A cell classification program recognizes patterns in the distributions of measurements of cellular parameters, whether or not such parameters are derived from image analysis; most such programs are designed to find discrete **clusters** corresponding to different cell types.

In the 1960's, much of what was known about the development and differentiation of blood cells had been learned from visual observation of normal and pathologic blood and marrow smears. An automated differential counter would have to do what a technician could. At a minimum, it should be able to distinguish among the mature leukocyte types present in normal peripheral blood, i.e., the **granulocytes**, including **neutrophils**, **eosinophils**, and **basophils**, and the **mononuclear cells**, i.e., **lymphocytes** and **monocytes**. The instrument would also have to flag "abnormal" cells, i.e., immature red and white cell types normally found in marrow, but not blood, and subdivide the neutrophils into the immature "**bands**," cells in which the nucleus had not completely segmented into lobes, and the mature "**segs**," cells in which segmentation was complete, the cell shown to the left being a seg.

If you looked at a hematology text, it would tell you that there were stages in the development of cells, all arising from a common **hematopoietic stem cell** that couldn't be described because nobody had ever seen one for sure. Then, there were supposedly discrete stages in the development of each lineage; in the case of neutrophils, the earliest recognizable progenitor cells were **myeloblasts**, large cells with large nucleolated nuclei and basophilic cytoplasm (both due to the presence of the relatively large amounts of RNA needed in protein synthesis), and without cytoplasmic granules. The next stage, **promyelocytes**, were, if anything, larger, and had large, immature cytoplasmic granules. Then came **myelocytes**, with cytoplasmic granules more or less identical to those in mature neutrophils but with round nuclei. **Metamyelocytes** had kidney-shaped nuclei; they matured into **bands**, which matured into **segs**.

This model was, as has been shown by a lot of elegant studies involving monoclonal antibodies and multiparameter flow cytometry, accurate in many particulars. However, there was one major problem with it, particularly as it was applied to differential counter design by people who knew a lot about engineering, mathematics, and or statistics, but not too much about biology. It appeared from the textbooks that a real hematologist should always be able to tell whether a cell was, for example, a promyelocyte or a myelocyte. The appearances of the cells, stained with Wright's or similar stains, were discussed in the books; one stage might have a purplish-pink cytoplasm, the next pinkish-purple (I am not making this up!)

What a real hematologist was more likely to tell you, at least if you were an aspiring hematologist, was that while there were "textbook examples" of each of the described cell types, there were also intermediate forms. You might also find out another little secret, namely, that hematologists couldn't always tell whether a very immature cell, or blast, was a myeloblast, or a lymphoblast (lymphocyte progenitor), or an erythroblast (erythrocyte progenitor); they made the calls by looking at the more mature surrounding cells. If these were myelocytes, the cell was a myeloblast, etc.

When you look at normal peripheral blood, you see different types of mature cells, which differ from one another in appearance in obvious ways. If you plot any of a number of descriptive parameters of these cells in a two-dimensional space, you get clusters. For example, lymphocytes, monocytes, and granulocytes form separate clusters in a plot of forward vs. orthogonal scatter values. Plots of the same parameters for cells in marrow, where there is a continuum of maturing cell types, feature not so much clusters as connected blobs, more dense in some places than in others, a pattern I was later to dub a **ginger root**. In the 1960's, there were a lot of people trying to make instruments find clusters that weren't there, because they were unaware of the continuous nature of many processes in cell differentiation. Unfortunately, despite our increased 21st-century level of knowledge and sophistication, there are still some people trying to find nonexistent clusters. We'll get back to this in several contexts later in the book.

Between the late 1960's and the mid-1970's, about ten different commercial differential counters based on slide scanning technology were introduced to the market, each claiming to identify more types of "abnormals" than the next. There was also competition to add other features, such as measurements of red cell size and hemoglobin content or reticulocyte counting. Although the engineers concentrated on refining an inadequate technology, a few individuals with more of a biological orientation began to examine other possible means of distinguishing leukocyte types.

Differential Leukocyte Counting: An Early Flow Systems Approach

One early alternative approach to leukocyte differentiation was taken by Hallermann et al[61]; this little-cited work of the early 1960's anticipates many of the later publications (and, possibly, some of the later patents) on flow cytometric differential counting. During the 1950's, blood cell counters based on flow cytometric detection of light scattering by cells were, as was mentioned previously, built by several manufacturers. These were entirely suitable for erythrocyte counting; since the number of leukocytes in blood was, in most cases, only about 0.1 percent of the number of erythrocytes, the inclusion of leukocytes in the erythrocyte count did not produce significant errors. The leukocyte count, however, was of at least as much interest as the erythrocyte count.

To count leukocytes in a hemocytometer, blood was diluted in a solution that lysed the erythrocytes. A similar procedure had to be used in early blood cell counters based on either light scattering or electronic (Coulter) volume measurement, because neither measurement could reliably discriminate leukocytes from erythrocytes. A measurement method that could make this distinction was suggested by the work of Kosenow[62] and others[63,64], who demonstrated characteristic staining of different types of leukocytes by **acridine orange**.

Optical cell counters used dark-field illumination, which was also a preferred technique for fluorescence excitation; the addition of a fluorescence detector to the scatter detector in such an instrument could allow detection of leukocytes based upon temporal coincidence of scatter and fluorescence pulses, while the nonfluorescent erythrocytes could be counted in the usual fashion by tallying scatter pulses. Since the leukocyte count is typically only about 0.1 % of the erythrocyte count, it might not, in practice, be necessary to discriminate between scatter pulses which were and which were not accompanied by fluorescence pulses in order to achieve an acceptably accurate erythrocyte count, but the leukocyte count could be corrected, if necessary.

The fluorescence approach, however, promised to go beyond discrimination of leukocytes from erythrocytes. Hallermann et al reported that granulocytes in acridine orange-stained cell suspensions could be distinguished from mononuclear cells on the basis of flow cytometric determination of the intensity of red cytoplasmic metachromatic fluorescence, which was greater in the granulocytes. Few other workers in analytical cytology seem to have been aware of this work at the time of its publication; I unearthed the reference to Hallermann et al when I was searching the literature on differential counting in the late 1970's.

Kamentsky's Rapid Cell Spectrophotometer

The individual who set the pace for the development of flow cytometry as an analytical cytologic tool is Louis Kamentsky, who began to study the problem of automating cervical cytology screening during the early 1960's, at which time he was working at IBM's Watson Laboratory at Columbia University; IBM's effort actually got its start from a back-fence conversation between neighbors in a New York suburb who were, respectively, a pathologist and an IBM manager (H. Derman, personal communication).

Kamentsky (L. Kamentsky, personal communication) had developed both instrumentation and statistical techniques for optical character recognition; even during the earliest stages of his studies on cell classification, his experience led him to anticipate having to use multiple parameters to develop a discriminant function to identify abnormal cells. His familiarity with the existing state of the art in hardware and software image analysis led him to doubt that high-resolution scanning and feature extraction by image processing could be done fast enough to serve as the basis for a clinical laboratory instrument.

Pathologists in New York, among them Herbert Derman, Leopold Koss, and Myron Melamed, taught Kamentsky that **nucleic acid content** and **cell size** were useful parameters for cervical cell classification; he learned how to measure these microphotometrically from Torbjörn Caspersson and Bo Thorell, in Stockholm. He then built the **Rapid Cell Spectrophotometer (RCS)**, a flow cytometer based on a transmitted-light microscope, with an arc lamp source and high-N.A. optics, allowing reasonably accurate absorption measurements on cells passing, without sheath flow, through a channel in a slide.

For work with cervical cells, the RCS measured nucleic acid content by absorption at 260 nm and cell size by light scattering at 410 nm[65]. The light scattering measurement was indirect. The absorption of cells (other than hemoglobin-containing erythrocytes) at 410 nm was known to be minimal; the parameter actually measured in the apparatus was light transmission at 410 nm. Since almost all of the light loss was due to scattering, high transmission signals were assumed to correspond to low scatter signals, and *vice versa.*

Kamentsky experimented with electrostatic and fluidic cell sorters, which could remove selected cells for examination by a pathologist and permit verification of the RCS's performance; a syringe pump-based sorter was described in 1967[66]. A refined version of the RCS, showing the computer, and the original prototype, which conveys more of the true flavor of a laboratory-built instrument, are illustrated in Figure 3-3.

Kamentsky and his colleagues also investigated the use of the RCS in leukocyte differential counting, with Leonard Ornstein providing expertise in histochemical staining. Figure 3-4, to the right, shows a contour plot of a blood cell population stained with the Feulgen stain for DNA and

Figure 3-3. ABOVE: Kamentsky's Rapid Cell Spectrophotometer, as shown in several publications; this is actually the third version of the instrument. BELOW: The real first RCS prototype, warts and all (courtesy of L. Kamentsky).

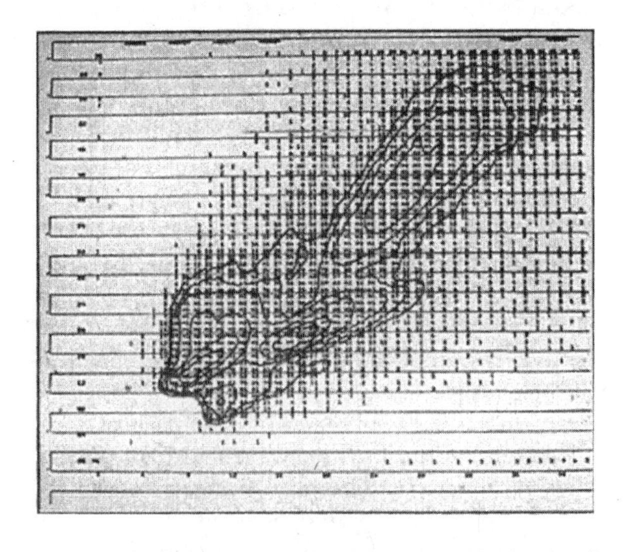

Figure 3-4. A two-parameter histogram of blood cells stained with Feulgen stain and naphthol yellow S, analyzed in the RCS, showing hand-drawn contour lines.

naphthol yellow S for protein; a number of cell clusters are visible. The contour lines are drawn by hand on a computer printout of numbers of cells corresponding to each pair of parameter values; the figure thus represents one of the first two-parameter histograms ever to be obtained from a flow cytometer. Incidentally, if you need to send somebody a 32 × 32 or 64 × 64 2-parameter histogram in a text format, this gimmick still works. You can probably beam a 32 × 32 to a Palm device. But I digress.

The RCS could actually measure four parameters; it was equipped with a dedicated digital computer (an IBM 1130) for acquisition and analysis of data from several hundred cells/second. Equivalent multiparameter analysis capabilities were not added to other flow cytometers for more than a decade after the RCS was built.

When the apparatus was first described, however, its most notable features were its speed and its inclusion of a sorter. The speed was achieved by substituting rapid microphotometric measurements of entire cells for pixel-by-pixel scans, eliminating the need for a laborious feature extraction process, and by using a fluidic specimen transport mechanism instead of a motorized microscope stage. This made it feasible to deal with much larger cell samples than could be processed by a computerized image analyzing microscope. Kamentsky viewed the sorter primarily as a necessity for verification of the instrument's performance in cytologic screening on a cell-by-cell basis; others would later exploit sorting's preparative uses.

An RCS prototype lent by IBM to Stanford was involved in the development of the **Fluorescence Activated Cell Sorter (FACS)** by Leonard Herzenberg and his colleagues; the RCS also influenced the subsequent development, by Leonard Ornstein and his colleagues, of Technicon Instruments Corporation's **Hemalog D**, the first of a series of flow cytometric differential leukocyte counters.

Fulwyler's Cell Sorter

The syringe pump sorter developed by Kamentsky and Melamed was not the first cell sorter described in the literature; Mack Fulwyler, then at Los Alamos Scientific Laboratory, reported using droplet deflection to separate cells on the basis of electronic cell volume in 1965[67], shortly after the publication of the first paper by Kamentsky et al[65]. The Fulwyler apparatus (M. Fulwyler, personal communication; M. Van Dilla, personal communication) was also, in a sense, developed to verify an instrument's performance.

Scientists at Los Alamos had been examining distributions of electronic cell volume measurements obtained with a Coulter counter, using a multichannel pulse height analyzer, a fairly common apparatus in a nuclear research establishment like Los Alamos, to accumulate distributions. It was noted that red blood cells frequently produced a **bimodal distribution**, i.e., one with two peaks. The question arose as to whether there was truly a bimodal distribution of cell volumes, the alternate hypothesis being that the bimodal distribution was artifactual, perhaps produced by differences in orientation and/or position of the asymmetric red cells as they passed through the Coulter orifice.

Fulwyler adapted the principle of the ink jet printer, then recently developed by Richard Sweet of Stanford[68], which used electrostatic deflection to deposit charged ink droplets in the desired pattern on paper. After cells were measured during passage through a Coulter orifice, the stream was broken up into droplets, which could be charged and then deflected into collection vessels as they passed between plates maintained at high voltages of opposing polarities. When cells from either peak of the bimodal distribution were sorted and reanalyzed, the original bimodal distribution was again observed, showing that it was due to an artifact. The Los Alamos group then turned its attention to the exploitation of real volume differences between cells; by 1967, they had successfully prepared highly (>95%) purified suspensions of blood granulocytes and lymphocytes[69].

3.3 MODERN HISTORY

The history of flow cytometry since 1967 has been discussed in some detail in references 1-9 and 1028. The remainder of this chapter describes events as I remember them happening.

Cell Cycle Analysis: Scanning versus Flow Systems

I spent some of my college years doing mathematical modeling of complex metabolizing systems[70] in the naive expectation that this would provide a rational approach to the design of anticancer drugs. While in medical school, I responded to suitable financial inducements, and put cell dynamics aside to work on computer statistical analysis of electrocardiograms[71]. In 1967, I was told to brush up on the literature of automated cell analysis in preparation for my impending stint as a "Yellow Beret" at the National Cancer Institute, where I was to work on methods for automating **cell kinetic studies** of acute leukemias.

Cancer had, for many years, been viewed as a consequence of disturbed cell growth patterns. The refinement in the 1950's of techniques for measuring cell growth had, by the mid-1960's, made it clear that cancer cells didn't just grow faster than normal cells. This meant that the simplest approach to cancer chemotherapy, i.e., giving a drug or drugs that killed the fastest growing cells, wouldn't work in all cases, although this would be, and still is, effective in treating those malignancies in which almost all of the cells grow rapidly.

In his *Growth Kinetics of Tumours*[382], Gordon Steel describes 1965 as marking the "end of the beginning." In the beginning, growth rate could be estimated only by watching a tumor increase in size or by counting cell concentrations or numbers in culture. In the 1920's, Payling Wright correlated the **frequency of mitosis** with cell growth rates[1094]. By the late 1940's, it was appreciated that DNA was the genetic material, and that its replication was therefore a central event in cellular reproduction.

Howard and Pelc[1095-6] used the radioactive isotope ^{32}P as a tracer, detecting its incorporation into cells' DNA by **autoradiography**. Slides containing the cells were coated with a photographic emulsion or film and left in the dark for some time, allowing radioactive decay of the isotope to expose the emulsion. After subsequent development, silver grains could be seen in the emulsion overlying portions of the cells into which the isotope had been incorporated. They showed that DNA synthesis does not occur continuously during cell growth; instead, there is a **cell cycle** that includes two **gaps**, one (G_1) preceding and one (G_2) following the DNA **synthetic (S)** period. The **mitotic (M)** phase of the cycle follows G_2 and precedes the G_1 phase of the next cycle.

Studies of cell kinetics were facilitated by the introduction in 1957 of **tritiated thymidine (^{3}H-TdR)** as a radioactive tracer[1097]. Under most conditions and in most cell types, tritiated thymidine is either incorporated into DNA or lost from cells, making it highly specific. Quastler and Sherman[1098] demonstrated the use of ^{3}H-TdR in analyses of growth kinetics of animal cells in 1959, and others quickly applied the material and autoradiographic techniques to studies of the growth kinetics of normal and malignant cells. One could, for example, estimate the fraction of cells in S phase from the **labeling index**, i.e., the percentage of cells over which grains could be seen in an autoradiograph. This was a reasonably tedious task, as was estimating the overall duration of the cell cycle from the **percentage of labeled mitoses (PLM)**. Obtaining a quantitative estimate of the amount of tracer incorporated in cells required **grain counting**, a procedure that far surpassed reticulocyte counting in the speed with which it could addle an observer.

The prevailing oncologic opinion in the mid-1960's was that, once the kinetics of normal and malignant cell growth were defined, clinicians could devise drug dosages and time schedules for administration which would exploit kinetic differences to kill maximal numbers of cancer cells with minimal host toxicity. This would require the collection of a large data base, a task made difficult by the necessity to rely on autoradiographic measurements of ^{3}H-TdR incorporation by cells as the principal means of determining cell growth rates. NCI wanted a system that would scan blood and marrow specimens, identify immature and mature, normal and malignant blood cells and determine DNA synthetic activity by grain counting.

This was obviously a considerably more complex task than automated differential leukocyte counting, which itself wasn't exactly easy. Feasibility studies were being done by Perkin-Elmer under Ken Preston's direction, with Marylou Ingram of the University of Rochester providing biological backup. Seymour Perry of NCI, who initiated the project, was also being advised by Marvin Zelen and other statisticians at NCI, by Mort Mendelsohn and Judy Prewitt, who, with Brian Mayall, were working on automated differential counting with the CYDAC system at the University of Pennsylvania, and by Lew Lipkin of the National Institute of Neurologic Diseases and Russ Kirsch of the National Bureau of Standards, who were attempting to use artificial intelligence to analyze and reconstruct the microscopic structure of the central nervous system[72].

To catch up with the latest developments in quantitative cytology and cytochemistry, I was told to attend a conference sponsored by the New York Academy of Sciences, held in June, 1967. Among the speakers was Lou Kamentsky, who by that time was measuring three parameters in the RCS, which already incorporated a dedicated computer as well as a cell sorter. I became an instant convert to flow cytometry.

When I got to NCI, I expressed the opinion that we should be trying to do cell kinetics by developing differential cell stains for the different cell types, running the cells through a sorter, and doing autoradiography on the sorted fractions. This did not sit well with my image processing colleagues, primarily because they couldn't envision a flow cytometer which measured the dozen or more parameters that they thought would be necessary for cell identification. Instead, we set up an image processing lab as a joint NCI-NIND-NBS venture. We mounted the mirror scanner built by Ken Preston on Lew Lipkin's microscope; Phil Stein, one of my partners in crime from high school, redesigned the scanner electronics and developed computer-controlled drives for the microscope stage and monochromator and an interface to Lew's LINC-8 computer. Phil also gave me a crash course on what had developed in electronics since we built amplifiers in our basements; this was necessary because the hiring freeze then on at NIH meant that I had to do some electronics construction and simple design work if I expected our instrument to be working by the time my hitch was up. We came up with a pretty sophisticated scanning optical microscope[73,74], as I mentioned on pp. 14-15; unfortunately, it took two minutes to scan a cell at a single wavelength, making it difficult to do many biologically relevant experiments and increasing my desire to work with flow cytometers.

Autoradiography was not quite the only game in town for analysis of DNA synthetic patterns in cell populations; one could measure DNA content on a cell-by-cell basis, using UV absorption or a Feulgen stain. This had actually been done by a few people; it hadn't caught on because microspectrophotometry required expensive and uncommon apparatus. Using Feulgen stain and our scanning microscope, we accumulated a 200-cell DNA content distribution in a mere 12 hours, taking up an hour or so of mainframe computer time in the process[74]. By the time the work was published, flow cytometry would have changed the rules of the game.

Cancer Cytology: Scanning versus Flow Cytometry

TICAS, a somewhat more practical image analysis system than our "Spectre II," was assembled in the late 1960's by George Wied of the University of Chicago, Gunter Bahr of the Armed Forces Institute of Pathology, and Peter Bartels of the University of Arizona in order to pursue automated interpretation of cervical smears. They organized a conference on "Automated Cell Identification and Cell Sorting," held in Chicago in 1968[75]. There were several presentations on flow cytometry, which were received with attitudes ranging from skepticism to hostility.

An industrial group reported good results in analysis of cervical cytology specimens using a device called the Cytoscreener[76], which performed medium-resolution image analysis of UV absorption of cells in a flow system. They got a chilly reception from the audience at this and subsequent meetings, although it was subsequently established by an NCI-sponsored study that their instrument worked at least as well as anything else developed at the time.

Kamentsky's RCS also showed some promise in tests on cervical specimens[77], using UV absorption an light scattering to measure nucleic acid and cell size. Kamentsky also discussed experiments on identification of leukocytes stained with a Feulgen reagent and with naphthol yellow S[78]. Mack Fulwyler described experiments in progress at Los Alamos on flow cytometric **fluorescence** measurement, using an **argon ion laser**, then inseparable in the public imagination from the "death ray" in the James Bond movie *Goldfinger*, for excitation[79].

Once it had been established, during the 1950's, that DNA and total nucleic acid content were useful parameters for discriminating between normal and malignant cells, the use of fluorescent reagents for measurement of these parameters had been suggested. In 1968, Walter Sandritter was one of the few vocal advocates of fluorescence flow cytometry of DNA as a basis for cancer screening, as indicated by his presentations in Chicago and at a subsequent symposium[80] in Cardiff.

During the late 1960's, Dittrich and Göhde, in Germany, developed a fluorescence flow cytometer using arc-lamp epiillumination, the **Impulscytophotometer (ICP)**[83], in which the cells flowed in a line extending along the axis of the high-N.A. microscope objective used as a condenser and collection lens. They also introduced **ethidium bromide** as a stoichiometric fluorescent stain for DNA, eliminating the need for the tedious process of Feulgen staining.

There was also interest in fluorescence flow cytometry at Stanford, where the Herzenbergs were developing a cell sorter which they eventually hoped to use to separate cells stained with fluorescent antibodies; they borrowed one of Kamentsky's prototypes to determine its efficacy for fluorescence measurements[81] and, in late 1969, described sorting of fluorescently stained cells by an instrument using a mercury arc lamp for excitation[82]. A paper by Van Dilla et al[79], which had appeared a few months earlier, described results with Feulgen-stained cells, showing a reasonably clean DNA content histogram, and also anticipated extension of the technology to work with fluorescent antibodies, to fluorescence detection at multiple wavelengths, and to multiparameter analysis using fluorescence, Coulter volume, and light scattering measurements, which were being investigated at Los Alamos by Paul Mullaney.

Figure 3-5. ABOVE: Publicity photo of the first Bio/Physics Systems Cytofluorograf. BELOW: The elusive Dr. Kamentsky in his natural habitat, shortly before the Cytofluorograf picture was taken (courtesy of L. Kamentsky).

Early Commercial Flow Cytometers

In 1970, Phywe AG of Göttingen began selling a commercial version of the ICP, built around a Zeiss fluorescence microscope. This instrument was rapidly applied by European workers to studies of tumor cell DNA content and of the effects of therapy on cell kinetics (see Ref. 232). Most people working with flow cytometry in the United States were unaware of the existence of the Impulscytophotometer until 1973 or 1974, when Barthel Barlogie brought an instrument to M. D. Anderson hospital in Houston.

Optical flow cytometers for research purposes also became available in the United States in 1970, when Lou Kamentsky, who left IBM to found Bio/Physics Systems, began to produce the **Cytograf** and **Cytofluorograf**, which, respectively, used helium-neon and argon ion laser light sources. The Cytograf measured forward scatter and extinction at 633 nm; it could be used to discriminate dead from live cells on the basis of uptake of Trypan blue. The Cytofluorograf, shown in Figure 3-5, on the previous page, measured forward scatter at 488 nm and green (about 530 nm) and red (above 640 nm) fluorescence excited by the 10-15 mW air-cooled laser. Both the Cytograf and Cytofluorograf used a meniscus-sensing arrangement to determine sample volume flow rates, facilitating cell counting; both allowed counting of cells in gating regions bounded on four sides. A 100-channel pulse height analyzer was available as an accessory; with this added option, the Cytofluorograf sold for just over $20,000.

Although the Cytofluorograf could be used for DNA content analysis, its fluorescence measurement sensitivity was not sufficient to permit measurement of immunofluorescence under ordinary circumstances. Much of the research effort at Bio/Physics Systems focused on development of differential counters and cytology apparatus using acridine orange as a stain. By 1971, Dick Adams and Lou Kamentsky had shown that lymphocytes, monocytes, and granulocytes in whole blood samples stained with acridine orange in isotonic saline could be discriminated by their progressively higher levels of red cytoplasmic fluorescence[588], these identifications were confirmed independently by sorting by the Los Alamos group in 1973[590]. While the differential counter never reached the market, Bio/Physics Systems did produce the **Hemac** hematology counter, which used scattering and extinction of red light from a helium-neon laser to count and size blood cells.

The first commercial flow cytometric differential counter, introduced in the early 1970's, was Technicon's Hemalog D[84,85] (Figure 3-6), which used light scattering and absorption measurements made at different wavelengths in three different flow cytometers to classify leukocytes. Chromogenic enzyme substrates were used to identify neutrophils and eosinophils by the presence of moderate to high and very high concentrations of peroxidase and, in another channel, to identify monocytes by their esterase content. Basophil identification was based on detection of glycosaminoglycans in basophil granules using Alcian blue. A single tungsten-halogen lamp served as light source for all three flow systems.

Although the Hemalog D employed cytochemical staining procedures that were well regarded by hematologists for such purposes as determination of lineage of leukemic cells, the apparatus, which worked pretty well, was initially regarded with a great deal of suspicion, at least in part due to the novelty of flow cytometry. The developers and manufacturers of image analyzing differential counters, which certainly didn't perform much better than the

Figure 3-6. The Hemalog D Differential Counter.

Hemalog D, did what they could to keep potential users suspicious of flow cytometry for as long as possible; the technology would eventually be legitimized by its dramatic impact on immunology.

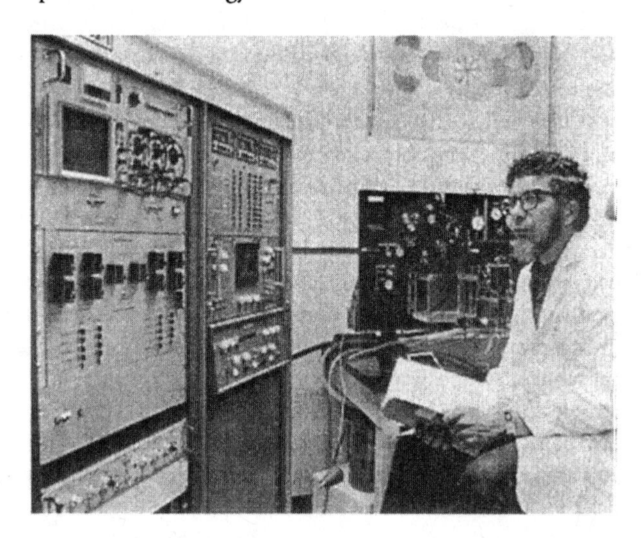

Figure 3-7: Leonard Herzenberg with B-D's first commercial version of the FACS, 1974 (NIH photo).

In 1972, Len Herzenberg's group at Stanford described an improved version of their **Fluorescence-Activated Cell Sorter (FACS)**, which used a fairly powerful argon ion laser instead of the arc lamp source used in the original, and which could detect the relatively weak fluorescence of cells stained with fluorescein- and rhodamine-tagged antibodies[86]. The instrument was produced commercially by Becton-Dickinson (B-D) two years later. The original version, the FACS-1 (Figure 3-7), measured forward scatter, which was used as a trigger signal, and fluorescence above 530 nm, and was equipped with a Nuclear Data pulse height analyzer for distribution analysis and with Tektronix event counters to keep track of the total number of cells counted and the cells in each of two gate or sort regions. Although flow cytometry had only gradually begun to attract the attention of cell biologists, cell sorting immediately caught the fancy of immunologists, and B-D placed instruments in a number of active and prestigious laboratories within a few years.

Coulter Electronics, which by 1970 had become a very large and successful manufacturer of laboratory hematology counters, pursued the development of fluorescence flow cytometers through a subsidiary, Particle Technology, under Mack Fulwyler's direction in Los Alamos. The **TPS-1 (Two Parameter Sorter)**, Coulter's first product in this area, reached the market in 1975. It used an air-cooled 35 mW argon ion laser source and could measure forward scatter and fluorescence.

Not Quite Commercial: The Block Projects

In 1972, I went to work for G. D. Searle & Co., a pharmaceutical firm that was then heavily, if not profitably, involved with medical instrumentation. Among other things, I evaluated instruments and instrument concepts that various people were trying to convince Searle to back or buy. We were moderately interested in getting into the differential leukocyte counter business, but hadn't seen anything we liked enough to get serious about.

Early in 1973, Myron Block and Tomas Hirschfeld of Block Engineering came to Searle with a proposal to develop a clinical blood cell counter which would use a flow cytometer to count and size erythrocytes, platelets, and leukocytes, do a differential leukocyte count and, for good measure, calculate the hemoglobin content of the blood by integration of the absorption of hemoglobin in the individual erythrocytes. The whole blood samples were to be fixed and stained with a mixture of three fluorescent dyes, and analyzed in an instrument which would use five separate illumination beams, separated in space, to derive measurements of absorption in the ultraviolet (indicating DNA content) and the Soret band (indicating hemoglobin content), of light scattering, and of four fluorescence parameters, three representing fluorescence of the dyes and the fourth representing nonradiative energy transfer between two of the dyes. A dedicated minicomputer would be used to process data in real time, using a multivariate discriminant function for leukocyte classification. I thought this was a wonderful idea. A group of us from Searle went to visit Block's plant and talk to the people who would be involved in the project, and came away convinced that they could build the instrument and make it work.

I had not been paying much attention to the details of analytical cytology since about 1970. A few weeks before hearing the presentation from Block, I had bumped into Judy Prewitt at a meeting and spent a few hours with her finding out what was, or was not, happening in the field of automated differential counting. Judy was responsible for a lot of the mathematical and computational methodology used in feature extraction and cell classification[87], and I had great respect for her opinions, possibly because they agreed with mine. Neither of us thought much of what was on the

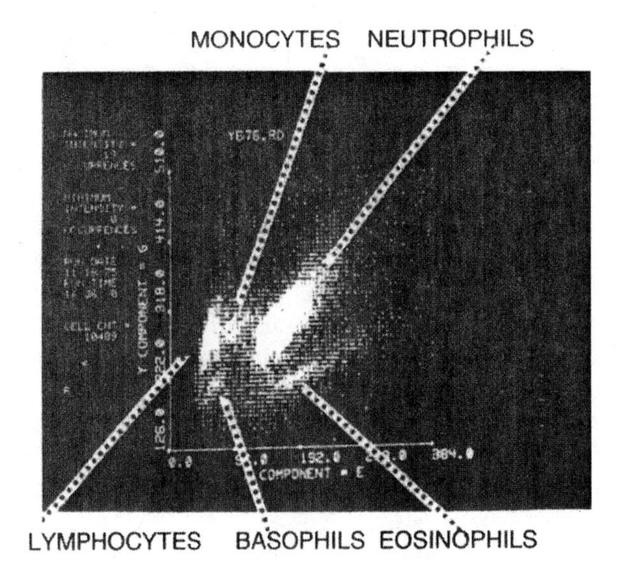

Figure 3-8. A two-parameter display from the Block differential counter showing five leukocyte clusters.

market or in development in the way of image analyzing systems; we reserved judgment on the Hemalog D. The conversation didn't cover any other flow cytometers, and I assumed that multiple illumination beam instruments with dedicated computers were already in use in research laboratories. I didn't really learn otherwise until the Engineering Foundation Conference on Automated Cytology held in December, 1973, at which point work on the first of our multiple illumination beam systems[88] was well underway at Block.

The trichrome fluorescent stain used in the system was developed by a physical chemist, Marcos Kleinerman, working in his basement. It was a mixture of ethidium bromide, a basic dye which was well known as a DNA fluorochrome and which imparted red fluorescence to cell nuclei, and of two acid dyes, brilliant sulfaflavine and a stilbene disulfonic acid derivative used as a laundry brightener and known by us as LN, for "long name."[88,89,90,91] The two dyes had different pKs, and hence had different affinity for cell proteins of different pKs; the result was that neutrophil granules stained primarily with LN, whereas eosinophil granules took up much more sulfaflavine. On slides, one could see blue granulocytes and green eosinophils; lymphocytes, monocytes, and basophils were distinguishable by cell size and nuclear and cytoplasmic morphology. The stain took some tweaking before it performed as well in flow systems; we had to scrap the fixatives and buffers originally proposed and start again from scratch. We also came to the realization that an instrument which derived five illuminating beams from a short-lived and highly explosive xenon arc lamp was not suitable for use in a clinical laboratory, and built a "simple" three-beam system using helium-cadmium and argon lasers[92]. Figure 3-8 shows a display from that system.

It was primarily due to the technical success of our approach that the Block differential counter never made it into production. We were able to develop algorithms for leukocyte classification using seven or eight measured parameters, and thus to discover that we could do as well using a single blue illuminating beam to excite cells stained with only two dyes, sulfaflavine and ethidium bromide, and measuring forward and orthogonal scatter and the dye fluorescences. This simplified system was no longer protectable by any of the patents for which Block had applied. We were also reasonably sure[93] that there could be even simpler systems, based on our success (unpublished experiments) in identifying lymphocytes, monocytes, neutrophils, and eosinophils in unfixed, unstained blood using multiple wavelength, multiple angle scatter measurements.

By 1976, automated differential counters had become the focus of a bureaucratic brouhaha at the Food and Drug Administration, requiring premarket clearance by that agency. The image analysis systems got around this through a "grandfather clause"; a new flow system couldn't, although Technicon's Hemalog D, which was "grandfathered," wasn't selling all that well anyway, because the hematologists hadn't yet come to trust flow cytometry. The end result was that nobody was very interested in pursuing this line of investigation further.

We had been doing other things with flow cytometry in addition to differential counting. An attempt was made to develop an instrument to detect bacterial growth, and a flow cytometer was developed that could detect fewer than 100 molecules of fluorescein-tagged antibody bound to a single virus particle[94]; one of these was actually sold to NASA. We also produced a system that would retrieve and store single cells after they went through a flow cytometer, to allow cell-by-cell validation of flow cytometric procedures for cervical cytology screening and other critical diagnostic tests[95].

The Evolution of Flow Cytometers in the 1970's

Although the commercial production of the Cytofluorograf and Impulscytophotometer in 1970 and the FACS in 1974 allowed laboratories which had not developed and built their own apparatus to pursue applications of fluorescence flow cytometry and sorting, advances in the technology itself during the 1970's occurred primarily in the relatively small community of labs in which instruments were developed and built. What got done in any given lab was determined by the biological problems and/or clinical applications under investigation, and also by the migration of instruments and/or investigators from one place to another. This process has recently received some attention from real historians of science, resulting in several publications by Alberto Cambrosio and Peter Keating[1099-1101,2427] and in a video history by Ramunas Kondratas[1102], which was funded by B-D and is available from the Smithsonian Institution Archives. Wallace Coulter and Lou Kamentsky, among others, were not interviewed.

As has already been mentioned, RCS prototypes and people who worked with them played a role in the development of both the Technicon Hemalog D blood cell counter and the Stanford/B-D FACS; the latter instrument represented a convergence of the RCS lineage and the lineage of the Los Alamos cytometer/sorters.

The Los Alamos instruments were oriented toward multiparameter analysis[655]; the lab received substantial funding from the National Cancer Institute for work on applications in cancer cytology[1103] and cell cycle analysis[228-9,1104] as it related to cancer chemotherapy. In the most elaborate of Los Alamos' cytometers, cells were analyzed in a rather elaborate quartz flow chamber with a built-in Coulter volume sensor. Optical access was available on four sides, permitting measurements of fluorescence in two spectral regions and of scatter at several angles[110,111,157]; multiangle scatter measurements proved invaluable for the identification of different types of leukocytes and were incorporated into commercial instruments by the late 1970's. Two clones of the Los Alamos multiparameter sorter were delivered to the National Cancer Institute in the early 1970's, accompanied by minicomputer-based data analysis systems, which had been developed to replace the less flexible two-parameter pulse height analyzer originally used. The Los Alamos cytometer designs were copied by investigators at other institutions, e.g., the Salk Institute, Colorado State University, the University of California at Los Angeles, where flow cytometry was first used to detect phagocytosis by uptake of fluorescent particles[1105], and the University of Houston, where the instrument was applied to flow cytometric analysis of bacteria[553].

Los Alamos also provided the inoculum for the subsequent growth of another major center for flow cytometer development, that at Lawrence Livermore Laboratory, where, from the mid-1970's on, flow sorting was perfected as a means for separating human chromosomes[276-7,1050,1106]. Other work done at Livermore related to cell cycle analysis[384,387], measurement of sperm cells[121], and detection of intracellular enzymes using fluorogenic substrates[363-4].

At Stanford, the emphasis remained on sorting on the basis of relatively weak fluorescence signals from bound antibody and antigen, with the aim of isolating morphologically indistinguishable viable lymphocytes with differences in antigen responsiveness and other functional characteristics[2,86,154,1100-2,1107]. This had two notable effects on instrument design. Droplet sorting was used because it allowed more cells to be processed and collected in a given time than would have been possible using a fluidic sorting mechanism. Placing the observation point, i.e., the intersection of the laser beam and the cell stream, in a jet in air, rather than in a flow chamber, shortened the distance between this point and the droplet breakoff point at which droplets containing selected cells had to be charged, decreasing the transit time between these points and making faster sorting possible.

Since B-D's commercial version of the FACS became available within two years of the appearance of the first publication[86] describing the instrument, it was easier for most large immunology labs to buy an instrument than it would have been to build one. One notable descendant of the Stanford instrument was the computer-controlled, multiparameter cytometer/sorter built at the Max Planck Institute for Biophysical Chemistry in Göttingen[119,147,239,314,405]. This apparatus used mirrors, rather than lenses, for laser beam focusing and light collection, allowing operation at short ultraviolet wavelengths. It was used to measure such parameters as intrinsic protein fluorescence, membrane fluidity (using fluorescence polarization), and receptor proximity (using energy transfer), and to establish the utility of Hoechst 33342 as a vital DNA stain and thioflavin T as an RNA stain.

At the University of Miami, work concentrated on simultaneous electronic measurement of Coulter volume and AC impedance (electrical opacity)[715,1108-9] of cells; the group there also first showed the feasibility of demonstrating DNA synthetic activity by using immunochemical detection of bromodeoxyuridine (BrUdR) incorporation[391] and investigated rare earth chelates as fluorescent labels for cytometry[1110].

Work done at the University of Rochester on slit-scanning static cytofluorometry[1111] was extended to flow systems, leading to the development of progressively more elaborate apparatus for processing pulse waveforms and for imaging cells in flow[171-4,1112].

Collaborations with investigators at Memorial Sloan-Kettering Cancer Center, which had originated during the RCS development program at IBM, provided new applications for the Bio/Physics Systems Cytograf and Cytofluorograf instruments during the 1970's. Most notable among these were techniques using acridine orange for simultaneous determination of DNA and RNA content and for analysis of chromatin structure and DNA synthesis[262-3,386,525]. Studies were also done on cell sizing measurements[97-8,160]. This work was facilitated by the addition of a minicomputer-based data analysis system and the development of software for multiparameter analysis[128]. While the Cytofluorografs available before 1976 were not sufficiently sensitive to be used for immunofluorescence analysis, the Memorial group did investigate lymphocyte activation using DNA and RNA measurements[597-600]; similar studies were done at Los Alamos[1113] and elsewhere[601]. This work apparently failed to excite the imaginations of immunologists, who were committed to using antibody reagents even in the pre-monoclonal era.

Other work on detection of lymphocyte activation using fluorescein fluorescence polarization measurements[414] was done on a laboratory-built instrument at the Ontario Cancer Institute[5,338]; polarization measurements were also used to detect early responses of hematopoietic cells to cytokines[417]. This work represented one of the earliest instances of the use of functional probes in flow cytometry. The Ontario group also studied changes in Hoechst dye uptake and retention during lymphocyte activation; this work played an important role in the identification of the efflux pump mechanisms now widely studied for their roles in anticancer drug resistance[248,253].

At the German Cancer Research Centre in Heidelberg, dual-beam fluorescence excitation capability was added to commercial instruments and then incorporated into a laboratory-built sorter[258,323] that also had a computerized data analysis system capable of producing three-parameter displays. The Heidelberg group also introduced **DAPI** as a DNA stain and the combination of DAPI and **sulforhodamine 101** for DNA and protein staining in flow cytometry[296].

While the Heidelberg instrument followed the pattern established at Los Alamos, Livermore, and Stanford of using multiwatt, water-cooled argon and krypton ion lasers for fluorescence excitation, other instruments built in Europe during the 1970's utilized smaller light sources. The simplest approach to flow cytometry involved the addition of a flow chamber and electronics to a fluorescence microscope[99], as had been done in the original Impulscytophotometer[83].

Kachel et al[656] combined fluorescence and Coulter volume measurement capability in the **Fluvo-Metricell**, which was marketed by **HEKA**, while Eisert and his coworkers[616-9] built instruments capable of highly precise optical size measurements using multiple small laser sources; one such system was eventually produced by **Kratel**.

The arc source instrument described by Lindmo and Steen[100-2] observed cells in sheath flow after a jet in air intersected the flat surface of a cover slip, making multiangle scatter and fluorescence measurements with sufficient sensitivity to characterize bacteria[103]. An early commercial version of this apparatus was produced by **Leitz**; a later version was made by **Skatron**, and an even later one, formerly available from **Bio-Rad** as the **Bryte HS™**, is now being produced by **Apogee** in the U.K.

By the mid-1970's, potential customers' interest in immunofluorescence measurement and sorting had increased to the point at which both Bio/Physics Systems and Coulter needed to develop new instruments to compete with Becton-Dickinson's FACS. Coulter's TPS-1 offered sorting capability, but its combination of relatively low-N.A. optics and a relatively low-power air-cooled argon laser source left it with limited fluorescence sensitivity. Bio/Physics introduced the **FC-200** flow cytometer, which substituted a flat-sided quartz flow cuvette for the thick-walled round capillary used in the original Cytofluorograf and replaced the original fluorescence collection lens with a higher-N.A. microscope objective. This instrument had sufficient sensitivity to measure immunofluorescence, but did not include a sorter. However, at the time, none of the manufacturers seemed to be in a great rush to add extra beams, more than one or two additional measurement parameters, and/or computers to flow cytometers as they were, for several reasons.

An instrument that could measure forward scatter and immunofluorescence could, using scatter as the trigger signal, do the single-parameter immunofluorescence measurements and, using fluorescence as the trigger signal, do the single-parameter DNA content measurements which, as far as most people were concerned, represented the state of the art. The addition of a second fluorescence parameter made it possible to measure two-color fluorescence from a dye such as acridine orange, or to measure DNA and total protein content using propidium iodide to stain DNA and fluorescein isothiocyanate as a covalent protein stain[217,295].

Two-parameter immunofluorescence measurements were desirable, but difficult. Monoclonal antibodies had been described in 1975[785], but would not become available as reagents, even to those in the vanguard of flow cytometry, for several years; obtaining reasonably specific staining of two cellular antigens using polyclonal antisera was nontrivial. Then, there was the question of fluorescent labels. While immunofluorescence microscopy using two or more different dye labels had been done, this was typically accomplished by manually switching illumination between excitation wavelengths for the two labels, e.g., blue for fluorescein and green for rhodamine. Adapting this technique to flow cytometry would have required a dual-laser apparatus.

Since their instrument didn't have dual-wavelength excitation capabilities, Loken, Parks, and Herzenberg[115] resorted to a compromise in order to do the first flow cytometric measurements of two-color immunofluorescence. Instead of using the 488 nm emission line of their argon ion laser for excitation, they used the 515 nm line to excite both fluorescein- and rhodamine-labeled antibodies. While 488 nm is very near the excitation maximum of fluorescein, rhodamine excitation is only about 5% of maximum at this wavelength. At 515 nm, rhodamine excitation is considerably improved, and, although fluorescein excitation is definitely suboptimal, the relative strengths of the fluorescein and rhodamine signals are reasonably well balanced. Fluorescence compensation circuits, which were used to reduce interference between the fluorescein and rhodamine signals, were described for the first time in this 1977 paper. From a practical point of view, however, two-color immunofluorescence remained in the "don't try this trick at home" category.

Multiangle scatter measurements had also not yet made it to prime time. Following the demonstration at Los Alamos that orthogonal scatter measurements could discriminate among lymphocytes, monocytes, and granulocytes[157], we had incorporated orthogonal as well as forward scatter measurements into our instruments at Block[92-3]. It was difficult for B-D (or Stanford) or Coulter to do this with their stream-in-air systems, because the light scattered from the small, round stream produced considerable interference, which got considerably worse when the stream was vibrated during droplet sorting. Bio/Physics Systems had a different problem; the mounting arrangement of the lasers in Cytofluorografs put the polarization of the beam in a direction that precluded making measurements of orthogonal scatter.

Going to three- or four- (or more-) parameter measurement capability also involved a major escalation in cost, because microprocessors, which had first appeared in 1973, had not developed to the point at which they might even be considered for use in data analysis. A minicomputer system was the only possible choice.

In late 1976, Myron Block and I tried to interest Bernie Shoor, of B-D, and Mack Fulwyler, who was just leaving Coulter's flow cytometry operations to join B-D, in pursuing commercial development of our computerized, multibeam, multiparameter system[92-3]. They told us they didn't think anybody would need all those beams and all those parameters. Things changed fast.

In 1976, Bio/Physics Systems was bought by **Ortho Diagnostics**, a subsidiary of Johnson and Johnson. By late 1977, Fred Elliott and others at Ortho were developing prototypes of the **System 50 Cytofluorograf**, a droplet sorter which incorporated a flat-sided flow chamber and high-efficiency collection optics, allowed measurements of forward and orthogonal scatter, extinction, and fluorescence at two or more wavelengths, and offered multiple laser excitation and a computer-based data analysis system as options.

By 1979, immunologists at NIH, with B-D's aid, had added a krypton laser emitting at 568 nm to the argon laser with which the FACS was normally equipped, and examined cells stained with antibodies labeled with FITC and with isothiocyanate (**XRITC**) and sulfonyl chloride (**Texas red**) derivatives of **rhodamine 101**[116,117]. FITC fluorescence was excited at 488 nm and measured at 510-550 nm; XRITC or Texas red fluorescence was excited at 568 nm and measured at 590-630 nm. Since the two measurements were made at different positions and at different times, there was essentially no crosstalk between the signals, therefore, no fluorescence compensation was needed. Work with the dual laser FACS was described in October 1979 at a meeting sponsored by B-D and NIH, at which B-D announced commercial availability of a dual-laser version of the **FACS IV**; the instrument also had computerized data analysis and sort control. The meeting precipitated a stampede of users, all of whom wanted to be first in their states with a dual laser cell sorter.

Coulter, under Bob Auer's direction, had also improved its breed of flow cytometers, introducing the **EPICS** series, droplet sorters that used large laser sources and that incorporated microprocessors into their data analysis systems. Although the first EPICS was intended as a single-beam instrument, the feeding frenzy underway in the user community led to the rapid addition of multiple-beam excitation capability.

The demand for dual-laser instruments was due primarily to the dissemination of monoclonal antibody methodology into the immunology community, which made immunofluorescence experiments with two antibodies relatively easy to do. This, in turn, led to the development of the covalent labels **XRITC** and **Texas red**[116-7], derivatives of **rhodamine 101**, which were designed for use in flow cytometry, and provided the impetus for people to acquire the dual-laser systems, which then provided the only practical approach to two-color immunofluorescence measurements. An illustration of the rapidity with which the field of multistation flow cytometry developed from 1979 on is given by the fact that papers describing conjugation procedures for XRITC and Texas red did not appear in print before 1982; for over two years, word-of-mouth and manufacturers' product information provided a large and growing community of users with the only data available.

Dog Days: The Genesis of Cytomutts

By 1976, I had talked to and visited a few people in the Boston area who were using flow cytometers, gone to a few meetings at which I met people who developed flow cytometers, and come to appreciate that the multistation multiparameter instruments built at Block had capabilities that didn't exist in any other flow cytometers. It also seemed that the apparatus was largely wasted on differential blood cell counting, and could be put to more productive use by biomedical researchers once they became familiar with the technology.

In 1977, with missionary zeal, I assumed part-time proprietorship of a flow cytometry service laboratory at the Sidney Farber (now Dana-Farber) Cancer Institute, hoping to spread the word around Harvard, which, where multiparameter flow cytometry was concerned, hardly deserved to be called "the Stanford of the East." I soon discovered that grants policies, at least as they were then, provided no mechanism by which the apparatus already built and lying idle at Block could be moved across the Charles River and used. I also discovered that if I wanted to attempt to duplicate high-sensitivity, multiple illumination beam, multiparameter flow cytometers, I would have to do it without the services of a machine shop or an electronics shop. I couldn't see that I had a choice. I had become a flow cytometry junkie; I didn't know how to do much of anything with fewer than two beams and none of the manufacturers would sell me an instrument with more than one beam for fluorescence excitation. Mort Mendelsohn dropped in for a visit, and told me that I was crazy to try to build a flow cytometer by myself, and that I would never get funded. He was half right; I got funded.

The lab I was in at the Farber contained an old-style Cytofluorograf. It was also the repository for the carcasses of a Feulgen microspectrophotometer and a scanning cytofluorometer built in Caspersson's lab at the Karolinska Institute, which had been used in the development of chromosome banding techniques by Caspersson in collaboration with Sidney Farber, George Foley, and Ed Modest in Boston[269-72]. The Zeiss microscope optics had long since disappeared from these instruments, but photomultipliers and housings, power supplies, and some other electronics remained. I was able to scrounge the fluidics system, flow chamber, illumination and collection optics, optical bench, and mounts from one of the Block prototypes. All I needed was a data analysis system and some lasers.

I wanted to avoid writing software at all costs. The best way to do this seemed to be to use a Data General minicomputer for data analysis; I could then use the software developed by Brough Turner at Block and/or the software written by Tom Sharpless at Memorial Sloan-Kettering, both of which ran on Data General hardware. When an unused Data General Eclipse minicomputer turned up in the basement, I figured I was all set.

When I started looking into lasers, I was given the impression I'd need hundreds of milliwatts of laser power to make decent measurements. This didn't completely square with my experience at Block; the most powerful laser on the multibeam systems there ran at about 10 mW, and even the laser on the instrument used for virus analysis was never operated above 100 mW. However, the people I knew who were running FACSes told me they used much higher powers, and I assumed that I'd be getting weaker signals from live cells stained with antibodies than I got from fixed cells stained with nucleic acid and protein stains. I ordered a 6-watt water-cooled argon ion laser, good for about 2 watts at 488 or 515 nm and 100 mW UV, and a 1-watt krypton ion laser, good for about 500 mW in the red (647 nm), 100 mW green (520 or 530 nm) or yellow (568 nm), and 50 mW UV. The optical bench from Block wasn't big enough to hold the lasers, so I got a 4 by 8 foot optical table. The big lasers necessitated the then customary ritual of bringing in 150 ampere, three-phase, 220 V electric current and plumbing and pumps to supply cooling water at a rate of 6 gallons/minute.

The first version of the instrument used beams from the argon and krypton lasers to illuminate a thin-walled capillary flow chamber from a Block Cytomat, which also provided the illumination and collection optics. The red/green fluorescence detector assembly from the Cytofluorograf, transplanted outside the chassis of that apparatus, provided two fluorescence detectors; pending completion of the computer system, the counters and pulse height analyzer from the Cytofluorograf were used for data analysis. The third fluorescence detector and the orthogonal scatter detector were photomultiplier assemblies removed with loving care from the remains of the Karolinska-built equipment. Good blood lines all, but a few too many for a pedigree. I accordingly acknowledged the new beast's mixed ancestry and its descent from the Cytomat, and named it "Cytomutt."

I was not entirely surprised to find that Cytomutt, with its high-N.A. collection optics, didn't seem to require a lot of laser power. I could get good DNA content measurements from unfixed cells stained with Hoechst 33342 using less than 10 mW of UV from the argon or krypton lasers. This allowed an arc lamp to be substituted as a UV source; although it made the optics a little trickier to align, this eliminated the almost certain need to spend $10,000 a year replacing the laser plasma tubes, which wore out much faster when operated in the ultraviolet.

When the 6-watt argon laser was cranked down to minimum power, it put out 200 mW at 488 nm, which was a lot more than was necessary to get strong immunofluorescence signals. It was therefore possible to insert a beamsplitter, taking off almost 100 mW to illuminate a second flow chamber, which, with minimal added detectors and electronics, was used for simple tasks such as screening monoclonal antibodies. The extra head on Cytomutt was dubbed "Cerberus." The system as it looked around 1980 is shown in Figure 3-9.

Figure 3-9. The author with Cytomutt and Cerberus.

The late 1970's and early 1980's were still very much the heyday of big lasers in flow cytometry; for much of that time, more multiwatt ion lasers were sold for flow cytometry than for any other use. Providing power and cooling water for three cytometer manufacturers and a laser manufacturer or two became a major logistic problem at Society for Analytical Cytology meetings, and the use of big lasers added to the prices and installation costs of the hundreds of instruments which came into use during this time. "Laser machismo" eventually became enough of a marketing gimmick that it was difficult to convince people to buy better-performing instruments with smaller lasers.

In 1978, Stuart Schlossman, at the Farber, began a collaboration with Ortho Diagnostics that led to the development of the first of many monoclonal antibodies reactive with cell surface antigens on human lymphoid cells[321-2]. He advised his then-colleagues at Ortho to purchase a B-D FACS, the same apparatus he used for immunofluorescence flow cytometry; this advice didn't exactly thrill Lou Kamentsky and his group, then manufacturing Ortho's own flow cytometers in the Boston area. A "gunfight at the OKT corral" was arranged, with the same samples being run on the FACS, which used a water-cooled argon laser emitting 200 mW at 488 nm, and on an Ortho FC-200, which used an air-cooled laser emitting 20 mW at 488 nm. The sensitivities of the two instruments, defined by the distance between peaks of histograms representing stained and unstained cells, were comparable, but the results of the test never found their way into either the scientific literature or Ortho's advertising.

Witnessing the shootout removed any doubts I had left that efficient optical design would make it possible to replace big, water-cooled, lasers with smaller, cheaper, air-cooled lasers and/or arc lamps as light sources for flow cytometry. Shortly thereafter, my colleagues and I built a dual-beam instrument with UV illumination from an arc lamp and laser illumination at 633 nm from a 7 mW He-Ne laser, allowing Hoechst dyes for DNA to be used in combination with oxazine 1 for RNA content measurement or with dicarbocyanine dyes for membrane potential estimation. The instrument also featured a hardwired data analyzer allowing the use of four parameters to define as many as eight gating regions. We brought this "Son of Cytomutt" to the 1981 Analytical Cytology meeting in nearby New Hampshire, and ran it, while the manufacturers' large systems sat idle due to the lack of electrical power and cooling water for the lasers.

The 1980's: Little Things Mean a Lot

Our demonstration at the New Hampshire meeting was not the only indication that small might be beautiful. At around the same time, B-D announced its **FACS Analyzer**, a benchtop system using an arc lamp source. It could measure fluorescence in two wavelength regions, light scattering at large angles, and (electronically) cell volume[152], and was offered with a microcomputer-based companion data analysis system. During the 1980's, other manufacturers also moved in the direction of somewhat smaller, more user-friendly instruments. Ortho, which had acquired rights to manufacture and distribute the Impulscytophotometer, also produced the **Spectrum III**[1114], an instrument designed for the clinical market, using an air-cooled argon laser source and measuring forward and orthogonal light scattering and two-color fluorescence. Both the B-D FACS analyzer and the Ortho Spectrum III employed closed fluidic systems, and did not offer sorting as an option; Coulter, in contrast to its competitors, chose to make its initial approach to the clinical market with a sorter, the **EPICS C**, which incorporated the optical bench and droplet sorter used in the research instruments of the EPICS series and placed virtually every function of the apparatus under computer control[873].

The FACS Analyzer and the EPICS C preserved some of the flexibility of research flow cytometers, at a price. The Analyzer's optics were, and had to be, very efficient, in order to permit immunofluorescence measurements to be done using the relatively weak blue-green excitation available from the arc lamp source. By changing excitation filters, however, one could use the arc lamp's strong UV, blue-violet, and green lines to excite dyes that could not be used with 488 nm argon lasers. Maintaining performance required maintaining optical alignment; this was clearly harder to do for some people than for others. The EPICS C achieved the capability for operation at one of several wavelengths by the simpler expedient of using a water-cooled argon ion laser source, which affected its size, price, and the logistics related to installation. By the time the second generation of clinical fluorescence flow cytometers were introduced in the late 1980's, the manufacturers had adopted Henry Ford's philosophy about color choices on the Model T; you could have any excitation wavelength you wanted, as long as it was 488 nm.

B-D's **FACScan**, the first of these benchtop instruments, used highly efficient optics, as had the FACS Analyzer, but substituted an air-cooled, 15 mW argon laser source for the arc lamp in the Analyzer. The FACScan flow chamber, very similar if not identical to what B-D has subsequently used in the FACSort, FACSCount, and FACSCAlibur, incorporates a high- N.A. "immersion" lens, with an optical coupling gel rather than immersion oil, to maximize light collection. The FACScan was also noteworthy for its introduction of a third fluorescence measurement channel; in addition to forward and orthogonal scatter at 488 nm, fluorescence could be measured in fixed emission ranges in the green, yellow-orange, and red. Data acquisition and analysis and much of the rest of the operation of the FACScan were originally controlled by a Hewlett-Packard microcomputer with a 68000-series processor. Coulter's **EPICS Profile**, originally introduced with capabilities for forward and orthogonal scatter and two fluorescence measurements, soon added a third fluorescence channel. This instrument incorporated a microprocessor-based controller, and could be interfaced to more elaborate data analysis systems built around IBM-compatible personal computers. The Profile achieved high light collection efficiency with a flow chamber design incorporating integral lenses and mirrors.

The 1980's also brought changes in the design of flow cytometers and sorters used for research, predominantly in the directions of using smaller lasers and more efficient light collection optics and the incorporation of microcomputer systems for instrument control as well as for data acquisition and analysis. The direction of the evolution of the apparatus was largely determined by the development of new parameters, reagents, and analytical methods, and the emergence of clinical applications of flow cytometry.

Measurements in the Main Stream

Fluorescence flow cytometry, since its inception, has been employed predominantly for measurements of **cell surface and intracellular antigens**, on the one hand, and of **cellular nucleic acid (DNA and sometimes RNA) content**, on the other. Qualitative and quantitative changes in these cellular parameters have been, and still are, used to define and characterize normal and abnormal cellular differentiation and function.

Immunofluorescence Comes of Age: Monoclonal Antibodies and Multiple Labels

The 488 nm argon ion lasers employed as light sources in most commercial flow cytometers are well suited for excitation of **fluorescein**, the popularity of which as a fluorescent label for antibodies antedated the introduction of both the laser and the cytometers. The subsequent development of labels such as **phycoerythrin** and its **tandem conjugates** was driven by the emergence of **monoclonal antibody reagents**, on the one hand (the subject of several articles and an entire book by Cambrosio and Keating [1099-1101,2428-9]), and the desirability of holding the cost and complexity of instruments down (at least in relative terms) by using only a single laser light source.

The major obstacle to progress in multicolor immunofluorescence between 1940 and the late 1970's was the difficulty of achieving specific staining with polyclonal antisera; as a result, little effort was expended during this time on discovery or development of fluorescent labels with emission spectra suitable for use in multicolor immunofluorescence measurements. Once monoclonal antibodies were developed as reagents, it became logical to look for new labels. Dual-laser flow cytometry using antibodies labeled with fluorescein and Texas red or XRITC gave better results than could be achieved with a single-laser instrument and fluorescein- and rhodamine-labeled antibodies, but greatly increased the cost and complexity of the apparatus required by adding a second water-cooled laser or a dye laser.

It was obviously desirable to have several labels that could be excited at a single wavelength, ideally by the 488 nm argon ion laser line prevalent in flow cytometers, and which emitted in different, reasonably well separated spectral regions; however, for a variety of reasons, it was, and is, not possible to simply design and synthesize molecules with the required characteristics. Nature, however, had provided a solution in the form of the phycobiliproteins, which are components of the photosynthetic apparatus of algae, and which, in their native configuration, nonradiatively transfer energy from blue-green and green light to chlorophyll, which could otherwise not utilize light from these spectral regions in photosynthesis. In the early 1980's, while Alex Glazer of Berkeley and Lubert Stryer of Stanford were collaborating on studies of the biochemistry of phycobiliproteins, Vernon Oi, an alumnus of the Herzenberg lab, moved to Stryer's department; it became apparent soon thereafter that these materials could be useful as fluorescent labels for antibodies, and in other circumstances as well[114]. Glazer and Stryer described the first tandem conjugate, in this case made from the phycobiliproteins phycoerythrin and allophycocyanin, in 1983[306]. A patent was secured by Stanford, which gave B-D several years' worth of exclusive rights to phycobiliproteins as labels for flow cytometry; by the late 1980's, numerous manufacturers were producing phycobiliprotein-labeled monoclonal antibodies.

Monoclonal antibodies to human lymphocyte surface antigens were among the first to come into widespread use, and were investigated as both diagnostic and therapeutic agents for conditions such as rejection of transplanted organs[1003,2430]. The analysis of lymphocyte subpopulations originally required an initial separation of the lymphocytes from granulocytes and other cells by density gradient centrifugation. It had been known for some time that a combination of forward and orthogonal scatter measurements could discriminate among lymphocytes, monocytes, and granulocytes[157]; it remained for Hoffman et al[175-6] to demonstrate the practicality of combining the scatter and immunofluorescence measurements for rapid analysis of immunologically defined lymphocyte subpopulations in whole blood. This, incidentally, introduced the concept of multiparameter gating to a substantial segment of the immunology community (recall the anecdote of p. 41).

Developments in DNA Content Analysis

The initial description of fluorescence flow cytometry by Van Dilla et al[79] in 1969 included a histogram of cellular DNA content, determined using a fluorescent Feulgen stain. The paper by Dittrich and Göhde describing the Impulscytophotometer[83], published the same year, described staining with **ethidium bromide**, which, although initially not offering the precision achieved with the Feulgen stain, greatly simplified sample preparation. In 1973, Crissman and Steinkamp[217] described the use of **propidium iodide**, a homolog of ethidium with a slightly longer emission wavelength, making it more suitable for use in combination with fluorescein in instruments with argon ion laser sources.

Early work with ethidium and propidium was done with fixed cells; treatment of samples with RNAse was required to eliminate fluorescence due to binding of the dyes to double-stranded RNA. In 1974, Crissman and Tobey[228] described a rapid staining procedure using **mithramycin**; although the DNA-specific fluorescence of this dye eliminated the need for RNAse treatment, the requirement for excitation at 457 nm or shorter wavelengths made the procedure usable only in systems using either large ion lasers or arc lamps for excitation. The first widely used rapid procedure for DNA staining was that reported by Krishan[218] in 1975, employing propidium iodide in a hypotonic sodium citrate solution, which rendered many cell types permeable to the dye. Subsequent modifications to this procedure by others[220-6] added low concentrations of nonionic detergent, which improved cell permeabilization and facilitated storage, and RNAse treatment.

The UV-excited, blue fluorescent **Hoechst dyes**, including compounds 33258 and 33342, were introduced by Latt[238]; in addition to offering the advantage of DNA specificity, these dyes provided the only reliable means of stoichiometrically staining DNA in living cells[239]. Another UV-excited, blue fluorescent, DNA-specific stain, **4'-6-diamidino-2-phenylindole (DAPI)**, was introduced by Stöhr et al[258], while Barlogie et al[233] described DNA-specific staining using a combination of ethidium bromide and mithramycin.

Flow Cytometry of RNA Content

In the late 1970's, Darzynkiewicz et al[262-3] developed flow cytometric methods for simultaneous measurement of RNA and DNA content using **acridine orange** as a metachromatic stain. This dye, applied to permeabilized cells under carefully controlled conditions, forms a green fluorescent complex with DNA and a red fluorescent complex with RNA. On the basis of analysis of such systems as mitogen-stimulated lymphocytes and leukemic cells undergoing chemical-induced partial differentiation *in vitro*, it was shown that patterns of DNA and RNA staining defined subcompartments of the cell cycle, distinguishing proliferating and quiescent cells. This is illustrated schematically in Figure 3-10.

The histogram shown at the top of the figure illustrates the distribution of DNA content in a population containing growing cells. Cells in the "first gap," or G_1, phase of the cell cycle (see p. 86) have a DNA content described as "diploid," or "2C," i.e., the amount of DNA contained in the 2 sets of chromosomes present before DNA replication begins. During the DNA synthetic, or S, phase, DNA content increases to twice this amount, the "tetraploid," or "4C" value. It remains at 4C through the "second gap," or G_2, phase, and during mitosis, the M phase. After mitosis, there are two daughter cells, each with a 2C DNA content.

It was recognized that many differentiated cells, such as resting peripheral blood lymphocytes, normally remained in a **quiescent state**, described as G_0 or G_{1Q}, characterized by a "diploid" (2C) DNA content; Darzynkiewicz and his co-workers showed that such cells had a low RNA content. Within 12 hours or so following exposure to mitogens, lymphocytes enter the G_1 phase and begin to synthesize RNA. RNA content continues to increase during the S phase, beginning about 30 hours after stimulation, in which DNA synthesis occurs.

Analysis of DNA content alone cannot discriminate cells in G_0 (G_{1Q}) from cells in the **proliferative** G_1 state, because the DNA content remains at 2C until the S phase begins. Measurements of RNA content can be used to make this distinction and, in addition, to define different stages within G_1. Cells pass from G_{1Q} through a brief transitional phase called G_{1T} (in which no cells are shown in the figure), in which RNA content is slightly increased, and then into G_{1A}, during which RNA content increases further, but remains lower than the RNA content of any S phase cell. They then enter G_{1B}, in which RNA content is at or above the lowest value seen in S phase cells. RNA content increases approximately linearly during S and G_2.

In exponentially growing cultures, which lack cells in G_{1Q}, cells appear to pass from S through G_2 and M back into G_{1A}. Normal cells, such as stimulated lymphocytes, when maintained in long-term culture, tend to revert back to a G_{1Q} state, although quiescent, low-RNA "S_Q" and "G_{2Q}"

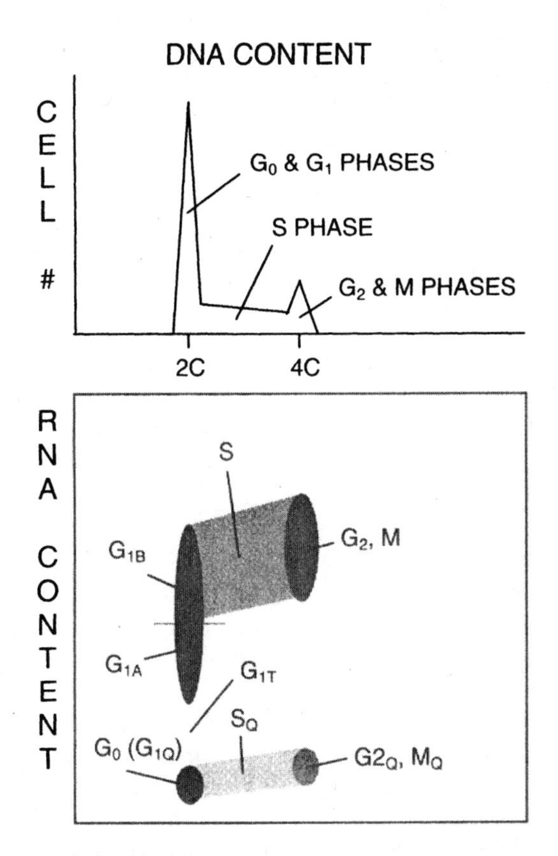

Figure 3-10: Cell cycle phases defined by DNA content (schematic histogram at top) and DNA/RNA content (schematic cytogram at bottom).

populations can appear transiently in cells deprived of nutrients or exposed to cold or to inhibitors of protein synthesis. Transition to quiescent (Q) states during S and G_2 appears to be somewhat more common in transformed and malignant cells than in normal cells.

Acridine orange staining for DNA and RNA content determination has been widely used in flow cytometry; excitation is feasible in both laser and arc source instruments. Using different preparative procedures, acridine orange can also be used to demonstrate differences in nuclear chromatin structure, evidenced by different sensitivities of DNA to heat or acid denaturation; this allows mitotic cells to be distinguished from G_2 cells, which cannot be done by DNA or DNA/RNA staining.

It is not feasible to do immunofluorescence measurements on cells stained for DNA and RNA with acridine orange, for two reasons. Many antigens are unlikely to emerge unchanged from the detergent permeabilization and acid treatment required to achieve specific staining of DNA and RNA. Also, the fluorescence of acridine orange interferes with the fluorescence of virtually all immunofluorescent labels that can be excited by blue or blue-green light. In 1981, I described a DNA/RNA staining procedure, using **Hoechst 33342** and **pyronin Y**, that could be used on intact or fixed cells also stained with fluorescein-labeled antibodies[113], employing a 488 nm beam to excite fluorescein (green fluorescence) and pyronin (orange fluorescence) and adding a second illuminating beam in the near UV (325 to 375 nm) to excite the Hoechst dye. The Hoechst/pyronin stain can be thought of as a modern fluorescent equivalent to the classical methyl green/pyronin stain discussed on p. 75.

Measurements of Functional Parameters

Intracellular enzyme activity, detected and quantified using **chromogenic** or **fluorogenic substrates**, was among the earliest parameters measured by flow cytometry. The original Stanford sorter[82] detected intracellular fluorescein fluorescence resulting from hydrolysis of **fluorescein diacetate (FDA)** (pp. 24-7); such staining was employed by numerous workers as a basis for tests of cell "viability," as defined by structural integrity of the plasma membrane. The Technicon Hemalog D differential counter[84-5] (p. 88) used absorption measurements and chromogenic substrates to identify neutrophils, eosinophils, and monocytes.

In 1983, Bass et al[362] described the use of another fluorogenic substrate, **2′,7′-dichlorodihydrofluorescein diacetate (H_2DCF-DA)**, for the detection of **oxidative enzyme activity**, in particular, the **respiratory burst** in activated neutrophils. H_2DCF-DA, like FDA, enters intact cells; once inside, it hydrolyzes to the colorless dihydro compound, which is oxidized to the fluorescent dichlorofluorescein in the presence of hydroperoxides. Assessment of cells' oxidative metabolism had previously been described by Thorell and others using an intrinsic parameter, i.e., the autofluorescence of the reduced forms of pyridine nucleotides[185,191-4], but this required UV excitation. The H_2DCF-DA technique has found much wider use.

Flow cytometric analyses of other functional parameters, such as **intracellular calcium ion concentration**, **intracellular pH**, and cytoplasmic and mitochondrial **membrane potential**, were developed for the analysis of a wide range of cell activation processes involving transmembrane signaling. These parameters, like oxidative metabolism, are only relevant when measured in live cells.

In 1977, Price et al, at the Ontario Cancer Institute[417] (p. 91), used fluorescein fluorescence polarization measurements[414] for flow cytometric detection of cell activation by mitogens and growth factors. Intracellular pH measurements, based on changes in the spectrum of fluorescein, were reported by Visser, Jongeling, and Tanke in 1979[519]; later that year, in collaboration with Lou Kamentsky and Peter Natale, I described flow cytometric methods for cell membrane potential estimation using **cyanine dyes**. Darzynkiewicz, Staiano-Coico, and Melamed observed increased mitochondrial uptake of **rhodamine 123** in activated lymphocytes in 1981[493], although they were not aware at the time that the dye uptake was driven by mitochondrial membrane potential.

In 1981, I used **chlortetracycline** for flow cytometric detection of changes in **membrane-bound calcium**[112]. Measurements with this probe shared a disadvantage with membrane potential measurements; they yielded broad fluorescence distributions reflecting differences in cell volume and other factors not related to the functional parameter under study. **Quin-2**[513], the first fluorescent probe described for flow cytometry of intracellular Ca^{++} concentration, had the same problem. The newer **indo-1**[858], a UV-excited probe that changed its emission spectrum on chelating a calcium ion, allowed intracellular [Ca^{++}] to be estimated by a **ratiometric measurement** (p. 47). The ratio of violet and blue-green emission intensities yielded a quantity proportional to the intracellular calcium concentration; distributions were sufficiently narrow to allow detection of relatively small responsive subpopulations among larger populations of cells unaffected by a given stimulus. Flow cytometric measurements using indo-1 were reported by Valet, Raffael, and Russmann in 1985[862].

Valet and his coworkers had previously (1981) reported **ratiometric pH measurement** by flow cytometry; they originally used a UV-excited dye[200], which was inaccessible to users of most cytometers. In 1986, Musgrove, Rugg, and Hedley showed that **2′,7′-*bis*-(2-carboxyethyl)-5-(and-6)-carboxyfluorescein (BCECF)** could be used for ratiometric pH measurement with 488 nm excitation. Calcium, pH, and membrane potential probes have since been used in combination with one another[872], and in combination with other labels, e.g., fluorescent antibodies[866].

Functional assays of enzyme activity, membrane potential, calcium ion, and pH usually involve measurements of the variation of these parameters over time following manipulation of cells; the concept of doing such **kinetic assays** by flow cytometry was articulated by Martin and Swartzendruber in 1980[436]. Kinetic, as well as static assays, are also used for determination of **drug uptake** and **efflux** by cells. Krishan and Ganapathi[352] used the intrinsic fluorescence of anthracyclines as a flow cytometric parameter in 1980, while Kaufman and Schimke[354], in 1981, described the use of a fluorescent analog of methotrexate to study amplification and loss of the dihydrofolate reductase gene.

Flow cytometric procedures for determination of levels of **glutathione** and **sulfhydryl** or **thiol groups** in cells were first discussed in 1983 by Durand and Olive[333-4]; much initial motivation for this work derived from the known role of thiols in the radiation resistance of tumor cells. Flow cytometry of intracellular glutathione has, more recently, become important in studies of tumor cells' drug resistance[801,1116] and of HIV infection and AIDS[1117-8]. Rice et al[801] (1986) described the use of monochlorobimane, a UV-excited, blue-fluorescent material now thought to be the most specific probe for glutathione measurement; a staining protocol is also given by Roederer et al[1118].

Clinical Uses of Fluorescence Flow Cytometry

As I mentioned previously, much of the motivation and support for the initial development of flow cytometry came from the shared beliefs of investigators and government and industrial funding organizations that the technology would lead to successful automation of cancer cytodiagnosis, on the one hand, and differential leukocyte counting and related tasks in hematology, on the other.

Although we still don't have the flow cytometric equivalent of a Pap smear, fluorescence flow cytometry, from its very beginnings, began to find applications in oncology. DNA content measurements were used clinically for diagnosis and for determination of the effects of drugs on tumor cell proliferation kinetics from the early 1970's on, with European workers taking the lead[1115]. By 1980, it had been made clear to the general medical community that DNA content abnormalities were common in cancer and leukemia[608]. Issues of sample preparation and storage and the use of chicken and rainbow trout erythrocytes as standards for DNA content measurements were addressed in a series of papers by Vindeløv et al[222-5] in 1982.

Interest in the clinical use of DNA content measurements increased markedly after 1983, when Hedley et al[610] described a method for measuring DNA in nuclei extracted from paraffin-embedded material. This allowed the prognostic significance of DNA content abnormalities in various tumor types to be determined in retrospective as well as prospective studies, and made flow cytometry vastly more attractive as a field of interest to young pathologists in a hurry, who would no longer have to wait five or more years to publish their data. Nomenclature for DNA content measurements was standardized in 1984 by a committee established by the Society for Analytical Cytology[741].

Although many more elaborate flow cytometric methods have been and are being used for analyses of cancer and leukemias, including DNA/RNA content analysis, studies of DNA synthetic patterns using BrUdR and anti-BrUdR antibodies, immunofluorescence and immunofluorescence-gated DNA content measurements, measurements of functional parameters related to drug resistance, and detection of oncogenes and their nucleic acid and protein products, DNA content analysis remains the clinical flow cytometric procedure most widely used in oncology.

When it came to automating the differential leukocyte count, flow cytometry was successful beyond our wildest dreams. Although there are still some slide-scanning, image-analyzing automated differential counters in service, three- and five-part differentials are now done predominantly by flow cytometers which don't even use fluorescence, but measure Coulter volume, AC impedance, light scattering, and/or absorption. Where the original goal in design of differential counters was simply to "flag" abnormal or immature cells in peripheral blood, fluorescence flow cytometry has instead redefined our concepts of normal and pathologic blood cell development, and can even be applied to marrow, once seemingly sacrosanct.

Clinical application of flow cytometric immunofluorescence measurements began almost immediately after the B-D FACS, the first instrument with the necessary measurement sensitivity, became commercially available. One of the first uses was in immunophenotyping human leukemias. By the late 1970's, groups led by Melvyn Greaves[1119-20] and Stuart Schlossman[1121], using polyclonal antisera, had shown that T cell acute lymphoblastic leukemia in children was unlikely to respond to chemotherapy. At this time, however, the lack of availability of standardized antibody reagents prevented widespread clinical use of immunofluorescence flow cytometry.

In the days of polyclonal antibodies, standardization for applications such as immunoassay depended primarily on reliable quantification of binding characteristics; this was emphasized in much of the early work from the Herzenberg lab at Stanford[154]. By the early 1980's, there were numerous commercial monoclonal reagents, making standardization imperative as much for researchers as for potential clinical users. However, the homogeneity of individual antibodies and the reproducibility with which new batches of reagent could be produced made it less important to develop quantitative standards than to achieve some consensus about which antibodies reacted with which antigens[1100].

This led to the first of a series of International Workshops on Human Leukocyte Differentiation Antigens[1122-7], which have defined **CD** or **"Cluster of Differentiation" Antigens** on leukocytes and other cell types based on experiments by hundreds of laboratories with hundreds of antibodies and hundreds of cell types. The 5th Workshop, held in Boston in November 1993[1126-7], was the first to provide data on quantitative expression of each of the tested antibodies. The printed volume from the 7th Workshop (Harrogate, UK, 2000) came out in 2002[3091]; information on the antigens (now up to CD247) is also available in "Protein Reviews on the Web (PROW)," an on-line journal from NIH[2431].

As was mentioned previously (p. 96), monoclonal antibodies to lymphocyte surface antigens were among the first to become available, and were used for quantitative analyses of T cell subsets in patients with such conditions as autoimmune diseases and graft rejection. In late 1981, it was reported that the T cell subset ratio was abnormal in an immunodeficiency state then newly described in male homosexuals[1128]. Within a short time, AIDS consciousness and fears of a heterosexual epidemic had become prevalent in the general public. In the few months that elapsed between the emergence of AIDS phobia and the discovery of the Human Immunodeficiency Virus, a lot of flow cytometers were peddled to a lot of clinical institutions on the basis that significant numbers of worried well people would be parting with several hundred dollars for T cell subset analyses once or twice a year. This was at best a questionable sales practice; subset analysis would never have been an appropriate screening test for AIDS. However, the buying frenzy left numerous consenting adults with flow cytometers in their labs and good economic reasons to find clinical applications, and determinations of the proportion and absolute count of CD4-positive T cells in peripheral blood have remained among the most useful predictors of the course of HIV infection, and among the most widely used immunofluorescence measurements in clinical flow cytometry. Cambrosio and Keating have recently focused their historical sights on phenotyping in general[2432-5].

Fluorescence flow cytometry has also provided the first practical method of automating the blood reticulocyte count, which (see p. 78) had previously been a tedious and imprecise procedure. The parameter of interest is the RNA content of immature red cells; Tanke et al, in 1981[286], showed that **pyronin Y** fluorescence could be used to identify reticulocytes; others developed procedures using **acridine orange**[764-5] and **cyanine dyes**[768]. Better discrimination of reticulocytes was achieved by Sage, O'Connell, and Mercolino, in 1983[288], using **thioflavin T**, but this dye was not usable with 488 nm excitation. A dye that was, **thiazole orange**, emerged from a study of structures related to thioflavin T carried out by Lee, Chen, and Chiu at B-D[769] and published in 1986.

An early conference on clinical cytometry was held in 1982 under the joint auspices of the Engineering Foundation and the Society for Analytical Cytology[587]; existing and projected applications of flow and image cytology in hematology, oncology, immunology, genetics and bacteriology were discussed. It was clear even then that fluorescence flow cytometry was being brought into the clinic via the back door by researchers who found their results clinically useful, and that, as a consequence, the calibrators and standards without which the instruments could never have been produced for the clinical market were largely unavailable. Although this deficiency has not been completely rectified, problems of standardization and quality control of instruments and procedures, and training and performance assessment of laboratory personnel, have been and are being addressed by numerous organizations involved in both cytometry and laboratory medicine.

The End of History?

I have to stop this discussion somewhere and get back to the technical details; I'll close by pointing you to the summary Table 3-1 on the next page and mentioning a few more significant firsts. The use of flow cytometry for detection of **specific nucleic acid sequences** was reported in 1985[902]. The ultimate in flow cytometric sensitivity, i.e., **single molecule detection**, was achieved in 1987[660]. In 1988, Nolan et al[1642] described a fluorescence flow cytometric procedure for detecting expression of a **β-galactosidase reporter gene** in transfected cells; since then, detection of gene expression has become much easier due to the introduction of **GFP** and its relatives[1648,2376-8].

Jumping back to "ancient history," Watson's recent "cytometry-oriented" historical surveys of the origins of the physics of fluidics and optics[2436] and of numbers and statistics[2437] are entertaining and informative. And, jumping forward again, a look at this week's journals should make it obvious that the range of applications of flow and static cytometry and cell sorting is still being extended. That should hold off the end of history for us, at least for a while.

A BRIEF OUTLINE OF FLOW CYTOMETRIC HISTORY

YEAR 1945 1950 1955 1960 1965 1970 1975 1980 1985 1990 1995 2000

PHYSICAL PARAMETER

Scatter Absorption Extinction Polarized fluorescence
Coulter volume Fluorescence Phase
Opacity Multiangle scatter Polarized scatter

CELLULAR PARAMETER

Presence/Size Nucleic acid content Antigen content Nucleic acid sequence
Protein content
DNA content RNA content
DNA base ratio
Chromatin structure
Membrane integrity pH Calcium
Enzyme activity Membrane potential
Endocytosis Apoptosis
Membrane and cytoplasmic viscosity
Drug uptake and efflux
Lectin binding sites Sulfhydryls/Glutathione
Redox state
Gene expression

REAGENTS

DNA Stains

Feulgen stains Ethidium Hoechst dyes
Propidium DAPI TOTO- and related dyes
Mithramycin

RNA Stains

Acridine orange
Pyronin Y
Thioflavin T
Thiazole orange

Antibodies/ Labels

Fluorescein Rhodamine Texas red/XRITC
Phycobiliproteins Tandem conjugates
Monoclonals

Functional Probes

Enzyme substrates Potential probes
pH probes
Indo-1

SPECIMENS

Bacteria Eukaryotic cells Viruses Molecules
Chromosomes Organelles

YEAR 1945 1950 1955 1960 1965 1970 1975 1980 1985 1990 1995 2000

Table 3-1. A brief outline of flow cytometric history.

4. HOW FLOW CYTOMETERS WORK

4.1 LIGHT AND MATTER

Introduction

In contemporary flow cytometry, measurements of light scattered, emitted, or absorbed by cells provide the values of almost all measurable parameters. Other techniques of analytical cytology, such as confocal microscopy and scanning cytometry, are just as dependent on optical principles and measurements as is flow cytometry. To understand cytometry, flow or otherwise, you need to know some basic facts about light and its interactions with matter.

Photometry versus Radiometry: What's in a Name?

When I just said that most of the information we get about cells from flow cytometry is derived from measurements of light, you probably knew what I meant. From a physicist's point of view, however, I was obviously in error. According to the precise definitions of physics, **light**, which is measured by **photometry**, is electromagnetic radiation perceptible to the human eye. The eye is most sensitive to wavelengths around 550 nm, and, in most of us, incapable of seeing much below 400 or above about 750 nm. Since, in our cytometric peregrinations, we may delve into the ultraviolet and/or the infrared, we are really measuring **radiant energy** or **radiation**, i.e., doing **radiometry**.

Lucky us. The International System of Units (SI Units)[2438] deals with radiant energy in **joules** and with **radiant flux** (energy per unit time) in **watts** (1 watt equals 1 joule/second), units with which we are apt to have at least some familiarity. If we were, instead, forced into using the physical units related to light, we'd be up to our eyeballs in candelas, lumens, lamberts, nits, and apostilbs. Under these circumstances, we had best not make light of radiation. We should, however, get the **quantities** and **units** we will be using on the table. The table, in this case, is Table 4-1, on the next page.

Physical Measurement Units

Of the quantities in this table, the one that is probably least familiar to you is the **solid angle**, measured in **steradians**. You may recall from high school geometry that the circumference of a circle of radius r is $2\pi r$, and that that angle which intercepts an arc of length r along the circumference is defined as one **radian**, which is approximately 57.3 degrees. A sphere of radius r has a surface area of $4\pi r^2$; one **steradian** is defined as that solid angle which intercepts an area equal to r^2 on the surface of the sphere. Figure 4-1 illustrates these concepts.

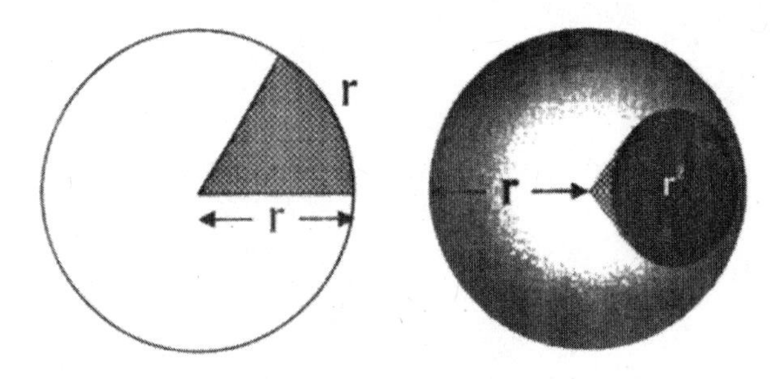

Figure 4-1. Radian and steradian.

The "pie wedge" (or is it a pi wedge?) with its apex at the center of the circle on the left side of the figure subtends a plane angle of 1 radian. The cone with its apex at the center of the sphere on the right side of the figure subtends a solid angle of 1 steradian. A "fisheye" lens of the type customarily described as having a 180-degree field of view collects light over a solid angle of 2π steradians, i.e., a hemisphere. The rest of the tabulated units are likely to be old acquaintances.

Then, there are a few other primary and derived quantities we will need.

Next stop is high school chemistry, where we first encounter the **mole**. One mole of an element or compound contains **Avogadro's number**, or 6.02×10^{23} molecules, of the substance, and has a mass in grams equal to the **molecular weight** of the material. You may notice from the table of prefixes that 1 yoctomole, or 10^{-24} mol, comes out to less than one molecule (it would be about 0.602 molecule); as far as I know, there is not yet an international convention for which corner to start from when slicing off a yoctomole. And, while we're down at the atomic level, we should mention that the **charge** of a single electron is about 1.6×10^{-19} **coulombs**. This means that an **electric current** of 1 **ampere** (A), which is 1 coulomb/second, represents a "flow" of 6.25×10^{18} electrons/second in the direction opposite to the direction in which the current is said to "flow." The electrons don't really travel very much, but I won't get into that now.

If you want definitive information on SI Units, you can get all you'll ever need from the National Institute of Science and Technology (NIST)'s Web site:

http://physics.nist.gov/cuu/Units/index.html

QUANTITY	UNIT
area (A)	square meter (m^2)
volume (V)	cubic meter (m^3)
mass (m)	kilogram (kg)
time (t)	second (s)
angle (θ)	radian (rad)
solid angle (Ω)	steradian (sr)
frequency (ν)	hertz (Hz)
wavelength (λ)	meter (m)
charge (q)	coulomb (C)

FACTOR	PREFIX	SYMBOL
10^{24} (= $(10^3)^8$)	yotta	Y
10^{21} (= $(10^3)^7$)	zetta	Z
10^{18} (= $(10^3)^6$)	exa	E
10^{15} (= $(10^3)^5$)	peta	P
10^{12} (= $(10^3)^4$)	tera	T
10^{9} (= $(10^3)^3$)	giga	G
10^{6} (= $(10^3)^2$)	mega	M
10^{3}	kilo	k
10^{2}	hecto	h
10^{1}	deka	da
10^{-1}	deci	d
10^{-2}	centi	c
10^{-3}	milli	m
10^{-6} (= $(10^{-3})^2$)	micro	μ
10^{-9} (= $(10^{-3})^3$)	nano	n
10^{-12} (= $(10^{-3})^4$)	pico	p
10^{-15} (= $(10^{-3})^5$)	femto	f
10^{-18} (= $(10^{-3})^6$)	atto	a
10^{-21} (= $(10^{-3})^7$)	zepto	z
10^{-24} (= $(10^{-3})^8$)	yocto	y

Table 4-1. SI units and prefixes.

Light in Different Lights

Since physicists from Newton's time to the middle of this century noted that light resembles waves in some aspects of its behavior and particles in others, both practitioners and teachers of physics have found it convenient to deal with light in whichever guise was more suitable to the context at hand. When I took high school physics, both the wave and the particle aspects of light were avoided in favor of an introduction to **geometrical optics**, which deals with the laws of reflection and refraction and with image formation by lenses and mirrors. In college physics, which had as a prerequisite the mathematics required to appreciate the properties of waves, the weightier topics of **physical optics** crept in, and interference, diffraction, and polarization were approached in terms of electromagnetic fields and waves, after which lip service was paid to such things as the photoelectric effect and the particle aspects of light. Further discussion of the concepts of **quantum mechanics**, and the interactions of light particles or **photons** with matter, was left for advanced courses in physics and/or chemistry.

It's All Done With Photons

In some respects, it made sense to deal with light in this schizophrenic fashion, provided you knew when to shift gears. In relation to flow cytometry, we can treat the light coming out of a laser as a "beam," and confidently predict what will happen when we bounce it off a plane mirror or two. When we want to focus that "beam" to a small spot to illuminate the cell stream, however, we have to go beyond geometrical optics to diffraction theory to calculate the focal length(s) of the lens(es) needed to achieve the desired spot size and geometry. Even after the light and the cells interact, we can cover most aspects of light scattering using a wave model of light. Once we get to fluorescence, however, we're forced to confront photons, whether we like it or not, and we find that photons behave in ways which seem strange to people who have learned "classical" physics. We also quickly discover that problems associated with the detection and measurement of light are often best treated in terms of photons.

The truth is that light isn't sometimes waves and sometimes photons, it's sometimes waves and always photons. Geometrical optics and wave formulas work because photons, in large numbers, and over large (compared to atomic dimensions) distances, behave in ways that are, on the average, predictable and well modeled by equations describing waves. However, when you take "wave" phenomena such as interference to their limits, and set up an experiment in which half of your light should go one way and half the other, once you get down to detecting very small amounts of light, one photon at a time, you find that photons don't split in half; they either go one way or the other. So, you might ask, why don't we forget about all this wave stuff?

If you'd really like an answer to that question, you should, as the English Lit instructors say, compare and contrast two books by Richard Feynman. In Volume I of *The Feynman Lectures on Physics*[640], written for undergraduates at CalTech, he gives elegant descriptions of the behavior of light as a wave in a language that might best be described as mathematics with English subtitles. Mathematics is, of course, a language in which any serious student of physics must become fluent, and it is a useful language for the description of physical phenomena because a few lines of succinct formulae tell the whole story – to the fluent reader. The material on light in the *Lectures* is hard going, even for the budding physicists, who will have to keep studying math through their undergraduate and graduate careers just to be able to keep learning physics, which gets mathematically more difficult as it gets more advanced. It looks a lot like physics, but, as Feynman confides in some relatively non-mathematical asides, it doesn't play at the single-photon level; what does is the theory of **quantum electrodynamics**, the mathematics of which are far too complex to be taught to undergraduate physics students.

In the second book, a small and remarkable work called *QED*[641], which I invoked on pp. 4-6, Feynman provides equally elegant descriptions in diagrams and English, intelligible (and entertaining) to interested laymen, rather than in mathematical terms, of photons and electrons and their interactions, which, as explained by quantum electrodynamics, account for most of what happens in the physical world, excluding gravitation and radioactivity but specifically including all of chemistry and biochemistry. This book may or may not help physicists. They have to plow through the mathematics of classical physical optics, which provide a good enough approximation to much of what goes on in the real world to be useful for everyday work; then, if they want to work in the areas in which only quantum electrodynamics gives them the right answers, they have to go into that area in a mathematically rigorous way. On the other hand, a physicist I know told me that even physicists appreciated Feynman's habit of explaining complex physical phenomena without resorting to complicated math. Unfortunately for those of us who would just like to get a little bit more of a handle on what's going on in the instruments we use in our biomedical work, most books about physics are written by people who teach physics, and most people who teach physics see their primary mission in life as teaching physicists, and don't write books like *QED*. We have to take what we can get; for now, what we will get is back to photons.

Photons are particles that, unlike electrons, protons, and neutrons, have no **rest mass**; they are composed of pure electromagnetic energy, and the absorption and emission of photons by atoms and molecules is the only mechanism by which the atoms and molecules can gain or lose energy. Absorption and emission are **quantized**, that is, each discrete process by which an atom or molecule gains or loses energy is always associated with the same energy gain or loss, and therefore involves a photon of the same energy every time it occurs.

Aggregates of photons are detectable as **electromagnetic radiation**, which behaves like a wave traveling at the **speed of light** (c). The speed of light is approximately 3×10^8 meters per second in a vacuum, and less in materials. Some physicists in the Boston area have recently succeeded in slowing light to a speed I can easily beat on my bicycle; this is not yet of much practical interest, but it makes me feel as if I'm in great physical shape.

From the **frequency** (ν), in hertz (Hz) (formerly called cycles per second), or from the **wavelength** (λ), in meters, of an electromagnetic "wave," we can calculate the **energy** [E, in joules (J)] of a single photon, using the formulas

$$E = h\nu \quad \text{and} \quad E = hc/\lambda \, .$$

In these, h is **Planck's constant**, which is roughly 6.63×10^{-34} joule-seconds. A single photon coming out of an argon ion laser emitting at a wavelength of 488 nm has an energy of approximately 4.07×10^{-19} J. To get a whole joule out of a 488 nm laser, you'd need 2.45×10^{18} photons. Since 1 watt (W) is equal to 1 J/s (that's joule/second), a laser emitting 10 mW at 488 nm is putting out 2.45×10^{16} photons per second. Photon energies are higher at shorter and lower at longer wavelengths; a 325 nm (UV) photon from a helium-cadmium laser has an energy of 6.12×10^{-19} J, so 1 J of photons at this wavelength contains only 1.63×10^{18} photons, while, at the 633 nm (red) emission wavelength of a helium-neon laser, the energy of a single photon is 3.14×10^{-19} J, and there are 3.18×10^{18} photons/joule.

A Few Warm Bodies

The photons most readily accessible to man have, throughout history, been derived from **incandescent** sources, that is, objects that produce light solely by virtue of their temperature. The distribution of energies of photons emitted from an incandescent object shifts toward higher modal values as the temperature increases. At relatively low temperatures, emission in the infrared predominates; this is perceptible as **heat** rather than as light. At about 1000 degrees Kelvin (K) (or "1000 Kelvins" in SI), the object appears dull red; at about 1750 K, yellow. By the time the surface temperature of the sun, about 6000 K, is reached, the color is what we have become conditioned to as "white." Stars hotter than the sun appear bluish white to our eyes.

The physical theory that has been developed to deal with incandescence describes the behavior of an ideal radiating source called a **black body**, which absorbs all radiation falling on it and emits radiation at all frequencies with 100 percent efficiency. According to the **Wien displacement law**, the wavelength λ_m (in µm) at which maximum power is radiated from a black body at temperature T is given by $\lambda_m = 2898/T$. The related wavelength λ_m', at which the maximal number of photons are emitted per unit time (remember, there are more photons per joule at longer wavelengths) is given by $\lambda_m' = 3670/T$. Working out the numbers for the sun, we get maximum power output at 0.483 µm, or 483 nm; that's blue-green, but the maximum photon output is at 612 nm, which is orange. For the cooler but more accessible 3400 K photoflood lamp, we come up with maximum power output at 852 nm and maximum photon output at 1079 nm; both of those wavelengths are in the infrared. In other words, incandescent sources of any kind are going to give us a lot of heat along with our light, and we're going to have to use a pretty hot incandescent source, such as an arc lamp, if we want to get much light out in the blue, violet, and ultraviolet.

The attractive alternative, in our modern world, is to dispose of the warm bodies and use a **laser**, which does not depend upon incandescence to produce its light. I said on p. 50 that a fairly puny laser is actually brighter than the sun; the numbers to follow will prove it. The total solar power reaching the earth (actually, reaching the outer layer of the atmosphere) is about 1400 W/m^2; the 0.8 mW (800 µW) lasers in bar code scanners have a beam diameter of 0.8 mm and, therefore, a beam area of $5 \times 10^{-7}\ m^2$, and thus produce 1600 W/m^2. Will wonders never cease?

Polarization and Phase; Interference

Two characteristics of light that are most easily understood in terms of the wave model are **polarization** and **phase**. Light behaves as a **transverse** wave or, more precisely, as two perpendicular transverse **electric** and **magnetic** waves, both perpendicular to the direction of propagation. We can start trying to clarify this with Figure 4-2.

Wave motion is **periodic**; it repeats itself in both space and time. An electromagnetic wave has associated with it periodic changes in both the **electric field** and the **magnetic field**. Electric and magnetic fields are **vector fields**; that is, each has both a **magnitude** and a **direction**. In a light wave, the electric field vector **E** and the magnetic field vector **B** are mutually perpendicular, or orthogonal, and both are also perpendicular or orthogonal to the axis of propagation.

Using a Cartesian coordinate system, with the light propagating along the z-axis, we have the electric field vector oscillating along the y-axis and the magnetic field vector oscillating along the x-axis; this keeps the electric and magnetic waves perpendicular to one another and perpendicular to the axis of propagation. This arrangement has actually overspecified the system somewhat; all of the **E** vectors are lined up in a neat row in a single plane, as are all of the **B** vectors. This is the way things are in what we call **plane** or **linearly polarized** light.

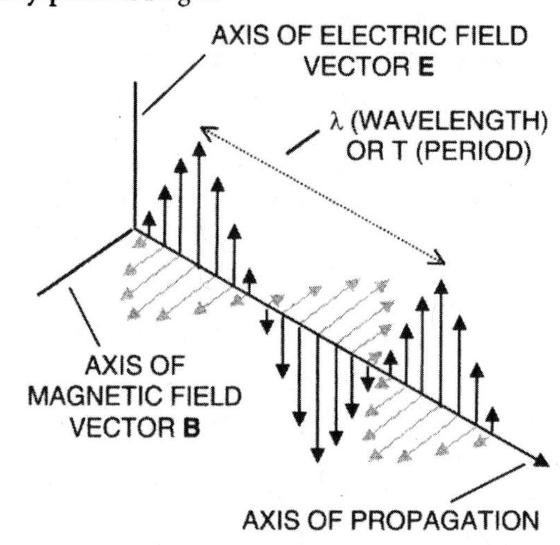

Figure 4-2. Light as an electromagnetic wave.

Since we know that the magnetic field vector is perpendicular to the electric field vector, and since most of the interesting things that happen when light interacts with matter relate directly to the electric field vector, we will, from this point on, keep track only of what's happening to the electric field or **E** vector to simplify things.

If you look at the axis of propagation in Figure 4-2, you might wonder why I identified the same dimension as both the **wavelength** and the **period**. I did because the wave can be considered as propagating in **space** or as propagating in **time**. Looking at the axis as a time axis, and the figure as representing successive values of the amplitudes of the **E** and **B** vectors at a fixed point in space, we can define the time interval, in seconds, between two successive occurrences of the maximum amplitude of the **E** vector as the **period** of the wave. The inverse of the period is the **frequency**, in Hertz, which indicates the number of complete **cycles** from maximum value to maximum value that occur in one second.

If you look at the axis of propagation as a space axis, the figure represents values of the amplitudes of the **E** and **B** vectors at different points in space at a fixed point in time. The distance between two points in space at which the vector is at maximum amplitude is the **wavelength**, measured in meters, or, more often, in **micrometers** (1 µm = 10^{-6} m; they used to be called **microns**), **nanometers** (1 nm = 10^{-9} m), or **Angstrom units** (1 Å = 10^{-10} m). The inverse of wavelength is also referred to as **frequency**; this, however, is a spatial, rather than a temporal frequency, and is measured in cycles per unit distance, rather than in cycles per second. The customary unit is the **wavenumber**; it denotes the number of cycles per centimeter, and is abbreviated cm^{-1}. Spatial frequency is often given the same symbol, ν, as is

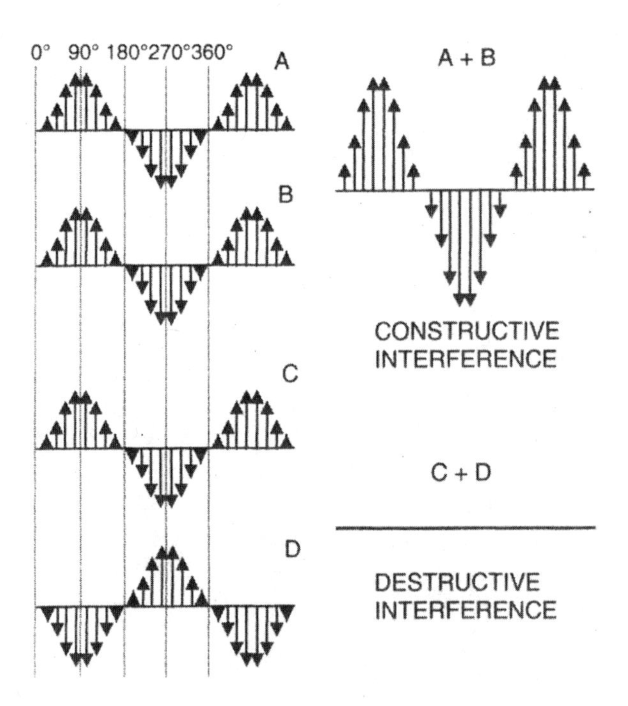

Figure 4-3. Constructive and destructive interference.

used for temporal frequency; as your high school physics teacher must have told you, check your units.

Each **cycle** of a wave can be represented as a full circle; a cycle doesn't have to be measured from peak to peak, but could be measured from trough to trough, or between any two places at which the amplitude and slope of the waveform are equal. The full circle corresponds to an angle of 360°, or 2π radians. Two waves of the same frequency, displaced from one another along the axis of propagation, are said to differ in **phase**. The phase difference is conveniently expressed as an **angle**, given in degrees or radians; phase differences are therefore restricted to values between 0° and 360° (or between -180° and +180°).

Two waves of identical amplitude with a phase difference of 0° (or 360°) are superimposable and add to one another, producing a wave with twice the amplitude; this is called **constructive interference.** Two waves with a phase difference of 180°, or π radians, cancel one another; this is **destructive interference.** The principle is illustrated in Figure 4-3; waves A and B are in phase, that is, they have a phase difference of 0°, and add; waves C and D have a phase difference of 180°, and cancel. Both **phase contrast** and **interference contrast** microscopy are based on interference.

Suppose we had two plane polarized waves of the same frequency and phase, but with different, orthogonal planes of polarization, such that the **E** vector of one wave oscillates along the x-axis, while the **E** vector of the other oscillates along the y-axis. The two **E** vectors would then add, as vectors add, to give us a new wave, with an **E** vector at a 45° angle ($\pi/4$ radians) to both the x- and y-axes. Note that this situation is different from the case illustrated in Figure 4-3, in which the waves were polarized in the same plane. If the two waves were equal in amplitude, but different in phase by 90° or $\pi/2$ radians, they would add to produce a wave with an **E** vector that would rotate around the z-axis; this resultant wave would be **circularly polarized.** Circularly polarized light is somewhat crudely illustrated in Figure 4-4.

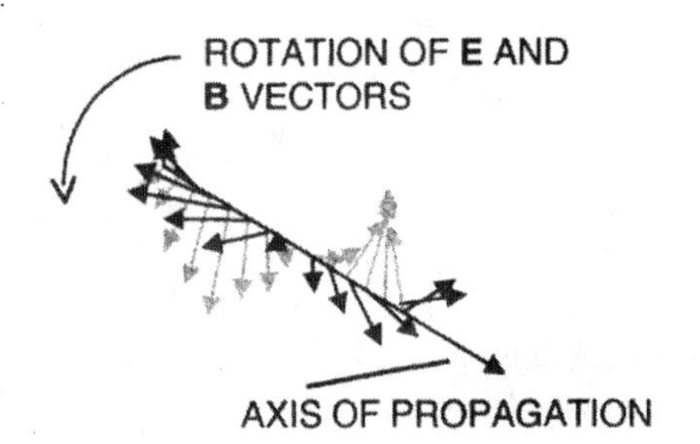

Figure 4-4. Circularly polarized light.

Two waves with perpendicular E-vectors that are different in amplitude and out of phase by 90° produce **elliptically polarized** light. Circular and elliptical polarization can be either **left-** or **right-handed**, depending upon the direction of rotation of the vectors. As long as the **E** and **B** vectors remain perpendicular to one another, they can rotate freely in their mutual plane. So, we could have a situation in which both the electric and the magnetic field vector, while remaining perpendicular to one another in the x-y plane, moved through a 360° range of angles with the x- and y-axes; this would give us **unpolarized** light.

Light Meets Matter: Rayleigh and Mie Scattering

In a vacuum, light, whether we're thinking of it as waves or photons, travels in straight lines at a velocity, *c*, of approximately 3×10^8 m/s, carrying with it its oscillating electric field. Just get some matter, even a few atoms worth, into the picture, though, and things change. Even in unperturbed atoms and molecules, there is some separation of positively charged protons and negatively charged electrons. When exposed to an oscillating electric field, these positively and negatively charged atomic constituents move in opposite directions in response to the field, becoming alternately closer together and farther apart, and thus giving rise to oscillating fields, i.e., electromagnetic waves, of their own, with the same frequency as the light which initiated the process, but not necessarily either of the same phase or propagating in the same direction. The resulting phenomenon is described as **light scattering**, although the "scattered" light is actually new photons, rather than old ones that have changed direction (p. 5). So far, so good, but where does the quantization I mentioned before fit in? The best answer is that it is conspicuous by its absence.

Among the interactions between light and matter, scattering is perhaps the most common and certainly the most casual. In more intimate encounters, in which photons are absorbed, atoms and molecules tend to be finickier about the energies of the photons involved, and quantization is more obvious. Even in these cases, however, we are dealing with probabilities. If we look at the absorption spectrum of a compound, we find one or more peaks corresponding to wavelengths, or frequencies, or photon energies at which the molecule involved is most likely to absorb photons. Although absorption of photons of other energies is less likely, it is not impossible, and the likelihood of absorption increases as the photon energy approaches regions near absorption peaks. The peaks can also be dealt with in a mechanical model as representing **resonant frequencies** of the molecule; transfer of energy to the molecule from an incident wave becomes more efficient as the frequency of the wave approaches a resonant frequency.

Scattering exhibits this characteristic as well. Materials scatter light at wavelengths at which they do not absorb. Atoms and small molecules (dimensions less than 1/10 wavelength) scatter, but do not absorb light in or near the visible region (let's consider this wavelength range to be 350-850 nm), and therefore appear transparent to the eye; they typically have absorption bands in the ultraviolet below 300 nm. As the wavelength of incident light decreases, approaching these absorption bands, the amplitude of oscillation induced in the intramolecular dipoles formed by charge separation increases, which means that the intensity of light scattering increases. The intensity of such **Rayleigh scattering** is directly proportional to a property of the scattering molecules called **molecular polarizability** (the term polarizability as used here relates to electric dipole formation, not to polarized light), and inversely proportional to the fourth power of the wavelength of the incident light. The cloudless sky appears blue in sunlight because gas molecules in the atmosphere scatter more light at shorter (i.e., blue and violet) than at longer wavelengths.

Polarization can be produced by Rayleigh scattering of unpolarized light. The **E** vectors of both the incident and the scattered light oscillate in planes perpendicular to the axis of propagation of the incident light. Light scattered at 90° to the incident light, however, itself propagates in a plane perpendicular to the axis of the incident light, i.e., in the same plane in which its **E** vector oscillates. If you are having trouble visualizing this, we can go back to Cartesian coordinates, and you can lend me a hand, let's say your right one. Cock your thumb and finger at right angles (go ahead, make my day), and then extend your middle finger perpendicular to the plane defined by your thumb and forefinger (the old hand has gotten a little stiff since high school physics, eh?).

Suppose the incident light propagates along the z-axis, represented by your middle finger. The **E** vectors of the incident and of the scattered light are oscillating in the plane defined by the your thumb and index finger, which respectively represent the x- and y-axes. Now imagine your thumb and index finger alternately lengthening and contracting, or move them back and forth, to represent the oscillation of the **E** vectors. Looking in the direction of the incident beam, i.e., end-on at your middle finger, you could readily perceive changes in length of either your thumb or index finger.

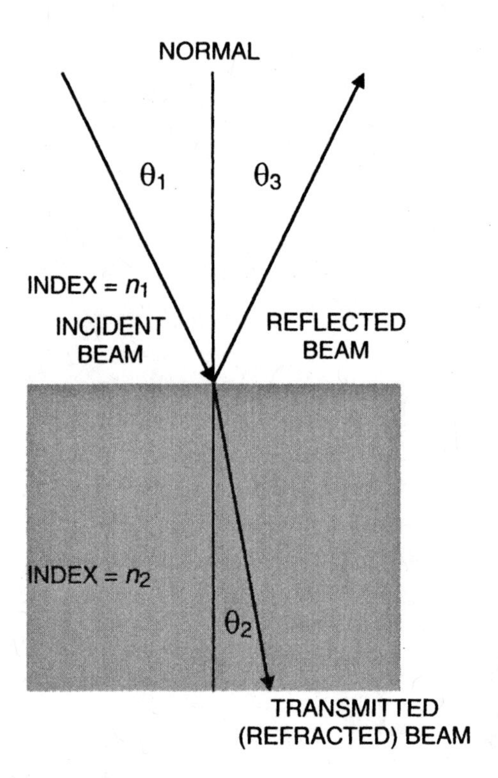

Figure 4-5. Reflection and refraction of light at a surface.

However, if you looked in the direction of light scattered at 90° from the incident beam, say, along the x-axis, or end-on at your thumb, you would be able to perceive a change in length of your index finger much more readily than a change in length of your thumb.

This is a handy way of getting you to appreciate one of the basic rules (of thumb?) of this silly game, which is that oscillations of the **E** vector in the direction of propagation don't count as light. In our coordinate system, with incident light propagating along the z-axis, when we look at 90° scattered light along the x-axis, we only see, or detect, light associated with oscillations of the **E** vector in the y-direction. This light, according to the definition in the previous section, is therefore linearly polarized. If you don't trust your right hand, you can see real polarization occurring via this mechanism by looking through a polarizing filter or polarized sunglasses at the blue sky at angles of 90° and 180° from the sun. At 90°, rotating the filter produces a noticeable change in brightness, because much of the light from this direction is polarized; at 180°, little or no effect is seen.

In Rayleigh scattering, the intensities of forward (near 0°) and back (near 180°) scattered light are nearly equal, because the scattering particles are so small that there can be only small differences in phase between light scattered from any two points within a particle. For larger particles, substantial phase differences may exist, leading to **interference**. Over a range of particle sizes from about 1/4 wavelength to tens of wavelengths, increasing amounts of light are scattered in the forward direction; this is **Mie scattering**, named for

Gustav Mie, who worked out the relevant theory. Flow cytometric forward scattering measurements for cell "sizing" are based on Mie's analysis; Mie scattering itself is a complex phenomenon influenced by numerous factors in addition to particle size, and we will discuss it and them in detail later.

A Time for Reflection – and Refraction: Snell's Law

Both **reflection** and **refraction** of light at surfaces result from light scattering, and both, like Mie scattering, involve interference. These phenomena can be described by either wave or photon models. The reflection and refraction of light at a plane surface are illustrated in Figure 4-5.

Incident light strikes the surface at an angle θ_1 to the normal. The **angle of reflection,** θ_3, is equal to the **angle of incidence,** θ_1, regardless of the material of which the surface is made. The angle θ_2 at which light is **transmitted** through the surface, does, however depend upon the composition of the material. According to **Snell's law of refraction,**

$$n_1 \sin \theta_1 = n_2 \sin \theta_2 ,$$

where n_1 and n_2 are the **refractive indices,** respectively, of the materials through which the incident and transmitted beams pass. In the figure, the incident beam is shown in air, which has a refractive index very close to 1 (about 1.0003).

According to wave models, the light reflected from a surface is light scattered backwards by the layers of molecules of the scattering material nearest the surface. Reflected light is observed at the angle $\theta_3 = \theta_1$ because light scattered at all other angles is removed by destructive interference. Light scattered forward, i.e., in the direction of transmission, by the surface layers is parallel to the incident beam, but lags behind the incident beam in phase, and adds to the now somewhat attenuated incident beam to produce a resultant wave with a slight phase lag. This wave is scattered from the next deeper layer of molecules, with the result that the light transmitted through this layer is still further retarded in phase, etc. The cumulative phase lags result in the light wave appearing to travel through the material at a velocity lower than the speed of light in a vacuum, c ($\approx 3 \times 10^8$ m/s); the velocity of light in a material of refractive index n is c/n. Fine, but why does the transmitted light travel at an angle different from the angle of incidence?

The wave model gives us both highly complicated and simple explanations, which come down to the same thing. The complicated explanation states that Maxwell's equations, which describe electromagnetic radiation beautifully if you're fluent in vector calculus, have to be satisfied at the surface as well as on either side of it; you then wade through three pages of formulas and find out that the transmitted light has to change direction. The simple explanation considers a wave entering the surface of the material.

As before, we'll have the wave coming in at the angle θ_1. This is a transverse wave; its "crests," or wavefronts, can be treated as planes that are parallel to one another and perpendicular to the direction in which the light is traveling. Suppose that the wavelength, i.e., the perpendicular distance between the wavefronts, is 600 nm in air (refractive index $\approx$1) and that the material has a refractive index of 1.2. The frequencies of the incident wave in air and the transmitted and scattered light in the material will be the same. The velocity of a wave is the product of the frequency and the wavelength. In air, the velocity is close to c, and the wavelength is 600 nm. In the material, the velocity is c/n, $= c/(1.2)$. Since the frequency remains constant, the wavelength in the material must be 600/(1.2) or 500 nm.

As the wave enters the surface, then, the wavefronts have to be 600 nm apart on the outside and 500 nm apart on the inside. Since the wave enters the surface, and is transmitted, at angles, the distance between wavefronts **along the surface, on the outside,** will be greater than 600 nm; it will be ($600/\sin \theta_1$) nm. The distance between wavefronts **along the surface, on the inside,** will be ($500/\sin \theta_2$) nm. Since these two distances must be equal, the angles of incidence and transmission will be unequal unless both are zero. We can bend the wavefronts, but we can't break them.

The explanations of reflection and refraction based on the photon model (see *QED*[641] for details) also, incidentally, strip some of the mystery from a well-known, but less well understood, law of optics known as **Fermat's principle of least time.** This principle can be used, without benefit of photons, to find the directions in which reflected and refracted light go, based on the light always picking the path that takes least time to traverse. There are two problems with Fermat's principle; one is that, just when you least expect it, light will turn out to take the path that takes the most time to traverse, the other is that it seems to require the light to have road maps or some other advance information.

When we're looking at light as photons, we can't say any given photon will go any given way, we can only compute the probability that a photon will go in any particular direction. As it turns out, when we consider reflection and refraction, we end up with a substantial probability that photons will go in the directions we have otherwise found reflected and refracted light to prefer, and a very low probability that they will go in other directions. The neat thing about this is that probabilities for most trajectories cancel out, leaving nonzero probabilities only for trajectories that take approximately the same time to traverse. This time is usually the shortest possible time, but occasionally the longest, in accordance with Fermat's principle. The photons don't need road maps; they just roll dice at every intersection and get there anyway.

Polarization by Reflection; Brewster's Angle

Light reflected from a surface is actually scattered from the material beneath the surface as well as from material at or near the surface. This light is linearly polarized, for the same reason that Rayleigh scattered light at 90° from the incident beam is linearly polarized; that is, oscillations of the **E** vector of the reflected light are only detectable in the direction perpendicular to the plane defined by the axis of propagation of the incident light and the normal to the surface. The maximum polarization occurs when the angle between the reflected light and the transmitted light is 90°. At that point, $\theta_3 + \theta_2 = 90°$, and, since sin (90°- x) = cos x, Snell's law gives us $(\sin \theta_3/\cos \theta_3) = n_2/n_1$, making $\theta_3 = \tan^{-1}(n_2/n_1)$. This value of θ_3 is known as **Brewster's angle**. Each time light strikes a surface at Brewster's angle (remember that $\theta_1 = \theta_3$), the reflected light is linearly polarized normal to the plane of incidence, and the transmitted light therefore contains a greater proportion of light polarized parallel to the plane of incidence. The end windows of a laser plasma tube are typically affixed at Brewster's angle to the optical axis of the tube. Since light emitted from the laser has effectively been reflected many times by the mirrors at opposite ends of the laser cavity, thus passing through these **Brewster windows** many times, the laser light is highly polarized.

Dispersion: Glass Walls May Well a Prism Make

Like the atmosphere, the "transparent" materials such as glass and quartz out of which we make optical gadgets do not absorb much visible light, but scatter increasing amounts of incident light as the wavelength gets nearer their absorption peaks in the ultraviolet. As the intensity of scattering increases, more phase lag is introduced into the transmitted beam; light thus appears to travel more slowly through the material. The refractive index therefore is higher at shorter wavelengths than at longer ones, meaning that the angles at which light of different wavelengths or colors are transmitted will differ, with blue and violet light more strongly deviated than orange and red. The change in refractive index with wavelength is called **dispersion**; it is the mechanism by which a **prism** forms a spectrum from incident white light. The degree of dispersion is different for different types of glass; high-dispersion glasses are used to make prisms. Dispersion is also the basis for **chromatic aberration**, an undesirable characteristic of lenses arising from different colors of light being focused at different distances from the lens.

Interference in Thin Films

A small fraction of the incident light is reflected from any interface between two materials of different refractive indices. When we pass monochromatic light through slabs of different thickness made from the same material, and measure the amount of light reflected, we find considerable variation with thickness, because the reflections from the front and back surfaces will, depending upon the thickness, interfere constructively or destructively. The difference in thickness over which a change from constructive to destructive interference occurs is less than a wavelength, so the effect is usually much more noticeable in thin layers of material than in thick slabs. A layer of material of a given thickness will reflect colors selectively, because there will be constructive interference at some wavelengths and destructive interference at others. Which wavelengths are maximally reflected will also depend upon the angle of incidence of the light, because this will determine the angle at which light reflected from the back surface is transmitted through the medium, and, therefore, the distance traveled by this reflected light.

Interference effects produce the iridescent colors reflected from thin films of oil on water; under much more carefully controlled conditions, thin layers of dielectric material can be deposited on optical components to produce **interference filters, antireflection coatings**, and **mirrors.** A typical antireflection coating for a camera lens, for example, is made by depositing a material such as magnesium fluoride, with a refractive index intermediate between the indices of air and glass, on the lens surface. Destructive interference at the wavelength λ will occur if the thickness of the coating is $\lambda/4$; the coating thickness used is generally about 125 nm, which is chosen for maximum efficacy against reflections in the green region of the spectrum, where the eye is most sensitive. The coating appears purple since, by design, it reflects very little green light (instead allowing it to be transmitted; energy is conserved) but does reflect somewhat more light at the violet and red ends of the spectrum.

Interference filters are usually made of several "sandwiches" of dielectric material separated by spacers; the structure of the filter is determined by the location and width of the **passband** and the degree of transmission of wanted light and reflection of unwanted light required. Dielectric thin films can also be used to make highly reflective **mirrors**, such as those used in lasers. Operation of lasers at some wavelengths requires that the mirrors reflect more than 99.5 percent of incident light. While such high reflectivity can be achieved fairly readily, it is generally possible to maintain it over only a small portion of the spectrum, and thus necessary to use different sets of mirrors when a laser is operated in different spectral regions. A spectrally selective mirror is said to be **dichroic**; in cytometry, the term is generally reserved for a spectrally selective mirror used at or near a 45° angle of incidence to separate light into two spectral bands.

Interference and Diffraction; Gratings

Diffraction, in its broadest sense, describes any departure from the predictions of geometric optics. Using geometric optics, for example, we would predict a sharp edge illuminated by a point source to cast a sharp shadow; instead, we see alternating bright and dark fringes at the periphery of the shadow of such an object. Light passing through one or more slits or apertures also produces fringe patterns. Diffraction results from interference, and it is diffraction that ultimately limits the resolution of optical systems; we will have more to say on this subject later.

We have already noted that selective transmission and reflection of different wavelengths can be achieved using thin layers of material. Similar effects can be obtained by passing light through an array of small slits or by reflecting light from a surface containing closely spaced (i.e., separated by a few wavelengths) grooves. Such structures are referred to as **gratings**; they are useful because they can provide greater spectral separation than can be obtained using prisms and because the spectra they produce are linear, whereas those produced by prisms are not. Precisely ruled gratings find application in spectrophotometers, while less precisely ruled ones are used in costume jewelry. Although both old-fashioned phonograph records and compact discs can act as gratings to disperse light, their primary esthetic appeal is, or should be, auditory.

The "smooth" surface of a crystal does not behave as a grating when illuminated with visible light, because the periodic structure of atoms or molecules in a crystal has dimensions much smaller than the wavelength of the light. Crystalline materials, however, will produce diffraction patterns in response to electromagnetic radiation at appropriately shorter wavelengths, e.g., x-rays, and this provides the basis for x-ray crystallographic determination of molecular structures.

Optical Activity and Birefringence

Although molecular dimensions are small compared to the wavelengths of visible and near-visible light, the scattering of light at these wavelengths is highly dependent upon structural characteristics of the molecules. The presence of asymmetric carbon atoms and/or helical structure in a small or large molecule, and the presence of oriented asymmetric macromolecules in a material, typically make the molecule or material scatter light of different polarizations differently.

These molecular properties account for **optical rotation**, a situation in which the plane of polarized light transmitted by a material is rotated with respect to the plane of polarization of incident light. They are also the basis for **birefringence**; a birefringent medium exhibits different values of refractive index for different polarizations of incident light. Both optical rotation and birefringence may, when measured in whole cells, provide some information about internal macromolecular structure.

Some molecules exhibit little or no birefringence when randomly oriented but become strongly birefringent when aligned, either by mechanical forces, e.g., in a flowing fluid, or by an electric field. This has provided a basis for a number of practical inventions, ranging from Edwin Land's original sheet polarizer (whence the Polaroid Corporation's name) to the ubiquitous liquid crystal display.

Matter Eats Light: Absorption

As you are now aware, a lot of optical phenomena can occur without benefit of absorption. Some of those in which we are most interested, however, such as fluorescence, cannot. Wave models don't add much to discussions of either absorption or fluorescence, both of which are conveniently understood only in terms of relations between molecules and photons.

According to quantum mechanics, molecules can only absorb energy as quanta, or photons, and any given type of molecule can only absorb photons with energies in specific ranges. Thus, a molecule can only exist in a countable number of discrete **energy states** (I said countable rather than finite because the number is potentially infinite, but we generally don't have to worry about more than a few hundred). Absorption of a photon by a molecule in a minimum-energy **ground state** raises the molecule to a higher-energy **excited state**. A change from one state to another is called a **transition**.

There is a hierarchy of energy states. The **total energy** content of a molecule is the sum of **electronic, vibrational, rotational, translational, electron spin orientation**, and **nuclear spin orientation** energies; the energies of photons associated with these different types of transitions differ by orders of magnitude. Electronic transitions involve absorption in the near infrared, visible, and ultraviolet, with photon energies ranging from about 2×10^{-19} J at 1 μm to about 10×10^{-19} J at 200 nm. Changes in vibrational energy states accompany absorption in the infrared; the photon energies range from about 2×10^{-21} J at 100 μm to about 7×10^{-20} J at 3 μm. Thus, the energies involved in vibrational transitions are 1-2 orders of magnitude smaller than those involved in electronic transitions.

Rotational transitions result from absorption of microwave radiation, with photon energies 2 to 3 orders of magnitude lower than those involved in vibrational transitions. Molecular translation requires even lower energies. We associate molecular motion with heat; while many of us now use quantized absorption of microwave energy to increase the motion of water molecules, many more of us rely on thermal energy to heat food. Even in microwave ovens, some of the cooking is accomplished by the transfer of thermal energy from water molecules to other molecules that do not absorb microwave energy.

A substantial amount of molecular motion can be produced solely by thermal agitation even at room temperature. At any given Kelvin or absolute temperature T, the ratio of the numbers of molecules, n_{upper} and n_{lower}, in upper and lower energy states associated with an energy difference ΔE can be calculated from the **Boltzmann distribution law**,

$$n_{upper} / n_{lower} = e^{-\Delta E/kT} ,$$

where k is **Boltzmann's constant** (1.38×10^{-23} J/K). Let's plug in some figures for different kinds of transitions. My microwave oven runs at 2450 MHz, or 2.45×10^9/s; the corresponding photon energy (p. 103) is 1.62×10^{-24} J. At a room temperature of 300 K, n_{upper}/n_{lower} for rotational transitions of this energy (i.e., $\Delta E = 1.62 \times 10^{-24}$ J) turns out to be 0.9996. In other words, heat energy at this ambient temperature raises almost as many molecules to the upper state as remain in the lower state.

Next, we'll consider a vibrational transition produced by another high-tech device sometimes used for heating water, in this case a carbon dioxide laser emitting at a wavelength of 10.6 μm. The photon energy is 1.88×10^{-20} J, and n_{upper} / n_{lower} at 300 K is .0107. Thus, the thermal agitation present at room temperature is sufficient to put only about one percent of molecules into an upper vibrational state.

Absorption of visible light produces electronic transitions. A (red) He-Ne laser operating at 633 nm emits photons with an energy of 3.14×10^{-19} J. At 300 K, the ratio n_{upper}/n_{lower} for this transition energy is 1.2×10^{-33}. In other words, at room temperature, it is close to impossible for thermal energy to result in electronic excitation. At 6000 K, the ratio is still only 0.023. Among other things, this means that if you're trying to get things to light up by heating them, you have to get them a lot hotter than room temperature. More to the point, however, it means that if we want to detect and quantify molecules based on their absorption and/or emission of light, we don't have to worry too much about the molecules getting into excited states due to thermal agitation.

Absorption: Counting the Calories

Since we've been talking about heating water, we'll resurrect the **calorie** (just call it cal), sometimes known as the **small calorie**, which is the amount of heat energy required to raise the temperature of 1 g (or 1 ml) of water by 1 K. There are 4.184 J in a calorie. For those of us who diet, there are no small calories, only large ones; the calories we count are really **kilocalories** (kcal), each of which is equal to 4184 J. This digression serves as more than a coffee break; it gives us the numbers we need to be able to relate energies in joules per photon to the kilocalories per mole commonly used to describe energies of chemical reactions.

There are 6.02×10^{23} molecules in 1 mol, and 4184 J in 1 kcal, so the energy change, in kcal/mol, resulting from the absorption of 6.02×10^{23} photons (a "mole" of photons, officially known as 1 **einstein**) by 1 mol of a substance can be calculated by multiplying the photon energy (in J) by $(6.02 \times 10^{23})/4184$, or 1.44×10^{20}.
The photon energy in J corresponding to an energy change given in kcal/mol can be found by multiplying by the reciprocal figure, 6.95×10^{-21}.

From this, we find that absorption in the microwave oven described above is producing an energy change of only 2.33×10^{-4} kcal/mol; that's low calorie cooking indeed. Absorption of 10.6 μm radiation from the CO_2 laser is good for 2.71 kcal/mol. How do you handle a hungry molecule? Try some visible light; at 633 nm we end up with 45.2 kcal/mol. We all live by converting ADP to ATP and vice versa, which only takes 7.3 kcal/mol; but we all live off plants, which accomplish this by photosynthesis, in which chlorophyll absorbs red (680 nm) light, good for about 42 kcal/mol, and, ever obedient to the laws of thermodynamics, wastes a chunk of it while providing us with our meal tickets.

A Selective Diet

Molecules are finicky eaters; those photon energies or wavelengths at which a given molecule will absorb are determined by the structure of the molecule, according to **selection rules** dictated by quantum mechanics. The electrons we're trying to excite reside in **orbitals**. Any given molecular orbital may contain up to two electrons, but, according to the **Pauli Exclusion Principle**, the two electrons must have opposite spins. Thus, an electron cannot go from a low-energy orbital ground state to a higher-energy orbital excited state if there is another electron of the same spin already present there.

Molecular symmetry, or the lack thereof, can affect absorption. Because transitions require a precise spatial relationship between the molecular dipole moment and the E vector of the light that is being absorbed, asymmetric or oriented molecules may show **dichroism**, absorbing light of different polarizations differently. Circular and linear dichroism in absorption are analogous to optical rotation and birefringence in scattering.

Speaking of scattering, in the discussion of that phenomenon, I mentioned that most small molecules absorb light in the ultraviolet. How, then, do we make molecules absorb at visible frequencies? Obviously, we have to arrange for electronic transitions to involve smaller changes of energy. One of the more tried and true methods involves making large systems of conjugated double bonds, typically including some atoms other than carbon. The π orbitals in such molecular structures are formed by hybridization of p orbitals from the individual constituent atoms; one consequence of this is that the energy difference between electronic excitation states becomes smaller, shifting absorption to longer wavelengths. Most of the dyes we use, fluorescent or otherwise, contain condensed and/or multiple heterocyclic rings. The people who design dyes have gotten quite good at tailoring their spectral characteristics.

Of course, if a material contained a whole lot of electrons free to run around loose, it would be apt to exhibit strong absorption over a broad range of wavelengths. This type of electronic structure is found in metals (and semiconductors), and accounts for both their broadband absorption and their high reflectivity; very high electric field intensities are produced in the surface layers of molecules of a metal object, resulting in intense scattering. Materials with discrete absorption bands also show changes in scattering characteristics near their absorption bands; this classically was referred to as **anomalous dispersion**.

The Chance of a Lifetime

In addition to being finicky eaters, molecules are fast eaters; absorption associated with electronic transitions occurs in about 1 **femtosecond**, i.e., 10^{-15} s. According to the **Franck-Condon Principle**, this time is too short for any nuclear displacement to occur. What does occur is charge displacement, or a change in the molecular dipole moment. Once absorption has occurred, things take on several aspects of a Roman orgy, because, depending upon how you look at it, molecules either don't keep down what they've absorbed or don't remain in excited states for long. The length of the average **lifetime** for a molecule in the excited state is dependent upon the processes by which the molecule can lose the absorbed energy, a subject we will consider further in our discussion of fluorescence.

The **uncertainty principle** of quantum mechanics tells us, among other things, that the more rapidly the energy of a system is changing, the less precisely we can define its energy. Thus, **excited states with long lifetimes are associated with narrow absorption peaks, whereas excited states with short lifetimes are associated with broad absorption peaks.** We will presently find out that this aspect of the uncertainty principle will limit how much we can do in the way of designing multiple fluorescent probes with minimal spectral overlap.

Spinning a Tale of Degeneracy

Discrete states of a molecule with the same energy content are said to be **degenerate**. In some cases, an external influence, such as an applied electric or magnetic field, can **resolve** degeneracy, producing **splitting**, i.e., the appearance of two or more spectral lines where there was only one before. The production of splitting by an applied magnetic field is known as the **Zeeman effect**. Applied magnetic fields increase the energy change associated with changes in orientation of electron and nuclear spins. Even in strong magnetic fields, however, we're talking about pretty low energies, down in the radio frequency bands between a few dozen and a few hundred MHz. This is the region in which **nuclear magnetic resonance** (NMR) and **electron spin resonance** (ESR) spectroscopy, which measure absorption related to spin orientation changes, are done.

Since the energies involved are so low, there are nearly equal numbers of molecules present in upper and lower states (consider the calculation for microwave absorption on the previous page). Thus, techniques such as NMR, which measure absorption in this energy region, can only detect absorption by a few molecules among many billions, and thus are relatively low in sensitivity. This is why NMR, which has become indispensable as a tool for qualitative analysis and structure determination, is not usable for the detection of very low concentrations of substances.

Facing Extinction: Cross Section and Optical Density

At this point, you are probably thinking that what you really needed to know about absorption was how to use it to measure things, and that all this business about molecules sucking up photons leaves you pretty far from that knowledge. It's closer than you may think. Suppose we have a slab of material, the thickness of which is d cm, containing n molecules/cm^3 of some absorbing substance. Moving closer to reality, we could make this slab a cuvette d cm across, filled with a solution of the absorber at a concentration of a mol/dm^3 (that's the the official SI unit for moles/liter; we'll use M). We would then have $n = \mathrm{N}a/10^3$, where N is Avogadro's number.

Beer's law, also called the **Beer-Lambert law**, relates the intensity of light entering the cuvette, I_0, and the intensity of light transmitted through the cuvette, I, by the formula

$$\ln (I_0/I) = \sigma n d,$$

where σ is a molecular property of the absorber called its **absorption cross section**. The cross section is expressed as an area (in this case, in cm^2); this area is not the actual physical area of the molecule but, rather, the area over which it will act as a barrier to the passage of light. Beer's law can also be stated as

$$\ln (I_0/I) = \alpha C d,$$

where C is the concentration and α is the **absorption coefficient**, which, like the cross section, reflects the capacity of the absorbing substance to absorb light.

The units here get a little sticky. If we have n in molecules/cm^3 and d in cm, we have to have σ in cm^2 to get things to work out. Since α is often represented in cm^2/mol, C must be in mol/cm^3 rather than in M (mol/dm^3), meaning that $C = a/10^3$. Converting to decimal logarithms gives

$$\log_{10} (I_0/I) = \varepsilon a d = A.$$

Here, A is the **absorbance**, or **optical density (O.D.)**, a dimensionless quantity, and ε is the **decadic molar extinction coefficient**, represented in units of dm^{3}mol^{-1}cm^{-1}. The preferred SI units for the extinction coefficient, concentration, and thickness are m^2/mol, mol/dm^3, and mm, but a confusing variety of units are used in the literature, and, if you're not careful, your calculations may go off by powers of ten.

Absorbance or O.D. units are convenient to use because, being logarithmic, they add instead of multiplying. An object with an O.D. of 1.0 absorbs 90% of the incident light. If another object with an O.D. of 1.0 is placed in the path of the 10% of light transmitted by the first object, 10% of this light, or 1% of the light incident on the first object, is transmitted by the second object. Thus, two objects with O.D. 1.0 are equivalent to one object with O.D. 2.0. The additive nature of absorbance units also makes it possible to express the absorbance of a mixture of substances at any given wavelength as the sum of the absorbances of the components of the mixture at that wavelength.

Absorption cross sections and absorption or extinction coefficients are, obviously, functions of wavelength, and are calculated from spectrophotometric measurements of known concentrations of absorbing materials in cuvettes of known path lengths. Note, however, that spectral curves often show the percentage of light absorbed, or the percentage of light transmitted, on a linear scale, rather than absorbance on a logarithmic scale.

The cross section can be explicitly related to the extinction coefficient by combining the several versions of Beer's law. Since $\ln x = 2.303 \log_{10} x$, we get

$$\sigma = 2.303\varepsilon a/n = 3.82 \times 10^{-21}\varepsilon,$$

which allows us to calculate the cross section from the extinction coefficient and vice versa. We can take an example

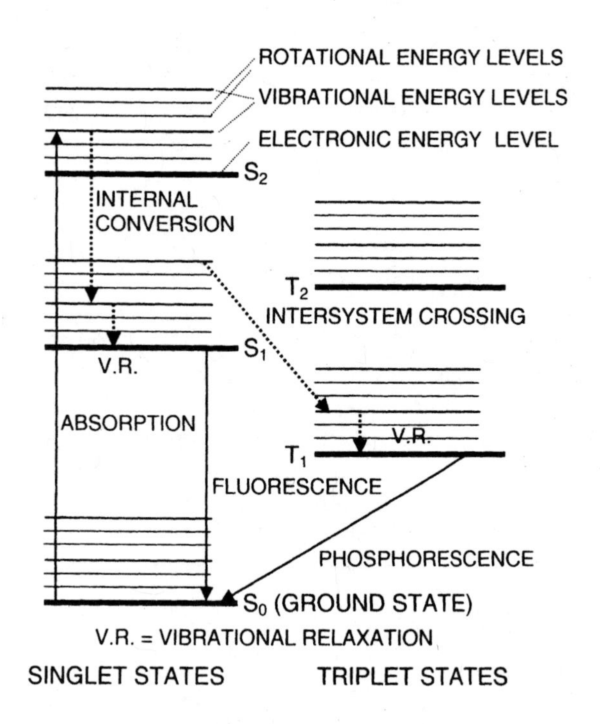

Figure 4-6. The Jablonski diagram of electronic energy levels, or states, and transitions. Solid arrows show radiative and dotted arrows nonradiative transitions.

from Lakowicz[23], who calculates, given that the value of ε for anthracene at 253 nm is 160,000 $dm^3mol^{-1}cm^{-1}$, that the absorption cross section at this wavelength is 6.1×10^{-16} cm^2 or, if you prefer, 6.1 $Å^2$; recall that 1 Å = 10^{-10} m = 10^{-8} cm. The actual area of an anthracene molecule is about 12 $Å^2$; thus, at 253 nm, an anthracene molecule absorbs about half the photons that come its way.

Unexciting Times: Emigrating from the Excited States

Absorption of UV or visible light leaves us with molecules in electronic excited states. There are a number of mechanisms by which they can get out of those excited states; they are often illustrated diagrammatically as was first done by Jablonski in 1935[1129]. A Jablonski diagram is shown in Figure 4-6. At a minimum, absorption involves a **transition** from the **electronic ground state,** known as S_0, to the first electronic **excited state,** S_1, but there's usually more to it than that. We have previously (pp. 109-10) discovered the hierarchy of energy states, and noted that vibrational transitions only require 1/100 to 1/10 as much energy as electronic transitions. One consequence of this is that each electronic state has associated with it a set of vibrational states. Immediately following absorption, the molecules involved are likely to be in higher vibrational excited states associated with S_1 or, if they have ordered New York cut photons, with S_2 (some of the molecules absorb enough energy to make it to the second electronic excited state).

In general, the excited molecules "shake off" their excess vibrational energy and, if they are in S_2, their excess electronic energy, in about 1 picosecond (10^{-12} s) by mechanisms called **internal conversion** and **vibrational relaxation,** which are **nonradiative transitions,** i.e., changes in energy level that are not accompanied by the emission of photons. The excess energy is lost either to other vibrational modes of the excited molecule or, through collision or radiationless transfer, to other molecules, thus ultimately being converted to heat.

This leaves the excited molecules in the lowest vibrational energy level of the first electronic excited state, S_1, trying to book passage back to the ground state S_0, which can be reached by several alternate routes. From our point of view, the route taken will determine whether getting there is all the fun, half the fun, or no fun at all.

What we would generally like the molecules to do is fluoresce, emitting all or some of their remaining excitation energy as photons that we can detect in our instruments. Fluorescence, however, is only one of several mechanisms by which **emission** can occur. **Luminescence** encompasses **fluorescence** and **phosphorescence,** both of which are types of **spontaneous emission.** There is also **stimulated emission,** which is the basis for laser operation. Then, there are nonradiative mechanisms through which the energy can be lost, including internal conversion, **resonance energy transfer,** various types of **quenching,** and **bleaching.**

Fluorescence: Working the Stokes Shift

If we get lucky, molecules that have returned, by internal conversion and vibrational relaxation, to the lowest vibrational state of S_1 will get themselves back to the ground state S_0 in relatively short order by purging themselves of photons, a process we have come to know and love as **fluorescence.** In almost all cases, the energy content of the photons emitted will be lower than the energy content of the photons originally absorbed, for two reasons. First, as we have just mentioned, some of the absorbed energy is lost by internal conversion before fluorescence occurs. Second, although fluorescence emission will get the molecules back to S_0, they will often be in higher vibrational states of S_0, and will lose a little bit more energy by vibrational relaxation in the 10^{-12} seconds following fluorescence emission, arriving back at the lowest vibrational state of S_0. Thus, fluorescence emission will occur at longer wavelengths than the absorption that preceded it.

The wavelength difference between the absorption or excitation maximum and the emission maximum is known as the **Stokes shift,** after George Stokes, who described and named fluorescence in the mid-1800's. The term fluorescence was derived from fluorspar, which is fluorescent, by analogy to opalescence, another optical effect named for a mineral[642]. The material Stokes observed was a solution of quinine; the light source was sunlight, the excitation filter a dark blue stained glass window, and the emission filter a glass of white wine. I am not making this up.

If we look at a fluorescence spectrum, such as that shown in Figure 4-7, we generally see that the emission spectrum is shaped like a mirror image of the absorption (or excitation)

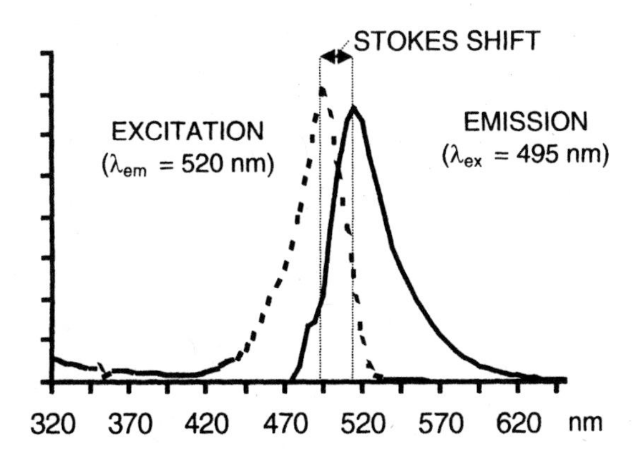

Figure 4-7. Fluorescence spectrum of fluorescein.

spectrum. Also, in general, the shape of the emission spectrum remains the same, irrespective of whether the material is excited at shorter or longer wavelengths within the absorption region. This is so because the energy difference between the longer and shorter wavelength photons is dissipated by nonradiative mechanisms following absorption.

Okay, you might say, but the excitation spectrum overlaps the emission spectrum. If the shape of the emission spectrum remains the same, no matter what the excitation wavelength is, wouldn't that mean that we could get 500 nm emission from 510 nm excitation, seemingly violating the Law of Conservation of Energy? Well, we could get 500 nm emission from 510 nm excitation, if the molecule in question was already in a vibrational excited state when the 510 nm excitation photon arrived. The cost of electronic excitation remains the same, but the molecule itself is coming up with some of the money.

The width and fine structure of the excitation spectrum reflect the energy differences between vibrational states of the electronic excited states; absorption peaks are typically skewed with long tails toward shorter wavelengths, because the transitions with the highest probabilities of occurrence are those to the first few vibrational levels of S_1. The width and fine structure of the emission spectrum reflect energy differences between the vibrational states of S_0; emission spectra usually have long tails toward longer wavelengths, because the most probable transitions in emission are those to the lowest few vibrational states of S_0.

In fluorescence, exit from the excited state generally obeys first order exponential kinetics. If fluorescence were the only process by which molecules returned to the ground state, the mean lifetime of molecules in the excited state would be $1/k_f$, where k_f is the rate constant for fluorescence emission. This quantity is defined as the **intrinsic lifetime**, τ_0. In reality, there are other processes competing with fluorescence to deexcite the molecule; the actual **excited state lifetime**, τ, will therefore be shorter. The actual mean excited state lifetimes for most fluorescent molecules are on the order of 10^{-8} s, or 10 ns. Since the decay kinetics are exponential, the mean lifetime is not the median lifetime; the fraction $1/e$, or about 37 percent, rather than half, of the molecules will remain in the excited state at $t = \tau$, while about 63 percent will have left the excited state by then.

It is often possible to use differences in fluorescence lifetimes between molecules to advantage for analytical purposes. The lanthanide rare earth elements, such as europium and terbium, exhibit atomic fluorescence due to emission from electrons in *f* orbitals; the excited state lifetime of such fluorescence is typically several microseconds. When rare earth chelates are used as fluorescent labels, it is possible to detect their fluorescence in the presence of much higher, but shorter-lived, background fluorescence by using a pulsed source such as a nitrogen laser and electronically gating the fluorescence measurement system so that 50 nanoseconds or so elapse before the measurement is made. This is an example of **time-resolved** fluorescence spectroscopy.

The **quantum efficiency**, or **quantum yield**, of fluorescence, Φ, is equal to the ratio of the number of photons emitted to the number of photons absorbed, and can also be represented as

$$\Phi = \tau/\tau_0 = k_f/(k_f + \Sigma k_i),$$

where Σk_i is the sum of rate constants for the competing nonradiative deexcitation processes. The total fluorescence emission obtained from a fluorescent material is the product of the number of photons or amount of light absorbed and the quantum efficiency. If you want to have a lot of light emitted, you've got to get a lot of light absorbed first; thus, good fluorescent probes need to have high extinction coefficients.

Phosphorescence

The S in S_0, S_1, etc. stands for **singlet**. In absorption, an electron is raised from a low-energy orbital to a higher-energy orbital, typically leaving behind another electron in the low-energy orbital. According to the Pauli exclusion principle, two electrons in the same orbital must have opposite spins, which are described as paired. A singlet excited state is one in which the electrons in the high- and low-energy orbitals have paired spins; absorption processes which result in formation of singlet excited states have relatively high transition probabilities, and emission leaving both electrons in the low-energy orbital also has a high probability of occurrence. Such transitions are described as **allowed**.

A **triplet** excited state is one in which the electrons in high- and low-energy orbitals have the same spin. Transitions to and from triplet states require changes of spin and therefore have a low probability of occurrence; they are called **forbidden**. The first triplet excited state, T_1, generally has a lower energy than the first singlet excited state, S_1. Following absorption, molecules can relax via a nonradiative transition to the T_1, rather than the S_1 state; this is called **intersystem crossing**. The transition to the ground state S_0 from T_1 is forbidden. This doesn't mean it can't happen; like a lot of other supposedly forbidden things, it does happen, but it takes longer to occur because it has a lower probability and because the energy must be passed through secret Swiss bank accounts. The associated emission is called **phosphorescence**. Lifetimes for phosphorescence are much longer than for fluorescence, typically milliseconds to seconds, thanks to which we have television and watch dials that glow in the dark. Since the energy difference between T_1 and the ground state S_0 is usually smaller than the energy difference between S_1 and S_0, phosphorescence typically occurs at longer wavelengths than fluorescence.

Fluorescence Polarization

I have already mentioned that, in order for absorption to occur, the **E** vector of the incident light must be aligned with the dipole moment of the absorbing molecule. If the incident light is linearly polarized, only those molecules that happen to be oriented properly with respect to the plane of polarization will absorb light. Since only those molecules absorb the incident light, only they are capable of fluorescence emission. Like absorption, emission can only occur in a direction determined by the orientation of the molecule.

Absorption occurs so rapidly (10^{-15} s) that the absorbing molecules have no time to move during the process. If fluorescence emission occurred as rapidly as absorption, or if the molecules involved were completely immobilized, the fluorescence emission occurring following excitation by linearly polarized light would be linearly polarized, although not necessarily polarized in the same plane as the exciting light. By now, we know that that isn't the way the world works; fluorescence is going to occur over a period of nanoseconds following absorption, and it's a cinch that at least some of the molecules are going to change their orientations (i.e., rotate) before they emit. This means that some **fluorescence depolarization** will occur. The more motion there is before emission, the more depolarization we can expect.

We can make use of this effect to determine the relative rotational freedom of fluorescent molecules, or the **fluidity** of their microenvironment; or, looking at the other side of the coin, we can determine the extent to which molecular movement is restricted, or the **viscosity** of the microenvironment. This is done by using appropriate polarization optics, which may be as simple as Polaroid filters in different orientations, and making measurements of fluorescence intensities polarized in the planes parallel and perpendicular to the plane of polarization of the excitation. These intensities are, respectively, denoted by $I_{\parallel}$ and $I_{\perp}$.

From the intensities, we can compute either the **fluorescence polarization**, p, as

$$p = (I_{\parallel} - I_{\perp})/(I_{\parallel} + I_{\perp}),$$

or the **fluorescence emission anisotropy**, r, where

$$r = (I_{\parallel} - I_{\perp})/(I_{\parallel} + 2\,I_{\perp}).$$

Values of both polarization and anisotropy increase as molecular rotation is increasingly restricted. The use of fluorescence polarization and anisotropy measurements to measure rotational diffusion of molecules in membranes and the cytosol will be discussed further in the chapter on parameters and probes.

As a general rule, when we are not trying to measure anisotropy or polarization in our cytometers, we pay little or no attention to the degree or direction of polarization of fluorescence. Most of the time, we get away with it. However, an article published in 2000 by Asbury, Uy, and van den Engh[2439] suggests that polarization effects may represent a fair-sized skeleton in our cytometric closet.

Since the light emitted by most lasers used as light sources in cytometry is polarized, both scattered light and fluorescence emission are typically polarized to some degree. This makes the intensity of detected signals more dependent on the angle and direction at which they are detected than would otherwise be the case. Differences from instrument to instrument in optical geometry, and in the polarization response of optical elements such as lenses, dichroics, and filters, may therefore lead to otherwise inexplicable differences in the intensities of signals measured from supposedly identical cells or particles. Further complications may be introduced by the fact that different fluorescent probes exhibit differing degrees of fluorescence polarization, some intrinsic to the molecular structure of the probes, and some dependent on binding to macromolecules and on other environmental characteristics.

The bottom line for most users is that polarization-related differences in the response of different instruments may interfere with the standardization of quantitative fluorescence measurement. The bottom line for those of us who develop and manufacture instruments is that we need to determine the nature and extent of those differences, in hopes of reconciling results from existing systems and improving the design of future systems. A simple solution was suggested by Asbury, Uy, and van den Engh; placing a polarizer at the so-called "**magic angle**" (54.7° for linearly polarized source emission) in the light path of each fluorescence detector removes the dependence of intensity measurements on polarization, with only a modest loss of overall signal intensity.

Stimulated Emission

One of the stranger things photons can do is make more photons just like themselves. I don't mean from nothing; there has to be some energy input to start with, but it's still pretty remarkable. It took Einstein to figure it out. We already know that a photon is likely to be absorbed by a molecule if the energy difference between the molecule's ground and excited states is equal to the energy of the photon. It turns out that just having photons of that energy around also increases the likelihood that molecules already excited will emit identical photons.

That, of course, is the catch. In general, when we're talking electronic excitation, there are a lot fewer excited molecules than molecules in the ground state, as our exercises with the Boltzmann distribution showed. On the other hand, we also calculated that, for less energetic transitions, we could end up with nearly equal numbers of molecules in the upper and lower energy levels. So, what happens if we put a whole lot of energy and some well-chosen photons into a small volume of molecules?

There are, as it turns out, a number of ways of doing this. One of them is to confine an incandescent ionized gas, or **plasma**, using a magnetic field, to the central portion of a tube placed in a mechanically rigid structure in which mirrors at both ends efficiently reflect light in some spectral region in which components of the plasma emit, with one of the mirrors transmitting a tiny bit of light. The plasma will contain a higher proportion of ions or molecules in an excited state than in a corresponding lower energy state, or a **population inversion.** With time, more and more light will bounce back and forth between the mirrors, and, wonder of wonders, the photons in the light transmitted out one end will all be of the same wavelength, and in the same phase, and going in the same direction. In other words, this light output will be **coherent**. We start with a little light of a given wavelength, and **stimulated emission** gives us more, so we are getting **gain** at that wavelength, or, to recoin a phrase, Light Amplification by Stimulated Emission of Radiation. In other words, we've made ourselves a **laser**.

Since actually building working lasers involves substantial amounts of both high tech and black art, I am tempted to say that reproduction of photons is an unnatural act. As it turns out, though, there is at least one known example of a natural laser; there is some gain at carbon dioxide emission lines in the atmosphere of Mars. Assuming we don't put enough laser light through flow cytometers to get the dyes in cells to lase (and I wouldn't swear we don't), there should not be competition between fluorescence and stimulated emission in flow cytometry.

Resonance Energy Transfer

There are a number of ways in which a molecule in an excited state can transfer energy nonradiatively to other molecules; collision, for example, is inelegant but effective. The transfer of energy, however, is more efficient when the energy differences between the S_0 and S_1 states of the donor and acceptor molecules are the same, in which case there is said to be a resonance between them. This condition holds when an emission peak of the donor species and an absorption peak of the acceptor species overlap substantially. Even then, however, significant **resonance energy transfer** can only occur when donor and acceptor molecules are within about 60 Å of one another in space, because, as Förster showed in the 1950's, the amount of energy transfer is inversely proportional to the sixth power of the distance between donor and acceptor. As will be discussed further in the chapter on parameters and probes, this sensitivity to intermolecular distances makes it possible to use measurements of the extent of energy transfer between molecular species as a "spectroscopic ruler" to measure those distances[315, 2347].

Energy transfer is **nonradiative**; that is, the donor is not emitting a photon that is then absorbed by the acceptor. Nonradiative transfer of absorbed light energy through one or more phycobiliproteins to chlorophyll is widespread in nature and essential to life; while these molecules are fluorescent to some extent in their natural environment, they are more fluorescent when separated from it. Energy transfer can be used for spectral shifting of fluorescence, since emission from the acceptor can result from absorption by the donor. This trick is commonly employed in scintillation counting, and is, as was noted in Chapter 1, also the basis for tandem conjugate labels.

Quenching, Bleaching, and Photon Saturation

More often than not, cytometry uses fluorescence as its medium of exchange for information; we use fluorescent probes or dyes to measure the amounts of various substances of interest on or in cells. We must therefore be concerned with processes that may, as it were, change the rate of exchange, making the fluorescence signal no longer proportional to the amount of probe or label used. Quenching, bleaching, and photon saturation can all, in some circumstances, interfere with our measurements; in other circumstances, we may be able to use these effects to advantage.

Quenching results when excited molecules relax to the ground state via any nonradiative pathways that provide alternatives to fluorescence. Loss of energy by vibration and collision, by energy transfer, and by intersystem crossing may all account for quenching. Molecular oxygen and paramagnetic molecules and heavy ions such as iodide quench by increasing the probability of intersystem crossing. Polar solvents such as water generally quench fluorescence to some extent because such molecules reorient around excited state dipoles. This may explain the popularity of cold showers as a means of getting out of excited states.

Bleaching occurs when the structures of fluorescent molecules are altered sufficiently to render them nonfluorescent. Bleaching can result directly from the action of light on the fluorescent molecules or from chemical reactions between the fluorescent species and other reactive molecules, such as oxidizing agents present in solution or free radicals produced by photochemical reactions.

In fluorescence excitation, one is shooting at a finite number of targets. We have a certain number of dye molecules, in each of which electrons can be raised to excited states, usually returning to the ground state after a few nanoseconds. However, there is always some chance that absorption of one or more photons will alter molecular electronic structure enough to break a chemical bond, yielding a nonfluorescent product(s); the process is called **photolysis**.

Bleaching can occur in flow cytometers at surprisingly low laser power levels, as was shown by Pinkel et al[96]. These authors set up a dual-beam flow cytometer, with excitation from two separated argon ion laser beams, both at 488 nm, each focused to an elliptical spot approximately 100 µm wide (perpendicular to flow) and 15 µm high (parallel to flow) at its intersection with the cell stream. The two observation points thus produced were separated in space by 10-100 µm. The power in the second "probe" beam was kept constant at 150 mW; power in the first beam was varied between zero (i.e., the beam was blocked) and 400 mW. This instrument was used to examine sperm cells stained for DNA by the acriflavine Feulgen method; distributions obtained from fluorescence measurements made in the probe beam were compared using different powers in the first beam. If passage through the first beam resulted in bleaching, i.e., destruction of dye molecules, fewer dye molecules would be present by the time the cells reached the probe beam, and the intensity of fluorescence detected in the probe beam would therefore be lower.

A slight decrease in fluorescence intensity was noticeable when the power in the first beam was set at 8 milliwatts (mW); when power was 400 mW, fluorescence intensity measured in the probe beam was decreased by about 75 percent, compared to values obtained with the first beam off. This indicated that most of the dye was destroyed, or at least temporarily converted to a nonfluorescent form, by exposure to argon ion laser emission at 488 nm at power levels commonly used for flow cytometry. Pinkel et al also showed that increasing excitation power above 15 mW in a single beam instrument did not result in a proportional increase in the peak channel number of the observed fluorescence distribution; this provided further evidence of significant bleaching. While most dyes used in flow cytometry are less sensitive to bleaching than is acriflavine, the message is clear; there is always a power level above which "throwing more photons at the problem" may be counterproductive.

My colleagues at Block and I were not at all surprised by these results. Tomas Hirschfeld had concluded, some years before the Pinkel paper came out, that one might get a better signal-to-noise ratio in some kinds of immunofluorescence analysis by taking advantage of bleaching. What we had been trying to do was to detect single viruses in serum isolates, using antibodies tagged with a macromolecule (polyethylenimine) to which several hundred molecules of fluorescein had previously been conjugated.

We encountered two major problems; the serum was highly fluorescent, due to its content of flavins and other naturally fluorescent molecules, and the highly fluoresceinated antibodies weren't fluorescent, because the fluorescein molecules were close enough to quench one another by energy transfer. This was an example of a common phenomenon called **concentration quenching**; one normally runs fluorescence spectra on micromolar or submicromolar concentrations of dyes to avoid it.

It occurred to Tomas that if he had several hundred fluorescein molecules quenching one another, it meant that they couldn't be spending a lot of time in the excited state, i.e., **quenching shortened the excited state lifetime**. He then reasoned that the quenched molecules would be less likely than unquenched molecules to bleach if he threw a lot more photons at them.

The interfering background substances weren't present in enough concentration to be quenched to anything like the extent to which the fluorescein molecules were. Therefore, clobbering the sample with a lot of laser power should rapidly bleach the molecules responsible for the background fluorescence. From that point on, photons would be squeezed out of the fluoresceins at a leisurely pace until they went to their reward. As long as the relative probabilities of fluorescence and bleaching remain the same, one can expect to get a certain number of photons out of a molecule before it gets bleached. Tomas concluded, and found, that he could get the same number of fluorescent photons out of a quenched fluorescein over a long time period than would be obtained from an unquenched fluorescein in a shorter time[108].

"Bleaching out the background" before making a cytometric measurement is analogous to letting the background fluorescence fade out when you are looking at cells under the fluorescence microscope (p. 12). It only works when the fluorescent material you want to measure is quenched to a significantly greater degree than are interfering fluorescent substances in the background. We had planned to use this strategy in the gadget we built to count single viruses[94], and Mike Hercher and others had done fairly extensive calculations of how much laser power we'd need. It turned out that, with 100 mW of excitation at 488 nm, we'd be able to see a few dozen molecules of fluorescein, and we wouldn't gain anything by increasing the power because we would have zapped every last fluorescein molecule with the 100 mW. This made us curiouser and curiouser about why almost everybody else who was building flow cytometers was using big lasers to get a watt or so of excitation power. It seemed that they wouldn't need that much power unless they were throwing away photons right and left everywhere in their instruments; to make a long story short, they were. But there are some justifications for the use of high laser power.

As the intensity of the incident light goes up, the probability of hitting dye molecules with photons increases. Up to a point, this will result in increased fluorescence emission. Once the number of dye molecules in the excited state becomes equal to the number of dye molecules in the ground state, however, a state of **photon saturation** is reached. Further net transition between the ground state and the first excited state is impossible; throwing more photons at the sample will not result in increased fluorescence.

A 1992 study by van den Engh and Farmer[1130] added considerably to our understanding of how of saturation and bleaching affect measurements made in flow cytometers. The theoretical section of this paper calculated the average time interval τ_p at which photons hit a dye molecule; this quantity is equal to $1/\sigma I$, where σ is the molecule's absorption cross section, in units of area, and I is the intensity of incident light, in units of photons per unit area per unit time. A typical dye might have an extinction coefficient of 3×10^4 $mol^{-1}cm^{-1}$; the formula on p. 111 would convert this to an absorption cross section of 1.15×10^{-20} m^2. A 1 W, 488 nm laser beam, focused to a 20 μm spot, provides an average illumination intensity, or **photon flux**, of 7.87×10^{27} photons $m^{-2}s^{-1}$; at the center of the Gaussian beam, the intensity is twice this. A dye molecule in the center of the beam will thus encounter a photon about every 6 ns. This time τ_p is of the same order of magnitude as the excited state lifetime τ.

Additional fluorescence cannot result from the interaction of photons with dye molecules that are already in the excited state; as van den Engh and Farmer put it, such molecules have an effective cross section of zero. The probability of a photon producing excitation of a molecule is $\tau_p/(\tau + \tau_p)$; this reaches a maximum value when $\tau = \tau_p$, and the excitation rate becomes $1/\tau$. Increasing excitation intensity beyond this point simply decreases the efficiency of excitation, and does not increase the total amount of emitted fluorescence. Van den Engh and Farmer's data also showed that bleaching is almost entirely due to absorption of single photons by molecules in the ground state, rather than to absorption of second photons by molecules in excited states.

A molecule is unlikely to go through an infinite number of cycles of excitation and emission; the probability of its undergoing bleaching while in an excited state is constant, and, therefore, as the number of cycles increases, the likelihood that the molecule will be destroyed increases. A molecule of Hoechst 33258 bound to DNA was found to have an effective life of about 100 excitation-emission cycles, whereas a molecule of DNA-bound propidium iodide would last for just over 200 cycles. The critical point here is that these figures are independent of excitation intensity. You get the same number of photons from a dye molecule whether you excite it at low intensity for a long time or high intensity for a short time, and you get the maximum number of photons out only by bleaching all of the dye molecules.

For any given observation period, there is a lower range of intensities at which neither bleaching nor saturation are significant; in this range, usually corresponding to laser power levels of no more than a few tens of milliwatts, fluorescence emission increases linearly with excitation intensity. Above this range, the relative increase in emission is less than the relative increase in intensity, and above the intensity at which $\tau = \tau_p$, there is no relative increase in emission. Thus, at least for dyes of the type studied by van den Engh and Farmer (i.e., DNA stains), there is some advantage to using very high excitation powers, i.e., hundreds of milliwatts.

Since a typical laser beam has a Gaussian profile, focusing the beam to a 20 μm spot produces a large variation in illumination intensity over the width of even a small (e.g., 2 μm) core stream. At low laser powers, this will result in unacceptably large CV's for fluorescence measurements. However, if the power is high enough, the intensity variation over the width of the core will produce only small variations in fluorescence emission, with minimal effects on the measurement CV; effects of fluctuations in laser output that would otherwise decrease precision will be minimized, as well.

One downside to measurements at high excitation intensity relates to the photochemical effects of dye bleaching on the biological and/or biochemical properties of the sample. Kissane et al[256] reported that UV laser powers above 100 mW substantially decreased viability of sorted cells stained with Hoechst dyes. It is likely that this is due to formation of photoadducts between the dye and DNA that are not repairable by normal mechanisms, and it seems probable that chromosomal DNA, or sperm, sorted after staining with Hoechst dye and exposure to very high laser powers, might be similarly damaged.

There are also other situations in which the use of high power levels may be counterproductive. Doornbos, de Grooth, and Greve[2440] carried out theoretical and experimental studies of bleaching and saturation in a flow cytometer, considering the behavior of fluorescein (FL), phycoerythrin (PE), and allophycocyanin (APC), and using a wide range of both laser powers and observation times. While APC fluorescence could be accurately fit to a model in which exit from the excited singlet state could only proceed via fluorescence emission, nonradiative decay, bleaching, and conversion to the triplet state, the fluorescence of FL and PE could not, indicating the involvement of other processes. For APC, it was found that the best signal-to-noise ratio for detection could be obtained with relatively low excitation power (30 mW), and a much longer observation time (1 ms) than is commonly used in conventional flow cytometers. Under these conditions, Doornbos, de Grooth, and Greve calculated it should be possible to detect a single molecule of cell-associated APC. In fact, a scanning system built by their group[2383], using an 8 mW, 635 nm laser diode for excitation, with an observation time of approximately 1 ms, resolves beads bearing small numbers of APC molecules better than do current commercial flow cytometers. Long observation times, and very slow flow rates, have also been used to advantage in flow cytometers specially designed for observation of DNA molecules[660,1144,2327,2333-4], viruses[94], and bacteria[2326].

Here's a riddle: when can a lifetime be a dead time? As was noted on p. 111, the uncertainty principle results in an inverse relationship between the widths of absorption peaks and excited state lifetimes. For most fluorophores, the excitation spectrum and the absorption spectrum are the same, or very nearly so, and emission spectra tend to be mirror images of excitation spectra. Thus, one would expect a fluorophore with a narrow emission peak to have a long excited state lifetime. The emission peaks of the lanthanide elements mentioned above are quite narrow, and, as would be predicted, their excited state lifetimes are very long, meaning microseconds. You would therefore not want to use a lanthanide label in a high-speed sorter, in which a cell spends less than a microsecond in the excitation beam; you wouldn't be able to excite each label molecule more than once. Well, you'd also have to look downstream some distance to detect the emission, and make the measurements over a long time period (it has been done[1385-6]), but that's another problem.

But let's say we have two dyes, one with a 3 ns lifetime and one with a 9 ns lifetime, and the excitation intensity is high enough so that a photon capable of exciting either comes by every 3 ns. You can get three excitation-emission cycles out of the dye with the 3 ns lifetime for every one you get out of the dye with the 9 ns lifetime. In other words, the 3 ns dye is just at photon saturation; the 9 ns dye is well past it, and its lifetime, i.e., the time it spends in the excited state, has become a "dead time" during which it cannot be excited, no matter how many eligible photons pay it court.

This may explain why some probes with short lifetimes yield higher intensity signals from flow cytometers than do other probes, with longer lifetimes that, according to spectrofluorimetric measurements, have higher absorption at the excitation wavelength and higher quantum efficiency. The average spectrofluorometer doesn't get its specimens anywhere near photon saturation, meaning that the time between excitations is very long, and the time molecules spend in the excited state is negligible, whereas photon saturation is not infrequently approached or reached in flow cytometers.

I was thinking that a cover blurb describing this book as being about excited states, vibrational relaxation, and stimulated emission might boost sales; "no emission without a quantum" seemed like a useful slogan. However, good taste won out, at least for the cover. Before I start to improve my image, I should cover a few odds and ends to complete this discussion of the interaction of light and matter complete.

Quantum Flotsam and Jetsam

Inelastic Scattering and Doppler Measurements

When we talked about scattering, I described what is more precisely defined as **elastic** scattering, which results in emission of photons at the same energy (or frequency, or wavelength) as the incident light. **Inelastic** scattering occurs when the scattering object is moving; the **Doppler shift** results in scattering at a higher frequency than that of the incident light if the object is moving toward the source and in scattering at a lower frequency if the object is moving away. The faster the object is moving, the larger the frequency difference between incident and scattered light. Since relatively low energies are involved in molecular motion, the frequency differences between the incident and the scattered radiation are relatively small. Inelastic scattering of microwave radiation has been widely and profitably employed by law enforcement agencies. Until the advent of lasers, it was difficult to use light to measure particle velocities, because small frequency differences were hard to detect. Now, **laser Doppler velocimetry** can be used for such purposes, and the cops can tell whether your particle is going over 65 before you can take any countermeasures.

Raman Scattering

There is a small probability that a molecule will undergo a vibrational transition at precisely the time at which scattering occurs, resulting in the emission of a photon differing in energy from the energy of the incident photon by the amount of energy involved in the vibrational transition. This is **Raman scattering**. The vibrational event involved in Raman scattering can be either a transition to a higher vibrational level or a transition to a lower level. In the former case, which is the likelier of two rather improbable events, the Raman emission is at a wavelength longer than that of the incident light, and is described as **Stokes Raman emission**. In the latter case, the Raman emission is at a wavelength shorter than that of the incident light, and is described as **Anti-Stokes Raman emission**. Nobody is getting something for nothing in either case. Until lasers came along, Raman spectroscopy was next to impossible, because the intensity of Raman scattering is only about 1/1000 the intensity of Rayleigh scattering. It is now fairly easy to do, and can provide useful information about molecular vibrations. Stokes Raman emission can also be a significant source of interference with some flow cytometric fluorescence measurements, notably that of phycoerythrin fluorescence, since the Raman emission from water illuminated at 488 nm is at about 590 nm, and at least some filters used for phycoerythrin measurement transmit this wavelength.

Nonlinear Optics and Harmonic Generation

At the high radiant flux densities achievable with lasers, electric field intensities are extremely high. Many physical effects are described by formulae containing terms that include higher powers of field intensity as well as the first power; the higher-order terms are usually negligible and what we observe are those effects that depend linearly on field intensity. Lasers have facilitated the study and application of **nonlinear optics**. One common application is in **harmonic generation**, in which nonlinear effects in crystals result in generation of light at two or more times the frequency of the incident light. Second harmonic generation, or **frequency doubling** of the 1064 nm YAG (yttrium aluminum garnet) laser line, for example, produces 532 nm, while third harmonic generation, or frequency tripling, produces 355 nm. Frequency-doubled YAG lasers have been used as sources in cytometry, and may be more widely used in the future as they become competitive in price with argon lasers. In the long run, solid-state lasers doubled from the 980 nm range to the 490 nm range will probably replace argon lasers in most fluorescence flow cytometers; Coherent introduced such a solid-state laser in 2001[2441].

Two-Photon and Multiphoton Excitation

At very high photon fluxes, it is possible to use two or more low-energy photons to induce a transition to an excited state that would normally require a single photon with two or more times the energy content. Denk, Strickler, and Webb[1131] accomplished this in the context of scanning fluorescence microscopy, using a tightly focused pulsed red laser beam to excite UV-absorbing fluorophores. Fluorescence emission increases quadratically with excitation intensity; since the depth of focus of the excitation beam is very small, the technique allows detection of fluorescence from a thin "slice" of the specimen. There are now at least a hundred multiphoton confocal microscopes in operation; they typically use very expensive pulsed Ti-sapphire lasers for excitation, although it has been claimed that there are cheaper ways to play the game. Multiphoton excitation for flow cytometry poses problems because of the small size of the excitation spot, which makes it difficult to get signals from more than a small region of a particle the size of a typical cell during transit through the instrument[2442].

4.2 OPTICAL SYSTEMS

Having sated our voyeuristic curiosities about the private lives of photons, we can now turn our attention to the building blocks of both the peeping Tom's and the analytical cytologist's hardware, i.e., the lenses and other elements that make up **optical systems**. Optical systems have only one basic function, which is diverting light from where it wants to go to where the user wants it to go. In order to learn how this is done, we must first consider where light wants to go to start with. This gets us away from quanta and back into the domain of geometric optics.

Light Propagation and Vergence

We usually consider light as originating from a mythical object called a **point source**, from which we imagine **rays** emanating in all directions, i.e., over a solid angle of 4π steradians. Rays go in straight lines unless reflected or refracted. Reflection is simple; the angle of incidence equals the angle of reflection. Refraction involves an important optical property of the material, i.e., the **index of refraction. Snell's law** and the indices of refraction of two materials allow us to determine the angle at which a ray will be "bent" at the interface between them, and provide much of what we need to model the behavior of optical systems (Fig. 4-5, p. 106, and pp. 107-8).

Rays from a point source get farther apart as they go along; they are **divergent**, or have **negative vergence**. Parallel rays have zero vergence; light consisting of parallel rays is said to be **collimated.** Rays that **converge**, or have **positive vergence**, get closer together as they go along. Convergent and divergent light in three-dimensional space can also be thought of as **waves** with spherical **wavefronts**; imagine something like an ice cream cone, with the tip of the cone being the point of origin or convergence of the rays and the hemispherical surface of the ice cream representing the wavefront. The wavefronts in collimated light are planar.

At a distance D from a point of origin (a point source; D here is, by convention, negative) or a point of convergence (in this case D is, by convention, positive), the **reduced vergence**, $V = n/D$. Vergence is measured in **diopters**, where 1 diopter $= 1\ m^{-1}$.

If a ray travels a distance L through a medium with a refractive index n, we say that it traverses an **optical path length** $S = Ln$. The optical path length reflects the distance light would have traveled in a vacuum in the same time it took to go the distance L in the medium. If many rays from one point travel the same optical path length to end up at another point, they will have the same phase (the peaks of their waves will coincide) and they will reinforce in such a way as to form a **real image**.

Image Formation by Optical Systems: Magnification

Images can be formed by mirrors, but we are concerned here with lenses. Lenses are pieces of material (usually glass, plastic, or quartz) with a shape such that all rays traversing them reach a distant point having traveled the same optical path. In Figure 4-8, that point is at infinity. The rays from the point source all emerge parallel to each other, so we say they are **collimated**. The distance from the point source to the lens is the **focal length**, f, of the lens.

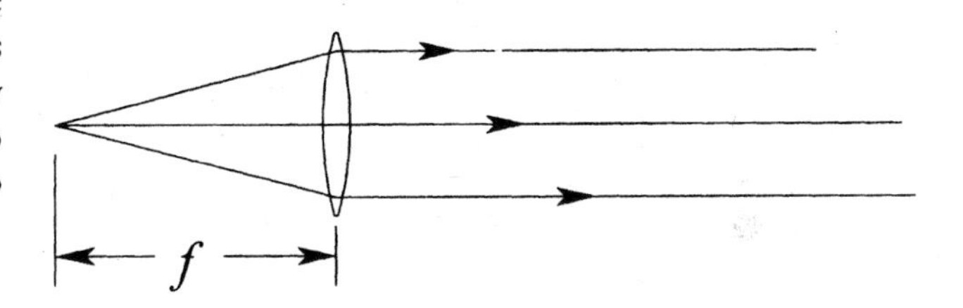

Figure 4-8 Light from a point source at the focal length of a lens is collimated. This defines the focal length, f.

Light rays are entirely reversible, so the arrow heads on the rays of Figure 4-8 could be reversed, bringing light from, say, a distant star to a focus at the focal length f of the lens. That's a lot easier way to measure a focal length than going a long way away to see if the light is really collimated. The best version of a real point source is a distant star, and since it is very far away, all the rays from it that go through the lens are parallel. That means we can evaluate our lens by looking at the focus to see if the rays all converge on one point, forming an image of the distant point source. If they do, there will be little rings around the point image, due to the wave nature of light. Those rings are **diffraction rings**, and a lens that can show them is said to be **diffraction limited** (as opposed to **aberration limited**).

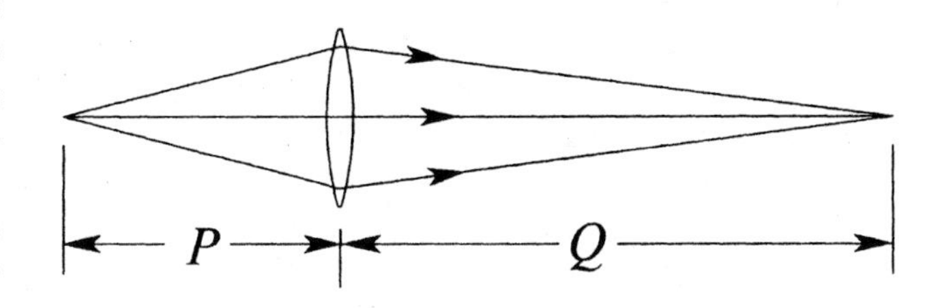

Figure 4-9. Rays from a point source can be focused to an image point.

In Figure 4-9, we see the formation of an image point that is not at infinity. If we put the point source, which we now refer to as the **object**, at a distance P, greater than the focal length f, from the lens, an image of the object will be formed at the distance Q on the other side of the lens; the distances P and Q and f are related by the **Lens Formula,**

$$1/P + 1/Q = 1/f \ .$$

Real point sources are hard to come by, and what we really want to make images of are things that are small, but bigger than points. So we use the lens of Figure 4-9 to make a real image from point images of multiple points in a real object, as shown in Figure 4-10 below.

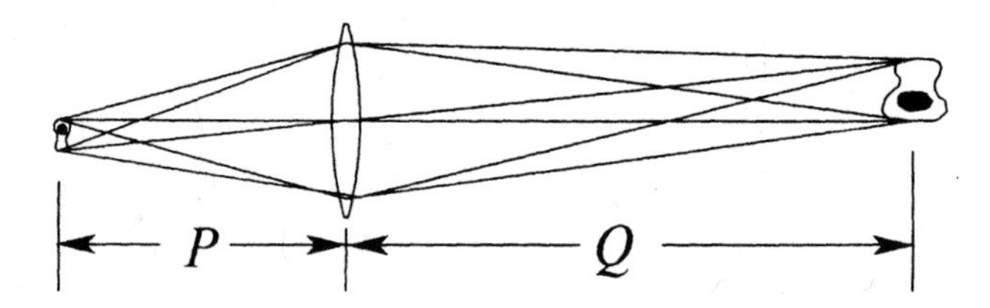

Figure 4-10. Rays from many points of an object add up to make an image.

We generally use lenses to make an image larger or smaller than the object. The **transverse**, or **lateral**, **magnification** of the lens shown in Figure 4-10 is M, where $M = P/Q$. But look closely at the object and the image in the figure, and you will see that their proportions are not the same; the ratio of width to height is larger for the image than for the object. This change of proportions has nothing to do with the fact that the image is inverted; it arises because the **axial magnification,** M_{axial} , of the lens is not the same as the lateral magnification, M; in fact, $M_{axial} = M^2$.

Lens Types and Lens Aberrations

The lens in Figures 4-8 to 4-10 is a **simple lens**, or **singlet**, made of a single material. Both surfaces are curved outward, or **convex**, making the lens a **bi-convex** or **double convex** lens. A lens with one flat surface and one surface curved outward is called **plano-convex**. Bi-convex and plano-convex lenses converge light, and are termed **positive** lenses. A lens with both surfaces curved inward is said to be **bi-concave** or **double concave**; a **plano-concave** lens has one flat side and one curved inward. Bi-concave and plano-concave lenses diverge light, and are termed **negative** lenses.

A **concavo-convex** lens, with greater curvature on the convex than on the concave side, converges light and is also known as a **converging**, or **positive, meniscus lens.** A **convexo-concave** lens, with greater curvature on the concave than on the convex side, diverges light and is therefore called a **diverging**, or **negative, meniscus lens.**

If the object is at some distance between f and $2f$ from a convex lens, a magnified, real, inverted image is formed at a distance between $2f$ and infinity. If the object is at $2f$, the image, again real and inverted, is at $2f$ and the same size as

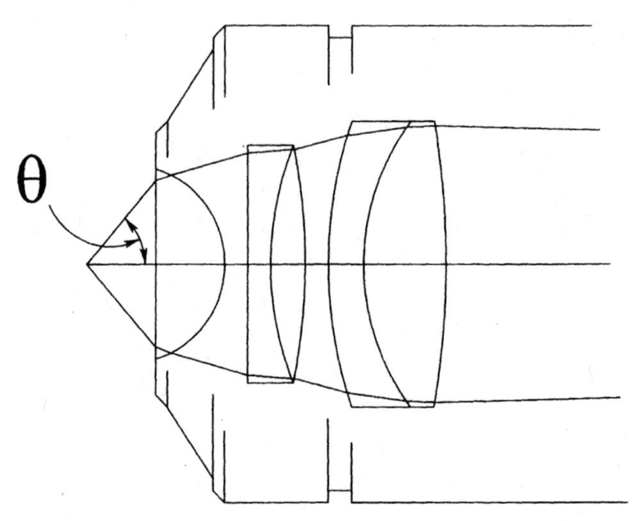

Figure 4-11. Elements of a typical microscope lens, showing the half angle θ that defines the acceptance cone and the numerical aperture (N.A.) of the lens.

the object. If the object is between $2f$ and infinity, a minified, real, inverted image is formed between f and $2f$.

When the object is at a distance less than f, a magnified, erect, **virtual image** appears. When real images are formed, light diverging from the object converges in the image plane; a real image can be seen on a screen placed in the image plane. When a virtual image is formed, light from the object appears to be diverging from the image plane, although neither the object nor a real image is located there. The eye contains the necessary optics to convert virtual images to real retinal images, and thus to make use of simple convex lenses as magnifying glasses and of mirrors for a variety of purposes. However, the detectors in our instruments need to be presented with real images. Simple lenses do not make the best images, due to the presence of **aberrations.**

Most lenses are **spherical**, that is, their curved surfaces are the surfaces of portions of a sphere. **Spherical aberration** results from rays passing through the outer portions of a lens coming to a focus at a different point from rays passing through the central portion. **Chromatic aberration** is a consequence of dispersion (p. 108); since the refractive index of a material is higher at shorter wavelengths, rays of light of different wavelengths come to foci at different distances from the lens. Other aberrations include **astigmatism**, resulting in different lateral magnifications in the vertical and horizontal directions perpendicular to the lens axis, **curvature of field**, resulting in the image of a rectangular object taking on a "barrel" or "hourglass" shape, and **coma**, resulting in images with comet-like "tails." Spherical aberration may be corrected by using **aspheric** rather than spherical surfaces in lenses. Although it is difficult for lens grinders to produce aspheric elements one at a time, it is simple and inexpensive to mold aspheric lenses once a prototype is made. Chromatic aberration is corrected by making lenses incorporating two or more elements made of materials with different dispersion characteristics. The simplest such lenses incorporate two elements, and are called **achromatic doublets**, or just **achromats;** they bring blue and red light to a focus at the same point. **Apochromatic** lenses have at least three elements, and focus blue, green, and red light to the same point. The **planapochromatic** lenses used in good microscopes also incorporate elements to correct spherical aberration, and to correct for curvature of field, providing an image that is in focus over most of the field of view, or flat, whence the "plan" in planapochromatic.

Numerical Aperture (N.A.) and Lens Performance

When a lens needs to collect rays over a very wide angle, multiple elements are needed to achieve diffraction-limited performance. Figure 4-11 shows a cross section of a typical microscope lens. The **half angle** θ shown in the figure is a measure of the **acceptance cone** of the lens. The **numerical aperture (N.A.)** of the lens is given by $\text{N.A.} = n \sin \theta$, where n is the refractive index of the medium between the subject and the front element of the lens. The N.A. of a lens is important in determining its resolution. The numerical aperture is important also because it measures the **resolution** of the lens; two self-luminous points can be resolved if they are a distance Δx apart, where $\Delta x = 0.61\lambda/(\text{N.A.})$, with λ representing the wavelength of the illuminating light in a vacuum. The wavelength in vacuum is specified here, because, as was noted on p. 107, the wavelength of light in material media is shorter than the wavelength in vacuum; in a medium of index n, $\lambda_{med} = \lambda_{vac}/n$. One consequence of this shortening is that cells in water ($n = 1.33$) can be resolved almost exactly 1.33 times better than cells in air ($n = 1.0003$). The use of **immersion lenses**, with water or an appropriate oil or gel with an index higher than that of air, increases both resolution and light collection.

Even the most exquisitely engineered microscope lenses impose some restraints on the user. For instance, they are designed to minimize aberrations when used at specified distances. Usually that means that Q, which, in this context, is called the **tube length**, is on the order of 150 mm; in many newer lenses, however, Q = infinity. Departure from the specified **conjugates**, i.e., the tube length and the **working distance**, which is the intended distance of the object from the front element of the lens, degrades the image quality. Also, most high-N.A. microscope objective lenses have large magnifications at their specified conjugates. That may mean that the axial magnification is so large that the objects in the field of view must all be in the same plane to be in focus. To meet the needs of such application areas as scanning microscopy, in which it is desirable to have a relatively large depth of focus, objective lenses with low magnification but relatively high N.A. have become available.

To be completely frank, though it is essential to have lenses with high resolution if you are doing confocal microscopy, or anything fancy in the way of scanning or image

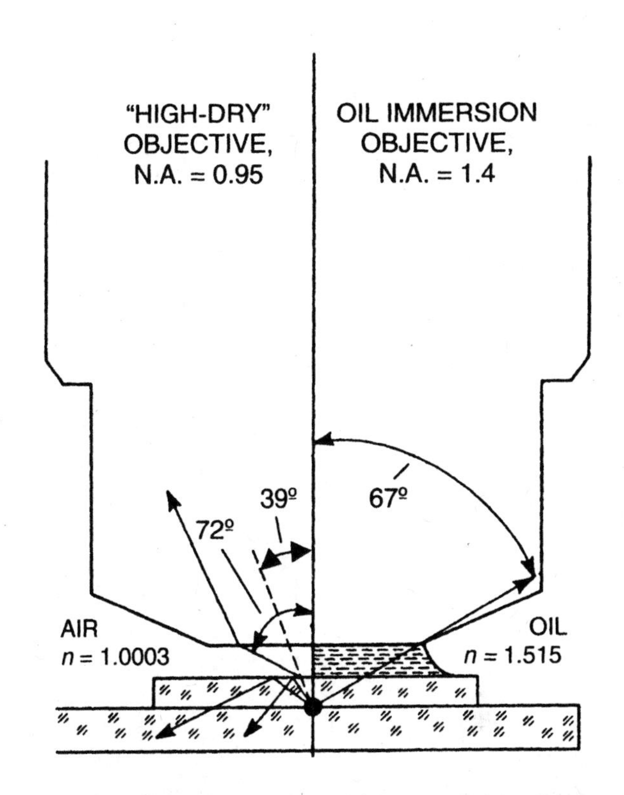

Figure 4-12. Showing the effect of N.A. on light collection. Modified from Murphy DB: *Fundamentals of Light Microscopy and Electronic Imaging*[2407], Copyright 2001 by Wiley-Liss, Inc. Used by permission.

analysis, or even just trying to look at a blood smear by eye, neither flow cytometry nor low-resolution scanning laser cytometry requires high-resolution optics. What we do need are optics that collect as much light as possible. That means a large acceptance cone, i.e., high N.A. However, there are a lot of high-N.A. lenses that have been optimized for high resolution in transmitted light microscopy by putting in a lot of lens elements, and, even with antireflection coatings on the elements, there is likely to be enough light lost at all the interfaces to reduce the light transmission of such lenses. The effects of N.A. on light gathering power can be appreciated from Figure 4-12.

The figure, modified from one in Murphy's admirable *Fundamentals of Light Microscopy and Electronic Imaging*[2407], shows the acceptance angles of a "high-dry" objective (N.A. = 0.95) and an oil-immersion objective (N.A. = 1.4); it is split to show the high-dry objective on the left and the immersion objective on the right. Rays of light defining the acceptance cone are shown coming from a "cell" depicted as a black dot between a slide and cover slip, with the line defining the split between the two objectives passing through the center of the cell and perpendicular to the slide, cover slip, and front surface of the lenses. We will assume that the cell is mounted in a medium with the same index as both the glass in these elements and the immersion oil, $n = 1.515$. We will first examine the high-dry objective.

Since the medium closest to the front element of this lens is air, with $n = 1.0003$, N.A. could not be any higher than this value. In fact, about the best one can do with a high-dry lens is N.A. 0.95. Let us consider the refraction of light coming from the cell at the interface between the cover slip and the layer of air between the cover slip and the front element of the lens. Snell's law tells us that light rays leaving the cell at angles larger than 41° to the perpendicular will not emerge from the cover slip at all; they will either be transmitted along the surface of the cover slip by **total internal reflection**, or be reflected back from the cover slip toward the slide. In practice, the half angle of the largest cone of light that will make it from the cell into the objective is 39°; this is the angle between the dotted line and the perpendicular through the cell. However, a ray of light coming from the cell at 39° and hitting the outer surface of the cover slip will be refracted according to Snell's law and emerge at an angle of 72° ($1.515 \sin 39° \approx 1.0003 \sin 72° \approx 0.95$, which is the N.A. of the lens). This is interesting, because it shows us that the quantity $n \sin \theta$, which defines N.A., remains the same from one medium to another. However, what does not remain the same is the half angle. We are trying to get light out of a cell and into our lens, and the biggest cone of light we can capture coming out of the cell has a half angle of 39°. So what fraction of the total amount of the light coming out of the cell does that cone represent?

Assuming that we don't have any anisotropy or other directional effects due to polarization, etc., the light should be coming out of the cell uniformly in all directions, i.e., over a solid angle of 4π steradians. We need to calculate the solid angle subtended by a 39°cone. It just so happens that the solid angle, in steradians, subtended by a cone with half angle θ is $2\pi(1 - \cos \theta)$; a 39° cone therefore subtends a solid angle of 1.40 steradians. Since 4π steradians = 12.57 steradians, we can only collect about 11 percent of the light from the cell using the N. A. 0.95 objective. Now let's turn our attention to the oil immersion objective on the right.

On this side, we have immersion oil, matching the index of refraction of the slide, mounting medium, cover slip, and front element of the lens, in between the cover slip and the lens. There is therefore no change in refractive index between the cell and the lens, so the half angle of the cone of light collected from the cell remains at 67° all the way along. This subtends a solid angle of 3.83 steradians; the oil immersion lens could, therefore, collect just over 30 percent of the light coming from the cell, or almost three times as much as the high-dry lens.

It is generally stated that the light gathering power of lenses increases as the square of N.A.; according to this formula, the oil immersion lens should collect 2.17 times as much as the high-dry lens. Of course, if we were using the high-dry lens to look at a cell on the surface of a slide, without the refraction at the surface of the cover slip, we would be able to collect light over a somewhat larger half angle. However, since, in flow cytometry, we are almost always looking at cells in either a stream of water in air or a stream of water in a quartz cuvette, we can expect to have the acceptance cones of lenses used without a coupling medium, such as immersion oil or a gel, restricted somewhat. When you can't afford to throw photons away, it pays to use the most efficient light collection system available, and the manufacturers of cytometric instruments have increasingly taken this lesson to heart.

Gradient Index, Fresnel, and Cylindrical Lenses

In a convex spherical or aspheric lens, the optical path length is made equal for rays passing through the center and edges of the lens by making the lens thicker at the center than at the edges. In a **gradient index**, or "**grin**" **lens** (the trade name "Selfoc" often used to describe such lenses properly refers only to those made by Nippon Sheet Glass), the thickness of the lens is constant, but the refractive index changes from the center to the outside. Such lenses can be made by diffusing various chemicals, which will change the index, into a cylinder of glass from the outside, and then heating the glass and drawing it out until a desired smaller diameter is reached. Small slices of the drawn material can then be cut and the ends polished, allowing large numbers of small lenses to be produced without the need for grinding curved surfaces. The process is virtually identical to that used in the production of fiber optics, and, in fact, grin lenses less than a millimeter in diameter and only a few millimeters long are often used in fiber optic systems.

A **Fresnel lens** is a flat surface with concentric trapezoidal grooves cut or, more commonly, molded into one surface; the angles of the grooves vary with the distance from the center, and, thus, rays entering the lens at different distances from the center are reflected and refracted at different angles. A Fresnel lens superficially resembles a grating, and you might guess it worked by diffraction, but it's all straight geometric optics. Just to confuse matters, there is also something called a Fresnel zone plate, which can focus light, but which does work by diffraction. Fresnel lenses with very high light gathering power can be made cheaply; they are most likely to be seen in overhead projectors, where they are used as condenser lenses.

Cylindrical lenses converge (if they are convex) or diverge (if they are concave) light in only one plane. Whereas a convex spherical lens will focus a circular collimated beam to a point, a convex cylindrical lens will focus the same beam to a line. Crossed convex cylindrical lenses with different focal lengths are typically used in the illumination optics of flow cytometers to focus a laser beam with a circular cross section to an elliptical spot. In a stream-in-air flow system such as is found in most high-speed cell sorters, the stream itself acts as a relatively strong cylindrical lens. **Toric lenses** have cylindrical surfaces with different curvatures in two perpendicular planes, each of which includes the optical axis. A simple toric lens is a piece of glass or plastic with a spherical surface on one face and a cylindrical surface on the other. Opticians use toric lenses to correct the astigmatism that results from the eye's own lens being somewhat toric.

The Helmholtz Invariant and Throughput

We can do some remarkable things with optics, but we can't beat the second law of thermodynamics and the law of conservation of energy. There are a few concepts that can help us keep out of trouble in this regard, and one of them is **throughput**. Figure 4-13 will get us started with this.

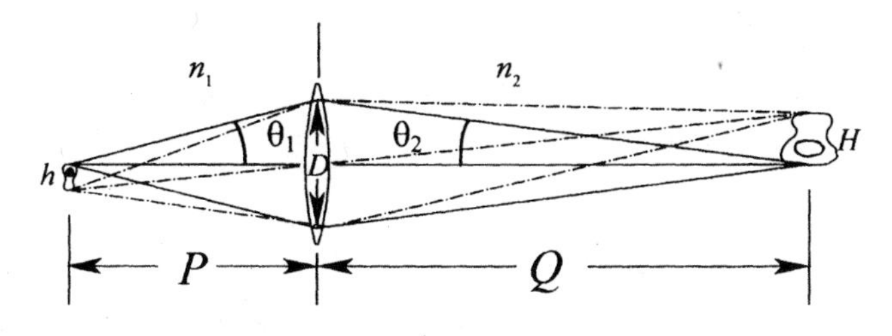

Fig 4-13. Multiple views of magnification.

Figure 4-13 largely reproduces Figure 4-10 (p. 120), which introduced us to magnification. We have provided for media of different indices, n_1 and n_2, on the object and image sides of the lens. We already know that the lateral magnification of the lens in the figure is *M*, and that $M = Q/P$. There are two similar triangles with apices at the center of the lens, one including perpendicular sides with the dimensions *h* (the height of the object) and *P*, and the other including perpendicular sides with the dimensions *H* (the height of the image) and *Q*. Because these triangles are similar, $H/h = Q/P = M$; this is the **magnification equation**. We can also see that $h/P = H/Q$, and, multiplying both sides of this equation by the lens diameter, *D*, we get $hD/P = HD/Q$.

The **Abbe sine condition**, derivable from geometric optics and also from thermodynamics[2443], tells us that

$$n_1 h \sin \theta_1 = n_2 H \sin \theta_2 .$$

We already know that the magnification, *M*, is equal to H/h; rearranging the terms in the equation thus tells us that

$$H/h = n_1 \sin \theta_1 / n_2 \sin \theta_2 = M .$$

But $n_1 \sin \theta_1$ and $n_2 \sin \theta_2$, respectively, are the numerical apertures (N.A.'s) of the lens from the object side and the image side, respectively. So we can express magnification not only in terms of the ratios of image size (height) to object height, H/h, and image distance to object distance, Q/P, but also in terms of the ratio of the N.A. for light collection and the N.A. for imaging. And we see that, for any given optical system, the product of the linear dimension of the object or image and the N.A. on the appropriate side of the system remains constant, or invariant. In the two-dimensional model we have just considered, the constant $h\ n \sin \theta$ is sometimes referred to as the **Helmholtz invariant** and sometimes as the **Lagrange** or **optical invariant**.

We can find a similar invariant quantity for the three-dimensional case. We consider an object, or source, of area A_1 in a medium of index n_1, with the lens collecting a cone of light in a cone with a half angle θ_1. The lens forms an image of area A_2 in a medium of index n_2, and the half angle of the cone on the image side is θ_2. We can actually use Figure 4-13 to play this game if we just assume that the areas of the object (with height *h*) and the image (with height *H*) in the figure are A_1 and A_2. The solid angles, Ω_1 and Ω_2, subtended by the cones with half angles θ_1 and θ_2 can be calculated from the formula $\Omega = 2\pi(1 - \cos \theta)$. And we end up with

$$n_1^2 A_1 \Omega_1 = n_2^2 A_2 \Omega_2 = \Theta.$$

The invariant quantity here, Θ, is variously known as the **optical extent**, **étendue**, or **throughput** and, as the last of these terms suggests, it tells you how much light can be transmitted through the system.

It is not the case that throughput can't be increased; we do it all the time when we switch from the high-dry to the oil immersion lens while using a microscope. However, once you've picked your optical components, you've defined the throughput of your system, and, while you can lose light, you can't get more than there was in the first place. The bottom line on all of this is that you get the most light into an optical system by collecting over the largest possible angle, or through the largest possible aperture. We'll get to a practical application of throughput a little further on, when we talk about light sources.

Photons in Lenses: See How They Run

To return to the peregrinations of photons for a moment, we might consider how lenses get photons to alter course. Let's keep it simple and think of photons leaving an object at a distance $2f$ from our simple convex lens. If the lens weren't there, they'd just go off in all directions. However, the formulation of quantum electrodynamics is that photons have the highest probability of going along paths from which any given deviation results in the smallest possible change in the time taken to traverse the path. A photon headed along the optical axis would normally get to the image plane a lot faster than a photon headed off toward the rim of the lens. That's why we make a convex lens convex. The lens is made of a material of refractive index greater than that of air, and light travels more slowly in such a material than in air. We make the center of the lens thickest, and make it thinner toward the rim, so that getting from the object plane to the image plane takes about the same time regardless of the angle at which the photon originally started.

If we really wanted to get the photons synchronized, we'd have to shape the lens so that its curvature was not spherical; in practice, spherical surfaces are easier to grind, so we usually put up with the **spherical aberration** that results from the lens not being quite the right shape. For some applications, molded or ground **aspheric** lenses are preferable.

The refractive index *n* for air is 1.0003, and sin θ is ≤ 1; therefore, the highest value of N.A. that could theoretically be achieved for an optical system in air would be 1.0003.

Higher N.A., and thus increased light gathering capacity, can, as Figure 4-12 shows, be achieved by filling the space between the optical system and the object with a medium of higher refractive index, such as **immersion oil**; the system must, of course, be designed with this in mind.

Aperture and Field Stops; The f Number

If you are familiar with photography, or, in this day and age, with video, you may recall that the **f number** (*f/#*) of a lens indicates the light gathering capacity of the lens. The iris diaphragm in a camera lens is an **aperture stop**; that is, it limits the cross-section and solid angle through which light can be collected. The f number is the ratio of focal length to aperture stop diameter; for small values of x,

$$f/\# \approx 1/[2 \times (\text{N.A.})] .$$

Unlike an aperture stop, a **field stop** limits the field of view, but does not affect the angle over which light is collected. Field stops are typically located in the image plane, and some of us (*mea culpa*) refer to them as image plane apertures. Field stops, usually in the form of "pinholes," are frequently used in flow cytometers to limit the area from which light can reach the detectors to the region immediately surrounding the observation point.

Depth of Field and Focus and Resolution of Lenses

It seems logical that as an optical system deals with light collected over an increasing range of angles, it becomes harder to get the light precisely where we want it. Thus, systems with high light gathering power should have relatively small depths of field and focus, while decreasing the aperture should increase depth of field and focus at the expense of getting less light through the system. This is exactly the way things work; you probably know that stopping down a camera lens increases depth of field, and you may remember that cheap cameras with f/11 lenses don't need a focus adjustment. You may also have been disappointed by the poor visual results you got using a high N.A. microscope lens because its depth of field and depth of focus were so small.

Depth of field, Δ, denotes the longitudinal range of object distances over which a "sharp" image of an object is obtained at a fixed point in the image plane, where the sharpness criterion is defined by the acceptable diameter of the "blur circle" in the image of an axial point object. For typical microscope objectives working at a wavelength λ, Δ is given by

$$\Delta = \lambda[(n^2-(\text{N.A.})^2)^{1/2}/(\text{N.A.})^2] .$$

Depth of focus, Δx, denotes the longitudinal range of distances in the image plane over which the image of an object at a fixed distance remains sharp; the relation between depth of field and depth of focus is expressed by $\Delta x = M\Delta$.

To determine the **limit of resolution** of a lens, we must take into account the phenomenon of **diffraction**. The diffracted image of a point shows a bright central spot, or **Airy disk,** surrounded by alternating dark and light circles. At a wavelength λ, the radius of the Airy disk is 0.61λ. The images of two points are just resolved when the distance between the images is equal to this radius; this is the case when the distance between the two points is 0.61λ/(N.A.).

4.3 LIGHT SOURCES

The Best and the Brightest

If we intend to do fluorescence measurements, the light will undoubtedly come from either a laser or a mercury arc lamp. Why are we restricted to these two sources? It's pretty obvious that the laser can readily be induced to put more photons than we really need through a very small area; we used to burn little holes through pieces of paper to impress visitors. Why can't we use a high-intensity quartz-halogen lamp? The microscope companies, after all, sell quartz-halogen illuminators for fluorescence microscopy, and even promote them for immunofluorescence work.

The characteristic we are looking for in a light source is commonly thought of as brightness, which was the old photometric term for it. What we're after is the analogous radiometric quantity, called **radiance** or **sterance.** We can describe radiant energy coming from anywhere, be it a source, an element of an optical system, a wall, etc., in terms of radiant **areance** or **exitance**, which measures the **power**, or **radiant flux** (φ) emitted per unit area (*A*), in units of W/m^2. A true point source can be described in terms of its **intensity** or **pointance**, which measures power per unit solid angle (Ω), in units of W/sr. There are no real point sources; arcs, filaments, **light emitting diodes (LEDs)**, and the good old sun are, instead, what are called **extended sources. Radiance** (L) measures power emitted from, transmitted through, or reflected by a surface, per unit of its area, per unit solid angle; that is, $L = \varphi/A\Omega$. The units of radiance are $Wm^{-2}sr^{-1}$. It is this quantity that is used to compare light outputs of extended sources.

We have already noted that the throughput of an optical system cannot be increased at any point in the system. Since throughput, $\Theta = n^2 A\Omega$, for a system in air, with $n \approx 1$, $\Theta = \varphi/L$. We are starting with whatever throughput we get to work with **at the source**, and, as we go through the system, things can only get worse. We can't take light from a system with low radiance and "squeeze it down" through a small cross-section. If we could, we'd probably be able to start a fire by holding a convex lens up to a lightning bug's rear end.

The optical system that transmits illumination light from an extended source to a specimen must, at a minimum, include a lens to collect light from the source and another to converge the collected light on the specimen. Both lenses are generally referred to as **condensers**; for purposes of this discussion, they will be distinguished as the **lamp condenser** and the **microscope condenser**. To add to the confusion, it should be noted that, in the epiillumination systems generally used in fluorescence microscopes, the microscope objec-

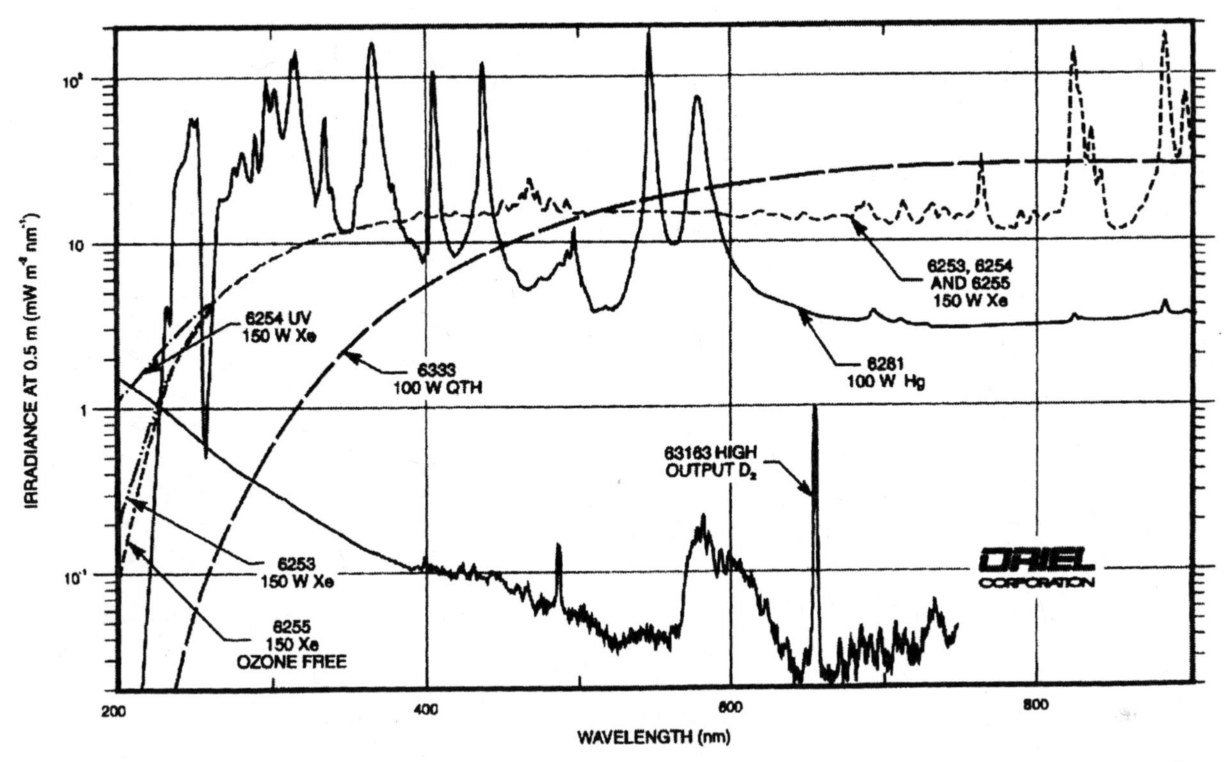

Figure 4-14. Output characteristics of arc (Hg, Xe), quartz-halogen (QTH), and deuterium (D_2) lamps (courtesy of Oriel Corp.).

tive, which collects the emitted fluorescence, also serves as the microscope condenser.

The most efficient illumination, i.e., that which produces maximum radiant flux through the specimen, can be obtained by making an image of some portion of the emitting surface in the plane of the specimen. In order to maximize light collection and specimen illumination, the condenser lenses should be of high N.A., to make the collection and illumination angles as large as possible. Consider, for the moment, that a single convex lens serves as both lamp condenser and microscope condenser. If the emitting surface of the lamp is located two focal lengths from the lens on one side, and the specimen plane is located two focal lengths away on the other, the lens will form a 1:1 image of the source in the specimen plane, and the solid angles for collection and illumination will be the same. If a magnified image of the source is to be formed, the lamp must be placed closer to the lens, and the specimen moved further away; the lens can then collect light over a larger solid angle, but the radiant flux per unit area in the specimen plane will not be increased. If a minified image of the source is formed, the lamp must be placed more than two focal lengths away from the source, and will therefore collect light over a smaller solid angle; again, the radiant flux through the specimen will not be increased.

The largest possible collection angle obtainable using a convex collecting lens will be achieved if the lens is placed one focal length from the source; it will then collect a collimated beam of light, requiring that a second lens be used to converge light on the specimen. It is also possible to collect light from a substantial fraction of the emitting surface of a source using ellipsoidal or parabolic reflectors. However, while this can allow substantial amounts of power to be directed through surface areas larger than the emitting surface of the source, it is still not possible to put more radiant flux per unit area of the specimen than can be collected per unit area from the source. I'll say it again: **you can't put any more light through a given area of the specimen than comes from the same area of an extended source.** And, if you don't collect all the light you can from the source, you can't get it to the specimen.

Thus, in illumination optics, as was the case for collection optics, it is obviously a good idea to use the **fastest** (highest N.A. or lowest *f*/#) condenser lens you can get to collect light from a lamp. The condenser lens doesn't have to produce a high quality image; it just has to be well enough corrected to do a decent job of "collimating" the light. Manufacturers of arc lamp systems, slide and movie projectors, and microscopes generally use aspheric lenses for this light collection job; it would otherwise require a lens shaped something like a marble, which would have intolerable spherical aberration. We can't improve on things, by the way, by using an ellipsoidal collector with an arc lamp; this will let you make a brighter image on a movie screen but not in a microscope or flow cytometer; it's a throughput thing. One might be able to collect and refocus the largest fraction of source emission from an arc using paraboloids, a notion suggested by Mike Hercher, to whom I am indebted for many explanations of throughput. The available paraboloids may not be good enough, and you would have to use humungous interference filters, but it might be possible. You still couldn't get around throughput; you could only collect more of the light to start with. Many people, including some who do optics for manufacturers of flow cytometers, do not seem to appreciate throughput; I kept hounding Mike, and also Rob Webb, until I got explanations in English, but I'll take the rap for this version. Whatever else you do, however, if you want to shine lots of light on a cell, you need a high-radiance source.

Harc, Harc, the Arc!

Now that we are all aware of the desirability of using the brightest lamp source available for flow cytometry, it would be nice to know which of the lamps available is the brightest. May I have the envelope, please? The lucky winner of the brightness sweepstakes is a **100 watt mercury short arc lamp**, for example, the Osram HBO 100W/2; the characteristics of this and some other lamps are shown in Figure 4-14. A 100 W Hg arc lamp has an average (photometric) brightness four times that of a 75 W xenon lamp, eleven times that of a 150 W xenon lamp, almost six times that of a 200 W mercury lamp, and about 100 times the brightness of a quartz-halogen lamp. The radiances go in pretty much the same order. The 100 W lamp has the highest radiance because the arc itself is smaller in size (about 0.25 by 0.25 mm, or 250 × 250 μm) than the arcs in more powerful lamps, and the difference in size more than offsets the lower total flux.

There is one lamp with higher average brightness than a 100 W mercury arc; this is a 500 W xenon arc lamp. However, the mercury lamp, as I calculate things, has a higher radiance in the region of its strong spectral lines at 365, 405, 436, 546, and 578 nm than does the xenon lamp, which has no strong lines in the near UV or visible spectrum. The 500 W xenon lamp, in other words, won't give you more usable power than a mercury lamp in those regions of the spectrum in which you're apt to be interested, and it has the undesirable characteristic of self-destructing without warning, doing a creditable imitation of a hand grenade in the process. This may have adverse consequences for your mental and physical health and for the physical health of any optical elements in the path of the fragments. If you still think you need a 500 W xenon lamp, you probably have space for it in the garage next to the assault weapons.

Is it just coincidental that a better and more peaceful approach to increasing excitation power in arc lamp flow cytometers has come from Norway? Steen and Sørensen[1132] modified the front end electronics and arc lamp power supply in a flow cytometer to increase the lamp current by a factor of ten for a few microseconds after the trigger signal goes above its threshold value, thus substantially increasing the excitation power while the cell is still in the observation region. The modification is inexpensive, increases the sensitivity of the instrument, does not substantially decrease the life of the lamp, and does not appear to increase the likelihood that the lamp will explode; whether anyone will attempt this trick in a commercial instrument remains to be seen.

Why, you may ask, does anyone bother using a 50 or a 200 W mercury lamp, a xenon lamp, or a quartz-halogen lamp in a fluorescence microscope if the 100 W mercury lamp is so good? There are several reasons. The radiance issue is germane to flow cytometry, where we want to illuminate a relatively small region of space. The 250 μm by 250 μm arc in the 100 W lamp is ideal for illuminating an area its own size (or smaller, using a field stop) in a flow cytometer, but we would benefit from the use of a larger arc size if we wanted to get uniform illumination over the entire field of view of a fluorescence microscope or an imaging cytometer, especially when using the lower magnifications. The other mercury lamps have larger arcs; in the 200 W lamp, arc dimensions are 600 by 2200 μm. The other lamps also have longer average lifetimes, e.g. 400 hours for a 200 W mercury and 1000 hours for a 200 W mercury-xenon lamp as opposed to 200 hours for the HBO100W/2.

Xenon lamps are generally chosen for their spectral characteristics, or lack thereof. Fluorescence microscopy pretty much grew up with mercury lamps. At first, only the ultraviolet emission from these lamps, principally the strong 365 nm line, was used for excitation. As a result, the dyes that were first found useful for fluorescence microscopy were those that could be excited in the near UV, either because they had absorption maxima in that spectral region or because they had high enough quantum efficiencies to emit appreciably when excited at wavelengths far from their absorption maxima.

Xenon lamps do not have strong spectral peaks in the visible region; they are preferred to mercury lamps as sources for spectrofluorometers for that reason, and are similarly useful for microspectrofluorometry. In principle, they allow use of a wider range of dyes; in practice, however, the continuum between the strong lines of a mercury arc is apt to be as bright a source in those spectral regions as is a xenon arc, and the mercury lamp is better near its strong UV, blue, and green lines.

A publication by Koper et al[107] on the addition of arc lamp illumination to a B-D FACS reported that ILC Technology (Sunnyvale, CA) had made xenon lamps with a zinc iodide additive, producing strong emission lines between 460 and 480 nm. Such lamps should be useful for excitation of fluorescein. ILC Technology told me they were no longer available, but some more were apparently made for use in the B-D FACS Analyzer; special lamps are also now available from Partec, which manufactures arc source cytometers.

Fluorescein and acridine orange, both of which have absorption maxima between 450 and 500 nm, in a region in which a mercury lamp has no strong lines, first became useful in fluorescence microscopy at a time when only UV excitation was used. These dyes are better excited by the blue mercury line at 436 nm than by the UV line, and thus became more useful when filters permitting use of blue excitation became widely available. Both fluorescein and acridine orange are ideally suited for excitation by the 488 nm line of an argon ion laser; this characteristic has made these dyes particularly useful for flow cytometry. There are other dyes that are easily used in fluorescence microscopy, but see only limited use in flow cytometry, e.g., brilliant sulfaflavine, which has an excitation maximum at about 420 nm. This dye is well excited by the UV and blue lines of a mercury lamp, but very poorly excited at 488 nm.

Arc lamps require relatively elaborate power supplies, including circuitry to generate a high-voltage pulse or RF (radio frequency) pulse train to ionize the gas, producing a conductive medium in which an arc can be started and maintained. Although arc lamps intended for spectrophotometry and microscopy are sold with power supplies that are supposed to provide light output regulated to better than 1%, I have measured 6-10% variations in light intensity, synchronized with the power line frequency, in arc lamp systems made by several manufacturers. A lamp, or any other light source, with such intensity fluctuations is apt to be unacceptable for cytometry.

Several of the largest manufacturers of lamps have, since the late 1980's, undertaken major efforts at developing arc lamps for use as automobile headlights, primarily to provide the stylists with headlights less than 1" high, and the status-conscious automobile buyer with yet another option to covet. It seems to have worked; I see a lot of cars with what the industry calls "discharge headlights" on the road these days, but I haven't heard of anybody using one in a flow cytometer yet. O Gucker, where art thou?

Peters[106] compared laser and mercury arc lamp illumination for flow cytometry, and found that arc lamps offered some advantages over ion lasers as UV sources. At the present state of the art, I would favor UV helium-cadmium lasers and/or violet diode lasers[2444] over both arc lamps and ion lasers for most applications; I will have more to say about this later.

Quartz-Halogen Lamps

In comparison to arc lamps, quartz-halogen lamps are less expensive and can be operated from simpler power sources; they employ incandescent filaments, but produce more green and blue light than conventional filament lamps. They can thus be used for fluorescence excitation in microscopy. The Technicon Hemalog D differential counter[84,85] successfully used a quartz-halogen lamp for flow cytometric absorption and scattering measurements; the newer hematology systems from Bayer Diagnostics, which absorbed Technicon, add a red laser, but keep the lamp as a source for absorption measurements.

I have detected brightly stained fluorescent objects (beads and cells stained with DNA fluorochromes) using a microscope-based flow cytometer with a quartz-halogen source. However, the quartz-halogen lamp is at best less than a tenth as bright as an arc lamp, and its low photon output severely limits the precision with which fluorescence can be measured at conventional flow rates. LEDs (see below) are better.

Light Emitting Diodes (LEDs)

LEDs are everywhere these days, not just on electronic gadgets. My bicycle tail light, and various auxiliary rear lights on automobiles, use red LEDs, and green LEDs are replacing incandescent sources in traffic lights. What has made this possible is the development of "high-brightness" LEDs by companies including Cree, LumiLeds (a joint venture of Agilent, formerly part of Hewlett-Packard, and Philips), and Nichia. Although an LED, like a lamp, is an extended source, it is not an incandescent source; when current is applied to the device, photons are emitted as electrons traverse the energy "band gap" in a semiconductor material. The precise composition of the material determines the energy range of these transitions, and emission is typically restricted to a relatively narrow spectral region (tens of nanometers). The lower end of the range of available LED emission wavelengths has moved steadily downward; there are now UV and violet LEDs in addition to blue, green, yellow, red, and infrared devices. Bob Hoffman and Eric Chase presented a poster at the 2000 ISAC meeting describing the use of UV and blue LEDs as illumination sources in flow cytometers[2445]. They measured DNA content in DAPI-stained calf thymocyte nuclei, with a CV of 2 percent, using a UV LED light source, and, using a blue LED source, could readily discriminate different levels of immunofluorescent staining of leukocytes by a PerCP-antiCD45 or a PE-antiCD4 antibody. I have recently compared a 100 W mercury arc lamp, a 50 W quartz-halogen lamp, and a high-brightness blue (460 nm) LED as illumination sources in a fluorescence microscope; the LED is at least as good as the quartz-halogen lamp in terms of its ability to produce visually detectable images of dimly stained objects, while consuming only a few hundred milliwatts of electrical power, rather than tens of watts.

Illumination Optics for Lamps and LEDs

Figure 1-21 and the related discussion on p. 51 show that two simple cylindrical lenses are all you need to get proper illumination for a flow cytometer using one (or more) laser sources; the monochromatic nature of laser light is, in general, all that is needed to define the excitation wavelength range.

The optics required for lamp illumination in flow cytometry are more complicated. Even when the same microscope objective is used for illumination and collection, as in an epiilluminated fluorescence microscope or a typical arc source flow cytometer, we also have to have a condenser lens for the lamp, and possibly a few additional lenses and diaphragms, in addition to some optical filters and a dichroic mirror (Figure 1-4, p. 9).

What we will consider first, however, is the plain, old-fashioned **transmitted light microscope** and its illumination optics. Separate lenses, i.e., the **condenser** and **objective**, are used for illumination and light collection. This configuration has also been used in flow cytometers, beginning with Kamentsky's[65]; the original Block system[88] used two opposed pairs of lenses to permit absorption and scatter measurements, and Technicon's Hemalog D differential counter[84,85] and the Steen/Lindmo instrument[100-103], now being resurrected by Apogee Flow Systems, also used or can use opposed lenses.

In transmitted light microscopy, one of the goals (if I said objectives that would just confuse things) is generally high spatial resolution. Resolution increases with increasing N.A., which is a function of the angle over which light is collected. Now, if we are concerned with scattered light or fluorescence, we can essentially treat our specimens as what optical theorists call **self-luminous objects**. It doesn't matter whether we illuminate the specimen over a large angle or a small one; light will be scattered, and fluorescence emitted, in all directions, i.e., over a solid angle of 4π steradians.

If what we are looking at, however, is light transmitted by the specimen, the situation is quite different. In order for the light to come out of the specimen at large angles, it has to go into the specimen at large angles. Thus, the condenser has to have an N.A. equal to that of the objective. It's amazing to me how many people either don't know that or have forgotten it. The N.A. value engraved on a lens simply defines the highest N.A. you can get; actually getting it is up to you. The immersion objectives on good microscopes typically have an N.A. of 1.3-1.4, as do the condensers. However, recall that the n in the formula that defines N.A. as $n \sin \theta$ is the refractive index, which is only 1.0003 for air. In other words, if the condenser is not immersed, you might as well not bother immersing the objective, because the effective N.A. can't be higher than 1. If the condenser is immersed, but its aperture diaphragm is stopped down, restricting the region through which light enters the condenser, you also won't be able to achieve the full N.A. of the microscope objective.

The condenser lens illuminates the specimen with light derived from the source. There are two basic arrangements used for specimen illumination; they are shown in Figure 4-15. In **Köhler illumination**, an image of the light source is formed in the back focal plane of the condenser lens. Thus, light rays from the source (image) leave the condenser lens and pass through the specimen as parallel bundles. Each point in the specimen plane is illuminated by light coming from all points of the source image. The alternative illumination setup is called **critical illumination**, for which an image of the source is formed in the specimen plane.

The position of the arc in an arc lamp will change with time; this phenomenon is referred to as **arc wander**, and is more of a problem with some lamps than with others. Significant arc wander may make it preferable to use Köhler illumination, which will average out intensity fluctuations at various points in the arc image. However, the overall level of illumination obtained from Köhler illumination is typically lower than that achieved when critical illumination is used. Critical illumination can be used satisfactorily with an arc source if a large enough image of the arc is formed in the specimen plane to keep the intensity distribution homogeneous over the observation region, even in the presence of arc wander.

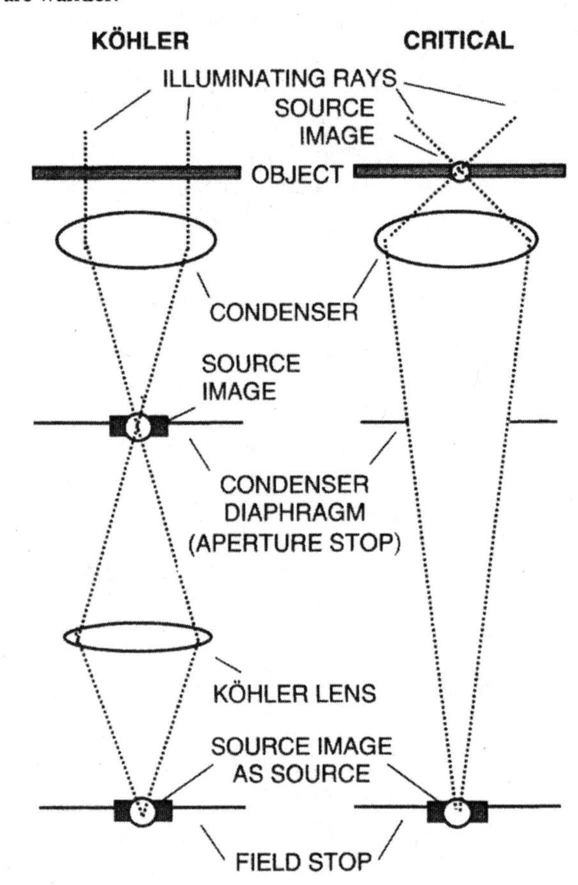

Figure 4-15. Köhler and critical illumination.

When Köhler illumination is employed, maximal illumination is achieved when the source image fills the entire aperture of the condenser in the condenser's back focal plane. The use of intermediate **relay lenses** relaxes the rather severe constraint on the physical geometry of the system that would be imposed by the use of a single lens to collect light from the source and image the source in the back focal plane of the condenser. In a typical transmitted light microscope with a substage illuminator, one or two lenses are used to collect light from the source, forming an image of the source in the plane of the substage diaphragm, which serves as a field stop. In critical illumination, the condenser is used to form an image of this image in the specimen plane; in Köhler illumination, an additional **Köhler lens** forms an image of this image in the back focal plane of the condenser.

Arc Source Epiillumination for Flow Cytometry

An optical system typical of those used in an arc source flow cytometer or an epiilluminated fluorescence microscope

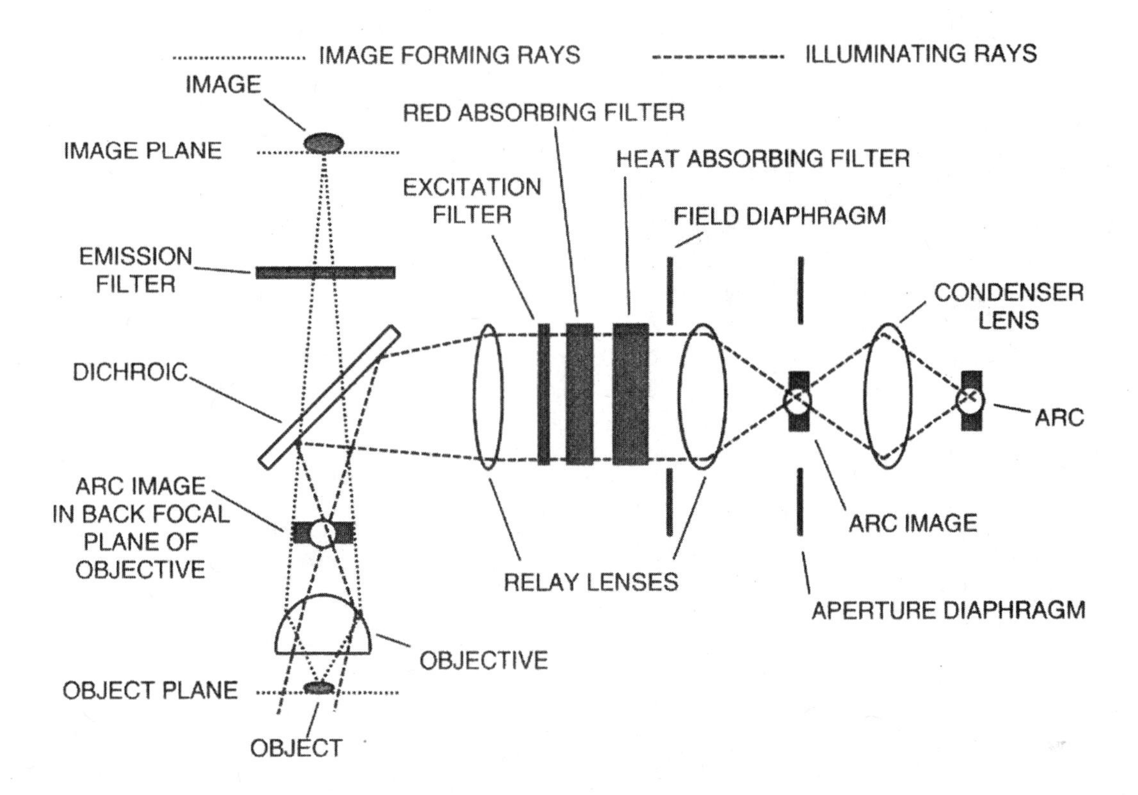

Figure 4-16. Optics for arc source epiillumination for fluorescence microscopy or flow cytometry.

is diagrammed in Figure 4-16. Infrared (heat) and red light are removed from the arc source by glass **heat absorbing** and **red absorbing filters**; since these are color glass rather than dielectric or interference filters, their transmission is not affected by angle of incidence. Their placement is not critical, but putting them between the source and the **excitation filter** used to select excitation wavelength reduces the heat load on that filter. In the past, color glass excitation filters, such as the UV-transmitting Schott UG 1, were generally used for UV and violet and sometimes used for blue illumination; interference filters were preferred for blue-green and green illumination. These days, whether you're doing fluorescence microscopy, image cytometry, or flow cytometry, it's more likely that you'll be using interference filters, and getting the right filter characteristics is critical to the quality of results.

In some instruments, a collimated beam is formed in the excitation optics to allow placement of interference filters normal to the beam; since the spectral transmission of interference filters changes with the angle of incidence, this arrangement maximizes the chance of the filter performing as specified. It is obviously difficult to avoid having converging beams impinging on the dichroic from both the illumination and collection sides of an epiilluminated fluorescence microscope, and the fluorescence collected by the objective will also pass through the **emission filter(s)** as a converging beam. However, this beam, being on the image side of a high-N.A. lens, forms a cone with a relatively small half angle; it is unlikely that any of its constituent rays strikes the emission filter at an angle far enough from normal to cause problems. It is also possible to reject even unwanted light that hits the filter at somewhat larger angle by keeping the passbands of excitation and emission filters well separated.

If critical illumination is used, images of both the arc and the object are formed in the image plane; with Köhler illumination, light from the arc is diverging as it passes through the image plane. The latter arrangement should result in less stray light from the arc getting through the filters and reaching the detector. However, if the emission filters contain a colored (absorptive) backing, filter fluorescence is more likely to occur and, because the light is diverging, one cannot simply use a field stop to eliminate the filter fluorescence, as is easily done in a flow cytometer with a laser source and orthogonal geometry.

Figure 4-16 shows a single emission filter, as would be used in a fluorescence microscope; a microscope-based flow cytometer making multicolor fluorescence measurements would use a series of filters and dichroics, similar to those shown in Figure 1-21 (p. 51), to separate the desired wavelength regions.

Lasers as Light Sources for Flow Cytometers

If arcs work for fluorescence microscopy and flow cytometry, why does anybody use lasers? Looking back over flow cytometric history, we see that Kamentsky's first instruments[65,66,78], the original Stanford sorter[82], and the ICP[83] all used arc sources. We have already seen that collecting the light from an arc source and putting into a small volume such as the observation region of a flow cytometer is a nontrivial task, just in terms of selecting lenses. We haven't really paid any attention to another major problem area, i.e., the selection of filters or other optical elements to define the excitation and emission wavelength regions that will be used. It is obviously much easier to get laser beams to go where you want them to than it is to ride herd on arc lamp emissions, and one doesn't have to tweak and/or realign laser focusing optics nearly as frequently as is required with arc lamp optics. Two-thirds of the optical filter problems associated with arc lamps are essentially eliminated when using laser sources, because you dispense with the excitation filter and dichroic, and only have to worry about emission filters for the detectors. The monochromaticity of laser light often permits you to relax the specifications on the emission filters, as well. Lasers are undeniably brighter than arc lamps; finally, before they became ubiquitous in CD players and as pointers, they were trendy and sexy and some people got (and may still get) a feeling of power from having to have the whole building replumbed and rewired so they could/can connect their instruments.

The Los Alamos group, which first reported the use of a laser for fluorescence flow cytometry[79], probably was influenced by the high radiance and the resulting ease with which an apparatus could be designed and built; since high technology tends to come to Los Alamos before it gets to a lot of other places, people there had the first crack at using lasers anyway. The use of an argon laser instead of an arc lamp in the second version of the Stanford sorter[86] was motivated by the improvement in the quality of weak (immuno)-fluorescence measurements possible using the brighter source.

There is no doubt (L. Kamentsky, personal communication) that the novelty and trendiness of lasers strongly influenced the decision in the late 1960's to use them as sources in the production Cytograf and Cytofluorograf. The latter was the first commercial product of any kind to incorporate an argon ion laser; its light source was a small, air-cooled device very much like those now used in most commercial flow cytometers, with a power output of about 10 mW. The laser machismo which subsequently developed among commercial and noncommercial builders of flow cytometers made it difficult, during the 1970's and early 1980's, to appreciate that bigger was not necessarily better. Fortunately, times have changed.

Laser Illumination: Going to Spot

Laser sources differ radically from extended sources in several respects; one is that the emission from lasers is confined to a very small solid angle, so it is generally possible (neglecting transmission losses) to focus all of the energy in the beam to a circular or elliptical spot. Most lasers used for flow cytometry emit a so-called **TEM_{00}** beam, in which the energy distribution is **Gaussian**. Spot "diameters," in the case of a circular spot, or "width" and "height," in the case of an elliptical spot, define the "**$1/e^2$ points,**" at which intensity is $1/e^2$ (or 0.135) times the value on-axis. Approximately 87.5 percent, or the fraction $[1 - (1/e^2)]$, of the total emission is contained in the region within the $1/e^2$ points; the area of this region corresponds to the area of a central elliptical region of a bivariate (two-dimensional) Gaussian (normal) distribution within two standard deviations of the bivariate mean.

Let's consider a typical laser illumination setup for a flow cytometer. The laser is assumed to be emitting a beam with a diameter of D mm. We will assume that the beam is collimated, i.e., not diverging at all, because its actual divergence is pretty small. If we took a convex spherical lens of focal length *f* mm and placed it in the beam, with its axis coincident with the beam axis, the beam would be focused to a round spot with diameter d µm at the focal distance *f* mm from the lens. The formula generally used for estimating the spot diameter d is

$$d \cong (4/\pi)(\lambda f/D) \cong 1.27(\lambda f/D),$$

with λ being the wavelength (in µm, **not** nm) of the laser emission. This formula neglects lens aberrations and assumes that the beam is focused to a **diffraction limited spot**; the formula is not, and cannot be, derived from geometric optics.

Loken and Stall calculated the spot size normally obtained in the original B-D FACS, which focused the beam with a 125 mm focal length spherical lens, as 55 µm with 515 nm illumination and 61 µm with 458 nm illumination[6], based upon laser beam diameters of 1.5 mm and 1.2 mm, respectively, at these wavelengths. Let us assume the center of a 20 µm wide core stream runs along a diameter of a 60 µm spot, and determine the variation in illumination over the width of the core. A table of the Gaussian distribution shows that the $1/e^2$ points, 30 µm off-axis, represent distances of 2 standard deviations (S.D.) on either side of the mean. A point 10 µm off-axis is thus 0.67 S.D. from the mean, and receives only 0.79 times peak beam irradiance.

What this means is that a small particle traveling near the outside of a 20 µm core stream receives only 79 percent as much illumination as an identical particle traveling down the core axis. If fluorescence emission is a linear function of illumination intensity, two identical particles following these different trajectories will produce signals differing in amplitude by over 20 percent. This does not make for low coefficients of variation. If one observes a low CV in a fluorescence distribution measured with this beam geometry and core size, the only possible explanation is that the illumination power used is at a level at which the change in fluorescence intensity for a given change in illumination intensity is relatively small (p. 117), minimizing the effects of uneven illumination. If you have enough laser power, you can be sloppy about illumination optics.

If the laser power level is lower, there are two obvious ways to improve performance of the system just described. One is to make the core smaller. Suppose we keep the 60 µm round spot and go to a 10 µm core diameter; the edges of the core are now 5 µm off axis, at which point the irradiance is almost 95 percent of the peak irradiance. That's better than 80 percent but it still isn't where we want to be. We'd have to have a 5 μm core to keep illumination variations to within 1 percent of the peak value with a 60 μm spot. This is acceptable if you're not processing 10 μm cells; if you want to look at bacteria, or chromosomes, you can probably get to a 5 μm core and keep the round spot as is. In fact, the unmodified B-D FACS didn't do badly at chromosome analysis[109], although high laser power, rather than smaller particle or core size, was probably responsible for the low CVs.

It's not that easy to use a really small core if we're looking at cells, and easiest way to get more even illumination over the width of the core is to form an elliptical focal spot, using **crossed cylindrical lenses**, as shown in Figure 4-17, decreasing the 60 μm dimension along the direction of flow and increasing the dimension perpendicular to the direction of flow. Going back to the Gaussian distribution, we find that we can keep illumination at over 98 percent of the peak value within the central 10 percent of the beam width. With a 100 μm beam width, we can handle a 10 μm core. By keeping the beam height a few times as large as a cell diameter, we avoid the "slit-scan" effect discussed on p. 51, and can use peak detectors instead of integrators, making the signal processing simpler, but losing the ability to derive information about cell size from pulse width or to discriminate doublets by their pulse height vs. pulse area. If we do want the additional information, we can simply use a shorter focal length lens, and add the extra electronics.

I should note that, in some of its instruments, B-D uses a wedge, or prism, and a spherical lens, rather than crossed cylindrical lenses, to form an elliptical spot. Either way, lasers make optical design much easier than do lamps.

If you want to focus down to extremely small spots, e.g. 5 μm or less, it may be advisable to expand the laser beam first. This is not usually done in commercial laser source flow cytometers. I have tried beam expansion several times, and find that it causes as many problems as it cures, so I've given up on it; commercial instruments, by and large, don't use beam expansion.

Suppose you want to use two or more laser beams? My approach, and that used in many commercial instruments, has generally been to use mirrors to get the unfocused beams lined up in space and then to put them through a single set of crossed cylindrical lenses. Others, such as the groups at Livermore and Los Alamos, have gone to more elaborate arrangements with the beams at small angles to one another going through different sets of lenses, primarily to get very precise control over focus and spot size for high-precision analyses of chromosomes. If there isn't a lot of power to spare, and/or if the beam diameters and wavelengths used are different enough so that one lens pair won't give you acceptable dimensions for the spots from both lasers, it may be easiest to place the lasers and their focusing optics on opposite sides of the flow cell; instruments from both Cytomation and Luminex follow this practice, and I've been doing it in my Cytomutts, when necessary, for years.

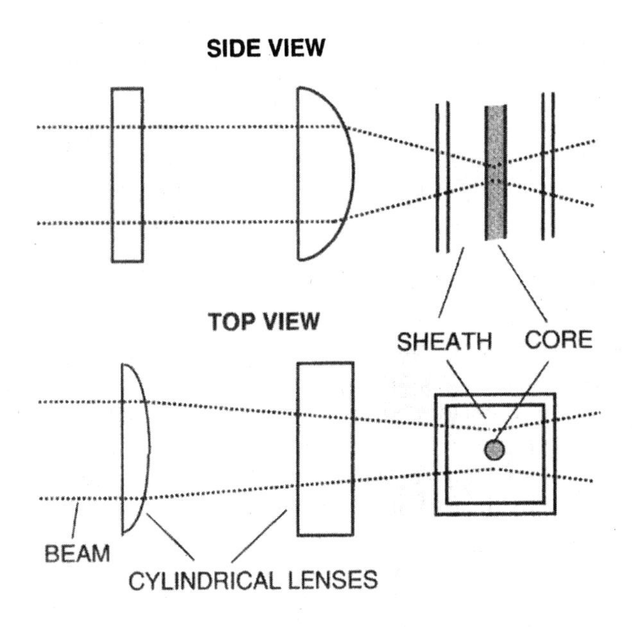

Figure 4-17. Use of crossed cylindrical lenses to focus a laser beam to an elliptical spot on the core stream.

Shedding Light on Cells: Lasers, Lamps, and LEDs

By now, it should be clear that lasers and arc lamps are usually the most realistic choices as light sources for flow cytometers. However, I have mentioned that LEDs have been used successfully as sources for two common types of flow cytometric analysis, namely, DNA content determination and immunofluorescence measurements. So what's the bottom line? How much light can each of these sources get through a cell? Let's do the math and find out.

If a laser beam were focused to an elliptical spot 100 μm wide by 20 μm high, a not unreasonable size to use in a flow cytometer, it can be calculated, from characteristics of the bivariate Gaussian distribution, that about 10 percent of the total power in the beam would illuminate a cell 10 μm in diameter at or near the center of the focal spot. That percentage figure will hold for any laser wavelength or power. If the laser in question is an air-cooled, 488 nm argon ion laser, emitting about 20 mW, probably the most widely used source in fluorescence flow cytometers at present, we get all of 2 mW through our 10 μm diameter cell. That is an illumination intensity of 2.55×10^7 W/m^2; since a 488 nm photon has an energy of 4.07×10^{-19} J, the figure is equivalent to 6.27×10^{25} photons·m^{-2}·s^{-1} (photons per square meter per second), and the photon flux through the 10 μm cell is 4.9×10^{15} photons·s^{-1} (photons per second).

There would be less power and fewer photons from a focused laser beam impinging on smaller targets. Only about 1 percent of the light (roughly 5×10^{14} photons·s^{-1}) would hit a 1 μm diameter bacterium; about 0.002 percent of the light (10^{12} photons·s^{-1}) would hit a 0.1μm virus particle.

The laser calculation is relatively easy; to get the corresponding numbers for lamps and LEDs, we have to dig up data on radiance from sources such as Figure 4-14 (p. 125). The figure originally appeared in an optics catalog from Oriel Corporation, now Thermo Oriel (Stratford, CT; www.oriel.com), a major supplier of light sources. It compares the **spectral irradiance** of mercury (Hg) and xenon (Xe) arc and quartz tungsten halogen (QTH) filament lamps. The irradiance values are given in $mW \cdot m^{-2} \cdot nm^{-1}$ (milliwatts per square meter area per nanometer wavelength) at a distance of 0.5 m; the corresponding radiance is calculated by determining the solid angle represented by a square meter at this distance and by integrating over an appropriate wavelength region.

The 100W Hg arc lamp has an irradiance of about 120 $mW \cdot m^{-2} \cdot nm^{-1}$ in a 10 nm band around the strong ultraviolet (UV) line at 366 nm, so total irradiance in this spectral band is approximately 1.2 $W \cdot m^{-2}$. The surface area ($4\pi r^2$) of a sphere with a radius of 0.5 m is $4\pi \cdot (0.25)$ m^2. Since this is the surface area subtended by a solid angle of 4π sr (steradians), 0.25 m^2 is the surface area occupied by 1 sr; the radiance is thus about 0.3 $W \cdot m^{-2} \cdot sr^{-1}$, or 300 $mW \cdot m^{-2} \cdot sr^{-1}$.

The arc in a 100W Hg lamp is 0.25 mm (250 μm) in diameter; its surface area is $4\pi \cdot (.000125)^2$ m^2; therefore the surface area of a 1 sr segment of the arc is $(.000125)^2$ m^2, or 15625 μm^2. The power radiated through 1 sr is 300 mW. A spot 10 μm in diameter on the surface has an area of $\pi \cdot (25)^2$ μm^2, or 78.5 μm^2; thus, about 1.5 mW [(78.5/15625)·300] of UV power in the bandwidth discussed would be emitted through this surface. This represents the maximum amount of power collected from the arc which can be directed back through the same area, thus, if the arc is used to illuminate a cell 10 μm in diameter, no more than 1.5 mW can impinge on the cell at any given time, no matter how efficiently light is collected from the arc and transmitted through the optics.

Mercury arc lamps have strong emission lines at several wavelengths other than 366 nm; peak radiance near that of the 366 nm line is also obtainable at 313 nm, farther in the UV, at 405 nm (violet), 436 nm (blue-violet), 546 nm (green), and 578 nm (yellow). Under the best conditions, neglecting transmission losses in lenses and in the filters used for wavelength selection, one would expect to be able to direct at most 1-1.5 mW in any of these wavelength regions from an Hg arc lamp through a 10 μm cell at any time. Since we have already figured out that we can get 2 mW (10 percent of the total beam power) through a cell with relatively little effort if we use a 20 mW laser source, we can calculate that the 100W Hg arc lamp will, at best, provide cytometrically usable light output equivalent to that from a 15 mW laser. That's more than adequate to do a lot of cytometry, especially when oil immersion lenses are used for illumination and light collection, but it's suboptimal for applications such as high speed sorting.

Illumination intensities obtainable from Hg arc lamps in the real world may be even lower than the above calculations suggest. Unger et al[2446] measured an illumination intensity of 4.8×10^5 $W \cdot m^{-2}$ in a fluorescence microscope using green (546 nm) illumination from a 100W Hg arc lamp, with a N.A. 1.4, 60 × oil immersion objective serving as condenser and collector lens. At 546 nm, a photon has an energy of 3.64×10^{-19} J; the photon flux through a 10 μm cell in this setup would therefore be about 10^{14} $photons \cdot s^{-1}$, or about 1/50 of what one could expect to get from a 20 mW laser. However, this level of illumination proved more than adequate to detect fluorescence from single molecules of a tetramethylrhodamine conjugate using a cooled CCD camera and an observation time of 100 ms.

A xenon arc lamp, as can be seen from Figure 4-14, has a relatively flat emission spectrum between the near ultraviolet and the near infrared (350-750 nm); this makes the Xe arc a desirable illumination source for spectrophotometers and spectrofluorometers. However, since the radiance of the Xe arc over this range is only about 1/10 the radiance of the Hg arc at its strong emission lines, the Xe arc is less desirable as a source for cytometry, except possibly in the region between 450 and 500 nm, where its radiance is slightly higher than that of Hg arc lamps. In any given 10 nm wavelength band, a Xe arc lamp probably won't put any more photons through a 10 μm cell than would a 200 μW laser.

The radiance of a quartz tungsten halogen filament lamp is substantially lower than that of a xenon arc lamp below 600 nm, and slightly higher between 600 and 800 nm. However, the area of the emitting surface of the filament lamp is much larger than the area of the arc in an arc lamp; thus, much less power – probably only a few tens of microwatts at most – can be collected from and directed through a small area. This makes filament lamps poorly suited for fluorescence excitation in flow cytometry, although they have been used quite successfully in flow cytometers which measure absorption and light scattering. For these purposes, they have the advantage that it is relatively easy to achieve precise regulation of output power.

What about LEDs? The current brightness champion among blue LEDs is Cree, Inc.'s XBright 470 nm device, which emits 150 mW from an 810 × 810 μm surface area. The emission from the 78.5 μm^2 area corresponding to the area of a cell 10 μm in diameter is about 18 μW, or less than 1/100 what one could get from a 20 mW laser. However, measurements made on my fluorescence microscope show that a blue LED should deliver about twice as much light to a cell as can be obtained from a quartz halogen lamp.

I'm sure LEDs will get brighter, but I don't think they will ever get hundreds of times brighter, so I don't see them replacing lasers for many flow cytometry applications. However, I will mention that Agilent is now producing a simple flow cytometer module that uses a red diode laser and a blue LED as light sources. It was designed for a restricted range of applications, and it appears to do the jobs it was designed to do. I have also seen a very impressive, inexpensive fluorescence imaging cytometer prototype with a blue LED source. The next few years should be interesting, but, for now, we need to focus, and focus on, lasers.

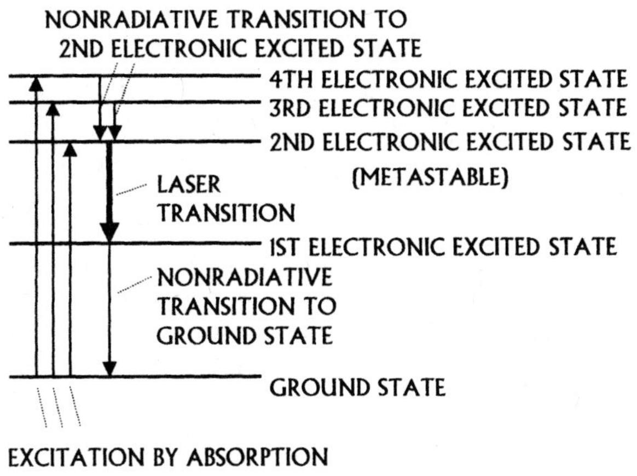

NONRADIATIVE TRANSITION TO
2ND ELECTRONIC EXCITED STATE
4TH ELECTRONIC EXCITED STATE
3RD ELECTRONIC EXCITED STATE
2ND ELECTRONIC EXCITED STATE
(METASTABLE)
LASER
TRANSITION
1ST ELECTRONIC EXCITED STATE
NONRADIATIVE
TRANSITION TO
GROUND STATE
GROUND STATE
EXCITATION BY ABSORPTION

Figure 4-18. Energy levels involved in laser action.

Lasers: The Basic Physics

Einstein on the Beam: Stimulated Emission

The word "**laser**," as previously noted, is an acronym for "Light Amplification by Stimulated Emission of Radiation." The physical process behind all lasers is **stimulated** or **induced emission**, described by Einstein in the early 1900's. In order for any kind of light emission to occur, the prospective emitter, an atom, ion, or molecule, must be excited by absorption of a photon, raising an electron to a higher energy level. After a brief period of time, the molecule typically returns to a lower energy state by emitting a photon with energy less than or equal to that of the absorbed photon. Under most circumstances, only a small fraction of the molecules in a material are in excited states, and the photons emitted from different excited molecules are different in wavelength, phase, and polarization; such emission as occurs is called **spontaneous emission**. However, as Einstein showed, once a molecule (or atom or ion) has been excited by absorption, the mere presence of a photon or photons of a particular energy in its vicinity increases the probability that it will emit a photon of the same energy (frequency or wavelength), phase, and polarization. Thus, photons can **induce** or **stimulate** the emission of like photons, and the light generated by stimulated emission is **monochromatic**, and **coherent**, i.e., the emitted radiation is at the same wavelength, in phase with, and propagating in the same direction as the stimulating radiation. No other mechanism can generate light with such uniform characteristics.

Stimulated emission becomes more likely as the fraction of the molecules in excited states increases, and can become self-sustaining when there is a **population inversion**, i.e., when the excited molecules outnumber those in the lower energy state. In general, it is difficult to create population inversions for energy transitions between the lowest excited state and the ground state of a molecule, because the ground state is more favorable on thermodynamic grounds according to the Boltzmann law (p. 109). Many practical lasers emit at a wavelength corresponding to the energy of a transition between a **metastable** higher energy excited state, i.e., one with a relatively long lifetime, and a lower energy excited state. The lasing medium is excited, or **pumped**, by electrical energy or by a high-intensity light source, causing the molecules in the medium to undergo transitions to excited states with energies equal to or higher than that of the metastable state; those at higher energies subsequently drop to the metastable state nonradiatively. Initially, spontaneous emission occurs at a particular laser wavelength as molecules drop from the metastable state to the lower excited state; thereafter, spontaneously emitted photons stimulate the emission of additional photons at that laser wavelength and the process continues. The population inversion required to sustain stimulated emission is maintained because molecules rapidly leave the lower energy state of the laser transition by thermodynamically favorable transitions to excited states of still lower energy or to the ground state. A diagram of the energy levels typically involved in laser action appears as Figure 4-18.

Although the acronym "laser" stands for "Light Amplification by Stimulated Emission of Radiation," an operating laser is more like an amplifier that has been driven into oscillation by application of positive feedback. One can drive an audio amplifier into oscillation in this fashion by placing a microphone in front of the speaker; the **resonant frequency** of oscillation is a function of the distance between the microphone and speaker.

Look, Ma, One Cavity: Optical Resonators

The initiation of stimulated emission in a volume of a suitable material will not in and of itself produce the concentrated, low-divergence light beams that characterize lasers and on which so much of their utility depends; it will, instead, result in light emission in all directions, i.e., over a solid angle of 4π steradians. This is so because, whereas the photons produced by stimulated emission travel in more or less the same direction as the stimulating photons, the spontaneously emitted photons responsible for the first round of stimulation do not have any directional preference. It is therefore necessary to perform some geometrical and optical manipulations in order to make a usable laser.

First, the volume of **lasing medium** in which stimulated emission occurs is shaped to produce some directionality of emission. As was just mentioned, spontaneously emitted photons are equally likely to be emitted in any given direction, and photons produced by stimulated emission, which follow the paths of these stimulating photons, will therefore also be equally likely to be emitted in any given direction. The probability that one photon will stimulate emission of others in the medium is proportional to the length of the path of the photon in the medium. If the medium were formed into a spherical shape, this average path length would be the same in all directions. The **gain** of the lasing medium, i.e., the number of stimulated photons emitted per unit distance per incident photon, is predominantly dependent on the quantum mechanical properties, i.e., energy levels and transition probabilities, of the medium. If the gain is high enough, and the intensity of excitation of the medium is sufficient, stimulated emission may be sustained in a spherical volume, but emission will be neither directional nor coherent.

In gas, ion, and solid-state lasers, the lasing medium is shaped into a long, thin cylinder or rod; photons emitted parallel to or at small angles to the axis of this cylinder are more likely to stimulate emission than photons emitted along or near the radius, because the path of the axial photons is substantially longer. Thus, the geometry of the medium will favor emission along the axis. Making the medium longer will, in general, increase the amount of power that can be obtained.

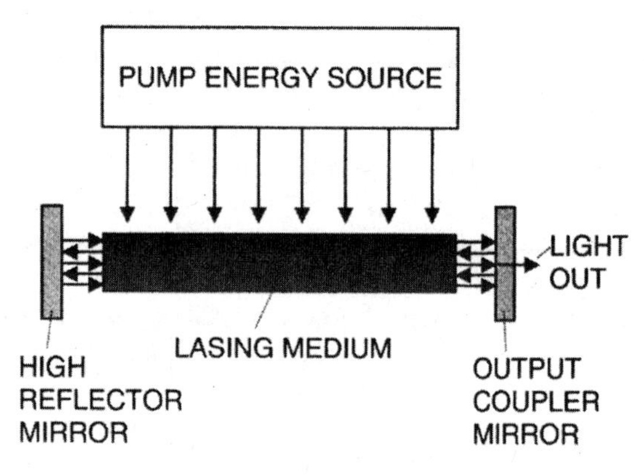

Figure 4-19. Schematic of a laser. The cavity is the region between the mirrors.

The directional property achieved by shaping the lasing medium is augmented by placing the medium inside a relatively rigid structure, called an **optical resonator**, with precisely aligned and spaced mirrors, highly reflective at the desired output wavelengths, mounted at opposite ends. Light emitted along the axis of the resonator is reflected back along the same path again and again; light at increasingly larger angles to the axis is less and less efficiently reflected back through the medium. Since light produced by stimulated emission is identical in wavelength, phase, and direction to the stimulating light, most of the emission confined within the **laser cavity**, i.e., the space between the mirrors, will be concentrated along or very near its axis.

Laser output is produced by making one of the mirrors, called the **output coupler**, able to transmit a small fraction of incident light; the amount of transmission permissible varies with the gain of the medium, which must be high enough to make up for the light lost by transmission outside the cavity and the light lost by absorption within the cavity. The mirror opposite the output coupler, called the **high reflector**, is made to reflect as much light as possible. The spacing between the mirrors is critical. If they are an even number of wavelengths apart, there will be constructive interference between the rays incident on and those reflected from the mirrors, maximizing output; if not, there will be destructive interference, which may be enough to prevent laser action entirely. A schematic of a laser is shown in Figure 4-19.

Laser Action à la Mode

The resonator can be thought of as analogous to an organ pipe; the length of the pipe, and the effective distance between the mirrors of the resonator, determine the frequency of the standing wave sustained by the structure. In the case of the resonator, this characterizes what is known as the **longitudinal mode** of the laser.

The energy profile of the beam itself, or the **transverse electromagnetic mode (TEM)** of the laser, is determined by the geometry of the medium as well as by the geometric optics of the mirrors. If stimulated emission is confined to a volume close to the axis of the resonator, the laser will operate in what is called TEM_{00} (pronounced "tee-ee-em-zero-zero"); the intensity profile associated with this mode is Gaussian. As the effective cross section of the medium increases, other transverse excitation modes, cartooned in Figure 4-20, are superimposed on TEM_{00}. These modes are individually undesirable in lasers designed for use in cytometry because they are, in general, not radially symmetric, but, rather, multilobed, and, therefore, are likely to produce nonuniform illumination.

Diode lasers are very different in structure from gas, ion, metal-vapor, and most solid state lasers. A diode laser is basically a light emitting diode fabricated so that polished facets on the semiconductor material itself or adjacent structures of differing refractive index reflect emitted light back into the active region of the diode, favoring stimulated emission and directional propagation. In the common **edge**

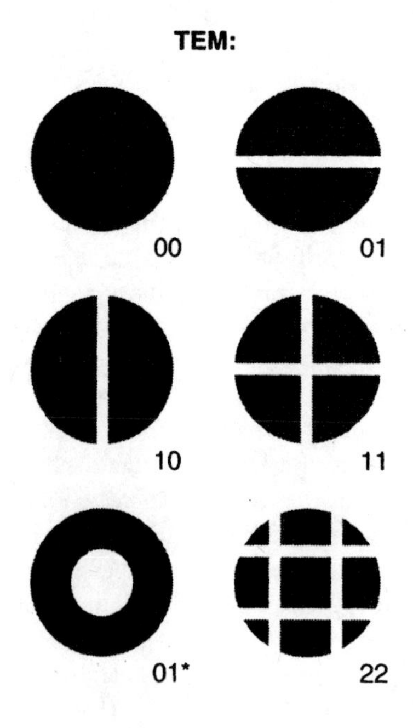

Figure 4-20. Laser transverse electromagnetic modes.

emitting diode laser designs, the emitting surface is a stripe about 1 µm high and 3 to 5 µm wide; the emission mode structure (see Figures 4-21 and 4-22) is substantially different from those shown in Figure 4-20.

Pumping Ions

To produce and maintain the population inversion necessary to sustain stimulated emission and laser action, energy must be injected from the outside. The method by which this **pumping** is done varies with the lasing medium used.

In **gas** [e.g., **helium-neon (He-Ne)**], **ion** (e.g., **argon** and **krypton**), and **metal vapor** [e.g., **helium-cadmium (He-Cd)**] lasers, an electric current is used to produce the **plasma** which serves as the lasing medium; in some of these lasers, particularly larger ion lasers, a magnetic field is used to confine the plasma to a region near the axis of the **plasma tube** in which the medium is contained.

Pulsed dye lasers, in which the lasing medium is a solution of fluorescent dye, and **pulsed solid-state** lasers, in which the lasing medium is a rod made of a material such as **ruby** or **yttrium aluminum garnet (YAG)**, are typically **optically** pumped; light from a flash lamp is often the pump energy source. In a **CW (continuous wave**, as opposed to pulsed) dye or solid-state laser, the CW output of a **pump laser**, typically an ion laser in the former case and a diode laser in the latter, is used. **Diode lasers** themselves are pumped by input of electric current. Since a substantial power density of excitation is typically necessary to produce a population inversion, all laser types have a **threshold level** of pump power, below which laser action cannot be achieved.

Laser Efficiency: Your Mileage May Vary

The **efficiency** of lasers varies greatly. An argon ion laser emitting a watt or so of light typically consumes about ten kilowatts of electrical power while in operation; the overall efficiency of this system is therefore on the order of 0.01 percent. A CW dye laser, optically pumped with the 1-watt output of the argon laser, might emit a few hundred milliwatts; the efficiency of the dye laser, neglecting the power consumption of the pump laser, is typically 20-30 percent. Diode lasers are also relatively efficient.

Less efficient lasers are more likely to require forced-air or water cooling, particularly when high power outputs are needed; this increases their size, complexity, and cost. Efficiency is strongly dependent on the gain of the laser, which may vary substantially for different laser lines. In the example of the argon laser given above, the same power input that could produce 1 watt of visible output might produce only 10-20 mW of UV output; efficiency would then drop to 0.00001-0.00002 percent.

Mirrors and Prisms for Wavelength Selection

In media that support laser action at different wavelengths, the **gain** at different wavelengths varies, profoundly affecting efficiency. When gain is low, higher mirror reflectivity is needed to maintain laser action, thus, less light can be allowed to pass through the output coupler mirror; the lower the transmission of this mirror, the lower the laser's output. Because of this, and because the combination of high reflectivity and controlled transmission of laser mirrors is achieved through the use of dielectric coatings, it is not generally possible to produce mirrors which will reflect and transmit appropriately in all of the spectral regions in which media such as argon and krypton exhibit laser action. It is thus generally necessary to use **interchangeable mirror sets**, each designed for emission over a wavelength range of no more than 100 nm. The mirrors with the broadest bandwidth are those typically installed in the krypton and argon-krypton ion lasers used for light shows, which allow simultaneous emission of light in the blue, green, yellow, and red spectral regions between 460 and 680 nm. In some instances, there is **competition** between two lasing processes during multiline operation; the power levels of the yellow and red lines in krypton lasers frequently exhibit seesaw behavior on this basis.

In medium and high power lasers, wavelength selection within the spectral range attainable with a single set of mirrors is generally done by insertion of a **Littrow prism** in the cavity between the mirrors. The **dispersion** of the prism results in light of different wavelengths being refracted at different angles on passage through the prism. At any given position of the prism, only a relatively narrow range of possible emission wavelengths will be reflected along the axis of the laser cavity between the high reflector and output coupler mirrors. Gain in this selected wavelength range will be sufficient to maintain laser action; gain at wavelengths above and below the selected range will not. The emission wavelength is changed by changing the orientation of the prism; this usually involves a vertical angular adjustment.

Dispersive elements other than prisms, e.g., **gratings**, can also be used for wavelength selection; it is also possible to insert an **optical filter** or another interference-based component, an **etalon**, in the laser cavity to restrict the range of emission wavelengths by reducing transmission, and, therefore, gain, outside of the desired narrow wavelength range. Although low power argon lasers of the type most commonly used in benchtop flow cytometers can, like their larger counterparts in sorters, be equipped with Littrow prisms, most are instead fitted with fixed narrow bandwidth mirrors that confine output to 488 nm. The He-Ne and diode lasers used for cytometry also do not make provision for the emission wavelength to be changed.

Brewster Windows for Polarized Output

In many types of lasers, **polarization** is introduced into the beam by putting windows between the medium and the end mirrors. The windows are placed at **Brewster's angle** (pp. 107-8) to the axis of the system. At this angle (about 57 degrees for glass), reflection from the window surface is minimized for light of one polarization, while a small percentage of light of the perpendicular polarization is reflected out of the cavity. The slight difference in transmissions of the two polarizations is magnified many times by the feedback characteristic of the optical resonator structure, with the result that the laser output in a system with such **Brewster windows** is highly polarized, typically in a ratio of at least 500:1.

Laser Power Regulation: Current and Light Control

Ion lasers, small or large, can generally be operated in either a **current control mode** or a **light control mode**. In the current control mode, the laser power supply is regulated to deliver a constant current; if the mechanical and optical characteristics of the laser do not change during operation, light output remains constant. If things change, e.g., if a mirror becomes slightly misaligned, light output decreases even though power supply current remains the same. In the light control mode, power supply output is regulated by a feedback circuit that samples the energy in the beam and adjusts the laser current to maintain constant light output. This works well when the laser is emitting at a single wavelength. When emission of several lines occurs simultaneously, it is more difficult to keep power constant, particularly if the gains differ considerably and/or if there is competition between lines.

The air-cooled argon lasers in benchtop flow cytometers are operated in the light control mode; so are diode lasers, which are usually built with a light-sensing photodiode in the same package. In the case of diode lasers, the incorporation of a light control feedback loop into the power supply is almost essential to prevent the laser from frying itself when it is turned on. He-Ne and He-Cd lasers typically do not incorporate light control circuits.

Beam Profiles and Beam Quality

The ion lasers widely used for cytometry are usually operated in the radially symmetric TEM_{00} or Gaussian mode, discussed on p. 134. He-Cd lasers emitting at 325 nm, used as UV sources in low-power systems, often emit in the TEM_{01^*} or "donut" mode, so-called because it produces a radially symmetric spot with a dark center. These and other modes were sketched in Figure 4-20 (p. 134). Actual Gaussian, "donut" and "yecchh" (loads of modes with nodes) intensity profiles, measured from He-Ne, CO_2, and diode lasers, are illustrated in Figure 4-21.

The intensity profiles in Figure 4-21 are drawn as isometric "peak and valley" plots; the shading lines in this figure were in different colors in the original, reflecting the tendency of laser beam profiling software to use overkill (chromatic *and* isometric representation) in data display. You can find some of that in flow cytometry software, too.

The Gaussian profile of the He-Ne laser beam in Figure 4-21 is almost as smooth as a computer-generated curve. The "donut" from the CO_2 laser is closer to a Bundt cake than a bagel, suggesting that the beam is not of the highest quality. Then we have the diode laser, which looks really

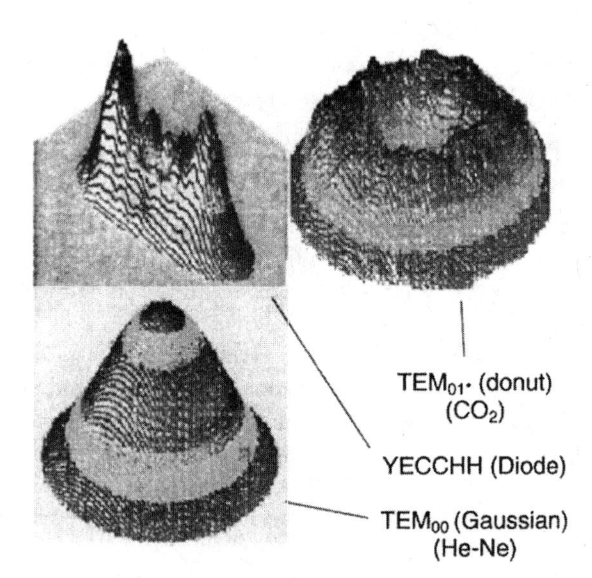

Figure 4-21. Beam intensity profiles of a CO_2 laser, a diode laser, and a He-Ne laser (courtesy of Excitech, Ltd.)

ugly, as most of them tend to do; the structure of diode lasers, as was mentioned on p. 135, is quite different from that of gas, ion, and most solid state lasers, and the ugly beam comes with the territory. But there is good news; it is possible, using fairly simple optics, to clean up a diode laser beam to a point at which it looks almost Gaussian, as shown in Figure 4-22.

BEAM FOCUSED

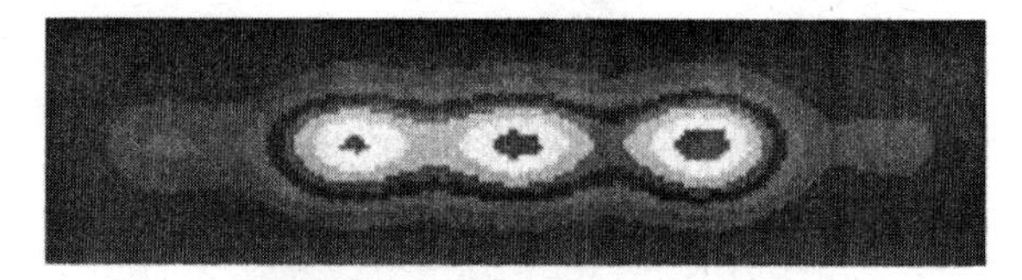

REAR CYLINDRICAL LENS SLIGHTLY DEFOCUSED

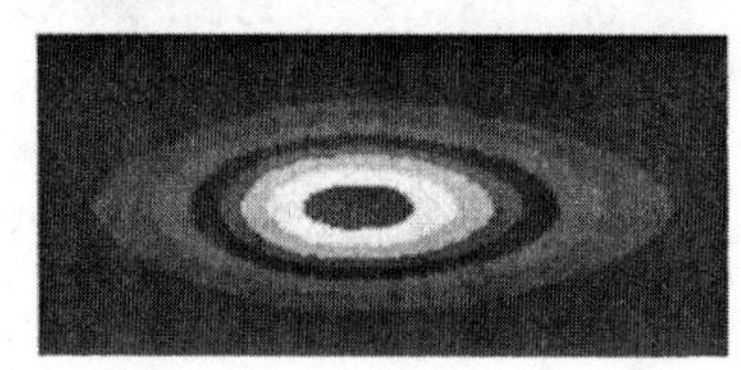

Figure 4-22. Profiles of the beam from a 397 nm (violet) laser diode, focused through crossed cylindrical lenses.

The displays of Figure 4-22 show chromatic plots of intensity at different points in the focused beam of a 397 nm violet diode laser. The top panel shows the profile obtained at the point of focus of both crossed cylindrical lenses; there are multiple modes similar to those that appear in the diode laser profile in Figure 4-21. The bottom panel of Figure 4-22 shows a much smoother, near Gaussian profile, achieved by defocusing the rear cylindrical lens slightly. We were shooting for this smooth beam profile in order to be able to do fluorescence measurements with reasonably good precision; with only 4 mW of available power, it was unlikely that we'd get help lowering the CV's from bleaching and saturation effects. And our optical manipulation worked; we measured DNA content in DAPI-stained nuclei of cultured lymphoblasts, with a 1.7% CV for the G_0/G_1 peak[2444].

Several manufacturers now offer violet, red, and infrared diode laser systems incorporating optics that produce a circular beam that is very close to Gaussian in its intensity profile. In Blue Sky Research's CircuLaser™ modules, this is accomplished by putting a high-N.A. (0.7) cylindrical microlens into the case housing the laser diode, capturing the entire beam and reducing its relatively large divergence in the vertical direction to match the smaller divergence in the horizontal direction. Although a circular beam can also be obtained using lenses external to the diode package, this generally involves the loss of more of the laser's output power.

A Gaussian beam should be focusable to a diffraction-limited spot; an approximation formula for calculating spot diameter was given on p. 130. Actually getting a spot of the size predicted by the formula would require both an ideal Gaussian beam and a lens that was diffraction limited, rather than limited by aberrations. There is also a throughput issue associated with focusing laser beams; to put it most simply, **the smaller the spot, the smaller the distance over which it stays small.**

Compared to just about anything else, the beam that comes out of a laser looks collimated; the narrow beam that emerges from your laser pointer makes a small spot on the screen on which your PowerPoint presentation is being projected. But, if you actually bothered to measure the spot diameters at the pointer output and the screen, you'd see that the spot on the screen was bigger. The beam coming out of the laser is diverging; it just isn't diverging very much.

In actuality, the output beam of a laser has been brought to a focus, or **waist**, somewhere near its point of exit from the laser package, courtesy of the laser mirrors. A typical beam diameter at this point is somewhere between 0.5 and 2.0 mm, and the beam divergence angle is usually less than 1 milliradian. If you put the beam through a **beam expander**, which works like a telescope in reverse, what comes out is a beam with a larger diameter and a smaller divergence. If you use the telescope in an observatory, you can produce a beam a few feet in diameter with a divergence so low that the beam will be less than 100 feet across when it is reflected from a mirror structure left by the Apollo astronauts on the moon in 1969. Ever since then, astrophysicists have been able to use the transit and return time of pulsed laser beams to measure the distance from the earth to the moon.

Biophysicists, or at least the cells they look at, need smaller beam diameters, and getting smaller beam diameters means putting up with higher beam divergences. If W_0 is the radius of the beam at the waist, and λ is the wavelength, the **divergence angle** $\theta_0 = \lambda/\pi W_0$. The **depth of focus**, or **confocal parameter**, of a Gaussian beam, is defined as the distance between the points at which the beam area is twice the area at the waist. The depth of focus is usually expressed as $2z_0$; the quantity z_0 is known as the **Rayleigh range**, and is equal to $\pi W_0^2/\lambda$. The divergence and depth of focus formulas also apply to the axis dimensions (height and width) of elliptical beams; an elliptical beam diverges more rapidly in the plane of its shorter axis than in the plane of its longer axis.

It is generally impractical to attempt to use a laser beam to fill the field of view of a microscope objective; even if the illumination profile were a near-ideal Gaussian, intensity over the field would vary unacceptably. A bigger problem arises from the coherent nature of laser light; interference effects produce an intolerable "speckle" in the image. For this reason, practical use of laser light sources for cell imaging applications generally involves scanning the field of view with a relatively small-diameter beam.

Scanning laser cytometers, which produce low-resolution images, generally use focal spots 5 µm or larger in diameter, permitting uniform illumination through the thickness of a layer of cells on a slide. At 488 nm (0.488 µm), a 5 µm focal spot has a depth of focus of 80 µm; a 10 µm focal spot has a depth of focus of 320 µm. A 2.5 µm spot, used at the highest magnifications in scanning laser cytometry, has a depth of focus of 20 µm.

In confocal and multiphoton confocal microscopy, it is advantageous to have spot diameters of 1 µm or less; this concentrates excitation light in a narrow layer of the specimen, improving resolution of fluorescence images. A 0.5 µm, 488 nm focal spot has a depth of focus of only 0.8 µm.

If we're trying to do a "slit-scan" (p. 51) of cells or particles to attempt to measure cell size (diameter) from pulse width, calculating the depth of focus, $2z_0$, won't really tell us what we need to know, because it gives us the points at which the beam area is twice the minimum area, meaning that the beam radius is $\sqrt{2}$ times W_0, or (1.414) W_0. If we want the pulse width measurement to be precise, we need to be sure that the beam radius W is constant to within a small percentage at any point in the core stream. There is a formula that can help us find out. The beam radius $W(z)$ at a distance z from the beam waist is

$$W(z) = W_0\,[1 + (z/z_0)^2]^{1/2}.$$

From this, we can calculate values of (z/z_0), and, therefore, z, for which $W(z)$ is within any given percentage of W_0. For example, if we want to keep variation to within 1 percent, we set $W(z) = (1.01)W_0$, and find that $z = 0.142z_0$. The corresponding values of z for 2 percent and 5 percent variation are, respectively, $0.201z_0$ and $0.32\,z_0$.

This tells us that it would not be very practical to use a beam 2 µm high at the waist for slit-scanning; even if we are willing to tolerate as much as 5 percent variation in beam height over the core, we'd have to work with a core diameter (and a particle diameter) of less than 4 µm. If we use a 5 µm beam waist, the distance over which variation in beam height is less than 5 percent is just over 25 µm, which is a reasonably manageable number, meaning that it is larger than either a typical cell diameter or a typical core diameter.

The divergence angle for an ideal Gaussian beam focused to a given spot size can be calculated from the formula given on p. 137. We can also measure both the spot size and the divergence angle of an actual laser beam. For a given spot size, the ratio of the observed to the calculated divergence angle, called M^2, serves as a measure of beam quality. A value of 1.2 or less is considered good. However, some manufacturers (and some knowledgeable users) put more stock in M^2 than do others.

Puttin' on My Top Hat?

While scanning and confocal microscopy put all of the energy in the laser beam through the cell of interest, most of the elliptical beam in a flow cytometer, by design, does not illuminate the core stream containing the cells. If we're not using a high power laser, the Gaussian beam profile in the TEM_{00} mode forces us to widen the focal spot to get uniform enough illumination over the width of the core stream to permit high precision in fluorescence measurement (p. 131). This means that much of the beam is wasted. We could get much more usable energy through the cells if we had a "top hat" beam profile with steep shoulders and a flat top. Unfortunately, it isn't that easy to get such a profile.

There are a few ways to flatten the top of the beam somewhat; one can widen the bore of the lasing medium to produce either a mixture of TEM_{00} and TEM_{01} modes or true **multimode** output, which adds in some of the higher, lobed modes as well. Alternatively, the beam can be put through an aperture and defocused slightly. This permits use of smaller focal spots, wasting less of the energy in the beam. It is possible, although tricky, to get a somewhat flat-topped profile out of a diode laser; as a general rule, however, you also end up with ugly pulse shapes, which can cause problems in signal processing.

The newest approach to getting uniform illumination from lasers involves the use of arrays of microlenses, which produce arrays of overlapping, small, focused beams. The profiles I have seen generated using this technique have fairly steep shoulders, but there is a lot of intensity variation across the top of the top hat. I'm not sure how much of this can be smoothed out, and I'm also not sure how much of the beam energy is lost in playing this game. I'm sure we'll find out in the next few years.

Harmonic Generation and Modulation

The light produced by lasers, like all other light, has associated electric and magnetic fields, and, since the radiance of a laser beam is substantially higher than that of an incandescent source, the associated electric field intensity may be high enough to produce nonlinear responses in certain materials. One notable application of these phenomena is in **harmonic generation**, in which nonlinear effects in crystals result in generation of light at two or more times the frequency of the incident light. **Second harmonic generation**, or **frequency doubling** of the 1064 nm YAG laser line, for example, produces 532 nm; **third harmonic generation**, or **frequency tripling**, produces 355 nm. The same nonlinear crystals may also be used to produce output at the sum and/or difference of the frequencies of two incident beams; in the case of YAG lasers, frequency summing can produce emission at 473 nm. Although a reasonably broad range of crystalline materials capable of harmonic generation is available, the range of wavelengths at which continuous (CW) output can be obtained is restricted.

It is sometimes desirable to vary the output power of a laser more rapidly than can be accomplished by adjustments to the power supply. **Modulation** at frequencies up to several hundred megahertz is possible using **electro-optic modulators**, which incorporate crystals that change their refractive index as a function of an applied voltage. **Acousto-optic modulators**, which use sound waves to produce changes in density that affect the light transmission characteristics of a substrate, work at lower frequencies, generally below 100 MHz. A light sensor and a modulator, connected by a feedback circuit, placed in the output path of a laser, can be used as a "noise eater," providing light regulated output, albeit at a relatively high price.

Lasers Used and Usable in Cytometry

NOTE: Discrete laser emission wavelengths are given to fractions of a nanometer in Table 4-2, on the next page; in most of the rest of the book, I'll stick to whole numbers with inconsistent rounding off, which is what everybody does most of the time.

Argon and Krypton Ion Lasers

The most popular lasers for fluorescence flow cytometry are **argon ion lasers**; they are usually operated at 488 nm, a wavelength useful for excitation of fluorescein, phycoerythrin and its tandem conjugates, propidium iodide or ethidium bromide, acridine orange, pyronin Y, various rhodamine and cyanine dyes, anthracycline drugs such as adriamycin, and various GFP variants. Argon ion lasers (and **krypton ion lasers**, which are less friendly but work on the same principles) are available from several manufacturers. At present, Coherent seems to be the most popular source for the large, water-cooled systems; smaller, air-cooled lasers are likely to come from Melles Griot (which absorbed Omnichrome), Spectra-Physics, and Uniphase.

The lasing medium in an ion laser is a plasma, which, in the larger lasers, is confined in a strong magnetic field generated by a solenoid. A high-voltage pulse is used to ionize the gas to start the plasma, a procedure similar to that used in starting an arc lamp. Ion lasers require a high current to maintain the plasma discharge; the bigger ones also put a high current through the solenoid, generally increasing power consumption enough to necessitate water-cooling.

Ar Ion	Kr Ion	He-Cd	He-Ne	Solid State
275.4				
300.3				
302.4				
305.5				
		325.0		
333.6				
	337.4			
	350.7			
351.1				
		354.5		355 (Nd:YAG x 3)
	356.4			
363.8				
	406.7			
	413.1			
	415.4			
				430 (Cr:LiSAF)
		441.6		
454.5				
				457 (Nd:YVO$_4$ x 2)
457.9				
				460 (semi)
	468.0			
465.8				
472.7				
	476.2			
476.5				
	482.5			
488.0				488.0 (semi)
496.5				
501.7				
514.5				514.5 (Yb:YAG x 2)
	520.8			
528.7				
	530.9			
				532 (Nd:YAG x 2)
		533.7		
		537.8		
			543.5	
	568.2			
			594.1	
			611.9	
			632.8	
		635.5		
		636.0		
	647.1			
	676.4			
	752.5			
	799.3			

Table 4-2. Emission wavelengths of argon and krypton ion, helium-cadmium, helium-neon, and solid-state lasers that emit at discrete wavelengths. Diode, dye, and some solid-state lasers are tunable over ranges of wavelengths. See text for details.

In addition to the strong blue-green and green lines at 488.0 and 514.5 nm, argon ion lasers emit at 454.5, 457.9 (violet-blue), 465.8, (blue) 472.7, 476.5, (blue-green) 496.5, and 501.7 (green) nm. Emission can also be obtained in the ultraviolet at 333.6, 351.1 and 363.8 nm, and in the green at 528.7 nm, using specially coated mirrors. In addition, the largest high power argon ion lasers can produce some deep ultraviolet lines between 275.4 and 305.5 nm. An infrared argon laser line at 1090 nm is not likely to be useful for cytometry in the near term.

Ion lasers are made out of relatively esoteric materials and are complex enough in their construction so that the relationship between price and power output is highly nonlinear. An argon laser that puts out 25 mW at 488 nm costs around $6,000; for less than ten times that amount, you can almost certainly get more than fifty times the power, from whichever manufacturer you like. The bigger laser is not necessarily a better investment. The best reason to use an argon ion laser is a heavy commitment to measurements of weak fluorescence, e.g. immunofluorescence, adriamycin uptake, etc.; the 15-25 mW air-cooled argon lasers now most common in flow cytometers, which have plasma tube lifetimes of 6,000 hours or more, are more than adequate for the job in benchtop systems with efficient light collection optics. Even for larger instruments, with less efficient light collection, one could still consider using an air-cooled argon laser; systems with power outputs as high as a few hundred mW are available, generally for less than $10,000.

Despite their many glamorous aspects, large ion lasers are basically big, expensive light bulbs. The hotter you run them, the faster they burn out. In order to be capable of putting out 100 mW in the UV, an argon laser generally has to be capable of 5-6 watts "all lines" power in the visible, and 1.3-2.0 watts at 488 nm. To get 100 mW of UV out of it, you need to run it near maximum rated current, and you're apt to need a plasma tube replacement at least once every couple of years. Run the same laser, or the next smaller model, at 200 mW output at 488 nm and you're close to idle current; the plasma tube is apt to last for several years. If you build your own instruments, you can use a beamsplitter to get two 100 mW 488 nm beams, and another one to get two 50 mW beams from one of the 100 mW beams, and run three flow cytometers from a single argon laser (as in "Cerberus"), if you have a large optical table on which to put it[105]. However, it may make more sense to use smaller, air-cooled argon lasers at a ratio of one laser per cytometer.

There are several reasons why you need big argon lasers to get UV output. You need high current, because the UV lines are emitted from a higher ionization state of argon than is needed to emit the visible wavelengths, and you need large size, because gain is relatively low (low-gain lines benefit from longer cavity lengths). So, don't expect that there will be any little air-cooled UV argon lasers for $6,000, soon or ever.

It is possible[6] to produce mirrors for big argon ion lasers that permit simultaneous emission at 351/363 nm in the ultraviolet and at 488 nm and other visible wavelengths; this provides the dual-wavelength source needed, for example, to do simultaneous analysis of DNA, using Hoechst 33342, and surface antigens, using antibodies labeled with fluorescein, phycoerythrin, and phycoerythrin-based tandem conjugates. In order to emit in the ultraviolet, the laser has to be run at very high current; since the laser is much more efficient at 488 nm than in the UV, power output at 488 nm is quite high. If the sensor in the light output regulator circuit responds to both UV and visible light, or to light at 488 nm alone, there may be considerable fluctuation in UV power output. If the sensor's optical bandwidth is restricted so that 488 nm light is blocked, the sensor, and the regulation electronics, then respond to fluctuations in UV power output. The relatively large changes in current that may be necessary to keep UV output stable are then likely to result in large fluctuations in power output at 488 nm. It is probably preferable to use separate sources for the UV and the visible wavelengths, and, at present, I would favor helium-cadmium over ion lasers as UV sources.

Coherent's "Enterprise" argon laser system is capable of simultaneous output of over 100 mW at 488 nm and a few dozen mW in the UV. This laser is water-cooled by a closed, recirculating system, and is fairly widely used in cell sorters. Since many localities now have environmental regulations requiring that conventional water-cooled lasers be equipped with recirculating systems, this makes some sense; personally, I'd rather stick to air-cooled lasers. My name is Howard, and I've been dry for over fourteen years... .

Why, you might wonder, did so many people spend so much money a few years back buying instruments equipped with large argon and krypton lasers? The krypton laser is a rude beast whose hour came round at last for light shows and two-color immunofluorescence measurements. Unlike argon lasers, which have high gain lines and low gain lines, krypton lasers have low gain lines and lower gain lines. Take the same size laser that would put out five watts in all-visible-lines mode with an argon tube, put in a krypton tube instead, and you're lucky if you get one watt of visible light out. But what visible light! Blue-green, at 468.0, 476.2, and 482.5 nm, green, at 520.8 and 530.9 nm, yellow, at 568.2 nm, and red, at 647.1 (the strongest krypton line) and 676.4 nm, all at once, explaining the popularity of krypton lasers for light shows, and also explaining why the multicolor shows tend to be indoors; the outdoor spectacles need the more powerful blue-green and green argon lasers. With different mirror sets, and appropriate adjustment of solenoid magnetic fields, krypton lasers can also emit multiple ultraviolet (337.4/350.7/356.4 nm), violet (406.7/413.1/415.4 nm) and infrared (752.5/799.3 nm) lines.

If all the colors of light are the good news, the bad news is that the optimum values for gas pressure and solenoid magnetic field for krypton laser operation are different for different lines; since the gains at all lines are low, these parameters need to be controlled. Then, if the alignment is really good and the optics are really clean, you'll get laser output. You can literally spit on the mirrors of an argon laser running at 488 nm and still get laser output, albeit less than before. Krypton laser (and UV argon laser) mirrors need to be squeaky clean or you might as well not bother trying to get the laser to run. If the optical alignment isn't near-perfect in the visible, as indicated by maximum all-lines power output near the manufacturer's specs, you'll never get an ion laser to work in the ultraviolet. Low gain at all lines means that krypton laser plasma tubes run hotter and die sooner than argon laser tubes. Last but not least, there is, as I mentioned previously, **competition** among the visible krypton lines; if, for example, you attempt to run yellow and red simultaneously, in the light regulated mode, you'll probably find that, while total power stays constant, the yellow and red power outputs alternately go up and down.

Krypton lasers got the call for flow cytometry primarily because, way back in 1979, when monoclonal reagents were beginning to make staining with two antibodies practical, but before there were phycobiliprotein labels, immunologists found it difficult to do studies of two surface antigens with argon lasers. Fluorescein was and is the obvious label for demonstrating one antigen. The dye most widely used as a second label was tetramethylrhodamine, which is poorly excited at 488 nm, absorbing maximally near 550 nm.

While it was possible to resolve signals from antibodies conjugated with fluorescein and tetramethylrhodamine by using the strong line at 515 nm for excitation, provided one went to offbeat optical filter combinations and introduced an electronic compensation circuit to take out the crosstalk between the signals[115], this didn't always work well.

XRITC and Texas Red[116,117], reactive derivatives of rhodamine 101 that have absorption and emission maxima at wavelengths 30 or 40 nm longer than those of tetramethylrhodamine, cannot be excited satisfactorily by any of the argon laser lines, but can be used with either a krypton laser at 568 nm or a CW dye laser operating near the absorption maximum of rhodamine 101 at about 590 nm[118]. In 1979, when these dyes and the related instrument modifications were unveiled, it seemed easier to use krypton lasers than dye lasers; thus, these were pushed by the manufacturers and you simply had to have a krypton laser in order to keep up to the state of the art, even if you never actually did two-color immunofluorescence.

I once put a prism in an all-visible-lines or "light show mode" krypton laser beam to break the beam up into seven beams of different colors, and directed them all through a flow cytometer, just to preserve the Cytomat/Cytomutt tradition of working with more beams than anybody else had. I actually could make measurements in three or four of the beams; unfortunately, power output in individual beams jumped all over the place. The krypton laser may be used to excite porphyrin fluorescence in blood cells (C. Stewart, personal communication) or in tumor cells labeled with porphyrin derivatives, using the violet lines. I have used the blue krypton lines to excite fluorescein and propidium, but only when my argon laser was out for repairs.

Keeping the optics of an ion laser clean and in alignment is a challenge best met by following the manufacturer's instructions. Mirrors and Brewster windows are generally cleaned with methanol; the slightest contamination with grease, from fingers or elsewhere, will cause your solvent to deposit a film of crud on optics, which can cripple or kill your laser output. Acetone is a good cleaner for some mirrors and destroys others; don't use it unless the manufacturer says so. Electronic or HPLC grade solvents are generally free enough of contaminants to be safe to use for cleaning optics.

Changing mirrors and getting the laser to lase again is a tedious procedure that must be done a little bit differently for each manufacturer's lasers; it is learned by doing and remains something of a black art to even experienced laser jockeys. The first time you put the UV optics in by yourself, you should set aside a day. When you become confident you can do it in one hour, you will budget one hour on a busy day and find the task takes three hours. This is another good reason for using smaller, simpler, single wavelength lasers with sealed optics, and, as of 2002, the major cytometer manufacturers have gotten this message.

Dye Lasers

With the advent of phycoerythrin[6,114] as a second label, it became practical to do two-color immunofluorescence measurements with single-beam excitation at 488 nm, using fluorescein- and phycoerythrin-labeled antibodies. The krypton laser was no longer necessary for two-color immunofluorescence, and was not useful for three-color measurements adding Texas red or XRITC as a third label because the 568 nm yellow line from the krypton laser, while it could excite the third dye, lay smack in the middle of the emission spectrum of phycoerythrin. At this point, interest turned to CW dye lasers as excitation sources.

The lasing medium in a dye laser is a fluorescent dye, usually dissolved in an organic solvent such as ethanol or ethylene glycol. Which dye is actually used depends on the wavelengths at which operation is desired; there are dyes now available for use with blue-green/green (457-515 nm) argon ion pump lasers that permit operation at wavelengths extending from 540 to over 900 nm. While ion lasers lase only at a few discrete wavelengths, a dye laser can emit at any of a wide range of wavelengths in the emission spectrum of the dye used. Wavelength selection is usually done using a wedge, grating, or filter rather than a prism.

The dye in a CW dye laser circulates through a nozzle, producing a flat-walled stream; continuous circulation is necessary to minimize bleaching of the dye and to allow for cooling by a heat exchanger. CW dye lasers require minimal electrical power, most of which is used to operate the circulator pump. In pulsed dye lasers, especially those with low duty cycles, the medium does not either heat or bleach very rapidly, so the dye need not be circulated; a pulsed dye laser has been made using Jell-O as the lasing medium.

Some dyes bleach faster than others, but they all need to be replaced after a few months' operation. This, and the fact that CW dye lasers are relatively hard to keep tweaked, and do not maintain output power as stably as do ion lasers, originally restricted their use as sources for flow cytometry to people who do relatively esoteric things with flow systems[119] and will put up with idiosyncratic apparatus. Improvements in dye laser design, and a perceived market for sources usable for excitation of Texas red or XRITC and allophycocyanin (which, while its maximum absorption is in the red, is reasonably well excited near 600 nm), led manufacturers to promote and users to buy flow cytometers with dye laser sources. Using a single argon laser as a 488 nm source and as a pump laser allowed multicolor immunofluorescence work to be done using only one large laser.

The dye most commonly employed as a lasing medium in flow cytometry is rhodamine 6G. The threshold power required from the pump laser to achieve output from a rhodamine 6G dye laser is usually about 700 mW. With this power input, over 100 mW of light can typically be obtained from rhodamine 6G at wavelengths between about 570 and 620 nm. For Texas red excitation, the dye laser is usually operated at 594 nm; for excitation of Texas red and allophycocyanin, a longer wavelength, 605-610 nm, is often used. The disadvantage of using dye lasers in dual-laser systems lies in the resulting inability to use either the dye laser or the pump laser as a UV source. In principle, one could get around this using dual-wavelength UV/visible mirrors in the argon pump laser, but this solution requires great technical skill on the part of the user and also is apt to involve frequent plasma tube replacements. In most cases, there are better alternatives. If you must use a dye laser, you're now probably best off using a 532 nm frequency-doubled YAG laser, which runs on house current and needs no water cooling, as a pump, rather than a big argon laser.

Helium-Neon Lasers

If you want to meet a nice laser, try a **helium-neon (He-Ne) laser**. They plug into the wall, they don't need water cooling, they're relatively small and very stable, the mirrors aren't even adjustable, and the plasma tubes last for years. Besides, they're relatively cheap. The most common He-Ne lasers emit red light at 632.8 nm; they are available with power outputs ranging from less than 1 mW to about 50 mW; other visible wavelengths at which He-Ne lasers are now available include 543.5 (green), 594.1 (yellow), and 611.9 (orange-red) nm.

Red He-Ne lasers were used at Los Alamos, beginning in the late 1960's, for scatter measurements at various[110,111] angles, and, in the 1970's, were incorporated into Ortho's instruments and Technicon's hematology systems for scatter and axial extinction measurements. At Los Alamos, they used a 5 mW laser; the 0.8 mW He-Ne supplied with the Ortho systems provided less than optimal fluorescence excitation, but it had very low noise and was therefore quite good for extinction measurements. I started using 5-7 mW red He-Ne lasers around 1980, to excite fluorescence of cyanine dye probes of membrane potential[112], for RNA measurements using oxazine 1[113], and for immuno-fluorescence excitation, using antibodies conjugated to allophycocyanin[8,114]. Mike Loken's group at B-D[644] and Bob Hoffman's at Ortho[645,646] examined red He-Ne lasers as sources for one- and two-color immunofluorescence measurements using phycocyanin and/or allophycocyanin as labels; there was general agreement that red He-Ne lasers were well suited for these applications, and they began to be offered in commercial flow cytometers. In recent years, some manufacturers have substituted diode lasers emitting in the 635 nm region for red He-Ne lasers.

Until 1985, 633 nm was the only visible wavelength available from commercial He-Ne lasers, and the 633 nm He-Ne lasers have been more widely used than the 543 nm, 594 nm, and 611 nm varieties. The 543 nm green He-Ne laser is useful for excitation of immunofluorescence from antibodies labeled with phycoerythrin or its tandem conjugates, and can also be used to excite DNA stains such as propidium[2471]. However, the gain of the green He-Ne line is quite low, and available power levels remain below 2 mW. This power level is sufficient to permit immunofluorescence measurement in an efficiently designed flow cytometer; B-D uses a green He-Ne laser in the FACSCount, a dedicated instrument for counting CD4- and CD8-positive T lymphocytes based upon staining with phycoerythrin and tandem-conjugate or PerCP-labeled antibodies[2472].

The 594 nm yellow He-Ne laser is usable for excitation of Texas red and of the Texas red/allophycocyanin combination; it operates at the same wavelength at which dye lasers are often used for these purposes. Maximum power output is still only about 2 mW, adequate for immunofluorescence work if a well-designed optical system is used. The major disadvantage of the 594 nm laser is the proximity of its lasing wavelength to the emission region of phycoerythrin; the availability of large numbers of labels excitable by 488 nm and 633 or 635 nm lasers has decreased demand for Texas red and other yellow-excited labels.

Helium-Cadmium and Helium-Selenium Lasers

Helium-cadmium (He-Cd) lasers, which can emit 5-200 mW in the blue (441.6 nm) and 1-100 mW in the UV (325.0 nm), depending on size, are relatively practical sources for flow cytometry. Lower power output in the UV at 354 nm is also available in some lasers. Like He-Ne lasers, He-Cd lasers plug into the wall and do not require water cooling; they need few or no adjustments and have relatively long plasma tube lifetimes. They are also cheaper than most ion lasers, but not by much. The Block Cytomat instruments used a single He-Cd laser with mirrors designed to permit simultaneous emission at 325 and 441 nm. Similar dual-wavelength models are now commercially available; there is little or no competition between the UV and blue wavelengths. The lasing medium is cadmium vapor; the pressure of both it and helium and the temperature of the medium must be carefully controlled to assure stable operation. Melles Griot has absorbed Liconix and Omnichrome, the American He-Cd manufacturers; Kimmon, in Japan, also makes He-Cd lasers.

I have used the 441 nm blue He-Cd line (at 10-100 mW) for ratiometric intracellular pH measurement using fluorescein derivatives[8] and for excitation of the chromomycin family of DNA stains (olivomycin, chromomycin A_3, and mithramycin), and the 325 nm UV line (at 1-35 mW) for excitation of the fluorescence of the DNA stains DAPI and Hoechst 33342 and of the calcium probes quin-2 and indo-1. I have found that 10 mW at 325 nm and 40 mW at 441 nm produce strong fluorescence signals from bacteria stained with a mixture of DAPI or Hoechst 33342 and mithramycin or olivomycin, and I[1133] and others[1134-5] have found similar or slightly higher power levels usable for chromosome analysis, using Hoechst 33342 or DAPI and chromomycin or mithramycin. Separate UV and blue He-Cd lasers or a single dual-wavelength laser should also work for measurement of BrUdR incorporation by a relatively simple, antibody-free staining technique developed at Los Alamos[647], using Hoechst 33342 and mithramycin; this is discussed in detail on p. 456.

He-Cd laser technology was originally pursued because a hollow cathode design, different from that commonly used, permits simultaneous emission of UV or blue, green (533.7 and 537.8 nm) and red (635.5 and 636.0 nm) beams. The "white-light" multiline He-Cd laser was intended for the graphics arts industry. Such lasers are now available from Cooke; their power levels are in the 10 mW range for each of the spectral regions. On the plus side, they are less noisy than conventional He-Cd lasers, but their size and cost are disadvantages; these days, combinations of diode and solid-state lasers may be a better bet.

The major disadvantage of He-Cd lasers is optical noise (light output fluctuations) at frequencies around 300 kHz; as plasma tubes age, helium pressure tends to build up, increasing the amplitude of these fluctuations and decreasing measurement precision. I will have more to say about this, and about light source noise in general, a little later on.

Helium-selenium lasers represent another type of **metal vapor laser**, similar to He-Cd lasers in construction but less tractable and with much shorter lifetimes (I heard of them dying after a few days in operation). They emit many lines ranging from blue-green through green, yellow, orange, and red, usually at very low power. They have not, as far as I know, been looked at as light sources for flow cytometry. I stopped trying to lay hands on one when the green and yellow He-Ne lasers first appeared.

Diode Lasers: Red, Infrared, Violet, and UV

Diode lasers began to tantalize flow cytometer designers at about the time commercial CD players, which were the first large-volume commercial products to incorporate diode lasers, appeared on the market. However, until the late 1980's, practical CW diode lasers had only been made to work in the near infrared (IR) (780 nm and above), and there were few fluorescent probes suitable for excitation at the available laser wavelengths.

Diode lasers are, like transistors, made of materials classed as **semiconductors**. The light emission from semiconductors is not from excited atoms or ions, as is the case in ion, He-Ne, and He-Cd lasers, and not, strictly speaking, from excited molecules, as is the case in dye lasers. The electrons that are excited in a diode laser are "free" in a crystalline material. Such "free electrons" also occur in metals; the nuclei in a metal are packed relatively close together, and the electrons in the outermost shells are not tightly held by any given nucleus, and may be excited from the so-called "valence band" to the so-called "conduction band" by ambient thermal energy. Energy transfer among electrons in a metal can occur fairly readily; this is what makes metals good conductors of electricity.

Semiconductors are so named because, though they do not conduct electricity well when in an unperturbed state, they may become conductive in the presence of an applied electric field or of incident light. Their electronic structure differs critically from that of metals in that there is a substantial energy difference, or "bandgap," between the valence and conduction bands, with almost all of the electrons lying in the valence band under normal conditions. The application of an electrical current to an appropriately configured semiconductor can result in light emission as electrons relax from the conduction to the valence band; this type of spontaneous emission is what occurs in light emitting diodes (LEDs).

A laser diode is basically an LED with its geometry tailored to provide a resonator structure that will support stimulated emission. The active regions of diode lasers typically have dimensions on the order of a few micrometers; they use either polished facets on the semiconductor material itself or adjacent structures of differing refractive index to perform the function of the mirrors used in larger lasers. Because the efficiency of diode lasers is extremely high, typically on the order of 20-30 percent, high reflectivity is not needed. Diode lasers are much less expensive than any other type of laser, because hundreds of lasers, if not more, can be produced from slices of a single semiconductor wafer. The down side of this is that small differences in semiconductor composition and dimensions of the finished chips affect the output wavelength; wavelength variation also occurs with changes in temperature. That's why Table 4-2 doesn't have a column listing diode laser emission wavelengths to a tenth of a nanometer.

The first practical diode lasers were, and many diode lasers still are, made of gallium aluminum arsenide (GaAlAs); the emission wavelength, other things being equal, is varied by changing the ratio of gallium to aluminum in the semiconductor material. The emission wavelengths theoretically achievable with GaAlAs lasers range from about 650 nm, at which point the material is almost pure AlAs, to about 900 nm, at which point the material is almost pure GaAs. However, GaAlAs lasers that emit below 750 nm are typically unstable, and immolate themselves within minutes to hours. The shortest wavelength GaAlAs lasers available in quantity emit at 750-780 nm; millions of them go into CD players, laser printers, and CD-ROM drives, and they cost at most a few dollars.

Gallium indium phosphide (GaInP) lasers go down to around 670 nm, providing up to 50 mW emission from small devices and hundreds of milliwatts from larger ones and arrays. The 635-640 nm diodes now used as alternatives to 633 nm He-Ne lasers in cytometric applications are made of aluminum gallium indium phosphide (AlGaInP); they are now available with output powers ranging from a few hundred microwatts to 35 mW.

The problem with long wavelength sources in flow cytometry stems primarily from the fact that most of the fluorescent dyes that are of any use as probes for cellular parameters require excitation at or below 650 nm. Indeed, many of the dyes usable with red excitation are, or are closely related to, substances long known and used as stains for transmitted light microscopy. It would probably be at least as difficult to devise new red-excited stains, especially specific ones, as to develop new types of diode lasers. The difference between 635 nm operation and 670 nm operation is, therefore highly significant.

Before 635 nm diode lasers became available, I managed to do DNA content analysis using rhodamine 800[737] with a 670 nm diode laser as a light source; this wavelength is also usable for excitation of dibenzodicarbocyanine dyes, used either alone as membrane potential probes or in reactive form (e.g., Cy5.5) as antibody labels, and for some aluminum phthalocyanine dyes which have also been investigated as antibody labels.

Working at 635 nm adds rhodamine 700, oxazine 750, Molecular Probes' TOTO-3 and TO-PRO-3 and some dyes of the SYTO series, and DRAQ5[2338-9] to the list of DNA dyes, Cy5 and allophycocyanin and its tandem conjugates to the antibody labels, oxazine 1 for RNA, and assorted cyanine and oxonol membrane potential probes. Doornbos et al[1136] were first to report using a 635 nm diode laser for DNA measurements with TO-PRO-3 and immunofluorescence measurements with allophycocyanin.

In an earlier paper, Doornbos et al[1137] discussed the use of 670 nm and 780 nm diode lasers as light sources for forward and right angle scatter measurements; they were able to discriminate lymphocytes, monocytes, and granulocytes in flow cytometers using either wavelength. However, CV's of fluorescence measurements were disappointingly large. My guess is that this may have had something to do with the typically ugly beam intensity profiles of diode lasers.

At this point, those of us who build flow cytometers are a lot farther along the learning curve; general- and special-purpose flow cytometers using 635 nm diode lasers, alone or in combination with other light sources, are now available from a number of manufacturers. There is a 635 nm diode laser in the lunchbox-sized OptoFlow Microcyte (Figure 1-25, p. 58), which is very likely the smallest production model flow cytometer, and you can put one on the Cytomation MoFlo sorter (Figure 1-24, p. 58), which is probably the largest. And, for many routine applications, e.g., counting $CD4^+$ T cells, an instrument with only a red diode laser source can do just as good a job as a conventional benchtop flow cytometer with a 488 nm laser[2447].

Diode lasers are used for laser printing and for compact disc recording and playback; in these applications, achieving a smaller focal spot size allows more information to be stored in and/or retrieved from the same area, either as more dots per inch on a page or, more importantly, as more bits per unit area on a disc. Recordable and CDs, written and read with 780 nm diode lasers, top out at around 700 MB per disc. A CD "burner" costs more than a reader because it needs a higher power diode laser to "burn" the CD by photobleaching spots in an IR-absorbing dye layer.

Using a shorter wavelength (650-660 nm) red diode laser instead of an IR laser allows a DVD to store 5.7 GB on a side. Getting 30 GB worth of data, music, or (where the money is) video onto one side of a disc would require a "blue" laser (matching some of the video). This prospect attracted the interest of some very large companies, with proportionally deep pockets, but the company that first succeeded was a smaller one, Nichia, in Japan, where Shuji Nakamura succeeded in making the first "blue" laser diodes in the late 1990's[2448]. These lasers are now available with emission in the range from 370 nm (UV) to 445 nm (blue-violet); while the newer devices at the far ends of the range still have short operating lifetimes (perhaps 3,000 hr), the 5 mW and 30 mW diodes with output between 395 and 415 nm, which have been in production for some time, are quoted as having lifetimes in excess of 15,000 hr.

Violet laser diodes are good excitation sources for a lot of materials that otherwise would require a krypton laser for excitation. The list[2444] includes Molecular Probes' labels Cascade Blue and Cascade Yellow, monobromo- and monochlorobimane, both used for detection of intracellular glutathione[1653], ECFP, the cyan-fluorescent reporter protein, and the DNA dyes mithramycin and chromomycin A_3. Laser diodes operating at the short (370-400 nm) end of the range are effective excitation sources for DAPI (38% of maximum excitation at 395 nm) and usable with the Hoechst dyes; they cannot be used with the calcium probe indo-1, but it may be possible to synthesize a similar calcium probe that would work with 370 nm excitation.

Diodes emitting at 405-425 nm could be used to measure intrinsic absorption and fluorescence of porphyrins in cells. The strong absorption of hemoglobin in erythrocytes in this wavelength region (the Soret band) influences the cells' light scattering properties, enabling discrimination of leukocytes from erythrocytes in dilute, unstained whole blood[2449]; zinc protoporphyrin fluorescence in erythrocytes may provide clinically relevant information in cases of iron-deficiency anemia and lead poisoning[1266]. Violet diode lasers could be used in practical clinical instruments; violet krypton lasers cannot.

When I wrote the last edition of this book, in 1994, 635 nm laser diodes were limited in power to about 5 mW and the diodes themselves cost several hundred dollars; in 2002, a packaged 635 nm, 35 mW laser system with a power regulator, temperature control to increase output stability, and beam circularizing optics costs less than a 633 nm He-Ne laser with the same power output. Although violet diode laser systems available now cost several thousand dollars, the major cost is the cost of the diodes, which can't stay high if millions of them are made for optical storage devices.

I mentioned the relative variability of diode laser emission wavelengths; this has advantages and disadvantages. Small variations in device dimensions and composition result in manufacturers' inventories including lasers with a range of emission wavelengths; for a price, the buyer (meaning the cytometer manufacturer or the laser system manufacturer; making a diode laser system is not a trick to try at home) can specify a selected wavelength range. If you're buying a diode laser system, you can also specify a wavelength range; again, it will cost you.

A red He-Ne laser will emit at 632.8 nm, not 632.85 or 632.75 nm, for all of its useful life. When cytometer manufacturers buy red diodes, they typically specify an emission range no narrower than, say, 635 to 640 nm; a diode in your instrument could emit anywhere in that range. If you measure fluorescence at 660 nm, your detector filter probably has a higher transmission at 640 nm than at 635 nm. More stray laser light leaking through the filter means higher fluorescence background, so which end of the range the diode emits at will affect measurement sensitivity.

The emission wavelength of a diode laser varies with temperature; this doesn't happen uniformly, but, rather, in jumps, and each wavelength change represents a different longitudinal mode of the laser. The resultant "**mode hopping**" may affect the stability of the intensity profile, even though the light control feedback circuitry needed to keep the laser from burning itself out keeps the total power output relatively constant. The best way to eliminate this problem is to control the temperature of the diode; this is absolutely essential for the violet diodes, and becoming more common in red diode systems designed for critical applications such as cytometry.

A diode laser is basically an LED, and it will emit a small amount of incoherent light, sometimes referred to as **LED glow**. You may not be all aglow yourself if you encounter this light, which is usually at wavelengths longer than the laser wavelength, meaning it may be able to get through the filters on one or more of your fluorescence detectors. Because it is incoherent, LED glow diverges fairly rapidly, but, if the laser diode is close to the flow chamber, enough LED glow may get in and be scattered from cells to show up as increased fluorescence background. The workaround for this problem is fairly simple; a bandpass or short pass filter is mounted between the laser diode and the flow chamber.

As I mentioned previously, diode lasers must be operated in a light control mode; as a result, their optical noise levels are typically substantially lower than those of most ion or He-Ne lasers. Diode lasers therefore offer an advantage

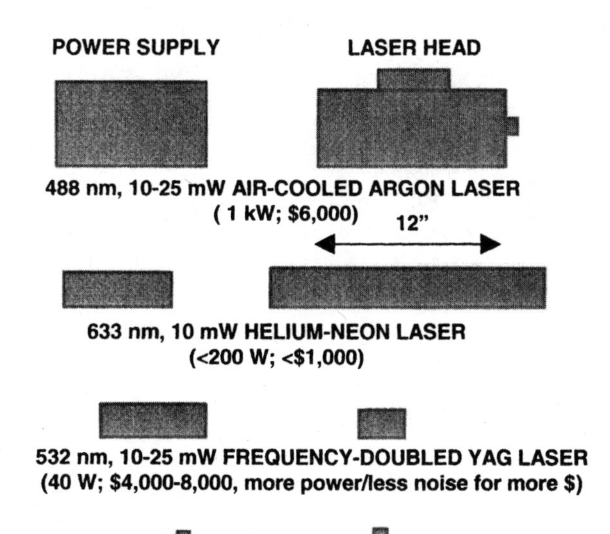

Figure 4-23. Sizes, power requirements, and approximate costs of some smaller lasers used for cytometry.

over these other sources for extinction measurements, which are more sensitive to optical noise than are scatter and fluorescence measurements. When putting a diode laser into an optical system, it is critical to position optical elements in such a way as to avoid reflecting light back into the laser, which can interfere with the operation of the light control circuitry and result in fluctuations in power output.

As Figure 4-23 shows, red diode lasers are 1/1000 the size (volume) of, consume 1/1000 as much power as, and cost less than 1/10 as much as, air-cooled argon lasers. This commends the diodes for use in portable equipment, and in applications and locations where cost and power consumption are an issue (e.g., $CD4^+$ cell counting in Africa, or (at least in mid-2001) California)[2447]. Violet diode lasers, and solid-state lasers, are also small and energy-efficient, if not inexpensive; the violet lasers should get considerably cheaper, and some of the solid-state sources may (see below). If it is ever possible to produce flow or image cytometers for a tenth of today's prices, the instruments will are likely to use laser diodes, or even LED's, as light sources.

Solid-State Lasers: Like, YAG Me!

Neodymium-YAG (yttrium aluminum garnet) lasers, in which the lasing medium is a solid rod of crystalline material pumped by a flashlamp or a diode laser, can produce power outputs of tens of watts at 1064 nm. Doubling or tripling the output of a diode-pumped YAG laser, using a crystal within the laser cavity, can yield green light (CW or pulsed) at 532 nm or UV light (pulsed only) at 355 nm. Frequency-doubled diode-pumped YAG lasers (green YAG lasers from here on) can be used for excitation of ethidium and propidium, rhodamine, phycoerythrin and its tandem conjugates, and the reporter protein dsRed[2450], among other things. And the green YAG laser does a better job of exciting most or all of these than does a 488 nm argon ion laser, and, unlike the argon laser, excites very little autofluorescence in most mammalian cell types.

In 1994, miniaturized green YAG lasers emitting 2 mW at 532 nm were offered for sale for $4,000. This was about twice the price of a 2 mW green (543 nm) He-Ne laser, which does an even better job of exciting phycoerythrin. Today, you can buy that 2 mW green YAG laser in the form of a laser pointer for about $200. Unfortunately, cheap green YAG lasers tend to be noisy, and when I say noisy, I mean they may turn themselves on and off. Most of these cheap lasers, by the way, come from Russia and China. A so-called single longitudinal mode green YAG laser, which may or may not actually have only one longitudinal mode, but which will have relatively low optical noise, can be had from several sources (I can speak for products from Coherent and Uniphase), but will set you back more than the cost of an argon laser. The small size and low power consumption are attractions; Guava and Luminex, among others, use green YAG lasers with about 10 mW power output in their instruments, which can be operated from batteries. If you'd like, you can buy a 10 W green YAG laser, and it will still run off house current. You're not likely to need to put that kind of power through cells, but, if you're still using a dye laser, a multiwatt 532 nm YAG makes a much more efficient pump than does a multiwatt ion laser.

The tripled UV lasers could be useful for excitation of DNA stains and calcium probes. They are at the right wavelength (355 nm), but they only operate in pulsed modes. However, a technique called **mode-locking** allows the lasers to emit regularly spaced pulses at 80-100 MHz; this repetition rate is high enough so that the laser behaves more or less as if it was a continuous light source. Both Lightwave Electronics and Spectra-Physics have introduced mode-locked 355 nm YAG lasers, and BD Biosciences is offering them as sources in both benchtop systems and sorters. The lasers are expensive, possibly even more expensive than the UV ion lasers that represent the only alternative sources offering power levels of tens to hundreds of milliwatts. However, mode-locked YAG lasers share the modest power and cooling requirements of their CW cousins, and therefore do not require the expensive infrastructure needed to support UV ion lasers.

An attempt to develop a solid-state CW UV laser for flow cytometry is ongoing at Light Age, a company that makes **alexandrite** lasers. Alexandrite, which is beryllium aluminum oxide containing chromium, emits between 700 and 850 nm, and can be pumped by 635-670 nm diode lasers. UV emission of 10-15 mW at approximately 370-380 nm has been achieved from a frequency-doubled, diode-pumped alexandrite laser; one of Light Age's prototypes was run for several months, and maintained a reasonably low optical noise level. The tunability of Alexandrite lasers is another point in their favor. The lasers have been expensive, although probably not any more expensive than UV ion lasers or mode-locked UV YAG lasers.

Alexandrite is an example of a **vibronic** crystalline laser material. Such materials behave similarly to laser dyes; laser transitions may occur between an excited electronic state and any vibrational state associated with a lower electronic energy state, allowing output to be tuned over a broad range. **Titanium-doped sapphire** lasers can operate in pulsed or CW mode between 660 and 1180 nm. Pump energy can be supplied by a high-power green YAG laser. Pulsed Ti-sapphire lasers are used for multiple-photon excitation of fluorescent dyes and intrinsically fluorescent cellular constituents; their high cost (typically over $100,000) has, to date, limited their use. **Cr:LiSAF** lasers, in which the laser transitions occur in Cr^{3+} ions in a matrix of $LiSrAlF_6$, have been doubled to 430 nm; a laser based on this technology was, and may still be, offered for sale by Melles Griot, but violet diodes may be a better bet for this wavelength range.

Melles Griot also offers 457 nm CW lasers made by frequency doubling the output of **neodymium yttrium vanadate ($Nd:YVO_4$)**. The biggest of these runs on house current, drawing 75 W at the plug, and puts out 400 mW; I am not clear why people aren't using these lasers instead of water-cooled argon lasers in chromosome sorters. Admittedly, the water-cooled lasers do put out other wavelengths besides 457.9 nm, but the solid-state jobs are about half the price and let you do without the plumbing and the cooling arrangements.

Although green Nd:YAG lasers are good excitation sources for a lot of dyes and labels used in cytometry, they can't quite reach fluorescein, which is not only still very popular as a label, but which also has been derivatized into probes for a large number of structural and functional parameters. Fluorescein can be excited at wavelengths as high as 515 nm; in fact, the first work on two-color immunofluorescence flow cytometry[115] was done using fluorescein- and rhodamine-labeled antibodies excited by the 514.5 nm line of an argon ion laser. As it happens, the primary emission wavelength of **ytterbium YAG (Yb:YAG)** lasers is 1029 nm; they could be doubled to 514.5 nm and used for fluorescein excitation. Yb:YAG lasers have some advantages over Nd:YAG in terms of ease of pumping and stability, and this suggests that green Yb:YAG lasers might have fewer problems than green Nd:YAG lasers, but the emergence of 488 nm solid-state lasers (see below) has probably discouraged development of a green Yb:YAG for cytometry, and is also likely to set back development of 473 nm, frequency-summed Nd:YAG lasers, which have been offered in some cytometers.

Some years back, Uniphase, a major supplier of argon and He-Ne lasers as well as green YAG lasers, described a 544 nm solid-state laser using an erbium-doped fluoride glass fiber as the lasing medium[1138]. This is not a frequency-doubled laser; pumping to the excited state by 971 nm light is accomplished by two-photon absorption (see pp. 118-9). The 544 nm wavelength is ideal for excitation of phycoerythrin and its tandem conjugates; since power levels of 10 mW should have readily been attainable, erbium fiber lasers might have been preferable to both green He-Ne and frequency-doubled YAG lasers as sources for flow cytometry. Uniphase's recent dramatic up-and-down performance in the financial markets suggests that the company's development programs may not be running at the level originally intended. Business aside, fiber lasers can be made to operate at other visible wavelengths; they also have excellent **pointing stability**, meaning the beam doesn't have as much of a tendency to undergo slight changes in direction as do beams from other types of lasers. This would also be advantageous for flow cytometry. Maybe Uniphase should have sold Enron moral fiber lasers.

But, getting back to (or away from) business, the big story in lasers for cytometry in recent years is about **frequency-doubled semiconductor (diode) lasers**. Doubling a semiconductor laser operating at 750-1000 nm will (do the math) yield UV, violet, or blue light. The process is inefficient; a laser capable of producing several hundred milliwatts is required to get a few milliwatts of visible light. The resulting laser system is typically considerably more complex and more expensive than a diode laser, because other components, notably, an external mirror and a crystal of the material used for harmonic generation, must be incorporated into the system. Before 2000, there were a few doubled-diode lasers on the market, but they were neither powerful nor cheap. Things have since changed, attracting much attention.

Coherent's "Sapphire" laser, apparently (and confusingly) named for its blue output, was introduced to the cytometry community at the May, 2000 ISAC meeting[2441]. The Sapphire isn't a Ti:sapphire laser and doesn't, as far as I know, have any sapphire in it at all; it is a frequency-doubled, diode-pumped semiconductor laser; Coherent currently sells a 10 mW, 460 nm model and both 20 mW and 200 mW models emitting at 488 nm; output wavelengths are guaranteed to ± 2 nm. BD Biosciences and DakoCytomation, among others, offer the 20 mW, 488 nm Sapphire in benchtop instruments; iCyt-Visionary Bioscience is retrofitting the 200 mW, 488 nm Sapphire to sorters. These lasers are not inexpensive, but they are small in size, run on small amounts of house current, and, reportedly, feature long lifetimes (>10,000 hr).

Another laser company, Novalux, has recently been making the rounds with another frequency-doubled semiconductor laser, the Protera; this will also deliver 10- 20 mW at 488 nm, and is priced competitively with argon lasers in the same power range.

The semiconductor lasers in both Coherent's Sapphire and Novalux's Protera are **surface emitting lasers**, and, unlike typical edge-emitting diodes, produce circular, near-Gaussian beams. The technology could yield relatively inexpensive laser sources at wavelengths ranging from UV to yellow. I would not be surprised to see the cytometer manufacturers and/or third parties produce drop-in solid-state replacements for the air-cooled 488 nm argon lasers now installed in most instruments, and soon.

Laser and Light Source Noise and Noise Compensation

Laser noise may originate from several sources. Poor power supply regulation and/or design typically result in light output fluctuations at the frequency of the line current used to run the power supply or at a multiple thereof; if a high-frequency switching circuit is used in the power supply, light noise due to power supply problems may also occur at the switching frequency. In some lasers, particularly He-Ne and He-Cd lasers, noise may be found at frequencies of a few hundred kilohertz, due either to radio frequency energy used to pump the lasing medium or to fluctuations in the medium itself.

In most circumstances, the level of light source noise determines the minimum detectable signal level for scatter measurement channels; although preamplifier baseline restoration circuitry (p. 55) removes the steady DC component of the background noise produced by stray illuminating light, the AC component of the background, representing fluctuations around the DC level, is amplified along with the signals produced by particles passing through the illumination beam. To be detectable, the signal from a particle must be substantially above the level of the fluctuations; therefore, the use of a light source with lower noise allows detection of smaller particles.

Noise also has a direct effect on the precision of both scatter and fluorescence measurements. Noise is usually specified as an average, **RMS** (**R**oot **M**ean **S**quare) value, and/or as a **peak-to-peak** value; both are given as percentages. The RMS noise value is actually the coefficient of variation (CV) of the power output level. If noise is normally distributed, the RMS value can be estimated from the peak-to-peak value as follows. Since more than 99 percent of the area of a normal distribution lies within three standard deviations of the mean, dividing the peak-to-peak noise percentage by 6 provides an approximation to the RMS noise percentage. RMS noise levels are generally on the order of 0.05% or less for diode and high-quality solid-state lasers, 0.2% to 0.5% for water-cooled ion lasers, and 0.5% to 1% for air-cooled argon and He-Ne lasers.

In an otherwise perfect cytometer, measuring particles that were absolutely identical, the CVs of scatter measurements could be no lower than the RMS noise of the light source. The same restriction on precision would hold for fluorescence measurements using low power illumination, where a given percentage change in excitation power produces the same percentage change in emission intensity; at higher power levels (p. 117), precision might be less affected.

It should be remembered that operation in the light control mode does not guarantee protection against noise. If the light output drops substantially enough for the supply current to rise to its maximum value, any further mechanical or optical deterioration cannot be compensated for by either the light control or the current control circuitry; thus, a change in light output will occur.

One solution to source noise problems is the incorporation of circuitry in the signal processing electronics that senses variations in illumination and corrects measured signal levels accordingly. This was done in Kamentsky's RCS, in the Block apparatus, and in several laboratory-built instruments, including mine. While commercial fluorescence flow cytometers do not now offer source noise compensation, it is relatively easy to implement using digital signal processing, and could thus become common in the future; this could help tame cheap, noisy YAG's.

The principal problem with He-Cd lasers is plasma noise, at frequencies between 300 and 400 kHz. It is difficult to keep RMS noise levels much below 1.5% even when the laser is new. Noise levels tend to increase thereafter, especially if the laser is left idle for long periods, because this leads to an irreversible buildup of helium in the plasma tube, which increases noise. This effect can be minimized, but not eliminated, by running the laser for at least several hours a week. That's not much help if the problem has already developed; I have had several He-Cd lasers become unusable in this fashion.

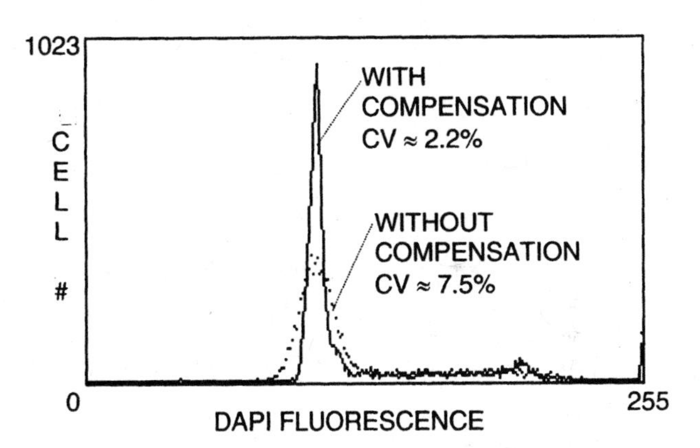

Figure 4-24. Effect of noise compensation circuitry on precision of fluorescence measurements using a noisy He-Cd laser for excitation.

If I measure DNA content with DAPI using these lasers and my standard electronics, I get atrocious precision; the corresponding distribution shown in Figure 4-24 has a CV of about 7.5%, because the illumination intensity reaching any given cell at any given time varies over such a large range due to noise. Things get better when I use the noise compensation circuit. A beam splitter reflects about 5% of the laser power to a photodiode detector, providing a reference signal. The fluorescence of dye in the cells is detected by the cytometer's PMT, and an analog circuit is used to divide the PMT signal by the reference signal, yielding a compensated fluorescence value. The DNA content distribution measured using the compensation circuit has a CV ≈2.2%. Incidentally, if I were making a pH or calcium measurement, using a ratio of fluorescence emission from one dye at two wavelengths, I wouldn't need to monitor laser output; taking the ratio compensates for the fluctuations in source power output because the numerator and denominator are equally affected by the noise.

There are other ways of compensating for laser noise. "Noise eaters" that you can put between the laser and the rest of the world sense laser output, and use the signal to control acousto-optic or electro-optic modulators, which adjust their light transmission rapidly in response to an applied voltage. These gadgets, however, cost thousands of dollars, as opposed to tens of dollars for a noise compensation circuit, and may require that the laser beam be polarized, which many He-Cd beams are not.

Some pretty sophisticated flow cytometry can be done by using a 325 nm He-Cd laser, a low-power 488 nm argon laser, and a 633 nm He-Ne laser. B-D's LSR uses those sources, although they haven't put in noise compensation. To get UV excitation for Hoechst dyes and 488 nm for immunofluorescence, it has always made more sense to me (and others[107,643]) to use an He-Cd laser or an arc lamp (and, these days, maybe a diode or a solid state UV laser) for the UV than to get UV and 488 nm out of one big, one big and one small, or two big ion lasers. In 2002, even manufacturers who have stuck with the big lasers for years are prepared to change. There are both flow and scanning system that use violet/UV and red diodes and green YAG or solid-state 488 nm lasers. I have seen the future, and it is air-cooled. To paraphrase an old standard, with apologies to Paul Simon:

Fifty Ways to Lose Your Laser

If you have thought of doing flow cytometry
With dyes requiring excitation by UV,
Using an ion laser makes it plain to see
There must be fifty ways to lose your laser.

The plasma tube goes, Mose,
The power supply's fried, Clyde,
The optics burn, Vern,
And the solenoid melts,
And, when they're replaced,
A hose comes unlaced,
So you want to cuss, Gus,
And try something else.

You can pick up an arc lamp, as has long been done
In places where a great big laser's hard to run,
But lining up your optics won't be much more fun
Than learning fifty ways to lose your laser.

Don't throw in the towel, Raul,
There's help on the way, Jay;
Just send the arc back, Jack,
And listen to me.
What will do the job, Bob,
Plugs into the wall, Paul;
An He-Cd, Lee,
Will get you UV.

Kit Snow, Scott Cram, Tom Frey et al and I have shown
You can sort chromosomes with HeCad light alone,
This might help when your budget's cut back
to the bone,
And there are fifty ways to lose your laser.

We've got diodes, too, Lou,
And alexandrite, Dwight,
Mode-locked YAG as well, Mel,
So listen to me.
With power costs high, Guy,
You'll shell out too much, Dutch,
Argon's not the way, Ray,
In this century.

Why pay to replumb and rewire? Don't be dense.
Small lasers can't be used for national defense,
But, for cytometry, they make a lot of sense,
And you'll find fifty ways to use your laser.

Danger!!! Laser!!! Hazards and Haze

Any laser that emits more than 800 μW anywhere from the UV to the near IR is classified as hazardous by the U.S. Bureau of Radiological Health. Once you get past the "Star Wars" notion of a laser as a handy tool for vaporizing enemies and competitors, you start to think of laser hazards primarily in terms of eye damage. This makes sense, up to a point, but there have only been a few dozen accidental injuries reported resulting from undiverged or focused beams getting into peoples' eyes.

Laser damage to the eye, or any other tissue, is caused when the absorption of light by the tissue causes enough heating to kill cells. The power density of the laser beam and the absorption coefficient of the tissue are the primary determinants of how much damage will be done. Argon lasers emitting a watt or so are routinely used for "spot welding" detached retinas, but you can shine a focused beam at lower power (say about 80 μW) into your eyeball without hurting your retina. In fact, Rob Webb, who taught me a lot about building and using flow cytometers, developed a highly useful gadget called the scanning laser ophthalmoscope, which produces a wide-field, high-resolution video image of the retina by scanning across it with a low-power laser. Incidentally, UV light doesn't penetrate into the eye very well; this means that it is unlikely to damage the retina. It also means we can't measure the UV-excited fluorescence of pyridine nucleotides in the retina, which might have given some useful information about metabolism in diabetes and other conditions. We can, however, measure flavin fluorescence with blue light from lamps or He-Cd or argon lasers.

Argon lasers are also used by dermatologists to treat pigmented lesions such as hemangiomas. These blood vessel tumors, which are reddish-purple in color, absorb more blue-green light than the surrounding normal tissue, and thus get hotter faster. The tolerance of normal skin for laser radiation is also a function of pigmentation; I am fair-skinned, and experience has taught me that I won't burn my finger by putting it into a 200 mW, 488 nm beam 2 mm in diameter, while colleagues of mine with darker complexions will quickly withdraw their fingers from the same beam. If I accidentally get my fingers into the beam at a point where it is converging toward the flow system, I will get a slight burn. Black construction paper, however, will catch fire if held in either the 2 mm or the focused beam.

People tend to worry most about invisible laser beams, meaning those in the UV and IR wavelength ranges, because it's harder to avoid what you can't see. UV light is of special concern because the photon energy levels are high enough to cause ionization; this can lead to mutation and carcinogenesis in exposed tissue. The probability of mutation increases considerably if UV-excited dyes, e.g., Hoechst dyes, are bound to DNA; mutation rates can also be increased by irradiation of acridine orange-treated cells with blue light or by irradiation of methylene blue-treated cells with red light. That said, I must point out that I know a lot of sun worshippers and former sun worshippers who haven't gotten skin cancer, and I've never heard of anybody picking up a tan in a flow lab, regardless of the state of undress of arc lamps, lasers, and (heh, heh!) people.

At the power levels that are, or should be, used in flow cytometry (a few hundred mW or less), widely diverged laser beams and diffuse reflections should not represent major hazards to users' eyes or skin. The potential problems come from concentrated (undiverged or converging) beams. It is fairly easy for users to identify the paths of beams from lasers emitting at visible wavelengths, and thus to avoid exposure. Most flow cytometers don't use either UV or IR beams, which might otherwise be more dangerous. Flow cytometers, like other systems that incorporate lasers, are required by law to be equipped with light shielding, and operators are required to be provided with safety goggles that prevent laser light from reaching the eye. Unfortunately, adjustments to the laser and illumination optics are best made, and often only possible, when one can see the beam, which requires removing the shields and the goggles.

Even if you were running one of the larger ion lasers used in flow cytometry at full power, you probably wouldn't be able to kill yourself directly using the laser beam; you'd be more likely to die from complications of the hernia you got trying to lift the power supply. You could very easily kill yourself, however, while fiddling with the optics of an ion laser by allowing a careless finger to brush past one of the many open electrical connections carrying several hundred volts at over fifty amperes. There have been a few dozen fatal accidents associated with laser use; in all cases, the victims were electrocuted. Many of them were far from novices at working with lasers. I got a very fast trip across my lab once while tweaking optics with sweaty fingers; since then, I have made it a point always to have somebody else, preferably trained in CPR, around the lab when working on a big laser.

I have already made mention of the fact that almost none of the applications of flow cytometry require use of more than about 50 mW of laser power when an efficiently designed optical system is used. Most of the small lasers run on house current and don't require the user to get into the laser head and tinker with the optics, and their lower-power beams pose less of a hazard to users' eyes and virtually no hazard to skin. Small lasers are therefore considerably safer than large ones, a point that probably helped speed the adoption of the former in commercial flow cytometers.

What about other possible effects of lasers that are, at present, inexplicable by science? There is a provocative body of literature on effects of low-power laser irradiation on cells and tissues; try doing a search on "laser acupuncture" in the MEDLINE database some time if you're interested. It has been claimed that shining a few hundred microwatts of red He-Ne laser light on the appropriate acupuncture points stimulates immune response, reduces gastroesophageal reflux (not in me, unfortunately), and relieves symptoms of arthritis, diabetes, and gallbladder disease, to name a few. Maybe a little laser exposure is good for you.

To sum up, big lasers, which few of us use anymore, are associated with big hazards; one should respect them. Most people who use benchtop flow cytometers with small visible lasers are in greater danger from their reagents and their samples than from their lasers. But watch out for the laser pointers; many of them put out more than the legal 800 μW, especially the green ones.

4.4 LIGHT COLLECTION

We now know what we need to do to illuminate cells, whether they are in a flow system or on a microscope slide. We next turn to the task of collecting at least some of the light emitted and/or scattered by the cells, and directing as much of it as possible, and as little other light as possible, to our photodetectors.

The optical elements of microscopes have been designed to do this job. The lenses provide **spatial resolution**, enabling us to collect a great deal of the light coming from a very small region of space and relatively little of the light coming from other regions a very small distance away. Optical filters provide **spectral resolution**, allowing discrimination between scattered, fluorescent, and background light.

Microscope Objectives

A microscope objective is designed to form a real, magnified image of a small object at a fixed distance, referred to as the **tube length**, from its mount. It is customary to describe objectives in terms of two principal parameters, magnification and N.A., and also to specify the working distance between the front element of the lens and the object. When placed at the working distance from the object, the objective forms a real image, magnified at its stated magnification, in a plane (the **image plane**) approximately one tube length behind the "shoulder" of the objective (that's where it screws into the nosepiece; you or I might have picked a different anatomical reference).

Looking at the Observation Point

In order to select a collection lens, we ought to look at a magnified image of the observation point in a flow cytometer; I've sketched one as Figure 4-25. What I've drawn is what you see if you use a decent microscope lens and make an image of the intersection of the laser beam with a round capillary flow chamber or with a stream in air. You find out where the core is and how big it is by running a fluorescent dye through the system; if your excitation is anywhere between UV and blue-green, Mercurochrome, which is a fluorescein derivative, is fine.

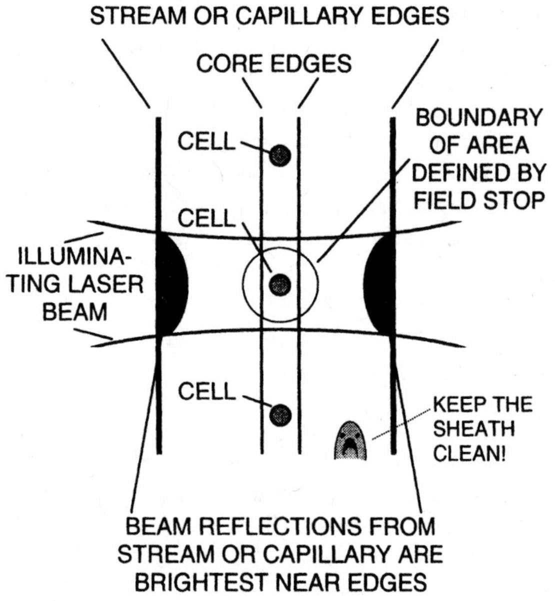

Figure 4-25. Looking at the observation point.

If the image is big enough, you'll see that, while there is a lot of reflection of the excitation beam off the capillary or stream, most of the reflection is from the edges, as long as the stream or capillary is, say, at least 50% wider than the beam. Since we want to measure light coming from the core, we almost instinctively think of using a field stop, i.e., a pinhole or slit in the image plane, so that the only light that gets through is from the region of the core. I've shown the region we want to look at in the figure.

What happens if we don't use a field stop? That depends upon what we're trying to measure. You might figure that you don't have to worry about all of the reflected light at 488 nm if you're measuring fluorescence above 515 nm, because you can get rid of it with optical filters. This is true, to some extent; however, as we will presently see, this approach places a considerably heavier burden on the filters, in terms of the performance required, and increases the likelihood of noise due to light leakage and/or filter fluorescence. If we're trying to measure light scattered at large angles to the excitation beam, we've got a problem; all of that 488 nm light is going to clobber our scatter signal, which is at the same wavelength as the excitation beam. In this case, we're forced into using some type of light barrier to insure that we collect light predominantly from the cells in the core rather than from elsewhere.

Stops vs. Blockers

I mentioned that what I've shown in the figure is what we see when the stream or capillary is wider than the beam by 50% or more. This is usually not the case in cell sorters in which measurements are made of cells flowing in a stream in air. If, for example, we have a 70 μm stream and a beam 80 μm wide, we will see a "ring" of light from the illuminating beam around the entire circumference of the stream, and we simply won't have the option of using a field stop to get rid of stray scattered light.

The alternative typically in stream-in-air cell sorters is the inclusion of an **obscuration** or **blocker bar**, the effect of which you can simulate by placing a finger over the drawing along the path of the illuminating beam. Particles in the core will emit fluorescence and scatter light in all directions; the obscuration bar will block a substantial amount of fluorescence and scattered light, but it will allow some of both to get into the collection system, while preventing almost all of the light reflected from the stream from going into the collection system.

Up to a point, this is a good idea. What we want to be able to do in a flow cytometer is to detect weak signals above background, and anything which increases the ratio of (photons collected from cells) to (photons collected from other sources) is helpful, unless it diminishes the number of signal photons to fewer than can be reliably detected above noise based upon sampling statistics. The use of a field stop in the image plane, which masks light coming from everywhere but the immediate vicinity of the intersection of the core stream with the illuminating beam, accomplishes much the same purpose as the use of an obscuration bar, but allows more signal photons to get through. While this generally isn't done in stream-in-air flow cytometers because of the relative sizes of beam and stream, it is done in flow cytometers in which observation occurs within a capillary or cuvette. When the stream in a stream-in-air system is wider than the beam, use of a field stop instead of an obscuration bar increases the signal intensity, and the signal-to-noise ratio is improved[105]. If the observation point is in a flat-sided cuvette, reflections of the laser beam off the cuvette walls are minimal; however, it still helps to use a field stop to reduce problems such as interference from Raman scattering.

Signal versus Noise: To See or Not to See

We have already found that we can only extract a certain number of photons from the sample, no matter how much excitation power we use, so it seems like a good idea to collect as many of those photons as we can get. We have just found out, however, that increasing light collection from the sample doesn't help us much if it is accompanied by a proportionate increase in light collection from the background.

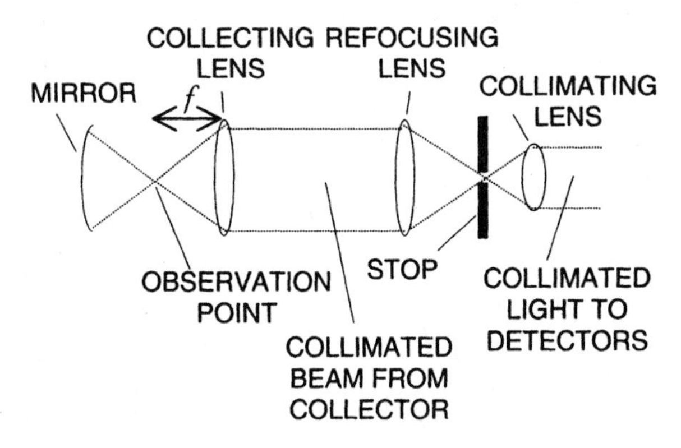

Figure 4-26. Maximizing light collection: Not always a good way to do things.

One can collect the largest possible amount of light by putting the collection lens at a distance from the observation region equal to its focal length, *f.* In such an arrangement, illustrated in Figure 4-26, no image is formed; the light rays emerge from the collector lens in parallel, in a collimated beam. The lenses used for collection in this fashion are typically not microscope objectives, but somewhat longer focal length, high-aperture aspheric lenses of the type normally used as condensers. The beam collected in this manner usually contains a substantial amount of light at the excitation wavelength. The burden of excluding this light from the fluorescence detector falls entirely on the optical filters used to limit detector bandwidth, unless the collimated beam of collected light is brought to a focus in a plane containing a pinhole as a field stop before being routed to the detector filters. A refocusing lens and field stop are shown in Figure 4-17. Behind them is an additional collimating lens; this is needed to reconverge the beam in order to relay light to the filters and detectors.

The setup in Figure 4-26 also illustrates another trick for increasing light collection; a **concave spherical mirror**, with a focus at the observation point, is placed opposite the lens, allowing the lens to capture some of the light emitted in this direction, which would otherwise be lost. Mirrors can converge and diverge beams as can lenses; a concave surface converges a beam, and a convex surface diverges one. Since refraction is not involved, mirrors, unlike lenses, do not exhibit wavelength-dependent variations in focal length. Most flow cytometers do not use the collection arrangement shown in Figure 4-26, but, instead, make a real image of the observation point with a high-N.A. lens, and employ field stops to prevent light from outside the observation region from reaching the detector. However, since the lens has to be farther than its focal length (between 1 and 2 focal lengths) away from the observation point in order to make an image, it must necessarily collect light over a smaller solid angle, and therefore get less light from the specimen. Are we missing a free lunch somewhere?

Of course not. In the first place, the imaging lens generally produces a fairly highly (20-50 ×) magnified image of the observation point. If placed two focal lengths from the observation point, it would produce an image the same size as the observation region, let's say about 50 by 50 μm. Geometric optics tells us that to produce a highly magnified image, the lens is going to be not much more than one focal length from the observation point, so it won't collect all that much less light than it would if it were exactly one focal length away.

Since the lens is almost one focal length from the observation point, the collected light emerging from it is going to converge to an image plane some distance between two focal lengths and infinity behind it; the closer the lens is to one focal length from the observation point, the closer the image is to infinity. In this case, "close to infinity" is about one tube length, or about 6 inches. The light is "close to collimated"; throughput considerations are at work here, and they tell us (p. 123) that, when we collect light over a large solid angle to make a magnified image, the image forming rays occupy a smaller solid angle in proportion to the amount of magnification. Thus, if we make a magnified image of the observation region, we don't need either a refocusing or recollimating lens, and we eliminate any transmission losses that might be associated with these optical elements. The dichroics used for spectral separation, which might have trouble with the divergent beam emerging from behind the stop in Figure 4-26, can easily handle the beam from the imaging lens. Also, because axial magnification goes as the square of lateral magnification, we collect much less interfering light from planes closer to and farther away from the observation region as we increase the magnification of the image.

Now, what about the stop in Figure 4-26 and the field stops used with imaging optics? If you're looking at a 50 by 50 μm observation region magnified 50 times, the corresponding region of the image is 2.5 by 2.5 mm, which is pretty macroscopic, and your field stop can be slightly larger than this and do a good job of blocking light collected from outside the observation region. With the optics in Figure 4-26, the refocusing lens is forming something pretty much equivalent to a 1:1 image in the plane of the stop. In order to block light from outside the observation region as effectively as could be done with an imaging system, you'd need a stop with an opening less than 100 μm across. These are available, but, in practice, you'd probably have to use a substantially bigger opening, because it would be difficult to keep the small one from drifting slightly out of alignment and blocking all the light. This means you let in more light from outside the immediate vicinity of the observation region. When the cell is in the observation region, most of what's in the neighborhood is water, and, with too big a stop, *you'll collect* signals from whatever is in the water. If you're using dyes with high background fluorescence, you may overload your baseline restoration circuitry. If you're using 488 nm excitation and trying to detect phycoerythrin fluorescence, you'll end up with higher background noise and lower fluorescence sensitivity due to Raman scattering from water than you would have if you had made a magnified image and used an appropriately sized field stop.

In principle, light can be collected over extremely large solid angles by **ellipsoidal**[120] or **parabolic**[648] **reflectors**; in practice, use of such elements for fluorescence collection in a flow cytometer requires higher performance optical filters than would be necessary in a system with better spatial resolution, and orthogonal scatter signals may be masked by noise. There are also the same problems with stops as were just discussed.

Flow cytometers built around microscopes, which use high-N.A. objectives[83,88,100-103], often show better precision than laser source systems, particularly in measurements of asymmetric cells, because the wider angles of illumination and collection in the former minimize effects of differences in particle orientation[121]. However, instrument performance is not necessarily improved simply by increasing the N.A. of the collecting lens. In order to make effective use of a high-N.A. objective, it may be necessary to compensate for the concomitant small depth of focus. In the Impulscytophotometer[9,83], this was done by directing sample flow along the axis of the objective, so that portions of cells passed sequentially through the focal plane. In the Steen/Lindmo instrument[100-103], the sheathed sample is directed against a flat surface, resulting in reduction of core thickness and in orientation of asymmetric cells, allowing most cells in the sample to be kept in focus even by a high-N.A. lens.

The spatial resolution required in flow cytometry is only that which is necessary to define and isolate the image of the intersection of the illuminating beam and the core. People tend to want to use the "best" microscope objectives; estimating quality from price, they choose an apochromat or planapochromat. These lenses do have high N.A.; without it, they would not be capable of high resolution. However, the resolution increase with increasing N.A. is not due to the collection of more light, but to the collection of light over a wider angle. A planapochromat is designed to produce flatness of field over a relatively large viewing area, and high resolution, for observation and photomicrography at high magnification. Its light gathering power may actually be sacrificed to achieve better performance in these areas. Although equal amounts of light go into the front elements of a simple achromat and a planapochromat with the same N.A., the light transmission of the achromat is likely to be higher, because the planapochromat achieves its flat field and high resolution through the use of more optical elements than are used in the achromat, and light is lost in the extra elements, even with the best of antireflection coatings. For flow cytometry, the achromat is almost always the better choice; it will certainly be cheaper, and very probably have a longer working distance and greater total light gathering capacity than will the more highly corrected lens. In recent years, I have been using, and recommending, inexpensive, single-element molded aspheric lenses with relatively high N. A. (0.62-0.68) as collection lenses, and these have found their way into some commercial instruments. And, after having had problems with low light transmission in some very expensive microscope lenses, I make it a practice to measure transmission of a low-power laser beam through any lens before I put down the money.

Spectral Selection: Monochromators versus Filters

With few exceptions, the definition of the spectral response of the detectors in flow cytometers is accomplished by the use of **optical filters**, although more sophisticated methods have been examined. Alternatives to filters may be desirable if it is necessary to make measurements in many wavelength regions. In the late 1970's and early 1980's, there were reports on the measurement of emission spectra in flow systems using polychromatic detectors[122] and on the use of monochromators to replace filters (Stöhr discussed this at a 1981 meeting, but apparently didn't publish on it). In the mid-1980's, Buican[649] described the use of Fourier transform methods for flow cytometric measurement of fluorescence in multiple spectral regions using a single detector PMT.

Monochromators and Polychromatic Detection

Spectral selection using **prism** or **grating monochromators** is accomplished by diverting different wavelengths of light in different directions and then positioning an aperture or slit between the prism or grating and the detector. The basic elements of a prism monochromator are shown in Figure 4-27. Incoming light is focused on an **entrance slit** at the focus of a **collimating lens**, which delivers a collimated beam to the prism, from which different wavelengths emerge at different angles. The **focusing lens** forms an image of the dispersed spectrum in the plane of the **exit slit**; the wavelength of light emerging from the exit slit is selected by rotating the prism about an axis perpendicular to the plane of the diagram.

The angular dispersion of a prism is not a linear function of wavelength; the angular dispersion of a grating is. Monochromators are more often built with gratings than with prisms as the dispersing element for this reason; the linear behavior makes it easier to motorize wavelength selection. Gratings also have higher dispersion than prisms, making monochromators more compact, but gratings are also less selective. A **spectrophotometer** typically uses a single monochromator to select the illumination wavelength for absorption measurements; a **spectrofluorometer** uses a monochromator to select excitation wavelength and one or two more monochromators to select emission wavelength.

Polychromatic detection is accomplished by placing multiple detectors in several regions of a spectral image. If a linear array of diode or **CCD (charge coupled device)**

detectors were placed where the exit slit appears in Figure 4-27, each element of the detector array would respond to radiation in a different spectral region. It would be more difficult to use conventional PMTs for polychromatic detection because of their size; an optical path length of several feet would be required to achieve the required spacing between desired spectral regions. Newer, multianode PMTs might be usable in this application. Experimental work with polychromatic detection using diode and CCD arrays in flow cytometry continues[1139], although this technology is not used in commercial systems.

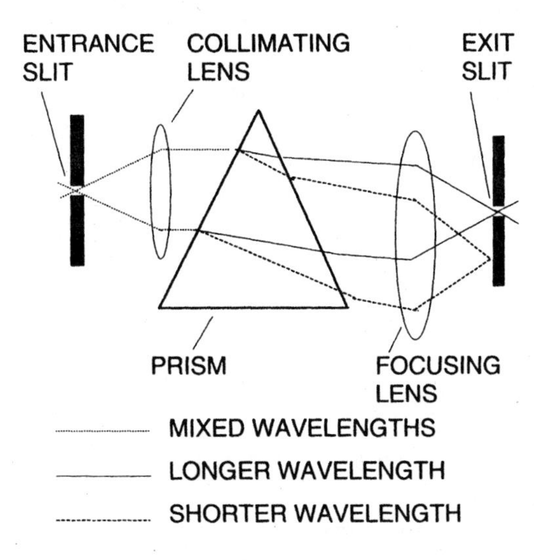

Figure 4-27. A prism monochromator. Output wavelength selection is accomplished by rotating the prism. Polychromatic detection could be done with a linear diode or CCD array at the position of the exit slit.

Buican[649,1140] built a flow cytometer in which fluorescence emission in eight spectral regions could be measured using a detector incorporating a Fourier transform or interferometric spectrometer. Such a detector system, however, is an order of magnitude more expensive than monochromators, which are themselves several times more expensive than a set of filters. The Fourier transform method has been widely adopted for infrared absorption spectroscopy because it offers some improvement (the so-called Fellgett advantage) in signal-to-noise ratio when detector noise is a limiting factor. This is not generally the case in flow cytometry; thus, it is not clear that interferometric detection offers either theoretical or practical advantages for most applications.

In general, the cell sample analyzed in a flow cytometer contains no more than three or four fluorescent materials, with well-known spectral characteristics, selected for efficient excitation by the illumination wavelength(s) used. Even in work with eight or more colors, optical filters remain the most cost-effective means for detector bandwidth limitation.

Interference Filters: Coatings of Many Colors

An **interference filter**, also called a **dichroic, dielectric,** or **reflective filter**, is composed of a transparent glass or quartz substrate(s) on which multiple thin layers of dielectric material, sometimes separated by spacer layers, are deposited. Constructive and destructive interference occur between reflections from the various layers, with the wavelength range(s) transmitted being determined primarily by the thickness of the layers and the selectivity by the number of layers. The unwanted wavelengths are reflected.

The layers of dielectric material on a substrate are collectively referred to as an **interference coating**, or, more commonly, simply as a **coating**. Coatings are used in optical elements other than filters; lenses, for example, are commonly coated to reduce reflections at their surfaces. Since, in the absence of a coating, about 5 percent of light is lost at each surface, the light transmission of the highly corrected multielement lenses used in microscopy and photography would be unacceptably low if antireflection coatings were not used.

Coatings are described as **hard** or **soft**; hard coatings are durable and moisture-resistant, but somewhat more difficult and more expensive to manufacture. They are typically used on surfaces that may be exposed to air and/or cleaned with tissues or swabs and solvents. Laser mirrors and dichroics are usually hard coated; most interference filters are made with soft coatings. The individual layers of coated substrates must be glued together, and a seal applied around the edges of the filter to prevent ambient moisture from degrading the coating.

Absorptive Filters versus Interference Filters

In an **absorptive filter**, such as a **color glass filter**, unwanted light is disposed of by absorption. It is possible to remove much more of the unwanted light by absorption than by interference; light transmission outside the **passband**, i.e., the region in which the filter is designed to transmit light, is commonly a few percent for a purely dielectric filter and only a few hundredths of a percent or less for a color glass filter. Transmission in the passband, however, is usually lower for interference filters than for color glass filters. Interference filters, as a general rule, allow sharper transitions between rejected regions and the passband(s), and permit greater selectivity; filters with a **FWHM (Full Width at Half Maximum** [transmission]) bandwidth of 3 nm are readily available.

In order to measure weak fluorescence, e.g., immunofluorescence, we really need to get rid of stray excitation light, and of excitation light scattered within the collection angle of the fluorescence collection optics. Letting 1 percent of that light through isn't good enough; we therefore can't use just a pure interference filter.

However, if we put a long pass color glass filter in a laser beam, we see that some of the absorption responsible for removal of excitation and scattered light is associated with fluorescence emission from the filter; this emission from the filter is at wavelengths that will pass through the filter, thus creating noise at the detector. So what do we do?

Well, we should, of course, use a field stop, as discussed in the preceding pages, to eliminate as much of the extraneous excitation light as possible, but we still need to specify what kind of filter goes between the field stop and the detector. And the winner is: both. The "interference filters" we use are made with an absorptive layer behind the coated layers. You can usually tell which side the absorptive layer is on because it appears more strongly colored; the coated side is more reflective, i.e., shinier. When you mount the filter in your instrument, the shiny side of the filter should face the collection lens; the other side should face the detector. When in doubt, reflect light at the unwanted wavelength from the filter onto a piece of paper (or a power meter); the side of the filter that reflects the most light should face the collection lens.

Interference filters specifically designed for use in cytometry frequently have low enough transmission outside the passband so that additional filters are not needed. They also have little arrows telling you which way to insert them in case you forget the "shiny side out" rule. If there is a need to further restrict bandwidth, the field stop (and the blocker bar, if one is used) and the interference filter get rid of enough of the stray excitation light to permit placing a color glass filter between them and the detector without worrying too much about filter fluorescence. This doesn't mean it can't happen; if it does happen, the absorptive layer of the interference filter should not be neglected as a source of such fluorescence. Loken and Stall discussed the evaluation of filters in some detail[6].

We generally neglect the fact that absorption results in the conversion of light into heat. A laser beam hitting a color glass filter may thus destroy dye molecules in that filter, changing its absorption, fluorescence, and transmission characteristics. Those dye molecules may fade more slowly during exposure to sunlight, arc lamps, etc. If one keeps it out of strong light, however, a color glass filter should have a relatively long lifetime. If you drop one and break it, the transmission characteristics of the pieces are the same as those of the intact filter; you can cover a lot of detectors with a 1" square filter if you don't mind getting glass splinters.

The companies that make interference filters have ways of cutting them; don't attempt this stunt at home. Interference filters have finite lifetimes even when they're not dropped; moisture eventually gets through their seals and causes the soft coatings to degrade. This is sometimes referred to as "delamination," which more properly describes the filter coming unglued, but the end result is the same: "Old interference filters never dye, they just lose their layers." This typically happens after a few years; if you hold the filter up to the light and see mottling around the edges, it's time to get a new filter. The old-timers didn't discover the problem until their cohorts of new filters aged.

Filter Transmission Characteristics

Figure 4-28 shows idealized transmission curves of **bandpass, short pass,** and **long pass** interference filters. The

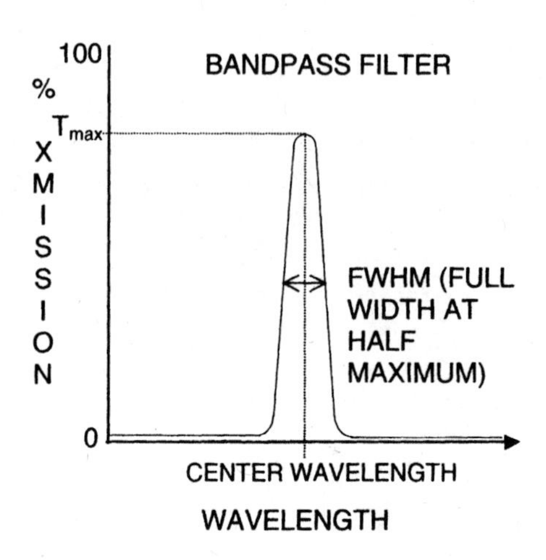

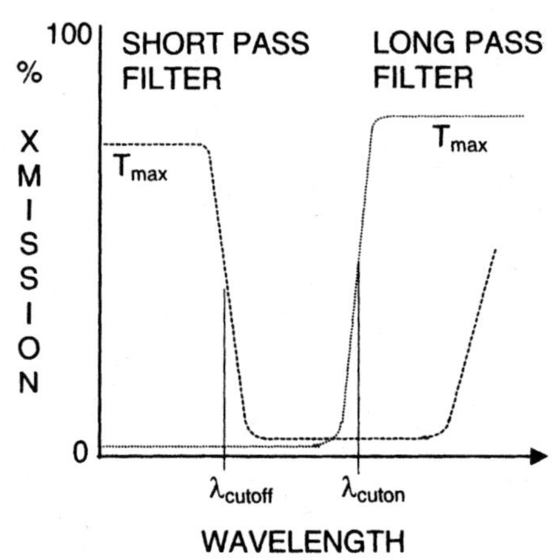

Figure 4-28. Light transmission characteristics of bandpass, short pass, and long pass interference filters.

pertinent characteristics of a bandpass filter are its **center wavelength**, usually specified in nanometers (nm), its **maximum transmission**, T_{max}, and its **bandwidth**, usually given by the FWHM in nanometers. The bandpass filters used in flow cytometry typically have a FWHM of 10-50 nm and a T_{max} of 65-80%.

If you turned the curve for the bandpass filter upside down, you'd have the transmission curve of a **notch filter**, which is not included in the figure. One of the more common notch filters is the magenta filter used for photographic color separation. A green bandpass filter blocks the red and blue/violet ends of the spectrum and transmits the green in the middle; a magenta notch filter passes red and blue/violet but blocks green. Narrow notch filters specifically designed to block laser lines may be useful for high-sensitivity fluorescence measurements.

Long pass filters are characterized by their T_{max} and by their **cut on wavelength**, λ_{cuton}, at which light transmission is 50% of T_{max}. Long pass color glass filter transmission curves resemble those of long pass interference filters, but typically show much lower transmission outside the passband.

Short pass filters are characterized by a T_{max} and a **cutoff wavelength**, λ_{cutoff}, at which transmission is 50% of T_{max}.

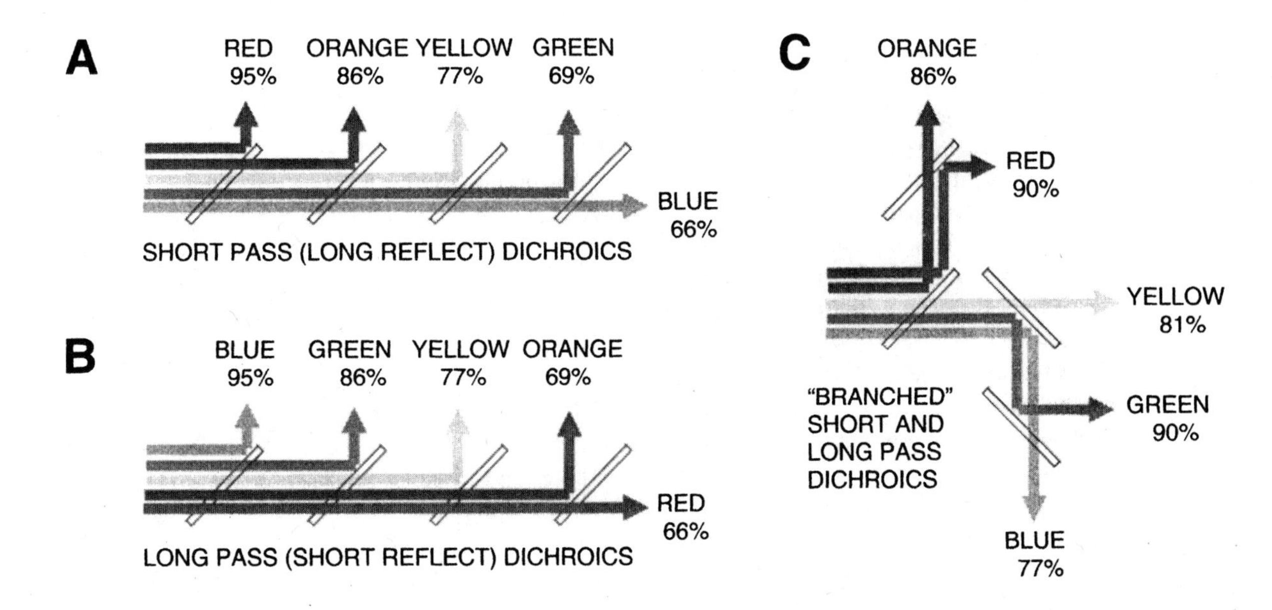

Figure 4-29. Transmission of several wavelength regions through different dichroic configurations.

They can be tricky to work with. The transmission of most short pass filters falls off sharply below about 420 nm due to absorption by some of the coating materials. However, the big problem is at at the long end; they're really notch filters. A 480 nm short pass filter passes blue light but also passes a lot of red light. To get rid of this, you can use either a color glass red absorbing filter or a longer wavelength (e.g., 580 nm) short pass filter, which blocks out to the near infrared. If you use short pass filters, keep your guard up.

You can measure transmission and reflectance characteristics of optical filters using a spectrophotometer[882], but, since most spectrophotometers don't reliably measure transmission below 1 percent, you can't tell whether a filter is truly "blocked" to 0.01% outside its passband. I have a few mW available from various laser sources at wavelengths from the UV to the near infrared, and a power meter that can detect a few nanowatts; this lets me measure transmission down to 0.01 percent reliably, in a well-darkened room. The filter manufacturers can do the same.

The transmission characteristics of an interference filter are determined by the distance incident light must travel between the layers of the filter; this is shortest for rays that are normal, or perpendicular, to the filter, and increases for rays at progressively larger angles to the normal. The transmission characteristics thus change with the angle of incidence of light, so users and manufacturers must specify the angle of incidence at which a filter is designed to work. If you hold a bandpass filter up to a white light source and vary the angle at which you look through it, you will see a change in the color of the transmitted light. Absorptive filters do not exhibit this angular dependence.

Dichroics

Epiilluminated fluorescence microscopes, and arc source flow cytometers, make use of what are termed **dichroics**, **dichroic mirrors**, or **dichroic beam splitters**; these are also used in light collection systems to direct light in different spectral regions to different detectors. Dichroics are interference filters; those used in fluorescence microscopy are usually the long pass type. Excitation light at a short wavelength, outside the passband, is reflected from a long pass dichroic; fluorescence emitted at longer wavelengths is transmitted through it. A short pass dichroic reflects longer wavelengths and transmits shorter ones. It is desirable to minimize the fluorescence of a dichroic, because any fluorescence excited in it by light at the illumination wavelength may be transmitted to the detector. Dichroics are, therefore, almost always made without the absorptive layers typically used in "interference" filters.

Filter sets for fluorescence microscopy have been developed that incorporate excitation filters and dichroics that, respectively, transmit and reflect in several discrete wavelength regions. This allows simultaneous viewing of (for example) blue, green, and red fluorescence excited by UV, blue-green, and yellow light. While these filter sets provide lovely visual images, their real purpose is to eliminate changing filter blocks during multicolor imaging experiments, minimizing the likelihood that the specimen will move slightly during that mechanical operation.

Managing light loss in dichroics was mentioned on p. 53; we now return to it with the aid of Figure 4-29. In a flow cytometer with 488 nm excitation, the blue side scatter signal is strongest, green and yellow (fluorescein and phycoerythrin) fluorescence signals are fairly strong, and orange and red (tandem conjugate) signals are weakest. We want to save as much of the weaker signals as possible.

It is reasonably easy to make a dichroic that reflects at least 95% of light in some region outside its passband, and relatively difficult to make a dichroic, or any other interference filter, with 95% transmission in its passband.

Thus, weak signals should be reflected from, rather than transmitted through, dichroics. Figure 4-29 assumes 95% reflectance and 90% transmission from the dichroics.

The configuration illustrated in (A) in Figure 4-29 is that shown in Figure 1-21 (p. 51); a linear array of short pass (long reflect) dichroics preserves better than 85% of the weaker red and orange signals, although 31% of the green signal is lost. This is preferable to the configuration in (B), which loses over 30% of both the orange and red signals. If the transmission of the filters were lower, say 85%, only 52% of the red signal would get through. For optimal light transmission, it may be best to have the dichroics and detectors in a branched configuration, such as that in (C), in which no signal must pass through more than two dichroics.

Most flow cytometers use a single lens, with its axis orthogonal to both the axis of flow and the axis of the illuminating laser beam, to collect fluorescence and side scatter signals. In some instruments, a spherical mirror is placed on the opposite side of the flow chamber from this lens, to reflect additional fluorescence emission and scattered light back into it. One can also get more light to the detectors by using two separate collection lenses on opposite sides of the chamber. I do this in most of my instruments; it allows collection of four-color fluorescence with none of the fluorescence signals being transmitted through more than one dichroic, reducing the light losses associated with multiple dichroics. One could also use relay lenses and/or fiber optics (see pp. 157-9) to direct light from both collection lenses to the same detector(s).

Neutral Density Filters

It is sometimes desirable to attenuate light without discriminating wavelengths; this can be done with **neutral density (N.D.) filters**. A neutral density filter can work by reflection or by absorption. A partially "silvered" (i.e., coated) mirror will reflect some portion of the light reaching it and transmit the remainder. Absorptive neutral density filters are color filters incorporating substances that absorb over the entire visible spectrum and which therefore appear gray or black in color; they are often seen on beaches. Neutral density filters are often used in front of the forward scatter detectors in flow cytometers, particularly in instruments that use high-power laser sources.

Beamsplitters; Ghosts and Ghostbusters

A reflective neutral density filter can serve as a **beamsplitter**, diverting the reflected portion of the incident beam in one direction and transmitting the remainder in another. Uncoated glass surfaces typically reflect about 5% of incident light; common cover slips thus can be used as beamsplitters to divert a small portion of an incident beam, e.g., for a side scatter measurement. To send 50% of the light one way and 50% of the light the other way, you need a reflective coated surface.

Beamsplitters (Figure 4-30) are frequently made in the form of cubes; the cube is formed by fusing two right triangular prisms with the coating along the "hypotenuse" face of one prism. Light therefore enters and leaves a cube beamsplitter normal to the surface, minimizing undesirable "ghost" reflections that are shown coming from the rear face of the plate beamsplitter.

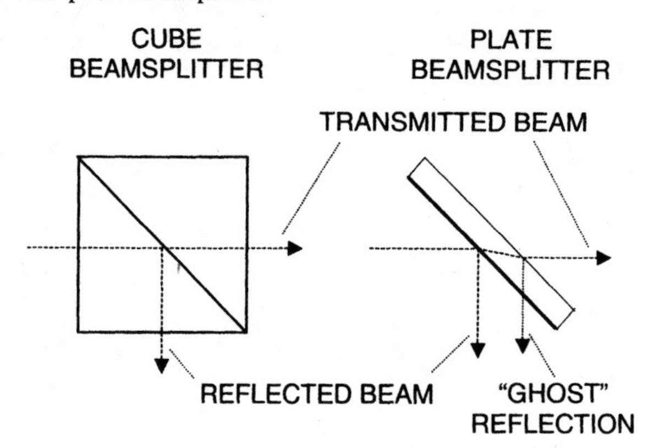

Figure 4-30. Cube and plate beamsplitters.

The more colors you measure in a flow cytometer, the more dichroics you need; and they also produce ghost reflections. The problem starts to get noticeable when there are a lot of them, so who(m) do you call? Well, you may have to call your optical filter supplier, and get dichroics and filters that you can place at angles other than 45 and 90 degrees, or you may need to use wedges to move the ghost reflections further apart so they don't go where you don't want them. But you probably won't have to call anybody; the manufacturer of your flow cytometer should take care of the problem for you. Until, that is, you decide to measure a different set of colors; then, you'll have to call both the manufacturer and the filter supplier.

It should be obvious that absorptive neutral density filters cannot be used as beamsplitters, because they are designed to absorb light, not reflect it. They, or at least the relatively inexpensive plastic ones you can buy in sheets in real camera stores, also tend to melt and/or burn when you try to use them to attenuate a few hundred milliwatts worth of laser beam.

Optics for Polarization Measurements

It is occasionally of interest to measure fluorescence or scattered light in two planes of polarization, respectively perpendicular and parallel to the plane of polarization of the excitation beam. This requires that light of different polarizations be separated and diverted to different detectors. Such measurements are relatively easy to do in flow cytometers with argon ion laser sources, because the laser source produces highly polarized illumination; you do have to be able to get at the detector filters. Two approaches can be used. The first makes use of a **polarizing beamsplitter**; this is an optical element that looks more or less like those shown in Figure 4-30, but which transmits light of one polarization and reflects light of the other.

Most beamsplitters polarize light to some extent; Asbury et al, in their extensive study of polarization in flow cytometry[2439], showed that beamsplitters caused strong, orientation-dependent polarization artifacts. Polarizing beamsplitters, which polarize light in well-defined and predictable ways, must be selected for the appropriate wavelength and tend to be fairly expensive.

If what you'd like to do is discriminate eosinophils from other granulocytes by their higher depolarized scatter signals[710,986,1137], you can use a simpler, less expensive approach employing **polarizing filters**. These are typically made of a plastic in which are embedded optically active molecules that have been oriented in one direction, e.g., by application of an electric field. This results in differential absorption of light of different polarizations. If you have a pair of polarized sunglasses, you have probably noticed that rotating one of the lenses about its axis changes its transmission, making a hazy blue sky appear darker blue.

Polarizing filters decompose the electric field vectors of incident light into parallel and perpendicular components; what comes out is light of one polarization. If you put two polarizing filters together, and rotate one, there will be one position at which light transmission through the pair is at a maximum, and another, at 90° to the first, at which no light gets through, because the second filter is then positioned to transmit only the polarization which has been excluded by the first filter. There is generally an orientation mark on such filters to facilitate finding the appropriate positions.

Setting up a cytometer for quick and dirty scatter polarization measurements requires, first, that two detectors be positioned and equipped with proper wavelength selection filters to measure scatter in the same spectral region. Using beads to provide a standard signal, the detector gains are adjusted to provide equal signal strengths. Then, a polarizing filter is placed in front of each detector. The filter on the "parallel" detector is rotated to produce maximum signal strength from the beads; the filter on the "perpendicular" detector is rotated to produce minimum signal strength. Gains are then readjusted.

If you're trying to measure fluorescence emission anisotropy or polarization against some standard, it may be necessary to play additional games, such as changing the laser polarization with a retardation plate, and you probably want to use higher quality components. See p. 114, then read Asbury et al[2439] and proceed from there.

Tunable Filters

It is now possible to make **tunable filters**, in which an electrical input changes the passband. These devices don't have conventional filters in them; some of them use acousto-optic modulators and gratings; others use liquid crystals. They can respond in milliseconds, and have been used in multispectral imaging and confocal microscopy, but not for flow cytometry. Because they polarize light, tunable filters have relatively low (35% or so) transmission; they are also on the pricey side, at least at present.

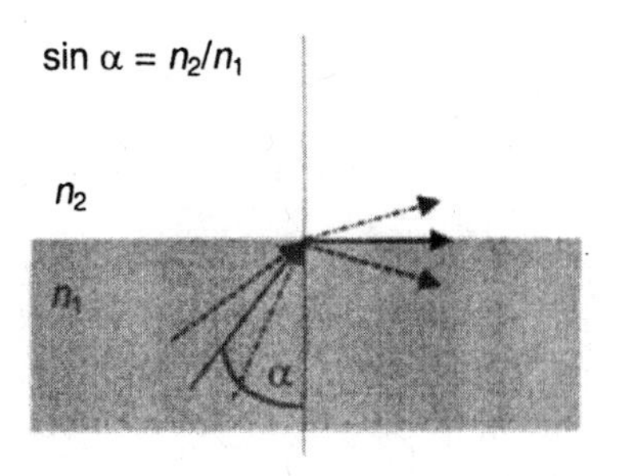

Figure 4-31. Total internal reflection.

Fiber Optics and Optical Waveguides

Lenses and mirrors aren't the only components useful for getting light from point A to point B; the communications industry, as the ads on TV remind us, gets a lot of light from one end of the country to another through **fiber optics**. These, and other **optical waveguides**, work by **total internal reflection**; the principle is illustrated in Figure 4-31. This shows three rays of light traveling in the direction of the boundary between regions of two different refractive indices n_1 and n_2, where $n_1 > n_2$. Light is refracted according to Snell's law (pp. 106-7); light striking the boundary at the **critical angle**, $\alpha = \sin^{-1} n_2/n_1$, will be refracted along the boundary, while light striking the boundary at larger angles will be reflected back into the medium from which it came. Light striking the boundary at angles smaller than the critical angle will be refracted at the boundary and emerge into the medium with the lower refractive index.

An optical fiber consists of a plastic, glass, or silica **core** surrounded by a **cladding** with a lower refractive index. The refractive indices of the core and cladding determine the **numerical aperture (N.A.)** of the fiber; this defines the maximum half angle θ at which light entering the fiber will propagate through it, and also the maximum angle at which light will emerge from the other end. The fiber illustrated in Figure 4-32, below, is a **step index** fiber; there are also **gradient index fibers**, in which refractive index decreases continuously with distance from the fiber axis.

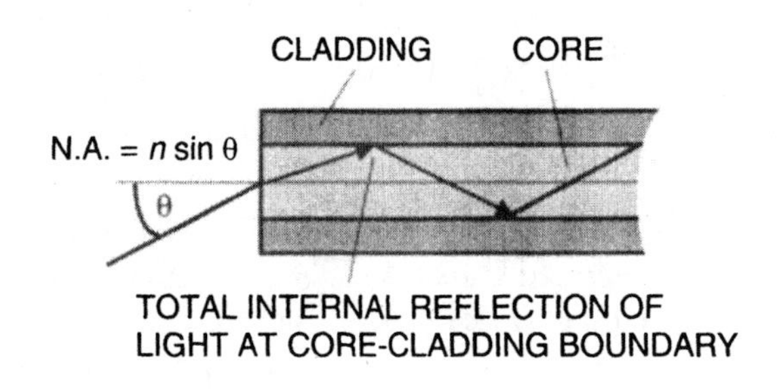

Figure 4-32. A fiber optical waveguide.

Single fibers used for optical communication typically have core diameters between 5 and 120 μm and outside diameters of 125-200 μm. They are thus, at least in principle, well suited to delivering light to, and collecting light from, regions of space roughly the size of the observation volume of a flow cytometer. In the mid-1980's, Mike Hercher and I[877] built a fluorescence flow cytometer in which fiber optics were used for both illumination and collection; the only lens in the whole thing was the one we used to focus the illuminating laser beam into the fiber optic that delivered light to the round capillary flow chamber.

This prompted a brief episode of euphoria in which we envisioned "flow cytometers on a chip," with the flow chamber, diode laser and detectors, and all on a single substrate, incorporating optical waveguides to move light to and from the sample stream. Waveguides of various geometries can be made on a substrate by depositing layers of material of the appropriate refractive index, or by doping the surface to change the refractive index. However, I now feel honor bound to point out that waveguide-based cytometers present a few problems without obvious solutions.

Light collection is limited by the N.A. of the fiber or waveguide. If you use a silica core, the maximum N.A. you can get is 0.4; for fluorescence collection, you'd be better off with a higher N.A. Plastic fibers with N.A. of 0.54 and glass fibers with N.A.'s of 0.66 or higher would be suitable, but, unfortunately, waveguide flow cytometers tend to have high levels of stray scattered light, and this induces fluorescence in plastic and glass fibers, which decreases measurement sensitivity in a big way. Beam shaping is at best hard to do, and collecting light selectively from the observation point is a lot easier to do using lenses and field stops. Finally, if you put the source, detectors, and all on one chip, you have to throw out the whole thing when it clogs. If waveguide flow cytometers ever do reach maturity, it will be because somebody finds an application for which their disadvantages don't disqualify them. I've been waiting for about 18 years now.

The Ortho System 30 and System 50 of the late 1970's used plastic fiber optics to relay light to the detectors; one end of the fiber was placed in an image plane, serving as a field stop; the other end went to the PMT housing. The fluorescence detector filters were placed in front of the fibers, forestalling problems with fluorescence that might have been induced in the fibers by stray laser light. BD Biosciences' new LSR II and FACSAria (Chapter 8) also use relay fibers.

Through a Glass Darkly: Light Lost (and Found) in Optical Components

I just mentioned problems with waveguide cytometers due to fluorescence induced in the optics by stray illuminating light; you should also know about some other instances in which the materials of which optical elements are made can cause severe problems.

The first set of headaches comes from the UV transmission characteristics of commonly used glasses such as BK-7 and optical crown glass; both lose very little light at wavelengths between 400 and 800 nm, but transmission falls off noticeably below 400 nm and rapidly below 350 nm. If you look in something like the Oriel Optics Catalog, you'll see transmission is usually tabulated for fixed thicknesses of glass, e.g., 10 mm. Obviously, the thicker the glass element through which light must pass, the lower the transmission.

The "long" UV wavelengths used for fluorescence excitation range from the 350 nm krypton laser line to the 366 nm mercury arc lamp line; a 10 mm thickness of BK-7 transmits over 85% of incident light in this wavelength range, but 10 mm of crown glass transmits only 72% of incident light. If you're not using a lot of thick lenses or cube beamsplitters in your system, you can generally live with these figures and get away with glass illumination optics instead of quartz or silica, which have high transmission even at 260 nm. You should also be okay with glass optics if you use a violet diode laser. If you want to use a UV He-Cd laser, which emits at 325 nm, you'll need quartz or silica optics, despite their premium prices; the transmission of 10 mm of crown glass at 325 nm is less than 30 percent.

When you do use quartz or silica optics for UV, you should avoid antireflection coatings designed for visible light, because these coatings have very high absorption in the UV, particularly down at 325 nm. The mirrors that are used to bounce laser beams around in cytometers can also pose problems here; they typically have coatings with very high reflectivity (>95%) above 420 nm or so and very low reflectivity (<50%) in the UV. Bounce an illuminating beam off two mirrors like that and you've lost 75% of your power. I tend to use uncoated lenses and minimize the number of mirrors in the system when I work with UV; where I do need a mirror, I use the same coating as is applied to UV laser mirrors. As a last resort in cases where I need reflectivity in the UV and at long wavelengths, I'll use an aluminum mirror, which has about 85% reflectivity across the board.

Since glass can absorb in the UV and violet, it should come as little surprise to you that glass can fluoresce when illuminated in that wavelength region. With illumination at 488 nm and longer wavelengths, you can probably use glass flow chambers; with UV illumination, and even with illumination at 441 nm, however, considerable fluorescence is observable in most types of optical glass. This does not generally interfere with measurements of strong fluorescence but is likely to impair or preclude measurements of weak fluorescence. In arc lamp systems, when the same lens is used for illumination and light collection, it is wise to use fluorite lenses because they have high UV transmission and low fluorescence, both of which are desirable.

Plastic optical elements, e.g., Fresnel lenses and fiber optics, are worse offenders than glass ones and tend to fluoresce noticeably even when illuminated at wavelengths of 500 nm and above. Problems with fluorescence from optical elements can generally be avoided by minimizing the amount of stray excitation light which gets into the collection optics, using field stops, blocker bars, and/or coatings.

I noted above that some Ortho instruments used plastic fiber optics to transmit light between the fluorescence emission filters and the PMTs; fiber fluorescence was not a problem. When Ed Luther and I (unpublished) used the same fibers to collect light from an argon laser source cytometer, collection was highly efficient, i.e., we got the same signal intensities as we would get from a lens of the same N.A. as the fiber, but fluorescence induced in the fibers limited the sensitivity with which we could measure immunofluorescence. More recently, when I was trying to measure some very weak autofluorescence with UV and blue excitation, I found problems with fluorescence in the lens; I'll have more to say about this later.

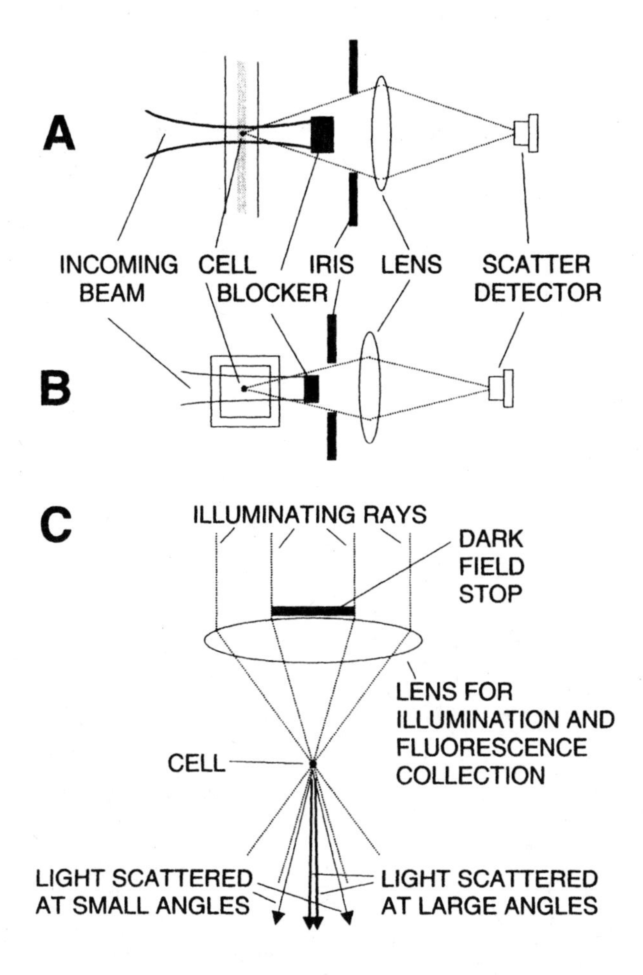

Figure 4-33. Optical arrangements for collection of forward scattered light.

Collection Optics for Forward Scatter Signals

Three basic optical arrangements are used for collection of forward scattered light; they are illustrated in Figure 4-33. Panel A of the figure shows a side view of the configuration used in flow cytometers in which cells are observed in a stream in air or in a round capillary. The stream and capillary walls act as cylindrical lenses, refracting light from the illuminating beam in the plane perpendicular to the axis of flow; in order to avoid interference from this source, it is necessary to collect light scattered at angles above and/or below this plane. A beam dump or blocker bar is placed with its long axis perpendicular to the plane of the drawing, i.e., perpendicular to the direction of flow. The illuminating beam diverges from its focal point on the core stream; the blocker must be large enough to intercept the diverging beam, and its height determines the minimum angle to the illumination beam at which forward scattered light can be collected. The maximum angle is determined by an iris diaphragm in front of the lens, which brings the collected light to a focus at the detector.

Panel B of the figure shows a top view of the configuration used in flow cytometers in which observation is done in flat-sided cuvettes. The arrangement is similar to that shown in A, except that the axis of the blocker bar is parallel to the direction of flow. The cylindrical lens effects of a round stream and/capillary are eliminated when a flat-sided flow chamber is used; it is therefore possible to measure light scattered in the plane of the long axis of the elliptical focal spot. This is advantageous because the divergence of the illuminating beam is lower in this plane, allowing the blocker bar to be made narrower, and permitting collection of light scattered at smaller angles to the beam.

Since crossed cylindrical lenses are used to produce the elliptical focal spot, it is possible to place the back focusing lens in such a way that the beam is brought to a focus on the blocker bar, rather than on the core stream; this is shown in the drawing, and its effect is to further decrease the minimum angle at which scattered light can be collected. The iris and lens function as in panel A. Ortho replaced the conventional blocker bar with a reflective strip, which, with a lens, diverted the transmitted beam to an extinction detector.

Panel C illustrates dark field illumination as used with arc or incandescent lamps; this configuration, used in the original Gucker apparatus[29] (see Figure 3-1, p. 74), was also adopted by Steen[102,1142], who added a valuable wrinkle.

In dark field illumination, a circular stop, coaxial with the illuminating lens, results in the illuminating rays forming hollow cones of light with central shadows, converging to and diverging from the plane of the specimen. It is customary to collect light scattered by the specimen into the shadow region on the side opposite the illuminating lens.

Steen[1142] noted that light scattered at larger angles emerges near the optical axis of the system, while light scattered at smaller angles emerges nearer the edge of the shadow cone. Diverting light from these two regions to different detectors allows separate measurement of small- and large-angle scatter signals. The characteristics of "orthogonal," "90°," or "side" scatter signals dominate scatter signals collected at angles as small as 18°, but the amplitudes are higher at the smaller angle. Steen's approach, utilized in instruments produced by Skatron of Norway, also variously made and distributed commercially by Leitz, Bruker, Ortho, and Bio-Rad, and currently being revived by Apogee, has, among other things, facilitated sensitive multiangle scatter measurements of bacteria.

In experimental systems, fiber optics have been used successfully for forward scatter signal collection, but noise levels tend to be high. My experience in recent years has been that forward scatter measurement sensitivity can be improved by using a relatively high N.A. collection lens, but I suspect I'm not collecting light at the really small angles.

4.5 DETECTORS

Once we have collected light from various directions, and selected the spectral region(s) from which we will allow light to reach the detector(s), we reach the point in our cytometer at which, Presto!, the light changes back into photons. The numbers of photons we collect from a cell vary over a wide range, depending upon whether we are measuring absorption or extinction (too many), light scattering (lots for forward scatter, enough for orthogonal scatter), strong fluorescence, e.g., of dyes bound to DNA (usually more than enough), or weak fluorescence, e.g., immunofluorescence (please, Sir, I want some more). The **detectors** we generally use are those best suited to the numbers of photons with which they will have to deal.

It is common practice to use **photodiodes** as detectors for absorption, extinction, and forward scatter signals and **photomultiplier tubes (PMTs)** as detectors for orthogonal scatter and fluorescence signals. In flow cytometers, whether the source is a laser or a lamp, extinction signals are detected by sampling the illumination beam on-axis. Since most light is transmitted through the sample, the light levels reaching the detector are high, and may be high enough to damage the detector unless attenuated by placing a neutral density (N.D.) filter in front of the detector. This filter should be reflective rather than absorptive so it doesn't melt or catch fire (no, I'm not kidding). The sensitivity of a PMT is rarely needed to detect extinction signals.

Objects the size of cells scatter light approximately as predicted by Gustave Mie's theory; the lion's share of the scattered light goes in the forward direction, at relatively small angles to the beam axis. This light can be detected by a photodiode if there is enough of it, i.e., if the excitation power is sufficient, and if it is in a spectral region to which a photodiode is sensitive.

Silicon Photodiodes

As was mentioned previously, a silicon photodiode produces current when photons impinge upon it; solar cells are silicon photodiodes with a large enough aggregate surface area to provide enough current to do something useful. A photodiode does not require an external power source in order to operate. The peak sensitivity of silicon photodiodes is at about 900 nm; at this wavelength, the **responsivity** of the devices, i.e., the current produced per unit of incident radiant power, is about 0.5 amperes/watt (A/W). At 500 nm, even in a "blue-enhanced" or "UV-enhanced" photodiode, responsivity is only about 0.28 A/W; at 350 nm, the figures are about 0.12 A/W for a UV-enhanced diode, 0.05 A/W for a blue-enhanced, and half that for the garden variety. What manufacturers' diodespeak calls "blue-enhanced" and "UV-enhanced" devices, incidentally, actually end up with improved performance over most of the visible spectrum compared to "standard" photodiodes.

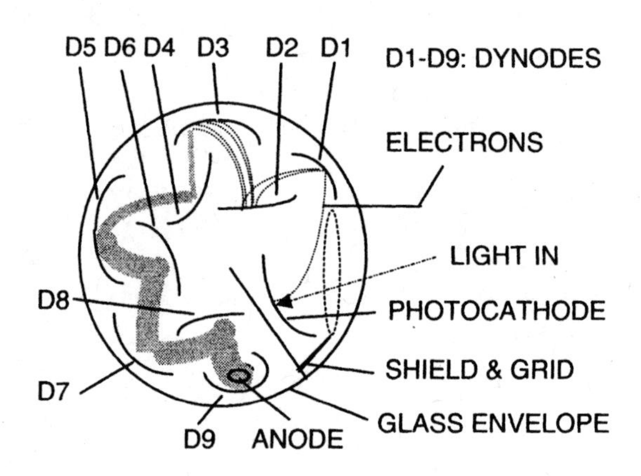

Figure 4-34. Elements of a photomultiplier tube (PMT).

To relate responsivity in A/W into what goes on at the photon level, we need to deal with photodiodes, and with other detectors, in terms of **quantum efficiency** (Φ); for detectors, or for the photosensitive elements of detectors, this is defined as

$$\Phi\ (\%) = 100 \times (\text{electrons out})/(\text{photons in})\ .$$

By definition, 1 A = 1 coulomb/s = 6.2×10^{18} electrons/s. Also by definition, 1 W = 1 J/s . However, although the number of electrons in a coulomb remains constant, the number of photons in a joule varies. At 350 nm, there are 1.76×10^{18} photons/J, at 500 nm, there are 2.52×10^{18} photons/J, and at 800 nm, there are 4.03×10^{18} photons/J. Thus, the responsivity in A/W at a quantum efficiency of 100 percent (one electron out for each photon in) will vary with wavelength. At 350 nm, this responsivity would be about 0.28 A/W, at 500 nm, about 0.41 A/W, and at 800 nm, about 0.65 A/W. The actual responsivities for a typical UV-enhanced silicon photovoltaic photodiode at these wavelengths are, respectively, about 0.12, 0.28, and 0.47 A/W; we can therefore calculate the respective quantum efficiencies as 44, 68, and 72 percent. And the quantum efficiencies are what's important.

Photodiodes are usually operated in the **photovoltaic** mode, in which no external voltage is applied across the diode; they can also be run in the **photoconductive** mode, with a **bias voltage** applied. This does not increase responsivity or quantum efficiency, but speeds up response time of the device, usually with some increase in noise. The response time of photodiodes increases with their intrinsic capacitance and therefore with their size; devices with an active area of 1 mm^2 or less usually have fast enough rise times for most cytometric applications.

Photovoltaic and photoconductive photodiodes have no gain. The small currents they produce in response to incident light must be amplified electronically. This is generally done by an active electronic preamplifier circuit that converts small input currents to proportional, but much larger, output voltages. The voltages may or may not be amplified further along in the signal processing electronics. The preamplifier and amplifier circuits I use for forward scatter measurement are probably fairly typical; they have an overall gain of about 5 million volts/ampere.

Knowing this, I know that if the 5 million V/A preamplifier's output pulse has a peak amplitude of 5 V, which is fairly typical when I measure light scattered by 2 μm plastic microspheres, using 20 mW excitation at 488 nm, the photocurrent generated in the detector was 1 μA.

Given the diode's responsivity of 0.25 A/W at 488 nm, this would mean that 4 μW of scattered light reached the detector. If I used a He-Ne laser at 633 nm, where the diode responsivity is 0.4 A/W, I'd need less laser power; when I use a big argon laser (200 mW at 488 nm) for excitation, I have to put a 1.0 neutral density filter in front of the detector to keep the signals on scale.

Photomultiplier Tubes (PMTs)

Like photodiodes, **photomultiplier tubes** produce current at their anodes when photons impinge upon their light-sensitive **photocathodes**. Unlike most photodiodes, PMTs do require external power sources; they also incorporate gain, which can be quite high (10^7 or more electrons out for each photon in). The gain mechanism of PMTs is one of the closest things in the physical world to a free lunch, which probably explains why PMTs are among the few types of **vacuum tubes** (without the vacuum, they're down the tubes) that survive in this solid state era. RCA, which developed the PMT, produced an excellent handbook on PMTs[123] which can be used should the reader want to flesh out the bare bones I have provided on page 54 and in what follows; Hamamatsu, which now supplies most of the PMTs used in flow cytometry, also has a handbook[2451].

A schematic top view of the inside of a side-window PMT is shown in Figure 4-34. The photocathode is the part of a PMT that responds directly to light; when photons hit it, it emits electrons. The photocathode, like a photodiode, therefore exemplifies the **photoelectric effect**, for studies of which Einstein was awarded his Nobel Prize in physics (relativity was regarded as too radical back then). The PMT also contains a series of electrodes called **dynodes**, to each of which is applied a potential slightly more positive than that on its neighbor dynode nearer the photocathode. There is also an **anode**, which is kept at a more positive potential than any of the other electrodes. The electrodes are arranged in space so that **photoelectrons**, i.e., electrons emitted from the photocathode in response to incident photons, are accelerated toward the first dynode by the electric field set up by the potential difference between these two electrodes.

When they hit a dynode, electrons cause **secondary emission** of more electrons from the dynode surface; these are accelerated toward the next dynode by the electrode field between the dynodes, and then produce further secondary emission, and so on. The higher the applied potential between dynodes, the more energy is imparted to the electrons between stages; this results in emission of increasing numbers of secondary electrons from each successive dynode surface and increases the gain of the tube. Anywhere between 10^3 and 10^8 electrons may reach the anode for every electron that left the cathode, and the gain mechanism is termed "noise-free" because the dynodes only emit electrons when struck by electrons. As I mentioned on page 54, in practice, the anode is usually kept near ground potential; the photocathode is at minus a few hundred to a couple of thousand volts, and the dynodes (D1-D9 in the diagram) are at progressively higher voltages. The shield and grid shown in the diagram prevent interactions between the anode and the photocathode.

There is some noise in PMTs; most of it results from **thermionic** (i.e., due to temperature alone) **emission** of electrons from the photocathode in the absence of incident light, and from DC leakage. These generate a **dark current** that, with photoelectron and secondary electron statistics, determines the signal-to-noise ratio obtainable from the PMT. When extremely low light levels must be detected, as in single photon counting (p. 164), it is common practice to refrigerate the detector to decrease thermionic emission, thereby reducing dark current. In flow cytometry, extraneous light reaching the detector, rather than dark current, is usually the principal noise source[8,124]; in this case, cooling the PMT won't help.

The **spectral response** of a PMT is determined by the composition of the photocathode. Tubes with **bialkali** photocathodes (some of the standard compositions are referred to as S-4 and S-5) have peak photocathode responsivity of about 40 mA/W at about 400 nm; their responsivity falls off sharply above 550 nm and is low enough at 600-650 nm to make these tubes largely unusable in this region.

Multialkali photo-cathodes, such as the S-20 type, extend the usable wavelength range to beyond 750 nm; even longer wavelength response is obtained using **gallium arsenide** (GaAs) cathodes, which provide high responsivity (50 mA/W) and a relatively flat responsivity curve between 300 and 850 nm, making them useful in spectrophotometers and spectrofluorometers. Tubes with GaAs photocathodes, however, tend to cost much more than other PMTs, and also typically have lower gain; in general, unless you are looking at fluorescence or orthogonal scatter beyond 800 nm, you shouldn't need one.

For fluorescence above 550 nm, you need multialkali tubes; they are noticeably better than bialkali tubes even for measurement of fluorescein fluorescence at 520 nm. However, you can get by with a bialkali tube for scatter measurements at 633 nm because scatter signals are so much stronger than fluorescence signals. For fluorescence or scatter measurements in the blue or blue-green, bialkali tubes, which are the cheapest, are just dandy. I'll go into more details shortly.

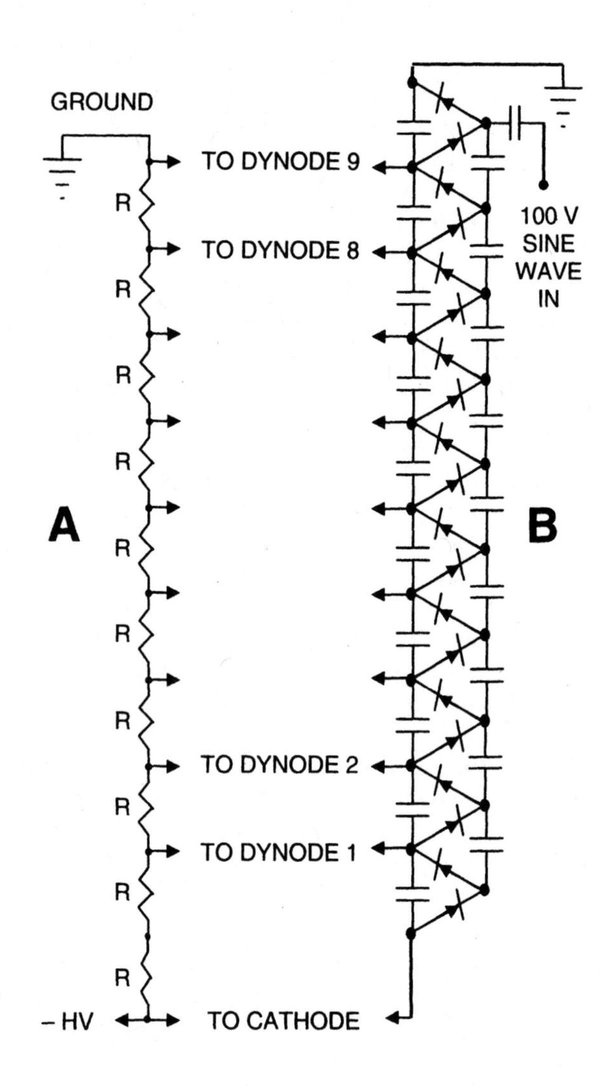

Figure 4-35. PMT electrode voltage supply circuits. Panel A: resistive dynode chain; Panel B: Cockcroft-Walton voltage multiplier.

To get signals out of a PMT, you have to plug it into a socket, which provides a way of connecting the photocathode, anode, and dynodes to the voltages required for each. Until recently, the customary way of powering a PMT involved connecting a very well regulated high voltage (300-2000 V, depending on the tube type and the gain desired) power supply to a network of resistors connected in series, known as a **dynode chain**; such a circuit is shown in panel A of Figure 4-35. The photocathode is connected to the end at the lowest potential, here indicated by –HV, and the dynodes are connected at every junction of resistors in between, up to the highest potential, which here is ground (0 V). Good regulation of PMT power supply voltage is essential because the relation between applied voltage and PMT gain is nonlinear (it is close to logarithmic); therefore, a change of one or two percent in applied voltage may change gain substantially.

The total resistance in the dynode chain is generally about 1 megohm (the value of the resistance R in this case is 100 KΩ), meaning the chain draws about 1 mA from a 1000 V supply. To keep PMT response linear within 1%, the anode current must be no greater than a small fraction of the current drawn by the dynode chain, say 10 μA or less. A 1994 publication[1142] from Hamamatsu pointed out that an alternative circuit, the Cockcroft-Walton voltage multiplier shown in panel B of Figure 4-35, offers several advantages over conventional dynode chains. This circuit was originally used to generate high voltages for particle accelerators, work with which earned Cockcroft and Walton the 1951 Nobel Prize in physics; it uses a network of diodes and capacitors to convert a sine wave input with amplitude of about 100 V into successively higher voltages. The multiplier circuit generates less heat than a resistive dynode chain, which results in lower noise; it also provides good linearity at anode currents well over 100 μA, and draws less input current because no resistive elements are used. Small detector modules incorporating a miniature PMT, shielded housing, and a power supply utilizing a Cockcroft-Walton multiplier are now available; one is shown in Figure 4-36 (F).

PMTs require **shielding** from stray light and magnetic fields; commercially available PMT housings provide both and may also incorporate a dynode chain, power supply, and even a preamplifier. When not in use, PMTs should be kept in the dark; exposure of an unconnected PMT to room light will affect the photocathode, increasing dark current, while exposure of a PMT to the same amount of room light while a power supply is connected will make the tube unusable, at least for purposes of cytometry.

PMT designs use a variety of physical layouts. The first PMTs were so-called **side-window** tubes, shown in Figures 4-34 and 4-36; their descendants are living and well today. A side-window tube usually has a wire grid, which blocks a small fraction of light, in front of the cathode; this **shadow effect** sometimes caused trouble in microspectrophotometry, and a lot of old hands in analytical cytology preferred **end-window** tubes (A in Figure 4-36), which don't have the obstruction. Grid shadows should not pose a problem in flow cytometry; there is no practical reason not to use side-window tubes, which tend to be less expensive than other types, and have largely replaced the end-window tubes used in earlier commercial and laboratory-built instruments. The primary selection criteria for PMTs are high gain and red sensitivity; the 1 1/8 inch diameter side-window R928, R1477, and R3896 tubes, all from Hamamatsu, are now widely used, and the last of these, which has very good red sensitivity, is made without a grid, eliminating any possibility of a grid shadow problem.

The commercial history of PMTs is enlightening. They were invented in the 1930's at RCA (the Radio Corporation of America, of which only the logo remains), and produced in small quantities until World War II, when it was discovered that the "white noise" produced by amplifying the dark current of a PMT was useful for jamming radar. This led to mass production of PMTs, making them affordable to scientific instrument builders after the war. They have been widely used in apparatus for nuclear medicine, and deeply

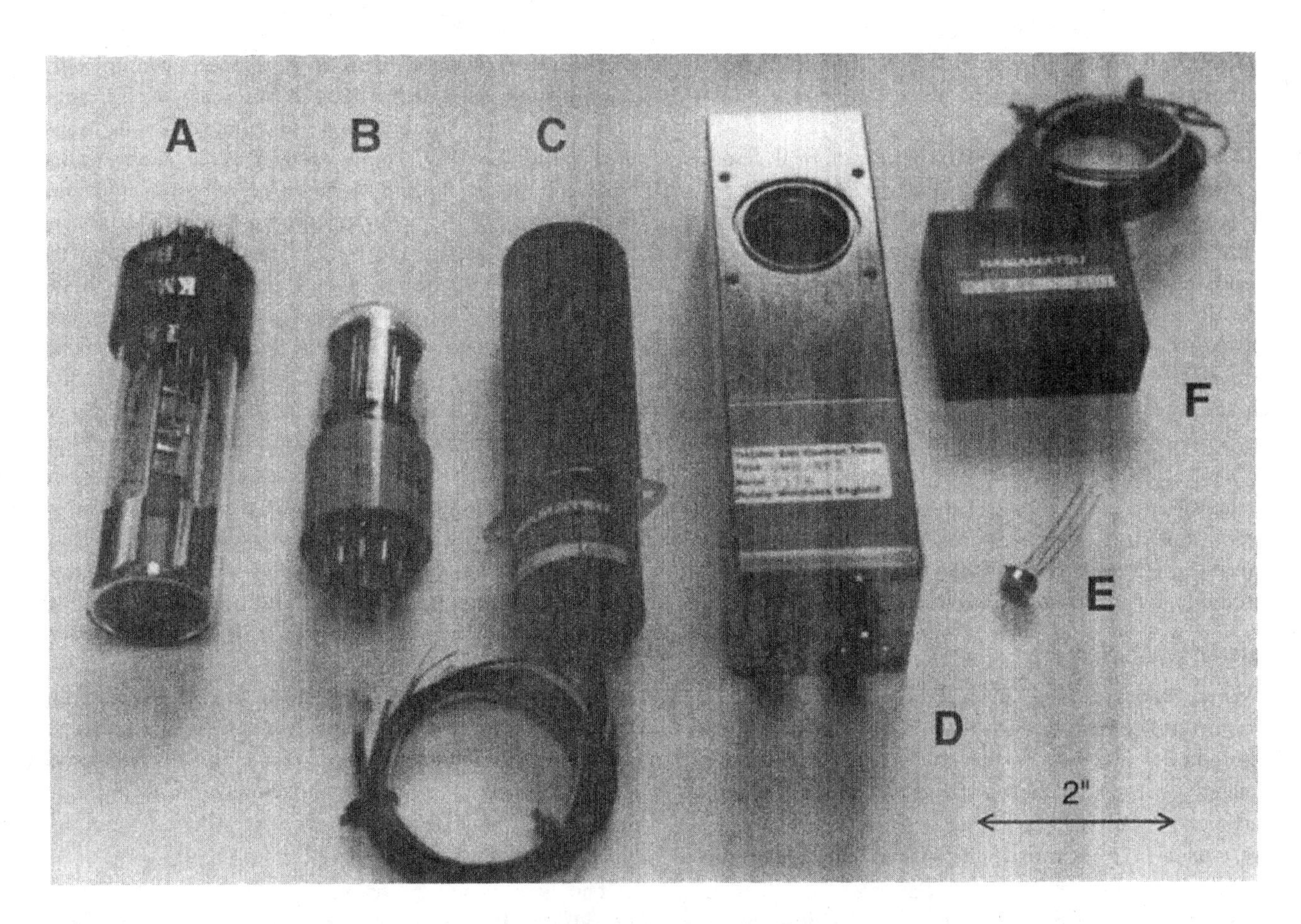

Figure 4-36. Detectors and housings. A: End-window PMT. B: 1 1/8" side-window PMT. C: 1 1/8" side-window PMT, with a magnetic shield, in a socket with a voltage multiplier. D: RF shielded housing for a side-window PMT, with a dynode chain. E: Silicon photodiode. F: Detector module with small side-window PMT and voltage multiplier power supply.

(underground) used by physicists hunting for neutrinos and various kinds of quarks. We're not the only ones analyzing rare events.

When RCA was phagocytosed by General Electric, the RCA division which made PMTs was bought out by its management, and became Burle Industries; Burle has chosen to focus on the relatively large market for PMTs for scintillation cameras and does not now make affordable red-sensitive PMTs, but it has at least kept the *Photomultiplier Handbook*[1143] alive (it can now be found at www.burle.com).

More recently, the PMT division of the British company Thorn EMI, which supplied the end-window tubes in the older B-D FACS instruments, was bought out by its management, becoming Electron Tubes, Ltd.; this may have staved off extinction but didn't do much for fluorescence. Luckily, Hamamatsu, which has become pretty much the only game in town, especially for red-sensitive tubes, appears committed to manufacturing PMTs usable for cytometry. The industry probably can't sell more than a few thousand tubes a year for flow cytometry, although other fluorescence instrumentation has the same requirements and expands the market somewhat; some larger companies obviously haven't thought the sales volume was worth the effort. If everybody thinks that way, we're in trouble.

Sensitivity Training: Photodiode versus PMT

The following arithmetic should provide you with some feel for why PMTs are used as detectors for relatively weak signals, e.g. fluorescence and orthogonal scatter. When I ran my big laser source instrument using 200 mW of excitation at 488 nm, with about 300 volts applied to a fluorescence detector PMT, and associated electronics with an overall gain of 1 million volts/ampere, I got 10 volt pulses from 3.7 µm fluorescent polystyrene spheres; the output (anode) current from the photomultiplier was therefore 10 µA. To find how much light was detected, I have to factor in the gain of the PMT. Operating at 300 V, the gain is about 1000; a PMT anode current of 10 µA thus corresponds to a cathode current of 10 nA. Since the PMT photocathode responsivity is 30 mA/W, this indicates that 0.333 µW of light reached the PMT cathode. This, incidentally, is about 8 percent of the forward scatter signal intensity calculated on p. 161.

While a photodiode is much more efficient at converting light to current than is a PMT cathode, I would only get about 80 nA of output current from a photodiode, even when it responded to the exceptionally bright fluorescence signals produced by the microspheres. I could certainly get away with using a photodiode as a detector in this case, and

I and a few manufacturers have used diodes with high-gain electronics to detect orthogonal scatter and bright fluorescence, as well as forward scatter.

Diodes are much harder to use to measure weak fluorescence signals, e.g. immunofluorescence. These signals typically require at least 100 times as much electron gain from the PMT to yield a 10 V output signal as do the bright fluorescence signals just discussed. Since a conventional photodiode doesn't have gain available, I'd have to work with <800 pA output from the diode if I wanted to use it as a fluorescence detector. That's practically impossible, even with the most meticulous electronics design; there are too many stray currents running around which will swamp 800 pA. Even dealing with 80 nA requires pretty careful electronics design and construction practice; while photodiodes cost less than PMTs and their associated electronics, if you add enough fancy electronics to a photodiode, it may be easier and cheaper to use a PMT.

Single Photon Counting

When you do need to squeeze every last photon out of a sample, you may have to bite the bullet and do the job with a photodiode, because the quality of your measurements ultimately depends on photon statistics. The more photons you detect, the better your measurement precision, signal-to-noise, sensitivity, etc., and photodiodes are better than PMTs at converting photons into electrons.

The ultimate light detection task is **single photon counting**, where you literally want to pick up the little buggers one at a time. This technique is used for bioluminescence measurements; it has also been employed[660,1144] by the Los Alamos group in detection of single phycoerythrin molecules and very small (≤ 5 kilobase) fragments of DNA labeled with fluorescent dyes. In single photon counting, each photoelectron released from the detector cathode generates a current pulse; sensitivity is limited by dark current, and is improved by refrigerating the detector, which decreases thermionic emission.

In flow cytometry, all cells of the same size spend, or at least should spend, the same amount of time in the illuminating beam; different numbers of fluorescence photons will therefore be collected from cells bearing different amounts of fluorescent dyes. Since the precision of any individual measurement is a function of the number of photons counted, measurements of lower fluorescence intensities are less precise than measurements of higher intensities. In a static cytometer, it is feasible to measure fluorescence with uniform precision by illuminating each cell until a preset number of photons have been counted; fluorescence intensity can be determined from the time required to accumulate the preset photon count. This technique has been used by Deutsch and Weinreb[1145] for fluorescence polarization measurements.

Avalanche Photodiodes (APDs)

The **avalanche photodiode (APD)** is a detector which combines some desirable properties of photodiodes and PMT's; like the former, it has high cathode quantum efficiency, like the latter, it has gain. When a bias voltage ranging between a few hundred and a few thousand volts is applied across an APD, some electron acceleration occurs within the device, leading to secondary electron emission. The resulting gain, typically on the order of 100-1,000, is considerably less than one can get with photomultipliers. However, since the cathode quantum efficiency of the diode is roughly an order of magnitude higher, an APD at a gain of 1,000 is as good as a PMT at a gain of 10,000. The APD's requirement for a bias voltage is something of a disadvantage, but the current required is lower than that needed by a PMT. APDs also suffer from relatively high dark currents, and their gain fluctuates considerably with slight variations in temperature. Both of these problems can be dealt with effectively by controlled cooling of the devices. However, once you add the bias supply and cooling circuitry to an already expensive APD, you end up paying about as much as you would for a PMT with its housing and power supply. Thus far, calculations indicate that the higher cathode quantum efficiency could make the APD assembly competitive with, or even superior to, a PMT as a sensitive detector for flow cytometry; experiments have not yet made the case. However, several manufacturers, including Luminex, Partec, and Optoflow, have used APDs for fluorescence detection in their products.

The "avalanche" in avalanche photodiode describes a phenomenon that occurs at relatively high applied bias voltages; liberation of a single photoelectron at the cathode leads to a massive electronic catharsis, followed by a refractory period. The response of the APD in this "Geiger" mode is markedly nonlinear, but well suited to single photon counting. The "Geiger" metaphor refers to the Geiger counter, in which ionization of gas by radiation produces a similar electrical breakdown phenomenon.

RCA had a division in Vaudreuil, Quebec, Canada (a suburb of Montreal) that developed single photon counting modules incorporating cooled APDs. These modules were and are used in the high sensitivity, low flow rate flow cytometers built at Los Alamos for sizing DNA fragments[1144]. The output of the devices is digital; detection of a photon results in output of a logic pulse, and the subsequent data processing hardware and software used at Los Alamos are substantially different from those used in conventional flow cytometers. Agronskaia et al[2452] described a circuit that allowed them to convert signals from an APD photon counting module to analog pulses, facilitating a comparison of conventional detection using a PMT and photon counting with an APD module for nucleic acid sizing in a slow flow system. Photon counting, as one might expect, gave better resolution of low intensity peaks.

Using APDs in "Geiger" mode for photon counting will generally not provide the dynamic range available from a PMT, because the APD itself becomes insensitive to photons for a brief period after each pulse is generated. There may be workarounds for this. If you're interested in laying out a few

	λ(nm)	300	350	400	450	500	550	600	650	700	750	800
Photons/J × 10^{18}		1.51	1.76	2.01	2.27	2.52	2.77	3.02	3.27	3.53	3.77	4.0
mA/W @ Φ = 100%		244	284	324	366	406	447	487	527	569	608	650
DETECTOR	GAIN	QUANTUM EFFICIENCY (Φ) (PERCENT) AT TABULATED WAVELENGTH										
UV-enhanced diode	1.0	41	44	46	61	68	73	77	76	75	74	72
931B PMT	8.0×10^6	3	18	18	14	10	6	2	0.2	—	—	—
9798B PMT	1.3×10^6	15	17	16	13	11	8	6	3	2	1	—
R928 PMT	1.0×10^7	25	22	21	17	15	12	9	7	6	4	2
R4457 PMT	1.2×10^7	22	20	19	17	15	13	11	9	3	0.6	0.1
R1477 PMT	1.0×10^7	22	22	22	21	20	16	13	11	9	6	4
R3896 PMT	9.5×10^6	29	28	26	23	21	19	15	12	10	8	6
R6357 PMT	4.0×10^6	32	30	29	27	25	22	18	15	11	7	4
R636 PMT	2.7×10^5	22	21	16	13	11	9	8	7	6	5	5
C 972 Channel PMT	2.0×10^8	3	4	4	5	6	7	8	7	5	3	1.5

Table 4-3. Cathode quantum efficiencies of diode and PMT detectors between 300 and 800 nm. The R928, R4457, R1477, R3896, R6357, and R636 are made by Hamamatsu. The 9798B is by Electron Tubes; the C972 is made by PerkinElmer Optoelectronics. Burle, Electron Tubes, and Hamamatsu all make the "industry standard" 931B.

thousand bucks for one of the APD modules from Vaudreuil, you'll need to contact PerkinElmer Optoelectronics; RCA was bought by GE, which sold the Vaudreuil division to E G & G, which changed its name to Perkin-Elmer after an acquisition and dropped the hyphen some time later.

PMTs: Picking a Winner

Table 4-3 shows quantum efficiencies at various wavelengths between 300 and 800 nm for a typical UV-enhanced silicon photovoltaic photodiode and for the photocathodes of a variety of PMTs; maximum available gains are also tabulated. The number of photons/J and the responsivity in mA/W for a quantum efficiency of 100% are also tabulated. I had similar tables in the second and third editions; some of the PMTs in them have become irrelevant, but there are some new and interesting varieties on the market.

The venerable 931B was developed by RCA, and is now made by Burle, Electron Tubes, Hamamatsu, and possibly others. It's a 1 1/8" side-window type with a bialkali photocathode and costs under $100. Thorn EMI's 9798B is an end-window type which was used as the red fluorescence detector in older B-D FACSes. The R928, R1477, and R3896, all from Hamamatsu, are 1 1/8" side-window PMTs with multialkali cathodes; the R3896, as noted on p. 162, does not have a grid in front of the photocathode. The Hamamatsu R4457 and R6357 are older and newer miniature (1/2") multialkali side-window tubes used in detector modules such as that shown in Figure 4-36(F); the R6357 is a gridless design. The 1 1/8" side-window Hamamatsu R636 has a gallium arsenide photocathode.

The C 972 channel photomultiplier is made by PerkinElmer Optoelectronics (sans hyphen; see left). It has a photocathode, but, in place of a series of dynodes, it substitutes a narrow semiconductive channel, across which a high voltage is applied, between the cathode and anode. Photoelectrons liberated from the cathode hit the walls of this channel, releasing secondary electrons, etc., etc., with all the free electrons getting accelerated toward the anode. The net effect is that the device behaves as if it had a whole bunch of dynodes; gains can be as high as 2×10^8. The concept is interesting, but, as can be seen from Table 4-3, the cathode quantum efficiency is pretty low, and you can't amplify photoelectrons unless you generate them at the cathode first. Noise-free gain is a great concept, but 10^6 times zero is still zero. I'll probably have to try a channel photomultiplier to satisfy my curiosity.

The table shows that a diode's quantum efficiency is only about twice that of the best PMT photocathodes at 300-400 nm, about 4 times at 550 nm, and over 10 times beyond 700 nm. At this point, the quantum efficiency of bialkali tubes, such as the 931B, has dropped below 0.1%.

The widely used R928 costs about $400; the souped-up R1477, a selected R928, doesn't cost much more, and is therefore preferable. Hamamatsu now pushes the R3896, a stellar performer; its initial equally stellar (about $1,000) price has come down since the last edition of the book came out, and I'd recommend it, particularly for measurements of really long wavelength labels, such as PE-Cy7 and APC-Cy7. Out in their territory, at 800 nm, APDs actually give PMTs more of a run for the money than is the case at shorter wavelengths.

The gallium arsenide R636 is useful in spectrofluorometers because it has a relatively flat response curve, but its maximum gain is quite low; unless you need to work at 900 nm, the R1477 and R3896 are better choices. For most work at or below 500 nm, or for scatter measurements out as far as 633/635 nm, the bargain-priced 931B will do a fine job, but, for measuring fluorescence anywhere above 500 nm, a tube with a higher quantum efficiency is worth its price. I found, when I excited propidium with less than 5 mW at 488 nm, that the extra quantum efficiency of an R928 (vs. a 931B) helped lower measurement CVs.

I have already mentioned the compact Hamamatsu detector modules that incorporate 1/2" PMTs. The current H7710-03 features an R6357 PMT; other, less expensive modules in the series are made with less spectacular tubes, which will probably be fine at 550 nm and shorter wavelengths.

Hamamatsu has also gone in some other interesting directions in PMT development. They have made ultraminiature PMTs that fit into the 16 mm diameter, 12 mm long TO- 8 "can" package normally used for transistors and diodes. The first generation of these tubes had neither high gain nor high fluorescence sensitivity, but the newest offerings, the R7400U series, include at least one tube with high red sensitivity; however, while the quantum efficiency of this tube is competitive, the gain (5×10^5) is still on the low side. These PMTs are also available in modules; I have been told that neither the tubes nor the modules are significantly cheaper than the larger varieties.

The other notable Hamamatsu offering is a **multianode** PMT, with a square or linear array of anodes and fine mesh dynodes. The different anodes respond to light impinging on different areas of the cathode, at least up to a point. The linear array multianode PMT can receive the light dispersed from a grating, with the outputs from the different anodes then providing spectral information. Zeiss has apparently used a multianode PMT in a spectral detector for its Meta confocal microscope system; I have also heard of one being used in an experimental flow cytometer.

Photomultipliers: Inexact Science

After all this discussion of PMT sensitivity, I am obliged to let you in on one of the dirty little secrets of electro-optics; the tabulated values are a rough guide. There is a lot of variation from device to device in most of the important parameters; cathode sensitivity and gain for a given applied voltage will vary over at least a 2:1 range, and individual variations in photocathode composition make for individual deviations from the spectral response curves of Table 4-3. The good news is that manufacturers test the sensitivity of individual PMTs and provide the results to the buyer. So, if you acquire two R3896's, you probably want to use the "hotter" one at the longer wavelength. My impression is that plain silicon photodiodes don't vary nearly as much as PMTs, although avalanche diodes may.

Charge Transfer Devices: CCDs, CIDs, Etc.

You are, by now, likely to have encountered the **charge coupled device**, or **CCD**, either in its low-cost form in your camcorder or digital camera, or in its rarer, cooled, more esoteric and expensive guise in imaging cytometers designed for low light level measurements. CCDs are one of a class of photodetectors described as **charge transfer devices**; there are also, for example, **charge injection devices**, or **CIDs**.

In all of these, exposure to light causes accumulation of electric charge in individual elements that are usually arranged in a linear or rectangular array; attached electronic circuitry senses the amount of stored charge in each element at regular intervals. Charge transfer devices are well suited for imaging; because they integrate over time, they are useful for measurement of low light intensities, especially when cooled. However, they tend to be relatively slow, and, on that account, they have not been widely used in flow cytometry. Newer, faster arrays may be useful in polychromatic detection for measurement of emission spectra in flow[1139].

I hear that there are now ways of getting gain out of CCDs, but I don't have either details or confirmation. New **CMOS image sensors** are starting to give CCDs a run for their money in the commercial camera markets; whether they will make inroads in science remains to be seen.

Intel, which joined forces with Mattel to produce the QX3 Computer Microscope, a cute toy that uses a CCD to provide 320 by 200 pixel images, decided in late 2001 to stop making the gadgets; they came on the market at $119.95, and I've snapped up a few for $49.95. There is still time to introduce your kids or grandkids to microscopy and cytometry via this route.

4.6 FLOW SYSTEMS

It has probably not escaped your notice that, to this point, in this book on flow cytometry, I have gone into great detail about light, optics, light sources, lenses, filters, and detectors and said very little indeed about **flow systems**, without which flow cytometry wouldn't be flow cytometry. I claim there has been a method to this madness. All the other stuff doesn't change just because you work in a flow system, and all the other stuff works when you don't work in a flow system. You can use the same light sources, and the same lenses, and the same filters and detectors, to illuminate and collect and detect light from cells on slides, or in culture dishes, or in microtiter (or nanotiter) plates, or in small capillaries, as you use to do the same jobs for cells in flow systems. A few chapters from now, we will discuss parameters and probes, virtually all of which can be measured in or applied to cells in static as well as in flow cytometers. There are some cytometric tasks for which flow cytometry is preferable, and some for which it is not, but most of the fundamentals of flow cytometry are the fundamentals of cytometry in general. Among those that are not are the theoretical and practical details of **fluid mechanics** or **hydrodynamics** and flow systems, to which we now turn.

In a flow cytometer, it is the task of the flow system to transport cells in the sample to and through the measurement station(s). In "static" microspectrophotometers or image analysis systems, the same job is usually delegated to precisely made and well-controlled mechanical hardware. However, while the mechanical transport system in a static cytometer may be inactive while actual measurements are being made, the flow system in a flow cytometer is continually active, and must move the entire cohort of cells in a sample past the measurement station(s) along almost identical trajectories at almost identical velocities if satisfactory data are to be obtained from the measurement process. This requires that a stable flow pattern be achieved and maintained, and both designers and users of flow cytometers must play active roles in this process.

Flow System Basics

The design of flow systems and the underlying physical principles have been discussed at length by Pinkel and Stovel[650] and by Kachel, Fellner-Feldegg, and Menke[1146]. If you feel a strong urge to design your own flow cytometer, you will probably want to refer to one or both of those publications. If you're willing to put up with what the manufacturers give you, and/or to do things my way, stick with me, and I will expand on the brief discussion of flow systems that appeared on pp. 55-57, hoping to hit the high points of the references just cited.

Almost all modern optical flow cytometer designs make use of **sheath flow,** or **hydrodynamic focusing,** to confine the sample or **core** fluid containing the cells to the central portion of a flowing stream of cell-free **sheath** fluid. Sheath flow improves the precision with which the cell sample can be positioned in the **observation region** of the cytometer by restricting cells to the central region of the stream, and reduces the likelihood of obstruction of the flow system. Stable, unobstructed flow minimizes variations in the position and velocity of the core stream; when flow becomes unstable or **turbulent,** due to obstruction or other causes, measurements are likely to become imprecise and inaccurate.

Figure 4-37 illustrates some aspects of fluid flow in a flow cytometer. Core (sample) and sheath inlet tubes are shown near the top of the flow chamber; near where these enter, there may also be a third port which can be connected to vacuum, allowing easy removal of air bubbles and back suction to clear clogs out of the orifice. Application of vacuum is more likely to be successful for the first purpose than for the second.

Gently Down the Stream: Laminar Flow

Flow must be stable from the region of the core injector tip downward if core velocity and position are to be maintained well enough to allow good measurements to be made. We want **stationary** or **streamline flow,** a condition characterized by the constancy over time of flow velocity at any given point in the system. Since the law of conservation of mass dictates that the same volume of fluid must pass in the same time through the narrow and wide portions of the capillary, the flow velocities at different points in the system will be different, i.e., higher in the narrow portions than in the wider ones. In fact, the product of cross-sectional area, A, and average flow velocity, v, remains constant and equal to the volume flow rate, Q, at any point along the flow system. But why are we talking about "average" velocity?

Water, which is the major component of both the sheath and core fluids and which therefore determines their flow characteristics, is not what physicists call an ideal liquid; it exhibits **viscosity,** which, in physical terms, means that some work must be done on a volume of fluid to get it to change its shape. While the everyday definition of viscosity conjures up fluids such as glycerin, which has a viscosity about 1000 times that of water, the effects of the viscosity of water on its pattern of flow are noticeable enough. In particular, we observe that the stationary flow of water through small tubes is **laminar.** If we look at a cylindrical tube of radius R containing flowing water, we find that the velocity of water at different distances from the axis or center of the tube varies. Velocity is highest along the axis; at the walls of the tube, there is actually a thin **boundary layer** of water that is not moving (i.e., it has zero velocity). At any intermediate point a distance r along the radius, the velocity is proportional to $(R - r)^2$. This produces a so-called **parabolic profile** of flow velocities, as if the water were broken up into thin cylindrical layers (**laminae** in Latin) that were sliding over one an-

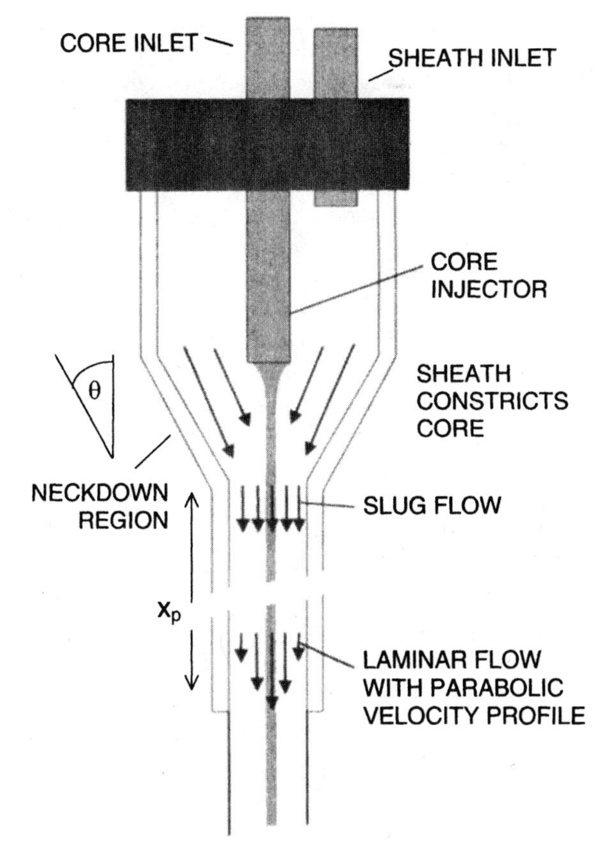

Figure 4-37. Fluid flow in a flow cytometer.

other. So, we can't assume that the velocity of the fluid will be constant across the entire cross section of the tube, but we can do our calculations based on average velocity and use the conservation law.

I can't think of a better illustration of a laminar flow profile than Figure 4-38. This shows a chain of diatoms (*Thalassiosira spp.*) in water flowing at 10 mL/min through the 3 mm wide, 300 μm deep flow chamber of the "Flow CAM," an apparatus developed by Sieracki et al[2453] at the Bigelow Laboratory for Ocean Sciences, Boothbay Harbor, ME, and now available commercially from Fluid Imaging Technologies. The instrument stores a digital image of each particle measured, in addition to fluorescence and size data.

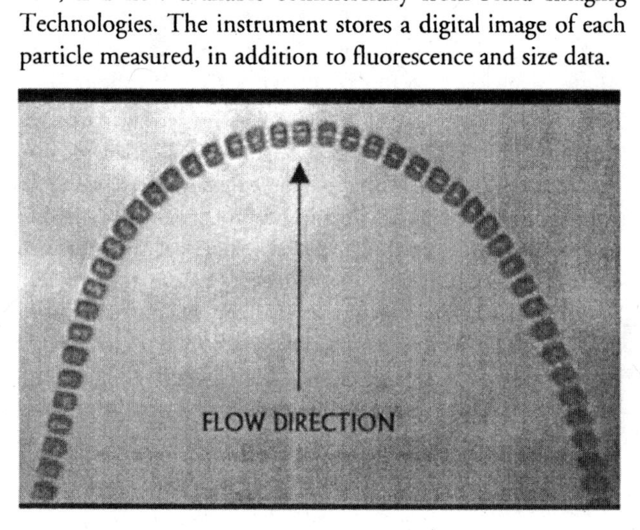

Figure 4-38. Laminar flow profile illustrated by diatoms in the "Flow CAM" imaging flow cytometer. Contributed by Chris Sieracki, Bigelow Laboratory for Ocean Sciences.

Returning to Figure 4-37, we see that at some point near the core injector tip, the cross section of the flow chamber is gradually decreased; the length of the chamber over which this happens has been called the "neckdown region." It's a grotesque name, with vaguely Rabelaisan overtones, but it will do. As the cross section decreases, the flow velocity increases; also, the ratio of core cross section (or diameter) to sheath cross section (or diameter) may be changed, depending upon the relative **volume flow rates** of sheath and core. What we are aiming for is a core of small enough diameter so that cells generally pass through the observation region one at a time; what we most want to avoid in the neckdown region is anything that will generate **turbulence**. Sharp edges and/or sudden changes in diameter will do that, and are to be avoided. In terms of design, a neckdown region with a gentle conical taper ($\theta \approx 30°$) is good. People have introduced various dodges such as tapered and/or eccentric injectors in order to orient asymmetric cells in flow[650,651]; for now, we will stick to the basics.

The figure shows an extension of the capillary past the tip of the neckdown region; this is the configuration used in Ortho's early Cytofluorografs and in my Cytomutts, with round capillaries, and in most modern benchtop instruments, with rectangular cuvettes. In stream-in-air systems, such as the B-D FACS and Coulter EPICS series sorters, the stream emerges through an orifice placed at the end of the neckdown region.

When the cross section of a tube through which a viscous fluid is flowing decreases, the velocity profile at the point of entrance to the constricted region is nearly constant across almost the entire cross section; this is referred to as **slug flow**. The fluid must flow for some distance x_p through the constricted portion of the tube before the parabolic flow profile reestablishes itself. Pinkel and Stovel[650] state that, for water at 20° C, this distance, in mm, is

$$x_p = 6 \times 10^{-5} d^2 v,$$

where d, in μm, is the diameter of the constricted portion of the tube, and v, in m/s, is the average fluid velocity. For a flow velocity of 10 m/s, which is a common value in flow cytometers, values of x_p for (constricted) tube diameters of 70, 100, and 250 μm, are, respectively, 3, 6, and 38 mm. This means that in a system in which cells are observed in a stream in air after exiting a 70 μm orifice, the parabolic velocity profile will be established at the observation point if it is 3 mm below the orifice. In an instrument in which cells are observed inside a 250 μm square quartz cuvette less than 38 mm long, the observation point is necessarily in a region where the flow velocity profile is not parabolic. This can be advantageous, because velocity differences between cells at different distances from the core axis will be minimized.

In laminar flow, the flow in the region of the boundary layer, i.e., the region near the tube walls, is inherently unstable, as a result of which any irregularities of the walls, or adherent particles or bubbles, may produce turbulence. If you happen to have a flow system handy, you can actually observe the flow pattern (sheath flow, I hope) by running an aqueous solution of a dye such as methylene blue through the core injector. By manipulating the core injection rate, you should be able to produce stable cores of varying diameters. If you see the core wiggling, or diffusing into the sheath, your flow system needs work. It is critical, by the way, to use an aqueous solution of dye for this exercise; watching the behavior of an ethanol or methanol core and an aqueous sheath can make you think your flow system needs an exorcist when absolutely nothing is wrong.

The crew I worked with at Block Engineering probably developed an appreciation of the importance of flow stability for good instrument performance earlier than did most other people, for the simple reason that we were always looking at measurements made at two or more observation points. Using an oscilloscope, we would observe waveforms from the detectors at two observation points, using the waveform from the first (upstream) detector to trigger the display. When the velocity was constant, the interval between pulses from the two detectors was constant; when flow was disturbed, the second pulse moved with respect to the first (the effect is called **jitter**), because cells took different times to traverse the distance between observation points. We built injector assemblies to allow changing the injector position

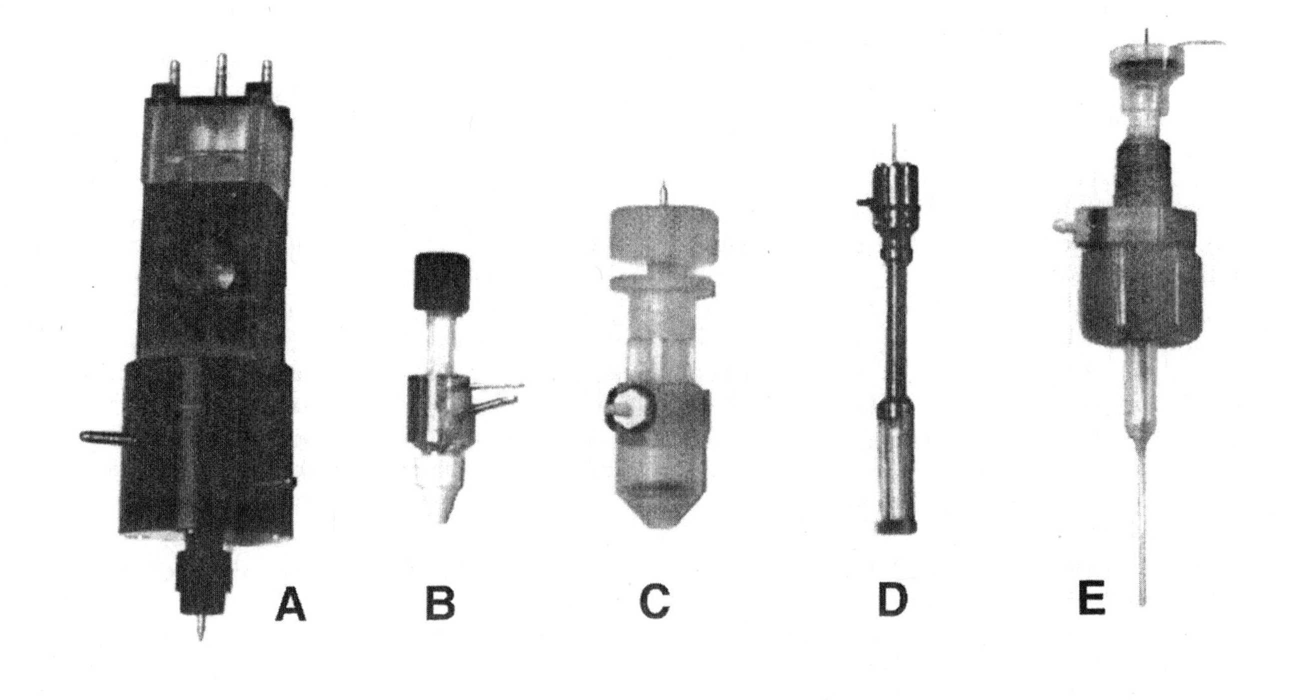

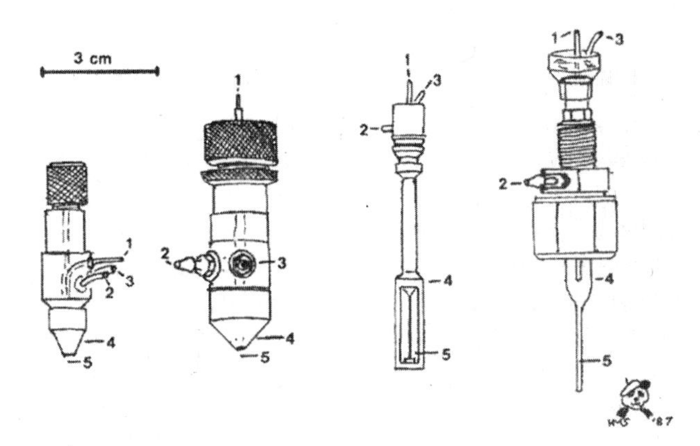

Figure 4-39. Flow chamber designs. A: Los Alamos type double sheath flow chamber. B: Stream-in-air nozzle from B-D FACS. C: Stream-in-air nozzle from Coulter EPICS. D: Sorting flow chamber (observation in cuvette) from Ortho Cytofluorograf. E: Flow chamber from a Cytomutt (observation in thin-walled capillary). Flow is from bottom to top in A, and from top to bottom in B-E. The sketch at left identifies various features of the flow chambers shown in B-E. 1: Core (sample) inlet tube. 2: Sheath inlet tube. 3: Vacuum connection (burp line). 4: Position of tip of core injector. 5: Position of observation point.

(**core steering**), allowing us to place the core in the position that gave the stablest flow pattern. This was not always in the geometric or optical center of the round capillary used in our flow chambers!

The adjustment mechanisms in the instruments I myself have built are cruder, although they do not get to the level of needles stuck through corks[102]. This makes it harder to arrive at the best core position, and harder to maintain it, but, when you are there, you can get good measurement precision (coefficients of variation 1.5 percent or less) even with needles stuck through corks (O. Bakke, personal communication; D. Pinkel, personal communication; L. Scherr, personal communication; H. Steen, personal communication).

Flow Chambers; Backflushes, Boosts, and Burps

Flow chambers, or **flow cells** (the latter term invites confusion with biological cells and I will avoid it), used in various instruments are shown in a photograph and scale drawing in Figure 4-39. The flow chambers shown were or are used only with laser source flow cytometers, except for the Cytomutt chamber, which has been used with both laser and arc lamp light sources. The chamber shown at A is similar to those built for apparatus at Los Alamos. Observation is done in a 1 cm diameter photometer cuvette; a second sheath, moving at very low velocity, is used to fill much of the volume of the flow chamber.

The B-D FACS and Coulter EPICS flow chamber designs at B and C are used to generate streams that are observed in air after they emerge from an orifice; both the flow chamber and the orifice are sometimes referred to as **nozzles**.

In the EPICS flow chambers and older B-D nozzles, the orifice is formed by a sapphire watch jewel; the shape of such an orifice is not a cylinder, but a truncated cone. The proper orientation for a watch jewel in a flow chamber is with the larger diameter toward the outside; this substantially reduces surface tension that would otherwise cause the emerging stream to spread out. The original Stanford apparatus used a quartz orifice; the B-D FACS flow chamber shown uses a ceramic orifice. It is also possible to sort cells emerging from thin-walled capillaries such as are used in the Cytomutt flow chamber shown at E. Sorting flow chambers are typically mounted in a bracket that also holds the transducer used to stabilize the droplet breakoff pattern.

A major difference between the stream-in-air nozzles and the other flow chambers shown in the figure is the much shorter distance between the injector tip and the stream exit from the flow chamber in the former. During sample changes, after a sample is removed, it is customary to let sheath fluid flow back through the sample inlet tube, or, in a system using a syringe pump for sample feed, to suck back on the feed syringe. These **backflushing** maneuvers will clear the sample line of cells from that sample, but will not remove any cells that had already left the injector but had not yet emerged from the chamber when the sample was removed. It is thus common practice to drive some of the new sample through the system at a higher than normal rate, either by briefly increasing the sample drive pressure or, if a syringe pump is used for sample feed, by increasing its delivery rate; the accelerated sample delivery is called a **boost**.

Both backflushing and boosting can introduce turbulence, i.e. disrupt laminar flow, which, among other things, may result in some of the cells from the old sample becoming mixed with the cells in the new sample. The shorter the distance between the injector tip and the exit or observation point, the less such **sample carryover** should occur. Stream-in-air flow chamber designs are therefore well suited for high throughput operations. Graves et al, who have been working for some time in this area, recently examined the relationships of various nozzle design parameters on the time taken for flow to stabilize between samples[2454]. They found that flow stabilized more rapidly when a large (150-200 µm) orifice diameter was used, because, when a smaller (50 µm) orifice was used, the boost generated higher back pressure, causing turbulence.

The original B-D sorting nozzle design lacked a connection through which air bubbles might be bled out of the system; removing bubbles required demounting the flow chamber and inverting it, which generally sprayed sample all over the place. In the HIV era, this doesn't even play in Peoria. Most newer flow chamber designs incorporate a bleed or "**burp**" line that can be connected to a vacuum line or a syringe.

Most stream-in-air designs have no provision for core steering, i.e., changing the core position laterally with respect to the sheath stream. People who used them told me that some of the older B-D nozzles consistently gave better measurement precision than others; I would take this to be an indication of how well the core was centered.

The Ortho flow chamber shown at D was used for observation of cells within a quartz cuvette with a square cross section; the channel is approximately 200 µm across. A 75 or 100 µm orifice was generally used for sorting; the observation point is positioned just above the orifice jewel, which is held, with the cuvette, in a steel cage. The flat surfaces in the cuvette minimize scatter of excitation light from the flow chamber walls; this reduces noise in scatter measurements.

The Cytomutt flow chamber shown at E is based on the designs used in the Block Cytomat systems[88,92]. Observation is done in a thin-walled quartz capillary, typically between 150 and 175 µm in inside diameter, which is held in a standard compression tube fitting. The same capillaries can be used for both analysis and sorting; for sorting, the observation point is usually moved down near the tip of the capillary to minimize the distance between it and the droplet breakoff point. The sample and sheath injection ports are fit into the tube fitting and sealed with epoxy glue. The core injector is a piece of 27 gauge stainless hypodermic tubing glued into the bottom portion of the barrel of a plastic Luer-lok syringe; core position is adjusted by twisting this fitting. Bubbles are removed through a vacuum connection. The large bore of Cytomutt flow chambers makes them less likely to clog than most sorter nozzles, and clogs (or bubbles) are usually readily dislodged by running a piece of stainless steel wire down through the injector, without removing the flow chamber. This largely eliminates the need to realign the system after a problem has been dealt with.

Cuvettes vs. Streams for Analysis and Sorting

Observation in a stream in air rather than in a cuvette offers potential advantages. There are two fewer interfaces from which light may be scattered in a stream in air system, and there are no cuvette or capillary walls that might get scratched or dirty. Proponents of stream-in-air interrogation for systems in which sorting is to be done argue that the greater distance between the observation point and the orifice in a design such as Ortho's leads to less accurate sorting. This need not be so; observation within the walls of the flow chamber allows more power to be used to drive the transducer than would be acceptable in a stream-in-air system, in which high power levels distort the stream at the observation point, degrading its optical quality and decreasing measurement precision. Sorting accuracy depends on maintaining the relative positions of the interrogation point and the droplet breakoff point; increased transducer drive may allow better control of this factor.

When the objective is high speed sorting, stream-in-air systems have been preferable, because they are compatible with higher analysis rates. The flow rate of sample through the observation point in the Ortho chamber was relatively low; velocity through a 75 µm orifice was 10 m/s, meaning that flow velocity in the 200 µm square cross section cuvette was only about 1.1 m/s. The Ortho system used a beam focused to a spot less than 10 µm high, allowing useful pulse width information to be derived, but cell transit time through the beam was around 9 µs, making it essentially impossible to process 100,000 cells/s. Beckman Coulter's Altra and BD Biosciences' FACSAria, newer sorters utilizing cuvettes, provide shorter transit times and can process tens of thousands of cells/s. However, the transit time through the beam in modern high speed stream-in-air sorters can be less than 1 μs, still potentially yielding a performance advantage for the stream-in-air systems.

On the minus side, stream-in-air systems generate aerosols, a concern when specimens contain potentially hazardous materials; the optical characteristics of the stream are also subject to change when the sheath is turned off and on. Light collection from streams in air is also usually less efficient than is light collection from flat-sided cuvettes. Some cytometer manufacturers allow the user to make the tradeoff.

Flow cuvettes with square or rectangular cross sections, designed for analysis rather than sorting, are now standard in commercial benchtop fluorescence flow cytometers. Such chambers were first used in Technicon's Hemalog D differential leukocyte counter. In the mid-1970's, Ortho introduced the FC-200 Cytofluorograf, featuring a chamber with a 200 μm square cross section; a spherical mirror on the side of the cuvette opposite to the collector lens for fluorescence and orthogonal scatter signals could be used in this system to increase light collection. Similar designs were used in Ortho's System 30, Spectrum, and Cytoron analyzers. Coulter offered a closed system analytical flow chamber for the EPICS sorters; the large cross section allowed cells to be analyzed at lower flow velocities for time-of-flight cell sizing measurements[104].

Observation in square cuvettes is standard in the Beckman Coulter Elite and Altra sorters and EPICS XL analyzers; the cuvettes have the front element of the collection lens built in. The new DakoCytomation CyAn analyzer uses a square cuvette without an integral lens. The flow chambers used in the new BD LSR series analyzers, and in the older FACSCalibur, FACScan, FACSort, FACSTrak, and FACSCount, have a rectangular cross section, roughly 180 by 400 μm, permitting use of a high-N.A. (1.2) lens, which increases light collection substantially. BD's new FACSAria sorter and the new Beckman Coulter FC500 analyzer also have rectangular cuvettes with coupled high-N.A. lenses.

Dan Pinkel, then at Lawrence Livermore Laboratory, designed a sorting flow chamber with an observation portion of square or rectangular cross section; the quartz cuvette (made by NSG Precision Cells) fits into a holder that allows it to be mounted to a B-D transducer mount. Although this chamber initially produced good results in analysis, there were some problems fitting watch jewels to it for sorting; these were eventually resolved[652]. A design similar in appearance to Pinkel's was offered by Coulter for the EPICS; this utilizes a cuvette only a few millimeters long with a square channel 76 μm across, attached where the tip of the nozzle is shown in Figure 4-39. This chamber design decreases noise in orthogonal scatter measurements, as compared to a stream-in-air system, and still permits sorting to be done at reasonably high speeds. The Beckman Coulter Elite and Altra sorters can be fitted with this cuvette, with 150 or 250 μm square flow cuvettes, or with stream-in-air nozzles. DakoCytomation sorters restrict observation to streams in air; all three manufacturers provide the same range of orifice diameters, i.e., 50 to 400 μm.

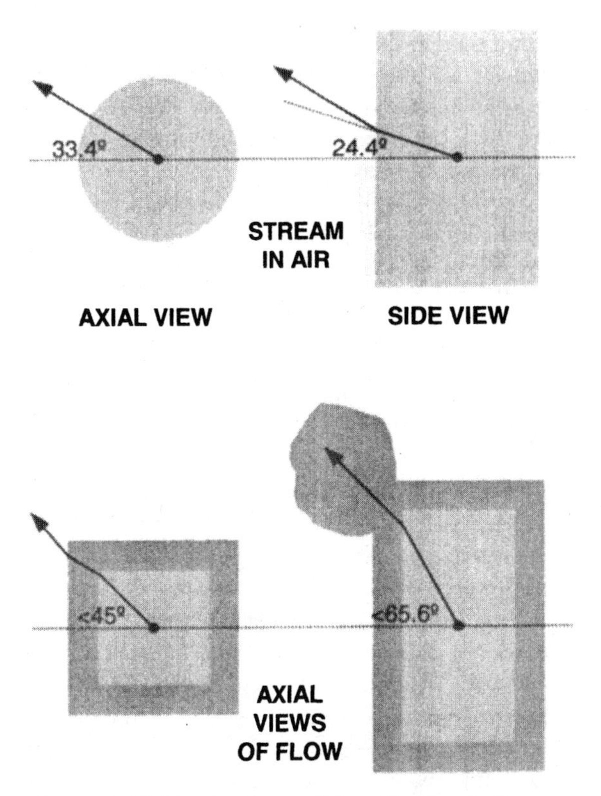

Figure 4-40. New angles on light collection in flow.

Light Collection from Streams and Cuvettes

To understand how the configuration and composition of a stream in air or a flow cuvette determine how much light can be collected, you don't need to know all the angles, but you do need to know a few of them; the old ones are shown in Figure 4-12 and discussed on pp. 121-2, and the new ones appear in Figure 4-40. You might also want to look back at p. 150 and Figure 4-25.

In flow cytometry, we are generally looking at particles in an aqueous medium, which I will assume for the purposes of the discussion here is water. Water has a refractive index n of 1.33 at 20 °C. Light coming from a particle in the stream will be refracted at the air-water interface in a stream-in-air system, and, assuming the cuvette is made of quartz (n = 1.46), at the water-quartz interface and the quartz-air interface in a system using a flow cuvette without an optically coupled lens. The angle that is of importance to us is the half angle of the cone of light coming from the particle in water that we can manage to coax into our collecting lens after all of the refractions have taken place.

The upper part of Figure 4-40 shows axial and side views of a stream in air. In most high-speed sorters, particles are interrogated in a stream in air; the lenses used for side scatter and fluorescence collection in these instruments are usually "ultra" long working distance (10-13 mm from the stream, so they won't get wet) microscope objectives, with an N.A. of 0.55. The formula N.A. = n sin θ tells us that the largest angle θ at which an N.A. 0.55 lens, working in air, will collect light is 33.4°. If we look at the axial view of the stream in air, we see that there is no problem collecting light that is emitted (or scattered) at that angle from a central particle along a radius of the circular cross section of the stream, because such light will strike the interface at normal incidence, and not be refracted.

However, as the side view shows, any light from the particle that travels outside the plane perpendicular to the axis of flow will be refracted at the air-water interface. Light that reaches the lens in air at an angle of 33.4° will have come from the particle at an angle of 24.4° in water. So the half angle θ of the cone from which we can collect light using an N.A. 0.55 lens is somewhere between 24.4° and 33.4°. The solid angle Ω corresponding to θ can be calculated from the formula $\Omega = 2\pi(1 - \cos\theta)$; it would be 0.562 sr (steradians) for a cone with a 24.4° half angle and 1.035 sr for a cone with a 33.4° half angle. To calculate the "right" solid angle, it is apparently necessary to use elliptic functions (don't ask); for the present purpose, after consultation with Rob Webb, I decided to approximate the right answer by taking the square root of the sum of the squares of the high answer and the low one. This yields 0.869 sr. By definition, a sphere surrounding the cell, representing 4π sr, or 12.57 sr, would receive 100% of the light from the particle; 0.869 sr represents a relatively unimpressive 6.9% of the total. In this calculation, and those that follow in this section, we neglect the fact that transmission of the lens is always less than 100%.

Up to a point, we could improve light collection from the stream by using a lens with a higher N.A., say 0.68 or 0.7. An N.A. 0.7 lens gets us to a 44.4° half angle for the radial rays and a 31.8° half angle for the refracted rays; the back-of-the-envelope calculation says it collects over a 1.529 sr solid angle, or about 12.2% of the total light, a 77% improvement over the N.A. 0.55 lens. An N.A. substantially higher than 0.7 won't help, because, as it turns out, 48.8° is the critical angle for an air-water interface, and light hitting the interface at that angle and larger angles is subject to total internal reflection (see pp. 121-2 and 157-8 and the figures on those pages). This means that most of the extra light that, say, an N.A. 0.95 lens would collect from a specimen in air will travel (gently?) down the stream as if the stream were a light pipe. In fact, Mariella et al[2455-6] have demonstrated that a fiber optic sharpened to a conical tip and inserted into the stream along the axis of flow downstream from the observation point will do a pretty good job of collecting light scattered at large angles. Of course, this precludes using the stream in air for sorting.

If you're using a stream in air because you want to sort at the absolute maximum rate, you'll probably need to use a 50 or 70 μm orifice to allow you to get a droplet rate of at least 50,000/s. Even if your beam is focused to a 20 μm spot, there is likely to be enough reflection from the stream (see p. 150) to make it necessary to use an obscuration bar in front of the fluorescence/side scatter collection lens. This will result in the loss of about 30% of the light that would otherwise be collected by the lens.

While the square and rectangular cuvettes used in flow cytometers minimize noise due to stray scattered light, allowing obscuration bars to be dispensed with and thereby facilitating high-precision measurements using relatively low-powered light sources, some problems with light collection arise from the refractive properties of the cuvettes. The bottom half of Figure 4-40 shows axial views of a square and a rectangular cuvette. The geometry of a square, as shown at the left, restricts collection to a half angle of less than 45°; any light leaving a particle centered in the flow stream at that angle will emerge from the cuvette into air at an angle of no more than 72°, limiting the maximum N.A. of a collection lens usable with a square cuvette to 0.95. In practice, it is difficult to obtain lenses with an N.A. that high and the long working distance of a few millimeters necessary for observation in cuvettes. An N.A. 0.65 "high dry" objective with a long working distance, or an aspheric lens with an N.A. between 0.62 and 0.68, would be a more realistic choice as a collection lens. The N.A. 0.65 lens would collect light emitted by a particle in water in a cone with a half angle of 29.3°, corresponding to a solid angle of only 0.801 sr, or about 6.4% of the total, a little less than is collected from a stream in air by an N.A. 0.55 lens.

An N.A. substantially higher than 1.0 can only be achieved if the lens is optically coupled to the specimen, in this case the cuvette, by a medium with a relatively high refractive index, matched to those of the materials of which the cuvette and lens are made. It is also necessary to use a rectangular cross section for the cuvette rather than a square one, as can be seen from the picture on the bottom right of Figure 4-40. Light leaving the target particle in water at an angle above 65.6° will be subject to total internal reflection; the maximum usable N.A., even for a coupled lens, is thus limited to 1.21 (recall from p. 122 that N.A. is invariant from medium to medium). The gray "blob" shown surrounding the ray emerging from the rectangular cuvette represents the coupling medium, which could be immersion oil if the chamber is horizontal, a gel if it is vertical (as in the B-D FACScan, etc.), or the substance of a lens attached directly to or built into the chamber. The solid angle over which an N.A. 1.2 lens collects light from a particle in the rectangular flow cuvette is 2.655 sr, or just over 21% of the total; this is more than three times as much light as is collected from a stream in air or a square cuvette by the lenses normally used with them.

Various authors[653-4,1147,2457] have described flow chamber designs incorporating optical elements that, in theory, permit the collection of two to six times as much light as can be collected from a square cuvette or stream in air. Goodwin et al[2458] measured the actual light collected by an N.A. 0.55 and an N.A. 0.85 objective as about 4% and about 9% of total emission, respectively. Considering that the transmission of the lenses was just above 80%, these figures are in reasonable agreement with the calculations just worked through.

In studies done during the course of designing the Cytomutt family[105], my colleagues and I found we could obtain essentially identical sensitivity and precision in fluorescence measurements using a stream-in-air system, a flat-walled flow chamber, or a round-walled chamber, provided that a stable flow pattern was achieved and maintained and that the beam and stream or capillary dimensions permitted use of a field stop, rather than an obscuration bar, to decrease collection of stray scattered light. The key to performance then resided in the design of illumination and collection optics, rather than in the geometry of the flow chamber.

For some time, I had better luck with my round capillary systems than with stream-in-air systems, which led me to conclude that observation in an enclosed space was inherently better. What I actually compared, however, were capillaries with outer diameters of about 300 µm and inner diameters of about 150 µm, on the one hand, and streams in air with diameters between 70 and 100 µm, on the other. Round streams, and round capillaries, behave as cylindrical lenses; the smaller the capillary, the stronger the lens. Therefore, the optical properties of the capillary and stream-in-air systems that I had compared were quite different.

Inspired by the large stream issuing from a broken capillary, Dick Adams and I (unpublished) did a brief comparison, and found that large (≥ 300 µm diameter) streams in air gave us fluorescence measurements at least as good as those obtained using round capillaries of the same outside diameter, and that scatter measurements made using the stream-in-air system were noticeably less noisy than those made using a capillary. We could detect forward scatter signals from small bacteria using a laser emitting less than 0.5 mW; when a capillary was substituted for the stream in air, signals were barely detectable above noise. There are practical problems with large streams in air; one has to use large volumes of sheath fluid per unit time, and streams in air in general, as mentioned previously, raise biohazard safety issues. Both of these objections can be eliminated by using a system in which the sheath flows through standing water and light is collected by a water immersion lens, with no interfaces between the lens and the core stream; such a flow chamber was used in the Block Engineering apparatus with which scatter measurements were made of single virus particles[94].

If there is a bottom line here, it is that you can't have it both ways. If you want to sort at the absolute maximum rate, you'll have to put up with inefficient light collection; if you want to hoard photons, you'll have to slow down. And, speaking of slowing down, Figure 4-41 shows flow chamber designs used in several flow cytometers with arc lamp sources. These instruments typically have highly efficient illumination and light collection optics, but, as we found on pp. 131-2, their lamps don't put out anywhere near as much usable light as do lasers, and both sample flow rates and cell acquisition rates are, accordingly, lower than in most laser-based instruments.

Panel A illustrates the arrangement in the Dittrich/Göhde Impulscytophotometer (ICP)[9,83]; this is an **axial flow** system, in which cells confined in a sheath are observed as they emerge from a tube oriented along the axis of a fluorescence microscope objective and make a 90° turn into a flowing stream. The Phywe ICP used suction to draw cells through the system; the later Ortho version forced them through with gas or air pressure. Wolfgang Göhde felt that axial flow of cells through the objective focal plane was an important determinant of the high precision obtained with the ICP under the best conditions. However, Göhde's newer flow system designs for Partec achieve equivalent precision with what amounts to a modified orthogonal geometry.

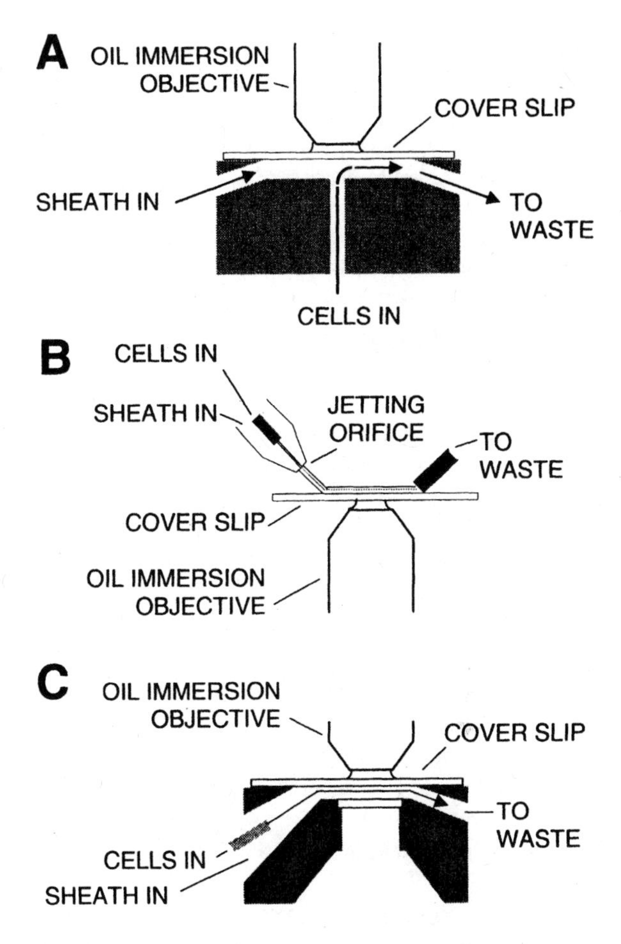

Figure 4-41. Flow chamber designs used with arc source flow cytometers. A: Impulscytophotometer. B: Lindmo/Steen system (Skatron). C: Closed system similar to some used by Bio-Rad, Heka, and Partec.

So does the instrument developed by Lindmo and Steen[100-103] (originally sold as the Leitz MPV-Flow; later by Skatron, Ortho (Europe), Bruker, Bio-Rad and (in the near future) Apogee). In this cytometer, exemplified in Panel B, cells confined in a sheath are ejected from a nozzle as they would be in a stream-in-air system. The stream then impinges on a cover slip at a relatively acute (about 20°) angle, and is observed with the high N.A. oil immersion lens of an epiilluminated fluorescence microscope downstream from the point of impact; fluid is removed from the cover slip by suction still further downstream.

If the flow pattern isn't stable before the cells leave the nozzle, it won't be stable after they hit the cover slip, and the excellent precision reported won't be achieved. If the flow pattern is stable, some improvement results from a further stabilization of the existing laminar sheath flow pattern by forces between the fluid and the cover slip. The stream flattens out, a phenomenon my colleagues and I noted when using a pen nib design to deposit cells from a flow cytometer onto movie film[95]. This tends to orient asymmetric cells and also to confine them to a narrow region of space, allowing a high aperture objective, with its small depth of focus, to be used without compromising performance.

A closed flow system was/will be available on the Bio-Rad/Apogee versions of the Lindmo/Steen instrument; it resembles that shown in Panel C. The Fluvo II, designed by Kachel[656], the newer Partec flow cytometers, designed by Göhde, and Thomas's RATCOM and NPE instruments are similar. All are basically orthogonal flow systems, although they are not symmetric about the axis of flow; some allow observation of the stream from two or more sides. The Partec instruments offer the options of multistation illumination, with two lamps or a lamp and a laser, and fluidic sorting in a closed system.

Since the same microscope objective serves as condenser lens and fluorescence collector in most arc source cytometers, increasing the N.A. of the lens improves both illumination and collection. Temsch et al[2459] reported that substitution of a gel-coupled N.A. 1.25 lens for an N.A. 0.8 lens in an arc source instrument increased fluorescence signal intensities approximately fourfold.

The flow chamber in B-D's FACS analyzer, an arc source instrument introduced in the early 1980's and soon eclipsed by the FACScan, was essentially a thick-walled round capillary, with an hourglass profile that allowed its use for electronic (Coulter) volume measurement; a second sheath was used largely to provide some fluid flow on both sides of the volume sensing orifice. Some of the old Los Alamos flow chambers (about the size and shape shown in Panel A of Figure 4-39), used with laser source cytometers, also incorporated electronic volume measurement capability[655]; the combination of electronic volume measurements with optical measurements in cytometry has probably been pursued for the longest period of time by Bob Leif and his colleagues[654,1108-9,2457]. The NPE/RATCOM instruments, and a variety of clinical hematology analyzers (including Beckman Coulter's), incorporate both optical and electronic volume measurement capabilities; an electronic volume measurement option is also offered by Partec.

When You've a Jet...

While observation of cells in a stream in air instead of in a closed system eliminates some of the complexity, and may lower the cost, of a flow chamber, more constraints on the hydrodynamics are, in general, associated with stream-in-air systems. Pinkel and Stovel[650] have covered some of the details.

One of the interesting things they mention is that the exit velocity v (in m/s) of fluid from an orifice or nozzle is, to a first approximation, directly proportional to the square root of the pressure used to propel the fluid through the orifice; i.e., the sheath pressure. If the sheath pressure, ΔP, is given in pounds per square inch (PSI),

$$v = 3.7\ (\Delta P)^{1/2}\ ;$$

if the pressure is given in atmospheres, the constant is 14, and if in pascals (N/m^2), it is 0.044. This relationship applies to typical stream-in-air nozzles, in which the orifice is placed immediately below the neckdown region; it does not hold for flow systems in which there is a capillary tube of any appreciable length, because of the effects of viscous drag.

A dimensionless constant called the **Reynolds number**, Re, is often used to characterize the stability of fluid flow. If d is the diameter of the stream, in cm, ρ is the density of the fluid, in gm/cm^3, η is the fluid viscosity, in poise (1 poise = 1 g/cm·s), and v_{av} is the average fluid flow velocity, in cm/s,

$$\mathrm{Re} = d\rho v_{av}/\eta.$$

Laminar flow is possible for values of Re < 2300; it may be possible at higher values, but turbulence is much more likely to occur. Thus, in flow cytometry, it makes sense to keep Re below 2300. For water at 20 °C, a temperature at which its density is very nearly 1 gm/cm^3 and its viscosity is 0.01 poise, we can relate stream diameter d, **in µm, not cm**, and v_{av}, **in m/s, not cm/s**, by

$$\mathrm{Re} = d v_{av}.$$

In other words, if v_{av} is 10, and d 230 or below, a person won't get turbulent flow. Unless you're trying to sort really big objects in a droplet sorter (the principle works for stream diameters as large as 1 mm (1000 µm)), you can work at 10 m/s. What you may have to worry about under more typical conditions, however, are **changes in temperature**; the viscosity of water is highly temperature dependent.

Turbulence is apt to be a problem at the inlets, rather than the outlets, of flow chambers, stream-in-air and otherwise. This happens because the sheath inlet is usually a relatively small diameter tube going fairly abruptly into a larger diameter chamber. If you are trying to squeeze the last few tenths of a percent out of an already very good measurement CV, as is the case in chromosome sorting, which stimulated Pinkel and Stovel's efforts, you need to devote some attention to this possible source of flow instability.

Figure 4-42 illustrates three configurations of core and sheath inlets in otherwise identical flow chambers. The "bad" configuration, at left, has the small sheath inlet entering the chamber at a right angle. The "better" configuration represents a compromise reached in many commercial systems; the sheath inlet tube is gently curved, and a perforated "spider" is put in to break up the flow pattern; empirically, this facilitates faster establishment of laminar flow downstream. The "best" configuration, at right, gently sneaks the core into a large bore sheath flow. I've seen chambers like this in action, and they give good CVs; I haven't yet bitten the bullet and made a Cytomutt flow chamber with a more hydrodynamically correct core and sheath inlet.

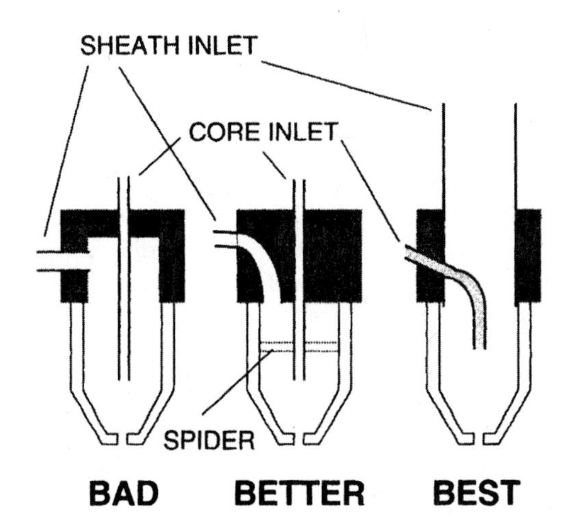

Figure 4-42. Minimizing turbulence generated at the sheath inlet to flow chambers.

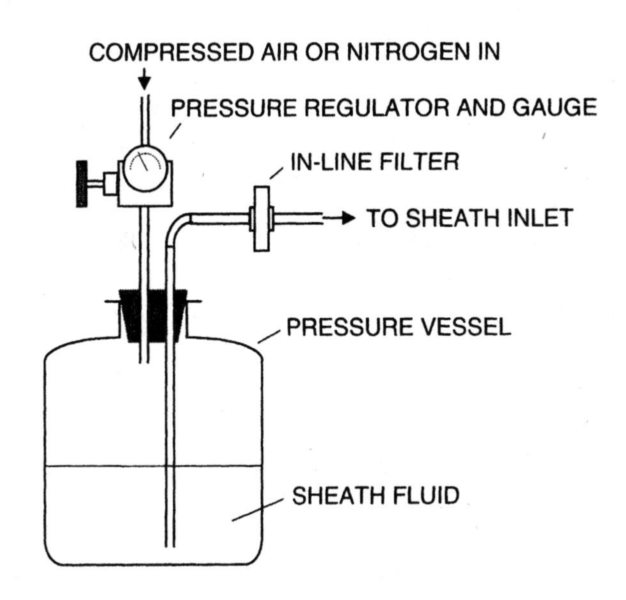

Figure 4-43. Sheath fluid supply plumbing.

Core and Sheath: Practical Details

If you looked at pp. 50-1 and 130-1, or otherwise have some acquaintance with flow cytometry, you've run across the notion that core sizes affect measurement precision. It thus may be useful to know how big a core you're actually running. Let's look at a numerical example. Suppose we have a stream-in-air system with an orifice 80 μm in diameter, through which we want cells to pass at a velocity of 10 m/s. When things are working properly, running the system for one second will squirt out a volume of fluid that would fill a cylinder 80 μm in diameter and 10 m high. The volume of this cylinder is π times the square of the radius times the height. Converting the radius and height measurements to cm, which will give us volume in cm^3, or mL, we find that, in one second, we run a total fluid volume of

$$\{\pi \times [(4 \times 10^{-3})^2] \times 1000\} \text{ mL} \approx 0.05 \text{ mL}.$$

In one minute, or 60 seconds, we'd get about 3 mL through the orifice. Knowing any two of the three variables, orifice diameter, flow velocity, and flow rate, we can calculate the third. If you have a commercial system, the manufacturer has provided you with the internal dimensions of the flow chamber (if you built your own, you should know!), and you can measure the amount of fluid which goes through the system per unit time; this will let you calculate the flow velocity.

Now, how can we figure out the core diameter or radius? If the instrument has a syringe pump, or some other means for measuring the sample or core volume flow rate, it's fairly easy; however, most flow cytometers are not so equipped. We therefore resort to subterfuge, and possibly centrifuge as well, and get and run a sample of cells, or beads, at a known concentration, determined by an automated counter, or, as a last resort, with a hemocytometer.

Suppose the concentration is 10^6 particles/ml, and we find we're counting 1000 particles/s (for the moment, assume we count every one that comes through). The count rate divided by the cell concentration gives us the rate of flow of the core fluid, which is 10^{-3} mL/s. The velocity is 10 m/s; from the core flow rate and velocity, we calculate the volume of fluid that would fill a cylinder with its height equal to 10 m and its radius equal to the core radius. The core radius is

$$[(\text{core flow rate})/(\pi \times \text{velocity})]^{1/2} .$$

The formula gives us a core radius of 5.64 μm; the core diameter is therefore 11.28 μm.

As was mentioned earlier, a parabolic velocity profile is established within a relatively short distance of the orifice in a stream-in-air system. Pinkel and Stovel report that, near the exit of an 80 μm nozzle, with flow at 10 m/s, fluid elements on axis and 10 μm off axis will differ in velocity by 1 m/s, or 10 percent. Particles traveling off-axis at lower velocities will, on that account, spend more time in the laser beam; this should result in higher signal intensities. However, we also have the Gaussian intensity distribution in the beam to consider here; the intensity of illumination diminishes with distance from the axis. In real stream-in-air instruments, the two effects undoubtedly counteract one another to some extent; the off-axis particles spend more time in a lower-intensity beam, and thus should receive, emit, and scatter more nearly the same amount of light than would be the case if the velocity profile were more sluglike than parabolic. In flow systems in which cells are observed in capillaries, the observation point may be at a level at which slug flow persists; widening the core might then be expected to degrade precision more noticeably.

Grace Under Pressure: Driving the Sheath and Core

The pressure-driven sheath used in most flow cytometers operates on the same principle as the wash bottle you used in chemistry if you took it as long ago as I did. The pressure bottle in the original Becton-Dickinson FACS looked exactly like one of those old wash bottles; a sheath bottle differs, however, in being connected to a pressurized gas supply with a regulator and a gauge so you don't have to blow into it. The setup is shown in Figure 4-43.

The range of drive pressures used for sheath fluid in various instruments is from about 3 PSI on the older Ortho Cytofluorografs to about 30 PSI in the Lindmo/Steen apparatus; the high speed sorters at Livermore and Los Alamos could go up to several hundred PSI, but the current commercial high speed sorters seem to top out near 100 PSI. For the lower pressures used in most commercial and lab-built systems, a variety of plastic bottles available from many laboratory equipment suppliers will do just fine; Ortho favored a plastic bottle held in a metal box to limit its expansion under pressure. I, and most of the manufacturers, now use stainless steel pressure vessels, originating at Alloy Products Corporation and acquired through as few intermediaries and with as few markups as possible. These are relatively expensive if you don't eliminate the middlemen when you buy them, but they are very unlikely to shatter. They typically have plastic on the outside around the bottom, which prevents them from making electrical contact with whatever they're sitting on; be advised that steel vessels containing saline can give you a nasty shock if you touch them while the drop charging circuit of your sorter is on.

The most common cause of turbulent flow in a previously "healthy" flow system is complete or partial **obstruction**. **Complete obstruction** of flow is relatively easy to diagnose, if not always equally easy to correct. **Partial obstruction** can be much more troublesome, because sample flow continues, but turbulence, caused by the obstruction, results in a broader range of cell trajectories and velocities, which typically alters pulse shapes and broadens distributions of parameter values. Inexperienced operators would frequently note the increase in CVs and promptly make unnecessary adjustments to the optics and electronics in an attempt to improve the quality of measurements; things then went from bad to worse, often requiring a visit from service personnel may be required to get signals to reappear. Since most flow cytometers in use these days don't permit the operator to make the adjustments, the misalignment problem has largely disappeared; the obstruction problem has not.

Most flow cytometer users don't get the opportunity to redesign their flow systems to minimize obstructions, and must therefore do what they can solely in terms of keeping apparatus, sheath fluids and samples as clean as possible. **Filtration** of sheath fluids and all diluents used in sample preparation through 0.22 or 0.47 μm filters will remove large particles and also minimize bacterial contamination. Even pharmaceutical grade stuff may have some particles in it; if you wouldn't inject it intravenously, don't use it as sheath fluid.

It's good practice to keep a 0.22 or 0.47 μm filter in the line between the reservoir and the flow chamber, to clean out any residual junk. However, if the sheath hasn't been filtered previously, such an in-line filter may clog rapidly enough to decrease the flow rate. This can happen when you follow the nearly universal practice of driving the sheath with gas (air or nitrogen) blown into the reservoir at constant pressure, which means that as the resistance of the filter goes up, the flow rate goes down. In theory, if the sheath were driven by a robust constant volume pump, instead of by pressure, flow velocity could be maintained. Don't bet on it. If you are working with really small particles, such as bacteria, it is advisable to prefilter sheath through a 0.11 μm filter; you probably don't want a filter with pores that small in your sheath line because it is likely to offer substantial resistance to fluid flow.

Generally speaking, flow cytometric analysis, as opposed to sorting, can be done using water instead of a saline solution as a sheath fluid; a prominent exception occurs in the case of electronic (Coulter orifice) cell volume measurements, which, like droplet cell sorting, require that cells be suspended in an electrically conductive medium. This usually means an isotonic sodium chloride solution, with or without a little buffering. Why not use saline solutions all the time? Well, all other things being equal, water is cheaper than saline. Water also doesn't leave salt deposits all over everything. For analysis, even using live cells, the interval between injection of the core (which does have to be isotonic to keep cells osmotically happy) into the sheath and the measurement is so brief that the cells won't have time to object to being in a yucky hypotonic medium until well after they've passed the observation point, unless there's an obstruction downstream which causes water to back up into the sample. For sorting and Coulter volume measurements, which require ionic solutions, normal saline for injection, which is fairly clean and sterile, is readily available in liter bottles at reasonable prices. "Injectable" or not, it does have to be filtered.

Nitrogen supposedly produces fewer bubbles than does compressed air when used as a sheath propellant. When house air supplies are non-existent or inadequate, and gas from tanks must be used for sheath drive, nitrogen has the advantage of being cheaper than air is. It remains to be determined whether substitution of nitrogen for air affects cell physiologic parameters such as membrane potential; when in doubt, use air and isotonic solutions.

When gas pressure is used to drive the core, or sample, as well as the sheath, the relative sheath and core flow rates are dependent upon the difference in pressures between the sheath container and the sample container. This determines the core size. If the pressure drive system for the sheath is a giant wash bottle, that for the core is a micro wash bottle; the pressure vessel is frequently the sample tube itself, with a volume of only a few mL. Separate regulators for core and sheath pressure are advisable; pressure in the sample vessel is adjusted to change the core flow rate. A **differential pressure gauge**, monitoring the difference in pressure between sheath and sample vessels, provides the best operator feedback for this process.

Unless valves in the fluid lines are closed, if there is no tube mounted for sample feed, sheath pressure forces sheath fluid back through the sample inlet tube; this is a relatively easy way of clearing out the residue of the last sample before running the next one, but it can also be a way of diluting the next sample with sheath fluid when you'd prefer not to have diluted it.

On the other side of that coin, most pressure feed systems also have switchable inlets for drive gas at both the regulated pressure and a higher boost pressure; when a tube is first put on the system, the higher pressure is applied to drive cells into the sample tubing. The input is then switched to the regulated pressure. Flow should then be allowed to stabilize before any measurements are made.

In order to avoid the consequences of Shapiro's First Law (p. 11), we'd like to have a filter in the sample path to prevent particles greater in diameter than the orifice (cell clumps and other junk) from clogging it. B-D used to use a small filter made of a bundle of glass microcapillary tubes, each with a 40 µm internal diameter; the filter was placed at the very tip of an aspirator assembly which was inserted into the tube containing the sample. Ortho used a hollow fiber filter that was much larger physically; most of the people I know replaced it with a few pieces of nylon mesh with a 40 µm pore size held over the aspirator tip with a piece of plastic tubing. You can now buy sample filters from various sources; Partec makes some nice ones. You can also buy nylon mesh and roll your own.

Just as is the case with a pressure-driven sheath, when a filter is used with a pressure-driven core, as the filter clogs, the core flow rate decreases. This can be a real problem when the cells being analyzed are treated with an equilibrium stain such as acridine orange or one of the cyanine dyes, because relative flow rates of core and sheath affect dye diffusion rates from core to sheath, and changes in these flow rates may affect staining intensity. Changes in flow rate and core size may also affect measurement precision.

If a pump, instead of gas pressure, is used to drive the sheath, the volume flow rate of the sample is directly adjustable, and less susceptible to the sources of variation just mentioned. The pump used to drive the core should be a **positive displacement pump**, e.g., a **syringe pump**, which produces a (relatively) constant volume flow rate even when driving a varying resistance. Dick Sweet's piece on flow sorters in reference 9, circa 1980, contained statements to the effect that syringe pumps weren't good for sorting because they produced pulsatile flow; opinions have changed. I have almost always used syringe pumps for sample feed; others are beginning to find them advantageous[5,100-103,649-650], and some newer commercial systems use them.

When really smooth drive is necessary, it can be obtained from a syringe pump that uses a feed screw instead of the more common rack and pinion drive. Pulsations can also be reduced by putting a small air bubble between the syringe piston and the sample; this gets away from true constant-volume operation, but remains closer to this ideal than would a constant pressure drive in the presence of increasing resistance.

If you're good enough at designing fluidics, you can even manage to make a peristaltic pump produce a stable flow pattern in a flow cytometer; I can't do it, but the people at OptoFlow seem to have managed, and I have heard of a variety of other pump types being used successfully.

Gravity, all by itself, can be used to drive fluids through flow systems; it can also effect whatever else you're using. Changes in the height at which the sheath tank is placed and/or in the liquid level in the sheath tank can affect flow. Gravity is often used to advantage when very slow flow rates are required, e.g., for the outer sheath of double-sheath flow systems, or for systems designed for molecular analysis.

In hematology analyzers, which are the prototypical clinical flow cytometers, sample and sheath may both be propelled by syringe pumps, which, in combination with valves and other mechanical and electronic components, allow blood samples to be withdrawn from the tubes in which they were collected, diluted, treated with appropriate reagents, and analyzed. This is a good way to handle samples that should all be subjected to the same treatment, and which can be analyzed in a minute or so. On the other hand, if you wanted to do a four-hour preparative sort on a droplet sorter, you'd need a syringe the size of a Bugatti Royale piston for the sheath; pressure drive from a big reservoir is the only practical way to go. You'd also probably have to use pressure feed for the sample, keeping it a container with appropriate temperature regulation and a magnetic stirrer or other mechanical device to keep the cells in suspension.

Figure 4-39 shows that the fluid connections in flow chambers are readily accessible; up until recently, it was fairly easy to get at those connections, at least on instruments designed for research purposes, and hook up whatever core and sheath drive mechanisms you might want. Third parties such as Cytek Development provide such gadgets, optimized for such tasks as kinetic experiments with calcium probes. However, if you're in the market for flow cytometer and contemplate playing games with the fluidics, it's a good idea to get confirmation from the manufacturer that the changes can be made without rendering the cytometer inoperative.

Perfect Timing: Fluidics for Kinetic Experiments

Kinetic experiments, in which the addition of one or more reagents to a cell sample and the introduction of the sample into the core stream must all be carried out at precisely timed intervals, are likely to need more elaborate external plumbing than the flow cytometer manufacturers normally supply.

The adaptation of the analytical chemical technique of **flow injection analysis**[1147-8] to cytometry was described by Lindberg, Ruzicka, and Christian[1149] shortly before the last edition of this book appeared. In addition to allowing rapid and flexible, automated, sample preparation and solution handling, flow injection analysis offers methods for analysis of data from samples in which chemical equilibrium has not been reached, which may be useful in both static and flow cytometric applications.

More recently, John Nolan, Larry Sklar, and their colleagues have developed instrumental approaches to both rapid kinetic experiments and high throughput flow cytometry[2454,2460-3]; these are discussed on pp. 364-6.

Oriented and Disoriented Cells

When velocities are different at different points in a flowing stream, an object in the stream is subjected to unequal forces at different points on its surface. These forces cause cells, particularly asymmetric cells, to tumble in the stream, and, in some cases, to assume one orientation rather than another. Kachel, Fellner-Feldegg, and Menke[1146] show photographs of the behavior of erythrocytes in flow, demonstrating the orienting effect of a rectangular core injector.

Johnson and Pinkel[651] made modifications to a commercial flow cytometer to permit high-resolution DNA analysis of sperm. Beveling the round core injector to a chisel point produced a ribbonlike stream; asymmetric sperm heads were oriented in the plane of the stream, and CVs of DNA measurements were substantially reduced. Orienting sperm and monitoring orientation are critical to selecting x- and y-chromosome enriched fractions by sorting, as will be discussed further in Chapter 10.

Matchmaker, Matchmaker, Make Me a(n) Index Match!

Differences in chemical composition between the core and sheath fluids (as in the instances when, for example, you use a water sheath and a saline core, or a saline sheath and a saline core with added protein) may be reflected in refractive index differences between the core and sheath. The refractive index mismatch usually isn't a problem for eukaryotic cells, but definitely will be for bacteria or really small (0.5 μm) beads and may be for platelets. The noise is predominantly in the scatter channels; fluorescence measurements should not be affected significantly although variable scattering of excitation and emission at the interface might be expected to increase CVs slightly. However, don't be too quick to assign blame to a refractive index mismatch when the noise is due to particles in the sheath fluid. The easiest way to tell the difference is to run the sheath without the core; if the noise persists, it can't be due to an index mismatch. Refractive indices of natural water samples may differ; Cucci and Sieracki have demonstrated the effects of differences in sheath and sample salinities on forward scatter signals from a variety of small marine organisms[2464].

Flow Unsheathed

Kamentsky's original Rapid Cell Spectrophotometer[65], and many of the early hematology counters that preceded it, did not use sheath flow, and there are still some flow cytometers that don't, including a fluidic sorter for very large particles (*Drosophila* embryos)[2325] and a microfluidic sorter for very small ones (bacteria; DNA fragments)[2326-7]. Among commercial systems, we have the Flow CAM, an imaging flow system for analysis of fairly large marine organisms[2453], and the **Guava PC™**, designed to count and analyze cells smaller than 60 μm[2465]. The principal problems with not using a sheath are that there are likely to be differences in flow velocities between particles traveling at different distances from the center of the stream and that there is a somewhat greater likelihood of the flow system becoming obstructed by large particles or aggregates. It would probably be difficult to obtain a distribution with a CV of 1% if you measured DNA content of nuclei in a flow cytometer without sheath flow, but it is evidently not a problem to get measurement CVs of 5-10%; if these are adequate for the application, you can safely dispense with the sheath.

Flow Systems: Garbage In, Garbage Out...

It should be obvious that spending a few hundred thousand bucks on an instrument won't do anything to relieve you of the burden of designing experiments. Flow cytometers are like computers in many respects; both are high-tech, often high-budget gadgets which can very rapidly and very precisely do exactly what you've told them to do, and neither has any capacity for doing what you meant instead of what you said.

If, for example, you are interested in finding cells so rare that they represent only one cell in every 10^8 cells in your sample, you should bear in mind that, even if your flow cytometer could zip through 100,000 cells/s, and could identify your cells of interest flawlessly, you would only encounter 3 or 4 cells of interest an hour. If all you wanted to do was get a count of your population of interest, with a precision of 10 percent, you'd have to count 100 cells to get the contribution to variance due to sample size down into the acceptable range. This would take 25 to 30 hours, not counting time to change samples, replenish sheath fluid, etc. It is all but impossible to do a single experiment of this kind by flow cytometry; it is beyond impossible and heading for irrational to contemplate doing such analyses for some routine purpose.

On the other hand, if you're looking for one cell in 10^5, you could get a good count in a few minutes; whether or not it is feasible do so depends on the efficacy of your cytometric identification procedure. It's easy to do if you can stain your cells of interest with one or, even better, two very bright fluorescent dyes; it's nearly impossible to do if you are relying on something like a dim immunofluorescent stain as the sole identification criterion. In general, you have to decide what measurement(s) you need to make of cells to answer the question(s) you're asking about them.

Recall that a flow cytometer, by itself, doesn't even have an intuitive way of telling what is a cell and what isn't; you basically provide a definition for it by picking a trigger channel or channels, setting a threshold value or values and, optionally, by defining selection regions in your measurement space using hardware and/or software gating. If no other gating is used, the machine treats every particle for which the amplitude of the trigger signal is above the set threshold value as a cell.

When you analyze clean samples from relatively homogeneous cell populations, you don't go too far wrong in using a forward light scatter measurement as a trigger parameter; immunofluorescence measurements have traditionally been done in this way, with the fluorescence measurement gated by the volume or scatter signal. In samples containing cell populations with greater variance in characteristics, thresholds and gates set on a fluorescence channel indicating the presence and amount of cellular DNA may provide more stringent criteria for distinguishing cells of interest from debris, on the one hand, and from aggregates, on the other. At present, the quality of both instruments and antibody reagents is high enough for it to be feasible to trigger on immunofluorescence signals to identify cells in relatively messy samples.

Given the prices so many people have to pay for flow cytometer time, it's wise to have some idea of what your sample looks like before you put it in the machine. The best way to get some idea is to look at the sample under a phase contrast and/or fluorescence microscope. It is true that the flow cytometer can pick up more subtle differences than you can discern by visual microscopic observation, but I'm not talking about subtle differences. If your immunofluorescent stained cells are lying on the slide in one long strand, you can cancel your appointment with the machine. If, unbeknownst to you, the cells are sitting in the tube in one long strand, putting them in the cytometer is apt to cancel other people's appointments as well. This can lead to stress, if it's your cytometer, and to physical harm, if it's someone else's.

Although in-line filtration is an obvious method of preventing large particles from entering and obstructing the flow system, filters may exacerbate the problem of sample carryover, which is encountered even in cytometers without them. Cells may accumulate in various places in the flow system during turbulent flow while samples are being changed, and subsequently pass into the measurement system.

Sample carryover is tolerable, if only because it is unnoticeable, in a lot of routine flow cytometry. If you are running a whole bunch of similar samples, each with approximately the same cell concentration, you probably won't know or care if 1 or 2 percent of the cells in the $(n+1)$st sample were really stragglers from the nth sample. Carryover can really kill you, however, when you are looking for rare cell subpopulations and/or when you run cell samples with very low cell densities after cell samples with very high densities.

Let's suppose we have a 1 percent carryover rate between two samples, each of which contains 10^6 cells/ml, and each of which is run at a rate of 1 µl/s, with data from 10,000 cells being collected. When the second sample is run, 1,010 cells will come through the machine each second; 1,000 of these will be from sample 2, and 10 will represent carryover from sample 1 (on the average). If we were now to put on a third sample with only 10^5 cells/ml, and run at the same sample flow rate, carryover, now from sample 2, would still give us 10 cells/s from sample 2, while we would only collect 100 cells/s from sample 3; for the third sample, carryover would be greater than 9 percent instead of 1 percent. You can generally get a good sense of the extent of carryover in your system from the persistence of fluorescent plastic beads, with which most of us start our day's runs, in subsequent cell samples. Reducing carryover may require some combination of backflushing, filter change or removal, and vigorous cleaning of the flow system.

Particles from the last sample aren't the only foreign agents which may introduce disinformation into data; microorganisms, particularly fungi, may grow in flow systems and in the sheath fluid supply, and surprisingly large amounts of hydrophobic dyes (acridine orange, the cyanines, and propidium are notorious in this regard) in stained samples may adhere to even short lengths of plastic sample inlet tubing, from which they are readily transferred to cells in subsequent samples. Chlorine bleach usually clears out both microorganisms and dyes; you do, however, have to be sure you've washed the bleach out thoroughly before you put samples back into the instrument, or you're apt to end up with yet another class of artifacts. I usually follow the bleach with water, then 70% ethanol, and then rinse well with water. If there's any liquid in the tubing and the flow cell when I shut down the system, I want it to be clean water.

In a posting (28 August 2001) to the Purdue Cytometry Mailing List, Mario Roederer recommended running 0.1 N NaOH (made in clean water and filtered through a small pore filter), CoulterCleanse solution, and distilled, deionized filtered water, in that order, through a stream-in-air sorter to solubilize any residual DNA, RNA, cells, and microorganisms. Mario also uses 70% ethanol for sterilization. I (and he) caution that it is probably a good idea to find out from the manufacturer whether there is any part of your instrument's fluidic system that won't stand up to the NaOH solution - or to any other solution or solvent you plan to run through the system.

The statistical distributions of cell arrival times and of the time intervals between successive cells' transits of the observation point may provide information about measurement quality. Lindmo and Fundingsrud[351] examined the latter distribution and its relation to various aspects of sample preparation. In some cases, even when samples did not contain cell clumps, the distribution did not fit the expected Poisson statistics; the authors hypothesized that cells that had stuck to the sample tubing were dislodged when hit by other cells passing through.

Watson[819] later extended this type of observation to permit the use of **time measurements for quality control** within individual sample runs. The interval between cell arrival or acquisition times is used to derive a real-time measurement of **sample flow rate**, which is displayed against time; data collected during periods of rapid fluctuation of flow rate are discarded. Gross et al[1150], using a combination of a computer algorithm for excluding such "burst" data, a rigorous cleaning procedure for the flow system, and staining and analysis techniques which minimized contributions from instrument noise and nonspecific fluorescence, reliably detected rare cells at frequencies of 1 per million using a benchtop flow cytometer; getting the garbage out, one way or another, definitely helps.

4.7 ELECTRONIC MEASUREMENTS

The ambiguous heading above will allow me to discuss electronic measurements of cells, i.e., **electronic** or **Coulter volume measurement**, which is based on **DC impedance**, and **electrical opacity**, which is based on **AC impedance**, and also to introduce a little bit about electronics in general.

Electricity and Electronics 101

Electrons and protons possess equal and opposite **charge**. The SI unit of charge (q), the **coulomb (C)**, is approximately 6×10^{18} times the charge on an electron, which is about 1.6×10^{-19} C. The electrons in an atom hang around the general vicinity of the protons, but the degree of mutual attraction varies, and may change with electronic energy levels. Electrons are, by convention, said to bear a **negative** charge; protons bear a **positive** charge. Particles of like charge repel one another; particles of opposite charge attract one another. **Coulomb's Law** states that the **electrostatic force** between two charges is proportional to the product of their magnitudes and inversely proportional to the square of the distance between them. The electrostatic force repelling two electrons spaced 1 mm apart is 10^{43} times as strong as the gravitational force attracting them.

Charge Separation, Electric Fields, and Current

According to the **Law of Conservation of Charge**, charge cannot be created, but it can be redistributed. If the electrons in a material are held relatively loosely, it is possible for charge to be transferred from one point in the material to another, and the material is called a **conductor**. If the electrons are more tightly held, charge transfer is not possible, and the material is called an **insulator**. There are also materials that normally behave as insulators, but which will conduct when some of their electrons are raised to appropriate energy levels; these materials are **semiconductors**. Atoms or molecules can lose or gain electrons to become positively or negatively charged **ions**; the Law of Conservation of Charge demands that ions be formed in pairs, with opposite charges. Ionic solutions and **plasmas**, which are essentially ionic gases, conduct electricity.

Separation of charges produces an **electric field**, which exerts a force on any charged particle in the field; the strength of the field is given in terms of force per unit charge. The **potential energy** of the particle varies with its position in the field. For example, if an electric field exists (never mind how it gets there) between two metal plates bearing opposite charges, an electron placed between the plates will be repelled by the negatively charged plate and attracted toward the positively charged one; the electron therefore has **potential energy**, which will decrease as it is accelerated toward the positive plate, being converted to **kinetic energy**. The potential energy of a charged particle in an electric field is proportional to its charge; the difference in potential energy, or **electrical potential difference**, between two points in the field at which the potential energy of a 1 coulomb charge changes by 1 joule, is defined as 1 **volt (V)**. The amount of potential energy lost, or kinetic energy acquired, as a single electron (charge 1.6×10^{-19} C) moves through a 1 volt potential difference, is called an **electron volt (eV)**; it is about 1.6×10^{-19} joules.

The mutual repulsion of like charges will cause redistribution of charge in a conductor, such that any excess charges are uniformly distributed on the surface of the conductor. The electric field "inside" the conductor is zero. Surrounding a volume with a conductor therefore produces **electrostatic shielding**, which is why low-level electrical signals are generally transmitted through **coaxial cable**, familiarly known as "coax" (pronounced "co-axe"). The signal is, in theory, carried in the central conductor, which is surrounded by an insulator, which is surrounded by braided or twisted wire and/or metal or metal coated foil. In principle, this outer conductive layer, which is connected to "ground," carries no current, and shields the inner conductor from electrical interference. In practice, the "shield" serves as a **return path** for the signal, and does carry current, which makes it less than perfect as a shield.

This has come to the attention of manufacturers of audio cables, who are now happy to sell you cables in which the signal and return signal are carried by a twisted pair of insulated wires surrounded by a foil shield that is only connected to ground at the signal source. This shield cannot carry current, and therefore works as advertised; the twisted pair configuration of the signal wires also reduces interference. Shielded twisted pair cables are probably not a necessity for a home audio system; I have put them in mine more as a matter of principle than because I think they'll make an audible difference.

The transfer of charge through conductors produces what is called an **electric current**; one **ampere (A)** of current represents the transfer of one coulomb of charge per second. It is totally incorrect to think of electric current "flow" in conductors in terms of an electron going into one end of a wire and coming out at the other end. An electric field produced at the "positive" end of the conductor accelerates a free electron toward that end, temporarily creating an excess positive charge "one atom over"; this generates a local field that accelerates a free electron from the vicinity of the next atom, etc., the process being repeated until an electron enters the "negative" end of the conductor. There is some analogy to the science toy made of a series of suspended metal balls, which is used to demonstrate transfer of momentum. You pick up the ball at one end, and let it swing into its static neighbor. The neighbor doesn't move, nor do any of the other balls except the one at the other end of the chain, which flies away from its neighbor to about the distance from which you released the first ball. In this instance, it is the local electric field which propagates along the conductor, moving at the speed of light; the current is, by convention, said to "flow" in the direction in which the field moves, which is opposite to that in which the electrons "move". The "speed of light" at which the field propagates in a conductor is not, by the way, the old familiar 3×10^8 m/s (186,000 miles/s) at which light travels in a vacuum; electricity, like light, moves more slowly through material media.

Resistance, Voltage, and Power; Ohm's Law

You could more or less guess, from the description of current flow in a conductor, that, just as friction and other real world effects prevent the metal balls from bouncing back and forth in perpetual motion, there must be some losses involved in all of those electrons jumping around. With the exception of a few esoteric materials that display the property of **superconductivity** at very low temperatures, there aren't any perfect conductors. Every material offers some **resistance** to the flow of electric current. According to **Ohm's Law**, the flow of a current of I amperes through a material with a resistance of R **ohms (Ω)** produces a drop in electrical potential, or a voltage difference, of E volts, across the resistance; that is:

$$E = IR .$$

It's amazing how much you can get done using just that formula.

Just as friction losses dissipate mechanical energy as heat, resistance losses dissipate electrical energy as heat. The amount of energy, in joules, lost per second is EI (remember that volts are joules/coulomb and amperes are coulombs/second), or I^2R. Joules per second, however, are **watts (W)**. A current flow of 1 A through a 1 Ω resistor produces a potential, or voltage, drop of 1 V across the resistor, and dissipates 1 W. The resistor gets hot, but not very; 4.184 J are equal to a "small" calorie, which is the amount of heat needed to raise the temperature of 1 gm of water by 1 °C.

Watt's a resistor made of? Generally speaking, resistors are made of conductive material, but not much of it. Again, as you'd probably expect, the less conductor you have, the harder it is to get current through it. It's much harder to move the same amount of traffic down a two-lane road than to move it down an eight-lane highway; similarly, the resistance of a given length of thin wire is higher than the resistance of the same length of thicker wire made of the same material. Many resistors are intentionally made to get hot, but, with exceptions, they are designed to dissipate the heat rather than to be destroyed by it.

Put 10 A through a 12 Ω resistor, which is what happens when you connect it to 120 V, and you dissipate 1200 W; the resistor is apt to get red hot, at least until the bell rings and the toast pops out. The filament of a 60 W light bulb is a 240 Ω resistor that gets white hot with 120 V across it, at which point it draws 0.5 A; the bulb must be filled with inert gas to prevent the filament from burning up. The active element in a **fuse** is a thin metal strip or wire that will melt once the current flowing through it gets above a certain value. And I oversimplified in all of the above cases; the resistance of materials typically increases with temperature. So, looking more carefully at the 60 W light bulb, we would find that the resistance of the cold filament is only about 1 Ω. The filament initially, and very briefly, draws a current much higher than 0.5 A (that's why bulbs are more likely to die just as they are turned on), increasing its resistance to 240 Ω as it reaches operating temperature.

Alternating and Direct Current; Magnetism

Where does current come from? We have already considered one current source in our discussion of the photovoltaic photodiode, or solar cell, which is a photoelectric source of current. The more light hits it, the more current the cell generates; however, the polarity of the current remains the same. The positive end of the cell remains positive. Current sources with this characteristic are said to generate **direct current**, or **DC**. DC is also produced by batteries, which convert chemical energy into electrical energy, and by some types of electromechanical generators.

Most of the current generated by electric utilities worldwide is not DC, but **alternating current**, or **AC**. The generators that produce either DC or AC make use of **magnetic fields** to convert mechanical energy into electrical energy. Magnetic fields result in part from the motion of charges; motion of a conductor in a magnetic field generates a current in the conductor. Passage of current through a conductor in a magnetic field creates a force on the conductor that may make it move if it is free to do so.

A microphone can be made from a very small coil of wire attached to a very small diaphragm and suspended in the field of a small permanent magnet; small currents are generated in the coil as the diaphragm is moved back and forth by interaction with sound waves in the air. These currents, electronically amplified, can be fed through a larger coil of wire, attached to a larger diaphragm and suspended in the field of a larger magnet; this device, a loudspeaker, converts the applied electrical energy back to mechanical energy, in the form of sound. The process of **transduction** between electrical and mechanical energy can work in both directions; a loudspeaker can function as a microphone, although its high mass makes it relatively insensitive, and a microphone can function as a loudspeaker, or at least as an earphone, although such a delicate device is likely to be destroyed by application of even a small amount of current.

The current generated in a microphone is an alternating current. No current is generated in the coil when the diaphragm is at rest; motion of the diaphragm away from its rest position in either direction causes an electrical potential difference across the coil, but the **polarity**, i.e., which side of the coil is positive and which negative, changes with the direction of motion. Alternating current is characterized by its **frequency**, *f*, measured in **hertz** (Hz), which used to be called **cycles per second**. Two alternating currents of the same frequency can additionally be related by their difference in **phase**; this is expressed by the same angular measure described in connection with the discussion of light on pp. 104-5.

Electric generators and electric motors are similar in their mechanical construction; they are comprised of substantial coils of wire suspended in the fields of relatively strong magnets. The magnets themselves are often electromagnets; passage of current through a coil of wire will create a magnetic field, and the field intensity is higher when the coil of wire is wrapped around a magnetic material such as an iron alloy. When a source of mechanical energy such as a steam turbine or internal combustion engine is used to move the coil, electric current generated in the coil can be used to operate other electrical devices. When current is applied to the coil, the coil moves, providing mechanical energy.

The alternating current output of generators is frequently in the form of a **sine wave**; that is, the voltage, V(t), at any time, t, is related to the maximum voltage, V_{max}, by the formula

$$V(t) = V_{max} \sin (2\pi f t).$$

The mathematical technique of **Fourier Analysis** often makes it convenient to deal with more complex alternating current waveforms as sums of series of sine (or cosine) waves.

Inductance, Reactance, Capacitance, Impedance

A conductor exhibits a property called **inductance (L)**, and responds to alternating current in a frequency-dependent fashion. The alternating current produces a changing magnetic field, which generates a voltage opposite in polarity to the applied voltage. In an inductance of 1 **henry (H)**, a voltage of 1 volt is induced by a current changing at the rate of 1 ampere/second. An inductor therefore has a property called **reactance**; reactance, like resistance, provides an impediment to the flow of current, but, unlike resistance, is dependent on the frequency of the current. The **inductive reactance** of an inductor, X_L, in ohms, is:

$$X_L = 2\pi f L \ ,$$

where *f* is the frequency in Hz and L is the inductance in henries. Inductive reactance increases with increasing frequency; an inductor conducts better at lower frequencies.

If the positive and negative sides of a DC current source are applied to a **capacitor**, a device consisting of two conductors separated by an insulator, there is a transient current flow, which is opposed by the accumulation of charges of opposite polarities on the conductors, and which stops when the potential difference between the conductors is equal to the potential of the source. If the current source is then removed, the potential difference between the conductors remains. The stored charge can be extracted as current. The amount of charge that can be stored in a capacitor increases with the surface area of the opposed conductors, and also varies with the **dielectric constant** of the insulator used. The **capacitance**, measured in **farads (F)**, is equal to the amount of charge on either electrode, in coulombs, divided by the potential difference between the electrodes, in volts; 1 farad equals one coulomb/volt. Real capacitors have capacitance values ranging from picofarads (pF) to thousand of microfarads (μF); some capacitance exists between any two nearby conductors.

Direct current will not flow "through" a capacitor; alternating current will, because the capacitor exhibits **capacitive reactance (X_C)**, which like inductive reactance, is a function of the frequency of the applied current. For X_C in ohms (Ω), C in farads, and *f* in Hz,

$$X_C = 1/(2\pi f C) \ .$$

A 1000 pF [or 1 nanofarad (nF)] capacitor has a reactance of about 159,000 Ω (159 kΩ) at 1,000 Hz (1 kHz), 15,900,000 Ω (15.9 MΩ) at 10 Hz, and 1,590 Ω (1.59 kΩ) at 100,000 Hz (100 kHz). Thus, a capacitor conducts better at higher frequencies.

When alternating current is applied to a resistor, the voltage and current vary together, in phase. In a capacitor, current and voltage are out of phase, with current 90° ahead of voltage. In an inductor, current and voltage are also out of phase, with current 90° behind voltage. Thus, in a circuit that contains both inductance and capacitance, the effects of one tend to cancel those of the other. The combined effect of resistance, inductive reactance, and capacitive reactance is referred to as the **impedance (Z)** of the circuit; this is, obviously, a frequency-dependent quantity.

Impedance is not simply the sum of resistance and reactance; it is expressed as

$$Z = [R^2 + (X_L - X_C)^2]^{1/2} .$$

It may seem as if we've gone a long way to get to a definition of impedance simply to explain the measurement made in a Coulter, or impedance, counter; the foregoing will, however, come in handy when we start talking about electronic circuits later on.

The Coulter Principle: Electronic Cell Sizing

Much of the stuff of which cells are made doesn't conduct electricity all that well. While intracellular fluid is a conductive ionic solution, the movement of most ions across the cell membrane, which is largely composed of lipid, an insulator, is restricted. Cells are therefore relatively poor conductors.

Blood is a suspension of cells in plasma, which is a reasonably good conductor. It was found in the late 19th century that the fraction of blood made up by the cells could be estimated from the conductance (conductance, measured in mhos, is the reciprocal of resistance) of blood; as the ratio of cells to plasma increases, the conductance of a given volume of blood decreases.

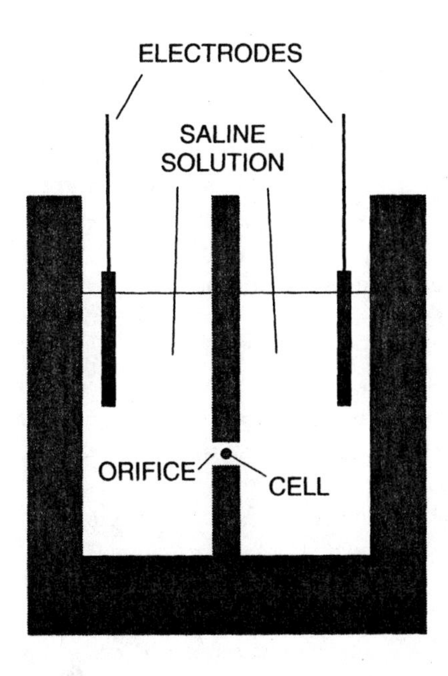

Figure 4-44. The Coulter orifice.

In the late 1940's, Wallace Coulter extended this type of measurement to the single cell level with his development of an electronic method for detecting, counting, and sizing cells based on their relatively low conductance. The principle of the **Coulter orifice** is illustrated in Figure 4-44.

Two chambers filled with a conductive saline solution are separated by a barrier containing a small orifice (typically 100 µm or less in diameter and no more than a few hundred µm in length) that provides the only fluid (and electrical) connection between the chambers. Most of the resistance, or impedance, in this arrangement is in the orifice. The electric circuit is analogous to that made up of two railroad rails connected by a thin wire; my friends and I used to enjoy demonstrating where the resistance was in that setup by dropping unwound wire coat hangers from an overpass onto the local subway tracks, briefly creating a conductive path between the third rail and one of the grounded rails on which the trains traveled. Coulter's analogue, while generating considerably less heat, light, and sound, has been vastly more useful and profitable.

A DC power supply that provides a constant current is connected to electrodes in each of the two chambers. As long as the orifice is entirely filled with saline, its impedance remains constant; we know from Ohm's law that, with a constant current flowing in the circuit, the voltage applied across the electrodes must also remain constant.

In operation, a stream of cells is passed through the orifice. A cell in the orifice displaces an equivalent volume of saline solution; the impedance of the orifice is therefore increased during the cell's transit, with the extent of the increase dependent on the volume of the cell. In order to keep the current through the system constant, the power supply must transiently apply a higher voltage between the electrodes. If the voltage output of the power supply is continuously monitored, a temporary increase in voltage output, or a **voltage pulse**, is observed whenever a cell passes through the orifice.

The details, which are discussed at length by Kachel[1151], are quite a bit more complicated. The electric field intensity varies considerably with distance from the axis of the orifice; as a result, the widths, amplitudes, and shapes of pulses produced by a cell's passage vary depending on its position in the stream. In order to deal with this problem, some instruments analyze individual pulse shapes, and reject pulses that appear to be due to cells close to the wall of the orifice. Some also employ **hydrodynamic focusing**, i.e., sheath flow, to confine the cells to a region of the stream near the axis of the orifice. Once the necessary corrections are made, cell volume can be derived from pulse amplitude.

Electrical Opacity: AC Impedance Measurement

Coulter and Hogg[715] established that the Coulter orifice could be operated using an AC, rather than a DC power supply. When the frequency used is in the radio frequency range, the cellular parameter measured is referred to as **electrical opacity**; this reflects the AC impedance of cells, and is more dependent on cellular structure and less dependent on size than is DC impedance. Some of Beckman Coulter's hematology instruments (e.g., the STK-S) and some from other manufacturers incorporate opacity measurements.

4.8 ANALOG SIGNAL PROCESSING

In optical flow cytometry, some light falls on the scatter and fluorescence detectors even when a particle is not passing through the observation region, producing a **background** current output from the detector; this background current fluctuates about some **baseline** value above zero. As a particle passes through the observation region, it produces a temporary increase in current output, i.e., a current **pulse**, at each detector. This is why flow cytometry has also been called "pulse cytophotometry." As is the case with Coulter volume and opacity measurements, information about the cells is derived from characteristics of the pulse.

Beam Geometry and Pulse Characteristics

Depending upon the illumination geometry, the **peak** amplitude, or **height**, and/or the **integral**, or **area**, of the fluorescence pulse will be proportional to the total amount of fluorescent material contained in the cell or particle. To understand this better, we need to look at Figure 4- 45.

The figure illustrates what happens when cells of different sizes pass through focused beams of different sizes. Suppose the 20 µm diameter particle contains twice as much dye as the 10 µm diameter particle. Since the volume of the 20 µm particle is eight times the volume of the 10 µm particle, the amount of dye per unit volume is higher in the smaller particle; if we looked at them under a fluorescence microscope, we would see a small, bright particle and a larger, dimmer one. At the top of the figure, we see the small particle in the middle of a 30 µm beam waist, and the larger par-

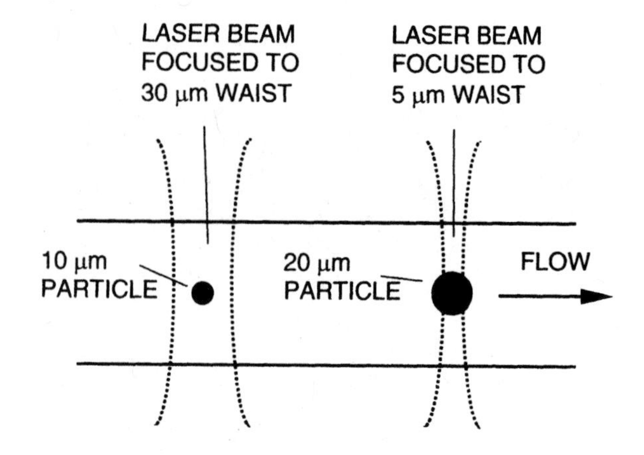

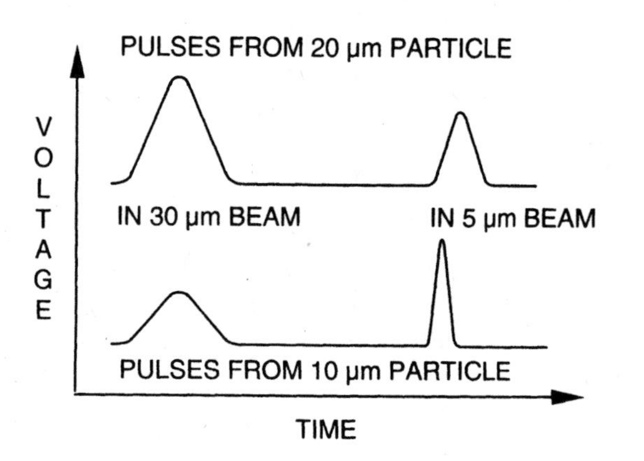

Figure 4-45. Effect of beam geometry on pulse shape.

ticle en route through a 5 µm beam waist. Note that we assume here that the power in the laser beams is low enough to prevent substantial bleaching of dye molecules, say 5-10 mW; this keeps the relationship between illumination and emission intensities linear.

The bottom portion of the figure illustrates the pulse shapes we may expect at the fluorescence detector preamplifier output as the particles traverse first the 30 µm and then the 5 µm beam. The scale is approximate. The pulses produced by both particles as they pass through the 30 µm beam will be Gaussian, because they both pass through the Gaussian intensity profile of the beam; the amplitude will reach a peak when the particles pass through the center of the beam.

Since the beam diameter is substantially larger than the particle diameter, the particles will receive approximately equal illumination, despite their different sizes. Thus, the peak amplitudes will be proportional to the amounts of dye contained in the two particles; the 20 µm particle, containing twice as much dye, produces a peak twice as high does the 10 µm particle.

The situation is different at the 5 µm beam. If both particles are traveling at the same velocity, the 20 µm particle is going to be illuminated for a longer time than the 10 µm particle. Since the beam waist is smaller than the diameter of either particle, neither particle will be entirely contained within the illuminated region at any one time. The larger particle will produce a longer pulse; the peak amplitude will be lower than the peak amplitude of the shorter pulse produced by the smaller particle because the smaller particle contains more molecules of dye per unit volume, i.e., it is brighter. Making the beam waist small compared to the particle diameter produces a **slit-scan**; one can get several types of information from pulses thus produced which cannot be derived using a larger beam waist. The **duration** of the pulse, or **pulse width**, yields size information[97,98]. The **peak height** gives information about **brightness**, or **fluorescence density**. In order to get a measure of the total amount of fluorescent material contained in the particles, however, you need the **integral**, or **area**, of the pulse, rather than the peak amplitude or pulse height. This makes the electronics somewhat more complicated.

Electronics 102: Real Live Circuits

As they come from the detectors, the current pulses we deal with in cytometry don't give us much to work with. The output current from the anode of a PMT or photodiode is only a few microamperes; let's say 10 µA (10^{-5} A) at most. To get the signal to the point where we can comfortably deal with it, we need an **amplifier**, something that will provide higher levels of current and/or voltage output than are available direct from the detector.

Circuits: Current Sources and Loads

It is customary to speak of electronic **circuits**; at a minimum, a circuit is composed of a **current source**, which supplies current, and a **load**, through which the current passes, eventually returning to the source, or completing the circuit. A circuit is not a perpetual motion machine; energy from the source is transferred to the load. In one of the simplest circuits, a resistance (which might be a light bulb) connected to a battery, the chemical energy in the battery is gradually lost by conversion into heat and light (an aside: only about 3% of the total wattage dissipated in an incandescent bulb goes into light; the rest is lost as heat).

Both current sources and loads have associated **impedances**. Ohm's law holds for impedances, that is, $E = IZ$; this means that, at a constant voltage, a low impedance load draws more current than a higher impedance load. As you may have noticed, the amount of power stereo amplifiers are rated to deliver is often specified separately for speaker impedances of 4 or 8 Ω; that's why. A low impedance source can supply more current than a high impedance source for a given voltage. PMTs and photodiodes are high impedance sources.

Figure 4-46 shows some circuit elements and the basic patterns in which they are interconnected; it features resistors and capacitors, which are known as **passive elements**. We'll leave inductors out of this discussion. Elements connected end to end are said to be **in series**; elements that have their corresponding terminals connected to the same point, or **node**, are said to be in parallel.

The resistances of resistors in series add, which makes sense. If we connect the series resistors R_1 and R_2 to a source, the same current, I, passes through both resistors. If E_1 is the voltage across R_1 and E_2 the voltage across R_2, Ohm's law gives us $E_1 = IR_1$ and $E_2 = IR_2$; the total voltage across the pair, $E = E_1 + E_2$. Resistors (and reactances) in series can therefore be used as **voltage dividers**; the voltage across a 10 kΩ resistor in series with a 90 kΩ resistor is 1/10 the voltage across the pair. A **variable resistor**, or **potentiometer**, contains a resistive element with a third connection which can be physically moved to make contact at any point between the two end terminals, providing a variable voltage divider usable as a light dimmer or volume control, among other things.

According to **Kirchhoff's Current Law**, the sum of currents entering any node in a circuit equals the sum of currents leaving that node. Circuit elements in parallel carry current in inverse proportion to their impedances; the voltage drops across any two elements in parallel are equal.

The charge storage capacity of a capacitor, all other things being equal, is a function of the surface area of its elements or plates; intuition correctly tells us that the capacitance of two capacitors in parallel is the sum of their individual capacitances. It is a little less obvious that two capacitors in series can store less charge than can either alone, but that's the way it works.

Ground Rules

If you are at all familiar with the electrical wiring of cars or bicycles, you've probably noticed that a lot of the loads have only a single wire connected to them. How is the circuit completed? The answer is that the other end of the load, and the other end of the battery providing the current, are both connected to the metal frame of the car or bicycle; they are said to be **grounded**, or, in British, **earthed**. The planet itself is conceived as representing the mother of all reference voltages, and, in fact, most power wiring is electrically connected at some point to a real ground, usually through the medium of a fairly substantial copper rod.

By convention, the "zero voltage" point in a circuit is defined as **ground**; this is represented in circuit diagrams by the symbol shown at the upper left in Figure 4-46. All points shown as connected to ground in a circuit diagram are electrically connected to one another; this is sometimes accomplished by a wire connection and sometimes accomplished by connecting the points to a metal (or otherwise conductive) enclosure in which the circuit is housed. Since the electric field inside a conductor is zero, such an enclosure provides **shielding** from stray electric fields.

In the real world, all grounds are not created equal. Let's go back to the car, where the positive terminal of the battery is connected to one terminal of various loads, and the negative terminal is connected to the car's chassis, as are the other ends of the loads. The current from a load is, therefore, being conducted back to the battery through the chassis, which has a low, but finite, resistance, which – Ohm's law again –

GROUND RESISTOR CAPACITOR

R_1 R_2 RESISTORS IN SERIES: $R = R_1 + R_2$

RESISTORS IN PARALLEL: $R = \frac{R_3 R_4}{R_3 + R_4}$ R_3 R_4

C_1 C_2 CAPACITORS IN SERIES: $C = \frac{C_1 C_2}{C_1 + C_2}$

CAPACITORS IN PARALLEL: $C = C_3 + C_4$ C_3 C_4

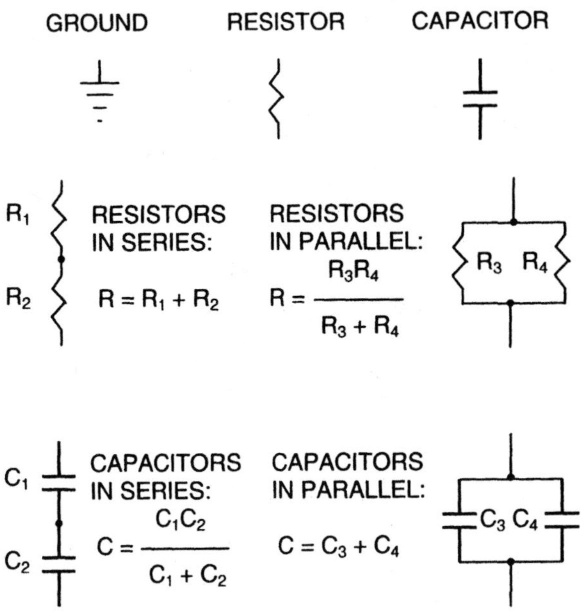

Figure 4-46. Some circuit elements.

means there is a potential difference between the point on the chassis at which the load makes contact and the point at which the battery is connected.

Most household AC power wiring in the United States delivers electricity to loads through two wires. One is called "hot," generally color-coded black (an unfortunate choice, since electronikers frequently use black as a color code for ground); the second, generally color-coded white, is called "neutral," and theoretically represents ground. Well, it is connected to ground, at the power station, at the transformer outside the house, and perhaps at points in between.

The nominal "110-115 V" value for household wiring is actually an RMS average; the peak values are higher. The voltage on the hot wire fluctuates (at 60 Hz), reaching extremes at 156 V above and 156 V below the voltage on the neutral wire. Inside the house, the wiring is divided up into parallel circuits, each with its own hot and neutral wires, and each with a fuse or circuit breaker limiting current in that circuit, generally to 15 or 20 A. If nothing at all is plugged into one of these parallel circuits, there should be no connection between its hot and neutral wires, and the neutral wire should, in theory, be at ground potential. You could touch it with wet bare hands without ill effect, again in theory. DO NOT ATTEMPT THIS TRICK AT HOME, IN THE LAB, OR ANYWHERE ELSE!!! The neutral wire is never at ground, and is usually hot enough to kill you, because there's usually current flowing in the circuits.

That's why modern wiring has a third, ground wire, color-coded green. This is connected to ground, either through the metal conduit and boxes in which the electric circuits are contained, or through a separate wire. You can safely connect yourself to the ground connection on a three-wire plug; in fact, it's a good idea to do so when you're working with solid-state electronics, which will be susceptible to static damage if you're not grounded. In a three-wire circuit, current goes from the hot wire through the load and back to the generating station via the neutral wire; the ground wire isn't carrying current. If the metal case of an electric or electronic appliance is connected to the third wire ground, and a **short circuit**, for example, an electrical connection between the hot or neutral wire and the case, develops due to mechanical damage, the ground connection provides a return path for the current. If the case isn't connected to ground, and you're holding it when the short develops, you become the ground connection, after which you may become the underground connection.

While the third-wire ground of power wiring is good enough to protect people from many of the potential lethal effects of 110 V AC, it may leave something to be desired as a "zero volt" reference. In many buildings in which AC wiring has been added over time, there may be substantial differences between the "ground" voltages at the third wire in different circuits. This seems to happen particularly often in hospitals, and seems to be exacerbated by the power surges and spikes which inevitably occur in such places. This makes it necessary to exercise extreme caution when plugging in electronic equipment. I and several people I know have had the experience of connecting a computer system to a flow cytometer, only to have some electronic components in one, the other, or both go up in smoke. It is often said that electronic devices work because they contain a small amount of smoke; when the smoke comes out, they stop working. In this instance, the proximate cause of the smoke coming out has been a measurable potential difference of tens of volts between the third wire grounds of two different circuits to which different components were connected.

There are several things you can do to minimize the likelihood of such disasters. If possible, plug any electronics you're connecting to your flow cytometer into the same circuit from which the cytometer gets power, i.e., into an outlet connected to the same circuit breaker. If you can't, use a voltmeter and/or the differential input of an oscilloscope to measure potential differences (AC and DC) between the grounds of the circuits to which you have access and which you might need to use in combination. If you don't see a big difference between "ground" potentials, you may be okay, but the problems tend to come from voltage transients. For this reason, you should always use surge suppressors with three-way protection (they block surges between hot and neutral and between either of those wires and the ground wire) and noise filtering on all of your electronics. Maybe now you know why you have had to replace all those boards in your instrument.

As I mentioned previously, low-level signals are typically transmitted through coaxial cable; the center conductor, in principle, carries the signal, while the outer conductor, usually connected to ground at least one point, provides shielding. By now, I hope you're becoming accustomed to the idea that, in a circuit, there has to be a conductive **return path** from the load back to the source as well as a path from the source to the load. Where's the return path in the coaxial cable? In the outer conductor, or shield. But wait a minute; doesn't that mean there's current flowing from the "ground" end of the load to the "ground" end of the source, which, as Ohm's law tells us, implies there is a potential difference between these two points? Yup. It probably doesn't amount to more than a few dozen microvolts, but there is one. When you start trying to measure signal levels around 1 millivolt, which is the low end of the bottom decade of a 4-decade log amp's input range, the details of ground connections become significant. In general, low-noise circuit design separates **power ground**, i.e., the point at which the return paths from the electronics to the power supplies converge, from **signal ground**, the "zero voltage" reference for signals. Grounds for analog signals are often separated from ground for digital signals, which tend to be noisier. The trick is to prevent what are called **ground loops**, i.e., current paths in the ground wiring which increase noise. This, as might be expected, requires both science and art; when you try to hunt down noise in circuits, you find your prey is not easily run to ground.

Couplings, Casual and Otherwise; Transformers

If you have any experience with audio equipment, you're probably familiar with hum at power line frequencies and multiples thereof (60 and 120 Hz, in the U.S.) as a major component of noise. How does this get into audio equipment? Primarily by **capacitive** and **inductive coupling**. I previously described a capacitor as being composed of two conductors separated by an insulator, and mentioned that AC could get from one side of a capacitor to another, while DC could not.

Maybe you thought, at that time, that the definition of a capacitor was too broad; it could for example, fit any two unshielded wires. So, indeed it could, and does; there is a small capacitance between any two conductors, and the closer they are together (while not actually in electrical contact) the higher the capacitance gets. A little bit of an AC signal in one can, and will, be transmitted to the other. This capacitive coupling is an electrostatic field effect, and is usually guarded against by **shielding**, which, as has already been mentioned, surrounds the conductor to be protected with another conductor, usually connected to ground.

Inductive coupling involves interactions between the magnetic fields surrounding two inductances; just as any old wire can be part of a capacitor, any old wire can be an inductor. The simplest defense against inductive coupling of an AC current to the outside world is to have the signal path and return path running through a **twisted pair** of wires; the associated magnetic fields pretty much cancel one another out. Putting a grounded shield around the twisted pair (p. 180) placates the capacitive demons, as well.

Inductive coupling is used intentionally and effectively in **transformers**, by which I mean not the metamorphosing toys, but the heavier and uglier devices used to convert low voltage to high voltage AC and *vice versa*. Suppose we had a coil with ten turns of wire placed next to a coil with twenty turns of wire, with a 10 V peak-to-peak AC signal running through the first coil; this would induce a 20 V peak-to-peak signal in the second coil, not only in principle, but pretty much in practice, too. Whether we could do anything effective with the 20 V signal, however, would depend upon the amount of current we were able to draw. If the coupling efficiency were 100%, putting 10 A at 10 V through the first coil, or **primary**, would allow us to get 5 A output current at 20 V from the **secondary**. Note that the products of voltage and current on both sides here are equal; in practice, there is always some loss. Coupling efficiency of transformers at relatively low frequencies, say below 100 kHz, is increased by winding both coils on a core of magnetic material, which is what makes transformers, particularly those used in power supplies, heavy.

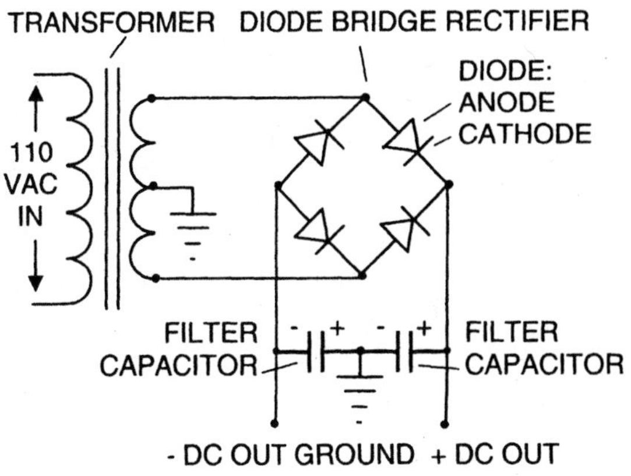

Figure 4-47. A line-powered DC power supply.

Power Supplies

Resistors, capacitors, inductors, and transformers are all **passive devices**; you can't get any more power or wattage out of them than you put into them. You can use a transformer to convert a low-voltage AC signal to a higher voltage signal, but you'll get proportionally less output current. Doing useful things with signals usually requires one or more stages of **power amplification**, in which the output voltage and/or current are increased while the temporal characteristics of the signal, or **waveform**, are preserved. This requires **active electronics**, incorporating devices such as transistors and vacuum tubes, and these need **power supplies**, which, for reasons we will get into shortly, must provide direct current. In principle, we could run our active electronics on battery power, which provides DC, but it's much more convenient to be able to plug things into the wall.

Edison originally envisioned household electricity as being provided by local generating stations as direct current. The drawback with this scheme is that the current has to be carried between the generating station and the users at a fixed voltage, say 110 V. If the ten houses on a block used 50 A apiece, the wire supplying the current just to that block would have to provide 500 A, which is a lot of current; you'd need a conductor more than a half inch in diameter to carry it without appreciable voltage drops and losses due to heating. If, instead, power is supplied as AC, as suggested by Westinghouse, the voltage on transmission lines can be thousands of volts, reducing the current transmitted, and allowing the use of thinner, cheaper conductors. A series of transformers placed at local power stations and in neighborhoods reduce voltage to the nominal 110 VAC required for household use, incidentally providing electromagnetic fields about which we have to decide whether to worry.

An active device, e.g., a power amplifier, has to be able to provide a (relatively) high-voltage, high-current replica of the input waveform. It doesn't know in advance how that waveform varies in time; it therefore needs to be able to call on its power supply for maximum power at any time, which means it needs DC power. As it turns out, most active devices run on voltages which are either substantially lower or substantially higher than 110 V; transistors typically require 25 V or less, while vacuum tubes need hundreds or, in some cases (e.g., TV picture tubes), thousands of volts.

A typical DC power supply, operated from AC line voltage, is shown in Figure 4-47. The first element in a power supply, after the on-off switch and the fuse or circuit breaker, which aren't shown here, is a transformer. This converts the 110 V AC from the power outlet into an AC voltage in the neighborhood of the DC voltage required by the electronics. The circuit symbol for a transformer is actually made up of back-to-back symbols for two inductors; the two vertical lines between them represent the metal core of the transformer.

The next element in the power supply is a **rectifier**, which converts the AC into DC; the rectifier shown here is composed of four **diodes** in what is known as a **bridge** configuration. A diode, as its name suggests, is a two-terminal electronic device; its two electrodes are a **cathode** and an **anode**. Current will flow from the anode to the cathode, but not from the cathode to the anode; the circuit symbol lets you know the direction in which current will flow. When current does flow from the anode to the cathode, there is a voltage drop across the diode; this is more or less fixed and its value depends on the materials of which the diode is made.

In the olden days, when we used vacuum tubes, which ran at high voltages and low currents, typical rectifiers were vacuum (tube) diodes, which are similar to PMTs without the dynodes. Instead of depending on light hitting the cathode to break electrons loose, which would have provided an extremely inefficient source of solar power, a high-resistance wire, or filament, running on relatively low-voltage AC, was used to heat the cathode, resulting in thermionic emission of electrons. Thermionic emission is a nuisance in a PMT, where it is the source of the unwanted dark current; in some applications, the PMT is cooled to reduce thermionic emission. In rectifiers and most other vacuum tubes, thermionic emission is an absolute necessity, hence the heated cathode.

Recall that in a PMT, the anode, thanks to a high-voltage DC power supply, is kept at a higher potential than the cathode, as are the dynodes; electrons accelerate from an electrode at one potential toward an electrode at a higher potential. When a diode is used as a rectifier, an AC potential is applied between the anode and cathode. When the anode is at a higher potential, electrons emitted from the cathode are accelerated toward the anode; current flows from anode to cathode. When the anode is at a lower potential than the cathode, there is no current flow. The diodes used in most power supplies these days are **semiconductor diodes**, usually made out of silicon; they generally run at lower voltages than vacuum tube diodes, but can handle larger currents, which is what the transistors in most modern electronic equipment want. Semiconductor diodes, while they do get warm due to their internal voltage drops, don't need heated filaments, which eventually burn out, nor do they need to be placed in evacuated glass bulbs; they have replaced vacuum diodes for almost all applications.

For this discussion, I will consider the diodes in the bridge as located at 1, 4, 7, and 10 o'clock. The AC voltage in the transformer secondary coil fluctuates; the top is alternately positive and negative with respect to the bottom. When the top is positive, the diodes at 1 and 7 conduct; current can flow out of the diode bridge through the wire connected at 3 o'clock, and return through the wire connected at 9. When the bottom of the secondary is positive, the diodes at 4 and 10 conduct, and the 3 o'clock point remains positive with respect to 9 o'clock, so current continues to flow in the same direction as before.

What's coming out of the rectifier at this point, however, isn't a constant DC voltage; it is, instead, the top halves of sine waves, one after the other, and the output voltage, while constant in polarity, fluctuates all the way down to zero. This isn't usable for running electronic circuits; we need to provide a DC output that remains at or above some fixed level. In order to do this, we put in the **filter capacitors** shown in the circuit diagram. A capacitor stores charge, which can later be extracted as current; the voltage across the capacitor will go as high as the voltage applied across its terminals, and then drop as the capacitor discharges. As the voltage difference between the 3 and 9 o'clock outputs of the rectifier increases from zero to its maximum value, the capacitor charges, approaching the maximum voltage; as the voltage returns toward zero, the capacitor discharges, maintaining some DC voltage across the rectifier output terminals. In order to supply high output currents, it is necessary that filter capacitors have high capacitances, typically thousands of microfarads. The capacitors used generally incorporate an electrolyte in their construction in order to achieve the high capacitance values required; such **electrolytic** capacitors are electrically polarized, and connecting them the wrong way will destroy them.

Once a capacitor is charged, completing a circuit between its positive and negative terminals will discharge it; the rate of discharge, and the discharge current (amperes = coulombs/second) depend on the resistance of the circuit. Short out a charged 10,000 μF capacitor with a screwdriver and you may melt the metal where it makes contact. On a more practical note, it used to be common practice to put a **bleeder resistor** across the output terminals of DC power supplies to gradually discharge the filter capacitors after power was turned off, thus preventing people from getting zapped while working on the electronics. These days, power supplies incorporate additional active devices called **voltage regulators**, which reduce the DC output fluctuations to a much lower level than could be accomplished by a Godzilla-sized filter capacitor.

In the power supply shown, the middle of the transformer secondary is shown connected to ground. The rectifier outputs at 3 and 9 o'clock provide DC voltages above and below ground. Power supplies with such symmetric positive and negative DC outputs, usually +/- 12 or 15 V, are typically required for analog electronic circuits. Digital or logic circuits more often require a unipolar power supply, typically + 5 V and ground. The electronic circuit of such a supply would omit the ground connection at the transformer secondary center tap; the 9 o'clock output would be grounded, the positive output would remain at 3 o'clock, and only a single filter capacitor would be needed.

The power supply of Figure 4-47 is what is known as a **linear** power supply. The current entering the transformer is 60 Hz line current. Many electronic devices, e.g., most computers, now use **switching** power supplies, in which a transformer input current at a frequency ranging from tens to hundreds of kilohertz is generated electronically. Inductive coupling is more efficient at these higher frequencies, making it possible to use smaller and lighter transformers and generate less heat for a given output power. Switching supplies, however, because of their high operating frequencies, are much more likely to generate noise which will interfere significantly with signal processing (at least in the context of flow cytometry) than are linear supplies.

Active Electronics: Tubes, Transistors, ICs

I don't know why I feel compelled to bring up vacuum tubes in a transistorized world. I know I'm not alone; there are rock musicians and audio nuts who insist that vacuum tube amplifiers sound better than transistor amplifiers. They have kept the vacuum tube industry alive, but I'm not sure I believe their claims. I do believe that it is somewhat easier to conceptualize the way vacuum tubes work than it is to do the same for transistors.

Let's go back to the vacuum diode. Following in the footsteps of Lee DeForest, we'll make a similar tube, but we'll put a third electrode between the anode and cathode, making it a wire **grid**, which won't mechanically prevent electrons from traveling from cathode to anode. We now have a three-electrode tube, or a **triode**. We'll heat the filament, and apply a fixed DC voltage between the anode and cathode, with the anode positive; a fixed current should then flow between the anode and cathode. Now, suppose we apply an AC signal to the grid. The area of the grid is small, so it won't act like a dynode; most of its effects will be exerted by its electric field. Depending on the potential difference between cathode and grid, the grid will either accelerate or decelerate electrons heading toward the anode from the cathode, or, equivalently, increase or decrease current flow from anode to cathode. Thus, there will be an AC current waveform analogous to that on the grid superimposed on the DC anode current. The AC anode current, however, is not derived from the grid input current; it comes from the DC power supply, and its voltage and current can be higher than the voltage and current applied to the grid. Thus, the triode can act as an **amplifier**. This is what made radio and television transmission practical, and what got rid of the big horns required by early phonographs and the even bigger horns needed for the acoustic phonograph recording process.

Vacuum tubes were improved on; further refinements were achieved by adding extra electrodes to make tetrodes, pentodes, etc., but they all used hot filaments and required sealed evacuated envelopes. When one of these components failed, as was inevitable, the tube ceased to function. By the 1950's, there were problems in making electronic computers larger than a certain size; even if it were possible to meet the enormous power and cooling requirements of systems containing thousands of tubes, the tubes' intrinsic failure rate would result in the computers' breaking down at progressively shorter intervals.

Working at Bell Laboratories in the 1940's, John Bardeen, Walter Brattain, and William Shockley developed a solid-state alternative to the vacuum tube, the **transistor**. The first transistors were made of germanium, which, like carbon and silicon, has four valence electrons. "Doping" germanium by adding impurities such as aluminum, with three valence electrons, produces a **p-type** (p for positive) semiconductor, a crystal structure with "holes," or relative deficiencies of electrons. Doping with impurities such as phosphorus, with five valence electrons, produces an **n-type** (n for negative) semiconductor, with extra electrons. A **p-n junction** between p- and n-type materials conducts electricity when a potential is applied to make the p-type material positive with respect to the n-type, but not when the potential is reversed; it therefore acts as a diode.

When fabricated with the appropriate geometry, a three-terminal semiconductor device containing two junctions (p-n-p or n-p-n) can be made to operate in such a way that fluctuations in a small current flowing across one junction are reproduced in the larger current flowing across the other; the device, a transistor, can, like the vacuum triode, be used as an amplifier. I told you it wasn't as easy to explain as a vacuum tube; maybe that's why its inventors got the Nobel Prize. For a lucid nonmathematical explanation of transistors and almost everything else, see Rodney Cotterill's *Cambridge Guide to the Material World*[152].

Vacuum tube electronics wasn't exactly intuitive; most engineers who had mastered it found transistor electronics even more difficult, because currents, rather than voltages, are what count in transistors, while tube electronics dealt primarily with voltages. I won't get into details. What is called a **bipolar transistor** has three electrodes, called the **emitter**, **base**, and **collector**; current input to the base modulates current flow between emitter and collector. A newer device, the **field effect transistor (FET)** has analogous electrodes called the **source**, **gate**, and **drain**; FETs are best dealt with in terms of voltages; the voltage applied to the gate controls the circuit between the source and drain.

Bipolar transistors have low impedances, and therefore tend to draw relatively high currents, although they don't get as hot as tubes. FETs have very high impedances; much of today's action in electronics depends on **complementary metal oxide silicon (CMOS)** devices, which incorporate FETs. These draw very little current, although their current requirements increase with operating frequency, which is why a microprocessor runs warmer at 2.2 GHz than at 1.1 GHz.

The microprocessor, however, is, as should be obvious, not a single transistor; it is an **integrated circuit (IC)**, fabricated on a single silicon substrate, containing millions of transistors. Transistors carrying small amounts of current don't need to be very large; they could, in theory, be reduced to molecular dimensions, and, in practice, the individual devices in a complex circuit such as a microprocessor may have conductive paths a fraction of a micrometer wide. Integrated circuits were first developed in the 1960's, also eventually garnering their inventors Nobel Prizes, and electronics hasn't been the same since. Except where a lot of power must be handled, almost all analog and digital signal processing is accomplished using integrated circuits; even some power devices, such as audio amplifiers, are now more likely to be built as monolithic structures than assembled from individual transistors.

Analog Nirvana: Operational Amplifiers

As I mentioned before, it was difficult to design real circuits using vacuum tubes, and, for me at least, harder to do so using transistors. If you were designing a tube amplifier, you generally wanted a **voltage gain**; for example, a 1 V peak-to-peak signal out for a 10 mV peak-to-peak signal in, which represents a voltage gain of 100. Transistors, however, were specified in terms of their **current gain**, a parameter known as beta (β). Like a lot of people, I never mastered beta. However, some people who did made it possible for the rest of us to do most of our own analog circuit design by inventing a truly wondrous device called an **operational amplifier**, or **op amp**.

An op amp is a three-terminal electronic device (neglecting, for the moment, its connections to the power supply and some adjustment points which may be added). It has two inputs, a **noninverting**, or **plus input**, and an **inverting**, or **minus input**, and one **output**. An ideal op amp has infinite input impedance, that is, it draws no current from a signal source, and zero output impedance, that is, it can supply as much voltage and current as are necessary. It also is capable of infinite voltage gain. However, the op amp is normally operated with its output connected to its minus input through a **feedback network**, which can be anything ranging from a simple wire to a complex circuit which itself contains other op amps. When the feedback network is connected, the output will do whatever it takes to minimize the voltage difference between the minus and plus inputs.

A real op amp may not have infinite input impedance, but those with FET inputs have impedances of about $10^{12}\ \Omega$, which is close enough. Running on a +/- 12 V power supply, a real op amp can put out voltages between -10 and + 10 V at currents of at least 10 mA; this is all that's needed for many purposes, and there are ways of boosting both voltage and current outputs to the range of hundreds of volts and tens of amperes. At low frequencies, the gain of real op amps may be in the range of 100,000; many newer devices can give you a voltage gain between 10 and 100 at frequencies of over 100 MHz.

Figure 4-48 shows some basic op amp circuits. The simplest is the **follower**, in which the output is connected directly to the minus input; the output voltage equals the input voltage. Followers are used with high impedance sources, such as pH electrodes; the very high input impedance of the op amp draws almost no current, while the output can provide the same voltage while boosting the current sufficiently to drive a chart recorder, A-D converter, etc.

In the **follower with gain**, the op amp has to make the voltage at the minus input equal to that at the plus input; the output is driving a voltage divider made up of the two resistors, and Ohm's law pretty much gets us the rest. The **inverting amplifier** is also relatively easy to figure out using Ohm's law. The plus input is grounded; the output therefore has to keep the minus input at ground, as well. In an inverting amplifier circuit, the input impedance is not the op amp's impedance, but R_1, which is the resistance between the input voltage and the **virtual ground** at the minus input.

The **differential amplifier**, as shown, subtracts one input voltage from another; it can also be built in fancier guises, to do sums and differences of more than two voltage inputs, with an additional input resistor required for each input. Gain can also be obtained. Differential amplifiers are useful in several contexts. I mentioned previously that ground isn't always ground, especially when you're trying to look at very low voltage signals. If you run such signals over a shielded twisted pair into a differential amplifier, you end up "bringing ground with you," so to speak. The sum and difference capability is used in audio mixers, and also in the analog fluorescence compensation circuits incorporated in most flow cytometers.

Remember how we got into this discussion? Trying to figure out how to convert the current output from a PMT or photodiode detector in a flow cytometer into a voltage?

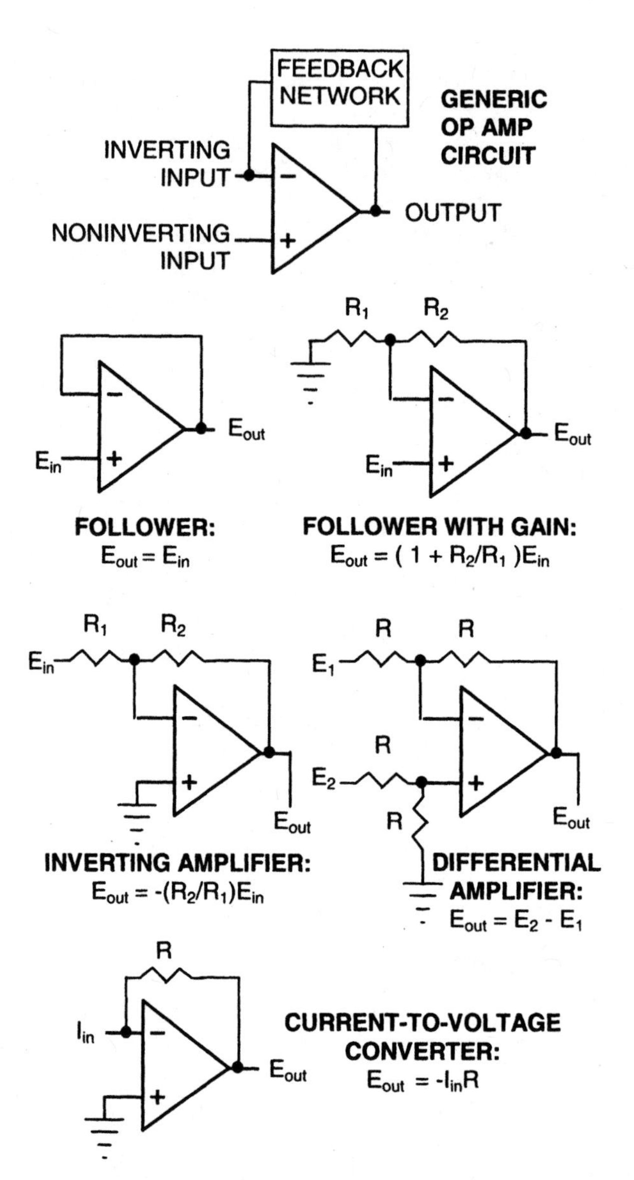

Figure 4-48. Basic operational amplifier circuits.

Guess what. I'm finally getting to the point. We use a **current-to-voltage converter** circuit, also known as a **transresistance amplifier** or a **transimpedance amplifier**. This is pretty much an inverting op amp circuit, minus the input resistor. But hold on a second. Doesn't that give us an input impedance of zero ohms, and an infinite voltage gain? Yes and no. The current is going into a virtual ground, but what we're concerned with is an input current, not an input voltage. In order to maintain the minus input at ground, the op amp has to produce an output voltage that draws as much current as is being put in, so there will be no current flow into or out of the minus input. Ohm's law tells it how to do that. So, on to how we process the signals from our photodetectors.

Detector Preamplifiers and Baseline Restoration

In a real flow cytometer, the PMT anode is generally near ground; the cathode is at minus a few hundred volts.

The current flow is from anode to cathode, so the PMT output will actually draw current from the minus input of the current-to-voltage converter. For reasons discussed previously, we'd like to keep the PMT output current below 100 μA (10^{-4} A); we also want to keep the op amp output below 10 V. If we use a 100 K resistor, we'll get 10 V out for 100 μA in. The first stage of our photodetector preamplifier, then, will be a current-to-voltage converter. This would, in fact, be all we needed if it weren't for the background light, which produces a fluctuating DC baseline on which the pulses we're trying to measure are superimposed. Figure 4-49 shows a photodetector preamplifier circuit incorporating baseline restoration.

The first stage of the circuit is the current-to-voltage converter just discussed. The 5 pF capacitor in parallel with the 100 K feedback resistor removes high frequency noise and prevents oscillation. The preamplifier output is taken from the output of the current-to-voltage converter, but this output also goes into a larger feedback network incorporating two additional op amps and associated circuitry. The first stage of this circuit is an inverting op amp with a gain of 2.2, but note that diodes of reversed polarities are connected in parallel with the 22 K resistor in the op amp feedback loop. Whatever the op amp output does, one or the other diode will conduct; they therefore **limit** or **clip** the voltage across the resistor, insuring it will be no more than the diode drop, which for silicon diodes is about 0.6 V. The net effect of the second op amp is that any pulses, either above or below the baseline, get "sliced off"; the amplified baseline goes into the second stage of the baseline restorer.

In this stage, there is an op amp set up as if it were an inverting amplifier, but there is a capacitor instead of a resistor between the output and the minus input. Such a circuit acts as an **integrator**; it can also be conceived of as a **low pass filter**, since the capacitor's reactance decreases with increasing frequency, reducing the effective gain of the circuit. The signal at the output of the third op amp, then, represents the baseline value, averaged over time. Since it has been through two inverting amplifiers, it has the same sign it started out with. We then complete the feedback network of the entire circuit by putting the amplified baseline through a 10K resistor into the minus input of the first stage current-to-voltage converter, using the summing property of the op amp illustrated in the differential amplifier circuit.

What we end up doing, overall, is subtracting a current proportional to the baseline current from the PMT current at the input stage. Feedback is a wonderful thing; you can vary the component values in the baseline restorer circuit over a substantial range without changing the performance characteristics very much. Trust me; I've built a lot of these circuits, and, under virtually all circumstances, I find that the DC output level when no cells are coming through is generally within a few millivolts of ground.

While the circuit illustrated isn't the only one that can be used for baseline restoration, it is reasonably typical, and has been employed in Cytomutts and in some commercial systems. Phil Stein and I have fiddled around with resistor and capacitor values and different op amps and tweaked performance some, but what's in the figure still works.

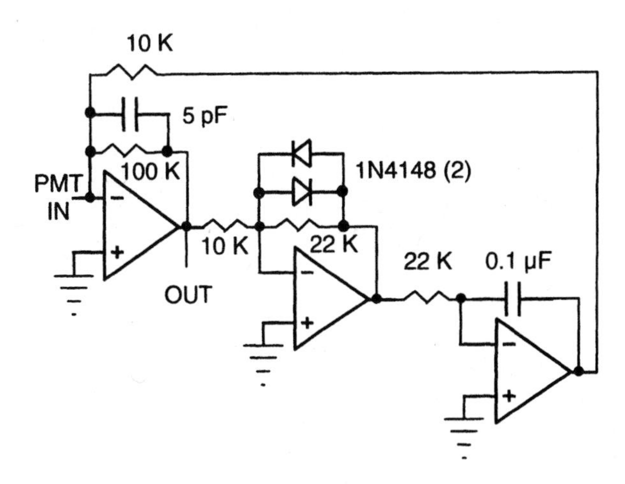

Figure 4-49. A photodetector preamplifier circuit.

Analog Pulse Processing: Front Ends and Triggering

The duration of the pulses produced by cells' passage through the illuminating beam is quite short. Velocities of 1-10 m/s are commonly used in flow cytometry; at 10 m/s, a particle traverses a 50 μm beam in 5 μs and a 5 μm beam in 500 nsec. Before the option of digital pulse processing became available, it was necessary to use analog **peak detectors**, **integrators**, and **pulse width measurement circuits** to provide short-term (microseconds to tens of microseconds) storage of the appropriate analog values for long enough to permit analog-to-digital conversion of the data.

Before the height, integral, or width of a pulse from a cell can be captured by the appropriate analog circuit, it is necessary to establish that a cell is present in the measurement system; this is the function of a **front end** or **trigger** circuit, which operates on analog inputs derived from one or more detectors. Figure 4-50 gives some idea of what happens in the front end circuit. The baseline-restored preamplifier output signal at the top of the figure, which we will assume is the **trigger signal**, includes a pulse. The trigger signal is fed to the plus input of a **comparator**; a device we have encountered, in company with some of the concepts of digital circuitry, on pp. 27-8. The minus input of the comparator is connected to a circuit that sets a **threshold voltage**, V_{th}. The threshold voltage is usually set with a **variable resistor** or **potentiometer**; one could also set it from a computer using a **digital-to-analog converter (DAC)**, especially considering that good DACs and the associated microprocessor interface circuitry now cost less than precision potentiometers, and have even displaced potentiometers from their traditional roles as audio volume and tone controls.

A comparator is related to an op amp; the signals applied to its inputs are **analog signals**, which can vary in voltage over at least a +/- 10 V range. The output produced by the comparator is a **digital signal** or **logic level**. When the

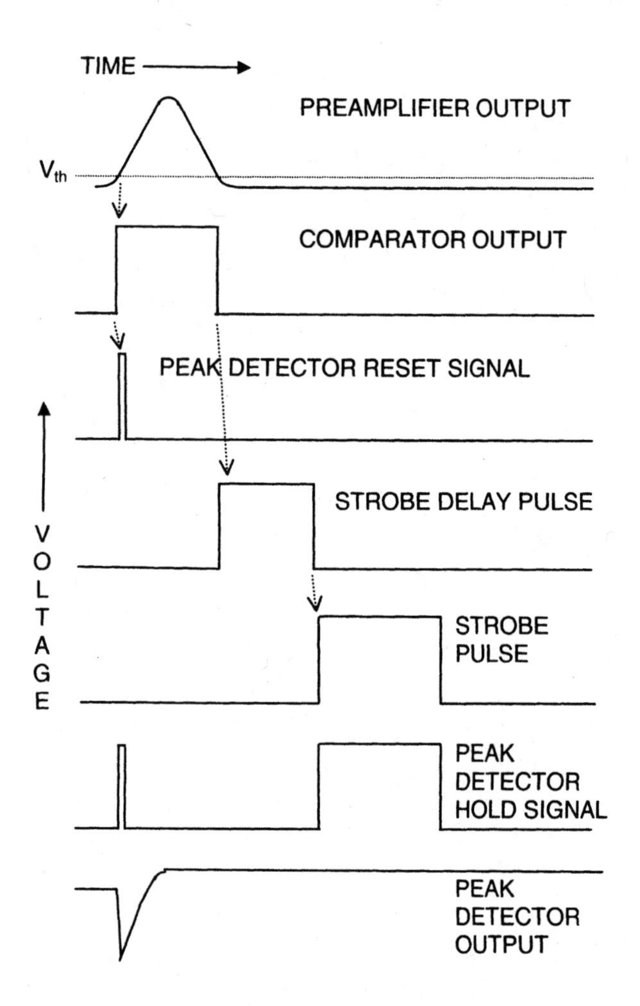

Figure 4-50. Waveforms in preamplifier, front end electronics, and peak detector circuits.

signal level on the plus input is less than or equal to the signal level on the minus input, the output is **logical zero** (typically 0 V, or ground). While the signal level on the plus input is greater than the signal level on the minus input, the output is **logical one** (typically 3.5 to 5 V).

When a cell is not passing through the observation point, the trigger signal is at baseline level, below V_{th}, and the comparator output is at logical zero. As a cell passes through the observation point, a pulse appears on the trigger signal, and, once the pulse voltage rises above V_{th}, the comparator output changes to logical one, remaining there until the trailing edge of the pulse, when the trigger signal again drops below threshold. In some instruments, it is possible to use logical or Boolean combinations (AND, OR, etc.) of the outputs of comparators connected to two or more input signals for triggering. When the comparator signal goes positive, the front end "knows a pulse is there."

Comparators take a finite time to **change output states,** i.e., to change their output levels from logical zero to logical one, after the input signal goes above V_{th}; the fastest ones may respond in 10 nsec, but they tend to be somewhat unstable, particularly with noisy input signals. The LM311 comparator, which is widely used, takes about 300 nsec to change state. We want to use this change of state to tell a peak detector circuit to start looking for a peak value to "memorize" and/or to tell an integrator and/or a pulse width measurement circuit to get on the job. Figure 4-50 deals only with a peak detector.

Peak Detectors

Peak detector, integrator, and pulse width measurement circuits all store their acquired signal values in **capacitors.** The circuits incorporate operational amplifiers and other active electronic components that prevent the capacitor from discharging unless a digital logic "**reset**" signal is applied, and allow the input to the capacitor to be disconnected, holding the voltage on the capacitor approximately constant, when a digital logic "**hold**" signal is applied.

The **peak detector reset signal, strobe delay pulse,** and **strobe pulse** shown in Figure 4-50 are all, at least in my "Cytomutt" front end electronics, generated by digital circuits called **monostable multivibrators,** or "**one-shots.**" These circuits can generate a logic pulse in response to either the rising edge or the falling edge of another digital signal. The duration of the output pulse of a one-shot is set by the values of an external resistor and capacitor attached to the device; a potentiometer can be used as the external resistor to provide a variable pulse duration.

In the configuration shown in Figure 4-50, the leading edge of the comparator output triggers a brief (< 200 ns) reset pulse, which, when applied to the peak detector, discharges its storage capacitor. During the reset pulse, a hold pulse is also applied to the peak detector, isolating its input and preventing the capacitor from charging. When a hold pulse is not present, the peak detector receives an input signal, which, in this case, is the same preamplifier output shown in the top trace of Figure 4-50. The peak detector's storage capacitor charges, following the input voltage as long as this voltage is increasing. When the input voltage falls below its maximum value, a diode in the peak detector circuit prevents the capacitor from discharging. The schematic diagram of a peak detector is shown in Figure 4-51.

In an ideal circuit, the voltage across the storage capacitor would remain that to which the capacitor was originally

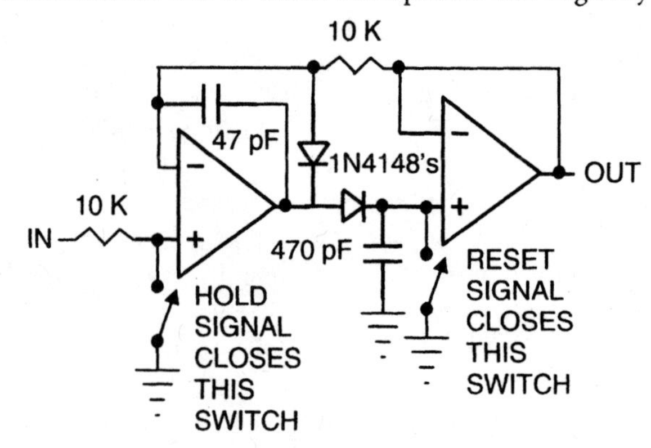

Figure 4-51. Schematic diagram of a peak detector.

charged, as in Figure 4-50; in the real world, charge is lost, resulting in a decrease in voltage, or **droop**, with time. This is observable in the output of a real peak detector, shown in Figure 4-52.

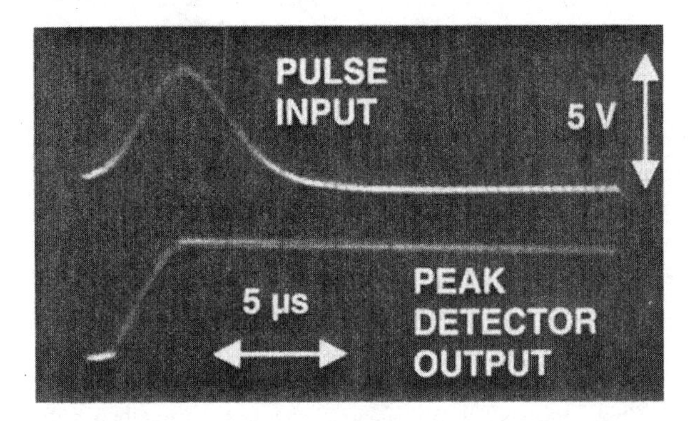

Figure 4-52. TOP: Preamp (upper trace) and peak detector (lower trace) outputs from a sample of propidium iodide-stained cell nuclei.

Peak detector performance is affected by the capacitance value chosen for the hold capacitor. Larger capacitors, other things being equal (some types of capacitors are "leakier" than others), decrease the droop, since they take longer to discharge, but this also makes the circuit take longer to reset, decreasing its fidelity of response to rapidly changing signals. Since the peak detectors used in flow cytometers only need to hold signal values for tens of microseconds, at most, i.e., until the values are transferred to a computer, to sort logic, or to other data analysis circuitry, it is generally possible to use capacitors of no more than a few hundred pF.

The reset and hold "switches" shown in the peak detector schematic of Figure 4-51 are electronic switches. The devices used for the task are usually field effect transistors, either discrete or incorporated in integrated circuits. It takes on the order of 100 nsec for the switches to respond, and about the same time for the capacitor to discharge through the resistance of the switch, which is about 100 ohms.

Reset and hold signals, which reach levels of at least 5 V in a matter of a few nanoseconds, are applied to circuit components in close physical proximity to those which carry the analog signals in peak detectors, and a small amount of crosstalk is inevitable. This results in the injection of small amounts of additional charge into the hold capacitor, introducing some inaccuracy into the held signal value. Charge injection errors are relatively inconsequential for signal levels above 100 mV, but seriously compromise the function of peak detectors at lower signal levels.

The strobe delay logic pulse shown in Figure 4-50 is, at least in my circuitry, adjustable in length. It is not strictly necessary in a single-beam flow cytometer, but is needed in a multistation instrument, where it allows the same reset pulse to reset peak detectors connected to signals coming from two or more illuminating beams separated in space. The strobe delay pulse is triggered by the falling edge of the comparator signal.

The falling edge of the strobe delay pulse triggers the strobe, or "data ready," logic pulse, which indicates that the outputs of peak detectors, integrators, and/or pulse width measurement circuits have reached stable values after the passage of a cell through the measurement system. Generally speaking, we know, or can find out with the aid of an oscilloscope monitoring the detector signals from the first and last observation points, how much time elapses between these signals. We can then set the length of the strobe delay so that the strobe signal will not be sent until the data from all peak detectors are valid. The strobe signal is transmitted, with the peak detector outputs, to the digital data acquisition system, where the rising edge of the strobe pulse is used to initiate the analog-to-digital conversion process.

The hold signal shown in Figure 4-50 is not generated by a one-shot; it is derived from the output of a digital circuit called an **OR gate**, with the reset and strobe signals connected to the inputs. The output of an OR gate is high when either or both inputs are high. When the hold signal is at logical one, the input to the peak detectors is connected to ground, preventing the output value from changing (neglecting the effects of droop). We have already noted that the hold signal is kept at logical one during the reset pulse, allowing the peak detector capacitor to discharge all the way to ground. Keeping the hold signal at logical one during the strobe pulse insures that the value being digitized by the data acquisition system will not change in the middle of the digitization process.

In the example of Figure 4-50, the interval between the point at which the preamplifier output pulse signal first rises above threshold and the point at which the strobe pulse returns to logical zero represents the minimum time required to process a pulse; this varies from instrument to instrument, but is typically between a few microseconds and a few tens of microseconds. The arrival of a second cell in the measurement region during this interval represents a **coincidence**, and the signal from the second cell can potentially interfere with that from the original cell to produce values of peak height, integral, and/or pulse width not accurately representative of either cell. Different instruments incorporate different circuit arrangements for dealing with coincidences; in simpler systems, a coincidence results in signals from both cells being aborted, while, in more sophisticated apparatus, "pipeline" processing of pulses allows resolution of almost all but the most closely separated cells. If you make the electronics more complicated, you can track as many cells as you want through as many beams as you want; book chapters by Hiebert and Sweet[639] and Hiebert[1153] provide additional details.

In the actual peak detector circuit of Figure 4-51, the output signal from the preamplifier goes to the positive input of the first op amp in the peak detector through a 10 KΩ resistor, to insure that the peak detector input impedance will be at least that high when the hold switch is closed.

When the trigger signal goes above threshold, the peak detector is reset; the voltage on the 470 pF hold capacitor drops to zero. The output of the first op amp goes to the ungrounded side of the hold capacitor through a diode; as long as the op amp output is higher than the voltage on the capacitor, the diode will conduct, charging the capacitor and increasing its voltage. This happens on the up side of the pulse.

The capacitor is also connected to the plus input of the second op amp, which is connected as a follower; its output is therefore the same voltage as is on the hold capacitor. The second op amp is typically a very high impedance FET-input type, and thus draws only minuscule current from the capacitor. The output is also connected, via a 10 KΩ resistor, to the minus input of the first op amp; the inner feedback loop of the first op amp contains a diode and a small (47 pF) capacitor. In this instance, the diode is connected so it will only conduct when the voltage at the minus input of the op amp is higher than the voltage at the op amp output. This happens after the pulse amplitude reaches its peak and begins falling back toward the baseline. The combined feedback loops compensate for the diode drop; the first op amp puts out extra voltage, so that the peak voltage on the hold capacitor is equal to the peak voltage of the input signal.

Pulse Integral or Area Measurements

When the beam dimension along the axis of flow is larger than a cell diameter, the peak height of a fluorescence pulse should be proportional to the total amount of fluorescent material in a cell. When this dimension is smaller, it is necessary to measure the integral, or area, of the pulse (Figure 4-45; p. 184). There are two basic approaches taken to this measurement in flow cytometers.

The first, and simpler, approach uses the preamplifier itself as an integrator. We have already seen that the pulses we're looking at last for only a few microseconds; the analog electronics used for processing such pulses therefore have to respond as rapidly as they would have to process periodic signals with periods similar to the pulse duration. Thus, the **frequency response** required of the preamplifier should extend at least to a few hundred kHz and possibly to above 1 MHz. This is an order of magnitude out of the league of stereo; we're right up in the frequency range of AM radio.

The electronics we use in this frequency range tend to become unstable and oscillate unless we add a little bit of capacitance to the circuit. This keeps the system stable, but decreases the high frequency response; what we have done is to insert a **low pass filter** (electronic, not optical) into the circuit. If we add more capacitance, we can slow the preamplifier down to the point at which the output pulse rises and falls at a considerably slower rate than does the input pulse; at this point, the preamplifier is behaving like an integrator. The peak value of the output pulse no longer faithfully represents the peak value of the input pulse, but approximates the area under the pulse. We can then use a peak detector to capture this peak value; its output will be the integral. This is a quick solution, although it gives us slow electronics that produce what are known as **long-tailed pulses**, decreasing the rate at which we can process cells without adding further circuitry or a farmer's wife. Because integration using low pass filters changes the shape of the pulse, it is also called integration by **pulse shaping**.

Integration by pulse shaping is generally necessary when logarithmic amplifiers (log amps) are used, because the output signal we need in that situation is the log of the integral of the pulse. If the output of a log amp is routed to an integrator, the integrator output will be the integral of the log, which is not the same as the log of the integral. However, if the input to the log amp is a low pass filtered signal, with the peak proportional to the integral of the original pulse, a peak detector operating on the log amp output will acquire the right signal, i.e., one proportional to the log of the integral.

What I will call the *de rigeur* electronic approach to integration, a circuit called a **gated integrator**, also uses a capacitor to store the integral value. In this circuit, however, speed need not be, and usually is not, sacrificed. The input signal to the integrator is turned on at the beginning of the pulse and off at the end, by a **gating signal** operating the same type of electronic switch that is used to reset a peak detector. The integrator also needs a reset switch to discharge the capacitor when it's time to process the next pulse.

Well, then, can we use the peak detector reset circuit of Figure 4-50 to reset the integrator and/or the comparator output to gate it? No, not if we want accurate values. The timing of reset signals for integrators and pulse width measurement circuits is more critical than the timing for peak detectors. In the sequence illustrated in Figure 4-50, the reset signal is not generated until the comparator changes output state, but this does not occur until the signal level exceeds threshold, which occurs some way up the leading edge of the pulse. An integrator reset at this time would "miss" the first portion of the pulse.

The problem is typically solved with the aid of an additional circuit element called an **analog delay line** to delay the input signal to the integrator by anywhere from half a microsecond up. What is a delay line? Very simple. Remember that current flows through a conductor at the speed of light, which, while it is slower in a conductor than in a vacuum, is pretty fast. A 1 µs delay line is basically a few hundred feet of wire. There are some tricks involved in dealing with the inductance created by packing all that wire into a reasonably sized package, but the finished product works. The real-time trigger signal is applied to the comparator; and the resultant reset pulse reaches each integrator just as the delayed pulse at its input "begins."

If hold signals for integrators are not applied precisely at the end of a pulse, some error is introduced into the output. A peak detector signal, however, can be held for an arbitrary time, because it reaches its maximum value in the middle of the pulse. Thus, as is also the case for reset signals, the timing of hold signals is more critical for integrators than for peak detectors.

Integrators, like peak detectors, store analog signal values in capacitors, and their dynamic characteristics, such as response time and output droop rate, are similarly affected by the capacitance of the hold capacitor. Charge injection introduces inaccuracies into the output signals of integrators in the same manner as occurs in peak detectors.

Pulse Width Measurement Circuits

A **pulse width measurement circuit** is essentially a **timer** that is turned on when the pulse starts, or rises above a certain threshold, and is held at whatever level it eventually reaches once the pulse amplitude falls back below threshold. In the classical analog implementation, the circuit is an integrator; its storage capacitor is charged not by the input signal, but by the input of a constant voltage that is applied after the circuit is reset and disconnected when the hold signal is applied. While it is charging, the voltage on the storage capacitor represents a **linear ramp**, increasing linearly with time; the voltage on the capacitor at the end of the measurement is therefore proportional to the duration of the pulse.

Pulse width measurement circuits, like integrators and unlike peak detectors, require precision timing to be accurate. A **constant threshold** pulse width measurement can be timed using a comparator output, with the rising edge of the comparator signal triggering reset and start signals, and the falling edge triggering a hold signal. Timing becomes trickier when the objective is to measure **constant fraction** pulse width, i.e., the interval between the times the pulse reaches a constant fraction of its eventual peak height.

A pulse width measurement circuit with a direct digital output can be made by replacing the integrator with a **digital counter**, to which the input is a **clock** signal. A clock signal is a logic signal that goes from positive to negative at a constant, known frequency; it is usually produced by a crystal-controlled oscillator. The reset signal sets the counter to zero and connects the clock circuit to its input; the value in the counter will then increase linearly with time until the hold signal disconnects the clock, leaving a digital value proportional to pulse duration in the counter.

This sounds like a great idea, but there is a problem. The pulse durations we are interested in measuring are only a few microseconds. In order to measure pulses with a precision of better than one percent, the counter has to end up with a stored value of several hundred. In order to get the counter to count to that value in a few microseconds, we need a fairly fast clock, operating at a frequency of around 100 MHz. It is not at all impossible to obtain clocks that fast, but we could get equivalent or better precision using the linear ramp analog circuit and digitizing the held value; even an inexpensive 12-bit ADC would give us a range of pulse widths between 0 and 4,096, allowing for greater precision than we would get with a digital circuit using a 100 MHz counter.

If all that is required is a relatively crude measure of pulse width, say, one good enough to tell the difference between a single cell and two or more coming through the system in close proximity, it may be feasible to do without a pulse width measurement circuit. In general, the area of a geometric figure is proportional to both its width and its height; the ratio of the peak and integral of a pulse can therefore be used as an approximation of pulse width.

Analog Pulse Processing: The Bottom Line

Peak detectors, integrators, and pulse width measurement circuits operate on baseline-restored signals, which may be on a linear or a logarithmic scale, with or without fluorescence compensation applied. The held signal outputs of peak detectors, integrators, and pulse width measurement circuits are sometimes referred to as **stretched** pulses, a term borrowed from nuclear instrumentation, which involves a lot of pulse processing. You could also call them aroused pulses, because they stay up, but let's not. They are also properly referred to as **flat-topped** pulses, and the flat-topped portions are known as **quasi-D.C.** signals.

Although we don't use these features explicitly in flow cytometry, pulses in other contexts are often characterized by their **rise times** and **fall times**, respectively defined as the times taken to get from 10% to 90% of peak value and from 90% to 10% of peak value. One can make more rectangular flat-topped pulses, i.e., pulses with shorter rise times, by using the strobe pulse to electronically switch the input of an amplifier between ground and the peak detector (or integrator, etc.) output. The output of that amplifier will then be at ground until the strobe pulse comes on, at which point it will go rapidly to the peak (or integral) value; when the strobe pulse goes off, the amplifier output will go back down to ground. The output of such a circuit may have a shorter fall time than would the output of a peak detector following a reset pulse, because the circuit has no hold capacitor to discharge. I used to use circuits of this type to generate real-time dot plots on oscilloscopes; I long ago abandoned both the oscilloscopes and the circuits in favor of digital computation.

In the real world, we have to be careful about choosing which of our measured parameters we use as **trigger signals**, particularly for integrators, when we are looking at low-level signals such as those obtained from immunofluorescence measurements. Figure 4-50 shows a robust trigger signal pulse rising above a baseline that is as straight a line as the laser printer can produce. Such pulses (and such quiet baselines) are easy to get from forward light scatter or DNA fluorescence measurements of most eukaryotic cells. However, if we look instead at immunofluorescence, the signals are considerably weaker, and the baseline may get noisy. If we try to use a noisy, weak signal as a trigger signal, lowering the threshold in an attempt to capture signals from cells generating low-amplitude pulses, the comparator starts to trigger on the baseline fluctuations and we start to measure background noise that we, or, more to the point, the instrument may count as cells. We also have to be careful, especially in the presence of a noisy baseline, to turn the integrator off (place it in a **hold** mode, not reset it) after the pulse ends; otherwise, we just keep adding noise to the integral.

Different flow cytometers have different triggering and thresholding schemes; some require that the operator choose a single signal as the trigger signal, while others allow thresholds to be set for all measured parameters. In the latter case, setting the threshold at zero for a particular parameter essentially eliminates that parameter's influence on triggering. I should mention that instruments that make measurements in multiple beams and use analog pulse processing generally require that the trigger signal be obtained from a measurement made in the first beam used. This should be obvious, but every so often I and other people I know have tried to trigger on a signal from a second or third beam and wondered why we weren't picking up any measurements from the first beam. Of course, if there's some pressing reason to trigger on signals from the second beam, it can be done without a time machine by running signals from the first beam through delay lines.

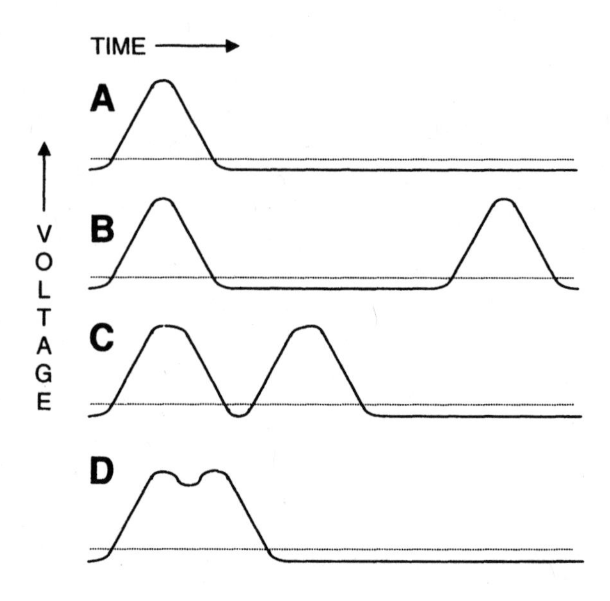

Figure 4-53. Telling two cells from one gets harder as the cells go through closer together.

Dead Times, Doublets, and Problem Pulses

The foregoing discussion of pulse height, area, and width measurement touched briefly on the subject of coincidences, but glossed over situations in which it may be difficult to tell that a coincidence is a coincidence. Figure 4-53 gives you the idea. Every flow cytometer, at least so far, has a **dead time**; i.e., a period after the arrival of a cell at the first measurement station during which the apparatus cannot successfully deal with a second cell. This is the hardware equivalent, if you will, of a neuron's refractory period following propagation of a nerve impulse. For the front end electronics described on the previous pages and in Figure 4-50, this period begins when the comparator output goes high and lasts until the end of the strobe pulse. In my single-beam instruments, this is around 50 µs; when I use two or more beams, it may be as long as 200 µs.

The dead time is frequently taken as indicating the number of cells per second that can be analyzed; e.g., with a 50 µs dead time, it should be possible to analyze 20,000 cells/s. It should be, provided the cells are well trained in synchronized swimming and arrive at 50 µs intervals. Real cells don't; they arrive at random times that generally fit a Poisson distribution. If you measure cell arrival times, and fit such a distribution, you can calculate the fraction of coincidences expected for any analysis rate, and select an analysis rate at which the fraction of coincidences will be no higher than you're willing to accept.

It is generally not a problem to determine that a coincidence has occurred when cells arrive separated enough in space and time so that the signal returns to the baseline. Neither panel A nor panel B in Figure 4-53 presents much of a problem in that regard.

Panel C shows the pulses that might be obtained from two cells passing through in rapid succession; the signal may not get all the way back to the baseline, but it does drop below the threshold level indicated by the dotted line; the comparator should go low and then high again. This won't happen when pulses are almost superimposed, as in panel D, which is one pattern which may be observed when a **doublet**, i.e., two cells either stuck together or very close in space, goes through the beam. If you get fancy, it is possible to pick up such occurrences in the front end electronics by measuring widths of comparator output pulses; the more conventional way of dealing with doublets is to wait until data are digitized and compare peak and integral values of signals. Wersto et al have a good recent paper[2466].

Trigger Happy?

In most flow cytometers, only one signal is used as a trigger signal; the important characteristic is that it have a relatively low background noise level. In immunofluorescence work, it has long been the custom to use forward scatter as a trigger signal, while in DNA analysis, it's as likely to be the fluorescence signal from the DNA stain, e.g., propidium. There are alternatives.

These days, signal-to-noise ratios of immunofluorescence signals are much better than they used to be, and it is often possible to trigger on an immunofluorescence signal. I have, for example, analyzed T-lymphocyte subsets in whole blood, without lysing the red cells, by triggering on fluorescence of a PE-Cy5 tandem-labeled anti-CD3 antibody, and immunofluorescence triggering is featured in at least one commercial instrument intended for subset analysis (The B-D FACSCount). It is advisable to use a log-amplified signal if the raw signal intensity is much below 100 mV, because comparators themselves are likely to behave unpredictably when their input voltages get too low.

In immunofluorescence analysis under other circumstances, I have found it advantageous to trigger on a nuclear (DNA) fluorescence signal; this is also a good way to deal with blood or marrow without red cell lysis, and, in addition, it allows you to discriminate single cells from doublets. Triggering on nuclear fluorescence isn't new; it was used in the original Cytofluorograf and the Block apparatus for differential leukocyte counting without red cell lysis, and later revived for the same purpose by Terstappen and Loken[1154]. This represents a simple but effective approach to rare event analysis. Nucleated cells account for approximately 1 of every 1,000 cells in blood; if you're interested in nucleated cells, and you trigger on scatter, running 10,000 cells/s, you'll only acquire data from 10 nucleated cells in that time. If you trigger on nuclear fluorescence, you can increase your analysis rate to several hundred nucleated cells a second, because, as far as the instrument is concerned, the other few hundred thousand cells aren't there. It's not difficult to detect populations accounting for a few tenths of a percent of the nucleated cells; these are a few cells per million in the input sample. There are some drawbacks; forward scatter signals are often messy enough to be unusable under these circumstances, and side scatter signals may not be as clean as you would expect, but fluorescence signals are not unacceptably compromised.

There are some circumstances under which there isn't a clean signal usable as a trigger; one of these is analysis of microorganisms. The instrument I use for bacterial analysis employs a technique known as **coincidence detection**. The "coincidence" here does not refer to two cells coming through the instrument in close proximity; what is detected is the coincidence of signals above threshold at two (or more) detectors. In this instance, fluorescence and forward scatter signals, both with relatively high background noise levels, are fed to separate comparators, with independently settable threshold levels; the logical AND of the comparators when both signals are above threshold initiates the reset pulse and other front end signals.

The logic game could be carried further; one might want to detect "live" cells stained green with calcein and "dead" cells recognizable by their red propidium-stained nuclei; in this case, it would be appropriate to use the logical OR of the signals from dual comparators to initiate front end response. Some commercial flow cytometers can implement coincidence detection and logical OR multiparameter triggering; some cannot (see Chapter 8).

Analog Linear, Log, and Ratio Circuits

The signal processing electronics in a flow cytometer have finished their job when they send the data analysis system flat-topped or held pulses, of amplitude 0-10 V or thereabouts, representing pulse peak or integral amplitudes or widths, and an accompanying logic pulse or strobe signal to indicate to the data analysis system that valid data are available.

Certain algebraic manipulations of either raw pulses or held levels are commonly accomplished using analog or hybrid analog and digital electronics, before signals are routed to the data analysis system, rather than digital computation thereafter. The electronic components employed include **linear sum and difference circuits**, used for fluorescence compensation, **logarithmic amplifiers,** used to facilitate analysis and display of data with values spanning a large dynamic range, and **ratio circuits**, used in fluorescence measurements of certain physiologic probes, compensation for power fluctuations in the light source, and calculation of quantities such as antigen surface density.

Linear Circuits; Fluorescence Compensation

Linear circuits generate outputs that are proportional to sums and/or differences of their inputs; they typically incorporate op amps in the differential amplifier configuration shown in Figure 4-48. They are most commonly used for **fluorescence compensation**[115], a practice made necessary by the overlap of emission spectra of fluorescent antibody labels, and previously discussed on pp. 36-8, with the aid of Figures 1-18 and 1-19. You might want to flip back there for a brief refresher.

Fluorescence compensation is, of necessity, an empirical process. The filter passbands shown indicate the range of wavelengths transmitted by each filter; the percentage of light transmitted varies with wavelength within the passband of each filter, and there may be noticeable variation between filters with the same overall specifications. There is some variation in emission spectra of labels; for example, fluorescein emission intensity, when excitation is at or near 488 nm, increases as pH rises from 7.0 toward 8.0. There are also variations from PMT to PMT in responsivity at different wavelengths. In short, there is no way to determine *a priori* what fraction of each signal must be subtracted from the others in order to yield reliable measurements of the emission intensities of each antibody-bound fluorescent label.

Figures 4-54 and 4-55, respectively, illustrate a simple case in which hardware compensation is used, and the hardware used for the compensation. When measurements are made of cells or, in this instance, of beads, bearing both fluorescein (FL)- (green) and phycoerythrin (PE)- (yellow) labeled antibodies, the substantial emission of FL in the yellow spectral region is detected by the yellow fluorescence detector intended for PE measurement. Similarly, the emission of PE in the green spectral region is detected by the green fluorescence detector intended for FL measurement. The spectral overlaps are evident in the left panel of Figure 4-54. The resulting positions of clusters in the dot plot make it impossible to use lines parallel and perpendicular to the axes to separate the display into quadrants containing what we would call dual negative, FL-positive/PE-negative, FL-negative/PE-positive, and dual-positive beads (or cells). Our partitioning of the display, of course, is for the purpose of setting gates that will allow us to count and/or sort the cells in each region. There is no law of nature that demands that the gating regions be rectangular; having rectangular gating regions does, however, let us use somewhat simpler hardware to set gates, and, for better or worse, it seems to be much easier for people to think in terms of rectangular gates.

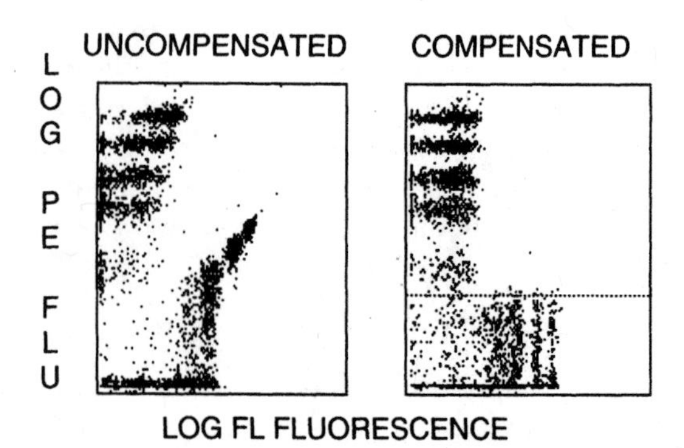

Figure 4-54. Uncompensated and compensated fluorescence signals from FL- and PE-labeled beads.

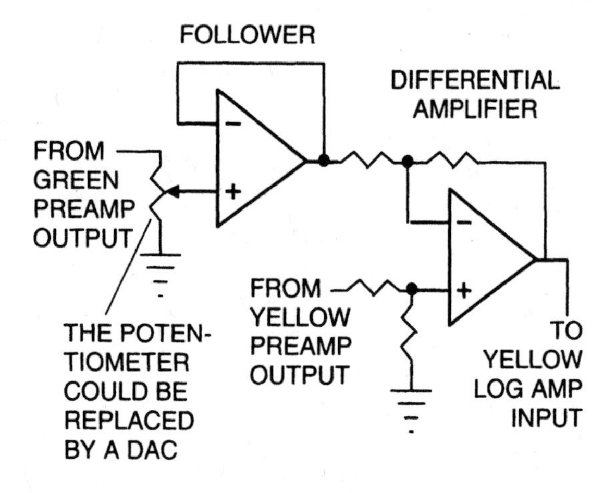

Figure 4-55. One side of the two-color compensation circuit used to generate the compensated data in Figure 4-54.

The data points on the left side of Figure 4-54 could be thought of as clustered along axes which are at an acute angle to one another; what we'd like to do is make the axes **orthogonal**. There are a number of well-known statistical procedures for doing this, including **factor analysis** and **principal component analysis.** They involve **linear transformation** of the data, which means computing values of a and b such that a data point that would be at (x,y) in the old coordinate system is at (x′,y′), where x′ = x - ay and y′ = y - bx, in the new coordinate system.

In the case of the yellow and green fluorescence, we have the old yellow signal, composed of a contribution from PE and a contribution from FL, and the old green signal, composed of a contribution from FL and a contribution from PE, and what we want to do is new yellow and green signals, respectively representing "pure" PE fluorescence and "pure" FL fluorescence.

We make the assumption that the yellow fluorescence from FL is a linear function of the amount of bound FL, which can be estimated from the green fluorescence signal, and the green fluorescence from PE is, similarly, a linear function of the amount of bound PE, which can be estimated from the yellow fluorescence signal. There are some cases in which this might not be true, for example, if there is substantial energy transfer from FL-labeled to PE-labeled antibodies, but we neglect this possibility.

The output from the green detector preamplifier is routed through the circuit of Figure 4-55, which subtracts a fraction of the output from the green detector from the yellow detector signal. You determine the amount subtracted by turning the knob on the variable resistor, or potentiometer, connected at the input to the follower; if compensation is mediated by the computer, a digital-to-analog converter (DAC) is used instead of a potentiometer to determine how much of the green signal is subtracted from the yellow signal. Another copy of the same circuit, and another potentiometer or DAC, are used to subtract a fraction of the yellow signal from the green signal.

The result is shown in the right-hand display of Figure 4-54. The operator, working by eye, twiddles knobs until the data clusters lie more or less parallel to the axes, so that the display can be partitioned into quadrants in the conventional manner. Since most people find twiddling knobs easier than doing algebra, there aren't many loud complaints about two-color compensation.

Things get somewhat more difficult when three colors, green, yellow, and red, are involved. In theory, it is necessary to subtract a little of both the yellow and the orange signals from the green signal, a little of the green and red from the yellow, and a little of the green and yellow from the red. There are now three circuits, each of which needs two potentiometers or DACs, for a total of six adjustments. Many instrument designs, with some justification provided by the spectra shown in Figure 1-18 (p. 37), omit subtracting red from green and green from red, dropping the number of adjustments to four.

For four colors, green, yellow, orange, and red, full compensation would require four circuits, each with three potentiometers or DACs. The circuit in the yellow channel, for example, would need to have some green, some orange, and some red subtracted; a total of twelve adjustments would be required. Looking to Figure 1-18 again for corners to cut, we find that the green channel should only need one adjustment, for correction from the yellow channel, while the yellow and orange channels would each need three. The red channel would have two adjustments, reducing the total number required from twelve to nine. Anybody who could actually get hired to operate a flow cytometer would be likely to find it easier to do the algebra required to solve the compensation problem mathematically than to adjust nine knobs to get a well-orthogonalized display. Bagwell and Adams[1155] have described the algebra, software, and control samples required for the mathematical approach, about which considerably more will be said in Chapter 5. To eliminate some of the suspense, let me say here that **neither hardware nor software compensation can fit clusters into quadrants across the full intensity range**; we'll see why later.

There is a more compelling reason than the anxiety induced by large numbers of knobs for switching from hardware to software compensation. Hardware compensation must be applied at the level of the linear signals coming out of the detector preamplifiers; the circuitry, as Figure 4-55 shows, is placed between the preamp outputs and the inputs to the logarithmic amplifiers that were almost universally used in immunofluorescence measurements. As was mentioned on pp. 35-37, a four-decade logarithmic amplifier must deal with inputs covering the voltage range between 1 mV and 10 V; the more circuitry there is connected to its inputs, the less likely it is that the noise level there will remain below 1 mV. If the noise level rises to just over 3 mV, a half-decade of dynamic range is lost; if it reaches 10 mV, a full decade of dynamic range is lost. In short, complex compensation circuitry almost inevitably decreases dynamic range. Recognizing this, Coulter, in designing its EPICS XL four-color, single-laser instrument, eliminated both linear compensation circuits and logarithmic amplifiers in favor of high-resolution analog-to-digital converters and software; other manufacturers have since followed suit.

In addition to their use in hardware compensation, linear circuits have been employed by the Los Alamos group for source noise compensation in extinction measurements[659] and in a cytochemical method for detection of BrUdR incorporation into DNA without the use of antibodies[647].

Logarithmic Amplifiers and Dynamic Range

In many applications of flow cytometry, the values of the measured variable span a restricted range. In DNA content analysis, for example, in the absence of clumping and aneuploidy, the brightest cells are twice as bright as the dimmest ones. In other cases, e.g., immunofluorescence analysis, the brightest cells may be hundreds of times brighter than the dimmer ones. In these circumstances, it has often made sense to use an electronic component called a **logarithmic amplifier** (familiarly, if not always contemptuously, known as a **log amp**) somewhere in the signal path, to increase **dynamic range**.

As its name suggests, a logarithmic amplifier produces an output signal, the amplitude of which is proportional to the logarithm of the input signal amplitude. The components typically used in flow cytometers have a range of at least three **decades**, i.e., they can accommodate signals varying in amplitude by at least a factor of 1000. Figure 4-56 illustrates the effect of the logarithmic transformation. The figure shows distributions of fluorescence from a population of beads (Quantum Simply Cellular® beads, Flow Cytometry Standards Corp., now available from Bangs Laboratories) bearing known numbers of molecules of phycoerythrin-labeled antibodies. The top plot shows data on a 256-channel linear scale; the bottom plot shows data on a 4-decade log scale mapped into a 256-channel linear scale. The logarithmic transformation was not, in this instance, done using a log amp; instead, data were digitized to 16 bits precision and the logarithmic values were obtained by digital computation. We couldn't have done any better using an ideal log amp; we might have done a lot worse using some real ones, for reasons that should shortly become clear.

Flow cytometers generally follow the standard practice of amplifying signals to values ranging between 0 and 10 V before they are converted to digital form. We will therefore assume that 10 V is the maximum output available from our

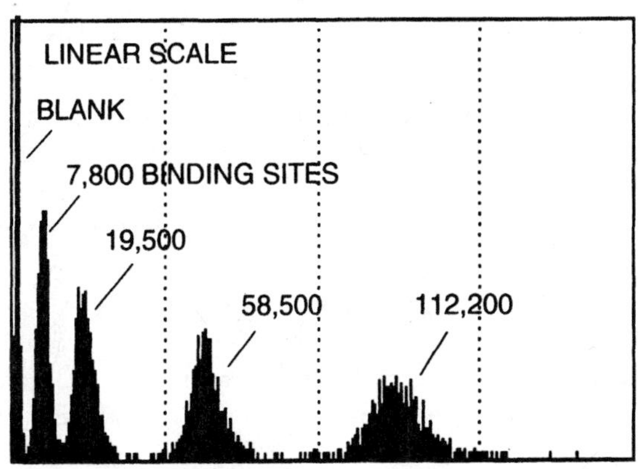

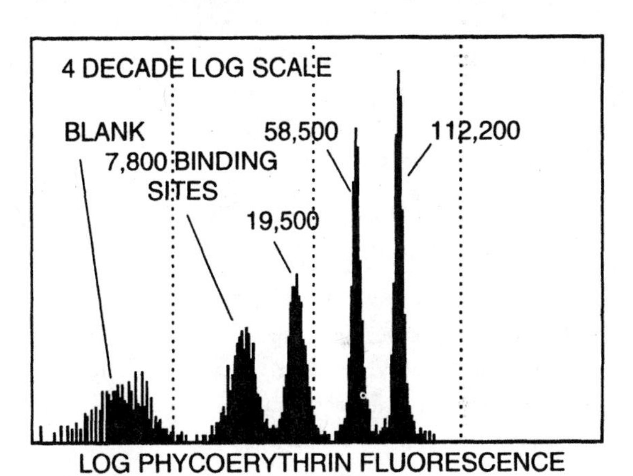

Figure 4-56. Signals from phycoerythrin-labeled anti-CD8 antibody bound to beads, displayed on 256-channel linear and 4-decade logarithmic scales.

preamplifier. If we want to keep our brightest cells on scale, we'll want them to produce pulses with a maximum height, after amplification, of 10 V. Cells which are dimmer by factors of 10, 100, 1,000 or 10,000 will, therefore, produce pulses with heights of 1 V, 100 mV, 10 mV, and 1 mV, respectively.

If we convert linear analog signals to digital signals, using an 8-bit analog-to-digital converter (ADC), we will have a 256 channel range (channels 0-255) corresponding to the range between 0 and 10 V; the difference between channels is just under 40 mV. A 10 V signal will be in channel 255, a 1 V signal in channel 25 or 26, a 100 mV signal in channel 2 or 3, and signals between 1 and 10 mV, or, for that matter, between 0 and 40 mV, will show up in channel 0. For most practical purposes, signals below 100 mV, which would correspond to "weakly positive" cells in an immunofluorescence analysis, will not be distinguishable from signals from "negative" cells.

If we run the linear signal through a logarithmic amplifier, however, we can manage to get everything on scale. Let's imagine a log amp for which 10 V in gives us 10 V out, and let's make the log amp's response 2.5 V/decade, i.e., for every tenfold increase in the input signal, we will get an increase of 2.5 V in the output signal. This should give us four decades worth of dynamic range. For a 10 V signal in, we get a 10 V signal out, for a 1 V signal in, 7.5 volts out, for a 100 mV signal in, 5 V out, for a 10 mV signal in, 2.5 V out, and for a 1 mV signal in, 0 V out. Under ideal conditions, such a log amp should, given the same linear input data, produce the log scale distribution shown in Figure 4-56.

If an 8-bit ADC is used to digitize the log amp output, signals with amplitudes of 1 mV, 10 mV, 100 mV, 1V, and 10 V, which, on a linear scale, would, respectively, be in channels 0, 0, 2 or 3, 25 or 26, and 255, end up in channels 0, 63, 127, 193, and 255. Sounds great. In fact, it is great, except that you may have more than a millivolt of noise running around in the electronics, which would prevent you from getting as many decades of log amplifier dynamic range as you might have paid for (three-and-a-half to four decades are generally provided in flow cytometers). Notice that the data in Figure 4-56 reflect a noise level fluctuating around 3 mV, which is around channel 31 on the 256-channel 4-decade log scale.

Twin Peaks: Distributions on Linear and Log Scales

When you use a linear intensity scale, and you obtain a distribution that has peaks near the low end and near the high end of the scale, the higher peak doesn't look as much like a peak as does the lower one, even when there are approximately the same number of cells in each. When the same data are displayed on a logarithmic scale, it's often easier to perceive the two peaks and to assess the relative sizes of the populations in them.

Figure 4-56 illustrates this point. There are 1,045 events in the peak representing beads with 58,500 antibody binding sites, and 1,180 events in the peak representing beads with 112,200 binding sites. It is not at all easy to guess this from the display on the linear scale; the relationship is more apparent when the data are displayed on a log scale. We can consider each peak in a multimodal distribution such as those in Figure 4-56 as a separate distribution. Suppose we had two peaks centered at channels 100-101 and channels 200-201, with the **coefficient of variation (CV)**, i.e., the standard deviation divided by the mean, 2% for each distribution. Assuming these peaks to be Gaussian, we should find 99% of the events in each within 3 standard deviations of the mean. Looking at the peak at channels 100-101, then, we should find 99% of the events in it between channels 95 and 106, because, if the mean is at channel 100, and the CV is 2%, one standard deviation is two channels. The peak at channels 200-201 is lower, and wider, because, if the mean is at channel 200, one standard deviation is four channels instead of two channels. Thus, to find 99% of the events in this peak, we have to count everything between channels 189 and 212. It's not easy to tell by eye whether there are the same number of events in the lower and the higher peak.

When the data are plotted on a 256-channel linear scale, each channel represents 1/256 of the range of values. When the data are plotted on a 4-decade log scale, each decade occupies 64 channels. Since $\log(a \times b) = \log a + \log b$, we can calculate that the linear value corresponding to channel $(n + 1)$ is $10^{1/64}$ (about 1.0366) times the linear value corresponding to channel n. On a logarithmic scale, distributions with the same CV occupy the same number of channels, regardless of mean channel position.

This property makes log scales particularly well suited, as Brian Mayall has pointed out (not, as far as I know, in print, but in conversations and at meetings), for looking at flow karyotypes, i.e., histograms obtained from flow cytometry of chromosomes. One of the first things you'd like to find out from a flow karyotype is whether there are extra or missing chromosomes. Counting chromosomes is not that hard to do when you look at slides, because all the chromosomes from a particular metaphase stay close to the same place. However, when you do a flow karyotype, you mix up the chromosomes from a few thousand metaphases, stain them with a DNA fluorochrome, and run them through the system to generate a histogram, leaving you with no way of keeping track of how many or which chromosomes came from each cell.

In a real flow karyotype, in which there are a couple of dozen peaks, it is difficult to compare them when they are displayed on a linear scale. On a log scale, the standard deviation ends up being the same number of channels wherever the mean may be, which means that peaks containing the same number of chromosomes should be the same height as well as the same width. This enables you to spot a trisomy or the lack of a chromosome (assuming it occurs in a substantial fraction of the cell population) at a glance.

The real world isn't always so kind. Looking back at the log scale display in Figure 4-56, we notice that the peak representing beads with 19,500 binding sites, which contains 1,002 events, is considerably shorter and wider than that representing beads with 58,500 sites, which contains 1,045 events; the peak representing beads with 7,800 binding sites, which contains 912 events, is shorter and wider still. If peaks with the same CVs and roughly the same numbers of events should be roughly the same height and width when displayed on a log scale, what's the problem?

Simple. The peaks don't have the same heights and widths because they don't have the same CVs. When you're measuring signals varying in intensity over a range of decades, it's almost a sure bet that the weaker signals will be associated with larger CVs. In the case of weak fluorescence signals, photon statistics may be involved, but noise also has its effects. If there are 3 mV of noise on a 10 V signal, or even a 1 V signal, the noise accounts for less than 1% of the signal value. The same 3 mV of noise on a 100 mV signal represents 3% of the signal value, while, on a 10 mV signal, 3 mV noise represents 30% of the signal value.

If you can measure the brightest and dimmest chromosomes with roughly the same CV (and you need a CV around 1% to be able to resolve the different peaks), Brian's suggestion of displaying data on a log scale to spot aneuploidy works. However, even when your CV varies, as in immunofluorescence measurements, log scales make peaks easier to recognize on displays. Unfortunately, you can't always assume that your log amp is generating an accurate log scale.

Falling Off a Log: Log Amps Behaving Badly

What are log amps made of? In the classical log amp, the central component is an op amp, with an element in the feedback circuit generating the log response characteristic. As it happens, the voltage drop across the base-emitter junction of a transistor is proportional to the log of the collector current; at 20 °C, the voltage changes by about 60 mV for each tenfold change in current. The simplest log amp is an op amp with a transistor and a few other components in its feedback loop; its output ranges between -100 and -400 mV.

As you may have guessed, the voltage change across the transistor junction used to generate logarithmic response varies with factors other than current, notably ambient temperature. In order to compensate for this, many log amps incorporate another op amp circuit, with another transistor; the two transistors are typically a matched pair made on a single substrate, and, sharing the same environment, respond identically to changes in ambient temperature. Some of the fancier log amp modules have built-in temperature regulation. With all of that, it is still difficult to get a log amp module incorporating a single op amp-based log circuit to respond accurately over a range of more than 3 decades. For this reason, several manufacturers make log amp modules which incorporate four or five stages of amplification; each stage is optimized for response in a restricted input range, and the sum of the outputs produces a reasonably accurate logarithmic response.

Whether a single-stage or multistage log amp module is used as the nucleus of a log amp, it is generally necessary to amplify the output to get signals into the 0-10 V range prior to digitization, and often necessary to condition the signal by adding or subtracting an offset voltage and/or putting a baseline restorer into the op amp design. Because the logarithm of zero is undefined (or "minus infinity"), even those log amp modules that respond to input signals that are more than a few mV on either side of ground act unpredictably at really low input levels. Other modules need added circuitry to restrict the input range to one polarity. Most log amp circuits incorporate several adjustments for trim, offset, and gain, and require some time on the bench with a skilled technician before they work properly; they don't always stay adjusted.

In 1983, Muirhead, Schmitt, and Muirhead[1156] described a procedure for measuring the response of log amps and converting data to linear values. Either test pulses from a light-emitting diode (LED) or fluorescence signals from calibration beads could be used as fluorescence standards. The log amp studied was a component of a commercial flow cytometer; while it was reasonably accurate at some points in its operating range, deviations of 10% or more from true logarithmic response were noted in other regions. Schmid, Schmid, and Giorgi[1157] reported similar results in 1988, using a calibration procedure in which channel differences between multiple peaks were plotted as PMT gain was varied to place the upper peak at various points in a 256-channel range. They found the characteristic response patterns of individual log amps to be relatively stable over time, allowing the calibration curve to be used to make corrections when transforming logarithmic to linear data.

Figure 4-57 illustrates typical response curves of different types of log amps as determined using a procedure described in 1988 by Parks, Bigos, and Moore[1158]. Measurements are made of bright fluorescent beads; the ratio of fluorescence intensities is determined on a linear scale. A ratio between 1.5 and 2 is desirable. Then, a series of fluorescence measurements are made with the log amp, with an acquisition gate for single beads set using scatter parameters. The gain is initially adjusted to place the median channel of the peak representing the brighter beads at or near the top channel of the ADC (255 in the illustration), and the number of channels' difference between the medians of the bright and dim peaks is recorded. The PMT gain is then lowered to move the median channel of the upper (bright) peak down 5-10 channels, and the number of channels between the peaks is again recorded. The procedure is continued, moving the peaks to lower channels. If PMT voltage must be lowered below 300 V to place the peaks at the lower end of the range, a neutral density filter should, if possible, be placed in front of the PMT, because PMT response may itself be nonlinear at low voltages. The response curve is a plot of median channel of the upper peak, along the horizontal axis, vs. the number of channels' difference between the medians of the bright and dim peaks, along the vertical axis. Since the ratio of intensities of the beads is constant, the difference between the medians of the two peaks, which is proportional to the logarithm of this ratio, should also be constant (log bright = log dim + log ratio). The ideal response curve, indicated by dotted lines in Figure 4-57, is a line parallel to the horizontal axis.

The first real amplifier response curve, at the top, is characteristic of the log amps built in the Herzenberg lab at Stanford[1159]; these incorporate a temperature-stabilized single stage module (Model 2531A, Optical Electronics, Inc., Tucson, AZ). I have also used this module in Cytomutt log amps, and have observed similar response. There is typically a substantial deviation from the ideal at high input signal values, then a response reasonably close to ideal over about a3-decade range, followed by an abrupt turn up or down at the low end as the noise catches up with the input signal. The amplifier is usable over a 4-decade range.

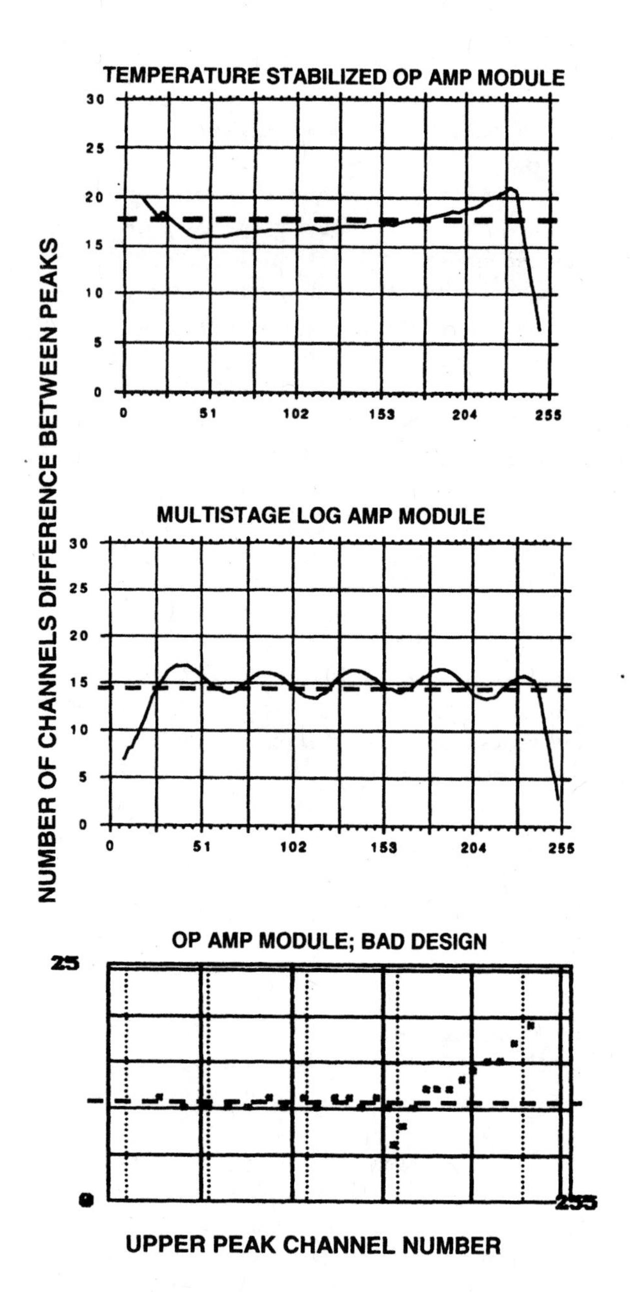

Figure 4-57. Response curves of different types of log amps as determined according to Parks, Bigos, and Moore[1158]. The horizontal dotted lines show ideal response.

The middle curve is characteristic of log amps built with multistage modules such as the Texas Instruments (Dallas, TX) TL441, used in some Becton-Dickinson instruments, or the Analog Devices (Norwood, MA) AD640 and its relatives, used in a design described by De Grooth et al[1160] in 1991. The bumps reflect the response characteristics of the individual stages in the module; there are relatively large deviations at high and low input signal values, with the latter due primarily to noise. Like the single-stage module above, the multistage amplifier illustrated is usable over a range of almost 4 decades.

The curve shown in the bottom panel, generated in my lab, illustrates some highly undesirable characteristics of a log amp, which were confirmed by Dave Parks and Marty Bigos when I sent them one of my early design efforts to test. The curve deviates wildly from ideal at the high end, and includes a relatively sharp notch in the third decade. This log amp would occasionally yield distributions with crimps in them; when it didn't, however, it seemed to be pretty good, because it was accurate at the low end and could handle an input range that appeared to be larger than 4 decades. However, it would have been an extremely bad idea to try to do log-to-linear conversions with this beast; it wasn't even stable over time. Bad log amps have also, unfortunately, turned up in some commercial instruments.

The take-home message here is: calibrate your log amps, even if you're not trying to do quantitative immunofluorescence, just so you know. And hope that the next flow cytometer you buy will have replaced the #@%$&^% log amps with trustworthy linear circuits and high-resolution ADCs. It will make flow cytometry in the rest of this century much easier.

Limits to Dynamic Range

If we had a perfect 5-decade log amp, it might not help much in extending the dynamic range of flow cytometric fluorescence measurements, for reasons best explained with the aid of Table 4-4. In previous discussions of PMTs and preamps, we established that we usually don't want to get more than 100 μA out of the PMT, and that the maximum signal we can put into the log amp is 10 V. Suppose our baseline-restored preamp gets 10 V out for 100 μA in, and drives a 5-decade log amp with an output slope of 2V/decade, and we measure 10 μs fluorescence pulses at 520 nm, with an R928 PMT. The tube's cathode quantum efficiency at 520 nm is 14%; the cathode current is 59 mA/W. We assume a PMT gain of 10,000.

The anode current of 100 μA yielding a 10 V preamp output at the top of the highest (fifth) decade results from amplification of a cathode current of 10 nA by the PMT gain of 10,000. Since a 1 A current is about 6.2×10^{18} electrons/s, 10 nA in a 10 μs pulse represents 620,000 photoelectrons. Dividing by the quantum efficiency, 0.14, we see that 4,430,000 photons reach the cathode during the pulse. The contribution to measurement CV from Poisson photoelectron statistics is $[(620{,}000)^{1/2}/620{,}000]$, or 0.13%, based on the number of photoelectrons emitted from the cathode. We use the number of photoelectrons because they are what we are actually counting; we are only inferring that the photons were there. If we used a 16-bit (65,536 channel) linear ADC, which has a difference between channels of about out 153 μV, the 10V signal would be in the highest channel, i.e., channel 65,535.

The top of the next (fourth) decade is a 1 V signal, corresponding to channel 6,553; the contribution to CV, from 62,000 electrons, would be 0.4%. The top of the third decade, at 100 mV, or channel 655, corresponds to 6,200 photoelectrons; the contribution to CV is 1.3%. The top of the fourth decade, at 10 mV, or channel 66, represents 620 elec-

DECADE	LOG AMP OUTPUT	LOG AMP INPUT	ANODE CURRENT	CATHODE CURRENT	PHOTONS/ 10 μs	ELECTRONS/ 10 μs	CV (%)	CHANNEL (16 BIT ADC)
5	10 V	10 V	100 μA	10 nA	4,430,000	620,000	0.13	65,535
4	8 V	1 V	10 μA	1 nA	443,000	62,000	0.4	6,554
3	6 V	100 mV	1 μA	100 pA	44,300	6,200	1.3	656
2	4 V	10 mV	100 nA	10 pA	4,430	620	4.0	66
1	2 V	1 mV	10 nA	1 pA	443	62	12.9	7
	0 V	100 nV	1 nA	100 fA	44	6	40.8	0

Table 4-4. Logarithmic amplifiers: What goes in, what comes out.

trons; the contribution to CV increases to 4%. At the bottom of the fourth decade, 1 mV, or channel 7, there are 62 electrons in the pulse, with the contribution to CV 12.9%, or almost one channel width (1/7 ≈ 14%). The bottom of the fifth decade, a pulse from 6 photoelectrons, ends up at channel 0, the contribution to CV would be 40.8%.

These are actually best case calculations. The numbers of electrons are calculated for 10 μs rectangular pulses; real pulses from a flow cytometer are apt to be shorter in duration and Gaussian in shape, and would probably represent something less than half the numbers of photons and photoelectrons used in the table. The PMT gain of 10,000 used in the calculations is conservative; it is typically attained with PMT voltages of 400-500 V. However, if the calculations are repeated assuming higher PMT gains, we end up with even fewer photons at the cathode and fewer photoelectrons emitted, so things only get worse. If the gain is increased to 100,000, the lowest decade gives us only 6 photoelectrons in 10 pulses.

In my experience, displays of flow cytometric immunofluorescence data on 4-decade log scales that place the brightest stained cells near the top of the top decade almost always have the unstained cells occupying at least half of the bottom decade, while, in displays that place the unstained cells near the bottom of the bottom decade, the brightest stained cells fall below the midpoint of the top decade. This suggests that the real dynamic range of most immunofluorescence data is closer to 3 1/2 than to 4 decades. We have established that 3 1/2 decades of dynamic range are fairly easy to get; with a 3 mV noise level, we can't do any better than that. If we reduce the noise, another 1/2 decade is probably within reach, especially if we settle for lower accuracy. Five-decade response from a single detector approaches the realm of fantasy.

We end up at pretty much the same point if we stop thinking about photons and electrons for a bit, and think instead about numbers of antibody molecules detectable on cell surfaces. If the high end of the top decade represents 1,000,000 molecules, the low end of the fourth decade is 101 molecules, while the fifth decade encompasses the range between 11 and 100 molecules. Calculations of the minimum number of fluorescent molecules detectable by conventional flow cytometers generally end up with figures above 100 molecules. Such calculations, which will be discussed in detail later on, are typically done using data collected with log amps, and may in themselves be suspect. I have produced differences of several hundred molecules in calculated detection limits, simply by interchanging log amps with slightly different DC offset levels, and therefore believe that some of the lowest reported threshold values may well be due to artifacts. This issue should be resolved within the next few years, but the only practical way to resolve it will be to determine sensitivity using linear amplification and high-resolution analog-to-digital conversion instead of log amps, to which good riddance.

None of the above should be taken as preventing you from making flow cytometric measurements over a dynamic range larger than 4 decades; you just can't do it with one detector and keep everything on scale. If you use one detector, at low gain, for the bright stuff, and another, at high gain, for the dim stuff, you can easily cover almost 8 decades. Marine biologists probably analyze samples encompassing the largest dynamic range; with five detectors, they could go from marine bacteria (10^{-12} g) to whales (10^{8} g). Even if the largest objects in the sample are only krill, the krill (and Shapiro's First Law), rather than detector dynamic range limitations, are the main obstacles to success.

Ratio Circuits

Ratio circuits are analog circuits that produce an output proportional to the ratio of two input signals. They are usually made from modules called **analog multipliers**, which can also be connected to produce signals proportional to the product, square, or square root of input signals. One application of ratio circuits has been in the derivation of signals proportional to **surface density of antigenic or receptor sites**, which can be calculated by dividing the number of bound ligand molecules by the cell surface area. An accurate approximation to cell surface area can be obtained from digital computation of the 2/3 power of volume[311]. Use of this technique is obviously feasible only with flow cytometers capable of making volume measurements. However, since both cell surface area and the "cell size" measurement obtained from small angle light scattering are proportional to the square of cell diameter, the ratio of immunofluorescence and scatter signals can serve acceptably as a parameter representative of antigen surface density for making comparisons[657,658].

Compensation for light source intensity fluctuations using a ratio circuit was discussed on pp.147-8 and illustrated in Figure 4-24. If the output of the light source increases or decreases, fluorescence and scatter signal amplitudes from identical particles will increase or decrease proportionally. If the source intensity is continuously monitored by a dedicated detector, the signal produced by dividing the raw fluorescence and scatter signals (i.e., the signals before baseline restoration has been applied) by the signal from this detector will be free of variations due to source fluctuations. Steen[102] has published a schematic for a source noise compensation circuit.

Ratio circuits are also useful for measurements of probes such as BCECF and indo-1, which change spectral characteristics with pH, cytoplasmic [Ca^{++}], etc. In some cases, a single excitation wavelength is used and the ratio of emission intensities at two wavelengths is computed; in others, two excitation beams are used and the ratio of emission intensities at one wavelength is computed. Ratiometric measurements have already been alluded to on p. 47; they will be discussed further in Chapters 7 (Parameters and Probes) and 10 (Using Flow Cytometers).

4.9 DIGITAL SIGNAL PROCESSING

The analog signal processing techniques we have just discussed leave the information from cells in the form of **stretched pulses,** i.e., slowly varying (constant in the absence of droop) voltage signals, stored in the short-term analog memories of peak detectors, integrators, and/or pulse width measurement circuits. In some cases, additional electronic circuitry may have been used to form linear combinations of signals, as in fluorescence compensation, to convert signals from a linear to a logarithmic scale, or to derive a signal representing the ratio of two other signals. The front end electronics, which include some rudimentary digital circuits, generate a digital logic signal, the strobe pulse, to inform a conventional data analysis system that all the analog voltage signals from a cell (or, more accurately, an event) are ready to be **digitized**, i.e., converted into numbers that can be further manipulated by the digital computer used for data analysis. We have already had a first look at digitization, and at the **analog-to-digital (A-D or A-to-D) converter (ADC)**, which is the hardware responsible for the task, in Chapter 1.

Most commercial and laboratory-built flow cytometers still use analog and hybrid circuits for pulse processing, but, over the years, more and more builders and manufacturers of instruments have introduced digital processing at progressively earlier stages. The Coulter EPICS XL[1183] and some of my newer Cytomutts[2466], for example, dispense with log amps, and use high-fidelity integrators or peak detectors and high-resolution ADCs for conversion between linear and logarithmic scales. This allows fluorescence compensation and ratio calculations to be done in software, eliminating the need for analog compensation and ratio circuits.

Analog-to-Digital Conversion

An analog-to-digital converter, as the name implies, has an **analog** input, usually a voltage in the range 0-10 V, 0-10.24 V, or something like that, and a **digital** output, which is a binary number with 0's and 1's represented by different logic voltage levels. For this and future discussions of logic circuits, I will stick to the ground-for-0, 5 V-for-1 "positive logic" convention which roughly represents the voltage levels encountered in older TTL and CMOS digital circuits. Just so you know, these days, microprocessor chips and their associated logic run on lower DC voltages, from below 2 V to 3.3 V, and even analog circuits tend to run on ± 5 V supplies, with ADC input voltage ranges typically ± 1 or 2 V. It doesn't really matter; the whole point of A-to-D conversion is to leave voltages and currents behind and convert the data to numbers.

Analog signal processing is pretty much instantaneous, and definitely continuous. The output of a current-to-voltage converter circuit, or an analog log amp, changes as the input changes. Some digital circuits also change outputs whenever their inputs change. By contrast, analog-to-digital converters, like digital computers, do not, by and large, change their outputs "instantaneously"; they take a finite time to do what they do, and the process, more often than not, involves a sequence of steps.

The rate at which the devices execute that sequence of steps is controlled by an internal or external **clock** signal. Clock signals were mentioned on p. 195, in the discussion of digital approaches to pulse width measurement; in the world of microprocessors and digital logic, clocks, overclocking, and clock envy are ubiquitous. You have undoubtedly noticed that AMD and Intel regularly compete for the fastest clock speed, with the current loser always claiming its processor is faster even if its clock is slower. And the fastest ADCs are no slouches in the clock department; some convert data at rates of several GHz. The ADCs we use in flow cytometry, even for digital pulse processing, are quite a bit slower, at least for now, but, as will soon become obvious, speed isn't everything.

Free Samples? Hold it!

If an ADC takes a finite time to operate, you might well ask how we can feed it a continuous input signal. The answer is that we can't. Instead, we take a series of **samples** of the input signal. In what now seem like ancient times, this required the use of a completely separate circuit called a **sample-and-hold**. A sample-and-hold operates in a fashion similar to a peak detector (pp. 192-4; Figures 4-51 and 4-52). It is controlled by "sample" and "hold" logic signals; the "sample" signal allows the input signal to charge a capacitor (called a "hold" capacitor), and the "hold" signal disconnects the input signal from the capacitor. The output of the circuit is the capacitor voltage, buffered by a high-impedance op amp.

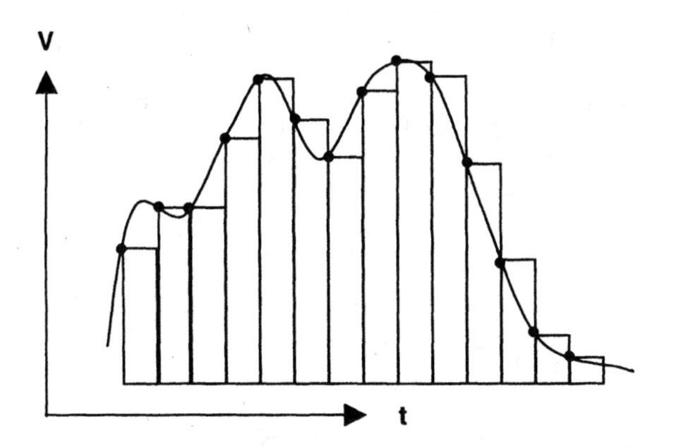

Figure 4-58. Continuous and sampled signals. The filled circles represent the sample points.

Figure 4-58 illustrates the effect of sampling an arbitrary voltage waveform. The dark filled circles represent sample points, taken at regular intervals; each sample point is an accurate representation of the value of the waveform at the point at which the sample was taken. If we used a sample-and-hold in the sampling process, the output of the sample and hold would, in the absence of droop, remain at the value of the last sample and then jump rapidly to the value of the next sample. Rapidly, not instantaneously, because it takes time for the hold capacitor to charge or discharge to the new sample voltage. But can we "trust" the series of samples to faithfully represent the waveform?

The short answer is yes, provided we take samples at short enough intervals. Using **Fourier analysis**, it is possible to represent any arbitrary waveform as the sum of sine and/or cosine waves of discrete frequencies. According to the **Nyquist sampling theorem**, it is possible to completely reconstruct a waveform from samples taken at twice the frequency of the highest frequency component of the waveform. If we look at the waveform in Figure 4-58, we see some sine wave-like "ups and downs," but the sample points appear to be taken frequently enough to reproduce those fluctuations. If the waveform had changed more rapidly than the one shown in the figure, or if we had taken samples less frequently, that might not have been the case; imagine what the "connect the dots" line, or the "staircase" sample and hold output, would look like if we only used every other sample point, or every third sample point.

The process of reconstructing a waveform from samples also typically involves Fourier analysis; it generates coefficients for sine and/or cosine waveforms at the various frequencies that make up the original waveform. If we attempt to reconstruct a waveform from samples taken at a rate below the **Nyquist frequency**, which is twice the frequency of the highest component of the waveform, the computer we're using won't suddenly emit sparks and smoke; instead, it will give us an initially reasonable-looking set of coefficients that gives substantial weight to frequencies that weren't actually of significance in the original waveform. This is described as **aliasing**, or **sample aliasing**, and, since it leads to totally incorrect results, should be avoided. But how?

The necessary fix is an **antialiasing filter**, a low pass (electronic, not optical) filter that is inserted into the signal path somewhere ahead of the sample-and-hold. That sounds easy enough, and, because engineers have devoted a great deal of time and effort over the years to the design of filters with appropriately smooth responses in their passbands and appropriately sharp high frequency cutoff rates, we can take it for granted that a suitable filter will be available. In fact, the filter, and the sample-and-hold, may well be built into the same module as the analog-to-digital converter itself.

The bottom line on sampling, then, is that samples are "free"; as long as we take enough of them, the discrete samples will contain all the information that was found in the original continuous signal. However, we will have to pay the piper when we actually do our analog-to-digital conversion.

Quantization: When Are Two Bits Worth a Nickel?

Since we now live in a digital world, we are likely to have opinions on how many bits we need for this or that. The 16-bit digital audio recording that seemed so awesome in the early days of compact discs is now looked at with disdain; even music hobbyists can now afford computer-based hardware with 24-bit ADCs. In deciding how many bits we need for acquisition and analysis of data from cytometry, we have to consider what we want to achieve in terms of precision and dynamic range, and determine how each of these will be affected by **quantization error**, which is an essential, necessary, unavoidable characteristic of the analog-to-digital conversion process.

The ADCs used to date for flow cytometry generally convert to somewhere between 8 and 20 bits' resolution. An n-bit converter has 2^n channels, and we typically consider its output in terms of channel numbers between 0 and $(2^n - 1)$. The corresponding channel numbers, and the voltages represented by the **least significant bit (LSB)**, i.e., the voltage difference between channel n and channel $(n + 1)$, assuming 10 V full scale, are shown below in Table 4-5.

Number of Bits	# of Channels	Voltage/LSB
8	256	39.1 mV
10	1,024	9.77 mV
12	4,096	2.44 mV
14	16,384	610 µV
16	65,536	153 µV
18	262,144	38.1 µV
20	1,048,576	9.54 µV
22	4,194,304	2.38 µV
24	16,777,216	596 nV

Table 4-5. Characteristics of analog-to-digital converters.

For an 8-bit converter, the lowest voltage that can make the output of an ADC register as channel 1 instead of channel 0 is just over 39.1 mV. Any signal level between 0 V (ground) and 39.1 mV at the input would produce a digital output of 0, and any signal greater than 39.1 mV and less than or equal to 78.2 mV would produce a digital output of 1. At the top end, a signal level greater than 9.961 V would produce a digital output of 255.

By convention, the voltage value representing a given channel is taken as the middle of the range of voltages that would be converted to that channel value. In the case of the 8-bit, 10 V full scale ADC, the voltage values would be 19.55 mV for channel 0, 58.65 mV for channel 1, 9.98045 V for channel 255, etc. For any given channel, the **quantization error** is constant in magnitude at one-half the channel width, or ± (1/2) LSB. In effect, digitizing the signal has the same effect as adding a small amount of **quantization noise**, with the noise having a mean of 0 and a uniform, or rectangular, distribution between - (1/2) LSB and + (1/2) LSB. The standard deviation of the noise is 0.29 LSB; this emerges from the properties of rectangular distributions, rather than from my hat. So, while sampling a signal does not degrade it, digitizing a signal does.

While its magnitude is constant, the relative importance of quantization error is quite different at opposite ends of the measurement scale. At the top end of the range, channel 255, the error is small; for the 8-bit converter, the peak-to-peak range of quantization noise is only 0.391% of full scale. However, at the bottom end of the range, the error is large; for channel 0, the peak-to-peak range of quantization noise represents 100% of the channel value for an 8-bit converter - or, for that matter, for a converter with any number of bits, since the channel value for channel 0 is always (1/2) LSB.

Suppose we are measuring DNA content in a tumor specimen, and want to be able to clearly resolve peaks from tumor and stromal cells differing in G_0/G_1 DNA content by 1% or less. If we use a linear scale, we need to place the G_0/G_1 peak of a DNA content distribution at a value less than one-half of full scale in order to get both the G_0/G_1 and G_2/M peaks on scale. If we use an 8-bit converter, and the G_0/G_1 peak of the stromal cells appears at channel 100, the corresponding voltage is 3.91 V. The peak-to-peak quantization noise at channel 100 is 1% of the signal value; the standard deviation of the noise is 0.29 percent of the signal value, and, if there are any other contributions to the variance of the measurement, other than quantization noise, we will probably not be able to resolve the tumor and stromal cell G_0/G_1 peaks.

As I have pointed out elsewhere[8], DNA is the only substance present in cells in which cell-to-cell variations in content are so small that measurement to a precision of 1 percent is warranted. Distributions of cellular DNA content represent most of the few "needles" among the many "haystacks" found in biology. Sure, if you run plastic beads through the instrument, you can get scatter distributions with CVs less than 1 percent, but you don't need anything like that kind of precision to measure light scattering from biological particles, which give distributions with CVs no less than 7-10%. In fact, the major contributors to variance in measurements of almost all cellular parameters are biology and photoelectron statistics.

The first data acquisition systems for flow cytometers dealt only with data reduced to stretched pulses representing peak heights, integrals, etc., and used 8-bit ADCs quite successfully. Most instruments weren't measuring DNA content with high (CVs of 1% or less) precision. Measurements of low intensity fluorescence, which had high CVs due to both photoelectron statistics and the biological characteristics of samples, were made using log amps, which placed signal levels high enough up on the 256-channel scale so that quantization error was not a concern. Almost all of the more recent instruments use 10-bit ADCs, which are more than adequate for high-precision DNA content measurements on a linear scale.

We run into problems, however, when we try to take logarithmically transformed data digitized with a 10-bit converter and interconvert between log values and linear values. This becomes necessary when we want to revise compensation settings after the fact using software. Plots of data that have been thus manipulated tend to have a "grainy" quality, although some software packages smooth out the transformed data by **dithering**, i.e., adding small amounts of additional noise in the form of random numbers. This subterfuge inevitably further degrades data, although usually not significantly, and intuition suggests that we could eliminate the need for it by using higher resolution ADCs. But how many bits do we need? Would 16 bits be enough?

A look back at Figure 4-56 (p. 199) tells us "almost, but not quite." The figure shows fluorescence distributions of beads bearing different, known amounts of fluorescent antibodies; data are displayed on a 256-channel (8-bit) linear scale in the top panel and on a 256-channel log scale in the bottom panel. However, the raw data were taken from one of my "Cytomutts"[2467], using a 16-bit (65,536-channel) ADC; conversion to a 256-channel log scale was done by a digital computer using a look-up table.

We note that the peak representing the blank beads in the log scale display shown in the bottom panel of Figure 4-56 looks different from the other peaks in the histogram; the lower half of the peak has what has lately come to be known as a "**picket fence**" appearance, with no events appearing in roughly every other channel. This tells us that we will "run out of bits" when we try to fit 16-bit data onto an 8-bit, 4-decade logarithmic scale.

The look-up table we use for the logarithmic conversion is a table of 65,535 bytes, with "addresses" 0-65,535. The byte at address *n* contains an 8-bit number between 0 and 255 representing the log value corresponding to the linear value *n*. When we compare data displayed on linear and log scales, which we can do with another backward glance at Figure 4-56, we see that the log scale makes peaks at the high end of the scale narrower and peaks at the low end of the scale wider than they appear on a linear scale. It is therefore not a surprise to find, when we calculate the numbers in the look-up table, that, at the high end, one value on the logarithmic scale represents a large range of linear values.

In fact, 2,316 linear values, between 63,220 and 65,535, all map to log value 255. And 2,234 values, between 60,986 and 63,219, map to log value 254. As we move toward the lower end of the scale, fewer and fewer linear values map to a given log value. By the time we get down to channel 22 linear (channel 34 log), there is only a single linear value for each log value. And, for linear values lower than 22, we actually have a choice of two or more log values; since there can only be one value at any address in a look-up table, it is necessary to select from the alternatives.

That is what I had to do when I worked out the look-up table used to generate the log values shown in Figure 4-56. For linear values of 18, where the choice of log values was 28 or 27, I chose 28, so there were no points plotted for channel 27. For linear values of 17, where the choice was 26 or 25, I chose 26; there were no points plotted for Channel 25. That's how the "picket fence" got built; I can whitewash it somewhat by pointing out that my instrument noise level was right around channel 21 linear, anyway, and there wasn't any point in trying to make data at that level and below look better. However, I would have been able to eliminate the picket fence effect using higher resolution ADCs. The problem I had was that, when I built my 16-bit data acquisition system, there pretty much weren't any higher resolution ADCs fast enough to handle the data at the rate at which they were generated.

When Bob Auer was designing the electronics for the Coulter EPICS XL[1183], in which the goal was to eliminate log amps and allow data to be displayed on a 10-bit (1,024-channel) log scale, he calculated that he would need 20-bit linear data to obtain a display without a picket fence. And he faced the same problem I did; there weren't any usable 20-bit ADCs. So he used a "divide-and-conquer" approach usually known in technical jargon as **subranging**.

The XL splits held signals from each of its channels; one is fed directly into a 15-bit converter, and the other is amplified, using analog circuitry, by a factor of 32, and then fed into a 15-bit converter. For the smallest signals, the unamplified signal converts to a digital value of zero; the digital value used for data analysis is the value obtained from the amplified signal. For the large signals, the digital value obtained from the unamplified signal is shifted left by five bits (equivalent to multiplication by 32), and this 20-bit value is used for data analysis. The large signals have 20-bit values, in which the lower 5 bits are all zeros, but they still have 15-bit precision; the 15-bit data from the small signals are treated as 20-bit values in which the upper 5 bits are all zeros. For the signals on the borderline, a switch controlled by a comparator is used to determine whether the digital value used will be from the amplified or the unamplified signal, and a small amount of noise is fed into the comparator to produce slight changes in the position of the crossover point; this avoids the generation of "spikes" or "notches" in histograms, which would otherwise be likely to occur at the crossover point because of slight differences in the gain of the analog and digital signal amplification. Most important, there are no picket fences to be seen. The design works.

Both my quick-and-dirty 16-bit data analysis system and the more elegant 20-bit data analysis system in the EPICS XL eliminate the need to use log amps, and allow fluorescence compensation and other procedures involving log-to-linear and linear-to-log conversion to be done by software without making the data look ratty. However, they still operate on held analog signals, and that poses a problem. If we want a 4-decade dynamic range, we have to be able to get accurate integrals or peak heights of pulses ranging in amplitude from 1 mV to 10 V. It is not at all easy to make peak detectors and/or integrators respond over such a large dynamic range. Designing the integrators for the EPICS XL to meet this specification was (R. Auer, personal communication) harder than designing the data acquisition and analysis system; I gave up on trying to design peak detectors with the necessary performance for my "Cytomutt" and, instead, bought them (ironically enough, from the same company from which I used to buy log amp modules). And the peak detectors only work in my 16-bit systems when they are performing well beyond the manufacturer's spec; the last few I bought are up to spec but not good enough to use.

The obvious solution to the problem would be to eliminate peak detectors, integrators, etc., in favor of **digital pulse processing**. It sounds easy enough; after all, that's the way we process all of our audio signals these days. In order to understand why digital pulse processing has taken as long as it has to find its way into commercial flow cytometers, we have to take a closer look at the innards of ADCs.

Analog-to-Digital Converters (ADCs) (and Digital-to-Analog Converters (DACs))

In the pulse height analyzer era, raising the issue of using general purpose computers for flow cytometric data analysis was likely to get the hardware freaks upset about a characteristic of ADCs known as **differential nonlinearity**. There are several different kinds of ADCs; you can see Horowitz and Hill[1056], newer books on digital audio[2467-8], and the websites of manufacturers such as Analog Devices (www.analog.com) for more thorough descriptions than I am about to give). Some types of ADCs have more differential nonlinearity than others.

Pulse height analyzers incorporated what are known as **Wilkinson** converters, which are a type of **dual slope** ADC. The dual slope ADC was conceived as an improvement on – what else? – the **single slope** ADC. A single slope ADC compares its input voltage with a **reference voltage** produced by a circuit called a **linear ramp generator** (see p. 195); this voltage starts at the low end of the range (say 0 V, or ground, or even a little bit below ground), and increases linearly with time. When a single slope ADC is used for flow cytometry, the **strobe pulse** from the flow cytometer (pp. 191-3) is used to generate a **start convert** signal, a logic pulse which resets the ramp generator, and also resets a **binary counter**, which is a digital circuit capable of counting up to, and storing, a value of (2^n - 1), where n is the number of bits to which the ADC converts. The reset leaves a value of zero in the counter.

When the ramp voltage goes above zero, a comparator turns on, or **enables**, the counter, which starts counting pulses from a **clock** circuit, i.e., an oscillator operating at a precisely controlled frequency, which puts out logic pulses at regular intervals. When the ramp voltage reaches the input voltage, another comparator stops the counter, which holds the count it reached. Since the ramp voltage is linear with time, and the clock operates at a constant frequency, the number stored in the counter at this point is proportional to the input voltage. The Wilkinson and other double-slope ADCs work by charging a capacitor to the input voltage, and then discharging it with a linear current drain, counting the time it takes for the voltage to reach zero, but it's the same basic principle.

The absolute accuracy of single- and dual-slope ADCs is not all that good, but, because of the linear characteristics of the reference voltage generator or current drain, they have very good **differential linearity**, that is, the voltage differences between any two adjacent channels are very nearly equal. These ADCs thus produce relatively smooth histograms. On the minus side, single-slope ADCs also incorporate a lot of parts, which tends to make them expensive, and they're fairly slow, with speed determined by the clock frequency. The **conversion time**, i.e., the time required to generate valid digital output data following a start convert pulse, is also a function of the input voltage. For example, if the ADC uses a 10 MHz clock, and the input signal corresponds to an output value of 127, the conversion takes some small setup time – a microsecond or so – plus 127 clock ticks, or 12.7 µs. If the digital signal value is 10, everything's done in the setup time plus 1 µsec. If it's 255, the conversion time is the setup time plus 25.5 µs, and so on. Even with a 100 MHz clock, converting a value of 1023 takes 10.23 µs plus the setup time.

Most of the ADCs available these days are faster, cheaper, and less complex than Wilkinson converters. Many of these are the so-called **successive approximation** type; they generate comparison voltages using a **digital-to-analog converter (DAC)**, which takes a digital input and generates an analog (voltage or current) output. A classical DAC incorporates a **voltage divider** made up of a string of resistors in series, so that points between any two resistors will be at 1/2, 1/4, 1/8, and successive fractions of the output voltage. This divider is the primary source of differential nonlinearity in a successive approximation ADC.

The reason for this is that, in order to get the voltage outputs needed from the divider, the resistors must have values R, R/2, R/4, etc. Resistors are real-world components. Most of the resistors used in electronic circuits are only specified to be within 5% or 10% of their stated resistance value; for slightly more money, you can get resistors made to a tolerance of 1%, and for substantially more than that, parts made to a tolerance of .01% or better. Let's suppose the smallest resistor in the voltage divider of a successive approximation ADC, i.e., the resistor in the "1's" bit, has a precision of 1%. In order to maintain the same voltage accuracy (not precision) in the resistor in the "128's" bit, the resistor has to be precise to better than .01%. That was a tough order when converter modules were fabricated from individual components. In the modern era of large scale integrated circuits, resistors in the DAC incorporated into a typical successive approximation ADC are "trimmed" to the necessary precision during fabrication of the circuit, usually by laser ablation of resistive material. The quality of the trim determines the differential linearity of the converter. In the interest of full disclosure, I will mention that DACs can now use calibrated current sources instead of calibrated resistors, and that many use combinations of the two; the state of the art of circuit fabrication allows differential linearity to be well controlled either way.

The differential linearity spec quoted for most successive approximation ADCs is 1/2 LSB, which means that, in an 8-bit ADC, the "widths" of channels, nominally 39.1mV, could vary between about 20 mV and about 60 mV. The differential nonlinearities in an 8-bit single- or dual-slope ADC arise from the slight deviations from true linear response in the ramp generator due to causes such as temperature fluctuations, and are very small; the channel widths generally vary by less than 1 mV. If we tried to generate histograms from the same data using an 8-bit Wilkinson ADC and an 8-bit successive approximation ADC, the latter would be apt to produce a pretty ragged looking display, because what would come out of a Wilkinson ADC as, say,

60 cells in channel 41 and 60 cells in channel 42 could come out of a successive approximation ADC as 90 cells in 41 and 30 in 42, or 30 and 90, or anywhere in between. Not what you'd like to send in to the journals.

As Tom Sharpless pointed out in the original Big Yellow Flow Cytometry Book[128], there's a quick fix for this problem. What we do is get a 12-bit successive approximation ADC, with output range 0-4095, and 2.5 mV between channels. This will be more expensive than an 8-bit ADC, in part because it has to have a much better trimmed resistive divider. If we use the upper (most significant) 8 bits' output of this converter, we can reduce the apparent differential nonlinearity. The nonlinearity will still be 1/2 LSB, but we are now taking the converter's 16's bit as our 1's bit, and thus, from our point of view, differential nonlinearity is reduced by a factor of 16.

Without doing statistical analyses on the data, you can't tell the difference between data taken with an 8-bit Wilkinson ADC and data taken to 8 bits' resolution using a 12-bit successive approximation ADC, and it's hard to tell the difference between data from the 8-bit Wilkinson ADC and data taken to 8 bits' resolution using a 10-bit successive approximation ADC. Successive approximation converters, which have a fixed conversion time, are also considerably faster than single- and dual-slope ADCs, another point in favor of the successive approximation devices.

When ADCs were more expensive, people would **multiplex** them, switching several different analog signals in succession to the input of a single ADC, and storing the output data in sequence. Some of the data acquisition boards available for personal computers, and used with older flow cytometers, still follow this practice. Multiplexing several input signals into one ADC obviously takes longer than using several ADCs in parallel, one for each signal, because you have to convert the first signal, store it, convert the second signal, store that, etc. By the time the previous edition of this book appeared, 16-bit successive approximation ADCs with conversion times of 10 μs were available for less than $25, and 12-bit converters were even cheaper, leading most of the cytometer manufacturers toward using a separate ADC for each data channel. This made it relatively simple to keep total conversion times for all channels (and few instruments had more than five) under about 20 μs, theoretically allowing collection of data on as many as 50,000 events/s.

The recent development of ADCs has, to a large extent, been driven by the needs of the digital audio and telecommunications markets. Newer ADCs incorporate technologies other than successive approximation, and achieve high-resolution conversion with better differential linearity.

The fastest converters, appropriately called **flash** ADCs, incorporate a voltage divider and a large number of comparators [(2^n - 1) comparators for an n-bit conversion], but, unlike the ADCs we have discussed previously, convert in a single step. **Digital storage oscilloscopes**, which are rapidly replacing the analog variety, use 8-bit or 10- bit flash converters with conversion rates of several GHz. While pure flash converters are typically relatively low-resolution devices (10 bits or less), hybrids that incorporate several flash stages and other circuitry have achieved higher (14-16 bits) resolution. The fastest 16-bit successive approximation converter requires 1 μs for a conversion; hybrid 16-bit converters can convert at 5-10 MHz rates (100-200 ns conversion time). BD Biosciences' FACSDiVa digital pulse processing electronics use 14-bit, 10 MHz converters, which, when the system was designed, were the fastest available devices with that resolution. Analog Devices, the manufacturer of those ADCs, now offers 14-bit, 105 MHz converters.

Digital Pulse Processing and DSP Chips

As long as the ADCs used in flow cytometry only have to digitize held analog signals such as those stored in peak detectors and integrators, conversion times of several μs are perfectly acceptable. Once we start doing digital pulse processing, conversion rates must be higher.

We have noted (p. 205) that, according to the Nyquist theorem, faithful reconstruction of a waveform in which the highest frequency component is at a frequency f requires that we sample and digitize the signal at a rate of at least $2f$. Now, let us imagine that the pulses we measure in flow cytometry are not Gaussian, but are, instead, the similarly shaped upper halves of sine waves. A 10 μs pulse would then be half of a sine wave with a period of 20 μs, and a corresponding frequency of 50 kHz. So, by the Nyquist theorem, we ought to be able to reconstruct the waveform if we sample at a rate of 100 kHz, meaning one sample every 10 μs. At this point, we should be questioning the validity of the conceptual "model" we have just made of a pulse.

The Nyquist theorem is perfectly valid, provided we are prepared to use Fourier analysis to reconstruct the waveform; this, unfortunately, is likely to require more computation, and more time, than we are likely to have available if we are trying to digitize and process eight or more channels worth of pulse trains coming out of a flow cytometer. If we want to get decent pulse information from a fast and relatively simple computational process, we have to take more samples of our pulses than the Nyquist theorem would suggest. Figure 4-59 may be of help.

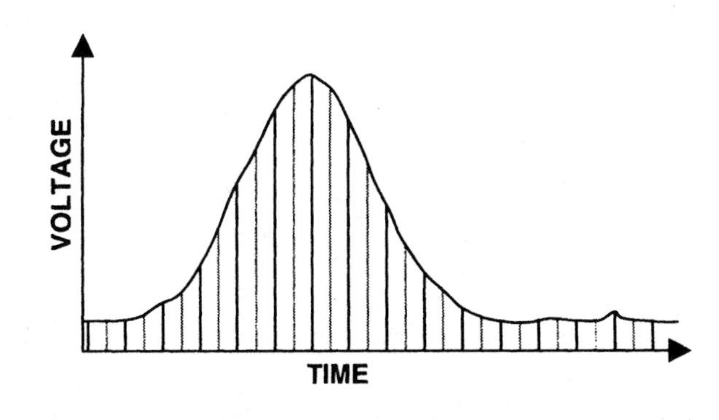

Figure 4-59. Digital pulse processing: "slicing" a slightly noisy Gaussian pulse with a baseline.

The trace in the figure represents a slightly noisy Gaussian pulse, with a baseline with DC and AC components. The vertical lines represent samples, or "slices," taken at regular intervals; I have shown alternate samples in black and gray.

At first glance, it appears that we might get a more than adequate characterization of the pulse by using every other sample, i.e., either all of the black lines or all of the gray lines. If you "connected the dots" at the point at which each set of lines intersects the trace, you'd get two curves that would follow the trace pretty closely. However, what we need to do by way of pulse processing is find the height, the width, and the area, or integral, of the pulse, and, if we start with the height, we see that the gray samples and the black samples aren't equivalent. As it happens, the sample that hits the trace closest to the true peak value is one of the black samples; the values represented by gray samples on either side of it appear to be at least 4 or 5 per cent lower than the peak value. On the other hand, if we're trying to find the pulse width, or where the pulse "begins" and "ends" by rising above the baseline, the gray samples may give us a more accurate idea than the black ones; as was the case with the peak height estimate, this is no more than the "luck of the draw." But we do get the sense that, if we tried to estimate the area or integral of the pulse, by taking the sum of a number of black or gray samples nearest the peak value, we'd end up with better agreement between the values from the black samples and the values from the gray samples than we had in our peak height and pulse width estimates.

Calculation confirms what intuition suggests; Phil Stein, Nancy Perlmutter, and I spent a couple of days working up a model, in a Microsoft Excel spreadsheet, that simulates digitization of Gaussian pulses with various random noise levels. Since we had simulated the pulses, we knew their true areas, peak heights, and widths; we then examined the accuracy and precision of 12-, 14-, 16-, and 18-bit measurements of peak height, width, and area, and compared measurements obtained by examining 8, 16, or 32 samples of a pulse. We found that area could be estimated to better than 1% using only 8 samples of the pulse; taking 16 samples increased precision considerably, while taking 32 samples yielded only a small additional improvement. As was expected, peak height and width measurement were less precise, although values were within a few percentage points of the "true" value; taking more samples substantially reduced the measurement error, again as expected.

These days, digital pulse processing is typically accomplished using a **Digital Signal Processor**, or **DSP Chip**. A DSP chip is a digital computer designed specifically for signal processing; it typically has a smaller instruction set than a general-purpose microprocessor, with emphasis on fast fixed- and/or floating-point arithmetic operations. Unlike general-purpose computers, a DSP chip is likely to have separate memories for program and data storage, and facilities to allow multiple computers and multiple analog-to-digital converters and other data sources to be interconnected. In particular, a DSP chip can very rapidly process a data stream, keeping track of a **running sum** of any given number of data values. Thus, for example, the chip could output a series of numbers representing the sum of input data points 1 through 8, the sum of data points 2 through 9, the sum of data points 3 through 10, etc. This would be an 8-point **unweighted** running sum; it would also be possible to multiply different data points in the input by different coefficients, producing a **weighted** running sum. In the case where the output stream of the DSP chip is an unweighted, 8-point running sum of input data, we say that an **8-point moving average filter** has been applied to the data.

Because each point in the output data stream of a moving average filter represents contributions from multiple points in the input data stream, the maximum change in values between two consecutive points in the output stream will be less than the maximum change in values between two consecutive points in the input stream. The highest frequency information in both the input and output streams is predominantly represented by the changes in values between consecutive points; the high frequency information in the output stream is therefore attenuated in comparison to the high frequency information in the input stream. Thus, the moving average filter is a low pass filter, and, like an analog low pass filter, behaves as an integrator.

When we try to capture the integral of a pulse using analog or hybrid electronic integrators, we have to select a trigger signal that will generate appropriately timed signals for the electronics that will hold the signal values. In digital pulse processing, the input signal streams are sampled at a relatively high rate, and the digital values are kept in a **buffer memory** until pulses have been identified and their areas and/or other characteristics have been determined. When we use a DSP chip to compute a running sum, we can use the points at which this sum reaches local maxima to represent the approximate centers of pulses, and the values of the sum at those maxima to represent the areas of those pulses. With a fast DSP chip and enough buffer memory, it is possible to identify a pulse in a signal from any detector at any of several observation points, and, if necessary, to work backward through the input data stream in the buffer to find and characterize low-level signals. The flexibility of this digital "triggering after the fact" is advantageous.

In principle, we could use DSP to eliminate analog baseline restoration by monitoring the baseline signal digitally and subtracting an averaged value from all signal values; this is the digital version of the analog computation performed by an analog baseline restorer. However, unless the DC baseline signal is very small, this approach may lose dynamic range. In a fluorescence channel, DC baseline is low, e.g., 1/1,000 of full scale, corresponding to a digital value of 65 on a 16-bit linear scale. If data are to be converted to an 8-bit or even a 10-bit logarithmic scale, there is ample room in the top channel of the scale, which encompasses many linear values (see p. 207). However, in a forward scatter channel, DC baseline is likely to be much higher, on the order of half of full scale, corresponding to a digital value of 32,767. Subtracting this value from the raw digitized signal values leaves only 32,767 channels available to describe pulse amplitudes, effectively turning the 16-bit ADC into a 15-bit ADC. This is not the direction in which we would like to go; we already know (p. 207 again) that we need 20-bit data to eliminate the picket fences that occasionally make bad neighbors in our cozy cottage industry.

For digital pulse processing in anything approaching a conventional flow cytometer, ADC conversion rates, as previously mentioned, must be at least 1 MHz; high-speed sorting and data processing demand higher conversion rates. As noted (p. 209), BD Biosciences elected to use a 14-bit (16,384-channel), 10 MHz ADC in their FACSDiVa system digital pulse processing system, because a higher-resolution part with the same conversion rate was not available. For the record, as of late 2002, you can get 14-bit converters with conversion rates above 100 MHz, but the fastest 16-bit converter I have seen advertised is a 5 MHz part, and the fastest 18-bit converter, introduced in mid-2002, has an 800 kHz conversion rate. The electronics required for high-speed digital pulse processing are way too complex to make subranging (as used in the Beckman Coulter XL, etc.; see p. 207) a viable option, and there are some tradeoffs involved in using 14-bit converters, the most obvious of which is a "picket fence" effect at the low end of the logarithmic scale.

A brief digression on audio recording may help us understand the tradeoffs currently involved in digital pulse processing. There is, in fact, a distinct possibility that slower, high-precision, high-resolution cytometric data analysis without digital pulse processing could make effective use of the **oversampling delta-sigma** ADCs now used for digital audio recording[2468-9]. These converters continuously sample signals at rates of a few MHz, and put out 24-bit data at rates as high as 192 kHz. You can find an easily understood audiovisual introduction, which also covers some of the basics of ADCs and sampling presented in previous pages, if you download the "Short Course in Digital Audio" from Syntrillium Software, at (http://support.syntrillium.com/tutorials.html).

The Screwy Decibel System

Modern audio components are likely to have a specified **signal-to-noise (S/N) ratio** of better than 80 **decibels (dB)**. Decibels (the **bel** derives from, and honors, Alexander Graham Bell) measure the ratio of signal amplitudes on a logarithmic scale, with a few tricks involved. When you're listening to speakers, or any other sound sources, your hearing is responding to the acoustic **power** delivered to your ears. The power that drives a loudspeaker with an impedance of Z ohms comes from an amplifier that puts a current I through the speaker, producing a voltage difference E across the load. Ohm's law tells us that E = IZ; the power, P, is EI, or E^2/Z.

The ratio, in decibels, of two **power** levels, P_1 and P_2, is

$$dB = 10 \log_{10} P_1/P_2 .$$

However, we have already seen that power varies as the square of voltage; therefore, the ratio in decibels of the two **voltage** levels, E_1 and E_2, corresponding to the power levels P_1 and P_2, is

$$dB = 20 \log_{10} E_1/E_2 .$$

In our discussions of signal processing, we are generally interested in the ratios of signal voltages, or signal and noise voltages, not in powers, so we use the second formula.

If we say that the (voltage) **signal-to-noise ratio (S/N)** of a circuit or system is 80 dB, it means that the ratio of (signal voltage amplitude)/(noise voltage amplitude) is 10,000:1. If the output is 10 V, the noise level is 1 mV. To those of us who grew up in the era of vinyl records and analog tape recording, a S/N of 80 dB looks pretty good, but, having previously looked at dynamic range in cytometry, we are now (I hope) aware that an 80 dB range just barely encompasses 4 decades.

I might add that the frequency response spec of 20-20,000 Hz ± 0.5 dB, which also sounds pretty good when it is hung on audio equipment, allows for about 6 percent variation in output level over the frequency range, and that ± 0.5 dB is about as close to ideal as the best log amp responses get. So there is plenty of motivation for moving toward digital signal processing and high-resolution analog-to-digital conversion in cytometry.

A stereo audio CD contains 16-bit data for each channel, which is played back at the rate of 44,100 data points per channel per second; it obviously must be recorded using a sampling rate of at least 44.1 kHz and 16-bit ADCs. However, most professional audio recording is now done at higher sampling rates, using ADCs with more bits, for a number of reasons.

One is that music, or at least most of the music I listen to, has a fairly wide dynamic range. The **SPL** or **Sound Pressure Level** scale used to describe audible sound defines the level of 0 dB SPL as the intensity of a 1 kHz tone barely audible to the normal human ear; at a rock concert, the level frequently reaches 120 dB SPL, which, before there were rock concerts, was observed to make most people uncomfortable, and thus came to be known as the **threshold of pain**. The SPL scale, incidentally, describes pressure levels rather than audio power levels; two sounds differing by a factor of 10 in amplitude are thus 20 dB apart, rather than 10 dB apart, on this scale. Microphones and their associated amplifiers convert pressures into voltages, which, as noted at left, also differ by 20 dB, rather than 10 dB, for each factor of 10 difference in amplitude.

Classical music covers a range from 40 dB SPL (*ppp*, very soft) to 100 dB SPL (*fff*, very loud). When recording it, we need to keep the *fff* signal at or below the top of the range of our ADC to avoid **clipping**, which would distort the sound when it was played back (this isn't usually a problem in rock music, in which clipping is frequently introduced intentionally). But back to Symphony Hall, where we have adjusted the gain of our microphone preamplifiers so that the loudest part of the piece, at 100 dB SPL, comes out at the high end of a 16-bit converter, i.e., channel 65,535. The softest (*ppp*) passages, at 40 dB SPL, with 1/1,000 the amplitude, will therefore end up at channel 65, and much of the music will come out somewhere in between, probably at channel numbers below 1,000.

Now, suppose we want to do our audio mixing and editing digitally. This is going to involve some multiplications and subtractions, and, the more of them we do, the more likely we are to end up with some "picket fence" effects at the low end of the digital scale in the output data. There are a number of fixes for this. One of the things we can do is use "companders" (compressor-expanders) in the electronic circuitry, transforming amplitudes according to a nonlinear curve so that sounds at the low and middle ends of the range get converted to higher channel values than they would otherwise.

The same trick has long been used in cytometry; the nonlinear curve with which we are already familiar is a logarithmic curve, and the input side of a compander acts pretty much like a log amp. And, yes, it may only be accurate to 6 percent or so, but that is within the ± 0.5 dB tolerance usually specified for high quality audio.

We could also add a little random noise to the signal, getting rid of the "picket fences," albeit also degrading signal quality slightly; this maneuver is also used in cytometry, as when data have been converted from relatively low resolution (10 bits or less) log scales to linear and back during after-the-fact software compensation.

The diehard digital audio purists, and most of the professionals in the recording business, even those who record rock, won't have any of this. They can now get 24-bit converters that sample at 192 kHz, giving them enough dynamic range to do all of the editing and add all of the effects they need, and then interpolate and round off to convert the finished product to a 16-bit, 44.1 kHz sample rate CD-quality audio file. Or they can leave it as a 24-bit, 192 kHz DVD-quality audio file.

You can now buy an 8- or 10-channel, 24-bit, 192 kHz digital audio recording accessory for your computer for under $2,000; the 24-bit converters used in such systems cost less than $10 apiece. Table 4-5 (p. 206) tells us that 1 LSB in a 24-bit converter with a 10 V full scale input range amounts to 596 nV; since the noise level in even the most carefully designed electronics is likely to be tens of microvolts, we can guess that we won't capture a real 24-bit signal. However, we are likely to be able to get 20 bits' worth of useful information, and that should eliminate the picket fence effect and the inaccuracy at the low end of the standard 4-decade scale. I speculate that one could feed the outputs of peak detectors and integrators into such a gadget, and get good data from cells at rates of a few thousand/s. That would be perfectly adequate for high-sensitivity, high precision fluorescence flow cytometry, which is likely to require relatively slow flow rates and thus not process tens of thousands of cells per second, anyway. I've already got the audio recorder; playing with it (for audio as well as for flow data) will have to wait until I finish writing this book. At worst, it should be useful for the sequel to my audio opus, *Songs for the Jaundiced Ear*[2470].

Pulse Slicing: Déjà Vu All Over Again

Moving from CD players back toward CD antigens, it seems inevitable that the detector signals in most flow cytometers will eventually be handled by DSP techniques, eliminating the analog and hybrid baseline restorers, threshold sensing and front end electronics, integrators, peak detectors, and pulse width measurement circuits on which we have just spent so much time. A fast enough DSP pulse processing system would even be able to sample light source power output and perform digital compensation for noise fluctuations, almost certainly performing better than the analog source noise compensation circuit I described on pp. 147-8. That's the future of digital pulse processing in cytometry; however, we should not forget that it also has a distinguished past.

Slit-scanning flow cytometers[171-4, 690-7, 1111-2, 1166-72] were pioneered by Leon Wheeless and his colleagues beginning in the 1970's, and investigated by a number of others then and since. A slit-scanning cytometer uses an extremely narrow region of illumination; beam heights under 5 μm are the norm. The resulting signals from cells can reveal morphologic information in the form of intensity variations along the length of a pulse. The original work in this field used relatively primitive computers for signal analysis, and was done at a time when it was only feasible to use 8- and 10-bit ADCs for data acquisition. This limited the dynamic range of data collection; however, many of the early applications of slit-scanning were to analysis of fairly strong fluorescence, low dynamic range linear signals from cells or chromosomes stained with nucleic acid dyes, and investigators' primary motivation for studying the technique was the desire to pursue applications of pulse shape information.

Other experimental systems[1173-5, 2473-4] digitized detector signals at rates up to tens of MHz in order to collect enough points (in some cases, over 100) to provide detailed information on the shapes of pulses as well as the more conventional measures of pulse height, width, and area. The RATCOM Personal Cytometer, the predecessor of today's NPE instruments (see Chapter 8), was probably the first commercial instrument to implement digital pulse processing, using 8-bit ADCs, around 1990; the objective in this case was to provide a highly accurate integral for DNA analysis, rather than high dynamic range. Earlier attempts at digital pulse processing used specially fabricated **application-specific integrated circuits (ASICs)** to perform the function of DSP chips, which were not yet available with the required speeds and processing capabilities; newer digital pulse processing systems may use both DSP chips and ASICs.

Those of us who design flow cytometers today typically want to be able to process sixteen or more detector signals, captured in three or more illumination regions, from tens of thousands of cells per second. We expect that the data will have a large dynamic range, that is, at least four decades, and that there will be substantial spectral overlap between signals, requiring fast software compensation and conversion between linear and logarithmic scales, and we want to do all of the processing in a short enough time to permit us to make sort decisions before the cells of interest are lost to us.

Getting pulse shape information is not a major concern, at least for the time being, and we can usually settle for relatively crude indicators of pulse width and height, which will be used primarily for discriminating between single cells and multiplets. What we care most about is getting accurate integral values representative of the amounts of fluorescence measured from each cell in each spectral region.

In theory, taking a large number of digital samples of each pulse should provide a very large dynamic range for the value of the integral. If a pulse is at least 6.4 µs long, an 8-bit converter sampling at 40 MHz (1 sample every 25 ns), will acquire at least 256 samples of the pulse, each with a value between 0 and 255. The integral, which is the sum of the 256 values, can theoretically range between 0 and 65,280; this appears to yield the same dynamic range as a 16-bit ADC. However, we are still dealing with an 8-bit converter, which, if it operates over a range of 0-10 V, has channels spaced 39 mV apart (Table 4-5, p.206). Unless the pulse amplitude exceeds 39 mV at some point, the converter output will remain at 0, and the integral will remain 0. The converter won't even notice 10 mV or 1 mV pulses. We're stuck in the upper 2 1/2 decades of a 4-decade dynamic range.

If we move to a 12-bit converter, with channels spaced about 2.4 mV apart, we won't have a problem with 10 mV pulses, or even 3 mV pulses; this gives us a 3 1/2 decade range, although the 1 mV pulses at the bottom of the fourth decade remain invisible. Since the 12-bit converter output ranges between 0 and 4,095, we could, in theory, get away with 16 samples of a pulse, dropping the sampling rate from 40 MHz to 2.5 MHz. With a 14-bit converter, output ranges between 0 and 16,383, and the channels are 600 µV apart. If we kept the sampling rate at 2.5 MHz, we'd get 16 samples of a 6.4 µs pulse, the values of the integral could range between 0 and 262,128, and the converter would respond to a 1 mV pulse. And, if we sampled at 10 MHz with the 14-bit ADC, as is done in BD's FACSDiVa electronics, we could get 16 slices of a 1.6 µs pulse, again giving us a maximum possible value of 262,128 for the integral. However, if we think about it, we will quickly find that we won't, and don't want to, ever get actual values close to that maximum. Here's why.

We expect to be looking at pulses that are more or less Gaussian in shape. In order to get a pulse integral value of 262,128, we have to record a value of 16,383 for each of the 16 slices of the pulse in question. Such a pulse would be rectangular, not Gaussian; if we encountered a real pulse that gave us this integral value, we could be sure that we were not capturing an accurate value of the integral. We would, presumably, be looking at the lower portion of a Gaussian pulse with amplitude exceeding the maximum acceptable input value for the ADC, and our signal would be **clipped**. Clipping (p. 211) is a well-known phenomenon in audio recording; if you overdrive preamplifiers and/or amplifiers, or put too large a signal into a digital recording device, you turn sine waves into roughly square waves, producing audible distortion.

In audio applications, clipping is often done intentionally, without penalty, to produce a distinctive sound effect; in digital pulse processing and football, there are penalties. On our playing field, the only way to avoid them is to keep the pulse amplitude sufficiently low so that no more than one point – the peak of the Gaussian pulse – produces the maximum output from the ADC. The other samples of the pulse will have lower values; we can easily calculate what they will be using a table of the Gaussian or normal distribution, and this will tell us what the practical attainable maximum value for the pulse integral should be.

In Chapter 5, we will delve into the anatomy of the Gaussian distribution in some detail. For the present, we will simply consider an integral of the Gaussian pulse computed by taking 16 14-bit slices evenly spaced over an interval in which theory tells us we will find a little more than 99 percent of the area of the pulse. The maximum value for the 16-slice integral, obtained when the true peak value is 16,383 (corresponding to a 10 V pulse if the full scale range is 0-10 V), is about 102,500. If 10 V represents the top of a 4-decade dynamic range, 1 mV represents the bottom; the digital value for the 16-slice integral of a 1 mV pulse is 10, that for the integral of a 10 mV pulse is 102, etc.

In theory, a digital value of 1 should correspond to the integral of a 100 nV pulse, allowing an integral of 16 14-bit slices to represent data spanning a five decade dynamic range. However, Table 4-5 (p. 206) tells us that the 14-bit converter will not produce an output above 0 unless the input signal level exceeds 610 nV. That means that there is little point in attempting to display integral data from a system such as the FACSDiVa on a five-decade logarithmic scale. In fact, since a four-decade scale would encompass digital values between 10 and 102,500, and since we know (p. 207) that we need 20-bit integral values (ranging between 104 and 1,048,575) to generate a smooth four-decade logarithmic scale occupying 1,024 channels, we should not be surprised when we see some "picket fence" effects at the lower end of the FACSDiVa's displays. But how much does it matter?

In Defense of de Fence

I feel obliged to make the same apologies for the FACSDiVa's picket fences as I made on p. 207 for my own, which result from my use of a 16-bit ADC, rather than a 20-bit converter, to convert held peak signal values. As already noted, digital pulse processing makes it possible to use almost any combination of signals from any illuminating beam(s) for triggering, and provides digital integral values covering a sufficiently large range to eliminate the need for logarithmic amplifiers and to permit fluorescence compensation to be done rapidly enough (by yet another DSP chip) for sort decisions to be based on compensated data values. That's the good news; the bad news is that, if we want to get 16-slice integrals from tens of thousands of cells per second, we're pretty much stuck with using 14-bit converters, and picket fences appear in the bottom half of the bottom decade on a four-decade logarithmic scale.

If we define that scale in terms of numbers of molecules of some substance measured in or on a cell, we can cover the range from 100 molecules to 1,000,000 molecules. Most of the CD antigens we detect on cells from the blood and immune system do not reach the level of 1,000,000 molecules per cell, and most of the flow cytometers we use to detect these antigens cannot reliably detect 100 molecules of antigen above background. So, we might want to regard the picket fences in our displays as if they bore the sign "beware of the data," and not try to play with what is behind them. I'll have more to say about this later (pp. 244 and 562-3).

Digitization: Tying It All Together

A few years back, a good friend of mine sent me a book called *The 85 Ways to Tie a Tie*[2475]. I knew three of them, and have added a fourth to my repertoire because it is useful for woolen and woven neckties. I don't think there are 85 ways to get high-resolution digital data out of a flow cytometer, but there are a couple of approaches I haven't mentioned, and now is the time to summarize existing methods.

As noted above, digital pulse processing is unique in offering highly flexible triggering schemes and the potential for really effective source noise compensation; if 18-bit, 20 MHz ADC's were available, there would be almost no downside, other than the complexity of the circuitry involved.

Digitizing held peak and/or integral values to 20 bits' precision, as is done in the Beckman Coulter EPICS XL, requires a high level of performance from peak detectors and integrators, and also, in its present form, introduces the complexity required for subranging analog-to-digital conversion. The electronics can be made fast enough to deal with tens of thousands of cells per second; although Beckman Coulter's current sorters still use log amps and hardware compensation, it seems likely that a switch will be made to high-resolution digitization to keep the product line competitive. There is a possibility that 24-bit ADCs designed for audio recording could be used for high-resolution capture of flow cytometric data from a few thousand cells per second; this approach might also permit the use of somewhat lower-performance peak detectors and integrators.

DakoCytomation's MoFlo line of high-speed sorters uses log amps, and digitizes log scale held peak and integral signal values to 16 bits' precision. Calibration curves for the log amps are stored in memory, and the digitized log data are converted to a linear scale to allow real-time software compensation to be accomplished by a DSP chip. The 16-bit digitization of the original log data provides enough resolution to allow compensated data to be displayed on a four-decade log scale with no picket fence effects.

The manufacturers' marketing and sales people are a good deal more partisan about the virtues of their respective methodologies than are the people who design and build the hardware and develop the firmware and software. At the present state of the art, while all of the approaches to high-resolution digital flow cytometric data processing involve tradeoffs, any of them is preferable to doing things the old-fashioned way with log amps and hardware compensation, especially when more than three fluorescence measurements are made of each cell.

4.10 PERFORMANCE: PRECISION, SENSITIVITY, AND ACCURACY

The performance of flow cytometers is generally discussed in terms of **precision**, **sensitivity**, and **accuracy**, often in that order. It is, perhaps not coincidentally, easiest to characterize precision, and hardest to characterize accuracy.

Precision describes the extent to which identical values are obtained from measurements of identical particles; the measure of precision most commonly used is the **coefficient of variation** or **CV**.

The definition of **sensitivity** has changed somewhat as cytometry has evolved; the measures originally proposed were the minimum size of particles or amount of a substance detectable above background by the apparatus, inferred from measurements of scatter or fluorescence signal intensities. Current practice aims to quantify the extent to which populations with very small signal intensities can be resolved from one another.

Accuracy describes the degree to which the measurement results produced by the flow cytometer conform to "true" values. Until recently, considerably more attention was paid to precision and to sensitivity than to accuracy, with good reason; until precision and sensitivity have been established, it is difficult to define a framework in which to characterize accuracy.

Precision; Coefficient of Variation (CV)

As was noted in the introductory discussion on pp. 18-22, the measurement precision of a cytometer is routinely characterized by accumulating a distribution of measured values of fluorescence or light scattering intensities from "nearly identical" particles, and computing the **coefficient of variation (CV)**, which, expressed as a percentage, is 100 times the **standard deviation (S.D.)** divided by the **arithmetic mean**, or **average**.

The particles may be artificial, e.g., fluorescent polystyrene microspheres, or biological, e.g., fixed, stained cell nuclei. Among cellular parameters measurable by flow cytometry, only DNA content is so precisely regulated as to vary by less than 2% from cell to cell in homogeneous, nondividing populations. Nuclei from quiescent or largely quiescent cell populations such as avian erythrocytes or mammalian peripheral blood lymphocytes, stained with appropriate fluorochromes, can therefore be used to test the precision of fluorescence measurements. When this is done, the CV is calculated for the G_0/G_1 or diploid peak of the DNA fluorescence distribution. Because inclusion of early S phase cells may affect calculations done by computation of the mean and variance, it is usually convenient to estimate the standard deviation as one-half the width of the peak at 60% of its maximum height or as (1/2.36) the **full width at half maximum** height **(FWHM)**.

When brightly stained fluorescent beads and cell nuclei stained for DNA are used to assess instrument precision, enough photons are typically collected from specimens so that photon statistics are not a major contributor to the variance of fluorescence distributions. In most cases, i.e., for most flow cytometric measurements, CVs below 5% are acceptable; for DNA measurements, CVs below 3% are preferable. Most modern instruments can achieve CVs between 1% and 2% in scatter and fluorescence measurements of beads.

Although beads are now more commonly used than nuclei to measure instrument precision, the intrinsic CV of nuclei is likely to be lower; the most precise instruments, using a highly DNA-selective stain such as DAPI or a mixture of mithramycin and ethidium, can achieve CV's on the order of 0.5 percent. The stains just mentioned, respectively, require UV and blue-violet excitation; DNA fluorochromes excitable at the 488 nm wavelength most commonly used in flow cytometers are less DNA-selective, and require that nuclei be treated with RNAse to eliminate fluorescence variations due to RNA staining.

Precision may be decreased by fluctuations in light source output, by poor alignment of optics, and by disturbances of fluid flow. If the optics of a flow cytometer are adjustable, it is customary to make such adjustments as are needed to maximize the signal from the fluorescent particles used for precision determination before making a definitive estimate of precision; brightly stained beads with a low intrinsic CV are often referred to as **alignment beads** because they are used for this purpose. In modern instruments that normally do not need alignment, fluid flow problems should be the primary suspect when a decrease in precision (i.e., an increase in CV) is noted, and the flow system should be checked and flushed and cleaned, if necessary, before a service call is initiated. A consistent decrease in laser output power points to laser light noise as the source of an otherwise explained loss of precision.

It should be obvious that it is pointless to attempt to determine either the sensitivity of an instrument or its measurement accuracy until it can be demonstrated that performance is up to the manufacturer's specifications in terms of precision.

Sensitivity I: Minimum Detectable Signal

The sensitivity of a flow cytometer may be defined in several different ways. Measurement of what should properly be called **instrument sensitivity** was originally based on determination of the minimum amount of fluorescent material, or minimum number of photons from a test light source, detectable by the instrument when no cells are present. This was originally done by determining the minimum concentration of dye which, when run through the system, increased the fluorescence detector signal above background, and calculating the number of dye molecules in the observation volume; detection limit experiments were also done by Loken and Herzenberg using cells stained with antibodies bearing both fluorescent and radioactive labels[154]. The Stanford and B-D FACS instruments of the mid-1970's were found to be capable of detecting about 3,000 molecules of fluorescein. By the late 1970's, the Block apparatus used for analyses of viruses[94] could detect a few dozen molecules of fluorescein; in the 1980's, work by the Los Alamos group culminated in the detection of single molecules of phycoerythrin in solution, using single photon counting techniques on a slow flow system[660].

In applications of flow cytometry such as immunofluorescence measurement, the paramount issue is generally how many molecules of a label such as fluorescein, attached to a ligand, are detectable on a cell surface; this provides a more practical definition of sensitivity, which, in this context, is more likely to be limited by cellular characteristics, in particular, autofluorescence, than by instrumental factors.

For example, the instrument sensitivity of 3,000 fluorescein molecules mentioned above for the 1975 vintage FACS instruments was not attainable in immunofluorescence measurements of lymphocytes, because the fluorescence of unstained lymphocytes, as then measured with those instruments, was broadly distributed in the range equivalent to the fluorescence of 5,000-10,000 fluorescein molecules. It was therefore necessary to have several times that amount of fluorescein bound to antibody on a cell surface before one could be sure that the cell was stained.

A third measure of sensitivity is useful in evaluating the efficiency of flow cytometer designs. If measurements are made of unstained cells (autofluorescence) and of weakly stained cells, two peaks representing brighter and dimmer cells will appear in the fluorescence distribution. Sequential measurements are made of a bimodal population, starting at low laser power; the ratio of the linear values representing the bright and dim cell peaks is plotted against laser power and power is increased and measurements repeated until the ratio reaches a plateau value; the laser power used is then noted. By this criterion, i.e., the amount of laser power at 488 nm needed to achieve maximum fluorescein fluorescence measurement sensitivity, the B-D FACScan (plateau at 8-10 mW) and Coulter Elite are among the most sensitive apparatus; the stream-in-air sorters (50 mW for the B-D FACStar, up to a few hundred mW for older B-D and Coulter EPICS sorters) are least sensitive. However, by the same criterion, the old arc source B-D FACS Analyzer, using 546 nm light to excite phycoerythrin-labeled antibodies, was approximately ten times as sensitive as any instrument using

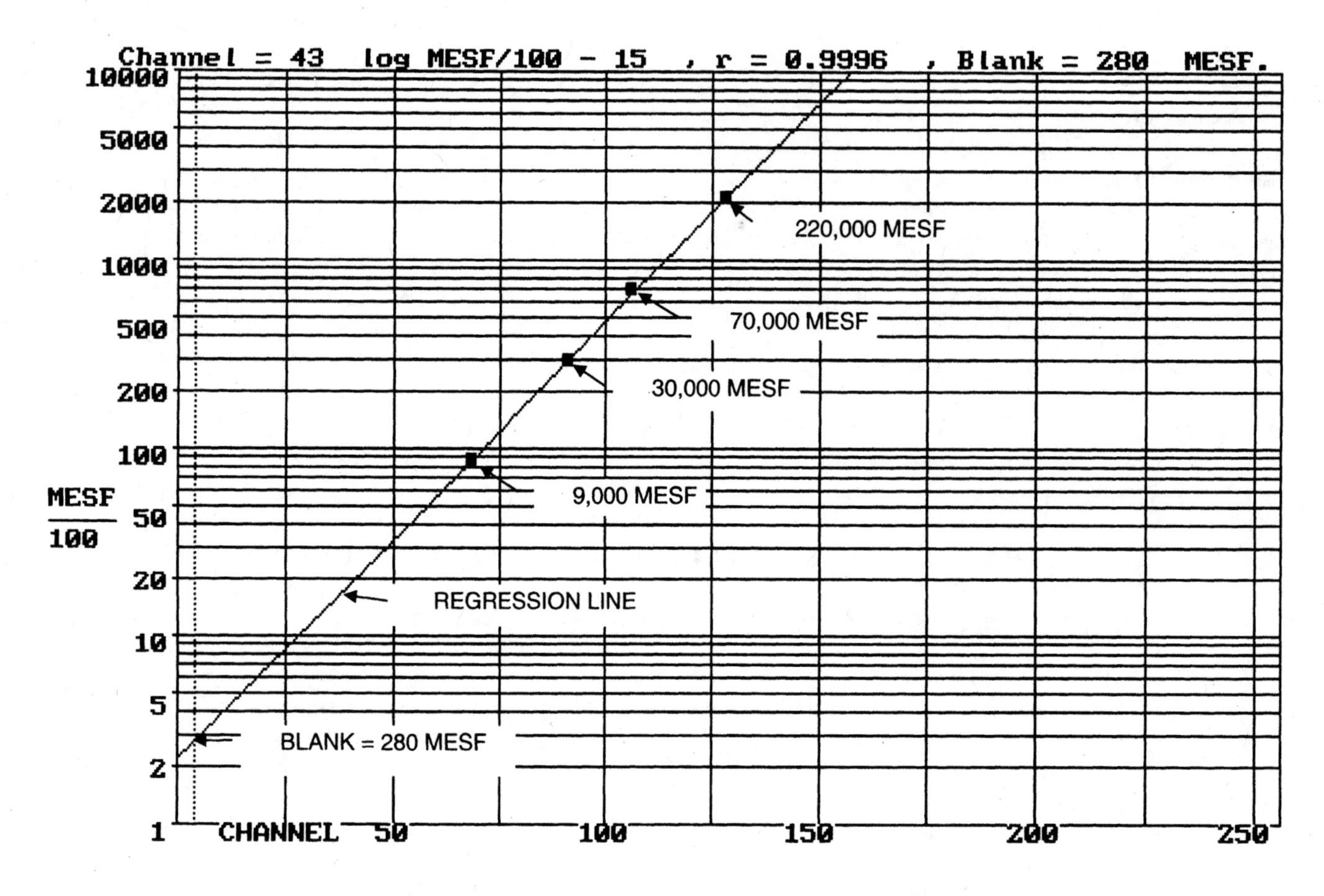

Figure 4-60. MESF threshold sensitivity determination using fluorescein-labeled beads.

488 nm light and fluorescein antibodies, in part because the highly efficient phycoerythrin is optimally excited at 546 nm and in part because autofluorescence, which in most cells is due predominantly to flavins, is much less with 546 nm excitation than with 488 nm excitation.

Sensitivity II: MESF Units

A technique still widely used for determination of flow cytometer sensitivity, originally developed by Dr. Abraham Schwartz[1161], expresses detection limits in terms of **Molecules of Equivalent Soluble Fluorochrome**, or **MESF Units**. The method is illustrated in Figure 4-60; it uses a mixture of beads, including some labeled with several different, known, numbers of molecules of the fluorescent dye of interest and some undyed blank beads. The expression of fluorescence in terms of equivalent soluble fluorochrome is necessary because the amount of dye on the beads is measured by comparing the bulk fluorescence of a bead suspension and a dye solution; in general, the quantum efficiency of a dye is decreased following its covalent linkage to a bead or other substrate.

The bead mixture is analyzed in the flow cytometer; a histogram of fluorescence from single beads, on a log scale, is acquired using a forward/orthogonal scatter gate. This histogram yields the channel numbers corresponding to the median values of the peaks representing the dyed and blank beads; the peak locations of the dyed beads and the mean numbers of dye molecules on each dyed bead population are used to derive the equation of a least-squares regression line relating the channel number to the log of the number of MESF units per bead. The resulting equation is then used to calculate the number of MESF units corresponding to the channel number of the blank bead peak. This **MESF threshold** represents the minimum number of molecules of the fluorochrome detectable by the instrument; in essence, it tells you how many molecules the cytometer sees when there aren't any molecules there. Figure 4-60 represents data from beads covalently labeled with fluorescein; the median fluorescence of the blank beads was calculated to be equivalent to the fluorescence of 280 molecules of soluble fluorescein. Surveys of several hundred laboratories[1161] in 1989-90 found median fluorescein detection thresholds near 900 MESF.

Abe Schwartz later suggested that the MESF threshold determination be modified by using the first channel number above the peak representing the blank beads, rather than the median channel of the peak; the calculation then yields the minimum number of dye molecules detectable above the background of blank particles. This number, obtained using the data shown in Figure 4-60, is 620 MESF.

When unstained cells are added to the bead mixture used for sensitivity determination, the level of cellular autofluorescence can be expressed in MESF units. The 1989-90 multilaboratory survey conducted by Schwartz and Fernández-Repollet[1161] found the lymphocyte autofluorescence peak corresponded to 657 ± 270 MESF of fluorescein; the 98th percentile value was 2552 ± 1748 MESF. Monocytes and granulocytes had somewhat higher autofluorescence. In comparing the value of 657 MESF for lymphocytes to the 5,000 fluorescein equivalent value reported by Loken and Herzenberg in 1975[154], we should note that the older measurement was made over a much wider spectral range; a 530 nm long pass filter was used, rather than the 520 or 525 nm bandpass filters now favored for fluorescein measurement. Since the peak of cellular autofluorescence emission is at around 550-560 nm, the larger bandwidth of the long pass filter will increase the autofluorescence signal to a greater extent than the fluorescein signal. In addition, it is likely that the older measurement was affected by both filter fluorescence and relatively high background light levels.

The major sources of error in sensitivity determination and autofluorescence measurement in MESF units as just described appear to be offsets and deviations from true logarithmic response in log amp circuits. Both of these have a much greater effect on the lowest decade, in which the blank generally lies, than on the higher decades; as a result, the fact that the labeled beads lie very close to the regression line provides little assurance that the line can be accurately extrapolated.

The problems likely to be introduced by log amps may be avoided by measuring the beads at high gain on a linear scale. Start with a low enough gain so you can see the brightest beads, then, increase the gain so that all but the one or two dimmest labeled beads and the blank go off scale. You can then calibrate your channel numbers in MESF units. If, for example, the peak representing the beads with 9,000 MESF is placed at channel 200, then each channel represents 45 MESF. If the peak representing the unlabeled beads is at channel 10, that corresponds to 450 MESF.

I should add that when you're looking for very small signals, e.g., weak immunofluorescence or nucleic acid probes, you're likely to be better off using a linear scale for two reasons. The dynamic range with which you have to deal under those circumstances is usually not very large, so you can keep everything on scale, and not have to deal with inaccuracies that log amps might introduce at the low end. Also, in order to be accurate, you should express the fluorescence intensity of stained cells in terms of their difference from unstained cells, and you can't do the subtraction on a log scale. However, whether you work on a linear or a log scale, you do need to establish the linearity of your instrument, which leads me to digress briefly from sensitivity to accuracy.

Accuracy I: Linearity (and Nonlinearity)

The **accuracy** of flow cytometers, i.e., the degree to which the measured values approach the true values of the variables measured, is less often discussed than the precision or sensitivity. Numerous instrumental factors can affect accuracy; one of those most often encountered is **nonlinearity**. One of the more common examples of nonlinearity is the situation in which G_2/S phase cells, which we know have twice the amount of DNA in G_0/G_1 phase cells, appear to have substantially more or less than that. This can be due to substantial offsets in the electronics, in which case dealing with the problem may require anything from turning a knob or trim potentiometer to redesigning the instrument. However, in older instruments, nonlinearity in DNA content measurements was more likely to be caused by the use of pulse height instead of pulse area or integral measurements with beam geometries in which the pulse height was not proportional to the integral.

Perhaps the easiest way to assess the linearity of a fluorescence channel in a flow cytometer is to analyze high-intensity fluorescent plastic beads, measuring the integral or area rather than the height of pulses on a linear scale. There will invariably be a small population in the fluorescence distribution representing doublets; the channel number at which the peak for this population is found should, ideally, be twice the channel number at which the main peak representing the singlet population is found. If the ratio of the doublet and singlet peak channel numbers is more than 2.02, or less than 1.98, it is probably advisable to adjust the instrument. For critical work, the linearity assessment should be repeated at different PMT gains to cover as much of the measurement range as possible.

Accuracy is also often estimated by adding internal standards to specimens, e.g., avian and/or fish erythrocytes, which contain known amounts of DNA, to cell nuclei stained for DNA, or by interspersing controls, such as beads bearing known amounts of fluorescent antibody, among immunofluorescence samples. These and other techniques for measuring and maintaining accuracy are essential for both quantitative measurements and quality control, and will be discussed in Chapter 10.

Sensitivity III: What's All the Noise About?

We are now approaching the end of our long strange trip (drip?) through the hardware of the flow cytometer. Like medical students, who develop the symptoms of each new disease they study, we are now acutely aware that things can go wrong with most parts of the instrument, and suspicious that, according to Murphy's Law, they will. And, like medical students, who are told that, under most circumstances, when they hear hoofbeats, the hooves are more likely horses' than zebras', we have to try to put what we've learned into perspective, and figure out what's likely to go wrong enough to bother us.

Most of what goes wrong enough to seriously compromise the sensitivity and precision of flow cytometric measurements falls under the heading of **noise**; since the early days of flow cytometry, the sources and effects of noise have been studied in detail from both theoretical and experimental viewpoints. In the 1970's, Holm and Cram[1162] and McCutcheon and Miller[124] established that fluorescence measurement precision decreased, i.e., that CV increased, as lower fluorescence intensities were measured; this was what would be expected on the basis of photon statistics.

Sensitivity IV: More Photons Give Better Precision

In 1983, Pinkel and Steen[1163] described a method for determination of sensitivity based on measurement of the CV of fluorescence distributions of pulses from a light-emitting diode (LED) placed in the light collection path. PMT voltage and amplifier gain were adjusted to keep the pulses in the same position as the intensity of the pulses was varied. Using LED pulses instead of test beads allowed sensitivity to be determined with the flow cytometer's laser turned off, eliminating most of the effects of background light; the magnitude of these effects was then estimated by repeating the sensitivity determination with the laser on. CV's obtained from plastic particles can be compared to standard curves obtained using the LED to determine whether photon statistics contribute significantly to variance, in which case precision may be increased by using higher laser powers.

Ubezio and Andreoni[1164] also investigated the relative contributions of photon statistics and instrumental factors to CV, by analysis of measurements made of propidium-stained nuclei at different laser power levels. Between 10 and 60 mW, the channel number of the diploid peak was exactly proportional to excitation power, i.e., results were not significantly affected by bleaching or saturation. A linear relationship was observed between $(CV)^2$ and the reciprocal of excitation power, with a fixed offset due to instrumental factors (contributions to CV from different sources add in a "root-mean-square" (RMS) manner; the total CV is the square root of the sum of the squares from individual contributions).

Sensitivity V: Background Effects

A 1992 article[1165] by Steen considered effects of signal strength, background, and detector quantum efficiency on sensitivity and precision; a pulsed LED was again used for sensitivity determination. When only photon statistics and background light noise are considered, maintaining instrument fluorescence sensitivity in the face of increasing background noise requires increasing the excitation intensity. When background noise is significant, sensitivity increases with the square root of excitation intensity; cutting the detection limit in half requires a fourfold increase in power. When noise is negligible, the same increase in sensitivity requires only a twofold increase in power. The increased power needed to maintain sensitivity as noise increases has an effect that, at first, seems paradoxical; given two instruments with equivalent sensitivity, the precision is higher in the instrument with the higher background.

This is so because the effect of background on precision diminishes at increasing signal levels. At the detection limit, the signal-to-noise ratio, i.e., the ratio of the number of photons coming from the sample particle to the number coming from the background, is 1. The more photons come from the background, the more are needed from the particle in order to detect it; increasing the excitation power gets more photons. All other things being equal, however, CV decreases, and precision increases, as the number of photons collected from the sample particle increases.

Near the detection limit, CV is near 100% for any instrument; well above the limit, CV is lower, and precision is higher, for the instrument with higher excitation power. This is not to say that you will improve performance of your instrument by taking the cover off and running it at the beach on a sunny day; decreasing the background while keeping excitation power constant will increase sensitivity and also improve precision at least slightly. Also bear in mind that virtually all dyes are susceptible to photodamage at some power level, and when you get past that level (which can be as low as 10 mW for a label such as PerCP), increasing excitation power only makes things worse.

A concrete illustration of the effect of background can be obtained from sensitivity values for two cytometers. Steen[1165] found the fluorescein fluorescence detection limit of a B-D FACScan to be 826 MESF; in the absence of background, the limit would have been 285 MESF. The corresponding figures for his arc source instrument[100-2], the prototype of the Skatron Argus/Bio-Rad Bryte HS, were 1800 MESF with background and 500 MESF without background.

In the arc source instrument, which uses the epiillumination system of a fluorescence microscope, much of the fluorescence background is due to fluorescence induced in the cover slip and lens, and possibly in the dichroic, by excitation light; this becomes a much more severe problem when using UV (366 nm) or blue (436) nm excitation than when excitation is at higher wavelengths. The use of quartz or silica in place of glass in optical components can reduce, but probably not eliminate, the fluorescence background; Partec, the major manufacturer of arc source flow cytometers, supplies quartz lenses and filters.

Sensitivity VI: Electrons Have Statistics, Too

In a PMT, the random processes don't stop after photons hit the cathode; Steen[1165] considers the effects of electron statistics in the PMT at various gain levels on precision. PMT gain results from secondary electron emission from successive dynodes; as the voltage between dynodes is increased, electrons acquire more energy before striking the next dynode, and release more secondary electrons. Electron statistics increase the contribution to CV; for an R928 PMT at 300 V, this factor is almost 1.4; with 1000 V applied to the tube, the factor is less than 1.1.

Most flow cytometers provide several stages of electronic gain, allowing the preamplifier output to be amplified by a selectable factor before signals are digitized. Steen's calculations and data provide excellent reasons for not using such electronic amplification; an amplifier gain of 10 and a low PMT voltage result in a larger contribution to CV from the PMT and 10 times as much amplifier noise as an amplifier gain of 1 and a higher PMT voltage that increases PMT gain by a factor of 10. If your preamp can get a full-scale output signal from a 10-100 µA input, you can avoid using additional gain stages; check with your manufacturer.

Source Noise Fluctuations and Performance

The background light reaching a flow cytometer's detectors, which are shielded from outside light, is dominated by light derived from the source. In the case of scatter detectors, background light levels are relatively high, and due primarily to stray illumination light scattered by interfaces in the flow chamber (or, in stream-in-air systems, by the air/stream interface) and by optical components. In the case of fluorescence detectors, the background light level is typically lower. Fluorescence from materials in the core and sheath and from optical elements, including flow chamber components, lenses, and filters, predominates; there may also be some Raman scattering from water molecules. The levels of stray light exhibit the same fluctuations as the source. The background light can be considered as composed of a constant (DC) component, which is, ideally, removed by baseline restoration, and a variable (AC) component, which is not.

I Blurred It Through the Baseline

Periodic fluctuations in the output of the light source result from such factors as power supply and laser plasma noise and arc wander. After baseline restoration, these fluctuations show up as noise or "ripple" on the baseline; higher source noise levels result in higher baseline noise. At illumination levels that do not produce photon saturation, the CV of fluorescence and scatter measurements cannot be lower than the RMS noise level of the source (which is the CV of source intensity) unless a source noise compensation circuit (pp. 147-8) is used. In the absence of background, source noise affects precision to at least this extent. When background is present, it also exerts proportionally larger effects on smaller signals.

Suppose the light source is an air-cooled argon ion laser, with 1% RMS noise. If no background is present, and the flow cytometer is otherwise ideal, measurements of completely uniform fluorescent particles will have a CV representing contributions from the source variation and from photon statistics. From the photon statistical CV's given for large signals in Table 4-3, and the contributions from electron statistics in the PMT previously discussed, we can guesstimate a contribution of 0.3%. The final CV, in percent, will be $[(1)^2 + (0.3)^2]^{1/2} = (1.09)^{1/2} \cong 1.044\%$.

If the same light source is used, with background increased to produce 1 µA at the PMT output, or 100 mV at the output of the 100,000 V/A current-to-voltage converter in the preamp, the 100 mV DC average will be removed by baseline restoration. The AC representing the 1% RMS noise, will show up as 1% of 1 V, or 1 mV, RMS, centered on the 0 V baseline. Assuming the noise is Gaussian, I use the quick-and-dirty rule of thumb given on p. 147 to estimate peak-to-peak voltage as 6 times RMS voltage. In this instance, with 1 mV RMS noise, the noise waveform should be within ± 3 mV of zero at least 99% of the time.

If you're looking at the highest decade (1-10 V), 3 mV is negligible at the top, and represents only 0.3% of signal at the bottom. In the next decade, when you get down to 100 mV, the 3 mV is 3% of signal value. By the bottom of the third decade, 3 mV looms very large on top of a 10 mV signal, and the 3 mV noise level prevents you from getting into the bottom half of the fourth decade. Even with perfectly identical 100 nA pulses going into the preamp, which should give you identical 10 mV pulses out, you'd see a very much broadened distribution due to the background noise. Thus, in the presence of background light, source noise fluctuations will decrease precision; the lower the signal value, the worse it gets. Fluctuations will also compromise sensitivity; with 3 mV noise, signals below 3 mV become undetectable. More background with the same RMS noise, or, alternatively, the same background with more RMS noise, will make things proportionally worse.

Good flow cytometer design should keep both the background and the source noise level low. Observation in flat-sided cuvettes instead of round streams or capillaries has been the principal means by which background has been reduced; background in fluorescence channels has also been reduced by the use of better designed filters, which do not themselves fluoresce and which transmit virtually no light at the excitation wavelength.

In stream-in-air instruments, in which some of the illuminating laser light is scattered in the direction of the fluorescence collection lens, an obscuration bar is typically used to keep most of this light from reaching the lens. Although fluorescence signal levels can be increased when the bar is removed, sensitivity may actually be decreased if the excitation light used induces substantial fluorescence in the lens elements. UV and blue-violet light are the worst offenders here; I have found fluorescence due to UV excitation light even in a quartz lens, although fluorescence in glass lenses is much higher. Where maximum light collection is the goal, it may be necessary to use catadioptric reflective lenses for light collection to eliminate fluorescence.

Since scatter channels operate at the illuminating wavelength, they cannot include filters that block this wavelength; background is therefore higher than in fluorescence channels. Baseline noise generally determines the sensitivity of a scatter channel; excessive noise may impede or prevent use of the scatter signal for triggering. A water-immersed flow system, in which light scattering interfaces are several millimeters away from the sample stream and well out of the field of view of the collection lens, provides the lowest scatter background. Block's instrument for virus detection[94], using such a flow system, could trigger on scatter signals from particles smaller than 0.1 µm, using a 100 mW water-cooled argon laser source, with 0.2% RMS noise. It was also possible to detect the same particles, with a predictably lower signal-to-noise ratio, using a 10 mW air-cooled argon laser with a similar RMS noise level, while such particles could not be detected in either a round capillary or a stream in air using either laser, due to the higher background. Illustrating the same principle, I found I could detect and trigger on scatter signals from 0.3 μm particles in a stream-in-air system using a He-Cd laser when the source noise was reduced to 3% peak-to-peak by an electro-optic modulator, but not when the same laser was running at slightly higher power with 15% peak-to-peak noise.

Source noise compensation (pp. 147-8; Figure 4-15) can be used to improve precision and, to some extent, sensitivity of measurements, although it is difficult to use with scatter detectors, because of the relatively high background noise levels involved. The minimum measurement CV calculated previously for measurements with an air-cooled argon laser with 1% RMS noise was 1.044%; instruments using such light sources achieve CVs in the range of 1.5%, suggesting that source noise may be the major contributor to variance and that CVs might be improved by noise compensation.

Restoration Comedy: The Case of the Disappearing Leukocytes

Baseline restoration is great stuff, up to a point, but we should remember that backgrounds, DC or AC, are never desirable. A 1 V DC background on the signal coming out of the preamp first stage, means that 10 μA of the PMT output current represented DC background; that only leaves us 90 μA to work with if we want to restrict output current to 100 μA to prevent nonlinear PMT response.

Under some circumstances, fluorescence backgrounds can get very high; with a 10 V background, even if you have a 10 V signal, you won't see it, because the op amps in the preamp and baseline restorer stop working like op amps when they reach **saturation**, i.e., get to the voltage or current limits imposed by their power supplies. Restorer failure on this basis was the culprit in one of my most puzzling cases, one I have called "The Case of the Disappearing Leukocytes" (I know, I should have gotten this into Dr. Watson's book[1030]).

My colleagues at Block and I used the first multiple excitation beam flow cytometers[88,92] to count blood cells stained with a mixture of three dyes[89,90,91,93]. Ethidium bromide, now well known in flow cytometry, produced red nuclear staining. Brilliant sulfaflavine (BSF), excited at 420 nm, stained eosinophil granules green, and LN ("Long Name"), excited by UV light, stained neutrophil and eosinophil granules blue. BSF and LN, both sulfonated acid dyes, produced intense staining, but there was a tremendous amount of background fluorescence in cell suspensions. The instruments[88,92] incorporated baseline restorers and noise compensation, and, in addition, used a very small core stream to minimize background fluorescence.

I was somewhat surprised when John Steinkamp (personal communication) told me in 1978 that he was not able to detect LN and BSF signals in blood cells stained with the dye mixture, when he examined the cells in Los Alamos' dual laser source instrument[125]; the multiwatt ion lasers used there for UV and violet illumination should have produced very strong fluorescence signals from the cells.

Shortly thereafter, I was able to run a sample of leukocytes stained according to our protocol on one of Ortho's ICP instruments[9,83], using UV illumination to excite LN and blue illumination to excite BSF. I was unable to get signals from cells that I had seen were brightly stained, having examined the sample under the fluorescence microscope.

This represented a clear violation of:

Shapiro's Second Law
of Flow Cytometry
[Flip) Wilson's Rule[126]]:
What You See
Is What You Get!

The provision in the ICP for observation of the sample stream allowed resolution of the mystery, if not of the cells. The background fluorescence from dye in the core stream was very strong.

Under staining conditions that result in high dye background, the cells may brighter than the core, per unit volume, but the volume of material from which the fluorescence signal is derived, which is larger than the cell volume, becomes critical. Taking typical numbers, suppose we have a 20 μm core illuminated by a beam 20 μm high. The **observation volume** thus defined will be roughly the volume of a cylinder with diameter equal to the core diameter and height equal to the beam height; this is $\pi \times 10^2 \times 20$ femtoliters (fL), or 6284 fL, where a femtoliter is 10^{-15} liters, or what used to be called a cubic micron (1 μm^3). The volume of a neutrophil is about 400 fl. Suppose the cell is 5 times as bright as the core; since the observation volume is about 15 times the cell volume, we can expect the fluorescence signal obtained during the passage of a cell through the observation volume to be only 1 1/3 times the background signal from an observation volume containing no cells.

Even when the ratio of core fluorescence to cell fluorescence is low and the observation volume is small, we can be prevented from resolving cell fluorescence over background (core) fluorescence if the background fluorescence is bright enough to drive some stage of the electronics into saturation. The instruments originally used to detect BSF and LN fluorescence[88,92] were designed to deal with the high background fluorescence associated with these dyes; a 5 μm core and 20 μm beam height were used, giving an observation volume of <400 fl, and, in the electronics, feedback was applied to the first stage of the preamplifier to subtract the DC baseline. The Los Alamos instrument[125] and the ICP had larger observation volumes and thus had to deal with more background fluorescence; while I am not certain about the resulting effects on the electronics in the Los Alamos system, it was clear that the ICP's electronics could not cope with the high background fluorescence. Thus, the mystery of the disappearing leukocytes was solved, proving that what you see is not always what you get, and that the flow cytometer is not always quicker than the eye.

Top 40 Noise Sources

To this point, I haven't said much about noise and related problems originating in flow cytometer electronics, although I have mentioned that there always seemed to be a few millivolts of noise about. Flow cytometer design, like everything else, faces economic constraints, and it doesn't pay to design components with better performance than you need. As long as people were using 8-bit A-D converters, in which the distance between channels is 39 mV, any noise level or offset below 39 mV was acceptable, especially when weak signals went into log amps. If you measure the outputs of older commercial flow cytometers, you'll find that "zero" can be anywhere within 30 mV of ground, and that peak detectors and integrators may be inaccurate in the range below a few hundred millivolts. This means you can pretty much forget about upgrading an old instrument to 4-decade digital log performance simply by hanging a 16-bit or higher resolution ADC on its outputs.

The manufacturers have fixed things in the design of newer instruments; recent work on the subject, to be discussed below, indicates that electronic noise is no longer a limiting factor for sensitivity, at least in commercial systems. In the course of trying to wring the last few millivolts of noise out of one of my Cytomutts[2467], Phil Stein and I hung an electronic (not optical) spectrum analyzer on the output of the prototype low noise electronics to see where the noise was coming from; in addition to the usual power line noise, the most significant contributions came from the horizontal sweep generator in the computer monitor and the power supply in a He-Ne laser. Things got noticeably quieter when we moved the offending laser supply further away from the electronics and used a notebook computer with an active-matrix screen to run the instrument, eliminating the monitor (this consideration later justified buying LCD displays for the lab long before they got as cheap as they are now).

The laser supply and monitor noise were in the frequency range below 100 kHz; when we looked up in the 100 MHz range, we found detectable noise spikes from most of the local FM stations. Just when we think we've gotten the grunge out of the flow system, it shows up in the electronics.

Sensitivity 007: Q and B (Dye Another Day?)

After the third edition of this book appeared, with its lengthy discussion of source noise fluctuation effects on sensitivity, Harald Steen took me to task; he didn't think source noise was nearly as important as photoelectron statistics as a limiting factor in sensitivity. I was offended until I realized that his light sources are quieter than mine. The dominant current approach to the definition and measurement of sensitivity is based on an amplification of Harald's earlier work[1163-5] by Jim Wood, formerly of Beckman Coulter, and Bob Hoffman, of BD Biosciences[2476-80], and I again acknowledge their help in putting this section together. The topic is obviously important if rival manufacturers have cooperated over a period of years to make better sense of things.

Q and B are neither mysterious secret agents nor (yet) Jim and Bob's new nicknames. Q is a measure of detection efficiency, with units of photoelectrons (or fractions of photoelectrons) emitted from the detector photocathode per molecule-equivalent (MESF) of fluorochrome measured. It is a function of many variables, potentially being affected by optical misalignment and disturbances of fluid flow, but more directly and obviously influenced by which fluorochrome is measured, by the power of the excitation source and by the efficiency of the light collection optics and the quantum efficiency of the detector photocathode.

B is a measure of background noise, usually expressed in MESF units as the number of fluorochrome molecule equivalents required to produce that level of background. B includes contributions from photoelectrons generated by photons emitted by fluorochrome molecules not associated with the particle (e.g., dye in solution), by photons from fluorochromes other than the one of interest (crosstalk due to spectral overlap), by photons emitted by cellular constituents (autofluorescence), by photons at the excitation wavelength from the light source that have passed through the detector filter, and by photons resulting from Raman scattering of the illuminating light by water. Electrons produced by thermionic emission from the photocathode (dark current) are also included in B, but usually not significant except in the far red and near infrared. The quantity F, also usually expressed in MESF units, denotes the number of particle-associated fluorochrome molecules.

It is possible to determine Q and B using only sets of beads. Measurements are made on the flow cytometer using a linear, rather than a logarithmic, scale. As a first step, the scale of the instrument must be calibrated in terms of MESF units of the fluorochrome of interest, using beads labeled with known amounts of that fluorochrome. Bangs Laboratories makes fluorescein- , phycoerythrin-, and PE-Cy5-labeled Quantum MESF beads, and BD Biosciences provides QuantiBRITE beads bearing known numbers of molecules of phycoerythrin[2481-2]; such beads can be used for a single-point MESF calibration, assuming the linearity of the instrument is within specification.

The next step uses beads from sets bearing various amounts of fluorescent dyes that emit in the same spectral region as the fluorochrome(s) of interest; these are available from a number of manufacturers, including Bangs Laboratories, Molecular Probes, and Spherotech. Dimly fluorescent or unstained beads will emit relatively few fluorescence photons, and the variances (i.e., the squares of the standard deviations, see pp. 19-22 and Chapter 5) of fluorescence distributions from such beads will therefore be dominated by contributions from photoelectron statistics, but will also contain "basis" contributions resulting from variations in bead size and dye loading and from instrumental factors. Since the brightest bead in such a bead set typically carries at least 100,000 molecules of dye, the variance of its fluorescence distribution will be dominated by the "basis" factors. The bright bead is first measured, and its fluorescence variance calculated. The measurement need not be made on the calibrated scale; in fact, it will usually be necessary to lower the gain to get the fluorescence distribution from the bright bead on scale. The gain is next (re)adjusted, if necessary, to the value used for MESF calibration, and fluorescence from three dim beads is measured; variances are calculated for each of the resulting bead peaks, while MESF units per bead are determined from the positions of the peaks.

"Photoelectron variances" for each of the three dim beads are calculated by subtracting the measured variance of the brightest bead from the measured variances of the dim beads. Next, the photoelectron variances are plotted (as the dependent variable, i.e., on the vertical axis) against the MESF values for the dim beads, and the equation of a regression line best fitting through the points is determined using the method of least squares (a spreadsheet for the calculation has been developed by Bob Hoffman and Jim Wood, and should be available on the Wiley web site associated with this book). The slope of the line will be 1/Q, and it will intersect the vertical axis at B/Q, allowing B and Q to be determined.

B and Q have been calculated for a reasonable number of working flow cytometers. Before I give you the details, however, I will ask you to look back at Table 4-4, on p. 203, in which I tabulated signal levels that might be expected to come out of a PMT in a flow cytometer. If we assume that the 5-decade scale used in constructing that table represents signals from between 10 and 1,000,000 cell-associated MESF, we can calculate Q without beads; using the numbers of photoelectrons in the sixth column of the table we get Q = 0.62 photoelectrons/MESF. I suggested that the calculations in the table might be optimistic, and they were. Hoffman and Kuhlmann[2482] observed that typical values for Q in BD FACScan and FACSCalibur cytometers were 0.25 photoelectrons/MESF for phycoerythrin (PE) measurements and only 0.012 photoelectrons/MESF for fluorescein measurements. They found it unusual to see a value of Q much higher than 0.4, even for PE, on these optically efficient benchtop systems.

B varies considerably more than does Q with operating conditions. Once the instrument is properly aligned, assuming there is no substantial variation in laser output, one would expect Q to remain relatively constant. Some factors that contribute to B, such as Raman scattering and illumination light leakage through filters, will also remain relatively constant; however, others, such as fluorescence from dye in solution and from other fluorochromes with spectral overlap, will vary. Typical values of B associated with bead measurements as just described were 2,000 MESF for FITC and 300 MESF for PE. However, background contributed by unbound fluorescent antibody in no-wash immunofluorescence analyses was 4,000 fluorescein MESF and 1,600 PE MESF. Spectral overlap from fluorescein-labeled anti-CD45 antibody on lymphocytes generated a 7,800 MESF background level in the detector used for PE.

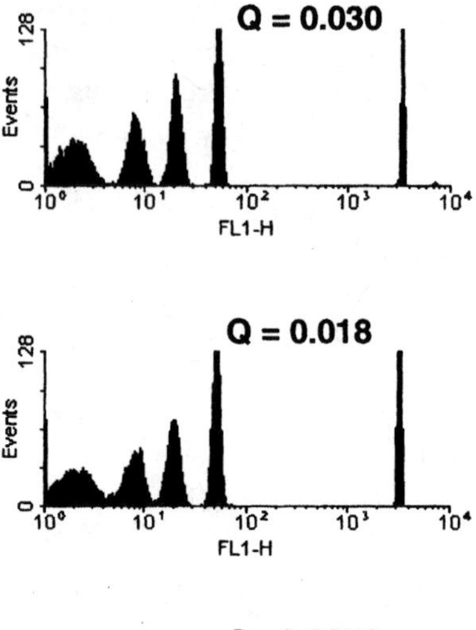

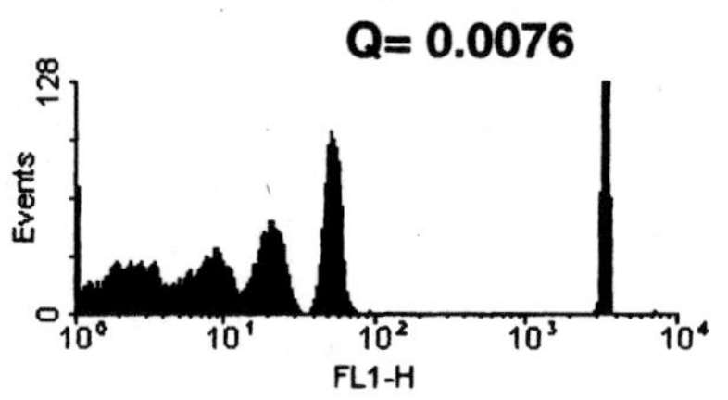

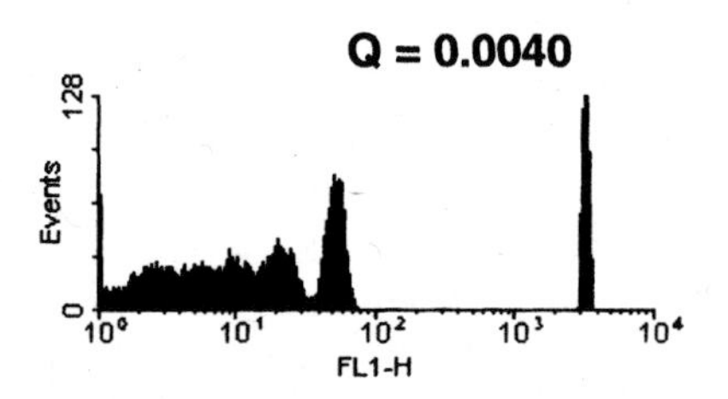

Figure 4-61. 488 → 525 nm fluorescence distributions for the four dimmest beads and the brightest bead of an 8-bead dyed bead set (Spherotech Rainbow Beads) measured with progressively lower values of Q. Courtesy of Bob Hoffman, BD Biosciences.

It's easy enough to figure out why Q is important in determining sensitivity. Figure 4-61 shows distributions of green fluorescence (the "fluorescein channel," FL1 in BD parlance) of a mixture of five of the eight beads in a Spherotech Rainbow Bead set. The measurements were all made in the same cytometer; Q was progressively decreased by lowering laser power, while B remained constant. Note that the peaks representing the dimmer beads become progressively broader as Q decreases, to the point at which the second dimmest peak is not clearly separable from either the dimmest or the third dimmest when Q = 0.0040. That makes sense; when fewer photoelectrons contribute to each measurement, the CV of the peak increases, and the peak gets wider, although its center (median or mode) remains in pretty much the same place, meaning that the degree of overlap between neighboring peaks keeps increasing as Q drops. Not incidentally, since it has already been noted that the typical value of Q for a fluorescein measurement channel is 0.012[2482], we can conclude that the top two curves show better separation of dim peaks than is likely to be observed in an average instrument, and hope that we will do better than the bottom two.

While decreasing Q has the effect of broadening all of the distributions, increasing B has the primary effect of broadening the dimmer peaks and pushing the distributions up the scale, and can be an equally effective way of preventing us from telling dim objects from dimmer ones, as can be seen from Figure 4-62.

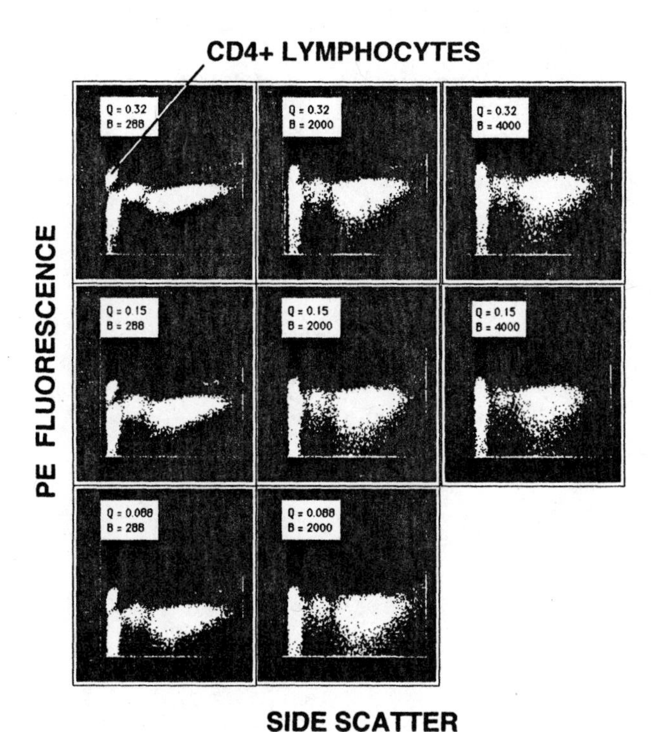

Figure 4-62. Separation (or lack thereof) of dimly stained CD4+ lymphocytes from unstained cells as observed with various values of Q and B. Courtesy of Bob Hoffman, BD Biosciences.

The cells shown in the figure were stained with a subsaturating concentration of PE-labeled anti-CD4 antibody in the presence of a saturating concentration of PerCP-labeled antibody, which exhibits little or no spectral overlap with the PE-antibody. CD4+ lymphocytes normally bind about 50,000 anti-CD4 antibody molecules per cell; it is probable that the CD4+ cells shown here are binding at most only a few thousand molecules of PE-labeled antibody. The CD4+ lymphocyte population (identified in the top left panel of the figure, with low side scatter and higher PE fluorescence than the other cell clusters) is readily distinguishable when background is low (B = 288 MESF), although separation from unstained lymphocytes gets worse as Q drops from 0.32 to 0.088. However, when B is 2000 or 4000, only the samples for which Q = 0.32 show even a hint of separation between CD4+ cells and other lymphocytes.

A "separation parameter," SP, where

$$SP = (Q \times F)^{1/2}/[1 + 2(B/F)]^{1/2},$$

may be calculated[2477]; it defines the difference, in normalized standard deviation units (see Chapter 5), between a stained population, with mean intensity F MESF units, and an unstained or blank population. As it turns out, decreasing B, if possible, can allow a constant degree of separation between populations to be maintained in the face of a decrease in Q. To look at things in another light, as it were, the sensitivity of the system, meaning its ability to separate dim fluorescence peaks, cannot be predicted from either Q or B alone; values of both must be known, and it is important to remember that B can change dramatically with experimental conditions, e.g., whether or not samples are washed before immunofluorescence measurements are made.

You may not want to get involved in the measurement of Q and B, but it is a good idea to keep sets of beads on hand and run them every now and then; if you find that you are less able to resolve dimly stained populations than you once were, it's time to adjust or fix the instrument.

That's it for now on how and how well flow cytometers work, except for a final message for you to take along while you read the next chapter on data analysis: Both the flow system and the data analysis system of a flow cytometer can give you garbage out in response to garbage in.

5. DATA ANALYSIS

Flow cytometry is an analytical technique for getting information about or, if you prefer, for answering questions about, cells, and an experimenter who wants to be successful in getting information or answering questions must be at least aware of, or, preferably, actively involved in, the selection of methods for both data acquisition and data analysis.

The **data acquisition** process in flow cytometry comprises all those operations which are required to make measurements of a specified **physical** characteristic(s) of cells in the sample, such as forward light scattering or green fluorescence intensity, and to convert the data to a numerical form suitable for manipulation by digital computers and long-term storage on magnetic or optical media.

The **data analysis** phase of flow cytometry includes any and all of the subsequent operations used to derive information about the **biological** characteristics of some or all of the cells in the sample from the measured values of the physical characteristics. Methods of flow cytometric data analysis may differ greatly in their complexity, depending primarily upon **what the experimenter wants to know about the cells.**

The successful application of flow cytometry to the characterization of cells in mixed populations requires that the instrument and/or the analysis software have the capacity to discriminate among the different cell types that may be present in a sample. This is provided by **multiparameter analysis**, which is what has made flow cytometry as useful as it is and should continue to be.

As multiparameter instruments and measurement techniques have proliferated, there have been calls from many quarters for the use of progressively more sophisticated statistical and numerical methods in flow cytometric data analysis and, indeed, some of these techniques may be highly informative when applied to the appropriate data by experimenters who understand both the biology and the statistics. Unfortunately, there seem to be a great many people spending a lot of time with sophisticated numerical methods in attempts to violate what I would call:

Shapiro's Seventh Law
of Flow Cytometry:
No Data Analysis Technique
Can Make Good Data
Out of Bad Data!!!

Data analysis methods share some of the characteristics of flow cytometers and lasers; bigger, or more complex, isn't necessarily better.

5.1 GOALS AND METHODS IN DATA ANALYSIS

Cell Counting

If a population of cells existed that were identical in all respects, one might still want to use a flow cytometer to count them, for example, to determine their concentration in a sample. Even in this idealized simple case, the detector output(s) would contain pulses not produced by the passage of single cells through the cytometer's observation region. In addition to pulses resulting from spatial and temporal coincidences of two cells, one could expect other pulses resulting from the presence of contaminating particles in the cell suspension and pulses due to optical and/or electrical noise in the system. Also, it would be necessary to correct the measurement time to take into account the dead time of the instrument. The measurement procedure would have to deal with both coincidences and dead time in order to produce an accurate count.

Tasks are typically divided between the instrument and its data analysis system; **triggering**, which eliminates low-level signals not representing cells, has traditionally been the province of the instrument, while **gating**, which may be used both to refine the trigger level and to eliminate cell multiplets, has been done with both hardware and software. With the advent of digital pulse processing, the boundary between the instrument "front end" and the analysis software has become less distinct; it has already been noted (p. 210) that software-based digital triggering can be much more flexible and sophisticated than the older, hardware-based process.

Characterization of Pure Cell Populations

Analysis of "homogeneous" populations of real cells usually has the purpose of defining and quantifying cellular characteristics which do vary from cell to cell, e.g., cell size, DNA content, or the amount of a particular antigen present on the cell surface; these parameters are usually analyzed in cell populations subjected to different manipulations. Studies of the effects of anticancer agents on DNA synthesis in cultured tumor cell lines provide a good example of this type of analysis; the analytical procedures used are typically concerned with comparing the **distributions** of DNA content in different samples, usually with the aid of mathematical models. However, even when distributions are to be collected from "homogeneous" populations, hardware and/or software gating, sometimes of a relatively elaborate nature (pp. 25-6), may be needed to eliminate debris and clumps from the analysis.

Identification of Cells in Mixed Populations

Differential leukocyte counting and T cell subset analysis exemplify the task of identification and counting of cell subpopulations in a mixed population. Accurate identification of cells generally requires measurements of several cellular parameters. Cells' intrinsic light scattering properties, which may give indications of size and surface or internal structure or granularity, and cellular content of DNA, RNA, lipid, various proteins, enzymes, receptors, and antigens, measured with extrinsic probes, have all proven to be useful parameters for cell classification. The analytical techniques used are generally interactive and empirical, and, at present, are typically based on definition of one or more **two-parameter gates**, even when more than two parameters are analyzed. By contrast, automated procedures developed for cell identification are more likely to make use of **multivariate statistics** and related methods.

Characterization of Cell Subpopulations

As one focuses attention on changes in structural or functional parameters in cell subpopulations which are found with diminishing frequency in a sample, e.g., relative frequencies of megakaryocytes with octoploid (8C), 16C, and 32C DNA content in bone marrow, it becomes more important to eliminate the contributions of noise, dead cells, debris, and cell multiplets to the data in order to obtain reliable information about the cells of interest. Comparison of distributions may be the end objective, but, without the development of appropriate gating techniques, there will be no distributions to compare.

Data Analysis Hardware and Software Evolve

In the early days of flow cytometry, multichannel pulse height analyzers were the most widely used means for accumulating and displaying signal intensity distributions. Thereafter, mini- and, later, microcomputers took over. Most flow cytometer builders were not exactly in the vanguard of that revolution; they continued to build their machines using pulse height analyzers, but, since the manufacturers of the pulse height analyzers were using minicomputers to build them, computers found their way into flow cytometers. After a while, people began to realize that it made more sense, and cost less, to equip a flow cytometer with a computer system designed to process data from a flow cytometer than it did to use a computer system designed to process data from nuclear spectroscopy.

Although the multichannel analyzer dealt reasonably adequately with distributions of single parameters, until computers came into wide use, few flow cytometers were equipped for simultaneous correlated analysis of two parameters. The simplest and, for most users, the only way of examining relationships between parameters was the production of a dot plot, or cytogram, using short term analog stored data (stretched pulses), the strobe signal, and the oscilloscope typically built into the flow cytometer. Hardwired additions to the basic system for dot plot generation, described on pp. 26-7, made it possible to set multiparameter gates, and, with the aid of a pulse height analyzer, to accumulate single-parameter distributions for cell subpopulations, but did not allow either accumulation of two-parameter distributions or data storage; photographs of the oscilloscope screen provided the only permanent records of two-parameter data.

Despite their impressive price tags, most flow cytometers sold before the 1980's did not have as much computer capability as Kamentsky's original apparatus. Since then, radical change has occurred as a result of the development of personal computers and the increased level of familiarity of flow cytometer users with computers in other contexts. Flow cytometer users, like everybody else, are now used to menus, mice, word processing, spreadsheets, and graphics software, and expect that either cytometer manufacturers or third party software developers will make it straightforward and simple not only to do elaborate analyses of flow cytometric data, but to move the results, in tabular and/or graphic form, into our clinical records, reports, slides, and scientific publications. Understanding what we're doing to the data, and deciding whether the selected analytical methods and the measurement values and results make sense, remain up to us, and will for the foreseeable future. For now, we'll consider computers and their use in cytometric data analysis.

5.2 COMPUTER SYSTEMS FOR FLOW CYTOMETRY

The title of the Chapter in the First Edition corresponding to this one in the present edition was "Data Analysis With and Without Computers." Now, there is no data analysis without computers. The original functions of digital computers in multiparameter flow cytometry were **data analysis**, **display**, and **storage**; because of the speed limitations of the first generation of computers which could practically be dedicated to flow cytometry, real-time control was more often done with hardwired, parallel-processing devices such as window comparators, which could be "supervised" by computers but which did not demand continuous attention from the computer. In the highest speed modern instruments, even real-time control is typically accomplished by hybrid electronics incorporating microprocessors, and personal computers are sufficiently fast to be used effectively for overall control.

The Beginning

When computers first became generally available to scientific researchers, in the 1960's, cytometry was heavily oriented toward image analysis. High-resolution image data from scanners were typically written on 9-track magnetic tape and carried to mainframe computers for analysis, which took minutes per cell. At that time, an IBM engineer named Lou Kamentsky considered the problems of cell image analysis at high resolution, concluded that there weren't powerful enough light sources, good enough sensors, fast enough processors, or enough memory available to make this practical, and developed optical flow cytometry. Although his original flow cytometer[65] incorporated a dedicated computer, this, in the mid-1960's, was a luxury affordable only by people who worked for IBM or the U.S. Government. The system was an IBM 1130, a small mainframe with less computing power than some of us now wear on our wrists.

When Kamentsky left IBM to found Bio/Physics Systems in 1970, his argon laser source flow cytometers were considerably less expensive (under $20,000) than the minicomputer systems of the time, and he did not offer customers the option of a computerized data analysis system. The reconciliation of flow cytometers and computers began at Los Alamos[1176-7], which, in the early 1970's, duplicated a few of its cell sorter systems, with dedicated DEC (Digital Equipment Corporation, since digested by Compaq, now merged with Hewlett-Packard) PDP-11 minicomputers, for use at the National Cancer Institute. Software for these systems was developed at both NIH and Los Alamos. At around the same time, the Jovins incorporated a PDP-11 into the cell sorter they built in Göttingen[147]. By the mid-1970's, Michael Stöhr and his colleagues[323], in Heidelberg, were interfacing various small computers to commercial flow cytometers, Tom Sharpless[128], at Memorial Hospital in New York, was developing a data acquisition and analysis system using a Data General (Data General Corp., Westborough, MA) Nova minicomputer, and Block Engineering was building Novas into prototypes of the Cytomat multibeam flow cytometers[88,92].

By the end of the 1970's, inflation and the inclusion of large lasers and sorting capability had driven the prices of flow cytometers well above $100,000; at the same time, demand arose among an increasing number of experienced users for more data analysis capacity than was available from pulse height analyzers. It thus became feasible for flow cytometer manufacturers to produce minicomputer-based data analysis systems. Becton-Dickinson had already incorporated Nuclear Data's PDP-11-based dual-parameter pulse height analyzer in its top-of-the-line systems, and continued development using Digital's PDP-11 and VAX computers. Ortho chose to base its computerized data analysis systems on Data General minicomputers.

The End of the Beginning

In a rapidly developing field, the last competitor to solidify a product design may gain the advantage of using the latest components; among commercial flow cytometer manufacturers, Coulter gained a clear advantage by basing its data analysis systems on microprocessors from the beginning, and consolidated it by shifting toward processors and software compatible with a *de facto* industry standard.

The standard was, of course, established by IBM (International Business Machines Corporation, Armonk, NY) with its Personal Computer (PC) family, introduced in the early 1980's. While IBM elected not to go into the cytometry business in the 1960's, we now find tens of thousands of descendants of IBM PCs in wide use for both flow cytometry and image analysis. They are, luckily, only a small minority of the hundreds of millions of PC-compatible computers installed worldwide, and can utilize both the software and add-on hardware developed for this large market for a wide range of applications.

In developing the PC, IBM took cognizance of the earlier success of Apple Computer (Cupertino, CA) with its Apple II line. Apple's personal computers incorporated a so-called **open architecture**; hardware could be connected directly to the microprocessor's **address**, **data**, and **control** lines, or **buses**, simply by plugging circuit boards into connectors, or **slots**, on the computer's **motherboard**. The electrical specifications, timing, and pin connections of the IBM PC and compatibles, like those of the Apple II, were made freely available by the manufacturers, facilitating development of peripherals by third parties, and thus giving rise to whole new industries. Both the Apple II and the IBM PC originally came into wide use primarily for business applications; however, once large-volume markets were established, the computers became accessible to other groups, including scientific researchers.

The Apple II was not enough computer to serve effectively as the data analysis system for a multiparameter flow cytometer system. Its major limitations lay in its restriction to 8-bit arithmetic operations and its small (64 KB; 1 KB = 1024 bytes of memory) address space. To be sure, this seemed like enough memory when the chips were developed; at the time, 64 KB of **random access memory (RAM)** cost several hundred dollars, and personal computers were shipped with only 8 or 16 KB of memory. And a few adventurous souls did hook 8-bit microcomputer systems up to flow cytometers. Kratel's flow cytometers originally came equipped with data analysis systems incorporating Radio Shack (Tandy Corp., Ft. Worth, TX) TRS-80 personal computers, and a number of individuals tried Apple IIs. Atari's 800 series ran almost twice as fast as an Apple II and had better graphics, but lacked the Apple II's open architecture and shared its addressing and arithmetic limitations. The Ataris were regarded as game machines, even after I showed that they could be made to capture multiparameter flow cytometric data faster than a PDP-11. The data acquisition and analysis systems I now use with 32-bit "Wintel" PCs was developed using 8-bit Atari computers; the lesson I learned, and now pass on to you, is that, no matter how good a computer is, it will never be widely used if it isn't taken seriously.

The IBM PC/XT and PC/AT and their clones did 16-bit arithmetic and could address a megabyte (1 MB = 1024 KB) of memory, including 640 KB of user RAM, with the drawback that they talked to memory in 64 KB segments; they also had separate instructions and addressing for **input-output (I/O) ports**, to which they could send, and from which they could receive, 8-bit data. The 640 KB RAM limit didn't seem like much of a problem in the early 1980's, when a little-known company called Microsoft developed the **Disk Operating System (DOS)** that still runs many IBM PC descendants. DOS hobbles the 80386, 80486s, and Pentium processors used in most PC's since the middle of that decade, which have the hardware capacity to address huge amounts of RAM but don't have an easy way to do it running under DOS. Microsoft's **Windows** operating environment, although its older versions (3.x, 95, 98 and Me) still run under DOS, provides for addressing gigabytes (GB; 1 GB = 1024 MB) of memory, and, these days, if you don't already have 1 GB of memory in your computer, you can add it for under $200.

The huge market represented by the large number of PC-compatible computers in use has generated a cornucopia of software, which includes at least one version of every important programming language, and a wide range of peripheral hardware for data acquisition, storage, and display. There are also numerous books dealing with PC family hardware and interfacing[627-8, 1059-62]. Coulter and I have continued to use IBM-compatible computers in flow cytometers, and most cytometer manufacturers have followed suit(s).

The major competition to Intel's 80x86/Pentium series of microprocessor chips now comes the Motorola/IBM PowerPCs, which have replaced the Motorola 68000 series chips used in earlier generations of the Apple Macintosh. The Mac preceded Windows in achieving a uniform user interface at the Mr. Rogers level of user-friendliness by imposing Gulag discipline on programmers and hardware developers[879-81]. From 1984, when they were introduced, until 1987, Macintosh computers had closed architectures. Since then, some Macs have offered expansion slots, along with progressively faster processors. Macs used to be relatively expensive, even with institutional discounts; prices have become somewhat more competitive over the years.

Becton-Dickinson's CONSORT 30 series of computer workstations used 68000-based computers made by Hewlett-Packard; this hardware was well designed for rapid data acquisition, but the computers, being specialty items rather than personal computers, were neither widely used nor inexpensive. Until 1994, B-D elected to deal with users who wanted to do data analysis on PC-compatible computers by facilitating data file transfer over Hewlett-Packard's (IEEE 488) instrumentation bus or via networks. B-D then introduced the FACStation, the first of a continuing series of Macintosh-based data analysis systems, and continues to use Macs with most of the FACS instrument line, although the latest FACSDiVa digital pulse processing systems use PC-compatible computers.

Even relatively sophisticated modern flow cytometers can now be run from laptop computers; one could probably use palmtops to run the simplest instruments. The credit goes not to the flow cytometer manufacturers, but to the computer industry; the original IBM PC had the processing power of a 1960's mainframe, and current PC-compatibles offer over 1,000 times the memory and 1,000 times the speed of that mainframe for 1/1,000 of the cost, while consuming les than 1/100 the power.

Data Rates and Data Acquisition Systems

Acquisition of data from a multiparameter flow cytometer can be a demanding job even for a minicomputer, but may be done as easily with a small microcomputer as with a large mini. This paradox is explained by considering the several ways in which data can be fed into a computer.

The **data rate**, i.e., the rate at which data are generated, in a flow cytometer is the product of the number of parameters measured and the rate at which cells are analyzed. However, the raw figures can be deceiving. If two parameters are being recorded for 1,000 cells each second, the average data rate is 2,000 points/s, suggesting that the computer has 500 μs in which to deal with each data point. In actuality, however, data from two channels arrive almost simultaneously, and, since the data are **asynchronous**, that is, the cells arrive at unpredictable times, the minimum interval between cells may be 50 μs or less, which would require that the computer process data at a rate of 25 μs per data point.

The electronics in what we now must call a "classic" flow cytometer provide **analog outputs**, in the form of stretched pulses representing pulse heights, areas, or widths, and the digital **strobe signal** which tells the computer that data need to be digitized and processed. It is preferable for the computer system to respond specifically to the strobe signal, rather than to check at regular intervals to see if there has been a strobe signal, since the latter strategy leaves open the possibility that there won't be enough time for the computer to deal with the data once the strobe signal has been detected.

The strobe signal could be used to generate an **interrupt** in the computer; this is one of the standard techniques for making computer systems responsive to asynchronous stimuli from the outside world. An interrupt generally causes the computer to stop execution of whatever program is in progress and to jump to a specific location in memory and begin executing the **interrupt servicing routine** stored there. The servicing routine has to accomplish two ends; it must store contents of any computer registers in use by the interrupted program that might be modified during the response to the interrupt, and restore those contents after responding, and it also must locate the source of the interrupt and take appropriate action.

Suppose we had a single A-D converter, with a multiplexer at its input, allowing it to handle input signals from multiple channels. If the rising edge of the strobe signal were used to generate an interrupt, the interrupt routine would have to save registers, generate a **start convert** pulse for the first A-D conversion, wait for the conversion to finish, read data from the ADC into memory, generate a start convert for the second conversion, and so on until all parameters were captured, restore registers, and go back to what it was doing. This uses a lot of machine time to do a rote operation.

Things get speeded up considerably if we use separate ADCs for each parameter; the computer still has to save and restore registers, but now it only has to generate one start convert pulse for all of the ADCs and then read data from all of them in succession. This is still inefficient. It only takes a few chips to generate a start convert pulse, and if we build a circuit to do this, and then use the **end-of-conversion (EOC)** signal from the slowest ADC to generate the interrupt, the data are ready and waiting, and the computer only has to read them into memory.

An even more expeditious way of handling such data involves the use of a **direct memory access (DMA) channel**, via which the computer interface can put data directly into memory without tying up the computer's central processor. When you acquire 8-parameter data from 5,000 cells each second, a liability that also makes it difficult to use a digital computer for real-time control comes to the fore. If your DMA channel has to get 8 data points out of the way in the 25 μs during which it is guaranteed (by front end electronics dead times) that another cell won't come along, the data have to be shoveled into memory in 3 μs per data point, because the computer or its DMA channel are still dealing with the data **serially**, i.e., one point at a time. The obvious solution to that problem is to use more DMA channels, one for each parameter, and stuff the data into a few different areas of memory in parallel. Then you have more time to deal with each data point. Of course, these days, 3 μs is plenty of time, but you get the point.

If you're building your own instruments, in small numbers, you may want and/or need to take advantage of commercially available hardware and software for data acquisition; if you're after the highest possible speed, and/or are going into production, you'll have to design your own hardware.

PC Data Acquisition Boards

Companies such as Data Translation, Measurement Computing (formerly ComputerBoards), and National Instruments, to name a few, produce plug-in **data acquisition boards** for personal computers. These typically incorporate one or more ADCs; 12-bit resolution is now more or less standard, but 16-bit boards have become more widely available in the past few years. Boards typically use between one and four ADCs; multiplexers are included to allow each ADC to handle more than one input, allowing a single board to be used to process as many as 16 signals. Some boards also have a channel or two of analog output from a DAC, and/or eight or more digital lines that can be used for input and/or output. Data acquisition rates above 1 MHz (1,000,000 conversions/s) are readily available for 12-bit boards; the rate is specified for conversions from a single input, meaning that a 200 kHz board collecting multiplexed data from 8 channels, e.g., the outputs of a 5, 6, 7 or 8-parameter flow cytometer, would have an upper limit of 25,000 cells/s. Data transfer can be accomplished via DMA; the boards can also be set up to generate interrupts or addressed directly by the computer to initiate conversions and/or read data. Most plug-in boards are designed for PCI sockets, making them usable in both "Wintel" computers and in those Apple Macintosh models that have PCI slots. There are still some ISA boards available for older PC's; there is also a recent trend in favor of building the data acquisition board into a freestanding box that connects to the PC using a high-speed serial connection, via either the Universal Serial Bus (USB) or Apple's FireWire, now standardized as IEEE 1394.

Once data acquisition boards became readily available for PC's, they provided a relatively easy means of interfacing personal computers to laboratory-built cytometers and to older instruments that either lacked computer interfaces or were equipped with computers that were less sophisticated, less standardized, or harder and more expensive to maintain than personal computers. Kachel, Messerschmidt, and Hummel[1178] described an interface for an 8-parameter IBM PC/AT-compatible using a 12-bit data acquisition board. In order to operate the computer with flow cytometers that did not make stretched pulse outputs available externally, these authors built external front end circuitry; it had the capacity for multiparameter triggering, a feature not then found on most commercial cytometers.

Preprocessors for Data Acquisition

The first computers used for flow cytometry were minicomputers, for which suitable data acquisition modules were not readily available. Special-purpose electronics controlled data acquisition by the IBM 1130 in Kamentsky's original apparatus[65]. Block's Cytomat instruments[88,92], which were equipped with dedicated Data General Nova minicomputers, had special hardwired preprocessors built in front of the computers to manage the 8-parameter data generated by multibeam measurements of 30,000 or more cells/s. The first generation of preprocessors involved relatively sophisticated and complex electronics; later on, the same trend toward large-scale integration of circuitry that gave us microprocessors also facilitated preprocessor design using more complex building blocks that were orders of magnitude smaller and less expensive than their predecessors.

When I gave up on trying to duplicate the Block computer system in the early 1980's, I ended up developing an interface using an 8-bit Atari computer. This gadget, which cost $399 when I first bought it, wasn't very fast, didn't have DMA, and could only address 64K of memory. At that, it was probably five to ten times faster than Kamentsky's IBM 1130, and had more memory, so I didn't think it would be impossible to use. I built a parallel processing interface, using a separate ADC for each data parameter to be recorded. A-D conversions were initiated by the strobe pulse, which triggered electronics that sent a start convert pulse to the 8-bit ADCs. The output of each ADC went to a 2,048 byte random-access memory (RAM) chip. The interface could acquire and store 4-parameter data from 2,048 cells without the computer being involved. Since there wasn't room in the computer's main memory for all the data, histograms had to be updated while the data were in the interface's memory chips. No problem; the A-D conversions took about 12 µs, and it only took a few microseconds more to get data into the memories. I used a few digital electronics tricks to enable the computer to talk to all 2,048 locations in a RAM using only one memory address.

By the time I could afford an IBM PC, which was around 1985, I found I could readily adapt the interface to talk to that computer. Since the PC could address 640K of RAM memory (gasp!), it was fairly simple to run the interface, fill up its RAM's, then quickly transfer the data to the PC and work on it while collecting the next 2,000 cells. I could have collected more cells using bigger RAM's; when I started, the 2K units were the biggest available. A few years later, Terry Fetterhoff and Bob McCarthy and their colleagues[1021] built a version of the interface with 8K RAM's, which could capture data from 8,192 cells at a time.

You might ask, as Bob Leif did in his review of the second edition[1179], why I didn't just use DMA instead of the on-board RAM's; Bob pointed out, quite rightly, that he had used DMA in his old instruments at Miami. Simple. While IBM even had adequate DMA capability for Kamentsky to utilize in the ancient 1130, the original IBM PC had klunky, slow DMA that couldn't have kept up with the data. This is one reason why B-D used Hewlett-Packard computers in the CONSORT 30. IBM personal computers from the PC/AT series on, and all of the 286/386/486 and Pentium clones, had better DMA capabilities. However, successive versions of Windows have made it more difficult to use either DMA or interrupts for real-time processing, and hardware advances have made it easier to build preprocessors and memory into data acquisition systems, the end result being that most flow cytometers now rely on external hardware for the first stage of data capture. I could also claim, with some justification, that my interface and its successors did use DMA, with their own dedicated controllers and private preserves of memory, but there isn't much point in fighting over semantics.

My venerable parallel processing interface was sold by my company as the **4Cyte™** Model I, and I still use the boards in a few of my instruments, although I have moved on to designs incorporating higher-resolution ADCs for critical applications.

Van den Engh and Stokdijk[1180] described a considerably more elegant parallel processing interface, which cranks out higher resolution multiparameter data at a rate of about 4 µs/cell. This was originally used on the high-speed "MoFlo" sorter built at Livermore; the sorter is now being produced commercially by DakoCytomation, which has its own long history of building parallel processing interfaces for fast flow cytometric data acquisition and sort control[1181-2]. Cytomation's original CICERO systems were built around minicomputers; their last models used Pentium class processors with parallel 16-bitA-D converters.

The processor designs created by Auer et al[1183] for Coulter's EPICS XL cytometer were among the most ambitious for their time (the early 1990's). I have previously (p. 207 and 214) mentioned that this apparatus does 4-decade log conversion and fluorescence compensation with software; the hardware on which the software originally ran was a system of interconnected InMos Transputer chips, microcomputers which were designed to be operated in parallel. A PC-compatible computer does the data analysis and display, but the time-critical, computationally intensive tasks involved in data acquisition are delegated to other processors. BD's FACSDiVa digital pulse processing electronics are also based on a multiprocessor system, and, as more manufacturers and developers of flow cytometers move toward the use of DSP at progressively earlier stages of signal processing, multiprocessor systems will continue to replace simpler data acquisition systems.

That's enough about flow cytometry hardware for now, and pretty much concludes what I will say about DSP and digital pulse processing. We will return in a bit to the [usually personal] computer hardware and software needed to store and maintain data and results in a usable form, but we will first consider the nature of data in general, and what we ought, and ought not, to do with them once they are in the computer.

5.3 PRIMARY DATA: FREQUENCY DISTRIBUTIONS

The first step in analysis of data obtained from flow cytometers usually involves examination of one- and two-dimensional **frequency distributions**, or **histograms**, of measured values of cellular parameters. When only pulse height analyzers were available for data analysis, frequency distributions were the only data recorded by the apparatus; these days, frequency distributions are almost always generated from **list mode** data stored in raw form in computer memory and/or on external storage media.

Until the 1980's, most data analysis in flow cytometry consisted of collecting and partitioning histograms of a single parameter. If this parameter was DNA content, inspection or relatively minimal mathematical manipulation was used as a basis for dividing the histogram into areas representing cells in the G_0/G_1, S, and G_2 + M phases of the cell cycle. If the parameter was immunofluorescence, the histogram was partitioned, by inspection, into areas representing "negative" and "positive" cells. In some cases, different histograms were compared on the basis of the channels in which peak values were to be found. Logarithmic amplifiers had not yet come into vogue; linear scales were used, even for immunofluorescence. People who did flow cytometry were up to their ears in frequency distributions, but rarely bothered to think about them as such.

From the 1980's on, two- and, later, multicolor immunofluorescence measurements found increasingly wider use, and logarithmic display scales and log amps became ubiquitous, changing the peak shapes of immunofluorescence distributions, but otherwise having little effect on the level of statistical sophistication applied to data analysis. Flow cytometer manufacturers and third parties provided improved software for DNA content distribution analysis, but the software was applied at the spinal reflex level by most users. Most of what was written about the formal statistics of distributions obtained from flow cytometry was intended for consumption by a relatively small audience of statistically adept users.

Notable early attempts to demystify data analysis were made by Bruce Bagwell, in a book chapter[1184], and by Jim Watson, who supplemented his book on flow cytometry[1030] with an equally admirable volume on data analysis[1185]. Both Jim's discussion of statistics, and mine, which follows, owe a great deal to M. J. Moroney's classic *Facts From Figures*[1186], which has provided a good nuts-and-bolts introduction to the subject for general readers since the early 1950's. While you may need a British-to-American Dictionary to get through parts of it, *Facts From Figures* is entertaining as well as informative, and Watson's recent historical survey of the origins of numbers and statistics[2437] is definitely worth a read.

You Say You Want a Distribution

Let's face it; you're not going out now after any of the works just mentioned, and I've already started to discuss distributions and where they come from, so I'll continue, with the aid of Figure 5-1.

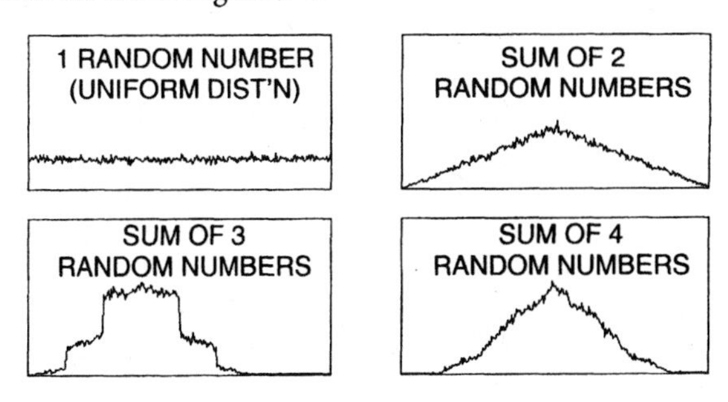

Figure 5-1. Distributions of sums of uniformly distributed random numbers approach the Gaussian or normal distribution.

We've already run across the Gaussian, or normal, distribution, also known as the "bell curve," on p. 22. It describes the intensity profile of a TEM_{00} laser beam (pp. 134-6); it also provides a pretty good approximation to what we get when we plot fluorescence histograms of calibration beads on a linear scale. In fact, measurements of many different characteristics of many different samples turn out to be normally distributed.

Now, most of us, even most statisticians, don't intuitively think, "Hey, if we go out and make a lot of measurements, and plot values on the *x* axis and frequencies on the *y* axis, we'll get a curve described by the equation

$$y = \frac{1}{\sigma\sqrt{2\pi}} e^{-\frac{(x-\mu)^2}{2\sigma^2}}$$

where μ is the **mean** and σ is the **standard deviation**, which just happens to be the equation of the normal or Gaussian distribution."

Gauss Out of Uniform

In my experience, the easiest distribution to deal with intuitively is the **uniform distribution**, in which all values have the same frequency of occurrence. You should get a uniform distribution with two possible values by flipping a fair coin a number of times and plotting the occurrences of heads and tails; throwing an unbiased die should give you a uniform distribution with six possible values. When you're looking for hundreds of values, you use a computer.

Computers can be programmed for **random number generation**, which means that, given a range of numbers, they are equally likely to spit out any number in the range. It turns out that the computers really generate **pseudorandom** numbers; values eventually repeat, but pseudorandom is close enough. The upper left panel of Figure 5-1 shows a computer-generated frequency distribution of 50,000 random numbers, and it's pretty uniform.

If, instead of plotting the distribution of individual random numbers, you plot that of the sums of pairs of random numbers, you get the triangular distribution in the upper right panel of the figure. The distribution of sums of three random numbers, at the lower left, looks like some rocks in Arizona and some ziggurats in Iraq, but nobody talks about things being distributed on a ziggurat curve. However, moving to the lower right panel, which shows the distribution of sums of four random numbers, we get a bell-shaped curve that looks pretty Gaussian. By the time you get to the sum of eight random numbers, even the computer can't tell the distribution from Gaussian, take my word for it.

Practically any measurement you can make has associated with it several sources of error, each of which will contribute some random variation to the measurement. What Figure 5-1 tells us is that, even if the distribution of the variation from each source were completely uniform, we could expect measured values to follow an approximately normal distribution. We don't have to rely on the figure, either; the **Central Limit Theorem** of statistics gets us to the same conclusion in a more formal, mathematical way.

About Binomial Theorem, I'm Teeming With a Lot o' News...

Another intuitive approach to the normal distribution comes from examination of the **binomial distribution**, which, among other things, describes the outcomes to be expected from multiple tosses of coins, fair or biased. In the case of the fair coin, the probability of the coin's coming up heads is equal to the probability of its coming up tails, both probabilities are taken as 1/2, eliminating consideration of the coin's landing on edge. For two throws of the coin, the possible outcomes are: two heads, heads followed by tails, tails followed by heads, and two tails. If we consider the outcomes only in terms of **combinations**, or numbers of heads and tails, and not in terms of **permutations**, which differentiate outcomes according to which comes first, the outcome of one head and one tail is twice as likely as either that of two heads or that of two tails. This very rudimentary distribution already has a peak.

For three throws, the outcomes are three heads, two heads and one tail, one head and two tails, and three tails. The mixed outcomes are each three times as likely as all heads or all tails; in the case of two heads and one tail, for example, the tail could come up on either the first, second, or third throw. For four tosses, the two heads and two tails are six times as likely as all heads or all tails; one head and three tails or one tail and three heads are four times as likely as all heads. Work it out with a pencil if you have problems.

For n throws, outcomes with r heads and $(n - r)$ tails, or r tails and $(n - r)$ heads, are $(n!)/[(r!)(n - r)!]$ times as likely as outcomes with all heads or with all tails. Note that $n!$, or ***n* factorial**, is $[(n)(n - 1)(n - 2)(n - 3)...(3)(2)(1)]$. The quantity $(n!)/[(r!)(n - r)!]$, abbreviated ${}^nC_r$ is the **number of combinations of *n* things taken *r* at a time.**

${}^nC_r$ is also known as the **binomial coefficient**. A binomial is an algebraic expression with two terms, e.g., $(p + q)$. When a binomial is raised to the nth power, the result is a series of terms, the rth term being expressed as ${}^nC_r p^r q^{n-r}$. For example,

$$(p + q)^4 = p^4 + 4p^3q + 6p^2q^2 + 4pq^3 + q^4.$$

You can verify the calculation, given that $0! = 1! = 1$. The binomial coefficients for expansions to increasing powers arrange themselves in **Pascal's Triangle**, shown in Figure 5-2.

1 1
1 2 1
1 3 3 1
1 4 6 4 1
1 5 10 10 5 1
1 6 15 20 15 6 1
1 7 21 35 35 21 7 1
1 8 28 56 70 56 28 8 1
• • •

Figure 5-2. Pascal's Triangle

The nth row of the triangle contains the coefficients for the nth power; the sum of these coefficients is 2^n, i.e., 2 for the 1st row, 16 for the 4th row, etc. You can also calculate the coefficient at any point in the $(n + 1)$st row by adding the coefficients to the left and right of it in the nth row. The triangle has endured for centuries as a way of getting children of all ages to play with numbers.

It also provides a good picture of the binomial distribution for any number n of coin tosses. Figure 5-3 illustrates binomial distributions for $n = 2, 4, 8$, and 16, assuming a fair coin, with the probabilities p and q $(= 1 - p)$ for heads and tails both equal to 0.5. The numbers of occurrences of various outcomes have been multiplied by appropriate constants to yield the same total number of events in each distribution.

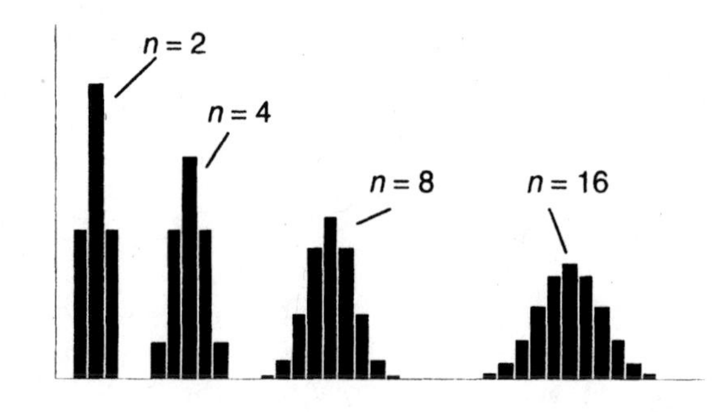

Figure 5-3. Binomial distributions for $n = 2, 4, 8$, and 16, with $p=q = 0.5$.

As n increases, binomial distributions get to look more and more like the normal distribution, except that binomial distributions are **discrete**, that is, probabilities are defined only for distinct values representing outcomes, while the normal distribution is **continuous**, with a probability assigned to any outcome between -∞ and ∞. The distribution for $n = 16$ in Figure 5-3 is hard to distinguish from a normal distribution, either by eye or mathematically. In fact, students of statistics have long been taught that a normal distribution can be used as an approximation to a binomial distribution for which $n > 10$.

The normal approximation to the binomial made life much easier in the precomputer days because one could always find tables of the normal distribution. On the other hand, one could also find tables of logarithms and slide rules, neither of which get much use these days. At the risk of being accused of statistical heresy, I will suggest that, in this era of digitized data with discrete values, we might more properly use binomial distributions as such. Don't worry about it yet.

If you'd like to see binomial distributions which don't look like normal distributions, take a look at Figure 5-4, which shows three binomial distributions for which $n = 16$. When $p = 0.5$, the distribution is symmetric, nearly Gaussian. When $p = 0.25$, the distribution is noticeably asymmetric, or **skewed**; when $p = 0.125$, the skewness is pronounced. This would be the distribution of heads obtained from tossing a decidedly biased coin (1 chance in 8 of coming up heads). The distributions shown are shifted to the left; distributions for values of p greater than 0.5 would be shifted to the right. As n increases, the skewness of distributions decreases.

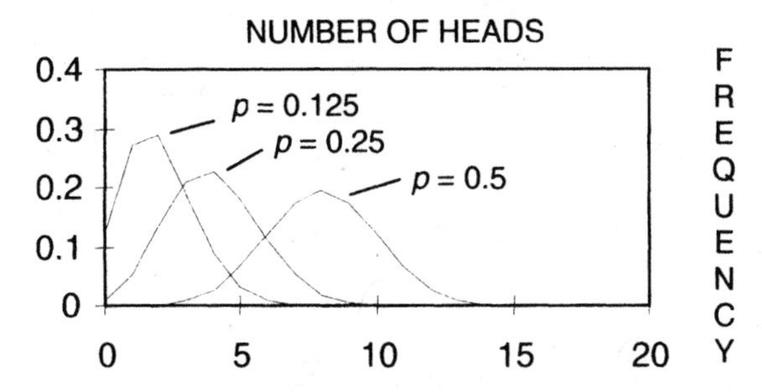

Figure 5-4. Binomial distributions for $n = 16$ and $p = 0.5$, 0.25, and 0.125. When $p \neq 0.5$, the distribution is asymmetric, or skewed.

While never likely to be mistaken for normal, skewed binomial distributions do approximate another well-known continuous distribution, the **Poisson** distribution. If you've ever encountered the Poisson distribution, you have undoubtedly heard it characterized as describing the statistics of rare occurrences; it thus might not come as a surprise that binomial distributions for which n and p are relatively small, which therefore also deal with rare occurrences, develop a resemblance to Poisson distributions.

Distributions Have Their Moments

Statistical versus Cytometric Parameters

We've already gotten used to talking about parameters in the context of flow cytometry, and now is the appropriate time to point out that the word "parameter" has a very well-defined meaning in a statistical context. Distributions are characterized by one or more **parameters**, which appear in the formula defining the distribution; in the case of the normal or Gaussian distribution, for which the formula appears on p. 231, the parameters are μ and σ.

In the case of the binomial distribution, the formula for which is

$$y(k) = {}^{n}C_{r}\,p^{k}(1-p)^{n-k},$$

where $y(k)$ denotes the probability of k occurrences, in n trials, of the outcome with probability p, the parameters are n and p.

The Poisson distribution (pp. 19-20), with the formula

$$y(k) = (e^{-\mu}\mu^{k})/k!,$$

where $y(k)$ denotes the probability of k occurrences of the specified outcome, is characterized by the single parameter μ.

Especially since I have already let it slip that, in the case of the normal distribution, μ is the **mean** and σ the **standard deviation**, you may be wondering how the parameters of a distribution relate to these better known descriptors.

Mean, Variance, and Standard Deviation

Mathematical statistics defines a series of **moments** of a distribution, which are very much analogous to the moments of physics. The ***m*th moment about the origin** of a discrete distribution such as the binomial, denoted by μ'_m, is defined as

$$\mu'_m = \Sigma\, x^m \cdot y(x);$$

it represents the sum of the mth powers of all possible outcomes, each weighted by its probability of occurrence. For continuous distributions, an integral replaces the sum. To relieve the math anxiety, let me reveal that the first moment about the origin is good old μ, better known as the **arithmetic mean** or **average**. As for the analogy between statistical and physical moments, a line perpendicular to the horizontal axis at μ passes through the center of gravity of the distribution.

The ***m*th central moment**, or ***m*th moment about the mean**, is denoted by μ_m, and defined as

$$\mu_m = \Sigma\,(x-\mu)^m \cdot y(x).$$

The second moment about the mean is better known in the guise of σ^2, the **variance**, which is the square of the **standard deviation (S.D.)**, σ.

The formulas you'll usually encounter for the mean and variance include a **number of observations**, N; the mean is $(\Sigma x)/N$ and the variance is $(\Sigma(x-\mu)^2)/N$; for small samples ($N < 20$), about which we needn't worry, the variance is generally more accurately expressed by $[\Sigma(x-\mu)^2]/(N-1)$. These formulas don't look quite like those for the moments, but the $y(x)$ terms in the moments include the factor $1/N$, which explains the apparent discrepancy. Different statistics books write the same formulas in different forms.

Parameters, or powers of parameters, aren't always moments, although they are in the case of the normal distribution. For the Poisson distribution, the mean is equal to the parameter μ; so is the variance. For the binomial, the mean is *np*; the variance is *np*(1 - *p*).

With Many Cheerful Facts About the Square of the Hypotenuse: Euclidean Distance

While the mean locates the physical center of a distribution, the standard deviation, and its square, the variance, provide an indication of the average distance from the center of points in the distribution. The "square root of squares" nature of the standard deviation relates to the time-honored definition of the **Euclidean distance** between two points in space, shown in Figure 5-5.

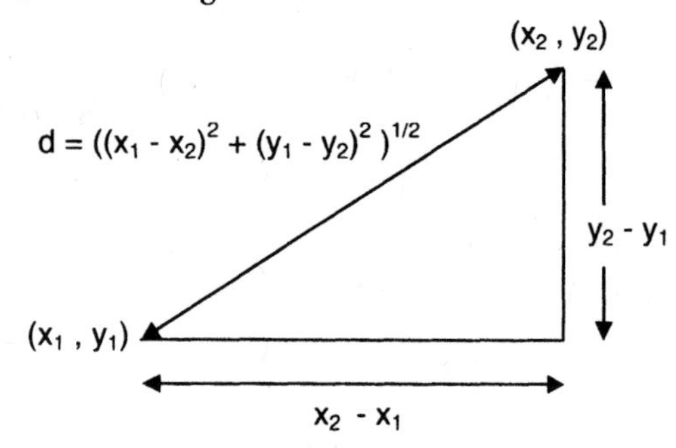

Figure 5-5. Euclidean distance between two points, with an assist from Pythagoras.

In the two-dimensional case, the line between two points (x_1,y_1) and (x_2,y_2) represents the hypotenuse of a right triangle with sides of length $(x_2 - x_1)$ and $(y_2 - y_1)$; from the Pythagorean Theorem, we know the length of the hypotenuse is the square root of the sum of the squares of the sides, or $[(x_2 - x_1)^2 + (y_2 - y_1)^2]^{1/2}$. The distance formula works in any number of dimensions; you can sketch it out for yourself in three if you have doubts.

Higher Moments: Skewness and Kurtosis

The **skewness**, γ_1, of a distribution is defined as:

$$\gamma_1 = \mu_3 / \sigma^3,$$

where μ_3 is the third moment about the mean and σ is the standard deviation. Since the normal distribution is symmetric about the mean, its skewness is zero. The skewness of the Poisson distribution is $\mu^{1/2}$; that of the binomial distribution is $(1 - 2p)/[n(p)(1 - p)]^{1/2}$.

If skewness is a relatively simple concept, **kurtosis**, and the related **excess**, are a little subtler. Both can be symbolized as γ_2; kurtosis is defined as

$$\gamma_2 = (\mu_4/\sigma^4) - 3,$$

and excess as $\gamma_2 = (\mu_4/\sigma^4)$. The excess of a normal distribution is zero, that of a Poisson distribution is 1/μ, and that of a binomial distribution is $(6p^2 - 6p + 1)/[np(1 - p)]$. Kurtosis and excess characterize the relative proportions of the central portion and tails of a distribution. A distribution with heftier tails than a normal distribution is called **platykurtic**; one with proportionally more in its central portion is **leptokurtic**.

Some Features of the Normal Distribution

If x is normally distributed with a mean μ and a standard deviation σ, the quantity z, where $z = (x - \mu)/\sigma$, is normally distributed with mean 0 and standard deviation 1. The standardized quantity is usually what appears in tables of the normal distribution. P(Z) is the probability that $z \leq Z$; it is the integral of, or the area under, the standardized normal curve between -∞ and Z. Y(Z) gives the height of the distribution at z = Z (or at z = -Z, since the distribution is symmetric). Some key values are shown in Table 5-1.

Z	P(Z)	Y(Z)	Notes
0.00	0.5000	0.3989	
0.68	0.7517	0.3166	quartile
1.00	0.8413	0.2420	~0.6 max
1.18	0.8810	0.1989	~1/2 max
1.41	0.9208	0.1476	1/e point
1.65	0.9505	0.1022	~5% > Z
1.96	0.9750	0.0584	~2.5% > Z
2.00	0.9772	0.0540	
2.05	0.9798	0.0487	~2% > Z
2.33	0.9901	0.0264	~1% > Z
2.58	0.9951	0.0143	~0.5% > Z
3.00	0.9987	0.0044	
3.10	0.9903	0.0033	~0.1% > Z
3.30	0.9995	0.0017	~.05% > Z
3.73	0.9999	0.0004	~.01% > Z
4.00	0.99997	0.000130	
4.27	0.99999	0.000048	
4.42	0.999995	0.000023	
4.75	0.999999	0.0000050	
4.90	0.9999995	0.0000024	
5.00	0.9999997	0.0000015	

Table 5-1. Some landmarks of the normal or Gaussian distribution.

Half of the symmetric distribution lies above the mean; three-quarters lies within 0.68 standard deviation (S.D.) of the mean, and it's a safe gamble that more than 99 and 44/100 percent of the distribution is less than 3 S.D. from the mean. When you deal with large numbers of observations, that may not be as much comfort as you need; note that 3 of every 10^7 points lie more than 5 S.D. above the mean. If you're really ambitious about rare event analysis, you might worry about the 1 event in 10^9 lying more than 6 S.D. above the mean, or even the 1 in 10^{12} lying more than 7 S.D. above. You're safe at 8 S.D., above which you can only expect 1 event in 10^{15}.

When working with the approximately normal distributions encountered in flow cytometric measurements of calibration beads, or of DNA content, we have gotten used to calculating the **coefficient of variation (CV)**, that is, the S.D. divided by the mean, to compare the relative dispersion or spread of two distributions. The **FWHM (Full Width at Half Maximum height)** of a normal distribution is 2.36 standard deviations; as noted on pp. 22-3, the CV of a flow cytometric near-normal distribution may be "conveniently" estimated by multiplying the difference in channel numbers between the left and right half-maximum points by (1/2.36) and dividing by the channel number of the peak. Or, if you'd prefer to do a really convenient calculation, i.e., one that you can do in your head, just find the difference in channel numbers between points at which the height is 0.6 maximum, where the width is 2 S.D. Divide by twice the peak channel number, and you've got the CV.

Many of the distributions observed in flow cytometry, e.g., immunofluorescence and right angle scatter distributions and distributions obtained from measurements of total cellular protein with stains such as fluorescein isothiocyanate, are, when measured on a linear scale, skewed toward lower values. Most of these distributions look more like the normal distribution when displayed on a logarithmic scale, and have been described as exhibiting **lognormal** distributions, i. e., distributions in which the log of the measured quantity is normally distributed. Other cellular characteristics, including generation time, also follow approximately lognormal distributions.

The "approximately" in the preceding sentence is really key here. There are billions of humans, bearing trillions of cells each, and when you do some rough and ready calculations using extreme values of lognormal distributions, you expect to find small numbers of cells with generation times of milliseconds and centuries, and lymphocytes the size of neutrons and filberts. Common sense steps in and tells us that distributions encountered in real life are **truncated** (the term is in boldface because it is used in statistics in exactly the sense in which I am now using it); values don't really range from minus to plus infinity. In addition, as can be appreciated from a paper by Coder, Redelman, and Vogt[1187], many distributions that appear lognormal really aren't.

Measures of Central Tendency: Arithmetic and Geometric Means, Median, and Mode

The mean represents what statisticians call a **measure of central tendency**; it indicates the position of a representative portion of a distribution. The **arithmetic mean**, or **average**, ($\Sigma x/N$), i.e., the sum of the individual values (x) of all of the observations in the distribution divided by the number, N, of observations is familiar to almost everybody. It is not the only mean there is; those of us who do flow cytometry are familiar, perhaps too familiar, with the **geometric mean**, which is the Nth root of the product of all of the observed values, or, in mathematical notation, $(\Pi x)^{1/N}$.

You might ask why so many flow cytometry software programs compute geometric means, since you aren't likely to find more than passing references to them in most statistics books. Well, I'll tell you. A lot of flow cytometry data is displayed on logarithmic scales and, essentially, stored in logarithmic form; the values we have to work with are the logarithms of the observed values. Adding the logarithms of two numbers gives you the logarithm of their product, and dividing the logarithm of a number by N gives you the logarithm of its Nth root. So, if you have N logarithms of observed values, and you take the arithmetic mean of the logarithms, adding them all up and dividing by N, what you get just happens to be the logarithm of the geometric mean of the observed values. It is easier to do this computation than it is to convert all the logarithmic values to linear values and take their arithmetic mean, especially when you are dealing with 8-bit or 10-bit data on the logarithmic scale, which will give you noticeable picket fence effects when you try to convert to a linear scale and back to a log scale. As it becomes the norm (a good name for a measure of central tendency if I ever heard one, but not a defined term) to store high-resolution list mode linear data, there should be little reason to bother with geometric means.

Other measures of central tendency include the **mode**, which is the value of the distribution that occurs most frequently, and the **median**, the value above and below which 50% of the distribution can be found. The arithmetic and geometric means are both susceptible to the influence of **outliers**, i.e., data points far from the center of the sample distribution. Small numbers of outliers can change the values of both means, in some cases substantially. The mode and median are largely immune from the effects of outliers.

Measures of Dispersion: Variance, Standard Deviation, CV, and Interquartile Range

The **variance**, its square root, the **standard deviation (S.D.)**, and the **coefficient of variation (CV)**, i.e., the S.D. divided by the arithmetic mean, are **measures of dispersion** for a distribution. Other measures of dispersion include the **range**, or difference between the highest and lowest values, and the **interquartile range (i.q.r.)**, the difference between the values within which the central 50% of the distribution lies.

Robustness in Statistics; the Robust CV

We need to worry at least a little about the actual form of the distributions with which we deal with because you can't just do a lot of statistical tests and expect accurate results. Statisticians speak of **parametric** tests, which involve quantities computed using estimated values of the parameters of distributions, and **nonparametric** tests, which don't. Many parametric tests are based on the assumption that the distribution being examined is normal, and may fail when it deviates substantially from normal. Tests that maintain their efficacy when their underlying assumptions are violated are called **robust**.

In traditional statistical work, a **sample** of data points is used to **estimate** the properties of the distribution represented by the sample. Some measures of central tendency and dispersion provide better **estimators** than others. For example, the range is not a particularly good measure of dispersion when the sample includes a reasonable number of points, because the value of the range obtained from the sample is determined solely by outliers. The arithmetic and geometric means are, as previously noted, affected by outliers; the S.D. is, to an even greater extent, because the variance, from which the S.D. is derived, includes terms representing the squares of the differences between outliers and the mean. The median and interquartile range, the values of which are determined by the locations of the central 50% of the distribution, are not greatly affected by outliers; they are **robust**.

In a symmetric distribution, the mean is equal to the median, in a symmetric **unimodal** distribution, i.e., one with a single peak, the mode is equal to the mean and median. In a skewed distribution, the mean, median, and mode all have different values. When we work with skewed distributions, e.g., when defining a "peak channel," we are more likely to focus on the mode of the distribution than on either the mean or median, and, estimating CVs by eye gets much harder than it is when we're dealing with relatively narrow normal distributions. The mode isn't a particularly robust statistic, anyway.

We'd like to have something analogous to the CV to use to compare skewed distributions, even if we can't do the calculation in our heads. The CV is obtained by dividing the S.D., a measure of dispersion, by the mean, a measure of central tendency. Something that may suit our needs admirably is the **robust CV (RCV)**.

The robust CV is obtained by dividing the interquartile range, a robust measure of dispersion, by the median, a robust measure of central tendency. In order to make it directly comparable to CV for normal distributions, we introduce a factor of 0.75. Since, as can be seen from Table 5-1, the interquartile range of a normal distribution encompasses 1.36 S.D., extending 0.68 S.D. above and below the mean, which is also the median, the adjusted value

$$\text{RCV} = 0.75 \times (\text{i.q.r.})/\text{median}$$

should be close to the value of the CV for a normal distribution.

The median, i.q.r., and RCV are also useful for dealing with distributions on logarithmic scales, which pose particular problems when attempts are made to calculate means and standard deviations using channel numbers. The mean of the logs isn't the log of the means, and vice versa; this is one reason some software calculates a geometric mean. An alternative approach is to convert everything from log back to linear and work from there; however, given the relatively poor agreement between log amps and log tables, this seems inadvisable. Eventually, we'll be keeping the data in 16-bit or higher precision linear form, and using the log scales primarily for displays; even then and thereafter, however, we'll run into less trouble if we use robust statistics.

"Box-and-Whiskers" Plots of Distributions

"Box-and-whiskers" plots, described by Tukey[1188], and shown in Figure 5-6, provide an easily appreciated graphical summary of a distribution. The box spans the interquartile range; the whiskers define the end points of the range unless they are more than 1.5 i.q.r. away, in which case dots along the axis may be used to show outliers. Tufte[1025] suggested a simpler, and even more informationally dense, version of the box-and-whiskers plot; this is also shown in the figure. Locating the median by a horizontal gap between lines may be a little tricky, but you can always make the gap bigger. Tufte's works on information display[1025,1189,2418-9] speak well of "small multiples," a type of display which might include those figures, shown in many papers involving flow cytometry, in which the reader is given the opportunity to compare 16 or 32 immunofluorescence histograms, each smaller than a postage stamp. However, box-and-whiskers plots, either in Tukey's original form or Tufte's minimalist style, can present the information more effectively. Nobody gets bragging rights for histograms any more, anyway.

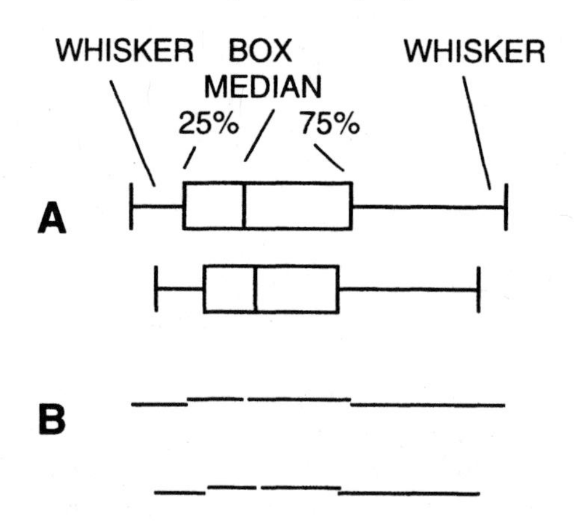

Figure 5-6. "Box-and-whiskers" plots showing medians and interquartile ranges of two distributions. A: As originally described by Tukey[1188]. B: In the minimalist interpretation suggested by Tufte[1026].

Calculating and Displaying Histograms

The computer algorithm for calculating histograms was described on p. 25. Today's software allows histogram computation to be stopped when the total number of cells or the number of cells in a given channel or range of channels reaches a preset value, or (in the case of data acquisition software) after a preset time has elapsed. These capabilities were anticipated by the pulse height analyzers (PHAs) used on the previous generation of flow cytometers, as was some limited capacity for gated analysis, which is erroneously considered a relatively "modern" technique in flow cytometry.

The gating capacities of PHAs were limited to the use of a gating pulse, which essentially provided a trigger signal for the analyzer, allowing computation of histograms from low-level and/or noisy signals, e.g., immunofluorescence signals, for which forward scatter signals served as gating pulses. However, the hardwired two-parameter analyzer incorporated into the original Bio/Physics Systems Cytofluorograf allowed the definition of a two-parameter gate bounded by a four-sided polygon; an associated gating pulse could be used to trigger the PHA to produce a histogram of a third parameter for cells within the gating region. Modern technology allows us to be much more creative about gating.

Figure 5-7 shows histograms of distributions plotted in various formats. PHAs usually produced a series of dots or a "bar graph"; when connected to chart recorders, they drew lines. The computer can easily reproduce all of these types of display and add a few. I find "odd" and "even" bar graphs, in which every other point of a distribution is displayed as a bar, to be useful for visual comparison of two histograms; they are particularly effective when the two histograms are displayed in different colors or shades of gray.

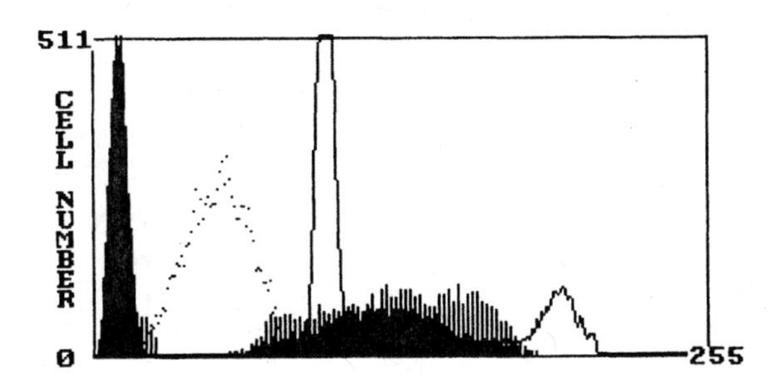

Figure 5-7. Histogram display formats.

Bivariate and Multivariate Distributions and Displays

When mixed cell populations are subjected to flow cytometric analysis, frequency distributions of any one cellular parameter generally show considerable overlap from one cell type to another. It is almost always possible, however, to distinguish different types of cells by looking at **multivariate frequency distributions** of two or more parameters in spaces of two or more dimensions.

Two-parameter **dot plots** or **cytograms**, and **two-dimensional frequency distributions**, shown as **gray scale**, **chromatic (color)**, **peak-and-valley ("isometric")**, or **contour plots**, all of which were discussed in Chapter 1, are the most commonly used forms of graphical presentation of bivariate flow cytometric data.

Dot Plots; Correlation and Covariance

Dot plots provide as good an indication as any other form of display of the **range** of bivariate data. They are also useful in determining **correlations** between parameters, i.e., the extent to which values of two parameters track one another. The **correlation coefficient** r_{xy} between variables x and y is defined as

$$r_{xy} = \Sigma[(x - \mu_x)(y - \mu_y)]/N\sigma_x \sigma_y,$$

where μ_x and μ_y are the mean values of x and y, and σ_x and σ_y are the associated standard deviations. The **covariance** of x and y, σ_{xy}, is

$$\sigma_{xy} = \Sigma[(x - \mu_x)(y - \mu_y)]/N.$$

Figure 5-8 shows dot plots of some computer-generated near-normal data illustrating various degrees of correlation. The panel at the left shows data that are essentially uncorrelated (r = -.002). The middle panel shows data with some correlation; r = .320. The x- and y-values in the right panel are highly correlated (r = .950). In general, if you plot two parameters that show a correlation this high, you might as well only measure one of them.

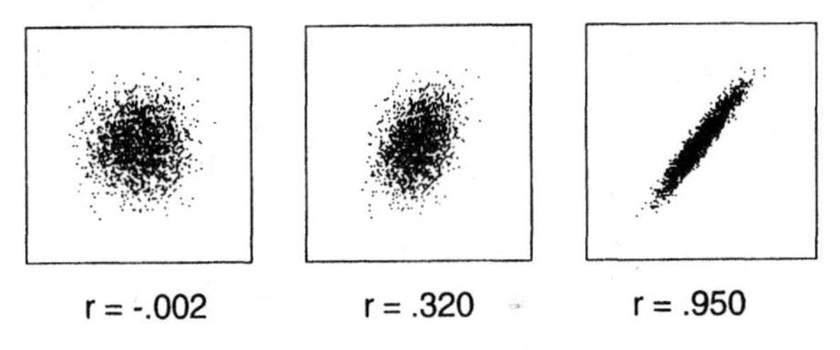

Figure 5-8: Dot plots of computer-generated data showing various degrees of correlation.

Linear Regression; Least-Squares Fits

When variables are well correlated, it is often useful to be able to express the value of one **dependent** variable as a function of the other **independent** variable. If y is the dependent, and x the independent, variable, it is possible to find values of the **slope** m and **y-intercept** b which best describe a linear relationship (y = mx + b) between the two. In the procedure called **least-squares linear regression**, values of m and b are selected which minimizes the sum of squares of the distances of actual data points (x,y) from the line y = mx + b. Euclidean distance comes into play again. The use of regression in calculating instrument sensitivity in MESF units was shown in Figure 4-60 (p. 216) and discussed on pp. 216-7.

When you do a regression calculation, the key critical item in the result is the minimized sum of squares, or **residual**. This determines the **confidence interval** for the value of the slope, m. If you look at the right panel of Figure 5-8, you can pretty much figure out that the regression line would run from the bottom left corner to the top right corner of the panel, meaning there wouldn't be much doubt about the value of m. There isn't anything that prevents you from deriving a least squares line for essentially uncorrelated data such as are shown in the left panel of Figure 5-8. However, if you look at the data, you won't (or shouldn't) be able to convince yourself that a regression line should run in any particular direction, and the residual will, in essence tell you that the slope could be anywhere between -∞ and ∞. In general, if you can't see a high degree of correlation in the data, regression, linear or otherwise, is a waste of time.

Linear regression can readily be extended to more than two dimensions through the magic of matrix algebra. There are also cases in which data clearly accumulate along a trajectory which doesn't happen to be a line; in such cases, it is possible to find least squares fits to polynomials, exponentials, etc. Such **nonlinear least-squares curve fitting** figures prominently in the methodology for deconvoluting DNA histograms, which was introduced on pp. 25-6.

Breaking off Undiplomatic Correlations

Correlation and regression are used, ubiquitously but inappropriately, in many studies comparing different methods of clinical measurement, e.g., comparison of results obtained by newer and older hematology counters. In general, a correlation coefficient becomes larger as the range of included data points increases; since evaluations of new methodology are typically designed to include extreme values at the high and low ends, and since nobody is willing to put up money for the study unless the new method seems to work reasonably well, it can be expected that there will be a high degree of correlation between the results obtained using the old and new method. This was pointed out in 1986 by Bland and Altman[2483], but, years later, when Rebecca Gelman designed a study using the Wilcoxon paired-sample (signed rank) test to evaluate volumetric capillary cytometry against flow cytometry for CD4+ lymphocyte counting, she noted that 6 of 7 comparisons of technologies for this purpose published between 1993 and 1995 still relied on correlation and regression. If you ever get involved in an instrument comparison, you should look at the paper by O'Gorman, Gelman et al[2484], and make sure the statistician responsible for the experimental design has read it.

Multivariate Measures of Central Tendency and Dispersion

In the matrix algebra used for multivariate statistics, individual observations are replaced by an **observation vector,** and means by a **means vector**; the function of the variance is subsumed by the **variance-covariance matrix**, which has the variances of the individual variables as its diagonal elements and the covariances (σ_{xy}) as its off-diagonal elements.

If you think about it, you will realize that it is not possible to come up with definitions for multivariate medians, quartiles, percentiles, etc. One can find the central positions of bivariate and multivariate Gaussian distributions (for an isometric plot of a bivariate Gaussian, look at the intensity distribution of a TEM_{00} laser beam, shown as an isometric plot in Figure 4-21 on p. 136), but these central positions are means. It is also possible, using fairly hairy mathematics, to find boundary contours for "rings," in the two-dimensional distribution, or "shells," in the three- or multi-dimensional distributions, that contain given percentile fractions of the population, but only because the Gaussian distribution is symmetric. That doesn't help us much in looking at most of the real multivariate distributions we encounter in cytometry, which are asymmetric in bizarre enough ways for me to have adopted the term "ginger root" to describe them.

Beyond Dot Plots: Two-Parameter Histograms

We found in chapter 1 (pp. 26-7) that dot plots can provide information we can't get from single-parameter histograms. However, in considering the broad area of discrimination and selection of cell populations, even in two dimensions, we quickly encountered a need for alternatives, and discovered that the display of a bivariate frequency distribution as a two-parameter histogram, which provides indications of the relative frequencies at which cells with different values of two measurement parameters occur within the sample population, could be much more informative.

Bivariate Distributions: Display's the Thing!

A useful discussion of techniques of graphical presentation of bivariate and multivariate data appeared in an article by Graedel and McGill[127], which shaped my original discussion of the subject; I have since been influenced by the monumental contributions made to the art and science of data visualization by Tufte[1026,1189,2418-9].

Not so long ago, those of us lucky enough to have computers attached to our flow cytometers were obsessed with minimizing the number of bytes of data we had to keep on hand, in order not to exceed the 640 KB capacity of our computer memories or the 360 KB capacity of a floppy disk. Even though seemingly gigantic 5 and 10 MB hard drives had become available to us, they did not offer storage space for as much data as we could fairly readily generate in a matter of days. We also were usually limited to 8-bit ADCs, which meant that our single-parameter histograms couldn't occupy more than 256 channels; the two-parameter distributions were typically 64 × 64, so as not to take up either too much space or too much computing time. This got us used to accumulating data from only 5,000 or 10,000 events in each sample, and, to tell the truth, the displays didn't look too bad.

Now that we can get 20-bit linear data, and have at least hundreds of MB RAM and tens of GB hard drive space in the laptop computers we use for off-line data analysis, it's tempting to go to 1,024-channel histograms and 256 × 256 2-parameter displays. However, this forces us to reconsider how many events we need to acquire and how to display them. Figure 5-9, on the next page, helps us see why; not coincidentally, the figure also appears, uncaptioned but in color, on the book's cover.

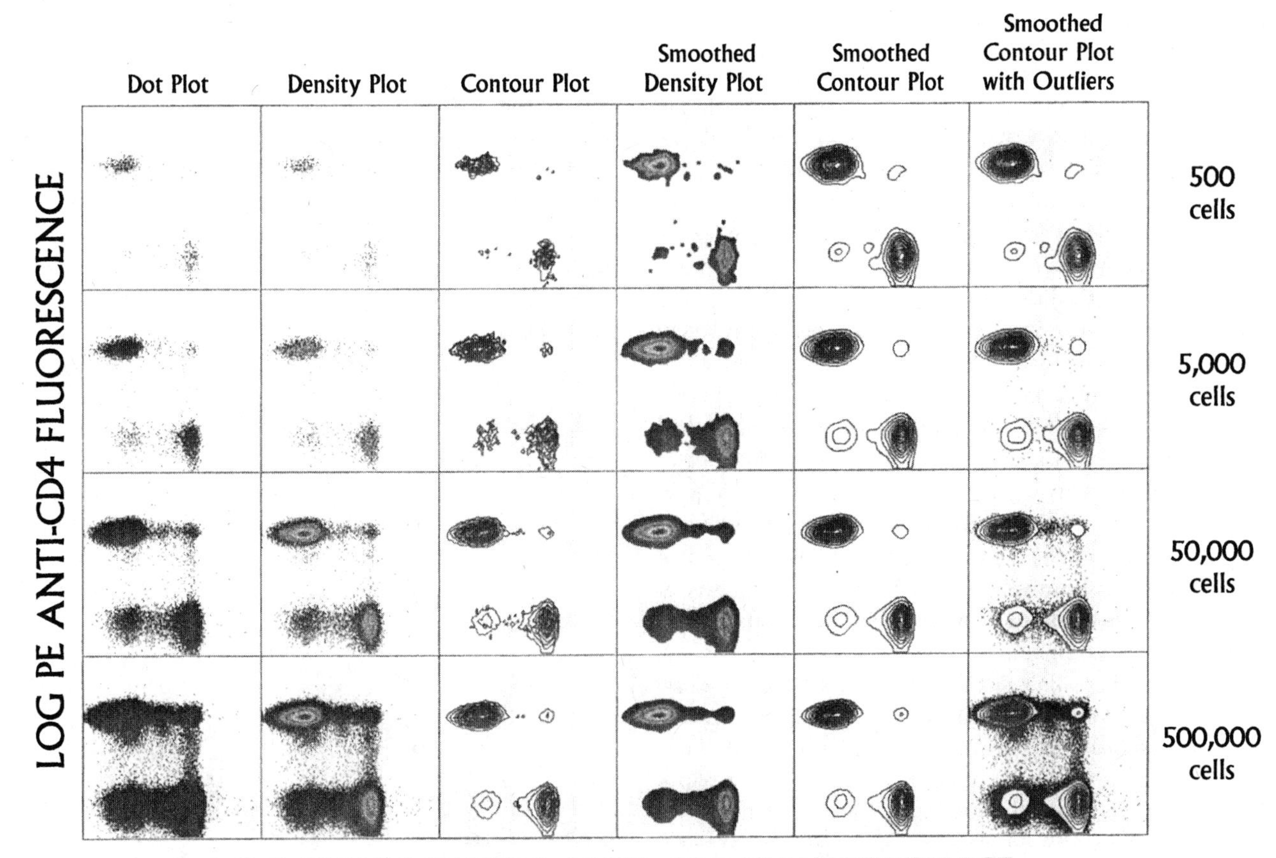

Figure 5-9. Varieties of two-parameter data display, with different numbers of cells represented in the plots in each row, and different plot types for the same numbers of cells in each column. The plots themselves are shown in color, without captions, on the front cover of the book. Plots by Jennifer Wilshire, using FlowJo software; data supplied by Mario Roederer.

Figure 5-9 shows several variations of the most popular types of two-parameter data display. There are dot plots, unsmoothed and smoothed density plots, and unsmoothed and smoothed contour plots; the smoothed contour plots are shown with and without outliers.

Displaying by the Numbers

When we look at the upper left corner of the display, what jumps out at us is that there aren't a whole lot of cells represented in the dot plot. Moving one square to the right, to the 500-cell density plot, we notice that there are so few cells that most of the plotted points still represent single occurrences. Actually, I'm not even sure you can tell the difference between dots of different shades of gray in the figure; you may need to look at the front cover to pick up the few dots that aren't colored blue.

Move to the 500-cell contour plot, and you'll notice that there are a lot more pixels in the contour lines than there are cells represented in the display. If you look at the smoothed density plot and the smoothed contour plot, and scan down the columns with those plots from top to bottom, you'll find it hard to tell that the data in the second and third rows came from ten and a hundred times the number of cells represented in the first row.

Is the glass in the first row 10% or 1% full or is it 90% or 99% empty? The smoothed plots look ready for a journal. Now, nobody will accuse you of faking data if you use flow software to produce smoothed plots; it can be argued that the software produces an accurate representation of the structure of the data, even if there are only 500 cells involved. But most of us would be leery about publishing flow data from only 500 cells, especially when we can analyze tens of thousands of cells a second. The smoothed plots simply look too good for the data, at least to my way of thinking.

If you look down the column of dot plots, you'll notice that, as has been previously mentioned, large areas of blackout accumulate as the number of cells increases. On the plus side, you can still see the single occurrences. Every event is represented; you just can't tell high-density areas from low ones, although I would argue that, even in the dot plot of 50,000 cells, you could pretty well guess where the high-density areas were.

I prefer to keep things as simple as possible, using dot plots, which are frequently all that's necessary, when I can, and only going to calculation and display of bivariate distributions when I can't get the information I want from a dot plot. Dot plots are fine when your primary purpose is defining windows around clusters and obtaining counts, at least if you do things my way.

The unsmoothed density plots are, like dot plots, completely honest; what you see is what you get. Every event appears. However, if you look at the smoothed density and contour plots, you'll notice that most of the single occurrences have disappeared, except, of course, in the smoothed contour plot with outliers, which is actually a dot plot of low-frequency events on top of a contour plot of higher-frequency events. And, as I play Goldilocks (moving toward gray scale, these days), it strikes me that the density plot of 500,000 cells is showing too many, that the density plot of 5,000 cells is showing too few, and that the density plot of 50,000 cells is just right.

Yet I said on p. 238 that the practice used to be to show no more than 5,000 or 10,000 cells in a two-parameter display. What's changed? Easy. On p. 238, I was talking about 64 × 64 display; when you move to a 128 × 128 or a 256 × 256 display, you have 4 times, or 16 times, as much real estate into which you can fit your events, and you need proportionally more events to make an unsmoothed display look as substantial.

If you're looking for rare events, and what you eventually want to show people is a two-parameter display with a gate drawn around however many (or few) of the elusive little buggers you manage to turn up, it's acceptable to have hundreds of thousands, or even millions, of cells represented overall. In theory, you will have worked your gating and chosen your display parameters so that none of the millions of non-rare events show up in the same area, and, if you can, it might be nice to show a plot of a lot of cells from a sample that doesn't have any of the rare events in it, just to firm up your case.

Of course, if you've done an experiment, presumably not involving rare event analysis, that, for whatever reason, only yielded data from a few thousand, or even a few hundred, events, you've got to go with what you've got. But you might consider leaving your two-parameter displays in a 64 × 64 format, and condensing your 1,024-channel histograms to 256 channels; the displays will look better.

While I'm on the subject of histograms, I should mention that, if you want to apply mathematical models to your data, for example, to estimate numbers of cells in various fractions of the cell cycle on the basis of DNA content, you may need a lot more cells than you think. The ISAC Guidelines for implementation of clinical DNA cytometry[1746] call for 10,000 cells in the S-phase region, exclusive of debris, etc.; it has, more recently, been suggested that there must be at least 200 cells per channel in the S-phase region for a model to produce reliable results[2485]. For a 256-channel histogram, with a G_0/G_1 peak at channel 100 and a G_2+M peak at channel 200, the requirement would be for almost 20,000 cells in the S-phase region; if 10% of the cell population were in S-phase, the sample would have to contain at least 200,000 events. While it would be of interest to have precise information on DNA synthesis in peripheral blood CD4+ lymphocytes in AIDS patients[2330] or in cancer cells circulating in blood[2331], finding enough cells to analyze is likely to be a problem in the former case and will definitely be a problem in the latter.

Economies of Scale

I have already complained (p. 31) that density plots, whether gray scale or color, are almost never accompanied by a scale or key indicating the numbers of events represented by each color or gray level. Figure 5-10 shows a 64 × 64 density plot with the scale I usually use.

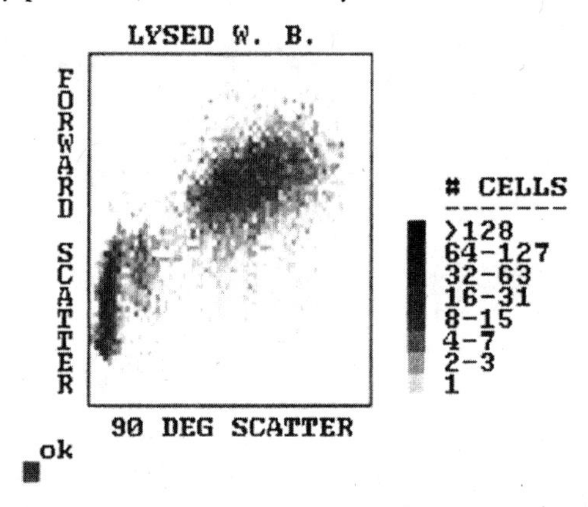

Figure 5-10. Density plot of 90° vs. forward scatter for leukocytes in lysed whole blood, showing a scale indicating the numbers of events represented by each gray level.

The density plots shown in gray scale in Figure 5-9, and in color on the book cover, are unaccompanied by a scale. Like most people who do cytometry these days, I use color printers for output. Edward Tufte[1026] correctly points out that there is no generally accepted scale which maps colors to values, and also that the number of people with one form or another of color blindness is significant. However, Tufte notwithstanding, there are several widely accepted color scales, including that of the spectrum, "Roy G. Biv," and the resistor color code (black, brown, red, orange, yellow, green, blue, violet, gray, white; the politically correct mnemonic for this goes something like: "Bad Boys Rudely Ogle Young Girls But Veritable Gentlemen Won't").

Increasingly lighter or darker shades of gray do provide an intuitive scale. In the gray scale displays of Figure 5-9, points representing increasing numbers of cells are shown in progressively lighter levels of gray, for the first three steps of the scale, but the fourth level is represented by a darker

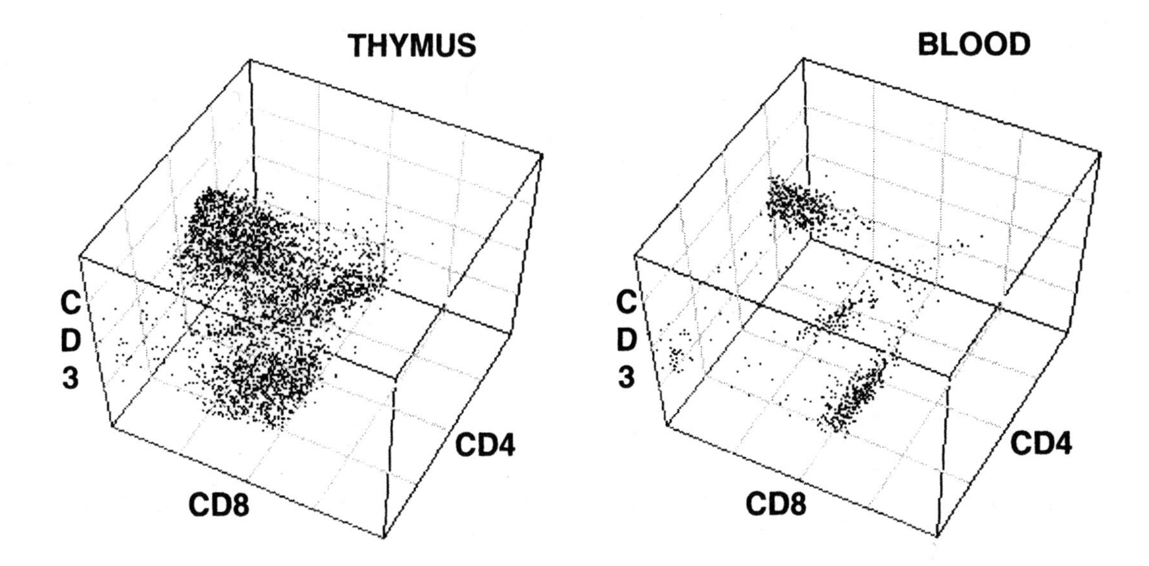

Figure 5-11. 3-Dimensional "cloud" plots of the distribution of CD3, CD4, and CD8 antigens on CD-3 positive cells from neonatal thymus and umbilical cord blood, stained with FITC-anti-CD8, PE-anti-CD4, and PerCP-anti-CD3. Data were provided by Janis Giorgi and Ingrid Schmid (U.C.L.A.); plots were generated with WinMDI software. All axes are on a log scale; there is nothing Gaussian about the multivariate distribution in thymus.

shade than the third. The scale for Figure 5-10 was chosen to represent increasing densities by progressively darker levels of gray, which I would argue is more intuitive. I will also reassert the preference I expressed on p. 32 for the binary logarithmic density scale used and shown in Figure 5-10; it would not be difficult to make this a standard for displays, much as the four-decade logarithmic scale for axes has become a standard.

I see three major problems with contour plots. The computations take time; while that's less of a problem as computers keep getting faster and cheaper, why waste time? In most of the contour plots you see published, all the lines are the same color, and, as is the case with most density plots, there is usually no accompanying scale. Contour plots convey a lot more information when lines at different heights, or the spaces between lines, are shown in different colors or in different shades of gray; displays in the latter format don't look all that different from smoothed density plots, which are easier to compute. Finally, contour plots generally don't show single occurrences, and, if you set them up to do so, they tend to get full of squiggles and look really ugly. If you're going to show single occurrences as dots overlaying a contour plot, you might as well use an unsmoothed density plot. I've been saying this for years, and I seem to have gotten some former contour plot addicts to agree with me. Maybe what we need is a twelve gray level program and some inspirational bumper stickers.

How Green Were My (Peaks and) Valleys

I discussed the display of two-parameter data as a **three-dimensional projection**, also known as a **peak-and-valley** or **isometric** plot, on pp. 31-32. People don't use such plots much these days, and I'm not going to say anything here that might promote a revival.

Clouds on the Horizon: 3-Dimensional Displays

There is some advantage to be able to look at data in three dimensions, particularly in dissecting the "ginger root" distributions found in tissues such as the bone marrow and thymus, in which cells express and lose antigens as they develop. Many flow software packages have the capacity to generate "cloud plots" of such distributions; the plots in Figure 5-11 above, contrasting CD3/CD4/CD8 antigen expression in the thymus and blood, were done with Joe Trotter's WinMDI freeware (see Chapter 11).

To make full use of cloud plots, it is necessary to be able to change the display angle to allow visualization of parts of a plot that might be hidden if the orientation were fixed. 3-dimensional displays have been described by Kachel and Schneider[1023], Stewart and Price[1024], Ormerod and Payne[1025], Kachel, Messerschmidt and Hummel[1178], and Greimers et al, who include an H-P Pascal program[1199]. Leary, Ellis, and McLaughlin[1200] described the use of a stereoscopic display unit for viewing 3-D displays; others, including Ed Luther (unpublished work at Bar Harbor and in my lab in the 1980's) have produced dual displays usable with less elaborate stereoscopic viewers or 3-D glasses.

Sloot and Figdor[1201] have taken a different approach to the display of 3-parameter data; they perform calculations that project the 3-dimensional data onto a plane in such a way that clusters are clearly resolved. This is mathematically correct, and useful, although it is somewhat less intuitive than 3-D display and probably won't draw as much of a crowd as the fancier display technology.

$$F_1 = f_1 \quad + k_{21}f_2 \quad + k_{31}f_3 \quad + k_{41}f_4 \qquad (1)$$

$$F_2 = k_{12}f_1 \quad + f_2 \quad + k_{32}f_3 \quad + k_{42}f_4 \qquad (2)$$

$$F_3 = k_{13}f_1 \quad + k_{23}f_3 \quad + f_3 \quad + k_{43}f_4 \qquad (3)$$

$$F_4 = k_{14}f_1 \quad + k_{24}f_4 \quad + k_{34}f_3 \quad + f_4 \qquad (4)$$

where F_1 = measured value of fluorescence in band 1 (e.g., green fluorescence),
F_2 = measured value of fluorescence in band 2 (e.g., yellow fluorescence)
F_3 = measured value of fluorescence in band 3 (e.g., orange fluorescence)
F_4 = measured value of fluorescence in band 4 (e.g., red fluorescence)

f_1 = actual fluorescence emitted by label 1 (e.g., fluorescein) in band 1(green)
f_2 = actual fluorescence emitted by label 2 (e.g., phycoerythrin) in band 2 (yellow)
f_3 = actual fluorescence emitted by label 3 (e.g., PE-Texas red) in band 3 (orange)
f_4 = actual fluorescence emitted by label 4 (e.g., PE-Cy5) in band 4 (red)

k_{ij} = fraction of fluorescence from label i in band j : derived from measurement of singly labeled specimens

Table 5-2. Equations (1-4) that must be solved to permit 4-color fluorescence compensation.

It is difficult, although not impossible, to set gates interactively using ellipsoids or planes in a simulated 3-dimensional space; the newer generation of flow cytometer operators, who have been exposed to video games since infancy, are probably a lot better at it than I am, but, while 3-D plots are widely available, 3-D gating doesn't seem to have found its way into most commercial software yet, and, since packages now allow for an almost limitless number of one- and two-dimensional gates, there really isn't much need for 3-D gating.

I discussed the difficulties of computing and displaying three-dimensional histograms with the introduction to cloud plots on pp. 32-3; there isn't any more that needs to be said about either subject here.

5.4 COMPENSATING WITHOUT DECOMPENSATING

The subject of fluorescence compensation was introduced on pp. 36-38, which included Figures 1-18 and 1-19; hardware compensation was further discussed on pp. 197-199, which included Figures 4-54 and 4-55. Instruments that measure no more than four fluorescence channels may still use hardware compensation, although most people who do four-color work are unhappy with hardware compensation. If you already have or are shopping for a flow cytometer that measures fluorescence in five or more spectral bands, you need software compensation, which is provided for in most newer instruments and which is also available in some third-party flow cytometry software packages. I should emphasize that there usually isn't much point in using the methodologies described on the pages immediately preceding this one to display uncompensated data.

In discussing software compensation, I will follow one of the approaches described in the now classic paper on the subject by Bagwell and Adams[1155]. Table 5-2 presents a set of equations relating measured values of fluorescence in four fluorescence bands, e.g., green, yellow, orange, and red, to actual fluorescence emission from four labels, e.g., fluorescein, phycoerythrin, PE-Texas red, and PE-Cy5. All we need to do to solve these equations, using matrix algebra, is find the numerical values of the coefficients k_{ij} .

The values of **spillover coefficients** k_{i1}, k_{i2}, k_{i3}, and k_{i4} are determined for each label i from measurements of the fluorescence of controls containing beads or cells stained only with that label. The value k_{ij} is defined as the ratio of (mean fluorescence of label i in band j) to (mean fluorescence of label i in band i); it is also possible to determine the coefficient values k_{ij} from the slopes of regression lines through the clusters representing the label i control on two-parameter dot plots. The measurements needed to derive the k_{ij} values, and the measurements to which compensation is applied, must be on a linear scale.

While it is possible to convert values from log to linear and back to apply software compensation, measurements made on an 8-bit or even a 10-bit logarithmic scale won't translate smoothly to an 8- or 10-bit linear scale, because the distance between two channels on the log scale may correspond to more than two channels in the linear scale, leaving gaps in distributions. I have already mentioned that the Coulter EPICS XL needs a 20-bit linear scale to translate to a 10-bit, 4-decade log scale, and that I use a 16-bit linear scale to translate to an 8-bit, 4-decade log scale, and can live with the small area of picket fence at the bottom of the low decade. If your instrument has software compensation built into its software, you probably have enough bits.

The compensation model I am using in this edition of the book is different from the one I used in the last one. What I have shown in Table 2 are equations that relate to **subtractive** compensation, for which the solution tells us how much of the fluorescence signals from interfering labels we need to subtract from each fluorescence signal to get the fluorescence contribution from the primary label in that channel. We will, for example, calculate and subtract green fluorescence contributions from phycoerythrin, phycoerythrin-Texas red, and phycoerythrin-Cy5 from the green signal, which should leave us with the value of green fluorescence due to fluorescein alone.

In the last addition of the book, I described **additive** compensation, also dealt with by Bagwell and Adams, which adds all of the fluorescence contributions from each of the labels. For example, the green, orange, and red contributions from phycoerythrin are added to the primary yellow fluorescence value to get a total phycoerythrin fluorescence value. This approach seemed attractive because it would theoretically capture more of the total fluorescence signal from each label. However, in practice, additive compensation will add more noise than signal, and should be avoided.

The additive compensation model I used in the last edition also included compensation for autofluorescence, which I have not included in the subtractive model I now recommend. It doesn't buy you all that much.

The bottom line on compensation is that existing hardware compensation circuits don't, and can't, compensate correctly across the whole measurement range; software can. However, you need to change your mindset about what compensated data should look like. Take a look at Figure 5-12.

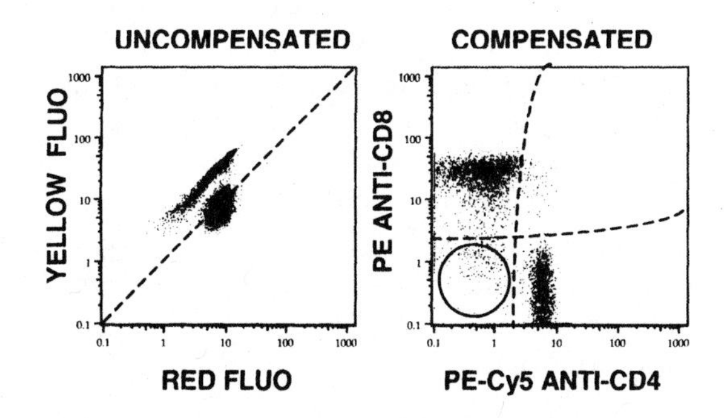

Figure 5-12. How compensation gets data to not quite fit into quadrants.

This figure reproduces the two rightmost panels of Figure 1-19 (p. 36), for which the caption was "How compensation gets data to fit into quadrants." I lied. The caption of Figure 5-12 is closer to the truth, and, having already mentioned (p. 198) that neither hardware nor software compensation can fit clusters into quadrants across the full intensity range, I can now tell you why.

The cells in the figure are those that fell into a T lymphocyte gate defined on the basis of low side scatter and staining with fluorescein anti-CD3 antibody. There are substantial numbers of CD4+CD8- and CD8+CD4- cells, and a scattering of double negatives and double positives. If you look at the left panel of the figure, which shows uncompensated data, the CD4-CD8+ cells form a very tight, elongated cluster just above the diagonal, while the CD4+CD8- cells form a cluster with a less extreme aspect ratio on the diagonal.

If we think of these clusters as ellipsoids, we can envision that each has two axes; the principal, or longer, axes of both clusters lie at substantial angles to the axes of the display. Having been introduced to correlation (pp. 237-8), we recognize that the data points in the uncompensated display of both clusters are correlated, with a higher correlation between the points in the CD4-CD8+ cluster than in the CD4+CD8- cluster. This reflects the fact both the yellow and red fluorescence signals are comprised of linear sums of PE and PE-Cy5 fluorescence, in different proportions. What compensation needs to do is allow us to create a new set of axes, representing PE and PE-Cy5 fluorescence; if we are successful in compensating, we expect there to be little correlation between the PE and PE-Cy5 fluorescence values in the clusters on the compensated display.

The right panel of Figure 5-12 tells us that we have largely succeeded. The principal axis of the compensated CD4+CD8- cluster, in which cells should not be exhibiting significant PE fluorescence, is essentially parallel to the PE axis, meaning that the PE-Cy5 fluorescence value is uncorrelated with the PE fluorescence value. The principal axis of the compensated CD4-CD8+ cluster, in which cells should not exhibit significant PE-Cy5 fluorescence, is almost parallel to the PE-Cy5 axis, meaning that the PE fluorescence value is at most only slightly correlated with the PE fluorescence value.

I have drawn a circle in the lower left corner of the right panel to indicate the position of a double negative cluster. If one broke up the display into classic quadrants using lines tangent to the upper and right edges of this cluster and perpendicular or parallel to the PE and PE-Cy5 axes, both the CD4-CD8+ and the CD4+CD8- clusters would extend across the quadrant boundaries. People who grew up with hardware compensation might conclude that the compensated display was **undercompensated**, and want to recalculate, changing k_{ij} values to subtract more PE-Cy5 from the yellow signal and more PE from the red signal. They would be wrong.

Let us consider, for the moment, that we are only doing two-color compensation between PE and PE-Cy5. A double negative cell has few bound molecules of either label, meaning that it has small values of both yellow and red fluorescence, and also that the variances of both fluorescences for cells in the double negative cluster are small in absolute terms. Subtracting a fraction of the red signal from the yellow signal to get the PE signal won't change the variance of the resulting value much, nor will the variance of the PE-Cy5 signal be changed much when that signal value is derived by subtracting a fraction of the yellow fluorescence from the red fluorescence.

Next, consider cells in the CD4+CD8- cluster. They have no more bound PE than do the double negatives, but they have many times as much bound PE-Cy5, meaning that both the yellow and the red fluorescence signals will be higher for cells in this cluster than for double negatives, and also that the variances of these signals will be higher. We will again subtract a fraction of the red signal from the yellow signal to get the PE signal, but, because we started out with higher variances, we will end up with higher variances, meaning that, even after compensation, the CD4+CD8- cluster will have a greater range of PE fluorescence values than will the double negative cluster. It will be impossible to separate "positive" from "negative" cells based on a fixed value of fluorescence, which is essentially what we do when we draw quadrants using lines perpendicular and parallel to the axes of a two-parameter display. We could do somewhat better by drawing lines with positive slopes, but, as Mario Roederer has shown in a comprehensive analysis of artifacts and limitations of compensation[2486], the correct separation lines would be curves looking more or less like those I have drawn in the right panel of Figure 5-12. In other words, if you want to get to compensation Heaven, you have to pass through the curly gates.

There's another problem that comes up with compensated data, which can be appreciated after a very careful look at Figure 5-12. What we are doing in compensation, whether in hardware or software, is subtracting a fixed fraction of one signal from another. That's deterministic, but because our photoelectron noise-limited data show so much random variation, we often end up with negative numbers as a result of compensation, and when we want to display data on log scales, we run into the problem that the logs of negative numbers are undefined. There are electronic and computational tricks for eliminating the negative numbers, but they leave us with a bunch of data points piled up at the very bottom of the scale. If you look carefully, you'll see the points just off the axes in both single positive clusters in the right panel of Figure 5-12. At best, this phenomenon makes it difficult to appreciate where the median of a negative cluster lies. At worst, we end up with an artifactual situation where it looks as if there's a tight cluster of negatives right at the axis and another one slightly off the axis, with a space in between. And, of course, once you start doing flow cytometry, your conditioned reflex response to seeing what looks like two separated clusters is to draw gates around each of them and count the numbers of events in the gates. These gates, however, are more surly than pearly. They are, as it were, gates in a picket fence, and, as I said on p. 214, we should look at picket fences as having "beware of the data" signs posted on them.

Mario Roederer and some of his former colleagues in the Herzenberg Lab at Stanford have been playing around with some variations on the good (or bad) old logarithmic data display scale that make it possible to eliminate the artifacts in the negative regions from displays of compensated data. What they came up with at first struck me as way too complicated for even relatively sophisticated users, but things have improved; see pp. 562-3 for more details and a dramatic illustration.

My modest counterproposal is that, since we can fairly easily put the math that defines and draws the curly gates into flow software packages, we do so, and crosshatch the areas on the low ends, allowing the numbers of cells in those crosshatched areas to be displayed, but preventing ourselves from being led astray by display artifacts. If you will, we'll be sending those infernally deluding data points to Hell in a handbasket.

To sum up: compensation is, or should be, based on algebra, and not a matter of taste. It should, wherever possible, be done with software, preferably, software incorporated into the instrument system or made to work with it relatively seamlessly. Compensation has artifacts and limitations; the more colors you measure, the more careful you need to be. See Mario's paper[2486] and the much less intimidating displays on his web site (http://www.drmr.com) for details.

5.5 DEALING WITH THE DATA

To this point, I have covered compensating the data, and displaying data values in one, two, and three dimensions. Three-dimensional displays don't lend themselves particularly well to data analysis, although statistical methods and classification techniques such as neural networks can be applied to data in spaces of any number of dimensions. In discussing methods of analysis, I'll start with one-dimensional data and work my way up.

Comparing and Analyzing Univariate Histograms

Visual inspection of multiple histograms or of the summarized data in box-and-whiskers plots only goes so far. Real statisticians do **tests of significance** to determine differences between distributions.

The first such test widely applied to flow cytometric histograms was the **Kolmogorov-Smirnov (K-S)** test, suggested for the purpose by Young[612] in 1977. This test calculates the **cumulative distribution**, which is the integral, of each of the two histograms to be compared. The algorithm for calculating a cumulative distribution is straightforward; channel n of the cumulative distribution is the sum of channels 0 through n of the original histogram. Examples using near-normal histograms are shown in Figure 5-13. I had an ulterior motive in using two near-Gaussian histograms for Figure 5-8; the curves incidentally illustrate how integration of pulses works.

Returning to the matter at hand, the K-S test calculates differences between the two cumulative distributions on a channel-by-channel basis; the maximum difference, or **D-value**, is examined to determine whether there is a statistically significant difference between the distributions.

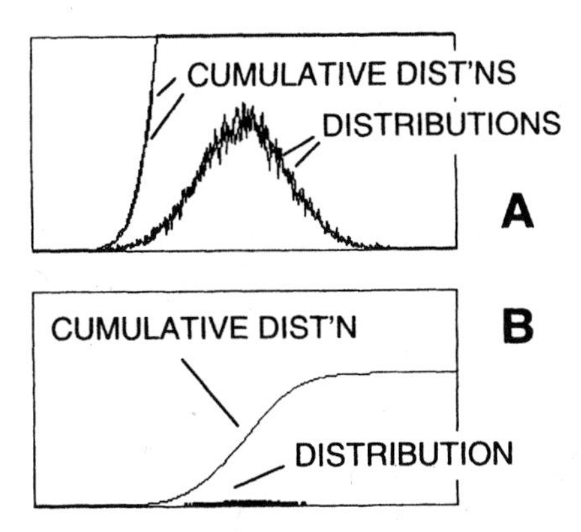

Figure 5-13. A: Two generated near-normal distributions and their cumulative distributions, or integrals (scale 0-256). B: One of the distributions and its integral, plotted on a different scale (0-16,384) to show the final integral value (10,000).

The K-S Test, Clonal Excess, and χ^2 Tests

The K-S test was used by Ault[611] in 1979 to detect **clonal excess** in peripheral blood lymphocytes in patients with B cell lymphomas. B cells bear surface immunoglobulins containing either κ or λ light chains, and different cells express different amounts of immunoglobulin. The cells in a B cell lymphoma represent a clone; all have the same light chain type and, in general, all express similar amounts of cell surface immunoglobulin. By applying the K-S test to histograms from aliquots of cells stained with fluorescent anti-κ and anti-λ antibodies, it was possible to detect contributions from circulating lymphoma cells making up a few percent of the B cell population.

The problem with the K-S test is that it is, if anything, too sensitive when it is done on a channel-by-channel basis. If you look hard, you'll see slight differences between the two distributions, and between the two cumulative distributions, in Figure 5-13. The maximum difference was 169; these distributions should, ideally, not be significantly different. A shift of a few channels due to vagaries of cytometer alignment can really screw up a channel-by-channel K-S test.

Things go better when data from a number of channels are combined, or **binned**. Cox et al[1190] used χ^2 (**chi-square**) statistics for histogram comparison, based on the assumption of Poisson variation within channels. This method is naturally applicable to binned data, and less susceptible to "false positives" than is the K-S test. Bagwell, Lovett and Ault[1011] used χ^2 tests to detect clonal excess. Many flow software packages include K-S test routines; the χ^2 tests of Cox et al are less common.

In recent years, Lampariello[2487-8] has discussed the use of both K-S and χ^2 tests as applied to the problem of discriminating weakly stained and unstained cells in histograms.

Roederer et al[2489] have described methods for comparison of univariate histograms in which the χ^2 test is applied after control data have been binned in such a way as to put equal numbers of events in each bin, which appears advantageous.

"Nonparametric" Histogram Comparison

Bagwell, Hudson, and Irvin[1191] developed a method of comparing histograms based on the application of Student's (see p. 19) t-test of the difference between two means on a channel-by-channel basis; they originally described the test as "nonparametric," an adjective which, while possibly applicable in the flow cytometric sense, was improperly used in the statistical sense, because the t-test is very much a parametric test. The method is now known as **average histogram comparison**.

Cumulative (Overton) Subtraction

Immunofluorescence histograms frequently show substantial overlaps between "negative" and "positive" populations; when measurements are made of antigens expressed at low density, one may see only a shift in position of the test histogram vs. that of a control rather than evidence of stained and unstained populations.

It has been fairly common practice to determine the "percent positive" in such histograms using the method of **cumulative subtraction** published by Overton in 1988[1192], which calculates reverse cumulative distributions on a percentage basis. Channel n of a cumulative distribution calculated on this basis would contain the percentage of the total number of events lying in channel n and all lower channels in the original histogram; channel n of a reverse cumulative distribution contains the percentage of the total number of events lying at channel n and all higher channels in the original. The "percent positive" is calculated from the maximum difference between the reverse cumulative distributions of the test histogram and a control histogram, e.g., an isotype control. Overton subtraction is included in numerous manufacturer-supplied and third party software packages.

That said, I should recall my argument on pp. 48 in favor of progressing beyond the concept of "positive" and "negative" when comparing immunofluorescence histograms. Since it is now not only possible, but relatively practical, to determine numbers of antibody-binding sites per cell when measuring weak immunofluorescence, it is appropriate to express differences between histograms in terms of properties of the distributions such as median and i.q.r., rather than to derive an illusory "percent positive."

Constant CV Analysis

Watson[1185] points out that the resolution of the ADC will have some effect on the CV of measurements, because the percentage difference between values in adjacent channels decreases with increasing channel numbers. He describes a procedure for histogram comparison that incorporates a correction to make CV constant. However, because the technique assumes underlying Gaussian distributions, its robustness, and, therefore, its suitability for analysis of immunofluorescence data, are not clear. Watson's book[1185] contains an extensive discussion of histogram comparison. He recently published an approach to discrimination of weakly stained from unstained cells using cumulative frequency subtraction and ratio analysis of means[2490].

Another Approach to Histogram Comparison

Zeng et al[2491,3322-3] have considered the problem of histogram comparison in the context of establishing molecular similarity between antibodies and, by inference, between the sites on cells to which they bind. Their eventual goal is to be able to use binding patterns for proteomic analysis.

Deconvoluting Single-Parameter Histograms

Single-parameter histograms are often obviously multimodal; good examples are immunofluorescence histograms containing separated peaks representing stained and unstained cells and histograms of cellular DNA content, which contain contributions from one or more populations of cells in G_0/G_1, S, and (G_2 + M) phases of the cell cycle and from nuclear fragments and other debris.

Considerable effort has been expended in the design of computer programs for identification and quantitative analysis of the components of DNA content distributions. Bagwell's book chapter[1184] provides a readable discussion of the underlying mathematical models and the workings of the programs; we will return to the subject when DNA analysis is discussed in later chapters.

Analysis of Two-Parameter Data

Two-Parameter Gating, Bitmap and Otherwise

A lot of flow cytometric analyses, e.g., immunofluorescence studies of human peripheral lymphocytes, are based on **gated analyses** from **two-dimensional selection regions** defined by **bitmaps**; the methodology was introduced in Chapter 1 and illustrated in Figures 1-17 (p. 34) and 1-19 (p. 38).

A **bitmap** is an area in computer memory set up to correspond to a selection region defined by the user in an interaction with the computer. Once the outline of the bitmap is drawn, a computer program sets bits corresponding to points on and inside the outline to 1, while bits corresponding to points outside the outline are set to 0.

Gated analysis using a bitmap is done by examining the bitmap parameters from all cells, and using data only from cells with bitmap bits set to 1 to compute one- and/or two-parameter histograms. It is also fairly common to produce color dot plots of parameters other than those used to set the bitmaps, in which cells in different bitmaps appear in different colors. B-D's "Paint-a-Gate™" software was among the first programs that did this, and, while almost all flow cytometric analysis software now available through instrument manufacturers or third parties includes such plotting capability, B-D's proprietary name seems to have found its way into the vernacular to describe the process.

Rectangles and Quadrilaterals: Hardware and Software Gating

Before computers became widely available, selection windows were set in hardware, as has been mentioned on pp. 27-9. Gating windows were restricted to a rectangular shape in most instruments, although Ortho's instruments could draw quadrilateral gates.

The B-D FACS II, which could measure two fluorescence parameters, had rectangular selection windows. When the first dual-label immunofluorescence work was done in the Herzenberg lab in the 1970's, using fluorescein and rhodamine labels and 515 nm excitation from a single argon laser, clusters representing cells stained with fluorescein-labeled antibody, rhodamine-labeled antibody, and both antibodies came out at acute angles to one another, and it was difficult or impossible to count or sort cells from only one of these populations. Since doing just that was the *raison d'être* of the place, it was necessary for Loken, Parks, and Herzenberg to develop fluorescence compensation[115]; a few op amps, potentiometers and resistors, and, *voilà!*, the fluorescein and rhodamine signals were back at right angles, and the sorting and counting could be done with the rectangular gates at hand.

One can implement rectangular gates in a digital computer program by comparing parameter values with set upper and lower limits, but the processing time per cell may be less if a rectangular bitmap is used instead. That consideration also applies in the case of some more sophisticated selection regions.

Ellipses and Beyond: Advantages of Bitmaps

Some people like to use **elliptical** selection windows, which offer an obvious advantage over rectangles or parallelograms because clusters are generally roughly elliptical, and because the corners of rectangles and parallelograms may get in the way when clusters are close together. Elliptical windows may be advantageous when the selection procedure must be automated, although, when the selection window is set by a user interacting with the system, a polygonal bitmap can get close enough to an ellipse to offer the same advantages over quadrilaterals[662]. There's no easy way to do elliptical windows in hardware. If you do them the obvious way in software, there is a computational liability in the sense that you need to compute quadratic functions (i.e., functions with terms including x^2 and y^2) to get the equations of ellipses; this takes a while.

If you use bitmaps, however, you don't have to compute quadratic functions for every cell coming through; you simply draw the ellipse, with the aid of the computer, and have the computer fill the elliptical bitmap. Thereafter, you determine whether the cell falls in the elliptical selection region by whether or not it fits in the bitmap.

Storage Requirements for Bitmaps

Once you define a bitmap, you'd like to be able to store it so you can use it again. To store a bitmap of arbitrary shape, possibly including disconnected regions, requires as many bits as there are points in the measurement region (i.e., 4,096 bits for a 64 × 64 bitmap, 16,384 bits for a 128 × 128 bitmap, etc.). However, if bitmap regions are restricted to closed polygons with an upper limit of *n* vertices, and a standard polygon fill routine is used in the computer, providing the coordinates of the vertices of the polygon, which only requires 2*n* storage locations, allows the bitmap to be reconstructed at run time. For elliptical regions, you only need to store the coefficients of the equation defining the ellipse, and so on. Since the average personal computer now has at least a hundred megabytes of memory, it's unlikely that anybody is still trying to economize on RAM for bitmap storage.

Analysis of Two-Parameter Distributions

See You Around the Quad

The most common form of data analysis applied to two-parameter distributions is the compilation of **quadrant statistics**, i.e., the derivation of counts or percentages of cells lying in quadrants of a two dimensional space. This can actually be done without bothering to compute the two-parameter histogram, by creating four rectangular gating regions that, like Arizona, New Mexico, Utah, and Colorado, have a corner in common. In this (these?) corner(s?), I should remind you that quadrants have their limits, especially when you work with multicolor compensated data.

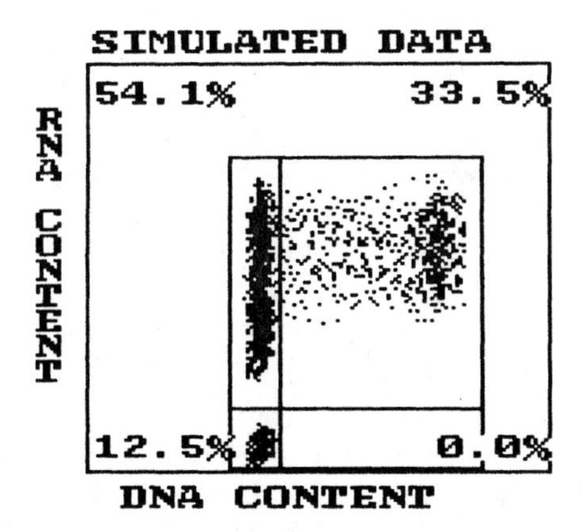

Figure 5-14. Quadrant statistics. Which quadrants are 1, 2, 3, and 4? Answer: Arizona, New Mexico, Utah, and Colorado. Other software may number them differently.

Quadrants themselves represent a throwback to the days when most flow cytometers only had rectangular selection regions for sorting and counting; this, as previously noted, was what got us fluorescence compensation. With today's computer-based data analysis systems, the shape of selection gates doesn't really matter, except that we humans like straight lines that meet at right angles, even when we lay out state borders.

I'm not sure whether all flow software packages number quadrants in the same order; you usually know the order yours uses, and might be surprised if and when you have to use or look at data from somebody else's. I am sure that most software packages only let you define the inner boundaries of the quadrant; the outer boundaries are the outer boundaries of the whole measurement space. Unfortunately, if you count stuff out to the outer boundaries, you increase the likelihood of including junk, which has boldly gone where no cell would go, in with cell counts; it is therefore advisable to have a way of moving the outer boundaries of the quadrant in, as shown in Figure 5-14. Score a point for my ancient **4Cyte™** software on this issue.

Bivariate Cell Kinetic Analysis

White and Terry[1193] have described a method for analysis of bivariate distributions of DNA content (propidium fluorescence) vs. fluorescence of antibodies to bromodeoxyuridine. This type of data is typically collected with propidium fluorescence (red) on a linear scale and antibody fluorescence (green) on a log scale; there is significant bleed of the propidium signal into the antibody fluorescence channel, with the result that unlabeled cells in the G_2 + M phases of the cell cycle, which have higher propidium fluorescence than cells in the G_0 and G_1 phases, also show higher green fluorescence. Correct estimation of the fractions of antibody labeled cells in different cell cycle phases requires a procedure equivalent to fluorescence compensation.

Additional discussion of this subject can be found in papers by Schmidt[2492], Cain and Chau[2493], Torricelli et al[2494], Johansson et al [2495], and White et al[2496].

Bivariate Karyotype Analysis

The most complete flow cytometric resolution of the human karyotype obtainable to date is achieved by staining with Hoechst 33342 and chromomycin A_3, which allow discrimination of differences in base composition between chromosomes of similar size. In bivariate displays of Hoechst dye vs. chromomycin fluorescence, the chromosomes typically appear as bivariate near-normal components superimposed on a continuous distribution of debris. Bivariate karyotypes are shown on pp. 318 (Figure 7-13) and 478 (Figure 10-15).

The complexity of a bivariate karyotype display, with dozens of peaks, makes automated or semiautomated analytical procedures essential if flow karyotype analysis is ever to become a routine procedure. Work in this direction is described by Dean, Kolla, and Van Dilla[1194], van den Engh, Hanson and Trask[1195], Boschman et al[1196], and Moore and Gray[1197]. However, it is probably fair to say that modern genetic analysis is oriented more toward molecular approaches.

Smooth(ing) Operators: When are Filters Cool?

Although I took a few slaps at data smoothing in my previous discussion of two-parameter displays, I didn't mean to imply there was no use for smoothing in general. Some form of filtering or smoothing is usually required before a curve fitting procedure can be applied to either univariate or multivariate data.

Sloot, Tensen, and Figdor[1198] describe a Fourier-transform based low-pass filter procedure which removes the higher frequency noise from distributions while preserving the underlying structure; this is more sophisticated than the more commonly used "moving average" filter, which simply subtracts a fraction of the values at adjacent points from each point in a distribution.

Bivariate Analysis in Hematology Analyzers

Hematology analyzers such as Abbott's Cell-Dyn series, Technicon (Bayer)'s H- and Advia series, and various Coulter systems do at least some of their differential counting via automated location of cell clusters in bivariate spaces. The algorithms used are not published, as far as I know.

5.6 MULTIPARAMETER DATA ANALYSIS

Multiparameter versus Multivariate Analysis

Multiparameter analysis is a term that covers a multitude of flow cytometric sins. As originally defined, multiparameter analysis involved collection and manipulation of data from measurements of more than one cellular characteristic. The definition was a bit fuzzy. Using an instrument such as the ICP, with a single fluorescence detector, to derive a DNA content distribution clearly was not multiparameter analysis. Using a Cytofluorograf with two fluorescence detectors to measure DNA and RNA in cells from the red and green fluorescence of acridine orange obviously was multiparameter analysis. Using a B-D FACS I, with a scatter detector and a fluorescence detector, to measure surface IgG on lymphocytes was a borderline case. The scatter measurement was used to provide a gating signal, enabling the fluorescence signal to be resolved above noise. Some people would consider this multiparameter analysis; some would not. Of course, as more people acquired two-parameter analysis capability, the growing community of flow cytometry snobs redefined multiparameter analysis to require measurement of at least three parameters, and so forth.

Some people refer to multiparameter analysis as if it were synonymous with **multivariate statistical analysis**; it is not. If one does statistical analysis of multiparameter data, it should probably be multivariate analysis, but one can do multiparameter flow cytometry without doing multivariate analysis, just as one can do single-parameter flow cytometry without doing any formal statistical analysis. In general, an exploratory phase of data analysis precedes any development of formal statistical techniques. Let me give you an example from a long time back.

Multiparameter Analysis of Leukocyte Types: 1974

We tend to think of multi-, now meaning at least four-parameter analysis, as a relatively recent addition to flow cytometry; that's a misconception While the technique didn't become widespread until the mid-1980's, when computer-based data analysis systems became generally available, it's been around since the 1960's.

Until the mid-1980's, however, nobody measured as many parameters or did as much statistical analysis of the data as was done between 1974 and 1977 in the Block differential counter project to which I have previously alluded[88-93]. We collected 6- and 7-parameter list mode data on over 20 million leukocytes from over 1000 normal and abnormal blood samples, using a 5-beam flow cytometer and, later, 3-beam and 2-beam instruments, and measuring forward and right angle scatter, extinction, and three or four fluorescence signals. This took several years, and left us with a few cabinets full of data on 9-track magnetic tape; it is now possible to accumulate 15-parameter data on 20 million leukocytes in an afternoon, and the data probably won't come close to filling up a CD-ROM.

Since the instrument was designed to count erythrocytes and platelets as well as leukocytes, we worked with unlysed whole blood samples, which were rapidly fixed in warm glutaraldehyde to make them permeable to the combination of ethidium bromide (EB) and the acid dyes LN and brilliant sulfaflavine (BSF) used for staining. Since the sample was unlysed, in order to be able to count 2,000 leukocytes in a minute, we had to measure the accompanying red cells, which outnumbered leukocytes by around 1,000:1, and platelets, which outnumbered leukocytes by about 50:1, giving a data rate in excess of 33,000 cells/s. We could not make multiparameter measurements in multiple beams at this rate; we therefore resorted to the same trick used by Hallermann et al[61] and others who attempted to develop leukocyte counters in the early 1960's, i.e., identification of leukocytes by nuclear fluorescence. Platelets don't have nuclei; erythrocytes with nuclei are abnormal and should be counted and identified as such, and any other nucleated cells in blood are either leukocytes or something the doctor wants to know about.

We thus designed the flow cytometers in our data collection system[88,92] so that the first beam through which cells passed would excite the red fluorescence of EB bound to nuclear DNA. Forward scatter and extinction measurements, also made in this beam, were used to identify, count, and size erythrocytes and platelets. The forward scatter signal was used to gate the red fluorescence detector; a red fluorescence signal above a set threshold indicated that a nucleated cell had come through the beam, triggering a series of logic signals which resulted in the fluorescence and right angle scatter signals from that cell being collected as the cell traversed the next beam(s). Coincidences of nucleated cells were unlikely, since only about 2,000 came by per minute; the electronics, however, would abort data collection in the

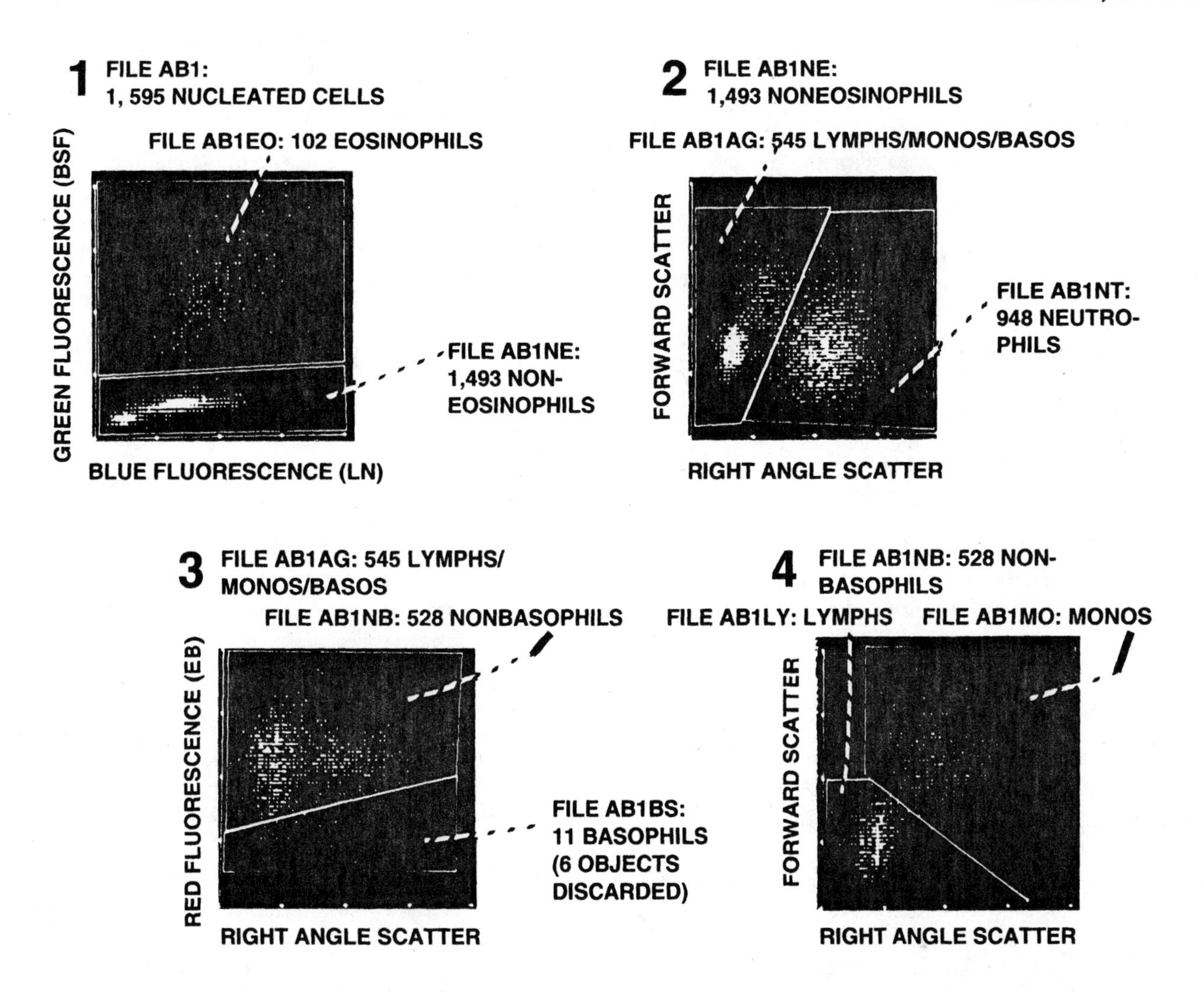

Figure 5-15. Multiparameter analysis of peripheral blood leukocyte types: Partitioning the population in two-dimensional subspaces to derive a training set for development of an automated algorithm.

event that two nucleated cells arrived at the first detector within too short a time period.

Automated Differentials via Discriminants

Using a minicomputer-based data collection system with a high-speed preprocessor, we could run blood samples through the machine and record data from the nucleated cells, and only from the nucleated cells, in list mode, so that we could display it and figure out how to do the differential count. I don't mean that we had to figure out which parameters to use, or whether they would work; we knew that different cells stained differently, and Wilson's Rule led us to expect that what we'd get from the flow system would correspond to what we saw under the fluorescence microscope. What we needed to do was to collect enough data so that we could use the multivariate statistical technique of **discriminant function analysis** to obtain an algorithmic procedure for leukocyte classification that would be implemented by a dedicated minicomputer incorporated in the instrument we intended to produce for clinical use.

If you feel the need for a digression into statistical analysis at this point, the subject, as it relates to image and flow cytometry, has been discussed in varying degrees of detail by Kamentsky[1], Prewitt[87], Sharpless[128], and Bartels[129-140,661]; a short book by Manly[1202] provides an accessible summary of multivariate statistical techniques, without cytometric examples. I will try to keep it short.

A **discriminant function** is a function of several variables (parameters, in the cytometric rather than the statistical sense) which, when evaluated for members of different classes (different cell types, in the cytometric case), assumes different ranges of values. We were looking for one or more **linear discriminant functions**, i.e., linear functions of our six or seven measured parameters that would enable us to tell different types of leukocytes apart. We restricted our search to linear functions because we couldn't compute quadratic functions or anything more complicated fast enough in real time to use them in a clinical instrument, even with a built-in minicomputer.

The procedure we used to find discriminant functions required that we start with a **training set**, i.e., that we provide the program with data from cells which we believed were lymphocytes, monocytes, neutrophils, etc. The computer could then come up with the functions of our measured variables which best distinguished each type of cell from all the other types. We would get the training set by using interactive computer graphics to extract clusters of cells from two-dimensional data displays. We selected the procedure for doing this by examining all of the possible two-dimensional displays (there are 20 if 6 parameters are measured) and using those on which clusters of cells were most widely separated. What we ended up with is shown in Figure 5-15, on the previous page.

Interactive Analysis: Finding the Training Set

We started out with a file called AB1, containing 1,595 nucleated cells. First, we looked at a bivariate distribution of blue fluorescence (LN) and green fluorescence (BSF) for all the cells in the file. We saw a diffuse cluster of cells with high BSF fluorescence standing out from everything else. We knew that eosinophils stained intensely with BSF, because we originally had looked at and marked the positions of Wright's stained blood cells on slides, then washed the slides, restained them with our fluorescent dyes, and reexamined the cells under the fluorescence microscope. We could thus say, with a high degree of confidence, that the cells with high measured values of BSF fluorescence were eosinophils, so we used the interactive graphics program to separate the file AB1 into an eosinophil file, AB1EO, containing 102 cells, and a "noneosinophil" file, AB1NE, containing the remaining 1,493 cells.

In step 2, we looked at a bivariate distribution of only those cells in the noneosinophil file AB1NE, with right angle scatter and forward scatter as the display parameters. We knew that neutrophils, with their cytoplasmic granules, scattered more light at wide angles than did lymphocytes and monocytes. This allowed us to separate the neutrophils (which we put in file AB1NT, with 948 cells) from the lymphocytes, monocytes, and basophils (in file AB1AG, with 545 cells).

In step 3, we looked at blue (LN) fluorescence vs. red nuclear (EB) fluorescence for the 545 cells in file AB1AG. From the restaining experiments mentioned previously, we knew that basophil nuclei did not appear as brightly stained as did nuclei of other cells; analysis of abnormal samples containing high percentages of basophils had also revealed large numbers of cells with low LN and low EB fluorescence in the location of the display we assigned to the basophil file, AB1BS, which contained 11 cells. In addition, we discarded 6 objects with values of EB fluorescence that suggested they were not basophils, but dye particles or pieces of cellular debris that got past our original EB thresholding logic. This left a "nonbasophil" file, AB1NB, with 528 cells.

In step 4, we examined extinction and right angle scatter values for the cells in AB1NB, partitioning it into files AB1LY (lymphocytes) and AB1MO (monocytes), relying on the known facts that monocytes are larger than lymphocytes, and therefore produce larger extinction signals, and that monocytes scatter more light orthogonal to the incident beam than do lymphocytes. The isolated files AB1EO, AB1NT, AB1BS, AB1LY, and AB1MO were then used as a training set of cells for a program that found a linear discriminant function.

Everything we did, up to this point, was multiparameter flow cytometric analysis; none of it was multivariate statistical analysis. Only after the training sets were defined did we begin to do the formal statistics involved in finding discriminant functions that would identify different cell types automatically.

Multiparameter Analysis of Leukocyte Types: 2002

Manual and interactive gating for identification of leukocyte subpopulations from flow cytometric data have progressed considerably over the past 28 years; a the title of a 2001 paper by DeRosa et al[2497], from the Herzenberg laboratory at Stanford, says it all: "11-color, 13-parameter flow cytometry: Identification of human naive T cells by phenotype, function, and T-cell receptor diversity."

In order to deal with data sets of this complexity, Roederer and his former colleagues at Stanford have extended their probability binning approach[2489] to permit comparisons of multivariate data[2498-9]; Baggerly[2500] has suggested some refinements to this methodology. In practice, however, cell clusters are still defined by skilled observers setting gates on multiple two-dimensional data displays.

The large volume of simpler T-cell subset analyses performed as a result of the worldwide HIV epidemic, and the need to maintain good quality control across multiple laboratories, have motivated flow cytometer manufacturers to develop automated software to eliminate subjective aspects of human-computer interaction from subset analysis. Some systems have reached the field; work in this area continues.

Procedures for Automated Classification

Discriminant functions represent one class of procedures usable for automated classification of cells, or other objects, based on measurements of multiple characteristics. Other procedures which have been used include **cluster analysis** and **principal component analysis**, which like discriminant functions, are based on multivariate statistics, and **genetic algorithms** and **neural network analysis**, which have less formal statistical bases.

Discriminant Functions and How They Work

What is a linear discriminant function, anyway? A two-dimensional or bivariate linear discriminant function is the equation of that line which, drawn across the bivariate distribution, best separates the clusters of data. A three-dimensional linear discriminant is the equation of a plane that partitions a three-dimensional data space to provide the best separation between one cluster and others. A four-dimensional linear discriminant is the equation of a hyperplane, etc. Knowing this, let's look at Figure 5-15 again.

In our interactive graphic analysis, we drew four-sided figures separating the cell clusters. In each of these cases, however, the boundary between the separated groups of cells

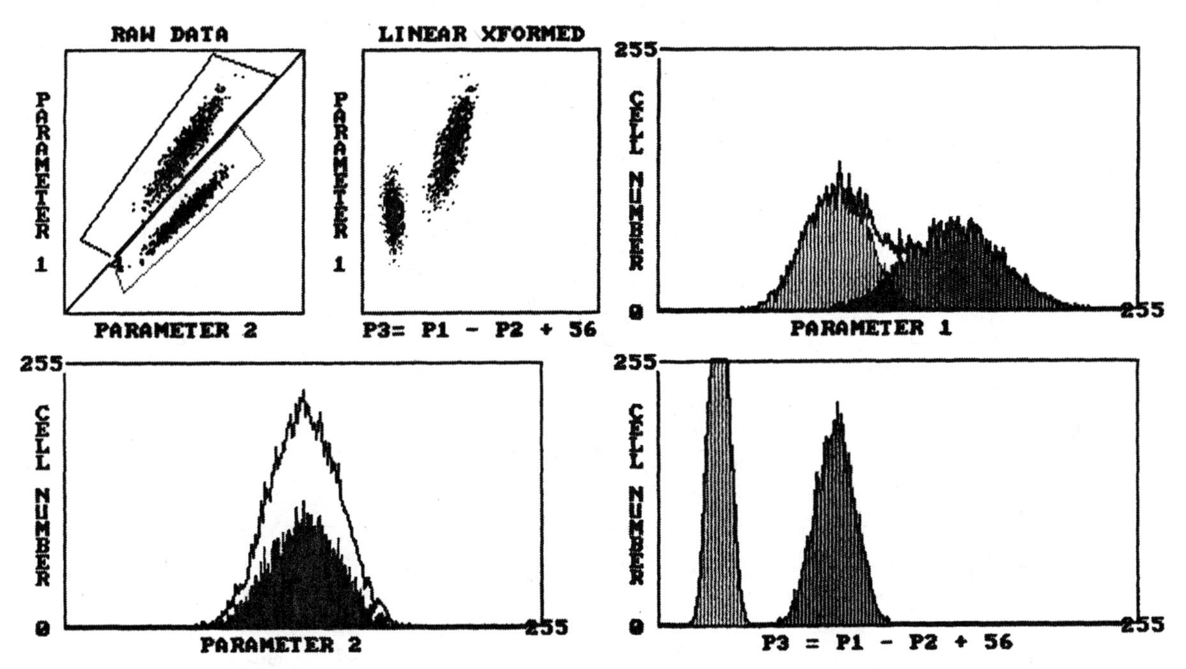

Figure 5-16. Linear transformation of data to allow separation of clusters as would be done using linear discriminant analysis or principal component analysis. P3 = P1 - P2 + 56 is a discriminant.

was defined by only one side of the four-sided figure, i.e., by a line. In 3 of 4 cases (steps 1,2, and 4), that line was not parallel to either the vertical or horizontal axis of the display.

In the fourth case (step 3), the line separating the basophils from the lymphocytes and monocytes was parallel to the horizontal axis; we could almost have used EB fluorescence alone to make the distinction, except that, as we noticed some pages ago, the second dimension (LN fluorescence) gave us a better indication of where to draw the line.

What we were doing with our interactive graphics, then, was linear discriminant analysis by eyeball. This got us about as good a separation of leukocyte classes as we could get from more formal discriminant function analysis in two dimensions; the reason for using six parameters in the differential leukocyte counter was that the additional parameters would further improve discrimination, at the expense of requiring us to work in a space of more than two dimensions. The real advantage of the discriminant function approach, however, is that it allows the leukocyte classification task to be done by a computer without operator intervention; this feature is absolutely necessary in an instrument for routine blood cell differential counting in the clinical laboratory, and just as necessary in a fluorescence based instrument which will automatically analyze T cell subsets, or one which will identify normal and abnormal cell types in bone marrow. While, in apparatus intended for research or for critical clinical uses, it may well be advisable for some human observer to keep a close watch on what is going on, it is anticipated, or at least hoped, that the easier samples can and will be handled entirely by the machine.

Since the empirical interactive partitioning of the two-parameter displays in Figure 5-15 could be done using straight lines, we were confident that linear discriminants would be adequate for automated leukocyte identification and counting in our system.

The procedure for finding linear discriminant functions has some similarities to that used in fluorescence compensation, as is illustrated in an informal way in Figure 5-16. The top left panel of Figure 5-16 is a two-dimensional dot plot showing two distinct clusters of data points that cannot be separated using rectangular windows; bitmap selection regions have been drawn around both. These clusters are separated by a diagonal line drawn across the screen. If you can face the trauma of bringing analytic geometry back to mind, you may recall that the equation of the 45° diagonal line is $y = x$; in this instance, given our axis labels, the equation is Parameter 1 = Parameter 2. For the cluster above the line, Parameter 1 > Parameter 2 , for all points in the cluster; similarly, for the cluster below the line, Parameter 1 < Parameter 2 , for all points in the cluster.

The panels at the upper right and lower left show distributions of Parameters 1 and 2 for the total population, drawn as curves. Distributions of these parameters for the two clusters, separated using the bitmaps shown, are plotted as bar graphs. The distributions of Parameter 1 for the two clusters overlap substantially; the distributions for Parameter 2 for the two clusters overlap almost completely. We want to compute a linear discriminant function, which, in this case, is a linear function of Parameters 1 and 2 that separates the two clusters. From the fact that the diagonal line separates the clusters, we know that we could use the function g = Parameter 1 - Parameter 2 as a discriminant. From a statistical point of view, we would have absolutely no problem using this function. However, if you think about it, you'll realize that g is negative for all points in one cluster and positive for all points in the other. In flow cytometry, we tend to work only in the one quadrant of two-dimensional Cartesian space in which both variables have values greater than or equal to zero. We'd like to be able to plot our transformed data on the same axes we use for raw data, and, when we do analog computation, using hardware, we like to keep voltages positive, or at least not negative.

We therefore want to add a medium-sized positive number to g to get a function we can plot more easily; I picked 56 by inspection, and computed a new Parameter 3, the value of which is determined by a function f, i.e., Parameter 3 = f = Parameter 1 - Parameter 2 + 56. A dot plot of Parameter 3 vs. Parameter 1 is shown in the upper middle panel of Figure 5-17; the panel at the lower right shows the distributions of Parameter 3 for the total population, as a line, and for the separated clusters, as bar graphs. You can drive a tank through the space between the two peaks on the Parameter 3 distribution. Also, as you probably noticed, the clusters in the dot plot of Parameter 3 vs. Parameter 1, unlike the clusters in the dot plot of Parameter 2 vs. Parameter 1, can be separated by rectangular windows. The take-home message, however, is that, having computed Parameter 3, we don't need to use rectangular windows.

One of the things we have just done is called **reducing the dimensionality of the data.** We've taken two-parameter data; if you will, a **vector** (just barely, in the two-parameter case) of observations, and computed a single number, a **scalar** quantity, which conveys what, to us at least, was the important information in the data set, i.e., which points belong to which clusters or, in the real world, how many enforcer T cells and how many informer T cells, or whatever, are in the sample. Sure, going from a two-dimensional vector to a number is no big deal, but the procedure works as well to go from three-, or four-, or from as-many-dimensions-as-your-computer-can-handle- dimensional vectors to one number, and that's the power of the discriminant function approach. Our two-dimensional exercise, with selection of a **training set** of clusters using bitmaps and subsequent semiempirical computation of a discriminant function, was a walkthrough version of the same procedure used for six-parameter data in the differential counter project, which operated in a harder to visualize six-dimensional space.

Hokanson et al[2501] reported that both discriminant function analysis and logistic regression were useful in discriminating breast cancer cells from hematopoietic precursors in bone marrow.

Principal Component Analysis

Principal component analysis[1200,1202-3], and the related technique of factor analysis, like discriminant analysis, are used to reduce the dimensionality of data. In principal component analysis new variables, which are linear functions of the old variables, are computed in such a way that the first new variable accounts for most of the variation in the data, the second variable, for the next most, and so on.

The choice of new variables is equivalent to a translation and rotation of the coordinate axes, again similar to what is done in fluorescence compensation. The new axes remain orthogonal to one another. A principal component analysis of the data shown in Figure 5-16 would place the axis representing values of one new variable along the diagonal line separating the clusters in the upper left-hand panel of the figure, with the axis of the other new variable perpendicular to it. The principal component is that linear function of the raw parameters that falls along the major axis of an elliptical cluster and which, therefore, accounts for the largest component of variance of the data.

Cluster Analysis

The technique of **cluster analysis**[1202] attempts to assign members of a population to different classes based on values of a number of measured variables. **Hierarchical** cluster analysis constructs a "tree" or **dendrogram** of relationships between clusters; this method was used in assigning antibodies to CD classes in the Fifth Leukocyte Differentiation Workshop[1126]. Hierarchical cluster analysis can proceed by **agglomeration** or **division**, which are the dignified analytical terms for "lumping" and "splitting." Agglomeration starts with each data point assigned to its own class and reduces the number of classes as similarities between data points emerge; division starts with all data points in one class and looks for differences.

In non-hierarchical cluster analysis, an initial number of clusters and their approximate locations in the measurement space are specified; assignments are refined based on a measure of the distance of data points from the cluster means,. using either Euclidean distance or the **Mahalanobis distance**, which incorporates terms from the variance-covariance matrix. Murphy[1204] described the application of non-hierarchical cluster analysis to flow cytometric data in 1985.

Principal component analysis may be applied to reduce the dimensionality of the data set before cluster analysis is done; this approach was followed by Kosugi et al[1203] in 1988. Other investigators who have applied cluster analysis to problems such as immunophenotyping[1205-12] have combined the technique with other methods, e.g., artificial intelligence techniques such as expert systems[1205] and principal components[2502]

Neural Network Analysis

Classification methods are generally described as **supervised**, meaning they require a training set, or **unsupervised**, meaning the algorithm finds the classes in the structure of the data. **Neural network analysis**[1213-4] might be described as a "latchkey" method; you provide a neural network with a training set, but you never seem to be sure how it does what it does.

A neural network is generally implemented as a mathematical model of stylized "neurons," connections between which are strengthened and weakened as the network "learns" from a training set, after which it can be used to classify objects in new data sets. The nature of the algorithm being implemented by the network is not, in general, obvious to the user. At present, neural networks are usually run as simulations on conventional digital computers; however, much of the interest in the methodology relates to the possibility of implementing neural networks in chips, some of which are already available.

Frankel et al[1215, 1217, 2503] have used neural networks for analysis of phytoplankton populations and blood cells; Boddy et al[1216] have also applied the method to phytoplankton; Redelman[1218] and Kim et al[3323] have also considered neural network analyses of blood cells. Godavarti et al[2474] found that neural networks performed better than conventional clustering algorithms in classifying cells based on pulse features measured using digital pulse processing, and Davey et al[2504] found neural networks superior to principal component analysis and several other multivariate statistical methods in classifying bacteria stained with dye mixtures.

Genetic Algorithms

Genetic algorithms[1219] are another class of nonclassical procedures said to be well suited for application to classification problems. The "genetic" name arises from the *modus operandi* of developing several algorithms, each of which operates on the training set, eliminating the poorer performing algorithms, and combining features of the better performers in the next generation. "Darwinian" or "evolutionary" might have been a more appropriate adjective; "genetic" probably sounds better when you're looking for venture capital. I haven't seen genetic algorithms used for flow data.

5.7 ANALYSIS OF COLLECTED DATA: HOW MUCH IS ENOUGH/TOO MUCH?

Nobody would bother building instruments as expensive as flow cytometers to collect and display data from cells if the data were simply going to be filed away and never looked at again. Sure, there are some applications of flow cytometry that require almost no formal data collection or analysis. The best example I can think of is screening monoclonal antibodies to see whether they react with all, none, or some of the cells in a cell population. People who do a lot of this run samples through in rapid succession; the flow pattern never gets a chance to stabilize as the operator accumulates all of the necessary information from a glance at the dot display of scatter vs. fluorescence on the monitor oscilloscope and hurries on to the next sample. This undeniably represents efficient and effective utilization of a flow cytometer, provided it's a bare bones instrument. I'm not sure you can even play this game on the modern benchtop instruments. I do know that anyone who would propose to purchase a high-end flow cytometer and use it 99 percent of the time for screening clones should, and probably will, have his or her budget examined.

On the other hand, if you can justify or are attempting to justify the acquisition of a high-end instrument, you almost undoubtedly have a problem which requires that you collect a lot of data and then chew on the data for a bit before you can answer the biological question you are asking. Telling the difference between immunofluorescence positive and immunofluorescence negative cells is easy, at least some of the time; telling the difference between cancer cells and normal cells must be hard because we don't know how to do it yet. Much massaging of stored flow cytometric data, past and present, was/is aimed at the development of cell classification criteria which could later be implemented in hardware and/or software for real-time decision making, either in diagnostic apparatus or for purposes of cell sorting. Whether you're trying to automate cervical cytology or the differential leukocyte count or to isolate pluripotent stem cells, you will end up in an iterative process of data collection, analysis, refinement of criteria, more collection, more analysis, etc. Getting the job done and the questions answered will, as often as not, depend at least as much on the proper choice of parameters to be measured as on the method(s) of data analysis employed.

Once you get flow cytometric data stored in list mode, there are all kinds of options for number crunching. You can apply any statistical test you like, and you will generally be working with large enough numbers of data points to make statisticians in other fields drool. If I were you, though, I wouldn't be too quick to disparage plain, old-fashioned dot plots. Those of us who deal with them day in and day out tend to forget that they allow us to determine the structure of large masses of data almost by inspection. An article some years back in *Science*[141] described a great advance in manipulation of multidimensional statistical data that was made possible by interactive computer graphics. The great advance turned out to be dot plots, which we in analytical cytology had had for years. While some of us have been lamenting a lack of rigorous statistical procedures for data analysis, statisticians have been aching to be able to reduce many of their problems to the level of most of ours. Now, even the snobs in our field can use humble but effective methodology without feeling embarrassed.

A good example of data extraction by dot plot inspection appears in Figure 5-17 (next page), another golden oldie from the Block differential counter project. The two-dimensional display of LN fluorescence vs. 90° scatter shows a high degree of correlation between measured values for all leukocyte classes. This told us we could do as good a differential count without LN fluorescence as we could with it, which would allow the use of a single-beam instrument with a two-dye stain instead of a two- or three-beam instrument with three dyes. We didn't need principal component analysis or even have to calculate correlation coefficients to reduce the dimensionality of our data.

The good news was that having a multiparameter instrument with which we could measure the same things in the same cells two different ways at the same time was really helpful in finding the most efficient ways to make measurements. The bad news was that, once we showed that the clinical task, i.e., differential leukocyte counting, for which Block hoped to sell a lot of fancy flow cytometers could be accomplished by a simpler, cheaper instrument that didn't infringe on any of the patents involved, the project lost economic viability. Such are the ways of science.

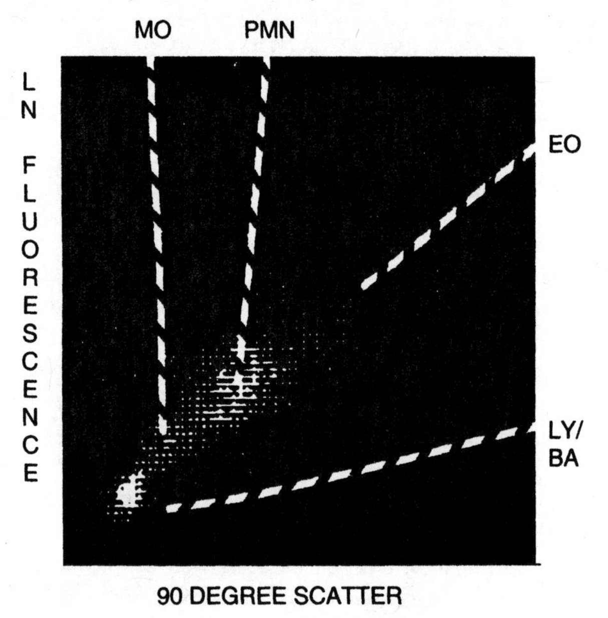

Figure 5-17. Calculation is not always necessary to reduce the dimensionality of data.

5.8 DATA ANALYSIS ODDS AND ENDS

Data Storage

The Flow Cytometry Standard (FCS) File Format

The Flow Cytometry Standard (FCS) format proposed by Murphy and Chused[1022] in 1984, and revised (to FCS2.0) by a committee of the International Society for Analytical Cytology in 1990[1220-1], has now been almost universally adopted by manufacturers of flow cytometry apparatus and by third party software developers.

The standard format makes it possible for users to read and manipulate data files acquired on a variety of instruments. If you happen to have data files that are not in FCS format, it makes sense to get a program that will convert them, so you can use the full range of software now available.

A revision of the standard to FCS3.0 was proposed in 1997[2505], but the full FCS3.0 standard has not been published, as its predecessors were, in *Cytometry*. It is available from ISAC's Web site at <http://www.isac-net.org/links/topics/FCS3.htm>.

The most notable additions to the FCS3.0 standard were a mechanism for handling data files larger than 100 MB and support for UNICODE text for keyword values. There has always been a considerable amount of flexibility in the standard; the ISAC Data File Standards Committee, which is responsible for maintaining and revising the standard, has been at work for several years on generalized FCS file parsing software, to be placed in the public domain.

An FCS file consists of a HEADER, identifying the file as an FCS file and specifying the version of FCS used, and containing numerical values identifying the position of the following TEXT segment, in which any of a large number of keywords and numerical values may be used to describe the specimen and the conditions under which the experiment was done, and the DATA segment, which contains numerical data in a format specified in the TEXT segment. The header and text portions of an FCS2.0 file collected using a B-D FACSCalibur are shown as Table 5-3 below.

FCS2.0 58 1536 1537 173946 0 0
/$BYTEORD/1,2,3,4/$DATATYPE/I/$NEXTDATA/0/
$SYS/Macintosh System Software
8.1.0/CREATOR/CELLQuestª 3.2.1/
$TOT/12315/$MODE/L/$PAR/7/
$P1N/FSC-H/$P1R/1024/$P1B/16/$P1E/0,0/
$P2N/SSC-H/$P2R/1024/$P2B/16/$P2E/0,0/
$P3N/FL1-H/$P3R/1024/$P3B/16/$P3E/4,0/
$P4N/FL2-H/$P4R/1024/$P4B/16/$P4E/4,0/
$P5N/FL3-H/$P5R/1024/$P5B/16/$P5E/4,0/
$P6N/FL2-A/$P6R/1024/$P6B/16/$P6E/0,0/
$P7N/FL2-W/$P7R/1024/$P7B/16/$P7E/0,0/
$CYT/FACSCalibur/CYTNUM/E2252/
$BTIM/18:04:11/$ETIM/18:04:29/
BD$ACQLIBVERSION/3.1/BD$NPAR/7/
BD$P1N/FSC-H/BD$P2N/SSC-H/BD$P3N/FL1-H/
BD$P4N/FL2-H/BD$P5N/FL3-H/BD$P6N/FL2-A/
BD$P7N/FL2-W/
BD$WORD0/104/BD$WORD1/376/BD$WORD2/444/
BD$WORD3/387/BD$WORD4/399/BD$WORD5/400/
BD$WORD6/400/BD$WORD7/400/BD$WORD8/400/
BD$WORD9/400/BD$WORD10/301/BD$WORD11/25
4/BD$WORD12/499/BD$WORD13/0/BD$WORD14/3
82/BD$WORD15/451/BD$WORD16/391/BD$WORD1
7/407/BD$WORD18/129/BD$WORD19/100/BD$WOR
D20/100/BD$WORD21/100/BD$WORD22/100/BD$W
ORD23/1/BD$WORD24/1/BD$WORD25/0/BD$WOR
D26/0/BD$WORD27/0/BD$WORD28/136/BD$WORD
29/32/BD$WORD30/52/BD$WORD31/52/BD$WORD
32/52/BD$WORD33/52/BD$WORD34/7/BD$WORD3
5/195/BD$WORD36/25/BD$WORD37/5/BD$WORD3
8/280/BD$WORD39/3/BD$WORD40/3/BD$WORD41
/100/BD$WORD42/100/BD$WORD43/0/BD$WORD4
4/1023/BD$WORD45/1023/BD$WORD46/1023/BD$W
ORD47/52/BD$WORD48/789/BD$WORD49/10/BD$
WORD50/6/BD$WORD51/52/BD$WORD52/0/BD$W
ORD53/0/BD$WORD54/0/BD$WORD55/0/BD$WOR
D56/0/BD$WORD57/0/BD$WORD58/0/BD$WORD59
/0/BD$WORD60/0/BD$WORD61/0/BD$WORD62/0/B
D$WORD63/0/
BD$LASERMODE/1/CALIBFILE/FALSE/
P7THRESVOL/52/
$FIL/02/$DATE/16-Nov-01/

Table 5-3. Header and text portion of an FCS 2.0 data file.

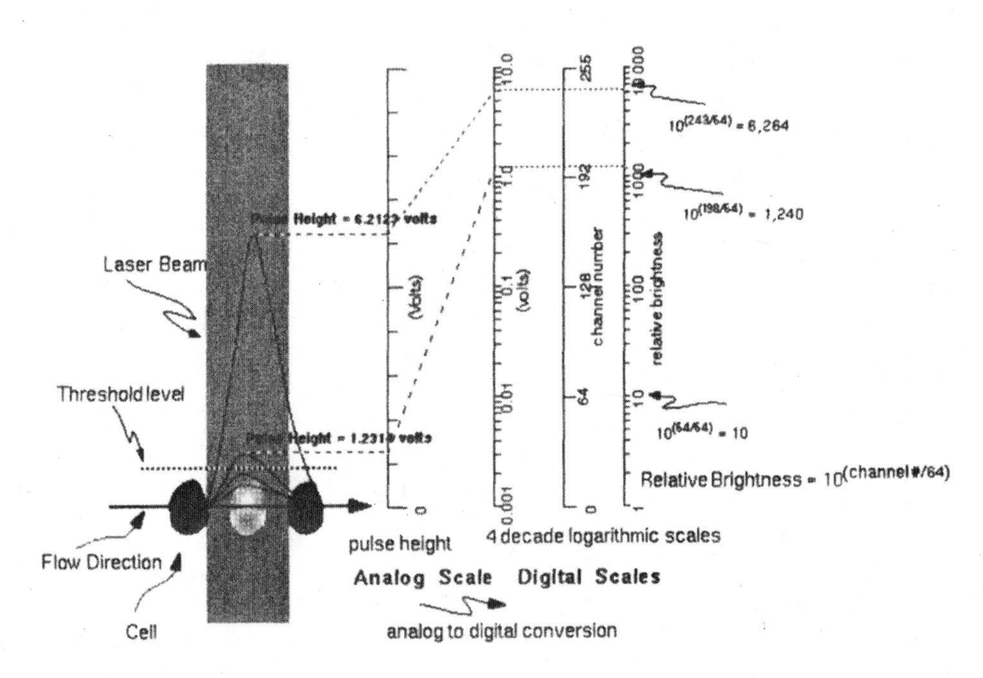

Figure 5-18. Linear and log scales revisited. Courtesy of David Coder.

You shouldn't have to worry much about the structure of your FCS files; the software you use for data acquisition will almost certainly make it easy for you to get information into the computer that the program will subsequently transfer to the file.

Magnetic/ Optical Tumors in the Digital Attic

By the time the Block differential counter project petered out in 1977, we had accumulated multiparameter list mode data from over 20,000,000 leukocytes in data files on 9-track tape, in the belief that all this stuff was going to have to be submitted to the Food and Drug Administration. The magnetic tumor metastasized from file cabinet to storage room to a warehouse in Cambridge, from which it was lost in the mid-1980's, well before the disease of accumulating flow cytometric data had become epidemic.

Nowadays, the magnetic tumor is being replaced by optical tumors, i.e., CD-ROMAs and DVD-ROMAs. A single 650 MB CD-ROM can supposedly hold all of classical Greek literature and everything that's been written about it until the present. In 1987, Mann[1222], who also considered other aspects of multiparameter flow cytometric analysis, raised the issue of data compression, but, these days, there isn't much point to it. People still do have problems converting data from old storage formats, e.g., 8" floppy discs, to new ones. If you want to save old data, it's best to save the hardware and software you need to read it with.

Linear and Log Scales and Ratios: Proceed with Care!

We are all used to collecting and displaying some data (forward and side scatter signals and fluorescence of DNA dyes) on linear scales and other data (fluorescence, much of the time) on log scales. However, things sometimes get confusing when we want to manipulate parameter values, and I have put up Figure 5-18, a really neat illustration from Dave Coder, which I hope will help me clarify things.

The left side of the figure shows a cartoon of cells passing through a laser beam and generating pulses of different voltages, which are then converted into numbers. There are a 10-volt linear and a 4-decade logarithmic voltage scale, covering the range from 1 mV (or, in the case of the linear scale, 0 V) to 10 V. There is also a 4-decade logarithmic relative brightness scale, covering the range from 1 to 10,000 arbitrary units. And, in between, there is a scale with channel numbers on it; the numbers range from 0 to 255, and the scale looks linear. And that's what can get us into trouble.

The 256-channel scale tells us that we are dealing with an 8-bit ADC. If this operates over a 0-10V input range, channel 0 corresponds to 0 V, channel 64 to 2.5 V, channel 128 to 5 V, channel 192 to 7.5 V, and channel 255 to 10 V. The numbers in the exponents to the right of the brightness scale tell us that the 1.231 V signal comes out at channel 198 and the 6.212 V signal comes out at channel 243. That means that the signals must have gone through a log amplifier before they got to the ADC, because the signal representing the 1.231 V pulse is a little over 7.5 V in amplitude (channel 198 vs. channel 192) at the ADC input.

Now, suppose you want to take the ratio of values of those two pulse amplitudes. If you go to the relative brightness scale, and convert from channel numbers back to the corresponding linear values there's no problem. The formula for doing this appears to the right of the brightness scale. The ADC scale we're using is a 256-channel, 4-decade scale, meaning that there are 64 channels/decade. Therefore, to convert from a channel number n on this log scale to a numerical value on the relative brightness scale, we raise 10 to the power (*n*/64). The application of this formula to the pulses at channels 243 and 198 gives us linear brightness values of 6,264 and 1,240; the ratio of these values, to two decimal places, is 5.05. You get the same value if you divide the pulse height voltages (6.212 V and 1.231 V).

What you do *not* want to do is divide the ADC channel numbers to get a "ratio." Just in case you're not good at doing arithmetic in your head, I'll tell you that 243/198 is 1.227, a number substantially different from 5.05. The catch is that the channel numbers don't represent the voltage values of the pulse heights; they represent the logs of the voltage values. What we want is a ratio a/b. Dividing log a by log b doesn't give us the ratio a/b; it doesn't even give us the log of a/b. What it gives us is the log of the (a)th root of b. While that number might be of some use for something (I can't think of what, but I'll be charitable), it won't help you if you're looking for a ratio, or for the log of a ratio.

There is, however, a way to get the log of the ratio using channel numbers and the log scale; log a – log b = log (a/b), so all we have to do is subtract the channel values.

In this instance, we are trying to get the ratio of brightnesses, or voltages, represented by channels 243 and 198; 243-198 = 45, and $10^{(45/64)}$ is 5.048, which is close enough to 5.046 and 5.051 to prove my point. (6,264 for the 6.212 V pulse and 1,240 for the 1.231 V pulse),

If you want to do the log ratio computation, and then plot the resulting value as a derived parameter, you may run into another snag. If the ratio you're after turns out to be less than 1, you'll end up with negative channel numbers. There's an easy fix; simply add a constant to bring the negative numbers back on scale. Since (log x + c) = log (10^cx), your adjusted channel numbers will remain proportional to the log of the ratio you were looking for.

This issue seems to come up on the Purdue Cytometry Mailing List a few times a year, so I thought I should expound on it here.

Ratios Only Help if Variables are Well Correlated

In what I have said above about ratios, I haven't considered why we might want to use a ratio of two parameter values instead of the parameter values themselves; the discussion of ratiometric measurements on p. 47 provides one example. We use the ratios of 405 nm and 480 nm emission intensities from the UV-excited probe Indo-1 as a measure of cytoplasmic [Ca^{++}], because the raw values of either don't give us the information we need. When complexed with Ca^{++}, the dye emits maximally at 405 nm; the free dye emits maximally at 480 nm. Fluorescence measurements at either wavelength depend to a greater extent on the amount of dye taken up and retained by a cell than on calcium concentration; the ratio, which tells us the proportions of complexed and free dye, is largely independent of the amount of dye in the cell, and also largely independent of other confounding influences, such as light source noise.

So far, so good. But suppose we want to separate two populations based on a ratio measurement. For this to work, it would be helpful to have the ratio of the two primary parameters be substantially different in the two populations. If we then performed separate **linear regression** analyses (pp. 237-8) on data points from the clusters representing the two populations, we would expect to get different slopes for the regression lines. We could tolerate similar slopes if the y-intercept values were substantially different.

But there's a catch, and it's pretty much the same one I mentioned when cautioning about the use of regression on pp. 237-8. Let's look back there, and again notice Figure 5-8, which shows clusters of data points with low and high degrees of **correlation**. If we calculated ratios of y- and x-values for the highly correlated data points in the right panel of the figure, for which the **correlation coefficient, r,** is .950, and then plotted the distribution of ratio values, we would expect the distribution to have a fairly low variance; the points all fall within a short distance of a line, and the mean of the ratio would be the slope of the line.

If we did the same ratio calculations for the data points in the left panel of Figure 5-8, where the data are essentially uncorrelated (r = -.002), we would find a relatively high variance for the ratio.

The ranges of x- and y-values are essentially the same for the data points in the left and right panels of Figure 5-8, and the variances of the distributions of x- and y-values for points in both panels are relatively high. However, the variance of the ratio of x- and y-values for the well-correlated data in the right panel is considerably lower than the variances of the x- and y-values themselves, while the variance of the ratio for the uncorrelated data in the left panel remains high.

To go back to the Indo-1 example, if you were looking at Indo-1 fluorescence measurements from two populations with different levels of cytoplasmic [Ca^{++}], it would be much easier to separate the populations if the data points for each were well correlated than it would if they were essentially uncorrelated.

Indo-1 ratios work well in mammalian cells, into which it is possible to load hundreds of thousands of dye molecules. If we try to do the same measurements, using the same observation time, in bacteria, into which we may only be able to load a few thousand molecules at most, we can expect the variances of the raw fluorescence measurements to be substantially higher than in mammalian cells because we will be collecting fewer photoelectrons; it will thus be more difficult to detect small changes. Keep that in mind for the discussions of ratiometric measurements of functional parameters on pp. 402 and 407.

Ratios, of course, have uses outside cytometry; among other things, they are used for figuring odds. So much for the odds; here's the end. On to cell sorting.

6. FLOW SORTING

The addition of cell sorting capability to a flow cytometer makes it possible to isolate highly purified populations of cells with precisely defined characteristics. Any parameter(s) measurable in a flow cytometer can provide a basis for selection of cells, and the limit on the degree of homogeneity that can be achieved in the selected population is set primarily by the precision with which the selection parameter(s) can be measured. To the uninitiated, flow sorting seems like something out of fantasy, almost an implementation of the Maxwell Demon (or, in this case, Mack's Swell Demon) at the cellular level. Perhaps because of this, sorters are widely coveted, and the desire to use them in experiments sometimes leads experimenters to overlook more expeditious methods of procedure.

The literature on the technical details of cell sorting, as opposed to flow cytometry, is manageable, to say the least; there are probably still not more than a few dozen papers. The well-illustrated 1976 *Scientific American* article by Herzenberg et al[2] remains a valuable source for historic perspective. Other older references covering technical details of the hardware are the chapter by Sweet[142] and its revision by Lindmo, Peters, and Sweet[1223] in the first and second editions of the Big Yellow Book[9,1028], and Pinkel and Stovel's contribution[650] in the volume edited by Van Dilla et al[624], which appeared about midway between them. However, if you're looking for more detail than appears here, the best place to go is to three consecutive chapters by Durack[2506], van den Engh[2507], and Leary[2508] in a relatively new book edited by Durack and Robinson, *Emerging Tools for Single-Cell Analysis: Advances in Optical Measurement Technologies*[2399].

Although flow sorting based on electroacoustic or electromechanical **fluidic switching** has been described[66,142,146,2325-7,2509], most commercial and laboratory-built instruments built before the 1990's, and many built since, employ **droplet sorting**, in which the fluid stream is broken up into droplets, and the droplets containing the selected cells are electrically charged and deflected into a collection vessel by passage through an electric field.

6.1 SORT CONTROL (DECISION) LOGIC

Whichever sorting mechanism is used, it is first necessary to determine whether each cell passing through the flow cytometer meets the selection criteria; when the criteria are met, it is then necessary to generate a logic signal which will activate the sorting mechanism. This activating signal must be delayed until the cell reaches the droplet breakoff point, in a droplet sorter, or the point at which the stream is diverted or captured, in a fluidic sorter. Depending upon the dimensions of the flow chamber, and upon whether cells are observed inside the chamber or in air, it may take anywhere from a few microseconds to a few hundred microseconds for a cell to traverse the distance between the observation point and the breakoff or diversion point.

One could, in principle, use one-shots (monostable multivibrators), such as are used in triggering circuits and discussed on p. 192, to generate the time delays required for sorting. This is undesirable, however, because it decreases the rate at which cells can be sorted. In a droplet sorter generating 40,000 drops/s, a droplet period is 25 µs; using typical figures of 10 m/s for velocity and 2.5 mm[142] for the distance from the nozzle to the droplet breakoff point, and assuming the observation point of a stream-in-air system is located 0.5 mm below the nozzle, 200 µs, or 8 droplet periods, will elapse between the time a cell passes the observation point and the time at which it reaches the breakoff point.

The use of a one-shot for timing would force a 200 µs wait between sort pulses, limiting the rate of analysis to less than 5,000 cells/s. Assuming that a cell's traverse of the observation point takes 5 µs, and that another 15 µs are required to arrive at a sort decision from the signal peak value(s) once the strobe pulse (see pp. 27-8 and 191-4) starts, the electronics could produce a sort decision within one droplet period after the arrival of a cell and would then end up idle for 7 droplet periods before they could process another cell. This is inefficient; if the decision electronics can respond in one droplet period, they can, in principle, make up to 8 sort decisions during the time required for a cell to reach the droplet breakoff point. If there were some way of keeping track of 8 droplets' worth of yea-or-nay for 200 μs, it would thus be possible to speed up the sorting rate.

There are, in fact, several ways. The mechanism used to queue sort decisions in many earlier commercial sorters is a digital circuit called a **shift register**. A shift register circuit accepts a stream of logic pulses ("1"s and "0"s) as input; the same sequence of logic pulses appears at the output after a set time delay. The time delay between the appearance of a given pulse sequence at the input and its appearance at the output is an adjustable, integral number of periods of a fixed-frequency **clock signal** that is also applied to the shift register. If this clock signal is synchronized with the transducer drive frequency, each delay period corresponds to one droplet period. Once the processing electronics respond to the signal and strobe pulses and reach a sort decision, the buck passes to the shift register. "Word" of the decision doesn't get to the drop charging circuits until they need to act upon it. By that time, the front end has made several other decisions, all of which are in the shift register pipeline en route to the drop charging circuit.

A sort decision, like a cell classification decision, is **binary**. A cell is either sorted, or not sorted, just as a cell is assigned or not assigned to a given class. The hardware and/or software used to make sort decisions can be identical to those used for data analysis, as discussed in the previous chapters. Complex decision functions can be realized in analog hardware that can form sums, differences, or ratios of log or linear signals, or in software, using digital computers to calculate the required functions and also to generate delayed sort pulses, eliminating the need for shift registers.

While hardware controllers can process multiparameter signals in parallel, most of the computers used in earlier cell sorters had to perform complex computations one step at a time. Computer-controlled sorting, as first reported by Arndt-Jovin and Jovin in 1974[147], then required a minicomputer. Early production instruments with computer-controlled sorting were not necessarily either better or capable of implementation of more complex decision functions than instruments in which sorting was not controlled by a digital computer, as was recognized by B-D, which replaced its original sort control electronics with a computer, and then replaced the computer with new sort electronics in three successive generations of FACS flow sorters.

At present, however, the issue is not whether computers should be used for sort control, but how many to use; almost all newer instruments incorporate multiple digital processors, using various combinations of PC hardware, DSP chips, and ASICs and other hardwired (or hardcoded) digital circuitry. Current products from all three major manufacturers of high-speed droplet sorters, BD Biosciences, Beckman Coulter, and DakoCytomation, can process between 50,000 and 100,000 sort decisions/s.

6.2 PRESELECTED COUNT CIRCUITS AND SINGLE CELL SORTING

Using a simple **digital counter** circuit, it is possible to arrange things so that a sorter will sort a preselected number of cells and then stop sorting. Among the things a digital counter can do is change an output from logical 1 to logical 0 after it receives a preset number of pulses as input. If the positive sort decision pulses are fed into the counter, and the logical AND of the sort pulses and the counter output drives the sorter activation mechanism, once the preset number of cells are sorted, cells will no longer be sorted, even if they meet the selection criteria.

If the preset number is one, and if a **two-axis positioner** (i.e., a stage with precise and reproducible X-Y motion control) bearing a slide, culture dish, or microtiter plate is placed in the location normally used for a collection vessel, individual cells meeting the selection criteria can be deposited in a predetermined pattern on a slide for subsequent analysis by microscopy or on an agar plate or other solid medium so that colony growth or failure to grow can be monitored, or inoculated into media in wells on the microtiter plate[145] and grown into clones (assuming they are not traumatized by the reagents and/or light to which they were exposed prior to sorting). Single cell sorting accessories are available from all major manufacturers of droplet sorters.

Due to mechanical constraints, single cell sorting systems often operate by deflecting all droplets except those containing a selected cell; the tube that normally conveys the undeflected stream to a waste container is removed. This setup is advantageous because it is easier to assess the precise location of an undeflected than of a deflected stream, and thus easier to determine where to position the microtiter well or the area on a slide or dish that is to receive the selected cell. It is also somewhat easier to collect an uncharged single droplet than a charged one.

Sorting single cells or very small numbers of cells may be impractical in fluidic sorters in which fluid flows continuously through the sort outlet, since it may be difficult to find a few cells in a large volume of fluid. BD Biosciences' FACSCalibur can be fitted with a filter apparatus that separates the sorted cells from the fluid; this is better adapted to sorting cells that do not need to be kept viable than to cells that do.

6.3 DROPLET SORTING, HIGH-SPEED AND LOW

Droplet sorting first requires **droplet generation**, i.e., breaking the stream containing the cell sample into droplets, and stabilizing the pattern of droplet formation to place the point at which droplets break off from the main stream at a fixed distance downstream from the observation point (or from the last observation point in a multistation instrument). Between the time a cell traverses the observation point(s) and the time the cell reaches the **droplet breakoff point**, the measured values of parameters used for selection are fed into the sort control logic, which generates a sort decision signal.

The decision signal is applied to **drop charging circuitry**, which, if a cell is to be sorted, applies a voltage to the stream at the time at which the cell should lie somewhere in the droplet breaking off from the stream. This leaves an electric charge on those droplets selected for sorting. The droplet stream, containing charged and uncharged droplets, passes through the electric field between two **deflecting plates** to which high voltages of opposite polarities have been applied. Charged droplets are deflected out of the main stream toward the deflection plate bearing an opposite charge, and the charged droplet streams are collected, while the uncharged main stream passes into an aspirated tube leading to the waste reservoir. The drop charging and deflection technique was originally developed by Sweet[68] for ink-jet graphic printing, and was first used for cell sorting by Fulwyler[67].

The hardware which must be added to turn a flow cytometer into a droplet flow sorter consists of electronics to facilitate droplet generation, hardware and/or software that can rapidly establish whether cells meet selection criteria, and the circuitry for charging and deflecting the droplets containing the selected cells

Droplet Generation

The high points of a droplet sorter are shown in caricature (definitely not to scale) in Figure 6-1. As Lord Rayleigh observed in the 1800's, a stream of fluid emerging from an orifice is hydrodynamically unstable and breaks up into a series of droplets that, in the aggregate, have a smaller surface area and lower surface tension. Droplet formation occurs in streams considerably larger than those found in flow cytometers, as any urologist can tell you. In the absence of external forces applied to the stream, the pattern of droplet formation varies unpredictably.

However, when the stream is subjected to vibration within a certain range of frequencies, the pattern of droplet formation becomes stabilized in time and in space, i.e., the droplet breakoff point remains at a fixed position, and the distance, s, between the observation point and the breakoff point remains constant. The vibration, which, in droplet sorters, is produced by a piezoelectric or electromagnetic transducer acoustically coupled to the flow chamber, quite literally "makes waves" in the fluid stream; the droplets, once separated from the stream, are spaced one wavelength λ apart.

A droplet formation pattern is obtained in a stream of diameter D only when $\lambda > \pi D$; the droplet breakoff point is closest to the orifice, minimizing the distance s, when the wavelength is 4.5 stream diameters, but the breakoff pattern can remain stable within the range of values of D such that $4D \leq \lambda \leq 8D$. The stream shown in Figure 6-1 is thus

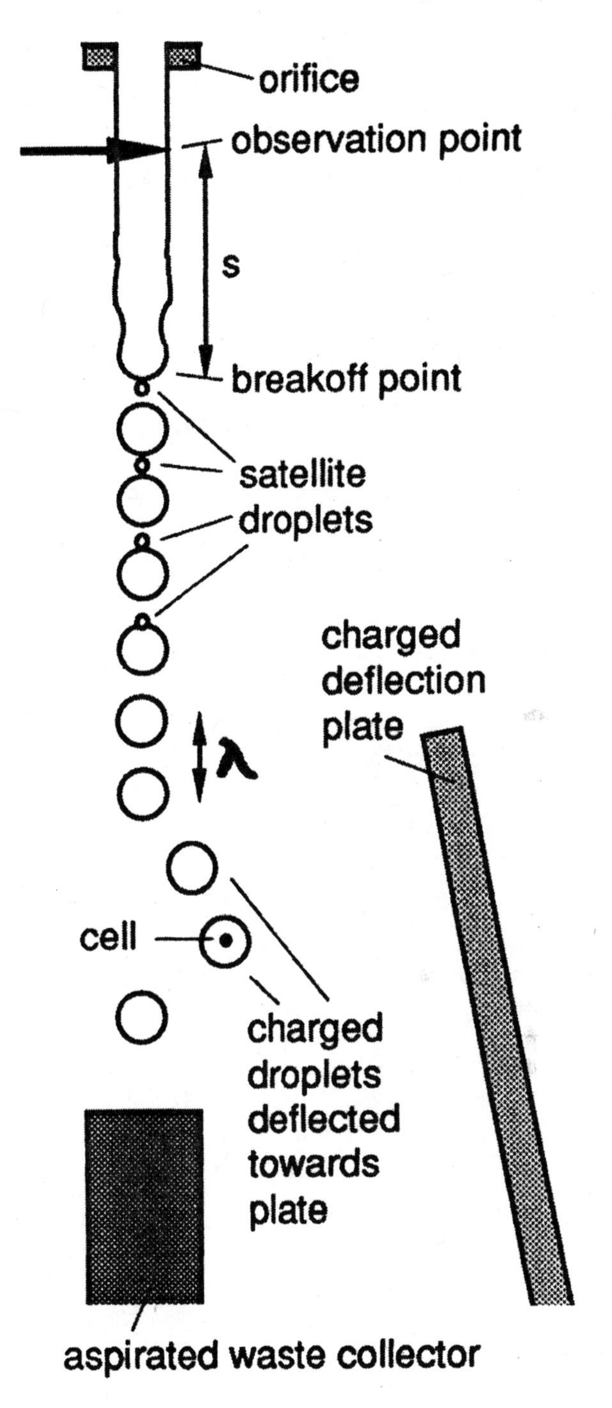

Figure 6.1 Droplet sorting.

fatter than in real life. The wavelength (or droplet spacing), λ, the frequency of the applied vibration, f, and the stream velocity, v, are related by the equation

$$v = f\lambda.$$

Let's look at the numbers for some real instruments. If we have a 50 µm stream diameter, and a velocity of 10 m/s, the 40 kHz drive frequency used with a 50 µm orifice in older sorters produces a droplet spacing of 250 µm, or 5 stream diameters, a near-ideal value. To get the same 5-

diameter spacing with a 70 µm orifice and the same 10 m/s velocity, we'd need to use a 28.6 kHz drive frequency. In some sorting experiments with a Cytomutt flow chamber, which produces a stream about 150 µm in diameter, stable droplet formation and good sorting accuracy were obtained with a 6 m/s stream velocity and an applied frequency of 10 kHz; I found the corresponding droplet spacing was 600 µm, or 4 stream diameters.

Cell sorting was developed using mammalian blood cells as samples; the technique can be more successfully applied to other cells and particles if the physical parameters of the system are modified. In the 1980's, Joe Gray and his colleagues at Lawrence Livermore Laboratory built instruments with very high stream velocities (50 m/s), permitting droplet generation frequencies as high as several hundred kHz[667]. Their initial studies were restricted to chromosomes; mammalian cells did not survive exposure to the high pressures (200 PSI) needed to generate high-velocity streams. The damage resulted from the use of air pressure to drive the sample; the cells, which contained dissolved air, decompressed explosively upon leaving the nozzle. When a syringe pump was substituted as the sample drive mechanism, it became possible to recover at least some cell types intact.

Once "the bends" had been prevented, there were a few other kinks left to work out in terms of optics[1130] and electronics[1180] suitable for dealing with observation times much shorter, and data acquisition rates much higher, than those associated with conventional flow sorting. This was done by Ger van den Engh, then also at Livermore, who developed a simpler and more robust high-speed sorter, the "MoFlo." High-speed sorters were used at Livermore and Los Alamos for preparative sorting of human chromosomes, from which DNA libraries were generated for use in the Human Genome Project. Although Fellner-Feldegg[665] had reported in 1984 that the B-D FACS could be operated with a 25 µm orifice and an 80 kHz droplet generation frequency for sorting small particles, the industry had been relatively slow to move toward higher-speed sorters for eukaryotic cells. The pace of development picked up when, in 1994, Cytomation produced a commercial version of the MoFlo under license from Livermore; Livermore also licensed Systemix to use the MoFlo technology for clinical applications.

We can bring our calculations on droplet generation up to date with some numbers from Ger van den Engh's recent book chapter[2507]. When a jet of fluid driven by a pressure P emerges from an orifice, the jet velocity is proportional to $(P)^{1/2}$, and independent of the orifice diameter. Allowing for some pressure drop in the fluidics upstream of the orifice, when the operating pressure is 100 PSI, saline emerges from a 70 µm orifice at a velocity of 37 m/s, for which the optimal droplet generation frequency is 118 kHz. It takes approximately 30 droplet periods to traverse the distance from the orifice to the droplet breakoff point.

Drop Charging And Deflection

The application of positive or negative voltage pulses to a stream of conductive liquid produces positive or negative surface charges on the stream. A droplet separated from the stream while the charging voltage was applied carries excess surface charge of the same polarity as existed on the stream at the time of separation. When a stream containing charged and uncharged droplets is passed through an electrostatic field established between a pair of parallel or near-parallel plates kept at high positive and negative voltages with respect to ground potential, the charged droplets are deflected, with positively charged droplets moving toward the negatively charged plate and *vice versa,* while the trajectories of uncharged droplets are not altered. It is thus possible to collect the streams of positively charged, negatively charged, and uncharged droplets, and the cells contained therein, in separate vessels. While 300 mOsm "normal" (0.85%) or buffered saline are often used as sheath fluids, salt concentrations as low as 10 mEq/L can provide sufficient conductivity for droplet sorting. Charging pulses are applied in fixed phase with the transducer drive signal.

Drop Deflection Test Patterns

There are actually two streams of "uncharged" droplets, as can be seen from Figure 6-2, which shows the stream pattern produced by a test circuit such as was used in Ortho's System 50 flow sorters; most current instruments have added the feature. When these sorters are operated in the "test" mode, the first droplet of each 4-droplet sequence is negatively charged, the third is positively charged, and the second and fourth are uncharged. The operator can then directly assess the stability of the droplet generation pattern by observing the diverging streams produced. Positively charged droplets are deflected toward the negatively charged plate; negatively charged droplets are deflected toward the positively charged plate. The bifurcation visible in the stream of "uncharged" droplets occurs because these droplets actually carry small charges. The separation of a positively charged droplet leaves a slight negative charge on the residual stream, which is acquired by the "uncharged" droplet that follows, while the separation of a negatively charged droplet similarly results in the next "uncharged" droplet carrying a slight positive charge. Thus, one sees four, rather than three, streams, because drops which should be in the sequence "left, center, right, center" are actually in the sequence "left, slightly right of center, right, slightly left of center."

Two- and Four-Way Sorts: How Much Voltage?

The separation between the streams can be increased, within limits, by increasing the droplet charging voltage, by increasing the potential difference between the deflection plates, and/or by making the deflection plates larger, i.e., longer in the direction of flow, so droplets spend more time in the field. It is also possible to apply more than one level of

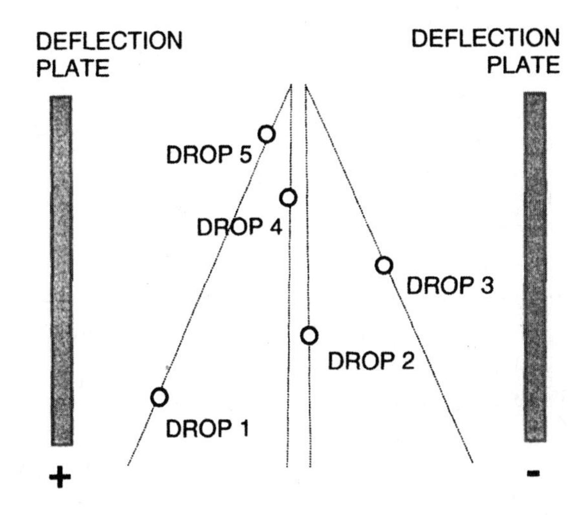

Figure 6-2. A droplet sorter test stream pattern. Drop 1 is negatively charged. Drop 2, which is uncharged, acquires a slight positive charge. Drop 3 is positively charged; drop 4, uncharged, acquires a slight negative charge. Drop 5 is positively charged, etc. The result is four streams, following trajectories shown by the dotted lines.

charging voltage to droplets. For example, a droplet charged by a 100 V pulse will carry more positive charge than a droplet charged by a 50 V pulse, and its trajectory will deviate more rapidly toward the negatively charged deflection plate. It will therefore be possible to collect two streams of positively charged drops in separate containers. This is the basis of four-way sorting, a relatively recent addition to commercial droplet sorters; but a feature of the multiparameter sorter described by Donna and Tom Jovin in 1974[147].

The potential difference between deflection plates in a droplet sorter can be anywhere from 2000 V to 6000 V. The voltage pulses used for droplet charging in flow sorters are usually somewhere between 100 and 200 volts on either side of ground in amplitude. However, excessive charge on droplets may interfere with droplet and jet stability, since surface charges counteract the effects of surface tension, which keeps drops intact and is a determinant of the location of the droplet breakoff point. Ger van den Engh has noted instability when voltages exceeding ±80 V are applied to a 70 μm stream[2507]. Zhou et al have described an inexpensive but effective low voltage charging circuit[2510].

Figure 6-1 shows **satellite droplets** in the stream; these are formed by fluid present in the neck or "ligament" regions of the stream, i.e., the places between the developing droplets, and typically travel at different velocities from the larger droplets in the stream. The pattern of satellite breakoff is determined by operating conditions, but its dependence on stream diameter, fluid velocity, frequency and waveform of applied vibration, timing of drop charging pulses, etc. is complex. You win the game if the satellites go faster than the larger droplets; they then merge with the main drops (which have the same charge polarity) after a few wavelengths, as shown in the picture. You lose if the satellites go slower and fuse with the drops behind them, which may not be similarly charged, or if the satellites form a separate charged stream that hits a deflection plate, screwing up the deflection pattern of the real sort stream.

How Many Drops Should be Charged?

It used to be common practice to allow one, two, or three droplets to be charged, to compensate for uncertainties in estimating the **droplet delay**, i.e., the time cells take to travel from the observation point (or, in a multistation sorter, the last observation point) to the breakoff point. As the distance cells must travel between these points increases, the accuracy with which their arrival time at the breakoff point can be predicted decreases. Typically, three drops were charged; the basic idea is that you have the best chance of getting the cell you want if you charge the droplet you think the cell is in, the one before it, and the one after it.

We have already noted that, when a charging voltage is applied as a droplet breaks off, the fluid that will form the next droplet carries a small charge of the opposite polarity; when this second droplet is charged, the fluid that will form the third carries a slightly higher charge of the opposite polarity. If the charging voltage applied to all three drops were the same, the residual charges on the second and third droplets would result in their having progressively less net charge than the first droplet, and thus in their being deflected less by the constant field between the deflection plates. This would produce multiple sort streams containing first, second, and third droplets, respectively, on the lines of Figure 6-2, only worse, making it less likely that all the selected cells would end up where they were supposed to, i.e., in the collection vessel. This is prevented by designing drop charging circuits so that, when more than one droplet is charged, the amplitudes (positive or negative) of voltages applied to successive droplets are increased slightly. A resistor-capacitor circuit accomplishes this very simply, acting as an integrator; the longer (i.e., the more drop periods) the drop charge signal is left on, the more charge is applied per unit time.

Guilty as Charged?

The charge on sorted droplets can pose problems, especially when one tries to collect sort streams in tubes or flasks that are (and most are) made of nonconductive materials. If there is no way to dissipate the excess charge, you will end up with a bunch of droplets repelling one another, which may decrease yields. It can be a great practical joke to run a test sort stream into a plastic container and watch people react to the electric shocks they get by dipping a finger into the liquid. When it's time for work, however, it's time to stop the collected cells from charging, not by taking away their plastic, but by sticking a grounded, sterilized (if appropriate) platinum or stainless steel wire into the collection vessel. The steel wires used as obturators in some intravenous catheters are ideal for this purpose, being cheaper than platinum, sterile, and disposable. I have suggested to a few of the manufacturers that they should sell collection tubes made of conductive plastic (let's face it, B-D makes a lot more money from plastic goods than from flow cytometers and antibodies), but, thus far, I haven't had any takers.

Determining Droplet Delay Settings

In the original stream-in-air flow sorter built at Stanford, droplet delay times were determined from the distance, s, between the observation point and the droplet breakoff point. This distance was measured using a low-power microscope fitted with an eyepiece micrometer, or estimated, in terms of a number of wavelengths, by inspection of the stream illuminated by a stroboscope operating at the droplet generation transducer drive frequency. The earlier flow sorters manufactured by B-D and Coulter also adopted this practice. Realizing that television cameras are cheap compared to cell sorters, various labs adopted the practice of observing droplet streams on TV monitors instead of by eye; newer commercial systems do the same, and some incorporate software that uses image processing algorithms to monitor the position of the breakoff point and automatically institute corrective measures should flow become unstable.

Since the frequency, f, at which the transducer is driven is known, and the fluid flow velocity v can be calculated from the known orifice diameter and the volume flow rate, the wavelength λ can be calculated relatively precisely. Droplet delay is measured in periods of the transducer drive frequency; in the system shown in Figure 6-1, a charging signal, if necessary, should be applied (s/λ) periods after the cell to be selected arrives at the (last) observation point.

The calculation becomes more complicated for instruments in which observation is done in a cuvette or capillary, because the fluid velocities v_i and v_e are different up- and downstream from the orifice. If s_i is the distance between the (last) observation point and the orifice, and s_e is the distance between the orifice and the droplet breakoff point, the number of droplet periods of delay required is $[s_e + (v_e/v_i)s_i]/\lambda$. The ratio of external and internal velocities, v_e/v_i, is essentially equal to the ratio of internal and external cross-sectional areas, a_i/a_e, which can be substituted in the formula.

The preselected count feature can also be used to determine the droplet delay needed for sorting. Fluorescent spheres are run through the instrument, and the selection windows are set to sort single spheres into one sort stream. A jig that holds a microscope slide in position to receive the sorted spheres is then inserted, and the instrument is set to sort 10 to 100 spheres; this process is repeated with different settings of the droplet delay control. This produces a number of drops of saline on the slide; each is examined to determine the droplet delay setting at which the largest number of spheres appear in a drop. Although it is generally recommended that this examination be done under a fluorescence microscope, the beads are generally recognizable in transmitted light; however, no lab that can afford a cell sorter should be too impoverished to put an inexpensive fluorescence microscope close by the lab.

That notwithstanding, I should mention a high-tech and a lower-tech variation on the above method for droplet delay determination. De Grooth et al[1224] used a beam splitter to divert a few percent of the power in the illuminating laser beam of their sorter into a fiber optic, which illuminates a capillary in a small stainless steel chamber. Another fiber is used to collect fluorescence emission from the capillary. The sorter is set up so that deflected droplets are collected into the chamber, and run at various droplet delay settings; the setting which produces the highest fluorescence reading from the capillary, i.e., the largest number of fluorescent beads diverted into the capillary, is chosen. These days, you could illuminate with an LED.

Lazebnik, Poletaev, and Zenin[1225] described a rapid and simple technique for droplet delay measurement using cells or beads coupled with horseradish peroxidase. A scatter gate is set to sort the particles, and 100 to 2,000 are sorted into a single well of an ELISA immunoassay strip at each of several droplet delay settings. The normal indicator is then added to the strip; the darkest color develops in the well containing the most particles, and the corresponding delay setting is then used.

In the past, possibly because the fluorescein fluorescence always looked greener in the other fellow's flow cytometer, many people modified their commercial instruments to incorporate features of the systems they didn't buy[663]; for example, old B-D and Coulter systems were fitted with the sort test circuits and counters incorporated in Ortho's sorters, and microscopes for stream observation were added to Ortho systems. As time passed, the manufacturers themselves built in the useful features they hadn't thought to include; this is a good thing, because newer systems are, in general, harder to modify than older ones due to the increased level of integration of hardware and software. Of course, if you build your own cytometer, you can put in whatever you like. Or so I tell myself.

Fractional Droplet Delays

In some systems, it is possible to select "fractional" droplet delay periods, e.g., 8 3/4 drops. This feature is included to allow for situations in which the distance, s, between the observation point and the droplet breakoff point is not an integral multiple of the wavelength λ. If an integral delay period is used, the effect is to apply the droplet charge pulse to part of one drop and part of the next. The fractional portion of the droplet delay setting should properly be thought of as an offset correcting for a phase difference.

Cells arrive at the observation point of a sorter in a random sequence, and therefore in a random phase relationship to the transducer drive and drop charging signals, which are generally synchronized. A cell might go through smack in the middle of a droplet period, at the very beginning, or at the very end; cells at the extreme points of the droplet period may perturb droplet formation and, potentially, affect the purity of a sort. If yield can be sacrificed in favor of purity, it is possible to sort only those cells that arrive near the middle of a droplet period.

Transducers and Transducer Drive Signals

The earliest sorter designs used sine wave generators to drive their piezoelectric transducers, and square or approximately square waves of the same frequency or a multiple to clock and drive the drop charging circuitry. The original rationale for the sine wave in the transducer drive (R. Sweet, personal communication) seems to have been a combination of the desire to prevent the transducer from overheating (which, as it turns out, doesn't happen) and the knowledge that Lord Rayleigh's theory dealt with sinusoidal perturbation of streams. The electronics are simplified considerably if a simple square wave can be used for transducer drive. Fellner-Feldegg[665] found that a square wave applied to a B-D sorter transducer actually produces sinusoidal vibrations, because the transducer mechanical assembly effectively acts as a low pass filter, and recommended the use of a square wave for transducer drive; it is not clear whether and which sorter manufacturers have taken his advice.

Optimal sorting performance requires good frequency stability in the transducer drive. In older instruments, crystal-controlled oscillators were used to generate fixed drive frequencies, while a variable drive frequency generator might incorporate less stable circuitry. Crystal-controlled digital frequency synthesizers now provide both a high degree of stability and a broad range over which drive frequencies may be adjusted.

Almost all manufacturers of droplet sorters use piezoelectric transducers for drop drive, although B-D has offered the option of a moving-coil electromagnetic transducer, essentially the guts of a loudspeaker, for use at the relatively low (< 10 kHz) drive frequencies needed with very large (up to 400 μm) sort orifices.

The sort nozzle assembly designed by Ger van den Engh for the MoFlo incorporates a drive transducer with an acoustic waveguide to focus energy toward the nozzle tip, and also includes a second transducer that can be operated as a sensor for a closed-loop drive control system[2507].

Improving Droplet Sorting

A lot of tinkering has been done with both commercial and laboratory-built droplet sorters to facilitate routine sterile sorting of multiple samples in one working day, and it was and is needed. Flow sorting, to date, has been something like heart surgery; the learning curve is steep, meticulous technique is required, and it's done very well only in places that do a lot of it all the time. The folks who work in those places generally have a better sense of how sorters should be designed than did the engineers who designed some sorters. The practical wisdom doesn't always get written down, but when it does, as in Phil Dean's article on "Helpful hints in flow cytometry and sorting"[664], it's well worth having.

In an instrument with multiple measurement stations, the distance between the first observation point and the breakoff point may be quite large, leading to considerable uncertainty in estimates of droplet delay period and thus to the requirement to charge large numbers of droplets to get reasonable yields, lowering throughput. Martin et al[143] described a system in which a scatter signal from cells, measured immediately upstream from the breakoff point, was used to resynchronize drop charging pulse timing.

Sorting Large Objects with Droplet Sorters

Over the years, progress in biology and biotechnology has increased interest in sorting objects substantially larger than mammalian cells, e.g., animal and plant cell hybrids, *C. elegans*, early embryos of *Drosophila* and other species, pancreatic islets, tumor cell spheroids, and beads substrates used for combinatorial chemistry.

According to Sweet[142], sorting by droplet generation and deflection is possible using streams as large as 1 mm in diameter. Although large stream diameters necessitate using lower stream velocities and droplet generation rates than are now prevalent, resulting in lower sample throughput, even rates of a few hundred objects sorted per second could provide a great improvement over the micromanipulation currently used to isolate some of the specimens just mentioned.

Although the rule of thumb for sorting smaller objects was to keep the orifice diameter about 5 times the average particle diameter, most people working with large objects have pushed the envelope, attempting, for example, to pass 100 μm particles through 200 μm orifices. I have no reports of oxytocin being added to the sheath fluid. Jett and Alexander[671] modified a B-D sorter for operation at a reduced (≈ 7 m/s) flow velocity with a 200 μm orifice at a droplet generation frequency of 4.5 kHz. This instrument could sort objects, e.g., tumor cell spheroids, as large as 100 μm in diameter[672]. Harkins and Galbraith sorted plant protoplasts and other materials as large as 95 μm diameter, using a modified Coulter EPICS instrument with a 204 μm nozzle and droplet generation frequencies as high as 8 kHz[673].

The analysis and sort rates that could be achieved using older droplet sorters, which had relatively long processing dead times, were limited to about one-tenth the droplet generation frequency, since higher analysis rates would result in the almost continuous occurrence of coincidences. At the droplet generation rates of 4-8 kHz used for sorting large objects, it was not possible to sort more than 800 cells/s. Since the volume, and, therefore, the mass of droplets increase with the cube of their diameter, substantially more work must be done to deflect large drops than is necessary to deflect small ones; this can be accomplished by increasing the drop charge voltage, increasing the voltage between the deflection plates, and/or increasing the length of the deflection plates.

Mack Fulwyler, the father of droplet sorting, was a diabetic, which undoubtedly provided some of the motivation for his attempts to sort pancreatic islets. Mack is shown in

Figure 6-3 with a B-D FACS 440 that he and Bill Hyun modified for islet sorting; it may not be easy to tell from the figure, but it seemed to me that the deflection plates in the apparatus were at least six inches long.

Figure 6-3. Mack Fulwyler with his islet sorter (courtesy of Bill Hyun, UCSF).

Although Ger van den Engh has done some work on droplet sorting large particles using instruments he has developed at his new company, **Cytopeia**, it is fair to say that most current efforts at sorting very large objects, and very small ones, utilize fluidic switching, to which we now turn, as a sorting mechanism.

6.4 FLUIDIC SWITCHING CELL SORTERS

When we start dealing with large objects and low sort rates, and/or with hazardous samples, we get into the areas of competence of fluidic switching sorters; the mechanisms of several of these are diagrammed in Figure 6-4.

In the course of his work at IBM in the 1960's, Kamentsky experimented with both droplet and mechanical sorting; in 1967, he and Melamed[66] described the addition of sorting capacity to the Rapid Cell Spectrophotometer in the form of a syringe pump which could withdraw cells from the sample stream. Friedman[1226], working with Kamentsky at Bio/Physics Systems in 1973, developed an improved fluid switching sorter design, shown in panel A of Figure 6-4, in which an acoustic transducer was coupled to the fluid stream. The transducer normally had a high-frequency signal applied, creating turbulent flow, which diverted cells into the waste stream; when a cell was to be sorted, the signal was turned off, and laminar flow was established within a few milliseconds, resulting in the selected cell being diverted into the collection tube. Although sort rates of several hundred cells/s could be achieved (L. Kamentsky, personal communication), the acoustic sorter was not added to instruments in the Bio/Physics Systems or Ortho product lines; faster droplet sorters were used instead.

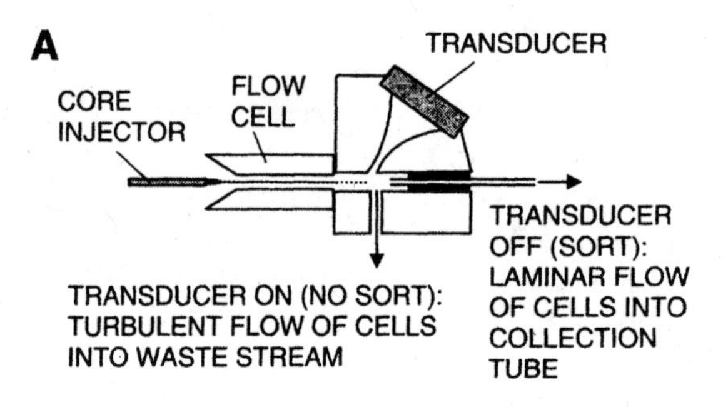

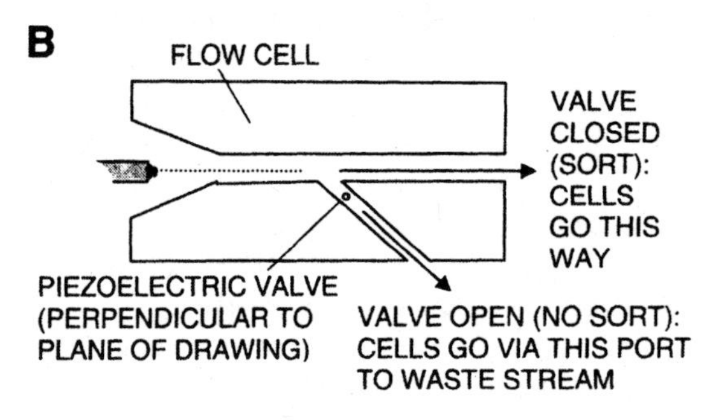

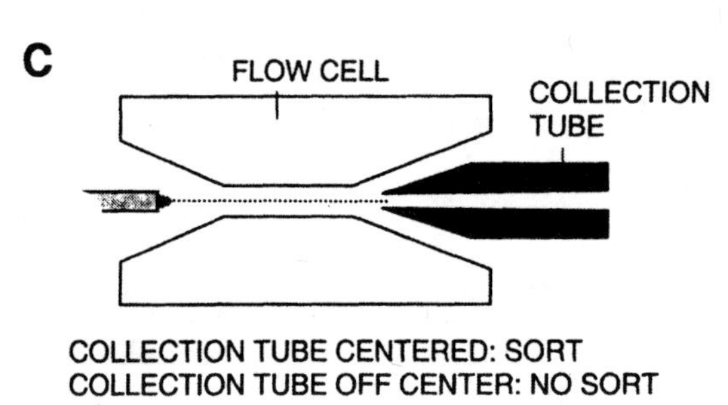

Figure 6-4. Fluidic sorter designs. A: Friedman's acoustic sorter. B: Fluidic switching sorter as used in some Partec instruments. C: Sorting arrangement in B-D FACSort and FACSCalibur.

The fluidic sorter mechanism shown schematically in Panel B of Figure 6-4 is incorporated into instruments now available from Partec; it features a closed fluidic system and uses gas controlled by piezoelectric valves to divert the fluid stream to sort cells. Sort rates of over 500 cells/s have been reported. In 1989, Gray et al[1227] reported successful sorting of pancreatic islets up to 300 μm in diameter using a modified Partec instrument.

In 1991, B-D introduced the FACSort™, a benchtop instrument combining the measurement capabilities of the FACScan flow cytometer with a closed fluidic sorting system, shown in Panel C of Figure 6-4. A collecting tube, normally placed eccentrically, is moved into the center of the stream downstream from the observation point to sort cells. Sort rates of several hundred cells/s are possible. In the other fluidic sorters just described, if no cells are being sorted, little or no fluid comes out of the collecting tube; there is continuous sheath fluid flow into the collecting tube in the FACSort mechanism. It was noted (p. 259) that this might pose a problem in recovering rare cell types, and that a filter system is offered as a solution. FACSort-style fluidic sorting is now optional on the FACSCalibur, which has replaced the FACScan and FACSort. Although the capabilities of the sorting mechanism are limited, people willing to work within its limitations have found the fluidic sorter easy to use.

In principle, a fluidic sorter should be more precisely controllable than a droplet sorter, because the sort decisions can be synchronized to cell arrival times rather than to droplet periods, which are asynchronous with arrival times. At relatively low sort rates, however, this may be of only academic interest. A fluidic switching sorter with a closed fluidic switching system would definitely be preferable to a droplet sorter for use with highly infectious materials; this is of more than academic interest.

Sorting Large Objects Using Fluidic Switching

In the mid-1990's, **Union Biometrica**, a small company then located in Cambridge, Massachusetts, developed an apparatus for sorting *C. elegans* at the request of a local researcher. This system has evolved into the **COPAS™** (Complex Object Parametric Analyzer and Sorter) series of instruments, variously optimized to sort embryonic and adult nematodes, *Drosophila* and zebrafish embryos, large beads, and *Arabidopsis* seedlings. The flow systems in the COPAS™ line feature fluid flow paths as large as 1 mm in diameter. The instruments vary in analytical capability; they can measure fluorescence and extinction, and, in some cases extract morphological information from time-of-flight measurements. All use the same sorting mechanism. The fluid stream containing specimens is normally intersected by a high-speed air jet shortly below the last observation point, diverting the stream to a waste collector or to a reservoir from which it can be recycled. When an object is to be sorted, the jet is turned off, allowing the desired particle to travel straight down into a collecting vessel. Thus, while the COPAS instruments are fluidic switching sorters, they are also stream-in-air systems. The sorting mechanism is said to be extremely reliable, although slow; sort rates range from a few dozen particles per second for the largest objects to a few hundred per second for smaller particles. It was suggested to me by one of the principals of Union Biometrica (P. Hansen, personal communication) that they might be able to sort as many as 1,000 particles/s from a small diameter stream.

An alternative fluidic switching instrument, capable of sorting dozens of *Drosophila* embryos/s on the basis of GFP fluorescence and autofluorescence, was described in 2001 by Furlong et al[2325] at Stanford. The technology has been licensed to Union Biometrica, but information for those hardy individuals who want to try building their own version of the apparatus is available at <http://www.stanford.edu/~profitt>.

Sorting Very Small Objects: Microfluidic Switching

Stephen Quake and his colleagues at CalTech have recently developed a series of microfluidic devices that have been used to sort bacteria[2326,2509] and that could be used to sort smaller objects such as viruses and DNA fragments[2327]. While earlier attempts at building microfluidic flow cytometers and cell sorters, including mine, used silicon, glass, ceramic, or metal substrates, Quake et al employ a technique called soft lithography, pioneered by George Whitesides at Harvard[2511], using silicon molds to fabricate microfluidic circuits from silicone elastomer. The CalTech group has developed devices that incorporate pumps and valves on-chip[2512-3], and manifolds that permit large numbers of fluidic circuits to be controlled by a small number of external pressure lines[2514]. A picture of a microfluidic sorter, built by Anne Fu and used for sorting bacteria, appears in Figure 6-5.

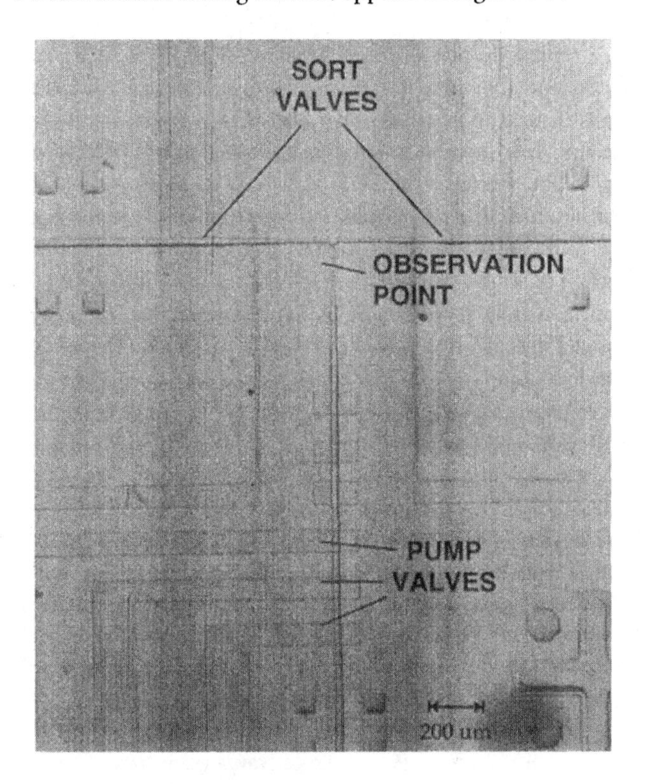

Figure 6-5. A microfluidic flow sorter for bacteria. Courtesy of Anne Fu and Stephen Quake (California Institute of Technology).

The sorter is built of two layers of elastomer. The top layer incorporates connections to a small air pressure manifold that operates the valves. The bottom layer, which rests atop a glass cover slip, contains the sample fluid inputs and outputs and the T-shaped sorting channel. The entire assembly is placed on the stage of an inverted epifluorescence microscope that incorporates optics for laser illumination and 2-color fluorescence detection.

Sample flows from the bottom of the "T" in the figure, propelled by the peristaltic action of the three pump valves, which are opened and closed in sequence. The sort valve on the right is normally kept closed and the one on the left is normally kept open, forcing the fluid stream down the left arm of the "T." When a particle of interest is detected, both valves are toggled, and a small amount of fluid containing the selected particle enters the right side of the "T."

The microfluidic sorters built by the Quake group to date have not used sheath flow, and, because different particles take different times to traverse the observation region, the CVs of fluorescence measurements are relatively large; the device shown also lacks a scatter channel for triggering on nonfluorescent or weakly fluorescent objects. It was used to separate bacteria bearing native and mutant strains of GFP, based on different ratios of green and yellow fluorescence; the ratio has a smaller CV than the individual fluorescence measurements (see p. 47).

The flow velocity through the microfluidic sorter is extremely low, typically 10-15 mm/s, or around 1/1,000 that in a typical droplet sorter. This allows particles to be illuminated and measured for a much longer time (i.e., milliseconds instead of microseconds), increasing measurement sensitivity. It is possible to analyze approximately 1,000 bacteria/s, but, because the fluid flow rate is so slow (tens of nanoliters/s), the concentration of bacteria in the input sample must be on the order of 10^8/mL to achieve that analysis rate. The observation region of the microfluidic sorter is approximately 6 × 10 μm in cross-section; devices with a smaller cross-section have been used to size DNA fragments. Devices with larger flow channels can be used to analyze and sort eukaryotic cells; flow rates, though somewhat faster, are still relatively slow.

Microfluidic sorters can sort no more than a few hundred cells/s. However, one can reverse and stop flow through the devices; neither droplet sorters nor other fluid switching cell sorters have this capability. Among other things, it allows static cells to be observed for a fixed period, eliminating measurement variances due to flow velocity variations, and permits sort decisions to be made on the basis of repeated observations of a single cell over time.

Commercial development of the technology developed by the Quake group is now underway at **Fluidigm Corporation** (formerly Mycometrix Corporation).

Shuichi Takayama, an alumnus of the Whitesides lab at Harvard now working at the University of Michigan, has built microfluidic flow cytometers using an air sheath[2515], which could potentially be used for sorting; fluid flow velocities are in the range of 1 m/s.

6.5 CELL MANIPULATION BY OPTICAL TRAPPING

Ashkin[1228] noted in 1970 that small particles could be captured and accelerated by the radiation pressure in intense light beams. He and his coworkers at Bell Laboratories subsequently used laser beams to manipulate single cells[1229], bacteria, and viruses[1230]. Work in this area was also done by Tudor Buican and others at Los Alamos[1231]; the instrument developed there is now available from **Cell Robotics** (Albuquerque, NM). A special section on optical trapping appeared in *Cytometry* (Vol. 12, No. 6, 1991). Cell Robotics now appears to be staking its future on a device that uses a laser instead of a lancet to draw small samples of peripheral blood.

6.6 CELL DAMAGE SELECTION ("CELL ZAPPING")

Photodamage Cell Selection

An alternative to sorting for removal of a relatively small population of cells from a sample was proposed by me[150], and, independently, by Martin and Jett[151] and also, probably, by a lot of other people. Photodamage cell selection, or "cell zapping," makes use of a high energy pulsed laser focused on a point downstream from the (last) observation point in a flow cytometer to kill cells selected on the basis of measurements made upstream; you kill the cells you don't want and keep the rest.

In the late 1970's, it seemed a zapper would be useful for removal of small numbers of cancer cells from patients' bone marrow prior to reinfusion following intensive chemotherapy. The lasers and electronics then available made it possible to process 50,000 or more cells/s with a "zapper," about ten times as many as could then be sorted. It might be practical to spend a day zapping cells to treat a patient; it would be virtually impossible to spend a week sorting for the same purpose.

Herweijer, Stokdijk, and Visser were the first to actually build a cell zapper[668]; they achieved sort rates of 30,000 cells/s, using an acousto-optic crystal to switch the killing beam, and photosensitizing cells with bromodeoxyuridine and Hoechst 33342. A 5-decade reduction in the number of viable cells was obtained using 400 mW of UV light at 351 and 363 nm from an argon ion laser killing; the system could process 30,000 cells/sec. Keij et al[2516] reported similar processing and kill rates using 20-100 mW of 275 nm UV from a doubled 514 nm argon ion laser, but noted problems with the doubling crystal and laser modulator.

High speed sorting can now achieve throughput rates equal to or better than those reported to date for cell zapping, but neither technology will play a major role in cancer treatment in the near future unless both the methods of identifying cancer cells and the available drugs are good enough to make the therapeutic strategy worthwhile. Even then, it will probably make more sense to identify and separate the stem cells needed to repopulate the marrow, a task made considerably easier over the years by advances in analytical flow cytometry, than it will to attempt to zap the cancer cells. Results of recent clinical trials of intensive chemotherapy and stem cell replacement seem to have dimmed enthusiasm for the overall concept, pushing clinicians in the direction of other therapeutic strategies.

Sorting (Zapping) Without Flow (Gasp!)

Although the cell zapping technique does not seem to offer much of an advantage for cell selection in flow systems, it makes excellent sense to use cell zapping to select cells which prefer to grow attached to solid substrates and/or to one another. Laser microbeams were used for precise cell killing within a few years of the demonstration of practical lasers in the early 1960's; Higgins et al[669] explored mechanisms and methodology in 1980.

It remained for Schindler, Olinger, and Holland[670] to develop a relatively practical, integrated system for selection of attached cells using laser photodamage. The apparatus, the ACAS 570, also incorporated a computer-based general-purpose microphotometer and microfluorometer, allowing several parameters to be measured to provide selection criteria. It was sold by **Meridian Instruments**, but not in large enough volume to keep the company from going out of business.

While Schindler originally described the operation of the instrument in saving the wanted cells from destruction as "Passover selection," it might now appropriately be called "Schindler's List Mode."

Oncosis is now attempting to develop a system for cell zapping on solid substrates for clinical use in removing Non-Hodgkin's lymphoma cells from marrow prior to reinfusion into patients. The claimed advantage of the system is that it is, to paraphrase company literature, more deterministic than sorting or zapping. We'll see.

Electrodamage Cell Selection in Flow

If someone in the United States had wanted to build a fancy cell zapper during the 1980's, it could probably have been paid for out of Strategic Defense Initiative research funds. Bakker Schut, de Grooth, and Greve[1232], in the less violent Netherlands, took an alternative "Star Wars" approach to cell selection; they used the Force- electrostatic force, that is.

Electroporation forms small pores in cell membranes by application of a strong electric field pulse for a few microseconds. When low field strengths and short pulse durations are used, the cell membranes reseal; such transient electroporation is used to transfect cells with genetic material[1233], and was also used by Berglund and Starkey[1234] for fluorescent antibody staining of oncogene products in live cells. At higher field strengths and/or longer pulse durations, permanent membrane damage leads to cell kill.

Electric pulses of 10 µs with field strengths near 3×10^6 V/m, applied across a Coulter orifice downstream from an optical observation point, were found to kill 99.9% of lymphocytes and cells from the K562 erythroleukemic cell line. The apparatus used operates at rates of 1,000 cells/s. Though not as fast as the photodamage flow sorter described above, it is a great deal less expensive.

While the ACAS 570 for a time found a niche for studies and selection of attached cells, flow-based cell damage selection systems, optical or electrical, have not; what they seem to need is a "killer application."

6.7 MEASURES OF CELL SORTER PERFORMANCE: PURITY, RECOVERY (YIELD), AND EFFICIENCY

Different aspects of sorter performance are important in different applications of sorting. Since it is generally not possible to maximize the **purity** of the sorted population and the **recovery (yield)**, i.e., the fraction of cells of interest, relative to the original number in the sample, collected by sorting, at the same time, it is necessary for the experimenter to choose the appropriate *modus operandi*.

If you are trying to get the highest possible purity in your sorted cell population, you have to work hard to keep cells you don't want out of the sorted fraction. Since cells come through in random sequence, the probability of getting an unwanted cell in your deflected fraction increases as you deflect more droplets per selected cell. If, on the other hand, you're trying to isolate cells that occur as 0.1 percent of the population, you almost have to be willing to settle for more contamination as long as you can get enough of the cells you need.

Coincidence Effects on Performance

It is customary to include **coincidence detection** circuitry in cell sorters to make it possible to determine whether unwanted cells are present in close enough proximity to selected cells to be deflected with them; the sorter can be set either to sort or to inhibit sorting of a cell in the event such a coincidence is detected. This can reduce contamination of sorted fractions; it will not eliminate it because there are always periods of dead time during which the processing electronics are "busy" and thus "blind" to approaching cells.

Pinkel and Stovel[650] calculate, based on Poisson statistics of cell arrival times, that, if coincidences are rejected. i.e., if no sort occurs when another cell passes through the system in proximity to a cell meeting selection criteria, the maximum percentage of the desired cell population which can theoretically be recovered is 37% if the fraction of cells of interest is near zero. This recovery rate is achieved when the product of the number of drops deflected, n, the cell processing rate, u, and the droplet period, T, is equal to 1. If the cells of interest make up 20% of the total population, maximum recovery rate, which is achieved when $n \times u \times T \approx 1.5$, is still only 43%. If the cells of interest are 50% of the total, the recovery rate is better than 75%, at $n \times u \times T = 2$.

The practical significance of all this is that two-pass sorting, with coincidences being neglected in the first pass, is virtually mandatory to improve recovery of cells that comprise a very small fraction of the total population being sorted. It is also advisable when working with rare cell types to use any bulk cell separation methods, e.g., centrifugal elutriation or immunomagnetic separation, that can be applied to increase the fraction of desired cells before undertaking flow sorting for final purification. McCoy et al[1235] point out that triggering on fluorescence instead of scatter may also facilitate sorting rare cells.

Papers by Keij et al[1236] and van Rotterdam, Keij, and Visser[1237] examined several models of coincidence and their effects on instrument performance in the context of high speed sorting and cell zapping. These authors differentiate between **beam coincidences** and **pulse processing** coincidences. In a beam coincidence, particles arrive together (or in rapid succession) at the observation point. The **pulse width**, or time taken for a single particle to traverse the beam, is equal to

(cell diameter + beam height)/flow velocity.

For a 10 μm cell, a 22 μm beam, and a 33 m/s velocity, pulse width is ≈1 μs. Reduction of beam coincidences can be achieved by making the beam smaller and increasing the flow velocity. However, in most systems, the dead time of the pulse processing electronics (tens of microseconds in conventional systems, and over 2 μs in high-speed systems) accounts for most of the coincidences; in most conventional instruments, the obvious, if not the only, way to improve sort performance is by using faster processing electronics. The CICERO™ system for computer-based analysis and sort control, the first product introduced by Cytomation, was designed to be retrofitted to older sorters to provide increased processing speed; while the product is no longer available, there are a number of devoted users (and systems) still going strong.

To increase total throughput in droplet sorters, it is generally necessary to increase droplet generation frequency as well as cell processing rate, because coincidences otherwise become intolerable. If flow velocity is kept constant, raising the frequency requires use of a smaller orifice. An arbitrarily small orifice, however, cannot be used for sorting because, as cell diameter increases relative to orifice and stream diameter, the passage of cells through the orifice and the presence of cells in neck regions of the stream causes progressively larger perturbations in the droplet pattern[666], resulting in erratic deflection trajectories which may result in contamination of selected fractions with unwanted cells. Perturbation is more apt to occur when a cell is at the front or back of a droplet than when it is in the middle. Where a cell will appear in a droplet can be established by sensing the phase of the transducer signal when the cell traverses the beam; Merrill et al[144] used this information to further constrain sort decisions; current instruments incorporate similar circuitry.

Conventional wisdom tells us to keep the droplet generation frequency as high as possible to maximize sort rate. However, if one must charge three drops to get reliable sorting with a 30 kHz droplet generation rate, and it is possible to get equivalent performance by charging only one drop at 10 kHz, the effective sort rates are the same. By using the lower frequency, one can use a larger orifice, minimizing stream perturbation by large cells and decreasing the likelihood of clogging. I been satisfied to do this in sorting with my Cytomutts. On the other hand, people who are doing routine sorting and charging three drops at 30 kHz could get better yields and about the same purity, most of the time, charging one drop or two.

It has already been noted (p. 260) that the optimal droplet generation frequency for a 70 μm orifice in a high-speed sorter with a 100 PSI operating pressure is 118 kHz[2507]. Poisson statistics predict that the fraction of droplets containing a single particle will be maximized when the average number of cells/droplet is 1; under these conditions, 37% of droplets will contain no particles, 37% will contain one, and 26% will contain more than one. Acceptable cell recovery levels can be attained only by sorting both those droplets containing a single wanted particle and those containing wanted particles in company with unwanted ones; purity can be increased by a second round of sorting. At occupancy rates approaching one particle/droplet, it becomes infeasible to charge and sort more than one droplet at a time. In practice, it is usually necessary to operate at occupancy rates of no more than one cell/3 droplets to minimize beam coincidences, meaning that the maximal sample throughput attainable in a modern high-speed sorter is unlikely to be more than 40,000 events/s.[2507] Note that, since a typical maximum sample flow rate for a sorter is 1.5 μL/s, this requires that there be 2.66×10^7 cells/mL in the input sample.

6.8 OTHER CONSIDERATIONS

Many technical difficulties associated with cell sorting arise not in sorting hardware and/or software, but in ancillary operations that must be performed to insure that the sorted cells (or other particles) can be put to the use for which they were intended. If you want to do biochemical analyses of cells with S phase DNA content, the cells don't need to be viable, and system won't need sterilization prior to operation. If sorted cells are to be grown in culture, however, they must be kept sterile.

Doing the Math

In the preliminary discussion of cell sorting on pp. 40-42, I mentioned that some people seem to avoid calculating how long it might actually take to acquire the number of cells they will need to successfully complete the projects in their high-ticket grant applications. It's important to do the math, not least because it is likely that the reviewers will do the math whether you do it or not.

James Leary and his colleagues, now at the University of Texas Medical Branch at Galveston, have addressed problems of high-speed sorting of very rare events[2501,2508,2517]. They point out that, among other things, you need to consider how good your measurement parameters are at distinguishing the cells you want from the cells you don't, and that Poisson statistics may be inaccurate when dealing with extremely low-frequency events. They also note that if, for example, you are attempting to sort cells present at a frequency of 1 cell in 10^6 in your sample population, you will have to process almost 3,000,000 events to be 95% sure that you will collect one wanted cell. You could get lucky and hit it on the first try, but you could also win the lottery and, if you are dedicated to science, fund your lab with the proceeds and not have to write more grant applications.

Win or lose, it is a bad idea to assume that you will be able to process 100,000 cells/s, and find and sort one of those one-in-a-million rare cells every ten seconds or so; see p. 42 for some real-life numerical details[2331].

Speed Limits: The Reynolds Rap

Back on p. 174, I discussed the Reynolds number, Re, a useful indicator of the stability of fluid flow. Laminar flow can be maintained if Re is below 2300; turbulent flow is likely to occur if Re is above 2300. For a saline stream at 20° C, Re is very nearly equal to the product of stream diameter in µm and flow velocity in m/s.

If your sorter is a stream-in-air system in which the stream diameter is reduced fairly rapidly between the sample injection point and the nozzle orifice, you may well be able to work at values of Re above 2300 without running into turbulence; if you're using a sorter in which cells are observed in a cuvette, or just don't like living on the edge, you probably want to keep Re below 2300. Table 6-1, on the next page, based on calculations done by Ruud Hulspas, provides some safe values of stream velocity for a range of orifice diameters, and includes drive pressures and optimal droplet generation frequencies corresponding to those velocities and orifice diameters.

There is some slop in the calculations. In reality, the actual final diameter of the stream formed by a jetting orifice is somewhat less than the diameter of the orifice, but the law of conservation of matter dictates that the velocity of the narrower stream is higher than the exit velocity from the orifice, so things tend to balance out.

Nozzle Diameter (µm)	Velocity (m/s)	Pressure (PSI)	Droplet Rate (kHz)
50	46.0	162.6	204.4
70	32.8	84.9	104.3
100	23.0	43.6	51.1
125	18.4	29.3	32.7
150	15.3	21.6	22.7
200	11.5	13.9	12.8
250	9.2	10.3	8.2
300	7.6	8.4	5.7
400	5.7	6.4	3.1

Table 6-1. Safe speed limits for sorting based on Reynolds number calculations (thanks to Ruud Hulspas).

For small orifice diameters, the Reynolds number is usually not what limits sorting speed; it is much easier to get fluidics hardware to work at 100 PSI than at 162.6 PSI, and high-speed sorters instruments are typically designed to work at the lower pressure. However, as you start contemplating using larger and larger orifices to sort larger and larger particles, the numbers, Reynolds and otherwise, catch up with you. If you use a 400 µm orifice, your maximum droplet generation rate is only 3.1 kHz, and, if you want to keep the coincidence rate manageable, you won't be able to sort more than 1,000 particles/s. This is only a few times the rate you could get from a fluidic sorter.

Calculations can be reassuring, but, in the real world, it's a good idea to check your stream patterns, which will tell you whether you have turbulence. If you have sample to spare, do that check with the cells or particles of interest running, because they may perturb flow, particularly when they are a little too big for your orifice. We don't want *West Side Story* in the sorter lab, so keep the sharks away from the jets.

Instrument Utilization

Multistation flow cytometry is harder to do than single station flow cytometry, because it is necessary to keep track of events at different points in time and space; cell sorting is harder than flow cytometric analysis for the same reason. It usually takes longer to set up for sorting than to set up for analysis, even if it is not necessary to sterilize the flow system. Also, since it is generally possible to get usable results from analyzing tens of thousands of cells, while sorting may require isolation of millions of cells, a sorting experiment can effectively stop dozens of analyses from being done on an instrument. That shouldn't stop you from designing or doing sorting experiments, but you should always think about whether there is a way to answer your question without sorting.

Monitoring versus Sorting for Cell Preparation

Some bulk separation procedures can yield larger cell populations than could be obtained by sorting, with comparable purity. For example, separation of lymphocytes based on cell surface phenotype can be done using columns, plates, or magnetic beads coated with monoclonal antibodies; this can produce 90-95% pure $CD4^+$ or $CD8^+$ T-cell populations. Monitoring the quality of bulk cell separation by flow cytometric analysis allows more cells to be harvested, and also frees up sorter time for other users.

De Mulder et al[149] used continuous flow cytometry to monitor monocyte purification by counterflow centrifugation. Zola et al[978] noted that adding known numbers of cells covalently labeled with FITC into mixtures being purified allowed the effects of successive steps to be established by determining the fraction of stained cells remaining. Immunomagnetic separation[1238] has become a widely used alternative to column and centrifugal methods for bulk cell separation, and is also useful for first-stage enrichment of desired populations; monitoring the process by flow is commonplace.

Collection Techniques: Life and Death Decisions

In most applications of sorting, why the cells are being sorted determines how they are sorted and collected. The ground rule of droplet sorting is that the sheath fluid must be a conductive ionic solution. Historically, normal saline (0.85% NaCl) has been the most commonly used sheath fluid; it preserves cell viability adequately for short sorting runs. However, for longer runs, buffered solutions (Michael Stöhr proposed HEPES; Dave Parks likes RPMI 1640 medium without pH indicator dye[674]) may keep cells happier. Collection tubes should contain added protein.

For viable cell sorting, it is generally advisable to keep cells cooled to 4° C. This can be done using ice, a recirculator, or thermoelectric cooling[675]. Whether or not cells are cooled, however, it is important that the temperature of sheath and sample fluids be kept constant during sorting runs because viscosity, which has profound effects on the hydrodynamics of sheath flow and droplet generation, changes dramatically with temperature. Sterile, viable cell sorting is usually done into tubes of cooled medium; a sterile ground wire may be inserted to discharge the deflected droplets.

A number of devices for cell collection have been described that are optimized for some particular purpose, e.g., demonstration of specific DNA sequences, extraction of various chemicals, isotope counting, morphologic analysis. Patrick and Keller described a simple gadget[676], made from intravenous tubing components and culture tubes, for collecting cells in a layer of protein over a base of Ficoll/Hypaque, allowing good preservation of morphology and viability. Métézeau et al[677] used nitrocellulose filters with an applied vacuum to obtain high local concentrations of sorted material for biochemical analysis or examination of morphology. Ger van den Engh has collected cells on "tapes" containing plastic microwells, into each of which a single cell can be sorted; the tape is dispensed from a reel, and the wells can be sealed to allow tape with collected cells to be stored on a take-up reel. This arrangement uses less elaborate positioning hardware than may be needed to sort into microtiter plates.

Cells may be collected on slides for such diverse purposes as morphologic analysis, immunofluorescent staining, and *in situ* hybridization of DNA. Cells generally adhere better when the slides are precoated with protein or with a synthetic peptide such as poly-L-lysine[678]. Alberti et al[679] found that preservation of cell structures was improved when cells were sorted onto slides coated with newborn calf serum from a sheath containing only 1/100 the salt concentration of normal saline. When normal saline was used, salt crystallized on and around cells as the slide dried, destroying much of their morphology.

Collecting cells that will subsequently be examined by electron microscopy requires attention to morphologic preservation. Sebring, Johnson, and Spall[1239] described a method for harvesting small numbers of cells; Penney et al[1240] reported improved preservation of cell morphology when fixative was added to the sheath fluid.

Dilutions of Grandeur

You can pretty well count on your sample fluid getting highly diluted by admixture with sheath. Let's return to our benchmark example of a sorter with a 70 µm orifice and a 100 PSI drive pressure, using a 118 kHz droplet generation frequency and a flow velocity of 37 m/s. If we calculate the total fluid flow rate as was done on p. 175, we find it is about 8.5 mL/min. The maximum sample (core) flow rate in a typical high-speed sorter is 1.5 µL/sec, or 90 µL/min, meaning that the ratio of the volumes of sheath fluid to sample fluid emerging from the nozzle, and the ratios of sheath and sample volumes in a droplet, or in a pool of sorted droplets, is almost 100:1.

Two things follow from this. First, the cell concentration in your collection vessel is going to be no more than about 1/100 the cell concentration in your input sample, even if you don't add any fluid to the collection vessel to start with. Second, if there's something your cells need badly enough for you to keep it in the sample fluid, you might want to add some to the sheath. You could put some of the good stuff in the collection vessel as well, although that will result in the concentration of collected cells being reduced still further.

What's the volume of a droplet? The most reliable easy way to calculate it is to divide the fluid flow rate by the droplet generation rate. If the flow rate is 8.5 mL/min, and the droplet generation rate is 118 kHz, the volume of a droplet comes out to about 1.2 nanoliters. The volume of a lymphocyte, incidentally, is about 200 femtoliters, so the cell doesn't occupy very much of the droplet volume. That's no surprise, given the dilution factors involved.

Can Getting Sorted Be Hazardous To Cells' Health?

Although cells of many types have been sorted and thereafter grown in culture, it is advisable to remember that there are no guarantees that every cell type will hold up under every manipulation. The Hoechst dyes, used to stain DNA prior to sorting live cells and chromosomes, also serve as photosensitizers for zapping[668]. Kissane et al[256] reported that plating efficiencies of Hoechst dye stained cells recovered from sorting decreased at laser powers above 100 mW; Libbus et al[1241] noted an increased incidence of chromosome aberrations in sorted vs. unsorted sperm. The high UV laser powers used in chromosome sorting, which produce extensive dye bleaching[1130], may damage DNA as well; this may favor chromosome sorting with lower-power lasers[1133-5,1242].

While the findings on DNA stains and high-power UV may not surprise us, we tend view cell surface staining as a benign procedure, and assume that a surface-labeled cell getting sorted isn't any worse than a human going bungee jumping. However, Chen, St. John, and Barker[1243] studied the electrical excitability of rat pituitary cells sorted after antibody labeling, and found that cells sorted using more than 10 mW laser power showed acutely altered electrical characteristics. Laser acupuncture?

Seidl et al[2518] investigated effects of magnetic separation (using the MACS system from **Miltenyi Biotec**) or droplet sorting on a B-D FACStarPLUS on membrane integrity (propidium exclusion), microviscosity, membrane potential, and Annexin-V staining in breast cancer cells and normal skin fibroblasts. Both separation techniques, as well as ancillary preparative steps such as enzymatic cell dissociation, affected membrane physiology, but the authors felt that neither magnetic separation nor droplet sorting was clearly preferable for cell preparation.

I have suggested that pressurization with a gas such as xenon might improve survival of cells otherwise liable to decompression injury after high-speed sorting, but I don't think anybody has actually tried this trick.

6.9 BIOHAZARD CONTROL AND BIOSAFETY IN FLOW CYTOMETERS AND SORTERS

Before the AIDS epidemic, it seemed that clinical labs were stampeding to buy cell sorters whether or not they needed to sort; many later stampeded to retrofit their sorters with closed fluidic systems, and gave up sorting. In the late 1970's and early 1980's, flow labs, which invariably featured low light levels and were frequently equipped with good stereo equipment, provided an attractive ambiance for impromptu social gatherings. The wine and cheese disappeared as soon as we started running samples we thought could kill us if we weren't careful.

In the early days of sorting, we were concerned more with preventing microorganisms from the laboratory air from getting into the sample than with preventing microorganisms in the sample from getting into the laboratory air, but, since the same measures can be used to accomplish both objectives, we had a leg up when we became seriously concerned with biohazards and biosafety.

In 1981, Merrill[148] examined methods used to minimize contamination in sorters, using culture plates containing confluent *Escherichia coli* placed around a modified B-D FACS II sorter to detect aerosolization of T4 bacteriophages introduced into the sample stream. He found that a major reduction in aerosol generation was achieved by collecting the undeflected stream in a vacuum-exhausted tube. Maintaining the area around the collection vessels at a slight negative pressure further reduced aerosol contamination.

A 1995 report by Ferbas et al[2519], who used bacteriophages to examine aerosol generation and contamination in a Coulter Elite sorter concluded that, if the sorting chamber door was kept closed and vacuum was maintained on the waste collection tube. Noting that their findings were consistent with Merrill's, they wrote "Our results argue strongly that the cytometers tested do not pose significant risk to the operator during sorting of infectious specimens."

Janis Giorgi reviewed sorting biohazardous (HIV-infected) specimens in 1994[2071]. She and others led ISAC to form a Biohazards Working Group to develop guidelines for sorting unfixed cells; these were published in June, 1997[2520-1]. The guidelines include recommendations for sample handling, operator training and protection, lab design, and instrument setup and maintenance, as well as details on methods for assessing aerosol generation. Activity in this area continues.

Recently, Sørensen et al[2522] described safety modifications to a FACSVantage sorter for sorting cells transduced with retroviral vectors, and Oberyszyn and Robertson[2323] described the use of "Glo Germ" fluorescent particles as an alternative to bacteriophages for determination of aerosol contamination.

Many institutions and organizations in which infectious disease research and treatment are carried on now have cell sorters in biosafety facilities; the sorter manufacturers, as can be seen from Chapter 8, all provide a variety of biohazard control options for their instruments.

6.10 CONCLUSIONS

It is still easier to analyze cells than to sort them, and somewhat harder to build a sorter, and keep it working, than it is to build and maintain a flow cytometer for analysis, but things are improving. Papers I cited as sophisticated in the Second Edition[974-7] seem old-fashioned now, and, while sorting is still best used when nothing else will do the job, in such cases, the user's hard work is likely to be amply rewarded. And, thanks to the manufacturers, the work isn't as hard as it used to be (Chapter 8 again). I used to say that whenever I got the urge to sort cells, I would lie down until it passed off, but I'll be dusting off my sorting hardware as soon as I finish writing.

If I do have one pet peeve about cell sorters, it is this: there are a lot of people who spend a lot of time doing sorting tasks that require minimal analysis, e.g., pulling out cells or bacteria that express a single surface antigen or fluorescent protein. As things now stand, you're likely to have to tie up a three-laser, twelve-parameter instrument to do a job that could be done in an instrument using a single laser (and probably a relatively inexpensive one), and measuring only fluorescence and scatter at one or two angles, using simple and inexpensive electronics. There are now some "low-end" cytometers, but there aren't any low-end sorters. I'd love to see one, but I'm not about to go into the business. On the other hand, if somebody wants a consultant, I'm here.

The free advice is this: **Don't sort when you don't have to.** If you are looking for rare events, and/or want to isolate large numbers of cells, take advantage of every available bulk separation technique as an alternative to sorting and/or to pre-enrich samples for the population(s) you want. Years ago, we were pretty much limited to centrifugal separation as best supporting actor; these days, **immunomagnetic separation**, for which **Dynal**, **Immunicon**, and **Miltenyi**, among others, provide a variety of reagents and apparatus, offers a wide range of options. Sorters are just like people; they'll do more for you when you make their work easier.

We now move from the hardware to the physical and cellular parameters we measure, and the probes with which we work when making the latter measurements.

7. PARAMETERS AND PROBES

We will now begin a detailed discussion of the applications of flow cytometry by considering the **parameters**, i.e., what can be measured, and the **probes**, i.e., the reagents that may be needed necessary to make the measurements.

In Chapter 1 (p. 2), I mentioned that the term **parameter** is used in several different senses in cytometric jargon. It is helpful to differentiate between the **physical parameters**, such as electrical impedance, light scattering and fluorescence, that are detected and quantified by the instrument, and the **cellular parameters**, such as cell size, cytoplasmic granularity, and DNA content, values of which we derive, or hope to derive, from measurements of the physical parameters. I will consider the physical parameters first.

7.1 PHYSICAL PARAMETERS

Electrical Parameters

DC Impedance (Coulter Volume)

Electronic volume measurement by the **Coulter principle**[49,50,51] is perhaps the most common flow cytometric technique for cell size determination; there are many more electronic cell counters in use worldwide than there are optical flow cytometers. The measurement is described on p. 10 and pp. 182-3; in brief, passage of a nonconductive particle, such as a cell, through the saline-filled orifice of a Coulter volume measurement circuit with a constant (DC) current applied across the orifice produces a voltage pulse with an amplitude that should, theoretically, be proportional to the volume of the particle.

Deviations from ideal behavior in Coulter orifice measurements are typically caused by particles passing through the orifice in close proximity to the walls, or through other regions of the orifice in which the electric field is nonuniform; problems may also occur when particles are asymmetric. Observation of a bimodal red blood cell volume distribution motivated Mack Fulwyler to build his first flow sorter, and thus to identify measurement artifacts as the source of the bimodality. The orifice must be longer than it is wide in order to produce a uniform electric field in the central region. The diameter of the orifice must be at least several times the particle diameter to keep the field in the region of a particle relatively uniform, but, if the orifice diameter is much larger than the particle diameter, the change in impedance produced by particles' traverse of the orifice will be small, decreasing signal-to-noise ratio and sensitivity.

Electronic volume measurements are sensitive to changes in the conductivity of cells; thus, cells with intact membranes, which have very low conductivity, will have a larger apparent volume than cells of the same physical size with damaged membranes, which have higher conductivity.

Volume sensing orifices have been and are incorporated into a number of laboratory-built and commercial flow cytometers, including current models from NPE and Partec. Wietzorrek et al[2524] have described a flow cytometer, capable of measuring DC impedance, fluorescence, and light scattering, that also includes provision for video imaging in flow.

Electronic volume measurement using appropriately sized orifices was employed some years ago to detect and size particles as small as bacteria[680] and viruses[681,2525-9]; a recent report by Saleh and Sohn[2530] describes the use of a microfabricated "Coulter counter on a chip" to detect and size colloidal particles as small as 87 nm in diameter using orifices with lateral dimensions of only a few micrometers. Detailed discussions of the theory and methodology of electronic cell volume measurement appear elsewhere[9,682-3,1151].

AC Impedance (Electrical Opacity); Capacitance

Cells' impedance has resistive and capacitive or dielectric components. When direct current (DC) or low frequency alternating current (AC) is applied across the orifice, cell impedance is determined primarily by the dielectric properties of the plasma membrane. As higher frequencies are applied, the dielectric properties of the cell interior become more important in determining impedance.

Coulter and Hogg[715] developed apparatus in which both DC resistance and radio frequency (34 MHz) AC impedance were measured, and defined **opacity** as the ratio of AC impedance to DC resistance. Leif et al[716], using the apparatus built by Coulter and Hogg, found that various fractions of human erythrocytes separated on the basis of buoyant density differed in opacity but not in volume; opacity measurements are also said to be capable of discriminating among different types of leukocytes. Hoffman and Britt[717] studied impedance of CHO cells at a lower AC frequency (1 MHz) and found the DC and AC impedance signals to be well correlated, although they did note substantial differences in the ratio of DC and AC signals between cells and plastic spheres.

Bulk physical separation of different cell types, presumably on the basis of dielectric properties, has been done using applied AC fields, and it seems likely that cell types separable in bulk by dielectrophoresis will also be found to be distinguishable on the basis of opacity measurements. If, however, the same underlying structural differences, e.g., in cytoplasmic granularity, underlie differences in cells' internal dielectric properties as are responsible for differences in light scattering behavior, the utility of opacity measurements will be considerably less than the utility of dielectrophoresis.

AC impedance measurements are not, as far as I know, provided for in any commercial fluorescence flow cytometers; however, both DC and AC impedance measurements are used in hematology instruments. Kraai et al[2531] reported that hematopoietic stem cells, normally detected by immunofluorescence measurements of CD34 antigen, appeared in the "immature" region defined by AC and DC impedance measurements in a Sysmex hematology counter.

Sohn et al[2532] measured **capacitance** of SP2/0 mouse myeloma cells in a microfluidic flow system, and reported that the histogram of capacitance values tracked the histogram of DNA content as measured by fluorescence flow cytometry. The CV of the capacitance histogram was much higher than one would expect or accept in a fluorescence measurement, but there is presumably some room for improvement.

Acoustic Measurements of Cells in Flow

Changes in cellular structure change the mechanical and, therefore, the acoustic properties as well as the optical and electrical properties of cells. In order to achieve the same spatial resolution with acoustic measurements as is achieved using visible light, equivalent wavelengths are required; these correspond to acoustic frequencies of several GHz. **Acoustic microscopy** on fixed and living cells has been done using apparatus operating at these high frequencies[718-9]. In 1984, Sweet, Fulwyler, and Herzenberg[720] reported making "zero-resolution" measurements of cells in a flow system at much lower acoustic frequencies (25-150 MHz). Apparently, the gain wasn't worth the pain; they thereafter remained with the stain.

Optical Parameters: Light Scattering

Almost all fluorescence flow cytometers, and many hematology counters, measure light scattering at small and larger angles to the incident beam, which we have come to know as **forward scatter** and **side scatter**. I have already cautioned (p. 5) against acceptance of the oversimplified and erroneous notion that forward scatter intensity is a measure of cell size. The simplified concept of side scatter intensity as an indicator of internal granular structure is somewhat more defensible, but the fact is that light scattering is a complex phenomenon.

Scattering: The Mueller Matrix Model

A general model for dealing with light scattering by particles, accounting for scattering of light of all polarizations at all angles, is based on the **Stokes vector** and the **Mueller matrix**[713,1244]. The Stokes vectors describe incident and scattered light; each has four components representing intensity and the degrees of linear and circular polarization. The Mueller matrix characterizes a scattering particle; it has sixteen elements, each of which is a function of the scattering angle. The Stokes vector for scattered light is the product of the Stokes vector for incident light and the Mueller matrix. In principle, if you know all the elements of the Mueller matrix, you have completely characterized the light scattering behavior of a particle but, in general, you can't compute the sixteen elements of the Mueller matrix simply by knowing the four elements of the Stokes vector for incident light and measuring the four elements of the Stokes vector for light scattered by the particle. However, the number of independent elements in the Mueller matrix is generally less than sixteen because of symmetry.

In the case of spherical particles, the Mueller matrix has four independent elements, which can be calculated as a function of particle radius according to Mie theory. For particles with a plane of symmetry, there are seven independent elements in the Mueller matrix, and you have a shot at finding them by making measurements of the intensity and polarization of scattered light. This, if you will, is the Holy Grail long sought by the knights of Los Alamos' round table, who, beginning in the late 1960's, measured scattering by plastic particles and by cells over a large range of angles, wavelengths, and linear polarizations. This took a lot more hardware than you'll find in the flow cytometers at the general store. The question remains as to whether the benefit, in terms of the information obtained about the biological particles under study, is worth the cost.

Knowing all 16 elements of the Mueller matrix, which tells you everything there is to know about scattering, leaves you with only half the information you'd get from Los Alamos' old 32-angle scatter measurement system, but that's still a lot of numbers to crunch. Those numbers also pose some problems because many don't relate in straightforward ways to observable morphological, biochemical, or functional parameters of cells. We use forward scatter measurements all the time because they're very reliable for telling us when there's a cell or other particle in our measurement system, and frequently useful for telling us roughly how big the cell is, and/or whether it's dead or alive. We use side scatter measurements all the time because they let us distinguish different cell types, but we're most comfortable when the distinctions made by side scatter measurements correspond to differences we can easily see by looking under a dark field or phase or interference contrast microscope. The failure of fancier scattering measurements to catch on may not lie so much in the difficulty of implementing them as in the difficulty of reconciling them with the rest of our experience. So let's get back to the simpler scattering measurements for a while, and then see what additions might or might not be worthwhile to pursue further.

Forward Light Scattering and Cell Size

Forward scatter measurements have been widely used for estimation of cell size since it was demonstrated by Mullaney et al[110] that the intensity of light scattered at small angles (0.5-2.0°) from an incident laser beam was roughly proportional to particle volume, as predicted by Mie's earlier theoretical work. So-called **forward scatter** or **small-angle scatter** measurements can be useful, provided one remains aware of their limitations. Most of these arise from the fact that many factors other than cell size influence cells' small-angle light scattering, as is attested to by a large body of literature[93,97-8,111,153-63,1244-5].

Forward light scatter signal intensity is strongly affected by the **wavelength** of light used and by the precise **range of angles** over which light is collected, the latter being determined by the focal lengths and numerical apertures of collecting lenses and the size, shape, and position of irises, slits, and obscuration bars in the optical system. Since no two manufacturers of flow cytometers use the same optical design for forward scatter measurements, it is unlikely that exactly the same results will be obtained from measuring the same cells in different instruments.

Among cellular properties other than size that influence forward scatter measurements are differences in **refractive index** between the cells and the suspending medium, cells' internal structure, and the presence within or upon cells of material with strong **absorption** at the illumination wavelength used. If the refractive indices of cells were the same as that of the suspending medium, the cells would not scatter incident light. The difference in index between cells and the medium is maintained, at least in part, by the action of the membrane as a permeability barrier to water and solutes. Cells with damaged membranes, i.e., those cells which are identified as "dead" by uptake of dyes such as trypan blue or ethidium bromide, have a lower refractive index, and thus produce smaller forward scatter signals. This characteristic was originally used in flow cytometric immunofluorescence work to discriminate damaged cells, which tend to stain nonspecifically with fluorescent antibodies, from the intact cells that were of interest to experimenters[154,164]. Such discrimination is less than perfect[165], particularly when the sample contains cells of different types and/or sizes.

The presence of strongly absorbing material in cells tends to decrease the amplitude of forward scatter signals; highly textured surface or internal structures, e.g., the specific granules of blood granulocytes, may have a similar effect. Electronic volume measurement indicates that blood granulocytes have a larger volume (about 350 fl) than lymphocytes (about 200 fl), but granulocytes produce smaller forward scatter signals in some flow cytometers. Thus, one cannot reliably estimate the relative sizes of two cells of different types from small-angle light scatter signals. Not only can't we compare apples and oranges; we can't, strictly speaking, always compare apples and apples, because **theory predicts, and experiments confirm, that, even for uniform particles, forward scatter amplitude will not be a monotonic function of particle size.**

I can't think of a better way to illustrate this than to include pictures from a poster presentation made by Kevin Becker et al at the ISAC XXI Congress in May, 2002[2533], which I have adapted into Figure 7-1.

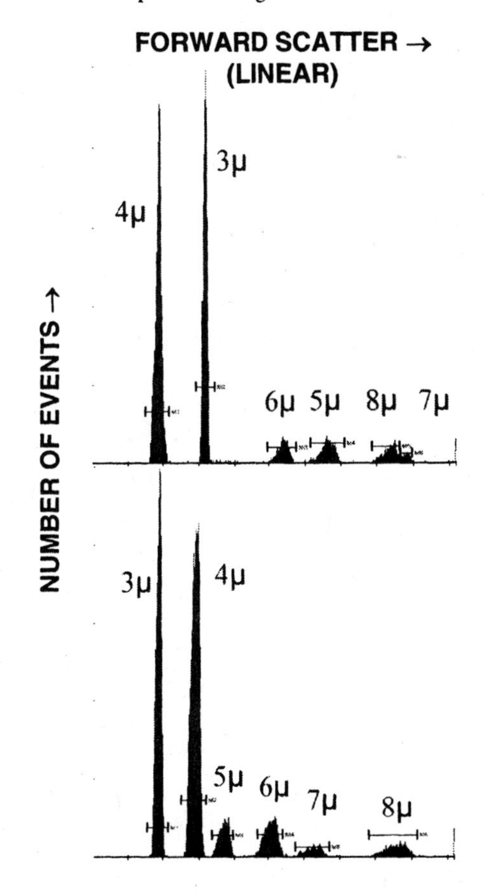

Figure 7-1. Forward scatter does not measure particle size (courtesy of Kevin Becker, Phoenix Flow Systems).

The histograms in Figure 7-1 show forward scatter intensities for polystyrene beads of various diameters, as measured using two popular benchtop flow cytometers. The trace on the top is from a BD Biosciences FACSCalibur, and the one on the bottom is from a Beckman Coulter EPICS XL. Both instruments use low-power, air-cooled 488 nm argon ion laser light sources. The relationship between forward scatter intensity and bead diameter indicated by the top histogram is not monotonic, that is, the signals from bigger beads are not necessarily bigger than the signals from smaller beads. The bottom histogram looks a little more promising, but I can tell you that there is no way to draw a calibration curve that relates forward scatter signal intensity to the first, second, or third power of particle size in any believable way, thus excluding the possibility that the forward scatter signal intensity is proportional to particle diameter, cross-sectional area, or volume. And these are beads; their diameters can be and have been measured accurately by electron microscopy, and their refractive indices are pretty much all the same, so we can't find any way to wiggle out of accepting the fact that forward scatter does not measure size.

This need not engender despair on the part of flow cytometer users. Like many of the other relatively rough measurements made by scientists, forward scatter measurements are good enough for many, and perhaps even most, of the purposes for which they are used. It is ignorance of and/or blind faith in techniques that gets experimenters into trouble.

After the first edition of this book came out, a reviewer of a manuscript cited my statement, printed in boldface on the previous page, as invalidating calculations of cell surface antigen density based upon ratios of fluorescence and forward scatter measurements. The measurements in question were made on cells in different stages of the cell cycle, and DNA content was also measured. In this case, knowing that S and G_2 phase cells are bigger than G_1 phase cells, it was not illogical to treat their forward scatter signals as roughly indicative of area, particularly since nobody has convincingly demonstrated that other parameters, such as refractive index, change drastically during the cell cycle. In other circumstances, e.g., granulocyte stimulation, changes in forward scatter signals may be influenced by changes in granule morphology as well as by swelling due predominantly to water movements, with accompanying changes in refractive index; it is thus hard to determine the extent to which a change in forward scatter signals reflects a change in size. It is always risky to assume that small (say 2 to 5%) differences in forward scatter signals precisely reflect small differences in size.

Forward Scatter and "Viability"

In general, "dead" cells, i.e., cells with sufficient membrane damage to render them permeable to dyes such as propidium, which normally does not enter intact cells, have lower forward scatter signals than live cells. McGann, Walterson, and Hogg[1257] measured forward and side scatter and Coulter volume of osmotically stressed and frozen-thawed cells. Osmotic swelling increased volume and decreased scatter signals, while membrane damage from freezing and thawing decreased scatter signals, but not cell volume. Scherer et al[2534] noted that changes in membrane lipid packing such as those that occur in apoptosis can affect forward scatter signals in the absence of apparent changes in cell volume or refractive index. In general, care is required when drawing conclusions about changes in scatter values. Care may also be required when drawing conclusions about relationships between dye uptake and cell death, but we'll open that can of worms later.

Side Scatter and Cytoplasmic Granularity

The group at Los Alamos set the pace for flow cytometric measurement of laser light scattering at multiple angles, as they had set the pace for forward light scattering measurements. The most elaborate of the instruments built at Los Alamos for this purpose[111] incorporated a sectoral solid state detector that measured 32 different signals, each representing the intensity of light scattered over a different range of angles from the incident beam. A dedicated minicomputer was required for data handling; while this apparatus was used to demonstrate differences in 32-parameter multiangle scatter signatures among different subpopulations in a variety of cell samples, it was difficult to determine whether a few and which few of the 32 scattering angle regions from which measurements were taken might contain the information most useful for cell discrimination[231,570,698].

In 1975, Salzman et al showed[157] that unfixed, unstained blood lymphocytes, monocytes, and granulocytes could be distinguished from one another by using measurements of forward and **orthogonal (90°) scatter** (now almost always called **side scatter**). Most of the information required for this discrimination is obtained from the side scatter signal, which is low for lymphocytes, higher for monocytes, and highest for granulocytes. The granular structures in the cytoplasm of granulocytes obviously present many more opportunities for scattering of incident light than does the more uniform cytoplasm of lymphocytes, and the higher intensity of light scattered at large angles to the incident beam probably represents some combination of multiple reflections and the summation of single scattering events from individual granules.

Whether or not theory can provide a complete explanation for the effect, it is reproducible and useful. After learning of the work at Los Alamos, I convinced my colleagues to incorporate side scatter measurement channels into Block's Cytomat-H differential counter prototypes and Cytomat-R research apparatus[92-3], where they proved useful enough in discriminating lymphocytes from monocytes and granulocytes to render our patented three-dye staining process largely unnecessary (see pp. 253-4).

Lymphocyte Gating: Forward Scatter Aside

When monoclonal antibody reagents became available, in the late 1970's and early 1980's, forward and side scatter measurements were used to identify lymphocytes and set a gate for immunofluorescence analyses in antibody-stained samples from lysed whole blood or buffy coat, following a procedure established by Hoffman et al[175,176,699]. **Two-angle scatter gating** of lymphocytes eliminated the time-consuming and labor-intensive process of enriching samples for mononuclear cells by centrifugation over a discontinuous density gradient using Ficoll-Hypaque or other separation media; it was also desirable because the separation procedure could lead to differential loss of lymphocyte subpopulations, e.g., the CD8+ (cytotoxic/suppressor) T cells[592-3]. It was not long, however, before trouble came to Paradise.

It was first noted that the advantages of two-angle scatter gating might not be realized if windows were improperly set[662,700]. In the mid-1970's, when the technique was first described, there was no better way to establish the extent to which lymphocytes and monocytes might cross-contaminate one another's gates. In 1990, Loken at al[1251] described the technique of **back-gating**, in which a sample stained with anti-CD14 and anti-CD45 antibodies (lymphocytes have high CD45, and monocytes high CD14) is used to allow determination of the fraction of the total lymphocyte population excluded from, and the fraction of other cell types included in, a particular lymphocyte gate. Widespread application of back-gating helped make it feasible for large numbers of laboratories to deal consistently with T cell subset analyses in HIV-infected patients.

While back-gating represented an improvement over simple two-angle scatter gating, it has substantial limitations. Once a scatter gate has been established by back-gating, subsequent analyses do not explicitly identify cells within the gate as lymphocytes or other cell types. If, for example, an experimenter wanted to determine levels of an antigen that was found on activated, but not resting, lymphocytes, and which was also present on other cell types, it would not be possible to know precisely which antigen-bearing cells in the scatter gate were and were not lymphocytes.

As more monoclonal antibodies and labels for them became available, making three-color immunofluorescence measurements feasible for routine clinical use, back-gating has largely been replaced by more specific gating procedures. **T-gating**, which identified T lymphocytes as lying in a gate drawn on a plot of side scatter vs. CD3 antibody immunofluorescence (as in Figures 1-14, 1-15, and 1-16, pp. 31-32) was described by Mandy et al[1027]. While it improved the accuracy with which CD4- and CD8-positive T lymphocytes could be identified, T-gating did not allow an estimate to be made of the size of the total lymphocyte population, which is often of clinical relevance.

Current practice, as described by Nicholson, Jones, and Hubbard[1252] defines a lymphocyte gate on the basis of high CD45 antigen expression and low side scatter, as illustrated in Figure 1-17 (p. 34). T cells in the lymphocyte gate are then identified by the presence of CD3 antigen. A four-color immunofluorescence measurement using antibodies to CD45, CD3, CD4, and CD8 allows $CD3^+4^+8^-$, $CD3^+4^-8^+$, $CD3^+4^-8^-$, and $CD3^+4^+8^+$ populations to be identified in a single tube (Figure 1-19, p. 38). An additional point in favor of more specific gating comes from the knowledge that different lymphocyte subsets are not uniformly distributed within the forward vs. side scatter gate but, rather, tend to have different median values of forward and side scatter[703,1247,1253]. A tightly set scatter gate will, therefore, selectively exclude one or more lymphocyte types.

Other Applications of Side Scatter

Changes in the side scatter signals of neutrophils occur in association with degranulation and membrane ruffling following activation[592,701]; side scatter signal amplitude also decreases on storage of anticoagulated blood samples, making it advisable to do gated immunofluorescence measurements on whole blood within 24 hrs or less following collection of samples.

The combination of forward and side scatter measurements was used, alone or in combination with measurements of lectin binding, autofluorescence, or immunofluorescence, in the earliest successful attempts to enrich **hematopoietic cell populations** by sorting[161,177-9], and to discriminate cell subpopulations in samples from a variety of sites[702-705].

Forward and side scatter signals have also been used to assess **nuclear morphology** of prostate[706] and bladder (J. Coon, personal communication) cancer cells. Papa et al[1254] reported that side scatter signal intensity from nuclei decreases as chromatin is decondensed by lowering ionic strength or by releasing histone H1 at low pH. Zucker et al[1255] and Nusse et al[1256] showed that, when certain preparative steps are employed, **mitotic nuclei** can be discriminated on bivariate displays of DNA content (propidium fluorescence) vs. side scatter.

Scatter signals may be profoundly influenced by inclusions in cells, particularly if these contain material with a refractive index markedly different from that of the cytoplasm. For example, Dubelaar et al[707] found marked differences in forward and side scatter signals between cyanobacteria containing and lacking **gas vacuoles**.

Nordström et al[2535] report that increased side scatter provides a good indication of whether or not insect cells have been infected with recombinant baculoviruses, eliminating the need for and cost of a fluorescent marker.

Extrinsic cellular parameters can be measured with appropriate reagents using side scatter signals. For example, Böhmer and King[708] labeled lymphocytes with **antibodies conjugated to colloidal 40-nm gold particles**; the gold label produced increased side scatter signal amplitudes more than tenfold and did not interfere with fluorescent antibody labeling.

The orthogonal design of most laboratory-built and commercial laser source flow cytometers has made it relatively easy to measure forward scatter, and to utilize the collection optics also used for fluorescence to measure side scatter. It requires at least a little bit of instrument modification to make measurements of light scattered over other angular ranges, and, while collection at two angles undoubtedly provides a great deal more information than collection at one angle, I am not yet convinced that collection at 32 angles provides an equivalent improvement over collection at two angles. The side scatter parameter has proven useful enough to have been incorporated into almost all commercial designs.

What is the Right Angle for "Right Angle" Scatter?

As the number of immunofluorescence signals measured increases, it may become difficult to make room in the orthogonal detector optical assembly for a scatter measurement detector. Stovel et al[709] coped with this problem by equipping the Stanford cell sorter with an outboard detector for large angle scatter signals, using a 2 mm diameter gradient index ("GRIN") lens coupled to the detector via a fiber optic. This lens collected light scattered at 130° from the beam, and its optical axis was not perpendicular to the direction of fluid flow. The 130° scatter signal was said to provide information equivalent to that in the side scatter signal for discriminating lymphocytes from other leukocytes; in addition, the positioning of the detector greatly reduced orientation artifacts normally found in side scatter signals from erythrocytes.

As was mentioned on p. 159, Steen[1142] noted that scatter signals collected at angles as small as 18° from the beam axis have the characteristics, and reflect the same cellular features, as signals collected orthogonal to the axis. In selecting a collection angle for large angle scatter, we therefore have a great deal of latitude. I have mentioned that Mariella et al collected very good large angle scatter signals by placing a fiber optic in the stream with its axis along the axis of flow[2455-6]; as a rule, however, there are practical reasons for not doing this.

Does Side Scatter = Total Protein?

I and others have observed[8,103,1255] that side scatter measurements correlate pretty well with flow cytometric measurements of **total protein** made using covalently and noncovalently bonded fluorescent dyes such as brilliant sulfaflavine, fluorescein isothiocyanate, sulforhodamine 101, and "LN" (also see Fig. 5-17, p. 254). I will discuss this further in the section on measurement of protein content.

Optimizing Side Scatter: Not as Easy as It Looks

We pretty much take it for granted that we can set up and align a flow cytometer using polystyrene beads with a very narrow size distribution, tweaking the optics to get the largest amplitude and the lowest CV we can in forward scatter and fluorescence measurements. However, as Doornbos et al noted[2312], it may be difficult to play this game with a side scatter channel; plots of forward vs. side scatter often show "Lissajous-like" patterns of wavy lines and loops, with relatively small CVs in forward scatter and large CVs in side scatter. This is explicable by Mie theory (which holds reasonably well for beads, if not for cells), and is due to extreme sensitivity of the side scatter signal to small variations in the size of highly symmetric particles. Such variations are typically found in small subpopulations of beads as a result of imperfections in the production process. The implication of all this for the average user: minimizing the CV of the side scatter distribution of beads doesn't always guarantee optimal alignment of your instrument. If forward scatter and fluorescence look good, and side scatter doesn't, don't futz with side scatter unless you can improve it without losing performance on the other channels.

Polarized 90° Scatter Measurements Reveal Eosinophils and Malaria Pigment-Containing Monocytes

Polarization measurements of orthogonal light scattering were introduced by de Grooth et al[710]. Under normal circumstances, randomly polarized light scattered orthogonal to the incident beam becomes linearly polarized (see pp. 78-82); linearly polarized light scattered at 90° remains polarized. In theory, some portion of the light collected orthogonal to the incident beam represents light scattered several times; the more times, the greater the likelihood that this light will be depolarized. **Eosinophil** granulocytes were found to be distinguishable from neutrophil granulocytes by higher values of depolarized side scatter. An as yet unidentified subpopulation of lymphocytes was also distinguished by higher depolarized scatter signals.

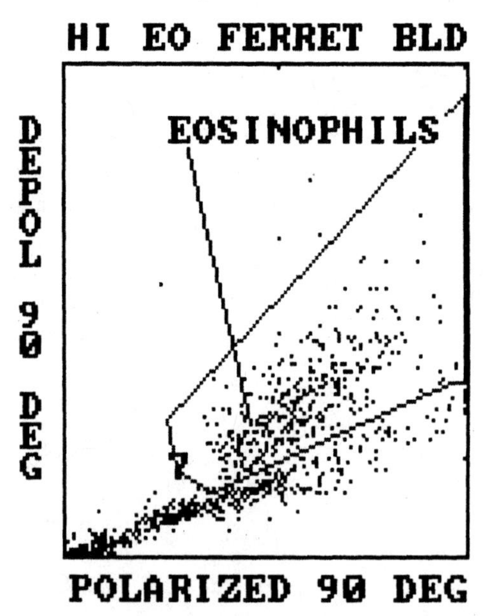

Figure 7-2. Depolarized 90° scatter signals can be used to identify eosinophil granulocytes.

Figure 7-2 shows polarized vs. depolarized 90° scatter values for leukocytes in lysed whole blood from a ferret with a high eosinophil count due to an experimental infection with microfilaria. The eosinophils are clearly separable from the other white cell types; a count of 100 particles sorted from a gate set as shown in the figure revealed 99 eosinophils and one worm.

It is relatively trivial to set up an instrument to make depolarized scatter measurements using inexpensive plastic polarizing filters (see p. 157). The technique requires no reagents, and represents the easiest way to count and sort eosinophils from a wide range of species (L. Terstappen, personal communication). The sorter manufacturers won't tell you about it, because it is the subject of a patent now held by Abbott and used in that company's Cell-Dyn hematology instruments. The birefringence of eosinophil granules, which is responsible for the cells' higher depolarized scatter signals, was noted decades ago; one could very likely use polarized and depolarized extinction signals to count or sort eosinophils without infringing on anybody's patent.

In 1999, Mendelow et al[2536] reported that a population of cells with relatively low polarized side scatter values and high depolarized side scatter values appeared when the peripheral blood of malaria patients was analyzed in a Cell-Dyn instrument. It was suggested that these cells were monocytes containing the malaria pigment hemozoin, which is known to be birefringent. Nordström et al[2537] subsequently confirmed this by cell sorting.

Many bacteria have birefringent cell walls, which suggests that polarized and depolarized scatter measurements might be useful in discriminating among genera or species. However, depolarized scatter signals, even from eukaryotic cells, are typically of much lower intensity than polarized scatter signals, suggesting that depolarized signals from bacteria could be too weak to permit precise measurements.

Multiple Wavelength Scattering Measurements

Any apparatus that incorporates a lamp instead of a laser as a light source for light scattering measurements will, obviously, measure light scattered over a larger range of wavelengths than will a laser source instrument. Technicon's older blood cell and differential counters[84,85], the Cytomat arc source instrument[88], the apparatus (Skatron Argus, Bio-Rad Bryte, etc.) developed by Lindmo and Steen[100-3], and the old B-D FACS analyzer all fell into this category.

The Technicon Hemalog D apparatus was noteworthy for its utilization of measurements of light scattering at different wavelengths for cell classification. In this system, basophil leukocytes were stained with a blue basic dye, which imparted a blue color to their specific granules. The granules therefore absorbed red light; however, their absorption was not high enough to permit unequivocal discrimination between basophils and other cells on the basis of absorption, especially since the number of granules per cell is highly variable. Satisfactory discrimination was achieved by making separate measurements of the scattering of red and near-infrared light by the cells. Since basophils, which contained the blue dye, absorbed more red light than infrared light, the ratio of red scatter amplitude to near-IR scatter amplitude was lower for the basophils, which could thus be discriminated from other cell types in the two-dimensional measurement space.

While developing the Block differential counter, we measured forward and 90° scatter signals from fixed and unfixed, unstained leukocytes at different laser wavelengths (325 and 441 nm from a helium-cadmium laser, 488 and 515 nm from an argon ion laser, and 633 nm from a helium-neon laser), and were somewhat surprised to find that ratios of scatter intensities at different wavelengths were different for different cell types, to an extent that allowed us to obtain differential counts of lymphocytes, monocytes, neutrophils, and eosinophils in unstained blood samples. Our enthusiasm for further work along these lines was tempered when we found that slight alterations in the geometry of the optical system used for forward scatter measurements could produce large changes in the relative positions of cell clusters. I was thus not surprised to read others' reports[156,161,180] that laser light scattering at different wavelengths (351/363, 457, 488, and 515 nm, all from argon lasers) provided information that could discriminate different types of blood cells.

A principal problem with studies of this type lies in the difficulty of making sure that the optical geometry is the same at the different wavelengths used. Even in the case in which UV and visible light are emitted in the same collinear beam by the same argon laser, the beam diameters are apt to be different, and the focusing lens, unless custom made for the purpose, will not have the same focal length at both wavelengths. This makes it hard to determine the extent to which apparent differences in scattering at different wavelengths may be due to small but significant differences in such factors as beam size and collection angle. Others have reported that UV scatter signals, obtained from the B-D FACS using an argon laser source, were less satisfactory than signals at 488 nm for cell discrimination. Using the Block instruments, we usually obtained better discrimination from the UV scatter signals than from signals at 488 nm. This may represent another instance among many in which each of several groups gets its best results when using the methodology with which it is most familiar.

From Russia with Lobes

A few pages back, I said we'd get back to the complex stuff about scatter; by now, you probably think the Mueller matrix has been sent to Siberia. Well, it has, quite literally, and I am happy to report that it is alive and well there, in Valeri Maltsev's lab at the Institute of Chemical Kinetics and Combustion.

Over the years, a few attempts have been made to apply relatively sophisticated multiangle scattering measurements to cell sizing and characterization. Since Mie's theory of scattering, which forms the basis for attempts to measure cell size measurement by forward scatter, uses a model in which particles are spherical and homogeneous, we might reasonably expect that, however bad our results might be with spherical cells, they would get worse when we tried to size asymmetric cells. This presented a problem to people concerned with the most common and most profitable type of cell volume measurements, i.e., the manufacturers of blood cell counters for clinical laboratories, because erythrocytes are normally not spherical. Technicon's H-1 hematology

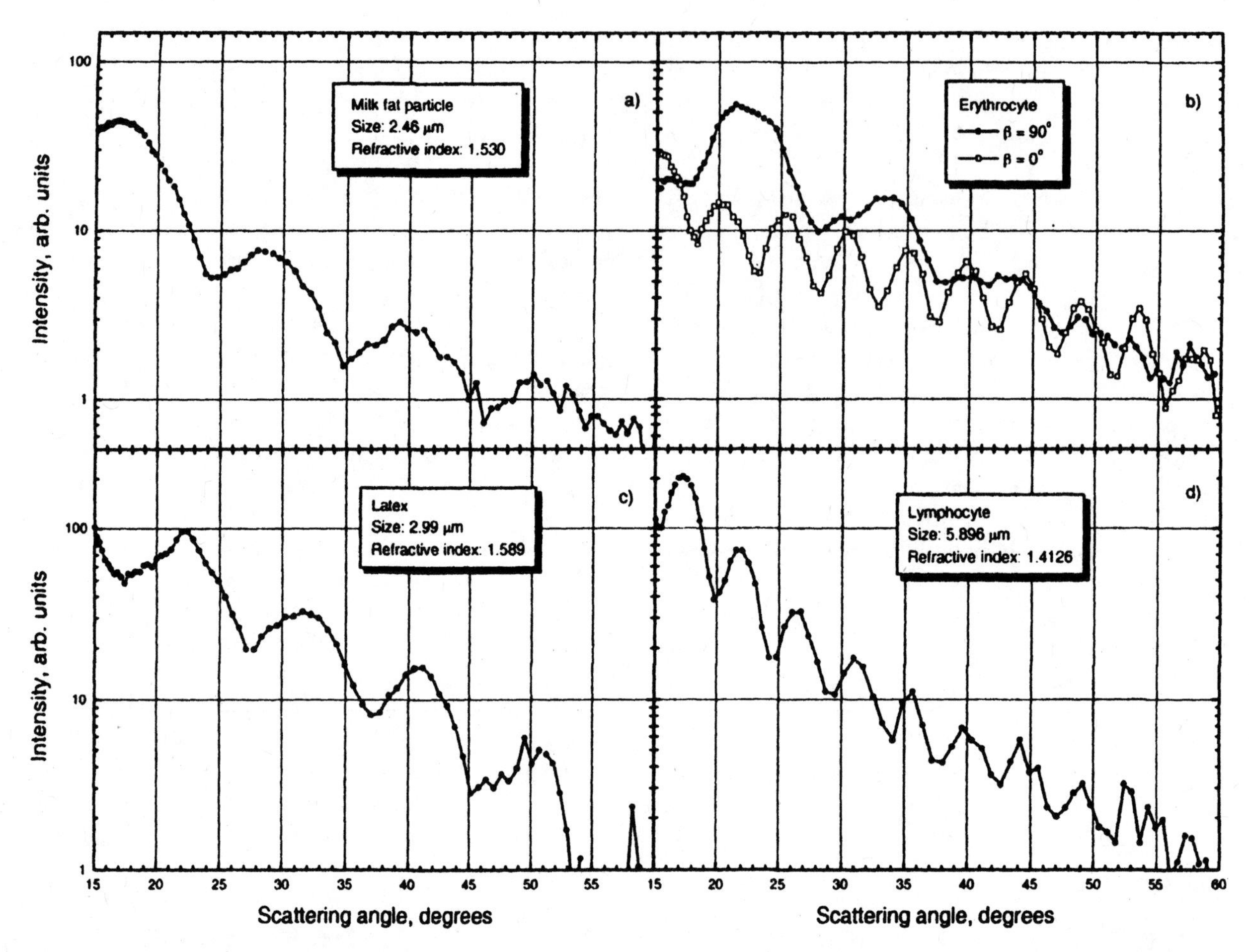

Figure 7-3. Indicatrices (plots of intensity vs. scattering angle) of a milk fat globule, a latex bead, a lymphocyte, and two erythrocytes. From: Shvalov AN et al, Cytometry 37:215-220, 1999 (Reference 2539), © John Wiley & Sons, Inc., used by permission.

instruments, and their successors (Technicon is now Bayer Diagnostics; see Chapter 8), employ a couple of ingenious maneuvers to deal with the problem. The discoid erythrocytes are converted, without changing their volumes, to spheres, using a procedure described by Kim and Ornstein[166]. This eliminates artifacts due to cell asymmetry; however, one still encounters substantial variation in the refractive indices of individual cells due to cell-to-cell differences in hemoglobin concentration. A measurement procedure devised by Tycko et al[684] meets this problem head-on; since the variation of refractive index with hemoglobin concentration is linear, it is possible, by making measurements of light scattered at two angles, to determine hemoglobin concentration and obtain volume from tables calculated using Mie theory. The system can be calibrated in absolute units of volume, using droplets of water-immiscible oils of different refractive indices as standards. This elegant methodology is demonstrably successful for its clinical hematologic purpose; it is unlikely to be adaptable to sizing other cell types.

Maltsev et al have come up with a simpler way of sizing sphered red cells in the course of extensive investigations on multiangle light scattering by cells and beads. They have devised a unique scanning flow cytometer in which particles are illuminated by a laser beam directed along the axis of flow; a fixed spherical mirror is used to collect scattered light. As the particle moves down the flow stream, the scattering angle from which light reaches the mirror changes. Thus, it is possible to derive a plot of intensity vs. scattering angle, known as an **indicatrix**, for each individual particle traversing the system. Indicatrices for a milk fat particle, a latex bead, a lymphocyte, and two (unsphered) differently oriented erythrocytes appear in Figure 7-3.

Indicatrices are uniquely characteristic of particle types; those for beads agree closely with calculations according to Mie theory. The sizes and refractive indices of spherical and near-spherical particles can be determined from their indicatrices, and, in at least some cases, it is possible to specify two or three angles for which light scattering measurements can be used for sizing. However, as Murphy's Law would predict, these angles are different for different cell types.

In cytometry, as in many other fields, there is little new under the sun; Loken, Sweet, and Herzenberg made measurements of multiangle light scattering in a modified FACS in the 1970's[155]; they did not, however, pursue the theoretical side of their investigations far enough to yield practical results. A commercial application of the Maltsev group's work to hematology is now being pursued, but it still seems unlikely that the high-speed cell sorter manufacturers will be rushing to put indicatrix measurement capability into their products any time soon.

Optical Parameters: Absorption

Since, even in this day and age, a lot of quantitative chemical and biochemical analyses are done with absorption measurements, we might consider why we don't make more use of absorption in cytometry. Absorption measurements are somewhat more demanding than scatter and fluorescence measurements. In any microscope, light scattered at larger angles than can be collected by the microscope objective used cannot readily be distinguished from light absorbed by the specimen; accurate absorption measurements therefore require light collection over a relatively large angle, necessitating use of a lens with a high N.A. It's not that hard to build a flow cytometer with high-N.A. lenses, oriented along the same axis, for illumination and collection, suitable for absorption measurements; this configuration is used in blood cell counters. However, most laser source fluorescence flow cytometers incorporate a high-N.A. collection lens placed at right angles to the illuminating beam, making it impossible to mount a high-N.A. lens along the beam axis, and use low-N.A. optics to form an illuminating beam, which is not effective for absorption measurements.

To maximize precision in absorption measurements, it is also necessary to match the refractive indices of the object under study and the medium in which it is suspended, this maneuver minimizes apparent absorption due to scattering. However, in work with intact cells, it is generally not possible to completely avoid scattering by internal cellular structures, e.g., cytoplasmic granules, even when the indices of sheath, sample, and cell membranes are matched.

The major problem with absorption measurements arises from the fact that, in most cases, cells, even when stained, do not absorb more than a small fraction of the light passing through them. Most of the light reaching the detector is unaffected by interaction with the specimen; it is, therefore, background, and, as Johnnie Cochran might summarize the Wood-Hoffman model discussed on pp. 221-3, "If there's too high a B, you just can't see." Fluctuations in the intensity of illumination therefore have larger effects on absorption measurements than on measurements of scattered light and fluorescence. In the latter, signal intensity is proportional to source intensity; a 1% change in illumination intensity produces a 1% change in a scatter or fluorescence signal. Suppose, however, that an absorption measurement is made of a particle that removes 10% of the light from the incident beam. A 1% increase in illumination intensity occurring while the particle passes through the observation region would result in an apparent light loss of 9% instead of 10%, while a 1% decrease in illumination intensity would result in an apparent light loss of 11%. In this case, a 1% change in illumination intensity produces a 10% change in the amplitude of an absorption signal; the less light the particle removes from the beam, the larger the effect of source intensity fluctuations. The intensity fluctuations of the laser sources most commonly used in fluorescence flow cytometers, while small (a few percent peak-to-peak), are sufficient to make absorption measurements unacceptably imprecise. Because of the problems just discussed, the use of absorption measurements in flow cytometry is somewhat restricted; they are most commonly employed for semiquantitative detection of strongly absorbing substances in apparatus designed for blood cell counting and sizing, employing lamps rather than lasers as light sources.

Absorption Effects on Light Scattering

I noted on p. 279 that the presence of an absorbing (colored) substance in cells may be inferred from differences in intensity of scattering signals at wavelengths at which the material does and does not absorb strongly. A dramatic illustration of this, shown in Figure 7-4, comes from the work of Ost et al[2449], who measured red and violet light scattering by leukocytes and erythrocytes in unstained dilute whole blood.

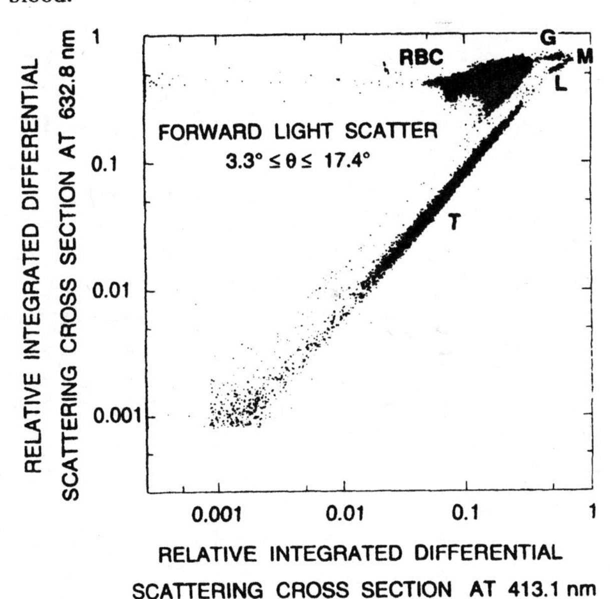

Figure 7-4. Erythrocytes scatter less light at a wavelength at which hemoglobin exhibits strong absorption. G: granulocytes; L: lymphocytes, M: monocytes, RBC: erythrocytes, T: thrombocytes. From: Ost V et al, Cytometry 32:191-197, 1998 (Reference 2449), © John Wiley & Sons, Inc., used by permission.

Erythrocytes contain high concentrations of hemoglobin, which absorbs strongly in the violet spectral region but only weakly in the red spectral region. If more of the photons incident on a cell are absorbed, fewer remain to be scattered. Their relatively low forward scatter intensities at 413.1 nm (violet) allow the erythrocytes (RBC) to be distinguished from leukocytes (G, M, L) and thrombocytes (T) on the two-dimensional dot plot of violet vs. red forward scatter.

Optical Parameters: Extinction

Extinction is a convenient term that describes **light loss** from the incident beam regardless of the mechanism involved. If a detector is placed where the beam stop used for a forward scatter measurement would normally be (see Fig. 1-21, p. 51), the passage of a cell through the measurement system will result in the detector signal decreasing in proportion to the total amount of light absorbed by the cell or scattered out of the field of view of the detector. Axial extinction or light loss signals are relatively easy to measure; photodiodes are typically used for the purpose. Like absorption signals, extinction signals are more sensitive to source noise fluctuations than are scatter and fluorescence signals. Although extinction measurements do not seem to be available in current commercial systems, Ortho's System 30 and System 50 instruments, made between the late 1970's and the late 1980's, featured an extinction measurement; the light source was a very low power (0.8 mW) He-Ne laser with very low RMS noise (0.05%). Steinkamp[659] described noise compensation circuitry that allowed fairly precise light loss measurements to be made using somewhat noisier argon lasers. It is now possible to achieve very low noise levels in diode lasers, which could be economical light sources for extinction measurements should anyone still be interested in doing them.

Returning for the moment to the subject matter of Figure 7-4 (previous page), Ost et al[2449] pointed out that, unlike forward scatter measurements, extinction measurements in the violet beam would not have been able to discriminate erythrocytes from other cell types, because, while the erythrocytes absorb more light and scatter less light than the other cell types, the total light losses from the beam are approximately equal for erythrocytes and other cells. To generalize, extinction measurements are not particularly good for quantifying amounts of stuff in or on cells or particles. So why should anyone still be interested in doing them?

The answer is that extinction measurements are useful for sizing particles; pulse width, rather than pulse height or area, is the critical characteristic. I will say more about this in the discussion of measurement of intrinsic cellular parameters (pp. 285-8).

Other Transmitted Light Measurements

Progress beyond van Leeuwenhoek's level in the microscopy of living, unstained biological materials has largely been made in this century with the development of dark field, polarized light, phase contrast, and interference contrast techniques. From the 1930's on, the newer methods of microscopy were, in their turn, adapted to quantitative measurement. The apparatus originally built[712] for **interferometric dry mass determinations** and **phase-based optical path length measurements** was cumbersome and difficult to use, but did yield accurate results, and was applied to a reasonable range of biological specimens.

Flow cytometry has benefited tremendously from the phenomenal pace of technological growth since the 1960's, which gave us lasers, vastly improved detectors and imaging devices, integrated circuits, microprocessors, and the electronics and computational methods required to store, manage, and process the bewildering mass of data we can generate. We can easily implement the entire range of microscopic measurement techniques with which our predecessors struggled. Turned on by technology, and surrounded by its cornucopia of gadgets, we tend to forget that, even with genetic engineering, we're basically looking at the same animalcules that fascinated van Leeuwenhoek.

By the 1920's, dark field microscopy could be used to visualize viruses and other particles below the resolution limit of transmitted light microscopy, as well as to examine the structural details of unstained cells; the obvious utility of the technique led to the development of the first flow cytometers, designed to detect and count particles based on light scattering, by the 1940's. The range of light scattering measurements available in modern flow cytometers essentially encompasses the subject material of dark field microscopy, and has expanded into polarized light microscopy.

Interference and Phase Measurements

The technology exists for making flow cytometric measurements of **optical path** and **dry mass** corresponding to those done with much greater difficulty by an older generation of **phase** and **interference microscopes**[714]. Again, it's harder to do these measurements than to do our bread-and-butter scatter, extinction, and fluorescence measurements. Work in this direction will probably yield a couple of parameters corresponding to things we can observe, at least some of which may be measurable more easily by other methods, and a whole bunch of data, which nobody will be able to interpret, but which will be just interesting enough to keep another generation of investigators funded. That's science. If we want to learn from history, however, we should note that biological microscopy has been advanced far more by the discovery, use, and refinement of staining methods than by the improvement of techniques for observation of unstained specimens, and that the primary path of development of analytical cytology has, similarly, been in the direction of measurement techniques using reagents, which have enabled us to characterize cells' chemistry and function with amazing sensitivity, specificity, and precision using relatively pedestrian measurement techniques.

There are instances, however, where the increased complexity of the instrumentation needed to make a measurement without using a reagent are offset by the potential disadvantages associated with use of the reagent. A good example is the sorting of sperm for sex selection in animals and humans. This was mentioned briefly on pp. 26 and 178 and will be discussed further in Chapter 10. Of note here is van Munster's work on enrichment of X- and Y-sperm by sorting based on interferometric volume measurement[2541-2].

Optical Parameters: Fluorescence

There is not much to add to what I have said to this point about basic fluorescence measurement in flow cytometry. Observation periods in conventional flow cytometers are on the order of microseconds, while the fluorescence lifetimes (pp. 112-8) of most probes and labels, and of fluorescent constituents of cells, are on the order of a few nanoseconds. This means that hundreds of photons are likely to be emitted from each fluorescent molecule in or on a cell during the cell's traverse of an illuminating beam at a wavelength at or near the molecule's excitation maximum.

While even the most efficient fluorescence collection optics collect no more than about 20% of the total emission from cells, the discussion of Q and B on pp. 221-3 tells us that, during the time a cell spends in the beam, the photocathode of a detector PMT will crank out photoelectron for every few molecules of our most efficient probes, and perhaps one photoelectron for every few hundred molecules of our least efficient probes. Using our conventional measurement technique, once we have determined the spectral bandwidth of a particular detector, we have no way of telling whether the photons reaching it to generate the photoelectron current we measure came from a dye we were trying to measure, from a fluorescent cellular constituent, or from stray excitation light. There are, however, ways of making at least some of those distinctions.

Fluorescence Lifetime Measurements

Pinsky et al[1274] and Steinkamp and Crissman[1275] used **phase-sensitive detection** to measure **fluorescence lifetimes** by flow cytometry. The phase referred to here relates to the electronic signal, not the phase of light waves. The laser used for excitation is modulated at a frequency between 10 and 50 MHz, meaning that the intensity fluctuates with a period of 20 to 100 nsec. The fluorescence emitted by cells also fluctuates at the modulation frequency, but there is a difference in phase between the excitation and emission waveforms, corresponding to the excited state lifetime. It is relatively simple to electronically dissect components of the fluorescence signal having different phase relationships to the excitation waveform, and therefore representing emission from fluorophores with different lifetimes. This enables the separation of signals from molecules with similar emission spectra.

Using the phase-sensitive flow cytometer built at Los Alamos, Steinkamp and his colleagues[2543-7] have demonstrated several possible uses of the fluorescence lifetime measurement, including separating weak signals from fluorescently labeled antibodies and/or gene probes from the autofluorescence background, discriminating signals from two probes with similar emission spectra but different lifetimes, and distinguishing emission signals from a single probe under different binding conditions. Capacity for lifetime measurement has not yet been incorporated into commercial flow cytometers. Lifetime measurements may also be made in imaging systems; van Zandvoort et al[2548] were able to distinguish signals from the dye SYTO-13 bound to DNA and the same dye bound to RNA using this method.

Fluorescence Polarization Measurements

As was noted on p. 114, fluorescence polarization measurements may be used to assess the mobility of fluorescent molecules. The laser sources used in most flow cytometers emit linearly polarized light. If the molecules in a sample excited by such a laser were completely immobile, the fluorescence emission would also be linearly polarized. However, any molecules that changed their orientations while in the excited state would emit in a different plane of polarization. Small molecules in solution are freer to move than are identical molecules bound to macromolecules or macromolecular assemblies; the degree of polarization of fluorescence emission can, therefore, be used to estimate the proportions of free and bound dye. Typical practice involves measurement of fluorescence emission in two planes of polarization, with analog or digital computation used to derive values of **fluorescence polarization** or **fluorescence anisotropy** using the formulae shown on p. 114.

Flow cytometric measurements of polarization, anisotropy, etc. are not difficult, in principle. Since the laser beams used for excitation in most instruments are already linearly polarized, it is only necessary to use polarizing filters, prisms, or beamsplitters ahead of the fluorescence detectors (PMTs) to make the intensity measurements in two planes of polarization[5,406,407-9]. Asbury et al[2439] have provided the most thorough and most current discussion of this subject.

Energy Transfer Measurements: Something to FRET About

Fluorescence Resonance Energy Transfer (FRET), also distinguished as **Förster energy transfer**, was introduced on pp. 44-5 and p. 115. Intramolecular energy transfer is responsible for the relatively large Stokes' shifts (p. 112-3) of phycobiliproteins and tandem conjugates, but the effect has many other interesting applications in cytometry, most notably in determining proximity between and/or colocalization of different molecules in or on cells.

For energy transfer between two (donor and acceptor) chromophores to occur, the absorption spectrum of the acceptor must overlap the emission spectrum of the donor. If the donor and acceptor are sufficiently close, absorption of excitation energy by the donor can result in nonradiative transfer of this excitation energy to the acceptor, with subsequent emission by the acceptor at wavelengths characteristic of the acceptor's emission spectrum.

The smaller the distance between donor and acceptor, the more likely it becomes that resonance energy transfer, rather than emission of fluorescence by the donor, will occur. Förster showed that the probability of energy transfer varies as the inverse sixth power of the distance between the chromophores, and this high-order dependence allows the intensity of energy transfer to be used to provide a sensitive measure of this distance, or what Stryer and Haugland[315] termed a "spectroscopic ruler."

Energy transfer measurements using single beam excitation at 488 nm can be made using fluorescein, which is optimally excited at this wavelength, as the donor, and tetramethylrhodamine, which has less than 5% of its maximal absorption at 488 nm, as the acceptor. Chan, Arndt-Jovin, and Jovin[314] used flow cytometric measurements of **energy transfer**[23] between fluorescein- and tetramethylrhodamine-labeled concanavalin A to estimate the proximity of receptors for this lectin on the surface of Friend mouse erythroleukemia cells.

The flow cytometric energy transfer measurement technique originally described by Chan et al is applicable to the study of many different types of ligand-receptor interactions; Szöllösi et al[316] reported that flow cytometry offered advantages over conventional steady-state fluorimetry for energy transfer measurements on cell surfaces.

Several groups have used energy transfer measurements to demonstrate associations between cell surface structures[1508-11,2549-50] and to investigate molecular assemblies on beads[2551]. Bene et al used both energy transfer and anisotropy measurements in a study of receptor clustering[2552]. Energy transfer between different lipid membrane labels has also been used to identify and sort cell hybrids[1512].

General principles and methodology have been reviewed by Szöllösi et al[783, 2347] and by Mátyus[1507]; two recent papers discuss the use of phycoerythrin and allophycocyanin as an energy transfer pair for FRET measurements[2553-4].

Quenching and Energy Transfer

Quenching of fluorescence (p. 115) occurs when energy is transferred from a fluorescent donor to a nonfluorescent acceptor. It can occur via the Förster mechanism or other mechanisms. Such quenching can be undesirable; for example, as one labels an antibody molecule with an increasing number of molecules of a fluorescent dye, each additional molecule adds less and less additional fluorescence because of quenching due to intramolecular interactions.

Parenthetically, while it is true that, if we were able to put dozens of molecules of a covalently bound label on an antibody, we would run the risk of changing the reactivity by getting one of the label molecules into the binding region, it is the quenching effect that prevents us from simply putting a lot of fluorescent dye on some other macromolecule and conjugating that to an antibody.

Quenching can be also desirable; when one is trying to distinguish between fluorescent particles in a cell and those bound to the surface, adding a spectrally appropriate cell-impermeant, nonfluorescent dye to the sample will quench the fluorescence of the extracellular, but not of the intracellular particles.

Matko et al[2555] used quenching by a long range electron transfer (LRET) mechanism to study protein clustering at cell surfaces. Packard and Komoriya[2556-7] have synthesized fluorescent indicators of protease activity in which two dye molecules attached to a peptide quench one another by an exciton transfer mechanism rather than a Förster mechanism; cleavage of the peptide by a protease renders the indicator molecule many times more fluorescent.

Measuring Fluorescence Spectra in Flow

Measurement of fluorescence emission spectra in flow was mentioned on pp. 152-3. Since current commercial instruments from BD Biosciences, DakoCytomation, and Partec permit measurement of fluorescence excited by a single laser (usually at 488 nm) in as many as 7 different spectral bands, using only dichroics and filters for spectral separation, one might say that low-resolution spectral measurement capability is now available off the shelf.

Earlier efforts[122,649,1139-40] and more ambitious later projects have used polychromatic detection. Asbury et al[2558] measured spectra of cells and chromosomes in flow using a monochromator that changed the wavelength detected by a PMT during the course of a run; the overall run yielded spectra of DNA dyes in the cells or chromosomes, although only a single wavelength was detected from any individual particle.

Fuller and Sweedler[2559] built a slow-flow system in which a grating dispersed collected fluorescence emission to a CCD, so that different pixel regions on the CCD responded to different wavelengths. This system was used to discriminate and size submicron synthetic lipid vesicles.

Gauci et al[2560] used a dispersing prism and an image-intensified diode array detector to capture fluorescence spectra of individual *Dictyostelium* spores stained with fluorescent antibodies bearing a variety of labels.

Two-Photon Fluorescence Excitation in Flow

It's not easy, and may not be all that's useful, but it's been done[2442]. See pp. 118-9.

Bioluminescence Detection in Flow?

People ask about this on the Purdue Cytometry Mailing List every now and then. Bioluminescence assays generally involve a very long observation time (some of them use photographic film as a detector), because you don't get a very high photon flux from most luminescent probes. It therefore seems unlikely that you'd collect a lot of photons from a cell bearing luminescent material if you ran it through the flow cytometer for a few microseconds with the laser turned off.

Lindqvist et al[1276] reported detection of luminescence from insect cells transfected with luciferase genes and exposed to luciferin; they did, however, use a 488 nm laser to illuminate the cells, and their paper does not make clear whether the measurements would work with the laser turned off.

So the bottom line of this little section (and of every paragraph in it) is that, when you ask me how I feel about the prospect of doing bioluminescence measurements in flow, I will tell you: turned off.

7.2 INTRINSIC CELLULAR PARAMETERS

I have called cellular parameters that can be measured without reagents **intrinsic**, and those that require reagents, or **probes**, for their measurement **extrinsic**. The contents and physical states of chemically defined entities in cells and organelles are **structural** parameters; **functional** parameters include biological properties and activities as well as a few specific chemical entities, such as intracellular pH and [Ca^{++}], which undergo rapid physiologic changes in living cells. The distinction between structural and functional parameters is more arbitrary than the distinction between intrinsic and extrinsic parameters; obviously, structural parameters such as DNA content provide considerable information about cell function. However, the classification I have used seems agreeable to most people's sensibilities.

Depending upon the composition of a cellular sample and upon the experimenter's goals, the same parameter can sometimes be used both to differentiate one cell type from others in a mixed population and to provide information about the cell's biological state. For example, tumor cells with near haploid DNA content are readily distinguished from normal host cells or stroma in the same sample, and can also be classified as being in the G_0/G_1 or in the S phase of the cell cycle, based solely upon a single measurement of DNA content. Note, however, that, in this instance, tumor cell G_2 and M phase and stromal cell G_0/G_1 phase DNA contents will overlap; distinguishing these populations will therefore require multiparameter measurements.

I will turn now to intrinsic cellular parameters, after which I will introduce probes and extrinsic cellular parameters. The list of measurable cellular **parameters** that appeared as Table 1-1 on p. 3 is reproduced as Table 7-1 on the next page for convenience.

Cell Size

From the preceding sections of this chapter, I hope you have concluded that DC impedance (Coulter volume) measurement is a reasonably good physical parameter to use for measuring cell or particle size (volume, in this case), that forward light scattering is an unreasonably bad one, and that extinction might be useful. It would, however, help if we had a "gold standard" for at least some form of size measurement; there is one, but it is not cytometric, or at least not totally cytometric.

Mean Cell Volume: The Cellocrit as Gold Standard

The Coulter orifice itself does not provide an absolute measurement of cell volume; this is best obtained, at least in terms of a mean value for a cell population, by first determining the **cellocrit**. The cellocrit is analogous to the familiar **hematocrit** measurement of clinical hematology, which determines the fraction of total blood volume occupied by red cells. The cellocrit is obtained by centrifuging a suspension containing a known number of cells per unit volume in a tube or capillary with a uniform cross-section along its length, and measuring the heights of the column of packed cells and of the entire column of fluid (i.e., the packed cells plus the supernatant). The cellocrit, expressed as a decimal fraction, is the ratio of the first of these to the second. If the cellocrit is to be reported as a percentage, the value is 100 times this ratio.

For cellocrit measurements to be accurate, the cells must occupy a fairly high fraction of the fluid volume; red cells are present at very high concentrations (about 5,000,000/µL) in blood, and the normal range of hematocrits runs from about 35% to about 50%. The best way to make cellocrit measurements on cells other than red blood cells is to pellet the cells and make the measurement after the pellet is resuspended in a very small volume of fluid. For critical work, isotope-labeled macromolecules can be added to the suspension to allow correction of cellocrit values for the volume of suspending medium trapped in the packed cell column.

In order to derive a measure of mean cell volume from the cellocrit, an accurate count of cells per unit volume must be obtained from another aliquot of the sample used for cellocrit measurements. You should know by now that you won't get an accurate or precise cell count using a microscope and a hemacytometer; the cell count will have to be done by a cytometer of one sort or another. The mean cell volume is then computed by dividing the cellocrit (as a fraction) by the cell count. Measurements of cell volume made in this fashion provide a primary standard for calibration of apparatus that is used for routine optical and/or electronic measurement of cell size.

Cell Volume, Area, and Diameter

Different cell size measurements may be needed for different applications. For example, if one wants to determine the **concentration** of a dye in cells, a value for cell **volume** is needed, while in calculations of **antigen density**, **surface area** is the desired quantity. The relationships between these are nonlinear; radius or diameter varies as the cube root of volume, cross-sectional area and surface area as the square of radius or diameter or the two-thirds power of volume. Schwartz et al described a simple method of using log amplifiers to derive cell diameter and surface area from electronic volume measurements[152]; ratio circuits have also been used by others for analog computation of receptor or antigen density from volume and presumptive cross-sectional area measurements[311,657-8]. These days, we'd use digital computers.

Cell Sizing: Slit Scans and Pulse Widths

A somewhat more practical general method of cell size measurement, available in some commercial flow cytometers, relies on the principle of **slit-scanning** (see pp. 183-4). The durations or widths of pulses produced at the detectors by particles passing through the illuminating beam will always vary with the size of the particles; particles with larger diameters or cross-sections will produce wider pulses. However, differences in pulse width between particles of different sizes become most apparent when the illuminating beam is

PARAMETER	MEASUREMENT METHOD AND PROBE IF USED
Intrinsic Structural Parameters (no probe)	
Cell Size	Electronic (DC) impedance, extinction, small angle light scattering; image analysis
Cell shape	Pulse shape analysis (flow); image analysis
Cytoplasmic granularity	Large angle light scattering, Electronic (AC) impedance
Birefringence (e.g., of blood eosinophil granules)	Polarized light scattering, absorption
Hemoglobin, photosynthetic pigments, porphyrins	Absorption, fluorescence, multiangle light scattering
Intrinsic Functional Parameter (no probe)	
Redox state	Fluorescence (endogenous pyridine and flavin nucleotides)
Extrinsic Structural Parameters (probe required)	
DNA content	Fluorescence (propidium, DAPI, Hoechst dyes)
DNA base ratio	Fluorescence (A-T and G-C preference dyes, e.g., Hoechst33258 and chromomycin A_3)
Nucleic acid sequence	Fluorescence (labeled oligonucleotides)
Chromatin structure	Fluorescence (fluorochromes after DNA denaturation)
RNA content (single and double-stranded)	Fluorescence (acridine orange, pyronin Y)
Total protein	Fluorescence (covalent- or ionic-bonded acid dyes)
Basic protein	Fluorescence (acid dyes at high pH)
Surface/Intracellular antigens	Fluorescence; scattering (labeled antibodies)
Surface sugars (lectin binding sites)	Fluorescence (labeled lectins)
Lipids	Fluorescence (Nile red)
Extrinsic Functional Parameters (probe required)	
Surface/Intracellular receptors	Fluorescence (labeled ligands)
Surface charge	Fluorescence (labeled polyionic molecules)
Membrane integrity (not always a sign of "viability")	Fluorescence (propidium, fluorescein diacetate [FDA]); absorption or scattering (Trypan blue)
Membrane fusion/turnover	Fluorescence (labeled long chain fatty acid derivatives)
Membrane organization (phospholipids, etc.)	Fluorescence (annexin V, merocyanine 540)
Membrane fluidity or microviscosity	Fluorescence polarization (diphenylhexatriene)
Membrane permeability (dye/drug uptake/efflux)	Fluorescence (anthracyclines, rhodamine 123, cyanines)
Endocytosis	Fluorescence (labeled microbeads or bacteria)
Generation number	Fluorescence (lipophilic or covalent-bonded tracking dyes)
Cytoskeletal organization	Fluorescence (NBD-phallacidin)
Enzyme activity	Fluorescence; absorption (fluorogenic/chromogenic substrates)
Oxidative metabolism	Fluorescence (dichlorofluorescein)
Sulfhydryl groups/glutathione	Fluorescence (bimanes)
DNA synthesis	Fluorescence (anti-BrUdR antibodies, labeled nucleotides)
DNA degradation (as in apoptosis)	Fluorescence (labeled nucleotides)
"Structuredness of cytoplasmic matrix"	Fluorescence (fluorescein diacetate [FDA])
Cytoplasmic/mitochondrial membrane potential	Fluorescence (cyanines, rhodamine 123, oxonols)
"Membrane-bound" Ca^{++}	Fluorescence (chlortetracycline)
Cytoplasmic $[Ca^{++}]$	Fluorescence ratio (indo-1), fluorescence (fluo-3)
Intracellular pH	Fluorescence ratio (BCECF, SNARF-1)
Gene expression	Fluorescence (reporter proteins)

Table 7-1. Some cellular parameters measurable by cytometry.

focused to a spot that has at least one axis (that parallel with the flow) no larger than, and, ideally, smaller than, the diameter of the particles to be measured, producing a slit- scan (pp. 50-1).

The major drawback of typical slit-scanning systems lies in their use of a small focal spot size; the smaller the spot to which a laser beam is focused, the more rapidly the beam converges to and diverges from a focus (pp. 137-8). A beam focused to a height of 2 μm maintains that height over a very small distance; one would have to use a core diameter of 4 μm or less to minimize variations in pulse width due to beam divergence. If the beam height is 5 μm, it is possible to work with a more practical core diameter, 20 μm.

Small beam heights also affect the range of angles over which forward scatter can be measured. A classical forward scatter "size" measurement collects light from angles smaller than 2°, using a beam stop to block the axial illuminating beam from reaching the scatter detector. For this to work, the illuminating beam must be diverging at an angle well below 2°, which can generally be arranged when relatively large focal spots are used. When the spot becomes only a few μm high, the illuminating beam diverges at larger angles, and the angle over which the "forward" scatter signals are collected is correspondingly increased, thus increasing the influence on the scatter signal of internal and surface structure relative to cell size. In a slit-scanning flow cytometer in which observation is done in a cuvette, the forward scatter measurement can be done around a vertical blocker bar; since the width of the illuminating beam is much larger than its height, its divergence in the horizontal plane is smaller than its divergence in the vertical plane, and it is possible to make forward scatter measurements at smaller angles than would be possible with a stream-in-air system, in which a horizontal blocker bar must typically be used in forward scatter measurements to prevent light scattered from the stream itself from reaching the scatter detector. My inclination would be to get rid of this problem by getting rid of forward scatter measurements, but, while I will lay out my rationale over the next few paragraphs, I don't expect the manufacturers to jump on this particular bandwagon any time soon.

I have already mentioned that slit-scanning, i.e., using a small beam height, increases the effect of differences in particle diameter on pulse width. However, Hammond et al[1248] have pointed out that one can, at least up to a point, get reasonable size measurements of particles as small as 1 μm in diameter from a beam with a 20 μm height, simply by subtracting an offset corresponding to the signal component due to beam width. They applied this transformation to pulse width signals from endosomes, and displayed the resultant values on an expanded scale, allowing better visualization of differences between signals from these small particles.

Pulse width measurements have been shown to be useful for cell sizing by several groups of investigators[97,98,104,160,168-9,617-9,2561]; extinction and side scatter as well as forward scatter signals may be used. Accuracy and precision may be increased by using a dual-beam system to monitor flow velocities of individual particles and making corrections to raw pulse width values[619]. However, how pulse width is measured is critical. One can measure the width between two points at a constant threshold level on the rising and falling edges of the pulse, which is referred to as **constant-threshold pulse width**. However, this has the effect of decreasing the pulse width value for lower-amplitude pulses. Hammond et al[1248] used the **constant-fraction pulse width** measurement advocated by Leary et al[104], in which the width is measured between points on the leading and trailing edges of the pulse at signal levels equal to a constant fraction of the pulse peak height.

There is a substantial difference between constant-threshold and constant-fraction pulse width for Gaussian pulses of different heights. However, as the rise and fall times of a pulse become shorter, the difference between constant-threshold and constant-fraction measurements decreases. If one has square rather than Gaussian pulses, there should be no practical difference between the two methods of width measurement.

We ordinarily prefer not to get square pulses out of our cytometers, because we are typically interested in the information contained in the area and/or the peak height of pulses as well as in pulse width information. However, if we split off some of the signal, and put it through a logarithmic amplifier and/or a fairly high-gain linear amplifier before going to a pulse width measurement circuit, we end up with a signal that will give us acceptable pulse width measurements using a constant threshold.

For particles at least several μm in diameter, using an extinction signal for pulse sizing makes a lot of sense. I have already (p. 282) mentioned that extinction is not particularly useful for quantifying amounts of cell-associated material, meaning that we won't usually need the peak height or area of an extinction pulse. A cheap diode laser can be made quiet enough to do precise extinction measurements, and we only need a photodiode detector. The drawback is that the background noise level catches up to us pretty quickly when we try to look at submicron particles. We can't readily limit the view of the optics, because we inevitably have a limited N.A. to work with on the axis of the illuminating beam, where we have to make our extinction measurements, due to the presence of a high-N.A. lens in the fluorescence collection system, and we won't get good enough spatial resolution from a low N.A. lens to look at a really small area of the core.

My proposed solution to the problems of on-axis and near on-axis measurements is to do away with them. The orthogonal collection optics in a flow cytometer are typically high-N.A. optics, meaning they have high spatial resolution. If, instead of focusing the beam to a 5 μm height, one uses a more tractable 10 or 20 μm beam height, and places a slit over the side scatter detector, it is possible to collect a side scatter signal from a region of the beam only a few μm high. If, for example, the orthogonal collection optics have 20 ×

magnification, the image of the intersection of a 20 µm beam with the core stream is 400 µm high, and a 40 µm high slit, readily obtainable from optics supply houses, will restrict the region of side scatter detection to 2 µm. I have already noted that it is impractical to use a 2 µm beam height because of problems with rapid beam divergence; the approach of slit-scanning using a real slit eliminates those problems. And, as usual, there's nothing much new in cytometry; Leon Wheeless told me he made width measurements using slits in the 1970's.

If orthogonal scatter is needed as a trigger signal, it is simple enough to use a beamsplitter to divert the full beam height side scatter signal to one PMT and the slit-height side scatter signal for pulse width sizing to another. Log and/or high-gain amplification produce the near-square pulses we need for the quick-and-dirty width measurement. And, best of all, we have a much better signal-to-noise ratio in side scatter measurement than we do in either forward scatter or extinction measurement.

Size Measurements in the Submicron Range

Scatter measurements for sizing, without benefit of pulse width, are actually more respectable when we're dealing with small stuff like bacteria, endosomes, and viruses, than when we're dealing with eukaryotic cells. Among other things, many bacteria are either spherical or ellipsoidal, or close enough so that they behave according to Mie theory. Maltsev's group will at least concede that they can find angles at which bacteria could be properly characterized by scattering in a standard flow cytometer[2540]. Koch, Robertson, and Button[2562-3] have developed a method for estimating bacterial cell volume and mass from forward scatter signals, using Rayleigh and Rayleigh-Gans theory (applicable to smaller particles than Mie theory) as well as Mie theory.

People who work routinely with bacteria and smaller particles tend to use smaller focal spot sizes than people who work with eukaryotes, although the beam divergence problem previously mentioned limits how small one can practically make the focal spot. Decreasing spot size concentrates the energy in the input laser beam in a smaller volume, increasing signal amplitude from particles in the sample. Noise due to light scattering by particulate contaminants in the sample and sheath fluids may not change appreciably, since, although a smaller volume is illuminated, the intensity of illumination increases. Thus, a substantial increase in signal-to-noise can only be achieved if the amount of stray light collected is decreased.

When cells are measured in a stream in air, decreasing spot diameter to less than half stream diameter greatly reduces stray scatter from the stream; changing to a larger stream or capillary further reduces stray scatter, and substituting a square cuvette makes an additional improvement. The increased beam intensity results in more photons being scattered from particles, which decreases variance of measurements due to photon statistics; however, unless a Gaussian beam is at least 5 times the core stream width, the intensity profile will itself will keep the theoretical minimum CV of measurements above 5 percent[685]. When you start with a square cuvette, making the beam smaller helps, up to a point[686] but, in practice, the decreased spot size starts to catch up with you as the increasing beam divergence causes light to spill around the beam stop. Small marine bacteria[686], small beads[687], and large viruses (F. Elliott and H. Shapiro, unpublished) have been detected using forward scatter in laser source systems using flat-sided cuvettes for observation; the practical particle size limit for detection seems to be about 150 nm.

In the submicron size range, scattering from particles is predominantly Rayleigh scattering, varying in intensity, all other things being equal, as the inverse sixth power of particle diameter[94]. While such scattering remains more intense in the forward direction than at 90°, the relative influences of size, internal and surface structure, and particle asymmetry and orientation no longer vary markedly with scattering angle. It may thus be more practical to use a 90° or large angle scattering measurement than a forward scatter measurement for detection and sizing of submicron particles[687], because, when observation is done in a square cuvette or a water-immersed system[94], and field stops are appropriately used in the detection optics, there is much less background noise from large angle scatter measurements than from forward scatter measurements.

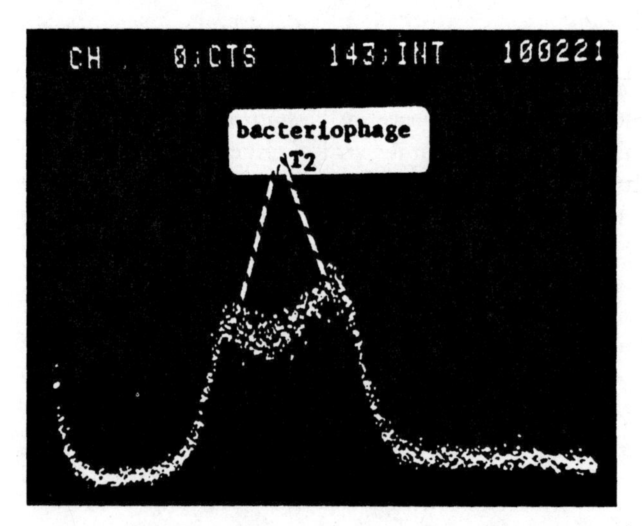

Figure 7-5. Side scatter signals from T2 bacteriophages.

Figure 7-5 shows a distribution of side scatter signals observed from T2 bacteriophages using the water-immersed flow system built at Block[94]; the bimodal distribution, which is well separated from background noise, may reflect the presence of virions with and without DNA. The sample was illuminated by a 100 mW, 488 nm argon laser beam. While the distribution shown is wide, extrapolating backward from the 6th power (Rayleigh) dependence of scattering intensity on particle diameter indicates that the CV of the corresponding diameter measurement should be no more than a few percent. The phages could still be resolved well above background using the same optics with an air-cooled argon laser source emitting 10 mW at 488 nm; since the limiting noise source in such scatter measurements is stray light, this is not surprising. Steen and Lindmo measured large angle scatter signals from large viruses in their arc source system[688,1246], clearly establishing that, in scattering measurements, problems are generally not solved by throwing more photons at them; Steen (personal communication) has also recently measured scatter signals from viruses in prototype instruments built by **Apogee** (see Chapter 8).

Other Size Measurement Techniques

Terstappen et al[1247] modified the angle of forward scatter collection and used a nonlinear transformation of the side scatter signal to improve resolution of peripheral blood leukocytes. In principle, improving signal-to-noise could further increase the sensitivity of scatter measurements.

Zarrin, Bornhop and Dovichi described an apparatus[689] in which **laser Doppler velocimetry** was used with a square cuvette-based flow cytometer to measure particles as small as 90 nm. In laser Doppler velocimetry, the scatter signal from moving particles is modulated, while the background noise is not; the signal is detected at the modulation frequency, increasing signal-to-noise.

A less radical sizing method was suggested by Gray, Hoffman, and Hansen[167]; samples were prepared in a solution containing a fluorescent compound of high molecular weight, which did not get into the cells. The core diameter was kept constant, resulting in a decrease in the baseline fluorescence signal when a cell passed the observation point. One disadvantage of this approach is that it reduces the number of fluorescence measurements of extrinsic parameters that could be made concomitantly with the cell size measurement. Another is that, when the core size is small, redistribution of dye in the core around cells produces increases in fluorescence over baseline. This "dye exclusion artifact" has recently been further investigated by Steen and Stokke[2564], who note its possible adverse effects on fluorescence measurements under some conditions.

In concluding the discussion of cell size measurements, I should mention an interesting paper by Bator et al[170], who used high speed photography of cells passing through an electronic volume measurement orifice to correlate differences in cell shape with differences in measured volume.

Cell Shape and Doublet Discrimination

The most elegant flow cytometric apparatus for determining cell shapes was, without doubt, the multidimensional high-resolution slit-scanning system developed for cancer cytology studies by Leon Wheeless and his colleagues at the University of Rochester[4,171-4,690], beginning in the 1970's. In this instrument, profiles of cells were obtained in the direction of flow and in two orthogonal directions, using the fluorescence of acridine orange excited by a sharply focused 488 nm laser beam. Rapid digitization allowed several dozen points on each pulse profile to be measured and captured in a dedicated minicomputer system. As I mentioned on p. 212, Wheeless' slit-scanning work represented the first application of digital pulse processing in flow cytometry; while the low resolution of available converters limited the dynamic range of measurements, the large number of points measured per pulse allowed fairly detailed morphologic information to be captured.

Slightly less fancy scanning flow cytometers were built and used for analysis of chromosomes[691-4] and phytoplankton[695], and for two-dimensional imaging of cells in flow[696,1249]. The analytical procedures used in high-resolution slit-scanning and imaging flow cytometers require optical and electronic hardware and software rather different from, and more complex than, that which is used in commercial and most laboratory-built flow cytometers. Progress in electronics has gradually made it easier and more affordable to do multiple point pulse profile processing on both laboratory-built and modified commercial instruments [697,2311,2565], and to incorporate imaging capability into flow systems. Commercial instruments which offer one or both of these features include **Fluid Imaging Technologies'** FlowCAM[2453] (p. 168; also see Chapter 8) and **CytoBuoy b.v.**'s CytoBuoy[2566](see Chapter 8), both designed for aquatic applications, the **Union Biometrica** COPAS systems for large particle sorting (p. 265; also see Chapter 8), and clinical urine analyzers from **International Remote Imaging Systems and Sysmex (Toa)**[2567] (Chapter 8). **Amnis** is developing a flow cytometer with hyperspectral imaging capability[2568]; see Chapter 11 for contact information.

David Galbraith and his colleagues, at the University of Arizona[2473-4] used high-speed digitization to collect multiple point pulse profiles of cells, and found that the skewness and kurtosis (these are higher moments of statistical distributions; see p. 234) of pulses provided useful information for discriminating among cell types.

More information about asymmetric cells than is available in conventional flow cytometers can be derived from making measurements of the same parameter at two different angles; this technique is used in sperm sorting, where the fluorescence of Hoechst 33342 is measured in both forward and orthogonal directions to improve resolution of X- and Y-chromosome bearing cells[651,2322,2569]. Sperm sorting will be discussed further in Chapter 10.

Zucker, Perreault, and Elstein[1250] used a combination of forward scatter and extinction signals to identify sperm of different refractive indices and orientations in an unmodified Ortho flow cytometer; the instrument is no longer available, and current research flow cytometers do not include extinction measurement capability.

However, some information about cell shape and orientation is available even to users of conventional flow cytometers, which have low-resolution slit-scan capability by virtue of using relatively small beam heights. Relationships between extinction pulse widths, peak heights, and integrals, tend to be different for symmetric and asymmetric cells, and for single particles and doublets or multiplets. Some shape information can also be obtained from the relationship between forward and large angle (orthogonal or 90°) light scatter signals; cells of different shapes will generally cluster in different regions of a two-dimensional distribution of these parameters, although the locations of these clusters may not always be predictable *a priori.*

Wolfgang Göhde, Sr., was one of the first to consider the problem of doublet discrimination. Flow cytometer manufacturers acknowledged his contribution by licensing his patent when they began to promote the use of a combination of pulse height and pulse area as a means of eliminating most doublets from DNA content analyses.

The same amount of fluorescence (i.e., the same value of the pulse integral) will be recorded when a single G_2 or M phase cell passes through the measurement system or when two G_0 or G_1 phase cells pass through in close proximity. However, the height of the fluorescence pulse will, in general, be lower for the doublets than for the singlets; while, in principle, two G_0 or G_1 cells passing through "side by side" should generate a pulse approximately as high as that produced by a single G_2 or M cell, this orientation of doublets is rare because the hydrodynamics of flow favor them being oriented along the axis of flow.

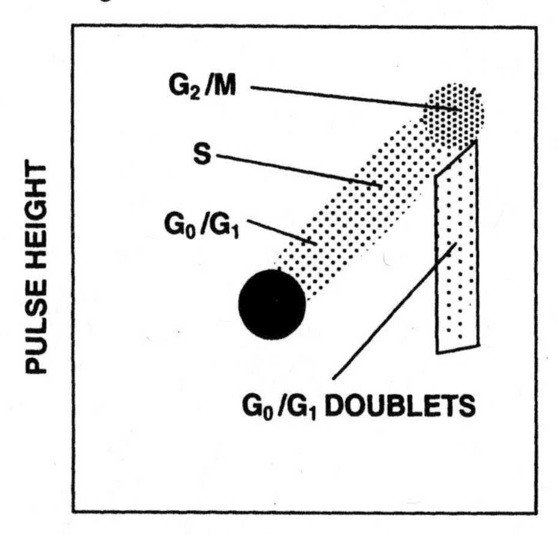

Figure 7-6. The principle of doublet discrimination using pulse height and pulse integral measurements.

Figure 7-6 illustrates the principle of doublet discrimination based on pulse height and integral measurements; the quadrilateral gate shown on the figure would remove most, but not all, G_0/G_1 doublets. Modern DNA histogram analyses procedures employ somewhat more sophisticated methods for doublet and multiplet discrimination, but the pulse height-vs.-pulse integral method is still better than nothing. It might be used more often for immunofluorescence analysis, in which unrecognized doublets may push experimenters to invalid conclusions.

If you go back to high school geometry, and recall that the areas (integrals) of plane figures such as rectangles and triangles are proportional to the product of base (width) times height, you will appreciate that dividing a pulse integral by a pulse height should give a quantity representative of pulse width, meaning that it is possible to obtain pulse width information even in instruments that do not have pulse width measurement circuitry or capability, as long as the beam is small enough so that pulse height and pulse integral do not contain identical information (see pp. 183-4).

Pulse width itself is useful for doublet discrimination[97]; in the context of DNA analysis, G_0/G_1 doublets will have larger pulse widths than single G_2 or M cells. Wersto et al[2466] provide a good recent reference on the topic. The **Luminex 100** instrument (see Chapter 8), designed for doing multiplexed ligand binding assays on color coded plastic beads, uses a pulse width measurement obtained by low-resolution digital pulse processing to eliminate doublets from analyses.

Measurement of Intrinsic Parameters Using Absorption or Extinction Signals

Microspectrophotometric absorption measurements of cells on slides can be used for quantitative analysis of substances with weak absorption, e.g., pararosaniline, used in the Feulgen staining procedure, as well as for measurements of substances with high extinction coefficients, e.g. nucleic acids (measured at 260 nm) and hemoglobin (measured in the Soret band around 415 nm). When static specimens are analyzed on slides, it is possible to select mounting and immersion media with refractive indices matched to those of the specimens, thus minimizing artifacts due to light scattering. The spatial resolution and the numerical apertures of illumination and collection optics are typically higher in microspectrophotometers than in flow cytometers, which makes it possible to do better absorption measurements in the former[34,52].

There has been some application of flow cytometers to measurement of nucleic acids and hemoglobin; Kamentsky's original instrument[65,66,77] and the Cytoscreener[76] were both equipped for measurement of nucleic acid absorption at 260 nm, while the Block arc source instrument[88] measured nucleic acid absorption at 260 nm and hemoglobin absorption at 420 nm. The nucleic acid measurements did not appear to be as precise as those obtained by microspectrophotometry; the hemoglobin measurements were adequate for discrimination of erythroid cells from leukocytes and produced estimates of hemoglobin content in reasonable agreement with those obtained by conventional hemoglobinometry.

Stewart, Stewart and Habbersett[1258], using the multiparameter sorter at Los Alamos, demonstrated that dead cells, i.e., cells with membranes permeable to propidium iodide, could be differentiated from other cells in leukocyte suspensions on the basis of lower extinction signals.

Fluorescence Measurements of Intrinsic Parameters

Autofluorescence: Pyridine and Flavin Nucleotides

The **autofluorescence** of most mammalian cells appears to be due primarily to the presence of **pyridine (NAD, NADP) and flavin (FMN, FAD) nucleotides** which, respectively, impart UV-excited blue and blue-excited green fluorescence to cells[181,183]. Fluorescence spectra of these materials are shown in Figure 7-7.

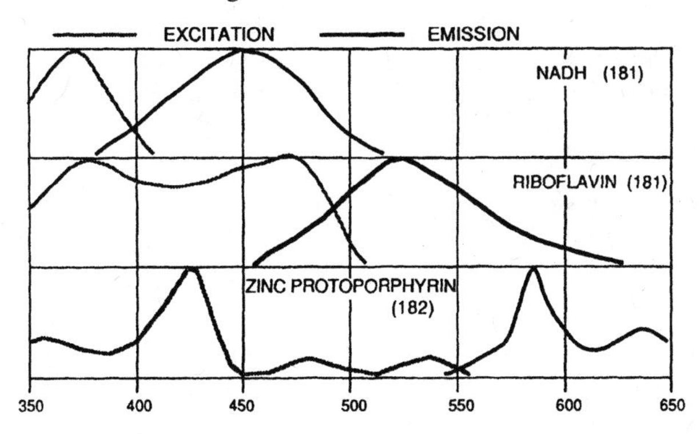

Figure 7-7. Fluorescence spectra of some materials implicated in mammalian cell autofluorescence.

Many investigators have been concerned with autofluorescence solely as a source of interference with measurement of weak fluorescence signals, e.g. fluorescein immunofluorescence. The fluorescence emission from unstained murine lymphocytes, excited at 488 nm and measured through the long pass filter combination used for single parameter measurements of fluorescein immunofluorescence, was originally reported to have an average intensity equal to that which would be measured from 10,000 molecules of antibody-bound fluorescein (reference 184; D. Parks, personal communication). The 10,000 molecule figure was obtained in the days when the FACS measured fluorescence between 520 and about 650 nm, using a long pass color glass filter, and was probably affected by filter fluorescence, Raman scatter, and other factors. Today's instruments use narrower band filters for green fluorescence emission, eliminating the Raman scatter contribution; improved filter design has essentially eliminated filter fluorescence.

Surveys conducted in 1989-90 by Abe Schwartz and his colleagues[1161] found a mean value of 657 fluorescein MESF for lymphocyte autofluorescence; Hoffman and Kuhlmann[2482] reported somewhat lower values, but, even at the level of a few hundred MESF, it is autofluorescence that limits the sensitivity of immunofluorescence measurements using fluorescein-labeled antibodies. Autofluorescence of lymphocytes in the region normally used for phycoerythrin measurement is only 50 MESF.

On the plus side, the high autofluorescence of blood neutrophils[161] and the higher autofluorescence of eosinophils[162] have, in conjunction with forward and side scatter measurements, been used for identification and sorting of these cells. The spectrum of eosinophil autofluorescence is nearly identical to that of riboflavin; Mayeno, Hamann, and Gleich[1259] have shown by chemical analysis of extracts that granule-associated flavin adenine dinucleotide (FAD) accounts for about 85% of eosinophil autofluorescence. In my experience, eosinophil autofluorescence is equivalent to a few thousand fluorescein MESF. However, I have also noted yellow autofluorescence (in the 575 nm spectral region) from eosinophils when using green (532 nm) excitation; I would not expect this to come from flavins (see Figure 7-7). Havenith et al[1260] have sorted alveolar macrophages from dendritic cells based on the high autofluorescence of the former; Njoroge et al[2570] have isolated autofluorescent macrophages from blood cell cultures.

Pyridine and Flavin Nucleotides and Redox State

Pyridine and flavin nucleotide fluorescence vary with the **oxidation-reduction** or **redox state** of cells, and NADH (and NADPH) fluorescence measurements have been used to monitor redox states of cells, tissues, and organs since the technique was described by Britton Chance and Bo Thorell in the late 1950's[185-90]. In 1979, Thorell described flow cytometric measurements of the redox state of liver cells based on NADH fluorescence[191]; by the time of his death in 1982, he had added the capacity to measure endogenous fluorochromes and flavin nucleotides to his apparatus as well[192,193]. Hafeman et al[194] used flow cytometry of NADH fluorescence in blood neutrophils to demonstrate that the respiratory burst induced in these cells by chemotactic stimuli and phorbol esters is an all-or-none event; Van De Winkel and Pipeleers[1261] sorted insulin-containing pancreatic cells based on changes in redox state following exposure to glucose.

Pyridine and Flavin Nucleotides and Cancer

Several groups[1262-5,2571-2] have described characteristic changes of the ratio of pyridine to flavin nucleotide fluorescence in bulk in tissue specimens from tumors and precancerous states and in tumors *in vivo*; this work has not provided a new diagnostic application for flow cytometry but has, instead, led to the development of devices for detection of tissue ischemia[2573] and for endoscopic cancer diagnosis[2574]. **Xillix**, a company in Richmond, British Columbia, Canada <http://www.xillix.com> sells instruments for autofluorescence bronchoscopy and other endoscopic applications.

Bacterial Autofluorescence Measurements

Jim Ho, of the Canadian Defence Research Establishment, has developed instruments that measure UV-excited bacterial autofluorescence and particle size in aerosols; these parameters may help distinguish among species. Like Gucker's original aerosol particle counter (p. 74), Ho's FLAPS (Fluorescence Aerodynamic Particle Sizer)[2575-6] is designed to detect biowarfare agents. A second-generation instrument (J Ho, personal communication; N. Dovichi, personal communication) resulted from collaboration between Jim and Norm Dovichi, then at the University of Alberta and now at the University of Washington. Norm passed through Los Alamos and played with chemical and nonchemical applications of flow cytometry[685,687,689,886-7] before he put his experience to use in developing the capillary electrophoresis-based gene sequencer[2576-7] and hit the big time. The improved bacterial aerosol analyzer measures fluorescence emission in 16 spectral regions and sizes bacteria by laser Doppler velocimetry[685]; particles are detected and sized in a red laser beam, and fluorescence is excited by a pulsed tripled YAG laser downstream. Mixed populations are resolved with the aid of principal component analysis (p. 252).

Porphyrin Fluorescence in Erythroid Cells

In the early development of erythroid cells in the bone marrow, **porphyrins** may accumulate in cells in amounts that outstrip the cells' capacity to synthesize **heme** from porphyrin and iron. Heme is nonfluorescent due to the quenching of the intrinsic fluorescence of porphyrin by iron, but both the porphyrin precursors of heme and the **zinc protoporphyrin** formed when iron incorporation is impaired, as occurs in **iron deficiency** and **lead intoxication**, fluoresce[182]. The Stokes shifts for porphyrin fluorescence are large; excitation maxima are in the violet (Soret band) and emission maxima in the orange or red (see Figure 7-7). The fluorescence of red cells in bulk has been used for diagnostic purposes and that of red cell precursors in marrow is detectable in a flow cytometer using excitation from the violet (406-422 nm) lines from a krypton laser (C. Stewart, personal communication). Zinc protoporphyrin has been measured in single cells using image cytometry[1266-7]; I have been able to measure its fluorescence in a flow cytometer using 100 mW excitation at 441 nm from a He-Cd laser (Figure 7-8), but only by taking great pains to reduce stray light and fluorescence from optical components. Using flow or image cytometry to follow the disappearance of fluorescent red cells from the circulation may be helpful in monitoring treatment of lead poisoning, which is a common clinical condition. Violet laser diodes could provide useful, and, ultimately, inexpensive light sources for clinical instruments intended for such measurements.

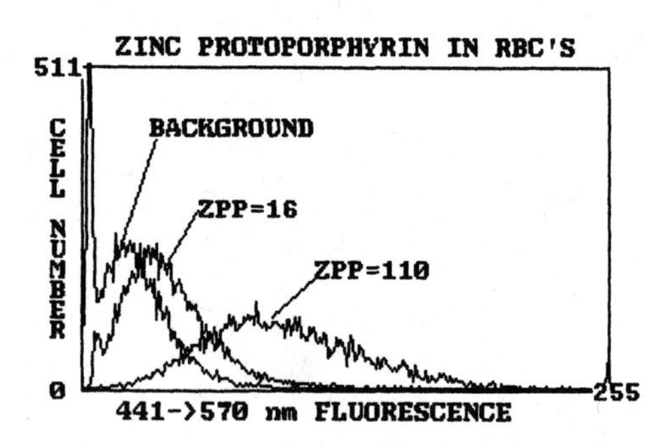

Figure 7-8. Zinc protoporphyrin fluorescence in human red cells; orange fluorescence (>570 nm) was excited with a 100 mW He-Cd laser at 441 nm.

Red cell porphyrin fluorescence is also encountered in **erythropoietic protoporphyria**, a relatively rare hereditary disorder, in which the bones and teeth are colored red by excess porphyrin deposition, and in which the effects of porphyrins on the nervous system lead to bizarre behavior. King George III of England is believed to have been a victim; other sufferers were persecuted as werewolves. Brun, Steen, and Sandberg[1268] have made flow cytometric measurements of porphyrin in patients with this disease, presumably not when the moon was full.

A potential clinical application of porphyrin metabolism in lymphocytes, rather than erythrocytes, is suggested by the fact that activated T cells accumulate substantial amounts of intracellular protoporphyrin IX in the presence of exogenous delta-aminolevulinic acid (ALA)[2578-9]. This may allow pathological proliferation of T cells to be at least partially arrested by photodynamic therapy.

Other Pigments

Autofluorescent pigments such as the UV-excited, blue fluorescent **lipofuscins** accumulate within mammalian cells as a function of age. Jongkind et al[195,196] used autofluorescence to sort cells of presumptively different ages for subsequent biochemical analysis. Hunt et al[1269-70] examined the accumulation of **ceroid**, a fluorescent pigment produced by oxidation of lipid/protein complexes, and found in the lipid-laden macrophages, or foam cells, which infiltrate atherosclerotic lesions. They found that a 424 nm bandpass filter was more selective than a 490 nm long pass filter for detection of the UV-excited fluorescence of ceroid in macrophages. Puppels et al[1271] measured **carotenoids** in single cells, but by Raman microspectroscopy rather than fluorescence flow cytometry.

The Jovins, in Göttingen, made flow cytometric measurements of the **fluorescence of aromatic amino acids** in proteins in differentiating cells, using a frequency-doubled argon laser operating in the UV at 257 nm for excitation[199].

Chlorophyll and Phycobiliproteins

Plant photosynthetic pigments, including **chlorophyll**, which contains a magnesium-porphyrin complex, and the **phycobiliproteins** present in algae, have a characteristic autofluorescence that has been studied *in vivo* by Trask et al[197] and Olson et al[198], among others. Chlorophyll itself has absorption regions in the ultraviolet and violet (below about 450 nm) and the far red (about 680 nm), and emits in the near infrared. Since most of the light penetrating past the surface layers of the ocean is in the blue and green regions of the spectrum, most algae contain relatively large amounts of **phycoerythrin**, **phycocyanin**, and **allophycocyanin**, which absorb blue-green, yellow-orange, and red light, respectively, and which, in concert, allow nonradiative transfer of incident light energy to chlorophyll. The phycobiliproteins have, to say the least, been exploited as fluorescent labeling reagents[114] and will be discussed further in this context.

Since different algal species contain different amounts of the various pigments, and since the spectra of the same types of phycobiliproteins (e.g., phycoerythrin) from different species may also differ, it is possible to use multistation flow cytometric measurements of fluorescence excited at different wavelengths to classify and count phytoplankton populations[711]. This has been pursued as an approach to environmental monitoring.

Flow cytometry and flow cytometric measurements of intrinsic pigment fluorescence in phytoplankton and marine bacteria have become widely used in oceanography[1051-2]; among other things, this led to the discovery of extremely abundant, but previously overlooked, chlorophyll-bearing marine bacteria[1272].

Xu, Auger, and Govindjee[1273] used a flow cytometer to measure chlorophyll fluorescence in isolated spinach thylakoids. The cytometer does measure the variable (light-dependent) component of fluorescence, although the high excitation intensity may perturb measurements under some circumstances.

Infrared Spectra and Cancer Diagnosis

Benedetti et al[1277-82] and Wong and Rigas and their collaborators[1283-8], using **Fourier transform infrared (FT-IR) spectroscopy**, described differences between normal cells and cells from leukemias and several solid tumor types. Differences were most consistently found in the nucleic acid phosphate stretch bands at 1080 cm^{-1} and in a protein band at 1540 cm^{-1}. Most of these studies involved bulk measurements, although leukemic cells have been examined by FT-IR microscopy[1281-2].

This work, like the work on pyridine and flavin nucleotide fluorescence mentioned previously, raises an important question. Since, in many cases, e.g., analyses of cervical specimens, actual malignant cells represent a very small fraction of the cells present in an abnormal specimen, the measured differences between specimens are more likely to reflect some alteration or premalignant state in a large fraction of the cells in a specimen than to represent signals from the malignant cells themselves. It is not clear that flow cytometry can contribute to the resolution of this issue.

7.3 PROBES, LABELS, AND [NOT] PROTOCOLS FOR EXTRINSIC PARAMETER MEASUREMENTS

As I said way back in Chapter 1, most of the measurements of extrinsic parameters done in most flow cytometers use fluorescent reagents. The discussion that follows will deal primarily with fluorescent staining and fluorescence measurements, covering other staining techniques when they are applicable in conventional flow cytometers or should be considered as part of an alternative approach to flow cytometry.

I have to introduce some ground rules here, for a number of reasons. There are a lot of probes now; some of them are dyes, and some of them are natural products, and big ones at that, e.g., the fluorescent proteins and the phycobiliproteins. I know a lot about some probes and not much about others, so I will only cover probes with which I have been fairly intimately involved in much detail.

If you need more details, there are two excellent sources. For the classicists, there is the new 10th Edition of *Conn's Biological Stains*, edited by Richard Horobin and John Kiernan[2580]. *Conn's* went 25 years (1977-2002) between the 9th and 10th Editions, so I don't feel too bad about the 8-year (1995-2003) gestation period for this 4th Edition of *Practical Flow Cytometry*.

But I digress. If you want the last word on fluorescent probes, their structure, their spectra, their uses, thousands of references, and (on the web), price and delivery information, you want to look at the 9th Edition of Molecular Probes' *Handbook of Fluorescent Probes and Research Products*[2332], edited by Dick Haugland. There's not much point in duplicating what's in there, especially since you can get the book for nothing (at least if you hurry), or pull up a regularly updated version online at <http://www.probes.com>.

As to protocols, *Practical Flow Cytometry* has always been more of a "why-to" book than a "how-to" book in terms of its discussions of which probes and parameters may be most useful for which applications. I cover how the probes work, just as I cover how the instruments work, but, in the majority of cases, I can't and don't provide the protocols, i.e., detailed information on the exact concentrations of reagents, compositions of buffers, incubation times and temperatures, etc., that you need to make the probes work in your lab with your instrument and your samples.

Although you can't just run out to The Heme Depot and get such information, there are sources. I mentioned several on pp. 62-5. I would turn first to *Current Protocols in Cytometry*[2389], which is available in looseleaf and/or CD-ROM versions, both frequently updated. When a significant methodological advance in flow or image cytometry is reported in the literature, it's a good bet that the editors of *Current Protocols* will hound one or more of the authors into producing a writeup of the procedure including all of the technical details that were omitted from the "Materials and Methods" section of the journal article. The various volumes of *Methods in Cell Biology* devoted to cytometry[1034,2385-8] may also be helpful.

Most people who publish good new methodology are anxious to proselytize; you can usually get experimental details from them via snail mail or e-mail, or from their lab or personal web sites. The Purdue Cytometry Mailing list is also a good place to go for help; you'll get more and better answers from the experts if your inquiries about procedures indicate that you've done your homework.

At this point, having told you what I don't do, I'll get back to what I do do before I get into deep...never mind.

Probes, Labels, and Dyes

I introduced a distinction between probes and labels on pp. 42-3, and I will now say a bit more about that subject. We use reagents to measure extrinsic cellular parameters because there aren't ways, or because we haven't found the ways, to estimate or measure all of the attributes of cells in which we might be interested.

There is a hierarchy of specificity, or of selectivity, of reagents, and we can appreciate it by considering the history of staining, some of which is covered in *Conn's*[2580], and some of which is in older books[24,32-3,1085]. If we look at stains classically used for blood cells, which date back to Paul Ehrlich's work in the 1880's[33], we find they are mixtures of an acid dye (typically eosin), which binds to basic elements within cells, meaning mostly proteins, and a basic dye(s) (typically one or more of the azure dyes, all thiazines), which bind(s) to acidic elements, notably nucleic acids, but also sulfonated glycosaminoglycans. We don't expect eosin to bind to acidic elements or azure dyes to bind to basic elements, but that's as specific as those dyes get. However, we can legitimately call those dyes **probes**, even if they are probes of some relatively nonspecific characteristics or constituents of cells.

At the other end of the hierarchy, we have antibodies and gene probes, large, or relatively large, molecules themselves, which, respectively, are exquisitely specific for macromolecular structure and sequence, at least when used under appropriate conditions. They, too, are probes, but, in order to detect them using an optical flow cytometer, we have to attach a **label**. Until the 1980's, almost all the labels we had were relatively low molecular weight dyes; since then, it has become increasingly likely that the label used for an antibody will be a macromolecule, specifically, a phycobiliprotein, which may also have small dye molecules covalently attached to it to modify its spectral characteristics.

Some of the dye probes we use are sensitive to changes in their chemical environment, and would be useless if they weren't. Measurements of intracellular pH are usually done using dyes that change their spectral characteristics as pH changes. That's an essential characteristic for a probe, but an undesirable one for a label, which we would like to be able to attach to a specific probe molecule without altering either the spectral characteristics of the label or the specificity of the probe, and which we would like to behave (i.e., fluoresce) pretty much the same way regardless of the environment in which it finds itself. Well, that's not the way the world works; just to cite one example, fluorescein, one of the all-time favorite fluorescent labels, is environmentally sensitive; its fluorescence increases with pH, enough so that derivatives of fluorescein are routinely used as pH probes.

I will start getting down to the specifics by discussing dyes and staining mechanisms in general, and the preparative techniques, such as **fixation** and **permeabilization**, which may be needed to get dyes into cells, and then cover the reagents used to measure the parameters in Table 7-1, starting with nucleic acid dyes and with labels.

Dyes and Quality Control: Gorillas in the NIST

In previous editions of this book, I said that if there were a **Shapiro's Third Law of Flow Cytometry**, it would probably be this:

Shapiro's Third Law of Flow Cytometry:
What's in the Bottle
Isn't Necessarily
What's on the Label!

This was particularly apt to be true for older bottles containing less common dyes; the Biological Stain Commission, which was set up to provide some degree of quality control over stains used routinely in research and clinical laboratories, has never gotten very far into the business of certifying fluorescent dyes, and even certified lots of dye may contain impurities. The problem of dye purity is hardly a new one in histochemistry and cytochemistry; Scott's piece on "Lies, damned lies – and biological stains"[721] and related correspondence[722] and Horobin's review[723] provide some historical background.

Spectroscopy does not provide much assistance when you're trying to find out whether your reagents are pure. You can only compare spectra of your lot with published spectra when you're working with relatively well-known stains; when what you've got definitely isn't the stain you wanted, the spectrum is likely to tell you, but, when what you've got is 45% what you wanted and 55% other junk, or when you're working up new compounds, spectra won't necessarily help.

A widely applicable, and perhaps the most useful, method for determining dye purity is **thin layer chromatography**[724], which can quickly tell you when you've got a mixture of compounds in the bottle. Most of the companies that specialize in dyes for analytical cytology will give you some idea of the purity of their products if you ask. If you're looking for something that isn't in their catalogs, try to find it in laser grade; laser grade dyes are apt to be reasonably clean. I know a lot of people who are happy that they didn't publish results that were obtainable only from a single bottle of dye, and a few who are unhappy that they did. I include myself in the former category.

When measuring fluorescence spectra of dyes, it is a good idea to make sure that the photodetector in your instrument will respond over the spectral range of emissions you intend to measure. Ideally, the fluorescence spectrum should be corrected for variations with wavelength in source emission intensity and detector response. Most spectrofluorometers use photomultiplier tubes with minimal response above 600 nm and negligible response above 650 nm; no amount of electronic "correction" can produce an accurate spectrum at longer wavelengths from such detectors. The Hamamatsu R928 and R1477 will respond to wavelengths as long as 800 nm; beyond that, you need an R3896 or a gallium arsenide tube s uch as an R636. It doesn't matter if the source and excitation monochromator in the instrument will provide excitation and select emission out to 900 nm if the tube won't respond past 650. When in doubt, look at the PMT.

In recent years, the manufacturers and distributors of dyes used in flow cytometry seem to have taken it upon themselves to supply high-quality products. You still need to be careful if you play the kinds of games I play, i.e., buying laser dyes, drugs, and other interesting chemical structures and trying to find out whether they are of any use in flow cytometry.

Since the last edition of this book emerged, several organizations, including the Centers for Disease Control, Food and Drug Administration, Health Canada, National Institutes of Health, and NCCLS have pursued the standardization of fluorescent reagents. The 800-pound gorilla of standardization, at least in the U.S., is the National Institute of Standards and Technology (NIST) (formerly the National Bureau of Standards), which has been recruited into producing some standardized fluorescent reference materials, starting with a certified fluorescein solution and expected to progress to fluorescein-labeled beads and to solutions and beads containing other fluorescent materials. Several publications have already emerged from NIST's collaborative efforts[2581-4].

The Dyes are Cast: An Overview

The spectral characteristics of a representative sample of fluorescent dyes, probes, and labels, and the excitation wavelengths available from various light sources, are shown in Figure 7-9, on the next page. Although there are thousands of compounds in Molecular Probes' catalog (the indispensable *Handbook*[2332]), and there are at least a few labels that are available conjugated to hundreds of different monoclonal antibodies, I'd guess that at least three-quarters of the samples run through flow cytometers contain at least one of the dyes/probes/labels shown in the figure.

A few selected parameters and applicable fluorescent probes are found in Table 7-2, on the page opposite the spectral chart. I've narrowed down the range of light sources in the chart to include four excitation wavelength ranges likely to be found in fluorescence flow cytometers, i.e., blue-green (488 nm), red (633-640 nm), UV (325-365 nm), and violet (395-415 nm). Probe emission maxima are indicated next to probe names.

Consideration of Figure 7-9 and Table 7-2 makes apparent the rationale for the use of multiple, spatially separated fluorescence excitation beams in multiparameter flow cytometry. Even when one can choose from a number of probes to select those with desired spectral characteristics, the use of several separated beams generally facilitates resolution of fluorescence signals from multiple probes. When a choice of probes is not available, multistation flow cytometry may provide the only means of making correlated measurements of two or more parameters of interest.

I've been beating this drum for over twenty years now; as I write this, I am looking at an article[8] called "Multistation multiparameter flow cytometry: A critical review and rationale," which I published in *Cytometry* (3:227-43) in January 1983. That article, which was adapted from a grant application (not funded the first time around), turned out to be the foundation for *Practical Flow Cytometry* and for this chapter. It included spectra of six nucleic acid dyes, five labels (no phycobiliproteins), six cyanine dyes (used as membrane potential probes; the reactive Cy2, Cy3, Cy5, etc. not having been developed at the time), and chlorotetracycline, a marginally useful probe of membrane-bound calcium. There were no usable diode lasers then, and He-Ne lasers only came in red, but, more to the point, there were no more than a few dozen multibeam flow cytometers in operation, and I'd guess that most of them used argon and krypton or dye lasers to excite fluorescein- and Texas red-labeled monoclonal antibodies. Today, it's hard not to do multiparameter flow cytometry, and getting harder not to do multistation multiparameter flow cytometry; we've got better hardware, better software, and, to get back to the topic immediately at hand, more and better reagents.

The first nineteen spectra shown in Figure 7-9 are those of labels, almost all widely used and readily available conjugated to antibodies and/or nucleic acid probes. The next twelve spectra are of nucleic acid dyes, including all five dyes (ethidium, Hoechst 33342, DAPI, mithramycin, and acridine orange) shown in my 1983 version of the figure. Then come four spectra of fluorescent reporter proteins, none of which had been described (although the original GFP had) when the last edition of this book was written. And, wrapping up, there are spectra of indo-1, a reliable and widely used calcium probe, in the presence and absence of calcium ions. To get 37 spectra onto one page, I had to change my spectrum display format slightly, but it is still essentially a streamlined version of the "box-and whiskers" plot shown in Figure 5-6 (p. 236).

The list of light sources in Figure 7-9 now includes Nd:YAG and semiconductor (diode) lasers as well as argon and krypton ion lasers, He-Cd and He-Ne lasers, and the HG arc lamp. I have drawn long vertical lines corresponding to the popular 488 and 633-635 nm excitation wavelengths through the spectra to facilitate orientation. If you look carefully, you'll notice that phycoerythrin and its tandem conjugates are excited much more efficiently by green light (doubled Nd:YAG lasers at 532 nm, green He-Ne lasers at 543 nm, or the Hg arc line at 546 nm) than by 488 nm light from an argon or semiconductor laser. However, since the use of a green source in place of a 488 nm source precludes excitation of fluorescein and its derivatives, and the use of a green source with a 488 nm source potentially puts stray green light into the fluorescein detector, green sources have not been widely used to date. That may change.

Table 7-2 differs from its predecessors in the article[8] and in previous editions of this book in two respects. They covered many more excitation wavelengths, and they listed dyes that had been used once or twice and weren't widely available. This time around, it made more sense to me to concentrate on materials that were widely used and widely available; there are, for example, UV-excited pH probes, but I didn't list them because they are relatively hard to find and not extensively described. Note that only one of the labels in Figure 7-9 did not make it into Table 7-2; the label is Texas red, which is not particularly well excited by 488 nm or red light, although I have heard claims that it can be excited in the UV and/or violet.

You won't find quantum efficiencies or relative fluorescence intensities of the dyes and labels in either Figure 7-9 or Table 7-2; those details will be covered in subsequent discussions of the individual materials. For now, we'll move on to a consideration of staining mechanisms.

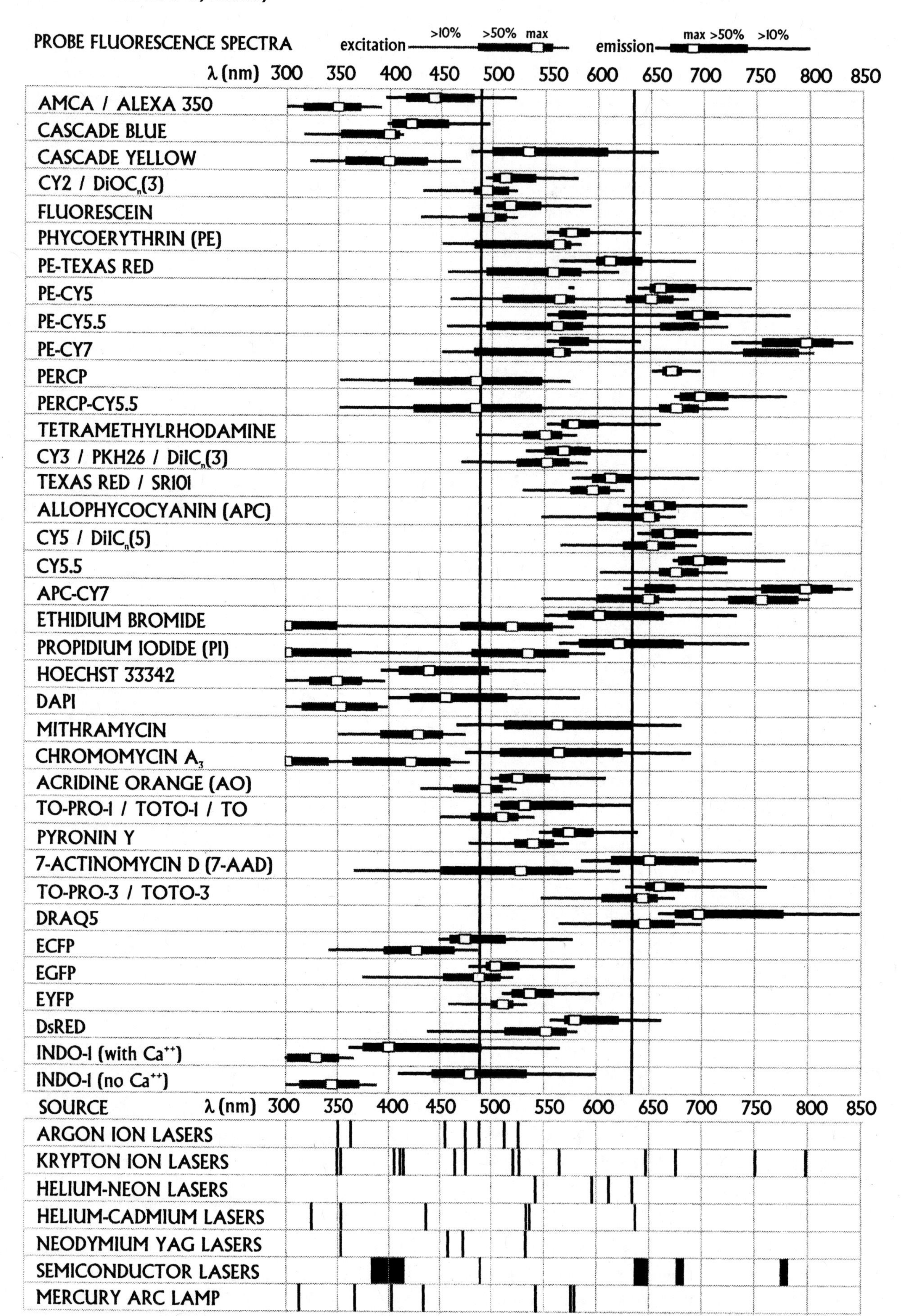

Figure 7-9. Probe fluorescence spectra and source emission wavelengths.

EXCITATION:	UV (325-355 nm)	VIOLET (395-415 nm)	BLUE-GREEN (488 nm)	RED (633-640 nm)
FLUORESCENT LABELS	AMCA, Alexa 350 (440)	Cascade Blue (420) Cascade Yellow (520)	Fluorescein, Cy2 (520) Cy3 (565) PE (575) PE-Texas Red (610) PE-Cy5 (660) PerCP (670) PerCP-Cy5.5, PE-Cy5.5 (700) PE-Cy7 (780)	APC, Cy5 (660) APC-Cy5.5, Cy5.5 (700) APC-Cy7 (780)
DNA-SELECTIVE DYES	Hoechst dyes (440) DAPI (455)	Hoechst Dyes (440), DAPI (455) Chromomycin, Mithramycin (560) ?7-AAD (660)	AO (520) 7-AAD (660) DRAQ5 (700)	DRAQ5 (700)
NONSELECTIVE NUCLEIC ACID DYES			TO-PRO-1, etc. (530) Pyronin Y (575) Ethidium (600) Propidium (615) AO (650)	TO-PRO-3, etc. (660)
REPORTER PROTEINS		ECFP (470)	EGFP (510) EYFP (535) dsRED (575)	
ENZYME SUBSTRATE FLUOROPHORES	7-amino-4-chloro-methylcoumarin (470) ELF 97 (530)	3-cyano-7-hydroxycoumarin (450)	Fluorescein, rhodamine 110 (520) resorufin (585)	
MEMBRANE POTENTIAL PROBES			DiBAC$_n$(3), DiOC$_n$(3), JC-1, Rhodamine 123 (520) JC-1 (585) DiOC$_n$(3) (610)	DiIC$_n$(5) (660)
pH PROBES			BCECF (520) Carboxy SNARF-1 (580) BCECF (620) Carboxy SNARF-1 (640)	
Ca^{++} PROBES	indo-1 (405) indo-1 (480)		fluo-3 (520) Fura Red (660)	

Table 7-2. Fluorescence spectral properties of a selection of reagents usable for common cytometric tasks. Emission maxima are indicated next to names of probes; probes for which two maxima are listed may be usable for ratiometric measurements.

Mechanisms of Staining by Fluorescent Dyes

The selective staining of different cellular constituents by fluorescent dyes is largely accounted for by two mechanisms.

The first of these is involved in staining by fluorescent and nonfluorescent dyes, and involves the development of contrast due to **differences in concentration** of dye from one region of the cell to another, which result from differences in the affinities of various cellular constituents for the dye. Thus, basic dyes are bound in relatively high concentrations to acidic materials such as nucleic acids and glycosaminoglycans, while dyes with high lipid solubility stain membranes and fat droplets, and so on.

The second mechanism of **fluorochroming** involves an increase in the quantum efficiency of a fluorescent dye when it is bound to a particular substance or in a particular environment (e.g., nonpolar vs. polar). The binding of the Hoechst dyes and DAPI to the outer groove of the DNA molecule results in approximately a hundredfold increase in fluorescence, as does the intercalative binding of ethidium and propidium; cyanine dyes such as thiazole orange (TO in Figure 7-9) and TO-PRO-1 increase fluorescence several thousandfold on intercalative binding to nucleic acid. However, acridine orange (AO), which also intercalates into DNA, exhibits slightly lower quantum efficiency when bound than when free in solution, i.e., the fluorescence is slightly quenched; thus, the bright nuclear staining produced by this dye must result almost exclusively from increased concentration of the dye in the nucleus. Fluorescent nuclear staining with AO or with dyes such as neutral red or safranin, which also decrease quantum efficiency on binding to DNA, improves on microscopic observation as the cell is left illuminated for a few seconds; the free dye, which initially produces high background fluorescence, is bleached more rapidly than the DNA-bound dye. It is hard to observe fluorescence from thiazole orange-stained nuclei under the microscope; there is little or no background fluorescence, and the dye in nuclei bleaches rapidly.

Binding of acid dyes to proteins usually is not associated with an increase in quantum efficiency, and background fluorescence from free dye tends to be relatively high as a result. For this reason, many investigators prefer to use **reactive derivatives of dyes**, e.g., fluorescein isothiocyanate (FITC), for protein staining. After incubation with FITC leaves some fluorescein covalently bound to protein, the unreacted dye is removed by washing, resulting in very low background fluorescence.

Even "specific" stains such as the Hoechst dyes and ethidium may bind nonspecifically to some materials in cells, particularly when the dye and/or the interfering material are present at high concentrations. **Nonspecific staining** may also occur when environmental factors such as salt concentration or pH are outside the range in which specific staining has been reported.

Environmental Sensitivity

Environmentally sensitive dyes are not dyes that recycle their garbage, but, rather, dyes that exhibit changes in quantum efficiency and/or spectral shifts upon binding to macromolecules or ions, changing ionization state, or moving from polar environments, i.e., ionic solutions such as the cytosol, to nonpolar environments, such as membrane lipid bilayers.

Environmental sensitivity is desirable, if not absolutely essential, for DNA stains, for example. You want the dye to fluoresce when bound to DNA and not otherwise. Environmental sensitivity is, in general, undesirable in dyes that you want to use to label antibodies or other ligands. Fluorescein, as I mentioned previously, is sensitive to the pH of its environment; as the pH drops, so does the quantum efficiency. As a result, if you have the same amount of fluorescein antibody bound to a tube of cells at pH 6.5 and to another tube of cells at pH 7, it will appear that there is more antibody bound to the cells at higher pH. Propidium fluorescence varies with salt concentration[227]; if you use a water sheath with a saline sample, you may find that your peak locations are unstable and/or your CVs are high.

Metachromasia

The term **metachromasia** is used to describe pronounced changes in color and/or fluorescence emission (and absorption) wavelength that occur when dyes such as toluidine blue or acridine orange bind to RNA or glycosaminoglycans. While the precise mechanisms of metachromasia have been debated since Paul Ehrlich coined the term[205,206], it is accepted that the effect results from interactions between dye molecules themselves, typically involving the formation of dimers, oligomers, or polymers, in which π orbitals of individual molecules interact one another with a resultant shift in the positions of electronic energy levels, rather than from effects of the environment or other molecular species on the electronic structure of the dye molecules.

The altered spectrum of the **metachromatic** molecular complex is contrasted with the normal **orthochromatic** spectrum of the dye in conditions under which complex formation does not occur. When acridine orange is used for differential staining of DNA and RNA (introduced on pp. 96-7 and discussed further on pp. 312 and 320-2), individual molecules of dye intercalated into the double helix of native DNA are normally far enough from one another to prevent complex formation, and therefore retain their orthochromatic green fluorescence. Dye molecules bound to denatured DNA or single-stranded RNA are free to form metachromatic aggregates.

A metachromatic shift of absorption, excitation or emission maxima to longer wavelengths is described as a **bathochromic shift**, or **red shift**; a shift to shorter wavelengths is described as a **hypsochromic shift**, or **blue shift**. If the absorptivity (i.e., the molar extinction coefficient) of the material increases, it is said to undergo a **hyperchromic shift**; a decrease in absorptivity is described as a **hypochromic shift**. In fluorescence metachromasia, it is relatively common for the excitation spectrum to shift to shorter wavelengths (hypsochromic shift) while the emission spectrum shifts to longer wavelengths (bathochromic shift). This is what happens in the case of acridine orange; the excitation maximum of the metachromatic red (>650 nm) fluorescence is near 460 nm, while the excitation maximum of the orthochromatic green (520 nm) fluorescence is near 490 nm .

In general, more careful control of dye (and cell) concentrations and staining conditions is needed to get reliable and reproducible metachromatic staining than might be necessary otherwise. Staining of DNA and RNA with acridine orange, for example, can be tricky; stabler staining patterns may be obtained by continuous addition of stain to the running sample[734].

Many dyes change their fluorescence spectra to some extent with changes in the polarity of the solvent; small changes are usually not described as metachromatic, although the underlying mechanism may be the same.

Spectral Changes and Ratiometric Measurements

In order to make ratiometric measurements of cytoplasmic [Ca^{++}] (introduced on p. 47) or pH, we need there to be substantial spectral shifts in the probes used. As can be seen from Figure 7-9, the excitation and emission spectra of indo-1 change substantially when Ca^{++} ions are bound to the probe, allowing the ratio of UV-excited 405 nm and 480 nm emission intensities to be used as an indicator of cytoplasmic [Ca^{++}]. Similarly, ratiometric pH measurement typically depends on differences in excitation and/or emission spectra of different protonation states of the same dye.

Ratiometric measurement of membrane potentials, which will be discussed further on, is more complex; a method described from my lab by Novo et al[2357] takes advantage of what is likely to be metachromasia resulting from formation of dye aggregates at high concentrations, but some other techniques may depend more on environmental sensitivity of probes.

Internal Energy Transfer in Probes and Labels

The phycobiliproteins, which nature has designed for efficient intramolecular energy transfer, and their tandem conjugates, representing early attempts to improve on nature in this regard, remind us that energy transfer may be intimately involved in determining the spectral characteristics of fluorescent reagents. Both phycobiliproteins and tandem conjugates are used almost exclusively as labels; we would prefer that their fluorescence not exhibit pronounced environmental sensitivity, and, in general, it does not.

Newer synthetic probes may use changes in intramolecular energy transfer characteristics to advantage; I have already mentioned probes for protease activity in which fluorescence of a fluorophore quenched by energy transfer in the intact probe molecule is markedly increased when the distance between fluorophore and quencher is changed by enzyme action on the probe[2556-7]. Some other examples will be discussed in subsequent sections.

"Vital" Staining

The availability of flow sorters stimulated interest in the development and application of stains that could permit viable cells with selected characteristics to be isolated for biochemical analysis or for further short- or long-term observation in culture. The stains used for such studies should, ideally, neither perturb the parameters measured nor compromise the viability of the cells being observed. Many of the probes now in use fall short of this ideal; this does not necessarily preclude their use but does require additional caution on the part of experimenters. In essence, we are now forced to reexamine the phenomenon of **vital staining**.

Staining and "Viability"

Dyes that can enter and stain living cells have long been described as **vital stains** even though most such dyes known to classical cytology are also known to be toxic to cells at the concentrations used for so-called vital staining. The semantic problems get worse as we go along; there is considerable disagreement about the definition of "viability" as it applies at the cellular level. People concerned with growing cells in culture generally take **clonogenicity** or **reproductive viability** as a criterion; this has two notable disadvantages. First, the definition of viability in terms of reproductive capacity excludes fully functional differentiated cells such as nerve and muscle cells, blood granulocytes, etc., and preservation of some specific cell function therefore seems to be a more suitable criterion of viability for such cells. Second, while reproduction of cells in culture provides unequivocal evidence of viability, failure to reproduce may be due to deficiencies in the medium and/or the experimenter rather than to cell damage.

If reproductive capacity is too stringent a criterion of viability, it is clear that some other criteria are not stringent enough, e.g. the capacity of the cell to exclude dyes such as trypan blue, eosin, and ethidium bromide. Cells that have been so damaged as to be incapable of performing their typical differentiated functions, e.g., chemotaxis and phagocytosis in the case of granulocytes, may still be classified as "viable" by a **dye exclusion test**.

"Intact" versus "Live" Cells

I find it convenient to use the term **intact cells** to describe cells that have not been treated with fixatives or lysing agents, and that do not show obvious morphologic damage or functional impairment. This avoids the issue of reproductive viability, but does preserve most of the distinction between cells in which studies of functional parameters are and are not appropriate. Among the extrinsic parameters listed in Table 7-1, membrane-bound and cytoplasmic [Ca^{++}], membrane integrity and permeability, endocytosis, membrane fluidity, structuredness of cytoplasmic matrix, membrane potential and pH are meaningful only as characteristics of intact cells. The validity of studies of responses of these parameters to biologic stimuli is best established by the inclusion of controls that demonstrate known responses to standard stimuli, e.g., changes in calcium distribution in cells treated with calcium ionophores.

Getting Dyes Into – and Out of – Intact Cells

In order to stain constituents not located on the surfaces of intact cells, a dye must be capable of crossing the cell membrane, either by **diffusion** or by some form of **active transport** or **carrier-mediated transport** (endocytosis can be considered as falling in the latter category). Most dyes that have been described as vital stains are small molecules that are relatively lipid soluble and are either positively charged or electrically neutral at physiologic pH. High lipid solubility favors partitioning of dyes from aqueous media into the lipid bilayer phase of the cell membrane and into membranous or lipid-containing intracellular structures. Positively charged dye molecules are attracted to negatively charged cell constituents such as glycosaminoglycans and nucleic acids; in addition, in living cells, positively charged molecules are concentrated from the medium into the cytosol and from the cytosol into mitochondria because there are interior-negative electrical potential gradients across both the cytoplasmic and the mitochondrial membranes.

Active and carrier-mediated transport both giveth and taketh away; there are a number of efflux mechanisms that may be active in cells. Perhaps the most notable is the glycoprotein efflux pump responsible for multiple drug resistance, which efficiently clears a wide variety of molecules from cells. Microorganisms exhibit an even broader range of transport mechanisms than do mammalian cells, making it risky to assume that dyes will interact in the same ways with prokaryotes and eukaryotes.

The cytoplasmic membranes of bacterial, fungal, and plant cells are surrounded by **cell walls**, which may offer barriers to dye entry not encountered in animal cells, which lack cell walls. Of particular note is the **outer membrane** found in Gram-negative bacteria, which, in its native state, prevents most lipophilic materials from entering the organisms[2078-9].

Permeancy and Permeability

Cytometry has picked up words from the lexicon of histology and histochemistry that describe the propensities of dyes and other chemicals for entering cells and the propensities of cells for taking up dyes and other chemicals. To repeat what I said when I introduced this topic on p. 43, materials that can readily cross the intact cytoplasmic membranes of cells are said to be **membrane-permeant**, or, more simply, **permeant**; materials that are excluded by intact cytoplasmic membranes are described as **membrane-impermeant**, or just **impermeant**. In order to stain living cells, i.e., to act as a **vital stain**, a dye must be permeant.

Cells that can take up a dye or other chemical are said to be **permeable** to the material; cells that cannot take up a material are said to be **impermeable** to it. The terms "permeability" and "impermeability" are used to characterize cell membranes in particular as well as cells in general.

A look at the structures of some nucleic acid dyes, shown in Figure 7-10, on the next page, may provide some help in understanding permeancy and impermeancy. Organic compounds bearing at least two positive charges, exemplified by fluorescent nucleic acid binding dyes such as DAPI, propidium iodide (PI) and TO-PRO-1, are impermeant.

Ethidium bromide (EB), not shown in the figure, shares the heterocyclic ring structure of PI, but carries only a single, delocalized positive charge; the ring nitrogen bears an ethyl (C_2H_5) group instead of the n-propyltrialkyl quaternary ammonium group (the "pro" in propidium) found on the ring nitrogen in PI.

Both dyes form complexes with DNA and RNA, and are toxic to cells once taken up. However, EB normally enters cells, albeit slowly, and is known to be pumped out of at least some bacteria[2585], while PI is normally excluded by its additional charge. Although EB is often used in lieu of PI for dye exclusion tests, it should not be because it is not truly impermeant. One can generalize from this; one should not assume that a dye is impermeant without determining whether or not it is being transported out of cells.

TO-PRO-1, a dye developed by Molecular Probes, has a delocalized positive charge on the heterocyclic ring structure and a propyltrialkyl (quaternary) ammonium on one ring nitrogen; it is, as was noted on the previous page, impermeant. Its parent compound, thiazole orange, has two N-methyl groups and lacks the quaternary ammonium substituent; it carries a single delocalized positive charge and is highly permeant. Thiazole orange was originally synthesized by Lee et al[769] at B-D as a stain for blood reticulocytes, making it highly desirable that it be permeant. Molecular Probes has produced a number of variants of the molecule, both permeant (the SYTO dyes) and impermeant (TO-PRO-1 and its homologues and some of the SYTOX dyes). The fluorescent properties of all of these dyes, which enhance fluorescence as much as several thousand times on binding to DNA or RNA, derive from the heterocyclic ring structure; the permeancy properties derive from the nature of various ring substituents, which have only minor influence on the fluorescence spectrum. Appropriate modification of ring substituents can influence not only whether or not a dye is permeant, but how rapidly it will enter, or leave, cells.

You might want to take some guesses as to the permeancy or impermeancy of the other dyes shown in Figure 7-10; the answers will appear in the subsequent discussion of nucleic acid stains.

There are two basic strategies for introducing impermeant dye molecules into living cells. The first, **permeabilization**, basically involves making holes in the membrane. This can be done physically, as in microinjection or membrane permeabilization by electrical breakdown (electroporation) or mechanical manipulation. Alternatively, membranes can be permeabilized by treatment with **fixatives**, or with

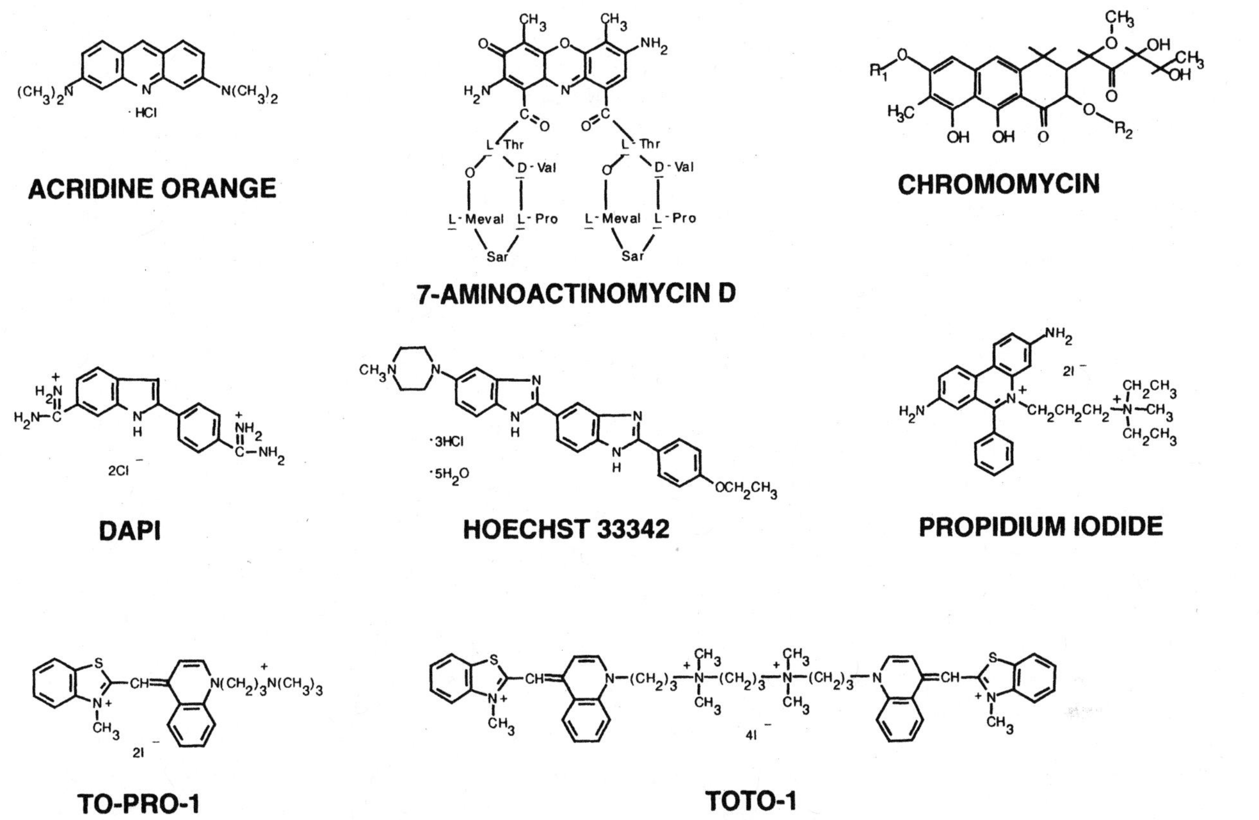

ACRIDINE ORANGE

7-AMINOACTINOMYCIN D

CHROMOMYCIN

DAPI

HOECHST 33342

PROPIDIUM IODIDE

TO-PRO-1

TOTO-1

Figure 7-10. A "rogues' gallery" of nucleic acid dyes. The chromomycin structure shown is the chromomycin nucleus; chromomycin A_3, mithramycin, and olivomycin each have different oligosaccharides as R1 and R2. Ethidium is similar to propidium, but with a ring N-ethyl group; thiazole orange (TO) is similar to TO-PRO-1, but has N-methyl groups on both rings. All structures except chromomycin were provided by Molecular Probes, Inc.

milder chemical agents such as lysolecithin. In either case, one can expect that solutes other than dye molecules will pass from cell to medium and *vice versa* during the time in which the membrane is permeable to dye. This in itself can produce alterations in cell physiology[207]; e.g., membrane potential drops to zero for a time. It then falls to the investigator to show that permeabilization didn't perturb the cells sufficiently to cast doubt on the validity of results obtained after staining and cytometry. I have been permeabilized by a 15 gauge hypodermic needle without ill effects; I would not expect to survive permeabilization by a 12 gauge shotgun.

The second strategy for getting dyes across membranes of living cells involves chemical modification of the dye molecules themselves. Acid dyes such as fluorescein do not cross cell membranes, but their electroneutral esters do. Diacetylfluorescein, which will forever be improperly called fluorescein diacetate (FDA), also called is uncharged and lipid soluble, and thus permeant; once inside the cell, ester molecules are readily and rapidly hydrolyzed to free fluorescein by nonspecific esterases, which are present in almost all cells. This effect serves as the basis for a test of cell "viability," or, more accurately, of **membrane integrity**; the ester is not fluorescent, while the free fluorescein anion, which is retained by intact cells, renders them highly fluorescent.

If one makes chromogenic or fluorogenic esters that require less ubiquitous enzymes than nonspecific esterases for hydrolysis, the accumulation of color or fluorescence in intact cells can be used to provide an indication of the presence and relative activity of various enzymes in those cells. Chemical modifications other than esterification, e.g., oxidation or reduction, may also be made to render dye molecules capable of crossing the cell membrane, and may also introduce a requirement for intracellular enzyme action to restore the staining characteristics of the unmodified molecule.

Vital Dye Toxicity and Photosensitization of Cells

It is important to remember that a permeant "vital" stain may eventually damage or kill cells; in this context, as in others, it is wise to establish that measurement conditions do not themselves perturb what one is attempting to measure.

Deleterious effects on cells other than those due to direct chemical toxicity of dyes may occur due to **photosensitization** when stained cells are exposed to high-intensity laser light. Many fluorescent dyes sensitize cells to photodynamic damage, although dyes that bind to DNA, such as the Hoechst dyes and the acridines, are probably best known in this regard. The probability of damage to cells stained with such dyes increases with the laser power used for analysis as well as with dye concentration.

When analyzing possible toxicity due to photosensitizing dyes, one should also bear in mind that a few hours' exposure of stained cells to fluorescent lighting in the laboratory may produce photodynamic damage equal to or greater than that resulting from a few microseconds in a low-power laser beam. To verify that cells are indeed being damaged by the dye and/or the laser, it is usually necessary to compare treated cells with cells kept in the dark.

Fixation – Why and How

Permeabilization can be, but is not always, accomplished in the context of **fixation** of cells. Most staining procedures in histology and histopathology, and many histochemical and cytochemical procedures, were originally developed using fixed specimens. The term "fixation" itself reminds us that one of the several original purposes of fixation was to make cells and tissues physically rigid enough to survive embedding and sectioning. Fixation was also necessary to make cells permeable to dye, thus allowing optimal staining, and to preserve specimens by inhibition of autolysis and microbial action. Baker[32] provides an excellent discussion of fixation mechanisms and of the classical fixatives; among these, **ethanol** and **formaldehyde** (and **paraformaldehyde**) are widely used for flow cytometry. More modern work on fixatives and mechanisms, with emphasis on the actions and use of **glutaraldehyde**, was summarized by Hopwood[208] and in a volume edited by Stoward[209]; Hopwood[725] provides a somewhat more recent large review.

Fixation is not an option when an experimenter's primary objective is to sort and retrieve living cells. However, when the objective is the demonstration and/or quantification of chemical constituents in cells, and there is no need for the cells to be kept alive during or after the process, it makes sense to work with fixed cells.

Fixation for Biohazard Control

Until the HIV epidemic emerged in the 1980's, most people who made flow cytometric measurements of unfixed cells preferred to work with unfixed material, which was at least perceived as being freer of cell aggregates, debris, etc., and exhibiting less autofluorescence than fixed specimens. While, statistically speaking, experimenters were far more likely to be exposed to such pathogens as the hepatitis B virus than to HIV, the perceived threat from the latter agent was what changed behavior, inside and outside the laboratory. Fixation with the objective of killing HIV and other viruses that may be present in specimens has now become routine in most flow cytometry involving cells or body fluids of human origin, and appropriate precautions against infection are widely, if not universally, observed when unfixed specimens must be analyzed.

The effects of fixatives on the infectivity of cells carrying HIV were examined by Cory, Rapp, and Ohlsson-Wilhelm[1303], who found that 30 min exposure to 0.5% paraformaldehyde, 1.85% formaldehyde, absolute methanol, or 1:1 methanol-acetone reduced infectivity by more than 99.99%.

Ericson et al[1304] studied the effects of commercial proprietary red cell lysing and fixing solutions on the HTLV-1 retrovirus. Coulter ImmunoPrep/QPrep and Immuno-lyse (Coulter Immunology, Hialeah, FL), FACS Lyse (B-D Immunocytometry Systems, San Jose, CA), and GenTrak whole blood lyse and fix solution (GenTrak, Plymouth Meeting, PA) all substantially reduced infectivity of infected cells seeded into blood and cultured after 5 min exposure. Ortho-lyse (Ortho Diagnostic Systems, Raritan, NJ), which had no added fixative, and ammonium chloride lysing solution, with 0.1 or 1.0 % paraformaldehyde added, did not reduce infectivity after 5 min but did after 60 min; other solutions worked better after 60 min than after 5 min.

This is good news and bad news; good news, because most people who do flow cytometry use commercial fixing and lysing solutions containing agents that are known to inactivate at least the two retroviruses studied in the references just cited, and bad news, because the development of new fixation techniques must now entail determining whether or not they inactivate various pathogens, a task for which most experimenters do not have much enthusiasm. If you remain undaunted, read on.

Fixation Mechanisms

Most of the effects of fixatives result from their action on **proteins**, which are denatured, precipitated, and/or **cross-linked** by fixatives at appropriate concentrations. The classification of fixatives as **coagulant** (e.g., acetone, ethanol, and methanol) or **non-coagulant** (e.g., acetic acid and formaldehyde) is based upon visual observation of their effects on solutions of albumin. However, both the coagulant alcohol and the non-coagulant aldehyde fixatives polymerize proteins. While the formation of protein aggregates is generally useful in stabilizing tissue specimens for embedding and sectioning, it is obviously only desirable up to a point when one wants to prepare suspensions of unclumped single cells for flow cytometric analysis. For many applications, it has been found useful to combine fixatives with **detergents**, which improve the speed and uniformity of reagent penetration into cells and tissues[725].

Permeabilization versus Fixation

Since specimens are usually not kept after they have been analyzed by flow cytometry, the preservative action of fixatives is generally unnecessary. In some cases, it may actually be desirable to remove portions of cells, as when flow cytometric measurements of DNA content are made on **isolated nuclei** from which the surrounding cytoplasm and extracellular matrix have been stripped by a combination of mechanical means, enzymes, hypotonic media, and **non-ionic detergents** such as Nonidet P-40 or Triton X-100. In the strictest sense of the term, specimens so treated can be said to consist of "unfixed cells," but they certainly cannot properly be described as "vitally stained." From a practical point of view, it makes sense to consider such cells as fixed; bear in mind, however, that while the cells are dead, accompanying pathogens may well not be.

Cells can be permeabilized without being fixed. Transient permeabilization with lysolecithin or electroporation, or addition of pore-forming peptide antibiotics such as nisin[2586-7], will allow cells to take up normally impermeant dyes such as propidium; at least some such treatments won't even kill the cells, much less fix them. Even the detergent treatment mentioned above, which makes the cells so permeable that everything but the nucleus washes away, doesn't do many of the things we expect fixation to do.

And cells can be fixed without being permeabilized, or at least exposed to fixatives without being permeabilized. The kicker here is that both "permeabilization" and "fixation" are terms covering a multitude of sins. As I wrote a quarter-century ago[93],

"Though I'm no linguist, I have heard
that every Eskimo
Learns sixteen different words describing
different kinds of snow.
I find it hard to understand how
histochemists live
And work with but a solitary word for fixative."

If we look at an intact cell as a plastic bag, we can envision permeabilization as turning it into a mesh bag. The problem is that it takes a finer mesh bag to hold marbles than to hold golf balls, and a finer mesh bag to hold golf balls than to hold baseballs. The antibiotic gramicidin A forms a 4.7 Å pore in cell membranes[2588]; this makes the membranes of cells exposed to gramicidin A permeable to most inorganic ions, with the result that the transmembrane electrical potential gradient is reduced to zero. However, the cells do not take up propidium. Nisin, which, as mentioned above, does make cells permeable to propidium, forms an 8 Å pore in membranes; bigger impermeant molecules need bigger holes to get through membranes. If you're trying to stain an intracellular antigen with a fluorescein-labeled IgG monoclonal antibody (MW 160,000), you'll probably need a much bigger hole than nisin makes, and, if you're working with a phycoerythrin-labeled monoclonal antibody (MW 400,000), you'll need a bigger hole than that.

If you need to expose large molecules or supramolecular structures in cells to antibodies or gene probes, but you want to keep at least some smaller molecules in or associated with the cells, you'll have to resort to fixation, and, in particular, to fixation with a cross-linking agent, which you will probably need to use in combination with a detergent or other permeabilizing agent. The cross-linking fixative will attach the small stuff you want to bigger stuff, which will keep it from getting through the big holes you made in the membrane.

Fixative Effects on Scatter Signals

The effects of alcohol and aldehyde fixation on proteins result in a marked increase in the **refractive index** of the cytoplasm, which rises from the value of approximately 1.35 observed in living cells to about 1.54. The refractive index of an isotonic saline solution is about 1.335. Since the amount of light scattered by cells is a function of the difference in refractive index between cells and the suspending medium, the amplitudes of light scattering signals are considerably higher for alcohol- or aldehyde-fixed than for unfixed cells. This may improve discrimination between cell types based on light scattering measurements. However, increased scattering may also affect the precision of fluorescence measurements because the effective amplitudes of both the excitation beam and the emitted fluorescence may be altered due to scattering by cytoplasmic structures. Scattering data may be difficult to interpret because various fixatives and postfixation treatments may in themselves change cell sizes, sometimes exerting different effects on different cells or on cells in different states. This topic was discussed in detail by Penttila, McDowell, and Trump[210], who also used flow cytometric measurements of light scattering and extinction to get a clearer description of cell volume changes during cell injury and fixation[211].

Carulli et al[2589] found that 10 minutes' exposure to BD FACS Lysing Solution, which contains a fixative as well as a lysing agent, allowed eosinophils in whole blood samples to be discriminated from other granulocytes on the basis of forward and side scatter signals.

Fixation for Surface Antigen Measurements

While some protein structures and enzyme activities remain unaltered after fixation, it is unwise to assume that the structure that particularly interests you will survive intact. For this reason, it has become common practice to stain cell surface antigens with fluorescent antibodies before fixing the cells to allow fluorescent staining of other constituents, e.g., DNA.

There are several reasons for doing immunofluorescence analyses on fixed cells. One is that appropriate fixation kills pathogens, but there are others. If, for example, you want to do simultaneous analyses of DNA content and surface antigens, using fluorescent antibodies and propidium[330-1], you'll need to fix the cells to permit RNAse treatment and nuclear staining while preserving the antigen-antibody complexes bound to the cell membrane. If you have only occasional access to a flow cytometer, and/or you collect samples over a period of days, you can't have everything stained fresh. And, of course, when you walk into the flow cytometer room with 100 freshly stained samples and the technician tells you the machine is down for two days, fixation is your only salvation.

Before you fix cells, it's a good idea to run them over a Percoll gradient, or do whatever else you like to do, to get rid of dead cells and debris. It's essential to wash the cells several times to get rid of any protein that may be in the medium; otherwise, you will end up with cell clumps at best and fluorescent-labeled aspic at worst.

Years ago, I used a modified version of Raul Braylan's time-honored procedure[331] for ethanol fixation of immunofluorescent stained cells. Washed cells, at a concentration of 2 x 10^6/ml or less, in phosphate buffered saline, pH 7.3, at 4 °C, are placed on a vortex mixer, and an equal amount of cold absolute ethanol is added slowly while vortexing. Cells fixed according to this procedure can be kept for at least several days, and often considerably longer, at 4 °C.

A frontal approach to fixation for immunofluorescent staining was taken by Van Ewijk et al[726], who compared immunofluorescence signals from fresh mouse and human lymphoid cells and from similar cells fixed before immunofluorescent staining. Paraformaldehyde, glutaraldehyde, acrolein, and osmium tetroxide at various concentrations were tested as fixatives; staining of Thy-1, T-2090, Lyt-1, Lyt-2, and Th-B antigens on murine and of B1, B2, β_2-microglobulin, HLA-DR, CALLA, OKT-3, and Leu-4 on human cells was quantified by flow cytometry. Fixation with 0.5% paraformaldehyde prior to immunofluorescent staining was found to preserve all mouse and human antigens studied; cells so fixed were stable and could be stored for at least one week prior to immunofluorescent staining without adverse effects.

Fixatives: Coming Out of Aldehyding Places

At present, **formaldehyde** and its supposedly less evil twin, **paraformaldehyde**, are probably the most popular fixative for immunofluorescent stained cells; **ethanol**, glutaraldehyde, and methanol have also been used. One might expect that alcohols would be preferable to aldehydes for fixing cells prior to immunofluorescence measurements, because aldehyde fixatives react with a variety of amines found in cells to produce fluorescent materials. Glutaraldehyde is notably worse than formaldehyde in this regard; on the positive side, this led to widespread use of chicken erythrocytes fixed in glutaraldehyde as low-intensity fluorescent particle "standards"[164] in the early days of immunofluorescence measurements. So-so standards are better than no standards at all.

Formaldehyde itself tends to polymerize in aqueous solution, making it difficult to control the effective concentration of fixative (at some point, the solution should become informaldehyde); methanol is added to many formaldehyde solutions to retard polymerization. Methanol-free formaldehyde is generally regarded as preferable for fixation. Relatively stable, methanol-free formaldehyde solutions are available; however, many people prefer to use paraformaldehyde, which is polymerized formaldehyde, available in solid form, to make up solutions containing a well-controlled concentration of formaldehyde. It is relatively common practice to heat solutions prepared from paraformaldehyde, ostensibly to facilitate depolymerization, to add buffer to prevent the formation of formic acid, and to make up fresh solutions frequently. However, Helander has shown[2590] that depolymerization occurs at room temperature, that only very small amounts of formic acid actually do occur in paraformaldehyde solutions, and that solutions are stable for at least a week. It may take considerably longer than that to dispel the urban legends.

We can certainly put the knock knock on formaldehyde; it's toxic, mutagenic, and potentially carcinogenic. Since detecting the slightest whiff of it in your house is likely to put you on the phone to a trial lawyer, why put up with it in your lab? Bostwick et al[2594] found they could switch to a mixture of 56% ethanol and 20% ethylene glycol and establish a "formalin-free surgical pathology lab." I guess we all have our pet causes.

It may be desirable to use non-aldehyde, cross-linking fixatives such as dimethylsuberimidate[212], 1-Ethyl-3-(3-Dimethylaminopropyl)-Carbodiimide Hydrochloride (EDC)[2591-2], N-Succinimidyl-3-(2-Pyridyldithio) Propionate (SPDP)[2591], or dimethyl 3,3'dithiobispropionimadate (DTBP)[2593] to minimize fluorescence induced by fixation, although these materials may be more expensive and less stable than more commonly used fixatives. Alternatively, Tagliaferro et al[2595] have described a procedure that uses Schiff base formation with fuchsin to quench fluorescence in glutaraldehyde-fixed material.

Over the years, a fair amount of effort has been devoted to producing stabilized blood cell controls for cytometry, which necessarily involves accomplishing at least some of the goals of fixation. Jani et al[2596] have tested the "TransFix" solution, developed by Barnett et al to prepare one such control[2597], as a fixative and transport medium for blood specimens intended for CD4+ cell counts. They found that counts remained stable in fixed specimens kept for 7 days under simulated tropical conditions expected to be found in resource-poor settings with a high prevalence of HIV infection. TransFix preserves most common surface antigens, and could find a wide range of applications; if it does, it is almost certain that flow cytometer manufacturers and third parties now producing fixing and lysing solutions will work toward producing their own improved products to permit longer specimen storage times under wider ranges of environmental conditions.

There are applications for which existing commercial fixatives are less than ideal; one of them is counting CD34+ stem cells in blood, which are consistently lower in specimens subjected to red cell lysis and fixation than in those subjected to lysis alone[2598]. It has been observed that different commercial fixing and lysing solutions have different effects on the forward and side scatter signals from leukocyte populations and on levels of expression of some antigens[2599,2600]; Macey et al[2600] advocate making immunophenotyping measurements on unfixed, unlysed whole blood using a permeant nuclear stain to identify nucleated cells. This approach can provide a good reference method for analyzing selective cell loss due to lysing reagents and fixatives; more studies in this area are badly needed. Manufacturers' tendency to keep the compositions of their preparative solutions secret does not facilitate evaluation of these products.

Fixation for Intracellular Antigen Measurements

When you're after **intracellular antigens**, with or without surface antigens, the cells have to be fixed before you can finish your immunofluorescent staining. Jacobberger et al[727] and Levitt and King[728] recommended methanol fixation for such purposes; 900 µl of ice-cold or colder 100% methanol are layered over 100 µl phosphate buffered saline (PBS), pH 7.4, containing 10^6 washed cells. Cells so fixed can also be stained with propidium iodide for simultaneous flow cytometry of intracellular immunofluorescence and DNA content. Clevenger, Bauer, and Epstein[1289] fixed cells at 4 °C with 0.5% paraformaldehyde and then permeabilized them with 0.1% Triton X-100 in PBS.

For staining both surface and intracellular antigens, one can either stain the surface antigen and then fix and permeabilize the cells[1290], or fix the cells in such a way that surface antigens are preserved. Other fixatives that have been used include buffered formaldehyde-acetone[1291], periodate-lysine-formaldehyde[1292], and paraformaldehyde followed by methanol, which was found in careful comparative evaluations by Pollice et al[1293] and Schimenti and Jacobberger[1294] to minimize loss of signal from antigens and to provide good DNA staining characteristics in terms of G1 peak location and CV and relative lack of clumps and debris. Schmid, Uittenbogaart, and Giorgi[1295] reported good results with fixation for 1 hr at 4 °C in 0.25% buffered paraformaldehyde followed by permeabilization for 15 min at 37 °C with 0.2% Tween 20; this treatment preserves relationships between leukocyte clusters on displays of forward vs. side scatter. Other permeabilizing agents that have been used for staining nuclear antigens include *n*-octyl-beta-D-glucopyranoside[1296], digitonin[1297], saponin[1298], and lysolecithin[2601].

There is no "one-size-fits-all" solution to the problem of preparing cells for such analyses; Koester and Bolton[2602] offer some guidelines and recommendations. However, there are fixative solutions available from numerous manufacturers that can be used for flow cytometric analysis of a variety of the more widely studied intracellular antigens.

Fixation for DNA Content Determination; Getting DNA (and Antigens) Out of a Tight Fix

DNA content analyses may be done on cells that have been liberated from tissues and tumors by mechanical and/or chemical treatment rough enough to render them permeable, in which case fixation is not necessary to permit staining with impermeant dyes. Fixation remains useful, however, when material needs to be transported before analysis, in which case the fixative serves the classical purpose of preserving the specimen. Howell et al[1299] found the very classical fixative mixture of 20:1 methanol:acetic acid useful in preparing cells from urine and bladder washing for both cytologic examination and DNA content analysis. Rouselle et al[2603] evaluated fixation with a variety of commonly used mixtures and found DNA content distributions with the lowest CVs were obtained using 68% ethanol and 85% methanol.

DNA content analysis may also be done on cells released from paraffin, which may have been fixed before the experimenter was born and which, in hindsight, may have been fixed a little more thoroughly than may be desirable for DNA analysis. Fixation is known to decrease accessibility of DNA stains to DNA, probably as a result of extensive cross-linking of histones to DNA[2604]. Luckily, cross-linking can be reversed; Overton and McCoy[2605] report that resuspending formalin-fixed cells in saline and heating them at 75 °C for at least an hour prior to staining restores the accessibility of the DNA to propidium.

The same general trick may be useful in restoring antigens in fixed tissue to enough of a semblance of their native state to permit or improve immunofluorescent staining. Boenisch[2606] examined **heat-induced antigen retrieval** of a number of tissue antigens using two popular techniques.

Catch the Wave: Fixation by the (Cook) Book

A lot of histologists and pathologists now seem to be using heat, with or without added chemical agents, for fixation. This isn't a completely new idea; in the 1970's, the Block group found that, when warmed to 40 °C, a glutaraldehyde fixative took less than 20 seconds to render red blood cells impervious to lysis by distilled water.

We didn't use a microwave oven back then, but that's what people are doing now. It has been established that the microwaves don't accomplish anything more subtle than heating the specimen, which decreases fixation time when the oven is used in conjunction with a chemical fixative[2607]. And, for those of you who have noticed warm and cold spots in different regions of the same microwaved entree, I will point out that Login et al[2608] have devised methods for calibrating and standardizing microwave ovens for various fixation procedures. I'll have to find out whether they work for day-old Hunan pan-fried noodles.

Microwave fixation does not seem to have been used much for flow cytometry, but a provocative paper by Grutzkau et al[2609] suggests that quantitative flow cytometry to determine optimal conditions for microwave and chemical fixation and for permeabilization could be useful in preparing samples for immunoelectron microscopy, and that microwave fixation might be useful for flow cytometric analysis of intracellular antigens.

Fixation Artifacts

If you are trying to reproduce a staining technique described in the literature, it's a good idea to use the fixation procedure employed by the author(s) of the paper in question. Even then, particularly if you're trying to do quantitative analyses, you may need to find out whether, and to what extent, variations in fixative composition and conditions affect staining. Holtfreter and Cohen[1300] noted that 50% ethanol fixation yielded the same DNA fluorescence from nucleated frog erythrocytes and leukocytes, while higher concentrations gave bimodal distributions, presumably due to the different density of chromatin in the two cell types. Haynes, Moynihan, and Cohen[1301] found that binding of an anti-human CD25 antibody to paraformaldehyde-fixed frog cells was artifactual, there being no evidence of reactive epitopes on unfixed cells. Jumping to humans, Cahill, Macey, and Newland[1302] reported that formaldehyde fixation of platelets can result in artifactual expression of antigens indicative of activation. The best way to avoid artifacts and other troubles with fixation is, of course, to work with unfixed cells.

Red Blood Cell Lysis: The Distilled Essence

On those occasions when you can get away without fixing, and want to make immunofluorescence measurements of leukocytes from blood or bone marrow without accompanying red cells, you resort to lysing agents, which are often either proprietary mixtures or ammonium chloride solutions (the two are not mutually exclusive). I have always preferred **hypotonic lysis**; I dilute the blood 1:10 with distilled water, wait for a few seconds until it clears, and bring the osmotic strength back to normal with concentrated saline. I was thus pleased to see that Terstappen, Meiners, and Loken[1305], who contributed quite a bit to the analysis of leukocyte antigens, found that hypotonic lysis preserves cellular immunofluorescence and scatter signals, and avoids cell loss encountered with ammonium chloride. This method also requires no washes, something else I favor in my lab, where flow cytometers outnumber people.

7.4 NUCLEIC ACID DYES AND THEIR USES

Fluorescent nucleic acid-binding dyes can used to measure a number of the extrinsic parameters shown in Table 7-1 (p. 286), including **DNA content, DNA base ratio**, single- and double-stranded **RNA content**, and also to characterize **chromatin structure** and to detect **DNA synthesis**. A "rogues' gallery" of nucleic acid stains appears in Figure 7-10 (back on p. 301).

DNA Content Measurement

Many, if not most, users of flow cytometers and sorters have at least a passing acquaintance with measurements of the **DNA content** of cells and chromosomes, which can be done rapidly and precisely by flow cytometry using a variety of fluorescent stains. I introduced DNA content measurements and some of the dyes used to make them on pp. 21-26, 43-4, and 96-7.

DNA content measurements were shown to be relevant to tumor pathology and chemotherapy in the late 1960's, and stimulated interest in fluorescence flow cytometry at a time when few, if any, instruments were sensitive enough to make immunofluorescence measurements.

The ideal dye for measurement of DNA content would be **DNA-specific**; that is, it would form a fluorescent complex with DNA, but not with RNA or with other macromolecular species. It would also not exhibit any **base** or **sequence preference**; in other words, the fluorescence from a given number of dye molecules bound to another given number of base pairs' worth of DNA would be the same, regardless of the proportions of A-T and G-C base pairs in the DNA.

Dick Haugland, of Molecular Probes, has taken me to task over the use of the term "DNA-specific" to describe real DNA dyes; he says there aren't any really DNA-specific dyes and suggests **DNA-selective** as a more appropriate adjective. I'll go with that from now on, and recommend you do the same. Some, but not all, of the improvements in DNA content measurement since the 1960's have resulted from the discovery or synthesis and use of dyes with greater DNA selectivity.

Feulgen Staining for DNA Content

By 1969, Van Dilla et al had used a modified **Feulgen procedure** with fluorescent stains (acriflavine and auramine O) and an argon laser source flow cytometer to produce DNA content distributions with a coefficient of variation of 6% for the diploid cell peak[79]. Feulgen staining is highly DNA-selective, but the procedure is also technically intensive, and the search for dyes that might be easier to use began almost immediately.

DNA Staining with Ethidium and Propidium

The first step in this direction was taken in 1969, when Dittrich and Göhde published a relatively sharp DNA content distribution obtained using their arc source flow cytometer to measure the fluorescence of fixed cells stained with **ethidium bromide**[83]. **Propidium iodide**[216] was introduced by Crissman and Steinkamp[217] in 1973 as a substitute for ethidium bromide in procedures for simultaneous quantitative analysis of DNA and protein content, in which fluorescein isothiocyanate (FITC) was used as a covalent stain for protein. Both fluorescein and propidium were excited at 488 nm; the rationale for the use of propidium lay in the fact that its emission maximum is 10-15 nm farther into the red region of the spectrum than that of ethidium, making it easier to separate red and green fluorescence signals from propidium and fluorescein using optical filters. It has generally been noted since that propidium produces fluorescence histograms with somewhat lower coefficients of variation (CVs) than are obtained using ethidium; this, rather than spectral characteristics, probably accounts for the greater popularity of propidium.

Ethidium and propidium form complexes with double-stranded DNA and RNA by intercalating between base pairs[213-5]. An intercalated dye molecule finds itself in a hydrophobic environment that results in a shift of its absorption spectrum and an increase in its fluorescence quantum efficiency. Excitation in the UV (320-360 nm) or blue-green (480-550 nm) spectral regions produces 20 to 30 times as much fluorescence emission from ethidium or propidium molecules bound to nucleic acid as would be emitted by the same number of dye molecules in solution. In addition, since binding to nucleic acid itself results in a local increase in the concentration of dye molecules, nuclei, chromosomes, and other structures containing double-stranded nucleic acid (DNA or RNA) are brightly stained by ethidium or propidium, provided cells are either fixed or permeabilized to allow the dye to enter.

Since ethidium does not rapidly cross the membranes of intact cells, and is likely to be pumped out when it does, it has widely, but erroneously, been regarded as impermeant, and used in dye exclusion tests; as I mentioned on p. 300, this is not a good idea. Propidium, which, by virtue of its double positive charge (see pp. 299-300) is impermeant, is a suitable alternative. The double charge also gives propidium a higher binding affinity for double-stranded nucleic acid than ethidium; the former dye will displace the latter from cells permeable to both[2610].

Neither ethidium nor propidium is DNA-selective; the dyes were originally used to stain fixed cells, and specimens were treated with RNAse to eliminate artifactual broadening of DNA content distributions that would otherwise result from the fluorescence of dye bound to double-stranded RNA. In 1975, Krishan described a simplified staining method that eliminated the steps of fixation and RNAse treatment; cells were suspended in a solution containing 0.1% sodium citrate and 50 μg/ml propidium iodide[218].

The cells first examined by Krishan were predominantly hematologic in origin; such cells are readily susceptible to lysis in hypotonic media. Substantial amounts of cytoplasm, and of cytoplasmic RNA, are lost as a result of hypotonic treatment, but only in cell types that undergo osmotic lysis. Look et al[221] (also A. T. Look, personal communication) found that the CVs of DNA content histograms obtained from propidium-stained cells were sharpened by the incubation of samples with 0.05 mg/ml ribonuclease (RNAse) for 30 min at room temperature before analysis; these days, almost everybody uses RNAse.

Fried, Perez and Clarkson, who studied hypotonic propidium staining in some detail[219], added 0.1% Triton X-100 to Krishan's propidium/citrate mixture and reported good results in staining cells that had been grown as attached monolayers[220]. The addition of the detergent at this relatively low concentration allows stained samples to be kept at room temperature for at least several days without significant change in the fluorescence histograms.

Neither ethidium nor propidium exhibits a strong base preference; this makes the dyes useful in determination of total DNA content of cells in such applications as plant taxonomy, which will be discussed in Chapter 10.

Ethidium and Propidium: Ionic Strength Effects

The binding affinity of ethidium to DNA is strongly dependent on the **ionic strength** of the solution[215]; fluorescence intensity of a solution of DNA in isotonic saline will be less than that of a solution with the same concentration of DNA and a lower ionic strength. Propidium has similar properties. Martens, van den Engh, and Hagenbeek[227] noted shifting peaks in propidium fluorescence distributions from cell samples in hypotonic solution run in a cell sorter using isotonic saline sheath fluid. This drift can be eliminated by the use of a sheath fluid with the same ionic strength as the sample, i.e., use a distilled water sheath with samples in hypotonic solution, or add salt to taste to the samples if you use a saline sheath in your instrument. This problem has been described to me by many people; the solution is in the solutions.

DNA Content: Sample Preparation and Standards

Because abnormalities in DNA content are commonly found in cancer, and may be relevant to prognosis, applications in clinical oncology motivated much of the work on preparation of tissue samples for DNA content determination.

Four papers[222-5] by Vindeløv and his coworkers summarized their extensive experience in preparation, storage, and analysis of propidium-stained samples; they were also among the first to consider the problem of **standardization of DNA content measurements**. Both **chicken erythrocytes**, with a DNA content of 35% of the human diploid value, and **rainbow trout erythrocytes**, with a DNA content of 80% of the human diploid value, were added to samples[224]. The use of two standards eliminates calibration errors due to **nonlinearity** in the instrument.

Taylor[226] investigated rapid methods for preparation of samples for DNA content analysis from cells grown as monolayers or in suspension, from solid tissues, and solid tumors, using propidium and other stains for DNA. He obtained histograms with good precision using an isotonic saline solution with approximately 1.0% Triton X-100 added. This concentration of detergent lyses cells and solubilizes cytoplasm; it also, unfortunately, tends to lyse nuclei after a few hours. RNAse (1 mg/ml, equivalent to 50-75 Kunitz units/ml) is used with propidium staining to eliminate RNA fluorescence artifacts. Staining according to Taylor generally requires less than 10 min incubation of cells with the staining solution; samples are then filtered through a 50 μm nylon mesh and analyzed.

Singh[2611] reports that passage through a tissue press followed by pipetting yields cleaner suspensions of isolated nuclei for DNA content analysis in less time than does mincing tissue.

Chromomycin A_3, Mithramycin, and Olivomycin

In 1974, before rapid DNA staining techniques employing propidium had been developed, Crissman and Tobey described a rapid DNA staining procedure using the antitumor antibiotic **mithramycin**[228], with argon laser excitation at 457 nm. These authors had used fluorescent Feulgen staining for analysis of drug effects on cell cycle progression[229]; the reduction of sample preparation time to 20 minutes from several hours made it feasible to monitor population kinetics almost continuously during experiments. Cells were stained with mithramycin in a solution containing ethanol, which, while not fully fixing the cells during the short staining time, did make them permeable to the drug.

The use of mithramycin was prompted by the report that complexes of the drug and DNA were fluorescent, while the drug failed to interact with RNA[230]; thus, no RNAse treatment was required to achieve selective DNA staining. Crissman et al[231] did detailed studies of staining with mithramycin and with the structurally related antibiotics **olivomycin** and **chromomycin A_3**, all of which are highly DNA-selective. They found that optimal staining of cells fixed in ethanol or electron microscopy grade glutaraldehyde was achieved with 50-100 μg/ml dye, with pH in the range 5-9, salt (NaCl) concentration 0.15-1.0 M, and with 15-200 mM added magnesium (Mg^{++}). Magnesium ion is required for formation of the complexes of chromomycin A_3, mithramycin and olivomycin with DNA, which also involve the 2-amino group of guanine[230]; the three dyes have a strong base preference, and act as fluorochromes for **G-C rich regions** of DNA. All are impermeant, at least in terms of short-term staining of intact cells; they would, obviously, be ineffective as antitumor antibiotics if they didn't eventually get in, but this is not the place to delve into pharmacology.

The emission maximum of the olivomycin-DNA complex is at approximately 545 nm, but there is substantial emission between 480 and 500 nm. The chromomycin complex has an emission maximum at about 555 nm, but its emission spectrum does not have the short-wavelength shoulder characteristic of the olivomycin complex. The mithramycin complex spectrum also lacks the shoulder and has an emission maximum at about 575 nm. The excitation maxima of the DNA complexes of all three dyes are at approximately 440 nm[231]. The quantum efficiency of the dye-DNA complex is relatively low; this may limit precision in work with chromosomes or bacteria due to photon statistics.

While chromomycin A_3, mithramycin, and olivomycin can all be excited by the 457 nm argon ion laser line, shorter wavelengths, e.g. the violet lines from a krypton or diode laser, 441 nm from a He-Cd laser or the 436 nm line from a mercury arc lamp, are preferable. The spectral characteristics just mentioned led directly to the development of another popular DNA stain. Until 1974, there was little interaction between American workers, almost all of who were using laser source flow cytometers, and European researchers in the field, who were using arc source instruments[232]. In mid-1974, Barthel Barlogie arrived in Houston, Texas with a Phywe ICP system, with which he began studies of tumor cell DNA content using ethidium bromide, as was then customary. The blue-violet excitation in the ICP seemed better suited to mithramycin than to ethidium, which first prompted a trial of DNA staining with mithramycin (B. Barlogie and H. Shapiro, unpublished), and next motivated the use of a mixture of **mithramycin and ethidium**[233].

Mithramycin Plus Ethidium: Do's and Don'ts

DNA-bound ethidium is not very efficiently excited by blue-violet light; the DNA-mithramycin complex is optimally excited, and energy transfer occurs between mithramycin and ethidium provided the molecules are in close proximity. Thus, in cells stained with the mithramycin-ethidium mixture, ethidium fluorescence comes primarily from ethidium bound to DNA, and broadening of fluorescence distributions due to RNA fluorescence is largely eliminated when violet or blue-violet (400-457 nm) excitation is used. I have run across people who used a mithramycin-ethidium mixture with 488 nm excitation; this makes little or no sense, because excitation of mithramycin at this wavelength is negligible, while excitation of ethidium is quite good. Most of the emitted fluorescence will, therefore, be due to direct emission from the ethidium rather than to energy transfer from mithramycin, and the DNA selectivity gained with blue-violet excitation will be lost.

Interest in chromomycin and mithramycin DNA staining, with or without ethidium, may increase as violet diode lasers, emitting near 405 nm, find their way into more commercial flow cytometers (see Chapter 8); excitation of these dyes at 405 nm is about 80% of maximum.

The Hoechst Dyes (33258, 33342, 34580?)

A series of highly DNA-selective, UV-excited, blue fluorescent bisbenzimidazole dyes, originally synthesized by Loewe[234] for Hoechst as antiprotozoal drugs, and designated as **Hoechst 33258**, **33342**, **33378**, and **33662**, were introduced to the flow cytometry community in 1974 by Latt[235-8], although they were not widely applied until somewhat later.

Detecting BrUdR by Hoechst Dye Quenching

The original motivation for studies of the Hoechst dyes lay in the fact that complexes of the dyes with DNA containing **bromodeoxyuridine** (variously abbreviated as BrdUrd, BUdr, BrUdR, etc.) were quenched, i.e., less fluorescent than complexes of the dyes with unsubstituted DNA. The fluorescence difference was sufficient to allow detection of **DNA synthesis** and of sister chromatid exchanges in cells incubated with BrUdR. Flow cytometric methods were subsequently developed that allow detection of BrUdR incorporation by Hoechst dye fluorescence quenching without the use of anti-BrUdR antibodies; these will be discussed further in the section on DNA synthesis.

Hoechst Dyes Have an A-T Base Preference

The Hoechst dyes act as DNA-selective fluorochromes, binding to sequences of three A-T base pairs in DNA[268]; this strong **A-T base preference** accounts for the popularity of Hoechst 33258 in combination with chromomycin A_3 for staining chromosomes and bacteria, which will be discussed further in the section on DNA base composition. The Hoechst dyes bind in the minor groove of the DNA helix rather than by intercalation. Their affinity for DNA is sufficiently strong that they will displace bound molecules of a variety of intercalating dyes[2612].

Hoechst 33342 as a Vital DNA Stain

Additional interest in Hoechst 33342 was aroused when Arndt-Jovin and Jovin showed in 1977 that at least some living cells could be stained with this dye, sorted on the basis of DNA content, and subsequently grown in culture[239]. At the time, this finding represented a source of frustration to most of the (lucky few) investigators who had cell sorters, since the argon lasers supplied with most sorters could not emit the UV lines necessary to excite the Hoechst dyes.

Fluorescence histograms in early publications on **vital staining with Hoechst 33342**[239-42] showed broad peaks (CV of 6% or more), which were commonly attributed to cell-to-cell differences in dye uptake. Conventional wisdom held that it was necessary to use at least 100 mW of laser power in the UV to excite Hoechst 33342 to produce acceptable histograms. It is now almost certain that the poor precision of the measurements was due to a combination of noise in the light sources and uneven illumination resulting from the laser beam being focused to too small a spot in the apparatus used by earlier authors. It now seems clear that, at UV power levels of 100 mW or more, dye saturation occurs, eliminating the effects of uneven illumination and reducing CVs.

In 1981, I reported that low-power UV excitation from a mercury arc or 10 mW from an argon or krypton laser could be used, with appropriate optics, to produce Hoechst 33342 fluorescence histograms with low CVs from vitally stained cells[113]. I and others[643] have since observed that 1-10 mW at 325 nm from a He-Cd laser provides more than adequate excitation for cells stained with Hoechst dyes in a flow cytometer with a properly designed optical system. In fact, the precision of DNA measurements I have made using air-cooled He-Cd lasers has been better than the precision of measurements I made using large, water-cooled argon and krypton lasers from several manufacturers.

While there are a few reports in the literature[243-6] that suggested that other dyes could be used for flow cytometric determination of DNA content in living cells, Hoechst 33342 is the only compound that has been extensively used for the purpose. It is clear that some cells can survive and have survived staining with the dye, passage through a UV laser beam, sorting, and washing with retention of reproductive integrity, and that other cells, under similar conditions, have failed to survive[239-42,247-55]. Exposure to high laser power levels (over 100 mW) appears to decrease survival[256]. It seems likely that the different toxicity of the dye to different cell types is related to differences in uptake and/or retention.

Hoechst Dyes In – And Out Of – Living Cells: The Drug Efflux Pump Discovered

Stoichiometric vital staining of DNA by Hoechst 33342 generally requires exposure of cells to 5-10 µM dye for at least 30 min. Lower concentrations (1-2 µM) and/or shorter incubation periods can be used to demonstrate differences between cell types in rates of dye uptake and/or efflux. This technique was originally used by Lalande and Miller[248-9], and was later adopted by Loken[241,250] and by me and my co-workers[257] to differentiate resting T lymphocytes (low uptake) from resting B cells and activated T cells (high uptake). Staining differences among lymphoid cell populations are more pronounced in the mouse than in the rat or man (makes you wonder, doesn't it).

The demonstration in 1981 by Lalande, Ling, and Miller[253] that the staining intensity of living cells exposed to Hoechst 33342 is determined by the operation of an **efflux pump** brought flow cytometry to bear on what has since become an active and important area of application, i.e., the study of drug transport and distribution in cells and of transport-related mechanisms of drug resistance. Work along these lines has also provided valuable new insights into the nature of vital staining.

Krishan[729] was one of the first to investigate mechanisms of Hoechst 33342 uptake and retention. He found that some live cells, e.g., cultured leukemic cells resistant to adriamycin, stained poorly, if at all, with Hoechst 33342 because the operation of the energy dependent pump produced rapid efflux of the dye; isolated nuclei from these cells, as expected, stained normally. When the pump was blocked by phenothiazines (e.g., 15 µM trifluoperazine added to mouse splenocytes incubated with 10 µM Hoechst 33342 for 60 min at 37 C) or Ca^{++} channel blockers (e.g., verapamil), Hoechst 33342 produced stoichiometric staining of live cells.

Baines and Visser[2613] reported in 1983 that, when mouse bone marrow was vitally stained with Hoechst 33342 and sorted on the basis of fluorescence intensity, stem cells were found in the lowest-fluorescence fraction. Others[1568,2614-5] have since noted that both Hoechst 33342 and rhodamine 123 fluorescence are low in primitive hematopoietic stem cells. Goodell et al have defined a **side population (SP)** of cells, in which primitive stem cells are found, on the basis of relatively low UV-excited blue and red fluorescence after Hoechst 33342 staining[2616]. The low Hoechst 33342 and rhodamine 123 fluorescence in stem cells result from the activity of one or more efflux pumps; the effects of pumps on staining with Hoechst dyes will be considered further in the discussions on membrane permeability and drug resistance and on stem cells in this chapter and in Chapter 10.

Hoechst Dye Staining Mechanisms

When the Hoechst dyes are used to stain fixed or permeabilized cells for DNA content analysis (see Taylor[226] for one such staining protocol), lower dye concentrations (3 µM or less) than are used for vital stoichiometric staining are mandatory to avoid **nonspecific fluorescence**.

Additional insight into the binding and staining mechanisms of the Hoechst dyes came following the report of Watson et al[730] that the ratios of Hoechst 33342 fluorescences in the 515-560 nm (green) and 390-440 nm (violet) bands were different for different subclasses of chicken thymocytes vitally stained with 5 µM dye. The green fluorescence in one cell type increased over a 2 hour incubation period; a second cell type reached an equilibrium fluorescence intensity in a few minutes.

Steen and Stokke[731], using an arc source flow cytometer with a grating monochromator fitted to the detector, measured emission spectra of fixed rat thymocytes stained with varying concentrations of Hoechst 33258, and found the emission shifted toward the green at increasing dye concentrations. Stokke and Steen[732] characterized two binding modes of the dye; at low dye/phosphate ratios, high-affinity binding with high quantum yield blue fluorescence is predominant; as dye concentration increases, there is lower affinity binding to secondary sites, with quenching, emission shifting toward the green, and DNA precipitation. Changes in Hoechst dye emission may help probe nuclear **chromatin structure** in different cell types, provided that differences in uptake and/or efflux patterns can be controlled.

Ellwart and Dormer[1310] exploited the spectral shift of Hoechst 33342 as a measure of **cell viability**, or membrane integrity. Live cells exposed to relatively low dye concentrations in the presence of propidium exclude propidium, but retain small amounts of the Hoechst dye, and shorter wavelength emission predominates. Cells that have completely lost membrane integrity take up propidium. Cells in transition exclude propidium, but accumulate higher concentrations of Hoechst 33342, and fluorescence emission shifts to longer wavelengths.

Stokke et al[1311] found that the fluorescence of Hoechst 33258 in fixed erythroid precursor cells was less than that in myeloid cells in the same sample; the fluorescence of the dye in nuclei from erythroid cells was the same as that in myeloid cells. This effect may be due to quenching and/or reabsorption of fluorescence by hemoglobin in the erythroid cells; it is not observed with dyes such as mithramycin.

Vinogradov and Rosanov[1312] have synthesized analogs of the Hoechst dyes lacking the piperazine ring found in the compounds made by Loewe; their DNA-binding selectivity and A-T preference reportedly exceed those of Hoechst 33258.

Hoechst 34580: Violet Time?

As you will find on perusal of Chapter 8, violet diode lasers are rapidly finding their way into commercial flow cytometers. While we normally think of the Hoechst dyes (and DAPI, discussed next) as "UV-excited," these dyes have some excitation cross section left (12% of maximum excitation for DAPI, 3% for Hoechst 33342) even at the 405 nm nominal emission wavelength of violet diodes[2444]. When the Hoechst dyes or DAPI are used to stain mammalian cell nuclei, one can usually expect to find millions of dye molecules in a cell. Under these circumstances, the variances of fluorescence distributions are not greatly broadened by photoelectron statistics when efficient collection optics and PMTs are used, even when only a few mW of excitation power at 405 nm are available.

Hoechst 34580 becomes interesting in this context because, when bound to DNA this dye exhibits an excitation maximum near 380 nm, about 30 nm longer than the absorption maxima of the other Hoechst dyes and 20 nm longer than that of DAPI. Its excitation is 15% of maximum at 405 nm, and 50% of maximum at 395 nm, a wavelength now available from selected violet diodes. With Nancy Perlmutter, I found that Hoechst 34580 was permeant, yielding a DNA histogram with a CV less than 5% from intact Jurkat cells excited by a 4 mW, 397 nm diode laser[2444]. However, the dye appeared to be unstable in solution; we were unable to obtain good histograms from either intact cells or isolated nuclei from day-old solutions. The dye is now available from Molecular Probes, and probably worth some further investigation; I'll get to it when I finish writing.

DAPI (and DIPI): Dyes Known for Precision

A highly DNA-selective, impermeant dye with properties similar to those of the Hoechst dyes, **4'-6-diamidino-2-phenylindole**, or **DAPI**, was introduced for flow cytometry of DNA content by Stöhr et al[258]. The dye had been previously used for visualization of virus and *Mycoplasma* infection of cultured cells[259]; its high DNA specificity and intense fluorescence have allowed visualization of single DAPI-stained virus particles under the fluorescence microscope[260]. Taylor[226] and others[261] have described nuclear isolation protocols for use with DAPI staining.

DAPI, like the Hoechst dyes, has a strong **A-T base preference**. Many people, myself included, believe that DAPI yields DNA histograms with CVs lower than are obtained using other dyes (I have noted some dramatic differences when analyzing cells removed from paraffin-embedded material). While well-controlled studies on whole cells or nuclei seem to be lacking, Otto and Tsou[733] did compare DAPI, the related compound DIPI, and Hoechst 33258 and 33342 as stains for flow karyotyping of chromosomes from CHO cells, using an ICP with arc lamp UV excitation at 366 nm. They found 5 μM to be the optimal concentration of DAPI for staining chromosomes. DIPI staining was very slightly brighter (103%) than DAPI; the Hoechst dyes were only 86% as bright. However, DAPI produced histograms with the lowest CVs (2.2% vs. 2.7% for DIPI, 2.8% for Hoechst 33258, 2.9% for Hoechst 33342).

Figure 7-11, on the facing page, illustrates the most dramatic example of high-precision DNA analysis of which I am aware. Lewalski, Otto, Kranert, and Wassmuth[1313], using a Partec PAS-II flow cytometer, analyzed enzymatically decondensed, DAPI-stained spermatozoa to confirm the suspected production of unbalanced spermatozoa in heterozygous rams carrying a 1; 20 chromosomal translocation. In the top panel, X- and Y-chromosome-bearing spermatozoa from a cytogenetically normal ram appear as two distinct peaks; the difference in DNA fluorescence intensity between the gonosomes averaged 4.8%. In the bottom panel, sperm samples from a heterozygous 1; 20 translocation carrier yield a histograms with five major and two minor (arrows) peaks, attributed to spermatozoa with a normal, balanced, and unbalanced chromosomal status, with CVs of 0.5-0.6%. Because the translocated chromosomal segment represents 2.4% of the total DNA content, as determined from the flow cytometric data, histograms with five instead of the expected six peaks are observed.

Darzynkiewicz and his colleagues[738-9] have found that staining with DAPI is less affected by the state of chromatin condensation than is staining with other DNA stains; this most probably accounts for the lower CVs generally observed with the dye.

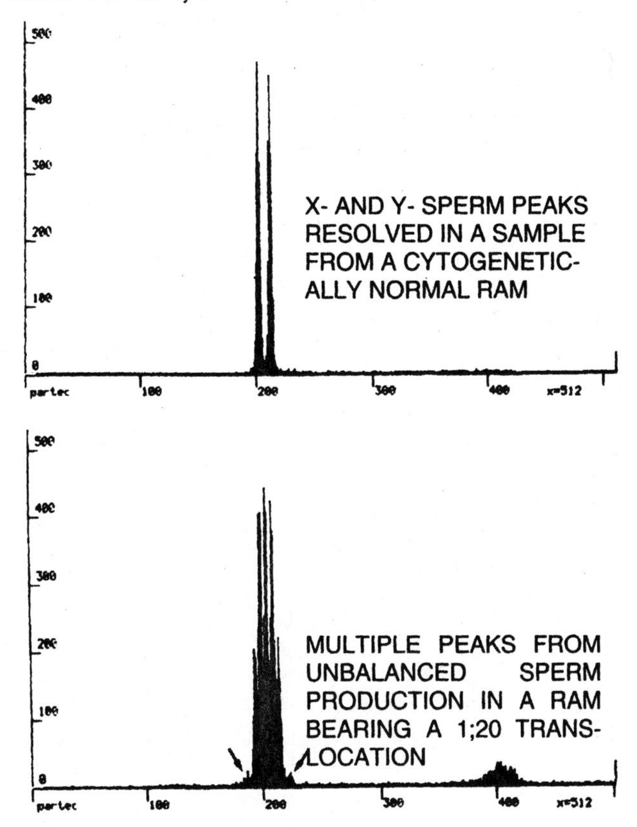

Figure 7-11. DNA content distributions in sperm from a normal ram and a ram bearing a 1;20 chromosomal translocation, stained with DAPI and analyzed on a Partec arc source flow cytometer; peak CV's are 0.5-0.8%. From H. Lewalski et al, Cytogenetics and Cell Genetics 64:286, 1993[1313]. The figure was kindly provided by Dr. Lewalski and is reprinted with the permission of S. Karger, publishers.

Determinants of High Precision in DNA Analysis

Because DNA content is so well regulated in most cells, measurements of DNA content are more likely than measurements of any other cellular parameter to result in fluorescence distributions with extremely low CVs. Even at that, the distributions shown in Figure 7-11 are remarkable. The level of precision represented in them is not readily achievable in laser source flow cytometers, for several reasons. Most benchtop laser source instruments use air-cooled argon ion lasers emitting 15-25 mW at 488 nm as light sources; the RMS optical noise level of such lasers is usually specified as 1%, and, since the power output is not sufficient to saturate the dyes used for DNA measurement, the CV of a fluorescence distribution obtained with such an instrument cannot be less than the RMS noise level of the source.

If the laser power were increased to the level needed for photon saturation, and/or the light source noise were reduced to a small fraction of its normal value, there might still be problems associated with the nature of the dyes available for DNA content analysis using 488 nm excitation. Propidium is the most common of these; it is not highly DNA-selective, and the RNAse treatment customarily used to minimize contributions to the fluorescence signal from dye bound to double-stranded RNA may not completely eliminate interfering fluorescence. We can therefore attribute some of the high precision achieved in DNA content measurement with DAPI, and, to a slightly lesser extent, with the Hoechst dyes, to the relatively high DNA-selectivity of the dyes, and to their relative insensitivity (notably that of DAPI) to the differences in chromatin condensation. However, that still isn't the whole story.

I should not need to mention that fluorescence CVs in the range of 1% are unlikely to be achieved in any flow cytometer in which the optics are even slightly misaligned, or in which there are any significant sources of turbulence in the fluidics. For purposes of the present discussion, however, we can assume that the optics and fluidics of whatever cytometer we use can be put into optimal operating condition.

At the end of the day, it is the optical design of the instrument, more than anything else, which makes it possible to achieve extremely low CVs in fluorescence measurement. Most instruments that can do it use arc lamp sources, and have relatively high-N.A. illumination optics, as well as relatively high-N.A. collection optics. As I mentioned on the previous page, photoelectron statistics are not a significant problem in measurements of the DNA content of eukaryotic cells; the relatively low UV power output of the arc lamp is more than adequate for excitation of DAPI or the Hoechst dyes. The high-N.A. (epi)illumination insures that excitation light reaches the cell from a substantially wider range of angles than would be the case in a typical laser source system. This minimizes effects of cellular asymmetry and orientation on illumination and light collection, resulting in lower fluorescence CVs[96]. We will revisit this issue in the discussion of sperm sorting in Chapter 10.

7-Aminoactinomycin D (7-AAD)

7-AAD, a fluorescent analog of the antitumor antibiotic actinomycin D, was synthesized for possible application to chromosome banding by Modest and Sen Gupta[1314], and was investigated as a cytometric DNA stain by Gill et al[1315] and, more recently, by Zelenin et al[735]. The complex of this dye with DNA has an absorption maximum at about 550 nm and an emission maximum at about 660 nm. 7-AAD is impermeant and highly DNA-selective, exhibiting a G-C base preference.

The quantum efficiency of 7-AAD is low, as was interest in the dye until Rabinovitch et al[736] demonstrated that, because of its long emission wavelength, the dye could be used in combination with fluorescein- and phycoerythrin-labeled antibodies for simultaneous flow cytometry of DNA content and two-color immunofluorescence using only a single 488 nm excitation beam. The DNA content distributions obtained using 7-AAD in this way have been disappointingly broad[736], probably because binding of the rather large dye molecule to DNA is affected by chromatin structure to a greater extent than is the case with other DNA stains. Stokke and Steen[740] utilized this property of 7-AAD for discrimination of different leukocyte types based on differences in chromatin structure.

Schmid et al[1316] have used 7-AAD to discriminate dead cells, i.e., cells that have lost membrane integrity, in unfixed samples stained with fluorescein- and phycoerythrin-labeled antibodies. Unlike propidium, which has also been used, 7-AAD emission is at a sufficiently long wavelength to minimize interference with phycoerythrin fluorescence. Schmid, Uittenbogaart, and Giorgi[1317] have also reported that 7-AAD uptake into cells can be used to demonstrate apoptosis in thymocytes also stained with fluorescein- and phycoerythrin-labeled antibodies.

Acridine Orange

Under carefully controlled conditions, the blue-excited green fluorescence of **acridine orange (AO)** molecules intercalated into DNA can be used to provide accurate and precise estimates of cellular DNA content. The techniques described by Darzynkiewicz and his coworkers[262-3], which were mentioned on pp. 44 and 96-7, and will be discussed more fully in the sections on probes of chromatin structure and RNA content, involve detergent treatment to permeabilize cells, acid denaturation to convert RNA to the single-stranded configuration, which forms a red (metachromatic) fluorescent complex with the dye, and careful adjustment of cell and dye concentrations to prevent formation of metachromatic fluorescent dye aggregates in association with DNA.

Many people who have tried to use acridine orange have been frustrated by the sensitivity of staining to very slight changes in operational parameters of the flow cytometer. A paper by Pennings et al[734] showed that DNA/RNA staining with acridine orange could be improved by using a roller pump and manifold to introduce dye and cells continuously into the flow cytometer sample stream, which stabilized the equilibria between free dye, dye intercalated into DNA, and dye complexed with RNA.

Styryl Dyes; LDS-751

Compounds containing a quinolinium, benzothiazole, or other heterocyclic ring system, linked to a substituted aminostyryl group, have been known as fluorescent nuclear stains since 1932, when von Jancso[1318] described staining of intraerythrocytic parasites by styrylquinolin. Perhaps because this brief paper is not widely known, Wang and Jolley[1319], then at Abbott Laboratories, received a patent on staining with styryl dyes in 1985.

In 1988, Terstappen and Loken and their colleagues[1320-1] described cell staining by the laser dye **LDS-751**, which predominantly stains DNA, can be excited effectively at 488 nm, and emits above 640 nm. LDS-751 enters intact cells, but stains cells with damaged membranes more intensely, and can therefore be used to discriminate these two classes of cells, even in fixed samples[1320]. It can also be used in combination with another nucleic acid stain and fluorescent antibodies to perform extended leukocyte differentials, platelet, and reticulocyte counts on unlysed, unfixed whole blood[1321]. LDS-751 was said to be the same dye as the laser dye Styryl-8; this is apparently not the case, at least for the LDS-751 sold by Molecular Probes.[2332].

I have had good luck staining mammalian cell nuclei with LDS-751, but have not succeeded in getting it to stain bacteria, which I find somewhat surprising. I also note a recent report by Snyder and Small[2517], to the effect that cultured murine fibroblasts and monocytes stained with two different samples of Molecular Probes' LDS-751 and examined by confocal microscopy showed membrane potential-dependent mitochondrial staining but no nuclear staining.

Latt et al[470] described fluorescent staining of DNA by three compounds supplied by Eastman Kodak, designated EK4, LL585, and VL772. VL772 forms fluorescent complexes with both DNA and RNA; the DNA complex has an absorption maximum at 510 nm and an emission maximum at about 565 nm. LL585 was said to be DNA-specific; the complex of this dye with DNA has an absorption maximum at 569 nm and an emission maximum at about 600 nm. Both VL772 and LL585 showed a strong A-T preference. I have looked at these dyes; I found that both VL772 and LL585 formed fluorescent complexes with RNA. Also, the dyes are lipophilic; LL585, in particular, stains mitochondria in intact cells. I didn't get very good DNA histograms even when I stained isolated nuclei, and the background fluorescence was too high to do good measurements of bacteria (and, I would presume, of chromosomes). LL585 is a styryl dye; EK4 and VL772 are cyanines.

Cyanine Dyes I: Thiazole Orange, etc.

Cyanine dyes[460-1], many of which were developed by Eastman Kodak as sensitizers for photographic film, are compounds in which two heterocycles (benzoxazoles, benzothiazoles, etc.) are joined by a conjugated polymethine chain. **Symmetric** cyanine dyes, in which the heterocycles are identical in structure and orientation, are well known as probes of **membrane potential**[460]. Cyanines have, however, been known for some time to act as nuclear stains, a property noted in the parent compound, **cyanine** (1,1′di-*iso*-amyl-4,4′-quinocyanine iodide) in the late 1800's. **Dicyanine A** (diethyl-2,4′-quinocarbocyanine iodide), an **asymmetric** cyanine, was described as an RNA stain in Kodak literature in the mid-1970's. Fluorescent nuclear staining by

diethyloxacyanine, used as an optical brightener for textiles, was illustrated by Paton and Jones[1322] in 1976, and Jacobberger, Horan, and Hare described fluorescent staining of DNA in intraerythrocytic malaria parasites and RNA in blood reticulocytes by **dimethyloxacarbocyanine** (**$DiOC_1(3)$** in the notation of Sims et al[460]; see Figure 7-12 below) in 1983[290] and 1984[768], respectively.

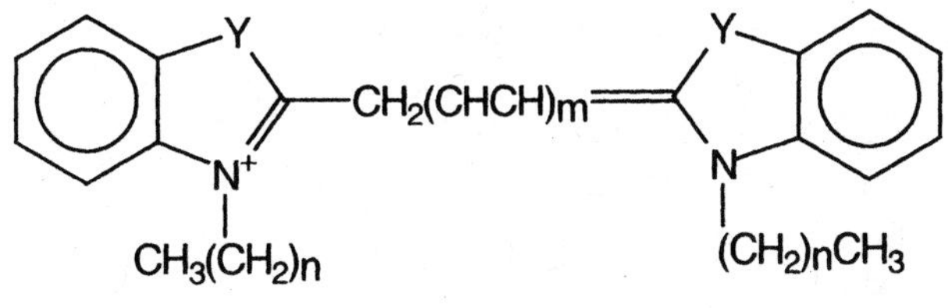

Figure 7-12. Structure of symmetric cyanine dyes given the formula "$DiYC_{n+1}(2m + 1)$" by Sims et al[460]. When Y is O or S, the substituent is oxygen or sulfur. When Y is I, the substituent is $C(CH_3)_2$. When Y is L or Q, the rings are isoquinolines joined at the 2 or 4 positions, respectively.

In 1983, Sage, O'Connell, and Mercolino[288], then at B-D, developed a reticulocyte analysis technique using **thioflavin T**, a basic thiazole dye mentioned as an RNA stain by Arndt-Jovin[289]. Thioflavin T is a true RNA (and DNA) fluorochrome, increasing its quantum efficiency many times when bound to nucleic acid. The excitation maximum for the dye is at approximately 440 nm, making it usable with an Hg arc lamp source emitting at 436 nm, such as was used in B-D's FACS Analyzer. Thioflavin T can also be excited by He-Cd (441 nm) or argon laser (457, but not 488 nm) sources. Emission is at 490-500 nm.

After B-D introduced the FACScan, a clinically oriented flow cytometer with an excitation wavelength fixed at 488 nm, it became desirable to find a dye with the characteristics of thioflavin T that could be used with that excitation wavelength. Lee, Chien, and Chiu[769,1323] took a frontal approach to this problem, studied the relationship between chemical structure and RNA fluorochrome activity in cyanine dyes, thioflavin, and ethidium, and, in 1987, described **thiazole orange (TO)**, which, chemically, is 1,3′-dimethyl-4,2′-quinothiacyanine, as most suited for reticulocyte analysis. Thiazole orange, when bound to RNA, has an absorption maximum at 509 nm and an emission maximum at 533 nm, and its fluorescence quantum efficiency is increased approximately 3,000 times over that of the free dye. Like thioflavin T, thiazole orange also behaves as a DNA fluorochrome; it and related dyes are useful for reticulocyte counting only because reticulocytes normally contain no DNA. Van Bockstaele and Peetermans[1324] noted in 1989 that 1,3′-diethyl-4,2′-quinothiacyanine iodide, a dye that had been widely available for many years, could be substituted for thiazole orange, which was difficult to get in at least some markets due to patent issues.

Lee and her coworkers at B-D also synthesized 1,3′-dimethyl-4,2′-quinothiacarbocyanine, which they named **thiazole blue**; this dye also behaves as a DNA and RNA fluorochrome and, when bound to nucleic acid, has an excitation maximum at about 640 nm, making it useful with red He-Ne or diode laser sources. As is the case with thiazole orange, the diethyl analog of thiazole blue has been commercially available for some time.

The ring structure of thiazole orange is the same as that of TO-PRO-1 (Figure 7-10, p. 301); the ring structure of thiazole blue has a trimethine bridge instead of a monomethine bridge between the heterocycles, and both thiazole orange and thiazole blue have methyl groups on both ring nitrogens (Molecular Probes' TO-PRO-3 shares the thiazole blue ring structure but has an n-propyltrialkylammonium side chain on the quinoline ring nitrogen, as does TO-PRO-1). Since thiazole orange and thiazole blue cyanine dyes are lipophilic and bear only a single delocalized positive charge; they are permeant (the TO-PRO dyes are impermeant), and are concentrated inside cells (and intracellular organelles) by any inside-negative potential difference across the cytoplasmic and/or organellar membrane. The dyes normally reach even higher concentrations in mitochondria than in the cytosol. Although mitochondrial staining by other cyanine dyes is pronounced, neither thiazole orange nor thiazole blue produces much mitochondrial staining.

This doesn't mean the dyes aren't concentrated in energized mitochondria in intact cells; they virtually have to be, according to chemical principles. However, the dyes aren't all that fluorescent in aqueous solution, and they aren't that much more fluorescent in nonpolar solvents such as butanol. Their fluorescence is enhanced dramatically by intercalation into double-stranded nucleic acid, be it DNA or RNA, and also, somewhat surprisingly, by binding to single stranded nucleic acid.

Symmetric cyanine dyes, such as $DiOC_1(3)$, are fluorescent in aqueous solution; their fluorescence is enhanced, typically two- to sixfold, in nonpolar solvents, and enhanced by approximately the same factor when they bind to nucleic acids. The fluorescence both of symmetric dyes and of asymmetric dyes such as thiazole orange and thiazole blue requires that all the ring structures in the dye molecule remain in the same plane, and that the probability of loss of excitation energy due to transfer to other molecules in the environment remain low.

The great degree of fluorescence enhancement noted when thiazole orange or thiazole blue bind to nucleic acid should not be taken as an indication that these dyes have higher quantum efficiencies than the symmetric cyanine dyes; what it does indicate is that thiazole orange and thiazole blue, when not bound to nucleic acid, have very low quantum efficiencies. This is probably so because there is some freedom for one of the rings in the unbound dye to rotate relative to the other; when the rings are not in the same plane, no fluorescence can occur. There may also be more nonradiative loss of excitation from unbound molecules to solvent and other molecules in the environment than there is when the dyes are intercalated into double-stranded nucleic acid.

The high quantum efficiency of bound dye makes thiazole orange and its relatives very difficult dye to use in microscopy; as was discussed on pp. 115-8, high quantum efficiency makes for rapid bleaching. It is relatively hard to do a reticulocyte count with a flow cytometer using acridine orange as the RNA stain, because the RNA-bound dye is quenched and background fluorescence is high. When you look under the microscope, however, the fluorescence of background dye bleaches out quickly and that of the quenched dye-RNA complexes persists, facilitating counting. When you look at thiazole orange-stained cells under a fluorescence microscope, there is very little fluorescence background, but the fluorescence from RNA-bound dye disappears before you have a chance to count the cells. In a flow cytometer, the cell doesn't get a chance to get bleached until it goes through the beam, at which time the detector can capture as much fluorescence as is emitted; thiazole orange works fine. It and its relatives also work fine in other instruments, for example, high-resolution gel scanners for detecting small amounts of DNA following electrophoresis[1325-6].

Cyanine Dyes II: TOTO and YOYO à GoGo

The gel scanning project undertaken by Glazer, Peck, and Mathies[1327] in the 1980's had two goals; the first was optimizing methods for detection of very small amounts of DNA on gels, and the second was developing dyes that could bind so strongly to DNA that dye-labeled fragments could be removed from gels and mixed with unlabeled nucleic acids without dye migrating from the labeled to the unlabeled molecules. The strategy adopted in pursuit of the second goal was the synthesis of homodimeric dyes, the first of which was **ethidium homodimer**. When DNA-ethidium homodimer complex is mixed with a 50-fold excess of unlabeled DNA, about 30% of the dye is retained within the original complex indefinitely.

The quantum efficiency of ethidium increases only 20- to 30-fold on binding to double-stranded nucleic acid; that of thiazole orange, as previously mentioned, increases 3,000-fold. The difference, as just discussed, lies in the lower quantum efficiency of unbound thiazole orange; the use of thiazole orange or a dye with similar fluorescence properties for staining DNA or RNA on gels should, therefore, result in lowered background fluorescence compared to ethidium homodimer.

In order to achieve the high binding affinity of ethidium homodimer and the fluorescence properties of thiazole orange, Rye et al[1328] synthesized compounds named **TOTO-1** and **YOYO-1**, in which two molecules of thiazole orange (TO), in the first case, and oxazole yellow (YO), its quinooxacyanine analog, in the second, were joined by a diazaundecamethylene linker. The binding affinity of these dyes for DNA is sufficiently high so that fragments labeled with TOTO-1 or YOYO-1 and with ethidium can be mixed and separated by electrophoresis. TOTO-1 has the spectral characteristics of thiazole orange; YOYO-1 has those of oxazole yellow (excitation maximum for nucleic acid-bound dye 489 nm; emission maximum 509 nm), and both are over 1,000 times more fluorescent in the nucleic acid-bound than in the free form. As little as 4 picograms of DNA-bound dye can be detected on a gel by the confocal scanner. **TOTO-3**, which combines two thiazole blue chromophores using the same linker as is used in TOTO-1 and YOYO-1, is red-excited. All three dyes and some related dimers are available from Molecular Probes.

Benson, Singh, and Glazer[1328] also synthesized **heterodimeric** DNA-binding dyes designed for energy transfer, including **TOTAB**, a thiazole orange-thiazole blue heterodimer for which the emission maximum of the DNA-bound dye lies at 662 nm. Fluorescence at this wavelength is enhanced 100-fold on binding to double-stranded DNA, while fluorescence of the thiazole orange donor chromophore at 532 nm is quenched by over 90%. The heterodimeric dyes, like the homodimers, are suitable for high-sensitivity detection of nucleic acids on gels[1329]. TOTAB, which, at this writing, is not yet commercially available, could be used for determination of DNA content using 488 nm excitation in cells also stained with fluorescein- and PE-labeled antibodies and, perhaps, PE-Texas Red tandem-labeled antibodies as well. There would, however, also be a catch.

Cyanine Dyes III: Alphabet Soup

Staining fixed cells or permanent cell nuclei for DNA content analysis tends to be a relatively casual procedure; we more or less just throw the dye in and go, without bothering to think that it may take more than a few seconds for dye and cells to come to equilibrium. When van den Engh, Trask, and Gray[1330] actually checked on this point, they found that mouse thymocyte nuclei permeabilized with Triton X-100 and stained with dyes at concentrations normally used for flow cytometry required about 5 min to come to equilibrium with Hoechst 33258, 20 min to equilibrate with propidium, and over an hour to reach equilibrium with chromomycin A_3. These dyes bounce on and off the DNA fairly readily, compared to the dimers just discussed, which stick like glue, and which therefore should take a longer time to equilibrate. I've tried staining permeabilized, RNAse-treated nuclei with TOTO-1 and YOYO-1 (the RNAse is essential because the dyes are not DNA-specific), and found that the DNA histogram peak positions and CVs did not stabilize even after several hours' incubation with the dyes; the monomeric dyes reach equilibrium much faster.

Hirons, Fawcett, and Crissman[1331] described use of TOTO-1 and YOYO-1 for flow cytometric DNA analysis, precipitating a brief stampede in that direction. In my view, however, it is important to separate the desirable spectral characteristics of the dyes from what are likely to be undesirable binding characteristics. Using thiazole orange or its cheaper diethyl analog instead of TOTO-1 to do DNA histograms should, in principle, realize the same advantages while requiring less incubation time.

Richard Haugland, Victoria Singer, Stephen Yue, and others at Molecular Probes have developed additional monomeric and dimeric cyanine DNA and RNA stains with some interesting properties[2332,2618]. The TO-PRO® family of impermeant monomeric dyes contains BO-PRO-1 and -3, JO-PRO-1, LO-PRO-1, PO-PRO-1, and -3, TO-PRO-1, -3, and -5, and YO-PRO-1 and -3, with excitation maxima ranging from 435 to 745 nm and emission maxima ranging from 455 to 770 nm. All TO-PRO series dyes have two positive charges and one intercalating moiety. The TOTO family of impermeant dimeric dyes contains corresponding BOBO, JOJO, LOLO, POPO, TOTO, and YOYO dyes, except that there is no TOTO-5; all TOTO series dyes have four positive charges and are **bis-intercalators**; that is, each has two intercalating moieties, which bind between nearby base pairs of a double-stranded nucleic acid helix, while the linker interacts with the minor groove. The nature of the strong interactions of TO-PRO and TOTO dyes with single-stranded nucleic acids remains less clear.

Molecular Probes' SYTO series of dyes are cell permeant relatives of the TO-PRO dyes. Their permeancy indicates that, like their parent thiazole orange (the TO in TO-PRO, TOTO, and SYTO), they lack the quaternary ammonium groups present in TO-PRO and TOTO dyes. The fact that their structures have not been revealed, at least in the Molecular Probes *Handbook*[2332], suggests that the dyes have extra ring side chains that improve their permeancy, and that the company does not want to make it easy for other people to learn its tricks. Some of the structures may be found in Molecular Probes' patents. The SYTO- dyes stain both DNA and RNA; some of them also stain other cellular structures, e.g., mitochondria.

The SYTOX dyes, are really, really impermeant, suggesting that they bear at least three positive charges (structures have not been published); they are promoted as viability indicators, but SYTOX Orange was found to be an optimal stain for DNA fragment sizing in a slow flow instrument with a green YAG laser source built at Los Alamos[2333-4].

Then there are PicoGreen, designed for quantitative fluorescence assay of double-stranded DNA in solution, and SYBR Green I and II and SYBR Gold, designed for demonstration and quantification of nucleic acids on gels. PicoGreen is reasonably selective for double-stranded DNA, increasing its fluorescence over 1,000 times on binding, while exhibiting a much smaller degree of fluorescence enhancement on binding to double-stranded RNA or to single-stranded nucleic acid. PicoGreen and the SYBR dyes have been used for DNA fragment sizing at Los Alamos[2333] and for detection of viruses in conventional flow cytometers[2336-7].

Base Preference in Cyanine and Styryl Dyes

I have observed (unpublished) that the fluorescence of TOTO-1, YOYO-1, TOTO-3, related monomeric cyanine dyes, and LDS-751 in solutions of a deoxyadenosine/deoxythymidine polymer (polydA-dT) is severalfold higher than the fluorescence of the same dyes in solutions of a deoxyguanosine/deoxycytosine polymer (polydG-dC) at the same dye and polymer concentrations, suggesting that the dyes have an A-T preference. Excitation and emission maxima are slightly different for dye in polydA-dT and in polydG-dC. Stephen Yue of Molecular Probes (personal communication, 1994) has made similar observations. However, it has also been established (R. A. Keller, personal communication, 1994; A. N. Glazer, personal communication, 1994, and references 1144 and 1325-9) that the intensity of TOTO-1, YOYO-1, and TOTAB fluorescence is proportional to the mass of DNA to which the dye is bound, regardless of the base composition of the dye. In this context, at least, the dimeric dyes do not show a base preference; the lack of one is critical for use of dimeric dyes in quantitative analysis of DNA in flow systems, in solution, or on gels.

The issue of base preference of cyanine dyes is further clouded by evidence from my lab (unpublished results and reference 1133), Frey et al[1134], and Hirons, Fawcett and Crissman[1331] that TOTO-1 and YOYO-1, thiazole orange, and thiazole blue produce differential staining of chromosomes with similar DNA content, in much the same way as do dyes with strong base preferences such as Hoechst 33258 and chromomycin A_3. In fact, something approximating the bivariate flow karyotype obtained from chromosomes stained with Hoechst 33258 and chromomycin (see the section on DNA base composition) can be produced by dual beam (UV and visible) excitation of chromosomes stained with TOTO-1 alone[1331]. It would certainly be useful to have a greater variety of dyes available for bivariate chromosome analysis, particularly dyes that could be excited at 488 nm; it is likely, based on studies of the existing cyanine and styryl dyes, that such molecules could be engineered, but there doesn't seem to be enough of a market to get Molecular Probes or other companies to make the effort.

Seeing Red: LD700, Oxazine 750, Rhodamine 800, TO-PRO-3, and DRAQ5 as DNA Stains

In the mid-1980's, while examining a number of dyes suitable for excitation by red (633 nm) helium-neon lasers, Sandra Stephens and I ran across three that stained DNA stoichiometrically[737]. They are LD700, oxazine 750 (OX750), and rhodamine 800 (R800), all developed as laser dyes. The dye-DNA complexes of these three dyes have absorption maxima at 660, 690, and 705 nm and emission maxima at 670, 700, and 715 nm. OX750 appears not to have a base preference; LD700 and R800 have a G-C preference. OX750 and R800 are usable with cheap 670 nm diode lasers.

Red-excited DNA stains aren't all that useful unless you can stain cells simultaneously with them and something else; it would be nice to do DNA/immunofluorescence measurements in a system with a diode laser using R800 and allophycocyanin or Cy5. However, at the concentrations (~25 µM) at which OX750 and R800 have to be used to produce good staining, they quench the fluorescence of allophycocyanin considerably.

Any of the TO-PRO or SYTOX dyes, and at least some of the SYTO and SYBR series and PicoGreen, should be usable for DNA content analysis in permeabilized or fixed cells treated with RNAse. It has been shown that TO-PRO-3 works well in this context when used with red excitation[2619]; this dye can also be used in combination with propidium in systems with only a single 488 nm laser, as energy transfer from propidium, which excites well at 488 nm, to TO-PRO-3, which does not, is fairly efficient[2620-1].

I mentioned **DRAQ5**[2338-9] on p. 44; this is a permeant anthraquinone dye which, when bound to DNA, has an excitation maximum near 650 nm and an emission maximum near 700 nm. DRAQ5 appears to be reasonably DNA-selective, and the combination of DNA selectivity and permeancy allows it to be used to produce reasonably good DNA content histograms when it is used as a vital stain, making it the only dye other than Hoechst 33342 usable for that purpose.

The fluorescence of DRAQ5 is not significantly increased when the dye is bound to DNA, and some manipulation of the relative concentrations of dye and cells may be necessary to obtain the best quality DNA content measurements. On the plus side, it is possible to excite the dye at 488 nm. DRAQ5 is available from **Biostatus, Limited.** This company also sells the somewhat less permeant DRAQ5NO[2622], also developed by Paul Smith's group, and said to be useful in discriminating cells in various stages of apoptosis, as APOPTRAK™.

Miscellaneous DNA-Selective Dyes

The diamidines **M&B 938**[2623] and **hydroxystilbamidine**[2332,2624] are structurally similar to DAPI; they are UV-excited, impermeant, and relatively highly DNA-selective, with an A-T base preference. M&B 938 was investigated as a drug by Rhône-Poulenc Rorer, now part of Aventis. Hydroxystilbamidine, now available from Molecular Probes, has been used as a viability indicator[2624]; since complexes of the dye with DNA exhibit an excitation maximum at 390 nm[2332], it should be better excited by violet diode lasers than DAPI or the Hoechst dyes.

3-amino-6-methoxy-9-(2-hydroxyethylamine) acridine (AMHA) is maximally excited at 375 and emits maximally at 510 nm; it is permeant and somewhat DNA-selective, probably with an A-T base preference. It is not commercially available

What Do DNA Stains Stain?

DNA content measurements demand high precision because DNA content is so nearly constant for substantial numbers of cells in a sample and because minute variations in DNA content may have considerable biologic significance.

All of the DNA stains just discussed can be used, on appropriately prepared samples and in properly aligned instruments, to give precise estimates of the total DNA content of cells that accurately reflect DNA content differences between different cells and cell types from the same organism, e.g., sperm cells, which are haploid, and somatic cells, which are predominantly diploid, or tumor cells and accompanying normal stromal cells, which may have different modal values of DNA content. This is really quite remarkable in light of the fact that different dyes bind in different fashions and, in some cases, to different chemical components of DNA molecules.

A little more care is required when attempts are made to determine the actual **mass of DNA**[1332-3], or **genome size**[1334-6], in cells from different species. Dyes with different base preferences will give different results in cells with different base compositions; in analyzing cells from 80 species of Tetrapoda, Vinogradov and Borkin[1335] found that ratios of DNA content values determined with different dyes were as high as 1.8 for this reason.

Darzynkiewicz and his colleagues[738-9] examined the accessibility to different dyes of DNA in nuclei under various conditions. During chemically induced differentiation of Friend erythroleukemia cells, and during spermatogenesis, chromatin condenses, and nuclear staining intensity with some DNA stains decreases; accessibility of DNA to different dyes is increased to different extents by acid extraction of nuclear histones. The increase in accessibility is lowest (45%) for DAPI, highest (1200%) for 7-AAD.

It seems logical that the dye least affected by chromatin structure, i.e., DAPI, would give the DNA histograms with the lowest CVs, since the effects of slight cell-to-cell differences in chromatin structure would be minimized, and DAPI indeed seems to produce the lowest CVs (Fig. 7-11, p 311). By contrast, it is hard to get even fair stoichiometric staining with 7-AAD, which has been shown to produce differential staining of different types of leukocytes based on chromatin structure differences[740].

If DAPI is the dye that gives best CVs, mithramycin may be the most DNA-selective dye; DAPI and the Hoechst dyes, while DNA-selective in mammalian cells, may stain other materials in plants and bacteria, often showing some spectral shift when they do.

Rundquist[1337] studied binding of DAPI and 7-AAD to fibroblast nuclei, and found that DAPI binding, but not 7-AAD binding, was increased following detergent treatment. 7-AAD binding was decreased by fixation with formaldehyde, but not with ethanol, following detergent extraction. Removal of basic protein with HCl resulted in an increase of about 100% in both DAPI and 7-AAD binding. Scatchard analysis suggested the existence of at least two classes of binding sites for both dyes; Bertuzzi et al[1338], analyzing propidium binding to fixed lymphocyte nuclei, also found evidence for two classes of binding sites with different affinities.

Beisker and Eisert[1339] used fluorescence depolarization of ethidium and propidium to study denaturation of DNA *in situ* by acridine orange, which transfers energy nonradiatively to both of the former.

DNA Ploidy and Aneuploidy: The DNA Index

When DNA staining is stoichiometric, we tend to refer interchangeably to cells DNA content and **ploidy**, a practice that was discouraged by the Society for Analytical Cytology's Committee on Nomenclature[741]. It was recommended that measurements of DNA content in abnormal or putatively abnormal "test" cells be made in conjunction with measurements of DNA content in normal diploid "reference" cells from the same individual, and that differences in DNA content from normal be expressed in terms of the **DNA index**, which is the ratio of the mean or modal channel numbers of G_0/G_1 peaks in the distributions from the test and reference cells. The term **DNA aneuploidy** can be used to describe a sample containing a stem line with abnormal DNA content; **aneuploidy** without the qualifier is a term reserved for describing samples with abnormal karyotypes.

Comparisons between karyotype and DNA content as measured by flow cytometry have been done in human tumors and tumor-derived cell lines by Tribukait et al[742], Petersen and Friedrich[743], and Bigner et al[744]. In general, there is good agreement between measured DNA content and chromosome number; discrepancies tend to occur in cases with very high (hypertriploid or more) chromosome numbers, and these tend to have higher DNA content than would be expected from chromosome number. It has been suggested that some abnormal chromosomes may contain increased amounts of DNA.

Sample Preparation for DNA Content Analysis

The "garbage in, garbage out" aspect of flow cytometry affects DNA content measurements in several ways. One typically wants to measure nuclear DNA content in a good representative sample of a cell population. This is easiest when the cell population is a homogeneous line growing in suspension culture. Even when a minimal complicating factor, such as the cells growing attached to a dish, is thrown in, the potential for differential loss during preparation arises. When the sample is from real tissue, or a real tumor, it is necessary to make sure the cells, or nuclei, that are measured come from the part of the specimen that is of interest, and that the method used to prepare a single cell suspension does not selectively destroy some cell types[745]. Debris interferes with measurements; if a piece of tissue is minced too well, or if a paraffin block is cut too thin[746-7], the real cells may get lost in the debris.

It is customary, and acceptable, to use mathematical models[1307-9] to determine fractions of cells in different cell cycle phases and to reduce the contributions of clumps and debris to a DNA histogram (Figure 1-10, pp. 25-6); it is unacceptable, although far from unheard of, to use mathematical models to attempt to derive the fraction of cells in S phase from a histogram with barely recognizable peaks. We will get back to this issue when we talk about applications of DNA content analysis in Chapter 10.

DNA Base Composition

The base-paired structure of double-stranded DNA places constraints on the base composition of properly assembled molecules; the number of adenine residues must be the same as the number of thymine residues, and the number of guanine residues must be the same as the number of cytosine residues. The ratio of **A + T (adenine + thymine)** to **G + C (guanine + cytosine)** is not similarly constrained. In human cells, this ratio happens to be close to one; in bacteria, the G + C content of DNA varies from less than 25% to more than 85%[264]. Less dramatic variations in DNA base composition are observable in human chromosomes.

If a fluorescent dye binds with different affinities to A-T and to G-C pairs (or sequences) in DNA, dye molecules will be concentrated in regions of the DNA molecule containing the bases for which the dye has higher affinity. On the other hand, if binding to A-T and binding to G-C pairs or sequences have different effects on the fluorescence yield of a dye, even if the same amount of dye is bound to A-T and to G-C, there will be differences in the amounts of fluorescence emitted by A-T and G-C bound dye. Both of these mechanisms have been observed in the interactions of fluorescent dyes with DNA[230,265-8].

The first practical application of differential fluorescent staining of DNA came in the late 1960's with the development of **chromosome banding techniques** by Caspersson et al[269-72]. Before this, it was often impossible to obtain accurate karyotypes by visual observation; attempts at automation of karyotyping by analysis of microscope images had reached a dead end because size and shape were the only parameters that were measured. Using quinacrine and other dyes, it became possible to produce banding patterns along chromosomes that facilitated discrimination between chromosomes of similar sizes and shapes. Studies of the mechanisms involved in producing banding led to the development of improved staining techniques that have been applied to flow cytometry as well as to microscopy and image analysis of chromosomes.

Quinacrine binds with similar affinities to A-T and G-C; however, the fluorescence of DNA-quinacrine complexes increases with A + T content[265,273]. Hoechst 33258, which binds preferentially to A-T[268], does not produce a banding pattern when it is used as a single dye to stain chromosomes. However, when a G-C binding molecule such as actinomycin D, which can serve as an acceptor for nonradiative energy transfer, is added in combination with the Hoechst dye, a banding pattern similar to that obtained with quinacrine appears. The fluorescence of Hoechst dye bound to A-T in close proximity to actinomycin bound to G-C is quenched due to energy transfer; the fluorescence of Hoechst dye in regions containing relatively long sequences of A-T pairs is not quenched[274]. The fluorescence of DNA stains is also affected by interactions between DNA and chromosomal proteins[275].

Since differential staining of chromosomes by combinations of fluorescent dyes typically involves energy transfer from molecules of one dye to molecules of the other, dyes that work well in combination almost always have excitation maxima in different spectral regions, because for energy transfer to occur, the absorption spectrum of the acceptor species must overlap the emission spectrum of the donor species. The difference in spectral characteristics makes it necessary to use a flow cytometer with two excitation beams at different wavelengths to fully characterize specimens stained with pairs of DNA fluorochromes.

The combination of Hoechst 33258 and chromomycin A_3, developed for chromosome analysis by the Livermore group[276-9], provides a good example of a dye pair. Separated UV (351/364 nm) and blue (458 nm) beams from argon lasers are used to excite the blue fluorescence of the A-T bound Hoechst dye and the yellow fluorescence of the G-C bound chromomycin. Figure 7-13 gives an example of the results obtained from dual laser flow cytometric analysis of chromosomes stained with the combination of dyes.

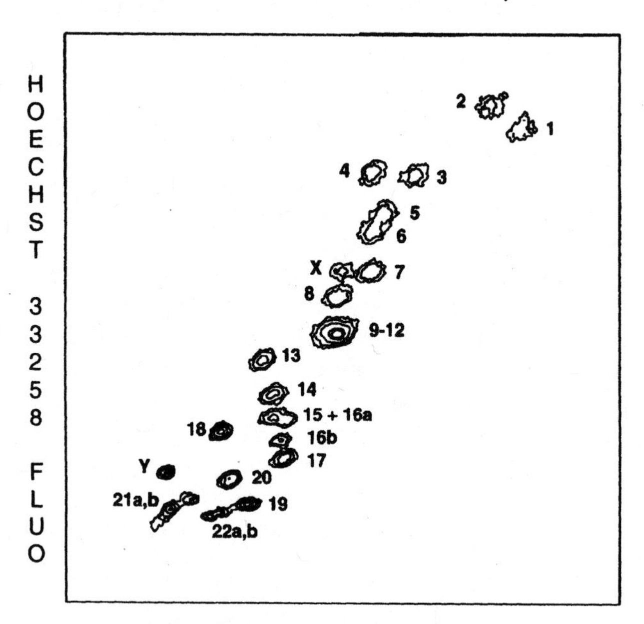

Figure 7-13. Flow karyotype of human chromosomes stained with Hoechst 33258 and chromomycin A_3. Provided by Ger van den Engh.

If the dyes had no base preference, the ratio of chromomycin fluorescence to Hoechst fluorescence would be the same for all chromosomes, and all of the data points would cluster along a straight line at 45° to the x and y axes, making it difficult to resolve chromosomes (e.g., 14-17) with similar DNA content. The differential staining gives greatly improved resolution in the two-dimensional space, because ratios of chromomycin fluorescence (G + C) to Hoechst fluorescence (A + T) do vary substantially among chromosomes of similar DNA content. This is true for humans; it is not for some other species.

Van Dilla et al[279] used the combination of Hoechst 33258 and chromomycin A_3 to differentiate bacterial species with different fractions of G + C; the principle is illustrated in Figure 7-14. The ratios of chromomycin fluorescence to Hoechst dye fluorescence for *Staphylococcus aureus*, *Escherichia coli*, and *Pseudomonas aeruginosa* reflect the marked differences in base composition (% G + C approximately 31, 50, and 67, respectively) among the three species. It is easy to separate the species, either by partitioning the two-dimensional space defined by the chromomycin and Hoechst dye fluorescence intensities, or by partitioning a histogram of fluorescence ratios.

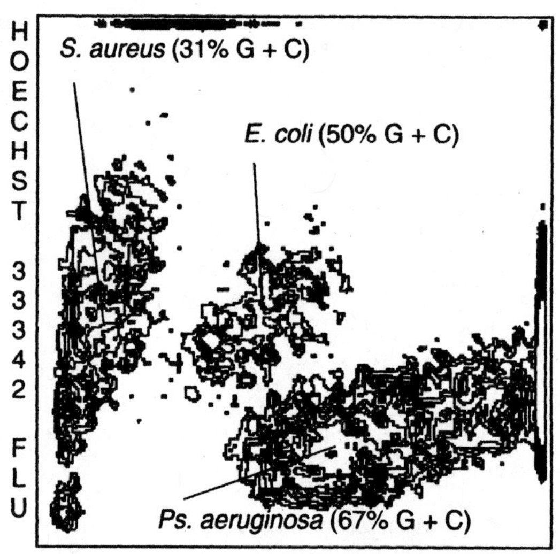

Figure 7-14. Hoechst/chromomycin fluorescence signatures of bacterial species with different DNA base compositions.

Sanders et al[1340] found a linear relationship (r = 0.99) between the log of the ratio of chromomycin A3 to Hoechst 33258 fluorescence and the log of % G + C over the range of 28-67% G + C; two cell populations could be identified in mixtures of two species that differed in base composition by as little as 4% G + C. This is a good example of the utility of a ratio as a single parameter in cases in which there are large variances in the absolute intensities of the two signals from which the ratio is derived; note that the clusters are markedly elongated, indicating that the ratio of chromomycin fluorescence to Hoechst dye fluorescence remains nearly constant for each of the bacterial species. In another paper, Sanders et al[1341] used changes in base composition to detect bacteriophage infection of *E. coli*.

It has been suggested that Hoechst/chromomycin staining could be useful in clinical laboratories for identification of bacteria in urine specimens. This might be so, provided the clinical instrument did not require the two 12-watt argon lasers used as light sources by van Dilla et al. In previous editions, I suggested that 5-10 mW at 325 nm and 20-40 mW at 441 nm from air-cooled He-Cd lasers might be adequate, when used with efficient optics, for flow cytometry of bacteria and chromosomes. I have already mentioned (p. 142; references 1133-5 and 1242) that He-Cd lasers have been used for bivariate analysis of chromosomes stained with Hoechst dyes and chromomycin. As for bacteria, Figure 7-14 was obtained using an instrument with a single He-Cd laser emitting 325 and 441 nm beams. Although the simpler instrumentation makes flow cytometry of bacterial base composition easier to do, the time required for fixation and staining and the relative lack of specificity of base composition for identification of bacterial species remain major obstacles to clinical use of the procedure.

In the past, I have used DAPI, which is A-T specific, in combination with olivomycin, which is G-C specific, to stain ethanol-fixed bacteria, choosing this dye pair because the absorption maxima of DAPI and olivomycin are closer to the 325 and 441 nm excitation wavelengths available from He-Cd lasers than are the absorption maxima of Hoechst 33258 and chromomycin A_3, which were used by Van Dilla et al[279]. I have also used mithramycin in place of chromomycin. When stained with DAPI/olivomycin or DAPI/mithramycin, *S. aureus*, *E. coli*, and *Ps. aeruginosa* are well resolved in a dual-beam flow cytometer with He-Cd laser excitation; however, the Hoechst dyes appear to give somewhat better separation than DAPI. Differences in Hoechst dye and DAPI staining of chromosomes have also been noted; Bernheim and Miglierina[1342] find that chromosomes 1 and Y stain more brightly with DAPI than with Hoechst 33258, presumably on the basis of different heterochromatin content.

As I mentioned on p. 315, dyes such as TOTO-1, YOYO-1, thiazole orange, and styryl-8, all excitable at 488 nm, appear to exhibit some sensitivity to base composition when used to stain chromosomes[1133-4,1331]. Since flow cytometers with 488 nm sources outnumber those with UV sources, many more laboratories would be able to do bivariate chromosome analysis and sorting if dye combinations were found that could be excited by a single 488 nm beam. Further work on cyanine and styryl dyes may provide the right dyes, at least for chromosome analysis; since these dyes also stain RNA, and since it is relatively difficult to get rid of RNA in bacteria, it may be harder to find alternatives to DAPI or Hoechst dyes and the chromomycins for determination of bacterial base composition. Structure-activity studies to improve the DNA specificity of cyanine and styryl dyes might help here.

Chromatin Structure; Identifying Cells in Mitosis

Differences in the staining and structure of nuclear chromatin in different cell types from the same organism were described by microscopists during the 1800's, well before the nature and roles of nucleic acids and nucleoproteins were understood. Such differences in chromatin structure are routinely used by hematologists and pathologists as aids in the identification and characterization of normal and abnormal cells, and it seems obvious that objective, quantitative, reproducible flow cytometric measures of chromatin structure should provide information of equivalent value.

The general approach to flow cytometry of chromatin structure is antithetical to the approach used in DNA content analysis. For determination of cellular DNA content with maximum accuracy and precision, it is necessary for the DNA in all cells examined to be equally accessible to the fluorochrome used, a condition best achieved by removal of most of the histones and other proteins that might interfere with staining[738]. For analyses of chromatin structure, it is essential that the conformation of nuclear material either be maintained or be modified in a predictable way.

The most extensive work on flow cytometry of chromatin structure is without doubt that done by Darzynkiewicz and his colleagues[262,1343-4]. They have employed a technique of **partial denaturation of DNA by acid or heat** treatment to demonstrate differences between different cell types and between cells in different phases of the cell cycle. Fixed cells are treated with RNAse to remove RNA, leaving DNA as the only nucleic acid present. Cells are then subjected to conditions that partially denature DNA, e.g., exposure to pH 1.5 for 30 seconds, and are then stained with **acridine orange (AO)**. The denatured DNA assumes a single-stranded conformation, and forms a polymeric complex with AO, shifting the absorption maximum of the dye to shorter wavelengths and the emission maximum to longer wavelengths, i.e., from the green (530 nm) spectral region to the red (>600 nm). AO monomers bind by intercalation to the remaining native helical DNA, retaining their normal absorption and fluorescence characteristics. When the stained cells are measured in a flow cytometer with blue or blue-green excitation (e.g., at 488 nm), the green fluorescence (515-575 nm) provides an estimate of the amount of DNA remaining in the native configuration, while the red fluorescence (600-700 nm) gives an estimate of the amount of denatured DNA.

Stokke and Steen[740] characterized the chromatin structure dependence of **7-aminoactinomycin D** binding to leukocytes. The level of 7-AAD binding is related to transcriptional activity, as indicated by DNA susceptibility to DNAse 1 digestion and by RNA synthesis. Stokke, Holte and Steen[1345] analyzed stimulated lymphocytes, and found that cells in G_1 bound almost twice as much 7-AAD as cells in G_0 ; dye binding increased almost linearly during the G_0 to G_1 transition, and correlated with expression of the early activation antigen 4F2 (now CD98). In diploid cells from non-Hodgkin's lymphomas, G_0/G_1 cell size, as measured by light scattering, was strongly correlated with 7-AAD binding. Using a dual-beam (UV and 488 nm) instrument, Stokke et al[1346] were able to identify G_0, G_1, S, and G_2 phase cells on a bivariate display of Hoechst dye vs. 7-AAD fluorescence and to discriminate live from dead cells; the mixture of Hoechst dye and 7-AAD can also be combined with fluorescein- and phycoerythrin-labeled antibodies for measurements of two cellular antigens.

Although it seems likely that some of the differences in chromatin structure between different cells could produce different patterns of staining by combinations of DNA fluorochromes, there is very little on this topic in the literature. Cowden and Curtis[280] compared staining of mouse thymocytes (condensed chromatin) and hepatocytes (more loosely organized chromatin) by various dyes, and noted different fluorescence intensities in the two types of nuclei when mithramycin and 7-aminoactinomycin D, among other dyes, were applied following RNAse treatment. Crissman (H. Crissman, personal communication, also reference 1347) has examined the use of combinations of DNA stains in combination with three-beam illumination to demonstrate differences in chromatin conformation by differences in DNA accessibility to different dyes; unfortunately, very few people have access to the three-beam flow cytometers needed to do this kind of work.

Chromatin Structure Identifies Mitotic Cells

The AO/acid denaturation technique provides more information about cells' progress through the cell cycle than can be obtained from DNA content measurements alone; it is particularly useful for discriminating **mitotic (M) cells** from cells in the G_2 phase, which have the same (4C) DNA content. As shown in Figure 7-15, cells in M phase (the identification of clusters was confirmed by flow cytometry of cultures in metaphase arrest) show a much higher red fluorescence, i.e., their DNA is less resistant to denaturation.

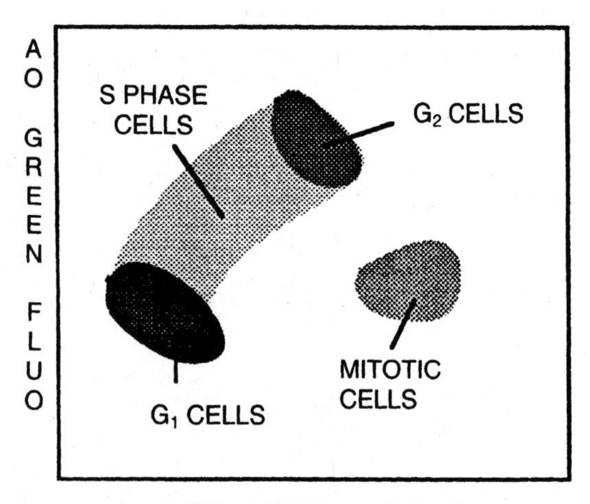

Figure 7-15. Flow cytometry of chromatin structure: AO staining patterns of an exponentially growing culture of Friend mouse erythroleukemia cells following partial DNA denaturation by acid (after Darzynkiewicz et al[262]).

Larsen et al[748] have developed an alternative procedure for **mitotic cell discrimination**; they report that, if nuclei isolated using nonionic detergents are fixed with formaldehyde, their fluorescence intensity following staining with ethidium, mithramycin, or propidium is quenched relative to unfixed isolated nuclei. Quenching is significantly less in the case of mitotic cells, causing these to form a separate cluster in a two-dimensional forward scatter vs. fluorescence measurement space.

It was previously mentioned (p. 277) that changes in **side scatter** of nuclei may reflect changes in chromatin structure[1254] and can be used to identify mitotic cells[1255-6]. Mitotic cells can also be identified by demonstration of certain nuclear antigens; this will be discussed further in subsequent sections.

RNA Content

The presence of relatively large amounts of **RNA** in growing cells was demonstrated many years ago. RNA is largely responsible for the **cytoplasmic basophilia**, or tendency of cytoplasm to stain intensely with basic dyes, which characterizes most immature and some leukemic blood cells. Histochemical demonstration and differentiation of RNA and DNA is largely based on the work of Brachet[36], who showed that a mixture of the basic dyes **methyl green** and **pyronin Y** stained nuclear DNA green and cytoplasmic RNA red. Pyronin Y, like methylene blue, is a basic, tricyclic heteroaromatic dye; both are homologues of **acridine orange**, which is employed in the best known flow cytometric technique for RNA content measurement, that described by Darzynkiewicz et al[262-3] and discussed on pp. 44 and 96-7. In this procedure, AO is used to stain both RNA and DNA. with results illustrated in Figures 3-10 (p. 97) and 7-16 (p. 321).

RNA/DNA Staining with Acridine Orange (AO)

In order to achieve good quantitation of DNA and RNA[1348-9], cells are first permeabilized with Triton X-100, and then stained with AO at a relatively low pH in the presence of EDTA. Under these conditions, DNA remains intact, i.e., in its native, double-stranded, helical form, while virtually all of the RNA present is converted to the single-stranded form. The low pH preserves the cytoplasm, which would otherwise be solubilized by the Triton X-100.

The AO concentration, which is critical, is adjusted so that DNA-bound dye is exclusively in the monomeric, intercalatively bound form, which fluoresces green. The AO bound to RNA is present in the form of a complex of RNA and dye polymers, and exhibits metachromatic red fluorescence. When the proper staining conditions are maintained, DNA content distributions estimated from AO green fluorescence show high precision (CVs less than 3%), while AO red fluorescence comes almost entirely from RNA, as shown by the loss of over 85% of red fluorescence following RNAse treatment.

Cell Cycle Compartments Defined on the Basis of RNA and DNA Content

Darzynkiewicz et al have used RNA and DNA content to define cell cycle compartments[262] on the basis of analysis

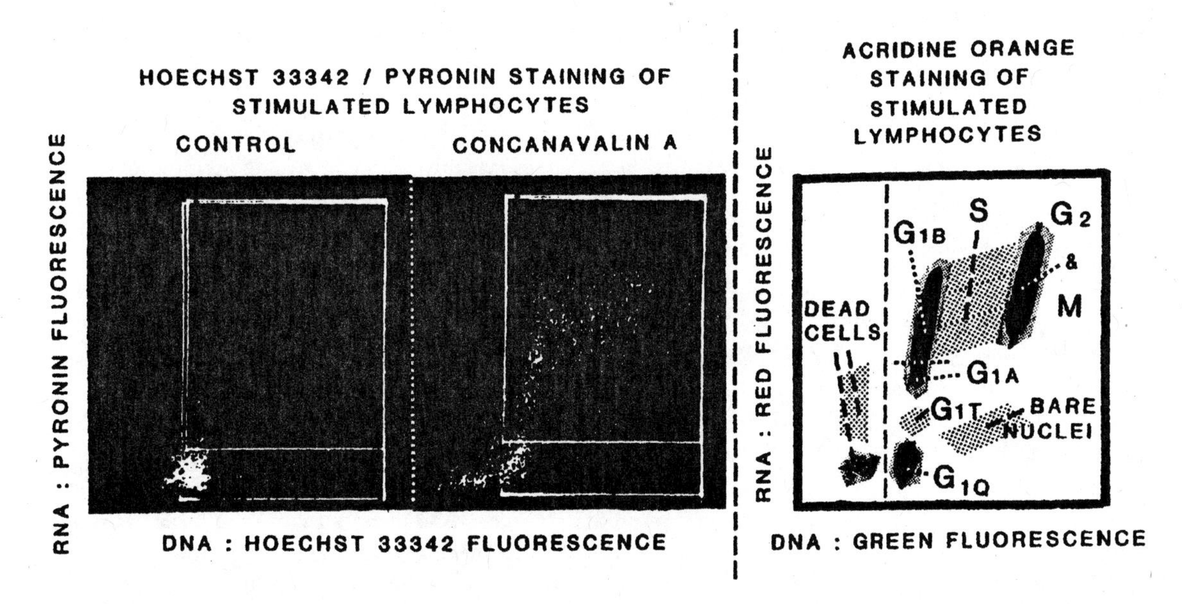

Figure 7-16. DNA and RNA content analysis of mitogen-stimulated lymphocytes. Left: typical results using the combination of Hoechst 33342 and pyronin Y described by the author[113]. Right: Typical results with acridine orange as described by Darzynkiewicz et al[262], showing cell cycle compartments defined on the basis of DNA and RNA content.

of such systems as mitogen-stimulated lymphocytes and leukemic cells undergoing chemical-induced partial differentiation *in vitro*. Typical DNA/RNA staining patterns of stimulated and unstimulated lymphocytes are shown in Figure 7-16, which also shows that other stains for DNA and RNA, in addition to AO, can be used to demonstrate the same cell cycle compartments.

The **quiescent state**, described as G_0 or G_{1Q}, in which cells such as peripheral blood lymphocytes normally remain, is characterized by a "diploid" (2C) DNA content and a low RNA content. Within 12 hours or so following exposure to mitogens, lymphocytes enter the G_1 phase and begin to synthesize RNA. RNA content continues to increase during the S phase, beginning about 30 hours after stimulation, in which DNA synthesis occurs.

Analysis of DNA content alone cannot discriminate cells in G_0 (G_{1Q}) from cells in the **proliferative** G_1 state, because the DNA content remains at 2C until the S phase begins. Measurements of RNA content can be used to make this distinction and, in addition, to define different stages within G_1. Cells pass from G_{1Q} through a brief transitional phase called G_{1T}, in which RNA content is slightly increased, and then into G_{1A}, during which RNA content increases further, but remains lower than the RNA content of any S phase cell. They then enter G_{1B}, in which RNA content is at or above the lowest value seen in S phase cells. RNA content increases approximately linearly during S and G_2.

In exponentially growing cultures, which lack cells in G_{1Q}, cells appear to pass from S through G_2 and M back into G_{1A}. Normal cells, such as stimulated lymphocytes, when maintained in long-term culture, tend to revert back to a G_{1Q} state, although quiescent, low-RNA "S_Q" and "G_{2Q}" populations can appear transiently in cells deprived of nutrients or exposed to cold or to inhibitors of protein synthesis. Transition to quiescent (Q) states during S and G_2 appears to be somewhat more common in transformed and malignant cells.

The pattern just described, in which RNA content increases during proliferation and decreases during quiescence, has been observed in human and animal blood, connective tissue, and epithelial cells of normal and malignant origin, and thus appears to be reasonably general. In many of the cell types examined, e.g., blood cells and leukemic cell lines, RNA content decreases with **differentiation** or maturation of cells. In other cell types, particularly those in which mature, nonproliferating cells are actively involved in protein synthesis, differentiated cells may contain more RNA than is found in their less mature progenitors, and high RNA content may not identify proliferative states.

As is illustrated in the left portion of Figure 7-16, the patterns of cellular DNA and RNA content observed using AO staining can also be demonstrated when other dyes, in this instance Hoechst 33342 and pyronin Y, are employed to stain DNA and RNA. Other parameters can substitute for RNA content for the definition of cell cycle compartments; it is possible, for example, to use **DNA and nuclear protein content** to identify proliferative and quiescent cell subpopulations[749-53].

AO: Problems and Some Solutions

I mentioned previously that, when staining conditions are properly adjusted, the red and green fluorescence of AO provide good estimates of RNA and DNA content. It has not been easy for many people, myself included, to duplicate the results obtained by Darzynkiewicz et al[262-3] using AO. This is due to the fact that the staining depends on a rather complex chemical equilibrium between AO bound to DNA, AO bound to RNA (and to other AO molecules), AO in solution in the sample (core) stream, and AO diffused into the sheath stream. The staining is therefore dependent, not only upon the concentration of dye and the concentration of cells used, but also upon the dimensions and flow rates of core and sheath fluid streams.

Most of the earlier work reported by Darzynkiewicz and his colleagues was done on Ortho Cytofluorograf flow cytometers. It was fairly easy to achieve similar results using these instruments, while adaptation of the AO staining procedures to other flow cytometers generally required some modifications of the staining protocol. As mentioned on p. 312, Pennings et al[734] reported that the stability and quality of staining were greatly improved when reagent was added continuously to the sample stream using a manifold and pump.

There are, however, other notable disadvantages associated with AO staining. The procedure used for DNA/RNA analysis requires membrane permeabilization by Triton X-100, and cannot, therefore, be used on living or intact cells. The orthochromatic (green) and metachromatic (red) fluorescence emissions of AO, between them, overlap the emission regions of fluorescein and almost every other material that can be used as a fluorescent antibody label with excitation near 488 nm. This makes it difficult to employ AO staining for DNA/RNA content analysis in conjunction with immunofluorescent staining to identify cells by phenotype or demonstrate expression of growth- or differentiation-related antigens. While UV-excited, blue fluorescent antibody labels could, in principle, be used in conjunction with AO, the antigens detected would have to retain their conformation at low pH following Triton X-100 treatment; this seems unlikely.

Finally, AO itself tends to adhere to the tubing and flow system components of flow cytometers and, unless great care is taken to remove residual dye from the instrument following analysis of AO-stained samples (e.g., by flushing with Clorox or another bleach), unstained or weakly stained cells (e.g., samples for immunofluorescence analysis) that are run thereafter will take up the dye and exhibit fluorescence, which may lead to misinterpretation of the data obtained.

Pyronin Y, Oxazine I, and Other Tricyclic Heteroaromatic Dyes as RNA Stains

In attempting to develop alternative methods for flow cytometry of DNA and RNA content that could overcome the disadvantages associated with AO[113], I examined the properties of a number of tricyclic heteroaromatic dyes that are similar in structure to AO and that had been reported to stain DNA and/or RNA. These are summarized in Table 7-3. Among the dyes described in this table are **pyronin Y** and **methylene blue**, the xanthene and thiazine homologs, respectively, of AO, both of which are well known as RNA stains. Also included are the thiazine dye **new methylene blue** and the oxazine dye **brilliant cresyl blue**, both widely used in laboratory hematology to demonstrate the RNA in blood reticulocytes. Some of the dyes produce metachromatic staining; these tend to have higher dimerization constants than the dyes that do not exhibit metachromasia. However, dye binding to and staining of RNA can occur without metachromasia. All of the dyes bind to DNA, and were found by Müller et al to exhibit varying degrees of preference for G-C regions[266-7].

In the **methyl green-pyronin** technique for DNA/RNA staining[36], nuclei are stained green, while the cytoplasm is red, suggesting that DNA is stained predominantly by methyl green, while RNA is stained predominantly by pyronin. Methyl green, like the Hoechst dyes, binds without intercalation to DNA and exhibits a strong A-T preference[268]. These facts led me to try the combination of **Hoechst 33342 and pyronin Y**[113] for staining DNA and RNA in intact cells. While this approach has the disadvantage of requiring the use of dual-wavelength (UV and blue-green or green) excitation, the Hoechst/pyronin technique does produce results comparable with those obtained using AO (Figure 7-16). DNA content may be estimated with excellent precision (CVs of the G_1 peak below 2%) in intact cells stained with Hoechst 33342 in combination with pyronin Y. That the intensity of pyronin Y and fluorescence is largely representative of RNA content was established by the loss of up to 85% of this fluorescence when ethanol-fixed cells were treated with RNAse.

There is now good evidence that the absence of fluorescence from pyronin bound to DNA in cells stained with Hoechst/pyronin results from blocking of pyronin binding to DNA by the Hoechst dye. In 1982, Pollack et al[281] reported that methyl green, which, like the Hoechst dyes, binds to A-T sequences in the minor groove, blocks pyronin binding to DNA, even though pyronin binds intercalatively and has a G-C preference. In 1990, Loontiens et al[2612] reported that Hoechst 33258 displaces intercalators from DNA, and, in 1995, Toba et al[2343] noted that pyronin Y alone produced fluorescent staining of both DNA and RNA, and that both Hoechst 33342 and 7-AAD decreased pyronin fluorescence in RNAse-treated cells in a concentration-dependent fashion.

Darzynkiewicz and his colleagues[754-9] studied the interactions of pyronin with both cells and nucleic acids. With the Los Alamos group, and using the Los Alamos three-beam flow cytometer, it was shown that the combination of Hoechst 33342, pyronin Y, and fluorescein isothiocyanate (FITC) could be used to stain DNA, RNA, and protein in fixed cells[754-6].

The question remained as to the specificity of pyronin Y as an RNA stain in intact cells. Cowden and Curtis[760] reported in 1983 that the dye stained mitochondria of some live cultured cells. In my 1981 paper on Hoechst/pyronin staining[113], I had shown that RNAse treatment largely abolished pyronin fluorescence in fixed cells, and noted that pyronin fluorescence intensities were similar in fixed and unfixed cells. This suggested that most of the pyronin fluo-

CLASS	COMPOUND NAME	λ_{max} (nm)	R1	R2	R3	R4	A	B	C	A_{DNA} ($\times 10^4$)	K_D ($\times 10^4$)	Meta-chro-matic?	Retic or RNA Stain?
ACRIDINES	proflavine	444	H	H	H	H	–	–	–	8.8			
X = CH	acridine orange (AO)	492	Me	Me	Me	Me	–	–	–	20.0	1.6	yes	yes
Y = NH	coriphosphine O	467	Me	Me	H	H	–	Me	–	3.2		yes	
AZINES	neutral red	540	Me	Me	H	H	–	Me	–	0.66		yes	yes
X = N													
Y = NH													
XANTHENES	3,6-diaminoxanthylium	496	H	H	H	H	–	–	–	1.8			
X = CH	acridine red		H	Me	H	Me	–	–	–				
Y = O	pyronin Y (G) (GS)	545	Me	Me	Me	Me	–	–	–	1.3	0.33		yes
	pyronin B	550	Et	Et	Et	Et	–	–	–				yes
THIOXANTHENES	3,6-diaminothioxanthylium	510	H	H	H	H	–	–	–	11.0			
X = CH	thiopyronin G	563	Me	Me	Me	Me	–	–	–	2.1	2.1		
Y = S													
OXAZINES	Capri blue GB (L)	643	Me	Me	Me	Me	–	–	–	0.28			
X = N	Capri blue GON		Me	Me	Et	Et	–	–	–			yes	
Y = O	oxazine 1	645	Et	Et	Et	Et	–	–	–				
	brilliant cresyl blue	636	Et	Et	H	H	Me	–	NH_2				yes
THIAZINES	thionin	598	H	H	H	H	–	–	–	3.7	2.2	yes	yes
X = N	azure C	609	Me	H	H	H	–	–	–				
Y = S	azure A	628	Me	Me	H	H	–	–	–			yes	yes
	azure B	647	Me	Me	Me	H	–	–	–	8.2	0.98	yes	yes
	methylene blue	665	Me	Me	Me	Me	–	–	–	8.1	0.71	yes	yes
	new methylene blue	636	Et	H	Et	H	Me	Me	–			yes	yes
	toluidine blue	636	Me	Me	H	H	–	Me	–	2.4		yes	yes

Table 7-3. Tricyclic heteroaromatic compounds usable for staining DNA and/or RNA. The general structure of these dyes is shown above the table. In the table, λ_{max} is the wavelength of maximum absorption, A_{DNA} the affinity for calf thymus DNA, in M^{-1}, and K_D the dimerization constant, also in M^{-1}. Affinities for DNA and dimerization constants are from Muller et al[2666-7]. The tabulated values for A_{DNA} and K_D should be multiplied by 10^4 to get the true values; for example, A_{DNA} for acridine orange is 8.8×10^4 M^{-1}.

rescence from intact cells was also accounted for by dye bound to RNA; it was impossible to prove the point, because I couldn't treat intact cells with RNAse.

What Does Pyronin Y Stain? Double-Stranded (Ribosomal) RNA and Sometimes Mitochondria

Darzynkiewicz et al[757-8] found that pyronin Y, at concentrations below 3.3 µM, produced primarily mitochondrial staining in live cells, while, at concentrations above 5 µM, the dye formed complexes with RNA; the interaction with RNA appears to be responsible for the irreversible toxicity of the dye which they noted at this concentration. They reported that the pyronin-DNA complex was not fluorescent, although a fluorescent complex with DNA could be formed when pyronin Y was added at high concentrations. They also found that complex of pyronin Y with single-stranded RNA was not fluorescent, and attributed the RNAse-sensitive fluorescence of pyronin Y in cells to dye bound to double-stranded RNA in polyribosomes[756,759,1350]. Schmid et al[2626] have since (1999) reported that 2 µg/mL pyronin Y produces stoichiometric staining of DNA in fixed, RNAse-treated cells.

Traganos, Crissman, and Darzynkiewicz[1350] found that the intensity of pyronin Y RNA staining varied with changes in conformation of RNA during mitosis and hyperthermia of CHO cells. Total RNA content detected after staining with AO increased in M as compared to G_2 phase cells, consistent with continued RNA synthesis during G_2. The content of double-stranded RNA, stained with ethidium (after DNAse treatment), was also somewhat higher in M cells. In contrast, stainability of RNA with pyronin decreased by 27% in M compared to G_2 cells, and stainability of RNA with pyronin was decreased in G_2 cells compared to cells in G_1. The effect was seen at a relatively narrow range of dye concentration (1.0-2.0 µg/ml), and was thought to involve selective denaturation and condensation of ribosomal RNA by pyronin in single ribosomes; this process does not occur in polyribosomes.

Since the ribosomal RNA content of cells generally parallels their total RNA content, at least in the context of representing proliferative vs. quiescent states, Darzynkiewicz et al gave pyronin Y a clean bill of health as an RNA stain in fixed cells – they use the dye – but cautioned against its indiscriminate use in intact cells[756].

The caution relates specifically to the possibility that pyronin Y fluorescence measured in intact cells may reflect contributions from dye bound to mitochondria as well as from dye bound to RNA. In the rat and human lymphoid cells with which I have used the dye, mitochondrial fluorescence does not seem to be significant in intact cells stained with 5 µM pyronin Y. Cellular and mitochondrial uptake of pyronin Y and other cationic dyes, such as the cyanines and rhodamine 123, is driven in large part by the interior-negative electrical potential gradients across the cytoplasmic and mitochondrial membranes, as will be discussed at great length in the section on probes of membrane potential. I have already mentioned that pyronin Y fluorescence intensities are similar in intact and ethanol-fixed cells; in the latter, the RNAse sensitivity of pyronin fluorescence establishes that it comes predominantly from pyronin-RNA complexes.

If a substantial portion of pyronin Y fluorescence in cells originated from mitochondria, fluorescence should be greatly decreased in cells treated with proton ionophores and/or other uncouplers of oxidative phosphorylation, which decrease or abolish potential gradients across the mitochondrial and, in some cases, the cytoplasmic membrane. I have not noted more than a 5-10% difference in fluorescence between intact lymphoid cells stained in the presence and absence of a "cocktail" of uncouplers and ionophores that demonstrably eliminate mitochondrial uptake and retention of rhodamine 123 in cells such as fibroblasts. Thus, in intact or fixed rat and human lymphoid cells, it appears safe to interpret pyronin Y fluorescence as primarily indicative of RNA content.

In other cell types, containing more mitochondria, fluorescence from dye in these organelles might interfere with RNA content determination in intact cells using pyronin Y. However, such interference would be eliminated by addition of uncouplers. The mitochondrial-vs.-RNA fluorescence issue may be moot if RNA staining is being done to discriminate proliferating and quiescent cell compartments, since, as will be discussed in later sections, both cytoplasmic and mitochondrial membrane potential can serve, in lieu of RNA content and in combination with DNA content, as indicators of proliferative activity.

Surviving Vital Staining with Pyronin Y

I found that cells "vitally" stained with 5 µM pyronin Y could not be grown in culture, and elected to settle for other benefits of vital staining, i.e., relative ease of sample preparation and the elimination of any fixation artifacts that might interfere with immunofluorescent staining done concomitantly with RNA staining with pyronin Y and/or RNA/DNA staining with pyronin and Hoechst 33342[113]. Pyronin Y can be used with fluorescein- and/or PE/Cy5 tandem- or PerCP-labeled antibodies; since the dye is not well excited at 488 nm, fluorescence signals, even from 5 µM dye, are relatively weak. The emission filters generally used for phycoerythrin are well suited for measurement of pyronin Y fluorescence. While the Hoechst/pyronin technique has the disadvantage of requiring dual wavelength (UV and 488 nm) excitation, RNA content measurement alone in cells stained with both dyes may provide useful information and only requires a single 488 nm excitation beam.

Edward Srour, of the University of Indiana, who was interested in studying the cell cycle in human hematopoietic stem cells, decided to find out whether the Hoechst/pyronin DNA/RNA staining procedure would work at lower dye concentrations without killing the cells. It did; he and his colleagues[2340-2] now routinely sort and culture cells stained with Hoechst 33342 at a concentration of 1.6 µM (1 µg/mL) and pyronin Y at a concentration of 3.3 µM (also 1 µg/mL), adding 50-100 µM verapamil to block efflux of the dyes. They have recently published a detailed protocol for the staining procedure[2627]. I will have more to say about this in the section on stem cells in Chapter 10.

In my original procedure, intact cells were incubated with 5-10 µM Hoechst 33342 for about 5-10 min at 37 °C, after which concentrated Pyronin Y (Sigma-Aldrich and Polysciences both produce satisfactorily pure preparations of the dye) was added to a concentration of 5 µM and incubation continued for 30-45 min. Cells stained with this mixture can subsequently be stained with labeled antibodies, provided that the concentrations of both Hoechst 33342 and pyronin Y are maintained in all staining and washing solutions. When cells are stained with fluorescent antibodies and subsequently fixed, Hoechst/pyronin staining is considerably easier; dyes are added after the last wash step to achieve concentrations of 1 µg/mL Hoechst 33258 or 33342 and 3.3-5 µM (1-1.6 µg/mL) pyronin Y.

In a dual-beam (UV/488 nm) instrument, the fluorescence of Hoechst 33342, excited in the UV and measured in the range between 440-480 nm, is indicative of DNA content; pyronin Y fluorescence, measured between 570-600 nm, is indicative of RNA content. I have recommended leaving the Hoechst dye in the staining solution even if only RNA measurements are to be made, to block any artifacts due to pyronin Y binding to DNA; this may be unnecessary.

Other DNA Dyes Usable with Pyronin Y

Toba et al[2343-5] showed that a combination of 7-AAD and pyronin Y could be used to measure DNA and RNA content in fixed cells in a flow cytometer with a single 488 nm excitation beam; this technique also accommodates immunofluorescent staining with fluorescein-labeled antibodies. Schmid et al[2346] refined the staining procedure, and used an additional red laser to excite allophycocyanin-labeled antibody.

According to a March, 2002 posting to the Purdue Cytometry Mailing List from Derek Davies, DRAQ5 can also be used in combination with pyronin Y. I haven't tried.

For the record, Fred Kasten, who has headed the Biological Stain Commission and whose experience as a histochemist far exceeds mine, pronounces the "pyro" in pyronin to rhyme with "hero," which is the way I pronounce the "piro" in Shapiro. I had always thought the "pyro" rhymed with "my roe," which is how some Shapiros from Philadelphia pronounce the "piro." Fred's pronunciation more accurately reflects the Greek origin of pyronin's name, although I'm finding it hard to switch, not being enough of a pyromaniac to burn my bridges.

Tips on Tricyclics (Don't Get Depressed)

In general, it should be preferable to use tricyclic dyes with little or no metachromatic tendency as RNA stains in combination with dyes such as Hoechst 33342, since the orthochromatic absorption or fluorescence of RNA-bound tricyclic dye can provide an adequate estimate of RNA content under these circumstances. The laser dye **oxazine 1**, for example, gives no evidence of being metachromatic, but orthochromatic oxazine 1 fluorescence (excited at 633 nm; measured above 665 nm) in fixed cells stained with 5-10 μM dye is predominantly RNAse-sensitive, although not as much so as pyronin Y, and therefore may provide an estimate of RNA content.

There are, however, reasons for considering the use of metachromatic dyes other than acridine orange for simultaneous DNA/RNA content estimation. Among these, dyes such as methylene blue and the related azure dyes and toluidine blue may be of particular interest because their absorption maxima suit them to use in instruments with inexpensive He-Ne or diode laser sources operating in the 630-650 nm range. Methylene blue and toluidine blue are strongly metachromatic[205]; the absorption bands of concentrated solutions shift to shorter wavelengths, and it would be expected that polymeric forms of such dyes bound to RNA would therefore show metachromatic fluorescein emission at wavelengths longer than the 700-725 nm emission from the monomeric form[282]. Under suitable conditions, thiazine dyes can be employed to produce blue staining of nuclear DNA and selective, metachromatic purple staining of cytoplasmic RNA[283]. However, when I looked for metachromatic fluorescence in cells stained with these dyes, using an instrument with a red He-Ne laser source, I did not find any. It might be easier to discover suitable dyes by spectrofluorometry of dye complexes with various forms of nucleic acids.

At least one lot of dye I have encountered that was labeled pyronin Y wasn't, at least by thin layer chromatographic comparison to material from Aldrich and Polysciences. The funny thing was that the phantom dye was better excited at 488 nm than was pyronin Y, and produced staining with a degree of RNAse-sensitivity indicating that it was at least as specific as pyronin Y. I couldn't get the structure determined; I wish I knew what it was. 3,6-diaminoxanthylium and 3,6-diaminothioxanthylium, both shown in Table 7-4, might be worth a look as RNA stains excitable at 488 nm.

I have played briefly with **thiopyronin G**, usable with 532 nm YAG laser, 546 nm arc lamp or 543 nm green He-Ne laser excitation; this should work as a metachromatic dye. Unfortunately, I can't find a source for the dye; John Spikes, of the University of Utah, sent me a few milligrams in exchange for scientific song lyrics, which he collects. He, and others, have used thiopyronin as a photosensitizer, but it seems to be handed down from photobiologist to photobiologist. My interest in thiopyronin has waned, because I expect it to suffer from many of the same problems as does acridine orange; with the availability of Hoechst/pyronin, 7-AAD/pyronin, and, potentially, DRAQ5/pyronin, there seem to be several choices of two dyes that are better than one, especially when one wants to measure antigens as well as DNA and RNA.

Most tricyclic dyes do not increase fluorescence on binding to either RNA or DNA; on a mole-for-mole basis, the fluorescence of dye bound to nucleic acid is usually quenched relative to the fluorescence of free dye. The high local concentration of dye in dye-nucleic acid complexes accounts for the fluorescence of nucleic acid-containing structures being detectable above background, but background fluorescence is typically high. What you see in the fluorescence microscope is not always what you get in the flow cytometer; quenched dye, in this instance, dye bound to nucleic acid, bleaches more slowly than unquenched dye. Thus, when you look at AO-stained material under the microscope, much of the background fluorescence has already disappeared due to bleaching by the time your eyes focus, the green fluorescence from DNA-bound AO monomer fades rapidly, and the red fluorescence and phosphorescence from dye-RNA polymer persists the longest.

Tricyclics Gag on Mucopolysaccharides

Polyanionic cell constituents other than nucleic acids, notably **glycosaminoglycans (GAGs)** (formerly called **mucopolysaccharides**) such as **heparin** and **chondroitin sulfates**, bind tricyclic dyes and stain metachromatically with those dyes that do exhibit metachromasia. The metachromatic staining of GAG-containing granules in **blood basophils** and **tissue mast cells** was, in fact, noted by Paul Ehrlich, who gave both of these cell types their names, and has been used as an identifying characteristic since his time. It is sometimes possible to use tricyclic dyes to selectively stain GAGs in cells that also contain large amounts of RNA, by lowering the pH to the region of 1.0, at which point most of the phosphate groups of nucleic acids are not ionized while most of the sulfate groups of GAGs remain charged; the basic dye therefore binds to GAGs but not to RNA. Saunders has described the use of acridine orange for identification of GAGs[284]. Timar et al[761] have described a flow cytometric procedure for staining GAGs with AO; RNA staining is eliminated not by manipulating pH, but by altering electrolyte concentration.

There does not seem to be any easy way to use tricyclic dyes to selectively stain RNA in the presence of GAGs;

Darzynkiewicz et al have noted that heparin in mast cells interferes with RNA content estimation using AO because heparin-bound AO makes a significant contribution to the red fluorescence signal[262]. In most other cell types, however, in which large amounts of interfering GAGs are not found, the absorption or fluorescence of tricyclic dyes can give reasonable estimates of RNA content. Bauer and Dethlefsen[285] compared RNA content estimates in HeLa and CHO cells made by UV microspectrophotometry and by flow cytometry using a modification of the Darzynkiewicz AO staining procedure. They found a correlation coefficient of 0.93 between paired measurements made by the two methods following various RNAse treatments. They also reported some background contribution to red fluorescence due to DNA; it is not clear whether this might have been eliminated by further adjustment of staining conditions.

Reticulocyte Counting: Cyanines Beat Tricyclics; RNA in Nucleated Cells: Cyanines Don't

One of the more important applications of RNA content determination is in the detection and counting of reticulocytes in blood; the procedure and its flow cytometric implementation have been discussed on pp. 78 and 99. Under normal circumstances, red cells lose their nuclei before being released from the bone marrow into the blood, and neither young nor mature red cells contain DNA. Some RNA, a residue of the protein synthetic apparatus, remains in the young cells, and is lost during the cells' first day or two in circulation. If sufficient RNA is present in a red cell, an insoluble complex can be formed between this RNA and (tricyclic) basic dyes such as methylene blue; the netlike appearance of this precipitated material gives the name "reticulocyte" to this class of immature red cells.

The first flow cytometric approaches to reticulocyte counting, in the late 1970's and early 1980's, were based on the use of tricyclic dyes. Ortho received approval from the U. S. Food and Drug Administration (USFDA) for clinical use of a procedure using acridine orange[764-5], but there were problems with this dye due to high background fluorescence, although Schmitz and Werner reported improved results using optimized dye concentrations and logarithmic amplification on the AO red fluorescence channel[766]. Tanke et al described a method using pyronin Y[286-7]. its major disadvantage was the requirement that cells be fixed. Later in the 1980's, the approach to reticulocyte counting shifted toward the use of basic dyes, different in structure from the tricyclics, which increase fluorescence by factors ranging from severalfold to several thousandfold on binding to RNA. These dyes, predominantly cyanines such as thiazole orange and related compounds such as thioflavin T and auramine O, stain both RNA and DNA. While they are nearly ideal for RNA detection in reticulocytes, from which DNA should be absent, they are poorly suited for demonstrating RNA in nucleated cells, unless the cells are treated with DNAse as described below.

Propidium Stains Double-Stranded RNA; What of Other Dyes?

Propidium iodide (PI) has been used by several investigators for analysis of RNA content. This dye binds by intercalation to double stranded nucleic acids, RNA as well as DNA. When PI is used for DNA content estimation in cells containing large amounts of RNA, broad fluorescence distributions are obtained due to the contribution of RNA-bound dye to the "DNA" fluorescence signal. The accuracy and precision of DNA measurements made with PI are improved when cells are treated with RNAse prior to analysis.

Frankfurt[293] described a complementary technique in which cells were treated with DNAse and stained with PI before flow cytometry was done; cellular fluorescence under these circumstances is due primarily to PI bound to double-stranded RNA. The contents of single- and double-stranded RNA in cells appear to follow one another; Wallen et al examined cultured mouse mammary tumor lines and reported good correlation between flow cytometric measurements of RNA using PI and AO and good correlation of both with RNA content estimated by UV microspectrophotometry[294].

Wright, Higashikubo, and Roti Roti[1351] combined propidium staining of double-stranded RNA with fluorescein isothiocyanate (FITC) staining of total protein (see the section on that) for flow cytometric studies of nuclear matrices prepared from nuclei by DNAse I digestion followed by NaCl extraction, and found that, while heat shock could protect some double-stranded RNA from RNAse digestion, propidium staining remained specific for double-stranded RNA.

Pyronin Y, as previously mentioned, forms fluorescent complexes primarily, if not exclusively, with double-stranded ribosomal RNA, and could presumably be used instead of propidium without the necessity for RNAse digestion. The monomeric and dimeric cyanines intercalate, and thus stain double-stranded RNA, but would require enzymatic removal of both DNA and single-stranded RNA to be useful.

An RNA-specific fluorochrome similar in structure and spectral characteristics (UV-excited, blue fluorescence) to DAPI was described by Wachtler and Musil[292]; I have not seen reports either of the use of this dye in flow cytometry or of whether it stains single- and/or double-stranded RNA.

7.5 FLUORESCENT LABELS AND PROTEIN DYES

More than half of the 37 spectra shown in Figure 7-9 (p. 296) are those of low and high molecular weight **labels** that are attached to other molecules to allow those molecules to be detected in or on cells or beads using cytometry. **Reactive** labels bind covalently to the molecules to be labeled; low molecular weight reactive labels can also be used as stains for cellular **proteins**, which can also be demonstrated using dyes, almost always acidic in character, which form ionic bonds with amino groups and other basic structures in proteins. Structures of a number of reactive labels are shown in Figure 7-17.

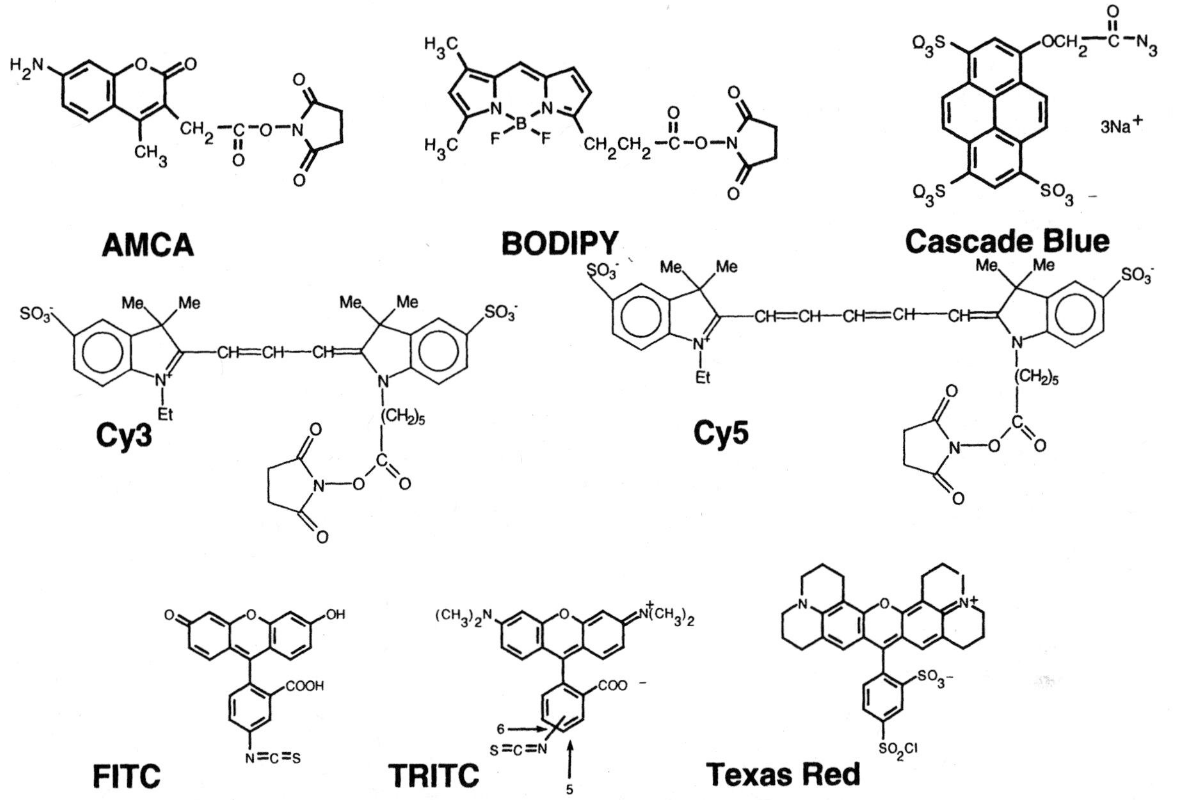

Figure 7-17. Chemical structures of some reactive labels. AMCA and BODIPY succinimidyl ester, Cascade Blue azide, FITC, TRITC, and Texas Red structures were provided by Molecular Probes, Inc.

Estimating Total and Basic Protein Content of Cells

Total protein content of fixed cells is estimated by staining with a variety of acid dyes that bind ionically or covalently to positively charged groups on proteins. It is difficult to establish the specificity of these dyes as protein stains, or how much dye is bound to materials other than proteins, because, while cells stained with DNA or RNA fluorochromes can be examined in the presence and absence of DNAse or RNAse, attempting to do the same trick with protein stains and proteolytic enzymes leaves you with samples containing stain in the presence and absence of cells. We thus have to be satisfied with a large body of indirect evidence that tells us that most of whatever isn't DNA or RNA in most cells is protein. By using acid dyes as protein stains, we increase the likelihood that what is being stained is actually protein, because these dyes are reacting predominantly with amino groups, which are considerably more abundant in proteins than in carbohydrates and lipids.

Total protein staining is useful both in monitoring growth and metabolism of otherwise homogeneous cell populations and in analysis of mixed populations containing cell types with different protein content, e.g., blood leukocytes. In most applications, protein stains are not used alone, but are combined with DNA fluorochromes. Most of the stains in common use have, therefore, been selected to have spectral characteristics complementary to those of ethidium or propidium on the one hand, or to those of DAPI or the Hoechst dyes, on the other.

Fluorescein Isothiocyanate (FITC)

Freeman and Crissman[295] examined several dyes suitable for argon ion laser excitation at 457 or 488 nm and selected fluorescein isothiocyanate (FITC) as a protein stain for use in combination with PI[217]. FITC was preferred to sulfonated dyes (e.g., brilliant sulfaflavine) because it binds covalently, leaving fluorescein moieties attached to cells after washing.

Because there is some overlap between the emission spectra of fluorescein and PI, fluorescence compensation may be required to eliminate contributions to the green fluorescence signal from PI and contributions to the orange or red fluorescence signal from fluorescein. When FITC is used for total protein staining, the fluorescein signal is usually relatively strong, and bleed into the orange or red channel is pronounced. This contrasts with the more common situation in which PI is combined with fluorescein-labeled antibodies; in this case, the effects of fluorescein spectral overlap on PI fluorescence are usually negligible, while those of PI spectral overlap on uncompensated fluorescein fluorescence are marked.

Sulforhodamine 101 (SR101)

Stöhr et al[296] examined various combinations of DNA and protein stains and selected one in which DAPI was used to stain DNA and sulforhodamine 101 (SR101) was used to stain protein. These were used in an instrument with UV and 488 nm beams exciting DAPI and SR101, respectively; while it is also possible to excite both dyes with UV light, Heiden, Göhde, and Tribukait[1352] report that dual-beam excitation with UV and green is superior. There is relatively little spectral overlap between the emission spectra of DAPI and SR101 to begin with; when dual-beam excitation is used, any bleed of DAPI fluorescence into the SR101 measurement channel is eliminated.

SR101 binds ionically, rather than covalently, to proteins; staining with it and with other sulfonated dyes is an equilibrium process, and there is generally a considerable amount of free dye remaining in the background. Under some circumstances, background fluorescence of acid dyes used to stain protein may interfere with or prevent measurement of protein content (see "The Case of the Disappearing Leukocytes," pp. 220-1). However, this fluorescence may often be reduced considerably by lowering the ionic strength of the staining mixture[8,91,93].

Although, in theory, SR101 should be usable in combination with fluorescein-labeled antibodies in an instrument with 488 nm excitation, the much higher concentration of SR101 relative to fluorescein in this situation may result in more spectral overlap than can be compensated for by some instruments.

Hematoporphyrin (HP) as a Protein Stain

Takahama and Kagaya[1353] described a simplified simultaneous one-step staining method for DNA and cell protein using 0.001% DAPI or Hoechst 33342 as the DNA stain and 0.03% hematoporphyrin (HP) to stain protein. The fluorescence emission of HP is at 670 nm; the excitation spectrum of porphyrins suggests that HP would definitely be usable as a total protein stain with blue-violet excitation (436 nm arc lamp, 441 nm He-Cd laser, or 457 nm argon laser) and might well work with 488 nm excitation. If 488 nm excitation could be used, the large Stokes' shift of porphyrins should insure that emission in the green spectral region would be sufficiently low to permit simultaneous use of fluorescein-conjugated antibodies and HP. Tanaka et al[1354] compared staining with the DAPI/HP combination by flow and image cytometry and reported that it yielded more accurate results than did PI and fluorescein.

Rhodamine 101 (or 640) as a Vital Protein Stain

Since neither the covalent bonding nor the ionic bonding acid dyes penetrate membranes of living or intact cells to any appreciable extent, they can only be used for protein staining in fixed or permeabilized cells, nuclei, etc. In 1982, Crissman and Steinkamp reported that rhodamine 640 (i.e. rhodamine 101, the uncharged dye that is the chromophore of SR101 and of the reactive labels XRITC and Texas Red) could be used at concentrations of 1-5 µg/ml to stain living or intact cells, yielding fluorescence distributions nearly identical to the total protein content distributions obtained from fixed cells stained with XRITC or FITC[297]. One would expect that other uncharged rhodamine dyes with absorption and fluorescence at shorter wavelengths, e.g., rhodamine B, might also be suitable for "total protein" staining in intact cells.

What Do "Total Protein" Stains Stain?

Holme et al[1355] compared forward scatter and Coulter volume signals from platelets with fluorescein fluorescence values obtained when platelets were stained for total protein and found a high degree of correlation; since platelets are predominantly composed of protein, this provides further evidence in support of FITC as a total protein stain.

In fixed cells, measurements of orthogonal (90 °) light scattering may provide information similar to that obtained using protein stains[8,103,1255]; this was alluded to on p. 278. Figure 5-17 (p. 254) shows that in glutaraldehyde- fixed peripheral blood leukocytes, total protein content, as estimated from the UV-excited blue fluorescence of the acid dye "LN"[91,93] is tightly correlated, on a cell-by-cell basis, with orthogonal light scatter intensity in the several different classes of leukocytes[8]. Others (H. Crissman, personal communication; M. Pallavicini, personal communication) have also noted that fluorescent protein stains and side scatter measurements yield similar distributions. When beginning studies on a cell population, it is therefore advisable to examine the correlation between side scatter and total protein content as determined with a fluorescent dye. If the side scatter measurement appears to provide the same information as the protein stain, the scatter signal can then be used in the absence of the protein stain, allowing the fluorescence channel that would have been used for protein content estimation to be devoted to analysis of a more specific parameter, e.g., a surface antigen.

Staining to Demonstrate Basic Protein

Protein staining by acid dyes is dependent on the **ionization constants** of the dyes and of amino groups and other charged sites on proteins. Thus, binding of dyes to cells is affected by the pH of the staining solution. At alkaline pH, only the more strongly basic groups in proteins remain ionized and thus able to bind acid dyes; as pH decreases, more basic groups become ionized. Thus, the same basic dye can be used to stain **basic proteins** and total proteins at different pHs; brilliant sulfaflavine, for example, is used as a basic protein stain at pH 8 and a total protein stain at pH 2.8[298].

Covalent Labels for Antibodies and Other Molecules

The covalent bond forming dyes such as FITC, which has just been discussed as a stain for total protein content, are probably better known, and undoubtedly more widely used, as **fluorescent labels** for a variety of large and small molecules that can be bound strongly and specifically to various cellular constituents. Specific ligands thus labeled can be used as reagents for a number of structural and functional parameters, including **surface sugars** (demonstrated using tagged lectins), **surface** and **intracellular antigens** (fluorescent antibodies), **surface charge** (fluorescent polycations), **surface** and **intracellular receptors** (fluorescent hormones, growth factors, neurotransmitters, viruses, etc.), **endocytosis** (fluorescent macromolecules, microorganisms, or plastic particles), **DNA synthesis** (fluorescent antibody to detect BrUdR in chromatin; fluorescent nucleotides), and **specific nucleic acid sequences** (fluorescent oligonucleotide probes). Spectral characteristics of the more important fluorescent labels are given in Figure 7-9 (p. 296) and Table 7-2 (p. 297).

Fluorescein Isothiocyanate (FITC) as a Label

Fluorescein, conjugated as its reactive isothiocyanate derivative **(FITC)**, is by far the most popular fluorescent label; its excitation maximum is very close to the 488 nm argon ion laser wavelength available in almost all flow cytometers, its quantum efficiency is high, and it had been in use long enough before flow cytometers became available for conjugation and staining procedures to have become well established, particularly in immunology.

As was mentioned on p. 77, the fluorescent label used in the first description of the fluorescent antibody technique by Coons et al in 1941[44] was anthracene, which fluoresces blue when excited by UV light. The fluorescent antibody technique itself was developed after attempts to visualize the absorption of dye-labeled cell-bound antibodies in transmitted light[1087-8] had failed. It was possible to see concentrations of anthracene-labeled antibodies against the darker background of unstained tissue, but the UV-excited blue autofluorescence of pyridine nucleotides and other tissue components (see Figure 7-7 and text on pp. 290-1) limited the sensitivity of the fluorescent antibody method as originally described.

In 1950, Coons and Kaplan[45] reported that the use of **fluorescein-labeled antibodies** improved sensitivity; the green fluorescence of fluorescein made it easier to discern concentrations of antibody against the tissue background. Two major barriers to widespread use of the fluorescent antibody method remained; few labs had fluorescence microscopes, and most investigators were neither equipped nor willing to handle the phosgene gas used to prepare fluorescein isocyanate for conjugation. In 1958, Riggs et al[46] described the use of fluorescein isothiocyanate for labeling. The synthesis of FITC involves thiophosgene, which is less noxious and more manageable than phosgene, and FITC is stable enough to be marketed commercially in a form ready for conjugation.

The intensity of fluorescein fluorescence emission varies with pH within the range (6.5-8.0) likely to be encountered in samples; pH must therefore be controlled when quantification of results is important.

Labeling with Lissamine Rhodamine B and Tetramethylrhodamine Isothiocyanate (TRITC)

It was also in 1958 that Chadwick et al[299] described the use of **Lissamine rhodamine B 200**, or **sulforhodamine B**, for antibody labeling. This compound was conjugated by converting its sulfonic acid group to a sulfonyl chloride. Since the red fluorescence of rhodamine dyes and the green fluorescence of fluorescein can be readily distinguished by an observer, particularly when excitation filters are switched to allow selective excitation of first one dye and then the other, it became possible to examine material stained with two different antibodies, each with its own distinct fluorescent label. Once relatively easy-to-use fluorescein and rhodamine labels became available, immunology laboratories began to acquire fluorescence microscopes, which had been improved considerably between the 1940's and the 1960's. The stage was thus set for the relatively rapid acceptance of flow cytometry by immunologists, and it was not until flow cytometers came into widespread use in the late 1970's that significant further progress in the development of fluorescent labels for antibodies was to occur. By this time, an isothiocyanate derivative of rhodamine, **tetramethylrhodamine isothiocyanate (TRITC)** had also become available.

Multicolor Fluorescence I: FITC and TRITC

As was just noted, by the time flow cytometers became available to immunologists, many of them were using fluorescein and rhodamine labels to do two-color work under the fluorescence microscope, switching from a combination of a blue primary (excitation) filter and a yellow or green secondary (observation) filter for detection of fluorescein-labeled antibody to a combination of a green primary filter and a red secondary filter for detection of rhodamine- labeled antibody. It was also possible, although difficult, to look at the fluorescence from both labels at once, using blue-green excitation and a yellow secondary filter; this relied heavily on the human observer's ability to discriminate weak red fluorescence from strong green fluorescence.

The microscopes used for immunofluorescence work typically have mercury or xenon arc lamp sources; mercury lamps have strong emission lines at 436 nm (blue) and 546 nm (green) and put out considerably less light at intermediate wavelengths, while xenon lamps have a relatively flat emission spectrum in the blue and green regions. The absorption maximum of fluorescein is at about 490 nm; that of tetramethylrhodamine is at about 555 nm. Using a conventional blue (centered near 436 nm) excitation filter for fluorescein gives suboptimal excitation, while using a green filter centered near 546 nm for rhodamine gives near-optimal excitation. Using a 500 nm short-pass filter to excite both dyes doesn't give very good excitation of either, allowing the dark-adapted eye and its associated brain to resolve the relatively weak emission from both.

The 488 nm argon laser line commonly used for excitation in flow cytometers is essentially at the absorption maximum of fluorescein, while the absorption of tetramethylrhodamine at this wavelength is only about 5% of its maximum value. This factor alone dictates that, if equal amounts of fluorescein- and tetramethylrhodamine-tagged antibodies were present on a cell, the total fluorescence emission from the fluorescein antibodies would be 20 times that from the tetramethylrhodamine antibodies. When bound to antibodies, both fluorescein and tetramethylrhodamine usually have quantum efficiencies somewhere between 0.2 and 0.5 (roughly on the order of 50% of the quantum efficiencies of the free dyes); these are not different enough to balance out the very large difference in absorption. The differences in molar extinction coefficients between fluorescein and tetramethylrhodamine are also not great enough to overcome the differences in absorption at 488 nm. Thus, one can realistically expect at least an order of magnitude more fluorescence from fluorescein-tagged antibody than from an equivalent amount of tetramethylrhodamine-tagged antibody when 488 nm excitation is used. The resolution of the signals from the two dyes therefore depends upon the use of detectors equipped with optical filters chosen to take advantage of the difference in emission spectra between fluorescein and tetramethylrhodamine.

The emission maximum of fluorescein (Figure 7-9) is at about 520 nm, while that of tetramethylrhodamine is at about 570 nm; emissions from the two dyes should be well resolved using a green (515-550 nm) filter for the fluorescein fluorescence detector and a red-orange (560 nm long pass) filter for the tetramethylrhodamine fluorescence detector. This, however, does not take into account the large absorption difference at 488 nm. To do this, we have to imagine spectra in which we either increase the height of the fluorescein emission spectrum by a factor of 10 or decrease the height of the tetramethylrhodamine emission spectrum by a like factor. In this view, what seemed to be an insignificant "red tail" on the fluorescein emission spectrum, going out to 600 nm, now overshadows most of the tetramethylrhodamine emission. For best resolution of the tetramethylrhodamine emission, we really should use a long pass filter that cuts on at a wavelength of 590 nm or longer on the tetramethylrhodamine fluorescence detector. At that, we would have to use electronic or software fluorescence compensation to get rid of crosstalk from the longer wavelength fluorescein emission, and, if there is a lot more fluorescein than tetramethylrhodamine on a cell, even this won't help.

Loken et al[115], in the Herzenberg lab at Stanford, first did two-color immunofluorescence flow cytometry as just described, and also evened the balance between fluorescein and tetramethylrhodamine absorption by operating the argon laser to produce a mixture of visible lines (mostly 488 & 515 nm) or a 515 nm beam. The 515 nm line is not great for tetramethylrhodamine, but it's no better for fluorescein, and using this excitation wavelength forces you to employ a fluorescein detector filter that cuts on at a longer wavelength than you would otherwise use. Before phycobiliproteins were introduced as labels, this was about the best that could be done in the way of two-color immunofluorescence analysis with a single-beam cytometer; on a historical note, the fluorescein/rhodamine labeling work forced the Herzenberg group to develop fluorescence compensation.

Multicolor Fluorescence II: Rhodamine 101 Dyes

The most obvious alternative to the less than satisfactory single beam excitation method for cells labeled with fluorescein and rhodamine involved the use of separated excitation beams at different wavelengths, in essence imitating the observer at the fluorescence microscope switching filter combinations. There was no simple way to do this using fluorescein as the first label and tetramethylrhodamine as the second, because the only lines available from ion lasers well suited to tetramethylrhodamine excitation (528 nm from argon and 530 nm from krypton lasers) are in the region of fluorescein emission. Instead, derivatives of **rhodamine 101** were substituted for tetramethylrhodamine. In rhodamine 101, the ring structure of other rhodamine dyes is extended (see the structures of TRITC and Texas Red in Figure 7-17), leading to shifts in the absorption and emission spectra, which have maxima at wavelengths about 30 nm longer than the absorption and emission maxima of tetramethylrhodamine.

Adding a second beam, derived from a krypton laser emitting at 568 nm or a dye laser at 590 nm, to the existing 488 nm argon laser beam in a flow cytometer made it possible to examined cells simultaneously stained with antibodies labeled by conjugation with FITC and with X-rhodamine isothiocyanate (**XRITC**, Research Organics, Cleveland, OH) or the sulfonyl chloride derivative **Texas red** (from Molecular Probes, which was then located in Plano, TX), both bearing the chromophore of rhodamine 101[116,117]. Fluorescein fluorescence excited at 488 nm was measured at 510-550 nm, and rhodamine 101 fluorescence excited at 568 nm was measured at 590-630 nm; there was essentially no crosstalk between the signals.

While the rhodamine 101 derivatives solved some problems involved in two-color immunofluorescence flow cytometry, they precipitated others. People who used XRITC found it hard to remove unbound dye from protein conjugates, to which the hydrophobic dye adhered. Texas red became more popular, but had a tendency to inactivate antibodies (rat monoclonal antibodies in particular) and, sometimes, to part company with antibodies to which it had been conjugated. Rhodamine 101 supposedly has a quantum efficiency of 1.0, but Texas red-labeled antibodies don't seem to give as much of a fluorescence signal as do fluorescein-labeled antibodies. The people I know who have the best luck with Texas red immunofluorescence originally used the label almost exclusively in the forms of **Texas red-avidin** or **-streptavidin**, which will form extremely high-affinity bonds with biotin-conjugated antibodies[300]. Excitation of Texas red requires a krypton laser or an argon laser-pumped tunable CW dye laser, neither of which is as easy to keep running as an argon laser. He-Ne lasers at 594 nm were not then powerful enough to present a viable option; the few milliwatts available at this wavelength might, however, do the job in modern instruments with efficient optics.

Early Problems with Multicolor Fluorescence

In addition to the difficulties presented by the rhodamine 101 derivatives, users faced some more general problems associated with two-color immunofluorescence measurements. As long as only a single antibody was used to stain cells, a significant enhancement of fluorescence signals could be obtained by using **indirect staining**, in which the fluorescent label is attached to a second **developing antibody** that binds to molecules of another antibody already bound to antigens on or in cells. It is, as a rule, not possible to put more than three molecules of fluorescein (or other dyes) directly on an IgG antibody molecule without either decreasing the specificity of the antibody reaction or decreasing the quantum yield of attached dye past the point of diminishing returns. It is, however, possible for more than three (usually 5 or 6) molecules of a developing antibody to bind to a single cell-bound antibody molecule. Thus, **direct immunofluorescent staining** will put no more than 3 dye molecules at each antigenic site, while indirect staining will put 15-18 dye molecules at each antigenic site, resulting in **amplification** of the fluorescence signal.

Most of the earlier single-color work with monoclonal antibodies was done using indirect staining, with something like a fluoresceinated goat anti-mouse Ig used as a developing antibody, since this *modus operandi* both eliminated the need for fluorochrome conjugation of each new monoclonal antibody and offered increased sensitivity as compared to direct fluorescent staining. Indirect staining of this kind, however, did not provide for labeling cells with two mouse monoclonal antibodies of the same immunoglobulin class.

When there were reasonably large numbers of antigenic sites present on cells, the obvious option in this instance was direct staining, using different labels (fluorescein and rhodamine 101) on each of the two antibodies. If some amplification was required, the use of **labeled avidin** and **biotinylated antibodies**[300], or of **hapten-conjugated antibodies** and differently labeled **anti-hapten antibodies**[301] resulted in more dye molecules being bound per antigenic determinant than would have been possible using direct staining.

What would have been most desirable, however, was a direct labeling technique that allowed more than three fluorescent molecules to be attached directly to each antibody molecule. The group at Block tried a "brute force" approach to such a method in which antibodies were conjugated to synthetic polymers bearing hundreds of fluorescein molecules (see p. 116); this didn't work, because the fluorescein fluorescence was quenched under those circumstances.

Shechter et al[302] had better luck with a procedure developed to produce highly fluorescent, physiologically active derivatives of insulin and other peptide hormones; they were able to attach 7 molecules of rhodamine to a molecule of **lactalbumin**, which was then conjugated to a molecule of hormone. As it turned out, however, when it came to designing fluorescent macromolecules suitable for antibody labeling, Nature had, as usual, outdone man.

Phycobiliproteins to the Rescue!

The **phycobiliproteins**[114,772,1356] are a family of macromolecules found in red algae and cyanobacteria (formerly called blue-green algae), in which they play critical roles in the function of the photosynthetic apparatus. Photosynthesis in green plants involves the direct interaction of light with chlorophyll, which has absorption maxima near 440 nm in the blue-violet and near 700 nm in the far red. Light in these spectral regions is not transmitted much beyond the surface layers of bodies of water, while green and blue light penetrate to greater depths. The survival of algae and cyanobacteria at these depths depends on the capacity of phycobiliproteins to absorb the shorter wavelength light and participate in a chain of nonradiative energy transfers that finally makes the light energy for photosynthesis available to chlorophyll.

The **phycoerythrins** absorb blue-green and green light, the **phycocyanins** green and yellow light, and the **allophycocyanins** orange and red light. While the function of phycobiliproteins in their natural environment is that of nonradiative energy transfer, these molecules are all highly fluorescent, and their particular fluorescence characteristics have been of great interest to users of flow cytometers since Oi, Glazer, and Stryer demonstrated the utility of phycobiliproteins as antibody labeling reagents[114].

The chromophores in phycobiliproteins are **bilins**, which are pyrrole pigments derived from the same building blocks as **porphyrins**; each phycobiliprotein molecule contains a large number of such chromophores. The extinction coefficients of phycobiliproteins are extremely high, and the quantum yields are also high. An antibody molecule directly labeled with fluorescein will have between 1 and 3 chromophores associated with it; an antibody molecule to which 6 fluoresceinated antibody molecules have been bound during an indirect staining procedure will have 6 to 18 associated chromophores. An antibody molecule directly labeled by conjugation with a phycobiliprotein may have as many as 34 associated chromophores, each with an absorbance and quantum yield roughly comparable to those of fluorescein. Spectra of R-phycoerythrin and allophycocyanin are shown in Figure 7-9 (p. 296); Figure 1-18 (p. 37) shows that, with excitation at 488 nm, a phycoerythrin-labeled antibody molecule will emit several times as much fluorescence as a fluorescein-labeled antibody molecule.

In addition to their high extinction coefficients and quantum yields, phycobiliproteins, and phycoerythrins in particular, are characterized by broad shoulders in their excitation spectra, allowing them to be excited effectively at wavelengths substantially below their emission maxima.

While the molecular weights of phycobiliproteins are sufficiently high so that conjugates may contain only one molecule of phycobiliprotein per antibody molecule, the high extinction coefficients and quantum efficiencies of the phycobiliproteins insure that one molecule will get the job done.

Phycoerythrins: R-PE, B-PE, and Others

While the peak absorption of **R-phycoerythrin (R-PE)** is at 565 nm, with the emission maximum at 578 nm, the absorption at 488 nm is approximately 50% of maximum. This allows R-PE to be used very effectively in combination with fluorescein for two-color immunofluorescence flow cytometry; only a single excitation beam (e.g., 488 nm from an argon laser) is required. Fluorescein fluorescence is measured, as customary, in the green region around 530 nm, while R-PE fluorescence is detected in the orange-red region above 570 nm. If a 590 nm long pass filter is used instead of a 570 nm long pass filter, interference from the red tail of fluorescein emission can generally be minimized to the point at which fluorescence compensation[115] is not necessary; this was more practical years ago, when most instruments measured two-color fluorescence, than it would be now.

B-phycoerythrin (B-PE), like R-PE, has a molecular weight of about 240,000, and contains 34 bilin chromophores per molecule; its absorption maximum is at 545 nm, and it is less well excited at 488 nm than is R-PE. PE-labeled antibodies have been widely available since the late 1980's, and the combination of fluorescein-labeled and PE-labeled antibodies has been standard for two-color immunofluorescence measurements since that time. B-D licensed Stanford University's patent on phycobiliproteins, which was based on the work by Oi, Glazer, and Stryer[114], and controlled the market on phycobiliprotein-labeled antibodies until the late 1990's; since that time, PE-labeled antibodies have become available from an increasingly large number of manufacturers.

Ong, Glazer, and Waterbury[305] suggested that the phycoerythrin from the marine cyanobacterium *Synechococcus* WH8103 **(S-PE)** might be useful as a label because it contains more phycourobilin groups, which are responsible for the shorter wavelength absorption of phycoerythrins, than does any other known phycoerythrin. The absorption maximum of S-PE is at 492 nm; the emission maximum is at 565 nm. A solution of S-PE excited at 488 nm has a fluorescence intensity 19 times higher than that of an equimolar solution of fluorescein, and about twice as high as an equimolar solution of R-PE. S-PE would, if it could be produced in quantity, be the ideal phycobiliprotein to use alone or in combination with fluorescein for immunofluorescence measurements with 488 nm excitation. Since most of S-PE emission is at wavelengths between 555 and 585 nm, it is also likely that fluorescein- and S-PE-labeled antibodies could be used in combination with propidium iodide for simultaneous DNA content and two-color immunofluorescence analyses. The theoretical advantages of S-PE, to which I referred in previous editions of this book, have not been realized at this writing, because S-PE is too hard to come by and/or because R-PE is good enough for most purposes.

Allophycocyanin (APC) and APC-B

Allophycocyanin (APC) is of particular interest as a single label because it exhibits high (about 75% of the maximum value) absorption in the 633-638 nm range in which red He-Ne and diode lasers operate. The absorption maximum of APC is at 650 nm; its emission maximum is at 660 nm. APC contains 6 bilin chromophores and has a molecular weight of about 100,000. **Allophycocyanin B (APC-B)** emits at slightly longer wavelengths (maximum near 680 nm). My colleagues and I measured fluorescence from human leukocytes stained with an APC-conjugated antibody in a flow cytometer using a 7 mW He-Ne laser source[303]. Although we used a suboptimal long pass filter (665 nm cut on, which transmits less than 50% of the APC fluorescence) on the fluorescence detector, we obtained results that compared favorably with those from another instrument in which cells stained with fluorescein-antibody were illuminated with 200 mW at 488 nm.

Loken et al[644] found that 25-50 mW from a large He-Ne laser mounted on a B-D FACS, which has less efficient fluorescence collection optics than the Cytomutt I used, provided near-optimal excitation of APC-labeled antibodies; Doornbos et al[1136] were able to discriminate cells stained with APC-labeled antibody from unstained cells using a flow cytometer with a diode laser source emitting 3 mW at 635 nm.

If excitation in the 590-650 nm range usable with APC is available, this label offers advantages over PE and fluorescein for use with antibodies reactive with low surface density antigens in cells such as macrophages[1357], which exhibit high autofluorescence when excited at 488 nm.

Glazer and his associates[773-4,1356] studied the physical chemistry and spectral properties of APC, in the process uncovering and solving a potential problem with its use as an antibody label. Native phycobiliproteins are composed of subunits; B-PE and R-PE each contain 13 subunits, while APC is made up of 3. Although phycobiliproteins are stable under most conditions associated with their use as labels, APC tends to dissociate into subunits at concentrations below 10^{-8} M, and APC concentrations might fall below this range in washed samples stained with APC-labeled antibody or ligand. Cross-linking the subunits while APC is in the trimeric state prevents dissociation, stabilizing the labeled antibody.

Phycocyanins

C-phycocyanin (C-PC), with an absorption maximum at 620 nm and an emission maximum at 650 nm, has been used as an antibody label by Hoffman et al[646], who were able to detect and resolve fluorescence from C-PC-labeled and APC-B-labeled antibodies, using a 642-657 nm bandpass filter for the former and a 675-695 nm bandpass filter for the latter, in an Ortho flow cytometer equipped with a 7 mW, 633 nm He-Ne laser source. Taking advantage of the small, but not insignificant absorption of C-PC at 488 nm, Daley et al[775] used a PC-avidin conjugate with biotinylated antibodies and fluorescein- and PE-labeled antibodies to do three-color immunofluorescence measurements on a Coulter EPICS instrument with a single 488 nm laser source.

R-phycocyanin II (R-PC-II)[776], a phycocyanin found in some species of *Synechococcus*, might be useful as a third label; it has absorption peaks at 533, 554, and 615 nm and an emission peak at 646 nm. However, as is the case with S-PE, it may be impractical to produce R-PC-II in quantity.

Phycobiliprotein Tandem Conjugates: PE-APC, PE-Texas Red, PE-Cy5, PE-Cy5.5, PE-Cy7, etc.

Glazer and Stryer[306] (1983) were the first to prepare a **tandem conjugate** of B-PE and APC in which energy transfer between these proteins, with the phycoerythrin molecule the donor and the allophycocyanin molecule the acceptor, results in strong emission at 660 nm on excitation at wavelengths between 470 and 560 nm. This material was shown by B-D to be usable as a third antibody label in combination with fluorescein and a phycoerythrin, permitting 3-color immunofluorescence measurements with 488 nm excitation. The PE-APC tandem conjugate, however, had two notable disadvantages; first, the chemistry involved in its preparation and conjugation was nontrivial, and, second, energy transfer was not complete, resulting in substantial emission from the conjugate in the same spectral range as phycoerythrin emission. In 2001, Tjioe et al[2628] found it somewhat easier to prepare and use PE-APC tandem conjugates.

The phycobiliprotein tandem conjugates now in widest use incorporate only a single phycobiliprotein molecule, namely, phycoerythrin, to which are conjugated several molecules of a lower molecular weight fluorochrome. The first widely used conjugates prepared in this fashion incorporated **phycoerythrin and Texas Red**; antibodies labeled with such conjugates, are available from a number of manufacturers, each using its own trade name for the conjugate. While most of the emission from PE-Texas Red conjugates is in the 610-620 nm emission region of Texas Red, incomplete energy transfer results in substantial emission from the conjugates in the PE emission region around 580 nm, and PE itself has substantial emission in the 610-620 emission range. As a result, a lot of fluorescence compensation must be applied to separate the fluorescence signals from a PE-labeled antibody and another antibody labeled with a PE-Texas Red tandem conjugate; while the spectra of PE and PE-Texas Red are shown in Figure 7-9 (p. 296); the spectral overlap problem is best appreciated from Figure 1-18 (p. 37) and the related discussion (pp. 36-7).

PE-Cy5 tandem conjugates, introduced by Waggoner et al[1359], incorporating a single phycoerythrin molecule and several molecules of the cyanine dye label **Cy5**, are now widely preferred over PE-Texas Red conjugates as a third label for immunofluorescence analyses using 488 nm excitation; they emit at the emission maximum of Cy5, near 660 nm.

Lansdorp et al[1360] took an intermediate step on the road to PE-Cy5 conjugates by preparing Cy5-labeled anti-PE antibodies, which could be complexed with PE-labeled antibodies; they observed that energy was transferred efficiently from PE to Cy5 in such complexes, enabling them to be used as a third label in combination with fluorescein- and PE-labeled antibodies, provided uncombined anti-PE antibody sites were blocked. Energy transfer is at least as efficient in PE-Cy5 conjugates as in complexes; Figure 7-9 (p. 296) and Figure 1-18 (p. 37) demonstrate that there is much less emission spectral overlap between PE and a PE-Cy5 conjugate than there is between PE and a PE-Texas Red conjugate. Accordingly, less fluorescence compensation is required. PE-Cy5-labeled monoclonal antibodies are widely available.

The list of phycoerythrin tandem conjugates has expanded considerably since the last edition of this book was written; it now includes PE-Cy5.5 (emission maximum near 700 nm) and PE-Cy7 (emission maximum near 770 nm). As you will find in Chapter 8, there are now flow cytometers on the market that will allow you to make simultaneous measurements of the fluorescence of fluorescein, PE, PE-Texas red, PE-Cy5, PE-Cy5.5, and PE-Cy7, and side scatter to boot, using a single 488 nm excitation beam. Don't look for hardware compensation in those puppies.

Allophycocyanin Tandem Conjugates: APC-Cy7 and APC-Cy5.5

In the last edition, I suggested that "It might...be desirable to prepare tandem conjugates of APC that could be used in conjunction with APC itself for multicolor immunofluorescence measurements employing a 633 nm He-Ne or 635 or 650 nm diode laser source." It didn't take long for this to happen; in 1996, Roederer et al[2629] and Beavis and Pennline[2630] contested priority for APC-Cy7, when the paper that had been accepted first was published second[2630]. APC-Cy5.5 seems to have just come along out of the blue (or the far red); I couldn't find a claimant on MEDLINE. APC-Cy7 emits maximally near 770 nm, and APC-Cy5.5 near 700 nm; both are now available from a number of companies, conjugated to a variety of monoclonal antibodies.

Mercy Me! PerCP!

The first new reference added to the Third Edition, reference 1027, by Mandy, Bergeron, Recktenwald, and Izaguirre, was a tandem in its own way. It introduced the concept of T cell gating for lymphocyte subset analysis (see pp. 30-4 and 277); it also described the use of a new third label, **peridinin chlorophyll protein**, or **PerCP**. This material is a component of the photosynthetic apparatus in a dinoflagellate; as illustrated in Figure 7-9, it has an absorption maximum near 490 nm and a relatively sharp emission peak at about 680 nm. The sharpness of the emission peak minimizes crosstalk between PerCP and PE, and therefore also minimizes the amount of fluorescence compensation needed.

So much for the good news. PerCP-labeled antibodies, available only from B-D, are, or at least were, on the pricey side. If you happen to be using a B-D FACScan or FACSCalibur, that may be the only bad news about PerCP; both Mandy et al[1027] and Nicholson, Jones, and Hubbard[1252] have reported excellent results with this label using that instrument. The other problem with PerCP, however, relates to its intolerance of high illumination power levels. The FACScan and FACSCalibur, which have very efficient light collection optics, use only 10-15 mW laser power for illumination, and the illuminating beam is on the order of 100 µm wide. If more power, or the same power in a narrower beam, is used for illumination, fluorescence signals diminish in intensity, instead of increasing in intensity. Many people found that their cytometers and PerCP didn't go well together, and favored PE-Cy5 as a third label.

Meanwhile, photon saturation and bleaching (pp. 115-8) had been rounded up as prime suspects in PerCP's "Case of the Disappearing Photons," but, as it turned out, that wasn't quite what was going on. A clue as to what might be happening came when Davis and Houck[2631], at B-D, described a **PerCP-Cy5.5** tandem conjugate, with maximum emission near 700 nm. The tandem exhibits even less fluorescence emission overlap with PE than does PerCP itself; PerCP-Cy5.5-labeled antibodies are now available from BD Biosciences. But what broke the case is that PerCP-Cy5.5 does not suffer from the same "saturation" problem as PerCP, even when laser powers as high as 200 mW are used.

Phycobiliproteins and Tandems: Dirty Little Secrets

PerCP is not completely unique in its response to high illumination levels; phycoerythrins and allophycocyanin exhibit similar behavior, to different extents. Two ISAC poster presentations from the Herzenberg lab[2632-3] described apparent saturation of phycoerythrin, with fluorescence emission intensities reaching a plateau when excitation energy (at 488 nm) was slightly less than 100 mW. An instrument with two 488 nm beams was set up, allowing fluorescence intensity to be measured in a 100 mW beam some 20 µs after a measurement was made in the first beam, which, during a series of runs, was operated at power levels ranging from a few mW to almost 1 W. It was noted that fluorescence intensity from cells bearing PE-labeled antibodies, measured in the second beam, decreased in a nonlinear manner as illumination intensity in the first beam increased. This appeared to indicate that PE had been bleached. Apparent bleaching was much less for PE-Cy5-labeled antibody, and still less for cells bearing fluoresceinated antibody.

However, the critical observation was that, if cells that had been saved after running through the instrument were reanalyzed after a few minutes, almost half of the PE fluorescence was recovered. It now seems clear that the culprit in the transient disappearance of fluorescence from PerCP and PE is accumulation of molecules in relatively long-lived (7 µs in the case of PerCP[2634]) triplet states. These molecules must return from the excited triplet state to the ground state before they can be excited again. Triplets (pp. 113-4) account for phosphorescence, a phenomenon in which emission may occur minutes, rather than nanoseconds, after excitation.

I once asked Alex Glazer whether phycoerythrin could be used to make a dye laser, and he explained that it couldn't, because, while one PE molecule is equipped with 34 chromophores, excitation of more than one of them at a time results in loss of energy by nonradiative mechanisms, rather than in fluorescence. For a dye laser to lase, photon saturation, i.e., the situation in which the number of chromophores in the excited state is equal to the number in the ground state (p. 116), must be achieved; the laser action is due to fluorescence emission from those excited molecules. This can't happen in PE, because multiple chromophore excitation doesn't lead to fluorescence, at least in part due to triplet conversion[2635].

Making a PE tandem creates a pathway whereby energy from one of the phycoerythrin chromophores can be transferred nonradiatively to an acceptor molecule such as Cy5, from which fluorescence emission will occur; this provides an alternative to triplet conversion and other means of depopulating the singlet excited state of the PE chromophore, decreasing the likelihood of triplet conversion. Thus, tandems should, and do, tolerate higher illumination power levels before beginning to manifest the saturation and bleaching syndrome noted in their parent macromolecules. This paradigm holds for PE and its tandems, and for PerCP and PerCP-Cy5.5, and I am sure that, if somebody goes looking, she or he will find that allophycocyanin and its tandems follow the same rules.

I don't mean for what I have just said to imply that phycobiliproteins and PerCP don't bleach; they do, and, if you don't believe me, look (by eye) at a small amount of phycoerythrin antibody solution before and after you leave it in sunlight for a few hours. The point is that triplet conversion, rather than bleaching, accounts for the relatively low illumination power tolerance of these materials.

For dyes for which triplet conversion is not a problem, e.g., DNA fluorochromes, maximum detection sensitivity is typically achieved by increasing illumination power to the point at which dye saturation occurs; in an observation period on the order of 1 µs, each dye molecule will go through one or two hundred excitation-emission cycles before being irreversibly bleached[1130]. However, optimizing detection of phycobiliproteins, which can allow detection at the single molecule level, typically requires much longer observation periods, on the order of a millisecond or more, during which time approximately 100 photons should be emitted by each phycobiliprotein molecule[660,888,2440].

I am indebted to Ken Davis of BD Biosciences and Dave Parks of the Herzenberg lab for providing me with details of the poster presentations for which abstracts appear in references 2631-4. I am assured that the details will be published at some time; for now, the manuscripts appear to be in long-lived triplet states.

Speaking of tandems following rules, the probability that energy transfer between a donor and an acceptor species will occur varies with the extent to which the donor emission spectrum and the acceptor excitation spectrum overlap. Look back at Figure 7-9 (p. 296), and you will note that this overlap diminishes pretty drastically as we move from PE-Texas red to PE-Cy5 to PE-Cy5.5 to PE-Cy7. As a result, energy transfer between donor and acceptor in this series is progressively less efficient, with the result that the longer wavelength emitting tandems also exhibit more and more emission in the spectral range in which PE normally emits. There's just a speck of PE emission in the PE-Cy5 spectrum; there's quite a bit in the PE-Cy5.5 and PE-Cy7 spectra.

If there's a lot of PE emission in the PE-Cy7 spectrum, it means that a fair fraction of the PE chromophores have not donated energy to Cy7, and, therefore, that there isn't as much emission from Cy7 as there would be from Texas red in PE-Texas red or from Cy5 in PE-Cy5. In some instances, the number of photons or photoelectrons your instrument actually collects and generates from PE-Cy7 can be less than 1/100 the number it would get from PE. If you're designing a multicolor immunofluorescence experiment involving a bunch of tandem labels, you'll want to have an idea of the numbers of molecules of various antigens you're likely to find on your cells, so you don't end up using PE-Cy7 or other inefficient labels to attempt to discriminate cells bearing small amounts of surface antigen from unstained cells.

To make matters worse, while we more or less glibly refer to "PE-Cy5," "PE-Cy7," etc., what we are really talking about is a bunch of PE molecules with different numbers of Cy5 or Cy7 molecules covalently attached to them in different places. A monoclonal antibody is molecularly homogenous; a tandem conjugate is not. That means that the degree to which donor emission will bleed through into the tandem's emission spectrum varies from lot to lot of tandem conjugate or tandem conjugate-labeled antibody. If you're setting up fluorescence compensation for an experiment involving tandem conjugates, you need a single-label compensation control for each antibody. That is, if you're using PE-Cy5.5-antiCD4 in one set of tubes, and PE-Cy-5.5-antiCD8 in another, you need to make sure you compensate them differently, using the appropriate control for each.

If you look hard, you'll find that there are some phycobiliprotein labeling products out there other than PE, APC, and the tandems I have already mentioned. In the phycobiliprotein department, **Martek Biosciences** makes available both some offbeat phycobiliproteins derived from cryptomonad algae, and reactive **phycobilisomes**, which can provide at least an order of magnitude more fluorescence signal than individual phycobiliprotein molecules. Telford et al[2636a,b] examined several of the cryptomonad phycobiliproteins; some can be excited by yellow or red light, and others require green (520-550 nm) excitation.

Molecular Probes offers tandem conjugates of PE and APC with its **Alexa dyes**[2348], which will be discussed in a subsequent section. They have also (R. Haugland, personal communication) investigated tandems in which the phycobiliprotein is the acceptor and the low molecular weight compound is the donor, e.g. Alexa-488-APC. Alexa 488 has excitation and emission characteristics similar to fluorescein; the tandem conjugate excites at 488 nm and emits at the 660 nm emission wavelength of APC.

And Alex Glazer and his colleagues at Berkeley have managed to get *E. coli* to produce recombinant phycobiliproteins[2637-9], complete with biotin for attaching to streptavidin, antibodies, etc. and an affinity tag to facilitate purification. If we carried that over to our species, we could just program our lymphocyte subsets to fluoresce in different colors, and phenotype without benefit of antibodies. The antibody and phycobiliprotein vendors are probably not too worried just yet.

Future Tandems: Heterocycles Built for Two?

There should be some advantages to tandem labels that do not include a phycobiliprotein molecule, but which couple two lower molecular weight chromophores in a way that facilitates energy transfer between them. Dimeric nucleic acid stains designed in this way[1328-9], also from Alex Glazer et al, have been mentioned on p. 314. One such dye, TOTAB, incorporates a thiazole orange and a thiazole blue chromophore; when bound to nucleic acid, the thiazole orange donor dye absorbs maximally around 500 nm and transfers most of its energy to the thiazole blue acceptor, which emits at about 660 nm. It should be possible to synthesize similar molecules incorporating two or more low molecular weight chromophores and the reactive groups necessary to conjugate them to proteins. An oxacarbocyanine and an indocarbocyanine dye, for example, in such a tandem compound would yield good absorption at 488 nm and emission around 575 nm, providing the desirable spectral characteristics of phycoerythrin in a lower molecular weight label.

The lower molecular weight of a completely synthetic tandem label should make it easier to preserve activity in smaller ligands, such as hormones and growth factors, after labeling; the other principal advantage of synthetic ligands is the relative ease with which absorption and emission spectral characteristics can be tailored to enable several different labels to be used with a single source. This prospect is particularly appealing because it could allow labels to be designed to fit inexpensive sources such as red diode lasers.

Glazer et al have, in fact, continued work on other relatively low molecular weight materials in which energy transfer is exploited; they are not antibody labels, but a new generation of labels and primers for DNA sequencing and sizing[2640].

There's a nice kind of symmetry to the phycobiliprotein/tandem story. Phycobiliproteins first came into use as labels because nature had done an excellent job of designing molecules to transfer fluorescence excitation energy. The PE-APC tandem, however, which combines two natural products, isn't as good for some human purposes as some synthetic and semisynthetic molecules have turned and may turn out to be. I expect that synthetic tandem labels will eventually come into wide use. I should also point out that nature is way ahead of us on the phycobilisomes in the photosynthetic apparatus, and that, if the dye designers turn their attention to an artificial photosynthetic molecule, they might accumulate considerably more wealth than can be realized by making slight improvements in cytometry.

Cyanine Dye Labels: From Cy-Fi to Hi5 for Cy5

During most of the 1980's, the phycobiliproteins were very much in the limelight, which was appropriate not only because lime is about the right excitation wavelength, but because they provided a means of attaching several efficient fluorescent labels, with spectra in regions in which cellular autofluorescence is minimal, to antibodies and other ligands. The fuss about the phycobiliproteins diverted many people's attention from the search for other, low molecular weight fluorescent labels with large Stokes' shifts and/or with excitation maxima in the green, red, and infrared spectral regions. In terms of the design of lower molecular weight labels, one of more successful efforts has been the development of **reactive cyanine dye labels**, largely due to the persistence of Alan Waggoner and his colleagues[1361-4].

Cyanines, like many of the compounds now in use as laser dyes, are not soluble to any appreciable extent in water, but are soluble in nonpolar solvents. This characteristic is essential to the use of cyanine dyes as probes of membrane potential, as it allows the dyes to pass freely through the lipid bilayer portion of cytoplasmic and organellar membranes, but is undesirable in labels because it increases the tendency of unconjugated label to stick to protein and the tendency of conjugates to bind nonspecifically by this mechanism. There is, in fact, considerable interest in the preparation of water-soluble analogues of existing laser dyes for use as laser dyes, since the substitution of aqueous dye solutions for the solutions in organic solvents now employed would make it easier to keep the dye medium cool.

The spectral characteristics of most fluorescent organic dyes are determined by the structure of the heterocyclic ring(s), and the conjugated double bonds connecting rings, if such connections are present (they are in cyanines), which form the backbone of the dye. Water solubility and reactivity can be modified by the addition of functional groups on the ring; many of the functional groups that can be added to increase water solubility may also be capable of being derivatized to produce dye molecules that can be covalently bound to proteins, etc. The art lies in adding the functional groups in ways that do not substantially alter spectra.

The first attempts of which I am aware to produce reactive derivatives of cyanines (A. Waggoner & H. Shapiro, unpublished), around 1981, failed to yield suitable antibody labels; the compounds synthesized shifted their absorption maxima and decreased their quantum efficiency on conjugation. Waggoner's group subsequently produced cyanines that could bind covalently to protein sulfhydryl groups[1361], and then isothiocyanate derivatives[1362]. The two newest series of labels[1363-4] use a succinimidyl ester group to link to proteins; the first of these[1363] employed carboxymethyl groups to increase solubility, while the second[1364] uses sulfonate substituents for the purpose.

The reactive **oxacarbocyanine** dye **Cy2** (see Figure 7-9) shares the spectrum of the $DiOC_N(3)$ series of dyes, which are well known as membrane potential probes. The absorption and emission spectral characteristics of these dyes are similar to those of fluorescein.

Cy3 and **Cy5** are, respectively, derived from **indocarbocyanine** and **indodicarbocyanine**; their spectral characteristics are shown in Figure 7-9. **Cy3** absorbs maximally at about 545 nm, and can be excited most effectively with a green He-Ne laser source at 543 nm or the mercury arc lamp line at 546 nm; however, the absorption of the material is high enough that it excites adequately at the 488 nm argon laser wavelength available in most cytometers. The 515 nm argon laser line is better, as are krypton lines at 520 and 530 nm and the emission from a frequency-doubled YAG laser at 532 nm. The emission peak of Cy3 is at about 565 nm; however, a substantial fraction of Cy3 emission is transmitted by the bandpass filters typically used for phycoerythrin detection. **Cy5** absorbs maximally near 640 nm; used by itself, it is very effectively excited by 633 nm He-Ne lasers, 635-650 nm diode lasers, or the 647 nm krypton laser line. For immunofluorescence work, Cy5 is now probably more widely used in the form of **PE-Cy5 tandem conjugates**[1359], which can be excited at 488 nm, than as a primary label.

Cy5.5 is a reactive derivative of **dibenzoindodicarbocyanine**, with maximal absorption near 675 nm and maximal emission at 695-700 nm. This material is optimally excited by 670 nm diode lasers, which are dirt cheap (think red laser pointers); its absorption at 633 nm is sufficient to make it possible to use Cy5- and Cy5.5-labeled antibodies for two-color immunofluorescence analyses in an instrument with a 633 nm He-Ne or 635- 650 nm diode laser source. The **IMAGN 2000**, a clinical instrument which did lymphocyte subset analysis by a low-resolution scanning technique dubbed **volumetric capillary cytometry**[1365,2484,2641], rather than flow cytometry, used Cy5 and Cy5.5 as antibody labels. **Biometric Imaging**, which produced the instrument, was acquired by Becton-Dickinson. **Cy3.5** , with maximum absorption near 580 nm and maximum emission near 600 nm, is to Cy3 as Cy5.5 is to Cy5.

Wessendorf and Brelje[1366], using microscopy and densitometry rather than flow cytometry, compared the brightness of immunofluorescent staining with the same antibody tagged with fluorescein, tetramethylrhodamine, Lissamine rhodamine, Texas Red, and Cy3.18, an early cyanine label; they concluded that the cyanine dye was brightest. This is not too surprising; cyanine dyes have higher extinction coefficients than most other dyes, and quantum efficiencies at least as high, and these authors used excitation wavelengths optimal for the individual dyes.

Cy7 is a reactive indotricarbocyanine dye; it absorbs in the near infrared (about 750 nm) and emits around 770 nm.

The indotricarbocyanine structure itself is not as stable chemically as are the indodicarbocyanine and -carbocyanine dyes fluorophores of Cy5 and Cy3; this makes it harder to prepare Cy7 labels and also results in a relatively short shelf life. Like Cy5, Cy7 is more widely used as an acceptor in PE- and APC-based tandems than as a primary immunofluorescent label; while Cy2, Cy3, and Cy5 are fairly popular as labels for nucleic acid probes used in **fluorescence in situ hybridization (FISH)**, Cy7 seems not to be.

Cy3, Cy5, and related dyes are available from **Amersham Biosciences** in kits that contain just about everything needed for antibody conjugation except the antibody.

There is a "sticking point" to cyanine dye labels; antibodies labeled with Cy3, Cy5, and/or PE-Cy5 and, possibly, other cyanines, seem to adhere to monocytes and, to a lesser extent, to granulocytes, resulting in low levels of nonspecific staining. I first heard about this from Carl Stewart and from Alan Waggoner, and the three of us (unpublished) demonstrated the effect using Cy3 and Cy5 antibodies in a Cytomutt with 543 and 633 nm He-Ne laser excitation. The effect came to general attention when PE-Cy5 tandem conjugate-labeled antibodies came into wider use and were found to bind to monocytes even when the antibody was not reactive with monocyte surface antigens. According to van Vugt et al[2642], this interaction involves the high-affinity IgG receptor CD64 present on the surface of monocytes. However, Stewart and Stewart[2643] report that some CD64-negative monocytic and myeloid leukemia cells also bind PE-Cy5 antibodies not reactive with their cell surface antigens. Tjioe et al[2628] noted that antibodies labeled with a PE-APC tandem conjugate bind to monocytes and granulocytes, although not to the same extent as do PE-Cy5-labeled antibodies. Antibody manufacturers have come up with various proprietary ways of minimizing such irrelevant binding.

Blue Notes: AMCA and Cascade Blue

A number of **UV-excited, blue fluorescent labels** have come into use for antibody labeling. After having been told at least twice that fluorescein was adopted as a label because the UV-excited blue autofluorescence present in most cells and tissues interfered with the UV-excited blue fluorescence of anthracene, which was the first label, you might well ask why. The simple answer is that people who had "maxed out" their instruments in terms of the number of detectors they could have looking at the 488 nm beam and the krypton, dye, or He-Ne laser beam wanted to measure still more antigens. If you had a UV laser available, you could use a UV-excited, blue fluorescent label; if you restricted its use to antibodies for which the cells of interest have relatively large numbers of binding sites, you wouldn't have a big problem with autofluorescence.

The **coumarins** are a family of compounds that exhibit UV-excited blue fluorescence and that, incidentally, were originally used therapeutically as anticoagulants and, in proportionally larger doses, as rodent poisons. Coumarin derivatives have been exploited as laser dyes, and may also act as pH indicators; the latter characteristic is usually an undesirable one for a label. The first popular coumarin label was **7-amino-4-methylcoumarin-3-acetic acid**, or **AMCA**. As shown in Figure 7-9, AMCA absorbs maximally at about 350 nm, and has an emission maximum near 455 nm; it can be excited optimally by argon or krypton UV lasers and by the mercury arc lamp line at 366 nm, and adequately by the 325 nm He-Cd laser line.

Aubry et al[1367] stated that, in addition to being usable for multicolor immunofluorescence analysis, AMCA was well suited for use in combination with acridine orange for simultaneous measurement of DNA and RNA content and cellular antigens, using UV and 488 nm excitation beams. My personal preference would be to use Hoechst 33342, pyronin Y, and fluorescein-antibodies[113] for the same purpose, in most cases; however, there are situations in which acridine orange may be more appropriate than pyronin for RNA content measurement, and also situations in which one would want to use acridine orange to measure something other than RNA content, e.g., DNA sensitivity to denaturation. In either of these cases, AMCA-labeled antibodies might be useful, provided the antigenic determinant of interest was preserved during the acridine orange staining procedure. Delia et al[1368] used AMCA as a third label in a flow cytometer with collinear UV and 488 nm illuminating beams. Molecular Probes' **Alexa 350**, a sulfonated aminocoumarin, has similar spectral characteristics, but a quantum yield almost twice as high as that of AMCA[2348].

Cascade Blue[1075,2332,2644], a reactive derivative of **pyrene**, was introduced by Molecular Probes; its spectrum is shown in Figure 7-9. The absorption maximum is near 390 nm. While absorption at 325 nm is probably too low to rely on a UV He-Cd laser for excitation, adequate excitation should be available from the 366 nm arc lamp line or the UV lines from argon or krypton or mode-locked tripled YAG lasers. Anderson et al[2644] excited both Cascade Blue and Molecular Probes' **Cascade Yellow** (emission maximum near 550 nm) using the violet krypton lines at 407 and 413 nm, and a violet diode laser should be usable as well. The emission maximum of Cascade Blue is at about 415 nm; emission above 500 nm is considerably less than is the case with AMCA and Alexa 350, facilitating compensation in multicolor immunofluorescence experiments.

Hey, BODIPY!

Molecular Probes also offers the **BODIPY** dyes[1075,2332]; which are **boron dipyrromethane** derivatives with long enough formula names to justify one of the manufacturer's many catchy abbreviations. The original member of the series, **BODIPY FL**, has spectral properties similar to those of fluorescein, with a somewhat narrower emission peak, less sensitivity to pH changes, and greater resistance to bleaching, all desirable characteristics. The excitation maxima of the various BODIPY dyes cover the range from 500 to 646 nm; emission maxima range from 506 to 660 nm. Five of the most popular BODIPY dyes are available with water-solubilizing spacers and amine-reactive ester groups attached to facilitate conjugation to antibodies or other proteins.

A variety of peptides, proteins, polysaccharides and other molecules with BODIPY dyes attached are also produced. **Bis-BODIPY-phosphatidylcholine**, localized in the inner leaflet of the plasma membrane, was used by Meshulam et al[1369] to measure phospholipase A activation in neutrophils by flow cytometry. Knaus et al[1370-1] and Martin et al[1372] have used BODIPY labeling with other small ligands; most recently, Haugland et al[1373] used phalloidin labeled with the new long-wavelength fluorophore **BODIPY 581/591** (581 nm excitation, 591 nm emission), which is spectrally similar to Texas Red, for fluorescence microscopic visualization of the intracellular distribution of F-actin. However, BODIPY dyes have still not moved as far up in the charts as might have been expected.

Alexa Dyes: Some Thoughts on Dyemographics

I write a new edition of *Practical Flow Cytometry* every few years and, if I'm lucky, sell a few thousand copies. Dick Haugland, who founded Molecular Probes, puts together a new edition of his *Handbook*[2332] every few years, with updates at intervals on CDs and or his web site, and gives away tens of thousands of copies. The rest of the time, he presides over a large and talented group of people who make dyes. They might be considered dye hackers, in the old, honorable sense of the word; they know the structures and the spectra, and how to tweak the structures to optimize the performance of the dyes. And the company isn't losing money.

The **Alexa dyes**[2332,2348] are a series of sulfonated coumarin- and rhodamine-based labels, with spectral characteristics similar to those of some of the more popular labels previously mentioned in this section: e.g., AMCA, fluorescein, Texas red, Cy3, Cy5, Cy5.5, and Cy7. However, the Alexa dyes have higher quantum yields, better photostability, and better charge characteristics (allowing more dye molecules to be put on a protein before the law of diminishing returns cuts in on fluorescence). So what happens? People who do their own dye conjugations have written testimonials for the Alexa dyes in the latest edition of the *Handbook*[2332], but the companies that make and sell antibodies in volume kept making conjugates with fluorescein, and tandems with Cy5, Cy5.5, and Cy7. The word on the street was that the antibody companies didn't want to pay the premium royalties Molecular Probes wanted to get paid for its premium dyes.

Just look at fluorescein, which is the Microsoft software equivalent in fluorescent labels. It has a lot of problems, e.g., its fluorescence yield varies markedly over the near-physiologic pH range, but, while everybody knows about, and many people complain about, the problems, the antibody manufacturers load fluorescein onto their antibodies, and almost everybody buys them and uses them. Molecular Probes has produced at least three green dyes – BODIPY FL, Oregon Green, and Alexa 488 – that are better labels than fluorescein, in most respects, including photostability, which keeps the dyes green longer – with fluorescein envy. The race is not always to the swift.

Luckily, Molecular Probes does well enough to keep the Alexa dyes and a few thousand other gourmet chemicals in the catalogue for its discerning customers. And, as of early 2003, at least a few antibody producers (BD/Pharmingen and Caltag, for sure) had started to offer Alexa dye labels.

Other Organic Fluorescent Labels: A Dye Named JOE, etc.

The DNA sequencing crowd uses fluorescent dyes, many of which are similar or identical to dyes used in cytometry, but is likely to know them by different names or nicknames. In the interest of nickname translation, I will mention that 6-JOE is 6-carboxy-4′,5′-dichloro-2′,7′-dimethoxyfluorescein, 5-FAM is 5-carboxyfluorescein, TAMRA is carboxytetramethylrhodamine, ROX is carboxy-X-rhodamine, and TET is c-carboxy-2′,4,7,7′tetrachlorofluorescein. The halogenated fluoresceins have excitation and emission maxima at somewhat longer wavelengths than those of fluorescein itself. Molecular Probes sells them in reactive form as succinimidyl esters for sequencing and probe labeling; there doesn't seem to be any compelling reason to use them in cytometry.

In the golden oldies department, I should mention **nitrobenzoxadiazole (NBD)** as a label; this material is usually applied to proteins in the form of its reactive chloride derivative, 4-chloro-7-nitrobenz-2-oxa-1,3-diazole (NBD chloride), which reacts with amino and thiol groups. The adduct with amino groups has spectral properties similar to fluorescein; Wallace et al[622] used **NBD-phallacidin** for flow cytometric studies of cytoskeletal structure. NBD fluorescence is strongly environmentally dependent; for details, see the Molecular Probes *Handbook*[2332].

Most dyes suitable for use in CW tunable dye lasers emitting in the red and infrared have absorption maxima in the orange and red spectral regions; reactive derivatives of these dyes should, in principle, be suitable labels for use in instruments with He-Ne or diode laser excitation in the red. Joel Wright and I (unpublished) fooled around with a derivative of the oxazine dye **Nile blue** without much luck in the early 1980's; Monsigny et al[1374] have, more recently, had better luck, producing both a **reactive benzoxazinone dye** with red emission that can be excited at 488 nm and another that apparently stains nucleic acids.

Aluminum phthalocyanine dyes have been derivatized for use as labels[1375]; these dyes themselves have been used as sensitizers for photodynamic treatment of cancer, and their uptake has been monitored by flow cytometry[1376-8]. The derivatized versions had strong absorption in the UV and red spectral regions with emission at longer red wavelengths. I was able (D. Schindele and H. Shapiro, unpublished) to resolve populations of beads labeled with two different aluminum phthalocyanine dyes in a cytometer with 670 nm diode laser illumination; I didn't have great luck with cells bearing antibodies labeled with phthalocyanine derivatives, but further work along this line would seem worthwhile, if a source of the dyes can be found.

Whitaker et al[1379] report that fluorescent **rhodol** derivatives can be made with fluorescence properties similar to fluorescein and rhodamine, but with better photostability and without the pH sensitivity of fluorescein.

In the development of laser dyes, attempts have been made to find and exploit chemical structures with fluorescence spectra showing large Stokes' shifts. Power levels of hundreds of milliwatts or more are required to pump CW dye lasers; at present, most dyes that emit in the red and infrared require a krypton ion laser (647/676 nm) for pumping. Since argon lasers are more tractable than their krypton counterparts, it would be desirable to use argon lasers as dye laser pumps for red- and IR-emitting dyes. Argon lasers, however, emit in the blue and green. Red and IR emission can be obtained with argon laser pumping when dyes with large Stokes' shifts, e.g., pyridyl and styryl dyes[307-9], are used as the lasing media. Derivatives of these dyes might make good labels; the pyridyl and styryl dyes might also be useful in low molecular weight tandem labels.

The IR diode lasers used in CD players emit at around 785 nm, which is above the emission maximum of the longest-wavelength label discussed up to this point (Cy7). Since relatively high power (tens of milliwatts) 785 nm diode lasers cost only a few dollars, the notion of using such a laser to excite at least one additional fluorescent label is attractive. One principal problem with this approach is that dyes with the large conjugated ring structures needed to produce IR absorption tend to be chemically unstable. Linda Lee, who synthesized thiazole orange while at B-D[769], made two relatively stable IR-excited dyes, **BHMP** and **BHDMAP**, during a brief stint at Biometric Imaging[2645]. The dyes, with emission maxima near 805 nm, were used successfully in a prototype volumetric capillary cytometer with dual (633 nm He-Ne and 785 nm diode) lasers. BHMP and BHDMAP are heptamethine cyanine dyes with a dialkylpyridinium ion at the central methine, which contributes to stability and makes the dyes water-soluble.

Quantum Dots

A fair amount of buzz has recently been devoted to the prospect of using **semiconductor nanocrystals**, better known as **quantum dots**, as labels for biological molecules or as tags for beads used in multiplex assays[2646-53].

In semiconductor light-emitting diodes (LEDs) and lasers, absorption of optical or electrical energy creates **electron-hole pairs**; one atom of the material temporarily loses an electron, while another somewhere in the vicinity temporarily gains one. When electron-hole pairs recombine, some of the absorbed energy is lost in the form of a photon. The emission wavelength is dependent on the **Bohr radius**, which is the average distance between electrons and holes in an excited-state pair. The Bohr radius is a function of the composition of the semiconductor material, but it normally on the order of tens of nanometers.

In semiconductor crystals with dimensions smaller than the Bohr radius, additional energy is required to create electron-hole pairs, and absorption and emission are thus shifted to shorter wavelengths. In such **nanocrystals**, typically no more than about 10 nm in diameter, the emission wavelength becomes more dependent on the size of the crystal than on its composition, since the energy required to keep an electron and a hole separated increases dramatically as the distance between them decreases. The emission wavelength of a CdSe crystal with a diameter of 2.1 nm is approximately 510 nm; that of a crystal with a 3.1 nm diameter crystal of the same material is approximately 560 nm.

The fluorescence spectrum of a nanocrystal is considerably different from that of an organic dye. Organic dyes (see Figure 4-7, p. 113, for the fluorescein spectrum) typically have small Stokes' shifts, i.e., their excitation maxima are within 20 nm of their emission maxima. The excitation spectrum and the emission spectrum of an organic dye often resemble mirror images of one another; both are substantially skewed, with a short-wavelength "shoulder" in the excitation spectrum and a long-wavelength "tail" in the emission spectrum.

The emission spectrum of a nanocrystal is typically symmetric, with a full width at half maximum of at most a few tens of nanometers, the emission bandwidth is thus substantially less than for a typical organic dye. This means that, given a pair of nanocrystals and a pair of dyes with the same emission peak wavelengths, a given combination of optical filters and dichroics will do a better job of separating fluorescence signals from the nanocrystals than of separating fluorescence signals from the dyes.

Moreover, the excitation spectra of nanocrystals are relatively independent of emission wavelength; progressively shorter wavelengths are increasingly effective for excitation. Excitation at 400 nm is typically at least twice that at 488 nm. This suggests that violet diode lasers will be economical and useful excitation sources for work with quantum dots, in either scanning or flow cytometers.

Being inorganic, nanocrystals are much less susceptible to photobleaching than are organic dyes; a nanocrystal is likely to be putting out over 75% of its original fluorescence output after an observation time sufficient to photobleach over 95% of the fluorescence emission from a dye. The relative chemical and photostabilities of nanocrystals should be particularly useful for work in the far red and infrared spectral regions (nanocrystals can be made with emission wavelengths above 1000 nm). Nanocrystals also have much higher absorption (and excitation) cross-sections than do dyes, meaning that, although the quantum efficiencies of nanocrystals and dyes are about the same, the fluorescence from a nanocrystal is typically equivalent to the fluorescence from a dozen or two dye molecules.

So why can't you buy 200 different monoclonal antibodies labeled with quantum dots? Well, as usual, there have been some practical problems, some of which appear to have been solved, and some of which remain.

Since both the emission wavelengths and the fluorescence intensities of nanocrystals depend on their size, preparative methods must yield crystals with highly homogenous size distributions in order to keep emission peaks confined to a small spectral range and maintain low emission bandwidths.

Earlier work with nanocrystals produced materials with fluorescence lifetimes in the range of 100 ns, which is definitely on the long side for flow cytometry, although acceptable for static cytometry and imaging. Newer materials have fluorescence lifetimes of approximately 15 ns, still a few times longer than the lifetimes of most commonly used dyes, but probably acceptable for flow cytometry.

Semiconductor materials don't much like hanging out in aqueous solutions. Practical quantum dot labels therefore must consist of a semiconductor core, a shell of another semiconductor material that confines excitation to the core, and an outer layer that allows the particle to remain dispersed in aqueous solution and that provides some means by which it may be attached to a biomolecule. CdSe crystals are typically made with ZnS or CdS shells; the hydrophilic coating may be silica or one of several sulfur-containing acids. A dihydrolipoic acid coating, in particular, readily binds avidin, allowing biotinylated antibodies to be bound to the coated crystals with very high affinity[2654].

However, once you've made a particle the size of a quantum dot and coated it with something hydrophilic that binds proteins, you've increased the likelihood that the particle will be more or less nonspecifically bound to cells, and/or scarfed up by phagocytes. While the effects of phagocytosis on any particular experiment may be difficult to predict, nonspecific binding is almost certain to decrease the ratio of fluorescence intensities you measure from cells that do and do not bear the material you are trying to detect with the aid of a quantum dot label.

Preliminary results on flow and image cytometry of cells labeled with quantum dot-tagged antibodies were presented by Bill Hyun of UCSF and collaborators from **Quantum Dot Corporation** at the 2000 ISAC meeting[2655], but I haven't yet seen a publication on flow cytometry using quantum dots. I guess people are still in the excited state over them, and need to relax and write a paper. You can now buy streptavidin-conjugated quantum dots with a 605 nm emission wavelength from Quantum Dot Corporation.

I put a brief writeup of green fluorescent protein in the last edition, guessing that while nobody had yet done flow cytometry of cells expressing GFP, it might be the next big thing; I have similar vibes about quantum dots. I won't tell you about all the predictions I made in earlier editions that didn't pan out, at least not here.

Getting Labels Onto Molecules of Interest

If there is one basic principle applicable to labeling, whether the label is a large or a small molecule, fluorescent or nonfluorescent, and whether what is being labeled is an antibody, drug, hormone, or other material, it is this: **It is absolutely critical to establish the effects of the fluorescent tagging procedure on the affinity and biological activity of the effector portion of the molecule being tagged.** Failure to do this can produce unpleasant surprises.

The more popular reactive forms of labels are those that bind to amino groups on proteins. Early in the game, fluorescein and tetramethylrhodamine isothiocyanates (FITC and TRITC) were among the most popular labeling reagents. Molecular Probes still sells isothiocyanate reagents, but recommends succinimidyl esters and sulfonyl halides because conjugates prepared using isothiocyanates reportedly deteriorate over time[2332]. **Research Organics** advocates dichlorotriazinylaminofluorescein (DTAF) as a replacement for fluorescein. Maleimide derivatives of labels react reasonably selectively with protein sulfhydryl groups.

The extremely high affinity of **biotin** for **avidin** or **streptavidin**[300] is widely exploited for labeling purposes. Labels may be covalently bound to avidin or streptavidin, and the labeled material used to detect biotinylated antibodies; alternatively, labels, especially larger molecules such as phycobiliproteins, may be biotinylated, and avidin or streptavidin then used to form a complex between the label and a biotinylated antibody. In some cases, at least, antibody-label complexes formed using a biotin-avidin reaction are stable enough to permit several complexes, each bearing a different antibody and a different fluorescent label, to be used simultaneously.

Maintaining activity of a labeled biomolecule may require the interposition of a third chemical structure, or **linker**, between the label and the molecule being labeled, particularly if the molecule itself is relatively small; I won't go into the details.

Having dwelt at length on the topic of labeling *per se*, we will now address the use of fluorescently labeled reagents for measurement of cell surface structures and properties.

7.6 IMPROVING SIGNALS FROM LABELS: AMPLIFICATION AND OTHER TECHNIQUES

In many cases, the cell- (or bead-) associated molecules we are trying to detect and quantify using a labeled reagent are abundant enough so that we can accomplish our task using **direct labeling**, incubating our cells or beads with a reagent bearing one or a few molecules of label, and then measuring the fluorescence intensity of the particles of interest in a cytometer, with or without a wash step to decrease the concentration of unbound reagent in the sample. When we are hunting rarer prey, we need to consider various ways of getting bigger signals from each molecule of interest than direct labeling can provide. In order to choose methodology appropriate to our measurement problems, we must appreciate several factors that potentially limit measurement sensitivity, and determine which of these are most relevant to our particular measurement problem.

The ultimate limit to sensitivity is set by the number of molecules of interest present in or on the particles we propose to analyze. Tomas Hirschfeld pointed out in the 1970's that, once we are detecting and counting single molecules, or small numbers of molecules, we can expect fluctuations in the number of molecules in the observation volume to become a major source of statistical variance in quantitative measurements[2657]. Some 20 years later, when the sensitivity of their capillary electrophoresis detectors had reached the single-molecule level, Chen and Dovichi[2658] demonstrated the effect experimentally.

In the discussion of Q and B and their effects on measurement sensitivity (pp. 221-3), it was noted that a FACScan or FACSCalibur, which, as flow cytometers go, has very efficient collection optics, will typically generate only a single photoelectron at the PMT photocathode for each four molecules (strictly speaking, for each four MESF) of phycoerythrin passing through the observation point; when fluorescein is detected, an average of 80 MESF must pass through the observation point to generate a single photoelectron[2482]. In these instances, measurements are limited by photoelectron statistics.

In a slow flow system such as the apparatus at Los Alamos used for DNA sizing and for single molecule detection of phycoerythrin, the observation time is much longer than in a conventional flow cytometer, and each MESF of phycoerythrin may generate a burst of 100 photoelectrons during its passage through the measurement system. If there are an average of 100 phycoerythrin molecules in the measurement system at any given time, they will give rise to 10,000 photoelectrons; the expected standard deviation would be 100 photoelectrons, and, in the absence of other sources of variance, one would be able to achieve a measurement CV of 1 percent. However, Poisson statistics would dictate that the standard deviation of the number of molecules in the measurement volume would be 10 molecules; the lowest possible CV is then not 1 percent, but 10 percent. Thus, what Chen and Dovichi[2658] define as "**molecular shot noise**" becomes the limiting factor in precision; in order to achieve 1% precision, one must measure or count at least 10,000 molecules.

These days, it is not uncommon to be able to detect a few hundred molecules of cell- or bead-associated antibodies or other target molecules in a conventional flow cytometer; while there may be fluorescence amplification techniques that let us collect hundreds of photoelectrons from thousands of molecules of a label for each target molecule present, they will not liberate us from molecular shot noise; we will, instead, come to recognize it as a major source of variance. Of particular interest is a recent paper by Elowitz et al[2659], which considers molecular noise in the context of gene expression, noting that low intracellular copy numbers of molecules can limit the precision of gene regulation.

Limits to Sensitivity: Autofluorescence

The spectral characteristics of covalent labels exert a significant effect on the sensitivity of flow cytometric measurements of weak fluorescence signals such as are obtained from cells stained with fluorescently tagged antibodies, hormones, lectins, etc. This is so because it is frequently not instrument performance, but **cellular autofluorescence** that limits the number of labeled molecules detectable on or in cells.

As mentioned on pp. 216-7, the intensity of lymphocyte autofluorescence in the green spectral region used for fluorescein fluorescence measurement is the same as would be obtained from over 600 MESF of fluorescein. In most newer laser source instruments, this level of autofluorescence should be detectable above background.

There is, however, no way to use the green fluorescence signal alone to distinguish the proportions contributed by antibody or ligand fluorescence and autofluorescence. Consider two cells, one bearing no bound antibody and emitting autofluorescence equivalent to the emission from 1,000 fluorescein MESF, and the other bearing 300 MESF of fluorescein on cell-bound antibody or ligand and emitting autofluorescence equivalent to 700 fluorescein MESF. Both cells produce 1,000 fluorescein MESF at the detector. Good discrimination of stained and unstained cells will, therefore, require that several times as many fluorescein molecules as are detectable against zero background be present on the surface of the stained cells.

There is a potential problem with measurements of both sensitivity and autofluorescence, in that both are (or were) generally done using log amps, and both involve extrapolating a reference line calculated from measured values of fluorescence obtained from beads bearing known amounts of label. These known amounts typically range from 5,000 to 500,000 MESF, and PMT and instrument gains are generally adjusted to place peaks from the fluorescence distributions of labeled beads in the upper decades of a log scale. The blank beads used for sensitivity determination and the unstained cells used to determine autofluorescence levels fall at the low end of the scale, usually below 1,000 MESF.

When the regression line (see Figure 4-60 on p. 216) is calculated, a correlation coefficient and, in some cases, other measures of the distance from the line of the points representing labeled beads, are provided. If the correlation coefficient is low (low, in this instance, probably means less than 0.98, because we're working on a log scale), and/or the distances of the labeled beads from the line are large, it is fairly obvious that the log amp doesn't work all that well; the measured sensitivity and/or autofluorescence values are definitely suspect in such cases. However, all log amps become unreliable at some point in the lower region of their range, and good correlations and small distances of labeled beads from the regression line unfortunately provide no guarantee that the log amp is behaving at the level of the blank bead or the unstained cell. I have produced differences of several hundred MESF in measured sensitivity simply by interchanging log amps with DC offset levels differing by 1 or 2 mV. That's why I suggested (p. 217) that sensitivity and autofluorescence levels should be determined using a **linear fluorescence scale**, and that high-sensitivity quantitative fluorescence measurements also be done on a linear scale. I'll make the point again, because it's very important; if you're trying to measure a few hundred molecules of bound antibody and you're not using a linear scale, you run the risk of discovering you've been kidding yourself in private by embarrassing yourself in public.

Improving Sensitivity: The New Wave(length)

In cases in which cellular autofluorescence is a factor limiting sensitivity, one obvious way of improving sensitivity is to work with excitation wavelengths at which cellular autofluorescence is lower than it is when 488 nm excitation is used.

The excitation and emission spectra of autofluorescence reported for several mammalian cell types[181,183] are close to the spectrum of riboflavin (Fig. 7-7, p.291). Relatively little cellular autofluorescence is excited by light at wavelengths above 515 nm[304], which are used for excitation of Cy3, tetramethylrhodamine, the rhodamine 101 derivatives XRITC and Texas red, Cy5, and the phycobiliproteins and tandem conjugates. Thus, the use of any of these labels can provide increased sensitivity for weak fluorescence measurements in mammalian cells.

The ratio of median emission from antibody-labeled cells to median emission from unstained cells, with both emission values expressed on a linear scale, can be taken as a measure of sensitivity; this figure was 10-15 times higher for cells stained with phycoerythrin- labeled antibodies and excited at 546 nm by the mercury arc lamp in an old B-D FACS Analyzer than for cells stained with fluorescein-antibodies and excited at 488 nm by an argon ion laser in a FACStar cell sorter (L. Lanier, M. Loken, N. Warner, personal communications and meeting presentations). While the FACScan has somewhat better PE fluorescence sensitivity than the FACStar, the analyzer, when using green excitation, remained superior[1380-1]. The 532 nm line from a doubled YAG laser and the 543 nm line from a green He-Ne laser are usable for high-sensitivity, low autofluorescence measurements, although the low power levels (1-2 mW) available from green He-Ne lasers may limit sensitivity in practice. Since many more current instruments have red He-Ne or diode lasers than have green YAG or He-Ne lasers, it will usually be most practical to use allophycocyanin as a label with a red excitation source to make sensitive fluorescence measurements in mammalian cells without interference from autofluorescence.

If, however, one were interested in measuring cell surface antigens on *Synechococcus* or other species of cyanobacteria, a phycobiliprotein would be an extremely poor choice of label; cyanobacteria contain phycobilisomes stuffed full of phycoerythrin, allophycocyanin, etc., with the result that their autofluorescence, which is considerable, would be likely to swamp the signal from a phycobiliprotein label.

Vesey et al[2660] provide a good example of how to select a label based on studies of the autofluorescence of particles of interest and of other particles and fluorescent solutes likely to be encountered in the sample; their target was *Cryptosporidium* oocysts in drinking water, and they found, using a Coulter EPICS Elite flow cytometer, that phycoerythrin-, Cy-3-, or tetramethylrhodamine-labeled antibodies excited by a low-power Green He-Ne laser provided the best signal-to-background ratio for detection.

Correcting and Quenching Autofluorescence

If your apparatus can discriminate between cellular autofluorescence and the fluorescence of your label, based on some difference in spectral characteristics, you may be able to pick out the label from the background.

If you happen to have a flow cytometer with two ion lasers, you can follow the lead of Steinkamp and Stewart[780], who, using the multilaser flow cytometer at Los Alamos, took advantage of the fact that autofluorescence is better excited in the violet than at 488 nm (Figure 7-7), while fluorescein is very poorly excited by violet light. They made dual-beam measurements of cells stained with fluorescein-antibodies using a 413 nm beam from a krypton laser and a 488 nm beam from an argon laser, and subtracted a portion of the violet-excited fluorescence signal from the fluorescence signal excited at 488 nm.

Roederer and Murphy[781] devised a method applicable to flow cytometers with only a 488 nm excitation beam; they noted that, while green (530 nm) fluorescence in cells stained with fluorescein-ligands represented the sum of autofluorescence and fluorescein fluorescence, red (625 nm) fluorescence resulted almost entirely from autofluorescence; thus, by subtracting a portion of the red signal from the green signal, they could compensate for autofluorescence. This was done by digital computer processing of list mode data; Alberti, Parks, and Herzenberg[782] used the fluorescence compensation electronics in the FACS to perform similar manipulations on-line.

The methods described in the preceding two paragraphs allow you to compensate for autofluorescence relatively well on a cell-by-cell basis. If the issue is detecting populations of weakly stained cells above autofluorescence background, it may be sufficient to do histogram subtraction, an approach described by Corsetti et al[1389] and Müller et al[1390] (also see pp. 245-6)

A direct chemical approach to autofluorescence on a cell-by-cell basis was taken by Hallden et al[1391] and Mosimann et al[2661], who, respectively used crystal violet and trypan blue to quench intracellular autofluorescence, allowing detection of surface antigens with fluorescein-labeled antibodies and of fluorescein-labeled nucleic acid probes.

The bottom line here, however, is that none of the methods just described deal as effectively with autofluorescence as does choosing a label and an excitation and emission wavelength at which autofluorescence is minimal.

Raman Scattering Effects on Sensitivity

Phycoerythrin offers an attractive alternative to fluorescein when extreme sensitivity is required, because it becomes possible to decrease the effect of cellular autofluorescence, as well as to place a large number of fluorescent chromophores on each antibody molecule. When PE is excited at 488 nm, rather than above 515 nm, cellular autofluorescence is not decreased. Moreover, there is the potential for interference from **Raman scattering** (p. 118), which occurs when a molecule undergoes a vibrational transition while scattering light. Raman emission is most likely to be at a frequency that is the sum of the frequencies of the incident light and the vibrational transition. The frequencies are expressed in **wave numbers (cm^{-1})**, which are the number of wavelengths per centimeter; the relationship between wavelength λ in nm and frequency ν in cm^{-1} is $\nu = 10^7/\lambda$. In flow cytometry, the dominant transition involved in Raman scattering comes from stretching of O-H bonds in water, at around 3600 cm^{-1}. At 488 nm, or 20492 cm^{-1}, the Raman emission is at 16892 cm^{-1}, i.e., $10^7/(20492 - 3600)$ nm, or 592 nm, a wavelength that may be transmitted by filters used for PE fluorescence measurement. When excitation at 532 nm is used, Raman emission is at 658 nm, which should not interfere with measurement of PE fluorescence but which could interfere with measurements of a PE-Cy5 tandem conjugate. Water Raman frequencies for various excitation wavelengths appear in Table 7-4 below.

EXCITATION (nm)	H_2O RAMAN EMISSION (nm)
325	368
350	400
365	420
408	478
415	488
436	517
441	524
458	548
476	574
488	592
515	632
520	640
532	658
543	675
546	680
568	714
594	756
611	783
633	820
650	849
675	892

Table 7-4. Raman emission from water.

In extremely high-sensitivity fluorescence measurements using single photon counting, broadband Raman emission, at wavelengths other than those tabulated above, may be the limiting noise source[1144].

Increasing Sensitivity: Amplification Techniques

Since experimenters can't readily change the number of molecules of the molecules they're trying to find in or on cells, they often attempt to shift the odds in their favor by increasing the amount of label that will be hung on each target molecule.

Some flow cytometer users have the option of increasing illumination power; few have the option of increasing light collection efficiency, and, while you can always try to find a PMT with more sensitivity than the one you have, there is a limit to how much measurement sensitivity can be increased by going that route. When increasing excitation power or fluorescence collection or detection efficiency will not improve measurements, increasing the number of fluorescent molecules bound per molecule of antibody or ligand using an amplification technique may be the only option available.

Amplification by Indirect Staining

The simplest way to do this in the context of immunofluorescence measurements is by using **indirect staining**; approximately six molecules of a polyclonal secondary **developing** antibody can be bound to each molecule of primary antibody used, putting six times as many molecules of label at each antigenic site as would be there if a directly labeled antibody were used. There are a number of variations on the basic theme.

Cohen et al[778] approached the problem of detecting small numbers of cell surface receptor sites by devising what they termed the "**super avidin-biotin system**" (**SABS**). Biotinylated antibodies bound to receptor sites are reacted, in succession, with phycoerythrin-streptavidin, with biotinylated anti-streptavidin antibody, and again with phycoerythrin-streptavidin. The SABS system reportedly allows quantification of antigenic sites within the range of 100-1300 sites per cell; as few as 50 sites/cell can be detected, a sensitivity approaching that of conventional radioimmunoassay.

Zola et al[1380-1] used a somewhat simpler technique, in which a primary mouse monoclonal antibody was developed with a horse anti-mouse antiserum that was then labeled with a phycoerythrin-streptavidin conjugate; they could reliably detect 100 primary binding sites, and applied their methodology to quantification of cytokine receptors.

Amplification Using Labeled Particles

A greater degree of amplification than is possible with indirect staining may be achieved by attaching antibodies (at other than their specific binding sites) to **larger particles**, among which are included fluorescently tagged **viruses**, **membrane microvesicles or liposomes**, **bacteria**, and **plastic beads** (typically coated with antibody, antigen, hormones, etc.) as cell surface labels. Quantum dots (pp. 339-40) also fit into the particle category. Problems with clumping, nonspecific binding, and endocytosis of labeled particulates can occur when this approach is taken, and caution is therefore advisable.

Truneh et al[777,1382] described the use of **unilamellar liposomes** containing **carboxyfluorescein** or sulforhodamine, and conjugated to **protein A**, for detection of small numbers of receptor sites on cell surfaces. Carboxyfluorescein diffuses out of liposomes extremely slowly, and therefore provides a relatively stable internal label for these structures. Gray et al[1383] also reported successful use of the carboxyfluorescein liposome technique. However, the sensitivities achieved are probably matched by modern instruments in analyses of samples stained only with directly labeled antibodies.

I have also already mentioned (p. 277) the use of a **nonfluorescent particle label, colloidal gold**[708], which can be conjugated to antibodies and other ligands and detected on cells by the change in their light scattering characteristics. Antibodies coupled to small (0.1-0.5 μm) nonfluorescent polystyrene beads can also be detected when bound to surfaces by strongly increased side scatter signals[2662]. Siiman et al[2663] were able to detect the binding of two different antibodies, one bound to a polystyrene bead coated with colloidal gold or silver and the other bound to an uncoated polystyrene bead, to cells using measurements of light scattering at 10-20° and at 20-65°; different wavelengths were needed for maximum discrimination of populations depending on whether colloidal gold or silver was used.

While nonfluorescent particles and/or colloidal silver/gold labeling probably do not provide a substantial increase in sensitivity over what can routinely be obtained using direct fluorescence measurements, they do make it possible to detect cell- or bead-bound antibodies in instruments using less expensive light sources and detectors than are needed in most fluorescence flow cytometers.

Phycobilisomes (p. 335), with molecular weights in excess of 10,000,000, might be thought of as biological particles; when directly or closely linked to antibodies, they appear to sterically hinder binding to cell surface antigens. However, Telford et al[2664] found that a red-excited stabilized phycobilisome coupled to streptavidin via a spacer (PBXL-3L, Martek) provided substantial amplification when compared to APC-streptavidin in labeling cell-bound biotinylated antibodies.

Amplification Using Enzymes as Labels: Playing the Hole CARD

A classical amplification technique applicable (originally with some difficulty) to flow cytometry is that of **enzyme-linked assay**. A ligand that reacts specifically with the substance sought by the experimenter (the chemists call the substance sought the **analyte**, which has always sounded kinky to me) is conjugated to an enzyme; after excess ligand-enzyme is removed, the analyte-ligand-enzyme complex is detected by reaction with a **chromogenic** or **fluorogenic substrate**, a colorless material that is acted upon by the enzyme to form a colored or fluorescent product. Once this reaction is complete, thousands of molecules of detectable product are present in the vicinity of each molecule of analyte.

Because of its large amplification factor, enzyme-linked assay offers high sensitivity, and is particularly useful for detection of small amounts of antigens and small numbers of copies of specific genetic sequences in cells; the obstacle to the application of the technique in flow systems is the necessity to trap the reaction product of the enzyme reaction in the cells being analyzed. Technicon (now Bayer Diagnostics) did this successfully in the Hemalog[84-5] and H-1 series of hematology analyzers, which use absorption measurements to identify neutrophil and eosinophil leukocytes by the presence of peroxidase in their specific granules. Kim et al adapted **immunoperoxidase**[779] and **alkaline phosphatase**[1384] labeling techniques for use on the Technicon instruments to subtype lymphocytes; the results based on forward scatter and absorption measurements were comparable in speed, sensitivity, and accuracy to those obtained by fluorescence flow cytometry.

In 1989, Bobrow et al[2665-6] described the technique of **catalyzed reporter deposition**, now familiarly known as **CARD**, which has since facilitated enzyme-based amplification in flow cytometry. The analyte of interest (e.g., an antigen) is reacted with an appropriate reagent (e.g., an antibody) conjugated with horseradish peroxidase (HRP). The specimen is then reacted with a phenolic HRP substrate (e.g., tyramine). HRP action on tyramine yields a highly reactive species that binds covalently to proteins. The tyramine may be labeled with biotin, as done in the original publication, and its presence demonstrated by subsequent reaction with fluorescent labeled streptavidin[2667-70]. However, it is becoming more common in current practice to label the tyramine directly with a fluorescent dye[2332,2671]. In either case, many fluorescent labeled molecules of the tyramine reaction product will accumulate, and stay, in the region of each reagent-analyte complex. Multiple rounds of amplification are possible, using HRP-labeled antibodies against the label deposited in the first CARD stage. Tyramide amplification reagents and kits for flow cytometry, using biotin, as developed by Kaplan et al[2669-70], are available from **Flow-Amp, Ltd.**; Molecular Probes offers a line of reagents including both biotinylated and fluorescent dye-labeled tyramine[2332]. Molecular Probes also offers an enzyme amplification technology described as **Enzyme-Labeled Fluorescence (ELF)**; antibodies or other reagents are tagged with alkaline phosphatase, either directly or via a biotin-streptavidin reaction, and the **ELF 97** phosphatase substrate is added, forming a bright green fluorescent, insoluble precipitate in the vicinity of the antigen or analyte molecules of interest[2332].

Amplifying the Analyte: The Polymerase Chain Reaction (PCR)

An enzyme-based amplification method to which we will return in a later section is the **polymerase chain reaction (PCR)**, which can be used to produce multiple copies of a known genetic sequence ***in situ*** for detection using labeled probes, amplified or otherwise.

Amplification Techniques: Pros and Cons; Fluorescent vs. Nonfluorescent Labels

Since the fairly simple and straightforward modifications of the **indirect staining** technique described above have permitted measurement of as few as 100 binding sites on cells, it is not clear that using larger labeled particles for amplification will further improve sensitivity.

Enzyme amplification is particularly useful in that it can allow qualitative analyses of small amounts of analytes in cells and tissues to be done with chromogenic substrates and a simple transmitted light microscope. As was mentioned on p. 77, the first approaches to antibody labeling involved conjugation of antisera with azo dyes; however, complexes of tissue antigens with such dyes, even in relatively high concentrations, exhibited only weak coloration under the light microscope. This led Coons, Creech, and Jones[44] to turn to fluorescent labels, which were more readily detectable at low concentrations against the darker background present in a fluorescence microscope. Enzyme-linked immunoassay achieves what simple labeling cannot.

Most fluorescence flow cytometers have neither absorption nor extinction measurement capabilities, and are therefore not well equipped for measurement of the reaction products of chromogenic substrates, although forward and right angle scatter measurements may be usable under some circumstances. Enzyme-linked methods have therefore, to date, been used principally with transmitted light microscopy and with static and image cytometry, and are probably in wider use than is fluorescence flow cytometry.

On one of my occasional descents from my ivory tower, I heard a real pathologist give a highly sophisticated talk on lymphomas, involving detection of multiple antigens, *in situ* hybridization with genetic probes, PCR, etc. Flow cytometers weren't involved; all the work was done using enzyme-linked reagents, and much of it was done on tissue slides, where the objective was to determine anatomical localization of various things, a task to which flow cytometers are, and will, for the most part, remain, unsuited. The semiquantitative analyses available from microscopic observation were all that was necessary to answer the biological questions at hand. I'm sure some of the work could have been done using fluorescence flow cytometry, but it wouldn't have been done any better, cheaper, or faster. Food for thought.

Coventry et al[2672] used video image analysis to compare the detection sensitivity of immunoperoxidase staining on slides with that of flow cytometry; they found that detection of 100-200 cell surface molecules in specimens on slides was possible only if heavy metal-enhanced immunoperoxidase methods were used, while high-sensitivity flow cytometric methods[1380-1] could, in some cases, detect as few as 50 molecules.

To some extent, amplification increases sensitivity at the expense of precision, because amplification factors aren't exact; you won't get exactly six molecules of developing antibody per molecule of primary antibody, or 100,000 molecules of product per molecule of enzyme-linked antibody. If you're looking for 100 copies of something, however, you have Poisson sampling statistics and molecular shot noise[2658] (p. 341) working against you, anyway. Even if you hung a 2 mW diode laser on each molecule, you'd still count only 100 molecules, and the measurement CV could be no less than 10 percent.

Improving Sensitivity: Time-Resolved Fluorescence

On p. 283, it was mentioned that phase-sensitive detection of a fluorescence signal in a flow cytometer using a modulated light source could discriminate fluorescence signals from different materials with different fluorescence lifetimes. The lifetimes of most of the materials responsible for autofluorescence are on the order of tens of nanoseconds; if one uses a label with a substantially longer lifetime, and a light source emitting brief (picosecond or shorter) pulses, and delays the fluorescence emission measurement for 50 nsec or so following the pulse, the contributions of autofluorescence and of other interferences, e.g., Raman scattering, stray scattered light, and filter fluorescence, are eliminated from this **time-resolved fluorescence** signal. Chelates of rare earth metals such as europium and terbium have fluorescence lifetimes on the order of microseconds (see p. 113), and Bob Leif suggested in the 1970's[1110], as did Tomas Hirschfeld (personal communication), that they might be useful as labels for time-resolved flow cytometry with pulsed sources. Condrau et al[1385-6] described measurements of cells bearing europium-chelate labeled antibodies in a modified flow cytometer in which acousto-optic modulation is used to deflect the illuminating beam away from the flow cell once a scatter signal is detected; their papers also include good bibliographies on time-resolved fluorescence immunoassays. Related work by Beverloo et al[1387-8] describes the use of **phosphor particles** as luminescent (in this case, phosphorescent rather than fluorescent) labels for time-resolved image cytometry. One drawback to both rare earth chelates and phosphors as labels in the context of flow cytometry is the long measurement time required to take advantage of long excited state lifetimes, which run to hundreds of microseconds; this would limit throughput to a few hundred cells/sec, but the tradeoff may be worthwhile if extreme sensitivity is required.

7.7 MEASURING CELL SURFACE AND INTRACELLULAR ANTIGENS

Almost all of the fluorescence flow cytometers now in use have been applied, at one time or another, to measurements of **cell surface immunofluorescence**. The desire to do immunofluorescence measurements motivated researchers and clinicians to learn about flow cytometry and buy flow cytometers; and this legitimized other flow cytometric techniques not based on immunofluorescence, such as DNA analysis and differential leukocyte counting.

The literature on flow cytometry of immunofluorescence is so vast that I will not even attempt to summarize the fraction of it that I have seen. Instead, I'll try to present the basics and to touch on some issues that I don't think have been well covered elsewhere.

History and Background

In the late 1960's, when most of the few people involved in flow cytometry were looking at DNA stains and trying to tell normal from abnormal cells in cervical cytology specimens, Len Herzenberg alone had the vision to appreciate that the combination of the apparatus and fluorescent antibody reagents could unravel the genealogy and developmental history of the lymphocytes.

Within a few years, flow cytometry, predominantly in the form of gated single-parameter immunofluorescence measurements, and cell sorting and subsequent functional analysis, had been used to define a taxonomy of lymphocytes, in mice and men, where none had existed previously. The methodology that Herzenberg and his associates and students had used to study murine cells was adapted and aggressively applied to human immunology by Stuart Schlossman in the United States, Melvyn Greaves in Great Britain and a rapidly growing crowd of others.

This first wave of activity was all the more remarkable because it occurred several years before the discovery of monoclonal antibodies; the identification and preparation of both reagents and samples were considerably more complicated than they are today. While immunologists, even in those days, seemed to have the opinion that fluorescent antibodies were the only specific reagents, and that neither cytochemistry nor multiparameter flow cytometry would have any use in future work on blood cell development, differentiation, and pathology, this viewpoint was not justified then; subsequent history has eroded any remaining basis for it.

Obviously, antibodies were exquisitely specific, but, until the advent of **monoclonal antibodies**[785], you could never be sure for what they were so specific, and never count on getting a serum with quite the same specificity next time you bled the same animal. For you physicists and electronikers out there (bear with me, immunologic sophisticates), I should amplify on this a bit.

Monoclonal Antibodies for the Uninitiated

For the past few hundred years, or since before anybody knew there were lymphocytes, we have been in the business of deliberately inducing **immunity** to various diseases in humans and animals, by injecting material from organisms that cause those diseases (or, sometimes from organisms closely related to the disease-causing organisms). The substance injected is called an **antigen**. If, a week or more after the antigen is injected, you extract some blood serum from the animal (or human) that received the antigen, you can (usually) demonstrate formation of a precipitate if antigen is added to the serum, due to the formation of complexes between molecules of antigen and molecules of serum proteins called **immunoglobulins (Igs)**, which are the **antibodies**.

Immunoglobulins are produced by **B lymphocytes**; in order for such production to occur, the concerted action and interaction of **T lymphocytes** and other blood cells called **macrophages** is also required. There are several different classes of Igs; B cells start out producing a very large molecule called IgM, and then typically switch to producing IgG. IgE, which is involved in allergic reactions, represents another class of Igs. The **gamma globulin** and **antitoxin** or **antiserum** that some of us remember being used in medicine and others have encountered in old novels are serum immunoglobulin fractions. All Igs of a given class have relatively similar "backbone" structures, but, at some points in these molecules, there exist **variable regions** that differ from molecule to molecule in amino acid composition and sequence; it is the variable regions that react with antigen. Also, Ig molecules are **multivalent**, i.e., each molecule contains at least two sites capable of binding antigen.

Antibodies react with antigens in a very specific fashion and, typically, with very high affinities, but a conventional antiserum contains not one type of Ig molecule with one unique structure, but a large number of antibodies, with different molecular structures, each reacting with different affinity with a different chemical group on the antigen. Lymphocytes, in the course of their development, undergo **gene rearrangement** of the genes that encode the variable regions of Igs; this "shuffling" of genetic material insures that there will be a great diversity of amino acid sequences, and therefore of reactivities, represented in the B cell population of any given individual, although any given B cell can produce only one specific molecular structure. B cells normally carry Igs around on their cell surfaces; when the **surface membrane immunoglobulin (SmIg)** on a B cell binds to an antigen, it stimulates the B cell to reproduce.

Thus, when an animal is **immunized**, or injected with antigen, B cells that produce, and bear on their surfaces, Igs that react with that antigen reproduce. Each B cell forms a **clone** of progeny, all of which produce Ig molecules with the same structure, which is different from the structures of the Igs produced by all the other clones. The antiserum is therefore described as **polyclonal**. Polyclonal antisera, by their biological nature, contain a small fraction of mixtures of Igs that react, with differing affinities, with different determinants on the same antigenic molecule, as well as a larger fraction of "irrelevant" antibodies which, while they may be very relevant to protecting the organism from infections, don't react at all with the antigen the experimenter injected to produce the antiserum, but may react with any of a tremendous variety of other proteins or macromolecules foreign to the animal in which the antiserum was produced.

If you extract the immunoglobulin fraction of an immunized animal, you end up with a number of Ig molecules that will react specifically with the antigen you're after and a much larger number of Ig molecules that will react with other things. **Adsorption**, or reaction with materials containing these irrelevant antigens, can be used to remove some of the irrelevant antibodies, and **affinity purification**, which traps desired antibodies on a column or other substrate containing immobilized antigen, can further increase the specificity of a polyclonal antiserum – to the point at which as many as a few percent of the Ig molecules will react with antigen! Even at this rather shocking level of impurity, specificity may be sufficient to permit definition of the chemical structures on or in cells with which a polyclonal antiserum is reactive.

In the days before monoclonal antibodies, this was what had to be done before one could even think of using an antiserum for immunofluorescent staining; the whole process seemed more art – many would say black art – than science. For those of us who worked in industry, the difficulties of obtaining and/or standardizing immune reagents were a major obstacle to the development and dissemination of procedures involving immunofluorescence. Very few people envisioned a need for more than two immunofluorescent labels; it was hard enough to do single-color fluorescence, and two colors represented a *tour de force*.

Things changed dramatically for the better in the late 1970's, with the dissemination and refinement of methods for **monoclonal antibody production**[785]. If you have the opportunity to look at a copy of the First Edition of *Flow Cytometry and Sorting*[9], which was published in 1979, you can try to find the single appearance of the words "monoclonal antibodies" in that book's 716 pages. They are not, by the way, found in any of the chapters that deal with applications of flow cytometry to immunology.

Monoclonal antibodies are produced by forming hybrids between B cells from animals immunized with an antigen and cultured cells derived from B cell tumors. The tumor cells, unlike normal B cells, are **immortal**, i.e., they can be grown in culture for generations on end. The hybrid cells acquire the property of producing one molecular species of Ig with one specific reactivity from their normal B cell parents, and the ability to grow indefinitely from their malignant parents. Cultures can be started from a single hybrid cell; these will thereafter continue producing identical Ig molecules as long as the culture is maintained.

Monoclonal antibodies are, thus, very well defined, in terms of their own structure, and one can be pretty well sure that all of the Ig molecules in a monoclonal will react with the same site on the antigen with the same affinity. However, if the structure of a given cellular antigen is unknown, simply having a monoclonal antibody instead of a polyclonal antiserum against the antigen doesn't tell you any more about the structure, although it may make it operationally easier to elucidate that structure.

Cell Surface Antigens: Structure versus Function

It was obvious to some people long ago that antigens weren't present on cell surfaces simply for decoration, or as Nature's bar code, but, rather, that each antigen had a function, so that, in most cases, **one person's surface antigen is another person's receptor**. This is much better and more widely appreciated in 2003 than it was in 1978. Monoclonal antibodies have, in fact, greatly facilitated studies of receptor-ligand interactions, in which considerations of specificity formerly demanded use of labeled natural ligands; the antibodies can be made with higher affinity and the structure on the receptor to which they bind can be selected to a great extent.

The advent of monoclonals also eliminated most of the difficulties associated with reaction of cells with two or more antisera to demonstrate two or more discrete antigens on or in the cells; this, in turn, awakened interest among immunologists in increasing the number of available antibody labels. The 1977 publication by Loken, Parks and Herzenberg on two-color immunofluorescence using fluorescein and tetramethylrhodamine as labels with argon ion laser excitation[115] made it clear that improvements in labels and reagents, as well as in equipment, would be necessary to make multicolor work generally feasible.

Moving Toward Multicolor Immunofluorescence

The improvement in labels began with the development of the long-wavelength rhodamine 101 dyes XRITC and Texas red, and continued with the introduction of the phycobiliproteins and, later, of tandem conjugates. While XRITC and Texas red forced some improvement in equipment, by mandating the use of dual-wavelength excitation, the improvement increased the cost and complexity of apparatus. The phycobiliproteins and tandem conjugates have since made multicolor fluorescence easily accessible to immunologists and other workers without the need for much specialized training in instrumentation.

The expanding armamentarium of monoclonal antibody reagents and labels precipitated some changes in immunologists' attitudes; they had enough and good enough reagents to do multiparameter analysis and, now that they had more than two parameters to keep track of, they had adequate motivation to use computers for data analysis and storage. Without computerized data analysis systems, it is unlikely that two-color, let alone three- and four-color immunofluorescence measurements, would have become commonplace.

In retrospect, I should not have expected that immunologists would take kindly to my glib suggestions in the mid-1970's that they run out and do 3-color and 4-color immunofluorescence measurements using multiple excitation beam, multiparameter flow cytometry. Most immunologists were unaware, as I was aware, of the existence of the hardware and software technology required, and painfully aware, as I was unaware, of how difficult it was to produce even good two-color immunofluorescent staining using conventional antisera and the flow cytometers that were then generally available. We literally could not give away a multiple beam apparatus in the mid-1970's; times have certainly changed.

While the demonstration in 1979 of two-color immunofluorescence using dual-laser excitation of fluorescein and XRITC generated a substantial demand for dual-laser instruments, a look at the literature reveals that the number of papers involving fluorescein/XRITC or fluorescein/Texas red immunofluorescence published during the 5 years after the apparatus and labels became available was considerably smaller than the number of dual-laser flow cytometers purchased with such studies in mind. Clearly, multicolor immunofluorescence was harder than it looked.

In the 1980's, history repeated itself, as people stampeded after flow cytometers with CW dye lasers, which could allow them to do 4-color immunofluorescence, using an argon laser at 488 nm to excite fluorescein and phycoerythrin, and a dye laser at 600 nm to excite Texas red and allophycocyanin, provided reagents were available. Some settled for 3-color immunofluorescence, which could be done with a single beam, using fluorescein, phycoerythrin, and phycocyanin[775] or a phycoerythrin-allophycocyanin[306] or phycoerythrin-Texas red tandem conjugate. The first couple of years' worth of publications and presentations actually demonstrating 3-[317-20] and 4-color (L. Herzenberg et al, personal communications and meeting presentations) immunofluorescence using the labels just mentioned came either from Herzenberg's laboratory at Stanford or from B-D's antibody production facility. This was strong evidence in favor of the view that **access to and skill in the use of reagents, and not access to hardware, were the most critical contributing factors to success with multicolor immunofluorescence measurements.**

Today, phycobiliprotein and tandem labels, an unbelievably wide selection of monoclonal antibodies, and instruments that can measure eight or more colors of fluorescence, are available and easy to use compared to what came before. Access is much less of a factor than skill.

Antibody Reagents and Staining Procedures

Most of us, myself included, don't make or conjugate our own **antibodies**; whatever isn't supplied by our collaborators has to be scrounged or bought. Once you get past the realization that there are few commodities you can legally buy that cost as much per small vial, you can appreciate the good points of commercial monoclonal antibodies. They are generally well-defined and well-controlled, you can get them with a variety of labels attached, and you generally can find out how much to use to stain how many cells by reading the label or package insert. Commercial polyclonal antisera, adsorbed and/or affinity-purified or otherwise enriched in antibodies for the antigens you're after, are also available and relatively convenient to use.

Susan Sharrow, who worked with both flow cytometers and monoclonals at NIH more or less since the inception of each, took me to task years ago for suggesting that it was easier to make flow cytometers than monoclonals. It's gotten easier to do both; hybridoma clones are available from places like the American Type Culture Collection, and you can buy kits for purifying the immunoglobulin fraction and labeling the antibodies. Given the range of monoclonals available, labeled, off the shelf, most people won't to have to roll their own. We did do it in my lab once, just for the experience.

Antibody Fragments versus Antibodies as Reagents

For some applications, it is preferable to use enzymatically cleaved antibody **fragments** instead of the whole immunoglobulin molecule. Certain cells, e.g., macrophages, monocytes, and granulocytes, have surface receptors that bind the **Fc** portion of antibodies more or less independently of the antigenic specificity of the antibodies; these cells may, by this mechanism, therefore be stained by labeled antibodies to antigens that are not present on the cell surface. Enzymes can be used to produce **monovalent (Fab)** or **bivalent ($F(ab'_2)$) antigen binding fragments** that will not bind to Fc receptors, eliminating the problem of nonspecific staining. The Fab fragments are useful in situations in which it is necessary to avoid crosslinking of antigenic sites and/or agglutination, although their binding affinities are lower than those of the multivalent antibodies from which they are derived. Fragments of some antibodies are available off-the-shelf, with or without fluorescent labels.

Engineered Antibodies: Phage Display and scFvs

It is now possible to produce monoclonal antibodies, human as well as murine, without benefit of either immunization or hybridoma technology, using **phage display**. It was shown in the late 1980's that proteins coded for by gene sequences introduced into the genome of filamentous bacteriophages could be expressed on the surface of the virions[2673]; it was also demonstrated[2674-5] that a functional antibody variable region (**Fv**) could be expressed in the periplasmic region of *E. coli* into which genes coding for the variable regions of both light and heavy immunoglobulin chains had been introduced. In 1990, McCafferty et al[2676] described production of active antibody molecules in filamentous phage.

The technology has since come a long way; there now exists a phage display library of 2×10^9 members, about three-fifths of which code for functional human **single chain variable fragments (scFvs)**; it is possible, with a minimum of additional genetic manipulation, to derive Fabs and immunoglobulins with the specificity of the scFvs[2677]. Clonal selection has become the province of biotech companies; using appropriate gimmicks (including cell sorting), you can grow up a potful of monoclonal antibody to the antigen of your choice, starting, if you like, with a single B cell[2678]. Since you can hold the light and heavy chain portions of an scFv together with a synthetic linker, you can add things on to the linker, producing such goodies as antibodies tagged with alkaline phosphatase[2679] or GFP[2680], or high-avidity tetrameric antibodies[2681].

Much of the current work on engineered antibodies is focused on development of agents that can be used to treat human disease, rather than on expanding the repertoire of reagents for cytometry, but it's a safe bet that we'll get some interesting things to play with during the next few years.

Molecular Probes' Zenon Antibody Labeling

Some lower-tech, although not less interesting, antibody engineering is involved in a new methodology for antibody labeling recently developed by Molecular Probes[2332], which they call Zenon technology. The Zenon One Mouse IgG_1 labeling kits incorporate purified Fab fragments of a goat antibody to the Fc portion of mouse IgG_1 ; the Fab fragments are available with any of a number of labels, including fluorescent dyes, phycobiliproteins and tandems, biotin, and enzymes, covalently attached. A Zenon One Fab reagent can be used to label a very small amount (1 µg or less) of a mouse monoclonal antibody; after 10 minutes incubation, any unbound labeling reagent is, in essence, inactivated by addition of an excess of indifferent mouse IgG_1 . Multiple antibodies labeled with different dyes can be used together, and quantitative measurements are possible, although the excess indifferent antibody added to remove unreacted labeling reagent will, over a period of hours, remove some reagent from the antibodies of interest. The Zenon methodology should be particularly helpful to investigators developing their own monoclonal antibodies, as it should facilitate simultaneous analysis of two or more antibodies on cells.

Antibody Shelf Life and Quality Control

Antibodies keep pretty well, but they don't have an infinite shelf life. You need to keep an eye out for crud accumulating in your vials of antibody, spinning down aggregates and so on when bad things happen to good antibodies. Periodic measurement of the fluorescence intensity of antibody bound to beads with known binding capacities allows you to monitor reagent quality. Manufacturers usually tell you how many milligrams of antibody protein are in the vial, which may be helpful; they don't usually tell you how many molecules of label are on each antibody molecule, which would also be helpful.

Direct Staining Using Monoclonal Antibodies for 2, 3, and More Colors with 488 nm Excitation

To my way of thinking, the best way to proceed when doing immunofluorescence is the way that requires the least work. I use monoclonal antibodies whenever I can get them, and, unless I'm looking for an antigen that is scarce enough to force me to employ some kind of amplification procedure (pp. 343-5), I prefer to stain using antibodies with directly conjugated labels. Most people, myself included, have tended to use fluorescein on the first antibody and phycoerythrin on the second. Fluorescein is available on more antibodies than phycoerythrin, and fluorescein-labeled antibodies are usually cheaper; however, phycoerythrin is a better label in the sense that it is brighter on a per-molecule-of-antibody basis (see Figure 1-18, p. 37). The figure also indicates that, for 3-color work with 488 nm excitation, PE-Cy5 tandem conjugates are preferable as labels to PE-Texas red tandem conjugates, because there are fewer spectral overlap problems with phycoerythrin. Four-color immunofluorescence with a single 488 nm beam once pretty much forced you to use PE-Texas red conjugates, although phycocyanin conjugates were reported usable if the antigen density was high enough[775]. These days, you have PE-Cy5.5, PerCP-Cy5.5, and PE-Cy7 to play with, and you can do 7-color immunofluorescence using a 488 nm beam if you've got the right instrument.

Multicolor Work Using Multiple Lasers; Biotin-Avidin Labeling

If you have a second laser emitting at 590 nm or above, i.e., a CW dye laser, a red He-Ne laser, a krypton ion laser, or a visible diode laser, in addition to a 488 nm argon laser, it may make sense to use allophycocyanin even as a third label, because compensation should not be necessary. Lanier, Loken, and others at B-D[787-8] did a lot of serious, big-league three-color immunofluorescence in this fashion with fluorescein, phycoerythrin, and allophycocyanin, using an argon laser emitting about 100 mW at 488 nm to excite the first two labels and a He-Ne laser emitting under 50 mW at 633 nm to excite the third. Hoffman et al[646] did 5-color measurements using phycocyanin and allophycocyanin B as labels with 633 nm excitation, and fluorescein, phycoerythrin, and a PE-Texas red conjugate excited by a second 488 nm beam; these days, you'd probably choose APC, APC-Cy7, and APC-Cy5.5, in that order, as red-excited labels.

Three-laser instruments are more common than they used to be, and, in the future, we can expect the most common three-beam systems to use 488 nm argon or solid-state lasers, red He-Ne or diode lasers, and violet diode lasers; Cascade blue and cascade yellow, or materials with similar spectra, are probably the best bets as labels for immunofluorescence when using violet excitation.

Mixing Colors: Do's and Don'ts

It helps to consider in advance which labels you'll want on which antibodies. One of the more general rules with which you contend in flow cytometry is that, when you have equal intensities of fluorescence coming from two materials excited at the same wavelength, it's easier to eliminate interference in the shorter wavelength signal from the longer wavelength fluorescent material than to eliminate interference in the longer wavelength signal from the shorter wavelength fluorescent material. This is so because the emission spectra of most materials are skewed toward longer wavelengths; the "red tail" of fluorescein emission, for example, overlaps a substantial part of the phycoerythrin emission spectrum, while there is very little overlap between the shorter wavelength portion of the phycoerythrin emission spectrum and the peak region of the fluorescein emission spectrum. Thus, you should use fluorescein to label the antibody that will give the weaker signal; this will also minimize crosstalk between fluorescein and phycoerythrin immunofluorescence signals, so you will need less fluorescence compensation.

The same argument applies with even greater force when you are considering combining immunofluorescent stains with other dyes that produce very strong fluorescence signals, e.g., nucleic acid dyes. The weak fluorescence of fluorescein-labeled antibodies can be discriminated fairly effectively from propidium fluorescence, although some overlap compensation is advisable; it would be much more difficult to discriminate phycoerythrin fluorescence from that of a nucleic acid dye such as TO-PRO-1. While phycoerythrin and propidium fluorescence emissions overlap considerably, Corver, Cornelisse, and Fleuren[1392] have shown that phycoerythrin antibody fluorescence can be discriminated from propidium fluorescence at levels above about 50,000 antibody binding sites per cell, making measurement of DNA and two antigens possible on a 3-color instrument.

Since the emission spectrum of fluorescein and the absorption spectrum of phycoerythrin overlap, **energy transfer** from fluorescein to phycoerythrin will occur and, if fluorescein and phycoerythrin molecules are in close proximity, may be significant. This will result in lower fluorescein fluorescence intensity than would otherwise be expected. Chapple, Johnson, and Davidson[1393] showed, by dual labeling of cells with pairs of monoclonal antibodies, that R-phycoerythrin could quench fluorescein emission when both were attached to the same cell. If quantitative or semiquantitative information is important, it is advisable to compare intensities of one- and two-antibody control samples to determine the extent of quenching. It is also important to note that fluorescence compensation won't work properly when there is substantial energy transfer; this will interfere with quantitative measurements, and, in the most critical cases, it may be desirable to use multiple illumination beams for accurate quantification.

Cocktails for Five: Multiplex Immunofluorescence

In practice, there are numerous situations in which the identification of multiple, discrete cell subpopulations is of primary importance, while quantification is either secondary or not necessary.

The relative ease with which cell samples can be stained using mixtures of monoclonal antibodies has led to the independent development by several groups of immunofluorescence staining techniques involving **multiplex labeling,** which may allow demonstration of three or more antigenically distinct cell populations in a single sample using only two fluorescent labels, offering considerable savings of time and labor compared to conventional staining techniques.

Definition and counting of subpopulations in mixed populations, as in lymphocyte subset analysis, is most easily and most commonly done by using antibodies that react with only one of the cell types to be counted. If only a single label is used, a separate sample is required for each subset to be demonstrated. As a rule, two-color immunofluorescence labeling readily allows visualization of three stained subpopulations (one double positive and two single positives) and one unstained population. Multiplex labeling allows for more economical use of the two-parameter measurement space.

In the simplest form of multiplex labeling, a procedure described by Saunders and Chang[792] and proposed by Mack Fulwyler[793] and, independently, by Tomas Hirschfeld and Myron Block (personal communication, 1981), a mixed cell population is stained with mixtures of antibodies, each of which should, in principle, react with only one cell subpopulation. Any given antibody present in the mixture may bear one or both of the two labels used, but a unique ratio of labels must be chosen for each antibody present. If phycoerythrin and fluorescein are the two labels chosen, cells reacting with an antibody present only in the phycoerythrin-labeled form will appear along the phycoerythrin fluorescence axis, cells reacting with an antibody present exclusively as the fluorescein conjugate will appear along the fluorescein fluorescence axis, and cells reacting with an antibody present as a 1:1 mixture of fluorescein and phycoerythrin conjugates will appear along a line at 45° to both axes.

A refinement of multiplex staining, described by Horan et al[791], uses mixtures of fluorescein- and phycoerythrin-labeled and unlabeled antibodies to produce fluorescence intensity differences of several orders of magnitude among the various cell subsets, with data display on a logarithmic scale. Liu et al[1394] demonstrated that helper and suppressor/cytotoxic T cells, NK and B cells, and monocytes could be identified in this fashion in whole blood samples from both normal and HIV-infected individuals, as shown in Figure 7-18.

Multiplex labeling has the potential to fail ungracefully when applied to certain types of samples. For example, small numbers of T cells bearing both CD4 and CD8 antigens may be found in peripheral blood[794]; these might be counted as other cell types. This behavior can, however, be identified by comparisons between multiplex labeled and conventionally labeled samples, allowing the prospective user to determine when the method is and is not suitable. In this context, the results reported with multiplex labeling in HIV-infected samples by Liu et al[1394] were encouraging.

When the previous edition of this book was written, there was a great deal of interest in possible applications of multiplex immunofluorescence to analyses of cells. Most laboratories had only two- or three-color fluorescence measurement capability, and multiplex analysis would have provided them the only means for investigating many phenomena then (and, in many cases, now) of great interest. For example, it was felt that determining the state of activation of various cell subpopulations in HIV-infected patients might provide indicators of both clinical course and response to therapy. Semiquantitative or quantitative analysis of activation antigens such as CD25 and CD71 (the IL-2 and transferrin receptors), and/or of parameters such as RNA content, were known to provide reliable evidence of lymphocyte activation[597-606].

If conventional staining procedures were used, it would be necessary to use two antibodies, and, therefore, two fluorescence channels, simply to define a single subpopulation of interest. The analysis of an activation antigen or RNA in this population would require a third antibody, or a dye such as pyronin Y, and a third fluorescence channel. In a 3-color instrument, analysis of activation in T-helper and suppressor/cytotoxic cells, B cells, and NK cells would require three tubes using conventional staining. If multiplex staining were used, a cocktail of antibodies labeled with fluorescein and

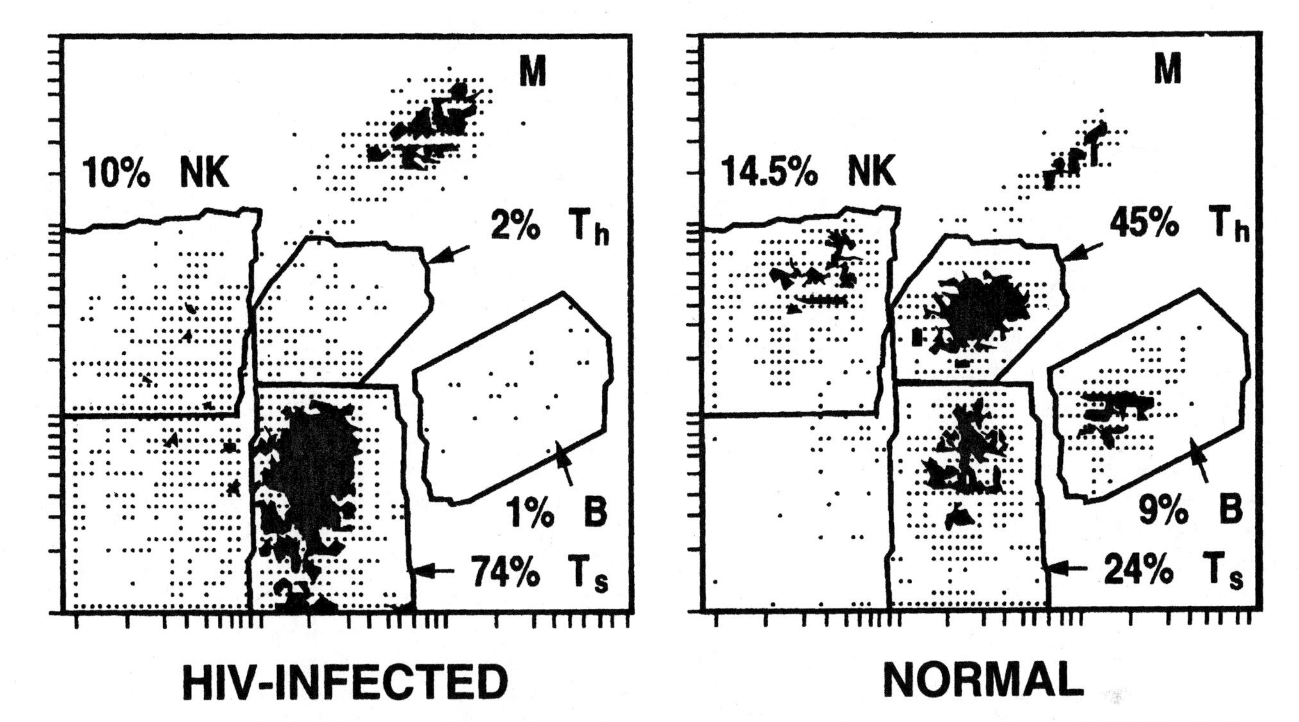

Figure 7-18. Multiplex immunofluorescent labeling to demonstrate helper (T_h) and suppressor/cytotoxic (T_s) T cells, B cells, NK cells, and monocytes (M) in peripheral blood from an HIV-infected and a normal individual[1394]. Erythrocytes were lysed after dextran sedimentation, and 2×10^6 washed leukocytes were stained with a cocktail containing anti-CD3 (T cells; B-D Leu-4: 10 µl unconjugated, 10 µl FITC-Ab), anti-CD4 (T_h; B-D Leu-3a: 20 µl PE-Ab), anti-CD8 (T_s; B-D Leu-2a: 19 µl unconjugated, 1 µl PE-Ab), anti-CD14 (monocytes; B-D Leu-M3: 20 µl FITC-Ab; Coulter My4: 5 µl PE-Ab), anti-CD16 (NK cells; B-D Leu-11c: 20 µl PE-Ab) anti-CD19 (B cells; Coulter B1: 5 µl FITC-Ab), and anti-CD20 (B cells; B-D Leu-12: 5 µl PE-Ab) antibodies at 4 C for 30 min. Cells were washed three times and fixed in 1.5% paraformaldehyde before analysis. Five leukocyte subclasses can be identified and counted in the two-dimensional measurement space. The figure was provided by Alan Landay of Rush-Presbyterian-St.Luke's Medical Center.

PE-Cy5 could be used to identify all four subclasses in one tube, with a third, spectrally distinct marker, e.g., PE-anti-CD25 or -CD71 or pyronin Y serving to define the activation state of all of the subclasses.

The multiplex labeling approach could be, and was, extended to analyses using cocktails containing more than two labels; Carayon, Bord, and Raymond did this to identify eight leukocyte subsets in a single tube[1395]. Buican and Purcell[329] reported that, by application of mathematical techniques similar to those used in computerized tomography, distributions of multiple antibody labeling could be uniquely reconstructed from two-color fluorescence analyses. Buican and Hoffmann[795] also explored the use of two-color measurements to identify multiple subpopulations based on analysis of multiple samples with different labeling ratios, and described apparatus for automated preparation of the required samples[796]. Multiplex labeling was also discussed by Wood[1396], Mansour et al[1397,1401], van Putten et al[1398-9], and Hunter et al[1400].

In a non-flow application, Ried et al[1402] used a 3-color multiplex fluorescence method to identify seven different DNA probes bound to chromosomes by digital imaging microscopy. However, a principal obstacle to the use of multiplex labeling with nucleic acid probes in flow cytometry lies in the low signal intensities expected due to low copy numbers of the nucleic acid sequences sought. Multiplex labeling cannot be expected to work well with very low intensity immunofluorescence signals, either; at levels at which one-color fluorescence is difficult to discriminate from noise, it is even more difficult to attempt to quantify intensity levels of two or more colors.

Where they were feasible, however, it was clear that multiplex labeling methods had great potential for saving time and money. Less sample and possibly less antibody is required, and only a single tube, instead of several, must be analyzed; this results in less preparation time, and in higher sample throughput for both flow cytometers and data analysis systems. However, interest in multiplex analysis of cells has waned as instruments capable of measuring four or more colors replaced two- and three-color systems in laboratories. Multicolor measurements offer most of the advantages of multiplex measurements; it is entirely feasible, for example, to derive counts of CD4+ and CD8+ T lymphocytes from a single tube of lysed whole blood using a mixture of anti-CD45, anti-CD3, anti-CD4, and anti-CD8 antibodies (see Figure 1-17, p.34), and this procedure, like a multiplex staining procedure, uses relatively little reagent and sample, and requires relatively little analysis time.

What has kept interest in multiplex fluorescence measurement very much alive and well since the previous edition appeared is the application of the technique to chemical analyses of ligands bound to beads[2370-5,2682], this subject was introduced on p. 48, and will be discussed further in Chapters 8 and 10.

Cocktail Staining Helps Identify Rare Cells

Cocktail staining has proven useful for the identification of rare cells. Stained cells representing less than 0.1 percent of a population are almost impossible to discriminate from unstained cells on a single-parameter histogram or on a dot plot of fluorescence versus an indifferent parameter such as forward scatter. However, if most or all of the unwanted cells are labeled with a contrasting color, using a cocktail of antibodies, it becomes much easier to find very small subpopulations. This approach was used by Ryan et al[883,1010] for detection of small numbers of leukemic cells following therapy; cocktail staining was also used by Gross et al[1150,2331] to identify cells present at frequencies as low as 1:10,000,000.

Immunofluorescence Staining Procedures

Due to limitations on reagent specificity and to what was available in the way of labels, earlier immunofluorescence staining methods had to incorporate multiple staining and washing steps, which were time-consuming and labor-intensive. Staining was done either in the cold (4 °C) or with an inhibitor such as sodium azide added to the cells to prevent **capping** or **stripping** of antibody. Even today, when one is using conventional antisera and/or indirect labeling, washing is necessary to prevent unacceptable levels of nonspecific background fluorescence, and the low temperature and/or inhibitor become necessary because of the length of time required. With directly labeled antibodies, things get about as easy as they can get; you throw all the antibodies in at once.

It has been argued[175,176] (also R. Hoffman, personal communication) that, when one uses directly labeled monoclonal antibodies, most or all of the wash steps can be omitted with few deleterious effects on results. Hoffman and Hansen[176] described a method for T cell subset analysis in which 50 µl of whole blood were incubated with 100 ng of monoclonal antibody for 10 minutes at room temperature, following which 2 ml of a lysing reagent (8.29 g NH_4Cl, 37 mg disodium EDTA, and 1 g $KHCO_3$ per liter, pH 7.3-7.5) was added. Samples were introduced into the flow cytometer within 10 minutes after the lysing reagent was added; immunofluorescence measurements were gated using a combination of forward and side scatter measurements so that, in theory, only lymphocytes were analyzed. This basic methodology, which worked as well with multiple antibodies as with a single antibody, was widely adopted.

Hoffman and his colleagues, then at Ortho, also worked out some quick staining procedures for using **biotin-avidin** reagents almost as if they were direct labels; since the affinity of avidin for biotin is so high, if you know the amount of biotin-antibody in your staining mixture, you can figure out how much labeled avidin to add so as to end up with slight antibody excess.

Hoffman et al also described a technique[645] for immunofluorescence analysis of leukocytes in whole blood without lysis; a bright fluorescent counterstain, which stains leukocyte nuclei and/or cytoplasm but not erythrocytes, is used to detect and classify leukocytes as well as to gate the fluorescence measurements; Terstappen and Loken[1154] also described triggering on the fluorescence of nuclear stains for analysis in unlysed blood. I used the same trick when I used Hoechst dye staining of nuclei to gate immunofluorescence measurements, but I usually didn't go out of my way to work with unlysed whole blood, since, at one leukocyte per 1,000 erythrocytes, you have to run for several minutes to get immunofluorescence signals from a few thousand leukocytes. If you're willing to accept the longer analysis time, you can even trigger on immunofluorescence signals, which facilitates doing things like just looking at the T cells and ignoring everything else in the sample.

In my view there are few excuses for not using the simpler immunofluorescent staining methods whenever possible, especially since there never seem to be enough hands available to stain all the samples you want to stain. There are, however, situations in which fast, no-wash immunofluorescent staining methods don't work so well, e.g., in the case when the number of antigenic sites per cell is extremely low. Hoffman and Kuhlmann[2482] found that unbound fluorescein- and PE-labeled antibodies, respectively, contributed 4,000 and 1,600 MESF of background fluorescence to the signal in no-wash immunofluorescence analysis, meaning that, if you are set on detecting a few hundred molecules of cell- or bead-bound antibody, you need to wash your cells, and you may need to resort to one or another of the amplification techniques discussed on pp. 343-5.

The simplest amplification technique is, of course, **indirect staining**; Parks et al[674] have discussed alternative methods of indirect staining and their advantages and disadvantages. The most commonly used indirect staining technique uses a fluorescent polyvalent or monoclonal **second antibody** specific for the immunoglobulin type of the primary antibody. After cells are exposed to the primary antibody and washed, they are incubated with the second antibody and washed again before being analyzed. Polyvalent antisera contain antibodies reactive with multiple sites on the primary antibody, and therefore can provide greater amplification than can monoclonal second antibodies; however, unless polyvalent sera are subjected to adsorption and/or affinity chromatography to enrich their content of anti-immunoglobulin antibodies, they are apt to produce high background fluorescence due to binding of labeled irrelevant antibodies to cells not bearing any primary antibody. Monoclonal second antibodies should not produce high background fluorescence; it should therefore be possible, albeit expensive, to use a mixture of monoclonal antibodies reactive with different sites on the primary antibody to achieve the amplification factor associated with polyclonal antisera and avoid the background fluorescence. I'm not sure whether or not anybody actually does this.

Protein A, which is produced by bacteria of the genus *Staphylococcus*, binds strongly to some IgG antibodies and, when fluorescently labeled, can be used as a second reagent for indirect staining. The principal disadvantage of using protein A results from its lack of reactivity with some primary antibodies and its predictable reactivity with host immunoglobulin already bound to cells.

Use of a **biotin-avidin** reaction[300], in which fluorescently labeled avidin or streptavidin is reacted with biotinylated primary antibody, typically produces brighter staining than would be obtained if the fluorochrome were directly bound to the antibody; this is especially likely to be the case when the fluorochrome is a large molecule such as a phycobiliprotein.

The combination of **hapten-conjugated antibodies** and **fluorescently labeled anti-hapten antibodies**[301] for indirect staining offers advantages in that this approach provides one of the more straightforward ways of indirectly labeling two or more antigens. For example, one could use primary antibodies conjugated with arsanilate and trinitrophenyl radicals, and fluorescein-anti-arsanilate and phycoerythrin-anti-trinitrophenyl secondary antibodies, to obtain amplified signals while minimizing interference between the two staining systems.

Putting antibodies and labels together seems to hold the same fascination for some of us as does playing with Legos and Erector Sets. Along this line, Wognum et al[789] described labeling cells with phycobiliproteins using a tetrameric complex of a mouse monoclonal antibody to a cell surface antigen and a mouse monoclonal antibody to a phycobiliprotein cross-linked with two molecules of a rat monoclonal antibody to mouse immunoglobulin.

Automated Sample Preparation

Several manufacturers offer devices for rapid lysis, fixation, and immunofluorescent staining of whole blood. You add the blood and antibodies; it does the rest, and, minutes later, you run the sample. Automation has been welcomed, especially by clinical users; more versatile preparation apparatus to meet the more varied needs of the research community is at an earlier evolutionary stage.

Fluorescence Measurements: Lurching Toward Quantitation

Calibration and Controls: Round One

Calibration and control procedures for immunofluorescence or ligand binding work differ substantially from those used for an application such as DNA analysis. In DNA analysis, the cells are typically brightly stained, and one calibrates the apparatus with beads and/or (preferably and) nuclei stained with the DNA stain being used, with precision being the primary concern and linearity next on the list. In measurements of weaker fluorescence, the small number of photons coming from stained cells may itself be the limit to precision; the emphasis is on **sensitivity and specificity**. We have to have some idea of how many fluorescent molecules we are detecting and of how much of the fluorescence is due to label specifically bound to the cells of interest.

I have already made the point that most of the argon ion laser source flow cytometers now in use have enough radiometric sensitivity to detect fewer than 1,000 molecules of antibody-bound fluorescein, and that the practical limit of sensitivity is set by the autofluorescence background. Instrumental approaches to discriminating fluorescein immunofluorescence or ligand fluorescence from autofluorescence were discussed on p. 342; in most cases, however, we don't have to carry things quite so far, because we can control for both autofluorescence and nonspecific fluorescence in properly designed **negative control samples**. These, ideally, should be cells that are known not to bear the antigen or ligand-binding entity being measured, to which labeled antibody or ligand has been added.

Isotype controls, containing cells and a labeled "irrelevant" antibody of the same isotype class as the reagent, were once almost universally used, and are still widely used; although some have deemed them unnecessary for many common clinical measurements[2683], others have not[2684]. If you are trying to do quantitative measurements, it's a good idea to have isotype controls, especially if the antigen you are looking for is not present in abundance. When an indirect staining procedure is employed, negative controls should also include samples in which the labeled second reagent is added to cells in the absence of primary antibody.

The ideal **positive control** is composed of cells that are known to bear approximately the same amount of antigen as the cells being analyzed; the objective in instrument setup is to use negative and positive controls to keep both negative and positive cells on scale. This is rarely a problem when you use log amplifiers, but it can be tricky when you're working on a linear scale. Through the years, people have come up with a number of nonfluorescent and not-very-fluorescent test objects to make this stage of the game easier. To answer an old and common question, in this instance, at least, the chicken came first.

Glutaraldehyde-fixed chicken erythrocytes become weakly fluorescent due to the presence of compounds formed by the aldehyde binding to amino groups in proteins, and were used since the early 1970's as controls approximating the fluorescence intensity of cells stained with fluorescent antibodies. **Osmium tetroxide-fixed cells**, which are essentially nonfluorescent, have been used to demonstrate spurious fluorescence signals due to fluorescence induced in color glass filters by scattered light.

As hard as it may be for some of you "younger" readers to believe, back in the 1970's, cell sorter operators in many prestigious immunology labs didn't bother with bead controls at all. By the mid-1980's, Coulter, Flow Cytometry Standards Corp., and Polysciences sold lightly labeled fluorescent beads with fluorescence in the same range as that of antibody-stained cells. Ortho distributed Fluorotrol, a mixture of fixed unstained thymocytes and thymocytes covalently labeled at two different intensities with fluorescein; Coulter supplied a preparation of fluorescent fixed cells under the name ImmunoSure. Flow Cytometry Standards and Ortho gave estimates of the number of fluorescein molecules bound to their materials. Caldwell et al[790] proposed the use of cellular fluorescence intensity itself as a quality control parameter in clinical flow cytometry, describing procedures for utilizing the relatively constant stainability of normal human peripheral lymphocytes to control for alterations in reagents as well as for instrumental sources of variability. However, while microscope photometry[1487] demonstrated consistency and reproducibility in fluorescence measurements of beads from Coulter, Covalent Technology, and Polysciences, these beads incorporated dyes other than fluorescein, and therefore could not readily be used to standardize measurements of fluorescein immunofluorescence.

In the late 1980's and 1990's, manufacturers shifted toward the production of controls bearing known amounts of common antibody labels, or known numbers of antibody binding sites, attempting to simplify the process of quantifying flow cytometric measurements of ligand binding or immunofluorescence. These materials were evaluated by relatively large groups of investigators in the United States, Canada, and Europe, and a series of meetings held in 1997 and 1998 brought most of the interested parties together and generated a special issue of *Cytometry*[2379] in which progress was summarized and critical issues were addressed.

Quantitative Fluorescence Cytometry: Definitions

Henderson et al[2685] defined **quantitative fluorescence cytometry (QFCM)** as "the calibrated measurement of fluorescence intensity from labeled particles...so as to determine the actual number of fluorescent ligands...labeling each particle." A cytometer measures **fluorescence intensity (FI)** of particles, but the measurements are initially reported as unitless histogram channel numbers. **Relative fluorescence intensities (RFIs)** express the relationship between the FIs of two measured particles; they can also be used to convert from histogram channel numbers on a logarithmic scale to linear units. The objective of QFCM as applied to immunofluorescence is determining **antibody-binding capacity (ABC)**, i.e., "the number of antibody molecules bound by a particle when specific binding sites are saturated." To convert from **MESF (Molecules of Equivalent Soluble Fluorochrome) Units** (see pp. 216-7) to ABC, it is necessary to know the labeled antibody's **effective fluorescence/protein ratio, F/P_{eff}**, where F/P_{eff} is defined as "the average number of MESF per functional ligand molecule in a fluorochrome-ligand conjugate." F/P_{eff} is analogous to, but more precise than, the older fluorochrome/protein, or F/P, ratio.

Schwartz et al[2686] established a taxonomy of **fluorescence standards** used for flow cytometry. A **Type 0 standard**, or **certified blank**, is a particle approximately the size of a cell (a lymphocyte is the typical cell in most discussions of QFCM). Certified blanks have a single, very low fluorescence intensity. **Type I standards** are better known as **alignment particles**; they are typically highly labeled and highly uniform in size, producing a very bright fluorescence signal with a low CV. Type Ia standards are smaller than cells (think of the 1-2 µm beads typically used to align an instrument and check its precision); Type Ib standards are cell-sized.

Type II standards, or **reference particles**, are the size of cells, and yield bright fluorescence signals; they need not be uniform in size. Type IIA particles are not spectrally matched to fluorochromes in the sample, and their fluorescence is not environmentally sensitive; Type IIB particles are environmentally sensitive and spectrally matched. Type IIC particles are not themselves labeled, but bind a relatively large amount of labeled antibody, obviously thereby sharing its spectrum and environmental sensitivity.

Type III standards, or **calibration particles**, are cell-sized, and not necessarily uniform in size. They come in sets including particles with several fluorescence intensity levels, ranging from dim to bright. Type IIIA, IIIB, and IIIC standards, respectively, share the properties of Type IIA, IIB, and IIC standards in terms of labeling, spectra and environmental sensitivity, and antibody binding. Type IIIC standards may be used to determine appropriate settings for **fluorescence compensation**; since they provide a known, wide range of fluorescence intensities using the same antibodies as are used to stain cell samples, they may be superior to single antibody-stained cell samples for this purpose.

Calibration Particles for QFCM

Successful quantitative fluorescence cytometry requires a well-aligned instrument; there has been some discussion of measurements of instrument performance on pp. 214-7, and additional coverage will be found in Chapter 10. I will deal here with the particles and methods needed to do QFCM with an apparatus in good working order.

"**Quantum Beads**" bearing known numbers of **Molecule Equivalents of Soluble Fluorochrome (MESF)** of fluorescein and phycoerythrin are Type IIIB standards. They were developed and produced by Flow Cytometry Standards Corporation[1161], which also developed **Quantum Simply Cellular® (QSC)** beads, Type IIIC standards capable of binding known numbers of antibody molecules (see pp. 48-9 and 199-201). The surface of a QSC bead is coated with a mixture of polyclonal goat antibodies to the Fc portions of various mouse Ig subtypes; the mechanism of binding of mouse monoclonal antibodies (MoAbs) to QSC beads is therefore quite unlike the mechanism by which the monoclonal antibodies bind to their targets. Fluorescent MoAbs may need to be incubated with QSC beads for a week or more in order for the fluorescence intensity of the beads to reach a plateau; the beads can then be used for some time thereafter[2687]. QSC beads, used with directly labeled antibodies, provide a standard curve spanning a range from fewer than 10,000 to over 100,000 molecules of bound antibody, allowing direct comparison with labeled cells (Figure 1-20, p. 49). Both Quantum Beads (PE-Cy5- and APC-labeled beads are now offered in addition to the fluorescein- and PE-labeled products) and Quantum Simply Cellular Beads (now available to bind human antibodies as well as mouse antibodies) may be purchased from **Bangs Laboratories**.

In 1985, Poncelet and Carayon[1488] developed a method they called **quantitative indirect immunofluorescence assay (QIFI)** to determine the absolute number of antibody binding sites on lymphoid cells by indirect immunofluorescence flow cytometry after saturation with monoclonal antibody, when binding is likely to be monovalent. The p67 T cell-associated antigen was quantified on cell lines by the binding of radiolabeled antibody. Then, saturating doses of unlabeled antibody followed by fluorescent anti-mouse antibody were applied to the cell lines to build a standard curve relating the mean fluorescence intensity of cells to the mean number of cell-bound primary antibody molecules. This curve (a straight line) was used to assess the absolute number of antibody molecules bound to other lymphoid cells. The technique was shown to be applicable to other IgG antibodies even when they were used in unpurified form.

The QIFI assay as originally described required the use of cell lines and radiolabeled antibodies to establish a standard curve; this limited its appeal, although, as Dux et al[1489] reported, the method allowed quantitative analysis even of antigens expressed at low densities. Poncelet et al[1490,1493] subsequently developed beads, now available as the **QIFIKIT™** (Biocytex, Marseille, France; available from **DakoCytomation**), which can be used as secondary standards, making quantitative indirect immunofluorescence assay generally accessible. The QIFIKIT beads are Type IIIC standards, containing known numbers of mouse monoclonal antibodies bound to an antigen (CD5) on the bead surface. A labeled secondary antibody therefore binds to a QIFIKIT bead by the same mechanism as if it were binding to a mouse monoclonal antibody in or on a cell. Using the QIFI methodology, Poncelet et al[1491] examined normal human lymphocytes, and found that expression of CD4 and CD45 antigens was relatively stable, at 48,000 ± 6,000 and 180,000 ± 17,000 binding sites/cell, respectively, while other antigens were more variable. In HIV-infected individuals, CD4 expression remained at 46,000 sites/cell, even with disease progression[1492]. Bikoue et al[2688], using QIFI technology, reported finding a mean CD4 expression level of 47,000 sites/cell on CD4+ lymphocytes.

BD Biosciences, which was just B-D at the time, took a serious plunge into quantitative immunofluorescence in the mid-1990's, developing the **QuantiBRITE** family of phycoerythrin-labeled beads and antibodies.[2481-2,2689]. QuantiBRITE beads are Type IIIB particles with four intensity levels of phycoerythrin labeling, ranging from approximately 1,500 PE molecules/bead to approximately 130,000 PE molecules/bead. QuantiBRITE kits include one antibody that is a 1:1 PE:IgG conjugate; antibodies to CD20, CD38, CD64, and HLA-DR are available, as well as custom 1:1 PE conjugates. Using the QuantiBRITE system, Davis et al[2481] reported the mean level of expression of CD4 on CD4+ lymphocytes to be 49,000 binding sites/cell; they found binding of their (B-D's) monoclonal antibody to be bivalent, indicating that there should be approximately 98,000 CD4 binding epitopes on the cell surface. Pannu et al, also at B-D, found a mean of 51,000 CD4 sites/CD4+ lymphocyte[2482].

Although it does not seem to be in the current catalogue, B-D made a 1:1 PE:CD4 antibody that was used in the CD4 studies they did and also in several studies of CD38 levels on CD8+ T cells in patients with HIV infection. Janis Giorgi had noted that elevated RFI of CD38 expression was an indicator of poor prognosis in HIV infection[2690], and initially developed a method for quantitation of cell surface CD38 on CD8+ T cells using the fluorescence intensity of CD4 on CD4+ T cells as a standard[2691]. While an initial study by Iyer et al[2689] found that measured CD38 fluorescence intensity values obtained using QuantiBRITE beads as intensity standards were essentially the same as those obtained using CD4 as a standard, a later multicenter study[2692] detected small but significant differences, and recommended that laboratories consistently use one standard or the other.

I will generalize from this; while, in principle, it would be desirable for flow cytometry labs everywhere to get the same numbers from the same samples, experience with quality assurance surveys over the years tells us that a great deal of effort is required simply to insure that critical clinical data such as CD4+ T cell counts remain reliable from lab to lab. Quantitative fluorescence measurements, and immunofluorescence measurements in particular, present more of a challenge. Recall that, while cellular DNA content can be measured with very high precision, the agreed upon definition of DNA aneuploidy is not based on a numerical value for DNA content, but, instead, on the DNA index, i.e., the ratio of DNA contents in tumor and stromal cells in the same sample (p. 317). DNA content measurement is based on analyses of strong signals, measured on linear scales, with minimal contributions to variance from photoelectron statistics, and the biological variance of DNA content, at least in G_0/G_1 phase stromal cells, is extremely small. Fluorescence compensation is not, as a general rule, used for DNA content measurement.

Immunofluorescence measurement is based on analyses of relatively weak signals, with fluorescence intensities typically no more than 1/100 of those typically encountered in DNA content measurement. Measurements are almost always made on logarithmic scales, and often represent the outputs of logarithmic amplifiers, which are likely to deviate substantially from true logarithmic response. Photoelectron statistics may contribute substantially to measurement variance, particularly at low signal levels, and the biological variance of the amounts of most antigens in or on most cell types is relatively large. Fluorescence compensation is likely to be applied to measurements used for quantitative immunofluorescence flow cytometry, and may, particularly if done in hardware, introduce significant inaccuracy into results. All of this makes it unlikely that a large number of labs will be able to get fluorescence intensity measurements that agree within one or two percent.

It is not unlikely, however, that we will be able to get agreement within ten percent, or even five percent, and that may be good enough. I mentioned above that Quantum Simply Cellular Beads incorporate antibodies to several classes of mouse immunoglobulins; they are therefore unlikely to have exactly the same number of binding sites for antibodies of each class, and might also bind differently to antibodies bearing different labels. This could make it possible to get different values for numbers of binding sites per cell when the beads are used with different antibodies to the same cell surface antigen, although the results would remain consistent for individual antibodies. However, as can be seen from Figure 7-19, the number of CD4 epitopes on CD4+ T lymphocytes determined using antibody-binding beads (R. Vogt, personal communication), about 47,000 sites/cell, is very close to those reported by Poncelet et al[1491-2] and Bikoue et al[2688] using the QIFI method, and by others using QuantiBRITE beads[2481-2].

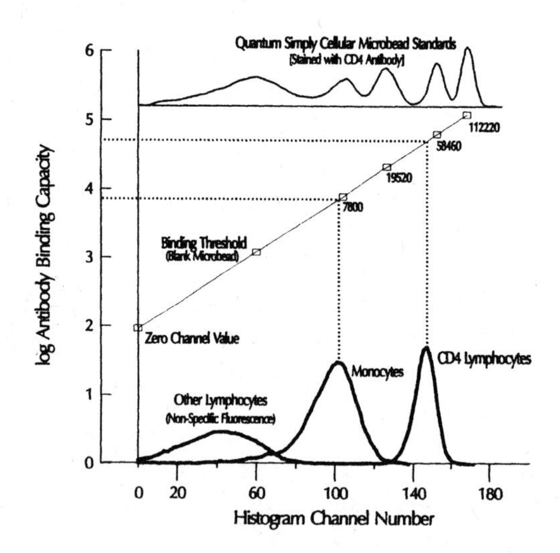

Figure 7-19. Quantitative determination of CD4 epitopes on peripheral blood lymphocytes and monocytes using QSC antibody-binding beads as intensity standards (provided by Robert Vogt, Centers for Disease Control).

In 1989 and 1991, Vogt et al[1505-6] examined intra- and interlaboratory variation in quantitative fluorescence measurement using both fluorescein-labeled beads and Fluorotrol as standards; while reproducibility was very good in a single laboratory, a study of 43 cytometers in 34 laboratories showed CVs in the 20-30% range for measurements of CD4 intensity of lymphocytes, although CVs dropped to 11% when a corrective factor based on Fluorotrol intensity values was applied. Those early results were encouraging. However, Lenkei et al[2693] reported in 1998 that ABCs determined using three different instruments in one laboratory, with different calibrators and with different fluorochrome conjugates of monoclonal antibodies, varied by 20-40 percent. Serke et al[2687], while noting that ABCs determined using the QIFIKIT and QuantiBRITE beads were similar, found large differences in titration curves between cells and QSC beads for a large number of MoAbs to different CD antigens, and differences in binding of fluorescein- and PE-labeled conjugates of the same MoAb to QSC beads. It is likely that the behavior reported for QSC beads relates to the fact that these beads bind antibodies via the Fc portion, while relevant cellular antigens bind the variable regions of the same antibodies.

Defining a Window of Analysis

Relating a measurement channel number to ABCs or MESF units requires the definition of a **Window of Analysis**, according to a procedure specified by Schwartz et al[2694]. While Figure 7-19 illustrates the results of such a procedure, assigning ABC values for anti-CD4 antibody to lymphocytes and monocytes based on fluorescence values obtained from QSC beads to which the antibody had been bound, I will work from Figure 4-60 (p. 216), which deals in MESF units, because it includes more of the "nuts and bolts."

The first item of business is measurement of a Type IIIB or IIIC standard; the former will give a scale calibrated in MESF units, and the latter a scale calibrated in ABCs. The Type IIIB standard used in Figure 4-60 was a set of Quantum beads labeled with fluorescein. The values we need are the histogram channel numbers for the peaks representing the four labeled beads and the blank bead in the bead kit. In this instance, measurements were recorded on a 256-channel (8-bit) logarithmic scale; Schwartz et al recommend converting measurements on higher resolution scales to a 256-channel scale, for reasons that will shortly become apparent.

Working with the known values for MESF units (or ABCs) assigned to the labeled beads by the vendor, we perform a linear regression analysis to get an equation of the form:

$$[\text{channel \#}] = a\,[\log \text{MESF}] + b.$$

In the example shown in Figure 4-60, I happen to have used the log of [MESF/100] which is [log MESF] - 2, so the right regression equation is actually:

$$[\text{channel \#}] = 43 \log \text{MESF} - 101,$$

rather than the equation shown at the top of the figure.

The slope, a, of the regression line defines the **coefficient of response**, indicating, at least in theory, the number of decades of dynamic range available in the measurement. If the 256-channel scale includes exactly four decades, the value of a should be 64 (i.e., 256/4); if it is a three-decade scale, a should be 85.33 (i.e., 256/3). Note that the value of a associated with Figure 4-60 is 43; this suggests that the measurement scale encompasses 5.9 decades. Don't believe it. What it actually indicates is that the maximum voltage output from the log amp used to process the signals was less than the full-scale input voltage of the analog-to-digital converter used to collect the data. That happened in my home-built Cytomutt; you are unlikely to have the same problem in your instrument, and, if you have what is nominally a four-decade logarithmic scale and end up with a value of *a* that is less than 59 or more than 69 (these values correspond to 4.3 and 3.7 decade ranges), you should probably have your system looked at. In case you haven't already guessed, we convert everything to a 256-channel scale so we can always use 64 as the ideal value of *a* for a four-decade logarithmic scale and 85.33 as the ideal value for a three-decade scale.

The points representing the measured values of the beads are pretty close to the regression line in both Figure 7-19 and Figure 4-60. Part of that is due to the fact that we are working on a semi-log scale; the deviations would be larger on a linear scale. However, we really can't rely on our eyeballs to give us a robust measure of how close our measured points are to the calculated line. The regression line is calculated using the method of least squares, meaning that the chosen line minimizes the sum of the squares of the differences between observed and calculated values of channel numbers. While I am on the subject of the regression line, I should point out something else that might otherwise confuse you.

Under most circumstances, we plot the value of a dependent variable in the vertical (Y) direction, and the value of an independent variable in the horizontal (X) direction, and calculate a regression equation: $Y = aX + b$. In the game we're playing here, we have the dependent variable, channel number, plotted on the X axis, and the independent variable, MESF units (or ABCs) plotted on the Y axis, so our regression equation is really: $X = aY + b$. Just so you know.

In order to determine the extent to which our logarithmic scale deviates from ideal response, we go back to our regression equation, and, taking the actual channel numbers we used to derive it, work backwards to calculate the MESF values that would fall on the line. For example, in Figure 4-60, the data point representing the bead with 70,000 MESF is at channel 108. The equation tells us that

$$108 = 43 \log[\text{MESF}] - 101,$$

so log [MESF] = (209/43), or 4.860, and the value of MESF that would fall on the line at channel 108 is $10^{4.860}$, or 72,521 MESF.

Having calculated all of the MESF (or ABC) values as just described, we find the percentage differences between them and the known MESF (or ABC) values supplied by the bead vendor. For example, 72,521/70,000 = 1.036; the known data value therefore deviates from the value calculated to lie on the regression line by 3.6 percent. Our quality measure is the **average residual percent**, which we get by doing the calculations for all the data points corresponding to known MESF or ABC values (we leave the blank out), adding them up, and dividing by the number of data points. The average residual percent should be less than 3 for MESF and less than 5 for ABCs; the line in Figure 4.60, which looks so close to the data points, actually has an unacceptably high value.

Getting back to the window of analysis, we next obtain the MESF numbers corresponding to channels 0 and 255, in the same fashion as we calculated values on the line to get the average residual percentage. For the data in Figure 4-60, the **zero channel value (ZCV)** is 223 MESF; the **maximum channel value** would be 190,138,370 MESF, but, as I noted a few paragraphs back, that's not a believable value because of a peculiarity of the electronics. The **window of analysis** runs from the zero channel value to the maximum channel value; it is generally suggested that instrument gain settings be adjusted to make the window of analysis run from 100-300 MESF or ABC (an acceptable range for ZCV) to 1,000,000-3,000,000 MESF or ABC.

Note that if the MESF or ABC value for the blank bead is higher than the ZCV, which it will be if some or all of the blank beads appear on scale, the instrument cannot actually detect a signal at the level of the ZCV. It used to be commonplace to define the **detection threshold** as the median value of the peak representing blank beads or unstained cells; it is advisable to use the median rather than the mean because low values tend to pile up in channel 0, which would skew the value of the mean, but not the value of the median. It makes more sense to define a threshold value to include no more than the upper few percent of events in the blank peak; that this number should be no higher than 1000 MESF or ABC, and, if it is that high, the bottom decade of the four-decade range (from 100 to 999 MESF/ABC) is essentially totally occupied by noise, making measurement values in this range meaningless.

The bottom line is that you can't expect to get accurate quantitation of values in peaks that overlap substantially with the blank; that concludes this discussion, which has gotten long enough so that it might be called "Window of Analysis Restaurant." You can get the programs to do the calculations just described from the instrument manufacturers and from third parties (see Software Sources in Chapter 11); just remember that you can't get anything you want from QFCM.

Type IIIA versus Type IIIB and IIIC Standards

While Type IIIB and IIIC standard beads provide accurate calibration of a cytometer's fluorescence intensity scale, and are spectrally matched to the labels used for QFCM, these types of beads, and particularly those bearing phycobiliproteins or antibodies, are somewhat less stable than Type IIIA beads. Type IIIA beads, which are hard-dyed, emit fluorescence in one or more of the same spectral regions in which common labels emit, but are not spectrally matched to the labels and are not environmentally sensitive. The relative fluorescence intensities of a Type IIIA and a

Type IIIB or IIIC bead may be influenced by the excitation wavelength and power used, and the detector bandwidth and sensitivity.

However, once a Type IIIB or Type IIIC bead has been used to establish a window of analysis, running a sample of Type IIIA beads with appropriate emission characteristics, using the same instrument gain settings, can provide a "Rosetta Stone" that will allow conversion between fluorescence values obtained from the Type IIIA beads and MESF or ABC values. As long as no significant changes are made in instrument configuration, setting up the instrument by adjusting gains to keep the Type IIIA beads at or very near the same channels in which they appeared in the original run will restore the window of analysis. Type IIIA beads are available from **Beckman Coulter, DakoCytomation, Molecular Probes, Polysciences,** and **Spherotech**.

Other Aspects of Fluorescence Quantitation

What is "Positive"? What is "Negative"?

Some of the harder-to-answer questions in analysis of immunofluorescence and ligand binding relate to how one defines **what is "negative" and what is "positive,"** this could be considered to be quantitation at the 1-bit level. Even at this level, there aren't always unequivocal answers. Generally speaking, if the cells in your sample have been taken directly from places in which they acquire or lose a particular antigen, you'll see a **continuous distribution** of immunofluorescence intensities going from some positive value all the way down to zero. This type of distribution is typically observed when cells such as stimulated peripheral blood lymphocytes are stained for activation antigens such as the IL-2 and transferrin receptors (CD25 and CD71, respectively), which emerge following exposure of cells to antigens or mitogens.

People typically consider cells from such continuous immunofluorescence distributions as "positive" for the antigen in question when their fluorescence intensity exceeds that of all but a small (usually 1-5%) fraction of a control population. It is not always clear what constitutes an appropriate control population. When stimulated lymphocytes are analyzed, unstimulated lymphocytes are generally used as a control. However, unstimulated lymphocytes may express low levels of activation antigens, and an unstimulated cell population may contain some cells that were activated before being drawn from the donor, which can be expected to bear the antigen being sought. Things get even more complicated when the antigen in question is one that is expressed on other cell types, such as monocytes, which may contaminate a lymphocyte gate.

When cells acquire or lose an antigen outside the compartment in which they are sampled, it is possible to obtain distributions containing clearly positive and clearly negative cells. T lymphocytes, for example, acquire both CD4 and CD8 antigens in the thymus, and generally lose one or the other before leaving the thymus. Thus, the distributions of these antigens in T cells from peripheral blood are typically bimodal, with few, if any, cells lying between the positive and negative peaks, while the distributions of the same antigens in thymocytes are continuous[787] (also see Figure 5-11, p. 241).

Chapter 5 devotes a fair amount of space to discussion of mathematical and statistical techniques that attempt to resolve weakly stained from unstained cells. Sladek and Jacobberger[1497] considered the problem of analysis of data from cells expressing low levels of intracellular antigen and, in a comparison of several methods, found mathematical modeling preferable to histogram subtraction (p. 245).

I am of the opinion that is advisable to rephrase unanswerable questions rather than to torture the data to extract the answers you want; when dealing with continuous distributions, it makes more sense to me to get accurate numerical values for medians and quartiles (p. 235-6) than to assign more arbitrary numbers to represent fractions of "positive" and "negative" cells. However, there may be a way of testing mathematical methods that purport to discriminate dim positives from negatives.

Making Weakly Fluorescent Beads and Cells: Do Try This Trick at Home!

The precision with which you can put only a few hundred molecules of a label or antibody on a cell or bead is limited by Poisson statistics (see the discussion of molecular shot noise on p. 341). You can't buy antibody-binding beads with a binding capacity of 100 antibodies. However, you can buy beads with a binding capacity of 5,000 antibodies, and, by incubating them with a 1:49 mixture of labeled and unlabeled antibodies, get an average of 100 labeled antibodies on each bead[2695]. The standard deviation, thanks to Poisson statistics, will, of course, be at least 10, meaning that the CV of the fluorescence distribution can be no lower than 10 percent, but you may not even be able to detect the fluorescence on your flow cytometer, and, if you can, it's a pretty good bet that photoelectron statistics will contribute more than molecular shot noise to the variance. You can play the same game with cells; a CD4+ T cell has about 50,000 binding sites for CD4 antibody, and, if you stain peripheral blood cells with a 1:99 mixture of labeled and unlabeled antibody, you'll get an average of 500 molecules of label on each CD4+ T cell[2696].

The stoichiometry described above only applies if the labeled and unlabeled antibody have equal shots at binding to cells. An unlabeled IgG antibody has a molecular weight of about 160,000, and, if you put 3 molecules of fluorescein on it, the molecular weight goes up to about 161,200. The difference in molecular weights between labeled and unlabeled antibody won't affect the binding characteristics much, so the ratio of labeled to unlabeled antibodies on beads, or cells, will be the same as the ratio in solution.

An IgG antibody labeled with 1 molecule of phycoerythrin has a molecular weight of about 400,000, and the much lighter unlabeled antibody will have an advantage in binding to beads or cells, with the result that the ratio of labeled to unlabeled bead- or cell-bound antibody will be lower than the ratio of labeled to unlabeled antibody in solution. You can fix this, and get nearly the same ratios for bead- or cell-bound antibodies as you start with in solution, by using phycoerythrin that has been bleached by exposure to strong light (an hour or two of sunlight will do) as the "unlabeled" antibody when you work with a phycoerythrin-labeled antibody.

A slight variation on this theme should provide a control for testing various computational methods of discriminating weakly labeled and unlabeled cells. It requires an instrument with 488 nm and red (633 or 635 nm) excitation beams. Stain beads or cells with a near-normal concentration of APC-labeled antibody to which you add a very small amount of fluorescein-labeled antibody. You should then be readily able to discriminate stained and unstained beads or cells on the basis of APC fluorescence, and to titrate the fluorescein fluorescence down so you have difficulty discriminating stained and unstained objects. If the computational method gives you the same numbers you get from gating and counting the APC-stained and unstained beads or cells, it's a winner. You don't need to know the exact stoichiometry of stained and unstained cells here, so it doesn't matter that conjugated APC adds at least 33,000 to the molecular weight of an antibody. You also don't have to worry about energy transfer from the fluorescein to the APC, because you can raise the ratio of fluorescein- and APC-labeled antibodies to get a weak, but detectable fluorescein signal.

Correlating Cytometry and Biochemistry: Studies of Antibody Binding Chemistry

Binding of antibodies, or any other ligands, to chemical structures on or in cells follows the rules of chemical kinetics and thermodynamics. Most of the antibodies used for cytometry are bivalent, but whether binding to their targets is bivalent, monovalent, or mixed depends on a number of factors, with the affinity or avidity of the antibody and the abundance and availability of the targets all exerting strong influences.

We generally attempt to saturate binding sites when performing immunofluorescent staining procedures, particularly when the objective is quantitative fluorescence measurement. However, Eric Martz pointed out a long time ago (personal communication, 1993) that if antibody and cell concentrations and incubation time are arranged so that binding is limited by the diffusion rate of antibody, rather than by the amount of antibody present in the sample, the fluorescence intensities of cells and beads should remain proportional to the number of binding sites on each, even at subsaturating antibody concentrations. In theory, we should be able to get away with using less antibody. In practice, most people will continue to titrate antibodies, adding more until the cells of interest don't get any brighter, and only take comfort from the theory when they find, as we all do from time to time, that an antibody straight out of the bottle is too dilute to saturate binding sites.

Flow cytometry can be, and has been, used to determine the binding parameters of antibodies. Bardsley et al[1496] developed a model and computer program for determining association constants of antibodies binding to cell surface antigens, which takes into account variations in antibody concentration as well as variations in antigen expression. Benedict et al[2697] described a flow cytometric assay and its use to determine the binding affinity of an anti-CD34 scFv antibody. Siiman and Burshteyn[2698] used competitive binding to determine binding constants for a variety of labeled and unlabeled MoAbs to cell surface markers, and to enumerate the numbers of target binding sites present on cells. A provocative paper by Lamvik et al[2699] reports that binding of unlabeled secondary antibody may increase binding of primary monoclonal antibodies (labeled in this instance, to allow the determination to be done) to cell surface antigens on fixed, permeabilized cells. Both equilibrium and kinetic measurements (about which more will be said in a later section) have been used in flow cytometric studies of antibody binding[2700].

Quantitative measurement of cellular constituents using immunofluorescence flow cytometry benefits from comparison and correlation of results obtained using flow cytometry and other analytical techniques. Sarin and Saxena[1500], for example, established a correlation between flow cytometric measurements of histocompatibility antigens and measurements by ELISA.

Correlating Cytometry and Biochemistry: Intracellular Antigen Measurements

Establishing correlations between flow cytometry results and results of more classical methods is of particular importance in analysis of intracellular antigens. The usual procedures for analysis of intracellular antigens require fixation and permeabilization of cells, in order to achieve the conflicting ends of allowing molecules as big as labeled antibodies to get into cells and retaining molecular targets which may be smaller in size.

Jacobberger and his colleagues[2701-2] have studied quantitative aspects of antibody binding to cellular constituents, with particular emphasis on intracellular antigens. A comprehensive review[2701] discusses basic chemistry and immunochemistry, fixation and sample preparation, and both equilibrium and kinetic measurement methods for flow cytometric analysis. Another paper[2702] describes correlation of quantitative measurements of SV40 T antigen with quantitative Western blots.

There are at least a few tricks that allow flow cytometry to be used to demonstrate and quantitate intracellular antigens in living cells, allowing cells bearing desired antigens in appropriate amounts to be sorted for culture and further analysis.

Detection of cytokine production by T cells (reviewed by Maino and Picker[2703]) has become a favored means of identifying cells participating in an immune reaction against a defined antigen. The usual procedure involves incubation of cells with the antigen, addition of brefeldin A to prevent cytokines from being secreted, staining of relevant surface antigens, fixation and permeabilization, and staining of intracellular cytokines. An alternative technology developed by Manz et al[2704] creates a **cell-surface affinity matrix** on the cell surface, first biotinylating cell surface proteins, and subsequently attaching streptavidin-tagged anti-cytokine antibodies to the biotin molecules. Cytokine secretion is not inhibited; instead, cytokines are secreted, and captured by the antibodies attached to the cell surface, on which they can be demonstrated following addition of fluorescent-labeled anti-cytokine antibodies. Pittet et al[2705] combined an anti-cytokine cell-surface affinity matrix with tetramer staining (introduced on pp. 47-8); this permits sorting of live, antigen-specific, cytokine-secreting cells.

Berglund and Starkey[1234] used **electroporation** to permit the introduction of labeled anti-oncogene antibodies into cells, at least some of which survived the procedure. Morris et al[2706] may have developed a kinder, gentler method; they made a 21-residue peptide carrier, pep-1, incorporating sequences known to be able to promote transport of proteins across membranes, and succeeded in introducing a number of proteins, including two different fluoresceinated antibodies, into living mammalian cells following the formation of complexes between the proteins and pep-1. The antibodies localized correctly to intracellular actin and to a lysosome-associated membrane protein. It remains to be seen whether carrier peptides can make it possible to do quantitative or semiquantitative immunofluorescence analyses of intracellular antigens in living cells.

Analyzing Immunofluorescence Data

We now turn our attention to what we can and can't learn from immunofluorescence measurements. Some aspects of this have come up in Chapter 5, others will come up in Chapter 10.

Some quantitative questions may actually be easier to answer than some qualitative ones. Like most histograms obtained from flow cytometry, immunofluorescence histograms typically contain data from 10,000 or more cells, and thus represent huge sample sizes when compared to most of the data sets with which statisticians normally have to contend. As a result, you can pretty much assume that a visible difference between two smoothed histograms is statistically significant; you cannot, however, assume that the significant difference is due to a significant biological difference between the cell populations from which the distributions were obtained, because instrumental variation and differences in reagents and sample preparation can also produce statistically significant differences. Intensity calibration makes histogram comparison easier than it might otherwise be.

If you are trying to demonstrate **differences between populations**, it is often useful to construct "envelopes" in which the area between the high and low values for each channel, taken from two or three replicates of control and experimental histograms done at different times, is shaded. Gaps between the control and experimental envelopes provide strong support for the hypothesis that control and experimental populations differ. Alternative methods for averaging and comparing histograms have been described by Marti et al[798] and Traill et al[799].

Mathematical models, while not applied to immunofluorescence to nearly the extent to which they have been used for DNA analysis, had been used by some authors. Takase et al[1494] used nonlinear least squares curve fitting to calculate the mean and standard deviation of immunofluorescence distributions, and Shabtai et al[1495] used a Euclidean distance vector to describe fluorescence intensity changes in two-color measurements.

When the previous edition of this book appeared, there were a few papers in the literature[1501-4] reporting fluorescence intensity measurements in terms of fluorescein MESF units. Others had used simpler, more empirical approaches; Terstappen et al[1498] represented amounts of surface antigens on myeloid cells by their mean fluorescence intensities in comparison with the background fluorescence of each cell type. Christopoulos et al[1499] used the same measure of intensity to quantify platelet-bound immunoglobulin.

Estimating Antigen or Receptor Surface Density

Surface density of antigenic or receptor sites can be estimated by dividing the number of bound ligand molecules by the cell surface area. An approximation to an absolute value for cell surface area can be obtained from an electronic measurement of cell volume by analog or digital computation of the 2/3 power of volume[311]. Use of this technique is obviously feasible only with flow cytometers capable of making volume measurements. However, since both cell surface area and the "cell size" measurement obtained from forward light scattering are (allegedly) approximately proportional to the square of cell diameter, the ratio of immunofluorescence and scatter signals can serve acceptably as a parameter representative of antigen surface density for making comparisons.

If we actually know how many antigenic or receptor sites there are on the cell, you might think we should aim for surface density in sites per square micrometer. However, while flow cytometry, with appropriate calibrators, can fairly readily give us cell volumes in femtoliters, we'd have to make some pretty rash assumptions to get a surface area from that. It's acceptable to assume the cell is roughly spherical, and calculate an approximate radius from the volume. In reality, we know that no cell surface is a completely smooth sphere, and that real cell surfaces differ in their degree of apparent roughness, as seen, for example, by scanning electron microscopy; this means that cells of roughly the same size can have substantially different true surface areas. Assuming cells are smooth and spherical so you can report "exact" values of surface antigen density doesn't make much sense.

Quantitative Fluorescence: Problems and Prospects

Shortly before the publication of the October 1998 *Cytometry* special issue on Quantitative Fluorescence Cytometry[2379], Nicholson and Stetler-Stevenson opined: "Although quantitative flow cytometry is currently drawing a lot of attention, much of the attention is focused in the wrong area. The greatest value that quantitation has provided to the clinical laboratory has been on focusing attention to properly setting up and calibrating flow cytometers. Being able to consistently perform semiquantitative measurements has the most benefit in most clinical settings." They noted that quantitation of CD38 expression on CD8+ T cells in HIV infection[2689-92] represented the only case in which measurements of antibody binding capacity might be important for clinical management.

I won't argue. I am also of the opinion that paying attention to quantitative aspects of immunofluorescence measurement is at least as important for quality assurance as for any other clinical purpose. However, had quantitative fluorescence cytometry not developed to the level at which it was by 1998, it would not have been possible to establish the clinical relevance of CD38 measurements, and we can only find additional instances in which quantitation may be necessary by making more quantitative measurements.

At the present time, we have more problems than solutions. The number of antibody binding sites on cells is known to be influenced by preparative methods, including lysis and fixation, and we will need to standardize the preparative techniques in order to permit widespread application of any quantitative fluorescence measurement procedure. Antibody affinity and binding valency can exert profound effects on ABC determinations, and antibody labels may influence these characteristics. Ideally, there should be little or no spectral overlap between the antibody label used for quantitative measurements and those used to define the gates in which the cells of interest lie, because spectral crosstalk into the channel used for quantitation will increase the background noise level and reduce measurement sensitivity. The label best established for quantitation is probably phycoerythrin; however, the temptation to attempt to do quantitative measurements using phycoerythrin, with a 488 nm laser also exciting antibodies labeled with fluorescein, PE-Texas red, PE-Cy5, PE-Cy5.5, and PE-Cy7, must be strongly resisted. We can, after all, now use three red-excited labels (APC, APC-Cy5.5, and APC-Cy7) for gating, minimizing interference with quantitative measurements using PE. And we still worry about calibrating the calibrators and standardizing the standards.

As I mentioned on pp. 294-5, there are some heavy hitters on the case. The National Institute of Standards and Technology (NIST) is getting into the business of standardizing fluorescent materials[2582-4], and will be producing and certifying standard solutions and particles, and an NCCLS (NCCLS used to be the National Committee for Clinical Laboratory Standards; having gone international, it is now simply NCCLS, usually pronounced "nickels") subcommittee is developing a guideline for Fluorescence Calibration and Quantitative Measurements of Fluorescence Intensity, which should help implement standardization not only in cytometry, but in other fields in which fluorescence quantitation is becoming of interest, e.g., microarray analysis. A brief prospectus was provided in 2000 by three key players in the NIST and NCCLS projects[2581].

7.8 NUCLEIC ACID SEQUENCE DETECTION

Figure 7-20 provides a relatively dramatic illustration of the combination of molecular biologic techniques with flow cytometry for detection of specific nucleic acid sequences. The figure shows a small population of CD4-positive lymphocytes containing HIV-1 viral nucleic acid, as detected by Patterson et al[1403] using fluorescein-labeled oligonucleotide probes after *in situ* PCR amplification.

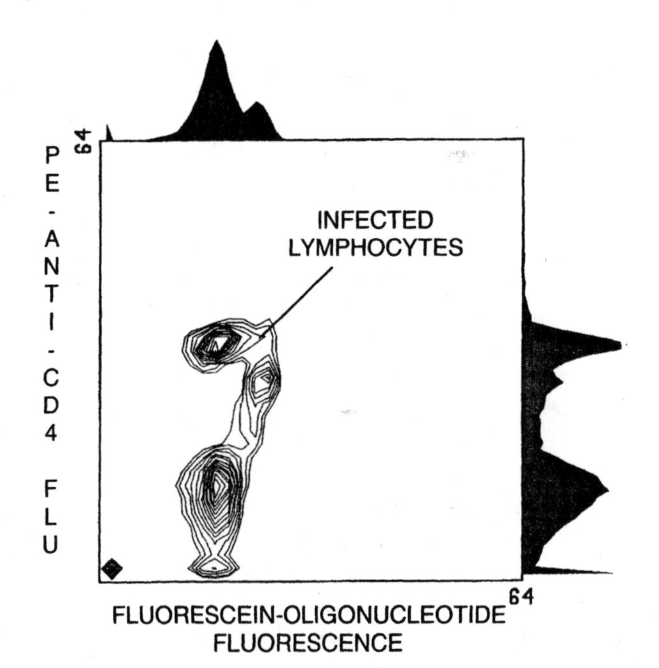

Figure 7-20. Two-dimensional contour plot of log PE-anti-CD4 fluorescence vs. log of fluorescence of fluorescein-labeled oligonucleotide probes specific for HIV-1 nucleic acids, showing oligonucleotide probe fluorescence in some of the CD4-positive lymphocytes after in situ PCR of viral sequences[1403]. The figure was provided by Charles Goolsby of Northwestern University.

Detection of sequences by flow cytometry after **Fluorescent *In Situ* Hybridization (FISH)** was described by Trask et al[902] in 1985. Between then and the time the previous edition of this book appeared, a variety of methods to attach fluorescent labels to probes were tried and described. Bauman, Bayer, and van Dekken[1404-5] detected poly-(A)+ RNA using a poly-biotin-d(U)-tailed oligo-d(T) probe and streptavidin-fluorescein; nuclei were stained with DAPI to provide a trigger signal. Amann et al[1406] used detection of specific ribosomal RNA sequences, present at the level of several thousand copies per cell, for bacterial identification, initially employing probes end-labeled with a single molecule of tetramethylrhodamine. Some signal amplification was obtained by substituting digoxigenin-labeled probes, which were rendered fluorescent by attaching fluoresceinated anti-digoxigenin antibodies[1407]. However, this made it somewhat more difficult to get probe and label through bacterial cell walls. Oligonucleotide probes end-labeled with single molecules of fluorescein and tetramethylrhodamine could be used for two-color analysis; the addition of more molecules of label per probe molecule increased signal, but decreased signal-to-noise due to increased nonspecific binding[1408].

Timm and Stewart[1409] discussed procedural aspects of "Fluorescent *In Situ* Hybridization En Suspension" (FISHES), including preparation of digoxigenin-labeled probes and staining and washing steps; they found that washing with a solution containing formamide and bovine serum albumin reduced cell loss. Lalli et al[1410] also discussed optimization of the preparative procedures. Bardin et al[1411] noted that biotin could be attached to the 5′ as well as to the 3′ end of oligonucleotide probes, offering some potential for increasing signal when fluorescent streptavidin is attached.

Bains et al[1412] described PRINS (Primed IN Situ labeling) to quantify poly-A and histone messenger RNA in cells. Poly(A) mRNA was labeled by incorporation of fluoresceinated nucleotides into cDNA chains generated by reverse transcription from the site of oligo(dT) primer-specific annealing on the poly(A) template.

While the sea of literature threatened to become filled with TOADFISH (Totally Outlandish Acronyms Describing Fluorescent *In Situ* Hybridization), this didn't stopped a lot of people for going after the big game, which certainly includes detection of low copy numbers of viral gene sequences in infected cells. When the previous issue of this book appeared, this required amplification of those sequences by *in situ* PCR[2708-9], as was done in the cells used to generate Figure 7-20. Surpassing even this feat of genetic engineering, Embleton et al[1413] reported that it was possible to separately produce and assemble amplified heavy and light chain DNA by PCR and subsequent primer additions in cells in suspension. Long et al[1414] examined *in situ* PCR on slides and in cell suspensions and found that results were comparable; nonspecific results were less frequent with cell suspensions.

Mirsky et al[1415], using conventional methods of detecting DNA after sorting single cells into microtiter plates, reported that the operational sensitivity of *in situ* PCR for detection of bovine leukemia virus was 90% when testing single infected cells. In addition, they were able to reliably amplify DNA from a single infected cell among as many as 10^5 uninfected cells and established that the sensitivity for detecting a single infected cell among 20, 100, or 1000 uninfected cells was at least 90%. This work put the concept of detecting small numbers of virus-infected cells directly by fluorescence flow cytometry following *in situ* PCR on a firm footing.

In a more mundane but possibly more practical application of genetic sequence detection by flow cytometry, van Dekken et al[1416] used labeled chromosomal probes to discriminate bladder cancer cells with monosomy-9 from stromal cells, enabling determination of the tumor cell kinetics by gated analysis of BrUdR incorporation.

If immunofluorescent staining has pretty much been reduced to cookbook procedures, nucleic acid probe staining may be said to still require both technical skill and FISHerman's luck (FISHerperson's luck, if you insist). Andreeff and Pinkel have edited a big book on basic principles and clinical applications of FISH[2710].

In recent years, it has become possible to use tyramide signal amplification (p. 344) in FISH; new probe labeling techniques[2671,2711-2] can reduce or eliminate nonspecific background fluorescence, making it possible to detect a few dozen copies of a sequence in cells.

Peptide Nucleic Acid (PNA) Probes

A new class of probes, **peptide nucleic acid (PNA)** probes[2713-4], appear to offer some significant advantages for cytometry, but clearly have a potential range of applications that extends far beyond this field.

A PNA molecule is comprised of the same purine and pyrimidine bases as are found in nucleic acids, attached to a pseudopeptide with amide or peptide bonds forming the backbone of the polymer, rather than sugars linked by phosphates. PNA molecules are insensitive to digestion by proteases and nucleases, and bind to DNA and RNA; a strand of PNA complementary to a region of double helical DNA will disrupt the double helix and bind to it, displacing its complementary strand. PNAs have been used in experiments on antisense therapy; some short sequences can enter intact cells, while others can be transported inside cells by carrier peptides. PNA probes tagged with thiazole orange or a homologue are nonfluorescent, but become fluorescent on hybridization with nucleic acid; PNA probes incorporating a fluorescent label and a quencher also become fluorescent on hybridization. Less stringent conditions are required for PNA probe hybridization than for hybridization with oligonucleotide probes.

I can leap into fantasy and envision using PNA probes to do FISH on living cells, which can subsequently be sorted. In the real world, PNA probes have been applied to identification of microorganisms based on rRNA sequences[2714] and to determination of telomere length[2715-6], both of which will be discussed in Chapter 10.

7.9 PROBES FOR VARIOUS CELL CONSTITUENTS

Surface Sugars (Lectin Binding Sites)

The specificity of binding of various **lectins** to different carbohydrate moieties of cell surface and other glycoproteins was appreciated well before the first fluorescence measurements were made in a flow cytometer. In the early 1970's, as flow cytometry began to become popular, the use of lectins as reagents for characterization of cell surfaces was logical and attractive; before the advent of monoclonal antibodies, no immune reagents were nearly as well defined, chemically or functionally, as were lectins. In addition, the mitogenic activity of lectins such as **concanavalin A (con A)**, **phytohemagglutinin (PHA)**, and **pokeweed mitogen (PWM)**, and reports of different patterns of lectin binding in normal and transformed cells, suggested that quantitative analysis of lectin binding to cells might provide valuable information in research and clinical laboratories.

The first reported flow cytometric analyses of lectin binding to cells were done by Kraemer et al[310-2] at Los Alamos, who examined binding of fluorescein-con A conjugates. In the initial study, fluorescence was the only cellular parameter measured, and it was difficult to resolve the weak fluorescence of fluorescein-con A above background noise. Later work was done with a multiparameter instrument, in which cell volume measurements were used to gate fluorescence measurements, improving discrimination of cell-associated fluorescence from noise. Analog electronics were used to compute the 2/3 power of cell volume, a quantity proportional to cell surface area, and to derive a measure of the cell surface density of lectin binding sites from the ratio of the fluorescence signal to the derived surface area.

Another relatively early paper[313], published by Bohn in 1976, described the use of a dye exclusion technique with two-color fluorescence measurement to discriminate between intact cells, to the surfaces of which lectin binding was measured, and cells with damaged membranes. In this paper, Bohn also pointed out that the capability of flow cytometry to precisely measure cell-associated ligand in cell suspensions from which unbound ligand had not been removed by washing could be useful in the analysis of many types of **ligand-receptor interactions**. Both Bohn's work and that of Kraemer et al were considerably more sophisticated than most of what was in print when they were published, and were probably not widely appreciated on that account.

While the present emphasis on analysis of cell surface structures with antibodies has decreased interest in the use of labeled lectins, the latter can still be used effectively as aids in enumeration and separation of cell subpopulations. Measurements of lectin binding have typically been used with measurements of other parameters for this purpose. For example, Nicola et al[179] used measurements of forward light scattering and of the fluorescence of fluorescein-PWM and rhodamine-labeled antineutrophil serum to separate hematopoietic progenitor cells from murine fetal liver, while Bauman et al[1417] purified murine stem cells and committed progenitors using a combination of wheat germ agglutinin and monoclonal antibodies. The anti-H lectin identifies nucleated and non-nucleated erythrocytes[1418]; other lectins have been described as markers for eosinophils[1419], and so on.

Lectin binding has also been applied to flow cytometric analysis of parasites[1420-1] and to subcellular organelles; Guasch, Guerri, and O'Connor[1422,2717] have examined lectin binding to Golgi fractions from rat liver. Others have examined changes in lectin binding before and after physiologic changes induced in cells, e.g., by cell-cell and cell-cytokine interactions[1423-6].

Lectin binding patterns change during immortalization and neoplastic transformation of cell lines[1427] and, conversely, during induced differentiation of tumor cells[1428]; heterogeneity of tumor cells[1429] may reflect different propensities of different cells for metastasis, which has been shown to be related to expression of cell surface glycoproteins that are themselves lectins[1430-2].

Because lectins are polyvalent, they have a somewhat greater tendency to agglutinate cells than do antibodies. McCoy et al [784] describe a method for blocking all but one active site of a fluoresceinated lectin to produce a monovalent label. The extent to which this technique may be generalized is not completely clear; however, monovalent, monomeric derivatives of lectins may be produced by other chemical means[1433]. Some formal chemistry can be done with lectins on cell surfaces; sugar competition assays can be used to define binding affinities[1434] and sites[1435].

Reagents for anyone interested in future work along these lines, in the form of fluorescent conjugates of a good selection of lectins, are available from a number of sources (e.g., Molecular Probes and Polysciences). I haven't found recent general review on the use of lectins in flow cytometry; there is a 1987 paper by McCoy[1436].

Analysis of Total Cellular Carbohydrate Content

While lectins bind specifically to particular carbohydrates, they cannot effectively be used as reagents for determination of total cellular carbohydrate content. Instead, Duijndam and van Duijn[1437] employed a technique well known in classical histochemistry, the **periodic acid-Schiff reaction**, to determine the carbohydrate content of erythrocytes. Periodic acid oxidation was followed by pararosaniline staining, providing a strong enough fluorescence signal to allow measurement by flow cytometry. Dimethylsuberimidate fixation was used and yielded low autofluorescence and only faint staining of unoxidized cells. This methodology should be generally applicable.

Specific Detection of Cellulose

Taylor et al[2718] have shown that a cellulase and an isolated bacterial cellulose binding domain (CBD) conjugated to fluorescent dyes can be used for specific detection of cellulose by flow cytometry or laser scanning cytometry.

A Probe for Cell Surface Aldehydes

In conjunction with studies on the mechanism of lymphocyte mitogenesis by compounds that produce oxidation at sites in the cell membrane, Roffman and Wilchek[806] synthesized a **fluorescein diaminobutyryl hydrazide** that could be used for quantitative determination of **aldehydes** in the membrane. They found, incidentally, that the mitogenic activity of different oxidizing agents was not correlated with the degree of aldehyde formation.

Probes for Lipids and Cholesterol

Nile Red

Greenspan et al[802-3] described **Nile red**, an oxazone prepared from the oxazine dye Nile blue, as a fluorescent stain for **neutral lipids** in cells and tissues. The dye fluoresces yellow when dissolved in neutral lipids, and red when in more polar lipids, e.g., phospholipids; its fluorescence is quenched in aqueous solution. Nile red can be excited at 488 nm; fluorescence of the lipid-bound dye is typically measured at about 550 nm. Dive et al[1438] used Fourier transform flow cytometry[1140] to detect emission spectral changes in Nile red in differentiating ovarian granulosa cells; the fluorescence component related to lipid droplets increased with maturation. While the Fourier transform technique is not accessible to most investigators, Smyth and Wharton[1439], examining differentiating adipocytes, found that the ratio of gold to red fluorescence from Nile red also reflects the accumulation of cytoplasmic lipid droplets, and can be used to define cells as being differentiated or undifferentiated. Brown, Sullivan, and Greenspan[1440] reported that flow cytometry of Nile red fluorescence could distinguish macrophages bearing orange-colored phospholipid inclusions from control alveolar macrophages, in which yellow-gold fluorescence from the neutral lipid droplets predominates. Nile red has lately been used to quantitate polyhydroxyalkanoic acids in bacteria[2719-20].

Filipin

Muller et al[804] reported in 1984 that the polyene antibiotic **filipin**, which was known to form fluorescent complexes with **membrane-associated cholesterol**, could be used for quantitative flow cytometric analysis of unesterified cholesterol in formaldehyde-fixed cells. Kruth et al[805] used filipin staining to detect and isolate cholesteryl ester-containing "foam" cells from atherosclerotic aortas by flow cytometry. Filipin is excited in the UV; fluorescence is measured between 510 and 540 nm. Hassall[1441] used a combination of cyanine dye-labeled low-density lipoprotein (LDL) to measure lipoprotein uptake, Nile Red to measure cholesteryl ester accumulation, and filipin to study free cholesterol homeostasis in a multiparameter analysis of human foam cell-forming macrophages. Hassall and Graham[2721] found that changes in filipin fluorescence correlated with changes in cholesterol biosynthesis in these cells.

Lipid Droplet Detection Using Scatter Signals

Suzuki et al[1442], studying triacylglycerol accumulation in cytoplasmic lipid droplets in the U937 macrophage-derived cell line, found that the extent of lipid droplet formation in each cell could be assessed in the absence of any staining by changes in the intensity of 90° light scatter; using the scatter measurement might save you a fluorescence channel if you're running short.

Probes for Cytoskeletal Organization / Actins

Wallace et al[622] used the fluorescent probe **NBD-phallacidin** for flow cytometric analysis of changes in **cytoskeletal organization** (in particular, changes in actin conformation) in blood neutrophil granulocytes following exposure to chemotactic peptides. This material is available from Molecular Probes. Phalloidin binds to F-actin; deoxyribonuclease I (DNAse I) binds to G-actin, and Haugland et al[1373] have shown that the two types of actin can be analyzed simultaneously in cells using **BODIPY-phalloidin** and **fluorescein-DNAse I**, both available from Molecular Probes.

7.10 TIME AS A PARAMETER: KINETIC MEASUREMENTS

Flow cytometry can be used for quantitative analyses of the degree to which almost any fluorescent substance associates with cells, whether the material is bound to the cell surface or is taken into the cell; it is often useful to examine the processes of binding, uptake, and efflux over time.

When the material is taken up over a period of hours, it is customary and sensible to examine cell samples at intervals. When uptake occurs over a period of seconds to minutes, the speed and throughput of flow cytometry can be utilized to best advantage if a single sample is analyzed continuously, and the time at which each cell arrived at the observation point is recorded with the values of scatter, fluorescence, and other more conventional cytometric parameters. I have alluded to the existence of **kinetic measurements** on p. 2, and, on pp. 177-8, mentioned the fact that the more sophisticated varieties may require nonstandard hardware.

I will stress here, as I did on p. 2, that, while kinetic measurements in flow cytometry are based upon measurements of single cells, the temporal patterns obtained represent data from successive cells in the same sample. We measure cell 1 at time 1, cell 2 at time 2, ... , cell *n* at time *n*, and make the assumption that cell 1, had we been able to bring it back into the flow cytometer and remeasure it at time *n*, would have looked more or less the same as cell *n* did. If that assumption holds, we can expect a flow cytometric kinetic measurement to provide more or less the same results as might be obtained from multiple sequential microphotometric measurements of the same cell. If not, we will have to resort to a technique such as scanning laser cytometry, which will allow us to observe the same cell repeatedly over time, to get the kinetic data we need.

Measurements of fluorescence versus time can, as should be obvious, be used to analyze the kinetics of efflux or dissociation of fluorescent material from cells[347], as well as the kinetics of binding and/or uptake. The time measurement technique is also applicable to studies of enzyme kinetics, in which development of color or fluorescence in cells as a result of enzyme action on chromogenic[348] or fluorogenic substrates is followed over time, and is essential in analyses of other rapidly changing functional parameters such as membrane potential and cytoplasmic calcium ion concentration.

The first publication on time measurement in flow cytometry was by Martin and Swartzendruber; it appeared in 1980[346].

Figure 7-21 is a typical display of data from a timed measurement; it shows changes in cytoplasmic [Ca^{++}] over a period of about 200 seconds following stimulation of T-lymphoblastoid cells with a mitogenic monoclonal antibody.

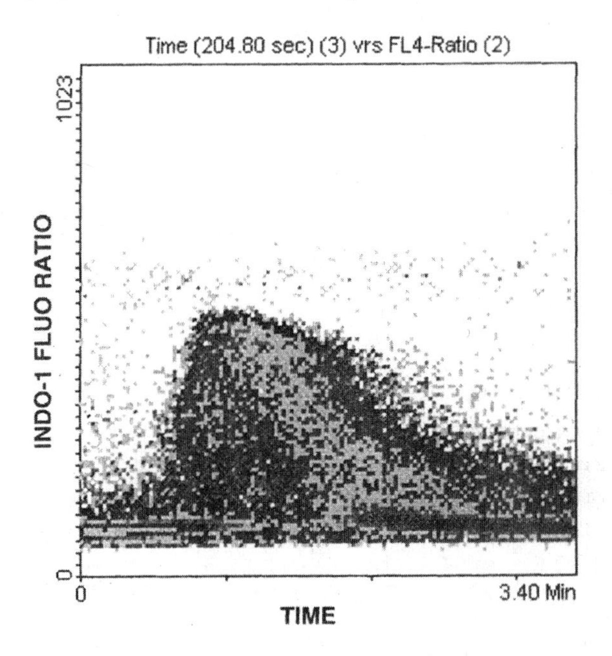

Figure 7-21. Plot of cytoplasmic [Ca^{++}], as indicated by indo-1 fluorescence ratio (pp. 47, 403-4), versus time in T lymphoblastoid cells following stimulation with OKT3 antibody to the T cell receptor. Data were provided by Keith Kelley (Miles Research Center).

Note that while most of the cells represented in Figure 7-21 appear to change their cytoplasmic calcium concentration in response to the applied stimulus, there is considerable heterogeneity in the degree of response. This pattern is not uncommon in kinetic measurements of functional parameters; data from fluorescent ligand binding to ostensibly homogeneous cell populations usually produces nicer looking curves.

The time measurement implemented by Martin and Swartzendruber[346] relied on a linear ramp generator to produce a voltage that increased as a linear function of time; this voltage was used in lieu of a detector signal as an input parameter for the data analysis system. It is now more common to use digital timers, the outputs of which can be read by the data acquisition hardware, for time measurements. Kachel et al[811] and Beumer et al[812] have described digital timing circuitry for collecting kinetic data; such circuits, some of which make use of the real-time clock incorporated in the computer system[813], are now available in most commercial flow cytometers.

A crude estimate of cells' arrival times in systems that lack a time parameter can be obtained by simply plotting events in the order in which they occurred; this makes the erroneous assumption that cells arrive at evenly spaced intervals, and Lindmo and Fundingsrud[351] showed that this was very much not the case (see pp. 144-5). Nonetheless, if your instrument hasn't got a time parameter, erroneous assumptions may be better than nothing at all. Some third-party data acquisition systems and add-ons for older instruments add time measurement capability.

Sample Handling for Kinetic Measurements

When the events being monitored in flow cytometric kinetic studies occur over very short time periods, it is desirable to be able to minimize the delay between the start of sample incubation and the passage of the first cells through the observation point(s) of the flow cytometer. Sample handling systems to permit rapid kinetic analysis by flow cytometry have evolved over the years[349-50,810,1532-4]. A commercial version of Kelley's system[1533], available from **Cytek Development**, provides mixing and temperature regulation in a unit that mounts close to the nozzle holder for reduced transit times, and allows stimuli to be added to a sample with observation of results within 1 second. An electronic circuit activated at the time of injection generates a data time stamp for direct correlation of injection and cellular response.

Dunne[1534] emphasized the importance of **time window analysis**, also called **fixed-time flow cytometry**, in which each cell reaches the observation point in a controllable, fixed time after a stimulus is added, in studies of rapid cellular responses; this can be achieved by allowing the cell sample and the solution containing the stimulus to mix in a T-junction upstream from the nozzle; the time between stimulation and observation is varied by adjusting the pressure and the length of tubing between the mixing junction and the observation point. Tárnok[2722-3] describes a relatively simple mixing device for time window analysis that he has used to study calcium transients in neuronal cells, and to sort cells with unusual patterns of calcium response. The constant pressure maintained by his mixing arrangement is essential for neuronal cells, because they undergo calcium shifts in response to pressure changes that may occur in syringe pump-driven mixing apparatus.

John Nolan, Larry Sklar, and their collaborators at Los Alamos and the University of New Mexico have been involved for some years in efforts to increase both the time resolution of kinetic measurements and the sample throughput of flow cytometers. They have optimized nozzle design parameters for stream-in-air instruments to stabilize flow in the shortest possible time (Graves et al[2454], p. 170), and refined mixing apparatus hardware and control software to reduce the interval between mixing and analysis from 300 ms to 55 ms[2460-1]. They have also developed a dynamic temperature regulation unit that provides more accurate maintenance of sample temperature than was previously possible, and also permits analysis of cellular and chemical responses to rapid temperature changes[2427].

With a view toward improving sample throughput as well as refining kinetic measurements, Durack et al[1535] developed the technique of **time interval gating** to allow data from kinetic experiments involving multiple samples, each of which is measured briefly (i.e., for a few seconds) at intervals of several minutes, to be collected in a single large list mode data file, facilitating subsequent analysis of the data.

The technology of **flow injection analysis**, a rapid analytical chemical method pioneered by Ruzicka[1147], has been applied by him and his colleagues to cytometry[1148-9,1536]. Their methodology uses computer-controlled systems of pumps and valves to handle samples, and software to facilitate non-equilibrium analysis.

Sklar et al[2462-3] have also developed systems for **high throughput flow cytometry**; the most advanced[2463] introduces 1-3 µL "slugs" of sample, interspersed with air bubbles to reduce sample carryover, into the cytometer at a rate of one sample every 1.3 seconds. The MoSkeeto™ sampler now offered by **DakoCytomation** appears to be a commercial version.

Time as a Quality Control Parameter

Watson[819] cleverly extended the earlier observations of cell arrival times[351] to permit the use of **time measurements for quality control** within individual sample runs. The interval between cell arrival or acquisition times is used to derive a real-time measurement of **sample flow rate**, which is displayed against time; in addition, a measured parameter of interest (DNA content, in the case illustrated in Figure 7-22) is shown (schematically) vs. time in a two-dimensional display.

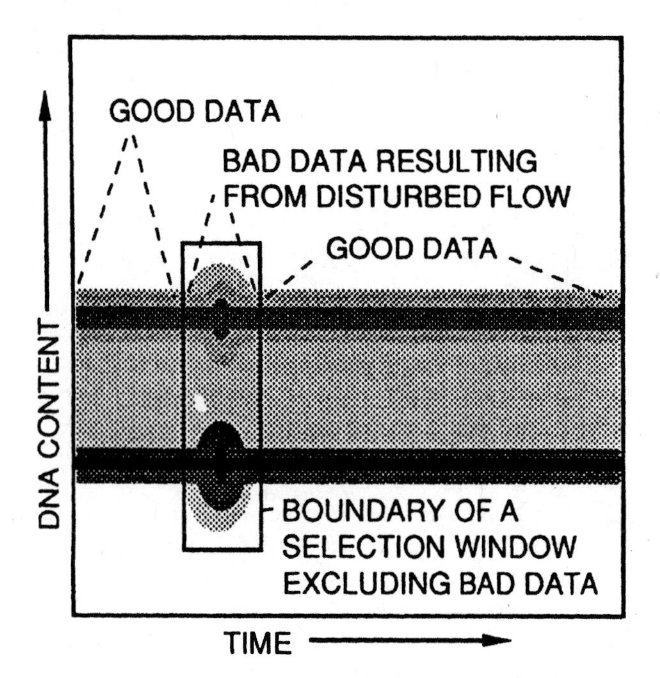

Figure 7-22. Time used as a quality control parameter in DNA content analysis (after Watson[819]).

The "lumps" in the middle of the display result from a temporary degradation in measurement precision caused by a transient disturbance of the flow pattern. Before and after the transient, the measurement precision is considerably better. However, if the DNA content distribution were calculated for the entire sample, the coefficient of variation would be increased by inclusion of the relatively small number of cells from which data were acquired during the flow disturbance. By defining a selection window as shown in the figure, it is possible to exclude the bad data points from the DNA content distribution. In many cases, e.g., when one analyzes cells from a very small sample of tumor tissue, being able to "rescue" data in this fashion may make the difference between getting and not getting enough information to characterize a specimen. Even when there is plenty of sample available, however, Watson's technique provides a convenient means of assessing the quality of measurements. Kusuda and Melamed[2725] used the chronological order of events in a list mode file as a time parameter and were able, in some cases, to correct for drift in data values.

Slooowww Flooowww

Conventional flow cytometers use flow velocities of 1-10 m/sec, making it almost impossible to examine the same cell twice at time intervals separated by more than a few hundred µsec. The relatively high flow velocities also result in shear stress on cells in the sample, making the use of flow cytometry for studies of cell adhesion suspect.

Amblard et al[1537] developed a chamber for flow cytometric analysis over an extended range of stream velocity which, with adaptations to the electronics, permitted standard analysis at velocities as low as 0.01 m/sec in a B-D FACS instrument. Conjugates formed by the adhesion between human B and resting T lymphocytes, disrupted in conventional flow cytometers, could be detected and precisely quantified provided analysis velocity was kept below 0.1 m/sec. A much longer chamber, operating at this velocity, with widely separated measurement stations, could allow multiple observations of an individual cell to be made over periods of tens to hundreds of seconds; nobody seems to have built one yet. However, slow flow systems are in use for such applications as DNA fragment sizing, which will be discussed in Chapter 10.

7.11 LABELED LIGAND BINDING

The flow cytometric analysis of ligand binding to cell surface and intracellular receptors, or to receptors on beads or other particles, can be approached in several ways, the most obvious of which is probably the use of labeled ligands. To produce labeled ligands, you need to have enough relatively pure unlabeled ligand on hand to yield an adequate supply of labeled material for your experiments. Acquiring the starting material used to be the first hard part of the project; the second part was verifying that labeling the ligand didn't drastically alter its binding characteristics. In this age of cloning, a respectable variety of labeled ligands such as cytokines and hormones are available off-the-shelf, and you probably won't have to worry about getting and labeling your ligand; if not, this isn't the place to look for help. This is a good place to mention that antibodies also count as ligands; their binding to cells can be analyzed using exactly the same techniques.

The production of labeled ligands, and the flow cytometric analysis of ligand binding, have become so common that I have given up trying to keep track of every ligand. Any listing or tabulation you see will be incomplete by the time it appears; I've run across several papers on flow cytometry of ligand binding that didn't turn up in my literature search on flow cytometry. As was the case with immunofluorescence, I'll have to stick to the basics with selective excursions into specifics.

Just as **cell surface receptors** for ligands ranging from small peptide hormones to viruses and bacteria can be demonstrated by flow cytometry using fluorescently labeled analogues, **cytosolic** and **nuclear receptors** for ligands such as steroid and thyroid hormones may also be demonstrable. However, since the ligands themselves are relatively small molecules, it is, in general, more difficult to find or synthesize analogues with binding characteristics similar to those of the natural ligands than to prepare usable fluorescent derivatives of larger molecules. In order to demonstrate receptors within intact cells, the fluorescent ligand used must obviously have the capability of crossing or being carried across the cell membrane, as well as the binding characteristics previously mentioned. The size and specificity constraints on analogues make it infeasible to attach more than one fluorochrome moiety to each molecule of ligand; since the number of intracellular receptor sites is generally small, one can therefore expect only weak fluorescence signals at best from cells stained with the labeled ligand analogue.

The flow cytometric literature in this area recounts numerous attempts at demonstration and quantitation of **estrogen receptors** in cells. The presence or absence of estrogen receptors, as determined by radioligand binding analysis of biopsy specimens, is of significance in assessing prognosis and planning therapy of breast cancer. Fisher et al[378], using fluoresceinated estrone (17-FE) and fluorescence microscopy, demonstrated heterogeneous binding patterns of the material in tissue samples; Tyrer et al[379], Kute et al[380], and Van et al[381] made flow cytometric measurements using 17-FE and other fluorescent analogues.

A study by Benz, Wiznitzer, and Lee[828], who utilized bovine serum albumin (BSA) as a carrier molecule to which an average of 25 estradiol and 4 fluorescein molecules were bound, generated a heated exchange of correspondence between Ashcroft[829] and Benz[830], which did not resolve the issue of how accurately one can study binding sites for small molecules using much larger ones. If you simply want to determine the number of binding sites, **anti-receptor antibodies** may be preferable to labeled ligand analogues as reagents, but, if the objective is to measure binding affinities, you may be literally and figuratively stuck with labeled ligands. On the bright side, Marchetti et al[1443] found comparable results in analyses of **glucocorticoid receptors** using flow cytometry with antireceptor antibodies and fluorescein-labeled steroid analogues and classical radioligand binding methods.

Labeling Strategies

The procedure described by Shechter et al[302], in which small peptide hormones such as insulin are fluorescently labeled by reaction with fluorescein- or tetramethylrhodamine-labeled **lactalbumin**, provides a way of getting a few more fluorescent molecules onto a ligand than can be done by direct labeling. This becomes important when the number of binding sites per cell is small; under these circumstances, flow cytometry may allow detection and quantitation of amounts of bound ligand too small to be seen under the fluorescence microscope. This degree of sensitivity is also associated with the technique of **video intensification microscopy (VIM)**[445-6], which offers the advantage of allowing the same cell to be observed over a period of time. A paper by MacInnes et al[447] illustrates the use of VIM and cell sorting in a complementary fashion to study luteinizing hormone releasing hormone binding sites on rat pituitary cells.

When the ligand for which receptors are to be demonstrated is a small molecule, covalent attachment of ligand molecules to a fluorescently tagged protein, or, alternatively, a phycobiliprotein, may yield a material that will still bind specifically to receptors, thereby enabling attachment of more than a single fluorochrome moiety to each receptor site. This technique was used by Osband et al[448] to show that a subpopulation of T cells bore H_2 (cimetidine-reactive) histamine receptors; the reagent employed was a conjugate of histamine with fluoresceinated albumin. Hallberg et al[449] used the same material to demonstrate H_1 and H_2 receptors on platelets; Muirhead et al[849], however, found only nonspecific binding of similar conjugates to lymphocytes.

Formal Analysis of Ligand Binding

Bohn[313,326-7] was probably the first to appreciate that flow cytometry could be used for **quantitative analyses of fluorescent ligand binding** to cells that yielded essentially the same data as were obtained from conventional radioligand binding assays; in some circumstances, bulk fluorimetry must be used in addition to flow cytometry for precise determination of the ratio of bound to unbound ligand, but, in other cases, information can be derived solely from flow cytometric data.

The flow cytometric method offers a particular advantage over radioisotope techniques in that the binding equilibrium is not disturbed because there is no need to wash away unbound ligand; the speed of the flow cytometric analysis and the elimination of the necessity to deal with and dispose of radioactive materials also weigh in its favor. Sklar and Finney[450] demonstrated the capabilities of flow cytometry in an analysis of steady-state binding of a fluoresceinated chemotactic peptide to neutrophils.

The study of Benz, Wiznitzer, and Lee[828] on estrogen receptors provides good examples of the calculation of binding affinities and the construction of the **Scatchard** and **Lineweaver-Burk plots** commonly used with radioligand assays. Other older examples of quantitative analysis, dealing with binding of antibodies to cell surface receptors, were published by Uckun et al[1444] and Krause et al[1445]; Ziegler et al studied labeled **insulin**[2313]. In addition, Chatelier and Ashcroft[797] proposed an alternate method, **isoparametric analysis**, for determining ligand-binding characteristics.

More recently, Gordon[2726] described a Scatchard analysis of **concanavalin A** binding to lymphocytes, noting that cytometric plots seemed more precise than those obtained using radiolabeled ligands, particularly at low ligand concentrations. Macho[2727] et al quantified **fatty acid** binding to cells using a BODIPY-labeled dodecanoic acid. Cherukuri et al[2728] used flow cytometry, fluorescence spectroscopy, and two-photon microscopy in an elaborate mechanistic study of binding and internalization of fluoresceinated poly-D-lysine by murine macrophages. Waller et al[2729] established the fluorescence characteristics of six fluorescently labeled ***N*-formyl peptides**, which bind to neutrophils, and developed and validated competitive binding protocols to determine binding constants of unlabeled ligands. Lauer et al[2730] performed real-time quantitative flow cytometric analysis of the interaction of fluoresceinated cholera toxin subunits with ganglioside receptors inserted into phospholipid membranes supported by glass beads. All of the reports just described may be helpful to anyone interested in doing quantitative work with other ligands.

Labeled Ligands versus Anti-Receptor Antibodies

In some cases, there is a choice of labeled ligands available that will bind to a particular receptor; for example, transferrin receptors can be demonstrated using either fluoresceinated transferrin[451] or a monoclonal antibody such as Ortho's OKT9, which reacts with the transferrin receptor[452] (CD71). Flow cytometry can, of course, be utilized to establish the fact that two different ligands do react with the same or closely spaced sites on cell surfaces. Fingerroth et al[453] used flow cytometric analysis to study the binding of fluorescently tagged viruses, antibodies, and complement components to human B cells, and showed that the Epstein-Barr virus binds to the type 2 complement receptor.

Ligand Binding Detected by Functional Changes

When the ligand of interest is not available in sufficient quantity or purity to permit ligand labeling, the only feasible way to identify cells that have bound and responded to ligand is by demonstrating induced changes in functional parameters. This approach was taken by Price et al[417], who isolated erythroid and myeloid precursor cells from bone marrow based upon SCM responses to impure preparations of growth stimulatory factors, and by Osband et al[454], who isolated H_2 receptor bearing T cells based on membrane potential responses to a combination of histamine and an H_1 receptor antagonist. For this approach to work, it is necessary to have stable and reproducible flow cytometric methods for the detection of physiologic changes induced by ligand binding; while such methods have improved over the years, the methods of preparing labeled ligands and the sensitivity of instruments have also improved, making it easier to detect binding *per se*. However, demonstration of functional changes remains useful for distinguishing between cells that merely bind ligand and cells that respond. Tordai et al[1446], for example, used binding of labeled antireceptor antibodies and induction of calcium fluxes by unlabeled thrombin to identify functional thrombin receptors on T-lymphoblastoid cells.

Fluorescent Ligand Binding: Some Examples

High-density lipoprotein (HDL) binding was studied by Schmitz et al[1447], using a rhodamine label; they compared fluorescent and radiolabeled ligand binding and estimated the number of binding sites on human lymphocytes, monocytes, and granulocytes. Traill et al[1448] used **"DiI" (dioctadecyl-indocarbocyanine)** to label HDL and quantified its binding to lymphocytes; they also[1449-50] studied binding of DiI-labeled **low-density lipoprotein (LDL)**. Corsetti et al[1451-2] labeled LDL with *N,N*-dipentadecylaminostyryl-pyridinium iodide, which has very flow fluorescence in aqueous solution and which therefore should yield lower fluorescence background in HDL-binding measurements.

Laborda et al[1453] and Torres et al[1454] examined the binding and endocytosis of fluorescently labeled **α-fetoprotein** and **transferrin** by lymphoid cells. Midoux, Roche, and Monsigny[1455] studied binding, uptake, and degradation of fluoresceinated **neoglycoproteins**. Others have used fluorescein as a label for the clotting **factor VIII**[1460], for **anaphylotoxin C5a**[1461], and for **low molecular weight heparin**[2731]. Richerson et al[1462] and Sumaroka et al[1463] examined binding of mycobacterial **muramyl dipeptides** to macrophages, while other groups studied binding of bacteria to platelets[1464-5] and gastrointestinal cells[1466]. Szabo and Damjanovich[1467] reported the used of a fluorescein-labeled, protein A-rich strain of *Staphylococcus* as a second-step reagent for immunofluorescence.

While some work has been done with fluoresceinated preparations of the **lipopolysaccharide (LPS)**[1456-9] derived from various species of Gram-negative bacteria, Triantafilou et al[2732] used Alexa 488 hydrazide to label the oligosaccharide core of the material, avoiding interaction of the label with lipid A, which is the portion of the molecule involved in binding. Their labeled preparation can be used at concentrations in the range of those reported to be present in the blood of patients with Gram-negative sepsis.

Indirect staining was used by Gabrilovich et al[1468] to quantify binding of the **HIV gp120 protein** to cells; the protein was labeled with biotin and subsequently reacted with Texas Red-streptavidin. Indirect staining also provides amplification to facilitate detection of small numbers of binding sites for **growth factors** and **cytokines**. Wognum et al used biotinylated ligands followed by phycoerythrin-streptavidin to detect receptors for **erythropoietin**[1469] and for

interleukin-6 (IL-6)[1470]; further amplification with biotin-anti-phycoerythrin antibody increases the signal for erythropoietin receptor detection[1471]. Biotinylated ligands have also been used to quantify receptors for **IL-2**[1472-3] and **IL-4**[1474-5], enabling demonstration of fewer than 100 receptor sites/cell in some cases. Lawrence et al[2733] used a fluorescein-labeled high-affinity opioid in studies of the kappa opioid receptor, amplifying signals by staining the bound ligand with a phycoerythrin-labeled anti-fluorescein antibody.

Both direct fluorescent labeling and biotinylation of small molecules may, as mentioned before, result in loss of activity and even in the labeled ligand acquiring the characteristics of an antagonist, as was shown by Helmreich et al[1476], who fluoresceinated **glucagon** at different sites, and by Newman et al[1477-8], who made biotinylated derivatives of **parathyroid hormone**. Jans et al[1479] used spacers to preserve biological activity when preparing biotinylated **vasopressin**. In other cases, even small molecules may be labeled without apparent loss of activity and binding specificity; active fluorescein derivatives of **transforming growth factor beta**[1480], of **chemotactic peptides**[701], and of the fibronectin peptide **arg-gly-asp-ser**[1481] have been prepared, as has an active fluorescent **NBD-phorbol ester**[1482]. Chianelli et al[1483] have used fluorescence flow cytometry with a labeled antireceptor antibody in competitive assays to determine whether derivatization of IL-2 to various nonfluorescent products affects binding.

Moving from very small ligands to very large ones, I should mention the use of labeled **aggregated human immunoglobulin** in a flow cytometric assay for **circulating immune complexes** based on competitive binding to surface receptors on Raji cells[1484], and a couple of papers on binding of fluorescein-labeled viruses to cells[1485-6].

Since ligand binding is one of the phenomena investigated in the context of **high throughput screening** of lead (I don't mean Pb!) compounds in drug development, I should point out that a high throughput flow cytometer could be gainfully employed in this area.

7.12 FUNCTIONAL PARAMETERS I

The distinction I have made between structural and functional parameters is, to some extent, arbitrary; it is, after all, the function of cells that determines, maintains, and/or changes their structure. In general, what I have called **structural parameters** describe cells' stable morphologic characteristics and their content of specific chemical constituents, while the classification of **functional parameters** has been reserved for physical and chemical properties of cells which are defined operationally and/or which change rapidly. Both of these characteristics of functional parameters tend to make them more difficult to measure, by flow cytometry or otherwise, than are structural parameters. The motivation to make such measurements is provided by the perceptions that **functional heterogeneity** within cell populations is of biological significance and that observations of their functional characteristics can aid in understanding and predicting cells' behavior. I'll start with the outside of the cell here and work my way in.

Cell Surface Charge

Valet et al[372,377] have estimated **cell surface charge** from the binding of **fluoresceinated polycations** to the cell surface. They believe their flow cytometric method can be somewhat more informative than can the conventional measurement of surface charge based upon cells' electrophoretic mobilities, since flow cytometric analysis can be done using labeled polycations of different molecular weights and structures to define the accessibility, as well as the number, of various charged sites on cell surfaces.

Cell Membrane Characteristics

Membrane Integrity versus "Viability": Dye Exclusion Tests

I have discussed **membrane integrity** in connection with vital staining on pp. 299-301. Most people who work with cells are familiar with tests of "viability" based on cells' capacity to exclude acid dyes such as **trypan blue**, **eosin**, **erythrosin**, **nigrosin**, and **primulin** and some basic dyes such as **propidium**. The basis of all of these **dye exclusion tests** is the same; the dyes used are **impermeant**, and do not normally cross intact cell membranes.

A slight variation on the same theme is provided by using **esters** of **fluorescein** or related compounds as the reagents; **fluorescein diacetate (FDA)**, introduced on pp. 24-7, is lipophilic, uncharged, and nonfluorescent, and thus readily crosses cell membranes. Once inside cells, FDA is hydrolyzed by nonspecific esterases to produce the fluorescent fluorescein anion, which is retained (for a few minutes, at least) by cells with intact membranes and lost by cells with damaged membranes[335].

When doing dye exclusion tests, bear in mind that **propidium and ethidium are not interchangeable**. Although the two dyes behave nearly identically as intercalating stains for double-stranded nucleic acid, ethidium will enter intact cells slowly, especially if the pH of the medium is high, but is likely to be pumped out by an efflux pump. Propidium, which has one more positive charge, will not enter normally enter intact cells; you should thus use propidium, not ethidium, to test membrane integrity.

In these days of multicolor immunofluorescence measurements, it is preferable to have a dye exclusion indicator that emits at a longer wavelength than propidium; Schmid et al[1316] have shown that **7-amino-actinomycin D (7-AAD)** is useful for this purpose, and the red-excited **TO-PRO-3**, which, like propidium, is doubly charged and impermeant (pp. 300-1), is also useful. In work with bacteria using a dual-laser instrument and adding both dyes to samples, I have noted that cells that take up propidium take up TO-PRO-3 and *vice versa*; I never observed cells that took up only one of the dyes. I have not had the opportunity to play the same game with either propidium or TO-PRO-3 and

SYTOX Green, which is a current favorite for membrane integrity testing, especially in bacteria[2734], and which I suspect has at least three positive charges.

Bhuyan et al[336] and Roper and Drewinko[337], in papers published in 1976, showed independently that cells treated with lethal doses of cytotoxic drugs remained "viable" by dye exclusion tests for several days, and that the fraction of viable cells estimated by dye exclusion or ^{51}Cr release[336] was considerably higher than the fraction of cells that retained clonogenicity in culture. **As a general rule, cells that let in trypan blue or propidium are dead, but cells that don't are not necessarily viable.**

Despite the limitations of dye exclusion tests, there have been a lot of papers published that describe various flow cytometric adaptations and applications[4, 165, 197, 240, 313, 326-7, 338-45]. As might be expected, dye exclusion provides a somewhat more accurate measure of viability when applied to situations in which cells are killed by punching holes in their membranes, e.g., freezing or immune lysis[4,344].

Cells that do have holes in the membrane usually admit molecules considerably bigger than dyes, and thus tend to stain nonspecifically with fluorescent antibodies, lectins, etc. If fluorescein is being used to label such ligands, the addition of erythrosin or ethidium (or, preferably, propidium) to the sample produces strong fluorescence in the dead cells, allowing them to be gated out of analyses[165, 313, 326-7]. Khaw et al[342] worked the other side of this street, detecting loss of membrane integrity in cardiac muscle cells by their binding of fluorescent spheres coated with an antimyosin antibody. Stöhr and Vogt-Schaden[339] used a mixture of Hoechst 33342 and propidium to stain DNA in unfixed cells; this allows discrimination of cells with intact (blue fluorescence) and damaged (red fluorescence) membranes. Wallen et al showed that this test, like other dye exclusion tests, does not give a good indication of reproductive viability[345], reinforcing the conclusions others reached earlier[336-7].

If you think about it, you'll probably guess that it is easier to see a cell lightly stained with eosin or propidium, under a fluorescence microscope, than it is to see a cell lightly stained with trypan blue under a transmitted light microscope. This would suggest that dye exclusion tests done by absorption and fluorescence might give different results. During the development of the Block differential leukocyte counter, some comparative studies of trypan blue and ethidium (we didn't appreciate the difference between ethidium and propidium at the time) exclusion were done; the proportion of damaged cells estimated by ethidium fluorescence was always higher than that estimated by trypan blue absorption, and the ratio of the two remained nearly constant from sample to sample (K. F. Mead, unpublished). Others[807] have made similar observations.

Berglund et al[808] examined the UV-excited, blue fluorescent sulfonated dye **Calcofluor White M2R (CFW)**, an optical brightener closely related to the "LN" used in the Block differential staining system, as a dye for assessing membrane integrity; it works well with animal cells but may stain the walls of plant cells even when the membranes are intact. There may be some advantage to using CFW or dyes with similar spectral characteristics when one is trying to measure other parameters by green and orange or red fluorescence using 488 nm excitation. However, as a general rule, one can use propidium as a membrane integrity indicator even when utilizing orange fluorescence measurements for other purposes, by adjusting the propidium concentration; if not, 7-AAD (or TO-PRO-3 if you have a red laser and a 488 nm laser) will almost certainly work.

If, for example, you are trying to measure pyronin Y fluorescence in intact cells, adding propidium at a concentration of 1-2 µg/ml to the sample (this is less than 1/20 the concentration at which the dye is commonly used for stoichiometric DNA staining) will produce strong red fluorescence in cells with damaged membranes. If the detector used for the pyronin measurement has a response extending to at least 600 nm, the cells with damaged membranes will show up off-scale in this channel; alternatively, a separate red (600-640 nm) detector could be used. When using cells stained with Hoechst 33342, bear in mind that, if it gets in, propidium will strongly quench the Hoechst dye fluorescence; if you are triggering on the Hoechst dye fluorescence signal, you may have to adjust the threshold accordingly. 7-AAD, in my experience, does not quench Hoechst dye fluorescence to the same extent.

Even if a cell does let propidium or the other "excluded" dyes in, you can't be absolutely sure it's nonviable. Cell membranes can be transiently **permeabilized**, not only to dyes, but to macromolecules, by physical or chemical means, as discussed on pp. 300-301, and then resealed with retention of viability - even reproductive viability. If you permeabilize cells in a solution of propidium, reseal them, and add FDA (to a final concentration between 500 ng and 1 µg/ml, from a 0.5-1 mg/ml working solution in DMSO), any cells that weren't permeabilized will show only green cytoplasmic fluorescence, any cells that didn't reseal or had damaged membranes to start with will show only red nuclear fluorescence, and cells that were permeabilized and resealed will show green cytoplasm and red nuclei (F. Tsang & H. Shapiro, unpublished). Jones and Senft[809] noted that air-dried smears could be made from cell suspensions stained with FDA and propidium, allowing the fraction of cells with intact membranes to be determined as long as a week after slides are made.

Two compounds closely related to FDA and originally developed for intracellular pH estimation also offer advantages for dye exclusion tests because the fluorescent materials formed from them in cells are lost from the cells much more slowly than is fluorescein. The compounds are **carboxyfluorescein (COF)** and **2',7'-bis'(carboxyethyl)-5-(6')-carboxyfluorescein (BCECF)**; they will be discussed in more detail in the section on pH measurement. However, when it comes to dye-exclusion tests, the real champion among fluorescein-based dyes is **calcein**, introduced into cells as the **acetoxymethyl (AM)** ester; once the ester is cleaved, the free dye stays inside cells much longer than the others (but see pp. 376-7). In general, however, people now seem to use nucleic acid dyes in preference to various fluorescein derivatives for dye exclusion testing, because the objective is often to eliminate dead cells from immunofluorescence analysis, and any of the fluorescein derivatives will produce enough of a green fluorescence signal to make it difficult or impossible to measure immunofluorescence using fluorescein, PE, or even PE tandems.

Detecting "Dead" Cells in Fixed Samples

Many of the samples now run through flow cytometers have been fixed before they are run. Identification of cells that had lost membrane integrity before fixation is useful, because when a sample is stained with fluorescent antibodies to a cell surface antigen or antigens prior to fixation, antibodies will enter cells with damaged membranes, and those cells will therefore exhibit high levels of nonspecific fluorescence. Riedy et al[1516] developed a method for identification of membrane-damaged cells using the irreversible binding of photoactivated **ethidium monoazide (EMA)**. EMA is a positively charged molecule containing the nucleic acid-binding fluorophore ethidium; it is excluded by "viable" cells with intact membranes and enters cells with damaged membranes. When added to a sample prior to fixation, EMA can be covalently bound to nucleic acids by photochemical crosslinking using visible light. When the samples are subsequently washed, stained with antibodies and fixed, ethidium is retained only in those cells that had damaged membranes prior to fixation; at appropriate levels, the ethidium fluorescence is distinguishable from fluorescein and phycoerythrin emission.

These days, there's an easier way to do things. Schmid et al[2735] found that, if cells were exposed to 4 μg/mL **7-AAD** before fixation, the dye could be kept in only those cells that had lost membrane integrity prior to fixation by addition of 4 μg/mL **actinomycin D (AD)** to subsequent fix, stain and wash solutions and buffers. AD is nonfluorescent, but competes with 7-AAD for binding sites on DNA. The 7-AAD/AD combination was used with fluorescein- and PE-labeled antibodies and TO-PRO-3 (with added RNAse) to permit two-color immunophenotyping and DNA content determination in cells that were not membrane-damaged prior to fixation.

Membrane Fusion and Turnover; Cell Tracking

It is often desirable to be able to attach permanent or relatively permanent fluorescent labels to cell membranes, for such purposes as detecting cell aggregation or hybridization, determining the localization and fate of cells isolated from and reinjected into an animal, and establishing the rate of turnover of various components of the membrane itself. A number of fluorescent compounds have been used for such membrane labeling.

Covalent labels such as FITC and XRITC bind to membrane proteins; Abernethy et al[1540] found that lymphocytes labeled with both dyes and reinjected into sheep could be followed over periods of days, and that FITC-labeled cells isolated from lymph could be double labeled with XRITC and recovered. Capo et al[1541] noted that there was some nonspecific transfer of fluorescein molecules (about 10,000 molecules) from FITC-labeled to unlabeled cells *in vitro*, suggesting that FITC is not the stablest of labels. Weston and Parish[1542] reported that Hoechst 33342-, BCECF-, and calcein-labeled cells could be followed for 2-3 days, while cells covalently labeled with **carboxyfluorescein succinimidyl ester (CFSE)** retained label for weeks.

Another class of labels are incorporated into the membrane lipid bilayer itself. Among the most widely studied are derivatives of cyanine dyes, including **"DiI" (dioctadecylindocarbocyanine)** and **"DiO" (dioctadecyloxacarbocyanine)**[1543], both of which can be excited at 488 nm, with DiO emitting green fluorescence at about 500 nm and DiI emitting yellow fluorescence at about 565 nm. Fluorescein and phycoerythrin emission filters are, respectively, well suited for measurement of DiO and DiI. The octadecyl (C_{18}) side chains of these probes reside in the lipid bilayer; the chromophores remain at the surface.

When used to label neuronal cells, which do not divide, DiI and DiO persist in the membrane for months; they were found to remain detectable for more than a week in cultured endothelial and smooth muscle cells[1544]. St. John[1545] found that DiI was toxic to embryonic rat neurons, and that the C_{12} derivative, which was not, could be used effectively for long-term labeling. Ledley et al[1546] used DiI as a marker to determine localization of transplanted hepatocytes and thyroid follicular cells by flow cytometry.

Paul Horan and his colleagues[1547-52] introduced the **PKH** series of tracking dyes, which includes both radioactively and fluorescently labeled compounds incorporating longer aliphatic chains (e.g., C_{24}) than are found in DiI and DiO. Presumably, these molecules incorporate into the outer layer of the lipid membrane bilayer and penetrate into the inner layer, providing stronger binding and, as a result, longer persistence. These dyes have been used to study peritoneal macrophages[1548-9], to discriminate target from effector cells in cytotoxicity assays[1550], to track lymphocytes used in adoptive immunotherapy[1552], and to label bacteria for studies of phagocytosis[1553]. I know of no comparative study of PKH dyes vs. DiI and DiO, but the PKH dyes demonstrably work.

Cell Proliferation Analyzed Using Tracking Dyes

As Horan et al pointed out[1547], if a cell labeled with a tracking dye subsequently proliferates, each daughter cell will get approximately half the label; in the next generation, each daughter cell will carry one-quarter of the label, and so on. Analysis of the labeling intensities of cells grown *in vivo* or *in vitro* after labeling will, therefore, allow determination of the number of division cycles through which each cell has progressed since the label was applied. While similar information can be obtained from bromodeoxyuridine labeling studies, it is much simpler to obtain using tracking dyes. Ashley et al[1554] used the yellow-fluorescent dye PKH26 GL to follow leukemic cells through several cell divisions; more recently, Yamamura et al[1555] developed a method for analysis of proliferation of stimulated lymphocytes using the same dye. Results are shown in Figure 7-23.

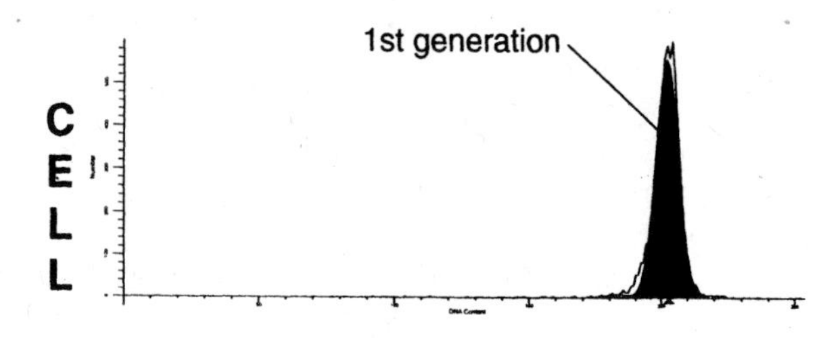

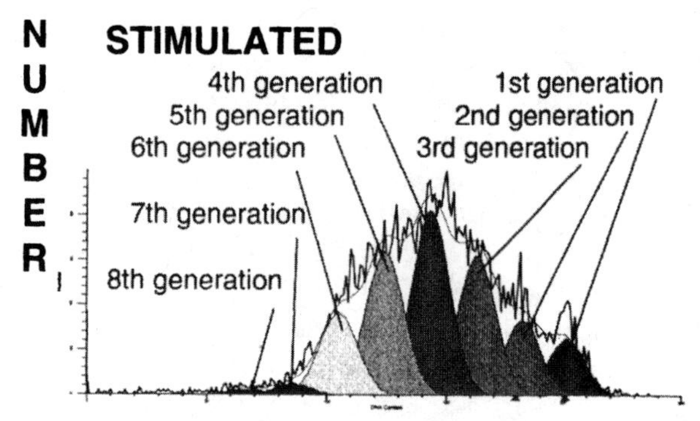

Figure 7-23. Cell proliferation indicated by dilution of fluorescence of the tracking dye PKH26 in mitogen-stimulated and unstimulated peripheral lymphocytes after 5 days in culture[1555]. The data and figures were provided by Yasuhiro Yamamura (Ponce (PR) School of Medicine), Abraham Schwartz (Flow Cytometry Standards Corp.), Bruce Bagwell, and Don Herbert (Verity Software House).

The raw histograms in the figure were analyzed using a variant of the ModFit DNA analysis program from **Verity Software House**, which does nonlinear least squares fits to decompose a histogram into a sum of distributions representing successive generations of daughter cells. The model can identify as many as ten successive generations of cells. This cannot be done using BrUdR or tritiated thymidine; the tracking dye technique also has the advantage of being usable with live cells, permitting sorting of cells in different generations. The PKH dyes and analysis software are available from **Sigma**.

Tracking dyes are particularly useful for analysis of responses of lymphocytes to antigens. When mitogens are used for lymphocyte stimulation, a majority of cells respond. In antigen stimulation, the fraction of responsive cells ranges from 1 or 2 percent, in mixed lymphocyte reactions, to one cell in several hundred thousand, which is typical when cells are exposed to viral antigens. In a lymphocyte population, there should be a spectrum of affinities for a particular antigen; cells that begin proliferating earlier may have different characteristics from those that begin proliferating later. This could be established by sorting of cells with different labeling intensities following stimulation by antigen, which might also provide clones useful for their cellular or humoral specificities. Sorting experiments along this line should also be able to determine to what extent late proliferative responses in antigen-stimulated cells are the result of recruitment, rather than of specific reaction to antigen.

A look at Figure 7-23, however, shows that the fluorescence distributions calculated by the model for successive generations of PKH26-labeled cells are rather broad and overlap considerably, meaning that sorting from any particular segment of the distribution would be likely to yield a mixture of cells from several generations. Since the previous edition of this book appeared, **carboxyfluorescein succinimidyl ester (CFSE)**[2349-50] has become widely used as a tracking dye for analysis of cell proliferation.

CFSE is, as Dick Haugland points out in the Molecular Probes *Handbook*[2332], an inappropriate term to describe the succinimidyl ester of 5(6)-carboxyfluorescein diacetate. He suggests CFDA SE; the name may not sell, but the sales receipts from the chemical itself should salve any wounds to his pride. CFSE is not properly a membrane label; it, like FDA and the esters of carboxyfluorescein and calcein, readily crosses cell membranes and enters cells, and is cleaved by intracellular esterases to a fluorescent form. Unlike fluorescein, carboxyfluorescein, and calcein, the fluorescent form of CFSE carries a reactive succinimidyl group, allowing it to bind covalently to amino groups in proteins; this accounts for the extremely long-term retention of the dye.

Both CFSE and the PKH dyes have been reported to be toxic to cells; this limits the concentrations at which they can be used to label a first generation, and, since tenth-generation cells will, on average, be only 1/1024 as bright as first generation cells, starting with lower concentrations of tracking dyes will usually decrease the number of generations discernible by flow cytometric measurements.

The PKH dyes and other long-chain fatty acid derivatives are more difficult to load into cells than is CFSE, but it is possible to make measurements of fluorescein- and PE-Cy5-labeled antibodies in cells stained with PKH26, while it is impossible to use fluorescein-labeled antibodies and difficult to use PE and some of its tandems in CFSE-stained cells. However, as can be seen from a comparison of Figures 7-23 and 7-24, CFSE clearly outdoes PKH26 in delineating generations. It is likely that if cells were sorted from a narrow middle region of one of the peaks shown in the bottom panel of Figure 7-24, the sorted population would come predominantly from a single generation.

Tracking dyes have had a substantial impact on studies of cellular immunology in recent years, and their applications to the study of lymphocyte development and activation will be discussed further in Chapter 10 (p. 501).

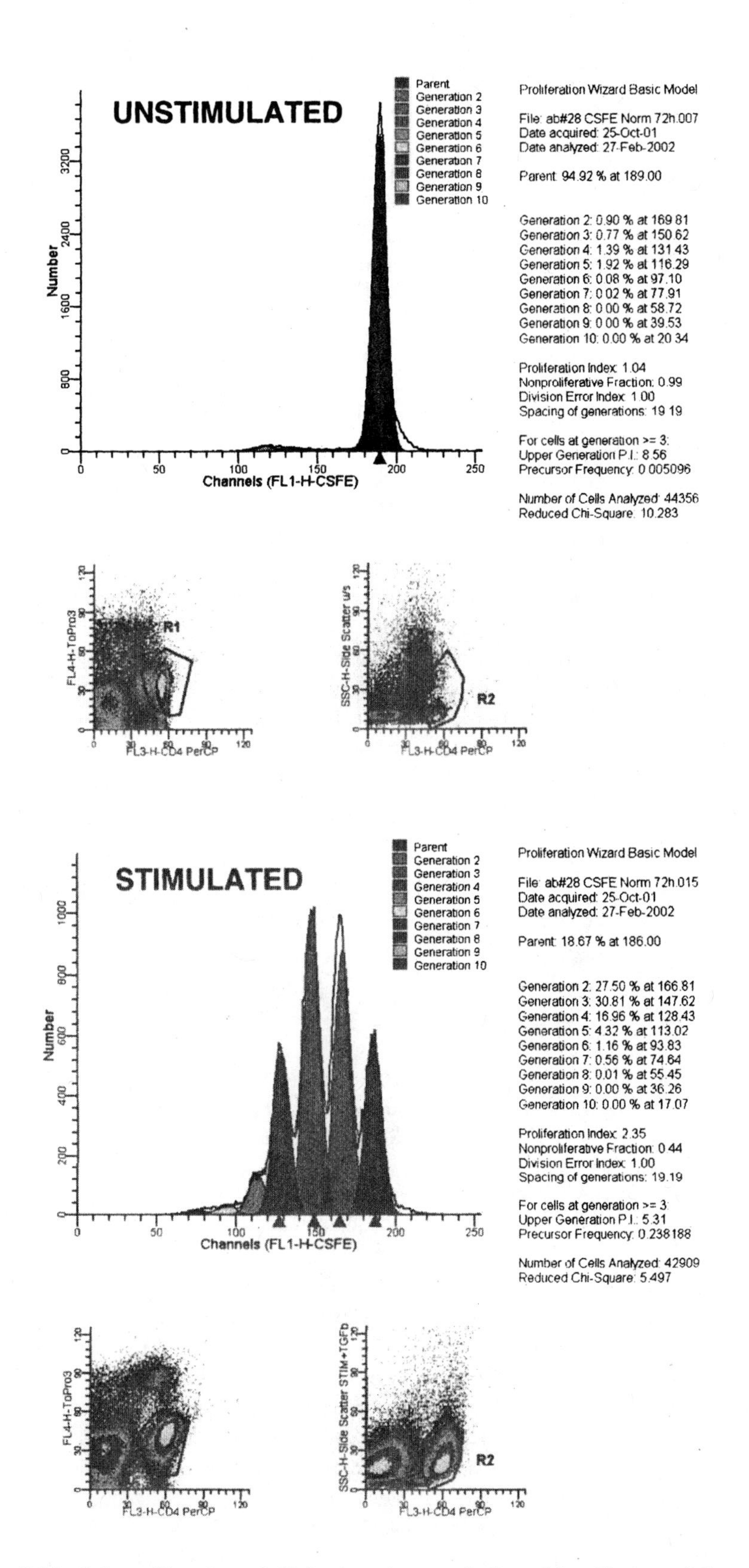

Figure 7-24. Cell proliferation of CD4+ lymphocytes indicated by dilution of fluorescence of carboxyfluorescein introduced into cells as the succinimidyl ester of carboxyfluorescein diacetate (CFSE). TO-PRO-3 was used to exclude dead cells. The data and figures were provided by Jonni Moore (U. of Pennsylvania) and Don Herbert (Verity Software House).

It would be desirable to have a tracking dye that did not emit fluorescence in the range of major popular antibody labels. There is nothing magic about the chromophores of either the PKH dyes or CFSE. The critical components of a PKH dye or of DiO, DiI, etc. are the long chain alkyl groups, which lock the compound into the lipid bilayer; the cyanine dye serves as a label. The chromophore in PKH26 is basically Cy3; Molecular Probes offers similar dyes with C_{18} alkyl groups and red-excited (Cy5-like) or near-infrared-excited (Cy7-like) chromophores.

The key characteristics of a CFSE-like dye are permeancy and a reactive group (amine-reactive, in the case of CFSE). Molecular probes sells a succinimidyl ester of SNARF-1 carboxylic acid, acetate, this compound, for which I'll accept any nickname with fewer than 10 letters, can be excited at 488 nm, but emits at longer wavelengths than does CFSE. Unfortunately, SNARF-1 itself was designed as a pH indicator; its emission characteristics change rather more dramatically with pH than do those of fluorescein, and any changes in intracellular pH with cell cycle stage or generation number might compromise the use of the SNARF-1 derivative for proliferation studies. A permeant reactive dye with spectral characteristics similar to Cy5.5 or rhodamine 800 (red-excited, emission >690 nm) would permit use of fluorescein, PE and almost all PE tandems for antibody labeling and gating, and could be very useful.

Membrane Organization and Fluidity/Viscosity

Lipid Packing Assessed with Merocyanine 540

Merocyanine 540 (MC540) is a green-excited, orange-fluorescent dye that, when it is applied to nerve axons, exhibits very small (0.3%) fluorescence changes in response to changes in membrane potential[457]. It can be excited at 488 nm, and its fluorescence is detectable through the 575 nm filters normally used for phycoerythrin measurement. MC540 was found by Valinsky, Easton, and Reich[467] to stain immature and leukemic but not normal leukocytes; the staining was independent of membrane potential. The dye was subsequently used to photosensitize leukemic[1556] and malaria-infected[1557] cells.

McEvoy et al[1558] noted that MC540 binds preferentially to membranes with loosely packed lipids, and provides a flow cytometric method for assessing lipid organization in individual cells. Analyses of cells stained simultaneously with MC540 and 1-[4-trimethylammo-niumphenyl]-6-phenyl-1,3,5-hexatriene, which gave a signal to normalize for surface area, showed that all leukocytes in peripheral blood bound equivalent amounts of dye per unit surface area, indicating that lipids of the plasma membranes of all cell types are organized similarly. Lymphocytes, monocytes, and neutrophils activated by appropriate stimuli all bound increased amounts of dye per unit surface area, indicating a change in lipid organization to a less-ordered state.

Bright staining with MC540 is often achieved by UV illumination of the cells during staining. Szabo et al[1559] showed that staining of mouse spleen cells was greatly enhanced by UV illumination before addition of the dye, and that UV treatment caused an increased permeability toward propidium iodide and, later, intracellular fluorescein as well as increased MC540 staining, although the increase in MC540 fluorescence preceded the other effects. They suggested that penetration of MC540 to the more fluid inner membrane structures explains the fluorescence increase.

Double-staining experiments by Belloc et al[1560], using MC540 and Hoechst 33342 on total bone marrow or peripheral blood cells, confirmed that the MC540-stained population included all the cycling cells, indicating that MC540 can be used as a marker for human hemopoietic cells. However, Pyatt[2736] et al found more total and committed progenitor cells in the CD34+MC540 dim population than in the CD34+MC540 bright population.

Around the time Annexin V[2353-4] was recognized as detecting membrane asymmetry in apoptotic cells, it was reported that MC540 was also effective for that purpose[2737-8,2970], staining apoptotic cells more brightly. The dye has been used only sporadically to detect apoptosis, but appears to be effective[2739-40]. Changes in MC540 fluorescence occur during platelet activation, and are apparently dependent on changes in accessibility of phosphatidylserine[2741-2].

I suspect that at least some oxonol dyes may also respond to membrane asymmetry; I will have more to say about this in the section on membrane potential dyes.

Membrane Fluidity and Microviscosity: Assessment Using Fluorescence Polarization

The concepts of **membrane fluidity** and **microviscosity** date from the early 1970's[397-8], by which time the Singer-Nicholson **fluid mosaic model of the cell membrane**[399] was generally appreciated, if not equally generally accepted. While it had become clear that many of the processes involved in the regulation of cell growth, differentiation, and differentiated function were mediated by receptor proteins extending through the cell membrane, it was also evident that the function of these proteins might be altered by local or global changes in the state of the **lipid bilayer** portion of the membrane.

Shinitzky and Inbar[397-8] investigated this possibility using measurements of the polarization of fluorescence of a hydrophobic probe bound to the membrane bilayer to derive an estimate of the physical state of membrane lipids. As was noted on pp. 114 and 283, when fluorescence is excited by polarized light, both the **polarization** and the **anisotropy** of fluorescence increase as the motion of chromophores becomes more restricted, i.e., as the "viscosity" of the medium increases, or as the "fluidity" decreases[400].

The probe Shinitzky and Inbar used for their measurements is **1,6-diphenyl-1,3,5-hexatriene (DPH)**. This is an uncharged hydrophobic material that is essentially nonfluorescent when coerced into aqueous media but which exhibits UV-excited blue fluorescence in nonpolar solvents. When a small amount of DPH in tetrahydrofuran (THF) or dimethyl sulfoxide (DMSO) is added to a cell suspension, examination of cells under the fluorescence microscope shows DPH in most membranous and lipid-containing structures, inside the cells as well as in the outer membrane. If you look at cells such as peripheral blood lymphocytes, in which there aren't a lot of intracellular membranous structures, you'll get the idea that DPH is only in the outer membrane, but it ain't so.

The original measurements of DPH fluorescence polarization were done on cuvettes full of cells in a modified spectrofluorometer; an instrument specially designed for such measurements was later manufactured by Elscint (Haifa, Israel). Differences in anisotropy were found between normal and leukemic cells[401-2], and between cells before and after exposure to mitogens[403] and viruses[404-5]. These findings stimulated Arndt-Jovin et al to implement anisotropy measurements on their cell sorter[400,406]; they found changes in anisotropy during DMSO-induced differentiation of Friend leukemia cells.

Several factors that affect membrane microviscosity have now been identified. The cholesterol content and the fractions of unsaturated and saturated fatty acids in membrane lipids exert strong influences; cholesterol depletion decreases viscosity while increasing cholesterol content has the opposite effect. Increasing the content of unsaturated fatty acids decreases microviscosity[398]. Changes in lipid composition could explain some of the slower changes in microviscosity observed in connection with interactions of various materials with the membrane; however, it is more difficult to explain rapid changes in microviscosity on this basis.

In experiments with phospholipid membrane vesicles, Lelkes[411] showed that measured microviscosity increases with the electrical potential difference across the cell membrane, whether the interior of the vesicles is made electrically negative with respect to the exterior (as is the case with cells) or vice versa. He suggested that this results from an electrostrictive effect, i.e., that the high electric field, proportional to the transmembrane potential difference, which exists only across the thickness of the membrane bilayer maintains orientation of the lipid molecules and that mobility is decreased as field strength increases.

Karnovsky et al[412] have attacked the concept of membrane microviscosity as being too simplistic; they have amassed abundant evidence that the mobility of lipids differs in different domains of the membranes of individual cells. One certainly cannot expect to derive detailed information from measurements of the average anisotropy of a few million cells in a cuvette. This doesn't mean that there may not be a place for flow cytometric measurements of DPH emission anisotropy in single cells, but nobody seems to have found a good use for them yet. Later studies, which have looked at fluorescence polarization using DPH[845] and other membrane probes[413,846] have not provided a major impetus for a resurrection in this area.

Details of flow cytometric measurement techniques have been given by Fox and Delohery[845], and, more recently, by Bock et al[1561-2] and Collins and Grogan[1563,1032]. Collins and Grogan[1563] compared bulk fluorometric and flow cytometric measurements and found them equivalent; they also[1564] examined a series of *n*-(9-anthroyloxy) fatty acid probes, where *n* = 2, 3, 6, 7, 9, 12, and 16, showing that the anthroyloxy moieties of the probes located at a graded series of depths in the outer leaflet of the plasma membrane of living HeLa cells. For different n, the efficiency of quenching with an aqueous phase quencher, Cu^{++}, decreased with increasing n. The probes should therefore be usable for measurements of dynamic parameters related to membrane fluidity at different depths in the plasma membrane. Vincent-Genod et al[2762] noted a decrease of 3-, 6-, and 9-(9-anthroyloxy) fluorescence anisotropy values in late apoptotic lymphocytes, compared to viable lymphocytes.

Seidl et al[2518] devised a method for estimating membrane microviscosity that does not require fluorescence polarization measurements. Pyrene decanoic acid (PDA), a UV-excited lipid probe, normally emits at 397 nm; collision between two molecules, one in the excited state and one in the ground state, forms an excimer that emits at 485 nm. Since the rate of excimer formation depends on the rate of diffusion of molecules in the membrane, a change in microviscosity results in a change of the excimer/monomer fluorescence emission ratio.

Lipid Peroxidation

Hedley and Chow[1565] described a method for measuring lipid peroxidation using time resolved flow cytometry. The naturally fluorescent fatty acid ***cis*-parinaric acid**, which is readily consumed in lipid peroxidation reactions, losing its fluorescence, was loaded into cells by exposure to a 5 µM concentration for 60 minutes at 37 °C. Fluorescence microscopy showed diffuse staining of surface and internal membranes. A 325 nm HeCd laser was used to excite parinaric acid fluorescence at 405 nm. Addition of the oxidant *t*-butyl hydroperoxide resulted in a burst of intracellular oxidation, demonstrated by simultaneously loading the cells with dichlorofluorescein, and in loss of parinaric acid fluorescence over time; this was followed by cell death, indicated by decreased forward light scatter and propidium uptake. Pretreatment of the cells with the antioxidant alpha-tocopherol (200 µM) reduced the rate of loss of parinaric acid fluorescence and delayed the onset of cell death. Simultaneous biochemical analysis of the lipid peroxidation breakdown product malondialdehyde revealed a close temporal correlation with loss of parinaric acid fluorescence, with and without alpha-tocopherol treatment, suggesting that the flow cytometric assay for lipid peroxidation is comparable to standard methods.

The problem with parinaric acid is that, while it is very well excited by a 325 nm He-Cd laser, it is very poorly excited at 350 nm and above, meaning that you will get almost no signal out of the probe even if you use a big argon or krypton laser for UV excitation. That didn't stop somebody I knew from publishing some essentially meaningless figures in which the distributions (mostly of noise) just happened to shift around in the right directions to support his hypothesis – and in a good single-name journal, too. As I have said before, bad flow happens to good journals.

If you only have a 488 nm laser, you can measure lipid peroxidation using a fluorescence emission ratio from hexadecanoyl-BODIPY-FL[2743], or by measuring decreasing fluorescence from 5-dodecanoylfluorescein[2744-5] or fluoresceinated phosphatidylethanolamine[2746-7].

Membrane Permeability to Dyes and Drugs: The Drug Efflux Pump(s) Revisited

Differences in the temporal pattern of uptake and retention of **Hoechst 33342** were used in the late 1970's for studies of lymphocyte activation[248-50,253,257]; this led to the demonstration in 1981 by Lalande, Ling, and Miller that dye retention was determined by the operation of what is now recognized as the **P-glycoprotein (Pgp) drug efflux pump.** This subject has already been discussed briefly on p. 309; we will return to it presently.

Flow cytometry of the kinetics of intracellular accumulation of **fluorescent drugs,** e.g., **anthracyclines** such as **doxorubicin** (daunomycin) and **adriamycin**[352-3,814], and **drug analogues,** such as **fluoresceinated derivatives of methotrexate**[354-6], can obviously be useful in pharmacology, e.g., for determination of mechanisms of drug resistance. The methotrexate analogues, in particular, have been valuable in studies of gene amplification[357]. Krishan, who has done extensive studies of anthracycline uptake and retention patterns in tumor cells and of their perturbation by phenothiazines and calcium channel blockers[352,815-7], has reviewed the general area of flow cytometric analysis of drugs and drug analogues on several occasions over the years[818,2748-50].

The cationic dye **rhodamine 123** had been shown by Lan Bo Chen and his colleagues in 1980 to stain mitochondria[489]; it was established in 1981 that the retention of the dye in mitochondria after washing was dependent on the existence of an interior-negative mitochondrial **membrane potential**[490]. By this time, it had already been noted that lymphocyte stimulation was accompanied by increased rhodamine 123 staining[493]. From 1981 to 1985, rhodamine 123 retention was thought to depend primarily on membrane potential. In 1985, Sonka et al[1566] found that in cells of an anthracycline-resistant tumor line, which was known to have an active efflux pump, rhodamine 123 retention was low. Substances such as verapamil, which were known to increase the sensitivity of anthracycline-resistant cells, inhibiting the pump, also shifted the cells from low to high rhodamine 123 retention. Tapiero et al[1567] reported similar findings in 1986.

At about the same time, Bertoncello, Hodgson, and Bradley[1568] and, slightly later, Mulder and Visser[1569] observed that bone marrow stem cells exhibited low rhodamine 123 retention. This characteristic remains useful for isolation of cells with marrow repopulating activity; it is now known to be due to the operation of an efflux pump, although it was originally attributed to decreased mitochondrial respiration.

Since the mid-1980's, various authors have described methods for demonstrating **multidrug resistance,** i.e., the presence and activity of the drug efflux pump, in tumor cells, based on uptake and or retention of Hoechst 33342[1570], anthracyclines[1571-3, 1576], rhodamine 123[1575, 1577], cyanine dyes[1577], and antibodies against P-glycoprotein[1573-4, 1576]. The pump has also been demonstrated in human peripheral lymphoid cells (B and NK, but not most T cells)[1578], trypanosomes[1579-80], and some bacteria[1581-2]. Kessel et al[1577], in an article well worth reading, demonstrated efflux from drug resistant cells of Hoechst 33342, rhodamine 123, and several cyanine dyes commonly used as membrane potential probes, and cautioned against drawing the conclusion that changes in rhodamine and cyanine dye fluorescence in cells are due to membrane potential changes. Oxonol dye probes of membrane potential, which are anionic and which apparently do not enter cells in large amounts, are largely unaffected by the operation of the pump. I will return to this issue in the discussion of membrane potential probes.

Assaraf and others[1583-5] further characterized methotrexate (MTX) resistance mechanisms, demonstrating that fluorescein-MTX is not transported by the carrier responsible for MTX uptake, but enters cells by passive diffusion. The labeled compound, however, remains useful for demonstration of increased intracellular dihydrofolate reductase, which causes one type of resistance to MTX. A second type appears to be due to decreased transport of MTX analogues into cells.

Fluorescent probes for specific transporters have been developed; Wiley et al[1586] described a fluorescein-labeled probe for the nucleoside transporter, while Knaus et al[1587-8] made BODIPY-labeled probes which bind to calcium channels. Using a somewhat less specific probe, the blue-violet-excited, green fluorescent anionic dye **lucifer yellow,** loaded into cells by electroporation, Dinchuk et al[1589] showed that an anion transport mechanism was responsible for efflux of the dye from lymphocytes. Aller et al[2751] and Natarajan and Srienc[2752-3], respectively, used **NBD-glucose** to study glucose uptake kinetics in rat brain cells and *E. coli.*

Dordal et al[2754] used a three-compartment mathematical model and kinetic measurements to assess the pharmacokinetics of doxorubicin.

We now know that there are multiple pumps that can mediate efflux of drugs and dyes from cells[2748-50]. In addition to P-glycoprotein, there are (at least) **multidrug resistance protein (MRP)**[2755-7], **breast cancer resistance protein (BCRP)**[2758], and **lung resistance protein (LRP)**[2748], and some tumors and cell lines contain more than one active pump. Whole-cell studies of one pump in such cells therefore require either the use of a substrate that is processed by only that pump or the use of a selective inhibitor(s) for the pump(s) not under study.

P-glycoprotein transports Hoechst 33342, rhodamine 123, and doxorubicin, among other compounds, and many acetoxymethyl (AM) esters, including indo-1 AM and fluo-3 AM, both used for cytoplasmic calcium measurements, and

calcein AM, used in tests of membrane integrity (pp. 370-1). If you are using these probes, it is advisable to determine whether or not the cells you are studying contain pumps that might interfere with your experiment. P-glycoprotein is inhibited by **cyclosporin A**, which competes with substrate molecules for binding sites, by the calcium channel blocker **verapamil**, which does not, by phenothiazines such as **trifluoperazine** and **prochlorperazine**, and, for those with a taste for political controversy, by **RU 486**.

MRP is a particularly effective transporter of glutathione S-conjugates of drugs; it also effluxes both **calcein AM** and free **calcein**, and carboxy-2',7'-dichlorofluorescein[2756], but not BCECF-AM[2757]. MRP is inhibited by **indomethacin**, **MK571**, and **probenecid**. **Mitoxantrone** was found by Minderman et al[2758] to be the only material that served as a substrate for BCRP in all of the cell lines they tested; fumitremorgin C inhibited mitoxantrone efflux by BCRP.

It is advisable to use antibodies[2755,2759], particularly antibodies that do not interfere with pump activity, as well as substrates to detect and quantitate protein pumps. Chen and Simon[2760] transfected cells with a Pgp-GFP fusion protein, allowing simultaneous quantitation of cells' content of protein and substrate, and showed that the degree of efflux depended only on the quantity of Pgp present in cells; Wang et al[2761] have standardized a flow cytometric assay for identification and evaluation of Pgp inhibitors.

There is a great deal more biochemical diversity among the prokaryotes than among the mammals, and we know that bacteria have a broader range of pumps than we do, and use some them to resist the actions of antibiotics. This is probably a good place to point out that all membranes are not created equal; the outer membrane of **Gram-negative bacteria** is, under normal conditions, impermeable to many lipophilic materials, such as cyanine dyes, which readily and rapidly enter almost all other types of cells, although it can be permeabilized, e.g., with EDTA. I have used this property as the basis of a "flow cytometric Gram stain" which will be discussed in Chapter 10; Molecular Probes advertises sets of nucleic acid dyes for the same purpose.

Endocytosis of Macromolecules and Particles

Fluorescently tagged macromolecules[368-70], plastic microparticles[371-3], and bacteria[374-5] have been used to study **endocytosis** and **phagocytosis**. Murphy et al[369] and Bassøe et al[375] have examined pH changes in the environment of endocytosed material, utilizing the pH-dependence of fluorescein fluorescence[202-4]. A refinement in this technique was introduced by Murphy et al[376]; cells are incubated with a mixture of fluorescein- and rhodamine B-labeled ligands, and excited at 488 nm, and the ratio of fluorescein and rhodamine fluorescence is calculated.

The emission intensity of fluorescein, when the dye is excited at 488 nm, decreases with decreasing pH over the range 8-4. The spectrum of rhodamine B does not change appreciably with pH over this range. The ratio of fluorescence signals from the two labels can thus be used to provide a measure of the ambient pH that does not vary with variations in the total amount of material in different cells. Additional material relating to this technique will appear in the later section on pH measurements.

As one would expect, work on flow cytometric measurement of endocytosis has concentrated heavily on analysis of blood mononuclear and polymorphonuclear phagocytic cells. Terstappen et al[825] developed a flow cytometric assay for circulating immune complexes in serum based upon phagocytosis of the complexes by granulocytes from healthy donors and subsequent immunofluorescent detection of intracellular complexes. Davis et al[826] have used fluorescein-labeled dextran to study fluid pinocytosis induced in polymorphonuclear cells by chemotactic peptides.

When microparticles, rather than macromolecules, are used to demonstrate endocytosis, it is possible to precisely specify both the size and spectral characteristics of the indicator used, and, if one cares, to obtain histograms of the numbers of particles taken up by cells, provided the particles are relatively bright and uniform. The particles themselves may also be coated with specific ligands. To permit simultaneous immunofluorescence analysis using fluorescein antibodies and measurement of phagocytosis, Rolland et al[827] employed 0.3 μm plastic particles labeled with ethidium as targets; the red fluorescence of the particles could readily be discriminated from fluorescein fluorescence.

Newer phagocytosis assays have added extra tricks. Using two-color analysis, orange- or red-labeled bacteria, and **2,7-dichlorofluorescin diacetate (DCFH-DA)**, an indicator of oxidative metabolism which will be further discussed shortly, it is possible to determine both phagocytosis and the subsequent occurrence of the respiratory burst[1590-1]; approximately the same trick can be done using green-labeled bacteria and **hydroethidine**, a derivative of ethidium that is oxidized to the red fluorescent form during the respiratory burst[1592]. Phagocytosis can also be assayed, with somewhat less certainty, with DCFH-DA and unlabeled bacteria, using the occurrence of the burst as an indicator of phagocytosis[1593]. Phagocytosis of immunoglobulin-coated sheep erythrocytes and respiratory burst can be assayed using the fluorescent product of DCFH-DA for the latter and changes in scatter signals for the former[1594].

Differentiating between green-labeled phagocytic targets attached to the cell surface or in solution and those that have actually been ingested can be accomplished by adding trypan blue[1595] or crystal violet[1596] to the sample. A somewhat more informative technique allows neutrophils to phagocytose fluorescein-labeled, heat-killed yeast cells, after which ethidium bromide is added to the solution. Ingested yeast cells are green; those attached to the cell surface are red[1597]. Similar distinctions can be made by allowing cells to ingest opsonized green fluorescent beads and then adding red fluorescent antibodies to stain the externally bound, but not the ingested beads[1598]; this requires lots of fluorescence compensation, and would probably work better with red beads and green antibodies. If all you want to measure is the ingested labeled bacteria, a commercial lysing reagent can be used to get rid of the rest[1599].

Ma et al[1600] used crystal violet to quench fluorescence from antibody in solution, which allowed them to demonstrate internalization of fluorescein-conjugated IgG from normal serum or serum containing anti-ribonucleoprotein by normal lymphocytes. The results showed that 54% of normal lymphocytes were penetrated by anti-RNP antibody and 23% by normal IgG, respectively. Suzuki et al[1601] employed DiI as a label to study endocytosis of lipid microspheres. Haynes et al[1602] studied intracellular digestion of endocytosed albumin by labeling the protein so heavily with rhodamine that the fluorescence of the dye was largely quenched; as protein was hydrolyzed within phagocytic vacuoles, small rhodamine-labeled fragments were released into solution, and became more fluorescent.

Wang Yang et al[1603] used pulse width and area measurements to analyze capping and endocytosis of fluorescein-immunoglobulin by mouse B cells; fluorescence decreased when the material entered acidic subcellular compartments. Chanh and Alderete[1604], however, observed decreases in fluorescence intensity associated with capping itself.

Bassøe et al[2763] have recently described the use of multiple probes, flow cytometry, and confocal microscopy to dissect the phagocytic process in neutrophils; they were able to demonstrate formation of reactive oxygen species (ROS) (pp. 379-80) in phagosomes.

Moving away from the mammals, flow cytometry and mathematical modeling have been applied to analysis of feeding by the ciliated protozoan *Tetrahymena pyriformis*[1605-7]. Flow cytometric analysis of the distribution of various types of microplankton before and after exposure to larger organisms has also been used to determine the dietary habits of the latter[1608]. Nature is red in cilium and pseudopod...and if it isn't, we can stain it so it is.

Enzyme Activity

Enzyme activity in single cells can be demonstrated and quantified by flow cytometry following incubation of the cells with **chromogenic** or **fluorogenic substrates**[358], which, respectively, yield colored and fluorescent products. Some products normally detected by their absorption are fluorescent, so the distinction between chromogenic and fluorogenic substrates can be blurry. The colored reaction products of chromogenic substrates are detected by scatter and absorption or extinction measurements; forward and side scatter as measured in conventional fluorescence flow cytometers are generally usable for this type of analysis. We have already encountered fluorogenic substrates in the contexts of membrane integrity determination (fluorescein diacetate, etc.) and signal amplification (p. 344).

Bayer's instruments for differential leukocyte counting (developed by Technicon, which was acquired by Bayer) measure **peroxidase** activity by scatter and absorption to discriminate among lymphocytes, monocytes, neutrophils, and eosinophils[84-5]. These systems have also been adapted to study staining kinetics[348] and, using immunoenzyme staining methods, to analyze T cell subsets in peripheral blood[779,1384]. Ross et al[1609] described a pararosaniline method for esterase staining and a naphthyl phosphate method for alkaline phosphatase staining, allowing the Technicon H-1 to be used for those relatively common cytochemical analyses. Kaplow, who collaborated in this work, had earlier used chromogenic substrates with the Bio/Physics Cytograf, which measured scattering and extinction of 633 nm laser light, to demonstrate esterase, peroxidase, and phosphatase activities in blood cells[359-60].

Dolbeare and Smith[358] discussed a number of **fluorogenic substrates** that can be used to demonstrate the activity of enzymes in, and in some cases on, cells; many such materials are available commercially. The majority of fluorogenic substrates are derivatives of **fluorescein**, of **coumarins** such as **4-methylumbelliferone (4-MU)**, or of **α-naphthol**. Substrates derived from fluorescein, as would be expected, yield blue-excited, green fluorescent products, while coumarin and α-naphthol-based substrates, the latter more often employed as chromogens rather than fluorogens, form products that require UV or violet excitation and emit in the blue and green regions of the spectrum. Derivatives of **resorufin**, which form green- to yellow-excited, orange-red fluorescent products, can be used to demonstrate esterases and oxidative enzymes. The excitation maximum of resorufin, however, is at about 570 nm, necessitating the use of a krypton laser at 568 nm, a dye laser, or, possibly, the 577 nm line of a mercury arc lamp for excitation.

Dolbeare and Vanderlaan[363] and Smith and Dean[364] have described fluorogenic reagents based upon other UV-excited, blue fluorescent and blue- excited, green fluorescent materials for flow cytometric demonstration of peptidases and acid phosphatases. Fluorogenic substrates suitable for use with red excitation are uncommon; Lee, Berry, and Chen[1610] described one candidate, **vita blue**, which could be derivatized for use as both an esterase substrate and a pH indicator, but the dye hasn't made it to the majors yet, as far as I know.

If cells are incubated with two (or more) fluorogenic substrates that yield products with different spectral characteristics, it is possible to demonstrate and monitor several enzymes in cells using multiparameter flow cytometric analysis[364-7]. Since the amount of detectable reaction product in a cell at any given time depends on many factors, including the rate of accumulation of substrate in the cell, the rate of entry of substrate to the cellular compartment containing enzyme, the rate of enzymatic reaction, and the rate of efflux of reaction product from the cell, it is not surprising that cellular fluorescence values, especially for different cell types, do not always correlate with enzyme activities[824].

Dick Haugland reviewed fluorogenic substrates and their uses in an 8-page article in 1995[2764]; since his current (2002) edition of the Molecular Probes *Handbook*[2332] devotes 56 pages to the subject and probably has at least 8 pages of references, you probably don't need to look for the article. Both article and handbook emphasize the desirability of trapping the fluorescent product in or on cells for flow cytometry and other assays done on cell suspensions.

Indicators of Oxidative Metabolism I: Tetrazolium Dye Reduction

Blairet al[820] used forward and side scatter measurements to demonstrate the development of oxidative enzyme activity in HL-60 human promyelocytic leukemia cells stimulated to granulocytic differentiation by dimethyl sulfoxide. The reagent employed was **nitroblue tetrazolium (NBT)**[361], widely used in histo- and cytochemistry for demonstration of **oxidative enzymes**. NBT, which is colorless, is reduced to a diformazan that forms an amorphous blue-black precipitate. **NBT reduction** is used as an indicator of the function of mature granulocytes and to determine whether morphologically immature cells belong to the myeloid series.

Blair et al found that the forward scatter signal was decreased, and the orthogonal light scatter signal was increased, in cells that reduced NBT. This work was noteworthy because it called attention to a general method for demonstrating color reactions in cells (or at least those which yield products with very high extinction) that requires neither absorption nor extinction measurements, and thus could be used with most flow cytometers.

Huet et al[1611] used side scatter signals for flow cytometric analysis of **3-(4,5-dimethylthiazolyl-2-yl)-2,5-diphenyltetrazolium bromide (MTT)** reduction; the color reaction produced when MTT is reduced is commonly used as a viability indicator in cell cultures. Fattorossi et al[1612] devised an alternate method for measurement of tetrazolium dye reduction by neutrophils; the cells were labeled with fluorescein-con A, the fluorescence of which was quenched by the formazan product of NBT reduction. Van Noorden, Dolbeare, and Aten[1613] proposed a more general method for detection of enzymatic reactions yielding colored formazan products; the formazan quenches glutaraldehyde-induced "autofluorescence" in cells.

Fluorescent tetrazolium dyes were investigated by Severin and Stellmach[821-2], who measured oxidative activity of living cells using **cyanoditolyl tetrazolium chloride (CTC)**, which is reduced to a water-insoluble fluorescent formazan that has an excitation maximum at 450 nm and emits in the range from 580 to 660 nm. CTC generates more product than NBT[1614]; Huang and Severin[1615] used it with lactate, fumarate, and other metabolic intermediates to characterize the activities of six different dehydrogenases in Ehrlich ascites cells, and it has become fairly popular for studies of bacteria[2252].

Indicators of Oxidative Metabolism II: 2,7-Dichlorofluorescin Diacetate (DCFH-DA), etc.

Bass et al[362] used another fluorogenic material to demonstrate oxidative metabolism of blood granulocytes. **2,7-dichlorodihydrofluorescein diacetate**, also known as **dichlorofluoresc*in* diacetate (DCFH-DA or H_2DCFDA)**, like FDA, is uncharged, nonfluorescent, and lipid soluble, and readily penetrates cells. Nonspecific intracellular esterases transform the ester into a nonfluorescent intermediate, DCFH. In the presence of **peroxidase**, and of H_2O_2 formed during the **respiratory burst** in activated granulocytes, DCFH is converted to the fluorescent **2,7-dichlorofluorescein (DCF)**. Duque and Ward[821] described the use of DCFH-DA and other dyes for quantitative assessment of neutrophil function. Cells are loaded with 5 μM DCFH-DA for 15 minutes at 37 °C, after which gated fluorescence analysis of the neutrophil population, identified by forward and side scattering characteristics, is done over time following administration of a stimulus such as a phorbol ester. Like fluorescein, DCF does leak out of cells; Molecular Probes offers **carboxy-H_2DCFDA**[2765], which responds to the respiratory burst by forming carboxydichlorofluorescein (carboxy-DCF), which is better retained by cells than is DCF[2332].

DCFH-DA has been used in combination with various particles for combined analyses of phagocytosis and respiratory burst in neutrophils[1590-1,1593-4]. The presence of peroxidase in neutrophils is evidently important for the development of the fluorescent product; monocytes, which have much less peroxidase, don't become as brightly fluorescent as neutrophils, even when H_2O_2 is added[1616].

Maresh and Monnat[1618], studying cells derived from myeloid leukemias, note the existence of a DCFH-DA oxidative pathway stimulated by fluoride, which appears distinct from that involved in the respiratory burst, and wonder whether a similar pathway in normal cells might confound some results obtained with this reagent.

Indicators of Oxidative Metabolism III: Hydroethidine (Dihydroethidium)

Dihydroethidium is probably more widely known under the name **hydroethidine (HE)**, which is a trademark of Prescott Laboratories. It is the product of reduction of ethidium by sodium borohydride; described in 1984 by Gallop et al[973]. The UV-excited, blue fluorescent HE, unlike ethidium, readily enters live cells, in which it may be oxidized to ethidium, which then exhibits its characteristic blue- or green-excited red fluorescence, enhanced by binding to double-stranded nucleic acids. HE has been used as an indicator of viability[1619]; when it is added at low concentrations, little ethidium accumulates in the nuclei of dead cells, while those of live cells become red fluorescent. HE has also been used as an indicator of the neutrophil respiratory burst[1592]; while DCFH-DA responds primarily to **hydrogen peroxide** generation, HE is more affected by **superoxide**, and the two dyes, used separately, allow discrimination of these two types of **reactive oxygen species (ROS)**[1620-1].

Indicators of Oxidative Metabolism IV: Dihydrorhodamine 123

In 1988, Rothe, Oser, and Valet[1622] introduced **dihydrorhodamine 123**, which is oxidized to the green fluorescent cationic dye **rhodamine 123**, to detect the respiratory burst in neutrophils. According to Henderson and Chappell[1623], rhodamine 123 fluorescence is produced only in cells containing peroxidase, meaning that the dye is relatively insensitive to superoxide production. Dihydrorhodamine is said to provide brighter signals than DCFH-DA.

Indicators of Oxidative Metabolism V: Detection of Hypoxic Cells

Some of us, e.g., radiobiologists and radiotherapists, are interested in finding cells that aren't getting and/or using their share of oxygen. Hypoxic cells in transplanted tumors and tumor cell spheroids had been distinguished by their pattern of uptake of Hoechst 33342[1624-6], but attempts were later made to develop more specific markers. Hodgkiss et al[1627-8] prepared 2-nitroimidazole derivatives with coumarin and indolizine chromophores that appeared useful for hypoxic cell detection. A more recent report[2766] utilized 2-nitroimidazole bound to theophylline (NITP). Under hypoxic conditions, this molecule is bioreductively bound to cellular macromolecules; it is then detected using anti-theophylline antibodies.

Detection of Caspase Activity

Caspases (cysteine-aspartic acid specific **proteases**) are activated by cell death inducing stimuli, and several reagents have been described for detection of caspase activity in apoptotic cells. **PhiPhiLux®** fluorogenic caspase substrates[2556-7] with either green or yellow emission, excitable at 488 nm, are available from **OncoImmunin**. Molecular Probes offers UV-excited, blue fluorescent (aminomethylcoumarin-based) and blue-green excited, green fluorescent (rhodamine 110-based) substrates for caspase-8, which is activated early in apoptosis, and caspase-3, which is activated later[2332]. Caspases can also be detected using fluorescent inhibitors. **Benzyloxycarbonyl-valinealanine-aspartic acid-fluoromethyl ketone (zVAD-FMK)** is a potent broad-spectrum inhibitor of caspases that binds irreversibly (covalently) to the enzyme active site and can block apoptosis. **Serologicals Corporation** offers a line of caspase inhibitors (originally Intergen's) usable for cytometry[2769]. **FAM-VAD-FMK** is a carboxyfluorescein (FAM) derivative of zVAD-FMK, and blocks apoptosis[2770]; there are also FAM-labeled inhibitors selective to various degrees for individual caspases, including caspase-3 and caspase-8, and yellow fluorescent sulforhodamine B-labeled (SR) derivatives of zVAD-FMK and of a caspase-3 inhibitor. **Immunochemistry Technologies** is an alternate source. Since pockets of caspase activity may be localized within cells, it may be advantageous to use scanning rather than flow cytometry for studies of caspases in apoptosis[2770-2].

Other Enzymes

Cathepsin L activity[1633-4] can be measured flow cytometrically in single viable cells by the intracellular cleavage of non-fluorescent (Z-Phe-Arg)2-rhodamine 110 to the green fluorescent monoamide Z-Phe-Arg-rhodamine 110 and rhodamine 110. (Z-Phe-Arg)2-R110 and (Z-Arg-Arg)2-R110 are several hundredfold more selective for cathepsin L than for cathepsin B, providing sensitivity for the former. Rhodamine-110-based substrates for nonspecific **aminopeptidases** have also been developed[2773].

For assessment of activities of **lysosomal enzymes**, van Noorden[1635] reports that the best results have been obtained with methods based on naphthol AS-TR derivatives and with methods for the demonstration of protease activity using methoxynaphthylamine derivatives as substrates and 5'-nitrosalicylaldehyde as coupling reagent.

Dive, Workman, and Watson[1636] used flow cytometry to measure the activity of **gamma-glutamyl transpeptidase**, by monitoring the conversion of gamma-glutamyl aminomethylcoumarin to aminomethylcoumarin. This was technically difficult because of the location of the enzyme on the cell exterior, resulting in rapid escape of the product.

Huang et al[1639-40] have developed fluorogenic substrates for **alkaline phosphatase**. Molecular Probes' **ELF-97 phosphate**[2774-5] produces a violet-excited, green-yellow fluorescent insoluble precipitate when cleaved by either alkaline or acid phosphatases; it can be used to demonstrate phosphatases in cells or for enzyme amplification using alkaline phosphatase-tagged antibodies or nucleotide probes (p. 344). This should be a good reagent to use with violet diode lasers.

Meshulam et al[1369] measured **phospholipase A** activation using bis-BODIPY-phosphatidylcholine, which localizes in the inner leaflet of the cell membrane. Sidhu et al[1641] used ethoxyresorufin and scanning laser cytometry to measure the activities of **cytochrome P-4501A1** and **NADPH DT-diaphorase**.

Detection of Enzymes and Products by Antibodies

Like other proteins, enzymes can be detected in or on cells using antibodies, although antibody reactivity doesn't absolutely indicate enzyme activity.

Tran-Paterson et al[1629] compared levels of the neutral endopeptidase-24.11, also known as the CD10 or CALLA antigen (Common Acute Lymphocytic Leukemia Antigen) on normal granulocytes, leukemic cells, and transfected COS-1 cells, as detected using fluorescent anti-CD10 antibody and by analysis of enzyme activity in cell suspensions; they found good correlation between results obtained by the disparate methods. Milhiet et al[1630] used a fluorescent inhibitor, *N*-[fluoresceinyl]-*N'*-[1-(6-(3-mercapto-2-benzyl-1-oxopropyl) amino-1-hexyl]thiocarbamide (FTI), for flow cytometric detection of CD10 as an enzyme.

Antibody detection of the products of enzymatic reactions is a more direct indicator of enzyme activity than is antibody detection of the enzyme protein itself. Antibodies to **phosphotyrosine**[1631-2] have been used for flow cytometric quantification of **tyrosine kinase** activity.

Enzyme Kinetics in Single Cells

By making appropriate timed multiparameter flow cytometric measurements, it is possible, as Dive, Workman, and Watson[813,1637-8] have shown, to characterize intracellular enzyme reactions in considerable detail, determining relevant parameters of reaction and inhibition kinetics. As is always the case when kinetic measurements are made using flow cytometry, it is necessary to validate the assumption that the time courses of reactions are similar in at least a substantial fraction of the cells in the sample. Scanning laser cytometry can be used to study enzyme kinetics in individual cells over time when this is necessary or desirable[2776-8].

Sulfhydryl (Thiol) Groups; Glutathione

Sulfhydryl or **thiol** groups in protein, and non-protein thiol compounds such as the tripeptide glutamyl-cysteinyl-glycine, more commonly known as **glutathione**, play an important role in cells' oxidative metabolism. When cells become hypoxic, the sulfhydryl/disulfide ratio increases, and the free radical scavenging properties of thiols are believed to play a pivotal role in increasing the resistance of hypoxic tumor cells to ionizing radiation[332] and to some chemotherapeutic agents. Flow cytometry of glutathione levels, in particular, has acquired new importance because of their relation to drug resistance in cancer cells[801,1116] and the progression of HIV infection[1117-8].

Olive and Durand[333-4] examined several maleimide and **bromobimane** derivatives as reagents for determination of cellular thiols. A maleimide derivative of coumarin emits blue fluorescence and can be used with UV excitation; **monobromobimane (MBB)** and didansylcysteine have similar spectral characteristics. **Fluorescein-5-maleimide** penetrates viable cells less well, but excites at 488 nm.

Treumer and Valet[800] and Rice et al[801], respectively, characterized ***o*-phthaldialdehyde (OPT)** and **monochlorobimane (MCB)**, both of which form UV-excited, blue fluorescent products with reduced glutathione (GSH), as rapid and specific vital stains for this substance. These reagents were used to demonstrate heterogeneity in GSH content in tumor cell populations. O'Connor et al[1649] advocated the use of **mercury orange**, and Poot et al[1650] the use of **chloromethylfluorescein**, both of which can be excited at 488 nm with, respectively, orange and green emission, for glutathione measurement.

While MCB is probably the most specific reagent for glutathione, because its reaction with the peptide is catalyzed by glutathione S-transferase, the form of this enzyme predominant in human cells (GST-pi) is much less effective in catalysis than is the form found in rodents[1651-2]. In 1994, Hedley and Chow[1653] evaluated a wide variety of methods for measurement of glutathione by flow cytometry, concluding that MBB was the reagent of choice for work with human cells. They found that MBB could be adequately excited by the 325 nm line of a He-Cd laser; its absorption maximum is 394 nm, making a violet diode laser a near ideal source for exciting the probe. MBB emission can be measured adequately at 450 nm; the emission maximum is at 490 nm, but one would normally want to cut off detection well below this wavelength to avoid interference from 488 nm laser light. Mercury orange was found usable, although results were less consistent and background fluorescence was higher. Chow and Hedley[2779] have measured GSH in clinical samples, attempting to correlate GSH concentration and drug resistance. MBB staining of intracellular glutathione is illustrated in Figure 7-25.

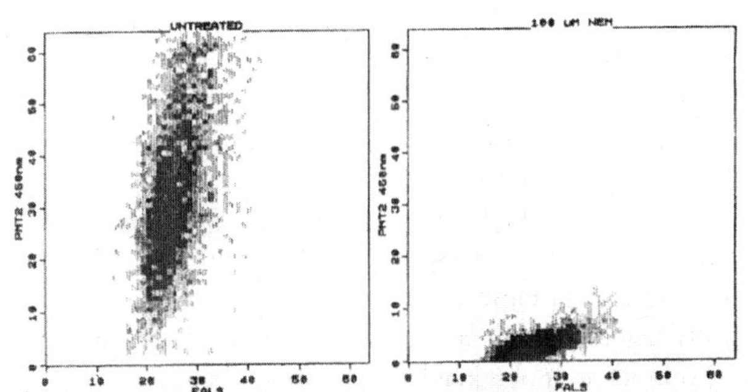

Figure 7-25. Specific staining of glutathione in cells by monobromobimane (MBB). The x-axis shows forward scatter, the y-axis MBB fluorescence, which is greatly reduced in the sample shown in the right panel, which was treated with N-ethylmaleimide (NEM) to deplete intracellular glutathione. The figure was provided by David Hedley (Princess Margaret Hospital, Toronto).

7.13 FUNCTIONAL PROBES II: INDICATORS OF CELL ACTIVATION

Introduction

The functional parameters I will emphasize in this section, i.e., **"structuredness of cytoplasmic matrix" (SCM)**, **cytoplasmic and mitochondrial membrane potential**, **"membrane-bound"** and **cytoplasmic calcium ion concentration**, and **intracellular pH**, and the effects of cell surface ligand-receptor interactions on these parameters, have been the subject of a good deal of research and a large number of publications.

Much of the earlier work, and of the earlier literature, did not deal with flow cytometry or with alternative methods of analysis of functional parameters in single cells. Most investigators now working in this area are well aware of the capabilities of flow cytometry, and are in possession of at least some of the information that has already been gained from flow cytometric studies of functional parameters and of the probes used for their measurement.

Cytoplasmic calcium ion concentration ($[Ca^{++}]$) and distribution, intracellular pH, and membrane potential ($\Delta\Psi$) have all been observed to change during the early stages of **surface receptor-mediated activation processes** related to the development, differentiated function, and pathology of a large number of cell types. Accordingly, $[Ca^{++}]$, pH, and $\Delta\Psi$ have, individually and collectively, been envisioned as **"second messengers"** mediating responses to cell surface ligand-receptor interactions.

The investigation of changes in the cellular ionic environment during activation has been actively pursued in hopes of clarifying the mechanisms of receptor function and of developing more precise tests of cell function for clinical and research use. Studies in this area have been facilitated by the development of fluorescent probes that allow estimation of [Ca^{++}], pH, and ΔΨ in cell populations and, in conjunction with instrumental methods such as flow cytometry, in individual cells.

While the current state of the art appears to allow fairly precise quantitation of [Ca^{++}], pH, and ΔΨ in terms of population averages, the use of flow cytometry to measure the same probes in single cells can reveal and has revealed marked heterogeneity even within supposedly homogeneous cell populations. Flow cytometric analysis of the effects of ligands, agonists, and antagonists on individual cells may yield otherwise unobtainable information about mechanisms of receptor function. It can also provide considerable insight into the sources of inaccuracy in measurements based on existing fluorescent probe technology.

Changes in the Cellular Ionic Environment Following Activation by Ligand Interaction With Cell Surface Receptors

In 1977, when I got interested in this subject, I read a then current review by Sonenberg and Schneider[421] dealing with physical and chemical changes in receptors, in the membrane, and in other cellular structures following ligand interactions with cell surface receptors. Few of the details had been worked out back then (for example, we didn't know JAK about kinases, and NO way did we suspect that a gas could act as a messenger), but the review and a few other articles written in that era did have the big picture:

The earliest detectable biochemical changes occur in milliseconds to seconds after the number of occupied receptor sites increases following exposure of cells to ligand. They include alterations in transmembrane ion fluxes, in intracellular [Ca^{++}], pH, and ΔΨ, and in activities of membrane-associated enzymes such as adenylate cyclase, protein kinases, and phospholipid methyltransferases[422].

Within seconds to minutes, there may be changes in cyclic nucleotide concentrations, increased protein phosphorylation, and alterations of membrane uptake of sugars, amino acids, and fatty acids. Ca^{++} redistributions within the cell and changes in the mobility of membrane lipids and proteins also may occur in this time frame. The biologic response to surface stimulation may occur in milliseconds, as happens in nerve impulse transmission, or after days, as in mitogenic activation of lymphocytes.

The observation that changes in [Ca^{++}], pH, and/or ΔΨ occurred early in the course of interaction of many different ligands with cell surface receptors in diverse cell types led to investigation of the role of ionic species in signal transduction between the cell surface and the nucleus and cytoplasmic organelles such as mitochondria. Ion flux changes were observed, and assigned a role in signal transduction, in the activation of lymphocytes[423-7], platelets[204,428-31], mast cells[432-5], and neutrophils[436-8], and in growth factor action[439-43], as well as in many other cell activation processes.

A single pattern of ionic events has been observed in most of the cell activation processes that have been examined to date. Following ligand binding to cell surface receptors, there is a rapid influx of Ca^{++}, usually accompanied by release of intracellular "membrane-bound" Ca^{++}, and resulting in a transient **rise in free cytoplasmic [Ca^{++}]**, which lasts for a few minutes at most. There is also typically an **increase in intracellular pH**, resulting from **Na^+/H^+ exchange** or **antiport**, which can be inhibited by amiloride. Changes in **membrane potential (ΔΨ)**, when they are observed, can go either in the direction of **depolarization (decrease in transmembrane potential)** or **hyperpolarization (increase in transmembrane potential)**, and appear to be due primarily to Na^+ and/or K^+ shifts across the membrane.

In some cases, **crosslinking of two or more receptors by a multivalent ligand** is required to induce ion flux changes; in others, binding of monovalent ligands to receptors is all that is necessary. When ion flux changes are initiated by crosslinking, bivalent antibodies to the receptor can produce the same effect as the natural ligand, while monovalent fragments cannot[432]. In most cases, materials that induce ion flux changes of the same nature and magnitude as those induced by natural ligands also induce the biologic responses induced by those ligands. For example, the **calcium ionophore A23187** is mitogenic for T lymphocytes, induces platelet aggregation, stimulates histamine release from mast cells, etc.[444].

If one monitors the average [Ca^{++}], ΔΨ, or pH of millions of cells in a suspension, it is possible to demonstrate **dose-response relationships** between the amounts of active ligand added and the magnitude of changes observed in the measured parameters. When the same parameters are followed in single cells, a correlation between ligand dose and degree of response may also be observed[362]. However, when the biologic responses of individual cells are examined, the usual pattern of response appears to be **all-or-none**[194]. Neurons exposed to neurotransmitters either generate a propagated action potential or they do not; lymphocytes either reproduce or do not, and so on.

It is generally accepted that an individual neuron is subject to the excitatory and inhibitory influences of various neurotransmitters, which are secreted by neighboring cells and which act on its dendrites and cell body, raising or lowering its resting membrane potential and, accordingly, changing its firing threshold. The membrane is a locus for summation of excitatory and inhibitory effects.

It is also known that the biologic responses of so-called "non-excitable" cells, such as lymphocytes, can be modulated by a great variety of cytokines, peptide hormones, neurotransmitters, and other ligands that react with cell surface receptors. It was hypothesized that, if the signals from these ligand-receptor interactions were transduced by ion fluxes across the cell membrane, the membranes of non-excitable cells, like those of neurons, might serve as "summing nodes" for excitatory and inhibitory stimuli.

The fact that ionic responses to ligand binding to surface receptors seem, even at the single cell level, to be dose-dependent, while biologic responses tend to be all-or-none, suggested that there might be [Ca^{++}] and/or $\Delta\Psi$ or pH threshold levels that must be reached for biologic responses to occur. Investigation of this possibility would require means to detect and quantify heterogeneity of cell populations with respect to ligand binding, [Ca^{++}], $\Delta\Psi$, pH, and changes in these parameters. Doing this nondestructively should allow one to make serial observations of cells, which could establish relationships between the nature and magnitude of cells' early responses to ligands and the same cells' subsequent biologic behavior.

It seemed to me at the time that cytometry, flow and otherwise, might usefully and profitably be applied to such studies. I knew there existed some fluorescent probe techniques for estimation of $\Delta\Psi$, [Ca^{++}], and pH; and I thought it would be fairly simple to adapt them for use with flow cytometry, and not much harder to answer all of the biological questions raised above. This view, needless to say, was excessively optimistic, for many reasons, some of which I will explain after the upcoming grand tour of functional parameters.

"Structuredness of Cytoplasmic Matrix" (SCM) and the Cercek Test for Cancer

"Structuredness of cytoplasmic matrix," or **SCM**[414], was probably the first functional parameter measured by flow cytometry. The measurement of SCM, like that of membrane fluidity, requires fluorescence polarization measurements. In this case, the fluorescence is emitted by **fluorescein**, produced by the action of nonspecific cellular esterases on **fluorescein diacetate (FDA)**, about which a fair amount has already been said (pp. 24-7, 369-71).

The original work in this area was done, and the term SCM was coined, in the early 1970's by Lea and Boris Cercek[415], then working in Manchester, England. They measured fluorescein fluorescence polarization in cell suspensions in cuvettes using a modified spectrofluorometer, and, in analyses of synchronized cell cultures, found changes during different phases of the cell cycle. The **fluorescence polarization**, calculated as described on p. 114, was taken as a measure of SCM.

In studies done over a period of several years (summarized in reference 414), the Cerceks found that SCM changed within a few hours after human lymphocytes were exposed to mitogenic lectins or antigens. They also reported that lymphocytes from cancer patients exhibited diminished SCM responses (i.e., smaller changes in polarization after exposure) to the lectin PHA when compared to cells from controls, and that lymphocytes from patients with cancer showed more marked SCM responses after exposure to preparations containing tumor antigens than did cells from controls. These results seemed to promise a rapid method for cancer diagnosis, and therefore excited considerable interest among cancer researchers and oncologists and in the diagnostics industry.

From about 1975 on, several groups of investigators attempted to duplicate the Cerceks' results with cancer patients, and at least as many failed as succeeded; this led, among other things, to everybody criticizing everybody else's methodology. It therefore seemed logical to try to put things on a more objective basis by making measurements of single cells, which has since been done by microfluorometry[416] as well as by flow cytometry[338,407-9,417-19]. Stewart et al[338] reported differences in the fluorescein polarization responses of lymphocytes from breast cancer patients and controls following exposure to PHA and pooled tumor antigen; their study was designed to exclude observer bias. Hartmann et al[418] noted diminished polarization responses of lymphocytes from cancer patients compared to cells from controls.

On another front, Price et al[417] demonstrated changes in fluorescence polarization of bone marrow cells within a few hours following addition of preparations of (granulocyte) colony stimulating factor and erythropoietin. By sorting cells with the most marked responses, they were able to obtain suspensions enriched in granulocytic or erythrocytic precursors dependent upon which cell growth factor was used as the stimulus. This suggested that changes in SCM, whatever their physicochemical explanation might be, could serve as general indicators of the effect of activators such as mitogens and growth factors on responsive cells.

Udkoff et al[419] analyzed the effects of the concentration of lectin, FDA, Ca^{++}, and K^{+} on the polarization responses of lymphocytes from normal donors, and found that all of the above might influence the shape of both fluorescence polarization and intensity distributions. To put it in lay terms, sometimes it works and sometimes it doesn't, and, when it doesn't, you can't always tell why (G. Price, R. Miller, L. Kamentsky, R. Udkoff, S. Chan, W. Eisert, W. Beisker, H. Steen, L. Scherr, and others, personal communications). I might add that SCM is not the only indicator of cell activation with which this problem has been encountered. This inhibited commercial development of an SCM-based cancer diagnostic test, but left the way open for further investigation and application of the effect.

Meisingset and Steen[420] investigated binding of fluorescein to proteins and lymphocytes, and noted both polarization and spectral changes on binding. Clearly, the rotational mobility of free cytoplasmic fluorescein will be greater than that of fluorescein bound to intracellular macromolecules, and changes in the ratio of free and bound dye will therefore change the polarization measured in a cell or cell suspension. Based upon studies using **time-resolved spectroscopy**, Kinoshita et al[847] concluded that the fraction of bound molecules and the anisotropy produced by binding are the primary determinants of intracellular fluorescein fluorescence anisotropy; they did not suggest a mechanism by which changes might occur in activation. Prosperi et al[848] found that various membrane-active agents changed rates of fluorescein influx and efflux; similar changes during activation might affect SCM by changing the ratio of free to bound dye.

The difficulty of measuring fluorescence polarization by flow cytometry has impeded investigation of the physicochemical basis of SCM response, although work with carefully modified commercial apparatus[1668-9] has established that polarization changes are routinely detectable in activated cells, and that such changes do not occur in the absence of calcium or in the presence of cytochalasin B.

The SCM test has been intensively studied by two Israeli physicists, Mordechai Deutsch and the late Aryeh Weinreb, and their colleagues at several hospitals and research institutions in Israel and elsewhere. Initial success with the SCM test using bulk fluorometry in the 1980's[1670-2] led Deutsch and Weinreb to develop a static cytometer, the Cellscan[1145], which uses a single photon counting technique (p. 164) to maximize the precision of measurements of fluorescence polarization of fluorescein and other dyes in single cells. A commercial version of this apparatus is available from **Medis Technology**; Deutsch et al have since developed a next generation apparatus, the Individual Cell Scanner[2780], which also makes high-precision polarization measurements.

Measurement precision is critical for SCM measurements because the largest changes in polarization tend to occur in those cells with the lowest intensity of intracellular fluorescence. The precision of a fluorescence intensity measurement made in a flow cytometer, in which observation time is constant, decreases as intensity decreases because fewer photoelectrons contribute to lower intensity measurements. The polarization values calculated from low intensity fluorescence measurements will also be less precise. The Cellscan and Individual Cell Scanner count a preset number of photons (usually 10,000 or 20,000) in each fluorescence measurement channel, and derive fluorescence intensity values from the time taken to reach the preset photon count; the contributions to variance from photoelectron statistics are therefore the same for both low and high intensity fluorescence measurements.

The polarization of intracellular fluorescein fluorescence in T cells decreases when the cells are exposed to mitogens; the average polarization of mitogen-treated cells is 9.7% lower than that of control cells at 15 min following addition of mitogen, and 13.5% less at 180 min[2781]. The coefficient of variation of the distribution of polarization values, presumably reflecting biologic variation, is large, approximately 75 percent, and, if the measurement technique itself is not highly precise, it may not be possible to distinguish differences between treated and control cells.

We now know a great deal about SCM and changes in SCM. The effect can be demonstrated using carboxyfluorescein[2782] and BCECF[2783] as well as fluorescein. Decreases in polarization, beginning at 40 min after exposure and lasting 24 hr, are induced by the T cell mitogens PHA, Con A, and anti-CD3 antibody, by phorbol esters, and by the calcium ionophore ionomycin, but not by pokeweed mitogen (PWM), a B cell mitogen[2781,2784]. The effects of ConA can be counteracted by addition of a competing sugar, methyl α-D-mannopyrannoside within 1 hr, but not after 5 hr[2781]. The mitogen-induced decrease in fluorescence polarization is inhibited by inhibitors of energy metabolism (NaN_3, NaF, KCN, and a proton ionophore), by proton kinase C inhibitors (H7, staurosporin) and by agents that disrupt cytoskeletal microtubules (colchicine, the *Vinca* alkaloid vinblastine) and microfilaments (cytochalasin B)[2781,2785].

Intracellular fluorescein fluorescence polarization values measured in cells from the Jurkat T-lymphoblastoid line were lower in S and G_2/M phase cells than in G_0/G_1 cells[2786]. Fluorescein fluorescence polarization in human lung-derived fibroblasts was decreased by IL 1-α and IL 1–β, and by TNF-α; the effects of these cytokines were inhibited by vinblastine[2787]. Increases in fluorescence polarization in mouse thymocytes and Jurkat cells occur in the early stages of spontaneous apoptosis and of apoptosis occurring in response to treatment with glucocorticoids, Fas ligand, or cancer chemotherapeutic agents; the increase in polarization is unaffected by treatment with caspase-3 inhibitors. The polarization changes in apoptosis appear to be associated with cell dehydration and shrinkage[2788-9].

In an earlier edition of this book, I suggested that intracellular pH changes related to cell activation might play a part in the SCM response. The excitation spectra of fluorescein, and of carboxyfluorescein and BCECF which can also serve as indicators of SCM response, are pH-sensitive, and changes in cytoplasmic pH occur in at least some cell activation processes, as will be discussed later, and could conceivably affect both the spectrum and the binding of the dye. However, after extensive discussions with Motti Deutsch and Reuven Tirosh, and consideration of recent findings[2780-9], I am persuaded that the effect is primarily due to changes in the organizational state and mobility of cytoplasmic proteins, and differences in free and bound amounts of different ionic forms of the dye. SCM changes are real; they're just hard to measure, which has given the whole procedure an undeservedly bad reputation.

Reports of clinical trials of the SCM test in cancer diagnosis continue to emerge[2790-8]; it seems to work overall, but is probably not quite sensitive or specific enough to be widely adopted. However, if you look at the data that have been obtained to date with single cell measurements, something very interesting emerges. Two phenomena are observed in the Cercek cancer test that supposedly differentiate between people with cancer and people without cancer. First, T cells from people with cancer don't show as much of a response to mitogens such as con A and phytohemagglutinin (PHA) as do T cells from people without cancer. That's old news; reports of diminished immune response in cancer, e.g., loss of skin reactions to tuberculin and other antigens, go back for generations. What isn't old news, however, is the observation that a substantial fraction, meaning 10 percent and sometimes several times that, of T cells from people with cancer show an SCM response, or evidence of activation, on exposure to tumor-derived antigens, which have little or no effect on T cells from people without cancer. This suggests that people with cancer either have a greatly expanded clone or clones of antigen-responsive T cells, or that some recruitment phenomenon is occurring in people with cancer but not in normals.

As I hope will be made clear by the end of this book, there are a lot of ways of measuring lymphocyte activation cytometrically, ranging from looking at calcium fluxes in the first few minutes to measuring activation antigens such as CD69 and CD98 (4F2) and cytokine production after 4-8 hours to measuring increased RNA and DNA synthesis after 24 and 36 hours. We also now have tetramers that allow us to identify cells that respond specifically to antigens.

The time course of SCM changes is roughly that of calcium changes; none of the proponents of the SCM test seem to have followed up and found out whether the activated cells go on to express activation antigens and reproduce. The studies done on SCM in blood cell precursors by Price et al[417] offer some encouragement, but we don't have the data for lymphocytes. Assuming that the cells from cancer patients that are activated by tumor antigens are doing what comes naturally for specifically activated lymphocytes, we ought to be able to detect activation by any of the several means just mentioned, all of which are much easier to implement than fluorescence polarization measurements, and most of which can be implemented on commercial flow cytometers. If the Cercek cancer test really works, this could be a big shot in the arm for clinical flow cytometry, because a lot of people will be getting tested every year or two. If the cancer test doesn't work, at least we might learn something about tumor immunology by looking further into patients' T cell responses to antigen.

The Cerceks themselves moved from England to Southern California around 1980, and went to work for Beckman. Shortly after they got there, I helped some folks at Beckman to build a flow cytometer that was supposed to be used for SCM work. The flow cytometer project stopped, and nothing further was heard until 1993, when the Cerceks published three papers[1673-5] describing the isolation of a cancer-associated, SCM-recognition, immunedefense-suppressing, and serine protease-protecting (CRISPP) peptide from the blood plasma of cancer patients. A consensus, synthetic 29 amino acid CRISPP peptide (CRISPPs) has the same cancer SCM-recognition (CR) activity and SCM-response modifying effects as the natural peptides; in other words, if you add it to normal T cells, they behave like T cells from people with cancer. The Cerceks subsequently attempted to determine the genetic origin of the CRISPP peptide[2799], and reported that CRISPP peptide increased DNA synthesis in cultured hepatocytes, while increasing DNA synthesis at low doses and decreasing DNA synthesis in higher doses when added to PHA-stimulated cultures of human lymphocytes[2800]. Before Beckman and Coulter merged, Coulter looked into the SCM test and decided not to pursue it; it doesn't look as if Beckman was any more interested in the CRISPP peptides than in the SCM test.

Thus far, neither the Cellscan nor the Individual Cell Scanner seem to have been used to make a multiparameter measurement of fluorescein fluorescence polarization and of another activation-related parameter (e.g., cytoplasmic Ca^{++}, pH, cytoplasmic or mitochondrial membrane potential, early activation antigen expression, or cytokine production), or even to detect phenotypic differences between responding and nonresponding cells. It would be nice to look into all of that. Every couple of years or so, somebody runs across the SCM literature and asks me about measuring fluorescence polarization of fluorescein in lymphocytes using a flow cytometer, and I have to say that my limited experience bears out Motti Deutsch's contention that flow cytometry isn't precise enough to be useful. Lou Kamentsky was trying to implement the SCM test using Ortho's flow cytometers in the late 1970's, without much luck, and I convinced him to look for another indicator of activation that might be easier to measure. I proposed membrane potential, which brings me (and you) to the next topic.

Optical Probes of Cell Membrane Potential ($\Delta\Psi$)

Membrane Potential and Its Physicochemical Bases

Electrical potential differences exist across the membranes of most prokaryotic and eukaryotic cells. These potential differences are due in part to the existence of **concentration gradients** of Na^+, K^+, and Cl^- ions across the cell membrane, and in part to the operation of various **electrogenic pumps**.

The potential differences across the cytoplasmic membranes of resting mammalian cells range in magnitude from about 10 to 90 mV, the cell interior being negative with respect to the exterior. There is also a potential difference of 100 mV or more across the membranes of **energized mitochondria**, with the mitochondrial interior negative with respect to the cytosol; this potential is dissipated when **energy metabolism** is inhibited.

In **prokaryotes**, the enzymes responsible for energy metabolism are located on the inner surface of the cytoplasmic membrane, and the potential difference across this membrane, which is typically 100-200 mV, depends largely on energy metabolism.

The resting potential across the cytoplasmic membrane of mammalian cells is frequently estimated from the **Goldman equation**[455]:

$$\Delta\Psi = \frac{RT}{F} \ln \frac{P_K[K^+]_i + P_{Na}[Na^+]_i + P_{Cl}[Cl^-]_o}{P_K[K^+]_o + P_{Na}[Na^+]_o + P_{Cl}[Cl^-]_i}$$

where $\Delta\Psi$ is the membrane potential, R is the gas constant, T is the temperature in degrees Kelvin, F is the Faraday, $[X]_i$ is the concentration of X ions inside the cell, $[X]_o$ is the concentration of X ions outside the cell, and P_X is the permeability of the membrane to X ions.

ΔΨ Measurement Using Microelectrodes

Membrane potentials can be measured directly using **implanted microelectrodes**, but this becomes increasingly difficult as the technique is applied to smaller cells. Changes in ΔΨ in response to ligand-receptor interactions have, however, been detected by such direct measurements. For example, Taki[456] reported that lymphocytes were electrically **depolarized** (i.e., the potential difference across the membrane decreased in magnitude) within 10 minutes following exposure to PHA; depolarization reached a maximum within a few hours, following which **hyperpolarization** (increase in magnitude of the potential difference) occurred over a period of several days. See p. 408 for a follow-up.

Indirect ΔΨ Measurement in Cell Suspensions Using Distributional Probes

Indirect estimates of ΔΨ can be obtained by monitoring the **distribution** of **radiolabeled lipophilic cationic indicators** (e.g., ^{3}H-triphenylmethylphos-phonium, or $TPMP^+$, which was described by Bakeeva et al in 1970[459]) or of **lipophilic cationic dyes**, such as cyanines and safranins, between cells and the suspending medium. Lipophilic indicators are used because this characteristic enables indicator molecules to pass freely through the lipid portion of the membrane; thus, the concentration gradient of an indicator species C^+ across the membrane is determined by the potential difference across the membrane according to the **Nernst equation:**

$$[C^+]_i/[C^+]_o = e^{-FE/RT},$$

where the notation is the same as was used previously. A ratio $[C+]_i/[C+]_o$ of 10 corresponds to a potential difference of 61 mV at 37 °C. Because their response to changes in ΔΨ is dependent upon their distribution across the cell membrane, lipophilic indicator cations are often described as **distributional probes.**

Once cells have been equilibrated with an indicator cation, **depolarization** of the cells will cause **release of indicator from cells into the medium**, while **hyperpolarization** will make **cells take up additional indicator from the medium.** The indicator distribution will not adequately represent the new value of ΔΨ until equilibrium has again been reached; this process requires periods ranging from a few seconds to several minutes. Thus, while distributional probes may be suitable for detection of slow changes in ΔΨ, they cannot be used to monitor the faster changes that occur during the propagation of action potentials in tissues such as nerve and muscle.

Most of the dyes now used as probes of ΔΨ were developed during a systematic search by Lawrence Cohen, of Yale University, and his coworkers[457] for materials that would exhibit rapid enough changes in optical characteristics to respond to action potentials in nerve cells. Among the dyes evaluated were many that proved unsuitable because of their tendencies to redistribute across cell membranes in response to slow potential changes. As might be expected, most lipophilic cationic dyes fell into this category. One of these, **3,3'-dihexyloxacarbocyanine [$DiOC_6(3)$]**, was studied by Hoffman and Laris[458], who were able to make estimates of ΔΨ in red blood cell suspensions based on the partitioning of the dye into the cells.

The use of a lipophilic cationic dye as a distributional probe of ΔΨ involves only a slight departure from the distributional probe technique for ΔΨ estimation using radiolabeled lipophilic cations The radioisotope method requires that a known volume of cells (or organelles, e.g., mitochondria) in suspension be equilibrated with the indicator; the intracellular and extracellular indicator concentrations are calculated from a determination of the amount of cell-bound indicator done by scintillation counting of the cells after washing to remove unbound indicator.

Estimation of ΔΨ of cells in suspension using symmetric cyanine dyes[458] is done in a spectrofluorometer. Addition of cells to micromolar solutions of dyes such as $DiOC_6(3)$ produces a suspension with lower fluorescence than that of the original solution, indicating that, **at micromolar external concentrations, the fluorescence of dye taken into cells is quenched.** When intracellular and extracellular ion concentrations are manipulated to hyperpolarize the cells, increasing cellular uptake of dye, the fluorescence of the suspension decreases further; when the cells are depolarized, releasing dye into the medium, the fluorescence of the suspension increases. The results obtained by Hoffman and Laris using this method to estimate ΔΨ in giant red cells from *Amphiuma* compared favorably with the results these authors obtained by microelectrode measurements.

This success led to a comparative study by Sims et al[460] in which 29 cyanine dyes were examined as indicators of ΔΨ in red cells and lipid membrane vesicles. This paper introduced the **abbreviated nomenclature for the cyanine dyes** that is now in general use, explained in Figure 7-12 (p. 313). The lipophilicity of cyanine dyes increases with the length of the alkyl side chains, i.e., as n in the formula "**$DiYC_{n+1}(2m+1)$**" increases. The wavelengths of maximum absorption and emission are essentially independent of the alkyl side chain length, denoted in the formula by (n + 1), but increase with the length of the polymethine bridge between the rings, i.e., as m in the formula increases[461]. The quantum efficiency of the dyes is increased, and the absorption and emission maxima are shifted to longer wavelengths, in nonpolar solvents. Spectra of some symmetric cyanine dyes and of tandem conjugates of phycobiliproteins and cyanine dyes appear in Fig. 7-9 (p. 296); the wide range of spectral characteristics available results in large part from the early development of the dyes by Eastman Kodak and their exploitation as photographic sensitizers[461].

The report of Sims et al[460] gave particular prominence to two dyes, **dipentyloxacarbocyanine [$DiOC_5(3)$]** and

dipropylthiadicarbocyanine [$DiSC_3(5)$], because they exhibited larger fluorescence changes than most of the other dyes in response to maximal hyperpolarization of the cells induced by the **potassium-selective ionophore valinomycin** (see reference 444 for an extensive discussion of ionophores), and because they were relatively stable and did not cause excessive lysis of cells.

Some people mistakenly assumed that $DiOC_5(3)$ and $DiSC_3(5)$ were the only cyanine dyes suitable for $\Delta\Psi$ estimation; this caused some problems because these dyes, and many of the others described by the Yale group, were not available commercially. The many investigators worldwide who became interested in using dyes for $\Delta\Psi$ measurements were supplied with them by Alan Waggoner[462], now at Carnegie Mellon University, who was responsible for most of the chemistry in the Yale project. Most people did not realize that dyes such as $DiOC_6(3)$ and $DiSC_2(5)$, which, respectively, can be substituted for $DiOC_5(3)$ and $DiSC_3(5)$, were readily available from Eastman and other companies (Eastman's product line is now sold by Acros Organics, but the dyes remain widely available). Results typical of those obtained in cuvette measurements with cyanine dyes are shown in Figure 7-26.

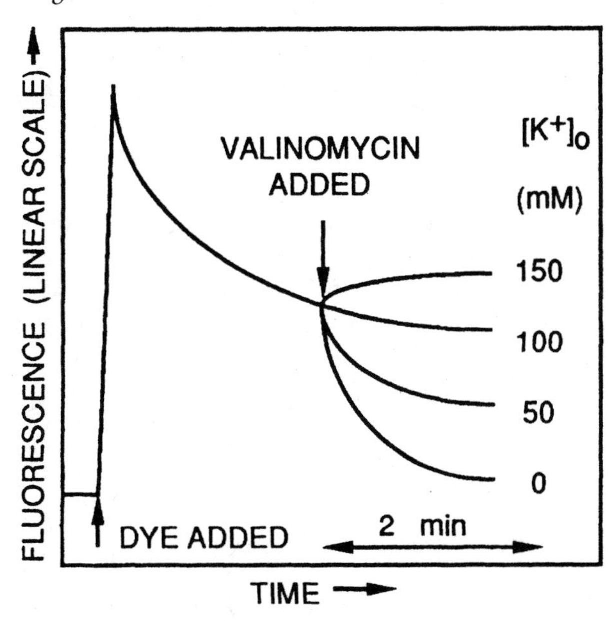

Figure 7-26. Valinomycin-induced changes in fluorescence intensity of red cell suspensions equilibrating with a cyanine dye at various external K^+ concentrations.

Under normal circumstances, the intracellular concentration of K^+, $[K^+]_i$, is considerably higher than the extracellular concentration $[K^+]_o$, while the intracellular concentration of Na^+, $[Na^+]_i$, is considerably lower than the extracellular concentration $[Na^+]_o$. The ionophore **valinomycin (VMC)**, which is lipophilic, forms a complex with K^+ ions and can thus readily transport them across cell membranes. Addition of VMC thus effectively increases the cells' potassium permeability (P_K in the Goldman equation), to the point at which membrane potential is determined almost entirely by the transmembrane $[K^+]$ gradient. If $[K^+]_o$ is low, VMC addition hyperpolarizes cells; if $[K^+]_o$ is high, VMC addition depolarizes cells, and, if $[K^+]_o = [K^+]_i$, VMC addition does not change $\Delta\Psi$.

Data such as those shown in Fig. 7-26 are obtained by normalizing curves so that the levels of fluorescence in each sample prior to VMC addition are identical. This normalization is necessary because even when the concentrations of cells and dye added to clean cuvettes are carefully controlled, equilibrium fluorescence readings obtained from cell suspensions vary due to such factors as dye adhesion to the cuvette walls. Despite this, people who worked with this technique for $\Delta\Psi$ estimation developed the notion that cyanine dye fluorescence could somehow be calibrated to read out $\Delta\Psi$ to the nearest millivolt. This was credible only as long as nobody looked at cyanine dye uptake at the single cell level. As far as I can determine (L. Cohen, personal communication; A. Waggoner, personal communication), cyanine dye-stained cells, other than nerve or muscle cells, weren't even looked at under a fluorescence microscope until I started playing with $\Delta\Psi$ probes in the mid-1970's.

At that time, the literature stressed the point that $\Delta\Psi$ estimation with cyanine dyes depended on **quenching of the fluorescence of intracellular dye**; in essence, the assumption was being made that all of the fluorescence measured in the cell suspension was coming from free dye in solution, or at least that the contribution of cell-associated dye to the total fluorescence signal remained constant. When I first looked at cells stained with micromolar concentrations of cyanine dyes, such as were used for $\Delta\Psi$ estimation in suspensions, I expected to see dark cells against a fluorescent background, which would certainly not make it easy to adapt cyanine dye $\Delta\Psi$ probes for flow cytometry. I was pleasantly surprised when I found brilliant fluorescence in the cells, and somewhat disappointed when I examined the broad fluorescence distributions (Figure 7-27, next page) obtained when they were run through a flow cytometer[424].

Single Cell Measurements with Distributional Probes

According to the Nernst equation (p. 386), $[C^+]_i/[C^+]_o$, i.e., the ratio of the intracellular and extracellular **concentrations** of a membrane-permeant cationic dye C^+, varies with $\Delta\Psi$. Since the extracellular concentration $[C^+]_o$ is the same for all cells in a suspension, the intracellular concentration $[C^+]_i$ should provide an indication of $\Delta\Psi$. If $\Delta\Psi$ is the same in all the cells, $[C^+]_i$ should be the same. However, the fluorescence distribution is not a distribution of dye concentration, i.e., of $[C^+]_i$, but of the **total fluorescence**, which we hope is proportional to the total amount of dye, per cell. To obtain a distribution of $[C^+]_i$, we need to divide the fluorescence value for each cell by the cell's volume, obtained by an electronic volume sensor or estimated from forward scatter or extinction; this is rarely done in practice.

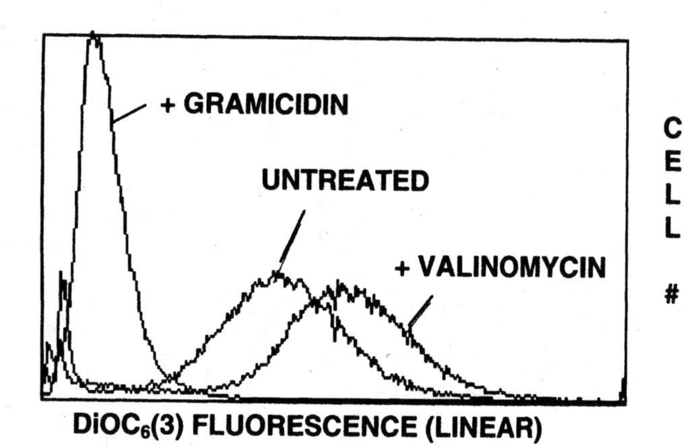

Figure 7-27. Distributions of the fluorescence of $DiOC_6(3)$ in CCRF-CEM T-lymphoblasts equilibrated with 50 nM dye for 15 min. A: untreated cells, B: cells depolarized with gramicidin, C: cells hyperpolarized with valinomycin. The residual peaks at the left of the distributions represent signals from debris.

While volume measurements were not made of the cells that provided the data of Fig. 7-27, experience tells us that the coefficient of variation (CV) of the cell volume distribution of human peripheral blood lymphocytes is no more than 15%, while the CV of the cyanine dye fluorescence distribution observed in these cells is about 30%. Thus, variance of cell volume cannot explain all of the variance of the fluorescence distribution. Does this mean that there is a broad distribution of membrane potentials in lymphocytes? Not necessarily. The cells treated with the ionophore **gramicidin (GRM)**, which forms channels in the membrane that permit most mono- and divalent ions to pass through, should be completely depolarized, yet their fluorescence distribution also shows a greater variance than could be explained by cell-to-cell differences in volume.

Recalling that there are large potential gradients across mitochondrial membranes, you might be tempted to think that **mitochondrial uptake** of dye was responsible for the variance in fluorescence in the distributions of Fig. 7-27. This possibility was investigated; lymphocytes in which the mitochondrial potential was abolished using a combination of antimycin, dinitrophenol, and oligomycin did show less cyanine dye fluorescence (by 15-20%) than cells that were not exposed to these inhibitors, but the variance of the fluorescence distribution was not significantly reduced.

In cells treated with GRM, $\Delta\Psi$ should be zero; and the concentrations of dye inside and outside the cells should therefore be the same. How is it, then, that we can manage to see the little buggers? Well, there are two reasons. The first is that the fluorescence of cyanine dyes is enhanced (approximately sixfold in the case of $DiOC_6(3)$, less for other dyes) when the dye is in a hydrophobic environment[460]. The second, which is quantitatively more important, is that the lipophilic character of the dye causes it to be concentrated in cells in the absence of a potential gradient because it binds to intracellular material, predominantly lipids and membranous structures (a 1997 paper[2801] actually describes $DiOC_6(3)$ as a "specific" stain for endoplasmic reticulum).

In the study of Sims et al[460], symmetric cyanine dyes with different alkyl side chain lengths were equilibrated with VMC-treated suspensions of red blood cells, which had been estimated to have a membrane potential of –40 mV. The Nernst equation would predict that the ratio of intracellular and extracellular dye concentrations should be less than 10; the ratio observed for diethylthiadicarbocyanine [$DiSC_2(5)$], the least lipophilic dye in the series, was over 100, and the ratios observed using other thiadicarbocyanine dyes increased with the length of the alkyl side chains, reaching a value above 10,000 for dihexylthiadicarbocyanine [$DiSC_6(5)$]. Thus, while uptake of cyanine dyes does provide an indication of membrane potential, the association of these dyes with cells is distinctly "Non-Nernstian."

Don't try to measure membrane potential with cyanine dyes with heptyl or longer side chains, e.g., $DiOC_7(3)$. In the series running from C_1 through C_6, the longer chain dyes, which are more lipophilic, get in faster, which is an advantage. At C_7 and above, the cyanines start to take on characteristics associated with the **cyanine tracking dyes** such as "DiI," "DiO," and the PKH series, discussed on pp. 45-6 and 371-4; they feel so warm and cozy in the lipid bilayer that they're not in any hurry to leave, even if the potential changes. The diheptyl (C_7) cyanines, which are not widely used, can be thought of as the "missing link" between membrane potential probes and tracking dyes, which typically have alkyl side chains containing at least 14 carbons.

Non-Nernstian binding of probe is also encountered when $\Delta\Psi$ is estimated with radiolabeled lipophilic cations such as $^3H\text{-}TPMP^+$. In order to get accurate values of cytoplasmic $\Delta\Psi$ using lipophilic cationic indicators, it is necessary to inhibit the mitochondria and to correct the results to account for uptake in the absence of a potential gradient. Felber and Brand[463-6] produced a few papers that resolved several controversies in the previous literature regarding lymphocyte membrane potentials. Their work, unfortunately, did not extend to single cell measurements.

This still leaves us looking for an explanation of the variance of cyanine dye fluorescence distributions. Further comparison of flow cytometric data[112] with data obtained from cuvette measurements can be helpful in this regard. When flow cytometry is done on cells exposed to different concentrations of cyanine dyes, a point is reached at which increasing the dye concentration does not increase fluorescence in the cells. If the dye used is $DiOC_6(3)$, this happens when cells at a concentration of 10^6/ml are incubated with 2 μM dye. The variance of the fluorescence distribution remains large, suggesting that cyanine dye fluorescence in cells results from fluorescence enhancement of dye bound to hydrophobic regions, and that the variance of fluorescence is due primarily to cell-to-cell variations in the number of binding sites.

Further evidence in support of this hypothesis comes from the observation that when the hydrophobic binding sites are saturated, at a dye concentration of 2 μM, the fluorescence distributions of cells are identical whether they are suspended in isotonic NaCl, in which they maintain normal ΔΨ, or in isotonic KCl, in which they are depolarized. In both cases, addition of VMC or GRM does not change cellular fluorescence. In other words, the cellular fluorescence no longer responds to changes in ΔΨ. However, if the fluorescence of cell suspensions equilibrated with 2 μM $DiOC_6(3)$ in NaCl and KCl is measured in a spectrofluorometer, the potential difference will be detectable, and hyperpolarization caused by addition of VMC to the cells in NaCl will produce a demonstrable fluorescence change. This means that the cells contain dye that is essentially nonfluorescent due to quenching, as well as dye bound to hydrophobic sites in which fluorescence is enhanced. VMC addition causes uptake of additional dye, which is also quenched once it enters the cells. Quenching of the fluorescence of symmetric cyanine dyes typically involves the formation of **aggregates**. So-called **H-aggregates** are nonfluorescent, while fluorescent **J-aggregates** typically emit at wavelengths substantially longer than do single dye molecules.

The dye concentration at which saturation of binding sites occurs is determined primarily by the lipophilicity of the dye; the fluorescence of cells (at 10^6/ml) equilibrated with 2 μM $DiOC_2(3)$, which is less lipophilic than $DiOC_6(3)$, is less than the fluorescence of cells equilibrated with the C_6 dye, and does change when ΔΨ is changed by ionophore addition or by manipulation of ion concentrations in the medium. In fact, if cells in 2 μM $DiOC_2(3)$ in NaCl (normal ΔΨ, higher fluorescence) are mixed with an equal volume of cells in 2 μM $DiOC_2(3)$ in KCl (depolarized, lower fluorescence), the cells and dye reequilibrate within a few minutes, as shown in Figure 7-28, yielding a fluorescence distribution that reflects the intermediate value of ΔΨ resulting from the ionic composition of the mixed suspending medium.

The distributions shown in Figure 7-28 are distributions of ΔΨ values estimated from the ratio of fluorescence and extinction signal amplitudes. This was done in an attempt to compensate for the effects of cell size variation in the cultured cell line used for the experiments, which results in broadening of the fluorescence distributions. The variance of the distributions of the fluorescence/extinction ratio remains fairly large, indicating that factors other than cell size contribute substantially to the variance in the number of sites to which dye binds with fluorescence enhancement.

From what has gone before, we can conclude that the cyanine dyes, under the best conditions, are not going to give us absolute values of ΔΨ to the nearest millivolt, whether we measure cell suspensions in cuvettes or individual cells in flow cytometers or microphotometers. However, while the fluorescence of cell-associated cyanine dye cannot provide an absolute measure of ΔΨ, it can provide a reasonably rapid indication of substantial changes in ΔΨ that occur over periods of seconds to hours, and allow us to examine correlations between ΔΨ and other parameters.

The breadth of the distributions obtained by flow cytometry means that cyanine dye fluorescence has only limited capability for detecting heterogeneity of ΔΨ in populations; if 50% of the cells have ΔΨ values that are approximately half those of the other 50%, you'll know it, while if 5% of the cells exhibit a 15% increase in ΔΨ, you won't. Are there, then, any better dye probes of ΔΨ than the cyanines?

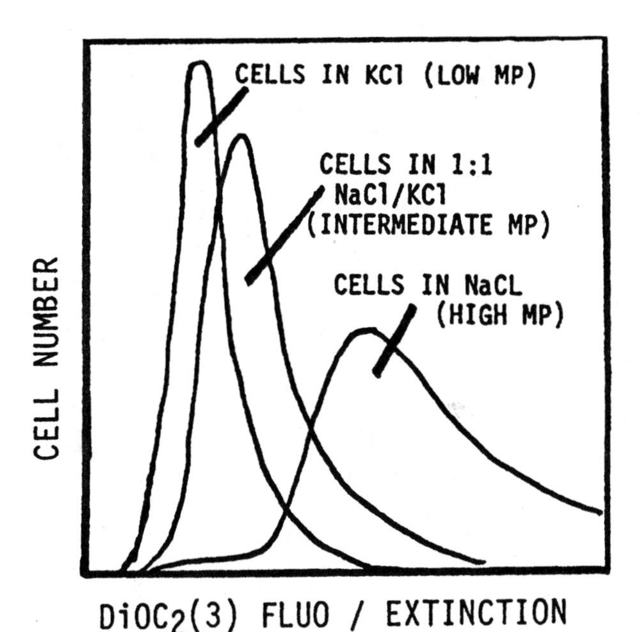

Figure 7-28. Distributions of MP (ΔΨ) (estimated from the ratio of $DiOC_2(3)$ fluorescence to extinction) in cultured CEM lymphoblasts suspended in isotonic NaCl, isotonic KCl, and a mixture of the two.

It has already been mentioned that any cationic dye that crosses cell membranes readily can serve as a distributional probe of ΔΨ. Among the classes of such dyes that have been investigated by Cohen, Waggoner et al[457,461,462] and by others are **cyanines** (e.g., the classical mitochondrial stain **pinacyanol** as well as the compounds discussed above), **acridines** (yes, acridine orange works as an ΔΨ probe if you use it at about 10 nM, but who needs it?), **oxazines** (e.g., Nile blue), **pyronins**, **rhodamines** (e.g., **rhodamine 123**), **safranins** (e.g., **Janus green**, well known as a mitochondrial stain, and **safranin O**) **styryl compounds** and **triarylmethane dyes** (e.g., **crystal violet**). I have done flow cytometry of cells stained with a reasonable number of cationic dyes from this list, and none of the dyes offers any obvious advantages from a metrologic point of view in flow cytometry, although some may be less toxic and/or more stable than the cyanines. Plášek and Sigler and their colleagues have refined calibration of cuvette measurements of ΔΨ using cyanine dyes, but have only obtained semiquantitative results using flow cytometry[2802-4].

Leslie Loew and his colleagues[1676-8] have described the use of the relatively hydrophilic methyl and ethyl esters of tetramethylrhodamine for "Nernstian" membrane potential measurement by imaging microspectrofluorometry; by measuring dye concentrations in various spatial regions, these probes allow simultaneous determination of cytoplasmic and mitochondrial membrane potential, once corrections are made for contributions for dye from out-of-focus regions and for non-Nernstian binding.

Oxonol Dyes as Membrane Potential Probes

Oxonols[426,850-1] (Figure 7-29), which are negatively charged, bind to the cytoplasmic membrane but do not accumulate in intact cells; probably because they do not enter cells at appreciable concentrations, they are much less toxic than cyanines and other cationic $\Delta\Psi$ probes. For the same reason, oxonol fluorescence, unlike fluorescence from cationic probes, is not greatly influenced by potential-dependent uptake of dye into mitochondria. These desirable characteristics of oxonols are offset somewhat by their weaker fluorescence, as compared to cyanines, and, since they don't produce distributions with any less variance, than do cyanines, I don't use oxonols much. I also find that the bright staining of damaged cells by oxonols, which, being negatively charged and highly lipophilic, stick like crazy to everything inside cells once they get in, is something of a nuisance. This may be dealt with by adding a dye such as crystal violet or trypan blue to quench the offending fluorescence, but it gets to be too much of a production for my taste.

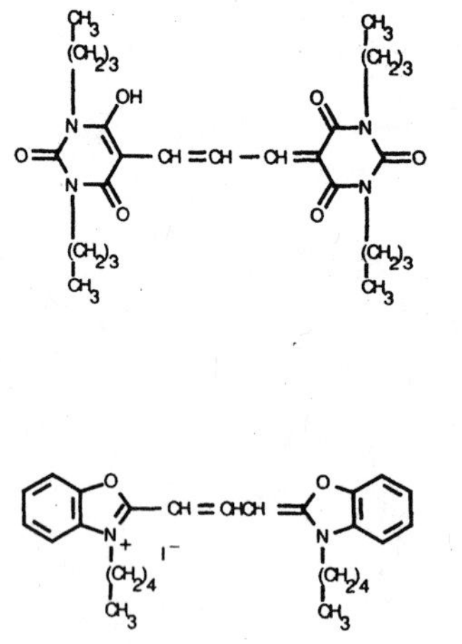

bis (1,3-dibutyl-barbituric acid) trimethine oxonol

$DiBAC_4(3)$

3,3′-dipentyloxa-carbocyanine iodide

$DiOC_5(3)$

Figure 7-29. Structures of two of the more popular membrane potential probes, an oxonol and a cyanine. Structures were provided by Molecular Probes.

When I have compared oxonols and cyanines in rat and human leukocytes, T cells, and human lymphoblastoid cell lines, I have gotten pretty much the same results in terms of being able to detect depolarization and hyperpolarization. (**oxonol-stained cells** are **brighter** when **depolarized** and **dimmer** when **hyperpolarized**, a mirror image, if you will, of cyanine dye response). This is probably so because I make measurements over short time periods, which, combined with the relative resistance of the cells I study to cyanine toxicity and the absence of significant mitochondrial interference with fluorescence, lets me get away with it. Ken Rosenthal and I grew cells in 2 μM $DiOC_6(3)$ for a few days, in the dark, of course. People whose cells won't stand up to that have good reason to use oxonol dyes.

I have confidence in the **estimates** (calling them **measurements** would imply more accuracy and precision than dyes and flow cytometry can give us) of $\Delta\Psi$ I get from cyanines, which do, after all correlate with microelectrode measurements[458,852], but oxonols work. Krasznai et al[2805] described a method for determination of "absolute membrane potential" of cells using the popular oxonol **$DiBAC_4(3)$**, a dye that is frequently referred to as "bis-oxonol" but should not be, because it is one member of a large family of bis-oxonols. A calibration curve is constructed from a plot of fluorescence intensity measured from stained cells vs. extracellular dye concentration; it is assumed that dye distribution is Nernstian, which is probably a risky assumption given the lipophilicity of oxonol dyes. However, Krasznai et al reported good agreement between their flow cytometric measurements and patch clamp measurements.

When cells are added to solutions of oxonol dyes, the fluorescence of the resulting suspensions is increased. Oxonols, being negatively charged, should tend to remain outside cells with interior-negative membrane potentials; a purely Nernstian distribution for cells with $\Delta\Psi = -61$ mV would make the internal oxonol concentration 1/10 of the external concentration. However, like most of the symmetric cyanines, the oxonols increase fluorescence in nonpolar solvents, with the result that any dye molecule bound to membranes or lipids becomes more fluorescent than it would be in aqueous solution. The lipophilicity of oxonols results in their being taken into cells against the electrochemical gradient across the cytoplasmic membrane, but the dyes are largely excluded from mitochondria by the electrochemical gradient across the mitochondrial membrane.

A careful choice of controls is required when oxonols are used. These dyes form complexes with valinomycin, complicating its use as a hyperpolarizing agent; monensin has been suggested as a replacement. The real danger, however, arises from using heat-killed or fixed cells as exemplars of cells with $\Delta\Psi = 0$, because both heat and fixation permeabilize cells and change the chemical structure of the membrane. The observed increased binding of oxonols is therefore likely to reflect influences of factors other than membrane potential, e.g., binding to intracellular proteins and changes in membrane asymmetry.

The binding of merocyanine 540, which can be envisioned as a chimerical combination of half a cyanine and half an oxonol, to cells is known to be influenced by changes in membrane asymmetry, hence the utility of that dye in detecting apoptotic cells. I have already (p. 374) noted my suspicion that at least some oxonol dyes may share the characteristics of their half-sibling, although I haven't done the experiments needed to prove the point. You're welcome to try and beat me to it using merocyanine 540 and $DiBAC_4(3)$ alternately with APC-annexin V in an apoptosis model system. In the meantime, stick with adjustments of the external potassium concentration or oxonol-friendly ionophores to get control cells with zero membrane potential.

Possible Alternatives to Distributional Probes for Flow Cytometry of Membrane Potentials

Dyes that work as Nernstian probes of membrane potential are obviously just dandy for monitoring membrane potential changes in single cells in **static** or image cytometers; even with non-Nernstian dyes, making repeated observations of the same cell over time eliminates the otherwise vexing problem of cell-to-cell variability in fluorescence intensity. However, under the best of circumstances, distributional dyes can only be used to monitor relatively slow $\Delta\Psi$ changes, occurring over periods of seconds to minutes. Better results would be expected using faster responding dyes, which sense membrane potential by different mechanisms.

The hardware technology for making static or imaging measurements is a lot simpler than it used to be, and it certainly seems easier than building a multistation flow system 100 meters long, which some of us have also thought of as a way to make repeated observations of the same cell several seconds apart. Because of the photodynamic toxicity of most membrane potential probes, however, you have to be careful, when making static measurements, about how you treat the cells between observations. For example, if you stain cells with a green-excited dye, you do all of the hunting for the cells under low-level red illumination using phase contrast and/or image intensification, thereby avoiding light and heat damage to the cells. When you want to measure the cells, you give them no more than a few milliseconds worth of green illumination at a time, using an arc or quartz lamp (with heat filters!!!) and a shutter, or else a xenon flash lamp. This minimizes photodynamic damage to the dye as well as to the cells. As long as you have several measurement values for the same cell, all you need to calculate is the percentage change in fluorescence in response to the stimulus applied, and you are largely unconcerned by the large cell-to-cell variations in fluorescence intensity that plague you when you do $\Delta\Psi$ estimation by flow cytometry.

As I mentioned previously, the distributional dye probes of $\Delta\Psi$ emerged as a spinoff of a project designed to find dyes that could respond within milliseconds to changes in nerve membrane potentials. The first dyes from which Cohen et al got fast responses typically showed fluorescence or absorption changes of about 0.1% in response to a 100 mV change in $\Delta\Psi$. One of these, merocyanine 540[467] (see above), was found to stain immature and leukemic but not normal leukocytes; the staining appeared to be independent of $\Delta\Psi$, but if the fluorescence had changed by 0.1% on VMC addition, nobody would have noticed. No need to worry about membrane potential affecting merocyanine fluorescence in the experiment suggested above.

Better **fast-response $\Delta\Psi$ probes** have since emerged from continued development efforts; one of the best dyes available a few editions back was a styryl compound that exhibited a 21% fluorescence change in response to a 100 mV potential change[468]. Believe it or not, this still wasn't good enough for flow cytometry. If we are looking for a 10 mV depolarization in mitogen-stimulated lymphocytes[465], we can expect to get a 2% fluorescence change in the best dye probe under the best circumstances, and we'd need an instrument CV of less than 1% to be able to detect it. That's one flow cytometer I can't tell you how to build. If you could build it, it wouldn't help, because cells stained with the fast **oxonol**, **merocyanine**, and **styryl** dye probes produce fluorescence distributions with the same large variances seen in cyanine dye fluorescence distributions.

Ratiometric Probes for Membrane Potential

Is there a way to make this a real measurement, i.e., to improve the accuracy or, more to the point, decrease the variance, of distributions observed in flow cytometric $\Delta\Psi$ estimation? Several possibilities suggest themselves. I have mentioned that the ratio of $[C^+]_i/[C^+]_o$ in cells varies with the lipophilicity of the dye C^+ as well as with $\Delta\Psi$. If two cyanine dyes of differing lipophilicities, e.g., a C_2 dye and a C_6 dye, are added to cells, the ratio of intracellular concentrations of the two dyes should vary with membrane potential. Since the dyes presumably bind to the same hydrophobic sites in cells, cells with more sites should take up more of both dyes. Thus, the cell-to-cell variation in numbers of binding sites should not affect the ratio of concentrations of the two dyes in cells. This ratio could therefore be used as an index of $\Delta\Psi$. In order to get this methodology to behave, however, you have to use a multistation system and two dyes with widely separated spectra; otherwise, energy transfer comes in and confuses things.

The **fast potential probes** respond to membrane potential changes by **changing their position and/or orientation in the membrane**; it is claimed that most do not penetrate to the cell interior. Since all of the transmembrane potential difference is developed across the thickness of the membrane, the electric field strength in the membrane itself can be quite high. Loew[469] has designed electrochromic $\Delta\Psi$ probes that undergo spectral changes in responses to changes in the applied electric field. The current champ is **di-4-ANEPPS** [1-(3-sulfonato-propyl)-4-[beta-[2-(di-*n*-butyl-amino)-6-naphthyl] vinyl] pyridinium betaine], which responds via a rapid (less than millisecond) spectral shift to membrane potential changes[1679]. The problem with this dye is that the electric field affects the excitation, rather than the emission, spectrum of the dye; you need a blue and a green excitation beam (441 nm from a He-Cd laser and 515 nm from an argon laser would do), and the membrane potential is calculated from the ratio of fluorescence intensities at 610 nm excited by the two sources. It may not be worth putting together a fancy flow cytometer to try the measurements; Chaloupka et al[2806] attempted cuvette measurements of

ΔΨ in *Saccharomyces cerevisiae* using di-4-ANEPPS, and found that the probe was not localized to the cell membrane and that the magnitudes of responses increased with time, and I suspect that taking a ratio won't cure all of those ills.

Loew's group[1680] also reported that the nonlinear optical phenomenon of **second harmonic generation**, which is responsible for the frequency doubling properties of crystals used in the laser industry, is sensitive to membrane potential; they have found a dye that gives a big signal, but it would definitely require more in the way of excitation power than most of us have in our flow cytometers.

Looking toward simpler solutions, we might reason that if the variance in distributions of cationic dye fluorescence in cells arises because of cell-to-cell variations in the number of dye binding sites, we ought to be able to do dandy ΔΨ measurements, at least on things like peripheral blood lymphocytes, if we can find a lipophilic cationic DNA fluorochrome. If we then add dye at lower concentrations than are needed to get stoichiometric staining of DNA, so as to be sure that some dye binding sites remain available, the fluorescence histogram we get should be a histogram of ΔΨ, shouldn't it? Yes, if the dye stains DNA and doesn't stain all kinds of other stuff in the cytoplasm. Most of the symmetric monomethine cyanine dyes, e.g., $DiOC_2(1)$, $DiSC_2(1)$, $DiQC_2(1)$, and $DiLC_5(1)$, stain cell nuclei, and show fluorescence enhancement when bound to DNA, but they stain cytoplasmic constituents as well. Some blue- and green-excited cationic DNA fluorochromes described by Latt et al[470] seem to present the same problems with nonspecific staining. The asymmetric cyanines, such as thiazole orange and thiazole blue, should be good candidates, except that they're not DNA-specific. I would bet that a suitable dye exists, but not on when someone will find it.

Farinas[2807] et al have managed to come up with a flow cytometric ratiometric technique for membrane potential measurement that uses an asymmetric cyanine dye; they measure fluorescence of $DiBAC_4(3)$ and SYTO-62 in a slow-flow microfluidic system. The ratio of fluorescence of the two dyes changes by approximately 2% for each 1 mV change in membrane potential. The measurement is not made at equilibrium; instead, mixing fluidics insure that the time interval between dye addition and observation is constant (see p. 365). I don't see why the measurement wouldn't work in a more conventional instrument with appropriate sample handling hardware.

Gonzalez and Tsien[2356,2808-9] have described (and, via **Aurora Biosciences**, now part of **PanVera**, patented) a high-sensitivity, fast-response ratiometric method for measuring cytoplasmic ΔΨ that also uses oxonol dye fluorescence, but in a far more specific way. Oxonols bind to both the inner and the outer face of the cytoplasmic membrane; as cells depolarize, dye shifts from the outer face to the inner face, while, as cells hyperpolarize, dye shifts from the inner face to the outer face. Gonzales and Tsien made energy transfer (FRET) measurements using fluorescently labeled lectins[2808] and, later, fluorescently labeled fatty acids[2809] as donors, and oxonols as acceptors, using the ratio of acceptor fluorescence to donor fluorescence as an indicator of ΔΨ. They achieved the best sensitivity with coumarin-labeled phosphatidylethanolamine as the donor and bis(1,3-dihexyl-2-thiobarbiturate)trimethine oxonol as the acceptor, with the fluorescence ratio changing 50% for a 100 mV change in ΔΨ. This is the highest sensitivity ever reported for a fast response (< 2 ms) method. An even faster response (< 0.4 ms, shorter than the duration of nerve action potentials) was achieved using a pentamethine oxonol acceptor.

I'll discuss another, lower-tech, ratiometric membrane potential measurement technique, useful for bacteria, later on.

Using Cyanine Dyes for Flow Cytometric ΔΨ Estimation, In Case You're Still Interested

Now, if things are as bad as I have said, why does anybody bother using cyanines – or other dyes – for flow cytometric estimation of cytoplasmic ΔΨ? I can tell you why I use them; I have a good idea of the limitations of the reagents and the technique and I only use them when I don't need a more precise estimation of ΔΨ than they can give me. You now know most of what I know about the limitations; if you're still interested in using the technique, here's how.

The dye I personally use most often is **$DiIC_1(5)$**, otherwise called **hexamethylindodicarbocyanine** and sometimes known as HIDC, which is available from Molecular Probes and (as a laser dye) from several other companies. I use $DiIC_1(5)$ because its fluorescence can be excited with a red He-Ne or diode laser and measured through 665 nm long pass glass filters (or 660-670 nm bandpass interference filters) using an R928 or other red-sensitive PMT. When I work with 488 nm excitation, I usually use **dihexyloxacarbocyanine [$DiOC_6(3)$]** or **hexamethylindocarbocyanine [$DiIC_1(3)$]**, the latter also known as acronol phloxine. $DiOC_6(3)$ is green fluorescent and can be used with the same detector/filter combination used for fluorescein; $DiIC_1(3)$ is orange fluorescent. I tend to use the C_6 dye at a final concentration of 50 nM and the others at 100 nM to 1 μm.

I make up 1 mM stock solutions of dye in DMSO and keep them in the dark at room temperature. I make working solutions by diluting the stock solution with ethanol so that I can reach the desired final dye concentration by adding 5 μl of working solution to each 1 ml of cell suspension. Since the dye really hates to hang around in aqueous solution, the cells tend to suck up most of it. As a result, you need to keep the cell concentration with which you work fairly constant; I generally shoot for 10^6/ml because that's a good concentration to use for flow cytometry.

If you are working with cells in protein-free media, dye equilibration is usually complete after 15 minutes at room temperature. If there's protein in the medium, I give the cells about 30 minutes at 37 °C. It's important to keep the incubation temperature and the interval between dye addition and introduction of the sample into the flow cytometer constant, and it's not a bad idea to run all samples in an experiment at the same flow rate, because this is an equilibrium staining procedure.

If the experiment you are doing extends over a period of a few hours, it's important to run controls during the course of the experiment as well as at the beginning and end. Controls should include an untreated cell sample, a sample of cells hyperpolarized by addition of 5 µM valinomycin (VMC), and another sample depolarized by addition of 10 µM gramicidin (GRM) (this assumes you are working with cells in a high-Na^+, low-K^+ medium). If you have any doubt about the health of your cells, make up a triple-dip sample of control cells, measure them after they have equilibrated with dye, and then add VMC to what's left of the sample. If the fluorescence of the cells doesn't increase, either the cells are dead or you've used too much dye. If it does increase, add GRM to the remaining cells; they should depolarize. If they don't, something's wrong. VMC, by the way, will not hyperpolarize cells after GRM addition has depolarized them, so make sure you do things in the right order.

I should also remind you that cyanines and other positively charged dye probes of membrane potential are subject to eviction from cells by the **glycoprotein drug efflux pump** (pp. 376-7). This is not as much of a problem when you're doing measurements of cytoplasmic membrane potential, and leaving cells in equilibrium with the dye, as it is when you're looking at mitochondria, and have to wash the cells. Washed cells with an active pump lose dye, period; equilibrated cells with an active pump come to equilibrium with lower concentrations of dye than they would if the pump weren't there, but the intracellular dye concentration still responds to membrane potential changes. If you're paranoid about the pump, use 100 nM $DiBAC_4(3)$; this oxonol dye is unaffected by it.

By following the procedures I have just described, you should be able to establish that the cells with which you are working have a nonzero potential difference across their cell membranes, and that the indicator dye you are using will respond to potential changes in either direction from the control value. You can then examine the effects of the biologic, chemical, or physical agents you are studying on the cells' $\Delta\Psi$.

It is advisable to keep the cells happy if you want them to continue responding over the course of a few hours. If the buffer capacity of the medium won't stand up to the metabolic output of the cells, you may get funny results because pH does have some ill-defined effects on $\Delta\Psi$ probes. If you are working with blood cells, avoid using NH_4Cl or a hypotonic medium to lyse red cells; the former depolarizes cells[471], and the latter may change cation contents (and thus presumably $\Delta\Psi$) and cell volumes of different cell types in different ways[472].

If you add practically any protein to a cell suspension that has been equilibrated with a cyanine or other lipophilic cationic indicator in a protein-free medium, the indicator concentration in cells will decrease because the indicator will bind to the protein in solution. When you do the measurement, it looks as if the cells have depolarized, but they probably haven't. If you're trying to find the effects of adding various proteins to cells, work in a medium with added protein, e.g., 1-10% albumin.

Similar artifactual "depolarization" may appear when you add appreciable concentrations of nonpolar solvents to a dyed cell suspension, which is why it's a good idea to keep the total amount of added DMSO, EtOH, etc. relatively constant from sample to sample and not add amounts of these solvents in excess of 1% of sample volume. If you are adding material dissolved in DMSO, make sure the sample is well mixed, because DMSO, left to its own devices, will sit at the bottom of an aqueous solution, and the hydrophobic molecules in it may be content to stay there unless prodded.

If you want to eliminate mitochondrial influences on the uptake of cyanines or other lipophilic cationic dyes, you can add an **uncoupler**. Felber and Brand[463] used 5-10 µM **carbonyl cyanide p-trifluoromethoxy-phenylhydrazone (FCCP)** in their $TPMP^+$ procedure; I have tried this a few times and it seems to work. Waggoner[462] cautions that the related uncoupler **carbonyl cyanide chlorophenylhydrazone (CCCP)** may interfere with cuvette measurements using $DiSC_3(5)$; I have used CCCP at 5-10 µM for flow cytometry without problems.

For the record, I have tried to measure $\Delta\Psi$ in cell suspensions in cuvettes using cyanine dyes on several occasions, and have only succeeded once; the flow cytometric procedure done on cells from the same samples at the same time has worked. Oxonol dyes, which are nontoxic and which respond to $\Delta\Psi$ changes more rapidly than do cyanines, have been advocated as superior $\Delta\Psi$ probes for cuvette measurements by Tsien, Pozzan and Rink[426]; I'll try them if I ever have occasion to attempt more cuvette measurements

Also for the record, there have been occasions when we shut off a flow cytometer after running cyanine dye-stained lymphocytes from one donor on one day, turned the system on again the next day, and found that the peak of the distribution of fluorescence from another donor's cells was within a few channels of the peak of the distribution obtained the day before. From my point of view, the flow cytometric technique is more reproducible than is the bulk measurement method.

There have been a lot of bad things said about cyanine dyes in the literature, many of which are true. At micromolar concentrations, cyanines have been observed to be toxic to bacteria[473] and mammalian cells[474-7,1683]. The dyes themselves may perturb $\Delta\Psi$ directly by altering membrane conductivity; their inhibition of energy metabolism may also result in $\Delta\Psi$ changes. When used to monitor neutrophil $\Delta\Psi$ responses to chemotactic peptides, $DiOC_6(3)$ and $TPMP^+$ have been reported to give contradictory results[478], while $DiSC_3(5)$ was found to be destroyed by oxidation following neutrophil activation[479].

The affinity of cyanine dyes for cell constituents is so high, and their solubility in water so low, that you can end up with inconclusive results when you use very low concentrations of dye and relatively high concentrations of cells, simply because the dye molecules grab on to the first cells they see[853]. Olive and Durand have taken advantage of this property and used $DiOC_7(3)$, which I told you not to use as an $\Delta\Psi$ probe, as an indicator of perfusion or penetration of materials into tumor cell spheroids[854].

The toxicity of the cyanine dyes is a liability shared in common with other families of cationic dyes and with lipophilic cations such as $TPMP^+$; when the cyanines are used in flow cytometry, at concentrations of 5-100 nM, toxicity is less than when radiolabeled cations or cationic dyes are used at micromolar concentrations for bulk measurements. Different cell types appear to have different degrees of susceptibility to cyanine dye toxicity; Crissman et al[855] found that simultaneous staining with $DiOC_5(3)$ improved both Hoechst 33342 staining of live CHO cells and cell survival following sorting, possibly because of effects on the efflux pump, while the supposedly less toxic rhodamine 123 affected neither staining nor survival. While this may not lead to the routine appearance of cyanine dyes in multivitamin pills, it does remind us that one cell's poison may be another cell's medium.

Cytoplasmic Membrane Potential: Summing Up

It is hardly feasible to make microelectrode measurements of $\Delta\Psi$ in large numbers of cells, and impalement with microelectrodes is probably more traumatic to lymphocytes than to squid axons. Despite this, cationic dyes such as the cyanines, radiolabeled cations, the anionic oxonol dyes, and microelectrodes have produced comparable results in most systems to which two or more measurement methods have been applied.

Tasaki and Byrne[1684] studied the time course of the intrinsic birefringence change of axons that coincides with action potentials, and attribute it to swelling of the nerve brought about by movement of water into the superficial layer of the axon. A close relationship was also demonstrated between nerve swelling and changes in light scattering and in dye absorbance. The intrinsic optical changes, by the way, are on the order of a few parts per ten thousand; you won't be noticing them on your scatter channel, and we'll probably have to stick with using dyes for flow cytometric measurements.

In looking at some studies[1685-93] on flow cytometry that emerged prior to the previous edition, I noticed that at least as many people seemed to use cyanines[1687-9] as oxonols[1690-1,1693]. Seamer and Mandler[1691] applied my trick of dividing by a size measure (Figure 7-28, p. 389) to lower the CVs from oxonol measurements. Damjanovich and Pieri[1692] correlated fluorescence flow cytometry with patch clamp measurements in lymphocytes. Radosevic et al[1693] used DiBAC4(3) to study membrane potential in conjugates of cytotoxic NK cells, which were labeled with a tracking dye, and K562 target cells. Conjugates could be identified on the basis of tracking dye fluorescence and light scattering signals; using a slit-scan technique, the membrane potential of each cell in a conjugate was measured separately, showing that depolarization of the K562 cell occurs as a consequence of the cytotoxic activity of the NK cell. In at least some of the studies in this group[1687-8], in which cyanine dyes were used, I suspect that the efflux pump had some effect on results. Watch out.

Mitochondrial Membrane Potential ($\Delta\Psi_m$)

Mitochondrial Staining with Rhodamine 123 Is Membrane Potential-Dependent

Since 1980, numerous investigators have used the lipophilic cationic dye **rhodamine 123** for investigations of mitochondrial structure and function; it is appropriate at this point to discuss the general issue of **estimation of mitochondrial membrane potential (which I will abbreviate as $\Delta\Psi_m$), as opposed to cytoplasmic membrane potential ($\Delta\Psi$)**, and to clear up some misconceptions that seem to have arisen concerning this particular dye.

As was mentioned on p. 385, there is normally a relatively large (100 mV or more, interior negative) gradient of electrical potential across mitochondrial membranes in cells. The electrical potential and pH gradients between the cytosol and the mitochondrial matrix are inextricably linked to the processes of **energy metabolism**[487-8]. When cells with normal mitochondrial function are allowed to equilibrate with lipophilic cations, potential differences alone generate gradients of cation concentration across mitochondrial and cytoplasmic membranes, with the concentration being highest in the mitochondria.

According to the **Nernst equation** (p. 386), the ratio of mitochondrial to cytosolic (free) cation concentrations should be 100:1 if the potential difference across the mitochondrial membrane is 120 mV, and 1000:1 if it is 180 mV. The presence of any high-affinity binding sites for the cation within mitochondria will result in the ratio of concentrations of (free + bound) cation in mitochondria and cytosol being higher than that predicted by the Nernst equation, unless there are much greater numbers of high-affinity binding sites within the cytosol.

If cell samples are allowed to come to equilibrium with equal concentrations of a series of cationic dyes of similar structure with increasing lipophilicity (e.g., cyanines), higher mitochondrial/cytosolic dye concentration ratios should be found for the dyes with higher lipophilicity (in the case of the cyanines, those dyes with longer alkyl side chains), due to increased binding to hydrophobic membrane structures, which are at least as abundant in mitochondria as in the cytoplasm. The ratio of intracellular and extracellular dye concentrations, $[C^+]_i/[C^+]_o$, should also increase with lipophilicity.

Under the equilibrium conditions just described, both cytoplasmic ($\Delta\Psi$) and mitochondrial ($\Delta\Psi_m$) membrane potential affect the total amount of dye taken up and bound by cells. The relative contributions of cytoplasm and mitochondria to the cytoplasmic "$\Delta\Psi$," as estimated by fluorescence of cells following equilibration with dyes such as cyanines, are, however, difficult to estimate *a priori*, because the total amount of dye in each compartment is the product of concentration and volume. In addition, one cannot expect to measure the same amounts of fluorescence emission from equal amounts of intra- and extramitochondrial dye, because quenching increases with dye concentration.

If cells are washed following a first equilibration with dye, the rate at which a new equilibrium between intramitochondrial, cytosolic, and extracellular dye concentrations is reached should be dependent upon mitochondrial affinity for the dye, because, after washing, the mitochondria serve as the primary source for redistribution of dye into the cytosol and thence into the extracellular medium. More and/or stronger mitochondrial binding of dye should be reflected in longer reequilibration times and, until equilibrium is reached, in higher ratios of intra- to extramitochondrial dye concentrations. Observation of such cells under a fluorescence microscope should show dye fluorescence to be confined to the mitochondria.

All that can be concluded, however, from the observation of fluorescence in the mitochondria of washed cells, or from microphotometric or flow cytometric measurement of cellular fluorescence under such conditions, is that the mitochondria were energized prior to dye addition and probably remained energized at the time of measurement; variations in $\Delta\Psi_m$ cannot be inferred from variations in cell fluorescence. This is so because the number and/or volume of mitochondria may vary from cell to cell and because mitochondrial fluorescence itself is likely not to be proportional to $\Delta\Psi_m$ due to concentration quenching. Recall (p. 388) that the ratio of concentrations of cyanine dyes inside and outside cells is typically at least several hundred, much higher than would be predicted by the Nernst equation; even if the ratio between cytosolic and mitochondrial concentrations is purely Nernstian, one can expect the concentration of dye in the mitochondria to be on the order of 10^4 times the external dye concentration. If the external dye concentration were only 10 nM, the mitochondrial concentration would be 100 μM, and one would expect dye fluorescence to be substantially quenched.

If the mitochondria are **deenergized**, i.e., if $\Delta\Psi_m$ is reduced to or near zero, before dye is initially added to the cells, e.g., by addition of a **metabolic inhibitor**, e.g., an **uncoupler** of oxidative phosphorylation, such as **CCCP**, no mitochondrial concentration of dye occurs in the first place, other than that which might result from dye binding to membranous structures in the absence of a potential gradient. Under these circumstances, the dye concentration in cells left in equilibrium with dye responds to differences in (cytoplasmic) $\Delta\Psi$, and flow cytometric fluorescence and cell size measurements can provide valid, semiquantitative estimates of this parameter provided dye concentrations are kept low, as discussed on p. 392, to minimize saturation and concentration quenching effects. Pretreatment with an inhibitor of mitochondrial energy metabolism similarly allows valid estimates of cytoplasmic $\Delta\Psi$ to be derived from uptake of $TPMP^+$ and similar substances.

If no inhibitor is added to cells, and they are examined while in equilibrium with dye, differences in fluorescence will be observed between cells with intact membranes and normal mitochondrial function, on the one hand, and damaged cells, on the other. Cells with deenergized mitochondria will, as just discussed, take up less dye than cells with intact energy metabolism; cells with cytoplasmic membrane damage sufficient to diminish or abolish the cytoplasmic $\Delta\Psi$ gradient will also take up less dye. Thus, equilibrium dye fluorescence measurements in cells can provide an indication of **cell viability** that is based upon **metabolic integrity** as well as membrane integrity.

Washing cells before measurement can be expected to produce some improvement in live-dead cell discrimination. If mitochondria are deenergized before equilibration with dye, very little fluorescence will be seen in washed cells even if the cytoplasmic membrane is intact. If mitochondria become deenergized after equilibration with dye, but before washing, some concentration of dye in mitochondria may be observed if high-affinity binding results in dye being lost relatively slowly from the mitochondria. If the mitochondria become deenergized after washing, dye should be lost from the mitochondria into the cytosol at a rate dependent upon the affinity of mitochondria for the dye.

Viewed against this background, the observed behavior of **rhodamine 123 (R123)** in mitochondria and cells is no different from what would be expected, and not markedly different from what would be expected and has been observed using other dyes. During the fall of 1979, when my initial publication on flow cytometric $\Delta\Psi$ estimation[424] and their first paper on R123 staining of mitochondria[489] were in press, Lan Bo Chen and Lincoln Johnson asked me if they could look at some R123-stained cells in one of my flow cytometers. We got to talking about membrane potentials; when I found out that R123 was a cationic dye I suggested that the mitochondrial staining they were seeing was potential-dependent, and we exchanged samples of dyes.

The R123 staining technique Johnson et al described[489] involved a 30 min equilibration of cells with 10 μg/ml (about 25 μM) R123, following which cells were washed and examined. Using that concentration of R123, I couldn't measure unwashed cells in the flow cytometer because of the intense background fluorescence. Backing off to 1 μM, I found (unpublished) that I could get R123 fluorescence to respond to the usual $\Delta\Psi$ manipulations involving VMC, GRM, and changes of (K^+) concentration in the same fashion as did the cyanines. Johnson and Chen[490] established that R123 staining was dependent upon the presence of the mitochondrial $\Delta\Psi$ gradient. They also observed that cells equilibrated with safranin and the cyanine dyes showed potential-dependent mitochondrial staining. In fact, cyanine dye-stained cells could be examined without washing, probably because cyanines are considerably more lipophilic than is R123, which should make the ratio of intra- and extramitochondrial dye concentrations at least somewhat higher for the cyanines.

A few weeks after these experiments were done, Paul Horan and coworkers exhibited pictures of cyanine dye-stained fibroblasts at the 1979 Asilomar conference on Analytical Cytology; these showed dye concentrated in mitochondria. Horan et al found, as did Johnson et al using R123, that stimulation of confluent fibroblasts by wounding caused an increase within minutes in mitochondrial fluorescence in cells adjacent to the damaged areas[491]. Until both groups published in 1981, however, most people who were aware of the capacity of R123 to stain mitochondria were unaware of the central role of mitochondrial $\Delta\Psi$ in R123 staining.

James and Bohman[492], for example, used R123 and flow cytometry to study proliferation of mitochondria during the cell cycle of HL-60 cells, and Darzynkiewicz et al[493] reported increased R123 staining of stimulated lymphocytes beginning approximately 24 hr following exposure to lectins, attributing most of the increase in fluorescence to an increase in the number of mitochondria per cell, in work submitted for publication before reports of the potential dependence of R123 staining appeared in print.

The R123 fluorescence response in lymphocyte activation was not surprising to me because I and others[425,456,484] (also M. Brand, personal communication), using $TPMP^+$, microelectrodes, $DiOC_6(3)$, and $TPMP^+$, respectively had found apparent hyperpolarization of lectin-stimulated T lymphocytes beginning at 5-12 hours, and usually preceding increases in RNA content as measured with pyronin Y. I also noted (unpublished) that R123 staining increased in stimulated T cells; an example of typical results obtained from lymphocytes stained with Hoechst 33342 in combination with various other probes appears as Figure 7-30.

In retrospect, some of the differences in R123 fluorescence between resting and stimulated lymphocytes may reflect behavior of the glycoprotein pump, which appears to be more active in resting than in stimulated cells, leading to observable differences in the fluorescence of Hoechst 33342 in when cells are equilibrated at lower concentrations of the dye than are typically used for stoichiometric DNA staining[248-50,253,257] (see pp. 309 and 376).

I had suggested[494] that flow cytometric $\Delta\Psi$ measurements might provide a more sensitive indicator of **cell viability** than was given by dye exclusion tests. This was of interest to several of my colleagues because it might provide a fast means for determining the effects of chemotherapeutic agents on cancer cells. Sam Bernal, Lan Bo Chen and I[495] looked at the effects of various drugs on R123 fluorescence in L1210 mouse leukemia cells, and concluded that loss of R123 fluorescence was a better and earlier predictor of loss of clonogenicity than were dye exclusion tests. However, later work (E. Adams, B. K. Bhuyan et al, presentations at the 1984 Cell Kinetics Society meeting and personal communications) showed that, while loss of R123 fluorescence generally indicates loss of clonogenicity, cells exposed to lethal doses of some drugs may not lose R123 fluorescence for some time. Bhuyan et al demonstrated by biochemical analyses that **R123 fluorescence parallels mitochondrial ATP content**, and that the changes in this parameter following lethal doses of different agents occur with different time courses.

When R123 first came into use, great stress was placed upon its relatively low toxicity as compared to dyes such as the cyanines. Darzynkiewicz et al[340], however, found that R123 inhibited the growth and cell cycle progression of cultured cells, while uncharged rhodamine dyes, which were not taken into mitochondria, were not inhibitory. Chen's group, which had reported different patterns of retention of the dye in normal and transformed cells[496], thereafter discovered a selective toxic effect of R123 on carcinoma cells *in vitro*[497] and *in vivo*[498]. They have since established that the dye interferes with mitochondrial energy metabolism[499].

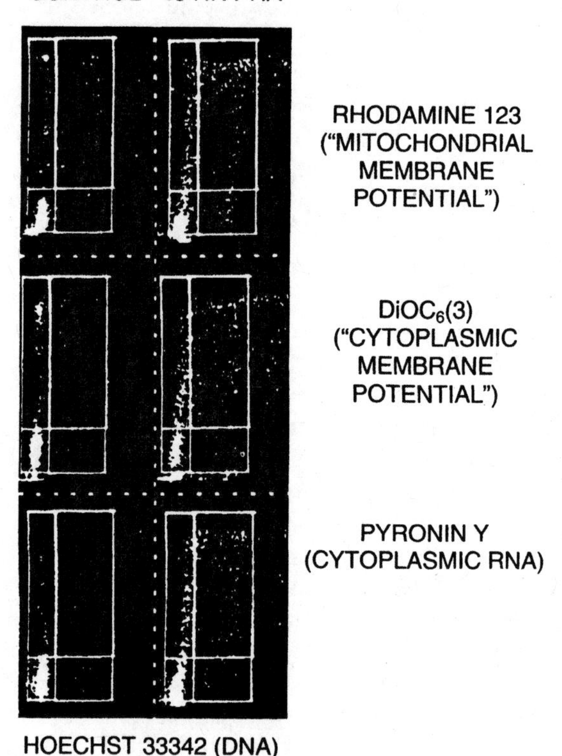

Figure 7-30. Two parameter flow cytometric analyses of lectin-stimulated human peripheral blood lymphocytes 48 hr after PHA addition.

The toxicity of the more lipophilic cyanine dyes, and of lipophilic cations such as $TPMP^+$, relative to R123 must be examined in the light of the fact that differences in lipophilicity exert profound effects on partitioning of the dye between mitochondria, cytosol, and medium. In cells equilibrated with equimolar concentrations of R123 and a lipophilic cyanine such as $DiOC_6(3)$, the intramitochondrial concentration of the cyanine dye may be hundreds of times higher than that of R123. Conversely, there exist other fluorescent probes of mitochondrial energization which, like R123, are relatively water-soluble, and which are also said to be nontoxic; e.g., the styryl dyes **DASPMI (dimethylaminostyrylpyridinium iodide)** and **DASPEI** described in 1976 by Bereiter-Hahn[500]. The studies of R123 as an anticancer drug recalled the therapeutic use of $DiSC_2(5)$ as an antihelminthic agent under the name Dithiazanine[501]. I have often wondered whether some of the unexplained sudden deaths associated with this dye/drug were related to the observation by uninformed patients of brilliant blue stools following its oral administration. In a more serious vein, it should be pointed out that Zigman and Gilman[475] have related the toxicity of cyanine dyes to redox potential; their work should be read by anyone interested in the pharmacologic, as opposed to the flow cytometric, aspects of dye probes of cell membrane potential. It is also evident from the work of Chen's group that thiacyanine dyes are more toxic than indo- and oxacyanines[1683].

Since the late 1980's, a lot of interesting things have happened in the mitochondrial area, with flow cytometry playing a pivotal role.

Borth, Kral, and Katinger[1699] established that R123 fluorescence of hybridomas was responsive primarily to glucose concentration, rather than to aging or growth; this was significant since R123 fluorescence was proposed as a useful parameter that has been proposed for use in monitoring cells in bioreactors. Myc et al[1700] used Hoechst 33342 and R123 to determine that there was no variation with phases of the cell cycle in R123 retention characteristics in either normal lymphocytes or leukemic cells.

Irion et al[1706] and Rottele and Zimmermann[1707] studied the kinetics of uptake of a variety of dyes with different lipophilicities into HeLa cell mitochondria *in situ*. The kinetics of uptake were found to be limited by diffusion; more lipophilic molecules were taken up more rapidly. Dye molecules accumulated in strongly lipophilic areas of the mitochondria, and both the potential gradient and hydrophobic interactions contribute to strong dye binding. Styrylpyridinium dyes are of interest because, unlike cyanines, they exhibit very low fluorescence in aqueous solution.

The Search for Better $\Delta\Psi_m$ Probes: Round One

Maftah et al[1696] and others[1697-8] described the use of **nonyl acridine orange (NAO)**, which they reported stained mitochondria in a membrane potential-independent fashion. One of the problems with attempting to measure $\Delta\Psi_m$ using the fluorescence of R123 or other dyes is that the number and size, or mass, of mitochondria vary from cell to cell; it is therefore not possible to determine to what extent differences in dye fluorescence between cells reflect differences in $\Delta\Psi_m$, as opposed to differences in these other factors. In theory, NAO fluorescence would provide a measurement of mitochondrial mass. Since the fluorescence emission maxima of NAO and R123 are quite close to one another, it would not be feasible to attempt ratiometric measurements. However, I suggested in the previous edition that the ratio of fluorescence of a suitably spectrally separated cationic dye, e.g., $DiIC_1(5)$ and NAO might provide an indicator of $\Delta\Psi_m$ normalized for differences in mitochondrial mass; I was wrong. I also speculated that the nonyl group locked NAO into the mitochondrial membrane, and it looks as if I was right about that; I can't finish the story without bringing in some other developments.

O'Connor et al used flow cytometry with rhodamine 123[1703] and other dyes[1704] to analyze isolated individual mitochondria; staining with R123, safranin O, and $DiOC_6(3)$ were potential dependent, while staining with NAO was not, as would be expected. Wolf and Kapatos[1705] were able to discriminate isolated neuronal synaptosomes from free mitochondria because, under the experimental conditions used, the former maintained a membrane potential and the latter did not.

Juan et al[2310] described a kinetic measurement procedure using rhodamine 123 to assess mitochondrial membrane potential; they found that, if dye and cell concentrations were kept constant, the rate of uptake of R123 by isolated rat hepatocytes was relatively constant, with increasing concentrations of added glucose or other substrates increasing the dye uptake rate and addition of uncouplers decreasing the dye uptake rate.

Hahn et al[1702] described a photocrosslinking fluorescent indicator of mitochondrial membrane potential; a carbocyanine dye was derivatized with a photoreactive nitrophenylazide group so that illumination would covalently attach it to nearby molecules. The dye, **PhoCy** (photofixable cyanine), specifically stained mitochondria in living fibroblasts; when stained cells were illuminated and fixed with formaldehyde, staining was retained owing to cross-linking, while fixation without illumination eliminated mitochondrial staining. If the mitochondria in cells were energized in the first place, one could, in theory, label them with PhoCy, wash out excess dye, lock the PhoCy into the mitochondria by illuminating the cells, apply some treatment, and restain with a spectrally distinct cationic dye to determine whether or not $\Delta\Psi_m$ had decreased. However, PhoCy itself apparently didn't make it to market, although some other reactive cationic mitochondrial dye probes eventually did.

$\Delta\Psi_m$, JC-1, and Apoptosis

Lan Bo Chen published two reviews, in 1988[1694] and 1989[1695], reflecting what he and his coworkers had done with R123 and other dyes up to that point. In 1991, his group described the use of the symmetric cyanine dye 5,5',6,6'-tetrachloro-1,1',3,3'-tetraethylbenzimidazolocarbocyanine iodide (**JC-1**), which forms fluorescent aggregates, as a probe of mitochondrial membrane potential[1681-2]. When excited at 490 nm, JC-1 monomers show an emission maximum at 527 nm, while aggregates emit maximally at 590 nm. In theory, the orange to green fluorescence ratio should give you a measure of membrane potential.

I've tried JC-1 in cells and bacteria; the biggest problems I've found with it is that it aggregates in solution and that it takes much longer to equilibrate to stable fluorescence values than do the more popular dyes such as $DiOC_5(3)$ and $DiOC_6(3)$. The fluorescence distributions in two dimensions are broad enough so that taking ratios doesn't help all that much. There are probably better dyes out there with similar properties if you want to do ratiometric measurements, but looking at two-color fluorescence from JC-1 has turned out to be very useful, even if it won't let you measure $\Delta\Psi_m$ to the nearest millivolt.

Chen's group and collaborators[1708], using NMR, found that R123 had different effects on intermediary metabolism of normal and cancer cells. On another front, the report by Chen's group[1682] that membrane potentials were different based on differences in JC-1 fluorescence at different points in mitochondria resonated with the conclusion drawn by Skulachev and his colleagues, based on analyses of fibroblast and cardiac myocyte mitochondria[1709] and on filamentous cyanobacteria[1710], that networks of mitochondria (or bacteria) act as power-transmitting protonic cables, with membrane potential changes propagating along the network[1711]. A cellular information superhighway? For what it's worth, Bedlack, Wei, and Loew[1712] observed localized membrane depolarizations that they believed to be involved in electric field-directed neurite growth. This gets us from subcellular power transmission lines to the kind a lot of people worry about these days; it's not at all illogical that electric field effects would be mediated via membrane potential changes, but I haven't seen anything along this line hit the newspapers yet. I haven't looked that hard at the journals. Chen's group published a methods paper and review on JC-1 in 1995[2810], but things didn't get really interesting until after that.

In 1993, Cossarizza et al[1701] reported using JC-1 for two-color flow cytometric measurement of mitochondrial membrane potential; while, as I said, it didn't give them values to the nearest millivolt, it did (and does) a much better job of discriminating cells with energized and deenergized mitochondria than did other popular dyes[2811].

In 1993, Richter[2812] suggested that maintenance of $\Delta\Psi_m$ was important in preventing **apoptosis**; in 1994, Cossarizza et al[2813] and Vayssiere et al[2814] reported that $\Delta\Psi_m$ was reduced in apoptosis, and, in 1995, Cossarizza et al[2815] reported that N-acetylcysteine stabilized $\Delta\Psi_m$ and prevented apoptosis in cells treated with agents that would normally induce the process. Richter and Cossarizza et al joined forces on a 1996 paper emphasizing the role of maintenance of $\Delta\Psi_m$ and ATP production in the prevention of apoptosis. The concept, and JC-1, started to get a lot of attention from that point on; I'll have more to say about the subject when I discuss apoptosis in Chapter 10. Meanwhile, I'll continue with the story of mitochondrial probes.

The Search for Better $\Delta\Psi_m$ Probes: Round Two

While, by 1997, JC-1 was recognized as more reliable in providing information on the state of energization of mitochondria than either $DiOC_6(3)$ or R123[2811], it did not provide a quantitative measure of $\Delta\Psi_m$, at least when applied to whole cells. Rottenberg and Wu[2817] noted in 1997 that $DiOC_6(3)$ itself induced apoptotic changes in lymphocytes from old mice, which led them to consider some attributes of this and other dyes that might affect their use as probes of $\Delta\Psi_m$, and, in 1998, to report that $DiOC_6(3)$ could be used for quantitative flow cytometric measurements of $\Delta\Psi_m$[2818].

It had been known for some time that DiOC6(3) and other oxacarbocyanine dyes inhibited NADH reductase in isolated mitochondria *in vitro*; Rottenberg and Wu reasoned that mitochondria *in situ* would be affected by substantially lower concentrations of dye in solution, because of the Nernstian and non-Nernstian concentration that occurs in cells. They found that the respiration of lymphocytes exposed to 40-100 nM $DiOC_6(3)$, a range of concentrations commonly used for assessment of was inhibited approximately 90 percent, suggesting that the probe itself was likely to alter the parameter they were trying to measure.

Apart from their effect on respiration and $\Delta\Psi_m$, extracellular dye concentrations in the 40-100 nM range create problems because, as mentioned on pp. 388 and 395, the concentration of dye in energized mitochondria can easily reach 10,000 times the extracellular dye concentration, and, at intramitochondrial concentrations in the 400-1000 µM range, the fluorescence of dye in mitochondria should be almost completely quenched.

I noted, in the discussion of measurements of cytoplasmic membrane potential, that $DiOC_6(3)$ fluorescence in cells, measured by flow cytometry, becomes independent of membrane potential at an extracellular dye concentration of 2 µM. I also noted that membrane potential changes in cells exposed to that concentration of dye can be measured in cuvettes, because the cuvette measurement measures the fluorescence of free dye in solution. In order to keep cellular fluorescence responsive to changes in cytoplasmic membrane potential, it is necessary to use lower concentrations of dye than might be used for cuvette measurements, and the 10-100 nM concentrations typically used for flow cytometry may be advantageous in terms of eliminating much of the effect of $\Delta\Psi_m$ on cytoplasmic membrane potential by virtue of extensive quenching of intramitochondrial dye.

When we are trying to measure $\Delta\Psi_m$ by flow cytometry, we cannot practically distinguish between fluorescence from intramitochondrial dye and fluorescence from dye in the extramitochondrial portions of the cell. We would prefer to have a monotonic relationship between measured fluorescence and $\Delta\Psi_m$; the fluorescence intensity measured from cells with energized mitochondria should be greater than that measured from cells with deenergized mitochondria, and cells with intermediate values of should yield intermediate values of fluorescence intensity. This is unlikely to happen when there is extensive quenching of dye in mitochondria; it is far more likely that a decrease in $\Delta\Psi_m$, resulting in release of dye into the cytosol, will lead to an increase in cellular fluorescence. Rottenberg and Wu[2818] found that 0.2 nM $DiOC_6(3)$ provided a monotonic fluorescence response across the range of $\Delta\Psi_m$.

Since the concentrations of JC-1 typically used for studies of $\Delta\Psi_m$ are in the range between 1 and 10 µM, it is not unlikely that perturbation by the dye will complicate any attempt to derive a quantitative measurement. Because the response of JC-1 to $\Delta\Psi_m$ is largely dependent on the concentration of dye being high enough to promote formation of J-aggregates, it is likely that reducing the concentration to the nanomolar range would eliminate the metachromatic fluorescence response that has provided most of the motivation for using the dye. Cossarizza et al[2819] did find a linear relationship between JC-1 orange (590 nm) fluorescence and $\Delta\Psi_m$ in isolated mitochondria, but it is likely that the intramitochondrial dye concentration in that experimental situation was lower than would be expected in mitochondria *in situ* due to the absence of the Nernstian and non-Nernstian concentration of dye into a cytoplasmic extramitochondrial compartment.

In 1996, Poot et al[2820], at Molecular Probes, described two reactive probes that responded to $\Delta\Psi_m$, **chloromethyl-X-rosamine (CMXRos)** and **CM-H_2XRos**, both given the **MitoTracker Red** trade name. Both probes are lipophilic and cationic, and are concentrated in mitochondria, where the chloromethyl groups on the probes bind covalently to thiols. CMXRos is fluorescent; its excitation maximum is at 578 nm, but some fluorescence is excitable at 488 nm. The emission maximum is at 599 nm. CM-H_2 XRos, the reduced form of CMXRos, is nonfluorescent; once inside cells with an active energy metabolism, it is oxidized to CMXRos.

The ability of these probes to bind covalently to constituents of mitochondria allows them to withstand fixation, as did the PhoCy dye described on p. 397, but there has been a running controversy in the literature as to how faithfully CMXRos measurements of fixed cells represent the state of mitochondrial energization before fixation. In dual staining experiments, Macho et al[2821] found CMXRos fluorescence to be highly correlated with $DiOC_6(3)$ fluorescence, and pursued immunofluorescence measurements of intracellular proteins such as Bcl-2 in cells that had been fixed and permeabilized following CMXRos staining. Ferlini et al[2822] questioned whether some of the correlation between CMXRos and cyanine dye fluorescence represented spectral crosstalk, but Macho et al[2823], in response, cited additional evidence that CMXRos responded to $\Delta\Psi_m$.

Molecular Probes offers several fixable mitochondrial probes in addition to CMXRos and CM-H_2XRos; these include **MitoTracker Green FM**, which has spectral characteristics similar to fluorescein but is almost nonfluorescent in aqueous solution, and **MitoTracker Orange CM-H_2TMRos** and **CMTMRos**, which emit at 576 nm[2332]. Hollinshead et al[2824], in a paper demonstrating the advantages of antibiotin antibodies, which react with endogenous biotinylated proteins in mitochondria, for specific labeling of those organelles in fixed material, reported that MitoTracker Green FM staining, representative of the state of mitochondrial energization before fixation, was preserved in fixed tissue. Poot and Pierce[2825] reported that, in cells stained with 200 nm each of MitoTracker Green FM and CMXRos, the ratio of CMXRos fluorescence to MitoTracker Green fluorescence provided better discrimination of apoptotic cells than did CMXRos fluorescence alone. A challenge by Gilmore and Wilson[2826] was followed by a response to Ferlini et al by Poot and Pierce[2827], maintaining that changes in CMXRos fluorescence are sensitive indicators of early apoptosis. Salvioli et al[2828], in a study of HL-60 cells, conceded that, while JC-1 and CMXRos did not always give identical results, CMXRos was a reliable probe for assessing $\Delta\Psi_m$, but Mathur et al[2829], studying cultured cardiomyocytes, found that JC-1 staining did reflect $\Delta\Psi_m$, but that CMXRos staining did not.

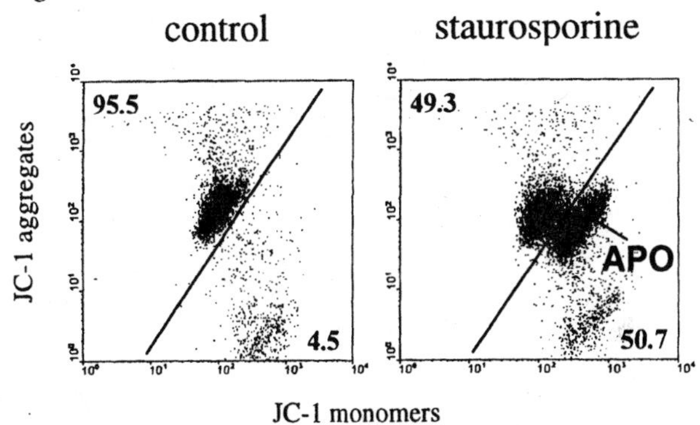

Figure 7-31. Demonstration of apoptotic HL-60 cells (APO) with deenergized mitochondria by JC-1 staining. Apoptosis was induced by staurosporine. The "JC-1 aggregates" axis displays 590 nm fluorescence; the "JC-1 monomers" axis displays 530 nm fluorescence. From: Salvioli et al, Cytometry 40:189-197, 2000 (Reference 2828), © John Wiley & Sons, Inc., used by permission.

If you decide to be conservative and rely on JC-1 as your probe for $\Delta\Psi_m$, Figure 7-31, above, gives you an idea of what you can expect. The apoptotic cells contain somewhat fewer JC-1 aggregates, as indicated by slightly lower 590 nm fluorescence, and substantially more JC-1 monomers, as indicated by 530 nm fluorescence, than do the unaffected or control cells. It would not be unusual to see lower values of 590 nm fluorescence in apoptotic cells than are shown in this figure.

In 2000, Keij et al[2830] reported that mitochondrial staining by NAO (p. 397), MitoTracker Green, and another Molecular Probes offering, **MitoFluor Green**, was affected by $\Delta\Psi_m$ changes induced by drugs prior to staining. A critique by Isola et al[2831] noted that most people stained first and applied treatments later. However, Keij et al have made the point that the mitochondrial uptake of all of the dyes they used is potential-dependent, which should bring us closer to the conclusion that neither NAO nor any of the MitoTracker dyes is a probe of "mitochondrial mass."

It had been alleged for many years that NAO binding to mitochondria was at least partly potential-independent, and reflected the affinity of the dye for **cardiolipin**, a polyunsaturated acidic phospholipid found exclusively in bacterial and mitochondrial membranes. NAO has, therefore, been used as a probe for cellular cardiolipin content as well as a probe for mitochondrial mass; Jacobson et al[2832], who have confirmed that NAO binding responds to $\Delta\Psi_m$, as Keij et al[2830] reported, may have slain that dragon. Or maybe not. Rhodamine 123 had magical properties attributed to it for years before JC-1 came along, but the mitochondrial staining properties of both dyes are largely deducible from their lipophilicities, structures, and association constants. We know that the MitoTracker dyes stay in mitochondria because they bind covalently once they are driven in by cytoplasmic and mitochondrial membrane potential gradients and by lipophilicity, and we can be pretty sure that NAO, once driven in by the same factors, is locked into mitochondrial membranes by its tracking dye-like nonyl side chain. If you're looking for a mitochondrial mass probe, anti-biotin antibodies[2824] may not be a bad bet.

That's about it for mitochondria; I should mention that Pham et al[2765] have developed a neat technique for looking at the activity of individual components of the respiratory chain in digitonin-permeabilized cells, using either 40 nM $DiIC_1(5)$ or 90 nM CMXRos as a probe of $\Delta\Psi_m$, and simultaneously detecting reactive oxygen species with carboxy-DCF (p. 379).

Bacterial Membrane Potentials

Bacteria, unlike eukaryotes, do not compartmentalize their respiratory enzymes inside mitochondria; the enzymes are, instead, located on the inner surface of the cytoplasmic membrane. As a result, the 100-200 mV, inside-negative, potential gradient generated by bacterial energy metabolism exists across the cytoplasmic membrane, and some bacteria stain in more or less the same potential-dependent fashion with rhodamine 123, cyanine dyes, and other fluorescent lipophilic cationic dyes[856-7]. Other bacteria, however, have to be coaxed into staining. In general, there's not much of a problem getting **Gram-positive** bacteria to take up lipophilic dyes, because the dyes readily get through the bacterial cell wall. The **Gram-negative** bacteria are more of a problem; the lipopolysaccharide-containing **outer membrane**, really a cell wall and not a membrane, efficiently excludes lipophilic compounds. It is possible to render the outer membrane permeable to lipophilic materials using any one of a number of chemical agents; the most widely used is probably **ethylene diamine tetraacetic acid (EDTA)**, which removes calcium from the structure.

Since bacterial membrane potential changes rapidly in response to the availability or lack thereof of suitable energy sources, and is rapidly dissipated when the organism is killed by drugs or other agents, it is possible to exploit this parameter and potential-sensitive dyes in rapid cytometric procedures for bacterial detection, identification, and antibiotic susceptibility testing. This subject will be further discussed in Chapter 10; here, I will consider some particular problems in bacterial membrane potential measurement, and describe at least one possible solution.

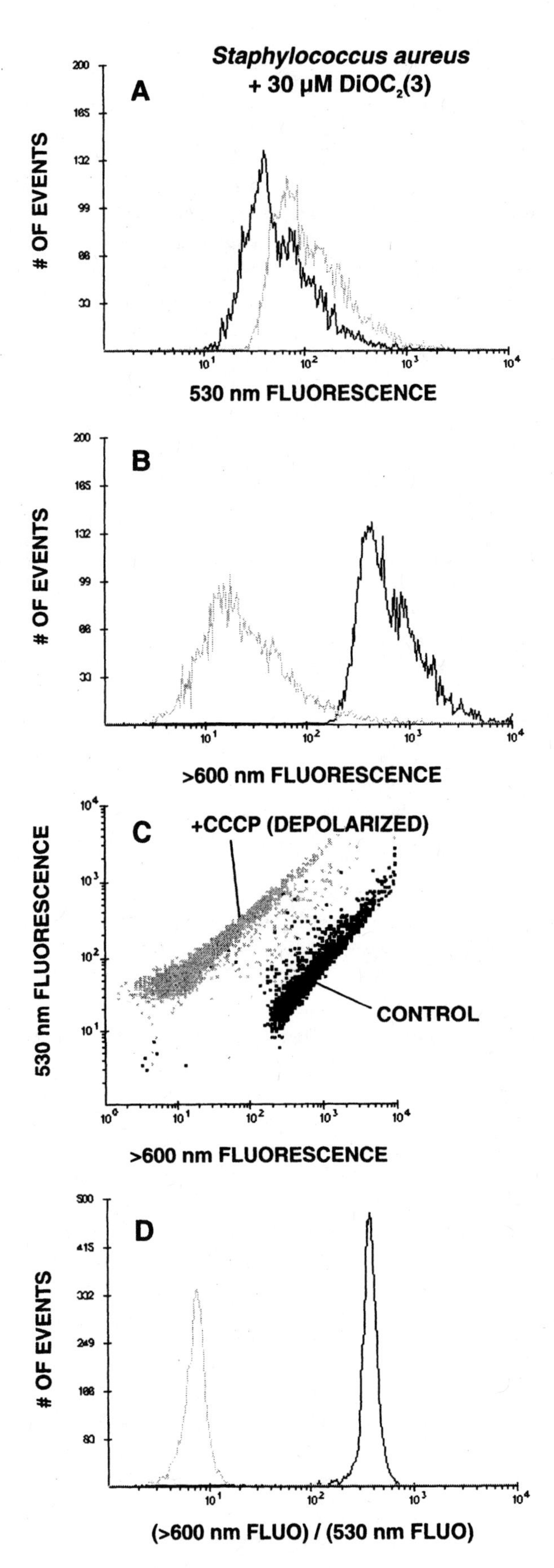

Figure 7-32 (legend on next page)

Figure 7-32 (opposite page). Measurement of the membrane potential of *Staphylococcus aureus* using $DiOC_2(3)$ and a ratiometric technique developed by Novo et al[2357]. Cells are loaded with 30 µM dye, and fluorescence, excited at 488 nm, is measured at 530 nm and at >600 nm. The curves and dots plotted in gray represent cells depolarized with CCCP; the curves and dots plotted in black represent control cells. Plots were created with FCS Express software (DeNovo Software).

Ratiometric ΔΨ Measurement in Bacteria

There are now a substantial number of papers in the literature (see Chapter 10 for references) describing flow cytometric measurement of ΔΨ in bacteria, both as a means of determining the metabolic state of bacteria in their natural environments and as a means of assessing response to antimicrobial agents or lack thereof. Rhodamine 123, $DiOC_6(3)$, and the oxonol dye $DiBAC_4(3)$ have all been used as probes, and they all share a disadvantage which can be appreciated from panels A and B of Figure 7-32 and from Figure 7-27 (p. 388). The distributions of dye fluorescence from cells that are maintaining normal membrane potential and from cells that have been completely depolarized by an ionophore (gramicidin A or CCCP) overlap enough so that one cannot clearly discriminate depolarized from control cells. This becomes a serious problem when one is attempting to determine whether bacteria will respond to a drug; killing 95 percent, or even 99 percent, of the population isn't good enough.

A few years back, David Novo, now known as the entrepreneurial genius behind **DeNovo Software** and its flagship product **FCS Express**, but then fresh out of college, came to my lab and did a creditable imitation of a very good postdoc. Dave wanted to develop a flow cytometric antimicrobial susceptibility test, and we started looking at cyanine dye fluorescence in *Staphylococcus aureus*, which, while otherwise fairly user-friendly, grows in clumps. This creates even more of a problem than there is with most other bacteria when you look at fluorescence distributions, because fluorescence from a clump of *n* cells is *n* times as bright as fluorescence from a single cell. In work with oxacarbocyanines some years back, I had noted that there was a fair amount of fluorescence detectable from these dyes at wavelengths above 580 nm, and Dave followed up on these observations with some rigorous spectrofluorimetric analyses that established that, at very high concentrations, the $DiOC_n(3)$ dyes developed a second emission peak at around 610 nm. The concentration dependence suggested that the long-wavelength fluorescence comes from J-aggregates.

In order to get substantial emission above 600 nm from microorganisms, we had to use much higher dye concentrations than we or anyone else had used previously; the concentration we found suitable for the most tractable of the dyes, $DiOC_2(3)$, was 30 µM. When we loaded bacteria with dye at this concentration, we found that 530 nm fluorescence from the organisms (panel A of Figure 7-32) no longer changed significantly with ΔΨ, but that fluorescence above 600 nm (panel B of Figure 7-32) did. Both 530 nm and >600 nm fluorescence signals appeared to be sensitive to cell (or clump) mass or size, and a dot plot (panel C of Figure 7-32, also shown in color on the back cover) showed high degrees of correlation between the two fluorescence values in clusters of depolarized and control cells, with the two clusters clearly separated. It thus came as no surprise to us that the ratio of

(>600 nm fluorescence)/(530 nm fluorescence) ,

multiplied by an appropriate constant to get it on the display scale, could serve as a measure of ΔΨ. This quantity is plotted in panel D of Figure 7-32, and it is clear that the degree of overlap between values of the ratio measured in depolarized and control cells is extremely small.

By examining valinomycin-treated organisms in buffers containing a range of potassium concentrations (distributions are shown on the back cover), the fluorescence ratio measurement can be calibrated to read out values of ΔΨ; we found that the usable range for *S. aureus* ran from approximately -30 mV to approximately -130 mV.

As far as I know, the ratiometric technique we described remains the most accurate and precise flow cytometric method for measurement (we've gone beyond estimation) of bacterial ΔΨ. The details may be found in reference 2357; I should, however, add a few things we have learned since that paper was published.

The technique can be applied to Gram-negative bacteria if EDTA is used to permeabilize the outer membrane, but, while it did provide the same large degree of separation of clusters of depolarized and control cells as we noted when working with Gram-positive organisms, we were unable to construct a calibration curve because the valinomycin-potassium buffers tended to lyse the EDTA-treated Gram-negative organisms.

We examined a number of indocarbocyanine, indodicarbocyanine, thiacarbocyanine, and thiadicarbocyanine dyes and did not find one that exhibited sufficient long wavelength fluorescence at high concentrations to permit its use for ratiometric measurements. The Molecular Probes dye **JC-9**[2332], a dibenzo homologue of $DiOC_2(3)$ that shares the tendency of JC-1 to form fluorescent J-aggregates, can be used for ratiometric measurements but the fluorescence ratio distributions obtained from control and depolarized bacteria stained with JC-9 are not as well separated as those obtained from bacteria stained with $DiOC_2(3)$. In our hands, JC-1 itself has not been usable for ratiometric measurements of ΔΨ in bacteria.

I suspect that $DiOC_2(3)$ could be used for ratiometric membrane potential measurement in isolated mitochondria, however, at the concentration necessary to produce a substantial fraction of J-aggregates, it is likely to perturb mitochondrial function[2818].

In the paper[2357], we reported that we were unable to grow bacteria after they had been exposed even briefly to 30 μM $DiOC_2(3)$. However, Imogen Wilding of GlaxoSmithKline reports (personal communication, 2001) that it is possible to culture organisms that have been exposed to the dye at 30 μM and at higher concentrations. I should have learned from my experience with pyronin Y (p. 324) that growing cells is not my strong point.

It is worth thinking about why this particular ratiometric method works so well in bacteria. I used to use cyanine dye fluorescence for $\Delta\Psi$ estimation, working with an external concentration of no more than 100 nM. Assuming that $\Delta\Psi$ = -122 mV, which would make the intracellular dye concentration 100 times the extracellular concentration according to the Nernst equation, and adding another factor of 10 for non-Nernstian effects that increase intracellular dye concentration, we come up with an intracellular concentration of 100 μM, which sounds very impressive, until we consider two other factors. First, the volume of the bacterial cell is only about 1 fL, and the word from Avogadro is that, given that intracellular concentration, we will only find about 60,000 molecules of dye in the organism. Second, most of those molecules are likely to be quenched, meaning that the quantum efficiency will not be on the order of 0.5, which is not atypical for cyanine dyes in nonpolar solvents, but is more likely to be a tenth of that value or less, so it will take at least ten molecules of intracellular dye to give us one MESF worth of fluorescence, leaving us with no more than 6,000 MESF. If we assume that the Q value (p. 222) for detecting a cyanine dye is similar to that for detecting fluorescein (.012 photoelectrons/MESF), we'll generate one photoelectron for every 80 MESF, or 75 photoelectrons for 6,000 MESF. If photoelectron statistics were the only source of variance in the fluorescence distribution, the CV would be no lower than 11.4 percent. The ratiometric measurement method uses a dye concentration 300 times higher, and, even though we can expect there to be even more quenching of the fluorescence of intracellular dye, the photoelectron statistics should improve substantially.

The ratiometric $\Delta\Psi$ measurement described by Gonzales and Tsien[2356,2808-9] (p. 392), using a labeled fatty acid or tracking dye and an oxonol, could, in principle, be applied to bacteria, but since one would expect there to be only a few thousand molecules of each of the dyes present per cell, it would probably be necessary to use either a slow flow system or a static or scanning cytometer to get good enough photoelectron statistics to make the exercise worthwhile.

$\Delta\Psi$ Measurement: Cautions and Conclusions

Whether you measure cytoplasmic, mitochondrial, or bacterial membrane potential, you will end up working with some extremely lipophilic dyes. Lipophilic implies hydrophobic; these materials are almost insoluble in water and would prefer to be almost anywhere else. That's what accounts for their non-Nernstian uptake into cells, and it also makes them stick to just about any surface to which they may be exposed while in aqueous solution. The dyes, and the ionophores used to prepare controls, which are also lipophilic, will bind to the tubing in your flow cytometer. If you don't flush the system carefully when you are finished, the dyes and ionophores will abandon the tubing for the next batch of cells that comes along. I use both chlorine bleach and alcohol to clean my system, and, when I think it is clean, I run some unstained cells through and see whether they become fluorescent during the run. If they do, I go through another cleaning cycle. As a last resort, I change all the sample tubing. That's easy to do on a Cytomutt; it may not be on your apparatus. If you use a shared instrument, it might be a good idea to check with your fellow users and the operator, if there is one, before you attempt to do ratiometric measurements in bacteria with 30 μM cyanine dye.

When you do immunofluorescence measurements on fixed cells, you don't find that your cells have good days and bad days; when you measure membrane potential, or pH, or calcium, in live cells, you do. If the cells (or at least the control cells) aren't happy, you won't be.

You do have to be fairly obsessive-compulsive about technique to get good results. The primary papers and the protocol books I mentioned in Chapter 2 and at the beginning of this chapter are a good place to start; if you need more detail or clarification, an e-mail to an author or a posting to the Purdue Cytometry Mailing list is likely to get you the information you need.

Optical Probes of Intracellular Calcium

The Bad Old Days

In 1984, when the First Edition of this book was written, it wasn't clear that flow cytometry had much to contribute in the area of intracellular calcium measurements; a lot of people were interested, and the existence and importance of calcium fluxes and redistributions was obvious, but the probes weren't there. Now, they are.

The older generation of optical and alternative techniques for measurement of intracellular Ca^{++} concentrations were discussed at length in a book by Thomas[502] and in review articles by Blinks et al[503] and Tsien[504]. Among the more classical optical probes of intracellular $[Ca^{++}]$[502-4] are **aequorin**, a luminescent protein now best known because of its connection to GFP, and **arsenazo III**, a metallochromic dye, neither of which enters intact cells. Arsenazo III response to changes in $[Ca^{++}]$ is measured by a spectral shift detected by two-wavelength absorption measurements; this technique could conceivably be adapted to flow cytometry but cells would have to be permeabilized or microinjected to get the dye in before they could be studied.

Chlortetracycline as a Probe of "Membrane-Bound" Calcium in Cells

Caswell and Hutchison[505-6] described the use of the antibiotic **chlortetracycline (CTC)** (not, I hope, likely to be confused with the tetrazolium dye with the same abbreviation (p. 379)) as a fluorescent probe of intracellular "membrane-bound" Ca^{++}. The concept of **"membrane-bound" Ca^{++}** arose from their finding that the fluorescence of CTC was increased when the compound was bound in hydrophobic regions, and that the fluorescence of Ca^{++} and Mg^{++} chelates of CTC was higher than that of CTC. Measurement of emission at 520 nm with excitation at 390 nm supposedly maximizes the discrimination of fluorescence emitted by molecules of CTC-Ca^{++} chelate bound in hydrophobic regions from fluorescence emitted by free CTC, bound CTC, and other chelates.

When cells equilibrated with 10-200 μM CTC are examined under a fluorescence microscope, fluorescence is observed in the same membranous and lipid-bearing structures that are stained by cyanine dyes. If the cells are then washed and the fluorescence of a cell suspension is measured in a cuvette, a gradual decrease in fluorescence is observed as CTC is released from cells into the medium. Treatment of the cells with **Ca^{++} ionophore A23187** or with **ligands reactive with cell surface receptors** (e.g., chemotactic peptides in the case of neutrophils[437], bethanechol in pancreatic acinar cells[507-8], con A in thymocytes[509]) causes a decrease in fluorescence, observable under the microscope as well as in a fluorometer. This is interpreted as indicating **release of Ca^{++} from hydrophobic (membrane) binding sites**.

Flow cytometry of CTC-stained cells[112], using UV or violet excitation from lasers or arc lamps, yields "haystack" fluorescence distributions with large variances, similar to those obtained from cells stained with dye $\Delta\Psi$ probes. The distributions shift following addition of A23187 or stimulating ligands, but one cannot discriminate subpopulations of stimulated cells. Thus, single cell studies with CTC are probably better done by methods that allow sequential observations of the same cells than by flow cytometry.

Spectral studies by Mathew and Balaram[510] suggest that two-wavelength emission measurements of CTC fluorescence excited in the UV might yield a ratio sensitive to shifts in intracellular Ca^{++}, which should have a narrower distribution than is observed for CTC fluorescence. Since the CTC spectrum is sensitive to pH changes in the physiologic range, however, which may also occur during cell activation, results might be difficult to interpret.

Probes for Free Cytoplasmic Calcium: Quin-2

The first probe to come into vogue for studies of **free cytoplasmic [Ca^{++}]** was **quin-2**, first introduced by Tsien and his associates.[504,511-8]. This material is an extremely selective calcium chelator that is introduced into cells as an inactive ester that yields reactive quin-2 after enzymatic hydrolysis. Quin-2 is UV-excited (334-337 nm is the favored wavelength region) and blue fluorescent (maximum emission from the intracellular Ca^{++} chelate is at about 490 nm). However, the fluorescence emission from the amount of quin-2 that can be gotten into a cell is pretty weak, generally no more than 10 times the level of cellular autofluorescence (in the spectral regions applicable to quin-2 excitation and emission measurement, this is generally pyridine nucleotide autofluorescence). The Ca^{++} chelate is about 7 times as fluorescent as the uncomplexed dye.

Quin-2 has been used for bulk measurements of cells in cuvettes and for single-cell measurements by microphotometry; it has been possible using either method to demonstrate rises in cytoplasmic [Ca^{++}] that occur within seconds following receptor stimulation or ionophore addition and which abate within a few minutes thereafter. The extremely high affinity of quin-2 for Ca^{++} results in considerable perturbation of Ca^{++} metabolism in quin-2-loaded cells and tissues; I have heard of the dye stopping hearts and triggering mitosis in lymphocytes. I also heard that some preparations of quin-2 yielded 10 or 12 spots on chromatography.

Even without its toxicity or impurities, quin-2 is not well suited for use with most flow cytometers because the absorption of the Ca^{++} chelate is extremely low at the wavelengths above 350 nm at which most ion laser and arc lamp sources are operated for UV excitation. Better excitation wavelengths are available from arc lamps, krypton lasers (a very weak line at 337 nm) and from He-Cd lasers. I have used the 325 nm line from a He-Cd laser for quin-2 fluorescence excitation, which allows demonstration of an ionophore-induced [Ca^{++}] transient by a shift in the position of a fluorescence distribution. The distribution itself, however, is a haystack, making it unlikely that a small number of activated cells within a heterogeneous population could be discriminated by flow cytometry using quin-2.

For microphotometric and cinefluorometric studies in which fluorescence from the same cells or tissues was measured repeatedly, quin-2 did not have the same disadvantages as for flow cytometry, and the probe provided investigators with a substantial amount of information about cytoplasmic [Ca^{++}] changes before Roger Tsien and his colleagues cooked up two much better reagents, **fura-2** and **indo-1**[858].

Fura-2 and Indo-1: Ratiometric Ca^{++} Indicators

Like quin-2, fura-2 and indo-1 are selective calcium chelators; since neither compound has as high an affinity for calcium as does quin-2, the newer probes do not perturb cellular calcium metabolism to anywhere near the extent to which quin-2 does. Also like quin-2, both fura-2 and indo-1 are typically introduced into cells in the form of acetoxymethyl (AM) esters, which are enzymatically hydrolyzed in cells to the free dyes. Cells are loaded by incubation with 1-15 μM concentrations of AM ester for 30-90 min at 37 °C; in some cases, an agent such as a Pluronic detergent may be needed to improve penetration of cells by the dye[2358].

Both fura-2 and indo-1 are at least ten times as fluorescent as quin-2, and the newer dyes also share the desirable characteristic, absent in quin-2, of undergoing substantial **spectral shifts** upon binding calcium. In the case of fura-2, the most pronounced change is in the excitation spectrum; while, **for indo-1, the emission spectra of the free dye and the calcium chelate differ substantially**, as shown in Figure 7-9 (p. 296) and Figure 7-33 (next page). Fura-2 and indo-1 can, therefore, both be used to make **ratiometric** measurements of cytoplasmic [Ca⁺⁺], which cancel out many extraneous factors, including the effect of cell-to-cell variations in dye content.

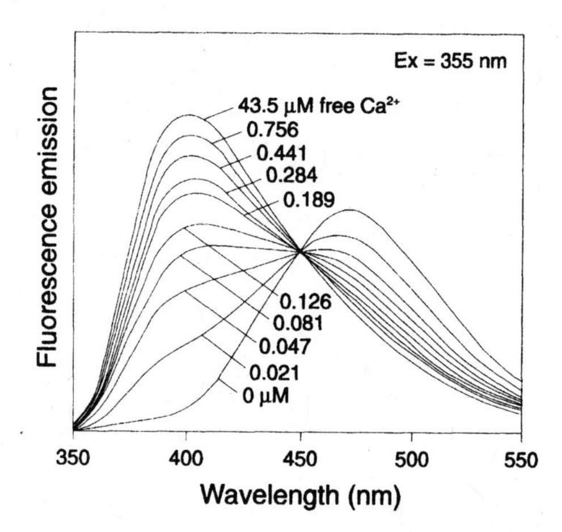

Figure 7-33. Emission spectra of indo-1 in solutions of increasing free calcium ion concentration. The figure was provided by Molecular Probes.

Fura-2 has an emission maximum at 510 nm; the absorption maximum for the calcium chelate is at about 335 nm and that for the free dye is at about 362 nm. If fluorescence at 510 nm is measured, first using 340 nm excitation and then using 380 nm excitation, the ratio ($F_{340\rightarrow510}/F_{380\rightarrow510}$) provides a measure of cytoplasmic [Ca⁺⁺]. The excitation wavelengths used for fura-2 measurements are readily available from xenon arc lamp sources, which do not have sharp spectral peaks in the near UV and visible regions, and fura-2 has been used for single cell [Ca⁺⁺] studies employing microspectrophotometers and video microscopes[859-61]. Fura-2 is less well suited for use in flow cytometers, because the excitation wavelengths are not readily obtained from the standard laser or arc sources.

Indo-1 fluorescence is excited by UV light; the free dye shows maximum emission at about 480 nm, and the calcium chelate emits maximally at about 405 nm. Any of the UV excitation sources commonly employed in flow cytometry can be used to excite indo-1, including He-Cd (325 nm), argon (351/363 nm) and krypton (350/356 nm) lasers and Hg arc lamps (366 nm). Fluorescence is measured through 10-20 nm bandpass filters centered near 405 and 480 nm, although I have been told by various people that they prefer 500 or even 520 nm filters to 480 nm filters for the longer wavelength measurement. The ratio, i.e., ($F_{UV\rightarrow405}/F_{UV\rightarrow480}$), is taken as a measure of cytoplasmic [Ca⁺⁺][862]. An example of flow cytometric measurement of cytoplasmic [Ca⁺⁺] changes in cell activation appears in Figure 7-21 (p. 365).

Since quin-2 provided only poor resolution of responding and nonresponding cell subpopulations[864], it was not widely used in flow cytometry. Indo-1, however, came into considerable vogue in a relatively short time after its properties were first described[853,865]. A particularly useful technique described by Rabinovitch et al[866-7] employs **indo-1 in combination with fluorescent antibodies** to follow cytoplasmic [Ca⁺⁺] responses in cell populations defined by the presence of a specific surface antigen.

While ratiometric [Ca⁺⁺] measurements made with either fura-2 or indo-1 eliminate the effect of cell-to-cell variations in dye content, measurements made with indo-1 have an additional advantage in that the effects of uneven illumination and of light source noise also are eliminated by virtue of their equal influences on the numerator and denominator of the ratio. This advantage, it should be noted, is common to other flow cytometric ratio measurements in which both parameters used in the ratio are measured at the same time in the same beam.

If aliquots of fura-2 or indo-1 loaded cells are placed in solutions with various known Ca⁺⁺ concentrations and treated with a **calcium ionophore** such as **A23187** or **ionomycin**, it is possible to calibrate the fluorescence ratio measurement to yield accurate molar values of cytoplasmic [Ca⁺⁺]. Kachel et al[1713] have described a general method for calibration of flow cytometric wavelength shift fluorescence measurements based on pulse modulation of the excitation source in a flow cytometer; it is not clear how readily this can be implemented in existing apparatus.

Indo-1 is widely used, at least by people with UV excitation sources in their flow cytometers[1714-8]. Since there are probably more than 10,000 fluorescence flow cytometers out there that don't have UV sources, that's small comfort. Luckily, there are alternatives.

Fluo-3 and Other Visible-Excited Ca⁺⁺ Probes

In 1989, Minta, Kao, and Tsien[1719] described a series of fluorescein- and rhodamine-based calcium indicators suitable for use with 488 nm excitation. The most widely used of these is **fluo-3**, which has the spectral characteristics of fluorescein, but which is almost nonfluorescent unless bound to calcium. The emission spectrum of fluo-3 at various values of [Ca⁺⁺] is shown in Figure 7-34 (next page).

Unlike fura-2 and indo-1, fluo-3 does not exhibit a spectral shift with changes in calcium concentration. A fluo-3 fluorescence distribution is a haystack; if you're stimulating a cell population, the haystack moves to the right when the cytoplasmic [Ca⁺⁺] goes up and back to the left when it goes back down. In 1990, Rijkers et al[1720] came up with a dodge that enables fluo-3 to be used in a more or less ratiometric mode; they load cells simultaneously with fluo-3 and with SNARF-1, a pH-sensitive dye that will be discussed in the next section. The amounts of the two dyes taken up by cells are roughly similar. The emission of SNARF-1 at 600 nm does not change appreciably during activation; as a result, the ratio of fluo-3/SNARF-1 fluorescence (520 nm emission to 600 nm emission) provides a measure of cytoplasmic [Ca⁺⁺] with considerably less variance than there is in the fluo-3 histogram.

In 1994, Novak and Rabinovitch[2833] introduced an improved ratiometric method combining fluo-3 with another calcium probe, **Fura Red**. Fura Red is excitable at 488 nm, emits maximally at 660 nm, and becomes less fluorescent on binding calcium; the ratio of fluo-3 to Fura Red fluorescence is therefore relatively sensitive to small changes in cytoplasmic [Ca^{++}].

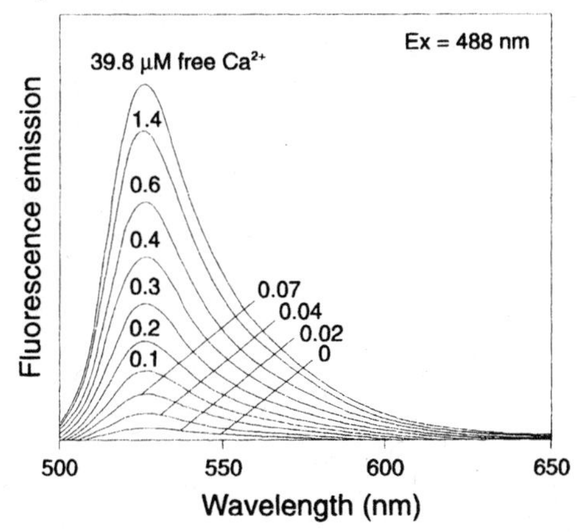

Figure 7-34. Emission spectrum of fluo-3 in solutions of increasing free Ca^{++} concentration. The figure was provided by Molecular Probes.

Molecular Probes (see the *Handbook*[2332]), which offers quin-2, indo-1, fluo-3, and Fura Red, has some other visible excited calcium indicators, such as the **Calcium Green** series.

Akkaya and Lakowicz[1722] reported they were developing a series of styryl-based, visible-excited, ratiometric indicators, but these have apparently not hit the market yet.

I am told that, while it might be possible to make an indo-1-like calcium probe that could be excited by a UV (370 nm) diode laser, it would be difficult, if not impossible, to tweak a similar molecule sufficiently to permit excitation by a violet diode.

AM esters and dyes loaded as AM esters may be pumped out of cells; Baus et al[2834] reported that the anion transport inhibitor sulfinpyrazone blocked the efflux of Fluo-3 from transformed murine T cells.

Flow Cytometric Probes of Intracellular pH

Since we have all been exposed, generally starting in junior high school (that might be middle school for some of you young ones), to dyes that change color with changes in pH, it should come as no surprise that some such materials can be used for single cell analysis. If a dye changes color with pH, one can find two wavelengths at which absorption, excitation, or emission will change differently as pH changes; the pH can then be estimated from the ratio of measurements made at those two wavelengths. As we have just noted while discussing the use of fura-2 and indo-1 as calcium probes, ratiometric measurements are highly desirable, especially in flow cytometry; as we shall shortly see, ratiometric measurements of intracellular pH by flow cytometry antedate ratio measurements of cell calcium.

The first description of a flow cytometric method for estimation of intracellular pH was published in 1979 by Visser, Jongeling and Tanke[519]; they measured the fluorescence of intracellular **fluorescein** produced by enzymatic hydrolysis of FDA. It is the excitation spectrum, rather than the emission spectrum, of fluorescein that shows the greatest dependence upon pH[202-4,369,376]. Since Visser et al did not have two-wavelength excitation capability in their flow cytometer, they estimated pH changes from changes in emission above 530 nm of cells excited at a single wavelength (488 nm).

When wavelengths above 465 nm are used for fluorescein excitation, emission intensity increases with pH; at excitation wavelengths below 465 nm, emission intensity decreases with pH. When excitation is at 465 nm, an **isosbestic point**, emission intensity is independent of pH. Visser et al noted that the average emission intensity of cells equilibrated with buffers of increasing pH increased, as would be expected, when 488 nm excitation was used; decreases in intensity with increasing pH were observed when the excitation wavelength was changed to 458 nm.

Valet et al[200] and Gerson[201] described ratiometric techniques for flow cytometric pH estimation in which the ratio of intensities of green and blue fluorescence emission from UV-excited dyes was used as a measure of pH. This ratio increases with pH. The probe used by Valet et al is **1,4-diacetoxy-2,3-dicyanobenzene (ADB)**, also called **2,3-dicyano-1,4-hydroquinone diacetate**; it is available from **Sigma.** ADB is enzymatically hydrolyzed inside cells to yield **2,3-dicyanohydroquinone (DCH)**. Gerson used the coumarin **4-methylumbelliferone (4-MU)** in free and esterified forms. These probes have an advantage over fluorescein derivatives in that they can be used with single beam excitation; 4-MU, however, also has the disadvantage of being able to cross cell membranes freely, and therefore tends to leak out of cells rapidly.

There exist **distributional radiolabel and dye probes of intracellular pH**. These are weak acids and bases that partition across membranes as a function of transmembrane pH gradients; 9-aminoacridine is an example. While they are usable for studies of cells in bulk, there is absolutely no reason to choose distributional pH probes for single cell studies, given the existence of dyes that can be used for ratiometric measurements.

The **calibration of ratiometric pH probes** is straightforward; fluorescence measurements are made of aliquots of cells suspended in high-potassium buffers at different, known pH's, and the cytoplasmic pH is made equal to the buffer pH by addition of the **proton ionophore nigericin** prior to flow cytometric measurement of emission or excitation ratios. This allows construction of a calibration curve. As is the case with the college chemistry variety of pH indicator techniques, cytometric measurements of pH will be most precise in the region of the pK_a of the indicator used.

An alternative calibration procedure was described by Chow et al[2835], who adapted a method described by Eisner et al[2836]. Cells are suspended in solutions containing mixtures of

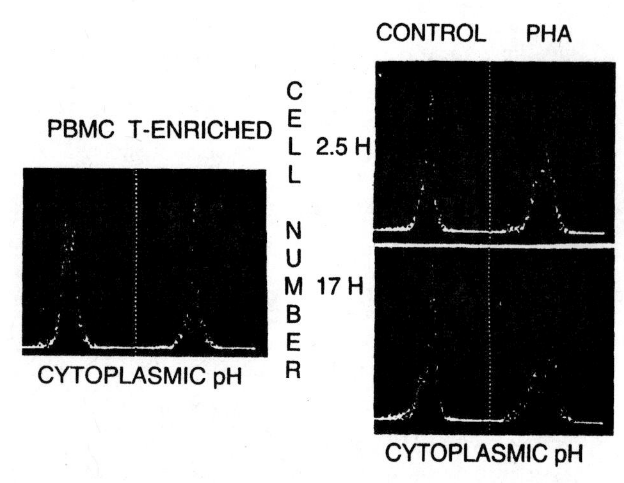

Figure 7-35. Estimation of cytoplasmic pH in human lymphocytes from carboxyfluorescein fluorescence ratio (520 nm emission; 488 and 441 nm excitation from argon and He-Cd lasers). The panel at left shows unstimulated peripheral blood mononuclear cells (PBMC) and a T cell enriched fraction. The panel at right shows control T-enriched cells and T-enriched cells exposed to PHA, 2.5 (upper pair) and 17 (lower pair) hours after exposure. PHA appears to increase pH in some cells and decrease it in others.

a weak acid (butyric acid) and a weak base (trimethylamine) in fixed ratios but different molar concentrations, allowing determination of a "pseudo null point" pH. The calibration curve is different for each cell type.

Valet et al[200] used ADB (DCH) primarily to demonstrate pH heterogeneity in tumor cell populations; Alabaster[520] employed the same probe for the same purpose. Gerson reported increases in pH within a short time following lymphocyte stimulation[427] and also with progression of stimulated lymphocytes through the cell cycle[521]; he has used bulk distributional measurements as well as flow cytometry with 4-MU and derivatives[201] for his analyses.

Musgrove, Rugg and Hedley[868] published a comparative study of **4-MU, DCH,** and **2′,7′-bis-(carboxyethyl)-5,6-carboxyfluorescein (BCECF)**[869] as flow cytometric pH probes; this provides a good reference on the use of these dyes and on the calibration procedure as well. UV excitation was used for 4-MU and DCH, with pH being estimated for 4-MU from the ratio $F_{UV\rightarrow 450}/F_{UV\rightarrow 560}$ and for DCH from the ratio $F_{UV\rightarrow 425}/F_{UV\rightarrow 540}$. BCECF was excited at 488 nm, and pH was estimated from the ratio $F_{488\rightarrow 520}/F_{488\rightarrow 620}$. 4-MU was poorly retained by cells; both DCH and BCECF were found to be usable over a pH range from 6.0 to 7.5. DCH gave the most sensitive indication of pH changes; BCECF was retained longest by cells. However, BCECF may be better suited to ratiometric measurements using dual-wavelength excitation than to emission ratio measurements.

Years ago, when I played the pH game briefly, I chose to pursue a flow cytometric pH estimation technique based on **fluorescein derivatives**, in preference to the probes described by Valet et al and Gerson, for several reasons. First, I wanted to be able to make **correlated measurements of pH and cell cycle position** in cells stained with Hoechst 33342, and, if possible, to measure pH and [Ca^{++}] together in quin-2 (or indo-1) loaded cells. Both Hoechst 33342 and the calcium probes are UV-excited and blue fluorescent, and thus cannot be used with ADB or 4-MU, which share their spectral characteristics. There are no substitutes presently available for either Hoechst 33342 or indo-1 for their respective purposes. However, I already knew[8] that it was possible to measure Hoechst dye fluorescence and pH independently in a three-beam flow cytometer if fluorescein derivatives were used for pH estimation.

Carboxyfluorescein (COF), at the time, was the probe best suited for such studies; while its spectrum is nearly identical to that of fluorescein[203-4], COF offers two significant advantages. The free dye produced intracellularly by enzymatic hydrolysis of COFDA is lost from cells much more slowly than is FDA-derived fluorescein. More important, **free COF does not enter the mitochondrial compartment**, and thus provides a good estimate of **cytoplasmic pH** rather than a composite of cytoplasmic and mitochondrial pH; BCECF shares this characteristic. In order for COFDA to enter cells, the buffer pH during loading must be kept relatively low ($\leq$ 7.0) to keep most of the material in an un-ionized form. Thereafter, pH can be restored to the physiologic (7.3-7.4) range used for most experiments. **BCECF is preferable to COF in that cells can be loaded at physiologic pH.**

The pH estimate obtained from **COF or BCECF fluorescence** is derived from the ratio of emission intensities at 520-550 nm obtained using **two excitation wavelengths** above and below 465 nm. The 488 nm argon ion laser line is nearly optimal as the longer excitation wavelength. For maximum sensitivity of the emission intensity ratio to pH, the shorter excitation wavelength used should be in the 420-450 nm range; the 441 nm He-Cd laser line and the 436 nm Hg arc lamp line are both suitable for this purpose. The ratio $F_{488\rightarrow 530}/F_{441\rightarrow 530}$ increases with increasing pH.

Nancy Allbritton and I (unpublished) used 488 and 441 nm excitation to make estimates of pH in human peripheral blood mononuclear cells, and in samples enriched in T cells, before and after PHA stimulation. Some of our results are shown as Figure 7-35.

The advantages of a ratio measurement are immediately apparent from the figure; this was the first "needle" distribution not representing DNA content that I had ever seen emerge from flow cytometry of real cells. The CVs of pH distributions are on the order of 5%; the displacement of peaks after lectin stimulation is at least 10%, suggesting that this parameter could be used to discriminate small subpopulations of activated lymphocytes within a few hours following stimulation. There remains some controversy[870-1] as to whether pH change can serve as a litmus test for lymphocyte activation. The indicators, however, keep getting better.

Figure 7-36 shows the emission spectra of another item from Molecular Probes' copious catalog[2332], the ratiometric pH indicator **carboxy SNARF-1**, which is a current favorite. This dye is well excited at 488 nm; the ratio of emission intensities at 640 and 580 nm provides a measure of pH.

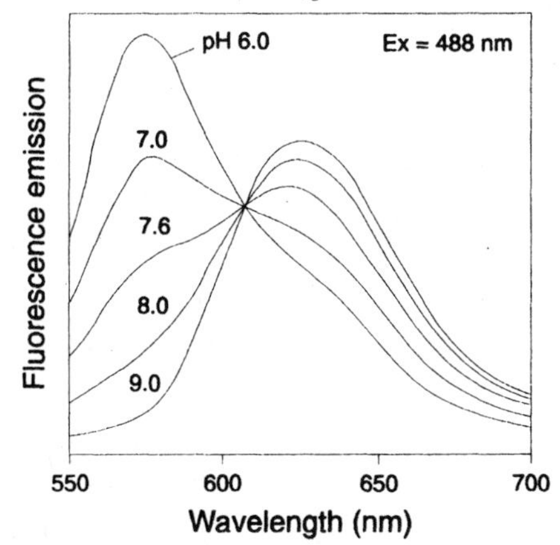

Figure 7-36. The pH-dependent emission spectra of carboxy SNARF-1 excited at 488 nm. The figure was provided by Molecular Probes.

Although some people got reasonably good results with ADB[1722-3], by the late 1980's, BCECF had become the indicator of choice. Hedley and Jorgensen[1724] found an 8% CV in the BCECF fluorescence ratio obtained from B16 melanoma cells, which would correspond to 0.4 pH units; they felt, however, that much of the variance was due to instrumental factors rather than pH variation. Musgrove, Seaman, and Hedley[1725] used BCECF to measure pH in samples enriched for different cell cycle phases; higher pH values were associated with S, G_2, and M phases of the cell cycle, with a corresponding increase in the percentage of G_1 cells at the lower pH range, suggesting cell-cycle dependence of pH.

Wang et al[1726] measured intracellular pH by the distribution of the weak acid, 5,5-dimethyl-2,4-oxazolidinedione-2-14C (^{14}C-DMO), a classical method, and by fluorescence flow cytometry with BCECF. The pH determined for CHO cells by the FCM method at external pH values of 6.0-8.1 agreed within 0.1 pH unit with that determined by the ^{14}C-DMO method.

Carboxy SNARF-1 is said to be more sensitive than either BCECF or ADB over the pH range 7.0-8.0[1727]; the dye is also better retained than ADB[1728]. Simultaneous detection of conjugate formation between cytotoxic and target cells during the cytotoxic process and determination of cytoplasmic [Ca^{++}] and pH are possible if one cell type is labeled with Fluo-3, and the other with SNARF-1; events positive for both dyes are identified as conjugates[1729].

In studies of rat basophil leukemia cells, Lee et al[1730] observed that external bicarbonate was necessary for pH regulation in these cells; it is thus advisable to have bicarbonate in your buffer when doing pH measurements.

Carboxy SNARF-1 and other dyes, e.g., BCECF and indo-1, are typically loaded into cells as acetoxymethyl (AM) esters; the hydrolysis of these esters results in a net acidification, and probe-induced perturbations, problematic in [Ca^{++}] studies with quin-2 (p. 403), may also need to be considered in critical analyses of pH[1731].

Measurement of pH in intracellular compartments is feasible if you can localize the pH indicator to those compartments. McNamara et al[2837] coated small polystyrene beads with phospholipids covalently labeled with fluorescein (pH-sensitive) and tetramethylrhodamine (pH-insensitive). The lipobeads are ingested by macrophages and directed to lysosomes; fluorescence measurements by flow or image cytometry can, respectively, provide values of average lysosomal pH or of the pH in an individual lysosome. Levitz et al[2838] covalently labeled the fungus *Cryptococcus neoformans* with fluorescein and 2',7'-difluorofluorescein (Oregon Green), which, respectively, have pK_as of 6.4 and 4.7, respectively, allowing sensitive pH detection over a broad range. They used the labeled fungi to study phagosomal pH.

Chitarra et al[2839] used ratiometric flow cytometric measurements of CFSE fluorescence to estimate internal pH in the bacterium *Clavibacter michiganensis*, taking the ability of cells to maintain a high pH in media of lower pH as an indicator of viability. This is, in effect, a rather fancy way of doing a dye exclusion test. It should not come as a surprise that the rather broad distributions of CFSE fluorescence ratio did not completely distinguish viable and killed cell populations; as was mentioned on pp. 256 and 402, if you only have a few thousand molecules of dye to work with, you probably won't get enough photoelectrons to do a precise measurement. The method would almost certainly work better in a slow flow, scanning or imaging system; others have reported making reasonably precise measurements of bacterial pH with the latter[2840-1].

The Hat Trick: Multiparameter Approaches to Ion Flux Measurements in Cell Activation

It has been my experience that you can use good parameters in multiparameter measurements to help make sense of bad ones. When I started working with membrane potential and immunofluorescence, using only forward scatter measurements for gating, I always suspected that I was measuring garbage along with my cells. I feel much more comfortable when I gate on Hoechst dye fluorescence, because I'm sure there's really a cell there.

Until I got ratiometric pH measurements working, I had no way of improving my conception of what was going on in flow cytometric estimation of $\Delta\Psi$ with cyanine dyes, [Ca^{++}] measurements with quin-2, etc. I thought I might be able to clarify things a little bit if I correlated signals from these probes with measurements of the amount of labeled ligand bound to the cell surface, but I couldn't work up much enthusiasm about this because I'd be looking at haystack vs. haystack. When ratiometric pH and [Ca^{++}] measurements became available, the picture changed considerably, and it became logical to reformulate some of the questions I originally tried to ask when there was finally some hope of answering them.

Since I wanted to look at stimulation of cells by ligands that react with cell surface receptors, I envisioned ways of getting some kind of dose-response curve, for which I would need a labeled ligand. If the ligand were bound covalently to a fluorescent bead, I could count the number of beads bound to a cell to determine the "dose" of ligand, and follow the time courses of [Ca^{++}], pH, and/or $\Delta\Psi$ changes in cells with various numbers of beads to determine the "response."

I figured that if I had the right dyes and the right excitation beams, I would to be able to get measurements of [Ca^{++}], pH, membrane potential, and ligand binding all at once, assuming that probe interactions wouldn't cause big problems. Since I and others[872, 1718,1729,1737-9] had done simultaneous measurements of multiple functional parameters, I thought (and think) the assumption was (and is) fairly safe. And, of course, I'd have at least one or two fluorescence channels available to gate out cell subpopulations

This is still about as fancy a flow cytometric experiment as I can conjure up; not surprisingly, it takes a fairly elaborate flow cytometer, with four beams. As it happens, all four beams (UV, violet, 488 nm, and red) can be derived from small, air-cooled lasers that plug into the wall. Nicer still, once we put the beast together, we can use it to do almost all of the other measurements that have been discussed in this chapter (I said all this in the previous edition, in 1994; in 2003, as you will find in Chapter 8, you can buy such an instrument).

I puckishly referred to the experiment described above as the hat trick, because it had three goals and might net me some valuable information about early events in cell activation if I could stick to it. However, it's on ice for now, and maybe for good. The reason is simple. A flow cytometer, no matter how fancy, is not very good at looking at individual cells over time, and it has become obvious that we need to do more of that in functional studies.

Wacholtz and Lipsky[1736] used single cell image analysis of changes in cytoplasmic [Ca^{++}] after mitogen stimulation of individual human T cells to determine the relation between the [Ca^{++}] signal and subsequent functional changes. Marked heterogeneity was observed in the magnitude of increase in [Ca^{++}], in the lag time of responses, and in the percentage of T cells that responded to anti-CD3 and to PHA. However, mitogenic stimuli that induced IL-2 production or DNA synthesis consistently generated increases in [Ca^{++}] that were sustained for 1 to 2 hours; elevations as small as 50 to 100 nM above control were associated with evidence of activation. Functional activation of T cells by PHA and anti-CD3 is thus correlated with the induction of small, but sustained increases in [Ca^{++}] that might or might not have been detectable by flow cytometry. Image cytometry worked fine.

And, these days, while you're doing the image cytometry, you can use the patch clamp technique and other modern electrophysiological methods to follow fast membrane potential changes and dissect the roles of the various calcium, chloride, and potassium channels in activation processes in lymphocytes and other cells[2842-69]. However, people who do that seem to use cell sorters to pick out interesting cells to work on, so you might want to finish the book before you order the new equipment.

NOsing Around for Nitric Oxide

Rao et al[1617] described a method for detection of the production of **Nitric Oxide ("Just Say NO")**, now recognized as a key intermediate in cellular signal transduction, by neutrophils, in which DCFH-DA fluorescence was measured in cells to which calmodulin inhibitors had been added. There are a number of probes for NO; Several authors[2870-2] have used 4,5-diaminofluorescein; Molecular Probes[2332] recommends DAF-FM (4-amino-5-methylamino-2′,7′-difluorofluorescein because it is unaffected by pH above 5.5 and more sensitive than other probes. These probes are all loaded into cells as "diacetates," i.e., as acetyl esters, in the manner of FDA.

Other Ions in the Fire

Amorino and Fox[2873] describe measurement of intracellular **sodium** ion concentration using **Sodium Green**. Balkay et al[2874] measured intracellular **potassium** ion, deriving the parameter from measurements of pH (using **BCECF**) made in the presence of **nigericin**, which keeps the ratio of potassium and hydrogen ions constant on both sides of the cell membrane. Pilas and Durack[2875] used **6-methoxy-*N*-(3-sulfopropyl) quinolinium (SPQ)**, a UV-excited, blue-green fluorescent probe quenched by halides, to measure intracellular **chloride** ion concentration. All of the probes mentioned come from – guess where – Molecular Probes, and the *Handbook*[2332] has much more information on them.

7.14 REPORTER GENES

Somebody Cloned My Gal: Enzymes as Reporter Genes

In this era of molecular biology, flow cytometric detection of enzyme activity has come into increasing use as a method for detecting and selecting cells expressing **reporter genes.** Reporter genes are genes that encode protein products that are relatively readily detectable; they are linked by standard recombinant DNA technology to a gene or genes of interest to an investigator, and the resulting construct is introduced into cells. Those cells in which the gene of interest is expressed can then be identified by the presence of the reporter gene product.

In the 1980's, genes encoding **surface antigens** not normally present on the cells of interest were used as reporter genes; expression of these genes was detected by immunofluorescence. This was unsatisfactory; what was generally observed were very broad distributions of relatively weak signals, and successful isolation of cells bearing the genes of real interest to investigators generally required multiple cycles of cell sorting, in which the brightest cells were sorted and clones, and the brightest progeny were resorted, etc.

In 1988, the Herzenberg lab[1642] reported the development of a new method, called **FACS-Gal**, in which the *lacZ* gene from *E. coli* was used as the reporter gene. This gene encodes the enzyme **β-D-galactosidase**, which is not normally present in many eukaryotic cell types. The activity of this enzyme can be detected by flow cytometry using the fluorogenic substrate **fluorescein di-β-D-galactoside (FDG)**, available from Molecular Probes, which is hydrolyzed to fluorescein by the enzyme. Since a few molecules of enzyme can produce a lot of molecules of fluorescein, the FACS-Gal technique made it much simpler to detect and sort cells expressing the reporter gene; the stronger signals also made it possible to distinguish cells expressing different amounts of the reporter gene and, therefore, of the gene(s) of interest, and to sort these separately for further analyses.

The FACS-Gal assay as originally published, although demonstrably usable (e.g., reference 1643) had some problems; loading the cells with FDG was difficult, and there were interferences from rare autofluorescent cells and from endogenous enzyme activity. Finally, it was difficult to do long sorting experiments because the enzymatic reaction continued over time. In 1991, several improvements were described by Fiering et al[1644]. Optimal loading conditions were determined, two-color fluorescence was found to identify interfering autofluorescent cells, chloroquine was used to inhibit endogenous mammalian enzyme, and **phenylethyl-β-D-thiogalactoside**, a competitive inhibitor, was added to stop the enzymatic reaction before sorting.

Jasin and Zalamea[1645] reported isolation of galactosidase-bearing sperm cells from transgenic mice using **5-dodecanoylaminofluorescein di-β-D-galactopyranoside**, a substrate that is more lipophilic than FDG; it may also be possible to use other substrates, e.g., **resorufin galactoside**[1646], which has been used for analyses of galactosidase activity in yeast.

Genes encoding enzymes other than β-D-galactosidase may be used as reporter genes; Puchalski et al[1646] transfected COS monkey kidney cells with a construct incorporating a recombinant **glutathione S-transferase (GST)** gene, which was demonstrable using **monochlorobimane**, using procedures that were discussed in the section on analysis of thiol groups and glutathione (p. 381). More recently, Lorincz et al[2876] developed a reporter system using *E. coli* **β-glucuronidase** as the enzyme and **fluorescein-di-β-D-glucuronide (FDGlcu)** as the fluorogenic substrate. Because selective inhibitors are available for β-glucuronidase and β-galactosidase, it is possible to detect both enzymes in transfected cells.

Puchalski and Fahl[1647] compared gene transfer by electroporation, lipofection, and DEAE-dextran transfection for compatibility with cell-sorting, with the requirements that the procedure used should generate morphologically homogeneous populations with at least a 5% colony-forming ability in which at least 10% of the cells expressed recombinant GST. Of the transfection techniques tested, only electroporation satisfied all requirements.

At this point, I'm supposed to say "But that's all ancient history..." because you can see GFP and its friends and relatives just a little bit down the page. I will, instead, point out that enzymes are still being used as reporter genes. One of the newer wrinkles was developed by Roger Tsien et al[2877], and made available by his company, **Aurora Biosciences**, which is now merged with **PanVera**. The enzyme is a **β-lactamase**, produced by the ampicillin resistance gene of *E. coli*; it is not found in untransfected mammalian or avian cells. The fluorogenic substrate used to detect it is called **CCF2**; it is loaded into cells as the acetoxymethyl ester, **CCF2-AM**. CCF2 contains a coumarin donor and a fluorescein acceptor attached to a cephalosporin β-lactam ring. When the ring is intact, the coumarin and fluorescein moieties of the probe are close enough together for energy transfer to occur, and excitation (the excitation maximum is 409 nm) produces green (530 nm) fluorescence. When the ring is cleaved by β-lactamase, the distance between the coumarin and the fluorescein is increased sufficiently to greatly reduce the efficiency of energy transfer, and the probe emits blue (450 nm) fluorescence. The intensity of blue emission from the hydrolyzed probe is about twice the intensity of green emission from the intact probe. CCF2 was not of great interest to the flow community when it was only feasible to excite it with a violet krypton laser; it may be a lot more interesting now that relatively inexpensive 405 nm violet diode lasers are available. Cavrois et al[2878] used the β-lactamase/CCF2 system to develop an assay for HIV-1 virion fusion with target cells.

Green Fluorescent Protein (GFP) et al

In early 1994, Chalfie et al[1648] described the use of a gene encoding an intrinsically fluorescent protein from the bioluminescent jellyfish *Aequorea victoria* as a reporter gene. The native **green fluorescent protein (GFP)** absorbs maximally at 395 nm, but can be excited moderately effectively at 488 nm, where its absorption is about one-third maximum; the emission spectrum has a sharp peak at 510 nm with a shoulder at 540 nm. I included this information in the previous edition; at the time, GFP expression in prokaryotic (*E. coli*) and eukaryotic (*Caenorhabditis elegans*) cells had been shown to render those cells fluorescent. I opined that "GFP should work for cell sorting, provided the cells used don't exhibit a lot of autofluorescence." Boy, did it. As I noted on p. 48, what I described in 1994 as a "mild mannered reporter" turned into Supermolecule. The first book[2376] and one of the big review articles[2377] are distinctly dated (1998), and it's hard to keep up with the literature. I won't even try. I don't do much GFP work, so I'll just try to cover some points that I think are important and point you elsewhere for more details.

Most of the probes we use in flow cytometry are things you add to cells after the fact. DNA dyes work on any cell that contains DNA, provided you can get the dye in. True, you wouldn't want to waste highly specific reagents, such as fluorescent antibodies and nucleic acid probes, on cells that you didn't think were carrying any detectable target material, but, in principle, you could throw any probe at any cell that comes through the door. Reporter genes, and fluorescent proteins in particular, are different; they have to be in or on the cells before they come through the door.

The protocols for labeling cells with fluorescent protein reporter genes are molecular biology protocols, not cytometry protocols; if your focus is on cytometry, what you need to know are the excitation and emission spectral characteristics of the proteins you're looking for and how much of them you expect to find in the cells. Figure 7-9 (p. 296) includes the spectra of four fluorescent proteins, **ECFP** (cyan; excitation maximum 434 nm, emission maximum 477 nm), **EGFP** (green; excitation maximum 489 nm, emission maximum 508 nm), **EYFP** (yellow-green, excitation maximum 514 nm, emission maximum 527 nm), and **DsRed** (orange; excitation maximum 558 nm, emission maximum 583 nm). They are all available, or, more accurately, the constructs you need to get the appropriate genes into cells are available, from **Clontech**, a division of **BD Biosciences.** Hawley et al[2879] recently described flow cytometric detection of all four at once using 458 nm excitation from an argon laser for ECFP, EGFP, and EYFP and 568 nm excitation from a krypton laser for DsRed.

All of these proteins have been engineered to give them more desirable fluorescence characteristics. ECFP, EGFP, and EYFP, produced by mutants of the *Aequorea* GFP gene[2377], are well-developed, as GFP variants go, while dsRed[2450], derived from a coral of the species *Discosoma*, is more of a work in progress. There are several variants, including dsRed Express[2884], which matures more rapidly and apparently yields more fluorescence than the original.

The longest wavelength fluorescent protein available from Clontech is HcRed, from the reef coral *Heteractis crispa*[2880]; its excitation and emission maxima are at 588 and 618 nm. HcRed could be excited by a 532 nm YAG laser, but the 568 nm krypton line, 590 nm from a dye laser, or 594 nm from a He-Ne laser (assuming you could get a He-Ne laser with enough power) would be a better bet.

As exploration of the animal kingdom[2881] and protein engineering[2377,2450,2882-5] have produced new and improved better fluorescent proteins, investigators have worked out optimal schemes to detect them by flow cytometry and put them to use[2886-96]. We can expect that to continue.

The brief review by Matz et al[2881] on GFPs packs more information into 7 pages than just about any other paper I have ever read; I heartily recommend it. Among other things, I learned that members of the GFP family are unique among pigment proteins in that they act as enzymes and synthesize their fluorophores from amino acids in their own polypeptide chains. Other pigment proteins, such as the phycobiliproteins, make their chromophores or fluorophores from small molecules and usually require several enzymes to get the job done. Thus, it would be a much harder job to make a phycobiliprotein reporter than it is to make a GFP reporter. The self-contained palette of GFP proteins also makes it much easier to modify their spectra by site-directed mutagenesis than it would be to change the spectrum of another type of pigment protein.

With a little help from their friends, GFPs themselves have "evolved" far beyond the point of merely marking transfection; there are fluorescent protein variants that are sensitive to pH[2897-9], calcium ion concentration (shades of aequorin!)[2900-2], and membrane potential[2903], and others that monitor cell surface receptor interactions[2378] and kinase activities[2904]. There is also a "timer" protein[2905] that gradually changes color from green to red. Of course, you do have to put genes into your cells to take advantage of all these benefits; I can't help you much there.

Minority Report(er)?

With fluorescent proteins and enzymes around, why does anybody still want to bother using proteins expressed on cell surfaces as reporters? Well, if your system works well enough so that only the cells expressing the reporter gene carry the product on their surfaces, you can then use antibodies on magnetic beads or a column to separate your transfectants instead of, or in addition to, a cell sorter. Christine et al[2906] describe a system for detecting recombinant switch activity in B cells based on expression of CD4 or a histocompatibility antigen normally not found on the cells.

Well, that does it for parameters and probes. If you decide to quit at this point and get into image analysis or confocal microscopy, you'll be using many of the same probes, detectors, electronics, computers, etc., so you haven't wasted too much time. If not, stick with me, and I'll tell you how to acquire a flow cytometer, and then go on to a survey of some of the applications of the technology.

8. BUYING FLOW CYTOMETERS

8.1 INTRODUCTION

Flow cytometry obviously cannot be done without flow cytometers, most of which are bought from manufacturers. Users' buying decisions are generally based on examination of manufacturers' literature, on demonstrations, and on information obtained from more experienced users. I get a lot of phone calls from people who want to know which instrument they ought to buy, and/or whether they ought to try building one. In this chapter, I will try to give you access to the information necessary to formulate rational answers to these questions.

Since the manufacturers will happily provide potential customers with literature containing photographs, diagrams, samples of data obtained from their instruments, and reasons why you should choose that particular system, I will not attempt to duplicate their copiously illustrated brochures. You may, therefore, want to arm yourself with promotional literature before you look at what I say about companies and apparatus; addresses and phone numbers of the manufacturers of flow cytometers and other cytometric apparatus appear in Chapter 11, "Sources of Supply," near the back of the book.

The origins of cytometry have been discussed in Chapters 1 and 3. Chapter 4 provides information on light sources, illumination and collection optics, detectors, and electronics that is equally relevant to static and flow cytometry, as well as information on fluidics and flow system design, augmenting introductory material in Chapter 1. Data analysis hardware and software are introduced in Chapter 1 and discussed in detail in Chapter 5, and Chapter 6 is devoted to sorting.

I have included information on all the commercial flow cytometers I know about in this chapter; a selection of other cytometric apparatus is discussed in Chapter 10.

8.2 HISTORY

Only a few companies make or have made cell sorters and/or optical flow cytometers intended for research and/or clinical use. By 1970, **Bio/Physics Systems**, which was founded by Lou Kamentsky, was selling its Cytograf and Cytofluorograf instruments, while **Phywe AG** marketed **Partec**'s commercial version of the Dittrich/Göhde ICP apparatus. Shortly thereafter, **Technicon** brought out the Hemalog D, the first of a series of flow cytometric leukocyte differential counters. **Becton-Dickinson** introduced the FACS cell sorter, based on the instrument developed in Len Herzenberg's lab at Stanford, in 1974; within a year, **Coulter** entered the cell sorter market with its TPS-1, later replacing this apparatus with the EPICS series, development of which was begun by Mack Fulwyler and continued by Bob Auer. In 1976, following its acquisition by Johnson & Johnson, Bio/Physics Systems became **Ortho Diagnostics Systems;** Ortho acquired marketing rights to the ICP from Phywe in 1978, and added cell sorters to its Cytofluorograf line at about the same time.

In the 1980's, alternatives to products of the "Big Three" American manufacturers, B-D, Coulter, and Ortho, began to emerge. At the beginning of the decade, these three companies offered laser source flow cytometers with cell sorting capability; sorting was optional on Ortho's instruments. Ortho also offered the arc source ICP, which could not be equipped for sorting, and the laser source Spectrum III, a highly automated system, without sorting, aimed at clinical users. In pursuit of the same market, B-D introduced the FACS analyzer, a small but sensitive analytical apparatus employing an arc lamp source, and Coulter produced the EPICS C[873], an ergonomically designed, computer-controlled "knobless" instrument that included sorting capability. During the 1980's, the arc source apparatus of

Lindmo and Steen[100-1] was commercialized as the MPV-Flow by **Leitz**; a refined version of this instrument, the Argus, was later produced by **Skatron**, and marketed in Europe by Ortho and in the U.S. by **Bruker**, which also introduced the Odam ATC 3000, a laser source sorting apparatus developed in France, which could make electronic volume measurements as well as optical measurements. A later version of the Lindmo/Steen instrument was made and sold by **Bio-Rad**; one is now being produced by **Apogee**, which also offers service for earlier versions.

An apparatus using arc lamp and/or laser sources, based on Eisert's designs[168,616-9], was sold for a time by **Kratel**, which also marketed data analysis hardware and software developed by Kachel. A later version of this data analysis system, and Kachel's Fluvo II flow cytometer, which made fluorescence and electronic volume measurements, became available from **HEKA. Partec** offered another arc source instrument, designed by Göhde and incorporating optical and electronic measurements and a closed fluidic sorter. Finally, three Japanese companies, **JASCO**, **Omron**, and **Showa Denko**, produced laser source instruments; Omron's was distributed in Europe by Ortho, and was the predecessor of the Cytoron Absolute, Ortho's last entry in the market.

After 1985, both B-D and Coulter brought out non-sorting benchtop instruments (the B-D FACScan and Coulter EPICS Profile) using low-power, air-cooled argon laser sources, and oriented at least as much toward the clinical as toward the research market. The efficient optical designs used in these instruments provided sensitivity at least equal to that obtained from these manufacturers' larger systems. Ortho, which, interestingly enough, had offered similar instrument configurations for years, ceased production of flow cytometers and sold its service operations to B-D in mid-1987. It reentered the American market in 1992, selling the Cytoron Absolute, built in Japan by Omron, and gradually withdrew again by the end of the decade.

Cytomation, a company founded in 1988, produced a series of add-on upgrades to improve data processing and sorting performance of existing cell sorters until 1994, when it began selling the MoFlo modular high speed cell sorter developed at Lawrence Livermore National Laboratory by Ger van den Engh and his coworkers[1180]. It was common to refer to B-D, Coulter, and Ortho as the "Big Three" of flow cytometry in the 1980's, if the term is used today, it denotes the trio of **BD Biosciences**, **Beckman Coulter**, and Cytomation, now **DakoCytomation**. I suspect that Becton-Dickinson decided on the "BD Biosciences" name when Coulter briefly jumped ahead of it in alphabetical order, becoming Beckman Coulter after being acquired by Beckman in 1997, but I can't prove it. The Big Three are the big makers of high speed sorters, but no longer the only makers; Ger van den Engh's new company, **Cytopeia**, arrived on the scene in 2000, and is now producing a relatively compact high speed sorter, the Influx, which, as of mid-2002, was to be sold by DakoCytomation, which has had an ongoing working arrangement with Cytopeia. **Systemix**, which had licensed the MoFlo technology from Livermore for clinical applications, decided not to pursue production of instruments.

I estimated in Chapter 1 that there were between 12,000 and 20,000 benchtop argon laser source flow cytometers in use worldwide, and fewer than 2,000 large systems with droplet sorting capability. It is therefore likely that several manufacturers have more instruments in the field than does Cytomation; **Luminex** and Partec come to mind. However, if we really want to get into a competition on eliminating fluid waste about this, perhaps we should do just that, and ask the manufacturers to supply us with the total volumes of sheath run through all their instruments per year. They don't know, and they probably wouldn't tell us if they did.

Six manufacturers returned detailed responses to a questionnaire I sent out in October 2001, asking for details about their instruments, and, as far as I am concerned, they are "The Big Six" for taking the time to do so. Their products are discussed in some detail in the sections of this chapter immediately following. Then comes a section that covers instruments from other manufacturers who did not return the questionnaire, but, in some cases, did send back some information. The rest was filled in from the manufacturers' web sites and literature, and/or from personal inspection. A separate section is devoted to clinical hematology instruments. Manufacturers are listed in alphabetical order within sections (with one exception, to be explained later).

I'll take the rap for the errors, but if some item of information about a particular instrument is of critical interest to you, check it out with the manufacturer. I tabulated the features of various instruments in the last edition of this book; I'm not going to do it this time around because there are too many instruments and too many features.

8.3 BD BIOSCIENCES

Background

The large cell sorter products from B-D began life differing only slightly in construction from the apparatus built by Herzenberg et al at Stanford[2,9,82,86]; more changes have since been made in the electronics and data analysis systems than in the optics.

The original FACS instruments used pulse height analyzers for data analysis; the FACS IV replaced a hardwired analyzer with the one that incorporated a dedicated minicomputer, allowing multiparameter analysis and sorting and data storage on magnetic media. In the next successor line, the FACS 400 had a minimal data analysis system that only generated dot plots; regions of interest for sorting could be defined on the two-parameter live display. The FACS 420 used an 8-bit microprocessor-based data analysis system that was also sold with B-D's arc source FACS Analyzer, the company's first entry into the non-sorting benchtop analyzer market. The FACS 440 had the same minicomputer-based data analysis system used in the FACS IV. The minicomputer was buttoned up so tightly within the system that it was literally impossible to take advantage of its computer capabilities; at least one user site installed a second computer to eliminate the drudgery of pushing all the buttons that had to be pushed to get the data analyzer to do anything useful.

Things were much improved when B-D elected to offer users a choice of micro- or minicomputer-based data analysis systems. The smaller CONSORT 30 and 32 systems were built around Hewlett-Packard Series 300 microcomputers using a 68000 series microprocessors running under a Pascal operating system. The larger CONSORT 40 incorporated a DEC MicroVAX minicomputer with a graphics terminal. A range of applications programs, and various combinations of floppy and hard disc storage, were offered for both computer systems, as was networking capability. In the mid-1990's, both the Consort 30 and Consort 40 series of data analysis systems were replaced by the FACStation series, which incorporated Apple Macintosh computers. The hardware and software have progressed from Motorola 680X0-based Macs to newer PowerPC-based current models.

BD Biosciences' current large Fluorescence Activated Cell Sorter (FACS™) is the FACSVantage SE™, which succeeded the FACStar, FACStar^{PLUS}, and FACSVantage. The company also makes three smaller instruments, the FACSCalibur™ and BD™ LSR II analyzers and the FACSAria™ high-speed sorter. These benchtop systems require neither special wiring nor water-cooling; the LSR II and FACSaria feature **FACSDiVa™** digital pulse processing, also available as an option on the FACSVantage SE™.

Except for the FACSDiVa option, which has a 6-month warranty, all BD Biosciences' instruments and additional components, including lasers, have a one-year warranty. Any options purchased after purchase of a new instrument have a 3-month warranty. A 4 1/2 day Key Operator Course is provided for 2 operators with each instrument purchase; an additional course is provided with the purchase of the FACSDiVa Option.

The BD FACSVantage SE™ Cell Sorter

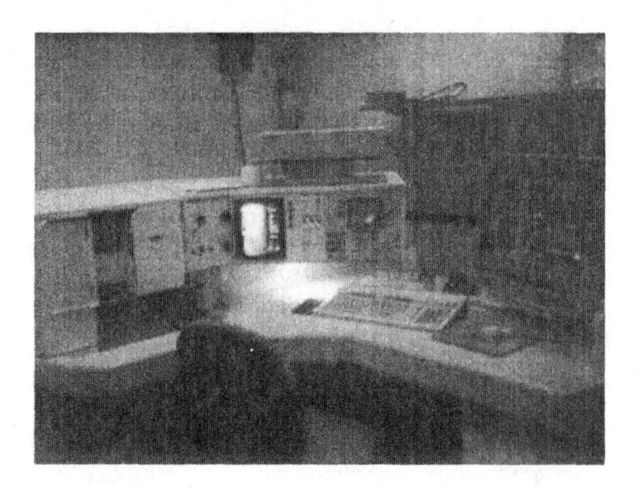

Figure 8-1. The BD FACSVantage cell sorter.

Light Sources and Illumination Optics:

The FACSVantage SE (Figure 8-1) can use as many as three laser beams to illuminate observation points separated in space. Available lasers include water- and air-cooled argon, krypton, or mixed-gas ion lasers, He-Cd, He-Ne, and 635 nm diode lasers. A dye laser may be added if another laser (water-cooled argon or frequency-doubled YAG) is available for optical pumping. Excitation optics are adjustable, allowing the user to steer the laser beam(s) and control the spacing between beams; an adjustable lens can be used to change the beam shape at the interrogation point.

Flow System:

The FACSVantage SE is a stream-in-air system, with ceramic orifice nozzles ranging in diameter from 50 to 400 μm available. Sheath and sample are pressure-driven; flow velocity is typically 10 m/s, varying with orifice size, and is adjustable by changing sheath pressure. Sample pressure is continuously adjustable from 2.0 PSI, for the 400 μm orifice, to 60 PSI, when the high speed TurboSort option is used with a 70 μm orifice. Fluid flows downward.

Forward Scatter:

An adjustable horizontal blocker bar and adjustable iris diaphragms allow the lower limit of collection to be set between 1° and 4°, and the upper limit to be set as high as 10°. A photodiode detector is standard; a PMT is optional.

Side Scatter and Fluorescence:

Signals are collected around a horizontal blocker bar by an N.A. 0.55 lens; an adjustable iris serves as a field stop. Detectors are Hamamatsu PMTs; three fluorescence detectors are standard, with additional detectors optional. BD Biosciences' Special Products Group can provide as many as twelve fluorescence detectors. Field stops are used in the optical system to limit the region of space from which light reaches the detectors; an image (≥ 40 ×) of each beam intersection point is formed in the plane of the corresponding field stop. When multiple illuminating laser beams are used, combinations of half mirrors and dichroics divert light from different observation points to different detectors. Filters and dichroics can be selected, and easily removed and changed, by the user.

Signal Processing:

BD offers two markedly different systems for signal processing and data acquisition; each has its own hardware, software, and computer platform. The standard FACSVantage SE uses analog and hybrid electronics for pulse processing; the FACSDiVa Option provides digital pulse processing.

Analog Electronics (standard):

Any parameter measured at the first observation point can be selected as a trigger signal. Current to voltage conversion of detected light pulses is done by a 20,000 V/A transimpedance amplifier with baseline restoration; further analog signal amplification is available. Linear signals are converted to a logarithmic scale by analog log amps with 4 decades of dynamic range. Fluorescence compensation for spectral overlap is done in hardware. Pulse integrals are captured by active integrators; peaks are captured by peak detectors. A pulse width measurement circuit that measures width at a constant fraction of pulse height provides information for doublet detection. The pulse peak, integral and width measurements from the analog electronics are digitized by 10-bit ADCs. The system allows a maximum acquisition rate greater than 30,000 events/s.

Digital Electronics (FACSDiVa Option):

Any signal measured at any observation point, or any Boolean combination of measured signals, can be used for triggering. Current-to-voltage conversion of detector signals is accomplished by transimpedance amplifiers with baseline restoration, but no additional analog amplification is available. Instead, all signal pulse trains are simultaneously sampled and digitized by 14-bit (16,384 channel) ADCs at 100 ns intervals (10 MHz conversion rate). An integral is obtained by taking the sum of at least 16 values collected during the pulse, providing a range of values running between 0 and 262,143 (18 bits). The maximum 14-bit value among these is taken to represent pulse peak height. Linear signals are converted to a logarithmic scale by digital processing, using an 18-bit look-up table. Both hardware and software fluorescence compensation can be implemented. The maximum data acquisition rate is 100,000 events/s, but may decrease with the number of signals being measured.

Sample Handling:

The standard sample tube used is 12 × 75 mm, with a minimum volume of 100 µl and a maximum of 4.5 mL. Temperature control for sample input and output is provided by circulating chilled or heated water through a water jacket surrounding the sample tube holder.

Sorting:

A piezoelectric transducer is used to generate droplets for sorting with the TurboSort Plus option; the MacroSorting option uses an electromagnetic transducer to provide the relatively high amplitudes and low frequencies needed for droplet generation when large orifices are used to sort correspondingly large particles. Sort rates vary depending on the nozzle size used, with larger nozzle sizes yielding slower sorting rates. Rates range from 35,000 to 2,000 events/s using the TurboSort Plus nozzles (50 to 100 µm diameter), and from 500 to 100 events/s using the MacroSorting nozzles (200 to 400 µm diameter). As many as four sort streams may be generated. Single cells can be sorted into multiwell plates; each cell's location can be stored in the list mode data file. Both analog and digital sort decision hardware are available.

Software:

For Use with Standard Analog Electronics:

CellQuest Pro software is provided for the analog electronics workstation, built around an Apple Macintosh computer with a G4 PowerPC processor. The FCS 2.0 data file format is used. The software allows protocol definition and batch analysis. As many as 8 parameters may be acquired; as many as 16 parameters, including parameters, such as ratios, derived from the acquired parameters, may be analyzed. Single parameter histograms may contain as many as 1,024 channels. The available 2-D display formats are dot, density, and contour plots. Rectangular, polygonal, elliptical, "Snap-to-Gate," and "Auto-polygon" regions can be defined, and Boolean combinations of regions can be used. Mean, median, peak channel (mode), SD, and CV are available for statistical analysis of data. Overlay of single parameter histograms may be used to compare histograms. There is provision for 3-D display of data, DNA histogram analysis (provided by third-party software), and export of data to graphics and spreadsheet programs.

For Use with FACSDiVa Option:

The FACSDiVa Option software is provided for the digital electronics workstation, built around a Hewlett-Packard X4000 (Intel/Windows) computer. Both FCS 3.0 and 2.0 file formats are supported. The software allows protocol definition and batch analysis. A maximum of 16 parameters can be acquired, with a total of 36 parameters for analysis; these include area and height for all 16 parameters acquired plus two ratio channels and a width and a time parameter. Linear to log transformation and fluorescence compensation are done in software. The maximum number of channels in a single parameter histogram is 262,141. Dot plots are the only available live 2-D displays. Rectangular, polygonal, elliptical, "Snap-to-Gate," and "Auto-polygon" regions can be defined, and Boolean combinations of regions can be used. Mean, median, SD, CV, minimum, and maximum are available for statistical analysis of data. DNA analysis is provided by third-party software and graphics and spreadsheet export are possible.

Other Details:

The Vantage SE (crate and frame) is approximately 100 cm wide × 182 cm long × 154 cm high, and weighs 286 kg. The DiVa option is 105 cm wide × 123 cm long × 94 cm high, and weighs 113.4 kg. Aerosol containment and biohazard controls include the application of a vacuum to the sort chamber to contain aerosols formed during sorting, and a drip containment system on the sample injection tube deals with drips from the sample introduction area. The instrument console operates on 110 VAC +/- 10%, 50/60 Hz, 20 A, or 230VAC, 50Hz. Each laser has its own specific power requirements.

Daily instrument optimization is needed to guarantee optimal performance. Sensitivity is configuration dependent. Fluorescence channels have detected <200 Molecules of Equivalent Soluble Fluorochrome (MESF) using Spherotech Rainbow RCP-30-5A particles. When using Bangs Laboratories' Quantum particles labeled with fluorescein, it has been possible to discriminate particles bearing <1,000 fluorescein MESF. The CV of fluorescence from propidium iodide-stained chicken erythrocyte nuclei is 3% or less.

The BD FACSCalibur™ Analyzer

The FACSCalibur, successor to the FACScan and FACSort, is a benchtop analyzer, with optional fluidic sorting. The primary excitation light source is a 488 nm air-cooled argon ion laser; a 635 nm diode laser is optional.

Forward and side scatter and three colors of fluorescence can be measured in the single 488 nm laser system, a fourth color is available with the 635 nm diode laser option. The cytometer and the computer system with which it is used take up about six running feet (less than two running meters) of bench space, as shown in Figure 8-2. The applications software includes numerous routines for automated setup, calibration, and data analysis.

Figure 8-2. The BD FACSCalibur benchtop cell sorter.

Light Sources and Illumination Optics:

The primary 488 nm laser beam is made elliptical by prisms and focused to a spot 20 μm high × 60 μm wide at the sample stream. The optics allow fine adjustments of the beam pathway. The optional 635 nm diode laser is mounted at right angles to the 488 nm laser; the 635 nm beam, which is already elliptical, bypasses the prisms, is reflected off a mirror, and passes through the same optics used to focus and steer the 488 nm beam. The 635 nm diode laser is focused upstream from the 488 nm laser; i.e., cells pass through the 635 nm beam first.

Flow System:

The quartz flow cuvette has inside dimensions approximately 180 × 430 μm. The sheath fluid is driven by pressure from a 4 liter pressurized tank; stream velocity through the flow cell is 6 m/s. Three fixed sample flow rates, 12, 35, and 60 μl/min, are available. Because the cuvette is rectangular, the core stream is elliptical; its size depends on the sample flow rate and the sheath pressure, with the slowest sample flow rate yielding the smallest core cross-sectional area. Fluid flows upward in the flow cuvette.

Forward Scatter:

A fixed vertical blocker bar is used to block the laser beam, and forward scatter signals are collected over a fixed range from 0.7° to 10°. In the dual laser 4-color setup, a 488 nm bandpass filter with a 10 nm bandwidth is placed in front of the scatter detector to eliminate scattered red laser light.

Side Scatter and Fluorescence:

The collection optics of the FACSCalibur follow the design originally used in the FACScan, providing efficient enough light collection to permit the use of low power lasers without sacrificing measurement sensitivity. Side scatter and fluorescence signals are collected using a custom-designed, long-working distance, N.A. 1.2 objective optically coupled to the flow chamber with a thixotropic gel. Magnification is approximately 13 ×. Field stops (1 mm wide in fluorescence channels and 1.5 mm wide for side scatter) are placed in the image planes in front of the PMT detectors. Three fluorescence detectors are used with the standard 488 nm laser; an additional fluorescence detector may be installed with the optional 635 nm diode laser. The filters normally used in the standard instrument provide for measurement of green (515-545 nm; "FL1"), yellow (564-606 nm; "FL2"), and red (> 670 nm; "FL3") fluorescence, excited at 488 nm. A fourth channel, measuring red (653-669 nm; "FL4") fluorescence excited at 635 nm, is added when the red laser is used.

Signal Processing:

Any signal or the Boolean AND of two signals may be used for triggering. A 20,000 V/A transimpedance amplifier with baseline restoration is used for current to voltage conversion of detector signals; additional analog signal amplification is available. Linear signals are converted to log using 4-decade analog log amps. Fluorescence compensation for spectral overlap is done in hardware. Pulse integrals are captured by active integrators; peaks are captured by peak detectors. Pulse width at a constant fraction of pulse peak height is available for one fluorescence detector. Data are digitized by 10-bit ADCs. The data acquisition system can process more than 30,000 events/s.

Sample Handling:

The FACSCalibur uses the same tube size and minimum and maximum volumes as the FACSVantage. However, the FACSCalibur also has an automated option for 12 × 75mm sample tube handling and 96 well microplates.

Sorting:

A closed system fluidic sorter is optional; the sort mechanism moves a collecting tube into the center of the flowing stream when a cell of interest passes by. The sort rate of 300 selected events/s is adequate for many tasks, and the closed system is advantageous for work with hazardous samples. However, since fluid is continuously collected by the sorting mechanism, it may be difficult to isolate rare cells, because they end up being highly diluted in the collection vessel. Although BD provides a concentrator module that continuously removes fluid from the sorted fraction using a filter that resides in a pressurized chamber, this filtration process has been shown to damage some types of live cells, and is considered most useful for sorting fixed cells or cells that do not need to remain viable after sorting.

Software:

CellQuest Pro software is provided on an Apple Macintosh G4 computer; the FCS 2.0 file format is used. A maximum of 6 parameters can be acquired, and 7 parameters, plus time, can be analyzed using CellQuest. The features of CellQuest software are described in the previous section on the FACSVantage.

Other Details:

The cytometer itself is 91.4 cm wide, 61.5 cm deep, and 67.3 cm high, and weighs approximately 109 kg. Aerosol containment and biohazard controls include closed fluidic sort paths and an automatic aspirator on sample inlet tubing. Power requirements are 120 VAC +/- 10%, 50/60 Hz, 20 A. Bangs Laboratories Quantum beads bearing <750 fluorescein MESF have been discriminated from unlabeled beads. The CV of propidium iodide-stained chicken erythrocyte nuclei is 3% or less.

The BD™ LSR II Analyzer

The BD LSR II is BD's newest benchtop analyzer, with as many as four fixed alignment laser sources and as many as fifteen fluorescence and two scatter channels. The older BD LSR was configurable with as many as three lasers and six fluorescence channels. The two systems use the same instrument enclosure (Figure 8-3); the LSR II replaces the LSR's analog signal processing with digital signal processing.

Figure 8-3. The BD LSR™ multi-beam benchtop analyzer.

Light Sources and Illumination Optics:

The LSR was available as a two or three laser source system that used a Coherent 488 nm, 20mW air-cooled argon ion laser, a Kimmon 325 nm, 8mW HeCd laser, and, optionally, a 633 nm, 17mW HeNe laser. The LSR II can use Coherent's "Sapphire" solid-state laser for 488 nm and Lightwave's mode-locked 355 nm YAG laser for UV; the Coherent VioFlame 405 nm, 25mW diode laser and a 638 nm diode laser are available as options.

Flow System:

The LSR and LSR II use the same flow cell and external fluidics setup as the FACSCalibur.

Forward Scatter:

The setup is apparently similar to that used in the FACSCalibur.

Side Scatter and Fluorescence:

The LSR and LSR II use a gel-coupled, N.A. 1.2 collection lens with a larger field of view than that of the lens used in the FACSCalibur. The LSR used field stops, mirrors, and dichroics similar to those in the FACSCalibur. In the LSR II, light from an image of each interrogation point is relayed through a multimode fiber optic, and signals from different spectral bands are separated by (highly) reflective dichroics placed in a polygonal pattern (Figure 8-4). Filters and lenses placed between the dichroics and the detector PMTs, correspondingly placed in a polygonal array, further restrict detector bandwidth. The older layout shown in Figure 8-4 allows seven detectors to collect a signal from one excitation beam; the current configuration provides for eight detectors. The dichroics are specially designed to work at the acute angles of incidence (not the customary 45°) necessitated by the octagonal layout. Light loss is minimized because each signal is required to pass through only a single optical filter, i.e., that placed immediately ahead of the detector. The relay fiber optics do not preserve polarization, likely precluding using the optical arrangement for polarized fluorescence or scatter measurements. On the plus side, I note that the use of fiber optic relay elements could allow light collected by a second lens placed 180° from the existing lens to be routed to the same detector, potentially doubling signal intensity; however, BD apparently does not now plan to offer this option.

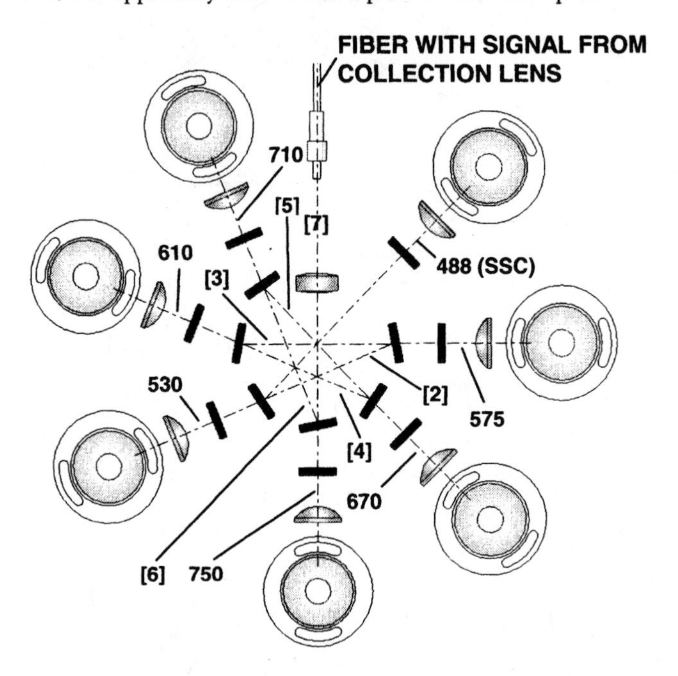

Figure 8-4. BD's "Octagon" collection optics. A number in square brackets associated with a path indicates the number of spectral bands traveling along that path; a number without brackets indicates the center wavelength (in nm; my guesses) of a single spectral band.

All fluorescence and side scatter detectors are Hamamatsu PMTs; all filters and dichroics can be changed as needed. Typical optical filter setups for the three laser source LSR II are: FL1 = 530 nm band pass (BP), FL2 = 575 nm BP, FL3 = 670 nm long pass (LP) or 682 nm BP, FL4= 500 nm BP, FL5 = 380 nm LP, 400 nm BP, or 424 nm BP, FL6 = 610 nm BP or 660 nm BP.

Signal Processing:

While the LSR used essentially the same signal-processing scheme as the FACSCalibur, the LSR II uses the FACSDiVa digital data acquisition system described in the previous section on the BD FACS Vantage system.

Sorting:

Not available.

Software

The LSR II uses the FACSDiVa Option software (see the previous section on the BD FACS Vantage). The standard BD LSR used CellQuest software, provided on an Apple Macintosh G4 computer. Data from as many as 6 fluorescence and 2 scatter parameters could be acquired, and as many as 8 parameters, selected from peak height, pulse width, area, and ratio, could be analyzed. All other features of the LSR's CellQuest software were as described for the FACSVantage. Software fluorescence compensation for LSR and LSR II systems is provided for by the included FlowJo™ analysis software (Tree Star, Inc.); DNA histogram analysis is provided for by the included Modfit LT™ software (Verity Software House).

Other Details:

The Sensor module is approximately 193 cm wide, 87 cm deep, 127 cm high, and weighs 239 kg. Aerosol containment and biohazard controls include closed fluidic paths and an automatic aspirator on the sample inlet tubing. The BD LSR requires two dedicated 120 VAC +/- 10%, 50/60 Hz, 20 A power lines.

The estimated detection limit is 200 fluorescein MESF, when using the Spherotech Rainbow RCP-30-5A Beads; the fluorescence CV of propidium iodide or Hoechst 33342-stained chicken erythrocyte nuclei is 3% or less.

The BD FACSAria™ Cell Sorter

The FACSAria (Figure 8-5), introduced in December, 2002, is a multilaser high-speed sorter that departs substantially from the design of other current commercial products of this type; to start with, it is a benchtop system requiring less than 1 kW of electrical power from standard 110 V, 60 Hz or 220 V, 50 Hz lines.

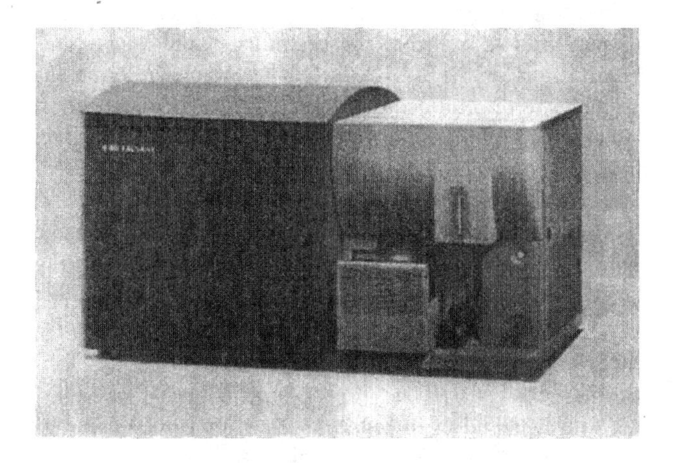

Figure 8-5. The BD FACSAria™ Cell Sorter

Light Sources and Illumination Optics:

The system uses two or three lasers; a 20 mW, 488 nm Coherent Sapphire solid state laser and a 17 mW, 633 nm He-Ne laser are standard, and a 20 mW, 407 nm violet diode laser is optional. The 407 nm laser can be used to excite usable fluorescence signals from Hoechst dyes used to stain DNA in living cells. Fiber optics (single-mode for the TEM_{00} sources) relay the illuminating laser beams to collimating lenses, from which they enter a series of four prisms that form elliptical spots 9 μm high × 65 μm wide, separated by 200 μm, on the sample stream inside a quartz cuvette. Optical alignment is fixed.

Flow System:

The quartz flow cuvette has a 160 × 250 μm rectangular cross section 15 mm in length. Both sheath and sample fluids are driven by pressure; sheath and cleaning fluid containers and a waste fluid tank are kept on an external cart with integral air pressure and vacuum supplies. Sheath fluid is transferred from the container on the cart into a reservoir inside the instrument, removing air bubbles and maintaining constant pressure independent of the fluid level in the external sheath container. The sample container accommodates a variety of sizes of sample tubes; once a tube is loaded, the container is sealed and pressurized, and the tube holder agitates the sample during the run to minimize settling of cells. Drive pressure can range from 2-75 PSI (14-517 kPa); typical stream velocity through the flow cell is 6 m/s. Because the cuvette is rectangular, the core stream is elliptical. Fluid flows downward; the stream exits the cuvette via a 70 μm or 100 μm nozzle.

Forward Scatter:

The setup is apparently similar to those used in the FACSCalibur and LSR II.

Side Scatter and Fluorescence:

Optical alignment is fixed; light is collected by a gel-coupled, N.A. 1.2 collection lens, similar if not identical to that used in the LSR II. Separate relay fiber optics transfer the signal from each beam to a polygonal array similar to that shown in Figure 8-4; an "octagon," more properly a nonagon, allows collection of as many as 8 signals (7 fluorescence signals plus side scatter) from the 488 nm beam intersections. As many as 3 signals (each) may be collected from the 633 nm beam and from the 407 nm beam. All detectors are Hamamatsu side-window PMTs; filters and dichroics may be changed as needed.

Signal Processing and Software:

The FACSAria uses the Windows-based FACSDiVa digital data acquisition system and software described in the section on the BD FACS Vantage cell sorter; a faster DSP chip is incorporated than was used in the original FACSDiVa, and additional software routines have been added to control sample handling and sorting.

Sample Handling:

The sample delivery system is, as mentioned, compatible with a variety of tube sizes; an arrangement for microplate sampling is likely to be made available in the future. The sampling tube is flushed inside and out with sheath fluid between samples to minimize carryover. There is also provision for filling the entire sample holder with ethanol for cleaning and disinfection.

Sorting:

Interchangeable 70 µm or 100 µm nozzle tips can be fitted to the cuvette with no adjustment required. Both two- and four-way sorting are possible; the system can process up to 60,000 compensated 8-parameter events/s and achieve a 100 kHz droplet generation rate at 75 PSI drive pressure. A camera monitors the position of the droplet breakoff point, and the transducer drive signal amplitude is controlled to maintain this position, keeping the droplet delay constant unless a clog or other substantial interruption of flow should occur. In these cases, fluid flow is shut off and the sample collection tubes are covered. During operation, both sample injection and sort collection chambers are closed; the sort chamber is sealed and operates under negative pressure. An additional Aerosol Management Option is available, which evacuates the sort collection chamber through a 0.01µm filter to trap aerosolized particles.

Other Details:

The benchtop portion of the FACSAria is 122 cm wide × 74 cm high × 71 cm deep and weighs 160 kg; space on the benchtop is also needed to accommodate the workstation keyboard, mouse, and LCD display. The fluidics cart, which can be placed on the floor, is 81 cm wide × 66 cm high × 66 cm deep and weighs 46 kg.

The estimated detection limit is <125 fluorescein MESF. CVs of 1.5% and 3.5% respectively, have been observed from the major fluorescence peaks of propidium- and Hoechst dye-stained chicken erythrocyte nuclei.

It is claimed that people with little or no technical training in flow cytometry or sorting can be taught to operate the FACSAria in substantially less time than is required to train operators of other high-speed cell sorters. This certainly sounds plausible.

The BD FACSCount

The FACSCount is a small (smaller than the FACScan) benchtop instrument dedicated to performing counts of CD4- and CD8-positive T cells[2472,3159]. It uses a 543 nm green He-Ne laser source, and measures fluorescence in two emission regions from phycoerythrin and a tandem conjugate. Prepackaged reagents incorporating a known concentration of fluorescent beads are used to allow derivation of absolute counts. The system is designed for use in parts of the world where HIV infection is common and facilities and money to support flow cytometers and cytometry are scarce.

8.4 BECKMAN COULTER, INC.

Background – and Signal-to-Background

By the 1970's, Wallace Coulter had already built a highly successful company that sold hematology counters utilizing his impedance-based cell detection and volume measurement method. He recognized both the scientific and commercial potential of fluorescence flow cytometry and cell sorting early enough to get his company into the cell sorter business by the end of the decade.

The original EPICS instruments collected fluorescence and side scatter using a pair of high-N.A. aspheric lenses, the first of which, placed at a distance equal to its focal length from the observation point, formed a "collimated beam," and the second of which brought that beam to a new focus, with a relatively large diameter field stop in the focal plane limiting the region of space from which light could reach the detectors. This system collected more light from cells than did some imaging systems, but the signal-to-background ratio was lower because background fluorescence and Raman scatter were collected from a much larger volume. While older stream-in-air EPICS cytometers could detect as few molecules of fluorescein as their competitors, there was more of a problem with phycoerythrin sensitivity due to Raman emission from water; this would have been largely avoided by the use of image forming optics. The interim solution was to use a smaller diameter field stop.

Collecting a collimated beam instead of forming an image also made it more difficult to use multiple excitation beams; imaging systems could separate optical signals from the two beam intersection points, while the EPICS optics, which could not, made it necessary to collect electronic signals using a **gated amplifier**, which turned the detectors on only when cells were traversing the appropriate beam. This, however, did not eliminate steady state fluorescence background from dye in the core stream; this could, in theory, be a problem in such applications as measurements of fluorescein immunofluorescence in cells also stained for DNA with DAPI. Although measurements I made on a Cytomutt set up with "nonimaging" optics indicated that interference between DAPI and fluorescein was insignificant in practice, there was some concern about other dye interactions. However, in this era of eight-color measurements, Coulter, now Beckman Coulter, has made the issues moot by using imaging optics for fluorescence and side scatter collection in its current instrument line.

Forward-looking from the beginning, Coulter used microprocessors in the EPICS line. The original research instruments were equipped with the MDADS data acquisition and analysis system; the main processor was an 8086 with an 8087 floating point coprocessor, and an 8080 was also used. Data storage was on 8" floppy or cartridge discs.

Coulter was also the first of the manufacturers to make the move to personal computers for data analysis, with the EASY-88 computer system, which used IBM PC-compatible hardware, including 5 1/4" floppy and hard discs and a removable-cartridge high density disc system, a floating point processor, and a 640 × 480, 4096-color graphics display. Instruments could be linked in an Ethernet-compatible network. The EASY-88 was said to be 2-3 times faster than a DEC PDP-11/23 minicomputer system.

Beckman Coulter's current analyzers and sorters are laser source flow cytometers with an orthogonal geometry. The EPICS® ALTRA™ with HyPerSort™ sorter (Figure 8-4) and Cytomics™ FC 500 analyzer (Figure 8-5) are designed primarily for research laboratory use, while the EPICS® XL and

XL-MCL analyzers are aimed at both clinical and research laboratories. All of the instruments come with a one-year warranty on all parts including the laser, with longer warranties available; the standard service option is five days during business hours. A five-day training course or optional training in customer's lab is provided.

The Beckman Coulter EPICS® ALTRA™ Cell Sorter

Figure 8-6. The Beckman Coulter EPICS® ALTRA™ sorter.

Light Sources:

All lasers on the ALTRA are optional and are customer-interchangeable on an industry standard optical table. A large variety of air-cooled and water-cooled lasers are available.

Illumination Optics:

The laser beam(s) are focused using crossed cylindrical lenses, and a variety of beam focusing lens assemblies are provided. The following focused beam spot sizes (height × width for a 488 nm laser beam) are available: 16 × 64, 16 × 85, 16 × 125, 6 × 100, 8.5 × 125, 16 × 33, and 11 × 125 µm. Inter-beam distances are set in time, not space, as 7, 20, 40 or 60 µs separation, meaning the beams are spaced not more than 100 µm apart at the sample stream.

Flow System:

A variety of flow chambers are available. For analysis with a closed fluidic system, a 250 µm square channel quartz cuvette with an integral N.A. 1.0 lens front element is used. For sorting, seven different jet-in-air nozzles, with orifice sizes ranging from 51-400 µm, are available, as are quartz flow chambers with a variety of internal and jetting orifice dimensions. Flow rates or velocities are customer adjustable and vary with the flow cell and sheath pressure; sheath flow velocity can be adjusted from approximately 1 m/s to about 30 m/s. Sample flow rate, sample and sheath pressure are all continuously adjustable by the operator. The sample and sheath pressure are obtained from electronically regulated compressed air or nitrogen, provided by a compressor, a nitrogen tank, or the house air supply. Sample pressure is adjustable from 0 to 15 PSI for the standard system, and from 0 to 100 PSI with the optional high speed sorting (HSS) option. Standard sheath pressure is 12 PSI; the range of adjustment is 1 to 100 PSI. Fluid flow is downward in sorting configurations.

Forward Scatter:

The forward scatter sensor, advertised as position-independent and using Fourier optics, incorporates two photodiodes as detectors; an optional PMT is available. The forward scatter collection lens N.A. varies depending on the collection half-angle used, which can be adjusted between 1.4° and 19°. For 2°, N.A. is 0.05; for 19°, N.A. is 0.40. The standard blocker bar is vertical; however, a variety of obscuration templates and different diameter field stops are available.

Side Scatter And Fluorescence:

The collection lens has an N.A. of 1.0. The first element is part of the flow cell, followed by an air gap; the other elements of the lens are contained in a brass cell, which is translatable to achieve best focus at the detectors. A 3 × magnified image of the interrogation point is formed inside the brass lens housing; a 500 µm diameter field stop is mounted in the image plane. No obscuration bar is used with the quartz cuvette flow chambers; a 2.5 mm wide horizontal blocker bar is used across the front of the fluorescence collection lens when running in the jet-in-air configuration. Six fluorescence channels, using Hamamatsu R1923 PMTs as detectors, are standard. An R1923 is also used to detect orthogonal scatter signals. PMT voltages are set from DACs.

All fluorescence filters are customer-removable without any need for re-alignment. The filters supplied are optimized to detect FITC, PE, ECD (PE-Texas Red), PC5 (PE-CY5), PC7 (PE-CY7), APC, and APC-CY7. Each fluorescence PMT can be used to collect signals from 2 laser intersection points using the AUX Channel feature. This feature allows a single PMT to collect signals excited by different lasers from fluorochromes having similar fluorescence emission properties (e.g., PE-CY5 and APC).

Signal Processing:

Single or multiple trigger signal(s) can be selected and the operator can set a separate threshold level for each signal used. Transimpedance amplifiers are used for current-to-voltage conversion of the signal pulses. A proprietary (patent pending) baseline restoration circuit is used. Analog signals for all non-log parameters are amplified with step gains. DAC-based attenuators are used on the forward scatter detector amplifier to provide gain adjustment, as these detectors use photodiodes instead of PMTs. Analog log amplifiers are used, and fluorescence compensation is done by software incorporating a 6 × 6 matrix inversion. Conventional peak detectors are used; integrators employ a switched capacitor circuit allowing the operator to select the time constant for either "conventional" or "high-speed" operation. The conventional mode allows the highest sensitivity to be obtained, using sheath pressure ≤ 15 PSI, with signal pulse widths 5-15 µs. For high-speed operation, at pressures up to 100 PSI, pulse width is less than 5 µs, and often 1-2 µs. Pulse peak and integral values are digitized to 10 bits.

The maximum acquisition rate depends on the number of parameters being collected, pulse width, and operator-selectable settings. The ALTRA processes the data pulses significantly faster than its predecessor, the Elite; conversion time is 1.1 μs per parameter and the ADC skips over unused parameters. Acquisition of one-parameter data with synchronous pulses could proceed at a rate of 130,000 events/s; real world numbers are, obviously, lower, but are compatible with acquisition of multiparameter data from 20,000 cells/s and sorting with a 100 kHz droplet generation frequency.

Sample Handling:

Tubes 12 × 75mm, 12 × 76mm and 17 × 100mm can be used, with no minimum sample volume. At present, there is no automated option for sampling from tubes or multiwell plates. An available module for sample delivery and sorting utilizes a re-circulating water bath providing temperature control from 4-40 °C.

Sorting:

The ALTRA generates droplets at approximately 25 kHz using the 100 μm tip and 90-100 KHz using the high-speed sorting 76 μm tip; it provides for two sort streams. A low abort rate and high purity are achieved when a cell is put into every 5th drop, corresponding to maximum analysis and sort rates of 5,000 events/s using the 100 μm nozzle and 20,000 events/s using the 76 μm tip. Users can sort faster based on the required specifications for purity and yield of their applications. Sorting of single cells into a large variety of multiwell plates (6, 12, 24, 48, 60 and 96 wells) or onto standard microscope slides is possible using the optional **AutoClone™** sorting module.

Software:

The ALTRA uses EXPO32, a Microsoft Windows based data acquisition and analysis software package utilizing the FCS 2.0 file format. Users can utilize "canned" protocols supplied with the software or can create unique protocols and combine them into a Worklist. Protocol definition includes parameters, plot definitions, gates, regions, and cytometer settings as well as disposition of the data, e.g., printout and archiving.

A maximum of eight parameters can be acquired from the instrument; derived parameters such as time, ratio, and time-of-flight (T-O-F) can also be computed and processed. The "PRISM" parameter allows identification and quantification of cell populations based on "positivity" or "negativity" for as many as six markers, in which case there are 64 possible categories of cells. Gated data acquisition is possible. The software does logarithmic/linear transformations and solves the matrix equation for fluorescence compensation. The maximum numbers of channels in single-parameter histogram and 2-parameter displays are, respectively, 1024 and 1024 × 1024. The display scale can be converted to a MESF or ABC (Antibodies Bound per Cell) scale after running appropriate bead controls. The following 2-D display formats are supported; dot, density, contour, and PRISM plots. Rectangular and polygonal regions can be created and combined using Boolean logic, with a maximum of 32 gates and 256 regions. Mean, median, mode, SD, CV, percentiles (%gated, %total) and counts of cells/μl can all be obtained. Comparison of data can be achieved by use of overlay plots of 1-D and 2-D data, as well as the calculation of Overton and K-S statistics. Isometric display of 1-D data and 3-D surface and tomogram plots are also available.

DNA analysis is provided by the 3rd party software package MultiCycle for Windows. Plots and histograms can be exported as PDF files, or into Microsoft Windows applications. Data can be exported in a text format or directly into Microsoft Excel.

Other Details:

The cytometer is approximately 132 cm high × 178 cm wide × 198 cm deep (including the tabletop) and weighs 500 kg, excluding water-cooled lasers. Aerosol and biohazard containment are provided by negative pressure in the sorting compartment that draws aerosols into the compartment, down a tubing vent and onto a filter that can be changed daily. The cytometer, workstation and air-cooled lasers require two 100/115 VAC, 20 A (16 A continuous) or two 220/240 VAC, 10 A 50/60 Hz dedicated power lines.

An on board, adjustable-rate, flashing LED provides a light source for testing detectors independent of the laser and flow chamber alignment; it can also be used to calibrate the gains of various signal pathways. Standard beads and bead sets must be run and recorded regularly to ensure proper alignment and calibration of the instrument.

In most configurations, <1,000 MESF of fluorescein and phycoerythrin can be detected; when the system is optimized for sensitivity, approximately 300 molecules of fluorescein and around 500 molecules of phycoerythrin can be detected. Fluorescence CVs of approximately 2% are achieved when measuring propidium iodide-stained peripheral blood lymphocytes. Using an optional PMT combined with a blocker optimized for low angle detection of forward scatter, particles as small as 0.14 μm diameter can be resolved above background.

The Beckman Coulter Cytomics™ FC 500 Analyzer

The FC 500 (Figure 8-7) is Beckman Coulter's newest benchtop analyzer, providing 5-color analysis from either single or dual laser excitation sources, with digital fluorescence compensation and log display using 20-bit linear list mode data.

Light Source:

The standard excitation source in the FC 500 is an air-cooled 20mW, 488 nm argon-ion laser; a 20mW, 633 nm He-Ne laser can be added. In the dual laser configuration, the laser beams are collinear, with no provision for multiple interrogation points.

Illumination Optics:

The beam(s) is/are focused by crossed cylindrical lenses to an elliptical spot 10 μm high × 80 μm wide at the sample.

Flow System:

A 150 × 450 μm rectangular channel quartz cuvette is used; sheath and sample flow upward. There are three user-

Figure 8-7. The Beckman Coulter Cytomics FC 500 analyzer.

selected sample flow rates, yielding data rates, on a sample of 10^6 cells/mL, of approximately 200, 600, or 1200 events/s. The sheath flow rate is fixed.

Forward Scatter:

As with the ALTRA, two solid-state detectors are used to collect forward scatter signals over a range of half-angles from 2° to 16°; the blocker bar has a "Maltese Cross" shape. An adjustable iris can be used to limit the half angle of collection to 8°. The forward scatter collection lens N.A. varies with the collection angle, ranging from 0.05 to 0.37.

Side Scatter and Fluorescence:

The front element of a N.A. 1.2 lens is built into the flow cell; a blocker bar is unnecessary. Additional, focusable lens elements form a 3 × magnified image; a 500 μm field stop is placed in the image plane. A photodiode is used to detect orthogonal scatter. Five fluorescence channels are standard; the detectors are Hamamatsu HC-120-32 PMT modules with enhanced red sensitivity. All optical filters are interchangeable by the user without the need for optical realignment. Bandpass filters and dichroics optimized for detection of fluorescein, phycoerythrin, PE-Texas red, PE-Cy5 or APC, and PE-Cy7 are provided with the instrument. All fluorescence filters are mounted on a removable single optical block, positioned on precision-machined dowel pins.

Signal Processing:

A single selectable trigger channel is used to acquire data, and the current pulses from the PMTs are converted to voltage pulses by a transimpedance amplifier utilizing a proprietary baseline restoration method. Analog circuits provide peak and integral values of the voltage pulses; these are digitized to 20 bits and stored, allowing the software to do log conversions and log scale displays and a full matrix inversion solution for fluorescence compensation, either in near real-time or after data have been stored in files.

Sample handling:

Sample input uses 12 × 75 mm tubes; sample volume typically is 0.5mL, with a minimum of 0.25mL and a maximum 3 mL. A 32 tube Multi Carousel Loader (MCL) with a presample bar code reader provides automated sample handling. An barcode reader wand is optional. Blood from closed Vacutainer tubes (B-D) or their equivalents can be automatically prepared for immunophenotyping using the optional Prep Plus II, followed by the optional T/Q Prep, which lyses and fixes the samples. If the samples need to be washed, the carousel can be placed into the optional Cell-Prep module, which aspirates each sample's cells into a hollow-fiber filter with a pore size adequate to retain the cells, but which allows hemoglobin and unbound reagents to escape. The carousel is then placed into the FC 500 for analysis. An optional module will allow aspiration from 96-well plates. The FC 500 has a maximum acquisition rate of 3,300 events/s, but the actual throughput rate depends on variables such as the flow rate, cell concentration, and the number of parameters acquired.

Sorting:

Not available on the FC 500.

Software:

The FC 500 uses RXP Software, a Windows 2000-based acquisition and analysis package utilizing the FCS 3.0 file format. The software allows for user-defined protocols; these can be combined to automatically process 32 tubes in a carousel. The RXP Software incorporates most analysis capabilities of the EXPO 32 software and adds additional features. As many as 16 parameters, including derived parameters such as ratios, time, and PRISM, can be acquired; as many as 24 parameters can be analyzed. Logarithmic display and fluorescence compensation are done in software and the 20-bit linear data can be saved in both compensated and uncompensated forms. Single parameter histograms have 1024-channel resolution and two parameter displays can have resolution ranging from 64 × 64 to 512 × 512. As many as 256 regions, of all types, can be created with as many as 32 regions available as gating regions; two to eight regions can be combined using Boolean logic for gating. Autogating, with user selectable levels, using elliptical and contour regions is available. All statistics are user definable on any histogram. Absolute cell counts using FlowCount™ fluorospheres are available. Display plots can be exported into Microsoft applications or converted to PDF files; data can also be exported directly into MS Excel or in text format. MultiCycle for Windows is provided for DNA analysis.

Other Details:

The current version of the FC 500 is approximately 61 cm high × 112 cm wide × 74 cm deep and weighs 85 kg. The power supply module is approximately 48 cm high × 41 cm wide × 51 cm deep and weighs 55 kg. The instrument and computer together require four dedicated 120V/20A or 240V/15A power lines. It is anticipated that a 20 mW, 635 nm diode laser may be substituted for the 23 mW, 633 nm He-Ne laser, and possible that a solid-state 488 nm source will be offered as well; this will allow the instrument to be housed in an enclosure only 90 cm wide.

To optimize instrument settings, the user can run the Auto Set-Up Wizard, which automatically adjusts PMT voltages, gains and color compensation settings while the appropriate bead sets and stabilized stained control cells are run on the instrument. Analyzing Spherotech™ Rainbow Calibration particles, the FC 500 can detect <600 MESF in the fluorescein channel and <300 MESF in the PE channel. Running Linear Flow Beads from Molecular Probes less than 600 MESF can be detected by the APC channel. Forward scatter signals from 0.5 µM diameter particles can routinely be resolved, and it is usually possible to resolve 0.3 µM diameter particles from background noise. All fluorescence channels except the 755 nm (PE-Cy7) channel have half-peak CVs of 1-1.5% when detecting the fluorescence from beads, with some beads yielding CVs of less than 1%. Fluorescence from beads detected by the PE-Cy7 channel has half-peak CVs of approximately 2%.

The EPICS® XL and XL-MCL Analyzers

The Beckman Coulter XL and XL-MCL (Figure 8-8) are non-sorting benchtop flow cytometers, designed for routine clinical laboratory applications as well as research use; they can measure as many as four fluorochromes excited by a single air-cooled laser. The XL-MCL includes a 32-tube multicarousel loader for automated sample handling.

Figure 8-8. The Beckman Coulter EPICS® XL-MCL analyzer.

Light Source:

The excitation light source is a JDS Uniphase 488 nm, 15 mW air-cooled argon ion laser.

Illumination Optics:

The laser beam is focused to an elliptical spot 10 µm high by 80 µm wide using crossed cylindrical lenses.

Flow System:

The flow system employs a 250 µm square channel quartz cuvette with upwards sheath and sample flow and an integral N.A. 1.0 lens front element. As with the FC 500, there are three user-selected sample flow rates and a fixed sheath flow rate.

Forward Scatter:

The XL uses the same forward scatter collection lens, diode detectors in the Fourier plane and cross-shaped blocker bar setup as the FC 500, but without the adjustable iris; a neutral density filter is available to reduce forward scatter signal intensity by a factor of 10 for samples containing large particles.

Side Scatter And Fluorescence:

In combination with the front element built onto the flow cell, a multielement focusable lens forms a 3 × magnified image; a 500 µm field stop is placed in the image plane. Like the FC 500, the XL uses a photodiode detector for orthogonal scatter. Three fluorescence channels are standard with an optional fourth channel; Hamamatsu R 1923 PMTs are used as fluorescence detectors. The standard four-color setup uses 525, 575, 620, and 675 nm bandpass filters and 488, 550, 600, and 645 nm dichroic long pass filters; a 488 nm blocking filter is also supplied. All optical filters are replaceable without need for realignment.

Signal Processing:

A single trigger channel is selectable. The XL uses the same combination of analog electronics and high-resolution digitization of the data pulses as the FC 500. Unlike the FC 500, the XL normally saves only 10-bit list mode data, although the 20-bit data are used internally for logarithmic conversion and fluorescence compensation. Linear, log and peak values of the pulse may be obtained from each fluorescence channel.

Sample Handling:

The XL uses the same size tubes and volumes as previously described with the FC 500. Samples can be prepared automatically with the XL-MCL using the optional Prep Plus II, T-Q Prep, and CellPrep stations as previously described with the FC 500, where the carousel is then placed into the XL-MCL for automated analysis. Throughput rates of greater than 100 tubes per hour can be achieved when acquiring 2,500 lymphocytes from a normal Q-Prep sample. The XL has the same maximum acquisition rate of 3,300 events/s as the FC 500.

Sorting:

Not available.

Software:

Two software packages are available, XL SYSTEM II software, version 3.0, and EXPO32 ADC software. Both are designed to accept industry standard barcodes when using the XL-MCL bar code reader and/or the optional wand, allowing for automated specimen identification. XL SYSTEM II software operates in a DOS environment under Windows 98SE and utilizes the FCS 2.0 file format. Batch processing and analysis of samples is possible. A maximum of 12 parameters, including time, ratio, and PRISM, can be processed, and gated acquisition is possible using as many as 8 regions from a total of 24 available. A variety of region types are available, and Boolean combination of regions is possible. A single parameter histogram has a maximum of 1024 channels, and a two-parameter display has a maximum of 256 × 256 channels' resolution. A variety of two-parameter display formats are available and overlays of both one and two-parameter data are possible. The software provides for fluorescence compensation using direct visual adjustment. Three-dimensional data display is not available. A large variety of statistics is available, and absolute cell counts can be calculated with the inclusion of appropriate beads in the sample. The software and available optional packages are designed to automate cell analysis and enumeration and report generation, all relevant for a clinical setting. Screen images can be captured in PCX format and ASCII file EPT format files can be generated. The software supports the industry standard SQL/ODBC database allowing for bidirectional connectivity to third-party software programs.

EXPO32 ADC software has the same features as the EXPO 32 software previously described with the ALTRA cell sorter, but includes Advanced Digital Compensation (ADC) for automated fluorescence compensation with up to four colors. A maximum of 16 parameters, including time, ratio, and PRISM can be processed. The software provides administrator tools that allow the operator to set user access levels, automate data archiving, and monitor usage for billing and accountability.

Other Details:

The XL cytometer is approximately 51 cm high × 61 cm wide × 57 cm deep and weighs 64 kg. The XL-MCL is approximately 25 cm wider and weighs 85 kg. The instrument requires space for ventilation: 31 cm from the back, 20 cm from the top, and 31 cm from each side. The Power Supply Module can be placed on the floor, is 48 cm high × 45.5 cm wide × 51 cm deep, weighs 55 kg, and requires 13 cm from the back for ventilation. The XL needs two dedicated 120V/20A or 240V/10A power lines; the XL-MCL needs an additional line.

Instrument settings, including compensation, can be set automatically using the software and running the appropriate bead sets and stabilized stained control cells. The user can change optical filters to optimize for fluorochromes other than FITC, PE, ECD, and PE-Cy5, and an optional PE-Cy7 optical kit is available.

The XL/XL-MCL can detect fewer than 1,000 MESF of FITC and PE on Bangs Laboratories' Quantum microbeads. The forward scatter channel can resolve 0.5 μm diameter plastic spheres from background noise. Precision is similar to that obtained using the FC 500.

8.5 DAKOCYTOMATION

Background

Cytomation got its start by manufacturing and distributing the CICERO system, an add-on to existing flow cytometers which provided improved, faster sort control and data analysis[1181-2]. The first versions of CICERO incorporated a DEC PDP-11 minicomputer system; later versions progressed through VAX minis, and the last, much smaller, less expensive versions, which worked even better, incorporated Pentium-based PC's running MS-DOS.

The original MoFlo high-speed cell sorter was designed and built by Ger van den Engh and his colleagues[1180], then at Lawrence Livermore Laboratory. It incorporated a parsimonious but flexible, bench-mounted optical design, allowing illumination of up to three observation points in a stream in air, and collection of forward and right angle scatter and fluorescence in as many as six wavelength regions. MoFlo, while preserving high-speed sorting capability (100,000 events/s and analysis), was considerably simpler in design than the original high-speed sorter[667] built at Livermore, and incorporated a sophisticated parallel-processing data analysis and sort control system run from a NeXT workstation.

I have great respect for Ger van den Engh's flow cytometer design, possibly because he borrowed and improved on a couple of features of my "Cytomutts." I also was impressed by Cytomation's computer interface when Brian Hall demonstrated it on the Cytomutt in my lab almost ten years ago. It had 16-bit data analysis, although the dynamic range of its front end electronics was not sufficient to allow it to do digital linear-to-logarithmic transformation with a dynamic range greater than three decades.

In the last decade, Cytomation has taken its own approaches to high-speed, high dynamic range data acquisition and analysis, melding the mechanical, optical, and analog portions of the front end electronics of the MoFlo instrument with its own sort control and data analysis system and software, initially running under MS-DOS and now under Windows NT, 2000, and XP. Modular design allowed for everything from tabletop analyzers to console sorters to be included in the MoFlo line, the most recent addition being the CyAn, a modular benchtop analyzer.

DakoCytomation resulted from the recent marriage of Cytomation, a company dedicated to production of high-speed research apparatus, and DAKO, an antibody manufacturer with a heavy clinical orientation. The happy couple could produce some interesting offspring.

The MoFlo® Cell Sorter

The MoFlo (see Figure 1-24, p. 58) is a modular, custom configured cell sorter built on an industry standard optical bench.

Light Source(s):

1-3 lasers; the system is user-configurable with any commercially available or prototype laser, emitting any wavelength(s) from ultraviolet to infrared, preferably in TEM_{00}.

Illumination Optics:

Separate cylindrical and spherical beam-shaping optics are used with each laser, forming as many as three spatially separated interrogation points, typically spaced 50-100 μm apart.

Flow System:

The jet-in-air CytoNozzle™ can be used with interchangeable ceramic tips providing 50, 70, 80, 90, 100, 120, 150, 200, or 400 μm orifice diameters. Sample (core) flow rates can be as high as 1.5 μL/s, with core diameters up to 8.5 μm; fluid flows downward at velocities up to 30 m/s. Sample and sheath pressure can be as high as 100 PSI.

Forward Scatter:

A horizontal obscuration bar is used; sizes ranging from 0.5-5.0 mm are available. Light at angles up to 10° is collected by a N.A. 0.15 lens. The detector can be a 25 mm

Advanced Photonics high-speed photodiode or an H957 Hamamatsu PMT module. Fluorescence collection from the forward direction is available as an option.

Side Scatter and Fluorescence:

A long working distance (13 mm), 0.55 N.A. microscope objective forms a 50 × magnified image of the interrogation point; a horizontal obscuration bar (available range 0.5-3.5 mm) is normally used. As many as 14 detectors may be fitted; the standard detectors are Hamamatsu H957 modules incorporating 1 1/8"side-window PMTs, available in a range of red sensitivities, but other detectors, including photodiodes, are available on request. An assembly of three stacked field stops, one for each beam intersection point, is placed in the image plane; from this point, lenses, mirrors, and dichroics are used to separate signals in different spectral bands and route them to the detectors, each of which is equipped with an additional filter(s) to define its response characteristics.

Signal Processing:

As many as four trigger channels can be selected; the Boolean OR as well as the Boolean AND combination can be used for triggering. A high gain-bandwidth transimpedance amplifier converts PMT output current signals (typically 0-100 μA) to the 0-10 V range. The preamplifier circuit also includes baseline restoration and output buffers. Hybrid digital/analog log amplifiers provide 80 dB (4-decade) dynamic range on each channel; deviations from ideal response across the range are later corrected by digital processing. The analog electronics, including switched-capacitor integrators and peak detectors, can operate on pulses less than 1 μs in duration; peak and integral are computed and digitized to 16 bits' precision within 5.4 μs. Fluorescence compensation for multi-color overlap of as many as 8 colors is implemented by a 1.6 giga-instruction/s DSP chip, performing matrix inversion on digitized data after any necessary correction for nonideal log amp response.

Sample Handling:

MoFlo options allow for automated sample delivery from 0.6-50 mL tubes, with temperature control for input and output. The MoSkeeto™ AutoSampler provides for delivery from 96-well plates.

Sorting:

Droplet formation rates up to 200 kHz and as many as four sort streams are available. The CyCLONE® option permits single-cell sorting into 96/384/1536-well plates and user-definable slide formats with data available on each sorted particle.

Software:

The Windows NT-based Summit™ Data Acquisition and Analysis Software is also available for offline analysis. It reads and writes all FCS formats, and includes capabilities for gated acquisition, postacquisition fluorescence compensation, and manipulation of "unlimited" numbers of parameters, including real-time computed parameters, and of rectangular, polygonal, and/or elliptical regions. Single-parameter histograms may contain as many as 4,096 channels; dual-parameter resolution can be as high as 1,024 × 1,024. The software also has full statistical capabilities, "publication-quality" graphics and spreadsheet/database export capability.

Other Details:

Footprint, dimensions, and power requirements of MoFlo systems vary over a wide range depending on the number and type of lasers installed.

The CytoShield™ product line includes: a Class I Biosafety Cabinet featuring formaldehyde decontamination, negative pressure over the complete work area, remote controls, clog detection, and redundant HEPA filters; an Aerosol Evacuation System, which provides negative pressure to the sort chamber, extracting aerosols to ULPA filters; and a Sort Integrity System, with positive pressure to the sort chamber using ULPA-filtered air.

Sensitivity has been measured at <200 MESF using DAKO 6-peak Fluorosphere Calibration Beads; the fluorescence CV from DAPI-stained trout erythrocytes is <2%. Particles as small as 0.2 μm can be detected using the side scatter signal for triggering.

DakoCytomation provides comprehensive support through a Technical Support Call Center, Regional Field Service Engineers, CytoLink™ Real-time Diagnostics and Solutions, and the MoFlo® Users Group, with full warranties and a wide range of service contracts available. A week-long training course is available on site or at DakoCytomation.

The CyAn™ Flow Cytometer

Light Source:

The CyAn (Figure 8-9) uses 1-3 fixed beams, typically deriving low-noise 351 nm, 50 mW and 488 nm, 150 mW beams, both TEM_{00}, from a Coherent Enterprise closed system water-cooled argon laser, and a 635 nm, 12.5 mW beam from a diode laser. A 20 mW Coherent Sapphire solid state laser may be substituted as a 488 nm source, and a 405 nm, 25 mW violet diode laser is also available.

Illumination Optics:

Beam-shaping optics include cylindrical and spherical elements to provide illumination over as much as the entire 250 μm width of the flow chamber. Three spatially separated interrogation points are typically spaced 200-400 μm apart.

Flow System:

Observation points are inside a UV-grade fused silica cuvette with a 250 μm square internal cross-section; fluid flow is upward. Sample flow rates may be as high as 1.5 μL/s, with core diameters up to 12.5 μm. Sample flow velocity may be as high as 12 m/s, at the maximum 10 PSI sheath pressure.

Forward Scatter:

Observation in a cuvette allows a vertical, rather than a horizontal, obscuration bar (0.5-5.0 mm widths available) to be used in the forward scatter channel. The maximum angle of light collection is 10°. A 0.15 N.A. collection lens is used; the detector may be a photodiode or a PMT.

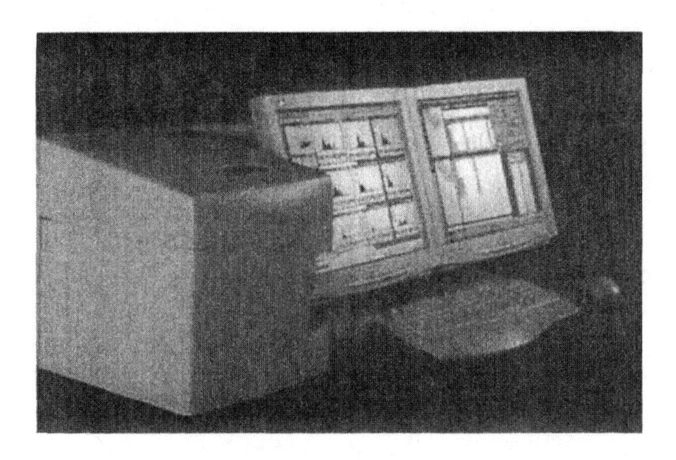

Figure 8-9. The CyAn benchtop flow cytometer.

Side Scatter and Fluorescence:

An aspheric lens, typically with 0.55-0.68 N.A., forms an image of the interrogation points, with a typical magnification of 12.5 ×; field stops similar to those used in the MoFlo are placed in the image plane. As many as 9 detectors may be mounted; these are Hamamatsu HC120 modules incorporating 1/2" side-window PMTs, with various red sensitivities available. Other detectors are available on request.

Signal Processing:

The CyAn uses essentially the same signal processing hardware and software as the MoFlo.

Sample Handling:

Manual loading from 5 mL tubes; automated, temperature-controlled delivery from tubes and multiwell plates should be available by the time this appears in print.

Sorting:

Not available.

Software:

Essentially the same as used with the MoFlo.

Other Details:

When it is equipped with the Enterprise laser the dimensions of the cytometer bench are 36 cm high × 116 cm wide × 57 cm deep. When all solid-state lasers (Coherent Sapphire 488 nm, violet and red diodes) are used, the bench is 36 cm high × 33 cm wide × 49 cm deep. The Enterprise laser requires 208-240V, 60 A, 3-phase power as well as the 110/220 V, 10/20 A, single phase power needed for the rest of the instrument. Various service, support, and training plans are available. Sensitivity has been measured at <200 MESF using DAKO 6-peak Fluorosphere calibration beads; the fluorescence CV from propidium iodide-stained chicken erythrocyte nuclei is <3%.

8.6 CYTOPEIA

You should, by now, have noticed that DakoCytomation and Cytopeia do not appear in the promised alphabetical order; I have two good reasons. The first is that DakoCytomation was Cytomation until quite recently; the second is that it was easier and more informative to discuss DakoCytomation first.

Ger van den Engh, who recently founded Cytopeia, shares my belief in Einstein's maxim that everything should be as simple as possible, but no simpler. However, while I tend to eliminate parts from my cytometers to save money, Ger's emphasis is on making his cytometers and sorters as efficient and reliable as possible. I have no doubt that he would be using diamond flow chambers if he thought they offered a performance advantage.

The MoFlo represented Ger's rethinking of the original Livermore high-speed sorter design in the late 1980's. A decade later, a good deal more had been learned about high speed sorting, and there had been considerable advances in computers, and some in electro-optics, as well. Since leaving Livermore for the University of Washington, Ger had built a series of sorters that incorporated new hardware, software, and knowledge as they became available, and he eventually got dragged into the business.

Cytopeia custom modifies instruments and prepares sorters for as many as three interrogation points, accommodating as many as four lasers, in which case two beams must be collinear. Fittings are provided to attach lasers selected by the customer to the optical bench. The first few instruments placed in the field have been designed to sort human cells for therapeutic use; the fluidics are placed in a clean room, with most of the bulky hardware and electronics behind a wall. However, the basic "InFlux" sorter design is modular, and in its simplest implementation, it provides an elegant, minimalist benchtop system. DakoCytomation will be selling one version of this instrument, with Summit software, under a preexisting cooperative agreement with Cytopeia.

The InFlux Cell Sorter

Light Sources:

User-selectable. The DakoCytomation version will probably use Coherent's Sapphire solid-state 488 nm laser.

Flow System:

The InFlux (Figure 8-10) is a stream-in-air system with sample event rates ranging from 0 to 100,000 particles/s, nozzle diameters ranging from 50-200 µm, and flow velocities from 5-25m/s. Sheath drive pressure can vary from 10-100 PSI; sample is driven at a 0-5 PSI pressure differential with respect to sheath. Fluid flows downward.

Forward Scatter:

Collected over angles from 1.5 to 10° using a horizontal blocker bar. The collection lens has an N.A. of 0.5 with 20 × or 50 × magnification; a 1 mm diameter field stop is used. The detector is a PMT.

Side Scatter and Fluorescence:

The collection lens (N.A. 0.5, 20 × or 50 ×) forms an image of the interrogation points; light from each intersection point passes through an 800 µm field stop. A 1.5 mm horizontal blocker bar is used. As many as 28 fluorescence channels can be implemented; standard detectors are Hamamatsu 957-12 or 957-06 PMT modules, but any

Figure 8-10 Cytopeia's InFlux cell sorter platform.

PMT specified by the customer can be supplied. Spectral separation is accomplished by mirrors, filters, and dichroics. Fluorescence channels can be used for DC measurements; the system may be configured to measure laser output, temperature, sheath pressure, etc. Options include a monochromator for spectral measurements and magic angle collection optics[2439] for precise polarization measurements.

Signal Processing:

The InFlux always uses parameter 1 as a trigger channel. Current-to-voltage conversion is done by transimpedance amplifiers; further signal amplification and logarithmic conversion are done by analog circuits, and both analog and digital baseline restoration can be implemented. Peak detectors and both switched-capacitor and pulse shaping integrators are available on all channels. The maximum data acquisition rate is 200 kHz, using as many as 28 parallel 16-bit ADCs.

Sample Handling:

Tube input: Tubes of a wide variety of sizes fit into the temperature-controlled sample holder. An automated sampler is not yet available.

Sorting:

Droplet sorting drive rates in the InFlux range from 25 kHz with the 200 μm orifice to 125 kHz with the 50 μm orifice. There can be as many as 6 sort streams. Indexed sorting into multiwell collection vessels, with all data stored for each sorted event, is available.

Software:

Cytopeia's InFlux software is Windows NT-based, using FCS file formats. The package is intended primarily for sort control and has limited analysis capabilities. Gated acquisition is possible with bitmaps for 12 parameters, and a maximum of 32 parameters. There is provision for 1024-channel single-parameter histograms and 256 × 256 dot plots. Free-form gating regions can be defined for any two parameters in each of 12 bitmap spaces; AND and OR combinations are possible. A live 3-D display can be rotated during data acquisition. The software also has statistical capabilities, including facilities for comparison of one- and two-parameter data. InFlux sorters sold by DakoCytomation will also be supplied with Cytomation's Summit software.

Other Details:

The instrument dimensions are determined by the lasers used; the sort module itself has a 61 × 61 cm footprint. Dual biohazard containment is standard; a laminar flow unit is available. There is a one-year warranty on the sorter itself; laser warranties depend on the laser manufacturer(s). One week of training is provided with instrument. The sensitivity is sufficient for measurements of chromosomes and bacteria; fluorescence CV for chromosomes is 1% or better.

Inquiries about all versions of the InFlux should be referred to Cytopeia rather than to DakoCytomation.

8.7 OPTOFLOW AS

Background

The Norwegian company Optoflow AS, a subsidiary of *Bio*DETECT, manufactures the MICROCYTE® family of flow cytometers. The version designed for field use (see Figure 1-25, p. 58) is probably the smallest instrument currently available; there are also a benchtop version and one intended specifically for water analysis. While the instruments only measure fluorescence in one spectral region (>650 nm), housings can be ordered in either red, green, blue, purple, or one of three shades of gray. These pioneering "designer cytometers" are available in the U.S. through *Bio*DETECT's Texas subsidiary.

The MICROCYTE® Flow Cytometer

Light Source:

635 nm, 5 mW diode laser.

Illumination Optics:

Polarizing optics are used to maximize scatter sensitivity and minimize interference of stray laser light with the fluorescence signal. The optics are built into a solid metal block.

Flow System:

The observation point is in a cuvette with a 250 μm square cross section. Both core and sheath are driven by pumps; the sample flow rate is fixed at 0.5 μL/s, facilitating calculation of particle numbers per unit volume.

Scatter and Fluorescence:

Two parameters are measured, with avalanche photodiodes (APDs) used as detectors for both. Scattered light is collected over a range of angles between 10° and 30°, intermediate between the ranges used for forward and side scatter in most other instruments. Fluorescence is collected at wavelengths above 650 nm (the detector response extends to 900 nm); sensitivity is reported as 1,000 MESF Cy5.

Signal Processing:

Either signal or both (in AND or OR combination) can be used for triggering. Preamplifiers incorporate current-to-voltage conversion and baseline restoration; analog log amplifiers are used. Data are converted to 8 bits' precision.

Sample Handling:

Tubes in a range of sizes, containing 0.1-1.0 mL of sample, may be used. Typical runs use fixed analysis times of 2, 20, or 200 s to analyze 1 to 100 µL of sample; at a sample concentration of 10^7 cells/mL, the data acquisition rate is 5,000 events/s.

Software:

The lab and field versions of the instrument have the capacity to generate 256 × 256 dot plots without an external computer. All versions can also be interfaced to a Windows-based PC; software provides for gated acquisition and two rectangular gating regions. Data can be stored in a proprietary format or as FCS 3.0 files, depending on the software used. Graphs can be saved as bitmaps and data exported to Microsoft Excel.

Other Details:

The field version of the MICROCYTE® is 33.3 cm wide × 33 cm high × 16 cm deep and weighs 12 kg; it can be operated from a 12 VDC source, drawing 2 A.

8.8 PARTEC GMBH

Background

Founded in the late 1960's, Partec has been producing fluorescence flow cytometers longer than any other company. In the original arc source Impulscytophotometer (ICP), marketed by Phywe AG and, later, by Ortho, cells flowed along the axis of illumination and light collection instead of perpendicular to it, and it was thought that this contributed to the extremely high fluorescence measurement precision achieved using this instrument. However, measurement precision of Partec's newer designs, sold under the Partec name since the 1980's, in which cells flow perpendicular to the illumination axis, has remained excellent, judging by published histograms of DNA content in cells stained with DAPI. Two spectacular examples appear in Figure 7-11 (p. 311).

Partec now offers instruments with arc lamp and/or laser sources, with optional fluidic sorting and electronic cell volume measurement. The new, modular CyFlow line (Figure 8-11) includes simple, single-laser, single-parameter instruments that can be run on batteries in resource-poor environments as well as multisource systems with digital processing, capable of measuring as many as 14 fluorescence parameters.

The CyFlow® and CyFlow® ML Flow Cytometers

Light Source:

The following sources are available: 100 watt long life Hg arc lamp (100 W), violet (nominal 407 nm) diode laser, blue (488 nm) solid state laser (20 or 200 mW), green (532 nm) Nd-YAG laser (up to 100 mW), or red (635 nm) diode laser (15 or 25mW, temperature stabilized). The CyFlow (Figure 8-11) uses a single source; the CyFlow-ML may be supplied with as many as 3 lasers plus the arc lamp.

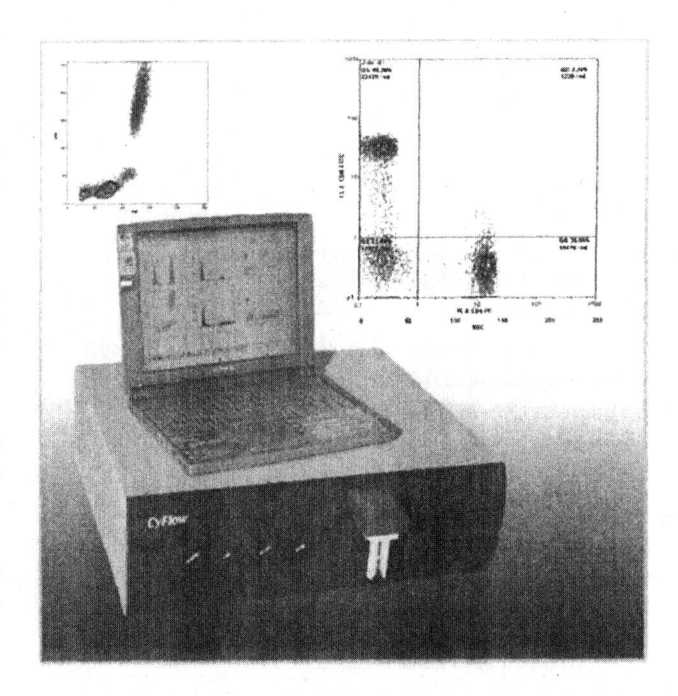

Figure 8-11. Partec's CyFlow flow cytometer.

Illumination Optics:

Köhler epiillumination optics through the microscope objective used for fluorescence light collection are used with the Mercury HBO arc lamp. Laser beams are normally focused with crossed cylindrical lenses, a separate set for each laser, to elliptical spots 10 µm high × 100 µm wide; other beam geometries are optional. A maximum of 3 interrogation points are spaced 30 µm apart.

Flow System:

Cells are analyzed in sheath flow in a synthetic quartz flow cell with a 250 µm square internal cross section. Sample flow rate is continuously adjustable by software between 0 and 3 mL/min. Flow velocity is typically 1 m/s, and is adjustable between 0 and 2 m/s. Typical core diameter is 5 µm; the adjustment range is 5-50 µm. Core (sample) is driven by a computer controlled air syringe pump; continuous steady flow is possible. The sheath is driven by air pressure, typically 200 mbar (2.9 PSI), generated by a pump; pressure is software adjustable from 50-800 mbar, providing flow velocities between 25 cm/s and 4 m/s. Sample flow rate can be measured using an optical encoder on the software calibrated syringe pump or by a patented method that measures the time taken for a sample meniscus to traverse the known distance between two sensing electrodes in an input tube of known diameter. An electronic particle volume sizing measurement is available as an option.

Forward Scatter:

Forward scatter signals are collected by a long working distance lens; a vertical blocker bar and an iris, respectively, limit lower and upper collection half-angles. The CyFlow typically collects light scattered at angles between 2° and 10° from the axis of the illuminating beam. The CyFlow-ML can be equipped with two "forward scatter" channels, the first collecting at angles between 2° and 6° and the second at angles between 6° and 14°. Angular ranges can be further restricted by substituting a wider blocker bar and/or a narrower iris. The standard detector is a PMT module.

Side Scatter And Fluorescence:

Objectives on opposite sides of the flow chamber are used to collect side scatter and fluorescence signals in the CyFlow®-ML. Partec has designed its own line of infinity-corrected Suprasil (quartz) microscope objectives with high UV light transmission. Both 20 ×, N.A. 0.65 and 40 ×, N.A. 0.8 dry objectives and a 40 ×, N.A. 1.25 glycerin immersion or gel-coupled objective are available, and may fairly readily be interchanged in the cytometers. Images are formed of the observation points, and, after passage through user-replaceable rectangular field stops 0.5-2 mm in width, light is recollimated before passing through or being reflected from dichroics or filters used for spectral separation. Standard detectors are Hamamatsu modules incorporating 1/2" side-window PMTs; a range of red sensitivities is available. The CyFlow is available with as many as 3 fluorescence channels and a side scatter channel; The CyFlow-ML can be fitted with 13 or 14 fluorescence channels and 2 or 1 side scatter channels. Filter sets for DAPI/Hoechst, fluorescein, propidium, PE, PE-CY5, APC, and a variety of other fluorochromes are available.

Signal Processing:

Trigger channel(s) are selectable; both upper and lower thresholds may be set, and both AND and OR combinations of signals used for triggering. Current-to-voltage conversion is accomplished by transimpedance amplifiers built into the PMT modules. Analog electronics provide baseline restoration and signal amplification with software-settable gains up to 1,000. The CyFlow utilizes analog log amplifiers selectable to produce linear, 3-decade log, or 4-decade log output; analog peak detectors, integrators, and pulse width measurement circuits process signals from each channel within 2 µs of the return of pulses to baseline, and output signals are digitized to 16 bits' precision. Electronics for all Partec flow cytometers are modular, with easily exchangeable boards in standard Euro-form sizes. Each channel is equipped with a card containing a high-speed, high-precision peak detector, integrator, and pulse width measurement circuit, with much of the circuitry digitally adjustable under the control of a custom designed application-specific integrated circuit (ASIC). Peak, integral, and pulse width are digitized by fast (1 µs) 16-bit ADCs, and transformation between log and linear scale data is accomplished by FloMax® software. The maximum data acquisition rate is > 10,000 cells/s.

Sample Handling:

The instrument accepts sample volumes from 0.5-2.2 mL in 2.5 mL Partec standard tubes. A mini-tube inset is available for sample volumes <0.5 mL.

Sorting:

Like many of the Partec cytometers, the CyFlow line can optionally incorporate a closed-system fluidic sorter, which diverts cells from the main stream into a sorting channel. Only minimal amounts of fluid emerge from the channel when cells are not being sorted. The sorter can be built with a channel diameter large enough to permit sorting of much larger particles (e.g., pancreatic islets) than can be sorted in droplet sorters, albeit at substantially lower rates (300 objects/s). Sort decisions are made in <10 µs based on any logical combination of free form gating regions; sort dead time is <50 µs. Sort delay is set by examining an image of the sorting system taken with an included CCD camera. The sorting module is compatible with acquisition rates greater than 10,000 events/s, and can sort up to 1,000 events/s.

Software:

Partec's FloMax® software runs under Windows, and incorporates capabilities for instrument control, single- and multiparameter data acquisition, and analysis. A desktop PC or a notebook computer with a >1.8 GHz CPU, CD-RW drive, and Ethernet connection is normally supplied by Partec. Data from more than 10,000,000 cells can be recorded in a single FCS 2.0 acquisition file. There are extensive capabilities for protocol definition, sample identification using barcodes, and operation of a sample autoloader; run-time features include automated cluster analysis and full matrix fluorescence compensation, and absolute cell counts may be generated. Batch processing of files according to predefined templates is possible, and automated multi-tube panel reports can be generated in Word or Excel formats. Acquired parameters can include pulse height, area, and width for as many as 16 channels; an additional 16 derived parameters, including ratios and event number or time, can be processed as well. Extensive logical gating capabilities are provided; as many as 32 regions can be defined, with rectangular, polygonal, elliptical, or free form boundaries. The maximum number of channels in a single-parameter histogram is 65,536; a subregion may be displayed and/or printed out with full resolution. Two-dimensional distributions can be as small as 32 × 32 or as large as 1024 × 1024; displays available include dot and density plots, and a zoom function permits enlarged display of subregions. Histogram overlays and 3-D displays are also available. There are extensive statistical capabilities, including mathematical models for DNA histogram analysis and curve fitting. Data and graphics may be exported in a variety of formats; the software can also generate high resolution PostScript files.

Other Details:

The CyFlow measures 43 cm × 37 cm × 16 cm and weighs 8 kg; the CyFlow-ML measures 50 cm × 50 cm × 32 cm and weighs 15-20 kg depending on the choice of light sources. The CyFlow may be operated from a 12 V or 24 V battery; both instruments are operable from a single 100-230 V, 50-60 Hz AC power line.

The PAS, PAS II, and PAS III Flow Cytometers

The Partec PAS series flow cytometers (Figure 8-12) were the first commercial flow cytometers to combine arc lamp and laser sources. Like the CyFlow line, the line includes simpler and more complicated instruments built from

Figure 8-12. The Partec PAS cytometer.

a basic kit of modules. PAS instruments are larger than CyFlow instruments, and can accept larger lasers (e.g., air-cooled argon ion lasers).

Light Source:

Standard: 488 nm air-cooled argon ion laser (20-50 mW), mercury arc lamp (100 W), and 635 nm red diode laser (25 mW); Optional: 532 nm green Nd:YAG laser, violet diode laser, blue (490 nm) solid-state laser.

Illumination Optics:

Köhler epiillumination optics are used with the arc lamp; the illuminated area is 50 µm × 100 µm. Laser beams are focused to elliptical spots 10 µm high × 100 µm wide by crossed cylindrical lenses; other beam geometries are optional. The PAS can be fitted with as many as 3 light sources, with at most of 2 interrogation points separated by 50 µm; The PAS III can use as many as 4 light sources, with at most 3 interrogation points separated by 30 µm.

Flow System:

Essentially the same as in the CyFlow cytometers.

Forward Scatter:

Essentially the same as in the CyFlow cytometers, except that the range of collection angles is 2°-12°.

Orthogonal Scatter And Fluorescence:

Essentially the same as in the CyFlow cytometers.

Signal Processing:

Either analog or digital processing; essentially the same as in the CyFlow cytometers.

Sample Handling:

Essentially the same as in the CyFlow cytometers; however, the PAS series can be fitted the ROBBY® Sample Automat and AutoLoader, a sample preparation and staining station which holds 36 sample tubes per carousel. Samples are prepared according to customized software defined protocols using Windows-based Partec Robby Prep software. The ROBBY can dispense antibodies, fluorochromes, lysing and fixation reagents out of a pool of 16 different chemicals, with adjustable sample incubation and mixing times. The ROBBY is also available as a stand-alone apparatus.

Sorting:

Essentially the same as in the CyFlow cytometers.

Software :

FloMax® software, as described in the section on the CyFlow instruments, is also available for the PAS line. Partec DPAC software, a simple single parameter analysis package running under Windows, is also available.

Other Details:

Dimensions of the PAS are 77 cm × 50 cm × 55 cm; the weight is 35 kg. Dimensions of the PAS III are 140 cm × 50 cm × 55 cm; the weight is 35-100 kg, dependent on which laser(s) is/are installed. The PAS series instruments operate on 110-240 V, 50/60 Hz AC current.

Warranty provisions: 12 months on all parts except filters, mirrors, other quartz or glass parts, disposables and cuvettes.

Service options: 1-3 year service contracts available.

Calibration materials: DNA Control UV for UV excitation or DNA Control PI for use with green or blue excitation are preparations of trout erythrocytes, used for precise alignment and target channel adjustments. Fluorescent beads and CountCheck beads are used for daily quality checks.

On-site training is provided with the instrument.

Sensitivity: <100 fluorescein molecules can be detected in the green fluorescence channel; forward and side scatter signals from submicron particles are measurable.

Precision: Fluorescence CV <1% on Partec DNA Control standards is guaranteed.

PA Ploidy Analyzer and CCA Cell Counter Analyzer

The Partec PA and CCA are one- or two-parameter push button operated desktop flow cytometers using a mercury arc lamp as a light source. The detectors are PMTs; in a two-parameter system, they can be configured to detect fluorescence in two spectral regions or fluorescence in one region and side scatter. The main applications are in cell biology and pathology for cell cycle analysis, apoptosis detection, micronucleus analysis, live/dead cell discrimination, and cell counting.

8.9 SOME OTHER FLOW CYTOMETER COMPANIES

Advanced Analytical Technologies, Inc. (AATI)

AATI makes the RBD2100, a small-footprint, benchtop flow cytometer, with a red diode laser source, intended for detecting and determining the viability of bacteria in environmental, food, and pharmaceutical samples and for characterizing microorganisms in fermenters.

Agilent Technologies, Inc.

Agilent, spun off from Hewlett-Packard, produces the 2100 Bioanalyzer, incorporating microfluidic "lab-on-a-chip" technology originally developed at Oak Ridge National Laboratory by Ramsey et al[2907-9] and brought to the product stage by Caliper Technologies Corporation. The instrument incorporates a red diode laser and a blue (470 nm) LED as light sources; samples are introduced in disposable "chip" cartridges containing an observation chamber and related fluidics. While the 2100 is intended primarily for capillary electrophoresis, chips are available that allow cell fluorescence to be measured in relatively slow sheath flow. Red fluorescence measurements are reasonably sensitive, with a detection limit of 5,000 MESF for Cy5 and related dyes; as might be expected, only relatively strong green fluorescence signals (>2,000,000 MESF of fluorescein) can be measured when the LED is used for excitation.

Apogee Flow Systems Ltd.

Apogee now provides service for the **Bryte HS**, the last commercial version of the arc source flow cytometer developed by Steen and Lindmo[100-3], and recently put improved versions of the instrument, which can measure small and large angle scatter and fluorescence in two to four wavelength regions, back into production.

Earlier versions of this apparatus were sold by Leitz as the MPV-Flow and by Skatron, a Norwegian company, as the Argus. The Argus was, for a time, distributed in Europe by Ortho and in the U.S. by Bruker. The rights to the instrument were then (in the 1990's) acquired by the Italian subsidiary of Bio-Rad Laboratories, a U.S. company headquartered in California. If that genealogy isn't complicated enough for you, here's another interesting wrinkle. Block Engineering, which, at the time, had some of the most advanced flow cytometry technology around, was acquired by Bio-Rad (the U.S. company) in 1977; at the time, Bio-Rad wasn't interested in flow cytometry, and lost an opportunity to become a leader in the field. Years later, they dipped a toe in the water, but, while the Bryte was equipped with an oil immersion lens, Bio-Rad never quite got totally immersed. As of 2002, they are, however, selling another flow cytometer; Bio-Rad is one of a number of companies developing multiplexed flow cytometric bead assays using Luminex's beads and software, and reselling the Luminex 100 cytometer, on which the assays are run.

The MPV-flow was built on an inverted fluorescence microscope, which could still be used as a microscope if you removed the flow chamber from the stage, and incorporated a photometer that could be used for static microphotometry as well as for flow cytometry. The Argus and other later versions of the instrument are assembled on something closer to an optical bench. A syringe pump is typically used for sample injection; cells in sheath flow are observed as they flow along a surface following extrusion from a nozzle. The flow cell in the Argus was open, with the sample stream exposed to air; the Bryte HS and Apogee's instruments can use a newer closed flow cell design.

All generations of this apparatus are demonstrably very precise and very sensitive; production models have shown CV's of less than 2% in measurements of bacterial DNA content, and done very well in characterizing bacteria by two-angle scatter. With a laser source added, the original laboratory-built system was capable of detecting light scatter signals from single virus particles; production instruments found a niche for work with bacteria. With an arc lamp source, the Argus and Bryte HS could reportedly detect a few thousand fluorescein molecules using blue-green excitation. Like other arc source systems, they could/can readily be changed from UV to blue-violet to blue-green to green excitation simply by changing optical filters; this provides an advantage over benchtop systems with fixed wavelength laser sources. Apogee's arc source **A10** replaces the Bryte.

In the late 1990's, Bio-Rad cooperated with Gary Salzman and his colleagues at Los Alamos National Laboratory, and with Harald Steen, in designing and producing two somewhat miniaturized versions of the Bryte apparatus, intended to be used by the U.S. Army for biowarfare agent detection (O, Gucker, where art thou?). One used an arc lamp source; the other retained the epiillumination optics but used a 532 nm green YAG laser source. Apogee now offers the **A20** for military use (you need a password to get the data sheet) and the **A30** for the rest of us. It uses a volumetric pump for sample feed, measures small and large angle scatter and two to four fluorescence parameters, and is available with a solid-state violet, blue, or green laser source. The A30 can detect scatter signals from medium to large viruses as well as bacteria.

Bentley Instruments

Bentley's Somacount and Bactocount flow cytometers are designed for somatic cell and bacteria counting in milk. They both rely on fluorescent DNA stains. The Somacount uses ethidium bromide, and doesn't draw enough current to have an argon laser (used in their earlier models), and the Bactocount is said to have a solid-state laser, so I'm guessing they use green YAGs. Food science marches on; there are several other companies in the same business as Bentley.

Chemunex SA

Chemunex manufactures both a flow cytometric apparatus (the D-Count) and a laser scanning system (the ChemScan RDI) for doing total viable counts of microorganisms in food, pharmaceuticals, cosmetics, drinking water, etc. Organisms are classified as viable if they produce and retain fluorescent material after incubation with fluorogenic substrates; many of Chemunex's test reagents appear to be esters of fluorescein or its derivatives, and both the scanning and flow systems apparently use argon lasers. Before they started producing their own flow cytometers, Chemunex was using small arc source instruments from Partec.

CytoBuoy b.v.

The CytoBuoy[2566,2910] is a flow cytometer with a fairly sophisticated optical and electronic design that is designed to sit in the water and count whatever drifts or swims by - in the size range of phytoplankton, anyway. The company is dedicated to marine biological applications, but the b.v. doesn't stand for "bon voyage"; it's Dutch.

Delta Instruments bv

Delta is another Dutch company; they make the SomaScope, an arc source fluorescence flow cytometer system for counting somatic cells in milk, having started out buying instruments from Partec. Delta's web site also lists a "BactoScope" for counting bacteria, but provides no further information on the technology. If it weren't for the web, I wouldn't know them from Edam.

Fluid Imaging Technologies, Inc.

From sea to shining sea - the FlowCAM[2453], from Maine, is an imaging flow cytometer for continuous monitoring of water; it was used to produce Figure 4-38 (p. 168).

FOSS Electric A/S

And back from water to milk again. Foss Electric was, as far as I know, the first producer of a cytometric instrument for counting somatic cells in milk. However, the original Fossomatic (not, as far as I know, advertised on Saturday Night Live) was not quite a flow cytometer; it stained cells with ethidium bromide and examined them on what might best be described as a rotating slide. The current version, the Fossomatic 5000, is a flow cytometer, as is Foss's BactoScan FC, designed for counting bacteria in milk.

Guava Technologies, Inc.

It would be almost poetic at this point if the Guava PC Personal Cytometer[2465] were designed for food analysis on tropical islands. Nope. The Guava is a very small (footprint not very much bigger than the laptop computer that sits on top and runs it) flow cytometer, sold with dedicated reagent kits and software for quantifying protein binding to cells or particles, total and viable cell counting, and detection of apoptotic cells. The flow chamber is micromachined; the system does not use sheath flow, but the chamber is big enough, and the illuminating beam wide enough, to permit cells to be measured at a reasonable rate without frequent clogs. The light source is a green (532 nm) YAG laser; the instrument measures forward scatter and fluorescence at about 575 and 675 nm. Protein binding measurements utilize phycoerythrin antibodies; viability is determined by exclusion of 7-aminoactinomycin D by nucleated cells stained with a permeant nucleic acid dye fluorescing at a shorter wavelength. I have seen the Guava PC in operation, but, so far, I haven't gotten hard answers to questions about its precision, which I wouldn't expect to be that good because there is no sheath. I'm guessing you probably wouldn't want to use the system for DNA histogram analysis, but it seems to do its assigned tasks reasonably well, and, while not inexpensive, it is easy to operate, and doesn't take up a whole lot of space.

Howard M. Shapiro, M.D., P.C.

Howard M. Shapiro, M.D., P.C., is my own corporate entity, in business since 1976. Although we have actually sold a few Cytomutts in the past, and can still probably find you one of our **4Cyte™** computer interfaces (if you have an ISA slot PC to plug it into) and software, instrument production is really not our thing. Most of the systems we have been built have been for research collaborators or for companies interested in developing and producing flow cytometric apparatus. If the next chapter gets your juices flowing, we'd be happy to help you build your own instrument.

iCyt–Visionary Bioscience

Gary Durack, most recently of the University of Illinois at Urbana, will be happy to build cytometers and/or parts for you, and/or modify your cytometer, and/or integrate it into a larger system, e.g., for high throughput screening. He's got good credentials and experience. We haven't worked together before, but I wouldn't be upset (and I hope he wouldn't) if you paid me for advice and him for hardware.

International Remote Imaging Systems

IRIS's Model 500 Urine/Fluids Workstation and Model 939UDx™ Urine Pathology System, are clinical urine analyzers employing real-time video flow imaging cytometry. Particles detected in the stained core stream are photographed by a computer-controlled microscope camera, using a triggered strobe flash. These products include a computer-controlled video camera and microscope assembly thatviews an optical flow cell. Particles are automatically classified based on size, shape, color, and staining intensity; images of particles are presented on a screen for visual confirmation. Samples are manually introduced into the Model 500; the Model 939UDx has automated sample handling.

Luminex Corporation

The Luminex 100 flow cytometer is a benchtop instrument designed specifically to perform multiplexed ligand binding analyses. I helped Luminex design it, and like to think of it, at least as far as optics go, as the first production instrument incorporating Cytomutt technology. The beads used for assays are about 6 μm in diameter, and are color-coded by staining with a mixture of two hydrophobic, red-excited fluorescent dyes, with emission maxima at about 660 and about 750 nm. The label used for protein assays is typically phycoerythrin (PE); Cy3 and some rhodamine dyes can be used as nucleic acid labels. Particles are observed at rates as high as 5,000/s as they flow upward in a chamber with a 200 μm square internal cross section, passing through separated 20 μm × 80 μm elliptical focal spots from a 5 mW, 635 nm diode laser and a 10 mW, 532 nm YAG laser. High-dry (N.A. 0.63) lenses on opposite sides of the flow cell respectively image the intersections of the 635 nm and 532 nm beams with the sample stream. Mirrors and dichroics separate side scatter, 660 nm fluorescence, and >715 nm fluorescence signals generated in the 635 nm beam, diverting them to avalanche photodiode (APD) detectors. Fluorescence at about 575 nm, excited by the 532 nm beam, is directed to a PMT module by a mirror and dichroic; a bandpass filter is placed directly in front of the PMT. Image plane field stops are used with all detectors. The side scatter channel is used as a trigger channel; pulse width information from this channel is used to exclude bead doublets from analysis.

The Luminex 100 was the first production cytometer to implement high dynamic range digital pulse processing. Signals from the APDs are digitized to 12 bits' precision, providing a usable dynamic range of almost three decades; the green-excited fluorescence signal used for assay readouts is converted by a 14-bit ADC, providing almost four decades of dynamic range. Luminex provides data capture software with relatively limited processing capability that can export data to FCS 2.0 files. I and others have found that the instrument can detect <500 PE molecules bound to beads; the red-excited fluorescence channels are less sensitive (detection limits of several thousand MESF), but good enough to do a variety of immunofluorescence measurements, e.g., CD4 lymphocyte counting using only the red laser for excitation, with side scatter triggering and cells bearing CD4-APC being counted within a lymphocyte gate defined by CD45-APC-Cy7 fluorescence and side scatter[2447]. At present, Luminex is working with Bayer Diagnostics toward extending the 100 for use cellular analyses, e.g., CD4 counting. As a general rule, the company sells instruments and beads to partner companies developing bead assays, rather than to end users. But things might change; get in touch with me if you're interested.

NPE Systems, Inc.

In the early 1980's, Rick Thomas and Jerry Thornthwaite played with a system designed primarily for DNA measurements, eventually forming a company named **RATCOM**, which, rather than heralding the era of punk marketing, acknowledged that Rick's middle initial is A. The RATCOM Personal Cytometer used an arc source and a photodiode detector, and made electronic volume and fluorescence measurements in a three-sided cuvette flow chamber. It was a benchtop system with an Intel-based personal computer for data analysis, ahead of its time because it used computer-controlled motorized optical mounts to maintain alignment and also implemented high-speed digital pulse processing, albeit with a limited dynamic range due to the relatively low precision of high speed ADCs available at the time.

In the early 1990's, RATCOM received a contract from NASA to develop a flow cytometer to be used on the Space Station, a project initiated by NASA and the Florida Division of the American Cancer Society, which was hoping that spinoff from the project would yield some improvements in cytometry relevant to cancer biology and oncology here on Earth.

Fast forward to 2001 (an appropriate year, what?), when an editorial and three papers in *Cytometry*[2911-4] described the use of an instrument, looking a lot like the pre-NASA RATCOM instrument but presumably benefiting some from the NASA project, for nuclear DNA and nuclear volume measurement in normal and cancer cells. In this work, the ratio of nuclear volume to DNA content, dubbed the Nuclear Packing Efficiency, or NPE, was shown to discriminate between nonmalignant and malignant cell populations with similar DNA content. Hence NPE systems. As I mentioned, the NPE analyzer appears to have many of the characteristics of Rick Thomas's older instruments, but, as I also mentioned, they were ahead of their time. The instrument is advertised as yielding DAPI fluorescence CVs of <1.4%, not a surprise for an arc-based system. The knock that has frequently been put on arc source systems is that they are harder to keep aligned than laser-based flow cytometers, and getting around that with computer-controlled mounts, which remain a feature of the NPE analyzer, was and still is a good idea. Whether the NPE concept will boost NPE as a company into orbit remains to be seen.

Union Biometrica, Inc.

Union Biometrica, now a unit of Harvard Biosciences, has developed and sells a series of large particle sorters incorporating their COPAS™ (Complex Object Parametric Analyzer and Sorter) technology. The company was started by Peter Hansen, whose experience in flow dates back to the 1970's, when he worked at Ortho, and Petra Krauledat, who worked for B-D for quite a while. Having already sold off another company, which developed a scatter/extinction based flow cytometer to do immunoassays, Peter and Petra capitalized Union Biometrica, which, in its early days, developed a prototype flow cytometric (again, extinction and scatter) veterinary hematology analyzer for another company. They also responded to request to build instruments that could analyze and sort *Caenorhabditis elegans* and *Drosophila* embryos. We're talking big stuff here; the COPAS stream diameters range from 250 µm to 1 mm.

Particles flow downward; they pass first through a 635 nm beam, derived from a diode laser, in which extinction and, optionally, fluorescence signals are measured, and then through a 488 nm beam (which, in the newest systems, comes from a Coherent Sapphire solid-state laser), in which fluorescence in one or more spectral regions is recorded. After emerging from the flow chamber, the stream is normally diverted sideways to a waste collector by a high-pressure air jet; when a particle is selected for sorting, the air jet is briefly turned off. The sorting speed depends on the stream size; with a small stream, it is possible to sort several hundred objects/s, but the sort rate for *Drosophila* embryos is only a few dozen/s. This rate appears to be somewhat higher than the rate achieved by Furlong et al[2325], who, independently, developed and demonstrated a *Drosophila* sorter at Stanford, and it is a great improvement on micromanipulation, which was all that the nematode, fly, and fish embryo biologists had previously had at their disposal for selecting organisms. Union Biometrica has licensed the Stanford technology, although how much of it is incorporated in the

COPAS instruments is unclear. However, if the price tag on the Union Biometrica instruments is too high for your budget, you can get plans to build your own instrument from Furlong et al at <http://www.stanford.edu/~profitt>.

When I last visited Union Biometrica, shortly after the acquisition by Harvard Biosciences, the plan seemed to be to refine their digital pulse processing technology to permit some degree of information about morphologic detail to be extracted, providing a much higher degree of sophistication in defining sorting criteria. Most of the people playing this game are putting one or more fluorescent reporter proteins into the organisms they work with, and looking for different patterns of expression is a lot more refined, and usually more appropriate, than simply looking for different levels of expression. There was also talk of developing a small (COPAS instruments are built up vertically, and share a small footprint that doesn't occupy much precious lab bench space), relatively inexpensive sorter for the small stuff with which the majority of flow cytometer and sorter users spend most of the time. Since my visit, things seem to have become less "COPASetic" at Union Biometrica; while a recent news article mentioned that Harvard Biosciences was looking to the COPAS line for big profits, it appears that Peter and Petra have taken their money and run.

8.10 HEMATOLOGY INSTRUMENTS, ETC.

I'm deviating from strict alphabetical order here because it makes sense to do so; I also won't guarantee that I've found all the manufacturers. If you want more details on these gadgets than appear here, try the manufacturers' web sites and/or the journal *Laboratory Hematology*. There is also a 1995 book by Groner and Simson[2915] on hematology analyzers; it was fairly comprehensive when it appeared, but could probably do with an update, because new and/or improved instruments keep on rolling out.

There used to be a relatively clear dividing line between the fluorescence flow cytometers and the flow cytometers used in clinical hematology labs for counting and classifying blood cells; the instruments in the latter group didn't use fluorescence. Not any more. Three things remain true. The hematology analyzers use a much wider range of physical parameters than you will find in fluorescence flow cytometers; i.e., DC and AC electrical impedance, absorption, extinction, forward, intermediate-angle, and side scatter (polarized and depolarized), and, now, fluorescence, usually for reticulocyte counting. There are also a lot more hematology analyzers than there are fluorescence flow cytometers; **ABX Diagnostics**, a French manufacturer, was recently reported on a trade publication to have a 20% share of the market, and to have sold over 5,000 instruments in 2000. Finally, while the software for hematology instruments does print out histograms and dot plots, the printout also contains the values for counts, sizes, percentages, etc. of various cell types, all obtained without operator intervention.

Technicon Instruments Corporation, now **Bayer Diagnostics**, offered the first flow cytometric differential leukocyte counter, the Hemalog D, to the market in 1971, at which time only two companies (Bio/Physics Systems and Partec) made fluorescence flow cytometers. The Hemalog D, and the H-series and later blood cell counters[84-5] from Bayer have been used for extinction and scatter measurements for research purposes[359-60,1609]; they could be used for T cell subset analysis using an immunoperoxidase staining procedure[779,1384], but were never widely promoted for that application. It is clear that Bayer has the know-how and the production and marketing capability to get into the fluorescence flow cytometer business, and it would not be a surprise to see them come out with an immunofluorescence analyzer.

Abbott Diagnostics, which makes the Cell-Dyn series of hematology instruments, remained poised on the brink of producing a fluorescence-based instrument for over a decade. The He-Ne laser-based Cell-Dyn 3000 and 3500 could do leukocyte differential counts based on measurements of multi-angle polarized and depolarized scatter[710,986] (pp. 278-9). A number of people involved in instrument design at B-D moved to Abbott in the days when Abbott was **Sequoia-Turner**, and made only impedance-based hematology counters; the low end of the Cell-Dyn line is still impedance-based. Sequoia-Turner was subsequently acquired by Unilever, which found it too small to provide much Unileverage, and divested it to Abbott. After an elephantine gestation period, Abbott came out with the Cell-Dyn 4000, which retains the scatter measurement capabilities of the lower-end 3200 and 3700 (model number inflation), but has an argon laser source and also measures fluorescence. The 4000 uses fluorescent nucleic acid stains to count reticulocytes and nucleated red cells, and can also do CD4 and CD8 counts with fluorescent antibodies, although, at present, the instrument is doing a "two-platform" analysis in a single box, using separate tubes for cell counts and immunofluorescence.

I mentioned ABX Diagnostics before; their instruments incorporate both impedance and optical measurements, including fluorescence. The same is true for **Sysmex**, which started out in Japan as **Toa Medical**, building impedance-based systems, and now makes instruments that incorporate optical measurements, including a dedicated reticulocyte counter and also the UF-100 flow cytometric urine analyzer, both of which measure fluorescence. As of 2001, Sysmex had some marketing arrangements with Roche Diagnostics, which, for a time, was selling its own COBAS line of hematology analyzers, developed, if I remember correctly, in France.

Then, of course there's **Beckman Coulter**, the original (at least from the Coulter side) 800 pound gorilla of hematology instrument manufacturers, and certainly the one most familiar to the majority of fluorescence flow cytometer users. They manufacture a veritable alphabet soup of hematology instruments, including the LH 700 Series, AC·T™ series, HmX, MAXM, STKS™ and Gen·S™. Their higher-end products measure AC and DC impedance and several optical parameters.

The industry is not exactly going to the dogs, but **IDEXX Laboratories** has just introduced a small but sophisticated hematology analyzer, the LaserCyte™, to the veterinary market. Maybe somebody will run some giant panda blood through one in time for my next edition.

8.11 LITTLE ORPHAN ANALYZERS (AND BIG ORPHAN SORTERS)

We now find the parent companies of many commercial flow cytometers, if not the hardware, among the dear departed. I have included descriptions of some of this apparatus for historical purposes and for the benefit of those of you who may become big siblings or adoptive parents. None of the instruments involved is apt to be left on your doorstep in a basket; any of them could make you a basket case. However, help is available (see Chapters 9 and 11).

Bio/Physics and Ortho: Cytofluorograf to Cytoron

If you can lay hands on one of the original Bio/Physics Systems Cytofluorografs (Model 4800, 4801, etc.), give it to the Smithsonian (they already have a B-D FACS). The later model FC-200, which had a rectangular flow cell and more efficient (N.A. 0.45) collection optics might have been worth keeping; it was more sensitive than most stream-in-air systems, if not quite up to the B-D FACScan.

Ortho's System 30 flow cytometer, the direct successor to the FC-200, and the System 50 cell sorter were both built on the same optical bench unit, which was more compact than those used in the early B-D FACS and Coulter EPICS sorters; the system was only slightly larger than the FACScan and Profile. The Ortho flow cytometer could be converted to a sorter, and vice versa, by changing flow chambers and adding (or removing) electronics. These instruments could be equipped with low- (20 mW air-cooled), medium- (100 mW water-cooled) or high-power (2-5 W) argon ion lasers and were always shipped with a second laser as well. In the standard versions, this was a 0.8 mW He-Ne laser, but the systems could be supplied with two ion lasers instead.

An elliptical focal spot(s) less than 10 μm high and about 130 μm wide was produced with crossed cylindrical lenses. Flat-sided quartz flow cells with a 200 μm square cross-section were used for observation in both the flow cytometers and the sorters. In the flow cytometer, the fluidic system was closed; in the sorter, a 75 or 100 μm watch jewel orifice mounted at the bottom of the flow cell, was used to define the jet diameter.

Both forward scatter and fluorescence signals were collected with relatively high-aperture aspheric lenses; on-axis extinction measurements were also made, using the He-Ne beam. Photodiodes were used as extinction and forward scatter detectors, but a PMT could be substituted as the forward scatter detector. PMTs were used to detect fluorescence and orthogonal scatter. The optical geometry permitted fluorescence measurements to be made in the forward direction as well as in the orthogonal direction.

The orthogonal collection lens formed an image of the observation point; field stops were introduced by using fiber optics of small diameter to transmit collected light to detectors (after some 15 years, B-D has, effectively, resurrected fiber optics as relay elements in its LSR II). The orthogonal collection lens could be replaced with the so-called "Ultrasense" optics, comprising a long working distance, high-dry microscope objective for improved light collection efficiency, and a minifying lens to reduce the image formed by this lens to a size compatible with the fiber optics in the detector system.

The Ortho instruments were the first to permit a choice of pulse peak, integral, and/or width measurements and included log amplifiers, but not very good ones. Kamentsky's hardwired analyzer (see p. 28) allowed definition of two parallelogram-shaped sort windows and counting of cells in selected windows. Ortho originally offered a Tracor pulse height analyzer, and later switched to computer-based data analysis systems. However, in their time, the capacity of the Ortho cytometers to measure lots of parameters was limited by what was (or wasn't) available in the way of computers.

The 2151 (upgraded from the 2150) was, in the late 1970's, the most elaborate and most versatile computer offered by any manufacturer of cell sorters; it was, unfortunately, also the most expensive. It used two Data General MP/200 microminicomputers to do a job now routinely done by personal computers or their equivalents. While it took Ortho's competitors some years to catch up with the 2151's capabilities, Ortho was stuck with a dinosaur by the time they did. Their 2140, a "simple" computer system also based on a Data General engine, didn't help things much, because of its high price and limited capabilities.

A few System 30's and 50's, upgraded with data analysis systems from Cytomation or Phoenix Flow Systems, or with my **4Cyte™** hardware and software, are still in productive operation. B-D supplied parts for the line for several years after the 1987 deal in which it bought out Ortho's flow cytometry business; these days, you have to contact Kevin Becker at Phoenix or scrounge if you need replacement components, and few people do.

I must admit that BD Biosciences' new FACSAria sorter, with its relatively small enclosure and efficient optics, including relay fiber optics, brings to mind – or to my mind, at least – the old Ortho System 50. I'm pretty sure the FACSAria will be a much better competitor in the market.

Ortho's Spectrum III flow cytometer was sold for clinical research applications (sold only to prevent disease?). It used a 100 mW argon laser source and incorporates the flow cytometer configuration used in the System 30, with Ultrasense optics. The focal spot was made relatively wide to minimize sensitivity to movement of the flow chamber. Forward and orthogonal scatter and red and green fluorescence could be measured.

Both core and sheath fluid were driven by positive displacement pumps; when a sample was presented by the operator, a controlled volume was aspirated through an external probe. The Cytoron Absolute (see below) continued that tradition. The Spectrum III incorporated a control computer which, like that in Coulter's old EPICS C, was used for most operator interaction with the instrument, controlling laser power, PMT gain settings, etc. In my limited exposure to Spectrum IIIs, I found them exasperatingly user-unfriendly, because the operator didn't have much control over what goes on.

Ortho's flow cytometer subsystems had the edge over those in other manufacturers' instruments, especially in regard to sensitivity and precision for measurements of very small particles (I've seen single virions in scatter in a System 50), until instruments such as the B-D FACScan came on the scene. Because they were designed for multibeam operation, Ortho's systems, in my opinion, did better at it than did other older systems. That opinion is not, by the way, influenced by the fact that I have had business dealings with Ortho; in the days when the older systems were designed, Ortho never asked my advice about hardware. One could demonstrably[643;645-6] get good results from Ortho systems using low-power light sources such as helium-cadmium lasers. The axial extinction pulse width measurement, also unavailable from surviving manufacturers, provided better sizing than did forward scatter measurements. On the other hand, Ortho's System 50 flow cell design was too complicated, leading to frequent clogging and making it hard to maintain good sorting performance. In the last analysis, it wasn't the quality of its instruments, but the quality of its management, that put Ortho out of the flow cytometer business in 1987.

But not for long. Around the time B-D bought Ortho out, Kevin Becker, who had worked for Ortho, founded Phoenix Flow Systems, which still provides rehabilitative services for older instruments. However, as it turned out, Ortho itself became the first flow cytometer company to rise from its ashes. Ortho's European component, based in Milan, continued to be involved with flow cytometry after 1987, distributing both Skatron's arc source instrument and a laser-based, four-parameter benchtop instrument manufactured in Japan by Omron. In 1992, Ortho brought a refined, five-parameter (two-angle scatter, three-color fluorescence) apparatus, the **Cytoron Absolute**, to the American market. The "Absolute" in the name reflected the instrument's capability for doing absolute $CD4^+$ T cell counts; it used a calibrated syringe pump for volumetric sample delivery.

The image-forming optics in the Cytoron were similar to those used in Ortho's older instruments; collection efficiency was increased by using a high-aperture (N. A. 0.9) lens. The flow cell was a flat-sided, square (200 µm I.D.) cuvette. An internal computer system controlled data acquisition and could perform some data analysis tasks; later versions of the firmware in the internal system allowed everything to be controlled by a Windows PC.

The Cytoron used an air-cooled argon ion laser for illumination; it was comparable in sensitivity to B-D's FACScan and Coulter's EPICS XL, but its maximum analysis rate was higher, i.e., 20,000 cells/s as compared to 5,000 or less for the FACScan and EPICS XL. During the mid-1990's, Ortho toyed with the idea of moving to 4-color fluorescence analysis capability and digital log transformation and compensation, even paying me to do some feasibility studies, but their flow operation, first to rise from its ashes, had fallen back by the end of the decade.

HEKA Elektronik GMBH: The FLUVO II Analyzer

HEKA sold the **FLUVO II** flow cytometer developed by Kachel[656], an arc source instrument that measured electronic cell volume and fluorescence in two or three wavelength regions. The optical system of the FLUVO II was basically that of an inverted fluorescence microscope; illumination and collection were done through an immersion objective. The flow system utilized a "tubeless transducer"[349] into which cells were introduced through a hole in the bottom of the sample container; the time taken for cells to reach the observation point was thus minimized, facilitating kinetic studies.

HEKA also sold Kachel's Z-80 based CYTOMIC 12 system[620] for one- and two-parameter data analysis; the later, PC-based "Cyto-Disp" computer supplied with the FLUVO II had a software suite that included Valet's "Diagnos 1" program for automated cell classification[874] in addition to more conventional programs.

The Kratel Partograph

The Partograph flow cytometer followed a design by Eisert[168, 616-9]. A flat-walled flow chamber with dual sheaths provides an extremely stable flow pattern. Two laser beams are focused through the same high-dry microscope objective which serves as a fluorescence collection lens, producing separated spots as small as 1 µm in diameter. When these spots are used for fluorescence measurement, integration of the fluorescence signal is required for quantification of total fluorescence in cells. However, considerable information is available from the pulse shape. The unique feature of the Eisert design lies in its capability to make absolute measurements of cell diameter from extinction pulse width; this is done by using the interval between a cell's traverse of the first and second illuminating beams to correct the raw value of pulse width for variations in cell velocity.

The Partograph FMP was available with low-power (15 mW) argon and/or He-Ne laser sources; an arc lamp source and sorting capability were also offered. The instrument fit on a benchtop. The computer systems used for data analysis were built around the Tandy (Radio Shack) TRS-80 (Z-80 chip) and, later, around a PC/AT compatible. Kratel also sold Kachel's CYTOMIC 12, later available from HEKA.

The ODAM ATC 3000

The **ODAM ATC 3000** was a multiparameter cell sorter developed in France[2916]. It could use one or two ion laser sources; the flow chamber incorporated an electronic cell volume measurement orifice in addition to a sorting orifice. The instrument's most unusual feature, however, was its use of a toroidal lens for light collection. As many as six signals (pulse height or integral, linear or log, and linear combinations or ratios) derived from the cell volume sensor and 3 PMT detectors could be processed by a 24-bit data analysis system built from bit-slice microprocessors. The computer system, available separately, could be programmed in Pascal by the user. The apparatus provided investigators in France with a native alternative to imported flow cytometers; I don't know of any being brought to the U.S. Bruker Spectrospin, S.A., distributed the ATC 3000 for a while; they were also, briefly, the U.S. distributor for the oft-abandoned Skatron Argus. Bruker eventually gave up flow cytometry to concentrate on their core business, which is high-ticket NMR and spectroscopy apparatus; I guess flow cytometry was just a sheath business.

Also Among the Missing

Three Japanese companies, **JASCO**, **Omron**, and **Showa Denko**, have made flow cytometers. I saw a paper in Japanese describing Showa Denko's apparatus, an argon laser source cell sorter. I don't know anything more about that instrument. JASCO's laser source benchtop analyzer was competitive with the EPICS Profile and FACScan; it was looked at by at least one American company, which decided not to pursue it for reasons not related to its performance. Omron's instrument became Ortho's Cytoron Absolute on the way to the orphanage.

Fred Elliott, who designed some of the innards of the Ortho instruments, started a company called **Cyto-Diagnostic Systems**, and developed a small laser source flow cytometer, with a data analysis system built around a PC/AT-compatible microcomputer, in the mid-1980's. The instrument was competitive at the time, but a deal for production and distribution was never made.

Flow Cytometer Rehabilitation; Used Instruments

There is a reasonably brisk trade in used flow cytometers and sorters. It is generally possible to get service and parts, at least for relatively recent models, from the manufacturers themselves, provided the manufacturers are still in business. **Phoenix Flow Systems** has made something of a specialty of rehabilitating old instruments, and they and Cytomation provide (past tense for Cytomation) add-on replacement computer systems for data analysis and, in Cytomation's case, for sort control, as well. **Applied Cytometry Systems** also provides replacement computers for some instruments. My (that is, **Howard M. Shapiro, M.D., P.C.'s) 4Cyte™** data acquisition hardware can also be used with older systems, if you can find a PC with an ISA slot into which to put the hardware. Once you've got a new computer system, and associated data acquisition software that can generate list mode and other data files in the Flow Cytometry Standard (FCS) format, you can use software from a variety of third-party providers for data analysis.

Following Suit

Bio/Physics Systems, later Ortho, was the first U.S. company to make fluorescence flow cytometers. As other manufacturers followed suit, suits followed. Ortho sued B-D, Coulter, and Technicon for infringing on various patents; B-D settled in connection with the 1987 deal. Technicon beat Ortho. It all took years. I wish I had five percent royalties on the legal fees.

These days, it seems as if everybody is suing everybody else. Judges and juries are being asked to decide complex technical issues; a dartboard might give better results. To make things worse, the patent examiners, who are overworked (and who, with one notable exception, are no Einsteins), are primarily interested in seeing that two individuals aren't both given patents for the same invention. To forestall this, they only have to search the patent literature; there's no need to read the journals.

As a result, Joe Schmo can get a patent on something that was published in the open literature in 1899 (even when patent examiners do literature searches, they only look at what's been computerized). This gets licensed to BioScam International. Sam Pull, at Ponzigen, then brings out a product based on the 1899 technology, which he figures is in the public domain. BioScam promptly sues, and the suits at Ponzigen decide to pay BioScam royalties, because it's cheaper than going to court. Meanwhile, in Milan, GenItalia, which had come up with a better version of the Ponzigen method, abandons it because it will be even harder to fight BioScam in court now that Ponzigen is paying royalties for a patent that shouldn't be worth diddly. No wonder my son went to law school.

The legal battles are starting to screw up science. I've told dozens of people who have inquired about sorting eosinophils that the easiest way to do it is to use polarized and depolarized scatter measurements, as described on pp. 278-9. However, as I mentioned back there, the patent on this technique is held by Abbott, which doesn't even manufacture sorters. If people who now work for BD Biosciences (and Leon Terstappen, who originally developed the method, did work for B-D for years) or Beckman Coulter or DakoCytomation tell their sorter customers to use polarized and depolarized scatter, they face suits by Abbott.

If you wait long enough (it's 20 years, now), the patents will expire and you can do what you want. That mode of operation, however, doesn't make for rapid progress. We may get some improvements in the system over the next few years because the software and biotech/pharmaceutical industries, which are vastly more economically significant than our little backwater of flow cytometry, have got the same problems we have and more. I can't wait.

One expired patent that may be of particular interest to the flow cytometry industry is Ortho's patent on Friedman's acoustic cell sorter[1226] (see pp. 264-5). This is a closed fluidic system sorter capable of isolating several hundred cells/s, a rate similar to that achieved by the fluidic sorter in B-D's

FACSCalibur. The acoustic sorter, however, unlike the mechanism used by B-D, only diverts fluid into the sort stream when cells are being sorted, meaning that rare cells can be sorted without being diluted by a large volume of extraneous fluid.

8.12 THIRD-PARTY SOFTWARE

The adoption of the FCS file format made it feasible for various third-party developers to write and sell, or, in some cases, give away, programs for analysis of flow cytometric data. A market for such software existed, and exists, because peoples' preferences in software tend to be idiosyncratic, and because the third-party programs generally have at least a few features which are not included, or not as well implemented, in software from the manufacturers.

This is not to knock the manufacturers' software; manufacturers, however, tend to have more people writing software than do the third-party companies, and breaking software projects into chunks, while sometimes necessary, almost always results in chinks between the chunks when the overall program is put together. If one good programmer can write the whole program, it's apt to run more smoothly.

I know most of the people who write third-party software, and almost all of them have a lot of experience with and knowledge of flow cytometry. A third-party program is generally written by one such person, and its features and performance characteristics are, as a result, tailored to meet the needs of a demanding user. The programmers writing for the larger companies may know a fair amount about flow cytometry, and may actually be better computer programmers than the third-party developers, but, even if they are working to specifications set by sophisticated, demanding users, it's not quite the same.

Third-party software can get pretty elaborate, especially with hardware added. Both Applied Cytometry Systems and Cytomation, for example, reverse engineered the relatively tight interface between the B-D FACScan and the Consort 32 computer system, which was normally essential for the cytometer's operation, and came up with replacement computer systems and software using Intel processors.

As flow cytometers tend increasingly toward digital signal processing and multiprocessor operation, and as hardware and software combinations replace hardware in the innards of the machines, there will be an increasing amount of software that will, of necessity, remain in the domain of the instrument manufacturers. For everything else, and, certainly, for any manipulation of data already collected in standard file formats, there will be third-party software. If you're satisfied with the software your manufacturer supplies, you may never need anything else. As it happens, though, even the manufacturers are beginning to realize that it takes time and costs them money to reinvent the wheel, and they are making deals with the third-party developers for versions of many of the better software packages which will interface smoothly with the manufacturers' software.

In the area of clinical instruments, the Food and Drug Administration (FDA, not to be confused with fluorescein diacetate) has regulatory authority over software; the software itself has to meet certain specifications. It is to be hoped that the FDA will recognize the benefits of the cooperative approach to software development mentioned above.

You can pretty much get the same data analysis capabilities in third-party software as are available from the manufacturers, and more. A list of third-party developers and some of their offerings appears in Chapter 11. Almost all of the producers offer demos, or at least showcase their products' capabilities on their web sites, so it's fairly easy to find out both whether a particular piece of software does the manipulation(s) you need to do and whether the way in which it does it/them fits with the way you think.

The committee that keeps up the FCS standard has been threatening to release some code for reading and writing FCS files, which may be (certainly is, when I wear my programmer's hat) the hardest part of a flow cytometry data analysis program to code. It would be kind of nice to have a public domain toolkit of program building blocks with which moderately sophisticated users could put together their own *ad hoc* programs. The DOS-based Forth software I wrote had some of this capability, but, at least in my limited experience, it has not been possible to duplicate the critical features in either a Windows or Macintosh GUI.

8.13 THE SELLING OF FLOW CYTOMETERS: HYPE AND REALITY

If you listen to the people who sell flow cytometers, you'll be forced to conclude that none of them works. Each manufacturer's people will tell you that the system they sell is the only one that really does work; a chorus from all the other manufacturers will deny this is so. The manufacturers may unite in denunciation of laboratory-built instruments.

If you listen to the people who use flow cytometers, you'll find that all of them work, and that none of them always works. You'll also find that one group of people thinks that one manufacturer's product is easiest to use and most trouble-free, and other groups argue just as passionately in favor of products from other manufacturers.

For example, people with a lot of experience with a B-D instrument always seem to be able to coax better performance out of it than can somebody who is used to other flow cytometers. I had an Ortho System 50 in my lab for a while; the Ortho people, or some of them, who worked on it could always get it to produce sharper DNA content distributions than I could, and I was always able to do better with my Cytomutt than they could. I look at the manufacturers' published precision specifications with a jaundiced eye, because my inquiries indicate that there is a substantial difference between the best that an instrument can do and its level of performance in routine use by the average operator.

If there weren't so many half-truths and misconceptions flying around as a result of manufacturers' and partisan users' efforts on behalf of their machines, it would be easier for people to choose the flow cytometers, bought or built, which they needed for their particular applications.

If there is one thing about flow cytometers that isn't the same the whole world over, it probably relates to user preferences in light sources. Arc source instruments have done much better in Europe than in the United States, where people prefer lasers. I would have to ascribe this to the fact that many users and potential users in the United States aren't as well trained in the use of fluorescence microscopes as are their European counterparts, and thus can't get the same good results with arc source systems.

If somebody gave me as much money as I wanted, and insisted that I spend it on one commercial flow cytometer, I'd have a problem. My decision would probably be influenced by the same factors that influence my purchases of bicycles, cars, computers, and stereo equipment. I always study the products in great detail, evaluate their performance features, and discover that none of them has all the features I want. I can't build cars; I end up buying the one that feels right. It's the same way with most software; I write my own software for flow cytometry data analysis, but, when it comes to word processors, spreadsheets, drawing packages, etc., I have to buy the one that feels right. With bicycles, computers, stereo equipment, and flow cytometers, I usually have more freedom of choice. I wouldn't like to spend a over a half a million dollars of taxpayers' or investors' money on a piece of equipment because it felt right, but, under the circumstances I just described, I guess I'd have to.

Over the years, we have seen convergent evolution in the development of commercial flow cytometers. If you can get some bell or whistle from one manufacturer, you can get its equivalent from another. Buying decisions tend to hinge on how well you like the user interface, but are also influenced by how well you like the salespeople and/or the local service representatives. Clear edges in performance get harder and harder to find, and manufacturers tend to bring instruments up to spec. That's nothing to complain about.

Over the next few years, there will probably be as many developments in the area of data analysis as in flow cytometric hardware. As more third parties become involved in the generation of hardware and software products for data analysis, users will have more options and more opportunities to tailor the user interface to their own tastes. That's also nothing to complain about. If you're not writing your own data analysis software, you definitely need to find software that feels right. That's why I haven't devoted more space and time to the particulars of this and that software package; what's important is it's not what the program does, but how comfortable you feel with the interface.

The one rule which is most important in deciding which flow cytometer to buy hasn't got anything to do with hardware *per se*. It is, instead,

Shapiro's Eighth Law of Flow Cytometry: Know Thy Cells!

You really don't want to lay out a lot of money, even if you, as a taxpayer, only contribute a minuscule share of it, for an instrument that won't let you do the analyses and/or experiments you need to do. The manufacturers will, almost without exception, let you run your samples through one of their instruments when you're in the process of making a buying decision. You shouldn't be making a buying decision unless you know enough about flow cytometry to know what cells you're going to be looking at and how they'll be prepared, and what kinds of data analysis you'll need to do. This is especially important if you're looking at something out of the ordinary. If you're trying to sort very big stuff, or very small stuff, be sure the machine can do it. If you're looking for very faint signals, try before you buy.

I have summarized the characteristics of a lot of general- and special-purpose flow cytometric equipment in the previous sections of this chapter; some of the manufacturers provided more information than others. I'd be surprised if I didn't make at least a few mistakes. As most of the manufacturers know, I have always been and still am willing to make corrections in public, in speech, prose, and/or verse. And, for this edition, the corrections should be posted on the web site.

8.14 APPLYING FOR A GRANT FOR A CYTOMETER

Unless you're a very wealthy amateur, have access to lots of industrial money, or hit the lottery, you're going to have to ask somebody for the wherewithal to purchase a flow cytometer. If you work in the United States, that somebody is usually the National Center for Research Resources at the National Institutes of Health, and the mechanism by which you ask for the cash is most often an application for a Shared Instrumentation Grant. To qualify, three or more NIH-funded Principal Investigators have to demonstrate a need for, a capability to use, and an institutional commitment to supporting a "single, commercially available instrument or integrated instrument system" that costs at least $100,000 (the maximum award is $500,000). The funding mechanism covers a lot of high-ticket gadgetry in addition to flow cytometers, e.g., NMR equipment, confocal microscopes, and electron microscopes, so you're not just competing with flow cytometry people.

As you might expect, I occasionally sit on panels that review applications requesting funding for cell sorters and flow cytometers; from what I hear from colleagues who sit on similar panels covering the other high-priced toys, the things that weigh for you and against you are pretty much the same there as well. In the rest of this chapter, I'll try to save you some grief if you're writing a grant application for an instrument, to NIH or elsewhere.

There are usually a few dozen applications in any given cycle, of which the four or five given highest priority, and maybe not even those, will get the money. All of the applications, as I mentioned above, are coming from investigators who have already been through the peer review process, and have gotten the funds to do their research. The review panel is not allowed to consider or question the underlying science; what is at issue is not whether the applicants know immunology, or genetics, or cell biology, or whatever their primary fields of interest may be, but whether they demonstrate an understanding of what to do, and how best to do it, with the equipment for which they are seeking funds. All of the applications are meritorious; however, they can't all be funded. What happens is that those that survive are those with the fewest obvious faults. I can tell you what some of those faults might be, and if you are writing or assisting in the writing of a grant application, you can at least try to avoid them.

The grant, if you get it, pays for the instrument; it doesn't pay for service contracts, operators' salaries, or any modifications to the site needed to install the instrument (it may cost thousands of dollars to bring in the power and cooling water connections for a big laser). Almost everybody who runs a shared instrument facility charges user fees, which are supposed to defray some of those costs. Almost everybody who has run a shared instrument facility knows that the facility is likely to run at a loss. If the administration at your institution doesn't come across as wildly enthusiastic about providing all funds necessary to make up the difference, you're apt to lose points, because somebody else's administration will look better.

It's advisable for the Principal Investigator (P.I.) on a cell sorter grant to be a person who knows a lot about flow cytometry. It is not required that he or she have an NIH grant, as long as there are three potential users who do have NIH grants and are Principal Investigators on those grants. NIH Policy is that the Principal Investigator on a Shared Instrumentation Grant doesn't even have to be a user of the instrument. However, in my experience, reviewers take a dim view of applications in which people with little or no expertise with a sophisticated instrument are listed as Principal Investigators.

In the old days, when there weren't user-friendly benchtop systems around, the graduate students and postdocs had to hang around the sorter lab, even to do simple analyses, and, at least on paper, it was easy to look knowledgeable. Now, a lot of people do a lot of respectable flow cytometry on benchtop systems, which is fine. However, if the P.I. on an application asking for a big sorter with four lasers is somebody whose experience appears limited to benchtop systems, it's a sure bet that that application won't look too strong in comparison with others.

The manufacturers are all too happy to write quotations according to which they will sell you a big sorter and a benchtop instrument (which will, of course, work together as an "integrated instrument system") for just under $500,000. I personally wouldn't ask for two instruments. You'll be very lucky if you get one instrument funded; leave some money for somebody else. It won't help your application if any of the reviewers gets even a fleeting impression that you're a pig.

If you have an instrument and are requesting an additional instrument, it helps to be able to show that what you've got is used a lot, which probably means nights and weekends; if you want to replace an old, obsolete instrument, bear in mind that most of us don't think a three year old cell sorter is obsolete because a new model has come out. Also, don't ask for more – or less – instrument than you need. If you're requesting an argon laser, a mixed gas laser, and a dye laser, it helps to include some details of the parameters that will be measured and the probes that will be used, because those lasers are expensive. The same goes for other add-ons. If you don't explicitly justify the need for them, they'll get axed from the budget. If you do get funded, you'll be short of cash, but, let's face it, if you do a poor job of justifying what you've asked for, you're less likely to get funded.

If you announce your intention to do experiments with Hoechst dyes, DAPI, or indo-1, but you have only asked for a single 488 nm argon laser, which doesn't emit the UV light needed to excite these dyes, you don't win points for trying to save money; it simply looks as if you don't know what you're doing.

Some instrument manufacturers will be happy to sell you various computer peripherals at substantially higher prices than you'd pay at a computer store. Having these items in your quote probably won't reduce your overall chance of getting funding if your application is otherwise perfect, but the budget is very likely to get cut, as it should be.

These days, applications are getting so thick, what with biographical sketches and information on other projects for all of the investigators, that many people seem to be skimping on the details of what will be done with instruments. Being concise can be a virtue; being concise to the point at which it becomes unclear that you need or can competently use the apparatus is a vice.

Finally, if you don't get funded the first time, and decide to resubmit the next time around, pay attention to the critique of your application. It's likely to call attention to some of the faults I have just mentioned. If you don't fix everything, your next application won't do any better than your first one. If you do, at least you've got a fighting chance. Good luck.

9. BUILDING FLOW CYTOMETERS

I am not going to tell you how to build flow cytometers. Not here, not now. The Third Edition of *Practical Flow Cytometry* represented a radical departure from the two previous editions and from their predecessor, *Building and Using Flow Cytometers: The Cytomutt Breeder's and Trainer's Manual,* which contained complete plans for the mechanical, optical, and electronic components of a multiparameter instrument using a 488 nm laser source. Wiley did the math and figured that, since only about one percent of the people who bought the First and Second Editions built machines, it made sense to kill the 50 pages of plans and devote more space in the Third (and this Fourth) Editions to how flow cytometers work and how to use them. I didn't argue.

In his Foreword to the Third Edition, Len Herzenberg suggested that biological scientists really shouldn't be thinking about building flow cytometers. That was easy enough for Len to say in 1994, but neither this book nor the previous versions would have been written if he hadn't felt differently in the mid-1960's. And I noted that the last issue of *Cytometry* I got before the Third Edition went to press in 1994 contained an article about a multidimensional computerized sort control method implemented in the Herzenberg lab[1740], and not on unmodified commercial equipment, either. Even now, Len always wants to measure more parameters than can be measured with the fanciest commercial flow cytometers, and he always has a few people in his lab building or modifying apparatus with that goal in mind.

9.1 WHY BUY A FLOW CYTOMETER?

In principle, you buy a flow cytometer because you want an instrument with a proven design, established reliability, and performance better than you could obtain from a system you might build yourself. You're willing to pay for this, and also willing to pay some more to insure that a trained service person will keep your instrument in good operating condition and fix it when it breaks.

A well-known analytical cytologist (Bas Ploem) told a bunch of people at a 1981 meeting that "Everybody who has bought a cell sorter up until this time has bought an instrument in development". I would tend to agree with him. It's been my experience that, during the first six months, at least, you're either having fits or having retrofits 90 per cent of the time. That may not be the case with modern benchtop systems, but it still seems to hold for the larger instruments.

Once the operating pattern gets a little stabler, you find that down time comes from annoying little things, such as clogs that require that you remove, clear, and replace the flow chamber, necessitating a fairly extensive realignment of the optics, and from real disasters, such as incineration of laser power supplies, which the laser manufacturer's people rather than the flow cytometer manufacturer's people usually end up fixing.

Even with those minor headaches, there's no question that if you are relatively new to this field, and you want to do something that has been done before, which a lot of people are now doing routinely with commercial flow cytometers, and you have the money to buy and maintain a commercial instrument, you should buy one. This is particularly true if you are interested in sorting; it is a good deal easier for an ingenue to build a flow cytometer than a cell sorter.

9.2 WHY BUILD A FLOW CYTOMETER?

Flow cytometers, like many other types of instruments, sell for anywhere from three to five times the cost of the parts which go into them. This is not to say that the manufacturers are making unconscionable profits; some of them aren't, or weren't, making any profits at all, at least on flow cytometers. It's just that when you factor in the labor costs, plant and equipment, marketing costs, etc., you end up with that multiplied figure. If you have yourself, or your graduate students, or a knowledgeable tech or two, or some other unsuspecting confederates, around as a source of labor, and some minimal shop facilities, like a small drill press and some hand tools, you can count on being able to build a facsimile, you should pardon the expression, of a commercial instrument for a parts cost no more than a third of the selling price. Should you try to do that?

Suppose you already have a sorter, and it's overloaded, and you'd like another instrument? You're no longer new to the field. If you've kept your sorter running, you've undoubtedly had to clear a bad clog and realign the system, and, if you can do that, you're not far away from being able to build your own flow cytometer or sorter. Again, if you can afford to buy another system, you might think about it. What you might conclude, though, is that what's on the market is pretty expensive.

If you'd like a five-beam, benchtop system just for analysis, there isn't a manufacturer making one. If you want a really slow flow system to analyze DNA fragments and viruses, ditto. Nobody blames the manufacturers for being in business to make money. If Len Herzenberg and his colleagues hadn't built the original fluorescence-activated cell sorter and shown that useful things could be done with it, generating a demand, nobody would be manufacturing sorters.

The problem is that today, when most of the companies in the business are big companies, fewer and fewer of them are willing to take a flyer on building something to meet the offbeat needs of a small user community. Most of the manufacturers are making their money selling benchtop systems and reagents, and, while they like to compete with one another, they don't like to compete with themselves, which usually means you have to buy one of their bigger, more expensive instruments if you want maximum flexibility. If you want to do unusual things that neither the benchtops nor the big machines can do, you have to think about building an instrument yourself, or of having someone (present company not excepted) build an instrument for you.

The funding situation being the way it is, maybe you don't have the money to buy a commercial system, even if you work in industry. Then, if you want a flow cytometer, or another flow cytometer, building your own may be the only option.

9.3 LEARNING TO BUILD YOUR OWN

Take my word for it, building flow cytometers isn't all that hard to do, and, at least until recently, it kept getting easier, as it became possible to buy bigger and bigger chunks of the do-it-yourself kit ready-made.

The annual research course on flow cytometry (see p. 67) usually features a hands-on lab in which three or four participants assemble a two-parameter fluorescence flow cytometer in an afternoon under the auspices of a couple of people from Los Alamos. It doesn't take much longer to do the same thing in your lab or basement. It can even be fun. Terry Fetterhoff and Bob McCarthy and their confederates described their experience with building and using a Cytomutt[1021] some years ago, when a lot of soldering was required; one of my more recent star pupils, Dennis Way[1741], got a four-parameter system running in a day and a half, and published on what he did with the instrument, not how he built it. That system has been run for at least ten years, and I've only been called about problems with it two or three times; the last time I checked, Dennis was likely to get lower CVs for beads and DNA standards than I did.

I had originally planned to update the do-it-yourself parts of the Second Edition and publish them on my own, as a sequel to *Building and Using Flow Cytometers*; so far, I haven't gotten around to doing that. The big problem is that I would have to revise the electronics and computer sections almost completely.

On the plus side, flow systems and optics haven't changed all that much since the 1980's; we have pretty much the same mounts, lenses, dichroics, filters, etc. now that we had then. There are a whole bunch of new lasers now available, including UV, violet, blue, green, and red diode or solid-state sources. There are also now PMTs on the market with higher red sensitivity than we used to be able to get, and they are available built into easy-to-use modules with integral high voltage supplies.

On the minus side, while electronics and computers have improved tremendously, it has gotten much harder to prototype and build your own circuits; most devices no longer come in packages that plug into sockets. Instead, you have to get somebody to build you a multilayer circuit board with all of the surface-mounted active electronic devices on it. That costs money. Computers and DSP chips and data acquisition systems are faster and more sophisticated, but software development has gotten harder rather than easier.

In the late 1960's, when I built my own audio systems, they were better than anything I could have bought; that was also true, at least to some extent, of the cytometers I built through the mid-1990's. Today, there's no way I can build audio equipment as good as what I can buy for a reasonable price, or flow cytometers as good as those you or I can buy for what may seem like an unreasonable price. I am still interested in building small, efficient, affordable instruments to do jobs that don't demand all the bells and whistles, and still trying to figure out which of the available component parts are best for doing the job. When I do, I'll probably post the plans on a web site and/or run a course; it's got to be easier than writing books. In the meantime, if you're looking for advice, I'm not hard to find.

10. USING FLOW CYTOMETERS: APPLICATIONS, EXTENSIONS, AND ALTERNATIVES

The successful application of flow cytometry in biological and medical research and in clinical medicine depends as much upon understanding of the biology of the systems being studied as upon familiarity with hardware, reagents, and methods of data analysis. Two editions ago, I wrote "It would be futile for this author and this book to attempt to provide even a rudimentary discussion of all of the areas of biology in which flow cytometry is used or of the biological problems to which the technique is applied"; the range of applications continues to expand.

The first flow cytometers were built and run in a few labs committed to instrument development. In the early days, almost any experiment that could be done with a flow cytometer would produce new information; in order to get the experiment done, it was frequently necessary for the experimenter to make a substantial refinement on existing technique.

During the 1980's and 1990's, flow cytometry became (not inappropriately) a mainstream technology, with one result being that most of today's users are trained in biology and/or medicine rather than in instrumentation. The number of papers and the range of journals in which they appear are large enough so that you can't just look under "Flow Cytometry" in MEDLINE or other databases and hope to find them all, as I pointed out in Chapter 2.

The range of parameters and probes in common use has also expanded considerably since the last edition was published. I've tried to cover as broad a range of parameters and probes as possible in Chapter 7, and to at least give some idea of the existing and potential applications of each. I therefore decided to approach this chapter as I did in the last edition, i.e., to consider the few basic tasks and procedures underlying almost all of the **applications** of flow cytometry, and then survey and comment on selected applications and point the reader to articles in which specific techniques and applications are discussed in detail. But, this time around, I also felt it was necessary to include some discussion of **extensions** of flow cytometry, involving instruments with characteristics and capabilities not yet available in commercial systems, and of **alternatives** to flow cytometry, notably scanning and imaging cytometry, which have become more practical and, in some cases, more affordable since the last edition appeared.

If we achieve our respective objectives in writing and reading, you should, by the end of this chapter, be able to make sense of, and sense nonsense in, anything in the literature which deals with flow cytometry.

10.1 THE DAILY GRIND

Keeping the Instrument Running: Diet and Exercise

In flow cytometry, it is not only "Garbage in, garbage out," but also "Garbage inside, garbage out." There are a few tricks to keeping flow cytometers running well, and there's more to calibration and standardization than running beads through the system. If the instrument isn't kept in shape, though, it is equally unsuited to any and all possible applications.

Generally speaking, the materials which go into a flow cytometer are air or gas, water or saline, plastic beads and/or samples which contain protein, carbohydrate, fat, and chemical additives such as coloring materials. We run on a similar mixture, beads generally excepted, and, like flow cytometers, are susceptible to pollutants in what we take in.

If your flow cytometer needs an air or gas line input, it's essential that the air be oil- and particle-free; if you're not sure, it's best to have a filter in the line. Even if you never plan to do sterile sorts, and only use water as sheath fluid, it's advisable to have the water go through a 0.47 µm filter before you put it into your sheath tank or bottle, and not a bad idea to keep a filter of the same pore size in the line between the sheath tank and the flow system.

If you work with microorganisms, platelets, or other small particles, your sheath fluid and reagents should be passed through a filter with a smaller pore size, e.g., 0.22 μm. If you run using an in-line filter with pores this small, either use a fairly large area cartridge filter or replace a smaller filter frequently, because the pores clog up, resulting in a substantial pressure drop across the filter, which will screw up your flow rate. For the most demanding work, e.g., trying to look at viruses or other really small stuff, you should consider preprocessing your fluids through an 0.1 μm filter; it is tricky to use these in-line, because the resistance is high enough to slow your fluid flow considerably.

Numerous problems may arise due to the tendencies of organic materials in samples to stick to metal and glass as well as plastic tubing in flow cytometers. On a within-day, sample-to-sample basis, dyes such as acridine orange, ethidium and propidium, and the cyanines may wash off the tubing and stain cells to which only fluorescent antibodies had intentionally been added, giving false positive staining. The dyes are usually removed effectively by rinsing the system with chlorine bleach; however, I have been told that the calcium ionophore ionomycin isn't always removed by a single chlorine bleach treatment.

When the bleach itself isn't washed out or neutralized, and you then try to do membrane potential or calcium probe studies on live cells, you find the cells dying instead of dyeing. When doing such experiments, I usually to follow the bleach with 50% ethanol and then saline; this may be superstition, but it generally avoids the cell kill I might see otherwise.

On a long-term basis, protein residue can screw up the optical quality of cuvette windows, cause flow disturbances, and provide a food source for microorganisms, which will happily grow in a flow cytometer if they are not frequently disturbed. The best preventive medicine for the ills just described is plenty of fluids. If you start up a flow system that hasn't been run for more than a few weeks, count on spending the first day unclogging it every five minutes. If they're painting the lab, run the flow system every day anyway, if you can, and never shut down at the end of a day without making sure that you've cleaned the flow system out with a purge of bleach, optionally followed by alcohol, and then by a lot of water, with backflushing so you get any residual bleach and/or alcohol out of the nooks and crannies.

Ion lasers, like flow systems, need their exercise; this may be truer of the big lasers than of the small ones. If you have a system with a big laser, you're apt to get into keeping mirrors tweaked and clean, checking on and fiddling with gas pressure, etc.; if you use a small argon laser, you're free from most of those headaches. It's advisable to monitor laser power output if you can; this is straightforward on big lasers, where you shift back and forth between current and light control mode. On smaller lasers, which are usually operated in a light control mode, what you need to do is let the laser warm up and check the operating current, which will creep up as the plasma tube ages and may scurry up if the optics get dirty or the alignment is knocked out of whack. You will also get clues that you're losing laser power from having to keep turning the PMT gains higher and higher to measure the same particles and/or the same cells stained with the same reagents, even though the system is well aligned and the CV's are low. Laser noise is an uncommon, or at least not a widely reported, problem in flow cytometers[876], you should be aware it exists but there's no need to go looking for it unless you're getting high CV's for which you can't find another explanation. I mentioned on pp. 142 and 147 that He-Cd lasers particularly need exercise; left idle, they develop increased helium pressure, which makes them get very noisy.

There isn't any exercise for the optics of your flow cytometer, which may get dirty, particularly if your lab is dusty or smoky. Clean living prevents problems; once they're there, you need live cleaning.

Particulars: Drawing a Bead on Flow Cytometer Alignment, Calibration, and Standardization

Flow cytometers can be calibrated either with synthetic standard particles or with cells or other biological material. It would be nice to have standardized calibration particles that would yield the same results for every user and every instrument, assuming proper alignment of each. We're not quite there, but we have reached a point at which it is possible for careful people using different instruments in different labs to analyze the same specimens and get both qualitative and quantitative results that are in good agreement.

This doesn't happen when the instruments are not in good working order, and there's not much point in analyzing samples on an instrument that isn't. So, as a general rule, it's a good idea to make sure that your flow cytometer is up to snuff before you invest time and money in preparing and running samples. If performance is substandard, you or somebody more skilled and knowledgeable will need to align and adjust the instrument. Since most modern flow cytometers make no provision for the user to adjust the mechanical, fluidic, and optical components, that may mean a service call, which will cost you money, and, even if you or somebody in your shop does the work, you really have to define what's broke before you or anybody else can fix it. As to how to do that, I have only one word for you – plastics.

It is, in principle, possible to produce spheres of polystyrene or other plastics that are extremely uniform in size, and to introduce fluorescent material into such spheres in a precisely controlled manner. It is also, in principle, possible to produce great sweet wines from Gewürztraminer, Riesling, and Semillon grapes. In practice, the sweet wines produced from those grapes are not all cast in the same mold. So it is with calibration particles. The good vineyards do not bottle the bad vintages; the manufacturers do not sell the really bad particles. A good vintage, or a good lot, depends upon factors not entirely under the vendor's control. People hoard small particles and great wines.

Fluorescent particles are available from several suppliers. I have generally had good luck with spheres from **BD Biosciences**, **Beckman Coulter**, **Flow Cytometry Standards Corporation** (their product line has been taken over by **Bangs Laboratories**), **Molecular Probes**, **Polysciences**, and **Spherotech**; I don't have much experience, good or bad, with beads from other suppliers. These days, practically everybody who sells beads intended for the flow cytometry market supplies reasonably complete information about their size, spectral characteristics, brightness, and uniformity, anyway; while the materials aren't cheap (p. 60), they generally perform as advertised.

I have discussed different types of particle standards on pp. 354-6, in the context of their application to quantitative immunofluorescence measurements. The process of determining whether or not a flow cytometer works begins with what are now called **Type I standards**, or **alignment particles**.

Alignment Particles: Fearful Asymmetry

Alignment particles are highly uniform and produce relatively strong fluorescence signals. When aligning an instrument, the routine is to run alignment particles, keeping core diameter relatively small to insure illumination uniformity, and adjust the positions of the flow cell and optical elements to maximize the fluorescence and forward scatter signals while minimizing the CV's of the distributions obtained. Data from alignment particles should be recorded on a linear scale.

When using alignment particles substantially smaller than cells (Type Ia), one should be aware that the effects of illumination inhomogeneity may be less evident with small particles than with large ones, as discussed in Chapter 4. A 2% fluorescence C.V. with 2 μm plastic spheres can translate into a 5% C.V. in measurements of cell nuclei stained for DNA content.

A well-aligned instrument operating on a linear scale should produce **symmetric peaks** when beads or nuclei are used as standards. This is particularly important when you are making DNA content measurements. The mathematical models commonly used to deconvolute DNA histograms and calculate numbers of cells in different cell cycle phases assume that the G_0/G_1 peak is symmetric. If the instrument is misaligned, and the bead peak is skewed with a tail to the right, it's a safe bet that the G_0/G_1 peak will be, also; the model will conclude that the cells in the tail are in S phase. In fact, if you run the model on the bead histogram, you'll be told there are S phase beads. Don't do the DNA analysis until you fix the alignment; if you've already collected the data, don't bother with the model, which will give worthless results.

Alignment particles provide the best measure of instrument performance in terms of **precision**, as was discussed on pp. 214-5. An instrument in good working order, running alignment particles, should produce fluorescence distributions with CVs no higher than 3 percent.

The beads from Beckman Coulter with which I am most familiar are intended for checking precision in connection with DNA measurements; their intrinsic CV is probably in the range of 1.5%.

Polysciences makes polystyrene particles in a range of sizes (from 0.1 to 10 μm) and colors. They have UV-excited, blue fluorescent and green-excited orange fluorescent spheres in addition to green fluorescent spheres usable with excitation wavelengths ranging from UV to blue-green; there are also spheres which are usable at other excitation wavelengths. Polysciences also supplies spheres with reactive groups to which other materials can be coupled and will make up spheres containing dyes of your choice, for a price. The uniformity of the custom spheres may vary. However, as a general rule, Polysciences 2 μm yellow-green fluorescent spheres are very uniform; I have observed scatter CV's as low as 1% and fluorescence CV's as low as 1.2% on well-aligned instruments. The yellow-green spheres are labeled during the polymerization process; this procedure is best undertaken using relatively hydrophobic dyes, and Polysciences uses a coumarin dye which mimics the emission spectrum of fluorescein fairly well, but which is excited better in the blue-violet (436, 441, or 457 nm) than at 488 nm.

Reference and Calibration Particles

For standardizing immunofluorescence measurements (pp. 354-6), it is advantageous to have spectrally matched **Reference (Type II)** and **calibration (Type III) particles** bearing either known numbers of bound molecules of fluorescein, phycoerythrin, and other antibody labels (Types IIB and IIIB particles) or known numbers of antibody binding sites (Types IIC and IIIC particles). This eliminates the effect of such factors as differences in PMT sensitivity curves and differences in emission filter bandwidths, facilitating comparison of results obtained with different instruments. However, for day-to-day assessment of instrument sensitivity (see the discussion of Q and B, pp. 221-3), it is usually more convenient to use Type IIA and IIIA particles, hard-dyed beads with fluorescence characteristics that are not environmentally sensitive and also are not closely spectrally matched to the probes used for immunofluorescence measurement. It is desirable for the set of particles to include a **Type 0 standard**, or **certified blank**. If you could resolve five bead peaks in the set last week, and can only resolve three this week (see Figure 4-61, p. 222), something's wrong.

Beads labeled with known numbers of MESF of fluorescein, phycoerythrin, and other labels, phycobiliproteins, and tandem conjugates are labeled in the aqueous phase; the fluorescent label is predominantly confined to the particle surfaces, and the fluorescence CVs of these particles are typically higher than those of hard-dyed beads. The hard dyed beads used as Type IIA standards, and those in Type IIIA bead sets, tend to have somewhat higher fluorescence CVs than do alignment particles; you should therefore avoid the temptation to use these beads for anything other than a first cut at instrument alignment. If you yield to temptation, be sure to use the brighter beads, which have lower CVs than the dimmer ones.

Compensation Standards

When you are doing multicolor immunofluorescence analysis, the actual signal intensities you get from different detectors will be a function of how efficiently the dyes you use are excited at your excitation wavelength(s) and of the spectral responses of your detectors, and even a relatively simple task such as determining fluorescence compensation settings to separate fluorescein and phycoerythrin immunofluorescence gets hard to do unless your particles are labeled with fluorescein and phycoerythrin rather than simply fluorescing green and yellow-orange. It is possible, in principle, to use beads labeled with spectrally matched mixtures of dyes other than the labels used to compensate for fluorescence overlap, but the manufacturers of those beads will have to make their cases for such products. However, when measuring fluorescence in more than three spectral bands, it is now common to use single-label and "every-label-but-one" cell samples to determine compensation settings.

Cells and Nuclei as Alignment Particles

I like to use **stained nuclei** for determination of both precision and linearity (p. 217), particularly in channels used for DNA content measurements. DAPI- or Hoechst dye-stained nuclei provide UV-excited fluorescence which is nominally blue, but which extends to almost 600 nm. Propidium iodide-stained cells, excited by light at wavelengths ranging from UV to yellow, emit red-orange fluorescence. If you are trying to measure DNA, you obviously want to use nuclei stained the same way you stain your samples. Peripheral lymphocyte nuclei usually have very low CV's; you can also use cell lines, but be aware that they can develop aneuploidy, which gradually increases CV's. I had that problem with CCRF-CEM cells; the ones I used originally had G_0/G_1 peak CV's below 1%, but they drifted up to just under 3% after a long time in culture.

When you're trying to measure precision in a fluorescence channel in which you use red excitation, whether or not you're doing DNA measurements, your best bet may be nuclei stained with DRAQ5 or with another red-excitable dye, e.g., TO-PRO-3, oxazine 750 or rhodamine 800. If you use a dual-beam instrument with 488 nm and red beams, DRAQ5 is advantageous, because, since the dye is excited by both 488 nm light and red light, you can use the same cells to align fluorescence detectors that look at two beams. You can play the same game in an instrument with UV and 488 nm beams using ethidium or propidium as the DNA stain, since both the UV and the 488 nm beam will excite these dyes. Figure 10-1 illustrates the principle.

Since the nuclei are stained with only one dye, all of the data points in the two-parameter dot plots of Figure 10-1

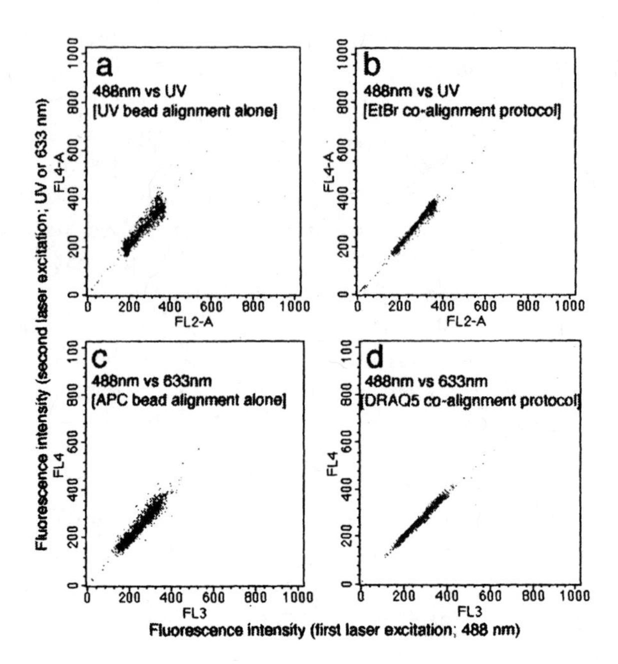

Figure 10-1. DNA histograms of ethidium- and DRAQ5-stained nuclei after alignment with beads alone (panels a and c) and after alignment using the nuclei as calibration particles (panels b and d). Courtesy of Paul Smith (University of Wales).

should, ideally, lie on a line. The plots in panels b and d in the figure are closer to this ideal than the plots in panels a and c, indicating that using the nuclei as alignment particles instead of beads results in better measurement precision.

Chicken and **rainbow trout erythrocytes** can serve as alignment particles; they are more commonly used as standards for DNA content estimation[224]. Procurement of trout cells can provide one of the many enjoyable experiences beside flowing streams that enrich the lives of flow cytometer users, at least those interested in DNA measurements.

Glutaraldehyde-fixed chicken erythrocytes have been popular with immunologists, because, without benefit of any staining at all, they fluoresce at about the level of brightness of cells stained with fluorescent antibodies, at least when measured in channels intended for measurement of fluorescein or phycoerythrin. Immunologists don't seem to use trout cells; I guess they only go fishing for complement.

Riese supplies fixed chicken and trout cells; **Orvis** (www.orvis.com) provides equipment for getting fresh trout cells. If you're after fresh chicken blood, and you live in a metropolitan area of any reasonable size, you can probably find a source of fresh-killed poultry; the old traditions held by many ethnic and religious minorities die hard.

Rose Colored Glasses: Optical Filter Selection

It helps to calibrate flow cytometers using particles labeled with the dyes you're going to use to stain cells. It's essential to make sure you fit your detectors (and, if you use arc lamps, your excitation path) with optical filters that transmit the wavelengths you want to measure and exclude the wavelengths you don't. Joe Trotter wrote a neat Web-based program that shows you how much light you will collect from a number of probes using various filters. You can find it online at (http://facs.scripps.edu/spectra).

These days, both the cytometer manufacturers and the filter manufacturers are well-informed about what is required of filters for flow cytometry, and they can generally tell you what you need for any particular probe combination and excitation wavelength(s). Most instruments use round interference filters, which are preferable to square ones because they are better sealed and therefore take longer to delaminate.

Filters do have a finite lifetime, and you should check your interference filters from time to time for evidence of delamination (pp. 153-4). Once the layers start to separate, you can generally see a pattern similar to that produced by light reflecting off oily water. It is then time to get a new filter. Although filters tend to delaminate after several years, they can die suddenly when dropped, even if the glass doesn't break, so be alert. It is also a good idea to check filter transmission, especially if you're having problems collecting enough light; this is easy to do if you have access to a spectrophotometer[882].

In some cases, you'll get too much light into a detector, particularly a forward scatter detector. This is curable with neutral density (N.D.) filters. If the N.D. filters in question are going to get into or near laser beams of any appreciable power, they should be the reflective type. Reflective N.D. filters are partially "silvered" (i.e., partially coated) mirrors that transmit some light reaching them and reflect the rest. Several companies (see Chapter 11) offer reflective filters in a wide range of optical densities at reasonable prices. For lower light level applications, it's generally fine to use absorptive N.D. filters; your local camera store probably carries Kodak's line of such products.

Be careful when using a filter not explicitly intended for the purpose as a **dichroic.** Most interference filters contain some absorptive components, which fluoresce, to decrease transmission outside the passband; good dichroics dispense with these, leaving that job to the detector filters. You also can't be sure what transmission (or reflection) of a filter is like at 45° unless you measure it. Newer instruments, such as BD Biosciences LSR, LSR II, and FACSAria, use dichroics designed to work at angles of incidence other than 45°; you'll need to consult the instrument manufacturer should you want to specify a dichroic for one of those instruments, and you may have to pay one of the filter manufacturers a substantial amount to custom-make the part.

Experimental Controls

What you will need in the way of experimental controls will vary depending on what you are trying to measure. If you suspect that somebody has run a cyanine dye, acridine orange, or another similarly sociopathic reagent through the instrument and not cleaned it afterwards, be sure to run some unstained cells before you start with stained samples; if the unstained cells start to fluoresce after they have been running for a while, the cytometer needs a cleaning. If you've already started running stained cells, and their fluorescence starts to increase, it's too late to retrieve the data.

If you're using familiar antibodies and familiar cells, e.g., for doing T cell subset analysis, you can probably dispense with isotype controls, but you should have some controls which, in the aggregate, contain all the cell types you are looking for, just to make sure they turn up where you expect them to in the measurement space. And it's a good idea to have either cells or antibody binding beads that will give you fluorescence signals near the upper end of your range, both for setting compensation and to make sure everything will stay on scale.

When you're exploring unfamiliar territory, you'll probably be better off starting off with too many controls than with too few.

Shake Well Before Using: When Controls Won't Help

Figure 10-2 provides an example of anomalous flow data that may perplex you in private and/or lead you to conclude in public that you have discovered some new biological phenomenon. The latter consequence is apt to be more embarrassing than the former.

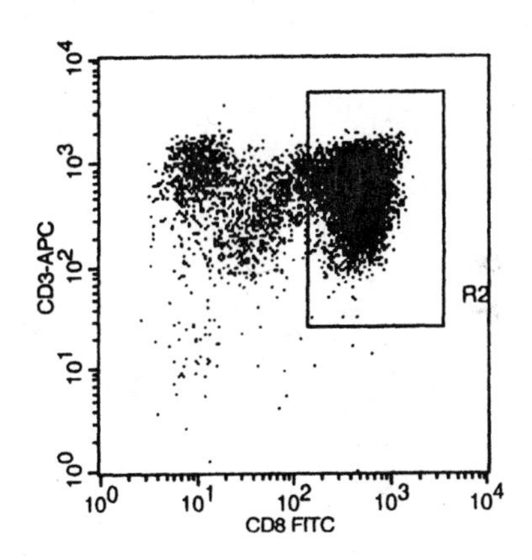

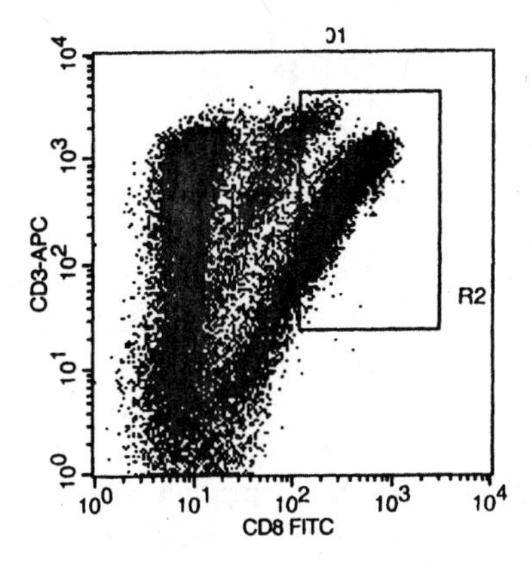

Figure 10-2. "Equal opportunity" (top panel) and "unequal opportunity" staining of T-cells (from Jörn Schmitz, Harvard Medical School)

The cells shown in Figure 10-2 were stained with the same antibodies; somebody forgot to mix the cells shown in the bottom panel, producing "unequal opportunity" staining instead of the desired "equal opportunity" staining illustrated in the top panel. It is relatively easy to identify this problem when the person preparing the cells is grossly negligent, and all or almost all of the samples, including the controls, yield results that resemble the bottom panel. The surviving lab personnel are apt to be more careful, and the occasional unmixed or incompletely mixed sample will therefore have more of an aura of mystery about it. Once you've seen the pattern, though, it's easy to recognize.

10.2 SIGNIFICANT EVENTS IN THE LIVES OF CELLS

Taking the Census: Cell Counting

Cell counting, discussed at length on pp. 18-21, is one of the most basic tasks in cytometry. Counting cells in a hemocytometer by visual observation is tedious and, unless one is willing to spend a long time on each sample, imprecise. The simple optical and electronic flow cytometers that became available for cell counting in the 1950's were widely adopted in both clinical and research laboratories, eventually giving rise to modern hematology counters. As was noted in the earlier discussion, these instruments feature volumetric sample delivery, allowing absolute cell counts to be derived from the number of cells counted per unit time.

Kamentsky's original Cytofluorograf incorporated a calibrated glass tube with photosensors at two levels in its pressure-fed sample delivery system, to detect the movement of the air-fluid interface, providing a measure of fluid flow per unit time. Fluid-level sensing volume measurement add-ons for flow cytometers have been developed [1538]; **Cytek** sells one such device. It is also relatively easy to adapt most instruments to receive samples from syringe pumps, which are readily available. However, most fluorescence flow cytometers do not use volumetric pumps or other calibrated means to introduce samples; absolute counts, when needed, are obtained by introducing beads into samples at known concentrations, and deriving cell counts from the ratio of the number of cells counted per unit time to the number of beads counted per unit time. The use of counting beads has facilitated performance of such clinical assays as CD4+ T cell counts using only fluorescence flow cytometers.

A recent study by Bergeron et al[2917] documents the sample delivery characteristics of several benchtop flow cytometers over a period of months. Sample flow rates, determined using counting beads, were relatively constant (CVs 5-7 percent). If you can accept this level of precision in cell counting for one or more of your applications, you can run counting beads at the beginning and end of sample runs to calculate the sample flow rate and check on its variation over time, and use the calculated rate to derive absolute cell counts for your experimental samples. Since you won't need to add counting beads to each sample, you'll save money; the beads aren't cheap.

A Counting Alternative: Image Analysis

It has probably not escaped your notice that digital cameras incorporating 2- and 3-megapixel CCD chips can be had for a few hundred dollars. That means the manufacturers are paying well under a hundred bucks for them. Such devices are beginning to be incorporated into a new generation of inexpensive image analysis systems.

A Danish company, **ChemoMetec**, recently introduced the **NucleoCounter**, a cell counter is built around a miniaturized low-power (1×) transmitted light fluorescence microscope illuminated by 8 green LEDs. There are no moving parts in the optical system (i.e., no focus adjustment), because the lens N.A. is low. The device uses a disposable cartridge containing propidium iodide; for a total count, cells are diluted with a permeabilizing solution and introduced into the cartridge, while, for a "nonviable" count, cells are introduced without prior permeabilization. The cartridge has a viewing area a few millimeters square, which contains a volume of 1.5 to 2 microliters; this is imaged onto a CCD through a red bandpass filter. Because the observed sample volume is relatively large, counts are more precise than would be obtained from a hemocytometer. The NucleoCounter takes about 30 seconds to do a count; it incorporates a USB connection to transfer output and images to a computer, but the device, which weighs 3 kg and measures 38 × 26 × 22 cm (W × H × D), is normally run as a self-contained system.

Beckman Coulter offers the **Vi-CELL™** viable cell counter, which uses image analysis to identify and count trypan blue-stained ("nonviable") and unstained ("viable") cells flowing through an observation chamber. This is also a relatively small, self-contained benchtop system.

A prototype image analyzing system, the **EasyCount**, which, like the NucleoCounter, incorporates a low-power, LED-illuminated fluorescence microscope, will be discussed in the section on CD4+ T cell counting.

The Doubled Helix: Reproduction

The Cell Cycle and Cell Growth

It is now possible, by using flow cytometric analysis of **DNA content, RNA content, cell size, cell cycle-related antigens, BrUdR incorporation**, and the fluorescence of **tracking dyes**, to define cells' position in the cell cycle with a precision previously unimaginable and, using cell sorting, to separate populations in different cell cycle phases for biochemical analysis or for studies in culture.

DNA Content Analysis

The determination of DNA content of cells or nuclei, previously discussed on pp. 21-6, 43-4, 96-7, and 301-17, is a common and widely used procedure in flow cytometry. In research laboratories, DNA content analyses are used to monitor the growth of eukaryotic and prokaryotic cells, and to detect perturbation of cellular growth patterns due to physical, chemical, and biological agents. In clinical applications, DNA content analyses are used to refine the diagnosis and estimate the probable biologic behavior of tumors and to monitor therapy.

Since the stains used for DNA content determination bind stoichiometrically, and produce very bright cellular fluorescence, it has, to some extent, been taken for granted that, as long as the instrument used is maintaining reasonable linearity and precision, DNA analysis will be simple and straightforward. This is true enough, provided cells are well prepared and well preserved, and these steps are not always simple and straightforward.

Nuclear isolation and staining and storage reagents and procedures, using various dyes, have been described by Vindeløv et al[222-5], Taylor[226], and Thornthwaite et al[261,889], among others; some of these have been discussed on p. 307. While these workers and others have expended some effort in trying to devise procedures suitable for application to a wide variety of tissues and tumors, it has been my experience that preparative and staining procedures have to be modified and optimized to suit the samples at hand. Clausen[613], for example, has provided an enlightening discussion of DNA analysis in keratinocytes. When analyzing DNA in solid tissues and tumors, some consideration must be given to sampling techniques; Greenebaum et al[890], for example, reported that needle aspiration gives better yields of abnormal cells than does standard excisional biopsy.

The development in 1983 by Hedley et al[610,891] of a technique for **DNA analysis of nuclei from paraffin-embedded material** represented a great advance in the application of DNA analysis to tumor pathology, since it made it possible to establish correlations between DNA aneuploidy and the biologic behavior of tumors by retrospective[892] as well as by prospective studies. The basic method involves deparaffinization in xylene or Histoclear and progressive rehydration in ethanol solutions, followed by acid pepsin treatment for nuclear isolation. The procedure works well on material originally fixed in formalin or a formaldehyde-acetone-acetic acid mixture; results have been poor with tissue fixed in Bouin's solution and unsatisfactory when mercury-based fixatives were used.

The thickness of sections taken from the paraffin block for processing will obviously have some effect on the proportion of nuclei which can be recovered intact. Although there has been some debate on the issue[747], it is reported[746] that 50 µm sections produce considerably less debris, and thus better histograms, than 30 µm or thinner sections. Hedley et al originally used DAPI as a stain, and reported[891] that, despite the DNA specificity of the stain, RNAse treatment improved precision. While, in my own limited experience with deparaffinized material, I have gotten much better results with DAPI than with propidium, I know other people who have obtained excellent precision using propidium.

Mathematical Models for DNA Analysis

Calculation of population fractions in different cell cycle phases has undoubtedly generated much too much debate over the years among people who do flow cytometric DNA analysis. When DNA content determination is based on a single parameter measurement, there will be artifacts and interferences to be dealt with under the best of circumstances, i.e., when the samples consists of freshly prepared cells from a population grown in suspension culture. Under less ideal circumstances, as in analysis of material from solid tumors, there may be no legitimate conceptual model for dealing with the complexity of the sample.

If the flow cytometer is linear, it can be expected that a population exhibiting a DNA diploid peak at channel n will have a DNA tetraploid peak at channel $2n$; however, minor nonlinearities (a few channels worth) are not uncommon in the flow cytometers used in the real world, and a noticeable fraction of older systems exhibited marked nonlinearity. A continuum of **debris** is frequently superimposed on at least the lower channels of a DNA histogram. **Doublets** due to **clumps** and/or **coincidences**, resulting from physical attachment or proximity of cells, may register as DNA tetraploid cells unless a two-parameter measurement scheme is used, in which a plot of pulse peak height vs. area (integral) or a plot of fluorescence vs. a pulse width measurement is used to permit discrimination between the doublets and the DNA tetraploid cells. Newer mathematical models[1184-5,1307-8] can deal with both debris and doublets.

Once one gets the debris and doublets out of the way, it becomes relatively easy to deal with a distribution which contains a single G_0/G_1 peak, an S phase distribution, and a G_2+M peak, and not too much harder to handle a highly DNA aneuploid distribution with proliferating cells and an isolated peak of normal DNA diploid stromal cells in G_0/G_1. Jerry Fried, who was one of the first people to do mathematical modeling of DNA histograms, opined to me years ago that, if you had a clean (little debris, few doublets) distribution with a CV of less than 3 percent, you might as well estimate cell cycle phases by eye. On the other hand, if you have overlapping DNA diploid and DNA aneuploid distributions, each with a visible G_0/G_1 and G_2+M peak, it's difficult for a model to decide which part of the S-phase belongs to which population, unless there are discernible jumps in that region.

Mathematical models for deconvoluting DNA histograms were originally intended principally for application to analysis of the perturbation of cell cycle kinetics by therapeutic agents, and they worked fine when applied to clean systems. What happened when people first started trying to use DNA analysis for clinical purposes was that they went running to mathematical models when they came up with distributions with CV's of 8 percent, or 12 percent, in the hope that the truth would come out of hiding if given enough computer time. This made precious little sense;

Shapiro's Seventh Law reminds us that bad data won't get better.

Generally speaking, the critical information sought from mathematical models, particularly by clinicians applying the models to distributions from tumor samples, is what is variously known as the **proliferative fraction**, **percentage of cells in S phase**, or **S-phase fraction (SPF)**, because this, like DNA aneuploidy, may be predictive of tumors' biologic behavior. To be sure, eyeball estimation and several different mathematical models may give several different numbers for SPF in a given sample[893-5].

In a previous edition, I said that it struck me as extremely unlikely that one method of SPF would turn out to yield a predictor of biologic behavior that was statistically significantly better than others. **I was wrong!** I didn't consider the effects of debris on SPF calculations. Kallioniemi et al[1742] used a background subtraction algorithm to compensate for the effects of slicing of tumor cell nuclei during preparation of paraffin-embedded specimens, and analyzed DNA histograms from breast and prostatic carcinomas. Median SPF's corrected for nuclear slicing were lower than uncorrected ones in both breast cancer (7.6% vs. 5.7%) and prostate cancer (6.7% vs. 4.2%). Corrected SPF levels resulted in a more significant survival difference between breast cancer patients with values above and below the median ($p = 0.0014$ vs. $p = 0.014$) and in a higher relative risk of death (4.5 vs. 3.1). The same was true for prostate cancer survival (p less than 0.0001 vs. $p = 0.002$) and relative risk of death (5.3 vs. 3.1). Also see pp. 25-6.

The two programs most widely used for DNA histogram analysis are Bruce Bagwell's **ModFit™** (Verity Software House) and Peter Rabinovitch's **Multicycle™** (Phoenix Flow Systems); both use modeling to minimize effects of clumps and debris. An example of the application of ModFit to deconvolution of a DNA histogram with DNA diploid and DNA aneuploid components is illustrated in Figure 1-10 (p. 25); the use of the same program, in a modified form, in studies with tracking dyes is shown in Figures 7-23 and 7-24 (pp. 372-3).

If you are hoping to get good data from a deconvolution program, you'll need a lot of cells in the sample[2940], perhaps as many as 200 cells per channel in S-phase (p. 240). That can mean 50,000 or more cells overall, even if you work with a 256-channel DNA histogram.

Clinical Application of DNA Content Analysis

As was noted on pp. 25-26, the quality of analysis of both ploidy and S-phase fraction has an impact on the prognostic relevance, or lack thereof, of the resultant data.

It has been reported[896] that **gating on forward and orthogonal scatter** may be useful in locating DNA aneuploid nuclei, since these parameters may differentiate tumor and stromal cell nuclei. It is also possible to stain tumor cell populations with fluorescent antibodies, e.g., anti-cytokeratin, which define tumor or stromal cell populations, and to use fluorescence gating[1743-4] to select the tumor cells for analysis. This facilitates detection of small subpopulations of DNA aneuploid nuclei in specimens.

In the mid-1980's, when it looked as if flow cytometry was going to go bounding into the clinical laboratory whether the clinical laboratory was ready for it or not, the National Cancer Institute set up a nationwide network of five institutions, all with considerable experience in the field (Memorial Sloan-Kettering Cancer Center, New York, NY, Montefiore Medical Center, New York, NY, Rush-Presbyterian-St. Luke's Medical Center, Chicago, IL, the University of California at Davis, Davis, CA, and the University of Rochester, Rochester, NY), to do cooperative studies on flow cytometry of bladder cancer. The good news is that these five institutions, which used different stains, hardware, and software, were able to obtain comparable results in analysis of deparaffinized sections from cases of human transitional cell carcinoma of the bladder[897-8]. The not-so-good news, or perhaps I should call it the sobering fact, is that it was much harder to get less of a consensus from analyses of specimens that were not previously in paraffin, even with all of that expertise being brought to bear on the problem. Preparation and preservation of samples to the point at which they can be sent around the country and produce reliable results was not then a *fait accompli*, which shouldn't have surprised me.

In October, 1992, a DNA Cytometry Consensus Conference[1745], bringing together 32 experts in the field from Europe and North America, was held under the auspices of the International Society of Analytical Cytology. Guidelines[1746] for implementation of clinical DNA cytometry, both flow and image, and consensus reviews of the role of DNA cytometry in bladder[1747], breast[1748], colorectal[1749], and prostate[1751] cancer and in neoplastic hematopathology[1750] were developed and published in *Cytometry* (Volume 14, Number 5) in 1993.

The Conference echoed the recommendations of the Society for Analytical Cytology's Committee on DNA Nomenclature[741], made in 1984, that the terms **DNA diploid** and **DNA aneuploid** be used to describe cells containing apparently normal and apparently abnormal amounts of DNA unless actual ploidy is established by cytogenetic studies, and that the degree of DNA content abnormality be given by the **DNA index (D. I.)**, which is the ratio of G_0/G_1 peak locations of the sample (tumor) cells and normal or reference cells. In order for a sample to be classified as DNA aneuploid, two distinct G_0/G_1 peaks must be present in the histogram.

It was stressed that the normal cell or stromal component present in tumor samples best represents DNA diploidy, and therefore provides the best standard. This is an important point because, while DNA content in normal cells from any given individual is quite uniform, sufficient polymorphism exists in the human race as a whole to account for a range of variation of about 7% in DNA content in DNA diploid cells (G. van den Engh, personal communication).

It was also recommended that details of the methods used to isolate and prepare cells and to analyze histograms be provided in publications dealing with DNA content analysis, and that mean CV's and ranges for both DNA diploid and DNA aneuploid populations be included. Samples from solid tumors should, in general, contain at least 20% tumor cells, and, particularly if SPF is to be determined, a minimum of 10,000 events should be analyzed. When S-phase populations of DNA diploid and DNA aneuploid cells overlap, a weighted average of the two SPF's should be reported; however, it has not been determined whether this practice yield prognostic information as good as or better than is obtained when all S-phase is attributed to the tumor cells.

It was further recommended that each laboratory should define its own ranges of DNA diploidy and aneuploidy, and of high, intermediate, and low SPF, for each tumor type, and that studies of outcomes be based on analyses of three or more groups (e.g., high, intermediate, and low SPF) rather than on analysis of patients falling above and below the median.

The consensus review on bladder cancer[1747] recommended analysis of both biopsy and bladder irrigation samples at diagnosis and of irrigation samples at follow-up; it was felt that flow cytometry should not be used either for bladder cancer screening or for work-up of microscopic hematuria.

According to the breast cancer consensus review[1748], while operable lesions with a D. I. of 1.0 have a favorable prognosis compared to DNA aneuploid tumors, DNA index does not achieve independent prognostic significance because it is strongly correlated with more powerful prognostic indicators. SPF, while strongly associated with tumor grade, retains independent predictive power (but see pp. 25-6). It was recommended that fresh frozen samples containing a substantial proportion of malignant cells be used for analysis.

The colorectal cancer consensus review[1749] concluded that prognostic significance of DNA aneuploidy has not been established, although there is strong evidence for it in Dukes' stage B and C lesions. SPF appears to be a more powerful prognostic factor.

According to the consensus review[1751], neither DNA aneuploidy nor SPF has been established as predictive in prostate cancer; few studies have been done on the latter, while studies on the former have yielded conflicting results.

In hematologic neoplasms[1750], DNA aneuploidy is associated with favorable outcomes in acute lymphocytic leukemia (ALL) in children, but not in adults; there is no prognostic value for ploidy analyses in acute myelocytic leukemia (AML) or non-Hodgkin's lymphoma (NHL), but SPF is prognostically significant in NHL.

The potential for clinical application of cytometry, whether flow or image, in cancer depends on the prevalence of the disease. According to the 1993 prediction of American Cancer Society[1752], there would be 1,208,000 new cancer cases in the U.S. in 1994; of these, 51,200 would represent bladder cancer, 183,000, breast cancer, 149,000, colorectal cancer, and 200,000, prostate cancer. Among the 94,200 hematologic neoplasms, there would be 45,000 cases of NHL, 12,500 of lymphocytic leukemia, and 11,400 of granulocytic leukemia, with the leukemia statistics representing both acute and chronic forms of the disease.

DNA Content Alternatives: Static Photometry and Scanning Laser Cytometry

Static photometry and **image analysis** can produce DNA content distributions from Feulgen-stained cells with precision almost comparable to that obtained from flow cytometry, at least when clinical specimens containing isolated nuclei are being analyzed. In theory, one can use image analysis to do DNA content determinations on nuclei in tissue, but the software remains a problem. While static methods are slower than flow cytometry, you do get to keep the cells on the slide, and to decide which are likely to be representative of tumor cells and stromal elements, and you don't need that many cells in a sample, which is advantageous when you are trying to do ploidy or SPF determinations on needle biopsy specimens.

The relatively new technique of **scanning laser cytometry** offers many of the advantages of both flow cytometry and image analysis. **CompuCyte**'s Laser Scanning Cytometer (**LSC**™)[2047,2380-1,2918] analyzes cells on a slide or other solid substrate, and produces low-resolution image data from which whole-cell measurements of DNA content, immunofluorescence, etc. can be derived. The LSC measures forward scatter and fluorescence in four or more spectral ranges using a 488 nm laser as its primary light source; most units also have a red laser and a violet diode laser is also offered as an option. The LSC is built around an upright microscope; the newer **iCyte**™, built around an inverted microscope, offers some additional features and the capability for analysis of specimens in microtiter plates and culture dishes. Both instruments are substantially faster than the older image analysis systems, making it feasible to analyze as many as a few thousand cells in a clinical sample. Although CVs obtained using the LSC in DNA analysis are typically higher than those obtained by flow cytometry, several reports have found the LSC and flow cytometry equally effective for ploidy and SPF measurement in solid tumors[2919-22]. Other applications of scanning laser cytometry in general and of the LSC in particular will be discussed elsewhere in this chapter.

The Mummy's GC/AT: DNA Content Analysis in Anthropology and Forensic Science

Cook[1753] reported the ability to detect intact nuclei in DAPI-stained cortical bone samples removed from individuals buried in Egypt approximately 2,000 years ago. When intact cells are found in a sample, it increases the likelihood that further investigations, such as extraction, amplification, and sequencing can be done successfully. Although the point was not specifically discussed, it seems likely that sorting intact nuclei from a sample containing large amounts of debris would yield a better starting preparation for further analysis.

This technique may be applicable to fresher material for forensic purposes as well, even if it won't quite get us to Jurassic Park. Schoell et al[2923-4] presented experimental evidence that flow sorting could be used to isolate sperm from vaginal samples taken from sexual assault victims for subsequent PCR amplification and identification; there are two major problems with this work. First, while cytometry allows the identification of sperm on the basis of haploid DNA content and other characteristics, flow sorting is not well suited to work with very small numbers of cells; a slide-based technique such as **micromanipulation**[2925] (relatively low-tech) or **laser capture microdissection**[2926-32] (relatively high-tech) would be preferable, assuming the authorities have the budget. Second, the authors appear to have published the same paper twice.

Cina[2933] and Di Nunno[2934-6] and their colleagues report that the degree of degradation of DNA in various tissues, as determined by flow cytometry, can provide a precise estimate of time of death; they recommend a needle biopsy of the liver as a source of cells. Maybe I should follow in my father's footsteps and set up a forensic cytometry lab.

Half a Genome is Better than None: Sperm Sorting

Brave New World aside, the separation of X- and Y-chromosome-bearing sperm is of great practical and economic importance for breeders of domestic animals[2324]. The obvious way in which to do this depends upon the demonstration of differences in DNA content between the two kinds of sperm. However, it is probably fair to say that earlier flow cytometric studies of sperm DNA[96,121,582-4] revealed as many instrumental differences between flow cytometers as DNA content differences in sperm.

The DNA in sperm heads is in a highly condensed state, and there are a variety of shapes of sperm heads, many of them asymmetric, in sperm from different animal species. This asymmetry causes artifacts in DNA content measurements in some flow cytometers in which sperm passing through the instrument in different orientations are unevenly illuminated or in which collection of fluorescence is perturbed by differences in orientation[121]. Slit-scanning[582,2311], and fluidics that align cells passing through the apparatus[583], have been used to improve measurement quality; Métézeau et al[1967] used a combination of axial extinction and forward scatter measurements to detect orientation, essentially compensating the DNA measurement. As Figure 7-11 (p. 311) dramatically illustrates, it is possible to build a flow cytometer optical system that can almost completely distinguish X- and Y-chromosome-bearing sperm, although whether such resolution can be achieved with preservation of viability remains in question.

It is now possible to discriminate and sort X- and Y-sperm from domestic animals[584,1968-78] and men[1979,2322-3] by flow cytometry. There was some question as to whether sperm stained with Hoechst 33342 for viable cell sorting DNA fluorochromes were desirable for use in insemination; Libbus et al[1241] noted an increased frequency of chromosome aberrations when sperm which had been stained and sorted were microinjected into hamster eggs, but Morrell and Dresser[1980] did not find abnormalities in the offspring of rabbits fertilized with sperm which had been stained with the Hoechst dye, whether or not the sperm had been sorted. There have been live births in animals following fertilization with sorted sperm[1973,1977]; Johnson, Flook and Hawk[1973] reported 94% female offspring in rabbits using X-enriched sperm and 81% male offspring using Y-enriched sperm.

Larry Johnson, of the U.S. Department of Agriculture, is the undisputed world leader in sperm sorting; after doing extensive work with sperm from domestic animals, he and his collaborators at the Genetics and IVF Instute (Fairfax, VA)[1979] sorted human sperm, achieving 75% purity of Y-sperm and 82% purity of X-sperm, and suggested that sorting might be a useful way of dealing with X-linked diseases in man. A report of this work inspired one of the science writers at *The Boston Globe* to produce a column suggesting that we might eventually be able to use gene probes to preselect "nice" offspring; I expect Larry got a good laugh out of the copy I sent him. Then he and his colleagues went back to work, and in 1998 they reported deliveries of normal babies after insemination or *in vitro* fertilization with X-enriched sorted sperm. Offspring were of the desired female gender in 92.9% of the pregnancies[2937-8].

I heard a rumor when I was writing the last edition that some success in isolation of X- and Y-sperm had been achieved using monoclonal antibodies raised by immunization of mice with sperm separated based upon DNA content, but I think I got a bum steer, which is just what cattle breeders are trying to avoid by separating sperm. Actually, one antibody that was thought to be preferentially expressed on Y-sperm has turned out not to be[1978,2939]. An attempt to separate X- and Y-sperm on a Percoll gradient also didn't work[1981].

Van Munster and colleagues[2541-2] found that sperm volume, measured by interferometric flow cytometry, was proportional to DNA content, and succeeded in obtaining fractions of bull semen enriched for X- and Y-sperm by sorting. This will undoubtedly stimulate somebody to build a sperm sorter incorporating a Coulter volume measurement; if it works as well as the current hardware and eliminates the need for Hoechst 33342 staining, the technology is likely to be widely used to assist both human and animal reproduction.

XY, Inc., established in 1996 as a jojnt venture between Cytomation (now DakoCytomation) and Colorado State University, runs several facilities for sorting animal sperm. They have licensed the patents based on Larry Johnson's work. Figure 10-3, on the next page, illustrates the procedure used at X-Y for separation of X- and Y-sperm from bull semen. The apparatus is a modified MoFlo cell sorter; a beveled injector needle is used to orient sperm, and Hoechst

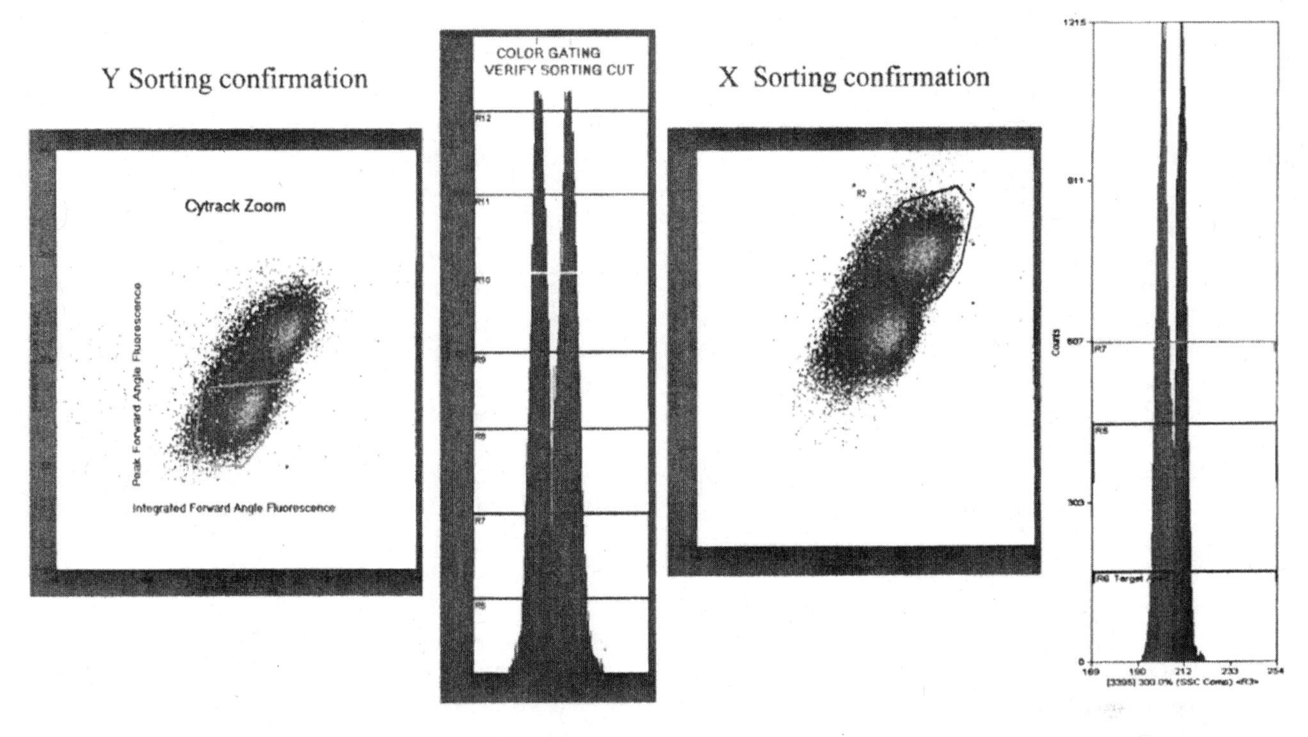

Figure 10-3. Separation of X- and Y- bull sperm based on forward and orthogonal scatter and forward and orthogonal Hoechst 33342 fluorescence measurements. Courtesy of Mike Evans, XY, Inc.

33342 fluorescence is measured in the forward direction as well as through conventional orthogonal collection optics. Properly oriented sperm are gated based on measurements of forward and side scatter and fluorescence; sort gates are set as shown on a magnified display of forward fluorescence peak vs. integral values.

The Widening G_0/G_1 re Detecting Mutation

The very first issue of *Cytometry* (July 1980) contained a paper by Otto and Oldiges[2941] on flow cytometric detection of chromosomal aberrations (clastogenic effects) induced in cultured cells by mutagens and x-rays. The authors used a high-resolution arc-source flow cytometer to analyze isolated chromosomes and cell nuclei stained with DAPI. The detection of clastogenic effects is a mainstay of toxicologic studies, and, until Otto and Oldiges proposed a flow cytometric method, required examination of metaphase chromosomes on slides. Their initial experiments correlated clastogenic effects with increases in the CV of peaks representing the largest chromosome. They then reasoned that, because cell division in cells containing chromosomal aberrations results in unequal distribution of DNA between daughter cells (i.e., mere aneuploidy is loosed upon the world), it should be possible to detect clastogenic effects by measuring DNA content distributions of whole nuclei and determining whether the CV of the G_0/G_1 peak was greater in cells exposed to clastogenic agents than in controls. The whole cell method has two obvious advantages; it eliminates the need to synchronize cells in mitosis to obtain chromosomes for analysis, and, because nuclei contain substantially more DNA than do chromosomes, it decreases the contribution of photoelectron statistics to measurement variance.

As Otto and Oldiges put it, "This effect is assumed to be measurable, using high resolution flow cytometry..." The DNA content distributions measured from their control cells had CVs of 1.2 and 2.0 percent, making it relatively easy to detect small, dose-dependent increases in cells exposed to clastogenic agents. It can safely be assumed that the method has been used by others who, for various reasons, have not obtained equivalent measurement precision. It is therefore logical to ask whether there is a reliable statistical procedure for analysis of CV data obtained from clastogenicity experiments. Misra and Easton[2942] have developed one that they claim is more robust than some that have been used previously; it's worth looking into if you work in this area.

Detecting DNA Synthesis: Cell Kinetics

Kinetics Before Flow Cytometry: Mitotic Indices, Doubling Times, and Radiolabel Studies

In order to precisely define the **kinetics** of a cell population (an excellent discussion of cell kinetics and the relevant radioisotope methodology can be found in Steel's book[382]), you really need to know both the cells' **DNA content**, and their **DNA synthesis rate**. I have given a little bit of the history of cell kinetic studies on pp. 85-6. Before 1950, it wasn't easy to measure either DNA content or synthetic rate, so the literature on kinetics was restricted to analysis of **mi-**

totic indices and **doubling times**. The **mitotic index** is obtained "simply" by counting the fraction of mitotic cells in the population, which isn't always as easy as it sounds. Under the best of circumstances, it isn't always easy to identify cells in early prophase and late telophase; in mixed cell populations, e.g., in bone marrow or tissues, you can't always be sure which of the cells you're looking at belong to the subpopulation of interest. Also, as I found out many years ago, unless you analyze, fix, or refrigerate samples within minutes after they are taken, you may get erroneously low mitotic indices; at least some mitotic cells in excised tissue at room temperature will proceed happily through mitosis. Finally, if the mitotic index is low, you have to count a lot of cells to get a precise value. Cytometry can help with that; see pp. 277, 320 and 462.

Doubling times for nonmotile cells can be determined unequivocally using the venerable method of **time-lapse photography**[2943-4], which allows an observer to monitor the development of clones from single cells. These days, it makes more sense to use a more sophisticated technique such as **laser scanning cytometry**, which allows multiparameter measurements to be done on each cell in a clone[2945]. Doubling times for cells in suspension can be estimated fairly reliably from visual or instrumental cell counts done at intervals. However, when we get back to the real world, we can't derive equally trustworthy information from such data as tumor volume measurements, because **cell loss** is at least as important as the kinetics of reproductively competent cells in determining tumor growth rates. Cell loss should be detectable in time-lapse studies of individual cells and their progeny, but may or may not be detectable in cell suspensions.

The introduction of **tritiated thymidine (^{3}H-TdR)** as a tracer in the late 1950's made it possible to estimate DNA synthesis rates in populations by bulk measurements of uptake, done in scintillation counters, and to determine heterogeneity in labeling patterns by **autoradiography**. Several assumptions underlie the use of ^{3}H-TdR for the study of DNA synthesis patterns in cell populations. It is assumed that the ^{3}H-TdR is either incorporated into DNA by cells or lost, i.e., that the tritium label is not transferred to cellular constituents other than DNA as a result of thymidine catabolism. It is also assumed that cells do not have large endogenous thymidine pools, and will thus incorporate exogenous ^{3}H-TdR into DNA if the labeled material is present during the period of DNA synthesis. This issue can be forced to some extent by adding a substance such as 5-fluorouracil during the period of ^{3}H-TdR exposure. It is assumed that the administered label is equally accessible to all cells in the population under study. Finally, it is assumed that the tritium label is not administering a heavy enough dose of radiation therapy to the cells to perturb their DNA synthesis. These same assumptions must be made when another labeled nucleotide, e.g., bromodeoxyuridine (BrUdR, BrdU, BrdUrd, etc.), is used in place of as ^{3}H-TdR as a DNA label, and it can sometimes be shown that one or more assumptions are invalid, depending upon the cell system chosen, the dose of tracer used, etc.

When incubated with ^{3}H-TdR, either for a brief period (a **pulse** of 5-60 minutes' duration) or for a longer time, cells synthesizing DNA, i.e., those in the S phase of the cell cycle, incorporate the tracer. The **labeling index (L.I.)**, i.e., the fraction of cells incorporating enough ^{3}H-TdR during incubation to produce exposed grains in an overlying autoradiographic emulsion, provides information about DNA synthesis patterns in a cell population. After pulse labeling, the L.I. should represent the percentage of cells in S phase; if there are no quiescent cells in the population, the L.I. should approach 100 percent as the labeling time increases.

Analysis of a cohort of cells labeled at a specific point in the cycle should provide an estimate of the average **generation** or **cell cycle time**, provided this does not exhibit tremendous cell-to-cell variance. The most common technique for deriving such information follows the **percentage of labeled mitoses (PLM)** over time following pulse labeling. The resulting **PLM curve** is a periodic function, with the time interval between the first and second relative maxima representing the generation time.

While some attempts were made to mechanize autoradiographic grain counting, the vast majority of studies involving quantitative autoradiography were done by human observers counting grains under transmitted light microscopes. This process was tedious, but the manpower and the microscopes were readily available at the time, whereas measurements of DNA content by Feulgen staining or UV absorption would have required microspectrophotometers, which were hard to come by. The few studies that combined Feulgen DNA content measurements and autoradiography took months to do.

Labeling Index versus DNA Content

Since the early 1970's, flow cytometers have provided us with cheap and plentiful DNA content analyses, while autoradiography has not gotten appreciably easier, despite the introduction of high specific activity tritiated thymidine and of other technical refinements that shorten exposure time from months to hours. As might be expected, this has led many people to employ flow cytometry almost to the exclusion of autoradiography in studies of cell kinetics. The distribution of **DNA content** in cell populations, like the labeling index, provides an estimate of the fraction of cells in S phase; in this case, the estimate is obtained from the fraction of cells with a measured DNA content between that of the G_0/G_1 (diploid or 2C) and G_2 + M (tetraploid or 4C) values (for the population in question; in tumor cells, the 2C value frequently differs from the 2C value for normal cells of the host).

Several circumstances can produce discordance between the S phase fractions estimated from labeling indices and from DNA content measurements. In principle, DNA repair on a massive scale might produce labeling indices higher than the fraction of cells with S phase DNA content, but the discrepancy, when one exists, is usually in the other direction. An "S_0" population of cells, which have stopped synthesizing DNA in the middle of S phase, will be characterized as in S phase by their DNA content but will not label with ^{3}H-TdR; the fraction of cells in S phase will therefore be higher than the labeling index.

The same result will occur if the cells being studied catabolize thymidine. Squirrel cells do this[831-2]; when looking at cell kinetics in hibernating ground squirrels[833], I found no labeling with thymidine in marrow cells from hibernating animals – and none in cells from animals that weren't hibernating, either. Using tritiated deoxycytidine, which squirrels don't catabolize, as a tracer, it was possible to demonstrate a greater than 10-fold decrease in DNA labeling during hibernation, although DNA histograms from hibernating and alert animals showed the same percentages of cells in S phase. This certainly convinced me that critical studies of cell kinetics require measurement of both DNA content and labeling; but it's easy to say things like that when you're not stuck with doing all the work.

Early Flow Cytometric Approaches to Labeling Using BrUdR and ^{3}H-TdR

A strategy proposed by Latt[236-8;383-4] (also see p. 308) was based on the observation that the fluorescence of dyes such as Hoechst 33258 and 33342 was partially quenched when the dyes bound to regions of DNA in which **bromodeoxyuridine (BrUdR)** was incorporated in place of thymidine. Latt, George and Gray[384] showed that Hoechst dye fluorescence in cells grown in the presence of BrUdR was decreased relative to the fluorescence of cells grown without BrUdR. Under these circumstances, however, Hoechst dyes can provide information about DNA content or information about BrUdR incorporation, but not both.

Other techniques for estimation of DNA synthesis rates based upon BrUdR effects on the fluorescence of a DNA fluorochrome were described by Swartzendruber[385] and by Darzynkiewicz et al[386]. Swartzendruber found that mithramycin fluorescence was enhanced in cells exposed to BrUdR, while Darzynkiewicz et al reported that the green DNA-specific fluorescence of AO was decreased approximately 40% in stimulated lymphocytes grown for a generation time in the presence of BrUdR. Maddox, Johnson, and Keating[1654] reported in 1989 that AO quenching could be used in a relatively routine fashion for detection of BrUdR incorporation in normal and leukemic marrow cells.

Gray et al[387] combined flow cytometry and isotope techniques to devise a rapid method analogous to the classical radioisotope technique of determining the **fraction of labeled mitoses** in a cell sample. Cells pulse labeled with tritiated thymidine (or another radioactive DNA precursor are harvested at different times and stained with a DNA fluorochrome, after which cells within a narrow window in mid-S phase (defined as "S_i") are sorted and the radioactivity per cell (RCS_i) determined by scintillation counting. The duration of S phase and of the total cell cycle are determined from fluctuations in RCS_i values with time.

Detection of Incorporated BrUdR with Hoechst Dyes and Propidium Iodide

Bohmer and Ellwart[388-9] and Noguchi et al[390] independently described essentially identical methods that estimated both BrUdR incorporation and total DNA content by combining ethidium or propidium with the Hoechst dyes for nuclear staining. Both dyes are excited by UV light; however, since Hoechst dye fluorescence is quenched by BrUdR and ethidium and propidium fluorescence are not, cells that have labeled will have relatively lower Hoechst dye fluorescence than unlabeled cells. By analysis of the Hoechst dye fluorescence distribution of cells with propidium fluorescence values corresponding to a window in S phase, it is possible to get the same information as would be obtained from cell sorting and scintillation counting in the RCS_i method[387], with less expenditure of time and effort.

Proponents of the method[1655-6] reported that flow cytometric analysis of BrUdR-quenched 33258 Hoechst fluorescence could be used to measure the G_1, S, and G_2 + M phases in each of three successive cell cycles after mitogen stimulation of peripheral blood lymphocytes. The data allow assessment of growth fraction, lag-time, compartment exit rate, compartment duration, and compartment arrest. Asynchronous cell populations can be analyzed as well[1657], and low-power He-Cd lasers are usable as UV excitation sources[1658-9].

Detection of BrUdR Incorporation with Anti-BrUdR Antibodies

Gratzner et al[391-2] proposed the use of fluorescent antibodies to detect BrUdR incorporation in the mid 1970's, at which time problems in the development and standardization of antisera limited application of the technique. Improved antisera[393] and, later, development of a monoclonal antibody to BrUdR[394] (available from B-D) increased sensitivity and specificity sufficiently to make immunofluorescence a practical and widely used method for estimation of DNA synthesis from BrUdR incorporation; it is possible to detect a much smaller amount of BrUdR with antibody than could reliably be detected by the quenching technique described above. The entire November 1985 issue of *Cytometry* was devoted to the development and use of monoclonal antibodies against BrUdR; this has also appeared in book form[834] and is an essential reference on the technique.

If you can use the anti-BrUdR antibody method, you can probably distinguish cells in early S phase from those in G_1 and cells in late S phase from those in G_2 well enough so you won't need to do elaborate mathematical analyses to deconvolute DNA histograms. The flow cytometric aspects of the procedure are straightforward, since the reagents used, normally propidium iodide and fluorescein-labeled antibodies, can be excited by a single blue-green beam, making the technique usable on most instruments.

Whether or not you can readily apply the anti-BrUdR immunofluorescence technique to the cell type(s) you want to study is another matter. The sample preparation methodology described by Dolbeare et al[395,835] is technically demanding, must be optimized for individual cell types, and works well with some cell types and not with others. Thus, although measurements of labeling indices may provide important information relevant to the prognosis and/or treatment of tumors, the immunofluorescence method for making such measurements, at its present state of development, is not readily adaptable to routine clinical use.

Difficulties in cell preparation arise because the "anti-BrUdR" antibody reacts only with BrUdR in single-stranded DNA; therefore, before the antibody can be used, nuclear DNA must be denatured *in situ* by treatment with acid and/or by heating[836-8]. In some cases, it is essentially impossible to get enough denaturation to bind antibody while leaving enough DNA to stain; anti-BrUdR antibodies cannot be used, for example, to analyze DNA synthesis in bacteria. A high-affinity monoclonal antibody to **iododeoxyuridine** in DNA has also been made[839]; this can offer increased sensitivity compared to anti-BrUdR antibodies, but still requires DNA denaturation before it can be used. According to Hoy, Seamer, and Schimke[1660], the cell density, volume of solution used, and pH at which denaturation is carried out are all critical. Larsen et al[1661] reported that DNA analysis was better, and that there was less cell aggregation and cell loss, when the wash steps were omitted following immunofluorescent staining. Still and all, it's not easy.

When technical considerations do permit its application, the anti-BrUdR antibody technique can be used with pulse and continuous labeling to estimate all cell cycle parameters that could be estimated from ^{3}H-TdR incorporation[840-3]. Bakker et al[1662-3] extended the methodology to permit double-labeling experiments to be done with chloro- and iododeoxyuridine and two different antibodies; Toba, Winton, and Bray[1664] reported that BrUdR incorporation and phenotype could be measured simultaneously, using fluorescein- and phycoerythrin-labeled antibodies, respectively, if 7-AAD instead of propidium were used to stain DNA. White and his colleagues[1193,1665] and others (see p. 247 and references 2492-6) have developed models for bivariate analysis of BrUdR incorporation data.

Cytochemical Detection of BrUdR Incorporation Using Difference and Ratio Signals

Crissman and Steinkamp[647] described a rapid cytochemical method for detecting BrUdR that does not use antibodies and does not require DNA denaturation; however, it calls for fancier hardware, once again dashing our hopes for a free lunch.

The quenching of Hoechst dye fluorescence by BrUdR serves as the basis for the new technique, as it does for the Hoechst/PI method described previously. In this instance, following culture in the presence of BrUdR, cells are fixed (in cold ethanol) and stained simultaneously for 1 hour with 0.5 µg/ml Hoechst 33342 and 5 µg/ml mithramycin in the presence of 5 mM $MgCl_2$. They are then analyzed in a dual-beam flow cytometer, using UV light to excite Hoechst dye fluorescence and violet or blue-violet light to excite mithramycin fluorescence.

In Crissman and Steinkamp's method, mithramycin fluorescence intensity indicates DNA content, while the **difference signal**, i.e., mithramycin fluorescence minus Hoechst fluorescence, provides a measure of the amount of BrUdR incorporated. The electronic subtraction magnifies the differences between fluorescence from the quenched and unquenched DNA fluorochromes; however, the results are probably cleaner than might be obtained by using a difference signal between Hoechst dye and propidium and ethidium fluorescence. This is so because ethidium and propidium themselves substantially quench Hoechst dye fluorescence, decreasing its intensity considerably and generally leading to lower measurement precision; mithramycin does not quench Hoechst dye fluorescence to nearly the same extent. The **ratio** of mithramycin fluorescence to Hoechst dye fluorescence may also be used to provide the measure of BrUdR incorporation. While Crissman and Steinkamp developed this technique with Los Alamos' multibeam flow cytometer, using very large argon and krypton ion lasers for excitation, it is perfectly feasible to measure the relatively bright fluorescence from Hoechst 33342 and mithramycin-stained cells in instruments using more modest light sources, e.g., a Cytomutt equipped with an air-cooled He-Cd laser emitting 325 nm and 441 nm simultaneously (I've tried this dye combination and light source with excellent results) or a multibeam (365 and 436 nm) Hg arc lamp system[88,844].

Breaking Up Is Easy To Do: SBIP, a Simpler Way to Detect BrUdR Incorporation into DNA

A 1994 publication by Li et al[1666] from Darzynkiewicz's lab described a greatly simplified method for detection of BrUdR incorporation. BrUdR acts as a photosensitizer for DNA; UV illumination of DNA containing the label leads to strand breaks at the sites of incorporation. After exposure to BrUdR, cell suspensions in Petri dishes are illuminated for 5 min with 300 nm light by placing them atop an analytical DNA transilluminator. The DNA strand breaks induced by illumination are then labeled by incorporation of digoxigenin- or biotin-labeled dUTP, with terminal deoxynucleotidyl transferase (TdT) added to catalyze the reaction. The labeled dUTP can then be detected using fluoresceinated anti-digoxigenin antibodies or fluorescein-straptavidin. Results obtained using the newer **SBIP (Strand Breaks Induced by Photolysis)** method and the standard technique employing anti-BrUdR antibodies are compared in Figure 10-4, on the next page.

The major advantage of the SBIP method is that it does not require DNA denaturation; this allows phycoerythrin antibodies to be used to detect cell surface or intracellular antigens, which might be destroyed during denaturation.

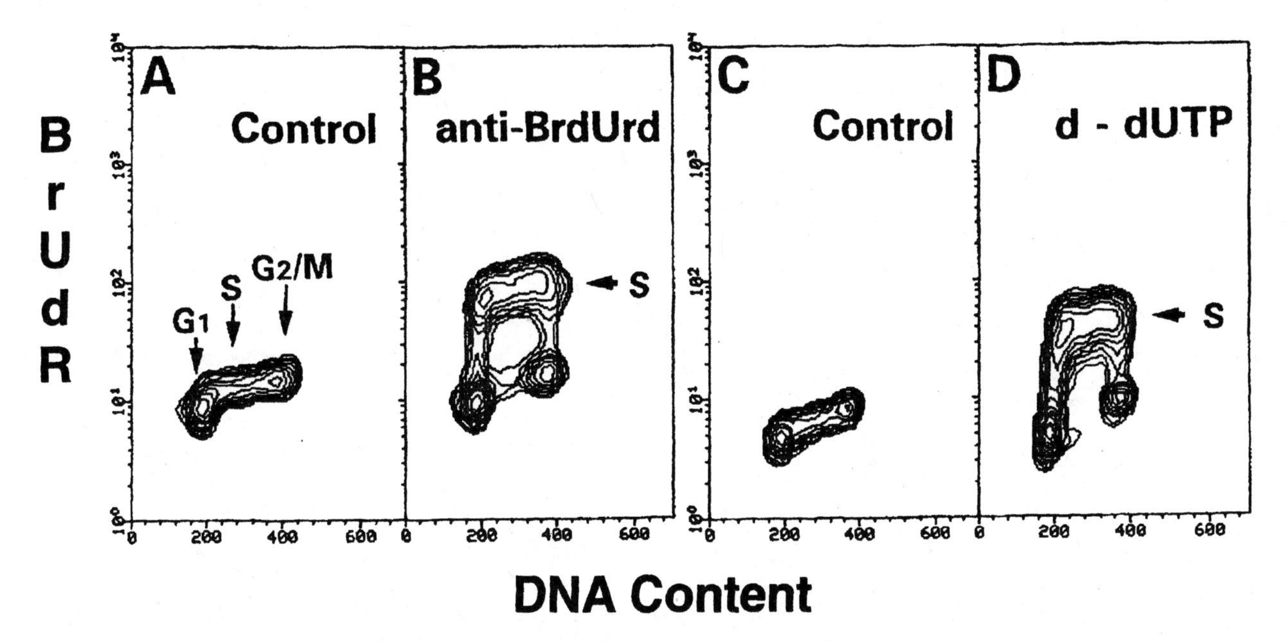

Figure 10-4. Detection of BrUdR incorporation using anti-BrUdR antibody (A and B) and the SBIP method (dUTP labeling of strand breaks induced by photolysis) (C and D) in HL-60 cells grown in the presence (B and D) and absence (A and C) of 50 µM BrUdR for 1 h. The x-axis shows DNA content, indicated by propidium fluorescence, on a linear scale; the y-axis shows BrUdR content, indicated by fluorescein fluorescence, on a logarithmic scale. The figure, reproduced with permission, is from X Li et al, Intl J Oncology 4:1157-61, 1994[1666], and was provided by Zbigniew Darzynkiewicz (New York Medical College).

Unlike the Hoechst/PI and Hoechst/mithramycin techniques, the SBIP method requires only 488 nm illumination. The method can also be adapted to distinguish between strand breaks due to apoptosis and to BrUdR, permitting simultaneous analysis of apoptosis and DNA synthesis[2074]. Further refinements of the SBIP technique were introduced in 1995. It was noted[2946] that prior incubation with Hoechst 33258 increased the number of strand breaks produced during a given period of UV illumination; it was also possible to incorporate BODIPY- or fluorescein-labeled dUTP rather than digoxigenin- or biotin-labeled dUTP at strand breaks, eliminating the need for the addition of fluorescently tagged anti-digoxigenin antibody or streptavidin to produce detectable labeling. However, in another paper published the same year[2947], the authors reported that BrdUTP, which is much less expensive than dUTP labeled with biotin, digoxigenin, BODIPY, or fluorescein, was efficiently incorporated at strand break sites and was readily detected by fluoresceinated anti-BrUdR antibodies, producing signals severalfold brighter than those obtained using directly or indirectly tagged nucleotides.

The application of the SBIP method to cells labeled with a tracking dye such as CFSE (see pp. 372-4) should allow derivation of cell cycle phase information in five or more successive generations of cells, surpassing the capacities of the Hoechst/PI method in this regard.

When I wrote the previous edition of this book, I was sure that the SBIP technique would replace methods that used anti-BrUdR antibody to detect BrUdR incorporation. However, within a year or two, the SBIP method was modified to use anti-BrUdR antibody, and the story doesn't end there.

Anti-BrUdR Antibody: Seeing the Light

The major objection to the original anti-BrUdR antibody methods arose because it was necessary to denature the DNA in order to get the antibody to react with incorporated BrUdR. A few years after the SBIP method was published, light bulbs lit up in the heads of Hammers, Kirchner, and Schlenke[2948], who reported in 2000 that the DNA denaturation step could be eliminated by incubating the cells with Hoechst 33258 and irradiating them with UV light, as was done in the SBIP method. They found that if the cells were then transferred to a hyoptonic (68 mOsm) buffer, anti-BrUdR antibodies would bind to the incorporated BrUdR. UV-A, UV-B, and UV-C sources were tested; light at 280-320 nm from an 8 W UV-B bulb yielded the best results.

Cytochemical Detection of BrUdR: Still Around

The anti-antibody folks haven't been resting on their laurels since the previous edition appeared, either. Frey[2620] noted in 1994 that the fluorescence of both **LDS-751** and **TO-PRO-3** was enhanced in the presence of DNA into which BrUdR had been incorporated. LDS-751 can be excited at 488 nm and emits at 670 nm; staining cells with a combination of LDS-751 and propidium can therefore be used to detect BrUdR incorporation using only a 488 nm excitation beam. As it happens, it is also possible to use a single 488 nm excitation beam for cells stained with a combination of propidium and TO-PRO-3, because there is enough energy transfer from propidium to TO-PRO-3 to produce a usable signal. Beisker et al[2949] used dual-laser instrument with a 488 nm beam exciting propidium fluorescence and red He-Ne or diode lasers exciting TO-PRO-3 fluorescence in isolated nuclei, and found that concentrations of 50 μM propidium and 0.3 μM TO-PRO-3 provided the best compromise between BrUdR detection sensitivity and DNA histogram resolution. Some of their results appear below in Figure 10-5.

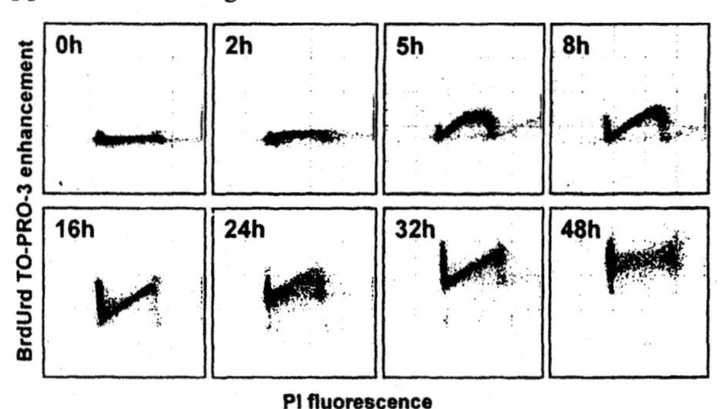

Figure 10-5. Use of a combination of propidium and TO-PRO-3 to detect bromodeoxyuridine incorporation. Propidium fluorescence insensitive to BrUdR incorporation) is shown on the X-axis; the Y axis shows the quotient of TO-PRO-3 and propidium fluorescence. From W. Beisker et al, Cytometry 37:221-9, 1999 (Reference 2949), © John Wiley & Sons, Inc., used by permission.

Cytochemical detection of BrUdR incorporation is somewhat less sensitive than detection using antibody and/or strand break techniques; cells are typically incubated with concentrations of 20-60 μM BrUdR, which may affect cell growth. However, cytochemical detection could, at least in principle, make it possible to assess BrUdR incorporation in viable cells, which is not possible when either antibody or strand break techniques are used.

Mozdziak et al[2950] found it possible to enrich cells with and without BrUdR to a purity of over 90% when sorting from mixtures of cells stained with Hoechst 33342 alone, but it was difficult to assess fractions of cells in different phases of the cell cycle in the BrUdR-containing cell population because fluorescence intensity was low. I am not aware of any reports in the literature on the interaction of **DRAQ5**[2338-9] with DNA containing BrUdR. DRAQ5 is the only dye other than Hoechst 33342 and a few other Hoechst dyes that has been reported to stain DNA stoichiometrically in living cells. If DRAQ5 fluorescence is either unaffected by the presence of BrUdR in DNA or enhanced by it, it should be possible to use a combination of Hoechst 33342 and DRAQ5 with UV and 488 nm or red excitation to measure BrUdR incorporation and cell cycle position simultaneously in viable cells.

If you have any experience with nucleic acid sequencers, you are likely to be aware that at least some of them work by incorporating fluorescently labeled nucleotides into oligonucleotides. Wouldn't it be nice if we could put those fluorescently labeled nucleotides into DNA *in vivo*? Yup. But we can't, at least not yet.

Label or Slander?

While we tend to regard established methods as having more of a ring of truth about them than newer ones, it is apparent that techniques based on incorporation of tritiated thymidine or other radioisotopes are not themselves without warts. High specific activity thymidine may administer enough radiation therapy to cells exposed to it to affect their rate of progress through the cell cycle. Also, the validity of many studies involving measurements of isotope uptake rests on assumptions that pool sizes remain constant and that the isotope used is not reutilized or metabolized. This has certainly not been established for every cell system studied with radioisotopes. One of the few published papers that compared the two kinetic methods found marked discrepancies between the fraction of stimulated thymocytes, as assessed by flow cytometry, in con A-treated and control cultures and the amounts of radiolabeled thymidine taken up[396]. Labeling experiments with BrUdR may be helpful in explaining results such as this. On the other hand, BrUdR can induce apoptosis in cells[1666], and, as noted above, it can inhibit growth at high concentrations, so it's not completely above suspicion, either.

Detecting RNA Synthesis Using Bromouridine

While I am still close to the subject of anti-BrUdR antibodies, I should point out that, according to Jensen, Larsen, and Larsen[1667], **RNA synthesis** can be detected relatively readily using these antibodies, which cross-react with **bromouridine** incorporated into RNA.

Generation Gaps: Tracking Dyes and Cell Kinetics

Tracking dyes have been introduced on pp. 45-6 and 371-4, and will be discussed further in the section on immunology later in this Chapter (p. 501). CFSE, in particular, allows viable cells in different generations to be identified and sorted, at least when used at relatively low concentrations. Tracking dyes offer an attractive alternative to BrUdR for labeling and following cells through generations, because, to data, getting the answers from BrUdR requires killing the cells. On the other hand, tracking dyes can't answer the question of whether or not a cell is synthesizing DNA over a short time period; we'll still need BrUdR or something like it to do that job, and, if we're working with pulse-labeled cells, we're likely to have to sacrifice the cells to get the answers.

Cell Cycle-Related Proteins: Cyclins, Etc.

Variations in many cellular characteristics during the cell cycle can be analyzed simply and directly by two-

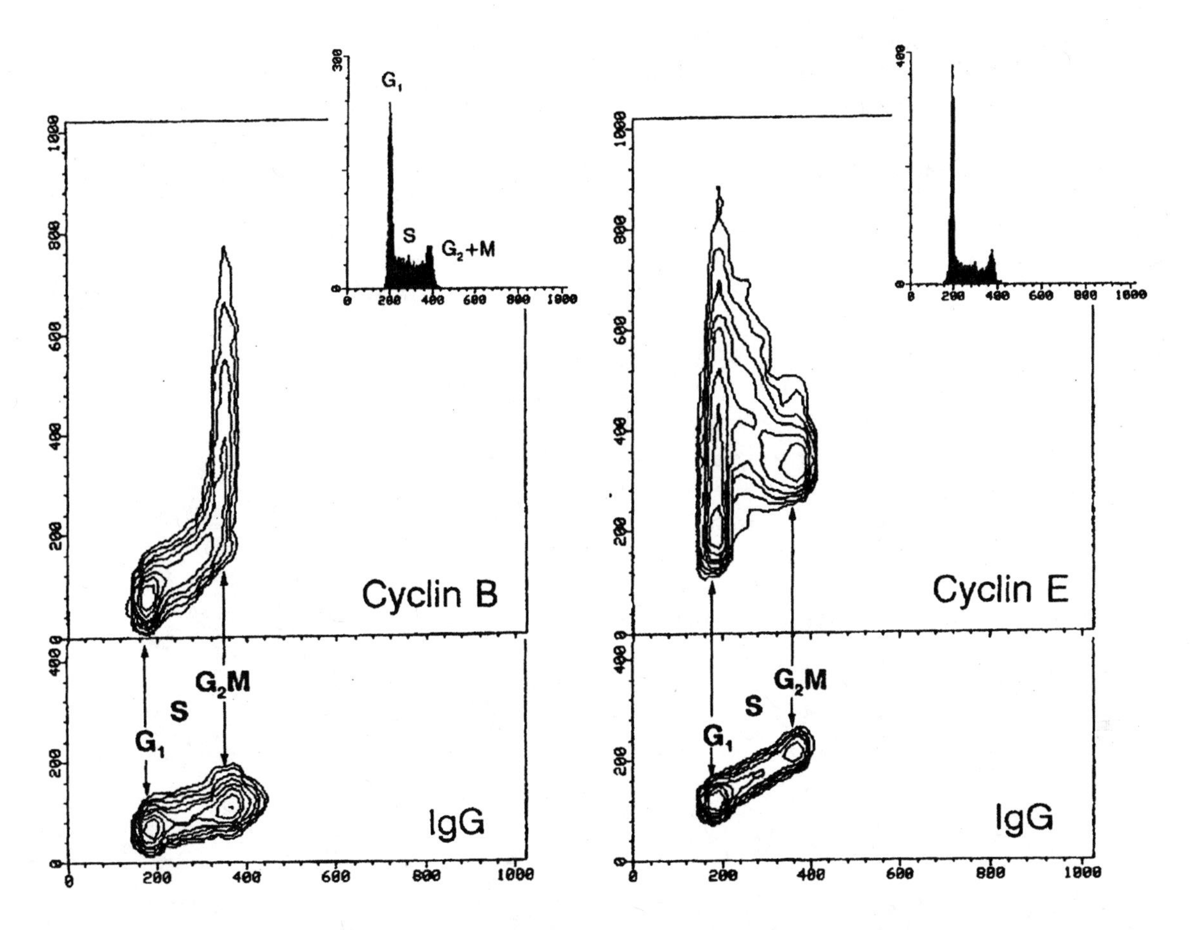

Figure 10-6. DNA content vs. expression of cyclins B and E in exponentially growing MOLT-4 cells; cells stained with nonspecific IgG instead of anticyclin antibodies are shown as controls. Cyclin B is expressed predominantly in late S, G_2, and M phases; cyclin E expression, while more heterogeneous, is predominantly in Gl and early S. The figure, provided by Zbigniew Darzynkiewicz and used by permission, is from a paper by Gong, Traganos, and Darzynkiewicz[1850] (Intl J Oncology 3:1037-42, 1993).

parameter flow cytometry of cells or nuclei stained simultaneously with a DNA fluorochrome and with a fluorescent probe that allows quantification of the other parameter being investigated.

In cells growing actively and continuously without maturation, the mass and volume of all cellular constituents must double over one cycle of growth. With the aid of DNA fluorochromes, we can observe the doubling of DNA content during the cell cycle. When we look at the joint distribution of any other cellular constituent and of DNA content in growing cells, we thus have to expect that, whatever the other constituent may be, there should be twice as much of it in M phase cells as there is in G_1 phase cells. This does not, however, always mean that the data points in the bivariate distribution lie along or about a straight line.

A strict linear relationship between cells' content of DNA and of another constituent can exist only if the constituent is synthesized at the same rate as is DNA. Deviations from such a relationship, such as are shown in Figure 10-6, indicate that a constituent is synthesized or expressed preferentially in one or more phases of the cell cycle, as is the case for **cyclin B** (predominantly expressed in late S, G_2, and M) and **cyclin E** (predominantly expressed in G_1 and early S, although the pattern of expression is more heterogeneous than that seen for cyclin B)[1850-1]. Note that the IgG controls in Figure 10-6, in which staining by antibody is nonspecific and should therefore be predominantly size-dependent, do show a linear relationship between antibody fluorescence and DNA content.

Darzynkiewicz et al[2951] reviewed the cytometry of cyclins in1996. Cyclins B and E have been mentioned above; **Cyclin A** is expressed in S and G_2, but is lost in mitosis just prior to metaphase, while **cyclin B_1** is lost between metaphase and anaphase[2952]. **D-type** cyclins are expressed early in G_1. It is inadvisable to use the presence or absence of various cyclins as an indicator of precise position in the cell cycle[2953]. This caution applies especially when working with tumor cells or with cells subjected to the influence of various agents that perturb the cycle; deviations from the "schedule" may occur in either case.

The distribution shown in Figure 3-10 (p. 97) and in Figure 7-16 (p. 321), to which I usually refer as the

"Darzynkiewicz Flag," is one that is commonly encountered in analyses of **RNA**[113,262-3,753-6], **total protein**[297,522,753-6], and **nuclear protein**[523,750-2] content during the cell cycle. Zbigniew Darzynkiewicz has, over the years, consistently been among the first to develop new methods for studying the birth and death of cells and to apply those methods to the study of cancer and other diseases, as should be obvious from the frequency with which his work is cited in this and previous editions of this book

While RNA and protein are not synthesized continuously during the cell cycle, the rates of synthesis of these two constituents do track one another[899]. Roti Roti et al[523] have used the distribution of nuclear protein vs. DNA content to divide the cell cycle into early and late G_1, early and late S, and early and late G_2 compartments; they were able to detect cell cycle-specific synthesis of nuclear proteins by measuring incorporation of radiolabeled precursors into sorted nuclei *in vitro*[1852].

Distributions obtained from simultaneous analysis of DNA content and of antigens that are expressed primarily during S phase[524,900] exhibit the pattern of panels B and D in Figure 10-5 (p. 457); these were actually obtained when anti-BrUdR antibody and dUTP, respectively, were used to detect cells in S phase[391-4,834-6,1666].

Since the previous edition of this book was written, our understanding of molecular mechanisms involved in progression through the cell cycle has continued to increase, with the 2001 Nobel Prize in Medicine being awarded to Leland Hartwell, Tim Hunt, and Sir Paul Nurse "for their discoveries of key regulators of the cell cycle." Analyses of DNA content and of antigens that are differentially expressed during different phases of the cell cycle remain of great interest to clinicians as well as to basic researchers.

An intracellular antigen defined by the **Ki-67** monoclonal antibody[1292,1661,1853-4] is preferentially expressed in proliferating cells (G_1, S, G_2, and M phases, but not G_0). The Ki-67 antibody came from the clone in the 67th well of a 96-well plate at the University of Kiel, which contained hybridomas resulting from immunization of mice with cells from Hodgkin's lymphomas. The Ki-67 antigen is not found in cells undergoing DNA repair or in tumors with unscheduled cyclin expression, and is considered a robust marker of cell proliferation; levels of Ki-67 expression are of prognostic value in at least some tumor types[2954].

The presence of **proliferating cell nuclear antigen (PCNA)**[1855-9] is also used to identify proliferating cells; this antigen is most prominently expressed during S phase. Landberg and Roos[1858] have developed no-wash staining techniques for unfixed cells facilitating demonstration of nuclear antigens; they have also shown that dual staining with Ki-67 and anti-PCNA antibodies identifies discrete G_0, G_1, S, G_2, and M phase compartments of the cell cycle. Teague and El-Naggar[1859] compared three monoclonal antibodies and two fixation protocols for analysis of PCNA, and found the most reliable determinations were obtained using methanol fixation and the PC10 antibody (DakoCytomation), which reacts with PCNA even in paraffin-fixed material. Larsen et al have recently reviewed the subject of PCNA detection[2955].

Transferrin is required for cell growth[2956], and proliferating cells express the **transferrin receptor (CD71)** on their cell surfaces. In well-defined experimental systems, e.g., mitogen- or antigen-stimulated lymphocytes, substantial CD71 expression is detectable on cells in the G_1, S, G_2, and M phases of the cell cycle, but not on G_0 cells (Figure 10-7).

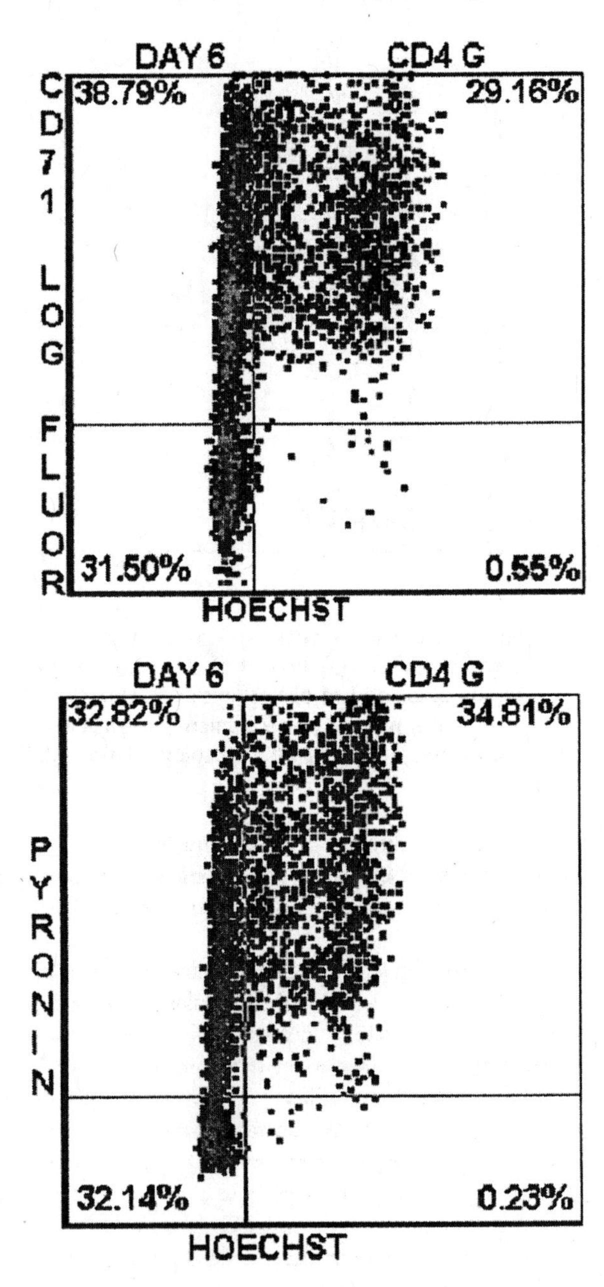

Figure 10-7. CD4 Cells from a 6-day-old mixed lymphocyte reaction (MLR), showing DNA content (Hoechst 33342) vs. transferrin receptor (top panel; CD71 fluorescence, log scale) and RNA content (bottom panel; pyronin Y fluorescence, linear scale). Data are from Lucia Vasconcellos and Edgar Milford (Brigham and Women's Hospital, Harvard Medical School).

Since it is, as a rule, easier to detect cell surface antigens than intracellular antigens, and since it is generally possible to preserve cell viability after detecting cell surface antigens, but not possible to do so after detecting intracellular antigens, it would seem to me that, if one wanted to demonstrate a proliferation-related antigen, CD71 would be a better choice than, say, Ki-67. However, while there seem to be numerous studies of Ki-67 expression on tumors[2954] and its relation to malignant potential, there are relatively few papers (reference 2957 is an example) that examine CD71 expression in this regard. There are also studies of Ki-67 in stimulated lymphocytes[2954], and, when I hear people who have elected to use Ki-67 as an indicator of lymphocyte proliferation complaining about the difficulty of doing the measurements, I always ask why they didn't try CD71 instead. However, there does not seem to be a published study of lymphocytes or any other cell type in which flow cytometry was used to look at Ki-67 and CD71 in and on the same cells at the same time. I'll do one if I can get around to it, but it would be nice if somebody else got me off the hook.

Flow cytometry has also been applied to studies of **oncogene expression** and its effects on the cell cycle. Some of the earliest work was done by Jacobberger, Fogleman, and Lehman[727], who had to develop and refine methods for demonstration of intracellular antigens as a first order of business. Lehman et al[1860] studied expression of the early tumor (T) and late viral (V) antigens in cells infected with simian virus 40 (SV40); the amount of T antigen per cell increased as cells entered successive stages of the cell cycle, while synthesis of V antigen began in late S and G_2 + M phases. Sladek and Jacobberger[1861] examined mouse fibroblasts after infection with a recombinant retrovirus encoding T antigen. As a result of T antigen expression, the duration of the G_1 phase was decreased and the duration of the G_2 + M phase was increased; the duration of S phase was unaffected by antigen expression. They subsequently established[1862] that T antigen is a concentration-dependent, positive cell cycle regulator in exponentially growing cells, and that endogenous negative control mechanisms responding to cell density override its effect. "Jake" Jacobberger and his colleagues continue to make significant contributions in this field[2701-2,2958-61].

Darzynkiewicz et al[262,525] defined cell cycle compartments as G_{1a} and G_{1b} according to whether RNA content has not or has reached the level associated with the earliest visible S phase cells. It was observed by Darzynkiewicz et al[526] that cell size (as estimated from light scatter measurements), RNA and protein content are all more heterogeneous (i.e., distributions of these parameters have higher CV's) in G_1 than in G_2 + M cell populations; this indicates an **unequal apportionment of cytoplasmic contents into daughter cells at cell division.** Most of the variance of the distributions of these parameters in cells in G_1 is contributed by cells in the G_{1a} compartment. This was taken to indicate that the RNA and protein content of cells equalizes prior to the transition to G_{1b}, in which compartment heterogeneity is minimal. The variance of distributions of fluorescence of a tracking dye such as PKH26 in cells (Figure 7-23, p. 372) can serve as an indicator of unequal distribution of membrane lipid between daughter cells.

Kimmel, Traganos, and Darzynkiewicz[527] established that the duration of G_{1a} is exponentially distributed and has a greater variance than the duration of G_{1b}, and that all or nearly all daughter cells enter G_{1a} following cell division. These findings support the transition probability model of cell kinetics proposed by Smith and Martin[528].

Relationships between cell size, DNA replication, and cell division have been the subject of speculation over many years; a brief review[529] by Baserga and a report by Zetterberg, Engstrom, and Dafgård[530] raised interesting questions best answered by cytometry. Zetterberg's group examined the effects of different types of **growth factors** on 3T3 cells starved to quiescence in low-serum media. Epidermal (EGF) and platelet-derived (PDGF) growth factors stimulated quiescent cells to undergo DNA synthesis and mitosis, but there was no growth (with EGF) or little growth (with PDGF) in cell size during the replication period, and the daughter cells would not undergo further division. When insulin was added, both growth in size and DNA synthesis occurred, although addition of insulin alone did not stimulate cells to enter mitosis. Thus, there may be separate control mechanisms associated with cytoplasmic growth and DNA replication.

When the previous edition of this book appeared, the determination of **relationships between surface antigen expression and cell cycle compartments defined by both DNA and RNA content,** while of interest to many investigators, was readily possible only for those who had the dual beam UV/488 nm excitation capability needed for Hoechst/pyronin/fluorescein staining, although similar studies could be accomplished, with more preparative effort, with single-beam instruments, using a technique described by Bauer et al[901], in which cells sorted on the basis of immunofluorescence are restained with acridine orange. It is now possible (pp. 324-5) to use the combination of 7-AAD and pyronin Y to determine DNA and RNA content in instruments with only a 488 nm illuminating beam, with fluorescein isothiocyanate or a fluoresceinated antibody, respectively, applicable to determination of total protein content or the content of a surface or intracellular antigen[2343-6]. If you have a 3-beam instrument with 6-color capability, and you want to study the relationship of two or three antigens to one another and to the cell cycle, you don't have much excuse not to use Hoechst 33342 and pyronin Y (and/or CD71) to define cell cycle phases and discriminate G_0 cells from G_1 cells. That's what the instrument is built for, and you might as well get your money's worth.

Sinnett, Flint, and Lalande[1863] described a procedure for **determining the point during S phase at which any single-copy DNA sequence is replicated.** Cells are synchronized at the G_1/S phase boundary, released and labeled with BrUdR, and sorted on the basis of DNA content at different times after release. Newly replicated BrUdR-substituted

DNA is then removed by UV irradiation and S1 nuclease treatment. PCR will thus not be able to amplify the target sequence in cells in which it has been replicated, while amplification, with production of detectable DNA, will occur in cells in which replication has not yet occurred.

Detecting Mitotic Cells

On p. 454, I promised you a follow-up on detecting mitotic cells; it was mentioned on p. 277 that a combination of side scatter and propidium fluorescence might be usable for the purpose, and the use of acridine orange staining to define differences in chromatin structure between interphase and mitotic cells was described and illustrated (Figure 7-15) on p. 320. While it looked for a while as if cyclins A and B1 might be helpful, a 1998 study by Juan et al[2962], from Darzynkiewicz's lab, established that an **antibody to phosphorylated histone H3 (anti-H3P)** reacted with mitotic cells, providing better discrimination than any of the anticyclin antibodies. In a 2001 review, Juan et al[2363] concluded that "the use of the H3-P antibody appears to provide the most advantages compared with the alternative methods of detection of mitotic cells"; this article provides protocols and caveats. By using antibodies to cyclins A and B1, in combination with anti-H3-P, it is possible, at least to some extent, to differentiate between early and late mitotic cells.

Figure 10-8, below, demonstrates identification of a mitotic population in cells stained with propidium iodide and anti-H3-P; the cells shown in the figure are those remaining after doublets were excluded by gating on a plot of peak vs. integral propidium fluorescence.

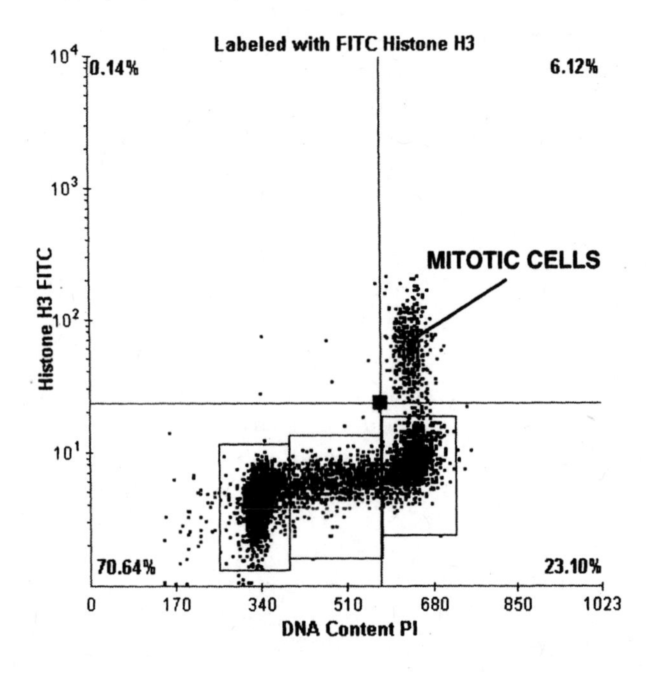

Figure 10-8. Identification of mitotic cells by staining with propidium iodide and antibody to phosphorylated histone H3. Courtesy of Jeffrey Scott and Dorothy Lewis (Baylor College of Medicine).

Memento Mori: Detecting Cell Death

Necrosis versus Apoptosis

There are two different mechanisms of cell death, **programmed cell death**, or **apoptosis**, and **accidental cell death**, or **necrosis**[2352]. **Necrosis** is identified primarily by **loss of membrane integrity**, as indicated by **dye exclusion tests, which have been discussed in detail on** pp. 46, 300-2, and 369-71.

Apoptosis seems to have aroused almost as much interest as sex, which is curious when you consider that, thanks to apoptosis, tadpoles lose a little tail. Apart from its role in embryonic development, apoptosis has been implicated in AIDS and cancer, and techniques for detecting apoptotic cells by cytometry, flow and otherwise, continue to be proposed by numerous investigators.

Comprehensive reviews of the subject by Darzynkiewicz et al appeared in 1992[1517] and 1997[2352]; in the latter, the term **cell necrobiology** was introduced "to comprise various modes of cell death; the biological changes which predispose, precede, and accompany cell death; as well as the consequences and tissue response to cell death. The term combines [the Greek] *necros* ("death") with *bios* ("life")."

Identifying Apoptotic Cells

The term "apoptosis" itself derives from a Greek word that describes phenomena such as the falling of leaves from trees. When, in 1997, I was inspired to wax lyrical about apoptosis, there was only one melody to use:

Les Feuilles Mortes (Autumn Leaves)

When outer leaflets of cell membranes
Let phosphatidylserine show,
Labeled annexin V will bind there,
And you can measure it in flow[2353-4,2963-4].

Mitochondria deenergize[2811-6,2965],
And superoxide levels rise[2966-8],
But the nuclear signs of apoptosis[1517,1521-4,1526]
Come later[2352]; then, the cell dies.

Did thymocytes get radiation?
Was dexamethasone to blame?
Or was it simply Fas ligation?
The end results are all the same.

Lytic enzymes in the cell are loosed[2557,2769-72];
Less glutathione stays reduced[2815,2967],
And, around the time the membrane's leaky[1520,1525,1527-31],
The vultures all leave their roost.

The journals publish three new assays
For apoptosis every week[2971-83];
Is it biology that varies
From cell to cell[2984-5], or just technique[2986]?

Cells can stay alive, as good as new,
If they can make bcl-2[2969],
But, if not, their DNA's in fragments[1517-9,2971,2982-3,2985]
When apoptosis is through.

(©Howard Shapiro; used by permission)

Apoptosis: Getting With the Program

In their 1992 review, Darzynkiewicz et al[1517] reported that the ATP-dependent lysosomal proton pump, demonstrated by supravital uptake of acridine orange, was preserved in apoptotic but not necrotic cells, and that bivariate analysis of cells for DNA and protein showed markedly diminished protein content in apoptotic cells, probably due to endogenous protease (what we now know to be caspase) activation. The sensitivity of DNA *in situ* to acid denaturation, probed by staining with acridine orange at low pH (see p. 320), was increased in both apoptotic and necrotic cells. They also reported that **mitochondrial membrane potential ($\Delta\Psi_m$)**, which they assayed by retention of rhodamine 123, was preserved in apoptotic, but not in necrotic cells; this is at variance with later findings by numerous other groups[2811-6,2965,2969]. The anti- and proapoptotic proteins in the bcl-2 family are associated with mitochondria, and it is now generally appreciated that **induction of mitochondrial permeability is one of the earliest events in apoptosis.** While there are disputes about which technique(s) is/are best to demonstrate loss of $\Delta\Psi_m$ (see pp. 394-400), this parameter is now widely used to detect apoptosis. Changes in **reactive oxygen species** occur[2966-8], but are not as widely used as indicators.

As was noted on p. 380, both fluorogenic substrates and inhibitors can be used to detect the activity of **caspases**, which are the prime enzymatic movers of events in apoptosis[2332,2556-7,2769-72].

ISNT there Light at the End of the TUNEL?

In apoptosis, but not necrosis, activation of an endonuclease results in fragmentation of DNA, which produces a characteristic "ladder" on gel electrophoresis. DNA fragmentation can be detected directly in cells using **in situ nick translation (ISNT)**[1518-9,2982] or **TdT(terminal deoxynucleotidyl transferase)-mediated biotin-dUTP nick-end labeling (TUNEL)**[1519,2971] assays. In both types of assay, cells are permeabilized, and dUTP labeled with fluorescein, digoxigenin, or biotin or, alternatively, BrdUTP is incorporated at the site of strand breaks. DNA polymerase I is used to catalyze incorporation in nick translation assays; however, incorporation is said to proceed more rapidly in TUNEL assays, in which TdT is used as catalyst[1519]. Fluoresceinated dUTP is detected directly; biotin- and digoxigenin-dUTP are, respectively, detected using labeled streptavidin and labeled anti-digoxigenin antibodies, and BrdU is detected with anti-BrdU antibodies. As was noted in the discussion of BrUdR labeling of DNA on p. 457, BrdUTP is a lot cheaper than other labeled nucleotides; it now seems to be very popular for TUNEL assays for the same reason.

Plasma membrane integrity is lost early in necrotic but not apoptotic cells; the former take up propidium iodide (PI) after a brief exposure, while the latter do not. However, apparent membrane permeability is slightly increased in apoptotic cells, resulting in different patterns of uptake of several dyes as compared to intact cells. Apoptotic cells stain faintly with ethidium[1527] and 7-AAD[1317], and take up propidium after a moderate incubation period[1528]. Hoechst dye uptake is increased, as is efflux of fluorescein produced by FDA hydrolysis[1529]. Changes in scatter signals occur in association with membrane permeability changes; apoptotic cells have lower forward and higher side scatter signals than do intact cells[1520-1,1523,1525,1530].

If you happen to have a flow cytometer with dual-beam UV and 488 nm excitation, the combination of PI and Hoechst 33342 provides good discrimination of live, necrotic, and apoptotic cells[1530]; live cells show low Hoechst dye fluorescence and high forward scatter, apoptotic cells show higher Hoechst dye fluorescence and lower forward scatter, and necrotic cells exhibit high PI fluorescence and low forward scatter. The use of a PI/7-AAD combination facilitates analysis of apoptosis in populations defined by fluorescein and phycoerythrin immunofluorescence[1531]. If you only have one beam, all is not lost; Schmid et al[1317] report that 7-AAD and scatter can distinguish apoptotic cells, and this technique can also be used with two-color immunofluorescence. Necrotic (or late apoptotic) cells take up a substantial amount of 7-AAD, intact cells take up little or none, and apoptotic cells take up a small amount.

Low molecular weight DNA is lost after cell permeability increases, resulting in decreased staining of apoptotic cells by DNA-specific fluorochromes, producing a **hypodiploid (sub G_0/G_1) peak**[1520-6]. This staining pattern is observed using a wide variety of DNA fluorochromes[1522]. However, in a flow cytometric analysis, one can never be sure that the sub G_0/G_1 events observed in a DNA histogram represent cells rather than apoptotic bodies (nuclear fragments); this makes it difficult to arrive at a reliable measure of the fraction of apoptotic cells by DNA analysis.

It should be appreciated that the features of apoptosis differ in at least some particulars from cell type to cell type (for example, see references 2984-6); this helps explain why assays developed using one or two cell types and one or two inducers of apoptosis may fail ingloriously when attempts are made to apply them to different model systems.

Apoptosis: The Case Against Flow Cytometry

Darzynkiewicz et al[2983,2986] point out that apoptosis is a dynamic process, sometimes resulting in the complete dissolution of cells within a few hours, and that many of the criteria required to precisely define the occurrence and extent of apoptosis in cell populations are morphological. For this reason, they now favor **laser scanning cytometry**[2382,2983,2986] over flow cytometry for studies of apoptosis, pointing out that this allows you to "have your cake and eat it too"[2986]. They do, however, have a favorite flow cytometric method, the **stathmo-apoptosis** method[2770,2987], in which a fluorescent irreversible caspase inhibitor, FAM-VAD-FMK, is added to cultures to arrest apoptosis, preventing cell disintegration. A **cumulative apoptotic index (CAI)** can be measured over periods as long as several days; cells are not prevented from undergoing the earliest stages of apoptosis, but, once they reach the stage of caspase activation, they bind the inhibitor, becoming fluorescent, and remain detectable.

Die Another Day: Cytometry of Telomeres

The maintenance of telomere length appears to be required for immortalization of tumors and cell lines[2988]. Methodology has been developed for measurement of telomere length by both flow cytometry[2715-6,222991-4] and image analysis[2989-90]. PNA probes (p. 362) are used. It is important to know the DNA index or chromosomal ploidy of the cells being studied in order to correct flow cytometric measurements of total telomere probe fluorescence for the number of chromosomes present in cells.

10.3 IDENTIFICATION OF CELLS IN MIXED POPULATIONS

Cell identification, in one form or another, is involved in almost all flow cytometry; indeed, it was the prospect of developing automated cell identification methods for **hematology** and **cancer cytology** that provided the impetus and the funding for the development of the first flow cytometers.

Automating a cell identification procedure, whether or not flow cytometry is the method used, requires, first, that there be a reasonably well-accepted **taxonomy** or **classification scheme** in place for the cells of interest, which allows cells to be classified on the basis of **cellular characteristics** or **parameters** which can be measured by the apparatus to be employed. In addition to being able to discriminate one type of cell from another, this apparatus must, of course, also be able to discriminate cells from other, irrelevant objects in samples.

Mixed Genotypes versus Mixed Phenotypes

The mixed cell populations encountered in cytometry are of two basic types. Marine biologists are likely to be working with samples containing **different genera and species** much of the time, and, not surprisingly, often rely on detection of differences in gene sequence to discriminate among cell types. Hematologists, on the other hand, are looking primarily at **phenotypic variation**. The fertilized ovum is the ultimate stem cell; while gametes, lymphocytes, and neoplastic cells may deviate significantly from the stem cell genotype, cytometric identification of different cell types is samples taken from blood, bone marrow, and the organs of the immune system is almost always based on differences in the amounts of gene products expressed.

No Parameter Identifies Cancer Cells

In the area of **cancer cytology**, we did not know when cytometry started, nor do we know now, **what distinguishes a malignant cell from a normal one**. After more than thirty years of intensive work in flow and image cytometry, cytochemistry, immunology, and molecular biology, we haven't found one constant difference, although we have described many characteristics in which some cancer cells differ from their normal counterparts. Where the difference has been defined, flow cytometry may help define whether it exists or not. As to the question, "What is a cancer cell?," it seems likely that the answer will be more in the form familiar to lawyers – "It is sometimes this, and sometimes that, and not this, and the other thing, or the third thing, in the presence of a fourth..." – than in the succinct form sought by physicians, medical scientists, the media and the public.

Many Parameters Identify Blood Cells

When it came to **classifying blood cells**, we knew, or thought we knew, many of the answers. Even before we were able to identify hematopoietic stem cells, there was a taxonomy of blood cells in place; this classification was based primarily on the appearance under the transmitted light microscope of cells stained with various eosin-methylene blue-azure dye mixtures such as **Giemsa's** and **Wright's stains**, and supported by subsequent demonstrations of cytochemical differences between different cell types, for example, differences in the content or activity of various enzymes.

Flow Cytometric Parameters Useful for Blood Cells

In order to develop automated instrumental methods of identifying blood cells, it was necessary to define **sets of cellular parameters** that could be measured by the apparatus and from which cells could be precisely identified using classification algorithms. Among the parameters found useful were **intrinsic parameters** such as **cell size**, **cytoplasmic granularity**, and **hemoglobin content**, and **extrinsic parameters**, including relatively nonspecific characteristics such as **affinities for various acidic and basic dyes** and more specific ones such as **glycosaminoglycan content** and **lipase and peroxidase activities**. By the early 1970's, when laser source flow cytometers were first used for cell surface immunofluorescence measurements, the other parameters just mentioned, singly or in combination, could be measured by image analysis or flow cytometry, and used to discriminate among **erythrocytes**, **lymphocytes**, **monocytes**, and **granulocytes**, and to distinguish **basophil**, **eosinophil**, and **neutrophil granulocytes**.

It was considerably more difficult to use any of the existing cell identification parameters to determine the **stage of maturity** of blood cells; it was similarly not possible to use these parameters to reliably distinguish any morphologic or biochemical differences between **lymphocyte subpopulations**, even when techniques such as rosetting were used to prepare samples enriched in one lymphocyte class or another. Within a few years of their introduction, **flow cytometric immunofluorescence measurements** were used to establish a taxonomy of lymphocytes; they have since been used to revise the genealogy of other blood cell types.

Specific Gene Products Identify Cell Types

The success of immunologic methods should not lead any of us to the conclusion that fluorescent antibodies are the only specific reagents. The general strategy for defining and identifying different cell types, whether or not the different cell types are from the same organism, is to demonstrate the presence, and/or measure the amount(s), of one or more **specific gene products** in or on cells, ideally with minimal perturbation of the cells, or, more to the point, with a minimum of effort expended in sample preparation. If the specific gene product happens to be intracellular hemoglobin in red cells, it can be demonstrated with just about as much specificity by its absorption of violet light as it can be by immunofluorescence. If the gene product in question is an enzyme, a chromogenic or fluorogenic substrate may be used to demonstrate its activity. If one wants to identify blood basophils, the metachromatic staining of glycosaminoglycans in their cytoplasmic granules can be demonstrated at least as easily as can the presence of IgE receptors on their cell surfaces. When the going gets tougher, for example, if one wants to discriminate between different types of hemoglobin in red cells, then antibodies certainly offer us one way of doing the job, but the molecular biologists would probably just as soon go looking for specific messenger RNAs.

If one is looking at **mixed populations of cells of different species**, as in flow cytometric analysis of phytoplankton, it is reasonably likely that a suitable combination of parameters can be found which will unequivocally distinguish one cell type from another; the multidimensional distribution of the set of parameters will be **multimodal**, showing **discrete clusters** representing the individual cell types. Multimodal distributions may sometimes, but not always, also be obtained from **mixed populations of mature cells from a single organism**, as in differential leukocyte counting in normal peripheral blood. By definition, cells from a single organism have a common precursor in the fertilized ovum; what we term **differentiation** describes the process by which different genes are activated in different cells, with each cell type producing its own specific set of gene products which make it **biochemically and/or morphologically different** from other cell types.

Maturation Processes and "Missing Links": The "Ginger Root" Model

Discrimination among cell types with a common precursor is particularly difficult when samples come from the anatomic compartment in which cell differentiation occurs. In peripheral blood, for example, it is trivial to distinguish red cells, which lack nuclei, from leukocytes, which have them, e.g., by measuring light scatter and the fluorescence of a DNA stain in nuclei. It is only slightly harder, using the right parameters, e.g., forward and side scatter, to separate monocytes and granulocytes, which have common precursors well beyond the primitive stem cell stage. In the marrow, enucleated red cells could clearly be told apart from their late nucleated precursors by a combination of hemoglobin absorption and nuclear fluorescence measurements, but the earliest red cell precursors, containing little hemoglobin and not much more of other characteristic gene products, could not easily be distinguished from primitive cells of the leukocytic series. Similarly, given the presence in the marrow of their common precursors, we have no parameters that can define separate clusters of monocytes and granulocytes. Finally, in the peripheral blood, although we can tell erythrocytes and granulocytes from other cell types, the vagaries of measurement make us unable, **using any parameter**, to make unequivocal distinctions between reticulocytes, which contain detectable RNA, and mature erythrocytes, which don't, or between stab (band) and segmented polymorphonuclear cells, where an arbitrary nuclear morphologic criterion is normally used.

This problem arises whenever we observe **continuous processes of maturation or transformation**. We conceptualized mature leukocytes into discrete categories, with some justification; we also conceptualized the process of differentiation into discrete stages, which was much more of an idealization. There aren't transitional stages between eosinophil and basophil granulocytes; there have to be transitional cells between metamyelocytes and mature granulocytes because no cell division occurs past the metamyelocyte stage. No matter which parameters we examine, continuous processes give us continuous distributions. If we do multiparameter flow cytometry of leukocytes in normal peripheral blood, we see clusters; if we look at marrow, we get something which I once described as resembling a **ginger root**, incidentally prompting Mike Loken to find and demonstrate a ginger root of suitable morphology to illustrate the point. Another appropriate analogy might be to a hand and fingers; the peripheral blood shows us five disconnected fingerprints, while the marrow yields a handprint, in which the fingertips are seen to be connected to the palm.

A similar problem can exist in numerical taxonomy of species, which has also relied heavily on multivariate statistics. If one looks only at modern forms, one is likely to find distributions of measurement parameters containing discrete clusters representing individual species, even when those species have a common ancestor. If the analysis is extended to include older and extinct species, connections between clusters appear; the distribution takes on a "ginger root" character.

The nodular appearance of ginger roots is what makes them particularly apt models for populations containing cells in different stages of development or species in different stages of evolution. A good example is given by the distributions obtained in **T cell subset analysis** of cells from thymus and peripheral blood, as illustrated in Figure 5-11 (p. 241). The relative numbers of helper/inducer and cytotoxic/suppressor T lymphocytes are determined using monoclonal antibodies directed against the CD4 antigen to identify the first of these cell types and antibodies directed against the CD8 antigen to identify the second[318,321-2]; both types also bear the CD3 antigen. Cell populations are illustrated in 3-dimensional "cloud" plots.

The plot on the right of Figure 5-11, representing cells taken from the blood, shows clear "antibody positive" and "antibody negative" populations; there are some dual positives, but few, if any, cells with intermediate fluorescence values. I claim that this pattern alone should tell us that **the CD4 and CD8 antigens emerge on cells during a developmental phase occurring outside the bloodstream.** Now how in blazes, you may ask, are we supposed to know that? Simple, I reply. It follows from:

Shapiro's Fourth Law of Flow Cytometry
(The Supermarket Theorem):
Most babies aren't born in supermarkets.

If you walk around a supermarket, you will see many women with children, some with infants and some with older children, and you will see many women without children, a few of whom look as if they are about to change their status at any moment. However, I certainly have never seen a baby born in the considerable time I have spent in supermarkets, and you probably haven't witnessed such a blessed event either. Since we both know where babies come from, we are forced to conclude that arrangements are usually made for them to be born in someplace other than a supermarket.

Now, to find out where T cells come from, we go not to Masters and Johnson, but to Reinherz and Schlossman[322], who, with others, have shown that the inducer T cells, which bear the CD4 but not the CD8 antigen, and the cytotoxic/suppressor T cells, which bear CD8 but not CD4, develop in the thymus, and that early and immature thymocytes bear neither the CD4 nor the CD8 antigen, while immature thymocytes bear both CD4 and CD8 and lose one or the other as they mature. If we look at a population of cells from the thymus, shown on the left side of Figure 5-11, we see a distribution in which, while there are areas of higher and lower cell density, there are continuous ranges of both CD4 and CD8 antigen expression. If it reminds you as much of a sliced ginger root as it does me, you're entitled to take a sushi break at this point.

The thymic cells with fluorescence values intermediate between those of the "positive" and "negative" clusters in the distribution from blood on the right side of the figure represent those cells that are acquiring and losing antigen. In the second edition of this book, not having either the data or the software needed to produce Figure 5-11, I constructed a simulated version of the distribution from 3-color fluorescence data published by Lanier, Allison, and Phillips[787] to illustrate the same point; I was relieved to find that it looked pretty much like the real distribution shown in the figure.

Since the processes of acquisition and loss of antigen during maturation do not occur instantaneously by quantum leaps, we would expect to find cells in all stages of maturation present in the tissue compartment (the thymus in this example) in which maturation occurs. Conversely, and more generally, if a sample of cells from a particular tissue compartment (the peripheral blood in this example) contains clearly distinguishable populations of cells bearing and lacking an antigen, and few cells which bear very little antigen, the acquisition and loss of the antigen in question must either occur outside the compartment from which the sample was taken, or be **rare events** in that compartment, like babies being born in a supermarket. The ginger root doesn't have the same density everywhere; it is thinner, or less dense, in the areas that represent rare events.

Figures 3-10 (p. 97) and 7-16 (p. 321), which depict DNA and RNA content in stimulated lymphocytes, illustrate a situation in which a rapid transition makes what should be a continuous distribution appear to contain discrete clusters. About 20 hours after exposure to mitogens, resting lymphocytes, which normally contain only small amounts of RNA, begin to make the larger amounts necessary for their subsequent multiplication. However, when you look at the distribution, there are a lot of resting (G_0 or G_{1Q}) cells, containing little RNA, and a lot of G_1 cells, containing noticeably more RNA, and few, if any, cells in between. Since we know that the transition from G_0 to G_1 takes place in the compartment (in this case, the culture vessel) from which the sample is taken, we have to conclude that cells don't take very long to gear up for RNA synthesis, or, in other words, that they don't stay in the transitional, or G_{1T}, state, for very long; if they did, we would expect to find a continuum of G_{1T} cells between the G_0 and G_1 clusters.

The evolutionary biologists had a similar problem for over a century after Darwin published his *Origin of Species.* When fossils were collected from sites in which different species had presumably evolved, the expected "missing links" between species were generally absent. This becomes readily understandable in terms of the theory of evolution in punctuated equilibria proposed in 1972 by Niles Eldredge and Stephen Jay Gould, according to which most evolutionary change occurs in brief bursts, resulting in the intermediary forms not being around very long. Finding one would, then, be expected to be a rare event.

In the absence of rare events generated by short-lived transitional states, the multidimensional distributions we get when we look at maturing cell populations in the environments in which they mature, no matter which parameters we look at, will contain continuously connected "ginger root" blobs rather than discrete clusters; unsupervised algorithms and expert systems can find clusters, but how you slice your ginger roots is a matter of taste. This can be restated as:

Shapiro's Fifth Law of Flow Cytometry
(The Barber Shop Theorem):
No man walks into a barber shop with a long beard who hasn't had a shorter one, and no barber can make a beard a lot shorter without cutting it.

Hair doesn't grow in quantum leaps, and as is the case with the ginger root, how you cut it is also a matter of taste.

The Barber Shop Theorem has some other practical applications in hematology, for example, in distinguishing between immature reticulocytes and mature red blood cells. Reticulocytes are identified by their content of RNA, which is lost within a day or so of the time these cells enter the blood from the marrow; the distribution of RNA in red cells is therefore continuous. To duplicate the reticulocyte count as performed by a human observer, you have to find some arbitrary point in the distribution at which to apply a razor.

As neutrophil granulocytes mature, they, like erythrocytes, have less and less use for their nuclei, but instead of turning the poor old things out in the cold, they let them have the spare room, in which the nuclei become progressively gnarled with age. The nuclei of mature granulocytes have several lobes connected by threads of chromatin; less mature "band" cells have nuclei that are twisted around in the fashion of crullers, but not really lobulated. Why and how do you tell the difference between the immature "bands" and the mature cells with segmented nuclei ("segs")? To answer the "why" first, finding an increase in the proportion of immature cells indicates that the bone marrow is responding to stress due to inflammation, infection, etc., so knowing the numbers and fractions of bands and segs is helpful in diagnosis. How? To coin a paraphrase, I don't know much about hematology, but I know what I like. Well, it's actually a little more precise than art criticism, but, again, it's an arbitrary distinction at its base because the process of nuclear involution is a continuous one. Nuclear morphology being kind of tough to assess in the average flow cytometer, various people have looked for various flow cytometric methods of telling bands from segs, because you need to be able to do it if you want to sell flow cytometric differential counters to a lot of clinical laboratories. My former colleagues and I ended up using the size distribution of the neutrophils; the immature cells are larger, and what you end up doing is applying the razor to a continuous distribution. Nonetheless, by cutting in the right place, we could get good band/seg ratios, just as one can and does get reasonable reticulocyte counts from RNA content distributions.

Back in 1980 or so, when everybody in the diagnostics business was mesmerized by monoclonals, the folks at Ortho decided to try their hand at distinguishing reticulocytes from mature erythrocytes[324,764], and at differentiating between bands and segs[325], using monoclonal antibodies. Did it work? Yes and no. Yes, you could produce distributions of antibody fluorescence which you could slice as other distributions are sliced to get a retic count or a band/seg ratio. And no, Virginia, they didn't find a surface antigen present on reticulocytes and absent on mature erythrocytes, or one present on bands and absent on segs, and I doubt that anyone else ever will.

When monoclonal antibodies were still a new technology, a lot of people had the idea that there would turn out to be a single surface antigen to identify each different cell type in the body. Experience has taught us that this is probably not the case. Surface antigens are not Nature's version of bar code; many of them have defined receptor functions and we can expect some function to be defined for most of the rest, because it's unlikely that structures that seem to be fairly well conserved during evolution serve a purely decorative purpose. This brings me to:

Shapiro's Sixth Law of Flow Cytometry:
There are some cell identification problems that even monoclonal antibodies can't solve.

Shave, if you must, but try to avoid splitting hairs.

Practical Multiparameter Gating: Color Wars

The most sophisticated current applications of flow cytometry involve the definition of clusters, or of nodes of ginger roots, in multidimensional spaces. This is often accomplished by defining a bitmap gate on the two-parameter display in which a particular cluster or node is best separated from other clusters or nodes. A color is assigned to each gate, making it possible to find the location of cells in any cluster on any two-parameter display by plotting cells falling within the gate in the assigned color. The procedure is illustrated in the color figures on the back cover showing normal and leukemic bone marrow; these were provided for the previous edition of this book by Leon Terstappen, then at B-D. The figures were was generated using B-D's Paint-a-Gate™ software; other manufacturers and third party developers have analogous programs.

In the late 1980's and early 1990's, the application of multiparameter flow cytometry to determining patterns of differentiation of myeloid leukocytes in normal and leukemic human bone marrow[1154,1212,1498,1760-71] represented the highest level of sophistication of the technology.

The multiparameter approach to cell differentiation in marrow was pioneered by Loken and Civin[990,1755-9] and their collaborators, who originally studied erythroid and B cell development, bringing B-D's resources in the areas of instruments, reagents, and software to bear on the problem. The work on phenotyping of myeloid leukemia, continued by Terstappen and his colleagues, epitomized the intelligent application of sophisticated cytometric methods to a clinical problem.

The figure depicting normal adult bone marrow shows the location of early leukocyte precursors, identifiable by their CD34 surface antigen, in a two dimensional display of forward (FLS) vs. polynomially transformed[1247] (ICLS) side scatter values, and in two-parameter displays showing expression of CD34, CD11b, and CD15. This pattern is consistently observed in normal marrow, which shows surprisingly little variation from individual to individual. The trajectories along which normal cells move as they mature, indicated by arrows on Panels A and C, were determined by examining cells sorted from various regions of the displays.

The analogous displays depicting bone marrow cells from a patient with acute myelocytic leukemia (AML) are

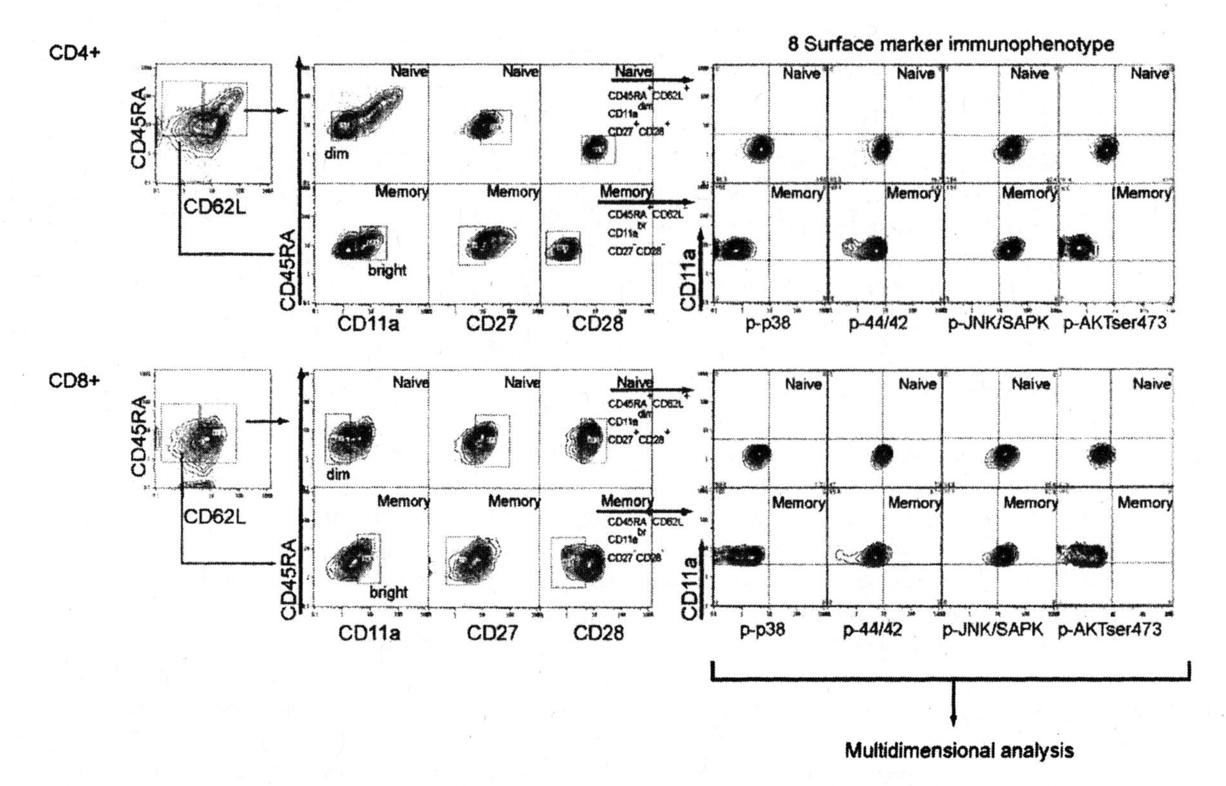

Figure 10-9. Subsetting of human T-cells into truly naive and memory classes and measurement of activated signaling kinases using 11-color fluorescence flow cytometry. Resting T-cells were prepared by density gradient centrifugation with other cells removed by adherence and magnetic separation, with purity assessed by measurement of forward and side scatter and CD3 antibody binding. From: Perez OD, Nolan GP: Nature Biotechnology 20:155-162, 2002 (reference 3000), © 2002 Nature Publishing Group; used by permission.

quite different. They are characterized by the appearance of leukemic cells in regions in which few or no cells appear in displays from normal marrow. As a group, however, leukemic patients are extremely heterogeneous; there are always cells where there shouldn't be cells, but no two leukemic patients show exactly the same pattern.

Interestingly enough, when my colleagues at Block and I[989] ran samples from patients with acute leukemia through the prototype differential counter system, without benefit of monoclonal antibodies, we also noted that leukemic cells showed up where there weren't normal cells, and that, while normal cells from different people showed up in pretty much the same places on displays, no two leukemic bloods looked alike. It may be more important to have a multiparameter measurement space, which we did have, than to have monoclonal antibodies or equivalently specific reagents. As I recall, some people from Technicon presented some data years ago on analyses of cells from patients with acute myelocytic leukemia using the peroxidase and esterase channels in their Hemalog D blood cell counter, and they found pretty much the same patterns we did, i.e., normals looked pretty much alike, and leukemias were different, and varied.

Verwer and Terstappen[1212] developed a computer procedure for automatic assessment of lineages of leukemias and Frankel et al[1215,1217] applied neural network methods to analysis of normal and leukemic marrow. Manual and automated multiparameter analysis make it possible to detect very small numbers of leukemic cells (between 1 in 10^3 and 1 in 10^6 cells) in marrow following the induction of clinical remission by combination chemotherapy, which falls under the rubric of **rare event analysis**, whether you are printing in color or black and white. I'll continue the story of leukemia phenotyping later on.

These days, complex gating strategies are more dependent on the number of colors you can measure than on the number of colors you can display. The state of the art in 1994, when the previous edition appeared, was 5-color fluorescence measurement[1306]; in the years since, colors have been added at the rate of about one a year, largely through the efforts of a large and creative group of people in the Herzenberg lab at Stanford[2497,2629,2644,2995-3000-3].

I can't think of a better illustration of the current (published) state of the art – 11 fluorescence colors plus forward and side scatter – than Figure 10-9, taken from a 2002 paper by Perez and Nolan[3000], for which the data were collected in the Herzenberg lab. The paper described the measurement of activated signaling kinases in human T cells separated by multiparameter gating into sub-subsets of truly naive ($CD45RA^+CD62L^+CD11a^{dim}CD27^+CD28^+$) and memory ($CD45RA^-CD62L^-CD11a^{bright}CD27^-CD28^-$) cells[2995-8,3000-3]. The input cell population of resting T cells was prepared and purified from peripheral blood by Ficoll-paque density gradient separation, followed by depletion of adherent cells and immunomagnetic separation using antibodies to CD14, CD16, CD19, CD44, and HLA-DR to remove macrophages, NK cells, B cells, and activated T cells. Cells were then stained with antibodies to CD4 (PE-Cy5.5), CD8 (Cascade Yellow), CD45RA (APC-Cy7), CD62L (PE-Cy7), CD11a (Cascade Blue), CD27 (Alexa 594), and CD28 (APC-Cy5.5), and with antibodies to phopshorylated or nonphosphorylated forms of one or more kinases. Antibodies to kinases were also, on occasion, labeled with fluorescein, PE, and APC, and, on occasion, with Alexa 488, Alexa 546, Alexa 568, Alexa 594, Alexa 633, Alexa 660, or Alexa 680. Discrimination between naive and memory cells reveals some differences in kinase activity between resting naive and memory cells; additional differences are noted when cells in different subsets are stimulated by kinase-activating agents (see the paper[3000] for details).

Finding Rare Cells

Flow cytometry has been applied, with varying degrees of success, to finding cell types that are present in samples at very low frequencies. How effectively this can be done is a function of how readily these rare events can be identified. I have already touched on some qualitative and quantitative aspects of **rare event analysis** and rare event sorting on pp. 19-20, 42, 178-80, 269, and 352.

If you look at rare event detection as basically a statistical problem, what you need to do is distinguish between samples from two distributions with different means. Whether or not such a distinction is possible, however, depends on the variances of the distributions as well as on the means.

One Parameter is Not Enough

Suppose, for example, you are trying to use a single-parameter immunofluorescence measurement to find rare positive cells among a population of negative cells. Let's say the positive cells have a mean fluorescence signal at channel 40, with a 30 percent coefficient of variation (CV), while the negative cells have a mean fluorescence signal at channel 20, also with a 30 percent CV; disregarding reality for now, we will assume linear scales and normal or Gaussian distributions.

The CV is equal to the standard deviation (S.D.) divided by the mean; S.D. for the positive cells is therefore 12 channels, and S.D. for the negative cells is 6 channels. Based on tables of the normal distribution and its integral (p. 234), about two/thirds of the positive cells will be found between channels 28 and 52, with half of them lying below channel 40, while almost 10 percent of the negative cells will be found between channels 28 and 52, although only about 13 negative cells in 10,000 should lie above channel 38. These numbers make it unlikely that positive cells present in a ratio of less than 1 per 1,000 negative cells can reliably be detected by single parameter measurements.

Now suppose we're looking at DNA content, and we want to find rare abnormal hypodiploid tumor cells (mean fluorescence at channel 20) among normal diploid cells (mean fluorescence channel 40). Thus, we're dealing with the same difference between population means we had in the last case. Let's use a realistic CV of 5 percent for the DNA measurements, and neglect S, G_2, and M phase cells for the moment. 99.8 percent of the abnormal cells will lie between channels 17 and 23; 99.9999 percent of the normal cells will lie above channel 30. Just going by the numbers, we could easily detect one hypodiploid cell per 10^6 normal cells.

In case you haven't noticed, though, I threw in a ringer in the DNA content example. In the real world, we are probably more likely go looking for tetraploid than for hypodiploid abnormal cells. So let's say we've got the normals at channel 20 and the abnormals at channel 40. Now, even if there are no S, G_2, or M phase cells in either population, if we only look at DNA content, without some way of telling doublets from single cells, we're going to get a substantial number of counts in the neighborhood of channel 40 resulting from two normal cells passing through the system together, either by virtue of being physically attached or by virtue of being in very close proximity as they go by the observation point. We can use some established tricks, such as comparing peak vs. integral of fluorescence signal or looking at scatter or extinction pulse width vs. fluorescence, to discriminate single abnormal cells from doublets, but, unless we go to a 3-D slit scanning system and/or sort all of our suspicious cells onto slides, we are very unlikely to be able to pick out one abnormal cell in a million normals.

Moving closer to the real world, and keeping the same 5% CV's, we'd think we'd be better off with abnormals at channel 30 and normals at channel 40 than with either abnormals at channel 40 and normals at channel 30, or abnormals at channel 40 and normals at channel 20. We already know that the last of these scenarios gets us into problems with doublets; the middle one will be troublesome if there are any normal S phase cells, because some of these can be expected to lie at channel 40 and, if the fraction of abnormal cells is too small to form a recognizable peak, we have no way to spot the abnormals. In theory, we may have doublets and S phase cells, but we shouldn't have 3/4 of a cell or nucleus going through, so we should readily spot the abnormals at channel 30 if the normals are at channel 40. We should, as long as there is no debris in the sample; in the real world, particularly when you go hunting for DNA aneuploidy in preps from solid tumors, you may have trouble finding nuclei but you can have all the debris you don't want. Well, fine, suppose you use a mathematical model to subtract out the debris distribution? No, that won't work either, because the model is only an approximation, and you're still going to need a substantial fraction – say 5 percent or more – of abnormal cells to be sure of finding them.

So far, then, having considered detection of rare cells using single-parameter measurements, we have discovered that we can tell from the numbers when there's no chance of our finding rare cells, but, also, that, when the numbers tell us there is a chance, factors other than statistics may prevent us from finding the rare cells. Does rare cell detection get any easier when we use multiparameter measurements? Would I ask if it didn't?

If we go back to the leukocyte differential counting procedure discussed on pp. 248-50, it becomes fairly obvious that multiparameter analysis improves things. That procedure started with a data file of 1595 leukocytes, obtained by gated analysis of whole blood, without red cell lysis, using nuclear fluorescence as the gating parameter. So, to count 1595 leukocytes, we also had to count about 1.5 million red cells and a few hundred thousand platelets, and we managed to pick out the leukocytes, which represented only about 0.1 percent of the total population analyzed, almost flawlessly (we decided after further analysis, that 6 of our 1595 objects weren't leukocytes after all). But it gets better. In the first step of breaking the leukocyte population into subpopulations, we manage to pick out the eosinophils, one of the less common leukocyte types; there were 102 of them, roughly 6 percent of the leukocytes and thus roughly 6 per 10^5 of the original cell population.

Ah, but that's not a fair test, you say, because the eosinophil stain was very bright and very specific. Objection sustained, but I'm ready for you. Let's look at the basophils, the rarest of the leukocytes. They were not picked out by using a specific stain for their IgE receptors or their glycosaminoglycans; instead, they were identified because the cluster in which they fell based on nuclear fluorescence, forward and right angle scatter, and extinction measurements lay far enough away from the other cell clusters to let us find them by blind luck. The hardest part of our job in identifying the basophils when the differential counting procedure was developed lay in finding a sample with 5 percent basophils, so we could be sure that the cluster was what we thought it was. Among 1595 (or 1589, if you really want to get picky) leukocytes, we identified 11 basophils, or about 7 cells per 10^6 cells in the original population. That suggested to me that multiparameter analysis, even using parameters that didn't seem to be all that specific for discrimination, was the most practical way to go when looking for rare events.

Cocktail Staining Can Help

This approach (p. 352) is often useful in rare cell detection problems involving immunofluorescence measurements. Ryan et al[883] considered the problem of detecting minimal residual disease in acute lymphoblastic leukemia based on finding and counting cells bearing the common acute lymphoblastic leukemia antigen (CALLA or CD10). They stained cells with a fluorescein-labeled anti-CALLA antibody combined with a "cocktail" of various phycoerythrin-labeled antibodies binding to different types of mature leukocytes, and performed analyses gated on forward and orthogonal scatter and pyhcoerythrin fluorescence to include only those mononuclear cells not stained by the phycoerythrin-labeled antibodies. This scheme allowed detection of CALLA-positive lymphoblasts at levels of 1 cell per 10^5 peripheral blood mononuclear cells, representing an improvement in sensitivity of better than a hundredfold over single-parameter immunofluorescence measurements.

Dirt, Noise, and Rare Event Detection

Gross et al[1150,2331] were able to reliably detect rare cells at frequencies as low as 1 in 10^6 by combining cocktail staining with a rigorous cleaning procedure for the flow system, which minimized sample carryover (see pp. 178-80), and a processing technique which excluded "bursts" of data likely to represent system noise and or debris, rather than cells, in the sample. Rare cell detection at this level is likely to be necessary in identifying fetal cells in maternal blood, or residual solid tumor cells, e.g., breast cancer cells, in bone marrow.

Really, Really Rare Events: Alternatives to Flow

When we start getting down to cells which represent 1 in 10^7 cells in the sample, we run across a new problem, which is fundamental to the flow cytometric method. If we count 10,000 cells/s, which is faster than a lot of us run routine analyses, we can run for an hour and find four of the cells we're looking for. To really be sure we're seeing them, we might want to count a few dozen, and that's a day's work. Unless we're talking about red cell variants, that's also a large total number of cells to have to take out of a patient.

Provided we have a way of staining the cells we're looking for intensely enough so that it's easy to find them, we may be better off using a static system like a wide-field microscope, and just counting the positive cells, rather than tying up a flow cytometer for days on end. This is certainly true when we're trying to do something else with very rare cells when we do find them, e.g., genetic analysis by *in situ* hybridization; using modern imaging and static systems, with computer-controlled stages and scanners, it is relatively simple, once a cell has been located, to find it again after the slide has been removed, processed, and replaced.

The **Automated Cellular Imaging System (ACIS)**[3004-5], from **Chromavision**, uses immunoenzyme staining and automated analysis of bright field microscope images to identify malignant cells in marrow at frequencies as low as 1 cell in 100 million, although an immunomagnetic enrichment step may be necessary to reach this detection level.

The **CellTracks™** [2383,3006-9], under development by **Immunicon**, features a small but capable scanning laser cytometer, with optical and mechanical components largely derived from CD player parts, that uses as many as three laser beams for measurement of fluorescently and immmunomagnetically labeled cells drawn by a magnetic field to the viewing surface of a disposable sample container. Immunicon has developed other immunomagnetic separation technology that has been used to enrich samples for residual tumor cells and other rare cell types prior to analysis using either flow or scanning cytometry[3010-3].

Oncosis's Photosis™ device[3014] is designed to scan a 250 cm^2 chamber containing as many as 500 million cells in a clinical specimen intended for stem cell transplantation, identify any non-Hodgkins lymphoma cells present, and destroy the tumor cells by "zapping" them with a pulsed laser beam. **Cyntellect**, a spinoff from Oncosis, offers the **LEAP**™ [3015] research platform for high throughput cell image analysis and manipulation.

10.4 TRICKS AND TWISTS: ODD JOBS FOR FLOW CYTOMETRY

Single Molecule Detection

The detection of single fluorescent molecules bearing multiple chromophores was proposed, and achieved using a static system, by Tomas Hirschfeld in the mid-1970's[884-5], when he was at Block Engineering. Mike Hercher, also at Block, then built a high-sensitivity flow cytometer[94] designed to perform assays for viral antigen based on the detection of small numbers of multiply labeled antibody molecules bound to a single virus particle. This was the apparatus used to measure the scatter signal distribution from bacteriophages shown in Figure 7-5 (p. 288); its fluorescence sensitivity was calculated to be sufficient for detection of 36 molecules of fluorescein, a number I hesitated to put in print until other people got into the same ballpark. When the original work with this apparatus was done, neither monoclonal antibodies nor phycobiliprotein labels were available; the use of both of these types of reagents would obviously make single-virus detection based on this approach easier, although factors other than instrument sensitivity influence feasibility.

Work on flow cytometric detection of small numbers of molecules continued at Los Alamos, where Dovichi et al[886-7] achieved sensitivities comparable to that of the Block apparatus[94] for detection of fluorescein and rhodamine dyes, and at Berkeley and Stanford, where Mathies and Stryer[888] established the feasibility of detecting single molecules of phycoerythrin. Such detection was actually achieved in 1987 by Nguyen et al at Los Alamos[660], using a slow flow system with a single photon counting detector.

While single molecule detection in flow systems represents an impressive achievement, flow cytometry is not the only technology, and is not necessarily either the most sensitive or the most efficient technology, for making fluorescence measurements at this level of sensitivity. Before the First Edition of this book was published, Coleman et al[260] had visualized DNA in individual DAPI-stained virions (and, I was told, in plasmids) by fluorescence microscopy and quantified it by microphotometry, using a photometer based on a Zeiss fluorescence microscope with a 100 W Hg arc lamp source. The single molecule detection limit for DAPI-stained DNA had also been attained by Morikawa and Yanagida[576] using video microscopy. Later, in preliminary experiments (A. Coleman and M. Block, personal communication), Coleman's microphotometer was used to detect bacteriophages based on the spatial coincidence of fluorescence signals from stained DNA and fluorescent antibodies.

Since, in many potential applications of high-sensitivity cytometry, e.g., virus detection in clinical specimens, the objects to be detected are not only of near-molecular dimensions but may be present at very low concentrations (< 10^6 particles/ml), detection in a flow system involves spending a lot of time running portions of the sample through the system which do not contain the objects of interest. In such cases, it is almost essential to concentrate the raw sample to increase throughput; it becomes logical to concentrate it to the point at which the analyte can be looked for on a slide in a static system. However, maximizing detection sensitivity in static or scanning systems requires that the illuminated region be made as small as possible.

Near-field optical microscopes, in which optical fibers of extremely small size (10 nm) are used to bring light to the specimen, allow selective illumination of regions approaching molecular dimensions, permitting not only the detection of single fluorescent molecules, but their characterization, e.g., by fluorescence lifetime. The Los Alamos group investigated this technology[1819]. However, in recent years, impressive spatial resolution in fluorescence microscopy has been achieved using conventional (far-field) optics, if not conventional illumination.

Stimulated Emission Depletion (STED) microscopy, described in a 2002 report by Dyba and Hell[3016], decreases the size of a detected fluorescent spot by quenching the excited molecules at the rim. Quenching is accomplished by stimulated emission. Two synchronized trains of laser pulses are used; the first pulse excites the fluorophores in the focal region, and a following red-shifted pulse, in a "donut" mode, quenches molecules at the rim of the focal spot, leaving fluorescence from the center largely unaffected. Using 760 nm excitation, Dyba and Hell achieved an effective spatial resolution of 33 nm, or 1/23 wavelength; they have been able to apply their technique to high-resolution imaging of structures in living cells and bacteria.

Single molecule detection has also been achieved, with less fanfare, in lower-tech, lower-budget apparatus. On p. 132, I noted that Unger et al[2446] were able to detect fluorescence of single molecules of tetramethylrhodamine-conjugated protein using a (relatively inexpensive) cooled CCD camera and a fluorescence microscope illuminated by a 100 W Hg arc lamp. Chiu et al[3017] used the same apparatus to develop a calibration procedure for quantitative measurements of GFP down to the single molecule level.

DNA Sizing, if not Sequencing, in Flow

In 1993, Goodwin et al, at Los Alamos, reported the use of a slow flow system, TOTO-1, and single photon counting to size DNA fragments as small as 5 kilobase pairs.

The Los Alamos group proposed the use of their methodology for **DNA sequencing**; a DNA fragment would be attached to a bead fixed in position upstream, and single molecules of the fluorescently labeled end base would detected downstream after enzymatic cleavage. This hasn't happened yet, but the DNA sizing work has continued.

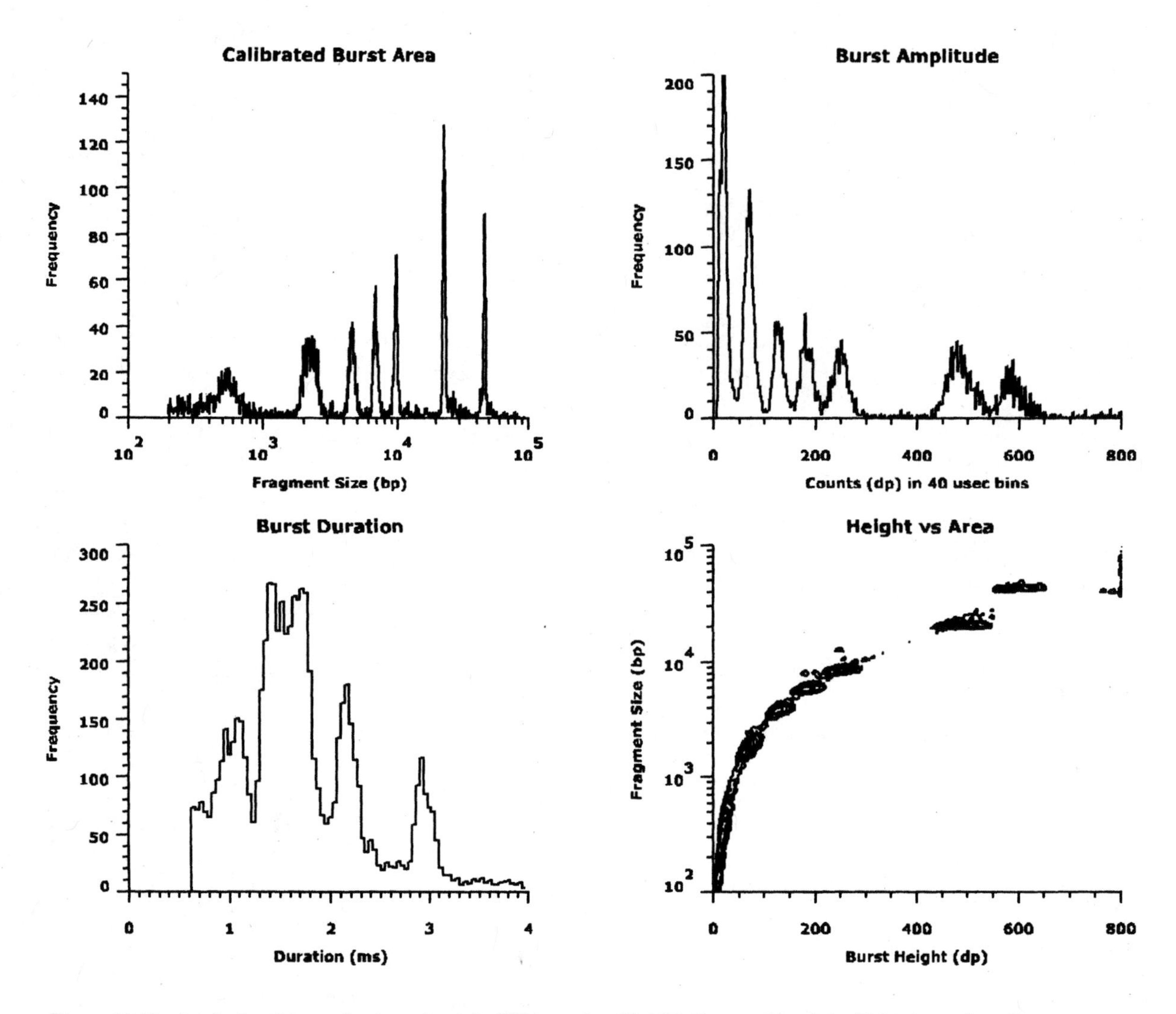

Figure 10-10. Analysis of bacteriophage lambda DNA and a *Hind* III digest of lambda DNA in a slow flow system. Courtesy of Rob Habbersett, Los Alamos National Laboratory.

Rob Habbersett, at Los Alamos, has reduced the DNA sizing apparatus to a small, benchtop system using 0.5 mW of 532 nm illumination from a YAG laser; samples flowing at 1 cm/sec are examined in a quartz cuvette with a 250 μm^2 flow channel, and fluorescence at 555-595 nm from SYTOX Orange (Molecular Probes)[2334] is collected using a 3.1 mm, 0.68 N.A. aspheric lens. A Perkin-Elmer single photon counting APD module is used as the detector, and data are processed by a multichannel scaling PC plug-in board that counts detector pulses in intervals ranging from 4 to 50 µs.

Figure 10-10 illustrates the result of analyzing a mix of DNA fragments from a restriction digest of bacteriophage lambda DNA, with a small amount of intact lambda DNA added. A total of 7800 individual fragments are represented. *Hind* III digestion produces 8 fragments from each completely cut lambda DNA molecule, with fragment lengths of 125, 564, 2027, 2322, 4361, 6557, 9416, and 23130 base pairs (bp). The 125 bp fragment was not detected in this example, and the two fragments at 2027 and 2322 bp (2nd peak from the left in the burst area histogram) are not separately resolved. Each individual detected event (i.e., fragment) comprises a "burst" of detected photons; the data processing software extracts the area, maximum amplitude, and duration of each burst, and stores them as list-mode parameters, from which all displays are generated. The burst-area scale was calibrated in terms of fragment length by fitting a Gaussian to each peak to find the center of the peak and performing a linear regression against the known fragment lengths.

While burst area is proportional to fragment size, burst amplitude, analogous to peak height, is not. The contour plot of amplitude vs. fragment length shows that, as fragment length increases, both amplitude and area increase until the fragment length is about 36 kbp. In the flow cell, the hydrodynamic forces extend the fragments as they accelerate off the end of the injector tip, and pass through the laser beam, which is focused to a spot approximately 11 µm in diameter. A 36 kbp fragment is long enough to fill the entire spot, placing a nominal maximum number of dye molecules in the probe volume and therefore [nearly] maximizing burst amplitude. Larger molecules extend to longer lengths (a fully extended 48.5 kbp lambda DNA molecule, for example, is roughly 16 μm long), and, while they produce only slightly higher burst amplitudes, they have substantially longer burst durations and larger burst areas. Circular DNA constructs (e.g. plasmids) or molecules that are kinked, looped and not fully extended in flow can produce higher amplitude bursts.

DNA fragment sizing apparatus has been built at the University of Twente[2440,2452] and at CalTech (a microfluidic system)[2326] as well as at Los Alamos. For a time, it appeared as if flow cytometric DNA fragment sizing would be competitive with agarose gel electrophoresis; the flow methods is faster, uses a smaller sample, and produces results on a linear scale. The picture has become somewhat clouded with the appearance of microfluidic capillary electrophoresis apparatus for DNA sizing, which also has speed and sample size advantages when compared to conventional gel electrophoresis. Of late, the Los Alamos group has advocated the use of DNA fragment sizing for rapid identification of bacterial pathogens[3018-20]; this is likely to attract more attention after the U.S. national experience with anthrax in 2001 than it might have before.

Solid Phase (Bead) Assays Using Flow Cytometry

By the 1980's, people at Los Alamos and elsewhere had become aware that flow cytometry might offer some advantages as a detection method for fluorescence immunoassays. Saunders, Martin, and Jett[1019] described a competitive binding assay in which fluorescently tagged antigen competed with an antigen analyte for binding sites on nonfluorescent, antibody-coated beads. Elsewhere, McHugh et al[1820] developed a sandwich assay for circulating immune complexes, in which the analyte was bound to beads coated with the complement component C1q, and rendered detectable by the addition of fluorescent anti-immunoglobulin antibodies.

Saunders et al[1834] developed a solid phase flow cytometric assay for the DNA-binding antibiotic actinomycin D, using mithramycin, which becomes fluorescent on binding to DNA, to compete with the analyte, which does not, for sites on DNA-coated spheres.

Flow cytometers can also be used to detect antigen-antibody reactions without making fluorescence measurements, for example, by using scatter or extinction signals to detect aggregation of small antigen-coated particles by soluble antibody. Sykulev, Cohen, and Eisen[1835-6] adapted this technique to determine the equilibrium constants of antigen-antibody reactions by examining the effects of addition of competing soluble antigen; they reported that this technique is usable with almost any soluble antigen and with antibodies exhibiting a very wide range of equilibrium constants.

By the time the previous edition of this book appeared, variations on the basic theme included assays for serum antibodies, employing antigen-coated beads and fluorescent anti-immunoglobulins[1821-6], sandwich assays for antigens, using antibody-coated beads and fluorescent antibodies[1827-30], and the use of small (0.1 μm), antibody-coated fluorescent beads instead of fluorescent antibodies to detect antigen bound to nonfluorescent beads coated with unlabeled antibody[1831].

Flow cytometric bead immunoassays are said to be equal to or better than conventional enzyme immunoassays in terms of sensitivity, specificity, and dynamic range; their major drawback to date is the requirement of a relatively expensive instrument for readout. Lindmo et al[1832] reported that the dynamic range of sandwich flow cytometric assays for antigens could be extended (from 2 decades to 3 decades) by employing two different sizes of particles coated with antibodies of substantially different affinities; Frengen et al[1833] considered ways of minimizing serum interference with flow cytometric solid phase assays.

Cocktails for 100: Multiplexed Bead Assays

Flow cytometric bead assays have now made it into the mainstream thanks to use of multiplexing schemes that permit separate analyses of a large number of ligands to be performed in a single aliquot of sample.[2370-5,3021].

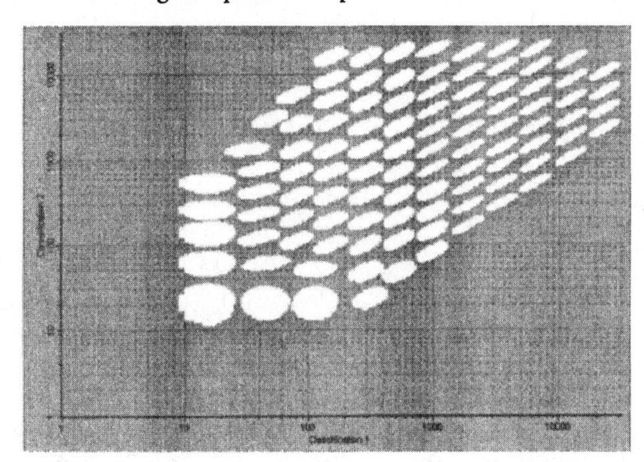

Figure 10-11. Positions of clusters representing 100 different color-coded beads used with the Luminex 100™ system for chemical analysis. The horizontal and vertical axes, respectively, indicate intensities of 635→658 and 635→>710 nm fluorescence. Courtesy of Luminex Corporation.

The Model 100 flow cytometer made by **Luminex** (Chapter 8) can perform as many as 100 ligand-binding assays per tube. Beads approximately 5.5 μm in diameter are color-coded with unique mixtures of two red-excited dyes; because large numbers of molecules of these dyes can be bound to beads, they provide strong signals, and fall into well-delineated clusters on the "bead map" illustrated in Figure 10-11 above.

A different color code is used for each analyte to be assayed; analyte is captured by an appropriate reagent (antibody, gene probe, etc.) bound to the bead, and detected by a "reporter" reagent labeled with phycoerythrin or another green-excited, yellow fluorescent label such as Cy3.

A mixture of beads appropriate for the desired measurements is mixed with a single aliquot of sample in a tube or in one well of a 96-well plate; the sample-bead mixture is then incubated with a mixture of all of the reporter reagents, and introduced into the flow cytometer, usually after a wash step. Each bead passes first through a red diode laser beam, in which measurements of side scatter and of the fluorescences (658 nm and >710 nm) of the color coding dyes are made, and then through a 532 nm green YAG laser beam, in which the fluorescence of the reporter reagent is measured. Side scatter signals are used to eliminate bead doublets and triplets from analysis; single beads are then classified by their color-coding, and distributions of reporter fluorescence associated with each bead type are analyzed to quantify the amount of each analyte of interest present in the sample.

Before producing the Model 100, Luminex made a series of beads that could be used for multiplex assays on a B-D FACScan; red- and orange-fluorescent dyes were used for color coding, and fluorescein or another dye with a similar spectrum was used as a reporter label. A PC-based data analysis system, attached to the FACScan, was used for calculating assay results. A number of publications describing multiplex analysis using Luminex beads have appeared in the literature[2370-5,3021].

Other companies, including **Bangs Laboratories** and **BD BioSciences**, have produced bead sets and assays using smaller numbers of beads, color coded by different intensities of a single fluorescent dye. There is a recent report of the use of the BD Biosciences kit for determining cytokines in tears[3022], and, no, I don't mean that the cytokines were emotionally upset.

Other coding schemes are possible; **Quantum Dot Corporation** has made beads tagged with mixtures of quantum dots of differing emission wavelengths[2648-9]; **3D Molecular Sciences** has produced plastic particless approximately 100 µm × 20 µm × 15 µm with "bar coding" based on width variations along the length of the particle, detectable by pulse shape analysis of a scatter or extinction pulse[3023]. These beads could presumably be used in a relatively inexpensive single-laser instrument.

Cells in Gel Microdroplets and on Microspheres

James Weaver and his colleagues at the Massachusetts Institute of Technology[1837-43] and, somewhat later, Eli Sahar and Raphael Nir and others at Tel Aviv University[1844-8] explored applications of the **gel microdroplet technique** originally described by Weaver et al[1837], in which single or multiple bacteria or eukaryotic cells are observed visually or cytometrically after encapsulation in agarose gel beads from 10-100 µm in diameter.

The method was originally envisioned as facilitating flow cytometric analyses of microorganisms, which, because of their small size, produce scatter signals which are difficult to detect in modern flow cytometers and which were impossible to detect in many older instruments. Since the nucleic acid content of microorganisms is approximately 1/1000 that of eukaryotic cells, fluorescence signals from microorganisms stained with propidium, DAPI, and other dyes which produce very strong signals from eukaryotic cells are also relatively weak; immunofluorescence signals would be correspondingly weaker. This made it infeasible to use fluorescence signals as trigger signals for flow cytometric analysis of microorganisms; however, the scatter signal from a 10 µm gel microbead could provide a robust trigger signal, allowing a fluorescence measurement to determine whether an organism (s) was(were) contained in the bead.

This rationale for the use of gel microdroplet technique has largely disappeared due to increases in the sensitivity of commercial flow cytometers; it is possible to quantify bacterial growth by direct analysis of cultures from liquid media, without the additional step of encapsulating the organisms in microdroplets. However, the method is potentially useful for other applications. When single organisms are encapsulated in microdroplets and allowed to grow, the entire clone produced by successive divisions remains in a single droplet (Figure 10-12), facilitating selection of colonies[1842,1845-8].

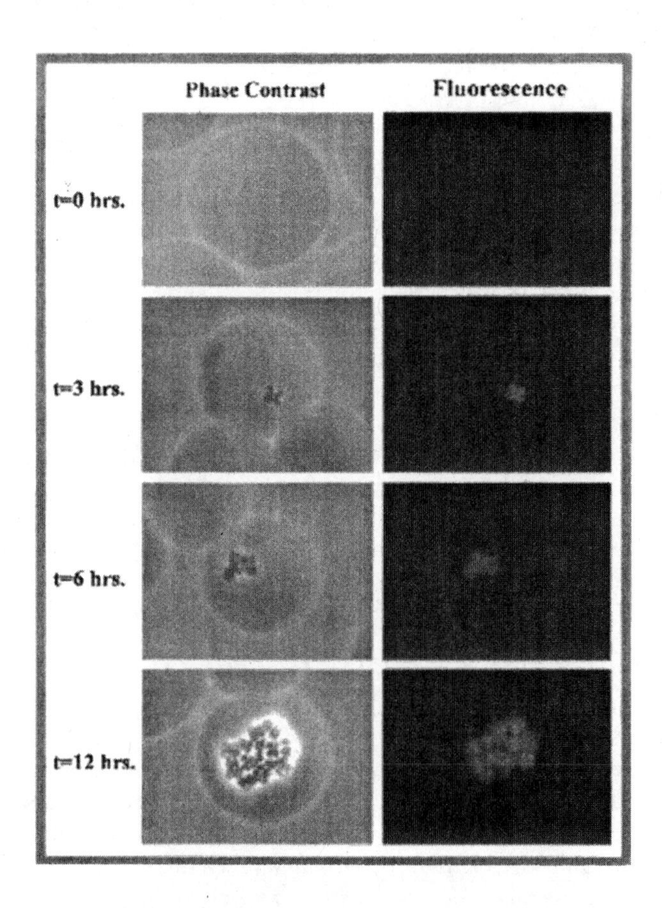

Figure 10-12. Growth of an encapsulated Gram-positive marine bacterium in gel microdroplets. Courtesy of Pat McGrath and Swee Kim Lin (One Cell Systems).

Such selection can be done on the basis of growth characteristics or on the basis of detection of metabolic products, which accumulate in the microdroplet[1839]. When applied to antibody-producing cells, the microdroplet technique represents a further diminution of scale over the microtiter plate[1840].

The production of microdroplets is relatively straightforward; cells in medium are added to an agarose mixture, which is added dropwise to dimethylpolysiloxane oil, in which it forms an emulsion. A kit[1843] is available from **One Cell Systems** (Cambridge, MA); McFarland and Durack have provided recipes "for the frugal investigator"[3024]. Cells can readily be released from microdroplets, e.g., by adding agarase, for growth in culture or further processing.

Under normal circumstances, most of the microdroplets (about 90 percent) will be unoccupied by cells, and those that are occupied will initially contain only a single cell. In most cases, forward and side scatter signals are sufficient to discriminate between occupied and unoccupied droplets, as shown in Figure 10-13; the fluorescence of dyes such as propidium can be used to identify dead cells. Viable cells can be stained for DNA with Hoechst 33342, providing a fairly precise estimate of the numbers of cells in colonies, or stained for total nucleic acid with SYTO-9 (this fluorescent dye was used in the cells shown in Figure 10-12), which provides a usable, if less precise, assessment of growth.

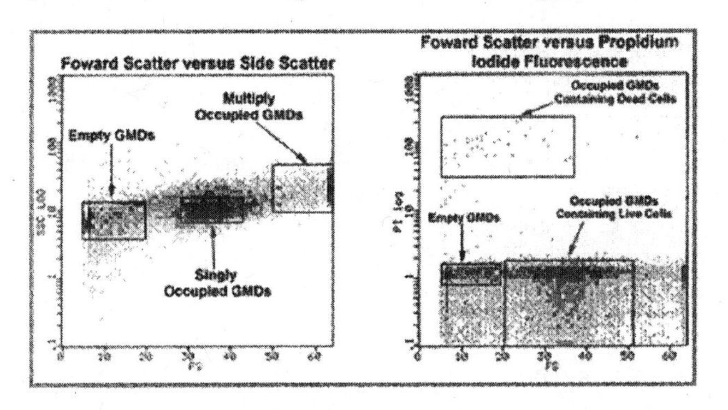

Figure 10-13. Identification of occupied and unoccupied gel microdroplets. Courtesy of Pat McGrath and Swee Kim Lin (One Cell Systems).

By incorporating appropriate antibodies in the gel, microdroplets can be made to capture one or more secreted products, which can then be detected using a second, fluorescently labeled, antibody[1840,3025-6]. However, the newer **cell-surface affinity matrix** technique[2704-5] may provide an equally effective means for detection of secreted products without the need for microdroplet generation or subsequent release of cells.

The use of gel microdroplets for such applications as antibiotic sensitivity testing, in which a dozen or more aliquots of sample subjected to different drug treatments must be analyzed, might be facilitated by multiplex color coding of the microdroplets; I am not aware that anyone has actually tried this.

Hanging Ten Pseudopodia?

In order to preserve characteristics of these normally adherent cells during flow cytometric analysis, Bloch, Smith, and Ault[1849] attached human monocyte-macrophages to 14-20 μm plastic microspheres; the cells retained phagocytic capacity, and could be maintained in long-term culture, stained for surface antigens, and examined and sorted while still in the adherent state. The technique should be applicable to other adherent cell types.

10.5 SINGLE CELL ANALYSIS: WHEN FLOW WON'T DO

From about the mid-1990's on, there has been an interesting interplay between high-sensitivity cytometry and high-sensitivity analytical chemistry. Richard Zare and his coworkers at Stanford[3027-8] exploited single cells as chemical detectors, using patch-clamp measurements of transmembrane electrical current and fluorescence measurements of intracellular calcium to quantify extremely small amounts of neurotransmitters and other ligands emerging from a capillary after electrophoresis.

Nancy Allbritton, who had been involved in that work, set up shop at the University of California at Irvine, and put the cells at both ends of the electrophoresis capillary; her group has used both conventional fluorescence detectors and detector cells to measure materials extracted directly from single cells into electrophoresis capillaries[3029-3032]. They emphasize determination of kinase activities in individual cells and in discrete regions of *Xenopus* oocytes.

Norm Dovichi et al[3033-8], now at the University of Washington, are doing what they describe as **chemical cytometry**, i.e., "the use of high-sensitivity chemical analysis to study single cells"[3033], and **metabolic cytometry**, "a form of chemical cytometry that monitors a cascade of biosynthetic and biodegradation products generated in a single cell"[3033]. The chemical analysis method they use is typically capillary electrophoresis. By using Hoechst 33342 fluorescence intensity, determined by image cytometry, as a criterion for cell selection, they have demonstrated variation of oligosaccharide metabolism during the cell cycle[3033]; they have also analyzed proteins from single cells[3038]. Their methodology has allowed them to detect as few as 50 molecules of a metabolic product. They have also made the point that, in addition to demonstrating cell-to-cell variation, analysis of single cells eliminates artifacts introduced in the preparation of extracts from cell populations for chemical analyses.

A special issue of *Current Opinion in Biotechnology* (Volume 14, Number 1, February, 2003) devoted to Analytical Biotechnology, and edited by Norm Dovichi and Dan Pinkel[3039], contains an article on kinase measurements by Sims and Allbritton[3032], some informative papers on analysis of single mitochondria[3040] and single cell electroporation[3041], and a review and update by Ibrahim and van den Engh on high speed sorting[3042].

10.6 APPLICATIONS OF FLOW CYTOMETRY

We now get into the area(s) in which I can't even attempt to keep up with the literature. Coverage from here on may be arbitrary and capricious at times. However, in the course of preparing this edition, I appealed far and wide for people working in the field of flow cytometry to send me references, reprints, figures, data, URLs, etc., so don't blame me if you didn't send me anything and I missed something you did, however neat it might have been.

I haven't excised the older references from this section, or, for that matter, from the rest of the book; choosing instead to put them into historical perspective as necessary.

Cell Differentiation, *Ab Ovo* and *De Novo*

A broad range of applications of flow cytometry in studies of cellular differentiation was envisioned by Donna and Tom Jovin[289] in the early 1970's, at which time their voices and those of a very few others seemed to be crying in the wilderness. So it goes in the prophecy business. The use of flow cytometry to detect and quantify **variations in antigens, nucleic acids, enzymes, and functional characteristics during normal and abnormal differentiation** is now commonplace. This is an area in which multiparameter analysis can be particularly useful. If one wants to determine, for example, whether different gene products are being produced simultaneously or sequentially, single cell analysis can provide information that is unavailable using more traditional biochemical approaches.

A study by Swartzendruber, Travis, and Martin[531] of BrUdR-induced differentiation in mouse teratocarcinoma cells provided a good illustration of the range of parameters that could be analyzed in the late 1970's. These authors studied multiangle light scattering behavior, DNA content, and enzyme kinetics in BrUdR-induced and spontaneously differentiating cells.

Interest in the role of cytoplasmic [Ca^{++}] and other functional parameters in the control of cell differentiation has also provided motivation for studies using flow cytometry, e.g., work by Levenson et al[532] on changes in mitochondrial membrane potential during the differentiation of Friend mouse erythroleukemia cells. Much of the earlier, and a fair amount of the later work in the area of blood cell differentiation has been done with leukemic cell lines such as HL-60 and K562, which, like Friend cells, undergo differentiation *in vitro* in response to a variety of chemical agents, including phorbol esters, DMSO, butyric acid, and retinoic acid. I'll cover more of blood cell development in the discussion of hematology and immunology.

As to development in general, it might be fair to say that, at least from a flow cytometric point of view, ontogeny has recapitulated hematology. We always knew that blood stem cells were in there somewhere; once enough monoclonal antibodies to leukocyte surface antigens had been raised, we managed to find the stem cells and clarify the developmental pathways. We now routinely harvest and transplant autologous and allogeneic blood stem cells, aided by cytometry at most steps along the way, and cytometry has also helped us define and mass produce some of the cytokines we use to extract an extra measure of performance from a hematopoietic system damaged by disease and/or drugs.

While we haven't come nearly as far in terms of taming embryonic stem cells, or stem cells that might help us regenerated damaged nerves or cardiac muscle, the model is now in place.

Embryonic stem cells have become a politically hot topic as well as a scientifically hot one. Flow cytometry has been used to analyze patterns of surface marker expression in stem cell differentiation[3043], and, with sorting, to detect and select cells that have maintained their embryonic characteristics[3044] and/or cells potentially useful for transplantation by virtue of differentiation along a specific pathway[3045]. **Multipotent adult progenitor cells (MAPC)** are of great scientific interest and may offer a less politically charged therapeutic alternative; their differentiation is studied using arrays to follow genotypic changes and flow cytometry to follow phenotypic ones[3046].

Differentiation in the Nervous System

I can't hope to cover the state of the art in every organ system, but I will call your attention of the work of Jeffery Barker, Dragan Maric et al at the National Institute of Neurologic Disorders and Stroke, National Institutes iof Health, on nerve cell differentiation[3047-53]. They have used analyses of the cell cycle, of protein expression, and of calcium and membrane responses to neurotransmitters to define and fractionate populations of stem and differentiated cells, and to determine patterns of expression of different receptors and ion channels during differentiation. Following on work by Morrison et al[3054], who sorted neural stem cells from emrbyonic peripheral nerves, and Rietze et al[3055], who sorted stem cells from the adult central nervous system, both using positive selection strategies, the Barker group developed negative selection criteria, isolating stem cells on the basis of failure to stain with any of a combination of markers for differentiation and apoptosis[3053].

Whole Embryo Sorting

Working with Herzenberg and others, Krasnow[1862] used the lacZ reporter gene (pp. 408-9) to identify cells from portions of bearing defined mutant genes known to be differentially expressed in different segments or regions of the embryo. With dissociated whole early embryos as the starting material, neuronal precursor cells were sorted; they differentiated into neurons with high efficiency in culture. Incipient posterior compartment cells that expressed the *en* and *wg* genes were also purified from early embryos; Cumberledge and Krasnow[1863], using such isolated cells, demonstrated that the *wg* protein provides a signal needed to maintain *en* expression. At the time at which these studies were done, and described as "whole embryo sorting," there were probably not a lot of people contemplating sorting whole embryos, as opposed to a whole embryo's worth of isolated cells. Since then, Furlong et al, based in the same building as the Herzenberg lab, have developed and built an instrument that sorts whole *Drosophila* embryos[2325]; you can build one from their plans (p. 266) or buy a similar device from **Union Biometrica** (p. 432).

Somatic Cell Genetics and Cell Hybridization

Reporter Genes Revisited

Much of the early work on the use of flow cytometry and sorting to isolate cells bearing transferred genes coding for specific human cell surface antigens from heterogeneous populations produced by cell hybridization and/or gene transfer was done in Ruddle's laboratory at Yale[533] and in Herzenberg's laboratory at Stanford[534]. In most cases, repeated sorting was needed to enrich cells bearing the desired antigens, which typically represent only a few hundredths of a percent of the cell population when the process is started. Le Bouteiller et al[535] described a method for quantitative analysis of expression of cloned HLA antigens in transformed cells.

I was somewhat surprised that repetitive sorting became a method of choice for isolation of small subpopulations of cells following gene transfer, especially since the procedure often began with the pooling of clones grown in isolation on solid or in semisolid media. It seemed to me that other assay methods, e.g., enzyme-linked immunoassay (ELISA), should have been sufficiently sensitive to identify the desired clones *in situ*, eliminating the pooling and sorting. This evidently reflected a more pronounced awareness of the shortcomings of flow cytometry than of the shortcomings of ELISA on my part.

The use of lacZ (p. 320)[1642-4] as a reporter gene detectable by fluorescence made it considerably easier to sort transfected cells, as the whole embryo sorting experiments discussed above indicate. And, in the previous edition of this book, it was noted that "the *Aequorea* green fluorescent protein[1648] (not yet used at the time) also looks promising for this purpose." The promise has been kept, and improved instrument sensitivity and sorting speed have made the job easier, although a few rounds of sorting may still be required.

Isolating and Characterizing Hybrid Cells

Sorting can also eliminate the need for selective media for **separating hybrid cells or cybrids following a cell fusion** experiment. Schaap et al[536] used dual labeling techniques to identify and sort cell hybrids; Jongkind and Verkerk[196] combined cell sorting with ultramicrochemical analysis of single sorted cultured human fibroblasts to permit **determination of enzyme activity in individual sorted cells** for genetic analysis. Tertov et al[1512] used **energy transfer between tracking dyes** with different spectral characteristics to identify and sort hybrid cells. Tracking dyes, particularly those that label membranes, e.g., "DiI," "DiO," and the PKH dyes (pp. 371-4) still provide the easiest approach to detecting hybrids.

Trask et al[1866] used bivariate flow karyotyping (see next section) and FISH (pp. 361-2) to characterize hybrid cells bearing human chromosomes, while Bouvet et al[1867] employed the former technique to identify pig chromosomes in pig-mouse hybrid cells. Some pig chromosomes could not be identified with certainty due to overlap with mouse chromosomes in the karyotypes; this could presumably be remedied by the use of fluorescent probes for the pig chromosomes, which should hybridize to regions containing oinkogenes. Sorry; my humor has peaks and troughs.

Chromosome Analysis and Sorting and Flow Karyotyping

It was noted in the mid-1970's[1106,3056] that individual chromosomes stained with a single DNA fluorochrome could be measured by flow cytometry, producing a **univariate karyotype** (Figure 10-14).

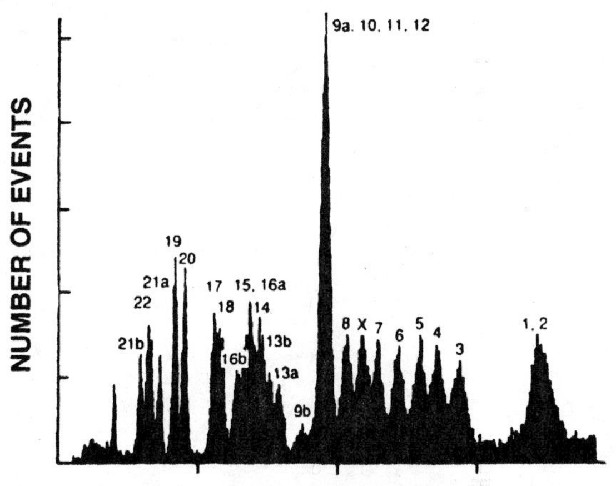

Figure 10-14. Univariate flow karyotype of human chromosomes stained with propidium iodide. From: Gray JW, Cram LS: Flow karyotyping and chromosome sorting. In: Melamed MR, Lindmo T, Mendelsohn ML (eds): *Flow Cytometry and Sorting* (2 Ed). New York, Wiley-Liss, 1990 (reference 1028), © John Wiley & Sons, Inc. Used by permission.

Although peaks representing most of the human chromosomes are discernible in the univariate karyotype of Figure 10-14, there is too much overlap between adjacent peaks for chromosomes to allow any given chromosome to be highly purified by sorting on the basis of propidium fluorescence alone. This reflects the fact that there are only small differences in DNA content between adjacent chromosomes.

The chromosomes shown in Figure 10-14 are stained with propidium, which is not highly selective for either A-T or G-C base pairs. By 1980, Langlois et al[277-8] had established that combinations of DNA fluorochromes that exhibited **base preferences** for A-T (e.g., Hoechst 33258 and DAPI) or G-C (e.g., chromomycin A_3 and mithramycin) could be used to produce a **bivariate karyotype** (pp. 317-9; Figures

7-13 (p. 318) and 10-15), allowing discrimination of chromosomes that are similar in total DNA content but contain different proportions of A-T and G-C base pairs.

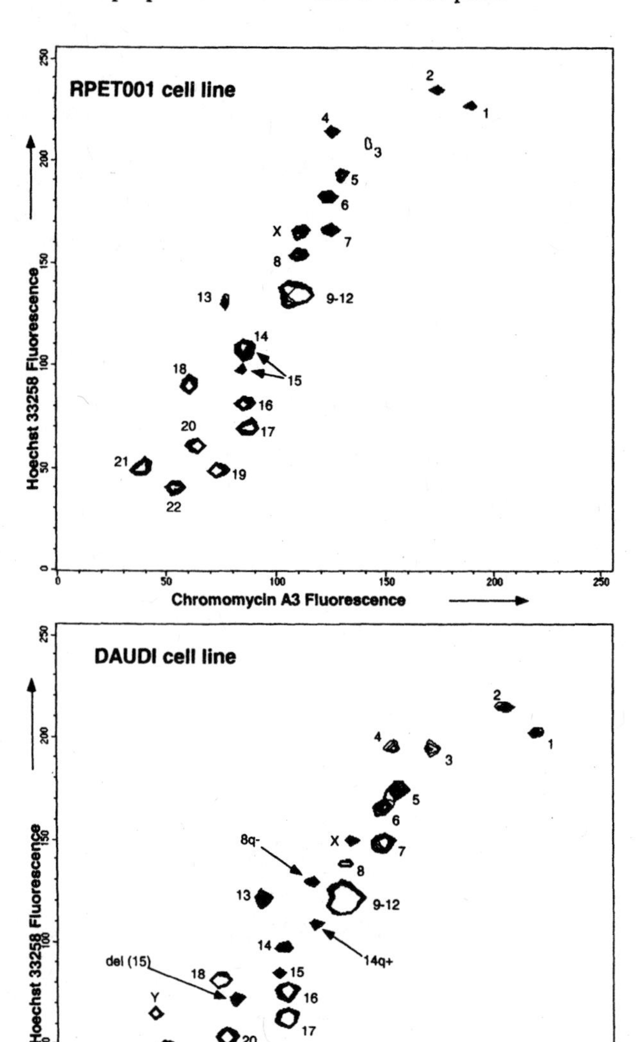

Figure 10-15. Bivariate karyotypes of RPET001 (top panel) and Daudi (bottom panel) human cell lines, based on Hoechst 33258 and chromomycin A_3 fluorescence. Courtesy of Simon Monard (Trudeau Institute).

A comparison of Figures 10-14 and 10-15 shows that the bivariate karyotype, unlike the univariate karyotype, places most of the human chromosomes in distinct clusters, making it relatively simple to purify individual chromosomes by sorting. **High-speed chromosome sorters**[667,904] with dual-beam illumination from high-power, water-cooled argon ion lasers (UV for Hoechst 3325, 457 nm for chromomycin A_3) were developed and used at Livermore and Los Alamos to do this, facilitating the task of mapping the human genome and yielding chromosome-specific probes now used for fluorescence in situ hybridization (FISH)[2710]. During the earlier stages of the genome project, when the first two editions of this book were written, it was common for experimenters to spend hours or even days sorting enough of a particular chromosome to build a genomic library[537]; amplification techniques have since made it possible to work with much less starting material, even as little as a single sorted chromosome[3057-60].

Even before high-speed sorters became commercially available, Los Alamos also kept at least one dual-beam commercial instrument busy sorting chromosomes with good results. I was told by Scott Cram that reasonably good resolution of major groups of ethidium bromide-stained chromosomes in univariate karyotypes could be achieved using instruments with low-power argon laser sources, e.g., the Coulter EPICS Elite and B-D FACScan; any skepticism I harbored on this subject vanished when I saw karyotyping done as a lab exercise at one of the annual flow cytometry courses.

While it has been demonstrated that bivariate flow karyotyping of chromosomes stained with a Hoechst dye or DAPI (with an A-T base preference) and chromomycin A_3 or mithramycin (with a G-C base preference) can be done using air- cooled He-Cd lasers instead of much larger water-cooled ion lasers[1133-5,1242] as light sources, the He-Cd sources do not put out quite enough power to achieve photon saturation of the dyes, and are noisier than ion lasers, with the result that CVs are higher and separation of chromosomes in the measurement space is not as good. However, it seems likely that the UV and 457 nm ion lasers now used in bivariate chromosome sorters could be replaced by mode-locked 355 nm YAG lasers and 457 nm ND:YVO_4 lasers, both available at power levels on the order of those achievable with large ion lasers.

Asymmetric cyanine[1133-4,1331] dyes of the TOTO and TO-PRO series (see p. 315) have also been shown to produce bivariate flow karyotypes, although resolution is much poorer than that obtained with Hoechst33258 and chromomycin A_3. Buoyed by preliminary results, I spent several years exploring cyanines and a variety of other dyes in an ultimately unsuccessful search for combinations that could produce bivariate karyotypes using 488 and 633 nm laser sources. Unless someone else has better luck, bivariate karyotyping will continue to depend on the dyes we know and on instruments with UV and blue-violet excitation; with luck, the instruments will get smaller.

Reviews and descriptions of preparative and analytical techniques for flow karyotyping and results appeared in a number of earlier papers[538-47,903, 1868-70]; the 1989 book *Flow Cytogenetics*[1050], edited by Joe Gray, remains an indispensable basic reference. Since the previous edition appeared, Lucretti and Doležel[3061] and Šimková et al [3062] have described methods for analysis and sorting of plant chromosomes, Ferguson-Smith has reviewed the use of chromosome sorting and painting in phylogenetics and diagnosis[3063], and Monard has discussed chromosome analysis and sorting[3064].

Unless chromosome suspensions are well prepared, which is nontrivial (see reference 1050), the best instrument in the world won't provide flow karyotypes with good resolution. Kuriki, Sonta, and Murata[1871] consider the effects of isolation buffers in karyotyping hamster chromosomes, and Telenius et al[1872] point out that contamination by chromatids can degrade the resolution of bivariate karyotypes.

It was suggested that flow cytometry would eventually be useful for karyotyping for clinical purposes; Arkesteijn et al[1873] analyzed karyotypes in various lineages of blood cells in cases of acute myelocytic leukemia. Figure 10-15 illustrates differences in bivariate karyotypes between the RPET001 cell line, derived from nonmalignant cells, and the Daudi cell line, derived from a leukemia. The Daudi karyotype in the bottom panel of the figure shows abnormalities due to a translocation between chromosomes 8 and 14 (a frequent occurrence in leukemias) and a deletion on chromosome 15. However, while chromosome sorting continues to be of some interest to researchers studying genetic diseases involving translocations and other gross chromosomal abnormalities, flow karyotyping, which requires both expensive apparatus and relatively labor-intensive sample preparation techniques, does not now seem slated for extensive clinical use.

Rabinovitch, Martin and Hoehn[548] have considered one possible alternative, i.e., the detection of small degrees of aneuploidy in interphase cells by DNA content determination, as is done in determining effects of clastogenic agents (p. 453). This also requires a well-standardized, reproducible preparative method and an instrument with good precision and reproducibility. It is also abundantly clear that, where it can be accomplished, DNA sequence detection, e.g., using FISH, provides better and more specific genetic information than is available from karyotyping.

The overall strategy for flow karyotyping a previously uninvestigated species involves isolation and preparation of metaphase chromosomes, followed by generation of a univariate flow karyotype. If the univariate karyotype reveals a number of chromosomes with DNA contents too close to one another to permit good resolution of individual peaks, a bivariate karyotype is analyzed; however, unless there are substantial differences in base composition between chromosomes with similar DNA contents, bivariate karyotyping will not improve resolution[3065]. When neither univariate nor bivariate karyotyping succeeds in resolving most or all of the chromosomes well enough to permit purification by sorting, it is common to analyze mutants in which substantial chromosome translocations and/or chromosome polymorphisms are found[3066-8]; this typically makes it easier to sort at least parts of some otherwise unobtainable chromosomes. It has also been possible to introduce single chromosomes from other species into cell lines by somatic cell hybridization, facilitating their purification by sorting[3069].

FISH[3070] and primed in situ labeling (PRINS)[3071] may be used to confirm identification of sorted chromosomes, usually based on identification of known repetitive sequences. Recently, Gygi et al[3072] reported the use of fluorescent sequence-specific **polyamide probes** (see Chapter 12) to distinguish human chromosome 9 from the other chromosomes with which it clusters on both univariate and bivariate histograms.

Rens et al[2311] described implementation of a slit-scan procedure for chromosome analysis and sorting which can be carried out on slightly modified commercial instruments. Stepanov et al[3073] described a flow cytometer in which mitotic cells are disrupted in the flow injector needle, releasing individual chromosomes and thus providing individual karyotypes on a cell-by-cell basis.

Probing Details of Cellular Structures and Inter- and Intramolecular Interactions

Dissection of Structures Using Antibodies, Ligands, and Genetic Methods

Since the late 1980's, flow cytometry has seen increasing use as a means of defining which regions of transmembrane and cell membrane associated proteins are actually exposed at the cell surface rather than embedded in the membrane. Studies of this type may make use of various combinations of monoclonal antibodies raised against different defined epitopes of isolated proteins, cells expressing mutant proteins, and ligands; flow cytometry provides quantitative answers as to which antibodies and/or lectins bind, and under what circumstances, and provides a powerful tool for structural analysis. This application is exemplified by work by Takahashi, Esserman, and Levy[1513], who showed that the transferrin receptor is exposed differently in low- and high-grade lymphoma cells, and by investigations by Klebba et al[1514-5] on transmembrane proteins in Gram-negative bacteria.

Intramolecular Interactions

A number of flow cytometric procedures for detecting inter- and intramolecular interactions are based on measurement of fluorescence resonance energy transfer (FRET)[2347] (pp. 283-4). Because the intensity of energy transfer is proportional to the inverse sixth power of the distance between the donor and acceptor fluorophores, slight changes in the distance between two suitably labeled interacting molecules result in large changes in energy transfer, providing a sensitive method for analyses of such phenomena as receptor-ligand interactions and changes in receptor subunit conformation. The first such studies were done using covalent labels on the molecules or submolecular units involved; more recently, it has become possible to clone suitable energy transfer pairs of fluorescent proteins into the structures of protein subunits between which interactions occur[2377-8].

Michnick and coworkers[3074-6] have taken an alternate approach. They have engineered subunits of enzymes such as dihydrofolate reductase and β-lactamase, and can produce cells containing signaling proteins in which each of the two moieties involved in interaction contains a cloned inactive subunit of the reporter enzyme. When the inactive subunits become sufficiently close to one another as a result of changes in the conformation of the signaling protein to which they are attached, they form an active binding site for the enzyme substrate. Active dihydrofolate reductase is detected by binding of fluorescently labeled methotrexate derivatives[354-6]; active β-lactamase is detected using the fluorogenic substrate CCF2[2877] (p. 409).

Clinical Flow Cytometry: Turf and Surf

The first review of clinical applications of cytometry was published in 1981 by Laerum and Farsund[586]. The Engineering Foundation and the Society for Analytical Cytology cosponsored conferences on Clinical Cytometry in April, 1982[587] and December, 1983. While the proceedings of the first of these meetings were not published, a symposium volume containing papers presented at the second did appear[979].

Beginning in September, 1986, a series of annual meetings on Clinical Applications of Cytometry were organized by Dr. Mariano La Via of the Medical University of South Carolina and others and held in Charleston, South Carolina; this led to the founding of the Clinical Cytometry Society, and, subsequently, to the inclusion of a volume on clinical cytometry (originally subtitled *Communications in Clinical Cytometry*, now simply called *Clinical Cytometry*) in each year's issues of the journal *Cytometry*. Numerous sessions, symposia, and courses on clinical cytometry, flow and otherwise, have been held at and/or in conjunction with meetings of a number of organizations concerned with those areas of laboratory medicine in which flow cytometry has had the greatest impact, i.e., hematology, immunology, and oncology (see Chapter 2 for details).

The earlier reviews[906,980-2] of clinical cytometry have been replaced by a number of books (also see Chapter 2). The most recent book dedicated to clinical flow cytometry is the Third Edition of *Flow Cytometry in Clinical Diagnosis*[2393], edited by Keren, McCoy, and Carey, published in 2001; I would also recommend a 2001 volume in *Clinics in Laboratory Medicine* on "New Applications of Flow Cytometry"[2398], edited by McCoy and Keren. Also of general interest is an online journal, *Case Studies in Clinical Flow Cytometry*, edited by Michael Borowitz and endorsed by both the Clinical Cytometry Society and ISAC. The journal is accessible at <http://www.flowcases.org>.

Earlier editions of this book made attempts at complete coverage of existing and projected clinical applications of flow cytometry according to a framework that Brian Mayall[983] and I[984] independently followed in surveying the field at the 1983 Engineering Foundation-SAC conference. In the previous edition and this one, I have made and make no pretense of trying to cover the entire field.

Clinical cytometry itself has changed over the years. In the 1970's, most clinical cytometry was done by simple hematology counters, on the one hand, and by image analyzing automated differential counters, on the other. The differential counters scanned Wright's-stained smears, and could not identify different lymphocyte types. Clinical cytometry today still involves flow cytometric hematology analyzers, which now perform differential counts as well as red and white cell, platelet, and reticulocyte counts, often without benefit of fluorescence measurements. Fluorescence flow cytometers are used for a wide range of clinical tests, predominantly involving immunophenotyping and DNA content analysis, none of which is done anywhere near as often as are differential leukocyte counts or Papanicoloaou smears.

A new breed of image analysis systems have come into clinical cytometry to facilitate and perform Pap smear screening for cervical and other gynecologic cancers, but, by and large, the people who deal with these systems publish in *Acta Cytologica* (http://www.acta-cytol.com) and *Analytical and Quantitative Cytology and Histology* (http://www.aqch.com), both of which are official journals of The International Academy of Cytology and The American Society of Cytopathology. The blood cell counter people tend to publish in *Laboratory Hematology* (http://labhem.cjp.com/), the official journal of the International Society for Laboratory Hematology, and the people who do clinical fluorescence flow cytometry are more likely to publish in *Cytometry* and *Clinical Cytometry*, and to go to meetings of the Clinical Cytometry Society and ISAC.

The turf issues in journals and societies mirror turf issues in hospitals and clinics. If the next successful clinical product incorporating a flow cytometer is built to do microbiology, or Pap smears, it is unlikely that the immunologists and/or pathologists who use their fluorescence flow cytometer to do immunophenotyping and DNA analysis will ever have much to do with it, and it is no more likely that the microbiologists or the cytopathologists would funnel a significant portion of their laboratory work load to the fluorescence flow cytometer in the clinical immunology lab, even if that instrument could do antimicrobial susceptibility tests or cervical cancer screening.

It was nicer when all of us went to the same meetings, but at least I (and you) can try to keep with lab hematology and cytopathology online. Hence my section title.

From this point on, I will deal with selected clinical applications of cytometry in the context of broader discussions of application areas. Phil McCoy has provided the bottom line for this section with a discussion of Medicaid and Medicare reimbursement for flow cytometry[3077]: you may be able to collect for doing reticulocyte counts, T cell counts, absolute CD4 and/or CD8 counts and ratios, single antigen [B27] HLA typing, immunophenotyping, and DNA content/cell cycle analysis.

Hematology

Clinical Application: Blood Cell Counting and Sizing

Before photoelectric[48] and electronic[49] cell counters were developed in the 1950's, the **counting and classification of blood cells** had to be done by human observers (pp. 77-9). Coulter's apparatus was introduced into laboratories in the

1960's and shown to be superior for **red cell counting**[50] and **sizing**[51]; the use of selective lysing agents to remove red cells made it feasible to do **white cell counting** with the same instruments. Acceptance came rapidly; the instrumentation was rapid, laborsaving, and produced more accurate and precise results than could be obtained by even the most dedicated, highly trained personnel.

It was slightly harder to make electronic counters count **platelets** because these cells were smaller in size, and thus produced smaller signals, which were harder to discriminate from noise. Nonetheless, electronic counting was perceived as the method of choice for platelet counting some time before it could actually be used effectively for the task. Modern hematology analyzers now count and size all formed elements in blood.

Fundamental issues and principles of cell counting were introduced on pp. 18-21. Commercial hematology instruments were covered briefly on pp. 433-4. Automated blood counts, and automated differential counts as well, are discussed in some detail in books by Bessman[985] and Groner and Simson[2915]. **NCCLS** (see p. 361) has issued an Approved Standard[3078] for quality control of hematology analyzers with recommended goals for accuracy and precision in measurements of hemoglobin concentration, erythrocyte, leukocyte, and platelet counts, and mean corpuscular volume.

Red Blood Cells (Erythrocytes)

Clinical Application: Reticulocyte Counts

Counting the relatively small fraction of RNA-containing immature cells, or reticulocytes (see pp. 78-9, 99, 312-4, and 326), which normally comprise about 1 percent of circulating red blood cells, gets into the realm of rare event analysis when red cell production is impaired, and the fraction of reticulocytes in the total red cell population decreases. The job gets easier when red cell turnover is increased, e.g., following bleeding or in hemolytic anemias, and the reticulocyte count (usually expressed as reticulocytes/1000 RBC) increases.

The biology of red cell maturation dictates that the distribution of RNA content in cells should be continuous; the distinction between reticulocytes and mature erythrocytes in visual counting is made only when enough RNA is present to form a visible precipitate after staining with basic dyes. Flow cytometric approaches to reticulocyte counting should ideally be based on the determination of the RNA content distributions in red cell populations, because shifts in the distribution to lower or higher mean fluorescence values provide the same clues to red cell kinetics as are obtained from decreases or increases in the reticulocyte count. However, what often happens in practice is that cells are determined to be "positive" or "negative" for RNA.

The reticulocyte count is normally in the neighborhood of 1%, or 10 reticulocytes per 1,000 red cells; thus, visual reticulocyte counts, generally based on counts of 1,000 red cells, are inherently imprecise due to Poisson sampling error[762-3]; if n reticulocytes are actually observed, the standard deviation is $n^{1/2}$. If, on examination of 1,000 red cells, 16 reticulocytes are seen, the standard deviation is 4 cells, and the CV is 4/16, or 25%. Taking the observed value plus 2 standard deviations as the **95% confidence limits**, i.e., the range within which there is 95% probability that the true value will be found, the true reticulocyte count is likely to lie between 8/1,000 and 24/1,000 red cells.

The only way in which to improve the precision of reticulocyte counts is to count more cells. If a flow cytometer is used to count 100,000 red cells, and 1,600 reticulocytes are counted, giving the same 16/1,000 count discussed above, the standard deviation is 40, the CV is a more acceptable 2.5%, and the 95% confidence limits are now 15.2/1,000 and 16.8/1,000 red cells, narrowing the range considerably. The hardest part of validating flow cytometric methods for reticulocyte counting has been evaluating agreement of the new techniques with visual counting, because visual counts are so abysmally imprecise.

Although some hematology counters identify reticulocytes by measuring the absorption of a cell-associated dye such as oxazine 750 or the fluorescence of acridine orange, the more recent trend has been toward using dyes which increase fluorescence by factors ranging from several dozenfold to several thousandfold on binding to RNA, including asymmetric cyanines such as thiazole orange and related compounds such as thioflavin T and auramine O.

The fact that all these dyes are positively charged and lipophilic raises some questions as to the mechanism of reticulocyte staining. The membrane potentials of reticulocytes appear to be higher[290-91] than those of mature red cells; this should cause vitally stained reticulocytes to take up basic dyes more avidly than would mature cells. Thus, the less mature cells should contain more dye as well as more RNA, and the staining difference between mature and immature cells would be enhanced. The experiments necessary to determine what contribution, if any, cell membrane potential differences make to reticulocyte staining (i.e., comparisons of fluorescence levels between unfixed and fixed or gramicidin-treated cells) have not been reported to date.

Corash et al[1783] compared thioflavin T, ethidium bromide, and thiazole orange as reticulocyte stains, and found the latter two preferable because they required less precise control of dye incubation time. Subsequent studies by Carter et al[1784], Van Hove et al[1785], Hansson et al[1786], and Schimenti et al[1787], among others, have established the validity and utility of flow cytometric reticulocyte counts done with thiazole orange; they are, as might be expected, more sensitive than manual counts[1786-7], in that they can detect cells with lower RNA levels. Thiazole orange reticulocyte counts can also be done on canine[1788-9] and feline[1790] blood. Van Petegem et al[1791] compared three benchtop flow cytometers, the B-D FACScan, Coulter EPICS Profile, and Ortho Cytoron Absolute, for reticulocyte enumeration using thiazole orange, and found the instruments gave equivalent results from counts of 30,000 cells. Guasch et al[1793] confirmed the report of Van Bockstaele and Peetermans[1792] that the diethyl analogue of thiazole orange, 1,3'-Diethyl-4,2'-quinolylthiacyanine iodide, yielded equivalent results.

Since thiazole orange stains DNA as well as RNA, red cells containing DNA due to the presence of intraerythrocytic parasites[771,1788], Howell-Jolly bodies[1785,1795], nuclei, or micronuclei become fluorescent when stained, and may be counted as reticulocytes; falsely high reticulocyte counts may also occur in the presence of red cell autofluorescence, giant platelets and high leukocyte or platelet counts[1785, 1796].

Dedicated flow cytometric reticulocyte counters, the R-1000 and R-3000, which measure RNA using argon ion laser excitation of **auramine O** fluorescence, have been developed by **Sysmex-TOA Medical Electronics** (Kobe, Japan)[1797], and validated in several studies[1798-1800]; Bowen et al[1801] found substantially equivalent performance in a comparison of reticulocyte counting using thiazole orange on a B-D FACS and auramine O on a Sysmex R-1000. More recent studies[3079-80] confirm that automated reticulocyte counting is far more precise than manual counting and that a variety of commercial instruments are efficiacious for the purpose. Riley et al[3081] have recently published an extensive review of reticulocyte counting methdology.

The obvious inadequacies of manual reticulocyte counts in terms of the number of cells counted and the difficulty of detecting cells with small amounts of RNA makes the manual count something less than a gold standard for comparison, creating a problem in validating flow cytometric counts. Oosterhuis et al[1794] have used a multivariate statistical model to compare manual and flow cytometric counts based on correlations of the results of each with hemoglobin concentration, mean cell volume, and erythrocyte density width, and found flow counts superior; their modeling method is generally applicable to tests in which reliable standard methods do not exist. NCCLS issued an Approved Guideline[3082] for reticulocyte counting in 1997, covering both flow cytometric and classical methods; Bruce Davis, who is working on a new guideline, tells me that a flow cytometric method based on thiazole orange will become the new gold standard.

There's just one little thing bothering me. I have done some work on reticulocyte counting over the years, aimed at finding red-excited dyes that would allow the use of inexpensive diode lasers instead of argon lasers in flow cytometric retic counters. As it happens, one of the newer counters from Sysmex does appear to use a red-excited dye. Now, the dyes used to stain retics are RNA dyes, and, if you stained blood with both a 488 nm-excited dye and a red-excited dye, and analyzed the cells in a dual-beam cytometer with 488 nm and red lasers, you would expect to see retics stained with both dyes. When I have tried this trick with thiazole orange and with a variety of putative red-excited retic stains, I find that there is some discordant staining; there are cells stained with both dyes, but there always seem to be cells stained with one dye and not the other. Maybe somebody out there would like to try the experiment using thiazole orange and a manufacturer's approved red-excited retic reagent. It's another one of those things that I haven't gotten around to while I've been writing this.

The Reticulocyte Maturity Index (RMI)

Since reticulocytes contain as much RNA as they ever will at the time at which they enter the circulation, and lose it over a period of a few days, the less mature reticulocytes will, on the average, contain more RNA than the more mature ones, and will therefore have higher fluorescence intensities when stained with dyes such as thiazole orange and auramine O.

Davis and Bigelow[1802-4] used either the mean fluorescence channel of the fluorescent cells or the fraction of highly fluorescent cells among the total number of reticulocytes as a **reticulocyte maturity index (RMI)**; the latter method was found to result in lower interlaboratory variability in a multicenter study[1805], which also demonstrated lower variability among labs using the Sysmex R-1000 than was found using the thiazole orange method. This is to be expected, since the R-1000 methodology is more standardized. The RMI, independent of reticulocyte count, has been found to be predictive of successful engraftment following marrow transplantation[1806-7] and of marrow recovery after intensive chemotherapy[1808], and a useful indicator of erythropoietic activity in anemia[3083].

Bayer Diagnostics' hematology analyzers, which count retics based on absorption of oxazine 750[737], calculate a reticulocyte maturity index from the fraction of cells with high absorption values. Other indicators of reticulocyte heterogeneity have been investigated. A monoclonal antibody which binds to relatively immature calls has been identified by Mechetner, Sedmak, and Barth[1809], while Bain and Cavill[1810], making use of the ability of a Technicon (now Bayer) hematology counter to measure both size and hemoglobin content of individual erythrocytes[684,1811], showed that while reticulocytes are often hypochromic macrocytes, the percentage of hypochromic macrocytes does not accurately predict the percentage of reticulocytes.

Erythrocyte Flow Cytometry: Other Clinical Uses

The detection of antibodies to cells is the bread and butter of **immunohematology**[1005]; a principal area of application is in **transfusion medicine**. Garratty and Arndt have discussed applications of flow cytometry in this field[3085-6]. Flow cytometric methods have been used[1006] to examine effects of regular blood component donation on donors, and for detection and quantification of specific cell-bound antibodies to platelets[1007,1939] and other cell types in **autoimmune cytopenias**. Changes in red cells[1963] during storage can also be monitored.

Davis[3084] has recently reviewed the use of flow cytometry in detecting **fetomaternal hemorrhage** by performing the equivalent of a **Kleihauer-Betke test**, and in detecting cells containing **hemoglobin F** and **intracellular parasites**.

White Blood Cells (Leukocytes)

Clinical Application: Differential Leukocyte Counting

While instrument developers anticipated that **flow cytometric differential white blood cell counters** would find the same rapid acceptance as had the earlier slide scanning systems, this was not to be. Differential counting, for better or worse, was intimately tied to the stained slide, and results obtained using fluorescent stains such as acridine orange[61,588-9] were discounted, even when confirmed by cell sorting[590]. **Technicon** (now **Bayer**)'s Hemalog instrument[84-5], based on **enzyme cytochemical stains** which were widely known in hematology, was not well accepted when it was introduced to the market in the early 1970's, although its performance was demonstrably superior to that of any competing image analyzing differential counter.

In the mid-1980's, there was a resurgence of interest and commercial activity in flow cytometric differential counting. Manufacturers of electronic (e.g., Coulter) and photometric (Ortho) instruments designed for blood cell counting in clinical laboratories began to offer **3-part (lymphocyte/monocyte/granulocyte) differential counts** based on differences in electronic cell volume[69] or in orthogonal light scattering[92-3,157,175-6], while other companies, recognizing the great success of Technicon's instrument, began to investigate the possibility of building inexpensive flow cytometric differential counters using new stains which could discriminate eosinophils and basophils from neutrophils, yielding a traditional **5-part differential count.**

It is apparent to most observers that there was a change in attitude toward flow cytometry on the part of clinical pathologists and laboratory hematologists in the mid-1980's; these clinicians were much more likely to accept a flow cytometric differential counter then than they had been in the mid-1970's, not because the technology of differential counting had improved dramatically, but because the highly visible applications of flow cytometry in research in hematology and immunology made the technology not only respectable, but desirable. Future developments in this area will depend not on science, but on the economics of health care. As long as it is decided that 5-part differential counts are still worth doing in volume, even in modern cost-constrained clinical laboratories, there will be new instruments. There are a lot of ways to do differential counts by flow cytometry[93], and the least complicated and least expensive will probably survive the natural selection process.

One of the more interesting approaches to the problem, implemented commercially by **Sequoia-Turner** (later **Unipath** and now **Abbott**)[986] in the Cell-Dyn 3000[1921], 3500, and 4000 instruments, is based on the observation by de-Grooth et al[710] that **polarized and depolarized orthogonal light scattering signals can discriminate eosinophils from neutrophils** in unstained blood (pp. 278-9). Two-parameter displays of a normal leukocyte population, obtained using one of these analyzers, are shown in Figure 10-16; the instrument measures scatter at smaller angles as well as orthogonal scatter.

Coulter's **VCS** (Volume/Conductivity/Scattering) technology uses measurements of DC and RF impedance and light scattering to perform a five-part differential count; it is implemented in the VCS[1922-3], STKS[1924], and MAXM hematology analyzers. Earlier instruments derived a differential count from the leukocyte volume distribution.

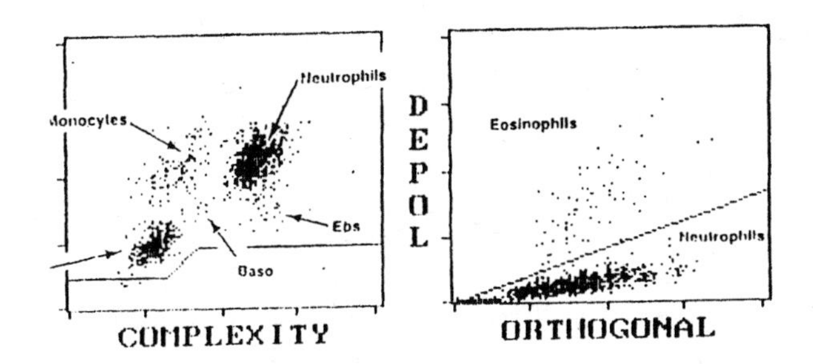

Figure 10-16. Clusters of peripheral blood leukocyte types in two-dimensional displays from an Abbott Cell-Dyn 3000 hematology analyzer. The figure was provided by Judy Andrews and Pamela Kidd (U. of Washington).

Bayer's **Advia** line, successors to the Technicon Hemalog and H-series instruments, retains the peroxidase staining used in the original Hemalog D, but have abandoned the Alcian blue basophil stain in favor of differential lysis, and identify monocytes based on their light scattering and low peroxidase content.

NCCLS (then the National Committee for Clinical Laboratory Studies) issued an Approved Standard on leukocyte differential counting in 1992[3087]. Since then, a number of comparisons of hematology analyzers have appeared in the literature[1926-31]. Bentley, Johnson, and Bishop[1931] did a comparative evaluation of the Unipath (now Abbott) Cell-Dyn 3000, Coulter STKS, Sysmex NE-8000, and Miles (Technicon) H-2, using the NCCLS protocol; they could not establish that any single instrument was clearly superior to the others. Buttarello et al[1930] reported similar results in a comparison of the same four instrument types.

Clinical hematology analyzers are designed to flag apparently abnormal bloods, because their capacities for detection of blasts, nucleated red cells, and immature granulocytes are limited. In general hospital and outpatient clinic settings, no more than 25% of samples are typically flagged, and thus require examination of a smear by a technologist or hematologist; in tertiary care institutions, over 60% of samples may be flagged. Nonetheless, automated differential counting still reduces technologists' and hematopathologists' workloads. If a more sophisticated apparatus that would flag fewer samples could be made cost-effective, there would probably be a substantial market for it.

The measurement parameters and methods and computer algorithms used in automated differential counters differ from instrument to instrument. Thus, while all instruments yield similar results, a careful comparison, such as the 1995 study of absolute lymphocyte counts published by Groner and Simson[3088], will reveal the biases of individual analyzers. Biases in lymphocyte counts may become problematic when leukocyte and lymphocyte counts from a hematology analyzer are combined with data from a fluorescence flow cytometer to obtain a "two-platform" absolute count of CD4+ cells.

The development of automated differential counters has been hampered to a considerable extent by the lack of a true "gold standard." Early in the game, it was thought that the results obtained by flow cytometric methods should agree with those obtained from visual analysis of Wright's or Giemsa-stained smears by a well-trained hematology technician, hematologist, or hematopathologist. It soon became apparent that most flow cytometric counters yielded higher monocyte and lower lymphocyte counts than were reported from smears, even when the methodologies used in the flow systems were quite different.

For example, the Block prototype, which identified monocytes primarily by forward and right angle light scattering, the Technicon Hemalog D, which identified monocytes by the presence of intracellular esterase, and the Ortho Cytofluorograf, using acridine orange staining, produced concordant results that did not agree with manual counts; results based on scatter and acridine orange staining were confirmed by sorting at Los Alamos[91]. Evidence that the flow systems were right, and the human observers (and slide-scanning differential counters) wrong, was provided by a 1974 study by Zucker-Franklin[3089], who examined peroxidase content, adherence to glass, and phagocytic capacity of mononuclear cells in human blood, and concluded that "the percentage of monocytes in normal blood is at least twice as high as is commonly recognized on routine smears." It is evidently not possible to tell small monocytes from lymphocytes on a Wright's-stained blood smear; this alone should disqualify it as a standard.

The cytochemistry and immunology of leukocytes are now considerably better understood than they were when most differential counters were developed, and it should be possible to identify the normal leukocytes (and any nucleated red cells) present in a peripheral blood specimen unambiguously using a combination of cytochemical stains, fluorescent antibodies, and light scattering characteristics.

Hübl et al[3090] proposed a reference method for the five-part differential count using a fluorescein-labeled anti-CD45 antibody, a PE-Cy5-labeled anti-CD14 antibody, and a cocktail of phycoerythrin-labeled anti-CD2, anti-CD16, and anti-HLA-DR antibodies; they reported good concordance with results of 500-cell manual differential counts, even for basophils, but sample preparation steps (lysis and washing) appeared to have some influence on counts.

To get a true "gold standard" count, it would be advisable to work with unfixed, unlysed whole blood samples and "no-wash" staining, using a vital nuclear stain to identify nucleated cells for triggering, to produce a differential count unbiased by any selective cell loss that might occur during lysis, fixation and/or washing. This seems entirely feasible.

Since all of the early development and most of the maturation of blood cells occur in the marrow, it is rare to have definitive diagnoses of hematologic diseases made on the basis of differential counts; what is usually required is an examination of a marrow aspirate or biopsy. This concept was difficult to get across to a lot of the engineering types involved in the development of the first generation of automated differential counters, who seemed to think peripheral blood differentials were the equivalent of a Pap smear for leukemia, and who also labored under the misconception that there were discrete categories of immature or "abnormal" cells which could be definitively identified by experienced hematologists. For at least some types of marrow examination, the flow cytometer can help the hematologist considerably more than the microscope can, and the hematologist may be able to drop out of the loop within a few years, or stay in it only to insure reimbursement.

CD Characters: Leukocyte Differentiation Antigens

International Workshops on Human Leukocyte Differentiation Antigens must be held to periodically keep track of an increasing number of defined human leukocyte cell surface antigens[905,1122-7]. The **CD (Cluster of Differentiation)** nomenclature now standard for the common leukocyte antigens runs to CD247; the results of the 7th workshop, held in Harrogate in 2000, are available in book form[3091] and also in an extensive online database via *Protein Reviews on the Web*[2431]. An online index of CD antigens is at <http://www.ncbi.nlm.nih.gov/prow/guide/45277084.htm>. The 8th International Workshop, chaired by Heddy Zola, will be held in Adelaide in December, 2004; information is available at <http://www.hlda8.org>.

For a more portable work on CD antigens, see *The Leucocyte Antigen FactsBook*[1874], edited by Barclay et al, now in its second (1997) edition[3092]. Academic Press's "*FactsBook*" series includes a few other titles relevant to leukocyte immunology as well. *The Adhesion Molecule FactsBook*[1875], with a 2000 second edition[3093], describes the chemical characteristics and cellular specificities of **adhesion molecules**. *The T Cell Receptor FactsBook*[3094], *The HLA FactsBook*[3095], and *The Cytokine FactsBook and WebFacts*[3096] are also handy references.

Granulocytes: Basophils

Basophils were originally so named by Paul Ehrlich on the basis of their propensity to accumulate cationic dyes in these granules at concentrations high enough to cause metachromasia, or shifts in the spectrum of dye. They are the rarest of the five major classes of peripheral blood leukocytes, usually comprising between 0.1% and 1.0% of cells in differential leukocyte counts, or fewer than 50 cells/μL blood. They can be identified on smears stained with Giemsa's, Wright's, and similar eosin-azure blood stains by the metachromatic absorption of azure dyes bound to glycosaminoglycans and other components of their cytoplasmic granules. Basophils can also be identified in suspension by their metachromatic staining with dyes such as Alcian blue, Astra blue and toluidine blue.

Basic Orange 21: The Best Basophil Stain Yet

Several manufacturers have investigated staining with a metachromatic dye described by Lawrence Kass, **Basic orange 21**[987,3097], which can be used either as an absorption stain or as a fluorescent stain. Dr. Kass, who is on the faculty at Case Western Reserve University, inherited a large collection of textile dyes from his father, and, in the tradition of Paul Ehrlich, proceeded to determine whether he could see anything interesting using the dyes. Perhaps unfortunately for those of us in the cytometry game, he is a very good morphologic hematologist, which means that he can usually see distinctions too subtle for low-resolution instruments like flow cytometers to pick up.

Basic orange 21, unlike most metachromatic dyes, has a metachromatic absorption peak at a longer wavelength than its orthochromatic absorption peak (560 versus 483 nm). The dye is rapidly absorbed by living cells, and is apparently nontoxic in the short term. It stains neutrophil granules orange, eosinophil granules a darker orange-brown, and basophils a striking purplish red; the metachromatic spectral shift can be demonstrated in solution if concentrated heparin sulfate is added to the dye. Basic orange 21 is the best stain I have ever come across for **visual counting of basophils** (or, for that matter, mast cells) by transmitted light microscopy. It is completely trivial to count basophils in unlysed whole blood in a hemocytometer using the dye.

Basic orange 21 also yields excellent basophil counts in a flow system in which green-excited red fluorescence can be measured; it works quite well with low-power green He-Ne laser excitation at 543 nm (L. R. Adams, H. Shapiro, and L. Kass, unpublished), and Crippen, Nilsson, and Matsson[2072-3] showed that basophils and mast cells could be distinguished from other cells using 515 nm excitation from an argon laser. If all you have is 488 nm, you're out of luck; this is much better at exciting the orthochromatic fluorescence of the dye than the metachromatic fluorescence.

Allergy Tests Using Basophil Degranulation

Basophils have receptors for IgE on their surfaces. Allergic individuals make IgE antibodies against substances to which they are allergic; these antibodies bind to the IgE receptors on basophil surfaces. Any allergen entering the circulation can cross-link IgE molecules bound to different receptors, initiating a signaling cascade that results in basophil degranulation, with the release of heparin, histamine, and a bunch of other stuff which, *in vivo*, leads to swelling, sneezing, etc., and which, *in vitro*, can be detected by chemical analysis and used as the basis of tests for allergens[1932-3].

Under most circumstances, the histamine release initiated by crosslinking of basophil cell surface IgE receptors is accompanied by loss of metachromatic staining. This forms the basis for the **basophil degranulation test (BDT)**[3098-101], in which reactivity with allergen is assessed by counting the number of basophils (i.e., cells with metachromatic granules) in an untreated control aliquot of a specimen and in aliquots exposed to putative allergens. Degranulation can also be initiated by anti-IgE antibody, which also cross-links receptors carrying bound IgE.

Milson et al[1934] and Nilsson[1935] demonstrated that the older Technicon hematology instruments, such as the H6000, which had a separate channel for identifying basophils by Alcian blue staining of their granules, could be used to do basophil degranulation tests for allergy on whole blood. Since the later Technicon H- series instruments and their Bayer successors detect basophils differently, they cannot be used for the same purpose.

Unless basophils are concentrated from peripheral blood by density gradient sedimentation or centrifugation, the number of cells present in each aliquot examined in a BDT is likely to be small enough to make results imprecise due to sampling error. This figured prominently in the so-called "Benveniste Affair" of 1988, in which results of BDT's were erroneously interpreted by Jacques Benveniste's group as demonstrating degranulation of basophils by solutions from which anti-IgE had presumably been totally removed by serial dilution, evidently lending some credence to a basic tenet of homeopathy and to the theory that water can "remember" what has been dissolved in it. Their paper was accepted and published in *Nature*[3102], with the condition that the Editor of Nature (John Maddox) be allowed to visit the Benveniste lab with colleagues of his choosing, scrutinize details of the original experiment, and observe a blinded attempt at repetition with rigorous controls. Maddox chose magician-debunker James ("The Amazing") Randi and NIH whistleblower/gadfly Walter Stewart to accompany him.

The BDT's in the Benveniste experiments were done by visual observation, and both sampling error and observer bias appear to have led the group to draw incorrect conclusions from the initial experiment. The experiment done with Maddox et al on hand found no apparent effect of high dilutions of anti-IgE, leading him and his colleagues to describe the original report as the result of a delusion[3103-4].

During the 1990's, a number of groups investigated flow cytometric assays as possible replacements for the BDT[3101]; in 1994, Sainte-Laudy et al reported that CD63 on basophil surfaces was upregulated after activation, and that the results of flow cytometric assays based on this characteristic correlated well (and substantially better than do results of assays measuring loss of metachromasia) with results of histamine release assays. The assay is commonly done using fluorescent anti-IgE antibodies to identify the basophils and anti-CD63 to detect activation; a kit (**BASOTEST**) is available from **ORPEGEN Pharma**.

In 1999, Belon et al[3105] published results of a multicenter study, based on BDTs done by visual observation, in which it was concluded that highly serially diluted histamine inhibited basophil degranulation. Since histamine at pharmacologic concentrations does exert this effect, the homeopaths (or at least the homeopath to Her Majesty Queen Elizabeth II) again claimed vindication. This led the producers of the BBC's "Horizon" science program to seek advice from a number of people involved in flow cytometry (they even posted a query to the Cytometry Mailing List) on the design of a controlled, blinded experiment in which effects of highly dilute histamine on basophil activation were measured by CD63 upregulation at two institutions in London. A transcript of the program in which results were revealed may be found at <http://www.bbc.co.uk/science/horizon/2002/homeopathytrans.shtml>.

Some years back, I attempted to develop a dedicated apparatus for performing BDTs using a cheap green He-Ne laser and Basic orange 21; today, I'd use a red laser and anti-IgE and anti-CD63 antibodies labeled with APC and APC-Cy7. However, there's no guarantee that allergists would run out and buy the gadget; some swear by the test and some don't[3101]. And some still seem to believe that homeopathy works.

Granulocytes: Eosinophils

People keep asking me about ways of identifying eosinophils. The objective is often to sort live eosinophils without activating them. There are several ways to do this.

Eosinophils are highly autofluorescent, producing strong blue (460 nm) fluorescence on UV excitation[3107] and strong green (525 nm) and yellow (575 nm) fluorescence on blue or blue-green excitation[162]. I have found that they also show some 575 nm autofluorescence on excitation at 532 nm; this is unusual. The original reports on identifying eosinophils based on high autofluorescence date back to the days when fluorescence filters were not very good and when side scatter measurements were not routine. Both of these problems have been corrected in modern flow cytometers and sorters. In my experience, when using 488 nm laser excitation, it is possible to define an eosinophil population in unstained samples by gating on a display of side scatter vs. green (same filter as used for fluorescein) or yellow (same filter as used for phycoerythrin) fluorescence. The green fluorescence is roughly the order of magnitude that would be expected from cells bearing a few thousand molecules of fluorescein, so you might want to try some low intensity fluorescein-labeled beads to make sure your instrument can detect signals in this range.

Once you have a cell sorter, sorting eosinophils can be less costly than sorting many other cell types because no antibody reagents are needed, but the costs of antibodies are, after all, rather low when compared to the cost of the sorter. Because they contain birefringent granules, eosinophils (at least those from mammalian and avian species) show higher depolarized side scatter signals than do other cells, and can be identified and sorted if one adapts an instrument to measure both polarized and depolarized side scatter (see pp. 278-9 and Figure 7-2)[710,986,3108].

If you can't live without monoclonals, I will mention that eosinophils, unlike other granulocytes, express CD4 weakly (about at the level of monocytes)[3007,3109-10]. However, some people seem to think that CD4 expression occurs only on activated eosinophils, while others think CD4 can be detected on all eosinophils, but that levels of expression vary with activation state. The truth could be established easily enough using a PE-Cy5- or APC-labeled anti-CD4 antibody and measuring polarized and depolarized scatter and 488 nm-excited autofluorescence. Unlike neutrophils, eosinophils lack CD16 and express CD49d[3111-2].

While eosinophils have higher (conventional or polarized) side scatter signals than neutrophils, there is enough overlap between the populations in unfixed preparations to preclude using side scatter as a sole parameter for identification. After aldehyde fixation, the forward and side scatter of eosinophils increases substantially, enabling them to be distinguished from neutrophils[3113]. Fixation or permeabilization also allows acid dyes (such as eosin or fluorescein) to enter the cells and bind to eosinophil basic protein in their granules, which is how the eosinophils got their name in the first place. Fluorescein isothiocyanate is an excellent stain for fixed or permeabilized eosinophils on that account[3114-5]; unfortunately, the basic protein also binds fluorescein and other acid dyes attached to antibodies, resulting in nonspecific staining. If you're interested in blocking nonspecific antibody binding, you might try an acidic molecule that is either nonfluorescent or will not be excited at the wavelength(s) you use for excitation.

Granulocytes: Neutrophils

(Clinical?) Tests of Neutrophil Function

Flow cytometry of appropriate parameters can be used for a variety of **functional studies** of blood cells. Provocative tests of **neutrophil function**[823] may use **tetrazolium dye reduction**[361,820,1611-5] (p.379), **dichlorofluorescein oxidation**[362,1616-8,1620] (p. 379), **dihydrorhodamine 123**[1622-3] (p. 379-80), **membrane potential probes**[823], or acridine orange[591] or orthogonal scatter measurement of **degranulation**[592]. **Phagocytosis assays**[374-5,1590-9] (pp. 377-8) employ fluorescent microspheres or bacteria[997].

Seligmann, Chused and Gallin[483] used flow cytometry and $DiOC_5(3)$ to demonstrate heterogeneity of **membrane potential response** to phorbol esters and chemotactic peptides in neutrophil granulocytes from normal controls and from patients with **chronic granulomatous disease**; their findings were comparable to those others obtained by bulk MP measurements. Studies of patients with chronic granulomatous disease and their families are still done using flow cytometry, with detection of respiratory burst rather than membrane potential change as the primary marker.

Moving from the granulocytes to their fellow phagocytes, the monocytes and macrophages, I note that Valet et al[372], using $DiOC_6(3)$, observed depolarization of guinea pig macrophages after incubation with lymphokines; similar results were reported by others who made microelectrode measurements of membrane potential.

Tests of neutrophil and monocyte/macrophage function are not universally used for clinical purposes. I haven't updated this section, instead electing to recommend a good and fairly recent compendium on the subject, *Phagocyte Function. A Guide for Research and Clinical Evaluation*, edited by Paul Robinson and George Babcock[2396].

Neutrophil CD64 in Inflammation and Sepsis

The high affinity Fc receptor for IgG, CD64, is upregulated on neutrophils in inflammation and infection[3116-20]. The fraction of neutrophils expressing CD64 and the level of expression are higher in patients with inflammatory syndromes and sepsis, than in patients with inflammation alone, in pregnant women with eclampsia or infection, or in controls. Bruce Davis <davisb@mmc.org> is pursuing commercial development of a point-of-care CD64 assay for sepsis through his company, **Trillium Diagnostics**. There was only one melody suitable for the following, which I wrote during one of Bruce's presentations on the subject.

When I'm Sick – CD64?

When you are healthy, blood PMNs
Don't need to express
Sites that tightly bind to Fc gamma, so
You won't see them when you do flow.
With inflammation, neutrophils show
What they lacked before.
As things get darker, what's the best marker?
CD64.

Its kinetics can let you know
When drugs take effect;
Whether you scan or flow,
This you can detect.

When tissue's hot, infected or not,
You will see the change;
Antigen will not be laid on quite as thick
If that's not what's making you sick.
For diagnosis, follow-up, too,
Who could ask for more?
It's so specific, it's just terrific-
CD64.

Platelets and Megakaryocytes

Ault[1939] has reviewed the general topic of flow cytometry of platelets on several occasions, most recently in 2001[1939,3121-2]; other recent reviews are by Michelson et al[3123] and Shankey et al[3124].

Viability and integrity of stored platelets may be assessed by **dye exclusion** or **membrane potential measurements**[998]. **Platelet activation** can be detected by measurement of **cytoplasmic $[Ca^{++}]$**[999] or by immunofluorescence analysis of **activation antigens**[1000,1936,1939] such as **P-selectin (CD62p)** and **gp53 (CD63)** or **bound fibrinogen**[1001,1939]. **Anti-platelet antibodies** can be detected using fluorescent antiimmunoglobulin antibodies[998,1939,1962].

In 1990, Kienast and Schmitz[1937] described staining of what appeared to be immature platelets by thiazole orange; increases in the numbers of such **reticulated platelets** in blood are indicators of thrombopoietic activity[1938-9]. In 1995, Ault and Knowles[3125] used *in vivo* biotinylation[3126] to label all circulating platelets in mice, and showed that reticulated platelets that appeared 24 hours later bore decreased levels of biotin (detected using fluorescent streptavidin), establishing that reticulated platelets are the youngest platelets.

Since the previous edition of this book appeared, increasing attention has been given to the presence and role of **activated platelets** in **cardiovascular disease**, including myocardial infarction and unstable angina[3122-4,3127-9]. There has also been work on **platelet microparticles**[3122] and on the interactions of platelets and leukocytes. Michelson et al[3130] report that circulating **monocyte-platelet aggregates** are a more sensitive indicator of platelet activation than is P-selectin expression. Platelet activation also results in alterations in **membrane phospholipid packing**, detectable using **annexin V**[3131] (which was used for this purpose some years before its application to the detection of apoptosis) or **merocyanine 540**[3132]; these reagents do not seem to have been widely used for platelet studies.

Thrombopoiesis is the result of the activity of **megakaryocytes**[992-6] in the bone marrow. These cells undergo **endomitosis** as the result of abortive mitosis[3133], and may have G_0/G_1 DNA content of 32C or higher, as compared to 2C for other G_0/G_1 phase cells. Megakaryocyte DNA content is increased in patients with coronary artery atherosclerosis[3134]. Platelets are formed by fragmentation of megakaryocyte cytoplasm; the induction of the fragmentation process appears to involve **nitric oxide**[3135].

Kuter and Rosenberg and their colleagues[1940-2] made incremental changes in platelet counts in rats by infusion of antiplatelet antibody or by platelet transfusion, and measured the ploidy response of megakaryocytes by flow cytometry, using propidium fluorescence corresponding to 4C or higher DNA content and fluorescein immunofluorescence from antiplatelet antibody as gating parameters. The flow cytometer involved in these studies was a Cytomutt, built at a cost of less than $20,000. Proportional changes in megakaryocyte ploidy were demonstrated; as the platelet count declined, ploidy increased. Using the megakaryocyte ploidy assay on bone marrow cultures, these workers identified and purified thrombopoietin; expending considerably less manpower and money than other groups that also isolated this elusive growth factor at about the same time.

The instrument Kuter and Rosenberg used was built shortly before the B-D FACScan became available, and I initially thought they would have been able to use a FACScan or one of the other commercial benchtop cytometers to do their work, but I was told that none of the instruments they tried could keep up with the data rate. Since only 1 of every 1,000 or so cells in the marrow is a megakaryocyte, even in the megakaryocyte-enriched cultures, you have to look at a few million cells to get a DNA histogram from a few thousand megakaryocytes, and the Mutt could evidently run faster than its pedigreed competitors. Shows you how much I know.

Hematopoietic Stem Cells

In the mid to late 1970's, a period in which most people were content to make single-parameter measurements, a considerable amount of effort and ingenuity was devoted, in a few centers, to the isolation of **hematopoietic stem cells** from animal marrow by multiparameter sorting on the basis of forward and orthogonal scatter, lectin binding, and autofluorescence [161,177-9]. The Geneva convention makes it impractical to do the same confirmatory spleen colony assays in man as can be done in the mouse, so flow cytometric approaches to human marrow and stem cell analysis came from different directions.

The emergence of marrow transplantation as a therapeutic procedure for treatment of leukemias, lymphomas, and solid tumors made normal donor marrow as well as pathological specimens more widely available, and flow cytometry has provided a visible and accessible means for the clinical researchers involved in transplantation programs to study the repopulation of the marrow and, at the same time, assess the efficacy of treatment. During the past decade, multicolor immunofluorescence flow cytometry has been used to provide further detail on normal blood cell development in the marrow[788,990,1154,1212,1498,1755-71,3136]; one aspect of this was discussed in detail in connection with leukemia phenotyping on pp. 467-8, which referred to two color illustrations on the back cover.

Over the same period in which flow cytometric immunofluorescence measurements of increasing sophistication were applied to marrow, different investigators extracted increasingly detailed cytokinetic information from marrow samples using DNA content, DNA/RNA, and chromatin structure analysis and BrUdR incorporation. There probably should be some more studies done than have been to date of the **kinetics of subpopulations, defined by immunofluorescence or other additional parameters**[991,1758], in marrow. The one cell type for which combined DNA and immunofluorescence analyses have been the rule rather than the exception seems to be the megakaryocyte, discussed in the previous section.

At present the isolation of human hematopoietic stem cells is big enough business to keep a number of biotechnology companies funded. I picked a few references[1762,1768,1943-61] on flow cytometry and human stem cells out of the then-recent literature for the previous edition; I have picked even fewer out for this one. If you look at any of the older references, you will quickly appreciate that, even a decade ago, stem cell research demanded the capacity to measure as many parameters as flow cytometer manufacturers are willing to make available. That still holds.

Human **CD34$^+$** hematopoietic stem cells are and lack lineage antigens characteristic of more mature cell types. Like rodent stem cells, they have an active glycoprotein pump[1949], which allows them to be identified by decreased rhodamine 123 retention (p. 376)[1954]. By keeping the concentration of pyronin Y low, and using verapamil to block dye efflux, Srour and his colleagues[2340-2,2627] have been able to use Hoechst 33342 and pyronin Y to define and sort CD34$^+$ stem cells from various stages of the cell cycle.

Cells capable of initiating hematopoiesis in long-term marrow culture lack CD45R; the primitive erythroid colony forming cells are CD45RO$^+$, while most granulopoietic colony-forming cells are CD45RO$^-$ [1947]. T cell precursors and possible lymphoid stem cells express terminal deoxynucleotidyl transferase[1950]. Hematopoietic stem cells can be found and counted in and isolated from peripheral blood[1944,1948,1951]; a higher proportion of peripheral blood stem cells than marrow stem cells express CD33[1948].

Although CD34$^+$HLA-DR$^-$ [1952-3,1957-8] stem cells are more primitive, CD34$^+$CD38$^-$HLA-DR$^+$ cells can give rise to multiple cell lineages in liquid culture[1961]. Pluripotent progenitor function has also been associated with a CD34$^+$Thy-1$^+$Rh123-dim cell subset[1954] and with expression of the c-kit ligand receptor[1953]. In 1992, it was reported[1957] that a CD34$^+$HLA-DR$^-$ population could give rise to mesenchymal stromal elements as well as multiple blood cell lineages; this was apparently not the case[1958], although even more primitive cell populations with such potential are now being studied.

Clinical Application: Monitoring CD34$^+$ Stem Cells According to the ISHAGE Protocol

During the past decade, increasing use has been made of CD34$^+$ **peripheral blood progenitor cells** in autologous and, to a lesser extent, in allogeneic transplantation[3137]. Because these cells normally represent less than 0.1% of the total nucleated cells in blood, it is necessary to use colony-stimulating factors such as G-CSF and GM-CSF to mobilize progenitor cells into the circulation. Counts of CD34$^+$ cells must then be done to determine the appropriate time for collection. Counts are also useful in following clinical progress of recipients of both marrow and peripheral stem cell transplants; multiparameter flow cytometry has become indispensable for conting CD34$^+$ cells, and may also assist in the process of detecting and removing residual cancer cells in marrow[1956] or material collected from peripheral blood.

A widely used set of guidelines for CD34$^+$ cell counting was developed by a committee of the International Society for Hematotherapy and Graft Engineering (**ISHAGE**, usually pronounced "ice age")[3138]. The Society has since changed its name to the International Society for Cellular Therapy (ISCT), but the ISHAGE guidelines persist. Stem cells are identified by their CD34$^+$CD45^{dim} phenotype and forward and side scatter signals in the same range as those of CD45^{bright} lymphocytes; 7-AAD staining is used to eliminate nonviable cells from analysis and beads are added to samples to derive an absolute count. Examples of the multiparameter gating procedure as implemented on benchtop analyzers from BD Biosciences and Beckman Coulter can be found online at ISCT's Web site: <http://www. celltherapy. org/committees/Committees/Graft_Evaluation/graft.htm>.

Side Population (SP) Stem Cells: Plastic, Fantastic

In 1996, Goodell et al[3139] reported isolation of a population of primitive stem cells from mouse marrow that were distinguished by low Hoechst 33342 fluorescence in the blue (440-460 nm) and red (>675 nm) spectral regions. They subsequently found cells with similar staining characteristics and properties in other species, including humans[3140]. These **side population (SP)** cells are CD34-negative and lineage negative, and appear to have the capacity to differentiate into cell types other than those found in the blood and immune system, e.g., liver and cardiac muscle cells[3141]. Preffer et al[3142] have found cells with the staining characteristics of SP cells in human peripheral blood; these cells, however, do not exhibit the extreme developmental plasticity just mentioned, and, thus far, have only been found to give rise to lymphoid and dendritic cell precursors in culture. SP cells in murine bone marrow are illustrated in Figure 10-17; Goodell has produced a definitive article on their detection[3143].

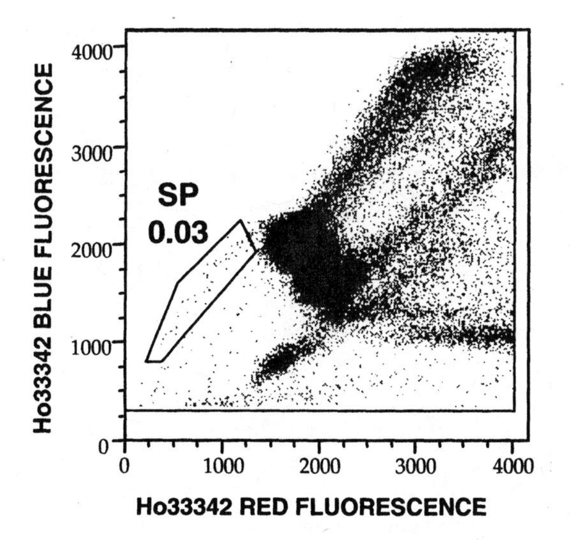

Figure 10-17. SP cells identified on a two-parameter display of blue vs, red Hoechst 33342 fluorescence of cells from murine bone marrow. Courtesy of Margaret Goodell, Baylor College of Medicine.

It was originally thought that Hoechst dye efflux from SP cells might be mediated by p-glycoprotein and/or the multidrug resistance protein MDR1 (p. 376). However, it was established in 2002 that the responsible transporter is from the **breast cancer resistance protein** family; the protein found in murine cells is **Bcrp1**[3144], while that found in human cells is **ABCG2** [3145-6]. This makes it possible to use fluorescent antibodies to ABCG2, or probes for the gene, rather than Hoechst dye fluorescence, to identify, or confirm the identity of, putative SP cells. The latter approach was taken by Lechner et al[3147], who described an apparent progenitor cell population in pancreatic islets.

Immunology

Immunologic applications have almost certainly provided the motivation for purchase of the majority of cell sorters and a substantial fraction of the flow cytometers now in use. The most common application of flow cytometry to immunology involves the detection and quantification, using fluorescently labeled monoclonal antibodies, of **surface antigens characterizing different stages of development of cells of the immune system**. My citation of only a tiny fraction of the tremendous number of publications on this topic[317-22,787-8,794] in previous editions did not stem from a lack of appreciation of the tremendous achievements represented by the development of the reagents and methodology and the elucidation of the pathways of differentiation of the immune system in mice and men. There were and are just too many good papers.

Immunologic Applications of Flow Cytometry: Still a Growth Industry

Flow cytometry aided in the identification of new surface antigens[907], in the demonstration of their specificity for particular cell lineages and states of maturation, and in the elucidation of their **receptor function**[908]. Originally, cells found to bear a single antigen were sorted and their functions defined in terms of cell-cell interactions *in vitro*; the trend toward multiparameter analysis and multicolor immunofluorescence studies[909-11] has continued, as has examination of cell function in terms of signaling-related ion fluxes and ion channels[912-3] as an adjunct to the more familiar demonstrations of cytotoxicity, help and suppression. Flow cytometry also continues to be applied to studies of **immune cell activation** by antigens and lymphokines. In fact, the technique is so commonly used in immunology that a significant fraction of papers in the literature in which multiparameter flow cytometry is used in nontrivial ways, but in which the term itself does not appear in the abstract or title, do not get assigned "flow cytometry" as a keyword, and never make it into the MEDLINE database.

The continuing development and commercialization of antibodies, labels, and instruments have made six- and eight-color immunofluorescence available to increasing numbers of users. Last time around, getting the reagents was the biggest problem, even for the relatively small number of people who had instruments capable of three-color fluorescence measurement. Now, with reagents and instruments in abundance, we have to worry most about making sense of all the data.

In the First Edition of this book, I made particular mention of the **use of cell sorting in isolation and selection of hybridomas and hybridoma variants**, which had been reviewed by Dangl and Herzenberg[549]. I used to see a lot of people use a lot of flow cytometer (and sorter) time screening hybridoma clones, and urged them to consider any possibly more efficient alternative[550]. There are now not only alternatives for hybridoma screening, but alternative sources of monoclonal antibodies, with viruses, bacteria, and fungi now standing in for the human immune system.

In the Second Edition, I singled out an article by Cantrell and Smith[551] on the regulation of T cell growth by the lymphokine interleukin-2 (IL-2), which dealt as much with cell biology as with immunology, as an example of extremely effective use of the technology of flow cytometry and sorting. I am gratified that a substantial fraction of current publications involving flow cytometry in immunology make similarly effective use of the technology, even if I didn't have anything to do with it.

In the Third Edition, I mentioned a nice study by Heagy et al[2289], on the inhibition of immune function by antiviral drugs, including AZT and some others that are routinely given to patients with HIV infection. The paper illustrated some good flow cytometric methodology for looking at lymphocyte activation, by 1994 standards, and, while we have a much more sophisticated repertoire of analytical methods with which to deal study that process today, we don't always make full use of what's available.

HIV Infection – The "Killer Application"

Clinical flow cytometry wouldn't be what it is today if it hadn't been for the AIDS epidemic. As soon as it was established that the disease was associated with low CD4$^+$ T cell counts, way too many people went out and bought flow cytometers. Although it has been pointed out that there are a lot of ways in which flow cytometry can be used in the study of HIV infection[2280-1,2397], the clinical focus, at least until recently, has been on CD4$^+$ T cell counting, and on getting the many labs involved with it to do it well enough to make the data usable[2282-5].

Shortly before I wrote the Second Edition of this book, Peter Duesberg[1018] proclaimed that HIV didn't cause AIDS. While very few people agreed with that, the ensuing debate did make it clearer that HIV infection is nothing at all like a typical lytic virus infection, which was the picture initially painted by some virologists who ought to have known better, and not even all that much like some other retroviral infections. As Lewis Thomas suggested, it can be helpful to find out that you didn't know things you thought you knew.

In addition to counting T cells, flow cytometry has proven indispensable for the studies of cellular immunology that are essential to understanding HIV infection. There have been demonstrations that flow cytometry can be used in a quasi-diagnostic way to detect serum antibodies to HIV[2287], or to HIV vaccine components[2288], or to find HIV antigens[2162-5] or mRNA[1403] in cells, but none of these initiatives has led to an accepted clinical test.

When the previous edition of this book appeared, it looked as if correcting altered glutathione metabolism in leukocytes[1117-8] might brighten the grim clinical picture in HIV infection, and some of us worried as to how glutathione measurements could be implemented given that most instruments didn't have the right light sources to excite the best reagents. I cited a then-recent paper by Pantaleo et al[2290] on the immune response to primary HIV infection, in which flow cytometry and PCR assays both demonstrated oligoclonal expansion of CD8$^+$ T cells, suggesting that cytotoxic cells are involved in clearing the initial viremia. Another interesting paper[2291] suggested that activated CD8$^+$ cells, i.e., cells expressing CD28 and HLA-DR, are responsible for anti-HIV activity later in the course of the disease. The paper on primary infection was pretty dramatic, not least because most, if not all of the patients studied had had the opportunity to avoid being infected with HIV.

In the years since, Highly Active Anti-Retroviral Therapy (**HAART**) has come on the scene, and patients with access to treatment are feeling better and living longer. We have learned a great deal more about the cellular dynamics of HIV infection, and the developed world seems to have been motivated by the combined forces of compassion, competition, and compulsion to bring programs for prevention, diagnosis, and treatment to the developing world. That's all good news; the bad news is that many people who could easily avoid becoming infected still opt to take their chances.

Clinical Application: T Cell Subset Analysis

The literature in this area is vast; a lot of the early clinical work with immunofluorescence flow cytometry literature described changes in the **ratios of B- and T cells**, and/or changes in the **ratios of helper/inducer T cells to cytotoxic/suppressor T cells** in various **disorders of immune function**, e.g., **autoimmune diseases** and **immunodeficiency syndromes**. It is clear that what some of us might consider mundane, single parameter immunofluorescence flow cytometry provided researchers and clinicians with a valuable tool for assessment of the function of the immune system.

Things got somewhat out of hand in the few months after the general public became conscious of the existence of AIDS (see p. 99), when a lot of flow cytcmeters were bought and sold on the faulty premise that subset analysis would be usable as a screening test. As the disease became better understood, it did become clear that the proportion and absolute count of CD4-positive T cells in peripheral blood were useful in following and predicting the clinical course of HIV infection, and attention turned to improving the precision and accuracy with which T cell subset analysis could be done for this clinical use.

As flow cytometry for lymphocyte subclassification first became popular, questions were raised about the equivalence of results obtained by different methods. In particular, it was reported that **Ficoll/Hypaque separation procedures used to prepare mononuclear cells for immunofluorescence analyses caused selective loss of CD8+ (cytotoxic/suppressor) T cells**[593-4,1002]. Such loss did not occur when leukocytes from whole blood were analyzed, after erythrocyte lysis, with **gating on forward and orthogonal scatter signals used to identify lymphocytes** for purposes of immunofluorescence measurement[175-6].

The first widely used guideline for immunophenotyping, published by the National Center for Clinical Laboratory Standards (now just NCCLS) in 1992[1876], recommended the procedure be done on lysed whole blood rather than separated lymphocytes. Subsequent guidelines from the Centers for Disease Control (CDC)[1877-8] and the Division of AIDS (DAIDS), National Institute of Allergy and Infectious Diseases, NIH[1879] also required the use of whole blood lysis.

By the late 1980's, it was recognized that gates set on the basis of forward and orthogonal scatter signals might both exclude some lymphocytes and include other cells, such as monocytes. The **back-gating** technique (see p. 277) was proposed by Loken et al[1251] as a means of determining the purity of the gate and the fraction of the lymphocytes included; the NCCLS standard (revised in 1998[3148]) recommended that the quality of the gate be determined, while the later CDC and DAIDS guidelines made this mandatory.

Many of the problems encountered with determination of **absolute CD4+ T cell counts** relate to the use of **"dual-platform"** methodology, in which total leukocyte counts and lymphocyte percentages derived from hematology analyzers are used in combination with percentages of CD4+ lymphocytes derived from fluorescence flow cytometers to calculate absolute counts. It has been established that **"single-platform"** methods, employing only one instrument, produce results with lower variability. Single-platform CD4+ cell counts can be done on conventional fluorescence flow cytometers if indicator beads at known concentrations are added to samples (see pp. 20-1) to provide cell counts[3149-50]; beads need not be used with instruments that incorporate volumetric sample delivery[3151-3]. A recent report by Bergeron et al[2917] (see p. 448) suggests that sample delivery rates of at least some benchtop flow cytometers are stable enough so that absolute counts could be done by adding beads to only the first and last sample in a run.

Most laboratories in affluent countries now routinely use three- and four-color fluorescence methodology for T cell subset analysis, with CD45/CD3/CD4 staining[3154] providing specific identification of lymphocytes. In a four-color instrument, CD45/CD3/CD4/CD8 staining allows accurate determination of all major T cell subsets in a single tube, as shown on pp. 33-5 and in Figure 1-17 (p. 34). It has been established that laboratories that use CD45 do better on quality assessment studies than laboratories that do not[3155]. The newest revisions of the DAIDS[3156] and CDC[3157] guidelines accommodate these recent advances in methodology, but also note that some further modifications in technique may reduce the costs of the procedure.

The April 15, 2002 issue of *Cytometry (Clinical Cytometry)* (Volume 50, Number 2, pp. 39-132) is a Special Issue on "CD4: 20 Years and Counting"[3158], edited by Frank Mandy.

T Cell Subsets: Alternative Technologies

A typical benchtop flow cytometer is an expensive instrument, and consumes a substantial amount of power, largely due to its use of an argon ion laser as a light source. Flow cytometer manufacturers have not been completely oblivious to the fact that millions of people with HIV infection live in areas where health care budgets are minimal and practical instruments may need to be run on battery power.

B-D's FACSCount (p. 418), a small, inexpensive flow cytometer intended exclusively for use in T cell subset analysis[2472,3159] was the industry's first effort at developing a CD4+ cell counter for resource-poor areas. The FACSCount uses a green He-Ne laser source, and reagents (antibodies to CD3/CD4 and CD3/CD8 and counting beads) come in prepackaged tubes. In recent years, a number of systems have been placed in Brazil, India, and Thailand.

George Janossy, in London, working with Debbie Glencross in South Africa and Ilesh Jani in Mozambique, among others, felt that flow cytometric CD4+ T cell counting could be made still more affordable by eliminating counting beads, minimizing reagent costs, and using cytometers with red diode laser sources. Progress along these lines can be assessed as detailed at **<http://www.affordcd4.com>** as well as in a number of scientific publications.

Sherman et al[3160] had shown in 1999 that CD4+ T cells could be identified accurately, and adequate counts obtained, by staining only with anti-CD4 antibody and determining the number of CD4+ cells in a "lymphocyte gate" defined on the basis of forward and side scatter signals. Janossy, Jani, and Göhde[3161], using a Partec prototype instrument, next demonstrated that CD4+ T cell counts obtained using **primary CD4 gating**, in which the cells of interest are identified in a two-parameter display of side scatter vs. antibody fluorescence, were equivalent to those obtained using conventional apparatus and multiple antibodies. Anticipating possible problems with sample deterioration in tropical sites, Jani et al[2596] showed that Transfix[2597], a fixative originally developed by Barnett et al for stabilizing whole blood controls, kept counts stable in samples maintained at 37 °C for as long as three days.

Glencross et al[3162] introduced the concept of **PanLeucogating**, using a combination of antibodies to CD45 and CD4 or CD8, to identify the total leukocyte population on a fluorescence flow cytometer, allowing a more accurate CD4+ cell count to be determined by a dual-platform method; Janossy et al[3153] established the utility of PanLeucogating in single-platform testing using a volumetric instrument.

The Luminex 100 flow cytometer (pp. 431-2), while designed for multiplexed biochemical analyses on beads, has been shown to be usable for CD4+ T cell counting[2447,3163]. The instrument has a volumetric sample delivery system, eliminating the need to use counting beads, and a PanLeucogated CD4+ cell count, using APC- and APC-Cy7-labeled anti-CD45 and anti-CD4 antibodies, can be implemented on a stripped version of the instrument using only the red diode laser and APD detectors.

Partec's Web site (http://www.partec/de) documents the incorporation of one of their CyFlow® flow cytometers (pp. 427-8) into a mobile laboratory in an SUV with off-road capability; the instrument uses a red diode laser source and identifies CD4+ cells in unlysed blood by single-parameter fluorescence measurements.

Other approaches to more affordable CD4+ cell counts and counters have used techniques other than flow cytometry. Since the number of CD4 epitopes per T cell in HIV-infected individuals is the same as in controls, and remains essentially constant, even with disease progression[1492], and since there are small or negligible amounts of CD4 present in serum and on monocytes and other blood cells, bulk determinations of CD4 content per unit volume of blood, especially when monocytes have been removed, should provide essentially the same information as counts of CD4-positive T cells. Such assays, which do not require the use of flow cytometers or other expensive instrumentation for readout, were developed by **Zynaxis** and by **T-Cell Sciences**; both the companies and the assays seem to have disappeared.

The IMAGN 2000 instrument[1365,2641] developed by **Biometric Imaging, Inc.** (since acquired by **BD Biosciences**) uses a scanning He-Ne laser and some other technology borrowed from bar-code readers. Sample cartridges contain anti-CD3 and anti-CD4 antibodies labeled with Cy5 and Cy5.5 (p. 336); rotation of the sample cartridge carousel mixes a known volume of unlysed blood with antibodies and introduces stained sample into a capillary of defined volume in which cells are measured; this gives the process the name "volumetric capillary cytometry," although "microvolume fluorometry" is also used. The IMAGN 2000 was found to be an effective single-platform system for CD4+ T cell counting[2484,3164], although precision for low counts was suboptimal. The IMAGN 2000 could be, and was, adapted fairly easily to do other cytometric chores, predominantly associated with **blood banking**[3165-76]. However, BD abruptly decided to stop selling (and, apparently, stop supporting) the system; this had little impact in the CD4+ counting world but left several blood bankers I know unhappy. Rumors that the IMAGN 2000 will be resurrected persist, but Elvis is still out of the building.

A couple of years ago, I prodded a number of people in academia and industry, all involved in the development of cytometry technology, to consider how today's technology could be applied to develop a small, rugged, inexpensive CD4+ T cell counter. The most interesting response came from Arjan Tibbe, then a doctoral student working at the University of Twente, in the Netherlands, under the direction of Jan Greve and the late Bart DeGrooth, and collaborating with Leon Terstappen of **Immunicon**. Arjan designed and tested a prototype instrument called the "**EasyCount**," based on imaging technology; in my opinion (and I do not have any financial interest in this instrument), it represents the most promising approach to doing reliable and economical CD4+ T cell counts and, perhaps, a number of other cell assays, in both resource-poor areas and the developed world, where health care costs could also do with some reduction.

Arjan's thesis work, also done in collaboration with Leon Terstappen, involved the development of a small, simple, laser scanning cytometer, the **CellTracks™** [2383,3006-9], now being commercialized by Immunicon for rare cell detection (p. 470). The CellTracks system uses disposable cartridges with an observation window featuring a series of deposited nickel lines. Cells tagged with antibodies bound to ferrofluid nanoparticles are introduced into the chamber, which is then placed in a magnetic yoke, aligning the antibody-tagged cells between the nickel lines, facilitating scanning (Immunicon's observation chambers and magnetic yokes are available from Molecular Probes in their Captivate™ product line, as is ferrofluid-conjugated streptavidin). While the CellTracks, which uses a red diode laser as a primary light source, can perform two- and three-color immunofluorescence analyses, including CD4+ counts, the objective of the EasyCount design exercise was to produce an even simpler system, ideally one with no moving parts.

The EasyCount is basically a low-power imaging fluorescence microscope, with illumination from a blue LED. A whole blood sample is mixed in a disposable plastic chamber with acridine orange, which renders all the nucleated cells intensely fluorescent, and with biotinylated anti-CD4 antibody and ferrofulid-conjugated streptavidin. The observation chamber (which, unlike the CellTracks chamber, does not have nickel lines on its surface) is then placed in the magnetic yoke, which draws cells bearing a relatively high density of antigen (meaning all of the CD4+ T cells and some monocytes) to the top of the chamber. The microscope, which operates at a low enough magnification so that focus adjustment is not needed, makes a fluorescence image of most of the viewing area on a CCD chip of the type used in relatively inexpensive digital cameras; software programmed into the chip allows the cells in the viewing area to be identified and counted, allowing the CD4+ T cell count to be calculated from the initial volume of blood used and the percentage of the chamber area in the field of view. The field of view is large enough to permit counting enough cells to keep the precision of the count at better than 10% when the number of circulating CD4+ T cells is as low as 200 cells/microliter. The count takes about a minute. The EasyCount is about the size of a toaster, but consumes much less power; it can be run for hours on batteries. It has no moving parts, requires minimal operator training (pipetting sample and reagent are about all that is required of the operator), and can be manufactured for around $1,000, meaning that it could be sold at a profit for $5,000, which is less than many hospitals in developing countries are paying for fluorescence microscopes which they now use to do substantially less accurate and precise CD4+ counts with **Dynal**'s magnetic beads[3177]. The consumable/reagent costs for EasyCount will probably be under $1. The EasyCount can also do a total leukocyte count, using an anti-CD45 antibody instead of the anti-CD4 antibody, and probably can also do a total lymphocyte count, using a cocktail of anti-CD3, anti-CD19 or CD20, and anti-CD56 antibodies, allowing calculation of the percentage of CD4+ T cells among lymphocytes.

What the EasyCount does is essentially a primary CD4-gated CD4+ T cell count; its results should therefore be expected to be comparable to those obtained using this methodology in flow. George Janossy and Frank Mandy are now looking retrospectively at some flow cytometric data to determine whether primary CD4-gating produces results equivalent to PanLeucogating or gating based on both CD45 and CD3 staining, and I suspect this will turn out to be the case. There may be some suggestion that this strategy will work in the forthcoming version of the CDC guidelines for CD4+ T cell counting, which Frank Mandy has helped develop. The World Health Organization is now funding construction of several EasyCount systems for clinical testing. For information, e-mail lterstappen@immunicon.com .

Because the EasyCount, like ChemoMetec's NucleoCounter (p. 448), examines cells that are brightly stained, relatively short observation times can be used even though the LED illumination sources are of relatively low intensity. A somewhat more adventurous approach to a simple CD4+ cell counter was recently described by William Rodriguez and Bruce Walker of Harvard Medical School, working in collaboration with John McDevitt at the University of Texas[3178]. This group developed a number of microfluidic devices intended to perform low-cost biochemical assays in HIV patients; an initial effort at extracting information on CD4+ counts by measuring soluble antigen has led to work on an imaging counter that identifies CD4+ cells based on the fluorescence of bound labeled anti-CD4 antibody.

Small, simple imaging counters with low associated instrument and reagent costs, such as those just described, should be usable in the many areas of the world in which HIV infection is prevalent and in which there is not a reliable supply of clean drinking water, let alone sheath fluid. If, as I expect, these instruments provide the same information as is now provided at higher cost by flow cytometers, it won't be long before activism and/or budgetary constraints bring the more cost-effective technology to the developed world. I'm sure we'll find other things to do with our flow cytometers; I'm also sure we'll find low-cost scanning and imaging methods for doing some of those other things. Ultimately, that should leave us with a few pennies more to spend on other areas of research and patient care.

Clinical Application: Transplantation

Horsburgh et al[3179] have recently reviewed the applications of flow cytometry to histocompatibility testing.

Before fluorescence flow cytometry existed, it was established that the presence of antibodies against donor cells in the serum of a kidney transplant recipient greatly increased the likelihood that the organ would be rejected. Crossmatching developed to detect such antibodies in potential recipients initially were based on the detection of lysis of donor cells by complement in the presence of recipient serum. However, this method failed to detect recipient antibodies that did not fix complement, and the presence of such antibodies was also associated with an increased likelihood of rejection.

In 1983, Garovoy et al[1880] developed the **flow cytometric crossmatch**, in which recipient serum is mixed with donor cells; antidonor antibodies are detected by staining the cells (typically, separated T cells, which do not normally bear immunoglobulins on their cell surfaces) with a fluorescent antihuman immunoglobulin reagent. Subsequent studies[1881-7] have shown that flow cytometric crossmatching can be more sensitive than the conventional method, particularly in high-risk recipients, e.g., those in whom grafts have previously failed, and continue to confirm the predictive power of the flow crossmatch[3180-].

Flow cytometric crossmatches are now done using two color fluorescence techniques to identify T or B or other specific donor cell types[1888]. Wang, Terashita, and Terasaki[1889] reported that the use of platelets as target cells reduced the false positive rate, but Terasaki platelets evidently did not become as popular as Terasaki plates. Wetzsteon et al[1890] developed flow cytometric methods to discriminate cytotoxic from noncytotoxic antibodies, improving sensitivity and specificity. Talbot et al[1891] recommend the addition of HLA-DR matching to the flow cytometric crossmatch, while Lazda[1892] reported that a strongly positive B cell crossmatch identifies a high-risk subset of patients. Scornik et al[1893], in a 1994 evaluation of flow crossmatching, noted needs for improved methods of measuring B cell antibodies and quantification of T cell antibodies; Berteli et al[1894] described a procedure to determine the specificity of anti-B cell antibodies. The need for quantification has been met, at least to some extent, using materials and methods discussed on pp. 354-9.

One Lambda, Inc., a company founded by Paul Terasaki, produces the **FlowPRA®** line of assays for **panel-reactive antibodies** that allow flow cytometry to detect the presence of antibodies to HLA Class I and/or II antigens; the reagents are beads, each of which bears an individual antigen. It should be fairly obvious that this type of analysis is ideally suited to multiplexing, and, indeed, One Lambda has licensed Luminex's technology and offers both color-coded antibody-bearing beads[3184] and its own private-labeled instrument, the **LABScan™ 100**. Flow crossmatching against

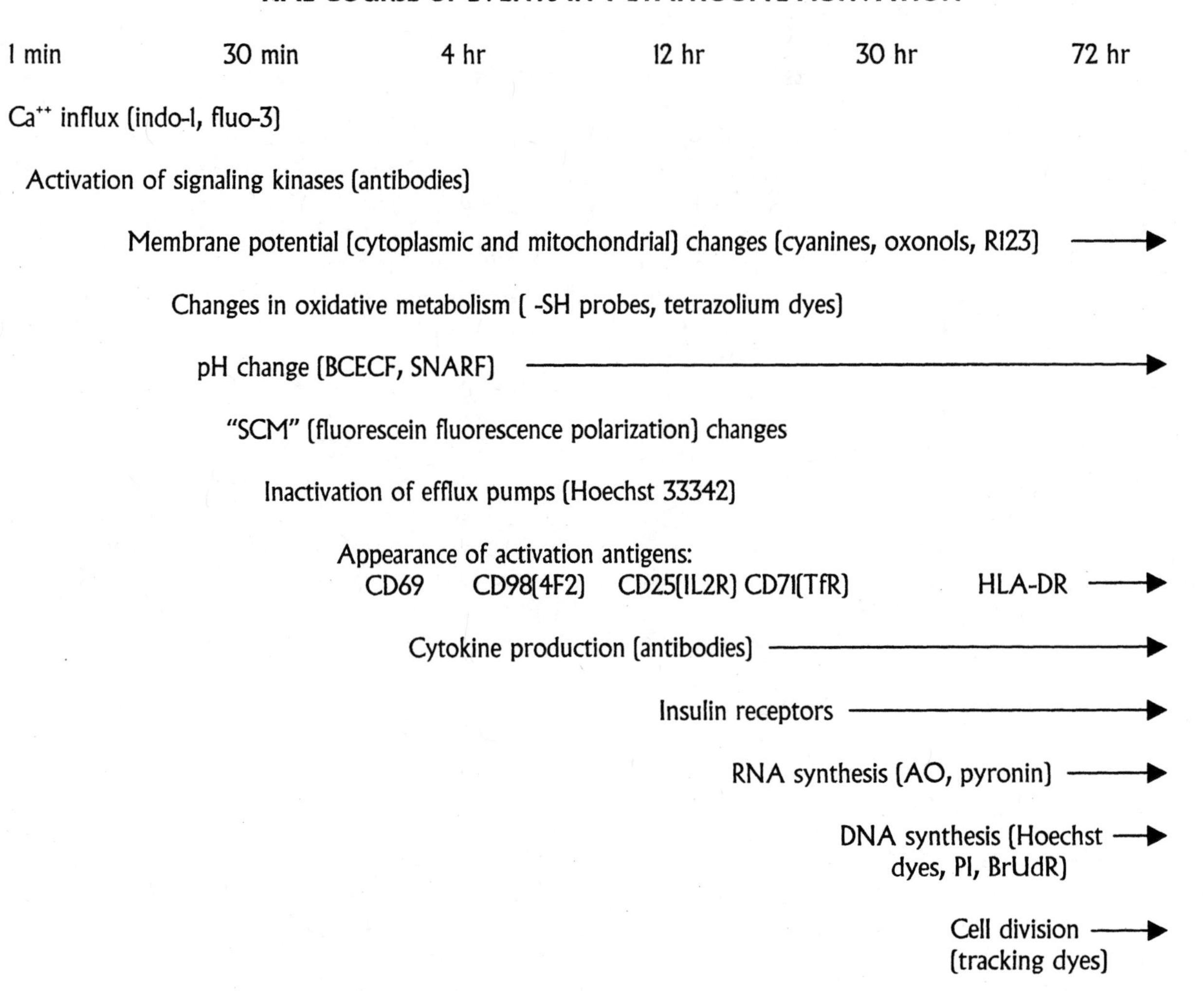

Figure 10-18. Time course of events in T lymphocyte activation and probes for their cytometric detection.

beads seems to have become popular; however, while most patients who reject transplanted kidneys have anti-HLA antibodies[3185], monitoring antibody levels by FlowPRA is not sensitive for diagnosis of early rejection[3186].

Other flow cytometric techniques have been used with success for diagnosis of **graft rejection**. In 1981, Cosimi et al[1003], at the Massachusetts General Hospital, reported that changes in T cell subset ratios could be used for monitoring renal transplant recipients, with a rise above 1.0 in the $(CD4^+)/(CD8^+)$ ratio indicating an impending rejection episode. This finding did not apply to some other populations, even in Boston; at the time, I was collaborating with Terry Strom, then at the Brigham and Women's Hospital, where a much less stringent immunosuppressive regimen was in use. The ratio of $(CD4^+)/(CD8^+)$ cells in patients at the Brigham was often greater than 1.0 in the absence of evidence of rejection; we accordingly looked at DNA synthesis in T cell subpopulations in an attempt to detect the lymphocyte activation we expected should occur during a rejection episode. We found[604] that an increase in the number of S/G_2 /M phase $CD4^+$ cells predicted a rejection episode, while an increase in the number of proliferative $CD8^+$ cells did not. We[605-6] and others[1895] thus turned our attention to means of detecting lymphocyte activation, for diagnosing rejection and for other purposes as well.

More recently, Yu et al[3187] have studied the leukocyte count and differential count in urine post-transplant as a predictor of renal graft rejection; Stalder et al[3188] proposed the use of markers of activation as a monitor of immunosuppressive therapy, after finding fewer indicators of activation in immunosuppressed graft recipients than in normal controls.

Detecting Lymphocyte Activation

There continues to be considerable interest in the development of **flow cytometric measures of lymphocyte acti-**

vation, both for basic research and for clinical purposes which include tissue typing and organ matching for transplantation, evaluation of cellular immune response, assessment of the activity of autoimmune disease processes, and monitoring of transplant recipients for early signs of graft rejection. Detection of activated T cells is perceived as providing a more precise indication of the dynamics of immune function, in these contexts, than could be obtained simply from counting absolute or relative numbers of different lymphocyte types in the blood. Figure 10-18 shows the time course of a number of events in T lymphocyte activation that may be observed by flow cytometry, and lists some probes that might be used in each case.

Foundations: From PHA (the Lectin) to PHA (the Pulse Height Analyzer)

When the previous editions of this book appeared, much of our understanding of lymphocyte activation was based not on the real thing, but on a convenient surrogate model, i.e., polyclonal cell activation by mitogenic lectins, such as phytohemagglutinin (PHA) and concanavalin A (con A) or antibodies, e.g., anti-CD3, which reacts with the constant region of the T cell receptor.

When I was a medical student at New York University in the early 1960's, the people who were most excited about PHA were the geneticists. When the lectin was added to peripheral blood, lymphocytes started dividing, providing numerous mitoses in which chromosomes could be counted and, to some extent, identified; chromosome banding, on which classification is now based, had not yet been developed. Cytogenetics itself was in such a sorry state that it was not established until after PHA came into use that the normal human chromosome complement was 46 and not 48 chromosomes.

Within a couple of years, it was discovered that certain antigenic preparations, such as tuberculin, could act as mitogens, but only when the lymphocytes to which they were added came from individuals who would exhibit delayed hypersensitivity skin responses to subcutaneously injected antigen. In 1964, Fritz Bach and Kurt Hirschhorn, then at N. Y. U., discovered that lymphocytes from two different individuals would, when mixed, proliferate; they noted that such a **mixed lymphocyte reaction (MLR)** could provide an *in vitro* test of histocompatibility[607].

Lymphocyte activation, whether by mitogens or antigens, has, since the 1960's, been detected and quantified by bulk measurements of the incorporation of **tritiated thymidine** (^{3}H-TdR), which is taken as an indication of the number of cells in a sample involved in **DNA synthesis**, i.e., the number of activated lymphocytes. Since DNA synthesis does not occur for some 30-36 h following lymphocyte stimulation, no indication of cell activation can be obtained either from ^{3}H-TdR uptake or from flow cytometric assessment of changes in DNA content at points earlier in the time course of the process. Lectin-activated and, later, mitogenic antibody-activated cells became widely used as a model for activation because the lectins commonly used stimulated most peripheral blood lymphocytes, that is, almost all of the T cells, making it possible to detect stimulation using ^{3}H-TdR within 48 hours, while, when lymphocytes were stimulated by antigens or in an MLR, a much smaller fraction of the population was activated, with the result that it was necessary to wait for many generation times, typically, 5-10 days, until the number of proliferating cells in the culture became large enough to be detectable by ^{3}H-TdR incorporation.

In the late 1960's, Fred Valentine and others at N.Y.U. followed proliferation of antigen-stimulated cells by **time-lapse cinematography** [2943-4]. It's hard to argue with these data; they revealed a mean doubling time of 12 h with the shortest time between 2 divisions of a given cell and its daughters being 8.5 h. ^{3}H-TdR uptake data demonstrated a doubling time in incorporation of about 10 to 12 hours, in the same ballpark. Under appropriate conditions, most cells entering the proliferative phase continued to divide for 3-5 days.

By the mid 1970's, it had been established, although it was not widely appreciated, that flow cytometry of **DNA content** at 30-36 h could detect smaller activated cell populations than would be detected at this time by isotope incorporation[595-6]. Also, Darzynkiewicz and his coworkers[262-3,597-600] and others[601] showed that **increases in RNA content in mitogen-stimulated lymphocytes**, detectable by increased metachromatic fluorescence of acridine orange (AO) (Figures 3-10, p. 97, and 7-16, p. 321), preceded increases in DNA content, occurring as early as 12 hours following exposure to mitogens. This established elevated RNA content as an indicator of lymphocyte activation, but only among the relatively small number of immunologically oriented users of flow cytometry who were amenable to looking at anything but immunofluorescence. It was not until 1980 that Noronha, Richman and Arnason[602] used DNA/RNA flow cytometry to demonstrate the presence of activated lymphocytes in cerebrospinal fluid (CSF) from patients with active multiple sclerosis; the number of activated cells they found increased with the activity of the disease, while no activated cells were found in CSF from control patients free of inflammatory neurologic disease. I assumed, at the time, that the publication of this work in *The New England Journal of Medicine* would get more attention than it seems that it did.

Functional Probes for Activation

The 1970's and 1980's saw the development of probes and methods, flow cytometric and otherwise, for measurement of **functional parameters**, such as membrane microviscosity, structuredness of cytoplasmic matrix, membrane-bound and **cytoplasmic [Ca^{++}]**, **pH**, and **membrane potential**. As was mentioned in the discussion of these areas in Chapter 7, changes in many functional parameters were found to occur within a few minutes to an hour following lymphocyte exposure to mitogens, which prompted me and my colleagues[112,424,484] and others[417-9,493,603,1896-7] to attempt to detect early T cell activation by flow cytometry using functional probes. Results to date have generally not been reliable enough to permit routine clinical application of such assays. At present, the most tractable functional parameter available for detection of lymphocyte activation seems to be **membrane potential**[484,493]; it is possible to demonstrate apparent cytoplasmic and mitochondrial hyperpolarization in activated cells at 5-12 hours, thus providing some improvement in speed of detection as compared to RNA fluorochromes.

I say "apparent" hyperpolarization because the increased staining with cyanine dyes or rhodamine 123 may also be affected by the decrease in **P-glycoprotein efflux pump activity** that occurs within a few hours of T cell activation (see p. 309). The role of P-glycoprotein in lymphocytes is discussed in several papers in a 2003 issue of *Clinical and Applied Immunology Reviews*[3189-94]. Although P-gp may provide an indicator of lymphocyte activation, it does not appear to play a key role, as evidenced by the fact that activation proceeds normally in P-gp-deficient mice.

Unfortunately, membrane potential/efflux, pH, and [Ca^{++}] measurements, have the distinct disadvantage of requiring cell samples that are not only viable, but also what I have called "happy," for analysis.

Despite the technical difficulties involved, flow cytometric studies of lymphocyte activation using physiologic probes kept some of us hooked for many years, on the basis that, when our ships came in, we would be able to detect stimulated lymphocytes in the first few minutes after they responded to surface ligand binding. This, in turn, should have given us all kinds of wonderful tests that we could use both to elucidate mechanisms of ligand-receptor interactions and to assess the quality of cell function for clinical purposes.

When I started playing around with membrane potential-sensitive dyes, I had the conviction that I was about to encounter blinding flashes of truth rather than blinding headaches. However, the more I (and others) worked in this area, the more confused things seemed to get. For example, Tsien et al[426], based on membrane potential estimation using oxonol dyes, reported that concanavalin A hyperpolarized mouse thymocytes; Taki[456], using microelectrodes, my colleagues and I[424], using DiOC$_6$(3) and flow cytometry, and Kiefer et al[425], using a radiolabeled cation, had described a depolarizing effect of lectins on mouse and human lymphocytes.

Felber and Brand[465] resolved this apparent discrepancy when they reported that hyperpolarization of thymocytes arises from activation of a Ca^{++}-dependent K$^+$ channel. This channel is fully activated in resting lymphocytes, and therefore cannot produce increased K$^+$ flux and the resulting hyperpolarization in response to lectin stimulation. They believed that the slight depolarization they observed in con A-stimulated lymphocytes was due to Na$^+$ influx, which occurs by nonelectrogenic as well as by electrogenic pathways[466]. This may also help explain the puzzling variations in the magnitude of the early lectin effects on MP in rat and human T cells which I have noted, and which have been observed by practically everyone to whom I have spoken who has looked at MP in such cells.

Further clarification of interactions between cytoplasmic [Ca^{++}], pH, and membrane potential came from the work of Grinstein et al[1910-5]. Thymic lymphocytes possess Ca^{++}-sensitive K$^+$ channels, which are activated by moderate increases in intracellular [Ca^{++}], resulting in hyperpolarization. At higher [Ca^{++}], nonselective cation channels open, producing depolarization. Variations in the levels of intracellular [Ca^{++}] in various earlier studies could explain some discrepancies.

The situation appears to be different in B cells. Monroe and Cambier[480-2], using flow cytometry with DiOC$_5$(3), consistently observed membrane depolarization in mouse B cells after reaction with multivalent (cross-linking) anti-immunoglobulin (anti-Ig), which is mitogenic for these cells. Monovalent Fab fragments of anti-Ig neither cross-link nor depolarize the cells; such fragments are not mitogenic. Depolarization of the cell membrane by raising [K$^+$]$_o$ does not stimulate B cells to reproduce, but does lead to increased expression of Ia antigen, an activation response also produce by cross-linking surface receptors with anti-Ig.

With Terry Strom, I used DiOC$_6$(3) and flow cytometry to demonstrate different responses of T and B cells to lectins[484] and cholinergic agents[485]. Ken Rosenthal and I demonstrated different patterns of MP change following Epstein-Barr virus addition to cells having and lacking receptors for the virus[486]. The problem is that you have to be careful and lucky to get experiments like this to work, and the same is true when you do pH and calcium studies, which may be one of the reasons there is so much controversy and inconsistency in the literature.

Membrane potential probes would not have been of great use to us for detection of early activation responses even if the magnitude and direction of potential changes had been consistent, because the fluorescence distributions obtained from cells labeled with cyanine or oxonol dye probes, even when normalized using a size-dependent scatter or extinction signal (Figure 7-28 p. 389), were too broad to permit discrimination of small subpopulations of activated cells. None of the membrane potential probes available today could improve the quality of flow cytometric membrane potential measurements sufficiently to solve this problem.

We had much better luck with membrane potential measurement later in activation, i.e., 5-12 hours and more after stimulation. We could consistently demonstrate apparent hyperpolarization, probably due in large part to increased mitochondrial activity, using either cyanine dyes or rhodamine 123 (see Figure 7-30 and pp. 395-7). However, these dyes provided essentially the same information as could be obtained using pyronin Y to stain RNA at 20 hours, and the latter dye, unlike the membrane potential probes, could be used on fixed cells.

We next turned our attention to probes that could be used for ratiometric measurements of functional parameters. While preliminary experiments with ratiometric pH measurement, using carboxyfluorescein as an indicator with dual-wavelength excitation (Figure 7-35, p. 406), suggested that early changes occurred in stimulated T cells, the literature of the mid-1980's contained conflicting reports of cytoplasmic alkalinization and acidification occurring in lymphocytes in response to mitogenic stimulation[427,521,870,871]. Later work by Grinstein indicated that calcium-dependent acidification is the dominant response[1912], but that alkalinization can occur as a result of activation of Na^+/H^+ by a calcium-independent mechanism[1913]. Intracellular pH, like membrane potential, turned out to be a sideshow; calcium remained in the center ring[1914-5].

DNA, RNA and Activation Antigens

By the time indo-1 became available, permitting ratiometric measurements of cytoplasmic [Ca^{++}] to be done by flow cytometry, my colleagues and I, having become increasingly frustrated in our attempts to tame flow cytometric functional probe measurements of cell activation, had, instead, started to collaborate with other investigators, who, using bulk methods and single-parameter immunofluorescence flow cytometry, were trying to characterize **activation antigens**, which appeared on, or increased dramatically on, lymphocytes within hours following mitogenic stimulation. This led to studies in which **combinations of DNA (and RNA, and sometimes membrane potential) stains and fluorescent antibodies were used to demonstrate early and late activated T cells** in clinical and experimental contexts.

We first approached the problem of detecting activated inducer and cytotoxic/suppressor cell populations, respectively, by examining DNA content in inducer and cytotoxic/suppressor populations defined by staining with anti-CD4 and anti-CD8 antibodies[604]. The clinical interest underlying these experiments related to prediction of graft rejection (p. 494).

We next looked at lectin-[605] and alloantigen-[606] activated T cells to determine the **kinetics of appearance and the distribution of several activation antigens**, i.e., antigens that appear on the surfaces of cells only after mitogenic stimulation. We found that all proliferating cells, including those in G_1, bore the **transferrin receptor (TfR, now CD71)**, the **4F2 antigen**[1898,1916-7] **(now CD98)**, and the **Tac antigen**, which is the **interleukin-2 receptor (IL2R, now CD25)**. The IL-2 receptor was detectable on stimulated cells at 8-12 hours; Redelman and Wormsley[1004] subsequently showed, using fluorescein immunofluorescence and pyronin Y staining, that IL-2 receptor appearance precedes increases in RNA content following lymphocyte stimulation, and is not affected by inhibitors of RNA synthesis.

The 4F2 (CD98) antigen was detectable on the surfaces of stimulated T cells as early as 4 h after exposure to lectin or antigen; the transferrin receptor (CD71) and HLA-DR were not detectable on the majority of activated cells until somewhat later. In the case of HLA-DR, simultaneous staining with Hoechst 33342, pyronin Y, and fluorescein-labeled antibody revealed that a substantial fraction of proliferating cells, identifiable by their increased RNA and DNA content, were not expressing HLA-DR. We found increased display of 4F2 antigens in peripheral blood lymphocytes taken from a transplant patient during an acute rejection episode; the number of cells bearing 4F2 diminished to normal levels following successful treatment with monoclonal antibodies.

We evaluated several parameters, including 4F2 and CD25 antigen expression, measured 12-24 h after initiation of mixed lymphocyte reactions (MLRs)[607], to determine whether these could serve as indicators of activation. The percentages of cells bearing either of these antigens at 12-24 h correlated well with measurements of thymidine incorporation made at 120 h. Also, removal of activated cells from mixed lymphocyte cultures by treatment with anti-CD25 antibodies and complement at 48-72 hours largely abolished proliferative responses otherwise detectable by thymidine incorporation at 96-120 hours; it thus seems likely that those T cells bearing 4F2 or CD25 activation antigens early in the activation process are the specifically activated T cells which are destined to proliferate later.

In order to demonstrate that flow cytometric immunofluorescence analysis could be made accessible and affordable, we showed[303] that 4F2 antigen on stimulated lymphocytes could be detected by measurements of the fluorescence of allophycocyanin (APC)-conjugated antibody in a flow cytometer using a 7 mW helium-neon laser source; this, however, failed to produce a massive demand for such instruments.

Biselli et al[1899] used dual color flow cytometry to study the kinetics of several activation antigens on $CD4^+$ and $CD8^+$ subsets after 24, 48, 72, 120, and 168 hours' incubation with PHA and Con A. Expression of these molecules followed a consistent time-course with no major differences between subsets. CD69[1918] expression peaked at 24 hours, whereas CD25 and CD71 expression peaked at 48 and 72 hours, respectively.

The expression of CD45RA remained stable for 72 h and then briskly decreased with no major differences between PHA and Con A activation; this presumably reflects the parallel increase in CD45RO expression which occurs with the conversion of T cells from the "naive" to the "memory" state following activation[1900-2].

As contrasted with functional probe assays of activation, measurements based on detection of activation antigens do not require live, let alone "happy," cells, and the physical presence of a newly synthesized or newly expressed protein on the cell surface provides a much more comfortable and substantial basis for clinical decision making than could ever be obtained from constant, let alone inconstant, ephemeral changes in ionic concentrations.

Since this work was done, a large number of studies (far too many to cite) have been published in which the state of activation of lymphocytes *in vivo* was assessed by measurement of a single activation antigen. I would suggest that this is not always a good idea, because different patterns of expression of activation antigens may be observed in response to different antigenic stimuli. For example, Santamaria et al[1919] found that CD69 was selectively expressed only on CD8$^+$ T cells infiltrating rejecting human heart allografts, while both CD4$^+$ and CD8$^+$ cells expressed CD25.

I would argue that **increased RNA content** and **transferrin receptor (CD71) expression** (see Figure 10-7 and pp. 460-1) are the parameters best suited to serve as "gold standards" for activation processes that can reasonably be expected to have been ongoing for 20 hours or more, *in vivo* or *in vitro*. Measurements of these parameters, unlike those of DNA content, identify cells in the G_1 as well as the S, G_2, and M phases of the cell cycle.

Pyronin Y, an effective stain for RNA, can be used in conjunction with Hoechst 33342 and, presumably, with DRAQ5 as vital DNA stains, and with antibodies labeled with fluorescein and with PE-Texas Red and/or PE-Cy5 tandem conjugates, allowing the state of activation to be delineated in immunologically defined cell subpopulations. Without such multiparameter experiments, it is not clear which activation antigens are the most trustworthy indicators in any particular situation; it would be foolish to expect that any given measurement, RNA content included, could be optimal in all contexts.

The transferrin receptor, CD71, is present on all proliferative phase cells, and can be detected on the surfaces of living cells by fluorescent antibody staining; I see little reason to use the Ki-67 antibody, which requires permeabilization and appears to stain the same cells.

When human peripheral blood mononuclear cells are stimulated with polyclonal mitogens such as PHA or sepharose-conjugated CD3 antibody, at least 60% of the T cells in both the CD4$^+$ and CD8$^+$ subpopulations typically express activation antigens; this degree of activation is readily detectable by ^{3}H-TdR incorporation after only 3-4 days, even though most cells do not initiate DNA synthesis for at least 30 hours after stimulation[1903]. Although as many as 2% of the lymphocytes in an unstimulated culture may be activated due to previous natural exposure of the host to antigens, this background activated population is negligible in the context of analysis of mitogen-stimulated cultures.

Mitogen Response versus Antigen Response

Studies of mitogen responsiveness may be useful in the context of demonstrating generalized immune deficiencies, such as occur in HIV infection[1904] and in some cancer patients (see pp. 383-5). However, in research applications, and in analyses of clinical situations in which successful therapy depends on restoration or generation of specific immune responses, notably treatment of AIDS[1905] and cytokine/cell therapy of cancer[1906-7], the cytometric assay strategy differs from that used in studies of activation induced by polyclonal mitogens.

In an alloantigen-stimulated culture or MLR, a much smaller fraction of lymphocytes, typically 0.5-1.5% of the population, is initially activated than is the case in a mitogen-stimulated culture. Detection of activation by incorporation typically requires at least 5-6 days in culture, by which time several doublings of the activated cell population have occurred; since the population of naturally activated cells is approximately equal in size to the population specifically stimulated by alloantigen, it is necessary to demonstrate a significant increase over background to detect specific activation, especially when a cytometric assay is done before proliferation of the alloantigen-stimulated population has occurred. We found that the chi-square test, applied to data from 5,000- to 10,000-cell samples, could discriminate between a match, a single-allele mismatch, and a two-allele mismatch in MLRs analyzed for 4F2 (CD98) or CD25 antigen 12-24 hours after stimulation.

The frequency of cells that recognize a specific bacterial or viral antigen, however, can be much lower than the frequency of responding cells in an MLR. In adults tested at least 20 years after primary infection with varicella-zoster virus, it was found[1904] that only 1/105,000 cytotoxic T lymphocytes (range 1/13,000 to 1/231,000) specifically recognized a viral protein. This suggests that proliferation would typically have to multiply the number of cells initially activated by antigen by a thousandfold to create a population of specifically responding cells roughly the same size as the background activated population. The mean generation time of cycling stimulated adult human lymphocytes is between 12[2943-4] and 20 hours[1903]; multiplying the original activated population by 1,024 could take as long as 10 days. The frequency of cells recognizing antigen in a recently immunized population may be considerably higher than the frequencies reported above. Waag et al[1909] reported that proliferation at 5 days induced by *Francisella tularensis* antigen in lymphocytes taken from human volunteers 14 days after immunization with a live vaccine strain exceeded that induced by con A.

Stimulation by **superantigens**, such as **Staphylococcal enterotoxin B (SEB)**, is restricted to those T cells with Vβ regions of the T cell receptor that can be crosslinked by the antigen; thus, while the fraction of responding cells in an SEB-stimulated culture will vary with the pattern of Vβ expression (SEB reacts with human Vβ3, 12, 14, 15, 17, and 20), this fraction will almost always be lower than the fraction of responding cells in a PHA-stimulated culture, and SEB is often used as a positive control stimulus in assays designed to measure stimulation by antigen.

In the situation in which immunization is expected to have occurred long ago, and in which there has not been recent challenge, frequencies of specifically activated cells on the order of 1/100,000 make it difficult for flow cytometry to detect specifically activated cells against the background activated population before substantial antigen-induced proliferation occurs. However, several advances made since the previous edition of this book was published facilitate the task, and have wider implications for the general field of immunology, as well.

Detecting Activation by CD69 Expression

The CD69 antigen[1918,3195-8], a surface receptor involved in kinase-mediated signaling and present in most blood cell types, including basophils, eosinophils, mast cells, and platelets, is expressed within 1-4 h of mitogenic or antigenic stimulation of T lymphocytes; expression in mitogen-stimulated cultures peaks at 24-48 h[1899], and the antigen is reported to be diluted by successive cell divisions.

In 1995, Maino et al[3199], at B-D, reported that early CD69 expression was detectable in lymphocytes stimulated with the comitogenic antibodies CD2 and CD2R, with pokeweed mitogen, with SEB, and with *Candida albicans*, and that the fraction of cells expressing CD69 at 4 h tracked ^{3}H-TdR uptake at 3 d in cultures stimulated with various doses of comitogenic antibodies. B-D later introduced the **FastImmune™ Activation System**, of reagents, which include an anti-CD69 antibody and gating antibodies appropriate for detection of activation in T, B, and NK cells. Lim et al[3200] suggest that quantitative, as well as quantitative , determination of CD69 expression may be useful in detecting activation.

Craston et al[3201] found that stimulation by mitogens stimulation resulted in the most rapid expression of CD69 on both T and NK cells, while alloantigen-stimulated cells responded more slowly. Caruso et al[3202] studied expression of CD25, CD69, CD71, and HLA-DR on T cells from healthy individuals stimulated with different mitogens and antigens; CD69 was the earliest expressed antigen, while HLA-DR was the latest. However, regardless of the stimulus used, lymphocytes expressing CD25 and CD71 were always more numerous than cells expressing CD69 and HLA-DR. Variations in the proportion of CD4+ and CD8+ T cells expressing each marker were observed with different stimuli. Activation marker expression showed overall agreement with ^{3}H-TdR uptake in discriminating between positive and negative responses, but the percentages of marker-positive cells were not correlated with amounts of ^{3}H-TdR uptake. Low doses of mitogens and antigens and/or short exposure times could induce activation antigen expression that was not followed by proliferation.

Hutchinson et al[3203], studying HIV-negative individuals with immunodeficiencies, found that most showed CD69 expression in response to mitogen, although ^{3}H-TdR uptake was diminished or absent. Sieg et al[3204] reported that, after anti-Vβ3 Ab stimulation, CD4+Vβ3+ cells from HIV-infected patients expressed CD69 and CD25, but demonstrated defects in expression of cell cycle-associated proteins, D-type cyclins, and cyclin A that precluded cell cycle progression. The proliferation defect was most apparent in patients with diminished CD4+ T cell numbers and higher plasma HIV RNA levels.

Thus, it appears that some caution is required in the use of CD69 expression as an indicator of T cell activation; I have said that CD69 is aptly named, because, while it may be sexy, it has little to do with reproduction.

Cytokines: Detecting Activation and More

CD69 expression is, however, a good marker of cells capable of **cytokine production**, which itself is an indicator of activation. Around the time the previous edition of this book was written, immunologists were seriously contemplating widespread use of ELISA measurements of cytokine production in 96-well plates in hopes of detecting a return of immune function in patients treated for HIV infection. The idea alone was enough to strike terror into the heart of anyone who remembered how hard it had been to get the dozens of labs involved in AIDS clinical trials to produce reasonably consistent CD4 counts.

Cytometry came to the rescue. Accumulation of cytokines in the Golgi apparatus of stimulated T cells permeabilized with paraformaldehyde and saponin had been observed by fluorescence microscopy in the late 1980's[3205-8]; by the early 1990's, several groups[3209-16] were developing and refining flow cytometric procedures for cytokine detection. The subject of cytokine detection[2703] was addressed briefly on pp. 357-8.

In 1993, Jung et al[3211] introduced the use of the ionophore **monensin** to block transport of cytokines out of the Golgi apparatus, increasing fluorescence signals. In 1994, Vikingsson et al[3213] reported that the number of interferon-γ (IFN-γ)-positive cells in cultures correlated with IFN-γ levels measured in culture supernatants. In 1995, Picker et al[3215] reported using **phorbol 12-myristate 13-acetate (PMA)** and **ionomycin** as an accessory cell-independent stimulus and **brefeldin A** to block transport out of the Golgi apparatus, and Prussin and Metcalfe[3216] demonstrated the specificity of intracytoplasmic cytokine staining by showing that staining could be blocked by excesses of cytokine or unlabeled antibody.

In 1997, Mehta and Maino[3217] combined staining for cytokines with staining for BrUdR incorporation to demonstrate that those cells that still synthesize cytokines at 48-72 h after stimulation have undergone cell division.

In 1997, Waldrop et al[3218] described modifications of the cytokine assay procedure that made it possible to identify **antigen-specific T cells.** Cells were incubated with antigen and accessory cells in slant tubes, and brefeldin was not added until 1 h after stimulation was initiated to prevent its possible interference with antigen processing. Anti-CD28 antibody was added as a costimulator, and anti-CD69 antibody was used to identify responding cells. The refined assay was used to study frequencies of response to various pathogens of CD4+ memory/effector cells from HIV+ patients. Further refinements described by Suni, Picker, and Maino[3219] in 1998 allowed the assay procedure to be performed on whole blood, eliminating the need for isolation of mononuclear cells. **BD Biosciences** now offers cytokine assays as part of its **FastImmune™** product line, which includes an antiCD28/antiCD49d costimulatory reagent.

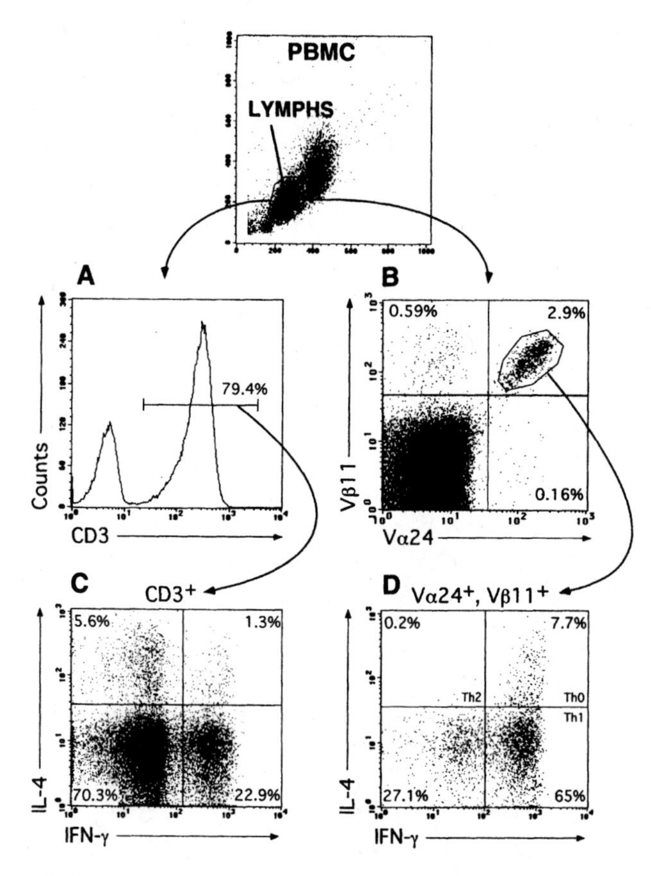

Figure 10-19. Intracellular cytokine staining. Courtesy of Calman Prussin, NIAID.

Intracellular cytokine staining is illustrated in Figure 10-19, provided by Calman Prussin of the National Institute of Allergy and Infectious Diseases. Human peripheral blood mononuclear cells were activated with PMA/ionomycin. Parallel samples were stained with either fluorescein-antiCD3 (panels A, C) or fluorescein-antiVβ24, PE/Cy5-anti Vβ11 (panels B, D) in addition to the cytokine antibodies, APC-antiIFN-γ and PE-anti IL-4. The populations shown in panels A and B are from the lymphocyte scatter gate in the top panel. CD3-positive T cells from panel A, represented in panel C, show a typical staining pattern for IFN-γ and IL-4 (C) demonstrating the simultaneous use of these mAbs to differentiate Th1, Th2, and Th0 cells. A small subset of T cells expressing the Vβ24, Vβ11 TCR, shown in the gate in panel B, are seen in panel D to have a unique pattern of cytokine expression. Numerical values in panels A and B represent positive cells as a percentage of total lymphocytes; values in panels C and D represent percentages of gated cells. Marker placement was determined from identical cell samples preincubated with unlabeled anti-cytokine antibodies.

Ins and Outs of Cytokine Staining

Following PMA/ionomycin stimulation, CD4 antigen is internalized and subsequently replaced. In the presence of brefeldin A, this does not occur; CD4+ cells typically lose about 90% of detectable CD4It is therefore standard procedure to stain cells for CD8, setting a gate around the more intensely stained CD3+ cells to define the CD8+ T cell population and another around the unstained CD3+ cells to define the CD4+ population. Hennessy et al[3220] examined several anti-CD4 antibodies and determined that BD's Leu3a/3b multiclone reagent alone allowed direct identification of the CD4+ subset in PMA-stimulated cells. Intracellular cytokine measurements have recently been discussed by Pala and Hussell[3221] and by Prussin and Foster[3222].

Although intracellular cytokine staining is probably adequate for assay purposes, it is desirable under other circumstances to be able to demonstrate cytokine production by living cells, allowing them to be sorted on the basis of their expression patterns. This was made possible in 1995 by Manz et al[2704], who developed a **cell surface affinity matrix** that allowed capture of secreted cytokines, which could subsequently be detected with fluorescent anti-cytokine antibodies. The original technology (p. 360) involved biotinylation of cell-surface proteins and attachment of streptavidin-conjugated anti-cytokine antibodies; subsequently, the procedure has been simplified by use of bispecific antibodies made by conjugating an anti-CD45 antibody with an anti-cytokine antibody[2705,3223-5]. These are commercially available from **Miltenyi Biotec.** Although CD69 and cytokine expression can be used, as described above, to identify antigen-specific cells, cell surface affinity matrix staining for cytokines can also be combined with **tetramer staining**, permitting even more precise definition of specificity[2705,3225].

Tetramer Staining: Talking the Talk; Walking the Walk?

Tetramer staining[2359-64,3225] to detect T cells reactive with defined peptide sequences was introduced on pp. 47-8. Although it has been possible to isolate at least some specifically reactive B cells by staining with fluorescently labeled antigen[1107,3226-7], T cells normally "see" antigen only after it has been processed by antigen-presenting cells, and antigen alone will not bind strongly to the T cell receptor. However, fluorescently labeled tetrameric complexes of antigenic peptides with MHC class I or II α chain and β_2-microglobulin mimic the processed antigen and its presenting cell well enough to be used to identify antigen-specific T cells. Since the initial description of tetramer staining of antigen-specific CD8+ T cells by Altman et al[2359] in 1996, the protein engineering technology has been refined[3228-30] and the application of the technique has yielded a great deal of previously unobtainable information. Since I don't have the time and space to devote to an extensive review, I will cite a short but informative one by Kelleher and Rowland-Jones[3231].

While tetramer staining is exquisitely specific, it defines a structural and not a functional characteristic of the cell[3232]. Virus-reactive T cells from patients with HIV infection may lack their expected cytolytic function, and fail to proliferate in response to antigenic or mitogenic stimulation, and similar functional defects have also been noted in tumor-specific T cells from cancer patients[3231]. In the era of 11-color flow cytometry, the obvious thing to do is combine tetramer staining with measurements of functional parameters, as noted in the previous section.

Tracking Dyes: Activation and Ontogeny

In the previous edition of this book, I predicted that tracking dyes such as PKH26 would be enormously useful in following cells through generations. I had noted the 1990 finding by Weston and Parish[1542] that cells covalently labeled with carboxyfluorescein succinimidyl ester (CFSE) retained label for weeks, but not the 1994 publication by Lyons and Parish[3233] in which CFSE was used to study division of lymphocytes. As can be seen from Figures 7-23 and 7-24 and pp. 371-4, there is a big difference between PKH26 and CFSE. Using a mathematical model, one can derive information about the proliferation of a cell population using either label, but, while it is usually possible to discern peaks representing different generations of CFSE-labeled cells by visual inspection of a one-or two-parameter histogram[2349], it is rarely or ever possible to do so in the case of PKH26-labeled cells. It is also possible to get sharper peaks in a CFSE fluorescence distribution by starting with an input cell population sorted from a narrow region of the fluorescence peak representing the first generation[2350,3234]. Lyons et al[2349] and Parish and Warren[3235] have recently reviewed CFSE methodology. Hasbold and Hodgkin[3237] have shown that isolated nuclei from CFSE-labeled cells retain enough of the dye to permit discrimination among different generations, allowing expression of nuclear antigens, BrUdR incorporation, etc. to be followed through successive generations of cells in otherwise homogeneous populations.

It should be obvious that tracking dye labeling provides a relatively easy method of isolating antigen-responsive T cells; label the cells, add antigen and whatever accessory cells and factors are necessary, and, after five or six days (or more), sort out the cells with very dim dye fluorescence; these will be the cells that have divided several times in response to the antigenic stimulus. Easy, and practical, but, as it turns out, that's not all. Keeping track of generations turns out to provide a great deal of interesting information about lymphocyte differentiation in a very graphical form. It is simple to determine, from two-parameter displays, whether the expression of a particular antigen or antigens or the production of a particular cytokine or cytokines differs from one generation of cells to another. This type of analysis has produced a substantial body of evidence that many developmental processes in both T and B lymphocytes are **division-linked**, or **division-regulated**[3237-48].

The original CFSE work was done in Australia, and the technique caught on there before it spread to the rest of the world. I wasn't paying much attention until I made my first visit to Australia in 1998. I've been keeping up since, but here I am busy writing while other people are doing the experiments I wanted to do.

What is "Early" Activation (Trick Question)?

It should now be obvious that, if we're willing to wait a few hours, we have a lot of different ways of detecting specific activation of lymphocytes. What can we do if we want to look at the first few minutes of the activation process?

There are the fast functional parameters; cytoplasmic [Ca^{++}] is one that comes to mind, as are pH change and "structuredness of cytoplasmic matrix" (pp. 383-5), which, because it is less well understood, harder to measure, and widely regarded as unreliable, is a definite dark horse. Carl June, who probably has had as much experience with flow cytometric calcium measurements as anyone[866-7,1714-5,1896-7,1920], made a case[1897] that they could be clinically useful, but it hasn't happened yet.

If we're willing to settle for not keeping the cells alive, we can look at **kinases**. Figure 10-9, on pp. 468-9, came from the *tour de force* paper by Perez and Nolan[3000] in which 11-color flow cytometry was used to determine the phosphorylation states of several different kinases in different T cell subsets; speed was not the object. However, Chow et al[3249] have shown that it can be, as illustrated by Figure 10-20.

Antibodies to phosphotyrosine[1631-2] were used by Far et al in 1994[1632] to define the overall level of protein phosphorylation in cells; since then, it has been possible to produce antibodies to both phosphorylated and nonphosphorylated forms of a number of kinases. Such antibodies are available from a number of sources; most of those used by Perez and Nolan[3000] and those used by Chow et al[3249] came from **Cell Signaling Technology**.

Figure 10-20 shows levels of phosphorylated **e**xtracellular signaling **r**elated **k**inase (**ERK1** and **ERK2**) in human peripheral T cells activated by various doses of PMA or (bottom right panel) by crosslinking anti-CD3 bound to the T cell receptor. Note that a substantial increase in phosphorylation, compared to control values, is detectable within 3 minutes after stimulation.

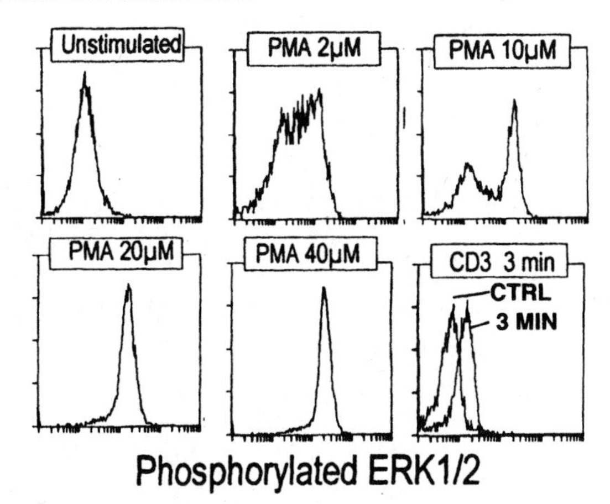

Figure 10-20. ERK1/2 phosphorylation in T cells exposed to various activation stimuli. From: Chow S et al, Cytometry (Commun Clin Cytom) 46:72-8, 2001 (reference 3249), © John Wiley & Sons, Inc. Used by permission.

So we can detect, or think we can detect, activation at 3 minutes by a change in kinase phosphorylation, although there is a hefty overlap between the histograms representing control and stimulated cells. So far, so good, but what flow cytometry can't tell us is whether there is temporal heterogeneity in kinase responses in cell populations.

Image analysis studies on calcium concentration changes in single lymphocytes over time done by Wacholtz and Lipsky[1735] revealed a considerable heterogeneity in both the magnitude of increases in intracellular [Ca++] and in the lag time before increases occurred, even when cells were stimulated with anti-CD3 or PHA, which are presumably reacting with more or less the same structures on all responsive cells.

If we define a proliferative state, or at least a state in which the cells tell us they're thinking about proliferating by synthesizing RNA, as a definitive indicator of activation, we have to determine whether the cells in or on which we find a putative "early" indicator of activation, such as a calcium flux change or kinase phosphorylation, get as far as the proliferative state. In the studies of CD25 and 4F2 expression on lectin stimulated cells and in MLRs to which I referred previously, this was done; antibody/complement lysis of cells expressing these antigens eliminated the normal proliferative response. In at least some situations, however, it is established that cells that respond to ligands with an increase in calcium flux do not necessarily go on to proliferate.

When cells react with real antigens, the reaction is between the antigen and a variable region of the cellular receptor, implying that different clones of cells with different binding constants for any particular antigen may behave differently. This is could conceivably translate into a broad spectrum of lag times in the small fraction of cells which do react with antigen to a great enough degree to manifest ion flux responses, display activation antigens, and proliferate, and it is not at all clear that all those that do the first thing do the second, etc. There is obviously some kind of threshold that must be crossed before specific activation occurs, but it may be crossed at different rates by different cells; in some recent talks I've given, I've compared the cells to personal computers, with the most reactive cells analogous to the Pentium and Power PC processors, the slightly slower ones, to the 486 and 68040, and so on, down to the slowly reacting 8088 and 68000 cells, and to the Z-80 and 6502 cells, which don't react. All but the last group run the program, but the different groups run it at different speeds.

From studies with activation antigens, we know that the most active cells will express CD25 and 4F2 within 12 hours following stimulation. However, if you lyse cells in an MLR, which exemplifies specific antigenic stimulation, 48 hours after initiation of the reaction, you don't eliminate all proliferation at 120 hours; you have to do the lysis at 72 hours to completely wipe out the culture. Since it is unlikely that cells make it all the way to DNA synthesis without at least transiently expressing CD25 and 4F2, this suggests that some cells in an MLR don't get to the "early" stage of activation at which they express CD25 and 4F2 until 60 hours or so after initiation of the culture. In other words, "early" isn't the same time for each cell.

It would be good to be able to separate the "early early" cells from the "late early" cells, in order to define both the chemical characteristics of their respective receptor-antigen interactions and any differences in their biologic behavior in immune responses *in vivo*. The most obvious way to do this may be by sorting proliferating cells from different generations using CFSE.

I realize that I haven't gotten into some other parameters that change during lymphocyte activation, such as oxidative metabolism (lymphocytes seem to do a mini-respiratory burst, complete with bioluminescence), or even the size change, or "blast transformation," which can be detected in flow cytometers just as it can under the microscope. But the bottom line is that we now have the hardware, the reagents, and the data analysis techniques needed to use flow cytometry productively to investigate real immune reactions instead of lectin stimulation, and we can go looking for activated cells in patients with HIV infection, or cancer, or chronic fatigue syndrome[2286], or post transplant.

Cancer Biology and Clinical Oncology

This represents another area in which the body of literature is so large that it is impossible to take a representative sample without writing another book. In the Second Edition, I cited a few then-recent reports[965-72] to try to exemplify the range of things being done with flow cytometry in the field. I itemized the following for the Third Edition:

Ornitz et al[966] used DNA flow cytometry to follow ploidy changes in the preneoplastic state and in neoplasia of pancreatic cells which occur in transgenic mice expressing SV40 T-antigen.

Tough and Chow[967] sorted high- and low-IgM natural antibody-binding populations from a heterogeneous lymphoma line and found that the level of antibody binding correlated inversely with tumorigenicity.

Kenter et al[968], using multiparameter flow cytometric analysis found differences in colcemid susceptibility between tumorigenic and nontumorigenic lymphocytes, in the process defining a cell cycle phase they called G1′, intermediate between G_0 and G_{1a} and absent in immortalized and tumorigenic cells.

Cook and Fox[969], using ADB, examined intracellular pH regulation in CHO cells subjected to hyperthermia at normal and acidic pH, and found that internal pH was regulated under both conditions, ruling out a disturbance of pH homeostasis as a mechanism for the thermosensitization observed at low pH.

Ota et al[970] observed that a glycoside from ginseng which inhibits growth of and induces melanogenesis in B16 melanoma cells, increased the number of peanut agglutinin binding sites on the cell membrane, while another glycoside, which induces melanogenesis but does not stimulate growth, does not change lectin binding patterns, although both compounds decreased membrane fluidity.

Other membrane effects of differentiating agents were studied by Fibach et al[971], who found that DMSO or hexamethylene-*bis*-acetamide decreased uptake of fluorescent fatty acid derivatives by murine erythroleukemia cells.

Bucana et al[972] studied uptake of hydroethidine in neoplastic cells; unoxidized hydroethidine could be distinguished from its oxidation product, ethidium, and quantified in cells not only by flow cytometry, but by bulk fluorimetry of microtiter plates.

When I wrote the last edition, the reviews on flow cytometry in oncology already written by others[586,608-9,906,980-2] covered the territory effectively and let me get off with my usual barbed commentary.

This time around, I have not done an exhaustive review of either the cancer research or the clinical oncology literature, so I will simply mention a few items I've run across, making no attempt at anything like thorough coverage of either field.

Cancer Diagnosis: Cervical Cytology

Considering that a good deal of the development of flow cytometry was paid for by agencies and companies interested in **automating the cytologic diagnosis of cervical cancer,** it may seem surprising that no flow cytometric clinical instrument for this purpose has yet emerged. However, pathologists, not entirely without reason, like to work with samples they can see and keep. As far as I know, there is only one FDA-approved automated cervical cytology screening instrument on the market; it is the **FocalPoint**™[3250] (formerly **AutoPap®**), made by **TriPath Imaging, Inc.** (Burlington, NC; http://www.tripathimaging.com). TriPath Imaging was formed by the merger of two companies active in the development of automated Pap smear screening, **NeoPath** and **AutoCyte,** and acquired the intellectual property of a third, **Neuromedical Systems.** Both TriPath Imaging and **Cytyc Corporation** (Boxborough, MA; http://www.cytyc.com) make apparatus for preparing high-quality slides from cervical cytology specimens; Cytyc is, or at least was, also working on an automated screening system. The predecessors of automated screening instruments were computerized microscope systems that facilitated presentation well-prepared of slides to cytotechnologists and cytopathologists.

The FocalPoint and its now-extinct competitors are/were imaging systems that examine(d) slides stained with Papanicolaou stain, and, whenever I have spoken or written about this subject in the past, I have lamented the fact that nobody seemed to be developing a system for cervical cytology screening that looked for some more specific markers associated with cell proliferation.

Well, such a system has finally been developed by **Molecular Diagnostics, Inc.** (Chicago, IL; http://www.molecular-dx.com) and is on the path toward FDA approval. The **InPath**™ system apparently uses fluorescently labeled antibodies to the **transferrin receptor (CD71)** and/or the **epidermal growth factor receptor (EGFR)** as an indicator of cellular dysplasia or neoplasia[3251-2], plus a nuclear stain (?DAPI) and a fluorescent antikeratin antibody selective for cervical cells. An additional test detects antigen(s) produced by the human papilloma virus (HPV), infection with which predisposes to the development of cervical cancer. Slides treated with the fluorescent reagents can later be stained with Papanicolaou stain. It will be interesting to see whether the system catches on.

Flow cytometry does, apparently, have some contribution to make to cervical screening; it is reported that forward and side scatter signatures can be used to assess the adequacy of a cervical specimen prior to slide preparation[3253]. Now, if you just throw the DNA stain and the antibodies into the mix, who knows what might happen?

Exfoliative cytology using flow cytometry has also been applied to monitoring the treatment of superficial bladder cancer by DNA or DNA/RNA analysis of bladder irrigation specimens[906]. In at least one urology department, an argon ion laser was used both to pump a dye laser for tumor phototherapy and as a light source for a flow cytometer to monitor the therapy.

DNA Content Measurements Yet Again

The earliest uses of flow cytometry in oncology dealt with analyses of **abnormalities of DNA content in tumor cells** and with **perturbation of the cell cycle by chemotherapeutic agents.** It was learned rather quickly that anything that kills cells will perturb the cell cycle sooner or later, and this line of research got less interesting with time. The demonstration of abnormalities in DNA content, on the other hand, provided and continues to provide prognostically relevant, objective bases for classification of tumors arising in many different tissues. At least one Cytomutt was used effectively and successfully[1008] in a surgical department in an institution in which the commercial instruments and their proprietors in the pathology department were absorbed by immunology and not very interested in DNA.

DNA analysis was improved considerably by mathematical modeling to minimize the effects of debris and clumps; Kallioniemi et al[2277] described a fully automated histogram analysis procedure which preserves the strong predictive power of the corrected aneuploid S-phase fraction. The issue of quality control of DNA analysis has also been addressed[2048,2278]; it was noted on pp. 25-6 that DNA ploidy and S phase fraction in breast cancer, once routinely measured and then abandoned[2320], can add prognostic information when appropriately determined[2321].

The development of a procedure permitting DNA content analysis of nuclei from **paraffin-embedded specimens**[610] made it possible to analyze large numbers of specimens from cases in which outcomes were already known; once they didn't have to wait for outcomes of 10-year prospective studies before they could publish, a lot more pathologists got interested in doing DNA flow cytometry. A few new wrinkles and improvements in the basic flow cytometric technique have been described[3254-7]. However, pathologists, as previously noted, still feel more comfortable with slides than with flow. John Crissman and his colleagues in Detroit (personal communication) put together their own static system, as did other groups of pathologists who probably would have been scared to build flow cytometers, and they could do DNA by fluorescence with precision almost as good as that obtainable from flow cytometers. Commercial systems for the purpose are now available; CompuCyte's Laser Scanning Cytometer, developed by Lou Kamentsky himself[2047] seems to have established a niche here[3258-60].

A Research News article by Rachel Nowak[2294] in the 24 June 1994 issue of *Science*, headlined "A New Test Gives Early Warning of a Growing Killer" described work by Brian Reid, Peter Rabinovitch and others at the University of Washington[2296-9] on factors that determine the likelihood that an individual with Barrett's esophagus will develop esophageal cancer. One is p53 gene mutation; a second is aneuploidy. According to Nowak, "More DNA means a brighter glow, which is detected by an instrument called a flow cytometer." The aneuploid cells are not necessarily malignant, but increasing aneuploidy suggests progression to cancer within a period of 18 months to 7 years. Survival of patients who have undergone surgery after early detection of progression is projected to be 80% at 5 years; 5-year survival in esophageal cancer has historically been under 5%. The Washington group has kept up its work on Barrett's esophagus[3261-4], identifying increased numbers of cells with 4N (G_2 or tetraploid) as an additional predictor of progression.

This brings to mind an interesting series of papers published in the late 1980's[2300-3] in which flow cytometric DNA analysis was adapted to detect hyperdiploidy in cultured dermal fibroblasts, which occurs in individuals genetically predisposed to cancer of the colon and nasopharynx, among other sites. The aneuploidy here isn't even a premalignant change, since it is occurring in cells of a completely different line from those that become malignant.

There isn't even universal agreement about how cells become aneuploid; they may undergo a second round of mitosis without cell division, becoming tetraploid and then losing chromosomes[2304-5], or they may go through cycles of unbalanced mitosis[2306]. Cell lines which undergo megakaryocytic differentiation make good model systems in which to study polyploidization[2307]; Mouthon et al[2308] published a study on one such line which included some very nice flow cytometry; also see p. 487.

Aneuploidy due to chromosomal instability precedes the development of colon cancer in patients with **ulcerative colitis**[3265]; Clausen et al[3266] sorted diploid and aneuploid cells from a colectomy specimen and subjected them to comparative genomic hybridization to analyze gene alterations.

Beyond DNA Content: Antigens, Oncogenes and Receptors, and Response to Therapy

DNA or DNA/RNA content measurements and analyses of chromatin structure have been applied to classification and estimation of prognosis of leukemias, lymphomas[3267], and solid tumors[979]. It is virtually certain that strong correlations exist between DNA (and RNA) content and proliferative activity of tumors, on the one hand, and surface antigen patterns, on the other; while, early in the game, different investigators emphasized different parameters, many more people now appreciate the utility of multiparameter measurements.

The appreciation that various **oncogenes** are expressed in many human cancers has led to the development of flow cytometric techniques for detecting oncogenes, initially by immunofluorescence of antibodies directed against gene products[1012-4], later with genetic probes. A paper by Stål et al[2279] on simultaneous analysis of DNA and c-*erb*B-2 expression in breast cancer exemplifies the technique; an article by Winter et al[3268] discusses the influence of the antiapoptotic **Bcl-2** gene on the proliferative activity of lymphomas[3268].

Jacobberger et al[3269-73] have applied multiparameter flow cytometry to analyses of interactions between oncogenes, growth factors, and receptors in several experimental systems.

Interest remains in determination of **drug sensitivity** of tumor cells. Detection of drug efflux using functional probes or antibodies, glutathione measurements, etc., at least for now, typically mean simply that chemotherapy, whatever the agent, is likely to be ineffective. *In vitro* drug sensitivity testing of bacterial pathogens is useful only because some drugs work and some don't; in most types of cancer, where no drugs work, drug sensitivity testing, cytometric or otherwise, is pointless. However, both flow and static cytometry may be useful in assessing response to various modalities of therapy; to chemotherapy, we must now add immunotherapy[3273] and, eventually, gene therapy[3274].

Chow et al[3249] (see pp. 501-2 and Figure 10-20) and Jacobberger et al[3300] have developed flow cytometric methods for determination of **kinase activities** in cells; these can be applied to detection of inhibition of kinases by drugs. The latter paper[3300] documents specific detection of the activity of **STAT5** in chronic myelocytic leukemia (CML) cells and lines expressing **Bcr/Abl**, and inhibition of STAT5 by **imatinib (STI-571; Gleevec)**, now a first line therapy for CML.

Immunophenotyping in Hematopathology

As immunologists adopted flow cytometry as a research tool for studying the differentiation and function of lymphocytes, immunofluorescence analysis was applied to the black sheep of the lymphocyte family, cells from leukemias and lymphomas, and such phenotyping was found to provide clinically relevant information. Although relatively few tumor specific antigens have been found, the demonstration on tumor cells of different antigens characteristic of various developmental stages in the tissue of origin provides information not only about the lineage, but also about the probable biologic behavior of the tumor[1009].

Immunophenotyping of leukemias, lymphomas, and other hematopoietic neoplasia is widely enough used for

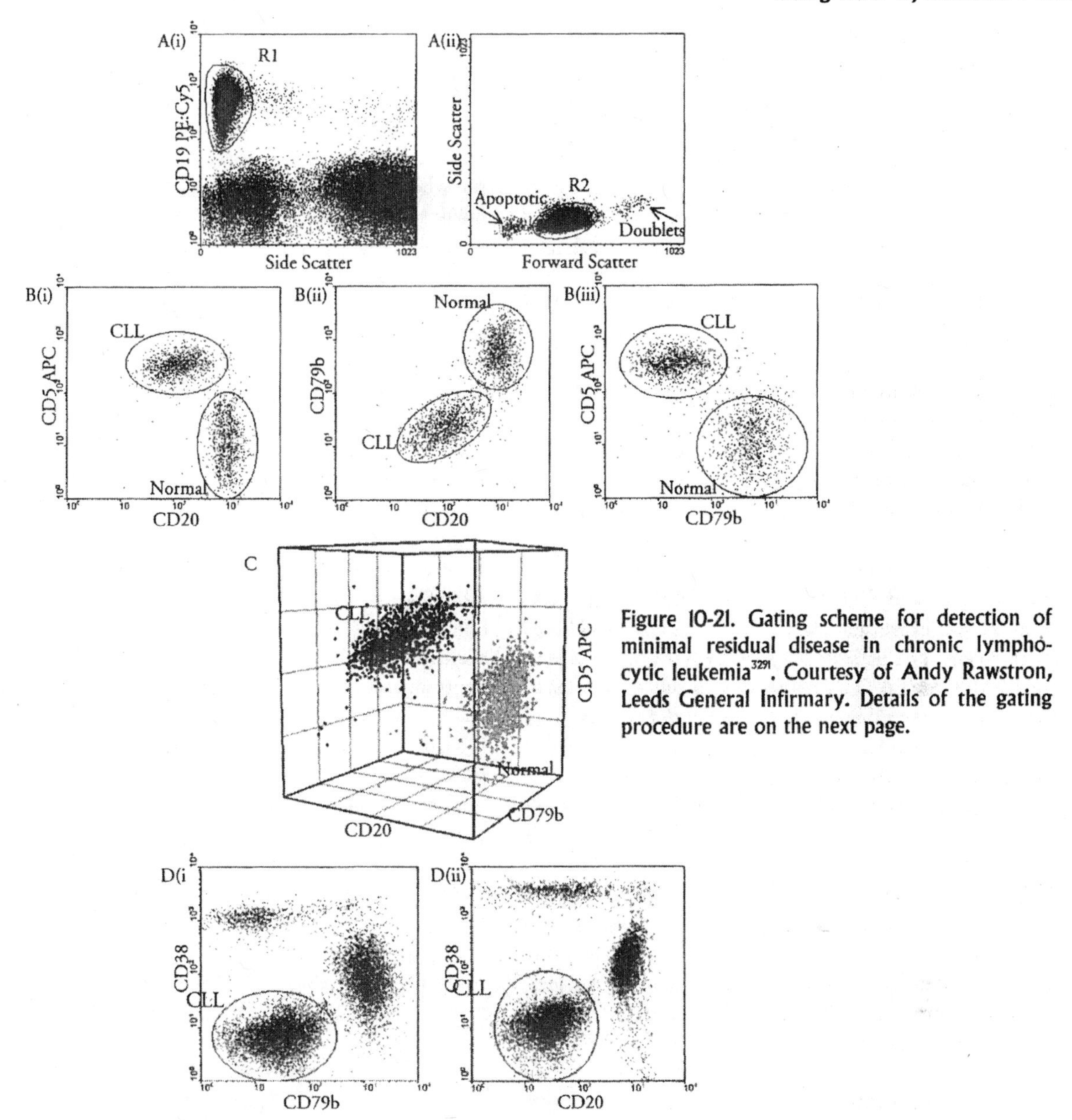

Figure 10-21. Gating scheme for detection of minimal residual disease in chronic lymphocytic leukemia[3291]. Courtesy of Andy Rawstron, Leeds General Infirmary. Details of the gating procedure are on the next page.

several organizations to have made recommendations about procedure. The conclusions of a U.S.-Canadian Consensus Conference held in Bethesda (MS) in 1995 were published in *Cytometry (Communications in Clinical Cytometry)* in 1997[3275-80], and the most recent NCCLS Guideline was issued in 1998[3281]. In 2001, Braylan et al[3282] summarized the consensus of a meeting of international experts held at the ISAC 2000 Congress on the minimum number of antibodies required to evaluate hematologic and lymphoid neoplasias; figures ranged from as many as 8 antibodies for plasma cell disorders to as many as 24 for acute leukemias. The authors cautioned that reducing the number of antibodies could significantly compromise diagnostic accuracy, monitoring, and/or therapy.

A 2003 book by Nguyen, Diamond, and Braylan, *Flow Cytometry in Hematopathology. A Visual Approach to Data Analysis and Interpretation*[3283], includes extensive examples of immunophenotyping in the text and on a CD-ROM. Phenotypic analysis of acute myeloid leukemia (AML) is discussed briefly on pp. 467-8; multiparameter phenotypes of normal marrow and marrow from an AML patient are shown in color on the back cover.

Detecting Minimal Residual Disease

In the leukemias and other neoplastic diseases for which effective therapy is available, it is useful to be able to detect **minimal residual disease** in blood or bone marrow specimens. While this is difficult in the absence of specific identifying characteristics, effective methodology has been described[611,883,1010-1,1150,3284-7].

Patients judged to be in clinical remission by standard methods of marrow examination may have as many as 5% residual leukemic cells in the marrow. However, while patients with no residual disease by the more stringent flow cytometric criteria tend to stay in remission for long periods of time, patients in whom multiparameter flow cytometry reveals residual leukemic cells at levels exceeding more than 0.01% of marrow cells usually relapse early and die unless alternative therapy is available. In recent years, it has been possible to make the argument that the undeniably elegant flow cytometric techniques used to detect minimal residual disease can also materially improve patient care in at least some hematologic malignancies[3288-96].

Figure 10-21, on the preceding page, illustrates the gating scheme developed by Rawstron et al[3291] for detection of minimal residual disease in chronic lymphocytic leukemia[3291]. In the first step, [panel A(i)], region R1 defines a putative B lymphocyte population based on forward scatter and CD19 expression and lack of CD3 expression (the CD3 gating process is not shown); in panel A (ii), the scatter gate R2 is defined to exclude apoptotic cells and doublets. Panels B(i), B(ii), and B(iii) show expression of CD5, CD20, and CD79b on normal lymphocytes and CLL cells, as does the 3-dimensional display of panel C. In this particular case, CD79b expression on the leukemic cells is low; this varies from case to case, and that is the reason for the inclusion of measurements of both CD20 and CD79b in the procedure. The combination of antibodies to these antigens with antibodies to CD5 and CD38 enables detection of CLL cells with all known expression patterns, at levels of 1 in 100,000 nucleated cells. The patients followed by Rawstron et al were treated with CAMPATH-1 and/or autologous transplant; all patients with detectable residual disease (CLL cells more than 0.05% of total leukocytes) had progressively increasing levels of leukemic cells on follow-up.

This article was published in *Blood*; the figure in the journal illustrating the gating scheme was of substantially poorer quality than the one Andy Rawstron donated for this volume. The axis labels were almost illegible, and one or two were incorrect, because, once Andy had reduced the figure to the size specified by the journal, he couldn't read them any better than I could without a magnifying glass. So the published version of the figure was definitely an example of bad flow happening to a good journal. If you looked at the materials and methods section, the sources of antibodies were specified, but no mention was made of which labels were used for which antibodies; it turns out there are a few tricks involved there. There's a lot of flow cytometry in *Blood*, but it is not always as critical to articles as it was in this case, and, for a variety of reasons, the article really didn't contain enough information to enable a skilled flow cytometer user to implement the procedure described. Yet people do want to implement procedures like this, for the clinically relevant purpose of detecting minimal residual disease. Getting it right is essential if clinicians are to obtain the information they want about prognosis and therapy.

The cytometry community probably needs yet another set of guidelines to deal with this issue. Conceding that it is unlikely that journal articles will contain complete procedural details, we have to define ways of getting them. If the authors managed to get approval from a Human Experimentation Committee to do the research, they must have submitted a detailed protocol that should contain all the necessary information, so the obvious thing to do would be to contact the corresponding author and ask for a copy of the protocol, or an equivalent detailed description of the procedure. As it happened, the e-mail address Andy Rawstron had given to *Blood* was inoperative due to various dot-complications, so I actually had to snail mail him to initiate a correspondence. He was an exceptionally good sport about the whole business, and I was pleased to say that the subsequent paper his group published in *Blood* about detecting neoplastic plasma cells in multiple myeloma patients[3294] included more legible figures and more procedural details.

But, getting back to my point, we all know that some antibodies work better than others, and that some labels work better than others, and that sample storage and preparation conditions may influence staining intensity, and we really need to have fairly rigid definitions of protocols to make sure everybody gets the right answers, especially when we need to quantify antigen expression, as may be necessary in analyzing cells from some leukemias[3297] and from patients with AIDS[2689-92].

I should also point out that, since there are fewer than 10,000 new cases of acute myelocytic leukemia in the U.S. each year[1752]; flow cytometric phenotyping at remission is therefore not likely to be applicable to more than 7,000 patients annually, and will not be able to generate more than a few million dollars in annual revenue for all parties involved. Perhaps we need an Orphan Diagnostic Act, much like the existing Orphan Drug Act, to make technology such as this available in situations in which there are too few patients to enable manufacturers or providers to bring products to market profitably via normal channels.

Flow cytometry, and cytometry in general, are not likely to get bailed out in this way unless they provide better (and, one hopes, more cost-effective) ways of doing things than do competing technologies. PCR and flow compete in the area of detection of minimal residual disease; flow is reported to be at least as sensitive[3298]. In the near future, we may see some competition between phenotyping and DNA array technology[3299] for classification and prognosis, but it looks as if the genomic technology, like flow, is better for establishing prognosis than for selecting among treatment options.

Biological Implications of Phenotyping Results

By now, marrow samples from thousands of patients with acute myelocytic leukemia, some with multiple leukemic clones, have been subjected to phenotypic analysis by multiparameter flow cytometry, and, as far as I know, no two clones are alike. This may tell us something about the biology of leukemias.

It is apparent that leukemia may result from various combinations of somatic mutations in hematopoietic cells. There may be dozens of possible mutations, yielding hundreds of genotypes, or perhaps even a few thousand genotypes. It doesn't seem all that likely, however, that there are many more genotypes than that.

We can then consider the likelihood of finding identical clones, given the number of genotypes we believe exist. This problem is similar to that posed in the mathematical puzzle that asks how many people must be in a room before there is a better than even chance that two of them have the same birthday.

Neglecting leap years, there are 365 possible birthdays. If there are two people in the room, the probability is (364/365) that they do not have the same birthday; if there are three people, the probability is (364/365) × (363/365). For four people, the probability that no two share a birthday is (364/365) × (363/365) × (362/365), and so on. When you get to 23 people, the probability that no two have the same birthday is less than 1/2; there is therefore a better than even chance that two of the 23 people have the same birthday.

If there were 365 leukemic genotypes, and if phenotype were completely determined by genotype, there would be a better than even chance that two identical phenotypes would have turned up by the time 23 marrow samples had been examined. However, many hundreds of samples have been examined, and no two phenotypes have been identical. This may simply mean that there are a lot more than 365 genotypes. If there were 1000 genotypes, I calculate that there would be even odds that two would be identical if 37 samples were examined. If no two phenotypes of over 300 are identical, either there are many thousand genotypes, or phenotype is not completely determined by genotype. The latter possibility is intriguing; it suggests that some host factor(s) might select for a particular leukemic phenotype.

This isn't all that farfetched. As I mentioned previously, a lot of reasonably smart people have been applying the methods of analytical cytology to normal and cancer cells for around forty years, and we still haven't found a consistent feature that differentiates all cancer cells from all normal cells. The whole definition of malignancy is, to some extent, an operational one; if it grows, spreads, and, when left untreated, eventually kills the patient, it's a cancer. I think it might be time to consider the medium in which cancer cells grow, i.e., the patient; maybe what we need to look for to define cancer in cytologic terms is a combination of characteristics of the tumor cells and characteristics of the host. This would certainly be compatible with what appears to be phenotypic selection.

The concept of **tumor stem cells** was recently revivified following a report by Al-Hajj et al[3301] that, among tumor cells isolated from nine human breast cancers, only those cells with a CD44+CD24- phenotype, which represented 1% or less of the population, readily induced tumors in immunocompromised mice. It made the newspapers.

Digression: A Slight Case of Cancer

A number of people, taking different routes, have arrived at viewpoints similar to that just expressed. I got there as a very interested spectator and patient, after undergoing a partial gastrectomy and radiation therapy for a mucosal ("MALT") lymphoma of the stomach that was diagnosed in 1991.

As an oncologist, I couldn't have asked for a better variety of cancer. Most gastric lymphomas are slow-growing, and prospects for long-term survival following treatment by surgery and irradiation are excellent[1772]. For many years, what are now known to be gastric lymphomas were classified as "pseudolymphoma," reflecting the fact that they looked like lymphomas, but apparently did not spread to lymph nodes or elsewhere, and the observation that patients might survive for ten years or more without any treatment other than antacids and ulcer drugs, which relieved the symptoms of pain and bleeding.

Modern thinking on gastric lymphomas is largely based on the work of Peter Isaacson, of University College and Middlesex School of Medicine, London, and his colleagues. Isaacson was largely responsible for the definition of the concept of **MALT lymphomas** (i.e., lymphomas of **M**ucosal **A**ssociated **L**ymphoid **T**issue)[1773]. Although gastric MALT lymphomas are among the most common tumors of this type, mucosal associated lymphoid tissue is not present in the normal stomach. MALT lymphomas of other sites in which lymphoid tissue is not normally found, e.g., the salivary gland and thyroid, arise following the acquisition of lymphoid tissue in these organs as a result of autoimmune disease (Sjogren's syndrome and Hashimoto's thyroiditis, respectively); in the stomach, lymphoid follicles develop in association with chronic gastritis.

The predominant cell in what were called pseudolymphomas is a B cell resembling the centrocytes seen in the lymphoid follicles normally present in the intestinal submucosa. It is phenotypically $CD21^+CD35^+CD5^-CD10^-$. Isaacson based his reclassification of pseudolymphomas as lymphomas in part on monoclonality, which is demonstrable by light chain staining and heavy chain gene rearrangement; there is no rearrangement of either bcl-1, bcl-2, or c-myc[1774]. Other evidence for the neoplastic nature of the cell comes from cytogenetics; rearrangements of chromosome 1p, trisomy 3, and trisomy 7 have been found[1775].

Intensive pathologic examination of gastrectomy specimens from a small number of patients showed that in all cases, small foci of lymphoma distinct from the original lesion were scattered throughout the specimen[1776]; the apparent multifocal nature of the tumor could explain the development of local relapse after a long disease-free interval, which is typical of the relatively infrequent recurrences that do occur.

It was thought that the propensity of the tumor to remain localized, rather than spreading to lymph nodes, marrow, etc., was due to the centrocyte-like nature of the neoplastic cells; normal centrocytes did not appear to recirculate. However, by using anti-idiotype antibodies, Isaacson's group demonstrated scattered MALT lymphoma cells in anatomically and otherwise histopathologically normal lymph nodes and spleen removed during surgery for gastric lesions.

The gastritis leading to the development of lymphoid tissue in the stomach occurs as a consequence of infection by *Helicobacter pylori*; this organism, which is also associated with peptic ulcer disease and gastric carcinoma, was identified in biopsy or surgical specimens from 101 of 110 patients (92%) with gastric MALT lymphomas[1777].

Isaacson hypothesized that gastric MALT lymphomas result from neoplastic transformation of cells involved in a normal immune response. In one test of this hypothesis, he placed cells from several gastric MALT lymphomas in culture with several strains of *H. pylori*, using other bacteria and phorbol ester as control stimuli. Phorbol ester induced proliferation of cells from all specimens; the neoplastic B cells proliferated otherwise only in the presence of both the specific strain of *H. pylori* found in the patient and T cells also taken from the tumor[1778]. This pattern of antigen- and T cell-dependent proliferation *in vitro* is consistent with the observation of microscopic foci of tumor, but not of larger lesions, in areas where antigen, in this case *H. pylori*, is absent. Interestingly enough, the immunoglobulins produced by the neoplastic B cells appear to react with autoantigens, rather than with *H. pylori*[1779-80].

Working with Isaacson's group, Claudio Doglioni had documented a high incidence of both gastric MALT lymphoma and *H. pylori* infection in northeastern Italy[1781]. They conducted a brief clinical trial there in which six patients with gastric MALT lymphoma demonstrated by endoscopic biopsy were treated with bismuth salts (i.e., Pepto-Bismol or its equivalent), metronidazole, and amoxicillin for two weeks to eradicate *H. pylori*. Follow-up biopsies were done 4 to 10 months later; in five of six patients, the lymphoma had disappeared, i.e., PCR failed to demonstrate residual disease in the stomach[1782].

I heard Isaacson speak in Boston in May, 1993, before his clinical study was published, and became a convert; I took the therapy for *H. pylori* ($30 worth of pills), just to be on the safe side. Since the lecture was delivered to an audience of pathologists, I prepared a summary, with references and circulated it to the gastroenterologist, oncologists, radiotherapist, and surgeon who had treated me. The results were interesting. I had asked my gastroenterologist about *Helicobacter* when I first had ulcerlike symptoms, before my diagnosis of lymphoma was established; at the time, he, like many others in Boston, was skeptical about the role of the organism in ulcer disease or anything else. After I told him about Isaacson's work, he had the pathologists pull out my specimen and look for *H. pylori*; they found the bacteria. I should have placed a bet.

There are big picture conclusions and little picture conclusions to be drawn from all this; let's look at the big picture first. I agree with Isaacson that MALT lymphomas probably aren't the only neoplasms in which the host environment provides some stimulus for tumor growth, and expect that this new paradigm may usefully be extended to other, more common cancers. Based on everything known about gastric MALT lymphomas, I am unlikely to have even a local recurrence of mine. However, there's another paradigm shift here. Where the prevailing wisdom was that the low rates of recurrence and spread were due to the localized, noncirculating nature of the neoplastic cells, the present picture, based on the anti-idiotype antibody studies, is that the cells do circulate. I got radiation therapy to the region of what's left of my stomach, which might have killed tumor cells remaining around there, but which almost certainly wouldn't have killed tumor cells in distant lymph nodes or my bone marrow. The likelihood is that those cells won't bother me. This goes completely against one of the central dogmas of cancer treatment, which is that curing the patient requires the eradication of every last cancer cell.

Taking *H. pylori* and the T cells responding to it out of the picture to eliminate the growth factors needed by the neoplastic B cells in a gastric MALT lymphoma isn't all that different from taking estrogen away from breast cancer cells which need it to grow. It also isn't that uncommon to find microscopic foci of breast cancer cells at autopsy in patients who were supposedly cured and died years later of something else. The fact is that most cancer cells are a hell of a lot harder to grow *in vitro* than they are to grow in their hosts; we might want to pay a little more attention to the role of host factors in looking at new modalities of treatment.

As for the smaller picture, being a dedicated analytical cytologist, I made sure that my cells got worked up; they were reported to be monoclonal for lambda light chain (MALT lymphomas are not classified as chocolate and vanilla), and, by flow cytometry, they were DNA diploid. However, since there is a history of lymphoma on both sides of my family, I thought it might be appropriate to look into the cytogenetics; the cells were found to have two extra chromosomes. If I tried hard, I could probably detect remaining cells in my marrow using DAPI and anti-lambda antibodies; for the time being, I prefer to let David Hedley remain as the only person I know who takes sternal marrow samples from himself.

From the point of view of a writer or a reader of a book on flow cytometry, one of the most sobering aspects of this whole story relates to Isaacson's methodology. There was no flow cytometry involved; no image analysis, either. All of the anti-idiotype and other antibody and gene probe work was done using enzyme-linked systems and transmitted light microscopy. If you go to the same meetings I do, you're probably used to seeing chromosome probes under fluorescence microscopes; in Isaacson's slides, they showed up as blue or brown dots on Giemsa-stained chromosomes. And, try as I might, I couldn't really think of any way in which flow or image cytometry would have enabled the work to be done better, or faster, or cheaper (and, almost nine years later, I still can't).

Analysis of Sperm

The use of flow cytometry and cell sorting for separation of X- and Y-sperm has been discussed on pp. 26, 282, 310-1, and 452-3.

Assessment of the quality of animal and human sperm by flow cytometry has been looked at in evaluation of infertility in animals and humans. Evenson, Darzynkiewicz and Melamed[341] approached this issue by correlating microscopic observations of sperm mobility with **dye exclusion** and **mitochondrial membrane potential** measurements, but found **chromatin structure**, as measured using acid denaturation and acridine orange, to be the most reliable indicator of sperm quality[941-3]. The **Sperm Chromatin Structure Assay (SCSA)** correlates well with alternate tests of sperm quality in bulls[944-5,1982]. Evenson et al[3302] compared SCSA measurements made on various arc lamp and laser source flow cytometers and found that, while there were differences in the appearance of displayed data from different instruments, all produced equivalent results. Evenson and Jost[3303] reviewed the SCSA in 2002.

Auger et al[1983] correlated mitochondrial membrane potential of sperm, as measured using rhodamine 123, with motility, and were able to use R123 fluorescence to sort highly motile sperm[1984]; more recently, Marchetti et al[3304] concluded that "Analysis of mitochondrial membrane potential is the most sensitive test by which to determine sperm quality." However, the sensitivity of membrane potential measurements does not appear to be related to their capacity to indicate apoptosis. Using annexin V staining, Ricci et al[3305] found no significant differences in the percentages of apoptotic sperm or of leukocytes in semen classified as normal and abnormal by WHO criteria.

Ericsson et al[1985] found carboxymethylfluorescein diacetate and hydroethidine useful in assessing viability of cryopreserved sperm, although they did not observe correlations between flow cytometric quality indicators and motility[1986]. The same group evaluated the toxicity of dyes to sperm cells; Hoechst 33342, hydroethidine, and a tracking dye decreased oxygen consumption, although R123 did not[1987].

Yeung et al[3306] studied 488 nm forward and side scatter and Coulter volume of sperm, and found that infertile sperm from knockout mice were larger than fertile sperm from heterozygous mice, suggesting that volume regulation may be critical for sperm function.

Haas and Cunningham[946] investigated another aspect of infertility using flow cytometry; they were able to detect and quantify **antibodies bound to sperm**. Haas and his colleagues[1988-92,1995] and others[1993-4,1996-8] have continued to use flow cytometry for this purpose; in comparative studies, it demonstrates advantages over alternative techniques[1992,1998].

The progress of the **acrosomal reaction** essential for sperm to become capable of fertilization can be monitored using fluorescent antibodies and lectins[1424,999-2005]. Harrison, Mairet, and Miller[1734], studying the effects of buffers used with sperm for *in vitro* fertilization, observed that bicarbonate in the buffer, which induced the calcium influx necessary to begin the acrosome reaction, also appeared to increase the number of cells with damaged membranes.

DNA content analysis of testicular aspirates and biopsies[2006-13] has proven useful in the evaluation of infertility, including that induced by cancer treatments; sperm are readily detected by their haploid DNA content, allowing their relative numbers to be estimated. Regarding DNA analysis in biopsies of undescended testes, in which cancer is likely to develop, Clausen et al[2013] note that DNA aneuploidy can occur in an undescended testis without any evidence of malignancy, and suggest it may indicate a preneoplastic state.

Commercial flow cytometers adapted for sperm counting are now available. BD Biosciences has produced a system of stains (SYBR-14 or propidium), diluent, and counting beads allowing the FACSCount (p. 418) to be used for sperm counting[3307]; Tsuji et al[3308] have reported on another device, the S-FCM.

Laser scanning cytometry and FISH were found useful by Baumgartner et al[3309] for detection of aneuploid sperm.

Isolating Fetal Cells from the Maternal Circulation for Prenatal Diagnosis

During pregnancy, both trophoblastic and blood cells from the fetus may enter the maternal circulation, in which they represent fewer than 1 of every 10^5 nucleated cells. The development of reliable methods for isolation of such fetal cells would provide a simple, low-risk alternative to amniocentesis and chorionic villus sampling for acquisition of cell samples for prenatal genetic diagnosis using *in situ* hybridization, PCR, and other molecular biologic methods. Bruch et al[1812] sorted trophoblast-like cells from maternal blood using three monoclonal antibodies against trophoblast, and, via PCR, found Y-specific sequences in two of three samples from mothers carrying male fetuses, although most of the sorted cells appeared to be maternal leukocytes. A higher purity of trophoblastic cells was obtained by combining immunomagnetic removal of maternal lymphocytes and sorting using antitrophoblast antibodies; fluorescence *in situ* hybridization with a Y-centromeric probe allowed detection of 47,XYY fetal cells[1813].

An alternative approach is based on isolation of fetal nucleated erythrocytes, which appear in the maternal circulation early in gestation. Bianchi et al[1814], using multiparameter sorting, isolated cells positive for the transferrin receptor (CD71) and negative for the CD3 and CD14 leukocyte antigens; Y chromosomal DNA sequences were detected in samples obtained at 11 and 12 weeks' gestation, and absent in subsequent samples at 16, 19, and 20 weeks, from women who delivered males, while Y DNA was not detected in two women who delivered females.

Price et al[1815] sorted fetal nucleated erythrocytes on the basis of a combination of forward and right angle scatter and the presence of CD71 (transferrin receptor) and glycophorin A on the cell surface. From such flow-sorted samples, Wachtel et al[1816] correctly identified fetal sex in 17/18 (94%) pregnancies of 10-21 weeks gestation. Their group also reported diagnosis of trisomy-21 in fetal cells from maternal blood[1817], as did Bianchi et al[1818].

Although specificity of fetal cell detection has been improved by use intracellular staining with antibodies to the gamma globin chain of hemoglobin F[3310-2], recent clinical studies have combined immunomagnetic separation, with negative selection for CD45+ cells[3310] and positive selection for cells containing gamma globin[3312], with flow sorting. The

NIFTY I multicenter trial[3313] were better using magnetic cell separation conducted by the National Institute of Child Health and Human Development found that target cell recovery and fetal cell detection were better using magnetic separation alone than with flow sorting with or without magnetic separation. All separation methods yield both fetal and maternal cells; discrimination of the two populations, which is essential for interpretation of FISH data used for prenatal diagnosis of trisomies, is most effectively accomplished using morphologic as well as biochemical criteria[3314], suggesting that the utility of flow cytometry in this field may be limited.

The March of Time: Circadian Rhythms, Aging, and Atherosclerosis

Periodic phenomena can only be effectively studied by methods that consume relatively little time compared to the period involved. Thus, it was really not possible to learn much about **circadian periodicity** of cell kinetics as long as kinetic studies required autoradiographic analysis with exposure times of weeks. Since DNA flow cytometry can be accomplished in minutes, it has been feasible to use the technique to discover circadian variations in cell proliferative activity in tissues[2028-36] and tumors[957,2027], and to analyze resulting circadian variations in drug effects[958], which may eventually be exploitable for scheduling kinetic-based therapy[959,2032].

Circadian fluctuations in counts of various blood cell types have been reported[2037]; this may have unwanted effects in following progression in such diseases as HIV infection[2038-9]. Circadian rhythms also affect the growth and metabolism of single-celled organisms such as the dinoflagellate *Gonyaulax*[2040-1,2049].

Changes at the cellular level related to **aging** can be investigated either by comparing cellular samples from young and old individuals or by examination of cells such as fibroblasts subjected to *in vitro* aging. Considering the popular hypothesis that aging is associated with impaired defenses against free radical reactions, Poot et al[960] examined glutathione recovery rates following oxidative stress in *in vitro* aged fibroblasts and in fibroblasts from patients with several metabolic diseases and found no impairment in comparison with normal cells. Martinez et al[961] quantified rhodamine 123 mitochondrial fluorescence in young (6 population doublings) and old (41 population doublings) fibroblasts, and found increased fluorescence in the older population. Cristofalo[2042] found the S-phase fraction and cell density to be independent biomarkers of aging in cell cultures. Differences in lymphocyte responses with age have been reported (e.g., references 1690 and 1692).

Aging has also been associated with reported increases in **DNA polyploidy** in several tissue types, including heart muscle and vascular smooth muscle[962,2043-4]. Similar increases in polyploidy in vascular smooth muscle occur following injury of the type used to induce experimental **atherosclerosis**. A methodological problem in this area of research lies in the necessity to discriminate tetraploid cells from doublets due to aggregation or proximity of diploid cells in the flow cytometer; Vliegen[963] et al have developed a mathematical procedure to correct for background and clumping. Independent confirmation that the observed polyploidy is not entirely artifactual was obtained when polyploid cell lines were isolated and cultured from rat aortas by Goldberg et al[964]. Black et al[2045] established that vascular smooth muscle polyploidy parallels inhibition, reversal, and redevelopment of **hypertension** in an experimental model, the spontaneously hypertensive rat. On another front, Stemme, Holm, and Hansson[2046] reported that T lymphocytes, which are found in abundance in atherosclerotic plaques, are memory cells (CD45RO$^+$); they also express the very late activation antigen VLA-1.

Although I have not done an extensive search of literature in this area since the previous edition, I suspect that the cytometric analysis of **platelet activation in atherosclerotic disease** (see p. 487) represents a significant contribution.

Clinical Application: Urine Analysis

International Remote Imaging Systems' "Yellow IRIS™" workstation[1019] was the first of a series of flow cytometers with imaging capability intended for analysis of urine specimens. Image data can be reviewed by an operator. The current instruments (p. 431) are the **Model 500** Urine/Fluids Workstation and the **939UDx™** Urine Pathology System; the former is manually fed, while the latter is completely automated. Both provide chemical as well as cytometric analyses of urine.

Sysmex manufactures the **UF-100**[3315-7] and **UF-50** urine analyzers; both combine fluorescence, light scattering, and impedance measurements to classify and count formed elements in blood. The numbers denote the number of samples the instruments can process per hour.

The Animal Kingdom

Lions and Pumas and Clams, Oh, My!

Flow cytometry has been and is being applied to a lot of species these days. Since, when the moon is full, I grow black and white fur and eat bamboo shoots, preferably with hoisin sauce, I decided to try to get flow cytometric data from giant panda cells to put into the Third Edition. I called Stephen O'Brien, of the National Cancer Institute, who used molecular genetic techniques to determine that the giant panda is a bear[1964], and asked whether he had ever done analyses of panda cells. No luck there, but, as it turned out, Eric Brown, a predoctoral student in his lab, sent me the contour plots of T cell subset analysis in a lion (*Panthera leo*) and a puma (or cougar, or mountain lion) (*Felis concolor*) shown in Figure 10-2. The antibodies used are the same ones used for T cell subset analysis in domestic cats, and the analyses are done for the same reason; the big cats can carry the same feline leukemia retrovirus as their smaller relatives. Bull et al [3318] recently studied polymorphic expression in the lion CD8alpha chain.

Dr. O'Brien's collaborators on the panda work, at the National Zoo, put me in touch with Dr. Marcia Etheridge, of the Department of Comparative Medicine at Johns Hopkins University School of Medicine, who has analyzed blood or its equivalent from a number of animal species on an Abbott Cell-Dyn 3500 hematology analyzer; Figure 10-23 shows clusters of clam cells on 2-dimensional histograms using essentially the same parameters as are shown in Figure 10-16 (p. 483), which depicts human blood.

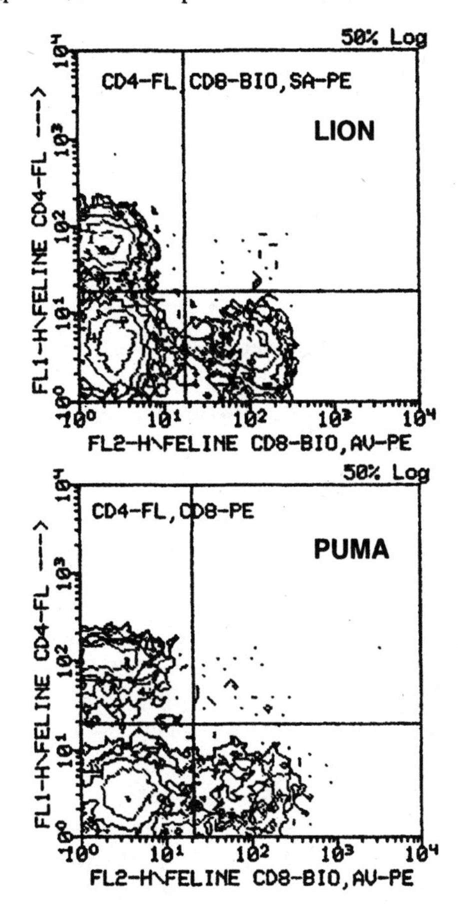

Figure 10-22. T cell subset analysis in lion (top) and puma (bottom) peripheral lymphocytes. The figure was provided by Eric Brown and Stephen O'Brien (National Cancer Institute).

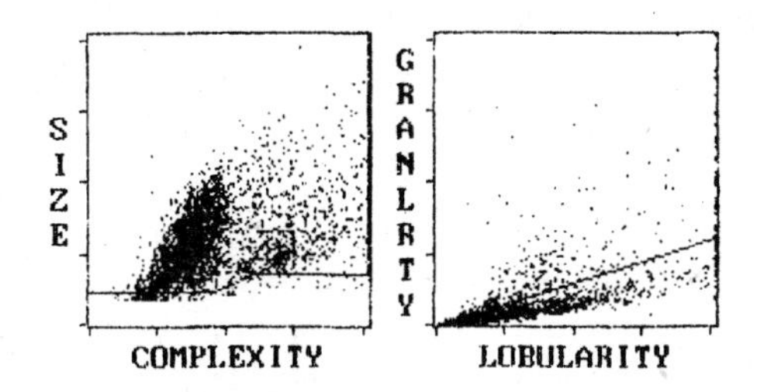

Figure 10-23. Analysis of clam cells in a Cell-Dyn 3500 hematology analyzer (see Figure 10-16, p. 483, for a comparison with human blood). The Figure was provided by Marcia Etheridge (Johns Hopkins University).

The phagocytic properties of clam hemocytes[3319] are relevant to pathology[3320] and toxicology[3321], and have been studied in those contexts using flow cytometry.

Flow cytometry can get the lion, or at least its cells, to lie down with the lamb, as well as the clam. Some interesting mathematical and statistical methods for extracting proteomic information from patterns of antibody binding to cell surface proteins[2491,3322-3] have emerged from my colleague Steve Mentzer's interest in the immunobiology of the sheep[3324], a favorite lab animal for thoracic and cardiac surgeons.

Blood from most mammals is closer to the human/lion/lamb model than the clam model. Wang et al[1965] showed that forward and right angle scatter could be used to perform three part differential leukocyte counts on porcine blood; Jain, Paape and Miller[1966] combined two-angle scatter and acridine orange fluorescence measurements for a differential count of bovine leukocytes which they reported discriminated eosinophils from other granulocytes.

Blottner et al[3325-9] have used flow cytometry of DNA content and vimentin expression to study spermatogenesis and seasonal variations in spermatogenesis in cattle, roe deer, brown hares, and mink. It's not clear whether they've gotten into a rut or just have a lot of seed money.

Starck and Beese[3330-1]combined flow cytometry with light and electron microscopy study cytological mechanisms involved in the rapid size changes of the small intestine and liver that are associated with feeding in snakes, using pythons and garter snakes as examples. They found that cell proliferation is not involved in the upregulation of organ size, but increases during downregulation, as cells worn out during the digestive process are replaced.

When I put out a call on the Purdue Cytometry Mailing List for flow cytometric data from giant pandas for this edition of the book, I got helpful e-mails from a lot of people, but the bottom line was that nobody seemed to have done flow cytometry on one. I found out that there are monoclonal antibodies that react with cell surface antigens in other bears and would therefore be potentially useful in studies of the immunological problems said to occur in pandas. E-mail me (hms@shapirolab.com) if you're interested.

Fish Story; FISH Story

I am informed that breeding sterile, triploid hybrids is a technique common in **fisheries** for preventing fish (vertebrates) and shellfish (invertebrates) from reproducing. While this sounds counterproductive, it isn't (look at beef); oysters, for example, survive (and stay marketable) longer if they don't bother to reproduce. Flow cytometry is, of course, a pretty good technique for determining whether you've achieved triploidy or not, and has been used for the purpose[951-2].

Genome size of diploid organisms, fish included, as indicated by DNA content, is useful for studies of classification and evolution. Ciudad et al measured DNA content of goldfish, tench, and zebrafish[3332]; Brainerd et al analyzed DNA content of different groups of pufferfish[3333], and suggested that the smooth pufferfish, which have the smallest genome sizes (0.7-1.0 pg diploid) measured to date in vertebrates, derive from ancestors with somewhat larger genomes.

While nobody seems to have done 11-color flow cytometry on fish cells, the ranges of probes used, parameters measured, and species subjected to cytometric analysis continue to expand. For example, Wong and Chan[3334] used rhodamine 123 staining to identify, and determine effects of cortisol on, chloride cells in the gills of Japanese eels.

The immune system supposedly evolved in fish; there is a journal called *Fish and Shellfish Immunology* in which you can expect to find a lot of flow cytometry, monoclonal antibodies and all, just as you would in almost any other immunology journal. In it, Inoue et al[3335] reported that staining with $DiOC_6(3)$ and measurement of forward and side scatter and fluorescence allowed resolution of five separate cell populations in fish blood, i.e., erythrocytes, thrombocytes mixed with lymphocytes, monocytes, neutrophils, and basophils. The same trick works with bird (quail) blood[3336].

Mayer et al[3337] isolated lymphocyte-like cells from sea lamprey intestine using forward and side scatter, and prepared a cDNA library with which they were able to determine that the lamprey cells contained many genes homologous to those found in human lymphocytes. Sequences with similarity to MHC, T cell receptor, or Ig genes were notable by their absence, suggesting to the authors that "the evolution of lymphocytes in the lamprey has reached a stage poised for the emergence of adaptive immunity."

The subject of genetic homologies easily gets us from fish to FISH, and the technique of "Zoo-FISH," or comparative chromosome painting, in which probes derived from flow sorted human chromosomes are used to detect homology with chromosomes of other mammalian species[3338-43]. Cat and human karyotypes are reported to resemble the putative ancestral mammalian karyotype[3339], while a comparison of human and harbor seal (*Phoca vitulina*) chromosomes shows resemblance to the putative ancestral carnivore[3341]. I guess you need to do Zoo-FISH on various seal species to find the karyotype of the average ancestral seal, or mean mother *Phoca*.

Flow Cytometry: For the Birds?

De Vita et al[2314] found DNA content potentially useful in sex identification in birds of prey, many of which are sexually monomorphic, which means it's hard to tell males from females by inspection. If you're trying to breed endangered falcons, and your pair contains two birds of the same sex, there won't be any falcon around. Sorry, I couldn't resist. It's a nice paper.

Having found flow cytometry applied to hawks, I thought I'd see if it had been applied to doves, as well. The closest I got was a paper by Itoh et al[3xxx] on binding of pigeon cytochrome C peptides to I-ab and the T cell receptor. There appear to be structural similarities between the pigeon peptides and MHC proteins, but probably just by cooincidence.

Big Stuff, Vegetable, Animal or Mineral

Since a good deal of the hardware technology of flow cytometry and sorting was developed by people interested in looking at lymphocytes and other small cells, most existing instruments require some modification to be usable for analysis and/or sorting of larger objects[671], be they **plant cells or protoplasts**[673], **tumor cell spheroids**[672], **pancreatic islets**[1227], or **megakaryocytes**. Large cell sorting was discussed in Chapter 6 (pp. 264-6). Both droplet sorters with large (> 200 µm) orifices and fluidic switching sorters have been used to sort a variety of appropriately sized specimens.

If there is a *Guinness Book of Records* in this area, the entry for largest objects subjected to flow cytometry might go to brine shrimp; in 1965, Linus Pauling and his colleagues counted these organisms by extinction measurements in a flow system[585] in conjunction with studies on the mechanism of anesthesia. Optical plankton analyzers[1167-8] may also be able to handle objects in this range. However, I also remember reading something years back about an optically activated fluidic potato sorter... .

Moving back to the animal world, I already mentioned (p. 265) that a modified Partec fluidic sorter had been used to sort pancreatic islets[1227]. A stain combining the dye neutral red and the chelating agent dithizone gave an 83% yield and 80% purity[2020]. On the other hand, a paper by Halban et al[2021], who used a conventional sorter to purify individual cell types from islets, and then mixed them in culture, reported that the cells will reaggregate to form structures resembling islets, so the cell transplant folks may not need to sort whole islets after all.

Large particles tend to settle faster than smaller ones, which can lead to problems during long analysis and sorting runs. Freyer, Fillak and Jett[2050], working with multicellular spheroids, found that adding 0.1% **xanthan gum** to suspensions greatly increased static viscosity, preventing settling of the particles in the sample tube. The gum does not increase viscosity in flow, and therefore does not interfere with sorter operation.

Flow Cytometry of Plant Cells and Chromosomes

A 1986 review by Brown et al[947] on flow cytometry in plant biology cited 44 references, remarking that the literature more than doubled in the two years since a previous review had appeared. There are now enough references to justify a book on applications of flow cytometry to plants, but, as far as I know, nobody has written one. However, Brown and Coba de la Pena[3344] have contributed a chapter on cytometry to a recent book on plant cell biology. Galbraith[3345] and Blackhall[3346] have reviewed flow cytometry and sorting of plant cells, and Lucretti and Doležel[3061] and Doležel et al[3347-8] have discussed methodology for flow cytometric analysis of plant chromosomes.

In general, people who do flow cytometry on materials from plants are measuring the same things the rest of us are used to measuring in animal cells, and facing many of the same problems.

Pollen is the most readily accessible single cell sample available from plants; however, since the objective in flow cytometry of plant materials is typically the study of some aspect of differentiated function, the samples commonly analyzed are cell suspensions or protoplasts[948,2016-8,3345-6,3349]. Chromosomes[949,3061-2,3065,3067,3071] and plant subcellular organelles such as chloroplasts and mitochondria have also been subjected to flow cytometric analysis.

Major areas of application of flow cytometry to plants include **measurement of DNA content** for taxonomic and other purposes[3350] and **analysis and sorting of plant chromosomes**. Figures 10-24, 10-25, and 10-26, illustrating these applications, were provided by Dr. Jaroslav Doležel, of the Laboratory of Molecular Cytogenetics and Cytometry, Institute of Experimental Botany, Olomouc, Czech Republic. A bibliography of the many publications of Dr. Doležel's group is available online at (http://www.ueb.cas.cz/olomouc1).

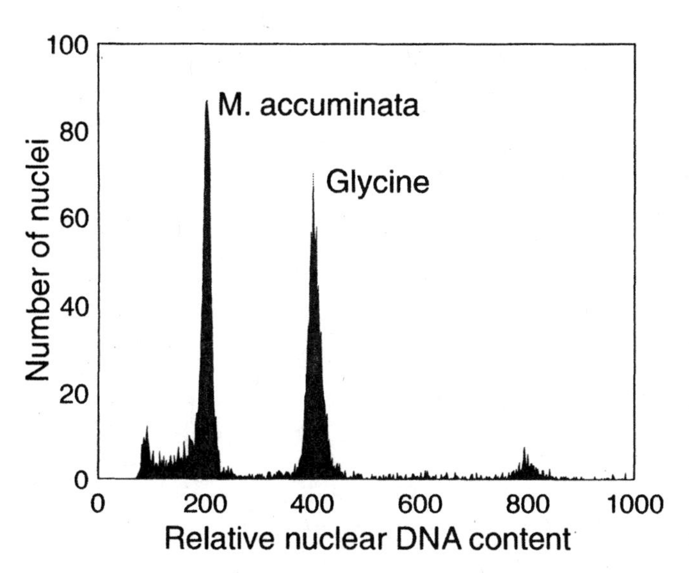

Figure 10-24. Determination of nuclear genome size in diploid banana (*Musa acuminata* "Calcutta 4"). Nuclei isolated from soybean (*Glycine max*, 2C = 2.50 pg DNA) were used as an internal reference standard. The small peak at channel 800 represents soybean nuclei in G_2 phase. Nuclear DNA content of *Musa* was estimated following the formula: *Musa* 2C DNA content = (*Musa* G_1 peak mean ÷ soybean G_1 peak mean) × 2.50 [pg DNA] and was found to be equal to 1.23 pg DNA. Courtesy of Jaroslav Doležel.

Measurement of Plant Cell DNA Content

Figure 10-24 illustrates determination of the 2C DNA content of a strain of bananas. A reference standard of known DNA content, soybean in this case, is used. Johnston et al[3351] recommended five species with a broad range of 2C DNA contents as standards: *Sorghum bicolor* cv. Pioneer 8695 (2C = 1.74 pg), *Pisum sativum* (pea) cv. Minerva Maple (2C = 9.56 pg), *Hordeum vulgare* (barley) cv. Sultan (2C = 11.12 pg), *Vicia faba* (broad bean) (2C = 26.66 pg), and *Allium cepa* (onion) cv. Ailsa Craig (2C = 33.55 pg). The reference standard should be one with 2C and 4C nuclear DNA content peaks close to, but not overlapping, the 2C and 4C peaks of the target species. Propidium iodide, which has no bias for A-T- or G-C-rich sequences within genomes, is recommended as the fluorochrome of choice for determination of plant DNA content; DAPI (A-T preference) should be used only if the estimated DNA value is corroborated using a second stain.

The Royal Botanic Gardens, Kew, UK, maintains a database of DNA content values; they are "C-values," i.e., values of DNA content in a haploid gamete nucleus, rather than the "2C-values" that would be obtained from somatic cell nuclei as shown in Figure 10-24. The URL is (http://www.rbgkew.org.uk/cval/homepage.html). The Plant DNA C-values Database contained data for 3,927 different Embryophyte plant species as of January, 2003; it combines data from Angiosperm, Pteridophyte, and Gymnosperm databases. If you would like to stop and smell the roses at this point (2C = 0.78-1.33 pg), see Yokoya et al[3352].

Vinogradov[1336,3353] suggested that plants with larger genome sizes should have higher G-C percentages; however, studies of a number of plant species[3354-5] have not found a significant correlation.

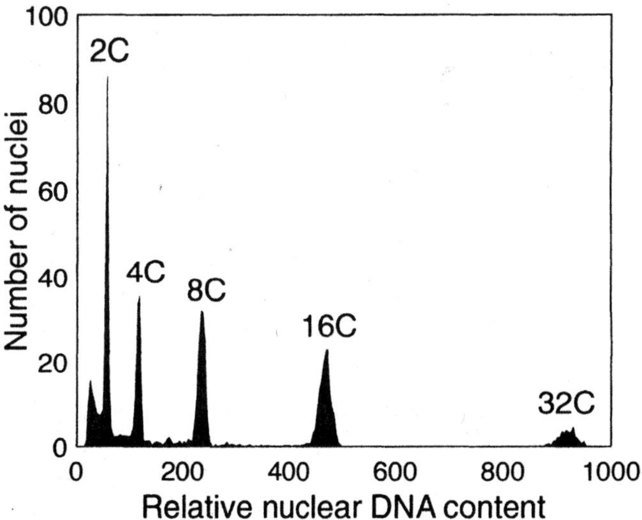

Figure 10-25. Histogram of relative nuclear DNA content of nuclei isolated from parenchymatic tissue of a young cactus plant (*Mammillaria san-angelensis*). Note the presence of peaks representing endopolyploid nuclei with DNA content reaching 32C. Courtesy of Jaroslav Doležel.

As Figure 10-25 shows, DNA content measurement may reveal cells with endopolyploid nuclei in plant tissue; this is a good reason to keep C-values in the database. DNA content measurement can also be used to determine sex in dioecious plants; however, since the difference in DNA content between male and female plants is typically only a few percent, it is necessary to use a high-resolution flow cytometer. Doležel and Göhde[3356] were able to resolve 3.7% differences in DNA content between male and female *Melandrium album* and *M. rubrum* using a Partec arc source flow cytometer and DAPI staining; they obtained G_0/G1 peak CVs of 0.53-0.70 % using a modification[3357] of the nuclear isolation technique originally published by Galbraith et al[3358] and an analysis rate of 20-50 nuclei/s. Pfosser et al[3359], using high-resolution flow cytometry, were able to detect aneuploidy resulting in differences in DNA content as small as 1.84 percent.

Houssa et al[2016] described a high-yield isolation procedure permitting flow cytometric analyses of DNA to be done using micrograms of shoot meristematic tissue.

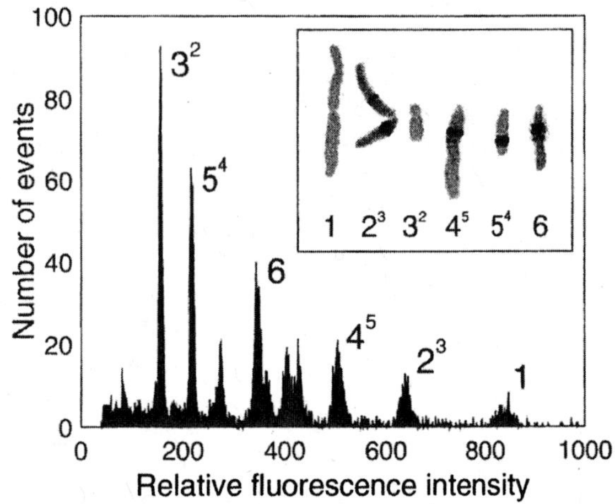

Figure 10-26. Flow karyotype of translocation line EF of broad bean (*Vicia faba* L., 2*n* = 12) obtained after analysis of DAPI-stained chromosome suspension. All chromosome types are resolved into well-discriminated peaks. Insert: images of flow-sorted chromosomes. Sorted chromosomes were identified after fluorescent labelling of *FokI* repeats (dark bands). Courtesy of Jaroslav Doležel.

Plant Chromosome Analysis and Sorting

Figure 10-26 shows a univariate karyotype of broad bean (*Vicia faba*). In the discussion of chromosome analysis and sorting on pp. 477-9, it was noted that, in many species in which chromosomes were similar in DNA content and base composition, it was necessary to work with lines containing chromosomal translocations to be able to sort material representing most of the karyotype. This procedure is frequently required in work with plants[3066-7,3360-3]; note that the chromosomes in Figure 10-26 come from a translocation line.

Other Flow Cytometric Applications in Plants

Some stains, particularly acid dyes, behave differently in plant and animal cells; Calcofluor white M2R, for example, works as an indicator of membrane integrity in animal cells but will stain cell walls of plant cells with intact cytoplasmic membranes, which lie inside the walls[808]. Bergonioux et al[2014] used acridine orange for DNA and RNA staining in nuclei isolated from plant protoplasts, and Ulrich[2015] described DAPI/sulforhodamine 101 staining for DNA and protein analyses in plant nuclei. Lucretti et al[3364] analyzed plant cell cycles using PI staining and antibodies to bromodeoxyuridine, and Taylor et al[2718] stained cellulose specifically (p. 363).

Measurements of the fluorescence of chlorophyll and other natural pigments as well as measurements of protein and nucleic acid content and specific antigens using extrinsic probes have been made in plant cells. Xu, Auger, and Govindjee[1273] measured chlorophyll fluorescence in isolated spinach thylakoids, and Pfündel and Meister[3365] used measurements of fluorescence above and below 710 nm to isolate maize mesophyll and bundle sheath chloroplast thylakoids.

Klock and Zimmermann[2018] used a flow cytometric assay to analyze the yield of fusion products in developing an improved method for electrofusion of plant protoplasts. Giglioli-Guivarc'h et al[3366] used BCECF to measure cytosolic pH of crabgrass mesophyll cell protoplasts, and O'Brien et al[3367] used Annexin-V and a TUNEL assay to monitor the progression of apoptosis in plants.

Wolters et al[2019] examined the origin of nuclear, chloroplast, and mitochondrial DNA in plant hybrids using a combination of flow cytometry and dot blot hybridization with DNA probes. Harkins et al[2017] examined gene expression in protoplasts isolated from transgenic tobacco plants by sorting; more recently, Galbraith[2887,3368-9] has led the move toward using GFP and its variants to study gene expression in plants.

Microbiology, Parasitology and Marine Biology

As I wrote elsewhere[3371], "Different people see microorganisms from different perspectives. To evolutionary and molecular biologists, microbes are relatives, with whom we set up correspondence. To biotechnologists, they are workers, to be employed and, perhaps, exploited. To environmental microbiologists, they may be merely scenery, or analogous to canaries in coal mines, but they are generally viewed as good neighbors if we have good fences. To clinical, food, and sanitary microbiologists, and to the defense establishment, microorganisms are enemies to be tracked, contained, and killed, and to leaders of rogue states and terrorist organizations, they are useful tools which are much easier to get through airports than are firearms and explosives."

It was van Leeuwenhoek's microscopy, in the 17th century, that first made us aware of the existence of the microbial world, but it was not until the advent of cytometry in the late 20th century that it became possible to carry out detailed studies of microorganisms at the single cell level. The motivation for doing this lies in the heterogeneity of microorganisms, which is far greater than the heterogeneity of perspectives from which they are contemplated by human observers. Even closely related species may exhibit marked differences in biochemistry and behavior, and mutation rates are such that a colony of more than a few hundred thousand cells almost certainly contains genetically different individuals, so that striking heterogeneity can exist within a clonal population of organisms that, in the aggregate, occupy too little space to be visible to the unaided human eye. The importance of single-cell analyses in microbiology was stressed by Davey and Kell in 1996 in a comprehensive, indispensable review[33].

In 2000, a special issue of *The Journal of Microbiological Methods* on "Microbial Analysis at the Single-Cell Level"[2403], featured papers on flow and image cytometry and other techniques; the complete text of all articles is available free of charge online. An issue of *Scientia Marina* devoted to "Aquatic Flow Cytometry: Achievements and Prospects"[2402], also appeared in 2000, and the July 1, 2001 issue of *Cytometry* was a special issue on "Flow Cytometry in the Marine Environment"[3372] (Volume 10, Number 5 of *Cytometry* (September, 1989) was a special issue on "Cytometry in Aquatic Sciences"). For those interested in alternative techniques, there is a 1998 book on *Digital Image Analysis of Microbes*[2420].

Measuring Microbes: Motivation

Microbiologists, whatever their perspective, subject specimens to cytometric analysis with well-defined goals in mind. In the simplest cases, it is necessary only that microorganisms be detected in a sample; at the next level of complexity, the organisms must be counted, explicitly or by the use of some surrogate indicator of the number present. Beyond this, an organism in pure culture may need to be identified and/or characterized as to its growth, metabolism, viability, and interaction with various chemical and physical agents. At the highest level of complexity, it may be necessary to detect, identify, count and characterize each of several organisms in a mixed population.

Even the simplest task, that of detection of microorganisms, may demand examination of several aspects of the specimen. As was mentioned in Chapter 3 (p. 74), **flow cytometric detection and counting of bacteria** in air[29-31] based on light scattering signals was demonstrated some time before similar technology was developed to deal with larger cells suspended in liquids. Detection becomes more difficult as the sample contains increasing amounts of inorganic or organic particulates in the microbial size range, making it necessary to use multiparameter approaches. As might be expected, gating measurements on the (relatively) strong scatter signal improve the signal-to-noise ratio in detection of immunofluorescently stained spores and bacteria[914-5], and Mansour et al[919], using a highly effective red cell lysing reagent and a centrifugation step, detected *Escherichia coli* seeded into human blood samples at concentrations of 10-100 organisms/mL by measuring forward scatter and ethidium fluorescence. Additional of parameters may further improve detection and identification; Donnelly and Baigent[916], for example, by combining forward scatter, propidium iodide, and fluorescein immunofluorescence, could detect *Listeria monocytogenes* in milk and discriminate this organism from several species of *Streptococcus* and *Staphylococcus*.

The first two generations of commercial fluorescence flow cytometers were not widely applied to studies of microorganisms. Early applications of flow cytometry to bacteria included fermentation process monitoring, industrial microbiology[553,558-9,562,569], clinical diagnosis[103,279,565-7], and environmental toxicology[560]. Preparative sorting was used to purify yeast basidiospores[572], and to isolate microbial subpopulations with desired metabolic characteristics[918] following gene transfer.

Although interest in all of these areas (and the last is certainly not least) has increased, even now, fewer than two thousand of the tens of thousands of papers in the cytometry literature deal with microbiological applications. However, interest in this area is increasing. The first book devoted to flow cytometric applications in microbiology[1052] and another that devoted considerable space to the use of flow cytometry in marine microbiology[1051] are now out of print and, more to the point, out of date. I don't have enough pages available to do justice to the subject here; it is time for a new book, and I intend to start work on *Microbiological Applications of Cytometry* shortly after I finish this tome.

Measuring Microbes: Instrument Issues

In principle, one can use a flow cytometer to measure the same parameters in bacteria or even viruses as are commonly measured in eukaryotic cells. In practice, since the size, mass, nucleic acid and protein content, etc. of bacteria are approximately 1/1,000 the magnitude of the same parameters in mammalian cells, it may be difficult to make good measurements, for two reasons. First, although modern instruments may reliably detect fewer than 1,000 molecules of fluorescent material in or on a cell, measurements at such low levels typically exhibit large variances due to photoelectron statistics; some microbial constituents may thus be undetectable or measurable with only limited precision. Second, flow cytometers designed with observation volumes large enough to permit precise measurements of eukaryotic cells have higher background levels of stray excitation light and Raman scatter than would an instrument designed explicitly for smaller particles.

However, when a properly designed or adapted instrument is used, sensitive and precise measurements can be made using light sources of modest power[94,103,198,561,567,2075-6]. Much of the earlier published work on flow cytometry of bacteria was done using custom-built or modified hardware; a lot of the rest was done on older Ortho systems, which, because observation was done in flat-sided cuvettes, had higher scatter sensitivity than did the stream-in-air instruments then produced by B-D and Coulter.

Among current commercial systems, the Lindmo/Steen arc source instrument[100-3], commercialized by Skatron, later sold as the Bryte HS by Bio-Rad, and now resurrected by Apogee, has excellent sensitivity for scatter measurements of microorganisms; the newest Apogee system, with a laser source, can easily detect medium- and large-sized viruses. Laser source instruments that make measurements in flat-sided cuvettes typically have somewhat better fluorescence sensitivity than arc source cytometers, but, while the cuvette decreases scatter noise compared to what would be observed in a stream-in-air system, the amplifiers for the signals produced by the diodes typically used to detect forward scatter may not provide enough gain to get signals from some smaller bacteria on scale.

Gant et al[2077], who studied antibiotic effects on scatter and propidium fluorescence signals from *E. coli*, found an unmodified B-D FACScan sufficiently sensitive for their needs. However, Dusenberry and Frankel[2076] reported that modifying a FACScan by focusing the laser beam to a smaller spot and substituting a PMT as a forward scatter detector increased sensitivity almost fivefold, and Button and Robertson (D. Button and B. Robertson, personal communication) have done most of their work on bacterial sizing and DNA content determination[686,2084-5,2092,2562-3] with an Ortho system modified to provide a smaller (narrower) focal spot.

In instruments I have built for analyses of bacteria, I have modified the electronics to provide threshold setting on both scatter and fluorescence channels, so that the signals from both must be above threshold to trigger the front end electronics (pp. 191-2); I find this often makes the difference between being able to detect small organisms and not being able to detect them.

Parameters Measured in Microorganisms

Forward and orthogonal light scattering, DNA and protein content, DNA base composition (A+T/G+C), lectin binding, membrane potential, and immunofluorescence measurements of **algae**, **bacteria**, **fungi**, and **protozoa** using flow cytometry had been reported in the literature by the mid-1980's[103,197-8,279,552-73,707,914-26], as had autofluorescence of photosynthetic and other pigments. By the end of that decade, flow cytometric measurements of genetic probes[1406-8] had also been made.

What we know about the behavior of most of the reagents now common use in cytometry was learned largely from observation of the interaction of those reagents with eukaryotic cells. **The most important fact about microbial cytometry is this: bacteria are not just little eukaryotes.** Uptake of, and efflux of, dyes, drugs, and other reagents by and from bacterial cells are affected by the structure of the cell wall, and by the presence of pores and pumps that may or may not be analogous to those found in eukaryotes. It is inadvisable to conclude that even well-characterized dyes behave in bacteria as they do in mammalian cells, and unwise to assume that staining procedures that were developed for mammalian cells can be "ported," as the computer types say, to protists and prokaryotes with no modification.

Bonaly et al[920] noted that stationary cells of *Euglena* did not stain stoichiometrically for DNA with intercalating dyes such as ethidium, because chromatin structure changed sufficiently to decrease the number of binding sites. Hoechst 33258 did stain stationary cells stoichiometrically; stoichiometric staining with ethidium could be achieved using acid treatment to extract nuclear basic proteins. Kuchenbecker and Braun[917] found that DNA in ethanol-fixed yeast cells was stained effectively after 12 minutes' incubation in a solution containing 100 µg/mL olivomycin, 40 mM $MgCl_2$, and 1 M NaCl; the DNA specificity of olivomycin eliminated the necessity for RNAse treatment, which is critical when staining yeast cells with ethidium or propidium because of their high RNA content (does anyone else remember that RNA was once called zymonucleic acid?). Bernstein et al reported that internal antigens in *Dictyostelium* cells were better visualized by immunofluorescence when the cells were permeabilized with 0.1 mM digitonin than when the cells were fixed. More recently, Walberg et al[3372] found substantial, unpredictable variability in patterns of uptake of nucleic acid binding dyes by three species of Gram-positive bacteria; under some circumstances, staining intensity was greater when organisms were vitally stained than when they were fixed. It is therefore likely that application of a published cytometric technique to a new species or strain of will require empirical optimization of preparative methods.

In the brief discussion of parameter measurements and probes that follows, I will emphasize the current state of the art; the later sections on microbiological applications will also feature quirks, but may mention outmoded methodology.

Flow Cytometric "Gram Stains"

The lipopolysaccharide **outer membrane of Gram-negative bacteria**, in its native state, presents a permeability barrier to most lipophilic materials[2078-9]; this is responsible for the resistance of the organisms to some antibiotics, and, from a cytometric point of view, also prevents staining of the intact organisms by many dyes. Chemical agents such as ethylene diamine tetraacetic acid (EDTA)[2081] may be used to permeabilize the outer membrane to drugs and dyes with at least transient retention of some metabolic function, although the characteristics of bacteria thus permeabilized are distinct from those of organisms in the native state. The properties of the outer membrane can be utilized in cytometric procedures analogous to Gram staining.

My initial approach to this[2075] was to use a single membrane potential dye, usually $DiIC_1(5)$, to stain multiple aliquots of a sample. The first two aliquots have both dye and EDTA added; the EDTA insures that dye can enter both Gram-positive and Gram-negative bacteria, and it will be concentrated in bacteria if a membrane potential is present. The second aliquot also contains the proton ionophore CCCP, which reduces membrane potential to near zero. If there are particles with similar scatter signatures in both aliquots that are more fluorescent in the aliquot without CCCP than in the aliquot with CCCP, I conclude that they represent bacteria, because they have membrane potentials. I then examine a third aliquot, which contains dye but neither EDTA nor CCCP; this will show dye uptake if the organisms are Gram-positive but not if they are Gram-negative. This procedure correctly determined the Gram staining characteristics of a number of common organisms, including *Staphylococcus aureus*, *Staphylococcus epidermidis*, *Streptococcus pyogenes* (Gram-positive), *Enterobacter cloacae*, *Escherichia coli*, *Klebsiella pneumoniae*, *Proteus vulgaris*, *Pseudomonas aeruginosa*, and *Salmonella typhimurium* (Gram-negative).

Molecular Probes has produced several kits containing fluorescent reagents for "Gram staining"; the LIVE *Bac*Light kit contains two dyes excitable at 488 nm; the orange fluorescent hexidium iodide is analogous to ethidium bromide, but has a hexyl group instead of an ethyl group on the ring nitrogen, making it highly lipophilic, and the green fluorescent SYTO 9 is much more water soluble. Both dyes enter Gram-positive bacteria; hexidium is excluded by Gram-negative organisms with intact outer membranes and cytoplasmic membranes. The result is that Gram-positive organisms fluoresce red, while Gram-negative organisms fluoresce green. Mason et al[3373] correctly determined Gram staining status of 45 species using hexidium iodide and SYTO 13, a close relative of SYTO 9. Molecular Probes also offers fluorescent conjugates of **wheat germ agglutinin (WGA)**. In 1990, this lectin was reported by Sizemore et al[3390] to stain Gram-positive but not Gram-negative cells, and Molecular Probes licensed their patent; however, I could not find other accounts of the use of WGA for cytometric Gram staining in more recent literature.

Detection and Sizing: Light Scattering

In the discussion of light scattering in Chapter 7 [pp. 275-6 (Figure 7-1) and 279-80], I emphasize the point that forward scatter signals from particles in the size range in which most eukaryotic cells lie are not precise indicators of cell size. However, I note on pp. 288-9 that scatter signals from bacteria and smaller particles can be used for sizing, and a number of groups have developed calibration procedures that allow cell volume to be computed from forward scatter signal intensity[2540,2562-3,3374-5]. For viruses and other particles in the size range below 0.2 μm, side scatter signals may be preferable to forward scatter signals for sizing[94,3376].

Scatter signal intensities depend on **refractive index** as well as on particle size; Green et al[3377] have described the use of flow cytometry to make measurements of both the size and the refractive index of phytoplankton organisms and marine particles.

Any mismatch of refractive index between the sheath fluid and the fluid in which cells are suspended adds noise to scatter measurements and may compromise accuracy[3378]. Even when scatter signals are only used for detection, it is advisable to correct index mismatches in order to maximize the chance of obtaining usable signals.

Detection and Sizing: Electrical Impedance

DC Impedance (Coulter volume) measurements (p. 273) were used to detect and size bacteria[680] in the 1950's and viruses[681,2525-9] in the 1970's. Saleh and Sohn[2530,3391] have recently described microfluidic flow cytometric impedance measurement system that can size colloidal particles in the viral size range.

Nucleic Acid (DNA and RNA) Staining

The number of ribosomes in a bacterial cell varies contain between a few hundred (in slowly growing organisms such as *Mycobacteria*) and tens of thousands (in metabolically active, exponentially growing *E. coli*). The content of double-stranded rRNA in most organisms is therefore several times the content of DNA. Dyes such as propidium and the SYTO and TO-PRO series, which form fluorescent adducts with both DNA and double-stranded RNA, will stain both; Guindulain and Vives-Rego[3379] found that at least two-thirds of SYTO-13 fluorescence in *E. coli* was due to dye binding to RNA.

While it is possible to treat fixed and permeabilized cells of at least some species with RNAse or DNAse, enabling nonselective dyes to be used for quantitative determination of DNA or the rRNA[3380], it is difficult to get the enzymes into other species, making it prudent to use more selective stains.

Steen[3381] recommends the mithramycin-ethidium combination, DAPI, or the Hoechst dyes for staining DNA; the former requires deep blue or violet excitation (now more readily available from violet diode lasers), while the latter need UV excitation, at least in the context of analysis of microorganisms. Bernander et al[3382] found that all of these stains yielded histograms of equivalent quality in both an arc source (Skatron Argus) and a laser source (B-D FACStar) flow cytometer. Stefan Andreatta[3383] noted in his Doctoral Thesis (and in a posting to the Purdue List) that precision of histograms obtained with Hoechst dyes and DAPI is improved if a small amount of Na-citrate/K-citrate buffer is added to the sample prior to staining in order to bring the pH to about 7.5; this eliminates the yellow nonspecific fluorescence that is often observed when DAPI is used to stain bacteria.

For RNA staining, Borth et al[3384] used pyronin Y (which stains double-stranded RNA) in combination with methyl green, to block DNA staining; these authors also used Hoechst 33258 to stain DNA, but neither they nor anyone else seems to have reported the use of a combination of a Hoechst dye and pyronin Y for simultaneous DNA/RNA staining in bacteria. I would expect it to work.

Müller et al[3385] took a more radical approach to quantification of bacterial rRNA content; they stained cells with fluorescently labeled oligonucleotide probes for 16S and 23S rRNA. These, as will be noted presently, are better known for their utility in species identification.

Total Protein Content: Scatter vs. Stains

Steen et al[103] noted in 1982 that large-angle light scatter signals from *E. coli* cells stained for total protein with FITC were linearly correlated with fluorescein fluorescence signals. However, since there are some circumstances under which materials other than protein (e.g., gas vacuoles[707]) influence side scatter signals from bacteria, it is advisable to use a stain for critical analyses of total protein content. Zubkov et al[3386] developed and validated a precise flow cytometric assay for measuring bacterial protein biomass using SYPRO dyes from Molecular Probes[2332], which were originally intended for sensitive detection of proteins on gels. The intensity of SYPRO-protein fluorescence of marine bacteria strongly correlated with their total protein content as measured by the bicinchoninic acid method; according to the authors' calibration, the mean biomass of planktonic bacteria from the North Sea in August 1998 was 24 fg of protein/cell.

Antibodies, Etc.: Labeling Strategies

If you happen to look in catalogs or compendia of antibodies, you will quickly discover a profound difference between bacteria and lymphocytes. Biotechnology has wiped out what used to be one of the bigger distinctions between the two; these days, bacteria can not only make antibodies, they can make better antibodies than lymphocytes can. But that's another story. I was hoping to call your attention to the fact that, while you can easily buy monoclonal antibodies to any of the many CD antigens found on lymphocytes, directly labeled with any one of a half-dozen or more tags, commercial monoclonal antibodies to bacterial and viral antigens are fairly scarce, and usually not available directly labeled. If you can find a directly labeled antibody, the odds are probably at least ten to one that the label is fluorescein.

There usually isn't a problem with fluorescein-labeled antibodies. However, if you happen to be looking at some species of *Pseudomonas*, you'll find they do a pretty good job of fluorescing at 525 nm in the absence of antibody. Similarly, you wouldn't want to use phycoerythrin to label an antibody against an antigen found in or on *Synechococcus*, which comes with an ample endogenous supply of the label. Vesey et al[2660] described a procedure they used to select labels for detection of pathogens in water samples; by taking into account both the autofluorescence of particles in the water, and background fluorescence in the water itself, they were able to optimize signal-to-noise for stained specimens. It's a good model to follow; you will almost certainly be looking for many fewer antibody-binding sites on microorganisms than you'd be apt to find on eukaryotic cells, and you'll also need to pick labels carefully for nucleotide probes and other ligands.

While it is not generally a problem to get a fluorescent antibody that will identify a particular species or even a particular strain of bacteria, there are not antibodies that differentiate among larger groups of organisms, making it infeasible to develop a manageable panel of antibodies that could be used to identify any of a large number of species. Ribosomal RNA probes offer an alternative.

Ribosomal RNA-Based Species Identification

Various probes complementary to different sequences in 16S rRNA can distinguish among the primary kingdoms (Eukaryotes, Eubacteria, Archaebacteria), and among smaller taxonomic groups, down to the species and strain level[3392-3]. Because many microorganisms cannot be grown in culture, it is desirable to be able to identify them without having to grow them; this can be accomplished using ribosomal probes. The several thousand ribosomes normally present per cell provide enough target sequences to make amplification unnecessary. Amann et al[1406], working with marine organisms, demonstrated the utility of 16S ribosomal RNA probes and flow cytometry for this purpose in 1990. The oligonucleotide probe technology has been improved since then, by optimizing probe sequences and binding conditions[1407-8,3387-9]; rRNA sequences have also been detected using peptide nucleic acid (PNA) probes[2714,3394] (p. 362), which allow cellular integrity to be better preserved and may provide stronger signals.

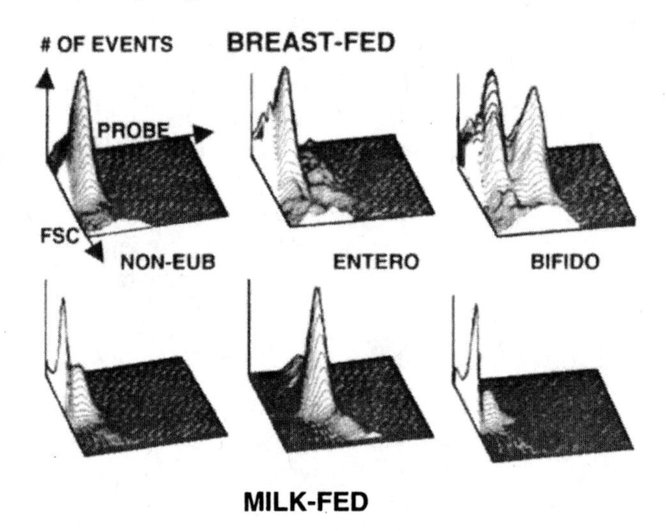

Figure 10-27. rRNA probe staining patterns demonstrate differences in enteric flora between breast-fed infants (top row) and infants fed reconstituted milk (bottom row). Left column: non-eubacterial probe; middle column: enterobacterial probe; right column: bifidobacterial probe. Courtesy of Vanya Gant (Hospital for Tropical Diseases, UCLH, London)

Figure 10-27 illustrates differences in the proportions of non-eubacterial organisms, enterobacteriaceae, and bifidobacteria in stool from breast-fed infants and from infants fed reconstituted milk. The same rRNA probe techniques work for the ecology of the gut[3395] and that of the ocean[1406,3394].

Functional Probes in Bacteria

Eukaryotes may maintain gradients of pH, membrane potential, and ion concentration across the membranes of internal organelles such as mitochondria, lysosomes, and plastids; at least some organelles are believed to be descended from free-living ancestors resembling bacteria[2255]. Bacteria do not have complex systems of internal compartments, much of their metabolic activity occurs at the cytoplasmic membrane, and there are typically gradients of both pH and membrane potential across this membrane in viable, active cells.

Membrane Potential, Permeability, "Viability," and Metabolic Activity

Kell et al[2250] presented a very thoughtful discussion of flow cytometry and its applications to the detection and quantification of metabolic, or functional, as well as structural heterogeneity in microbial cultures. They published a number of papers dealing with the flow cytometric determination of bacterial **viability**[2251], **injury**[2252], and **dormancy**[2253-4] using probes of **membrane potential ($\Delta\Psi$)** and of **oxidative metabolism**.

I have already discussed membrane potential and its estimation by flow cytometry, in more detail than you probably wanted, on pp. 385-402; here, I will summarize points relevant and, in some cases, peculiar to membrane potentials in bacteria.

The outer membrane of Gram-negative bacteria, in its native state, presents a permeability barrier to the uptake of dyes, particularly lipophilic dyes such as cyanines and oxonols; however, when the cells are permeabilized with EDTA under conditions that preserve viability[2081], these dyes will partition across the membrane as they do in other cells.

Unlike most of us, bacteria do not store large amounts of carbon and energy sources; this is one reason why nobody tries to sell them exercise equipment. As a consequence of this metabolic characteristic, bacteria lose their membrane potentials as soon as they use up the energy source in the medium. If an organism can grow on glucose and not on rhamnose, and you take its glucose away and give it rhamnose, the membrane potential drops to near zero within a few minutes. Put back glucose, or something else that will support growth, and the membrane potential comes back. Metabolically active bacteria have membrane potentials, more or less in the same way that metabolically active people have electroencephalographic waves.

Bacteria that are "dead" by **dye exclusion** criteria, that is, which let in propidium or the TO-PRO or SYTOX dyes, and/or won't retain fluorescein after FDA hydrolysis, are said to have lost **membrane integrity**. If one takes this to indicate the presence of a big hole or holes in the cytoplasmic membrane, loss of membrane integrity should result in the dissipation of ion concentration gradients across the membrane, with loss of the membrane potential. Membrane integrity, permeability, and "viability" have been discussed on pp. 299-303 and 369-371, and some relatively recent findings in bacteria may necessitate some rethinking of these concepts. Kell et al[3454] now equate "viability" and "culturability"; cytometry can measure "activity," but not "viability."

A great deal of bacteriology is dependent on being able to grow bacteria in culture. If we want to know whether there are bacteria in a supposedly sterile fluid, we put some into broth or onto an agar plate, incubate for some hours, and look for turbidity in the broth or colonies on the plate. Metabolically active bacteria will grow, when provided with the right nutrients, dead bacteria won't. However, there are also bacteria that may be **sublethally injured**, which basically means they look intact, but won't grow under normal conditions, and there are also bacteria that are **dormant**. Many bacteria and fungi are formal enough about dormancy so that they form **spores**, which exhibit little or no metabolic activity, but which are equipped to last for some time in inhospitable environments. Other bacteria, such as the *Micrococcus luteus* shown in the top panel of Figure 10-28, which came from a stationary phase culture in which their energy source was exhausted, simply cease metabolic activity; they lose membrane potential, but not membrane integrity. After nutrients are restored, such dormant organisms may regain metabolic activity, membrane potential, and the capacity to grow; the restoration of membrane potential following such resuscitation is evident in the bottom panel of Figure 10-28.

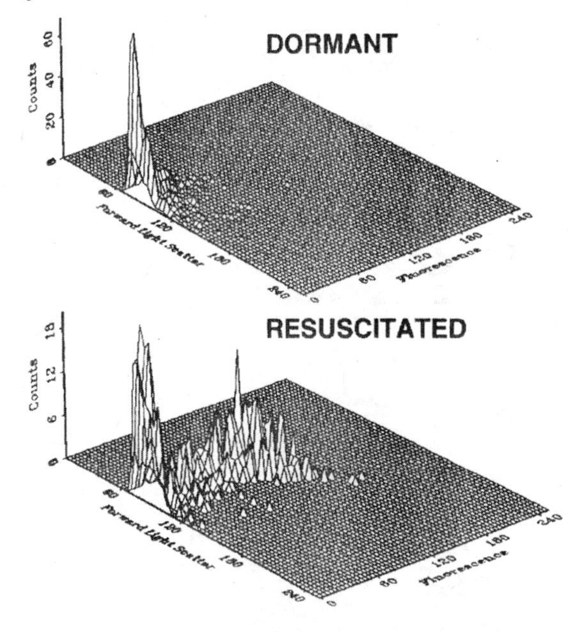

Figure 10-28. Membrane potential, as indicated by rhodamine 123 fluorescence, vs. forward scatter in dormant and resuscitated cultures of *Micrococcus luteus*. The figure was provided by Douglas Kell (University College of Wales, Aberystwyth; now at the University of Manchester Institute of Science and Technology).

Metabolic activity in the form of **respiration** in aerobic bacteria can be detected using an **indicator of oxidative metabolism** such as a **tetrazolium dye** (p. 379); Kaprelyants and Kell[2252,2254] used **cyanoditolyl tetrazolium chloride (CTC)**, a fluorescent indicator with spectral characteristics well matched to the arc lamp source in their Skatron instrument, for the purpose. They demonstrated that *M. luteus* cells that were frozen and thawed, causing a leakage of NADH from the cells, could regain respiratory activity, as indicated by reduction of CTC to an insoluble fluorescent formazan, when exogenous NADH was added to the medium. CTC was also used by Schaule, Flemming, and Ridgway[2256] to quantify respiring bacteria in drinking water with image analysis. Chromogenic tetrazolium dyes are widely used as indicators in microtiter plate assays of metabolic viability of both bacteria and mammalian cells.

Figure 10-29 illustrates the use of a combination of PNA rRNA probes and CTC to determine the fraction of metabolically *Salmonella typhimurium* in a mixed population of bacteria. A cocktail of non-*Salmonella Enterobacteriaceae* containing approximately equal numbers of *E. coli, C. freundii, P. vulgaris* and *S. dysenteriae* was prepared through volumetric mixing of cultures of known cell concentration. The mixture was divided in half and one portion killed with incubation in 10% buffered formalin for 30 min. Washed, formalin-killed cells were resuspended to their original volume in broth. Live and formalin-killed portions of *Salmonella typhimurium* were also prepared, and equal volumes of each preparation were combined to yield a final mixture containing four distinct populations: live (metabolically active) *Salmonella* (C), dead *Salmonella* (D), live non-*Salmonella Enterobacteriaceae* (B) and dead non-*Salmonella Enterobacteriaceae* (A). This mixture was incubated with CTC, fixed with formalin, hybridized with a Cy5-labeled *Salmonella*-specific PNA rRNA probe, and examined by flow cytometry.

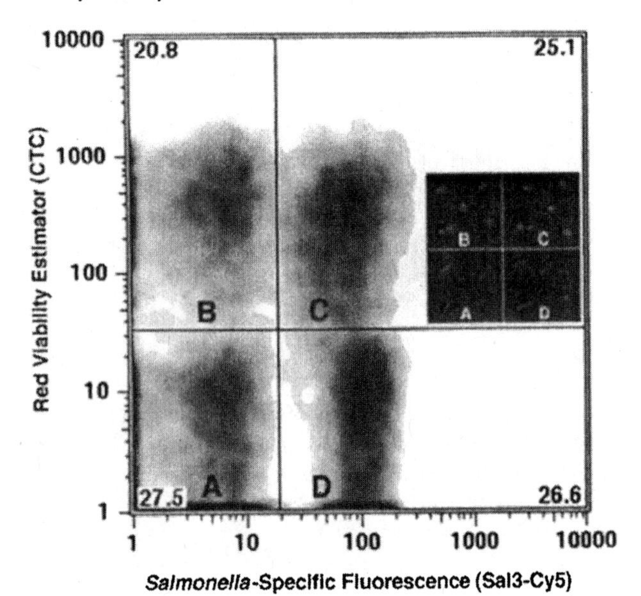

Figure 10-29. Combined rRNA probe and CTC staining for analysis of a genetically and metabolically complex cell mixture. Courtesy of Byron F. Brehm-Stecher and Eric A. Johnson, Food Research Institute, University of Wisconsin-Madison.

Believe it or not, a two-color analysis such as that shown in Figure 10-29 is fairly high-tech, as microbial cytometry goes. Two-angle scatter and one-color fluorescence was pretty much the norm for "multiparameter" cytometry of bacteria well into the 1990's, meaning that most people were only looking at one probe at a time.

While the CTC fluorescence shown in Figure 10-29 provides an indication that cells are, or at least were, viable, there wouldn't have been any way to confirm this, because the fixation required for rRNA probe hybridization killed the cells. Nebe von Caron[2610] and Hewitt[3396] and their collaborators have used single cell sorting of cells stained with multiple functional probes to confirm identification of cells as viable or nonviable, as illustrated in Figure 10-30.

Salmonella typhimurium stored for 25 days on nutrient agar at 4°C was re-suspended in buffered saline containing peptone, sodium succinate and glucose. Cells were treated with EDTA to allow dye penetration of the outer membrane, sonicated briefly to break up aggregates, and incubated with $DiBAC_4(3)$, ethidium bromide and propidium iodide for 30 min at 25°C. Cells were then sorted onto agar plates.

The lipophilic anionic oxonol dye $DiBAC_4(3)$ is used as a membrane potential probe; it stains electrically depolarized cells more brightly than cells that maintain membrane potential, and "permeabilised" cells, i.e., cells that have lost membrane integrity, brighter still, probably because of binding to hydrophobic regions in intracellular proteins. Ethidium enters the bacteria, but is actively pumped out by cells that maintain membrane potential. Intracellular ethidium enhances fluorescence on binding to double-stranded DNA or RNA. Propidium enters permeabilized cells and, because of its higher affinity for double-stranded nucleic acid (due at least in part to the extra positive charge; se pp. 306-7), displaces ethidium. $DiBAC_4(3)$ fluorescence is measured at 525 nm, and ethidium fluorescence at 575 nm.

The dye combination allows identification of cells in different functional stages. Active pumping cells do not stain significantly with any of the dyes. De-energized cells take up ethidium, but not $DiBAC_4(3)$ or propidium. Depolarized cells take up ethidium and $DiBAC_4(3)$, but not propidium, and permeabilized cells and "ghosts," with damaged membranes, take up both $DiBAC_4(3)$ and propidium.

Sorting of bacteria from different functional stages reveals that, in most cases, all but the permeabilized cells are capable of recovery. In the experiment shown in Figure 10-30, 35% of the electrically depolarized cells grew on agar plates; recovery was even better when cells were sorted onto special resuscitation media or into liquid media. Depolarization therefore is clearly not an indicator of indicates decline in cell functionality, but certainly not cell death. Recovery of actively pumping and de-energized cells approaches 100% (de-energized cells lose pump activity but maintain membrane potential at least briefly); however, fewer than 1% of events sorted from the regions containing permeabilized cells and ghosts form colonies on agar.

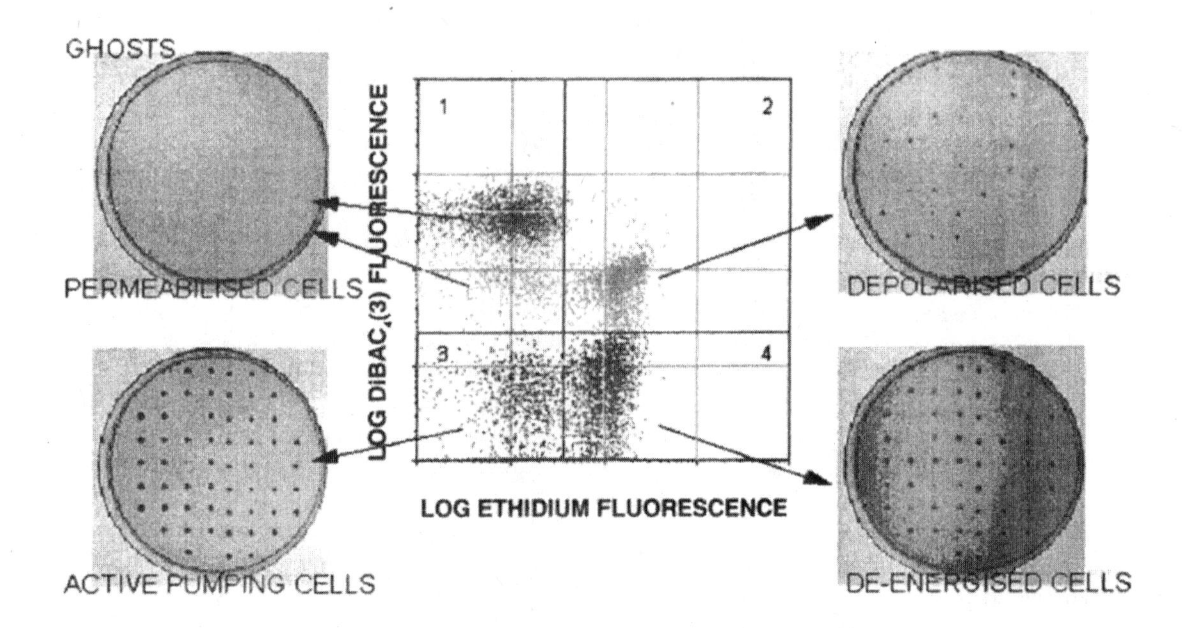

Figure 10-30. Recovery of *Salmonella typhimurium* in different functional states determined by staining with DiBAC$_4$(3), ethidium bromide, and propidium iodide. Courtesy of Gerhard Nebe-von-Caron (Unipath, Ltd.).

Life (and death) get still more complicated, as can be appreciated from Figure 10-31, below.

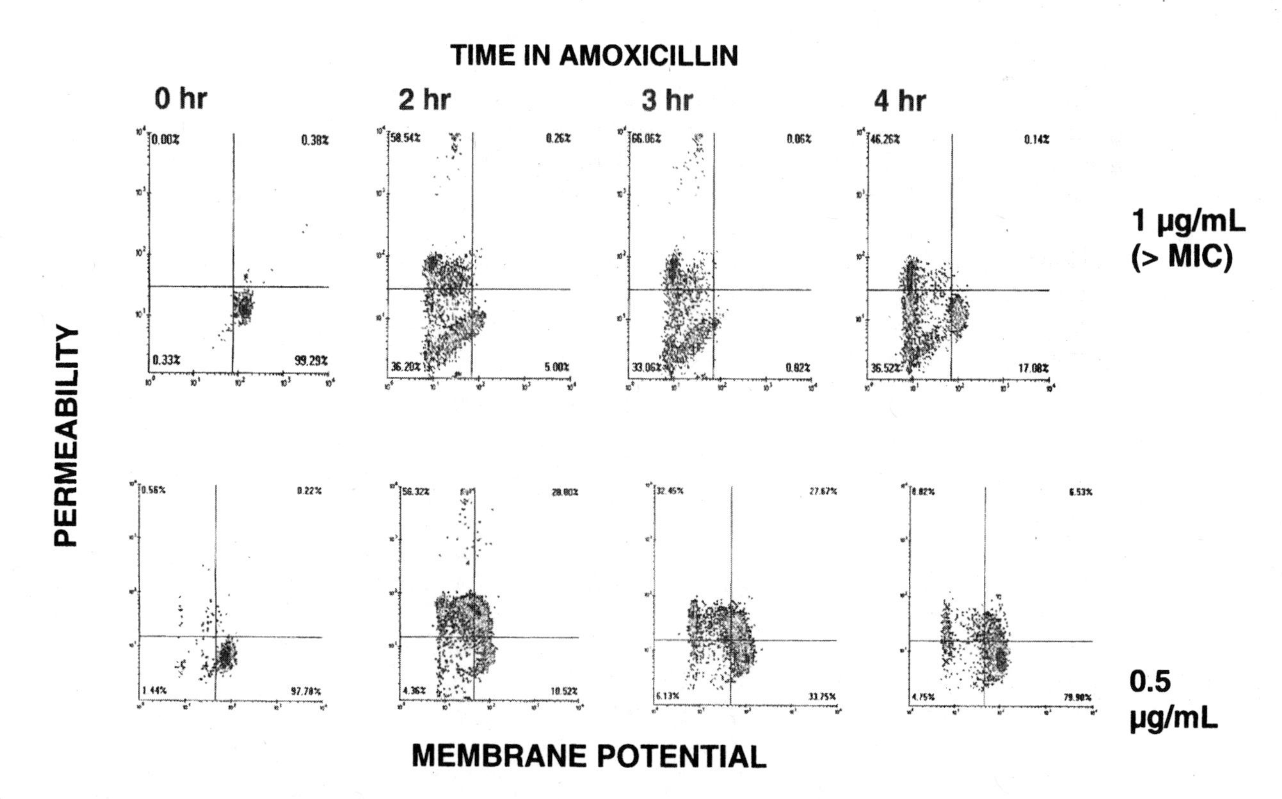

Figure 10-31. Effects of amoxicillin on membrane potential ($\Delta\Psi$) and permeability of *Staphylococcus aureus*, measured ratiometrically using DiOC$_6$(3) and TO-PRO-3[2351,2357].

After we developed the ratiometric membrane potential measurement method[2357] described on pp. 400-2, David Novo, Nancy Perlmutter, Richard Hunt and I turned our attention to multiparameter measurements of membrane potential and permeability in bacteria exposed to antibiotics[2351]. We used a cytometer with 488 and 633 nm lasers for measurements of bacteria exposed simultaneously to our membrane potential indicator dye, $DiOC_2(3)$, and TO-PRO-3. Figure 10-31 plots the ratiometric membrane potential (red/green DiOC2(3) fluorescence) of amoxicillin-treated *Staphyloccus aureus* against a ratiometrically normalized permeability measure, i.e., the ratio of TO-PRO-3 fluorescence to green $DiOC_6(3)$ fluorescence. The ratiometric permeability measurement removes most of the effects of cell size variation and clumping in the same way as is accomplished by the ratiometric $\Delta\Psi$ measurement.

The strain of *S. aureus* we used was amoxicillin sensitive, but we exposed it to concentrations of amoxicillin above (1 µg/ml) and below the minimal inhibitory concentration (MIC). In cultures treated with either dose, at time zero, most cells show low values of permeability and relatively high values of $\Delta\Psi$, appearing in the lower right quadrant of the display. After 2 hr at an amoxicillin concentration above MIC (top strip), many cells have lost $\Delta\Psi$ completely, and almost all have lost $\Delta\Psi$ to some extent, appearing in the lower and upper left quadrants; over 58% of the total have become permeable (upper left quadrant). By 4 hr, some regrowth has occurred; about 17% of the events measured show normal $\Delta\Psi$ and no permeability. The situation is quite different at an amoxicillin concentration below MIC (bottom strip). At 2, 3, and even 4 hr, a substantial fraction of events (as high as 28%) are in the upper right quadrant, indicating a membrane potential greater than zero with permeability to TO-PRO-3. By 4 hr, most cells (over 79%) have regained normal $\Delta\Psi$ and lost permeability. Bacterial counts were followed over this time period, which was too short to have allowed a high-$\Delta\Psi$, impermeable population to appear due to growth of the small population of such cells present after 2 hr. Although it is possible that some intermediate-$\Delta\Psi$, permeable events represent aggregates of high-$\Delta\Psi$, impermeable viable cells and permeable, low-$\Delta\Psi$ dead cells, it appears that many of these events are accounted for by transiently permeabilized viable cells.

The phenomenon of transient membrane permeability induced by sublethal injury thus appears to be real, and may be quite general; we have since observed it in *Pseudomonas aeruginosa* and *Escherichia coli* as well as in *S. aureus*[3371,3397], and others (S. Barbesti, personal communication; G. Nebe-von-Caron, personal communication) have also encountered it. Amor et al[3398] noted propidium uptake in the presence of carboxyfluorescein retention in sublethally injured *Bifidobacterium* species; their paper also described an effective ratiometric $\Delta\Psi$ measurement technique using the ratio of $DiBAC_4(3)$ fluorescence to side scatter, but they apparently did not examine cells simultaneously stained with $DiBAC_4(3)$ and propidium.

Digression: A Therapeutic Approach Based on Transient Permeabilization

Our primary conclusion, based on substantial differences in patterns of membrane potential and permeability responses of different microorganisms to different antibiotics[2351], was that a simple, rapid, "one-size-fits-all" flow cytometric test for bacterial antibiotic susceptibility, long sought by ourselves and others, was likely to be unattainable, although multiparameter cytometry seemed well suited to dissecting these distinct responses in the context of development of new antimicrobial agents.

Several months later, it occurred to me that, if multiply resistant bacteria that could not be killed by an existing combination of agents could be transiently permeabilized, it would be possible to derivatize a large number of generally toxic heterocyclic compounds [e.g., by adding the quaternary ammonium groups that differentiate propidium from ethidium (p. 300)] to forms that would be taken up by, and kill, the permeabilized bacteria, while not entering host cells. This formed the basis for a successful patent application (U.S. 6,562,785, May 13, 2003); I now have some evidence (unpublished) that TO-PRO-3 itself can kill bacteria if transported in, and am looking for the right pharmaceutical industry partner for an intensive development project.

Investigators who previously used flow cytometry to study the interactions of antimicrobial agents and bacteria[2263,3371,3395-403,3479,3488] typically measured only a single parameter, usually membrane permeability or membrane potential in attempts to determine "viability"; although their results correlated fairly well with classical susceptibility tests, they did not establish cytometric criteria for viability on a cell-by-cell basis, and could not have detected transient permeabilization. One take-home message is that it is always a good idea to measure multiple parameters when you're not sure what's going on. Another is that, even now, drug leads don't always come from combinatorial chemistry, genomics, or proteomics; the promising therapeutic approach described above may be one of the first to emerge from **cytomics**, and making it work, or finding out that it won't, will require some fairly sophisticated cytometry. I'm ready.

So Few Molecules, So Little Time: Viruses

Even with three beams and twelve colors, conventional flow cytometers cannot give us as much information as we would like to have about microorganisms. When we are trying to detect signals from only a few thousand, or a few hundred, or a few dozen fluorescent molecules, the typical observation time of a few microseconds is simply not long enough for us to collect enough photons to get precise measurements.

If we stain DNA with a Hoechst dye, we can expect at most 1 dye molecule per 4 A-T base pairs at a saturating dye concentration. If that DNA is in *E. coli* (roughly 50% A-T), an organism with 2 copies of the chromosome (roughly 9.2 million base pairs) will bind at most 1.15 million dye

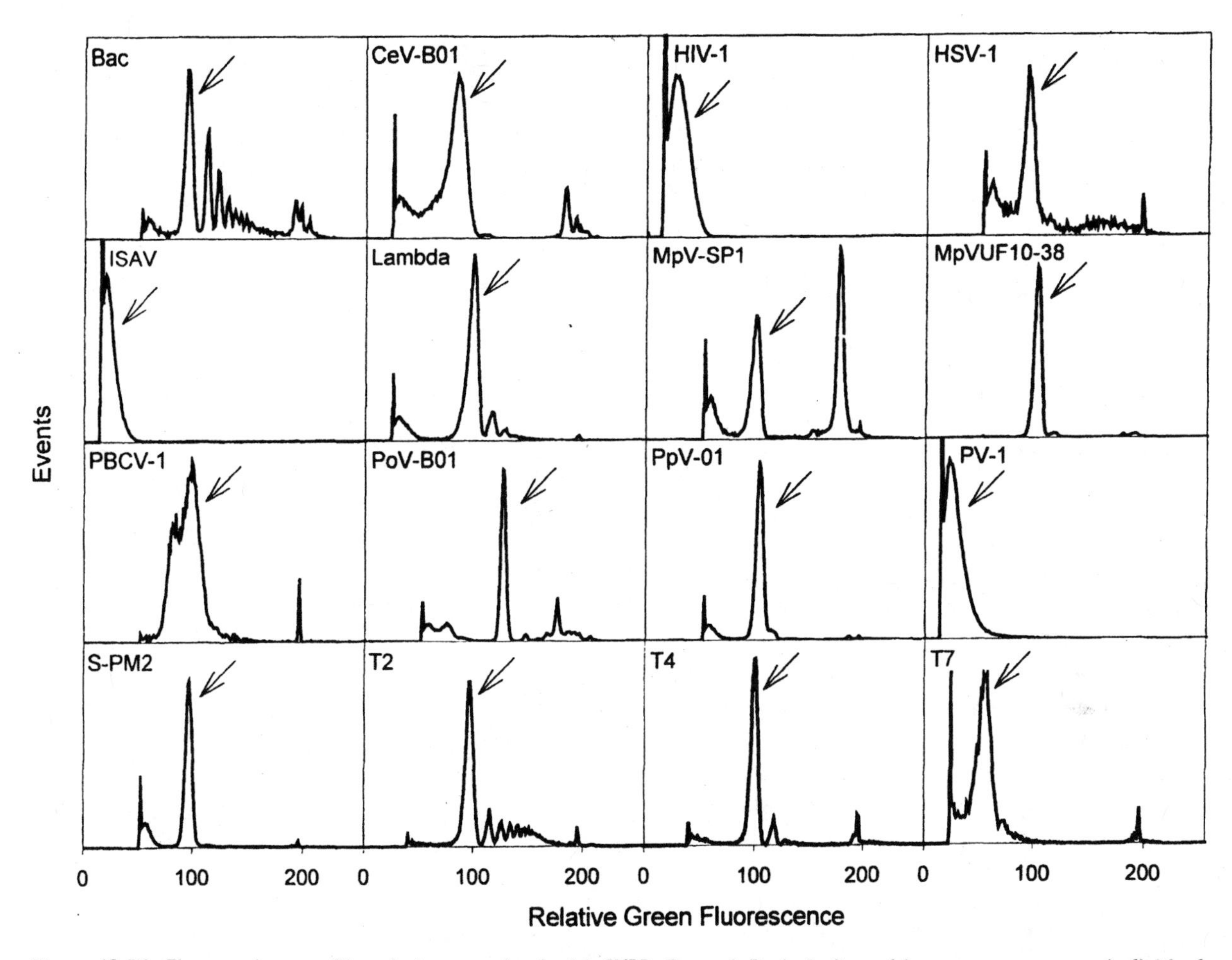

Figure 10-32. Fluorescence profiles of viruses stained with SYBR Green I. Peaks indicated by arrows represent individual virions; the highest intensity peaks (clearest in the panel showing PBCV-1) represent 0.95 μm fluorescent beads. Key: Bac, a baculovirus; CeV-B01, MpV-SP1, MPVUF10-38, PoV-B01, PpV-01, algal viruses; HIV-1, human immunodeficiency virus; HSV-1, Herpes simplex virus; ISAV, infectious salmon anemia virus; Lambda, T2, T4 and T7, bacteriophages; PBCV-1, a virus infecting *Paramecium* and *Chlorella*; PV-1, poliovirus; S-PM2, a virus infecting *Synechococcus*. Reprinted from Brussaard CPD, Marie D, Bratbak G: Flow cytometric detection of viruses. Journal of Virological Methods 85:175-82 (reference 2337), Copyright 2000, with permission from Elsevier Science.

molecules, all of which should have the relatively high quantum efficiency they acquire on binding. If, in our brief observation time, we can generate one photoelectron at the fluorescence detector PMT cathode for each 100 dye molecules, we'll get 10,000 photoelectrons from each cell; the CV of the measurement in an otherwise perfect instrument will therefore be no lower than 1 percent. CV values reported by Bernander et al[3382] for Hoechst 33258-stained *E. coli* with 2 copies of the chromosome were around 6 percent, indicating that measurement precision was not limited by photoelectron statistics. CVs for cells with 8 chromosomes, stained with Hoechst 33342 or DAPI, were under 3 percent.

Most bacteria contain several times as much double-stranded rRNA as DNA; we should therefore get strong, if not specific, signals from bacteria stained with TO-PRO-, SYTO-, or SYTOX dyes. However, when we attempt to detect nucleic acid in viruses using conventional flow cytometers[2337] (Figure 10-32), we can expect problems, for a number of reasons.

The genome sizes of the viruses studied by Brussaard et al[2337] range from 9.2 to 300 kbp, but it is probable that, in many cases, the compact structure of the virion will limit access of dye to the viral nucleic acid, reducing the number of available binding sites. Also, while dye should bind to accessible double-stranded viral DNA and RNA more or less as it binds to DNA in prokaryotes and eukaryotes, it is not clear how well dye will bind to single-stranded viral nucleic acid, nor how much fluorescence enhancement will occur as a result of binding. If we extracted the naked DNA of a dsDNA virus with a 50 kbp genome, such as bacteriophage lambda, we would expect to be able to bind no more than about 6,000 molecules of a Hoechst dye to the nucleic acid; this would generate about 60 photoelectrons at the fluorescence detector of a conventional flow cytometer.

Hoechst dyes typically increase fluorescence by a factor of about 100 on binding to nucleic acid. In a conventional flow cytometer, fluorescence is collected from a volume of approximately 10,000 fL (10^{-11} L). The virion will occupy a volume less than 1 fL. If the concentration of dye used is 3 μM, there will be 1.8×10^7 molecules of unbound dye in the observation volume. Since the quantum efficiency of unbound dye is 1/100 that of bound dye, we should generate 1 photoelectron for every 10^4 molecules of free dye. We are then trying to find the 60 photoelectrons generated by the virion in the company of 1,800 photoelectrons resulting from detection of fluorescence of free dye. Poisson statistics tell us that 1,800 photoelectrons are really $1{,}800 \pm (1{,}800)^{1/2}$, or $1{,}800 \pm 42$ photoelectrons, so we will, at best, barely be able to detect a molecule of viral DNA above background. This gives us a signal to noise ratio near 1, which translates to a CV near 100%.

The data in Figure 10-32 could not have been collected using a Hoechst dye; the dye used was SYBR Green I, which increases fluorescence almost a thousandfold on binding to nucleic acid, and has a higher quantum efficiency when bound than do the Hoechst dyes. Using this dye, we'd probably get 120 photoelectrons from the DNA molecule, and no more than a few hundred from the background, making the nucleic acid easily detectable. Brussaard et al[2337] did not find a strong linear relationship between staining intensity and genome size in the viruses they worked with, suggesting that dye access to nucleic acid is somewhat restricted, but enough dye apparently bound to viral nucleic acid to allow fluorescence signals from all but the small-genome RNA viruses (HIV-1, ISAV, and polio) to be clearly discriminated from background.

Now turn your attention back to pp. 471-3 and Figure 10-10. The panel in the upper left corner of that figure shows the intensities of fluorescence signals from bacteriophage lambda DNA, and from fragments of that DNA as small as 564 bp. The peaks from the intact DNA and the larger (> 5 kbp) fragments) appear to have fairly low CVs. What's the trick? The trick is that the observation time on Rob Habbersett's slow-flow instrument at Los Alamos, from which the data were obtained, is a few milliseconds, rather than a few microseconds, meaning that, all other things being equal, a given chunk of DNA generates about 1,000 times as many photoelectrons in the slow-flow instrument as in a conventional flow cytometer.

We can measure DNA adequately in bacteria in a conventional instrument; we can also, as the calculation on p. 402 shows, get reasonable, if not spectacular, statistics in the ratiometric membrane potential measurement, in which we pack a whole lot of dye into the bacteria. If we're interested in measuring immunofluorescence, or nucleic acid probes (even rRNA probes), or in ratiometric measurements of pH (see p. 407) or calcium, we'll have to use longer observation times if we want decent precision. The manufacturers don't seem to be in a big rush to make instruments with the appropriate performance characteristics.

Applications in Marine Microbiology

Since life started in the ocean, it seems appropriate to start my detailed review with marine microbiology. A rapidly growing body of cytometric literature deals with **marine microorganisms**, which include **bacteria**, **phytoplankton**, **zooplankton**[570-1,686,707,927-32,1051], and, most recently, **viruses**[2336-7,3404-6].

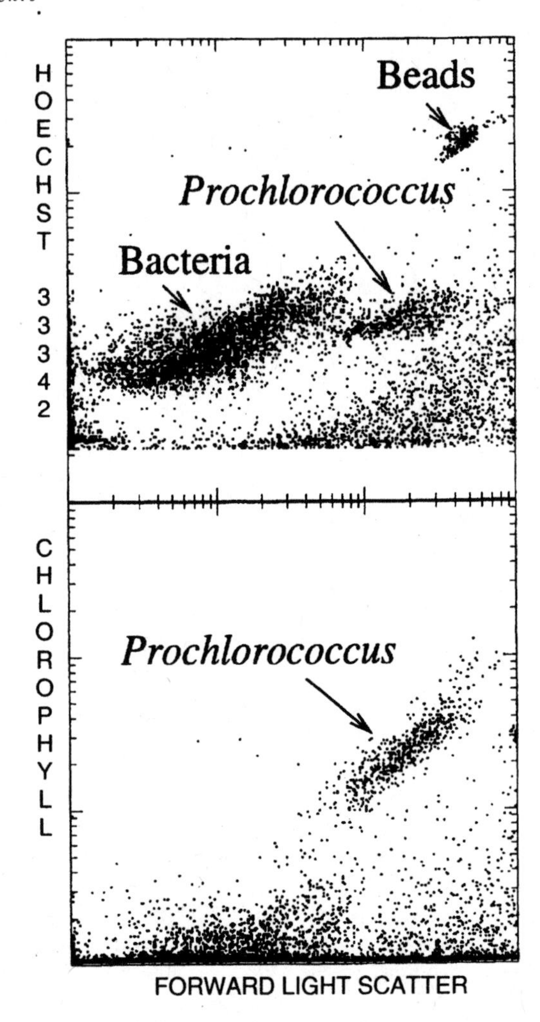

Figure 10-33. Forward scatter, Hoechst 33342 (DNA), and chlorophyll fluorescence (all on log scales) of the recently discovered marine bacterium *Prochlorococcus*[1272] measured with a dual-beam flow cytometer. The figure was provided by Brian Binder, Sallie Chisholm, Sheila Frankel, and Robert Olson (M.I.T. and Wood's Hole Oceanographic Institution).

Figure 10-33 shows the flow cytometric signature of the marine bacterium *Prochlorococcus*, discovered with the aid of a flow cytometer by Chisholm et al in 1988[1272]. The UV-excitable beads in the top panel are 0.46 μm diameter spheres from Polysciences; 0.57 μm yellow-green fluorescent beads from the same supplier are off scale on the bottom panel. There's a lot of *Prochlorococcus* in the ocean[3407-9]; according to Zubkov et al[3410], almost a third of the total bacterioplankton turnover of amino acids can be attributed to the species.

Prochlorococcus isn't the only new species flow cytometry has helped discover. In July, 1994, Courties et al[2083] reported the discovery of the smallest known eukaryotic organism, a green alga named *Ostreococcus tauri* which is the main component of the phytoplankton in the Mediterranean Thau lagoon of France. *Ostreococcus*, at 0.97 by 0.70 µm, may be bigger than *Prochlorococcus*, but, for a eukaryote, it's tiny. It has one plastid and one mitochondrion, and its DNA content is about 33 fg.

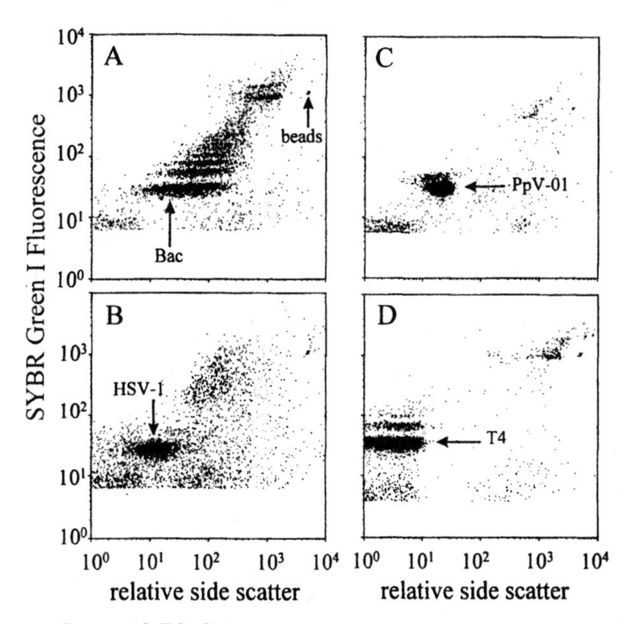

Figure 10-34. Side scatter and fluorescence signatures of four viruses shown in Figure 10-32. Reprinted from Brussaard CPD, Marie D, Bratbak G: Flow cytometric detection of viruses. Journal of Virological Methods 85:175-82 (reference 2337), Copyright 2000, with permission from Elsevier Science.

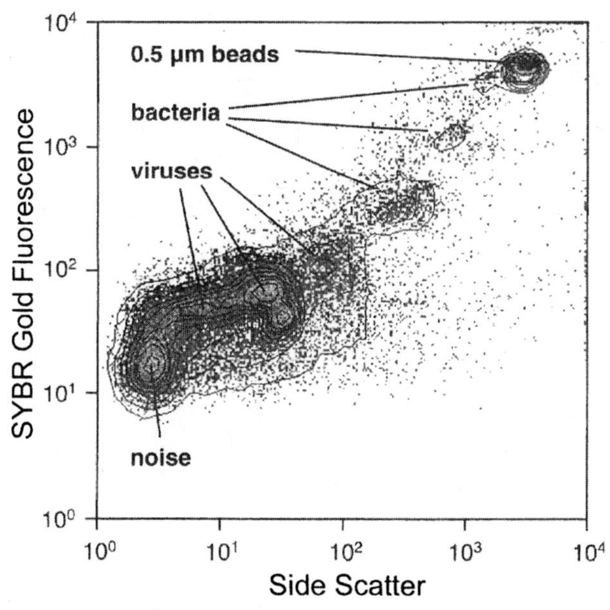

Figure 10-35. Side scatter and fluorescence signatures of particles in a water sample from a small Alpine pond. Courtesy of Stefan Andreatta (University of Innsbruck).

The new mystery guests in the aquasphere are viruses. Figures 10-34 and 10-35 show scatter signals from cultured viruses[2337], including the algal virus PpV-01, and from a sample of pond water that contains bacteria and viruses[3406]. The data in Figure 10-35 are particularly impressive; they were collected from a stream-in-air (MoFlo) sorter with the drop drive on!

Before the previous edition appeared, Troussellier, Courties, and Vaquer[2086] reviewed applications of flow cytometry in marine microbial ecology, which include identification of organisms[2087-8], determination of biomasses and productions[2089-92] and the measurements *in vitro* of bacterial and phytoplanktonic growth[2093-8] and metabolism[2099-100]. Monfort and Baleux[2101] found flow cytometry reliable as compared to epifluorescence microscopy for counting bacteria in aquatic ecosystems. Balfoort et al[2140] compared the performance of a laboratory-built and a commercial flow cytometer in phytoplankton analysis. Monger and Landry[2102] compared Hoechst 33342 and DAPI for staining DNA in marine bacteria, and found lower background and lower CV's with the Hoechst dye. Other authors[2103-4] considered aspects of scatter and absorption by phytoplankton cells, and applications of flow cytometry to characterizing organic[2105] particles in water. Flow cytometry has also been used to study larger organisms' feeding on smaller ones[1605-8]. At the high end of the size scale, Le Gall et al[2106] measured DNA content and base composition of macroalgal nuclei. Intracellular toxin was determined flow cytometrically in the marine dinoflagellate *Gonyaulax*, which causes "red tide"[571], and flow cytometric quantification of the autofluorescence of luciferin in this organism was used to study the circadian cycle of its bioluminescence[2049]; DNA content analyses[2040-1] established the periodicity of the *Gonyaulax* cell cycle.

Good collections of papers on cytometry and marine biology appear in recent special issues of *Scientia Marina*[2402] and *Cytometry*[3372].

Multiangle light scattering, autofluorescence, and DNA content[197-8,570] have all been used as identifying parameters for **phytoplankton**, with rRNA probes added to the mix as the technology improved[3394,3408,3411-3] (the fishing grounds may be alarmingly depleted, but there's still plenty of FISH in the sea), and flow cytometry has become a standard method for monitoring phytoplankton population dynamics. Although several Coulter EPICS instruments made ocean voyages[928] (I haven't heard them called EPICS Seas), they and other large instruments were/are not really optimized for phytoplankton analysis; most groups now use benchtop flow cytometers.

Quantification of the types and numbers of microflora and -fauna in the oceans should provide a sensitive means of monitoring changes in the quality of the environment. Figure 10-36 is directly related to this issue; it was provided by Don Button and Betsy Robertson of the University of Alaska, who used a modified Ortho instrument[2084]. Panel A shows a mixed bacterial population obtained from Prince William Sound, site of the 1989 *Exxon Valdez* oil spill[2085]; Panel B shows a seawater culture inoculated with gravel

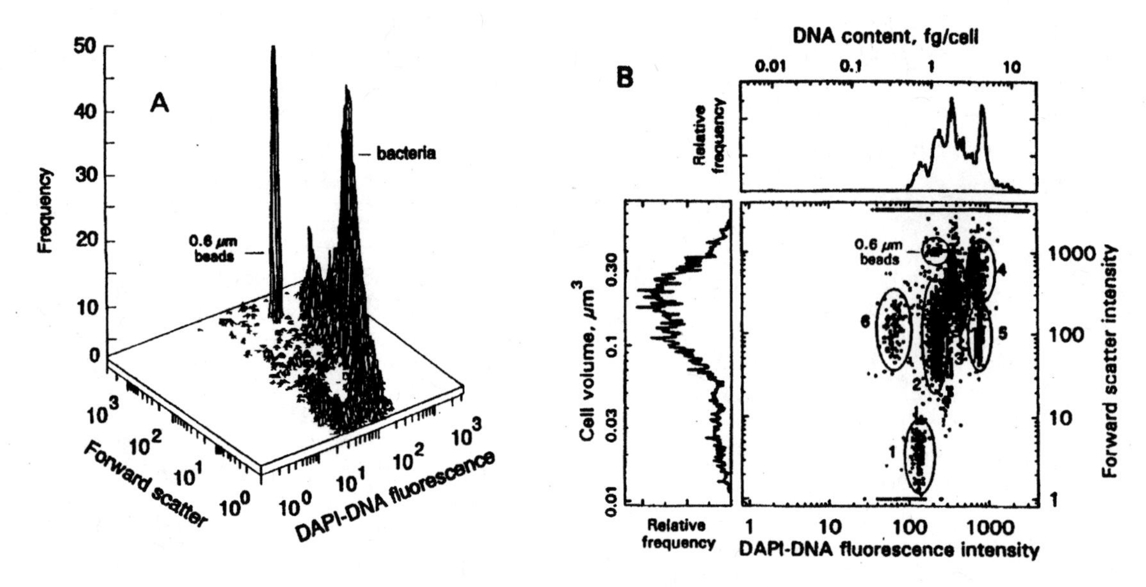

Figure 10-36. Size and DNA content of bacteria in seawater from Prince William Sound, Alaska (A), and of bacteria in a culture growing on a crude oil/terpene mixture (B). The figure was provided by Don Button and Betsy Robertson (U. of Alaska). Note that the scales express parameter values, not arbitrary numbers.

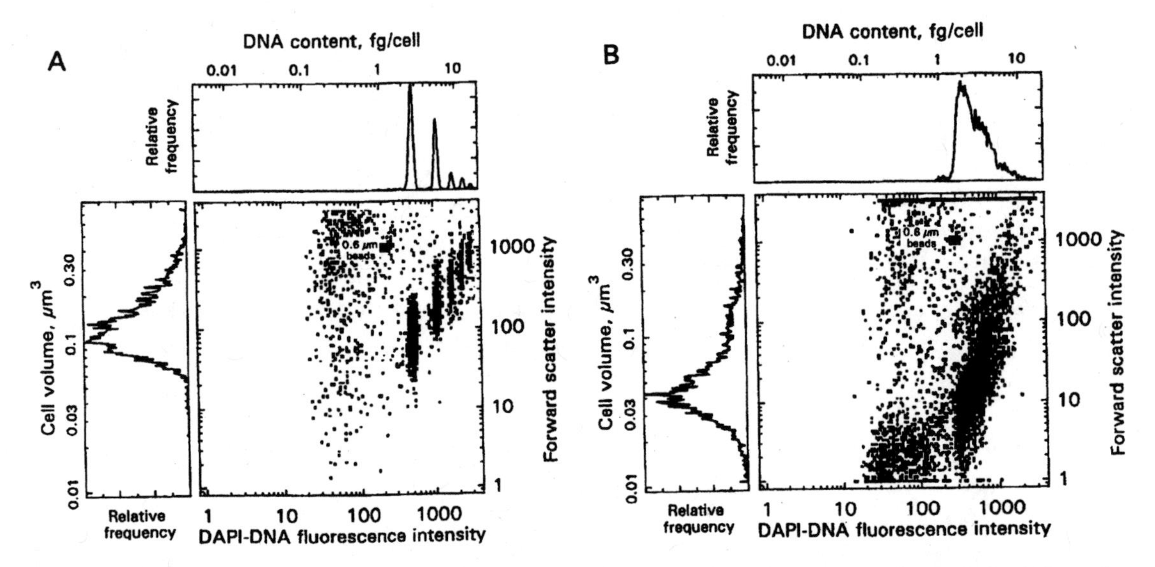

Figure 10-37. Size and DNA content of *Oligobacterium* RB1 isolated from Resurrection Bay, Alaska (A), and of the indigenous population from which RB1 was isolated (B). From Don Button and Betsy Robertson.

using a crude oil/terpene mixture as a carbon source. Some organisms will metabolize anything.

All kidding aside, Figure 10-36 and Figure 10-37, which came from the same place, exemplify a practice that ought to be more widespread in flow cytometry. Note that the scales on the histograms and dot plots are given in fg/cell DNA content and cell volume in μm^3 (fl), not channel numbers or arbitrary powers of ten. The volumes are derived from Mie theory[2562-3], and normalized with respect to the forward scatter signal and known Coulter volume for *E. coli*; the DNA scale was obtained from a standard curve of *E. coli* treated with rifampicin to produce cells with different integral numbers of genomes. The cells were stained with DAPI, which is A-T specific; the DNA content scale assumes the 50% G-C content characteristic of *E. coli*. If the organisms involved have a substantially different G-C content, the number is off; however, the scale remains consistent, and G-C content variations can be corrected for[3416]. Maybe if the journals gave notice that scales such as these would be mandatory in two years, the rest of us would get motivated.

Figure 10-37 shows DNA content and cell volume of microorganisms isolated from Resurrection Bay, near Seward, Alaska. Panel A shows the growth pattern at fast growth rates of *Oligobacterium* RB1 isolated by extinction culture; there are cell subpopulations with different numbers of copies of the genome. Panel B shows the signature of the indigenous population of microorganisms from which RB1 was isolated. I really appreciate these figures, because I wrote parts of the previous edition on board ship in both Prince William Sound and Resurrection Bay, on what was supposed to be a vacation to celebrate finishing the manuscript. I did get to observe some interesting marine microbiological phenomena; the toilets ran on seawater, and, late at night, the bioluminescent organisms would put on their own little light show. When you flushed, they flashed; it was like the Fourth of July.

Flow cytometry of marine microorganisms involves very long runs and accumulates huge amounts of data. Multivariate data collection and analysis[2107-8] are essential for discrimination of multiple species in samples; it is generally felt that automated on-line analysis, e.g., by neural networks[1215-6,2503-4,3417-20] will ultimately be necessary to keep up with the amount of data generated, particularly as more parameters are measured.

Extensions: Cytometers for Marine Applications

In the 1980's, Dubelaar et al[695,1167-8] described the Optical Plankton Analyzer, a multibeam instrument with slit-scanning capability optimized for analysis of a wide size range of marine organisms; also around that time, I helped Penny Chisholm, Rob Olson, Sheila Frankel et al, at M.I.T. and the Woods Hole Oceanographic Institution, build somewhat simpler cytometers[198,2082].

The latest instruments from Dubelaar are the CytoBuoy[2566,2910] series, commercially available from **CytoBuoy, b.v.** (see p. 430); they are designed to be mounted on buoys or in submersible vessels, and incorporated sophisticated, miniaturized optics and electronics with low power consumption. Detector signals may be telemetered to a remote monitoring computer. **Fluid Imaging Technologies** (pp. 168 and 431) offers the FlowCam[2453] imaging flow cytometer, which, like the CytoBuoys, can accommodate larger particles than most flow cytometers.

Cavender-Bares et al[3414] modified a Coulter EPICS V flow cytometer to allow rapid switching between single- and dual-sheath fluidic systems, making it possible to achieve optimal conditions for analysis of a wider range of size classes and concentrations of plankton organisms than would be possible using a conventional instrument.

Olson et al[3415] describe a microfluorimeter and a flow cytometer using the "pump-during-probe" technique, which measure chlorophyll fluorescence yield of phytoplankton cells following excitation by a pulsed argon laser. They have also built a submersible flow cytometer; for details, see: http://www.whoi.edu/science/B/Olsonlab/

Trask et al[197,711] pointed out that the excitation spectrum of autofluorescence from algal phycobiliprotein photosynthetic pigments[927,929] is extremely useful for identification; a multistation flow cytometer, with excitation beams at blue, green, yellow, and red wavelengths, suitable for excitation of algal phycobiliprotein pigments would be the ideal apparatus to use on phytoplankton analysis, because it should improve discrimination of species in unstained samples. This could now be implemented with small enough lasers to fit the apparatus comfortably on board a ship, but I haven't heard of anybody doing it.

References: Flow Cytometry and Oceanography

Claude Courties, Marc Troussellier and Louis Legendre, of the Observatoire Océanologique, Banyuls-sur-Mer, France, maintain an online list of references on flow cytometric applications in oceanography. The list is regularly reviewed, and can be consulted and downloaded at http://www.obs-banyuls.fr/FCM-Oceanography/Database.html; as of February 24, 2003, it contained 447 references.

General Microbiology

Previously Noted

As reported in the previous edition, Madelin and Johnson[2109] used an aerosol particle counter to size fungal and actinomycete spore aerosols at different humidities; this was a good way to commemorate the Golden Anniversary of the Gucker experiments.

Numerous authors[1514-5,2127-39] have used flow cytometry to measure binding of labeled antibodies to bacterial cell surfaces, for purposes as varied as determining levels of antibodies in serum and analyzing the structure of transmembrane proteins (p. 479). Nelson, Neill, and Poxton[2122] used flow cytometry to observe binding of anti-lipopolysaccharide monoclonal antibodies to whole bacteria under physiologic conditions, and compared this method with whole cell ELISA and immunoblotting. Only flow cytometry could demonstrate subpopulations of cells with different binding characteristics. Eitzman, Hendrick, and Srienc[2148] described a procedure for measuring quantitative immunofluorescence in single yeast cells, using a transfected *E. coli* lacZ gene as the test antigen.

Allman et al[2142] examined forward and orthogonal scatter signals, Coulter volume, and ultrastructure of *Azotobacter vinelandii* under various culture conditions and observed that forward scatter tracked Coulter volume better than did orthogonal scatter. In another paper[2143], they found that the addition of DNA content, measured with mithramycin-ethidium bromide, to forward and orthogonal scatter measurements gave much better resolution of different species than did the two scatter signals. Miller and Quarles[2144] reported some success in bacterial identification using FITC/PI staining.

Hiraga et al[2145] used flow cytometry to monitor morphologic effects of plasmid genes in *E. coli*; Poulsen and Jensen[2146] detected morphologic effects of chromosomal genes in the same organism in the same way. Sazer and Sherwood[2147] stained the fission yeast *Schizosaccharomyces pombe* with Hoechst dye and chromomycin, noting the different base ratios of mitochondrial and nuclear DNA, and reported that mitochondrial growth and DNA synthesis could occur in the absence of DNA synthesis. However, Carlson et al[3439] subsequently questioned whether DNA fluorescence measured from the cytoplasmi was due to mitochondria.

Héchard et al[2149] used flow cytometry to monitor the interaction between cocultured *Listeria monocytogenes* and an antagonistic *Leuconostoc* strain producing an anti-*Listeria* bacteriocin. Pinder et al[2150] examined flow cytometric counting of bacteria in suspension using forward scatter and ethidium fluorescence and found excellent correlation with plate counts. Russo-Marie et al[2151] applied the FACS-Gal reporter gene technique, using fluorescein galactopyranoside (pp. 408-9), to differentiating *Myxococcus xanthus* cells. Manafi and Kneifel[2152] provided a discussion of fluorogenic substrates in general and their applications in microbiology.

Cell Cycles and Cell Division

Boye, Lobner-Olesen, and Skarstad and numerous others[2110-26,2141,3381-2,3421-33] continue to apply DNA flow cytometry to determining characteristics and control mechanisms of the **bacterial cell cycle**. Ueckert et al[3434] combined propidium staining for loss of cell membrane integrity and CFSE labeling to study the effects of mild heat treatment and low concentrations of nisin on division of *Lactobacillus plantarum*; they were able to discriminate as many as eight generations of daughter cells. Porro et al[3435-7] used DNA and protein and/or cell wall labeling to follow cohorts of daughter cells of *Saccharomyces cerevisiae*.

Fluorescent Protein Methods in Microbes

Valdivia and Falkow[3440-5] and coworkers were among the first to apply flow cytometry and GFP expression in studies of bacteria and yeasts, with emphasis on analysis of genes related to bacterial virulence. Others[3446-53] have used GFP and other fluorescent proteins in a wide range of applications.

Microbial Communities: Will Flow Work?

Few of us believe that eukaryotic cells can't tell the difference between life in suspension culture and life in the urban cellular environment of a multicellular organism. As we learn more about microorganisms, it is becoming more difficult to maintain the notion that even free-living autotrophs go through life uninfluenced by their peers.

Bacteria differentiate; the process, exemplified by sporulation in *Bacillus subtilis*[3455] and the generation of swarmer and stalked cells in *Caulobacter crescentus*, involves asymmetric cell division under the control of signal transduction proteins[3456-7]. Flow cytometry and fluorescent probes have been applied in studies of both phenomena just mentioned.

A variety of chemical signal molecules mediate **quorum sensing**[3458-9], the process by which groups of bacteria coordinate their behavior, taking on some of the functions of multicellular organisms. Flow cytometric studies of dormancy and resuscitation in *Micrococcus luteus*[2253-4] aided in the discovery of resuscitation promoting factor (Rpf), a "bacterial cytokine"[3460] that stimulates the growth of *M. luteus* and of other high G-C Gram-positive organisms, including *Mycobacterium tuberculosis*, which make similar proteins.

Quorum sensing regulates bioluminescence, mating, sporulation, and virulence factor expression, and the formation of **biofilms**[3461] in a variety of bacterial species. An established biofilm comprises microbial cells and an **extracellular polymeric substance (EPS) matrix**; we can expect that flow cytometry of bacteria removed from a biofilm will omit some relevant particulars, as we have come to expect of flow cytometry of cells removed from solid tissues and tumors. If we want to analyze bacteria in biofilms *in situ*, we need only keep the same probes and switch to scanning, image, or static cytometry.

Bad Guys Don't All Wear Black Hats: Microbial Detection/Identification in Health-Related Contexts

There are a lot of microorganisms out there; Elmer Pfefferkorn of Dartmouth, one of my daughter's microbiology professors, described the world as coated with a "fecal veneer." We and our ancestors have coexisted with a huge number of genera and species for millions of years, only learning within the past two centuries that a few among the invisible multitudes with which we share the planet are our implacable enemies, and that a few others can do us harm when they get into the wrong parts of our bodies and/or when our defenses are down.

We pay sanitary and food microbiologists, clinical microbiologists, and the microbiologists working in our macroscopic defense establishment to look for the bad guys in a large number of samples from a wide variety of sources, and a quick look at the literature tells us that most of these microbiologists are still searching for faster, cheaper, more sensitive detection methods. The cytometric strategies for detection are pretty much the same in all areas.

The Basic Questions

Neglecting the details for the moment, the microbiology laboratory can be envisioned as a black box to which the inputs are material ("atoms" in today's information technology parlance) and from which the outputs are items of information ("bits"). The input is a solid or liquid sample; the outputs are the answers to one or more questions:

1) Is evidence of the activity of microorganisms present in the sample?

2) If so, what microorganisms?

And, for clinical and defense purposes:

3) To which antimicrobial agents, if any, are the microorganisms susceptible?

The evidence of microbial activity required for an affirmative answer to the first question might include the presence of microbial antigens or gene sequences, or the presence of host antibodies to microorganisms, or the presence of viable microorganisms; if the organisms themselves cannot be found, it may be more difficult to obtain definitive answers to the second and third questions.

Moving from the ivory tower to the real world of the laboratory, still minus the details, it is clear that somewhat different processes are in place for dealing with bacteria, fungi, and protozoan parasites, on the one hand, and viruses, on the other. For the former, finding evidence of microbial activity almost always means eventually finding viable (i.e., culturable) organisms in the input sample; determining the identities and antimicrobial susceptibilities of these organisms typically requires the establishment of pure cultures.

Routine culture of viruses is much less common; detection is typically accomplished using PCR and other nucleic acid amplification techniques, which also allow identification and, in some cases, susceptibility testing. Clinical laboratory diagnosis of many viral infections is based on demonstration of antiviral antibodies,

There is little question that cytometry, flow and otherwise, can answer the questions listed above, and that, in at least some cases, the required information can be obtained faster by cytometric methods than by classical bulk culture-based technology. However, it is almost certain that cost-effectiveness will be the principal determinant of whether cytometry will become widely used for detection and identification of microbial pathogens.

Bacteriology laboratories haven't changed a lot during the past century; Pasteur would feel pretty much at home in a "modern" lab. The first task facing the lab when a new sample comes in is **bacterial detection**, that is, determining whether there are microorganisms, or pathogenic microorganisms, in the sample at all. In some cases, this can be done by looking at the sample under a microscope, perhaps using Gram's or another specialized staining technique, or staining with a fluorescent dye such as acridine orange and using a fluorescence microscope. This procedure is referred to as **direct detection**; it works best on samples that normally contain few or no bacteria, or samples in which the pathogens you're after can readily be distinguished from other bacteria normally present in that type of specimen.

Direct detection, however, is labor-intensive. It takes ten or fifteen minutes to take a urine sample, for which most microbiology texts recommend an initial attempt at direct detection, spin down the sediment, make a slide, and stain it. It also takes a well-trained, skilled technologist, because bacteria are near the limit of resolution of optical microscopes, and the tech has to be able to keep the scope clean and in good alignment and know how to use it. Otherwise, the whole business is a waste of time. It only takes a minute, however, for a minimally trained person to put a small aliquot of the urine specimen on a plate containing blood agar or some other nutrient-rich *cordon bleu* medium, and wait until the next day, or at least the next shift, to see if anything grows. By that time, a more trained individual is likely to be able to look at all of the plates on which there are colonies, and determine that some of them don't need to be looked at further because whatever is on them is likely to be insignificant or a contaminant.

Detection: Intrinsic Parameters are Not Enough

At its simplest, the flow cytometer mimics the microscope. Measurements of forward and/or orthogonal light scattering signals provide the information a technician would obtain by visual examination of an unstained slide. An alternative cytometric measurement of size, capable of discriminating particles in the bacterial size range from larger and smaller objects is obtained from measurements of electrical (DC) impedance made using a Coulter orifice. It was shown more than twenty years ago[1015-7] that particle counting by this technique was effective in detecting clinically significant bacteriuria (the criterion in use at the time was 100,000 organisms/mL). The false negative rate for such analyses was low, i.e., bacteria, whenever present, were detectable. The false positive rate was unacceptably high, because the single-parameter electronic volume measurement could not discriminate between bacteria and other organic and inorganic particles in the same size range.

Two bacterial species differing substantially in size, shape, and/or refractive index may be distinguished by their forward and orthogonal scatter signatures, but two species chosen at random are far too likely to occupy the same region of the measurement space to make two-parameter scatter measurements alone generally useful either for distinction between species or for discrimination of microorganisms from other small particles. Fungi, which are, in general, larger than bacteria, can be distinguished from them on the basis of scatter signals, but, again, scatter signals alone may not be adequate to discriminate between fungi and other particles in a specimen.

Detection: Fluorescence Improves Accuracy

Improvements in cytometric detection of microorganisms typically rely on fluorescent staining; nucleic acid stains, fluorogenic substrates, and membrane potential-sensitive dyes have all been used for this purpose. These reagents all produce stronger signals, and require less sample preparation time, than would be necessary to detect organisms using labeled antibodies, lectins, or oligonucleotide probes. Although the specificity of antibodies and oligonucleotides may be useful for identifying organisms once they have been detected, it limits the utility of the reagents in the detection process unless the sample is being screened for one or a very few pathogenic species.

Detection: When the Tough Get Going

When you are looking very hard for one pathogenic species, e.g., *Bacillus anthracis*, flow cytometry, notwithstanding fifty plus years' work on detecting that particular bug, may go right out the window. In late 2001, after anthrax-by-mail had caused a number of fatal and near-fatal infections[3462], I got a call from a *Wall Street Journal* reporter interested in why the Army didn't seem to be willing to release its flow cytometric technology to the people then screening federal offices and postal facilities for anthrax spores. I told him that it was unlikely that flow was going to get anywhere near the level of sensitivity that those screeners seemed to be achieving. At the time, I thought they were using PCR; they weren't. They would turn on a mail sorting machine and suck air out of it for ten or twenty minutes, blowing the air over a plain old agar plate[3463]. Pretty low tech, but they could reliably detect a single colony of *B. anthracis* on a plate, and I doubt that any fancier method could have done better.

Similar detection problems arise under less dramatic circumstances. If you are trying to find a dozen or so colonies of *E. coli* O157:H7 on a side of beef, the only way you can hope to do it is to hose down the carcass, filter the wash water through a 0.22 μm pore filter, and plate the filter. You might try to use appropriate antibodies and/or rRNA probes and a fluorescence microscope or a scanning cytometer of some kind, but it would be a lot easier and a lot cheaper to just incubate the plate for a few hours and then apply specific reagents.

A blood culture can turn out positive if there is one viable bacterial cell in the 10 mL of blood originally drawn. If we had a fast enough flow cytometer to run all the blood through in a reasonable time, and we wanted to make sure we had a positive culture, by counting around 50 organisms, we would have to draw and process a unit of blood from the patient. We don't do that, of course; the machines that process blood cultures detect products of bacterial metabolism in the blood, after it has been incubated for a while, and will generally not respond unless there are at least 10^5 metabolically active organisms/mL of medium. I noted on p. 515 that a flow cytometric technique[919] had detected *E. coli* seeded into human blood samples at concentrations of 10-100 organisms/mL; while cytometry is impractical for detecting that one viable cell in 10 mL without an incubation period, it should be able to provide a definitive answer after an incubation period six to ten generation times shorter than is necessary for competitive technology.

Identification: Too Many Broths...

The next step in the process, for samples in which organisms have been detected, is **bacterial identification**. As a first step, it is traditional to look at a Gram-stained slide of the organism. What happens next depends on how many suspects there are.

In 1988, Phillips and Martin[2258], who had done several studies on immunofluorescence analyses of bacteria at the U.K. Biological Defence Establishment at Porton Down, using an Ortho flow cytometer[568,914-5], concluded that the low scatter sensitivity then available limited what could be done in terms of bacterial identification. Without enough of a scatter signal to trigger on, detecting the immunofluorescence signals was hopeless. Obernesser, Socransky, and Stashenko[2259], doing immunofluorescence analyses on oral flora in a decidedly civilian environment, reached a similar conclusion in 1990. The conclusions of both reports might have been different had the investigators been using instruments better suited to bacterial analysis.

There is, however, a fundamental problem with focusing on the use of flow cytometry for bacterial identification, as opposed to bacterial detection, or even to sensitivity testing, and it is this: there are too many kinds of bacteria, and the nature of flow cytometry makes it very difficult for this technology to be competitive in a situation where there are thousands of species about which you need to be concerned.

It has been more or less traditional for people involved in a project to identify bacteria by flow cytometry to start out with a few well-known organisms, and demonstrate the ability to tell them apart. A good example can be found in Figure 7-14 (p. 318), which shows separated clusters of *Staphylococcus aureus*, *Escherichia coli*, and *Pseudomonas aeruginosa* in a two-parameter display of Hoechst 33342 and chromomycin A_3 fluorescence. Van Dilla et al[279], in 1984, applied this dye combination, which is sensitive to DNA base composition, and which had been used to stain and sort human chromosomes, to bacterial analysis; the three organisms shown, which, respectively, contain 31%, 50%, and 67% G+C, are well separated. Further work in this area[1340-1] made it possible to resolve a few more species and to detect bacteriophage infection in *E. coli* by the attendant change in base composition.

This approach is fine, if you're trying to identify one of half a dozen or so organisms. If you were looking at cerebrospinal fluid from a patient with signs of bacterial meningitis, you'd know that the infection was likely to be caused by one of four or five organisms, and you could probably say that it was one of them or something else using base composition, or multiplex antibody staining with two-color immunofluorescence, or FITC/PI, or even forward and orthogonal scatter. However, if you happen to be looking at urine, or blood, both of which account for much higher percentages of the average clinical bacteriology lab's workload than does spinal fluid, even if you consider only the common pathogens found in your institution during the past year, your list is likely to have dozens or even hundreds of organisms on it.

This is most commonly done using a battery of selective media, in a 96-well plate or equivalent, each well of which is inoculated with organisms from the same colony. An intermediate step is often required, in which cells picked from a single colony on a plate are grown in broth culture to provide enough inoculum for identification. It helps to know the Gram-staining characteristics of the organism before inoculating it into selective media, because different sets of media are frequently used for Gram-positive and Gram-negative bacteria.

After another few hours, it becomes apparent which energy sources will and will not support the growth of the organism, and which chemical reactions it can and cannot perform. There are a lot of possible patterns of growth and metabolism in 96 wells, and the developers and manufacturers of bacterial identification tests have made sure that they will get different patterns for any of at least several hundred species. Well, okay. Why can't you just do the same tests by flow cytometry? You could, of course; just look at the cell growth, or the cell metabolism, or whatever, in each well, with the flow cytometer. The problem is that when you try to do that, one well at a time, even with high-throughput flow cytometry, it may take you ten minutes or more to read one 96-well test. The conventional 96-well test is read in under a minute in a plate reader that uses 96 LED's as light sources and 96 photodiodes as detectors. Even with microfluidic flow cytometers, there's no way you'll get a system with 96 flow cells at any reasonable cost. Conventional flow cytometry for bacterial identification is only likely to be practical when you're looking for very small numbers of one or a few different organisms; you can use a gene probe or fluorescent antibody for specific identification, and scatter parameters and perhaps a nucleic acid stain to narrow down the gating region, treating the identification problem as rare event analysis (pp. 469-71).

If you happen to be looking for biological warfare agents, you've really got a problem. In the age of genetic engineering, what the other side is likely to throw at you may be an organism for which you haven't got the right fluorescent antibody or gene probe. I'll have some suggestions about what to do in this situation in a later section.

Identification: Can Multiplexing Help?

On p. 475, I suggested several possible solutions to the multiple aliquot problem involving multiplex fluorescence. Cells in media containing different nutrients or inhibitors (for identification), or different drugs (for susceptibility testing) could be "color coded," either by embedding them in gel microdroplets containing mixtures of dyes in different proportions, or by directly labeling the cells with mixtures of tracking dyes or high-affinity nucleic acid dyes with different spectral characteristics. The direct labeling approach would probably require use of a high-sensitivity, slow-flow cytometer to detect more than four or five differently labeled populations; the gel microdroplet approach could probably be run on a commercially available instrument such as a Luminex 100.

Ye et al[3464] may have come up with a better idea. Their group at Glaxo Wellcome, which has been using Luminex's technology for detection of single nucleotide polymorphisms[2373-4], has adapted the same technology to bacterial identification using genus- and species-specific rRNA sequences. They report that assays can be done at a rate of almost 100/h, starting with DNA extracted from a few thousand organisms, with a cost per test of about one dollar.

We can now look at some of what has been attempted in the areas of environmental and sanitary microbiology, food microbiology, biowarfare agent detection, and clinical microbiology, remembering that hindsight is usually 20/20.

Environmental and Sanitary Microbiology

Several groups[2153-5] have used flow cytometry to study growth of food-borne pathogens and agents causing fish diseases under various environmental conditions. Volsch et al[2156] combined propidium staining with immunofluorescence to detect two different serotypes of ammonia-oxidizing bacteria, present at concentrations of 0.1 to 2%, in activated sludge from sewage plants.

Porter et al[2157], using fluorescent antibodies and sorting mixtures of *Staphylococcus aureus* and *E. coli*, found that *S. aureus*, when mixed in different proportions with *E. coli*, could be selectively recovered at a purity in excess of 90%, even when *S. aureus* composed less than 1% of the total cells. Sorting with fluorescently labeled antibodies specific for *E. coli* was also tested for the ability to recover E. coli from natural lake water populations and sewage; sample purities of greater than 70% were routinely achieved, as determined by colony counts. Populations of E. coli released into environmental samples were recovered at greater than 90% purity.

Water That Made Milwaukee (and Sydney) Infamous

In 1993 and 1994, the drinking water supply in Milwaukee became contaminated with the protozoan parasite *Cryptosporidium parvum*, which caused many cases of diarrhea and some deaths. I noted in the previous edition that a flow cytometric technique for monitoring drinking water for this and other parasites, developed by Vesey et al[2223-4], was being used in Australia, and suggested that we ought to think about it in the United States, too. "Maybe I'll stick to Foster's for a while," said I.

In 1998, while I was visiting Sydney (and being escorted by Hopi Yip, who was involved in the water analyses, to visit Duncan Veal, Belinda Ferrari, and others at Macquarie University, where the methodology was developed), the flow cytometric monitoring system detected *Cryptosporidium* in drinking water from several sources in town. Nobody seemed to know where the parasite came from; the grisly end of a dead kangaroo was one suggestion. Interestingly enough, there did not appear to be a widespread outbreak of diarrhea due to *Cryptosporidium* in Sydney at the time, although it might have been somewhat harder to detect due in the presence of another GI bug, presumably viral (I dubbed it the "Anzac two-step"), that had been making the rounds for some weeks before parasites were found in the water supply.

Cryptosporidium is hard to stain, and doesn't have a lot of surface antigens; the procedure used for detection in 1998 involved presumptive flow cytometric identification based on scatter signals and single-color immunofluorescence, followed by fluidic sorting onto filters, and subsequent visual examination. The Macquarie methodology for detection of both *Cryptosporidium* and *Giardia* has been improved considerably by the use of dual color immunofluorescence and immunomagnetic separation[3465-70], but I am told that flow cytometry is no longer in routine use for water monitoring in Sydney, and I've gone back to American beer (Sam Adams isn't from Milwaukee, anyway).

For what it's worth, in 2001, Lindquist et al[3471] compared four fluorescent antibody-based methods for detection of *Cryptosporidium parvum* in surface water, and reported that solid-phase cytometry had the highest presumptive and confirmed detection rates. Although flow cytometry had the next highest presumptive detection rate in reagent water, it placed third in analyses of spiked surface and tap waters, and, as used by these authors, lacked a confirmation procedure.

Food Microbiology

Garcia-Armesto et al[2158] and Laplace-Builhe et al[2159] considered applications of flow cytometry to food microbiology. Flow cytometry has been used in conjunction with viability markers such as FDA for rapid counting of yeast, mold, and bacterial cells in food products using **Chemunex**'s single-parameter flow cytometer[2160]; this has allowed rapid detection of low numbers of microbial contaminants. The correlation between flow cytometric results and product shelf life was excellent. Chemunex's instrument line (p. 430), now includes flow and solid phase cytometers; they have been used for fermentation monitoring as well as for detection of microbial contaminants in food and water.

I have already (pp. 430-1) mentioned that instruments for counting bacteria in milk are available from **Bentley Instruments**, **Delta Instruments bv**, and **FOSS Electric A/S**. Using a conventional flow cytometer with fluorescent antibodies and propidium to detect *Listeria monocytogenes* in milk, Donnelley, Baigent, and Briggs[2161] reported that flow cytometry yielded a 5.86% false positive rate and a 0.53% false negative rate when compared with standard culture procedures. McClelland and Pinder[3472] were able to detect *Salmonella typhimurium* in dairy products at levels of 10^4/ml using flow cytometry and fluorescent monoclonal antibodies.

Hennes et al[2335] showed that bacteriophages parasitic on Cyanobacteria retained dimeric cyanine dyes such as YOYO-1 sufficiently well to be used to identify their target species in mixed populations by fluorescence microscopy. Goodridge et al[3473] adapted this technique for detection of *E. coli* O157:H7, using a bacteriophage shown to attack most known varieties of this strain in combination with immunomagnetic separation. They estimated that detection limits in food products could be as low as 10^2-10^3 organisms/ml, a level of sensitivity equivalent to that obtainable using PCR.

Bioterrorism and Bioopportunism

As I have pointed out before (p. 74), flow cytometry of biological specimens began with attempts at rapid detection of bacteriological warfare agents by Gucker et al under U.S. Army auspices in the 1940's. Various American, British, and Canadian defense agencies, and their counterparts elsewhere, have continued to provide support for, and do, significant work on flow cytometry of bacteria[2423,2574-5,3474-6].

For most military applications of flow cytometry, the specimen is likely to be the output of a device that concentrates particulate matter in aerosols; in the early 1990's, the U.S. Army build several dozen mobile chemical and biological agent detection labs into Humvees, equipping each with a Coulter EPICS XL flow cytometer into which were fed samples from aerosol concentrators, stained with the acridine dye coriphosphine O. The Army next contracted with Los Alamos and Bio-Rad to develop a smaller instrument with lower power consumption; the result was what is now sold by Apogee (to qualified buyers; otherwise they have to kill you) as the A20 (see p. 430). A parallel effort attempted to develop a detection system based on the B-D FACSCount, the notion being that several hundred of these would be mounted on top of buildings at military bases around the world. I can just see the KC-135 modified to change the sheath water.

It's gotten hard to keep track of what's going on on the military side of the biowarfare game, but business seems to be picking up in the homeland security area since the anthrax attacks. A company called **PointSource Technologies, LLC** (http://www.safewater.tv) has been getting some press recently, having received a patent on what appears to be a flow cytometric method for detecting individual bacterial pathogens in water supplies using scatter signals in sixteen different angular ranges. Their Vigilant X-3 Microbiometer was used to monitor the water supply at Qualcomm Stadium during the 2003 Super Bowl (see the Web site for details). Not that the biggies in the flow cytometry business aren't looking for a piece of the action; they're just being more discreet.

If I seem to lack enthusiasm for much of the effort that has been made of late to apply cytometry to biowarfare agent detection, it is because I think many of the decision makers involved in the process, as opposed to the people actually doing the work, have gotten overly enthusiastic about the technology and lost sight of what it can and can't do. Airborne Instruments Laboratory built a bacterial detection system for the Army in the 1960's that was essentially a static cytometer in a somewhat bulky backpack; it collected cells on a tape and stained them with an acridine dye. I suspect its performance was not much worse than that of a Humvee-based flow cytometer. I also believe that some really compact, small, inexpensive cytometric instruments could now be developed for use by the military if and when realistic specs are arrived at, and that flow may not be the best way to go.

Meanwhile, recent past and current events have occasioned some concern about whether what is published in the open literature might give aid and comfort to bioterrorists and rogue states, with a group of journal editors and other interested parties concluding that "Journals and scientific societies can play an important role in encouraging investigators to communicate results of research in ways that maximize public benefits and minimize risks of misuse."[3477]

Viruses and Other Intracellular Pathogens

Direct detection of viruses by flow cytometry is relatively new, but the technology has been used for decades to detect **infection of cells by viruses and other intracellular pathogens**, such as chlamydia, by changes in nucleic acid content and/or by expression of pathogen-derived or pathogen-related sequences or antigens[574,614,933-5,1403,2230]. A report showed that cytokine-induced expression of **human immunodeficiency virus** in a chronically infected line was accompanied by increased expression of membrane-bound IL-1β[936]; **HIV P24 antigen**[2162-5] and mRNA[1403] can also be detected in peripheral blood mononuclear cells. **Virus-cell interactions** can also be studied using membrane probes[486,575] or covalently labeled virus particles[453,1485-6].

McSharry et al[2166] described a rapid method for detection of **herpes simplex virus** in clinical samples following amplification in tissue culture. Qvist et al[2167-8] developed a method for flow cytometric detection of bovine viral diarrhea virus in blood cells from persistently infected cattle, and found it to be at least as sensitive as virus isolation in cell culture. Others applied flow cytometry to interactions of cells with rickettsiae[2169-70,2187] and to analysis of antibody binding to chlamydia elementary bodies[2171]. McSharry reviewed the uses of flow cytometry in virology in 1994[3478]; more recent work will be discussed in the section on clinical microbiology.

Scatter signals from medium-sized and large viruses are detectable in laboratory-built and in some commercial flow cytometers[94,688,1246] (see Figure 7-5, p. 288), and we have already noted (pp. 522-4) that detectable fluorescence signals can be obtained from viruses treated with cyanine nucleic acid dyes. Autofluorescence is not a problem when measuring individual virions, so it is likely that a measurable immunofluorescence signal could be obtained from a single virion using a slow flow system and, if necessary, appropriate amplification techniques.

It is now feasible to build a flow cytometer that could characterize phenotypic and, possibly, genotypic heterogeneity in virus populations, and, when operated in an environment designed for biohazard containment, allow sorting of single virions with different characteristics for subsequent study in culture. The instrument is not likely, in itself, to be any more expensive than the fancier cell sorters now on the market; we're not talking about the superconducting supercollider. However, it seems to me that these gadgets should be installed at the CDC, NIH, and Fort Detrick, because virus sorting will, as cell sorting did, allow questions to be answered that could not be answered any other way.

Clinical Microbiology

Álvarez-Barrientos et al[3479] produced an extensive and well-referenced review on applications of flow cytometry to clinical microbiology in 2000. I wrote a shorter piece, echoing much of what is said here, in 2001[3480].

Clinical microbiology is regarded as an area in which flow cytometry looks very promising. It's looked very promising for a generation, and millions of dollars in venture capital have been lost on what, in retrospect, weren't outstandingly sensible approaches.

When most people contemplate the gold in the hills – or ills – of clinical microbiology, they are thinking about **diagnostic bacteriology**, on which over a billion dollars are spent annually. The detection and identification phases of diagnosis have been introduced on pp. 528-31.

The ideal clinical specimen for flow cytometric analysis is one in which relatively few particles and few bacteria are found in the absence of an infection, and in which large numbers of bacteria are found when an infection is present. Blood is hardly ideal, because there are too many large objects such as platelets and red and white cells getting in the way, and, even when the patient is suffering from overwhelming septicemia, there may be no more than 1 organism in each mL of blood (see p. 530).

While spinal fluid, even in a case of meningitis, typically contains far fewer cells than blood, and while you may only need to look for a few different kinds of bacteria or fungi in it, there are still problems; the sample volume submitted for analysis is usually much less than 1 mL, and specimens from patients with bacterial meningitis may contain only a few dozen microorganisms.

Urine is an excellent specimen. Between a third and half the specimens submitted for bacteriologic workup nationwide are urine specimens, so any cost-effective technology that could improve analysis in this area should find widespread application. And urine is, believe it or not, relatively clean. In uninfected individuals, urine is sterile and contains few particles; in patients with urinary tract infections, urine typically contains at least tens and, more often, hundreds of thousands of bacteria per mL. This means that urine samples could be analyzed by flow cytometry without much prior processing; there's no need for centrifugation or filtration, or for lysis of cells. In fact, if your instrument can count both the white cells and the bacteria in a urine specimen, it will do much better at diagnosing urinary tract infections.

I have already mentioned (p. 529) that a Coulter counter fitted with a small orifice can detect almost all cases of significant bacteriuria, but that it also identifies other noninfectious particles in urine as bacteria. The capacities of Sysmex's UF-50 and UF-100 flow cytometric urine analyzers (p. 510) for detection of bacteriuria have gotten mixed reviews[3485-7]. I don't doubt that an instrument using a suitable combination of nucleic acid stains, fluorogenic substrates such as FDA, and membrane potential probes would be able to identify bacteria in samples, determine their metabolic activity, and discriminate them from noninfectious inorganic and organic particles in the same size range.

In the previous edition, I waxed almost lyrical about the capabilities of small, red laser source flow cytometers and red-excited cyanine membrane potential dyes for bacterial detection and identification. In this edition, I have outlined an approach to a flow cytometric "Gram stain" using these instruments and reagents (pp. 516-7); last time around, I pointed out that the instrument could also identify fungi, and leukocytes (which, in urine, provide independent confirmation of the existence of bacterial infections) in the same samples used for bacterial detection. I also mentioned that, since bacteria will maintain membrane potentials in media in which they can grow, and lose membrane potential in media in which they cannot or in the presence of compounds that inhibit their metabolism, it would be possible to do identification steps beyond the "Gram stain" using the flow cytometric membrane potential measurement. For example, metabolic inhibition by crystal violet will distinguish *Staphylococcus* from *Streptococcus* species. I am less enthusiastic about this approach now; I suspect that it makes more sense to do growth- and inhibition-based identification using scanning cytometry and "nanowell plates" containing small numbers of organisms, preferably incorporating microfluidics on-chip[2512-4] to facilitate sample preparation, and the same approach may be in order for susceptibility testing. I am sobered by what I hear from the people I know in the diagnostic microbiology business; they tell me that all of this has to be done for a quarter, and that may be a tall order.

Antimicrobial Susceptibility Testing: One Size Does Not Fit All

What the clinicians generally care most about is not so much detection and identification as **antimicrobial susceptibility testing**, i.e., determining which drugs kill or arrest the growth of the organism, and what concentrations of the effective agents are required. This can be initiated from the same inoculum used for the identification process, sometimes making the clinically relevant information about which drug to use available before identification of the organism has been completed.

From the 1980's on, an increasing percentage of clinical bacteriology laboratories have installed automated systems for identification and susceptibility testing, which provide results in 2 to 7 hours, rather than in 15 to 24 hours, as is the case when traditional culture methods are used. The automated systems, although originally conceived as labor- and money-saving, as well as timesaving, have turned out to increase laboratory costs. However, an older study by Doern et al[2257] and more recent work by Barenfanger et al[3481-4] document the clinical impact of rapid identification and susceptibility testing; patients for whom diagnoses were made more rapidly with the aid of automated systems had significantly lower mortality rates and hospital costs and shorter hospital stays. The motivation for developing cytometric techniques for rapid susceptibility testing is stronger than ever. Now, all we have to figure out is how to do them.

In many clinical situations, e.g., in the context of managing urinary tract infections, knowing whether the pathogen is bacterial or fungal and, if bacterial, whether it is Gram-positive or Gram-negative, is usually enough to allow a physician to decide which antibiotic(s) to prescribe, at least initially. The use of existing systems for sensitivity testing would permit a prescription to be changed, if necessary, within 24 hours. Other situations, e.g., septicemia and meningitis, may require more rapid answers.

In considering responses to biological warfare, the defense agencies have traditionally though in terms of administering antitoxins in the short run and immunizing personnel against the agent, in the longer run. It is probably more sensible to consider antimicrobial agents as the short-run defense; it is poor strategy for your adversary to release an agent against which no antimicrobial is effective, and also poor strategy to use an agent against which you are likely to have stockpiled antitoxin and vaccine. Unfortunately, the major reason we now have to think about defense against biological warfare is that it has been, and may again be, used by less rational individuals than we once thought might use it[3462], so all bets are off.

When I first took the position that, faced with serious bacterial infections and/or biowarfare agents, we should determine bacterial antimicrobial susceptibility first, initiate treatment, and then wait a day or two for definitive identification, it seemed to be heresy; it now appears that other people are willing to consider what I call the "Andromeda Strain" approach. The precise species and strain of the organism are of epidemiologic and/or forensic significance; getting the patient on the right drug is time-critical, and, if you don't have specific reagents for an organism, you can still figure out how to do that.

Over the years, a lot of people have considered the use of flow cytometry for **antimicrobial susceptibility testing**. There are two main classes of antimicrobial agents. **Bactericidal agents**, e.g., penicillin and streptomycin, kill bacteria; **bacteriostatic agents**, e.g., chloramphenicol, keep organisms from growing, but don't kill them. **The most obvious way in which to determine the effects of either type of drug on cultures is to compare bacterial counts over time in treated and control cultures**, and, after all these years, there doesn't seem to be a better indicator or set of indicators of antibiotic sensitivity that will work in all types of bacteria.

In early work, Cohen and Sahar[2260] used light scatter and ethidium fluorescence to identify and count bacteria in cultures from body fluids and exudates, and detected susceptibility to amikacin within 1 h in 12 of 13 positive specimens. Gant et al[2077], using a FACScan, reported being able to discriminate between the effects of different antibiotics based on morphologic (i.e., scatter signal) changes. Raponi et al[2261] noted changes in *E. coli* capsular morphology in response to low doses of antibiotics.

I found years ago[2075] that bactericidal antibiotics take away membrane potentials; in addition, some antibiotics, such as aminoglycosides, diminish membrane potential in susceptible organisms within 5-10 min. Using scatter and membrane potential probe fluorescence measurements **and counts** from my little red laser system, I could determine antibiotic sensitivity of several common species within 60-90 min. Other investigators reported similar results in sensitivity testing with commercial flow cytometers; Bercovier et al[857] used rhodamine 123 as the membrane potential probe, while Ordonez and Wehman[2263] used $DiOC_5(3)$.

Early enthusiasm for the technique was tempered by the realization that the need for multiple aliquots greatly increases the time needed to test a single clinical sample. If you are testing ten antimicrobial agents at two concentrations, you have to look at twenty aliquots of sample plus a control.

A flow cytometric system can get susceptibility information for one sample in an hour or so, but ten minutes of the hour will be used to run the multiple aliquots through the instrument. If you have a couple of dozen broth cultures, flow will get the first one done faster than the 96-well sensitivity tester, but the 96-well apparatus will finish the lot well ahead.

Enthusiasm was tempered, but not extinguished. Work on flow cytometric susceptibility tests continued, as summarized in reviews by Álvarez-Barrientos et al[3479], Davey et al[3488], and Walberg and Steen[3489].

Bacteria: Confusion Reigns

Until recently, most reported studies on bacteria and fungi measured only a single fluorescence parameter, assessing "viability" by membrane integrity, determined using fluorogenic substrates such as FDA or dyes such as propidium and Sytox Green, or by the presence of a membrane potential, determined with cyanines, oxonols, or rhodamine 123. For what it's worth, while the results of flow cytometric susceptibility tests of bacteria correlate with culture results, estimates of the fraction of viable bacteria by cytometry and from culture are typically different by at least a factor of five. The best correspondence I have ever seen was obtained in my lab, in studies initiated by Jared Silverman of Cubist Pharmaceuticals on the mechanism of action of the lipopeptide antibiotic daptomycin[3490]; viable counts estimated by ratiometric membrane potential were within a factor of two of plate counts. Not coincidentally, daptomycin's antimicrobial action appears to be intimately related to its rapid depolarizing action on the cell membrane.

As was noted on pp. 519-22, once people started looking at multiple functional parameters[2351,2610,3396-8], things became more complicated. Different antibacterial agents may have different effects on various functional parameters in different bacteria against which they are effective, meaning that no single functional parameter is optimal for susceptibility determination for all drugs and all species.

Counts still work. In most cases, one can get adequate counts of organisms in 96-well plates by determining turbidity (which is affected by the size as well as the number of organisms[3488]), or by using nucleic acid stains such as Pico green or tetrazolium dyes, and a flow cytometer is an expensive alternative. The "nanowell plate" approach I suggested on p. 534 may make for shorter incubation times than are needed using 96-well plates.

If one considers the plate count to be a "gold standard" for susceptibility testing and other viability determinations, it is important to remember that what is counted on a plate are colonies, and that **a colony forming unit (CFU) may be a single organism or an aggregate**. When nucleic acid stains or indicators of metabolism are used for the readout of a microwell assay, the signal is proportional to the number of organisms, not to the number of aggregates. If nothing is done to break up aggregates, the "events" detected by flow cytometry (and those sorted into plates, as illustrated in Figure 10-30, p. 521) may represent single cells or aggregates, and, if an appropriate cytometric indicator of "viability" is selected, flow cytometric data should agree well with plate counts, as was the case in our daptomycin studies[3490]. These were done primarily with *Staphylococcus aureus*, which forms clumps of cells. Other organisms, such as *Streptococcus pyogenes*, form long chains; disaggregation may be necessary before flow cytometry can be done. Mild sonication is the preferred technique[2610,3518].

The molecular biologists would like to do away with growing bacteria altogether, relying on detection of resistance genes in amplified DNA. Flow cytometry could play a part in that if multiplexed bead assays are used for the readout[3464], or if slow flow is used to read "fingerprints" of strains with known resistance characteristics based on DNA fragment size in digests[3018-20].

I still believe that multiparameter flow cytometry can play an important role in analyses of the bases of antibiotic resistance and in the development of new antimicrobial agents (p. 522). This has taken on a new urgency with the emergence of drug-resistant strains of organisms causing tuberculosis and pneumonia, among other diseases, and with increasing concern about bioterrorism. Flow cytometry can provide both direct and indirect measures of drug uptake by microorganisms, facilitating understanding of the permeability barriers and efflux mechanisms involved in drug resistance[3548-9]. This work will require running a lot of samples, and a lot of hazardous samples, at that.

Mycobacteria: Down for the Count

Speaking of hazardous samples, *Mycobacterium tuberculosis* is high on my list of organisms to avoid. A single individual working at Maine's Bath Iron Works managed to infect over 400 contacts with a fairly nasty strain of TB, and the bug is readily spread by coughing, sneezing, etc., unlike, say, HIV or Ebola virus.

While tuberculosis is a bad disease, the organism itself grows slowly enough *in vitro* so that it takes weeks to determine whether a culture is positive. This makes conventional susceptibility testing take a long time. Ronald Schell and his coworkers at the Wisconsin State Laboratory of Hygiene have developed several flow cytometric assays for antimicrobial susceptibility in mycobacteria[3491-4]. They initially[3491] used FDA as an indicator; this made it necessary to stain organisms, which is more of a problem in mycobacteria than in most other species, and to work with unfixed specimens, which scared most potential users off. However, they subsequently established[3494] that organisms in fixed specimens could be identified and counted based on their scatter characteristics, enabling the test to be done safely in most laboratories. Results are available in 72 hours, which classifies as rapid in this context.

Antifungal Susceptibility: Flow Does the Job

Previously, Green et al[2262] used scatter and propidium fluorescence to determine susceptibility of *Candida albicans* and other fungi to a variety of agents within 3 h. Much subsequent work on flow cytometric antifungal susceptibility tests has relied primarily on dye exclusion tests using propidium[3495-7]; other studies[3498-500] utilize Molecular Probes' dye **FUN-1**[2332], which localizes in vacuoles and imparts red fluorescence to metabolically active cells with intact membranes, while cells with damaged membranes exhibit diffuse yellow-green fluorescence. Acridine orange[3501] and membrane potential dyes[3502-3] have also been used as viability indicators. Perhaps because many antifungal agents act primarily on the cytoplasmic membrane, results of flow cytometric antifungal susceptibility tests of *Aspergillus* and *Candida* species and of *Cryptococcus neoformans* and *Torulopsis glabrata* correlate well with both the NCCLS M-27standard method[3497,3499-500,3502] and with clinical outcomes[3499]; results are available several days earlier using flow.

Antiviral Susceptibility by Flow Cytometry

Changes in the fraction of cells synthesizing viral antigens and in the patterns of expression of viral antigens have been used as the basis for tests of susceptibility of herpes simplex virus type 1[3504], cytomegalovirus[3505-7], and human herpesvirus 6[3508] to a variety of antiviral drugs; the flow cytometric assays are substantially less labor-intensive than conventional alternative methods.

From my point of view, if flow cytometry can make it anywhere in clinical microbiology, it will be in the area of bacterial detection, using inexpensive instruments and nucleic acid and/or membrane potential dyes. The technology is particularly well suited to urine microbiology, but the same machine that can work on urine can also handle cultures in liquid media. I've been saying all this for twenty years, and not much has happened. There are at least 10,000 potential sites in the U.S. in which such instruments could be placed, which represents a nice market; when you consider that there's a chance the instrument could be made inexpensive enough to put in a doctor's office, the market is even more attractive.

Or should be. BD has had a big share of the flow cytometry market and an even bigger share of the diagnostic microbiology market for decades, and never produced a flow-based microbiology product. They have invited me to pitch one to them on two occasions fifteen years apart; no sale. There are some smaller companies pursuing environmental and food microbiology (see Chapter 8), but it will probably take a big one to make a dent in the clinical area.

Cytometry in Vaccine Development

The earlier section on lymphocyte activation documented the present ability of flow cytometry to detect antigen-specific T lymphocytes and to determine their functional capacities. Prevention via immunization remains one of the best ways of dealing with infectious diseases, and active efforts are underway to develop and refine vaccines for relatively new agents, such as HIV, and for the granddaddy of them all, smallpox, extinct in the wild but still putting the terror in bioterrorism.

Using intracellular cytokine production as an indicator of CD8+ T lymphocyte response, Frelinger and Garba[3519] stimulated lymphocytes from unvaccinated and previously vaccinated individuals; while the levels of response diminished with time since vaccination, even those patients vaccinated more than 35 years prior to challenge showed some response.

Letvin et al[3509] discuss the prospects for developing HIV vaccines; their work to date with animal models indicates that vaccines can elicit protective antigen-specific cytotoxic T cells[3510], but also that a single nucleotide mutation in the virus can result in the death of an immunized animal[3511]. This outcome notwithstanding, papers by the Letvin group provide an excellent example of how to assess immune responses to vaccines.

Reed et al[3512] developed a flow cytometric bead assay for measuring the antibody response to the "protective" antigen of *Bacillus anthracis*; multiplexed immunoassays may facilitate demonstration of antibodies to multiple antigens from multiple pathogens, whether induced by prophylactic immunization or by infection[3513]. Cirino et al[3514] developed a flow cytometric assay to identify scFv antibody fragments that disrupted anthrax toxin binding, which are potentially useful in the treatment of the infection.

Geoffroy et al[3449] used GFP to enable monitoring of lactic acid bacteria used as live oral vaccine vectors, while Haidinger at el[3450] used GFP to facilitate removal of live bacteria from vaccines made from bacterial ghosts.

Microbiology Odds and Ends

Fouchet et al[2264] reviewed flow cytometric applications in the broad area of microbiology before the previous edition of this book was published. Some other older references[2265-7] show that dental researchers jumped on the flow microbiology bandwagon ahead of most of the rest of us. Then, there are some older and newer papers on opsonizing antibodies[2268-72] and bacterial interactions with cells[2273-4,3514-], including a few on my old friend *Helicobacter pylori* (yes, I've run some through my machine).

I still heartily recommend two excellent papers[2275-6] which together combine quantitative immunofluorescence, image analysis, and flow cytometry in direct demonstrations of anaerobic bacteria in stool. The subject matter may stink, but the methodology is superb. I might add that, thanks to papers like these and the work shown in Figure 10-27 (p. 518)[3395], it is no longer technically correct to look at bad immunophenotyping (Figure 10-2, p. 447) and say "This looks like crap"; we now know what crap looks like.

Parasitology

As long as flow cytometry remains an expensive technology, it is unlikely to find much use in clinical diagnosis of parasitic diseases, whatever its utility in research in this field.

There are probably nearly a billion people with malaria; most of them are poor and don't live in the United States, which has kept interest in malaria in this country focused on being able to treat the disease in American military personnel sent elsewhere.

By the time the previous edition of this book was written, flow cytometry had been used by several groups of investigators to demonstrate the presence in red blood cells of **intracellular parasites**, including various species of *Babesia* (babesiosis), *Plasmodium* (malaria), and *Trypanosoma* (African sleeping sickness and Chagas' disease)[290,577-81], and to study the cell cycle of the malaria parasite *Plasmodium falciparum*[937-8]. DNA fluorochromes had been shown to permit discrimination of parasitized from uninfected cells and, in some cases, purification of viable infected cells by sorting. Cells containing some parasites also could be identified by their higher apparent membrane potentials[290-1,939]. Jackson et al[940] applied fluorescence microscopy with an FDA/ethidium stain to determining "viability," i.e. membrane integrity, of intra- and extracellular *Leishmania*; the flow cytometric version of this assay has since been used with numerous species of parasites.

Flow cytometry can detect malaria parasitemia, using fluorescent dyes[2172-4], a cytochemical (Technicon H-1) hematology analyzer[2175], and polarized scatter measurements[2536-7] (p. 279). It has also been used to characterize the DNA content and growth kinetics[2176-9] of *Plasmodium* species and their effects on red cell antigens[2180-3], and to analyze the effects of drugs and other treatments[1557,1619,2184-6].

Flow cytometry was also used to study the DNA content[2188-92], growth[2193-7], surface characteristics[1481,2198-202,2214], and effects on the host[2203-5] of, and the effects of drugs on[1579-80,2206], various species of *Trypanosoma*. Growth, surface antigens, and host cell interactions of *Leishmania* species were also investigated[1421,2207-13].

Moving on to organisms which, at least occasionally, worry Americans, there was some flow cytometry done on intestinal parasites of the genera *Giardia*[2215-21,2224] and *Cryptosporidium*[2222-4]. Alderete and coworkers[2225-8] investigated the surface antigens of *Trichomonas vaginalis* and their relationship to infectivity; it appears that virulence of this organism is mediated by infection with a double-stranded RNA virus. Humphreys, Allman, and Lloyd[2229] used FDA/PI and an oxonol membrane potential dye in viability tests for *T. vaginalis*. *Pneumocystis carinii*, the causative organism of the pneumonia fatal to many people with HIV infection, also came under flow cytometric scrutiny[1420,2231-4].

There were a bunch of other papers on flow cytometry in parasitology[2235-49] that I included in the references to the Third Edition; at that point, I declined to "open that can of worms and protozoa." I will again beg off on a detailed discussion, and point you to some more recent references dealing with *Acanthamoeba*[3520-1], *Babesia*[3522], *Cryptosporidium*[3523-6] (also see p. 528), *Eimeria*[3527], *Encephalitozoon*[3528-9], *Giardia*[3530-1], *Leishmania*[3532-6], *Theileria*[3537], and *Trypanosoma*[3538] species.

Work on the biology and chemotherapy of malaria has continued[3539-45], and some progress seems to have been made toward development of a preventive vaccine[3546-7].

The first papers on detection of malaria parasitemia used the UV-excited Hoechst dyes and relatively large instruments[578-9]; Makler et al[771] took a step in the right direction by demonstrating that thiazole orange and the FACScan could be used to detect and analyze *Plasmodium* species in blood, and other recent papers used ethidium and propidium. That still keeps things in the realm of instruments that use blue-green or green excitation. However, I (unpublished) have managed to detect *P. falciparum* in parasitized red cells using oxazine 750 and a Cytomutt with a 7 mW red He-Ne laser, and this dye and others could be used in an inexpensive cytometer with a red diode laser as a light source[2447]. I don't know whether the billion victims are going to be any better off if such an instrument exists, but one can be made if it's needed. It can also be used for malaria research, for which funding doesn't seem to be lavish anywhere.

Meanwhile, you might want to run a copy of this section by the parasitologists and/or tropical medicine specialists in your institution, especially if they don't use your machine. Take them to lunch. Buy them acceptable beverages. You might gain friends and/or paying customers.

Pharmacology and Toxicology

Flow cytometric analysis of drug effects on cells *in vitro* has, for some time, been considered as an alternative to animal testing[2026].

Drugs and the Life and Death of Cells

Since some of the first users of fluorescence flow cytometry were involved in cancer diagnosis, research, and treatment, it is not surprising that much early work in both the United States and Europe used **DNA content analysis** to determine **effects of anticancer drugs on the cell cycle**. In the early 1970's, it seemed there was an unending litany of papers entitled "Effects of (your drug here) on cell cycle traverse in (your cell type)"; while, from today's point of view, few of those papers seem all that informative, flow cytometry has become steadily more valuable in the analysis of interactions between drugs and cells. Testing has grown more specific; where once one might have simply followed DNA content distributions over time to detect DNA damage, there are now, for example, tests based on dual-parameter DNA/protein[953] or DNA/RNA[954] analysis, or on detection of DNA denaturation[955-6]. Recent discussions of drugs and the cell cycle have been published by Shackney and Shankey[3550] and Traganos et al[3551].

In many cases, it is possible to study **drug uptake** directly, using fluorescent drugs such as the anthracyclines[352-3,814-8] or labeled analogs such as fluoresceinated derivatives of methotrexate[354-7]; this has been discussed on pp. 376-7. Charcosset et al[623] used flow cytometry to demonstrate that the uptake of a cationic drug, was as expected, dependent upon membrane potential; the particular drug studied was the fluorescent DNA intercalator *N*-methylellipticinium, which has been studied as an anticancer agent.

Even when direct assessment of drug uptake on a cell-by-cell basis is not possible, flow cytometry can readily provide quantitative analyses of the **effects of different drug doses on cells' metabolism** over time courses ranging from seconds to weeks, using the full range of parameters available.

The results of flow cytometric assays of drug effects on mammalian cell viability have not always correlated well with results obtained from culture. Poot et al[3552] report that culture of cells prior to drug treatment followed by assessment of proliferative survival by Hoechst 33342/ethidium detection of BrUdR incorporation after 72 h exposure to the label yields cytometric results comparable to those obtained from colony-forming assays.

In recent years, it has become appreciated that many cytotoxic drugs induce **apoptosis** in susceptible cells; Darzynkiewicz and his coworkers[3553-5] have described both drug effects and the methods for their study using both flow cytometry and laser scanning cytometry.

There is also interest in replacing established methods in **toxicology** by faster, more automated, flow cytometric methods. Zucker et al[2058-9] studied membrane effects of tributyltin, which were analyzed by both biochemical and flow cytometric assays. Yurkow et al[3556-7] used a specific antibody to monitor cellular levels of the protein of metallothionein, which increase following exposure to heavy metals. Effects of toxic agents on **spermatogenesis** can also be detected by flow cytometry[2060-2].

Erythrocyte Micronucleus Assays

Several flow cytometric adaptations have been described for the **bone marrow** and **blood erythrocyte micronucleus assays**[2063-8,3558-68]. These are standard tests for effects of **clastogenic agents.** It was noted on p. 453 that DNA damage due to such compounds or to radiation could be detected by increases in the CV of DNA histograms, but this requires carefully controlled, high-resolution flow cytometry[2941-2,3569]. While the nuclei of mammalian erythrocytes are normally extruded during maturation, cells exposed to clastogenic agents retain chromosome fragments, or micronuclei, which can be detected by appropriate staining.

Since the objective is to detect DNA-containing micronuclei in relatively immature red cells (reticulocytes), which normally contain RNA, it is necessary to discriminate between the two types of nucleic acid. The earlier versions of flow cytometric micronucleus assays used Hoechst 33342 or DAPI as DNA selective stains[2063-5], identifying reticulocytes by thiazole orange staining, and therefore required dual-laser flow cytometers with UV and 488 nm excitation, making it impractical to implement the assays on most fluorescence flow cytometers.

Beginning in the early 1990's, Torous and her colleagues[2066-8,3561-8] worked toward a robust micronucleus assay that can now be run on a single (488 nm)-laser flow cytometer. Reticulocytes are identified by staining with fluorescein-labeled anti-CD71 antibody; cells are fixed and stained with propidium after RNAse treatment, and malaria-parasitized erythrocytes are used as a DNA standard[3565]. This test proved to be reproducible in a multilaboratory study[3567]. A further improvement has been reported[3568] in which SYBR Green or SYTOX Green replaces propidium as the DNA stain and PE-labeled antibodies are used to detect CD71; a PE-Cy5 anti-glycophorin antibody is added to improve discrimination of erythroid cells.

Micronucleus assays for laser scanning cytometers have been described by Styles et al[3570] and Smolewski et al[3571].

Toxic Waste and B cell Proliferation

During the 1980's and early 1990's, the Centers for Disease Control determined peripheral blood lymphocyte immunophenotypes of approximately 900 individuals living near toxic waste cleanup (Superfund) sites and 600 controls; and discovered what appeared to be a higher incidence of B cell monoclonal lymphocytosis[3580] (BCML), which may or may not be a precursor of B cell chronic lymphocytic leukemia[3581], in the group near Superfund sites. A follow-up investigation confirmed the findings[3582]. The difference in incidence of BCML between the two populations was small enough so that it would not have been noticed in a study of fewer than 100 individuals; this should remind us that looking for small effects in small groups can be a (nontoxic) waste of time and money.

Radiation Dosimetry

Snopov et al[3572] used an antibody against thymine dimers to estimate the dose of radiation received by mononuclear leukocytes in irradiated blood. An older technique for estimating radiation dose from the frequency of mutations at the erythrocyte glycophorin A locus[3573-5] has been used to study survivors of the atomic bombing of Japan[3576-7] and the nuclear accident at Chernobyl[3578-9].

Food Science

Somatic Cell Counts in Milk

Counts of somatic (nucleated) cells in milk serve as the principal criterion for determining whether a cow's udder is healthy and, therefore, whether its milk is suitable for consumption. In most countries, such analyses are required by law; if the presence of mastitis is indicated by an elevated cell count, the milk cannot be sold.

The major problem in somatic cell counting in milk is discrimination of cells from fat droplets; Breer et al[950] showed that this could be done by adding acridine orange to a milk sample and analyzing the sample on a flow cytometer. A respectable number of Partec's smaller flow cytometers were sold to dairy concerns for somatic cell counting. When I first heard about this, it seemed to me that a flow cytometer was a bit pricey for this application; I was wrong. As it happens, the instruments then most widely used for somatic cell counting cost at least as much as flow cytometers. They were the **Fossomatic** series, manufactured by **Foss Electric A/S** (p. 431). In these instruments, a milk sample was diluted in buffer containing ethidium bromide, which stained nuclei; the suspension was then applied to a rotating disc, forming a thin layer of cells which passed under the objective of a fluorescence microscope with a PMT, fitted with an appropriate emission filter, placed in its image plane. Foss's newer Fossamatic 5000 instruments, and their competitors from **Bentley Instruments** in the U.S. (p. 430) and Delta Instruments bv in the Netherlands (p. 431), are flow cytometers. They are not intended to get 1 percent CV's on ethidium bromide fluorescence; all they have to do is discriminate stained nuclei from other stuff in the sample; and they work well enough so that there are hundreds in use.

Ostensson[2022] used a more conventional flow cytometer to follow total and differential leukocyte counts in blood and milk in experimental endotoxin-induced bovine mastitis, and Saad and Astrom[2023] examined estrogen effects on blood and milk leukocytes. Redelman et al[2024], using carboxydimethylfluorescein diacetate to label intact cells in milk, could identify five or more cell clusters on two-parameter displays of fluorescence vs. orthogonal scatter. Pillai et al[3583] and Rivas et al[3584] report that differential count information provides somewhat more information about udder health status than simple somatic cell counts; D'Haese et al[3585] examined a solid phase cytometer as an option for somatic cell counting and found it less effective than a Fossomatic instrument.

Bentley, Delta, and Foss also manufacture flow cytometric apparatus for counting bacteria in raw milk; this testing is, as far as I know, currently not mandatory in the U.S.

Seo et al[3586] reported that a rapid method combining flow cytometry and immunomagnetic bead separation followed by 6 h enrichment in culture could detect very small numbers of *E. coli* O157:H7 inoculated into raw milk, ground beef, and apple juice samples; as few as four organisms/g of ground beef could be found; Goodridge et al[3473] estimate detection sensitivity of 10^2 organisms for ml in milk using their flow cytometric bacteriophage assay.

Brewhaha

Jespersen, Lassen, and Jakobsen[2025] applied flow cytometry to detect wild yeast infections in breweries. After selective enrichment in culture for 48-72 h and staining with a fluorogenic substrate, they could find one wild yeast cell per 10^6 culture yeast cells. They probably used lagerithmic amplifiers.

A Loaf of Bread, a Jug of Wine...

Attfield et al[3587] showed that flow cytometry with oxonol dyes and propidium was useful for estimating the activity of reconstituted dry yeasts, which are widely used in both bread baking and winemaking. If your sourdough starter is a non-starter, check it out in your lab.

Malacrino et al[3588] found flow cytometry useful for monitoring populations of both yeast and the bacterium *Oenococcus oeni*, which improves taste by converting malic acid to lactic acid, in wine; Graca da Silveira et al[3589] used flow cytometry to monitor membrane integrity of *O. oeni*.

Resveratrol, a polyphenolic antioxidant found in red wine[3590-7], is reported to inhibit cell growth, including tumor cell growth, induce apoptosis in cultured leukemia cells, and, at high levels, decrease cell responses associated with inflammation. To your health!

Since cheese goes so well with wine, this is a good place to mention that Bunthof et al[3598] used Molecular Probes LIVE/DEAD BacLight kit[2332] with fluorescence and confocal microscopy and flow cytometry to monitor permeabilization of cheese starter bacteria.

Seeing the Blight

Phytophthora infestans, an important plant pathogen originally classified as a lower fungus but now thought to be closer to algae and diatoms, caused the great potato famine in Ireland in the mid-19th century; other species of the same genus cause root rot in squashes, cucumbers, peppers, and other species. Day et al[3599] report that Calcofluor White staining, flow cytometry and a sophisticated data analysis algorithm allow *Phytophthora infestans* sporangia to be discriminated from other airborne biological particles. We don't want fries with that.

Major Food Group: Chocolate

Last but not least in the food section, I come to a major reason I have kept such sanity as I have while generating this volume. According to Rein et al[3600-1], cocoa polyphenols inhibit platelet activation *in vitro*, with the effect also being detected after cocoa consumption. Time for a chocolate break.

Biotechniques and Biotechnology

This section is oriented toward industrial uses of flow cytometry, but includes a few tricks that didn't seem to fall under other headings I had. If you're interested, you might want to look at a 1996 book edited by Al-Rubeai and Emery, *Flow Cytometry Applications in Cell Culture*[2400]; a new book on *Flow Cytometry in Biotechnology*, edited by Larry Sklar, should be published by Oxford University Press in 2003 or 2004. There is a 2001 review article by Rieseberg et al on "Flow cytometry in biotechnology"[3602], but it has a lot of introductory material on flow cytometry and little detail on the applications.

In the 1970's, when people first got the notion that flow cytometry might be useful in the design and analysis of industrial processes involving prokaryotic and eukaryotic cells, the metabolic reactions involved were those that had evolved with the cells; today, the influence of recombinant DNA technology is pervasive. James Bailey[3603], who published some of the earliest papers on multiparameter flow cytometry of bacteria[553,558], was perhaps the first chemical engineer to appreciate what genetic manipulation could accomplish in the field he called "metabolic engineering," and combined flow cytometry and other experimental techniques with mathematical modeling to study the dynamics of bacterial and yeast populations. His students have continued to use flow cytometry for such applications as on-line monitoring of bioreactors[3604] and selection of yeast cells[3605] expressing antibodies with higher affinities than can be generated in the immune system[3606].

There seem to be only a few basic approaches to flow cytometry in biotechnology. Flow cytometers without sorting capability are used for **high throughput screening**, the objective being to look at a relatively simple interaction, such as ligand binding, in a maximum number of samples in a minimum time. Fast sample handling hardware (p. 365-6) has reduced the time needed to process a single sample to under two seconds; multiplexing may allow samples to be mixed before analysis, further increasing throughput.

Sorters, preferably high-speed sorters, are used to isolate the "best and the brightest" fraction of a percent of cells or beads from highly heterogeneous mixed populations, based on ligand binding and/or metabolic activity.

While the applications just mentioned typically do not require measurement of a large number of parameters, it is likely that successful on-line process monitoring will, although, once the right measurement parameters have been selected, it should be possible to get by with a lower analysis rate. A number of papers relevant to theoretical and practical aspects of process monitoring, dealing with bacteria, yeast, and eukaryotic cells, have appeared since the previous edition was published[3384,3396,3604,3607-20].

Protein and Gene Expression on Cells and Beads

While phage display (p. 348) provided the first widely used technology for generating and expressing combinatorial libraries of antibodies and other proteins, the more recent trend has been toward expression of engineered enzymes, antibodies, and receptors in bacteria and yeast[3621-32]. The antibodies, in particular, are expected to be useful in therapy as well as in diagnosis and research. Cell sorting seems to be the preferred mode for selection in this business, but simpler technologies have been reported to work[3606,3633-4].

Sydney Brenner and his colleagues[2328-9] have used sorting to identify differentially expressed genes by two-color analysis of oligonucleotide-bearing beads.

Getting Big Molecules into Small Cells

There are several ways of getting large molecules such as dextrans, proteins and DNA, or impermeant molecules, such as acid dyes, into cells; Lee et al[2051] used flow cytometry to evaluate three. The first was **cell fusion**, in this case, with red blood cell ghosts loaded with a fluorescent reporter molecule. The second technique, **osmotic lysis of pinosomes**, involved a brief exposure of cells to a hypertonic solution containing the reporter molecule; subsequently, a hypotonic media was added which lysed the pinosomes formed during hypertonic treatment. The third technique was **scrape loading**, which creates transient holes in the cell membrane, allowing reporter molecules to enter cells, by application of mechanical force. Osmotic lysis of pinosomes offers several advantages; it is simple, efficient (virtually all cells became fluorescently labeled), and enables larger amounts of material to be loaded more uniformly into cells while maintaining excellent viability.

Electroporation, in which pores in the membrane are transiently created by brief application of a strong electric field, is another widely used method for loading cells[1233-4,2052-6]. Weaver et al[2052-4,3635-7] have used flow cytometry and fluorescence microscopy with fluorescent reporter molecules of various sizes to analyze the dynamics of the electroporation process itself and the changes which occur in membrane transport characteristics.

Staying Alive, Staying Alive

Rodriguez and Lodish[3638] improved survival of attached cells following selection for gene expression and sorting into microwell plates by providing a drug-sensitive feeder layer of cells in each well which could be disposed of once sufficient growth of the sorted cells had taken place.

Katsuragi et al[3639] were searching for bacteria with high thiamine production, but the only assay to which they had access killed many of the bacteria. They were able to recover and culture high-producing strains by establishing colonies in microdroplets before performing the assay, since a small fraction of cells in the sorted microdroplets survived the procedure.

Zengler et al[3640] utilized the gel microdrop technique to culture and isolate "unculturable" bacteria from seawater and soil that require low nutrient concentrations to grow and would otherwise be overgrown by other organisms in the sample; their overall objective is identification of new sources of microbial metabolites.

Et Cetera

Kavanagh et al[965] developed flow cytometric assays to quantify **cell-to-cell communication** to facilitate studies of alterations in this process induced by tumor promoters. Cells are scrape-loaded with Lucifer yellow with or without rhodamine-labeled dextran; transfer of dye(s) between donor and recipient cells can be assessed by one- and two-color fluorescence flow cytometry.

Wilson, Mulligan, and Raison[2057] described an assay technique for antibodies to membrane-associated antigens using polyacrylamide microspheres coupled with cell (or organelle) membranes, which provide an easily stored, standardized antigen source usable in subsequent flow cytometric assays.

Nardelli, McHugh, and Mage[2069] developed a soluble macromolecular conjugation reagent, polyacrylamide-streptavidin (PASA), for the simplified preparation of multivalent protein-protein conjugates. Soluble linear polyacrylamide, with a molecular weight of approximately 10^6, is activated with carbodiimide and conjugated to approximately 20 streptavidin residues per molecule. PASA can bind biotinylated proteins to produce homo- or heteroconjugates of known composition; it was used to prepare multivalent antibody conjugates that could bind either of two antigenically distinct cell lines. The technique is also potentially useful for making immunotoxins, tumor labeling conjugates, and complex immunogens.

Vlieger et al[2070] described a flow cytometric assay for **quantitative analysis of polymerase chain reaction (PCR) products.** Magnetic beads coated with streptavidin are used to capture biotinylated PCR fragments, and analyzed by flow cytometry after hybridization with a hapten-labeled probe or by immunoenzymatic reactions. The method can detect fractions of femtomoles of product. An image analysis method using chemiluminescence was also described.

Alternatives: Microfluidic Cytometers, Flow and Static

In the early 1980's, I got the notion that it might be possible to build a "flow cytometer on a chip," incorporating integrated optics, fluidics, and electronics, and, with help from Mike Hercher, got as far as building a cytometer which used fiber optics (without lenses) for illumination and light collection[877].

My original inspiration was a gadget built by Jonathan Briggs et al[878], then at Syva Corporation. Their simple cytometer was essentially an epiilluminated system in which light from a He-Cd laser is coupled into an optical fiber with a 50 µm core. The fiber is dipped into a suspension of cells, or beads. The beam diverges from the end of the fiber, at an angle dictated by the N.A. of the fiber (pp. 157-8), resulting in the intensity falling off sharply with distance from the end of the fiber. The N.A. also limits the angle at which the fiber can collect light; the closer a particle emitting fluorescence is to the end of the fiber, the more light is collected. These two effects combine to limit the region in which substantial fluorescence can be excited in, and detected from, particles in the suspension to a volume within 50 µm or so of the tip of the fiber. When the solution is stirred, every time a fluorescent cell or bead passes through this volume, a fluorescence pulse can be detected, using a dichroic at the other end of the fiber to separate the excitation and emission wavelengths and appropriate optics to divert the emission signal to a PMT. Because of variability in signal intensities from particles at different distances from the fiber, the CV's were fairly large, around 8%; the single-fiber system also couldn't measure scatter. However, by bundling three fibers together, putting excitation down the middle fiber and collecting from the two outside, it is possible to restrict signal processing to events in which the two detector fibers produce almost equal signals, which reduces the CV. Since fibers can carry light over distances of miles without substantial losses, it would, for example, be feasible to keep the laser, PMT, electronics, etc., comfortably housed on shipboard, and (almost) deep six the business end of a fiber cytometer. Rob Olson was talking about building such a gadget, but he apparently went straight to a submersible cytometer.

Just before the previous edition was published, flow cytometry on a chip, if not on a ship, was resuscitated by Dan Sobek, Martha Gray, and Steve Senturia[2309], of MIT, who described experiments with a sheath flow system microfabricated in fused silica. Luckily, they didn't try to build in integrated optics; it has since become apparent that microfluidic cytometers work a lot better with conventional optics (see p. 158).

A newer generation of microfluidic flow cytometers have been built in plastic. **Agilent**'s commercial instrument (pp. 429-30) appears to incorporate molded parts. **Micronics** builds their cytometers by gluing layers of plastic together, while the Quake group at Caltech[2326-7,2509,2512-4] (and its commercial offshoot, **Fluidigm**), Whitesides et al at Harvard[2511], and Takayama et al at the University of Michigan[252=15] mold systems from silicone elastomers. Microfluidic cytometers are small, and can be inexpensive; they are not fast.

On the plus side, the Quake/Fluidigm systems can incorporate hundreds of pumps and valves on a single chip, and control them with no more than a few dozen inputs[2514]. Flow can be stopped and reversed. These devices might better be thought of as "intelligent microtiter (or nanotiter) plates" than as flow cytometers. They are not suited for high throughput screening, but, if cells selected in a high throughput screening process are sorted into a microfluidic chip, they can be grown on the chip, subjected to repeated perturbation and "high content" analysis using imaging, confocal, or laser scanning microscopy. While microfluidic sorting is slow, it doesn't generate aerosols, and may be the best way to deal with hazardous bacteria and viruses.

I don't know about cheap confocal microscopy, but the small, inexpensive imaging devices described on pp. 448 and 492-3, and scanning hardware such as is used in **Immunicon**'s "CellTracks[TM»2383,3006-9]" (p. 492) seem well matched to the intelligent microfluidic chip. I suspect that a book called *Practical Slow Cytometry* is waiting to be written.

Cytometry Afield

The Lymphocytes of the Long Distance Runner

Haq et al[2292] reported reductions in $CD3^+$ cells following the running of a marathon; they felt the changes were attributable to increases in serum cortisol. Gore et al[3685] can detect illicit use of erythropoietin by runners and other athletes using a combination of flow cytometric erythrocyte measurements and biochemical assays.

War and Peace

Not far from Marathon itself, prisoners of war liberated from a camp in Bosnia were found by Dekaris et al[2293] to have increases in activated T cells, a decrease in $CD4^+/CD8^+$ ratio, and lowered serum cortisol, along with general nutritional changes. The same group[3641] also reported increases in activated T, B, and NK cells in women displaced from their homes by war. Skarpa et al[3642] found increased numbers of

CD16+ and perforin-containing lymphocytes in patients with post-traumatic stress disorder, with the highest levels in those who had been tortured while in concentration camps.

Griffiths et al[3643], at the U.K.'s Biological Defence establishment at Porton Down, examined the effects of combinations of vaccines and the nerve gas antagonist pyridostigmine bromide on the general health and cellular immune status of guinea pigs, attempting to determine whether the treatment might cause symptoms of what will undoubtedly soon be described as "Gulf War I Syndrome." Although animals did show immune responses to the vaccines, neither health nor immune function appeared to be adversely affected.

Biselli et al[3644] reported a stress-related decrease in the fraction of CD29+ CD8+ lymphocytes in military student pilots. Gruzelier et al[3645], studying medical students at exam time, found that self-hypnosis mitigated the effects of stress on lymphocyte subpopulations, while Ruzyla-Smith et al[3646] reported that hypnosis only altered immune response significantly in highly hypnotizable subjects. I'm mesmerized.

Blood, Sweat, and Tears?

In addition to blood, semen, and urine, fluorescence flow cytometers have been used to analyze cells from nasal secretions[2292], saliva[2293], cerebrospinal fluid[2294], effusions[2295], stool[2296], and tears[3022], by people who looked for and got relevant information from these specimens. No sweat yet, although there are papers on effects of exercise on the immune system[3647] and on DNA content of benign and malignant tumors of sweat glands[3648].

Pulp Nonfiction

A few years back, somebody sent me an e-mail expressing a great deal of enthusiasm for using flow cytometry to characterize particles in process waters from the paper industry. I can't find it. However, just a couple of weeks ago, I got another e-mail on the same topic from Lari Vähäsalo in Finland, who reached the same conclusion and, to judge from a recently published paper[3649], an even greater level of enthusiasm.

Flow Cytometry On the Rocks

Acritarchs are organic-walled microfossils widely distributed in sedimentary rocks. Although different types of acritarchs have been classified on the basis of morphology, it has been difficult to perform chemical and biochemical analysis on individual subpopulations. Moldowan and Talyzina and colleagues[3674-5] used a conventional droplet sorter, measuring forward and side scatter and fluorescence, to separate different classes of acritarchs for subsequent biochemical analysis and electron and fluorescence microscopy. Until recently, geologists and paleontologists may have thought "FACS" stood for "Fluorescent Acritarchs in Cambrian Sediments"; now they've got the real dirt on the subject, and may find more uses for the technology.

To Boldly Go Where No Cytometer Has Gone Before

As I noted on p. 432, the flow cytometers from **NPE Systems** trace their ancestry to the RATCOM instrument that was supposed to have been used on the International Space Station[2911-4]. Changes in behavior of various leukocyte populations observed after space flight[3650-54], and concern that conditions encountered in space could not be simulated adequately in an earthbound laboratory[3655], provided the rationale for placing a cytometer in space; work on the design of such instruments and the necessary support systems has continued[3656-9].

Partec has just announced that its new **CyFlow Space** instrument is to be used on the Space Station. I was involved in the original cytometer-in-space project, and suggested that a scanning laser cytometer might be more appropriate than a flow cytometer; I haven't changed my mind. I should have designed and built the machine, but I just wasn't enterprising enough.

Meanwhile, back on the home planet, the Defense Advanced Research Projects Agency (DARPA) decided that it would be possible to detect early effects of biowarfare agents by having soldiers in the field wear small flow cytometers, which would perform leukocyte counts at appropriate intervals, and funded **Honeywell** and **Micronics** to develop a microfluidic system for the purpose. This could give "present arms!" a whole new meaning.

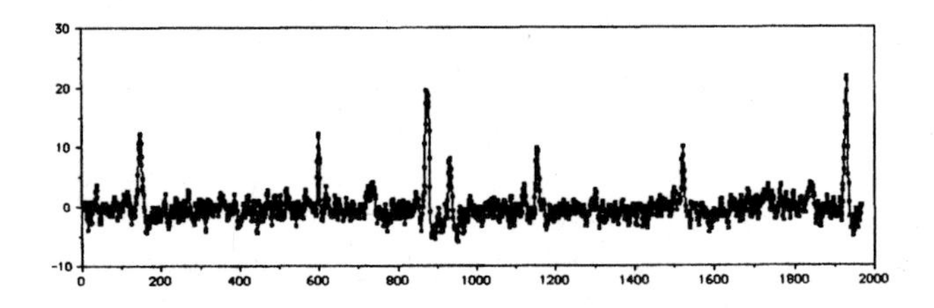

Figure 10-38. Flow cytometry *in vivo*. Courtesy of Charles Lin (Massachusetts General Hospital/Harvard Medical School).

From my point of view, however, the boldest new venture in flow cytometry is one that deals with inner, not outer, space. Figure 10-38 shows a trace of fluorescence pulses from leukocytes stained with a Cy5-labeled antiCD45 antibody and excited with a red laser. It looks pretty ordinary; the detector used was a conventional PMT, and cells were observed in a capillary. Well, not exactly a capillary, it was a venule in a mouse ear, and the mouse was alive, although not kicking very hard. Charles Lin and his coworkers at the Massachusetts General Hospital built the apparatus (for more information see <http://www.massgeneral.org/wellman/faculty/lin/profile.asp>), and we may all be able to learn something from it. *In vivo veritas.*

11. SOURCES OF SUPPLY

This chapter lists firms and organizations that sell or provide various services, supplies and items of equipment for which users and builders of cytometers have recurring needs. I have dealt personally with many of the listed companies; I haven't intentionally excluded anybody. The list that follows here picks up items in approximately the order in which they are discussed in the book. Addresses of U. S. branches, if any, are given for companies headquartered outside the U.S.

11.1 RESOURCES, SOCIETIES, JOURNALS

National Flow Cytometry Resource
Bioscience Division, M-888
Los Alamos National Laboratory
Los Alamos, NM 87545
Telephone (505) 667-1623
Fax (505) 665-3024
http://lsdiv.lanl.gov/NFCR/

International Society for Analytical Cytology (ISAC)
60 Revere Drive, Suite 500
Northbrook, IL 60062-1577
Phone (847) 205-4722
Fax (847) 480-9282
www.isac-net.org
E-mail: ISAC@isac-net.org

Cytometry (incorporating *Bioimaging*)
Editor-in-Chief: Charles L. Goolsby, Ph.D.
Northwestern University Medical School
Department of Pathology
Ward Building 6-204
303 East Chicago Avenue
Chicago, IL 60611-3008
Phone (312) 503-1847
Fax (312) 503-1848
E-mail: cytometry@nwu.edu

Clinical Cytometry Society (CCS)
www.cytometry.org
P.O. Box 25456
Colorado Springs, CO 80936-5456
Shipping Address:
5610 Towson View
Colorado Springs, CO 80918
Phone (719) 590-1620
Fax (719) 590-1619
Business E-mail: admin@cytometry.org

NCCLS
940 West Valley Road, Suite 1400
Wayne, PA 19087-1898
Phone (610) 688-0100
Fax (610) 688-0700
http://www.nccls.org
NCCLS used to stand for "National Committee for Clinical Laboratory Standards," but it is now a global organization that develops consensus documents for audiences beyond the clinical laboratory community.

11.2 OPTICAL SUPPLY HOUSES

The following supply tables, mounts, light sources, lenses, mirrors, filters, photometric apparatus, fiber optics, and other stuff besides. Their catalogs are educational as well as informative.

Coherent, Inc. (formerly Ealing Electro-Optics Inc.)
5100 Patrick Henry Drive
Santa Clara, CA 95054
Phone (408) 764-4000
Fax (408) 764-4800
http://www.coherentinc.com
A full line of optics; good buys on cylindrical lenses.

Edmund Industrial Optics
101 E. Gloucester Pike
Barrington, NJ 08007-1380
Phone (800) 363-1992
Fax (856) 573-6295
http://www.edmundoptics.com
A comprehensive line of optics and optical instruments.

Kinetic Systems, Inc.
20 Arboretum Road
Boston, MA 02131
Phone (617) 522-8700 or (800) 992-2884
Fax (617) 522-6323
http://www.kineticsystems.com
Optical tables and other vibration isolation equipment.

Linos (formerly Spindler & Hoyer)
459 Fortune Blvd.
Milford, MA 01757
Phone (508) 478-6200
Fax (508) 478-5980
http://www.linos.comportal/en/index.html

Melles Griot
16542 Millikan Avenue
Irvine, California 92606
Phone (949) 261-5600 or (800) 835-2626
Fax (949) 261-7589
http://www.mellesgriot.com
I particularly like their (inexpensive!!) microscope objectives, mirrors, glass lenses, and beamsplitters.

Newport Corporation
1791 Deere Ave.
Irvine, CA 92606
Phone (949) 863-3144 or (800) 222-6440
Fax (949) 253-1680
http://www.newport.com
A complete line of optical tables, breadboards, and mounts.

OptoSigma
2001 Deere Avenue
Santa Ana, CA 92705
Phone (949) 851-5881
Fax (949) 851-5058
http://www.optosigma.com

Rolyn Optics Company
706 Arrowgrand Circle
Covina, CA 91722
Phone (626) 915-5707
Fax (626) 915-1379
http://www.rolyn.com

Siskiyou Design Instruments
110 S.W. Booth Street
Grants Pass, OR. 97526
Phone (877) 313-6418
Fax (541) 479-3314
http://www.sd-instruments.com
Optical platforms, micromanipulators, other hardware for life sciences applications and [confocal] microscopy

Technical Manufacturing Corporation
15 Centennial Drive
Peabody, MA 01960
Phone (800) 542-9725 or (978) 532-6330
Fax (978) 531-8682
http://www.techmfg.com
Optical tables, breadboards, and isolators.

Thermo Oriel
150 Long Beach Blvd.
Stratford, CT 06615
Phone (203) 377-8282
Fax (203) 378-2457
http://www.oriel.com
Oriel is a good source for arc lamp systems and power supplies, among other things.

Thorlabs, Inc.
435 Route 206 North
Newton, NJ 07860
Phone (973) 579-7227
Fax (973) 300-3600
http://www.thorlabs.com
Small breadboards and mounts, optics, diode detectors, and lasers. Will also do reasonably priced custom machining with a short turnaround time.

11.3 PROBES AND REAGENTS

Linscott's Directory (http://www.linscottsdirectory.com), available in hard copy and online, can be useful for finding vendors of antibodies and other biological reagents.

Accurate Chemical and Scientific Corp.
300 Shames Dr.
Westbury, NY 11590
Phone (516) 333-2221 or (800) 645-6264
Fax (516) 997-4948
http://www.accuratechemical.com
Some monoclonal antibodies; good collection of photosensitizing dyes (cyanines, oxonols, etc.).

AdvanDx, Inc.
222 Partridge Lane
Concord, MA 01742
Phone (781) 405-1654
http://www.advandx.com
PNA probes for microorganisms.

Aldrich Chemical Company, Inc.
1001 West Saint Paul Ave.
Milwaukee, WI 53233
Phone (800) 325-3010 or (314) 771-5765
http://www.sigmaaldrich.com
Dyes and other chemicals.

Amersham Biosciences, Inc.
800 Centennial Avenue
P.O. Box 1327
Piscataway, NJ 08855-1327
Phone (732) 457-8000
Fax (732) 457-0557
http://www.apbiotech.com
Biochemicals, CyDyes, and more.

Applied Biosystems
850 Lincoln Centre Drive
Foster City, CA 94404
Phone (650) 638-5800 or (800) 327-3002
Fax (650) 638-5884
http://www.appliedbiosystems.com
Custom PNA Probe Service.

Beckman Coulter
11800 S.W. 147th Avenue
Miami, FL 33196
Phone (800) 526-3821
Fax (800) 232-3828
http://www.beckmancoulter.com
Monoclonal antibodies, flow cytometry controls, beads, immunoassay products.

BD Biosciences/BD Immunocytometry Systems
2350 Qume Dr.
San Jose, CA 95131-1807
Phone (877) 232-8995 (Ordering information)
Phone (800) 223-8226
Fax (408) 954-2347
http://www.bdbiosciences.com
Monoclonal antibodies, flow cytometry controls, beads, bead assays.

BioErgonomics, Inc.
4280 Centerville Road
St. Paul, MN 55127
Phone (800) 350-6466
Fax (888) 810-7189
http://www.bioe.com
Reagents for stem cell and T-cell isolation, cell stabilization and activation reagents.

Biomeda Corp.
P. O. Box 8045
Foster City, CA 94404
Phone (800) 341-8787
Fax (510) 783-2299
http://www.biomeda.com
Biomeda are specialists in phycobiliprotein probes.

Biosource International
542 Flynn Road
Camarillo California USA 93012
Phone (800) 242-0607
http://www.biosource.com
Antibodies, cytokines, peptides and custom products.

Boehringer Mannheim (see Roche Diagnostics Corporation)

Calbiochem-Novabiochem Corp.
10394 Pacific Center Ct.
San Diego, CA 92121
Phone (800) 854-3417
Fax (800) 776-0999
http://www.calbiochem.com
Antibodies, Biochemicals.

Caltag Laboratories
1849 Bayshore Blvd. #200
Burlingame, CA 94010
Phone (650) 652-0468 or (800) 874-4007
Fax (650) 652-9030
http://www.caltag.com
Monoclonal and polyclonal antibodies; avidin and streptavidin conjugates utilizing tandem dyes, "fix and perm" kits.

Cell Signaling Technology, Inc.
166B Cummings Center
Beverly, MA 01915
Phone (978) 921-6216 or (877) 616-CELL
Fax (978) 922-7069
http://www.cellsignal.com
Antibodies to kinases and phosphoproteins.

CHEMICON International, Inc.
28820 Single Oak Drive
Temecula, CA 92590
Phone (800) 437-7500 or (909) 676-8080
Fax (800) 437-7502 or (909) 676-9209
http://www.chemicon.com
Monoclonal antibodies and custom services.

Chromaprobe, Inc.
400 Brooktree Ranch Rd.
Aptos, Califonia 95003
Phone (888) 964-1400
Fax (831) 688-3600
http://www.chromaprobe.com
Monoclonal antibodies and custom antibody services.

Clontech (now part of BD Biosciences)
BD Biosciences Clontech
1020 East Meadow Circle
Palo Alto, CA 94303
Phone (877) 232-8995
http://www.clontech.comindex.shtml
Fluorescent protein vectors.

DakoCytomation California Inc.
6392 Via Real
Carpinteria, CA 93013
Phone (805) 566-6655 or (800) 235-5743
Fax (805) 566-6688
http://us.dakocytomation.com
Monoclonal antibodies, lysing reagent kits, DNA kits, beads, Medimachine for tissue dissagregation.

Diatec.com AS
Gaustadalleen 21
0349 Oslo
Norway
Phone (47) 22 95 86 25
Fax (47) 22 95 86 49
http://www.diatec.com
Monoclonal antibodies.

EBioscience
5893 Oberlin Dr. Suite #106
San Diego, CA 92121
Phone (888) 999-1371 or (858) 642-2058
Fax (858) 642-2046
http://www.ebioscience.com
Mouse and human monoclonals, cytokines and kits.

Enzyme Systems Products
486 Lindbergh Avenue
Livermore, CA 94550
Phone (888) 449-2664 or (925) 449-2664
Fax (925) 449-1866
http://www.enzymesys.com
Chromogenic and fluorogenic substrates; their catalog contains a large bibliography on this topic.

Exalpha Biologicals, Inc.
86 Rosedale Road
Watertown, MA 02472
Phone (800) 395-1137 or 617-924-3400
Fax (866) 924-5100 or (617) 924-5100
http://www.exalpha.com/biologicals/
Monoclonal antibodies and assay kits for flow cytometry.

Exciton Chemical Company, Inc.
P. O. Box 31126
Overlook Station
Dayton, OH 45437
Phone (937) 252-2989
Fax (937) 258-3937
http://www.exciton.com
Laser dyes.

Flow-Amp Systems, Ltd.
11000 Cedar Avenue
Cleveland, OH 44106
Phone (216) 721-0590
Fax (216) 721-1917 http://www.flow-amp.com
Tyramide amplification technology for flow cytometry.

Immunochemistry Technologies, LLC
9401 James Avenue South, Suite 155
Bloomington, MN 55431
Phone (952) 888-8788 or (800) 829-3194
Fax (952) 888-8988
http://www.immunochemistry.com
FAM and SR fluorescent caspase inhibitors.

IQ Products
IQ Corporation NV
Rozenburglaan 13a
9747 AN Groningen
The Netherlands
Phone (31) 0 50 5757 000
Fax (31) 0 50 5757 001
http://www.iqproducts.nl
Antibodies, cytokines, kits for flow cytometry.

Jackson ImmunoResearch Laboratories, Inc.
P.O. Box 9
872 W. Baltimore Pike
West Grove, PA 19390-0014
Phone (610) 869-4024 or (800) 367-5296
Fax (610) 869-0171
http://www.jacksonimmuno.com
Secondary antibodies conjugated to a wide range of fluorophores.

Martek Biosciences Corporation
6480 Dobbin Road
Columbia, MD 21045
http://www.martekbio.com
Superfluors™, SensiLight™, and CryptoLight™ dyes from algae, also fluorescent-labeled antibodies and streptavidin conjugates.

Molecular Probes, Inc.
4849 Pitchford Ave.
Eugene, OR 97402
Phone (541) 465-8300
Customer Service (541) 465-8338
Technical Assistance (541) 465-8353
Fax (541) 344-6504
http://www.molecularprobes.com

If it fluoresces, Molecular Probes is apt to sell it. Home of Texas red (they moved since it was synthesized). Dick Haugland keeps and supplies extensive bibliographies on just about every molecule he stocks, and the *Handbook of Fluorescent Probes and Research Chemicals*[2332], Molecular Probes' catalog, available at their Web site, is must reading.

Novocastra Laboratories – distributed by Vector Laboratories in the U.S., U.K., and Canada
30 Ingold Rd.
Burlingame, CA 94010
Phone (650) 697-3600 or (800) 227-6666
Fax (650) 697-0339
http://www.novocastra.co.uk
Antibodies.

OncoImmunin, Inc.
207A Perry Parkway, Suite 6
Gaithersburg, MD 20877
Phone (301) 987-7881
Fax (301) 987-7882
http://www.phiphilux.com
PhiPhiLux® fluorogenic caspase substrates for apoptosis and other fluorogenic enzymatic assays.

ORPEGEN Pharma
Gesellschaft für biotechnologische Forschung, Entwicklung und Produktion m.b.H.
Czernyring 22
D-69115 Heidelberg
Germany,
Phone +49 6221 9105-0
Fax +49 6221 9105-10
http://www.orpegen.com
BASOTEST for basophil degranulation

PanVera LLC
Discovery Center
501 Charmany Drive
Madison, WI 53719 USA
Phone (608) 204-5000 or (800) 791-1400
Fax (608) 204-5200
http://www.panvera.com
Aurora Biosciences has merged into PanVera, a subsidiary of Vertex Pharmaceuticals. Beta-lactamase reporter gene, antibodies and fluorescence based assays.

Pharmingen (now part of BD Biosciences)
BD Biosciences
10975 Torreyana Road
San Diego, CA 92121
Phone (877) 232-8995
Fax (858) 812-8888
http://www.bdbiosciences.compharmingen/
Monoclonal antibodies.

Phoenix Flow Systems
11575 Sorrento Valley Rd., Suite 208
San Diego, CA 92121
Phone (800) 886-FLOW or (858) 453-5095
Fax (858) 259-5268
http://www.phnxflow.com
Apotosis and cell proliferation kits.

Pierce Biotechnology, Inc.
P. O. Box 117
Rockford, IL 61105
Phone (800) 874-3723
Fax (800) 842-5007
http://www.piercenet.com
Cross-linking reagents.

Prozyme
1933 Davis Street, Suite 207
San Leandro, CA 94577-1258
Phone (800) 457-9444 or (510) 638-6900
Fax (510) 638-6919
http://www.prozyme.com
R-PE and APC conjugation kits and reagents

Polysciences, Inc.
400 Valley Rd.
Warrington, PA 18976-9990
Phone (800) 523-2575 or (215) 343-6484
Fax (800) 343-3291
http://www.polysciences.com
Dyes, calibration particles, enzyme substrates, fixatives, EM supplies, lectins, and contract R&D.

Quantum Dot Corp.
26118 Research Road
Hayward, CA 94545
Phone (510) 887-8775
Fax (510) 783-9729
http://www.qdots.com
Quantum dot nanocrystals label biomolecules and beads.

R&D Systems
614 McKinley Place N.E.
Minneapolis, MN 55413
Phone (612) 379-2956 or (800) 343-7475
Fax (612) 656-4400
http://www.rndsystems.com
Kits for fCM analysis of cytokine receptors.

Research Organics
4353 East 49th St.
Cleveland, OH 44125
Phone (800) 321-0570
Fax (216) 883-1576
http://www.resorg.com
Buffers, enzyme substrates, fluorescent labels.

Riese Enterprises
BioSure Division
12301 Loma Rica Drive, Suite G
Grass Valley, CA 95945-9355
Phone (916) 273-5095 or (800) 345-2267
Fax (916) 273-5097
http://www.biosure.com
BioSure carries flow cytometry sheath and staining solutions.

Roche Diagnostics Corporation
Roche Applied Science
P.O. Box 50414
9115 Hague Road
Indianapolis, IN 46250-0414
Phone (800) 428-5433
Fax (800) 428-2883
http://biochem.roche.com
Biochemicals, antibodies, reagents for apoptosis, cytotoxicity and cell proliferation.

Serologicals Corporation
5655 Spalding Drive
Norcross, GA 30092
Phone (678) 728-2000 or (800) 842-9099
Fax (678) 728-2247
http://www.serologicals.com
Intergen fluorescent (Martek) labels and caspase inhibitors.

Serotec Inc.
3200 Atlantic Ave., Suite 105
Raleigh, NC 27604
Phone (919) 878-7978 or (800) 265-7376
Fax (919) 878-3751
http://www.serotec.com
Antibodies.

Sigma-Aldrich Chemical Co.
P. O. Box 14508
St. Louis, MO 63178
Phone (800) 325-3010 or (314) 771-5765
Fax (800) 325-5052
http://www.sigmaaldrich.com
Biochemicals and organics.

Vector Laboratories
30 Ingold Road
Burlingame, CA 94010
Phone (650) 697-3600
Phone (800) 227-6666 (Ordering Information)
Fax (650) 697-0339
http://www.vectorlabs.com
Antibodies, lectins, biotin-avidin reagents.

Worthington Biochemical Corporation
730 Vassar Ave
Lakewood, NJ, 08701
Phone (732) 942-1660 or (800) 445-9603
Fax (800) 368-3108
http://www.worthington-biochem.com
Enzymes.

Zymed Laboratories, Inc.
561 Eccles Avenue
South San Francisco, CA 94080
Phone (800) 874-4494
Fax (650) 871-4499
http://www.zymed.com
Antibodies and immunochemicals.

11.4 CALIBRATION PARTICLES/CYTOMETRY CONTROLS

Bangs Laboratories, Inc.
9025 Technology Drive
Fishers, IN 46038-2886
Phone (317) 570-7020 or (800) 387-0672
Fax (317) 570-7034
http://www.bangslabs.com
MESF and antibody-binding beads, beads.

BioCytex
140, Ch. de l'Armee d'Afrique
13010 Marseille France
Phone (33) 4 96 12 20 40
Fax (33) 4 91 47 24 71
http://www.biocytex.fr
Quantitative flow cytometry kits utilizing calibrated bead suspensions.

BD Biosciences (see section 11.3 for address)
Flow cytometry control particles.

Beckman Coulter Corporation (see section 11.3 for address)
Flow cytometry control particles.

DakoCytomation (see section 11.3 for address)
Quantitative ImmunoFluorescence Indirect flow cytometry assay kit (QIFIKIT®) developed by BioCytex.

Duke Scientific Corporation
2463 Faber Place
Palo Alto, CA 94303
Phone (650) 424-1177 or (800) 334-3883
Fax (650) 424-1158
http://www.dukescientific.com

Flow Cytometry Standards Corporation – now part of Bangs Laboratories (see Bangs Laboratories, Inc. above)

Molecular Probes, Inc. (see section 11.3 for address)
Polysciences, Inc. (see section 11.3 for address)
Particles for flow cytometry alignment.

Riese Enterprises, BioSure Division (see section 11.3)
Fixed nuclei and red blood cells for flow cytometry controls.

Spherotech, Inc.
1840 Industrial Dr. Suite 270
Libertyville, IL 60048-9817
Phone (800) 368-0822 or (847) 680-8922
Fax (847) 680 8927
http://www.spherotech.com
Rainbow quantitative fluorescent beads; other dyed beads.

Streck Laboraties, Inc.
7002 S. 109th St.
La Vista, NE 68128
Phone (800) 228-6090
Fax (402) 333-6017
http://www.streck.com
Hematology and immunology stabilized control cells.

Seradyn
7998 Georgtown Road, Suite 1000
Indianapolis, IN 46268
Phone (800) 428-4007
Fax (317) 610-3888
http://www.seradyn.com
Beads.

11.5 FLOW CYTOMETERS (see Chapter 8)

Advanced Analytical Technologies, Inc.
2901 S. Loop Drive, Suite 3300
Ames, IA 50010
Phone (515) 296-6600
http://www.aati-us.com
RBD2100 flow cytometer for bacterial detection in industrial applications.

Agilent Technologies, Inc.
2850 Centerville Road
Wilington, Delaware 19808
Phone (800) 227-9770
http://www.agilent.comchem/labonachip
Model 2100 Bioanalyzer "Lab-on-a-Chip" performs cell fluorescence assays as well as chemical assays.

Apogee Flow Systems
Head Office:
25 Ross Way
Northwood, Middlesex HA6 3HU
U. K.
Phone +44 1923 842340
Fax +44 1923 842797
E-mail: sales@ApogeeFlow.com
http://www.ApogeeFlow.com
Factory and Laboratory:
Butlers Land Farm
Mortimer, ReadingRG7 2AG
U.K.
Phone +44 1923 842340
Supports Bio-Rad's Bryte-HS instruments, and makes the A10, A20, and A30 flow cytometers

BD Biosciences (see section 11.3 for address)
FACSVantage SE and FACSAria cell sorters; FACSCalibur BD LSR II, FACSCount, FACSArray flow cytometers.

Beckman Coulter (see section 11.3 for address)
Cytomics™ FC 500, XL, XL-MCL flow cytometers and ALTRA™ cell sorter.

*Bio*DETECT AS
Olav Helsets vei 6
P.O.B.150 Oppsal
N-0619 Oslo
Norway
Phone +47 22 62 70 80
Fax +47 22 62 72 75
http://www.biodetect.biz
*Bio*DETECT, Inc.
2500 City West Blvd,
Suite 300,
Houston TX 77042
Phone (713) 267-2300
Fax (713) 267-2267
E-mail: USA@biodetect.biz
MICROCYTE® compact flow cytometer for detection and identification of microorganisms (developed by Optoflow AS (http://www.optoflow.com)).

Bentley Instruments
4004 Peavey Road
Chaska, MN 55318 USA
Phone (952) 448-7600
Fax (952) 368-3355
http://www.bentleyinstruments.com
Somacount and Bactocount flow cytometers for somatic cell and bacteria counting in milk.

Chemunex SA
Immeuble 'Paryseine'
3 allee de la Seine
94854 Ivry-sur-Seine Cedex
Paris, France
Phone 33 (0) 1 49 59 20 00
Fax 33 (0) 1 49 59 20 01
Chemunex USA
1 Deer Park Drive
Suite H2
Monmouth Junction
NJ 08852
Phone (732) 329-1153 or (800) 411-6734
Fax (732) 329-1192
http://chemunex.com

Chemunex makes the D-count® flow cytometer and ChemScan RDI® imaging cytometer systems for microbial detection in industrial applications.

CytoBuoy b.v.
Zeelt 2
2411 DE Bodegraven
The Netherlands
Phone 31 (0) 348 688 101
Fax 31 (0) 348 688 707
http://www.cytobuoy.com
Cytobuoy flow cytometers, designed for *in situ* analysis of phytoplankton in natural waters.

Cytopeia
12730 28th Ave NE
Seattle, WA 98125
Phone (206) 364- 3400
http://www.cytopeia.com
Custom built cell sorters (InFlux platform).

DakoCytomation
4850 Innovation Drive
Fort Collins, CO 80525
Phone (800) 822-9902
Fax (970) 226-0107
http://www.cytomation.com
MoFlo® cell sorter, CyAn™ flow cytometer.

Delta Instruments bv
P.O. Box 379
9200 AJ Drachten
The Netherlands
Phone (+31) 512 54 30 13
Fax (+31) 512 51 33 79
http://www.deltainstruments.com
SomaScope™ for somatic cell detection in milk.

Fluid Imaging Technologies, Inc.
P.O. Box 350
211 Ocean Point Road
East Boothbay, Maine, 04544
Phone/Fax (207) 882-1100
http://www.fluidimaging.com
FlowCAM, an imaging flow cytometer for continuous monitoring of water.

FOSS Electric A/S
Slangerupgade 69, Postbox 260
DK-3400 Hillerød
Denmark
Phone +45 7010 3370
Fax +45 7010 3371
FOSS in North America
7682 Executive Drive
Eden Prairie, MN 55344, USA
Phone (952) 974-9892
Fax (952) 974-9823
http://www.foss.dk
Fossomatic and BactoScan FC systems for somatic cell and bacteria counting in milk.

Guava Technologies, Inc.
25801 Industrial Boulevard
Hayward, CA 94545-2991
Phone (866) 448-2827
Fax (510) 576-1500
http://www.guavatechnologies.com
Guava PC™ flow cytometer with dedicated software and reagents for absolute cell counting and viability tests.

Howard M. Shapiro, M.D., P.C.
(Howard M. Shapiro, M.D., P. C. is a company, not a person. I'm just Howard M. Shapiro, M.D.)
283 Highland Ave.
West Newton, MA 02465-2513
Phone (617) 965-6044
Fax (617) 244-7110
Cytometry Laboratory:
119 Braintree Street, Suite 102
Allston, MA 02134-1641
Phone (617) 783-8392
Fax (617) 783-4750
E-mail: hms@shapirolab.com (that's me as a person)
Cytomutt components, software, and consultation.

iCyt – Visionary Bioscience
1816 South Oak Street
Champaign, IL 61820
Phone (217) 328-9396
Fax (217) 328-9692
http://www.i-cyt.com
Custom cytometry instrumentation and data management.

International Remote Imaging Systems
9162 Eton Avenue
Chatsworth, CA 91311-5874
Phone (818) 709-1244 or (800) PRO-IRIS
Fax (818) 700-9661
http://www.proiris.com
Clinical urinalysis, video flow imaging flow cytometer.

Luminex Corporation
12212 Technology Blvd
Austin, TX 78727
Phone (512) 219-8020 or (888) 219-8020
Fax (512) 258-4173
http://www.luminexcorp.com
LX-100 flow cytometer for bead-based assays. See the Luminex website for strategic partners that develop kits and provide services for the Luminex technology.

NPE Systems, Inc.
7620 SW 147 Court
Miami, FL 33193
Phone (866) NPE-4567
Fax (305) 382-3947
http://www.npesystems.com
NPE Analyzer flow cytometer, combining electronic volume and fluorescence measurements.

One Lambda, Inc.
21001 Kittridge Street
Canoga Park, CA 91303-2801
Phone (818) 702-0042 or (800) 822-8824
Fax (818) 702-6904
http://www.onelambda.com
FlowPRA® tissue typing technology using the LABScan™ 100 flow cytometer.

Partec GmbH
Otto-Hahn-Str. 32
D-48161 Münster, Germany
Phone +49 2534 8008-0
Fax +49 2534 8008-90
http://www.partec.de
http://www.partec.usa
CyFlow, PAS and PAS-III, CCA-I and CCA-II flow cytometers, PA-I and PA-II ploidy analyzers, PPCS particle analyzer and cell sorter.

Union Biometrica, Inc. (Division of Harvard Biosciences)
35 Medford Street, Suite 101
Somerville, Massachusetts 02143
Phone (617) 591-1211
Fax (617) 591-8388
http://www.unionbio.com
COPAS™ flow cytometers for analysis, sorting and dispensing of multi-cellular organisms and large objects.

Hematology Instruments

Abbott Diagnostics
5440 Patrick Henry Dr.
Santa Clara, California 95054
Phone (408) 982-4800
http://www.abbottdiagnostics.com
CELL-DYN series of hematology analyzers.

ABX Diagnostics (U.S. Branch of French company)
34 Bunsen
Irvine, CA 92618-4210
Phone (949) 453-0500
Fax (949) 453-0300
http://www.abx.com
Pentra series of hematology analyzers.

Bayer Corporation
Diagnostics Division
511 Benedict Avenue
Tarrytown, NY 10591
Phone (914) 631-8000
Fax (914) 524-2132
http://www.bayerdiag.com
ADVIA® series and Technicon H* series of hematology analyzers.

Beckman Coulter (see section 11.3 for address)
LH 700 Series, AC·T™ series, HmX, MAXM, STKS™ and Gen·S™ hematology analyzers.

IDEXX Laboratories, Inc.
One IDEXX Drive
Westbrook, ME 04092
Phone (207) 856-0300
Fax (207) 856-0346
http://www.idexx.com
Idexx makes the LaserCyte™ veterinary hematology instrument.

Sysmex Corporation
6699 Wildlife Way
Long Grove, IL 60047
Phone (800) 3-SYSMEX
Fax (708) 726-3505
http://www.sysmex.com
KX-21, K-4500, SF-3000, SE-Series, XE-2100 hematology analyzers; UF-100 flow cytometric urine analyzer.

11.6 DATA ANALYSIS SOFTWARE/SYSTEMS

Hardware and Software

Applied Cytometry Systems, Ltd.
Unit 2 Brooklands Way
Brooklands Park Industrial Estate
Dinnington, Sheffield S25 2JZ
South Yorkshire, England
Phone 44 1909 566982
http://www.appliedcytometry.com
Applied Cytometry Systems, Inc. North America
3453 Ramona Ave., Suite 10
Sacramento, CA 95826
Phone (800) 500-FLOW
http://www.appliedcytometry.biz
Developer of EXPO32 software for Beckman Coulter cytometers; WinFCM, a Windows-based data acquisition and control system for the BD FACS series; and StarStation, PC-based software for the Luminex 100 cytometer.

Beckman Coulter (see section 11.3 for address)

BD Biosciences (see section 11.3 for address)

DakoCytomation (see section 11.5 for address)

Partec GmbH (see section 11.5 for address)

Commercial Software Sources

De Novo Software
64 McClintock Crescent
Thornhill, Ontario
L4J 2T1 Canada
Phone (905) 738-9442
Fax (905) 738-5126
http://www.denovosoftware.com
FCS Express and FCS Express Lite are data analysis and presentation programs running under Windows.

Ray Hicks
Phone +44 0797 4538647
Fax +44 0870 740 8595
E-mail: Sales@FCSPress.com
http://www.fcspress.com
FCSPress, a Macintosh program for FCM data analysis;
FCS Assistant, shareware utility for editing and exporting FCS data files.

Management Sciences Associates
6565 Penn Avenue
Pittsburgh, PA 15206-4490
Phone (412) 362-2000 or (800) MSA-INFO
Fax (412) 363-8878
E-mail: info@msa.com
http://www.msa.com
MacLAS® and WinLAS® list mode data analysis software for Macintosh and Windows platforms, designed to integrate with clinical data management systems.

Phoenix Flow Systems (see section 11.3 for address)
MultiCycle AV (DNA analysis), Win-FCM (acquisition for FACScan), MultiTime (kinetic analysis), WinReport, QC Tracker (QC software), Apo-Soft™ and MultiPlus (complete data analysis package) software for MS/DOS or MS Windows; Mac versions of some packages are available.

Tree Star, Inc.
20 Winding Way
San Carlos, CA 94070
Phone (800) 366-6045 or (650) 591-2854
Fax (650) 508-9186
http://www.flowjo.com
FloJo data analysis software for Macintoshes; a PC-compatible version should be available soon. ProJo, available free of charge, is a set of utilities that report, edit and administer FCS data files.

Verity Software House
P. O. Box 247
45A Augusta Road
Topsham, ME 04086
Phone (207) 729-6767
Fax (207) 729-5443
http://www.vsh.com
ModFit (DNA analysis), WinList (list mode data analysis), IsoContour, and ReticFit Software, running under Microsoft Windows. Macintosh versions of some programs are available.

Noncommercial Software Sources

A comprehensive catalogue of free flow cytometry software is maintained by Steve Kelley at the Purdue University Cytometry Laboratories website:
http://flowcyt.cyto.purdue.edu/flowcyt/software.htm

AUTOKLUS is cluster analysis software available for PC's, Cytometry 14:649-659, 1993. Author: Tom Bakker Schut, Netherlands. Richard Allen Cox is enhancing AUTOKLUS. The original AUTOKLUS software is free from the Purdue cytometry laboratories website.

Cylchred, developed by Terry Hoy, is cell cycle analysis software based on algorithms by Watson *et. al.* (1987) Cytometry 8:1-8 and Ormerod *et. al.* (1987) Cytometry 8:637-641 with modifications by Ormerod (1991) and Hoy (1996-99). The package accepts histograms in FCS single parameter binary format with a maximum of 1024 channels. Cylchred has operated in a DOS environment since 1996 and has been transposed into C++ by Nigel Garrahan and compiled as a 32 bit product for Windows 95. By implication it will not operate under Windows 3.1.
hoy@cardiff.ac.uk
http://www.uwcm.ac.uk/study/medicine/haematology/cytonetuk/documents/software.htm

CYTOWIN, developed by Daniel Vaulot with Jeff Dusenberry, is a windows 3.1 program designed to analyze single parameter histograms, two-parameter cytograms or list mode data. It can analyze data generated by the EPICS 5, Profile, Elite, FACScan, FACSort, and CICERO.
http://www.sb-roscoff.fr/Phyto/cyto.html#cytowin

Flow Explorer 4.0 is "Postcardware" developed by Ron Hoebe (see the Web site). The program scans directories for [FCS] flow data list files and displays parameters, notes, histograms, and bivariate density, dot, and contour plots.
AMC, Celbiology CMO
R.A. Hoebe, Room 351.2
1105 AZ Amsterdam
The Netherlands
http://wwwmc.bio.uva.nl/~hoebe/Welcome.html

IDLYK is a software package that does neural net clustering, rudimentary cell cycle and multivariate analysis. It should run on any computer platform with the IDL graphics language (IDLYK is freeware, IDL is very much not). IDLYK was developed by:
Robert Habbersett

LS-1 MS888
Los Alamos National Lab
Los Alamos, NM 87545
E-mail: robb@beatrice.lanl.gov.

MFI is a program that calculates median fluorescence intensities from list mode data; it was developed by:
Eric Martz
Department of Microbiology
University of Massachusetts
Amherst, MA 01003
E-mail: emartz@microbio.umass.edu
http://www.umass.edu/microbio/mfi/

RFlowCyt, is a basic R package for flow cytometry data analysis. R is `GNU S' - A language and environment for statistical computing and graphics; it was developed by:
Tony Rossini
Research Assistant Professor of Biostatistics
University of Washington
rossini@u.washington.edu
http://software.biostat.washington.edu/wikis/front/RFlowCyt

Soft Flow Hungary, Ltd.
Kedves u. 24
H-7628 Pecs, Hungary
Phone (36) 72 240064
Fax (36) 72 240065
Soft Flow, Inc. (North American office)
11513 Galtier Drive
Burnsville, MN 55337
Phone (800) 956-0100
Fax (612) 895-0900
http://www.visi.com~soft-flow/
Soft Flow's FCAP-list is flow cytometry analysis software for Macintoshes. HPtoMac disk conversion software allows Macintoshes to read 3.5" Hewlett Packard diskettes. FCB Applications are flow cytometry BASIC programs. FCAP-list includes a code generator that automatically creates FCB program code of any manually performed data analysis process, where FCB is flow cytometry BASIC code. The products were originally sold commercially and are now available free of charge.

WinMDI is list mode data analysis and display software running under Microsoft Windows developed by Joseph Trotter when he was at the Salk Institute. It is available free of charge at http://facs.scripps.edu/software.html

11.7 CYTOMETER REHABILITATION/ADD-ONS

Alternative Biomedical Services, Inc.
2326 West 78th Street
Hialeah, FL 33016
Phone (877) 227-1687 or (305) 558-4996
Fax (305) 558-6511
http://www.absbiomed.com
Service, parts, reagents, reconditioned cytometers from Beckman Coulter and Abbott (CellDyn series).

Automation Laboratory Technology
P. O. Box 255
Mossyrock, WA 98564-0255
Phone (800) 932-6883
E-mail: flow@atds.net
Service and maintenance of BD benchtop cytometers.

Cytek Development
46560 Fremont Blvd., Unit 116
Fremont, CA 94538
Phone (510) 657-0102
Fax (510) 657-0151
http://www.cytekdev.com
Volumetric sample delivery systems; sample delivery module for fast kinetic studies; sample preparation unit.

Spectron Corporation
11025 118th Place NE
Kirkland, WA 98033
Phone (425) 827-9317 or (800) 747-8624
Fax (425) 827-6942
http://www.spectroncorp.com
Reconditioned BD and Beckman Coulter cytometers

Laser upgrades (or downgrades to air-cooled systems) and the like are available from iCyt – Visionary Bioscience and Phoenix Flow Systems (see section 11.3 for address).

11.8 FLOW CYTOMETER PARTS

Flow System Plumbing

Alloy Products Corp.
1045 Perkins Avenue, PO Box 529
Waukesha, WI 53187-0529
Phone (800) 236-6603
Fax (262) 542-5421
http://www. alloyproductscorp.com
Alloy Products makes the stainless steel sheath tanks used by most flow cytometer manufacturers and by lab supply houses. Markups are lowest at the source.

Sigmund Cohn Corp.
121 South Columbus Ave.
Mt. Vernon, NY 10553
Phone (914) 664-5300
Fax (914) 664-5377
http://www.sigmundcohn.com
Fine stainless steel wire, indispensable for removing 76 μm particles from 75 μm orifices.

Cook, Inc.
P. O. Box 489
Bloomington, IN 47402
Phone (812) 339-2235 or (800) 457-4500
Fax (800) 554-8335
http://www.cookgroup.com
Cook carries stopcocks and other medical plumbing.

Hellma Cells, Inc.
80 Skyline Drive
Plainview, NY 11803
Phone (516) 939-0888
Fax (516) 939-0555
http://www.hellmausa.com
Hellma makes cuvette-type custom flow chambers.

NSG Precision Cells, Inc.
195G Central Ave.
Farmingdale, NY 11735
Phone (631) 249-7474
Fax (631) 249-8575
http://www.nsgpci.com
NSG also makes cuvette-type custom flow chambers.

Research Developments
3150-B Villa St.
Los Alamos, NM 87544
Phone (505) 662-4721
Jim Coulter, formerly of Los Alamos National Laboratory, custom makes flow chambers and other parts.

Specialty Glass Products, Inc.
2885 Terwood Rd.
Willow Grove, PA 19090
Phone (800) 850-4747
Fax (215) 659-7217
E-mail: sales@sgpinc.com
http://www.sgpinc.com
Custom capillaries for stream-in-air systems.

Value Plastics, Inc.
3325 Timberline Road
Ft. Collins, CO 80525
Phone (970) 223-8306 or (888) 404-5837
Fax (970) 223-0953
http://www.valueplastics.com
Fittings and couplings to get from hypodermic components to the water mains; a good catalog.

Vita Needle Co.
919-T Great Plain Avenue
Needham, MA 02492
Phone (781) 444-8629
Fax (781) 444-3956
http://vitaneedle.com
Stainless hypodermic tubing cut to your specs, or mine.

Photodetectors

Advanced Photonix, Inc.
1240 Avenida Acaso
Camarillo, CA 93012
Phone (805) 987-0146
Fax (805) 484-9935
http://www.advancedphotonix.com
Silicon photodiodes, avalanche photodiodes (APDs) and detector/preamplifier assemblies.

Burle Industries, Inc.
1000 New Holland Ave.
Lancaster, PA 17601-5688
Phone (717) 295-6888 or (800) 366-2875
Fax (717) 295-6096
http://www.burle.com
Photomultiplier tubes (PMTs) and accessories. Burle used to be RCA's PMT manufacturing operation.

Electron Tubes Inc.
100 Forge Way, Unit 5
Rockaway, NJ 07866
Phone (201) 575-5586
Fax (201) 586-9771
http://www.electron-tubes.co.uk
Previously part of Thorn EMI, Electron Tubes is the US affiliate of Electron Tubes Ltd in the UK and makes PMTs and accessories. I have used their PMT housings.

Hamamatsu Corp.
360 Foothill Rd.
Bridgewater, NJ 08807-0910
Phone (800) 524-0504 or (908) 231-0960
Fax (908) 231-1218
http://www.hamamatsu.com
Hamamatsu makes PMTs and accessories and pretty much has a lock on the flow cytometry market; they also supply photodiodes, APDs, and CCD camera chips.

PerkinElmer Optoelectronics (Headquarters)
44370 Christy St.
Fremont, CA 94538-3180
Phone (510) 979-6500 or (800) 775-6786
Fax (510) 687-1140
http://opto.perkinelmer.comindex.asp
E-mail: opto@perkinelmer.com
PerkinElmer Optoelectronics
2175 Mission College Blvd.
Santa Clara, CA 95054
Phone (408) 565-0850
Fax (408) 565-0793
Channel photomultipliers and photon counting modules
PerkinElmer Optoelectronics
22001 Dumberry Rd.
Vaudreuil, Quebec J7V 8P7

Canada
Phone (450) 424-3300
Fax (450) 424-3411
APD's, photon counting APD modules

RMD (Radiation Monitoring Devices), Inc.
44 Hunt Street
Watertown, MA 02472
Phone (617) 926-1167
Fax (617) 926-9743
http://www.rmdinc.com
Large area APDs and micro-APD arrays.

DC-DC Converter Modules for PMT HV Power Supplies

DEL Electronics Corp.
1 Commerce Park
Vahalla, NY 10595
Phone (914) 686-3600
Fax (914) 686-5424
http://www.delpower.com

MIL Electronics, Inc.
15O Dow St. Tower Two,
Manchester, NH 03101
Phone (603) 647-9201
Fax (603) 647-9201
http://www.milelectronics.com

Power Supplies (Low Voltage)

Power-One power supplies or their equivalents are available from local electronics distributors.

Other Electronics

Douglas Electronics
2777 Alvarado St.
San Leandro, CA 94577
Phone (510) 483-8770
Fax (510) 483-6453
http://www.douglas.com
Printed circuit boards for wire wrap and solder construction, some fitting standard buses and slots; schematic capture/ PC layout software for the Macintosh.

Global Specialties Corp.
1486 Highland Avenue, Unit 2
Cheshire, CT 06410
Phone (203) 272-3285
Fax (203) 468-0060
http://www.globalspecialties.com
Breadboarding sockets, test and design instruments.

11.9 LASERS

Laser Trade Publications

Laser Focus World
98 Spit Brook Road
Nashua, NH 03062
Phone (603) 891-0123
Fax (603) 891-0574
http://www.laserfocusworld.com

Lasers & Optronics
301 Gibraltar Dr., Box 650
Morris Plains, NJ 07950-0650
Phone (973) 292-5100
Fax (973) 292-0783
http://www.lasersoptrmag.com

Photonics Spectra
Laurin Publishing Co., Inc.
Berkshire Common
P. O. Box 4949
Pittsfield, MA 01202
Phone (413) 499-0514
Fax (413) 442-3180
http://www.photonicsspectra.com

Laser Manufacturers

Blue Sky Research Inc.
537 Centre Pointe Drive
Milpitas, CA 95035
Phone (408) 941-6068
Fax (408) 941-6069
http://www.blueskyresearch.com
Blue Sky's µLens™ technology generates, low-divergence, circular, diffraction-limited beams from laser diodes; the diode modules thus produced are called CircuLasers™.

Coherent, Inc.
Laser Group
5100 Patrick Henry Dr.
Santa Clara, CA 95054
Phone (800) 527-3786 or (408) 764-4983
Fax (800) 362-1170 or (408) 764-4800
http://www.coherentinc.com
Large argon and krypton lasers, CW dye lasers, solid state lasers, red and violet diode laser systems.

The COOKE Corporation
1091 Centre Road, Suite 100
Auburn Hills, MI 48326-2670
USA
Phone (248) 276-8820
Fax (248) 276-8825
http://www.cookecorp.com
Multicolor hollow-cathode He-Cd lasers.

Cyonics (now part of JDS Uniphase, see below)

Evergreen Laser Corp.
9G Commerce Circle
Durham, CT 06422
Phone (860) 349-1797
Fax (860) 349-3873
http://www.evergreenlaser.com
Laser re-tubing and re-manufacturing.

JDS Uniphase Corp.
163 Baypointe Pkwy
San Jose, CA 95134
Phone (408) 434-1800
Fax (408) 433-3838
http://www.jdsu.com
He-Ne, air-cooled argon, diode and solid-state lasers.

Laser Innovations
668 Flinn Avenue, #22
Moorpark, CA 93021
Phone (805) 529-5864
Fax (805) 529-6358
http://www.laserinnovations.com
Sales, service and support for Coherent ion lasers.

LiCONiX (now part of Melles Griot, see below)

Light Age, Inc.
Two Riverview Drive
Somerset, NJ 08873
Phone (732) 563-0600
Fax (732) 563-1571
http://www.light-age.com
The nUVo™ is a patented diode pumped alexandrite laser. The doubled nUVo™ operates between 360 and 400 nm with 1-10mW of power.

Lightwave Electronics
2400 Charleston Road
Mountain View, CA 94043
Phone (650) 962-0755
Fax (650) 962-1661
http://www.lwecorp.com
High power (up to 300mW) diode pumped 532nm lasers; mode-locked UV YAG lasers.

Melles Griot
Laser Group
2051 Palomar Airport Road, 200
Carlsbad, California 92009
Phone (800) 645-2737 or (760) 438-2131
Fax (760) 438-5208
http://www.mellesgriot.com
Argon, krypton and mixed gas ion lasers; He-Ne, He-Cd, diode and diode pumped solid state lasers.

Newport Corp. (see section 11.2 for address)
Red He-Ne, air cooled argon ion and diode lasers.

Novalux, Inc.
1170 Sonora Court
Sunnyvale, CA 94086
Phone (408) 736-0707
Fax (408) 735-0395
http://www.novalux.com
Protera solid-state 488 nm lasers.

Omnichrome (now part of Melles Griot, see above)

Power Technology Inc.
16302 Alexander Road
Alexander, AR 72002
Phone (501) 407-0712
Fax (501) 407-0036
http://www.powertechnology.com
Diode laser and diode pumped solid state laser modules; power supplies for diode lasers.

Research Electro Optics
1855 S. 57th Ct.
Boulder, CO 80301
Phone (303) 938-1960
Fax (303) 447-3279
http://www.reoinc.com
Green, yellow, orange, and red He-Ne lasers.

Spectra-Physics
1335 Terra Bella Ave.
P. O. Box 7013
Mountain View, CA 94039
Phone (650) 961-2550 or (800) 775-5273
Fax (650) 968-5215
http://www.splasers.com
Large and small argon and krypton lasers, CW dye lasers, He-Ne lasers, CW green and mode-locked UV YAG lasers.

Thorlabs (see section 11.2 for address)
Diode lasers.

11.10 OPTICAL FILTERS

Color Glass Filters

All the color glass filter manufacturers sell through distributors, e.g., the optical supply houses listed in 11.2.

Corning Glass Works's color glasses are now made by:
Kopp Glass, Inc.
2108 Palmer Street
Pittsburgh, PA 15218
Phone (412) 271-0190
Fax (412) 271-4103
http://www.koppglass.com

Hoya Corp. USA
101 Metro Drive, Suite 500,
San Jose, CA 95110
Phone (408) 441-3305
Fax (408) 451-9562
http://www.hoya.co.jp

Thermo Corion Optical Filters
(formerly Ditric Optics, Inc.)
8 East Forge Parkway
Franklin, MA 02038-3148
Phone (508) 528-4411 or (800) 598-6783
Fax (508) 520-7583
http://www.corion.com

Schott Glass Technologies, Inc.
400 York Ave.
Duryea, PA 18642
Phone (570) 457-7485
Fax (570) 457-6921
http://www.schottglasstech.com

Newport Industrial Glass, Inc.
10564 Fern Avenue
Stanton, CA 90680
Phone (714) 484-7500
Fax (714) 484-7600
http://www.newportglass.comhomefil.htm
Distributor for all of the color glass filter manufacturers.

Interference Filters

Chroma Technology Corp.
74 Cotton Mill Hill, Unit A-9
Brattleboro VT 05301
Phone (800) 824-7662
Fax (802) 257-9400
http://www.chroma.com

Thermo Corion Optical Filters (see address above)

Omega Optical, Inc.
210 Main Street
Brattleboro, VT 05301
Phone (802) 254-2690 or (866) 488-1064
Fax (802) 254-3937
http://www.omegafilters.com

The optical houses of section 11.2 also make and sell interference filters. While Omega's and Chroma's stock filters may cost a bit more than some of their competitors', each Omega and Chroma filter comes with a transmission curve. Omega and Chroma will also make custom and semicustom (i.e., with additional coatings) filters.

Neutral Density Filters

Many of the same companies that make interference filters also offer reflective neutral density (N.D.) filters. The color glass filter makers offer absorptive N.D. filters. Your local photo supply store probably has Kodak plastic neutral density filters, which may be all you need.

Polarizing Filters and Optics

For cheap plastic polarizers, try your local photo shop; for fancier stuff, try the optical supply houses.

Tunable Filters

Cambridge Research & Instrumentation, Inc.
35-B Cabot Road
Woburn, MA 01801
Phone (888) 372-1242 or (781) 935-9099
Fax (781) 935-3388
http://www.cri-inc.com
Liquid crystal tunable imaging filters.

11.11 AIDS TO TROUBLESHOOTING FLOW CYTOMETERS WHEN ALL ELSE FAILS

See reference 621.

11.12 PROFICIENCY TESTING

FAST Systems, Inc.
8-5 Metropolitan Ct.
Gaithersburg, MD 20878-4013
Phone (301) 977-0536
Fax (301) 977-7023
http://www.fastsys.com

11.13 SEX SELECTION (the book needed more sex)

MicroSort
Division of the Genetics & IVF Institute
3015 Williams Drive, Suite 101
Fairfax, Virginia 22031 USA
Phone (703) 876-3897 or (800) 277-6607
Fax (703) 995-4928
Division at Huntington Reproductive Center
23961 Calle de la Magdalena, Suite 541
Laguna Hills, CA 92653
microsort@givf.com
http://www.microsort.net/
Sperm sorting for gender selection in humans.

XY, Inc. at Moondrift
1108 North Lemay Avenue
Fort Collins, Colorado 80524
Phone (970) 493-3113
Fax (970) 493-3114
http://www.xyinc.com

For inquiries related to XY, Inc.'s research & development, sperm-sorting facility, sperm-evaluation laboratory and bull stud:
XY, Inc. at ARBL
3801 West Rampart Road
ARBL Building
CSU Foothills Research Campus
Fort Collins, Colorado 80523
Phone (970) 491-4764
Fax (970) 491-4374

11.14 ALTERNATIVE TECHNOLOGY

3D Molecular Sciences Ltd
Harston Mill
Harston, Cambridge CB2 5GG
UK
Phone +44 (0)1223 875 280
Fax +44 (0)1223 875 269
http://www.3d-molecularsciences.com
Developing multiplex bead assays using optically readable microfabricated encoded particles.

Amnis Corporation
2025 First Ave, Suite PH-B
Seattle, WA 98121
Phone (206) 374-7000
http://www.amnis.com
Developing its Image Stream technology for multispectral imaging of cells in flow.

Arcturus Engineering, Inc.
400 Logue Avenue
Mountain View, California 94043, USA
Phone (888) 446 7911 or (650) 962 3020
Fax (650) 962 3039
http://www.arctur.com
Arcturus's PixCell® is a laser capture microdissection system on a microscope platform.

ChemoMetec A/S
Gydevang 43
DK-3450 Allerød
Denmark
Phone (+45) 48 13 10 20
Fax (+45) 48 13 10 21
http://www.chemometec.com
NucleoCounter image analyzing cell counter.
(Distributor for U. S. and Canada:
New Brunswick Scientific Co., Inc.
P. O. Box 4005, 44 Talmadge Rd.
Edison, NJ 08818-4005
Phone (732) 287-1200 or (800) 631-5417
Fax (732) 287-4222
http://www.nbsc.com)

Chemunex SA (see section 11.5 for address)
The ChemScan RDI (known as Scan RDI™ in North America) analyzer uses laser scanning to detect bacteria on a filter membrane.

ChromaVision Medical Systems, Inc.
33171 Paseo Cerveza
San Juan Capistrano, CA 92675
Phone (888) 443-3310 or (949) 443-3355
Fax (949) 443-3366
http://www.chromavision.com
Automated Cellular Imaging System (ACIS®).

CompuCyte Corporation
12 Emily Street
Cambridge, MA 02139
Phone (800) 840-1303 or (617) 492-1300
Fax (617) 577-4501
http://www.compucyte.com
CompuCyte's LSC™ scanning laser cytometer[2047,2380-1,2918] was developed by Kamentsky and Kamentsky (and you know two heads are better than one). It has found and continues to find applications in many areas of basic and clinical cell analysis. The newer iCyte™ offers additional features.

Cyntellect, Inc. (spin-off from Oncosis, Inc)
6199 Cornerstone Court, Suite 111
San Diego, CA 92121-4740
Phone (858) 450-7079
Fax (858) 550-1774
http://www.cyntellect.com
Laser-Enabled Analysis and Processing (LEAP™) platform utilizes optical scanning, image analysis and targeting laser for cell analysis and manipulation[3015].

Dynal Biotech
P.O.Box 114
Smestad
N-0309 Oslo
Norway
Phone + 47 22 06 10 00
Fax + 47 22 50 70 15
http://www.dynal.no
Separation technology using magnetic Dynabeads®.

Fluidigm Corporation (formerly Mycometrix Corporation)
7100 Shoreline Court
South San Francisco, CA 94080
Phone (650) 266-6000
Fax (650) 871-7152
http://www.fluidigm.com
Fluidigm is commercializing microfluidic technology developed by Stephen Quake et al at CalTech for manipulation and sorting of cells and macromolecules[2326-7].

Immunicon Corporation
3401 Masons Mill Rd.
Huntingdon Valley, PA. 19006
Phone (215) 830-0777
Fax (215) 830-0751
http://www.immunicon.com
Immunicon's magnetic cell separation technology, some elements of which are available from Molecular Probes, is also the basis for the "CellTracks™" scanning laser cytometer[2383] and a simple cell counter.

Medis Technology
805 Third Avenue, 15th Floor
New York, NY 10022
Phone (212) 935-8484
Fax (212) 935-9216
http://www.medistechnologies.com
Medis Technology sells the Cellscan[1145] instrument developed at Bar-Ilan University for fluorescence polarization measurements.

Micronics, Inc.
8463 154th Avenue NE, Building F
Redmond, WA 98052
Phone (425) 895-9197
Fax (425) 895-1183
http://www.micronics.net
Micronics, with Honeywell, is developing the BioFlips wearable cytometer to detect a biowarfare attack by changes in a soldiers white blood cell count (funded by DARPA)

Miltenyi Biotec Inc.
12740 Earhart Avenue
Auburn, CA 95602, USA
Phone (530) 888-8871 or 800 FOR MACS
Fax (530) 888-8925
http://www.miltenyibiotec.com
Magnetic Cell Sorting Technology (MACS)

Oncosis, Inc.
6199 Cornerstone Court, Suite 111
San Diego, CA 92121-4740
Phone (858) 550-1770
Fax (858) 550-1774
http://www.oncosis.com
Photosis™ system for removing tumor cells from marrow[3014].

PE Biosystems
850 Lincoln Centre Drive
Foster City, CA 94404 USA
Phone (650) 638-5800 or (800) 345-5224
Fax (650) 638-5884
http://www.pebiosystems.com
FMAT™ 8100 microvolume laser scanning HTS system.

StemCell Technologies Inc.
777 West Broadway, suite 808
Vancouver BC Canada V5Z 4J7
Phone (800) 667-0322 or (604) 877-0713
Fax (800) 567-2899 or (604) 877-0704
http://www.stemcell.com
Cell separation, enrichment, expansion and evaluation technologies.

SurroMed, Inc.
2375 Garcia Avenue
Mountain View, CA 94043
Phone (650) 230 1961
Fax (650) 230 1960
http://www.surromed.com
SurroScan™, microvolume laser scanning cytometry system.

12. AFTERWORD

This is the space I reserved ahead of time in which to dot i's, cross t's, report late breaking news, and get in a few more opinions.

12.1 DOTTING i'S AND CROSSING t'S

As in the transitions between previous editions, I didn't renumber the old references. The Second Edition references started at 624, but reference 749 duplicated reference 523. The references in the Third Edition started at number 1027; those in this Fourth Edition are numbers 2315-3685. There are at least two duplicates; 3021 = 2682 and 3139 = 2616.

12.2 LATE BREAKING NEWS

New Book

Well, not that new, but I seem to have missed *Cytometric analysis of cell phenotype and function*[3660], edited by McCarthy and Macey, which appeared in late 2001.

New Protein Stain

Ferrari et al[3661] have used a new stain, **Beljian red**, a metabolite of the fungus *Epicoccum nigrum*, to label cysts of *Giardia*. The material has excitation peaks near 400 nm and 525 nm; it is non-fluorescent until it interacts with proteins, producing a fluorescent complex emitting at 605 nm.

Caveat on Fluorescent Caspase Inhibitors

Zbigniew Darzynkiewicz recently reported, in postings to the Purdue Cytometry Mailing List, that fluorescent caspase inhibitors (p. 380) may bind to other constituents of apoptotic cells, as well as to caspases. With Brian Lee and Gary Johnson, of Immunochemistry Technologies, he concludes that "While these reagents are good markers of apoptosis and very likely report activation of caspases or serine proteases, we currently suspect that mechanism of their retention in apoptotic cells may not be a strict function of their covalent interaction with caspase or protease enzymes." Further work on the additional binding sites should be published later in 2003.

Polyamide Probes

Gygi et al[3072] used polyamide probes to discriminate chromosome 9 from other chromosomes with which it normally clusters (p. 479). The probes came from Peter Dervan and his coworkers at Caltech, who have investigated whether synthetic organic chemistry can improve on nature in terms of designing sequence-specific DNA binding polymers[3662-4]. Polyamides can bind to native double-stranded DNA, and, in some cases, can enter intact cells and modulate gene function[3665].

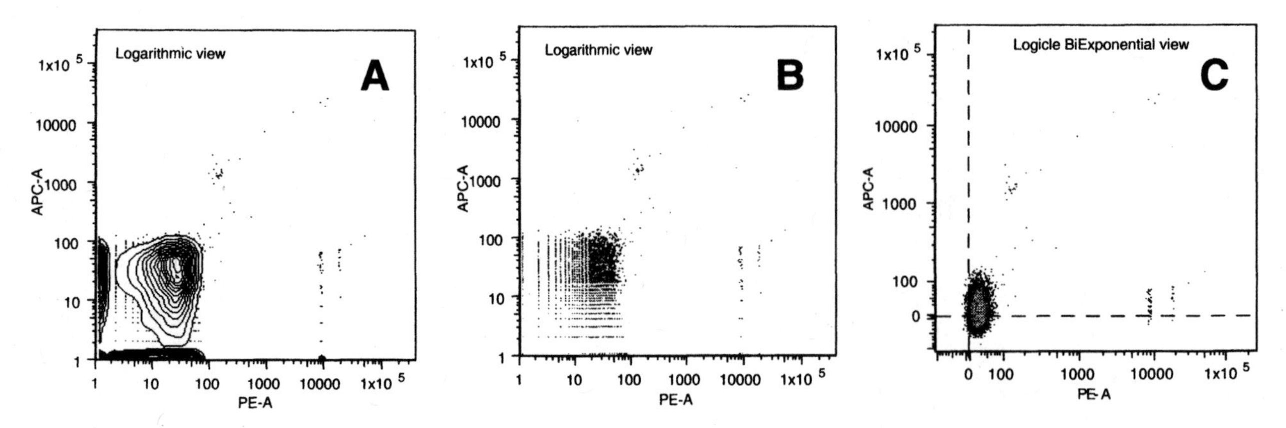

Figure 12-1. Tearing down the "picket fence" and reuniting the negatives using a BiExponential data transform instead of a logarithmic scale. Courtesy of David Parks and Wayne Moore (Herzenberg Lab, Stanford).

Tearing Down the (Picket) Fences

My description of logarithmic amplifiers and fluorescence compensation as "Deals With the Devil" (p. 35) seems to resonate with many flow cytometer users, and it seems that the more colors they measure, the more frustrated they get about the whole business. High-resolution digitization lets us get rid of the logarithmic amplifiers, but the large dynamic range of immunofluorescence data demands that we keep the logarithmic scale, or at least something like it. When our high-resolution digitization is not quite as high-resolution as it should be, we get "picket fences" at the bottom of the scale (p. 207), whether or not compensation is applied to the data. But it gets worse.

After baseline restoration is applied, signals from particles with little detectable fluorescence, i.e., those that are "negatives" in a given fluorescence channel, produce signals with maximum amplitudes near zero volts (ground). However, there is always some noise on the baseline, which means that the value we capture might be slightly below ground rather than slightly above ground. Whether we use a log amp or do our conversions from linear to log scales digitally, we have a problem; as numbers approach zero, their logarithms approach negative infinity, and the logarithms of negative numbers are undefined. It is therefore common practice to apply a "fudge factor," using additional circuitry at the input to a log amp or adding an appropriate constant if we are doing digital conversions, to keep the linear data values at least a little bit above ground.

Once we start compensating data, we are likely to end up with more negative values, and with a broader range of negative values, because the numbers we subtract from small signal values during compensation, whether we use analog compensation circuitry or digital compensation, are fixed fractions of large signal values, representing "ideal" compensation, while the small signal values, representing contributions from small numbers of photoelectrons, typically have a large variance. When we look at displays of compensated data on log scales, they almost invariably show one cluster of "negatives" plastered up against an axis, and another, discrete cluster just off the axis. Panels A and B of Figure 12-1 provide a good illustration. The data in all three panels of the figure represent signals from blank particles (and a few from contaminating fluorescent particles); they were taken from a FACSDiVa digital pulse processing system, and processed by Dave Parks and Wayne Moore of the Herzenberg lab using a beta version of Tree Star's FlowJo software incorporating a **BiExponential data transform**.

The log scale contour plot in Panel A shows three clusters of "negatives," one along each axis and one occupying most of the first decade. The contour lines conceal much of the "picket fence" appearance of the data, which is much more obvious in the log scale color density plot in Panel B.

There isn't anything sacred about the logarithmic scale; it just happens that it provides a convenient way of displaying data with a large dynamic range. Convenient, that is, as long as the data don't include zero and negative values. When we were stuck with using log amps, we really didn't have much choice as to what scale to use. Given the freedom made possible by high-resolution digital processing, Dave, Wayne, Mario Roederer et al decided to investigate other possible data transforms, looking for something that behaved like a log transform for large signal values, but was better behaved for small, zero, and negative values. What they came up with is a hyperbolic sine function that can have different coefficients in its positive and negative exponential components. Don't worry about it; you won't have to do the calculation. The bottom line is that the BiExponential function implemented in Wayne's prototype "Logicle" program and in the FlowJo beta is close to linear at the low end and close to log at the high end, and, as can be seen from Panel C of Figure 12-1, it puts all the "negatives" into the single cluster in which our brains expect them to be found.

If there is a down side to all of this, it is that the function has a lot of coefficients, which may take some tweaking to get the data into the right shape. Most users don't want to go there, and some who want to probably shouldn't. So, the challenge for the programmers is to make the process automatic. Otherwise, I'll stick with my counterproposal (p.

244) that we define the low end of the log scale as off limits for gating and just live with the artifactual multiple clusters. I have to admit, Panel C looks pretty nice.

New Instrument: The BD FACSArray™

BD Biosciences has recently introduced the FACSArray, a benchtop analyzer designed for both multiplexed bead assays and cellular fluorescence measurements. It has 532 nm (green) YAG and red diode laser sources and a FACSCalibur-type flow cell, and takes samples from microtiter plates, measuring forward and side scatter and fluorescence at (I'm guessing) 575 and >660 nm (green-excited) and 660 and >700 nm (red-excited), using APDs and PMTs as detectors. The FACSArray seems to be intended to go head to head with Guava and Luminex.

Science Special Section: Biological Imaging

The 4 April 2003 issue of *Science* (Volume 300, pp. 1-196) contains a special section on biological imaging, introduced by Hurtley and Helmuth[3676]. Notable articles in the issue include a comparative review of light microscopy techniques for live cell imaging by Stephens and Allan[3677] and a report by Scorrano et al[3678] demonstrating the roles of BAX and BAK in regulating Ca^{++} transport from endoplasmic reticulum to mitochondria, a process without which apoptotic death cannot occur.

Cytomics in Predictive Medicine: a *Clinical Cytometry* Special Issue and Other Recent Citings and Sightings

The May 2003 issue of *Clinical Cytometry* (*Cytometry*, Volume 53B, pp. 1-85) is a special issue on "Cytomics in Predictive Medicine," edited by Günter Valet and Attila Tárnok[3679], including articles on the predictive utility of immuno-phenotyping in acute myeloid leukemia[3680-1] and chronic fatigue syndrome[3682].

While those of us in the flow game may know and love *Clinical Cytometry*, I think we all have to admit that *The New England Journal of Medicine* has a considerably higher impact factor. Well, the May 1, 2003 issue of *NEJM* has an article[3683] and an editorial[3684] on the use of flow cytometric quantification of ZAP-70 kinase levels in chronic lymphocytic leukemia (CLL) cells to assess probable clinical course and outcome. The underlying biology is this: although the 50 to 70 percent of patients with hypermutated immunoglobulin heavy-chain variable regions usually have a good prognosis, those patients with unmutated genes do not. Using flow cytometry, Crespo et al[3683] found unmutated genes in all patients in whom at least 20% of B-CLL cells expressed ZAP-70 at levels equal to those found in T and NK cells; ZAP-70 expression itself was also determined to be of prognostic significance. Molecular methods that allow direct detection of the unmutated gene are "not [now] widely available in clinical practice[3683]," but Rai and Chiorazzi[3684] noted that ZAP-70 can be detected "by a relatively convenient and clinically available technology (multiparameter immunofluorescence flow cytometry)[3684]."

Back on pp. 468-9, I used Figure 10-9, from a 2002 paper by Perez and Nolan[3000] in which flow measurements of intracellular signaling kinases in T cell subsets were described, as an illustration of the current state of the art in flow cytometry. A little more than a year after this paper appeared, we have a high-profile article establishing clinical relevance of flow cytometric kinase measurements in leukemic cells and a high-profile editorial pointing out the advantages of the technique. Perhaps not as dramatic as the progress from the recognition of SARS as a new disease to the identification and sequencing of the virus in three months, but still pretty impressive.

In April, 2003, Maryalice Stetler-Stevenson of NIH, Jerry Marti of FDA, and Bob Vogt of CDC rounded up a few of the usual suspects for a meeting on "Identifying the Optimal Methods for Clinical Quantitative Flow Cytometry." Renewed interest in this topic has followed from flow cytometric analyses done on patients undergoing treatment with monoclonal antibodies; data suggest that the number of antibody binding sites present on the target cell type(s) may be of relevance in determining response. If this turns out to be true at the most simple-minded level, it may be possible to improve the therapeutic index of monoclonal antibody therapy just by picking appropriate mixtures of antibodies, but that's getting way ahead of the game; what we need to do now is figure out how to get multiple laboratories to do reliable, reproducible quantitative measurements. Stay tuned.

12.3 ANALYTICAL BIOLOGY, SUCH AS IT ISN'T: IS THIS ANY WAY TO RUN A SCIENCE?

Flow cytometers are, unfortunately, too often used by, or under the direction of, scientists who neither know nor care to know the details of their operation. In the long run, that attitude cannot contribute to scientific progress.

Scientific instrument development follows a traditional pattern in which new principles in physics and chemistry are first applied by physicists and physical chemists, who domesticate them to the point at which they can be used by analytical chemists. The resulting technology then diffuses into the organic chemical, biochemical, and biomedical communities over a period of several years. Chemists and physicists are expected to build their own instruments if they can't buy the apparatus they need to answer the questions they are asking; biologists are not.

Thus, while there is an established discipline of analytical chemistry, there is not an established discipline of analytical biology. Moreover, biologists and medical researchers are generally not exposed to general principles of instrumentation and measurement during their training. They certainly don't learn how to build instruments. Clinicians may be knowledgeable about apparatus they use, but are generally unaware of instruments and methodology not already established in their fields of specialization.

While clinical medicine has been unflatteringly described as a cottage industry, the epithet applies to medical research as well. Researchers are encouraged to formulate complex hypotheses and ask intricate questions, but are typically given no guidance to speak of regarding the choice of methodology that must be used to test the hypotheses and answer the questions. Most senior researchers will opine that graduate course work in instrumentation, measurement, and methodology is unnecessary; the students will learn from books, from articles, from their advisers, etc.

Just where are all these books and articles? Who taught the advisers, and how much attention have the advisers paid to methodology since they were young investigators? The imagined teaching resources don't exist. People learn how to build and use instruments from hanging around other people who build and use instruments, and the rest of their colleagues don't understand their methodology, may not believe their results, and, worst of all, waste a lot of time and money doing experiments the hard way due to a lack of understanding of methodology.

There seems to be room for improvement in a lot of the methodology with which biomedical research propels itself along these days. At one time, I wanted to set up an academic department to train people to develop instruments and methods, but I'm probably too old now. If you want to learn, read (or re-read) the book, and find a mentor. If you're setting up such a department and would like to have an old dog help teach you and your students some new tricks, get in touch.

12.4 COLOPHON

A colophon typically goes at the end of a book and gives details on the typesetting and production. I'm putting mine here so I can finish with some other stuff.

The reviewers of the First Edition of *Practical Flow Cytometry* were uniformly negative about the typesetting, which was done with a Centronics dot-matrix printer and an Atari 800 computer (total system cost just over $1,000). Many people assumed I was being chintzy about personal computer systems, and that I could have done much better using a $3,000 (yes!) Apple II or a $4,500 (yes! again) IBM PC for my word processing. Not true; when the book was typeset in late 1984, the setup I used and the Atari word processors provided the only means of getting a personal computer to produce the two-column, proportionally spaced text requested by the publisher. If I'd waited another two months, I could have produced nicer two-column proportional text using a letter-quality printer with the Atari. It took almost another year before the Apple and IBM word processors caught up.

Naturally, when the time came to consider putting out a Second Edition of the book, priorities one and two went to improving the typesetting and upgrading the sloppily scrawled and critically unacclaimed drawings that had illustrated the First Edition. By mid-1986, when I started evaluating the computer typesetting options, I was hearing all of these wonderful things about Desktop Publishing, and how it was already here for the Apple Macintosh and Coming Real Soon for the IBM PC. It sounded as if I could buy a single computer and a single software package to do everything I needed; set the type, draw illustrations, and do the layout, merging text and graphics. When I couldn't wait any longer for that ideal software, which was still not available for either the Mac or the PC when the second edition went to the printer, I had to wing it.

I first had to convert the files containing the text of the First Edition from single- to double-density Atari floppies, so I could read them on an IBM PC drive with the aid of shareware that translated them, more or less, into IBM format. The IBM format files were stripped of control codes and edited, and new text generated, using the XyWrite III Plus word processor, which formatted and set the text in double columns, using 10 point Times Roman as the body type. Text pages were printed on an Apple LaserWriter Plus printer (over $4,000!), which I lent to XyQuest for six weeks so they could fix the XyWrite printer drivers.

Merging of text and graphics was done by leaving space in the text page layouts for graphics and printing graphics separately. Graphics in IBM PC format were captured with SymSoft's HOTSHOT and printed on the LaserWriter Plus. Computer-generated drawings were done on a Macintosh SE using Cricket Draw and Apple's MacDraw, and also printed on the printer. Line art and photographs were copied on Canon copiers. The pages containing graphics were then inserted into the printer, and the text was printed in its proper location. This, believe it or not, was the most expeditious way to get the job done at the time.

Since I waited six years to do the Third Edition, I expected to be up to my armpits in fantastic page layout and document processing software when I started this one. Ha. I tried FrameMaker, PageMaker, and Ventura Publisher, on Mac and Windows platforms, and couldn't live with any of them. XyWrite had appeared in new DOS and Windows incarnations, which I didn't like either.

As luck would have it, Microsoft's version 6.0 of Word for Windows hit the stores just about the time I started typing. I had played around with version 2.0, which didn't hack it, but, luckily, I had translated all of the old files into Microsoft's Rich Text Format, so it was fairly painless to read them.

As we all think we know, cross-platform compatibility between Macintoshes and PC-compatibles is a piece of cake. Ha again. All of the drawings done on the Mac were done with old enough versions of the drawing programs so the new versions either wouldn't print them or would only convert them to PICT files much uglier than the originals. I ended up having John Brandes redo some of the old graphics using the Word for Windows drawing package; other old figures and photos and line art were scanned in by Chris Spychalski using a Hewlett-Packard ScanJet IIC, connected to an Apple Macintosh IIci. Image files in TIFF (Tagged Image File Format) were transferred to the PC-compatibles via floppy disks and sneakernet.

A few drawings for the Third Edition were made with Corel Draw 4, Micrografx Windows Draw, and ChemWindow, a chemical structure drawing program. PC screen capture and image format translation were done with Pizazz Plus for Windows (Application Techniques, Pepperell, MA) My objective was to have the whole book, artwork and all, on the hard disk, so I wouldn't need to run paper through the printer twice, and I was very impressed that I could do the whole job with a word processor, rather than having to resort to a page layout program. So I went along with the rest of the world and switched all of my word processing to Word for Windows.

The Second Edition was printed on the Apple LaserWriter Plus, a 300 dpi PostScript printer. The Third Edition was printed on 600 dpi Hewlett-Packard LaserJet 4M and LaserJet 4MP printers, making it feasible to print halftones from photographs. These printers had PostScript capability, but I used them in their native (PCL) mode, because, at least under Windows 3.1, they printed a lot faster that way.

By the time I started work on this Fourth Edition, I thought the layout part would be a piece of cake. In case you haven't heard, the typesetters used to print most books these days can work directly from Adobe Acrobat .pdf files, which are easy to generate from PostScript. I tried to use FrameMaker to do the book, but the available documentation was so atrocious that, after several months of screwing around, I still couldn't manage to get a single page of text to lay out properly, and decided to stick with Word. No picnic there, either. The computers at the lab and at the house and my laptop all had what was supposed to be the same version of Word for Office 2000 (separate copies of the program, of course; having written and sold software, I am against piracy). They also had the same printer drivers and the same fonts, but different computers would give me different layouts of the same text. Things improved when I moved from Windows Millennium Edition (hard to tell whether that is an operating system or just a resident virus) to Windows XP. It then turned out that the Acrobat Distiller printer driver would choke on my Word files and not produce .pdf files, and that each of the three PostScript drivers available for my 1200 DPI Lexmark Optra T610 printer was flaky in its own inimitable way. I settled on the least flaky one, which usually generated PostScript files that could be made into .pdf's.

After two major laptop hard drive crashes, I got smart and started keeping all of the files in duplicate on two external IEEE 1394 (FireWire) hard drives. And here I am. I need a new laptop; I was so desperate I even tried a titanium Mac for a while. Meanwhile, the folks at Wiley decided to play it safe and photograph my printed pages instead of trying to print the book from .pdf files. I hope what you see is what I got.

The body text for the Fourth Edition is 10 point Adobe Garamond; headings and captions are in Adobe Albertus MT. Many of the older figure labels are in Arial; the newer ones are in Helvetica. Shapiro's Laws of Flow Cytometry are set in Zapf Chancery. I'm still not sure whether symbols are coming out in Adobe's Symbol PostScript font or in Microsoft's Symbol TrueType font. There are a few other fonts scattered here and there. I hope you like the layout, but maybe I'll just send the publisher a double-spaced typed manuscript for my next book.

Since they were intended in part as revolutionary tracts, I thought it was appropriate that the First Edition and its predecessor should be Little Red Books. The Second Edition, with its redesigned cover, was, like Chinese pandas and newspapers, red all over as well as black and white. The blue cover design for the Third Edition was a little subdued, but Denise Papania has done a fantastic job with the cover for this one.

12.5 UNFINISHED BUSINESS

AIDS and Infectious Disease in the Third World

As those of you who have been involved in cytometry for many years, or have read through this book, know, it was the emergence of AIDS in the U. S. and Europe in the early 1980's that led to the widespread clinical use of fluorescence flow cytometry and the improvement of the previously shaky economic status of most of the instrument manufacturers. To borrow a phrase from the computer industry, HIV was, indeed, our "killer application." To their credit, many instrument manufacturers have attempted to produce smaller, more rugged, less expensive cytometric apparatus for clinical use in the Third World, but this has remained largely inaccessible to most of the African and Asian countries most affected by the epidemic.

I have described some approaches to affordable, practical cytometric systems for CD4+ T-cell counting on pp. 491-3. This subject is now a topic of discussion at the annual Cytometry Development Workshop, held every Fall at the Asilomar Conference Grounds in Pacific Grove, California, and attended by technical personnel from academic institutions, national laboratories, and industry. The Workshop has always had a strong ecumenical spirit, with information exchanged freely among competing laboratories and companies, and the session held in 2000 helped stimulate the development of the "EasyCount" instrument. More remains to be done; inexpensive cytometric apparatus originally developed for CD4+ cell counting could be adapted to meet the needs of other price-sensitive emerging markets in developed countries, e.g., food and water microbiology, and for other uses, e.g., clinical research on malaria and other parasitic diseases. The cytometry industry may need to have its arm twisted, as the pharmaceutical industry has had its arm twisted, to get the right products to resource-poor nations at the right price; if you have any leverage, use it. I'll be happy to help.

A Center for Microbial Cytometry

The pace of development of instruments, reagents, and techniques optimized for cytometry of microorganisms, i.e., bacteria, nano- and picoplankton, fungi, viruses, and small unicellular eukaryotes – has been glacial.

For example: My colleagues and I published the first paper on flow cytometric detection of single virions in 1979[94], and, although Harald Steen reproduced our results within a few years[1246], the next published paper on detection of single virions appeared in 1999[2336]. As I pointed out on pp. 522-4, no current commercial instrument has sufficient sensitivity to make precise multiparameter measurements of substances present at levels below a few thousand molecules in individual microorganisms. We need an instrument optimized for that purpose; once we have it, it can be used to look at a lot of interesting small stuff, e.g., some strange bacteria[3666] and bacterial nucleic acids[3667-8] that seem to circulate in blood, circulating immune complexes[825,1484,1820,3545], and platelet microparticles[1460,3121-2,3132,3669-70], as well as at viruses and bacteria from various sources.

As I said on p. 516, bacteria are not just little eukaryotes. Many of the dyes with which we have become familiar as stains for eukaryotic cells behave differently in bacteria; some behave differently in different bacterial species. Relatively few monoclonal antibodies against microbial antigens are commercially available, and the choice of labels for those few that are labeled is limited. Other specific reagents, e.g., rRNA probes, are also difficult to find. The immunologists and hematologists, even the most competitive ones, have cooperated over a generation to define the CD antigens, antibodies to which have become profitable for a number of companies; a similar cooperative effort will have to be undertaken to make better reagents available for microbiology.

It's not that there aren't a reasonable number of people who want and could benefit from improved instruments, reagents, and techniques for cytometry of microorganisms; the obstacle appears to be something like an activation energy barrier. In a word, the development process needs to be catalyzed; I am volunteering to be one of the enzymes. In 2002, I established a nonprofit corporation called **The Center for Microbial Cytometry.** The first mission of the center will be to get interested parties (from academia, government, and industry) together, by e-mail at first and by conference call or face-to-face meeting as appropriate later, to define (and catalog) the principal problems and currently available solutions. We can expect to find that, in many cases, one group of investigators has some solutions to another group's problems, which should lead to productive collaborations. Ideally, this would increase manufacturers' level of interest in providing the needed systems and materials, but, if it did not, we would at least have a data base and a clearing house that would facilitate do-it-yourself instrument modification and construction and exchange of reagents and techniques. The free market hasn't worked very well in this field over the past twenty or thirty years.

I don't expect the Center to accumulate a large staff and a building full of labs; we would only need people and resources on-site to do what none of our participating members was willing or able to do. I therefore also don't expect a large budget to be required, and it seems to me that the necessary funds could be obtained from some combination of government agencies, private foundations, and/or industrial donors. I'll start scrounging as soon as I finish writing. If you would like to be added to my list for further future mailings on the topic, e-mail me at <hms@shapirolab.com>.

A Nobel Prize for Herzenberg and Kamentsky?

We all know that the developers of really useful scientific methods are at least as likely to get Nobel Prizes as those who discover new scientific facts and principles. If you've read this far in this book, I shouldn't have to convince you that cell sorting is a really useful scientific method. It would not have become one without the combined efforts of Mack Fulwyler, Len Herzenberg, and Lou Kamentsky.

Mack, then at Los Alamos, described cell sorting by droplet generation and electrostatic deflection in 1965[67]. Lou set the pace for the development of flow cytometry as an analytical cytologic tool[1] during the 1960's, when he worked IBM's Watson Laboratory at Columbia University, and had added a sorter to his original instrument[65] by 1967[66]. Len recognized the potential of flow cytometry and sorting as a preparative method, and his lab at Stanford has been central in the development and refinement of the technology since the late 1960's, when the first fluorescence-activated sorter was described[82]. Since that time, log amps, fluorescence compensation, reporter genes, new fluorochromes, and many other innovations have come out of the lab (reread this book!), which always seems to be measuring one more color than anybody else, and which also functions as a highly successful core lab, honing the cytometric skills of both experienced users and novices.

Mack is gone now, but there is still time for Len and Lou to win that trip to Stockholm. If you share my sentiments about this, and have clout with anyone eligible to nominate them for the Nobel Prize (see <http://www.nobel.se/medicine/nomination/nominators.html>), let that person know.

12.6 FLOW AND THE HUMAN CONDITION

The psychologist Mihaly Csikszentmihalyi wrote a book called *Flow: The Psychology of Optimal Experience* (Harper & Row, 1990). Of course, he wasn't talking about flow cytometry, but some flow experiences can be pretty optimal. Nobody has won a Nobel Prize for flow cytometry, but Len Herzenberg and Lou Kamentsky still might (they may not be old enough). Several people have made millions out of flow cytometry; Wallace Coulter made hundreds of millions. I know of marriages, and fistfights, in which flow cytometry played a role.

I haven't gotten rich doing flow, but I've been invited to a lot of interesting places, and met a lot of nice people. Since it's unlikely that anybody but the people I mentioned could win a Nobel Prize for cytometry or sorting itself, as opposed to winning one for using it intelligently, flow people tend to be more cooperative and less competitive than workers in some other fields. It is more obvious than ever that flow can't bring world peace, but it can occasionally bring individual people from places or groups with opposing philosophies together in a common cause, which is a start. I think I'll keep doing it.

There's No Business Like Flow Business

There's no business like flow business
When show business is slow.
Where else can you play around with high-tech
Lasers, and computers, and cell clones?
I go in the lab each day and try tech-
Niques out with my tech;
It's in my bones.

There's no people like flow people;
They smile where lights are low.
Even when there's *Mycoplasma* in your cells,
Your laser smokes and emits bad smells,
Still, you wouldn't trade it for a stack of gels –
Well, I wouldn't, I know.
Let's go on with the flow!

12.7 ONE MORE THING

I almost forgot this, and it's very important: When your flow system clogs, whatever you do, don't smite it with your staff.

REFERENCES

1. Kamentsky LA: Cytology automation. Adv Biol Med Phys 14:93, 1973
2. Herzenberg LA, Sweet RG, Herzenberg LA: Fluorescence activated cell sorting. Sci Amer 234(3):108, Mar 1976
3. Braylan RC: Flow cytometry. Arch Pathol Lab Med 107:1, 1983
4. Horan PK, Wheeless LL Jr: Quantitative single cell analysis and sorting. Science 198:149, 1977
5. Miller RG, Lalande ME, McCutcheon MJ et al: Usage of the flow cytometer-cell sorter. J Immunol Meths 47:13, 1981
6. Loken MR, Stall AM: Flow cytometry as an analytical and preparative tool in immunology. J Immunol Meths 50:R85, 1982
7. Kruth HS: Flow cytometry: Rapid biochemical analysis of single cells. Analyt Biochem 125:225, 1982
8. Shapiro HM: Multistation multiparameter flow cytometry: a critical review and rationale. Cytometry 3:227-43, 1983
9. Melamed MR, Mullaney PF, Mendelsohn ML (eds): *Flow Cytometry and Sorting*. New York, Wiley, 1979
10. Hecht E, Zajac A: *Optics*. Reading (MA), Addison-Wesley, 1974
11. Hoenig SA: *How to Build and Use Electronic Devices without Frustration, Panic, Mountains of Money, or an Engineering Degree* (2 ed). Boston; Little, Brown, 1980
12. Horowitz P, Hill W: *The Art of Electronics*. New York, Cambridge University Press, 1980
13. Mims FG III: *Getting Started in Electronics*. Fort Worth, Radio Shack, 1983
14. Mims FG III: *Engineer's Notebook II*. Fort Worth, Radio Shack, 1982
15. Jung WC. *IC Op-amp Cookbook* (2 ed). Indianapolis, Sams, 1980
16. Lancaster D: *TTL Cookbook*. Indianapolis, Sams, 1974
17. Lancaster D: CMOS Cookbook. Indianapolis, Sams, 1977
18. Stein PG, Shapiro HM: *The Joy of Minis and Micros*. Rochelle Park (NJ), Hayden, 1981
19. Coffron JW: *Practical Hardware Details for 8080, 8085, Z80, and 6800 Microprocessor Systems*. Englewood Cliffs (NJ), Prentice-Hall, 1981
20. Zumchak EM: *Microcomputer Design and Troubleshooting*. Indianapolis, Sams, 1982
21. Foster CC: *Real Time Programming- Neglected Topics*. Reading (MA), Addison-Wesley, 1981
22. Scanlon LJ: *FORTH Programming*. Indianapolis, Sams, 1982
23. Lakowicz JR: *Principles of Fluorescence Spectroscopy*. New York, Plenum, 1983.
24. Gurr E: *Synthetic Dyes in Biology, Medicine, and Chemistry*. New York, Academic Press, 1971
25. Lillie RD (ed): *H. J. Conn's Biological Stains* (9 ed). Baltimore, Williams & Wilkins, 1977
26. Darzynkiewicz Z, Traganos F, Melamed MR: Detergent treatment as an alternative to cell fixation for flow cytometry. J Histochem Cytochem 29: 329, 1981
27. Derjaguin BV, Vlasenko GJ: Flow-ultramicroscopic method of determining the number concentration and particle size analysis of hydrosols and aerosols. J Colloid Sci 17:605, 1962
28. Moldavan A: Photo-electric technique for the counting of microscopical cells. Science 80:188, 1934
29. Gucker FT Jr, O'Konski CT, Pickard HB, Pitts JN Jr: A photoelectronic counter for colloidal particles. J Am Chem Soc 69:2422-31, 1947
30. Gucker FT Jr, O'Konski CT: Electronic methods of counting aerosol particles. Chem Revs 44:373, 1949
31. Ferry RM, Farr LE Jr, Hartman MG: The preparation and measurement of the concentration of dilute bacterial aerosols. Chem Revs 44:389, 1949
32. Baker JR: *Principles of Biological Microtechnique*. London, Methuen, 1958
33. Ehrlich P, Lazarus A: *Histology of the Blood*. Cambridge (UK), Cambridge University Press, 1900
34. Caspersson TO: *Cell Growth and Cell Function*. New York, Norton, 1950
35. Feulgen R, Rossenbeck H: Mikroskopisch-chemischer Nachweis einer Nucleinsäure von Typus der Thymonucleinsäure und auf die darauf beruhende elektive Färbung von Zellkernen in mikroskopischen Präparaten. Hoppe-Seyler's Z Physiol Chem 135:203, 1924
36. Brachet J: La localisation des acides pentosenucléiques pendant la développement des Amphibiens. C R Soc Biol 133:90, 1940
37. Caspersson T, Schultz J: Nucleic acid metabolism of the chromosomes in relation to gene reproduction. Nature 142:294, 1938
38. Avery OT, MacLeod CM, McCarty M: Studies on the chemical nature of the substance inducing transformation of pneumococcal types. Induction of transformation by a deoxyribonucleic acid fraction isolated from *Pneumococcus* Type III. J Exp Med 79:137, 1944
39. Papanicolaou GN, Traut HF: The diagnostic value of vaginal smears in carcinoma of the uterus. Am J Obst Gynec 42:193, 1941
40. Papanicolaou GN: Some improved methods for staining vaginal smears. J Lab Clin Med 26:1200, 1941
41. Friedman HP Jr: The use of ultraviolet light and fluorescent dyes in the detection of uterine cancer by vaginal smear. Am J Obst Gynec 59:852, 1950

42. Mellors RC, Silver R: A microfluorometric scanner for the differential detection of cells: application to exfoliative cytology. Science 114:356, 1951
43. Mellors RC, Keane JF Jr, Papanicolaou GN: Nucleic acid content of the squamous cancer cell. Science 116:265, 1952
44. Coons AH, Creech HJ, Jones RN: Immunological properties of an antibody containing a fluorescent group. Proc Soc Exptl Biol Med 47:200, 1941
45. Coons AH, Kaplan MH: Localization of antigen in tissue cells. II. Improvements in a method for the detection of antigen by means of fluorescent antibody. J Exp Med 91:1, 1950
46. Riggs JL, Seiwald RJ, Burckhalter JH et al: Isothiocyanate compounds as fluorescent labeling agents for immune serum. Am J Pathol 34:1081, 1958
47. Goding JS: Conjugation of antibodies with fluorochromes: modifications to the standard methods. J Immunol Meths 13:215, 1976
48. Crosland-Taylor PJ: A device for counting small particles suspended in fluid through a tube. Nature 171:37, 1953
49. Coulter WH: High speed automatic blood cell counter and cell size analyzer. Proc Natl Electronics Conf 12:1034, 1956
50. Brecher G, Schneiderman M, Williams GZ: Evaluation of electronic red blood cell counter. Am J Clin Path 26:1439, 1956
51. Mattern CFT, Brackett FS, Olson BJ: Determination of number and size of particles by electrical gating: blood cells. J Appl Physiol 10:56, 1957
52. Mellors RC (ed): *Analytical Cytology* (2 ed). New York, McGraw-Hill, 1959
53. Miner RW, Kopac MJ (eds): *Cancer Cytology and Cytochemistry.* Ann N Y Acad Sci 63:1033, 1956
54. von Bertalanffy L, Bickis I: Identification of cytoplasmic basophilia (ribonucleic acid) by fluorescence microscopy. J Histochem Cytochem 4:481, 1956
55. Armstrong JA: Histochemical differentiation of nucleic acids by means of induced fluorescence. Exptl Cell Res 11:640, 1956
56. von Bertalanffy L, Masin F, Masin M: Use of acridine-orange fluorescence technique in exfoliative cytology. Science 124:1024, 1956
57. Ingram M, Preston K Jr: Importance of automatic pattern recognition techniques in the early detection of altered blood cell production. Ann N Y Acad Sci 113:1066, 1964
58. Ingram M, Preston K Jr: Automatic analysis of blood cells. Sci Amer 223(5):72, Nov 1970
59. Mendelsohn ML, Kolman WA, Bostrom RC: Initial approaches to the computer analysis of cytophotometric fields. Ann N Y Acad Sci 115:998, 1964
60. Prewitt JMS, Mendelsohn ML: The analysis of cell images. Ann N Y Acad Sci 128: 1035, 1966
61. Hallermann L, Thom R, Gerhartz H: Elektronische Differentialzählung von Granulocyten und Lymphocyten nach intravitaler Fluochromierung mit Acridinorange. Verh Deutsch Ges Inn Med 70:217, 1964
62. Kosenow W: Die Fluorochromierung mit Acridinorange, eine Methode zur Lebendbeobachtung gefärbter Blutzellen. Acta Haemat 7:217, 1952
63. Jackson JR: Supravital blood studies, using acridine orange fluorescence. Blood 17:643, 1961
64. Schiffer LM: Fluorescence microscopy with acridine orange: a study of hemopoietic cells in fixed preparations. Blood 19:200, 1962
65. Kamentsky LA, Melamed MR, Derman H: Spectrophotometer: new instrument for ultrarapid cell analysis. Science 150:630, 1965
66. Kamentsky LA, Melamed MR: Spectrophotometric cell sorter. Science 156:1364, 1967
67. Fulwyler MJ: Electronic separation of biological cells by volume. Science 150:910, 1965
68. Sweet RG: High frequency recording with electrostatically deflected ink jets. Rev Sci Instrum 36:131, 1965
69. Van Dilla MA, Fulwyler MJ, Boone IU: Volume distribution and separation of normal human leukocytes. Proc Soc Exptl Biol Med 125:367, 1967
70. Shapiro HM: Input-output models of biological systems: formulation and applicability. Comput Biomed Res 2:430, 1969
71. Shapiro H, Simonson E, Cady LD Jr: Electrocardiographic correlations in normal patients. Am J Med Electronics 3:41, 1964
72. Lipkin LE, Watt RC, Kirsch RA: The analysis, synthesis, and description of biological images. Ann N Y Acad Sci 128:984, 1966
73. Stein PG, Lipkin LE, Shapiro HM: Spectre II: general-purpose microscope input for a computer. Science 166:328, 1969
74. Shapiro HM, Bryan SD, Lipkin LE, Stein PG, Lemkin PF: Computer-aided microspectrophotometry of biolgical specimens. Exptl Cell Res 67:81, 1971
75. Wied GL, Bahr GF (eds): *Automated Cell Identification and Cell Sorting.* New York, Academic Press, 1970
76. Finkel GC, Grand S, Ehrlich MP et al: Cytologic screening automated by Cytoscreener. J Assn Adv Med Instrum 4:106, 1970
77. Koenig SH, Brown RD, Kamentsky LA et al: Efficacy of a rapid cell spectrophotometer in screening for cervical cancer. Cancer 21:1019, 1968
78. Kamentsky LA, Melamed MR: Instrumentation for automated examinations of cellular specimens. Proc IEEE 57:2007, 1969
79. Van Dilla MA, Trujillo TT, Mullaney PF et al: Cell microfluorometry: a method for rapid fluorescence measurement. Science 163:1213, 1969
80. Evans DMD (ed): *Cytology Automation.* London, Livingstone, 1970
81. Saunders AM, Hulett HR: Microfluorometry: comparison of single measurements to a rapid flow system. J Histochem Cytochem 17:188, 1969
82. Hulett HR, Bonner WA, Barrett J et al: Automated separation of mammalian cells as a function of intracellular fluorescence. Science 166:747, 1969
83. Dittrich W, Göhde W: Impulsfluorometrie bei Einzelzellen in Suspensionen. Z Naturforsch 24b:360, 1969
84. Ornstein L, Ansley HR: Spectral matching of classical cytochemistry to automated cytology. J Histochem Cytochem 22:453, 1974
85. Mansberg HP, Saunders AM, Groner W: The Hemalog D white cell differential system. J Histochem Cytochem 22:711, 1974
86. Bonner WA, Hulett HR, Sweet RG et al: Fluorescence activated cell sorting. Rev Sci Instrum 43:404, 1972
87. Prewitt JMS: Parametric and nonparametric recognition by computer: an application to leukocyte image processing. Adv Computers 12:285, 1972
88. Curbelo R, Schildkraut ER, Hirschfeld T et al: A generalized machine for automated flow cytology system design. J Histochem Cytochem 24:388, 1976
89. Kleinerman M: Differential counting of leukocytes and other cells. U S Patent 3,916,205, 1975
90. Shapiro HM, Schildkraut ER, Hirschfeld T et al: White blood cell differential counting: a multiparameter flow cytophotometric technique (abstract in heptameter). Med Instrum 9:58, 1975
91. Shapiro HM, Schildkraut ER, Curbelo R et al: Combined blood cell counting and classification with fluorochrome stains and flow instrumentation. J Histochem Cytochem 24:396, 1976
92. Shapiro HM, Schildkraut ER, Curbelo R et al: Cytomat-R: a computer-controlled multiple laser source multiparameter flow cytophotometer system. J Histochem Cytochem 25:836, 1977
93. Shapiro HM: Fluorescent dyes for differential counts by flow cytometry: does histochemistry tell us much more than cell geometry? J Histochem Cytochem 25:976-89, 1977
94. Hercher M, Mueller W, Shapiro HM: Detection and discrimination of individual viruses by flow cytometry. J Histochem Cytochem 27:350, 1979
95. Schildkraut ER, Hercher M, Shapiro HM et al: A system for storage and retrieval of individual cells following flow cytometry. J Histochem Cytochem 27:289, 1979
96. Pinkel D, Dean P, Lake S et al: Flow cytometry of mammalian sperm: progress in DNA and morphology measurement. J Histochem Cytochem 27:353, 1979

97. Sharpless TK, Traganos F, Darzynkiewicz Z, Melamed MR: Flow cytofluorimetry: discrimination between single cells and cell aggregates by direct size measurements. Acta Cytol 19:577-81, 1975
98. Sharpless TK, Melamed MR: Estimation of cell size from pulse shape in flow cytofluorometry. J Histochem Cytochem 24:257, 1976
99. von Sengbusch G, Hugemann B: A fluorescence microscope attachment for flow-through cytofluorometry. Exptl Cell Res 86:53, 1974
100. Steen HB, Lindmo T: Flow cytometry: a high resolution instrument for everyone. Science 204:403, 1979
101. Lindmo T, Steen HB: Characteristics of a simple, high resolution flow cytometer based on a new flow configuration. Biophys J 28:33, 1979
102. Steen HB: Further developments of a microscope-based flow cytometer: light scatter detection and excitation intensity compensation. Cytometry 1:26, 1980
103. Steen HB, Boye E, Skarstad K, Bloom B, Godal T, Mustafa S: Applications of flow cytometry on bacteria: cell cycle kinetics, drug effects, and quantitation of antibody binding. Cytometry 2:249-57, 1982
104. Leary JF, Todd P, Wood JCS et al: Laser flow cytometric light scatter and fluorescence pulse width and pulse rise-time sizing of mammalian cells. J Histochem Cytochem 27:315, 1979
105. Shapiro HM, Feinstein DM, Kirsch A et al: Multistation multiparameter flow cytometry: some influences of instrumental factors on system performance. Cytometry 4:11, 1983
106. Peters DC: A comparison of mercury arc lamp and laser illumination for flow cytometers. J Histochem Cytochem 27:241, 1979
107. Koper GJM, Bonnet J, Christiaanse JGM et al: An epiilluminator/detector unit permitting arc lamp illumination for fluorescence activated cell sorters. Cytometry 3:10, 1982
108. Hirschfeld T: Fluorescence background discrimination by prebleaching. J Histochem Cytochem 27:96, 1979
109. Davies KE, Young BD, Elles RG et al: Cloning of a representative genomic library of the human X chromosome after sorting by flow cytometry. Nature 293:374, 1981
110. Mullaney PF, Van Dilla MA, Coulter JR et al: Cell sizing: a light scattering photometer for rapid volume determination. Rev Sci Instrum 40:1029, 1969
111. Salzman GC, Crowell JM, Goad CA et al: A flow-system multiangle light-scattering instrument for cell characterization. Clin Chem 21:1297, 1975
112. Shapiro HM: Flow cytometric probes of early events in cell activation. Cytometry 1:301, 1981
113. Shapiro HM: Flow cytometric estimation of DNA and RNA content in intact cells stained with Hoechst 33342 and pyronin Y. Cytometry 2:143-50, 1981
114. Oi VT, Glazer AN, Stryer L: Fluorescent phycobiliprotein conjugates for analyses of cells and molecules. J Cell Biol 93:981, 1982
115. Loken MR, Parks DR, Herzenberg LA: Two-color immunofluorescence using a fluorescence-activated cell sorter. J Histochem Cytochem 25:899, 1977
116. Hoffman PM, Davidson WF, Ruscetti SK et al: Wild mouse ecotropic murine leukemia virus infection of inbred mice: dual-tropic virus expression precedes the onset of paralysis and lymphoma. J Virol 39:597, 1981
117. Titus JA, Haugland R, Sharrow SO et al: Texas red, a hydrophilic red-emitting fluorophore for use with fluorescein in dual parameter flow microfluorimetric and fluorescence microscopic studies. J Immunol Meths 50:193, 1982
118. Parks DR, Stovel RT, Herzenberg LA: A tunable dye laser in a two laser FACS system. Cytometry 2:119, 1981
119. Arndt-Jovin DJ, Grimwade BG, Jovin TM: A dual laser flow sorter utilizing a CW pumped dye laser. Cytometry 1:127, 1980
120. Skogen-Hagenson MJ, Salzman GC, Mullaney PF et al: A high efficiency flow cytometer. J Histochem Cytochem 25:784, 1977
121. Van Dilla MA, Gledhill BL, Lake S et al: Measurement of mammalian sperm deoxyribonucleic acid by flow cytometry. Problems and approaches. J Histochem Cytochem 25:763, 1977
122. Wade CG, Rhyne RH Jr, Woodruff WH et al: Spectra of cells in flow cytometry using a vidicon detector. J Histochem Cytochem 27:1049, 1979
123. Engstrom RW: Photomultiplier Handbook. Lancaster (PA), RCA Corporation, 1980
124. McCutcheon MJ, Miller RG: Fluorescence intensity resolution in flow systems. J Histochem Cytochem 27:246, 1979
125. Steinkamp JA, Orlicky DA, Crissman HA: Dual-laser flow cytometry of single mammalian cells. J Histochem Cytochem 27:273, 1979
126. Shapiro HM: Discussion on flow systems. In: Wied GL, Bahr GF, Bartels PH (eds): *The Automation of Uterine Cancer Cytology*. Chicago, Tutorials of Cytology, 1976, p. 101
127. Graedel TE, McGill R: Graphical presentation of results from scientific computer models. Science 215:1191, 1982
128. Sharpless TK: Cytometric data processing. In reference 9, p.359
129. Bartels PH: Numerical evaluation of cytologic data. I. Description of profiles. Anal Quant Cytol 1:20, 1979
130. Bartels PH: Numerical evaluation of cytologic data. II. Comparison of profiles. Anal Quant Cytol 1:77, 1979
131. Bartels PH: Numerical evaluation of cytologic data. III. Selection of features for discrimination. Anal Quant Cytol 1:153, 1979
132. Bartels PH: Numerical evaluation of cytologic data. IV. Discrimination and classification. Anal Quant Cytol 2:19, 1980
133. Bartels PH: Numerical evaluation of cytologic data. V. Bivariate distributions and the Bayesian decision boundary. Anal Quant Cytol 2:77, 1980
134. Bartels PH: Numerical evaluation of cytologic data. VI. Multivariate distributions and matrix notation. Anal Quant Cytol 2:155, 1980
135. Bartels PH: Numerical evaluation of cytologic data. VII. Multivariate significance tests. Anal Quant Cytol 3:1, 1981
136. Bartels PH: Numerical evaluation of cytologic data. VIII. Computation of the principal components. Anal Quant Cytol 3:83, 1981
137. Bartels PH: Numerical evaluation of cytologic data. IX. Search for data structure by principal components transformation. Anal Quant Cytol 3:167, 1981
138. Bartels PH: Numerical evaluation of cytologic data. X. Introduction to multivariate analysis of variance. Anal Quant Cytol 3:251, 1981
139. Bartels PH: Numerical evaluation of cytologic data. XI. Nested designs in multivariate analysis of variance. Anal Quant Cytol 4:81, 1982
140. Bartels PH: Numerical evaluation of cytologic data. XII. Curve fitting. Anal Quant Cytol 4:241, 1982
141. Kolata G: Computer graphics comes to statistics. Science 217:919, 1982
142. Sweet RG: Flow sorters for biologic cells. In reference 9, p.177
143. Martin JC, McLaughlin SR, Hiebert RD: A real-time delay monitor for flow-system cell sorters. J Histochem Cytochem 27:277, 1979
144. Merrill JT, Dean PN, Gray JW: Investigations in high-precision sorting. J Histochem Cytochem 27:280, 1979
145. Stovel RT, Sweet RG: Individual cell sorting. J Histochem Cytochem 27:284, 1979
146. Duhnen J, Stegemann J, Wiezorek C et al: A new fluid switching flow sorter. Histochemistry 77:117, 1983
147. Arndt-Jovin DJ, Jovin TM: Computer-controlled multiparameter analysis and sorting of cells and particles. J Histochem Cytochem 22:622, 1974
148. Merrill JT: Evaluation of selected aerosol-control measures on flow sorters. Cytometry 1:342, 1981
149. De Mulder PHM, Wessels JMC, Rosenbrand DA et al: Monocyte purification with counterflow centrifugation monitored by continuous flow cytometry. J Immunol Meths 47:31, 1981
150. Shapiro HM: Apparatus and method for killing unwanted cells. U S Patent 4,395,397, 1983
151. Martin JC, Jett JH: Photodamage, a basis for super high speed cell selection. Cytometry 2:114, 1981

152. Schwartz A, Sugg H, Ritter TW et al: Direct determination of cell diameter, surface area, and volume with an electronic volume sensing flow cytometer. Cytometry 3:456, 1983
153. Kerker M, Chew H, McNulty PJ et al: Light scattering and fluorescence by small particles having internal structure. J Histochem Cytochem 27:250, 1979
154. Loken MR, Herzenberg LA: Analysis of cell populations with a fluorescence-activated cell sorter. Ann N Y Acad Sci 254:163, 1975
155. Loken MR, Sweet RG, Herzenberg LA: Cell discrimination by multiangle light scattering. J Histochem Cytochem 24:284, 1976
156. Loken MR, Houck DW: Light scattered at two wavelengths can discriminate viable lymphoid cell populations on a fluorescence-activated cell sorter. J Histochem Cytochem 29:609, 1981
157. Salzman GC, Crowell JM, Martin JC et al: Cell identification by laser light scattering: identification and separation of unstained leukocytes. Acta Cytol 19:374, 1975
158. Salzman GC, Mullaney PF, Price BJ: Light-scattering approaches to cell characterization. In reference 9, p.105
159. Salzman GC, Wilder ME, Jett JH: Light scattering with stream-in-air flow systems. J Histochem Cytochem 27:264, 1979
160. Sharpless T, Bartholdi M, Melamed MR: Size and refractive index dependence of simple forward angle scattering measurements in a flow system using sharply focused illumination. J Histochem Cytochem 25:845, 1977
161. Watt SM, Burgess AW, Metcalf D et al: Isolation of mouse bone marrow neutrophils by light scatter and autofluorescence. J Histochem Cytochem 28:934, 1980
162. Weil GJ, Chused TM: Eosinophil autofluorescence and its use in isolation and analysis of human eosinophils using flow microfluorometry. Blood 57:1099, 1981
163. Kerker M: Elastic and inelastic light scattering in Flow Cytometry (Paul Mullaney Memorial Lecture). Cytometry 4:1, 1983
164. Loken MR, Stout RD, Herzenberg LA: Lymphoid cell analysis and sorting. In reference 9, p.505
165. Dangl J, Parks D, Oi V et al: Rapid isolation of cloned isotype switch variants using fluorescence activated cell sorting. Cytometry 2:395, 1982
166. Kim YR, Ornstein L: Isovolumetric sphering of erythrocytes for more accurate and precise cell volume measurement by flow cytometry. Cytometry 4:419, 1983
167. Gray ML, Hoffman RA, Hansen WP: A new method for cell volume measurement based on volume exclusion of a fluorescent dye. Cytometry 4:428-34, 1983
168. Eisert WG: High resolution optics combined with high spatial reproducibility in flow. Cytometry 1:254, 1981
169. Monroe JG, Havran WL, Cambier JC: Enrichment of viable lymphocytes in defined cell cycle phases by sorting on the basis of pulse width of axial light extinction. Cytometry 3:24, 1982
170. Bator JM, Groves MR, Price BJ et al: Erythrocyte deformability and size measured in a multiparameter system that includes impedance sizing. Cytometry 5:34, 1984
171. Wheeless LL Jr: Slit-scanning and pulse width analysis. In reference 9, p.125
172. Cambier JL, Kay DB, Wheeless LL: A multidimensional slit-scan flow system. J Histochem Cytochem 27:321, 1979
173. Wheeless LL, Cambier MA, Kay DB et al: False alarms in a slit-scan flow system. Causes and occurrence rates. Implications and potential solutions. J Histochem Cytochem 27:596, 1979
174. Wheeless LL, Patten SF, Berkhan TK et al: Multidimensional slit-scan prescreening system: preliminary results of a single-blind clinical study. Cytometry 5:1, 1984
175. Hoffman RA, Kung PC, Hansen P et al: Simple and rapid measurement of human T lymphocytes and their subclasses in peripheral blood. Proc Natl Acad Sci USA 77:4914, 1980
176. Hoffman RA, Hansen WP: Immunofluorescent analysis of blood cells by flow cytometry. Int J Immunopharmacol 3:249, 1981
177. van den Engh GJ, Visser JWM: The morphology of the hematopoietic stem cell derived from scatter measurements on a light activated cell sorter. Annual Report, Radiobiological Laboratories, Amsterdam, 1977, p.47
178. Scott RB, Grogan WM, Collins JM: Separation of rabbit marrow precursor cells by combined isopycnic sedimentation and electronic cell sorting. Blood 51:1137, 1978
179. Nicola NA, Metcalf D, von Melchner H et al: Isolation of murine fetal hemopoietic progenitor cells and selective fractionation of various erythroid precursors. Blood 58:376, 1981
180. Otten GR, Loken MR: Two color light scattering identifies physical differences between lymphocyte subpopulations. Cytometry 3:182, 1982
181. Aubin J: Autofluorescence of viable cultured mammalian cells. J Histochem Cytochem 27:36, 1979
182. Lamola AA, Piomelli S, Poh-Fitzpatrick MB et al: Erythropoietic protoporphyria and lead intoxication: the molecular basis for difference in cutaneous photosensitivity. J Clin Inves 56:1528, 1975
183. Benson HC, Meyer RA, Zaruba ME et al: Cellular autofluorescence-is it due to flavins? J Histochem Cytochem 27:44, 1979
184. Ledbetter JA, Rouse RV, Micklem HS et al: T cell subsets defined by expression of Lyt-1,2,3 and Thy-1 antigens. Two-parameter immunofluorescence and cytotoxicity analysis with monoclonal antibodies modifies current views. J Exp Med 152:280, 1980
185. Chance B, Thorell B: Localization and kinetics of reduced pyridine nucleotide in living cells by microfluorometry. J Biol Chem 234:3044, 1959
186. Chance B, Cohen P, Jobsis F et al: Intracellular oxidation- reduction states *in vivo*. Science 137:499, 1962
187. Kohen E, Kohen C, Thorell B: A comparative study of pyridine-nucleotide metabolism in yeast and mammalian cells by microfluorometry-microelectrophoresis. Histochemie 12:95, 1968
188. Kobayashi S, Kaede K, Nishiki K et al: Microfluorometry of oxidation-reduction states of the rat kidney *in situ*. J Appl Physiol 31:693, 1971
189. Shapiro HM, Bier M, Zukoski CF: Continuous redox state monitoring of preserved organs. Proc Annu Conf Eng Med Biol 15:274, 1973
190. Apstein CA, Ahn J, Briggs L et al: Role of decrease in wall thickness in causing ischemic cardiac failure. Clin Res 27:436a, 1971
191. Thorell B: Intracellular red-ox steady states as basis for cell characterization by flow cytofluorometry. Blood Cells 6:745, 1980
192. Thorell B: Flow cytometric analysis of cellular endogenous fluorescence simultaneously with emission from exogenous fluorochromes, light scatter, and absorption. Cytometry 2:39, 1981
193. Thorell B: Flow-cytometric monitoring of intracellular flavins simultaneously with NAD(P)H levels. Cytometry 4:61, 1983
194. Hafeman DG, McConnell HM, Gray JW et al: Neutrophil activation monitored by flow cytometry: stimulation by phorbol diester is an all-or-none event. Science 215:673, 1982
195. Jongkind JF, Verkerk A, Visser WJ et al: Isolation of autofluorescent 'aged' human fibroblasts by flow sorting. Exp Cell Res 138:409, 1982
196. Jongkind JF, Verkerk A: Cell sorting and microchemistry of cultured human fibroblasts: applications in genetics and aging research. Cytometry 5:182, 1984
197. Trask BJ, van den Engh GJ, Elgershuizen JHBW: Analysis of phytoplankton by flow cytometry. Cytometry 2:258, 1982
198. Olson RJ, Frankel SL, Chisholm SW, Shapiro HM: An inexpensive flow cytometer for the analysis of fluorescence signals in phytoplankton: Chlorophyll and DNA distributions. J Exp Mar Biol Ecol 68:129, 1983
199. Arndt-Jovin DJ, Jovin TM: Automated cell sorting with flow systems. Ann Rev Biophys Bioeng 7:527, 1978
200. Valet G, Raffael A, Moroder L et al: Fast intracellular pH determination in single cells by flow cytometry. Naturwissenschaften 68:265, 1981
201. Gerson DF: Determination of intracellular pH changes in lymphocytes with 4-methylumbelliferone by flow microfluorometry. In: *In-*

tracellular pH: Its Measurement, Regulation, and Utilization in Cellular Functions, Nuccitelli R, Deamer DW (eds.), Alan R. Liss, New York, 1982, p.125

202. Heiple JM, Taylor DL: Intracellular pH in single motile cells. J Cell Biol 86:885, 1980
203. Thomas JA, Kolbeck PC, Langworthy TA: Spectrophotometric determination of cytoplasmic and mitochondrial pH using trapped pH indicators. In: *Intracellular pH: Its Measurement, Regulation, and Utilization in Cellular Functions*, Nuccitelli R, Deamer DW (eds.), Alan R. Liss, New York, 1982, p.10
204. Simons ER, Schwartz DB, Norman NE: Stimulus response coupling in human platelets: thrombin-induced changes in pH. In: *Intracellular pH: Its Measurement, Regulation, and Utilization in Cellular Functions*, Nuccitelli R, Deamer DW (eds.), Alan R. Liss, New York, 1982, p.463
205. Michaelis L, Granick S: Metachromasy of basic dyestuffs. J Am Chem Soc 67:1212, 1945
206. Sculthorpe HH: Metachromasia. Med Lab Sci 35:365, 1970
207. Miller DS, Lau Y-T, Horowitz SB: Artifacts caused by cell microinjection. Proc Natl Acad Sci USA 81:1426, 1984
208. Hopwood D: Recent advances in fixation of tissues. In: *Electron Microscopy and Cytochemistry*, Wisse E, Daems WTh, Molenaar I, van Duijn P (eds.), North Holland, Amsterdam, 1973, p.367
209. Stoward PJ (ed.): *Fixation in Histochemistry*. Chapman and Hall, London, 1973
210. Penttila A, McDowell EM, Trump BF: Effects of fixation and postfixation treatments on volume of injured cells. J Histochem Cytochem 23:251, 1975
211. Penttila A, McDowell EM, Trump BF: Optical properties of normal and injured cells. Application of cytographic analysis to cell viability and volume studies. J Histochem Cytochem 25:9, 1977
212. Hassell J, Hand AR: Tissue fixation with diimidoesters as an alternative to aldehydes. I. Comparison of cross-linking and ultrastructure obtained with dimethylsuberimidate and glutaraldehyde. J Histochem Cytochem 22:223, 1974
213. LePecq J-B, Yot P, Paoletti C: Interaction du bromhydrate d'ethidium (BET) avec les acides nucleiques (A.N.). Etude spectrofluorometrique. C R Acad Sci 259:1786, 1964
214. LePecq J-B, Paoletti C: A new fluorometric method for RNA and DNA determination. Anal Biochem 17:100, 1966
215. LePecq J-B, Paoletti C: A fluorescent complex between ethidium bromide and nucleic acids. Physical-chemical characterization. J Mol Biol 27:87, 1967
216. Hudson B, Upholt WB, Divinny J et al: The use of an ethidium bromide analogue in the dye-buoyant density procedure for the isolation of closed circular DNA: the variation of the superhelix density of mitochondrial DNA. Proc Natl Acad Sci USA 62:813, 1969
217. Crissman HA, Steinkamp JA: Rapid simultaneous measurement of DNA, protein and cell volume in single cells from large mammalian cell populations. J Cell Biol 59:766, 1973
218. Krishan A: Rapid flow cytofluorometric analysis of mammalian cell cycle by propidium iodide staining. J Cell Biol 66:188, 1975
219. Fried J, Perez AG, Clarkson BD: Flow cytofluorometric analysis of cell cycle distributions using propidium iodide. J Cell Biol 71:172, 1976
220. Fried J, Perez AG, Clarkson BD: Rapid hypotonic method for flow cytofluorometry of monolayer cell cultures. J Histochem Cytochem 26:921, 1978
221. Look AT, Melvin SL, Williams DL et al: Aneuploidy and percentage of S-phase cells determined by flow cytometry correlate with cell phenotype in childhood acute leukemia. Blood 60:959, 1982
222. Vindeløv LL, Christensen IJ, Keiding N et al: Long-term storage of samples for flow cytometric DNA analysis. Cytometry 3:317, 1982
223. Vindeløv LL, Christensen IJ, Nissen NI: A detergent-trypsin method for the preparation of nuclei for flow cytometric DNA analysis. Cytometry 3:323, 1982
224. Vindeløv LL, Christensen IJ, Nissen NI: Standardization of high-resolution flow cytometric DNA analysis by the simultaneous use of chicken and trout red blood cells as internal reference standards. Cytometry 3:328, 1982
225. Vindeløv LL, Christensen IJ, Jensen G et al: Limits of detection of nuclear DNA abnormalities by flow cytometric DNA analysis. Results obtained by a set of methods for sample-storage, staining and internal standardization. Cytometry 3:332, 1982
226. Taylor IW: A rapid single step staining technique for DNA analysis by flow microfluorimetry. J Histochem Cytochem 28:1021, 1980
227. Martens ACM, van den Engh GJ, Hagenbeek A: The fluorescence intensity of propidium iodide bound to DNA depends on the concentration of sodium chloride. Cytometry 2:24, 1981
228. Crissman HA, Tobey RA: Cell-cycle analysis in 20 minutes. Science 184:1297, 1974
229. Tobey RA, Crissman HA: Use of flow microfluorometry in detailed analysis of effects of chemical agents on cell cycle progression. Cancer Res 32:2726, 1972
230. Ward DC, Reich E, Goldberg IH: Base specificity in the interaction of polynucleotides with antibiotic drugs. Science 149:1259, 1965
231. Crissman HA, Stevenson AP, Orlicky DJ et al: Detailed studies on the application of three fluorescent antibiotics for DNA staining in flow cytometry. Stain Technol 53:321, 1978
232. Göhde W, Schumann J, Buchner T et al: Pulse cytophotometry: application in tumor cell biology and in clinical oncology. In reference 9, p.599
233. Barlogie B, Spitzer G, Hart JS et al: DNA histogram analysis of human hemopoietic cells. Blood 48:245, 1976
234. Loewe H, Urbanietz J: Basisch substitutierte 2,6-bis-benzimidazolderivate, eine neue chemotherapeutisch active korperklasse. Arzneimforsch 24:1927, 1974
235. Latt SA: Microfluorimetric detection of deoxyribonucleic acid replication in human metaphase chromosomes. Proc Natl Acad Sci USA 70:3395, 1973
236. Latt SA: Microfluorimetric analysis of deoxyribonucleic acid replication kinetics and sister chromatid exchanges in human chromosomes. J Histochem Cytochem 22:478, 1974
237. Latt SA: Detection of DNA synthesis in interphase nuclei by fluorescence microscopy. J Cell Biol 62:546, 1974
238. Latt SA, Stetten G: Spectral studies on 33258 Hoechst and related bisbenzimidazole dyes useful for fluorescent detection of deoxyribonucleic acid synthesis. J Histochem Cytochem 24:24, 1976
239. Arndt-Jovin DJ, Jovin TM: Analysis and sorting of living cells according to deoxyribonucleic acid content. J Histochem Cytochem 25:585, 1977
240. Hamori E, Arndt-Jovin DJ, Grimwade BG et al: Selection of viable cells with known DNA content. Cytometry 1:132, 1980
241. Loken MR: Simultaneous quantitation of Hoechst 33342 and immunofluorescence on viable cells using a fluorescence activated cell sorter. Cytometry 1:136, 1980
242. Lydon MJ, Keeler KD, Thomas DB: Vital DNA staining and sorting by flow microfluorometry. J Cell Physiol 102:175, 1980
243. Nicolini C, Kendall F, Desaive C et al: Physical-chemical characterization of living cells by laser-flow microfluorometry. Cancer Treatment Reps 60:1819, 1976
244. Burns VV: Studies with a fluorescent vital probe for DNA in mammalian cells. Exp Cell Res 107:459, 1977
245. Nicolini C, Belmont A, Parodi S et al: Mass action and acridine orange staining: Static and flow cytofluorometry. J Histochem Cytochem 27:102, 1979
246. Johnson TS, Swartzendruber DE, Martin JC: Nuclear size of G1/S transition cells measured by flow cytometry. Exp Cell Res 134:201, 1981
247. Pallavicini MG, Lalande ME, Miller RG et al: Cell cycle distribution of chronically hypoxic cells and determination of the clonogenic potential of cells accumulated in G_2 + M phases after irradiation of a solid tumor *in vivo*. Cancer Res 39:1891, 1979

248. Lalande ME, Miller RG: Fluorescence flow analysis of lymphocyte activation using Hoechst 33342 dye. J Histochem Cytochem 27:394, 1979
249. Lalande ME, McCutcheon MJ, Miller RG: Quantitative studies on the precursors of cytotoxic lymphocytes. VI. Second signal requirements of specifically activated precursors isolated 12 h after stimulation. J Exp Med 151:12, 1980
250. Loken MR: Separation of viable T and B lymphocytes using a cytochemical stain, Hoechst 33342. J Histochem Cytochem 28:36, 1980
251. Green DK, Malloy P, Steel M: The recovery of living cells by flow sorting machine. Acta Pathol Microbiol Scand, Sect A, Suppl 274:103, 1980
252. Visser JWM: Vital staining of haemopoietic cells with the fluorescent bis-benzimidazole derivatives Hoechst 33342 and 33258. Acta Pathol Microbiol Scand, Sect A, Suppl 274:86, 1980
253. Lalande ME, Ling V, Miller RG: Hoechst 33342 dye uptake as a probe of membrane permeability changes in mammalian cells. Proc Natl Acad Sci USA 78:363-7, 1981
254. Fried J, Doblin J, Takamoto S et al: Effects of Hoechst 33342 on survival and growth of two tumor cell lines and on hematopoietically normal bone marrow cells. Cytometry 3:42, 1982
255. Van Zant G, Fry CG: Hoechst 33342 staining of mouse marrow: effects on colony-forming cells. Cytometry 4:40, 1983
256. Kissane RJ, Tobey RA, Crissman HA et al: Detailed FCM and cell sorting studies of dye-binding kinetics, viability and cell growth of cells following DNA staining with Hoechst 33342. Cell Tissue Kinet 15:105, 1982
257. Williams JM, Shapiro HM, Milford EL, Strom TB: Multiparameter flow cytometric analysis of lymphocyte subpopulation activation in lectin-stimulated cultures. J Immunol 128:2676-81, 1982
258. Stöhr M, Eipel H, Goerttler K et al: Extended application of flow microfluorometry by means of dual laser excitation. Histochemistry 51:305, 1977
259. Russell WC, Newman C, Williamson DH: A simple cytochemical technique for demonstration of DNA in cells infected with mycoplasmas and viruses. Nature 253:461, 1975
260. Coleman AW, Maguire MJ, Coleman JR: Mithramycin- and 4'-6-diamidino- 2-phenylindole (DAPI)-DNA staining for fluorescence microspectrophotometric measurement of DNA in nuclei, plastids, and virus particles. J Histochem Cytochem 29:959, 1981
261. Thornthwaite JT, Sugarbaker EV, Temple WJ: Preparation of tissues for DNA flow cytometric analysis. Cytometry 1:229, 1980
262. Darzynkiewicz Z, Traganos F, Melamed MR: New cell cycle compartments identified by multiparameter flow cytometry. Cytometry 1:98, 1980
263. Traganos F, Darzynkiewicz Z, Sharpless T et al: Simultaneous staining of ribonucleic and deoxyribonucleic acids in unfixed cells using acridine orange in a flow cytofluorometric system. J Histochem Cytochem 25:46, 1977
264. Holländer R, Pohl S: Deoxyribonucleic acid base composition of bacteria. Zbl Bakt Hyg I Abt Orig A 246:236, 1980
265. Latt SA: Fluorescent probes of DNA microstructure and synthesis. In reference 9, p. 263
266. Müller W, Crothers D: Interactions of heteroaromatic compounds with nucleic acids. 1. The influence of heteroatoms and polarizability on the base specificity of intercalating ligands. Eur J Biochem 54:267, 1975
267. Müller W, Bünnemann H, Dattagupta N: Interactions of heteroaromatic compounds with nucleic acids. 2. Influence of substituents on the base and sequence specificity of intercalating ligands. Eur J Biochem 54:267, 1975
268. Müller W, Gautier F: Interactions of heteroaromatic compounds with nucleic acids: A-T- specific non-intercalating DNA ligands. Eur J Biochem 54:385, 1975
269. Caspersson T, Farber S, Foley GE et al: Chemical differentiation along metaphase chromosomes. Exp Cell Res 49:219, 1968
270. Caspersson T, Zech L, Modest EJ et al: Chemical differentiation with fluorescent alkylating agents in *Vicia faba* metaphase chromosomes. Exp Cell Res 58:141, 1969
271. Caspersson T, Zech L, Johansson C et al: Identification of human chromosomes by DNA binding fluorescent agents. Chromosoma 30:215, 1970
272. Caspersson T, Lomakka G, Zech L: The 24 fluorescence patterns of the human metaphase chromosomes- distinguishing characters and variability. Hereditas 67:89, 1971
273. Latt SA, Brodie S, Munroe SH: Optical studies of complexes of quinacrine with DNA and chromatin: Implications for the fluorescence of cytological chromosome preparations. Chromosoma 49:17, 1974
274. Latt SA, Sahar E, Eisenhard ME: Pairs of fluorescent dyes as probes of DNA and chromosomes. J Histochem Cytochem 27:65, 1979
275. Brodie S, Giron J, Latt SA: Estimation of accessibility of DNA in chromatin from fluorescence measurements of electronic excitation energy transfer. Nature 253:470, 1975
276. Jensen RH, Langlois RG, Mayall BH: Strategies for choosing a deoxyribonucleic acid stain for flow cytometry of metaphase chromosomes. J Histochem Cytochem 25:954, 1977
277. Langlois RG, Carrano AV, Gray JW, Van Dilla MA: Cytochemical studies of metaphase chromosomes by flow cytometry. Chromosoma 77:229-51, 1980
278. Langlois RG, Yu L-C, Gray JW, Carrano AV: Quantitative karyotyping of human chromosomes by dual beam flow cytometry. Proc Natl Acad Sci USA 79:7876-80, 1982
279. Van Dilla MA, Langlois RG, Pinkel D, Yajko D, Hadley WK: Bacterial characterization by flow cytometry. Science 220:620-2, 1983
280. Cowden RR, Curtis SK: Microfluorometric investigations of chromatin structure. I. Evaluation of nine DNA-specific fluorochromes as probes of chromatin organization. Histochemistry 72:11, 1981
281. Pollack A, Prudhomme DL, Greenstein DB et al: Flow cytometric analysis of RNA content in different cell populations using pyronin Y and methyl green. Cytometry 3:28, 1982
282. Lewis GN, Goldschmid O, Magel TT et al: Dimeric and other forms of methylene blue: Absorption and fluorescence of the pure monomer. J Am Chem Soc 65:1150, 1943
283. Bennion PJ, Horobin RW, Murgatroyd LB: The use of a basic dye (azure A or toluidine blue) plus a cationic surfactant for selective staining of RNA: A technical and mechanistic study. Stain Technol 50:307, 1975
284. Saunders AM: Histochemical identification of acid mucopolysaccharides with acridine orange. J Histochem Cytochem 12:164, 1964
285. Bauer KD, Dethlefsen LA: Total cellular RNA content: Correlation between flow cytometry and ultraviolet spectroscopy. J Histochem Cytochem 28:493, 1980
286. Tanke HJ, Niewenhuis IAB, Koper GJM et al: Flow cytometry of human reticulocytes based on RNA fluorescence. Cytometry 1:313, 1981
287. Tanke HJ, Rothbarth PH, Vossen JMJJ et al: Flow cytometry of reticulocytes applied to clinical hematology. Blood 61:1091, 1983
288. Sage BH Jr, O'Connell JP, Mercolino TJ: A rapid, vital staining procedure for flow cytometric analysis of human reticulocytes. Cytometry 4:222, 1983
289. Arndt-Jovin DJ: Cellular differentiation. In reference 9, p. 453
290. Jacobberger JW, Horan PK, Hare JD: Analysis of malaria parasite-infected blood by flow cytometry. Cytometry 4:228, 1983
291. Mikkelsen RB, Kazuyuki T, Wallach DFH: Membrane potential of Plasmodium-infected erythrocytes. J Cell Biol 93:685, 1982
292. Wachtler F, Musil R: Nucleoli visualized by silver staining combined with a new RNA-specific fluorochrome. Stain Technol 54:265, 1979
293. Frankfurt OS: Flow cytometric analysis of double-stranded RNA content distributions. J Histochem Cytochem 28:663, 1980
294. Wallen CA, Higashikubo R, Dethlefsen LA: Comparison of two flow cytometric assays for cellular RNA- acridine orange and propidium iodide. Cytometry 3:155, 1982

295. Freeman DA, Crissman HA: Evaluation of six fluorescent protein stains for use in flow microfluorometry. Stain Technol 50:279, 1975
296. Stöhr M, Vogt-Schaden M, Knobloch M et al: Evaluation of eight fluorochrome combinations for simultaneous DNA-protein flow analysis. Stain Technol 53:205, 1978
297. Crissman HA, Steinkamp JA: Rapid, one step staining procedures for analysis of cellular DNA and protein by single and dual laser flow cytometry. Cytometry 3:84, 1982
298. Leeman U, Ruch F: Cytofluorometric demonstration of basic and total proteins with sulfaflavine. J Histochem Cytochem 20:659, 1972
299. Chadwick CS, McEntegart MG, Nairn RC: Fluorescent protein tracers. A simple alternative to fluorescein. Lancet 1:412, 1958
300. Bayer E, Wilchek M: The avidin- biotin complex as a tool in molecular biology. Trends Biochem Sci 3:N257, 1978
301. Wofsy L, Henry C, Cammisuli S: Hapten- sandwich labeling of cell surface antigens. Contemp Top Mol Immunol 7:215, 1978
302. Shechter Y, Schlessinger J, Jacobs S et al: Fluorescent labeling of hormone receptors in viable cells: Preparation and properties of highly fluorescent derivatives of epidermal growth factor and insulin. Proc Natl Acad Sci USA 75:2135, 1978
303. Shapiro HM, Glazer AN, Christenson L et al: Immunofluorescence measurement in a flow cytometer using low-power helium-neon laser excitation. Cytometry 4:276, 1983
304. Abraham JL, Etz ES: Molecular microanalysis of pathological specimens *in situ* with a laser- Raman microprobe. Science 206:716, 1979
305. Ong LJ, Glazer AN, Waterbury JB: An unusual phycoerythrin from a marine cyanobacterium. Science 224:80, 1984
306. Glazer AN, Stryer L: Fluorescent tandem phycobiliprotein conjugates. Emission wavelength shifting by energy transfer. Biophys J 43:383, 1983
307. Kato K: High-efficiency, high-power difference- frequency generation at 2-4 μm in $LiNbO_3$. IEEE J Quant Elect QE-16:1017, 1980
308. Marason EG: Energy transfer dye mixture for argon pumped dye laser operation in the 700 to 800 nm region. Optics Commun 40:212, 1982
309. Hoffnagle J, Roesch LPh, Schlumpf N et al: CW operation of laser dyes styryl-9 and styryl-11. Optics Commun 42:267, 1982
310. Kraemer PM, Tobey RA, VanDilla MA: Flow microfluorimetric studies of lectin binding to mammalian cells. I. General features. J Cell Physiol 81:305, 1973
311. Steinkamp JA, Kraemer PM: Flow microfluorimetric studies of lectin binding to mammalian cells. II. Estimation of the surface density of receptor sites by multiparameter analysis. J Cell Physiol 84:197, 1974
312. Steinkamp JA, Kraemer PM: Quantitation of lectin binding by cells. In reference 9, p. 497
313. Bohn B: High-sensitivity cytofluorometric quantitation of lectin and hormone binding to surfaces of living cells. Exptl Cell Res 103:39, 1976
314. Chan SS, Arndt-Jovin D, Jovin TM: Proximity of lectin receptors on the cell surface measured by fluorescence energy transfer in a flow system. J Histochem Cytochem 27:56, 1979
315. Stryer L, Haugland RP: Energy transfer: a spectroscopic ruler. Proc Natl Acad Sci USA 58:719, 1967
316. Szöllösi J, Tron L, Damjanovich S et al: Fluorescence energy transfer measurements on cell surfaces: A critical comparison of steady- state fluorimetric and flow cytometric methods. Cytometry 5:210, 1984
317. Hardy RR, Hayakawa K, Parks DR et al: Demonstration of B-cell maturation in X-linked immunodeficient mice by simultaneous three-colour immunofluorescence. Nature 306:270, 1983
318. Lanier LL, Loken MR: Human lymphocyte subpopulations identified by using three- color immunofluorescence and flow cytometry analysis. Correlation of Leu-2, Leu-3, Leu-7, Leu-8 and Leu-11 cell surface antigen expression. J Immunol 132:151, 1984
319. Loken MR, Lanier LL: Three-color immunofluorescence analysis of Leu antigens on human peripheral blood using two lasers on a fluorescence-activated cell sorter. Cytometry 5:151, 1984
320. Parks DR, Hardy RR, Herzenberg LA: Three-color immunofluorescence analysis of mouse B-lymphocyte subpopulations. Cytometry 5:159, 1984
321. Kung PC, Goldstein G, Reinherz E, Schlossman SF: Monoclonal antibodies defining distinctive human T cell surface antigens. Science 206:347, 1979
322. Reinherz E, Schlossman S: The differentiation and function of human T lymphocytes. Cell 19:821, 1980
323. Stöhr M, Futterman G: Visualization of multidimensional spectra in flow cytometry. J Histochem Cytochem 27:560, 1979
324. Kirchanski SJ, Natale PJ: Reticulocyte detection via an automated flow cytometer: Histochemical versus immunochemical results. Cytometry 2:108, 1981
325. Li CKN, Hansen WP, Rubin RH et al: Immunofluorescence determination of neutrophil maturity. Cytometry 2:112, 1981
326. Bohn B: Flow cytometry: a novel approach for the quantitative analysis of receptor-ligand interactions on surfaces of living cells. Molec Cell Endocrinol 20:1, 1980
327. Bohn B, Manske W: Application of flow cytofluorometry to ligand binding studies on living cells. Practical aspects and recommendations for calibration and data processing. Acta Pathol Microbiol Scand, Sect A, Suppl 274:227, 1980
328. de Bruin HG, de Leur-Ebeling I, Aaij C: Quantitative determination of the number of FITC- molecules bound per cell in immunofluorescence flow cytometry. Vox Sang 45:373, 1983
329. Buican TN, Purcell A: 'Many-color' flow microfluorimetric analysis by multiplex labelling. Surv Immunol Res 2:178, 1983
330. Vyth-Dreese FA, Kipp JBA, DeJong TAM: Simultaneous measurement of surface immunoglobulins and cell cycle phase of human lymphocytes. Acta Pathol Microbiol Scand, Sect A, Suppl 274:207, 1980
331. Braylan RC, Benson NA, Nourse V et al: Correlated analysis of cellular DNA, membrane antigens and light scatter of human lymphoid cells. Cytometry 2:337, 1982
332. Shapiro HM: Redox balance in the body: an approach to quantitation. J Surg Res 3:138, 1972
333. Olive PL, Biaglow JE, Varnes ME et al: Characterization of the uptake and toxicity of a fluorescent thiol reagent. Cytometry 3:349, 1983
334. Durand RE, Olive PL: Flow cytometry techniques for studying cellular thiols. Rad Res 95:456, 1983
335. Rotman B, Papermaster BW: Membrane properties of living cells as studied by enzymatic hydrolysis of fluorogenic esters. Proc Natl Acad Sci USA 55:134, 1966
336. Bhuyan BK, Loughman BE, Fraser TJ et al: Comparison of different methods of determining cell viability after exposure to cytotoxic compounds. Exptl Cell Res 97:275, 1976.
337. Roper PR, Drewinko B: Comparison of *in vitro* methods to determine drug- induced cell lethality. Cancer Res 36:2182, 1976
338. Stewart S, Pritchard KI, Meakin JW et al: A flow system adaptation of the SCM test for detection of lymphocyte response in patients with recurrent breast cancer. Clin Immunol Immunopathol 13:171, 1979
339. Stöhr M, Vogt-Schaden M: A new dual staining technique for simultaneous flow cytometric DNA analysis of living and dead cells. Acta Pathol Microbiol Scand, Sect A, Suppl 274:96, 1980
340. Darzynkiewicz Z, Traganos F, Staiano- Coico L et al: Interactions of rhodamine 123 with living cells studied by flow cytometry. Cancer Res 42:799, 1982
341. Evenson DP, Darzynkiewicz Z, Melamed MR: Simultaneous measurement by flow cytometry of sperm cell viability and mitochondrial membrane potential related to cell motility. J Histochem Cytochem 30:279, 1982
342. Khaw BA, Scott J, Fallon JT et al: Myocardial injury: Quantitation by cell sorting initiated with antimyosin fluorescent spheres. Science 217:1050, 1982
343. Frankfurt OS: Assessment of cell viability by flow cytometric analysis using DNase exclusion. Exptl Cell Res 144:478, 1983
344. Jacobs DB, Pipho C: Use of propidium iodide staining and flow cytometry to measure antibody- mediated cytotoxicity: Resolution of

complement - sensitive and resistant target cells. J Immunol Meths 62: 101, 1983

345. Wallen CA, Higashikubo R, Roti Roti JL: Comparison of the cell kill measured by the Hoechst- propidium iodide flow cytometric assay and the colony formation assay. Cell Tissue Kinet 16:357, 1983
346. Martin JC, Swartzendruber DE: Time: A new parameter for kinetic measurements in flow cytometry. Science 207:199, 1980
347. Muirhead KA, Steinfeld RC, Severski MC et al: Anion transport hetrogeneity detected by flow cytometric measurement of NBD-taurine efflux kinetics. Cytometry 5:268, 1984
348. Beumer T, Pennings A, Beck H et al: Continuous measurement and analysis of staining kinetics by flow cytometry. Cytometry 4:244, 1983
349. Kachel V, Glossner E, Schneider H: A new flow cytometric transducer for fast sample throughput and time resolved kinetic studies of biological cells and other particles. Cytometry 3:202, 1982
350. Finney DA, Sklar LA: Ligand/receptor internalization: A kinetic, flow cytometric analysis of the internalization of N-formyl peptides by human neutrophils. Cytometry 4:54, 1983
351. Lindmo T, Fundingsrud K: Measurement of the distribution of time intervals between cell passages in flow cytometry as a method for the evaluation of sample preparation procedures. Cytometry 2:151, 1981
352. Krishan A, Ganapathi R: Laser flow cytometric studies on the intracellular fluorescence of anthracyclines. Cancer Res 40:3895, 1980
353. Nooter K, van den Engh G, Sonneveld P: Quantitative flow cytometric determination of anthracycline content of rat bone marrow cells. Cancer Res 43:5126, 1983
354. Kaufman RJ, Schimke RT: Amplification and loss of dihydrofolate reductase genes in a Chinese hamster ovary cell line. Mol Cell Biol 1:1069, 1981
355. Mariani BD, Slate DL, Schimke RT: S phase specific synthesis of dihydrofolate reductase in Chinese hamster ovary cells. Proc Natl Acad Sci USA 78:4985, 1981
356. Rosowsky A, Wright J, Shapiro H et al: A new fluorescent dihydrofolate reductase probe for studies of methotrexate resistance. J Biol Chem 257:14162, 1982
357. Schimke RT: Gene amplification, drug resistance, and cancer. Cancer Res 44:1735, 1984
358. Dolbeare FA, Smith RE: Flow cytoenzymology: Rapid enzyme analysis of single cells. In reference 9, p. 317
359. Kaplow LS, Lerner E: Computer-assisted monocyte esterase assay by flow-cytophotometry. J Histochem Cytochem 25:590, 1977
360. Kaplow LS: The application of cytochemistry to automation. J Histochem Cytochem 25:990, 1977
361. Nachlas MM, Tsou KC, DeSouza E et al: Cytochemical demonstration of succinic dehydrogenase by the use of a new p-nitrophenyl substituted ditetrazole. J. Histochem Cytochem 5:420, 1957
362. Bass DA, Parce JW, DeChatelet LR et al: Flow cytometric studies of oxidative product formation by neutrophils: A graded response to membrane stimulation. J Immunol 130:1910, 1983
363. Dolbeare F, Vanderlaan M: A fluorescent assay of proteinases in cultured mammalian cells. J Histochem Cytochem 27:1493, 1979
364. Smith RE, Dean PN: A study of acid phosphatase and dipeptidyl aminopeptidase II in monodispersed anterior pituitary cells using flow cytometry and electron microscopy. J Histochem Cytochem 27:1499, 1979
365. Watson JV: Enzyme kinetic studies in cell populations using fluorogenic substrates and flow cytometric techniques. Cytometry 1:143, 1980
366. Malin-Berdel J, Valet G: Flow cytometric determination of esterase and phosphatase activities and kinetics in hematopoietic cells with fluorogenic substrates. Cytometry 1:222, 1980
367. Haskill S, Becker S, Johnson T et al: Simultaneous three color and electronic cell volume analysis with a single UV excitation source. Cytometry 3:359, 1983
368. Goldstein JL, Brown MS, Krieger M et al: Demonstration of low density lipoprotein receptors in mouse teratocarcinoma stem cells and description of a method for producing receptor- deficient mutant mice. Proc Natl Acad Sci USA 76:2843, 1979
369. Murphy RF, Powers RS, Verderame M et al: Flow cytofluorimetric analysis of insulin binding and internalization by Swiss 3T3 cells. Cytometry 2:402, 1982
370. Métézeau P, Djavadi-Ohaniance L, Goldberg ME: The kinetics and homogenity of endocytosis of a receptor- bound ligand in a heterogeneous cell population studied by flow cytofluorometry. J Histochem Cytochem 30:359, 1982
371. Dunn PA, Tyrer HW: Quantitation of neutrophil phagocytosis using fluorescent latex beads. Correlation of microscopy and flow cytometry. J Lab Clin Med 98:374, 1981
372. Valet G, Jenssen HL, Krefft M et al: Flow cytometric measurements of the transmembrane potential, the surface charge density and the phagocytic activity of guinea pig macrophages after incubation with lymphokines. Blut 42:379, 1981
373. Steinkamp JA, Wilson JS, Saunders GC et al: Phagocytosis: Flow cytometric quantitation with fluorescent microspheres. Science 215:64, 1982
374. Bassøe C-F, Solsvik J, Laerum OD: Quantitation of single cell phagocytic capacity by flow cytometry. Acta Pathol Microbiol Scand, Sect A, Suppl 274:170, 1980
375. Bassøe C-F, Laerum OD, Glette J et al: Simultaneous measurement of phagocytosis and phagosomal pH by flow cytometry: Role of polymorphonuclear neutrophilic leukocyte granules in phagosome acidification. Cytometry 4:254, 1983
376. Murphy RF, Powers S, Cantor CR: Endosome pH measured in single cells by dual fluorescence flow cytometry: Rapid acidification of insulin to pH 6. J Cell Biol 98:1757, 1984
377. Valet G, Bamberger S, Hofmann H et al: Flow cytometry as a new method for the measurement of electrophoretic mobility of erythrocytes using membrane charge staining by fluoresceinated polycations. J Histochem Cytochem 27:342, 1979
378. Fisher B, Gunduz N, Zheng S et al: Fluoresceinated estrone binding by human and mouse breast cancer cells. Cancer Res 42:540, 1982
379. Tyrer HW, Pipho CJ, Mitra R et al: Studies to detect estrogen receptors in intact cells using cytofluorometric techniques. Cytometry 2:133, 1981
380. Kute TE, Linville C, Barrows G: Cytofluorometric analysis for estrogen receptors using fluorescent estrogen probes. Cytometry 4:132, 1983
381. Van NT, Raber M, Barrows GH, Barlogie B: Estrogen receptor analysis by flow cytometry. Science 224:876, 1984
382. Steel GG: *Growth Kinetics of Tumours*. Oxford, Oxford University Press, 1977
383. Latt SA: Fluorometric detection of deoxyribonucleic acid synthesis; possibilities for interfacing bromodeoxyuridine dye techniques with flow fluorometry. J Histochem Cytochem 25:913, 1977
384. Latt SA, George YS, Gray JW: Flow cytometric analysis of bromodeoxyuridine- substituted cells stained with 33258 Hoechst. J Histochem Cytochem 25:927, 1977
385. Swartzendruber DE: A bromodeoxyuridine (BUdR)- mithramycin technique for detecting cycling and non-cycling cells by flow microfluorometry. Exptl Cell Res 109:439, 1977
386. Darzynkiewicz Z, Andreeff M, Traganos F et al: Discrimination of cycling and noncycling lymphocytes by BUdR-suppressed acridine orange fluorescence in a flow cytometric system. Exptl Cell Res 115:31, 1978
387. Gray JW, Carver JH, George YS et al: Rapid cell cycle analysis by measurement of the radioactivity per cell in a narrow window in S phase (RCS_i). Cell Tissue Kinet 10:97, 1977
388. Böhmer R-M, Ellwart J: Combination of BUdR- quenched Hoechst fluorescence with DNA-specific ethidium bromide fluorescence for cell cycle analysis with a two-parametrical flow cytometer. Cell Tissue Kinet 14:653, 1981

389. Böhmer R-M, Ellwart J: Cell cycle analysis by combining the 5-bromodeoxyuridine/33258 Hoechst technique with DNA- specific ethidium bromide staining. Cytometry 2:31, 1981
390. Noguchi PD, Johnson JB, Browne W: Measurement of DNA synthesis by flow cytometry. Cytometry 1:390, 1981
391. Gratzner HG, Leif RC, Ingram DJ et al: The use of antibody specific for bromodeoxyuridine for the immunofluorescent determination of DNA replication in single cells and chromosomes. Exptl Cell Res 95:88, 1975
392. Gratzner HG, Pollack A, Ingram DJ: DNA replication in cells and chromosomes by immunological techniques. J Histochem Cytochem 24:34, 1976
393. Gratzner HG, Leif RC: An immunofluorescence method for monitoring DNA synthesis by flow cytometry. Cytometry 1:385, 1981
394. Gratzner HG: Monoclonal antibody to 5-bromo- and 5-iododeoxyuridine: A new reagent for detection of DNA replication. Science 218:474, 1982
395. Dolbeare F, Gratzner HG, Pallavicini MG et al: Flow cytometric measurement of total DNA content and incorporated bromodeoxyuridine. Proc Natl Acad Sci USA 80:5573, 1983
396. Betel I, Martijnse J, Van der Westen G: Mitogenic activation and proliferation of mouse thymocytes. Comparison between isotope incorporation and flow- microfluorometry. Exptl Cell Res 124:329, 1979
397. Shinitzky M, Inbar M: Microviscosity parameters and protein mobility in biological membranes. Biochim Biophys Acta 433:133, 1976
398. Shinitzky M, Henkart P: Fluidity of cell membranes- current concepts and trends. Int Rev Cytol 60:121, 1979
399. Singer SJ, Nicholson GL: The fluid mosaic model of the structure of cell membranes. Science 175:720, 1972
400. Jovin TM: Fluorescence polarization and energy transfer: Theory and application. In reference 9, p.137
401. Ben-Bassat H, Polliak A, Mitrani Rosenbaum S et al: Fluidity of membrane lipids and lateral mobility of concanavalin A receptors in the cell surface of normal lymphocytes and lymphocytes from patients with malignant lymphomas and leukemias. Cancer Res 37:1307, 1977
402. Inbar M, Goldman R, Inbar L et al: Fluidity difference of membrane lipids in human normal and leukemic lymphocytes as controlled by serum components. Cancer Res 37:3037, 1977
403. Inbar M, Shinitzky M: Decrease in microviscosity of lymphocyte surface membrane associated with stimulation induced by concanavalin A. Eur J Immunol 5:166, 1975
404. Levanon A, Kohn A, Inbar M: Increase in lipid fluidity of cellular membranes induced by adsorption of RNA and DNA virions. J Virol 22:353, 1977
405. Levanon A, Kohn A: Changes in cell membrane microviscosity associated with adsorption of viruses. FEBS Lett 85:245, 1978
405. Arndt-Jovin D, Ostertag W, Eisen H et al: Studies of cellular differentiation by automated cell separation. Two model systems: Friend virus-transformed cells and *Hydra attenuata*. J Histochem Cytochem 24:332, 1976
407. Lindmo T, Steen HB: Flow cytometric measurement of the polarization of fluorescence from intracellular fluorescein in mammalian cells. Biophys J 18:173, 1977
408. Epstein M, Norman A, Pinkel D et al: Flow-system fluorescence polarization measurements on fluorescein-diacetate stained EL4 cells. J Histochem Cytochem 25:821, 1977
409. Keene JP, Hodgson BW: A fluorescence polarization flow cytometer. Cytometry 1:118, 1980
410. Stewart SS, Miller RG, Price GB: A design for a real-time fluorescence polarization computer. Cytometry 1:204, 1980
411. Lelkes PI: Potential dependent rigidity changes in lipid membrane vesicles. Biochem Biophys Res Commun 90:656, 1979
412. Karnovsky MJ, Kleinfeld AM, Hoover RL et al: The concept of lipid domains in membranes. J Cell Biol 94:1, 1982
413. Schaap GH, de Josselin de Jong JE, Jongkind JF: Fluorescence polarization of six membrane probes in embryonal carcinoma cels after differentiation as measured on a FACS II cell sorter. Cytometry 5:188, 1984
414. Cercek L, Cercek B: Application of the phenomenon of changes in the structuredness of cytoplasmic matrix (SCM) in the diagnosis of malignant disorders: a review. Europ J Cancer 13:903, 1977
415. Cercek L, Cercek B, Ockey CH: Structuredness of the cytoplasmic matrix and Michaelis-Menten constants for the hydrolysis of FDA during the cell cycle in Chinese hamster ovary cells. Biophysik 10:187, 1973
416. Cercek L, Cercek B: Changes in the structuredness of cytoplasmic matrix (SCM) in human lymphocytes induced by phytohaemagglutinin and cancer basic protein as measured on single cells. Brit J Cancer 33:359, 1976
417. Price GB, McCutcheon MJ, Taylor WB et al: Measurement of cytoplasmic fluorescence depolarization of single cells in a flow system. J Histochem Cytochem 25:597, 1977
418. Hartmann W, Beisker W, Eisert R et al: Fluorescence polarization measurements on human lymphocytes from patients with and without neoplasia. Acta Pathol Microbiol Scand, Sect A, Suppl 274:183, 1980
419. Udkoff R, Chan S, Norman A: Identification of mitogen responding lymphocytes by fluorescence polarization. Cytometry 1:265, 1981
420. Meisingset KK, Steen HB: Intracellular binding of fluorescein in lymphocytes. Cytometry 1:272, 1981
421. Sonenberg M, Schneider AS: Hormone action at the plasma membrane: Biophysical approaches. In: ***Receptors and Recognition***, Series A, Vol 4, Cuatrecasas P, Greaves MF (eds), Chapman and Hall, London, 1977, p.1
422. Hirata F, Axelrod J: Phospholipid methylation and biological signal transmission. Science 209:1082, 1980
423. Freedman MH, Raff MC, Gomperts B: Induction of increased calcium uptake in mouse T lymphocytes by concanavalin A and its modulation by cyclic nucleotides. Nature 255:378, 1975
424. Shapiro HM, Natale PJ, Kamentsky LA: Estimation of membrane potentials of individual lymphocytes by flow cytometry. Proc Natl Acad Sci USA 76:5728-30, 1979
425. Kiefer H, Blume AJ, Kaback HR: Membrane potential changes during mitogenic stimulation of mouse spleen lymphocytes. Proc Natl Acad Sci USA 77:2200, 1980
426. Tsien RY, Pozzan T, Rink TJ: T-cell mitogens cause early changes in cytoplasmic free Ca^{+2} and membrane potential in lymphocytes. Nature 295:68, 1972
427. Gerson D, Kiefer H, Eufe W: Intracellular pH of mitogen-stimulated lymphocytes. Science 216:1009, 1982
428. Horne WC, Simons ER: Probes of transmembrane potentials in platelets: Changes in cyanine dye fluorescence in response to aggregation stimuli. Blood 51:741, 1978
429. Friedhoff LT, Kim E, Priddle M et al: The effect of altered transmembrane ion gradients on membrane potential and aggregation of human platelets in blood plasma. Biochem Biophys Res Commun 102:832, 1981
430. Rink TJ, Smith SW, Tsien RY: Cytoplasmic free Ca^{+2} in human platelets: Ca^{+2} thresholds and Ca- independent activation for shape change and secretion. FEBS Lett 148:21, 1982
431. Feinstein MB, Egan JJ, Sha'afi RI et al: The cytoplasmic concentration of free calcium in platelets is controlled by stimulators of cyclic AMP production (PGD_2, PGE_1, forskolin). Biochem Biophys Res Commun 113:598, 1983
432. Ishizaka T, Hirata F, Ishizaka K et al: Stimulation of phospholipid methylation, Ca^{2+} influx, and histamine release by bridging of IgE receptors on rat mast cells. Proc Natl Acad Sci USA 77:1903, 1980
433. Lewis RA, Austen KF: Mediation of local homeostasis and inflammation by leukotrienes and other mast cell-dependent compounds. Nature 293:103, 1981
434. Hesketh TR, Beaven MA, Rogers J et al: Stimulated release of histamine by a rat mast cell line is inhibited during mitosis. J Cell Biol 98:2250, 1984

435. White JR, Ishizaka T, Ishizaka K et al: Direct demonstration of increased intracellular concentration of free calcium as measured by quin-2 in stimulated rat peritoneal mast cell. Proc Natl Acad Sci USA 81:3978, 1984
436. Korchak HM, Weissmann G: Changes in membrane potential of human granulocytes antecede the metabolic responses to surface stimulation. Proc Natl Acad Sci USA 75:3818, 1978
437. Naccache PH, Volpi M, Showell HJ et al: Chemotactic factor- induced release of membrane calcium in rabbit neutrophils. Science 203:461, 1979
438. Grinstein S, Furuya W: Amiloride- sensitive Na^+/H^+ exchange in human neutrophils: Mechanism of activation by chemotactic factors. Biochem Biophys Res Commun 122:755, 1984
439. Boynton AL, McKeehan WL, Whitfield (eds): *Ions, Cell Proliferation, and Cancer.* New York, Academic Press, 1982
440. Schuldiner S, Rozengurt E: Na^+/H^+ antiport in Swiss 3T3 cells: Mitogenic stimulation leads to cytoplasmic alkalinization. Proc Natl Acad Sci USA 79:7778, 1982
441. Rothenberg P, Reuss L, Glaser L: Serum and epidermal growth factor transiently depolarize quiescent BSC-1 epithelial cells. Proc Natl Acad Sci USA 79:7783, 1982
442. Cassel D, Rothenberg P, Zhuang Y-X et al: Platelet-derived growth factor stimulates Na^+/H^+ exchange and induces cytoplasmic alkalinization in NR6 cells. Proc Natl Acad Sci USA 80:6224, 1983
443. Burns P, Rozengurt E: Serum, platelet-derived growth factor, vasopressin and phorbol esters increase intracellular pH in Swiss 3T3 cells. Biochem Biophys Res Commun 116:931, 1983
444. Gomperts B: *The Plasma Membrane: Models for Structure and Function.* London, Academic Press, 1977
445. Willingham MC, Pastan I: The visualization of fluorescent probes in living cells by video intensification microscopy (VIM). Cell 13:501, 1978
446. Schlessinger J, Shechter Y, Willingham MC et al: Direct visualization of binding, aggregation, and internalization of insulin and epidermal growth factor on living fibroblastic cells. Proc Natl Acad Sci 75:2659, 1978
447. MacInnes DG, Green DK, Harmar A et al: Neuroendocrine receptor-ligand binding using quantitative video-intensification microscopy and fluorescence- activated cell sorting. Quart J Exptl Physiol 68:463, 1983
448. Osband ME, Cohen EB, McCaffrey RP et al: A technique for the flow cytometric analysis of lymphocytes bearing histamine receptors. Blood 56:923, 1980
449. Hallberg T, Dohlsten M, Baldetorp B: Demonstration of histamine receptors on human platelets by flow cytometry. Scand J Hematol 32:113, 1984
450. Sklar L, Finney DA: Analysis of ligand-receptor interactions with the fluorescence activated cell sorter. Cytometry 3:161, 1982
451. Steiner M: Fluorescence microscopic studies of the transferrin receptor in human erythroid precursor cells. J Lab Clin Med 96:1086, 1980
452. Goding JW, Burns JF: Monoclonal antibody OKT-9 recognizes the receptor for transferrin on human acute lymphocytic leukemia cells. J Immunol 127:1256, 1981
453. Fingerroth JD, Weis JJ, Tedder TF et al: Epstein-Barr virus receptor of human B lymphocytes is the C3d receptor CR2. Proc Natl Acad Sci USA 81:4510, 1984
454. Osband M, McCaffrey R, Shapiro H: Cell sorting of histamine H_2-receptor bearing T-cells based upon changes in membrane potential following histamine binding. Blood 54(Supp 1):90a, 1979
455. Goldman DE: Potential, impedance, and rectification in membranes. J Gen Physiol 27:37-60, 1944
456. Taki M: Studies on blastogenesis of human lymphocytes by phytohemagglutinin, with special reference to changes of membrane potential during blastoid transformation. Mie Med J 19:245, 1970
457. Cohen LB, Salzberg BM: Optical measurement of membrane potential. Rev Physiol Biochem Pharmacol 83:35-88, 1978
458. Hoffman JF, Laris PC: Determination of membrane potentials in human and *Amphiuma* red cells by means of a fluorescent probe. J Physiol 239:519, 1974
459. Bakeeva LE, Grinius LL, Jasaitis AA et al: Conversion of biomembrane-produced energy into electric form. II. Intact mitochondria. Biochim Biophys Acta 216:13, 1970
460. Sims PJ, Waggoner AS, Wang C-H et al: Studies on the mechanism by which cyanine dyes measure membrane potential in red blood cells and phosphatidylcholine vesicles. Biochemistry 13:3315-30, 1974
461. Fisher NI, Hamer FM: A comparison of the absorption spectra of some typical symmetrical cyanine dyes. Proc Roy Soc A 154:703, 1936
462. Waggoner AS: Dye indicators of membrane potential. Ann Rev Biophys Bioeng 8:47, 1979
463. Felber SM, Brand MD: Factors determining the plasma-membrane potential of lymphocytes. Biochem J 204:577, 1982
464. Felber SM, Brand MD: Valinomycin can depolarize mitochondria in intact lymphocytes without increasing plasma membrane potassium fluxes. FEBS Lett 150:122, 1982
465. Felber SM, Brand MD: Early plasma-membrane-potential changes during stimulation of lymphocytes by concanavalin A. Biochem J 210:885, 1983
466. Felber SM, Brand MD: Concanavalin A causes an increase in sodium permeability and intracellular sodium content of pig lymphocytes. Biochem J 210:893, 1983
467. Valinsky JE, Easton TG, Reich E: Merocyanine 540 as a fluorescent probe of membranes: Selective staining of leukemic and immature hemopoietic cells. Cell 13:487, 1978
468. Grinvald A, Fine A, Farber IC et al: Fluorescence monitoring of electrical responses from small neurons and their processes. Biophys J 42:195, 1983
469. Loew LM: Design and characterization of electrochromic membrane probes. J Biochem Biophys Meths 6:243, 1982
470. Latt SA, Marino M, Lalande M: New fluorochromes, compatible with high wavelength excitation, for flow cytometric analysis of cellular nucleic acids. Cytometry 5:339, 1984
471. Lichtman MA, Weed RI: The monovalent cation content and adenosine triphosphatase activity of human normal and leukemic granulocytes and lymphocytes: Relationship to cell volume and morphologic age. Blood 34:645, 1969
472. Gelfand EW, Cheng RKK, Ha K et al: Volume regulation in lymphoid leukemia cells and assignment of cell lineage. N Engl J Med 311:939, 1984
473. Miller JB, Koshland DE Jr: Effects of cyanine dye membrane probes on cellular properties. Nature 272:83, 1978
474. Montecucco C, Pozzan T, Rink T: Dicarbocyanine fluorescent probes of membrane potential block lymphocyte capping, deplete cellular ATP and inhibit respiration of isolated mitochondria. Biochim Biophys Acta 552:552, 1979
475. Zigman S, Gilman P Jr: Inhibition of cell division and growth by a redox series of cyanine dyes. Science 208:188, 1980
476. Johnstone RM, Laris PC, Eddy AA: The use of fluorescent dyes to measure membrane potentials: A critique. J Cell Physiol 112:298, 1982
477. Smith TC: The use of fluorescent dyes to measure membrane potentials: A response. J Cell Physiol 112:302, 1982
478. Korchak HM, Rich A, Wilkenfeld C et al: Membrane potential changes during neutrophil (PMN) activation- A comparison of probes. J Cell Biol 95:243a, 1982
479. Whitin JC, Clark RA, Simons ER et al: Effects of the myeloperoxidase system on fluorescent probes of granulocyte membrane potential. J Biol Chem 256:8904, 1981
480. Monroe JG, Cambier JC: B cell activation. I. Anti-immunoglobulin-induced receptor cross-linking results in a decrease in the plasma membrane potential of murine B lymphocytes. J Exp Med 157:2073, 1983

481. Monroe JG, Cambier JC: B cell activation. III. B cell plasma membrane depolarization and hyper-Ia antigen expression induced by receptor immunoglobulin cross-linking are coupled. J Exp Med 158:1589, 1983
482. Cambier JC, Monroe JG: B cell activation. V. Differentiation signaling of B cell membrane depolarization, Increased I-A expression, G_0 to G_1 transition, and thymidine uptake by anti-IgM and anti-IgD antibodies. J Immunol 133:76, 1984
483. Seligmann B, Chused TM, Gallin JI: Human neutrophil heterogeneity identified using flow microfluorometry to monitor membrane potential. J Clin Inves 68:1125, 1981
484. Shapiro HM, Strom TB: Lectin effects on lymphocyte membrane potentials: B and T cells respond differently. Blood 54 (Supp 1):92a, 1979
485. Shapiro HM, Strom TB: Electrophysiology of T-lymphocyte cholinergic receptors. Proc Natl Acad Sci USA 77:4317, 1980
486. Rosenthal KS, Shapiro HM: Cell membrane potential changes follow Epstein-Barr virus binding. J Cell Physiol 117:39, 1983
487. Mitchell P: Keilin's respiratory chain concept and its chemiosmotic consequences. Science 206:1148, 1979
488. Skulachev VP, Hinkle PC (eds): *Chemiosmotic Proton Circuits in Biological Membranes*. Reading (MA), Addison-Wesley, 1981.
489. Johnson LV, Walsh ML, Chen LB: Localization of mitochondria in living cells with rhodamine 123. Proc Natl Acad Sci USA 77:990, 1980
490. Johnson LV, Walsh ML, Bockus BJ et al: Monitoring of relative mitochondrial membrane potential in living cells by fluorescence microscopy. J Cell Biol 88:526, 1981
491. Cohen RL, Muirhead KA, Gill JE et al: A cyanine dye distinguishes between cycling and non-cycling fibroblasts. Nature 290:593, 1981
492. James TW, Bohman R: Proliferation of mitochondria during the cell cycle of human cell line (HL-60). J Cell Biol 89:256, 1981
493. Darzynkiewicz Z, Staiano-Coico L, Melamed MR: Increased mitochondrial uptake of rhodamine 123 during lymphocyte stimulation. Proc Natl Acad Sci USA 77:6696, 1981
494. Shapiro HM. Cytological assay procedure. U S Pat No 4, 343, 782, Aug 1982
495. Bernal SD, Shapiro HM, Chen LB: Monitoring the effect of anticancer drugs on L1210 cells by a mitochondrial probe, rhodamine-123. Int J Cancer 30:219, 1982
496. Johnson LV, Summerhayes IC, Chen LB: Decreased uptake and retention of rhodamine 123 by mitochondria in feline sarcoma virus-transformed mink cells. Cell 28:7, 1982.
497. Lampidis TJ, Bernal SD, Summerhayes IC et al: Selective toxicity of rhodamine 123 in carcinoma cells *in vitro*. Cancer Res 43:716, 1983
498. Bernal SD, Lampidis TJ, McIsaac RM et al: Anticarcinoma activity *in vivo* of rhodamine 123, a mitochondria-specific dye. Science 222:169, 1983
499. Modica-Napolitano S, Weiss MJ, Chen LB et al: Rhodamine 123 inhibits bioenergetic function of isolated rat liver mitochondria. Biochem Biophys Res Commun 118:717, 1984
500. Bereiter-Hahn J: Dimethylaminostyrylmethylpyridiniumiodine (DASPMI) as a fluorescent probe for mitochondria in situ. Biochim Biophys Acta 423:1, 1976
501. Goodman LS, Gilman A: *The Pharmacologic Basis of Therapeutics*, 3rd Ed, New York, Macmillan, 1965, pp. 1064-6
502. Thomas MV: Techniques in Calcium Research. London, Academic Press, 1982
503. Blinks JR, Wier WG, Hess P et al: Measurement of Ca^{2+} concentrations in living cells. Prog Biophys Molec Biol 40:1, 1982
504. Tsien RY: Intracellular measurements of ion activities. Ann Rev Biophys Bioeng 12:91, 1983
505. Caswell AH, Hutchison JD: Visualization of membrane bound cations by a fluorescent technique. Biochem Biophys Res Commun 42:43, 1971
506. Caswell AH, Hutchison JD: Selectivity of cation chelation to tetracyclines: Evidence for special conformation of calcium chelate. Biochem Biophys Res Commun 43:625, 1971
507. Chandler DE, Williams JA: Intracellular divalent cation release in pancreatic acinar cells during stimulus-secretion coupling. I. Use of chlorotetracycline as fluorescent probe. J Cell Biol 76:371, 1978
508. Chandler DE, Williams JA: Intracellular divalent cation release in pancreatic acinar cells during stimulus-secretion coupling. II. Subcellular localization of the fluorescent probe chlorotetracycline. J Cell Biol 76:386, 1978
509. Mikkelsen RB, Schmidt-Ullrich R: Concanavalin A induces the release of intracellular Ca^{2+} in intact rabbit thymocytes. J Biol Chem 255:5177, 1980
510. Mathew MK, Balaram P: A reinvestigation of chlortetracycline fluorescence: Effect of pH, metal ions, and environment. J Inorg Biochem 13:339, 1980
511. Tsien RY: New calcium indicators and buffers with high selectivity against magnesium and protons: design, synthesis and properties of prototype structures. Biochemistry 19:2396, 1980
512. Tsien RY: A non-disruptive technique for loading calcium buffers and indicators into cells. Nature 290:527, 1981
513. Tsien RY, Pozzan T, Rink TJ: Calcium homeostasis in intact lymphocytes: Cytoplasmic free calcium monitored with a new, intracellularly trapped fluorescent indicator. J Cell Biol 94:325, 1982
514. Pozzan T, Arslan P, Tsien RY et al: Anti-immunoglobulin, cytoplasmic free calcium, and capping in B lymphocytes. J Cell Biol 94:335, 1982
515. Weiss A, Imboden J, Shoback D et al: Role of T3 surface molecules in human T-cell activation: T3-dependent activation results in an increase in cytoplasmic free calcium. Proc Natl Acad Sci USA 81:4169, 1984
516. Ohsako S, Deguchi T: Receptor-mediated regulation of calcium mobilization and cyclic GMP synthesis in neuroblastoma cells. Biochem Biophys Res Commun 122:333, 1984
517. Powell T, Tatham PER, Twist VW: Cytoplasmic free calcium measured by quin2 fluorescence in isolated ventricular myocytes at rest and during potassium-depolarization. Biochem Biophys Res Commun 122:1012, 1984
518. Kruskal BA, Keith CH, Maxfield FR: Thyrotropin-releasing hormone-induced changes in intracellular $[Ca^{2+}]$ measured by microspectrofluorometry on individual quin2-loaded cells. J Cell Biol 99:1167, 1984
519. Visser JWM, Jongeling AAM, Tanke HJ: Intracellular pH-determination by fluorescence measurements. J Histochem Cytochem 27:32, 1979
520. Alabaster O: Tumor cell metabolic heterogeneity: An adaptive survival response. Proc Amer Assn Cancer Res 24:6, 1983
521. Gerson DF, Kiefer H: Intracellular pH and the cell cycle of mitogen-stimulated murine lymphocytes. J Cell Physiol 114:132, 1983
522. Crissman HA, Van Egmond J, Holdrinet RS et al: Simplified method for DNA and protein staining of human hematopoietic cell samples. Cytometry 2:59, 1981
523. Roti Roti J, Higashikubo R, Blair OC, Uygur N: Cell cycle position and nuclear protein content. Cytometry 3:91-6, 1982
524. Kufe DW, Nadler L, Sargent L et al: Biological behavior of human breast carcinoma-associated antigens expressed during cell proliferation. Cancer Res 43:851, 1983
525. Darzynkiewicz Z, Sharpless T, Staiano-Coico L, Melamed MR: Subcompartments of the G_1 phase of cell cycle detected by flow cytometry. Proc Natl Acad Sci USA 77:6696, 1980
526. Darzynkiewicz Z, Crissman H, Traganos F et al: Cell heterogeneity during the cell cycle. J Cell Physiol 113:465, 1982
527. Kimmel M, Traganos F, Darzynkiewicz Z: Do all daughter cells enter the "indeterminate" ("A") state of the cell cycle? Analysis of stathmokinetic experiments on L1210 cells. Cytometry 4:191, 1983
528. Smith JA, Martin L: Do cells cycle? Proc Natl Acad Sci USA 70:1263, 1973

529. Baserga R, Growth in size and cell DNA replication. Exptl Cell Res 151:1, 1984
530. Zetterberg A, Engström W, Dafgård E: The relative effects of different types of growth factors on DNA replication, mitosis, and cellular enlargement. Cytometry 5:368, 1984
531. Swartzendruber DE, Travis GL, Martin JC: Flow cytometric analysis of the effect of 5-bromodeoxyuridine on mouse teratocarcinoma cells. Cytometry 1:238, 1980
532. Levenson R, Macara IG, Smith RL et al: Role of mitochondrial membrane potential in the regulation of murine erythroleukemia cell differentiation. Cell 28:855, 1982
533. Kamarck ME, Barbosa JA, Kuhn L et al: Somatic cell genetics and flow cytometry. Cytometry 4:99, 1983
534. Kavathas P, Herzenberg LA: Stable transformation of mouse L cells for human membrane T-cell differentiation antigens, HLA, and β_2-microglobulin: Selection by fluorescence-activated cell sorting. Proc Natl Acad Sci USA 80:524, 1983
535. Le Bouteiller PP, Mishal Z, Lemonnier FA et al: Quantitation by flow cytofluorimetry of HLA class I molecules at the surface of murine cells transformed by cloned HLA genes. J Immunol Meths 61:301, 1983
536. Schaap GH, Verkerk A, Van Der Kamp AWM et al: Selection of proliferating cybrid cells by dual laser flow sorting. Exptl Cell Res 140:299, 1982
537. Lebo RV: Chromosome sorting and DNA sequence localization. Cytometry 3:145, 1982
538. Sillar R, Young BD: A new method for the preparation of metaphase chromosomes for analysis. J Histochem Cytochem 29:74, 1981
539. Stöhr M, Hutter KJ, Frank M et al: A reliable preparation of monodispersed chromosome suspensions for flow cytometry. Histochemistry 74:57, 1982
540. van den Engh G, Trask B, Cram S et al: Preparation of chromosome suspensions for flow cytometry. Cytometry 5:108, 1984
541. Lebo RV, Bastian AM: Design and operation of a dual-laser chromosome sorter. Cytometry 3:213, 1982
542. Collard JG, Philippus E, Tulp A et al: Separation and analysis of human chromosomes by combined velocity sedimentation and flow sorting applying single-and dual-laser flow cytometry. Cytometry 5:9, 1984
543. Young BD, Ferguson-Smith MA, Sillar R et al: High resolution chromosome analysis of human peripheral lymphocyte chromosomes by flow cytometry. Proc Natl Acad Sci USA 78:7727, 1981
544. Disteche CM, Kunkel LM, Lojewski A et al: Isolation of mouse X-chromosome specific DNA from an X-enriched lambda phage library derived from flow sorted chromosomes. Cytometry 2:282, 1982
545. Krumlauf R, Jeanpierre M, Young BD: Construction and characterization of genomic libraries from specific human chromosomes. Proc Natl Acad Sci USA 79:2971, 1982
546. Fantes JA, Green DK, Cooke HJ: Purifying human Y chromosomes by flow cytometry and sorting. Cytometry 4:88, 1983
547. Lalande M, Kunkel LM, Flint A et al: Development and use of metaphase chromosome flow-sorting methodology to obtain recombinant phage libraries enriched for parts of the human X chromosome. Cytometry 5:101, 1984
548. Rabinovich PS, Martin GM, Hoehn H: Interphase flow-cytogenetics: Correlation of DNA fluorescence to aneuploidy in human fibroblast cultures. Hum Genet 61:246, 1982
549. Dangl JL, Herzenberg LA: Selection of hybridomas and hybridoma variants using the fluorescence-activated cell sorter. J Immunol Meths 52:1, 1982
550. Lo MMS, Tsong TY, Conrad MK et al: Monoclonal antibody production by receptor-mediated electrically induced cell fusion. Nature 310:792, 1984
551. Cantrell DA, Smith KA: The interleukin-2 T-cell system: A new cell growth model. Science 224:1312, 1984
552. Falchuk KH, Krishan A, Vallee BL: DNA distribution in the cell cycle of *Euglena gracilis*. Cytofluorometry of zinc deficient cells. Biochemistry 14:3439, 1974
553. Bailey JE, Fazel-Madjlessi J, McQuitty DN et al: Characterization of bacterial growth by means of flow microfluorometry. Science 198:1175, 1977
554. Paau AS, Cowles JR, Oro J: Flow-microfluorometric analysis of *Escherichia coli*, *Rhizobium meliloti*, and *Rhizobium japonicum* at different stages in the growth cycle. Can J Microbiol 23:1165, 1977
555. Paau AS, Lee D, Cowles JR: Comparison of nucleic acid content in free-living and symbiotic *Rhizobium meliloti* by flow microfluorometry. J Bacteriol 129:1156, 1977
556. Slater ML, Sharrow SO, Gart JJ: Cell cycle of *Saccharomyces cerevisiae* in populations growing at different stages. Proc Natl Acad Sci USA 74:3850, 1977
557. Hutter K-J, Eipel HE: Flow cytometric determinations of cellular substances in algae, bacteria, moulds and yeasts. Antonie van Leeuwenhoek 44:269, 1978
558. Bailey JE, Fazel-Madjlessi J, McQuitty DN et al: Measurement of structured microbial population dynamics by flow microfluorometry. J Am Inst Chem Eng 24:570, 1978
559. Fazel-Madjlessi J, Bailey JE: Analysis of fermentation processes using flow microfluorometry: Single-parameter observations of batch bacterial growth. Biotechnol Bioeng 21:1955, 1979
560. Hutter K-J, Stöhr M, Eipel HE: Simultaneous DNA and protein measurements of microorganisms. Acta Pathol Microbiol Scand, Sect A, Suppl 274:100, 1980
561. Steen HB, Boye E: Bacterial growth studied by flow cytometry. Cytometry 1:32, 1980
562. Agar DW, Bailey JE: Cell cycle operation during batch growth of fission yeast populations. Cytometry 3:123, 1982
563. Murphy RF, Daban J-R, Cantor CR: Flow cytofluorimetric analysis of the nuclear division cycle of *Physarum polycephalum* plasmodia. Cytometry 2:26, 1981
564. Bonaly J, Mestre JC: Flow fluorometric study of DNA content in nonproliferative *Euglena gracilis* cells and during proliferation. Cytometry 2:35, 1981
565. Martinez OV, Gratzner HG, Malinin TI et al: The effect of some β-lactam antibiotics on *Escherichia coli* studied by flow cytometry. Cytometry 3:129, 1982
566. Ingram M, Cleary TJ, Price BJ et al: Rapid detection of Legionella pneumophila by flow cytometry. Cytometry 3:134, 1982
567. Boye E, Steen HB, Skarstad K: Flow cytometry of bacteria: A promising tool in experimental and clinical microbiology. J Gen Microbiol 129:973, 1983
568. Phillips AP, Martin KL: Immunofluorescence analysis of Bacillus spores and vegetative cells by flow cytometry. Cytometry 4:123, 1983
569. Betz JW, Aretz W, Härtel W: Use of flow cytometry in industrial microbiology for strain improvement programs. Cytometry 5:145, 1984
570. Price BJ, Kollman VH, Salzman GC: Light-scatter analysis of microalgae. Correlation of scatter patterns from pure and mixed asynchronous cultures. Biophys J 22:29, 1978
571. Yentsch CM: Flow cytometric analysis of cellular saxitoxin in the dinoflagellate *Gonyaulax tamarensis* var. *excavata*. Toxicon 19:611, 1981
572. Cohen J, Perfect JR, Durack DT: Method for the purification of *Filobasidiella neoformans* basidiospores by flow cytometry. Sabouraudia 20:245, 1982
573. Muldrow LL, Tyndall RL, Fliermans CB: Application of flow cytometry to studies of pathogenic free-living amoebae. Appl Environ Microbiol 44:1258, 1982
574. Gershey EL: SV40-infected muntjac cells: Cell cycle kinetics, cell ploidy, and T-antigen concentration. Cytometry 1:49, 1980
575. Leary JF, Notter MFD: Kinetics of virus adsorption to single cells using fluorescent membrane probes and multiparameter flow cytometry. Cell Biophys 4:63, 1982
576. Morikawa K, Yanagida M: Visualization of individual DNA molecules in solution by light microscopy: DAPI staining method. J Biochem 89:693, 1981

577. Jackson PR, Winkler DG, Kimzey SL et al: Cytofluorograf detection of *Plasmodium yoelii*, *Trypanosoma gambiense*, and *Trypanosoma equiperdum* by laser excited fluorescence of stained rodent blood. J Parasitol 63:593, 1977
578. Howard RJ, Battye FL:*Plasmodium bergheii*-infected red cells sorted according to DNA-content. Parasitology 78:263, 1979
579. Howard RJ, Battye FL, Mitchell GF: Plasmodium-infected blood cells analyzed and sorted by flow fluorimetry with deoxyribonucleic acid binding dye 33258 Hoechst. J Histochem Cytochem 27:803, 1979.
580. Howard RJ, Rodwell BJ: *Babesia rodhaini*, *Babesia bovis*, and *Babesia bigemina*. Analysis and sorting of red cells from infected mouse or calf blood by flow fluorimetry using 33258 Hoechst. Exp Parasitol 48:421, 1979
581. Whaun JM, Rittershaus C, Ip SHC: Rapid identification and detection of parasitized human red cells by automated flow cytometry. Cytometry 4:117, 1983
582. Benaron DA, Gray JW, Gledhill BL et al: Quantification of mammalian sperm morphology by slit-scan flow cytometry. Cytometry 2:344, 1982
583. Pinkel D, Lake S, Gledhill BL et al: High resolution DNA content measurements of mammalian sperm. Cytometry 3:1, 1982
584. Garner DL, Gledhill BL, Pinkel D et al: Quantification of the X-and Y-chromosome bearing spermatozoa of domestic animals by flow cytometry. Biol Reprod 28:312, 1983
585. Robinson AB, Manly KF, Anthony MP et al: Anesthesia of artemia larvae: Method for quantitative study. Science 149:1255, 1965
586. Laerum OD, Farsund T: Clinical application of flow cytometry: A review. Cytometry 2:1, 1981
587. Shapiro HM: Conference report: Cytometry in the clinical laboratory. Cytometry 3:312, 1983
588. Adams LR, Kamentsky LA: Machine characterization of human leukocytes by acridine orange fluorescence. Acta Cytol 15:289, 1971
589. Adams LR, Kamentsky LA: Fluorimetric characterization of six classes of human leukocytes. Acta Cytol 18:389, 1974
590. Steinkamp JA, Romero A, Van Dilla MA: Multiparameter cell sorting: Identification of human leukocytes by acridine orange fluorescence. Acta Cytol 17:113, 1973
591. Abrams WR, Diamond LW, Kane AB: A flow cytometric assay of neutrophil degranulation. J Histochem Cytochem 31:737, 1983
592. Sklar LA, Oades ZG, Finney DA: Neutrophil degranulation detected by right angle light scattering: Spectroscopic methods suitable for simultaneous analyses of degranulation or shape change, elastase release, and cell aggregation. J Immunol 133:1483, 1984
593. De Paoli P, Villalta D, Battistin S et al: Re: Selective loss of OKT8 lymphocytes on density gradient separation of blood mononuclear cells. J Immunol Meths 61:259, 1983
594. Ritchie AWS, Gray RA, Micklem HS: Right angle light scatter: A necessary parameter in flow cytofluorimetric analysis of human peripheral blood mononuclear cells. J Immunol Meths 64:109-17
595. Costa J, Cassidy M, Yee C: Evaluation of the lymphocyte blastogenic response by rapid flow analysis. J Immunol Meths 8:339, 1975
596. Utsinger PD, Yount WJ, Fallon JG et al: Cytofluorometric analysis of the kinetics of lymphocyte transformation after phytohemagglutinin stimulation: Comparison with the kinetics of thymidine incorporation. Blood 49:33, 1977
597. Braunstein JD, Melamed MR, Darzynkiewicz Z et al: Quantitation of transformed lymphocytes by flow cytofluorimetry. I. Pnytohemagglutinin response. Clin Immunol Immunopathol 4:209, 1975
598. Braunstein JD, Melamed MR, Sharpless TK et al: Quantitation of lymphocyte proliferative response to allogeneic cells and phytohemagglutinin by flow cytofluorometry. II. Comparison with [^{14}C] thymidine incorporation. Clin Immunol Immunopathol 5:326, 1976
599. Darznykiewicz Z, Traganos F, Sharpless T et al: Lymphocyte stimulation: A rapid multiparameter analysis. Proc Natl Acad Sci USA 73:2881, 1976
600. Traganos F, Gorski AJ, Darzynkiewicz Z et al: Rapid multiparameter analysis of cell stimulation in mixed lymphocyte culture reactions. J Histochem Cytochem 25:881, 1977
601. Gill C, Fischer CL, Wilkins B et al: Lymphocyte blastoid transformation assay by cytofluorography. Med Instrum 10:9, 1976
602. Noronha ABC, Richman DP, Arnason BGW: Detection of in vivo stimulated cerebrospinal-fluid lymphocytes by flow cytometry in patients with multiple sclerosis. N Engl J Med 303:713, 1980
603. Nairn RC, Rolland JM: Fluorescent probes to detect lymphocyte activation. Clin Exp Immunol 39:1, 1980
604. Williams JM, Christenson L, Araujo JL Carpenter CB, Milford EL, Shapiro HM, Strom TB: A new approach to the monitoring of kidney transplant patients via flow cytometric analysis of T-cell subsets, activation antigens, and DNA content. Transpl Proc 15:1957-61, 1983
605. Cotner T, Williams JM, Christenson L Shapiro HM, Strom TB, Strominger JL: Simultaneous flow cytometric analysis of human T cell activation antigen expression and DNA content. J Exp Med 157:461-72, 1983
606. Williams JM, Loertscher R, Cotner T Reddish M, Shapiro HM, Carpenter CB, Strominger JL, Strom TB: Dual parameter flow cytometric analysis of DNA content, activation antigen expression, and T cell subset proliferation in the human mixed lymphocyte reaction. J Immunol 132:2330-7, 1984
607. Bach F, Hirschhorn K: Lymphocyte interaction: A potential histocompatibility test in vitro. Science 143:813, 1964
608. Barlogie B, Drewinko B, Schumann J et al: Cellular DNA content as a marker of neoplasia in man. Amer J Med 69:195, 1980
609. Barlogie B, Raber MN, Schumann J et al: Flow cytometry in clinical cancer research. Cancer Res 43:3982, 1983
610. Hedley DW, Friedlander ML, Taylor IW et al: Method for analysis of cellular DNA content of paraffin-embedded pathological material using flow cytometry. J Histochem Cytochem 31:1333, 1983
611. Ault KA: Detection of small numbers of monoclonal B lymphocytes in the blood of patients with lymphoma. N Engl J Med 300:25, 1979
612. Young IT: Proof without prejudice: Use of the Kolmogorov-Smirnov test for the analysis of histograms from flow systems and other sources. J Histochem Cytochem 25:935, 1977
613. Clausen OPF: Flow cytometry of keratinocytes. J Cutan Path 10:33, 1983
614. Aurelian L: Herpes simplex virus diagnosis. Antigen detection by ELISA and flow microfluorometry. Diag Gynec Obstet 4:375, 1982
615. Koper GJM, Christiaanse JGM: The look-up table: A classifier for cell sorters. Cytometry 1:394, 1981
616. Eisert WG, Ostertag R, Niemann E-G: Simple flow microphotometer for rapid cell population analysis. Rev Sci Instrum 46:1021, 1975
617. Eisert WG: Fast cell size distribution analysis by laser flow microphotometry-applications to ciliate populations. Microsc Acta 78:228, 1976
618. Eisert WG: Cell differentiation based on absorption and scattering. J Histochem Cytochem 27:404-9, 1979
619. Eisert WG, Nezel M: Internal calibration to absolute values in flow-through particle size analysis. Rev Sci Instrum 49:1617, 1978
620. Kachel V, Schneider H, Schedler K: A new flow cytometric pulse height analyzer offering microprocessor controlled data acquisition and statistical analysis. Cytometry 1:175, 1980
621. Kunitskaya-Peterson C: *International Dictionary of Obscenities*. Oakland, Scythian Books, 1981
622. Wallace PJ, Wersto RP, Packman CH et al: Chemotactic peptide-induced changes in neutrophil actin conformation. J Cell Biol 99:1060, 1984
623. Charcosset J-Y, Jacquemin-Sablon A, LePecq J-B: Effect of membrane potential on the cellular uptake of 2-N-methyl-ellipticinium by L1210 cells. Biochem Pharmacol 33:2271, 1984
624. Van Dilla MA, Dean PN, Laerum OD, Melamed MR (eds): *Flow Cytometry: Instrumentation and Data Analysis*. Orlando, Academic Press, 1985

625. Meyer-Arendt JR: *Introduction to Classical and Modern Optics* (2 Ed). Englewood Cliffs, Prentice-Hall, 1984
626. Falk D, Brill D, Stork D: *Seeing the Light: Optics in Nature, Photography, Color, Vision, and Holography.* New York, Harper & Row, 1986
627. Eggebrecht LC: *Interfacing to the IBM Personal Computer.* Indianapolis, Sams, 1983
628. Sargent M III, Shoemaker RL: *The IBM Personal Computer™ from the Inside Out* (Revised Ed). Reading (MA), Addison-Wesley, 1986
629. Kelly MG, Spies N: *FORTH: A Text and Reference.* Englewood Cliffs, Prentice-Hall, 1986
630. Nassau K: *The Physics and Chemistry of Color.* New York, Wiley, 1983
631. Campbell ID, Dwek RA: *Biological Spectroscopy.* Menlo Park (CA), Benjamin/Cummings, 1984
632. Taylor DL, Waggoner AS, Murphy RF, Lanni F, Birge RR (eds): *Applications of Fluorescence in the Biological Sciences.* New York, Alan R. Liss, 1986
633. Spencer M: *Fundamentals of Light Microscopy.* Cambridge, Cambridge University Press, 1982
634. Inoué S: *Video Microscopy.* New York, Plenum, 1986.
635. De Duve C: *A Guided Tour of the Living Cell* (in two volumes). New York, Scientific American Books, 1984
636. Alberts B, Bray D, Lewis J, Raff M, Roberts K, Watson JD: *Molecular Biology of the Cell.* New York, Garland, 1983
637. Roitt I, Brostoff J, Male D: *Immunology.* St. Louis, Mosby, 1985
638. Male D: *Immunology: An Illustrated Outline.* St. Louis, Mosby, 1986
639. Hiebert RD, Sweet RG: Electronics for flow cytometers and sorters. In reference 624, p. 129
640. Feynman RP, Leighton RB, Sands M: *The Feynman Lectures on Physics,* Volume I (Mainly Mechanics, Radiation, and Heat). Reading (MA), Addison-Wesley, 1963
641. Feynman RP: *QED.* Princeton, Princeton University Press, 1985
642. Stokes GG: On the change of refrangibility of light. Philos Trans 143:385, 1852
643. Bigler RD: A comparison of low-power helium-cadmium and argon ultraviolet lasers in commercial flow cytometers. Cytometry 8:441, 1987
644. Loken MR, Keij JF, Kelley KA: Comparison of helium-neon and dye lasers for the excitation of allophycocyanin. Cytometry 8:96, 1987
645. Hoffman RA: Immunofluorescence analysis of leukocytes without red cell lysis. Cytometry Supp 1:36, 1987
646. Hoffman RA, Reinhardt BN, Stevens FE Jr: Two color immunofluorescence using a red helium neon laser. Cytometry Supp 1:103, 1987
647. Crissman HA, Steinkamp JA: A new method for rapid and sensitive detection of bromodeoxyuridine in DNA replicating cells. Exptl Cell Res 173:256, 1987
648. Tokita N, Skogen-Hagenson MJ, Johnson TS, Raju MR, Belli J: Flow cytometric measurement of adriamycin fluorescence for determining drug cytotoxicity. Abstracts, Automated Cytology VII, Asilomar, California, II-7, 1979
649. Buican TN: An interferometer for spectral analysis in flow. Cytometry Supp 1:101, 1987
650. Pinkel D, Stovel R: Flow chambers and sample handling. In reference 624, p. 77
651. Johnson LA, Pinkel D: Modification of a laser-based flow cytometer for high-resolution DNA analysis of mammalian spermatozoa. Cytometry 7:268, 1986
652. Pinkel D: A square quartz channel as nozzle tip on a FACS II. In reference 624, p. 232
653. Watson JV: A method for improving light collection by 600% from square cross section flow cytometry chambers. Br J Cancer 51:433, 1985
654. Leif RC, Wells M: Optical analysis of the AMAC IIIS transducer. Appl Optics 26:3244, 1987
655. Steinkamp JA, Fulwyler MJ, Coulter JR, Hiebert RD, Horney JL, Mullaney PF: A new multiparameter separator for microscopic particles and biological cells. Rev Sci Instrum 44:1301, 1973
656. Kachel V, Glossner E, Kordwig E, Ruhenstroth-Bauer G: Fluvo-Metricell, a combined cell volume and cell fluorescence analyzer. J Histochem Cytochem 25:804, 1977
657. Matsui Y, Staunton DE, Shapiro HM, Yunis EJ: Comparison of MHC antigen expression on PHA- and MLC-induced T cell lines with that on T and B lymphoblastoid cell lines by cell cycle dependency. Human Immunol 15:285, 1986
658. Matsui Y, Shapiro HM, Sheehy MJ, Christenson L, Staunton DE, Eynon EE, Yunis EJ: Differential expression of T cell differentiation antigens and major histocompatibility antigens on activated T cells during the cell cycle. Eur J Immunol 16:248, 1986
659. Steinkamp JA: A differential amplifier circuit for reducing noise in axial light loss measurements. Cytometry 4:83, 1983
660. Nguyen DC, Keller RA, Jett JH, Martin JC: Detection of single molecules of phycoerythrin in hydrodynamically focused flows by laser-induced fluorescence. Anal Chem 59:2158-61, 1987
661. Bartels PH: Numerical evaluation of cytologic data: XIII. Curve fitting and curvilinear regression. Analyt Quant Cytol 5:229, 1983
662. Vitale M, Papa S, Mariani AR, Facchini A, Rizzoli R, Manzoli FA: Use of poligonal [sic] windows for physical discrimination among mononuclear subpopulations in flow cytometry. J Immunol Methods 96:63, 1987
663. Visser JWM, Tanke HJ: Local modifications to commercial instruments. In reference 624, p.223
664. Dean PN: Helpful hints in flow cytometry and sorting. Cytometry 6:62, 1985
665. Fellner-Feldegg H: Do we really understand the hydrodynamics of flow sorting? Abstracts, Analytical Cytology X, Asilomar, California, D3, 1984
666. Stovel RT: The influence of particles on jet breakoff. J Histochem Cytochem 25:813, 1977
667. Peters D, Branscomb E, Dean P, Merrill T, Pinkel D, Van Dilla M, Gray JW: The LLNL high-speed sorter: Design features, operational characteristics, and biological utility. Cytometry 6:290, 1985
668. Herweijer H, Stokdijk W, Visser JWM: High speed photodamage cell selection using bromodeoxy-uridine/Hoechst 33342 photosensitized cell killing. Cytometry 9:143-9, 1988
669. Higgins ML, Smith MN, Gross GW: Selective cell destruction and precise neurite transection in neuroblastoma cultures with pulsed ultraviolet laser microbeam irradiation: An analysis of mechanisms and transection reliability with light and scanning electron microscopy. J Neurosci Methods 3:83, 1980
670. Schindler ML, Olinger MR, Holland JF: Automated analysis and survival selection of anchorage-dependent cells under normal growth conditions. Cytometry 6:368, 1985
671. Jett JH, Alexander RG: Droplet sorting of large particles. Cytometry 6:484, 1985
672. Freyer JP, Wilder ME, Jett JH: Viable sorting of intact multicellular spheroids by flow cytometry. Cytometry 8:427, 1987
673. Harkins KR, Galbraith DW: Factors governing the flow cytometric analysis and sorting of large biological particles. Cytometry 8:60, 1987
674. Parks DR, Lanier LL, Herzenberg LA: Flow cytometry and fluorescence activated cell sorting (FACS). In: Weir DM (ed): *Handbook of Experimental Immunology,* 4th ed. Volume 1: Immunochemistry. p.29.1, Oxford, Blackwell, 1986
675. Göttlinger C, Meyer KL, Weichel W, Müller W, Raftery B, Radbruch A: Cell-cooling in flow cytometry by Peltier elements. Cytometry 7:295, 1986
676. Patrick CW, Keller RH: A simple device for the collection of cells sorted by flow cytometry. Cytometry 5:308, 1984
677. Métézeau P, Bernheim A, Berger R, Goldberg ME: A simple device to obtain high local concentrations of material sorted by flow cytometry for biochemical or morphological analysis. Cytometry 5:550, 1984
678. Kanz L, Bross KJ, Mielke R, Löhr GW, Fauser AA: Fluorescence-activated sorting of individual cells onto poly-L-lysine coated slide areas. Cytometry 7:491, 1986

679. Alberti S, Stovel R, Herzenberg LA: Preservation of cells sorted individually onto microscope slides with a fluorescence-activated cell sorter. Cytometry 5:644, 1984
680. Kubitschek HE: Electronic counting and sizing of bacteria. Nature 182:234, 1958
681. De Blois RW, Mayyasi SA, Schidlovsky G, Wesley R, Wolff JS: Virus counting and analysis by the resistive pulse (Coulter counter) technique. Proc Amer Assn Cancer Res 15:104, 1974
682. Grover NB, Ben-Sasson N-A, Naaman J: Electrical sizing of cells in suspension. In: Catsimpoolas N (ed) *Cell Analysis.* Volume I, p. 93, New York, Plenum, 1982
683. Kachel V: Sizing of cells by the electrical resistance pulse technique. In: Catsimpoolas N (ed) *Cell Analysis.* Volume I, p. 195, New York, Plenum, 1982
684. Tycko DH, Metz MH, Epstein EA, Grinbaum A: Flow-cytometric light scattering measurement of red blood cell volume and hemoglobin concentration. Appl Optics 24:1355, 1985
685. Zarrin F, Dovichi NJ: Effect of sample stream radius upon light scatter distributions generated with a Gaussian beam light source in the sheath flow cuvette. Anal Chem 59:846, 1987
686. Button DK, Robertson BR: Methodology for analysis of a small marine bacterium by flow cytometry. Cytometry Supp 1:103, 1987
687. Zarrin F, Risfelt JA, Dovichi NJ: Light scatter detection within the sheath flow cuvette for size determination of multicomponent submicrometer particle suspensions. Anal Chem 59:850, 1987
688. Steen HB, Lindmo T: Differential light scattering detection in an arc lamp based flow cytometer. Cytometry Supp 1:24, 1987
689. Zarrin F, Bornhop DJ, Dovichi NJ: Laser Doppler velocimetry for particle size determination by light scatter within the sheath flow cuvette. Anal Chem 59:854, 1987
690. Wheeless LL Jr, Kay DB: Optics, light sources, filters, and optical systems. In reference 24, p.22
691. Cram LS, Bartholdi MF, Wheeless LL Jr, Gray JW: Morphological analysis by scanning flow cytometry. In reference 624, p. 164
692. Lucas JN, Pinkel D: Orientation measurements of microsphere doublets and metaphase chromosomes in flow. Cytometry 7:575, 1986
693. Lucas JN, Gray JW: Centromeric index versus DNA content flow karyotypes of human chromosomes measured by means of slit-scan flow cytometry. Cytometry 8:273, 1987
694. Weier H-Ul, Eisert WG: Two-parameter data acquisition system for rapid slit-scan analysis of mammalian chromosomes. Cytometry 8:83, 1987
695. Dubelaar GBJ, Groenewegen AdC, Stokdijk W, Visser JWM: The OPA (optical plankton analyzer): A flow cytometer for algae. Cytometry Supp 1:25, 1987
696. Ong S-H, Horne D, Yeung C-K, Nickolls P, Cole T: Development of an imaging flow cytometer. Analyt Quant Cytol Histol 9:375, 1987
697. Weier H-Ul, Lucas JN, Mullikin JC, van den Engh G: Affordable two-parameter slit-scan data acquisition. Cytometry Supp 1:102, 1987
698. Schafer IA, Jamieson AM, Petrelli M, Price BJ, Salzman GC: Multiangle light scattering flow photometry of cultured human fibroblasts: comparison of normal cells with a mutant line containing cytoplasmic inclusions. J Histochem Cytochem 27:359, 1979
699. Hansen WP, Hoffman RA: Method and apparatus for automated identification and enumeration of specified blood cell subclasses. U S Patent 4,284,412, 1981
700. Thompson JM, Gralow JR, Levy R, Miller RA: The optimal application of forward and ninety-degree light scatter in flow cytometry for the gating of mononuclear cells. Cytometry 6:401, 1985
701. McNeil PL, Kennedy AL, Waggoner AS, Taylor DL, Murphy RF: Light-scattering changes during chemotactic stimulation of human neutrophils: kinetics followed by flow cytometry. Cytometry 6:7, 1985
702. Nielsen O, Larsen JK, Christensen IJ, Lernmark Å: Flow sorting of mouse pancreatic B cells by forward and orthogonal light scattering. Cytometry 3:177, 1982
703. Terstappen LWMM, de Grooth BG, Nolten GMJ, ten Napel CHH, van Berkel W, Greve J: Physical discrimination between human T-lymphocyte subpopulations by means of light scattering, revealing two populations of T8-positive cells. Cytometry 7:178, 1986
704. van Bockstaele DR, Berneman ZN, Peetermans ME: Flow cytometric analysis of hairy cell leukemia using right-angle light scatter. Cytometry 7:217, 1986
705. Ward GK, Stewart SS, Price GB, Mackillop WJ: Cellular heterogeneity in normal human urothelium: an analysis of optical properties and lectin binding. J Histochem Cytochem 34:841, 1986
706. Benson MC, McDougal DC, Coffey DS: The application of perpendicular and forward light scatter to assess nuclear and cellular morphology. Cytometry 5:515, 1984
707. Dubelaar GBJ, Visser JWM, Donze M: Anomalous behaviour of forward and perpendicular light scattering of a cyanobacterium owing to gas vacuoles. Cytometry 8:405, 1987
708. Böhmer R-M, King NJC: Flow cytometric analysis of immunogold cell surface label. Cytometry 5:543, 1984
709. Stovel RT, Parks DR, Nozaki T Jr: A 130 degree light scatter detection system. Abstracts, Analytical Cytology X, Asilomar, California, B23, 1984
710. de Grooth BG, Terstappen LWMM, Puppels GJ, Greve J: Light-scattering polarization measurements as a new parameter in flow cytometry. Cytometry 8:539, 1987
711. van den Engh GJ, Trask BJ, Visser JWM: Flow cytometer for identifying algae by chlorophyll fluorescence. U S Patent 4,500,641, 1985
712. Meyer-Arendt JR: Optical instrumentation for the biologist: microscopy. Appl Optics 4:1, 1965
713. Perry RJ, Hunt AJ, Huffman DR: Experimental determinations of Mueller scattering matrices for nonspherical particles. Appl Optics 17:2700, 1978
714. Barer R, Dick DAT: Interferometry and refractometry of cells in tissue culture. Exptl Cell Res Suppl 4:103, 1957
715. Coulter WH, Hogg WR: Signal modulated apparatus for generating and detecting resistance and reactive changes in a modulated current passed for particle classification and analysis. U S Patent 3,502,974, 1970
716. Leif RC, Schwartz S, Rodriguez CM, Pell-Fernandez L, Groves M, Leif SB, Cayer M, Crews H: Two-dimensional impedance studies of BSA buoyant density separated human erythrocytes. Cytometry 6:13, 1985
717. Hoffman RA, Britt WB: Flow-system measurement of cell impedance properties. J Histochem Cytochem 27:234, 1979
718. Johnston RN, Jipson V, Atalar A, Heiserman J, Quate CF: Acoustic microscopy: resolution of subcellular detail. Proc Natl Acad Sci U S 76:3325, 1979
719. Bereiter-Hahn J: Scanning acoustic microscopy of living cells. J Microsc 146:29, 1987
720. Sweet RG, Fulwyler MJ, Herzenberg LA: Acoustic Sensing in Flow. Abstracts, Analytical Cytology X, Asilomar, California, P1.2, 1984
721. Scott JE: Lies, damned lies - and biological stains. Histochem J 4:387, 1972
722. Lillie RD [Scott JE]: Biological stains: a comment on J. E. Scott's editorial [and Scott's reply]. Histochem J 5:487, 1973
723. Horobin RW: The impurities of biological dyes: their detection, removal, occurrence and histological significance - a review. Histochem J 1:231, 1969
724. Proctor GB, Horobin RW: A widely applicable analytical system for biological stains: reverse-phase thin layer chromatography. Stain Technol 60:1 ,1985
725. Hopwood D: Cell and tissue fixation, 1972-1982. Histochem J 17:389, 1985
726. Van Ewijk W, Van Soest PL, Verkerk A, Jongkind JF: Loss of antibody binding to prefixed cells: fixation parameters for immunocytochemistry. Histochem J 16:179, 1984
727. Jacobberger JW, Fogleman D, Lehman JM: Analysis of intracellular antigens by flow cytometry. Cytometry 7:356, 1986

728. Levitt D, King M: Methanol fixation permits flow cytometric analysis of immunofluorescent stained intracellular antigens. J Immunol Methods 96:233, 1987
729. Krishan A: Effect of drug efflux blockers on vital staining of cellular DNA with Hoechst 33342. Cytometry 8:642, 1987
730. Watson JV, Nakeff A, Chambers SH, Smith PJ: Flow cytometric fluorescence emission spectrum analysis of Hoechst-33342-stained DNA in chicken thymocytes. Cytometry 6:310, 1985
731. Steen HB, Stokke T: Fluorescence spectra of cells stained with a DNA-specific dye, measured by flow cytometry. Cytometry 7:104, 1986
732. Stokke T, Steen HB: Binding of Hoechst 33258 to chromatin in situ. Cytometry 7:227, 1986
733. Otto F, Tsou KC: A comparative study of DAPI, DIPI, and Hoechst 33258 and 33342 as chromosomal DNA stains. Stain Technol 60:7, 1985
734. Pennings A, Speth P, Wessels H, Haanen C: Improved flow cytometry of cellular DNA and RNA by on-line reagent addition. Cytometry 8:335, 1987
735. Zelenin AV, Poletaev AI, Stepanova NG, Barsky VE, Kolesnikov VA, Nikitin SM, Zhuze AL, Gnutchev NV: 7-amino-actinomycin D as a specific fluorophore for DNA content analysis by laser flow cytometry. Cytometry 5:348-54, 1984
736. Rabinovitch PS, Torres RM, Engel D: Simultaneous cell cycle analysis and two-color surface immunofluorescence using 7-amino-actinomycin D and single laser excitation: applications to study of cell activation and the cell cycle of murine Ly-1 B cells. J Immunol 136:2769, 1986.
737. Shapiro HM, Stephens S: Flow cytometry of DNA content using oxazine 750 or related laser dyes with 633 nm excitation. Cytometry 7:107, 1986
738. Darzynkiewicz Z, Traganos F, Kapuscinski J, Staiano-Coico L, Melamed MR: Accessibility of DNA in situ to various fluorochromes: relationship to chromatin changes during erythroid differentiation of Friend leukemia cells. Cytometry 5:355, 1984
739. Evenson D, Darzynkiewicz Z, Jost L, Janca F, Ballachey B: Changes in accessibility of DNA to various fluorochromes during spermatogenesis. Cytometry 7:45, 1986
740. Stokke T, Steen HB: Distinction of leukocyte classes based on chromatin-structure-dependent DNA-binding of 7-aminoactinomycin D. Cytometry 8:576, 1987
741. Hiddemann W, Schumann J, Andreeff M, Barlogie B, Herman CJ, Leif RC, Mayall BH, Murphy RE, Sandberg AA: Convention on nomenclature for DNA cytometry. Cytometry 5:445, 1984
742. Tribukait B, Granberg-Öhman I, Wijkström H: Flow cytometric DNA and cytogenetic studies in human tumors: A comparison and discussion of the differences in modal values obtained by the two methods. Cytometry 7:194, 1986
743. Petersen SE, Friedrich U: A comparison between flow cytometric ploidy investigation and chromosome analysis of 32 human colorectal tumors. Cytometry 7:307, 1986
744. Bigner S, Bjerkvig R, Laerum OD, Muhlbaier LH, Bigner DD: DNA content and chromosomes in permanent cultured cell lines derived from malignant gliomas. Analyt Quant Cytol Histol 9:435, 1987
745. Chassevent A, Daver A, Bertrand G, Coic H, Geslin J, Bidabe M-Cl, George P, Larra F: Comparative flow DNA analysis of different cell suspensions in breast carcinoma. Cytometry 5:263-7, 1984
746. Stephenson RA, Gay H, Fair WR, Melamed MR: Effect of section thickness on quality of flow cytometric DNA content determinations in paraffin-embedded tissues. Cytometry 7:41, 1986
747. Camplejohn RS, Macartney JC: Comments on "effect of section thickness on quality of flow cytometric DNA content determinations in paraffin-embedded tissues". Cytometry 7:612, 1986
748. Larsen JK, Munch-Petersen B, Christiansen J, Jørgensen K: Flow cytometric discrimination of mitotic cells: resolution of M, as well as G_1, S and G_2 phase nuclei with mithramycin, propidium iodide, and ethidium bromide after fixation with formaldehyde. Cytometry 7:54, 1986
749. This reference duplicated reference 523.
750. Auer G, Ono J, Caspersson TO: Determination of the fraction of G_0 cells in cytologic samples by means of simultaneous DNA and nuclear protein analysis. Analyt Quant Cytol 5:1, 1983
751. Auer G, Ono J, Caspersson TO: Cytochemical identification of quiescent and growth-activated tumor cells. Analyt Quant Cytol 5:5, 1983
752. Pollack A, Moulis H, Prudhomme DL, Block NL, Irvin GL III: Quantitation of cell kinetic responses using simultaneous flow cytometric measurements of DNA and nuclear protein. Cytometry 5:473, 1984
753. Darzynkiewicz Z, Traganos F, Staiano-Coico L: Cell and nuclear growth during G_1: kinetic and clinical implications. Ann N Y Acad Sci 468:45, 1986
754. Crissman HA, Darzynkiewicz Z, Tobey RA, Steinkamp JA: Correlated measurements of DNA, RNA and protein in individual cells by flow cytometry. Science 228:1321, 1985
755. Crissman HA, Darzynkiewicz Z, Tobey RA, Steinkamp JA: Normal and perturbed Chinese hamster ovary cells: correlation of DNA, RNA and protein content by flow cytometry. J Cell Biol 101:141, 1985
756. Darzynkiewicz Z, Kapuscinski J, Traganos F, Crissman HA: Applications of pyronin Y in cytochemistry of nucleic acids. Cytometry 8:138, 1987
757. Darzynkiewicz Z, Kapuscinski J, Carter SP, Schmid FA, Melamed MR: Cytostatic and cytotoxic properties of pyronin Y: relation to mitochondrial localization of the dye and its interaction with RNA. Cancer Res 46:5760, 1986
758. Kapuscinski J, Darzynkiewicz Z: Interactions of pyronin Y(G) with nucleic acids. Cytometry 8:129, 1987
759. Kapuscinski J, Traganos F, Crissman HA, Darzynkiewicz Z: Application of pyronin Y as a probe of conformation of RNA. Cytometry Supp 1:89, 1987
760. Cowden RR, Curtis SK: Supravital experiments with pyronin Y, a fluorochrome of mitochondria and nucleic acids. Histochemistry 77:535, 1983
761. Timar J, Boldog F, Kopper L, Lapis K: Flow cytometric measurements and electron microscopy of cell surface glycosaminoglycans using acridine orange. Histochem J 17:71, 1985
762. Greenberg ER, Beck JR: The effects of sample size on reticulocyte counting and stool examination: the binomial and Poisson distributions in laboratory medicine. Arch Pathol Lab Med 108:396, 1984
763. Savage RA, Skoog DP, Rabinovitch A: Analytic inaccuracy and imprecision in reticulocyte counting: a preliminary report from the College of American Pathologists Reticulocyte Project. Blood Cells 11:97, 1985
764. Seligman PA, Allen RH, Kirchanski SJ, Natale PJ: Automated analysis of reticulocytes using fluorescent staining with both acridine orange and an immunofluorescence technique. Am J Hematol 14:57, 1983
765. Vaughan WP, Hall J, Johnson K, Dougherty C, Peebles D: Simultaneous reticulocyte and platelet counting on a clinical flow cytometer. Am J Hematol 18:385, 1985
766. Schmitz FJ, Werner E: Optimization of flow-cytometric discrimination between reticulocytes and erythrocytes. Cytometry 7:439, 1986
767. Ryan D, Laczin J, Mitchell S, Kossover S: Determination of the reticulocyte count by flow cytometry: importance of the 90° light scatter (90° LS) parameter. Blood 64: Supp 1:46a, 1984
768. Jacobberger JW, Horan PK, Hare JD: Flow cytometric analysis of blood cells stained with the cyanine dye $DiOC_1(3)$: reticulocyte quantification. Cytometry 5:589, 1984
769. Lee LG, Chen C-H, Chiu LA: Thiazole orange: a new dye for reticulocyte analysis. Cytometry 7:508. 1986
770. Davis BH, Bigelow N: Flow cytometric quantitation of reticulocytes using thiazole orange. Cytometry Supp 1:37, 1987
771. Makler MT, Lee LG, Recktenwald D: Thiazole orange: a new dye for *Plasmodium* species analysis. Cytometry 8:568, 1987

772. Kronick MN, Grossman PD: Immunoassay techniques with fluorescent phycobiliprotein conjugates. Clin Chem 29:1582, 1983
773. Yeh SW, Glazer AN, Clark JH: Control of bilin transition dipole moment direction by macromolecular assembly: energy transfer in allophycocyanin. J Phys Chem 90:4578, 1986
774. Yeh SW, Ong LJ, Clark JH, Glazer AN: Fluorescence properties of allophycocyanin and a cross-linked allophycocyanin trimer. Cytometry 8:91, 1987
775. Daley JF, Woronicz J, Levine H: Three color immunofluorescence using a single laser flow cytometer. Cytometry Supp 1:103, 1987
776. Ong LJ, Glazer AN: R-phycocyanin II, a new phycocyanin occurring in marine *Synechococcus* species. Identification of the terminal energy acceptor bilin in phycocyanins. J Biol Chem 262:6323, 1987
777. Truneh A, Machy P: Detection of very low receptor numbers on cells by flow cytometry using a sensitive staining method. Cytometry 8:562, 1987
778. Cohen JHM, Aubry JP, Jouvin MH, Wijdenes J, Bancherau J, Kazatchkine M, Revillard JP: Enumeration of CR1 complement receptors on erythrocytes using a new method for detecting low density cell surface antigens by flow cytometry. J Immunol Methods 99:53, 1987
779. Kim YR, Martin G, Paseltiner L, Ansley H, Ornstein L, Kanter RJ: Subtyping lymphocytes in peripheral blood by immunoperoxidase labeling and light scatter/absorption flow cytometry. Clin Chem 31:1481, 1985
780. Steinkamp JA, Stewart CC: Dual-laser, differential fluorescence correction method for reducing cellular background autofluorescence. Cytometry 7:566, 1986
781. Roederer M, Murphy RF: Cell-by-cell autofluorescence correction for low signal-to-noise systems: application to epidermal growth factor endocytosis by 3T3 fibroblasts. Cytometry 7:558, 1986
782. Alberti S, Parks DR, Herzenberg LA: A single laser method for subtraction of cell autofluorescence in flow cytometry. Cytometry 8:114, 1987
783. Szöllösi J, Mátyus L, Trón L, Balázs M, Ember I, Fulwyler MJ, Damjanovich S: Flow cytometric measurements of fluorescent energy transfer using single laser excitation. Cytometry 8:120, 1987
784. McCoy JP Jr, Shibuya N, Riedy MC, Goldstein IJ: *Griffonia simplicifolia* I isolectin as a functionally monovalent probe for use in flow cytometry. Cytometry 7:142, 1986
785. Kohler G, Milstein C: Continuous cultures of fused cells secreting antibody of predefined specificity. Nature 256:495, 1975
786. Festin R, Björklund B, Tötterman TH: Detection of triple antibody-binding lymphocytes in standard single-laser flow cytometry using colloidal gold, fluorescein, and phycoerythrin as labels. J Immunol Methods 101:23, 1987
787. Lanier LL, Allison JP, Phillips JH: Correlation of cell surface antigen expression on human thymocytes by multi-color flow cytometric analysis: implications for differentiation. J Immunol 137:2501, 1986
788. Loken MR, Shah VO, Dattilio KL, Civin CI: Flow cytometric analysis of human bone marrow. II. Normal B lymphocyte development. Blood 70:1316, 1987
789. Wognum AW, Thomas TE, Lansdorp PM: Use of tetrameric antibody complexes to stain cells for flow cytometry. Cytometry 8:366, 1987
790. Caldwell CW, Maggi J, Henry LB, Taylor HM: Fluorescence intensity as a quality control parameter in clinical flow cytometry. Am J Clin Pathol 88:447, 1987
791. Horan PK, Slezak SE, Poste G: Improved flow cytometric analysis of leukocyte subsets: simultaneous identification of five cell subsets using two-color immunofluorescence. Proc Natl Acad Sci USA 83:8361, 1986
792. Saunders AM, Chang C-H: A new immune monitoring system for the determination of lymphoid cell subsets. Ann N Y Acad Sci 468:128, 1986
793. Fulwyler MJ: Apparatus for distinguishing multiple subpopulations of cells. U S Patent 4,499,052, 1985
794. Blue M-L, Daley JF, Levine H, Schlossman S: Coexpression of T4 and T8 on peripheral blood T cells demonstrated by two-color fluorescence flow cytometry. J Immunol 134:2281, 1985
795. Buican TN, Hoffmann GW: Immunofluorescent flow cytometry in *N* dimensions. The multiplex labeling approach. Cell Biophys 7:129, 1985
796. Buican TN, Hoffmann GW: An automated device for the preparation of complex reagent mixtures. The immunofluorescence tomograph. Cell Biophys 7:157, 1985
797. Chatelier RC, Ashcroft RG: Calibration of flow cytometric fluorescence standards using the isoparametric analysis of ligand binding. Cytometry 8:632, 1987
798. Marti GE, Schuette W, Magruder L, Gralnick HR: A method to average immunofluorescent histograms. Cytometry 7:450, 1986
799. Traill KN, Böck G, Winter U, Hilchenbach M, Jürgens G, Wick G: Simple method for comparing large numbers of flow cytometry histograms exemplified by analysis of the CD4 (T4) antigen and LDL receptor on human peripheral blood lymphocytes. J Histochem Cytochem 34:1217, 1986
800. Treumer J, Valet G: Flow-cytometric determination of glutathione alterations in vital cells by *o*-phthaldialdehyde (OPT) staining. Exptl Cell Res 163:518, 1986
801. Rice GC, Bump EA, Shrieve DC, Lee W, Kovacs M: Quantitative analysis of cellular glutathione by flow cytometry utilizing monochlorobimane: some applications to radiation and drug resistance *in vitro* and *in vivo*. Cancer Res 46:6105, 1986
802. Greenspan P, Mayer EP, Fowler SD: Nile red: a selective fluorescent stain for intracellular lipid droplets. J Cell Biol 100:965, 1985
803. Fowler SD, Greenspan P: Aopplication of Nile red, a fluorescent hydrophobic probe, for the detection of neutral lipid deposits in tissue sections: comparison with oil red O. J Histochem Cytochem 33:833, 1985
804. Muller CP, Stephany DA, Winkler DF, Hoeg JM, Demosky SJ Jr, Wunderlich JR: Filipin as a flow microfluorometry probe for cellular cholesterol. Cytometry 5:42, 1984
805. Kruth HS, Cupp JE, Khan MA: Method for the detection and isolation of cholesteryl ester-containing "foam" cells using flow cytometry. Cytometry 8:146, 1987
806. Roffman E, Wilchek M: The extent of oxidative mitogenesis does not correlate with the degree of aldehyde formation of the T lymphocyte membrane. J Immunol 137:40, 1986
807. Krause AW, Carley WW, Webb WW: Fluorescent erythrosin B is preferable to trypan blue as a vital exclusion dye for mammalian cells in monolayer culture. J Histochem Cytochem 32:1084, 1984
808. Berglund DL, Taffs RE, Robertson NP: A rapid analytical technique for flow cytometric analysis of cell viability using Calcofluor White M2R. Cytometry 8:421, 1987
809. Jones KH, Senft JA: An improved method to determine cell viability by simultaneous staining with fluorescein diacetate-propidium iodide. J Histochem Cytochem 33:77, 1985
810. Omann GM, Coppersmith W, Finney DA, Sklar LA: A convenient on-line device for reagent addition, sample mixing, and temperature control of cell suspensions in flow cytometry. Cytometry 6:69, 1985
811. Kachel V, Schneider H, Bauer J, Malin-Berdel J: Application of the CYTOMIC 12 flow cytometric compact analyzer for automatic kinetic measurements. Cytometry 3:244, 1983
812. Beumer T, Lenssinck H, Pennings A, Haanen C: An easy-to-build timer for kinetic measurements in flow cytometry. Cytometry 5:648, 1984
813. Dive C, Workman P, Watson JV: Improved methodology for intracellular enzyme reaction and inhibition kinetics by flow cytometry. Cytometry 8:552, 1987
814. Speth PAJ, Linssen PCM, Boezemann JBM, Wessels HMC, Haanen C: Quantitation of anthracyclines in human hematopoietic cell subpopulations by flow cytometry correlated with high pressure liquid chromatography. Cytometry 6:143, 1985

815. Krishan A, Sauerteig A, Wellham LL: Flow cytometric studies on modulation of cellular adriamycin fluorescence. Cancer Res 45:1046, 1985
816. Krishan A, Sauerteig A, Gordon K, Swinkin C: Flow cytometric monitoring of cellular anthracycline accumulation in murine leukemic cells. Cancer Res 46:1768, 1986
817. Krishan A: Flow cytometric monitoring of anthracycline transport in tumor cells. Ann N Y Acad Sci 468:80, 1986
818. Krishan A: Laser flow cytometric studies on intracellular drug fluorescence. In: Gray JW, Darzynkiewicz Z (eds): *Techniques in Cell Cycle Analysis.* Clifton (NJ), Humana Press, 1986
819. Watson JV: Time, a quality-control parameter in flow cytometry. Cytometry 8:646, 1987
820. Blair OC, Carbone R, Sartorelli AC: Differentiation of HL-60 promyelocytic leukemia cells monitored by flow cytometric measurement of nitro blue tetrazolium (NBT) reduction. Cytometry 6:54, 1985
821. Severin E, Stellmach J: Flow cytometry of redox activity of single cells using a newly synthesized fluorescent formazan. Acta Histochem 75:101, 1984
822. Stellmach J, Severin E: A fluorescent redox dye. Influence of several substrates and electron carriers on the tetrazolium salt-formazan reaction of Ehrlich ascites tumor cells. Histochem J 19:21, 1987
823. Duque RE, Ward PA: Quantitative assessment of neutrophil function by flow cytometry. Analyt Quant Cytol Histol 9:42, 1987
824. Jongkind JF, Verkerk A, Sernetz M: Detection of acid-β-galactosidase activity in viable human fibroblasts by flow cytometry. Cytometry 7:463. 1986
825. Terstappen LWMM, de Grooth BG, Nolten GMJ, ten Napel CHH, van Berkel W, Greve J: Flow cytometric detection of circulating immune complexes with the indirect granulocyte phagocytosis test. Cytometry 6:316, 1985
826. Davis BH, McCabe E, Langweiler M: Characterization of f-met-leu-phe-stimulated fluid pinocytosis in human polymorphonuclear leukocytes by flow cytometry. Cytometry 7:251, 1986
827. Rolland A, Merdrignac G, Gouranton J, Bourel D, LeVerge R, Genetet B: Flow cytometric quantitative evaluation of phagocytosis by human mononuclear and polymorphonuclear cells using fluorescent nanoparticles. J Immunol Methods 96:185, 1987
828. Benz C, Wiznitzer I, Lee SH: Flow cytometric analysis of fluorescein-conjugated estradiol (E-BSA-FITC) binding in breast cancer cell suspensions. Cytometry 6:260, 1985
829. Ashcroft RG: Measurement of ligand binding, E-BSA-FITC. Cytometry 7:298, 1986
830. Benz CC: Reply to Dr. Ashcroft. Cytometry 7:299, 1986
831. Adelstein SJ, Lyman CP, O'Brien RC: Variations in the incorporation of thymidine into the DNA of some rodent species. Comp Biochem Physiol 12:223, 1964
832. Adelstein SJ, Lyman CP: Pyrimidine nucleoside metabolism in mammalian cells: an *in vitro* comparison of two rodent species. Exptl Cell Res 50:104, 1968
833. Shapiro HM, Lyman CP, Sullivan JD: Cell kinetics in hibernating animals: Implications for metrology and therapy. Cell Tissue Kinet 11:687, 1978
834. Gray JW, Mayall BH: *Monoclonal Antibodies Against Bromodeoxyuridine.* New York, Alan R. Liss, 1985 (Also published as Cytometry 6(6):499-673, 1985)
835. Dolbeare F, Beisker W, Pallavicini MG, Vanderlaan M, Gray JW: Cytochemistry for bromodeoxyuridine/DNA analysis: Stoichiometry and Sensitivity. Cytometry 6:521, 1985
836. Moran R. Darzynkiewicz Z, Staiano-Coico L, Melamed MR: Detection of 5-bromodeoxyuridine (BrdUrd) incorporation by monoclonal antibodies: Role of the DNA denaturation step. J Histochem Cytochem 33:821, 1985
837. Beisker W, Dolbeare F, Gray JW: An improved immunocytochemical procedure for high-sensitivity detection of incorporated bromodeoxyuridine. Cytometry 8:235, 1987
838. Schutte B, Reynders MMJ, van Assche CLMVJ, Hupperets PSGJ, Bosmann FT, Blijham GH: An improved method for the immunocytochemical detection of bromodeoxyuridine labeled nuclei using flow cytometry. Cytometry 8:372, 1987
839. Vanderlaan M, Watkins B, Thomas C, Dolbeare F, Stanker L: Improved high-affinity monoclonal antibody to iododeoxyuridine. Cytometry 7:499, 1986
840. Yanagisawa M, Dolbeare F, Todoroki T, Gray JW: Cell cycle analysis using numerical simulation of bivariate DNA/bromodeoxyuridine distributions. Cytometry 6:550, 1985
841. Begg AC, McNally NJ, Shrieve DC, Kärcher H: A method to measure the duration of DNA synthesis and the potential doubling time from a single sample. Cytometry 6:620, 1985
842. Sasaki K, Murakami T, Ogino T, Takahashi M, Kawasaki S: Flow cytometric estimation of cell cycle parameters using a monoclonal antibody to bromodeoxyuridine. Cytometry 7:391, 1986
843. Sasaki K, Murakami T, Takahashi M: A rapid and simple estimation of cell cycle parameters by continuous labeling with bromodeoxyuridine. Cytometry 8:526, 1987
844. de Grooth BG, van Dam M, Swart NC, Willemsen A, Greve J: Multiple wavelength illumination in flow cytometry using a single arc lamp and a dispersing element. Cytometry 8:445, 1987
845. Fox MH, Delohery TM: Membrane fluidity measured by fluorescence polarization using an EPICS V cell sorter. Cytometry 8:20, 1987
846. Masuda M, Kuriki H, Komiyama Y, Nishikado H, Egawa H, Murata K: Measurement of membrane fluidity of polymorphonuclear leukocytes by flow cytometry. J Immunol Methods 96:225, 1987
847. Kinoshita S, Fukami T, Ido Y, Kushida T: Spectroscopic properties of fluorescein in living lymphocytes. Cytometry 8:35, 1987
848. Prosperi E, Croce AC, Bottiroli G, Supino R: Flow cytometric analysis of membrane permeability properties influencing intracellular accumulation and efflux of fluorescein. Cytometry 7:70, 1986
849. Muirhead K, Bender P, Hanna W, Poste G: Binding of histamine and histamine analogs to lymphocyte subsets analyzed by flow cytometry. J Immunol 135:4121, 1985
850. Wilson HA, Seligmann BE, Chused TM: Voltage-sensitive cyanine dye fluorescence signals in lymphocytes: Plasma membrane and mitochondrial components. J Cell Physiol 125:61, 1985
851. Wilson HA, Chused TM: Lymphocyte membrane potential and Ca^{2+}-sensitive potassium channels described by oxonol dye fluorescence measurements. J Cell Physiol 125:72, 1985
852. Jenssen H-L, Redmann K, Mix E: Flow cytometric estimation of transmembrane potential of macrophages - A comparison with microelectrode measurements. Cytometry 7:339, 1986
853. Chused TM, Wilson HA, Seligmann BE, Tsien RY: Probes for use in the study of leukocyte physiology by flow cytometry. In reference 632, p. 531
854. Olive PL, Durand RE: Characterization of a carbocyanine derivative as a fluorescent penetration probe.
855. Crissman HA, Hofland MH, Stevenson AP, Wilder ME, Tobey RA: Use of DiO-C_5-3 to improve Hoechst 33342 uptake, resolution of DNA content, and survival of CHO cells. Exptl Cell Res 174:388-96, 1988
856. Resnick M, Schuldiner S, Bercovier H: Bacterial membrane potential analyzed by spectrofluorocytometry. Curr Microbiol 12:183, 1985
857. Bercovier H, Resnick M, Kornitzer D, Levy L: Rapid method for testing drug-susceptibility of *Mycobacteria* spp. and gram-positive bacteria using rhodamine 123 and fluorescein diacetate. J Microbiol Methods 7:167, 1987
858. Grynkiewicz G, Poenie M, Tsien RY: A new generation of Ca^{2+} indicators with greatly improved fluorescence properties. J Biol Chem 260:3440, 1985
859. Tsien RY, Rink TJ, Poenie M: Measurement of cytosolic free Ca^{2+} in individual small cells using fluorescence microscopy with dual excitation wavelengths. Cell Calcium 6:145, 1985

860. Poenie M, Alderton J, Tsien RY, Steinhardt RA: Changes of free calcium levels with stages of the cell division cycle. Nature 315:147, 1985
861. Williams DA, Fogarty KE, Tsien RY, Fay FF: Calcium gradients in single smooth muscle cells revealed by the digital imaging microscope using fura-2. Nature 318:558, 1985
862. Valet G, Raffael A, Russmann L: Determination of intracellular calcium in vital cells by flow cytometry. Naturwissenschaften 72:600, 1985
863. Ransom JT, DiGiusto DL, Cambier J: Flow cytometric analysis of intracellular calcium mobilization. Meths Enzymol 141:53-63, 1987
864. Ransom JT, DiGiusto DL, Cambier JC: Single cell analysis of calcium mobilization in anti-immunoglobulin-stimulated B lymphocytes. J Immunol 136:54, 1986
865. Chused TM, Wilson HA, Greenblatt D, Ishida Y, Edison LJ, Tsien RY, Finkelman FD: Flow cytometric analysis of murine splenic B lymphocyte cytosolic free calcium response to anti-IgM and anti-IgD, Cytometry 8:396, 1987
866. Rabinovitch PS, June CH, Grossmann A, Ledbetter JA: Heterogeneity among T cells in intracellular free calcium responses after mitogen stimulation with PHA or anti-CD3. Simultaneous use of indo-1 and immunofluorescence with flow cytometry. J Immunol 137:952, 1986
867. June CH, Ledbetter JA, Rabinovitch PS, Martin PJ, Beatty PG, Hansen JA: Distinct patterns of transmembrane calcium flux and intracellular calcium mobilization after differentiation antigen cluster 2(E rosette receptor) or 3 (T3) stimulation of human lymphocytes. J Clin Inves 77:1224, 1986
868. Musgrove E, Rugg C, Hedley D: Flow cytometric measurement of cytoplasmic pH: A critical evaluation of available fluorochromes. Cytometry 7:347, 1986
869. Rink TJ, Tsien RY, Pozzan T: Cytoplasmic pH and free Mg^{2+} in lymphocytes. J Cell Biol 95:189, 1982
870. Rogers J, Hesketh TR, Smith GA, Metcalfe JC: Intracellular pH of stimulated lymphocytes measured with a new fluorescent indicator. J Biol Chem 258:5994, 1983
871. Grinstein S, Cohen S, Lederman HM, Gelfand EW: The intracellular pH of quiescent and proliferating human and rat thymic lymphocytes. J Cell Physiol 121:87, 1984
872. Lazzari KG, Proto PJ, Simons ER: Simultaneous measurement of stimulus-induced changes in cytoplasmic Ca^{2+} and in membrane potential of human neutrophils. J Biol Chem 261:9710, 1986
873. Leif SB, Leif RC, Auer R: The EPICS C analyzer: An ergometrically designed flow cytometer computer system. Analyt Quant Cytol Histol 7:187, 1985
874. Valet G: The Diagnos1 program system for the automated identification and classification of abnormal cells by flow-cytometry. Cytometry Supp 1:17, 1987
875. Hecht J: *The Laser Guidebook.* New York, McGraw-Hill, 1986
876. Shapiro HM: Technical tutorial: Laser noise and news. Cytometry 8:248, 1987
877. Shapiro HM, Hercher M: Flow cytometers using optical waveguides in place of lenses for specimen illumination and light collection. Cytometry 7:221, 1986
878. Briggs J, Fisher ML, Ghazarossian VE, Becker MJ: Fiber optic probe cytometer. J Immunol Methods 81:73, 1985
879. Apple Computer, Inc.: *Technical Introduction to the Macintosh Family.* Reading (MA), Addison-Wesley, 1987
880. Apple Computer, Inc.: *Programmer's Introduction to the Macintosh Family.* Reading (MA), Addison-Wesley, 1988
881. Apple Computer, Inc.: *Designing Cards and Drivers for Macintosh II and Macintosh SE.* Reading (MA), Addison-Wesley, 1988
882. Kelley KA, McDowell JL: Technical Tutorial: Practical considerations for the selection and use of optical filters in flow cytometry. Cytometry 9:277-80, 1988
883. Ryan DH, Mitchell SJ, Hennessy LA, Bauer KD, Horan PK, Cohen HJ: Improved detection of rare CALLA-positive cells in peripheral blood using multiparameter flow cytometry. J Immunol Methods 74:115, 1984
884. Hirschfeld T: Optical microscopic observation of single small molecules. Appl Optics 15:2965, 1976
885. Hirschfeld T: Quantum efficiency independence of the time integrated emission from a fluorescent molecule. Appl Optics 15:3135, 1976
886. Dovichi N, Martin JC, Jett JH, Keller RA: Attogram detection limit for aqueous dye samples by laser-induced fluorescence. Science 219:845, 1983
887. Dovichi N, Martin JC, Jett JH, Trkula M, Keller RA: Laser-induced fluorescence of flowing samples as an approach to single-molecule detection in liquids. Anal Chem 56:348, 1984
888. Mathies RA, Stryer L: Single-molecule fluorescence detection: A feasibility study using phycoerythrin. In reference 632, p. 129
889. Thornthwaite JT, Thomas RA, Russo J, Ownby H, Malinin GI, Hornicek F, Woolley TW, Frederick J, Malinin TI, Vazquez DA, Seckinger D: A review of DNA flow cytometric preparatory and analytical methods. In: Russo J (ed): *Immunochemistry in Tumor Diagnosis,* Boston, Martinus Nijhoff, 1985, p. 380
890. Greenebaum E, Koss LG, Sherman AB, Elequin F: Comparison of needle aspiration and solid biopsy technics in the flow cytometric study of DNA distributions of surgically resected tumors. Am J Clin Path 82:559, 1984
891. Hedley DW, Friedlander ML, Taylor IW, Rugg CA, Musgrove EA: DNA flow cytometry of paraffin-embedded tissue. Cytometry 5:660, 1984
892. Hedley DW, Friedlander ML, Taylor IW: Application of DNA flow cytometry to paraffin-embedded archival material for the study of aneuploidy and its clinical significance. Cytometry 6:327, 1985
893. Baisch H, Beck HP, Christensen IJ, Hartmann NR, Fried J, Dean PN, Gray JW, Jett JH, Johnston DA, White RA, Nicolini C, Zeitz S, Watson JV: A comparison of mathematical methods for the analysis of DNA histograms obtained by flow cytometry. Cell Tissue Kinet 15:235, 1982
894. Dean PN: Methods of data analysis in flow cytometry. In reference 624, p. 195
895. Gray JW, Darzynkiewicz Z (eds): *Techniques in Cell Cycle Analysis.* Clifton (NJ), Humana Press, 1987
896. Banner BF, Chacho MS, Roseman DL, Coon JS: Multiparameter flow cytometric analysis of colon polyps. Am J Clin Path 87:313, 1987
897. Coon JS, Weinstein RS: Technology assessment of urinary flow cytometry. Human Pathol 18:1195, 1987
898. Coon JS, Deitch AD, de Vere White RW, Koss LG, Melamed MR, Reeder JE, Weinstein RS, Wersto RP, Wheeless LL: Interinstitutional variability in DNA flow cytometric analysis of tumors: The National Cancer Institute's Flow Cytometry Network experience. Cancer 61:126, 1988
899. Skog S, Tribukait B: Discontinuous RNA and protein synthesis and accumulation during cell cycle of Ehrlich ascites tumour cells. Exptl Cell Res 159:510, 1985
900. Kurki P, Vanderlaan M, Dolbeare F, Gray JW, Tan EM: Expression of proliferating cell nuclear antigen (PCNA)/cyclin during the cell cycle. Exptl Cell Res 166:209, 1986
901. Bauer KD, Clevenger CV, Williams TJ, Epstein AL: Assessment of cell cycle-associated antigen expression using multiparameter flow cytometry and antibody-acridine orange sequential staining. J Histochem Cytochem 34:245, 1986
902. Trask B, van den Engh G, Landegent J, Jansen in de Wal N, van der Ploeg M: Detection of DNA sequences in nuclei in suspension by in situ hybridization and dual beam flow cytometry. Science 230:1401, 1985
903. Gray JW, Langlois RG: Chromosome classification and purification using flow cytometry and sorting. Annu Rev Biophys Biophys Chem 15:195, 1986

904. Gray JW, Dean PN, Fuscoe JC, Peters DC, Trask BJ, van den Engh GJ, Van Dilla MA: High-speed chromosome sorting. Science 238:323, 1987
905. Shaw S: Characterization of human leukocyte differentiation antigens. Immunol Today 167:1, 1987
906. Coon JS, Landay AL, Weinstein RS: Biology of disease: Advances in flow cytometry for diagnostic pathology. Lab Inves 57:453, 1987
907. Morimoto C, Rudd CE, Letvin NL, Schlossman SF: A novel epitope of the LFA-1 antigen which can distinguish killer effector and suppressor cells in human CD8 cells. Nature 330:479, 1987
908. Meuer SC, Cooper DA, Hodgdon JC, Hussey RE, Fitzgerald KA, Schlossman SF, Reinherz EL: Identification of the receptor for antigen and major histocompatibility complex on human inducer T lymphocytes. Science 222:1239, 1983
909. Uckun FM, Jaszcz W, Ambrus JL, Fauci AS, Gajl-Peczalska K, Song CW, Wick MR, Myers DE, Waddick K, Ledbetter JA: Detailed studies on expression and function of CD19 surface determinant by using B43 monoclonal antibody and the clinical potential of anti-CD19 immunotoxins. Blood 71:13, 1988
910. Fowlkes BJ, Kruisbeek AM, Ton-That H, Weston MA, Coligan JE, Schwartz RH, Pardoll DM: A novel population of T-cell receptor ••-bearing thymocytes which predominantly expresses a single V• gene family. Nature 329:251, 1987
911. Crispe IN, Moore MW, Husmann LA, Smith L, Bevan MJ, Shimonkevitz RP: Differentiation potential of subsets of $CD4^-8^-$ thymocytes. Nature 329:336, 1987
912. Havran WL, Poenie M, Kimura J, Tsien R, Weiss A, Allison JP: Expression and function of the CD3-antigen receptor on murine $CD4^+8^+$ thymocytes. Nature 330:170, 1987
913. Lewis RS, Cahalan MD: Subset-specific expression of potassium channels in developing murine T lymphocytes. Science 239:771, 1988
914. Phillips AP, Martin KL: Dual-parameter scatter-flow immunofluorescence analysis of bacillus spores. Cytometry 6:124-9, 1985
915. Phillips AP, Martin KL, Capey AJ: Direct and indirect immunofluorescence analysis of bacterial populations by flow cytometry. J Immunol Methods 101:219, 1987
916. Donnelly CW, Baigent GJ: Method for flow cytometric detection of *Listeria monocytogenes* in milk. Appl Envir Microbiol 52:689, 1986
917. Kuchenbecker D, Braun G: Rapid yeast DNA staining method for flow cytometry. J Basic Microbiol 8:509-12, 1985
918. Srienc F, Campbell JL, Bailey JE: Flow cytometry analysis of recombinant *Saccharomyces cerevisiae* populations. Cytometry 7:132, 1986
919. Mansour JD, Robson JA, Arndt CW, Schulte TH: Detection of *Escherichia coli* in blood using flow cytometry. Cytometry 6:186, 1985
920. Bonaly J, Bre MH, Lefort-Tran M, Mestre JC: A flow cytometric study of DNA staining in situ in exponentially growing and stationary *Euglena gracilis*. Cytometry 8:42, 1987
921. Lefort-Tran M, Bre MH, Pouphile M, Manigault P: DNA flow cytometry of control *Euglena* and cell cycle blockade of vitamin B12-starved cells.
922. Bernstein RL, Browne LH, Yu SC, Williams KL: Detergent treatment of *Dictyostelium discoideum* cells allows examination of internal cell type-specific antigens by flow cytometry. Cytometry 9:68, 1988
923. Fry J, Matthews HR: Flow cytometry of the differentiation of *Physarum polycephalum* myxamoebae to cysts. Exptl Cell Res 168:173, 1987
924. Kubbies M, Wick R, Hildebrandt A, Sauer HW: Flow cytometry reveals a high degree of genomic size variation and mixoploidy in various strains of the acellular slime mold *Physarum polycephalum*. Cytometry 7:481, 1986
925. Brunk CF, Bohman RE: Analysis of nuclei from exponentially growing and conjugated *Tetrahymena thermophila* using the flow microfluorimeter. Exptl Cell Res 162:390, 1986
926. Bloodgood RA, Salomonsky NL, Reinhart FD: Use of carbohydrate probes in conjunction with fluorescence-activated cell sorting to select mutant cell lines of *Chlamydomonas* with defects in cell surface glycoproteins. Exptl Cell Res 173:572, 1987
927. Wood AM, Horan PK, Muirhead K, Phinney DA, Yentsch CM, Waterbury JM: Discrimination between types of pigments in marine *Synechococcus* spp. by scanning spectroscopy, epifluorescence microscopy, and flow cytometry. Limnol Oceanog 30:1303, 1985
928. Olson RJ, Vaulot D, Chisholm SW: Marine phytoplankton distributions measured using shipboard flow cytometry. Deep Sea Res 32:1273, 1985
929. Olson RJ, Chisholm SW, Zettler ER: Dual-beam flow cytometry for distinguishing between phytoplankton pigment types. EOS 67:973, 1986
930. Vaulot D, Olson RJ, Chisholm SW: Light and dark control of the cell cycle in two marine phytoplankton species. Exptl Cell Res 167:38, 1986
931. Gerritsen J, Sanders RW, Bradley SW, Porter KG: Flow cytometric investigations of zooplankton feeding. EOS 66:1305, 1985
932. Rivkin RB, Phinney DA, Yentsch CM: Effects of flow cytometric analysis and cell sorting on photosynthetic carbon uptake by phytoplankton in cultures and from natural populations. Appl Envir Microbiol 52:935, 1986
933. Levitt D, Zable B, Bard J: Binding, ingestion, and growth of *Chlamydia trachomatis* (L_2 serovar) analyzed by flow cytometry. Cytometry 7:378, 1986
934. Waldman FM, Hadley WK, Fulwyler MJ, Schachter J: Flow cytometric analysis of *Chlamydia trachomatis* interaction of L cells. Cytometry 8:55, 1987
935. Rosenthal KS, Hodnichak CM, Summers JL: Flow cytometric evaluation of anti-herpes drugs. Cytometry 8:392, 1987
936. Folks TM, Justement J, Kinter A, Dinarello CA, Fauci AS: Cytokine-induced expression of HIV-1 in a chronically infected promonocyte cell line. Science 238:800, 1987
937. Hare JD, Bahler DW: Analysis of *Plasmodium falciparum* growth in culture using acridine orange and flow cytometry. J Histochem Cytochem 34:215, 1986
938. Hare JD: Two-color flow-cytometric analysis of the growth cycle of *Plasmodium falciparum* in vitro: Identification of cell cycle compartments. J Histochem Cytochem 34:1651-8, 1986
939. Midgley M: The interaction of a fluorescent probe with *Trypanosoma brucei brucei*: Evidence for the existence of a membrane potential. FEMS Microbiol Lett 18:203, 1983
940. Jackson PR, Pappas MG, Hansen BD: Fluorogenic substrate detection of viable intracellular and extracellular pathogenic protozoa. Science 227:435, 1985
941. Evenson DP, Darzynkiewicz Z, Melamed MR: Relation of mammalian sperm chromatin heterogeneity to fertility. Science 210:1131, 1980
942. Evenson DP, Melamed MR: Rapid analysis of normal and abnormal cell types in human semen and testis biopsies by flow cytometry. J Histochem Cytochem 31:248, 1983
943. Evenson DP, Higgins PJ, Grueneberg D, Ballachey BE: Flow cytometric analysis of mouse spermatogenic function following exposure to ethylnitrosourea. Cytometry 6:238, 1985
944. Ballachey BE, Hohenboken WD, Evenson DP: Heterogeneity of sperm nuclear chromatin structure and its relationship to bull fertility. Biol Reprod 36:915, 1987
945. Ballachey BE, Evenson DP, Saacke RG: The sperm chromatin structure assay: Relationship with alternate tests of semen quality and heterospermic performance of bulls. J Androl 9:109, 1988
946. Haas GG Jr, Cunningham ME: Identification of antibody-laden sperm by cytofluorometry. Fertil Steril 42:606, 1984
947. Brown SC, Jullien M, Coutos-Thevenot P, Muller P, Renaudin J-P: Present developments of flow cytometry in plant biology. Biol Cell 58:173, 1986
948. Galbraith DW, Harkins KR, Jefferson RA: Flow cytometric characterization of the chlorophyll contents and size distributions of plant protoplasts. Cytometry 9:75, 1988

949. Conia J, Bergounioux C, Perennes C, Muller P, Brown S, Gadal P: Flow cytometric analysis and sorting of plant chromosomes from *Petunia hybrida* protoplasts. Cytometry 8:500, 1987
950. Breer C, Lutz H, Super BS: Counting somatic cells in milk with a rapid flow-through cytophotometer.
951. Chaiton JA, Allen SK JR: Early detection of triploidy in the larvae of Pacific oysters, *Crassostrea gigas*, by flow cytometry. Aquaculture 48:35, 1985
952. Allen SK Jr, Thiery RG, Hagstrom NT: Cytological evaluation of the likelihood that triploid grass carp will reproduce. Trans Am Fisheries Soc 115:841, 1986
953. Maier P, Schawalder HP: A two-parameter flow cytometry protocol for the detection and characterization of the clastogenic, cytostatic, and cytotoxic activities of chemicals. Mutation Res 164:369, 1986
954. Darzynkiewicz Z, Williamson B, Carswell EA, Old LJ: Cell cycle-specific effects of tumor necrosis factor. Cancer Res 44:83, 1984
955. Darzynkiewicz Z, Traganos F, Xue S, Staiano-Coico L, Melamed MR: Rapid analysis of drug effects on the cell cycle. Cytometry 1:279, 1981
956. Frankfurt OS: Flow cytometry analysis of DNA damage and the evaluation of alkylating agents. Cancer Res 47:5537, 1987
957. Klevecz RR, Shymko RM, Blumenfeld D, Braly PS: Circadian gating of S phase in human ovarian cancer. Cancer Res 47:6267, 1987
958. Møller U, Larsen JK, Keiding N, Christensen IJ: Circadian-stage dependence of methotrexate in a keratinized epithelium. An *in-vivo* study using flow cytometry on the hamster cheek pouch epithelium. Cell Tissue Kinet 17:483, 1984
959. Hayes FA: Cell kinetic-based scheduling of chemotherapy: Hypothesis versus reality. J Clin Oncol 5:1713, 1987
960. Poot M, Verkerk A, Koster JF, Jongkind JF: De novo synthesis of glutathione in human fibroblasts during in vitro ageing and in some metabolic diseases as measured by a flow cytometric method. Biochim Biophys Acta 883:580, 1986
961. Martinez A, Vigil A, Vila JC: Flow-cytometric analysis of mitochondria-associated fluorescence in young and old human fibroblasts. Exptl Cell Res 164:551, 1986
962. Goldberg ID, Shapiro H, Stemerman MB, Wei J, Hardin D, Christenson L: Frequency of tetraploid nuclei in the rat aorta increases with age. Ann N Y Acad Sci 435:422, 1984
963. Vliegen HW, Vossepoel AM, van der Laarse A, Eulerdink F, Cornelisse CJ: Methodological aspects of flow cytometric analysis of DNA polyploidy in human heart tissue. Hostochemistry 84:348, 1986
964. Goldberg ID, Rosen EM, Shapiro HM, Zoller LC, Myrick K, Levenson SE, Christenson L: Isolation and culture of a tetraploid subpopulation of smooth muscle cells from the normal rat aorta. Science 226:559, 1984
965. Kavanagh TJ, Martin GM, El-Fouly MH, Trosko JE, Chang C-C, Rabinovitch PS: Flow cytometry and scrape-loading/dye transfer as a rapid quantitative measure of intercellular communication *in vitro*. Cancer Res 47:6046, 1987
966. Ornitz DM, Hammer RE, Messing A, Palmiter RD, Brinster: Pancreatic neoplasia induced by SV40 T-antigen expression in acinar cells of transgenic mice. Science 238:188, 1987
967. Tough DF, Chow DA: Tumorigenicity of murine lymphomas selected through fluorescence-detected natural antibody binding. Cancer Res 48:270,1988
968. Kenter AL, Watson JV, Azim T, Rabbitts TH: Colcemid inhibits growth during early G1 in normal but not in tumorigenic lymphocytes. Exptl Cell Res 167:241, 1986
969. Cook JA, Fox MH: Effects of acute pH 6.6 and 42.0°C heating on the intracellular pH of Chinese hamster ovary cells. Cancer Res 48:496, 1988
970. Ota T, Fujikawa-Yamamoto K, Zong Z-P, Yamazaki M, Odashima S, Kitagawa I, Abe H, Arichi S: PLant glycoside modulation of cell surface related to control of differentiation in cultured B16 melanoma cells. Cancer Res 47:3863, 1987
971. Fibach E, Nahas N, Giloh H, Gatt S: Uptake of fluorescent fatty acids by erythroleukemia cells. Exptl Cell Res 166:220, 1986
972. Bucana C, Saiki I, Nayar R: Uptake and accumulation of the vital dye hydroethidine in neoplastic cells. J Histochem Cytochem 34:1109, 1986
973. Gallop PM, Paz MA, Henson E, Latt SA: Dynamic approaches to the delivery of reporter reagents into living cells. Biotechniques 1:32, 1984
974. Cordier G, Dezutter-Dambuyant C, Lefebvre R, Schmitt D: Flow cytometry sorting of unlabeled epidermal Langerhans cells using forward and orthogonal light scatter properties. J Immunol Methods 79:79, 1985
975. Salari H, Takei F, Miller R, Chan-Yeung M: Novel technique for isolation of human lung mast cells. J Immunol Methods 100:91, 1987
976. Baron B, Métézeau P, Hatat D, Roberts C, Goldberg ME, Bishop C: Cloning of DNA libraries from mouse Y chromosomes purified by flow cytometry. Somat Cell Molec Genet 12:289, 1986
977. Larsen JK, Byslov AG, Christensen IJ: Flow cytometry and sorting of meiotic prophase cells of female rabbits. J Reprod Fert 76:587, 1986
978. Zola H, Krishnan R, Bradley J: A simple technique for evaluation of methods of cell separation. J Immunol Methods 76:383, 1985
979. Andreeff M (ed): *Clinical Cytometry*. Ann N Y Acad Sci 468:1-408, 1986
980. Lovett EJ III, Schnitzer B, Keren DF, Flint A, Hudson JL, McClatchey KD: Applications of flow cytometry to diagnostic pathology. Lab Inves 50:115, 1984
981. Quirke P, Dyson JED: Flow cytometry: Methodology and applications in pathology. J Pathol 149:79, 1986
982. Ryan DH, Fallon MA, Horan PK: Flow cytometry in the clinical laboratory. Clin Chim Acta 171:125-73, 1988
983. Mayall B: Cytometry in the Clinical Laboratory: Quo vadis? Ann N Y Acad Sci 468:1, 1986
984. Shapiro HM: The little laser that could: Applications of low power lasers in clinical flow cytometry. Ann N Y Acad Sci 468:18, 1986
985. Bessman JD: *Automated Blood Counts and Differentials: A Practical Guide*. Baltimore, Johns Hopkins University Press, 1986
986. Terstappen LWMM, de Grooth BG, Visscher K, van Kouterik FA, Greve J: Four-parameter white blood cell differential counting based on light scattering measurements. Cytometry 9:39, 1988
987. Kass L: Individual leukocyte determination by means of differential metachromatic dye sorption. U S Patent 4,581,223, 1986
988. Shapiro HM: White cell indices: An observer-independent, objective approach to the classification of normal and abnormal leukocytes. Blood 50:Supp 1:160, 1977
989. Shapiro HM, Young RE, Webb RH, Wiernik PH: Multiparameter flow cytometric characterization of cell populations in acute leukemia. Blood 50:Supp 1:209, 1977
990. Loken MR, Shah VO, Dattilio KL, Civin CI: Flow cytometric analysis of human bone marrow. I. Normal erythroid development. Blood 69:255, 1987
991. Pallavicini M, Summers LJ, Giroud FJ, Dean PN, Gray JW: Multivariate analysis and list mode processing of murine hemopoietic subpopulations for cytokinetic studies. Cytometry 6:539, 1985
992. Jackson CW, Brown LK, Somerville BC, Lyles SA, Look AT: Two-color flow cytometric measurement of DNA distributions of rat megakaryocytes in unfixed, unfractionated cell suspensions. Blood 63:768, 1984
993. Worthington RE, Nakeff A, Micko S: Flow cytometric analysis of megakaryocyte differentiation. Cytometry 5:501, 1984
994. Corash L, Mok Y, Levin L, Baker G, Chen H: Serial studies of megakaryocyte DNA content and platelet volume in response to variable degrees of thrombocytopenia. Blood 68:Supp 1:157a, 1986
995. Rabellino EM, Russel JB: Ploidy distribution profile of megakaryocytes in normal marrow and in patients with ITP. Blood 68:Supp 1:311a, 1986
996. Tomer A, Harker LA, Burstein SA: Purification of human megakaryocytes by fluorescence-activated cell sorting. Blood 70:1735, 1987
997. Bassøe C-F, Bjerknes R: Phagocytosis by human leukocytes, phagosomal pH, and degradation of seven species of bacteria measured by flow cytometry. J Med Microbiol 19:115, 1985

998. Kim BK, Chao FC, Shapiro HM, Kenney DM, Surgenor DM, Jacobsen MS, Button LN, Kevy SV: Membrane potential in stored platelets. Blood 68:Supp 1:299a, 1986
999. Davies TA, Drotts D, Weil GJ, Simons ER: Flow cytometric measurements of cytoplasmic calcium changes in human platelets. Cytometry 9:138, 1988
1000. Shattil SJ, Cunningham M, Hoxie JA: Detection of activated human platelets in whole blood using an activation-dependent monoclonal antibody and flow cytometry. Blood 68:Supp 1:312a, 1986
1001. Jackson CW, Ashmun RA, Jennings LK: Flow cytometric examination of fibrinogen binding to platelets activated in native plasma. Blood 68:Supp 1:318a, 1986
1002. Renzi P, Ginns LC: Analysis of T-cell subsets in normal adults. Comparison of whole blood lysis technique to Ficoll-Hypaque separation by flow cytometry. J Immunol Methods 98:53, 1987
1003. Cosimi AB, Colvin RB, Burton RC, Rubin RH, Goldstein G, Kung PC, Hansen WP, Delmonico FL, Russell PS: Use of monoclonal antibodies to T-cell subsets for immunologic monitoring and treatment in recipients of renal allografts. N Eng J Med 305:308, 1981
1004. Redelman D, Wormsley S: The induction of the human T-cell growth factor receptor precedes the production of RNA and occurs in the presence of inhibitors of RNA synthesis. Cytometry 7:453, 1986
1005. Nance SJ, Garratty G: Application of flow cytometry to immunohematology. J Immunol Methods 101:127, 1987
1006. Matsui Y, Martin-Alosco S, Doenges E, Christenson L, Shapiro HM, Yunis EJ, Page P: Effects of frequent and sustained plateletapheresis on peripheral blood mononuclear cell populations and lymphocyte functions of normal volunteer donors. Transfusion 26:446, 1986
1007. Rosenfeld CS, Nichols G, Bodensteiner DC: Flow cytometric measurement of antiplatelet antibodies. Am J Clin Pathol 87:518, 1987
1008. Mankin HJ, Connor JF, Schiller AL, Perlmutter N, Alho A, McGuire M: Grading of bone tumors by analysis of nuclear DNA content using flow cytometry. J Bone Joint Surg 67-A:404, 1985
1009. Greaves MF: Differentiation-linked leukemogenesis in man. Science 234:697, 1986
1010. Ryan DH, van Dongen JJM: Detection of residual disease in acute leukemia using immunological markers. In: Bennett JM, Foon KA (eds): *Immunologic Approaches to the Classification and Management of Leukemias and Lymphomas.* Boston, Martinus Nijhoff, 1988
1011. Bagwell CB, Lovett EJ III, Ault KA: Localization of monoclonal B-cell populations through the use of D-value and R-value contours. Cytometry 9:469-76, 1988
1012. Watson JV: Oncogenes, cancer and analytical cytology. Cytometry 7:400, 1986
1013. Watson JV, Sikora K, Evan GI: A simultaneous flow cytometric assay for *c-myc* oncoprotein and DNA in nuclei from paraffin-embedded material. J Immunol Methods 83:179, 1985
1014. Andreeff M, Slater DE, Bressler J, Furth ME: Cellular *ras* oncogene expression and cell cycle measured by flow cytometry in hematopoietic cell lines. Blood 67:676, 1986
1015. Human RP, Rowe GD: Automated screening for bacteriuria. Med Lab Sci 35:223, 1978
1016. Dow CS, France AD, Khan MS, Johnson T: Particle size distribution analysis for the rapid detection of microbial infection of urine. J Clin Pathol 32:386, 1979
1017. Alexander MS, Khan MS, Dow CS: Rapid screening for bacteriuria using a particle counter, pulse-height analyser, and computer. J Clin Pathol 34:194, 1981
1018. Duesberg PH: Retroviruses as carcinogens and patho-gens: Expectations and reality. Cancer Res 47:1199, 1987
1019. Saunders GC, Jett JH, Martin JC: Amplified flow-cytometric separation-free fluorescence immunoassays. Clin Chem 31:2020, 1985
1020. Deindoerfer FH, Gangwer JR, Laird CW, Ringold RR: "The Yellow IRIS" urinalysis workstation - The first commercial application of "automated intelligent microscopy". Clin Chem 31:1491, 1985
1021. Fetterhoff TJ, Hammer JE, Luckey DW, McCarthy RC: Construction of a flow cytometer based on the design by Shapiro: Modifications and practical considerations. Cytometry 8:340, 1987
1022. Murphy RF, Chused TM: A proposal for a flow cytometric data file standard. Cytometry 5:553, 1984
1023. Kachel V, Schneider H: On-line three-parameter data uptake, analysis and display device for flow cytometry and other applications. Cytometry 7:25, 1986
1024. Stewart SS, Price GB: Realtime acquisition, storage and display of correlated three-parameter flow cytometric data. Cytometry 7:82, 1986
1025. Ormerod MG, Payne AWR: Display of three-parametric data acquired by a flow cytometer. Cytometry 8:240, 1987
1026. Tufte ER: *The Visual Display of Quantitative Information.* Cheshire (CT), Graphics Press, 1983
1027. Mandy FF, Bergeron M, Recktenwald D, Izaguirre CA: A simultaneous three-color T cell subsets analysis with single laser flow cytometers using T cell gating protocol. Comparison with conventional two-color immunophenotyping method. J Immunol Methods 156:151-62, 1992
1028. Melamed MR, Lindmo T, Mendelsohn ML (eds): *Flow Cytometry and Sorting* (2 Ed). New York, Wiley-Liss, 1990, xii + 824 pp.
1029. Givan AL: *Flow Cytometry: First Principles.* New York, Wiley-Liss, 1992, xiv + 202 pp.
1030. Watson JV: *Introduction to Flow Cytometry.* Cambridge, Cambridge University Press, 1991, xvi + 443 pp.
1031. Ormerod MG (ed): *Flow Cytometry. A Practical Approach.* Oxford, IRL Press at Oxford University Press, 1990, xxiv + 279 pp.
1032. Grogan W McL, Collins JM: *Guide to Flow Cytometry Methods.* New York, Marcel Dekker, 1990, x + 228 pp.
1033. Radbruch A (ed): *Flow Cytometry and Cell Sorting.* Berlin, Springer, 1992, x + 223 pp.
1034. Darzynkiewicz Z, Crissman HA (eds): *Methods in Cell Biology,* Vol 33, *Flow Cytometry,* San Diego, Academic Press, 1990, xviii + 716 pp.
1035. Robinson JP (ed): *Handbook of Flow Cytometry Methods.* New York, Wiley-Liss, 1993, xii + 246 pp.
1036. Keren DF (ed): *Flow Cytometry in Clinical Diagnosis.* Chicago, American Society of Clinical Pathologists Press, 1989, xviii + 343 pp.
1037. Keren D, Hanson C, Hurtubise P (eds): *Flow Cytometry and Clinical Diagnosis.* Chicago, American Society of Clinical Pathologists Press, 1994, xi + 665 pp.
1038. Bauer KD, Duque RE, Shankey TV (eds): *Clinical Flow Cytometry. Principles and Application.* Baltimore, Williams & Wilkins, 1993, xvi + 635 pp.
1039. Riley RS, Mahin EJ, Ross W (eds): *Clinical Applications of Flow Cytometry.* New York, Igaku-Shoin, 1993, x + 914 pp.
1040. Coon JS, Weinstein RS (eds): *Diagnostic Flow Cytometry* (*Techniques in Diagnostic Pathology,* No 2). Baltimore, Williams and Wilkins, 1991, xii + 199 pp.
1041. Vielh P (ed): *Flow Cytometry. Guides to Clinical Aspiration Biopsy.* New York, Igaku-Shoin, 1991, xvi + 173 pp.
1042. Landay AL, Ault KA, Bauer KD, Rabinovitch PS (eds): *Clinical Flow Cytometry.* Ann N Y Acad Sci 677:1-468, 1993
1043. Macey MG (ed): *Flow Cytometry: Clinical Applications.* Oxford, Blackwell Scientific, 1994, xii + 308 pp.
1044. Laerum OD, Bjerknes R (eds): *Flow Cytometry in Hematology.* London, Academic Press, 1992, xii + 272 pp.
1045. Sun T: *Color Atlas-Text of Flow Cytometric Analysis of Hematologic Neoplasms.* New York, Igaku-Shoin, 1993, viii + 221 pp.
1046. Yen A (ed): *Flow Cytometry: Advanced Research and Clinical Applications.* Boca Raton, CRC Press, 1989, Vol I: 346 pp; Vol II: 293 pp.
1047. Jacquemin-Sablon A (ed): *Flow Cytometry. New Deve-lopments* (NATO Advanced Study Institute Series H, Cell Biology, Vol 67). Berlin, Springer, 1993, x + 473 pp.

1048. Métézeau P, Ronot X, Le Noan-Merdrignac G, Ratinaud MH (eds): *La Cytométrie en Flux pour l'Étude de la Cellule Normale ou Pathologique.* Paris, MEDSI/McGraw-Hill, 1988, 407 pp.
1049. Brugal G: Book Review: *La Cytométrie en Flux pour l'Étude de la Cellule Normale ou Pathologique.* Cytometry 9:512, 1988
1050. Gray JW (ed): *Flow Cytogenetics.* San Diego, Academic Press, 1989, xvi + 312 pp.
1051. Demers S (ed): *Particle Analysis in Oceanography* (NATO Advanced Study Institute Series G, Ecological Sciences, Vol 27). Berlin, Springer, 1991, xii + 416 pp.
1052. Lloyd D (ed): *Flow Cytometry in Microbiology.* London, Springer, 1993, xii + 188 pp.
1053. Saleh BEA, Teich MC: *Fundamentals of Photonics..* New York, John Wiley & Sons, 1991, xvii + 966 pp.
1054. Hecht J: *The Laser Guidebook* (2 Ed). Blue Ridge Summit, Tab Books/McGraw-Hill, 1992, xiv + 498 pp.
1055. Smith RF: *Microscopy and Photomicrography. A Working Manual.* Boca Raton, CRC Press, 1990, viii + 162 pp.
1056. Horowitz P, Hill W: *The Art of Electronics* (2 Ed). Cambridge, Cambridge University Press, 1989, xxiii + 1125 pp.
1057. Pease RA: *Troubleshooting Analog Circuits.* Boston, Butterworth-Heinemann, 1991, xii + 219 pp.
1058. Williams J (ed): *Analog Circuit Design: Art, Sciece, and Personalitiess.* Boston, Butterworth-Heinemann, 1991, x + 389 pp.
1059. Eggebrecht LC: *Interfacing to the IBM Personal Com-puter* (2 Ed). Indianapolis, Howard W. Sams and Company, 1990, xvi + 345 pp.
1060. Forst G: *PC Principles.* Cambridge, The MIT Press, 1990, xxviii + 560 pp.
1061. Sanchez J, Canton MP: *IBM Microcomputers. A Programmer's Handbook.* New York, McGraw-Hill, 1990, xii + 503 pp.
1062. Jourdain R: *Programmer's Problem Solver* (2 Ed). New York, Brady Publishing, 1992, xii + 596 pp.
1063. Apple Computer, Inc.: *Technical Introduction to the Macintosh Family* (2 Ed). Reading, Addison-Wesley, 1992, xxxii + 410 pp.
1065. Leonhard W, Simon B: *CD-MOM: The Mother of All Windows Books.* Reading, Addison-Wesley, 1993, xxxi + 1037 pp.
1066. Brodie L (and FORTH, Inc.): *Starting Forth* (2 Ed). Englewood Cliffs, Prentice-Hall, 1987, xviii + 346 pp.
1067. Zech R: *Forth for Professionals.. A Practical Programming Language for Research and Deve-opment.* New York, Ellis Horwood, 1990, 306 pp.
1068. Woehr J: *Forth: The New Model.* San Mateo, M&T Books, 1992, xx + 316 pp.
1069. Horobin RW: *Understanding Histochemistry. Selection, Evaluation, and Design of Biological Stains.* New York, John Wiley & Sons, 1988, 172 pp.
1070. Rost FWD: *Fluorescence Microscopy.* Volume I. Cambridge, Cambridge University Press, 1991, xiv + 236 pp.
1071. Rost FWD: *Quantitative Fluorescence Microscopy.* Cambridge, Cambridge University Press, 1992, xiii + 253 pp.
1072. Kohen E, Hirschberg JG (eds): *Cell Structure and Function by Microspectrofluorometry.* San Diego, Academic Press, 1989, xxiv + 465 pp.
1073. Wang Y, Taylor DL (eds): *Fluorescence Microscopy of Living Cells in Culture.* San Diego, Academic Press, 1989; Part A, xiv + 333 pp., Part B, xiv + 503 pp.
1074. Mason WT (ed): *Fluorescent and Luminescent Probes for Biological Activity a Practical Guide to Technology for Quantitative Real-Time Analysis.* London, Academic Press, 1993, xxi + 433 pp.
1075. Haugland RP, Larison KD: *Handbook of Fluorescent Probes and Research Chemicals* (5 Ed) Eugene (OR), Molecular Probes, Inc., 1992, x + 421 pp.
1076. Lacey AJ (ed): *Light Microscopy in Biology.* Oxford, IRL Press/Oxford University Press, 1989, xviii + 329 pp.
1077. Darnell J, Lodish H, Baltimore D: *Molecular Cell Biology* (2 Ed). New York, Scientific American Books, xl + 1105 pp.
1078. Watson JD, Gilman M, Witkowski J, Zoller M: *Recombinant DNA* (2 Ed). New York, W. H. Freeman and Company, 1992, xiv + 626 pp.
1079. Roitt I, Brostoff J, Male D: *Immunology* (3 Ed). St. Louis, Mosby-Year Book Europe, xii + 400 pp.
1080. Golub ES, Green DR: *Immunology a Synthesis* (2 Ed). Sunderland, Sinauer Associates, 1991, xxviii + 744 pp.
1081. Weer PD, Salzberg BM (eds): *Optical Methods in Cell Physiology.* New York, John Wiley & Sons, 1986, xv + 480 pp.
1082. Russ JC: *Computer-Assisted Microscopy. The Measurement and Analysis of Images.* New York, Plenum Press, 1990, xii + 453 pp.
1083. Pawley JB (ed): *Handbook of Biological Confocal Microscopy.* New York, Plenum Press, 1990, xiii + 232 pp.
1084. Cherry RJ (ed): *New Techniques of Optical Microscopy and Microspectroscopy.* Boca Raton, CRC Press, 1991, xii + 279 pp.
1085. Clark G, Kasten FH: *History of Staining* (3 Ed). Baltimore, Williams & Wilkins, 1983, x + 304 pp.
1086. Coons AH: The beginnings of immunofluorescence. J Immunol 87:499-503, 1961
1087. Reiner L: On the chemical alteration of purified antibody-proteins. Science 72:483-4, 1930
1088. Marrack J: Nature of antibodies. Nature 133:292-3, 1934
1089. Creech HJ, Jones NR: The conjugation of horse serum albumin with 1,2-benzanthryl isocyanates. J Am Chem Soc 42:1970-5, 1940.
1090. Coons AH, Creech HJ, Jones RN, Berliner E: The demonstration of pneumococcal antigen in tissues by the use of fluorescent antibody. J Immunol 45:159-70, 1942.
1091. Wintrobe MM, Lee GR, Boggs DR, Bithell TC, Athens JW, Foerster J (eds): *Clinical Hematology* (7 Ed). Philadelphia, Lea & Febiger, 1974, xxiv + 1896 pp.
1092. Wintrobe MM: Variations in the size and hemoglobin content of erythrocytes in the blood of various vertebrates. Folia Hematol 51:32, 1933
1093. Wintrobe MM: Erythrocyte in man. Medicine 9:195, 1930
1094. Wright GP: The relative duration of the various phases of mitosis in chick fibroblasts *in vitro.* J Roy Microsc Soc :414-7, 1925
1095. Howard A, Pelc SR: Nuclear incorporation of ^{32}P as demonstrated by autoradiographs. Exp Cell Res 2:178-87, 1951
1096. Howard A, Pelc SR: Synthesis of desoxyribonucleic acid in normal and irradiated cells and its relation to chromosome breakage. Heredity Supp 6:261-73, 1953
1097. Taylor JH, Woods PS, Hughes WL: The organization and duplication of chromosomes using tritiated thymidine. Proc Natl Acad Sci USA 43:122-8, 1957
1098. Quastler H, Sherman FG: Cell population kinetics in the intestinal epithelium of the mouse. Exp Cell Res 17:420-38, 1959
1099. Cambrosio A, Keating P: Between fact and technique; The beginnings of hybridoma technology. J Hist Biol 25:175-230
1100. Cambrosio A, Keating P: A matter of FACS: Constituting novel entities in immunology. Med Anthrop Quart 6:362-84, 1992
1101. Keating P, Cambrosio A: "Ours is an engineering approach": Flow cytometry and the constitution of human T cell subsets. J Hist Biol, 27:449-79, 1994
1102. Kondratas R: Smithsonian Video History Collection: The History of the Cell Sorter. Washington, Smithsonian Institution Archives, 1992 (www.si.edu/archives/ihd/videocatalog/9554.htm)
1103. Steinkamp JA, Crissman HA: Automated analysis of deoxyribonucleic acid, protein, and nuclear to cytoplasmic relationships in tumor cells and gynecologic specimens. J Histochem Cytochem 22:616-21, 1974
1104. Dean PN, Jett JH: Mathematical analysis of DNA distributions derived from flow microfluorometry. J Cell Biol 60:523-7, 1974
1105. Kwan D, Epstein MB, Norman A: Studies on human monocytes with a multiparameter cell sorter. J Histochem Cytochem 24:355, 1976

1106. Gray JW, Carrano AV, Steinmetz LL, Van Dilla MA, Moore DH II, Mayall BH, Mendelsohn ML: Chromosome measurement and sorting by flow systems. Proc Natl Acad Sci USA 72:1231-4, 1975
1107. Julius MH, Masuda T, Herzenberg LA: Demonstration that antigen-binding cells are antibody producing cells after purification using a fluorescence-activated cell sorter. Proc Natl Acad Sci USA 69:1934-8, 1972
1108. Thomas RA, Cameron BF, Leif RC: Computer-based electronic cell volume analysis with the AMAC II transducer. J Histochem Cytochem 22:626-41, 1974
1109. Thomas RA, Yopp TA, Watson BD, Hindman DHK, Cameron BF, Leif SB, Leif RC, Roque L, Britt W: Combined optical and electronic analysis of cells with the AMAC transducers. J Histochem Cytochem 25:827-35, 1977
1110. Leif RC, Clay SP, Gratzner HG, Haines HG, Vallarino LM: Markers for instrumental evaluation of cells of the female reproductive tract: Existing and new markers. In: Wied GL, Bahr GF, Bartels PH (eds): *The Automation of Uterine Cancer Cytology.* Chicago, Tutorials of Cytology, 1976, pp 313-44
1111. Wheeless LL, Patten SF: Slit-scan cytofluorometry. Acta Cytol 17:333, 1973
1112. Wheeless LL, Hardy JA, Balasubramanian N: Slit-scan flow system for automated cytopathology. Acta Cytol 19:45, 1975
1113. Cram LS, Gomez ER, Thoen CO, Forslund JC, Jett JH: Flow microfluorimetric quantitation of the blastogenic response of lymphocytes. J Histochem Cytochem 24:383-7, 1976
1114. Merle-Beral H, Klatzman D, Blanc C, Debre P: Evaluation des performances du Spectrum III. Nouv Rev Fr Hematol 27:193-9, 1985
1115. Andreeff M (ed): *Impulscytophotometrie.* Berlin, Springer, 1975, xvi + 182 pp.
1116. Hedley DW: Flow cytometric assays of anticancer drug resistance. Ann N Y Acad Sci 677:341-53, 1993 (in reference 1042)
1117. Roederer M, Staal FJT, Osada H, Herzenberg LA, Herzenberg LA: CD4 and CD8 T cells with high intracellular glutathione levels are selectively lost as the HIV infection progresses. Int Immunol 3:933-7, 1991
1118. Roederer M, Staal FJT, Anderson M, Rabin R, Raju PA, Herzenberg LA, Herzenberg LA: Disregulation of leukocyte glutathione in AIDS. Ann N Y Acad Sci 677: 113-25, 1993 (in reference 1042)
1119. Chessells JM, Hardisty RM, Rapson NT, Greaves MF: Acute lymphoblastic leukemia in Children: Classification and prognosis. Lancet 2:1307-9, 1977
1120. Janossy G, Roberts MM, Capellaro D, Greaves MF, Francis GE: Use of the fluorescence activated cell sorter in human leukemia. In: *Immunofluorescence and Related Staining Techniques,* Knapp W et al (eds), Amsterdam, Elsevier, 1978, pp 111-22
1121. Sallan SE, Chess L, Frei E 3rd, O'Brien C, Nathan DG, Strominger JL, Schlossman SF: Utility of B- and T-cell specific antisera in the classification of human leukemias. In: *Differentiation of Normal and Neoplastic Hematopoietic Cells,* Clarkson B et al (eds), Cold Spring Harbor, Cold Spring Harbor Laboratory, 1978, pp 479-83
1122. Bernard A, Boumsell L, Dausset J, Milstein C, Schlossman SF (eds): *Leukocyte Typing: Human Leucocyte Differentiation Antigens Detected by Monoclonal Antibodies.* Berlin, Springer, 1984
1123. Reinherz EL, Haynes BJ, Bernstein ID, Nadler LM (eds): *Leucocyte Typing II.* (2 Volumes). New York, Springer, 1985
1124. McMichael AJ, Beverley PCL, Cobbold S, Crumpton MJ, Gilks W, Gotch FM, Hogg N, Horton M, Ling N, MacLennan ICM, Mason DY, Milstein C, Spiegelhalter D, Waldmann H (eds): *Leucocyte Typing III. White Cell Differentiation Antigens.* Oxford, Oxford University Press, 1987, xxxiv + 1050 pp.
1125. Knapp W, Dörken B, Gilks WR, Rieber EP, Schmidt RE, Stein H, von dem Borne AEGKr (eds): *Leucocyte Typing IV. White Cell Differentiation Antigens.* Oxford, Oxford University Press, 1989, xxiv + 1182 pp.
1126. Schlossman SF, Boumsell L, Gilks W, Harlan JM, Kishimoto T, Morimoto C, Ritz J, Shaw S, Silverstein RL, Springer TA, Tedder TF, Todd RF (eds): *Leucocyte Typing V. White Cell Differentiation Antigens.* Oxford, Oxford University Press, 1995
1127. Schlossman SF, Boumsell L, Gilks W, Harlan JM, Kishimoto T, Morimoto C, Ritz J, Shaw S, Silverstein RL, Springer TA, Tedder TF, Todd RF: Update: CD Antigens 1993. J Immunol 152:1-2, 1994
1128. Gottlieb MS, Schroff R, Schanker HM, Weisman JD, Fan PT, Wolf RA, Saxon A: *Pneumocystis carinii* pneumonia and mucosal candidiasis in previously healthy homosexual men: evidence of a new acquired cellular immunodeficiency. N Engl J Med 305:1425-31, 1981
1129. Jablonski A: Über den Mechanismus des Photolumi-nescenz von Farbstoffphosphoren. Z Phys 94:38-46, 1935
1130. van den Engh G, Farmer C: Photo-bleaching and photon saturation in flow cytometry. Cytometry 13:669-77, 1992
1131. Denk W, Strickler JH, Webb WW: Two-photon laser scanning fluorescence microscopy. Science 248:73-6, 1990
1132. Steen H, Sørensen OI: Pulse modulation of the excitation light source boosts the sensitivity of an arc lamp-based flow cytometer. Cytometry 14:115-22, 1993
1133. Shapiro HM, Perlmutter NG: Bivariate chromosome flow cytometry using single-laser instruments. Cytometry Supp 6:71, 1993
1134. Frey T, Stokdijk W, Hoffman RA: Bivariate flow karyotyping with air-cooled lasers. Cytometry Supp 6:71, 1993
1135. Snow C, Cram LS: The suitability of air-cooled helium cadmium (HeCd) lasers for two color analysis and sorting of human chromosomes. Cytometry Supp 6:20, 1993
1136. Doornbos RMP, De Grooth BG, Kraan YM, Van Der Poel CJ, Greve J: Visible diode lasers can be used for flow cytometric immunofluorescence and DNA analysis. Cytometry 15:267-71, 1994
1137. Doornbos RMP, Hennink EJ, Putman CAJ, De grooth BG, Greve J: White blood cell differentiation using a solid state flow cytometer. Cytometry 14:589-94, 1993
1138. Piehler D: Upconversion process creates compact blue/green lasers. Laser Focus World 29(11):95-102, 1993
1139. Vesey G, Narai G, Ashbolt N, Veal D: Detection of specific microorganisms in environmental samples using flow cytometry. Methods Cell Biol 42: 489-522, 1994
1140. Buican T: Real-time Fourier transform spectrometry for fluorescence imaging and flow cytometry. Proc SPIE 1205:126-33, 1990
1141. Steen H: Light scattering measurement in an arc lamp-based flow cytometer. Cytometry 11:223-30, 1990
1142. Mitchell P: PMTs benefit from improved high-voltage supplies. Laser Focus World 28(1):135-40, 1992
1143. Burle Industries: *Photomultiplier Handbook.* Lancaster, PA, 1989
1144. Goodwin PM, Johnson ME, Martin JC, Ambrose WP, Marrone BL, Jett JH, Keller RA: Rapid sizing of individual fluorescently stained DNA fragments by flow cytometry. Nucleic Acids Res 21:803-6, 1993
1145. Deutsch M, Weinreb A: An apparatus for high-precision repetitive sequential optical measurement of living cells. Cytometry 16:214-26,1994.
1146. Kachel V, Fellner-Feldegg H, Menke E: Hydrodynamic properties of flow cytometry instruments. In reference 1028, pp 27-44
1147. Ruzicka J, Hansen E: *Flow Injection Analysis* (2 Ed). New York, Wiley-Interscience, 1988, xxii + 498 pp.
1148. Ruzicka J, Lindberg W: Flow injection cytoanalysis. Anal Chem 64:537-41, 1992
1149. Lindberg W, Ruzicka J, Christian GD: Flow injection cytometry: a new approach for sampling and solution handling in flow cytometry. Cytometry 14:230-6, 1993
1150. Gross H-J, Verwer B, Houck D, Recktenwald D: Detection of rare cells at a frequency of one per million by flow cytometry. Cytometry 14:519-26, 1993

1151. Kachel V: Electrical resistance pulse sizing: Coulter sizing. In reference 1028, pp 45-80
1152. Cotterill R: *The Cambridge Guide to the Material World.* Cambridge, Cambridge University Press, 1985, v + 352 pp.
1153. Hiebert R: Electronics and Signal Processing. In reference 1028, pp 127-44
1154. Terstappen LWMM, Loken MR: Five-dimensional flow cytometry as a new approach to blood and bone marrow differentials. Cytometry 9:548-56, 1988
1155. Bagwell CB, Adams EG: Fluorescence spectral overlap compensation for any number of flow cytometry parameters. Ann N Y Acad Sci 677:167-84, 1993 (in reference 1042)
1156. Muirhead KA, Schmitt TC, Muirhead AR: Determination of linear fluorescence intensities from flow cytometric data accumulated with logarithmic amplifiers. Cytometry 3:251-6, 1983
1157. Schmid I, Schmid P, Giorgi JV: Conversion of logarithmic channel numbers into relative linear fluorescence intensity. Cytometry 9:533-8, 1988
1158. Parks DR, Bigos M, Moore WA: Logarithmic amplifier transfer function evaluation and procedures for logamp optimization and data correction. Cytometry Supp 2:27, 1988
1159. Sweet R, Parks D, Nozaki T, Herzenberg L: A 3 1/2 decade logarithmic amplifier for cell fluorescence data. Cytometry 2:130, 1981
1160. De Grooth BG, Doornbos RMP, Florians A, Greve J: A fast logarithmic amplifier for flow cytometry. Cytometry Supp :117, 1991
1161. Schwartz A, Fernández-Repollet E: Development of Clinical Standards for Flow Cytometry. Ann N Y Acad Sci 677:28-39, 1993 (in reference 1042)
1162. Holm DM, Cram LS: An improved flow microfluorimeter for rapid measurement of cell fluorescence. Exp Cell Res 80:105-10, 1973
1163. Pinkel D, Steen HB: Simple methods to determine and compare the sensitivity of flow cytometers. Cytometry 3:220-3, 1982
1164. Ubezio P, Andreoni A: Linearity and noise sources in flow cytometry. Cytometry 6:109-15, 1985
1165. Steen HB: Noise, sensitivity, and resolution of flow cytometers. Cytometry 13:822-30, 1992
1166. Wheeless LL Jr: Slit-Scanning. In reference 1028, pp 109-25
1167. Peeters JC, Dubelaar GB, Ringelberg J, Visser JW: Optical plankton analyser: a flow cytometer for plankton analysis, I: Design considerations. Cytometry 10:522-8, 1989
1168. Dubelaar GB, Groenewegen AC, Stokdijk W, van den Engh GJ, Visser JW: Optical plankton analyser: a flow cytometer for plankton analysis, II: Specifications. Cytometry 10:529-39, 1989
1169. Robinson RD, Wheeless DM, Hespelt SJ, Wheeless LL: System for acquisition and real-time processing of multidimensional slit-scan flow cytometric data. Cytometry 11:379-85, 1990
1170. Zuse P, Hauser R, Manner R, Hausmann M, Cremer C: Real-time multiprocessing of slit scan chromosome profiles. Comput Biol Med 20:465-76, 1990
1171. Beisker W, Nüsse M: Optical design of a combined integral-light and slit-scanning multiparameter flow cytometer. Cytometry Supp 5:52, 1991
1172. Bakker Schut TC, Florians A, Radosevic K, de Grooth BG, Greve J: Signal processing in slit-scan flow cytometry of cell conjugates. Cytometry 14:459-64, 1993
1173. Buican T: A multiple bus data acquisition, processing, and control system for flow cytometry. Cytometry Suppl1:101, 1987
1174. Thomas RA, Thornthwaite J, Eggleston R: NASA inflight cytometry project. Cytometry Supp 5:117, 1991
1175. Parson JD, Olivier TL, Habbersett RC, Martin JC, Wilder ME, Jett JH: Characterization of digital signal processing in the DiDAC data acquisition system. Cytometry Supp 6:40, 1993
1176. Salzman GC, Wilkins SF, Whitfill JA: Modular computer programs for flow cytometry and sorting: the LACEL system. Cytometry 1:325-36, 1981
1177. Hiebert RD, Jett JH, Salzman GC: Modular electronics for flow cytometry and sorting: the LACEL system. Cytometry 1:337-41, 1981
1178. Kachel V, Messerschmidt R, Hummel P: Eight-parameter PC-AT based flow cytometric data system. Cytometry 11:805-12, 1990
1179. Leif RC: Book review: Practical Flow Cytometry, Second Edition. Cytometry 10:490-1, 1989
1180. van den Engh G, Stokdijk W: Parallel processing data acquisition system for multilaser flow cytometry and cell sorting. Cytometry 10:282-93, 1989
1181. Hall B, Ashcroft R, Malachowski G: CICERO's flexible software control of sorting eliminates contaminating "dawdler" cells. Cytometry Supp 1:97, 1987
1182. Ashcroft R, Hall B, Malachowski G: High flow rates and conventional sorters. Cytometry Supp 1:104, 1987
1183. Auer RE, Starling D, Weber B, Wood JCS: A data acquisition system for flow cytometry with wide dynamic range analog to digital conversion and digital signal processing. Cytometry Supp 6:18, 1993
1184. Bagwell CB: Theoretical aspects of flow cytometry data analysis. In reference 1038, pp 41-70
1185. Watson JV: *Flow Cytometry Data Analysis. Basic Concepts and Statistics.* Cambridge, Cambridge University Press, 1992, viii + 288 pp.
1186. Moroney MJ: *Facts From Figures* (3 Ed). London, Penguin Books, 1956, 472 pp.
1187. Coder DM, Redelman D, Vogt RF Jr: Computing the central location of immunofluorescence distributions: Logarithmic data transformations are not always appropriate. Cytometry (Commun Clin Cytometry) 18:75-8, 1994
1188. Tukey J : *Exploratory Data Analysis.* Reading (MA), Addison-Wesley, 1977, xvi + 688 pp.
1189. Tufte ER: *Envisioning Information.* Cheshire (CT), Graphics Press, 1990, 126 pp.
1190. Cox C, Reeder JE, Robinson RD, Suppes SB, Wheeless LL: Comparison of frequency distributions in flow cytometry. Cytometry 9:291-8, 1988
1191. Bagwell CB, Hudson JL, Irvin GL III: Nonparametric flow cytometry analysis. J Histochem Cytochem 27:293, 1979
1192. Overton RW: Modified histogram subtraction technique for analysis of flow cytometry data. Cytometry 9:619-26, 1988
1193. White RA, Terry NHA: A quantitative method for evaluating bivariate flow cytometric data obtained using monoclonal antibodies to bromodeoxyuridine. Cytometry 13:490-5, 1992
1194. Dean PN, Kolla S, Van Dilla MA: Analysis of bivariate flow karyotypes. Cytometry 10:109-23, 1989
1195. van den Engh G, Hanson D, Trask B: A computer program for analyzing bivariate flow karyotypes. Cytometry 11:173-83, 1990
1196. Boschman GA, Manders EMM, Rens W, Slater R, Aten JA: Semi-automated detection of aberrant chromosomes in bivariate flow karyotypes. Cytometry 13:469-77, 1992
1197. Moore, DH II, Gray JW: Derivative domain fitting: A new method for resolving a mixture of normal distributions in the presence of a contaminating background. Cytometry 14:510-8, 1993
1198. Sloot PMA, Tensen P, Figdor CG: Spectral analysis of flow cytometric data: Design of a special-purpose low-pass digital filter. Cytometry 8:545-51, 1987
1199. Greimers R, Rongy AM, Schaaf-Lafontaine N, Boniver J: CUBIC: A three-dimensional colored projection of Consort 30 generated trivariate flow cytometric data. Cytometry 12:570-8, 1991
1200. Leary JF, Ellis SP, McLaughlin SR: 3-D autostereoscopic viewing of multidimensional data for principal component/biplot analysis and sorting. Cytometry Supp 5:134-5, 1991
1201. Sloot PMA, Figdor CG: Ternary representation of trivariate data. Cytometry 10:77-80, 1989
1202. Manly BFJ: *Multivariate Statistical Methods. A Primer.* London, Chapman and Hall, 1986, x + 159 pp.
1203. Kosugi Y, Sato R, Genka S, Shitara N, Takakura K: An interactive multivariate analysis of FCM data. Cytometry 9:405-8, 1988

1204. Murphy, RF: Automated identification of subpopulations in flow cytometric list mode data using cluster analysis. Cytometry 6:302-9, 1985
1205. Salzman G, Krall R, Beckman R, Pederson S, Stewart C: A knowledge-based system as a cluster analysis assistant. Cytometry Supp 1:12, 1987
1206. Bierre P, Mickaels R, Thiel D: Multidimensional visualization and autoclustering of flow cytometric data. Cytometry Supp 5:64, 1991
1207. Valet G, Valet M, Tschope D, Gabriel H, Rothe G, Lellermann W, Kahle H: White cell and thrombocyte disorders. Standardized, self-learning flow cytometric list mode data classification with the CLASSIF1 program system. Ann N Y Acad Sci 677:233-51, 1993 (in reference 1042)
1208. Kelley S, Durack G, Maguire D, Ragheb K, Robinson JP: An automated system for immunophenotype analysis. Cytometry Supp 6:40, 1993
1209. Salzman GC, Parson JD, Beckman RJ, Stewart SJ, Stewart CC: Autogate: A Macintosh cluster analysis program for flow cytometry data. Cytometry Supp 6:43, 1993
1210. Bierre P, Thiel D, Mickaels R: Multiparameter cluster analysis using attractors. Cytometry Supp 6:44, 1993
1211. Bakker Schut TC, De Grooth BG, Greve J: Cluster analysis of flow cytometric list mode data on a personal computer. Cytometry 14:649-59, 1993
1212. Verwer BJH, Terstappen LWMM: Automatic lineage assignment of acute leukemias by flow cytometry. Cytometry 14:862-75, 1993
1213. Pao Y-H: *Adaptive Pattern Recognition and Neural Networks*. Reading (MA), Addison-Wesley, 1989, xviii + 309 pp.
1214. Aleksandr I, Morton H: *An Introduction to Neural Computing*. London, Chapman and Hall, 1990, xix + 240 pp.
1215. Frankel DS, Olson RJ, Frankel SL, Chisholm SW: Use of a neural net computer system for analysis of flow cytometric data of phytoplankton populations. Cytometry 10:540-50, 1989
1216. Boddy L, Morris CW, Wilkins MF, Tarran GA, Burkill PH: Neural network analysis of flow cytometric data for 40 marine phytoplankton species. Cytometry 15:283-93, 1994
1217. Frankel DS, Loken MR, Stelzer GT, Shults KE, Bagwell CB: Neural network analysis of flow cytometric data for normal and leukemic bone marrow. Cytometry Supp 6:44, 1993
1218. Redelman D: Improved procedures for training neural networks to analyze flow cytometric data. Cytometry Supp 6:43, 1993
1219. Goldberg DE: *Genetic Algorithms in Search, Optimization, and Machine Learning*. Reading (MA), Addison-Wesley, 1989, xiii + 412 pp.
1220. Dean PN, Bagwell CB, Lindmo T, Murphy RF, Salzman GC: Introduction to flow cytometry data file standard. Cytometry 11:321-2, 1990
1221. Data file standard for flow cytometry. Data File Standards Committee of the Society for Analytical Cytology. Cytometry 11:323-32, 1990
1222. Mann RC: On multiparameter data analysis in flow cytometry. Cytometry 8:184-9, 1987
1223. Lindmo T, Peters DC, Sweet RG: Flow sorters for biological cells. In reference 1028, pp 145-69
1224. De Grooth BG, Doornbos RM, Van Der Werf KO, Greve J: Simple delay monitor for droplet sorters. Cytometry 12:469-72, 1991
1225. Lazebnik YA, Poletaev AI, Zenin VV: Drop-delay measurement using enzyme-coated particles. Cytometry 13:649-52, 1992
1226. Friedman M: Digital fluidic amplifier particle sorter. U S Patent No 3,791,517, 1973
1227. Gray DW, Göhde W, Carter N, Heiden T, Morris PJ: Separation of pancreatic islets by fluorescence-activated sorting. Diabetes 38:133-5, 1989
1228. Ashkin A: Acceleration and trapping of particles by radiation pressure. Phys Rev Lett 24:156-9, 1970
1229. Ashkin A, Dziedzic JM, Yamane T: Optical trapping and manipulation of single cells using infrared laser beams. Nature 330:769-71, 1987
1230. Ashkin A, Dziedzic JM: Optical trapping and manipulation of viruses and bacteria. Science 235:1517-20, 1987
1231. Buican TN, Smyth MJ, Vcrissman HA, Salzman GC, Stewart CC, Martin JC: Automated single-cell manipulation and sorting by light trapping. Appl Optics 26:5311-6, 1987
1232. Bakker Schut TC, de Grooth BG, Greve J: A new principle of cell sorting by using selective electroporation in a modified flow cytometer. Cytometry 11:659-66, 1990
1233. Puchalski RB, Fahl WE: Gene transfer by electroporation, lipofection, and DEAE-dextran transfection: compatibility with cell-sorting by flow cytometry. Cytometry 13:23-30, 1992
1234. Berglund DL, Starkey JR: Isolation of viable tumor cells following introduction of labelled antibody to an intracellular oncogene product using electroporation. J Immunol Methods 125:79-87, 1989
1235. McCoy J Jr, Chambers WH, Lakomy R, Campbell JA, Stewart CC: Sorting minor subpopulations of cells: use of fluorescence as the triggering signal. Cytometry 12:268-74, 1991
1236. Keij JF, van Rotterdam A, Groenewegen AC, Stokdijk W, Visser JW: Coincidence in high-speed flow cytometry: models and measurements. Cytometry 12:398-404, 1991
1237. van Rotterdam A, Keij J, Visser JW: Models for the electronic processing of flow cytometric data at high particle rates. Cytometry 13:149-54, 1992
1238. Miltenyi S, MüW, Weichel W, Radbruch A: High gradient cell separation with MACS. Cytometry 11:231-8, 1990
1239. Penney DP, Leary JF, Cooper R Jr, Paxhia A: Electron microscopic identification and morphologic preservation of enriched populations of lung cells isolated by laser flow cytometry and cell sorting: a new technique. Stain Technol 65:165-77, 1990
1240. Sebring RJ, Johnson NF, Spall WD: Transmission electron microscopy of small numbers of sorted cells. Cytometry 9:88-92, 1988
1241. Libbus BL, Perreault SD, Johnson LA, Pinkel D: Incidence of chromosome aberrations in mammalian sperm stained with Hoechst 33342 and UV-laser irradiated during flow sorting. Mutat Res 182:265-74, 1987
1242. Frey T, Houck DW, Shenker BJ, Hoffman RA: Bivariate flow karyotyping with air-cooled lasers. Cytometry 16:169-74, 1994
1243. Chen GG, St John PA, Barker JL: Rat lactotrophs isolated by fluorescence-activated cell sorting are electrically excitable. Mol Cell Endocrinol 51:201-10, 1987
1244. Salzman GC, Singham SB, Johnston RG, Bohren CF: Light scattering and cytometry. In reference 1028, pp 81-107
1245. Ulicny J: Lorenz-Mie light scattering in cellular biology. Gen Physiol Biophys 11:133-51, 1992
1246. Steen H: Flow cytometric studies of microorganisms. In reference 1028, pp 605-22
1247. Terstappen LW, Mickaels RA, Dost R, Loken MR: Increased light scattering resolution facilitates multidimensional flow cytometric analysis. Cytometry 11:506-12, 1990
1248. Hammond TG, Majewski RR, Morre DJ, Schell K, Morrissey LW: Forward scatter pulse width signals resolve multiple populations of endosomes. Cytometry 14:411-20, 1993
1249. Ong SH, Nickolls PM: Optical design in a flow system for imaging cells. Australas Phys Eng Sci Med 14:74-80, 1991
1250. Zucker RM, Perreault SD, Elstein KH: Utility of light scatter in the morphological analysis of sperm. Cytometry 13:39-47, 1992
1251. Loken MR, Brosnan JM, Bach BA, Ault KA: Establishing optimal lymphocyte gates for immunophenotyping by flow cytometry. Cytometry 11:453-9, 1990
1252. Nicholson JK, Jones BM, Hubbard M: CD4 T-lymphocyte determinations on whole blood specimens using a single-tube three-color assay. Cytometry 14:685-9, 1993
1253. Rabinowitz R, Granot E, Deckelbaum R, Schlesinger M: Antigenic differences between subsets of peripheral blood lymphocytes differing in their right angle light scatter in flow cytometric analysis. Int Arch Allergy Immunol 97:200-4, 1992

1254. Papa S, Maraldi NM, Matteucci A, Santi P, Vitale M, Galanzi A, Manzoli FA: Chromatin organization in isolated nuclei: flow cytometric characterization employing forward and perpendicular light scatter. Cell Biochem Funct 6:31-8, 1988

1255. Zucker RM, Elstein KH, Easterling RE, Massaro EJ: Flow cytometric discrimination of mitotic nuclei by right-angle light scatter. Cytometry 9:226-31, 1988

1256. Nusse M, Julch M, Geido E, Bruno S, Di Vinci A, Giaretti W, Ruoss K: Flow cytometric detection of mitotic cells using the bromodeoxyuridine/DNA technique in combination with 90 degrees and forward scatter measurements. Cytometry 10:312-9, 1989

1257. McGann LE, Walterson ML, Hogg LM: Light scattering and cell volumes in osmotically stressed and frozen-thawed cells. Cytometry 9:33-8, 1988

1258. Stewart CC, Stewart SJ, Habbersett RC: Resolving leukocytes using axial light loss. Cytometry 10:426-32, 1989

1259. Mayeno AN, Hamann KJ, Gleich GJ: Granule-associated flavin adenine dinucleotide (FAD) is responsible for eosinophil autofluorescence. J Leukoc Biol 51:172-5, 1992

1260. Havenith CE, Breedijk AJ, van Miert PP, et al.: Separation of alveolar macrophages and dendritic cells via autofluorescence: phenotypical and functional characterization. J Leukoc Biol 53:504-10, 1993

1261. Van De Winkel M, Pipeleers D: Autofluorescence-activated cell sorting of pancreatic islet cells: purification of insulin-containing B-cells according to glucose-induced changes in cellular redox state. Biochem Biophys Res Commun 114:835-42, 1983

1262. Kluftinger AM, Davis NL, Quenville NF, Lam S, Hung J, Palcic B: Detection of squamous cell cancer and pre-cancerous lesions by imaging of tissue autofluorescence in the hamster cheek pouch model. Surg Oncol 1:183-8, 1992

1263. Liu CH, Das BB, Sha Glassman WL, et al.: Raman, fluorescence, and time-resolved light scattering as optical diagnostic techniques to separate diseased and normal biomedical media. J Photochem Photobiol B 16:187-209, 1992

1264. Schantz SP, Alfano RR: Tissue autofluorescence as an intermediate endpoint in cancer chemoprevention trials. J Cell Biochem Suppl 1:199-204, 1993

1265. Silberberg MB, Savage HE, Tang GC, Sacks PG, Alfano RR, Schantz SP: Detecting retinoic acid-induced biochemical alterations in squamous cell carcinoma using intrinsic fluorescence spectroscopy. Laryngoscope 104:278-82, 1994

1266. Sassaroli M, da Costa R, Vaananen H, Eisinger J: Distribution of non-heme porphyrin content of individual erythrocytes by fluorescence image cytometry and its application to lead poisoning. Cytometry 13:339-45, 1992

1267. Markowitz SB, Nunez CM, Klitzman S, Munshi AA, Kim WS, Eisinger J, Landrigan PJ: Lead poisoning due to hai ge fen. The porphyrin content of individual erythrocytes. JAMA 271:932-4, 1994

1268. Brun A, Steen HB, Sandberg S: Erythropoietic protoporphyria: a quantitative determination of erythrocyte protoporphyrin in individual cells by flow cytometry. Scand J Clin Lab Invest 48:261-7, 1988

1269. Hunt JV, Carpenter KL, Bottoms MA, Carter NP, Marchant CE, Mitchinson MJ: Flow cytometric measurement of ceroid accumulation in macrophages. Atherosclerosis 98:229-39, 1993

1270. Hunt JV, Bottoms MA, Skamarauskas J, Carter NP, Mitchinson MJ: Measurement of ceroid accumulation in macrophages by flow cytometry. Cytometry 15:377-82, 1994

1271. Puppels GJ, Garritsen HS, Kummer JA, Greve J: Carotenoids located in human lymphocyte subpopulations and natural killer cells by Raman microspectroscopy. Cytometry 14:251-6, 1993

1272. Chisholm SW, Olson RJ, Zettler ER, Goericke R, Waterbury JB, Welschmeyer NA: A novel free-living prochlorophyte abundant in the ocean euphotic zone. Nature 326:655-61, 1988

1273. Xu C, Auger J, Govindjee: Chlorophyll a fluorescence measurements of isolated spinach thylakoids obtained by using single-laser-based flow cytometry. Cytometry 11:349-58, 1990

1274. Pinsky BG, Ladasky JJ, Lakowicz JR, Berndt K, Hoffman RA: Phase-resolved fluorescence lifetime measurements for flow cytometry. Cytometry 14:123-35, 1993

1275. Steinkamp JA, Crissman HA: Resolution of fluorescence signals from cells labeled with fluorochromes having different lifetimes by phase-sensitive flow cytometry. Cytometry 14:210-6, 1993

1276. Lindqvist C, Karp M, Åkerman K, Oker-Blom C: Flow cytometric analysis of bioluminescence emitted by recombinant baculovirus-infected insect cells. Cytometry 15:207-212, 1994

1277. Benedetti E, Papineschi F, Vergamini P, Consolini R, Spremolla G: Analytical infrared spectral differences between human normal and leukemic cells (CLL) - I. Leukemia Res 8:483-9, 1984

1278. Benedetti E, Palatresi MP, Vergamini P, Papineschi F, Spremolla G: New possibilities of research in chronic lymphatic leukemia by means of Fourier transform-infrared spectroscopy-II. Leukemia Res 9:1001-8, 1985

1279. Benedetti E, Palatresi MP, Vergamini P, Papineschi F, Andreucci MC, Spremolla G: Infrered characterization of nuclei isolated from normal and leukemic (B-CLL) lymphocytes. Appl Spectrosc 40:39-43, 1986

1280. Spremolla G, Benedetti E, Vergamini P, Andreucci MC, Macchia P: An investigation of acute lymphoblastic leukemia (ALL) in children by means of infrared spectroscopy. Part IV. Haematologica 73:21-4, 1988

1281. Benedetti E, Vergamini P, Spremolla G: FT-IR analysis of single human normal and leukemic lymphocytes. Mikrochim acta 1:139-41, 1988

1282. Benedetti E, Teodori L, Trinca ML, Vergamini P, Salvati F, Mauro F, Spremolla G: A new approach to the study of human solid tumor cells by means of FT-IR microspectroscopy. Appl Spectrosc 44:1276-80, 1990

1283. Wong PTT, Rigas B: Infrared spectra of microtome sections of human colon tissues. Appl Spectrosc 44:1715-8

1284. Rigas B, Morgello S, Goldman IS, Wong PTT: Human colorectal cancers display abnormal Fourier-transform infrared spectra. Proc Natl Acad Sci USA 87:8140-4, 1990

1285. Wong PTT, Papavassiliou ED, Rigas B: Phosphodiester stretching bands in the infrared spectra of human tissues and cultured cells. Appl Spectrosc 45:1563-7, 1991

1286. Wong PTT, Wong RK, Caputo TA, Godwin TA, Rigas B: Infrared spectroscopy of exfoliated human cervical cells: Evidence of extensive structural changes during carcinogenesis. Proc Natl Acad Sci USA 88:10988-92,1991

1287. Rigas B, Wong PT: Human colon adenocarcinoma cell lines display infrared spectroscopic features of malignant colon tissues. Cancer Res 52:84-8, 1992

1288. Wong PT, Goldstein SM, Grekin RC, Godwin TA, Pivik C, Rigas B: Distinct infrared spectroscopic patterns of human basal cell carcinoma of the skin. Cancer Res 53:762-5, 1993

1289. Clevenger CV, Bauer KD, Epstein AL: A method for simultaneous nuclear immunofluorescence and DNA content quantitation using monoclonal antibodies and flow cytometry. Cytometry 6:208-14, 1985

1290. Hayden GE, Walker KZ, Miller JF, Wotherspoon JS, Raison RL: Simultaneous cytometric analysis for the expression of cytoplasmic and surface antigens in activated T cells. Cytometry 9:44-51, 1988

1291. Slaper-Cortenbach IC, Admiraal LG, Kerr JM, van Leeuwen EF, von dem Borne AE, Tetteroo PA: Flow-cytometric detection of terminal deoxynucleotidyl transferase and other intracellular antigens in combination with membrane antigens in acute lymphatic leukemias. Blood 72:1639-44, 1988

1292. Drach J, Gattringer C, Glassl H, Schwarting R, Stein H, Huber H: Simultaneous flow cytometric analysis of surface markers and nuclear Ki-67 antigen in leukemia and lymphoma. Cytometry 10:743-9, 1989

1293. Pollice AA, McCoy JP Jr, Shackney SE, Smith CA, Agarwal J, Burholt DR, Janocko LE, Hornicek FJ, Singh SG, Hartsock RJ: Se-

quential paraformaldehyde and methanol fixation for simultaneous flow cytometric analysis of DNA, cell surface proteins, and intracellular proteins. Cytometry 13:432-44, 1992
1294. Schimenti KJ, Jacobberger JW: Fixation of mammalian cells for flow cytometric evaluation of DNA content and nuclear immunofluorescence. Cytometry 13:48-59, 1992
1295. Schmid I, Uittenbogaart CH, Giorgi JV: A gentle fixation and permeabilization method for combined cell surface and intracellular staining with improved precision in DNA quantification. Cytometry 12:279-85, 1991
1296. Hallden G, Andersson U, Hed J, Johansson SG: A new membrane permeabilization method for the detection of intracellular antigens by flow cytometry. J Immunol Methods 124:103-9, 1989
1297. Anderson P, Blue M-L, O'Brien C, Schlossman SF: Monoclonal antibodies reactive with the T cell receptor ζ chain: Production and characterization using a new method. J Immunol 143:1899-1904, 1989
1298. Jacob MC, Favre M, Bensa JC: Membrane cell permeabilization with saponin and multiparametric analysis by flow cytometry. Cytometry 12:550-8, 1991
1299. Howell LP, Deitch AD, Andreotti VA, Westrick LA, White RD: Fixation method useful for cytologic examination and DNA flow cytometry of exfoliated bladder cells. Urology 41:472-5, 1993
1300. Holtfreter HB, Cohen N: Fixation-associated quantitative variations of DNA fluorescence observed in flow cytometric analysis of hemopoietic cells from adult diploid frogs. Cytometry 11:676-85, 1990
1301. Haynes L, Moynihan JA, Cohen N: A monoclonal antibody against the human IL-2 receptor binds to paraformaldehyde-fixed but not viable frog (*Xenopus*) splenocytes [published erratum appears in Immunol Lett 273:257, 1991]. Immunol Lett 26:227-32, 1990
1302. Cahill MR, Macey MG, Newland AC: Fixation with formaldehyde induces expression of activation dependent platelet membrane glycoproteins, P selectin (CD62) and GP53 (CD63). Br J Haematol 84:527-9, 1993
1303. Cory JM, Rapp F, Ohlsson-Wilhelm BM: Effects of cellular fixatives on human immunodeficiency virus production. Cytometry 11:647-51, 1990
1304. Ericson JG, Trevino AV, Toedter GP, Mathes LE, Newbound GC, Lairmore MD: Effects of whole blood lysis and fixation on the infectivity of human T-lymphotropic virus type 1 (HTLV-I). Cytometry (Commun Clin Cytom) 18:49-54, 1994
1305. Terstappen LW, Meiners H, Loken MR: A rapid sample preparation technique for flow cytometric analysis of immunofluorescence allowing absolute enumeration of cell subpopulations. J Immunol Methods 123:103-12, 1989
1306. Beavis AJ, Pennline KJ: Simultaneous measurement of five cell surface antigens by five-color immunofluorescence. Cytometry 15:371-6, 1994
1307. Rabinovitch PS: Practical considerations for DNA content and cell cycle analysis. In reference 1038, pp 117-42
1308. Bagwell CB, Mayo SW, Whetstone SD, Hitchcox SA, Baker DR, Herbert DJ, Weaver DL, Jones MA, Lovett EJ 3d: DNA histogram debris theory and compensation. Cytometry 12:107-18, 1991
1309. Dean PN: Data Processing. In reference 1028, pp 415-44
1310. Ellwart JW, Dormer P: Vitality measurement using spectrum shift in Hoechst 33342 stained cells. Cytometry 11:239-43, 1990
1311. Stokke T, Holte H, Davies CD, Steen HB, Lie SO: Quenching of Hoechst 33258 fluorescence in erythroid precursors. Cytometry 11:686-90, 1990
1312. Vinogradov AE, Rosanov JM: Some properties of new DNA-specific bisbenzimidazole fluorochromes without a piperazine ring. Biotech Histochem 68:265-70, 1993
1313. Lewalski H, Otto FJ, Kranert T, Wassmuth R: Flow cytometric detection of unbalanced ram spermatozoa from heterozygous 1; 20 translocation carriers. Cytogenet Cell Genet 64:286-91, 1993
1314. Modest EJ, Sen Gupta S: 7-substituted actinomycin D (NSC-3053) analogs as fluorescent DNA-binding and experimental antitumor agents. Cancer Chemother Rep 58:35-48, 1974
1315. Gill JE, Jotz MM, Young SG, Modest EJ, Sen Gupta S: 7-amino-actinomycin D as a cytochemical probe. I. Spectral properties. J Histochem Cytochem 23:793-9, 1975
1316. Schmid I, Krall WJ, Uittenbogaart CH, Braun J, Giorgi JV: Dead cell discrimination with 7-amino-actinomycin D in combination with dual color immunofluorescence in single laser flow cytometry. Cytometry 13:204-8, 1992
1317. Schmid I, Uittenbogaart CH, Keld B, Giorgi JV: A rapid method for measuring apoptosis and dual-color immunofluorescence by single laser flow cytometry. J Immunol Methods 170:145-157, 1994
1318. von Jancso N: Beobachtung chemotherapeutischer vorgäge im fluorescenzmikroscop. Klin Wochenschr 11:689, 1932
1319. Wang CJ, Jolley ME: Fluorescent nucleic acid stains. U S Pat 4, 544, 546, October, 1985
1320. Terstappen LW, Shah VO, Conrad MP, Recktenwald D, Loken MR: Discriminating between damaged and intact cells in fixed flow cytometric samples. Cytometry 9:477-84, 1988
1321. Terstappen LW, Loken MR: Five-dimensional flow cytometry as a new approach for blood and bone marrow differentials. Cytometry 9:548-56, 1988
1322. Paton AM, Jones SM: Techniques involving optical brightening agents. In: Norris JR, Ribbons DW (eds): *Methods in Microbiology*, Vol 5a, New York, Academic Press, 1971, pp 135-44
1323. Lee LG, Chen C-H: Detection of reticulocytes, RNA or DNA. U S Pat 4, 883, 867, November, 1989
1324. Van Bockstaele DR, Peetermans ME: 1,3′-diethyl-4,2′-quinothiacyanine iodideas a "thiazole orange" analogue for nucleic acid staining. Cytometry 10:214-8, 1989
1325. Rye HS, Quesada MA, Peck K, Mathies RA, Glazer AN: High-sensitivity two-color detection of double-stranded DNA with a confocal fluorescence gel scanner using ethidium homodimer and thiazole orange. Nucleic Acids Res 19:327-33, 1991
1326. Mathies RA, Peck K, Stryer L: Optimization of high-sensitivity fluorescence detection. Anal Chem 62:1786-91, 1990
1327. Glazer AN, Peck K, Mathies RA: A stable double-stranded DNA-ethidium homodimer complex: application to picogram fluorescence detection of DNA in agarose gels. Proc Natl Acad Sci USA 87:3851-5, 1990
1328. Benson SC, Mathies RA, Glazer AN: Heterodimeric DNA-binding dyes designed for energy transfer: stability and applications of the DNA complexes. Nucleic Acids Res 21:5720-6, 1993
1329. Benson SC, Singh P, Glazer AN: Heterodimeric DNA-binding dyes designed for energy transfer: synthesis and spectroscopic properties. Nucleic Acids Res 21:5727-35, 1993
1330. van den Engh GJ, Trask BJ, Gray JW: The binding kinetics and interaction of DNA fluorochromes used in the analysis of nuclei and chromosomes by flow cytometry. Histochemistry 84:501-8, 1986
1331. Hirons GT, Fawcett JJ, Crissman HA: TOTO and YOYO: New very bright fluorochromes for DNA content analysis by flow cytometry. Cytometry 15:129-40, 1994
1332. Rasch EM: DNA "standards" and the range of accurate DNA estimates by Feulgen absorption microspectrophotometry. Prog Clin Biol Res 196:137-66, 1985
1333. Lee GM, Thornthwaite JT, Rasch EM: Picogram per cell determination of DNA by flow cytofluorometry. Anal Biochem 137:221-6, 1984
1334. Capriglione T, Olmo E, Odierna G, Improta B, Morescalchi A: Cytofluorometric DNA base determination in vertebrate species with different genome sizes. Basic Appl Histochem 31:119-26, 1987
1335. Vinogradov AE, Borkin LJ: Allometry of base pair specific-DNA contents in Tetropoda. Hereditas 118:155-63, 1993
1336. Vinogradov AE: Measurement by flow cytometry of genomic AT/GC ratio and genome size. Cytometry 16:34-40, 1994

1337. Rundquist I: Equilibrium binding of DAPI and 7-aminoactinomycin D to chromatin of cultured cells. Cytometry 14:610-7, 1993

1338. Bertuzzi A, D'Agnano I, Gandolfi A, Graziano A, Starace G, Ubezio P: Study of propidium iodide binding to DNA in intact cells by flow cytometry. Cell Biophys 17:257-67, 1990

1339. Beisker W, Eisert WG: Denaturation and condensation of intracellular nucleic acids monitored by fluorescence depolarization of intercalating dyes in individual cells. J Histochem Cytochem 37:1699-704, 1989

1340. Sanders CA, Yajko DM, Hyun W, Langlois RG, Nassos PS, Fulwyler MJ, Hadley WK: Determination of guanine-plus-cytosine content of bacterial DNA by dual-laser flow cytometry. J Gen Microbiol 136:359-65, 1990

1341. Sanders CA, Yajko DM, Nassos PS, Hyun WC, Fulwyler MJ, Hadley WK: Detection and analysis by dual-laser flow cytometry of bacteriophage T4 DNA inside Escherichia coli. Cytometry 12:167-71, 1991

1342. Bernheim A, Miglierina R: Different Hoechst 33342 and DAPI fluorescence of the human Y chromosome in bivariate flow karyotypes. Hum Genet 83:189-93, 1989

1343. Darzynkiewicz Z: Acid-induced denaturation of DNA *in situ* as a probe of chromatin structure. Methods Cell Biol 33:337-52, 1990 (in reference 1034)

1344. Darzynkiewicz Z: Acid-induced denaturation of DNA *in situ* as a probe of chromatin structure. Methods Cell Biol 41:527-41, 1994

1345. Stokke T, Holte H, Steen HB: In vitro and in vivo activation of B-lymphocytes: a flow cytometric study of chromatin structure employing 7-aminoactinomycin D. Cancer Res 48:6708-14, 1988

1346. Stokke T, Holte H, Erikstein B, Davies CL, Funderud S, Steen HB: Simultaneous assessment of chromatin structure, DNA content, and antigen expression by dual wavelength excitation flow cytometry. Cytometry 12:172-8, 1991

1347. Saito S, Crissman HA, Nishijima M, Kagabu T, Nishiya I, Cram LS: Flow cytometric and biochemical analysis of dose-dependent effects of sodium butyrate on human endometrial adenocarcinoma cells. Cytometry 12:757-64, 1991

1348. Darzynkiewicz Z: Differential staining of DNA and RNA in intact cells and isolated cell nuclei with acridine orange. Methods Cell Biol 33:285-98, 1990 (in reference 1034)

1349. Darzynkiewicz Z: Simultaneous analysis of cellular RNA and DNA content. Methods Cell Biol 41:402-20, 1994

1350. Traganos F, Crissman HA, Darzynkiewicz Z: Staining with pyronin Y detects changes in conformation of RNA during mitosis and hyperthermia of CHO cells. Exp Cell Res 179:535-44, 1988

1351. Wright WD, Higashikubo R, Roti Roti JL: Flow cytometric studies of the nuclear matrix. Cytometry 10:303-11, 1989

1352. Heiden T, Göhde W, Tribukait B: Two-wavelength mercury arc lamp excitation for flow cytometric DNA-protein analyses. Anticancer Res 10:1555-62, 1990

1353. Takahama M, Kagaya A: Hematoporphyrin/DAPI staining: simplified simultaneous one-step staining of DNA and cell protein and trial application in automated cytological screening by flow cytometry. J Histochem Cytochem 36:1061-7, 1988

1354. Tanaka N, Ohtsuka S, Matsuyama M, et al.: [UV-microspectrophotometric and flow cytometric analysis of the same samples using DAPI/HP staining]. Gan To Kagaku Ryoho 20:731-6, 1993

1355. Holme S, Heaton A, Konchuba A, Hartman P: Light scatter and total protein signal distribution of platelets by flow cytometry as parameters of size. J Lab Clin Med 112:223-31, 1988

1356. Glazer AN: Phycobiliproteins - a family of valuable, widely used fluorophores. J Appl Phycology 6:105-12, 1994

1357. Fuchs HJ, McDowell J, Shellito JE: Use of allophycocyanin allows quantitative description by flow cytometry of alveolar macrophage surface antigens present in low numbers of cells. Am Rev Respir Dis 138:1124-8, 1988

1358. Festin R, Bjorkland A, Totterman TH: Single laser flow cytometric detection of lymphocytes binding three antibodies labelled with fluorescein, phycoerythrin and a novel tandem fluorochrome conjugate. J Immunol Methods 126:69-78, 1990

1359. Waggoner AS, Ernst LA, Chen CH, Rechtenwald DJ: PE-CY5. A new fluorescent antibody label for three-color flow cytometry with a single laser. Ann N Y Acad Sci 677:185-93, 1993

1360. Lansdorp PM, Smith C, Safford M, Terstappen LW, Thomas TE: Single laser three color immunofluorescence staining procedures based on energy transfer between phycoerythrin and cyanine 5. Cytometry 12:723-30, 1991

1361. Ernst LA, Gupta RK, Mujumdar RB, Waggoner AS: Cyanine dye labeling reagents for sulfhydryl groups. Cytometry 10:3-10, 1989

1362. Mujumdar RB, Ernst LA, Mujumdar SR, Waggoner AS: Cyanine dye labeling reagents containing isothiocyanate groups. Cytometry 10:11-9, 1989

1363. Southwick PL, Ernst LA, Tauriello EW, Parker SR, Majumdar RB, Majumdar SR, Clever HA, Waggoner AS: Cyanine dye labeling reagents--carboxymethyl-indocyanine succinimidyl esters. Cytometry 11:418-30, 1990

1364. Mujumdar RB, Ernst LA, Mujumdar SR, Lewis CJ, Waggoner AS: Cyanine dye labeling reagents: sulfoindocyanine succinimidyl esters. Bioconjugate Chem 4:105-11, 1993

1365. Manian BS, Dubrow B, Hartz T: Clinical evaluation of volumetric capillary cytometry - a new method for absolute cell cell counts in homogeneous format. Presented at the ESCAP Meeting, Grenoble, France, May, 1994

1366. Wessendorf MW, Brelje TC: Which fluorophore is brightest? A comparison of the staining obtained using fluorescein, tetramethylrhodamine, lissamine rhodamine, Texas red, and cyanine 3.18. Histochemistry 98:81-5, 1992

1367. Aubry JP, Durand I, De Paoli P, Banchereau J: 7-amino-4-methylcoumarin-3-acetic acid-conjugated streptavidin permits simultaneous flow cytometry analysis of either three cell surface antigens or one cell surface antigen as a function of RNA and DNA content. J Immunol Methods 128:39-49, 1990

1368. Delia D, Martinez E, Fontanella E, Aiello A: Two- and three-color immunofluorescence using aminocoumarin, fluorescein, and phycoerythrin-labelled antibodies and single laser flow cytometry. Cytometry 12:537-44, 1991

1369. Meshulam T, Herscovitz H, Casavant D, Bernardo J, Roman R, Haugland RP, Strohmeier GS, Diamond RD, Simons ER: Flow cytometric kinetic measurements of neutrophil phospholipase A activation. J Biol Chem 267:21465-70, 1992

1370. Knaus HG, Moshammer T, Friedrich K, Kang HC, Haugland RP, Glossman H: In vivo labeling of L-type Ca2+ channels by fluorescent dihydropyridines: evidence for a functional, extracellular heparin-binding site. Proc Natl Acad Sci U S A 89:3586-90, 1992

1371. Knaus HG, Moshammer T, Kang HC, Haugland RP, Glossmann H: A unique fluorescent phenylalkylamine probe for L-type Ca2+ channels. Coupling of phenylalkylamine receptors to Ca2+ and dihydropyridine binding sites. J Biol Chem 267:2179-89, 1992

1372. Martin RJ, Kusel JR, Robertson SJ, Minta A, Haugland RP: Distribution of a fluorescent ivermectin probe, bodipy ivermectin, in tissues of the nematode parasite *Ascaris suum*. Parasitol Res 78:341-8, 1992

1373. Haugland RP, You W, Paragas VB, Wells KS, Du Bose DA: Simultaneous visualization of G- and F-actin in endothelial cells. J Histochem Cytochem 42:345-50, 1994

1374. Monsigny M, Midoux P, Le Bris MT, Roche AC, Valeur B: Benzoxazinone derivatives: new fluorescent probes for two-color flow cytometry analysis using one excitation wavelength. Biol Cell 67:193-200, 1989

1375. Schindele D, Renzoni GE, Fearon KL, Vandiver MW, Ekdahl RJ, Pepich BV: Novel fluorescent probes for helium/neon or laserdiode excitation. Cytometry Supp 4:4, 1990

1376. Chan WS, Marshall JF, Lam GY, Hart IR: Tissue uptake, distribution, and potency of the photoactivatable dye chloroaluminum sulfonated phthalocyanine in mice bearing transplantable tumors. Cancer Res 48:3040-4, 1988
1377. Roberts WG, Berns MW: In vitro photosensitization I. Cellular uptake and subcellular localization of mono-L-aspartyl chlorin e6, chloro-aluminum sulfonated phthalocyanine, and photofrin II. Lasers Surg Med 9:90-101, 1989
1378. Korbelik M: Distribution of disulfonated and tetrasulfonated aluminum phthalocyanine between malignant and host cell populations of a murine fibrosarcoma. J Photochem Photobiol B 20:173-81, 1993
1379. Whitaker JE, Haugland RP, Ryan D, Hewitt PC, Haugland RP, Prendergast FG: Fluorescent rhodol derivatives: versatile, photostable labels and tracers. Anal Biochem 207:267-79, 1992
1380. Zola H, Neoh SH, Mantzioris BX, Webster J, Loughnan MS: Detection by immunofluorescence of surface molecules present in low copy numbers. High sensitivity staining and calibration of flow cytometer. J Immunol Methods 135:247-55, 1990
1381. Zola H, Flego L, Sheldon A: Detection of cytokine receptors by high-sensitivity immunofluorescence/flow cytometry. Immunobiology 185:350-65, 1992
1382. Truneh A, Machy P, Horan PK: Antibody-bearing liposomes as multicolor immunofluorescence markers for flow cytometry and imaging. J Immunol Methods 100:59-71, 1987
1383. Gray A, Huchins ER, Morgan J, Jaswon M: Signal enhancement in fluorescence microscopy and flow cytometry using fluorescent liposome-antibody conjugates as second layer reagents. Ann Biol Clin 50:169-74, 1992
1384. Kim YR, Paseltiner L, Kling G, Yeh CK: Subtyping lymphocytes in peripheral blood by direct immunoalkaline phosphatase labeling and light scatter/absorption flow cytometric analysis. Am J Clin Pathol 97:331-7, 1992
1385. Condrau MA, Schwendener RA, Niederer P, Anliker M: Time-resolved flow cytometry for the measurement of lanthanide chelate fluorescence: I. Concept and theoretical evaluation. Cytometry 16:187-94, 1994
1386. Condrau MA, Schwendener RA, Zimmermann M, Muser MH, Graf U, Niederer P, Anliker M: Time-resolved flow cytometry for the measurement of lanthanide chelate fluorescence: II. Instrument design and experimental results. Cytometry 16:195-205, 1994
1387. Beverloo HB, van Schadewijk A, van Gelderen-Boele S, Tanke HJ: Inorganic phosphors as new luminescent labels for immunocytochemistry and time-resolved microscopy. Cytometry 11:784-92, 1990
1388. Beverloo HB, van Schadewijk A, Bonnet J, van der Geest R, Runia R, Verwoerd NP, Vrolijk J, Ploem JS: Preparation and microscopic visualization of multicolor luminescent immunophosphors. Cytometry 13:561-70, 1992
1389. Corsetti JP, Sotirchos SV, Cox C, Cowles JW, Leary JF, Blumburg N: Correction of cellular autofluorescence in flow cytometry by mathematical modeling of cellular fluorescence. Cytometry 9:539-47, 1988
1390. Müller MR, Lennartz K, Nowrousian MR, Dux R, Tsuruo T, Rajewsky MF, Seeber S: Improved flow-cytometric detection of low P-glycoprotein expression in leukaemic blasts by histogram subtraction analysis. Cytometry 15:64-72, 1994
1391. Hallden G, Skold CM, Eklund A, Forslid J, Hed J: Quenching of intracellular autofluorescence in alveolar macrophages permits analysis of fluorochrome labelled surface antigens by flow cytofluorometry. J Immunol Methods 142:207-14, 1991
1392. Corver WE, Cornelisse CJ, Fleuren GJ: Simultaneous measurement of two cellular antigens and DNA using fluorescein-isothiocyanate, R-phycoerythrin, and propidium iodide on a standard FACScan. Cytometry 15: 117-28, 1994
1393. Chapple MR, Johnson GD, Davidson RS: Fluorescence quenching of fluorescein by R-phycoerythrin. A pitfall in dual fluorescence analysis. J Immunol Methods 111:209-18, 1988
1394. Liu C-M, Muirhead KA, George SP, Landay AL: Flow cytometric monitoring of human immunodeficiency virus-infected patients. Simultaneous enumeration of five lymphocyte subsets. Am J Clin Path 92:721-8, 1989
1395. Carayon P, Bord A, Raymond M: Simultaneous identification of eight leucocyte subsets of human peripheral blood using three-colour immunofluo-rescence flow cytometric analysis. J Immunol Methods 138:257-64, 1991
1396. Wood GS: A Venn diagram model that allows triple-label immunophenotypic analysis of cells based upon double-label measurements. Am J Clin Pathol 92:73-7, 1989
1397. Mansour I, Jarraya MA, Gane P, Roquin H, Rouger P, Doinel C: Triple labeling with two-color immunofluorescence using one light source: a useful approach for the analysis of cells positive for one label and negative for the other two. Cytometry 11:636-41, 1990
1398. van Putten WL, Kortboyer J, Bolhuis RL, Gratama JW: Three-marker phenotypic analysis of lymphocytes based on two-color immunofluorescence using a multinomial model for flow cytometric counts and maximum likelihood estimation. Cytometry 14:179-87, 1993
1399. van Putten WL, de Vries W, Reinders P, Levering W, van der Linden R, Tanke H J, Bolhuis RL, Gratama JW: Quantification of fluorescence properties of lymphocytes in peripheral blood mononuclear cell suspensions using a latent class model. Cytometry 14:86-96, 1993
1400. Hunter SD, Peters LE, Wotherspoon JS, Crowe SM: Lymphocyte subset analysis by Boolean algebra: A phenotypic approach using a cocktail of 5 antibodies and 3 color immunofluorescence. Cytometry 15:258-66, 1994
1401. Mansour I, Jarraya MA, Gane P, Reznikoff MF: Multiple labeling using two-color immunofluorescence with only one light source, two fluorescence photomultiplier tubes, and two light scatter detectors. Cytometry 15:272-6, 1994
1402. Ried T, Baldini A, Rand TC, Ward DC: Simultaneous visualization of seven different DNA probes by in situ hybridization using combinatorial fluorescence and digital imaging microscopy. Proc Natl Acad Sci USA 89:1388-92, 1992
1403. Patterson BK, Till M, Otto P, Goolsby C, Furtado MR, McBride LJ, Wolinsky SM: Detection of HIV-1 DNA and messenger RNA in individual cells by PCR-driven in situ hybridization and flow cytometry. Science 260:976-9, 1993
1404. Bauman JG, van Dekken H: Flow cytometry of fluorescent in situ hybridization to detect specific RNA and DNA sequences. Acta Histochem Suppl 37:65-9, 1989
1405. Bauman JG, Bayer JA, van Dekken H: Fluorescent in-situ hybridization to detect cellular RNA by flow cytometry and confocal microscopy. J Microsc 157:73-81, 1990
1406. Amann RI, Binder BJ, Olson RJ, Chisholm SW, Devereux R, Stahl DA: Combination of 16S rRNA-targeted oligonucleotide probes with flow cytometry for analyzing mixed microbial populations. Appl Environ Microbiol 56:1919-25, 1990
1407. Zarda B, Amann R, Wallner G, Schleifer KH: Identification of single bacterial cells using digoxigenin-labelled, rRNA-targeted oligonucleotides. J Gen Microbiol 137:2823-30, 1991
1408. Wallner G, Amann R, Beisker W: Optimizing fluorescent in situ hybridization with rRNA-targeted oligonucleotide probes for flow cytometric identification of microorganisms. Cytometry 14:136-43, 1993
1409. Timm EA Jr, Stewart CC: Fluorescent in situ hybridization en suspension (FISHES) using digoxigenin-labeled probes and flow cytometry. Biotechniques 12:362-7, 1992
1410. Lalli E, Gibellini D, Santi S, Facchini A: In situ hybridization in suspension and flow cytometry as a tool for the study of gene expression. Anal Biochem 207:298-303, 1992
1411. Bardin PG, Pickett MA, Robinson SB, Sanderson G, Holgate ST, Johnston SL: Comparison of 3' and 5' biotin labelled oligonucleotides for in situ hybridisation. Histochemistry 100:387-92, 1993

1412. Bains MA, Agarwal R, Pringle JH, Hutchinson RM, Lauder I: Flow cytometric quantitation of sequence-specific mRNA in hemopoietic cell suspensions by primer-induced in situ (PRINS) fluorescent nucleotide labeling. Exp Cell Res 208:321-6, 1993

1413. Embleton MJ, Gorochov G, Jones PT, Winter G: In-cell PCR from mRNA: amplifying and linking the rearranged immunoglobulin heavy and light chain V-genes within single cells. Nucleic Acids Res 20:3831-7, 1992

1414. Long AA, Komminoth P, Lee E, Wolfe HJ: Comparison of indirect and direct in-situ polymerase chain reaction in cell preparations and tissue sections. Detection of viral DNA, gene rearrangements and chromosomal translocations. Histochemistry 99:151-62, 1993

1415. Mirsky ML, Da Y, Lewin HA: Detection of bovine leukemia virus proviral DNA in individual cells. PCR Methods Appl 2:333-40, 1993

1416. Van Dekken H, Schervish EW, Pizzolo JG, Fair WR, Melamed MR: Simultaneous detection of fluorescent in situ hybridization and in vivo incorporated BrdU in a human bladder tumour. J Pathol 164:17-22, 1991

1417. Bauman JG, de Vries P, Pronk B, Visser JW: Purification of murine hemopoietic stem cells and committed progenitors by fluorescence activated cell sorting using wheat germ agglutinin and monoclonal antibodies. Acta Histochem Suppl 36:241-53, 1988

1418. Fibach E, Rachmilewitz EA: Flow cytometric analysis of the ploidy of normoblasts in the peripheral blood of patients with beta-thalassemia. Am J Hematol 42:162-5, 1993

1419. Lee MC, Turcinov D, Damjanov I: Lectins as markers for eosinophilic leukocytes. Histochemistry 86:269-73, 1987

1420. De Stefano JA, Cushion MT, Trinkle LS, Walzer PD: Lectins as probes to *Pneumocystis carinii* surface glycocomplexes. J Protozool 36:65S-66S, 1989

1421. Jacobson RL, Schnur LF: Changing surface carbohydrate configurations during the growth of *Leishmania major.* J Parasitol 76:218-24, 1990

1422. Guasch RM, Guerri C, O'Connor JE: Flow cytometric analysis of concanavalin A binding to isolated Golgi fractions from rat liver. Exp Cell Res 207:136-41, 1993

1423. Vasquez JM, Magargee SF, Kunze E, Hammerstedt RH: Lectins and heparin-binding features of human spermatozoa as analyzed by flow cytometry. Am J Obstet Gynecol 163:2006-12, 1990

1424. Miyazaki R, Fukuda M, Takeuchi H, Itoh S, Takada M: Flow cytometry to evaluate acrosome-reacted sperm. Arch Androl 25:243-51, 1990

1425. Grillon C, Monsigny M, Kieda C: Cell surface lectins of human granulocytes: their expression is modulated by mononuclear cells and granulocyte/macrophage colony-stimulating factor. Glycobiology 1:33-8, 1990

1426. Cerdan D, Grillon C, Monsigny M, Redziniak G, Kieda C: Human keratinocyte membrane lectins: charac-terization and modulation of their expression by cytokines. Biol Cell 73:35-42, 1991

1427. Rak JW, Basolo F, Elliott JW, Russo J, Miller FR: Cell surface glycosylation changes accompanying immortalization and transformation of normal human mammary epithelial cells. Cancer Lett 57:27-36, 1991

1428. Notter MF, Leary JF: Surface glycoproteins of differentiating neuroblastoma cells analyzed by lectin binding and flow cytometry. Cytometry 8:518-25, 1987

1429. Ward GK, Stewart SS, Dotsikas G, Price GB, Mackillop WJ: Cellular heterogeneity in human transitional cell carcinoma: an analysis of optical properties and lectin binding. Histochem J 24:685-94, 1992

1430. Gabius HJ, Engelhardt R, Hellmann T, et al.: Characterization of membrane lectins in human colon carcinoma cells by flow cytofluorometry, drug targeting and affinity chromatography. Anticancer Res 7:109-12, 1987

1431. Yeatman TJ, Bland KI, Copeland E3, Kimura AK: Tumor cell-surface galactose correlates with the degree of colorectal liver metastasis. J Surg Res 46:567-71, 1989

1432. Alam SM, Whitford P, Cushley W, George WD, Campbell AM: Flow cytometric analysis of cell surface carbohydrates in metastatic human breast cancer. Br J Cancer 62:238-42, 1990

1433. Kaku H, Mori Y, Goldstein IJ, Shibuya N: Monomeric, monovalent derivative of Maackia amurensis leukoagglutinin. Preparation and application to the study of cell surface glycoconjugates by flow cytometry. J Biol Chem 268:13237-41, 1993

1434. Whitehurst CE, Day NK, Gengozian N: Sugar competition assays reveal high affinity receptors for Erythrina cristigalli lectin on feline monocytes. J Immunol Methods 131:15-24, 1990

1435. Thurnher M, Clausen H, Sharon N, Berger EG: Use of O-glycosylation-defective human lymphoid cell lines and flow cytometry to delineate the specificity of Moluccella laevis lectin and monoclonal antibody 5F4 for the Tn antigen (GalNAc alpha 1-O-Ser/Thr). Immunol Lett 36:239-43, 1993

1436. McCoy JP Jr: The application of lectins to the characterization and isolation of mammalian cell populations. Cancer Metastasis Rev 6:595-613, 1987

1437. Duijndam WA, van Duijn P: Flow cytometric determination of carbohydrates in human erythrocytes. Histochemistry 88:263-5, 1988

1438. Dive C, Yoshida TM, Simpson DJ, Marrone BL: Flow cytometric analysis of steroidogenic organelles in differentiating granulosa cells. Biol Reprod 47:520-7, 1992

1439. Smyth MJ, Wharton W: Differentiation of A31T6 proadipocytes to adipocytes: a flow cytometric analysis. Exp Cell Res 199:29-38, 1992

1440. Brown WJ, Sullivan TR, Greenspan P: Nile red staining of lysosomal phospholipid inclusions. Histochemistry 97:349-54, 1992

1441. Hassall DG: Three probe flow cytometry of a human foam-cell forming macrophage. Cytometry 13:381-8, 1992

1442. Suzuki K, Sakata N, Hara M, Kitani A, Harigai, M, Hirose W, Kawaguchi Y, Kawagoe M, Nakamura H: Flow cytometric analysis of lipid droplet formation in cells of the human monocytic cell line, U937. Biochem Cell Biol 69:571-6, 1991

1443. Marchetti D, Van NT, Gametchu B, Thompson EB, Kobayashi Y, Watanabe F, Barlogie B: Flow cytometric analysis of glucocorticoid receptor using monoclonal antibody and fluoresceinated ligand probes. Cancer Res 49:863-9, 1989 [published erratum appears in Cancer Res 49:3142, 1989]

1444. Uckun FM, Fauci AS, Chandan-Langlie M, Myers DE, Ambrus JL: Detection and characterization of human high molecular weight B cell growth factor receptors on leukemic B cells in chronic lymphocytic leukemia. J Clin Invest 84:1595-608, 1989

1445. Krause D, Shearman C, Lang W, Kanzy EJ, Kurrle R: Determination of affinities of murine and chimeric anti alpha/beta-T-cell receptor antibodies by flow cytometry. Behring Inst Mitt 87:56-67, 1990

1446. Tordai A, Fenton J 2d, Andersen T, Gelfand EW: Functional thrombin receptors on human T lymphoblastoid cells. J Immunol 150:4876-86, 1993

1447. Schmitz G, Wulf G, Bruning T, Assmann G: Flow-cytometric determination of high-density-lipoprotein binding sites on human leukocytes. Clin Chem 33:2195-203, 1987

1448. Traill KN, Jurgens G, Bock G, Wick G: High density lipoprotein uptake by freshly isolated human peripheral blood T lymphocytes. Immunobiology 175:447-54, 1987

1449. Traill KN, Jurgens G, Bock G, Huber L, Schonitzer D, Widhalm K, Winter U, Wick G: Analysis of fluorescent low density lipoprotein uptake by lymphocytes. Paradoxical increase in the elderly. Mech Ageing Dev 40:261-88, 1987

1450. Huber LA, Bock G, Jurgens G, Traill KN, Schonitzer D, Wick G: Increased expression of high-affinity low-density lipoprotein receptors on human T-blasts. Int Arch Allergy Appl Immunol 93:205-11, 1990

1451. Corsetti JP, Weidner CH, Cianci J, Sparks CE: The labeling of lipoproteins for studies of cellular binding with a fluorescent lipophilic dye. Anal Biochem 195:122-8, 1991

1452. Corsetti JP, Sparks JD, Sikora B, Sparks CE: Cellular heterogeneity in binding and uptake of low-density lipoprotein in primary rat hepatocytes. Hepatology 17:645-50, 1993
1453. Laborda J, Naval J, Allouche M, Calvo M, Georgoulias V, Mishal Z, Uriel J: Specific uptake of alpha-fetoprotein by malignant human lymphoid cells. Int J Cancer 40:314-8, 1987
1454. Torres JM, Esteban C, Aguilar J, Mishal Z, Uriel J: Quantification of alpha-fetoprotein and transferrin endocytosis by lymphoid cells using flow cytometry. J Immunol Methods 134:163-70, 1990
1455. Midoux P, Roche AC, Monsigny M: Quantitation of the binding, uptake, and degradation of fluoresceinylated neoglycoproteins by flow cytometry. Cytometry 8:327-34, 1987
1456. Heumann D, Gallay P, Barras C, Zaech P, Ulevitch RJ, Tobias PS, Glauser MP, Baumgartner JD: Control of lipopolysaccharide (LPS) binding and LPS-induced tumor necrosis factor secretion in human peripheral blood monocytes. J Immunol 148:3505-12, 1992
1457. Bochsler PN, Maddux JM, Neilsen NR, Slauson DO: Differential binding of bacterial lipopolysaccharide to bovine peripheral-blood leukocytes. Inflammation 17:47-56, 1993
1458. Corrales I, Weersink AJ, Verhoef J, van Kessel KP: Serum-independent binding of lipopolysaccharide to human monocytes is trypsin sensitive and does not involve CD14. Immunology 80:84-9, 1993
1459. Weersink AJ, Van Kessel KP, Torensma R, Van Strijp JA, Verhoef J: Binding of rough lipopolysaccharides (LPS) to human leukocytes. Inhibition by anti-LPS monoclonal antibody. J Immunol 145:318-24, 1990
1460. Gilbert GE, Sims PJ, Wiedmer T, Furie B, Furie BC, Shattil SJ: Platelet-derived microparticles express high affinity receptors for factor VIII. J Biol Chem 266:17261-8, 1991
1461. Werfel T, Oppermann M, Schulze M, Krieger G, Weber M, Gotze O: Binding of fluorescein-labeled anaphylatoxin C5a to human peripheral blood, spleen, and bone marrow leukocytes. Blood 79:152-60, 1992
1462. Richerson HB, Adams PA, Iwai Y, Barfknecht CF: Uptake of muramyl dipeptide fluorescent congeners by normal rabbit bronchoalveolar lavage cells: a study using flow cytometry. Am J Respir Cell Mol Biol 2:171-81, 1990
1463. Sumaroka MV, Litvinov IS, Khaidukov SV, Golovina TN, Kamraz MV, Komal'eva RL, Andronova TM, Makarov EA, Nesmeyanov VA, Ivanov VT: Muramyl peptide-binding sites are located inside target cells. Febs Lett 295:48-50, 1991
1464. Sullam PM, Payan DG, Dazin PF, Valone FH: Binding of viridans group streptococci to human platelets: a quantitative analysis. Infect Immun 58:3802-6, 1990
1465. Yeaman MR, Sullam PM, Dazin PF, Norman DC, Bayer AS: Characterization of Staphylococcus aureus-platelet binding by quantitative flow cytometric analysis. J Infect Dis 166:65-73, 1992
1466. Clyne M, Drumm B: Adherence of Helicobacter pylori to primary human gastrointestinal cells. Infect Immun 61:4051-7, 1993
1467. Szabo G Jr, Damjanovich S: Fluorescent staphylococci as microbeads. Cytometry 10:801-2, 1989
1468. Gabrilovich DI, Kozich AT, Moshnikov SA, Pokrovsky VV: The direct binding of an HIV fragment with granulocytes from healthy subjects and infected patients. Scand J Immunol 35:369-72, 1992
1469. Wognum AW, Lansdorp PM, Humphries RK, Krystal G: Detection and isolation of the erythropoietin receptor using biotinylated erythropoietin. Blood 76:697-705, 1990
1470. Wognum AW, van Gils FC, Wagemaker G: Flow cytometric detection of receptors for interleukin-6 on bone marrow and peripheral blood cells of humans and rhesus monkeys. Blood 81:2036-43, 1993
1471. Wognum AW, Krystal G, Eaves CJ, Eaves AC, Lansdorp PM: Increased erythropoietin-receptor expression on CD34-positive bone marrow cells from patients with chronic myeloid leukemia. Blood 79:642-9, 1992
1472. Harel-Bellan A, Mishal Z, Willette-Brown J, Farrar WL: Detection of low and high affinity binding sites with fluoresceinated human recombinant interleukin-2. J Immunol Methods 119:127-33, 1989
1473. Taki S, Shimamura T, Abe M, Shirai T, Takahara Y: Biotinylation of human interleukin-2 for flow cytometry analysis of interleukin-2 receptors. J Immunol Methods 122:33-41, 1989
1474. Zuber CE, Galizzi JP, Harada N, Durand I, Banchereau J: Interleukin-4 receptors on human blood mononuclear cells. Cell Immunol 129:329-40, 1990
1475. Law CL, Armitage RJ, Villablanca JG, Le Bien TW: Expression of interleukin-4 receptors on early human B-lineage cells. Blood 78:703-10, 1991
1476. Heithier H, Ward LD, Cantrill RC, Klein HW, Im MJ, Pollak G, Freeman B, Schiltz E, Peters R, Helmreich EJ: Fluorescent glucagon derivatives. I. Synthesis and characterisation of fluorescent glucagon derivatives. Biochim Biophys Acta 971:298-306, 1988
1477. Newman W, Beall LD, Randhawa ZI: Biotinylation of peptide hormones: structural analysis and application to flow cytometry. Methods Enzymol 184:275-85, 1990
1478. Newman W, Beall LD, Levine MA, Cone JL, Randhawa ZI, Bertolini DR: Biotinylated parathyroid hormone as a probe for the parathyroid hormone receptor. Structure-function analysis and detection of specific binding to cultured bone cells by flow cytometry. J Biol Chem 264:16359-65, 1989
1479. Jans DA, Bergmann L, Peters R, Fahrenholz F: Biotinyl analogues of vasopressin as biologically active probes for vasopressin receptor expression in cultured cells. J Biol Chem 265:14599-605, 1990
1480. Durham LA 3d, Krummel TM, Cawthorn JW, Thomas BL, Diegelmann RF: Analysis of transforming growth factor beta receptor binding in embryonic, fetal, and adult rabbit fibroblasts. J Pediatr Surg 24:784-8, 1989
1481. Quaissi MA, Kusnierz JP, Gras-Masse H, Drobecq H, Velge P, Cornette J, Capron A, Tartar A: Fluorescence-activated cell-sorting analysis of fibronectin peptides binding to *Trypanosoma cruzi* trypomastigotes. J Protozool 35:111-4, 1988
1482. Balazs M, Szöllösi J, Lee WC, Haugland RP, Guzikowski AP, Fulwyler MJ, Damjanovich S, Feuerstein BG, Pershadsingh HA: Fluorescent tetra-decanoylphorbol acetate: a novel probe of phorbol ester binding domains. J Cell Biochem 46:266-76, 1991
1483. Chianelli M, Signore A, Hicks R, Testi R, Negri M, Beverley PC: A simple method for the evaluation of receptor binding capacity of modified cytokines. J Immunol Methods 166:177-82, 1993
1484. Kingsmore SF, Crockard AD, Fay AC, McNeill TA, Roberts SD, Thompson JM: Detection of circulating immune complexes by Raji cell assay: comparison of flow cytometric and radiometric methods. Diagn Clin Immunol 5:289-96, 1988
1485. Harabuchi Y, Koizumi S, Osato T, Yamanaka N, Kataura A: Flow cytometric analysis of Epstein-Barr virus receptor among the different B-cell subpopulations using simultaneous two-color immunofluorescence. Virology 165:278-81, 1988
1486. Stocco R, Sauvageau G, Menezes J: Differences in Epstein-Barr virus (EBV) receptors expression on various human lymphoid targets and their significance to EBV-cell interaction. Virus Res 11:209-25, 1988
1487. Kaplan DS, Picciolo GL: Characterization of instrumentation and calibrators for quantitative microfluorometry for immunofluorescence tests. J Clin Microbiol 27:442-7, 1989
1488. Poncelet P, Carayon P: Cytofluorometric quantification of cell-surface antigens by indirect immunofluorescence using monoclonal antibodies. J Immunol Methods 85:65-74, 1985
1489. Dux R, Kindler-Rohrborn A, Lennartz K, Rajewsky MF: Calibration of fluorescence intensities to quantify antibody binding surface determinants of cell subpopulations by flow cytometry. Cytometry 12:422-8, 1991
1490. Poncelet P, Mutin M, Burnet O, George F, Ambrosi P, Sampol J: Quantification of cell membrane antigenic sites in immuno-

cytometry with indirect IF: the QIFI assay. Cytometry Supp 5:82, 1991

1491. Poncelet P, Bikque A, Lavabre T, Poinas G, Parant M, Duperray O, Sampol J: Quantitative expression of human lymphocytes membrane antigens: definition of normal densities measured in immunocytometry with the QIFI assay. Cytometry Supp 5:82-3, 1991

1492. Poncelet P, Poinas G, Corbeau P, Devaux C, Tubiana N, Muloko N, Tamalet C, Chermann JC, Kourilsky F, Sampol J: Surface CD4 density remains constant on lymphocytes of HIV-infected patients in the progression of disease. Res Immunol 142:291-8, 1991

1493. Poncelet P, George F, Lavabre-Bertrand T: Immunological detection of membrane-bound antigens and receptors. In: Masseyeff R, Albert W, Staines NA (eds): *Methods of Immunological Analysis*, Vol 6, Weinheim, VCH Publishers, 1995

1494. Takase K, Iwaki K, Gunji T, Yata J: Fluorescence intensity analysis through simplex optimization in flow cytometry. J Immunol Methods 118:129-38, 1989

1495. Shabtai M, Malinowski K, Waltzer WC, Pullis C, Raisbeck AP, Rapaport FT: Quantitative analysis of surface marker densities after exposure of T-cells to concanavalin A (Con A): a sensitive early index of cellular activation. Cell Immunol 133:519-25, 1991

1496. Bardsley WG, Wilson AR, Kyprianou EK, Melikhova EM: A statistical model and computer program to estimate association constants for the binding of fluorescent-labelled monoclonal antibodies to cell surface antigens and to interpret shifts in flow cytometry data resulting from alterations in gene expression. J Immunol Methods 153:235-47, 1992

1497. Sladek TL, Jacobberger JW: Flow cytometric titration of retroviral expression vectors: comparison of methods for analysis of immunofluorescence histograms derived from cells expressing low antigen levels. Cytometry 14:23-31, 1993

1498. Terstappen LW, Hollander Z, Meiners H, Loken MR: Quantitative comparison of myeloid antigens on five lineages of mature peripheral blood cells. J Leukoc Biol 48:138-48, 1990

1499. Christopoulos CG, Kelsey HC, Machin SJ: A flow-cytometric approach to quantitative estimation of platelet surface immunoglobulin G. Vox Sang 64:106-15, 1993

1500. Sarin A, Saxena RK: Quantitative estimation of major histocompatibility complex antigens on live tumour cells. Indian J Exp Biol 28:1017-20, 1990

1501. Christensen J, Leslie RG: Quantitative measurement of Fc receptor activity on human peripheral blood monocytes and the monocyte-like cell line, U937, by laser flow cytometry. J Immunol Methods 132:211-9, 1990

1502. Ritzi EM: Quantitative flow cytometry of mouse mammary tumor virus envelope glycoprotein (gp52): alternative measures of hormone-mediated change in a viral cell surface antigen. J Virol Methods 40:11-30, 1992

1503. Pallis M, Robins A, Powell R: Quantitative analysis of lymphocyte CD11a using standardized flow cytometry. Scand J Immunol 38:559-64, 1993

1504. Witzig TE, Li CY, Tefferi A, Katzmann JA: Measurement of the intensity of cell surface antigen expression in B-cell chronic lymphocytic leukemia. Am J Clin Pathol 101:312-7, 1994

1505. Vogt RF Jr, Cross GD, Henderson LO, Phillips DL: Model system evaluating fluorescein-labeled microbeads as internal standards to calibrate fluorescence intensity on flow cytometers. Cytometry 10:294-302, 1989

1506. Vogt RF Jr, Cross GD, Phillips DL, Henderson LO, Hannon WH: Interlaboratory study of cellular fluorescence intensity measurements with fluorescein-labeled microbead standards. Cytometry 12:525-36, 1991

1507. Mátyus L: Fluorescence resonance energy transfer measurements on cell surfaces. A spectroscopic tool for determining protein interactions. J Photochem Photobiol B 12:323-37, 1992

1508. Trón L, Szöllösi J, Damjanovich S: Proximity measurements of cell surface proteins by fluorescence energy transfer. Immunol Lett 16:1-9, 1987

1509. Szöllösi J, Damjanovich S, Goldman CK, Fulwyler MJ, Aszalos AA, Goldstein G, Rao P, Talle MA, Waldmann TA: Flow cytometric resonance energy transfer measurements support the association of a 95-kDa peptide termed T27 with the 55-kDa Tac peptide. Proc Natl Acad Sci U S A 84:7246-50, 1987

1510. Szöllösi J, Damjanovich S, Balázs M, Nagy P, Tron L, Fulwyler MJ, Brodsky FM: Physical association between MHC class I and class II molecules detected on the cell surface by flow cytometric energy transfer. J Immunol 143:208-13, 1989

1511. Harel-Bellan A, Krief P, Rimsky L, Farrar WL, Mishal Z: Flow cytometry resonance energy transfer suggests an association between low-affinity interleukin 2 binding sites and HLA class I molecules. Biochem J 268:35-40, 1990

1512. Tertov VV, Sayadyan HS, Kalantarov GF, Molotkovsky JG, Bergelson LD, Orekhov AN: Use of lipophilic fluorescent probes for the isolation of hybrid cells in flow cytometry. J Immunol Methods 118:139-43, 1989

1513. Takahashi S, Esserman L, Levy R: An epitope on the transferrin receptor preferentially exposed during tumor progression in human lymphoma is close to the ligand binding site. Blood 77:826-32, 1991

1514. Klebba PE, Benson SA, Bala S, Abdullah T, Reid J, Singh SP, Nikaido H: Determinants of OmpF porin antigenicity and structure. J Biol Chem 265:6800-10, 1990

1515. Rutz JM, Abdullah T, Singh SP, Kalve VI, Klebba PE: Evolution of the ferric enterobactin receptor in gram-negative bacteria. J Bacteriol 173:5964-74, 1991

1516. Riedy MC, Muirhead KA, Jensen CP, Stewart CC: Use of a photolabeling technique to identify nonviable cells in fixed homologous or heterologous cell populations. Cytometry 12:133-9, 1991

1517. Darzynkiewicz Z, Bruno S, Del Bino G, Gorczyca W, Hotz MA, Lassota P, Traganos F: Features of apoptotic cells measured by flow cytometry. Cytometry 13:795-808, 1992

1518. Gold R, Schmied M, Rothe G, Zischler H, Breitschopf H, Wekerle H, Lassmann H: Detection of DNA fragmentation in apoptosis: application of in situ nick translation to cell culture systems and tissue sections. J Histochem Cytochem 41:1023-30, 1993

1519. Gorczyca W, Gong J, Darzynkiewicz Z: Detection of DNA strand breaks in individual apoptotic cells by the in situ terminal deoxynucleotidyl transferase and nick translation assays. Cancer Res 53:1945-51, 1993

1520. Swat W, Ignatowicz L, Kisielow P: Detection of apoptosis of immature CD4+8+ thymocytes by flow cytometry. J Immunol Methods 137:79-87, 1991

1521. Nicoletti I, Migliorati G, Pagliacci MC, Grignani F, Riccardi C: A rapid and simple method for measuring thymocyte apoptosis by propidium iodide staining and flow cytometry. J Immunol Methods 139:271-9, 1991

1522. Telford WG, King LE, Fraker PJ: Comparative evaluation of several DNA binding dyes in the detection of apoptosis-associated chromatin degradation by flow cytometry. Cytometry 13:137-43, 1992

1523. Pellicciari C, Manfredi AA, Bottone MG, Schaack V, Barni S: A single-step staining procedure for the detection and sorting of unfixed apoptotic thymocytes. Eur J Histochem 37:381-90, 1993

1524. Afanasyev VN, Korol BA, Matylevich NP, Pechatnikov VA, Umansky SR: The use of flow cytometry for the investigation of cell death. Cytometry 14:603-9, 1993

1525. Chrest FJ, Buchholz MA, Kim YH, Kwon TK, Nordin AA: Identification and quantitation of apoptotic cells following anti-CD3 activation of murine G0 T cells. Cytometry 14:883-90, 1993

1526. Zamai L, Falcieri E, Zauli G, Cataldi A, Vitale M: Optimal detection of apoptosis by flow cytometry depends on cell morphology. Cytometry 14:891-7, 1993

1527. Lyons AB, Samuel K, Sanderson A, Maddy AH: Simultaneous analysis of immunophenotype and apoptosis of murine thymocytes by single laser flow cytometry. Cytometry 13:809-21, 1992
1528. Vitale M, Zamai L, Mazzotti G, Cataldi A, Falcieri E: Differential kinetics of propidium iodide uptake in apoptotic and necrotic thymocytes. Histochemistry 100:223-9, 1993
1529. Ormerod MG, Sun XM, Snowden RT, Davies R, Fearnhead H, Cohen GM: Increased membrane permeability of apoptotic thymocytes: a flow cytometric study. Cytometry 14:595-602, 1993
1530. Sun XM, Snowden RT, Skilleter DN, Dinsdale D, Ormerod MG, Cohen GM: A flow-cytometric method for the separation and quantitation of normal and apoptotic thymocytes. Anal Biochem 204:351-6, 1992
1531. Schmid I, Uittenbogaart CH, Giorgi JV: Sensitive method for measuring apoptosis and cell surface phenotype in human thymocytes by flow cytometry. Cytometry 15:12-20, 1994
1532. Kelley KA: Sample station modification providing on-line reagent addition and reduced sample transit time for flow cytometers. Cytometry 10:796-800, 1989
1533. Kelley KA: Very early detection of changes associated with cellular activation using a modified flow cytometer. Cytometry 12:464-8, 1991
1534. Dunne JF: Time window analysis and sorting. Cytometry 12:597-601, 1991
1535. Durack G, Lawler G, Kelley S, Ragheb K, Roth RA, Ganey P, Robinson JP: Time interval gating for analysis of cell function using flow cytometry. Cytometry 12:701-6, 1991
1536. Lindberg W, Scampavia LD, Ruzicka J, Christian GD: Fast kinetic measurements and on-line dilution by flow injection cytometry. Cytometry 16:324-30, 1994
1537. Amblard F, Cantin C, Durand J, Fischer A, Sekaly R, Auffray C: New chamber for flow cytometric analysis over an extended range of stream velocity and application to cell adhesion measurements. Cytometry 13:15-22, 1992
1538. Schweppe F, Hausmann M, Hexel K, Barths J, Cremer C: An adapter for defined sample volumes makes it possible to count absolute particle numbers in flow cytometry. Anal Cell Pathol 4:325-34, 1992
1539. Stewart CC, Steinkamp JA: Quantitation of cell concentration using the flow cytometer. Cytometry 2:238-43, 1982
1540. Abernethy NJ, Chin W, Lyons H, Hay JB: A dual laser analysis of the migration of XRITC-labeled, FITC-labeled, and double-labeled lymphocytes in sheep. Cytometry 6:407-13, 1985
1541. Capo C, Mege JL, Benoliel AM, Mishal Z, Bongrand P: Quantification of the nonspecific intercellular transfer of fluorescent molecules between labeled and unlabeled rat thymocytes. Cytometry 8:468-73, 1987
1542. Weston SA, Parish CR: New fluorescent dyes for lymphocyte migration studies. Analysis by flow cytometry and fluorescence microscopy. J Immunol Methods 133:87-97, 1990
1543. Honig MG, Hume RI: Fluorescent carbocyanine dyes allow living neurons of identified origin to be studied in long-term cultures.
1544. Ragnarson B, Bengtsson L, Haegerstrand A: Labeling with fluorescent carbocyanine dyes of cultured endothelial and smooth muscle cells by growth in dye-containing medium. Histochemistry 97:329-33, 1992
1545. St John PA: Toxicity of "DiI" for embryonic rat motoneurons and sensory neurons in vitro. Life Sci 49:2013-21, 1991
1546. Ledley FD, Soriano HE, O'Malley BW Jr, Lewis D, Darlington GJ, Finegold M: DiI as a marker for cellular transplantation into solid organs. Biotechniques 13:584-7, 1992
1547. Horan PK, Melnicoff MJ, Jensen BD, Slezak S: Fluorescent cell labeling for *in vivo* and *in vitro* cell tracking. Methods Cell Biol 33:469-90, 1990 (in reference 1034)
1548. Melnicoff MJ, Horan PK, Breslin EW, Morahan PS: Maintenance of peritoneal macrophages in the steady state. J Leukoc Biol 44:367-75, 1988
1549. Melnicoff MJ, Morahan PS, Jensen BD, Breslin EW, Horan PK: In vivo labeling of resident peritoneal macrophages. J Leukoc Biol 43:387-97, 1988
1550. Slezak SE, Horan PK: Cell-mediated cytotoxicity. A highly sensitive and informative flow cytometric assay. J Immunol Methods 117:205-14, 1989
1551. Jensen BD, Schmitt TC, Slezak SE: Labeling of mammalian cells for in vivo cell tracking by a fluorescence method. Prog Clin Biol Res 355:199-207, 1990
1552. Wallace PK, Palmer LD, Perry-Lalley D, Bolton ES, Alexander RB, Horan PK, Yang JC, Muirhead KA: Mechanisms of adoptive immunotherapy: Improved methods for *in vivo* tracking of tumor-infiltrating lymphocytes and lymphokine-activated killer cells. Cancer Res 53:2358-67, 1993
1553. Raybourne RB, Bunning VK: Bacterium-host cell interactions at the cellular level: fluorescent labeling of bacteria and analysis of short-term bacterium-phagocyte interaction by flow cytometry. Infect Immun 62:665-72, 1994
1554. Ashley DM, Bol SJ, Waugh C, Kannourakis G: A novel approach to the measurement of different in vitro leukaemic cell growth parameters: the use of PKH GL fluorescent probes. Leuk Res 17:873-82, 1993
1555. Yamamura Y, Eyler E, Rodriguez N, Yano N, Bagwell B, Schwartz A: A new proliferation analysis model for lymphocyte mitogenic response. Cytometry Supp 7, "in press", 1994 – NOTE: the abstract referred to in the original reference 1555 was never published, but the material appeared in Yamamura Y, Rodriguez N, Schwartz A, Eylar E, Bagwell B, Yano N: A new flow cytometric method for quantitative assessment of lymphocyte mitogenic potentials. Cell Mol Biol (Noisy-le-grand) 41 Suppl 1:S121-32, 1995
1556. Sieber F, Spivak JL, Sutcliffe AM: Selective killing of leukemic cells by merocyanine 540-mediated photosensitization. Proc Natl Acad Sci USA 81:7584-7, 1984
1557. Smith OM, Traul DL, McOlash L, Sieber F: Evaluation of merocyanine 540-sensitized photoirradiation as a method for purging malarially infected red cells from blood. J Infect Dis 163:1312-7, 1991
1558. McEvoy L, Schlegel RA, Williamson P, Del Buono BJ: Merocyanine 540 as a flow cytometric probe of membrane lipid organization in leukocytes. J Leukoc Biol 44:337-44, 1988
1559. Szabo G Jr, Redai I Jr, Bacso Z, Hevessy J, Damjanovich S: Light-induced permeabilization and merocyanine 540 staining of mouse spleen cells. Biochim Biophys Acta 979:365-70, 1989
1560. Belloc F, Lacombe F, Bernard P, Dachary D, Boisseau MR: Selective staining of immature hemopoietic cells with merocyanine 540 in flow cytometry. Cytometry 9:19-24, 1988
1561. Bock G, Huber LA, Wick G, Traill KN: Use of a FACS III for fluorescence depolarization with DPH. J Histochem Cytochem 37:1653-8, 1989
1562. Bock G: [Measuring plasma membrane viscosity on-line using flow cytometry]. Biomed Tech 35:203-5, 1990
1563. Collins JM, Grogan WM: Comparison between flow cytometry and fluorometry for the kinetic measurement of membrane fluidity parameters. Cytometry 10:44-9, 1989
1564. Collins JM, Grogan WM: Fluorescence quenching of a series of membrane probes measured in living cells by flow cytometry. Cytometry 12:247-51, 1991
1565. Hedley D, Chow S: Flow cytometric measurement of lipid peroxidation in vital cells using parinaric acid. Cytometry 13:686-92, 1992
1566. Sonka J, Stohr M, Vogt-Schaden M, Volm M: Anthracycline resistance and consequences of the in situ-in vitro transfer. Cytometry 6:437-44, 1985
1567. Tapiero H, Sbarbati A, Fourcade A, Cinti S, Lampidis TJ: Effect of verapamil on rhodamine 123 mitochondrial damage in adriamycin resistant cells. Anticancer Res 6:1073-6, 1986
1568. Bertoncello I, Hodgson GS, Bradley TR: Multiparameter analysis of transplantable hemopoietic stem cells: I. The separation and enrich-

ment of stem cells homing to marrow and spleen on the basis of rhodamine-123 fluorescence. Exp Hematol 13:999-1006, 1985
1569. Mulder AH, Visser JW: Separation and functional analysis of bone marrow cells separated by rhodamine-123 fluorescence. Exp Hematol 15:99-104, 1987
1570. Morgan SA, Watson JV, Twentyman PR, Smith PJ: Flow cytometric analysis of Hoechst 33342 uptake as an indicator of multi-drug resistance in human lung cancer. Br J Cancer 60:282-7, 1989
1571. Herweijer H, van den Engh G, Nooter K: A rapid and sensitive flow cytometric method for the detection of multidrug-resistant cells. Cytometry 10:463-8, 1989
1572. Nair S, Singh SV, Krishan A: Flow cytometric monitoring of glutathione content and anthracycline retention in tumor cells. Cytometry 12:336-42, 1991
1573. Gheuens EE, van Bockstaele DR, van der Keur M, Tanke HJ, van Oosterom AT, De Bruijn EA: Flow cytometric double labeling technique for screening of multidrug resistance. Cytometry 12:636-44, 1991
1574. Krishan A, Sauerteig A, Stein JH: Comparison of three commercially available antibodies for flow cytometric monitoring of P-glycoprotein expression in tumor cells. Cytometry 12:731-42, 1991
1575. Ludescher C, Thaler J, Drach D, et al.: Detection of activity of P-glycoprotein in human tumour samples using rhodamine 123. Br J Haematol 82:161-8, 1992
1576. Van Acker KL, Van Hove LM, Boogaerts MA: Evaluation of flow cytometry for multidrug resistance detection in low resistance K562 cells using daunorubicin and monoclonal antibodies. Cytometry 14:736-46, 1993
1577. Kessel D, Beck WT, Kukuruga D, Schulz V: Characterization of multidrug resistance by fluorescent dyes. Cancer Res 51:4665-70, 1991
1578. Chaudhary PM, Mechetner EB, Roninson IB: Expression and activity of the multidrug resistance P-glycoprotein in human peripheral blood lymphocytes. Blood 80:2735-9, 1992
1579. Frommel TO, Balber AE: Flow cytofluorimetric analysis of drug accumulation by multidrug-resistant *Trypanosoma brucei brucei* and *T. b. rhodesiense.* Mol Biochem Parasitol 26:183-91, 1987
1580. Sutherland IA, Peregrine AS, Lonsdale-Eccles JD, Holmes PH: Reduced accumulation of isometamidium by drug-resistant *Trypanosoma congolense.* Parasitology 103:245-51, 1991
1581. Miyauchi S, Komatsubara M, Kamo N: In archaebacteria, there is a doxorubicin efflux pump similar to mammalian P-glycoprotein. Biochim Biophys Acta 1110:144-50, 1992
1582. Molenaar D, Bolhuis H, Abee T, Poolman B, Konings WN: The efflux of a fluorescent probe is catalyzed by an ATP-driven extrusion system in *Lactococcus lactis.* J Bacteriol 174:3118-24, 1992
1583. Assaraf YG, Seamer LC, Schimke RT: Characterization by flow cytometry of fluorescein-methotrexate transport in Chinese hamster ovary cells. Cytometry 10:50-5, 1989
1584. Assaraf YG, Slotky JI: Characterization of a lipophilic antifolate resistance provoked by treatment of mammalian cells with the antiparasitic agent pyrimethamine. J Biol Chem 268:4556-66, 1993
1585. Assaraf YG: Characterization by flow cytometry and fluorescein-methotrexate labeling of hydrophilic and lipophilic antifolate resistance in cultured mammalian cells. Anticancer Drugs 4:535-44, 1993
1586. Wiley JS, Brocklebank AM, Snook MB, Jamieson GP, Sawyer WH, Craik JD, Cass CE, Robins MJ: A new fluorescent probe for the equilibrative inhibitor-sensitive nucleoside transporter. 5'-S-(2-aminoethyl)-N6-(4-nitrobenzyl)-5'-thioadenosine (SAENTA)-chi 2-fluorescein. Biochem J 273:667-72, 1991
1587. Knaus HG, Moshammer T, Friedrich K, Kang HC, Haugland RP, Glossman H: In vivo labeling of L-type Ca2+ channels by fluorescent dihydropyridines: evidence for a functional, extracellular heparin-binding site. Proc Natl Acad Sci U S A 89:3586-90, 1992
1588. Knaus HG, Moshammer T, Kang HC, Haugland RP, Glossmann H: A unique fluorescent phenylalkylamine probe for L-type Ca2+ channels. Coupling of phenylalkylamine receptors to Ca2+ and dihydropyridine binding sites. J Biol Chem 267:2179-89, 1992
1589. Dinchuk JE, Kelley KA, Callahan GN: Flow cytometric analysis of transport activity in lymphocytes electroporated with a fluorescent organic anion dye. J Immunol Methods 155:257-65, 1992
1590. Trinkle LS, Wellhausen SR, McLeish KR: A simultaneous flow cytometric measurement of neutrophil phagocytosis and oxidative burst in whole blood. Diagn Clin Immunol 5:62-8, 1987
1591. Hasui M, Hirabayashi Y, Kobayashi Y: Simultaneous measurement by flow cytometry of phagocytosis and hydrogen peroxide production of neutrophils in whole blood. J Immunol Methods 117:53-8, 1989
1592. Perticarari S, Presani G, Mangiarotti MA, Banfi E: Simultaneous flow cytometric method to measure phagocytosis and oxidative products by neutrophils. Cytometry 12:687-93, 1991
1593. Burow S, Valet G: Flow-cytometric characterization of stimulation, free radical formation, peroxidase activity and phagocytosis of human granulocytes with 2,7-dichlorofluorescein (DCF). Eur J Cell Biol 43:128-33, 1987
1594. Casado JA, Merino J, Cid J, Subira ML, Sanchez-Ibarrola A: Simultaneous evaluation of phagocytosis and Fc gamma R-mediated oxidative burst in human monocytes by a simple flow cytometry method. J Immunol Methods 159:173-6, 1993
1595. Hed J, Hallden G, Johansson SG, Larsson P: The use of fluorescence quenching in flow cytofluorometry to measure the attachment and ingestion phases in phagocytosis in peripheral blood without prior cell separation. J Immunol Methods 101:119-25, 1987
1596. Cantinieaux B, Hariga C, Courtoy P, Hupin J, Fondu P: Staphylococcus aureus phagocytosis. A new cytofluorometric method using FITC and paraformaldehyde. J Immunol Methods 121:203-8, 1989
1597. Fattorossi A, Nisini R, Pizzolo JG, D'Amelio R: New, simple flow cytometry technique to discriminate between internalized and membrane-bound particles in phagocytosis. Cytometry 10:320-5, 1989
1598. Ogle JD, Noel JG, Sramkoski RM, Ogle CK, Alexander JW: Phagocytosis of opsonized fluorescent microspheres by human neutrophils. A two-color flow cytometric method for the determination of attachment and ingestion. J Immunol Methods 115:17-29, 1988
1599. White-Owen C, Alexander JW, Sramkoski RM, Babcock GF: Rapid whole-blood microassay using flow cytometry for measuring neutrophil phagocytosis. J Clin Microbiol 30:2071-6, 1992
1600. Ma JA, Chapman GV, Chen SL, Penny R, Breit SN: Flow cytometry with crystal violet to detect intracytoplasmic fluorescence in viable human lymphocytes. Demonstration of antibody entering living cells. J Immunol Methods 104:195-200, 1987
1601. Suzuki K, Takahashi K, Matsuki Y, Kawakami M Kawaguchi Y, Hidaka T, Sekiyama, Y Mizukami Y, Kawagoe M: Fluorescent probe-labeled lipid micro-sphere uptake by human endothelial cells: a flow cytometric study. Jpn J Pharmacol 60:349-56, 1992
1602. Haynes AP, Fletcher J, Garnett M, Robins A: A novel flow cytometric method for measuring protein digestion within the phagocytic vacuole of polymorphonuclear neutrophils. J Immunol Methods 135:155-61, 1990
1603. Wang Yang MC, Harvey NE, Cuchens MA, Buttke TM: Pulse profile analyses of endocytosis in capped B lymphocytes and BCL1 cells. Cytometry 9:131-7, 1988
1604. Chanh TC, Alderete BE: A rapid method for quantitating lymphocyte receptor capping: capping defect in AIDS patients. J Virol Methods 29:257-65, 1990
1605. Lavin DP, Fredrickson AG, Srienc F: Flow cytometric measurement of rates of particle uptake from dilute suspensions by a ciliated protozoan. Cytometry 11:875-82, 1990
1606. Hatzis C, Sweeney PJ, Srienc F, Fredrickson AG: A discrete, stochastic model for microbial filter feeding: a model for feeding of ciliated protists on spatially uniform, nondepletable suspensions. Math Biosci 102:127-81, 1990

1607. Fredrickson AG, Hatzis C, Srienc F: A statistical analysis of flow cytometric determinations of phagocytosis rates. Cytometry 13:423-31, 1992
1608. Cucci TL, Shumway SE, Brown WS, Newell CR: Using phytoplankton and flow cytometry to analyze grazing by marine organisms. Cytometry 10:659-69, 1989
1609. Ross DW, Bishop C, Henderson A, Kaplow L: Whole blood staining in suspension for nonspecific esterase and alkaline phosphatase analyzed with a Technicon H-1. Cytometry 11:552-5, 1990
1610. Lee LG, Berry GM, Chen CH: Vita Blue: a new 633-nm excitable fluorescent dye for cell analysis. Cytometry 10:151-64, 1989
1611. Huet O, Petit JM, Ratinaud MH, Julien R: NADH-dependent dehydrogenase activity estimation by flow cytometric analysis of 3-(4,5-dimethylthiazolyl-2-yl)-2,5-diphenyltetrazolium bromide (MTT) reduction. Cytometry 13:532-9, 1992
1612. Fattorossi A, Nisini R, Le Moli S, De Petrillo G, D'Amelio R: Flow cytometric evaluation of nitro blue tetrazolium (NBT) reduction in human polymorphonuclear leukocytes. Cytometry 11:907-12, 1990
1613. Van Noorden CJ, Dolbeare F, Aten J: Flow cytofluorometric analysis of enzyme reactions based on quenching of fluorescence by the final reaction product: detection of glucose-6-phosphate dehydrogenase deficiency in human erythrocytes. J Histochem Cytochem 37:1313-8, 1989
1614. Kuhlmann U, Severin E, Stellmach J, Wiezorek C, Echsler K: [Fluorescent formazans in flow cytometry. Studies of their oxygen sensitivity]. Acta Histochem Suppl 37:221-30, 1989
1615. Huang CJ, Severin E: Enzyme activities of six different dehydrogenases in Ehrlich ascites cells measured by flow cytometry. Acta Histochem 94:33-45, 1993
1616. Robinson JP, Bruner LH, Bassoe CF, Hudson JL, Ward PA, Phan SH: Measurement of intracellular fluorescence of human monocytes relative to oxidative metabolism. J Leukoc Biol 43:304-10, 1988
1617. Rao KM, Padmanabhan J, Kilby DL, Cohen HJ, Currie MS, Weinberg JB: Flow cytometric analysis of nitric oxide production in human neutrophils using dichlorofluorescein diacetate in the presence of a calmodulin inhibitor. J Leukoc Biol 51:496-500, 1992
1618. Maresh GA, Monnat RJ Jr: Novel fluoride-stimulated dichlorofluorescin dye oxidation pathway in human leukemia cell lines. Biochem Biophys Res Commun 194:869-75, 1993
1619. Davis WC, Wyatt CR, Hamilton MJ, Goff WL: A rapid, reliable method of evaluating growth and viability of intraerythrocytic protozoan hemoparasites using fluorescence flow cytometry. Mem Inst Oswaldo Cruz 87:235-9, 1992
1620. Rothe G, Valet G: Flow cytometric analysis of respiratory burst activity in phagocytes with hydroethidine and 2',7'-dichlorofluorescin. J Leukoc Biol 47:440-8, 1990
1621. Carter WO, Narayanan PK, Robinson JP: Intracellular hydrogen peroxide and superoxide anion detection in endothelial cells. J Leukoc Biol 55:253-8, 1994
1622. Rothe G, Oser A, Valet G: Dihydrorhodamine 123: a new flow cytometric indicator for respiratory burst activity in neutrophil granulocytes. Naturwissen-schaften 75:354-5, 1988
1623. Henderson LM, Chappell JB: Dihydrorhodamine 123: a fluorescent probe for superoxide generation? Eur J Biochem 217:973-80, 1993
1624. Siemann DW, Keng PC: Characterization of radiation resistant hypoxic cell subpopulations in KHT sarcomas. (II). Cell sorting. Br J Cancer 58:296-300, 1988
1625. Chaplin DJ, Trotter MJ, Durand RE, Olive PL, Minchinton AI: Evidence for intermittent radiobiological hypoxia in experimental tumour systems. Biomed Biochim Acta 48:S255-9, 1989
1626. Minchinton AI, Durand RE, Chaplin DJ: Intermittent blood flow in the KHT sarcoma--flow cytometry studies using Hoechst 33342. Br J Cancer 62:195-200, 1990
1627. Hodgkiss RJ, Jones GW, Long A, Middleton RW, Parrick J Stratford MR, Wardman P, Wilson GD: Fluorescent markers for hypoxic cells: a study of nitroaromatic compounds, with fluorescent heterocyclic side chains, that undergo bioreductive binding. J Med Chem 34:2268-74, 1991
1628. Hodgkiss RJ, Middleton RW, Parrick J, Rami HK, Wardman P, Wilson GD: Bioreductive fluorescent markers for hypoxic cells: a study of 2-nitroimidazoles with 1-substituents containing fluorescent, bridgehead-nitrogen, bicyclic systems. J Med Chem 35:1920-6, 1992
1629. Tran-Paterson R, Boileau G, Giguere V, Letarte M: Comparative levels of CALLA/neutral endopeptidase on normal granulocytes, leukemic cells, and transfected COS-1 cells. Blood 76:775-82, 1990
1630. Milhiet PE, Dennin F, Giocondi MC, Le Grimellec C, Garbay-Jaureguiberry C, Boucheix C, Roques BP: Detection of neutral endopeptidase-24.11/CD10 by flow cytometry and photomicroscopy using a new fluorescent inhibitor. Anal Biochem 205:57-64, 1992
1631. Dhar A, Shukla SD: Electrotransjection of pp60v-src monoclonal antibody inhibits activation of phospholipase C in platelets. A new mechanism for platelet-activating factor responses. J Biol Chem 269:9123-7, 1994
1632. Far DF, Peyron J-F, Imbert V, Rossi B: Immunofluorescent quantification of tyrosine phosphorylation of cellular proteins in whole cells by flow cytometry. Cytometry 15:327-34, 1994
1633. Banati RB, Rothe G, Valet G, Kreutzberg GW: Detection of lysosomal cysteine proteinases in microglia: flow cytometric measurement and histochemical localization of cathepsin B and L. Glia 7:183-91, 1993
1634. Assfalg-Machleidt I, Rothe G, Klingel S, Banati, R, Mangel WF, Valet G, Machleidt W: Membrane permeable fluorogenic rhodamine substrates for selective determination of cathepsin L. Biol Chem Hoppe Seyler 373:433-40, 1992
1635. van Noorden CJ: Assessment of lysosomal function by quantitative histochemical and cytochemical methods. Histochem J 23:429-35, 1991
1636. Dive C, Workman P, Watson JV: Can flow cytoenzymology be applied to measure membrane-bound enzyme kinetics? Assessment by analysis of gamma-glutamyl transpeptidase activity. Biochem Pharmacol 46:643-50, 1993
1637. Dive C, Workman P, Watson JV: Novel dynamic flow cytoenzymological determination of intracellular esterase inhibition by BCNU and related isocyanates. Biochem Pharmacol 36:3731-8, 1987
1638. Dive C, Cox H, Watson JV, Workman P: Polar fluorescein derivatives as improved substrate probes for flow cytoenzymological assay of cellular esterases. Mol Cell Probes 2:131-45, 1988
1639. Huang Z, Terpetschnig E, You W, Haugland RP: 2-(2'-phosphoryloxyphenyl)-4(3H)-quinazolinone derivatives as fluorogenic precipitating substrates of phosphatases. Anal Biochem 207:32-9, 1992
1640. Huang Z, You W, Haugland RP, Paragas VB, Olson NA, Haugland RP: A novel fluorogenic substrate for detecting alkaline phosphatase activity in situ. J Histochem Cytochem 41:313-7, 1993
1641. Sidhu JS, Kavanagh TJ, Reilly MT, Omiecinski CJ: Direct determination of functional activity of cytochrome P-4501A1 and NADPH DT-diaphorase in hepatoma cell lines using noninvasive scanning laser cytometry. J Toxicol Environ Health 40:177-94, 1993
1642. Nolan GP, Fiering S, Nicolas JF, Herzenberg LA: Fluorescence-activated cell analysis and sorting of viable mammalian cells based on β-D-galactosidase activity after transduction of *Escherichia coli lacZ*. Proc Natl Acad Sci USA 85:2603-7, 1988
1643. Saalmuller A, Mettenleiter TC: Rapid identification and quantitation of cells infected by recombinant herpesvirus (pseudorabies virus) using a fluorescence-based beta-galactosidase assay and flow cytometry. J Virol Methods 44:99-108, 1993
1644. Fiering SN, Roederer M, Nolan GP, Micklem DR, Parks DR, Herzenberg LA: Improved FACS-Gal: flow cytometric analysis and sorting of viable eukaryotic cells expressing reporter gene constructs. Cytometry 12:291-301, 1991

1645. Jasin M, Zalamea P: Analysis of *Escherichia coli* beta-galactosidase expression in transgenic mice by flow cytometry of sperm. Proc Natl Acad Sci USA 89:10681-5, 1992
1646. Wittrup KD, Bailey JE: A single-cell assay of beta-galactosidase activity in *Saccharomyces cerevisiae*. Cytometry 9:394-404, 1988
1647. Puchalski RB, Manoharan TH, Lathrop AL, Fahl WE: Recombinant glutathione S-transferase (GST) expressing cells purified by flow cytometry on the basis of a GST-catalyzed intracellular conjugation of glutathione to monochlorobimane. Cytometry 12:651-65, 1991
1648. Chalfie M, Tu Y, Euskirchen G, Ward WW, Prasher DC: Green fluorescent protein as a marker for gene expression. Science 263:802-5, 1994
1649. O'Connor JE, Kimler BF, Morgan MC, Tempas KJ: A flow cytometric assay for intracellular nonprotein thiols using mercury orange. Cytometry 9:529-32, 1988
1650. Poot M, Kavanagh TJ, Kang HC, Haugland RP, Rabinovitch PS: Flow cytometric analysis of cell cycle-dependent changes in cell thiol level by combining a new laser dye with Hoechst 33342. Cytometry 12:184-7, 1991
1651. Cook JA, Iype SN, Mitchell JB: Differential specificity of monochlorobimane for isozymes of human and rodent glutathione S-transferases. Cancer Res 51:1606-12, 1991
1652. Ublacker GA, Johnson JA, Siegel FL, Mulcahy RT: Influence of glutathione S-transferases on cellular glutathione determination by flow cytometry using monochlorobimane. Cancer Res 51:1783-8, 1991
1653. Hedley D, Chow S: Evaluation of methods for measuring cellular glutathione content using flow cytometry. Cytometry 15:349-58, 1994
1654. Maddox AM, Johnson DA, Keating MJ: 5-bromodeoxyuridine (BUdR) quenching of acridine orange fluorescence distinguishes cycling and non-cycling normal and malignant bone marrow cells in vitro. Leuk Res 13:781-90, 1989
1655. Rabinovitch PS, Kubbies M, Chen YC, Schindler D, Hoehn H: BrdU-Hoechst flow cytometry: a unique tool for quantitative cell cycle analysis. Exp Cell Res 174:309-18, 1988
1656. Poot M, Schmitt H, Seyschab H, Koehler J, Chen U, Kaempf U, Kubbies M, Schindler D, Rabinovitch PS, Hoehn H: Continuous bromodeoxyuridine labeling and bivariate ethidium bromide/Hoechst flow cytometry in cell kinetics. Cytometry 10:222-6, 1989 [published erratum appears in Cytometry 10:670, 1989]
1657. Ormerod MG, Kubbies M: Cell cycle analysis of asynchronous cell populations by flow cytometry using bromodeoxyuridine label and Hoechst-propidium iodide stain. Cytometry 13:678-85, 1992
1658. Goller B, Kubbies M: UV lasers for flow cytometric analysis: HeCd versus argon laser excitation. J Histochem Cytochem 40:451-6, 1992
1659. Kubbies M, Goller B, Van Bockstaele DR: Improved BrdUrd-Hoechst bivariate cell kinetic analysis by helium-cadmium single laser excitation. Cytometry 13:782-6, 1992
1660. Hoy CA, Seamer LC, Schimke RT: Thermal denaturation of DNA for immunochemical staining of incorporated bromodeoxyuridine (BrdUrd): critical factors that affect the amount of fluorescence and the shape of BrdUrd/DNA histogram. Cytometry 10:718-25, 1989
1661. Larsen JK, Christensen IJ, Christiansen J, Mortensen BT: Washless double staining of unfixed nuclei for flow cytometric analysis of DNA and a nuclear antigen (Ki-67 or bromodeoxyuridine). Cytometry 12:429-37, 1991
1662. Bakker PJ, Stap J, Tukker CJ, van Oven CH, Veenhof CH, Aten J: An indirect immunofluorescence double staining procedure for the simultaneous flow cytometric measurement of iodo- and chlorodeoxyuridine incorporated into DNA. Cytometry 12:366-72, 1991
1663. Bakker PJ, de Vries RJ, Tukker CJ, Hoebe RA, Barendsen GW: Application of a DNA double labelling method for the flow cytometric analysis of recruitment of non-cycling cells in a mixed population of P and Q cells. Cell Prolif 26:89-100, 1993
1664. Toba K, Winton EF, Bray RA: Improved staining method for the simultaneous flow cytofluorometric analysis of DNA content, S-phase fraction, and surface phenotype using single laser instrumentation. Cytometry 13:60-7, 1992
1665. White RA, Fallon JF, Savage MP: On the measurement of cytokinetics by continuous labeling with bromodeoxyuridine with applications to chick wing buds. Cytometry 13:553-6, 1992
1666. Li X, Traganos F, Melamed MR, Darzynkiewicz Z: Detection of 5-bromo-2-deoxyuridine incorporated into DNA by labeling strand breaks induced by photolysis (SBIP). Intl J Oncol 4:1157-61, 1994
1667. Jensen PO, Larsen J, Larsen JK: Flow cytometric measurement of RNA synthesis based on bromouridine labelling and combined with measurement of DNA content or cell surface antigen. Acta Oncol 32:521-4, 1993
1668. Rolland JM, Dimitropoulos K, Bishop A, Hocking GR, Nairn RC: Fluorescence polarization assay by flow cytometry. J Immunol Methods 76:1-6, 1985
1669. Dimitropoulos K, Rolland JM, Nairn RC: Analysis of early lymphocyte activation events by fluorescence polarization flow cytometry. Immunol Cell Biol 66:253-60, 1988
1670. Deutsch M, Weinreb A: Validation of the SCM test for the diagnosis of cancer. Eur J Cancer Clin Oncol 19:187-93, 1983
1671. Chaitchik S, Asher O, Deutsch M, Weinreb A: Tumor specificity of the SCM test for cancer diagnosis. Eur J Cancer Clin Oncol 21:1165-70, 1985
1672. Chaitchik S, Deutsch M, Asher O, Krauss G, Lebovich P, Michlin H, Weinreb A: An evaluation of the SCM test for the diagnosis of cancer of the breast. Eur J Cancer Clin Oncol 24:861-7, 1988
1673. Cercek L, Cercek B: Cancer-associated SCM-recognition, immunedefense suppression, and serine protease protection peptide. Part I. Isolation, amino acid sequence, homology, and origin. Cancer Detect Prev 16:305-19, 1992
1674. Cercek L, Cercek B: Cancer-associated SCM-recognition, immunedefense suppression, and serine protease protection peptide. Part II. Immunedefense suppressive effects of the CRISPPs peptide. Cancer Detect Prev 17:433-45, 1993
1675. Cercek L, Cercek B: Cancer-associated SCM-recognition, immunedefense suppression, and serine protease protection peptide. Part III. CRISPP peptide protection of serine proteases against inhibition. Cancer Detect Prev 17:447-54, 1993
1676. Ehrenberg B, Montana V, Wei MD, Wuskell JP, Loew LM: Membrane potential can be determined in individual cells from the Nernstian distribution of cationic dyes. Biophys J 53:785-94, 1988
1677. Gross D, Loew LM: Fluorescent indicators of membrane potential: microspectrofluorometry and imaging. Methods Cell Biol 30:193-218, 1989
1678. Farkas DL, Wei MD, Febbroriello P, Carson JH, Loew LM: Simultaneous imaging of cell and mitochondrial membrane potentials. Biophys J 56:1053-69, 1989 [published erratum appears in Biophys J 57:following 684, 1990]
1679. Montana V, Farkas DL, Loew LM: Dual-wavelength ratiometric fluorescence measurements of membrane potential. Biochemistry 28:4536-9, 1989
1680. Bouevitch O, Lewis A, Pinevsky I, Wuskell JP, Loew LM: Probing membrane potential with nonlinear optics. Biophys J 65:672-9, 1993
1681. Reers M, Smith TW, Chen LB: J-aggregate formation of a carbocyanine as a quantitative fluorescent indicator of membrane potential. Biochemistry 30:4480-6, 1991
1682. Smiley ST, Reers M, Mottola-Hartshorn C, et al.: Intracellular heterogeneity in mitochondrial membrane potentials revealed by a J-aggregate-forming lipophilic cation JC-1. Proc Natl Acad Sci U S A 88:3671-5, 1991
1683. Anderson WM, Delinck DL, Benninger L, Wood JM, Smiley ST, Chen LB: Cytotoxic effect of thiacarbocyanine dyes on human colon carcinoma cells and inhibition of bovine heart mitochondrial

NADH-ubiquinone reductase activity via a rotenone-type mechanism by two of the dyes. Biochem Pharmacol 45:691-6, 1993
1684. Tasaki I, Byrne PM: The origin of rapid changes in birefringence, light scattering and dye absorbance associated with excitation of nerve fibers. Jpn J Physiol 43:S67-75, 1993
1685. Aszalos A, Damjanovich S, Colombani P, Hess A: Lymphocyte populations with different sensitivity to cyclosporin have different plasma membrane potentials. J Med 18:351-74, 1987
1686. Aszalos A, Tron L, Paxton H, Shen S: Lymphocyte subpopulation with low membrane potential in the blood of cyclosporin- and prednisone-treated patients: in vivo selectivity for T4 subset. Biochem Med Metab Biol 41:25-9, 1989
1687. Vayuvegula B, Slater L, Meador J, Gupta S: Correction of altered plasma membrane potentials. A possible mechanism of cyclosporin A and verapamil reversal of pleiotropic drug resistance in neoplasia. Cancer Chemother Pharmacol 22:163-8, 1988
1688. Hasmann M, Valet GK, Tapiero H, Trevorrow K, Lampidis T: Membrane potential differences between adriamycin-sensitive and -resistant cells as measured by flow cytometry. Biochem Pharmacol 38:305-12, 1989
1689. Tanner MK, Wellhausen SR, Klein JB: Flow cytometric analysis of altered mononuclear cell transmembrane potential induced by cyclosporin. Cytometry 14:59-69, 1993
1690. Witkowski JM, Micklem HS: Transmembrane electrical potential of lymphocytes in ageing mice. Flow cytometric analysis of mitogen-stimulated cells. Mech Ageing Dev 62:167-79, 1992
1691. Seamer LC, Mandler RN: Method to improve the sensitivity of flow cytometric membrane potential measurements in mouse spinal cord cells. Cytometry 13:545-52, 1992
1692. Damjanovich S, Pieri C: Electroimmunology: membrane potential, ion-channel activities, and stimulatory signal transduction in human T lymphocytes from young and elderly. Ann N Y Acad Sci 621:29-39, 1991
1693. Radosevic K, Schut TC, van Graft M, de Grooth BG, Greve J: A flow cytometric study of the membrane potential of natural killer and K562 cells during the cytotoxic process. J Immunol Methods 161:119-28, 1993
1694. Chen LB: Mitochondrial membrane potential in living cells. Annu Rev Cell Biol 4:155-81, 1988
1695. Chen LB: Fluorescent labeling of mitochondria. Methods Cell Biol 29:103-23, 1989
1696. Maftah A, Petit JM, Ratinaud MH, Julien R: 10-N nonyl-acridine orange: a fluorescent probe which stains mitochondria independently of their energetic state. Biochem Biophys Res Commun 164:185-90, 1989
1697. Ratinaud MH, Leprat P, Julien R: In situ flow cytometric analysis of nonyl acridine orange-stained mitochondria from splenocytes. Cytometry 9:206-12, 1988
1698. Benel L, Ronot X, Mounolou JC, Gaudemer F, Adolphe M: Compared flow cytometric analysis of mitochondria using 10-n-nonyl acridine orange and rhodamine 123. Basic Appl Histochem 33:71-80, 1989
1699. Borth N, Kral G, Katinger H: Rhodamine 123 fluorescence of immortal hybridoma cell lines as a function of glucose concentration. Cytometry 14:70-3, 1993
1700. Myc A, De Angelis P, Kimmel M, Melamed MR, Darzynkiewicz Z: Retention of the mitochondrial probe rhodamine 123 in normal lymphocytes and leukemic cells in relation to the cell cycle. Exp Cell Res 192:198-202, 1991
1701. Cossarizza A, Baccarani-Contri M, Kalashnikova G, Franceschi C: A new method for the cytofluorimetric analysis of mitochondrial membrane potential using the J-aggregate forming lipophilic cation 5,5',6,6'-tetra-chloro-1,1',3,3'-tetraethylbenzimidazolcarbocyanine iodide (JC-1). Biochem Biophys Res Commun 197:40-5, 1993
1702. Hahn KM, Conrad PA, Chao JC, Taylor DL, Waggoner AS: A photocross-linking fluorescent indicator of mitochondrial membrane potential. J Histochem Cytochem 41:631-4, 1993
1703. O'Connor JE, Vargas JL, Kimler BF, Hernandez-Yago J, Grisolia S: Use of rhodamine 123 to investigate alterations in mitochondrial activity in isolated mouse liver mitochondria. Biochem Biophys Res Commun 151:568-73, 1988
1704. Petit PX, O'Connor JE, Grunwald D, Brown SC: Analysis of the membrane potential of rat- and mouse-liver mitochondria by flow cytometry and possible applications. Eur J Biochem 194:389-97, 1990
1705. Wolf ME, Kapatos G: Flow cytometric analysis of rat striatal nerve terminals. J Neurosci 9:94-105, 1989
1706. Irion G, Ochsenfeld L, Naujok A, Zimmermann HW: The concentration jump method. Kinetics of vital staining of mitochondria in HeLa cells with lipophilic cationic fluorescent dyes. Histochemistry 99:75-83, 1993
1707. Rottele J, Zimmermann HW: Transport and accumulation of lipophilic dye cations at the mitochondria of HeLa cells in situ. Cell Mol Biol 39:739-56, 1993
1708. Singer S, Neuringer LJ, Thilly WG, Chen LB: Quantitative differential effects of rhodamine 123 on normal cells and human colon cancer cells by magnetic resonance spectroscopy. Cancer Res 53:5808-14, 1993
1709. Amchenkova AA, Bakeeva LE, Chentsov YS, Skulachev VP, Zorov DB: Coupling membranes as energy-transmitting cables. I. Filamentous mitochondria in fibroblasts and mitochondrial clusters in cardiomyocytes. J Cell Biol 107:481-95, 1988
1710. Severina II, Skulachev VP, Zorov DB: Coupling membranes as energy-transmitting cables. II. Cyanobacterial trichomes. J Cell Biol 107:497-501, 1988
1711. Skulachev VP: Power transmission along biological membranes. J Membr Biol 114:97-112, 1990
1712. Bedlack R Jr, Wei M, Loew LM: Localized membrane depolarizations and localized calcium influx during electric field-guided neurite growth. Neuron 9:393-403, 1992
1713. Kachel V, Kempski O, Peters J, Schodel F: A method for calibration of flow cytometric wavelength shift fluorescence measurements. Cytometry 11:913-5, 1990
1714. June CH, Rabinovitch PS, Ledbetter JA: CD5 antibodies increase intracellular ionized calcium concentration in T cells. J Immunol 138:2782-92, 1987
1715. June CH, Rabinovitch PS: Flow cytometric measurement of cellular ionized calcium concentration. Pathol Immunopathol Res 7:409-32, 1988
1716. Griffioen AW, Rijkers GT, Keij J, Zegers BJ: Measurement of cytoplasmic calcium in lymphocytes using flow cytometry. Kinetic studies and single cell analysis. J Immunol Methods 120:23-7, 1989
1717. Jennings LK, Dockter ME, Wall CD, Fox CF, Kennedy DM: Calcium mobilization in human platelets using indo-1 and flow cytometry. Blood 74:2674-80, 1989
1718. Oda A, Daley JF, Kang J, Smith M, Ware JA, Salzman EW: Quasi-simultaneous measurement of ionized calcium and alpha-granule release in individual platelets. Am J Physiol 260:C242-8, 1991
1719. Minta A, Kao JP, Tsien RY: Fluorescent indicators for cytosolic calcium based on rhodamine and fluorescein chromophores. J Biol Chem 264:8171-8, 1989
1720. Rijkers GT, Justement LB, Griffioen AW, Cambier JC: Improved method for measuring intracellular Ca++ with fluo-3. Cytometry 11:923-7, 1990
1721. Akkaya EU, Lakowicz JR: Styryl-based wavelength-ratiometric probes: a new class of fluorescent calcium probes with long wavelength emission and a large Stokes' shift. Anal Biochem 213:285-9, 1993
1722. Gillies RJ, Cook J, Fox MH, Giuliano KA: Flow cytometric analysis of intracellular pH in 3T3 cells. Am J Physiol 253:C121-5, 1987
1723. Cook JA, Fox MH: Intracellular pH measurements using flow cytometry with 1,4-diacetoxy-2,3-dicyanobenzene. Cytometry 9:441-7, 1988

1724. Hedley DW, Jorgensen HB: Flow cytometric measurement of intracellular pH in B16 tumors: intercell variance and effects of pretreatment with glucose. Exp Cell Res 180:106-16, 1989
1725. Musgrove E, Seaman M, Hedley D: Relationship between cytoplasmic pH and proliferation during exponential growth and cellular quiescence. Exp Cell Res 172:65-75, 1987
1726. Wang ZH, Chu GL, Hyun WC, Pershadsingh HA, Fulwyler MJ, Dewey WC: Comparison of DMO and flow cytometric methods for measuring intracellular pH and the effect of hyperthermia on the transmembrane pH gradient. Cytometry 11:617-23, 1990
1727. van Erp PE, Jansen MJ, de Jongh GJ, Boezeman JB, Schalkwijk J: Ratiometric measurement of intracellular pH in cultured human keratinocytes using carboxy-SNARF-1 and flow cytometry. Cytometry 12:127-32, 1991
1728. Wieder ED, Hang H, Fox MH: Measurement of intracellular pH using flow cytometry with carboxy-SNARF-1. Cytometry 14:916-21, 1993
1729. Van Graft M, Kraan YM, Segers IM, Radosevic K, De Grooth BG, Greve J: Flow cytometric measurement of [Ca2+]i and pHi in conjugated natural killer cells and K562 target cells during the cytotoxic process. Cytometry 14:257-64, 1993
1730. Lee RJ, Oliver JM, Deanin GG, Troup CD, Stump RF: Importance of bicarbonate ion for intracellular pH regulation in antigen- and ionomycin-stimulated RBL-2H3 mast cells. Cytometry 13:127-36, 1992
1731. Worthington RE, Aubry J-P: Studies of changes in cytoplasmic pH and membrane potential. In reference 1047, pp. 17-27
1732. Wacholtz MC, Cragoe E Jr, Lipsky PE: A Na(+)-dependent Ca2+ exchanger generates the sustained increase in intracellular Ca2+ required for T cell activation. J Immunol 149:1912-20, 1992
1733. Demaurex N, Grinstein S, Jaconi M, Schlegel W, Lew DP, Krause KH: Proton currents in human granulocytes: regulation by membrane potential and intracellular pH. J Physiol 466:329-44, 1993
1734. Harrison RA, Mairet B, Miller NG: Flow cytometric studies of bicarbonate-mediated Ca2+ influx in boar sperm populations. Mol Reprod Dev 35:197-208, 1993
1735. Mason MJ, Grinstein S: Ionomycin activates electrogenic Ca2+ influx in rat thymic lymphocytes. Biochem J 296:33-9, 1993
1736. Wacholtz MC, Lipsky PE: Anti-CD3-stimulated Ca2+ signal in individual human peripheral T cells. Activation correlates with a sustained increase in intracellular Ca2+1. J Immunol 150:5338-49, 1993
1737. Mitsumoto Y, Mohri T: Dual-fluorescence flow cytometric analysis of membrane potential and cytoplasmic free Ca2+ concentration in embryonic rat hippocampal cells. Cell Struct Funct 14:669-72, 1989
1738. Bernardo J, Newburger PE, Brennan L, Brink HF, Bresnick SA, Weil G, Simons ER: Simultaneous flow cytometric measurements of cytoplasmic Ca++ and membrane potential changes upon FMLP exposure as HL-60 cells mature into granulocytes: using [Ca++]in as an indicator of granulocyte maturity. J Leukoc Biol 47:265-74, 1990
1739. Lund-Johansen F, Olweus J: Signal transduction in monocytes and granulocytes measured by multiparameter flow cytometry. Cytometry 13:693-702, 1992
1740. Bigos M, Parks DR, Moore WA, Herzenberg LA, Herzenberg LA: Pattern sorting: A computer-controlled multidimensional sorting method using K-D trees. Cytometry 16:357-363
1741. Way DL, Witte MH, Fiala M, Ramirez G, Nagle RB, Bernas MJ, Dictor M, Borgs P, Witte CL: Endothelial transdifferentiated phenotype and cell-cycle kinetics of AIDS-associated Kaposi sarcoma cells. Lymphology 26:79-89, 1993
1742. Kallioniemi OP, Visakorpi T, Holli K, Heikkinen A, Isola J, Koivula T: Improved prognostic impact of S-phase values from paraffin-embedded breast and prostate carcinomas after correcting for nuclear slicing. Cytometry 12:413-21, 1991
1743. Braylan RC, Benson NA, Nourse VA, Kruth HS: Cellular DNA of human neoplastic B cells measured by flow cytometry. Cancer Res 44:5010-6, 1984
1744. Oud PS, Henderik JBJ, Beck HLM, Veldhuizen JAM, Vooijs GP, Herman CJ, Ramaekers FCS: Flow cytometric analysis and sorting of human endometrial cells after immunocytochemical labeling for cytokeratin using a monoclonal antibody. Cytometry 6:159-64, 1985
1745. Hedley DW, Shankey TV, Wheeless LL: DNA cytometry consensus conference. Cytometry 14:471, 1993
1746. Shankey TV, Rabinovitch PS, Bagwell B, Bauer KD, Duque RE, Hedley DW, Mayall BH, Wheeless L: Guidelines for implementation of clinical DNA cytometry. International Society for Analytical Cytology. Cytometry 14:472-7, 1993
1747. Wheeless LL, Badalament RA, de Vere White RW, Fradet Y, Tribukait B: Consensus review of the clinical utility of DNA cytometry in bladder cancer.Report of the DNA Cytometry Consensus Conference. Cytometry 14:478-81, 1993
1748. Hedley DW, Clark GM, Cornelisse CJ, Killander D, Kute T, Merkel D: Consensus review of the clinical utility of DNA cytometry in carcinoma of the breast.Report of the DNA Cytometry Consensus Conference. Cytometry 14:482-5, 1993
1749. Bauer KD, Bagwell CB, Giaretti W, Melamed MR, Zarbo RJ, Witzig TE, Rabinovitch PS: Consensus review of the clinical utility of DNA flow cytometry in colorectal cancer. Cytometry 14:486-91, 1993
1750. Duque RE, Andreeff M, Braylan RC, Diamond LW, Peiper SC: Consensus review of the clinical utility of DNA flow cytometry in neoplastic hematopathology. Cytometry 14:492-6, 1993
1751. Shankey TV, Kallioniemi OP, Koslowski JM, Lieber ML, Mayall BH, Miller G, Smith GJ: Consensus review of the clinical utility of DNA content cytometry in prostate cancer. Cytometry 14:497-500, 1993
1752. Boring CC, Squires TS, Tong T, Montgomery S: Cancer statistics, 1994. CA 44 (1):7-26, 1994
1753. Cook M: Detection of DNA in ancient skeletal remains using DNA flow cytometry. Biotech Histochem 68:260-4, 1993
1754. Eldredge N: *Time Frames. The Evolution of Punctuated Equilibria.* Princeton, Princeton University Press, 1985, 240 pp.
1755. Civin CI, Loken MR: Cell surface antigens on human marrow cells: dissection of hematopoietic development using monoclonal antibodies and multiparameter flow cytometry. Int J Cell Cloning 5:267-88, 1987
1756. Loken MR, Shah VO, Hollander Z, Civin CI: Flow cytometric analysis of normal B lymphoid development. Pathol Immunopathol Res 7:357-70, 1988
1757. Shah VO, Civin CI, Loken MR: Flow cytometric analysis of human bone marrow. IV. Differential quantitative expression of T-200 common leukocyte antigen during normal hemopoiesis. J Immunol 140:1861-7, 1988
1758. Hollander Z, Shah VO, Civin CI, Loken MR: Assessment of proliferation during maturation of the B lymphoid lineage in normal human bone marrow. Blood 71:528-31, 1988
1759. Le Bien TW, Wormann B, Villablanca JG, Law CL, Steinberg LM, Shah VO, Loken MR: Multiparameter flow cytometric analysis of human fetal bone marrow B cells. Leukemia 4:354-8, 1990
1760. Terstappen LW, Safford M, Loken MR: Flow cytometric analysis of human bone marrow. III. Neutrophil maturation. Leukemia 4:657-63, 1990
1761. Terstappen LW, Loken MR: Myeloid cell differentiation in normal bone marrow and acute myeloid leukemia assessed by multi-dimensional flow cytometry. Anal Cell Pathol 2:229-40, 1990
1762. Terstappen LW, Huang S, Safford M, Lansdorp PM, Loken MR: Sequential generations of hematopoietic colonies derived from single nonlineage-committed CD34+CD38- progenitor cells. Blood 77:1218-27, 1991
1763. Terstappen LW, Safford M, Konemann S, Loken MR, Zurlutter K, Buchner T, Hiddemann W, Wormann B: Flow cytometric characterization of acute myeloid leukemia. Part II. Phenotypic heterogeneity at diagnosis. Leukemia 5:757-67, 1991

1764. Terstappen LW, Konemann S, Safford M, Loken MR, Zurlutter K, Buchner T, Hiddemann W, Wormann B: Flow cytometric characterization of acute myeloid leukemia. Part 1. Significance of light scattering properties. Leukemia 5:315-21, 1991
1765. Terstappen LW, Safford M, Konemann S, Loken MR, Zurlutter K, Buchner T, Hiddemann W, Wormann B: Flow cytometric characterization of acute myeloid leukemia. Part II. Phenotypic heterogeneity at diagnosis. Leukemia 5:757-67, 1991
1766. Terstappen LW, Safford M, Konemann S, et al.: Flow cytometric characterization of acute myeloid leukemia. Part II. Phenotypic heterogeneity at diagnosis. Leukemia 6:70-80, 1992
1767. Terstappen LW, Safford M, Unterhalt M, Konemann S, Zurlutter K, Piechotka K, Drescher M, Aul C, Buchner T, Hiddemann W, et al: Flow cytometric characterization of acute myeloid leukemia: IV. Comparison to the differentiation pathway of normal hematopoietic progenitor cells. Leukemia 6:993-1000, 1992
1768. Terstappen LW, Buescher S, Nguyen M, Reading C: Differentiation and maturation of growth factor expanded human hematopoietic progenitors assessed by multidimensional flow cytometry. Leukemia 6:1001-10, 1992
1769. Terstappen LW, Levin J: Bone marrow cell differential counts obtained by multidimensional flow cytometry. Blood Cells 18:311-30, 1992
1770. Reading CL, Estey EH, Huh YO, Claxton DF, Sanchez G, Terstappen LW, O'Brien MC, Baron S, Deisseroth AB: Expression of unusual immunophenotype combinations in acute myelogenous leukemia. Blood 81:3083-90, 1993
1771. Wormann B, Safford M, Konemann S, Buchner T, Hiddemann W, Terstappen LW: Detection of aberrant antigen expression in acute myeloid leukemia by multiparameter flow cytometry. Recent Results Cancer Res 131:185-96, 1993
1772. Cogliatti SB, Schmid U, Schumacher U, Eckert F, Hansmann M-L, Hedderich J, Takahashi H, Lennert K: Primary B-cell gastric lymphoma, a clinicopathologic study of 145 patients. Gastroenterology 101:1159-70, 1991
1773. Isaacson PG: Extranodal lymphomas: the MALT concept. Verh Dtsch Ges Pathol 76:14-23, 1992
1774. Wotherspoon AC, Pan L, Diss TC, Isaacson PG: A genotypic study of low grade B-cell lymphomas, including lymphomas of mucosa associated lymphoid tissue (MALT). J. Pathol. 1990; 162:135-40
1775. Wotherspoon AC, Pan LX, Diss TC, Isaacson PG: Cytogenetic study of B-cell lymphoma of mucosa-associated lymphoid tissue. Cancer Genet Cytogenet 58:35-8, 1992
1776. Wotherspoon AC, Doglioni C, Isaacson PG: Low-grade gastric B-cell lymphoma of mucosa-associated lymphoid tissue (MALT): a multifocal disease. Histopathology 20:29-34, 1992
1777. Wotherspoon AC, Ortiz-Hidalgo C, Falzon MR, Isaacson PG: *Helicobacter pylori*- associated gastritis and primary B-cell gastric lymphoma. Lancet. 1991; 338:1175-6.
1778. Hussell T, Isaacson PG, Crabtree JE, Spencer J: The response of cells from low-grade B-cell gastric lymphomas of mucosa-associated lymphoid tissue to *Helicobacter pylori* [see comments]. Lancet 342:571-4, 1993
1779. Hussell T, Isaacson PG, Crabtree JE, Dogan A, Spencer J: Immunoglobulin specificity of low grade B cell gastrointestinal lymphoma of mucosa-associated lymphoid tissue (MALT) type. Am J Pathol 142:285-92, 1993
1780. Hussell T, Isaacson PG, Spencer J: Proliferation and differentiation of tumour cells from B-cell lymphoma of mucosa-associated lymphoid tissue in vitro. J Pathol 169:221-7, 1993
1781. Doglioni C, Wotherspoon AC, Moschini A, de Boni M, Isaacson PG: High incidence of primary gastric lymphoma in northeastern Italy [see comments]. Lancet 339:834-5, 1992
1782. Wotherspoon AC, Doglioni C, Diss TC, Pan L, Moschini A, de Boni M, Isaacson PG: Regression of primary low-grade B-cell gastric lymphoma of mucosa-associated lymphoid tissue type after eradication of *Helicobacter pylori* [see comments]. Lancet 342:575-7, 1993
1783. Corash L, Rheinschmidt M, Lieu S, Meers P, Brew E: Enumeration of reticulocytes using fluorescence-activated flow cytometry. Pathol Immunopathol Res 7:381-94, 1988
1784. Carter JM, McSweeney PA, Wakem PJ, Nemet AM: Counting reticulocytes by flow cytometry: use of thiazole orange. Clin Lab Haematol 11:267-71, 1989
1785. Van Hove L, Goossens W, Van Duppen V, Verwilghen RL: Reticulocyte count using thiazole orange. A flow cytometry method. Clin Lab Haematol 12:287-99, 1990
1786. Hansson GK, Andersson M, Jarl H, Stemme S: Flow cytometric analysis of reticulocytes using an RNA-binding fluorochrome. Scand J Clin Lab Invest 52:35-41, 1992
1787. Schimenti KJ, Lacerna K, Wamble A, Maston L, Iaffaldano C, Straight M, Rabinovitch A, Lazarus HM, Jacobberger JW: Reticulocyte quantification by flow cytometry, image analysis, and manual counting. Cytometry 13:853-62, 1992
1788. Uemura T, Suzuki S, Ohnishi T: Flow cytometric enumeration of reticulocyte in the peripheral blood from canine infected with Babesia gibsoni. Zentralbl Veterinarmed [b] 37:468-72, 1990
1789. Abbott DL, McGrath JP: Evaluation of flow cytometric counting procedure for canine reticulocytes by use of thiazole orange. Am J Vet Res 52:723-7, 1991
1790. Reagan WJ, Vap LM, Weiser MG: Flow cytometric analysis of feline reticulocytes. Vet Pathol 29:503-8, 1992
1791. Van Petegem M, Cartuyvels R, de Schouwer P, van Duppen V, Goossens W, van Hove L: Comparative evaluation of three flow cytometers for reticulocyte enumeration. Clin Lab Haematol 15:103-11, 1993
1792. Van Bockstaele DR, Peetermans ME: 1,3'-Diethyl-4,2'-quinolylthiacyanine iodide as a "thiazole orange" analogue for nucleic acid staining. Cytometry 10:214-6, 1989
1793. Guasch R, Juan G, Carretero F, O'Connor JE: Flow-cytometric enumeration of reticulocytes with the new fluorochrome 1,3'-diethyl-4,2'-quinolylthiacyanine. Ann Hematol 65:184-7, 1992
1794. Oosterhuis WP, Zwinderman AH, Modderman TA, Dinkelaar RB, van der Helm HJ: Multivariate statistical modeling: alternative approach to test evaluation, applied to counting reticulocytes by flow cytometry. Clin Chem 38:1706-11, 1992
1795. Lofsness KG, Kohnke ML, Geier NA: Evaluation of automated reticulocyte counts and their reliability in the presence of Howell-Jolly bodies. Am J Clin Pathol 101:85-90, 1994
1796. Pappas AA, Owens RB, Flick JT: Reticulocyte counting by flow cytometry. A comparison with manual methods. Ann Clin Lab Sci 22:125-32, 1992
1797. Tatsumi N, Tsuda I, Kojima K, Niri M, Setoguchi K: An automated reticulocyte counting method: preliminary observations. Med Lab Sci 46:157-60, 1989
1798. Kojima K, Niri M, Setoguchi K, Tsuda I, Tatsumi N: An automated optoelectronic reticulocyte counter. Am J Clin Pathol 92:57-61, 1989
1799. Laharrague P, Corberand JX, Fillola G, Marcelino N: [Evaluation of an automatic analyzer of reticulocytes: the Sysmex R-1 000]. Ann Biol Clin 48:253-8, 1990
1800. Dalbak LG, Theodorsen L, Aune MW, Sandberg S: [Reticulocytes--new possibilities with automated counting]. Tidsskr Nor Laegeforen 113:709-12, 1993
1801. Bowen D, Bentley N, Hoy T, Cavill I: Comparison of a modified thiazole orange technique with a fully automated analyser for reticulocyte counting. J Clin Pathol 44:130-3, 1991
1802. Davis BH, Bigelow NC: Flow cytometric reticulocyte quantification using thiazole orange provides clinically useful reticulocyte maturity index. Arch Pathol Lab Med 113:684-9, 1989
1803. Davis BH, Bigelow NC: Clinical flow cytometric reticulocyte analysis. Pathobiology 58:99-106, 1990
1804. Davis BH, Bigelow NC: Flow cytometric reticulocyte analysis and the reticulocyte maturity index. Ann N Y Acad Sci 677:281-92 (in reference 1042)

1805. Davis BH, Di Corato M, Bigelow NC, Langweiler MH: Proposal for standardization of flow cytometric reticulocyte maturity index (RMI) measurements. Cytometry 14:318-26, 1993

1806. Davis BH, Bigelow N, Ball ED, Mills L, Cornwell G 3d: Utility of flow cytometric reticulocyte quantification as a predictor of engraftment in autologous bone marrow transplantation. Am J Hematol 32:81-7, 1989

1807. Sakairi K, Miyachi H, Tanaka Y, et al.: [Frequency of the development of RNA-rich reticulocytes in allogeneic bone marrow transplant recipients]. Rinsho Ketsueki 33:791-5, 1992

1808. Kuse R: The appearance of reticulocytes with medium or high RNA content is a sensitive indicator of beginning granulocyte recovery after aplasiogenic cytostatic drug therapy in patients with AML. Ann Hematol 66:213-4, 1993

1809. Mechetner EB, Sedmak DD, Barth RF: Heterogeneity of peripheral blood reticulocytes: a flow cytometric analysis with monoclonal antibody HAE9 and thiazole orange. Am J Hematol 38:61-3, 1991

1810. Bain BJ, Cavill IAJ: Hypochromic macrocytes: are they reticulocytes? J Clin Pathol 46:963-4, 1993

1811. Mohandas N, Kim YR, Tycko DH, Orlik J, Wyatt J, Groner W: Accurate and independent measurement of volume and hemoglobin concentration of individual red cells by laser light scattering. Blood 68:506-13, 1986

1812. Bruch JF, Metezeau P, Garcia-Fonknechten N, Richard Y, Tricottet V, Hsi BL, Kitzis A, Julien C, Papiernik E: Trophoblast-like cells sorted from peripheral maternal blood using flow cytometry: a multiparametric study involving transmission electron microscopy and fetal DNA amplification. Prenat Diagn 11:787-98, 1991

1813. Cacheux V, Milesi-Fluet C, Tachdjian G, Druart L, Bruch JF, Hsi BL, Uzan S, Nessmann C: Detection of 47,XYY trophoblast fetal cells in maternal blood by fluorescence in situ hybridization after using immunomagnetic lymphocyte depletion and flow cytometry sorting. Fetal Diagn Ther 7:190-4, 1992

1814. Bianchi DW, Stewart JE, Garber MF, Lucotte G, Flint AF: Possible effect of gestational age on the detection of fetal nucleated erythrocytes in maternal blood. Prenat Diagn 11:523-8, 1991

1815. Price JO, Elias S, Wachtel SS, Klinger K, Dockter M, Tharapel A, Shulman LP, Phillips OP, Meyers CM, Shook D et al: Prenatal diagnosis with fetal cells isolated from maternal blood by multiparameter flow cytometry. Am J Obstet Gynecol 165:1731-7, 1991

1816. Wachtel S, Elias S, Price J, Wachtel G, Phillips O, Shulman L, Meyers C, Simpson JL, Dockter M: Fetal cells in the maternal circulation: isolation by multiparameter flow cytometry and confirmation by polymerase chain reaction. Hum Reprod 6:1466-9, 1991

1817. Elias S, Price J, Dockter M, Wachtel S, Tharapel A, Simpson JL, Klinger KW: First trimester prenatal diagnosis of trisomy 21 in fetal cells from maternal blood. Lancet 340:1033, 1992

1818. Bianchi DW, Mahr A, Zickwolf GK, Houseal TW, Flint AF, Klinger KW: Detection of fetal cells with 47,XY,+21 karyotype in maternal peripheral blood. Hum Genet 90:368-70, 1992

1819. Ambrose WP, Goodwin PM, Martin JC, Keller RA: Alterations of single molecule fluorescence lifetimes in near-field optical microscopy. Science 265:364-7, 1994

1820. McHugh TM, Stites DP, Casavant CH, Fulwyler MJ: Flow cytometric determination and quantitation of immune complexes using human C1q-coated microspheres. J Immunol Methods 95:57-61, 1986

1821. McHugh TM, Miner RC, Logan LH, Stites DP: Simultaneous detection of antibodies to cytomegalovirus and herpes simplex virus by using flow cytometry and a microsphere-based fluorescence immunoassay. J Clin Microbiol 26:1957-61, 1988

1822. McHugh TM, Wang YJ, Chong HO, Blackwood LL, Stites DP: Development of a microsphere-based fluorescent immunoassay and its comparison to an enzyme immunoassay for the detection of antibodies to three antigen preparations from *Candida albicans.* J Immunol Methods 116:213-9, 1989

1823. Presani G, Perticarari S, Mangiarotti MA: Flow cytometric detection of anti-gliadin antibodies. J Immunol Methods 119:197-202, 1989

1824. Scillian JJ, McHugh TM, Busch MP, Tam M, Fulwyler MJ, Chien DY, Vyas GN: Early detection of antibodies against rDNA-produced HIV proteins with a flow cytometric assay. Blood 73:2041-8, 1989

1825. Elkhalifa MY, Kiechle FL, Gordon SC, Chen J, Poulik MD: A flow cytometric method to detect anti-pyruvate dehydrogenase antibody in primary biliary cirrhosis. Am J Clin Pathol 97:202-8, 1992

1826. Best LM, Veldhuyzen van Zanten SJ, Bezanson GS, Haldane DJ, Malatjalian DA: Serological detection of *Helicobacter pylori* by a flow microsphere immunofluorescence assay. J Clin Microbiol 30:2311-7, 1992

1827. Lisi PJ, Huang CW, Hoffman RA, Teipel JW: A fluorescent immunoassay for soluble antigens employing flow cytometric detection. Clin Chim Acta 120:171-9, 1982

1828. Kim KY, Han MY, Yoon DY, Cho BY, Choi MJ, Choe IS, Chung TW: Solid-phase immunoassay using a flow cytometer: quantitative and qualitative determination of protein antigens and a hapten. Immunol Lett 31:267-72, 1992

1829. Labus JM, Petersen BH: Quantitation of human anti-mouse antibody in serum by flow cytometry. Cytometry 13:275-81, 1992

1830. Frengen J, Schmid R, Kierulf B, Nustad K, Paus E, Berge A, Lindmo T: Homogeneous immunofluorometric assays of alpha-fetoprotein with macroporous, monosized particles and flow cytometry. Clin Chem 39:2174-81, 1993

1831. Renner ED: Development and clinical evaluation of an amplified flow cytometric fluoroimmunoassay for *Clostridium difficile* toxin A. Cytometry (Commun Clin Cytometry) 18:103-8, 1994

1832. Lindmo T, Bormer O, Ugelstad J, Nustad K: Immunometric assay by flow cytometry using mixtures of two particle types of different affinity. J Immunol Methods 126:183-9, 1990

1833. Frengen J, Kierulf B, Schmid R, Lindmo T, Nustad K: Demonstration and minimization of serum interference in flow cytometric two-site immunoassays. Clin Chem 40:420-5, 1994

1834. Saunders GC, Martin JC, Jett JH, Perkins A: Flow cytometric competitive binding assay for determination of actinomycin-D concentrations. Cytometry 11:311-3, 1990

1835. Sykulev YK, Cohen RJ, Eisen HN: Particle counting by flow cytometry can determine intrinsic equilibrium constants for antibody-ligand interactions in solution. Mol Immunol 30:101-4, 1993

1836. Sykulev YK, Sherman DA, Cohen RJ, Eisen HN: Quantitation of reversible binding by particle counting: hapten-antibody interaction as a model system. Proc Natl Acad Sci U S A 89:4703-7, 1992

1837. Weaver JC, Seissler PE, Threefoot SA, Lorenz JW, Huie T, Rodrigues R, Klibanov AM: Microbiological measurements by immobilization of cells within small volume elements. Ann N Y Acad Sci 434:363-72, 1984

1838. Williams GB, Threefoot SA, Lorenz JW, Bliss JG, Weaver JC, Demain AL, Klibanov AM: Rapid detection of *E. coli* immobilized in gel microdroplets. Ann N Y Acad Sci 501:350-3, 1986

1839. Weaver JC, Williams GB, Klibanov A, Demain AL: Gel microdroplets: rapid detection and enumeration of individual microorganisms by their metabolic activity. Bio/Technology 6:1084-9, 1988

1840. Powell KT, Weaver JC: Gel microdroplets and flow cytometry: rapid determination of antibody secretion by individual cells within a cell population. Bio/Technology 8:333-7, 1990

1841. Weaver JC, Bliss JG, Harrison GI, Powell KT, Williams GB: Microdrop technology: A general method for separating cells by function and composition. Methods 2:234-47, 1991

1842. Weaver JC, Bliss JG, Powell KT, Harrison GI, Williams GB: Rapid clonal growth measurements at the single-cell level: gel microdroplets and flow cytometry. Bio/Technology 9:873-6, 1991

1843. Goguen B, Kedersha N: Product Review: Clonogenic cytotoxicity testing by microdrop encapsulation. Nature 363:189-90, 1993

1844. Rosenbluh A, Nir R, Sahar E, Rosenberg E: Cell density dependent lysis and sporulation of *Myxococcus xanthus* in agarose microbeads. J Bact 171:4923-9, 1989
1845. Nir R, Lamed R, Gueta L, Sahar E: Single-cell entrapment and microcolony development within uniform microspheres amenable to flow cytometry. Appl Environ Microbiol 56:2870-5, 1990
1846. Nir R, Yisraeli Y, Lamed R, Sahar E: Flow cytometry sorting of viable bacteria and yeasts according to beta-galactosidase activity. Appl Environ Microbiol 56:3861-6, 1990
1847. Nir R, Lamed R, Sahar E, Shabtai Y: Flow cytometric isolation of growth rate mutants: A yeast model. J Microbiol Methods 14:247-56, 1992
1848. Sahar E, Nir R, Lamed R: Flow cytometric analysis of entire microbial colonies. Cytometry 15:213-21, 1994
1849. Bloch DB, Smith BR, Ault KA: Cells on microspheres: a new technique for flow cytometric analysis of adherent cells. Cytometry 3:449-52, 1983
1850. Gong J, Traganos F, Darzynkiewicz Z: Expression of cyclins B and E in individual MOLT-4 cells and in stimulated human lymphocytes during their progression through the cell cycle. Intl J Oncology 3:1037-42, 1993
1851. Gong J, Li X, Traganos F, Darzynkiewicz Z: Expression of G_1 and G_2 cyclins measured in individual cells by multiparameter flow cytometry: a new tool in the analysis of the cell cycle. Cell Prolif 27:357-71, 1994
1852. Holland JM, Wright WD, Higashikubo R, Roti Roti JL: Effects of irradiation on nuclear protein synthesis in G_2 phase of the cell cycle. Radiation Res 122:197-208, 1990
1853. Gerdes J, Schwab U, Lemke H, Stein H: Production of a monoclonal antibody reactive with a human nuclear antigen associated with cell proliferation. Int J Cancer 31:13-20, 1983
1854. Gerdes J, Lemke H, Baisch H, Wacker H, Schwab U, Stein H: Cell cycle analysis of cell proliferation-associated human nuclear antigen defined by the monoclonal antibody Ki-67. J Immunol 133:1710-5, 1984
1855. Celis JE, Bravo R, Larsen PM, Fey S: Cyclin: A nuclear protein whose level correlates directly with the proliferating state of normal as well as transformed cells. Leukemia Res 8:143-57, 1984
1856. Bravo R, Frank R, Blundell PA, MacDonald-Bravo H: Cyclin/PCNA is the auxiliary protein of DNA polymerase-delta. Nature 326:515-7, 1987
1857. Bolton WE, Mikulka WR, Healy CG, Schmittling RJ, Kenyon NS: Expression of proliferation associated antigens in the cell cycle of synchronized mammalian cells. Cytometry 13:117-26, 1992
1858. Landberg G, Roos G: Flow cytometric analysis of proliferation associated nuclear antigens using washless staining of unfixed cells. Cytometry 13:230-40, 1992
1859. Teague K, El-Naggar A: Comparative flow cytometric analysis of proliferating cell nuclear antigen (PCNA) antibodies in human solid neoplasms. Cytometry 15:21-7, 1994
1860. Lehman JM, Laffin J, Jacobberger JW, Fogleman D: Analysis of simian virus 40 infection of CV-1 cells by quantitative two-color fluorescence with flow cytometry. Cytometry 9:52-9, 1988
1861. Sladek TL, Jacobberger JW: Simian virus 40 large T-antigen expression decreases the G1 and increases the G2 + M cell cycle phase durations in exponentially growing cells. J Virol 66:1059-65, 1992
1862. Sladek TL, Jacobberger JW: Dependence of SV40 large T-antigen cell cycle regulation on T-antigen expression levels. Oncogene 7:1305-13, 1992
1863. Sinnett D, Flint A, Lalande M: Determination of DNA replication kinetics in synchronized human cells using a PCR-based assay. Nucleic Acids Res 21:3227-32, 1993
1864. Krasnow MA, Cumberledge S, Manning G, Herzenberg LA, Nolan GP: Whole animal cell sorting of *Drosophila* embryos. Science 251:81-5, 1991
1865. Cumberledge S, Krasnow MA: Intercellular signalling in *Drosophila* segment formation reconstructed in vitro. Nature 363:549-52, 1993
1866. Trask BJ, van den Engh G, Christensen M, Massa HF, Gray JW, Van Dilla M: Characterization of somatic cell hybrids by bivariate flow karyotyping and fluorescence in situ hybridization. Somat Cell Mol Genet 17:117-36, 1991
1867. Bouvet A, Konfortov BA, Miller NG, Brown D, Tucker EM: Identification of pig chromosomes in pig-mouse somatic cell hybrid bivariate flow karyotypes. Cytometry 14:369-76, 1993
1868. Cram LS: Flow cytogenetics and chromosome sorting. Hum Cell 3:99-106, 1990
1869. Green DK: Analysing and sorting human chromosomes. J Microsc 159:237-44, 1990
1870. Métézeau P, Schmitz A, Frelat G: Analysis and sorting of chromosomes by flow cytometry: new trends. Biol Cell 78:31-9, 1993
1871. Kuriki H, Sonta S, Murata K: Flow karyotype analysis and sorting of the Chinese hamster chromosomes: comparing the effects of the isolation buffers. J Clin Lab Anal 7:119-22, 1993
1872. Telenius H, de Vos D, Blennow E, Willat LR, Ponder BA, Carter NP: Chromatid contamination can impair the purity of flow-sorted metaphase chromosomes. Cytometry 14:97-101, 1993
1873. Arkesteijn GJ, van Dekken H, Martens AC, Hagenbeek A: Clinical applications of flow karyotyping in myelocytic leukemia by stimulation of different subpopulations of cells in blood or bone marrow samples. Cytometry 11:196-201, 1990
1874. Barclay AN, Birkeland ML, Brown MH, Beyers AD, Davis SJ, Somoza C, Williams AF: *The Leucocyte Antigen Facts Book.* London, Academic Press, 1993, x + 424 pp.
1875. Pigott R, Power C: *The Adhesion Molecule Facts Book.* London, Academic Press, 1993, viii + 190 pp.
1876. National Committee for Clinical Laboratory Standards: *Clinical Applications of Flow Cytometry: Quality Assurance and Immunophenotyping of Peripheral Blood Lymphocytes.* NCCLS Publication H24-T. Villanova (PA), NCCLS, 1992
1877. *Guidelines for the Performance of CD4+ T-Cell Determinations in Persons with Human Immunodefi-ciency Virus Infection.* Morbidity and Mortality Weekly Report, Vol. 41, No. RR-8, Atlanta, U. S. Public Health Service, Centers for Disease Control, 1992
1878. Nicholson JKA: Immunophenotyping specimens from HIV-infected persons: laboratory guidelines from the Centers for Disease Control and Prevention. Cytometry (Commun Clin Cytometry) 18:55-9, 1994
1879. Calvelli T, Denny TN, Paxton H, Gelman R, Kagan J: Guideline for flow cytometric immunophenotyping: a report from the National Institute of Allergy and Infectious Diseases, Division of AIDS. Cytometry 14:702-15, 1993
1880. Garovoy MR, Rheinschmidt M, Bigos M, Perkins H, Colombe B, Feduska N, Salvatierra O Jr: Flow cytometry analysis: A high technology crossmatch technique facilitating transplantation. Transplant Proc 15:1939-43, 1983
1881. Chapman JR, Deierhoi MH, Carter NP, Ting A, Morris PJ: Analysis of flow cytometry and cytotoxicity crossmatches in renal transplantation. Transplant Proc 17:2480-1, 1985
1882. Cook DJ, Terasaki PI, Iwaki Y, Terashita G, Fujikawa J, Gera J, Takeda A, Danovitch G, Rosenthal JT, Fine R et al: Flow cytometry crossmatching for kidney transplantation. Clin Transpl 1:375-80, 1988
1883. Talbot D, Givan AL, Shenton BK, Stratton A, Proud G, Taylor RM: The relevance of a more sensitive crossmatch assay to renal transplantation. Transplantation 47:552-5, 1989
1884. Kerman RH, Van Buren CT, Lewis RM, De Vera V, Baghdahsarian V, Gerolami K, Kahan BD: Improved graft survival for flow cytometry and antihuman globulin crossmatch-negative retransplant recipients. Transplantation 49:52-6, 1990
1885. Mahoney RJ, Ault KA, Given SR, Adams RJ, Breggia AC, Paris PA, Palomaki GE, Hitchcox SA, White BW, Himmelfarb J, Leeber DA: The flow cytometric crossmatch and early renal transplant loss.
1886. Bou-Habib JC, Krams S, Colombe BW, Lou C, Bubar OT, Yousif B, Amend WJ, Salvatierra O Jr, Melzer J, Garovoy MR: Impaired

kidney graft survival in flow cytometric crossmatched positive donor-specific transfusion recipients. Transplant Proc 23:403-4, 1991

1887. Ogura K, Terasaki PI, Johnson C, Mendez R, Rosenthal JT, Ettenger R, Martin DC, Dainko E, Cohen L, Mackett T, Berne T, Barba L, Lieberman E: The significance of a positive flow cytometry crossmatch test in primary kidney transplantation. Transplantation 56:294-8, 1993
1888. Bray RA, Lebeck LK, Gebel HM: The flow cytometric crossmatch. Dual-color analysis of T cell and B cell reactivities. Transplantation 48:834-40, 1989
1889. Wang GX, Terashita GY, Terasaki PI: Platelet crossmatching for kidney transplants by flow cytometry. Transplantation 48:959-61, 1989
1890. Wetzsteon PJ, Head MA, Fletcher LM, Lye WC, Norman DJ: Cytotoxic flow-cytometric crossmatches (flow-tox): a comparison with conventional cytotoxicity crossmatch techniques. Hum Immunol 35:93-9, 1992
1891. Talbot D, Cavanagh G, Coates E, Givan AL, Shenton BK, Lennard TW, Proud G, Taylor RM: Improved graft outcome and reduced complications due to flow cytometric crossmatching and DR matching in renal transplantation. Transplantation 53:925-8, 1992
1892. Lazda VA: Identification of patients at risk for inferior renal allograft outcome by a strongly positive B cell flow cytometry crossmatch. Transplantation 57:964-9, 1994
1893. Scornik JC, Brunson ME, Schaub B, Howard RJ, Pfaff WW: The crossmatch in renal transplantation. Evaluation of flow cytometry as a replacement for standard cytotoxicity. Transplantation 57:621-5, 1994
1894. Berteli AJ, Daniel V, Terness P, Opelz G: A new method for determining anti-B cell antibodies and their specificity using flow cytometry. J Immunol Methods 164:21-5, 1993
1895. Tötterman TH, Hanås E, Bergström R, Larsson E, Tufveson G: Immunologic diagnosis of kidney rejection using FACS analysis of graft-infiltrating functional and activated T and NK cell subsets. Transplantation 47:817-23, 1989
1896. Rabinovitch PS, June CH, Kavanagh TJ: Measurements of cell physiology: Ionized calcium, pH, and glutathione. In reference 1038, pp. 505-34
1897. June CH, Linette GP, Pierce PF, Jin N-R, Lum LG: Potential clinical applications of signal transduction measurements in marrow transplantation and HIV-1 infection. Ann N Y Acad Sci 677:225-32, 1993 (in reference 1042)
1898. Haynes BF, Hemler ME, Mann DL, Eisenbarth GS, Shelhamer J, Mostowski HS, Thomas CA, Strominger JL, Fauci AS: Characterization of a monoclonal antibody (4F2) that binds to human monocytes and to a subset of activated lymphocytes. J Immunol 126:1409-14, 1981
1899. Biselli R, Matricardi PM, D'Amelio R, Fattorossi A: Multiparametric flow cytometric analysis of the kinetics of surface molecule expression after polyclonal activation of human peripheral blood T lymphocytes. Scand J Immunol 35:439-47, 1992
1900. Akbar AN, Terry L, Timms A, Beverley PC, Janossy G: Loss of CD45R and gain of UCHL1 reactivity is a feature of primed T cells. J Immunol 140:2171-8, 1988
1901. Akbar AN, Timms A, Janossy G: Cellular events during memory T-cell activation in vitro: the UCHL1 (180,000 MW) determinant is newly synthesized after mitosis. Immunology 66:213-8, 1989
1902. Merkenschlager M, Beverley PC: Evidence for differential expression of CD45 isoforms by precursors for memory-dependent and independent cytotoxic responses: human CD8 memory CTLp selectively express CD45RO (UCHL1). Int Immunol 1:450-9, 1989
1903. Kubbies M: Alteration of cell cycle kinetics by reducing agents in human peripheral blood lymphocytes from adult and senescent donors. Cell Prolif 25:157-66, 1992
1904. Schroff RW, Gottlieb MS, Prince HE, Chai LL, Fahey JL: Immunological studies of homosexual men with immunodeficiency and Kaposi's sarcoma. Clin Immunol Immunopathol 27:300-14, 1983
1905. Cohen J: AIDS research shifts to immunity. Science 257:152-3, 1992
1906. Hermann GG, Geertsen PF, von der Maase H, Zeuthen J: Interleukin-2 dose, blood monocyte and CD25+ lymphocyte counts as predictors of clinical response to interleukin-2 therapy in patients with renal cell carcinoma. Cancer Immunol Immunother 34:111-8, 1991
1907. Hermann GG, Geertsen PF, von der Maase H, Steven K, Andersen C, Hald T, Zeuthen J: Recombinant interleukin-2 and lymphokine-activated killer cell treatment of advanced bladder cancer: clinical results and immunological effects. Cancer Res 52:726-32, 1992
1908. Arvin AM, Sharp M, Smith S, Koropchak CM, Diaz PS, Kinchington P, Ruyechan W, Hay J: Equivalent recognition of a varicella-zoster virus immediate early protein (IE62) and glycoprotein I by cytotoxic T lymphocytes of either CD4+ or CD8+ phenotype. J Immunol 146:257-86, 1991
1909. Waag DM, Galloway A, Sandstrom G, Bolt CR, England MJ, Nelson GO, Williams JC: Cell-mediated and humoral immune responses induced by scarification vaccination of human volunteers with a new lot of the live vaccine strain of *Francisella tularensis*. J Clin Microbiol 30:2256-63, 1992
1910. Grinstein S, Dixon SJ: Ion transport, membrane potential, and cytoplasmic pH in lymphocytes: Changes during activation. Physiol Revs 69:417-81, 1989
1911. Grinstein S; Smith JD: Ca2+ induces charybdotoxin-sensitive membrane potential changes in rat lymphocytes. Am J Physiol 257:pC197-206, 1989
1912. Gelfand EW, Cheung RK, Grinstein S: Calcium-dependent intracellular acidification dominates the pH response to mitogen in human T cells. J Immunol 140:246-52, 1988
1913. Grinstein S, Smith JD, Rowatt C, Dixon SJ: Mechanism of activation of lymphocyte Na+/H+ exchange by concanavalin A. A calcium- and protein kinase C-independent pathway. J Biol Chem 262:15277-84, 1987
1914. Gelfand EW, Cheung RK, Mills GB, Grinstein S: Uptake of extracellular Ca2+ and not recruitment from internal stores is essential for T lymphocyte proliferation. Eur J Immunol 18:917-22, 1988
1915. Roifman CM, Mills GB, Stewart D, Cheung RK, Grinstein S, Gelfand EW: Response of human B cells to different anti-immunoglobulin isotypes: absence of a correlation between early activation events and cell proliferation. Eur J Immunol 17:1737-42, 1987
1916. Wells RG, Lee WS, Kanai Y, Leiden JM, Hediger MA: The 4F2 antigen heavy chain induces uptake of neutral and dibasic amino acids in *Xenopus* oocytes. J Biol Chem 265:15285-8, 1992
1917. Freidman AW, Diaz LA Jr, Moore S, Schaller J, Fox DA: The human 4F2 antigen: evidence for cryptic and noncryptic epitopes and for a role of 4F2 in human T lymphocyte activation. Cell Immunol 154:253-63, 1994
1918. Lopez-Cabrera M, Santis AG, Fernandez-Ruiz E, Blacher R, Esch F, Sanchez-Mateos P, Sanchez-Madrid F: Molecular cloning, expression, and chromosomal localization of the human earliest lymphocyte activation antigen AIM/CD69, a new member of the C-type animal lectin superfamily of signal-transmitting receptors. J Exp Med 178:537-47, 1993
1919. Santamaria M, Marubayashi M, Arizon JM, Montero A, Concha M, Valles F, Lopez A, Lopez F, Pena J: The activation antigen CD69 is selectively expressed on CD8+ endomyocardium infiltrating T lymphocytes in human rejecting heart allografts. Hum Immunol 33:1-4, 1992
1920. Linette GP, Hartzman RJ, Ledbetter JA, June CH: HIV-1 infected T cells show a selective signaling defect after perturbation of CD3 antigen/receptor. Science 241:573-6, 1988
1921. van Leeuwen L, Eggels PH, Bullen JA: A short evaluation of a new haematological cell counter--the Cell-Dyn 3000--following a modified tentative NCCLS-procedure. Eur J Clin Chem Clin Biochem 29:105-10, 1991

1922. Jouault H, Imbert M, Mary JY, Ade P, Herpin J, Sultan C: [Leukocyte differential of the Coulter VCS. Evaluation and modalities of use in comparison with the Coulter STKR and manual count]. Ann Biol Clin 48:247-52, 1990
1923. Picard F, Terroux N, Levy JP: Use of Coulter VCS for differential leukocyte counts. Nouv Rev Fr Hematol 32:211-6, 1990
1924. Robertson EP, Lai HW, Wei DCC: An evaluation of leucocyte analysis on the Coulter STKS. Clin Lab Haemat 14:53-68, 1992
1925. Picard F, Guesnu M, Levy JP, Flandrin G: Use of the new Coulter MAXM for leucocyte differentials. Nouv Rev Fr Hematol 34:309-14, 1992
1926. Nelson L, Charache S, Wingfield S, Keyser E: Laboratory evaluation of differential white blood cell count information from the Coulter S-plus IV and Technicon H-1 in patient populations requiring rapid "turnaround" time. Am J Clin Pathol 91:563-9, 1989
1927. van Wersch JW, Bank C: A new development in haematological cell counting: the Sysmex NE-8000, automaton for cell count and physical five-part leukocyte differentiation. J Clin Chem Clin Biochem 28:233-40, 1990
1929. Swaim WR: Laboratory and clinical evaluation of white blood cell differential counters. Comparison of the Coulter VCS, Technicon H-1, and 800-cell manual method. AM J Clin Pathol 95:381-8, 1991
1929. Drayson RA, Hamilton MS, England JM: A comparison of differential white cell counting on the Coulter VCS and the Technicon H1 using simple and multiple regression analysis. Clin Lab Haematol 14:293-305, 1992
1930. Buttarello M, Gadotti M, Lorenz C, Toffalori E, Ceschini N, Valentini A, Rizzotti P: Evaluation of four automated hematology analyzers. A comparative study of differential counts (imprecision and inaccuracy). AM J Clin Pathol 97:345-52, 1992
1931. Bentley SA, Johnson A, Bishop CA: A parallel evaluation of four automated hematology analyzers. Am J Clin Pathol 100:626-32, 1993
1932. Dirscherl P, Grabner A, Buschmann H: Responsiveness of basophil granulocytes of horses suffering from chronic obstructive pulmonary disease to various allergens. Vet Immunol Immunopathol 38:217-27, 1993
1933. Du Buske LM: Introduction: basophil histamine release and the diagnosis of food allergy. Allergy Proc 14:243-9, 1993
1934. Milson TJ, Patrick CW, Sohnle PG, Patrick LC, Keller RH: Flow cytochemical analysis of atopic reactions. Diagn Immunol 3:182-6, 1985
1935. Nilsson TA: Allergy testing using degranulation of basophils and flow cytometry. Eur J Haematol Suppl 53:50-3, 1990
1936. Corash L: Measurement of platelet activation by fluorescence-activated flow cytometry. Blood Cells 16:97-106, 1990
1937. Kienast J, Schmitz G: Flow cytometric analysis of thiazole orange uptake by platelets: a diagnostic aid in the evaluation of thrombocytopenic disorders. Blood 75:116-21, 1990
1938. Ault KA, Rinder HM, Mitchell J, Carmody MB, Vary CP, Hillman RS: The significance of platelets with increased RNA content (reticulated platelets). A measure of the rate of thrombopoiesis. Am J Clin Pathol 98:637-46, 1992
1939. Ault KA: Flow cytometric measurement of platelet function and reticulated platelets. Ann N Y Acad Sci 677:293-308, 1993 (in reference 1042)
1940. Kuter DJ, Greenberg SM, Rosenberg RD: Analysis of megakaryocyte ploidy in rat bone marrow cultures. Blood 74:1952-62, 1989
1941. Kuter DJ, Rosenberg RD: Regulation of megakaryocyte ploidy in vivo in the rat. Blood 75:74-81, 1990
1942. Kuter DJ, Beeler D, Rosenberg RD: The purification of megapoietin: a physiological regulator of megakaryocyte growth and platelet production. Proc Natl Acad Sci USA 91:11104-8., 1994
1943. Andrews RG, Singer JW, Bernstein ID: Precursors of colony-forming cells in humans can be distinguished from colony-forming cells by expression of the CD33 and CD34 antigens and light scatter properties. J Exp Med 169:1721-31, 1989
1944. Siena S, Bregni M, Brando B, Ravagnani F, Bonadonna G, Gianni AM: Circulation of CD34+ hematopoietic stem cells in the peripheral blood of high-dose cyclophosphamide-treated patients: enhancement by intravenous recombinant human granulocyte-macrophage colony-stimulating factor. Blood 74:1905-14, 1989
1945. Ema H, Suda T, Miura Y, Nakauchi H: Colony formation of clone-sorted human hematopoietic progenitors. Blood 75:1941-6, 1990
1946. Andrews RG, Singer JW, Bernstein ID: Human hematopoietic precursors in long-term culture: single CD34+ cells that lack detectable T cell, B cell, and myeloid cell antigens produce multiple colony-forming cells when cultured with marrow stromal cells. J Exp Med 172:355-8, 1990
1947. Lansdorp PM, Sutherland HJ, Eaves CJ: Selective expression of CD45 isoforms on functional subpopulations of CD34+ hemopoietic cells from human bone marrow. J Exp Med 172:363-6, 1990
1948. Bender JG, Unverzagt KL, Walker DE, Lee W, Van Epps DE, Smith DH, Stewart CC, To LB: Identification and comparison of CD34-positive cells and their subpopulations from normal peripheral blood and bone marrow using multicolor flow cytometry. Blood 77:2591-6, 1991
1949. Chaudhary PM, Roninson IB: Expression and activity of P-glycoprotein, a multidrug efflux pump, in human hematopoietic stem cells. Cell 66:85-94, 1991
1950. Gore SD, Kastan MB, Civin CI: Normal human bone marrow precursors that express terminal deoxynucleotidyl transferase include T-cell precursors and possible lymphoid stem cells. Blood 77:1681-90, 1991
1951. Ravagnani F, Siena S, Bregni M, Brando B, Belli N, Lansdorp PM, Notti P, Pellegris G, Gianni AM: Methodologies to estimate circulating hematopoietic progenitors for autologous transplantation in cancer patients. Haematologica 76:46-9, 1991
1952. Srour EF, Brandt JE, Briddell RA, Leemhuis T, van Besien K, Hoffman R: Human CD34+ HLA-DR- bone marrow cells contain progenitor cells capable of self-renewal, multilineage differentiation, and long-term in vitro hematopoiesis. Blood Cells 17:287-95, 1991
1953. Briddell RA, Broudy VC, Bruno E, Brandt JE, Srour EF, Hoffman R: Further phenotypic characterization and isolation of human hematopoietic progenitor cells using a monoclonal antibody to the c-kit receptor. Blood 79:3159-67, 1992
1954. Baum CM, Weissman IL, Tsukamoto AS, Buckle AM, Peault B: Isolation of a candidate human hematopoietic stem-cell population. Proc Natl Acad Sci U S A 89:2804-8, 1992
1955. Lansdorp PM, Dragowska W: Long-term erythropoiesis from constant numbers of CD34+ cells in serum-free cultures initiated with highly purified progenitor cells from human bone marrow. J Exp Med 175:1501-9, 1992
1956. Lebkowski JS, Schain LR, Okrongly D, Levinsky R, Harvey MJ, Okarma TB: Rapid isolation of human CD34 hematopoietic stem cells--purging of human tumor cells. Transplantation 53:1011-9, 1992
1957. Huang S, Terstappen LWMM: Formation of haematopoietic microenvironment and haematopoietic stem cells from single human bone marrow stem cells. Nature 360:745-9, 1992
1958. Huang S, Terstappen LWMM: Formation of haematopoietic microenvironment and haematopoietic stem cells from single human bone marrow stem cells (correction). Nature 368:664, 1994
1959. Lansdorp PM, Schmitt C, Sutherland HJ, Craig WH, Dragowska W, Thomas TE, Eaves CJ: Hemopoietic stem cell characterization. Prog Clin Biol Res 377:475-84, 1992
1960. Lansdorp PM, Dragowska W: Maintenance of hematopoiesis in serum-free bone marrow cultures involves sequential recruitment of quiescent progenitors. Exp Hematol 21:1321-7, 1993
1961. Huang S, Terstappen LWMM: Lymphoid and myeloid differentiation of single human CD34+, HLA-DR+, CD38- hematopoietic stem cells. Blood 83:1515-26, 1994

1962. Rinder HM, Murphy M, Mitchell JG, Stocks J, Ault KA, Hillman RS: Progressive platelet activation with storage: evidence for shortened survival of activated platelets after transfusion. Transfusion 31:409-14, 1991
1963. Long KE, Yomtovian R, Kida M, Knez JJ, Medof ME: Time-dependent loss of surface complement regulatory activity during storage of donor blood [see comments]. Transfusion 33:294-300, 1993
1964. O'Brien SJ, Nash WG, Wildt DE, Bush ME, Benveniste RE: A molecular solution to the riddle of the giant panda's phylogeny. Nature 317:140-4, 1985
1965. Wang FI, Williams TJ, el-Awar FY, Pang VF, Hahn EC: Characterization of porcine peripheral blood leukocytes by light-scattering flow cytometry. Can J Vet Res 51:421-7, 1987
1966. Jain NC, Paape MJ, Miller RH: Use of flow cytometry for determination of differential leukocyte counts in bovine blood. Am J Vet Res 52:630-6, 1991
1967. Métézeau P, Cotinot C, Colas G, Azoulay M, Kiefer H, Goldberg ME, Kirszenbaum M: Improvement of flow cytometry analysis and sorting of bull spermatozoa by optical monitoring of cell orientation as evaluated by DNA specific probing. Mol Reprod Dev 30:250-7, 1991
1968. Johnson LA, Flook JP, Look MV: Flow cytometry of X and Y chromosome-bearing sperm for DNA using an improved preparation method and staining with Hoechst 33342. Gamete Res 17:203-12, 1987
1969. Johnson LA, Flook JP, Look MV, Pinkel D: Flow sorting of X and Y chromosome-bearing spermatozoa into two populations. Gamete Res 16:1-9, 1987
1970. Johnson LA, Clarke RN: Flow sorting of X and Y chromosome-bearing mammalian sperm: activation and pronuclear development of sorted bull, boar, and ram sperm microinjected into hamster oocytes. Gamete Res 21:335-43, 1988
1971. Gledhill BL: Selection and separation of X- and Y- chromosome-bearing mammalian sperm. Gamete Res 20:377-95, 1988
1972. Morrell JM, Keeler KD, Noakes DE, Mackenzie NM, Dresser DW: Sexing of sperm by flow cytometry. Vet Rec 122:322-4, 1988
1973. Johnson LA, Flook JP, Hawk HW: Sex preselection in rabbits: live births from X and Y sperm separated by DNA and cell sorting. Biol Reprod 41:199-203, 1989
1974. Ali JI, Eldridge FE, Koo GC, Schanbacher BD: Enrichment of bovine X- and Y-chromosome-bearing sperm with monoclonal H-Y antibody-fluorescence-activated cell sorter. Arch Androl 24:235-45, 1990
1975. Otto FJ, Hettwer H, Hofmann N: [Differentiation of human sperm cells using flow cytometry]. Urologe [a] 29:46-8, 1990
1976. Otto FJ, Hettwer H: Flow cytometric discrimination of human semen cells. Cell Mol Biol 36:225-32, 1990
1977. Cran DG, Johnson LA, Miller NG, Cochrane D, Polge C: Production of bovine calves following separation of X- and Y-chromosome bearing sperm and in vitro fertilisation. Vet Rec 132:40-1, 1993
1978. Hendriksen PJ, Tieman M, Van der Lende T, Johnson LA: Binding of anti-H-Y monoclonal antibodies to separated X and Y chromosome-bearing porcine and bovine sperm. Mol Reprod Dev 35:189-96, 1993
1979. Johnson LA, Welch GR, Keyvanfar K, Dorfmann A, Fugger EF, Schulman JD: Gender preselection in humans? Flow cytometric separation of X and Y spermatozoa for the prevention of X-linked diseases. Hum Reprod 8:1733-9, 1993
1980. Morrell JM, Dresser DW: Offspring from inseminations with mammalian sperm stained with Hoechst 33342, either with or without flow cytometry. Mutat Res 224:177-83, 1989
1981. Upreti GC, Riches PC, Johnson LA: Attempted sexing of bovine spermatozoa by fractionation on a Percoll density gradient. Gamete Res 20:83-92, 1988
1982. Karabinus DS, Evenson DP, Jost LK, Baer RK, Kaproth MT: Comparison of semen quality in young and mature Holstein bulls measured by light microscopy and flow cytometry. J Dairy Sci 73:2364-71, 1990
1983. Auger J, Ronot X, Dadoune JP: Human sperm mitochondrial function related to motility: a flow and image cytometric assessment. J Androl 10:439-48, 1989
1984. Auger J, Leonce S, Jouannet P, Ronot X: Flow cytometric sorting of living, highly motile human spermatozoa based on evaluation of their mitochondrial activity. J Histochem Cytochem 41:1247-51, 1993
1985. Ericsson SA, Garner DL, Redelman D, Ahmad K: Assessment of the viability and fertilizing potential of cryopreserved bovine spermatozoa using dual fluorescent staining and two-flow cytometric systems. Gamete Res 22:355-68, 1989
1986. Kramer RY, Garner DL, Bruns ES, Ericsson SA, Prins GS: Comparison of motility and flow cytometric assessments of seminal quality in fresh, 24-hour extended and cryopreserved human spermatozoa. J Androl 14:374-84, 1993
1987. Downing TW, Garner DL, Ericsson SA, Redelman D: Metabolic toxicity of fluorescent stains on thawed cryopreserved bovine sperm cells. J Histochem Cytochem 39:485-9, 1991
1988. Haas G Jr, D'Cruz OJ, De Bault LE: Assessment by fluorescence-activated cell sorting of whether sperm-associated immunoglobulin (Ig)G and IgA occur on the same sperm population. Fertil Steril 54:127-32, 1990
1989. D'Cruz OJ, Haas G Jr: Lack of complement activation in the seminal plasma of men with antisperm antibodies associated in vivo on their sperm. Am J Reprod Immunol 24:51-7, 1990
1990. D'Cruz OJ, Haas G Jr, Wang BL, De Bault LE: Activation of human complement by IgG antisperm antibody and the demonstration of C3 and C5b-9-mediated immune injury to human sperm. J Immunol 146:611-20, 1991
1991. D'Cruz OJ, Haas G Jr: Flow cytometric quantitation of the expression of membrane cofactor protein as a marker for the human sperm acrosome reaction. Fertil Steril 58:633-6, 1992
1992. Haas G Jr, D'Cruz OJ, De Bault LE: Comparison of the indirect immunobead, radiolabeled, and immunofluorescence assays for immunoglobulin G serum antibodies to human sperm. Fertil Steril 55:377-88, 1991
1993. Rasanen ML, Hovatta OL, Penttila IM, Agrawal YP: Detection and quantitation of sperm-bound antibodies by flow cytometry of human semen. J Androl 13:55-64, 1992
1994. Sinton EB, Riemann DC, Ashton ME: Antisperm antibody detection using concurrent cytofluorometry and indirect immunofluorescence microscopy. Am J Clin Pathol 95:242-6, 1991
1995. D'Cruz OJ, Haas G Jr: The expression of the complement regulators CD46, CD55, and CD59 by human sperm does not protect them from antisperm antibody- and complement-mediated immune injury. Fertil Steril 59:876-84, 1993
1996. Nikolaeva MA, Kulakov VI, Ter-Avanesov GV, Terekhina LN, Pshenichnikova TJ, Sukhikh GT: Detection of antisperm antibodies on the surface of living spermatozoa using flow cytometry: preliminary study. Fertil Steril 59:639-44, 1993
1998. Rasanen M, Lahteenmaki A, Saarikoski S, Agrawal YP: Comparison of flow cytometric measurement of seminal antisperm antibodies with the mixed antiglobulin reaction and the serum tray agglutination test. Fertil Steril 61:143-50, 1994
1999. Fenichel P, Hsi BL, Farahifar D, Donzeau M, Barrier-Delpech D, Yehy CJ: Evaluation of the human sperm acrosome reaction using a monoclonal antibody, GB24, and fluorescence-activated cell sorter. J Reprod Fertil 87:699-706, 1989
2000. Engh E, Clausen OP, Purvis K: Acrosomal integrity assessed by flow cytometry in men with variable sperm quality. Hum Reprod 6:1129-34, 1991
2001. Fenichel P, Donzeau M, Farahifar D, Basteris B, Ayraud N, Hsi BL: Dynamics of human sperm acrosome reaction: relation with in vitro fertilization. Fertil Steril 55:994-9, 1991

2003. Graham JK, Kunze E, Hammerstedt RH: Analysis of sperm cell viability, acrosomal integrity, and mitochondrial function using flow cytometry. Biol Reprod 43:55-64, 1990
2004. Uhler ML, Leung A, Chan SY, Schmid I, Wang C: Assessment of human sperm acrosome reaction by flow cytometry: validation and evaluation of the method by fluorescence-activated cell sorting. Fertil Steril 60:1076-81, 1993
2005. Purvis K, Rui H, Scholberg A, Hesla S, Clausen OP: Application of flow cytometry to studies on the human acrosome. J Androl 11:361-6, 1990
2006. Hellstrom WJ, Deitch AD, de Vere White RW: Evaluation of vasovasostomy candidates by deoxyribonucleic acid flow cytometry of testicular aspirates. Fertil Steril 51:546-8, 1989
2007. Hellstrom WJ, Tesluk H, Deitch AD, de Vere White RW: Comparison of flow cytometry to routine testicular biopsy in male infertility. Urology 35:321-6, 1990
2008. Kaufman DG, Nagler HM: Aspiration flow cytometry of the testes in the evaluation of spermatogenesis in the infertile male. Fertil Steril 48:287-91, 1987
2009. Ring KS, Burbige KA, Benson MC, Karp F, Hensle TW: The flow cytometric analysis of undescended testes in children. J Urol 144:494-8, 1990
2010. Fossa SD, Melvik JE, Juul NO, Pettersen EO, Amellem O, Theodorsen L: DNA flow cytometry in sperm cells from testicular cancer patients. Impact of different treatment modalities on spermatogenesis. Eur Urol 19:125-31, 1991
2011. Lee SE, Choo MS: Flow-cytometric analysis of testes in infertile men: a comparison of the ploidy to routine histopathologic study. Eur Urol 20:33-8, 1991
2012. Kenney RM, Kent MG, Garcia MC, Hurtgen JP: The use of DNA index and karyotype analyses as adjuncts to the estimation of fertility in stallions. J Reprod Fertil Suppl 44:69-75, 1991
2013. Clausen OP, Giwercman A, Jorgensen N, Bruun E, Frimodt-Moller C, Skakkebaek NE: DNA distributions in maldescended testes: hyperdiploid aneuploidy without evidence of germ cell neoplasia. Cytometry 12:77-81, 1991
2014. Bergounioux C, Perennes C, Brown SC, Gadal P: Nuclear RNA quantification in protoplast cell-cycle phases. Cytometry 9:84-7, 1988
2015. Ulrich W: Simultaneous measurement of DAPI-sulforhodamine 101 stained nuclear DNA and protein in higher plants by flow cytometry. Biotech Histochem 67:73-8, 1992
2016. Houssa C, Bomans J, Greimers R, Jacqmard A: High-yield isolation of protoplasts from microgram amounts of shoot meristematic tissues and rapid DNA content determination by flow cytometry. Exp Cell Res 197:153-7, 1991
2017. Harkins KR, Jefferson RA, Kavanagh TA, Bevan MW, Galbraith DW: Expression of photosynthesis-related gene fusions is restricted by cell type in transgenic plants and in transfected protoplasts. Proc Natl Acad Sci U S A 87:816-20, 1990
2018. Klock G, Zimmermann U: Facilitated electrofusion of vacuolated x evacuolated oat mesophyll protoplasts in hypo-osmolar media after alignment with an alternating field of modulated strength. Biochim Biophys Acta 1025:87-93, 1990
2019. Wolters AA, Vergunst AC, van der Werff F, Koornneef M: Analysis of nuclear and organellar DNA of somatic hybrid calli and plants between *Lycopersicon* spp. and *Nicotiana* spp. Mol Gen Genet 241:707-18, 1993
2020. Jiao L, Gray DW, Göhde W, Flynn GJ, Morris PJ: In vitro staining of islets of Langerhans for fluorescence-activated cell sorting. Transplantation 52:450-2, 1991
2021. Halban PA, Powers SL, George KL, Bonner-Weir S: Spontaneous reassociation of dispersed adult rat pancreatic islet cells into aggregates with three-dimensional architecture typical of native islets. Diabetes 36:783-90, 1987
2022. Ostensson K: Total and differential leukocyte counts, N-acetyl-beta-D-glucosaminidase activity, and serum albumin content in foremilk and residual milk during endotoxin-induced mastitis in cows. Am J Vet Res 54:231-8, 1993
2023. Saad AM, Astrom G: Effects of exogenous estrogen administration to ovariectomized cows on the blood and milk-leukocyte counts and -neutrophil phagocytosis measured by flow cytometry. Zentralbl Veterinarmed [b] 35:654-63, 1988
2024. Redelman D, Butler S, Robison J, Garner D: Identification of inflammatory cells in bovine milk by flow cytometry. Cytometry 9:463-8, 1988
2025. Jespersen L, Lassen S, Jakobsen M: Flow cytometric detection of wild yeast in lager breweries. Int J Food Microbiol 17:321-8, 1993
2026. Zbinden G: Reduction and replacement of laboratory animals in toxicological testing and research. Interim report 1984-1987. Biomed Environ Sci 1:90-100, 1988
2027. Klevecz RR, Braly PS: Circadian and ultradian rhythms of proliferation in human ovarian cancer. Chronobiol Int 4:513-23, 1987
2028. Sletvold O, Laerum OD: Alterations of cell cycle distribution in the bone marrow of aging mice measured by flow cytometry. Exp Gerontol 23:43-58, 1988
2029. Laerum OD, Sletvold O, Riise T: Circadian and circannual variations of cell cycle distribution in the mouse bone marrow. Chronobiol Int 5:19-35, 1988
2030. Smaaland R, Abrahamsen JF, Svardal AM, Lote K, Ueland PM: DNA cell cycle distribution and glutathione (GSH) content according to circadian stage in bone marrow of cancer patients. Br J Cancer 66:39-45, 1992
2031. Smaaland R, Laerum OD, Lote K, Sletvold O, Sothern RB, Bjerknes R: DNA synthesis in human bone marrow is circadian stage dependent. Blood 77:2603-11, 1991
2032. Sletvold O, Smaaland R, Laerum OD: Cytometry and time-dependent variations in peripheral blood and bone marrow cells: a literature review and relevance to the chronotherapy of cancer. Chronobiol Int 8:235-50, 1991
2033. Frentz G, Moller U, Holmich P, Christensen IJ: On circadian rhythms in human epidermal cell proliferation. Acta Derm Venereol 71:85-7, 1991
2034. Carbajo-Perez E, Carbajo S, Orfao A, Vicente-Villardon JL, Vazquez R: Circadian variation in the distribution of cells throughout the different phases of the cell cycle in the anterior pituitary gland of adult male rats as analysed by flow cytometry. J Endocrinol 129:329-33, 1991
2035. Aarnaes E, Clausen OP, Kirkhus B, De Angelis P: Heterogeneity in the mouse epidermal cell cycle analysed by computer simulations. Cell Prolif 26:205-19, 1993
2036. Kirkhus B, Clausen OP: Circadian variations in cell cycle phase progression of mouse epidermal cells measured directly by bivariate BrdUrd/DNA flow cytometry. Epithelial Cell Biol 1:32-8, 1992
2037. Wesemann W, Clement HW, Gemsa D, Hasse C, Heymanns J, Pohlner K, Schafer F, Weiner N: Immobilization and light-dark cycle-induced modulation of serotonin metabolism in rat brain and of lymphocyte subpopulations: in vivo voltammetric and FACS analyses. Neuropsychobiology 28:91-4, 1993
2038. Bourin P, Mansour I, Doinel C, Roue R, Rouger P, Levi F: Circadian rhythms of circulating NK cells in healthy and human immunodeficiency virus-infected men. Chronobiol Int 10:298-305, 1993
2039. Malone JL, Simms TE, Gray GC, Wagner KF, Burge JR, Burke DS: Sources of variability in repeated T-helper lymphocyte counts from human immunodeficiency virus type 1-infected patients: total lymphocyte count fluctuations and diurnal cycle are important. J Acquir Immune Defic Syndr 3:144-51, 1990
2040. Vicker MG, Becker J, Gebauer G, Schill W, Rensing L: Circadian rhythms of cell cycle processes in the marine dinoflagellate *Gonyaulax polyedra.* Chronobiol Int 5:5-17, 1988
2041. Homma K, Hastings JW: The s phase is discrete and is controlled by the circadian clock in the marine dinoflagellate *Gonyaulax polyedra.* Exp Cell Res 182:635-44, 1989

2042. Cristofalo VJ: Cellular biomarkers of aging. Exp Gerontol 23:297-307, 1988

2043. Hariri RJ, Hajjar DP, Coletti D, Alonso DR, Weksler ME, Rabellino E: Aging and arteriosclerosis. Cell cycle kinetics of young and old arterial smooth muscle cells. Am J Pathol 131:132-6, 1988

2044. Capasso JM, Bruno S, Li P, Zhang X, Darzynkiewicz Z, Anversa P: Myocyte DNA synthesis with aging: correlation with ventricular loading in rats. J Cell Physiol 155:635-48, 1993

2045. Black MJ, Adams MA, Bobik A, Campbell JH, Campbell GR: Vascular smooth muscle polyploidy in the development and regression of hypertension. Clin Exp Pharmacol Physiol 15:345-8, 1988

2046. Stemme S, Holm J, Hansson GK: T lymphocytes in human atherosclerotic plaques are memory cells expressing CD45RO and the integrin VLA-1. Arterioscler Thromb 12:206-11, 1992

2047. Kamentsky LA, Kamentsky LD: Microscope-based multiparameter laser scanning cytometer yielding data comparable to flow cytometry data. Cytometry 12:381-7, 1991

2048. Danesi DT, Spano M, Altavista P: Quality control study of the Italian group of cytometry on flow cytometry cellular DNA content measurements. Cytometry 14:576-83, 1993

2049. Johnson CH, Inoue S, Flint A, Hastings JW: Compartmentalization of algal bioluminescence: autofluorescence of bioluminescent particles in the dinoflagellate *Gonyaulax* as studied with image-intensified video microscopy and flow cytometry. J Cell Biol 100:1435-46, 1985

2050. Freyer JP, Fillak D, Jett JH: Use of xantham gum to suspend large particles during flow cytometric analysis and sorting. Cytometry 10:803-6, 1989

2051. Lee G, Delohery TM, Ronai Z, Brandt-Rauf PW, Pincus MR, Murphy RB, Weinstein IB: A comparison of techniques for introducing macromolecules into living cells. Cytometry 14:265-70, 1993

2052. Weaver JC, Harrison GI, Bliss JG, Mourant JR, Powell KT: Electroporation: high frequency of occurrence of a transient high-permeability state in erythrocytes and intact yeast. Febs Lett 229:30-4, 1988

2053. Bartoletti DC, Harrison GI, Weaver JC: The number of molecules taken up by electroporated cells: quantitative determination. Febs Lett 256:4-10, 1989

2054. Prausnitz MR, Lau BS, Milano CD, Conner S, Langer R, Weaver JC: A quantitative study of electroporation showing a plateau in net molecular transport. Biophys J 65:414-22, 1993

2055. Chakrabarti R, Wylie DE, Schuster SM: Transfer of monoclonal antibodies into mammalian cells by electroporation. J Biol Chem 264:15494-500, 1989

2056. Graziadei L, Burfeind P, Bar-Sagi D: Introduction of unlabeled proteins into living cells by electroporation and isolation of viable protein-loaded cells using dextran-fluorescein isothiocyanate as a marker for protein uptake. Anal Biochem 194:198-203, 1991

2057. Wilson MR, Mulligan SP, Raison RL: A new microsphere-based immunofluorescence assay for antibodies to membrane-associated antigens. J Immunol Methods 107:231-7, 1988

2058. Zucker RM, Easterling RE, Ting-Beall HP, Allis JW, Massaro EJ: Effects of tributyltin on biomembranes: alteration of flow cytometric parameters and inhibition of Na^+, K^+-ATPase two-dimensional crystallization. Toxicol Appl Pharmacol 96:393-403, 1988

2059. Zucker RM, Elstein KH, Easterling RE, Massaro EJ: Flow cytometric analysis of the cellular toxicity of tributyltin. Toxicol Lett 43:201-18, 1988

2060. Evenson DP, Janca FC, Jost LK, Baer RK, Karabinus DS: Flow cytometric analysis of effects of 1,3-dinitrobenzene on rat spermatogenesis. J Toxicol Environ Health 28:81-98, 1989

2061. Evenson DP, Baer RK, Jost LK: Long-term effects of triethylenemelamine exposure on mouse testis cells and sperm chromatin structure assayed by flow cytometry. Environ Mol Mutagen 14:79-89, 1989

2062. Hoover DM, Hoyt JA, Seyler DE, Abbott DL, Hoffman WP, Buening MK: Comparative effects of disulfiram and N-methyltetrazolethiol on spermatogenic development in young CD rats. Toxicol Appl Pharmacol 107:164-72, 1991

2063. Hayashi M, Norppa H, Sofuni T, Ishidate M Jr: Flow cytometric micronucleus test with mouse peripheral erythrocytes. Mutagenesis 7:257-64, 1992

2064. Hayashi M, Norppa H, Sofuni T, Ishidate M Jr: Mouse bone marrow micronucleus test using flow cytometry. Mutagenesis 7:251-6, 1992

2065. Cao J, Beisker W, Nusse M, Adler ID: Flow cytometric detection of micronuclei induced by chemicals in poly- and normochromatic erythrocytes of mouse peripheral blood. Mutagenesis 8:533-41, 1993

2066. Tometsko AM, Torous DK, Dertinger SD: Analysis of micronucleated cells by flow cytometry. 1. Achieving high resolution with a malaria model. Mutat Res 292:129-35, 1993

2067. Tometsko AM, Dertinger SD, Torous DK: Analysis of micronucleated cells by flow cytometry. 2. Evaluating the accuracy of high-speed scoring. Mutat Res 292:137-43, 1993

2068. Tometsko AM, Torous DK, Dertinger SD: Analysis of micronucleated cells by flow cytometry. 3. Advanced technology for detecting clastogenic activity. Mutat Res 292:145-53, 1993

2069. Nardelli B, McHugh L, Mage M: Polyacrylamide-streptavidin: a novel reagent for simplified construction of soluble multivalent macromolecular conjugates. J Immunol Methods 120:233-9, 1989

2070. Vlieger AM, Medenblik AM, van Gijlswijk RP, Tanke HJ, van der Ploeg M, Gratama JW, Raap AK: Quantitation of polymerase chain reaction products by hybridization-based assays with fluorescent, colorimetric, or chemiluminescent detection. Anal Biochem 205:1-7, 1992

2071. Giorgi JV: Cell sorting of biohazardous specimens for assay of immune function. Methods Cell Biol 42: 359-69, 1994

2072. Crippen T, Nilsson K, Matsson P: Analysis of monoclonal antibodies generated against the basophilic leukemia cell line, KU812. Cytometry Supp 4:73, 1990

2073. Crippen T, Nilsson K, Matsson P: A metachromatic dye which discriminates basophil granules. Cytometry Supp 4:73, 1990

2074. Li X, Traganos F, Darzynkiewicz Z: Simultaneous analysis of DNA replication and apoptosis during treatment of HL-60 cells with camptothecin and hyperthermia and mitogen stimulation of human lymphocytes. Cancer Res 54:4289-93, 1994

2075. Shapiro HM: Flow cytometry in laboratory microbiology: New directions. ASM News 56:584-8, 1990

2076. Dusenberry JA, Frankel SL: Increasing the sensitivity of a FACScan flow cytometer to study oceanic phytoplankton. Limnol Oceanogr 39:206-9, 1994

2077. Gant VA, Warnes G, Phillips I, Savidge GF: The application of flow cytometry to the study of bacterial responses to antibiotics. J Med Microbiol 39:147-54, 1993

2078. Nikaido H, Nakae T: The outer membrane of Gram-negative bacteria. Adv Microb Physiol 20:163-250, 1979

2079. Nikaido H: Permeability of the outer membrane of bacteria. Angew Chem Int Ed Engl 18:337-50, 1979

2080. Leive L: Release of lipopolysaccharide by EDTA treatment of *E. coli*. Biochem Biophys Res Commun 21:290-6, 1965

2081. Leive L, Kollin V: Controlling EDTA treatment to produce permeable *Escherichia coli* with normal metabolic processes. Biochem Biophys Res Commun 28:229-36, 1967

2082. Frankel SL, Binder B, Chisholm SW, Shapiro HM: A high-sensitivity flow cytometer for studying picoplankton. Limnol Oceanogr 35:1164-9, 1990.

2083. Courties C, Vaquer A, Troussellier M, Lautier J, Chrétiennot-Dinet MJ, Neveux J, Machado C, Claustre H: Smallest eukaryotic organism. Nature 370:255, 1994

2084. Robertson BR, Button DK: Characterizing aquatic bacteria according to population, cell size, and apparent DNA content by flow cytometry. Cytometry 10:70-6, 1989

2085. Button DK, Robertson BR, McIntosh D, Juttner F: Interactions between marine bacteria and dissolved-phase and beached hydrocarbons after the Exxon Valdez oil spill. Appl Environ Microbiol 58:243-51, 1992

2086. Troussellier M, Courties C, Vaquer A: Recent applications of flow cytometry in aquatic microbial ecology. Biol Cell 78:111-21, 1993

2087. Li WKW, Lewis MR, Lister A: Flow cytometric detection of prochlorophytes and cyanobacteria in the Gulf of Policastro, Italy. Arch Hydrobiol 124:309-16, 1992

2088. Campbell L, Vaulot D: Photosynthetic picoplankton community structure in the subtropical North Pacific Ocean near Hawaii (station ALOHA). Deep-Sea Res 40:2043-60, 1993

2089. Li WKW, Dickie PM: Relationship between the number of dividing and nondividing cells of cyanobacteria in North Atlantic picoplankton. J Phycol 27:559-65, 1991

2090. Li WKW, Dickie PM, Harrison WG, Irwin BD: Biomass and production of bacteria and phytoplankton during the springblook in the western North Atlantic Ocean. Deep-Sea Res 40:307-27, 1993

2091. Li WKW, Zohary T, Yacobi YZ, Wood AM: Ultraphytoplankton in the eastern Mediterranean Sea: towards deriving phytoplankton biomass from flow cytometric measurements of abundance, fluorescence, and light scatter. Mar Ecol Prog Ser 102:79-87, 1993

2092. Button DK, Robertson BR: Kinetics of bacterial processes in natural aquatic systems based on biomass as determined by high-resolution flow cytometry. Cytometry 10:558-63, 1989 [published erratum appears in Cytometry 11:451, 1990]

2093. Yentsch CM, Campbell JW: Phytoplankton growth: perspectives gained by flow cytometry. J Plankton Res 13: Supp 83-108, 1991

2094. Binder BJ, Chisholm SW: Relationship between DNA cycle and growth rate in *Synechococcus* sp, strain PCC 6301. J Bact 172:2313-9, 1990

2095. Partensky F, Vaulot D: Growth and cell cycle of two closely related red tide-forming dinoflagellates: *Gymnodinium nagasakiense* and *G.* cf. *nagasakiense*. J Phycol 27:733-42, 1991

2096. Vaulot D, Partensky F: Cell cycle distributions of prochlorophytes in the north western Mediterranean Sea. Deep-Sea Res 39:727-42, 1992

2097. Boucher N, Vaulot D, Partensky F: Flow cytometric determination of phytoplankton DNA in cultures and oceanic populations. Mar Ecol Prog Ser 71:75-84, 1991

2098. Subba Rao DV, Partensky F, Wohlgeschaffen G, Li WKW: Flow cytometry and microscopy of gametogenesis in *Nitzschia pungens*, a toxic bloom-forming marine diatom. J Phycol 27:21-6, 1991

2099. Demers S, Roy S, Gagnon R, Vignault C: Rapid light-induced changes in cell fluorescence and in xanthophyll-cycle pigments of *Alexandrium excavatum* (Dinophyceae) and *Thalassiostra pseudonana* (Bacillariophyceae): a photo-protection mechanism.

2100. Furuya K, Li WKW: Evaluation of photosynthetic capacity in phytoplankton by flow cytometric analysis of DCMU-enhanced chlorophyll fluorescence. Mar Ecol Prog Ser 88:279-87,1992

2101. Monfort P, Baleux B: Comparison of flow cytometry and epifluorescence microscopy for counting bacteria in aquatic ecosystems. Cytometry 13:188-92, 1992

2102. Monger BC, Landry MR: Flow cytometric analysis of marine bacteria with Hoechst 33342. Appl Envir Microbiol 59:905-11, 1993

2103. Agustí S: Allometric scaling of light absorption and scattering by phytoplankton cells. Can J Fish Aquat Sci 48:763-7, 1991

2104. Cunningham A, Buonnacorsi GA: Narrow-angle forward light scattering from individual algal cells: implications for size and shape discrimination in flow cytometry. J Plank Res 14:223-234, 1992

2105. Longhurst AR, Koike I, Li WKW, Rodriguez J, Dickie P, Kepay P, Partensky F, Bautista B, Ruiz J, Wells M, Bird DF: Sub-micron particles in northwest Atlantic shelf water. Deep-Sea Res 39:1-7, 1992

2106. Le Gall Y, Brown S, Marie D, Mejjad M, Kloareg B: Quantification of nuclear DNA and G-C content in marine macroalgae by flow cytometry of isolated nuclei. Protoplasma 173:123-32, 1993

2107. Li WKW: Bivariate and trivariate analysis in flow cytometry: phytoplankton size and fluorescence. Limnol Oceanogr 35:1356-68, 1990

2108. Demers S, Kim J, Legendre P, LegendreL: Analyzing multivariate flow cytometric data in aquatic sciences. Cytometry 13:291-8, 1992

2109. Madelin TM, Johnson HE: Fungal and actinomycete spore aerosols measured at different humidities with an aerodynamic particle sizer. J Appl Bacteriol 72:400-9, 1992

2110. Boye E, Lobner-Olesen A, Skarstad K: Timing of chromosomal replication in *Escherichia coli*. Biochim Biophys Acta 951:359-64, 1988

2111. Skarstad K, Boye E, Steen HB: Timing of initiation of chromosome replication in individual *Escherichia coli* cells. EMBO J 5:1711-7, 1986 [published erratum appears in EMBO J 5:3074, 1986]

2112. Skarstad K, von Meyenburg K, Hansen FG, Boye E: Coordination of chromosome replication initiation in *Escherichia coli*: effects of different dnaA alleles. J Bacteriol 170:852-8, 1988

2113. Skarstad K, Boye E: Perturbed chromosomal replication in recA mutants of *Escherichia coli*. J Bacteriol 170:2549-54, 1988

2114. Skarstad K, Lobner-Olesen A, Atlung T, von Meyenburg K, Boye E: Initiation of DNA replication in *Escherichia coli* after overproduction of the DnaA protein. Mol Gen Genet 218:50-6, 1989

2115. Lobner-Olesen A, Skarstad K, Hansen FG, von Meyenburg K, Boye E: The DnaA protein determines the initiation mass of *Escherichia coli* K-12. Cell 57:881-9, 1989

2116. Bernander R, Merryweather A, Nordstrom K: Overinitiation of replication of the *Escherichia coli* chromosome from an integrated runaway-replication derivative of plasmid R1. J Bacteriol 171:674-83, 1989

2117. Bernander R, Nordstrom K: Chromosome replication does not trigger cell division in *E. coli*. Cell 60:365-74, 1990

2118. Boye E, Lobner-Olesen A: The role of dam methyltransferase in the control of DNA replication in *E. coli*. Cell 62:981-9, 1990

2119. Jaffe A, Boye E, D'Ari R: Rule governing the division pattern in *Escherichia coli* minB and wild-type filaments. J Bacteriol 172:3500-2, 1990

2120. Hansen FG, Atlung T, Braun RE, Wright A, Hughes P, Kohiyama M: Initiator (DnaA) protein concentration as a function of growth rate in *Escherichia coli* and *Salmonella typhimurium*. J Bacteriol 173:5194-9, 1991

2121. Allman R, Schjerven T, Boye E: Cell cycle parameters of *Escherichia coli* K-12. J Bacteriol 173:7970-4, 1991

2122. Boye E, Lobner-Olesen A: Bacterial growth control studied by flow cytometry. Res Microbiol 142:131-5, 1991

2123. Dasgupta S, Bernander R, Nordstrom K: In vivo effect of the tus mutation on cell division in an *Escherichia coli* strain where chromosome replication is under the control of plasmid R1. Res Microbiol 142:177-80, 1991

2124. Lobner-Olesen A, Boye E: Different effects of mioC transcription on initiation of chromosomal and minichromosomal replication in *Escherichia coli*. Nucleic Acids Res 20:3029-36, 1992

2125. Lobner-Olesen A, Boye E, Marinus MG: Expression of the *Escherichia coli* dam gene. Mol Microbiol 6:1841-51, 1992

2126. von Freiesleben U, Rasmussen KV: The level of supercoiling affects the regulation of DNA replication in *Escherichia coli*. Res Microbiol 143:655-63, 1992

2127. Nelson D, Neill W, Poxton IR: A comparison of immunoblotting, flow cytometry and ELISA to monitor the binding of anti-lipopolysaccharide monoclonal antibodies. J Immunol Methods 133:227-33, 1990

2128. Cordery MC, Smith IM, Mackenzie NM, Parker DJ: Quantification by flow microfluorimetry of specific binding of antibody to some serotypes of *Haemophilus pleuropneumoniae*. Res Vet Sci 41:277-8, 1986

2129. Kravtsov AL, Korovkin SA, Naumov AV: [Quantitative determination of specific plague fluorescent immunoglobulin activity using pulse flow cytofluorometry]. Lab Delo 12:59-62, 1988

2130. Minas W, Sahar E, Gutnick D: Flow cytometric screening and isolation of *Escherichia coli* clones which express surface antigens of the oil-degrading microorganism *Acinetobacter calcoaceticus* RAG-1. Arch Microbiol 150:432-7, 1988
2131. Evans ME, Pollack M, Hardegen NJ, Koles NL, Guelde G, Chia JK: Fluorescence-activated cell sorter analysis of binding by lipopolysaccharide-specific monoclonal antibodies to gram-negative bacteria. J Infect Dis 162:148-55, 1990
2132. Cooper LJ, Schimenti JC, Glass DD, Greenspan NS: H chain C domains influence the strength of binding of IgG for streptococcal group A carbohydrate. J Immunol 146:2659-63, 1991
2133. Halstensen A, Lehmann AK, Guttormsen HK, Vollset SE, Bjune G, Naess A: Serum opsonins to serogroup B meningococci after disease and vaccination. Niph Ann 14:157-65, 1991
2134. Lutton DA, Patrick S, Crockard AD, Stewart LD, Larkin MJ, Dermott E, McNeill TA: Flow cytometric analysis of within-strain variation in polysaccharide expression by *Bacteroides fragilis* by use of murine monoclonal antibodies. J Med Microbiol 35:229-37, 1991
2135. Nelson D, Bathgate AJ, Poxton IR: Monoclonal antibodies as probes for detecting lipopolysaccharide expression on *Escherichia coli* from different growth conditions. J Gen Microbiol 137:2741-51, 1991
2136. Nelson JW, Barclay GR, Micklem LR, Poxton IR, Govan JR: Production and characterisation of mouse monoclonal antibodies reactive with the lipopolysaccharide core of *Pseudomonas aeruginosa*. J Med Microbiol 36:358-65, 1992
2137. Srikumar R, Dahan D, Gras MF, Ratcliffe MJ, van Alphen L, Coulton JW: Antigenic sites on porin of *Haemophilus influenzae* type b: mapping with synthetic peptides and evaluation of structure predictions. J Bacteriol 174:4007-16, 1992
2138. Srikumar R, Chin AC, Vachon V, Richardson CD, Ratcliffe MJ, Saarinen L, Kayhty H, Makela PH, Coulton JW: Monoclonal antibodies specific to porin of *Haemophilus influenzae* type b: localization of their cognate epitopes and tests of their biological activities. Mol Microbiol 6:665-76, 1992
2139. Siegel SA, Evans ME, Pollack M, Leone AO, Kinney CS, Tam SH, Daddona PE: Antibiotics enhance binding by human lipid A-reactive monoclonal antibody HA-1A to smooth gram-negative bacteria. Infect Immun 61:512-9, 1993
2140. Balfoort HW, Berman T, Maestrini SY, Wenzel A, Zohary T: Flow cytometry: instrumentation and application in phytoplankton research. Hydrobiologie 238:89-97, 1992.
2141. Boye E, Lobner-Olesen A: Flow cytometry: illuminating microbiology. New Biol 2:119-25, 1990
2142. Allman R, Hann AC, Phillips AP, Martin KL, Lloyd D: Growth of *Azotobacter vinelandii* with correlation of Coulter cell size, flow cytometric parameters, and ultrastructure. Cytometry 11:822-31, 1990
2143. Allman R, Hann AC, Manchee R, Lloyd D: Characterization of bacteria by multiparameter flow cytometry. J Appl Bacteriol 73:438-44, 1992
2144. Miller JS, Quarles JM: Flow cytometric identification of microorganisms by dual staining with FITC and PI. Cytometry 11:667-75, 1990
2145. Hiraga S, Jaffe A, Ogura T, Mori H, Takahashi H: F plasmid ccd mechanism in *Escherichia coli*. J Bacteriol 166:100-4, 1986
2146. Poulsen P, Jensen KF: Three genes preceding pyrE on the *Escherichia coli* chromosome are essential for survival and normal cell morphology in stationary culture and at high temperature. Res Microbiol 142:283-8, 1991
2147. Sazer S, Sherwood SW: Mitochondrial growth and DNA synthesis occur in the absence of nuclear DNA replication in fission yeast. J Cell Sci 97:509-16, 1990
2148. Eitzman PD, Hendrick JL, Srienc F: Quantitative immunofluorescence in single *Saccharomyces cerevisiae* cells. Cytometry 10:475-83, 1989
2149. Héchard Y, Jayat C, Letellier F, Julien R, Cenatiempo Y, Ratinaud MH: On-line visualization of the competitive behavior of antagonistic bacteria. Appl Environ Microbiol 58:3784-6, 1992
2150. Pinder AC, Purdy PW, Poulter SA, Clark DC: Validation of flow cytometry for rapid enumeration of bacterial concentrations in pure cultures. J Appl Bacteriol 69:92-100, 1990
2151. Russo-Marie F, Roederer M, Sager B, Herzenberg LA, Kaiser D: Beta-galactosidase activity in single differentiating bacterial cells. Proc Natl Acad Sci U S A 90:8194-8, 1993
2152. Manafi M, Kneifel W: Fluorogenic and chromogenic substrates--a promising tool in microbiology. Acta Microbiol Hung 38:293-304, 1991
2153. Pace J, Chai TJ: Comparison of *Vibrio parahaemolyticus* grown in estuarine water and rich medium. Appl Environ Microbiol 55:1877-87, 1989
2154. Morgan JA, Cranwell PA, Pickup RW: Survival of *Aeromonas salmonicida* in lake water. Appl Environ Microbiol 57:1777-82, 1991
2155. Thorsen BK, Enger O, Norland S, Hoff KA: Long-term starvation survival of *Yersinia ruckeri* at different salinities studied by microscopical and flow cytometric methods. Appl Environ Microbiol 58:1624-8, 1992
2156. Volsch A, Nader WF, Geiss HK, Nebe G, Birr C: Detection and analysis of two serotypes of ammonia-oxidizing bacteria in sewage plants by flow cytometry. Appl Environ Microbiol 56:2430-5, 1990
2157. Porter J, Edwards C, Morgan JA, Pickup RW: Rapid, automated separation of specific bacteria from lake water and sewage by flow cytometry and cell sorting. Appl Environ Microbiol 59:3327-33, 1993
2158. Garcia-Armesto MR, Prieto M, Garcia-Lopez ML, Otero A, Moreno B: Modern microbiological methods for foods: colony count and direct count methods. A review. Microbiologia 9:1-13, 1993
2159. Laplace-Builhe C, Hahne K, Hunger W, Tirilly Y, Drocourt JL: Application of flow cytometry to rapid microbial analysis in food and drinks industries. Biol Cell 78:123-8, 1993
2160. Brailsford M, Gatley S: Rapid analysis of microorganisms using flow cytometry. In reference 1052, pp. 171-80.
2161. Donnelly CW, Baigent GJ, Briggs EH: Flow cytometry for automated analysis of milk containing *Listeria monocytogenes*. J Assoc Off Anal Chem 71:655-8, 1988
2162. Cory JM, Ohlsson-Wilhelm BM, Brock EJ, Sheaffer NA, Steck ME, Eyster ME, Rapp F: Detection of human immunodeficiency virus-infected lymphoid cells at low frequency by flow cytometry. J Immunol Methods 105:71-8, 1987
2163. McSharry JJ, Constantino R, Robbiano E, Echols R, Stevens R, Lehman JM: Detection and quantitation of human immunodeficiency virus-infected peripheral blood mononuclear cells by flow cytometry. J Clin Microbiol 28:724-33, 1990
2164. Ohlsson-Wilhelm BM, Cory JM, Kessler HA, Eyster ME, Rapp F, Landay A: Circulating HIV p24 antigen-positive lymphocytes: a flow cytometric measure of HIV infection. J Infect Dis 162:1018-24, 1990
2165. Gadol N, Crutcher GJ, Busch MP: Detection of intracellular HIV in lymphocytes by flow cytometry. Cytometry 13:359-70, 1994.
2166. McSharry JJ, Costantino R, McSharry MB, Venezia RA, Lehman JM: Rapid detection of herpes simplex virus in clinical samples by flow cytometry after amplification in tissue culture. J Clin Microbiol 28:1864-6, 1990
2167. Qvist P, Aasted B, Bloch B, Meyling A, Ronsholt L, Houe H: Flow cytometric detection of bovine viral diarrhea virus in peripheral blood leukocytes of persistently infected cattle. Can J Vet Res 54:469-72, 1990
2168. Qvist P, Houe H, Aasted B, Meyling A: Comparison of flow cytometry and virus isolation in cell culture for identification of cattle persistently infected with bovine viral diarrhea virus. J Clin Microbiol 29:660-1, 1991
2169. Li H, Walker DH: Characterization of rickettsial attachment to host cells by flow cytometry. Infect Immun 60:2030-5, 1992
2170. Roman MJ, Crissman HA, Samsonoff WA, Hechemy KE, Baca OG: Analysis of *Coxiella burnetii* isolates in cell culture and the ex-

pression of parasite-specific antigens on the host membrane surface. Acta Virol 35:503-10, 1991

2171. Hall RT, Strugnell T, Wu X, Devine DV, Stiver HG: Characterization of kinetics and target proteins for binding of human complement component C3 to the surface-exposed outer membrane of *Chlamydia trachomatis* serovar L2. Infect Immun 61:1829-34, 1993
2172. Wernli M, Tichelli A, von Planta M, Gratwohl A, Speck B: Flow cytometric monitoring of parasitaemia during treatment of severe malaria by exchange transfusion [letter]. Eur J Haematol 46:121-3, 1991
2173. Kadjoian V, Gasquet M, Delmas F, Guiraud H, De Meo M, Laget M, Timon-David, P: Flow cytometry to evaluate the parasitemia of *Plasmodium falciparum.* J Pharm Belg 47:499-503, 1992
2174. van Vianen PH, van Engen A, Thaithong S, van der Keur M, Tanke HJ, van der Kaay HJ, Mons B, Janse CJ: Flow cytometric screening of blood samples for malaria parasites. Cytometry 14:276-80, 1993
2175. Bunyaratvej A, Butthep P, Bunyaratvej P: Cytometric analysis of blood cells from malaria-infected patients and in vitro infected blood. Cytometry 14:81-5, 1993
2176. Janse CJ, van Vianen PH, Tanke HJ, Mons B, Ponnudurai T, Overdulve JP: *Plasmodium* species: flow cytometry and microfluorometry assessments of DNA content and synthesis. Exp Parasitol 64:88-94, 1987
2177. Janse CJ, Boorsma EG, Ramesar J, van Vianen P, van der Meer R, Zenobi P, Casaglia O, Mons B, van der Berg FM :*Plasmodium berghei*: gametocyte production, DNA content, and chromosome-size polymorphisms during asexual multiplication in vivo. Exp Parasitol 68:274-82, 1989
2178. Ponzi M, Janse CJ, Dore E, Scotti R, Pace T, Reterink TJ, van der Berg FM, Mons B: Generation of chromosome size polymorphism during in vivo mitotic multiplication of *Plasmodium berghei* involves both loss and addition of subtelomeric repeat sequences. Mol Biochem Parasitol 41:73-82, 1990
2179. Jacobberger JW, Horan PK, Hare JD: Cell cycle analysis of asexual stages of erythrocytic malaria parasites. Cell Prolif 25:431-45, 1992
2180. Udomsangpetch R, Webster HK, Pattanapanyasat K, Pitchayangkul S, Thaithong S: Cytoadherence characteristics of rosette-forming *Plasmodium falciparum.* Infect Immun 60:4483-90, 1992
2181. Pattanapanyasat K, Webster HK, Udomsangpetch R, Wanachiwanawin W, Yongvanitchit K: Flow cytometric two-color staining technique for simultaneous determination of human erythrocyte membrane antigen and intracellular malarial DNA. Cytometry 13:182-7, 1992
2182. Pattanapanyasat K, Udomsangpetch R, Webster HK: Two-color flow cytometric analysis of intraerythrocytic malaria parasite DNA and surface membrane-associated antigen in erythrocytes infected with *Plasmodium falciparum.* Cytometry 14:449-54, 1993
2183. Wiser MF, Faur LV, Lanners HN, Kelly M, Wilson RB: Accessibility and distribution of intraerythrocytic antigens of *Plasmodium*-infected erythrocytes following mild glutaraldehyde fixation and detergent extraction. Parasitol Res 79:579-86, 1993
2184. Scheibel LW, Colombani PM, Hess AD, Aikawa M, Atkinson CT, Milhous WK: Calcium and calmodulin antagonists inhibit human malaria parasites (*Plasmodium falciparum*): implications for drug design. Proc Natl Acad Sci U S A 84:7310-4, 1987
2185. van Vianen PH, Klayman DL, Lin AJ, Lugt CB, van Engen AL, van der Kaay HJ, Mons B: *Plasmodium berghei*: the antimalarial action of artemisinin and sodium artelinate in vivo and in vitro, studied by flow cytometry. Exp Parasitol 70:115-23, 1990
2186. van Vianen PH, Thaithong S, Reinders PP, van Engen A, van der Keur M, Tanke HJ, van der Kaay HJ, Mons B: Automated flow cytometric analysis of drug susceptibility of malaria parasites. Am J Trop Med Hyg 43:602-7, 1990
2187. Baca OG, Crissman HA: Correlation of DNA, RNA, and protein content by flow cytometry in normal and *Coxiella burnetii*-infected L929 cells. Infect Immun 55:1731-3, 1987
2188. Wells JM, Prospero TD, Jenni L, Le Page RW: DNA contents and molecular karyotypes of hybrid *Trypanosoma brucei.* Mol Biochem Parasitol 24:103-16, 1987
2189. Muhlpfordt H, Berger J: Characterization and grouping of *Trypanosoma brucei brucei*, *T.b. gambiense* and *T.b. rhodesiense* by quantitative DNA-cytofluorometry and discriminant analysis. Trop Med Parasitol 40:1-8, 1989
2190. Kooy RF, Hirumi H, Moloo SK, et al.: Evidence for diploidy in metacyclic forms of African trypanosomes. Proc Natl Acad Sci U S A 86:5469-72, 1989
2191. Nozaki T, Dvorak JA: *Trypanosoma cruzi*: flow cytometric analysis of developmental stage differences in DNA. J Protozool 38:234-43, 1991
2192. Dvorak JA: Analysis of the DNA of parasitic protozoa by flow cytometry. Methods Mol Biol 21:191-204, 1993
2193. Finley RW, Dvorak JA: *Trypanosoma cruzi*: analysis of the population dynamics of heterogeneous mixtures. J Protozool 34:409-15, 1987
2194. McDaniel JP, Dvorak JA: Identification, isolation, and characterization of naturally-occurring *Trypanosoma cruzi* variants. Mol Biochem Parasitol 57:213-22, 1993
2195. Nozaki T, Dvorak JA: Intraspecific diversity in the response of *Trypanosoma cruzi* to environmental stress. J Parasitol 79:451-4, 1993
2196. Thompson CT, Dvorak JA: Quantitation of total DNA per cell in an exponentially growing population using the diphenylamine reaction and flow cytometry. Anal Biochem 177:353-7, 1989
2197. Rolin S, Paindavoine P, Hanocq-Quertier J, et al.: Transient adenylate cyclase activation accompanies differentiation of *Trypanosoma brucei* from bloodstream to procyclic forms. Mol Biochem Parasitol 61:115-25, 1993
2198. Mutharia LM, Pearson TW: Surface carbohydrates of procyclic forms of African trypanosomes studied using fluorescence activated cell sorter analysis and agglutination with lectins. Mol Biochem Parasitol 23:165-72, 1987
2199. Vincendeau P, Daeron M: *Trypanosoma musculi* co-express several receptors binding rodent IgM, IgE, and IgG subclasses. J Immunol 142:1702-9, 1989
2200. Brickman MJ, Balber AE: *Trypanosoma brucei rhodesiense* bloodstream forms: surface ricin-binding glycoproteins are localized exclusively in the flagellar pocket and the flagellar adhesion zone. J Protozool 37:219-24, 1990
2201. Jacobson KC, Fletcher RC, Kuhn RE: Binding of antibody and resistance to lysis of trypomastigotes of *Trypanosoma cruzi.* Parasite Immunol 14:1-12, 1992
2202. Jacobson KC, Washburn RG, Kuhn RE: Binding of complement to trypomastigotes of a Brazil strain of *Trypanosoma cruzi*: evidence for heterogeneity within the strain. J Parasitol 78:697-704, 1992
2203. Assoku RK, Gardiner PR: Detection of antibodies to platelets and erythrocytes during infection with haemorrhage-causing *Trypanosoma vivax* in Ayrshire cattle. Vet Parasitol 31:199-216, 1989
2204. Powell MR, Rowland EC, Sidner RA: *Trypanosoma cruzi*: flow cytometric analysis of lymphocyte subsets in susceptible and protected C3H/He mice. Exp Parasitol 73:197-202, 1991
2205. Lopez HM, Tanner MK, Kierszenbaum F, Sztein MB: Alterations induced by *Trypanosoma cruzi* in activated mouse lymphocytes. Parasite Immunol 15:273-80, 1993
2206. Connelly MC, Ayala A, Kierszenbaum F: Effects of alpha- and beta-adrenergic agonists on *Trypanosoma cruzi* interaction with host cells. J Parasitol 74:379-86, 1988
2207. Cruz AK, Titus R, Beverley SM: Plasticity in chromosome number and testing of essential genes in *Leishmania* by targeting. Proc Natl Acad Sci U S A 90:1599-603, 1993
2208. Doyle PS, Engel JC, Pimenta PF, da Silva PP, Dwyer DM: *Leishmania donovani*: long-term culture of axenic amastigotes at 37 degrees C. Exp Parasitol 73:326-34, 1991

2209. Darcy F, Torpier G, Kusnierz JP, Rizvi FS, Santoro F: *Leishmania chagasi*: in vitro differentiation of promastigotes monitored by flow cytometry. Exp Parasitol 64:376-84, 1987
2210. Jaffe CL, Perez ML, Sarfstein R: *Leishmania tropica*: characterization of a lipophosphoglycan-like antigen recognized by species-specific monoclonal antibodies. Exp Parasitol 70:12-24, 1990
2211. Rodriguez de Cuna C, Kierszenbaum F, Wirth JJ: Binding of the specific ligand to Fc receptors on *Trypanosoma cruzi* increases the infective capacity of the parasite. Immunology 72:114-20, 1991
2212. Bertho AL, Cysne L, Coutinho SG: Flow cytometry in the study of the interaction between murine macrophages and the protozoan parasite *Leishmania amazonensis*. J Parasitol 78:666-71, 1992
2213. Butcher BA, Sklar LA, Seamer LC, Glew RH: Heparin enhances the interaction of infective *Leishmania donovani* promastigotes with mouse peritoneal macrophages. A fluorescence flow cytometric analysis. J Immunol 148:2879-86, 1992
2214. Shaw KT, Shaw IT, Ryan P, Stevenson MM, Kongshavn PA: Identification of immunodominant *Trypanosoma musculi* antigens recognized by monoclonal antibody and curative immunoglobulin G2a antibody. Int J Parasitol 22:603-12, 1992
2215. Ward HD, Alroy J, Lev BI, Keusch GT, Pereira ME: Biology of *Giardia lamblia*. Detection of N-acetyl-D-glucosamine as the only surface saccharide moiety and identification of two distinct subsets of trophozoites by lectin binding. J Exp Med 167:73-88, 1988
2216. Erlandsen SL, Sherlock LA, Januschka M, Schupp DG, Schaefer FW 3d, Jakubowski W, Bemrick WJ: Cross-species transmission of *Giardia* spp.: inoculation of beavers and muskrats with cysts of human, beaver, mouse, and muskrat origin. Appl Environ Microbiol 54:2777-85, 1988
2217. Hoyne GF, Boreham PF, Parsons PG, Ward C, Biggs B: The effect of drugs on the cell cycle of *Giardia intestinalis*. Parasitology 99:333-9, 1989
2218. Heyworth MF, Pappo J: Use of two-colour flow cytometry to assess killing of *Giardia muris* trophozoites by antibody and complement. Parasitology 99:199-203, 1989
2219. Heyworth MF, Ho KE, Pappo J: Generation and characterization of monoclonal antibodies against *Giardia muris* trophozoites. Immunology 68:341-5, 1989
2220. Heyworth MF, Pappo J: Recognition of a 30,000 MW antigen of *Giardia muris* trophozoites by intestinal IgA from *Giardia*-infected mice. Immunology 70:535-9, 1990
2221. Heyworth MF: Relative susceptibility of *Giardia muris* trophozoites to killing by mouse antibodies of different isotypes. J Parasitol 78:73-6, 1992
2222. Cozon G, Cannella D, Biron F, Piens MA, Jeannin M, Revillard JP: *Cryptosporidium parvum* sporozoite staining by propidium iodide. Int J Parasitol 22:385-9, 1992
2223. Vesey G, Slade JS, Byrne M, Shepherd K, Dennis PJ, Fricker CR: Routine monitoring of *Cryptosporidium* oocysts in water using flow cytometry. J Appl Bacteriol 75:87-90, 1993
2224. Vesey G, Hutton P, Champion A, Ashbolt N, Williams KL, Warton A, Veal D: Application of flow cytometric methods for the routine detection of *Cryptosporidium* and *Giardia* in water. Cytometry 16:1-6, 1994
2225. Alderete JF: *Trichomonas vaginalis* NYH286 phenotypic variation may be coordinated for a repertoire of trichomonad surface immunogens. Infect Immun 55:1957-62, 1987
2226. Wang A, Wang CC, Alderete JF: *Trichomonas vaginalis* phenotypic variation occurs only among trichomonads infected with the double-stranded RNA virus. J Exp Med 166:142-50, 1987
2227. Alderete JF: Alternating phenotypic expression of two classes of *Trichomonas vaginalis* surface markers. Rev Infect Dis 10:S408-12, 1988
2228. Alderete JF, Newton E, Dennis C, Engbring J, Neale KA: Vaginal antibody of patients with trichomoniasis is to a prominent surface immunogen of *Trichomonas vaginalis*. Genitourin Med 67:220-5, 1991
2229. Humphreys MJ, Allman R, Lloyd D: Determination of the viability of *Trichomonas vaginalis* using flow cytometry. Cytometry 15:343-8, 1994
2230. Messick JB, Rikihisa Y: Presence of parasite antigen on the surface of P388D1 cells infected with *Ehrlichia risticii*. Infect Immun 60:3079-86, 1992
2231. De Stefano JA, Trinkle LS, Walzer PD, Cushion MT: Flow cytometric analyses of lectin binding to *Pneumocystis carinii* surface carbohydrates. J Parasitol 78:271-80, 1992
2232. De Stefano JA, Sleight RG, Babcock GF, Sramkoski RM, Walzer PD: Isolation of *Pneumocystis carinii* cysts by flow cytometry. Parasitol Res 78:179-82, 1992
2233. Armstrong MY, Koziel H, Rose RM, Arena C, Richards FF: Indicators of *Pneumocystis carinii* viability in short-term cell culture. J Protozool 38:88S-90S, 1991
2234. Lapinsky SE, Glencross D, Car NG, Kallenbach JM, Zwi S: Quantification and assessment of viability of *Pneumocystis carinii* organisms by flow cytometry. J Clin Microbiol 29:911-5, 1991
2235. Flores BM, Garcia CA, Stamm WE, Torian BE: Differentiation of *Naegleria fowleri* from *Acanthamoeba* species by using monoclonal antibodies and flow cytometry. J Clin Microbiol 28:1999-2005, 1990
2236. Montfort I, Ruiz Arguelles A, Perez Tamayo R: Phenotypic heterogeneity in the expression of a 30 kDa cysteine proteinase in axenic cultures of *Entamoeba histolytica*. Arch Med Res 23:99-103, 1992
2237. Hamelmann C, Foerster B, Burchard GD, Horstmann RD: Lysis of pathogenic and nonpathogenic *Entamoeba histolytica* by human complement: methodological analysis. Parasite Immunol 14:23-35, 1992
2238. Flores-Romo L, Bacon KB, Estrada-Garcia T, Shibayama M, Tsutsumi V, Martinez-Palomo A: A fluorescence-based quantitative adhesion assay to study interactions between *Entamoeba histolytica* and human enterocytes. Effect of proinflammatory cytokines. J Immunol Methods 166:243-50, 1993
2239. Raether W, Mehlhorn H, Hofmann J, Brau B, Ehrlich K: Flow cytometric analysis of *Eimeria tenella* sporozoite populations exposed to salinomycin sodium in vitro: a comparative study using light and electron microscopy and an in vitro sporozoite invasion-inhibition test. Parasitol Res 77:386-94, 1991
2240. Charif H, Darcy F, Torpier G, Cesbron-Delauw MF, Capron A: *Toxoplasma gondii*: characterization and localization of antigens secreted from tachyzoites. Exp Parasitol 71:114-24, 1990
2241. Burgess DE, McDonald CM: Analysis of adhesion and cytotoxicity of *Tritrichomonas foetus* to mammalian cells by use of monoclonal antibodies. Infect Immun 60:4253-9, 1992
2242. Baldwin CL, Black SJ, Brown WC, Conrad PA, Goddeeris BM, Kinuthia SW, Lalor PA, Mac Hugh ND, Morrison WI, Morzaria SP et al: Bovine T cells, B cells, and null cells are transformed by the protozoan parasite *Theileria parva*. Infect Immun 56:462-7, 1988
2243. Fell AH, Preston PM: Growth of *Theileria annulata* and *Theileria parva* macroschizont-infected bovine cells in immunodeficient mice: effect of irradiation and tumour load on lymphocyte subsets. Int J Parasitol 22:491-501, 1992
2244. Goddeeris BM, Dunlap S, Innes EA, McKeever DJ: A simple and efficient method for purifying and quantifying schizonts from *Theileria parva*-infected cells. Parasitol Res 77:482-4, 1991
2245. Peterson N, Liu JA, Shadduck J: *Encephalitozoon cuniculi*: quantitation of parasites and evaluation of viability. J Protozool 35:430-4, 1988
2246. Lunney JK, Murrell KD: Immunogenetic analysis of *Trichinella spiralis* infections in swine. Vet Parasitol 29:179-93, 1988
2247. Hillyer GV: *Fasciola hepatica*:Sp2/0 (helminth: myeloma) hybridoma expressing parasite antigen. Am J Trop Med Hyg 41:674-9, 1989
2248. Amen RI, Aten JA, Baggen JM, Meuleman EA, de Lange-de Klerk ES, Sminia T: *Trichobilharzia ocellata* in *Lymnaea stagnalis*: a flow

cytometric approach to study its effects on hemocytes. J Invertebr Pathol 59:95-8, 1992

2249. Fuller AL, McDougald LR: Analysis of coccidian oocyst populations by means of flow cytometry. J Protozool 36:143-6, 1989
2250. Kell DB, Ryder HM, Kaprelyants AS, Westerhoff HV: Quantifying heterogeneity: flow cytometry of bacterial cultures. Antonie Van Leeuwenhoek 60:145-58, 1991
2251. Kaprelyants AS, Kell DB: Rapid assessment of bacterial viability and vitality by rhodamine 123 and flow cytometry. J Appl Bact 72:410-22, 1992
2252. Kaprelyants AS, Kell DB: The use of 5-cyano-2,3-ditolyl tetrazolium chloride and flow cytometry for the visualisation of respiratory activity in individual cells of *Micrococcus luteus*. J Microbiol Methods 17:115-22, 1993
2253. Kaprelyants AS, Gottschal JC, Kell DB: Dormancy in non-sporulating bacteria. FEMS Microbiol Revs 104:271-86, 1993
2254. Kaprelyants AS, Kell DB: Dormancy in stationary-phase cultures of *Micrococcus luteus*: Flow cytometric analysis of starvation and resuscitation. Appl Envir Microbiol 59:3187-96, 1993
2255. Margulis L: *Symbiosis in Cell Evolution. Microbial Communities in the Archean and Proterozoic Eras* (2 Ed). New York, W. H. Freeman, 1993, xxvii + 452 pp.
2256. Schaule G, Flemming H-C, Ridgway HF: Use of 5-cyano-2,3-ditolyl tetrazolium chloride for quantifying planktonic and sessile respiring bacteria in drinking water. Appl Envir Microbiol 59:3850-7, 1993
2257. Doern GV, Vautour R, Gaudet M, Levy B: Clinical impact of rapid in vitro susceptibility testing and bacterial identification. J Clin Microbiol 32:1757-62, 1994
2258. Phillips AP, Martin KL: Limitations of flow cytometry for the specific detection of bacteria in mixed populations. J Immunol Methods 106:109-17, 1988
2259. Obernesser MS, Socransky SS, Stashenko P: Limit of resolution of flow cytometry for the detection of selected bacterial species. J Dent Res 69:1592-8, 1990
2260. Cohen CY, Sahar E: Rapid flow cytometric bacterial detection and determination of susceptibility to amikacin in body fluids and exudates. J Clin Microbiol 27:1250-6, 1989
2261. Raponi G, Keller N, Overbeek BP, Rozenberg-Arska M, van Kessel KP, Verhoef J: Enhanced phagocytosis of encapsulated *Escherichia coli* strains after exposure to sub-MICs of antibiotics is correlated to changes of the bacterial cell surface. Antimicrob Agents Chemother 34:332-6, 1990
2262. Green L, Petersen B, Steimel L, Haeber P, Current W: Rapid determination of antifungal activity by flow cytometry. J Clin Microbiol 32:1088-91, 1994
2263. Ordonez JV, Wehman NM: Rapid flow cytometric antibiotic susceptibility assay for *Staphylococcus aureus*. Cytometry 14:811-8, 1993
2264. Fouchet P, Jayat C, Hechard Y, Ratinaud MH, Frelat G: Recent advances of flow cytometry in fundamental and applied microbiology. Biol Cell 78:95-109, 1993
2265. Barnett JM, Cuchens MA, Buchanan W: Automated immunofluorescent speciation of oral bacteria using flow cytometry. J Dent Res 63:1040-2, 1984
2266. Kornman KS, Patters M, Kiel R, Marucha P: Detection and quantitation of *Bacteroides gingivalis* in bacterial mixtures by means of flow cytometry. J Periodont Res 19:570-3, 1984
2267. Fine DH, Mandel ID: Indicators of periodontal disease activity: an evaluation. J Clin Periodontol 13:533-46, 1986
2268. Sachsenmeier KF, Schell K, Morrissey LW, et al.: Detection of borreliacidal antibodies in hamsters by using flow cytometry. J Clin Microbiol 30:1457-61, 1992
2269. Tertti R, Eerola E, Lehtonen OP, Stahlberg TH, Viander M, Toivanen A: Virulence-plasmid is associated with the inhibition of opsonization in *Yersinia enterocolitica* and *Yersinia pseudotuberculosis*. Clin Exp Immunol 68:266-74, 1987
2270. Tertti R, Eerola E, Granfors K, Lahesmaa-Rantala R, Pekkola-Heino K, Toivanen A: Role of antibodies in the opsonization of *Yersinia* spp. Infect Immun 56:1295-300, 1988
2271. Tertti R, Granfors K, Lahesmaa-Rantala R, Toivanen A: Serum opsonic capacity against *Yersinia enterocolitica* O:3 in yersiniosis patients with or without reactive arthritis. Clin Exp Immunol 76:227-32, 1989
2272. Tosi MF, Czinn SJ: Opsonic activity of specific human IgG against *Helicobacter pylori*. J Infect Dis 162:156-62, 1990
2273. Dunn BE, Altmann M, Campbell GP: Adherence of *Helicobacter pylori* to gastric carcinoma cells: analysis by flow cytometry. Rev Infect Dis 13:S657-64, 1991
2274. Summersgill JT, Raff MJ, Miller RD: Interactions of virulent and avirulent *Legionella pneumophila* with human monocytes. J Leukoc Biol 47:31-8, 1990
2275. Apperloo-Renkema HZ, Wilkinson MHF, van der Waaij D: Circulating antibodies against fecal bacteria assessed by immunomorphometry: Combining quantitative immunofluorescence and image analysis. Epidemiol Infect 109:497-506, 1992
2276. van der Waaij LA, Mesander G, Limburg PC, van der Waaij D: Direct flow cytometry of anaerobic bacteria in human feces. Cytometry 16:270-9, 1994
2277. Kallioniemi O-P, Visakorpi T, Holli K, Isola JJ, Rabinovitch PS: Automated peak detection and cell cycle analysis of flow cytometric DNA histograms. Cytometry 16:250-5, 1994
2278. Silvestrini R (and the SICCAB Group for Quality Control of Cell Kinetic Determinations): Quality control for evaluation of the S-phase fraction by flow cytometry: A multicentric study. Cytometry (Commun Clin Cytometry) 18:11-6, 1994
2279. Stål O, Sullivan S, Sun X-F, Wingran S, Nordenskjöld B: Simultaneous analysis of c-*erb*B-2 expression and DNA content in breast cancer using flow cytometry. Cytometry 16:160-8, 1994
2280. Landay A, Ohlsson-Wilhelm B, Giorgi J: Application of flow cytometry to the study of HIV infection. AIDS 4:479-97, 1990
2281. Giorgi JV, Landay AL: HIV infection: diagnosis and disease progression evaluation. Methods Cell Biol 42: 437-55, 1994
2282. Waxdal MJ, Monical MC, Fleisher T, Marti GE: Inter-laboratory survey of lymphocyte immunophenotyping. Pathol Immunopathol Res 7:345-56, 1988
2283. Gelman R, Cheng S-C, Kidd P, Waxdal M, Kagan J: Assessment of the effects of instrumentation, monoclonal antibody, and fluorochrome on flow cytometric immunophenotyping: A report based on 2 years of the NIAID DAIDS flow cytometry quality assessment program. Clin Immunol Immunopathol 66:150-62, 1993
2284. Rickman WJ, Monical C, Waxdal MJ: Improved precision in the enumeration of absolute numbers of lymphocyte phenotypes with long-term monthly proficiency testing. Ann N Y Acad Sci 677:53-8, 1993 (in reference 1042)
2285. Homburger HA, Rosenstock W, Paxton H, Paton ML, Landay AL: Assessment of interlaboratory variability of immunophenotyping. Results of the College of American Pathologists Flow Cytometry Survey. Ann N Y Acad Sci 677:43-9, 1993 (in reference 1042)
2286. Landay AL, Jessop C, Lennette ET, Levy JA: Chronic fatigue syndrome: clinical condition associated with immune activation. Lancet 338:707-12, 1991
2287. Sligh JM, Roodman ST, Tsai CC: Flow cytometric indirect immunofluorescence assay with high sensitivity and specificity for detection of antibodies to human immunodeficiency virus (HIV). Am J Clin Path 91:210-4, 1989
2288. Gorse GJ, Frey SE, Newman FK, Belshe RB (and the AIDS Vaccine Clinical Trials Network): Detection of binding antibodies to native and recombinant human immunodeficiency virus type 1 envelope glycoproteins following recombinant gp160 immunization measured by flow cytometry and enzyme immunoassays. J Clin Microbiol 30:2606-12, 1992
2289. Heagy W, Crumpacker C, Lopez PA, Finberg RW: Inhibition of immune function by antiviral drugs. J Clin Invest 87:1916-24, 1991

2290. Pantaleo G, Demarest JF, Soudeyns H, Graziosi C, Denis F, Adelsberger JW, Borrow P, Saag MS, Shaw GM, Sekaly RP, Fauci AS: Major expansion of CD8+ T cells with a predominant Vβ usage during the primary immune response to HIV. Nature 370:463-7, 1994
2291. Landay AL, Mackewicz CE, Levy JA: An activated CD8+ T cell phenotype correlates with anti-HIV activity and asymptomatic clinical status. Clin Immunol Immunopathol 69:106-16, 1993
2292. Haq A, al-Hussein K, Lee J, al-Sedairy S: Changes in peripheral blood lymphocyte subsets associated with marathon running. Med Sci Sports Exerc 25:186-90, 1993
2293. Dekaris D, Sabioncello A, Mazuran R, Rabatic S, Svoboda-Beusan I, Racunica NL, Tomasic J: Multiple changes of immunologic parameters in prisoners of war. Assessments after release from a camp in Manjaca, Bosnia. JAMA 270:595-9, 1993
2294: Nowak R: A new test gives early warning of a growing killer. Science 264:1847-8, 1994
2295. Rabinovitch PS, Reid BJ, Haggitt RC, Norwood TH, Rubin CE: Progression to cancer in Barrett's esophagus is associated with genomic instability. Lab Invest 60:65-71, 1989
2296. Reid BJ, Blount PL, Rubin CE, Levine DS, Haggitt RC, Rabinovitch PS: Flow-cytometric and histological progression to malignancy in Barrett's esophagus: prospective endoscopic surveillance of a cohort [see comments]. Gastroenterology 102:1212-9, 1992
2297. Blount PL, Ramel S, Raskind WH, Haggitt RC, Sanchez CA, Dean PJ, Rabinovitch PS, Reid BJ: 17p allelic deletions and p53 protein overexpression in Barrett's adenocarcinoma. Cancer Res 51:5482-6, 1991
2298. Ramel S, Reid BJ, Sanchez CA, Blount PL, Levine DS, Neshat K, Haggitt RC, Dean PJ, Thor K, Rabinovitch PS: Evaluation of p53 protein expression in Barrett's esophagus by two-parameter flow cytometry. Gastroenterology 102:1220-8, 1992
2299. Reid BJ, Sanchez CA, Blount PL, Levine DS: Barrett's esophagus: cell cycle abnormalities in advancing stages of neoplastic progression. Gastroenterology 105:119-29, 1993
2300. Danes BS, Boyle PD, Traganos F, Melamed MR: A standardized assay to identify colon cancer genotypes by in vitro tetraploidy in human dermal fibroblasts. Dis Markers 4:271-82, 1986
2301. Danes BS, Boyle PD, Traganos F, Ringborg U, Melamed MR: In vitro hyperdiploidy in dermal fibroblasts: evidence for genetic predisposition in aerodigestive tract cancer. Clin Genet 31:25-34, 1987
2302. Danes BS, De Angelis P, Traganos F, Melamed MR: Tetraploidy in cultured dermal fibroblasts from patients with heritable colon cancer. Dis Markers 6:151-61, 1988
2303. Svendsen LB, Larsen JK, Christensen IJ: Human skin fibroblast in vitro tetraploidy. Flow cytometric DNA assay used to confirm metaphase assay in patients with various colonic diseases. Cancer Genet Cytogenet 39:245-51, 1989
2304. Okuda A, Kimura G: Commitment to ploidy conversion of 3Y1 cells during metaphase arrest by colcemid. Cell Tissue Kinet 21:21-31, 1988
2305. Shackney SE, Smith CA, Miller BW, Burholt DR, Murtha K, Giles HR, Ketterer DM, Pollice AA: Model for the genetic evolution of human solid tumors. Cancer Res 49:3344-54, 1989
2306. Dooley WC, Allison DC: Non-random distribution of abnormal mitoses in heteroploid cell lines. Cytometry 13:462-8, 1992
2307. Long MW, Heffner CH, Williams JL, Peters C, Prochownik EV: Regulation of megakaryocyte phenotype in human erythroleukemia cells. J Clin Invest 85:1072-84, 1990
2308. Mouthon M-A, Freund M, Titeux M, Katz A, Guichard J, Bréton-Gorius J, Vainchenker W: Growth and differentiation of the human megakaryoblastic cell line (ELF-153): A model for early stages of megakaryocytopoiesis. Blood 84:1085-97, 1994
2309. Sobek D, Senturia SD, Gray ML: Microfabricated fused silica flow chambers for flow cytometry. Presented at the 1994 Solid State Sensors and Actuators Workshop, Hilton Head, SC, June, 1994
2310. Juan G, Cavazzoni M, Sáez GT, O'Connor J-E: A fast kinetic method for assessing mitochondrial membrane potential in isolated hepatocytes with rhodamine 123 and flow cytometry. Cytometry 15:335-42, 1994
2311. Rens W, Van Oven CH, Stap J, Jakobs ME, Aten JA: Slit-scanning technique using standard cell sorter instruments for analyzing and sorting nonacrocentric human chromosomes, including small ones. Cytometry 16:80-7, 1994
2312. Doornbos RMP, Hoekstra AG, Deurloo KGI, De Grooth BG, Sloot PMA, Greve J: Lissajous-like patterns in scatter plots of calibration beads. Cytometry 16:236-42, 1994
2313. Ziegler O, Cantin C, Germain L, Dupuis M, Sekaly RP, Drouin P, Chiasson JL: Insulin binding to human cultured lymphocytes measured by flow cytometry using three ligands. Cytometry 16:339-45, 1994
2314. De Vita R, Cavallo D, Eleuteri P, Dell'Omo G: Evaluation of interspecific DNA content variations and sex identification in *Falconiformes* and *Strigiformes* by flow cytometric analysis. Cytometry 16:346-50, 1994
2315. Janeway CA, Travers P, Walport M, Shlomchik M: *Immunobiology*, 5th Ed. New York, Garland, 2001, xx + 732 pp (+ CD)
2316. Givan AL: *Flow Cytometry: First Principles*, 2nd Ed. New York, Wiley-Liss, 2001, xviii + 273 pp
2317. "Student" [Gossett WS]: On the error of counting with a haemacytometer. Biometrika 5:351-60, 1907
2318. Poisson SD: *Recherches sur la Probabilité des Jugements*. Paris, Bachelier, 1837.
2319. Soper HE: Tables of Poisson's exponential binomial limit. Biometrika 10:25-35, 1914
2320. American Society of Clinical Oncology. Clinical guidelines for the use of tumor markers in breast and colorectal cancer. J Clin Oncol 14:2843-77, 1996
2321. Bagwell CB, Clark GM, Spyratos F, Chassevent A, Bendahl P-O, Stol O, Killander D, Jourdan M-L, Romain S, Hunsberger B, Wright S, Baldetorp B. DNA and cell cycle analysis as prognostic indicators in breast tumors revisited. Clin Lab Med 21:875-95, 2001
2322. Johnson LA, Welch GR: Sex preselection: high-speed flow cytometric sorting of X and Y sperm for maximum efficiency. Theriogenology 52:1323-41, 1999
2323. Levinson G, Keyvanfar K, Wu JC, Fugger EF, Fields RA, Harton GL, Palmer FT, Sisson ME, Starr KM, Dennison-Lagos L, et al: DNA-based X-enriched sperm separation as an adjunct to preimplantation genetic testing for the prevention of X-linked disease. Hum Reprod 10:979-82, 1995
2324. Garner DL: Sex-Sorting mammalian sperm: concept to application in animals. J Androl 22:519-26, 2001
2325. Furlong EE, Profitt D, Scott MP: Automated sorting of live transgenic embryos. Nat Biotechnol 19:153-6, 2001
2326. Fu AY, Spence C, Scherer A, Arnold FH, Quake SR: A microfabricated fluorescence-activated cell sorter. Nature Biotech. 17:1109-11, 1999
2327. Chou H-P, Spence C, Scherer A, Quake S: A microfabricated device for sizing and sorting DNA molecules. Proc Natl Acad Sci USA. 96:11-3, 1999
2328. Brenner S, Williams SR, Vermaas EH, Storck T, Moon K, McCollum C, Mao JI, Luo S, Kirchner JJ, Eletr S, DuBridge RB, Burcham T, Albrecht G: In vitro cloning of complex mixtures of DNA on microbeads: physical separation of differentially expressed cDNAs. Proc Natl Acad Sci U S A 97:1665-70, 2000
2329. Brenner S, Johnson M, Bridgham J, Golda G, Lloyd DH, Johnson D, Luo S, McCurdy S, Foy M, Ewan M, Roth R, George D, Eletr S, Albrecht G, Vermaas E, Williams SR, Moon K, Burcham T, Pallas M, DuBridge RB, Kirchner J, Fearon K, Mao J, Corcoran K: Gene expression analysis by massively parallel signature sequencing (MPSS) on microbead arrays. Nat Biotechnol 18:630-4, 2000 (Comment in: Nat Biotechnol 18:597-8, 2000; Erratum in: Nat Biotechnol 18:1021, 2000)

2330. Shapiro HM, Lederman M, Connick E, Kessler H, Kuritzkes DR, Landay AL: Small differences in CD4+ T-cell production may go unnoticed [Letter]. AIDS 13:290-1, 1999

2331. Gross HJ, Verwer B, Houck D, Hoffman RA, Recktenwald D: Model study detecting breast cancer cells in peripheral blood mononuclear cells at frequencies as low as 10(-7). Proc Natl Acad Sci U S A 92:537-41, 1995

2332. Haugland RP: *Handbook of Fluorescent Probes and Research Products*, 9th Ed. Molecular Probes, Inc., Eugene, OR, 2002, 966pp (online at www.probes.com)

2333. Yan X, Grace WK, Yoshida TM, Habbersett RC, Velappan N, Jett JH, Keller RA, Marrone BL: Characteristics of different nucleic acid staining dyes for DNA fragment sizing by flow cytometry. Anal Chem 71:5470-80, 1999

2334. Yan X, Habbersett RC, Cordek JM, Nolan JP, Yoshida TM, Jett JH, Marrone BL: Development of a mechanism-based, DNA staining protocol using SYTOX orange nucleic acid stain and DNA fragment sizing flow cytometry. Anal Biochem 286:138-48, 2000

2335. Hennes KP, Suttle CA, Chan AM: Fluorescently labeled virus probes show that natural virus populations can control the structure of marine microbial communities. Appl Envir Microbiol 61: 3623-3627, 1995

2336. Marie D, Brussaard, CPD, Thyrhaug R, Bratbak G, Vaulot, D: Enumeration of marine viruses in culture and natural samples by flow cytometry. Appl Environ Microbiol 65: 45-52, 1999

2337. Brussaard CPD, Marie D Bratbak G: Flow cytometric detection of viruses, J Virol Methods 85: 175-82, 2000

2338. Smith PJ, Wiltshire M, Davies S, Patterson LH, Hoy T: A novel cell permeant and far red-fluorescing DNA probe, DRAQ5, for blood cell discrmination by flow cytometry. J Immunol Methods 229:131-9, 1999

2339. Smith PJ, Blunt N, Wiltshire M, Hoy T, Teesdale-Spittle P, Craven MR, Watson JV, Amos WB, Errington RJ, Patterson LH: Characteristics of a novel deep red/infrared fluorescent cell-permeant DNA probe, DRAQ5, in intact human cells analyzed by flow cytometry, confocal and multiphoton microscopy. Cytometry 40:280-91, 2000

2340. Ladd AC, Pyatt R, Gothot A, Rice S, McMahel J, Traycoff CM, Srour EF: Orderly process of sequential cytokine stimulation is required for activation and maximal proliferation of primitive human bone marrow CD34+ hematopoietic progenitor cells residing in G0. Blood 90:658-68, 1997

2341. Gothot A, Pyatt R, McMahel J, Rice S, Srour EF: Functional heterogeneity of human CD34(+) cells isolated in subcompartments of the G0/G1 phase of the cell cycle. Blood 90:4384-93, 1997

2342. Gothot A, van der Loo JC, Clapp DW, Srour EF: Cell cycle-related changes in repopulating capacity of human mobilized peripheral blood CD34(+) cells in non-obese diabetic/severe combined immune-deficient mice. Blood 92:2641-9, 1998

2343. Toba K, Winton EF, Koike T, Shibata A: Simultaneous three-color analysis of the surface phenotype and DNA-RNA quantitation using 7-amino-actinomycin D and pyronin Y. J Immunol Methods 182:193-207, 1995

2344. Toba K, Kishi K, Koike T, Winton EF, Takahashi H, Nagai K, Maruyama S, Furukawa T, Hashimoto S, Masuko M, Uesugi Y, Kuroha T, Tsukada N, Shibata A: Profile of cell cycle in hematopoietic malignancy by DNA/RNA quantitation using 7AAD/PY. Exp Hematol 24:894-901, 1996

2345. Toba K, Koike T, Watanabe K, Fuse I, Takahashi M, Hashimoto S, Takahashi H, Abe T, Yano T, Shibazaki Y, Itoh H, Aizawa Y: Cell kinetic study of normal human bone marrow hematopoiesis and acute leukemia using 7AAD/PY. Eur J Haematol 64:10-21, 2000

2346. Schmid I, Cole SW, Korin YD, Zack JA, Giorgi JV: Detection of cell cycle subcompartments by flow cytometric estimation of DNA-RNA content in combination with dual-color immunofluorescence. Cytometry 39:108-16, 2000

2347. Szöllösi J, Damjanovich S, Mátyus L: Application of fluorescence resonance energy transfer in the clinical laboratory: routine and research. Cytometry (Comm Clin Cytometry) 34:159-179, 1998

2348. Panchuk-Voloshina N, Haugland RP, Bishop-Stewart J, Bhalgat MK, Millard PJ, Mao F, Leung WY, Haugland RP: Alexa dyes, a series of new fluorescent dyes that yield exceptionally bright, photostable conjugates. J Histochem Cytochem 47:1179-88, 1999

2349. Lyons AB, Hasbold J, Hodgkin PD: Flow cytometric analysis of cell division history using dilution of carboxyfluorescein diacetate succinimidyl ester, a stably integrated fluorescent probe. Methods Cell Biol 63:375-398, 2001 (in reference 2385)

2350. Nordon RE, Ginsberg SS, Eaves CJ: High resolution cell division tracking demonstrates the Flt3 ligand dependence of human marrow CD34+CD38- cell production in vitro. Br J Haematol 98:528-39, 1997

2351. Novo D, Perlmutter NG, Hunt RH, Shapiro HM: Multiparameter flow cytometric analysis of antibiotic effects on membrane potential, membrane permeability, and bacterial counts of *Staphylococcus aureus* and *Micrococcus luteus*. Antimicrob Agents Chemother 44:827-34, 2000

2352. Darzynkiewicz Z, Juan G, Li X, Gorczyca W, Murakami T, Traganos F: Cytometry in cell necrobiology: analysis of apoptosis and accidental cell death (necrosis). Cytometry 27:1-20, 1997

2353. Koopman G, Reutelingsperger CP, Kuijten GA, Keehnen RM, Pals ST, van Oers MH: Annexin V for flow cytometric detection of phosphatidylserine expression on B cells undergoing apoptosis. Blood 84:1415-20, 1994

2354. Vermes I, Haanen C, Steffens-Nakken H, Reutelingsperger C: A novel assay for apoptosis. Flow cytometric detection of phosphatidylserine expression on early apoptotic cells using fluorescein labelled Annexin V. J Immunol 184:39-51 1995

2355. Shapiro HM: Membrane potential estimation by flow cytometry. Methods 21:271-9, 2000 (in reference 2402)

2356. Gonzalez, JE, Tsien, RY: Improved indicators of cell membrane potential that use fluorescence resonance energy transfer. Chem Biol 4:269-77, 1997

2357. Novo D, Perlmutter NG, Hunt RH, Shapiro HM: Accurate flow cytometric membrane potential measurement in bacteria using diethyloxacarbocyanine and a ratiometric technique. Cytometry 35:55-63, 1999

2358. Burchiel SW, Edwards BS, Kuckuck FW, Lauer FT, Prossnitz ER, Ransom JT, Sklar LA: Analysis of free intracellular calcium by flow cytometry: Multiparameter and pharmacologic applications. Methods 21:221-30, 2000

2359. Altman JD, Moss PA, Goulder PJ, Barouch DH, McHeyzer-Williams MG, Bell JI, McMichael AJ, Davis MM: Phenotypic analysis of antigen-specific T lymphocytes. Science 274:94-6, 1996

2360. Dunbar PR, Ogg GS, Chen J, Rust N, van der Bruggen P, Cerundolo V: Direct isolation, phenotyping and cloning of low-frequency antigen-specific cytotoxic T lymphocytes from peripheral blood. Curr Biol 8:413-6, 1998

2361. Burrows SR, Kienzle N, Winterhalter A, Bharadwaj M, Altman JD, Brooks A: Peptide-MHC class I tetrameric complexes display exquisite ligand specificity. J Immunol 165:6229-34, 2000

2362. Hoffmann TK, Donnenberg VS, Friebe-Hoffmann U, Meyer EM, Rinaldo CR, DeLeo AB, Whiteside TL, Donnenberg AD: Competition of peptide-MHC class I tetrameric complexes with anti-CD3 provides evidence for specificity of peptide binding to the TCR complex. Cytometry 41:321-8, 2000

2363. Novak EJ, Liu AW, Nepom GT, Kwok WW: MHC class II tetramers identify peptide-specific human CD4(+) T cells proliferating in response to influenza A antigen. J Clin Invest 104:R63-7, 1999; Comment in: J Clin Invest 104:1669-70, 1999

2364. Meyer AL, Trollmo C, Crawford F, Marrack P, Steere AC, Huber BT, Kappler J, Hafler DA: Direct enumeration of Borrelia-reactive CD4 T cells ex vivo by using MHC class II tetramers. Proc Natl Acad Sci U S A 97:11433-8, 2000

2365. Schena M, Shalon D, Davis RW, Brown PO: Quantitative monitoring of gene expression patterns with a complementary DNA microarray. Science 270:467-70, 1995; Comment in: Science 270:368-9, 371, 1995
2366. Lipshutz RJ, Morris MS, Chee M, Hubbell E, Kozal MJ, Shah N, Shen N, Yang R, Fodor SPA: Using Oligonucleotide Probe Arrays to Access Genetic Diversity. BioTechniques 19:442-447, 1995
2367. Hacial JG, Brody LC, Chee MS, Fodor SPA, Collins FS: Detection of heterozygous mutations in BRCA1 using high density oligonucleotide arrays and two-colour fluorescence analysis. Nature Genetics 14:441-7, 1996
2368. Kononen J, Bubendorf L, Kallioniemi A, Bärlund M, Schraml P, Leighton S, Torhorst J, Mihatsch MJ, Sauter G, Kallioniemi OP: Tissue microarrays for high-throughput molecular profiling of tumor specimens. Nat Med 4:844-7, 1998; Comment in: Nat Med 4:767-8, 1998
2369. Fulton RJ, McDade RL, Smith PL, Kienker LJ, Kettman JR Jr: Advanced multiplexed analysis with the FlowMetrix system. Clin Chem 43:1749-56, 1997
2370. Oliver KG, Kettman JR, Fulton RJ: Multiplexed analysis of human cytokines by use of the FlowMetrix system. Clin Chem 44:2057-60, 1998
2371. Iannone MA, Consler TG, Pearce KH, Stimmel JB, Parks DJ, Gray JG: Multiplexed molecular interactions of nuclear receptors using fluorescent microspheres. Cytometry 44:326-37, 2001
2372. Smith PL, WalkerPeach CR, Fulton RJ, DuBois DB: A rapid, sensitive, multiplexed assay for detection of viral nucleic acids using the FlowMetrix system. Clin Chem 44:2054-6, 1998
2373. Iannone MA, Taylor JD, Chen J, Li MS, Rivers P, Slentz-Kesler KA, Weiner MP: Multiplexed single nucleotide polymorphism genotyping by oligonucleotide ligation and flow cytometry. Cytometry 39:131-40, 2000
2374. Taylor JD, Briley D, Nguyen Q, Long K, Iannone MA, Li MS, Ye F, Afshari A, Lai E, Wagner M, Chen J, Weiner MP: Flow cytometric platform for high-throughput single nucleotide polymorphism analysis. Biotechniques 30:661-6, 668-9, 2001
2375. Yang L, Tran DK, Wang X: BADGE, Beads Array for the Detection of Gene Expression, a high-throughput diagnostic bioassay. Genome Res 11:1888-98, 2001
2376. Chalfie M, Kain S (eds): *Green Fluorescent Protein: Properties, Applications, and Protocols.* New York, Wiley-Liss, 1998, xiv + 385 pp
2377. Tsien RY: The green fluorescent protein. Annu Rev Biochem 67:509-44, 1998
2378. Chan FKM, Siegel RM, Zacharias D, Swofford R, Holmes KL, Tsien RY, Lenardo MJ: Fluorescence resonance energy transfer analysis of surface receptor interactions and signaling using spectral variants of the green fluorescent protein. Cytometry 44:361-8, 2001
2379. Lenkei R, Mandy F, Marti G, Vogt R (eds): Quantitative fluorescence cytometry: An emerging consensus. Cytometry 33:93-287, 1998
2380. Shapiro HM: Scanning laser cytometry. Unit 2.10. In: Robinson JP, Darzynkiewicz Z, Dean P, Dressler L, Rabinovitch P, Stewart C, Tanke H, Wheeless L (eds): *Current Protocols in Cytometry,* New York, John Wiley & Sons, 1999, pp 2.10.1-2.10.8.
2381. Darzynkiewicz Z, Bedner E, Li X, Gorczyca W, Melamed MR: Laser-scanning cytometry: A new instrumentation with many applications. Exp Cell Res 249:1-12, 1999
2382. Li X, Darzynkiewicz Z. The Schrodinger's cat quandary in cell biology: integration of live cell functional assays with measurements of fixed cells in analysis of apoptosis. Exp Cell Res 249:404-12, 1999
2383. Tibbe AGJ, de Grooth BG, Greve J, Dolan GJ, Rao C, Terstappen LWMM: Cell analysis system based on compact disk technology. Cytometry 47:173-182, 2002
2384. Ormerod MG: *Flow Cytometry,* 2d Ed. RMS Microscopy Handbooks, Volume 44, Oxford (UK), BIOS Scientific Publishers, 1999, 128 pp
2385. Darzynkiewicz Z, Crissman HA, Robinson JP (eds): *Methods in Cell Biology,* Vol 63, *Cytometry,* 3d Ed. Part A. San Diego, Academic Press, 2001, xxii + 650 pp
2386. Darzynkiewicz Z, Crissman HA (eds): *Methods in Cell Biology,* Vol 64, *Cytometry,* 3d Ed, Part B. San Diego, Academic Press, 2001, xxx + 614 pp
2387. Darzynkiewicz Z, Robinson JP, Crissman HA (eds): *Methods in Cell Biology,* Vol 41,*Flow Cytometry,* 2d Ed, Part A. San Diego, Academic Press, 1994, xxxii + 591 pp
2388. Darzynkiewicz Z, , Crissman HA (eds): *Methods in Cell Biology,* Vol 42, *Flow Cytometry,* 2d Ed, Part B. San Diego, Academic Press, 1994, xxxiv + 695 pp
2389. Robinson JP, Darzynkiewicz Z, Dean PN, Hibbs AR, Orfao A, Rabinovitch PS, Wheeless LL (eds)): *Current Protocols in Cytometry,* New York, John Wiley & Sons (continuing series in looseleaf and CD-ROM)
2390. Diamond RA, DeMaggio S (eds): *In Living Color. Protocols in Flow Cytometry and Cell Sorting.* Berlin, Springer, 2000, xxvi + 800 pp
2391. Ormerod MG (ed): *Flow Cytometry. A Practical Approach,* 3rd Ed. Oxford (UK), Oxford University Press, 2000, xx + 276 pp
2392. Jaroszeski M, Heller R: *Flow Cytometry Protocols.* Methods in Molecular Biology, Volume 91, Totowa (NJ), Humana Press, 1998, x + 274 pp
2393. Keren D, McCoy JP Jr, Carey JL (eds): *Flow Cytometry in Clinical Diagnosis,* 3rd Ed. Chicago, American Society for Clinical Pathology Press, 2001, xii + 739 pp
2394. Owens MA, Loken MR: *Flow Cytometry Principles for Clinical Laboratory Practice. Quality Assurance for Quantitative Immunophenotyping.* New York, Wiley-Liss, 1995, xiv + 224 pp
2395. Stewart CC, Nicholson JKA(eds): *Immunophenotyping* (Cytometric Cellular Analysis Series). New York, Wiley-Liss, 2000, xiii + 448 pp
2396. Robinson JP, Babcock GF (eds) *Phagocyte Function. A Guide for Research and Clinical Evaluation.* (Cytometric Cellular Analysis Series). New York, Wiley-Liss, 1998, xiv + 385 pp
2397. Cossarizza A, Kaplan D (eds): *Cellular Aspects of HIV Infection* (Cytometric Cellular Analysis Series). New York, Wiley-Liss, 2002, xiv + 458 pp
2398. McCoy JP Jr, Keren DF (eds): New applications of flow cytometry. Clin Lab Med 21:697-932, 2001
2399. Durack G, Robinson JP (eds): *Emerging Tools for Single-Cell Analysis. Advances in Optical Measurement Technologies* (Cytometric Cellular Analysis Series). New York, Wiley-Liss, 2000, x + 359 pp
2400. Al-Rubeai M, Emery AN (eds): *Flow Cytometry Applications in Cell Culture,* New York, Marcel Dekker, 1996, xii + 331 pp
2401. Zola H (ed): Special Issue: Flow Cytometry. J Immunol Methods 243:1-262, 2000
2402. Weaver JL (ed): Flow cytometry: measuring cell populations and studying cell physiology. Methods 21:199-312, 2000
2403. Reckermann M, Colijn F (eds): Aquatic flow cytometry: Achievements and prospects. Scientia Marina 64:121-268, 2000
2404. Alberghina L, Porro D, Shapiro H, Srienc F, Steen H (eds): Microbial analysis at the single-cell level. J Microbiol Methods 42:1-114, 2000. Full text of all articles is available at www.elsevier.com/ locate/jmicmeth
2405. Hecht J: *Understanding Lasers: An Entry-Level Guide.* New York, IEEE Press, 1994
2406. Harbison JP, Nahory RE: *Lasers: Harnessing the Atom's Light.* New York, Scientific American Library, 1997
2407. Murphy DB: *Fundamentals of Light Microscopy and Electronic Imaging.* New York, Wiley-Liss, 2001, xii + 368 pp
2408. Rost F, Oldfield R: *Photography with a Microscope.* Cambridge (UK), Cambridge University Press, 2000, x + 278 pp
2409. Herman B: *Fluorescence Microscopy,* 2d Ed. New York, BIOS Scientific Publishers/Springer (in association with the Royal Microscopical Society) xiv + 170 pp
2410. Petzold C: *Code. The Hidden Language of Computer Hardware and Software.* Redmond (WA), Microsoft Press, 1999

2411. Petzold C: *Programming Windows*, 5th Ed. Redmond (WA), Microsoft Press, 1999
2412. Smith SW: *The Scientist and Engineer's Guide to Digital Signal Processing*, 2d Ed. San Diego, California Technical Publishing, 1998, xiv + 650 pp (the entire book may be downloaded in .pdf format gratis at www.DSPguide.com)
2413. Lakowicz JR: *Principles of Fluorescence Spectroscopy*, 2d Ed. New York, Kluwer Academic/Plenum Publishers, 1999, xxvi + 698 pp
2414. Sharma A, Schulman SG: *Introduction to Fluorescence Spectroscopy*. New York, Wiley-Interscience, 1999 + 173 pp
2415. Valeur B: *Molecular Fluorescence*. Weinheim, Wiley-VCH, 2002, xiv + 387 pp
2416. Mason WT (ed): *Fluorescent and Luminescent Probes for Biological Activity. A Practical Guide to Technology for Quantitative Real-Time Analysis*, 2d Ed. London, Academic Press, 1999, xxviii + 647 pp
2417. Inoué S, Spring KR: *Video Microscopy. The Fundamentals*, 2d Ed. New York, Plenum Press, 1997, xxviii + 742 pp
2418. Tufte E: *The Visual Display of Quantitative Information*, 2d Ed. Cheshire (CT), Graphics Press, 2002, 156 pp
2419. Tufte E: *Visual Explanations. Images and Quantities, Evidence and Narrative*. Cheshire (CT), Graphics Press, 1997, 156 pp
2420. Wilkinson MHF, Schut F (eds): *Digital Image Analysis of Microbes. Imaging, Morphometry, Fluorometry and Motility Techniques and Applications*. Chichester, John Wiley & Sons, 1998, xxvi + 551 pp
2421. Wang XF, Herman B (eds): *Fluorescence Imaging Spectroscopy and Microscopy*. Volume 137 in Chemical Analysis: A Series of Monographs on Analytical Chemistry and Its Applications, (Winefordner JD, Series Ed). New York, John Wiley & Sons, 1996. xxxii + 483 pp
2422. Pawley JB (ed): *Handbook of Biological Confocal Microscopy*, 2d Ed. New York, Plenum Press, 1995, xxiv + 632 pp
2423. Stopa PJ: The flow cytometry of Bacillus anthracis spores revisited. Cytometry 41:237-44, 2000
2424. Garfield S: *Mauve. How One Man Invented a Color that Changed the World*. New York, W. W. Norton, 2000, 222 pp
2425. de Kruif, P: *Microbe Hunters*. With a new introduction by F. Gonzalez-Crussi. San Diego, Harcourt, Brace, 1996
2426. Udenfriend S: Development of the spectrophotofluorometer and its commercialization. Protein Sci 4:542-51, 1995
2427. Keating P, Cambrosio A: Fluorescence-Activated Cell Sorter. In: Bud R, Warner DJ (eds), *Instruments Of Science. An Historical Encyclopedia*. New York, Garland, 1998, pp 247-9
2428. Cambrosio A, Keating P: *Exquisite Specificity. The Monoclonal Antibody Revolution*. New York, Oxford University Press, 1995, xxii + 243 pp
2429. Cambrosio A, Keating P: Monoclonal Antibodies: From Local to Extended Networks. In: Thackray A (ed): *Private Science: Biotechnology and the Rise of the Molecular Sciences*. Philadelphia, University of Pennsylvania Press, 1998; pp165-81
2430. Cambrosio A, Keating P, Guttmann RD: New Medical Technologies and Clinical Practice: A Survey of Lymphocyte Subset Monitoring. Clinical Transpl 8:532-40, 1994
2431. Shaw S, Turni LA, Latz KS (eds): *Protein Reviews on the Web (PROW)*. On -line at: www.ncbi.nlm.nih.gov/prow
2432. Keating P, Cambrosio A: Interlaboratory Life: Regulating Flow Cytometry. In: Gaudillière J-P, Löwy I (eds): *The Invisible Industrialist: Manufacturers and the Construction of Scientific Knowledge*. London, Macmillan/New York, St. Martin's Press, 1998; pp 250-95
2433. Cambrosio A, Keating P: Of Lymphocytes and Pixels: The Techno-Visual Production of Cell Populations. Stud Hist Philos Biol BiomedSci 31:233-70, 2000
2434. Keating P, Cambrosio A: Real Compared to What?: Diagnosing Leukemias and Lymphomas. In: Lock M, Young A, Cambrosio A (eds): *Living and Working with the New Medical Technologies. Intersections of Inquiry*. Cambridge (UK), Cambridge University Press, 2000; pp 103-34
2435. Keating P, Cambrosio A: Biomedical Platforms. Configurations 8: 337-87, 2000 (a book with the same title has been submitted for publication)
2436. Watson JV: The early fluidic and optical physics of cytometry. Cytometry (Comm Clin Cytom) 38:2-14, 1999
2437. Watson JV: A brief history of numbers and statistics with cytometric applications. Cytometry (Comm Clin Cytom) 46:1-22, 2001
2438. Taylor BN: *Guide for the Use of the International System of Units (SI)*. NIST Special Publication 811, 1995 Edition. United States Department of Commerce, National Institute of Standards and Technology. See: http://physics.nist.gov/cuu/Units/index.html
2439. Asbury CL, Uy JL, van den Engh G: Polarization of scatter and fluorescence signals in flow cytometry. Cytometry 40:88-101, 2000
2440. Doornbos RMP, de Grooth BG, Greve J: Experimental and model investigations of bleaching and saturation of fluorescence in flow cytometry. Cytometry 29:204-14, 1997
2441. Gitin M, Ginouves P, Schulze M, Seelert W, Rosperich J, Pfaff J, Spinelli L: A compact solid-state 488 nm laser for cell analysis applications. Cytometry Supp 10:71, 2000
2442. Hänninen PE, Soini JT, Soini E: Photon-burst analysis in two-photon fluorescence excitation flow cytometry. Cytometry 36:183-8, 1999
2443. Boyd RW: Radiance Theorem. *Radiometry and the Detection of Optical Radiation*, New York, John Wiley & Sons, 1983, p 74ff
2444. Shapiro HM, Perlmutter NG: Violet laser diodes as light sources for cytometry. Cytometry 44:133-6, 2001
2445. Hoffman R, Chase E: Light emitting diodes as light sources for flow cytometry. Cytometry Supp 10:163, 2000
2446. Unger M, Kartalov E, Chiu C-S, Lester HA, Quake SR: Single-molecule fluorescence observed with mercury lamp illumination. BioTechniques 27:1008-14, 1999
2447. Janossy G, Jani IV, Kahan M, Barnett D, Mandy F, Shapiro H: Precise CD4 T-cell counting using red diode laser excitation: for richer, for poorer. Cytometry (Clinical Cytometry) 50:78-85, 2002
2448. Nakamura S, Pearton S, Fasol G: *The Blue Laser Diode. The Complete Story*. Berlin, Springer, 2000, 368 pp
2449. Ost V, Neukammer J, Rinneberg H: Flow cytometric differentiation of erythrocytes and leukocytes in dilute whole blood by light scattering. Cytometry 32:191-197, 1998
2450. Baird GS, Zacharias DA, Tsien RY: Biochemistry, mutagenesis, and oligomerization of DsRed, a red fluorescent protein from coral. Proc Natl Acad Sci U S A 97:11984-9, 2000
2451. "Photomultiplier Tube" Editorial Committee (Kume H, Chief Ed): *Photomultiplier Tube. Principle to Application*. Hamamatsu, Hamamatsu Photonics K. K., 1994, 244 pp
2452. Agronskaia A, Florians A, van der Werf KO, Schins JM, de Grooth BG, Greve J: Photon-counting device compatible with conventional flow cytometric data acquisition electronics. Cytometry 32:255-9, 1998
2453. Sieracki CK, Sieracki ME, Yentsch CS: An imaging-in-flow system for automated analysis of marine microplankton. Mar Ecol Prog Ser 168:285-96, 1998
2454. Graves SW, Nolan JP, Jett JH, Martin JC, Sklar LA: Nozzle design parameters and their effects on rapid sample delivery in flow cytometry. Cytometry 47:127-37, 2002
2455. Mariella R Jr, van den Engh G, Masquelier D, Eveleth G: Flow-stream waveguide for collection of perpendicular light scatter in flow cytometry. Cytometry 24:27-31, 1996
2456. Mariella RP Jr, Huang Z, Langlois RG: Characterization of the sensitivity of side scatter in a flow-stream waveguide flow cytometer. Cytometry 37:160-3, 1999
2457. Leif RC, Cayer ML, Dailey W, Stribling T, Gordon K. Use of a spherical multiparameter transducer for flow cytometry. Cytometry 20:185-90, 1995

2458. Goodwin PM, Ambrose WP, Martin JC, Keller RA. Spatial dependence of the optical collection efficiency in flow cytometry. Cytometry 21:133-44, 1995

2459. Temsch EM, Obermayer R, Doležel J, Greilhuber J: Application of an optical immersion-gel in a flow cytometer with horizontally oriented objective. Biotechnic Histochem 76:11-4, 2001

2460. Nolan JP, Posner RG, Martin JC, Habbersett R, Sklar LA. A rapid mix flow cytometer with subsecond kinetic resolution. Cytometry 21:223-9, 1995

2461. Seamer LC, Kuckuck F, Sklar LA: Sheath fluid control to permit stable flow in rapid mix flow cytometry. Cytometry 35:75-9, 1999

2462. Edwards BS, Kuckuck F, Sklar LA: Plug flow cytometry: An automated coupling device for rapid sequential flow cytometric sample analysis. Cytometry 37:156-9, 1999

2463. Kuckuck FW, Edwards BS, Sklar LA: High throughput flow cytometry. Cytometry 44:83-90, 2001

2464. Cucci TL, Sieracki ME: Effects of mismatched refractive indices in aquatic flow cytometry. Cytometry 44:173-8, 2001

2465. Phi-Wilson J, Harvey J, Goix P, O'Neill R: A technology for the rapid acquisition of cell number and viability. American Biotechnol Lab May, 2001, pp 35-6

2466. Wersto RP, Chrest FJ, Leary JF, Morris C, Stetler-Stevenson MA, Gabrielson E: Doublet discrimination in DNA cell-cycle analysis. Cytometry (Commun Clin Cytom) 46:296-306, 2001

2467. Shapiro HM, Perlmutter NG, Stein PG: A flow cytometer designed for fluorescence calibration. Cytometry 33:280-7, 1998 (in reference 2379)

2468. Baert L, Theunissen L, Vergult G, Maes J, Arts J (eds): *Digital Audio and Compact Disc Technology*, 3d Ed. Oxford, Focal Press, 1995, ix + 305 pp

2469. Pohlmann KC: *Principles of Digital Audio*, 4th Ed. New York, McGraw-Hill, 2000, xvi + 736 pp

2470. Shapiro HM: *Songs for the Jaundiced Ear* (audio CD). West Newton, MA, Shapiro Productions, 1998.

2471. Hudson JC, Porcelli RT, Russell TR. Flow cytometric immunofluorescence and DNA analysis: using a 1.5 mW helium-neon laser (544 nm). Cytometry 21:211-7, 1995

2472. Strauss K, Hannet I, Engels S, Shiba A, Ward DM, Ullery S, Jinguji MG, Valinsky J, Barnett D, Orfao A, Kestens L: Performance evaluation of the FACSCount System: a dedicated system for clinical cellular analysis. Cytometry (Commun Clin Cytom) 26:52-9, 1996

2473. Zilmer NA, Godavarti M, Rodriguez JJ, Yopp TA, Lambert GM, Galbraith DW. Flow cytometric analysis using digital signal processing. Cytometry 20:102-17, 1995

2474. Godavarti M, Rodriguez JJ, Yopp TA, Lambert GM, Galbraith DW: Automated particle classification based on digital acquisition and analysis of flow cytometric pulse waveforms. Cytometry 24:330-9, 1996

2475. Fink T, Mao Y: *The 85 Ways to Tie a Tie: The Science and Aesthetics of Tie Knots*. London, Fourth Estate, 1999, 144 pp

2476. Wood JCS, Hoffman RA: Evaluating fluorescence sensitivity on flow cytometers: an overview. Cytometry 33:256-9, 1998 (in reference 2379)

2477. Wood JCS: Fundamental flow cytometer properties governing sensitivity and resolution. Cytometry 33:260-6, 1998 (in reference 2379)

2478. Chase ES, Hoffman RA: Resolution of dimly fluorescent particles: a practical measure of fluorescence sensitivity. Cytometry 33:267-79, 1998 (in reference 2379)

2479. Hoffman RA, Wood JCS: Definition and characterization of flow cytometer fluorescence sensitivity. Cytometry Supp 11:125-6, 2002

2480. Hoffman RA, Kuhlmann CW: Effect of flow cytometer detection efficiency, background light, and spectral overlap on the resolution of dimly fluorescent populations. Cytometry Supp 11:121-2, 2002

2481. Davis KA, Abrams B, Iyer SB, Hoffman RA, Bishop JE: Determination of CD4 antigen density on cells: role of antibody valency, avidity, clones, and conjugation. Cytometry 33:197-205, 1998 (in reference 2379)

2482. Pannu KK, Joe ET, Iyer SB: Performance evaluation of QuantiBRITE phycoerythrin beads. Cytometry 45:250-8, 2001

2483. Bland JM, Altman DG: Statistical methods for assessing agreement between two methods of clinical measurement. Lancet i:307-10, 1986

2484. O'Gorman MRG, Gelman R, Site Investigators, NIAID New CD4 Technologies Focus Group: Inter- and intrainstitutional evaluation of automated volumetric capillary cytometry for the quantitation of CD4- and CD8-positive T-lymphocytes in the peripheral blood of persons infected with human immunodeficiency virus. Clin Diagn Lab Immunol 4:173-9, 1997

2485. Herbert DJ, Bagwell CB, Munson ME, Hunsberger BC: Examination of some DNA guidelines using simulation data. Cytometry Supp 9:96, 1988; also Herbert DJ, personal communication, 2002, and Purdue Web Site posting 3/28/02

2486. Roederer M: Spectral compensation for flow cytometry: visualization artifacts, limitations, and caveats. Cytometry 45:194-205, 2001

2487. Lampariello F, Aiello A: Complete mathematical modeling method for the analysis of immunofluorescence distributions composed of negative and weakly positive cells. Cytometry 32:241-54, 1998

2488. Lampariello F: On the use of the Kolmogorov-Smirnov statistical test for immunofluorescence histogram comparison. Cytometry 39:179-88, 2000

2489. Roederer M, Treister A, Moore W, Herzenberg LA: Probability binning comparison: a metric for quantitating univariate distribution differences. Cytometry 45:37-46, 2001

2490. Watson JV: Proof without prejudice revisited: immunofluorescence histogram analysis using cumulative frequency subtraction plus ratio analysis of means. Cytometry 43:55-68, 2001

2491. Zeng Q, Greenes RA, Young AJ, Boxwala A, Rawn J, Long W, Wand M, Salganik M, Milford EL, Mentzer SJ: Molecular identification using flow cytometry histograms and information theory. Proc AMIA Symp 2001:776-80, 2001

2492. Schmidt W: Comment on R.A: White's v function. Cytometry 24:289-91, 1996 - re: doubling time

2493. Cain SJ, Chau PC: A transition probability cell cycle model simulation of bivariate DNA/bromodeoxyuridine distributions. Cytometry 27:239-49, 1997

2494. Torricelli A, Bisiach M, Spinelli L, Ubezio P: From flow cytometric BrdUrd data to cell population growth and doubling time. Cytometry 29:222-32, 1997

2495. Johansson MC, Johansson R, Baldetorp B, Oredsson SM: Comparison of different labelling index formulae used on bromodeoxyuridine-flow cytometry data. Cytometry 32:233-40, 1998

2496. White RA, Meistrich ML, Pollack A, Terry NH: Simultaneous estimation of T(G2+M), T(S), and T(pot) using single sample dynamic tumor data from bivariate DNA-thymidine analogue cytometry. Cytometry 41:1-8, 2000

2497. DeRosa SC, Herzenberg LA, Herzenberg LA, Roederer M: 11-color, 13-parameter flow cytometry: Identification of human naive T cells by phenotype, function, and T-cell receptor diversity. Nature Med 7:245-8, 2001

2498. Roederer M, Moore W, Treister A, Hardy RR, Herzenberg LA: Probability binning comparison: a metric for quantitating multivariate distribution differences. Cytometry 45:47-55, 2001

2499. Roederer M, Hardy RR: Frequency difference gating: a multivariate method for identifying subsets that differ between samples. Cytometry 45:56-64, 2001

2500. Baggerly KA: Probability binning and testing agreement between multivariate immunofluorescence histograms: extending the chi-squared test. Cytometry 45:141-50, 2001

2501. Hokanson JA, Rosenblatt JI, Leary JF: Some theoretical and practical considerations for multivariate statistical cell classification useful in autologous stem cell transplantation and tumor cell purging. Cytometry 36:60-70, 1999

2502. Reyes C, Adjouadi M: A directional clustering technique for random data classification. Cytometry 27:126-35, 1997

2503. Frankel DS, Frankel SL, Binder BJ, Vogt RF: Application of neural networks to flow cytometry data analysis and real-time cell classification. Cytometry 23:290-302, 1996
2504. Davey HM, Jones A, Shaw AD, Kell DB: Variable selection and multivariate methods for the identification of microor-ganisms by flow cytometry. Cytometry 35:162-8, 1999
2505. Seamer LC, Bagwell CB, Barden L, Redelman D, Salzman GC, Wood JCS, Murphy RF: Proposed new data file standard for flow cytometry, version FCS 3.0. Cytometry 28:118-22, 1997
2506. Durack G: Cell-sorting technology. In reference 2399, pp 1-19
2507. van den Engh G: High-speed cell sorting. In reference 2399, pp 21-48
2508. Leary JF: Rare-event detection and sorting of rare cells. Inreference 2399, pp 49-72
2509. Fu AY, Chou HP, Spence C, Arnold FH, Quake SR: An integrated microfabricated cell sorter. Anal Chem 74:2451-7, 2002
2510. Zhou H, Lin B, Wu W, Zhang Y, Wang L: A low-voltage droplet charging circuit with simulative cell-sorting function for flow cytometer-cell sorter. Cytometry 39:306-9, 2000
2511. Whitesides GM, Ostuni E, Takayama S, Jiang X, Ingber DE: Soft lithography in biology and biochemistry. Annu Rev Biomed Eng 3:335-73, 2001
2512. Unger MA, Chou H-P, Thorsen T, Scherer A, Quake SR: Monolithic microfabricated valves and pumps by multilayer soft lithography. Science 288:113-6, 2000
2513. Quake SR, Scherer A: From micro to nano fabrication with soft materials. Science 290:1536-40, 2000
2514. Thorsen T, Maerkl SJ, Quake SR: Microfluidic large scale integration. Science 298:580-4, 2002
2515. Huh D, Tung Y-C, Wei H-H, Grotberg JB, Skerlos SJ, Kurabayashi K, Takayama T: Use of air-liquid two-phase flow in hydrophobic microfluidic channels for disposable flow cytometers. Biomed Microdevices 4:141-9, 2002
2516. Keij JF, Groenewegen AC, Dubelaar GB, Visser JW: High-speed photodamage cell selection using a frequency-doubled argon ion laser. Cytometry 19:209-16, 1995
2517. Rosenblatt JI, Hokanson JA, McLaughlin SR, Leary JF: Theoretical basis for sampling statistics useful for detecting and isolating rare cells using flow cytometry and cell sorting. Cytometry 27:233-8, 1997
2518. Seidl J, Knuechel R, Kunz-Schughart LA: Evaluation of membrane physiology following fluorescence activated or magnetic cell separation. Cytometry 36:102-11, 1999
2519. Ferbas J, Chadwick KR, Logar A, Patterson AE, Gilpin RW, Margolick JB. Assessment of aerosol containment on the ELITE flow cytometer. Cytometry (Commun Clin Cytom) 22:45-7, 1995
2520. Schmid I, Dean PN: Introduction to the biosafety guidelines for sorting of unfixed cells. Cytometry 28:97-8, 1997
2521. Schmid I, Nicholson JK, Giorgi JV, Janossy G, Kunkl A, Lopez PA, Perfetto S, Seamer LC, Dean PN: Biosafety guidelines for sorting of unfixed cells. Cytometry 28:99-117, 1997
2522. Sorensen TU, Gram GJ, Nielsen SD, Hansen JE: Safe sorting of GFP-transduced live cells for subsequent culture using a modified FACS vantage. Cytometry 37:284-90, 1999
2523. Oberyszyn AS, Robertson FM: Novel rapid method for visualization of extent of aerosol contamination during high-speed sorting of potentially biohazardous samples. Cytometry 43:217-22, 2001
2524. Wietzorrek J, Plesnila N, Baethmann A, Kachel V: A new multiparameter flow cytometer: optical and electrical cell analysis in combination with video microscopy in flow. Cytometry 35:291-301, 1999
2525. DeBlois RW, Bean CP: Counting and sizing of submicron particles by the resistive pulse technique. Rev Sci Instrum 41:909-16, 1970
2526. DeBlois RW, Bean CP, Wesley RKA: Electrokinetic measurements with submicron particles and pores by the resistive pulse technique. J Colloid Interface Sci 61:323-35, 1977
2527. DeBlois RW, Wesley RKA: Sizes and concentrations of several type C oncornaviruses and bacteriophage T2 by the resistive pulse technique. J Virol 23:227-33, 1977
2528. DeBlois RW, Uzgiris EE, Cluxton DH, Mazzone HM: Comparative measurements of size and polydispersity of several insect viruses. Anal Biochem 90:273-88, 1978
2529. Feuer BI, Uzgiris EE, DeBlois RW, Cluxton DH, Lenard J: Length of glycoprotein spikes of vesicular stomatitis virus and Sindbis virus measured *in situ* using quasi elastic light scattering and a resistive-pulse technique. Virology 90:156-61, 1978
2530. Saleh OA, Sohn LL: Quantitative sensing of nanoscale colloids using a microchip Coulter counter. Rev Sci Instrum 72:4449-51, 2001
2531. Kraai R, Reymer AG, Brouwer-Mandema GG, van Beckhoven JM, Hoogeboom M, le Cessie S, Kluin-Nelemans JC: Hemopoietic stem and precursor cell analysis in umbilical cord blood using the Sysmex SE-9000 IMI channel. Cytometry (Commun Clin Cytom) 46:114-8, 2001
2532. Sohn LL, Saleh OA, Facer GR, Beavis AJ, Allan RS, Notterman DA: Capacitance cytometry: measuring biological cells one by one. Proc Natl Acad Sci U S A 97:10687-90, 2000
2533. Becker CK, Parker JW, Hechinger MK, Leif R: Is Forward Scatter Monotonic On Commercial Flow Cytometers? Poster presented at the ISAC XXI Congress, San Diego, CA, May, 2002
2534. Scherer JM, Stillwell W, Jenski LJ: Anomalous changes in forward scatter of lymphocytes with loosely packed membranes. Cytometry 37:184-90, 1999
2535. Nordström T, Willamo P, Arvela M, Stenroos K, Lindqvist C: Detection of baculovirus-infected insect cells by flow cytometric side-scatter analyses. Cytometry 37:238-42, 1999
2536. Mendelow BV, Lyons C, Nhlangothi P, Tana M, Munster M, Wypkema E, Liebowitz L, Marshall L, Scott S, Coetzer TL: Automated malaria detection by depolarization of laser light. Br J Haematol. 104:499-503, 1999
2537. Kramer B, Grobusch MP, Suttorp N, Neukammer J, Rinneberg H: Relative frequency of malaria pigment-carrying monocytes of nonimmune and semi-immune patients from flow cytometric depolarized side scatter. Cytometry 45:133-40, 2001
2538. Soini JT, Chernyshev AV, Hanninen PE, Soini E, Maltsev VP: A new design of the flow cuvette and optical set-up for the scanning flow cytometer. Cytometry 31:78-84, 1998
2539. Shvalov AN, Surovtsev IV, Chernyshev AV, Soini JT, Maltsev VP: Particle classification from light scattering with the scanning flow cytometer. Cytometry 37:215-20, 1999
2540. Shvalov AN, Soini JT, Surovtsev IV, Kochneva GV, Sivolobova GF, Petrov AK, Maltsev VP: Individual Escherichia coli cells studied from light scattering with the scanning flow cytometer. Cytometry 41:41-5, 2000
2541. van Munster EB, Stap J, Hoebe RA, te Meerman GJ, Aten JA: Difference in volume of X- and Y-chromosome-bearing bovine sperm heads matches difference in DNA content. Cytometry 35:125-8, 1999
2542. van Munster EB: Interferometry in flow to sort unstained X- and Y-chromosome-bearing bull spermatozoa. Cytometry 47:192-9, 2002
2543. Deka C, Cram LS, Habbersett R, Martin JC, Sklar LA, Steinkamp JA. Simultaneous dual-frequency phase-sensitive flow cytometric measurements for rapid identification of heterogeneous fluorescence decays in fluorochrome-labeled cells and particles. Cytometry 21:318-28, 1995
2544. Deka C, Lehnert BE, Lehnert NM, Jones GM, Sklar LA, Steinkamp JA: Analysis of fluorescence lifetime and quenching of FITC-conjugated antibodies on cells by phase-sensitive flow cytometry. Cytometry 25:271-9, 1996
2545. Sailer BL, Valdez JG, Steinkamp JA, Crissman HA: Apoptosis induced with different cycle-perturbing agents produces differential changes in the fluorescence lifetime of DNA-bound ethidium bromide. Cytometry 31:208-16, 1998

2546. Keij JF, Bell-Prince C, Steinkamp JA: Simultaneous analysis of relative DNA and glutathione content in viable cells by phase-resolved flow cytometry. Cytometry 35:48-54, 1999

2547. Steinkamp JA, Lehnert NM, Keij JF, Lehnert BE: Enhanced immunofluorescence measurement resolution of surface antigens on highly autofluorescent, glutaraldehyde-fixed cells analyzed by phase-sensitive flow cytometry. Cytometry 37:275-83, 1999

2548. van Zandvoort MA, de Grauw CJ, Gerritsen HC, Broers JL, oude Egbrink MG, Ramaekers FC, Slaaf DW: Discrimination of DNA and RNA in cells by a vital fluorescent probe: Lifetime imaging of SYTO13 in healthy and apoptotic cells. Cytometry 47:226-35, 2002

2549. Bene L, Szöllösi J, Balazs M, Mátyus L, Gaspar R, Ameloot M, Dale RE, Damjanovich S: Major histocompatibility complex class I protein conformation altered by transmembrane potential changes. Cytometry 27:353-7, 1997

2550. Nagy P, Bene L, Balazs M, Hyun WC, Lockett SJ, Chiang NY, Waldman F, Feuerstein BG, Damjanovich S, Szöllösi J: EGF-induced redistribution of erbB2 on breast tumor cells: flow and image cytometric energy transfer measurements. Cytometry 32:120-31, 1998

2551. Buranda T, Lopez GP, Keij J, Harris R, Sklar LA: Peptides, antibodies, and FRET on beads in flow cytometry: A model system using fluoresceinated and biotinylated beta-endorphin. Cytometry 37:21-31, 1999

2552. Bene L, Fulwyler MJ, Damjanovich S: Detection of receptor clustering by flow cytometric fluorescence anisotropy measurements. Cytometry 40:292-306, 2000

2553. Batard P, Szöllösi J, Luescher I, Cerottini JC, MacDonald R, Romero P. Use of phycoerythrin and allophycocyanin for fluorescence resonance energy transfer analyzed by flow cytometry: Advantages and limitations. Cytometry 48:97-105, 2002

2554. Sebestyen Z, Nagy P, Horvath G, Vamosi G, Debets R, Gratama JW, Alexander DR, Szöllösi J. Long wavelength fluorophores and cell-by-cell correction for autofluorescence significantly improves the accuracy of flow cytometric energy transfer measurements on a dual-laser benchtop flow cytometer. Cytometry 48:124-35, 2002

2555. Matko J, Jenei A, Wei T, Edidin M: Luminescence quenching by long range electron transfer: a probe of protein clustering and conformation at the cell surface. Cytometry 19:191-200, 1995

2556. Packard BZ, Toptygin DD, Komoriya A, Brand L: Profluorescent protease substrates: intramolecular dimers described by the exciton model. Proc Natl Acad Sci U S A 93:11640-5, 1996

2557. Komoriya A, Packard BZ, Brown MJ, Wu M-L, Henckart PA: Assessment of caspase activities in intact apoptotic thymocytes using cell-permeable fluorogenic caspase substrates. J Exp Med 191:1819-28, 2000

2558. Asbury CL, Esposito R, Farmer C, van den Engh G: Fluorescence spectra of DNA dyes measured in a flow cytometer. Cytometry 24:234-42, 1996

2559. Fuller RR, Sweedler JV: Characterizing submicron vesicles with wavelength-resolved fluorescence in flow cytometry. Cytometry 25:144-55, 1996

2560. Gauci MR, Vesey G, Narai J, Veal D, Williams KL, Piper JA: Observation of single-cell fluorescence spectra in laser flow cytometry. Cytometry 25:388-93, 1996

2561. Manogaran PS, Kausalya S, Pande G. Flow cytometric measurement of NK cell immunoconjugates by pulse width processing. Cytometry 19:320-5, 1995

2562. Koch AL, Robertson BR, Button DK: Deduction of the cell volume and mass from forward scatter intensity of bacteria analyzed by flow cytometry. J Microbiol Methods 27:49-61, 1996

2563. Robertson BR, Button DK, Koch AL: Determination of the biomasses of small bacteria at low concentrations in a mixture of species with forward light scatter measurements by flow cytometry. Appl Environ Microbiol 64:3900-9, 1998

2564. Steen HB, Stokke T: Dye exclusion artifact in flow cytometers. Cytometry 47:200-5, 2002

2565. Huller R, Glossner E, Schaub S, Weingartner J, Kachel V: The Macro Flow Planktometer: a new device for volume and fluorescence analysis of macro plankton including triggered video imaging in flow. Cytometry 17:109-18, 1994

2566. Dubelaar GB, Gerritzen PL, Beeker AE, Jonker RR, Tangen K: Design and first results of CytoBuoy: a wireless flow cytometer for in situ analysis of marine and fresh waters. Cytometry 37:247-54, 1999

2567. Kubota F, Kusuzawa H, Kosaka T, Nakamoto H. Flow cytometer and imaging device used in combination. Cytometry 21:129-32, 1995

2568. Basiji D, Ortyn B, Finch L: High sensitivity multispectral imaging of cells in flow. Cytometry Supp 11, 52, 2002

2569. Rens W, Welch GR, Houck DW, van Oven CH, Johnson LA: Slit-scan flow cytometry for consistent high resolution DNA analysis of X- and Y-chromosome bearing sperm. Cytometry 25:191-9, 1996

2570. Njoroge JM, Mitchell LB, Centola M, Kastner D, Raffeld M, Miller JL: Characterization of viable autofluorescent macrophages among cultured peripheral blood mononuclear cells. Cytometry 44:38-44, 2001

2571. Zhang JC, Savage HE, Sacks PG, Delohery T, Alfano RR, Katz A, Schantz SP: Innate cellular fluorescence reflects alterations in cellular proliferation. Lasers Surg Med 20:319-31,1997

2571. Papadopoulos AJ, Zhadin NN, Steinberg ML, Alfano RR: Fluorescence spectroscopy of normal, SV40-transformed human keratinocytes, and carcinoma cells. Cancer Biochem Biophys 17:13-23, 1999

2572. Cordeiro PG, Kirschner RE, Hu QY, Chiao JJ, Savage H, Alfano RR, Hoffman LA, Hidalgo DA: Ultraviolet excitation fluorescence spectroscopy: a noninvasive method for the measurement of redox changes in ischemic myocutaneous flaps. Plast Reconstr Surg 96:673-80, 1995

2573. Weigel TL, Yousem S, Dacic S, Kosco PJ, Siegfried J, Luketich JD: Fluorescence bronchoscopic surveillance after curative surgical resection for non-small-cell lung cancer. Ann Surg Oncol 7:176-80, 2000

2574. Hairston P, Ho J, Quant FR: Design of an instrument for real-time detection of bioaerosols using simultaneous measurement of particle aerodynamic size and intrinsic fluorescence. J Aerosol Sci 28:471-482, 1997

2575. Ho J, Spence M, Hairston P: Measurement of biological aerosol with a fluorescent aerodynamic particle sizer (FLAPS): correlation of optical data with biological data. Aerobiologia 15:281-91, 1999

2576. Dovichi N: Development of DNA sequencer. Science 285:1016, 1999

2577. Dovichi NJ, Zhang J: How capillary electrophoresis sequenced the human genome. Angew Chem Int Ed Engl 39:4463-8, 2000

2578. Eleouet S, Carre J, Vonarx V, Heyman D, Lajat Y, Patrice T: Delta-aminolevulinic acid-induced fluorescence in normal human lymphocytes. J Photochem Photobiol B 41:22-9, 1997

2579. Hryhorenko EA, Rittenhouse-Diakun K, Harvey NS, Morgan J, Stewart CC, Oseroff AR: Characterization of endogenous protoporphyrin ix induced by delta-aminolevulinic acid in resting and activated peripheral blood lymphocytes by four-color flow cytometry. Photochem Photobiol 67:565-72, 1998

2580. Horobin RW, Kiernan JA (eds): *Conn's Biological Stains. A Handbook of Dyes, Stains, and Fluorochromes for Use in Biology and Medicine.* Published for the Biological Stain Commission. Oxford, BIOS Scientific Publishers, 2002, xvi + 555 pp

2581. Marti GE, Gaigalas A, Vogt RF Jr: Recent developments in quantitative fluorescence calibration for analyzing cells and microarrays. Cytometry (Commun Clin Cytom) 42:263, 2000

2582. Gaigalas AK, Li L, Henderson O, Vogt R, Barr J, Marti G, Schwartz A: The development of fluorescence intensity standards. J Res Nat Inst Stand Technol 106:381-9, 2001

2583. Schwartz A, Wang L, Early A, Gaigalas A, Zhang Y-Z, Marti GE, Vogt RF: Quantitating fluorescence intensity from fluorophore: The definition of MESF assignment. J Res Nat Inst Stand Technol 107:83-91, 2002

2584. Wang L, Gaigalas AK, Abbbasi F, Marti GE, Vogt RF, Schwartz A: Quantitating fluorescence intensity from fluorophores: Practical use of MESF values. J Res Nat Inst Stand Technol 107:339-53, 2002
2585. Jernaes MW, Steen HB: Staining of *Escherichia coli* for flow cytometry: Influx and efflux of ethidium bromide. Cytometry. 17:302-9, 1994
2586. McAuliffe O, Ross RP, Hill C: Lantibiotics: structure, biosynthesis and mode of action. FEMS Microbiol Rev 25:285-30, .2001
2587. Moll GN, Clark J, Chan WC, Bycroft BW, Roberts GC, Konings WN, Driessen AJ: Role of transmembrane pH gradient and membrane binding in nisin pore formation. J Bacteriol 179:135-40. 1997
2588, Kovacs F, Quine J, Cross TA: Validation of the single-stranded channel conformation of gramicidin A by solid-state NMR. Proc Natl Acad Sci U S A 96:7910-5, 1999
2589. Carulli G, Sbrana S, Azzara A, Minnucci S, Angiolini C, Marini A, Ambrogi F: Detection of eosinophils in whole blood samples by flow cytometry. Cytometry 34:272-9, 1998
2590. Helander KG: Formaldehyde prepared from paraformaldehyde is stable. Biotech Histochem 75:19-22, 2000
2591. Rubbi CP, Qiu J, Rickwood D: An investigation into the use of protein cross-linking agents as cell fixatives for confocal microscopy. Eur J Histochem 38:269-80, 1994
2592. Tymianski M, Bernstein GM, Abdel-Hamid KM, Sattler R, Velumian A, Carlen PL, Razavi H, Jones OT: A novel use for a carbodiimide compound for the fixation of fluorescent and non-fluorescent calcium indicators in situ following physiological experiments. Cell Calcium 21:175-83, 1997
2593. Campbell TA, Ware RE, Mason M: Detection of hemoglobin variants in erythrocytes by flow cytometry. Cytometry 35:242-8, 1999
2594. Bostwick DG, al Annouf N, Choi C: Establishment of the formalin-free surgical pathology laboratory. Utility of an alcohol-based fixative. Arch Pathol Lab Med 118:298-302, 1994
2595. Tagliaferro P, Tandler CJ, Ramos AJ, Pecci Saavedra J, Brusco A: Immunofluorescence and glutaraldehyde fixation. A new procedure based on the Schiff-quenching method. J Neurosci Methods 77:191-7 1997; Erratum in: J Neurosci Methods 82:235-6, 1998
2596. Jani V, Janossy G, Iqbal A, Mhalu FS, Lyamuya EF, Biberfeld G, Glencross DK, Scott L, Reilly JT, Granger V, Barnett D: Affordable CD4+ T cell counts by flow cytometry. II. The use of fixed whole blood in resource-poor settings. J Immunol Methods 257:145-54, 2001
2597. Barnett D, Granger V, Mayr P, Storie I, Wilson GA, Reilly JT: Evaluation of a novel stable whole blood quality control material for lymphocyte subset analysis: Results from the UK NEQAS immune monitoring scheme. Cytometry (Commun Clin Cytom) 26:216-22, 1996
2598. Brocklebank AM, Sparrow RL: Enumeration of CD34+ cells in cord blood: a variation on a single-platform flow cytometric method based on the ISHAGE gating strategy. Cytometry (Commun Clin Cytom) 46:254-61, 2001
2599. Bossuyt X, Marti GE, Fleisher TA: Comparative analysis of whole blood lysis methods for flow cytometry. Cytometry (Commun Clin Cytom) 30:124-33, 1997 (Letter and Authors' Reply: Cytometry (Commun Clin Cytom) 30:324-5, 1997)
2600. Macey MG, McCarthy DA, Milne T, Cavenagh JD, Newland AC: Comparative study of five commercial reagents for preparing normal and leukaemic lymphocytes for immunophenotypic analysis by flow cytometry. Cytometry (Commun Clin Cytom) 38:153-60, 1999
2601. Mikulka WR, Bolton WE: Methodologies for the preservation of proliferation associated antigens PCNA, p120, and p105 in tumor cell lines for use in flow cytometry. Cytometry 17:246-57, 1994
2602. Koester SK, Bolton WE: Intracellular markers. J Immunol Methods 243:99-106, 2000 (in reference 2401)
2603. Rousselle C, Robert-Nicoud M, Ronot X: Flow cytometric analysis of DNA content of living and fixed cells: a comparative study using various fixatives. Histochem J 30:773-81, 1998
2604. Linden E, Skoglund P, Rundquist I: Accessibility of 7-aminoactinomycin D to lymphocyte nuclei after paraformaldehyde fixation. Cytometry 27:92-5, 1997
2605. Overton WR, McCoy JP Jr: Reversing the effect of formalin on the binding of propidium iodide to DNA. Cytometry 16:351-6, 1994
2606. Boenisch T: Formalin-fixed and heat-retrieved tissue antigens: a comparison of their immunoreactivity in experimental antibody diluents. Appl Immunohistochem Mol Morphol 9:176-9, 2001
2607. Leonard JB, Shepardson SP: A comparison of heating modes in rapid fixation techniques for electron microscopy. J Histochem Cytochem 42:383-91, 1994
2608. Login GR, Leonard JB, Dvorak AM: Calibration and standardization of microwave ovens for fixation of brain and peripheral nerve tissue. Methods 15:107-17, 1998
2609. Grutzkau A, Kruger-Krasagakes S, Kogel H, Moller A, Lippert U, Henz BM: Detection of intracellular interleukin-8 in human mast cells: flow cytometry as a guide for immunoelectron microscopy. J Histochem Cytochem 45:935-45, 1997
2610. Nebe-von-Caron G, Stephens PJ, Hewitt CJ, Powell JR, Badley RA: Analysis of Bacterial Function by Multi-Colour Fluorescence Flow Cytometry and Single Cell Sorting. J Microbiol Methods 42:97-114, 2000 (in reference 2404; Full text is available at www.elsevier.com/locate/jmicmeth)
2611. Singh NP: A rapid method for the preparation of single-cell suspensions from solid tissues. Cytometry 31:229-32, 1998
2612. Loontiens FG, Regenfuss P, Zechel A, Dumortier L, Clegg RM. Binding characteristics of Hoechst 33258 with calf thymus DNA, poly[d(A-T)], and d(CCGGAATTCCGG): multiple stoichiometries and determination of tight binding with a wide spectrum of site affinities. Biochemistry 29:9029-39, 1990
2613. Baines P, Visser JW: Analysis and separation of mouse bone marrow stem cells by H33342 fluorescence-activated cell sorting. Exp Hematol 11:701-8, 1983
2614. Wolf NS, Koné A, Priestley GV, Bartelmez SH: In vivo and in vitro characterization of long-term repopulating primitive hematopoietic stem cells isolated by sequential Hoechst 33342-rhodamine 123 FACS selection. Exp Hematol 21:614- 22, 1993
2615. Bradford GB, Williams B, Rossi R, Bertoncello I: Quiescence, cycling, and turnover in the primitive hematopoietic stem cell compartment. Exp Hematol 25:445-53, 1997
2616. Goodell MA, Brose K, Paradis G, Conner AS, Mulligan RC: Isolation and functional properties of murine hematopoietic stem cells that are replicating in vivo. J Exp Med 183:1797-806, 1996
2617. Snyder DS, Small PLC: Staining of cellular mitochondria with LDS-751. J Immunol Methods 257: 35–40, 2001
2618. Singer VL, Jones LJ, Yue ST, Haugland RP: Characterization of PicoGreen reagent and development of a fluorescence-based solution assay for double-stranded DNA quantitation. Anal Biochem 249:228–38, 1997
2619. Van Hooijdonk CA, Glade CP, Van Erp PE: TO-PRO-3 iodide: a novel HeNe laser-excitable DNA stain as an alternative for propidium iodide in multiparameter flow cytometry. Cytometry 17:185-9, 1994
2620. Frey T: Detection of bromodeoxyuridine incorporation by alteration of the fluorescence emission from nucleic acid binding dyes using only an argon ion laser. Cytometry 17:310-8, 1994
2621. Corver WE, Fleuren GJ, Cornelisse CJ: Improved single laser measurement of two cellular antigens and DNA-ploidy by the combined use of propidium iodide and TO-PRO-3 iodide. Cytometry 28:329-36, 1997
2622. Wiltshire M, Patterson LH, Smith PJ: A novel deep red/low infrared fluorescent flow cytometric probe, DRAQ5NO, for the discrimination of intact nucleated cells in apoptotic cell populations. Cytometry 39:217-23, 2000
2623. Stockert JC, Trigoso CI, Cuellar T, Bella JL, Lisanti JA: A new fluorescence reaction in DNA cytochemistry: microscopic and spec-

troscopic studies on the aromatic diamidino compound M&B 938. J Histochem Cytochem 45:97-105, 1997

2624. Barber L, Prince HM, Rossi R, Bertoncello I: Fluoro-Gold: An alternative viability stain for multicolor flow cytometric analysis. Cytometry 36:349-54, 1999

2625. Jayat-Vignoles C, Ratinaud MH: Nucleic acid specificity of an acridine derivative permits its use for flow cytometric analysis of the cell cycle. Cytometry 27:153-60, 1997

2626. Schmid I, Ferbas J, Uittenbogaart CH, Giorgi JV: Flow cytometric analysis of live cell proliferation and phenotype in populations with low viability. Cytometry 35:64-74, 1999

2627. Srour EF, Jordan CT: Isolation and Characterization of Primitive Hematopoietic Cells Based on Their Position in Cell Cycle. In: Klug CA, Jordan CT (eds): *Hematopoietic Stem Cell Protocols (Methods in Molecular Medicine, Vol. 63)*, Totowa (NJ), Humana Press, 2001, pp. 93-111

2628. Tjioe I, Legerton T, Wegstein J, Herzenberg LA, Roederer M: Phycoerythrin-allophycocyanin: a resonance energy transfer fluorochrome for immunofluorescence. Cytometry 44:24-9, 2001

2629. Roederer M, Kantor AB, Parks DR, Herzenberg LA: Cy7PE and Cy7APC: bright new probes for immunofluorescence. Cytometry 24:191-7, 1996

2630. Beavis AJ, Pennline KJ: Allo-7: a new fluorescent tandem dye for use in flow cytometry. Cytometry 24:390-4, 1996; also see Publisher's Notice, Cytometry 24:395, 1996

2631. Davis KA, Houck DW: A Novel Red Dye for Flow Cytometry [Note: this title is omitted from the printed volume]. Cytometry Supp 9:141. 1998

2632. Parks DR, Bigos M, Herzenberg LA: A practical and theoretical examination of dye saturation and related phenomena. Cytometry Supp 5:117, 1991

2633. Parks DR, Bigos M, Moore W, Herzenberg LA: Short-term and long-term loss of fluorescent dye activity as a function of exposure to laser excitation. Cytometry Supp 6:85, 1993

2634. Bishop JE, Davis KA, Abrams B, Houck DW, Recktenwald DJ, Hoffman RA: Mechanism of higher brightness of PerCP-Cy5.5. Cytometry Supp 10:162-3, 2000

2635. Hofkens J, Schroeyers W, Loos D, Cotlet M, Köhn F, Vosch T, Maus M, Herrmann A, Müllen K, Gensch T, De Schryver FC: Triplet states as non-radiative traps in multichromophoric entities: single molecule spectroscopy of an artificial and natural antenna system. Spectrochim Acta 57A:2093-107, 2001

2636. Telford WG, Moss MW, Morseman JP, Allnutt FC: Cryptomonad algal phycobiliproteins as fluorochromes for extracellular and intracellular antigen detection by flow cytometry. Cytometry 44:16-23, 2001

2637. Cai YA, Murphy JT, Wedemayer GJ, Glazer AN: Recombinant phycobiliproteins. Recombinant C-phycocyanins equipped with affinity tags, oligomerization, and biospecific recognition domains. Anal Biochem 290:186-204, 2001

2638. Tooley AJ, Cai YA, Glazer AN: Biosynthesis of a fluorescent cyanobacterial C-phycocyanin holo-alpha subunit in a heterologous host. Proc Natl Acad Sci U S A 98:10560-5, 2001

2639. Tooley AJ, Glazer AN: Biosynthesis of the cyanobacterial light-harvesting polypeptide phycoerythrocyanin holo-α subunit in a heterologous host. J Bact 184:4666-71, 2002

2640. Glazer AN, Mathies RA: Energy-transfer fluorescent reagents for DNA analyses. Curr Opin Biotechnol 8:94-102, 1997

2641. Dietz LJ, Dubrow RS, Manian BS, Sizto NL: Volumetric capillary cytometry: a new method for absolute cell enumeration. Cytometry 23:177-86, 1996

2642. van Vugt MJ, van den Herik-Oudijk IE, van de Winkle JG: Binding of PE-Cy5 conjugates to the human high-affinity receptor for IgG(CD64). Blood 88:2358-61, 1996

2643. Stewart CC, Stewart SJ: Cell preparation for the identification of leukocytes. Methods Cell Biol 63:217-51, 2001 (in reference 2385)

2644. Anderson MT, Baumgarth N, Haugland RP, Gerstein RM, Tjioe T, Herzenberg LA, Herzenberg LA: Pairs of violet-light-excited fluorochromes for flow cytometric analysis. Cytometry 33:435-44, 1998

2645. Lee LG, Woo SL, Head DF, Dubrow RS, Baer TM: Near-IR dyes in three-color volumetric capillary cytometry: cell analysis with 633- and 785-nm laser excitation. Cytometry 21:120-8, 1995

2646. Bruchez M, Jr., Moronne M, Gin P, Weiss S, Alivisatos AP: Semiconductor nanocrystals as fluorescent biological labels. Science 281:2013-6, 1998

2647. Chan WC, Nie S: Quantum dot bioconjugates for ultrasensitive nonisotopic detection. Science 281:2016-8, 1998

2648. Han M, Gao X, Su JZ, Nie S: Quantum-dot-tagged microbeads for multiplexed optical coding of biomolecules. Nat Biotechnol 19:631-5, 2001

2649. Gao X, Chan WC, Nie S: Quantum-dot nanocrystals for ultrasensitive biological labeling and multicolor optical encoding. J Biomed Opt 7:532-7, 2002

2650. Chan WC, Maxwell DJ, Gao X, Bailey RE, Han M, Nie S: Luminescent quantum dots for multiplexed biological detection and imaging. Curr Opin Biotechnol 13:40-6, 2002

2651. Jovin TM: Quantum dots finally come of age. Nat Biotechnol 21:32-3, 2003

2652. Wu X, Liu H, Liu J, Haley KN, Treadway JA, Larson JP, Ge N, Peale F, Bruchez MP: Immunofluorescent labeling of cancer marker Her2 and other cellular targets with semiconductor quantum dots. Nat Biotechnol 21:41-6, 2003

2653. Jaiswal JK, Mattoussi H, Mauro JM, Simon SM: Long-term multiple color imaging of live cells using quantum dot bioconjugates. Nat Biotechnol 21:47-51, 2003

2654. Goldman ER, Balighian ED, Mattoussi H, Kuno MK, Mauro JM, Tran PT, Anderson GP: Avidin: A natural bridge for quantum dot-antibody conjugates. J Am Chem Soc 124:6378-82, 2002

2655. Hyun W, Daniels RH, Hotz CZ, Bruchez M: Nanocrystals as multicolor, single excitation probes for flow and image cytometry. Cytometry Supp 10:182, 2000

2656. Dale GL: Rapid production of quasi-stable antibody-phycoerythrin conjugates for use in flow cytometry. Cytometry 33:482-6, 1998

2657. Hirschfeld T: Limits of analysis. Anal Chem 48:16A-31A, 1976

2658. Chen DY, Dovichi NJ: Single-molecule detection in capillary electrophoresis: Molecular shot noise as a fundamental limit to chemical analysis. Anal Chem 68:690-6, 1996

2659. Elowitz MB, Levine AJ, Siggia ED, Swain PS: Stochastic gene expression in a single cell. Science 297:1183-6, 2002 (Comment in Science 297:1129-31, 2002)

2660. Vesey G, Deere D, Gauci MR, Griffiths KR, Williams KL, Veal DA: Evaluation of fluorochromes and excitation sources for immunofluorescence in water samples. Cytometry 29:147-54, 1997

2661. Mosiman VL, Patterson BK, Canterero L, Goolsby CL: Reducing cellular autofluorescence in flow cytometry: an in situ method. Cytometry (Commun Clin Cytom) 30:151-6, 1997

2662. Fortin M, Hugo P: Surface antigen detection with non-fluorescent, antibody-coated microbeads: an alternative method compatible with conventional fluorochrome-based labeling. Cytometry 36:27-35, 1999

2663. Siiman O, Gordon K, Burshteyn A, Maples JA, Whitesell JK: Immunophenotyping using gold or silver nanoparticle-polystyrene bead conjugates with multiple light scatter. Cytometry 41:298-307, 2000

2664. Telford WG, Moss MW, Morseman JP, Allnutt FCT: Cyanobacterial stabilized phycobilisomes as fluorochromes for extracellular antigen detection by flow cytometry. J Immunol Methods 254:13-30, 2001

2665. Bobrow MN, Harris TD, Shaugnessy KJ, Litt GJ: Catalyzed reporter deposition, a novel method of signal amplification. J Immunol Methods 125:279-85, 1989

2666. Bobrow MN, Shaugnessy KJ, Litt GJ: Catalyzed reporter deposition, a novel method of signal amplification. II. Application to membrane immunoassays. J Immunol Methods 137:103-12, 1991

2667. Earnshaw JC, Osbourn JK: Signal amplification in flow cytometry using biotin tyramine. Cytometry 35:176-9, 1999
2668. Chao J, DeBiasio R, Zhu Z, Giuliano KA, Schmidt BF: Immunofluorescence signal amplification by the enzyme-catalyzed deposition of a fluorescent reporter substrate (CARD). Cytometry 23:48-53, 1996
2669. Kaplan D, Smith D: Enzymatic amplification staining for flow cytometric analysis of cell surface molecules. Cytometry 40:81-5, 2000
2670. Kaplan, D, Smith D, Meyerson H, Pecora N, Lewandowska K: CD5 expression by B lymphocytes and its regulation upon Epstein-Barr virus transformation, Proc Natl Acad Sci U S A 98:13850-3, 2001
2671. van Gijlswijk RP, Zijlmans HJ, Wiegant J, Bobrow MN, Erickson TJ, Adler KE, Tanke HJ, Raap AK: Fluorochrome-labeled tyramides: use in immunocytochemistry and fluorescence in situ hybridization. J Histochem Cytochem 45:375-82, 1997
2672. Coventry BJ, Neoh SH, Mantzioris BX, Skinner JM, Zola H, Bradley J: A comparison of the sensitivity of immunoperoxidase staining methods with high-sensitivity fluorescence flow cytometry-antibody quantitation on the cell surface. J Histochem Cytochem 42:1143-7, 1994
2673. Smith GP: Filamentous fusion phage: Novel expression vectors that display cloned antigens on the virion surface. Science 228:1315-7, 1985
2674. Skerra A, Pluckthun A: Assembly of a functional immunoglobulin Fv fragment in Escherichia coli. Science 240:1038-41, 1988
2675. Better M, Chang CP, Robinson RR, Horwitz AH: Escherichia coli secretion of an active chimeric antibody fragment. Science 240:1041-3, 1988
2676. McCafferty J, Griffiths AD, Winter G, Chiswell DJ: Phage antibodies: Filamentous phage displaying antibody variable domains. Nature 348:552-4, 1990
2677. Krebs B, Rauchenberger R, Reiffert S, Rothe C, Tesar M, Thomassen E, Cao M, Dreier T, Fischer D, Hoss A, Inge L, Knappik A, Marget M, Pack P, Meng XQ, Schier R, Sohlemann P, Winter J, Wolle J, Kretzschmar T: High-throughput generation and engineering of recombinant human antibodies. J Immunol Methods 254:67-84, 2001
2678. de Wildt RM, Steenbakkers PG, Pennings AH, van den Hoogen FH, van Venrooij WJ, Hoet RM: A new method for the analysis and production of monoclonal antibody fragments originating from single human b cells. J Immunol Methods 207:61-7, 1997
2679. Kohl J, Ruker F, Himmler G, Razazzi E, Katinger H: Cloning and expression of an HIV-1 specific single-chain Fv region fused to Escherichia coli alkaline phosphatase. Ann N Y Acad Sci 646:106-14, 1991
2680. Griep RA, van Twisk C, van der Wolf JM, Schots A: Fluobodies: Green fluorescent single-chain Fv fusion proteins. J Immunol Methods 230:121-30, 1999
2681. Cloutier SM, Couty S, Terskikh A, Marguerat L, Crivelli V, Pugnieres M, Mani JC, Leisinger HJ, Mach JP, Deperthes D: Streptabody, a high avidity molecule made by tetramerization of in vivo biotinylated, phage display-selected scFv fragments on streptavidin. Mol Immunol 37:1067-77, 2000
2682. Earley MC, Vogt RF Jr, Shapiro HM, Mandy FF, Kellar KL, Bellisario R, Pass KA, Marti GE, Stewart CC, Hannon WH: Report from a workshop on multianalyte microsphere assays. Cytometry (Clin Cytom) 50:239-42, 2002
2683. Keeney M, Gratama JW, Chin-Yee IH, Sutherland DR: Isotype controls in the analysis of lymphocytes and CD34+ stem and progenitor cells by flow cytometry--time to let go! Cytometry: (Commun Clin Cytom) 34:280-3, 1998
2684. O'Gorman MR, Thomas J: Isotype controls--time to let go? Cytometry: (Commun Clin Cytom) 38:78-80, 1999
2685. Henderson LO, Marti GE, Gaigalas A, Hannon WH, Vogt RF Jr: Terminology and nomenclature for standardization in quantitative fluorescence cytometry. Cytometry 33:97-105, 1998 (in reference 2379)
2686. Schwartz A, Marti GE, Poon R, Gratama JW, Fernández-Repollet E: Standardizing flow cytometry: a classification system of fluorescence standards used for flow cytometry. Cytometry 33:106-14, 1998 (in reference 2379)
2687. Serke S, van Lessen A, Huhn D: Quantitative fluorescence flow cytometry: a comparison of the three techniques for direct and indirect immunofluorescence. Cytometry 33:179-87, 1998 (in reference 2379)
2688. Bikoue A, George F, Poncelet P, Mutin M, Janossy G, Sampol J: Quantitative analysis of leukocyte membrane antigen expression: normal adult values. Cytometry (Commun Clin Cytom) 26:137-47, 1996
2689. Iyer SB, Hultin LE, Zawadzki JA, Davis KA, Giorgi JV: Quantitation of CD38 expression using QuantiBRITE beads. Cytometry 33:206-12, 1998 (in reference 2379)
2690. Liu Z, Hultin LE, Cumberland WG, Hultin P, Schmid I, Matud JL, Detels R, Giorgi JV: Elevated relative fluorescence intensity of CD38 antigen expression on CD8+ T cells is a marker of poor prognosis in HIV infection: results of 6 years of follow-up. Cytometry (Commun Clin Cytom) 26:1-7, 1996
2691. Hultin LE, Matud JL, Giorgi JV: Quantitation of CD38 activation antigen expression on CD8+ T cells in HIV-1 infection using CD4 expression on CD4+ T lymphocytes as a biological calibrator. Cytometry 33:123-32, 1998
2692. Schmitz JL, Czerniewski MA, Edinger M, Plaeger S, Gelman R, Wilkening CL, Zawadzki JA, Wormsley SB: Multisite comparison of methods for the quantitation of the surface expression of CD38 on CD8(+) T lymphocytes: The ACTG Advanced Flow Cytometry Focus Group. Cytometry (Commun Clin Cytom) 42:174-9, 2000
2693. Lenkei R, Gratama JW, Rothe G, Schmitz G, D'hautcourt JL, Arekrans A, Mandy F, Marti G: Performance of calibration standards for antigen quantitation with flow cytometry. Cytometry 33:188-96, 1998 (in reference 2379)
2694. Schwartz A, Fernández Repollet E, Vogt R, Gratama JW: Standardizing flow cytometry: Construction of a standardized fluorescence calibration plot using matching spectral calibrators. Cytometry (Commun Clin Cytometry) 26:22-31, 1996
2695. Solajic Z, Shapiro H, Bergeron M, Mandy F: A calibration method for flow cytometric detection at ultra low levels of phycoerythrin fluorescence. Cytometry Supp 10:174. 2000
2696. Mandy F., Solajic Z, Bergeron M, Perlmutter N, Shapiro H: Quantification of small amounts of cell-bound labeled antibody using a high-precision, high-sensitivity flow cytometer. Cytometry 34:295, 1998
2697. Benedict CA, MacKrell AJ, Anderson WF: Determination of the binding affinity of an anti-cd34 single-chain antibody using a novel, flow cytometry based assay. J Immunol Methods 201:223-31, 1997
2698. Siiman O, Burshteyn A: Cell surface receptor-antibody association constants and enumeration of receptor sites for monoclonal antibodies. Cytometry 40:316-26, 2000
2699. Lamvik J, Hella H, Liabakk NB, Halaas O: Nonlabeled secondary antibodies augment/maintain the binding of primary, specific antibodies to cell membrane antigens. Cytometry 45:187-93, 2001
2700. Boulla G, Randriamampita C, Raposo G, Trautmann A: Binding kinetics of soluble ligands to transmembrane proteins: Comparing an optical biosensor and dynamic flow cytometry. Cytometry 40:76-80, 2000
2701. Jacobberger JW: Stoichiometry of immunocytochemical staining reactions. Methods Cell Biol 63:271-98, 2001(in reference 2385)
2702. Frisa PS, Lanford RE, Jacobberger JW: Molecular quantification of cell cycle-related gene expression at the protein level. Cytometry 39:79-89, 2000
2703. Maino VC, Picker LJ: Identification of functional subsets by flow cytometry: Intracellular detection of cytokine expression. Cytometry 34:207-15, 1998

2704. Manz R, Assenmacher M, Pfluger E, Miltenyi S, Radbruch A: Analysis and sorting of live cells according to secreted molecules, relocated to a cell-surface affinity matrix. Proc Natl Acad Sci U S A 92:1921-5, 1995
2705. Pittet MJ, Zippelius A, Speiser DE, Assenmacher M, Guillaume P, Valmori D, Lienard D, Lejeune F, Cerottini JC, Romero P: Ex vivo IFN-gamma secretion by circulating CD8 T lymphocytes: Implications of a novel approach for T cell monitoring in infectious and malignant diseases. J Immunol 166:7634-40, 2001
2706. Morris MC, Depollier J, Merty J, Heitz F, Divita G: A peptide carrier for the delivery of biologically active proteins into mammalian cells. Nat Biotechnol 19:1173-6, 2001
2707. Nicholson JKA, Stetler-Stevenson M: Quantitative fluorescence. To count or not to count, Is that the question? Cytometry (Commun Clin Cytom) 34:203-4, 1998
2708. Yang G, Olson JC, Pu R, Vyas GN. Flow cytometric detection of human immunodeficiency virus type 1 proviral DNA by the polymerase chain reaction incorporating digoxigenin- or fluorescein-labeled dUTP. Cytometry 21:197-202, 1995
2709. Timm EA Jr, Podniesinski E, Duckett L, Cardott J, Stewart CC. Amplification and detection of a Y-chromosome DNA sequence by fluorescence in situ polymerase chain reaction and flow cytometry using cells in suspension. Cytometry (Commun Clin Cytom) 22:250-5, 1995
2710. Andreeff M, Pinkel D (eds): *Introduction to Fluorescence In Situ Hybridization: Principles and Clinical Applications.* New York, Wiley-Liss, 1999, xii + 455 pp
2711. van de Corput MP, Dirks RW, van Gijlswijk RP, van Binnendijk E, Hattinger CM, de Paus RA, Landegent JE, Raap AK: Sensitive mRNA detection by fluorescence in situ hybridization using horseradish peroxidase-labeled oligodeoxynucleotides and tyramide signal amplification. J Histochem Cytochem 46:1249-59, 1998
2712. van Gijlswijk RP, van de Corput MP, Bezrookove V, Wiegant J, Tanke HJ, Raap AK: Synthesis and purification of horseradish peroxidase-labeled oligonucleotides for tyramide-based fluorescence in situ hybridization. Histochem Cell Biol 113:175-80, 2000
2713. Nielsen PE: Applications of peptide nucleic acids. Curr Opin Biotechnol 10:71-5, 1999
2714. Stender H, Fiandaca M, Hyldig-Nielsen JJ, Coull J: PNA for rapid microbiology. J Microbiol Methods 48:1-17, 2002
2715. Rufer N, Dragowska W, Thornbury G, Roosnek E, Lansdorp PM: Telomere length dynamics in human lymphocyte subpopulations measured by flow cytometry. Nat Biotechnol 16:743-7, 1998
2716. Hultdin M, Grönlund E, Norrback KE, Eriksson-Lindström E, Just T, Roos G: Telomere analysis by fluorescence *in situ* hybridization and flow cytometry. Nucleic Acids Res 26:3651-6, 1998
2717. Guasch RM, Guerri C, O'Connor JE: Study of surface carbohydrates on isolated Golgi subfractions by fluorescent-lectin binding and flow cytometry. Cytometry 19:112-8, 1995
2718. Taylor JG, Haigler CH, Kilburn DG, Blanton RL: Detection of cellulose with improved specificity using laser-based instruments. Biotech Histochem 71:215-23, 1996
2719. Gorenflo V, Steinbuchel A, Marose S, Rieseberg M, Scheper T: Quantification of bacterial polyhydroxyalkanoic acids by Nile red staining. Appl Microbiol Biotechnol 51:765-72, 1999
2720. Vidal-Mas J, Resina P, Haba E, Comas J, Manresa A, Vives-Rego J: Rapid flow cytometry--Nile red assessment of PHA cellular content and heterogeneity in cultures of Pseudomonas aeruginosa 47T2 (NCIB 40044) grown in waste frying oil. Antonie Van Leeuwenhoek 80:57-63, 2001
2721. Hassall DG, Graham A. Changes in free cholesterol content, measured by filipin fluorescence and flow cytometry, correlate with changes in cholesterol biosynthesis in THP-1 macrophages. Cytometry 21:352-62, 1995
2722. Tárnok A: Improved kinetic analysis of cytosolic free calcium in stress sensitive cells by fixed time flow-cytometry. Cytometry 23:82-9, 1996
2723. Tárnok A.: Rare event sorting based on changes in intracellular free calcium by fixed time flow-cytometry. Cytometry 27:65-70, 1997
2724. Graves SW, Habbersett RC, Nolan JP: A dynamic inline sample thermoregulation unit for flow cytometry. Cytometry 43:23-30, 2001
2725. Kusuda L, Melamed MR Display and correction of flow cytometry time-dependent fluorescence changes. Cytometry 17:340-2, 1994
2726. Gordon IL. Scatchard analysis of fluorescent concanavalin A binding to lymphocytes. Cytometry 20:238-44, 1995
2727. Macho A, Mishal Z, Uriel J: Molar quantification by flow cytometry of fatty acid binding to cells using dipyrrometheneboron difluoride derivatives. Cytometry 23:166-73, 1996
2728. Cherukuri A, Frye J, French T, Durack G, Voss EW Jr: FITC-poly-D-lysine conjugates as fluorescent probes to quantify hapten-specific macrophage receptor binding and uptake kinetics. Cytometry 31:110-24, 1998
2729. Waller A, Pipkorn D, Sutton KL, Linderman JJ, Omann GM: Validation of flow cytometric competitive binding protocols and characterization of fluorescently labeled ligands. Cytometry 45:102-14, 2001
2730. Lauer S, Goldstein B, Nolan RL, Nolan JP: Analysis of cholera toxin-ganglioside interactions by flow cytometry. Biochemistry 41:1742-51, 2002
2731. Harenberg J, Malsch R, Piazolo L, Huhle G, Heene DL: Analysis of heparin binding to human leukocytes using a fluorescein-5-isothiocyanate labeled heparin fragment. Cytometry 23:59-66, 1996
2732. Triantafilou K, Triantafilou M, Fernandez N: Lipopolysaccharide (LPS) labeled with Alexa 488 hydrazide as a novel probe for LPS binding studies. Cytometry 41:316-20, 2000
2733. Lawrence DM, el-Hamouly W, Archer S, Leary JF, Bidlack JM: Identification of kappa opioid receptors in the immune system by indirect immunofluorescence. Proc Natl Acad Sci U S A 92:1062-6, 1995
2734. Roth BL, Poot M, Yue ST, Millard PJ: Bacterial viability and antibiotic susceptibility testing with SYTOX Green™ nucleic acid stain. Appl Env Microbiol 63:2421-31, 1997
2735. Schmid I, Hausner MA, Cole SW, Uittenbogaart CH, Giorgi JV, Jamieson BD: Simultaneous flow cytometric measurement of viability and lymphocyte subset proliferation. J Immunol Meth 247:175-86, 2001
2736. Pyatt RE, Jenski LL, Allen R, Cornetta K, Abonour R, Traycoff CM, Srour EF: Use of merocyanine 540 for the isolation of quiescent, primitive human bone marrow hematopoietic progenitor cells. J Hematother 8:189-98, 1999
2737. Mower DA, Jr., Peckham DW, Illera VA, Fishbaugh JK, Stunz LL, Ashman RF: Decreased membrane phospholipid packing and decreased cell size precede DNA cleavage in mature mouse B cell apoptosis. J Immunol 152:4832-42, 1994
2738. Ashman RF, Peckham D, Alhasan S, Stunz LL: Membrane unpacking and the rapid disposal of apoptotic cells. Immunol Lett 48:159-66, 1995
2739. Chiu L, Cherwinski H, Ransom J, Dunne JF: Flow cytometric ratio analysis of the Hoechst 33342 emission spectrum: Multiparametric characterization of apoptotic lymphocytes. J Immunol Methods 189:157-71, 1996
2740. Laakko T, King L, Fraker P: Versatility of merocyanine 540 for the flow cytometric detection of apoptosis in human and murine cells. J Immunol Methods 261:129-39, 2002
2741. Watala C, Waczulikova I, Wieclawska B, Rozalski M, Gresner P, Gwozdzinski K, Mateasik A, Sikurova L: Merocyanine 540 as a fluorescent probe of altered membrane phospholipid asymmetry in activated whole blood platelets. Cytometry 49:119-33, 2002
2742. Waczulikova I, Rozalski M, Rievaj J, Nagyova K, Bryszewska M, Watala C: Phosphatidylserine content is a more important contributor than transmembrane potential to interactions of merocyanine 540 with lipid bilayers. Biochim Biophys Acta 1567:176-82, 2002

2743. Makrigiorgos GM: Detection of lipid peroxidation on erythrocytes using the excimer-forming property of a lipophilic BODIPY fluorescent dye. J Biochem Biophys Methods 35:23-35, 1997
2744. Makrigiorgos GM, Kassis AI, Mahmood A, Bump EA, Savvides P: Novel fluorescein-based flow-cytometric method for detection of lipid peroxidation. Free Radic Biol Med 22:93-100, 1997
2745. Chung WY, Benzie IF: Probe-assisted flow cytometric analysis of erythrocyte membrane response to site-specific oxidant stress. Cytometry 40:182-8, 2000
2746. Maulik G, Kassis AI, Savvides P, Makrigiorgos GM: Fluoresceinated phosphoethanolamine for flow-cytometric measurement of lipid peroxidation. Free Radic Biol Med 25:645-53, 1998
2747. Maulik G, Salgia R, Makrigiorgos GM: Flow cytometric determination of lipid peroxidation using fluoresceinated phosphoethanolamine. Methods Enzymol 352:80-91, 2002
2748. Krishan A: Monitoring of cellular resistance to cancer chemotherapy: drug retention and efflux. Methods Cell Biol. 64:193-209, 2001 (in reference 2386)
2749. Krishan A, Arya P: Monitoring of cellular resistance to cancer chemotherapy. Hematol Oncol Clin North Am 16:357-72, 2002
2750. Krishan A, Fitz CM, Andritsch I: Drug retention, efflux, and resistance in tumor cells. Cytometry 29:279-85, 1997
2751. Aller CB, Ehmann S, Gilman-Sachs A, Snyder AK: Flow cytometric analysis of glucose transport by rat brain cells. Cytometry 27:262-8, 1997 - NBD-Glucose
2752. Natarajan A, Srienc F: Dynamics of glucose uptake by single Escherichia coli cells. Metab Eng 1:320-33, 1999
2753. Natarajan A, Srienc F: Glucose uptake rates of single E. coli cells grown in glucose-limited chemostat cultures. J Microbiol Methods 42:87-96, 2000
2754. Dordal MS, Ho AC, Jackson-Stone M, Fu YF, Goolsby CL, Winter JN. Flow cytometric assessment of the cellular pharmacokinetics of fluorescent drugs. Cytometry 20:307-14, 1995
2755. Minderman H, Vanhoefer U, Toth K, Yin MB, Minderman MD, Wrzosek C, Slovak ML, Rustum YM: DiOC2(3) is not a substrate for multidrug resistance protein (MRP)-mediated drug efflux. Cytometry 25:14-20, 1996
2756. Laupeze B, Amiot L, Courtois A, Vernhet L, Drenou B, Fauchet R, Fardel O: Use of the anionic dye carboxy-2',7'-dichlorofluorescein for sensitive flow cytometric detection of multidrug resistance-associated protein activity. Int J Oncol 15:571-6, 1999
2757. Olson DP, Taylor BJ, Ivy SP: Detection of MRP functional activity: calcein AM but not BCECF AM as a Multidrug Resistance-related Protein (MRP1) substrate. Cytometry (Commun Clin Cytom) 46:105-13, 2001
2758. Minderman H, Suvannasankha A, O'Loughlin KL, Scheffer GL, Scheper RJ, Robey RW, Baer MR: Flow cytometric analysis of breast cancer resistance protein expression and function. Cytometry 48:59-65, 2002
2759. Muller MR, Lennartz K, Baack B, Heim MM, Seeber S, Scheulen ME: Simultaneous measurement of cellular P-glycoprotein content and function by multiparametric flow-cytometry. Int J Clin Pharmacol Ther 38:180-6, 2000
2760. Chen Y, Simon SM: In situ biochemical demonstration that P-glycoprotein is a drug efflux pump with broad specificity. J Cell Biol 148:863-70, 2000
2761. Wang EJ, Casciano CN, Clement RP, Johnson WW: In vitro flow cytometry method to quantitatively assess inhibitors of P-glycoprotein. Drug Metab Dispos 28:522-8, 2000
2762. Vincent-Genod L, Benderitter M, Voisin P: Micro-organisation of the membrane after radiation-induced apoptosis: A flow cytometry study. Radiat Environ Biophys 40:213-9, 2001
2763. Bassøe C-F, Li N, Ragheb K, Lawler G, Sturgis J, Robinson JP: Investigations of phagosomes, mitochondria, and acidic granules in human neutrophils using fluorescent probes. Cytometry 51B:21-9, 2003
2764. Haugland RP: Detecting enzymatic activity in cells using fluorogenic substrates. Biotech Histochem 70:243-51, 1995
2765. Pham NA, Robinson BH, Hedley DW: Simultaneous detection of mitochondrial respiratory chain activity and reactive oxygen in digitonin-permeabilized cells using flow cytometry. Cytometry 41:245-51, 2000
2766. Webster L, Hodgkiss RJ, Wilson GD. Simultaneous triple staining for hypoxia, proliferation, and DNA content in murine tumours. Cytometry 21:344-51, 1995
2767. Larison KD, BreMiller R, Wells KS, Clements I, Haugland RP: Use of a new fluorogenic phosphatase substrate in immunohistochemical applications. J Histochem Cytochem 43:77-83, 1995
2768. Telford W, Cox W, Singer V: Detection of endogenous and antibody-conjugated alkaline phosphatase with ELF-97 phosphate in multicolor flow cytometry applications. Cytometry 43:117-25, 2001
2769. Amstad PA; Yu G; Johnson GL; Lee BW; Dhawan S; Phelps DJ: Detection of Caspase Activation In Situ by Fluorochrome-Labeled Caspase Inhibitors. BioTechniques 31:608-16, 2001
2770. Smolewski P, Grabarek J, Phelps DJ, Darzynkiewicz Z: Stathmo-apoptosis: Arresting apoptosis by fluorochrome-labeled inhibitor of caspases. Int J Oncol 19:657-63, 2001
2771. Smolewski P, Bedner E, Du L, Hsieh TC, Wu JM, Phelps DJ, Darzynkiewicz Z: Detection of caspases activation by fluorochrome-labeled inhibitors: Multiparameter analysis by laser scanning cytometry. Cytometry 44:73-82, 2001
2772. Telford WG, Komoriya A, Packard BZ: Detection of localized caspase activity in early apoptotic cells by laser scanning cytometry. Cytometry 47:81-8, 2002
2773. Ganesh S, Klingel S, Kahle H, Valet G. Flow cytometric determination of aminopeptidase activities in viable cells using fluorogenic rhodamine 110 substrates. Cytometry 20:334-40, 1995
2774. Larison KD, BreMiller R, Wells KS, Clements I, Haugland RP: Use of a new fluorogenic phosphatase substrate in immunohistochemical applications. J Histochem Cytochem 43:77-83, 1995
2775. Telford W, Cox W, Singer V: Detection of endogenous and antibody-conjugated alkaline phosphatase with ELF-97 phosphate in multicolor flow cytometry applications. Cytometry 43:117-25, 2001
2776. Bedner E, Melamed MR, Darzynkiewicz Z: Enzyme kinetic reactions and fluorochrome uptake rates measured in individual cells by laser scanning cytometry. Cytometry 33:1-9, 1998
2777. Deutsch M, Kaufman M, Shapiro H, Zurgil N: Analysis of enzyme kinetics in individual living cells utilizing fluorescence intensity and polarization measurements. Cytometry 39:36-44, 2000
2778. Sunray M, Zurgil N, Shafran Y, Deutsch M: Determination of individual cell Michaelis-Menten constants. Cytometry 47:8-16, 2002
2779. Chow S, Hedley D. Flow cytometric determination of glutathione in clinical samples. Cytometry 21:68-71, 1995
2780. Zurgil N, Shafran Y, Fixler D, Deutsch M: Analysis of early apoptotic events in individual cells by fluorescence intensity and polarization measurements. Biochem Biophys Res Commun 290:1573-82, 2002
2781. Kaplan MR, Trubniykov E, Berke G: Fluorescence depolarization as an early measure of T lymphocyte stimulation. J Immunol Methods 201:15-24, 1997
2782. Cohen-Kashi M, Deutsch M, Tirosh R, Rachmani H, Weinreb A: Carboxyfluorescein as a fluorescent probe for cytoplasmic effects of lymphocyte stimulation. Spectrochim Acta A Mol Biomol Spectrosc 53A:1655-61, 1997
2783. Gelman-Zhornitsky E, Deutsch M, Tirosh R, Yishay Y, Weinreb A, Shapiro HM: 2', 7'- bis-(carboxyethyl)-5-(6')-carboxyfluorescein (BCECF) as a probe for intracellular fluorescence polarization measurements. J Biomed Optics 2:186-94, 1997
2784. Eisenthal A, Marder O, Dotan D, Baron S, Lifschitz-Mercer B, Chaitchik S, Tirosh R, Weinreb A, Deutsch M: Decrease of intracellular fluorescein fluorescence polarization (IFFP) in human peripheral blood lymphocytes undergoing stimulation with phytohaemag-

glutinin (PHA), concanavalin a (ConA), pokeweed mitogen (PWM) and anti-CD3 antibody. Biol Cell 86:145-50, 1996

2785. Eisenthal A, Marder O, Lifschitz-Mercer B, Skornick Y, Tirosh R, Weinreb A, Deutsch M: Inhibition of mitogen-induced changes in intracellular fluorescein fluorescence polarization of human peripheral blood lymphocytes by colchicine, vinblastine and cytochalasin B. Cell Struct Funct 21:159-66, 1996

2786. Zurgil N, Deutsch M, Tirosh R, Brodie C: Indication that intracellular fluorescence polarization of T lymphocytes is cell cycle dependent. Cell Struct Funct 21:271-6, 1996

2787. Marder O, Shoval S, Eisenthal A, Fireman E, Skornick Y, Lifschitz-Mercer B, Tirosh R, Weinreb A, Deutsch M: Effect of interleukin-1 alpha, interleukin-1 beta and tumor necrosis factor-alpha on the intracellular fluorescein fluorescence polarization of human lung fibroblasts. Pathobiology 64:123-30, 1996

2788. Zurgil N, Schiffer Z, Shafran Y, Kaufman M, Deutsch M: Fluorescein fluorescence hyperpolarization as an early kinetic measure of the apoptotic process. Biochem Biophys Res Commun 268:155-63, 2000

2789. Schiffenbauer YS, Trubniykov E, Zacharia BT, Gerbat S, Rehavi Z, Berke G, Chaitchik S: Tumor sensitivity to anti-cancer drugs predicted by changes in fluorescence intensity and polarization in vitro. Anticancer Res 22:2663-9, 2002

2790. Deutsch M, Ron I, Weinreb A, Tirosh R, Chaitchik S: Lymphocyte fluorescence polarization measurements with the cellscan system: Application to the SCM cancer test. Cytometry 23:159-65, 1996

2791. Ron IG, Deutsch M, Tirosh R, Weinreb A, Eisenthal A, Chaitchik S: Fluorescence polarisation changes in lymphocyte cytoplasm as a diagnostic test for breast carcinoma. Eur J Cancer 31A:917-20, 1995

2792. Birindelli S, Colnaghi MI, Pilotti S: New SCM (structuredness of the cytoplasmatic matrix)-based approach in breast cancer detection. Tumori 82:550-3, 1996

2793. Merimsky O, Deutsch M, Tirosh R, Wohl I, Weinreb A, Chaitchik S: Detection of colon cancer by monitoring the intracellular fluorescein fluorescence polarization changes in lymphocytes. Cancer Detect Prev 20:300-7, 1996

2794. Rahmani H, Deutsch M, Ron I, Gerbat S, Tirosh R, Weinreb A, Chaitchik S, Lalchuk S: Adaptation of the Cellscan technique for the SCM test in breast cancer. Eur J Cancer 32A:1758-65, 1996

2795. Birindelli S, Colnaghi MI, Pilotti S: Comments on adaptation of the Cellscan technique for the SCM test in breast cancer (Rahmani et al., Eur. J. Cancer, 32A:1758-65, 1996). Eur J Cancer 33:1333-5, 1997

2796. Merimsky O, Kaplan B, Deutsch M, Tirosh R, Weinreb A, Chaitchik S: Detection of melanoma by monitoring the intracellular fluorescein fluorescence polarization changes in lymphocytes. Cancer Detect Prev 21:167-77, 1997

2797. Avtalion N, Avtalion R, Tirosh R, Sheinberg A, Weinreb A, Avinoach I, Deutsch M: Preparation of a diagnostic antigen of human melanoma based on lymphocyte activation as measured by intracellular fluorescein fluorescence polarization. Cancer Detect Prev 23:64-71, 1999

2798. Klein O, Lin S, Embon O, Sazbon A, Zidan J, Kook AI: An approach for high sensitivity detection of prostate cancer by analysis of changes in structuredness of the cytoplasmic matrix of lymphocytes specifically induced by PSA-act. J Urol 161:1994-6, 1999

2799. Cercek L, Siaw M, Cercek B: A DNA probe study on the origin of the cancer recognition, immunedefense suppression and serine protease protection peptide. Cancer Detect Prev 19:325-30, 1995

2800. Cercek L, Carr BI, Siaw M, Cercek B: Effect of the cancer recognition, immunedefense suppression, and serine protease protection peptide on DNA synthesis in rat hepatocytes and human lymphocytes. Cancer Detect Prev 19:206-9, 1995

2801. Sabnis RW, Deligeorgiev TG, Jachak MN, Dalvi TS: DiOC6(3): a useful dye for staining the endoplasmic reticulum. Biotech Histochem 72:253-8, 1997

2802. Plášek J, Dale RE, Sigler K, Laskay G: Transmembrane potentials in cells: a diS-C_3(3) assay for relative potentials as an indicator of real changes. Biochim Biophys Acta 1196:181-90, 1994

2803. Plášek J, Sigler K: Slow fluorescent indicators of membrane potential: A survey of different approaches to probe response analysis. J Photochem Photobiol B 33:101-24, 1996

2804. Denksteinová B, Sigler K, Pl Plášek J: Three fluorescent probes for the flow-cytometric assessment of membrane potential in Saccharomyces cerevisiae. Folia Microbiol (Praha) 41:237-42, 1996

2805. Krasznai Z, Márián T, Balkay L, Emri M, Trón L: Flow cytometric determination of absolute membrane potential of cells. J Photochem Photobiol B 28:93-9, 1995

2806. Chaloupka R, Plášek J, Slavík J, Siglerová V, Sigler K: Measurement of membrane potential in Saccharomyces cerevisiae by the electrochromic probe di-4-ANEPPS: Effect of intracellular probe distribution. Folia Microbiol 42:451-6, 1997

2807. Farinas J, Chow AW, Wada HG: A microfluidic device for measuring cellular membrane potential. Anal Biochem 295:138-42, 2001

2808. Gonzalez JE, Tsien RY: Voltage sensing by fluorescence resonance energy transfer in single cells. Biophys J 69:1272-80, 1995

2809. Cacciatore TW, Brodfuehrer PD, Gonzalez JE, Jiang T, Adams SR, Tsien RY, Kristan WB Jr, Kleinfeld D: Identification of neural circuits by imaging coherent electrical activity with FRET-based dyes. Neuron 23:449-59, 1999

2810. Reers M, Smiley ST, Mottola-Hartshorn C, Chen A, Lin M, Chen LB: Mitochondrial membrane potential monitored by JC-1 dye. Methods Enzymol 260:406-17, 1995

2811. Salvioli S, Ardizzoni A, Franceschi C, Cossarizza A: JC-1, but not DiOC6(3) or rhodamine 123, is a reliable fluorescent probe to assess $\Delta\Psi$ changes in intact cells: Implications for studies on mitochondrial functionality during apoptosis. FEBS Lett 411:77-82, 1997

2812. Richter C: Pro-oxidants and mitochondrial Ca2+: Their relationship to apoptosis and oncogenesis. FEBS Lett 325:104-7, 1993

2813. Cossarizza A, Kalashnikova G, Grassilli E, Chiappelli F, Salvioli S, Capri M, Barbieri D, Troiano L, Monti D, Franceschi C: Mitochondrial modifications during rat thymocyte apoptosis: A study at the single cell level. Exp Cell Res 214:323-30, 1994

2814. Vayssiere JL, Petit PX, Risler Y, Mignotte B: Commitment to apoptosis is associated with changes in mitochondrial biogenesis and activity in cell lines conditionally immortalized with simian virus 40. Proc Natl Acad Sci U S A 91:11752-6, 1994

2815. Cossarizza A, Franceschi C, Monti D, Salvioli S, Bellesia E, Rivabene R, Biondo L, Rainaldi G, Tinari A, Malorni W: Protective effect of *N*-acetylcysteine in tumor necrosis factor-α-induced apoptosis in U937 cells: The role of mitochondria. Exp Cell Res 220:232-40, 1995

2816. Richter C, Schweizer M, Cossarizza A, Franceschi C: Control of apoptosis by the cellular ATP level. FEBS Lett 378:107-10, 1996

2817. Rottenberg H, Wu S: Mitochondrial dysfunction in lymphocytes from old mice: Enhanced activation of the permeability transition. Biochem Biophys Res Commun 240:68-74, 1997

2818. Rottenberg H, Wu S: Quantitative assay by flow cytometry of the mitochondrial membrane potential in intact cells. Biochim Biophys Acta 1404:393-404, 1998

2819. Cossarizza A, Ceccarelli D, Masini A: Functional heterogeneity of an isolated mitochondrial population revealed by cytofluorometric analysis at the single organelle level. Exp Cell Res 222:84-94, 1996

2820. Poot M, Zhang YZ, Kramer JA, Wells KS, Jones LJ, Hanzel DK, Lugade AG, Singer VL, Haugland RP: Analysis of mitochondrial morphology and function with novel fixable fluorescent stains. J Histochem Cytochem 44:1363-72, 1996

2821. Macho A, Decaudin D, Castedo M, Hirsch T, Susin SA, Zamzami N, Kroemer G: Chloromethyl-X-Rosamine is an aldehyde-fixable potential-sensitive fluorochrome for the detection of early apoptosis. Cytometry 25:333-40, 1996

2822. Ferlini C, Scambia G, Fattorossi A: Is chloromethyl-X-rosamine useful in measuring mitochondrial transmembrane potential? Cytometry 31:74, 1998

2823. Macho A, Decaudin D, Castedo M, Hirsch T, Susin SA, Zamzami N, Kroemer G: Chloromethyl-X-Rosamine - a fluorochrome for the determination of the mitochondrial transmembrane potential. Cytometry 31:75, 1998

2824. Hollinshead M, Sanderson J, Vaux DJ: Anti-biotin antibodies offer superior organelle-specific labeling of mitochondria over avidin or streptavidin. J Histochem Cytochem 45:1053-7, 1997

2825. Poot M, Pierce RH: Detection of changes in mitochondrial function during apoptosis by simultaneous staining with multiple fluorescent dyes and correlated multiparameter flow cytometry. Cytometry 35:311-7, 1999

2826. Gilmore K, Wilson M: The use of chloromethyl-X-rosamine (Mitotracker Red) to measure loss of mitochondrial membrane potential in apoptotic cells is incompatible with cell fixation. Cytometry 36:355-8, 1999

2827. Poot M, Pierce RC: Detection of apoptosis and changes in mitochondrial membrane potential with chloromethyl-X-rosamine. Cytometry 36:359-60, 1999

2828. Salvioli S, Dobrucki J, Moretti L, Troiano L, Fernandez MG, Pinti M, Pedrazzi J, Franceschi C, Cossarizza A: Mitochondrial heterogeneity during staurosporine-induced apoptosis in HL60 cells: Analysis at the single cell and single organelle level. Cytometry 40:189-97, 2000

2829. Mathur A, Hong Y, Kemp BK, Barrientos AA, Erusalimsky JD: Evaluation of fluorescent dyes for the detection of mitochondrial membrane potential changes in cultured cardiomyocytes. Cardiovasc Res 46:126-38, 2000

2830. Keij JF, Bell-Prince C, Steinkamp JA: Staining of mitochondrial membranes with 10-nonyl acridine orange, MitoFluor Green, and MitoTracker Green is affected by mitochondrial membrane potential altering drugs. Cytometry 39:203-10, 2000

2831. Isola R, Falchi AM, Diana A, Diaz G: Probing mitochondrial probes. Cytometry 41:148, 2000

2832. Jacobson J, Duchen MR, Heales SJR: Intracellular distribution of the fluorescent dye nonyl acridine orange responds to the mitochondrial membrane potential: implications for assays of cardiolipin and mitochondrial mass. J Neurochem 82:224-33, 2002

2833. Novak EJ, Rabinovitch PS. Improved sensitivity in flow cytometric intracellular ionized calcium measurement using fluo-3/Fura Red fluorescence ratios. Cytometry 17:135-41, 1994

2834. Baus E, Urbain J, Leo O, Andris F: Flow cytometric measurement of calcium influx in murine T cell hybrids using fluo-3 and an organic-anion transport inhibitor. J Immunol Methods 173:41-7, 1994

2835. Chow S, Hedley D, Tannock I: Flow cytometric calibration of intracellular pH measurements in viable cells using mixtures of weak acids and bases. Cytometry 24:360-7, 1996

2836. Eisner DA, Kenning NA, O'Neill SC, Pocock G, Richards CD, Valdeolmillos M: A novel method for absolute calibration of intracellular pH indicators. Pflugers Arch 413:553-8, 1989

2837. McNamara KP, Nguyen T, Dumitrascu G, Ji J, Rosenzweig N, Rosenzweig Z: Synthesis, characterization, and application of fluorescence sensing lipobeads for intracellular pH measurements. Anal Chem 73:3240-6, 2001

2838. Levitz SM, Nong SH, Seetoo KF, Harrison TS, Speizer RA, Simons ER: Cryptococcus neoformans resides in an acidic phagolysosome of human macrophages. Infect Immun 67:885-90, 1999

2839. Chitarra LG, Breeuwer P, Van Den Bulk RW, Abee T: Rapid fluorescence assessment of intracellular pH as a viability indicator of Clavibacter michiganensis subsp. Michiganensis. J Appl Microbiol 88:809-16, 2000

2840. Rechinger KB, Siegumfeldt H: Rapid assessment of cell viability of Lactobacillus delbrueckii subsp. Bulgaricus by measurement of intracellular pH in individual cells using fluorescence ratio imaging microscopy. Int J Food Microbiol 75:53-60, 2002

2841. Hornbaek T, Dynesen J, Jakobsen M: Use of fluorescence ratio imaging microscopy and flow cytometry for estimation of cell vitality for Bacillus licheniformis. FEMS Microbiol Lett 215:261-5, 2002

2842. Fukushima Y, Hagiwara S: Voltage-gated Ca2+ channel in mouse myeloma cells. Proc Natl Acad Sci U S A 80:2240-2, 1983

2843. Matteson DR, Deutsch C: K channels in T lymphocytes: A patch clamp study using monoclonal antibody adhesion. Nature 307:468-71, 1984

2844. DeCoursey TE, Chandy KG, Gupta S, Cahalan MD: Voltage-gated K+ channels in human T lymphocytes: A role in mitogenesis? Nature 307:465-8, 1984

2845. DeCoursey TE, Chandy KG, Gupta S, Cahalan MD: Voltage-dependent ion channels in T-lymphocytes. J Neuroimmunol 10:71-95, 1985

2846. Gallin EK: Ionic channels in leukocytes. J Leukoc Biol 39:241-54, 1986

2847. Moingeon P, Chang HC, Sayre PH, Clayton LK, Alcover A, Gardner P, Reinherz EL: The structural biology of CD2. Immunol Rev 111:111-44, 1989

2848. Gardner P, Alcover A, Kuno M, Moingeon P, Weyand CM, Goronzy J, Reinherz EL: Triggering of T-lymphocytes via either T3-ti or T11 surface structures opens a voltage-insensitive plasma membrane calcium-permeable channel: Requirement for interleukin-2 gene function. J Biol Chem 264:1068-76, 1989

2849. Deutsch C, Lee SC: Modulation of K+ currents in human lymphocytes by pH. J Physiol 413:399-413, 1989

2850. Gardner P: Patch clamp studies of lymphocyte activation. Annu Rev Immunol 8:231-52, 1990

2851. Amigorena S, Choquet D, Teillaud JL, Korn H, Fridman WH: Ion channel blockers inhibit B cell activation at a precise stage of the G1 phase of the cell cycle. Possible involvement of K+ channels. J Immunol 144:2038-45, 1990

2852. Freedman BD, Price MA, Deutsch CJ: Evidence for voltage modulation of IL-2 production in mitogen-stimulated human peripheral blood lymphocytes. J Immunol 149:3784-94, 1992

2853. Grissmer S, Lewis RS, Cahalan MD: Ca(2+)-activated K+ channels in human leukemic T cells. J Gen Physiol 99:63-84, 1992

2854. Nagy P, Panyi G, Jenei A, Bene L, Gaspar R, Jr., Matko J, Damjanovich S: Ion-channel activities regulate transmembrane signaling in thymocyte apoptosis and t-cell activation. Immunol Lett 44:91-5, 1995

2855. Lewis RS, Cahalan MD: Potassium and calcium channels in lymphocytes. Annu Rev Immunol 13:623-53, 1995

2856. Verheugen JA, Vijverberg HP, Oortgiesen M, Cahalan MD: Voltage-gated and Ca(2+)-activated K+ channels in intact human T lymphocytes. Noninvasive measurements of membrane currents, membrane potential, and intracellular calcium. J Gen Physiol 105:765-94, 1995

2857. Phipps DJ, Branch DR, Schlichter LC: Chloride-channel block inhibits T lymphocyte activation and signalling. Cell Signal 8:141-9, 1996

2858. Baricordi OR, Ferrari D, Melchiorri L, Chiozzi P, Hanau S, Chiari E, Rubini M, Di Virgilio F: An ATP-activated channel is involved in mitogenic stimulation of human T lymphocytes. Blood 87:682-90, 1996

2859. Varga Z, Bene L, Pieri C, Damjanovich S, Gaspar R, Jr.: The effect of juglone on the membrane potential and whole-cell K+ currents of human lymphocytes. Biochem Biophys Res Commun 218:828-32, 1996

2860. Gulbins E, Szabo I, Baltzer K, Lang F: Ceramide-induced inhibition of T lymphocyte voltage-gated potassium channel is mediated by tyrosine kinases. Proc Natl Acad Sci U S A 94:7661-6, 1997

2861. Hoth M, Fanger CM, Lewis RS: Mitochondrial regulation of store-operated calcium signaling in t lymphocytes. J Cell Biol 137:633-48, 1997

2862. Jensen BS, Odum N, Jorgensen NK, Christophersen P, Olesen SP: Inhibition of T cell proliferation by selective block of Ca(2+)-

activated K(+) channels. Proc Natl Acad Sci U S A 96:10917-21, 1999

2863. Freedman BD, Liu QH, Somersan S, Kotlikoff MI, Punt JA: Receptor avidity and costimulation specify the intracellular Ca2+ signaling pattern in CD4(+)CD8(+) thymocytes. J Exp Med 190:943-52, 1999
2864. Franco-Obregon A, Wang HW, Clapham DE: Distinct ion channel classes are expressed on the outer nuclear envelope of t- and b-lymphocyte cell lines. Biophys J 79:202-14, 2000
2865. Lepple-Wienhues A, Szabo I, Wieland U, Heil L, Gulbins E, Lang F: Tyrosine kinases open lymphocyte chloride channels. Cell Physiol Biochem 10:307-12, 2000
2866. Hoth M, Button DC, Lewis RS: Mitochondrial control of calcium-channel gating: A mechanism for sustained signaling and transcriptional activation in T lymphocytes. Proc Natl Acad Sci U S A 97:10607-12, 2000
2867. Fanger CM, Neben AL, Cahalan MD: Differential Ca2+ influx, KCa channel activity, and Ca2+ clearance distinguish Th1 and Th2 lymphocytes. J Immunol 164:1153-60, 2000
2868. Lewis RS: Calcium signaling mechanisms in T lymphocytes. Annu Rev Immunol 19:497-521, 2001
2869. Liu QH, Fleischmann BK, Hondowicz B, Maier CC, Turka LA, Yui K, Kotlikoff MI, Wells AD, Freedman BD: Modulation of Kv channel expression and function by Tcr and costimulatory signals during peripheral CD4(+) lymphocyte differentiation. J Exp Med 196:897-909, 2002
2870. Qiu W, Kass DA, Hu Q, Ziegelstein RC: Determinants of shear stress-stimulated endothelial nitric oxide production assessed in real-time by 4,5-diaminofluorescein fluorescence. Biochem Biophys Res Commun 286:328-35, 2001
2871. Navarro-Antolin J, Lamas S: Nitrosative stress by cyclosporin a in the endothelium: Studies with the NO-sensitive probe diaminofluorescein-2/diacetate using flow cytometry. Nephrol Dial Transplant 16 Suppl 1:6-9, 2001
2872. Havenga MJ, van Dam B, Groot BS, Grimbergen JM, Valerio D, Bout A, Quax PH: Simultaneous detection of NOS-3 protein expression and nitric oxide production using a flow cytometer. Anal Biochem 290:283-91, 2001
2873. Amorino GP, Fox MH. Intracellular Na+ measurements using sodium green tetraacetate with flow cytometry. Cytometry 21:248-56, 1995
2874. Balkay L, Marian T, Emri M, Krasznai Z, Tron L: Flow cytometric determination of intracellular free potassium concentration. Cytometry 28:42-9, 1997
2875. Pilas B, Durack G: A flow cytometric method for measurement of intracellular chloride concentration in lymphocytes using the halide-specific probe 6-methoxy-N-(3-sulfopropyl) quinolinium (SPQ). Cytometry 28:316-22, 1997
2876. Lorincz M, Roederer M, Diwu Z, Herzenberg LA, Nolan GP: Enzyme-generated intracellular fluorescence for single-cell reporter gene analysis utilizing Escherichia coli beta-glucuronidase. Cytometry 24:321-9, 1996
2877. Zlokarnik G, Negulescu PA, Knapp TE, Mere L, Burres N, Feng L, Whitney M, Roemer K, Tsien RY: Quantitation of transcription and clonal selection of single living cells with β-lactamase as reporter. Science 279:84-8, 1998
2878. Cavrois M, de Noronha C, Greene WC: A sensitive and specific enzyme-based assay detecting HIV-1 virion fusion in primary T-lymphocytes. Nat Biotechnol 20:1151-4, 2002
2879. Hawley TS, Telford WG, Ramezani A, Hawley RG: Four-color flow cytometric detection of retrovirally expressed red, yellow, green, and cyan fluorescent proteins. BioTechniques 30:1028-34, 2001
2880. Gurskaya NG, Fradkov AF, Terskikh A, Matz MV, Labas YA, Martynov VI, Yanushevich YG, Lukyanov KA, Lukyanov SA: GFP-like chromoproteins as a source of far-red fluorescent proteins. FEBS Lett 507:16-20, 2001
2881. Matz MV, Lukyanov KA, Lukyanov SA: Family of the green fluorescent protein: Journey to the end of the rainbow. Bioessays 24:953-9, 2002
2882. Heim R, Prasher DC, Tsien RY: Wavelength mutations and post-translational autoxidation of green fluorescent protein. Proc Natl Acad Sci U S A 91:12501-4, 1994 Dec 20;
2883. Matz MV, Fradkov AF, Labas YA, Savitsky AP, Zaraisky AG, Markelov ML, Lukyanov SA: Fluorescent proteins from nonbioluminescent Anthozoa species. Nat Biotechnol 17:969-73, 1999
2884. Bevis BJ, Glick BS: Rapidly maturing variants of the Discosoma red fluorescent protein (dsRed). Nat Biotechnol 20:83-7, 2002
2885. Fradkov AF, Verkhusha VV, Staroverov DB, Bulina ME, Yanushevich YG, Martynov VI, Lukyanov S, Lukyanov KA: Far-red fluorescent tag for protein labelling. Biochem J 368(Pt 1):17-21, 2002
2886. Ropp JD, Donahue CJ, Wolfgang-Kimball D, Hooley JJ, Chin JY, Hoffman RA, Cuthbertson RA, Bauer KD. Aequorea green fluorescent protein analysis by flow cytometry. Cytometry 21:309-17, 1995
2887. Galbraith DW, Lambert GM, Grebenok RJ, Sheen J: Flow cytometric analysis of transgene expression in higher plants: Green-fluorescent protein. Methods Cell Biol 50:3-14, 1995
2888. Ropp JD, Donahue CJ, Wolfgang-Kimball D, Hooley JJ, Chin JY, Cuthbertson RA, Bauer KD: Aequorea green fluorescent protein: simultaneous analysis of wild-type and blue-fluorescing mutant by flow cytometry. Cytometry 24:284-8, 1996
2889. Anderson MT, Tjioe IM, Lorincz MC, Parks DR, Herzenberg LA, Nolan GP, Herzenberg LA: Simultaneous fluorescence-activated cell sorter analysis of two distinct transcriptional elements within a single cell using engineered green fluorescent proteins. Proc Natl Acad Sci U S A 93:8508-11, 1996
2890. Valdivia RH, Hromockyj AE, Monack D, Ramakrishnan L, Falkow S: Applications for green fluorescent protein (GFP) in the study of host-pathogen interactions. Gene 173:47-52, 1996
2891. Lybarger L, Dempsey D, Franek KJ, Chervenak R: Rapid generation and flow cytometric analysis of stable GFP-expressing cells. Cytometry 25:211-20, 1996
2892. Lybarger L, Dempsey D, Patterson GH, Piston DW, Kain SR, Chervenak R: Dual-color flow cytometric detection of fluorescent proteins using single-laser (488-nm) excitation. Cytometry 31:147-52, 1998
2893. Zhu J, Musco ML, Grace MJ: Three-color flow cytometry analysis of tricistronic expression of eBFP, eGFP, and eYFP using EMCV-IRES linkages. Cytometry 37:51-9, 1999
2894. Beavis AJ, Kalejta RF: Simultaneous analysis of the cyan, yellow and green fluorescent proteins by flow cytometry using single-laser excitation at 458 nm. Cytometry 37:68-73, 1999
2895. Stull RA, Hyun WC, Pallavicini MG: Simultaneous flow cytometric analyses of enhanced green and yellow fluorescent proteins and cell surface antigens in doubly transduced immature hematopoietic cell populations. Cytometry 40:126-34, 2000
2896. Richards B, Zharkikh L, Hsu F, Dunn C, Kamb A, Teng DH. Stable expression of Anthozoa fluorescent proteins in mammalian cells. Cytometry. 2002 Jun 1;48(2):106-12.
2897. Elsliger MA, Wachter RM, Hanson GT, Kallio K, Remington SJ: Structural and spectral response of green fluorescent protein variants to changes in pH. Biochemistry 38:5296-301, 1999
2898. Hanson GT, McAnaney TB, Park ES, Rendell ME, Yarbrough DK, Chu S, Xi L, Boxer SG, Montrose MH, Remington SJ: Green fluorescent protein variants as ratiometric dual emission pH sensors. 1. Structural characterization and preliminary application. Biochemistry 41:15477-88, 2002
2899. Olsen KN, Budde BB, Siegumfeldt H, Rechinger KB, Jakobsen M, Ingmer H: Noninvasive measurement of bacterial intracellular pH on a single-cell level with green fluorescent protein and fluorescence ratio imaging microscopy. Appl Environ Microbiol 68:4145-7, 2002
2900. Romoser VA, Hinkle PM, Persechini A: Detection in living cells of Ca2+-dependent changes in the fluorescence emission of an indicator composed of two green fluorescent protein variants linked by a

calmodulin-binding sequence. A new class of fluorescent indicators. J Biol Chem 272:13270-4, 1997
2901. Miyawaki A, Llopis J, Heim R, McCaffery JM, Adams JA, Ikura M, Tsien RY: Fluorescent indicators for Ca2+ based on green fluorescent proteins and calmodulin. Nature 388:882-7, 1997; Comment in: Nature 388:834-5. 1997
2902. Miyawaki A, Griesbeck O, Heim R, Tsien RY: Dynamic and quantitative Ca2+ measurements using improved cameleons. Proc Natl Acad Sci U S A 96:2135-40, 1999
2903. Siegel MS, Isacoff EY: A genetically encoded optical probe of membrane voltage. Neuron 19:735-41, 1997
2904. Ting AY, Kain KH, Klemke RL, Tsien RY: Genetically encoded fluorescent reporters of protein tyrosine kinase activities in living cells. Proc Natl Acad Sci U S A 98:15003-8, 2001
2905. Terskikh A, Fradkov A, Ermakova G, Zaraisky A, Tan P, Kajava AV, Zhao X, Lukyanov S, Matz M, Kim S, Weissman I, Siebert P: "Fluorescent timer": protein that changes color with time. Science 290:1585-8, 2000; Comment in: Science. 290:1478-9, 2000
2906. Christine R, Siebenkotten G, Radbruch A: Sensitive analysis of recombination activity using integrated cell surface reporter substrates. Cytometry 37:205-14, 1999
2907. Ramsey JM, Jacobson SC, Knapp MR: Microfabricated chemical measurement systems. Nat Med 1:1093-6, 1995
2908. Ramsey JM: The burgeoning power of the shrinking laboratory. Nat Biotechnol 17:1061-2, 1999
2909. McClain MA, Culbertson CT, Jacobson SC, Ramsey JM: Flow cytometry of Escherichia coli on microfluidic devices. Anal Chem 73:5334-8, 2001
2910. Dubelaar GBJ, Gerritzen PL: CytoBuoy: a step forward toward using flow cytometry in operational oceanography. Scientia Marina 64:255-65, 2000 (in reference 2403)
2911. Cram LS: Spin-offs from the NASA space program for tumor diagnosis. Cytometry 43:1, 2001
2912. Thomas RA, Krishan A, Robinson DM, Sams C, Costa F: NASA/American Cancer Society High-Resolution Flow Cytometry Project-I. Cytometry 43:2-11, 2001
2913. Wen J, Krishan A, Thomas RA: NASA/American Cancer Society High-Resolution Flow Cytometry Project - II: Effect of pH and DAPI concentration on dual parametric analysis of DNA/DAPI fluorescence and electronic nuclear volume. Cytometry 43:12-5, 2001
2914. Krishan A, Wen J, Thomas RA, Sridhar KS, Smith WI Jr: NASA/American Cancer Society High-Resolution Flow Cytometry Project - III: Multiparametric analysis of DNA content and electronic nuclear volume in human solid tumors. Cytometry 43:16-22, 2001
2915. Groner W, Simson E: Practical Guide to Modern Hematology Analyzers. New York, John Wiley & Sons, 1995, 258 pp
2916. Gaucher JC, Grunwald D, Frelat G: Fluorescence response and sensitivity determination for ATC 3000 flow cytometer. Cytometry 9:557-65, 1988
2917. Bergeron M, Lustyik G, Ding T, Nicholson J, Janossy G, Shapiro H, Barnett D, Mandy F: The stability of currently used cytometers facilitates both the identification of pipetting errors and their volumetric operation: 'time' can tell all. Cytometry (Clin Cytom) 52B:37-9, 2003
2918. Kamentsky LA, Burger DE, Gershman RJ, Kamentsky LD, Luther E: Slide-based laser scanning cytometry. Acta Cytol 41:123-43, 1997
2919. Clatch RJ, Walloch JL, Zutter MM, Kamentsky LA: Immunophenotypic analysis of hematologic malignancy by laser scanning cytometry. Am J Clin Pathol 105:744-55, 1996
2920. Sasaki K, Kurose A, Miura Y, Sato T, Ikeda E: DNA ploidy analysis by laser scanning cytometry (LSC) in colorectal cancers and comparison with flow cytometry. Cytometry 23:106-9, 1996
2921. Clatch RJ, Walloch JL, Foreman JR, Kamentsky LA: Multiparameter analysis of DNA content and cytokeratin expression in breast carcinoma by laser scanning cytometry. Arch Pathol Lab Med 121:585-92, 1997
2922. Gorczyca W, Darzynkiewicz Z, Melamed MR: Laser scanning cytometry in pathology of solid tumors. A review. Acta Cytol 41:98-108, 1997
2923. Schoell WM, Klintschar M, Mirhashemi R, Pertl B: Separation of sperm and vaginal cells with flow cytometry for DNA typing after sexual assault. Obstet Gynecol 94:623-7, 1999
2924. Schoell WM, Klintschar M, Mirhashemi R, Strunk D, Giuliani A, Bogensberger G, Pertl B: Separation of sperm and vaginal cells based on ploidy, MHC class I-, CD45-, and cytokeratin expression for enhancement of DNA typing after sexual assault. Cytometry 36:319-23, 1999
2925. Jiang PZ, Shen XM, Huang H, Shi YM, Yao KT: Rapid isolation of cancer cells from tumor tissue by micromanuiplator and extraction of tiny amount of rna. Di Yi Jun Yi Da Xue Xue Bao 22:551-3, 2002
2926. Conn M (ed): *Methods in Enzymology*, Vol 356, *Laser Capture in Microscopy and Microdissection.* San Diego, Academic Press, 2002, 775 pp
2927. Emmert-Buck MR, Bonner RF, Smith PD, Chuaqui RF, Zhuang Z, Goldstein SR, Weiss RA, Liotta LA: Laser capture microdissection. Science 274:998-1001, 1996
2928. Suarez-Quian CA, Goldstein SR, Pohida T, Smith PD, Peterson JI, Wellner E, Ghany M, Bonner RF: Laser capture microdissection of single cells from complex tissues. Biotechniques 26:328-35, 1999
2929. DiFrancesco LM, Murthy SK, Luider J, Demetrick DJ: Laser capture microdissection-guided fluorescence in situ hybridization and flow cytometric cell cycle analysis of purified nuclei from paraffin sections. Mod Pathol 13:705-11, 2000
2930. Webb T: Laser capture microdissection comes into mainstream use. J Natl Cancer Inst 92:1710-1, 2000
2931. Fend F, Specht K, Kremer M, Quintanilla-Martinez L: Laser capture microdissection in pathology. Methods Enzymol 356:196-206, 2002
2932. Luzzi V, Mahadevappa M, Raja R, Warrington JA, Watson MA: Accurate and reproducible gene expression profiles from laser capture microdissection, transcript amplification, and high density oligonucleotide microarray analysis. J Mol Diagn 5:9-14, 2003
2933. Cina SJ: Flow cytometric evaluation of DNA degradation: A predictor of postmortem interval? Am J Forensic Med Pathol 15:300-2, 1994
2934. Di Nunno NR, Costantinides F, Bernasconi P, Bottin C, Melato M: Is flow cytometric evaluation of DNA degradation a reliable method to investigate the early postmortem period? Am J Forensic Med Pathol 19:50-3, 1998
2935. Di Nunno N, Costantinides F, Melato M: Determination of the time of death in a homicide-suicide case using flow cytometry. Am J Forensic Med Pathol 20:228-31, 1999
2936. Di Nunno N, Costantinides F, Cina SJ, Rizzardi C, Di Nunno C, Melato M: What is the best sample for determining the early postmortem period by on-the-spot flow cytometry analysis? Am J Forensic Med Pathol 23:173-80, 2002
2937. Fugger EF, Black SH, Keyvanfar K, Schulman JD: Births of normal daughters after microsort sperm separation and intrauterine insemination, in-vitro fertilization, or intracytoplasmic sperm injection. Hum Reprod 13:2367-70, 1998
2938. Belkin L: Getting the girl. The New York Times Magazine, July 25, 1999, pp 26-31, 38, 54-5
2939. Sills ES, Kirman I, Colombero LT, Hariprashad J, Rosenwaks Z, Palermo GD: H-Y antigen expression patterns in human X- and Y-chromosome-bearing spermatozoa. Am J Reprod Immunol 40:43-7, 1998
2940. Brown RD, Linden MD, Mackowiak P, Kubus JJ, Zarbo RJ, Rabinovitch PS: The effect of number of histogram events on reproducibility and variation of flow cytometric proliferation measurement. Am J Clin Pathol 105:696-704, 1996

2941. Otto F, Oldiges H: Flow cytogenetic studies in chromosomes and whole cells for the detection of clastogenic effects. Cytometry 1:13-17, 1980.
2942. Misra RK, Easton MDL: Comment on analyzing flow cytometric data for comparison of mean values of the coefficient of variation of the G1 peak. Cytometry 36:112-6, 1999
2943. Marshall WH, Valentine FT, Lawrence HS: Cellular immunity in vitro. Clonal proliferation of antigen-stimulated lymphocytes. J Exp Med 130:327-43, 1969
2944. Valentine FT: The transformation and proliferation of lymphocytes in vitro. In: Revillard JP (ed): *Cell-Mediated Immunity: In Vitro Correlates*. Basel/New York, S. Karger, 1971, pp. 6-43
2945. Bedner E, Ruan Q, Chen S, Kamentsky LA, Darzynkiewicz Z: Multiparameter analysis of progeny of individual cells by laser scanning cytometry. Cytometry 40:271-9, 2000
2946. Li X, Traganos F, Melamed MR, Darzynkiewicz Z. Single-step procedure for labeling DNA strand breaks with fluorescein- or BODIPY-conjugated deoxynucleotides: detection of apoptosis and bromodeoxyuridine incorporation. Cytometry 20:172-80, 1995
2947. Li X, Darzynkiewicz Z: Labelling DNA strand breaks with BrdUTP. Detection of apoptosis and cell proliferation. Cell Prolif 28:571-9, 1995
2948. Hammers HJ, Kirchner H, Schlenke P. Ultraviolet-induced detection of halogenated pyrimidines: simultaneous analysis of DNA replication and cellular markers. Cytometry 40:327-35, 2000
2949. Beisker W, Weller-Mewe EM, Nüsse M: Fluorescence enhancement of DNA-bound TO-PRO-3 by incorporation of bromodeoxyuridine to monitor cell cycle kinetics. Cytometry 37:221-9, 1999
2950. Mozdziak PE, Pulvermacher PM, Schultz E, Schell K: Hoechst fluorescence intensity can be used to separate viable bromodeoxyuridine-labeled cells from viable non-bromodeoxyuridine-labeled cells. Cytometry 41:89-95, 2000
2951. Darzynkiewicz Z, Gong J, Juan G, Ardelt B, Traganos F: Cytometry of cyclin proteins. Cytometry 25:1-13, 1996
2952. Gong J, Traganos F, Darzynkiewicz Z: Discrimination of G2 and mitotic cells by flow cytometry based on different expression of cyclins A and B1. Exp Cell Res 220:226-31, 1995
2953. Widrow RJ, Rabinovitch PS, Cho K, Laird CD: Separation of cells at different times within G2 and mitosis by cyclin B1 flow cytometry. Cytometry 27:250-4, 1997
2954. Endl E, Hollmann C, Gerdes J: Antibodies against the Ki-67 protein: assessment of the growth fraction and tools for cell cycle analysis. Methods Cell Biol63:399-418, 2001 (in reference 2385)
2955. Larsen JK, Landberg G, Roos G: Detection of proliferating cell nuclear antigen. Methods Cell Biol63:419-31, 2001 (in reference 2385)
2956. Barnes D, Sato G: Serum-free cell culture: a unifying approach. Cell 22:649-55, 1980
2957. Rahman SA, Yokoyama M, Nishio S, Takeuchi M: Flow cytometric evaluation of transferrin receptor in transitional cell carcinoma. Urol Res 25:325-9, 1997
2958. Sladek TL, Jacobberger JW: Cell cycle analysis of retroviral vector gene expression during early infection. Cytometry 31:235-41, 1998
2959. Sramkoski RM, Wormsley SW, Bolton WE, Crumpler DC, Jacobberger JW: Simultaneous detection of cyclin B1, p105, and DNA content provides complete cell cycle phase fraction analysis of cells that endoreduplicate. Cytometry 35:274-83, 1999
2960. Jacobberger JW, Sramkoski RM, Wormsley SB, Bolton WE: Estimation of kinetic cell-cycle-related gene expression in G1 and G2 phases from immunofluorescence flow cytometry data. Cytometry 35:284-9, 1999
2961. Sladek TL, Laffin J, Lehman JM, Jacobberger JW: A subset of cells expressing SV40 large T antigen contain elevated p53 levels and have an altered cell cycle phenotype. Cell Prolif 33:115-25, 2000
2962. Juan G, Traganos F, James WM, Ray JM, Roberge M, Sauve DM, Anderson H, Darzynkiewicz Z: Histone H3 phosphorylation and expression of cyclins A and B1 measured in individual cells during their progression through G2 and mitosis. Cytometry 32:71-7, 1998
2963. van Engeland M, Nieland LJ, Ramaekers FC, Schutte B, Reutelingsperger CP: Annexin V-affinity assay: a review on an apoptosis detection system based on phosphatidylserine exposure. Cytometry 31:1-9, 1998 (Review)
2964. Stuart MC, Reutelingsperger CP, Frederik PM: Binding of annexin V to bilayers with various phospholipid compositions using glass beads in a flow cytometer. Cytometry 33:414-9, 1998
2965. Zamzami N, Marchetti P, Castedo M, Zanin C, Vayssiere JL, Petit PX, Kroemer G: Reduction in mitochondrial potential constitutes an early irreversible step of programmed lymphocyte death in vivo. J Exp Med 181:1661-72, 1995
2966. Zamzami N, Marchetti P, Castedo M, Decaudin D, Macho A, Hirsch T, Susin SA, Petit PX, Mignotte B, Kroemer G: Sequential reduction of mitochondrial transmembrane potential and generation of reactive oxygen species in early programmed cell death. J Exp Med 182:367-77, 1995
2967. Zamzami N, Hirsch T, Dallaporta B, Petit PX, Kroemer G: Mitochondrial implication in accidental and programmed cell death: Apoptosis and necrosis. J Bioenerg Biomembr 29:185-93, 1997
2968. Hildeman DA, Mitchell T, Teague TK, Henson P, Day BJ, Kappler J, Marrack PC: Reactive oxygen species regulate activation-induced T cell apoptosis. Immunity 10:735-44, 1999
2969. Vaux DL, Korsmeyer SJ: Cell death in development. Cell 96:245-54, 1999
2970. Schlegel RA, Stevens M, Lumley-Sapanski K, Williamson P: Altered lipid packing identifies apoptotic thymocytes. Immunol Lett 36:283-8, 1993
2971. Gavrieli Y, Sherman Y, Ben-Sasson SA: Identification of programmed cell death in situ via specific labeling of nuclear DNA fragmentation. J Cell Biol119:493-501, 1992
2972. Frey T. Nucleic acid dyes for detection of apoptosis in live cells. Cytometry 21:265-74, 1995
2973. Reynolds JE, Li J, Eastman A: Detection of apoptosis by flow cytometry of cells simultaneously stained for intracellular pH (carboxy SNARF-1) and membrane permeability (Hoechst 33342). Cytometry 25:349-57, 1996
2974. Koester SK, Roth P, Mikulka WR, Schlossman SF, Zhang C, Bolton WE: Monitoring early cellular responses in apoptosis is aided by the mitochondrial membrane protein-specific monoclonal antibody APO2.7. Cytometry 29:306-12, 1997
2975. Koester SK, Schlossman SF, Zhang C, Decker SJ, Bolton WE: APO2.7 defines a shared apoptotic-necrotic pathway in a breast tumor hypoxia model. Cytometry 33:324-32, 1998
2976. Mentz F, Baudet S, Blanc C, Issaly F, Binet JL, Merle-Beral H: Simple, fast method of detection apoptosis in lymphoid cells. Cytometry 32:95-101, 1998
2977. Taga K, Yoshida M, Kaneko M, Asada M, Okada M, Taniho M, Tosato G: Contribution of automated hematology analysis to the detection of apoptosis in peripheral blood lymphocytes. Cytometry (Commun Clin Cytom) 42:209-14, 2000
2978. Lertworasirikul T, Bunyaratvej A: A rapid measurement of apoptosis-associated light scatter changes using a hematology analyzer. Cytometry (Commun Clin Cytom) 42:215-7, 2000
2979. Lizard G: Changes in light scatter properties are a general feature of cell death but are not characteristic of apoptotically dying cells. Cytometry (Commun Clin Cytom) 46:65-6, 2001
2980. Strebel A, Harr T, Bachmann F, Wernli M, Erb P: Green fluorescent protein as a novel tool to measure apoptosis and necrosis. Cytometry 43:126-33, 2001
2981. Steff AM, Fortin M, Arguin C, Hugo P: Detection of a decrease in green fluorescent protein fluorescence for the monitoring of cell death: an assay amenable to high-throughput screening technologies. Cytometry 45:237-43, 2001
2982. Span LF, Pennings AH, Vierwinden G, Boezeman JB, Raymakers RA, de Witte T: The dynamic process of apoptosis analyzed by flow

cytometry using Annexin-V/propidium iodide and a modified in situ end labeling technique. Cytometry 47:24-31, 2002
2983. Bedner E, Li X, Gorczyca W, Melamed MR, Darzynkiewicz Z: Analysis of apoptosis by laser scanning cytometry. Cytometry 35:181-95, 1999
2984. Frey T: Correlated flow cytometric analysis of terminal events in apoptosis reveals the absence of some changes in some model systems. Cytometry 28:253-63, 1997
2985. Del Bino G, Darzynkiewicz Z, Degraef C, Mosselmans R, Fokan D, Galand P: Comparison of methods based on annexin-V binding, DNA content or TUNEL for evaluating cell death in HL-60 and adherent MCF-7 cells. Cell Prolif 32:25-37, 1999
2986. Darzynkiewicz Z, Bedner E, Traganos F: Difficulties and pitfalls in analysis of apoptosis. Methods Cell Biol 63:527-46, 2001(in reference 2385)
2987. Smolewski P, Grabarek J, Lee BW, Johnson GL, Darzynkiewicz Z: Kinetics of HL-60 cell entry into apoptosis during treatment with TNF-α or camptothecin assayed by the stathmo-apoptosis method.
2988. Counter CM, Hahn WC, Wei W, Caddle SD, Beijersbergen RL, Lansdorp PM, Sedivy JM, Weinberg RA: Dissociation among in vitro telomerase activity, telomere maintenance, and cellular immortalization. Proc Natl Acad Sci U S A 95:14723-8, 1998
2989. Poon SS, Martens UM, Ward RK, Lansdorp PM: Telomere length measurements using digital fluorescence microscopy. Cytometry 36:267-78, 1999
2990. Poon SS, Lansdorp PM: Measurements of telomere length on individual chromosomes by image cytometry. Methods Cell Biol 64:69-96, 2001
2991. Law HKW, Lau YK: Validation and development of quantitative flow cytometry-based fluorescence in situ hybridization for intercenter comparison of telomere length measurement. Cytometry 43:150-3, 2001
2992. Roos G, Hultdin M: Flow cytometric determination of telomere length (letter). Cytometry 45:79, 2001
2993. Law HKW, Lau YK: DNA index and Q flow FISH measurement of telomere length (reply to Roos and Hultdin). Cytometry 45:80, 2001
2994. Baerlocher GM, Mak J, Tien T, Lansdorp PM: Telomere length measurement by fluorescence in situ hybridization and flow cytometry: tips and pitfalls. Cytometry 47:89-99, 2002
2995. Roederer M, Bigos M, Nozaki T, Stovel RT, Parks DR, Herzenberg LA: Heterogeneous calcium flux in peripheral T cell subsets revealed by five-color flow cytometry using log-ratio circuitry. Cytometry 21:187-96, 1995
2996. Roederer M, De Rosa S, Gerstein R, Anderson M, Bigos M, Stovel R, Nozaki T, Parks D, Herzenberg L: 8 color, 10-parameter flow cytometry to elucidate complex leukocyte heterogeneity. Cytometry 29:328-39, 1997
2997. Watanabe N, De Rosa SC, Cmelak A, Hoppe R, Herzenberg LA, Roederer M: Long-term depletion of naive T cells in patients treated for Hodgkin's disease. Blood 90:3662-72, 1997
2998. Bigos M, Baumgarth N, Jager GC, Herman OC, Nozaki T, Stovel RT, Parks DR, Herzenberg LA: Nine color eleven parameter immunophenotyping using three laser flow cytometry. Cytometry 36:36-45, 1999
2999. Baumgarth N, Roederer M: A practical approach to multicolor flow cytometry for immunophenotyping. J Immunol Methods 243:77-97, 2000 (in reference 2401)
3000. Perez OD, Nolan GP: Simultaneous measurement of multiple active kinase states using polychromatic flow cytometry. Nat Biotechnol 20:155-62, 2002
3001. Mitra DK, De Rosa SC, Luke A, Balamurugan A, Khaitan BK, Tung J, Mehra NK, Terr AI, O'Garra A, Herzenberg LA, Roederer M: Differential representations of memory T cell subsets are characteristic of polarized immunity in leprosy and atopic diseases. Int Immunol 11:1801-10, 1999
3002. De Rosa SC, Roederer M: Eleven-color flow cytometry. A powerful tool for elucidation of the complex immune system. Clin Lab Med 21:697-712, vii, 2001
3003. De Rosa SC, Brenchley JM, Roederer M: Beyond six colors: A new era in flow cytometry. Nat Med 9:112-7, 2003
3004. Bauer KD, de la Torre-Bueno J, Diel IJ, Hawes D, Decker WJ, Priddy C, Bossy B, Ludmann S, Yamamoto K, Masih AS, Espinoza FP, Harrington DS: Reliable and sensitive analysis of occult bone marrow metastases using automated cellular imaging. Clin Cancer Res 6:3552-9, 2000
3005. Witzig TE, Bossy B, Kimlinger T, Roche PC, Ingle JN, Grant C, Donohue J, Suman VJ, Harrington D, Torre-Bueno J, Bauer KD: Detection of circulating cytokeratin-positive cells in the blood of breast cancer patients using immunomagnetic enrichment and digital microscopy. Clin Cancer Res 8:1085-91, 2002
3006. Tibbe AG, de Grooth BG, Greve J, Liberti PA, Dolan GJ, Terstappen LW: Optical tracking and detection of immunomagnetically selected and aligned cells. Nat Biotechnol 17:1210-3, 1999
3007. Tibbe AG, de Grooth BG, Greve J, Liberti PA, Dolan GJ, Terstappen LW: Cell analysis system based on immunomagnetic cell selection and alignment followed by immunofluorescent analysis using compact disk technologies. Cytometry 43:31-7, 2001
3008. Tibbe AG, de Grooth BG, Greve J, Dolan GJ, Terstappen LW: Imaging technique implemented in celltracks system. Cytometry 47:248-55, 2002
3009. Tibbe AG, de Grooth BG, Greve J, Dolan GJ, Rao C, Terstappen LW: Magnetic field design for selecting and aligning immunomagnetic labeled cells. Cytometry 47:163-72, 2002
3010. Racila E, Euhus D, Weiss AJ, Rao C, McConnell J, Terstappen LW, Uhr JW: Detection and characterization of carcinoma cells in the blood. Proc Natl Acad Sci U S A 95:4589-94, 1998
3011. Terstappen LW, Rao C, Gross S, Kotelnikov V, Racilla E, Uhr J, Weiss A: Flow cytometry--principles and feasibility in transfusion medicine. Enumeration of epithelial derived tumor cells in peripheral blood. Vox Sang 74 Suppl 2:269-74, 1998
3012. Moreno JG, O'Hara SM, Gross S, Doyle G, Fritsche H, Gomella LG, Terstappen LW: Changes in circulating carcinoma cells in patients with metastatic prostate cancer correlate with disease status. Urology 58:386-92, 2001
3013. Hayes DF, Walker TM, Singh B, Vitetta ES, Uhr JW, Gross S, Rao C, Doyle GV, Terstappen LW: Monitoring expression of her-2 on circulating epithelial cells in patients with advanced breast cancer. Int J Oncol 21:1111-7, 2002
3014. Hanania EG, O'Neal, RA Eisfeld TM, Palsson BO, Koller MR: High-throughput cell purification for autologous stem cell transplantation of Non-Hodgkin's lymphoma (NHL) patients on the Photosis™ clinical platform. Cytometry Supp 10:105, 2002
3015. Koller MR, Hanania EG, Sasaki GC, Palsson BO: High-throughput imaging and laser-based manipulation of individual cells on the LEAP™ research platform. Cytometry Supp 10:62, 2002
3016. Dyba M, Hell SW: Focal Spots of Size λ/23 Open Up Far-Field Florescence Microscopy at 33 nm Axial Resolution. Phys Rev Lett 88:163901, 2002
3017. Chiu C-S, Kartalov E, Unger M, Quake S, Lester HA: Single-molecule measurements calibrate green fluorescent protein surface densities on transparent beads for use with 'knock-in' animals and other expression systems. J Neurosci Methods 105:55-63, 2001
3018. Huang Z, Jett JH, Keller RA: Bacteria genome fingerprinting by flow cytometry. Cytometry 35:169-75, 1999
3019. Kim Y, Jett JH, Larson EJ, Penttila JR, Marrone BL, Keller RA: Bacterial fingerprinting by flow cytometry: bacterial species discrimination. Cytometry 36:324-32, 1999
3020. Larson EJ, Hakovirta JR, Cai H, Jett JH, Burde S, Keller RA, Marrone BL: Rapid DNA fingerprinting of pathogens by flow cytometry. Cytometry 41:203-8, 2000
3021. See Reference 2682

3022. Cook EB, Stahl JL, Lowe L, Chen R, Morgan E, Wilson J, Varro R, Chan A, Graziano FM, Barney NP: Simultaneous measurement of six cytokines in a single sample of human tears using microparticle-based flow cytometry: allergics vs. non-allergics. J Immunol Methods 254:109-18, 2001

3023. Evans M, Sewter C, Hill E: An encoded particle array tool for multiplex bnioassays. ASSAY and Drug Devel Technol 1:199-208, 2003

3024. McFarland DC, Durack G: Gel microdrop encapsulation for the frugal investigator. In: Diamond RA, DeMaggio S (eds): *In Living Color. Protocols in Flow Cytometry and Cell Sorting*. Berlin, Springer, 2000, pp 184-93 (in reference 2390)

3025. Gray F, Kenney JS, Dunne JF: Secretion capture and report web: Use of affinity derivatized agarose microdroplets for the selection of hybridoma cells. J Immunol Methods 182:155-63, 1995

3026. Kenney JS, Gray F, Ancel MH, Dunne JF: Production of monoclonal antibodies using a secretion capture report web. Biotechnol 8:333-7, 1995

3027. Shear JB, Fishman HA, Allbritton NL, Garigan D, Zare RN, Scheller RH: Single cells as biosensors for chemical separations. Science 267:74-7, 1995

3028. Fishman HA, Orwar O, Allbritton NL, Modi BP, Shear JB, Scheller RH, Zare RN: Cell-to-cell scanning in capillary electrophoresis. Anal Chem 68:1181-6, 1996

3029. Sims CE, Meredith GD, Krasieva TB, Berns MW, Tromberg BJ, Allbritton NL: Laser-micropipet combination for single-cell analysis. Anal Chem 70:4570-7, 1998

3030. Meredith GD, Sims CE, Soughayer JS, Allbritton NL: Measurement of kinase activation in single mammalian cells. Nat Biotechnol 18:309-12, 2000

3031. Li H, Sims CE, Wu HY, Allbritton NL: Spatial control of cellular measurements with the laser micropipet. Anal Chem 73:4625-31, 2001

3032. Sims CE, Allbritton NL: Single-cell kinase assays: Opening a window onto cell behavior. Curr Opin Biotechnol 14:23-8, 2003

3033. Krylov SN, Zhang Z, Chan NW, Arriaga E, Palcic MM, Dovichi NJ: Correlating cell cycle with metabolism in single cells: Combination of image and metabolic cytometry. Cytometry 37:14-20, 1999

3034. Le XC, Tan W, Scaman CH, Szpacenko A, Arriaga E, Zhang Y, Dovichi NJ, Hindsgaul O, Palcic MM: Single cell studies of enzymatic hydrolysis of a tetramethylrhodamine labeled triglucoside in yeast. Glycobiology 9:219-25, 1999

3035. Krylov SN, Arriaga EA, Chan NW, Dovichi NJ, Palcic MM: Metabolic cytometry: Monitoring oligosaccharide biosynthesis in single cells by capillary electrophoresis. Anal Biochem 283:133-5, 2000

3036. Krylov SN, Arriaga E, Zhang Z, Chan NW, Palcic MM, Dovichi NJ: Single-cell analysis avoids sample processing bias. J Chromatogr B Biomed Sci Appl 741:31-5, 2000

3037. Krylov SN, Starke DA, Arriaga EA, Zhang Z, Chan NW, Palcic MM, Dovichi NJ: Instrumentation for chemical cytometry. Anal Chem 72:872-7, 2000

3038. Hu S, Lee R, Zhang Z, Krylov SN, Dovichi NJ: Protein analysis of an individual Caenorhabditis elegans single-cell embryo by capillary electrophoresis. J Chromatogr B Biomed Sci Appl 752:307-10, 2001

3039. Dovichi NJ, Pinkel D: Analytical biotechnology. Tools to characterize cells and their contents. Curr Opin Biotechnol 14:3-4, 2003

3040. Fuller KM, Arriaga EA: Advances in the analysis of single mitochondria. Curr Opin Biotechnol 14:35-41, 2003

3041. Olofsson J, Nolkrantz K, Ryttsén F, Lambie BA, Weber SG, Orwar O: Single cell electroporation. Curr Opin Biotechnol 14:29-34, 2003

3042. Ibrahim SF, van den Engh G: High-speed cell sorting: fundamentals and recent advances. Curr Opin Biotechnol 14:5-12, 2003

3043. Ling V, Neben S: In vitro differentiation of embryonic stem cells: immunophenotypic analysis of cultured embryoid bodies. J Cell Physiol171:104-15, 1997

3044. Eiges R, Schuldiner M, Drukker M, Yanuka O, Itskovitz-Eldor J, Benvenisty N: Establishment of human embryonic stem cell-transfected clones carrying a marker for undifferentiated cells. Curr Biol 11:514-8, 2001

3045. Andressen C, Stocker E, Klinz FJ, Lenka N, Hescheler J, Fleischmann B, Arnhold S, Addicks K: Nestin-specific green fluorescent protein expression in embryonic stem cell-derived neural precursor cells used for transplantation. Stem Cells 19:419-24, 2001

3046. Jiang Y, Vaessen B, Lenvik T, Blackstad M, Reyes M, Verfaillie CM: Multipotent progenitor cells can be isolated from postnatal murine bone marrow, muscle, and brain. Exp Hematol30:896-904, 2002

3047. Maric D, Maric I, Barker JL: Buoyant density gradient fractionation and flow cytometric analysis of embryonic rat cortical neurons and progenitor cells. Methods 16:247-59, 1998

3048. Maric D, Maric I, Smith SV, Serafini R, Hu Q, Barker JL: Potentiometric study of resting potential, contributing K+ channels and the onset of Na+ channel excitability in embryonic rat cortical cells. Eur J Neurosci 10:2532-46, 1998

3049. Maric D, Liu QY, Grant GM, Andreadis JD, Hu Q, Chang YH, Barker JL, Joseph J, Stenger DA, Ma W: Functional ionotropic glutamate receptors emerge during terminal cell division and early neuronal differentiation of rat neuroepithelial cells. J Neurosci Res 61:652-62, 2000

3050. Maric D, Maric I, Chang YH, Barker JL: Stereotypical physiological properties emerge during early neuronal and glial lineage development in the embryonic rat neocortex. Cereb Cortex 10:729-47, 2000

3051. Maric D, Maric I, Barker JL: Developmental changes in cell calcium homeostasis during neurogenesis of the embryonic rat cerebral cortex. Cereb Cortex 10:561-73, 2000

3052. Maric D, Liu QY, Maric I, Chaudry S, Chang YH, Smith SV, Sieghart W, Fritschy JM, Barker JL: GABA expression dominates neuronal lineage progression in the embryonic rat neocortex and facilitates neurite outgrowth via GABA(A) autoreceptor/Cl- channels. J Neurosci 21:2343-60, 2001

3053. Maric D, Maric I, Chang YH, Barker JL: Prospective cell sorting of embryonic rat neural stem cells and neuronal and glial progenitors reveals selective effects of basic fibroblast growth factor and epidermal growth factor on self-renewal and differentiation. J Neurosci 23:240-51, 2003

3054. Morrison SJ, White PM, Zock C, Anderson DJ: Prospective identification, isolation by flow cytometry, and in vivo self-renewal of multipotent mammalian neural crest stem cells. Cell 96:737-49, 1999

3055. Rietze RL, Valcanis H, Brooker GF, Thomas T, Voss AK, Bartlett PF: Purification of a pluripotent neural stem cell from the adult mouse brain. Nature 412:736-9, 2001

3056. Stubblefield E, Cram S, Deaven L: Flow microfluorometric analyses of isolated Chinese hamster chromosomes. Exp Cell Res 94:464-8, 1975

3057. VanDevanter DR, Choongkittaworn NM, Dyer KA, Aten J, Otto P, Behler C, Bryant EM, Rabinovitch PS: Pure chromosome-specific PCR libraries from single sorted chromosomes. Proc Natl Acad Sci U S A 91:5858-62, 1994

3058. Hui SM, Trask B, van den Engh G, Bartuski AJ, Smith A, Flint A, Lalande M, Silverman GA: Analysis of randomly amplified flow-sorted chromosomes using the polymerase chain reaction. Genomics 26:364-71, 1995

3059. Silverman GA, Schneider SS, Massa HF, Flint A, Lalande M, Leonard JC, Overhauser J, van den Engh G, Trask BJ: The 18q- syndrome: Analysis of chromosomes by bivariate flow karyotyping and the PCR reveals a successive set of deletion breakpoints within 18q21.2-q22.2. Am J Hum Genet 56:926-37, 1995

3060. Xiao Y, Slijepcevic P, Arkesteijn G, Darroudi F, Natarajan AT: Development of DNA libraries specific for chinese hamster chromosomes 3, 4, 9, 10, X, and Y by DOP-PCR. Cytogenet Cell Genet 75:57-62, 1996

3061. Lucretti S, Doležel J: Cell cycle synchronization, chromosome isolation, and flow-sorting in plants. Methods Cell Biol 50:61-83, 1995

3062. Šimková H, Cíhalíková J, Vrána J, Lysák MA, Doležel J: Preparation of HMW DNA from plant nuclei and chromosomes isolated from root tips. Biol Plant 46:369-73, 2003
3063. Ferguson-Smith MA: Genetic analysis by chromosome sorting and painting: Phylogenetic and diagnostic applications. Eur J Hum Genet 5:253-65, 1997
3064. Monard SP: Chromosome sorting and analysis by FACS. Methods Mol Biol 91:239-54, 1998
3065. Rabbitts P, Impey H, Heppell-Parton A, Langford C, Tease C, Lowe N, Bailey D, Ferguson-Smith M, Carter N: Chromosome specific paints from a high resolution flow karyotype of the mouse. Nat Genet 9:369-75, 1995
3066. Macas J, Gualberti G, Nouzova M, Samec P, Lucretti S, Doležel J: Construction of chromosome-specific DNA libraries covering the whole genome of field bean (Vicia faba L.). Chromosome Res 4:531-9, 1996
3067. Lucretti S, Doležel J: Bivariate flow karyotyping in broad bean (Vicia faba). Cytometry 28:236-42, 1997
3068. Burkin DJ, O'Brien PC, Broad TE, Hill DF, Jones CA, Wienberg J, Ferguson-Smith MA: Isolation of chromosome-specific paints from high-resolution flow karyotypes of the sheep (Ovis aries). Chromosome Res 5:102-8, 1997
3069. Lee JY, Koi M, Stanbridge EJ, Oshimura M, Kumamoto AT, Feinberg AP: Simple purification of human chromosomes to homogeneity using muntjac hybrid cells. Nat Genet 7:29-33, 1994
3070. Nguyen BT, Lazzari K, Abebe J, Mac I, Lin JB, Chang A, Wydner KL, Lawrence JB, Cram LS, Weier HU, et al. In situ hybridization to chromosomes stabilized in gel microdrops. Cytometry 21:111-9, 1995
3071. Kubaláková M, Lysák MA, Vrána J, Šimková H, Cíhalíková J, Doležel J: Rapid identification and determination of purity of flow-sorted plant chromosomes using C-PRINS. Cytometry 41:102-8, 2000
3072. Gygi MP, Ferguson MD, Mefford HC, Lund KP, O'Day C, Zhou P, Friedman C, van den Engh G, Stolowitz ML, Trask BJ: Use of fluorescent sequence-specific polyamides to discriminate human chromosomes by microscopy and flow cytometry. Nucleic Acids Res 30:2790-9, 2002
3073. Stepanov SI, Konyshev VN, Kotlovanova LV, Roganov AP: Karyotyping of individual cells with flow cytometry. Cytometry 23:279-83, 1996
3074. Remy I, Michnick SW: Clonal selection and in vivo quantitation of protein interactions with protein-fragment complementation assays. Proc Natl Acad Sci U S A 96:5394-9, 1999
3075. Remy I, Wilson IA, Michnick SW: Erythropoietin receptor activation by a ligand-induced conformation change. Science 283:990-3, 1999
3076. Galarneau A, Primeau M, Trudeau LE, Michnick SW: Beta-lactamase protein fragment complementation assays as in vivo and in vitro sensors of protein protein interactions. Nat Biotechnol 20:619-22, 2002
3077. McCoy JP, Jr.: Medicaid and medicare reimbursement for flow cytometry. Am J Clin Pathol 115:631-41, 2001
3078. Richardson Jones A et al: *Performance Goals for the Internal Quality Control of Multichannel Hematology Analyzers; Approved Standard.* Publication H26-A. Wayne (PA), NCCLS, 1996
3079. Buttarello M, Bulian P, Farina G, Temporin V, Toffolo L, Trabuio E, Rizzotti P: Flow cytometric reticulocyte counting. Parallel evaluation of five fully automated analyzers: An NCCLS-ICSH approach. Am J Clin Pathol 115:100-11, 2001
3080. Davis BH, Bigelow NC, Koepke JA, Borowitz MJ, Houwen B, Jacobberger JW, Pierre RV, Corash L, Ault KA, Batjer JD: Flow cytometric reticulocyte analysis. Multiinstitutional interlaboratory correlation study. Am J Clin Pathol 102:468-77, 1994
3081. Riley RS, Ben-Ezra JM, Goel R, Tidwell A: Reticulocytes and reticulocyte enumeration. J Clin Lab Anal 15:267-94, 2001
3082. Koepke JA et al: *Methods for Reticulocyte Counting (Flow Cytometry and Supravital Dyes); Approved Guideline.* Publication H44-A. Wayne (PA), NCCLS, 1997
3083. Davis BH, Ornvold K, Bigelow NC: Flow cytometric reticulocyte maturity index: A useful laboratory parameter of erythropoietic activity in anemia. Cytometry 22:35-9, 1995
3084. Davis BH: Diagnostic utility of red cell flow cytometric analysis. Clin Lab Med 21:829-40, 2001
3085. Garratty G, Arndt P: Applications of flow cytometry to transfusion science. Transfusion 35:157-78, 1995
3086. Garratty G, Arndt P: Applications of flow cytofluorometry to red blood cell immunology. Cytometry 38:259-67, 1999
3087. Koepke JA et al: *Reference Leukocyte Differential Count (Proportional) and Evaluation of Instrumental Methods; Approved Standard.* Publication H20-A. Wayne (PA), NCCLS, 1992
3088. Simson E, Groner W: Variability in absolute lymphocyte counts obtained by automated cell counters. Cytometry 22:26-34, 1995
3089. Zucker-Franklin D: The percentage of monocytes among "mononuclear" cell fractions obtained from human blood. J Immunol 112:234-40, 1974
3090. Hübl W, Wolfbauer G, Andert S, Thum G, Streicher J, Hubner C, Lapin A, Bayer PM: Toward a new reference method for the leukocyte five-part differential. Cytometry (Commun Clin Cytom) 30:72-84, 1997
3091. Mason D, Simmons D, Buckley C, Schwartz-Albiez R, Hadam M, Saalmuller A, Clark E, Fabio F, Morrissey JA, Vivier E et al (eds): *Leucocyte Typing VII.* New York, Oxford University Press, 2002, 1504 pp
3092. Barclay A, Brown M, Law S, McKnight A, Tomlinson M, van der Merwe P: *The Leucocyte Antigen FactsBook,* 2 Ed. Academic Press, 1997, 624 pp
3093. Isacke C, Horton M: *The Adhesion Molecule FactsBook,* 2 Ed. Academic Press, 2000, 328 pp
3094. Lefranc M-P, Lefranc G: *The T Cell Receptor FactsBook.* Academic Press, 2001, 397 pp
3095. Marsh S, Parham P, Barber L: *The HLA FactsBook.* Academic Press, 1999, 416 pp
3096. Fitzgerald K, O'Neill L, Gearing A, Callard R: *The Cytokine FactsBook and WebFacts,* 2 Ed. Academic Press, 2001, 528 pp
3097. Kass L: Metachromatic dye sorption means for differential determination of leukocytes. U. S. Patent 4,400,370. 1983.
3098. Benveniste J: The human basophil degranulation test as an in vitro method for the diagnosis of allergies. Clin Allergy. 11:1-11, 1981
3099. Yeung Laiwah AC, Patel KR, Seenan AK, Galloway E, McCulloch W: Evaluation of the human basophil degranulation test using the commercially available Baso-kit as a test of immediate-type hypersensitivity in hay-fever sufferers. Clin Allergy. 14:571-9. 1984
3100. Sainte-Laudy J: Standardization of basophil degranulation for pharmacological studies. J Immunol Methods 98:279-82, 1987
3101. Crockard AD, Ennis M: Basophil histamine release tests in the diagnosis of allergy and asthma. Clin Exp Allergy 31:345-50, 2001
3102. Davenas E, Beauvais F, Amara J, Oberbaum M, Robinzon B, Miadonna A, Tedeschi A, Pomeranz B, Fortner P, Belon P, Sainte-Laudy J, Poitevin B, Benveniste J: Human basophil degranulation triggered by very dilute antiserum against IgE. Nature 333:816-8, 1988
3103. Maddox J, Randi J, Stewart WW: "high-dilution" experiments a delusion. Nature 334:287-90, 1988
3104. Maddox J: Waves caused by extreme dilution. Nature 335:760-3, 1988
3105. Belon P, Cumps J, Ennis M, Mannaioni P, Sainte-Laudy J, Roberfroid M, Wiegant F: Inhibition of human basophil degranulation by successive histamine dilutions: Results of a european multi-centre trial. Inflamm Res 48 Suppl 1:S17-8, 1999
3106. Fisher P: The end of the Benveniste affair? [see comments]. Br Homeopath J 88: p186-7, 1999

3107. Vadas MA, Varigos G, Nicola N, Pincus S, Dessein A, Metcalf D, Battye FL: Eosinophil activation by colony-stimulating factor in man: Metabolic effects and analysis by flow cytometry. Blood 61:1232-41, 1983
3108. Lavigne S, Bosse M, Boulet LP, Laviolette M: Identification and analysis of eosinophils by flow cytometry using the depolarized side scatter-saponin method. Cytometry 29:197-203, 1997
3109. Lucey DR, Dorsky DI, Nicholson-Weller A, Weller PF: Human eosinophils express CD4 protein and bind human immunodeficiency virus 1 gp120. J Exp Med 169:327-32, 1989
3110. Rand TH, Cruikshank WW, Center DM, Weller PF: CD4-mediated stimulation of human eosinophils: lymphocyte chemoattractant factor and other CD4-binding ligands elicit eosinophil migration. J Exp Med 173:1521-8, 1991
3111. Thurau AM, Schylz U, Wolf V, Krug N, Schauer U: Identification of eosinophils by flow cytometry. Cytometry 23:150-8, 1996
3112. Gopinath R, Nutman TB: Identification of eosinophils in lysed whole blood using side scatter and CD16 negativity. Cytometry 30:313-6, 1997
3113. Carulli G, Sbrana S, Azzara A, Minnucci S, Angiolini C, Marini A, Ambrogi F: Detection of eosinophils in whole blood samples by flow cytometry. Cytometry 34:272-9, 1998
3114. Efthimiadis A, Hargreave FE, Dolovich J: Use of selective binding of fluorescein isothiocyanate to detect eosinophils by flow cytometry. Cytometry 26:75-6, 1996
3115. Bedner E, Halicka HD, Cheng W, Salomon T, Deptala A, Gorczyca W, Melamed MR, Darzynkiewicz Z: High affinity binding of fluorescein isothiocyanate to eosinophils detected by laser scanning cytometry: A potential source of error in analysis of blood samples utilizing fluorescein-conjugated reagents in flow cytometry. Cytometry 36:77-82, 1999
3116. Qureshi SS, Lewis SM, Gant VA, Treacher D, Davis BH, Brown KA: Increased distribution and expression of CD64 on blood polymorphonuclear cells from patients with the systemic inflammatory response syndrome (SIRS). Clin Exp Immunol 125:258-65, 2001
3117. Hirsh M, Mahamid E, Bashenko Y, Hirsh I, Krausz MM: Overexpression of the high-affinity Fcgamma receptor (CD64) is associated with leukocyte dysfunction in sepsis. Shock 16:102-8, 2001
3118. Naccasha N, Gervasi MT, Chaiworapongsa T, Berman S, Yoon BH, Maymon E, Romero R: Phenotypic and metabolic characteristics of monocytes and granulocytes in normal pregnancy and maternal infection. Am J Obstet Gynecol 185:1118-23, 2001
3119. Allen E, Bakke AC, Purtzer MZ, Deodhar A: Neutrophil CD64 expression: Distinguishing acute inflammatory autoimmune disease from systemic infections. Ann Rheum Dis 61:522-5, 2002
3120. Barth E, Fischer G, Schneider EM, Moldawer LL, Georgieff M, Weiss M: Peaks of endogenous G-CSF serum concentrations are followed by an increase in respiratory burst activity of granulocytes in patients with septic shock. Cytokine 17:275-84, 2002
3121. Ault KA, Mitchell J: Analysis of platelets by flow cytometry. Methods Cell Biol 42 Pt B:275-94, 1994
3122. Ault KA: The clinical utility of flow cytometry in the study of platelets. Semin Hematol 38:160-8, 2001
3123. Michelson AD, Barnard MR, Krueger LA, Frelinger AL, 3rd, Furman MI: Evaluation of platelet function by flow cytometry. Methods 21:259-70, 2000
3124. Shankey TV, Jeske WP, Walenga JM: Flow cytometric analysis of platelets and platelet function. In: Stewart CC, Nicholson JKA (eds): *Immunophenotyping* (Cytometric Cellular Analysis Series). New York, Wiley-Liss, 2000, pp 333-59 (in reference 2395)
3125. Ault KA, Knowles C: In vivo biotinylation demonstrates that reticulated platelets are the youngest platelets in circulation. Exp Hematol 23:996-1001, 1995
3126. Heilmann E, Friese P, Anderson S, George JN, Hanson SR, Burstein SA, Dale GL: Biotinylated platelets: A new approach to the measurement of platelet life span. Br J Haematol 85:729-35, 1993
3127. Schultheiss HP, Tschoepe D, Esser J, Schwippert B, Roesen P, Nieuwenhuis HK, Schmidt-Soltau C, Strauer B: Large platelets continue to circulate in an activated state after myocardial infarction. Eur J Clin Invest 24:243-7, 1994
3128. Becker RC, Tracy RP, Bovill EG, Mann KG, Ault K: The clinical use of flow cytometry for assessing platelet activation in acute coronary syndromes. TIMI-III thrombosis and anticoagulation group. Coron Artery Dis 5:339-45, 1994
3129. Nurden AT, Macchi L, Bihour C, Durrieu C, Besse P, Nurden P: Markers of platelet activation in coronary heart disease patients. Eur J Clin Invest 24 Suppl 1:42-5, 1994
3130. Michelson AD, Barnard MR, Krueger LA, Valeri CR, Furman MI: Circulating monocyte-platelet aggregates are a more sensitive marker of in vivo platelet activation than platelet surface p-selectin: Studies in baboons, human coronary intervention, and human acute myocardial infarction. Circulation 104:1533-7, 2001
3131. Watala C, Waczulikova I, Wieclawska B, Rozalski M, Gresner P, Gwozdzinski K, Mateasik A, Sikurova L: Merocyanine 540 as a fluorescent probe of altered membrane phospholipid asymmetry in activated whole blood platelets. Cytometry 49:119-33, 2002
3132. Thiagarajan P, Tait JF: Collagen-induced exposure of anionic phospholipid in platelets and platelet-derived microparticles. J Biol Chem 266:24302-7, 1991
3133. Vitrat N, Cohen-Solal K, Pique C, Le Couedic JP, Norol F, Larsen AK, Katz A, Vainchenker W, Debili N: Endomitosis of human megakaryocytes are due to abortive mitosis. Blood 91:3711-23, 1998
3134. Bath PM, Gladwin AM, Carden N, Martin JF: Megakaryocyte DNA content is increased in patients with coronary artery atherosclerosis. Cardiovasc Res 28:1348-52, 1994
3135. Battinelli E, Willoughby SR, Foxall T, Valeri CR, Loscalzo J: Induction of platelet formation from megakaryocytoid cells by nitric oxide. Proc Natl Acad Sci U S A 98:14458-63, 2001
3136. Loken MR, Wells DA: Normal antigen expression in hematopoiesis: Basis for interpreting leukemia phenotypes. In: Stewart CC, Nicholson JKA (eds): *Immunophenotyping* (Cytometric Cellular Analysis Series). New York, Wiley-Liss, 2000, pp 133-60 (in reference 2395)
3137. Gee AP, Lamb LS Jr: Enumeration of CD34-positive hematopoietic progenitor cells. In: Stewart CC, Nicholson JKA (eds): *Immunophenotyping* (Cytometric Cellular Analysis Series). New York, Wiley-Liss, 2000, pp 291-319 (in reference 2395)
3138. Sutherland DR, Anderson L, Keeney M, Nayar R, Chin-Yee I: The ISHAGE guidelines for $CD34^{+}$ cell determination by flow cytometry. J Hematother 6:85-9, 1996.
3139. See Reference 2616
3140. Goodell MA, Rosenzweig M, Kim H, Marks DF, DeMaria M, Paradis G, Grupp SA, Sieff CA, Mulligan RC, Johnson RP: Dye efflux studies suggest that hematopoietic stem cells expressing low or undetectable levels of CD34 antigen exist in multiple species. Nat Med 3:1337-45, 1997
3141. Goodell MA: Stem cells: Is there a future in plastics? Curr Opin Cell Biol 13:662-5, 2001
3142. Preffer FI, Dombkowski D, Sykes M, Scadden D, Yang YG: Lineage-negative side-population (SP) cells with restricted hematopoietic capacity circulate in normal human adult blood: Immunophenotypic and functional characterization. Stem Cells 20:417-27, 2002
3143. Goodell MA: Stem cell identification and sorting using the Hoechst 33342 side population (SP). Unit 9.18. In: Robinson JP, Darzynkiewicz Z, Dean P, Hibbs AR, Orfao A, Rabinovitch P, Wheeless L, (eds): *Current Protocols in Cytometry*, New York, John Wiley & Sons, 2002, pp 9.18.1-9.18.11.
3144. Zhou S, Morris JJ, Barnes Y, Lan L, Schuetz JD, Sorrentino BP: Bcrp1 gene expression is required for normal numbers of side population stem cells in mice, and confers relative protection to mitoxantrone in hematopoietic cells in vivo. Proc Natl Acad Sci U S A 99:12339-44, 2002
3145. Kim M, Turnquist H, Jackson J, Sgagias M, Yan Y, Gong M, Dean M, Sharp JG, Cowan K: The multidrug resistance transporter

ABCG2 (breast cancer resistance protein 1) effluxes hoechst 33342 and is overexpressed in hematopoietic stem cells. Clin Cancer Res 8:22-8, 2002

3146. Scharenberg CW, Harkey MA, Torok-Storb B: The ABCG2 transporter is an efficient Hoechst 33342 efflux pump and is preferentially expressed by immature human hematopoietic progenitors. Blood 99:507-12, 2002

3147. Lechner A, Leech CA, Abraham EJ, Nolan AL, Habener JF: Nestin-positive progenitor cells derived from adult human pancreatic islets of Langerhans contain side population (SP) cells defined by expression of the ABCG2 (BCRP1) ATP-binding cassette transporter. Biochem Biophys Res Commun 293:670-4, 2002

3148. Borowitz M et al: *Clinical Applications of Flow Cytometry: Quality Assurance and Immunophenotyping of Lymphocytes; Approved Guideline.* Publication H42-A. Wayne (PA), NCCLS, 1998

3149. Reimann KA, O'Gorman MR, Spritzler J, Wilkening CL, Sabath DE, Helm K, Campbell DE: Multisite comparison of CD4 and CD8 T-lymphocyte counting by single- versus multiple-platform methodologies: Evaluation of Beckman Coulter flow-count fluorospheres and the tetraONE system.The NIAID DAIDS new technologies evaluation group. Clin Diagn Lab Immunol 7:344-51, 2000

3150. Schnizlein-Bick CT, Spritzler J, Wilkening CL, Nicholson JK, O'Gorman MR: Evaluation of TruCount absolute-count tubes for determining CD4 and CD8 cell numbers in human immunodeficiency virus-positive adults. Site investigators and the NIAID DAIDS new technologies evaluation group. Clin Diagn Lab Immunol 7:336-43, 2000

3151. Mercolino TJ, Connelly MC, Meyer EJ, Knight MD, Parker JW, Stelzer GT, DeChirico G. Immunologic differentiation of absolute lymphocyte count with an integrated flow cytometric system: a new concept for absolute T cell subset determinations. Cytometry (Commun Clin Cytom) 22:48-59, 1995

3152. Connelly MC, Knight M, Giorgi JV, Kagan J, Landay AL, Parker JW, Page E, Spino C, Wilkening C, Mercolino TJ. Standardization of absolute CD4+ lymphocyte counts across laboratories: an evaluation of the Ortho CytoronAbsolute flow cytometry system on normal donors. Cytometry (Commun Clin Cytom) 22:200-10, 1995

3153. Janossy G, Jani IV, Bradley NJ, Bikoue A, Pitfield T, Glencross DK: Affordable CD4(+)-T-cell counting by flow cytometry: CD45 gating for volumetric analysis. Clin Diagn Lab Immunol 9:1085-94, 2002

3154. Nicholson JK, Hubbard M, Jones BM: Use of CD45 fluorescence and side-scatter characteristics for gating lymphocytes when using the whole blood lysis procedure and flow cytometry. Cytometry (Commun Clin Cytom) 26:16-21, 1996

3155. Gelman R, Wilkening C: Analyses of quality assessment studies using CD45 for gating lymphocytes for CD3(+)4(+)%. Cytometry (Commun Clin Cytom) 42:1-4, 2000

3156. Schnizlein-Bick CT, Mandy FF, O'Gorman MR, Paxton H, Nicholson JK, Hultin LE, Gelman RS, Wilkening CL, Livnat D: Use of CD45 gating in three and four-color flow cytometric immunophenotyping: Guideline from the National Institute of Allergy and Infectious Diseases, Division of AIDS. Cytometry 50:46-52, 2002

3157. Mandy FF, Nicholson JKA, McDougal JS: Guidelines for Performing Single-Platform Absolute CD4' T-Cell Determinations with CD45 Gating for Persons Infected with Human Immunodeficiency Virus. MMWR Recomm Rep 52(RR-2):1-13, 2003 (also available online at <http:// www.cdc.gov/mmwr/preview/mmwrhtml/rr5202a1.htm>)

3158. Mandy F, Nicholson J, Autran B, Janossy G: T-cell subset counting and the fight against AIDS: Reflections over a 20-year struggle. Cytometry 50:39-45, 2002

3159. Lopez A, Caragol I, Candeias J, Villamor N, Echaniz P, Ortuno F, Sempere A, Strauss K, Orfao A: Enumeration of CD4(+) T-cells in the peripheral blood of HIV-infected patients: an interlaboratory study of the FACSCount system. Cytometry (Commun Clin Cytom) 38:231-7, 1999

3160. Sherman GG, Galpin JS, Patel JM, Mendelow BV, Glencross DK: CD4+ T cell enumeration in HIV infection with limited resources. J Immunol Methods 222:209-17, 1999

3161. Janossy G, Jani I, Göhde W: Affordable CD4(+) T-cell counts on 'single-platform' flow cytometers I. Primary CD4 gating. Br J Haematol 111:1198-208, 2000

3162. Glencross D, Scott LE, Jani IV, Barnett D, Janossy G: CD45-assisted panleucogating for accurate, cost-effective dual-platform CD4+ T-cell enumeration. Cytometry 50:69-77, 2002

3163. Perlmutter N, Solajic Z, Mandy F, Shapiro H: Cellular analysis using the Luminex 100 system. Cytometry Supp 11:120, 2002.

3164. Glencross D, Scott L, Aggett H, Sonday S, Scott CS: Microvolume fluorimetry for the determination of absolute CD4 and CD8 lymphocyte counts in patients with HIV: A comparative evaluation. Clin Lab Haematol 21:391-5, 1999

3165. Dzik W, Sniecinski I, Fischer J: Toward standardization of CD34+ cell enumeration: an international study. Biomedial Excellence for Safer Transfusion Working Party. Transfusion. 39:856-63, 1999

3166. Krailadsiri P, Seghatchian MJ: III-7 evaluation of Imagn 2000: A new system for absolute leucocyte count. Transfus Sci 19:405-7, 1998

3167. Barbosa IL, Sousa ME, Godinho MI, Sousa F, Carvalhais A: Single- versus dual-platform assays for human CD34+ cell enumeration. Cytometry 38:274-9, 1999

3168. Chapple P, Prince HM, Wall D, Filshie R, Haylock D, Quinn M, Bretell M, Venter D: Comparison of three methods of CD34+ cell enumeration in peripheral blood: Dual-platform ishage protocol versus single-platform, versus microvolume fluorimetry. Cytotherapy 2:371-6, 2000

3169. Seghatchian J, Beard M, Krailadsiri P: The role of in process qualification in quality improvement of the Haemonetics MCS plus leucodepleted platelet concentrate. Transfus Sci 22:165-9, 2000

3170. Seghatchian J, Krailadsiri P, Chandegra B, Beard M, Beckman N, Bissett L, Morris B, Booker W: National evaluation of Imagn 2000 for quality monitoring of leucodepleted red cell and platelet concentrates: Comparison with flow and nageotte. Transfus Sci 22:77-9, 2000

3171. Brecher ME, Wong EC, Chen SE, Vampola C, Rocco RM: Antibiotic-labeled probes and microvolume fluorimetry for the rapid detection of bacterial contamination in platelet components: A preliminary report. Transfusion 40:411-3, 2000

3172. Noga SJ, Vogelsang GB, Miller SC, Meusel S, Loper K, Case R, Myers B, Rogers L, Flinn I, Borowitz M, O'Donnell P: Using point-of-care CD34 enumeration to optimize pbsc collection conditions. Cytotherapy 3:11-8, 2001

3173. Valbonesi M, Bruni R, Florio G, Zanella A, Bunkens H: Cellular contamination of plasma collected with various apheresis systems. Transfus Apheresis Sci 24:91-4, 2001

3174. Barnett D, Goodfellow K, Ginnever J, Granger V, Whitby L, Reilly JT: Low level leucocyte counting: A critical variable in the validation of leucodepleted blood transfusion components as highlighted by an external quality assessment study. Clin Lab Haematol 23:43-51, 2001

3175. Seghatchian J, Krailadsiri P, Scott CS: Counting of residual wbcs in wbc-reduced blood components: A multicenter evaluation of a microvolume fluorimeter by the United Kingdom Mational Blood Service. Transfusion 41:93-101, 2001

3176. Krailadsiri P, Seghatchian J, Rigsby P, Bukasa A, Bashir S: A national quality assessment scheme for counting residual leucocytes in unfixed leucodepleted products: The effect of standardisation and 48 hour storage. Transfus Apheresis Sci 26:73-81, 2002

3177. Didier JM, Kazatchkine MD, Demouchy C, Moat C, Diagbouga S, Sepulveda C, Di Lonardo AM, Weiss L: Comparative assessment of five alternative methods for CD4+ T-lymphocyte enumeration for implementation in developing countries. J Acquir Immune Defic Syndr 26:193-5, 2001

3178. Rodriguez W, Mohanty M, Christodoulides N, Goodey A, Romanovicz D, Ali M, Floriano P, Walker B, McDevitt J: Development of Affordable, Portable CD4 Counts for Resource-Poor Settings Using Microchips. Presented at the 10th Conference on Retroviruses and Opportunistic Infections, Boston, February 2003

3179. Horsburgh T, Martin S, Robson AJ: The application of flow cytometry to histocompatibility testing. Transpl Immunol 8:3-15, 2000

3180. El Fettouh HA, Cook DJ, Bishay E, Flechner S, Goldfarb D, Modlin C, Dennis V, Novick AC: Association between a positive flow cytometry crossmatch and the development of chronic rejection in primary renal transplantation. Urology 56:369-72, 2000

3181. Bishay ES, Cook DJ, Starling RC, Ratliff NB, Jr., White J, Blackstone EH, Smedira NG, McCarthy PM: The clinical significance of flow cytometry crossmatching in heart transplantation. Eur J Cardiothorac Surg 17:362-9, 2000

3182. O'Rourke RW, Osorio RW, Freise CE, Lou CD, Garovoy MR, Bacchetti P, Ascher NL, Melzer JS, Roberts JP, Stock PG: Flow cytometry crossmatching as a predictor of acute rejection in sensitized recipients of cadaveric renal transplants. Clin Transplant 14:167-73, 2000

3183. Scornik JC, Clapp W, Patton PR, Van der Werf WJ, Hemming AW, Reed AI, Howard RJ: Outcome of kidney transplants in patients known to be flow cytometry crossmatch positive. Transplantation 71:1098-102, 2001

3184. Pei R, Lee JH, Shih NJ, Chen M, Terasaki PI: Single human leukocyte antigen flow cytometry beads for accurate identification of human leukocyte antigen antibody specificities. Transplantation 75:43-9, 2003

3185. El-Awar N, Terasaki P, Lazda V, Nikaein A, Arnold A: Most patients who reject a kidney transplant have anti-HLA-antibodies. Tissue Antigens 60:553, 2002

3186. Muller-Steinhardt M, Fricke L, Kirchner H, Hoyer J, Kluter H: Monitoring of anti-HLA class I and II antibodies by flow cytometry in patients after first cadaveric kidney transplantation. Clin Transplant 14:85-9, 2000

3187. Stalder M, Birsan T, Holm B, Haririfar M, Scandling J, Morris RE: Quantification of immunosuppression by flow cytometry in stable renal transplant recipients. Ther Drug Monit 25:22-7, 2003

3188. Yu DS, Sun GH, Lee SS, Wu CJ, Ma CP, Chang SY: Flow-cytometric measurement of cellular changes in urine: A simple and rapid method for perioperatively monitoring patients after kidney transplantation. Urol Int 62:143-6, 1999

3189. Donnenberg AD, Donnenberg VS: P-gp expression and function in T lymphocytes. Clin Appl Immunol Revs 4: in press, 2003

3190. Donnenberg VS, Burckart GJ, Donnenberg AD: P-glycoprotein (P-gp) function in T cells: Implications for Organ Transplantation. Clin Appl Immunol Revs 4: in press, 2003

3191. Gupta S: P-glycoprotein expression in the cells of the immune system during aging. Clin Appl Immunol Revs 4: in press, 2003

3192. Pendse SS, David M. Briscoe DM, Frank MH: P-Glycoprotein and Alloimmune T Cell Activation. Clin Appl Immunol Revs 4: in press, 2003

3193. Ruefli AA, Johnstone R: A Role for P-glycoprotein in Regulating Cell Growth and Survival. Clin Appl Immunol Revs 4: in press, 2003

3194. Eisenbraun MD: Development, function and maintenance of T lymphocyte populations in P-glycoprotein-deficient mice. Clin Appl Immunol Revs 4: in press, 2003

3195. Cebrian M, Yague E, Rincon M, Lopez-Botet M, de Landazuri MO, Sanchez-Madrid F: Triggering of T cell proliferation through AIM, an activation inducer molecule expressed on activated human lymphocytes. J Exp Med 168:1621-37, 1988

3196. Lanier LL, Buck DW, Rhodes L, Ding A, Evans E, Barney C, Phillips JH: Interleukin 2 activation of natural killer cells rapidly induces the expression and phosphorylation of the Leu-23 activation antigen. J Exp Med 167:1572-85, 1988

3197. Testi R, Phillips JH, Lanier LL: Leu 23 induction as an early marker of functional CD3/T cell antigen receptor triggering. Requirement for receptor cross-linking, prolonged elevation of intracellular [Ca++] and stimulation of protein kinase C . J Immunol 142:1854-60, 1989

3198. Testi R, D'Ambrosio D, De Maria R, Santoni A: The CD69 receptor: a multipurpose cell-surface trigger for hematopoietic cells. Immunol Today 15:479-83, 1994

3199. Maino VC, Suni MA, Ruitenberg JJ: Rapid flow cytometric method for measuring lymphocyte subset activation. Cytometry 20:127-33, 1995

3200. Lim LC, Fiordalisi MN, Mantell JL, Schmitz JL, Folds JD: A whole-blood assay for qualitative and semiquantitative measurements of CD69 surface expression on CD4 and CD8 T lymphocytes using flow cytometry. Clin Diagn Lab Immunol 5:392-8, 1998

3201. Craston R, Koh M, Mc Dermott A, Ray N, Prentice HG, Lowdell MW: Temporal dynamics of CD69 expression on lymphoid cells. J Immunol Methods 209:37-45, 1997

3202. Caruso A, Licenziati S, Corulli M, Canaris AD, De Francesco MA, Fiorentini S, Peroni L, Fallacara F, Dima F, Balsari A, Turano A: Flow cytometric analysis of activation markers on stimulated T cells and their correlation with cell proliferation. Cytometry 27:71-6, 1997

3203. Hutchinson P, Divola LA, Holdsworth SR: Mitogen-induced T-cell CD69 expression is a less sensitive measure of T-cell function than [(3)H]-thymidine uptake. Cytometry 38:244-9, 1999

3204. Sieg SF, Harding CV, Lederman MM: Hiv-1 infection impairs cell cycle progression of CD4(+) T cells without affecting early activation responses. J Clin Invest 108:757-64, 2001

3205. Andersson U, Sander B, Andersson J, Moller G: Concomitant production of different lymphokines in activated T cells. Eur J Immunol 18:2081-4, 1988

3206. Sander B, Cardell S, Heremans H, Andersson U, Moller G: Detection of individual interleukin 4- and gamma interferon-producing murine spleen cells after activation with t-cell mitogens. Scand J Immunol 30:315-20, 1989

3207. Andersson U, Sander B: Detection of individual interleukin 2-producing cells after anti-CD3 antibody activation. Immunol Lett 20:115-20, 1989

3208. Sander B, Andersson J, Andersson U: Assessment of cytokines by immunofluorescence and the paraformaldehyde-saponin procedure. Immunol Rev 119:65-93, 1991

3209. de Caestecker MP, Telfer BA, Hutchinson IV, Ballardie FW: The detection of intracytoplasmic interleukin-1 alpha, interleukin-1 beta and tumour necrosis factor alpha expression in human monocytes using two colour immunofluorescence flow cytometry. J Immunol Methods 154:11-20, 1992

3210. Chikanza IC, Corrigal V, Kingsley G, Panayi GS: Enumeration of interleukin-1 alpha and beta producing cells by flow cytometry. J Immunol Methods 154:173-8, 1992

3211. Jung T, Schauer U, Heusser C, Neumann C, Rieger C: Detection of intracellular cytokines by flow cytometry. J Immunol Methods 159:197-207, 1993

3212. Assenmacher M, Schmitz J, Radbruch A: Flow cytometric determination of cytokines in activated murine T helper lymphocytes: Expression of interleukin-10 in interferon-gamma and in interleukin-4-expressing cells. Eur J Immunol 24:1097-101, 1994

3213. Vikingsson A, Pederson K, Muller D: Enumeration of IFN-gamma producing lymphocytes by flow cytometry and correlation with quantitative measurement of IFN-gamma. J Immunol Methods 173:219-28, 1994

3214. McIntyre CA, Horne CJ, Lawry J, Rees RC: The detection of intracytoplasmic interleukin-2 in Jurkat E6.1 and human peripheral blood mononuclear cells using direct conjugate, two-colour, immunofluorescent flow cytometry. J Immunol Methods 169:213-20, 1994

3215. Picker LJ, Singh MK, Zdraveski Z, Treer JR, Waldrop SL, Bergstresser PR, Maino VC: Direct demonstration of cytokine synthesis heterogeneity among human memory/effector T cells by flow cytometry. Blood 86:1408-19, 1995
3216. Prussin C, Metcalfe DD: Detection of intracytoplasmic cytokine using flow cytometry and directly conjugated anti-cytokine antibodies. J Immunol Methods 188:117-28, 1995
3217. Mehta BA, Maino VC: Simultaneous detection of DNA synthesis and cytokine production in staphylococcal enterotoxin B activated CD4+ T lymphocytes by flow cytometry. J Immunol Methods 208:49-59, 1997
3218. Waldrop SL, Pitcher CJ, Peterson DM, Maino VC, Picker LJ: Determination of antigen-specific memory/effector CD4+ Tcell frequencies by flow cytometry: Evidence for a novel, antigen-specific homeostatic mechanism in HIV-associated immunodeficiency. J Clin Invest 99:1739-50, 1997
3219. Suni MA, Picker LJ, Maino VC: Detection of antigen-specific t cell cytokine expression in whole blood by flow cytometry. J Immunol Methods 212:89-98, 1998
3220. Hennessy B, North J, Deru A, Llewellyn-Smith N, Lowdell MW: Use of Leu3a/3b for the accurate determination of CD4 subsets for measurement of intracellular cytokines. Cytometry 44:148-52, 2001
3221. Pala P, Hussell T, Openshaw PJ: Flow cytometric measurement of intracellular cytokines. J Immunol Methods 243:107-24, 2000
3222. Prussin C, Foster B: Detection of intracellular cytokines by flow cytometry. Unit 6.24. In: Coligan JE, Kruisbeek AM, Margulies DH, Shevach EM, Strober W, Coico R, (eds): *Current Protocols in Immunology*. New York, John Wiley & Sons, 2002, pp 6.24.1-6.24.11
3223. Brosterhus H, Brings S, Leyendeckers H, Manz RA, Miltenyi S, Radbruch A, Assenmacher M, Schmitz J: Enrichment and detection of live antigen-specific CD4(+) and CD8(+) T cells based on cytokine secretion. Eur J Immunol 29:4053-9, 1999
3224. Bitmansour AD, Douek DC, Maino VC, Picker LJ: Direct ex vivo analysis of human CD4(+) memory T cell activation requirements at the single clonotype level. J Immunol 169:1207-18, 2002
3225. Meidenbauer N, Hoffmann TK, Donnenberg AD:Direct visualization of antigen-specific T cells using Peptide-MHC Class I Tetrameric Complexes. Methods, in press, 2003
3226. Hayakawa K, Ishii R, Yamasaki K, Kishimoto T, Hardy RR: Isolation of high-affinity memory B cells: Phycoerythrin as a probe for antigen-binding cells. Proc Natl Acad Sci U S A 84:1379-83, 1987
3227. McHeyzer-Williams MG, Nossal GJ, Lalor PA: Molecular characterization of single memory B cells. Nature 350:502-5, 1991
3228. Sato Y, Sahara H, Tsukahara T, Kondo M, Hirohashi Y, Nabeta Y, Kawaguchi S, Ikeda H, Torigoe T, Ichimíya S, Tamura Y, Wada T, Yamashita T, Goto M, Takasu H, Sato N: Improved generation of HLA class I/peptide tetramers. J Immunol Methods 271:177-84, 2002
3229. Cunliffe SL, Wyer JR, Sutton JK, Lucas M, Harcourt G, Klenerman P, McMichael AJ, Kelleher AD: Optimization of peptide linker length in production of MHC class II/peptide tetrameric complexes increases yield and stability, and allows identification of antigen-specific CD4+t cells in peripheral blood mononuclear cells. Eur J Immunol 32:3366-75, 2002
3230. Brophy SE, Holler PD, Kranz DM: A yeast display system for engineering functional peptide-MHC complexes. J Immunol Methods 272:235-46, 2003
3231. Kelleher AD, Rowland-Jones SL: Functions of tetramer-stained HIV-specific CD4+ and CD8+ T cells. Curr Opin Immunol 12:370-4, 2000
3232. Maecker HT, Maino VC, Picker LJ: Immunofluorescence analysis of T-cell responses in health and disease. J Clin Immunol 20:391-9, 2000
3233. Lyons AB, Parish CR: Determination of lymphocyte division by flow cytometry. J Immunol Methods 171:131-7, 1994
3234. Nordon RE, Nakamura M, Ramirez C, Odell R: Analysis of growth kinetics by division tracking. Immunol Cell Biol 77:523-9, 1999
3235. Parish C, Warren HS: Use of the intracellular fluorescent dye 5-(and -6)-carboxyfluorescein diacetate succinimidyl ester (CFSE) to monitor lymphocyte migration and proliferation. Unit 4.9. In: Coligan JE, Kruisbeek AM, Margulies DH, Shevach EM, Strober W, Coico R, (eds): *Current Protocols in Immunology*. New York, John Wiley & Sons, 2002, pp 4.9.1-4.9.11
3236. Hasbold J, Hodgkin PD: Flow cytometric cell division tracking using nuclei. Cytometry 40:230-7, 2000
3237. Wells AD, Gudmundsdottir H, Turka LA: Following the fate of individual T cells throughout activation and clonal expansion. Signals from T cell receptor and CD28 differentially regulate the induction and duration of a proliferative response. J Clin Invest 100:3173-83, 1997
3238. Gett AV, Hodgkin PD: Cell division regulates the T cell cytokine repertoire, revealing a mechanism underlying immune class regulation. Proc Natl Acad Sci U S A 95:9488-93, 1998
3239. Hasbold J, Gett AV, Rush JS, Deenick E, Avery D, Jun J, Hodgkin PD: Quantitative analysis of lymphocyte differentiation and proliferation in vitro using carboxyfluorescein diacetate succinimidyl ester. Immunol Cell Biol 77:516-22, 1999
3240. Wells A, Gudmundsdottir H, Walsh M, Turka LA: Single-cell analysis of T-cell responses in vitro and in vivo: Role of costimulatory signals. Transplant Proc 31:814-5, 1999
3241. Fazekas de St Groth B, Smith AL, Koh WP, Girgis L, Cook MC, Bertolino P: Carboxyfluorescein diacetate succinimidyl ester and the virgin lymphocyte: A marriage made in heaven. Immunol Cell Biol 77:530-8, 1999
3242. McCall MN, Hodgkin PD: Switch recombination and germ-line transcription are division-regulated events in B lymphocytes. Biochim Biophys Acta 1447:43-50, 1999
3243. Deenick EK, Hasbold J, Hodgkin PD: Switching to IgG3, IgG2b, and IgA is division linked and independent, revealing a stochastic framework for describing differentiation. J Immunol 163:4707-14, 1999
3244. Hasbold J, Hong JS, Kehry MR, Hodgkin PD: Integrating signals from IFN-gamma and IL-4 by B cells: Positive and negative effects on CD40 ligand-induced proliferation, survival, and division-linked isotype switching to IgG1, IgE, and IgG2a. J Immunol 163:4175-81, 1999
3245. Gudmundsdottir H, Wells AD, Turka LA: Dynamics and requirements of T cell clonal expansion in vivo at the single-cell level: Effector function is linked to proliferative capacity. J Immunol 162:5212-23, 1999
3246. Tangye SG, Ferguson A, Avery DT, Ma CS, Hodgkin PD: Isotype switching by human B cells is division-associated and regulated by cytokines. J Immunol 169:4298-306, 2002
3247. Tangye SG, Avery DT, Hodgkin PD: A division-linked mechanism for the rapid generation of Ig-secreting cells from human memory B cells. J Immunol 170:261-9, 2003
3248. Tangye SG, Avery DT, Deenick EK, Hodgkin PD: Intrinsic differences in the proliferation of naive and memory human B cells as a mechanism for enhanced secondary immune responses. J Immunol 170:686-94, 2003
3249. Chow S, Patel H, Hedley DW: Measurement of MAP kinase activation by flow cytometry using phospho-specific antibodies to MEK and ERK: potential for pharmacodynamic monitoring of signal transduction inhibitors. Cytometry (Commun Clin Cytom) 46:72-8, 2001
3250. Vassilakos P, Carrel S, Petignat P, Boulvain M, Campana A: Use of Automated Primary Screening on Liquid-Based, Thin-Layer Preparations. Acta Cytol 46:291-5, 2002
3251. Patterson B, Domanik R, Wernke P, Gombrich M: Molecular biomarker-based screening for early detection of cervical cancer. Acta Cytol 45:36-47, 2001

3252. Keesee SK, Domanik R, Patterson B: Fully automated proteomic detection of cervical dysplasia. Anal Quant Cytol Histol 24:137-46, 2002

3253. Grundhoefer D, Patterson BK: Determination of liquid-based cervical cytology specimen adequacy using cellular light scatter and flow cytometry. Cytometry (Commun Clin Cytom) 46:340-4, 2001

3254. Ormerod MG, Titley JC, Imrie PR. Use of light scatter when recording a DNA histogram from paraffin-embedded tissue. Cytometry 21:294-9, 1995

3255. Seamer LC, Altobelli KK. Fluorescence drift detection as a novel QC procedure for DNA cell-cycle analysis. Cytometry (Commun Clin Cytom) 22:60-4, 1995

3256. Glogovac JK, Porter PL, Banker DE, Rabinovitch PS: Cytokeratin labeling of breast cancer cells extracted from paraffin-embedded tissue for bivariate flow cytometric analysis. Cytometry 24:260-7, 1996

3257. Corver WE, Fleuren GJ, Cornelisse CJ: Software compensation improves the analysis of heterogeneous tumor samples stained for multiparameter DNA flow cytometry.

3258. Gorczyca W, Darzynkiewicz Z, Melamed MR: Laser scanning cytometry in pathology of solid tumors. A review. Acta Cytol 41:98-108, 1997

3259. Gorczyca W, Davidian M, Gherson J, Ashikari R, Darzynkiewicz Z, Melamed MR: Laser scanning cytometry quantification of estrogen receptors in breast cancer. Anal Quant Cytol Histol 20:470-6, 1998

3260. Gorczyca W, Deptala A, Bedner E, Li X, Melamed MR, Darzynkiewicz Z: Analysis of human tumors by laser scanning cytometry. Methods Cell Biol 64:421-43, 2001(in reference 2386)

3261. Galipeau PC, Cowan DS, Sanchez CA, Barrett MT, Emond MJ, Levine DS, Rabinovitch PS, Reid BJ: 17p (p53) allelic losses, 4N (G2/tetraploid) populations, and progression to aneuploidy in Barrett's esophagus. Proc Natl Acad Sci U S A 93:7081-4, 1996

3262. Reid BJ, Levine DS, Longton G, Blount PL, Rabinovitch PS: Predictors of progression to cancer in Barrett's esophagus: Baseline histology and flow cytometry identify low- and high-risk patient subsets. Am J Gastroenterol 95:1669-76, 2000

3263. Reid BJ, Prevo LJ, Galipeau PC, Sanchez CA, Longton G, Levine DS, Blount PL, Rabinovitch PS: Predictors of progression in Barrett's esophagus II: Baseline 17p (p53) loss of heterozygosity identifies a patient subset at increased risk for neoplastic progression. Am J Gastroenterol 96:2839-48, 2001

3264. Rabinovitch PS, Longton G, Blount PL, Levine DS, Reid BJ: Predictors of progression in Barrett's esophagus III: Baseline flow cytometric variables. Am J Gastroenterol 96:3071-83, 2001

3265. Rabinovitch PS, Dziadon S, Brentnall TA, Emond MJ, Crispin DA, Haggitt RC, Bronner MP: Pancolonic chromosomal instability precedes dysplasia and cancer in ulcerative colitis. Cancer Res 59:5148-53, 1999

3266. Clausen OP, Andersen SN, Strømkjaer H, Nielsen V, Rognum TO, Bolund L, Kølvraa S: A strategy combining flow sorting and comparative genomic hybridization for studying genetic aberrations at different stages of colorectal tumorigenesis in ulcerative colitis. Cytometry 43:46-54, 2001

3267. Winter JN, Andersen J, Variakojis D, Gordon LI, Fisher RI, Oken MM, Neiman RS, Jiang S, Bauer KD: Prognostic implications of ploidy and proliferative activity in the diffuse, aggressive non-Hodgkin's lymphomas. Blood 88:3919-25, 1996

3268. Winter JN, Andersen J, Reed JC, Krajewski S, Variakojis D, Bauer KD, Fisher RI, Gordon LI, Oken MM, Jiang S, Jeffries D, Domer P: Bcl-2 expression correlates with lower proliferative activity in the intermediate- and high-grade non-Hodgkin's lymphomas: An Eastern Cooperative Oncology Group and Southwest Oncology Group cooperative laboratory study. Blood 91:1391-8, 1998

3268. Frisa PS, Goodman MN, Smith GM, Silver J, Jacobberger JW: Immortalization of immature and mature mouse astrocytes with SV40 T antigen. J Neurosci Res 39:47-56, 1994

3269. Jacobberger JW, Sizemore N, Gorodeski G, Rorke EA: Transforming growth factor beta regulation of epidermal growth factor receptor in ectocervical epithelial cells. Exp Cell Res 220:390-6, 1995

3270. Rorke EA, Jacobberger JW: Transforming growth factor-beta 1 (TGF beta 1) enhances apoptosis in human papillomavirus type 16-immortalized human ectocervical epithelial cells. Exp Cell Res 216:65-72, 1995

3271. Frisa PS, Walter EI, Ling L, Kung HJ, Jacobberger JW: Stepwise transformation of astrocytes by simian virus 40 large T antigen and epidermal growth factor receptor overexpression. Cell Growth Differ 7:223-33, 1996

3272. Rorke EA, Zhang D, Choo CK, Eckert RL, Jacobberger JW: Tgf-beta-mediated cell cycle arrest of HPV16-immortalized human ectocervical cells correlates with decreased E6/E7 mRNA and increased p53 and p21(Waf-1) expression. Exp Cell Res 259:149-57, 2000

3273. Clay TM, Hobeika AC, Mosca PJ, Lyerly HK, Morse MA: Assays for monitoring cellular immune responses to active immunotherapy of cancer. Clin Cancer Res 7:1127-35, 2001

3274. Jacobberger JW, Sramkoski RM, Zhang D, Zumstein LA, Doerksen LD, Merritt JA, Wright SA, Shults KE: Bivariate analysis of the p53 pathway to evaluate ad-p53 gene therapy efficacy. Cytometry 38:201-13, 1999

3275. Braylan RC, Borowitz MJ, Davis BH, Stelzer GT, Stewart CC: U.S.-Canadian Consensus recommendations on the immunophenotypic analysis of hematologic neoplasia by flow cytometry. Cytometry (Commun Clin Cytom) 30:213, 1997

3276. Stelzer GT, Marti G, Hurley A, McCoy P Jr, Lovett EJ, Schwartz A: U.S.-Canadian Consensus recommendations on the immunophenotypic analysis of hematologic neoplasia by flow cytometry: standardization and validation of laboratory procedures. Cytometry (Commun Clin Cytom) 30:214-30, 1997

3277. Stewart CC, Behm FG, Carey JL, Cornbleet J, Duque RE, Hudnall SD, Hurtubise PE, Loken M, Tubbs RR, Wormsley S: U.S.-Canadian Consensus recommendations on the immunophenotypic analysis of hematologic neoplasia by flow cytometry: selection of antibody combinations. Cytometry (Commun Clin Cytom) 30:231-5, 1997

3278. Borowitz MJ, Bray R, Gascoyne R, Melnick S, Parker JW, Picker L, Stetler-Stevenson M: U.S.-Canadian Consensus recommendations on the immunophenotypic analysis of hematologic neoplasia by flow cytometry: data analysis and interpretation. Cytometry (Commun Clin Cytom) 30:236-44, 1997

3279. Braylan RC, Atwater SK, Diamond L, Hassett JM, Johnson M, Kidd PG, Leith C, Nguyen D: U.S.-Canadian Consensus recommendations on the immunophenotypic analysis of hematologic neoplasia by flow cytometry: data reporting. Cytometry (Commun Clin Cytom) 30:245-8, 1997

3280. Davis BH, Foucar K, Szczarkowski W, Ball E, Witzig T, Foon KA, Wells D, Kotylo P, Johnson R, Hanson C, Bessman D: U.S.-Canadian Consensus recommendations on the immunophenotypic analysis of hematologic neoplasia by flow cytometry: medical indications. Cytometry (Commun Clin Cytom) 30:249-63, 1997

3281. Borowitz M et al: *Clinical Applications of Flow Cytometry: Immunophenotyping of Leukemic Cells; Approved Guideline.* Publication H43-A, Wayne (PA), NCCLS, 1998

3282. Braylan RC, Orfao A, Borowitz MJ, Davis BH: Optimal number of reagents required to evaluate hematolymphoid neoplasias: Results of an international consensus meeting. Cytometry 46:23-7, 2001

3283. Nguyen D, Diamond LW, Braylan RC: *Flow Cytometry in Hematopathology. A Visual Approach to Data Analysis and Interpretation.* Totowa (NJ), Humana Press, 2003, xiv + 220 pp (+ CD-ROM)

3284. Wells DA, Sale GE, Shulman HM, Myerson D, Bryant EM, Gooley T, Loken MR: Multidimensional flow cytometry of marrow can differentiate leukemic from normal lymphoblasts and myeloblasts after chemotherapy and bone marrow transplantation. Am J Clin Pathol 110:84-94, 1998

3285. Weir EG, Cowan K, LeBeau P, Borowitz MJ: A limited antibody panel can distinguish B-precursor acute lymphoblastic leukemia from normal B precursors with four color flow cytometry: Implications for residual disease detection. Leukemia 13:558-67, 1999

3286. Weir EG, Borowitz MJ: Flow cytometry in the diagnosis of acute leukemia. Semin Hematol 38:124-38, 2001

3287. Baer MR, Stewart CC, Dodge RK, Leget G, Sule N, Mrozek K, Schiffer CA, Powell BL, Kolitz JE, Moore JO, Stone RM, Davey FR, Carroll AJ, Larson RA, Bloomfield CD: High frequency of immunophenotype changes in acute myeloid leukemia at relapse: Implications for residual disease detection (Cancer and Leukemia Group B Study 8361). Blood 97:3574-80, 2001

3288. Campana D, Coustan-Smith E: Detection of minimal residual disease in acute leukemia by flow cytometry. Cytometry 38:139-52, 1999

3289. Szczepanski T, Orfao A, van der Velden VH, San Miguel JF, van Dongen JJ: Minimal residual disease in leukaemia patients. Lancet Oncol 2:409-17, 2001

3290. San Miguel JF, Vidriales MB, Lopez-Berges C, Diaz-Mediavilla J, Gutierrez N, Canizo C, Ramos F, Calmuntia MJ, Perez JJ, Gonzalez M, Orfao A: Early immunophenotypical evaluation of minimal residual disease in acute myeloid leukemia identifies different patient risk groups and may contribute to postinduction treatment stratification. Blood 98:1746-51, 2001

3291. Rawstron AC, Kennedy B, Evans PA, Davies FE, Richards SJ, Haynes AP, Russell NH, Hale G, Morgan GJ, Jack AS, Hillmen P: Quantitation of minimal disease levels in chronic lymphocytic leukemia using a sensitive flow cytometric assay improves the prediction of outcome and can be used to optimize therapy. Blood 98:29-35, 2001

3292. Coustan-Smith E, Sancho J, Behm FG, Hancock ML, Razzouk BI, Ribeiro RC, Rivera GK, Rubnitz JE, Sandlund JT, Pui CH, Campana D: Prognostic importance of measuring early clearance of leukemic cells by flow cytometry in childhood acute lymphoblastic leukemia. Blood 100:52-8, 2002

3293. Venditti A, Tamburini A, Buccisano F, Del Poeta G, Maurillo L, Panetta P, Scornajenghi KA, Cox C, Amadori S: Clinical relevance of minimal residual disease detection in adult acute myeloid leukemia. J Hematother Stem Cell Res 11:349-57, 2002

3294. Rawstron AC, Davies FE, DasGupta R, Ashcroft AJ, Patmore R, Drayson MT, Owen RG, Jack AS, Child JA, Morgan GJ: Flow cytometric disease monitoring in multiple myeloma: The relationship between normal and neoplastic plasma cells predicts outcome after transplantation. Blood 100:3095-100, 2002

3295. Bjorklund E, Mazur J, Soderhall S, Porwit-MacDonald A: Flow cytometric follow-up of minimal residual disease in bone marrow gives prognostic information in children with acute lymphoblastic leukemia. Leukemia 17:138-48, 2003

3296. Krampera M, Vitale A, Vincenzi C, Perbellini O, Guarini A, Annino L, Todeschini G, Camera A, Fabbiano F, Fioritoni G, Nobile F, Szydlo R, Mandelli F, Foa R, Pizzolo G: Outcome prediction by immunophenotypic minimal residual disease detection in adult T-cell acute lymphoblastic leukaemia. Br J Haematol 120:74-9, 2003

3297. Borowitz MJ, Shuster J, Carroll AJ, Nash M, Look AT, Camitta B, Mahoney D, Lauer SJ, Pullen DJ: Prognostic significance of fluorescence intensity of surface marker expression in childhood B-precursor acute lymphoblastic leukemia. A Pediatric Oncology Group study. Blood 89:3960-6, 1997

3298. Malec M, Bjorklund E, Soderhall S, Mazur J, Sjogren AM, Pisa P, Bjorkholm M, Porwit-MacDonald A: Flow cytometry and allele-specific oligonucleotide PCR are equally effective in detection of minimal residual disease in ALL. Leukemia 15:716-27, 2001

3299. Shipp MA, Ross KN, Tamayo P, Weng AP, Kutok JL, Aguiar RC, Gaasenbeek M, Angelo M, Reich M, Pinkus GS, Ray TS, Koval MA, Last KW, Norton A, Lister TA, Mesirov J, Neuberg DS, Lander ES, Aster JC, Golub TR: Diffuse large B-cell lymphoma outcome prediction by gene-expression profiling and supervised machine learning. Nat Med 8:68-74, 2002

3300. Jacobberger JW, Sramkoski RM, Frisa PS, Peng Ye P, Gottlieb MA, Hedley DW, Shankey TV, Smith BL, Paniagua M, Goolsby CL: Immunoreactivity of STAT5 phosphorylated on tyrosine 694 as a cell-based measure of Bcr/Abl kinase activity. Cytometry, in press, 2003

3301. Al-Hajj M, Wicha MS, Benito-Hernandez A, Morrison SJ, Clarke MF: Prospective identification of tumorigenic breast cancer cells. Proc Natl Acad Sci U S A 100(7):3983-8, 2003

3302. Evenson D, Jost L, Gandour D, Rhodes L, Stanton B, Clausen OP, De Angelis P, Coico R, Daley A, Becker K, Yopp T: Comparative sperm chromatin structure assay measurements on epiillumination and orthogonal axes flow cytometers. Cytometry 19:295-303, 1995

3303. Evenson D, Jost L: Sperm chromatin structure assay is useful for fertility assessment. Methods Cell Sci 22:169-89, 2000

3304. Marchetti C, Obert G, Deffosez A, Formstecher P, Marchetti P: Study of mitochondrial membrane potential, reactive oxygen species, DNA fragmentation and cell viability by flow cytometry in human sperm. Hum Reprod 17:1257-65, 2002

3305. Ricci G, Perticarari S, Fragonas E, Giolo E, Canova S, Pozzobon C, Guaschino S, Presani G: Apoptosis in human sperm: Its correlation with semen quality and the presence of leukocytes. Hum Reprod 17:2665-72, 2002

3306. Yeung CH, Anapolski M, Cooper TG: Measurement of volume changes in mouse spermatozoa using an electronic sizing analyzer and a flow cytometer: Validation and application to an infertile mouse model. J Androl 23:522-8, 2002

3307. Hansen C, Christensen P, Stryhn H, Hedeboe AM, Rode M, Boe-Hansen G: Validation of the FACScount AF system for determination of sperm concentration in boar semen. Reprod Domest Anim 37:330-4, 2002

3308. Tsuji T, Okada H, Fujisawa M, Hamaguchi Y, Kamidono S: Automated sperm concentration analysis with a new flow cytometry-based device, S-FCM. Am J Clin Pathol 117:401-8, 2002

3309. Baumgartner A, Schmid TE, Maerz HK, Adler ID, Tarnok A, Nuesse M: Automated evaluation of frequencies of aneuploid sperm by laser-scanning cytometry (LSC). Cytometry 44:156-60, 2001

3310. Lewis DE, Schober W, Murrell S, Nguyen D, Scott J, Boinoff J, Simpson JL, Bischoff FZ, Elias S: Rare event selection of fetal nucleated erythrocytes in maternal blood by flow cytometry. Cytometry 23:218-27, 1996

3311. DeMaria MA, Zheng YL, Zhen D, Weinschenk NM, Vadnais TJ, Bianchi DW: Improved fetal nucleated erythrocyte sorting purity using intracellular antifetal hemoglobin and Hoechst 33342. Cytometry 25:37-45, 1996

3312. Wang JY, Zhen DK, Falco VM, Farina A, Zheng YL, Delli-Bovi LC, Bianchi DW: Fetal nucleated erythrocyte recovery: fluorescence activated cell sorting-based positive selection using anti-gamma globin versus magnetic activated cell sorting using anti-CD45 depletion and anti-gamma globin positive selection. Cytometry 39:224-30, 2000

3313. Bianchi DW, Simpson JL, Jackson LG, Elias S, Holzgreve W, Evans MI, Dukes KA, Sullivan LM, Klinger KW, Bischoff FZ, Hahn S, Johnson KL, Lewis D, Wapner RJ, de la Cruz F: Fetal gender and aneuploidy detection using fetal cells in maternal blood: Analysis of NIFTY I data. National institute of Child Health and Development Fetal Cell Isolation Study. Prenat Diagn 22:609-15, 2002

3314. Cha D, Hogan B, Bohmer RM, Bianchi DW, Johnson KL: A simple and sensitive erythroblast scoring system to identify fetal cells in maternal blood. Prenat Diagn 23:68-73, 2003

3315. Delanghe JR, Kouri TT, Huber AR, Hannemann-Pohl K, Guder WG, Lun A, Sinha P, Stamminger G, Beier L: The role of automated urine particle flow cytometry in clinical practice. Clin Chim Acta 301:1-18, 2000

3316. Roggeman S, Zaman Z: Safely reducing manual urine microscopy analyses by combining urine flow cytometer and strip results. Am J Clin Pathol 116:872-8, 2001

3317. Regeniter A, Haenni V, Risch L, Kochli HP, Colombo JP, Frei R, Huber AR: Urine analysis performed by flow cytometry: Reference range determination and comparison to morphological findings, dipstick chemistry and bacterial culture results--a multicenter study. Clin Nephrol 55:384-92, 2001

3318. Bull ME, Gebhard DG, Tompkins WA, Kennedy-Stoskopf S: Polymorphic expression in the CD8alpha chain surface receptor of African lions (Panthera leo). Vet Immunol Immunopathol 84:181-9, 2002

3319. Brousseau P, Pellerin J, Morin Y, Cyr D, Blakley B, Boermans H, Fournier M: Flow cytometry as a tool to monitor the disturbance of phagocytosis in the clam Mya arenaria hemocytes following in vitro exposure to heavy metals. Toxicology 142:145-56, 2000

3320. Allam B, Ashton-Alcox KA, Ford SE: Haemocyte parameters associated with resistance to brown ring disease in Ruditapes spp. Clams. Dev Comp Immunol 25:365-75, 2001

3321. Fournier M, Pellerin J, Clermont Y, Morin Y, Brousseau P: Effects of in vivo exposure of Mya arenaria to organic and inorganic mercury on phagocytic activity of hemocytes. Toxicology 161:201-11, 2001

3322. Zeng Q, Wand M, Young AJ, Rawn J, Milford EL, Mentzer SJ, Greenes RA: Matching of flow-cytometry histograms using information theory in feature space. Proc AMIA Symp:929-33, 2002

3323. Kim EY, Zeng Q, Rawn J, Wand M, Young AJ, Milford EL, Mentzer SJ, Greenes RA: Using a neural network with flow cytometry histograms to recognize cell surface protein binding patterns. Proc AMIA Symp:380-4, 2002

3324. Zhao T, He C, Su M, West CA, Swanson SJ, Young AJ, Mentzer SJ: Cell adhesion molecule expression in the sheep thymus. Dev Comp Immunol 25:519-30, 2001

3325. Blottner S, Hingst O, Meyer HH: Seasonal spermatogenesis and testosterone production in roe deer (Capreolus capreolus). J Reprod Fertil 108:299-305, 1996

3326. Blottner S, Roelants H: Quantification of somatic and spermatogenic cell proliferation in the testes of ruminants, using a proliferation marker and flow cytometry analysis. Theriogenology 49:1275-87, 1998

3327. Blottner S, Roelants H: Calculation of spermatogenic transformations based on dual-flow cytometric analysis of testicular tissue in seasonal breeders. Andrologia 30:331-7, 1998

3328. Blottner S, Roelants H, Wagener A, Wenzel UD: Testicular mitosis, meiosis and apoptosis in mink (Mustela vison) during breeding and non-breeding seasons. Anim Reprod Sci 57:237-49, 1999

3329. Roelants H, Schneider F, Goritz F, Streich J, Blottner S: Seasonal changes of spermatogonial proliferation in roe deer, demonstrated by flow cytometric analysis of c-kit receptor, in relation to follicle-stimulating hormone, luteinizing hormone, and testosterone. Biol Reprod 66:305-12, 2002

3330. Starck JM, Beese K: Structural flexibility of the intestine of Burmese python in response to feeding. J Exp Biol 204:325-35, 2001

3331. Starck JM, Beese K: Structural flexibility of the small intestine and liver of garter snakes in response to feeding and fasting. J Exp Biol 205:1377-88, 2002

3332. Ciudad J, Cid E, Velasco A, Lara JM, Aijon J, Orfao A: Flow cytometry measurement of the DNA contents of G0/G1 diploid cells from three different teleost fish species. Cytometry 48:20-5, 2002

3333. Brainerd EL, Slutz SS, Hall EK, Phillis RW: Patterns of genome size evolution in tetraodontiform fishes. Evolution Int J Org Evolution 55:2363-8, 2001

3334. Wong CK, Chan DK: Effects of cortisol on chloride cells in the gill epithelium of Japanese eel, Anguilla japonica. J Endocrinol 168:185-92, 2001

3335. Inoue T, Moritomo T, Tamura Y, Mamiya S, Fujino H, Nakanishi T: A new method for fish leucocyte counting and partial differentiation by flow cytometry. Fish Shellfish Immunol 13:379-90, 2002

3336. Moritomo T, Minami A, Inoue Y, Nakanishi T: A new method for counting of quail leukocytes by flow cytometry. J Vet Med Sci 64:1149-51, 2002

3337. Mayer WE, Uinuk-Ool T, Tichy H, Gartland LA, Klein J, Cooper MD: Isolation and characterization of lymphocyte-like cells from a lamprey. Proc Natl Acad Sci U S A 99:14350-5, 2002

3338. Scherthan H, Cremer T, Arnason U, Weier HU, Lima-de-Faria A, Fronicke L: Comparative chromosome painting discloses homologous segments in distantly related mammals. Nat Genet 6:342-7, 1994

3339. Rettenberger G, Klett C, Zechner U, Bruch J, Just W, Vogel W, Hameister H: Zoo-FISH analysis: Cat and human karyotypes closely resemble the putative ancestral mammalian karyotype. Chromosome Res 3:479-86, 1995

3340. Solinas-Toldo S, Lengauer C, Fries R: Comparative genome map of human and cattle. Genomics 27:489-96, 1995

3341. Fronicke L, Muller-Navia J, Romanakis K, Scherthan H: Chromosomal homeologies between human, harbor seal (Phoca vitulina) and the putative ancestral carnivore karyotype revealed by zoo-FISH. Chromosoma 106:108-13, 1997

3342. Breen M, Thomas R, Binns MM, Carter NP, Langford CF: Reciprocal chromosome painting reveals detailed regions of conserved synteny between the karyotypes of the domestic dog (Canis familiaris) and human. Genomics 61:145-55, 1999

3343. Nash WG, Menninger JC, Wienberg J, Padilla-Nash HM, O'Brien SJ: The pattern of phylogenomic evolution of the Canidae. Cytogenet Cell Genet 95:210-24, 2001

3344. Brown S, Coba de la Pena, T: Cytometry. In: Hawes C, Satiat-Jeunemaitre B (eds): *Plant Cell Biology : A Practical Approach.* Oxford Oxford, University Press, 2001, pp 85-107

3345. Galbraith DW: Flow cytometry and sorting of plant protoplasts and cells. Methods Cell Biol 42 Pt B:539-61, 1994

3346. Blackhall NW: Flow Cytometry of Plant Cells. In: Spier RE (ed) *The Encyclopedia of Cell Technology*, New York, John Wiley & Sons, 2000 (2nd Edition in preparation)

3347. Doležel J, Macas J, Lucretti S: Flow analysis and sorting of plant chromosomes. Unit 5.3. In: Robinson JP, Darzynkiewicz Z, Dean P, Hibbs AR, Orfao A, Rabinovitch P, Wheeless L, (eds): *Current Protocols in Cytometry*, New York, John Wiley & Sons, 1999, pp 5.3.1-5.3.33.

3348. Doležel J, Lysák MA, Kubaláková M, Šimková H, Macas J, Lucretti S: Sorting of plant chromosomes. Methods Cell Biol 64:3-31, 2001(in reference 2386)

3349. Hammatt N, Lister A, Blackhall NW, Gartland J, Ghose TK, Gilmour DM, Power JB, Davey MR, Cocking EC: Selection of plant heterokaryons from diverse origins by flow cytometry. Protoplasma 154:34-44, 1990

3350. Galbraith DW, Lambert GM, Macas J, Doležel J: Analysis of nuclear DNA content and ploidy in higher plants. Unit 7.6. In: Robinson JP, Darzynkiewicz Z, Dean P, Hibbs AR, Orfao A, Rabinovitch P, Wheeless L, (eds): *Current Protocols in Cytometry*, New York, John Wiley & Sons, 1998, pp 7.6.1-7.6.22.

3351. Johnston JS, Bennett MD, Rayburn AL, Galbraith DW, Price HJ: Reference standards for determination of DNA content of plant nuclei. Am J Bot 86:609, 1999

3352. Yokoya K, Roberts AV, Mottley J, Lewis R, Brandham PE: Nuclear DNA amounts in roses. Ann Bot 85:557-61, 2000

3353. Vinogradov AE: Genome size and GC-percent in vertebrates as determined by flow cytometry: the triangular relationship. Cytometry 31:100-9, 1998

3354. Cerbah M, Coulaud J, Brown SC, Siljak-Yakovlev S: Evolutionary DNA variation in the genus Hypochaeris. Heredity 82 (Pt 3):261-6, 1999

3355. Barow M, Meister A: Lack of correlation between AT frequency and genome size in higher plants and the effect of nonrandomness of base sequences on dye binding. Cytometry 47:1-7, 2002
3356. Doležel J, Göhde W: Sex determination in dioecious plants Melandrium album and M. rubrum using high-resolution flow cytometry. Cytometry 19:103-6, 1995
3357. Ulrich I, Ulrich W: High-resolution flow cytometry pf nuclear DNA in higher plants. Proroplasma 165:212-5, 1991
3358. Galbraith DW, Harkins KR, Maaddox JM, Ayres NM, Sharma DP, Firoozabady E: Rapid flow cytometric analysis of the cell cycle in intact plant tissues. Science 220: 1049-51, 1983
3359. Pfosser M, Amon A, Lelley T, Heberle-Bors E. Evaluation of sensitivity of flow cytometry in detecting aneuploidy in wheat using disomic and ditelosomic wheat-rye addition lines. Cytometry 21:387-93, 1995
3360. Lysák MA, Cíhalíková J, Kubaláková M, Šimková H, Kunzel G, Doležel J: Flow karyotyping and sorting of mitotic chromosomes of barley (Hordeum vulgare L.). Chromosome Res 7:431-44, 1999
3361. Vrána J, Kubaláková M, Šimková H, Cíhalíková J, Lysák MA, Doležel J: Flow sorting of mitotic chromosomes in common wheat (Triticum aestivum L.). Genetics 156:2033-41, 2000
3362. Kubaláková M, Vrána J, Cíhalíková J, Šimková H, Doležel J: Flow karyotyping and chromosome sorting in bread wheat (Triticum aestivum L.). Theor Appl Genet 104:1362-72, 2002
3363. Neumann P, Pozárková D, Vrána J, Doležel J, Macas J: Chromosome sorting and pcr-based physical mapping in pea (Pisum sativum L.). Chromosome Res 10:63-71, 2002
3364. Lucretti S, Nardi L, Nisini PT, Moretti F, Gualberti G, Doležel J: Bivariate flow cytometry DNA/BrdUrd analysis of plant cell cycle. Methods Cell Sci 21:155-66, 1999
3365. Pfündel E, Meister A: Flow cytometry of mesophyll and bundle sheath chloroplast thylakoids of maize (Zea mays L.). Cytometry 23:97-105, 1996
3366. Giglioli-Guivarc'h N, Pierre JN, Vidal J, Brown S: Flow cytometric analysis of cytosolic pH of mesophyll cell protoplasts from the crabgrass Digitaria sanguinalis. Cytometry 23:241-9, 1996
3367. O'Brien IE, Reutelingsperger CP, Holdaway KM: Annexin-V and TUNEL use in monitoring the progression of apoptosis in plants. Cytometry 29:28-33, 1997
3368. Sheen J, Hwang S, Niwa Y, Kobayashi H, Galbraith DW: Green-fluorescent protein as a new vital marker in plant cells. Plant J 8:777-84, 1995
3369. Galbraith DW, Anderson MT, Herzenberg LA: Flow cytometric analysis and FACS sorting of cells based on GFP accumulation. Methods Cell Biol 58:315-41, 1999
3370. Davey HM, Kell DB: Flow cytometry and cell sorting of heterogeneous microbial populations-the importance of single-cell analyses. Microbiol Revs 60:641-96, 1996
3371. Shapiro HM: Microbial analysis at the single-cell level. Tasks and techniques. J Microbiol Methods 42:3-16, 2000 (in reference 2404; full text is available at www.elsevier.com/locate/jmicmeth)
3372. Walberg M, Gaustad P, Steen HB: Uptake kinetics of nucleic acid targeting dyes in S. aureus, E. faecalis and B. cereus: A flow cytometric study. J Microbiol Methods 35:167-76, 1999
3373. Mason DJ, Shanmuganathan S, Mortimer FC, Gant VA: A fluorescent gram stain for flow cytometry and epifluorescence microscopy. Appl Environ Microbiol 64:2681-5, 1998
3374. Julia O, Comas J, Vives-Rego J: Second-order functions are the simplest correlations between flow cytometric light scatter and bacterial diameter. J Microbiol Methods 40:57-61, 2000
3375. Bouvier T, Troussellier M, Anzil A, Courties C, Servais P: Using light scatter signal to estimate bacterial biovolume by flow cytometry. Cytometry 44:188-94, 2001
3376. Vorauer-Uhl K, Wagner A, Borth N, Katinger H: Determination of liposome size distribution by flow cytometry. Cytometry 39:166-71, 2000
3377. Green RE, Sosik HM, Olson RJ, DuRand MD: Flow cytometric determination of size and complex refractive index for marine particles: Comparison with independent and bulk estimates. Appl Opt 42:526-41, 2003
3378. Cucci TL, Sieracki ME: Effects of mismatched refractive indices in aquatic flow cytometry. Cytometry 44:173-8, 2001
3379. Guindulain T, Vives-Rego J: Involvement of RNA and DNA in the staining of Escherichia coli by SYTO 13. Lett Appl Microbiol 34:182-8, 2002
3380. Garcia-Ochoa F, Santos VE, Alcon A: Intracellular compounds quantification by means of flow cytometry in bacteria: Application to xanthan production by Xanthomonas campestris. Biotechnol Bioeng 57:87-94, 1998
3381. Steen HB: Staining and measurement of DNA in bacteria. Methods Cell Biol 64:539-51, 2001
3382. Bernander R, Stokke T, Boye E: Flow cytometry of bacterial cells: Comparison between different flow cytometers and different DNA stains. Cytometry 31:29-36, 1998
3383. Andreatta S: Cytometry of Aquatic Bacteria: Analyses at the community and subgroup level. Doctoral Thesis, University of Innsbruck, 2001
3384. Borth N, Mitterbauer R, Mattanovich D, Kramer W, Bayer K, Katinger H: Flow cytometric analysis of bacterial physiology during induction of foreign protein synthesis in recombinant Escherichia coli cells. Cytometry 31:125-9, 1998
3385. Müller S, Bley T, Babel W: Adaptive responses of Ralstonia eutropha to feast and famine conditions analysed by flow cytometry. J Biotechnol 75:81-97, 1999
3386. Zubkov MV, Fuchs BM, Eilers H, Burkill PH, Amann R: Determination of total protein content of bacterial cells by SYPRO staining and flow cytometry. Appl Environ Microbiol 65:3251-7, 1999
3387. Fuchs BM, Wallner G, Beisker W, Schwippl I, Ludwig W, Amann R: Flow cytometric analysis of the in situ accessibility of Escherichia coli 16S rRNA for fluorescently labeled oligonucleotide probes. Appl Environ Microbiol 64:4973-82, 1998
3388. Fuchs BM, Glockner FO, Wulf J, Amann R: Unlabeled helper oligonucleotides increase the in situ accessibility to 16s rRNA of fluorescently labeled oligonucleotide probes. Appl Environ Microbiol 66:3603-7, 2000
3389. Fuchs BM, Syutsubo K, Ludwig W, Amann R: In situ accessibility of Escherichia coli 23S rRNA to fluorescently labeled oligonucleotide probes. Appl Environ Microbiol 67:961-8, 2001
3390. Sizemore RK, Caldwell JJ, Kendrick AS: Alternate Gram staining technique using a fluorescent lectin. Appl Environ Microbiol 56:2245-7, 1990
3391. Saleh OA, Sohn LL: Direct detection of antibody-antigen binding using an on-chip artificial pore. Proc Natl Acad Sci U S A 100:820-4, 2003
3392. Giovannoni SJ, DeLong EF, Olsen GJ, Pace NR: Phylogenetic group-specific oligodeoxynucleotide probes for identification of single microbial cells. J Bacteriol 170:720-6, 1988
3393. DeLong EF, Wickham GS, Pace NR: Phylogenetic stains: Ribosomal RNA-based probes for the identification of single cells. Science 243:1360-3, 1989
3394. Worden AZ, Chisholm SW, Binder BJ: In situ hybridization of Prochlorococcus and Synechococcus (marine cyanobacteria) spp. with rRNA-targeted peptide nucleic acid probes. Appl Environ Microbiol 66:284-9, 2000
3395. Chmielewski V, Basset C, Holton J, Bloom S, Seman M, Drupt F, Gant V: Use of multiparameter flow cytometry for studying the composition of human colonic flora in health and inflammatory bowel disease using 16S rRNA in situ hybridisation. Cytometry Supp 11:102, 2002
3396. Hewitt CJ, Nebe-Von-Caron G: An industrial application of multiparameter flow cytometry: Assessment of cell physiological state and its application to the study of microbial fermentations. Cytometry 44:179-87, 2001

3397. Shapiro HM: Multiparameter flow cytometry of bacteria: Implications for diagnostics and therapeutics. Cytometry 43:223-6, 2001

3398. Amor KB, Breeuwer P, Verbaarschot P, Rombouts FM, Akkermans AD, De Vos WM, Abee T: Multiparametric flow cytometry and cell sorting for the assessment of viable, injured, and dead Bifidobacterium cells during bile salt stress. Appl Environ Microbiol 68:5209-16, 2002

3399. Mason DJ, Power GM, Talsania H, Phillips I, Gant VA: Antibacterial action of ciprofloxacin. Antimicrob Agents Chemother 39:2752-8, 1995

3400. Roth BL, Poot M, Yue ST, Millard PJ: Bacterial viability and antibiotic susceptibility testing with Sytox green™ nucleic acid stain. Appl Env Microbiol 63:2421-31, 1997

3401. Suller MT, Stark JM, Lloyd D: A flow cytometric study of antibiotic induced damage and evaluation as a rapid antibiotic susceptibility test for methicillin-resistant Staphylococcus aureus. J Antimicrob Chemother 40:77-83, 1997

3402. Jepras RJ, Paul FE, Pearson SC, Wilkinson MJ: Rapid assessment of antibiotic effects on Escherichia coli by bis-(1,3-dibutylbarbituric acid) trimethine oxonol and flow cytometry. Antimicrob Agents Chemother 41:2001-5, 1997

3403. Suller MT, Lloyd D: Flow cytometric assessment of the postantibiotic effect of methicillin on Staphylococcus aureus. 42:1195-9, 1998

3404. Shen CF, Meghrous J, Kamen A: Quantitation of baculovirus particles by flow cytometry. J Virol Methods 105:321-30, 2002

3405. Chen F, Lu JR, Binder BJ, Liu YC, Hodson RE: Application of digital image analysis and flow cytometry to enumerate marine viruses stained with SYBR gold. Appl Environ Microbiol 67:539-45, 2001

3406. Andreatta S, Hofer JS, Sommaruga R, Psenner R: Analysis and high speed sorting of free viruses by flow cytometry. Cytometry Supp 11:69, 2002

3407. Partensky F, Hess WR, Vaulot D: Prochlorococcus, a marine photosynthetic prokaryote of global significance. Microbiol Mol Biol Rev 63:106-27, 1999

3408. West NJ, Schonhuber WA, Fuller NJ, Amann RI, Rippka R, Post AF, Scanlan DJ: Closely related prochlorococcus genotypes show remarkably different depth distributions in two oceanic regions as revealed by in situ hybridization using 16S rRNA-targeted oligonucleotides. Microbiology 147:1731-44, 2001

3409. Gregori G, Colosimo A, Denis M: Phytoplankton group dynamics in the Bay of Marseilles during a 2-year survey based on analytical flow cytometry. Cytometry 44:247-56, 2001

3410. Zubkov MV, Fuchs BM, Tarran GA, Burkill PH, Amann R: High rate of uptake of organic nitrogen compounds by prochlorococcus cyanobacteria as a key to their dominance in oligotrophic oceanic waters. Appl Environ Microbiol 69:1299-304, 2003

3411. Simon N, LeBot N, Marie D, Partensky F, Vaulot D: Fluorescent in situ hybridization with rRNA-targeted oligonucleotide probes to identify small phytoplankton by flow cytometry. Appl Environ Microbiol 61:2506-13, 1995

3412. Casamayor EO, Pedros-Alio C, Muyzer G, Amann R: Microheterogeneity in 16s ribosomal DNA-defined bacterial populations from a stratified planktonic environment is related to temporal changes and to ecological adaptations. Appl Envir Microbiol 68:1706-14, 2002

3413. Suzuki MT, Taylor LT, DeLong EF: Quantitative analysis of small-subunit rRNA genes in mixed microbial populations via 5'-nuclease assays. Appl Envir Microbiol 66:4605-14, 2000

3414. Cavender-Bares KK, Frankel SL, Chisholm SW: A dual sheath flow cytometer for shipboard analyses of phytoplankton communities from the oligotrophic oceans. Limnol Oceanog 43:1383-8, 1998

3415. Olson RJ, Sosik HM, Chekalyuk AM: Photosynthetic characteristics of marine phytoplankton from pump-during-probe fluorometry of individual cells at sea. Cytometry 37:1-13, 1999

3416. Button DK, Robertson BR: Determination of DNA content of aquatic bacteria by flow cytometry. Appl Envir Microbiol 67:1636-45, 2001

3417. Wilkins MF, Boddy L, Morris CW, Jonker RR: Identification of phytoplankton from flow cytometry data by using radial basis function neural networks. Appl Environ Microbiol 65:4404-10, 1999

3418. Gregori G, Colosimo A, Denis M: Phytoplankton group dynamics in the bay of marseilles during a 2-year survey based on analytical flow cytometry. Cytometry 44:247-56, 2001

3419. Wilkins MF, Hardy SA, Boddy L, Morris CW: Comparison of five clustering algorithms to classify phytoplankton from flow cytometry data. Cytometry 44:210-7, 2001

3420. Boddy L, Wilkins MF, Morris CW: Pattern recognition in flow cytometry. Cytometry 44:195-209, 2001

3421. Seror SJ, Casaregola S, Vannier F, Zouari N, Dahl M, Boye E: A mutant cysteinyl-tRNA synthetase affecting timing of chromosomal replication initiation in B. cubtilis and conferring resistance to a protein kinase C inhibitor. Embo J 13:2472-80, 1994

3422. Boye E, Stokke T, Kleckner N, Skarstad K: Coordinating DNA replication initiation with cell growth: Differential roles for DnaA and SeqA proteins. Proc Natl Acad Sci U S A 93:12206-11, 1996

3423. Botello E, Nordstrom K: Effects of chromosome underreplication on cell division in Escherichia coli. J Bacteriol 180:6364-74, 1998

3424. Atlung T, Hansen FG: Effect of different concentrations of H-NS protein on chromosome replication and the cell cycle in Escherichia coli. J Bacteriol 184:1843-50, 2002

3425. Li Y, Sergueev K, Austin S: The segregation of the Escherichia coli origin and terminus of replication. Mol Microbiol 46:985-96, 2002

3426. Sergueev K, Court D, Reaves L, Austin S: E. coli cell-cycle regulation by bacteriophage lambda. J Mol Biol 324:297-307, 2002

3427. Bach T, Krekling MA, Skarstad K: Excess SeqA prolongs sequestration of oriC and delays nucleoid segregation and cell division. Embo J 22:315-23, 2003

3428. Jacquet S, Partensky F, Marie D, Casotti R, Vaulot D: Cell cycle regulation by light in Prochlorococcus strains. Appl Environ Microbiol 67:782-90, 2001

3429. Holtzendorff J, Partensky F, Jacquet S, Bruyant F, Marie D, Garczarek L, Mary I, Vaulot D, Hess WR: Diel expression of cell cycle-related genes in synchronized cultures of Prochlorococcus sp. Strain PCC 9511. J Bacteriol 183:915-20, 2001

3430. Winzeler E, Shapiro L: Use of flow cytometry to identify a Caulobacter 4.5 s RNA temperature-sensitive mutant defective in the cell cycle. J Mol Biol 251:346-65, 1995

3431. Jensen RB, Shapiro L: The Caulobacter crescentus smc gene is required for cell cycle progression and chromosome segregation. Proc Natl Acad Sci U S A 96:10661-6, 1999

3432. Kahng LS, Shapiro L: The CcrM DNA methyltransferase of Agrobacterium tumefaciens is essential, and its activity is cell cycle regulated. J Bacteriol 183:3065-75, 2001

3433. Hung DY, Shapiro L: A signal transduction protein cues proteolytic events critical to Caulobacter cell cycle progression. Proc Natl Acad Sci U S A 99:13160-5, 2002

3434. Ueckert JE, Nebe von-Caron G, Bos AP, ter Steeg PF: Flow cytometric analysis of Lactobacillus plantarum to monitor lag times, cell division and injury. Lett Appl Microbiol 25:295-9, 1997

3435. Porro D, Ranzi BM, Smeraldi C, Martegani E, Alberghina L: A double flow cytometric tag allows tracking of the dynamics of cell cycle progression of newborn Saccharomyces cerevisiae cells during balanced exponential growth. Yeast 11:1157-69, 1995

3436. Porro D, Srienc F: Tracking of individual cell cohorts in asynchronous Saccharomyces cerevisiae populations. Biotechnol Prog 11:342-7, 1995

3437. Porro D, Martegani E, Ranzi BM, Alberghina L: Identification of different daughter and parent subpopulations in an asynchronously growing Saccharomyces cerevisiae population. Res Microbiol 148:205-15, 1997

3438. Alberghina L, Smeraldi C, Ranzi BM, Porro D: Control by nutrients of growth and cell cycle progression in budding yeast, analyzed by double-tag flow cytometry. J Bacteriol 180:3864-72, 1998
3439. Carlson CR, Grallert B, Bernander R, Stokke T, Boye E: Measurement of nuclear DNA content in fission yeast by flow cytometry. Yeast 13:1329-35, 1997
3440. Valdivia RH, Falkow S: Bacterial genetics by flow cytometry: Rapid isolation of Salmonella typhimurium acid-inducible promoters by differential fluorescence induction. Mol Microbiol 22:367-78, 1996
3441. Cormack BP, Valdivia RH, Falkow S: FACS-optimized mutants of the green fluorescent protein (GFP). Gene 173:33-8, 1996
3442. Valdivia RH, Hromockyj AE, Monack D, Ramakrishnan L, Falkow S: Applications for green fluorescent protein (GFP) in the study of host-pathogen interactions. Gene 173:47-52, 1996
3443. Cormack BP, Bertram G, Egerton M, Gow NA, Falkow S, Brown AJ: Yeast-enhanced green fluorescent protein (yEGFP) a reporter of gene expression in Candida albicans. Microbiology 143 (Pt 2):303-11, 1997
3444. Cirillo DM, Valdivia RH, Monack DM, Falkow S: Macrophage-dependent induction of the Salmonella pathogenicity island 2 type iii secretion system and its role in intracellular survival. Mol Microbiol 30:175-88, 1998
3445. Lee AK, Falkow S: Constitutive and inducible green fluorescent protein expression in bartonella henselae. Infect Immun 66:3964-7, 1998
3446. Unge A, Tombolini R, Molbak L, Jansson JK: Simultaneous monitoring of cell number and metabolic activity of specific bacterial populations with a dual gfp-luxab marker system. Appl Environ Microbiol 65:813-21, 1999
3447. De Wulf P, Brambilla L, Vanoni M, Porro D, Alberghina L: Real-time flow cytometric quantification of GFP expression and GFP-fluorescence generation in saccharomyces cerevisiae. J Microbiol Methods 42:57-64, 2000 (in reference 2404; full text is available at www.elsevier.com/locate/jmicmeth)
3448. Lowder M, Unge A, Maraha N, Jansson JK, Swiggett J, Oliver JD: Effect of starvation and the viable-but-nonculturable state on green fluorescent protein (GFP) fluorescence in GFP-tagged Pseudomonas fluorescens A506. Appl Environ Microbiol 66:3160-5, 2000
3449. Geoffroy MC, Guyard C, Quatannens B, Pavan S, Lange M, Mercenier A: Use of green fluorescent protein to tag lactic acid bacterium strains under development as live vaccine vectors. Appl Environ Microbiol 66:383-91, 2000
3450. Haidinger W, Szostak MP, Beisker W, Lubitz W: Green fluorescent protein (GFP)-dependent separation of bacterial ghosts from intact cells by FACS. Cytometry 44:106-12, 2001
3451. Attfield PV, Choi HY, Veal DA, Bell PJ: Heterogeneity of stress gene expression and stress resistance among individual cells of Saccharomyces cerevisiae. Mol Microbiol 40:1000-8, 2001
3452. Maksimow M, Hakkila K, Karp M, Virta M: Simultaneous detection of bacteria expressing GFP and dsRed genes with a flow cytometer. Cytometry 47:243-7, 2002
3453. Abd H, Johansson T, Golovliov I, Sandstrom G, Forsman M: Survival and growth of Francisella tularensis in Acanthamoeba castellanii. Appl Environ Microbiol 69:600-6, 2003
3454. Kell DB, Kaprelyants AS, Weichart DH, Harwood CR, Barer MR: Viability and activity in readily culturable bacteria: a review and discussion of the practical issues. Antonie Van Leeuwenhoek 73:169-87, 1998
3455. Chung JD, Conner S, Stephanopoulos G. Flow cytometric study of differentiating cultures of Bacillus subtilis. Cytometry 20:324-33, 1995
3456. Jacobs C, Shapiro L: Microbial asymmetric cell division: Localization of cell fate determinants. Curr Opin Genet Dev 8:386-91, 1998
3457. Jensen RB, Wang SC, Shapiro L: Dynamic localization of proteins and DNA during a bacterial cell cycle. Nat Rev Mol Cell Biol 3:167-76, 2002
3458. Bassler BL: Small talk. Cell-to-cell communication in bacteria. Cell 109:421-4, 2002
3459. Miller MB, Bassler BL: Quorum sensing in bacteria. Annu Rev Microbiol 55:165-99, 2001
3460. Mukamolova GV, Kaprelyants AS, Young DI, Young M, Kell DB: A bacterial cytokine. Proc Natl Acad Sci U S A 95:8916-21, 1998
3461. Donlan RM: Biofilms: Microbial life on surfaces. Emerg Infect Dis 8:881-90, 2002 (online at: http://www.cdc.gov/ ncidod/EID/vol8no9/02-0063.htm)
3462. Atlas RM: Bioterrorism: from threat to reality. Annu Rev Microbiol 56:167-85, 2002
3463. Dull PM, Wilson KE, Kournikakis B, Whitney EA, Boulet CA, Ho JY, Ogston J, Spence MR, McKenzie MM, Phelan MA, Popovic T, Ashford D: Bacillus anthracis aerosolization associated with a contaminated mail sorting machine. Emerg Infect Dis 8:1044-7, 2002
3464. Ye F, Li MS, Taylor JD, Nguyen Q, Colton HM, Casey WM, Wagner M, Weiner MP, Chen J: Fluorescent microsphere-based readout technology for multiplexed human single nucleotide polymorphism analysis and bacterial identification. Hum Mutat 17:305-16, 2001
3465. Vesey G, Deere D, Weir CJ, Ashbolt N, Williams KL, Veal DA: A simple method for evaluating Cryptosporidium-specific antibodies used in monitoring environmental water samples. Lett Appl Microbiol 25:316-20, 1997
3466. Deere D, Vesey G, Ashbolt N, Davies KA, Williams KL, Veal D: Evaluation of fluorochromes for flow cytometric detection of Cryptosporidium parvum oocysts labelled by fluorescent in situ hybridization. Lett Appl Microbiol 27:352-6, 1998
3467. Deere D, Vesey G, Milner M, Williams K, Ashbolt N, Veal D: Rapid method for fluorescent in situ ribosomal RNA labelling of Cryptosporidium parvum. J Appl Microbiol 85:807-18, 1998
3468. Ferrari BC, Vesey G, Davis KA, Gauci M, Veal D: A novel two-color flow cytometric assay for the detection of Cryptosporidium in environmental water samples. Cytometry 41:216-22, 2000
3469. Veal DA, Deere D, Ferrari B, Piper J, Attfield PV: Fluorescence staining and flow cytometry for monitoring microbial cells. J Immunol Methods 243:191-210, 2000
3470. Ferrari BC, Veal D: Analysis-only detection of Giardia by combining immunomagnetic separation and two-color flow cytometry. Cytometry 51A:79-86, 2003
3471. Lindquist HD, Ware M, Stetler RE, Wymer L, Schaefer FW, 3rd: A comparison of four fluorescent antibody-based methods for purifying, detecting, and confirming cryptosporidium parvum in surface waters. J Parasitol 87:1124-31, 2001
3472. McClelland RG, Pinder AC: Detection of Salmonella typhimurium in dairy products with flow cytometry and monoclonal antibodies. Appl Environ Microbiol 60:4255–62, 1994
3473. Goodridge L, Chen J, Griffiths M: The use of a fluorescent bacteriophage assay for detection of Escherichia coli O157:H7 in inoculated ground beef and raw milk. Int J Food Microbiol 47:43-50, 1999
3474. Bruno JG, Sincock SA, Stopa PJ: Highly selective acridine and ethidium staining of bacterial DNA and RNA. Biotech Histochem 71:130-6, 1996
3475. Emanuel PA, Dang J, Gebhardt JS, Aldrich J, Garber EA, Kulaga H, Stopa P, Valdes JJ, Dion-Schultz A: Recombinant antibodies: A new reagent for biological agent detection. Biosens Bioelectron 14:751-9, 2000
3476. Stopa PJ, Mastromanolis SA: The use of blue-excitable nucleic-acid dyes for the detection of bacteria in well water using a simple field fluorometer and a flow cytometer. J Microbiol Methods 45:143-53, 2001
3477. Journal Editors and Authors Group: Uncensored exchange of scientific results. Proc Natl Acad Sci U S A 100:1464, 2003
3478. McSharry JJ: Uses of flow cytometry in virology. Clin Microbiol Rev 7:576-604, 1994

3479. Álvarez-Barrientos A, Arroyo J, Cantón R, Nombela C, Sánchez-Pérez M: Applications of flow cytometry to clinical microbiology. Clin Microbiol Rev 13:167-95, 2000

3480. Shapiro HM: Microbiology. Clin Lab Med 21:897-909; x-xi, 2001

3481. Barenfanger J, Drake C, Kacich G: Clinical and financial benefits of rapid bacterial identification and antimicrobial susceptibility testing. J Clin Microbiol 37:1415-8, 1999

3482. Barenfanger J: Significantly reduced variable costs, mortality, and length of stay with timely microbiologic procedures. Am Clin Lab 20:26-31, 2001

3483. Barenfanger J, Short MA, Groesch AA: Improved antimicrobial interventions have benefits. J Clin Microbiol 39:2823-8, 2001

3484. Barenfanger J, Drake CA, Lawhorn J, Kopec C, Killiam R: Outcomes of improved anaerobic techniques in clinical microbiology. Clin Infect Dis 35:S78-83, 2002

3485. Okada H, Sakai Y, Miyazaki S, Arakawa S, Hamaguchi Y, Kamidono S: Detection of significant bacteriuria by automated urinalysis using flow cytometry. J Clin Microbiol 38:2870-2, 2000

3486. Zaman Z, Roggeman S, Verhaegen J: Unsatisfactory performance of flow cytometer UF-100 and urine strips in predicting outcome of urine cultures. J Clin Microbiol 39:4169-71, 2001

3487. Manoni F, Valverde S, Antico F, Salvadego MM, Giacomini A, Gessoni G: Field evaluation of a second-generation cytometer UF-100 in diagnosis of acute urinary tract infections in adult patients. Clin Microbiol Infect 8:662-8, 2002

3488. Davey HM, Weichart DH, Kell DB, Kaprelyants AS: Estimation of microbial viability using flow cytometry. Unit 11.3. In: Robinson JP, Darzynkiewicz Z, Dean P, Hibbs AR, Orfao A, Rabinovitch P, Wheeless L, (eds): *Current Protocols in Cytometry*, New York, John Wiley & Sons, 1999, pp 11.3.1-11.3.20

3489. Walberg M, Steen HB: Flow cytometric monitoring of bacterial susceptibility to antibiotics. Methods Cell Biol 64:553-66, 2001(in reference 2386)

3490. Silverman JA, Perlmutter NG, Shapiro HM: Correlation of Daptomycin Bactericidal Activity and Membrane Depolarization in Staphylococcus aureus. Antimicrob Agents Chemother, submitted

3491. Norden MA, Kurzynski TA, Bownds SE, Callister SM, Schell RF: Rapid susceptibility testing of Mycobacterium tuberculosis (H37RA) by flow cytometry. J Clin Microbiol 33:1231-7, 1995

3492. Bownds SE, Kurzynski TA, Norden MA, Dufek JL, Schell RF: Rapid susceptibility testing for nontuberculosis mycobacteria using flow cytometry. J Clin Microbiol 34:1386-90, 1996

3493. Kirk SM, Schell RF, Moore AV, Callister SM, Mazurek GH: Flow cytometric testing of susceptibilities of Mycobacterium tuberculosis isolates to ethambutol, isoniazid, and rifampin in 24 hours. J Clin Microbiol 36:1568-73, 1998

3494. Moore AV, Kirk SM, Callister SM, Mazurek GH, Schell RF: Safe determination of susceptibility of Mycobacterium tuberculosis to antimycobacterial agents by flow cytometry. J Clin Microbiol 37:479-83, 1999

3495. Ramani R, Ramani A, Wong SJ: Rapid flow cytometric susceptibility testing of Candida albicans. J Clin Microbiol 35:2320-4, 1997

3496. Green LJ, Marder P, Mann LL, Chio LC, Current WL: LY303366 exhibits rapid and potent fungicidal activity in flow cytometric assays of yeast viability. Antimicrob Agents Chemother 43:830-5, 1999

3497. Ramani R, Chaturvedi V: Flow cytometry antifungal susceptibility testing of pathogenic yeasts other than Candida albicans and comparison with the NCCLS broth microdilution test. Antimicrob Agents Chemother 44:2752-8, 2000

3498. Wenisch C, Linnau KF, Parschalk B, Zedtwitz-Liebenstein K, Georgopoulos A: Rapid susceptibility testing of fungi by flow cytometry using vital staining. J Clin Microbiol 35:5-10, 1997

3499. Wenisch C, Moore CB, Krause R, Presterl E, Pichna P, Denning DW: Antifungal susceptibility testing of fluconazole by flow cytometry correlates with clinical outcome. J Clin Microbiol 39:2458-62, 2001

3500. Balajee SA, Marr KA: Conidial viability assay for rapid susceptibility testing of Aspergillus species. J Clin Microbiol 40:2741-5, 2002

3501. Kirk SM, Callister SM, Lim LC, Schell RF: Rapid susceptibility testing of Candida albicans by flow cytometry. J Clin Microbiol 35:358-63, 1997

3502. Ordonez JV, Wehman NM: Amphotericin B susceptibility of Candida species assessed by rapid flow cytometric membrane potential assay. Cytometry (Commun Clin Cytom) 22:154-7, 1995

3503. Peyron F, Favel A, Guiraud-Dauriac H, El Mzibri M, Chastin C, Dumenil G, Regli P: Evaluation of a flow cytofluorometric method for rapid determination of amphotericin B susceptibility of yeast isolates. Antimicrob Agents Chemother 41:1537-40, 1997

3504. Pavic I, Hartmann A, Zimmermann A, Michel D, Hampl W, Schleyer I, Mertens T: Flow cytometric analysis of herpes simplex virus type 1 susceptibility to acyclovir, ganciclovir, and foscarnet. Antimicrob Agents Chemother 41:2686-92, 1997

3505. McSharry JM, Lurain NS, Drusano GL, Landay A, Manischewitz J, Nokta M, O'Gorman M, Shapiro HM, Weinberg A, Reichelderfer P, Crumpacker C: Flow cytometric determination of ganciclovir susceptibilities of human cytomegalovirus clinical isolates. J Clin Microbiol 36:958-64, 1998

3506. McSharry JJ, Lurain NS, Drusano GL, Landay AL, Notka M, O'Gorman MR, Weinberg A, Shapiro HM, Reichelderfer PS, Crumpacker CS: Rapid ganciclovir susceptibility assay using flow cytometry for human cytomegalovirus clinical isolates. Antimicrob Agents Chemother 42:2326-31, 1998

3507. Kesson AM, Zeng F, Cunningham AL, Rawlinson WD: The use of flow cytometry to detect antiviral resistance in human cytomegalovirus. J Virol Methods 71:177-86, 1998

3508. Manichanh C, Grenot P, Gautheret-Dejean A, Debre P, Huraux JM, Agut H: Susceptibility of human herpesvirus 6 to antiviral compounds by flow cytometry analysis. Cytometry 40:135-40, 2000

3509. Letvin NL, Barouch DH, Montefiori DC: Prospects for vaccine protection against HIV-1 infection and AIDS. Annu Rev Immunol 20:73-99, 2002

3510. Barouch DH, Santra S, Schmitz JE, Kuroda MJ, Fu TM, Wagner W, Bilska M, Craiu A, Zheng XX, Krivulka GR, Beaudry K, Lifton MA, Nickerson CE, Trigona WL, Punt K, Freed DC, Guan L, Dubey S, Casimiro D, Simon A, Davies ME, Chastain M, Strom TB, Gelman RS, Montefiori DC, Lewis MG, Emini EA, Shiver JW, Letvin NL: Control of viremia and prevention of clinical AIDS in rhesus monkeys by cytokine-augmented DNA vaccination. Science 290:486-92, 2000

3511. Barouch DH, Kunstman J, Kuroda MJ, Schmitz JE, Santra S, Peyerl FW, Krivulka GR, Beaudry K, Lifton MA, Gorgone DA, Montefiori DC, Lewis MG, Wolinsky SM, Letvin NL: Eventual AIDS vaccine failure in a rhesus monkey by viral escape from cytotoxic T lymphocytes. Nature 415:335-9, 2002

3512. Reed DS, Smoll J, Gibbs P, Little SF: Mapping of antibody responses to the protective antigen of bacillus anthracis by flow cytometric analysis. Cytometry 49:1-7, 2002

3513. Jani IV, Janossy G, Brown DW, Mandy F: Multiplexed immunoassays by flow cytometry for diagnosis and surveillance of infectious diseases in resource-poor settings. Lancet Infect Dis 2:243-50, 2002

3514. Cirino NM, Sblattero D, Allen D, Peterson SR, Marks JD, Jackson PJ, Bradbury A, Lehnert BE: Disruption of anthrax toxin binding with the use of human antibodies and competitive inhibitors. Infect Immun 67:2957-63, 1999

3515. Logan RP, Robins A, Turner GA, Cockayne A, Borriello SP, Hawkey CJ: A novel flow cytometric assay for quantitating adherence of Helicobacter pylori to gastric epithelial cells. J Immunol Methods 213:19-30, 1998

3516. Stapleton AE, Fennell CL, Coder DM, Wobbe CL, Roberts PL, Stamm WE: Precise and rapid assessment of Escherichia coli adherence to vaginal epithelial cells by flow cytometry. Cytometry 50:31-7, 2002

3517. Alugupalli KR, Michelson AD, Barnard MR, Robbins D, Coburn J, Baker EK, Ginsberg MH, Schwan TG, Leong JM: Platelet activation by a relapsing fever spirochaete results in enhanced bacterium-platelet interaction via integrin alphaIIbbeta3 activation.Mol Microbiol. 39:330-40, 2001
3518. Braga PC, Bovio C, Culici M, Dal Sasso M: Flow cytometric assessment of susceptibilities of Streptococcus pyogenes to erythromycin and rokitamycin. Antimicrob Agents Chemother 47:408-12, 2003
3519. Frelinger JA, Garba ML: Responses to smallpox vaccine (letter). N Engl J Med 347:689-90, 2002
3520. Khunkitti W, Avery SV, Lloyd D, Furr JR, Russell AD: Effects of biocides on Acanthamoeba castellanii as measured by flow cytometry and plaque assay. J Antimicrob Chemother 40:227-33, 1997
3521. Borazjani RN, May LL, Noble JA, Avery SV, Ahearn DG: Flow cytometry for determination of the efficacy of contact lens disinfecting solutions against Acanthamoeba spp. Appl Environ Microbiol 66:1057-61, 2000
3522. Fukata T, Ohnishi T, Okuda S, Sasai K, Baba E, Arakawa A: Detection of canine erythrocytes infected with Babesia gibsoni by flow cytometry. J Parasitol 82:641-2, 1996
3523. Valdez LM, Dang H, Okhuysen PC, Chappell CL: Flow cytometric detection of Cryptosporidium oocysts in human stool samples. J Clin Microbiol 35:2013-7, 1997
3524. Cole DJ, Snowden K, Cohen ND, Smith R: Detection of Cryptosporidium parvum in horses: Thresholds of acid-fast stain, immunofluorescence assay, and flow cytometry. J Clin Microbiol 37:457-60, 1999
3525. Delaunay A, Gargala G, Li X, Favennec L, Ballet JJ: Quantitative flow cytometric evaluation of maximal Cryptosporidium parvum oocyst infectivity in a neonate mouse model. Appl Envir Microbiol 66:4315-7, 2000
3526. Neumann NF, Gyurek LL, Gammie L, Finch GR, Belosevic M: Comparison of animal infectivity and nucleic acid staining for assessment of Cryptosporidium parvum viability in water. Appl Envir Microbiol 66:406-12, 2000
3527. Fuller AL, Golden J, McDougald LR: Flow cytometric analysis of the response of Eimeria tenella (coccidia) sporozoites to coccidiocidal effects of ionophores. J Parasitol 81:985-8, 1995
3528. Moss DM, Croppo GP, Wallace S, Visvesvara GS: Flow cytometric analysis of Microsporidia belonging to the genus Encephalitozoon. J Clin Microbiol 37:371-5, 1999
3529. Santillana-Hayat M, Sarfati C, Fournier S, Chau F, Porcher R, Molina J-M, Derouin F: Effects of chemical and physical agents on viability and infectivity of Encephalitozoon intestinalis determined by cell culture and flow cytometry. Antimicrob Agents Chemother 46:2049-51, 2002
3530. Lloyd D, Harris JC, Maroulis S, Biagini GA, Wadley RB, Turner MP, Edwards MR: The microaerophilic flagellate Giardia intestinalis: Oxygen and its reaction products collapse membrane potential and cause cytotoxicity. Microbiology 146 Pt 12:3109-18, 2000
3531. Biagini GA, Lloyd D, Kirk K, Edwards MR: The membrane potential of Giardia intestinalis. FEMS Microbiol Lett 192:153-7, 2000
3532. Azas N, Di Giorgio C, Delmas F, Gasquet M, Timon-David P: Assessment of amphotericin B susceptibility in Leishmania infantum promastigotes by flow cytometric membrane potential assay. Cytometry 28:165-9, 1997
3533. Maarouf M, de Kouchkovsky Y, Brown S, Petit PX, Robert-Gero M: In vivo interference of paromomycin with mitochondrial activity of Leishmania. Exp Cell Res 232:339-48, 1997
3534. Guinet F, Louise A, Jouin H, Antoine JC, Roth CW: Accurate quantitation of Leishmania infection in cultured cells by flow cytometry. Cytometry 39:235-40, 2000
3535. Di Giorgio C, Ridoux O, Delmas F, Azas N, Gasquet M, Timon-David P: Flow cytometric detection of Leishmania parasites in human monocyte-derived macrophages: Application to antileishmanial-drug testing. Antimicrob Agents Chemother 44:3074-8, 2000
3536. Kamau SW, Hurtado M, Muller-Doblies UU, Grimm F, Nunez R: Flow cytometric assessment of allopurinol susceptibility in Leishmania infantum promastigote. Cytometry 40:353-60, 2000
3537. Yagi Y, Shiono H, Kurabayashi N, Yoshihara K, Chikayama Y: Flow cytometry to evaluate Theileria sergenti parasitemia using the fluorescent nucleic acid stain, SYTO16. Cytometry 41:223-5, 2000
3538. Martins-Filho OA, Eloi-Santos SM, Carvalho AT, Oliveira RC, Rassi A, Luquetti AO, Rassi GG, Brener Z: Double-blind study to evaluate flow cytometry analysis of anti-live trypomastigote antibodies for monitoring treatment efficacy in cases of human Chagas' disease. Clin Diagn Lab Immunol 9:1107-13, 2002
3539. van der Heyde H, Elloso M, vande Waa J, Schell K, Weidanz W: Use of hydroethidine and flow cytometry to assess the effects of leukocytes on the malarial parasite Plasmodium falciparum. Clin Diagn Lab Immunol 2:417-25, 1995
3540. Reinders PP, van Vianen PH, van der Keur M, van Engen A, Janse CJ, Tanke HJ: Computer software for testing drug susceptibility of malaria parasites. Cytometry 19:273-81, 1995
3541. Pattanapanyasat K, Yongvanitchit K, Heppner DG, Tongtawe P, Kyle DE, Webster HK: Culture of malaria parasites in two different red blood cell populations using biotin and flow cytometry. Cytometry 25:287-94, 1996
3542. Pattanapanyasat K, Thaithong S, Kyle DE, Udomsangpetch R, Yongvanitchit K, Hider RC, Webster HK: Flow cytometric assessment of hydroxypyridinone iron chelators on in vitro growth of drug-resistant malaria. Cytometry 27:84-91, 1997
3543. Barkan D, Ginsburg H, Golenser J: Optimisation of flow cytometric measurement of parasitaemia in Plasmodium-infected mice. Int J Parasitol 30:649-53, 2000
3544. Kumaratilake LM, Ferrante A: Opsonization and phagocytosis of Plasmodium falciparum merozoites measured by flow cytometry. Clin Diagn Lab Immunol 7:9-13, 2000
3544. Waitumbi JN, Opollo MO, Muga RO, Misore AO, Stoute JA: Red cell surface changes and erythrophagocytosis in children with severe plasmodium falciparum anemia. Blood 95:1481-6, 2000
3545. Stoute JA, Odindo AO, Owuor BO, Mibei EK, Opollo MO, Waitumbi JN: Loss of red blood cell-complement regulatory proteins and increased levels of circulating immune complexes are associated with severe malarial anemia. J Infect Dis 187:522-5, 2003
3546. Ballou WR, Kester KE, Stoute JA, Heppner DG: Malaria vaccines: Triumphs or tribulations? Parassitologia 41:403-8, 1999
3547. Hoffman SL, Goh LM, Luke TC, Schneider I, Le TP, Doolan DL, Sacci J, de la Vega P, Dowler M, Paul C, Gordon DM, Stoute JA, Church LW, Sedegah M, Heppner DG, Ballou WR, Richie TL: Protection of humans against malaria by immunization with radiation-attenuated plasmodium falciparum sporozoites. J Infect Dis 185:1155-64, 2002
3548. Levy SB: Active efflux mechanisms for antimicrobial resistance. Antimicrob Agents Chemother 36:695-703, 1992
3549. Nikaido H: Prevention of drug access to bacterial targets: permeability barriers and active efflux. Science 264: 382-8, 1994
3550. Shackney SE, Shankey TV: Cell cycle models for molecular biology and molecular oncology: exploring new dimensions. Cytometry 35:97-116, 1999
3551. Traganos F, Juan G, Darzynkiewicz Z: Cell-cycle analysis of drug-treated cells. Methods Mol Biol 95:229-40, 2001
3552. Poot M, Silber JR, Rabinovitch PS: A novel flow cytometric technique for drug cytotoxicity gives results comparable to colony-forming assays. Cytometry 48:1-5, 2002
3553. Bhatia U, Traganos F, Darzynkiewicz Z: Induction of cell differentiation potentiates apoptosis triggered by prior exposure to DNA-damaging drugs. Cell Growth Differ 6:937-44, 1995
3554. Deptala A, Li X, Bedner E, Cheng W, Traganos F, Darzynkiewicz Z: Differences in induction of p53, p21waf1 and apoptosis in relation to cell cycle phase of MCF-7 cells treated with camptothecin. Int J Oncol 15:861-71, 1999

3555. Darzynkiewicz Z, Juan G, Traganos F: Assaying drug-induced apoptosis. Methods Mol Biol 95:241-54, 2001

3556. Yurkow EJ, Makhijani PR: Flow cytometric determination of metallothionein levels in human peripheral blood lymphocytes: Utility in environmental exposure assessment. J Toxicol Environ Health A 54:445-57, 1998

3557. Burger J, Lord CG, Yurkow EJ, McGrath L, Gaines KF, Brisbin IL, Jr., Gochfeld M: Metals and metallothionein in the liver of raccoons: Utility for environmental assessment and monitoring. J Toxicol Environ Health A 60:243-61, 2000

3558. Grawé J, Zetterberg G, Amnéus H: Flow-cytometric enumeration of micronucleated polychromatic erythrocytes in mouse peripheral blood. Cytometry 13:750-8, 1992

3559. Grawé J, Abramsson-Zetterberg L, Zetterberg G: Low dose effects of chemicals as assessed by the flow cytometric in vivo micronucleus assay. Mutation. Res. 405:199-208, 1998

3560. Abramsson-Zetterberg L, Zetterberg G, Bergqvist M, Grawé J: Human cytogenetic biomonitoring using flow-cytometric analysis of micronuclei in transferrin-positive immature peripheral blood reticulocytes. Environ Mol Mutagen 36:22-31, 2000

3561. Tometsko AM, Dertinger SD, Torous DK: Analysis of micronucleated cells by flow cytometry. 4. Kinetic analysis of cytogenetic damage in blood. Mutat Res 334:9-18, 1995

3562. Dertinger SD, Torous DK, Tometsko KR: Simple and reliable enumeration of micronucleated reticulocytes with a single-laser flow cytometer. Mutat Res 371:283-92, 1996

3563. Dertinger SD, Torous DK, Tometsko KR: Flow cytometric analysis of micronucleated reticulocytes in mouse bone marrow. Mutat Res 390:257-62, 1997

3564. Torous DK, Dertinger SD, Hall NE, Tometsko CR: Enumeration of micronucleated reticulocytes in rat peripheral blood: A flow cytometric study. Mutat Res 465:91-9, 2000

3565. Dertinger SD, Torous DK, Hall NE, Tometsko CR, Gasiewicz TA: Malaria-infected erythrocytes serve as biological standards to ensure reliable and consistent scoring of micronucleated erythrocytes by flow cytometry. Mutat Res 464:195-200, 2000

3566. Weaver JL, Torous D: Flow cytometry assay for counting micronucleated erythrocytes: Development process. Methods 21:281-7, 2000

3567. Torous DK, Hall NE, Dertinger SD, Diehl MS, Illi-Love AH, Cederbrant K, Sandelin K, Bolcsfoldi G, Ferguson LR, Pearson A, Majeska JB, Tarca JP, Hewish DR, Doughty L, Fenech M, Weaver JL, Broud DD, Gatehouse DG, Hynes GM, Kwanyuen P, McLean J, McNamee JP, Parenteau M, Van Hoof V, Vanparys P, Lenarczyk M, Siennicka J, Litwinska B, Slowikowska MG, Harbach PR, Johnson CW, Zhao S, Aaron CS, Lynch AM, Marshall IC, Rodgers B, Tometsko CR: Flow cytometric enumeration of micronucleated reticulocytes: High transferability among 14 laboratories. Environ Mol Mutagen 38:59-68, 2001

3568. Dertinger SD, Torous DK, Hall NE, Murante FG, Gleason SE, Miller RK, Tometsko CR: Enumeration of micronucleated CD71-positive human reticulocytes with a single-laser flow cytometer. Mutat Res 515:3-14, 2002

3569. Taets C, Aref S, Rayburn AL: The Clastogenic Potential of Triazine Herbicide Combinations Found in Potable Water Supplies. Environ Health Perspect 106:197-201, 1998

3570. Styles JA, Clark H, Festing MF, Rew DA: Automation of mouse micronucleus genotoxicity assay by laser scanning cytometry. Cytometry 44:153-5, 2001

3571. Smolewski P, Ruan Q, Vellon L, Darzynkiewicz Z: Micronuclei assay by laser scanning cytometry. Cytometry 45:19-26, 2001

3572. Snopov SA, Berg RJ, van Weelden H, Samoilova KA, van der Leun JC, de Gruijl FR: Molecular dosimetry by flow cytometric detection of thymine dimers in mononuclear cells from extracorporally UV-irradiated blood. J Photochem Photobiol B 28:33-7, 1995

3573. Langlois RG, Bigbee WL, Jensen RH: Flow cytometric characterization of normal and variant cells with monoclonal antibodies specific for glycophorin A. J Immunol 134:4009-17, 1985

3574. Jensen RH, Langlois RG, Bigbee WL: Determination of somatic mutations in human erythrocytes by flow cytometry. Prog Clin Biol Res 209B:177-84, 1986

3575. Langlois RG, Nisbet BA, Bigbee WL, Ridinger DN, Jensen RH: An improved flow cytometric assay for somatic mutations at the glycophorin A locus in humans. Cytometry 11:513-21, 1990

3576. Langlois RG, Bigbee WL, Kyoizumi S, Nakamura N, Bean MA, Akiyama M, Jensen RH: Evidence for increased somatic cell mutations at the glycophorin A locus in atomic bomb survivors. Science 236:445-8, 1987

3577. Kyoizumi S, Akiyama M, Cologne JB, Tanabe K, Nakamura N, Awa AA, Hirai Y, Kusunoki Y, Umeki S: Somatic cell mutations at the glycophorin A locus in erythrocytes of atomic bomb survivors: Implications for radiation carcinogenesis. Radiat Res 146:43-52, 1996

3578. Jensen RH, Langlois RG, Bigbee WL, Grant SG, Moore D, 2nd, Pilinskaya M, Vorobtsova I, Pleshanov P: Elevated frequency of glycophorin A mutations in erythrocytes from Chernobyl accident victims. Radiat Res 141:129-35, 1995

3579. Jones IM, Galick H, Kato P, Langlois RG, Mendelsohn ML, Murphy GA, Pleshanov P, Ramsey MJ, Thomas CB, Tucker JD, Tureva L, Vorobtsova I, Nelson DO: Three somatic genetic biomarkers and covariates in radiation-exposed Russian cleanup workers of the Chernobyl nuclear reactor 6-13 years after exposure. Radiat Res 158:424-42, 2002

3580. Marti GE, Vogt RF, Zenger VE (eds): *Proceedings of the USPHS Workshop on Laboratory and Epidemiologic Approaches to Determining the Role of Environmental Exposures as Risk Factors for B-cell chronic lymphocytic leukemia and other B-cell lymphoproliferative disorders.* Atlanta, U. S. Department of Health and Human Services, 1997

3581. Marti GE, Carter P, Abbasi F, Washington GC, Jain N, Zenger VE, Ishibe N, Goldin L, Fontaine L, Weissman N, Sgambati M, Fauget G, Bertin P, Vogt RF, Jr., Slade B, Noguchi PD, Stetler-Stevenson MA, Caporaso N: B-cell monoclonal lymphocytosis and B-cell abnormalities in the setting of familial B-cell chronic lymphocytic leukemia. Cytometry 52B:1-12, 2003

3582. Slade BA: *Follow-up investigation of B-cell abnormalities identified in previous ATSDR health studies.* Publication PB99-138331. Bethesda, U. S. Department of Health and Human Services, Public Health Serrvice, 1999

3583. Pillai SR, Kunze E, Sordillo LM, Jayarao BM: Application of differential inflammatory cell count as a tool to monitor udder health. J Dairy Sci 84:1413-20, 2001

3584. Rivas AL, Quimby FW, Blue J, Coksaygan O: Longitudinal evaluation of bovine mammary gland health status by somatic cell counting, flow cytometry, and cytology. J Vet Diagn Invest 13:399-407, 2001

3585. D'Haese E, Nelis HJ, Reybroeck W: Determination of somatic cells in milk by solid phase cytometry. J Dairy Res 68:9-14, 2001

3586. Seo KH, Brackett RE, Frank JF: Rapid detection of escherichia coli O157:H7 using immuno-magnetic flow cytometry in ground beef, apple juice, and milk. Int J Food Microbiol 44:115-23, 1998

3587. Attfield PV, Kletsas S, Veal DA, van Rooijen R, Bell PJ: Use of flow cytometry to monitor cell damage and predict fermentation activity of dried yeasts. J Appl Microbiol 89:207-14, 2000

3588. Malacrino P, Zapparoli G, Torriani S, Dellaglio F: Rapid detection of viable yeasts and bacteria in wine by flow cytometry. J Microbiol Methods 45:127-34, 2001

3589. Graca da Silveira M, Vitoria San Romao M, Loureiro-Dias MC, Rombouts FM, Abee T: Flow cytometric assessment of membrane integrity of ethanol-stressed Oenococcus oeni cells. Appl Environ Microbiol 68:6087-93, 2002

3590. Ragione FD, Cucciolla V, Borriello A, Pietra VD, Racioppi L, Soldati G, Manna C, Galletti P, Zappia V: Resveratrol arrests the cell division cycle at S/G2 phase transition. Biochem Biophys Res Commun 250:53-8, 1998

3591. Hsieh TC, Juan G, Darzynkiewicz Z, Wu JM: Resveratrol increases nitric oxide synthase, induces accumulation of p53 and p21(WAF1/CIP1), and suppresses cultured bovine pulmonary artery endothelial cell proliferation by perturbing progression through S and G2. Cancer Res 59:2596-601, 1999
3592. Sgambato A, Ardito R, Faraglia B, Boninsegna A, Wolf FI, Cittadini A: Resveratrol, a natural phenolic compound, inhibits cell proliferation and prevents oxidative DNA damage. Mutat Res 496:171-80, 2001
3593. Surh YJ, Hurh YJ, Kang JY, Lee E, Kong G, Lee SJ: Resveratrol, an antioxidant present in red wine, induces apoptosis in human promyelocytic leukemia (HL-60) cells. Cancer Lett 140:1-10, 1999
3594. Carbo N, Costelli P, Baccino FM, Lopez-Soriano FJ, Argiles JM: Resveratrol, a natural product present in wine, decreases tumour growth in a rat tumour model. Biochem Biophys Res Commun 254:739-43, 1999
3595. Kampa M, Hatzoglou A, Notas G, Damianaki A, Bakogeorgou E, Gemetzi C, Kouroumalis E, Martin PM, Castanas E: Wine antioxidant polyphenols inhibit the proliferation of human prostate cancer cell lines. Nutr Cancer 37:223-33, 2000.
3596. Rotondo S, Rajtar G, Manarini S, Celardo A, Rotillo D, de Gaetano G, Evangelista V, Cerletti C: Effect of trans-resveratrol, a natural polyphenolic compound, on human polymorphonuclear leukocyte function. Br J Pharmacol 123:1691-9, 1998
3597. Falchetti R, Fuggetta MP, Lanzilli G, Tricarico M, Ravagnan G: Effects of resveratrol on human immune cell function. Life Sci 70:81-96, 2001
3598. Bunthof CJ, van Schalkwijk S, Meijer W, Abee T, Hugenholtz J: Fluorescent method for monitoring cheese starter permeabilization and lysis. Appl Environ Microbiol 67:4264-71, 2001
3599. Day JP, Kell DB, Griffith GW: Differentiation of Phytophthora infestans sporangia from other airborne biological particles by flow cytometry. Appl Environ Microbiol 68:37-45, 2002
3600. Rein D, Paglieroni TG, Pearson DA, Wun T, Schmitz HH, Gosselin R, Keen CL: Cocoa and wine polyphenols modulate platelet activation and function. J Nutr 130:2120S-6S, 2000
3601. Rein D, Paglieroni TG, Wun T, Pearson DA, Schmitz HH, Gosselin R, Keen CL: Cocoa inhibits platelet activation and function. Am J Clin Nutr 72:30-5, 2000
3602. Rieseberg M, Kasper C, Reardon KF, Scheper T: Flow cytometry in biotechnology. Appl Microbiol Biotechnol 56:350-60, 2001
3603. Reardon KF, Lee KH, Wittrup KD, Hatzimanikatis V: Jay Bailey as mentor--the students' perspective. Biotechnol Bioeng. 79:484-9, 2002
3604. Abu-Absi NR, Zamamiri A, Kacmar J, Balogh SJ, Srienc F: Automated flow cytometry for acquisition of time-dependent population data. Cytometry 51A:87-96, 2003
3605. Feldhaus MJ, Siegel RW, Opresko LK, Coleman JR, Feldhaus JM, Yeung YA, Cochran JR, Heinzelman P, Colby D, Swers J, Graff C, Wiley HS, Wittrup KD: Flow-cytometric isolation of human antibodies from a nonimmune Saccharomyces cerevisiae surface display library. Nat Biotechnol 21:163-70, 2003
3606. Foote J, Eisen HN: Breaking the affinity ceiling for antibodies and T cell receptors. Proc Natl Acad Sci U S A 97:10679-81, 2000
3607. Hewitt CJ, Nebe-von Caron G, Nienow AW, McFarlane CM: The use of multi-parameter flow cytometry to compare the physiological response of Escherichia coli W3110 to glucose limitation during batch, fed-batch and continuous culture cultivations. J Biotechnol 75:251-64, 1999
3608. Zhao R, Natarajan A, Srienc F: A flow injection flow cytometry system for on-line monitoring of bioreactors. Biotechnol Bioeng 62:609-17, 1999
3609. Muller S, Bley T, Babel W: Adaptive responses of Ralstonia eutropha to feast and famine conditions analysed by flow cytometry. J Biotechnol 75:81-97, 1999
3610. Shaw AD, Winson MK, Woodward AM, McGovern AC, Davey HM, Kaderbhai N, Broadhurst D, Gilbert RJ, Taylor J, Timmins EM, Goodacre R, Kell DB, Alsberg BK, Rowland JJ: Rapid analysis of high-dimensional bioprocesses using multivariate spectroscopies and advanced chemometrics. Adv Biochem Eng Biotechnol 66:83-113, 2000
3611. Hewitt CJ, Nebe-Von Caron G, Axelsson B, McFarlane CM, Nienow AW: Studies related to the scale-up of high-cell-density E. Coli fed-batch fermentations using multiparameter flow cytometry: Effect of a changing microenvironment with respect to glucose and dissolved oxygen concentration. Biotechnol Bioeng 70:381-90, 2000
3612. Harding CL, Lloyd DR, McFarlane CM, Al-Rubeai M: Using the microcyte flow cytometer to monitor cell number, viability, and apoptosis in mammalian cell culture. Biotechnol Prog 16:800-2, 2000
3613. Porro D, Venturini M, Brambilla L, Alberghina L, Vanoni M: Relating growth dynamics and glucoamylase excretion of individual saccharomyces cerevisiae cells. J Microbiol Methods 42:49-55, 2000 (in reference 2404; full text is available at www.elsevier.com/locate/jmicmeth)
3614. Muller S, Ullrich S, Losche A, Loffhagen N, Babel W: Flow cytometric techniques to characterise physiological states of Acinetobacter calcoaceticus. J Microbiol Methods 40:67-77, 2000
3615. Bunthof CJ, Bloemen K, Breeuwer P, Rombouts FM, Abee T: Flow cytometric assessment of viability of lactic acid bacteria. Appl Environ Microbiol 67:2326-35, 2001
3616. Vidal-Mas J, Resina P, Haba E, Comas J, Manresa A, Vives-Rego J: Rapid flow cytometry--Nile red assessment of pha cellular content and heterogeneity in cultures of pseudomonas aeruginosa 47t2 (NCIB 40044) grown in waste frying oil. Antonie Van Leeuwenhoek 80:57-63, 2001
3617. Amanullah A, Hewitt CJ, Nienow AW, Lee C, Chartrain M, Buckland BC, Drew SW, Woodley JM: Application of multi-parameter flow cytometry using fluorescent probes to study substrate toxicity in the indene bioconversion. Biotechnol Bioeng 80:239-49, 2002
3618. Bunthof CJ, Abee T: Development of a flow cytometric method to analyze subpopulations of bacteria in probiotic products and dairy starters. Appl Environ Microbiol 68:2934-42, 2002
3619. Abu-Absi NR, Srienc F: Instantaneous evaluation of mammalian cell culture growth rates through analysis of the mitotic index. J Biotechnol 95:63-84, 2002
3620. Patkar A, Vijayasankaran N, Urry DW, Srienc F: Flow cytometry as a useful tool for process development: Rapid evaluation of expression systems. J Biotechnol 93:217-29, 2002
3621. Boder ET, Midelfort KS, Wittrup KD: Directed evolution of antibody fragments with monovalent femtomolar antigen-binding affinity. Proc Natl Acad Sci U S A 97:10701-5, 2000
3622. Shusta EV, Holler PD, Kieke MC, Kranz DM, Wittrup KD: Directed evolution of a stable scaffold for T-cell receptor engineering. Nat Biotechnol 18:754-9, 2000
3623. Holler PD, Holman PO, Shusta EV, O'Herrin S, Wittrup KD, Kranz DM: In vitro evolution of a T cell receptor with high affinity for peptide/MHC. Proc Natl Acad Sci U S A 97:5387-92, 2000
3624. VanAntwerp JJ, Wittrup KD: Fine affinity discrimination by yeast surface display and flow cytometry. Biotechnol Prog 16:31-7, 2000
3625. Daugherty PS, Iverson BL, Georgiou G: Flow cytometric screening of cell-based libraries. J Immunol Methods 243:211-27, 2000
3626. Olsen M, Iverson B, Georgiou G: High-throughput screening of enzyme libraries. Curr Opin Biotechnol 11:331-7, 2000
3627. Wittrup KD: Protein engineering by cell-surface display. Curr Opin Biotechnol 12:395-9, 2001
3628. Kieke MC, Sundberg E, Shusta EV, Mariuzza RA, Wittrup KD, Kranz DM: High affinity T cell receptors from yeast display libraries block T cell activation by superantigens. J Mol Biol 307:1305-15, 2001
3629. Hayhurst A, Georgiou G: High-throughput antibody isolation. Curr Opin Chem Biol 5:683-9, 2001
3630. Levy R, Weiss R, Chen G, Iverson BL, Georgiou G: Production of correctly folded Fab antibody fragment in the cytoplasm of Es-

cherichia coli trxb gor mutants via the coexpression of molecular chaperones. Protein Expr Purif 23:338-47, 2001

3631. Chen G, Hayhurst A, Thomas JG, Harvey BR, Iverson BL, Georgiou G: Isolation of high-affinity ligand-binding proteins by periplasmic expression with cytometric screening (PECS). Nat Biotechnol 19:537-42, 2001
3632. Chen W, Georgiou G: Cell-surface display of heterologous proteins: From high-throughput screening to environmental applications. Biotechnol Bioeng 79:496-503, 2002
3633. Cohen N, Abramov S, Dror Y, Freeman A: In vitro enzyme evolution: The screening challenge of isolating the one in a million. Trends Biotechnol 19:507-10, 2001
3634. Yeung YA, Wittrup KD: Quantitative screening of yeast surface-displayed polypeptide libraries by magnetic bead capture. Biotechnol Prog 18:212-20, 2002
3635. Prausnitz MR, Corbett JD, Gimm JA, Golan DE, Langer R, Weaver JC: Millisecond measurement of transport during and after an electroporation pulse. Biophys J 68:1864-70, 1995
3636. Gift EA, Weaver JC: Observation of extremely heterogeneous electroporative molecular uptake by saccharomyces cerevisiae which changes with electric field pulse amplitude. Biochim Biophys Acta 1234:52-62, 1995
3637. Gift EA, Weaver JC: Simultaneous quantitative determination of electroporative molecular uptake and subsequent cell survival using gel microdrops and flow cytometry. Cytometry 39:243-9, 2000
3638. Rodriguez C, Lodish HF: Enhanced efficiency of cloning FACS-sorted mammalian cells. Biotechniques 24:750-2, 1998
3639. Katsuragi T, Tanaka S, Nagahiro S, Tani Y: Gel microdroplet technique leaving microorganisms alive for sorting by flow cytometry. J Microbiol Methods 42: 81-6, 2000 (in reference 2404; full text is available at www.elsevier.com/locate/jmicmeth)
3640. Zengler K, Toledo G, Rappe M, Elkins J, Mathur EJ, Short JM, Keller M: Cultivating the uncultured. Proc Natl Acad Sci U S A 99:15681-6, 2002
3641. Sabioncello A, Kocijan-Hercigonja D, Rabatic S, Tomasic J, Jeren T, Matijevic L, Rijavec M, Dekaris D: Immune, endocrine, and psychological responses in civilians displaced by war. Psychosom Med 62:502-8, 2000
3642. Skarpa I, Rubesa G, Moro L, Manestar D, Petrovecki M, Rukavina D: Changes of cytolytic cells and perforin expression in patients with posttraumatic stress disorder. Croat Med J 42:551-5, 2001
3643. Griffiths GD, Hornby RJ, Stevens DJ, Scott LA, Upshall DG: Biological consequences of multiple vaccine and pyridostigmine pretreatment in the guinea pig. J Appl Toxicol 21:59-68, 2001
3644. Biselli R, Farrace S, D'Amelio R, Fattorossi A: Influence of stress on lymphocyte subset distribution--a flow cytometric study in young student pilots. Aviat Space Environ Med 64:116-20, 1993
3645. Gruzelier J, Smith F, Nagy A, Henderson D: Cellular and humoral immunity, mood and exam stress: The influences of self-hypnosis and personality predictors. Int J Psychophysiol 42:55-71, 2001
3646. Ruzyla-Smith P, Barabasz A, Barabasz M, Warner D: Effects of hypnosis on the immune response: B-cells, T-cells, helper and suppressor cells. Am J Clin Hypn 38:71-9, 1995
3647. Mitchell JB, Dugas JP, McFarlin BK, Nelson MJ: Effect of exercise, heat stress, and hydration on immune cell number and function. Med Sci Sports Exerc 34:1941-50, 2002
3648. Hagler J, Trattner A, Nativ O, Hauben DJ, David M: Benign and malignant eccrine poroma--a flow cytometric comparison. Isr J Med Sci 32:1151-3, 1996
3649. Vähäsalo L, Degerth R, Holmbom B: Use of flow cytometry in wet end research. Paper Technology 44:45-9, 2003
3650. Meehan RT, Neale LS, Kraus ET, Stuart CA, Smith ML, Cintron NM, Sams CF: Alteration in human mononuclear leucocytes following space flight. Immunology 76:491-7, 1992
3651. Smolen JE, Fossett MC, Joe Y, Prince JE, Priest E, Kanwar S, Smith CW: Antiorthostatic suspension for 14 days does not diminish the oxidative response of neutrophils in mice. Aviat Space Environ Med 71:1239-47, 2000
3652. Crucian BE, Cubbage ML, Sams CF: Altered cytokine production by specific human peripheral blood cell subsets immediately following space flight. J Interferon Cytokine Res 20:547-56, 2000
3653. Mills PJ, Meck JV, Waters WW, D'Aunno D, Ziegler MG: Peripheral leukocyte subpopulations and catecholamine levels in astronauts as a function of mission duration. Psychosom Med 63:886-90, 2001
3654. Mehta SK, Kaur I, Grimm EA, Smid C, Feeback DL, Pierson DL: Decreased non-MHC-restricted (CD56+) killer cell cytotoxicity after spaceflight. J Appl Physiol 91:1814-8, 2001
3655. Pecaut MJ, Simske SJ, Fleshner M: Spaceflight induces changes in splenocyte subpopulations: Effectiveness of ground-based models. Am J Physiol Regul Integr Comp Physiol 279:R2072-8, 2000
3656. Crucian BE, Sams CF: The use of a spaceflight-compatible device to perform WBC surface marker staining and whole-blood mitogenic activation for cytokine detection by flow cytometry. J Gravit Physiol 6:P33-4, 1999
3657. Sams CF, Crucian BE, Clift VL, Meinelt EM: Development of a whole blood staining device for use during space shuttle flights. Cytometry 37:74-80, 1999
3658. Crucian B, Norman J, Brentz J, Pietrzyk R, Sams C: Laboratory outreach: Student assessment of flow cytometer fluidics in zero gravity. Lab Med 31:569-73, 2000
3659. Smolewski P, Bedner E, Gorczyca W, Darzynkiewicz Z: "liquidless" cell staining by dye diffusion from gels and analysis by laser scanning cytometry: Potential application at microgravity conditions in space. Cytometry 44:355-60, 2001
3660. McCarthy DA, Macey MG (eds): *Cytometric analysis of cell phenotype and function*. Cambridge (UK), Cambridge University Press, 2001, 430 pp
3661. Ferrari B,C, Attfield DA, Veal DA, Bell PJ: Application of the novel fluorescent dye Beljian red to the differentiation of Giardia cysts. J Microbiol Methods 52:133-6, 2003
3662. Dervan PB: Molecular recognition of DNA by small molecules. Bioorg Med Chem 9:2215-35, 2001
3663. Rucker VC, Foister S, Melander C, Dervan PB: Sequence specific fluorescence detection of double strand DNA. J Am Chem Soc 125:1195-202, 2003
3664. Suto RK, Edayathumangalam RS, White CL, Melander C, Gottesfeld JM, Dervan PB, Luger K: Crystal structures of nucleosome core particles in complex with minor groove DNA-binding ligands. J Mol Biol 326:371-80, 2003
3665. Mapp AK, Ansari AZ, Ptashne M, Dervan PB: Activation of gene expression by small molecule transcription factors. Proc Natl Acad Sci U S A 97:3930-5, 2000
3666. McLaughlin RW, Vali H, Lau PC, Palfree RG, De Ciccio A, Sirois M, Ahmad D, Villemur R, Desrosiers M, Chan EC: Are there naturally occurring pleomorphic bacteria in the blood of healthy humans? J Clin Microbiol 40:4771-5, 2002
3667. Nikkari S, McLaughlin IJ, Bi W, Dodge DE, Relman DA: Does blood of healthy subjects contain bacterial ribosomal DNA? J Clin Microbiol 39:1956-9, 2001
3668. Nikkari S, Lopez FA, Lepp PW, Cieslak PR, Ladd-Wilson S, Passaro D, Danila R, Relman DA: Broad-range bacterial detection and the analysis of unexplained death and critical illness. Emerg Infect Dis 8:188-94, 2002
3669. Kim HK, Song KS, Lee ES, Lee YJ, Park YS, Lee KR, Lee SN: Optimized flow cytometric assay for the measurement of platelet microparticles in plasma: Pre-analytic and analytic considerations. Blood Coagul Fibrinolysis 13:393-7, 2002
3670. Tocchetti EV, Flower RL, Lloyd JV: Assessment of in vitro-generated platelet microparticles using a modified flow cytometric strategy. Thromb Res 103:47-55, 2001
3671. Greve J: In memoriam: Bart de Grooth (1951-2001). Cytometry 47:i-ii, 2002

3672. Plaeger S: A tribute to Janis Giorgi, Ph. D. Cytometry (Commun Clin Cytom) 46:69-70, 2001
3673. Cohen J: Uninfectable. The New Yorker, July 6, 1998, pp 34-9
3674. Moldowan JM, Talyzina NM: Biogeochemical evidence for dinoflagellate ancestors in the Early Cambrian. Science 281:1168-70, 1998
3675. Talyzina NM, Moldowan JM, Johannison A, Fago FJ: Affinities of Early Cambrian acritarchs studied by using microscopy, fluorescence flow cytometry, and biomarkers. Review of Paleobotany and Palynology 108:37-53, 2000
3676. Hurtley SM, Helmuth L: The future looks bright... Science 300:75, 2003
3677. Stephens DJ, Allan VJ: Light microscopy techniques for live cell imaging. Science 300:82-6, 2003
3678. Scorrano L, Oakes SA, Opferman JT, Cheng EH, Sorcinelli MD, Pozzan T, Korsmeyer SJ: BAX and BAK regulation of endoplasmic reticulum Ca^{++}: A control point for apoptosis. Science 300:135-9, 2003.
3679. Valet G, Tárnok A: Cytomics in predictive medicine. Cytometry 53B:1-3, 2003
3680. Valet G, Repp R, Link H, Ehninger A, Gramatzki M, and SHG-SML study group: Pretherapeutic identification of high-risk acute myeloid leukemia (AML) patients from immunophenotypic, cytogenetic, and clinical parameters. Cytometry 53B:4-10, 2003
3681. Repp R, Schaekel U, Helm G, Thiede C, Soucek S, Pascheberg U, Wandt H, Aulitzky W, Bodenstein H, Sonnen R, Link H, Ehninger G, Gramatzki M, and SHG-SML study group: Immunophenotyping is an independent risk factor for risk stratification in AML. Cytometry 53B:11-19, 2003
3682. Stewart CC, Cookfair DL, Hovey KM, Wende KE, Bell DS, Warner CL: Predictive immunophenotypes: Disease-related profile in chronic fatigue syndrome. Cytometry 53B:26-33, 2003
3683. CrespoM, Bosch F, Villamor N, Bellosillo B, Colomer D, Rozman M, Marcé S, López-Gilllermo A, Campo E, Montserrat E: ZAP-70 expression as a surrogate for immunoglobulin-variable-region mutations in chronic lymphocytic leukemia. N Engl J Med 348:1764-75, 2003
3684. Rai KR, Chiorazzi N: Determining the clinical course and outcome in chronic lymphocytic leukemia. N Engl J Med 348:1797-9, 2003
3685. Gore CJ, Parisotto R, Ashenden MJ, Stray-Gundersen J, Sharpe K, Hopkins W, Emslie KR, Howe C, Trout GJ, Kazlauskas R, Hahn AG: Second-generation blood tests to detect erythropoietin abuse by athletes. Haematologica 88:333-44, 2003

INDEX

NOTE: Entries referring to pages containing relevant tables and figures are in *italic* type.

Encyclopaedic Companion to Medical Statistics

Encyclopaedic Companion to Medical Statistics

Second Edition

Edited by

Brian S. Everitt

Professor Emeritus, King's College, London, UK

and

Christopher R. Palmer

Director of the Centre for Applied Medical Statistics,

University of Cambridge, UK

With a Foreword by Richard Horton

A John Wiley and Sons, Ltd, Publication

This edition first published 2011

Registered office
John Wiley & Sons Ltd, The Atrium, Southern Gate, Chichester, West Sussex, PO19 8SQ, United Kingdom

For details of our global editorial offices, for customer services and for information about how to apply for permission to reuse the copyright material in this book please see our website at www.wiley.com.

Library of Congress Cataloging-in-Publication Data

The encyclopaedic companion to medical statistics / edited by Brian S. Everitt and Christopher R. Palmer ; with a foreword by Richard Horton. – 2nd ed.
p. ; cm.
Includes bibliographical references.
Summary: "The Encyclopaedic Companion to Medical Statistics, contains readable accounts of almost 400 statistical topics central to current medical research. Each entry has been written by an individual chosen for both their expertise in the field and their ability to communicate statistical concepts successfully to medical researchers. Real examples from the biomedical literature and relevant illustrations feature in many entries, and extensive cross–referencing signposts the reader to related entries"–Provided by publisher.
ISBN 978-0-470-68419-1 (h/b)
1. Medical statistics–Encyclopedias. I. Everitt, Brian. II. Palmer, Christopher Ralph.
[DNLM: 1. Statistics as Topic–methods–Encyclopedias–English. 2. Models, Theoretical–Encyclopedias–English. WA 13 E562 2010]
RA409.E527 2010
610.7203–dc22

2010018741

A catalogue record for this book is available from the British Library.

Print ISBN: 978-0-470-68419-1
ePDF ISBN: 978-0-470-66974-7

Set in 9/11pt in Times Roman by Thomson Digital, Noida, India
Printed in Singapore by Markono Print Media Pte Ltd

To Mary-Elizabeth
Brian S. Everitt

To Cathy-Joan, Laura, Carolyn and David
Christopher R. Palmer

Contents

Foreword

This encyclopaedia contains no entry for 'Peer review'. In my small corner of the medical statistical universe, this seems like a gross sin of omission. Instead, the process of evaluating research papers is discussed under 'Critical appraisal'. Are these two procedures synonymous? And, irrespective of whether they are or are not, should anybody care?

I believe that peer review and critical appraisal do differ, that these differences matter a great deal when considering the ways in which readers should interpret the medical literature, and that an understanding of these differences helps to place medical statistics in its proper context when surveying the wide horizon of clinical and public health research.

The editors of this quite wonderfully rewarding treatise on statistical terms have defined critical appraisal as 'the process of evaluating research reports and assessing their contribution to scientific knowledge'. This statement follows naturally from the meaning of the words 'criticism' (the art of judging) and 'appraisal' (the estimation of quality). That is to say, critical appraisal is an estimation of worth followed by some kind of judgment – a judgment that leans more towards an art than a science. As a non-statistician, I rather warm to the precise imprecision of this definition.

Now consider the more commonly embedded term 'peer review' and look how inferior it is! Who is this anonymous idealised peer? Generally, one would consider a peer to be an equal, somebody who comes from a group comparable to that from which the person under scrutiny has emerged. This intellectual egalitarian is subsequently set the task of viewing again (to take 'review' at its most literal meaning) the work under consideration. But to view with what purpose? None is specified.

Despite these practical shortcomings, editors of biomedical journals remain wedded to 'peer review'. We feel uncomfortable with the notion of critical appraisal. The embodiment of peer review as a distinct scientific discipline is the series of international congresses devoted to peer review in biomedical publication, organised jointly by *JAMA* and the *BMJ*. These congresses have spawned hundreds of abstracts, dozens of research papers, and four theme issues of *JAMA*. They are entirely commendable in every way. For the editors of *JAMA* and the *BMJ*, peer review encompasses a broad range of activities: mechanisms of editorial decision making, together with their quality, validity, and practicality, online peer review and publication, pre-publication posting of information, quality assurance of reviewers and editors, authorship and contributorship, conflicts of interest, scientific misconduct, peer review of grant proposals, economic aspects of peer review, and the future of scientific publication.

In other words, peer review is a tremendously elastic concept, allowing editors to stretch it to mean whatever interests them at a given (whimsical) moment in time and place. Indeed, its elasticity is seen by many of us as its great strength. The concept grows in richness and understanding as our own appreciation of its complexity and nuance soars. The impenetrable nature of peer review, and the obscure and hard-to-learn expertise it demands, feeds our brittle egos. The notion of critical appraisal, by contrast, is far thinner in meaning, with much less room for editorial manipulation and aggrandisement.

Even if peer review and critical appraisal do differ, should anyone actually care? Yes, they should, and for a very simple reason: the idea of peer review is now bankrupt. Its retention as an operation within the biomedical sciences reflects the interests of those who wish to preserve their own power and position. Peer review is fundamentally anti-democratic. It elevates the mediocre. It asphyxiates originality and it kills careers. How so?

Peer review is not about intelligent engagement with a piece of research. It is about defining the margins of what is acceptable and unacceptable to the reviewer. The mythical 'peer' is being asked to view again, after the editor, the work in question and to offer a comment about the geographical location of that work on the map of existing knowledge. If there is space on this map, and provided the work does not disrupt (too much) the terrain established by others, its location can be secured and marked by sanctioning publication. If the disruption is too great, the work's wish to seek a place of rest must be vetoed. Peer review is about the agency of power to preserve established orthodoxy. It has nothing to do with science. It has everything to do with ideology – and the maintenance of a quiet life of privilege and mystique.

Instead, critical appraisal is about incrementally working one's way towards truth[1]. It can never be about truth itself. The essence of biomedical research is estimation. Our world resists certainty. Critical appraisal is about transparent, measurable analysis that cuts a path towards greater precision.

1. Horton, R. 2002: Postpublication criticism and the shaping of clinical knowledge. *JAMA* 287, 2843–7.

Critical appraisal refuses to veil itself in the gaudy adornments that editors pin to peer review in order to embellish their own importance in the cartography of scientific inquiry. A far more robust instrument critical appraisal is for that refusal.

What do these differences tell us about the proper place of medical statistics in biomedicine today? In my view, as a lapsed doctor and a now wrinkled editor, medical statistics is the most important aspect of our critical appraisal of any piece of new research. The evaluations by so-called peers in the clinical specialties that concern a particular research paper provide valuable insight into how that work will be received by a community of practitioners or scholars. However, as an editor I am less interested in reception than I am in meaning[2].

I want a tough interrogation of new work before its publication, according to commonly agreed standards of questioning – standards that I can see and evaluate for myself. To return to my personal definition of critical appraisal, I want an estimation of quality combined with a judgment. I do not want a view from the club culture of one particular academic discipline. The rejection of peer review by the editors of this encyclopaedia is therefore a triumph of liberty against the forces of conformity.

Yet still today, too much of medicine takes medical statistics for granted. Time and again, we see research that has clearly not been within a hundred miles of a statistical brain. Physicians usually make poor scientists, and physicians and scientists together too often play the part of amateur statistician – with appalling consequences. The future of a successful biomedical research enterprise depends on the flourishing of the discipline we call medical statistics. It is not at all clear to me that those who so depend on medical statistics appreciate either that dependence or the fragility of its foundation.

If this magnificent encyclopaedia can be deployed in the ongoing argument about the future of twenty-first century academic medicine, then not only the research enterprise but also the public's health and well-being will be far stronger tomorrow than it is today.

Richard Horton
Editor, *Lancet*

2. Horton, R. 2000: Common sense and figures: the rhetoric of validity in medicine. *Satist Med* 19, 3149–64.

Preface to the Second Edition

In this second edition of the *Encyclopaedic Companion to Medical Statistics* there are over 30 completely new entries and the majority of entries from the first edition from 2005 have been revised and updated. The aim of this new edition remains the same as before and that is, quintessentially, to aid communication between medical researchers and statisticians.

We hope this aim has been met by providing fully cross-referenced articles encompassing a wide range of statistical topics likely to be encountered in today's medical literature, of suitable breadth and depth to give sufficient detail, but not to overwhelm with too much technical background and to provide helpful further references where needed for those wishing to explore topics more deeply. We believe a key strength of this single-volume reference work also remains from the first edition, namely the accessibility of the articles. Their readability is enhanced because contributors are not only experts in their respective fields but also adept at communicating statistical concepts to non-statisticians, including those who may admit to having a certain fear of handling data and not knowing how to deal with all the numbers arising from their medical research!

Another aspect of the aim to enhance communication between medical and statistical disciplines concerns encouraging timeliness of seeking statistical advice. It is our hope too that this *Encyclopaedic Companion* will serve to encourage medical researchers to consult with statisticians at the earliest opportunity within the life cycle of a research project. Relevant entries herein might be read before, and reviewed after, such consultations to enable clearer understanding and, ultimately, help facilitate better quality medical research.

Once again our thanks are due to a large number of people: the contributors for their sterling efforts in producing such excellent entries, the team at Wiley, in particular Richard Davies and Heather Kay and, of course, to our families.

Brian S. Everitt, Dulwich, London
Christopher R. Palmer, Cambridge, UK

Preface

Statistical science plays an important role in medical research. Indeed a major part of the key to the progress in medicine from the 17th century to the present day has been the collection and valid interpretation of evidence, particularly quantitative evidence, provided by the application of statistical methods to medical investigations. Current medical journals are full of statistical material, both relatively simple (for example, t-tests, p-values, linear regression) and, increasingly, more complex (for example, generalised estimating equations, cluster analysis, Bayesian methods). The latter material reflects the vibrant state of statistical research with many new methods having practical implications for medicine being developed in the last two decades or so. But why is statistics important in medicine? Some possible answers are:

(1) Medical practice and medical research generate large amounts of data. Such data are generally full of uncertainty and variation, and extracting the 'signal' from the 'noise' is usually not trivial.
(2) Medicine involves asking questions that have strong statistical overtones. How common is the disease? Who is especially likely to contract a particular condition? What are the chances that a patient diagnosed with breast cancer will survive more than five years?
(3) The evaluation of competing treatments or preventative measures relies heavily on statistical concepts in both the design and analysis phase.

Recognition of the importance of statistics in medicine has increased considerably in recent years. The last decade, in particular, has seen the emergence of evidence-based medicine, and with it the need for clinicians to keep one step ahead of their patients, many of whom nowadays have access to virtually unlimited information (much of it being virtual, yet some of it being limited in its reliability). Compared with previous generations of medical students, today's pre-clinical undergraduates are being taught more about statistical principles than their predecessors. Furthermore, today's clinical researchers are faced (happily, in our view) with growing numbers of biomedical journals utilising statistical referees as part of their peer review processes (see CRITICAL APPRAISAL and STATISTICAL REFEREEING). This enhances the quality of the papers journal editors select, although from the clinical researcher's perspective it has made publication in leading journals more challenging than ever before.

So statistics is (and are) prevalent in the medical world now and is set to remain so for the future. Clearly, clinicians and medical researchers need to know something about the subject, even if only to make their discussion with a friendly statistician more fruitful. The article on consulting a statistician quotes one of the forefathers of modern statistics, R.A. Fisher who, back in 1938, observed wryly: '*To consult the statistician after an experiment is finished is often merely to ask him to conduct a post-mortem examination. He can perhaps say what the experiment died of.*' Thus, one of our hopes for the usefulness and helpfulness of the *Encyclopaedic Companion to Medical Statistics* is that it may serve to encourage both productive and timely interactions between medical researchers and statisticians. Another sincere hope is that it fills a gap between, on the one hand, textbooks that delve into possibly too much theory and, on the other hand, shorter dictionaries that may not necessarily focus on the needs of medical researchers, or else have entries that are tantalisingly succinct. To meet these ends, the present reference work contains concise, informative, relatively non-technical, and hence, we trust, readable accounts of over 350 topics central to modern medical statistics.

Topics are covered either briefly or more extensively, in general, in accordance with the subject matter's perceived importance, although we acknowledge there will be disagreement, inevitably, about our choice of article lengths. Many entries benefit from containing real-life, clinical examples. Each has been written by an individual chosen not only for subject-matter expertise in the field but, just as importantly, also by ability to communicate statistical concepts to others.

The extensive cross-referencing supplied using SMALL CAPITALS to indicate terms that appear as separate entries should help the reader to find his or her way around and also serves to point out associated topics that might be of interest elsewhere within the *Encyclopaedic Companion*. All but the shortest entries contain references to further resources where the interested reader can learn in greater depth about the particular topic.

Thus, while hoping this work is found to be mostly comprehensible we do not claim it to be fully comprehensive. As co-editors we take joint responsibility for any errors ('sins of commission') and would positively welcome suggestions for possible new topics to consider for future inclusion to rectify perceived missing entries ('sins of omission').

Our thanks are due to numerous people – first, to all of the many contributors for providing such excellent material, mostly on time (mostly!) with particular gratitude extended to those who contributed multiple articles or who handled requests for additional articles so gracefully. Next, we appreciated the tremendous and indispensable efforts of staff at Arnold, especially Liz Gooster and Liz Wilson, and not least for their remaining calm during an editor's moments of anxiety and neurosis about the entire project. In addition we would like to thank Harriet Meteyard for her constant support and encouragement throughout the preparation of this book. Finally, our family members deserve especial thanks for having been extra tolerant of our time spent on developing and executing this extensive project from beginning to end. It is our hope that the *Encyclopaedic Companion* proves all these efforts and sacrifices to be well worthwhile, becoming a useful, regularly-thumbed reference added to the bookshelf of many of those involved in contemplating, conducting or contributing to medical research.

Brian S. Everitt and Christopher R. Palmer
January 2005

Biographical Information on the Editors

Brian S. Everitt – Professor Emeritus, King's College London. After 35 years at the Institute of Psychiatry, University of London, Brian Everitt retired in May 2004. Author of approximately 100 journal articles and over 50 books on statistics, and also co-editor of *Statistical Methods in Medical Research*. Writing continues apace in retirement but now punctuated by tennis, walks in the country, guitar playing and visits to the gym, rather than by committees, committees and more committees.

Christopher R. Palmer, founding Director of Cambridge University's Centre for Applied Medical Statistics, regularly teaches and collaborates with current and future doctors. His first degree was from Oxford, while graduate and postdoctoral studies were in the USA (at UNC-Chapel Hill and Harvard). He has shifted from mathematical towards applied statistics, with particular interest in the ethics of clinical trials and the use of flexible designs whenever appropriate. Fundamentally, he likes to promote sound statistical thinking in all areas of medical research and hopes this volume might help towards that end. Chris served as Deputy or Acting Editor for *Statistics in Medicine*, 1996–2000, and is a long-standing statistical reviewer for *The Lancet*. He and his wife have three children they consider to be more than statistically significant.

List of Contributors

K. R. Abrams (KRA), Centre for Biostatistics and Genetic, Epidemiology, Department of Health Sciences, University of Leicester, Leicester, LE1 7RH, UK

Colin Baigent (CB), Clinical Trial Service Unit and Epidemiological Studies Unit (CTSU), Richard Doll Building, Old Road Campus, Roosevelt Drive, Oxford OX3 7LF, UK

Alun Bedding (AB), Quantitative Sciences, GlaxoSmith-Kline, Medicines Research Centre, Gunnels Wood Road, Stevenage, Hertfordshire SG1 2NY, UK

Tijl De Bie (TDB), ISIS Research Group, Building 1, University of Southampton, Southampton SO17 1BJ, UK

Kathe Bjork (KB), Primetrics, Inc. Arvada, Colorado, USA (kathe@primetrics.net)

J. Martin Bland (JMB), Professor of Health Statistics, Department of Health Sciences, University of York, Heslington, York YO10 5DD, UK

Matteo Bottai (MtB), Division of Biostatics, Arnold School of Public Health, University of South Carolina, 800 Sumter Street, Columbia, SC 29208, USA, and also Unit of Biostatistics, Institute of Environmental Medicine, Karolinska Institutet Nobels väg 13, Stockholm, Sweden

Michelle Bradley (MMB), Health Information and Quality Authority, George's Court, George's lane, Dublin 7, Ireland

Sara Brookes (SB), Department of Social Medicine, University of Bristol, Canynge Hall, Whiteladies Road, Clifton, Bristol BS8 2PR, UK

Marc Buyse (MB), International Drug Development Institute (IDDI), 30 avenue provinciale, 1340 Louvain-la-Neuve, Belgium

M.J. Campbell (MJC), School of Health and Related Research, University of Sheffield, Regent Court, 30 Regent Street, Sheffield S1 4DA, UK

James R. Carpenter (JRC), Medical Statistics Unit, London School of Hygiene and Tropical Medicine, Keppel Street, London WC1E 7HT, UK

Lucy M. Carpenter (LMC), Department of Public Health, University of Oxford, and Nuffield College, Oxford OX1 1NF, UK

Susan Chinn (SC), Respiratory Epidemiology and Public Health, Imperial College, Emmanuel Kaye Building, Manresa Road, London SW3 6LR, UK

Tim Cole (TJC), MRC Centre of Epidemiology for Child Health, UCL Institute of Child Health, 30 Guilford Street, London WC1N 1EH, UK

Chris Corcoran (CCo), Department of Mathematics and Statistics, Utah State University, Logan, UT 84322-3900, USA

Nello Cristianini (NC), UC Davis Department of Statistics, 360 Kerr Hall, One Shields Avenue, Davis, CA 95616, USA

Sarah Crozier (SRC), MRC Lifecourse Epidemiology Unit, University of Southampton, Southampton General Hospital, Southampton SO16 6YD, UK

Carole Cummins (CLC), The University of Birmingham, Department of Public Health, Epidemiology and Biotatistics, 90 Vincent Drive, Edgbaston, Birmingham, B15 2TH

George Davey-Smith (GDS), School of Social and Community Medicine, University of Bristol, Oakfield House, Oakfield Grove, Bristol BS8 2BN, UK

Simon Day (SD), Roche Products Limited, Welwyn Garden City, Hertfordshire. AL7 1TW, UK.

Daniela De Angelis (DDA), Statistics, Modelling and Economics Department, Health Protection Agency, Centre for Infections, London and MRC Biostatistics Unit, Institute of Public Health, University Forvie Site, Robinson Way, Cambridge CB2 0SR, UK

Jonathan Deeks (JD), Public Health, Epidemiology and Biostatistics, University of Birmingham, Edgbaston, Birmingham B15 2TT, UK

Graham Dunn (GD), Health Sciences Research Group, School of Community Based Medicine, University of Manchester, Jean McFarlane Building, Oxford Road, Manchester M13 9PL, UK

Doug Easton (DE), Department of Public Health and Primary Care, University of Cambridge, Strangeways Research Laboratory, Worts Causeway, Cambridge CB1 8RN, UK

Jonathan Emberson (JE), Clinical Trial Service Unit and Epidemiological Studies Unit (CTSU), Richard Doll Building, Old Road Campus, Roosevelt Drive, Oxford OX3 7LF, UK

Richard Emsley (RE), Health Sciences Research Group, School of Community Based Medicine, University of Manchester, Jean McFarlane Building, Oxford Road, Manchester M13 9PL, UK

Brian S. Everitt (BSE), Biostatistics Department, Institute of Psychiatry, Denmark Hill, London SE5 8AF, UK

David Faraggi (DF), Department of Statistics, University of Haifa, Haifa 31905, Israel

W. Harper Gilmour (WHG), Section for Public Health and Health Policy, Division of Community Based Sciences, University of Glasgow, Glasgow G12 8RZ, UK

Els Goetghebeur (EG), Department of Applied Mathematics and Statistics, Ghent University, Krijgslaan 281-S9, 9000 Ghent, Belgium

Andrew Grieve (AG), Division of Health & Social Care Research, Department of Primary Care and Public Health Sciences, School of Medicine, King's College, Floor 7, Capital House, 42 Weston St, London SE1 3QD, UK

Julian P. T. Higgins (JPTH), MRC Biostatistics Unit, Institute of Public Health, University Forvie Site, Robinson Way, Cambridge CB2 0SR, UK

Theodore R. Holford (TRH), Division of Biostatistics, Yale School of Public Health, Yale, New Haven, CT 06520, USA

Sally Hollis (SH), AstraZeneca, Parklands, Alderley Park, Macclesfield, Cheshire SK10 4TF, UK

Torsten Hothorn (TH), Institut für Statistik, Ludwig-Maximilians-Universität München, Ludwigstrasse 33, DE-80539 München, Germany

Hazel Inskip (HI), MRC Lifecourse Epidemiology Unit, University of Southampton, Southampton General Hospital, Southampton SO16 6YD, UK

Tony Johnson (TJ), MRC Biostatistics Unit, Institute of Public Health, University Forvie Site, Robinson Way, Cambridge CB2 0SR, UK & MRC Clinical Trials Unit, 222 Euston Road, London, NW1 2DA

Karen Kafadar (KKa), Department of Mathematics, University of Colorado at Denver, PO Box 173364, Campus Box 170, Denver, CO 80217-3364, USA

Kyungmann Kim (KK), Department of Biostatistics and Medical Informatics, University of Wisconsin Medical School, 600 Highland Ave., Madison, WI 53792-4675, USA

Ruth King (RK), School of Mathematics and Statistics, Mathematical Institute, University of St Andrews, Fife KY16 9SS, UK

Wojtek Krzanowski (WK), School of Engineering, Mathematics and Physical Science, University of Exeter, Harrison Building, North Park Road, Exeter EX4 4QF, UK

Ranjit Lall (RL), Warwick Emergency Care and Rehabilitation, Division of Health in the Community, Warwick Medical School, University of Warwick, The Farmhouse, Gibbet Hill Campus, Coventry CV4 7AL, UK

Sabine Landau (SL), Biostatistics Department, Institute of Psychiatry, King's College, Denmark Hill, London SE5 8AF, UK

Andrew B. Lawson (AL), Division of Biostatistics and Epidemiology, College of Medicine, Medical University of South Carolina, Charleston, SC 29425, USA

Morven Leese (ML), Health Service and Population Research Department, Institute of Psychiatry, King's College, Denmark Hill, London SE5 8AF, UK

Andy Lynch (AGL), Department of Oncology, University of Cambridge, Li Ka Shing Centre, Robinson Way, Cambridge, CB2 0RE, UK

Cyrus Mehta (CM), President, Cytel Software Corporation, 675 Massachusetts Avenue, Cambridge, MA 02139, USA

Richard Morris (RM), Department of Primary Care and Population Health, UCL Medical School, Royal Free Campus, London NW3 2PF, UK

Paul Murrell (PM), Department of Statistics, The University of Auckland, Private Bag 92019, Auckland, New Zealand

Christopher R. Palmer (CRP), Department of Public Health and Primary Care, Institute of Public Health, University Forvie Site, Robinson Way, Cambridge CB2 0SR, UK

Max Parmar (MP), MRC Clinical Trials Unit, 222 Euston Road, London NW1 2DA, UK

Nitin Patel (NP), Cytel Software Corporation, 675 Massachusetts Avenue, Cambridge, MA 02139-3309, USA

John Powles (JP), Department of Public Health and Primary Care, Institute of Public Health, University Forvie Site, Robinson Way, Cambridge CB2 0SR, UK

P. Prescott (PP), Faculty of Mathematical Studies, University of Southampton, Southampton SO17 1BJ, UK

Sophia Rabe-Hesketh (SRH), Graduate School of Education and Graduate Group in Biostatistics, University of California, Berkeley, 3659 Tolman Hall, California 94720, USA and Institute of Education, University of London

Ben Reiser (BR), Department of Statistics, University of Haifa, Haifa 31905, Israel

Shaun Seaman (SRS), MRC Biostatistics Unit, Institute of Public Health, University Forvie Site, Robinson Way, Cambridge CB2 0SR, UK

Mark Segal (MRS), Division of Biostatistics, University of California, 185 Berry Street, Suite 5700, San Francisco, CA 94107, USA

Pralay Senchaudhuri (PSe), Cytel Software Corporation, 675 Massachusetts Avenue, Cambridge, MA 02139-3309, USA

Stephen Senn (SS), Department of Statistics, The University of Glasgow, Glasgow G12 8QQ, UK

Pak Sham (PS), Department of Psychiatry, The University of Hong Kong, Queen Mary Hospital, 102 Pokfulam Rd, Hong Kong

Charlie Sharp (CS), Computing Department, Institute of Psychiatry, Denmark Hill, London SE5 8AF, UK

Arvid Sjolander (AJS), Department of Medical Epidemiology and Biostatistics, Karolinska Institutet, Nobels Väg 12A, 171 77 Stockholm, Sweden

Anders Skrondal (AS), Division of Epidemiology, Norwegian Institute of Public Health, PO Box 4404 Nydalen, N-0403, Oslo, Norway

Nigel Smeeton (NCS), King's College London, Department of Primary Care and Public Health Sciences, Division of Health and Social Care Research, 7th Floor Capital House, 42 Weston Street, London SE1 3QD, UK

Nigel Stallard (NS), Warwick Medical School, University of Warwick, Coventry CV4 7AL, UK

Jonathan Sterne (JS), School of Social and Community Medicine, University of Bristol, Canynge Hall, 39 Whatley Road, Bristol BS8 2PS, UK

Elisabeth Svensson (ES), Swedish Business School/ Statistics, Örebro University, Örebro, Sweden

Matthew Sydes (MS), MRC Clinical Trials Unit, 222 Euston Road, London NW1 2DA, UK

Jeremy Taylor (JMGT), Department of Biostatistics, University of Michigan, 1420 Washington Heights, Ann Arbor, MI 48109-2029, USA

Kate Tilling (KT), School of Social and Community Medicine, University of Bristol, Canynge Hall, 39 Whatley Road, Bristol BS8 2PS, UK

Brian Tom (BT), MRC Biostatistics Unit, Institute of Public Health, University Forvie Site, Robinson Way, Cambridge CB2 0SR, UK

Rebecca Turner (RT), MRC Biostatistics Unit, Institute of Public Health, University Forvie Site, Robinson Way, Cambridge CB2 0SR, UK

Andy Vail (AV), Biostatistics Group, University of Manchester, Oxford Road, Manchester M13 9PL, UK

Stijn Vansteelandt (SV), Ghent University, Dept. of Applied Mathematics and Computer Science, Krijgslaan 281, S9, B-9000 Ghent, Belgium

Sarah L. Vowler (SLV), Bioinformatics Core, Cancer Research UK, Cambridge Research Institute, Robinson Way, Cambridge CB2 0RE, UK

Stephen J. Walters (SJW), Medical Statistics Group, School of Health and Related Research, University of Sheffield, Regent Court, 30 Regent Street, Sheffield S1 4DA, UK

J.G. Wheeler (JGW), Quanticate Ltd, Bevan House, 9-11 Bancroft Court, Hitchin, Herts, SG5 1LH, UK

Brandon Whitcher (BW), Clinical Imaging Centre, GlaxoSmithKline, Hammersmith Hospital, Du Cane Road, London W12 0HS, UK

Ian White (IW), MRC Biostatistics Unit, Institute of Public Health, University Forvie Site, Robinson Way, Cambridge CB2 0SR, UK

Janet Wittes (JW), 1710 Rhode Island Ave. NW, Suite 200, Washington DC 20036, USA

Mark Woodward (MW), The George Institute for International Health, PO Box M201, Missenden Road, Sydney NSW 2050, Australia

Ru-Fang Yeh (RFY), University of California San Francisco, Campus Box Number 0560, 500 Parnassus, 420 MU-W, San Francisco, CA 94143-0560, USA

Abbreviations and Acronyms

ACES	Active control equivalence study
ACET	Active control equivalence test
AD	Adaptive design
AI	Artificial intelligence
AIC	Akaike's information criterion
ANCOVA	Analysis of covariance
ANOVA	Analysis of variance
AR	Autoregressive
ARMA	Autoregressive moving average
AUC	Area under curve
BIC	Bayesian information criterion
BUGS	Bayesian inference Using Gibbs Sampling (software)
CACE	Complier average causal effect
CART	Classification and regression tree
CAT	Computer-adaptive testing
CBA	Cost-benefit analysis
CEA	Cost-effectiveness analysis
CI	Confidence interval
CONSORT	Consolidation of standards of reporting trials
COREC	Central Office for Research Ethics Committees
CPMP	Committee for Proprietary Medicinal Products
CPO	Conditional predictive ordinate
CrI	Credible interval
CRM	Continual reassessment method
CSM	Committee on Safety of Medicines
CUE	Cost-utility analysis
CV	Coefficient of variation
CWT	Continuous wavelet transform
DAG	Directed acyclic graph
DALY	Disability adjusted life-year
DAR	Dropout at random
DCAR	Dropout completely at random
DDD	Data-dependent design
DE	Design effect
DoF	Degrees of freedom
DIC	Deviance information criterion
DM	Data mining
DMC	Data monitoring committee
DSMC	Data and safety monitoring committee
DWT	Discrete wavelet transform
DZ	Dizygotic
EBM	Evidence-based medicine
EDA	Exploratory data analysis
EM	Expectation-maximisation
EMEA	European Medicines Evaluation Agency
FDA	Food and Drug Administration
GAM	Generalised additive model
GEE	Generalised estimating equations
GFR	General fertility rate
GIS	Geographical information system
GLIM	Generalised linear interactive modelling (software)
GLIMM	Generalised linear mixed model
GLM	Generalised linear model
GLMM	Generalised linear mixed model
GRR	Gross reproduction rate
GWAS	Genome-wide association studies
HALE	Health-adjusted life expectancy
HMM	Hidden Markov model
HPDI	Highest posterior density interval
HREC	Human research ethics committee
HRQoL	Health-related quality of life
IBD	Identity-by-descent
ICC	Intraclass (or intracluster) correlation coefficient
ICER	Incremental cost-effectiveness ratio
ICH	International Conference on Harmonization
IRB	Institutional review board
ITT	Intention-to-treat
IV	Instrumental variable
KDD	Knowledge discovery in databases
KM	Kaplan–Meier
kNN	k-nearest neighbour
LDF	Linear discriminant function
LR	Likelihood ratio
LREC	Local research ethics committee
LS	Least squares
LST	Large simple trial
MA	Moving average
MANOVA	Multivariate analysis of variance
MAR	Missing at random
MCA	Medicines Control Agency
MCAR	Missing completely at random

MCMC	Markov chain Monte Carlo
MHRA	Medicines and Healthcare Products Regulatory Agency
MLE	Maximum likelihood estimate (or estimation)
MREC	Multicentre research ethics committee
MSE	Mean square error
MTD	Maximum tolerated dose
MZ	Monozygotic
NI	Nonignorable (or noninformative)
NMB	Net monetary benefit
NNH	Number needed to harm
NNT	Number needed to treat
NPV	Negative predictive value
NRES	National Research Ethics Service
NRR	Net reproduction rate
OLS	Ordinary least squares
OR	Odds ratio
PCA	Principal component analysis
PDF	Probability density function
PEST	Planning and Evaluation of Sequential Trials (software)
PGM	Patient generated measure
PH	Proportional hazards
PK/PD	Pharmacokinetics/pharmacodynamics
POP	Persuade-the-optimist probability
PP	Per protocol
P-P	Percentile–percentile
PPP	Persuade-the-pessimist probability
PPV	Positive predictive value
QALY	Quality adjusted life-year
QoL	Quality of life
Q-Q	Quantile–quantile
QTL	Quantitative trait loci
RCT	Randomised controlled trial
REB	Research ethics board
REC	Research ethics committee
REML	Restricted maximum likelihood
ROC	Receiver operating characteristic
ROI	Region of interest
RPW	Randomised play-the-winner
RR	Relative risk
SD	Standard deviation
SE	Standard error
SEM	Standard error of the mean; structural equation model
SMR	Standardised mortality ratio
SPM	Statistical parametric map
SPRT	Sequential probability ratio test
SS	Sum of squares
SSE	Sum of squares due to error
SVM	Support vector machine
TDT	Transmission distortion test
TFR	Total fertility rate
TSM	Tree-structured method
TT	Triangular test
VAS	Visual analogue scale
WLSE	Weighted least squares estimate (or estimation)

A

accelerated factor See SURVIVAL ANALYSIS

accelerated failure time models See SURVIVAL ANALYSIS, TRANSFORMATION

active control equivalence studies The classic randomised CLINICAL TRIAL seeks to prove superiority of a new treatment to an existing one and a successful conclusion is one in which such proof is demonstrated. The famous MRC trial of streptomycin is a case in point (Medical Research Council Streptomycin in Tuberculosis Trials Committee, 1948). The trial concluded with a significant difference in outcome in favour of the group given streptomycin compared to the group that was not. In recent years, however, there has been an increasing interest in trials whose objective is to show that some new therapy is no worse as regards some outcome than an existing treatment. Such trials have particular features and difficulties that were described in an important paper by Makuch and Johnson (1989) in which they used the term 'active control equivalence studies' (ACES).

Actually, the term is not ideally chosen since, unlike bioequivalence studies, where the object is to show that the bioavailability of a new formulation is not only at least 20 % *less* than that of an existing formulation, but also at most 25 % *more*, and hence where *equivalence* to some degree is genuinely the aim, in ACES it is almost always the case that only noninferiority is the goal. It may be questioned as to why the rather modest goal of noninferiority should be of any interest in drug regulation. There are several reasons. The first is that the new drug may have advantages in terms of tolerability. Second, the new drug, while showing no net advantage to the existing one, may increase patient choice and this can be useful. For example, many people have an aspirin allergy. Hence, it is desirable to have alternative analgesics, even if no better on average than aspirin. Third, it may become necessary to withdraw treatments from the market and one can never predict when this may happen. There are now several statins on the market. The fact that this is so means that withdrawal of cerivastatin does not make it impossible for physicians to continue to treat their patients with this class of drug. Fourth, introduction of further equivalent therapies before patent expiry of an innovator in the class may permit price competition to the advantage of reimbursors (although such competition is probably not particularly effective; Senn and Rosati, 2003). However, the fifth reason is probably the most important. Drug regulation is designed to satisfy some minimum requirements for pharmaceuticals: that they are of sufficient quality, are safe and efficacious. Efficacy is demonstrated if the treatment is better than placebo, even if it is not as good as some other treatments. The comparison of a new drug to an active treatment may be dictated by ethics but the object of the trial may simply be an indirect proof that the treatment is better than placebo through comparison to an agent whose efficacy is accepted.

Recently the issue of the indirect comparison to placebo has been taken more seriously. Consider the case where we have a single effective treatment on the market, say A, whose efficacy has been demonstrated in a series of trials comparing it to placebo. We now run some new trials comparing a further treatment, B, to A. Taking all these trials together, they then have the structure of an incomplete blocks design. The effect of B compared to placebo can then be estimated using the double contrast of B compared to A and A compared to placebo. This approach has been examined in detail by Hasselblad and Kong (2001). A consequence of taking this particular view of matters is that the precision with which the effect of A was established compared to placebo cannot be exceeded by the indirect comparison of B to placebo, since the variance of this indirect contrast is the sum of the variances of the two direct contrasts.

This is, however, not the only difficulty with such studies. The following are some of those that apply.

Establishing a clinically irrelevant difference. If the route of a formal analysis compared to placebo via an indirect contrast is taken, this particular difficulty may be finessed. The new treatment is shown to be 'significantly' better than placebo, albeit using an indirect argument, and the extent of its inferiority to the comparator is only of relevance to the extent that it impinges on the proof of efficacy compared to placebo. If this proof is provided, then the comparison to the active comparator is 'water under the bridge'. If this particular approach is *not* taken, however, then any proof of efficacy of the new treatment rests on a demonstration that it is not 'substantially inferior' to the comparator, which comparator is accepted as being efficacious. This raises the issue as to what it means for a drug to be not substantially inferior to another one. This appears to require that some margin Δ, $\Delta > 0$, be adopted such that if τ is the extent by which the new treatment is inferior to the standard (where $\tau < 0$ indicates inferiority) then it is judged *substantially* inferior if $\tau \leq -\Delta$ and not substantially inferior or 'equivalent' if $\tau > -\Delta$.

Encyclopaedic Companion to Medical Statistics: Second Edition Edited by Brian S. Everitt and Christopher R. Palmer

Technical statistical aspects. In a Neyman–Pearson framework (see Salsbury, 1998) the test of noninferiority requires one to use a shifted NULL HYPOTHESIS. One might, therefore, adopt $H_0{:}\tau \leq -\Delta$. The situation is not as controversial as that for true bioequivalence, where the fact that two hypotheses have to be rejected, that of inferiority and that of superiority, means that an intuitive approach of seeing that the confidence limits for the difference lie within the limits of equivalence is not 'optimal' (Berger and Hsu, 1996), although the 'optimal' test may in practice be worse (Perlman and Wu, 1999; Senn, 2001). In practice, in the case of ACES if the lower conventional $1-\alpha$ *two-sided* CONFIDENCE INTERVAL for τ exceeds $-\Delta$, the hypothesis of substantial inferiority may be rejected at the level α and noninferiority asserted. It might be thought that a one-sided confidence interval would be sufficient for this purpose. However, the general regulatory convention is that all tests designed to show superiority are two-sided (despite apparent purpose) and, since such tests are a special case of a noninferiority test with $\Delta = 0$, use of one-sided tests for noninferiority would lead to inconsistencies (Senn, 1997; Committee for Proprietary Medicinal Products, 2000). In a Bayesian framework (see BAYESIAN METHODS) one might require that the posterior probability of noninferiority were less than some specified amount. Alternatively, use of a loss function would permit a decision analytic method, such as has been proposed for bioequivalence (Lindley, 1998), to be used.

Power of trials. Note that the reason one does not employ a value of $\Delta = 0$ in practice is that unless it is expected that the new treatment really is better than the standard, the power of the resulting test could never exceed 50 %. However, the clinically irrelevant difference is likely to be less than the clinically relevant difference used in conventional trials. Hence, if the new treatment is actually no better than the standard treatment, then, for a given sample size, the noncentrality parameter, $\delta = \Delta / SE(\hat{\Delta})$ is likely to be smaller for ACES than for trials designed to show superiority. Consequently, ACES either have lower power or higher sample sizes than conventional trials.

Assay sensitivity. A problem with ACES is that if the trial appears to show noninferiority of the new treatment, then there are three plausible explanations. The first, that of chance, is one that statistical analysis is designed to address. The second, that the new treatment is indeed noninferior, is what was desired to prove. However, a third possibility, that the experiment was not sensitive to find a difference, is difficult to exclude. This issue has been referred to as one of 'competence' (Senn, 1993) and affects whatever inferential framework one decides to use. An analogy may be useful here. In a game of hunt the thimble, a found thimble renders the quality of the strategy used for finding it irrelevant. It is no more 'found' if a good strategy were used than if a bad one were. However, a failure to find a thimble does not automatically justify the conclusion that the room does not contain one and the quality of the search employed is a crucial consideration in any judgement that it does not.

The effect of DROPOUTS, NONCOMPLIANCE *and the role of* INTENTION-TO-TREAT *analysis.* It is plausible that in many circumstances in conventional superiority trials if noncompliance or dropouts are a problem an intention-to-treat analysis will give a more modest estimate of the treatment effect than will a PER PROTOCOL analysis. In ACES, it is at least plausible that this may not be the case.

Conflict of requirements of additivity and clinical relevance. It may be that the clinically irrelevant difference is most meaningfully established on a scale that is not additive. For example, in a trial of an anti-infective, it could be most appropriate to establish that the difference in cure rate on the probability scale was not greater than some specified amount. Contrariwise, the log-odds scale might lend itself more readily to statistical modelling. This can lead to considerable difficulties (Holmgren, 1999), in particular because a trial does not recruit a random sample from the target population. It may be that further modelling using additional data may be necessary (Senn, 2000).

A common circumstance likely to make regulatory authorities ask questions is that a trial that was designed with optimism to show superiority to an active comparator fails to do so, but then is used to attempt to demonstrate noninferiority. This particular set of circumstances has become the subject of one of the European Medicine Evaluation Agency's 'points to consider' (Senn, 1997; Committee for Proprietary Medicinal Products, 2000). This stresses the desirability of establishing the trial's purpose pre-performance and also warns against establishing the clinically irrelevant difference, Δ, after the trial is complete. It regards putting a trial that was designed to show superiority to the purpose of noninferiority as an unacceptable use but accepts the converse. The guideline recognises that there are no issues of multiple testing involved with such switches (Bauer and Kieser, 1996) but that establishing values of Δ retrospectively may be biasing. Thus, it is preferable for investigators to specify in advance (e.g. by means of formal change to the CLINICAL TRIALS PROTOCOL) their intended switch of purpose and to fix the value of Δ prior to data unblinding. This, however, raises the issue as to whether the value of Δ is not something the regulator should declare for given indications rather than relying on the sponsor to do so. Otherwise, a regulator could be faced with the following position. Drug B is registered on the basis of comparison to a standard treatment A because the lower confidence interval for the treatment effect, τ_{B-A}, exceeds some pre-specified value Δ. However, a further drug, C, which has also been compared to A, is not granted a licence because a superiority trial was planned. Although superiority to A was not proven, the lower confidence interval for the treatment effect τ_{C-A} excludes a smaller possible difference between C and A than is excluded for the difference between B and A by the trial that has led to registration of B. *SS*

Bauer, P. and Kieser, M. 1996: A unifying approach for confidence intervals and testing of equivalence and difference. *Biometrika* 83, 4, 934–7. **Berger, R. L. and Hsu, J. C.** 1996: Bioequivalence trials, intersection–union tests and equivalence confidence sets. *Statistical Science* 11, 4, 283–302. **Committee for Proprietary Medicinal Products** 2000: Points to consider on switching between superiority and non-inferiority. **Hasselblad, V. and Kong, D. F.** 2001: Statistical methods for comparison to placebo in active-control studies. *Drug Information Journal* 35, 435–49. **Holmgren, E. B.** 1999: Establishing equivalence by showing that a specified percentage of the effect of the active control over placebo is maintained. *Journal of Biopharmaceutical Statistics* 9, 4, 651–9. **Lindley, D. V.** 1998: Decision analysis and bioequivalence trials. *Statistical Science* 13, 2, 136–41. **Makuch, R. and Johnson, M.** 1989: Issues in planning and interpreting active control equivalence studies. *Journal of Clinical Epidemiology* 42, 6, 503–11. **Medical Research Council Streptomycin in Tuberculosis Trials Committee** 1948: Streptomycin treatment for pulmonary tuberculosis. *British Medical Journal* ii, 769–82. **Perlman, M. D. and Wu, L.** 1999: The emperor's new tests. *Statistical Science* 14, 4, 355–69. **Salsbury, D.** 1998: Hypothesis testing. In Armitage, P. and Colton, T. (eds), *Encyclopedia of biostatistics*. Chichester: John Wiley & Sons, Ltd. **Senn, S. J.** 1993: Inherent difficulties with active control equivalence studies. *Statistics in Medicine* 12, 24, 2367–75. **Senn, S. J.** 1997: *Statistical issues in drug development*. Chichester: John Wiley & Sons, Ltd. **Senn, S. J.** 2000: Consensus and controversy in pharmaceutical statistics (with discussion). *The Statistician* 49, 135–76. **Senn, S. J.** 2001: Statistical issues in bioequivalence. *Statistics in Medicine* 20, 17–18, 2785–99. **Senn, S. J. and Rosati, N.** 2003: Editorial: Pharmaceuticals, patents and competition – some statistical issues. *Journal of the Royal Statistical Society Series A – Statistics in Society* 166, 271–7.

adaptive designs CLINICAL TRIALS that are adaptive are modified in some way by the data that have already been collected within that trial. The most common way the designs adapt is in the allocation of treatment, as a function of the response. For example, we may be interested in a dose that gives a 20 % chance of toxicity, where excesses to this level of toxicity would be harmful. Therefore, we may want to design the trial in such a way that, as more information is gathered, doses are allocated to optimise the estimate of that dose. If we were to use a traditional fully randomised approach to running the trial, which is not adaptive, we would probably not look at the data until the end of the trial, thereby risking exposing subjects to toxic doses and also possibly failing to produce an optimal estimate of the required dose. Another such example of an adaptive design is given in Rosenberger and Lachin (1993), whereby there are two treatments in the study, A and B, and as information emerges from the trial the treatment assignment probabilities are adapted in an attempt to assign more patients to the treatment performing better thus far. Therefore, when a patient enters the study, if treatment A appears to be better than treatment B, a patient has a greater than 50 % chance of being allocated treatment A – and vice versa.

Because adaptive designs modify the allocation of treatment on an ongoing basis, and thus protect patients from ineffective or toxic doses, they can be said to be more ethical than traditional designs. Rosenberger and Palmer (1999) consider the ethical dilemma between collective and individual ethics (see ETHICS AND CLINICAL TRIALS) and argue that in a clinical trial setting individual ethics should be uppermost; i.e. consideration should be towards doing what is best for patients in the current trial as opposed to doing what is best for future patients who stand to benefit from the results of current trial. The Declaration of Helsinki of October 2000 outlines the tension between these two types of ethics by stating: 'Considerations related to the well-being of the human subject should take precedence over the interests of science and society.' It is adaptive designs that address the individual ethics, as opposed to fully randomised designs, which address those collective ethics.

We will be dealing primarily with response adaptive designs here, such as those just outlined, and will not be describing those designs that attempt dynamically to balance the randomisation for covariate information, such as outlined by Pocock and Simon (1975) (see DATA-DEPENDENT DESIGNS, MINIMISATION).

The randomised play winner (RPW) design attempts to allocate treatments to patients sequentially based on a simple probability model. Rosenberger (1999) emphasises that the RPW design specifically applies to the situation where the outcome from a trial is binary, i.e. either 'success' or 'failure' and where there are only two treatments, e.g. drug A and drug B. At the start of the trial there is an assumed urn of α balls of type A (which relate to drug A) and β balls of type B (which relate to drug B). When a subject is recruited, a ball is drawn from the urn and then replaced. If the ball is type A then the subject is allocated to drug A, if type B then the subject is allocated to drug B. When the subject's outcome is available (and we assume that the outcome is available before the next subject is randomized), the urn is updated. If the response is a success on drug A, then a ball of type A is put into the urn, and similarly for a success on drug B. If the outcome is a failure on drug A, then a ball of type B is put into the urn, and again similarly for a failure on drug B. In this way, the balls build up such that a new subject has a better chance of being allocated to a better treatment.

Rosenberger (1999) concludes with a table of conditions under which the RPW rule is reasonable and provides a realistic alternative to the standard clinical trial design. These are given in the table.

adaptive designs *Conditions under which the RPW is reasonable (Rosenberger, 1999)*

- The therapies have been evaluated previously for toxicity
- The response is binary
- Delay in response is moderate, allowing adapting to take place
- Sample sizes are moderate (at least 50 subjects)
- Duration of the trial is limited and recruitment can take place during the entire trial
- The trial is carefully planned with extensive computations done under different models and initial urn compositions
- The experimental therapy is expected to have significant benefits to public health if it proves effective

Traditional dose–response studies, where patients are allocated to a limited number of doses along an assumed dose–response curve, are limited and, some would say, wrong. For example, if the assumed dose–response model is incorrect then patients may be allocated to ineffective or unsafe doses. One answer could be to increase the number of doses. However, this would result in many patients allocated to wasted doses. It would be much better to increase the number of doses and allocate doses to a subject based on current knowledge of the dose–response curve, which best optimises some pre-specified criteria. This is precisely what Bayesian response adaptive designs attempt to do, by employing Bayesian DECISION THEORY to a utility function. Thus, the dose that most optimally addresses the utility is allocated to the next available subject or cohort of subjects.

One of the first BAYESIAN METHODS described was the continual reassessment method (CRM), introduced by O'Quigley, Pepe and Fisher (1990), and originally devised for dose-escalation studies in oncology. Whitehead *et al.* (2001a) suggest that the method could also be used for applications in other serious diseases. The CRM envisages a study whereby human volunteers are treated sequentially, in order to detect a dose with a probability of toxicity of 20 %, i.e. TD20. The response is a binary response, 'toxicity' or 'no toxicity'. Before the study starts, investigators are asked to provide what their best guess is of a probability of toxicity at each of the series of doses. The first patient is then treated with the dose that is considered to be the closest to the TD20. Once the outcome is observed the PROBABILITY of toxicity at each of the doses is recalculated using the Bayesian method of statistics. The procedure continues in this way until it settles on a single dose. Whitehead *et al.* (2001a) point out that the CRM does home in on the TD20 quickly and efficiently, but there has been concern that early on in the trial subjects could be allocated to too high a dose, leading to potential toxicity problems. This has led to a number of modifications, such as starting at the lowest dose and never skipping a dose during the escalation.

Whitehead *et al.* (2001b) suggest practical extensions to the CRM for pharmacokinetic data, employing the use of Bayesian decision theory to allocate treatments optimally to subjects. They argue that conventional dose-escalation studies carried out in healthy volunteers do not normally employ statistical methodology or formal guidelines for dose escalation. As such the studies can take a long time to complete with little opportunity to skip doses. The methods proposed allocate doses in order to maximise the information about the dose–response curve, given a pre-specified safety constraint. They use two simple utility or gain functions, one that allocates the highest allowable dose under the safety constraint and the other that allocates doses in order to optimise the shape of the dose–response curve.

Krams *et al.* (2003) also use a Bayesian decision theory approach with sequential dose allocation to a Phase II study in acute stroke therapy by inhibition of neutrophils (ASTIN), which employs up to 15 dose levels. They use a response-adaptive procedure in order to find a dose that gives an improvement over that of placebo in the primary ENDPOINT, allocating the next subject either to the optimal dose or PLACEBO. Stopping rules were employed by which if the posterior probability of an effective drug or ineffective drug were greater than 0.9 then the decision would be made either to go on to a confirmatory trial (effective drug) or to stop development (ineffective drug). In this way, they were able to stop development of a compound more quickly than would have been possible under the traditional paradigm.

In 2006, the Pharmaceutical Research and Manufacturers of America (PhRMA) Adaptive Design Working Group published a series of papers in an issue of the Drug Information Association (DIA) journal detailing various aspects of these trials. Topics included terminology and classification; implementation; confidentiality and trial integrity; adaptive dose response; seamless Phase II/III; and sample size reestimation (see *Drug Information Journal* 40, 425–84, 2006). In addition, and reflecting the growing interest in adaptive designs, there have been numerous special editions of other journals devoted to these trials, including *Journal of Statistical Planning and Inference*, issue 136(2), 2006; *Journal of Biopharmaceutical Statistics*, issues 16(5),

2006 and 17(6), 2007; and *Statistics in Medicine,* issue 27(10), 2008. *AB*

Krams, M., Lees, K., Hacke, W., Grieve, A. P., Orgogozo, J.-M. and Ford, G. A. 2003: Acute stroke therapy by inhibition of neutrophils (ASTIN). An adaptive dose–response study of UK-279,276 in acute ischemic stroke. *Stroke* 34, 2543–8. **Pocock, S. and Simon, R.** 1975: Sequential treatment assignment with balancing of prognostic factors in controlled clinical trials. *Biometrics* 31, 103-15. **O'Quigley, J., Pepe, M. and Fisher, L.** 1990: Continual reassessment method: a practical design for Phase I clinical trials in cancer. *Biometrics* 46, 33–48. **Rosenberger, W. F.** 1999: Randomized play-the-winner clinical trials: review and recommendations. *Controlled Clinical Trials* 20, 328–42. **Rosenberger, W. F. and Lachin, J. M.** 1993: The use of response-adaptive designs in clinical trials. *Controlled Clinical Trials* 14, 471–84. **Rosenberger, W. F. and Palmer, C. R.** 1999: Ethics and practice: alternative designs for Phase III randomised clinical trials. *Controlled Clinical Trials* 20, 172–86. **Whitehead, J., Yinghui, Z., Patterson, S., Webber, D. and Francis, S.** 2001a: Easy-to-implement Bayesian methods for dose-escalation studies in healthy volunteers. *Biostatistics* 2, 47–61. **Whitehead, J., Zhou, Y., Stallard, N., Todd, S. and Whitehead A.** 2001b: Learning from previous responses in Phase I dose-escalation studies. *British Journal of Clinical Pharmacology* 52, 1–7.

adaptive randomisation See ADAPTIVE DESIGNS, RANDOMISATION

adjustment for noncompliance in randomised controlled trials In clinical medicine, 'noncompliance' occurs when a patient does not fully follow a prescribed course of treatment. The alternative terms 'adherence' and 'concordance' attempt to avoid the authoritarian overtones of 'compliance'. In randomised CLINICAL TRIALS, we are concerned with any departure from a randomised treatment, whether due to noncompliance or a treatment change agreed with medical staff. In a trial to compare two types of medication (drug A and drug B, say) for the treatment of heart disease, for example, patients may refuse or forget to take any of their medication or forget to take it some of the time (partial compliance). Patients allocated to receive drug A might switch to drug B, and vice versa. Some of the patients may even take another medication altogether (drug C, say) or, particularly if the therapy appears to be failing, receive a much more radical intervention such as surgery. A further complication for the estimation of treatment effects arises when patients who fail to comply with their prescribed treatment are also those who are more likely to be lost to follow-up.

Rationale. Conventionally, trials with departures from randomised treatment are analysed by INTENTION-TO-TREAT. This directly compares the *effectiveness* of the different treatment policies as actually implemented in the trial – e.g. 'drug A plus changes' versus 'drug B plus changes'. Unlike effectiveness, *efficacy* relates to the effects of the treatments themselves, and is not estimated by an intention-to-treat analysis. Researchers may also be interested in the effectiveness of an intervention in other circumstances, e.g. if public suspicion of the intervention had been reduced by the positive results of a clinical trial. In these circumstances, the rates of compliance may be improved and adjustment for this change may be attempted.

It is important to define the aim of adjustment for noncompliance. For example, in a trial of immediate versus deferred zidovudine in asymptomatic HIV infection, the initial plan was to defer zidovudine until the onset of symptomatic disease. However, following a protocol amendment, some individuals started zidovudine before the onset of symptomatic disease (White *et al.*, 1997). There was interest in estimating the effect that would have been observed under the original protocol. Zidovudine before the onset of symptomatic disease was therefore regarded as 'noncompliance'. Other individuals stopped zidovudine treatment because of adverse events. Additional adjustment for stopping treatment would not answer a clinically relevant question, so the analysis did not aim to estimate efficacy.

Adjustment for noncompliance is useful in a variety of situations. Patients may be most interested in treatment efficacy. Differences in compliance may help to explain variation of a treatment effect with time, between subgroups in a trial or between trials in a META-ANALYSIS. Reconciling trial data with observational data may require adjustment for noncompliance in the trial. Policy analysis may require projections for situations with improved compliance.

Most attempts to allow for noncompliance use on-treatment analysis or PER PROTOCOL analysis. This only provides a valid comparison of the treatments themselves (efficacy) if compliers and noncompliers do not differ systematically in their disease state or prognosis. In practice this is unlikely to be the case, so selection bias occurs. Heart disease patients who comply with their prescribed medication, for example, are also those who are likely to improve their diet or take more exercise and these changes, in turn, are likely to lead to a better outcome. SELECTION BIAS may often be reduced by adjustment for baseline covariates, but there is still no guarantee of an unbiased analysis. For example, in the Coronary Drug Project, 5-year mortality of poor compliers was 28.2 % compared with 15.1 % in good compliers, and adjustment for 40 baseline factors only reduced the difference to 25.8 % versus 16.4 % (The Coronary Drug Project Research Group, 1980).

Newer 'randomisation-based' methods can estimate efficacy while avoiding selection bias by directly comparing the groups as randomised as in an intention-to-treat analysis (White, 2005). This is made possible by considering the subgroup of 'compliers' who would have received their randomised treatment, whichever group they were randomised to. For example, a trial in Indonesian children compared vitamin A supplementation with no intervention, the outcome being 12-month mortality. Vitamin A supplementation was actually received by only 80% of the intervention arm and by none of the control arm. Sommer and Zeger (1991) considered the subgroup who did not receive vitamin A in the intervention arm and a corresponding subgroup of the control arm who *would not have received vitamin A if they had been allocated to receive it*. These 'noncomplier' subgroups were assumed to be unaffected by allocation to vitamin A. It is then straightforward to estimate the number of noncompliers in the control arm and their mean outcome, and hence the risk difference, risk ratio or odds ratio in compliers. This is often called the 'complier average causal effect' (CACE) estimate (Little and Rubin, 2000). This approach is a special case of PRINCIPAL STRATIFICATION.

A more general approach requires a model relating *potential outcomes* for each individual under different counterfactual treatments. A simple model might say that each individual would have blood pressure *b* mmHg lower if they took the drug with perfect compliance than if they did not take the drug, with proportional blood pressure reductions for partial compliance. Such a model may be fitted by observing that untreated blood pressure must have the same distribution in each randomised group (Fischer-Lapp and Goetghebeur, 1999). An important advantage of these methods is that no assumption is required about the relationship between compliance and potential outcomes. They are closely related to the use of INSTRUMENTAL VARIABLES methods (Dunn and Bentall, 2007).

The approaches just described are generally only able to estimate one treatment effect in a two-arm trial. They tend to be hopelessly imprecise in situations such as EQUIVALENCE STUDIES where patients may stop all treatment during the trial, so that the analysis requires estimation of the effect of both treatments. In this case it is possible to adjust the randomised comparison using observational estimation of one or more treatment effects – i.e. assuming there are no unmeasured confounders for treatment. Methods such as *marginal structural modelling* can work even when actual treatment is both a consequence of symptomatic deterioration and a cause of slower disease progression (see Little and Rubin, 2000, for references to this literature).

A trial with noncompliance has less POWER than one with perfect compliance, as a result of the reduced effect size as estimated in an intention-to-treat analysis, and it is natural to want to recover the lost power. However, many of the new procedures preserve the intention-to-treat SIGNIFICANCE LEVEL and therefore do not affect power. In some cases, it is impossible to regain power without making some assumption about comparability of noncompliers and compliers. In other situations, some gain in power is theoretically possible, but this is unlikely to be appreciable in practice (Becque and White, 2008). Significance testing should therefore rely on intention-to-treat analysis even when other methods are used to estimate efficacy. *IW/GD*

Angrist, J. D., Imbens, G. W. and Rubin, D. B. 1996: Identification of causal effects using instrumental variables (with discussion). *Journal of the American Statistical Association* 91, 444–72. **Becque, T. and White, I. R.** 2008. Regaining power lost by non-compliance via full probability modelling. *Statistics in Medicine* 27, 5640–63. **Dunn, G. and Bentall, R.** 2007: Modelling treatment-effect heterogeneity in randomized controlled trials of complex interventions (psychological treatments). *Statistics in Medicine* 26, 4719–45. **Fisher-Lapp, K. and Goetghebeur, E.** 1999: Practical properties of some structural mean analyses of the effect of compliance in randomized trials. *Controlled Clinical Trials* 20, 531–46. **Little, R. and Rubin, D. B.** 2000: Causal effects in clinical and epidemiological studies via potential outcomes: concepts and analytical approaches. *Annual Review of Public Health* 21, 121-45. **The Coronary Drug Project Research Group** 1980: Influence of adherence to treatment and response to cholesterol on mortality in the Coronary Drug Project. *New England Journal of Medicine* 303, 1038–41. **White, I. R** 2005: Uses and limitations of randomization-based efficacy estimators. *Statistical Methods in Medical Research* 14, 327–47. **White, I. R., Walker, S., Babiker, A. G. and Darbyshire, J. H.** 1997: Impact of treatment changes on the interpretation of the Concorde trial. *AIDS* 11, 999–1006.

age-period cohort analysis To understand the effect of time on a particular outcome for an individual it is essential to realise the relevant temporal prospective. Age affects many aspects of life, including the risk of disease, so this is an essential component of any analysis of time trends. Period denotes the date of the outcome and if the outcome varies with period it is likely to be due to some underlying factor that affects the outcome and varies in the same way for the entire population under study. Cohort, contrariwise, refers to generational effects caused by factors that only affect particular age groups when their level changes with time.

An example of a period effect would be a potential effect of an air contaminant that affected all age groups in the same way. If the level of exposure to that factor increased/decreased with time, exerting a change in the outcome in all age groups, then we would expect a related pattern across all age groups in the study. In studies that take place over long periods of time, the technology for measuring the outcome may change, giving rise to an artifactual effect that was not due to change in exposure to a causative agent. For example, intensive screening for disease can identify disease cases that would not previously have been identified, thus artificially increasing the disease rate in a population that has had no change in exposure over time.

Cohort (also called birth cohort) effects may be due to factors related to exposures associated with the date of birth, such as the introduction of a particular drug or practice during pregnancy that was brought in at a particular point in time. For example, a pregnancy practice associated with increased risk and adopted by the population of mothers during a particular time period could affect the risk during the lifespan of the entire generation born during that period. While it is common to refer to these effects as being associated with year of birth, they could also be the result of changes in exposure that occurred after birth. In many individuals, lifestyle factors that may affect disease risk over a lifetime are fixed as they approach adulthood. A quantification of these effects on such a generation would give rise to a comparison of these cohort or generational effects.

An inherent redundancy among these three temporal factors arises from the fact that knowing any two factors implies the value of the third. For example, if we know an individuals age (a) at a given date or period (p), then the cohort is the difference ($c = p - a$). This linear dependence gives rise to an identifiability problem in a formal regression model that attempts to obtain quantitative estimates of regression parameters associated with each temporal element:

$$E[Y] = \beta_0 + a\beta_a + p\beta_p + c\beta_c$$

Using the linear relationship between the temporal factors gives rise to:

$$\begin{aligned} E[Y] &= \beta_0 + a\beta_a + p\beta_p + (p-a)\beta_c \\ &= \beta_0 + a(\beta_a - \beta_c) + p(\beta_p + \beta_c) \end{aligned}$$

which has only two identifiable parameters besides the intercept instead of the expected three. Another way of visualising this phenomenon is that all combinations of age, period and cohort may be displayed in the LEXIS DIAGRAM (see the figure), which is obviously a representation of a two-dimensional plane instead of the three dimensions expected for three separate factors.

In general, these analyses are not limited to linear effects applied to a continuous measure of time, but instead they are applied to temporal intervals, such as disease rates observed for 5- or 10-year intervals of age and period. When the widths of these intervals are equal, the model may be expressed as:

$$E[Y_{ijk}] = \mu + \alpha_i + \pi_j + \gamma_k$$

where μ is the intercept, α_i the effect of age for the ith ($i = 1, \ldots, I$) interval, π_j the effect of period for the jth ($j = 1, \ldots, J$) interval and γ_k the effect of the kth cohort ($k = i - j + I = 1, \ldots, K = I + J - 1$). The usual constraints in this model imply that $\sum \alpha_i = \sum \pi_j = \sum \gamma_k = 0$. The identifiability problem manifests itself through a single unidentifiable parameter (Fienberg and Mason, 1979), which can be more easily seen if we partition each temporal effect into components of overall linear trend and curvature or departure from linear trend. For example, age can be given by $\alpha_i = \overleftrightarrow{i}\beta_a + \breve{\alpha}_i$, where $\overleftrightarrow{i} = i - 0.5(I + 1)$, β_a is the overall slope and $\breve{\alpha}_i$ the curvature. The overall model can be expressed as:

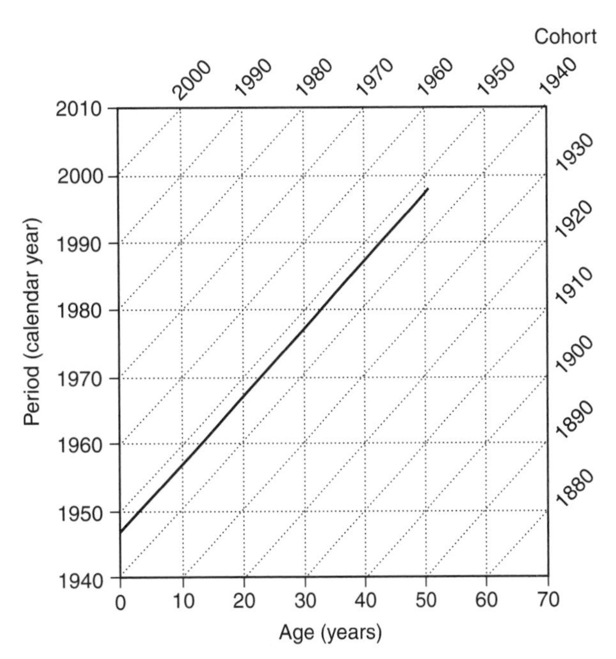

age-period cohort analysis *Lexis diagram showing the relationship between age, period and cohort. The diagonal line traces age-period lifetime for an individual born in 1947*

$$\begin{aligned} E[Y_{ijk}] &= \mu + (\overleftrightarrow{i}\beta_a + \breve{\alpha}_i) + (\overleftrightarrow{j}\beta_\pi + \breve{\pi}_i) \\ &\quad + (\overleftrightarrow{k}\beta_\gamma + \breve{\gamma}_k) \\ &= \mu + \overleftrightarrow{i}(\beta_a + \beta_\gamma) + \overleftrightarrow{j}(\beta_\pi + \beta_\gamma) + \breve{\alpha}_i \\ &\quad + \breve{\pi}_i + \breve{\gamma}_k \end{aligned}$$

because $\overleftrightarrow{k} = \overleftrightarrow{j} - \overleftrightarrow{i}$. Thus, each of the curvatures can be uniquely determined, but the overall slopes are hopelessly entangled so that only certain combinations can be uniquely estimated (Holford, 1983).

The implication of the identifiability problem is that the overall direction of the effect for any of the three temporal components cannot be determined from a regression analysis. Thus, we cannot even determine whether the trends are increasing or decreasing with cohort, for instance. The second figure on page 8 displays several combinations of age, period and cohort parameters, each set of which provides an identical set of fitted rates. Notice that as the period parameters are rotated clockwise, the age and cohort parameters are comparably rotated in the counterclockwise direction. Each of these parameters can be rotated a full 180°, but it is important also to realise that they cannot be rotated one at a time, only all together. Thus, even though the specific trends cannot be uniquely estimated, certain combinations of the overall trend can be uniquely determined, such as $\beta_\pi + \beta_\gamma$, which is called the *net drift* (Clayton and Schifflers, 1987a, 1987b). Alternative drift estimates covering shorter timespans can also be determined and these have practical significance in that they describe the experience of following a particular age group in time, because both period and cohort will advance together. Curvatures, by way of contrast, are completely determined, including polynomial parameters for the square and higher powers, changes in slopes and second differences. The significance test for any one of the temporal effects in the presence of the other two will generally be a test of the corresponding curvature and not the slope. Holford provides further detail on how software can be set up for fitting these models (Holford, 2004). *TRH*

Clayton, D. and Schifflers, E. 1987a: Models for temporal variation in cancer rates I: Age-period and age-cohort models. *Statistics in Medicine* 6, 449–67. **Clayton, D. and Schifflers, E.** 1987b: Models for temporal variation in cancer rates II: Age-period cohort models. *Statistics in Medicine* 6, 469–81. **Fienberg, S. E. and Mason, W. M.** 1979: Identification and estimation of age-period-cohort models in the analysis of discrete archival data. *Sociological Methodology* 1978, 1-67. **Holford, T. R.** 1983: The estimation of age, period and cohort effects for vital rates. *Biometrics* 39, 311–24. **Holford, T. R.** 2004: Temporal factors in public health surveillance: sorting out age, period and cohort effects. In Brookmeyer, R. and Stroup, D. F. (eds), *Monitoring the health of populations*. Oxford: Oxford University Press, pp. 99–126.

age-related reference ranges These are ranges of values of a measurement that identify the upper and lower limit of normality in the population, where the range varies according to the subject's age. Reference ranges are an important part of medical diagnosis, where a continuous measurement (e.g. blood pressure) needs converting to a binary variable for decision-making purposes. If the patient's value lies outside the measurement's reference range it is treated as abnormal and the patient is investigated further. The construction of reference ranges involves estimating the range of values that covers a specified percentage of the reference population, often 95 %. Usually this is the central part of the distribution with equal tail area probabilities, although in some cases the reference range is bounded at zero or infinity. For normally distributed data the range can be derived from the population MEAN and STANDARD DEVIATION (SD), the 95 % range, for example, being the mean plus or minus 2 SDs. For nonnormal data the simplest approach is to use quantiles, i.e. rank and count the data, then the 2.5% and 97.5% points are the lower and upper limits of the 95 % reference range. However, this is inefficient and requires a large sample. If the data are skew they can be transformed, e.g. to logarithms, and then the reference range can be calculated from the mean and SD on the transformed scale

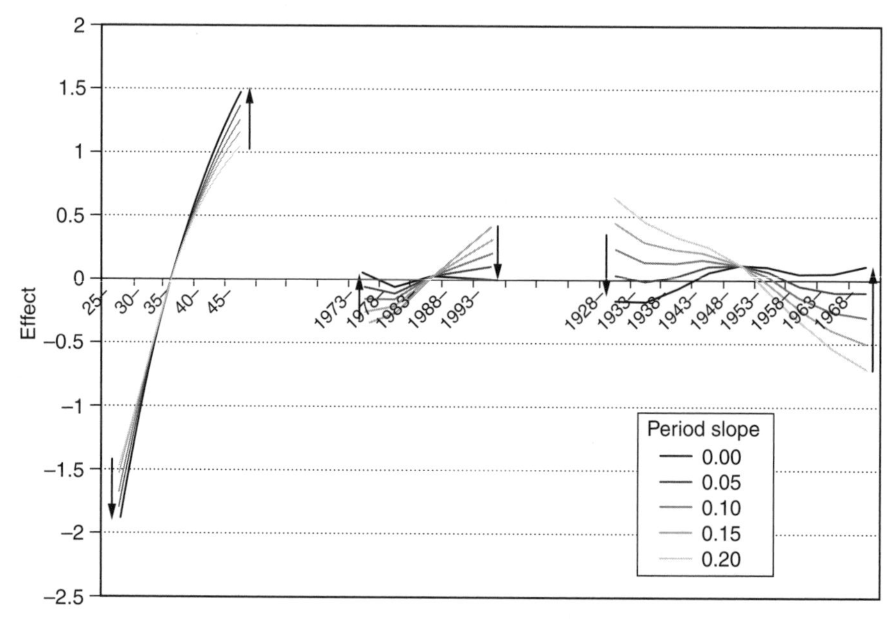

age-period cohort analysis *Age, period and cohort effects for pre-menopausal breast cancer incidence for SEER, 1973–1997*

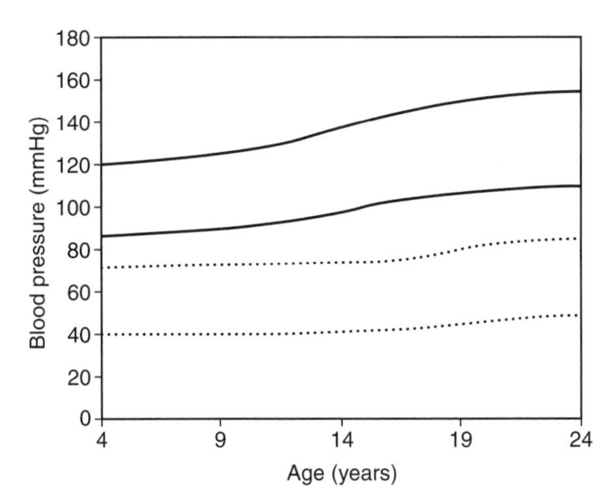

age-related reference ranges *Age-related 95 % reference ranges for blood pressure in boys: systolic (solid lines) and diastolic (dotted lines)*

and transformed back to the original scale. A more flexible variant is to use a Box–Cox power transformation (of which the logarithm is a special case), which adjusts for skewness more precisely (see TRANSFORMATIONS).

Age-related reference ranges are reference ranges that depend on age. They arise most commonly in paediatrics, notably for age-related measures of body size like height and weight, which can be displayed as GROWTH CHARTS. The principles of reference range estimation are essentially the same when they are age related, except that the ranges for adjacent age groups need to be consistent. To avoid discontinuities at the age group boundaries requires the summary statistics to define the reference range (e.g. the mean and SD) and to change smoothly with age but imposing this constraint complicates the fitting process. For normally distributed homoscedastic data, where the SD is constant across age, the age-related mean can be estimated by LINEAR REGRESSION and the reference range constructed around the regression curve using the residual SD. The regression curve is estimated using a smoothing regression function, e.g. a polynomial, fractional polynomial or generalised additive (cubic spline) curve. If the SD changes with age, as is often the case, a curve of the age-related SD also needs to be estimated by the regression methods of Aitkin (1987) or Altman (1993) and the age-related mean obtained using weighted linear regression with weights corresponding to the inverse square of the age-related SD. The age-related reference range is again constructed around the regression curve using the SD curve.

When the data are skew it may be possible to adjust for the skewness using a single, e.g. logarithmic, transformation at all ages. However, often the degree of skewness is itself age related, although this needs a large sample to show it. In this case an age-related summary statistic for the skewness has to be estimated, along with the age-related mean and SD. The LMS METHOD is a popular way to do this, or alternatively the EN method of Royston and Wright (1998). For more extreme nonnormal data, a nonparametric approach based on QUANTILE REGRESSION is needed, a form of least absolute errors regression, where smooth curves are constructed for the age-related upper and lower limits of the reference range. The figure gives age-related reference ranges for systolic and diastolic blood pressure in boys aged 4–24, estimated by the LMS method.

There are two advantages of reference ranges based on an underlying frequency distribution, as opposed to those derived using quantile regression. The first is efficiency – the standard errors of the reference range limits are smaller. The second is analytical convenience – data for individuals can be converted to z-SCORES, indicating how many SDs they are above or below the median of the distribution, which is a convenient way of adjusting for age prior to further analysis. *TJC*

[See also GROWTH CHARTS]

Aitkin, M. 1987: Modelling variance heterogeneity in normal regression using GLIM. *Applied Statistics* 36, 332–9. **Altman, D. G.** 1993: Construction of age-related reference centiles using absolute residuals. *Statistics in Medicine* 12, 917–24. **Cole, T. J. and Green, P. J.** 1992: Smoothing reference centile curves: the LMS method and penalized likelihood. *Statistics in Medicine* 11, 1305–19. **Koenker, R. W. and D'Orey, V.** 1987: Computing regression quantiles. *Applied Statistics* 36, 383–93. **Royston, P. and Wright, E. M.** 1998: A method for estimating age-specific reference intervals ('normal ranges') based on fractional polynomials and exponential transformation. *Journal of the Royal Statistical Society Series A*, 161, 79–101.

age-specific rates These are rates calculated within a number of relatively narrow age bands. A crude rate is the number of events occurring in a population during a specified time period divided by an estimate of the size of the population. However, when comparing rates between populations with different age distributions, it is necessary to consider rates at specific ages separately.

In the table, death rates are presented for Costa Rica and the United Kingdom for 1999, derived from data from the United Nations (2002). The final column gives the age-specific rates for broad age bands and the crude (total) rate. The age-specific rate is calculated as the number of deaths in the particular age group. In Costa Rica, the death rate at ages 0–5 is calculated as 1296/1 070 000. The rate is expressed per 1000 persons so the rate is multiplied by 1000 to give the rate of 1.2 per 1000 in the final column of the table.

age-specific rates *Population, number of deaths and death rates from all causes for Costa Rica and the United Kingdom for the year 1999*

Costa Rica				
Age group	*Population (100 000s)*	*% in age group*	*Deaths*	*Death rate/1000*
0–15	10.7	32 %	1296	1.2
15–49	17.4	52 %	2766	1.6
50–69	3.9	12 %	3447	8.8
70+	1.3	4 %	7523	56.6
Total	33.4		15032	4.5
United Kingdom				
Age group	*Population (100 000s)*	*% in age group*	*Deaths*	*Death rate/1000*
0–15	113.9	19 %	5850	0.5
15–49	288.0	48 %	31 228	1.1
50–69	126.1	21 %	120 759	9.6
70+	67.0	11 %	474 225	70.8
Total	595.0		632 062	10.6

The crude (total) rate for Costa Rica is less than half that for the UK. However, at no age is the rate in the UK double that for Costa Rica and for some age groups the rate is higher in Costa Rica than in the UK. Note that the percentages of the population in each age group (third column) differ markedly. The UK population is much older (11 % of the population are over 70 compared with 4 % in Costa Rica). The different age structure explains the misleading comparison between the crude rates.

Age-specific rates are cumbersome to compare across a number of populations. Standardisation methods are often used to provide an age-adjusted summary rate for each population.

Many countries publish age-specific rates for all cause and specific causes of death, e.g. the annual publications of the Office of National Statistics (ONS) in England and Wales (ONS, 2002). Age-specific disease incidence rates are also published in various countries, most notably cancer incidence, for which international data are compiled by the International Agency for Research on Cancer (Parkin *et al.*, 2003). Age-specific prevalence rates for exposures such as smoking can also be derived, but are more usually obtained from specific surveys such as the General Household Survey (Walker *et al.*, 2001). *HI*

[See also CAUSE-SPECIFIC DEATH RATE, STANDARDISED MORTALITY RATIO]

Office for National Statistics 2002: *Mortality statistics: cause. Review of the Registrar General on deaths by cause, sex and age, in England and Wales, 2001*. London: Office for National Statistics. **Parkin, M., Whelan, S., Ferlay, J., Teppo, L. and Thomas, D. B.** 2003: *Cancer incidence in five continents*, Vol. VIII. Lyon: IARC Scientific Publications. **United Nations** 2002: *2000 demographic yearbook*. New York: United Nations. **Walker, A., Maher, J., Coulthard, M., Goddard, E. and Thomas, M.** 2001: *Living in Britain: results from the 2000 General Household Survey*. London: The Stationery Office.

agreement Agreement in repeated assessments is a fundamental criterion for quality of assessments on rating scales. The use of rating scales and other kinds of ordered classifications of complex qualitative variables is interdisciplinary and unlimited. Rating scale assessments produce *ordinal data*, the ordered categories representing only a rank order of the intensity of a particular variable and not a numerical value in a mathematical sense, although the use of numerical labelling could give a false impression of quantitative data. (see RANK INVARIANCE). The main quality concepts of scale assessments are reliability and validity. Reliability (see MEASUREMENT PRECISION AND RELIABILITY) refers to the extent to which repeated measurements of the same object yield the same result, which means agreement in repeated assessments of various designs. In interrater reliability (see MEASUREMENT PRECISION AND RELIABILITY) studies are made of the level of agreement between observers that classify the same object or individual, and intrarater reliability (see INTRACLASS CORRELATION COEFFICIENT) studies refer to agreement in test–retest scale assessments by the same rater.

The frequency distribution of pairs of ordinal data is described in a square CONTINGENCY TABLE (see the figure with parts I, II and III on page 11), and in the case of continuous assessments on a visual analogue scale, VAS, by a scatter plot. The percentage agreement (PA) is a basic agreement measure. When the agreement is unsatisfactory small reasons for disagreement can be evaluated by a statistical method that takes account of the rank-invariant properties of ordinal data and that makes it possible to identify and measure systematic disagreement, when present, separately from disagreement caused by individual variability in assessments. Systematic disagreement is population based and reveals a systematic change in conditions or memory bias between test–retest assessments, or between raters who interpret the scale categories differently. Large individual variability, on the other hand, is a sign of poor quality of a rating scale as it allows for uncertainty in repeated assessments. The presence of systematic disagreement in the use of the scale categories between the two assessments is revealed by different frequency distributions, which means marginal distributions (parts I and II). A systematic disagreement regarding the categorical levels and in the way of concentrating the assessments on the categories are measured by the relative position (RP) and the relative concentration (RC) respectively. The RP expresses the extent to which the marginal distribution of assessments Y is shifted towards higher categories than the marginal distribution of X, rather than the opposite. A theoretical description is the difference between the probabilities $P(X<Y)-P(Y<X)$. Possible values of RP range from (-1) to 1, and RP is positive when higher scale categories are more frequently used in the assessments Y than in X when compared with the opposite. Correspondingly, the RC expresses the extent to which the marginal distribution of Y assessments is more concentrated to central scale categories than is the marginal distribution of X, theoretically described by the difference in probabilities $P(X_i<Y_k<X_j)-P(Y_i<X_k<Y_j)$. Possible values range from (-1) to 1, and a positive RC indicates that the assessments Y are more concentrated than X. Zero or very small values of both RP and RC mean that the systematic part of an observed disagreement paired assessments is negligible.

Systematic disagreement is evident by the marginal heterogeneity, and by pairing off the two sets of marginal frequencies, the so-called rank-transformable pattern of agreement (RTPA) is constructed. The RTPA describes the expected pattern in the case of systematic disagreement only. All pairs of observations of the RTPA will have the same rank ordering in the two assessments provided that the ranks are tied to the cells, which is the definition of the augmented ranking procedure (aug-ranks) (see RANKING).

Part II in the figure is the RTPA of the pattern in part I. The observed distribution of pairs in part I deviates from this RTPA, which means that some of the pairs of aug-ranks given to the observations differ. The relative rank variance (RV) is a rank-based measure of this observed individual variability, i.e. unexplained by the measures of systematic disagreement:

$$\mathrm{RV}=\frac{6}{n^3}\sum_{i=1}^{m}\sum_{j=1}^{m}x_{ij}(\varDelta\bar{R}_{ij})^2$$

where n is the number of paired assessments and $(\varDelta\bar{R}_{ij})^2$ is the square of the mean aug-rank difference of the ijth cell, and the summation is made over all cells ij of the $m\times m$ square table, $0\leq\mathrm{RV}\leq 1$ (Svensson *et al.*, 1996; Svensson, 1998a). The Cohen's coefficient kappa (κ) is a commonly used measure of agreement adjusted for the chance expected agreement (see KAPPA AND WEIGHTED KAPPA).

The calculations of Cronbach's alfa and other so-called reliability coefficients are based on the assumption of quantitative, normally distributed data, which is not achievable in

I		Rater X					II	Rater X					III	Rater X				
		A	B	C	D	tot		A	B	C	D	tot		A	B	C	D	tot
Rater Y	D			1	1	2	D				2	2	D		1	4	14	19
	C		2	2	14	18	C			1	17	18	C	1	2	10	4	17
	B	1	1	11	3	16	B			16		16	B	1	6	3	1	11
	A	2	8	3	1	14	A	3	11			14	A	1	2			3
	tot	3	11	17	19	50	tot	3	11	17	19	50	tot	3	11	17	19	50

I: PA, 12 %; RP, −0.49; RC, 0.16; RV, 0.08

II: PA, 12 %; RP, −0.49; RC, 0.16; RV, 0

III: PA, 62 %; RP = RC = 0; RV, 0.05

agreement *Examples of paired ordinal data from interrater assessments on a four-point scale with the ordered categories labelled A < B < C < D. The rank-transformable pattern of agreement (RTPA) is shaded. The measures of percentage agreement (PA), the relative position (RP), the relative concentration (RC) and the relative rank variance (RV) are given*

data from rating scales. There is also a widespread misuse of the correlation coefficient as a reliability measure. The correlation coefficient (see CORRELATION) measures the degree of association between two variables and does not measure the level of agreement. In part I of the figure the PA is 12 %, and the observed disagreement is mainly explained by a systematic disagreement in position. The negative RP value (−0.49) and the RTPA (part II) shows that the assessments Y systematically used lower categories than X. A slight additional individual variability, RV = 0.08 is observed. SPEARMAN'S RANK CORRELATION COEFFICIENT, r_s, is 0.66 in part I of the figure and 0.97 in part II, ignoring the fact that the assessments are systematically biased and unreliable. The same holds for the coefficient kappa (−0.14). In part III the marginal homogeneity and the zero RP and RC values confirm that the disagreement (39 %) is entirely explained by slight individual dispersion (RV = 0.05) from the RTPA, which is the main diagonal in this case. The r_s is 0.61 and the κ is 0.45.

Besides reliability studies, the level of disagreement is of main interest in paired assessments 'before and after' treatment for analysing change in outcome or treatment effect. In this application of the disagreement measures, nonzero RP and RC values indicate the level of common group change in outcomes, and the heterogeneity in changes among the individuals is measured by the RV (Svensson, 1998b). *ES*

Svensson, E. 1998a: Application of a rank-invariant method to evaluate reliability of ordered categorical assessments. *Journal of Epidemiology and Biostatistics* 3, 403–9. **Svensson, E.** 1998b: Ordinal invariant measures for individual and group changes in ordered categorical data. *Statistics in Medicine* 17, 2923–36. **Svensson, E., Starmark, J.-E., Ekholm, S., von Essen, C. and Johansson, A.** 1996: Analysis of inter-observer disagreement in the assessment of subarachnoid blood and acute hydrocephalus on CT scans. *Neurological Research* 18, 487–94.

Akaike's information criterion Akaike's information criterion (AIC) is an index used to discriminate between competing models. It is widely used when there is the issue of model choice where we wish to find the most parsimonious model (see Akaike, 1974). Often there may be a number of possible models that can be fitted to the data, from which parameters can be estimated using, for example, the MAXIMUM LIKELIHOOD ESTIMATION. Generally, complex models are more flexible, but contain a relatively large number of parameters, whereas simpler models with fewer parameters may compromise the fit of the model to the data. Essentially, the AIC statistic compares competing models by considering the trade-off between the complexity of the model and the corresponding fit of the model to the data. The AIC statistic is widely used, particularly as it can be used to compare even nonnested models when likelihood ratio tests cannot be applied.

Let $\boldsymbol{x}$ denote the data and $\hat{\theta}$ the corresponding maximum likelihood estimates (MLEs) of the parameters. Then, the AIC for a given model is denoted by:

$$\text{AIC} = -2\log \text{L}(\hat{\theta}; x) + 2p$$

where p denotes the number of parameters in the given model being fitted to the data and $\log \text{L}(\hat{\theta}; \boldsymbol{x})$ the corresponding log-likelihood evaluated at the MLEs of the parameters. The AIC statistic is calculated for each possible model being considered. The model deemed optimal is the one with the smallest AIC value, i.e. a model with a relatively small number of parameters that adequately fits the data. The AIC is generally easy to calculate given the maximum of the likelihood function and is very versatile, allowing us to compare, for example, nonnested models. We note that corrections have been suggested to the AIC statistic to allow for data with overdispersion (denoted by QAIC) and small sample sizes (AIC_c). See, for example, Burnham and Anderson (2002), Sections 2.4–5.

The AIC statistic has also been used to compare the performance of different models, relative to each other (Buckland, Burnham and Augustin, 1997; Burnham and Anderson, 2002, Section 2.6). It is not the absolute values of the AIC statistics that are important but their relative values, in particular their difference. For each model the term $\Delta\text{AIC} = \text{AIC} - \min \text{AIC}$ is calculated, where min AIC is the value of the AIC statistic for the model deemed optimal. Clearly, $\Delta\text{AIC} = 0$ for the model deemed optimal; the larger the value of ΔAIC the poorer the model. The relative penalised likelihood weights w_i can also be calculated for each model $i = 1, \ldots, m$, where:

$$w_i = \frac{\exp(-\Delta\text{AIC}_i/2)}{\sum_{j=1}^{m} \exp(-\Delta\text{AIC}_j/2)}$$

and AIC_i denotes the corresponding AIC value associated with model i. The weights provide a scale to interpret the difference in values for the models. Finally, these model weights can be used to obtain a (weighted) model-averaged estimate of parameters of interest. *RK*

[See also DEVIANCE, LIKELIHOOD RATIO]

Akaike, H. 1974: A new look at the statistical model identification. *IEEE Transactions on Automatic Control AC* 19, 716–72. **Buckland, S. T., Burnham, K. P. and Augustin, N. H.** 1997: Model selection: an integral part of inference. *Biometrics* 53, 603–18. **Burnham, K. P. and Anderson, D. R.** 2002: *Model selection and multimodel inference*, 2nd edition. Heidelberg: Springer Verlag.

allelic association This is an association between two alleles (at two different loci), or between an allele and a phenotypic trait, in the population. Since humans are diploid a more technical definition of the former is necessary: two alleles are associated if their frequency of co-occurrence in the same haplotype (i.e. the genetic material transmitted from one parent) is greater than the product of the marginal frequencies of the two alleles.

Association between two alleles is also known as *linkage disequilibrium*. The reason is that, in a large population under random mating, the extent of association between two alleles (as measured by the difference between the frequency of the haplotype containing the two alleles and the product of the frequencies of the two alleles) decreases by a factor equal to one minus the recombination fraction (see GENETIC LINKAGE) between the two loci, per generation. Thus allelic association represents a state of disequilibrium that tends to dissipate at a rate determined by the strength of linkage between the two alleles, towards the state of equilibrium when the frequency of the haplotype is equal to the produce of the frequencies of the two constituent alleles.

Associations between two alleles can arise in a population for a number of reasons. The mutation that gave rise to the more recent allele may have occurred on a chromosome that happened to contain the other allele. Random genetic drift during a population bottleneck may have led to the overrepresentation of some haplotypes. The mixing of two populations with different allele frequencies may have resulted in associations between alleles in the overall population. When, for any of these reasons, such allelic associations arose many generations ago, only those occurring between tightly linked loci are likely to have persisted to the current generation. We would therefore expect an imperfect inverse relationship between the extent of association between two alleles and the distance between them.

An association between an allele and a disease may be the result of a direct causal relationship. In other words, the allele is a causal variant that is functional and increases the risk of the disease. However, it could also be indirect, with the allele being in linkage disequilibrium with a causal variant. The presence of linkage disequilibrium between tightly linked loci means that it is possible to screen a chromosomal region for a causal variant without examining all the alleles, only a sufficient number to ensure that any causal variant in the region is likely to be in linkage disequilibrium with one or more of the alleles examined. The polymorphisms chosen to represent itself and associated polymorphisms in its vicinity in an association study are called TAG polymorphisms. The International HapMap Project (www.hapmap.org) has characterised the pattern of allelic associations among over 3 million single nucleotide polymorphisms (SNPs) in the human genome in three major populations (Europeans, Africans and Asians).

Classical epidemiological designs (CASE-CONTROL STUDIES, COHORT STUDIES, CROSS-SECTIONAL STUDIES) are readily applicable to the study of disease–allele associations, as are the statistical methods developed for these designs (e.g. LOGISTIC REGRESSION, SURVIVAL ANALYSIS). These designs are potentially susceptible to the problem of hidden population stratification, which can lead to spurious associations or mask true associations. Family-based association designs are robust to population stratification and usually consist of the use of either parental or sibling controls. Methods for the analysis of matched samples, such as the MCNEMAR'S TEST (also called the transmission disequilibrium test in the context of parental controls) and CONDITIONAL LOGISTIC REGRESSION are applicable to these designs.

The study of disease–allele associations is a complementary strategy to linkage analysis, in the localisation and identification of genes that increase the risk of disease. In general, allelic association is unlikely to be detected when the marker locus is quite far (>1 megabase) from the disease locus, but can be much more powerful than linkage when the marker locus is close enough to the disease locus to be in substantial linkage disequilibrium with it, particularly when the effect size of the disease locus is small. For this reason, allelic association is particularly appealing for searching regions that demonstrate linkage to the disease or to the investigation of specific candidate genes. However, technological developments have enabled the efficient genotyping of up to 1 million SNPs in a single array, and this has led to association studies on the whole-genome scale (called genome-wide association studies, or GWAS) that have coverage of over 90 % of common variants (allele frequency > 5 %) in the genome. *PS*

all subsets regression A form of regression in which *all* possible models are compared using some appropriate criterion for indicating the 'best' models. If there are p explanatory variables in the data, there are a total of $2^p - 1$ possible regression models because each explanatory variable can be in or out of the model and the model containing no explanatory variables is excluded. One possible criterion for comparing models is the MALLOWS C_p STATISTIC and to illustrate its use we will apply it to data that arise from a study of 25 patients with cystic fibrosis reported in O'Neill *et al.* (1983), and also given in Altman (1991). Data for the first three patients are given in the first table. The dependent variable in this case is a measure of malnutrition (PE_{max}). Some of the models considered in the all subsets regression of these data are shown in the second table, together with their associated C_p values, where p refers to the number of parameters in a particular model, i.e. a model that includes a subset of $p - 1$ of the explanatory variables plus an intercept. If C_p is plotted against p, the subsets of explanatory

all subsets regression *Cystic fibrosis data; first three subjects*

Sub	*Age*	*Sex*	*Height*	*Weight*	*BMP*	*FEV*	*RV*	*FRC*	*TLC*	PE_{max}
1	7	0	109	13.1	68	32	258	183	137	95
2	7	1	112	12.9	65	19	449	245	134	85
3	8	0	124	14.1	65	22	441	268	147	100

Sub: subject number
Sex: 0 = male, 1 = female
BMP: body mass (weight/height^2) as a percentage of the age-specific median in normal individuals
FEV: forced expiratory volume in one second
RV: residual volume
FRC: functional residual capacity
TLC: total lung capacity
PE_{max}: maximal statistic expiratory pressure (cmH_2O)

all subsets regression *Some of the models fitted in applying the all subsets regression to the cystic fibrosis data (size is one more than the number of variables in a model, to include the intercept)*

Model	*Size*	*Terms*	C_p
7	2	Sex	17.24
14	3	Sex, weight	4.63
21	4	Age, FEV, RV	2.62
28*	4	Age, BMP, FEV	4.5
35	5	Sex, weight, BMP, FEV	2.95
42	6	Age, weight, BMP, FEV, RV	2.8
49	6	Age, sex, height, FEV, TLC	6.99
56*	7	Age, sex, height, FEV, RV, TLC	7.06
63	8	Sex, weight, BMP, FEV, RV, FRC, TLC	6.49
70	9	Age, height, weight, BMP, FEV, RV, FRC, TLC	8.06
77	9	Age, sex, height, BMP, FEV, RV, FRC, TLC	10.29

* Models close to the line $C_p = p$.

variables most worth considering in trying to find a parsimonious model are those lying close to the line $C_p = p$.

All subsets regression has been found to be particularly useful in applications of COX'S REGRESSION MODEL (see Kuk, 1984). *BSE*

[See also MULTIPLE LINEAR REGRESSION]

Altman, D. G. 1991: *Practical statistics for medical research*. London: CRC/Chapman & Hall. **Kuk, A. Y. C.** 1984: All subsets regression in a proportional hazards model, *Biometrika*, 71, 587–92. **O'Neill, S., Leahy, F., Pasterkamp, H. and Tal, A.** 1983: The effects of chronic hyperfunction, nutritional status and posture on respiratory muscle strength in cystic fibrosis. *American Review of Respiratory Disorders* 128, 1051–4.

alternative hypothesis See HYPOTHESIS TESTS

AMOS See STRUCTURAL EQUATION MODELLING SOFTWARE

analysis of covariance (ANCOVA, ANOCOVA) This is an extension of the analysis of variance (ANOVA) that incorporates a continuous explanatory variable. Where ANOVA aims to detect if there is a change in the mean value of a variable across two or more groups, ANCOVA (or rarely ANOCOVA) does the same but adjusts for a continuous covariate.

Most commonly this covariate will be a baseline measurement, allowing the analysis to adjust for initial variation between participants and isolate the effects due to the treatment factor. However, sometimes a different covariate is used. For example, Karhune *et al.* (1994) consider the association between alcohol intake (divided into four categories) and numbers of Purkinje cells. In doing so they introduce age as a continuous covariate in order to 'control' or 'adjust' for the effects of age on cell numbers.

Under other circumstances the authors could have been interested in the effects of age and wanting to adjust for alcohol intake. Despite being the same analysis computationally, this is not typically what is thought of as analysis of covariance and might more commonly be presented as a 'regression'. Indeed the various analysis of variance methods can all be viewed from within a regression framework, which demonstrates that ANCOVA can be extended to cope with much more than one continuous covariate.

Mathematically, ANCOVA follows a similar path to that for ANOVA and the output is usually summarised in a similar table, although the details may vary.

The promised benefits of the analysis of covariance are clear. If one has an unbalanced observational study, then ANCOVA can adjust for differences in baseline values and remove a potential bias from the results. By the same token, if one has a randomised trial that is naturally balanced, then

ANCOVA reduces the amount of unexplained variation in the data and thus increases the power of the test.

However, ANCOVA can only be employed if the appropriate assumptions are met. These include those of ANOVA (i.e. normality of residuals, homoscedasticity) as well as the appropriateness of the ANCOVA model. Is the relationship with the covariate truly linear? Does the effect of the covariate vary between groups? Failing to meet these assumptions can lead to the introduction of important but subtle biases. It is a frequent concern that medical research papers report a covariate as having been 'controlled' or 'adjusted' for, with no evidence that the control or adjustment was appropriate. For further details see Altman (1991), Owen and Froman (1998), Miller and Chapman (2001) and Vickers and Altman (2001). *AGL*

[See also GENERALISED LINEAR MODEL]

Altman, D.G. 1991: *Practical statistics for medical research.* London: Chapman & Hall. **Karhune, P. J., Erkinjutti, T. and Laippala, P.** 1994: Moderate alcohol consumption and loss of cerebellar Prikinje cells. *British Medical Journal* 308, 1663–7. **Miller, G. A. and Chapman, J. P.** 2001: Misunderstanding analysis of covariance. *Journal of Abnormal Psychology* 110, 40–8. **Owen, S. V. and Froman, R. D.** 1998: Uses and abuses of the analysis of covariance. *Research in Nursing and Health* 21, 557–62. **Vickers, A. J. and Altman, D. G.** 2001: Analysing controlled trials with baseline and follow-up measurements. *British Medical Journal* 323, 123–4.

analysis of variance (ANOVA) Often referring to the one-way analysis of variance, it is a test for a common MEAN in multiple groups that we describe in detail here. Analysis of variance frequently arises in the comparison of more complicated models, but the same logical arguments apply. In all cases, the underlying concept is to partition the observed variance into quantities attributable to specific explanatory sources, and then consider important those sources that explain 'more than their fair share' of the variance.

Despite the confusion sometimes caused by the name, the one-way analysis of variance is a method for testing to see whether multiple samples come from populations that share the same mean. In this respect it can be viewed as an extension to the *t*-test, which assesses whether samples from two populations share a common mean. An analysis of variance performed on two samples is equivalent to performing a *t*-test.

ANOVA assumes that all the samples come from populations with a NORMAL DISTRIBUTION that share the same VARIANCE. It can be viewed in a number of ways, but essentially compares the estimate of the variance obtained within samples (that makes no assumption that the populations have a common mean) with an estimate of the variance from the sample means (which will require the assumption that the populations have the same mean). If the two estimates of the variance are different, then this is evidence that our assumption of equality failed and, therefore, that the populations do not all have the same mean.

Note that the variance of a single sample is estimated as the sum of squared differences from the mean divided by the sample size minus one. The sum of squared differences term is interpretable as a measure of the total variation in the sample. In the analysis of variance, by combining all groups together, one can calculate this measure for all the data. This is termed the 'total sum of squares' or 'total SS'.

Variation in the data is either 'between' or 'within' the samples. The 'within groups sum of squares' or 'within SS' can be calculated as the sum of squared differences from the individual sample means (rather than the differences from the overall mean that produced the total SS). 'Between groups sum of squares' or 'between SS' can be calculated directly, but is most easily calculated by subtraction of the within SS from the total SS.

The two estimates of the variance (or 'mean square' as it is often termed in this context) can then be calculated. The between groups mean square is equal to the between SS divided by the number of groups minus one. The within groups mean square is equal to the within SS divided by the number of observations minus the number of groups.

An *F*-statistic is then calculated as the between groups variance divided by the within groups variance. Under the assumptions of normality and homoscedasticity (common variance) this statistic will be an observation from an *F*-DISTRIBUTION if the groups come from populations with a common mean. The DEGREES OF FREEDOM of the *F*-distribution are the number of groups minus one and the number of observations minus the number of groups.

From the *F*-distribution, we can calculate the probability of observing such an extreme value of the *F*-statistic if the populations have a common mean. This is a one-tailed test. If the value is unusually small, this suggests the between groups variance is unusually small and so is not evidence of variation between the groups. Therefore, the test is to find the probability, if the populations do have a common mean, of observing a value greater than that observed.

A natural way of presenting ANOVA is the ANOVA table. Given N observations that fall into k groups, it is necessary to calculate the total SS and the within SS as described earlier and then the analysis can be completed as presented in the first table. Murphy *et al.* (1994) conducted an analysis of variance to see if milk consumption before the age of 25 affects bone density of the hip in later life. A total of 248 women participated in this part of their study ($N = 248$) and were divided into groups that represent low, medium and high milk consumptions ($k = 3$). The samples had similar variances and so at least one of the assumptions for ANOVA was

analysis of variance (ANOVA) *The analysis of variance table*

Source of variance	*Degrees of freedom*	*Sums of squares*	*Mean squares*	*F*	*P*
Between groups	k − 1	Between SS = Total SS − Within SS	Between MS = Between SS/(k − 1)	Between MS / Within MS	p
Within groups	N − k	Within SS	Within MS = Within SS/(N − k)		
Total	N − 1	Total SS			

analysis of variance (ANOVA) *Approximate reconstruction of the analysis of variance table from Murphy et al.* (1994).

Source of variance	*Degrees of freedom*	*Sums of squares*	*Mean squares*	*F*	*P*
Between groups	**2**	**0.15**	0.08	3.8	0.23
Within groups	**245**	**4.4**	0.02		
Total	**247**	4.6			

(Entries in bold were inferred from the paper, the rest simply follow from the calculations)

satisfied. As is common for reasons of space, the ANOVA table was not presented in the published paper, just the *P*-value, but enough data were presented for an approximate reconstruction.

We can infer that the within SS is approximately 4.4 and the between SS is approximately 0.15. This leads to an *F*-statistic of approximately 4. From the reported *P*-value (0.023), it can be calculated from the *F*-distribution (with 2 and 245 respectively for numerator and denominator degrees of freedom) that the *F*-statistic was 3.8. The conclusion then is that there is evidence that these samples do not come from populations that share a common mean. The reconstructed table is presented in the second table (entries in bold in this table were inferred from the paper, the rest simply follow from the calculations).

It is preferable to conduct an analysis of variance rather than to conduct *t*-tests between all pairs of groups. ANOVA avoids problems of multiple testing and thus keeps control of the SIGNIFICANCE LEVEL. Having conducted an ANOVA and rejected the hypothesis of common means, it may then be desired to test to see which groups are responsible (although a plot of the data might be as informative). In this case, care must be taken to correct for the problems of making MULTIPLE COMPARISONS.

It is important to take note of the assumptions being made, rather than simply ignoring them. ANOVA can be quite robust to variations from normality, but heteroscedasticity can be a serious problem. Residual plots can be used to help assess the normality and BOXPLOTS can be used to help assess the heteroscedasticity. Possible formal tests for the assumptions are the KOLMOGOROV–SMIRNOV TEST and LEVENES TEST respectively.

If the assumptions do not hold, then TRANSFORMATION of the data might correct this. Otherwise a number of nonparametric alternatives to ANOVA exist, the most commonly used being the KRUSKAL–WALLIS TEST and the FRIEDMAN TEST.

The one-way analysis of variance is appropriate when our data are simply divided into a number of groups. There are many other forms of analysis of variance. The TWO-WAY ANALYSIS OF VARIANCE should be used when the groups are defined by two factors. Suppose, for example, we had six groups: the three groups of women in Murphy *et al.* (1994) and three groups of men at the same levels of milk consumption. Rather than a one-way analysis of variance, a two-way analysis of variance with gender and milk consumption as the two factors would be appropriate in this instance.

If the data are multiple observations from the same subjects, perhaps measurements of cholesterol levels 0, 7, 14, 21 and 28 days after starting a new diet on several individuals, then a REPEATED MEASURES ANALYSIS OF VARIANCE would be appropriate. This is a special case of the two-way ANOVA and can be viewed as an extension of the paired sample *t*-test.

If there are observations of more than one characteristic from the individuals in several groups, i.e. measures of both the diastolic and systolic blood pressure, then a multivariate analysis of variance (MANOVA) can be used. If, however, it is desired to correct for a measured baseline covariate, such as body mass index, in the analysis, then an ANALYSIS OF COVARIANCE (ANCOVA) may be used.

All these techniques could be implemented through a regression framework, in most cases MULTIPLE LINEAR REGRESSION. The advantages of doing so would be the transition from the use of a HYPOTHESIS TEST to an actual estimate of effect sizes. This approach would also allow more flexibility; for instance in the case of Murphy *et al.* (1994) we could account for the natural ordering of the levels of milk consumption that ANOVA ignores. As a general principle, estimation and modelling are usually preferred to testing of hypotheses. For further details see Altman (1991) and Altman and Bland (1996). *AGL*

Altman, D. G. 1991: *Practical statistics for medical research.* London: Chapman & Hall. **Altman, D. G. and Bland, J. M.** 1996: Statistics notes: comparing several groups using analysis of variance. *British Medical Journal* 312, 1472–3. **Murphy, S., Khaw, K.-T., May, H. and Compston, J. E.** 1994: Milk consumption and bone mineral density in middle aged and elderly women. *British Medical Journal* 308, 939–41.

area under the curve (AUC) This is a simple and useful method of obtaining a summary measure from plotted data. Medical research is frequently concerned with serial data, as in repeated measurements (see REPEATED MEASURES ANALYSIS OF VARIANCE) on a subject over time, e.g. blood aspirin concentration measured at various times over a 2-hour interval (Matthews *et al.*, 1990). Say we have n measurements y_i taken at times t_i ($i = 1, \ldots, n$). Such data are frequently exhibited by plotting y_i versus t_i and joining the resulting points by straight-line segments resulting in a 'curve'. The resulting area under the curve (AUC) is often used as a single-number summary measure for the individual subject. Further analysis of the subjects or comparison of groups of subjects is carried out based on the summary measures. The AUC for the set of points (y_i, t_i) $i = 1, \ldots, n$ is typically calculated by the trapezium rule:

$$\mathrm{AUC} = \frac{1}{2}\sum_{i=1}^{n-1}(t_{i+1} - t_i)(y_i + y_{i+1})$$

The AUC is used as a summary measure in many areas of medical research, including bioequivalence and pharmacokinetics. It plays an especially important role in the analysis of RECEIVER OPERATING CHARACTERISTIC (ROC) CURVES. The area under the ROC curve of a diagnostic marker (test) measures the ability of the marker to discriminate between healthy and diseased subjects. It is the most commonly used measure of performance of a marker. We use the convention that larger marker values are more indicative of disease. Then if we randomly pick one subject from the healthy population and one from the diseased population we would 'expect' that the value of the marker for the healthy subject would be smaller than the corresponding value for the diseased subject. AUC is the probability that this, in fact, occurs. The larger the AUC, the better the overall discriminatory accuracy of the marker. An area of 1 represents a perfect test while an area of 1/2 represents a worthless test having a discriminatory ability, which is the equivalent of differentiating between healthy and diseased subjects by a fair coin toss. Consider the example discussed in the entry for the ROC curve. The points on the curve are given in the table.

area under the curve *Summary data used in an ROC curve*

Specificity	(y_i)	0	0.56	0.84	0.94	0.98	1.00
1-Sensitivity	(t_i)	0	0.04	0.12	0.32	0.60	1.00

The data presented result in an AUC as follows:

$$\begin{aligned}\mathrm{AUC} &= 0.5[(0.04-0) \times (0 + 0.56) + (0.12-0.04) \\ &\quad \times (0.56 + 0.84) + (0.32-0.12) \times (0.84 \\ &\quad + 0.94) + (0.60-0.32) \times (0.94 + 0.98) \\ &\quad + (1.00-0.60) \times (0.98 + 1.00)] = 0.91\end{aligned}$$

An area of 0.91 indicates the high discriminatory ability of the marker.

For the ROC curve, estimating the area by the trapezium rule is equivalent to computing the Wilcoxon or Mann–Whitney statistic divided by the products of the sample sizes on the healthy and diseased populations. For smoothed ROC curves, alternative estimates of the AUC are available (Faraggi and Reiser, 2002). The effectiveness of alternative diagnostic markers is usually studied by comparing their AUCs (Wieand *et al.*, 1989). Adjustments of these areas for covariate information, selection bias and pooling effects are discussed in the references given in the entry for the ROC curve. Schisterman *et al.* (2001) consider corrections of the AUC for measurement error. For further details see Hanley and McNeil (1982). *DF/BR*

Faraggi, D. and Reiser, B. 2002: Estimation of the area under the ROC curve. *Statistics in Medicine* 21, 3093–106. **Hanley, J. A. and McNeil, B. J.** 1982: The meaning and use of the area under the receiver operating characteristic (ROC) curve. *Radiology* 143, 29–36. **Matthews, J. N. S., Altman D. G., Campbell, M. J. and Royston, P.** 1990: Analysis of serial measurements in medical research. *British Medical Journal* 300, 230–5. **Schisterman, E., Faraggi, D., Reiser, B. and Trevisan, M.** 2001: Statistical inference for the area under the ROC curve in the presence of random measurement error. *American Journal of Epidemiology* 154, 174–9. **Wieand, S., Gail, M. H., James, B. R. and James, K. L.** 1989: A family of non-parametric statistics for comparing diagnostic markers with paired or unpaired data. *Biometrika* 76, 585–92.

artificial intelligence (AI) This branch of computer science is devoted to the simulation of intelligent behaviour in machines. Traditional focus areas of AI are machine vision, MACHINE LEARNING, natural-language processing and speech recognition. Historically an interdisciplinary field, and hence characterised by the presence of several competing paradigms and approaches, recently AI has started developing a more unified conceptual framework, based largely on the convergence of statistical and algorithmic ideas.

A constant theme of AI throughout its history has been 'pattern recognition', the crucial task of detecting 'patterns' (regularities, relations, laws) within data. This task has emerged as a roadblock in all the traditional areas mentioned earlier and hence has attracted significant attention. Since most current approaches to pattern recognition involve significant use of statistics, this has become an important tool in AI in general.

Recently, AI has been applied to a new series of important problems and this, in turn, has heavily affected general AI research. Important applications of modern AI include: intelligent data analysis (see also DATA MINING IN MEDICINE); information retrieval and filtering from the web; bioinformatics; and computational biology. Traditional application areas, by way of contrast, included the design of EXPERT SYSTEMS for medical or industrial diagnosis, methods for scheduling in logistics and creation of other decision-making assistant software.

The imprecise definition of what AI actually is has made it harder in time to gauge the impact of this research field on everyday applications. A number of widely used computer programs would have met early definitions of artificial intelligence, e.g. popular web-based recommendation systems or air travel planning advisors.

Popular techniques for pattern recognition such as NEURAL NETWORKS, decision trees and cluster analysis (see CLUSTER ANALYSIS IN MEDICINE) have made their way into the standard toolbox of data analysis and are commonly found in the toolbox of any biology lab. Machine vision methods are routinely used in analysing medical images, as well as parts of systems such as microarray machines for collecting gene expression data. Web retrieval and email filtering software also incorporate several ideas from natural-language processing and pattern recognition and the modern sequence analysis of genomic data heavily relies on techniques originally developed for speech recognition. Intelligent web agents exist to find, assess and retrieve relevant information for the user and speech-recognition systems are routinely used in automatic phone information systems. The field of artificial intelligence has clearly produced a number of practical applications, but – the critics say – these have been achieved without solving the general problem of building intelligent machines. Maybe for this reason, generally the main success story of AI is reported to be the defeat of the chess world champion Gary Kasparov by an IBM algorithm in 1997.

The origin of the field of AI is often identified with a paper by A. M. Turing, which appeared in 1950 in the journal *Mind*, and with a workshop held at Dartmouth College in the summer of 1956, although many key ideas had already been debated before, during the early years of cybernetics.

Modern techniques of artificial intelligence include Bayesian belief networks, part of the more general field of probabilistic graphical models; pattern-recognition algorithms such as SUPPORT VECTOR MACHINES, which represent the convergence of ideas from classical statistics and from neural networks analysis; statistical analysis of natural language text and machine vision algorithms; reinforcement learning algorithms which represent a connection with control theory; and many other methods. *NC/TDB*

Bishop, C. 1996: *Neural networks for pattern recognition*. Oxford: Oxford University Press. **Mitchell, T.** 1995: *Machine learning*. Maidenhead: McGraw-Hill. **Russell, S. and Norvig, P.** 2002: *Artificial intelligence: a modern approach*, 2nd edition. Harlow: Prentice Hall. **Shawe-Taylor, J. and Cristianini, N.** 2004: *Kernel methods for pattern analysis*. Cambridge: Cambridge University Press.

association This is the statistical dependence between two variables. Measures of association, unlike descriptive statistics of a single variable, summarise the extent to which one variable increases or decreases in relation to a change in a second variable. The basic graphical analysis of two variables is the SCATTERPLOT, which provides evidence of association in the shape and direction of the scatter of points. In the example given here, there appears to be an association between body mass index and systolic blood pressure values in a sample of a few thousand middle-aged men and women: higher values of body mass index tend to be associated with higher values of systolic blood pressure, suggesting a 'positive' association. A 'negative' association, in contrast, would describe a situation where an increase in one variable tends to be related to a decrease in the second variable.

Various statistical measures can be used to interpret the degree of association.

Correlation coefficient. This specifically measures the degree of *linear* association between two quantitative variables on a scale from negative one to positive one. A value of zero indicates a total absence of linear association, while a value of positive or negative one indicates a perfect linear relationship. The correlation coefficient between body mass index and systolic blood pressure in our example was 0.25, indicating a positive association that is less than perfectly linear. However, adherence to a linear relationship is only one form of association and it is easy to imagine other plausible patterns of association, such as a parabolic scatter, in which the change in one variable may be perfectly reflected in the change in the second variable, but the correlation coefficient might be close to zero.

Regression coefficient. In the case of simple linear regression, there is a complete correspondence between the correlation coefficient and the regression coefficient for the slope (β). The regression coefficient, therefore, also measures association, but its value is interpreted as the magnitude of change in the dependent variable that arises, on average, from a unit change in the independent variable. In our example, an estimate of $\hat{\beta} = 1.37$ indicated that a 1 kg/m^2 increase in the body mass index was associated with an average increase of 1.37 mmHg systolic blood pressure. However, in more complex regression models, the regression coefficient can measure other forms of association beyond linear dependence. For example, either the dependent or independent variable may be mathematically transformed, such as raising to a higher power, taking logarithms, etc., and the association measured by the regression coefficient would express a nonlinear change in one variable in response to a change in the second variable.

Relative risk. In the special case of two binary variables, various ratio measures are often used to quantify the degree of association. For example, one variable might be a measure of disease occurrence, the other a biological or environmental quantity. Most commonly the ratio would compare probability of disease expressed as an odds, a risk or some other relevant approximation to the risk. A relative risk value of 1, indicating equal risks in both groups, suggests that no association exists between the biological or environmental quantity and disease.

If a statistical measure suggests positive or negative association, this should not immediately be taken to imply that the association is valid and generalisable. Several considerations might lead us to question the importance of an observed statistical association.

First, consideration of the STANDARD ERROR of the measure of association, generally reflecting the size of the sample, places the magnitude of association in perspective with the magnitude of random error. Apparently strong associations may in fact be poorly estimated and fall short of statistical significance.

Second, an apparent association may be entirely spurious (i.e. 'confounded') due to the influence of other measured, or unmeasured, variables that have not been accounted for in the analysis. For example, in a preliminary statistical enquiry, risk of coronary heart disease may appear to be associated with watching television, although consideration of the underlying relationship with obesity and physical exercise would probably suggest that the preliminary finding was spurious. An association may alter after adjustment for the interdependence of other variables and the general validity of a measure of association would often depend on the extent to which such potential interdependencies have been taken into account. Studies measuring several variables often utilise multiple regression models to estimate adjusted regression coefficients and partial correlation coefficients by including all relevant variables in the model. However, even after allowing for such interdependencies, the much stronger claim of CAUSALITY between two variables would generally require examination of more stringent criteria.

Third, an observed association may be specific to the chosen range of the variables or to the particular group of subjects studied and any inference beyond the range of the data to hand would require careful consideration of the method of sample selection. Various forms of selection bias may limit the generalisability of the association. *JGW*

as treated See INTENTION-TO-TREAT

attenuation due to measurement error This is a bias reducing the size of a correlation or a regression coefficient due to imprecision of data measurement. Consider an analytical epidemiological study in which the aim is to estimate the CORRELATION between true average consumption of alcohol (mg per day) and true average systolic blood pressure (mmHg). Blood pressure measurements are well-known to be variable within individuals and a single measurement is likely to be rather imprecise (see MEASUREMENT PRECISION AND RELIABILITY). Such a statement is even more true of a single day's intake of alcohol as a measure of the true average daily intake of alcohol (even if that day's intake were found to be measured without error). Now, in the epidemiological study we chose, for each participant, to measure systolic blood pressure once and then ask them to recall their alcohol intake the previous day. If we now calculate the Pearson product–moment correlation between the two measures we are likely to get a positive value that may be statistically significant (assuming we have a large enough sample) but will not be particularly high (i.e. not far above zero). Suppose, for the sake of argument that we have found a value of this correlation to be 0.20. It should be fairly obvious that as the measures of systolic blood pressure and alcohol get less precise (equivalent for a fixed population to lowering their reliabilities) the correlation will tend to zero. This is attenuation due to measurement error.

Let the observed measurement of blood pressure for the ith participant be Y_i and the corresponding true average blood pressure be τ_i. Similarly, let the measured alcohol intake be X_i with a true average of η_i. We have estimated the correlation between Y and X, ρ_{YX}, when we are really interested in the correlation between the true values, $\rho_{\eta\tau}$. If the errors of measurement for blood pressure are uncorrelated with those for alcohol consumption then it can be shown that the following relationship holds:

$$\rho_{YX} = \rho_{\eta\tau}\sqrt{(\kappa_Y \kappa_X)} \qquad (1)$$

Here, κ_Y and κ_X are the reliabilities of the blood pressure and alcohol consumption measurements respectively. It follows that:

$$\rho_{\eta\tau} = \rho_{YX}/\sqrt{(\kappa_Y \kappa_X)} \quad (2)$$

Provided we know the reliabilities for the two measurements, this equation can be used to adjust the observed correlation between Y and X to obtain the required correlation between their true average values. If we know that $\kappa_X = 0.3$ and $\kappa_X = 0.7$, for example, the required correlation is $0.2/\sqrt{(0.3 \times 0.7)} = 0.44$.

If, instead of a correlation, the linear regression coefficient for the effect of blood pressure on alcohol consumption were of key interest then:

$$\beta_{YX} = \beta_{\eta\tau}\kappa_X \quad (3)$$

and, again, the required adjustment is straightforward. Equation (3) also holds approximately if we were to use a logistic regression to predict the presence/absence of hypertension.

These calculations are fine as long as we have valid estimates of the reliabilities. However, they are only valid in these very simple situations as described. Epidemiologists almost always wish to adjust their estimates to allow for confounding and some of these confounders are inevitably going to be prone to MEASUREMENT ERROR. Under these circumstances life is considerably more complicated! We cannot even be certain that the estimate of the required parameter will be attenuated, never mind being attenuated in a way described by equation (3). Readers are referred elsewhere to these much more challenging but more realistic situations (Carroll, Ruppert and Stefanski, 1995; Cheng and Van Ness, 1999; Gustafson, 2003). *GD*

Carroll, R. J., Ruppert, D. and Stefanski, L. A. 1995: *Measurement error in nonlinear models*. London: Chapman & Hall. **Cheng, C.-L. and Van Ness, J. W.** 1999: *Statistical regression with measurement error*. London: Arnold. **Gustafson, P.** 2003: *Measurement error and misclassification in statistics and epidemiology*. London: Chapman & Hall/CRC.

attributable risk As a measure of the public health significance of exposure to a risk factor for disease, the attributable risk provides an estimate of the proportion of diseased subjects that may be attributed to the exposure. It is defined by:

$$\lambda = \Pr\{D\} - \Pr\{D|\bar{E}\}$$

where Pr{D} is the probability that an individual develops disease and E and $\bar{E}$ represent whether an individual is exposed or not exposed to the factor of interest (Levin, 1953). Ideally, one would like to know both Pr{D} and $\Pr\{D|\bar{E}\}$ for the population under study, but for some study designs this is not possible, so if one wishes to use the measure, care is needed to design a study that will provide as good an estimate as possible, especially when one employs an observational study. Using BAYES THEOREM and rearranging the equation, we can obtain an expression expressed in terms of the relative risk (RR):

$$\lambda = \Pr\{E\}(RR-1)$$

where $\Pr\{E\}$ is the prevalence of exposure in the population at large. This is a convenient way of expressing the measure of association, because RR is often estimated using alternative study designs, including CASE-CONTROL, COHORT AND CROSS-SECTIONAL STUDIES.

Attributable risk is most easily interpreted when the factor of interest increases risk, i.e. $RR > 1$, and in these cases the possible range of the measure is from 0 to 1. An attributable risk of zero can occur when no individuals in the population are exposed to the factor of interest, or if the factor is not related to risk of disease, $RR = 1$. The measure is not easily interpreted when the exposure is protective, $RR < 1$, so it is generally not used in this case. By redefining the reference group, one can always express the results of a study in a form in which RR is greater than 1, so this is not a serious limitation. In addition, the measure is often expressed as a percent. As RR become large, λ goes to 1, but λ goes to zero either as the proportion exposed, $\Pr\{E\}$, becomes small or as the relative risk, RR, approached the null value of 1. If an entire population is exposed to a particular factor, $\Pr\{E\} = 1$, then the second equation (above) reduces to $\lambda = (RR - 1)/RR$.

The table shows a typical 2×2 table that can be used to display the results from an epidemiological study. In a case-control study, the column totals are generally regarded as being fixed by design and the odds ratio or cross-product ratio is used as a good approximation to the estimate of RR when the disease is rare. In addition, the exposure distribution in the controls, $\Pr\{E\} = \Pr\{E|\bar{D}\}$, is considered to be representative of the exposure distribution in the overall population. Substituting in the sample estimates of these quantities gives rise to what is the maximum likelihood estimate of λ:

$$\hat{\lambda} = \frac{ad-bc}{d(a+c)}$$

attributable risk *Results from an epidemiological study with two levels of exposure and disease status*

	Disease status		
Exposed	*D*	$\bar{D}$	*Total*
E	a	b	$a+b$
$\bar{E}$	c	d	$c+d$
Total	$a+c$	$b+d$	N

When setting a confidence interval about the estimate, Walter (1975) suggests using the normal approximation on the log transformation of the complement of the estimate:

$$\mathrm{Var}\left[\log\left(1-\hat{\lambda}\right)\right] = \frac{a}{c(a+c)} + \frac{b}{d(b+d)}$$

Alternatively, Leung and Kupper (1981) have suggested using a logit transformation in which:

$$\mathrm{Var}\left[\log\frac{\hat{\lambda}}{1-\hat{\lambda}}\right] = \left[\frac{a}{c(a+b)} + \frac{b}{d(b+d)}\right]\left[d(b+d)\right]^2$$

In a cohort study, the row totals in the table are regarded as fixed; therefore such a study does not provide a good internal estimate of the exposure distribution and neither does it provide a good estimate of the unconditional estimate of the probability of disease. In this case, the proportion exposed is usually derived from another study, perhaps an earlier case-control study or a survey of the entire population. A cross-sectional study provides both an estimate of the relative risk and the overall population distribution, so in that sense it is ideal for estimating attributable risk. However, a cross-sectional study suffers in other ways (see CROSS-SECTIONAL STUDIES). Walter (1976) discusses the properties of estimates of attributable risk using these alternative study designs.

Methods for estimating attributable risk for a particular exposure while adjusting for potential confounding factors depends on whether the effect is constant over the levels of the covariates under consideration. When the effect is constant, it can be represented as having a common relative risk over the strata when using a stratified approach, such as the Mantel–Haenszel method, or it can be represented by a main effect only in a model, such as the linear logistic model. In these situations, one can directly use the adjusted estimator or the relative risk, along with an estimate of the exposure distribution in the diseased group in the second equation (above) to obtain an estimate of the adjusted attributable risk (Walter, 1976; Greenland, 1987). However, the assumption that the association can be described without the inclusion of an interaction term is a strong one and it is critical in that a seriously biased estimate can result if it is not true.

An estimate of attributable risk that can be used either in a stratified analysis in which the effect is not homogeneous across strata or in a generalised linear model that includes interaction terms can be expressed as:

$$\lambda = 1 - \Sigma_{i,j}\frac{p_{ij}}{RR_{i|j}}$$

where *j* represents the levels of the factor(s) being adjusted, *i* represents the levels of exposure, p_{ij} is the proportion of diseased individuals in *(i,j)* and $RR_{i|j}$ the relative risk for exposure level *i* for individuals with level *j* of the covariates being adjusted (Walter, 1976; Benichou, 1993). *TRH*

Benichou, J. 1993: Methods of adjustment for estimating the attributable risk in case-control studies: a review. *Statistics in Medicine* 10, 1753–73. **Greenland, S.** 1987: Variance estimators for attributable fraction estimates, consistent in both large strata and sparse data. *Statistics in Medicine* 6, 701–8. **Leung, H. K. and Kupper, L. L.** 1981: Comparison of confidence intervals for attributable risk. *Biometrics* 37, 293–302. **Levin, M. L.** 1953: The occurrence of lung cancer in man. *Acta Unio Internationalis contra Cancrum* 9, 531–41. **Walter, S. D.** 1975: The distribution of Levin's measure of attributable risk. *Biometrika* 62, 371–4. **Walter, S. D.** 1976: The estimation and interpretation of attributable risk in health research. *Biometrics* 32, 829–49.

AUC See AREA UNDER THE CURVE

autocorrelation See CORRELATION

automatic selection procedures These are procedures for identifying a parsimonious model in regression in general and MULTIPLE LINEAR REGRESSION in particular. Such methods are needed because in regression analysis an underfitted model can lead to severely biased estimation and prediction. In contrast, an overfitted model can seriously degrade the efficiency of the resulting parameter estimates and predictions. Consequently a variety of techniques all with the aim of selecting the most important explanatory variables for predicting the response variable and thereby obtaining a parsimonious and effectively predictive model have been developed. Perhaps the three most commonly used methods are *forward selection*, *backward elimination* and a combination of both of these, known as *stepwise regression*.

The forward selection approach begins with an initial model that contains only an intercept and successively adds explanatory variables to the model from the pool of candidate variables until a stage is reached where none of the candidate variables, if added to the current model, would contribute information that is statistically important concerning the expected value of the response.

The backward elimination method begins with an initial model that contains all the explanatory variables being used in the study and then first identifies the single variable that contributes the least information about the expected value of the response; if this is deemed not to be 'significant' then the variable is eliminated from the current model. Successive steps of the method result in a 'final' model from which no further variables can be eliminated without adversely affecting, in a statistical sense, the predicted value of the expected response.

The stepwise regression method combines elements of both forward selection and backward elimination. The initial model considered is one that contains only an intercept. Explanatory variables are then considered for inclusion in the current model, as described previously for forward selection, but now in each step of the procedure variables included previously are also considered for possible elimination as in the backward method, and they might be removed if the presence of new variables in the model make their contribution to predicting the expected response no longer significant.

In multiple linear regression the criterion used for assessing whether or not a variable should be added to an existing model in forward selection or removed from an existing model in backward elimination is, essentially, the change in the residual sum-of-squares produced by the inclusion or exclusion of the variable. Specifically in forward selection an '*F*-statistic', known as the *F-to-enter*, is calculated as:

$$F = \frac{RSS_m - RSS_{m+1}}{RSS_{m+1}/(n-m-2)}$$

where $RSS_m$and RSS_{m+1} are the residual sums of squares when models with m and $m+1$ explanatory variables have been fitted. The *F*-to-enter is then compared with a preset term; calculated *F*s greater than the preset value lead to the variable under consideration being added to the model. In backward selection a calculated *F* less that a corresponding *F*-to-remove leads to a variable being removed from the current model. In the stepwise procedure variables are entered as with forward selection, but after each addition of a new variable those variables currently in the model are considered for removal by the backward elimination process. (For more details see Petrie and Sabin, 2005.) In other types of regression, for example, LOGISTIC REGRESSION, other criteria are used for judging whether or not a variable should be entered into or removed from the current model. When applying regression techniques to HIGH-DIMENSIONAL DATA more sophisticated variable selection techniques are needed (see, for example, Francois, 2008).

None of the automatic procedures for selecting subsets of variables is foolproof and it is possible for them to be seriously misleading in some circumstances (see Agresti, 1996). That said, at least one can be more confident in a chosen model if all three procedures converge on to the same set of variables, as occurs quite frequently, but not always, in practice. When different subsets of variables are indicated, judgement is necessary to decide on a preferred model, such judgement being based on the desire to create a parsimonious model that is likely to be generalisable, not overly complex as if modelling mere quirks of the particular dataset on which it is based, and yet including important or standard parameters deemed to be of clinical relevance. *BSE*

[See also ALL SUBSETS REGRESSION]

Agresti, A. 1996: *Introduction to categorical data analysis*. New York: John Wiley & Sons, Inc.. **Francois, D.** 2008: *High-dimensional data analysis: from optimal metrics to feature selection*. VDM Verlag. **Petrie, A. and Sabin, S.** 2005: *Medical statistics at a glance*, 2nd edition, Wiley-Blackwell, Chichester.

available case analysis This is an approach to multivariate data containing missing values on a number of variables, in which MEANS, VARIANCES and covariances (see COVARIANCE MATRIX) are calculated from all available subjects with nonmissing values on the variable (means and variances) or pair of variables (covariances) involved. Although this approach makes use of as much of the observed data as possible, it does have disadvantages. For example, the summary statistics for each variable may be based on different numbers of observations and the calculated variance–covariance matrix may now not be suitable for methods of multivariate analysis such as PRINCIPAL COMPONENTS ANALYSIS and FACTOR ANALYSIS for reasons described in Schafer (1997). *BSE*

[See also MISSING DATA, MULTIPLE IMPUTATION]

Schafer, J. L. 1997: *Analysis of incomplete multivariate data*. Boca Raton, Florida: Chapman & Hall/CRC.

average age at death This flawed statistic is sometimes used for summarising life expectancy and other aspects of mortality. For example, Andersen (1990) comments on a study that compared average age at death for male symphony orchestra conductors and for the entire US male population and showed that, on average, the conductors lived about 4 years longer. The difference is, however, largely illusory because as age at entry was birth, those in the US male population who died in infancy and childhood were included in the calculation of the average lifespan, whereas only men who survived long enough to become conductors could enter the conductor cohort. The apparent difference in longevity disappeared after accounting for infant and perinatal mortality.

In the other direction, a study in the USA that used average age at death of rock stars (which, on the basis of 321 such deaths, they found to be 36.9 years) to warn of the perils of rock music also got it wrong. It took no account of the rock stars still alive.

Proper analysis of mortality involves the determination of AGE-SPECIFIC RATES for mortality, which requires denominator data on the age distribution of the population (see Colton, 1974). *BSE*

Andersen, B. 1990: *Methodological errors in medical research*. Oxford: Blackwell Scientific. **Colton, T.** 1974: *Statistics in medicine*. Boston: Little, Brown and Co.

average treatment effect on the treated (ATT) See PROPENSITY SCORES

average treatment effect on the untreated (ATU) See PROPENSITY SCORES

B

back-calculation Also known as back-projection, this is a means of estimating, for example, past HIV infection rates and predicting the number of new AIDS cases in the future and was first proposed in the mid-1980s (Brookmeyer and Gail, 1986). The essence of the method is contained in the equation:

$$d(t) = \int_0^t h(s)\, p(t-s) \mathrm{d}s$$

where $d(t)$ and $h(s)$ denote the disease diagnosis rate at time t and the infection rate at time s, and $p(.)$ indicates the probability distribution (density) of the incubation time (or INCUBATION PERIOD). This expression states that the rate of disease diagnosis at time t depends on the rate of new infections at time s and on the distribution of the incubation time $t-s$. Therefore, if any two of these three components are known, the third can be *inferred*. Typically, the disease diagnosis rate for t up to the current time T and the distribution of the incubation time are assumed known and the infection rate is estimated.

The figure explains the idea in a discrete time framework using the HIV epidemic as an example. Here the interest is in estimating HIV incidence and in predicting future AIDS cases. Suppose data on new AIDS cases over time up to the current time T are available together with the information on the distribution of the incubation time. It is then possible to reconstruct the number of past infections that have resulted in the observed AIDS cases. The estimated incidence of HIV can be used in conjunction with the distribution of the incubation time to produce short-term projections of new AIDS diagnoses. Note that in this particular case, the MEDIAN length of the incubation time is of the order of 10 years, with very few individuals developing AIDS within a short time period from infection. The observed AIDS cases therefore provide information on infections that occurred in the distant past, rather than in recent years. Estimates of incidence of infection for the years just preceding T will necessarily be quite inaccurate, as they are based on little information. Care should then be taken in the interpretation of recent trends in the number of infections. However, this problem will not affect projections of AIDS cases as long as they are short term.

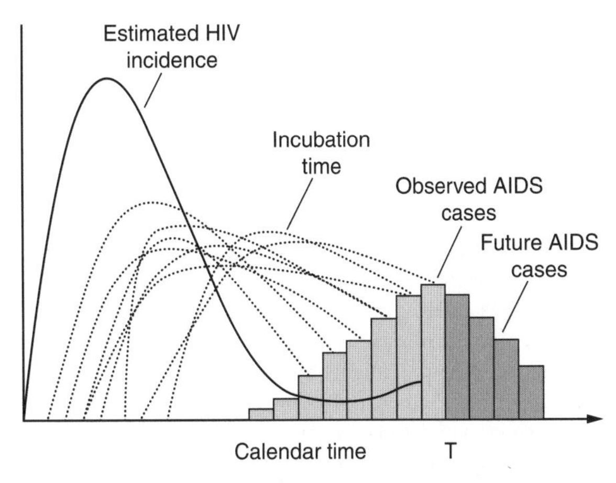

back-calculation *Back-calculation using HIV incidence and prediction of future AIDS cases*

A number of formulations of the back-calculation equation have been proposed. To give a flavour of the estimation problem, it is convenient to use a discrete version of our first equation. Let t_0 be the beginning of the epidemic and y_k the number of individuals that develop the disease endpoint of interest (e.g. AIDS in an HIV context) in the kth time interval $[t_{k-1}, t_k)$ for $k = 1, \ldots, K$. Suppose that f_{ij}, the probability of developing the disease endpoint in the jth time interval given infection in the ith interval, is also known. Then the expected number of new disease cases in $[t_{k-1}, t_k)$ can be expressed as:

$$E(y_k) = \sum_{i=1}^{k} E(h_i) f_{ik}$$

where h_i is the unobserved number of new infections in the ith time interval. Assuming that the h_i are independently distributed according to a POISSON DISTRIBUTION with parameter $E(h_i)$, then the y_k are also Poisson distributed with parameter $E(y_k)$. From this the likelihood for the observed data can be constructed and maximised to obtained estimates of the number of new infections over time (see MAXIMUM LIKELIHOOD DISTRIBUTION). In practice, estimation of $\mathbf{h} = (h_1, \ldots, h_k)$ is not so straightforward. The high dimensionality of $\mathbf{h}$ can lead to unstable estimates. In order to avoid lack of identifiability,

Encyclopaedic Companion to Medical Statistics: Second Edition Edited by Brian S. Everitt and Christopher R. Palmer

some structure needs to be imposed on the shape of **h**. This has typically been achieved by choosing fully parametric models for $\mathbf{h} = \mathbf{h}(\mathbf{q})$. The problem is then reduced to an estimation of **q**, conveniently chosen to be of a lower dimension than **h**. Alternatively, to retain some flexibility, weakly parametric models (i.e. step functions constant over a long period of time) have been specified or smoothness constraints on **h** have been introduced. This has created a rich literature, especially in the HIV field (see Brookmeyer and Gail, 1994).

Attractive in principle, given the simplicity of the idea, the method does require precise knowledge of at least two of the three components introduced already. However, perfect information is rarely available. For example, as in HIV, the incidence of the disease endpoint, typically acquired from surveillance schemes, might be affected by reporting delay or underreporting. Further, the distribution of the incubation time may also be imprecisely known. Results can be highly sensitive to misspecification of the inputs. It is therefore important that data are appropriately adjusted for delay in reporting before they are used in the back-calculation. Equally, it is essential that sensitivity analyses to the model chosen for the distribution of the incubation time are carried out. One more limitation of the method is the inability to provide precise estimates of the incidence of infection in recent times. This is a particularly serious problem for diseases with long incubation times, as seen in the HIV example.

These limitations notwithstanding, the back-calculation method has been widely used and developed in various ways, especially in the HIV area. Notably, the original methodology assumed a fixed distribution for the incubation time, independent of calendar time or age at infection. However, therapeutic changes over time and the discovery of a clear dependence of HIV progression on age at infection have made the time–age independence assumption untenable. This has led to the development of age–time specific versions of back-calculation. Equally, the need to estimate the number of individuals at different stages of the development of HIV has resulted in the development of 'staged' back-calculations, where the incubation time is divided into stages according to the value of markers of HIV disease. A final example is given by the need to refine estimation of HIV incidence, especially in recent years, and AIDS projections. This has resulted in a further development of the method, now able to incorporate external information on the disease spread as well as other surveillance data, in addition to AIDS diagnoses (see De Angelis, Gilks and Day, 1998; Becker, Lewis and Li, 2003).

The method and its developments have found important application in other contexts besides HIV. Examples include the assessment of the bovine spongiform encephalopathy epidemic in cattle and the consequent Creutzfeldt–Jakob disease epidemic in humans in Great Britain, the estimation of the Hepatitis C virus epidemic in France and the estimation of the number of new injecting drug users in Australia. *DDA*

Bacchetti, P. 1998: Back-calculation. In Armitage, P. and Colton, T. (eds), *Encyclopedia of biostatistics*, Vol. 1. Chichester: John Wiley & Sons, Ltd, pp. 235–42. **Becker, N. G., Lewis, J. J. C. and Li, Z. F.** 2003: Age-specific back-projection of HIV diagnosis data. *Statistics in Medicine* 22, 2177–90. **Brookmeyer, R. and Gail, M. H.** 1986: Minimum size of the acquired immunodeficiency syndrome (AIDS) epidemic in the United States. *Lancet* 2(8519), 1320–2. **Brookmeyer, R. and Gail, M. H.** 1994: *AIDS epidemiology: a quantitative approach*. New York: Oxford University Press. **De Angelis, D., Gilks, W. R. and Day, N. E.** 1998: Bayesian projection of the acquired immune deficiency syndrome epidemic. *Journal of the Royal Statistical Society C – App* 47, 449–81.

back-projection See BACK-CALCULATION

backwards regression See LOGISTIC REGRESSION, MULTIPLE LINEAR REGRESSION

balance See RANDOMISATION

bar chart A graphical display of data classified into a number of (usually unordered) categories. Equal width rectangular bars are used to represent each category, with the heights of the bars being proportional to the observed frequency in the corresponding category. An example is shown in the figure.

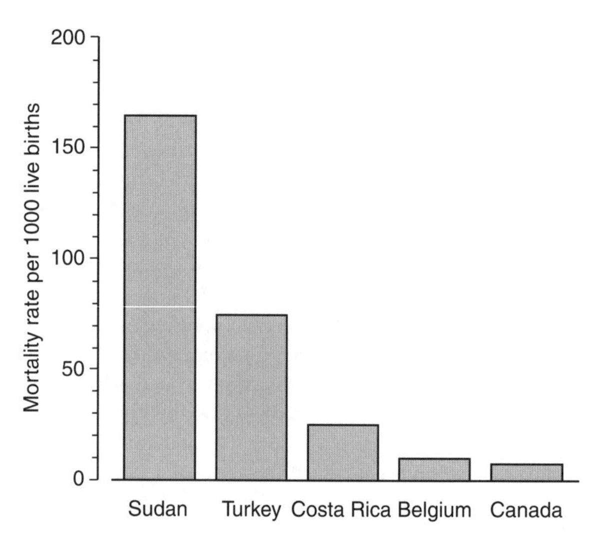

bar chart *Mortality rates per 1000 live births for children under five in five different countries*

An extension of the simple bar chart is the component bar chart (also known as the stacked bar chart) in which particular lengths of each bar are differentiated to represent a number of frequencies associated with each category forming the chart. Shading or colour can be used to enhance the display. An example is given in the second figure; here the numbers of patients in the four categories of a response variable for two treatments (BP and CP) are displayed.

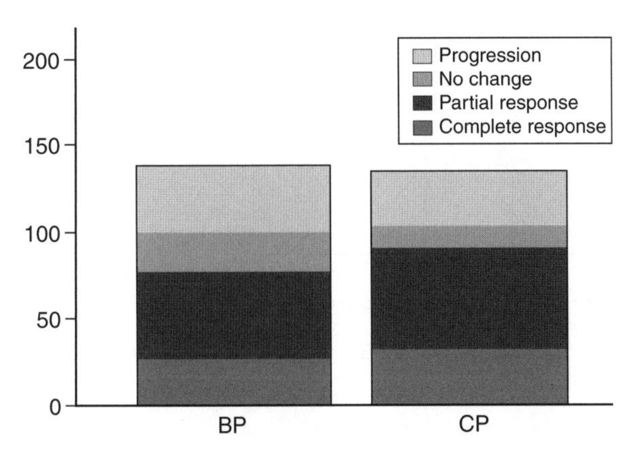

bar chart *Response to treatment*

The basic bar chart is often of little more help in understanding categorical data than the numerical data themselves. However, sophisticated adaptations of the graphic can become an extremely effective tool for displaying a complex set of categorical data. That this is so can be illustrated by an example taken from Sarkar (2008) that uses data summarising the fates of the 2201 passengers on the *Titanic*. The data are categorised by economic status (class of ticket, first, second or third, or crew), sex (male or female), age (adult or child) and whether they survived or not (the data are available on Sarkar's website, http://lmdv.r-forge.r-project.org/). The first diagram produced by Sarkar is shown in the third figure. This plot looks impressive but is dominated by the third 'panel' (adult males) as heights of bars represent counts and all panels have the same limits. Sadly, all the plot tells us is that there were many more males than females aboard (particularly among the crew, which is the largest group) and that there were even fewer children. The plot becomes more illuminating about what really happened to the passengers if the proportion of survivors is plotted and by allowing independent horizontal scales for the different

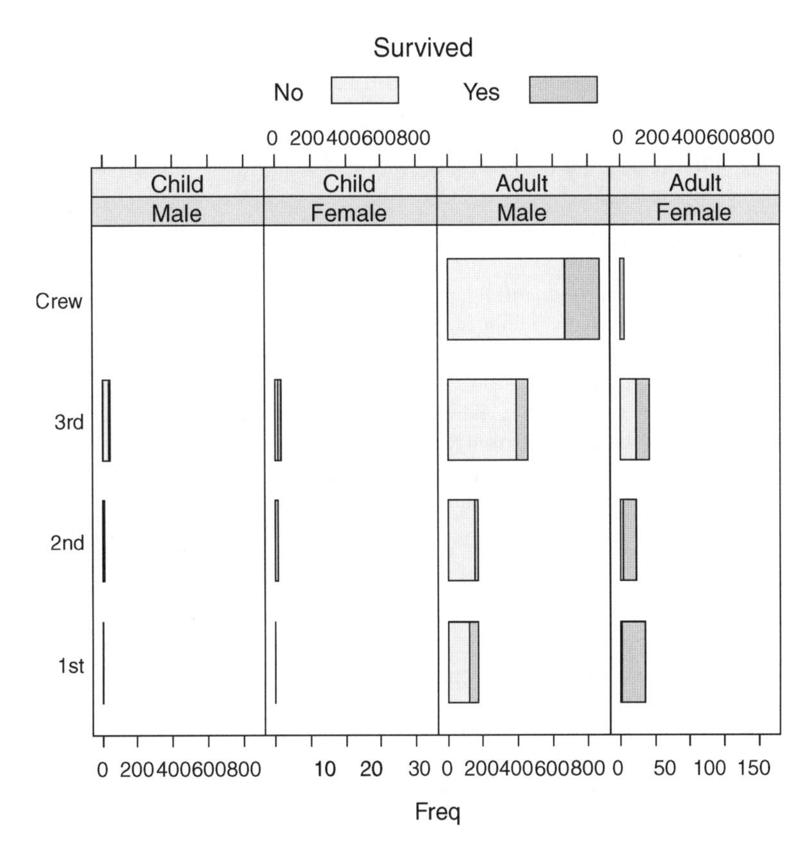

bar chart *Summary of the fate of passengers of the Titanic, classified by sex, age and class (used with the permission of Springer)*

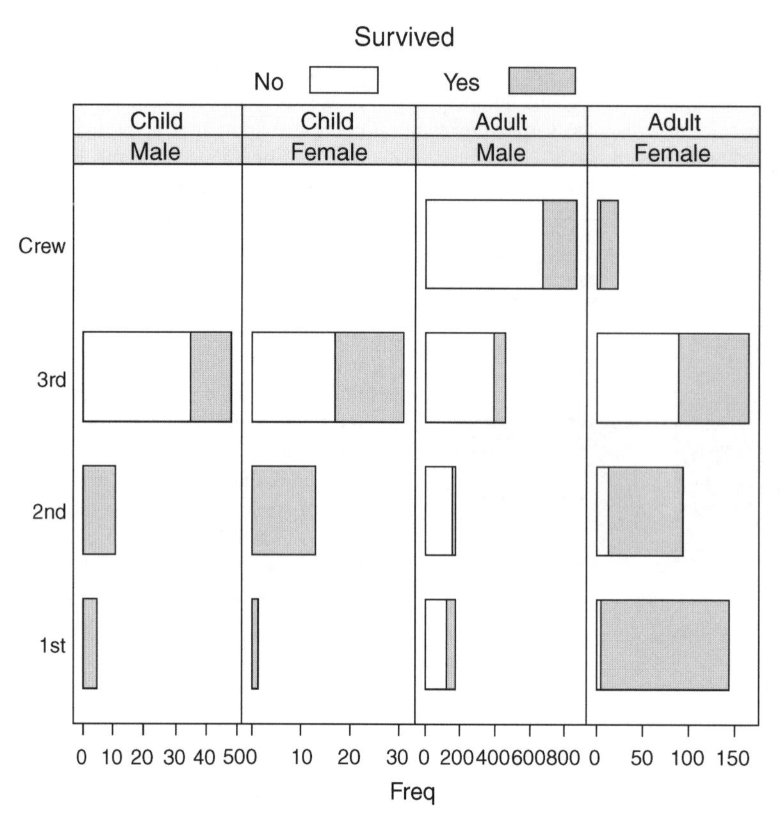

bar chart *Survival among different subgroups of passengers on the Titanic, with a different horizontal scale in each panel (used with the permission of Springer)*

'panels' in the plot; this plot is shown in the fourth figure, which emphasises the proportion of survivors within each subgroup rather than the absolute numbers. The proportion of survivors is lowest among third-class passengers and the diagram makes it very clear that the 'women and children first' policy did not work very well for this class of passenger. (I am grateful to Dr Sarkar and to Springer for allowing me to reproduce the two diagrams.) *BSE*

[See also HISTOGRAM, PIE CHART]

Sarkar, D. 2008: *Lattice: multivariate data visualization with R*. New York: Springer.

baseline measurements These are measurements taken at the beginning of a study. (This section, however, concentrates on their role within the context of a CLINICAL TRIAL.) Baseline measurements come in different varieties and may also have a variety of purposes (Senn, 1998). First are *demographic characteristics* of the patient, which either do not change (such as, for example, sex), change slowly, if at all (such as height), or change at the same rate for all patients (such as age). The second and simplest sort is a measurement of the same type as the outcome variable, but taken one or more occasions prior to RANDOMISATION: these might be referred to as *true baselines*. Third, one has *baseline correlates*, measurements taken before randomisation on variables other than the true outcome variable but predictive of it and which may vary during the trial.

Some such measurements are invariably collected as part of the process of deciding which patients may enter the trial. For example, it may be required in a trial of asthma that patients be aged 18–65, have a baseline forced expiratory volume in 1 second (FEV_1) no more than 75 % of that predicted by age, sex and height, can demonstrate 20 % reversibility when given a bronchodilator and are normotensive. Simply fulfilling these requirements will necessitate taking at least two FEV_1 measurements prior to randomisation, as well as diastolic and systolic blood pressures and recording height, sex and age of patients, the last two being variables that are always recorded anyway. In practice, many other things, such as, for example, concomitant medication and the centre in which the patient is treated, will also be recorded and such things are also potential candidates for any model.

All three kinds of baseline may be used for four further common purposes. First, to help characterise the patients in the trial; second, to compare the groups in various arms of the trials; third, to provide conditional estimates, which will generally have greater precision and will be conditionally unbiased given the observed baseline measurements; and fourth, to investigate the constancy or otherwise of the treatment effect as patients vary. There is a fifth purpose to which true baselines can be put: as part of a general repeated-measures framework, such as, for example, the random slopes approach of Laird and her co-writers (Laird and Ware, 1982; Laird and Wang, 1990).

The following are some issues that arise.

Generalising results. Using the actual baseline measurements observed is clearly superior to using the inclusion criteria to characterise the trial, as the latter simply define values the patients might have had, rather than values they did have. It is unclear, however, to what extent the baseline characteristics measured can be used as a basis for generalising the results, since many things that might be important have not been measured. In any case, the logic of clinical trials is comparative rather than representative. In many areas it is accepted that the patients in a clinical trial will be unrepresentative. It is hoped rather that an additive scale of measurement may permit useful application of the results. This may require use of additional covariate information from the target population (Lane and Nelder, 1982).

Comparing groups. This suggests perhaps that the second use of comparing the groups is more valuable. However, from another point of view, the comparison of groups is simply an unimportant resting point on the road to adjustment. If the baseline measurement is prognostic a superior inference will be made by conditioning on it, whether or not it is imbalanced. Particularly questionable is the common practice of comparing groups at baseline in terms of significance tests (Altman, 1985; Senn, 1989, 1994). This has no useful role as part of a general strategy for analysis but could possibly have some limited use as a test of the randomisation process itself (to detect fraud, for example) as part of some general quality control of the trial. That being so, however, the commonly employed significance levels of 5 % seem inappropriate.

Covariate adjustment. This in turn suggests that the third use of baseline measurements, to provide conditional estimates by stratification or adjustment using ANALYSIS OF COVARIANCE, is the most important of these purposes. However, many trialists appear to have a strong (one might say unreasonably strong) preference for simple analyses over more complicated ones, preferring simple *t*-TESTS to analysis of covariance and the log-rank test to PROPORTIONAL HAZARDS regression.

Effects in subgroups. The fourth use of baselines is also controversial. An issue of bias VARIANCE trade-off arises. Some, the 'splitters', are more worried about bias and consider that it is important to report treatment effects by subgroups defined by baseline measurements; others, the 'poolers', regard variance as being the bigger concern and point to the unreliability of inferences based on small groups.

Use of true baselines in repeated measures analysis. A controversial matter here is that the baselines are sometimes explicitly measured as part of the outcome despite the fact that, obviously, the treatments cannot affect the baselines. If the baselines, or some function of them, are also included as covariates this may lead to causally acceptable inferences. The simplest example is where change scores are used and baselines are fitted as covariates. Inferences as to the effect of treatment are then identical to those that would be made using raw outcomes and baselines as covariates (Laird, 1983).

Nonlinear models. For the general linear model, the expected value of the estimator conditioning on the covariate is the same as the unconditional estimator. This is not generally true for nonlinear cases. Gail *et al.* (1984) have considered where this does and does not hold for a variety of models. Robinson and Jewel have concentrated on the case of LOGISTIC REGRESSION (Robinson and Jewell, 1991) and Ford *et al.* on the proportional hazards model (Ford *et al.*, 1995). It is usually the case where nonlinear models are involved that fitting covariates leads to an increase in variance. However, there is a biasing of the treatment effect towards the null if prognostic covariates are not fitted, so it does not follow that fitting such covariates necessarily leads to a loss of power. There are many arguments, in fact, as to why the conditional estimators should be preferred (Lindsey and Lambert, 1998), but so-called marginal approaches using working correlations matrices have also become extremely popular, in particular via the GENERALISED ESTIMATING EQUATION approach of Liang and Zeger (1986).

Measurement error. It is well known that where a covariate is measured with error, the estimate of its effect on outcome is attenuated (see ATTENUATION DUE TO MEASUREMENT ERROR). The false conclusion is sometimes drawn that under such circumstances, analysis of covariance does not yield conditionally unbiased estimators (Chambless and Roeback, 1993). What has been overlooked is a second attenuation: that of the true baseline difference on the observed baseline difference (Senn, 1994, 1995). The variance of the observed covariate will exceed that of the 'true' covariate. The covariance of the two can be shown to be the variance of the true covariate. Hence, since the regression of observed on true is the covariance divided by the variance of true this regression is 1. However, the regression of true on observed is the covariance divided by the variance of observed and hence is less than 1. On average the true baseline difference is closer to zero than the observed baseline difference. The two attenuations exactly cancel out and so it turns out that correcting for an imbalance in observed covariates using the observed covariates is the right thing to do.

Correcting for true baselines. It has also been claimed that, in the case where the covariate is of the same kind as the outcome measure, in other words is a true baseline, analysis of covariance is only appropriate if the baselines are balanced and that unadjusted change scores provide an unbiased estimate in the more general case (Liang and Zeger, 2000). This is incorrect as the following counterexample shows. Imagine a trial in hypertension in which, quite irrationally, but as is theoretically possible, we include only patients who have diastolic blood pressures (DBP) of either 95 mmHg or 105 mmHg (to the nearest mmHg). Forty patients of each sort are recruited and are allocated, otherwise at random, but in proportions 3:1 in the first stratum and 1:3 in the second stratum. In the absence of any further knowledge, a perfectly reasonable estimate under these unreasonable circumstances would be obtained by subtracting mean DBP under the active treatment from mean DBP under placebo separately in each of the two strata and averaging the results. What would be misleading would be to take DBP at baseline from DBP at outcome and compare the average over both strata under active treatment with that under placebo. Yet, since we can recode 95 and 105 to a dummy variable with values 0 and 1 by dividing by 10 and subtracting 9.5, the first approach is formally equivalent to analysis of covariance and the second approach is simply that of change-scores.

Choice of covariates. Regulatory authorities are naturally nervous that sponsors may unfairly manipulate results by choosing the model most favourable to them and journal editors ought to have similar fears about their authors. One way of protecting the Type I error rate is to pre-specify the model, which is common practice with the pharmaceutical industry and recommended by various guidelines (International Conference on Harmonisation, 1999). A model-checking approach based on randomisation tests not using the treatment information is an alternative (Edwards, 1999). From the Bayesian point of view, however, this is simply a formal and pointless game. Having a variable in a model is equivalent to saying that one knows nothing about its effects. An excluded variable is one for which the effect is known to be zero. Other positions are possible. If users cannot agree which model is appropriate, then even given a fiction of prior ignorance about the treatment effect, different posterior opinions will obtain. This may give a sort of justification for frequentist sensitivity analysis. However, it should not be forgotten that whereas little may be known about the effect of treatment in advance of a trial, the same cannot be said for covariates, and such prior knowledge is an important guide in the choice of model (Senn, 2000).

Choice of true baseline to fit. It sometimes happens that in a run-in period a number of measurements are made of the eventual outcome variable. Often the last only is used as a covariate, although this is in fact wasteful as it implicitly assumes, which is unlikely, an autoregressive process. It is better either to fit the mean baseline or, more generally, each of the baseline measurements (Senn, 1997).

Subgroup analysis and treatment by covariate interactions. Such analyses are sometimes undertaken to examine the constancy of the treatment effects. The former is the natural extension of the stratification approach and the latter of analysis of covariance. Once such analyses have been undertaken a problem of combining results from individual strata arises. This issue arises frequently in the analysis of multicentre trials where centre-specific effects may be examined. If such effects are weighted by precision, this is equivalent to using SAS Type II sums of squares. If unweighted averages are used, the equivalence is to Type III sums of squares (Gallo, 2001). From one point of view, once an interaction has been fitted an overall effect is no longer of interest, but it can also be maintained that this attitude naively maintains the claims of reducing bias against reducing variance. Compromise positions involving random effects are sometimes used but not nearly as commonly in combining effects from various centres as in the analogous problem in meta-analysis of combining effects from various trials and hardly ever when other sorts of covariate are involved. Great care must be taken in interpreting effects when interactions are involved (Chuang-Stein and Tong, 1996) and it may be wise to fit a model without interactions as a check (Senn, 2000). This is recommended by international guidelines (International Conference on Harmonisation, 1999). Note that it is the interaction of covariates with the treatment effect that causes problems in this way. The interaction of covariates with each other is not an issue of the same importance, since it is merely the joint effect of these for which one seeks adjustment.

Of course, there are also very many technical issues to be confronted when considering adjustment for baseline measurements. A particular instructive case to consider is that where a number, say k, of binary covariates have been measured. One approach is that of stratification. One creates 2^k strata based on these covariates, forms a treatment contrast within each and then a combination, usually weighted, of them all. In the case of a linear model, if within-stratum variances are assumed constant and combined efficiently, this is equivalent to carrying out analysis of covariance using ordinary least squares having formed a factor for each binary covariate and fitting all interactions between covariates up to and including the highest. Note that if we characterise an interaction by the number of covariates r involved, with $r=1$ corresponding to main effects and $r=0$ to the general intercept, then there are in general $\binom{k}{r}$ such terms and that since $\sum_{r=0}^{k}\binom{k}{r}=2^k$ one is fitting the same number of degrees of freedom in both cases, as is indeed necessary, since the two are equivalent. If there are q treatments in the trial, fitting the treatment effect as a main effect only removes a further $q-1$ degrees of freedom. Most trials have two treatments and so $q=2, q-1=1$ and one further degree of freedom is removed. Going further and considering interactions between covariates and the treatment is moving a step along the road to splitting the treatment effect by subgroup. Interactions between treatment and covariates will not be considered here and instead the issue of interactions of covariates among themselves is considered.

The approach via analysis of covariance has greater flexibility than that using strata, since it is possible to fit the main effect of covariates only or a limited degree of interactions between them. Furthermore, if we move to covariates with more than two levels, models with reduced degrees of freedom are possible. Suppose, for example, that we have measured baseline severity on a three-point scale as 1, 2, 3. We can code this using two dummy variables. This is equivalent to fitting a linear and a quadratic covariate. Possible schemes for both approaches are illustrated in the table. Note that these are equivalent since $Z_1=X_2+2X_3-1$ and $Z_2=1-3X_2$. The circumstances under which one would wish to fit X_3 alone and especially X_2 alone are fewer than those where one might choose to fit Z_1 alone. Thus, there is an attractive flexibility and economy of the analysis of covariance approach. Of course, where truly continuous measures are involved the advantages for the covariate modelling approach increase as arbitrary cut-points have to be used if stratification is employed. *SS*

baseline measurements *Coding schemes for a three-point severity scale*

		Dummy variables		*Covariates*	
	Intercept	*Severity level 2*	*Severity level 3*	*Linear*	*Quadratic*
Severity		X_2	X_3	Z_1	Z_2
1	1	0	0	−1	1
2	1	1	0	0	−2
3	1	0	1	1	1

Altman, D. G. 1985: Comparability of randomized groups. *Statistician* 34, 1, 125–36. **Chambless, L. E. and Roeback, J. R.** 1993: Methods for assessing difference between groups in change when initial measurements is subject to intra-individual variation. *Statistics in Medicine* 12, 13, 1213–37. **Chuang-Stein, C. and Tong, D. M.** 1996: The impact of parametrization on the interpretation of the main-effect terms in the presence of an interaction. *Drug Information Journal* 30, 421–4. **Edwards, D.** 1999: On model prespecification in confirmatory randomized studies. *Statistics in Medicine* 18, 7, 771–85. **Ford, I. *et al.*** 1995: Model inconsistency, illustrated by the Cox proportional hazards model. *Statistics in Medicine* 14, 735–46. **Gail, M. H. *et al.*** 1984: Biased estimates of treatment effects in randomized experiments with nonlinear regressions and omitted covariates. *Biometrika* 71, 431–44. **Gallo, P.** 2001: Center-weighting issues in multicenter clinical trials. *Journal of Biopharmaceutical Statistics* 10, 2, 145–63. **International Conference on Harmonisation** 1999: Statistical principles for clinical trials (ICH E9). *Statistics in Medicine* 18, 1905–42. **Laird, N.** 1983: Further comparative analyses of pre-test post-test research designs. *The American Statistician* 37, 329–30. **Laird, N. M. and Wang, F.** 1990: Estimating rates of change in randomized clinical trials. *Controlled Clinical Trials* 11, 6, 405–19. **Laird, N. M. and Ware, J. H.** 1982: Random-effects models for longitudinal data. *Biometrics* 38, 4, 963–74. **Lane, P. W. and Nelder, J. A.** 1982: Analysis of covariance and standardization as instances of prediction. *Biometrics* 38, 3, 613–21. **Liang, K. Y. and Zeger, S. L.** 1986: Longitudinal data-analysis using generalized linear models. *Biometrika* 73, 1, 13–22. **Liang, K. Y. and Zeger, S. L.** 2000: Longitudinal data analysis of continuous and discrete responses for pre-post designs. *Sankhya – the Indian Journal of Statistics Series B* 62, 134–48. **Lindsey, J. K. and Lambert, P.** 1998: On the appropriateness of marginal models for repeated measurements in clinical trials. *Statistics in Medicine* 17, 4, 447–69. **Robinson, L. D. and Jewell, N. P.** 1991: Some surprising results about covariate adjustment in logistic regression models. *International Statistical Review* 58, 227–40. **Senn, S. J.** 1994: Testing for baseline balance in clinical trials. *Statistics in Medicine* 13, 17, 1715–26. **Senn, S.** 1995: In defence of analysis of covariance: a reply to Chambless and Roeback. *Statistics in Medicine* 14, 20, 2283–5. **Senn, S. J.** 1997: *Statistical issues in drug development*. Chichester: John Wiley & Sons, Ltd. **Senn, S. J.** 1998: Baseline adjustment in longitudinal studies. In Armitage, P. and Colton, T. (eds), *Encyclopedia in biostatistics*, Vol. 1. New York: John Wiley & Sons, Inc., pp. 253–7. **Senn, S. J.** 2000: The many modes of meta. *Drug Information Journal* 34, 535–49.

basic reproduction number A term used in the theory of infectious diseases for the average number of secondary cases that an infectious individual produces in a completely susceptible population. The basic reproduction number (R) of an infectious agent is a key factor determining the rate of spread and the proportion of the host population affected. The number depends on the duration of the infectious period, the probability of infecting a susceptible individual during one contact and the number of new susceptible individuals contacted per unit time; consequently, it may vary considerably for different infectious diseases and also for the same disease in different populations. The value of R has implications for whether there is a positive probability that an epidemic may occur and the proportion of the population infected were an epidemic to take place. The larger the value of R, the larger the fraction of the population that must be immunised to prevent an epidemic. A recent account of the use of the basic reproduction number is given in Wesley and Allen (2009). *BSE*

Wesley, C. L. and Allen, L. J. S. 2009: The basic reproduction number in epidemic models with periodic demographies. *Journal of Biological Dynamics* 3, 116–29.

Bayes' theorem Bayes' theorem is a method by which conditional probabilities (see CONDITIONAL PROBABILITY) may be manipulated. In particular, it provides a means of reversing the conditioning in order to obtain probability statements regarding specific events of interest. Bayes' theorem itself was described originally in 'An essay . . .' and published two years after the death of the Reverend Thomas Bayes in 1761 (Bayes, 1763). The use of Bayes' theorem in manipulating conditional probabilities is used widely, even if users may not be aware of it, but its use for more general quantities, for example relative risks, gives rise to considerable controversy (see BAYESIAN METHODS).

Following Spiegelhalter, Abrams and Myles (2004), consider two events a and b, using the *multiplication rule* of PROBABILITY; the probability of both a and b occurring, denoted '$a \wedge b$', is given by:

$$P(a \wedge b) = P(a|b) \times P(b) = P(b|a) \times P(a) \qquad (1)$$

Rearranging this equation yields an expression for $P(b|a)$:

$$P(b|a) = \frac{P(a|b) \times P(b)}{P(a)} \qquad (2)$$

Considering the events '$a \wedge b$' and '$a \wedge \bar{b}$', where $\bar{b}$ represents the event 'not b', then these are mutually exclusive and using the *addition rule* of probability (see probability), we can 'extend the argument' for a to include b:

$$P(a) = P(a \wedge b) + P(a \wedge \bar{b}) = P(a|b) \times P(b) + P(a|\bar{b}) \times P(\bar{b}) \qquad (3)$$

Combining equations (2) and (3) yields Bayes' theorem, which expresses $P(b|a)$ in terms of conditional probabilities for a and the probability of b:

$$P(b|a) = \frac{P(a|b) \times P(b)}{P(a|b) \times P(b) + P(a|\bar{b}) \times P(\bar{b})} \qquad (4)$$

Conversely, this equation could be considered in terms of $\bar{b}$, i.e.:

$$P(\bar{b}|a) = \frac{P(a|\bar{b}) \times P(\bar{b})}{P(a|b) \times P(b) + P(a|\bar{b}) \times P(\bar{b})} \qquad (5)$$

Dividing equation (4) by (5) yields:

$$\frac{P(b|a)}{P(\bar{b}|a)} = \frac{P(a|b) \times P(b)}{P(a|\bar{b}) \times P(\bar{b})} = \frac{P(a|b)}{P(a|\bar{b})} \times \frac{P(b)}{P(\bar{b})} \qquad (6)$$

Hence, equation (6) is Bayes' theorem in terms of the odds of event b, in which the prior odds of b, i.e. $P(b)/P(\bar{b})$, are modified in the light of the data, i.e. the LIKELIHOOD RATIO, to yield the posterior odds of b, conditional on knowing a. In the case when b is not a simple event, i.e. b or $\bar{b}$ and $b_1, \ldots, b_n$ are in fact n mutually exclusive events, equation (1.5) can be generalised so that the probability of $b_1|a$ is given by:

$$P(b_1|a) = \frac{P(a|b_1) \times P(b_1)}{\sum_j P(a|b_j) \times P(b_j)} \qquad (7)$$

Consider the case of wishing to determine whether a patient has a particular disease, D, that the background prevalence of the disease in the population is 30 %, but a test is available. The characteristics of the test are such that a patient who has disease D will test positive with probability 0.8, i.e. the SENSITIVITY, while the probability of a positive test result for nondiseased patients is 0.2, i.e. one minus the SPECIFICITY. Using Bayes' theorem we can calculate the probability that a patient who has tested positive does have disease D as:

$$\begin{aligned} P(D|T+) &= \frac{P(T+|D) \times P(D)}{P(T+|D) \times P(D) + P(T+|\bar{D}) \times P(\bar{D})} \\ &= \frac{0.8 \times 0.3}{0.8 \times 0.3 + 0.2 \times 0.7} = 0.63 \end{aligned} \qquad (8)$$

In terms of odds, the prior odds of having the disease are 0.3/0.7 = 0.43, i.e. just under 1 in 2, while the likelihood ratio is 4 for a positive test result, i.e. 0.8/0.2, and the posterior odds of having the disease, having tested positive, is therefore 1.72. *KRA*

Bayes, T. 1763: An essay towards solving a problem in the doctrine of chances. *Philosophical Transactions of the Royal Society* 53, 418.
Spiegelhalter, D. J., Abrams, K. R. and Myles, J. P. 2004:

Bayesian approaches to clinical trials and health-care evaluation. Chichester: John Wiley & Sons, Ltd.

Bayesian methods The use of Bayes' theorem for manipulating conditional probabilities of specific events of interest is used widely without controversy (see BAYES' THEOREM). However, Bayes' theorem may also be applied to more general quantities, e.g. relative risks, and in such settings the inclusion of *external* information in the form of the unconditional probability distribution (see CONDITIONAL PROBABILITY) for the quantity of interest, the prior distribution, rather than the prevalence as in diagnostic testing, *is* controversial and has attracted considerable debate (Spiegelhalter, Abrams and Myles, 2004).

In short, a Bayesian approach (generally) has been described as 'the explicit quantitative use of external evidence in the design, monitoring, analysis, interpretation and reporting of a health-care evaluation' (Spiegelhalter, Abrams and Myles, 2004). As such, it has been argued that a Bayesian approach is often more *flexible* than traditional methods as it can adapt to each unique situation; is more *efficient* in that it uses all available evidence thought to be relevant; is more *useful* in providing predictions and inputs for making decisions about individual patients and summarising evidence regarding a problem, e.g. making direct probability statements that are clinically relevant; and more *ethical* in both clarifying the basis for randomisation and fully exploiting the experience provided by past patients.

There are three elements of a Bayesian approach to medical statistics: subjective probability, assessment of evidence and decision theory. While the second is the one that is most often thought of as a Bayesian approach per se, i. e. the use of external evidence, the first underpins many of the purported advantages of a Bayesian approach, while the third illustrates the wider perspective that a Bayesian approach can give. A frequentist view of probability relies on a long-run view of the world, with probability being defined as the long-run frequency of events occurring (see PROBABILITY). While such a view is entirely consistent with replicable events, when considering unique events, such as the probability that a patient has a particular disease, such a reliance on repeatability makes little sense. A Bayesian approach views probability as a degree of belief in an event occurring, which does not rely only on repeatability but also encompasses a subjective nature of probability, as we all bring own experiences and background information in making probability assessments (Lindley, 1985). The use of a decision theoretic approach to statistical inference places the decision regarding a parameter within the context of the potential loss/gain in utility associated with making decisions (Lindley, 1985).

Fundamental in both frequentist and Bayesian approaches to statistical inference is the likelihood function (see LIKELIHOOD). From a frequentist perspective the likelihood function summarises how plausible different values of a parameter are by using an inverse argument, i.e. for a given value of the unknown parameter how plausible are the data that have been observed. A Bayesian approach uses the likelihood function, $P(y|\theta)$, in the same manner, i.e. as a summary of the relationship between data observed (y) and unknown parameter (θ), but using Bayes' theorem reverses the conditioning to obtain the probability distribution for the unknown parameter conditional on both the data and any background information summarised in the prior distribution, $P(\theta)$. Thus:

$$P(\theta|y) = \frac{P(\theta)P(y|\theta)}{P(y)} = \frac{P(\theta)P(y|\theta)}{\int P(\theta)P(y|\theta)d\theta} \quad (1)$$

Although this equation is applicable whether the model contains a single unknown parameter or multiple unknown parameters, the specification of $P(\theta)$ (see PRIOR DISTRIBUTIONS) and the computation of $P(y)$ (see COMPUTATIONAL METHODS) can be more difficult, but an added complexity is that we are often only interested in certain key parameters, e.g. a treatment effect, and wish to consider the other parameters as nuisance parameters. Thus, in addition to obtaining the joint posterior distribution, we often obtain the *marginal posterior distribution* for one or more parameters, say $\theta=(\delta, \phi)$; then the marginal posterior distribution for δ is given by:

$$P(\delta|y) = \int P(\theta|y)\mathrm{d}{\rightarrow}\varphi \quad (2)$$

As with the computation of $P(y)$ in equation (1), the integration out of the remaining model parameters in equation (2) is very rarely analytically tractable.

The prior distribution does not necessarily have to be temporally prior to the study in question, but rather is a summary of the pertinent external information, i.e. either based on other studies, subjective beliefs or a combination of the two. When there are multiple sources of external evidence in the form of other study results, then the prior distribution may be based on a synthesis of such evidence using meta-analysis or generalised evidence synthesis techniques, which may downweight some sources of external evidence, e.g. observational studies, or may adjust the results for potential confounders or in order to make the synthesis of more relevance to the study in question (Spiegelhalter, Abrams and Myles, 2004). In terms of using subjective prior beliefs, there have been a number of methods advocated for the elicitation of such beliefs using a variety of methods, ranging from informal discussion, through the use of structure questionnaires possibly using a 'trial roulette' format to the use of interactive computer elicitation techniques (Chaloner *et al.*, 1993; Spiegelhalter, Freedman and Parmar, 1994). When the beliefs of multiple individuals are elicited then consideration has to be given as to whether these should be pooled in a formal manner or used independently (Genest and Zidek, 1986).

A particular type of prior distribution that is often used is what is termed a 'noninformative' or 'vague/prior' distribution. Such a distribution is deemed to be 'vague' relative to the likelihood so that the data from the study in question dominate the analysis. While such an analysis appeals to analysts wishing to maintain a sense of objectivity but nevertheless take advantage of other aspects of adopting a Bayesian approach, e.g. the ability to make direct probability statements, when considering prior distributions for parameters in complex models other than main effects, e.g. variance components, careful consideration has to be given to what 'vague' really means, and this should be assessed as part of a sensitivity analysis (Spiegelhalter, Abrams and Myles, 2004). A related issue is that of whether a 'vague' prior distribution is invariant to transformations, i.e. what is vague on one scale may in fact be informative on another, and in such circumstances Jeffreys' priors may be considered, which, although not necessarily 'vague', are invariant to transformations (Bernardo and Smith, 1994). In complex multiparameter models the specification of a joint prior distribution can be a difficult task in itself, since assuming independence between all parameters, and thus being able to specify a series of univariate prior distributions, is usually unreasonable. A consequence is that we often have to specify conditional prior distributions.

In summary, there is no such thing as a 'correct' or single prior distribution, and consideration of a range (or 'community') of prior distributions is advocated (Spiegelhalter, Freedman and Parmar, 1994; Spiegelhalter, Abrams and Myles, 2004). Such a 'community' could contain a 'vague' prior distribution, a 'sceptical' prior distribution, i.e. one that places only a small probability on an intervention being beneficial, an 'enthusiastic' prior distribution and a prior essentially based at the null (Spiegelhalter, Freedman and Parmar, 1994; Spiegelhalter, Abrams and Myles, 2004).

Having obtained the posterior distribution using Bayes' theorem (1) all subsequent inference is based on it. Standard measures of location and uncertainty may be obtained, e.g. posterior mean and variance, and the posterior density itself may be plotted, which is especially important when it exhibits unusual behaviour, e.g. multimodal. CREDIBLE INTERVALS (CrIs) can also be calculated, which are analogous to CONFIDENCE INTERVALS but which have the interpretation often incorrectly ascribed to CIs, namely that they are intervals in which the unknown parameter lies with a specific posterior probability. CrIs can be obtained in a number of ways, either as equal-tail area intervals or as highest posterior density intervals (HPDIs), which have the property that no point outside the interval has a higher point probability than a point inside the interval and are particularly informative when the posterior distribution is either skew or multimodal (Spiegelhalter, Abrams and Myles, 2004). In addition to obtaining CrIs, a particularly appealing advantage of a Bayesian approach is that direct probability statements can be made that are of direct clinical relevance, e.g. the posterior probability that a relative risk is above a certain value or is within a certain specified range (Spiegelhalter, Abrams and Myles, 2004).

Another advantage of adopting a Bayesian approach is the ability to make predictive statements regarding future data by obtaining the posterior predictive distribution. The posterior predictive distribution for future data is obtained by integrating the likelihood function for the future data over the posterior distribution, i.e. current state of knowledge regarding the parameter, so that the predictive distribution for future data, x, having observed data y is given by:

$$P(x|y) = \int P(x|\theta)P(\theta|y)\mathrm{d}\theta \quad (3)$$

This equation can be used specifically in the monitoring of studies, since having obtained the posterior predictive distribution, direct probability statements can therefore be made regarding the eventual 'observed' study result and thus decisions made as to whether to continue or not (see CLINICAL TRIALS).

An alternative form of the predictive distribution is to use the prior distribution rather than the posterior distribution and so the resulting predictive distribution is in fact that for the data observed. Comparison of this with the observed data has been advocated a means by which prior-data conflict can be assessed, although this raises fundamental questions when subjective beliefs are used (Spiegelhalter, Abrams and Myles, 2004).

In many biomedical settings data accumulates sequentially over time and an important advantage in the use of a Bayesian approach is the ability of the Bayes' theorem to naturally accommodate such scenarios (Bernardo and Smith, 1994). Essentially, the posterior distribution at one time point becomes the prior distribution for the subsequent time point, assuming that the data can be considered to be conditionally independent. Thus, if data y_1 are observed first, followed by data y_2, then:

$$P(\theta|y_1, y_2) \propto P(y_1|\theta)P(y_2|\theta)P(\theta) \propto P(y_2|\theta)P(\theta|y_1) \quad (4)$$

Of fundamental importance to the practical application of Bayesian methods in a medical setting are the assumptions made regarding model parameters. In many situations specific model parameters may represent subgroups of individuals within a single study (see SUBGROUP ANALYSIS), studies within a meta-analysis or units within an institutional comparison setting. Such multiplicity of parameters requires assumptions to be made regardless of whether a frequentist or Bayesian approach is adopted. From a

Bayesian perspective three possibilities exist: the parameters can be thought to be identical and therefore all the data pooled and the common parameter estimated; the parameters can be thought to be independent and therefore each subgroup/study/unit analysed separately (specifying an independent prior distribution for each); or the parameters can be thought to be 'similar' in the sense that we thought them not to be systematically different, in which case they are termed 'exchangeable'. If the assumption of exchangeability a priori is thought to be a reasonable one then the parameters are assumed to be drawn from a common distribution (with unknown hyperparameters) – this specifies a hierarchical or multilevel model (see MULTILEVEL MODELS). Consequently, in estimating a specific parameter, i.e. the underlying effect in a subgroup/study/unit, we 'borrow strength' from the other parameters via the common distribution. In practical terms, this means that a Bayesian approach to problems of multiplicity ensures that individual parameters are shrunk towards some overall common effect and that the 'borrowing of strength' ensures that there is less uncertainty surrounding the underlying effect within an individual subgroup/study/unit than had been originally observed in the data (Spiegelhalter, Abrams and Myles, 2004). Specification of prior distributions for the unknown hyperparameters in the model then encompass the *degree* to which we believe individual subgroups/studies/units may be different to one another.

As with statistical modelling generally, model criticism can take the form of answering these questions: If a different statistical model were used would different conclusion be reached? How well does the model perform, i.e. how well does it model the data? In terms of different statistical models, obviously different models could be used and results compared or some form of model selection process may be used (see later). Regarding 'model fit', one approach is to consider prediction of the observed data based on the model and to compare this with the actual observed data using a cross-validation approach to produce the conditional predictive ordinate (CPO) (Gilks, Richardson and Spiegelhalter, 1996). Alternatively, an overall assessment of model performance can be calculated deviance information criterion (DIC) (Spiegelhalter *et al.*, 2002). In addition, the use of specific prior distributions raises the question of whether different conclusions would be drawn, legitimately, by individuals holding different prior beliefs. However, sometimes equally important is the specific specification of PRIOR DISTRIBUTIONS even though they may be intended to be 'vague'. Consequently, the use of a Bayesian approach dictates the need for careful and conscientious sensitivity analyses and this may appear daunting to the uninitiated analyst.

Model selection, whether relating to the specific parametric form or covariates included in a model, can be achieved either by qualitatively comparing aspects of model fit, e.g. CPOs and DIC discussed earlier or quantitatively via the use of Bayes' factors (Bernardo and Smith, 1994). Bayes' factors provide a means of assessing the relative plausibility of the two competing models, in an analogous manner to a LIKELIHOOD RATIO, but having integrated over the prior distributions for model hyperparameters. Consequently, the specification of improper prior distributions, which often arise when attempting to represent 'vague' beliefs, causes computational difficulties (Bernardo and Smith, 1994). While Bayes' factors themselves can be used to compare competing models directly, and which do not have to be nested, they can also be used in conjunction with prior model probabilities to obtain the posterior model probabilities, i.e. the plausibility of the competing models based on both data and subjective prior beliefs, and which can, in turn, be used to average across models, so that the estimation of a treatment effect, for example, takes into account both the within and between model uncertainty present (Kass and Raftery, 1995).

As has already been mentioned, the application of Bayesian methods to realistic biomedical problems can be computationally intensive, with only highly stylised examples being analytically tractable. In order to evaluate integrals such as those in equations (1) and (2) three broad techniques have been considered: asymptotic approximations, quadrature (numerical integration) techniques and simulation methods (Bernardo and Smith, 1994). The development of MARKOV CHAIN MONTE CARLO (MCMC) simulation methods together with user-friendly software such as WinBUGS (see BUGS AND WINBUGS) has enabled the use of a Bayesian approach to be a realistic choice for many analysts regardless of philosophical credence.

The table summarises the differences between a frequentist and a Bayesian approach to many of the issues that arise in the design, monitoring, analysis and interpretation of RCTs and which are now discussed briefly.

Although in practice Bayesian methods have been applied more frequently in the analysis of RCTs, use of Bayesian methods in specifically the design of early phase trails in which decisions as to the appropriate dose level or whether to initiate a confirmatory trial have to be taken as data accumulates has received attention (Gatsonis and Greenhouse, 1992; Stallard, 1998). The role that elicitation of prior beliefs and demands from various stakeholders (clinicians, patients and policymakers) has to play in confirming (or refuting) the need for a randomised trial on the basis of equipoise has also been advocated, whether or not these are used in a formal assessment of whether a proposed RCT is likely to lead to a definitive answer given the resources available and uncertainty in, for example, the event rate in the control group (Spiegelhalter, Abrams and Myles, 2004).

Bayesian methods *Comparison of frequentist and Bayesian approaches to design, monitoring and analysis/interpretation of RCTs (adapted from Spiegelhalter, Abrams and Myles, 2004)*

Issue	*Frequentist*	*Bayesian*
External information	Informally used in design	Used formally to specify prior
Sample size	Required to detect minimum clinically significant difference at pre-specified level of Type I and II errors	Assumed fixed, but assessment of probability of final CrI excluding clinically significant difference, allowing for uncertainty in inputs
Parameter of interest	Fixed state of nature	Unknown quantity
Randomisation	Justifies hypothesis testing	Not necessary due to subjective nature of probability
Basic question	How likely are data given value of parameter?	How likely is value of parameter given data?
Presentation of results	Likelihood functions, *P*-values and CIs	Plots of posterior, posterior probabilities of quantities of interest, CrIs, posterior used in decision model
Interim analyses	*P*-values and estimates adjusted for number of analyses	Inference not affected by number of analyses
Interim predictions	Conditional power	Use posterior predictive distribution
Subsets	Adjusted *P*-values, e.g. Bonferroni	Subset effects 'shrunk' using 'sceptical' prior

A crucial aspect of conducting large-scale PHASE III TRIALS is the issue of monitoring the trial as data accumulate in order to minimise exposure of patients to less effective (or even harmful) interventions. From a frequentist perspective such monitoring raises issues of multiplicity and for which methods to adjust for this to exist. The use of a Bayesian approach to accumulating evidence is entirely natural, in that at various stages during a trial the posterior distribution for the outcome is an assessment of the current state of knowledge and on which decisions regarding continuation/termination should be based without the need for adjustment (Fayers, Ashby and Parmar, 1997). An additional advantage is the ability to predict, using the posterior predictive distribution at interim inspections, what the consequences of continuation would be in terms of the eventual posterior distribution, conditional on the data so far (Abrams, 1998). An alternative approach that has been suggested extends the consideration of the posterior distribution to incorporate the potential losses of making (Spiegelhalter, Abrams and Myles, 2004).

One key question, however, is what prior to use in such monitoring situations. As regards the situation in which a difference in favour of one intervention has been detected, then a 'sceptical' prior (see PRIOR DISTRIBUTION) has been advocated, on the grounds that if the data so far are sufficient to convince a sceptic of the merits of a particular intervention then continuation would appear inappropriate (Fayers, Ashby and Parmar, 1997). Similarly, when no difference has been detected at an interim analysis, an enthusiastic prior distribution could be used to assess whether there is sufficient evidence for a proponent of an intervention to rule out a benefit.

Having conducted an RCT, how should the results be analysed and interpreted from a Bayesian perspective and what advantages do they confer? Ultimately, a Bayesian approach allows an exploration of how and why individuals interpreting the same RCT evidence may reach differing conclusions – namely that they held different a priori beliefs, although in the light of substantial evidence even 'sceptics' and 'enthusiasts' should converge to a consensus. The use of Bayesian methods also focuses attention on estimation and/or decision making and enables direct probability statements to be made that are of clinical relevance. It also enables the inclusion of pertinent external information, which in the case of RCTs that are relatively small, but which have produced large effects and appear to be 'too good to be true', provide a means by which such results can be ameliorated (Spiegelhalter, Abrams and Myles, 2004). An alternative approach in such circumstances is to ask the question: What prior beliefs would I have to hold in order not to accept the findings of an RCT? If the prior beliefs required to overturn such findings are so 'extreme' that it is unlikely for them to be held by a rational individual then the RCT results are accepted at 'face value'.

Frequently in RCTs, interest focuses on subgroups of patients (see SUBGROUP ANALYSIS) and interpretation of the effects of an intervention within such subgroups raises issue of multiplicity. A Bayesian approach to subgroup analyses considers the simultaneous analysis of the subgroups within a hierarchical model, in which a 'sceptical' prior distribution is placed on the degree to which the estimates of effectiveness within individual subgroups differ from one another – a consequence of such an approach is that aberrant effects in relatively small subgroups of patients are 'shrunk' towards a common overall effect, the degree of shrinkage depending on both the size of the subgroup and the degree of scepticism expressed. Such an approach thus reduces the possibility that spurious findings are accepted unwittingly.

Bayesian approaches to the analysis of RCTs other than two-group parallel designs have also been advocated, including CROSS-OVER TRIALS, FACTORIAL DESIGNS and CLUSTER RANDOMISED TRIALS (Spiegelhalter, 2001).

The growth of EVIDENCE-BASED MEDICINE and healthcare is based on the systematic searching for and synthesis of research evidence. Meta-analysis, the quantitative pooling of evidence from 'similar' studies, raises a number of methodological issues for which a Bayesian approach has been advocated.

The most fundamental issue in meta-analysis is *heterogeneity* – statistical, clinical and methodological. Statistical heterogeneity refers to the study-to-study variability in terms of the estimates associated with each study. When excessive statistical heterogeneity exists attempts should be made to explain it in terms of study- and patient-level covariates, but this is not always possible and so random effects models, which allow for such heterogeneity, are often used (Spiegelhalter, Abrams and Myles, 2004). Estimation of the variance components within such models can be problematic, especially when the number of studies is small and Bayesian methods have the advantage of not only allowing for the uncertainty in variance component estimates but also allow for the possibility of informative prior distributions on variance components based on other external evidence.

Clinical heterogeneity refers to the fact that different studies may have used different doses, may have had different patient populations, e.g. in terms of age, and may have considered different comparators. In particular, studies that compare different interventions only provide indirect evidence for other comparisons and the use of multiparameter evidence synthesis methods within a Bayesian framework have been advocated in order that the appropriate correlation and uncertainty is taken into account. A specific issue for which Bayesian methods are advantageous is when *baseline risk* is considered as a possible treatment modifier, i.e. the event rate in the control group. Clearly regression techniques have to allow for the correlation induced between the control group event rate and treatment effect, which is most easily accomplished by using effectively a multivariate meta-analysis model. Such multivariate models can also be used when multiple or surrogate outcomes are considered.

Methodological heterogeneity often refers to study design and Bayesian methods for the synthesis of evidence from a variety of disparate sources have been developed, e.g. randomised and observational studies, epidemiological and toxicological and qualitative and quantitative studies. These methods can allow for both heterogeneity between different sources of evidence and can be extended to allow for differing levels of bias associated with different study designs and quality.

Specific methodological issues in EPIDEMIOLOGY for which Bayesian methods have been advocated are: MEASUREMENT ERROR, MISSING DATA and pharmacoepidemiology, when assessing evidence on potentially rare but serious adverse events. One specific area of epidemiology for which Bayesian and empirical Bayes' methods have been used for some considerable time is SPATIAL EPIDEMIOLOGY, in which interrelationships between geographical areas are considered.

The comparison of institutions in terms of health outcomes, often referred to as profiling, raises a number of methodological issues, most notably multiplicity and issues concerned with interpreting the outcome in individual 'units' that appear aberrant and for which Bayesian methods have been applied.

In evaluating healthcare interventions interest often focuses not only on clinical effectiveness but also on cost-effectiveness (see COST-EFFECTIVE ANALYSIS), with both clinical outcomes and resource use/cost data collected as part of the study. Methodological issues arise when analysing both outcomes simultaneously, most notable of which is the correlation between the two and for which Bayesian methods have been advocated.

Although collection of both clinical and cost data within an RCT is highly desirable, such studies are often of relatively short duration and extrapolation to the longer term and to include other outcomes is often required. Such extrapolation is most frequently achieved within a decision-modelling framework, which decomposes the intervention/disease pathway into a finite number of transitions or states between which patients can move (see DECISION THEORY and MARKOV CHAIN MONTE CARLO). Decision models can assess either clinical or cost-effectiveness of competing interventions or policies, with different parts of the model being populated by either different sources of evidence or the same source, e.g. study, by using a common metric for different health states, usually a utility or quality of life outcome (see QUALITY OF LIFE MEASUREMENT). The key advantages that a Bayesian approach confers on such models are the ability to infer indirectly key model inputs on which there may be no direct evidence and allow for appropriate sources of uncertainty and correlation in the model inputs. The development of economic decision models can also play an important role in identifying aspects of the model (and therefore intervention/disease process) about which there is considerable uncertainty and on which further research may need to be commissioned.

While the areas of application above have concentrated on epidemiological and evaluation studies, Bayesian methods are beginning to be developed in other areas of biomedical research, most notably image analysis, time series and genetics, especially the analysis of gene expression data.

While the use of Bayesian methods in many areas of biomedical research conveys numerous advantages, their use requires careful and conscientious application, which places considerable emphasis on the role of sensitivity analyses with respect to the statistical model, prior distributions and computational methods (see MARKOV CHAIN MONTE CARLO and BUGS and WINBUGS). In order to improve and harmonise the reporting of analyses using Bayesian methods a checklist *BayesWatch* (Spiegelhalter, Abrams and Myles, 2004) has been developed. *KRA*

Abrams, K. R. 1998: Monitoring randomised controlled trials – Parkinson's disease trial illustrates the dangers of stopping early. *British Medical Journal* 316, 7139, 1183–4. **Bernardo, J. M. and Smith, A. F. M.** 1994: *Bayesian theory*. Chichester: John Wiley & Sons, Ltd. **Chaloner, K., Church, T., Louis, T. A. and Matts, J. P.** 1993: Graphical elicitation of a prior distribution for a clinical trial. *Statistician* 42, 341–53. **Fayers, P. M., Ashby, D. and Parmar, M. K. B.** 1997: Tutorial in biostatistics: Bayesian data monitoring in clinical trials. *Statistics in Medicine* 16, 1413–30. **Gatsonis, C. and Greenhouse, J. B.** 1992: Bayesian methods for Phase I clinical trials. *Statistics in Medicine* 11, 1377–89. **Genest, C. and Zidek, J.** 1986: Combining probability distributions: a critique and an annotated bibliography (with discussion). *Statistical Science* 1, 114–48. **Gilks, W. R., Richardson, S. and Spiegelhalter, D. J.** 1996: *Markov chain Monte Carlo methods in practice*. New York: Chapman & Hall. **Kass, R. and Raftery, A.** 1995: Bayes' factors and model uncertainty. *Journal of the American Statistical Association* 90, 773–95. **Lindley, D. V.** 1985: *Making decisions*, 2nd edition. Chichester: John Wiley & Sons, Ltd. **Parmar, M. K. B., Spiegelhalter, D. J. and Freedman, L. S.** 1994: The CHART trials: Bayesian design and monitoring in practice. *Statistics in Medicine* 13, 1297–312. **Spiegelhalter, D.** 2001: Bayesian methods for cluster randomized trials with continuous responses. *Statistics in Medicine* 20, 435–52. **Spiegelhalter, D. J., Abrams, K. R. and Myles, J. P.** 2004: *Bayesian approaches to clinical trials and health-care evaluation*. Chichester: John Wiley & Sons, Ltd. **Spiegelhalter, D. J., Freedman, L. S. and Parmar, M. K. B.** 1994: Bayesian approaches to randomised trials (with discussion). *Journal of the Royal Statistical Society Series A* 157, 357–87. **Spiegelhalter, D. J., Best, N. G., Carlin, B. P. and van der Linde, A.** 2002: Bayesian measures of model complexity and fit (with discussion). *Journal of the Royal Statistical Society B* 64, 583–640. **Stallard, N.** 1998: Sample size determination for Phase II clinical trials based on Bayesian decision theory. *Biometrics* 54, 279–94.

Bayesian networks See GRAPHICAL MODELS

Bayesian persuasion probabilities These are posterior probabilities that a new treatment being tested in a Phase II clinical trial is better than or no better than a standard treatment. In a Phase II trial INTERIM ANALYSES are carried out to determine whether or not to stop the trial early because, on the basis of the data already accrued, the new treatment appears either unlikely to be better than the standard treatment or unlikely not to be better than it.

One method of determining whether or not to stop the trial is that of persuasion probabilities. The persuade-the-pessimist probability (PPP) is defined as the posterior probability that the new treatment is better than the standard treatment. The persuade-the-optimist probability (POP) is the posterior probability that the new treatment is no better than the standard.

Prior to commencement of the trial, two pairs of prior distributions for the effectiveness of the standard and new treatments are chosen. One pair is that of an investigator who is optimistic that the new treatment is better than the standard. The other pair is that of someone who is pessimistic (or sceptical) about the effectiveness of the new treatment. Also pre-specified are thresholds PPP_{CRIT} and POP_{CRIT}.

At each interim analysis PPP and POP are calculated using the data collected so far.

If $POP > POP_{CRIT}$, the trial is stopped, because even an optimist should be persuaded that the new treatment is no better than the standard. Similarly, if $PPP > PPP_{CRIT}$, the trial is stopped because even a pessimist should be persuaded that the new treatment is better. For further details see Heitjan (1997). *SRS*

[See also BAYESIAN METHODS]

Heitjan, D. F. 1997: Bayesian interim analysis of Phase II cancer clinical trials. *Statistics in Medicine* 16, 1791–802.

benchmarking This is a procedure for adjusting a less reliable series of observations to make it consistent with more reliable measurements known as *benchmarks*. For example, data on hospital bed occupation collected monthly will not necessarily agree with figures collected annually and the monthly figures (which are likely to be less reliable because the annual figures will probably originate from a census, exhaustive administrative records or a larger sample) may be adjusted at some point to agree with the more reliable annual figures. Benchmarking is often used to adjust time-series data to annual benchmarks while preserving as far as possible the month-to-month movement of the original series (see, for example, Cholette and Dagum, 1994). *BSE*

Cholette, P. A. and Dagum, E. B. 1994: Benchmarking time series with autocorrelated survey errors. *International Statistics Review* 62, 365–77.

Berkson's fallacy Sometimes a spurious relationship can be concluded because the data from which the conclusion was derived came from a special source, which is not representative of the general population. Such bias is known as Berkson's fallacy and it can only be avoided by careful study design (Walter 1980; Feinstein, Walter and Horwitz, 1986; Woodward, 2005).

A classic example of this bias is the study of autopsies by Pearl (1929). Fewer autopsies than expected found both tuberculosis and cancer to occur together; the frequency of cancer was thus lower among tuberculosis victims than others. This led Pearl to the erroneous conclusion that tuberculosis might be offering people some kind of protection against cancer, even leading to the suggestion that cancer patients might be treated with the protein of the tuberculosis bacterium. The problem with this line of thinking is that not every death is autopsied; in this case it turned out that people who died with both diseases were less likely to be autopsied, leading to an artificial lack of numbers with both diseases in Pearl's autopsy series.

Berkson's fallacy is a particular problem with case-control studies. For example, suppose that both the case and control series are derived from hospitals. If it happened that anyone with both the 'case' disease and some other disease were more likely to be hospitalised than someone with only one of the pair, we may well see a relationship between the prevalence of the two diseases in the case-control study, even when there is really no such relationship in the general population. Exactly the same situation may also give rise to spurious relationships between any risk factor for the 'second' disease and the disease that defines cases. For instance, consider a hospital-based case-control study of coffee drinking and angina among the elderly. Suppose that coffee drinking is a risk factor for Parkinson's disease. If someone has Parkinson's disease she or he is unlikely to be hospitalised unless she or he develops a potentially life-threatening condition, such as angina. Most individuals with angina will be treated in the community, the exception, perhaps, being when there is a disabling co-morbidity. The result of these hypothetical conditions might be a disproportionate number with Parkinson's disease (who tend to drink coffee) among the angina cases in hospital than among the controls (people with other illnesses). The case-control study would thus find coffee drinking to be a risk factor for angina, even if this were not actually true. *MW*

[See also BIAS IN OBSERVATIONAL STUDIES]

Feinstein, A. R., Walter, S. D. and Horwitz, R. I. 1986: An analysis of Berkson's bias in case-control studies. *Journal of Chronic Diseases* 39, 495–504. **Pearl, R.** 1929: Cancer and tuberculosis. *American Journal of Hygiene* 9, 97–159. **Walter, S. D.** 1980: Berkson's bias and its control in epidemiological studies. *Journal of Chronic Diseases* 33, 721–5. **Woodward, M.** 2005: *Epidemiology: study, design and data analysis*, 2nd edition. Boca Raton: Chapman & Hall/CRC Press.

beta distribution This is a flexible PROBABILITY DISTRIBUTION, commonly used to describe a proportion. Whereas many of the distributions we encounter are nonzero over an infinite range of values, the beta distribution is nonzero only in the range 0 to 1. By rescaling, it can be useful any time that a distribution is required over a finite range. The distribution is defined by two parameters, r and s, and has the density function:

$$f(x) = x^{r-1}(1-x)^{s-1}/\beta(r,s)$$

where the $\beta(r,s)$ term can be viewed as a constant to ensure that the total probability is equal to 1.

The MEAN of the beta distribution is $r/(r+s)$ and the VARIANCE is $rs/([(r+s)^2(r+s+1)]$. The parameters r and s define the shape of the distribution. This shape can be wide ranging, with u-shaped curves, n-shaped curves, strictly increasing/decreasing curves and triangular distributions all possible. Some of the possible distributions are illustrated in the figure (see page 38). If r and s are equal then the distribution will be symmetric.

Note the similarities to the BINOMIAL DISTRIBUTION. Where the binomial models the distribution of the number of successes, when given the probability of a success, the beta can model the probability of a success given the number of successes. Indeed, in a Bayesian analysis (see BAYESIAN METHODS), the beta distribution is the conjugate prior for the binomial distribution.

The beta distribution is related to a number of other distributions. It contains the uniform distribution over [0,1] as a special case (when $r=1$ and $s=1$), it is increasingly well approximated by a NORMAL DISTRIBUTION as r and s increase and it can result from constructions of the form $A/(A+B)$ where A and B are both random variables with GAMMA DISTRIBUTIONS. For further details on how the beta distribution relates to other distributions, see Leemis (1986).

The beta distribution is most commonly used to model proportions. Suppose that we wish to estimate the specificity of a test that in trials correctly identifies 50 of the 52 participants that do not have a condition. The usual normal approximation will not suffice since it leads to an interval from 0.91 to 1.01, and a value greater than 1 makes no sense. There are a number of ways to use the beta distribution in estimating the interval (see Brown, Cai and DasGupta, 2001). *AGL*

Brown, L. D., Cai, T. T. and DasGupta, A. 2001: Interval estimation for a binomial proportion. *Statistical Science* 16, 101–33. **Leemis, L. M.** 1986: Relationships among common univariate distributions. *The American Statistician* 40, 2, 143–6.

bias Any experiment, study or measuring process is said to be biased if it produces an outcome that differs from the 'truth' in a systematic way. Bias can occur at any stage of the research process from the literature review through to the publication of the results (Armitage and Colton, 2005).

It is important to distinguish between bias or systematic error, on the one hand, and random error, on the other hand. For example, suppose that we had a population of subjects with a MEAN weight of 80 kg and a STANDARD DEVIATION of

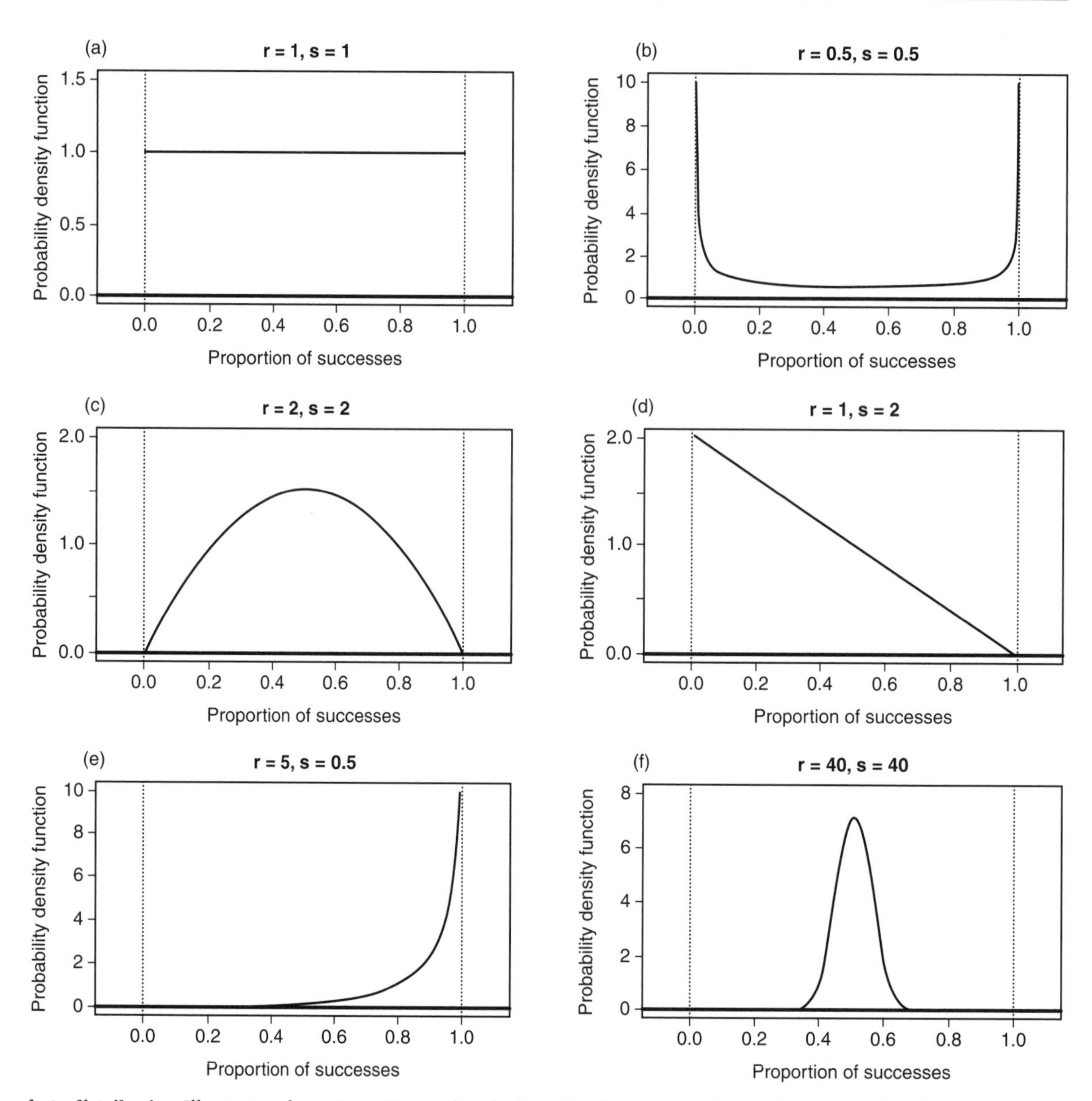

beta distribution *Illustrating the variety of forms that the beta distribution can take: (a) the uniform distribution over (0,1), (b) a bimodal concave distribution (in this case the Jeffrey's prior), (c) a curve with a single mode, (d) a linear function of the proportion, (e) a nonlinear but still strictly increasing distribution, (f) an example that is well approximated by the normal distribution*

10 kg. If we select a simple random sample of 25 subjects from this population and measure their weights using a well-calibrated set of scales, then it is possible that the mean weight for this sample will be substantially different from 80 kg. In fact, there is about a 1 in 20 chance that the sample mean will be more than 4 kg below or 4 kg above the true mean of 80 kg.

However, simple random sampling produces an unbiased estimate of the true mean weight because, if the process of selecting a simple random sample of 25 subjects and computing the sample mean weight were repeated a large number of times, the distribution of the sample means would be centred around the true mean of 80 kg. The larger the sample size, the closer the sample means will be clustered around the true population mean. In other words, the expected value of the sample mean equals the population mean. In this scenario, there is no bias and any deviation of the observed sample mean from the true value can be accounted for by pure chance, known as random variation or random error.

If, however, the weights of a random sample of subjects were measured using a *poorly* calibrated set of scales that weighed each subject as being 2 kg heavier than the actual weight, this would lead to a biased estimate of the true population mean weight. The size of this bias, or systematic error, would not be reduced by increasing the sample size and the distribution of the sample mean will be centred around 82 kg rather than 80 kg. The systematic error in this example is a measurement bias due to a faulty measuring instrument.

More generally, measurement bias could be due to such diverse causes as poor questionnaire design, faulty equipment, observer error or respondent error (Silman and Macfarlane, 2002). Examples of observer error include misreading the scale on an instrument, bias in reporting results by an unblinded evaluator in a clinical trial or bias in eliciting information about the exposure history of cases and controls in a CASE-CONTROL STUDY. Examples of respondent error include biased reporting of symptoms by unblinded patients in a clinical trial, bias in recall of exposure history by cases and controls in a case-control study (see BIAS IN OBSERVATIONAL STUDIES).

All types of study are susceptible to design bias. This can arise from many sources, such as SELECTION BIAS (when the subjects selected for study are not representative of the target population), NONRESPONSE BIAS (when there is a systematic difference between the characteristics of those who choose to participate and those who do not), noncomparability bias (when groups of subjects chosen for comparison in, for example, a case-control study are not in fact comparable). Randomised trials (see CLINICAL TRIALS) are generally regarded as being least susceptible to design biases. The scope for BIAS IN OBSERVATIONAL STUDIES, especially case-control studies, is much greater. Armitage and Colton (2005), Ellenberg (1994), Porta and Last (2008) and Sackett (1979) all provide a comprehensive description of sources of design bias.

Analysis bias arises from errors in the analysis of data. This covers such issues as confounding bias (in which confounding factors have not been appropriately adjusted for in the analysis), analysis method bias (including inappropriate assumptions about the distribution of variables, faulty strategies for handling MISSING DATA or OUTLIERS, unplanned SUBGROUP ANALYSIS and *data dredging*) (Armitage and Colton, 2005; Davey Smith and Ebrahim, 2002).

Ensuring that the interpretation of data is unbiased is just as important as ensuring that the processes of design, measurement and analysis are unbiased. Bias in the interpretation of data can be conscious or unconscious and is particularly difficult to address because it involves subjective judgements on the part of the researchers. Kaptchuk (2003) provides an overview of the issues involved.

There is some evidence to suggest that the source of funding for drug studies is related to the outcome. A systematic review by Lexchin *et al.* (2003) demonstrated a systematic bias in favour of the products made by the company funding the research. The main sources of this bias were thought to be inappropriate selection of treatments to compare against the product being investigated and publication bias. Potential sources of investigator bias are reviewed in detail by Greenland (2009).

Finally, publication bias (see SYSTEMATIC REVIEWS AND META-ANALYSIS) can arise from two main sources. First, researchers are more likely to submit papers for publication if the research produces a statistically and clinically significant result rather than an inconclusive result. Second, journal editors are more likely to publish papers reporting statistically and clinically significant results (Dubben, 2009). *WHG*

[See also NONRESPONSE BIAS, SELECTION BIAS]

Armitage, P. and Colton, T. (eds) 2005: *Encyclopaedia of biostatistics*, 2nd edition. New York: John Wiley & Sons, Inc. **Davey Smith, G. and Ebrahim, S.** 2002: Data dredging, bias or confounding. *British Medical Journal* 325,1437–8. **Dubben, H.** 2009: New methods to deal with publication bias. *British Medical Journal* 339, b3272. **Ellenberg, J. H.** 1994: Selection bias in observational and experimental studies. *Statistics in Medicine* 13, 557–67. **Greenland, S.** 2009: Accounting for uncertainty about investigator bias: disclosure is informative. *Journal of Epidemiology and Community Health* 63, 593-8. **Kaptchuk, T. J.** 2003: Effect of interpretive bias on research evidence. *British Medical Journal* 326, 1453–5. **Porta, M. and Last, J. M.** 2008: *A dictionary of epidemiology*, 5th edition. Oxford: Oxford University Press. **Lexchin, J., Bero, L. A., Djulbegovic, B. and Clark, O.** 2003: Pharmaceutical industry sponsorship and research outcome and quality: systematic review. *British Medical Journal* 326, 1167–76. **Sackett, D. L.** 1979: Bias in analytic research. *Journal of Chronic Diseases* 32, 51–63. **Silman, A. J. and Macfarlane, G. J.** 2002: *Epidemiological studies: a practical guide*, 2nd edition. Cambridge: Cambridge University Press.

bias in observational studies In an ideal study, an investigator seeks to estimate the effect of an exposure to a factor on an outcome of interest. We might like to be able to look at what happens to a population when the factor is at one level and then turn back time and rerun things at the second level; but that is impossible, of course.

Very often it is not even possible or practical to conduct an experiment in which the levels of exposure are controlled, so that one is left with analysing observational data that occur naturally. Bias is any systematic departure from this idealised construct, which is distinct from purely random error, which is zero on average. The latter can be dealt with by reducing variability in the measure of association, which can be accomplished in a variety of ways, including the increase of the overall sample size. However, bias cannot be reduced by increasing the sample size and it can only be controlled through carefully conducted research by an investigator. There have been attempts to catalogue the types of bias that can occur and these broadly fall into three sources: the selection of study subjects, errors in the information collected and confounding or entangling the effects with other causes of the outcome (Hill and Kleinbaum, 1998).

In order to discuss the sources of bias in an observational study in more concrete terms, consider a hypothetical epidemiological study in which the results are summarised in a 2×2 table (shown in the table). We are interested in studying the association between exposure and disease in a manner that avoids bias. Among the choices of study design from which data for this 2×2 table may have arisen are a CROSS-SECTIONAL STUDY, a COHORT STUDY or a CASE-CONTROL STUDY. In a cross-sectional study, N subjects are sampled and the four cell frequencies determined, but in a cohort study, a groups of exposed and unexposed subjects are chosen, essentially fixing the row totals, and then the column frequencies are determined by what transpires during the course of follow-up. For a case-control study, the column totals are regarded as fixed and subjects distributed to each row within a column depending on their exposure history, which would usually be gleaned by interview. Fundamental to each of these study designs is the realisation of a random sample, either overall or within the rows or columns.

bias in observational studies *Tabulated results from an epidemiological study with two levels of exposure and disease status*

	Diseased		
Exposed	*Yes*	*No*	*Total*
Yes	a	b	$a+b$
No	c	d	$c+d$
Total	$a+c$	$b+d$	N

Selection bias occurs when the proportion recruited from the target population that is counted in a cell of the 2×2 table depends on both the row and the column. One way in which this can occur in a cohort study is if there are differential diagnoses depending on the exposure status. For example, suppose that an exposure of interest occurs in a manufacturing plant that provides health insurance for its employees, but among the unexposed are substantial numbers who are uninsured. If the insured receive regular checkups from their physicians, this may increase the likelihood of a correct diagnosis among those exposed, while similar cases may have been missed for the unexposed that are uninsured. Clearly, this would bias an estimate of the odds ratio that would be calculated from such a study. Another potential source of such bias in a cohort study may arise from loss to follow-up, e.g. if instead of exposure the investigator is interested in whether a person is using a particular type of treatment. However, suppose that the treatment is not only ineffective but it also causes unpleasant symptoms in patients who are related to the occurrence of the disease outcome. If the individuals so affected drop out of the study, this would artificially lower the count in this cell of the 2×2 table and bias the estimate. Notice that the magnitude of the effect of this selection bias may be substantial, even if the number lost represents a small proportion of the total. This is especially true when the proportion that develops the disease is small, so that the portion lost in a cell of the table is relatively high, even though the proportion lost represents a small proportion of the overall sample.

In a case-control study, a common source of bias when selecting cases can occur when subjects with a prevalent disease are enrolled into the study, some of whom may have had the disease for some time. Those who have been ill for a long period of time will be more likely to be enrolled if such a study design is used, a phenomenon known as LENGTH-BIASED SAMPLING. If the primary aims of the study are to study the association between exposure and the occurrence of the disease, this will clearly lead to a biased estimate of association, but this could have been avoided by only enrolling newly diagnosed cases instead.

The choice of appropriate controls in a case-control study can be an especially common source of bias. If the cases are selected from among those who are diagnosed at a collaborating set of hospitals, then the controls should ideally be a representative sample of those who are healthy in the catchment areas of those hospitals. If all hospitals in an area are cooperating with a study, then this could be accomplished by recruiting a random sample of the overall population in the geographic area. Random digit dialling is one approach that has been useful in populations well covered by telephones, but it is becoming more difficult to employ the method with the increasing use of current technologies such as cell phones, caller ID and no-call lists. In some studies, controls are selected using subjects who have been admitted into the same hospital for a disease that is unrelated to the exposure of interest. This would result in a group of subjects from the same catchment area as the cases, thus avoiding one source of potential selection bias. The estimate of association in such a study would be the difference between the effect of exposure on the disease of interest and its effect on the 'control disease' (Breslow, 1978, 1982). If one has chosen a control disease that is not related to exposure, i.e. the effect is zero, then the estimate of association will be an unbiased estimate of the effect on disease risk. However, it is often difficult to be certain that this is the case because the assumption may just be the result of a lack of knowledge about the aetiology of disease affecting the controls.

A cross-sectional study can be a useful way of obtaining a snapshot of the association between two or more variables at a single point in time, especially if the population chosen for study is of broad interest and a carefully planned method for drawing a random sample has been put in place. Some national health surveys are good examples of such studies, such as those conducted by the National Center for Health Statistics. However, if the aim is to study disease aetiology or other outcomes that evolve over time, then the single snapshot in time can be a serious limitation. For example, in an epidemiological study, subjects with a disease who have been identified by a survey conducted at a single point in time would necessarily be a prevalent case, which is a potential source of bias here as it is in a case-control study.

Information bias in an observational study arises from error in the variables that have been collected as part of the data for each subject in a study. Such errors can either be differential or nondifferential, i.e. random. Differential error in reporting values summarised in a 2×2 table would arise if the error rate for reporting the variable in the column depended on the row or vice versa. This would obviously be a potentially important source of bias when estimating an association. However, bias can also arise when the error is nondifferential or purely random.

Case-control studies can be prone to information bias because someone with a serious illness may remember their history of exposure to the factor of interest quite differently from a healthy control. This RECALL BIAS can be especially significant when other studies of the exposure of interest have entered the public's consciousness or been reported in the news. One technique for minimising its effect is to use a well-structured interview in which the questions have been clearly and unambiguously phrased and posed in an identical manner to all subjects in the study. This requires considerable effort on the part of an investigator, in that the questionnaire would need to be pre-tested and the interviewers well trained.

Information bias can potentially also affect a study by subconsciously influencing evaluations by interviewers, professional diagnosticians or even laboratory technicians. This could happen if the individual has a preconceived idea of what the results of a study will be or of the way the results are going. Thus, it is generally preferred that the study hypotheses not be known to those responsible for collecting the data or that the status of a subject be masked, a procedure in which the person recording the data is said to be blind with respect to the outcome. These measures should reduce the possibility for differential errors, but not nondifferential errors.

While it is intuitively easy to appreciate that differential error of measurement can bias the results of an observational study, nondifferential error can also have an effect as well. If only a single variable is affected by nondifferential error, then the effect is generally to attenuate the effect, i.e. to bias the estimated association towards the null value of no association. This would tend to make the results of a study with nondifferential error in one of the variables conservative in the sense that it would make it more difficult to establish that an estimated association was not due to chance alone. Contrariwise, it would also result in an underestimate of an effect, which can be important when trying to determine the public health significance of exposure to a particular factor.

It is most desirable to minimise information bias during the design and data collection phase of a study by minimising measurement error, but it is generally not possible to be entirely successful in these efforts. One approach to correcting for bias at the data analysis phase is to introduce a correction factor that takes into account the measurement error. In the case of a 2×2 table, formulae have been provided for this (Barron, 1977; Copeland *et al.*, 1977) and similar approaches are also available for use in LOGISTIC REGRESSION (Rosner, Spiegelman and Willett, 1990). There is now a rich variety of statistical techniques for dealing with errors in variables, many of which are described in the text by Carroll, Ruppert and Stefanski (1995).

Confounding arises when the estimated effect for an association of interest is entangled with another factor, perhaps one that is well known to be associated with the outcome. It is conceptually related to aliasing in design of experiments, in which two effects are completely entangled, and *collinearity* in other contexts. The potential for confounding in an observational study of two variables exists when each is associated with a third variable, the confounder, in the presence of the factor of interest. Precise definitions of confounding go to the heart of the objectives of observational studies and various models have been proposed as a theoretical basis for its effect (Rubin, 1974; Wickramaratne and Holford, 1987). Alternatively, *collapsability* is sometimes used as a simple and practical alternative to more formal definitions of confounding (Bishop, Fienburg and Holland, 1973). An association is collapsible with respect to a putative confounder if the estimated association is unchanged when adjusting for the confounder in the analysis.

Approaches for dealing with a potential confounder are in essence to estimate the association holding the value of the confounder constant. In a designed experiment, this would be accomplished by selecting strata or blocks of subjects with identical values of the confounder and only vary the exposure of interest within the strata. One way of accomplishing a similar effect in an observational study is to stratify the data by the potential confounder and then combine information across the strata, if the effect is constant, using the MANTEL–HAENSZEL METHOD or something similar (Mantel and Haentszel, 1959). Alternatively, one can adjust for one or more putative confounders by including them in a model, such as the linear logistic model (Hosmer and Lemeshow,

1989) or an alternative GENERALISED LINEAR MODEL (McCullagh and Nelder, 1989) for a binary response.

It is entirely possible that an observational study will not be able to separate out the effect of an exposure of interest from the effect of another exposure that it thought to be a confounder. This is not unlike the more general problem of collinearity that arises in the context of regression analysis. In these situations, it may only be possible to conduct a new study in which the design has been carefully constructed so that one can tease apart the separate contributions of a factor of interest from its confounder. *TRH*

Barron, B. A. 1977: The effects of misclassification on the estimate of relative risk. *Biometrics* 33, 414–18. **Bishop, Y. M. M., Fienburg, S. E. and Holland, P. W.** 1973: *Discrete multivariate analysis: theory and practice*. Cambridge, MA: MIT Press. **Breslow, N.** 1978: The proportional hazards model: applications in epidemiology. *Communications in Statistics – Theory and Mathematics* A7, 4, 315–32. **Breslow, N.** 1982: Design and analysis of case-control studies. *Annual Review of Public Health* 3, 29–54. **Carroll, R. J., Ruppert, D. and Stefanski, L. A.** 1995: *Measurement error in nonlinear models*. London: Chapman & Hall. **Copeland, K. T., Checkoway, H., McMichael, A. J. and Holbrook, R. H.** 1977: Bias due to misclassification in the estimation of relative risk. *American Journal of Epidemiology* 105, 488–95. **Hill, H. A. and Kleinbaum, D. G.** 1998: Bias in observational studies. In Armitage, P. and Colton, T. (eds), *Encyclopedia of biostatistics*. Chichester: John Wiley & Sons, Ltd. **Hosmer, D. W. and Lemeshow, S.** 1989: *Applied logistic regression*. New York: John Wiley & Sons, Inc. **Mantel, N. and Haenszel, W.** 1959: Statistical aspects of the analysis of data from retrospective studies of disease. *Journal of the National Cancer Institute* 22, 719–48. **McCullagh, P. and Nelder, J. A.** 1989: *Generalised linear models*. London: Chapman & Hall. **Rosner, B., Spiegelman, D. and Willett, W. C.** 1990: Correction of logistic regression relative risk estimates and confidence intervals for measurement error: the case of multiple covariates measured with error. *American Journal of Epidemiology* 132, 734–45. **Rubin, D. B.** 1974: Estimating causal effects of treatments in randomized and nonrandomized studies. *Journal of Educational Psychology* 66, 688–701. **Wickramaratne, P. J. and Holford, T. R.** 1987: Confounding in epidemiologic studies: the adequacy of the control group as a measure of confounding. *Biometrics* 43, 751–65.

bimodal distribution

bimodal distribution This is a PROBABILITY DISTRIBUTION or a FREQUENCY DISTRIBUTION with two modes. Often the two modes in the distribution correspond to the data arising from two distinct populations. The first figure shows a bimodal density function arising from a weighted sum of two NORMAL DISTRIBUTIONS (a FINITE MIXTURE DISTRIBUTION). An example of a histogram with two distinct modes is shown in the second figure. The data here correspond to the sizes of myelinated lumbosacral ventral root fibres taken from a kitten of a particular age. The first mode is associated with axons of gamma neurons and the second with alpha neurons. Other examples of medical bimodal distributions are the age of incidence of Hodgkin's lymphoma and the speed of inactivation of the drug ioniazid in US adults. An account of the use of bimodal distributions in a medical setting is given in Hagberg *et al.* (2001). *BSE*

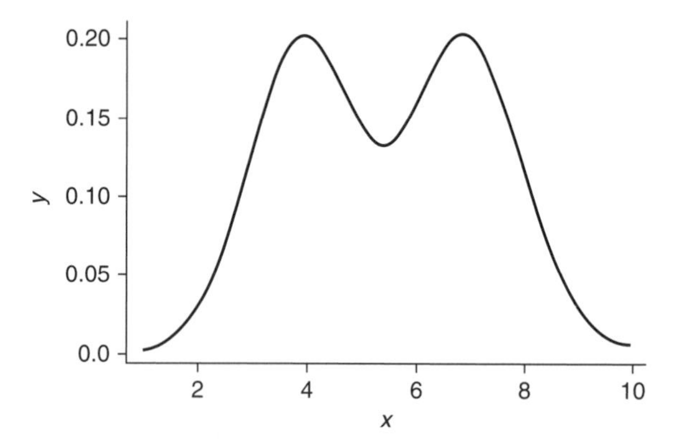

bimodal distribution *Finite mixture distribution*

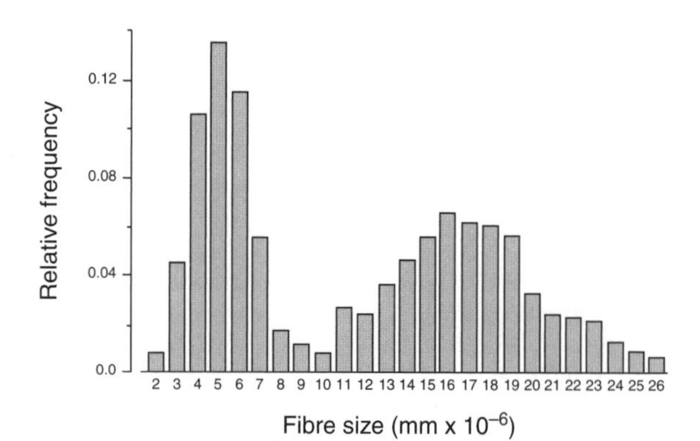

bimodal distribution *Histogram with two distinct modes*

Hagberg, G. E., Zito, C., Patria, F. and Saries, J. N. (2001) Improved detection of event-related functional MRI signals using probability functions *Neuroimage* 14, 1193–205.

binomial distribution

binomial distribution This is the PROBABILITY DISTRIBUTION of the number of 'successes', X, in a series of n independent trials, where the probability of a success is p for each trial. Specifically the distribution is given by:

$$\Pr(X = x) = \frac{n!}{x!(n-x)!} p^x (1-p)^{n-x}, x = 0, 1, 2, \ldots, n$$

where $n!$ (factorial n) is the product of all the integers up to and including n and 0! is defined to be 1. The mean of the distribution is np and its variance $np(1 - p)$. Some binomial distributions with $n = 10$ and different values of p are shown in the figure (see page 43). The distribution often occurs in medicine as the basis for testing the hypothesis that the

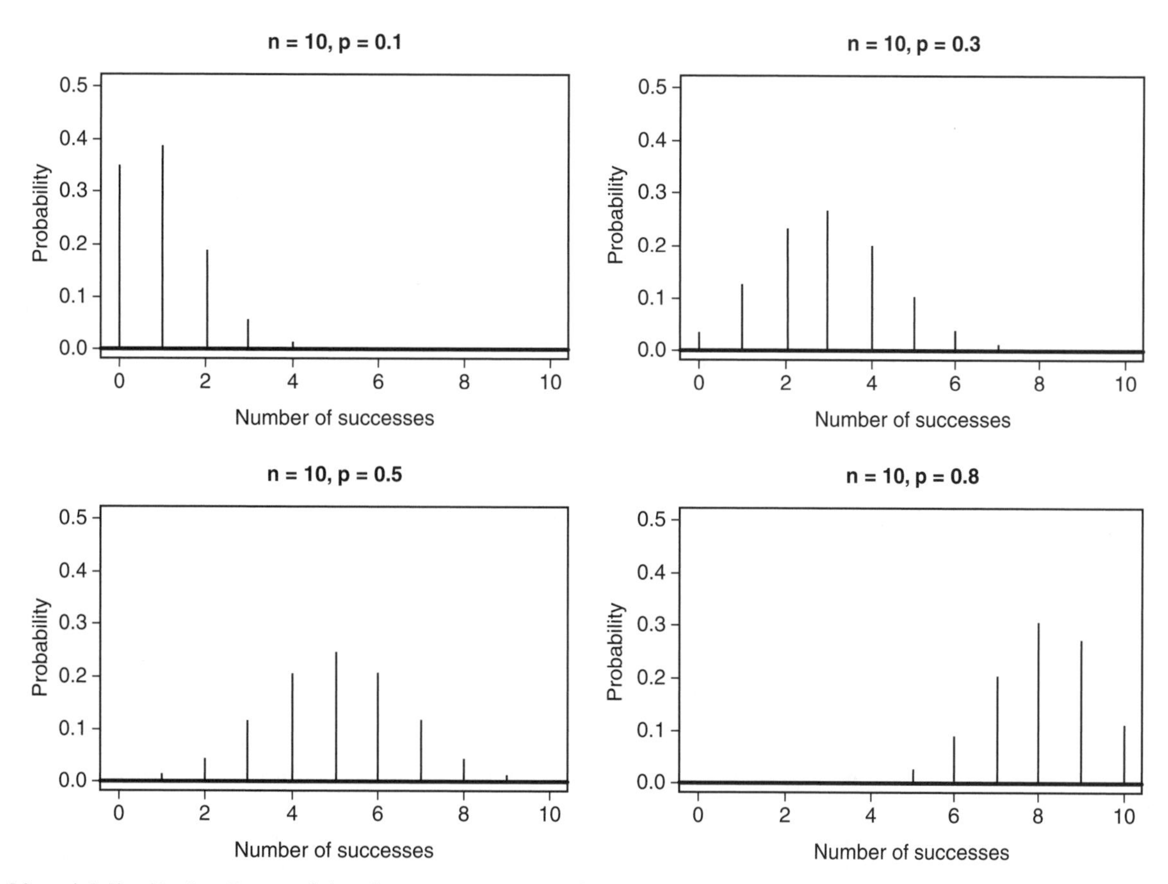

binomial distribution *Binomial distributions for various values of* n *and* p

probability of some event of interest takes a particular value. For example, a researcher may postulate that 10 % of a population is infected with a virus and, on sampling 20 people at random from the population, finds that 6 people have the virus. Is there any evidence that the infection rate is higher than the hypothesised value of 10 %? To answer this question a *P*-value can be computed from the binomial distribution as the probability that 6 or more people in the 20 sampled have the virus when the probability that a person is infected is 0.1, i.e. the sum:

$$\sum_{x=6}^{20} \frac{20!}{x!(20-x)!}(0.1)^x(0.9)^{20-x}$$

The resulting value is 0.01, giving strong evidence that the infection rate is larger than 10 %. As well as testing for a specific proportion, the binomial distribution can be used in calculating CONFIDENCE INTERVALS for a proportion. Villanueva *et al.* (2003) use the binomial distribution to estimate confidence intervals for the proportion of adverts in medical journals with inaccurate claims. More details of the binomial distribution can be found in Altman (1991). *BSE/AGL*

Altman, D. G. 1991: *Practical statistics for medical research.* London: Chapman & Hall. **Villanueva, P., Peiró, S., Librero, J. and Pereiró, I.** 2003: Accuracy of pharmaceutical advertisements in medical journals. *Lancet* 361, 27–32.

bioinformatics This is a term given for the coming together of molecular biology, computer science, mathematics and statistics to deal with the ever-expanding genomic and proteomic databases, which are themselves the result of rapid technological advances in DNA sequencing, gene expression measurement and macromolecular structure determination. In many cases such techniques give rise to HIGH DIMENSIONAL DATA. A comprehensive account of bioinformatics is given in Zvelebil and Baum (2007). *BSE*

Zvelebil, M. and Baum, J. 2007: *Understanding bioinformatics.* Garland Science.

birth cohort studies These are studies established to examine growth, development and health of children from birth. However, given sufficient follow-up they also provide insights into influences on adult disease that operate throughout the life course.

In principle, a birth cohort study is one in which all study participants are recruited at birth and then followed over time. The cohort is defined by the location and the time period in which the participants were born, which may be those born in one week or over a period of a year of more. The members of the cohort are then followed up at various time points to ascertain risk factors and health outcomes. As the cohort ages the focus of the research tends to shift. In the early years, the emphasis tends to be on childhood growth and development and risk of childhood illness, but as the cohort matures adult risk factors such as smoking and obesity and health measures such as blood pressure start to be of greater interest. Outcome variables in childhood such as height can later be considered as risk factors when assessing chronic disease later in life. Intergenerational and genetic factors are of interest and information on family members is included in many birth cohorts (Lawlor, Andersen and Batty, 2009).

The first cohort to be established using recruitment at birth was the National Survey of Health and Development, which studied babies born in Britain in the first week of March 1946 (Wadsworth *et al.*, 2006). The cohort has been followed up on more than 20 occasions since birth, the latest being at age 60-64 years. Contacts with the participants have been by postal questionnaire, home and clinic visits, and through schools and links with health and educational professionals.

Britain has two other birth cohort studies conducted on similar lines. They comprise those born between the 3 and 9 March 1958 (Power and Elliott, 1992) and between 5 and 11 April 1970 (Elliott and Shepherd, 2006) respectively. Both studies have included a number of follow-ups that have given insights into the growth and development of these cohorts through childhood, adolescence and into adulthood. Cross-cohort comparisons have also been possible and have allowed examination of secular trends, e.g. into Crohn's disease, ulcerative colitis and irritable bowel syndrome (Ehlin *et al.*, 2003). The Millennium Cohort was recruited in a different way (Smith and Joshi, 2002), and a further cohort study is planned for births in 2012.

Birth cohort studies are not of course confined to Britain, though until recently comprehensive national coverage has rarely been attempted elsewhere. The Scandinavian countries have well-developed linkage systems and in Norway and Denmark studies of over 100 000 births have been launched (Olsen *et al.*, 2001; Magnus *et al.*, 2006), and the United States has embarked on a study of a similar scale (Branum *et al.*, 2003, and see http://www.nationalchildrensstudy.gov).

Frequently, birth cohorts are located in one town or city. For example, the Pelotas Birth Cohort Study in Brazil recruited all births born in the city of Pelotas during 1982. It represents a good example of a birth cohort study with long-term follow-up in a developing country (Victora and Barros, 2006).

Many birth cohorts have been defined retrospectively. Thus births in a defined geographical area during a specified time period are identified from established records. The data can then be linked to other standard records such as death indices, or the study population can be traced and those still alive can be assessed by post or by interview. An example of this is the birth records of the 1920s and 1930s from the English county of Hertfordshire that were extracted in the 1980s. The population was traced through the National Health Service Central Register and details of deaths and current general practitioner addresses obtained. This allowed not only an analysis of mortality in relation to birth and infant weight but also enabled follow-up of the survivors to examine them for risk factors for chronic conditions such as cardiovascular disease (Syddall *et al.*, 2005).

Some retrospectively defined birth cohorts have focused on particular events that gave rise to extreme living conditions. For example, those born in Amsterdam in 1944–1945 around the time of the famine imposed by the German occupation have been followed up to assess the impact of famine at key stages of pregnancy and early life (Roseboom *et al.*, 2001). Similarly, a cohort of men born in 1916–1935 were identified from one district in Leningrad, a third of whom had experienced starvation during the siege of Leningrad in 1941–1944 when they were around the age of puberty (Sparén *et al.*, 2003). The whole cohort was followed up and invited to take part in health examinations to assess the long-term effects of the famine.

There is also interest in defining birth cohorts at an earlier time point than birth. A child's growth and development begins before birth and so characterisation of aspects of pregnancy is considered important in determining the long-term influences on the offspring's health. The Avon Longitudinal Study of Parents and Children (ALSPAC) recruited 14 000 pregnant women resident in the English county of Avon whose expected dates of delivery were between 1 April 1991 and 31 December 1992. The women and their offspring have been followed up by means of postal questionnaires on many occasions and a subsample known as the Children in Focus was seen at clinics 10 times before the age of seven years. From that age onwards clinics began for the entire cohort (Golding *et al.*, 2001).

Taking this one step further, with an increasing focus on the very early origins of life, two cohort studies have recruited women before pregnancy. The first of these recruited some 2500 women in six villages near Pune in India. Of these, over 1000 became pregnant and full data were obtained on nearly 800 births. This cohort has now been followed up into adolescence (Rao *et al.*, 2001). In the UK, the Southampton Women's Survey recruited over 12 500 women aged 20 to

34 years when they were not pregnant and over 3000 of them were studied throughout subsequent pregnancies and the children are being followed up (Inskip *et al.*, 2006). The recently launched US National Children's Study is mainly recruiting women in the first trimester of pregnancy but is also including some women before conception (http://www.nationalchildrensstudy.gov).

Birth cohort studies have many strengths. Usually they capture a cross-section of the population and they have all the advantages of longitudinal studies. However, the weakness is that over the life course a large percentage of prospectively defined birth cohorts tends to drop out. Many DROPOUTS are due to death as the cohort ages or to migration out of the region or country of study. Persistent questioning and requests to attend clinics or be visited at home adds to the attrition, as some participants feel that they have contributed enough and their motivation wanes. The remaining cohort may no longer represent the general population. Retrospectively defined cohorts can suffer less from this problem but then they often lack sufficient data on the early years. *HI*

[See also COHORT STUDIES]

Branum, A. M., Collman, G. W., Correa, A. *et al.* 2003: National Children's Study of environmental effects on child health and development. *Environmental Health Perspective* 111, 642–6. **Ehlin, A. G. C., Montgomery, S. M., Ekbom, A., Pounder, R. E. and Wakefield, A. J.** 2003: Prevalence of gastrointestinal diseases in two British national birth cohorts. *Gut* 52, 1117–21. **Elliott, J. and Shepherd, P.** 2006: Cohort profile: 1970 British Birth Cohort (BCS70). *International Journal of Epidemiology* 35, 836–43. **Golding, J., Pembrey, M., Jones, R. and The ALSPAC Study Team** 2001: ALSPAC – The Avon Longitudinal Study of Parents and Children 1. Study methodology. *Paediatric and Perinatal Epidemiology* 15, 74–87. **Inskip, H. M., Godfrey, K. M., Robinson, S. M., Law, C. M., Barker, D. J. and Cooper, C.** 2006: Cohort profile: The Southampton Women's Survey. *International Journal of Epidemiology* 90, 42–8. **Lawlor, D. A., Andersen, A. M. and Batty, G. D.** 2009: Birth cohort studies: past, present and future. *International Journal of Epidemiology* 38, 897–902. **Magnus, P., Irgens, L. M., Haug, K., Nystad, W., Skjaerven, R. and Stoltenberg, C.** 2006: Cohort profile: The Norwegian Mother and Child Cohort Study (MoBa). *International Journal of Epidemiology* 35, 1146–50. **Olsen, J., Melbye, M., Olsen, S. F. *et al.*** 2001: The Danish National Birth Cohort – its background, structure and aim. *Scandinavian Journal of Public Health* 29, 300–7. **Power, C. and Elliott, J.** 2006: Cohort profile: 1958 British Birth Cohort (National Child Development Study). *International Journal of Epidemiology* 35, 34–41. **Rao, S., Yajnik, C. S., Kanade, A., Fall, C. H. D., Margetts, B.M., Jackson, A. A., Shier, R., Joshi, S., Rege, S., Lubree, H. and Desai, B.** 2001: Intake of micronutrient-rich foods in rural Indian mothers is associated with the size of their babies at birth: Pune Maternal Nutrition Study. *Journal of Nutrition* 131, 1217–24. **Roseboom, T. J., van der Meulen, J. H. P., Osmond, C., Barker, D. J. P., Ravelli, A. C. J. and Bleker, O. P.** 2001: Adult survival after prenatal exposure to the Dutch famine 1944–45. *Paediatric and Perinatal Epidemiology* 15, 220–5. **Smith, K.** and **Joshi, H.** 2002: The Millennium Cohort Study. *Popular Trends* 107, 30–4. **Sparén, P., Vågerö, D., Shestov, D. B., Plavinskaja, S., Parfenova, N., Hoptiar, V., Paturot, D. and Galanti, M. R.** 2004: Long term mortality after severe starvation during the siege of Leningrad: prospective cohort study. *British Medical Journal* January, 328, 11. **Syddall, H. E., Aihie Sayer, A., Dennison, E. M., Martin, H. J., Barker, D. J. P., Cooper, C. and The Hertfordshire Cohort Study Group** 2005: Cohort profile: The Hertfordshire Cohort Study. *International Journal of Epidemiology* 34, 1234–42. **Victora, C. G. and Barros, F. C.** 2006: Cohort profile: The 1982 Pelotas (Brazil) Birth Cohort Study. *International Journal of Epidemiology* 35, 237–42. **Wadsworth, M., Kuh, D., Richards, M. and Hardy, R.** 2006: Cohort profile: The 1946 National Birth Cohort (MRC National Survey of Health and Development). *International Journal of Epidemiology* 35, 49–54.

biserial correlation See CORRELATION

bivariate boxplot This is a two-dimensional analogue of the BOXPLOT for univariate data, which is based on calculating 'robust' measures of location, scale and correlation. It consists essentially of a pair of concentric ellipses, one of which (the 'hinge') includes 50 % of the data and the other (called the 'fence') which delineates potential troublesome outliers. In addition, resistant regression lines of both y on x and x on y are shown, with their intersection showing the bivariate locations estimator. The acute angle between the regression lines will be small for a large absolute value of correlations and large for small absolute values. Details of the construction of the bivariate boxplot are given in Goldberg and Iglewicz (1992). This type of boxplot may be useful in

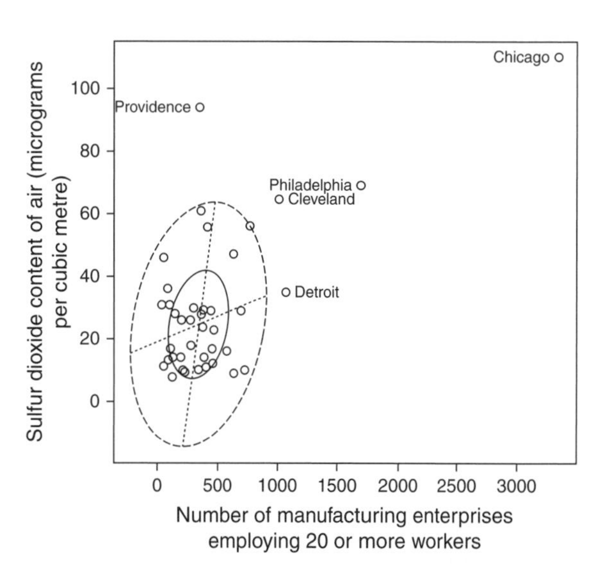

bivariate boxplot *Scatterplot of sulfur dioxide concentration against number of manufacturing enterprises for cities in the USA, showing the bivariate boxplot of the data* indicating the distributional properties of the data and in identifying possible outliers. An example for a SCATTERPLOT of the number of manufacturing enterprises employing more than 20 people against the pollution level as measured by sulfur dioxide concentration for a number of US cities is shown in the figure (see page 45); five cities are indicated as outliers, four that have more than a thousand manufacturing enterprises, but one, Providence, that has only a relatively small number of manufacturing enterprises. *BSE*

[See also SCATTERPLOT MATRICES]

Goldberg, K. M. and Iglewicz, B. 1992: Bivariate extensions of the boxplot. *Technometrics* 34, 307–20.

bivariate distribution

For each and every pair of feasible values, the probability that a pair of variables will take those values. This then is a natural extension of the idea of a univariate probability distribution, applicable when we are measuring two paired variables, e.g. a person's height and weight.

If the two variables are independent then the bivariate distribution will not be of particular interest. When they are correlated, as with height and weight, however, it becomes important to consider the bivariate distribution. If we were to look at a sample we might see that 20 % of the sample are over 6 feet tall and 20 % of the sample are under 50 kg in weight, but only by looking at the bivariate distribution would we know that there were no (or few) people who were both over 6 feet tall and under 50 kg in weight.

Whereas univariate distributions can usually be depicted in a BAR CHART or HISTOGRAM, bivariate distributions, due to their additional dimension, cannot. If the two variables are categorical then a simple cross-tabulation will probably be most informative, while if the two variables are continuous a SCATTERPLOT will probably be appropriate.

For example, whereas in previous univariate work McLaren *et al.* (2000) have separately looked at the distributions of red blood cell volume and haemoglobin levels when identifying anaemia, in a more sophisticated approach, McLaren *et al.* (2001) employ a bivariate distribution of red cell volume and haemoglobin.

The most commonly encountered bivariate distribution is the BIVARIATE NORMAL DISTRIBUTION, a special case of the MULTIVARIATE NORMAL DISTRIBUTION. *AGL*

McLaren, C. E., Kambour, E. L., McLachlan, G. J., Lukaski, H. C., Li, X., Brittenham, G. M. and McLaren, G. D. 2000: Patient-specific analysis of sequential haematological data by multiple linear regression and mixture distribution modelling. *Statistics in Medicine* 19, 1, 83–98. **McLaren, C. E., Cadez, I. V., Smyth, P. and McLachlan, G. J.** 2001: Classification of disorders of anemia on the basis of mixture model parameters. Technical Report 01–56. Irvine: Information and Computer Science Department, University of California.

bivariate normal distribution

This is a special case of the multivariate normal distribution with two variables and the most common example of a bivariate distribution. The bivariate normal distribution is worthy of mention because of all the multivariate normal distributions it is the most commonly used, the easiest to illustrate and the easiest to write out in mathematical notation.

Given two variables, X and Y, the probability density function of the bivariate normal distribution is defined by the means of X and Y (here denoted μ_X and μ_Y respectively), the STANDARD DEVIATIONS of X and Y (here denoted σ_X and σ_Y respectively) and the CORRELATION of X and Y (denoted ρ). Given these values, the probability that X and Y take values x and y respectively, $f(x,y)$, is:

$$f(x,y) = \frac{1}{2\pi\sigma_X\sigma_Y\sqrt{1-\rho^2}}\exp\left\{-\frac{1}{2(1-\rho^2)}\left[\left(\frac{x-\mu_X}{\sigma_X}\right)^2 - 2\rho\left(\frac{x-\mu_X}{\sigma_X}\right)\left(\frac{y-\mu_Y}{\sigma_Y}\right)+\left(\frac{y-\mu_Y}{\sigma_Y}\right)^2\right]\right\}$$

This formula may not appear particularly pleasant, but is easier to handle than for higher variate normal distributions because of the single correlation involved. For further details of the distribution see Chatfield and Collins (1980) and Grimmett and Stirzaker (1992).

When graphed as a SCATTERPLOT, data from this distribution will appear as a cluster of points in an approximately elliptical shape with the density of the points being greatest at the centre of the ellipse. The location of the ellipse will be dependent on the two means, while the standard deviations and correlation determine the angle and spread of the ellipse. The ellipse gets 'narrower' as the magnitude of the correlation increases and approaches a straight line at $\rho = 1$ or -1.

Suryapranata *et al.* (2001) use the bivariate normal distribution in order to compare simultaneously the clinical effectiveness and cost-effectiveness of two treatments for patients with acute myocardial infarction. By drawing a graph of the difference in effect against the difference in cost they were able to illustrate a CONFIDENCE INTERVAL for the differences between the two treatments as an ellipse.

A convenient property of the bivariate normal distribution is the fact that the marginal distributions of the two variables are univariate normal; i.e. if Y is ignored, then X by itself has a NORMAL DISTRIBUTION (and vice versa). Also, the conditional distributions of X and Y are normal. To put this another way, if Y is observed to take a particular value, then the unknown value of X still has a normal distribution given this knowledge. *AGL*

Chatfield, C. and Collins, A. J. 1980: *Introduction to multivariate analysis*. London: Chapman & Hall. **Grimmett, G. R. and Stirzaker, D. R.** 1992: *Probability and random processes*, 2nd edition. Oxford: Clarendon Press. **Suryapranata, H., Ottervanger, J. P., Nibbering, E., van't Hof, A. W. J., Hoorntje, J. C. A., de Boer, M. J., Al, M. J. and Zijlstra, F.** 2001: Long-term outcome and cost-effectiveness of stenting versus balloon angioplasty for acute myocardial infarction. *Heart* 85, 667–71.

Bland–Altman plot
See LIMITS OF AGREEMENT

blinding
See CLINICAL TRIALS, CRITICAL APRAISAL

blocked randomisation
See RANDOMISATION

Bonferroni correction
This correction is used when performing multiple significance tests in order to avoid an excess of false positives (Schaffer, 1995). Suppose, for example, STUDENT'S t-TEST is to be applied to sample data on six variables to assess mean differences in two populations of interest. If the NULL HYPOTHESIS of no difference in means holds for each of the six variables, and each of the six tests is performed at the 5 % SIGNIFICANCE LEVEL, the probability of falsely rejecting the equality of at least one pair of means is 0.26 (this assumes the variables are independent), a fivefold increase over the nominal significance level. The Bonferroni correction approach to this problem involves using a significance level of α/n rather than α for each of the n tests to be performed. For a small number of multiple tests (up to about 5) this method provides a simple and acceptable answer to the problem of inflating the Type I error. The correction is, however, highly conservative and not recommended if large numbers of tests are to be applied, particularly since its use can lead to the rather unsatisfactory situation where many tests are significant at the α level but none at level α/n (Perneger, 1998). In addition, the Bonferroni correction ignores the degree to which the variables may be correlated, which again leads to conservatism when such correlations are substantial. *BSE*

[See also MULTIPLE COMPARISON PROCEDURES]

Perneger, T. V. 1998: What's wrong with Bonferroni adjustments? *British Medical Journal* 316, 1236–8. **Schaffer, J. P.** 1995: Multiple hypothesis testing. *Annual Review of Psychology* 46, 561–84.

boosting
This is a class of optimization algorithms that can be applied to fit a number of classical and modern statistical models. Its origins come from machine learning and computer science (Meir and Rätsch, 2003; Schapire, 2003) but have been adopted in statistics as well. From a statistical point of view, boosting works by iteratively fitting residuals obtained from rather simple regression models (Bühlmann and Hothorn, 2007) (see MULTIPLE LINEAR REGRESSION). These models are called base-learners and determine the structure of the final model which, in essence, is the sum of all base-learners. The method is attractive because it can be applied to multiple linear regression, LOGISTIC REGRESSION, classification, SURVIVAL ANALYSIS, robust regression (see ROBUSTNESS), QUANTILE REGRESSION, etc. Furthermore, the regression relationship can be restricted to linear or additive functions, which facilitates interpretation of the final model. Unlike RANDOM FORESTS, boosting is sensitive to the most important hyperparameter, the number of iterations of the algorithms. Too large values will cause overfitting. Thus, cross-validation techniques have to be applied to determine an appropriate number of iterations. The algorithm is especially useful for model fitting for HIGH-DIMENSIONAL DATA, i.e. when the number of observations is smaller than the number of exploratory variables. Models fitted by boosting algorithms have been successfully applied to weight estimation for foetuses by three-dimensional ultrasound imaging or for predicting cancer subtypes based on gene expression and single nucleotide polymorphisms (SNPs) data. *TH*

Bühlmann, P. and Hothorn, T. 2007: Boosting algorithms: regularization, prediction and model fitting. *Statistical Science*, 22(4), 477–505. **Meir, R. and Rätsch, G.** 2003: An introduction to boosting and leveraging. In *Advanced lectures on machine learning (LNAI2600)*. **Schapire, R. E.** 2003: The boosting approach to machine learning: an overview. In Denison, D. D., Hansen, M. H., Holmes, C., Mallick, B. and Yu, B. (eds), *Nonlinear estimation and classification*. New York: Springer.

bootstrap
The bootstrap is a computationally intensive technique for statistical inference, which can be used when the assumptions that underpin much of classical statistical inference are questionable. This may be because the data are not normally distributed or the dataset is small so that theoretical results based on large sample theory are inapplicable. For example, the bootstrap can be used to estimate the BIAS and STANDARD ERROR of parameter estimates together with CONFIDENCE INTERVALS.

In effect, as we illustrate in the figure, the bootstrap is a data resampling technique. It was formally introduced by Efron (see the discussion in Efron and Tibshirani, 1993) and, although it has a sound theoretical basis, the idea there is something magical about it is reflected in its name. The term bootstrap derives from the phrase to pull oneself up by one's bootstrap, widely thought to be based on one of the 18th century adventures of Baron Munchausen. The Baron found himself at the bottom of a deep lake and saved himself by hauling himself up by his bootstraps.

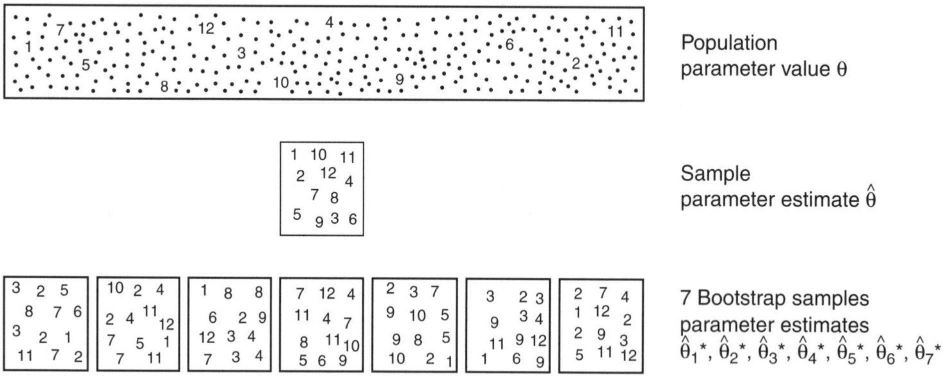

Population parameter value θ

Sample parameter estimate $\hat{\theta}$

7 Bootstrap samples parameter estimates $\hat{\theta}_1^*, \hat{\theta}_2^*, \hat{\theta}_3^*, \hat{\theta}_4^*, \hat{\theta}_5^*, \hat{\theta}_6^*, \hat{\theta}_7^*$

bootstrap *Schematic illustration of bootstrapping*

We will describe the idea using the figure. Suppose we have a population in which the true value of a quantity of interest, say adult height, is denoted by θ. We wish to estimate θ and take a sample of 12 individuals from this population. In the figure, the population is denoted by the large rectangle in the first row (note that the numbers identify population members and are not their adult heights). In this population, the 12 individuals to be included in the sample are numbered. They comprise the actual sample, which is shown in the second row. Our estimate of adult height, calculated from this sample, is denoted by $\hat{\theta}$.

In order to quantify how close $\hat{\theta}$, the estimate of adult height in our sample, is likely to be to θ, the actual adult height in the population, we need at the very least to estimate the variance of $\hat{\theta}$. Imagine doing this in the following way. Take a large number, say B, of samples of size 12 from the population. In each of these samples, calculate an estimate of adult height. Call these estimates $\hat{\theta}_1, \ldots, \hat{\theta}_B$. Then estimate the variance of $\hat{\theta}$ by the sample variance of $(\hat{\theta}_1, \ldots, \hat{\theta}_B)$. Of course, this approach is impossible in practice; if we could afford to draw B extra samples of size 12, we would have drawn a much larger sample initially! However, an approximation to it can be achieved as follows.

Suppose we sample with replacements from the 12 observations in the data (second row in the figure) to form a 'subsample', also of size 12. Seven possible such 'subsamples' are shown in the third row of the figure. For example, the first subsample, shown in the first rectangle in the third row, consists of the following observations (note some observations will occur more than once, and some not at all): $\{1, 2, 2, 2, 3, 3, 5, 6, 7, 7, 8, 11\}$. These 'subsamples' are known as *bootstrap samples*.

Using each of these bootstrap samples we calculate an estimate of adult height. By convention, these are denoted with a '*', to indicate they have been calculated from a bootstrap sample. From the seven bootstrap samples in the third row of the figure, we therefore get $\hat{\theta}_1^*, \ldots, \hat{\theta}_7^*$. Now we simply estimate the variability of the estimate of adult height calculated from the actual data, $\hat{\theta}$, by the sample variance of the bootstrap estimates $\hat{\theta}_1^*, \ldots, \hat{\theta}_7^*$. Of course, in practice we would need many more than seven bootstrap estimates.

Another way of looking at this is as follows. We wish to learn about the relationship between the true population parameter value, θ, and estimates of θ obtained from samples from the population, denoted $\hat{\theta}$. To do this, we pretend the observed data are the population and repeatedly sample from the data to learn about the relationship between $\hat{\theta}$ and estimates obtained from the resampled data, denoted $\hat{\theta}^*$. In other words, we say:

$$\text{Distribution of estimates } \hat{\theta} \text{ given } \theta \text{ is approximated by Distribution of estimates } \hat{\theta}^* \text{ given } \hat{\theta} \tag{1}$$

This is known as the *bootstrap principle*. It is important to separate this principle from simulation, which is used to estimate the distribution of estimates $\hat{\theta}^*$ given $\hat{\theta}$. In fact, there are two potential sources of error in bootstrap procedures. The first arises because the bootstrap principle does not hold true, i.e. the two distributions in equation (1) are not equal. The second arises because we only use a finite number of bootstrap samples, B, to estimate the distribution of the $\hat{\theta}^*$s. However, this error can be made as small as we like by simply increasing B, whereas the bootstrap error is fixed. One of the arts of bootstrapping is to consider simple functions of θ, such as $(\hat{\theta}-\theta)/\hat{\sigma}$ (where $\hat{\sigma}$ is the sample standard error of $\hat{\theta}$), for which the bootstrap principle is more nearly true.

To make things more concrete, we illustrate how to use the bootstrap to estimate VARIANCE. Consider the data in the table. We are interested in estimating the average change in the carbon monoxide transfer factor. The obvious estimate is the mean: $(33 + 2 + 24 + 27 + 4 + 1 - 6)/7 = 12.14$. Suppose we were able to draw a large number, B, samples of the same size as that in the table from the 'population' of smokers with chickenpox and estimate the average change on each. Denote the resulting estimates by $\hat{\theta}_1, \ldots, \hat{\theta}_B$ and recall that the true value in the population is called θ. Then an estimate of the variance would be:

bootstrap *Data on the carbon monoxide transfer factor for seven smokers with chickenpox, measured on admission to hospital and after a stay of one week (Davison and Hinkley, 1997, p.67)*

Patient	*Entry*	*Week*	*Change = (Week − Entry)*
1	40	73	33
2	50	52	2
3	56	80	24
4	58	85	27
5	60	64	4
6	62	63	1
7	66	60	−6

$$\frac{1}{B}\sum_{i=1}^{B}(\hat{\theta}_i - \theta)^2 \qquad (2)$$

Using the bootstrap principle (1), we estimate this by (a) replacing θ by its estimate from the data, $\hat{\theta}$, and (b) replacing the $\hat{\theta}_i$ by $\hat{\theta}_i^*$, where each $\hat{\theta}_i^*$ is the mean carbon monoxide transfer in the ith bootstrap sample.

The second table shows the bootstrap in action. The first row shows the observed differences, corresponding to the fourth column of the first table. The second row shows the frequency of these observations in the data in the first table; they all occur once. Rows 3–11 show the frequency of the observations in bootstrap samples 1–9. Thus, in the first bootstrap dataset, observation 1 does not appear, observation 2 appears three times, observation 3 appears twice, observation 4 once, observation 5 does not appear, observation 6 appears once and observation 7 does not appear: the mean is then 11.71. The table shows $B = 9$ bootstrap samples. We thus have $\hat{\theta}_1^*, \ldots, \hat{\theta}_9^*$, each of which stands in approximately the same relationship to $\hat{\theta}$ as $\hat{\theta}$ does to the true parameter θ. We can use these to learn about the relationship between $\hat{\theta}$ and θ.

Specifically, the bootstrap estimate of variance is:

$$\frac{1}{B}\sum_{i=1}^{B}\left(\hat{\theta}_i^* - \hat{\theta}\right)^2 \qquad (3)$$

Comparing with (2), we see the bootstrap version (3) is derived by (a) putting '*'s next to everything with a 'hat' and (b) putting a 'hat' on what is left. This rule of thumb is very useful in practice.

Substituting the bootstrap estimates from the second table gives:

$$\frac{1}{9}\Big[(11.71-12.14)^2 + (7.00-12.14)^2 + (18.57-12.14)^2 + (15.43-12.14)^2 + (15.71-12.14)^2 + (26.43-12.14)^2 + (13.29-12.14)^2 + (24.29-12.14)^2 + (16.43-12.14)^2\Big] = 7.17^2$$

bootstrap *Frequencies with which each difference from the original data in the first table appear in each of nine nonparametric bootstrap samples*

								Statistic
Observed differences	33	2	24	27	4	1	−6	(mean)
Frequency in observed data	1	1	1	1	1	1	1	$\hat{\theta} = 12.14$
1st bootstrap sample		3	2	1		1		$\hat{\theta}_1^* = 11.71$
2nd bootstrap sample	1	1	1			2	2	$\hat{\theta}_2^* = 7.00$
3rd bootstrap sample	3	1		1		2		$\hat{\theta}_3^* = 18.57$
4th bootstrap sample	1	1	1	2		1	1	$\hat{\theta}_4^* = 15.43$
5th bootstrap sample		2	2	2	1			$\hat{\theta}_5^* = 15.71$
6th bootstrap sample	4	1	1	1				$\hat{\theta}_6^* = 26.43$
7th bootstrap sample	1	2	1	1	1	1		$\hat{\theta}_7^* = 13.29$
8th bootstrap sample	2	1	2	2				$\hat{\theta}_8^* = 24.29$
9th bootstrap sample	1	1	1	2		2		$\hat{\theta}_9^* = 16.43$

However, $B = 9$ is not nearly enough. Typically we may need around $B = 800$ bootstrap samples to estimate the variance accurately (Booth and Sarkar, 1998). Taking $B = 1000$, we find that the bootstrap variance of the mean is 5.39^2, which compares with the maximum likelihood estimate of 5.38^2. The bootstrap estimate of the standard error of the mean, 12.14, is thus 5.39. Of course, this example is only illustrative; we know the answer anyway. However, in many circumstances we may not, e.g. if the data are not normally distributed and we want the standard error of the median or some other nonstandard measure of the data's 'centre'.

The bootstrap principle can clearly be applied much more widely. It is probably most often used to calculate confidence intervals (Carpenter and Bithell, 2000), where it avoids the need to rely on large sample theory or assumptions concerning the distribution of the data. For example, the distribution of individual patients' hospital costs is usually very skew and the bootstrap has been applied to calculate confidence intervals for the average cost of hospitalisation. Other applications include hypothesis tests, power calculations and estimating the predictive performance a statistical model will have when applied to a new dataset that was not used in formulating or estimating the model.

In order for the bootstrap principle (1) to hold, it is necessary for the bootstrap sampling to mimic the actual data sampling. Therefore, if we are bootstrapping a clinical trial with two treatments, we should sample with replacement within each treatment group, to preserve the randomisation. Other situations require different approaches.

The bootstrap resampling illustrated here does not depend on any statistical model and is an example of the *no-parametric bootstrap*. An alternative, the parametric bootstrap, is less widely used. This samples data from a parametric statistical model, such as a regression model, rather than with replacement from the observed data.

Lastly, note that the bootstrap, although it uses simulation, is characterised by the bootstrap principle (1). It is thus quite distinct from two other common uses of simulation, randomisation tests and MARKOV CHAIN MONTE CARLO (for fitting BAYESIAN MODELS). *JRC*

[**Acknowledgement:** James R. Carpenter was supported by ESRC Research Methods Programme grant H333250047, titled 'Missing data in multilevel models'.]

Booth, J. G. and Sarkar, S. 1998: Monte-Carlo approximation of bootstrap variances. *The American Statistician* 52, 354–7. **Carpenter, J. and Bithell, J.** 2000: Bootstrap confidence intervals: when, which, what? A practical guide for medical statisticians. *Statistics in Medicine* 19, 1141–64. **Davison, A. C. and Hinkley, D. V.** 1997: *Bootstrap methods and their application.* Cambridge: Cambridge University Press. **Efron, B. and Tibshirani, R.** 1993: *An introduction to the bootstrap.* New York: Chapman & Hall.

boxplot This is a graphical display useful for highlighting important distributional features of a variable. The diagram is based on the five-number summary of a dataset, the numbers being the minimum, the lower quartile, the median, the upper quartile and the maximum. The boxplot is constructed by first drawing a 'box' with ends at the lower and upper quartiles of the data, next a horizontal line (or some other feature) is used to indicate the position of the median within the box and then lines are drawn from each end of the box to the most remote observations. One convention modifies this last step by truncating the lines to within (unmarked) points given by the upper quartile plus 1.5 times the interquartile range (the difference between the upper and lower quartiles) and the lower quartile plus 1.5 times the interquartile range. In this case, any observations outside these limits are represented individually by some means in the finished graphic. Different computer packages may employ slightly different conventions for displaying extreme or outlying values.

The resulting diagram schematically represents the body of the data minus the extreme observations. Particularly useful for comparing the distributional features of a variable in different groups as illustrated in the figure, which shows the birthweights of infants with severe idiopathic respiratory disorder, classified by whether or not the infant survived. For other examples see Altman (1991). *BSE*

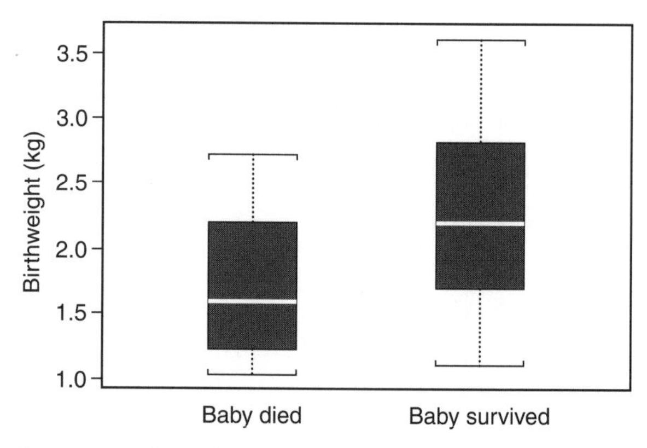

boxplot *Birthweights (kg) of infants with severe idiopathic respiratory disease syndrome*

[See also HISTOGRAM, STEM-AND-LEAF PLOT]

Altman, D. G. 1991: *Practical statistics for medical research.* London: Chapman & Hall.

Box–Cox transformation See TRANSFORMATIONS

Bradford Hill criteria Guidelines for drawing conclusions about causal relationships were proposed by Sir Austin Bradford Hill, Professor Emeritus of Medical

Statistics at the London School of Hygiene and Tropical Medicine, in his address to the Section of Occupational Medicine of the Royal Society of Medicine in 1965 (Bradford Hill, 1965). Bradford Hill's guidelines drew on his many contributions to chronic disease research in the post-war era, including the groundbreaking work with Richard Doll on the link between smoking and lung cancer. The following nine aspects were proposed for deciding whether a statistical association might be causal: strength – magnitude of the association, as observed by measures such as the ratio of incidence rates; consistency – repeated observation of the association in different populations and circumstances and by different researchers; specificity – whether a cause leads to a single effect in a given population; temporality – whether the cause precedes the effect in time; biological gradient – existence of a trend or dose–response curve between the cause and effect; plausibility – whether the association is consistent with current biological knowledge; coherence – ensuring that the interpretation of cause and effect does not conflict with what is known of the natural history of the disease; experiment – existence of experimental rather than observational evidence, such as through conducting a randomised trial or by introduction of a preventive measure; analogy – comparison with previous research that identified similar effect mechanisms.

Bradford Hill did not intend these guidelines to be philosophically rigorous 'criteria' for causal inference, rather a basis for decision making that could lead to timely action for the good of public health. With further consideration, some of the guidelines (e.g. 'specificity') are less than universal in their utility and some commentators have proposed alternative criteria more firmly rooted in deductive logic (Weed, 1986). However, in proposing these guidelines, Bradford Hill advocated an approach of making the best use of the totality of available evidence: 'All scientific work is incomplete – that does not confer upon us the freedom to ignore the knowledge we already have, or to postpone the action it demands.' *JGW*

Bradford Hill, A. 1965: The environment and disease: association or causation? *Proceedings of the Royal Society of Medicine* 58, 295.
Weed, D. L. 1986: On the logic of causal inference. *American Journal of Epidemiology* 123, 965–79.

bubbleplot This is a graphical display for three variables in which two variables are used to form a SCATTERPLOT and then the values of the third variable are represented by circles with radii proportional to these values and centred on the appropriate point in the scatterplot. An example is shown in the figure; here the data are for 41 cities in the USA and the two variables forming the scatterplot are average annual temperature and average annual wind speed, with the 'bubbles' representing the pollution level as measured by the concentration of sulfur dioxide in the air. The plot suggests that higher pollution levels are associated with a combination of lower annual temperature and higher average wind speed. More details of bubbleplots can be found in Everitt (2003). *BSE*

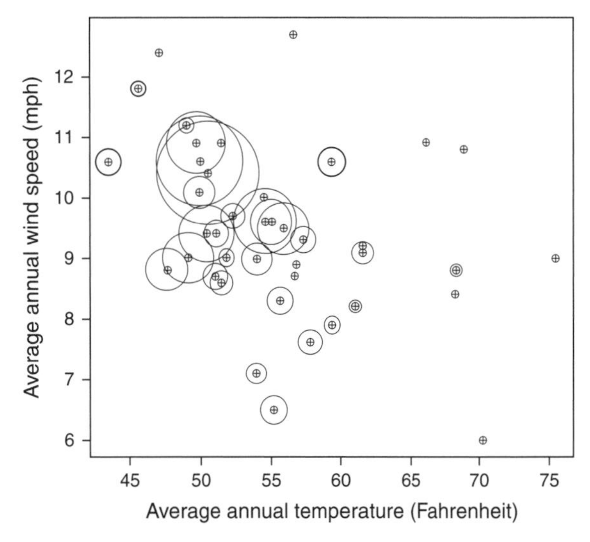

bubbleplot *Bubbleplot of annual temperature and wind speed against pollution level as measured by sulphur dioxide concentration in the air for 41 cities in the USA*

[See also BIVARIATE BOXPLOT and SCATTERPLOT MATRICES]

Everitt, B. S. 2003: *Modern medical statistics.* London: Arnold.

BUGS and WinBUGS The use of BAYESIAN METHODS in practical problems in medical statistics and other substantive areas of application has been hindered until relatively recently by computational aspects. In particular, the evaluation of integrals in order to obtain posterior marginal, conditional and predictive distributions in many multiparameter problems are not usually analytically tractable and asymptotic, numerical integration techniques or simulation-based methods are required (Bernardo and Smith, 1994). In many practical problems in medical statistics the structure and nature of the models used have made parameter estimation particularly amenable to the use of MARKOV CHAIN MONTE CARLO (MCMC) simulation methods and it is these that the software packages Bayesian inference Using Gibbs Sampling (BUGS) and WinBUGS (Windows version of BUGS) implement. (Latest versions at the time of writing are BUGS 0.5 and WinBUGS 1.4 are freely available from www.mrc-bsu.cam.ac.uk.)

BUGS and WinBUGS use the BUGS syntax, which is similar to that of S-PLUS & R, to specify the likelihood and prior distributions for the statistical model in question, together with initial starting values for the sampler (Gilks,

Thomas and Spiegelhalter, 1994). Within WinBUGS the specification of models may also be in terms of directed acyclic graphs (DAGs) using the Doodle feature (see GRAPHICAL MODELS), with the appropriate code being produced automatically.

Additions to the most recent version of WinBUGS (1.4) are the ability to use scripts so that WinBUGS may be used in 'batch mode' and improved graphics capabilities, together with calculation of the deviance information criterion (DIC) to assess model complexity and fit (see BAYESIAN METHODS). In addition, the suite of S-PLUS functions CODA (Best, Cowles and Vines, 1995) can be used to explore convergence issues with output from BUGS and WinBUGS.

Specific developments of WinBUGS are PKBUGS, which allows MCMC methods to be used for complex population *pharmacokinetic/pharmacodynamic* (PK/PD) *models* and GeoBUGS, which is an add-on to WinBUGS that fits spatial models and produces a range of maps as output.

Since BUGS and WinBUGS require the user to specify statistical models in terms of the LIKELIHOOD and PRIOR DISTRIBUTIONS (see BAYESIAN METHODS), using MCMC methods in order to evaluate the model is only recommended for users skilled at undertaking Bayesian analyses and must therefore be used with considerable care – the manual even comes with a 'health warning'! *KRA*

Bernardo, J. M. and Smith, A. F. M. 1994: *Bayesian theory*. Chichester: John Wiley & Sons, Ltd. **Best, N. G., Cowles, M. K. and Vines, S. K.** 1995: *CODA convergence diagnosis and output analysis software for Gibbs Sampler output: Version 0.3*. Cambridge: MRC Biostatistics Unit. **Gilks, W. R., Thomas, A. and Spiegelhalter, D. J.** 1994: A language and program for complex Bayesian modelling. *The Statistician* 43, 169–78. **Spiegelhalter, D. J., Thomas, A. and Best, N. G.** 2001: *WinBUGS version 1.4 user manual*. Cambridge: MRC Biostatistics Unit.

C

calibration Consider a situation in which we wish to measure serum concentrations of hormones, enzymes and other proteins, for example, using such methods as radioimmunoassays (RIA) and enzyme-linked immunosorbent assays (ELISA). Three key questions in the development of such assays are (a) how does the expected value (average) of the assay response change as a function of the true amount of the target material in the serum samples, (b) how does the VARIANCE (or STANDARD DEVIATION) of the assay results change with the average assay result and, subsequently, (c) how might we use a particular assay result to determine the amount of the target material in a new sample of serum? We leave question (c) for the time being and concentrate on questions (a) and (b). Let the assay response be Y and let the true level of the target material be X. We wish to determine the form of the functions F and G in the following two equations:

$$E(Y|X) = F(X) \qquad (1)$$

and

$$\mathrm{Var}(Y|X) = G(E(Y|X)) \qquad (2)$$

Here we assume that the values of X are known without MEASUREMENT ERROR. We are concerned with what is often referred to as *absolute calibration*. If we do not have access to the truth, but only have measurements using alternative assays, Y_1 and Y_2, say, then we are concerned with the problem of comparative calibration (for the latter see METHOD COMPARISON STUDIES). Typically, such a univariate calibration study involves performing the assay procedure (ideally with full, independent, replications) on each of N training samples or specimens with known values of X, and then using various data analytic and modelling procedures to evaluate the form of F and G. The statistical methods might be fully parametric (fitting linear or nonlinear models, for example, with an assumed parametric model for the variance) or nonparametric (essentially fitting an arbitrarily shaped smooth dose–response curve).

Suppose an analytical chemist wishes to use some form of absorption spectroscopy to study the composition of, say, certain body fluids. He or she is likely to use measurements of many peak heights from such spectra to measure several substances simultaneously. This activity is the multivariate analogue of the univariate case; i.e. multivariate calibration. Technically, multivariate calibration is much more difficult than the simpler univariate problem, but the ultimate aims and logic are similar. We start with the latter and then briefly discuss the former.

Instead of dealing with the technical complexities of fitting nonlinear models with heterogeneous error distributions, we will consider an example that, by comparison, appears to be quite simple. Suppose we have a simple colorimetric assay for urinary glucose. We obtain a series of specimens with known glucose concentrations (X) and then measure the absorbance (Y) using the relevant assay procedure. We assume that the calibration function F is a straight line and that the variance of the Y measurements is independent of X (i.e. the 'error' variance is constant). Fitting a simple linear regression model for Y using ordinary least squares gives us estimates of the intercept (α) and slope (β) of the straight line relating X to Y. Having answered questions (a) and (b) using the simple regression analysis, we now move on to question (c). Suppose we are presented with a new urine specimen and are asked to determine its glucose content.

The classical method of estimating the unknown X from our measurement, Y, involves using information from the above regression of Y on X. The required estimate is given by:

$$X = (Y - \alpha)/\beta \qquad (3)$$

An alternative is the so-called inverse estimator suggested by Krutchkoff (1967). This involves using the original X, Y data to regress X on Y to obtain estimates of the intercept (γ) and slope (λ), and then simply using these parameter estimates to predict X given a new Y_1, i.e.:

$$X = \hat{\gamma} + \hat{\lambda} Y \qquad (4)$$

For details of the properties of these two estimators, see the review by Osborne (1991).

To illustrate the ideas of multivariate calibration, consider a relatively simple example. Suppose we wish to measure the concentration of a particular metabolite in the blood (X) but we are now able to use, say, three different colorimetric assay procedures to obtain values Y_1, Y_2 and Y_3. Assuming that the three corresponding calibration curves (F_1, F_2 and F_3), as before, are all straight lines (but with different intercepts, slopes and 'error' variances) we can use MULTIVARIATE LINEAR REGRESSION (or three separate regressions) in order to estimate the parameters of the three calibration curves. The classical approach to the use of a new set of three measurements (Y_1, Y_2 and Y_3) on a new specimen to predict an unknown X is the multivariate generalisation of the univariate problem. Details of multivariate calibration are well beyond the scope of the present article, however, and readers are referred to Thomas (1994) and Naes *et al.* (2002) for further information. Considering our present example, one simple approach

Encyclopaedic Companion to Medical Statistics: Second Edition Edited by Brian S. Everitt and Christopher R. Palmer

(particularly if we are prepared to assume conditional independence of the Y values) might involve estimating the unknown X using each of the Y_1, Y_2 and Y_3 values separately (in each case using equation (3) above) and then producing a weighted average of these three estimates, with weights proportional to their estimated precision. An example of the inverse approach would be to produce a multiple regression to predict the unknown X from the three Y measurements. This has obvious technical drawbacks, however, because of MULTICOLLINEARITY (high correlations between the three Y values). One possible solution involves the use of principal components regression. A PRINCIPAL COMPONENTS ANALYSIS is carried out on the Y values and then one or more of the resulting components are used to predict the unknown X. Further details of principal components regression and alternative analytical strategies can be found in Thomas (1994) and Naes *et al.* (2002). Whatever method of prediction is used, however, it is important in both univariate and multivariate calibration problems that the performance of the predictions are adequately evaluated. This might involve validation using a test set of new X, Y values or internal cross-validation (use of the LEAVE-ONE-OUT CROSS-VALIDATION approach, for example) using the original training set. *GD*

Krutchkoff, R. G. 1967: Classical and inverse regression methods of calibration. *Technometrics* 9, 425–39. **Naes, T., Isaksson, T., Fearn, T. and Davies, T.** 2002: *A user-friendly guide to multivariate calibration and classification.* Chichester, UK: NIR Publications. **Osborne, C.** 1991: Statistical calibration: a review. *International Statistical Review* 59, 309–36. **Thomas, E. V.** 1994: A primer on multivariate calibration. *Analytical Chemistry* 66, 795A–804A.

caliper matching See MATCHING

canonical correlation analysis This technique establishes whether relationships exist between a priori groups of variables in a study. For example, in a study of heart disease, we might ask if there is a connection between personal physical characteristics such as age, weight and height, on the one hand, and the systolic and diastolic blood pressures of the individuals, on the other. Alternatively, in chronic depression, a study might be aimed at uncovering relationships between personal social and financial variables such as gender, age, educational level, income and a range of health variables including various indicators of depression. In another example, a public health survey might be conducted to explore connections between housing quality variables and indicators of different illnesses.

A first attempt at analysing the strength of association between two *groups* of variables (e.g. between housing quality and illness) might involve examination of all correlations between pairs of variables, one from each group. However, if each group contains more than just a few variables, such an approach is bound to lead to confusion. Ideally, one would like to replace each set of original variables by a new set, in such a way that the new variables were mutually uncorrelated within sets and just a few of them exhibited correlation between sets. Canonical correlation analysis takes just such an approach, and finds optimal sets of *linear transformations* of the original variables, one for each original group of variables. Suppose that $u_1, u_2, \ldots, u_s$ are the transformed variables for one set (say, the housing quality variables), while $v_1, v_2, \ldots, v_s$ are the transformed variables for the other set (say, the illness variables).

'Optimality' is defined by requiring the correlation between u_1 and v_1 to be as large as possible among all linear combinations of the original variables, that between u_2 and v_2 to be the next largest, that between u_3 and v_3 the third largest and so on, subject to the following constraints: $u_1, u_2, \ldots, u_s$ are mutually uncorrelated; $v_1, v_2, \ldots, v_s$ are mutually uncorrelated; and any u_i, v_j pair is uncorrelated when $i \neq j$.

It is clearly not possible to have more (uncorrelated) transformed variables than there were original variables in a set, so the number s of pairs that can be derived is equal to the *smaller* of the numbers of original variables in the two groups.

The effect of canonical correlation analysis is thus to channel all the association between the two groups of variables through the resulting pairs of linear combinations (u_1,v_1), (u_2,v_2), ... These derived variables are known as *canonical variates*. The only nonzero correlations remaining in the correlation matrix of the new variables are those between corresponding pairs of canonical variates, i.e. between u_i and v_i for $i = 1, \ldots, s$; they are known as the *canonical correlations* of the system. Most computer software packages that contain multivariate statistical procedures will conduct such an analysis. They will also quote a significance level against each canonical correlation, appropriate for testing the NULL HYPOTHESIS that all succeeding population canonical correlations are zero. Such significance levels should be treated with some caution, as they rely on the assumption that the data follow a MULTIVARIATE NORMAL DISTRIBUTION. Nonetheless, the number of significant canonical correlations is usually taken to indicate the number of (independent) connections that exist between the two groups of variables.

Inspection of the coefficients of each original variable in each canonical variate may also provide an interpretation of the canonical variate in the same manner as interpretation of principal components, which may help to identify the nature of the connection between the groups (see PRINCIPAL COMPONENT ANALYSIS). However, again a cautionary note is in order, because such interpretation is not quite as straightforward as for principal components. The reason for the complication is that there may be very diverse VARIANCES and covariances (see COVARIANCE MATRICES) among the original variables in the two groups, which affects the sizes of the coefficients in the canonical variates, and there is no convenient normalisation to place all coefficients on an equal footing. This drawback can be alleviated to some extent by restricting interpretation to the *standardised* coefficients, i.e. the coefficients that are appropriate when the original variables have been standardised, but nevertheless the problem still remains.

To illustrate the technique, consider a canonical correlation analysis between the 'health' variables and the 'personal' variables in the Los Angeles depression study of 294 respondents presented by Afifi and Clark (1984, Chapter 15). The four 'personal' variables were: gender, age, income, education level (numerically coded from the lowest, 'less than high school', to the highest, 'finished doctorate'), while the two 'health' variables were: CESD (the sum of 20 separate numerical scales measuring different aspects of depression) and health (a numerical score measuring 'general health'). The correlation matrix between these variables for the sample is shown in the table.

canonical correlation analysis *Correlation matrix for two health and four personal variables in the LA depression study*

	CESD	*Health*	*Gender*	*Age*	*Education*	*Income*
CESD	1.0	0.212	0.124	−0.164	−0.101	−0.158
Health		1.0	0.098	0.308	−0.270	−0.183
Gender			1.0	0.044	−0.106	−0.180
Age				1.0	−0.208	−0.192
Education					1.0	0.492
Income						1.0

Here the maximum number of canonical variate pairs is $s = \min(2, 4) = 2$. The first canonical correlation turns out to be 0.405 and this gives a significance level $P < 0.000\,01$. It might be argued that gender and education are unlikely to have normal distributions, so this significance level should not be taken too literally. Nevertheless, there does seem to be strong evidence that the first canonical correlation is significant. The corresponding canonical variates, in terms of standardised original variables, are:

$$u_1 = -0.490 CESD + 0.982 Health$$

$$v_1 = 0.025 Gender + 0.871 Age - 0.383 Education + 0.082 Income$$

High coefficients correspond to CESD (negatively) and health (positively) for the perceived health variables and to age (positively) and education (negatively) for the personal variables. Thus relatively older and uneducated people tend to score low in terms of depression, but perceive their health as relatively poor, while relatively younger but educated people have the opposite health perception.

The second canonical correlation is 0.266, which has a significance level $P < 0.001$ so also carries interpretative worth. The corresponding canonical variates are:

$$u_2 = 0.899 CESD + 0.288 Health$$

$$v_2 = 0.396 Gender - 0.443 Age - 0.448 Education - 0.555 Income$$

Since the higher value of the gender variable is for females, the interpretation here is that relatively young, poor and uneducated females are associated with higher depression scores and, to a lesser extent, with poor perceived health. Thus there are two interpretable 'dimensions' of connection between the two sets of variables. A SCATTERPLOT of the scores of respondents against each pair of canonical variates would help to identify any anomalous individuals in the sample.

A further interesting application of canonical correlation analysis occurs when there is just a single set of variables, but they are measured on individuals in a number of a priori distinct groups or populations. For example, a set of signs or symptoms $x_1, x_2, \ldots, x_p$ is observed on a sample of patients suffering from jaundice and each patient is classified into one of g illnesses that have the external manifestation of jaundice. We can thus define a set of indicator variables $y_1, y_2, \ldots, y_g$ that specify a patient's illness, by setting the values $y_i = 1$, $y_j = 0$ $(j \neq i)$ for a patient suffering from illness i. A canonical correlation analysis with the x values as one set of variables and the y values as the other set of variables will then produce the linear combinations of the x values that are most highly correlated with linear combinations of the group indicator variables. Since the latter define the best way to view group differences, the former are just the canonical variables that best discriminate between the g groups of individuals (see DISCRIMINANT FUNCTION ANALYSIS).

Finally, the variables in a study may fall into more than two *a priori* sets and some general between-set measure of association is required. Various possible definitions of such association may be made and, consequently, the ideas of canonical correlation analysis may be generalised in various ways. However, such generalisation is quite complicated and interpretation of the results becomes much more problematic. Gnanadesikan (1997) provides a brief overview and further references. *WK*

Afifi, A.A. and Clark, V. 1984: *Computer-aided multivariate analysis*. California: Wadsworth. **Gnanadesikan, R.** 1997: *Methods for statistical analysis of multivariate observations*, 2nd edition. New York: John Wiley & Sons, Inc.

canonical variates See CANONICAL CORRELATIONS ANALYSIS

capture–recapture methods This is an alternative approach to a census for estimating population size that operates by sampling the population several times, identifying individuals who appear more than once. Capture–recapture methods have a long history dating back to 1786, when Laplace used such a technique to estimate the size of the total population of France. Traditionally, capture–recapture methodology was primarily focused on wildlife populations, but has increasingly been applied to human populations, particularly within epidemiological situations. Within the ecological field, capture–recapture experiments involve observers going into the field and recording all animals that are observed (either visual sightings or trappings) at a sequence of capture events. On the initial capture event, all animals that are observed are recorded, uniquely marked and released back into the population. At each subsequent capture event, all unmarked animals are recorded and uniquely marked, all marked animals are recorded and all animals released. The data from such an experiment are simply the record of capture histories for each individual animal observed within the study. Each individual capture history is typically represented by a series of zeros and ones, where the 0 and the 1 denote the absence or presence, respectively, of the individual at each capture event.

There are generally two forms of models for capture–recapture data: closed and open, for which there have been a series of models proposed. Closed models assume that the population is constant throughout the study period, with no births, deaths or migrations, whereas open models allow for these transitions in the population. Generally speaking, the parameters of interest differ between the two models. For example, within closed populations, the total population size is generally of particular interest; conversely, for open populations, parameters of interest may include birth rates, death rates, migration rates and/or productivity rates. We initially, briefly, consider the capture–recapture methods often considered for wildlife data before considering in further detail epidemiological models.

For closed populations, Otis *et al.* (1978) described a series of different capture–recapture models, relating to possible heterogeneity in the capture rates as a result of time, trap response or individual effects. King and Brooks (2008) have incorporated these models into a Bayesian framework. Mixture models have become increasingly popular to model individual heterogeneity (Pledger, 2000; Morgan and Ridout, 2008). Additionally, there have been a series of models proposed for open populations, dependent on the parameters of interest, with perhaps the most widely used being the Cormack–Jolly–Seber model, where the survival rates are of primary interest, and the Arnason–Schwarz model, which incorporates multistrata data. Recent advances include the generalisation to multievent models (Pradel, 2005), where the state of an individual may only be partially observed.

Within the epidemiological literature, closed populations are usually modelled, with the total population size of particular interest – and this is what we shall focus on here.

capture–recapture methods *Example of an incomplete contingency table, with three sources: A, B and C. The entries n_{ijk} denote the number of individuals observed in the given cell, where 0/1 represents absence/presence on the given list. The cell n_{000} is unobserved and hence unknown*

		C=1	*C=0*
A=1	B=1	n_{111}	n_{110}
A=1	B=0	n_{101}	n_{100}
A=0	B=1	n_{011}	n_{010}
A=0	B=0	n_{001}	n_{000}

For example, many areas of scientific research focus on the estimation of population size: from the number of susceptibles to a given disease, to the number of drug addicts in a particular area or the number of injuries sustained in the workplace. However, it is usually impossible to enumerate each member of a population, possibly due to their number (e.g. the number of web pages on the internet) or when the population is 'hidden' (such as the number of injector drug users). Thus, data are often collected in the form of a series of incomplete population counts using a variety of sources or lists. Each source corresponds to a capture event and an individual being recorded by a given source corresponds to being observed at that capture event. It is assumed that each individual is uniquely identifiable by each source. Then, the data are simply the capture histories of all individuals observed. The data are usually summarised in the form of a 2^k contingency table, where k is the number of sources, and the cell entries correspond to the number of individuals that are observed by each combination of sources (i.e. the number of individuals observed with the same capture history). Clearly, the contingency table is incomplete since the number of individuals belonging to the population but not observed within the study is unknown (see the table, for example).

Unlike the ecological application the sources do not usually have a temporal sequencing as for the capture events and so different models have been developed within the epidemiological application. Within the epidemiological field the capture–recapture approach is sometimes called 'multiple record systems' and the corresponding estimate of the total population referred to as 'Bernoulli census estimates' or 'ascertainment corrected rates'.

The most common approach to analysing epidemiological data of this sort is via the use of LOG-LINEAR MODELS, introduced by Fienberg (1972). In these models, the logarithm of the expected cell count is expressed as a linear function of parameters. These parameters represent main effect terms for individual sources and associations between two or more sources. Thus, these models allow for interactions between different sources. The model assuming independence between each source is simply a special case with no interactions present. There are usually a number of possible log-linear models that can be fitted to the data, each specifying different sets of interactions between the sources.

Traditionally, classical analyses consist of initially finding the model which provides the best fit to the data, using, for example, LIKELIHOOD RATIO TESTS and/or information criteria, such as AKAIKE'S INFORMATION CRITERION (AIC) or *Bayesian information criterion* (BIC). Once the given model has been selected, the total population is estimated using MAXIMUM LIKELIHOOD ESTIMATION for the missing cell, combined with the observed number of individuals (see, for example, Hook and Regal, 1995). Recently, Bayesian approaches have also been developed for fitting log-linear models to the data, and in particular the issue of model choice (King and Brooks, 2001; King *et al.* 2005). This approach also allows the calculation of a model-averaged estimate of the total population, removing the model-dependence problem that may arise when only a single model is chosen on which to base inference. Alternative approaches to using log-linear models include use of the Rasch model (see Carriquiry and Fienberg, 1998) in order to model possible heterogeneity in the population and also latent class models when the individuals can be categorised into different subpopulations. *RK*

Carriquiry, A. L. and Fienberg, S. E. 1998: Rasch models. In Armitage, P. and Colton, T. (eds), *Encyclopedia of biostatistics*. Chichester: John Wiley & Sons, Ltd. **Fienberg, S. E.** 1972: The multiple recapture census for closed populations and incomplete 2^k contingency tables. *Biometrika* 59, 591–603. **Hook, E. B. and Regal, R. R.** 1995: Capture-recapture methods in epidemiology: methods and limitations. *Epidemiological Reviews* 17, 243–64. **King, R. and Brooks, S. P.** 2001: On the Bayesian analysis of population size. *Biometrika* 88, 317–36. **King, R. and Brooks, S. P.** 2008: On the Bayesian estimation of a closed population size in the presence of heterogeneity and model uncertainty. *Biometrics* 64, 816–24. **King, R., Bird, S. M., Brooks, S. P., Hay, G. and Hutchinson, S.** 2005: Prior information in behavioural capture–recapture methods: demography influences injectors' propensity to be listed on data sources and their drugs-related mortality. *American Journal of Epidemiology* 162, 694-703. **Otis, D. L., Burnham, K. P., White, G. C. and Anderson, D. R.** 1978: Statistical inference from capture data on closed animal populations. *Wildlife Monographs* 62, 1–135. **Morgan, B. J. T. and Ridout, M. S.** 2008: A new mixture model for capture heterogeneity. Journal of the Royal Statistical Society: Series C **57,** 433–46. **Pledger, S.** 2000: Unified maximum likelihood estimates for closed capture–recapture models using mixtures. *Biometrics* 56, 434–42. **Pradel, R.** 2005: Multievent: an extension of multistate capture–recapture models to uncertain states. *Biometrics* 61, 442–7.

carry-over See CROSSOVER TRIALS

CART This is an acronym for classification and regression tree. See TREE-STRUCTURED METHODS

cartogram This is a diagram in which descriptive statistical information is displayed on a geographical map by means of shading, by using a variety of different symbols or by some more involved procedure. Two examples in figures are given. A description of how cartograms may be constructed is given in Gusein-Zade and Tikanov (1993). *BSE*
[See also DISEASE CLUSTERING]

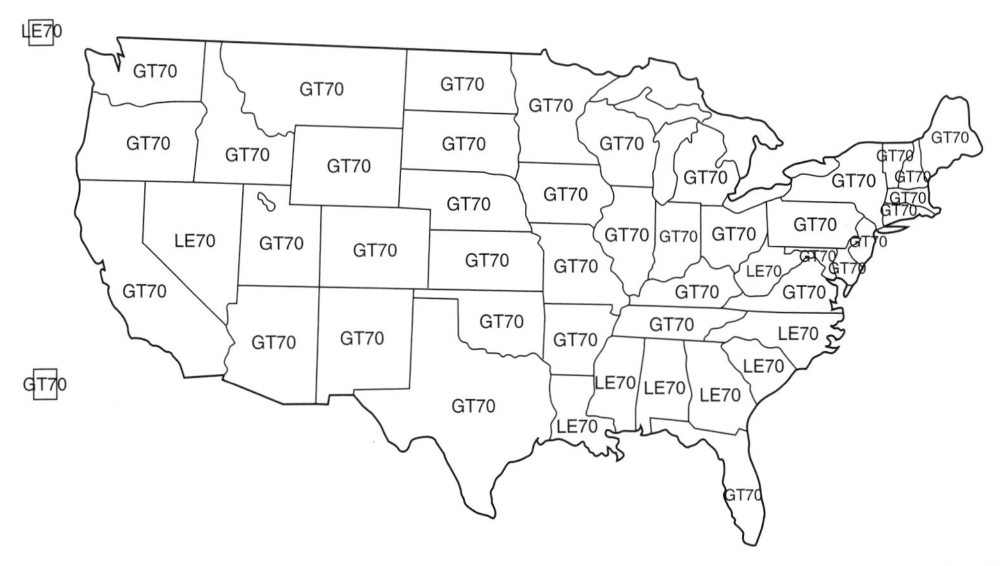

cartogram *Cartogram of life expectancy in the USA by state. <70 = 70 years or less, >70 = more than 70 years*

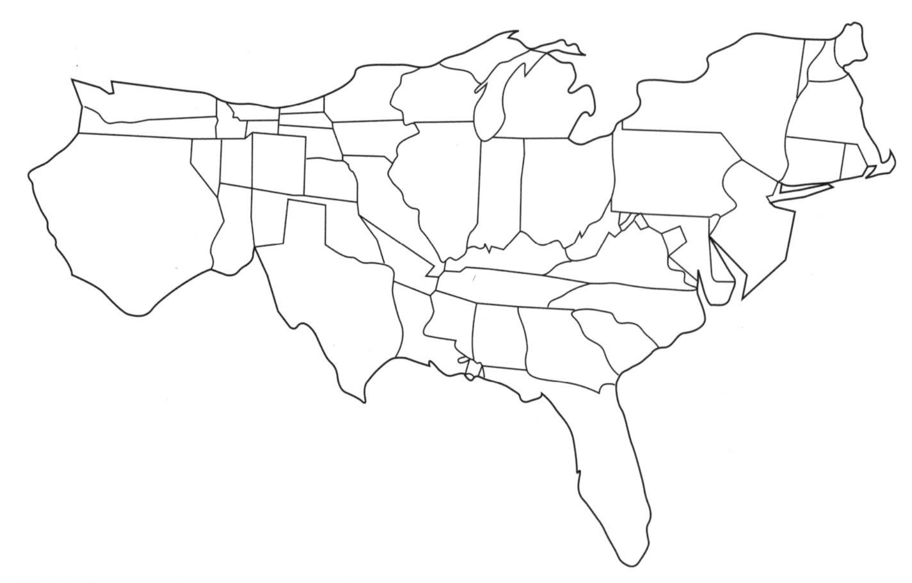

cartogram *1996 US population cartogram (all states are resized relative to their population)*

Gusein-Zade, S. and Tikanov, V. 1993: A new method for constructing continuous cartograms, *Geography and Geographic Information Systems* 20, 167–73.

case-cohort studies See CASE-CONTROL STUDIES

case-control studies These are studies in which a group of people with the disease of interest (the cases) is compared with a group without the disease (the controls). Exposure to the risk factor of interest is then ascertained in all those recruited to the study and the exposure levels of the cases are compared to those of the controls. Differing levels of exposure between the cases and controls indicate that the exposure is associated with the disease. Such studies are particularly appropriate for rare diseases for which follow-up studies are inappropriate, as insufficient cases of the disease would arise to provide sufficient statistical POWER.

It is important at the outset to have a clear definition of the type of person who is eligible to be a case. Issues that need to be considered include:

1. *The definition of the disease.* Can this be broadly defined or is a specific subtype of the disease the focus of interest? For example, in a study of leukaemia a decision needs to be made as to whether all cases of leukaemia will be included or only a specific subtype such as chronic myeloid.
2. *The age range of the cases.* Some diseases are likely to have different causes at different ages.
3. *The sex of the cases.* For some disorders, it would be appropriate to restrict to one or other sex. For example, while cases of breast cancer do arise in men, the aetiology of these is likely to be very different from that of the disease in women.
4. *Incident, prevalent or deceased cases.* It is usually preferable to recruit cases when they are diagnosed (incident cases). However, for diseases that are very rare it can take too long to recruit sufficient cases. They can therefore be supplemented by existing (prevalent) cases. However, for diseases that can lead to death, the prevalent cases will only represent the survivors. To avoid survivor bias, deceased cases can be included if the next-of-kin can provide the appropriate information on the exposure of interest.
5. *Hospital or community cases.* A decision needs to be taken as to where the cases are to found. If the disease is of such a nature that all cases are likely to come to hospital (e.g. breast cancer), then the hospital may be an appropriate recruitment location. For diseases that are often treated in the community, such as back pain, recruitment from hospital will only include cases at the most severe end of the disease spectrum.

Equally, besides the choice of cases, there are important considerations for the selection of controls, who should be drawn from the population at risk of becoming cases. In theory, the cases and controls can be considered as being part of a large hypothetical COHORT STUDY. Those who develop the disease are the cases, and those who have not acquired it form the pool from which the controls are drawn. Thus controls should be drawn carefully from the group of people who would be classified as cases if they happened to develop the disease.

Controls should be within the same age range as those chosen for the cases, and if only one sex is being considered for the cases the same restriction should apply to the controls. If the cases are recruited in hospital then the controls may also be recruited from the hospital or from the community defined by the catchment area of the hospital. Any exclusion criteria applied to the cases must also be applied to the controls, and the controls must be at risk of developing the disease. For example, in a case-control study of endometrial cancer, it was important that women who have had a hysterectomy, and thus no longer had an endometrium, were not included as controls (Barbone, Austin and Partridge, 1993).

Increasingly, matched case-control studies are being conducted (see MATCHED SAMPLES). One or more controls are chosen for each case, matched as closely as possible to the case for various factors that are not of intrinsic interest to the study. Common matching factors are age and sex. Thus for each case, a control of the same age (to perhaps within one year) and the same sex would be chosen. One-to-one matching gives rise to a matched pair study. To increase the statistical power of the study, more than one control can be chosen for each case. This is particularly useful if the disease is rare and it is hard to find sufficient cases. However, it is rarely worth studying more than four controls per case, as the effort spent in collecting the data on the extra controls tends to outweigh the minimal increase in power.

Frequently, the exposures will be assessed by QUESTIONNAIRES. This can present logistical difficulties if the cases are very sick and sometimes (as with deceased cases) the next-of-kin need to be questioned instead. For example, in a case-control study to examine the risk of sudden infant death syndrome in relation to used infant mattresses, all the exposure data were collected by questionnaire (Tappin *et al.*, 2002). Exposure to a used infant mattress was assessed by asking parents about routine night and day sleeping places and ascertaining the state of the mattress and whether it was new for this baby. Other studies have been able to link records with data held on the individuals. For example, a study of Alzheimer's disease in relation to levels of aluminium in drinking water was able to utilise records from the water companies to ascertain the aluminium levels in water piped to each address at which the cases and controls had lived during their lives (Martyn *et al.*, 1997).

A particular concern in case-control studies is the possibility of RECALL BIAS, particularly when obtaining information by questionnaire from the cases and controls. The cases are likely to have thought about the possible causes of their disease, whereas the controls will not. Thus the level of recall of past exposures may differ between cases and controls, which may lead to spurious differences in exposure between the two groups.

One of the most difficult aspects of a case-control study is ensuring that the choice of controls is appropriate. Controls that are not representative of the population at risk of the disease will lead to biased findings (see BIAS IN OBSERVATIONAL STUDIES). An example is when the cases are asked to choose a friend to act as the control. Such an approach tends to maximise the response rate in the controls (as the controls are usually willing to help their sick friends). However, if the exposure of interest is related to work or leisure activities or lifestyle, then the friends will be more similar to the cases than the average person in the population at risk. Thus *overmatching* of controls to cases takes place and little difference may be found between the exposures of the cases and the controls. Any association between the disease and the exposure can therefore be missed. Issues of BIAS need to be addressed at the design stage of a case-control study as no adjustment can be made in the analysis to take account of it.

In common with all epidemiological studies, *confounding* is an issue that has to be considered in case-control studies. A confounding factor is one that is related both to the disease under study and the exposure of interest. Account can be taken of confounding factors at the analysis stage, but it is important at the design stage to identify and collect as much information as possible on all putative confounding factors. Common confounding factors are age and sex as these two factors are invariably related to any disease and to most exposures. Matching at the design stage on one or more confounders provides a way of removing the effect of these confounders and so adjustment for them is not needed in the analysis.

The appropriate method of analysis depends on the type of case-control study employed. The analytical methods for matched case-control studies differ from those for unmatched studies, and it is important that the appropriate methods are employed to avoid bias in the results. Basic methods of analysis in unmatched studies are described next.

Estimates of risk of disease in the unexposed and exposed groups cannot be obtained from case-control studies, because the case-control ratio has been fixed in the design. However, it is possible to obtain an estimate of the relative risk of disease from the odds ratio. The simplest form of a case-control study arises when exposure is classified into two groups, namely exposed and unexposed. The data can then be presented in a 2×2 table.

case-control studies *A 2×2 table for comparing $a + c$ cases with $b + d$ controls*

	Cases	*Controls*
Exposed	a	b
Unexposed	c	d

The ODDS RATIO is the ratio of the odds of exposure in the cases (a/c) to the odds of exposure among the controls (b/d) and thus is calculated as ad/bc. Odds ratios above 1 imply that the exposure is associated with an increased risk of the disease whereas a value below 1 indicates that the exposure may be protective.

As an example, consider a study of cod liver oil in infancy in relation to child-onset Type 1 diabetes (Stene and Joner, 2003). The results are given in the second table The odds ratio is $197 \times 834/(777 \times 318) = 0.66$, indicating that cod liver oil appears to protect against child-onset diabetes.

case-control studies *Results of a case-control study in Type I diabetes*

Cod liver oil in 1st year of life	*Cases*	*Controls*
Yes	197	777
No	318	834

Standard analysis of a 2×2 table using a CHI-SQUARE TEST can be performed to test whether the odds ratio differs significantly from 1. The 95 % CONFIDENCE INTERVALS can be derived using a variety of approximate methods. The most sophisticated methods require iterative solutions but simple methods such as that proposed by Woolf in 1955 (see Breslow and Day, 1980; Schlesselman, 1982) provide reasonable approximations. His method involves calculating the approximate VARIANCE of the natural logarithm of the odds ratio as:

$$\frac{1}{a} + \frac{1}{b} + \frac{1}{c} + \frac{1}{d}$$

and then obtaining the 95 % confidence interval of the natural logarithm of the odds ratio from:

$$\log(\text{OR}) \pm 1.96 \times \sqrt{\text{var}}[\log(\text{OR})]$$

Taking exponentials of the two resulting values gives the 95 % confidence interval for the odds ratio.

In the example above the chi-square test gives a value of 15.3, indicating $p < 0.0001$ and a 95 % confidence interval for the odds ratio of 0.54 to 0.81. Thus this simple analysis indicates a strong association between cod liver oil and diabetes. However, odds ratios and their confidence intervals are usually derived using computer packages. One that is freely available and provides ready access to these forms of analysis is EPIINFO (http://www.cdc.gov/epiinfo/). This can be downloaded via the internet and is in the public domain.

More often we are interested in examining different levels of exposure. A trend in odds ratios across different levels of exposure provides more convincing evidence of a relationship between the exposure and the disease than results from the simple dichotomy of exposed versus unexposed.

Odds ratios can be calculated at each level of exposure compared with the baseline exposure level (usually the unexposed). A chi-square test for trend can be performed to test for linear trend in the odds ratios. In the third table there is an apparent trend with more frequent consumption of cod liver oil leading to lower odds ratios. The chi-square test for trend leads to a significance level of $p < 0.001$.

case-control studies *Case-control study data according to an ordinal exposure*

Cod liver oil in 1st year of life	*Cases*	*Controls*	*Odds ratio*
No	318	834	1 (reference)
Yes, 1–4 times per week	60	224	0.70
Yes, $\geq$5 times per week	137	553	0.65

More sophisticated analyses are possible using LOGISTIC REGRESSION. This allows assessment of exposure on a continuous scale (rather than requiring grouping into levels) as well as allowing for the effects of confounding factors. The coefficients obtained in the logistic regression model are the logarithms of the odds ratios, and the actual odds ratios and their confidence intervals can be readily obtained.

In a matched study, the controls have deliberately been chosen to be more similar to the cases than those generally at risk of the disease. This has to be recognised in the analysis.

The analysis of matched studies is more complex than for unmatched studies and in the discussion of the basic analysis, only matched pairs analyses will be considered. The analysis focuses on the pairs rather than the individuals contributing to the pairs and the standard presentation of the data is as follows:

	Control	
Case	*Exposed*	*Unexposed*
Exposed	r	s
Unexposed	t	u

Thus each pair contributes to one of the cells and the total $(r + s + t + u)$ is the total number of pairs rather than the total number of individual cases and controls in the study. Interest focuses on the pairs that are discordant for exposure. Thus a comparison is made between those pairs where the case is exposed but the control is not and the pairs where the control is exposed and the case is not. The odds ratio is calculated as the ratio of the discordant pairs s/t. The odds ratio will therefore be high if the pairs with case exposed and control unexposed greatly outnumber the pairs where the control is exposed but the case is not.

As an example, consider a study on hip osteoarthritis by Cooper *et al.* (1998), which examined the risk associated with previous hip injury to the affected hip (or to the hip on the same side – right or left – for the matched control). The data from the study are given in the fourth table The odds ratio is $46/11 = 4.2$.

case-control studies *Data from a matched case-control study in hip osteoarthritis*

	Control	
Case	*Previous hip injury*	*No previous injury*
Previous hip injury	1	46
No previous injury	11	553

Basic analysis of matched studies becomes difficult in anything other than the matched paired analysis of a simple dichotomous exposure. Nowadays, matched case-control studies are usually analysed using CONDITIONAL LOGISTIC REGRESSION. This is not the same as unconditional logistic regression mentioned above, as it takes account of the matching structure in the design of the study. Conditional logistic regression enables us to test the odds ratio calculated above, giving $p < 0.0001$ and a 95 % confidence interval around the odds ratio of 2.2 to 8.1. This indicates that hip injury is associated with an increased risk of hip osteoarthritis later in life.

Unmatched case-control studies can be analysed in most standard statistical computing packages. Logistic regression is used in many forms of analysis not confined to case-control studies. Conditional logistic regression is somewhat more specialised and is not available in all packages. It is most readily available in STATA for which there is a specific routine for this. It can be performed in SAS, S-plus and SPSS (see STATISTICAL PACKAGES) but these packages require use of a PROPORTIONAL HAZARDS model for which the likelihood function is identical, so cannot be used quite so readily. EPIINFO, mentioned above, does enable much of the basic analyses and is also convenient to use if one is examining data that are already tabulated (such as in a published paper), rather than the raw data on individuals. *HI*

[See also NESTED CASE-CONTROL STUDIES, OBSERVATIONAL STUDIES, SAMPLE SIZE DETERMINATION IN OBSERVATIONAL STUDIES]

Barbone, F., Austin, H. and Partridge, E. E. 1993: Diet and endometrial cancer: a case-control study. *American Journal of Epidemiology* 137, 393–403. **Breslow, N. E. and Day, N. E.** 1980: *Statistical methods in cancer research,* Vol. 1: *The analysis of case-control studies.* Lyon: International Agency for Research on Cancer. **Cooper, C., Inskip, H., Croft, P., Campbell, L., Smith, G., McLaren, M. and Coggon, D.** 1998: Individual risk factors for hip osteoarthritis: obesity, hip injury, and physical activity. *American Journal of Epidemiology* 147, 516–22. **Martyn, C., Coggon, D., Inskip, H., Lacey, R. and Young, W.** 1997: Aluminium concentrations in drinking water and risk of Alzheimer's disease. *Epidemiology* 8, 281–6. **Schlesselman, J. J.** 1982: *Case-control studies. Design, conduct and analysis.* Oxford: Oxford University Press. **Stene, L. C. and Joner, G.** 2003: Norwegian Childhood Diabetes Study Group. Use of cod liver oil during the first year of life is associated with lower risk of childhood-onset Type 1 diabetes: a large, population-based, case-control study. *American Journal of Clinical Nutrition* 78, 1128–34. **Tappin, D., Brooke, H., Ecob, R. and Gibson, A.** 2002: Used infant mattresses and sudden infant death syndrome in Scotland: case-control study. *British Medical Journal* 325, 1007–12.

categorising continuous variables This is the process of converting a continuous variable such as age into a categorical variable with a number of categories, e.g. 'young' (<40 years), 'middle aged' (40–60 years) and 'old' (>60). The practice is very common in medical research where clinicians appear to have a general preference for categorising individuals (see Altman, 1991). When used to simplify or improve the presentation and description of the data, grouping individuals into categories may often be useful, although the choice of category boundaries and number of categories may not be easy. When, however, the categorical variables created by the process are used in data analysis, rather than the original continuous variables, problems arise (see, for example, Hunter and Schmidt, 1990; Streiner, 2002).

Categorisation introduces an extreme form of measurement error; splitting a continuous variable into categories results in lost information and an inevitable loss of power in analysis. The apparent simplicity of the categorical variables and the ability to use proportions and odds ratios, which are more familiar to many clinicians, are unlikely to compensate for the lost power. Retaining the continuous variable and analysing the data using the appropriate statistical methodology will always be a far better strategy. *BSE*

Altman, D. G. 1991: Categorising continuous variables. *British Journal of Cancer* 64, 975. **Hunter, J. E. and Schmidt, F. L.** 1990: Dichotomisation of continuous variables: the implications for meta-analysis. *Journal of Applied Psychology* 75, 334–49. **Streiner, D. L.** 2002: Breaking up is hard to do: heartbreak of dichotomizing continuous data. *Canadian Journal of Psychiatry* 47, 262–6.

causal diagram See CAUSAL MODELS

causal effect (direct and indirect) This refers to the change in outcome distribution, or some feature thereof, that would arise under a specific intervention. It is commonly formalised using so-called counterfactual or potential outcomes $Y(a)$, which represent the outcome that one would – possibly contrary to fact – have observed for a given subject had the exposure A been set to the value a through some intervention or manipulation. The (average) causal effect of exposure a on the outcome can then be defined as the population-averaged difference $E\{Y(a) - Y(0)\}$ between the counterfactual outcomes under exposure level a and some reference exposure level 0. This is to be contrasted with the more usual expected difference $E(Y|A=a) - E$

$(Y|A=0)$, where the expected observed outcome Y is considered over *different* subpopulations defined by levels of observed exposure $A=a$ and $A=0$. This may not carry the interpretation of a causal effect when the subgroups of exposed and unexposed subjects are not inherently comparable. When the outcome is dichotomous (with $Y=1$ encoding 'disease' and $A=0$ encoding 'no exposure'), a causal effect can alternatively be quantified in terms of the causal relative risk (see RELATIVE RISK AND ODDS RATIO), $P\{Y(a)=1\}/P\{Y(0)=1\}$, the causal odds ratio (see RELATIVE RISK AND ODDS RATIO), $\text{odds}\{Y(a)=1\}/\text{odds}\{Y(0)=1\}$, or the causal population attributable fraction, $[P(Y=1)-P\{Y(0)=1\}]/P(Y=1)$. The latter expresses by what percentage of diseased subjects the disease could have been prevented/avoided by taking away the exposure (assuming that the exposure can only cause, but never prevent, disease). Causal effects are sometimes defined within strata of given values of pre-exposure covariates C, e.g. $E\{Y(a)-Y(0)|C\}$, within strata defined by the observed exposure, e.g. $E\{Y-Y(0)|A\}$ (in which case one refers to them as 'causal effects in the exposed' or 'treatment effects among the treated'), or within so-called principal strata defined by joint counterfactual outcomes (Robins, 1986, Section 12.2; Frangakis and Rubin, 2002). For instance, in the context of randomised experiments with treatment noncompliance, Frangakis and Rubin (2002) focus on the local average treatment effect (LATE) $E\{Y(1)-Y(0)|A(1)=1,A(0)=0\}$, where $Y(r)$ and $A(r)$ denote the counterfactual outcome and received treatment under randomised assignment to treatment $r=0$ or 1. The LATE thus encodes the effect of assignment to treatment 0 versus 1 within the principal stratum $\{A(1)=1,A(0)=0\}$ composed of subjects who would take the treatment if assigned to it (i.e. $A(1)=1$), but not otherwise (i.e. $A(0)=0$).

By extending the previous concepts to a joint exposure (A, M), where M is a mediator or intermediate variable on the causal path from exposure to outcome, definitions of direct and indirect causal effects can be constructed. These express the causal effect that manifests besides or through the change caused in that intermediate variable. Consider, for instance, the Methods for Improving Reproductive Health in Africa Trial, which investigated the effect of diaphragm and lubricant gel use (A) in reducing infection by HIV (Y) among susceptible women (Rosenblum *et al.*, 2009). Women received intensive condom counselling and provision, and were then randomly assigned to either the active treatment arm ($A=1$) or not ($A=0$). Because there was much lower reported use of condoms (M) in the intervention arm than in the control arm, special interest lies in the direct effect of assignment to the diaphragm arm, other than through its effect on condom use (Rosenblum *et al.*, 2009).

To formalize the concept of a direct causal effect, define for each subject $Y(a, m)$ to be the counterfactual outcome under exposure level a and mediator level m, and $M(a)$ to be the counterfactual mediator under exposure level a. The controlled direct effect of exposure level a versus reference exposure level 0, controlling for M, can then be defined as the expected contrast $E\{Y(a, m)-Y(0, m)\}$ (Robins and Greenland, 1992; Pearl, 2001). It expresses the exposure effect that would be realised if the mediator were controlled at level m uniformly in the population. In the example, with M expressing the percentage of sexual acts where the male condom was used, $E\{Y(1,0)-Y(0,0)\}$ expresses the effect of assignment to the diaphragm arm if in truth no male condoms were used.

There are a number of limitations to the concept of a controlled direct effect, in view of which alternative definitions have been proposed. First, it is often not realistic to imagine forcing the mediator to be the same for all subjects in the population. Second, indirect effects cannot be defined in a similar manner as controlled direct effects because it is impossible to hold a set of variables fixed, such that the effect of exposure on outcome would circumvent the direct pathway. In particular, the total causal effect, say $E\{Y(a)-Y(0)\}$, minus the controlled direct effect, say $E\{Y(a, m)-Y(0, m)\}$, may not represent an indirect effect, unless the exposure and mediator have linear effects on the outcome (VanderWeele and Vansteelandt, 2009). Both limitations can be overcome by considering so-called natural or pure direct effects (Robins and Greenland, 1992; Pearl, 2001), which are defined as the expected contrast $E\{Y(a, M(0))-Y(0, M(0))\}$ or total direct effects (Robins and Greenland, 1992), which are defined as $E\{Y(a, M(a))-Y(0, M(a))\}$. The natural direct effect essentially expresses what would be realised if the exposure was administered, but its effect on the mediator was somehow blocked. In the example, $E\{Y(1, M(0))-Y(0, M(0))\}$ expresses the effect of assignment to the diaphragm arm as it would have been observed if women had not changed their frequency of male condom use following randomised assignment.

The difference between the total causal effect and a pure natural direct effect, $E\{Y(a)-Y(0)\}-E\{Y(a, M(0))-Y(0)\}=E\{Y(a, M(a))-Y(a, M(0))\}$, measures an indirect effect as it expresses how much the outcome would change on average if the exposure were controlled at level a but the mediator were changed from level $M(0)$ to $M(a)$. It is termed the total indirect effect (Robins and Greenland, 1992). In the example, $E\{Y(1, M(1))-Y(1, M(0))\}$ would express the change in HIV incidence that would have been observed for women on the diaphragm arm if they had gone back to their original use of the male condom. Likewise, the difference between the total effect and the total direct effect gives the natural indirect effect $E\{Y(a)-Y(0)\}-E\{Y(a)-Y(0, M(a))\}=E\{Y(0, M(a))-Y(0, M(0))\}$. This expresses how much the outcome would change on average if the exposure were controlled at level 0 but the mediator were changed from its natural level $M(0)$ to the level $M(a)$, which it would have taken at exposure level a.

Direct effects are sometimes defined using concepts of principal stratification (Frangakis and Rubin, 2002); for instance, as the exposure effect within the principal stratum of individuals whose mediator level was not affected by the exposure, i.e. $E\{Y(a) - Y(0)|M(0) = M(a)\}$. Unlike the forgeoing definitions of direct effect, principal stratum direct effects do not conceptualise the possibility of manipulating the mediator. However, they have a more limited utility because of the inability to identify which individuals fall into which principal strata, because the principal strata are sparsely populated in many realistic applications and because they do not correspond to a natural definition of indirect effect (VanderWeele and Vansteelandt, 2009).

All foregoing causal effect definitions require the same population to be evaluated under different interventions/exposures. Because of this, they require specialised estimation techniques (see MARGINAL STRUCTURAL MODELS, INVERSE PROBABILITY WEIGHTING) that work under specific untestable assumptions, such as about the exchangeability of subjects under different exposure levels (e.g. assumptions that all confounders of the association between exposure and outcome have been measured). Current estimation techniques for the causal effect $E\{Y(a) - Y(0)\}$ (and likewise all other considered causal effect measures) infer the effect of 'non-invasive' interventions or manipulations that allow for setting the exposure at level a versus 0. Here, 'noninvasive' interventions that set A to the value a are those that have no effect among those for whom exposure level $A = a$ was naturally observed (VanderWeele and Vansteelandt, 2009). The use of counterfactual outcomes makes clear the precise meaning of vague statements such as 'causal effect', 'direct effect' and 'indirect effect', and makes clear, in particular, that there are different definitions of direct effect and indirect effect. Whether estimates of the considered effect measures can effectively be interpreted as 'causal' effects in a specific application depends on whether the required untestable exchangeability assumptions are met. *SV, EG*

Frangakis, C. and Rubin, D. 2002: Principal stratification in causal inference. *Biometrics* 58, 21–9. **Pearl, J.** 2001: Direct and indirect effects. In *Proceedings of the Seventeenth Conference on Uncertainty and artificial intelligence*. San Francisco: Morgan Kaufmann, pp. 411–20. **Robins, J. M.** 1986: A new approach to causal inference in mortality studies with sustained exposure periods – application to control of the healthy worker survivor effect. *Mathematical Modelling* 7, 1393–512. **Robins, J. M. and Greenland, S.** 1992: Identifiability and exchangeability for direct and indirect effects. *Epidemiology* 3, 143–55. **Rosenblum, M., Jewell, N. P., van der Laan, M., Shiboski, S., Van Der Straten, A. and Padian, N.** 2009: Analysing direct effects in randomized trials with secondary interventions: an application to human immunodeficiency virus prevention trials. *Journal of the Royal Statistical Society – Series C* 172, 443–65. **VanderWeele, T. J. and Vansteelandt, S.** 2009: Conceptual issues concerning mediation, interventions and composition. *Statistics and its interface*, 2, 457–468.

causality The two fundamental principles for establishing evidence for a cause and effect relationship are deduction and induction. The first, with its roots in early Greek philosophy, encompasses mathematical reasoning: starting with a premise formulated without reference to the outside world, a general theory is developed through logical reasoning and subsequently confirmed by statistical observation. Induction, contrariwise, aims to establish a general principle by observation of the natural world and seeks to confirm this principle through prediction and further observation. In contrast to deductive reasoning, inductive reasoning proceeds from the particular to the general. While there can be no doubt that much of modern science strives to make progress through inductive methods, the deductive principle has a firmer grounding in logic. As Hume pointed out, simply observing that one event follows another, no matter how many times it occurs, does not establish that one event *caused* the other.

The concept of what a cause is needs to be set in context with the knowledge of the time. In 1855 John Snow established the link between a 'morbid matter' carried in water and acute, often fatal, diarrhoea in London (Snow, 1855). Snow's evidence consisted of temporal statistical associations, geographical associations in relation to the water companies supplying the city and consideration of a plausible route of infection through ingestion of drinking water. It was the assembly of these various strands of evidence that led to the premise that the water source caused the diarrhoea and a subsequent outbreak in Central London gave Snow the chance to establish deductive proof, by removing the Broad Street water pump, with the immediate effect of stopping the epidemic (see HISTORY OF MEDICAL STATISTICS). Snow postulated a cause of cholera 30 years before the bacterium was isolated. As science delves deeper into genetic and molecular mechanisms, we alter our level of definition of causal factors. In reality, few cause–effect associations are simple two-factor problems. Many of our prominent chronic diseases are multifactorial in nature and many of the factors that contribute to causation are yet to be discovered.

Despite Snow's example, much scientific work progresses without deductive proof; empirical knowledge seems to accumulate through methods that may lend support to but do not actually guarantee correct conclusions. Statistical procedures are employed in observational sciences, where a host of unknown sources of error, including MEASUREMENT ERROR and random fluctuations, must be considered before even progressing towards consideration of causality. Formal 'proof' is generally not attainable in observational sciences and it may be fruitless to try to demonstrate causality in a formal way. More often, a number of associations must be considered together to assemble the totality of evidence. The BRADFORD HILL CRITERIA offer guidelines for considering the totality of evidence. Some of these criteria, such as establishing temporality and demonstrating reversibility, are powerful tools in determining causality, whereas others may depend on specific circumstances.

Bradford Hill's aim was primarily to provide a more structured approach to informing decisions for preventive action. Other writers have advocated decisions based on Popper's concept of refutation. Refutationists would suggest that science can advance more rapidly by formulating radical hypotheses and discarding erroneous theories than it can by fruitless accumulation of supporting evidence. We could make many statements in favour of a hypothesis and still fail to establish it, but we could take just *one* example that disproves the hypothesis to conclude it is false (e.g. one black swan is proof against the assertion that all swans are white). However, even the process of refutation depends on observation and is therefore itself not devoid of uncertainty. There remains a crucial role for statistics in any science based on observation: to set probabilistic limits on whether an observed association exists or should be ignored.

Many widely accepted statistical practices appear to adhere to deductive reasoning. The NULL HYPOTHESIS is a statement formulated prior to any knowledge of the data and therefore a test of the null is an inference from the hypothetical general population to the sample: 'If there is no association in the general population a result like the one seen in the sample would have probability P.' In the LIKELIHOOD approach to estimation, computations proceed from consideration of the hypothetical value for the parameter to support for that value in the particular data observed. However, frequentist theory would appear to be embedded in inductive reasoning: we imagine an experiment repeating an infinite number of times, under the assumption that given enough repetitions we will ultimately arrive at the truth. While this reasoning seems logical if conditions remain unaltered, some would argue that if conditions are likely to change in different populations or under the influence of additional variables, then a general principle cannot be asserted on this basis.

As with any discipline, our ability to determine meaningful results in statistical analysis depends on the approach we adopt. The discovery of a statistically significant result is grounds for rejecting a null hypothesis, but is not evidence for a previously unformulated hypothesis on the strength of the current data. Hence, dredging data for significant results takes us no nearer to establishing causality and will at best raise possibilities for future investigation. Likewise, if we are seeking 'explanation' in a statistical model, the decision to include a term in the model, guided only by considerations of statistical significance (refutation of the null), would not respect deductive principles (Maclure, 1985). If scientific explanation relies on testing specified prior hypotheses, a more consistent modelling approach would be to include model terms, even if they were not significant in the current data. For further details see Weed (1986). *JGW*

Maclure, M. 1985: Popperian refutation in epidemiology. *American Journal of Epidemiology* 121, 343. **Snow, J.** 1855: *On the mode of communication of cholera*, 2nd edition. London: Churchill. **Weed, D. L.** 1986: On the logic of causal inference. *American Journal of Epidemiology* 123, 965–79.

causal models These models attempt to discover whether an observed association between an exposure (e.g. cigarette smoking) and an outcome (e.g. high blood pressure) arises because the exposure causes the disease or whether it is a *spurious association*. A spurious association can arise if both the exposure and the disease have one (or more) common cause – e.g. if socioeconomic status is a cause of both high blood pressure and cigarette smoking. There are four major types of causal model (see Greenland and Brumback, 2002) and many definitions of 'causal' in an epidemiological context (see Parascandola and Weed, 2001).

One type of causal model is the *causal diagram* (Greenland and Brumback, 2002; Hernan *et al.*, 2002). Such diagrams link exposure, outcome and other variables by arrows representing direct causal effects. A hypothetical example for smoking and high blood pressure is shown in the first figure.

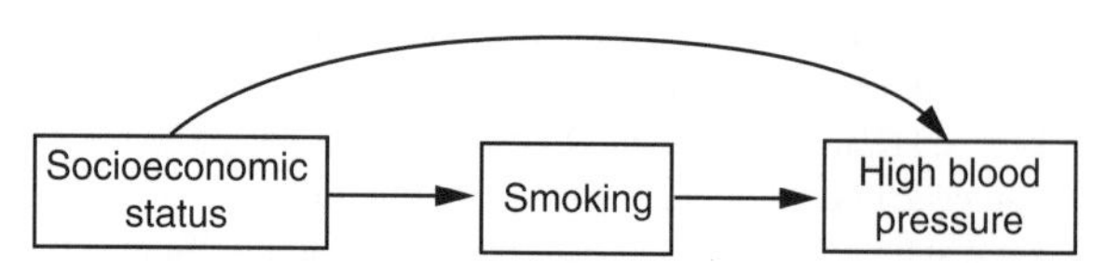

causal models *Causal diagram showing a possible relationship between smoking, high blood pressure and socioeconomic status*

Such diagrams can be drawn from prior knowledge of causal mechanisms. Uses include deciding whether adjusting for confounding variables would increase or decrease bias. For example, from the figure, one would adjust for socioeconomic status in analysing the relationship between smoking and high blood pressure (as it is a common cause). (For more details, see Hernan, *et al.*, 2002.)

A second type of causal model, the *counterfactual model*, arises from a definition of CAUSALITY: 'Exposure makes a difference in joutcome (or the probability of an outcome) when it is present, compared with when it is absent' (Parascandola and Weed, 2001). Thus the 'counterfactual' arises because the definition hypothesises about what 'might have been' if conditions had been other than those actually observed (Maldonado and Greenland, 2002). The definition also specifies that all other conditions should have remained constant; i.e. it is only the exposure that was (hypothetically) changed. For example, we could say that smoking causes high blood pressure if blood pressure is higher when a person smokes compared to what it would be if that same person were a nonsmoker. The causal effect of smoking on blood pressure for an individual could then be estimated by the difference between their blood pressure when they smoked and their blood pressure if they did not smoke. However, only one of these outcomes for each person is actually observed. The unobserved quantities are estimated using substitutes (Maldonado and Greenland, 2002). This could be by using unexposed individuals to estimate the outcomes if exposed individuals had in fact been unexposed (Greenland and Brumback, 2002) or by imputing the outcome from observed covariates. In a randomised trial, we assume that treatment and control groups are balanced in all respects, except that the exposure and therefore the outcome in the exposed group can be used as a substitute for the outcome in the control group were it to have been exposed (Mandolado and Greenland, 2002).

The third type of causal model discussed in the 2002 review, the *sufficient-component cause* model, assumes that the outcome is caused by several causes, none of which is necessary and sufficient alone to cause the disease (Greenland and Brumback, 2002). Such models are often shown as PIE CHARTS, with each slice representing one of the components of the overall cause. For example, high blood pressure could be caused by the presence of any two of smoking, low socioeconomic status and being overweight (see the second figure).

One disadvantage of such models is that in order to explain relationships such as that of smoking with lung cancer, it is necessary to assume that 'smoking is one element in a sufficient cause and that the other elements simply have not been identified yet' (Parascandola and Weed, 2001). This is clearly a strong assumption to make in many areas of medical statistics.

The fourth type of causal model, the STRUCTURAL EQUATION MODEL, may be thought of as a parameterisation of the causal graph (Greenland and Brumback, 2002). Here, each relationship on the graph is expressed as an equation, with each variable occurring as an outcome in only one of the equations. All parameters are estimated simultaneously and can include error terms and underlying LATENT VARIABLES (as unobserved variables) and correlation between any of the variables.

Singh-Manoux, Richards and Marmot (2003) used causal diagrams and structural equation models to analyse the causal relationship between leisure activities and cognitive function. The causal diagram showed eight measured variables (including the outcome) and seven unobserved variables, and included correlation between leisure activities entailing low and high cognitive effort. Participation in leisure activities, particularly social or high cognitive effort activities, was positively associated with cognitive function. *KT*

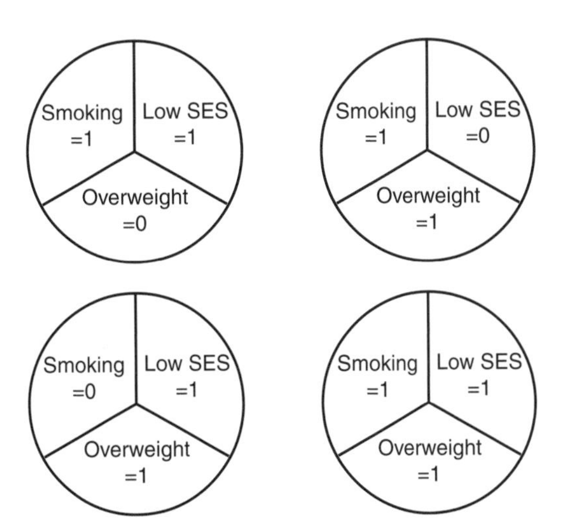

causal models *Sufficient-component cause model showing hypothetical causes of high blood pressure*

Greenland, S. and Brumback, B. 2002: An overview of relations among causal modelling methods. *International Journal of Epidemiology* 31, 1030–7. **Hernan, M. A., Hernandez-Diaz, S., Werler, M. M. and Mitchell, A. A.** 2002: Causal knowledge as a prerequisite for confounding evaluation: an application to birth defects epidemiology. *American Journal of Epidemiology* 155, 176–84. **Maldonado, G. and Greenland, S.** 2002: Estimating causal effects. *International Journal of Epidemiology* 31, 422–9. **Parascandola, M. and Weed, D. L.** 2001: Causation in epidemiology. *Journal of Epidemiology and Community Health* 55, 905–12. **Singh-Manoux, A., Richards, M. and Marmot, M.** 2003: Leisure activities and cognitive function in middle age: evidence from the Whitehall II study. *Journal of Epidemiology and Community Health* 57, 907–13.

cause-specific death rate This is a death rate calculated for people dying from a particular disease. All-cause mortality rates provide a summary of the overall mortality of a population, but the distribution of causes of death can vary considerably between populations. Classically, mortality from infectious diseases is higher in countries that are less developed, whereas diseases that are chronic, and primarily affect the elderly, predominate in the developed world. In studying a population's mortality experience, it is therefore necessary to consider the specific causes of death rather than simply the all-cause death rate.

The table gives the cause-specific rates for a variety of diseases for different countries. The rates are calculated by dividing the number of deaths from the cause in question during the year by an estimate of the mid-year population. Thus, the Romanian rate for tuberculosis is derived from 2130 (the deaths from tuberculosis occurring in the year 2000), divided by 22,435,000 (the number in the population for that year). The result is then multiplied by 100,000 to give the rate of 9.5 per 100,000 population.

cause-specific death rate Cause-specific death rates per 100 000 from selected causes from various countries (United Nations, 2002)

Country	*Tuberculosis*	*HIV/ AIDS*	*Lung cancer*	*Cardiovascular disease*	*Accidents and violence*	*All causes*
Argentina (1996)	29.5	–	23.2	297.3	52.2	863.9
Australia (1999)	0.2	0.4	17.9	135.2	21.9	337.9
Bahamas (1997)	0.7	50.2	3.6	81.9	28.8	264.8
Japan (1999)	2.1	0.0	20.6	122.2	29.7	740.8
Panama (1997)	4.8	15.6	6.0	118.3	56.9	418.7
Romania (2000)	9.5	2.2	37.9	701.9	64.2	1140.3
South Africa (1995)	32.2	13.2	9.7	103.6	119.4	606.2
United Kingdom (1999)	0.9	0.3	58.3	426.2	33.4	1062.3
United States (1998)	0.5	5.0	57.1	349.3	55.6	863.9

The table shows that accidents and violence are high in South Africa, whereas cardiovascular disease is of concern in Romania. It must be noted that these are crude rates and the different age structures of the populations have not been considered; the comparisons can therefore be misleading (see AGE-SPECIFIC RATES).

Cause-specific death rates are published by many countries and often are presented as age- and sex-specific rates for each cause. The rates are presented by cause of death as classified by the International Classification of Diseases (World Health Organisation, 1992). Regular updates to the classification are required to account for the changing definitions of disease and, in particular, to recognise new diseases. For example, acquired immunodeficiency syndrome (AIDS) did not feature in the 9th revision but has been included in the 10th. For analysis of disease time trends, 'bridging' across the revisions can present some difficulties due to the changing definitions.

Cause-specific rates can also be obtained for new cases of specific diseases. Many countries now have cancer registries that publish national and/or regional cancer incidence rates. These are collated internationally by the International Agency for Research (Parkin *et al.*, 2003), which publishes age- and sex-specific incidence rates for various types of cancer from many countries across the world. Similarly, rates for over 200 conditions have been compiled by the World Health Organisation (Murray and Lopez, 1996). *HI*

Murray, C. J. L. and Lopez, A. D. 1996: *Global health statistics. A compendium of incidence, prevalence and mortality estimates for over 200 conditions.* Harvard: World Health Organisation. **Parkin, M., Whelan, S., Ferlay, J., Teppo, L. and Thomas, D. B.** 2003: *Cancer incidence in five continents*, Vol. VIII. Lyon: IARC Scientific Publications. **United Nations** 2002: *2000 Demographic yearbook*. New York: United Nations. **World Health Organisation** 1992: *International statistical classification of diseases and related health problems*, 10th revision. Geneva: World Health Organisation.

censored observations This is a distinguishing characteristic of time-to-event data (see SURVIVAL ANALYSIS). Censored observations contain only partially observed information about the time to the event of interest; i.e. the exact time of the event is unknown as it may not yet have occurred or be known to have occurred. For example, in a study of time to recurrence of a particular medical condition, say leukaemia after 'successful' bone marrow transplantation, some patients may not experience a recurrence at the end of the study, some may drop out or be lost to follow-up, some may experience the event of interest during successive medical visits, while yet others may experience a 'competing' event, say death, which prevents further follow-up of these subjects.

There are different forms of censoring. The most common in medical studies is *right censoring*. This occurs when by the end of a subject's follow-up the event of interest has not been observed. In this situation all that is known is that the true unobserved 'survival' time exceeds the observed censored time. Most of the examples already seen are of this form.

Left censoring occurs when the true survival time of a subject is less than the actual time observed. For example, in the leukaemia study just examined, subjects may relapse before their first medical visit. Thus, only the incomplete information that the true recurrence times are less than the times to their first medical visit is available. Both forms of censoring are special cases of *interval censoring*, where a subject is known only to have experienced the event within a specific time interval.

Furthermore, the censoring mechanism can be independent or dependent. In the former, the true survival time of a subject, whose observation has been censored, is independent of the mechanism that brought about this censoring. Alternatively, the survival prospects for a homogeneous group of subjects are the same in those censored and in those continued to be followed up. Dependent censoring can occur when subjects are withdrawn from a study because of their apparent high or low risk of experiencing the event. This type of censoring makes standard survival analysis techniques invalid. Thus it is important when censoring occurs to collect as much information as possible on those subjects with censored observations, in order to decide whether the mechanisms behind the censoring types encountered are independent or dependent.

Censoring should not be 'ignored' as valuable information is contained in those subjects with censored times. Therefore, any analysis performed must take account of censoring to make valid inferences. For further details see Collett (2003). *BT*

Collett, D. 2003: *Modelling survival data in medical research*, 2nd edition. London: Chapman & Hall/CRC.

central limit theorem

See NORMAL DISTRIBUTION

Central Office for Research Ethics (COREC)

See ETHICAL REVIEW COMMITTEES

chi-square distribution

This is the PROBABILITY DISTRIBUTION of the sum of the squares of independent normally distributed random variables (denoted χ^2 and sometimes referred to as chi-squared). If we have a variable, X_1, that has a NORMAL DISTRIBUTION with MEAN 0 and VARIANCE 1, then the square of X_1 will have a chi-square distribution with one DEGREE OF FREEDOM, often denoted $\chi^2(1)$ or χ_1^2 depending on the text. Similarly, if we have n independent observations, each from a normal distribution with mean 0 and variance 1, say $X_1, X_2, \ldots, X_n$, then the sum $X_1^2 + X_2^2 + \cdots + X_n^2$ will have a chi-square distribution with n degrees of freedom, here denoted χ_n^2.

The chi-square distribution can arise from many other circumstances, but this definition makes it clear that since it is the distribution of a sum of squared numbers, the distribution's density function can only be positive for non-negative numbers. It can also be seen that a variable taking a chi-square distribution with, for example, seven degrees of freedom is the sum of seven independent variables each with a χ_1^2 distribution. Since the χ_1^2 distribution has a mean of 1 and variance of 2, this result tells us that the χ_n^2 distribution will have a mean of n and variance of $2n$.

In addition to these relationships, we note that the χ_2^2 distribution is identical to the EXPONENTIAL DISTRIBUTION with parameter 0.5 and that, in general, the χ_n^2 distribution is identical to the GAMMA DISTRIBUTION, with parameters $n/2$ and 2. As n becomes large, the χ_n^2 distribution is better approximated by a normal distribution (with mean n and variance $2n$). It has a slightly more complicated relationship with the F-DISTRIBUTION, for which it is the limiting distribution after scaling.

The shapes that the distribution can take and some of these relationships are illustrated in the figure (see page 68). For further details of the relationships with other distributions, see Leemis (1986). As can be seen, with one or two degrees of freedom the mode of the density function is at zero and is strictly decreasing with the value of the random variable. With more degrees of freedom, the density function takes the 'whale' shape, which is more usually associated with the chi-square distribution.

The chi-square distribution is most obviously used in the CHI-SQUARE TEST, but also appears when testing hypotheses about variances, when performing LIKELIHOOD RATIO tests and in some NONPARAMETRIC METHODS (e.g. the KRUSKAL–WALLIS TEST).

Here the central chi-square distribution has been discussed. There is also a non-central chi-square distribution that arises when the normal random variables that define the distribution have a nonzero mean. For further details see Altman (1991). *AGL*

Altman, D. G. 1991: *Practical statistics for medical research*. Boca Raton: CRC Press/Chapman & Hall. **Leemis, L. M.** 1986: Relationships among common univariate distributions. *The American Statistician* 40, 2, 143–6.

chi-square test

This is a statistical significance test used to assess a variety of hypotheses of categorical data, particularly when interest centres on the distribution of observations across categories. For example, a study might record the ethnic group of individuals and interest might centre on the proportion of individuals in each ethnic group.

Categorical data are often presented in FREQUENCY TABLES (or CONTINGENCY TABLES), where each cell of the table shows the number of observations (counts) for a particular combination of categories for the variables of interest. Chi-square tests are a form of HYPOTHESIS TEST where the hypothesis specifies the expected distribution of observations across the cells of the frequency table (the expected number of individuals in each cell). The chi-square test statistic, χ^2, provides a measure of how much the counts recorded in the study (the 'observed counts') deviate from counts predicted by the hypothesis (the 'expected counts'). A small P-VALUE is evidence that the hypothesis is false. (The formula for the chi-square test statistic is given in the entry for CONTINGENCY TABLES.)

Freeman *et al.* (2002) provide an example of the use of chi-square tests. This re-audit of hip fracture in East Anglia involved seven hospitals and analyses were performed to establish whether there were any differences between hospitals in terms of patient demographics. One of the variables tested was patient gender. In this case, the hypothesis to test was that the distribution of patients across male and female categories was the same at all of the hospitals. In other words,

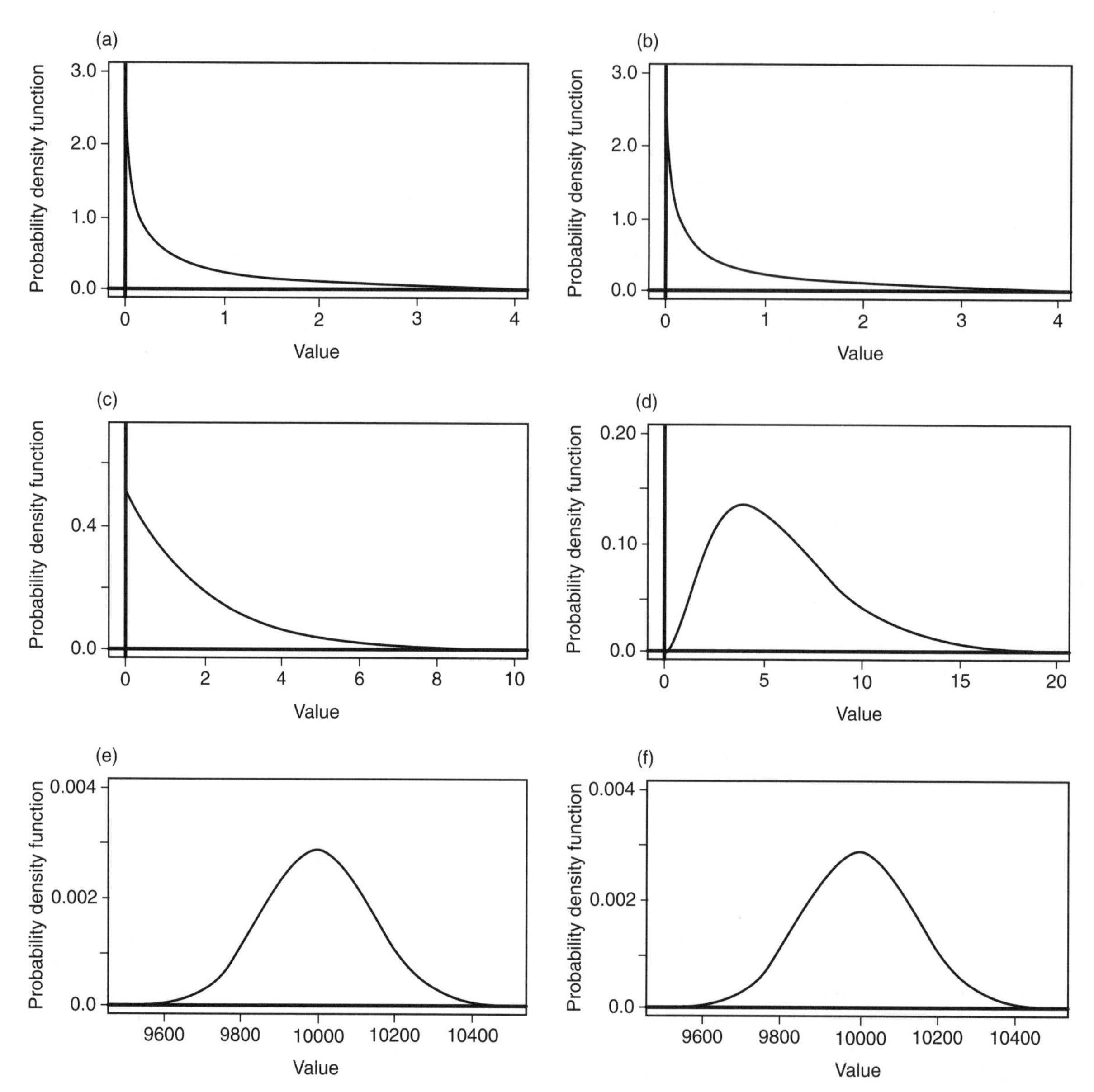

chi-square distribution *Illustrating the form of the chi-square probability density function and its relationships with other distributions: (a) chi-square distribution with one degree of freedom; (b) the F-distribution with one and one thousand degrees of freedom; (c) chi-square distribution with two degrees of freedom (equivalently the exponential distribution with a parameter of 0.5); (d) chi-square distribution with six degrees of freedom; (e) chi-square distribution with ten thousand degrees of freedom; (f) normal distribution with mean of ten thousand and variance of twenty thousand*

the hypothesis was that the proportion of hip fracture patients who were male was the same at all of the hospitals. The result of a chi-square test was $\chi^2 = 7.50$ and $P = 0.48$, indicating that there was no evidence of a difference between hospitals in the proportion of hip fracture patients who were male.

There are three common types of chi-square test: the goodness-of-fit test, the test for independence and the test for homogeneity.

The chi-square *goodness-of-fit* test is appropriate when there is a single group of subjects and a single variable of interest. The hypothesis in this sort of test specifies a precise distribution of subjects across categories. For example, this test could be used to determine whether the ethnic mix of a patient group from a study is different from the (known) ethnic mix in the general population.

The chi-square test for *independence* is appropriate when there is a single group of subjects and two variables of interest. In this case, the hypothesis is that the distribution of subjects across categories in one variable is the same for (i. e. independent of) the categories of the other variable. For example, this test could be used to determine whether the incidence of a disease is related to the ethnic mix of a patient group – in other words, whether the proportion of patients with a disease is the same for all ethnic groups.

The chi-square test for *homogeneity* is appropriate when there are several groups of subjects and a single variable of interest. This test is used to determine whether the distribution of a variable across categories is the same for all groups. The study by Freeman *et al.* (2002) provides an example of this sort of test where the groups are hospitals and interest centres on whether the distribution of male and female patients is the same at all hospitals.

Each of these tests is only appropriate when there is no natural ordering to the categories (i.e. the data are nominal). If the categories are ordered (i.e. the data are ordinal), e.g. when the variable of interest is age group, then other analyses should be used, such as the chi-square test for trend (see, for example, Altman, 1991).

The chi-square test is a parametric test; it is only a valid test when the expected number of observations in each cell is not too small. (A rough rule of thumb is that the expected number of observations in each cell should be at least 5. However, a less stringent criterion is commonly used and has been shown to yield satisfactory results; namely no more than 25 % of cells should have an expected cell count of less than 5, provided none is less than 1.) The test can become invalid when the expected number of observations in a cell is very small and/or when the total number of observations is small. In such cases it can be prudent to use a continuity correction (see YATES' CORRECTION) or, even better, employ a NONPARAMETRIC or EXACT METHOD alternative (e.g. FISHER'S EXACT TEST).

The sampling method should be considered when determining what sort of chi-square test to use. It is only valid to test a hypothesis that makes sense in relation to the way in which the data were gathered. In the hospital example, there are several independent groups of patients (from the seven hospitals) and a single variable of interest (gender). In this case, it is valid to perform a test for homogeneity, but it is not valid to perform a test for independence.

Chi-square tests only provide an overall test of a hypothesis. If a chi-square test is significant, it may be necessary to perform further post hoc tests in order to determine in more detail where significant deviations from the hypothesis exist. In the example from Freeman *et al.* (2002), had the test been significant, this would have indicated that the proportion of hip fracture patients who were male was *not* the same at all of the hospitals in the audit. This does not specify precisely those hospitals that were significantly different from each other. In this case, a *t*-test of the difference between proportions for each pair of hospitals could be used, but it should be noted that such post hoc tests would be subject to the usual problems of multiple comparisons (see MULTIPLE COMPARISON PROCEDURES).

When a variable has only two categories, e.g. gender, the data may be analysed in terms of proportions rather than frequencies. This situation may be analysed using logistic regression, which allows more complex hypotheses to be tested. An alternative and more flexible analysis for any number of variables with any number of categories is POISSON REGRESSION.

Chi-square tests are described in most introductory statistics texts (e.g. Altman, 1991; Wild and Seber, 2000; Agresti, 2002). *PM*

Agresti, A. 2002: *Categorical data analysis*, 2nd edition. New York: John Wiley & Sons, Inc. **Altman, D. G.** 1991: *Practical statistics for medical research*. London: Chapman & Hall. **Freeman, C., Todd, C., Camilleri-Ferrante, C., Laxton, C., Murrell, P., Palmer, C. R., Parker, M., Payne, B. and Rushton, N.** 2002: Quality improvement for patients with hip fracture: experience from a multi-site audit. *Quality and Safety in Health Care* 11, 3, 239–45. **Wild, C. J. and Seber, G. A.** 2000: *Chance encounters: a first course in data analysis and inference*. New York: John Wiley & Sons, Inc.

chi-square test for trend See CASE-CONTROL STUDIES

Christmas tree adjustment See INTERIM ANALYSIS

classification and regression trees (CART) See TREE-STRUCTURED METHODS

classification function See DISCRIMINANT FUNCTION ANALYSIS

clinical equipoise See ETHICS AND CLINICAL TRIALS

clinical trial protocols See PROTOCOLS FOR CLINICAL TRIALS

clinical trials Also known as randomised controlled trials (RCTs), these are studies of a medical intervention in which the allocation of patients to the various experimental groups, at least one of which is a control group, occurs by an *aleatory*, or chance, mechanism. Such a study has a number of essential features in addition to randomisation. It is hypothesis-driven with an unambiguous ENDPOINT assessed in a way that assures unbiased measurement; it has secondary endpoints that add credibility and interpretability to the primary outcome; it defines its study group in a way that allows logical inference to some definable population; it uses an ethically and scientifically defensible control group; and a clearly written protocol governs its procedures. Because the experimental units are humans, these trials require a formal process of informed consent as well as assurance that the safety of the participants is monitored during the course of the study.

Randomised controlled trials may enrol tens, hundreds or even thousands of participants. While the general structure of the design of such a trial is independent of its size, the number of participants enrolled affects many aspects of the conduct of a trial. Small single-centre trials tend to collect a lot of data. Multicentre trials typically collect somewhat fewer data on each patient, but the methods of collection tend to be highly structured to ensure comparability across centres. A so-called 'large simple trial' recruits thousands of people and collects a parsimonious amount of data directed at a few pointed questions (Yusuf, Collins and Peto, 1984).

A classification of trials of drugs or biologics relevant in the regulatory setting categorises them according to their phase. PHASE I TRIALS, the first trials of the product in humans, aim to gain a preliminary understanding of the safety of the product and to select doses for further study. PHASE II TRIALS define the dose more precisely and perhaps collect data germane to a preliminary glimpse at efficacy, often through the use of surrogate endpoints or biologic markers. In many therapeutic areas, Phase I and II trials are not randomised, and sometimes not controlled, because their goals can be achieved with simpler designs. Confirmatory, or PHASE III TRIALS aim to test the efficacy of the product. Typically, a Phase III trial is randomised. Trials performed after the product has been approved for licensing, whether they are randomised or not, are sometimes called PHASE IV TRIALS.

This section touches on the early history of the modern randomised controlled trial, discusses its salient features and describes some of its limitations. Several textbooks provide thorough introduction to these trials (Pocock, 1997: Friedman, Furberg and DeMets, 1998; Meinert, 1998).

The earliest modern randomised clinical trial studied streptomycin, the first truly active drug for the treatment of tuberculosis. Streptomycin, discovered in 1944, had shown very promising results in uncontrolled pilot studies in the USA. At the end of the Second World War, limited supplies of the drug were made available in Great Britain for clinical use. Because of the belief that the drug had promising therapeutic activity, a portion of the available supply was reserved for patients with the two most lethal forms of tuberculosis – meningeal and miliary disease. The rest could be used for the large majority of patients, who had pulmonary disease. Since it was manifestly impossible to treat everyone with pulmonary tuberculosis, the Medical Research Council decided that the best use of the small amount available would be to study its effects in a controlled setting. The study that followed (Marshall *et al.*, 1948) was a multicentre controlled trial comparing streptomycin to standard bed rest. Random numbers placed in sealed envelopes governed the treatment assignment of the 109 patients entered. Patients were eligible if they had 'acute progressive bilateral pulmonary tuberculosis of presumably recent origin, bacteriologically proved, unsuitable for collapse therapy, age group 15 to 25 (later extended to 30)'. The narrow age group was chosen to limit the number of eligible patients. Physicians responsible for evaluating radiological change with treatment were unaware of the treatment assignment. Results showed a clear benefit of streptomycin; the published report refers to a few of the observed treatment differences as 'statistically significant'.

Two anonymous editorials accompanied the publication of this trial. One dealt with the implications of the results for tuberculosis therapy. The second, entitled 'The controlled therapeutic trial' (Anon, 1948), commented on the noteworthy aspects of this study, including the precise definition of the patient population, the advantages of multicentre collaboration to assure adequate numbers of study subjects, the need for screening potential subjects so that they conformed to the eligibility requirements and the value of the 'ingenious system of sealed envelopes' to ensure that advanced knowledge of the treatment assignment would not influence the decision about patient eligibility.

Within six years of publication of the streptomycin trial, another trial dramatically exhibited the unprecedented power of the new methodology. The field trial of the newly developed Salk vaccine for the prevention of poliomyelitis was an enormous effort. Centrally coordinated at the University of Michigan by a group convened for that purpose, it involved public health departments in 44 of the 48 states of the USA. Within about a 6-week period, from 26 April to 15 June 1954, 402 000 children received either the Salk vaccine or placebo by random assignment. The very large numbers were required because of the relatively low attack rates of the poliovirus in normal populations of children.

The trial showed that the vaccine effectively reduced the incidence of polio and, by providing rough estimates of effectiveness for certain subgroups, suggested that the effectiveness of vaccination varied somewhat according to the primary manifestations of the disease (bulbospinal versus spinal) and according to the type of virus recovered. The comprehensive report describing this heroic effort and its results (Francis *et al.*, 1955) was published less than 2 years after the National Foundation for Infantile Paralysis first announced its decision to sponsor a formal trial of the vaccine.

Since then, use of the randomised prospective methodology has burgeoned. Bolstered by subsequent developments in statistical methodology, the randomised clinical trial is widely acknowledged as the 'gold standard' of evidence for evaluating therapies. However, because the randomised trial can only be employed to address a small fraction of the unanswered questions relating to medical interventions, a variety of methodologies – experimental and observational, prospective and retrospective – have been and continue to be widely used.

The statistical framework of a randomised clinical trial involves two sets of basic notions: TYPE I ERROR rate, *P*-VALUE and validity; POWER and precision. The *Type I error rate* is the probability that if the treatments under study do not differ (i.e. the NULL HYPOTHESIS is true), the study will show a statistically significant difference between them. The *P-value* is the probability under the null hypothesis that data would show by chance a difference as large as the difference observed. *Validity* in clinical trials refers not to the correctness of the answer in any particular trial but to the expectation that, under the null hypothesis, the data would behave in a way consistent with the pre-specified Type I error rate and that if the sample size were large enough, the estimated treatment effect would be the true effect.

Power, by the same token, is the probability that the study will show a statistically significant effect of treatment if the true effect of treatment is not zero. Power is then a function of the true effect: the larger the effect, the higher the power. *Precision* is a measure of the variability of the estimated effect of treatment. The higher the sample size and the lower the underlying variability of the measurements, the higher the precision.

Results from studies of interventions may help physicians decide on the best therapeutic option for particular patients. They permit regulatory agencies to approve products for marketing and widespread commercial distribution. Increasingly, governments, insurance companies and managed-care organisations use such studies to help decide which therapies to reimburse. The use of formal clinical data as the basis for these decisions rests on two obvious assumptions. The first is that the studies that form the database have in fact yielded correct results; i.e. any differences in outcome can with confidence be attributed to differences in the therapy administered. To the extent that studies possess this quality, they are said to have *internal validity*. The second is that the study results are, to some degree at least, generalisable to a population of subjects that is broader than the study population from which the data are derived. Studies having this property are said to have *external validity* and the broadness of a study's external validity is one critical measure of the study's overall importance.

A study lacking internal validity has no redeeming value. The mere presence of internal validity, however, does not guarantee its general usefulness if the study has very limited external validity. For example, selecting a very tightly defined study population that does not represent the larger universe of patients with the same condition may render the results at best narrowly applicable. Alternatively, perhaps the intervention under study is of such technical complexity that it is only available in a few selected centres and cannot be applied broadly. Examples of such problems have occurred in virtually every field of medicine.

The major barriers to internal validity relate mainly to the extent to which three major sources of confusion are permitted to interfere with unambiguous inference – bias, confounding and chance (Hennekens and Buring, 1987).

Used in the context of analytic clinical studies, 'bias' does not connote moral opprobrium but simply refers to any systematic error in the design or execution of a study that distorts the true relationship between intervention and outcome. Bias can originate with either the investigators or the study subjects and is of several types. Sometimes the very manner of selecting a study's participants introduces bias (*selection bias*); this is a particular problem, for example, in CASE-CONTROL STUDIES when the selection of cases, or agreement of the patient to participate, is not independent of the chance that the patient will have been exposed to the intervention of interest. In retrospective studies, the ability of patients to recall interventions in the past may be influenced by whether or not they have experienced certain medical outcomes (RECALL BIAS). The history obtained from a patient in any study may be skewed by knowledge on the part of either the interviewer (*observer bias*) or the study subject (*subject bias*) of the nature of the study or what interventions have occurred or what outcomes they have experienced. In prospective studies, losses to follow-up often occur; nonrandom losses that occur differentially in one group or the other may introduce serious bias. By far the best insurance against the various types of bias in clinical studies is the use of a prospective, randomised study design that keeps both the investigators and the study subjects unaware of which intervention the individual subjects are receiving (the so-called *double-blind* or *double-masked* design). Certain study designs, such as those that compare a surgical to a medical intervention or those where the side-effects of treatment frequently reveal the nature of the therapy, preclude blinding of either subjects or investigators. These cases require the most objective endpoints available (e.g. death from any cause, if this is an appropriate measure of treatment effectiveness). Where less objective endpoints are the most meaningful medically, evaluators who are blind to the therapy received should assess them. In designs that are not prospective and randomised, minimisation of bias poses a major challenge because there are no good analytic tools to correct for it. Once bias is present, the extent to which it has been eliminated by sound design and conduct is always open to question. The primary advantage of a well-conducted randomised clinical trial is that it removes many sources of bias.

Confounding refers to the distortion of the association between intervention and outcome by the association of another factor (the *confounder*) with both. A factor A is a confounder of an intervention I and an effect E if A has an effect on E independently of I and the use of I depends in some way on A. For example, if one were to try to compare two regimens for leukaemia (say, a vigorous and a more gentle one) by reviewing past series of patients treated with each, one might find that the aggressive regimen performed much better than the other. A more detailed analysis, however, might reveal that the aggressive regimen was used preferentially in younger patients, while the gentle treatment was reserved for older patients. Age would confound any attempt to compare the two regimens in this simple manner, since it is well known that younger leukaemia patients have a better prognosis than elderly ones when treated with virtually any regimen.

Note that, in contrast to bias, confounding is nobody's fault. It does not represent errors of commission or omission, but rather is a natural consequence of the often complex relationships among the many factors that determine clinical outcome following an intervention (or, more generally, following exposures of any kind). In OBSERVATIONAL STUDIES, when a variable is suspected of being a source of confounding, one attempts to correct for the confounding in either the design or the analysis. A prospective randomised design provides the strongest protection against confounding because randomisation tends to equalise the distribution of potential or actual confounders among the various treatment arms. Randomisation does this with both known and unknown confounders. While statistical methods are available for adjusting for known confounders, nothing other than randomisation can even approach the control of unknown confounders.

The very nature of clinical investigation permits the play of chance to deal the investigator a misleading answer. The reliability of a result is in part a function of the number of patients studied; increasing the sample size of any randomised study, no matter what the design, will decrease the probability that the patients studied are peculiar in some identified or unidentified manner. Moreover, an inadequate sample size clearly increases the likelihood that chance alone can deal a false or misleading result. A trial that is too small may end up showing a difference where none really is present or, more commonly, may fail to show a difference where a medically important one really does exist. The statistical procedure that evaluates the degree to which the observed result is consistent with chance is called a test of *statistical significance*.

The great virtue of randomisation is that it removes selection bias from the allocation of patients to therapy and that it tends to equalise baseline patient characteristics (actual or potential confounders) in the various arms of the study whether these characteristics are known or unknown. Thus, if the sample size is sufficiently large, any statistically significant differences in outcome can be attributed to the differences in therapy with much greater confidence than with other methodologies. Finally, randomised treatment assignment provides theoretical support for the inference that permits calculation of the probability that the observed differences might have arisen by chance (the P-value).

The conclusions of a clinical trial are usually stated in probabilistic terms; therefore, even when a trial has shown a statistically significant difference between two treatments, there remains the possibility that the results might have been due to chance. If, however, several independent trials show a consistent effect, the probability that the result can be ascribed to chance is enormously reduced. In such a case, chance plays a far smaller role in interfering with a strong conclusion that it does in most other aspects of daily life.

Various features of the randomised controlled trial, when taken together, form the most reliable route towards the documentation of a causal relation between an intervention and a clinical outcome. The prospective collection of data, with suitable measures to ensure correctness and completeness, serves to minimise MISSING DATA, misclassified data or erroneous data elements. The use of blinding, whenever feasible, minimises the presence of observer or subject bias. The use of a PLACEBO, when medically appropriate, allows isolation of the true effects of the test therapy from those attendant on the therapeutic setting in general. The contemporaneous relationship of test and control groups eliminates secular changes in patient selection or ancillary therapy that might affect results. Finally, as already described, the use of randomisation tends to equalise the distribution of confounding variables, known and unknown, among the intervention options and provides a formal basis for the application of statistical inference to the data.

Of course, a randomised controlled trial is not the only route to truth. Few would quibble with the claim that appendectomy cures acute appendicitis, that penicillin cures pneumococcal pneumonia, that radiotherapy cures early Hodgkin's disease or that the use of parachutes saves people who jump from planes (Smith and Pell, 2003), although no randomised data exist to support any one of these claims. In the presence of striking clinical effects, formal testing of a clinical hypothesis with a randomised trial may be unnecessary. Neither does a randomised trial necessarily always give the right answer. Like any other scientific experiment, a clinical trial is fallible. Careless data collection, inaccurate observations, inadequate sample size, faulty techniques of inference or simply the play of chance can result in misleading or frankly erroneous conclusions.

Adherence to high standards of design, conduct and analysis, as best one can judge from the published report, tends to bolster the credibility of the trial. The larger the difference in outcomes between control and test therapies, the larger the sample size (or in trials that count events, the more events) and the smaller the P-value of the comparison and the more precisely estimated the differences, the more credible the results.

In randomised controlled trials, as in other experiments, the design should address the question at hand. A clear medical question leads to a crisp design, while a fuzzy purpose may lead to an inadequate amount of data, concentration on irrelevant details and failure to answer useful questions. In practice, many trials have foundered because their vaguely articulated hypotheses have spawned inadequate designs. One should think the scientific hypothesis underlying the clinical experiment makes sense, for the more biologically plausible the hypotheses the more likely the results will be believable. There is often much subjectivity here, however, and what makes eminent sense to some may be implausible to others. One may, from time to time, be faced with results in clinical trials that seem to have no plausible scientific support.

Many of the most important clinical trials have been founded on hypotheses about the putative mechanism of action of the intervention. It is gratifying when such trials turn out positive, since then the result is consistent with the proposed mechanism. In the real world, however, it is often extraordinarily difficult to use clinical trials productively as probes for the underlying mechanism.

A randomised controlled clinical trial has a protocol that discusses the purpose of the trial, describes the procedures to be used and defines the endpoints as well as the criteria used to define 'success'. It justifies the sample size, the study group, the strata – if any – and plan for follow-up and the method to be used for monitoring safety during the course of the trial. Regulatory bodies often require that they review protocols for trials performed for the purpose of licensing a product or extending the indication in the label.

The complexity of the study and of the protocol depends on a number of factors. A small single-centre trial that is studying a short-term intervention for symptomatic relief may have a brief protocol, while a large, multicentre study measuring a complicated endpoint in subjects followed for several years may require a much more detailed protocol. Whether the protocol is simple or complicated, it should be written clearly and essential parts should be easily identified. Other documents may provide a further description of the study. For example, a data analysis plan may present the details of the statistical methods to be used. If the study has a data safety monitoring committee or an endpoint committee, their charters will describe them and their roles. The International Conference on Harmonization, in its guidelines for Good Clinical Practice, describes elements of a well-constructed protocol.

While randomisation produces study groups with equal expected distributions of both measured and unmeasured characteristics at baseline, randomisation by itself cannot ensure that the results of a study will be unbiased. The unbiasedness conferred by randomisation requires a statistical analysis that classifies people according to the treatment to which they were randomised, not the treatment they actually received. Analysis that preserves the randomisation is called *intent-to-treat* analysis.

A perfectly conducted clinical trial would have no problem analysing the groups as they were randomised, for each person would receive the assigned therapy, each would adhere to the protocol and each would provide a measurement for the primary endpoint. Even rigorously executed clinical trials, however, rarely meet this ideal. Many participants in trials adhere incompletely or not at all with their assigned regimen. The primary endpoint may be missing for some participants. Thus, while the principle of performing statistical analysis according to the randomised assignment is central to producing an unbiased result, in practice one is often forced to violate those principles. One needs reasonable approaches to assigning outcomes when actual observations are unavailable. Often investigators are tempted to analyse data from the subset of participants who complete the study according to the protocol. Such analyses can be subject to severe selection bias (Lamm *et al.*, 1981).

Sometimes an analysis other than the intent-to-treat analysis is appropriate. For example, in studies of infectious disease where treatment is given presumptively before determination of the infecting organism, the primary analysis may include only those patients who have an organism against which the agent being studied is likely to be effective. Such an analysis leads to an unbiased assessment of the effect of the intervention on people infected with the target organism.

A clear objective along with focused hypotheses will drive the choice of endpoints. When the endpoint is unambiguous, its definition and measurement is simple. The more subjective the endpoint, the more need for independent assessment. For example, in unblinded trials of cancer therapies where the endpoint is a measurement of change in tumour size as assessed by a CT scan, independent readers who do not know whether the subjects are in the treated or control group may be required to ensure unbiased measurement of size of the tumours. If the endpoint is a report by the subject (e.g. a score on a scale measuring pain), the trial may use several outcomes to support and corroborate the primary endpoint.

Control groups are essential to the design of a randomised controlled trial and the inferences that can be drawn from them.

The control can be a placebo, usual care, a different therapy or usual care plus a placebo. In selecting a control therapy, the investigators should choose one that is relevant to the question at hand and should attempt, insofar as possible, to choose a control that allows blinding.

The study group in a randomised controlled trial is necessarily heterogeneous. Sometimes, investigators want to ensure equal allocation of treatment and control in specific subgroups. In such cases, they define strata and randomise within strata. For example, in a study of prevention of heart attack, one might stratify by occurrence of a prior heart attack. The choice of whether or not to stratify should depend on the size of the study and the relationship of the particular variable to the outcome of interest. If a variable has the potential to be a strong confounder and the study is small, then stratification may be useful.

At the end of a study, one might want to analyse the data by subgroup, either those that defined the randomisation strata or others. The purpose of such analyses is sometimes to assess whether the treatment is effective within specific subgroups of the population and sometimes to assess whether the effect of treatment varies by subgroup. Such analyses should be undertaken and interpreted with great caution because trials are rarely large enough to support reliable inference within subgroups (Yusuf *et al.*, 1991).

While randomised controlled trials have great advantages over other designs in terms of producing validity inference about the effect of treatment, they have severe limitation in the actual setting of clinical investigation. A prospective trial is often large, complex and expensive, requiring the coordination of a small army of participants and many sites spread over counties, countries or continents. There is no realistic prospect, therefore, of employing the randomised clinical trial in more than a small proportion of the unresolved questions in therapy. Within specific disease categories, the restrictive ELIGIBILITY CRITERIA of many trials mean that the results of the study will be directly applicable only to a narrow segment of the total patient population with that disease. Sometimes, a trial would need to be very long to yield clinically meaningful results. The polio vaccine trial realised its scientific goals quickly because the effect of the vaccine could be determined over the few months following immunisation. It is another matter entirely to assess the impact of a screening intervention or preventive drug on the incidence of heart attacks or cancer; here the necessary follow-up time may be measured in decades. Even for trials that determine the endpoint of individual treatments quickly, slow accrual may make the study take so long to complete that the therapeutic question it poses is no longer of interest by the time the answer is available. Some subspecialties have particular problems with the use of the randomised controlled trials; for surgery, in particular, the acceptance and application of the randomised trial has been slow. Finally, a randomised controlled trial may not be usable at all if the competing interventions have been in use for enough time that the attitudes of physicians are fixed; physicians are understandably reluctant to randomise patients if they think they already know the answer. The decision to perform a randomised controlled trial requires a sizeable commitment in time, money and effort; the decision not to perform one entails some sacrifice in the quality of evidence provided to physicians when they are making choices for therapies for their patients. A combination of medical, societal and financial forces determines those questions that become the subject of randomised controlled trials. *JW*

Anon 1948: The controlled therapeutic trial. *British Medical Journal* 2, 791–2. **Francis Jr, T.*et al.*** 1955: An evaluation of the 1954 poliomyelitis vaccine trials. Summary report. *American Journal of Public Health* 45, 1–51. **Friedman, L. M., Furberg, C. D. and DeMets, D. L.** 1998: *Fundamentals of clinical trials*, 3rd edition. Heidelberg: Springer Verlag. **Hennekens, C. H. and Buring J. E.** 1987: *Epidemiology in medicine*. Boston and Toronto: Little, Brown and Co. **Lamm, G.*et al.*** 1981: Influence of treatment adherence in the coronary drug project. *New England Journal of Medicine* 304, 612–13. **Marshall, G., Blacklock, J. W. S., Cameron, C.*et al.*** 1948: Streptomycin treatment of pulmonary tuberculosis. A Medical Research Council investigation. *British Medical Journal* 2, 769–82. **Meinert, C.** 1998: *Clinical trials*. New York: Oxford University Press. **Pocock, S.** 1996: *Clinical trials: a practical approach*. New York: John Wiley & Sons, Inc. **Smith, G. C. and Pell, J. P.** 2003: Parachute use to prevent death and major trauma related to gravitational challenge: systematic review of randomised controlled trials. *British Medical Journal* 327, 1459–61. **Yusuf, S., Collins, R. and Peto, R.** 1984: Why do we need some large, simple randomized trials? *Statistics in Medicine* 3, 409–20. **Yusuf, S., Wittes, J., Probstfield, J. and Tyroler, H.** 1991: Analysis and interpretation of treatment effects in subgroups of patients in randomized clinical trials. *Journal of the American Medical Association* 266, 93–8.

clinical versus statistical significance This pair of terms is often confused, being mistakenly considered as interchangeable, when, in reality, neither implies the other necessarily. One of the unnecessary difficulties of statistics is that we are using ordinary English words, such as 'normal', 'confidence' and 'population', in a technical way. If only our founders had followed the example of the anatomists and named everything in Latin. It is too late now and we are stuck with our English terminology. Of all the words our predecessors appropriated, the one that must cause the greatest confusion is 'significant'.

The *Shorter Oxford English Dictionary* gives two definitions of 'significance': 'the meaning or import of something, meaning, suggestiveness' and 'importance, consequence'. The statistical usage relates to the first interpretation. If a difference is significant in a sample, there is evidence that the difference exists in the population that the sample represents. Hence the difference has meaning beyond the individuals who make up the sample. By clinical significance, we mean that the difference we have observed is important; that, for example, it implies that we should change clinical practice. Thus, this usage relates to the second interpretation, meaning that the significant difference is important.

If a difference or relationship is statistically significant, this implies that we have evidence that it is real, existing in the larger population, but not that it is important, having implications for clinical practice. Concluding that a difference or relationship is important depends on its magnitude, together with nonstatistical factors, so that we can decide whether it is big enough to influence clinical decisions. For example, in a large clinical trial with 2000 subjects in each arm, a difference of 1 mmHg in mean diastolic blood pressure would be statistically significant. As this was a trial, it would be reasonable to conclude that the difference was real and that the treatments had slightly different effects on blood pressure. Yet it is unlikely that such a difference would influence treatment decisions. It would not be important and so not clinically significant. Contrariwise, a small study might produce a nonsignificant difference that is quite large. We could not conclude that the difference was real, but we might think it important enough to carry out another trial.

Statisticians cannot expect to appropriate ordinary English words and then demand that their use be restricted. However, the use of 'significant' in research reports to mean something other than statistical significance can be potentially misleading and it makes sense to avoid it. In its instructions to authors, the *Lancet* asks that authors 'avoid non-technical uses of technical terms in statistics, such as ... "significant"'. *JMB*

[See also CRITICAL APPRAISAL]

cluster analysis in medicine The term cluster analysis covers a very wide range of methods for discovering groups in multivariate data. It is distinct from classification techniques such as DISCRIMINANT FUNCTION ANALYSIS or *classification and regression tree* (CART) (see TREE-STRUCTURED METHODS) analysis. These classify individuals into groups that have already been identified whereas cluster analysis looks for groups within the data. General texts on cluster analysis are by Everitt, Landau and Leese (2001) and Gordon (1999) and technical developments are often published in the *Journal of Classification*. However, there is a vast literature dealing with cluster analysis in various guises and this is necessarily a very brief review of the more widely used methods. Many specialised methods dealing with particular subject matters have become separated from mainstream cluster analysis, often using their own terminology, and they may be classified under other headings such as pattern recognition, ARTIFICIAL INTELLIGENCE or DATA MINING IN MEDICINE.

In medicine, the cases to be clustered are generally people and the multivariate data describe various aspects of their clinical, psychological or service use status. However, other units of analysis can of course be clustered, e.g. hospitals or health authorities, and cluster analysis can also be used to group variables, although this is less common. The data may be in the form of attributes, such as ethnic group, or continuous measurements such as blood pressure, or a mixture of both types may be analysed (mixed mode data). The objective of the analysis is to find subgroups of people who are relatively homogeneous with respect to these characteristics. The reason for performing the analysis might be administrative (e.g. to define strata in a survey sample) or, more usually, it might be related to a research question (e.g. to identify groups of people with a common gene structure). (A review of methods used in medicine is given by McLachlan, 1992.)

In the most widely used methods, one typically proceeds by choosing a measure of the proximity between cases in terms of the multivariate data (the term 'proximity' covers both similarity and distance, either of which may be calculated). Next, an algorithm for forming clusters is applied to the matrix of proximities. The investigator usually has to decide on the number of clusters forming the 'solution', i.e. a partition of the data based on a particular choice of proximity measure, algorithm and number of clusters. The choice of number of clusters is particularly difficult, since there are few reliable formal tests and the investigator may have to consider a range of solutions that seems reasonable based on the subject matter. The robustness of the final solution can be tested in a number of ways. These include using alternative clustering methods, using *split-half* methods on the data, the detection and exclusion of outliers or influential cases and validation against external data. Formal hypothesis tests of the NULL HYPOTHESIS of absence of cluster structure are theoretically possible but are rarely applied.

The definition of proximity between individuals depends on the type of data and the relative weight to be placed on different variables (Everitt, Landau and Leese, 2001). For example, binary attribute data may be coded as series of ones and zeros, denoting presence or absence of an attribute. In the case where each category is of equal weight, such as gender or white/nonwhite ethnic group, a *simple matching coefficient* (the proportion of matches between two individuals) could be used. However, if the attributes were the presence of various symptoms, proximity might be more appropriately measured using the asymmetric *Jaccard coefficient*, based on the proportion of matches where there is a positive match (i.e. ignoring joint negative matches). For continuous data, the *Euclidean distance* between individuals i and j,

$\left[\sum_{k=1}^{p}\left(x_{ik}-x_{jk}\right)^2\right]^{1/2}$, where p is the number of variables, is often used (applied to binary data it is the same as the simple matching coefficient). For mixed mode data, *Gower's coefficient* can be used to combine components of distance from either attribute or continuous data, after first scaling the continuous data to a 0–1 range. Many alternative proximity measures have been proposed to deal with specialised types of data; e.g. in genetics binary matches may be assigned different weights depending on the part of the genetic sequence from which they arise.

Hierarchical algorithms are possibly the most widely used of general purpose clustering methods and are included in most STATISTICAL PACKAGES. They use a heuristic algorithm successively to join or divide clusters on the basis of their proximity (thus being referred to as agglomerative or divisive methods respectively). The methods differ in the way in which the intercluster proximities are calculated from the interindividual distances. *Single linkage*, for example, uses the proximity between the closest individuals in two clusters to be joined, whereas *complete linkage* uses the most distant. *Ward's method* is another popular agglomerative method that joins clusters on the basis of an error sum-of-squares criterion; unlike the two linkage methods mentioned it requires the raw data (rather than just the proximity matrix) to be available during the clustering process. Divisive methods are less commonly used than agglomerative methods, except perhaps those specifically designed for attribute data. An example of the latter is the *monothetic divisive* method, which divides the sample according to the value of a single attribute at each stage, the attribute being chosen so as to create the most homogeneous groups at that stage. *Polythetic divisive* methods divide according to a number of variables considered together.

Both agglomerative and polythetic divisive methods generally produce a tree diagram, or *dendrogram*, which shows the process by which cases have been joined or divided, and this can be used to suggest the number of clusters present (by examining the jumps in the proximity at which clusters are joined or divided). Here we can see that a small dendrogram is formed using single linkage from the following matrix of proximities (the proximity between case 1 and itself is, of course, 0.0, and between cases 1 and 2 it is 2.0, etc) (the figure shows the relationship in diagrammatic fashion):

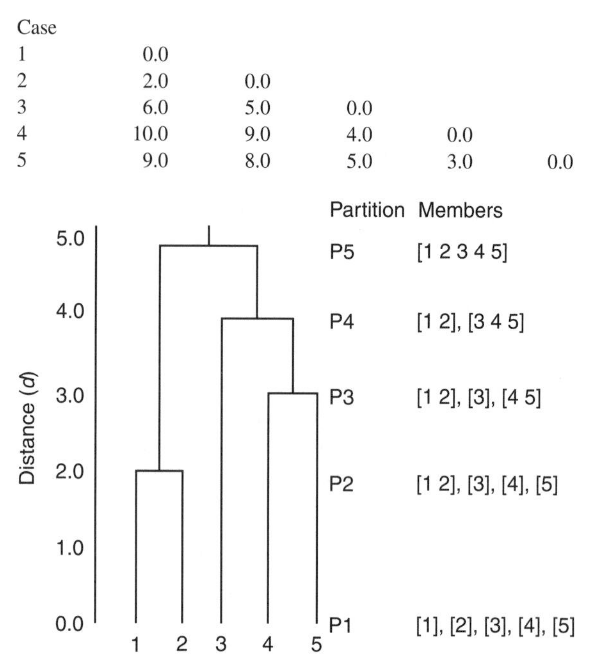

Case					
1	0.0				
2	2.0	0.0			
3	6.0	5.0	0.0		
4	10.0	9.0	4.0	0.0	
5	9.0	8.0	5.0	3.0	0.0

cluster analysis in medicine *A dendrogram produced using single linkage applied to a matrix of pairwise distances. A sequence of partitions P1–P5 is produced according to the minimum distance between cases in clusters to be joined. Cases 1 and 2 join first, then 4 and 5; case 3 joins the 4–5 cluster and finally all cases are joined*

A study of people with eating disorders made by Hay, Fairburn and Doll (1996) illustrates the use of two of the standard hierarchical methods mentioned and some of the robustness checks that can be made. The rationale for their study was that: 'Clinical experience, however, indicates that a substantial number of those who present for treatment of an eating disorder do not fulfil diagnostic criteria for either of these disorders. ... The aim of the present study was to derive an empirically based scheme for classifying those with recurrent binge eating.' The data were the first seven principal components based on 22 items from a QUESTIONNAIRE measuring eating disorder behaviours and attitudes. Ward's method was used, with complete linkage as a check. Clinical judgement, inspection of the dendrogram and formal tests were used to determine the number of clusters. The analysis was repeated without two OUTLIERS and robustness was examined using a 75 % subsample. Tests of construct validity were performed using variables external to clustering and predictive validity was assessed in terms of its success in predicting the time course of the illness compared to using standard diagnostic criteria.

Partitioning methods divide the data into a single partition (rather than a series as in hierarchical methods), the number of clusters being specified in advance. They iteratively reassign cases to clusters and recompute cluster centres, so as to optimise an objective function such as within-cluster variance. One popular method is *k-means*; a non-parametric method is *partitioning around medoids* (PAM). These methods usually need to have an initial partition from which to start the process and require the whole dataset to be available during the process. The *Kohonen* or *self-organising map* (SOM), a type of NEURAL NETWORK that successively reassigns cases to clusters, is an example of an 'online' method, i.e. one where cases are taken one at a time and 'presented' to cluster centres. The 'winning' (closest) centre is moved towards the case and the process continues with the next case, recycling cases until the system is stable. This method is quite similar to k-means, but does not need all cases to be available and can therefore cope with much larger datasets.

In addition to methods using heuristic algorithms, a number of so-called *model-based* methods has been developed. The model that underlies such methods is usually that the data are a sample from a FINITE MIXTURE DISTRIBUTION (see McLachlan and Peel, 2000). For categorical data, the populations could be multinomial: a method based on this assumption is *latent class analysis* (see Everitt, Landau and Leese, 2001). For continuous data, a mixture of MULTIVARIATE NORMAL DISTRIBUTION may be appropriate. Estimating the multivariate normal parameters and the mixing proportions by MAXIMUM LIKELIHOOD ESTIMATION can be a difficult computational problem, especially with small samples. However, the use of classification likelihood methods (which involve estimating cluster memberships treated as indicator variables rather than estimating mixing proportions) has simplified this task and made clustering based on multivariate normal models more widely available in standard software packages. Implementation of this method requires a specification of how size and shape (spherical or ellipsoidal) are assumed to vary. The second figure shows a three-component mixture identified by fitting a mixture of multivariate normals using classification likelihood and illustrates the results of making these choices for a particular dataset.

There are many methods that do not fall into the categories mentioned. For example, in *fuzzy* methods individuals have a grade or weight of cluster membership for different clusters as opposed to *crisp* methods, where cases are definitely assigned to one cluster. Model-based methods that produce probabilities of membership of the clusters for each case can be regarded as fuzzy, but there are also fuzzy methods that do not rely on a probabilistic model. Some methods can allow for *overlapping* clusters (overlap being a different concept to fuzziness, in that cases can belong to more than one cluster simultaneously).

Other methods can cluster cases and variables simultaneously. An example is *hierarchical classes* (not to be

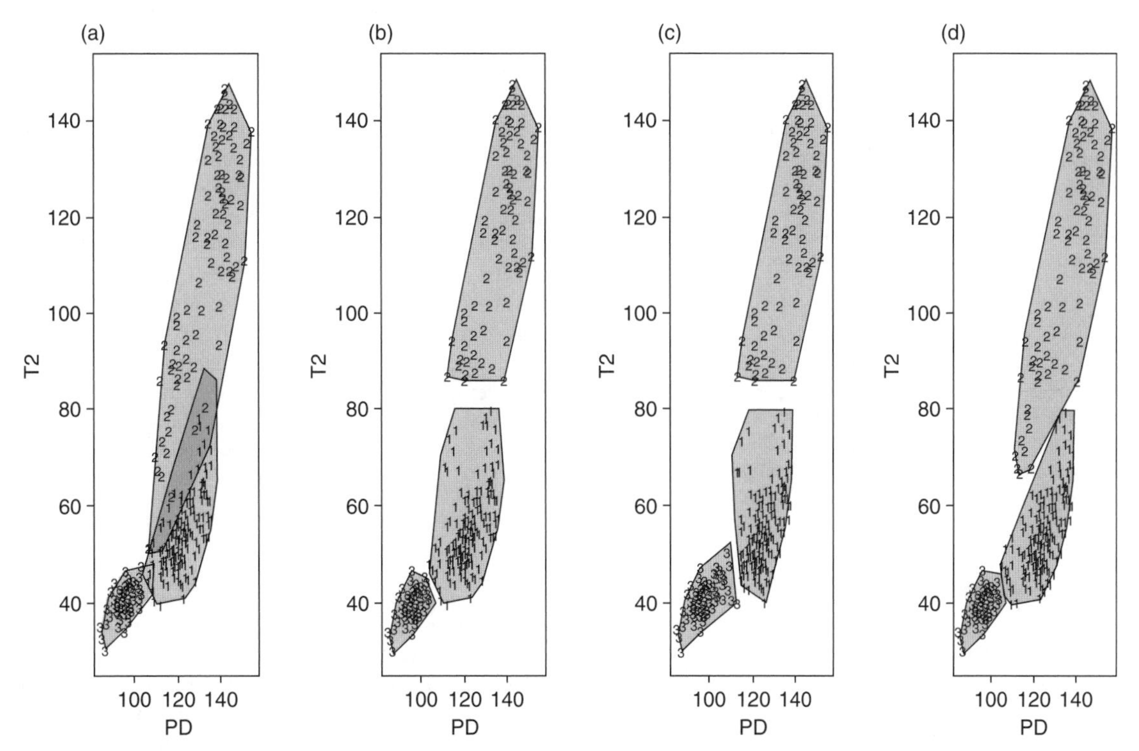

cluster analysis in medicine *Three-group solutions from applying various forms of classification maximum likelihood to a sample of 500 voxels from fMRI imaging data: (a) an a priori classification; (b) clusters assuming same-size spherical clusters; (c) clusters assuming same-size ellipsoidal clusters; (d) clusters allowing ellipsoidal clusters of different sizes*

confused with the more general term 'hierarchical methods' as described earlier). This is a method appropriate for attribute data and relatively often used in psychological or psychiatric applications. Another method for clustering both cases and variables is *direct data clustering*, which involves rearranging the rows and columns of the data matrix so that cases that are similar in terms of variables (and vice versa) appear next to each other. Arabie, Hubert and De Soete (1996) give a general review of standard and nonstandard methods and individual entries in this volume describe some of these in more detail.

Two types of medical data that may need special treatment, because of the size and complexity of their typical datasets, are genetic and imaging data. Gene expression data produced by *microarrays* (solid surfaces containing many, often thousands of, target genes against which genetic samples are compared) are characterised by very large datasets and also include a temporal dimension if the samples from the same person are taken at different time points (see MICROARRAY EXPERIMENTS). The correlation that this induces is sometimes modelled by including an *autoregressive* component in the analysis. Medical images, e.g. from functional magnetic resonance imaging, are also often characterised by large datasets. Furthermore, in addition to a temporal dimension, they may also exhibit *spatial autocorrelation* due to the physical contiguity of the measurements (see STATISTICS IN IMAGING). Genetic and imaging datasets often require methods not available in standard software and a number of websites are devoted to this type of specialised analysis. *ML*

Arabie, P., Hubert, L. J. and De Soete, G. (eds) 1996: *Clustering and classification*. Singapore: World Scientific. **Everitt, B. S., Landau, S. and Leese, M. and Stahl, D.** 2011: *Cluster analysis*, 5th edition. Wiley, Chichester. **Gordon, A. E.** 1999: *Classification*, 2nd edition. New York: Chapman and Hall. **Hay, P. J., Fairburn, C. G. and Doll, H. A.** 1996: The classification of bulimic eating disorders: a community-based cluster analysis study. *Psychological Medicine* 26, 801–12. **McLachlan, G. J.** 1992: Cluster analysis and related techniques in medical research. *Statistical Methods in Medical Research* 1, 27–48. **McLachlan, G. J. and Peel, D.** 2000: *Finite mixture models*. New York: John Wiley & Sons, Inc.

cluster randomised trials These are CLINICAL TRIALS in which groups or clusters of individuals are randomly allocated to treatments. The difference between cluster randomised trials and individually randomised trials is that in a cluster trial the main unit of randomisation is not the same as the unit on which the analysis is carried out. Thus the unit of randomisation may be as a group of people such as in a town but the outcome will be the behaviour of people in the town. The intervention is often aimed at and delivered by healthcare professionals, such as education to modify their treatment of patients, but the effectiveness of the intervention is assessed in terms of the outcome for the patient. In contrast, in individually randomised trials, both intervention and outcome are aimed at the same person. Useful references are Donner and Klar (2000), Campbell, Donner and Klar (2007) and Hayes and Moulton (2009).

The main reason for using a cluster trial is fear of *contamination*. This occurs because subjects in the same unit or treated by the same healthcare professional are likely to receive the same intervention. Thus it can be very difficult to ensure that subjects in a control group do not receive at least some of the intervention if they are physically in the same unit as the treated subjects. It may be difficult for healthcare professionals to switch from one style of treatment to another or subjects may compare notes on the treatments they have received. There are many different features associated with cluster randomised trials and some of the statistical aspects were first discussed by Cornfield (1978). The main feature is that patients treated by one healthcare professional tend to be more similar than those treated by different healthcare professionals. If we know by which doctor a patient is being treated, we can predict slightly better than chance the performance of the patient and thus the observations for one doctor are not completely independent, which is the usual assumption for analysis. What is surprising is how even a small correlation can greatly affect the design and analysis of such studies.

Cluster trials can be divided into those with a cohort design, in which patients are followed up over time, and cross-sectional design, in which the patients at baseline are not the same as those in the follow-up. A cohort design would follow up patients after treatment, but a public health campaign might adopt a cross-sectional design in which different individuals are questioned before and after the intervention, say a local radio campaign to reduce drink driving.

It is helpful, when planning and analysing a study, to have a model in mind from the start. We will consider a cohort design in general practice in which the same patients are followed up over time. For continuous outcomes y_{ij} for an individual j in practice i we assume that:

$$y_{ij} = \mu + z_i + \tau\delta_i + \beta x_{ij} + \varepsilon_{ij} \qquad (1)$$

where $j = 1, \ldots, n_i$ = the number of patients in practice i and $i = 1, \ldots, N$ = total number of practices. Here z_i is assumed to be a random variable with $E(z_i) = 0$, $\text{Var}(z_i) = \sigma_B^2$, and reflects the overall effect of being in practice i, τ is the additional effect of being in one of the treatment arms relative to the other where δ_i takes the value 1 for one treatment arm and zero otherwise and x_{ij} is a vector of the individual level (or practice level) covariates with regression coefficients β. We assume $\text{Var}(\varepsilon_{ij}) = \sigma^2$ and that z_i and ε_{ij} are independent and thus $\text{Var}(y_{ij}) = \sigma^2 + \sigma_B^2$. It can be shown that when a model is fitted that ignores z_i the STANDARD ERROR of the estimate of β

is too small and thus in general one is likely to increase the TYPE I ERROR RATE.

One feature of the model as written is that both σ^2 and σ_B^2 are assumed constant and independent of the treatment effect, but clearly this can be investigated and the model modified if necessary. For some models, we need to assume also that z_i and ε_{ij} are normally distributed. The model can be extended to counts or binary dependent variables using appropriate generalized linear models.

The INTRACLUSTER CORRELATION (ICC) is given by:

$$\rho = \frac{\sigma_B^2}{\sigma^2 + \sigma_B^2} \tag{2}$$

With cluster trials there are two sample size issues: how many clusters and how many patients per cluster. The basic principles for a completely randomised design have been discussed by Donner and Klar (2000). The idea is to obtain the sample size for an individually randomised trial and inflate the sample size by the design effect (DE), where $\text{DE} = 1 + (\bar{n}-1)\rho$ and $\bar{n}$ is a measure of the average cluster size. Values of the ICC up to about 0.05 are found in practice in primary care. Even with such a small ICC, with 20 patients per practice the sample size has to be doubled for a fixed POWER compared to an individual randomised trial.

Cornfield (1978) states that one should 'analyse as you randomise'. Since randomisation is at the level of the practice, a simple analysis would be to calculate 'summary measures', such as the mean value for each practice, and analyse these as the primary outcome variable.

Omitting the covariates from the model for simplicity it is easy to show that:

$$\bar{y}_i = \mu + \tau\delta_i + \bar{\varepsilon}_i \tag{3}$$

where $\bar{y}_i$ is the mean value for y_{ij} for practice i and:

$$\text{Var}(\bar{y}_i) = \sigma_B^2 + \frac{\sigma^2}{n_i} \tag{4}$$

Equation (3) is a simple model with independent errors, which are homogeneous if n_i's are always of similar size. An ORDINARY LEAST SQUARES estimate at practice level of τ is unbiased and the standard error of estimate is valid provided the error term is independent of the treatment effect.

Thus a simple analysis at the cluster level would be the following: if n_i's are the same or not too different, then carry out a two sample t-test on the practice level means; if the n_i's are different, then carry out a weighted two-sample t-test using the estimated inverse of the variance for weight. It is worth noting that if σ^2 is zero (all values from a practice are the same) then practice size does not matter in the analysis and if σ_B^2 is zero, then the weight is equivalent to the number of patients per practice.

The advantage of a practice-level approach is that it is simple and intuitive. It works for both continuous and binary data; for the latter, one analyses the proportions in each cluster and treats the proportions as continuous measures (as, for example, using a paired t-test).

However, there are a number of problems with a cluster-level approach. The main one is that it does not properly allow for patient-level covariates. It is unsatisfactory to use cluster averaged values of the patient-level covariates. A two-stage approach would be to use conventional methods to adjust for covariates, ignoring clustering, and then apply the cluster methods to the adjusted outcomes. However, fitting equation (1) does both stages simultaneously. The cluster-level method is also possibly inefficient since the number of DEGREES OF FREEDOM for any practice-level comparison is constrained by the number of practices and the method takes no account of the number of patients per practice.

The model given in equation (1) is termed a RANDOM EFFECTS MODEL, a two-stage multilevel model or a mixed model since it contains both random effects (z_i) and fixed effects (δ_i). The main method for fitting these types of models is by MAXIMUM LIKELIHOOD and this is available in a number of packages such as MlWin, SAS Proc Mixed, STATA, R and Splus. Some of the methods require distributional assumptions such as normality of the between-cluster random effect z_i, which can be difficult to verify empirically, particularly with small numbers of clusters. A further refinement to model fitting is to use a technique known as restricted maximum likelihood (REML). This method is useful for estimating variance components because the usual maximum likelihood estimates are biased (see COMPONENTS OF VARIANCE). This procedure is available in SAS Proc Mixed and MlWin. These methods estimate the parameters from a *cluster-specific* model and try to estimate the effect of the intervention *within* clusters.

A rather different method of estimating the parameters uses GENERALISED ESTIMATING EQUATIONS (GEE), which provide valid estimates of treatment effects even if the intracluster correlation is not precisely specified. Since it is an approximate method it requires more than 20 clusters to give valid estimates. GEE estimates parameters from a *population* or *marginal* model that tries to estimate the effects *on average* over clusters.

To explain this it is easier to select an example outside clinical trials. Suppose we had patients, clustered in some way, e.g. in families, and we were interested in the risk of high blood pressure for stroke. A marginal model looks at the risk of intervention people with high blood pressure, compared to low blood pressure, *on average*. In contrast, a cluster specific looks at the average of risk of people with high versus low pressure *within a cluster* such as a family. For a linear model, the marginal and cluster specific methods are estimating the same *population parameter*, although different methods of estimation may give differing *estimated* results. For a non-linear model such as a logistic regression, the population parameters are different, to a degree related to the difference in the mean levels of the clusters. (Further details are given by Neuhaus, Kalbfleisch and Hauck, 1991).

Some packages have an option to estimate a 'robust' standard error for a large number of procedures such as multiple and LOGISTIC REGRESSION under clustering, also known as the Huber–White estimate, for which no distributional assumptions are required. The method to avoid is the FIXED EFFECTS approach in which one fits DUMMY VARIABLES to each cluster. This removes the cluster-level variability, but gives estimates that are biased (Murray, 1998). *MJC*

[See also CLUSTERED BINARY DATA]

Campbell, M. J., Donner, A. and Klar, N. 2007: Developments in cluster randomized trials and Statistics in Medicine. *Statistics in Medicine* 26, 2–19. **Cornfield, J.** 1978: Randomization by group: a formal analysis. *American Journal of Epidemiology* 108, 100–2. **Donner, A. and Klar, N.** 2000: *Design and analysis of cluster randomisation trials in health research*. London: Arnold. **Hayes, R. J. and Moulton, L. H.** 2009: *Cluster randomised trials*. Boca Raton: Chapman & Hall/CRC. **Murray, D. M.** 1998: *Design and analysis of group randomized trials*. Oxford: Oxford University Press. **Neuhaus, J. M., Kalbfleisch, J. D. and Hauck, W. W.** 1991: A comparison of cluster-specific and population-averaged approaches for analysing correlated binary data. *ISI Review* 59, 25–35.

clustered binary data These are binary responses on units that are nested in clusters. Examples include repeated responses where occasions are nested in subjects, twin data where twins are nested in twin pairs and responses on children nested in doctors. Here the units are said to be at level 1 and the clusters at level 2. In three-level data, the 'level-2' clusters are themselves nested in 'level-3' clusters. For instance, children may be nested in doctors who are nested in hospitals, as shown in the figure.

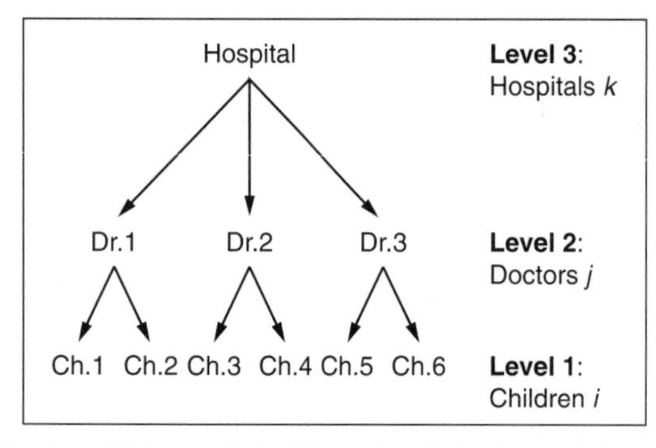

clustered binary data *Three-level clustered data*

Units within a cluster are expected to be more similar to each other than to units in different clusters and there is hence between-cluster heterogeneity and within-cluster dependence. Three types of statistical methods, accommodating the dependence induced by clustering, have been suggested for analysing clustered binary data.

1. *Cluster-specific models* where each cluster has its own effect(s):
   (a) RANDOM EFFECTS MODELS, where dependence is explicitly modelled by including cluster-specific random intercepts (and possibly coefficients) that are drawn from a distribution and hence vary over clusters. The random effects are assumed to be uncorrelated with the included covariates. These models are typically estimated using maximum marginal likelihood, where the random effects are 'integrated out' (see RANDOM EFFECTS MODELS FOR DISCRETE LONGITUDINAL DATA). The random effects are sometimes specified as categorical latent classes, leading to mixture regression.
   (b) Fixed intercept models, where dependence is explicitly modelled by including fixed intercepts that vary over clusters. In this case it is not necessary to assume that the cluster-specific effects are uncorrelated with the included covariates and the regression parameters can be interpreted as within-cluster effects. These models are typically estimated using maximum conditional likelihood, where the cluster-specific intercepts are 'conditioned out'.
2. *Marginal approaches* for marginal or population-averaged effects.
   (a) GENERALISED ESTIMATING EQUATIONS (GEE), where dependence is treated as a nuisance. GEE is an estimation algorithm that need not correspond to any statistical model.
   (b) Marginal statistical models, for instance the Bahadur, Dale and George–Bowman models. These models are usually estimated using MAXIMUM LIKELIHOOD ESTIMATION.
3. *Transition models*, where effects are conditional on responses of other units in the cluster. These models require that the units within a cluster are not 'interchangeable', the canonical example being longitudinal data where responses are time ordered. Transition models are sometimes called autoregressive models, lagged response models or dynamic models.

AS/SR-H

Fahrmeir, L. and Tutz, G. 2001: *Multivariate statistical modelling based on generalized linear models*. New York: Springer. **Molenberghs, G.** 2002: Model families. In Aerts, M., Geys, G., Molenberghs, G. and Ryan, L. M. (eds), *Topics in modelling of clustered data*. Boca Raton: Chapman & Hall/CRC, pp. 47–75. **Skrondal, A. and Rabe-Hesketh, S.** 2004: *Generalized latent variable modelling: multilevel, longitudinal and structural equation models*. Boca Raton: Chapman & Hall/CRC.

Cochran *Q*-test This test is used to see if the proportion of positive dichotomous outcomes varies between sets of matched data. It is used, for example, to test whether there is heterogeneity between people rating subject data or to see if there is a difference between treatments for trials using matched patients.

When McNEMAR'S TEST is applied to two paired groups, the Cochran Q-test seeks to identify whether the proportions of positive responses vary among many matched groups. The Cochran Q-test can be viewed as an extension of McNemar's test and are equivalent when the number of groups is two.

To calculate Cochran's Q-statistic, one must identify for each of the N samples or subjects the number of groups in which its response is positive and denote these values $S_1, S_2, \ldots, S_N$. One must also identify for each of the c groups (e.g. raters or time points) the total number of samples or subjects that are given a positive response and denote these values $T_1, T_2, \ldots, T_c$. These two sets of values will both sum to the total number of positive responses, which we denote T.

The Cochran Q-statistic is then calculated as:

$$Q = (c-1) \times \frac{c\sum_{j=1}^{c} T_j^2 - T^2}{cT - \sum_{n=1}^{N} S_n^2}$$

and compared to the CHI-SQUARE DISTRIBUTION with $c-1$ DEGREES OF FREEDOM.

Cochran (1950) illustrates this with an example where four different media are investigated for effectiveness in growing a bacterium when 69 matched specimens were grown in each medium. Four of the specimens had S_n values of 4 (i.e. bacteria grew in all four media), five had S_n values of 3, one had an S_n value of 2 and the rest S_n values of 0, giving:

$$\sum_{n=1}^{N} S_n^2 = 113,\ c = 4,\ T = 33$$

The total numbers of successful specimens by medium (the T_j) were 6, 10, 7 and 10, giving:

$$\sum_{j=1}^{c} T_j^2 = 285,\ Q = 8.052$$

When compared to the chi-square distribution with 3 degrees of freedom this gives a P-value of 0.045, indicating that there is evidence that the media do not all perform alike.

It may be apparent from these calculations that since only the sums of positive responses are used, any specimen (or row of matched data in the more general case) that contributed no positive responses can be ignored. By arguments of symmetry, one can see that specimens that give a positive response in every case are similarly uninformative and can be ignored. Note that this is akin to McNemar's test, where the concordant pairs do not contribute to the test statistic. For further details see Fleiss (1981). *AGL*

Cochran, W. G. 1950: The comparison of percentages in matched samples. *Biometrika* 37, 256–66. **Fleiss, J. L.** 1981: *Statistical methods for rates and proportions*. New York: John Wiley & Sons, Inc.

Cochrane Collaboration The Cochrane Collaboration is an international organisation that aims to help people make well-informed decisions about healthcare by preparing, maintaining and promoting the accessibility of systematic reviews of the effects of healthcare interventions. Systematic reviews produced by the Collaboration are published in The Cochrane Database of Systematic Reviews as part of The Cochrane Library, available online and on CD-ROM on a subscription basis (The Cochrane Database of Systematic Reviews, 2004). The Cochrane Collaboration is currently the largest organisation in the world engaged in the production and maintenance of systematic reviews. In 2003 more than 9000 contributors from 80 countries were involved and the second issue of the database in 2004 contained 1999 completed reviews and 1441 protocols for reviews.

The Cochrane Collaboration was named after Archie Cochrane, the British epidemiologist who, in his influential text, *Effectiveness and efficiency*, promoted the use of evidence from randomised controlled trials to inform the provision of healthcare services (Cochrane, 1972). He went on to emphasise the importance of systematic reviews, when in 1979 he wrote: 'It is surely a great criticism of our profession that we have not organised a critical summary, by specialty or subspecialty, adapted periodically, of all relevant randomised controlled trials' (Cochrane, 1979). This challenge led to the establishment during the 1980s of an international collaboration to develop the *Oxford database of perinatal trials* (Chalmers, 1989–1992). In 1987, the year before Cochrane died, he referred to a systematic review of randomised controlled trials (RCTs) of care during pregnancy and childbirth as 'a real milestone in the history of randomised trials and in the evaluation of care' and suggested that other specialties should copy the methods used (Cochrane, 1989). His encouragement, and the endorsement of his views by others, led to the opening of the first Cochrane Centre by Iain Chalmers (in Oxford, UK) in 1992 and the founding of the Cochrane Collaboration in 1993.

The Collaboration produces reviews through its Collaborative Review Groups, which are supported by fields, Cochrane centres and methods groups (The Cochrane Collaboration website). There are currently 50 Cochrane collaborative review groups, each being responsible for reviews in a particular area of healthcare. The 12 regional Cochrane centres support review activity and dissemination of the library around the world, while the nine fields provide links between the Collaboration and particular areas of healthcare (e.g. primary care), types of consumer (e.g. older people) or types of intervention (e.g. vaccines). The 10 methods groups undertake statistical and methodological research related to systematic reviews, advise the Collaboration on how systematic reviews should be undertaken and reported, monitor the quality of reviews and assist in the development of software and training materials.

Cochrane reviews aim to minimise bias and therefore reviews of healthcare interventions attempt to locate all randomised trials, whether or not they have been published. The Collaboration has worked to improve the identification of randomised controlled trials in the literature by systematically handsearching journals and conference proceedings and by working with the National Library of Medicine to improve indexing of randomised trials on Medline and PubMed. The resulting collection of citations, The Cochrane Central Register of Controlled Trials, is available as a second database on The Cochrane Library. In September 2010 it contained over 400 000 citations.

Publication of Cochrane reviews as electronic rather than paper documents has advantages that include the ability to update reviews when new trials are completed, full reporting of standardised details from all trials, including forest plots and data, and the ability for users of the Cochrane Library to reanalyse reviews using alternative summary statistics and statistical models, as well as viewing the analyses chosen by the author. Comments and criticisms of reviews can also be made online and published alongside the original review.

In its second decade the Cochrane Collaboration is continuing to register and publish new reviews of healthcare interventions, as well as tackling the challenges of how to obtain better systematic evidence of the harmful effects of interventions and how to ensure that systematic reviews are updated in a timely manner. The Collaboration is also now developing plans for the publication of Cochrane Reviews of diagnostic test accuracy. *JD*

Chalmers, I. (ed.) 1989–1992: *The Oxford database of perinatal trials*. Oxford: Oxford University Press. (Contents were subsequently transferred to and maintained in The Cochrane Database of Systematic Reviews.) **Cochrane, A. L.** 1972: *Effectiveness and efficiency. Random reflections on health services*. London: Nuffield Provincial Hospitals Trust. **Cochrane, A. L.** 1979: 1931–1971: a critical review, with particular reference to the medical profession. In *Medicines for the year 2000*. London: Office of Health Economics, pp. 1–11. **Cochrane, A. L.** 1989: Foreword. In Chalmers, I., Enkin, M. and Keirse, M. J. N. C. (eds), *Effective care in pregnancy and childbirth*. Oxford: Oxford University Press. **The Cochrane Collaboration website**: www.thecochranelibrary.com and www.cochrane.org. **The Cochrane Database of Systematic Reviews** 2004: Issue 2. The Cochrane Library. Chichester: John Wiley & Sons, Ltd.

coefficient of determination See CORRELATION

coefficient of variation This is a measure of dispersion defined as the STANDARD DEVIATION divided by the MEAN. Because the standard deviation and the mean share the same units, these units cancel out and leave the coefficient of variation (CV) as a dimensionless number. Because it is independent of measurement units, the coefficient can be used to compare the amount of dispersion for two sets of values – hence its alternative name, relative variability. Such comparison can be useful in some cases but it has to be remembered that the CV can only be used for ratio scale variables that have a true zero point, e.g. height and weight. However, the CV cannot be used on variables on an interval scale, e.g. temperature measured in degrees Centigrade, because it would have a different value from temperature measured in degrees Fahrenheit due to the different mean temperatures. *BSE*

cohort studies Also called medical follow-up studies, cohort studies are considered to be any epidemiological study in which the study population is identified before the occurrence of the disease event of interest and then followed in time until the first occurrence of the disease event or the end of the study, whichever comes first. These may also be referred to as survival studies, in which the outcome is death. Typically, subjects are classified as exposed or not exposed to one or more putative risk factors at the beginning of the study or, alternatively, they may provide more detailed information on exposure.

Because exposure is determined prior to an illness, this study design avoids bias due to selective recall by patients who have been recently diagnosed as in a CASE-CONTROL STUDY, especially when there may be rumours or preconceptions regarding the association between disease and the putative risk factor. Nevertheless, the potential for bias always deserves considerable thought when designing a study, especially for an observational study (Kleinbaum, Kupper and Morgenstern, 1982; Kelsey, Thompson and Evans, 1986; Prentice, 1995; Rothman and Greenland, 1998). The strongest evidence of the effect of an exposure on a disease event or death is provided by a study in which the level of exposure is assigned at random, as in a randomised controlled clinical trial. However, for factors that may be harmful, this would not be feasible in a human population due to ethical concerns.

In a typical cohort study, subjects are recruited for a period of time and then followed until a specified date, when the status of the subject is recorded and the results analysed. The figure (see page 83) presents a diagram showing a chronological representation for four hypothetical subjects. The date of enrolment is represented by a circle (•) and the date at which the disease is diagnosed or the subject dies is represented by a diamond (♦). A complete history from enrolment to the outcome is available for the first two subjects, but for the last two, only incomplete information is available because they are not observed until the outcome. In the third subject, follow-up continues until the study ends, when they are withdrawn alive. However, the fourth subject is lost to follow-up during the period represented by the dotted line and the fact that the outcome had actually occurred before the end of the study was not known to the investigators. The last two subjects are said to be *right censored* because only partial information on time to the outcome is available, i.e. it is known that outcome occurred sometime to the right of the last date of observation.

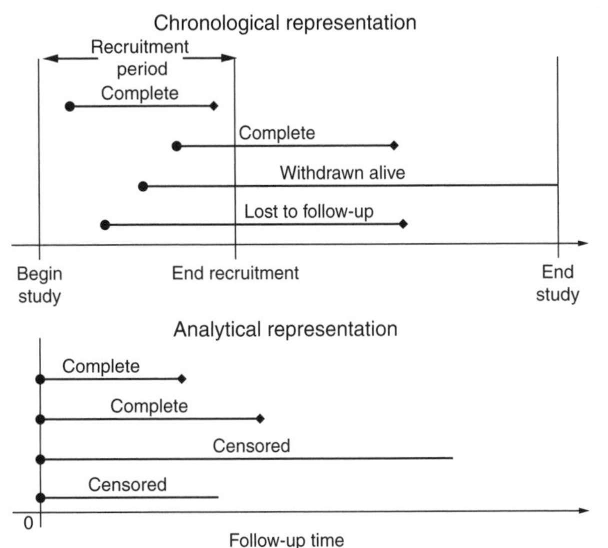

cohort studies *Chronological and analytical representations of a cohort study (• = start time, ♦ = time of outcome)*

For analysis, the time of follow-up and an indicator of whether the outcome was observed are used, as shown in the analytical representation in the figure. The hazard function, which is also called the incidence rate for a disease outcome or a mortality rate for a death outcome, is the basic quantity of interest. A proportional hazards or Cox model is commonly employed:

$$\lambda(\mathbf{X}; t) = \lambda_0(t)\exp\{\mathbf{X}\chi\}$$

in which **X** is a row vector of covariates, χ a column vector of corresponding parameters to be estimated and λ_0 (t) the underlying hazard that may depend on time (Prentice, 1995; Holford, 2002). Among the elements in the vector of covariates are indicators of the exposure level for each subject.

The number of disease cases or deaths usually exerts the greatest impact on the statistical power of a cohort study. Therefore, a rare disease will typically require a huge and expensive effort to accomplish. This may be due to the need to enrol a very large population that will be followed in time or it may be due the need for a long period of follow-up, especially if there is a long incubation period between the time when the exposure of interest occurs and the disease process begins. For example, in studies of cardiovascular disease, e.g. the Framingham Study and the MRFIT Study, a sample size of 5000 to 20 000 was used to obtain results of interest. However, for studies of diseases such as cancer, e.g. the Nurses' Health Study and the Iowa Women's Study, the outcome is less common, so sample sizes in the range of 50 000 to 100 000 or even larger may be used.

At one time, cohort studies were identified as prospective studies, but the current usage of retrospective and prospective refers to the temporal identification of the study population in relation to the study itself. Thus, a prospective cohort study would start by recruiting a study population that would subsequently be followed in time. However, in some circumstances, a more efficient design strategy would be retrospectively to identify a population in which records are available that will allow an investigator to reconstruct the cohort experience that would have been observed had the study population been enrolled in the study for the entire time period. For example, in a study of factors affecting occupational safety, company records might allow an investigator to go back in time and thus reconstruct the disease history of cohorts exposed to different factors of interest, i.e. a retrospective cohort. *TRH*

Holford, T. R. 2002: *Multivariate methods in epidemiology*. New York: Oxford University Press. **Kelsey, J. L., Thompson, W. D. and Evans, A. S.** 1986: *Methods in observational epidemiology*. New York: Oxford University Press. **Kleinbaum, D. G., Kupper, L. L. and Morgenstern, H.** 1982: *Epidemiologic research: principles and quantitative methods*. Belmont: Lifetime Learning Publications. **Prentice, R. L.** 1995: Design issues in cohort studies. *Statistical Methods in Medical Research* 4, 273–92. **Rothman, K. J. and Greenland, S.** 1998: *Modern epidemiology*. Philadelphia: Lippincott-Raven.

coefficient of determination See CORRELATION

coincidences These are surprising concurrencies of events, perceived by some as meaningfully related, with no causal connection. Carl Jung was fascinated by coincidences and even introduced the term *synchronicity* for what he saw as an *acausal* connecting principle needed to explain the phenomenon, arguing that such events occur far more frequently than chance allows. However, Jung gets very little support from Fisher who commented thus on coincidences:

'The one chance in a million will undoubtedly occur, with no more and no less than its appropriate frequency, however surprised we may be that it should occur to us.' Most statisticians would agree with Fisher and put down coincidences to the 'law' of truly large numbers: With a large enough sample, any outrageous thing is likely to happen. Some examples are given in Everitt (1999).

Those interpreting medical research studies should be aware of this when faced with an extremely small P-value if such did not arrive from a pre-planned analysis, perhaps arriving from a data-dredging exercise (or 'fishing expedition'). *BSE*

[See also PITFALLS IN MEDICAL RESEARCH, POST HOC ANALYSES]

Everitt, B. S. (2008) *Chance rules*, 2nd edition. Springer, New York.

collective ethics See ETHICS AND CLINICAL TRIALS

common errors in statistical analyses in medicine Statistical errors can occur at any stage in a study including the planning, design and execution stages. However, the most common errors typically occur during the analysis, presentation and interpretation of the collected data (see the first figure).

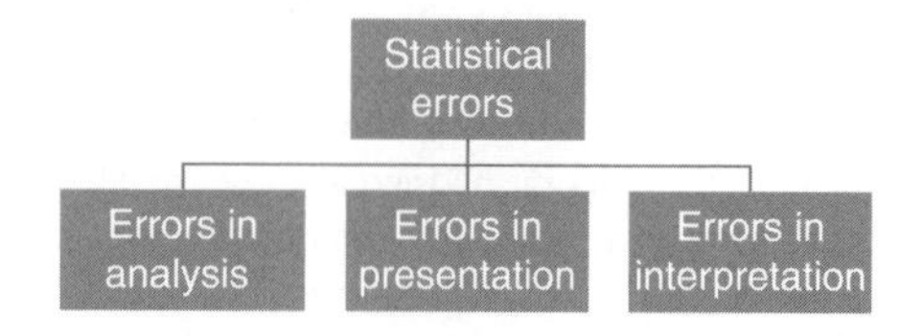

common errors in statistical analyses in medicine *Diagram of the most common errors occurring in a statistical analysis*

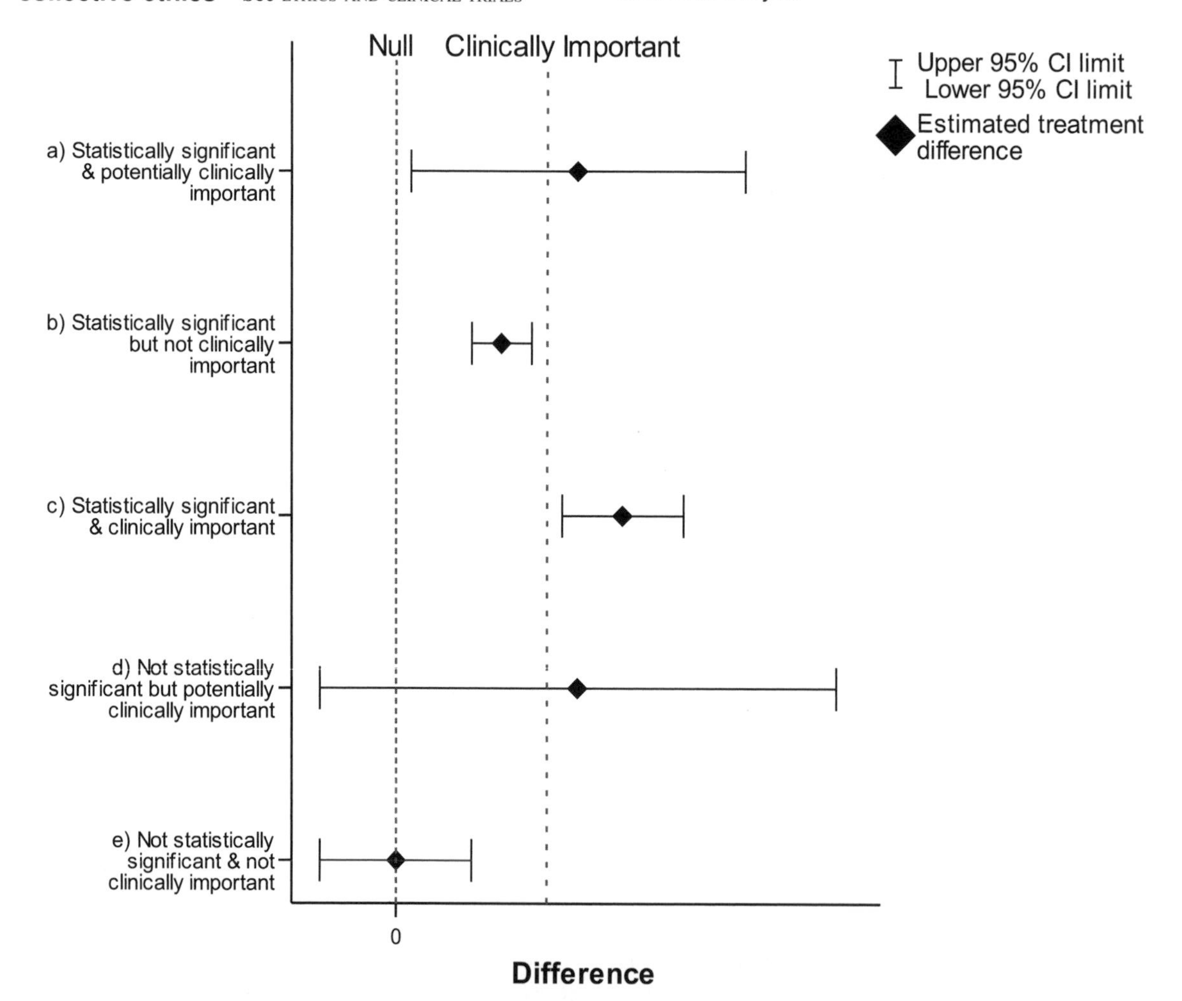

common errors in statistical analyses in medicine *Use of confidence intervals to help distinguish statistical significance from clinical importance (Walters, 2009)*

Box 1 *Basic errors in statistical analyses in medicine (from Altman, 1991)*

1. Using methods of analysis when the assumptions are not met.
2. Analysing paired data ignoring the pairing.
3. Failing to take account of ordered categories.
4. Treating multiple observations on one subject as independent.
5. Using multiple paired comparisons instead of analysis that considers all groups.
6. Performing within group analyses and then comparing groups by comparing P-VALUES or CONFIDENCE INTERVALS.
7. Quoting confidence intervals that include impossible values.

Box 2 *More advanced errors in statistical analyses in medicine (from Altman, 1991)*

1. Using correlation in method comparison studies.
2. Using correlation to compare to sets of time-related observations.
3. Assessing the comparability of two or more groups by means of hypothesis tests.
4. Evaluating a diagnostic test solely by means of SENSITIVITY and SPECIFICITY.

Altman (1991) describes seven 'basic' errors and four more 'advanced' errors in the statistical analysis of datasets (see Boxes 1 and 2). With the presentation of the results of a statistical analysis there are again several common errors (see Box 3).

The common misleading errors in graphical presentation are (Altman, 1991):

1. Lack of true zero on the vertical axis.
2. Change of scale in the middle of an axis.

Box 3 *Errors in the presentation of results of statistical analyses in medicine (from Altman, 1991)*
Box 3

1. Using standard errors (or confidence intervals) for descriptive information.
2. Presenting means (or medians) of continuous data without any indication of variability.
3. Presenting the results of a statistical analysis solely as a P-value.
4. Presenting results with spurious numerical precision.
5. Misleading graphical presentation (see GRAPHICAL DECEPTION).

Box 4 *Errors in the interpretation of the results of statistical analyses in medicine*

1. Association/relationship ≠ causality.
2. Nonsignificant ≠ no effect.
3. Statistically significant important.
4. Extrapolation from sample to population of interest when sample is not representative.

3. Three-dimensional effects.
4. Failure to show coincident points in a scatter diagram.
5. Showing a fitted regression line without a scatter of the raw data.
6. Superimposing two (or more) graphs with different vertical scales (especially when they do not start at zero).
7. Plotting means without any indication of variability.

Further discussion of these and many other issues can be found in Freeman, Walters and Campbell (2008).

There are several common errors in the interpretation of statistical analyses (see Box 4). However, the majority of errors in the interpretation of statistical analyses relate to understanding the meaning of hypothesis tests and P-values. The P-value from a hypothesis test is frequently interpreted as the probability that the observed effect is due to chance. This is incorrect. The P-value *is not* the probability that the observed effect is due to chance. It is the probability of obtaining the observed effect (or a more unlikely one) when the NULL HYPOTHESIS is true. The P-value assesses how likely it is to observe such an effect in a sample when there is no such difference in the population. Another false interpretation is the belief that $P=0.001$ implies a stronger effect that $P=0.01$. This may be so, but the P-values alone do not demonstrate this; we need a confidence interval for the observed effect.

Statistical significance is often used as the sole basis of the interpretation. A common mistake is that any significant effect, however small or implausible, is taken as real, and any nonsignificant effect is taken as indicating no difference. A statistically significant result does not imply that a result is important. Real, nonrandom effects may be very small and unimportant. Conversely, a nonsignificant result does not imply there is no effect. Big effects may not be significant if the sample size is low or the variability of the data is high. Again the estimation of a confidence interval for the effect may reduce the difficulties and help interpretation (see the second figure) (Walters, 2009).

Another frequent error in the interpretation of the results of statistical analysis is to equate a relationship or association and causation. An observed association between variables does not necessarily imply a causal relationship (see CAUSALITY). Most research is based on the principle of extrapolating findings from a sample to the population of interest (Campbell, Machin and Walters, 2007). In order to do this the sample must be representative of the population. This extrapolation can be compromised by studies with high dropout or refusal rates. There is also another error of extrapolation by estimating unknown data values outside the known limits of the data.

Further discussion of these common errors in the statistical analysis of medical data can be found in Altman (1991), Campbell, Machin and Walters (2007), Freeman, Walters and Campbell (2008) and Good and Hardin (2009). *SJW*

Altman, D. G. 1991: *Practical statistics for medical research*. London, Chapman & Hall. **Campbell, M. J., Machin, D. and Walters,S. J.** 2007: *Medical statistics: a text book for the health sciences*, 4th edition. Chichester: John Wiley & Sons, Ltd. **Freeman, J. V., Walters, S. J. and Campbell, M. J.** 2008: *How to display data*. Oxford: BMJ Books, Blackwell. **Good, P. I. and Hardin, J. W.** 2009: *Common errors in statistics (and how to avoid them)*, 3rd edition. Chichester: John Wiley & Sons, Ltd. **Walters, S. J.** 2009: Consultants' forum: should post hoc sample size calculations be done? *Pharmaceutical Statistics* 8, 2, 163–9.

competing risks This term is used particularly in SURVIVAL ANALYSIS to indicate that the event of intent (e.g. death) may occur from more than one cause. Survival analysis is concerned with the time (T) to the occurrence of some event of interest, such as remission, death due to a specific disease, discontinuation of use of a contraceptive device, etc. For some individuals T may not be observable due to the occurrence of some competing event. For example, if death from prostate cancer is of interest, then death from cardiovascular disease is a competing risk, as is death from old age. There may be multiple competing failure types from which each subject is at risk. We assume that there is only one failure time per study subject. For more than one time observed for each subject multistate models as discussed by Kalbfleisch and Prentice (2002) are recommended. The survival function of any specific failure type of interest is typically estimated by the product-limit estimator (KAPLAN–MEIER ESTIMATOR (KM)), treating the observed times of the other failure types as CENSORED OBSERVATIONS. The complement of the Kaplan–Meier estimator (1-KM) is often used to estimate the probability of failure due to a specified failure cause even in the presence of competing risks. The KM approach is only reasonable under the assumption that all competing risks are independent.

Independence can be considered to mean that the probability of the occurrence of some competing event is independent of the occurrence of any of the other competing events. In most situations this assumption is not valid. For example, patients with local relapse of breast cancer may have a higher probability of distance recurrence. In fact it has been argued that in the presence of competing risks a cause-specific survival function has no biological meaning since 'elimination' of some of the competing risks must influence the others. For example, Hougaard (2000) points out that treatment of stroke owing to thromboses by dissolving blood clots would increase the probability of haemorrhage. In order to avoid unrealistic assumptions on the relationships between the various competing risks the KM method should not be used but rather the cumulative incidence function needs to be estimated. The cumulative incidence function for a specific cause of interest (often called the subdistribution function) is the probability of the event of interest occurring before time t from an individual subject to all of the competing causes. To illustrate, suppose that 10 patients, subject to two competing risks (A and B) die at the times shown in the table.

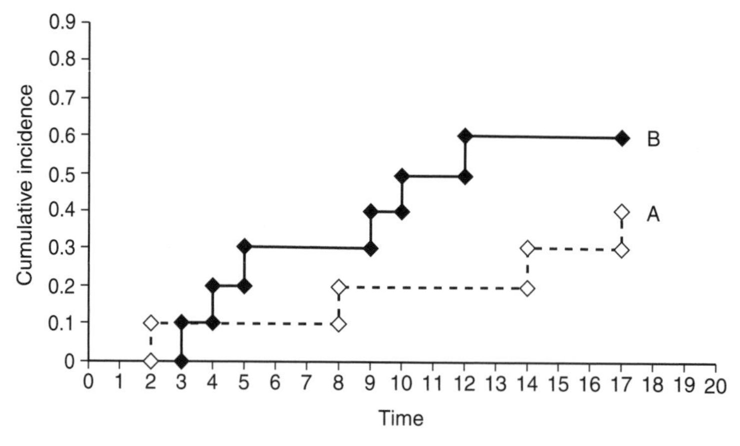

competing risks *Cumulative incidences of death for causes A and B for data in the competing risks table*

competing risks *Survival experience of 10 patients subject to two competing risks of death (A and B)*

Patients	*Time to death*	*Cause of death*
1	2	A
2	17	A
3	3	B
4	14	A
5	5	B
6	9	B
7	4	B
8	8	A
9	10	B
10	12	B

Corresponding cumulative incidences for causes A and B are presented in the first figure (see page 86). In order to illustrate the cumulative incidence computations consider time 5. The probability of death from cause A at time 5 or before is estimated by the number of cause A deaths occurring by time 5 divided by the total number of subjects in the study or 1/10; for cause B the corresponding probability is 3/10. Note that the sum of these two probabilities 4/10 is the overall probability of death at time 5 or before and is equivalent to 1-KM computed on all the deaths ignoring cause.

However, when computed for a specific cause 1-KM, the cumulative incidence estimate can differ substantially, as illustrated for cause A in the second figure. This shows that 1-KM tends to overestimate the probability of interest. The presence of censoring and/or explanatory variables complicates the computations (Kalbfleisch and Prentice, 2002). Statistical inference, both for a single cumulative incidence function and for the comparison of several cumulative incidences can be quite complex.

Crowder (2001) discusses in detail parametric modelling for competing risks – an approach not often taken for biomedical research. Green, Benedetti and Crowley (2003) focus on applications in oncology, although their methods are more generally applicable. In many situations it may not be clear which of the possible competing causes resulted in death. Flehinger, Reiser and Yashchin (2001) review the analysis of such *masked* data. *DF/BR*

Crowder, M. 2001: *Classical competing risks*. New York: Chapman & Hall. **Flehinger, B. J., Reiser, B. and Yashchin, E.** 2001: Statistical analysis for masked data. In Balakrishnan, N. and Rao, C. R. (eds), *Handbook of statistics*, Vol. 20: *Advances in reliability*. London: Chapman & Hall, pp. 499–522. **Green, S., Benedetti, J. and Crowley, J.** 2003: *Clinical trials in oncology*, 2nd edition. New York: Chapman & Hall. **Hougaard, P.** 2000: *Analysis of multivariate survival data*. New York: Springer Verlag. **Kalbfleisch, J. D. and Prentice, R. L.** 2002: *The statistical analysis of failure time data*, 2nd edition. New York: John Wiley & Sons, Inc.

complementary log-log model This is another commonly used model, besides the LOGISTIC REGRESSION model and the PROBIT MODEL, for investigating the relationship between a binary (or binomial) response, Y say, and explanatory or predictor variables. It is used in a variety of settings, e.g. in the analysis of data from toxicology studies (dose–response data) where interest lies in determining the effect on subjects' survival (e.g. mice mortality) of the exposure to different doses of a 'toxic' chemical compound. It is also used in serological studies in which serological tests are performed to detect the presence (i.e. the seropositivity) or absence of antibodies produced in response to an infectious disease such as malaria, so as to be able to

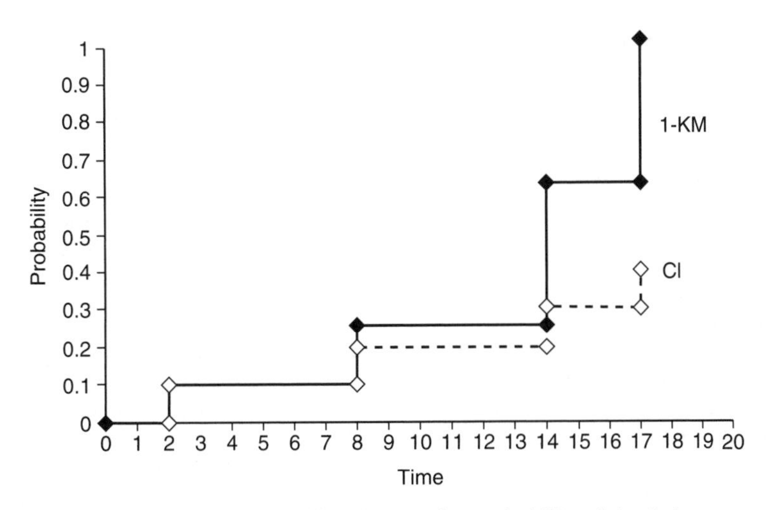

competing risks *Cumulative incidence and 1-KM estimates for probability of death by cause A*

calculate infection rates. Further examples are in dilution studies, where an estimate of the number of infective organisms present in a solution is required but can only be obtained through applying different dilutions of the solution to a number of plates that contains a growth medium and recording whether any growth has occurred after a fixed incubation period; in ageing studies where interest lies in self-reported mobility disability; and in the analysis of grouped (or interval) survival data (see SURVIVAL ANALYSIS), where the presence or occurrence of an event is known only to within a specific time interval.

Mathematically, the model relates the probability, P say, of a 'positive' response ($Y=1$) to a linear combination of the explanatory variables through the complementary log-log link function (see GENERALISED LINEAR MODELS); i.e.:

$$\log(-\log(1-P)) = \beta_0 + \beta_1 x_1 + \cdots + \beta_k x_k$$

where $x_1, \ldots, x_k$ are the k explanatory variables and the β values are the corresponding regression coefficients.

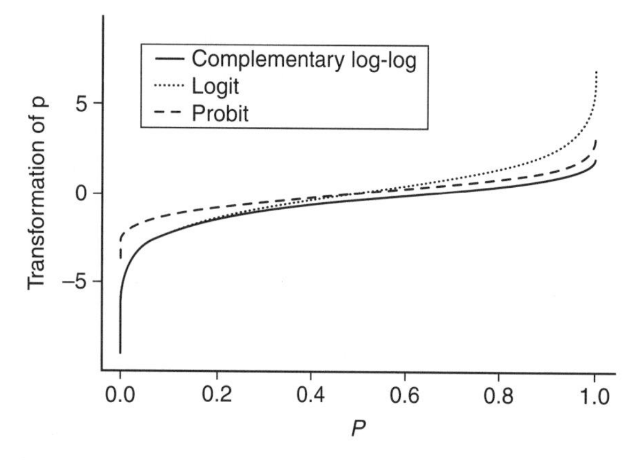

complementary log-log model *Three transformations of probability*

The figure presents a plot of the complementary log-log transformation of the probability P against the probability P itself. Also included on the graph are plots of the logit and probit transformations of P against P. Each of these three transformations converts a probability in the unit interval (0, 1) to any value whatsoever, thus eliminating the need to impose any restrictions on the regression coefficients. However, unlike the logit and probit transformations that are both (180° rotationally) symmetric about $P=1/2$, the complementary log-log transformation is asymmetric (see the figure). Thus this transformation is found to be more suitable where it is appropriate to deal with the probability of a positive response in an asymmetric manner, i.e. when the probability increases from 0 fairly slowly but approaches 1 quite suddenly. Observe also that the complementary log-log transformation does not differ appreciably from the logit transformation when P is small, say less than 0.2.

The justification for using this type of model in the analysis of data from many studies comes from assuming that each subject has an underlying, continuous latent or unobservable tolerance or threshold variable, Y^*, which is assumed to come from the Gumbel or extreme value distribution (see Davison, 1998). If a subject's tolerance variable, Y^*, exceeds a certain threshold θ (i.e. $Y^* > \theta$), then a positive response, $Y=1$, is observed. For example, in a toxicological study investigating the effect of different doses of an experimental drug on mice, a mouse may die if the exposure dosage exceeds the underlying tolerance the mouse has for the drug. In studies concerning mobility disability in the elderly, the underlying latent response variable for the self-reporting inability to walk a quarter of a mile may be the subject's true mobility level. Hence each individual's response to the question: 'Are you able to walk a quarter of a mile?' will depend on his or her cut-off point, which is the threshold level on this latent scale at which he or she will move from $Y=0$ to $Y=1$. Thus coefficients in the regression model above may be interpreted as the effects of the covariates on the latent variable, Y^*.

The complementary log-log model can also be derived from noting the relationship between the probability of a positive response in a time interval of length, T say (or an analogous measure, say volume), and the response rate, μ say, for this time interval, under the Poisson assumption (see POISSON DISTRIBUTION). For example, this relationship is utilised in the development of models for dilution and serological studies, where the probabilities of growth occurring on a plate at a particular dilution and of a person living in a particular disease endemic area being infected with this disease in one year, respectively, are of interest.

This model also follows naturally from the application of the proportional hazards assumption to grouped survival data. The regression coefficients in this case are interpreted as log hazard ratios (see SURVIVAL ANALYSIS) or log relative risks. However, if P is small ($P < 0.2$), the regression coefficients can also be interpreted as log odds. For further details see Collett (2002). *BT*

Collett, D. 2002: *Modelling binary data*, 2nd edition. London: Chapman & Hall/CRC. **Davison, A. C.** 1998: Extreme values. In Armitage, P. and Colton, T. (eds), *Encyclopedia of biostatistics*. Chichester: John Wiley & Sons, Ltd.

complete case analysis This is an analysis that uses only individuals who have a complete set of the intended measurements included in a study. An individual with a missing value on one or more variables will not be included in the analysis. When there are many individuals with missing values this approach can considerably reduce the effective sample size. In most circumstances complete case analysis is not to be recommended since other approaches such as multiple imputation can be used in order to retain as full a dataset as possible, thereby improving efficiency and reducing bias. *BSE*

[See also DROPOUTS]

complex interventions These are interventions that contain several interacting components and/or multiple outcomes. Most CLINICAL TRIALS focus on a single intervention with a primary outcome measure (see ENDPOINTS); the standard trial compares a group of patients who receive a new medication with a group who do not, and the allocation to groups is done at random. Increasingly, however, we wish to test more complex interventions. Multifaceted interventions or those that have multiple outcomes are hard to develop and their testing presents a range of challenges. Disciplines such a public health and preventive medicine need to assess complex packages of measures that are tailored to individuals or groups, rather than a single medication given in standard doses to recipients. Interventions targeted at behaviour change can involve various components and be particularly complex. Sometimes the intervention continues to develop after the initial assessment, and ongoing evaluation is needed, though this can be hard to accomplish.

There is no specific definition of what makes an intervention complex; even standard 'simple' interventions may have elements of complexity. CLUSTER RANDOMIZED TRIALS have been developed to deal with the complexity that arises when interventions are applied to groups such as hospital wards, surgeons conducting operations and community-based groups. However, the complexity can go much further. Multifaceted interventions to alter diet or physical activity, for example, can be extremely complex.

The British Medical Research Council has developed guidance for the development and evaluation of complex interventions (Craig *et al.*, 2008). Some of the key points that have emerged over the years relating to these interventions include the need to develop the intervention carefully, drawing on all the evidence to date using systematic reviews, understanding the theory underlying the intervention, modelling the processes and outcomes, and assessing feasibility and piloting the methods, before embarking on a full evaluation of the intervention. A clear description of the intervention is vital both for those involved to understand it fully, but also in reporting the findings so that the intervention can be introduced elsewhere, or tested further. Various approaches to developing and evaluating interventions have been developed, such as a multiple optimisation strategy (MOST) (Collins *et al.*, 2005) and the Reach, Effectiveness, Adoption, Implementation and Maintenance (RE-AIM) framework (Glasgow, Vogt and Boles, 1999). The National Institute for Health and Clinical Excellence (2007) has also produced guidance for planning, delivering and evaluating public health activities aimed at behaviour change.

The choice of study design depends on the intervention being assessed. Randomisation should be employed wherever possible and the gold standard is the randomised controlled trial, perhaps with variations on that methodology, such as preference trials, and randomised consent, stepped wedge and N-OF-1 TRIALS. However, sometimes randomised assessment is not possible. An example of this is the introduction of the Sure Start Local Programmes in England, which was a government initiative to provide care and support to families and children in the most deprived areas in the country. The intervention was undoubtedly complex in that services offered in each Sure Start area varied according to locally identified priorities, and no control groups were monitored contemporaneously. The evaluation team used a quasi-experimental approach to compare the data from Sure Start with similar data from children in the Millennium Cohort Study who lived in similarly deprived parts of the country but who did not yet have access to Sure Start programmes (Melhuish *et al.*, 2008). This falls short of the ideal of a randomised assessment but was a serious attempt to evaluate a large, widespread and costly intervention.

Public health assessments often need to maximise the information available from 'natural experiments' (Petticrew *et al.*, 2005). Thus the introduction of a new supermarket, traffic calming measures, the building of a major road may have health effects that may or may not be the intended consequence of the intervention. Randomised assessments are rarely possible and although less than ideal a before-and-after comparison, preferably contrasted with one or more comparable areas, may be the best that can be achieved. Particularly difficult is the assessment of interventions affecting entire populations introduced by national or local government. Examples are the introduction of water fluoridation (see http://www.southamptonhealth.nhs.uk/publichealth/fluoridation) or smoking bans in various countries (Haw *et al.*, 2006; Pell *et al.*, 2008). A prerequisite for such assessments is collection of data before the intervention commences and this can be politically difficult, as it may necessitate delaying the implementation of the new measures.

Complex interventions require imaginative approaches but they also need systematic evaluation and an understanding of the true nature of the intervention under study. All components of the intervention and its consequences, intended and unintended, have to be assessed and researchers need to understand the theoretical and practical background to the intervention that they are investigating. Campbell *et al.* (2000) offer the sensible advice that a mixture of RCTs and other research designs are needed fully to assess complex interventions. *HI*

Campbell, M. *et al.* 2000: Framework for the design and evaluation for complex interventions to improve health, *British Medical Journal* 321, 694–6. **Collins, L. M., Murphy, S. A., Nair, V. N. and Strecher, V. J.** 2005: A strategy for optimizing and evaluating behavioral interventions. *Annals of Behavioural Medicine* 30(1), 65–73. **Craig, P., Dieppe, P., Macintyre, S., Michie, S., Nazareth, I. and Petticrew, M.** 2008: Developing and evaluating complex interventions: the new Medical Research Council guidance. *British Medical Journal* 337, a1655. Full guidance available at: www.mrc.ac.uk/complexinterventionsguidance. **Glasgow, R. E., Vogt, T. M. and Boles, S. M.** 1999: Evaluating the public health impact of health promotion interventions: the RE-AIM framework. *American Journal of Public Health* 89, 1322–7. **Haw, S. J., Gruer, L., Amos, A., Currie, C., Fischbacher, C., Fong, G. T. *et al.*** 2006: Legislation on smoking in enclosed public places in Scotland: how will we evaluate the impact? *Journal of Public Health* 28, 24–30. **Melhuish, E., Belsky, J., Leyland, A. H., Barnes, J. and the National Evaluation of Sure Start** 2008: Research Team Effects of fully-established Sure Start Local Programmes on 3-year-old children and their families living in England: a quasi-experimental observational study. *Lancet* 372, 1641–7. **National Institute for Health and Clinical Excellence (NICE)** 2007: Behaviour change at population, community and individual levels. In *NICE Public Health Guidance*. London: NICE. **Pell, J. P., Haw, S., Cobbe, S., Newby, D. E., Pell, A. C. H., Fischbacher, C. *et al.*** 2008: Smoke-free legislation and hospitalizations for acute coronary syndrome *New England Journal of Medicine* 359, 482–91. **Petticrew, M., Cummins, S., Ferrell, C., Findlay, A., Higgins, C., Hoy, C. *et al.*** 2005: Natural experiments: an underused tool for public health? *Public Health* 119, 751–7.

complier average causal effect (CACE) See ADJUSTMENT FOR NONCOMPLIANCE IN RANDOMISED CONTROLLED TRIALS

component bar chart See BAR CHART

components of variance These are variance parameters that quantify the variation attributable to random effect terms included in a regression model. For example, a simple RANDOM EFFECTS MODEL for diastolic blood pressure measurements on patients recruited from a number of clinics includes random effects to represent the variability between clinics and random residual effects to represent the variability between patients. If no further random effect terms are added to this model, the model is said to include two components of variance. The VARIANCE of the random clinic effects in the model is the between-clinic variance component and the variance of the random residual effects is the between-patient variance component in this example. Under this model, the total variance of the individual patient measurements is assumed to be equal to the sum of the variance components.

Suppose in this example of blood pressure measurements on patients within clinics that the overall mean value is estimated as 80 mmHg, with the between-clinic variance component estimated as 7 and the between-patient variance component estimated as 135. The estimated between-clinic variance component allows construction of a 95% range for the mean blood pressure values at the different clinics, using the approach for calculating a reference interval. Here, values that are within approximately two (between-clinic) standard deviations of the overall mean are $80-1.96\sqrt{7} = 74.8$ mmHg and $80 + 1.96\sqrt{7} = 85.2$mmHg. It is therefore estimated that the majority of mean blood pressure values for different clinics lie between 74.8 mmHg and 85.2 mmHg.

Estimation of variance components is relevant in a number of application areas. In HEALTH SERVICES RESEARCH, variance components can be used to describe the variability between administrative or geographical units such as clinics, hospitals or towns and, separately, the variability between patients within units. In LONGITUDINAL DATA, variance components can be used to describe the variability between patients and, separately, the variability between measurements within patients.

When the data of interest are from a balanced design, there is a standard approach for estimation of variance components that is based on ANALYSIS OF VARIANCE. As an example, consider some data representing six repeated measurements of the peak expiratory flow rate (PEFR) for 10 patients with asthma. A simple random effects model for the PEFR measurements includes a between-patient variance component σ_b^2 and a within-patient variance component σ_w^2. Because the same number of observations is available for every patient, the dataset is balanced and the variance components can be estimated using an analysis of variance table for the data. The table presents the observed sums of squares and mean squares, as in a conventional analysis of variance. Under the random effects model assumed here, the expected values for the mean squares can be expressed in terms of the variance components σ_w^2 and σ_b^2. By equating the observed mean squares with their expected values, estimates for σ_w^2 and σ_b^2 are obtained as 191.41 for σ_w^2 and (11903.83 – 191.41)/6 = 1952.07 for σ_b^2.

components of variance *Observed sums of squares and mean squares*

Source of variation	*Degrees of freedom*	*Sums of squares*	*Mean squares*	*Expected mean squares*
Between patients	9	107134.51	11903.83	$\sigma_w^2 + 6\sigma_b^2$
Within patients	50	9570.70	191.41	σ_w^2
Total	**59**	**116705.21**		

Many study designs produce unbalanced data; e.g. health services research studies that include a number of hospitals or clinics commonly recruit varying numbers of patients from these and longitudinal studies need not collect equal numbers of measurements from all subjects. Several methods are available for estimation of variance components in unbalanced datasets. Extensions of the analysis of variance approach to the unbalanced case have been proposed, but these are not now commonly used. Estimation of variance components using the method of MAXIMUM LIKELIHOOD ESTIMATION can be achieved within many statistical software packages. However, maximum likelihood estimates of variance components are biased downwards in general. The preferred method for estimation of variance components in unbalanced data is RESTRICTED MAXIMUM LIKELIHOOD ESTIMATION (REML), which is also available within many software packages. In balanced datasets, REML estimation gives the same results as the analysis of variance approach as just described, whereas maximum likelihood estimation does not.

By definition, a component of variance is nonnegative, since it corresponds to the variance of a set of random effects. However, the methods for estimation of variance components can produce negative values. Usually, this occurs when the true value of the variance component is small and nonnegative. One approach to proceeding is to set the negative estimate to zero. Estimation and reporting should be handled with care for data in which a negative variance estimate has been obtained (Brown and Prescott, 1999). For further accounts of variance components, see Goldstein (1995), Searle (1971) and Snijders and Bosker (1999). *RT*

Brown, H. and Prescott, R. 1999: *Applied mixed models in medicine.* Chichester: John Wiley & Sons, Ltd. **Goldstein, H.** 1995: *Multilevel statistical models.* London: Arnold. **Searle, S. R.** 1971: *Linear models.* New York: John Wiley & Sons, Inc. **Snijders, T. and Bosker, R.** 1999: *Multilevel analysis.* London: Sage.

composite endpoint

See ENDPOINTS

compound symmetry

This term is used to describe the structure of a covariance matrix that has all its diagonal elements equal to the same value (say σ_1^2) and all its off-diagonal elements equal to another value (say σ_{12}), i.e. a covariance matrix with the form:

$$\Sigma = \begin{bmatrix} \sigma_1^2 & \sigma_{12} & \dots & \sigma_{12} \\ \sigma_{12} & \sigma_1^2 & \dots & \sigma_{12} \\ \vdots & \vdots & & \vdots \\ \sigma_{12} & \sigma_{12} & \dots & \sigma_1^2 \end{bmatrix}$$

Such a structure is assumed by some approaches to the analysis of longitudinal date, e.g. the random intercept model, although it is generally unrealistic since, in practice, variances often increase with time and covariances frequently increase with the time interval between two measurements. An account of testing for compound symmetry is given in Votaw (1948). *BSE*

[See also LINEAR MIXED EFFECTS MODELS]

Votaw, D. F. 1948: Testing compound symmetry in a normal multivariate distribution. *Annals of Mathematical Statistics* 19, 447–73.

conditional independence graphs

See GRAPHICAL MODELS

conditional logistic regression

This is a form of logistic regression that can be applied to matched datasets, particularly data from matched CASE-CONTROL STUDIES (see MATCHED PAIRS ANALYSIS). For such data the usual logistic regression model cannot be used since the number of parameters increases at the same rate as the sample size with the consequence that MAXIMUM LIKELIHOOD ESTIMATION is no longer viable. The problem is overcome by regarding particular parameters as a 'nuisance' that do not need to be estimated (see NUISANCE PARAMETERS). A conditional likelihood function can then be created that will yield maximum likelihood estimators of the parameters of most interest, i.e. the regression coefficients of the EXPLANATORY VARIABLES involved. The mathematics of the procedure are described, for example, in Collett (2003). The conditional logistic regression models can be applied using standard logistic regression software as follows: first, set the sample size to the number of matched pairs; next, use as explanatory variables the differences between the values for each case and control; then, set the value of the response variable to one for all observations; and, finally, exclude the constant term from the model. *BSE*

Collett, D. 2003: *Modelling binary data*, 2nd edition. Boca Raton: Chapman & Hall/CRC Press.

conditional probability

A conditional probability is the probability of an event given that another event has occurred. For example, consider two events a and b. The probability of both a and b occurring, denoted $P(a \wedge b)$, using the *multiplication rule* (see PROBABILITY) can be expressed as:

$$P(a \wedge b) = P(a|b) \times P(b) \quad (1)$$

Rearranging equation (1) yields the conditional probability of a given b as:

$$P(a|b) = \frac{P(a \wedge b)}{p(b)} \quad (2)$$

If a and b are independent, then $P(a \wedge b) = P(a)\,P(b)$ and hence from equation (2) $P(a|b) = P(a)$. Frequently, we wish to reverse the conditioning; i.e. rather than $P(a|b)$ we want $P(b|a)$ and this can be achieved using BAYES' THEOREM.

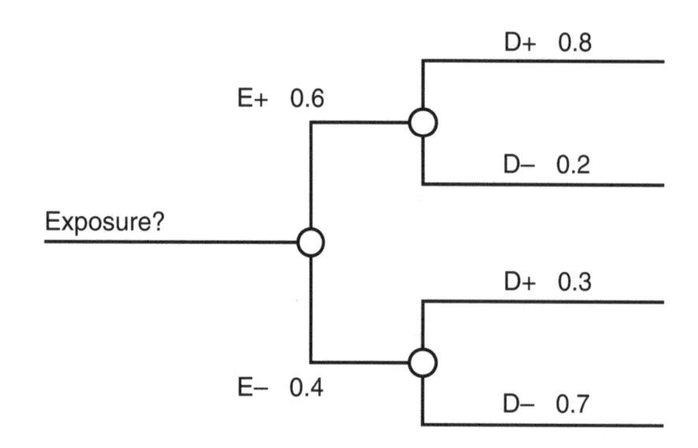

conditional probability *Decision tree*

Conditional probabilities are frequently used in epidemiology (Clayton and Hills, 1993). The figure shows a typical situation is which individuals can develop a disease or not denoted D+ and D− respectively, having been exposed or not, denoted E+ and E− respectively. The conditional probability that individuals develop the disease given that they were exposed, i.e. $P(D+|E+)$, is 0.8. *KRA*

Clayton, D. G. and Hills, M. 1993: *Statistical models in epidemiology*. Oxford: Oxford University Press.

confidence intervals This is a range of values calculated from a sample so that a given proportion of intervals thus calculated from such samples would contain the true population value. In research, we collect data on our research subjects so we can draw conclusions about some larger population. For example, in a randomised controlled trial comparing two obstetric regimes, the relative risk of Caesarean section for active management of labour compared to routine management was 0.97, with a confidence interval 0.60 to 1.56 (Sadler, Davison and McCowan, 2000). This trial was carried out in one obstetric unit in New Zealand, but we are not specifically interested in this unit or in these patients. We are interested in what they can tell us about what would happen if we treated future patients with active management of labour rather than routine management. We want to know not the relative risk for these *particular* women but the relative risk for *all* women.

The trial subjects form a sample that we use to draw some conclusions about the population of such patients in other clinical centres in New Zealand and other countries, now and in the future. The observed relative risk of Caesarean section, 0.97, provides an estimate of the relative risk we would expect to see in this wider population. It is called a *point estimate* because it is a single number. If we were to repeat the trial, we would not get exactly the same point estimate. Other similar trials cited by Sadler, Davison and McCowan (2000) have reported different relative risks: 0.75, 1.01 and 0.64. Each of these trials represents a different sample of patients and clinicians and there is bound to be some variation between samples. Hence we cannot conclude that the relative risk in the population will be the same as that found in our particular trial sample. The relative risk that we get in any

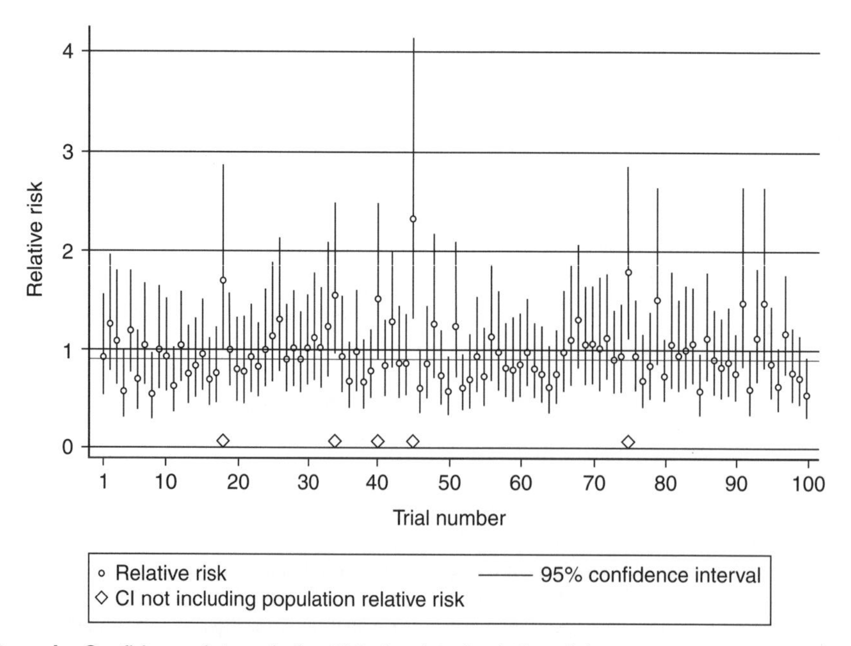

confidence intervals *Confidence intervals for 100 simulated relative risks*

particular sample would be compatible with a range of possible differences in the population.

We estimate this range of possibilities in the population with the confidence interval. A 95% confidence interval is defined in such a way that, if we were to repeat the trial many times and calculate a confidence interval for each, 95% of these intervals would include the relative risk for the population. Thus if we estimate that the population value is within the 95% confidence interval, we will be correct for 95% of samples.

This is a pretty difficult concept to get to grips with. The figure (see page 92) shows a computer simulation of relative risks and confidence intervals for 100 studies where the relative risk in the population is 0.90 and the sample size and Caesarean rate similar to those in the New Zealand study (Sadler, Davison and McCowan, 2000). Of these 100 confidence intervals, 5 include the population value (chosen to be 0.90).

Many researchers misunderstand confidence intervals and think that 95% of samples will produce point estimates within this confidence interval. This is simply not true. In the simulation, the first sample confidence interval is 0.46 to 1.15, and only 83% of sample relative risks are within these limits.

Such intervals are not unique and indeed many intervals with this property could be chosen. We usually choose the interval so that, of those intervals that do not include the population value, half will be wholly greater than that value and half wholly less. This often leads to intervals that are symmetrical about the point estimate, although in the case of RELATIVE RISKS AND ODDS RATIOS this symmetry usually occurs on the logarithmic rather than the natural scale.

In principle, a confidence interval can be found for any quantity estimated from a sample. There are several different methods for doing this, some simple and some not. First, we shall show how confidence intervals can be found for two of the simplest statistics, MEAN and proportion for continuous and categorical data respectively, and then see what they show about confidence intervals in general.

In the St George's Birthweight Study (Brooke *et al.*, 1989) data on birth weight and gestational age on 1749 pregnancies were obtained. For the 1603 births at 37 weeks' gestation or more the mean birth weight was 3384 g and the STANDARD DEVIATION was 449 g. This is a large sample and the sample mean will be an observation from a NORMAL DISTRIBUTION whose mean is the unknown mean birth weight in the population and whose standard deviation is well estimated by the standard error $449/\sqrt{1603} = 11.2$. For a normal distribution, 95% of observations are less than 1.96 standard deviations from the mean, so 95% of sample means will be less than 1.96 standard errors from the population mean. The 95% confidence interval has as a lower limit the sample mean minus 1.96 standard errors and as an upper limit the sample mean plus 1.96 standard errors, $3384 - 1.96 \times 11.2$ to $3384 + 1.96 \times 11.2 = 3362$ to 3406 g.

Similar methods can be used for many large sample estimates. We need the estimate to be from an approximately normal distribution and the standard error to be well estimated.

We can estimate a confidence interval for a proportion p using the standard error formula for a BINOMIAL DISTRIBUTION $\sqrt{p(1-p)/n}$. For example, in the St George's Birthweight Study 146 of 1749 births occurred at less than 37 completed weeks' gestation. The proportion is thus $146/1749 = 0.08348$ or 8.3%. The standard error is estimated by $\sqrt{0.08348(1-0.08348)/1749} = 0.006614$. The 95% confidence interval is thus $0.08348 - 1.96 \times 0.006614 = 0.07052$ to $0.08348 + 1.96 \times 0.006614 = 0.09644$. Rounding this, we get 0.071 to 0.096, which is from 7.1% to 9.6%.

For small samples things get much more complicated. We cannot assume that the estimate follows a normal distribution or that the standard error is a good estimate of the standard deviation of whatever distribution it does follow. For means, we can use a method based on the standard error if we assume that the data themselves follow a normal distribution. If we make this assumption then for a sample of n observations the difference between the sample mean and the unknown population mean divided by the standard error follows a t-DISTRIBUTION with $n - 1$ DEGREES OF FREEDOM. Rather than 95% of samples having means within 1.96 standard errors of the population mean, they have means within $t_{0.05}$ standard errors of the population mean, where $t_{0.05}$ is the two-sided 5% point of the t-distribution with degrees of freedom. In the birth weight study there were 11 babies born at 34 weeks' gestation. Their mean birth weight was 2477 g with a standard deviation of 531 g, giving the standard error $531/\sqrt{11} = 160.1$ g. There were $11 - 1 = 10$ degrees of freedom and the 5% point of the t-distribution is 2.228. The 95% confidence interval for the mean birth weight of babies born at 34 weeks was therefore $2477 - 2.228 \times 160.1$ to $2477 + 2.228 \times 160.1$, namely from 2120 to 2834 g.

For a proportion estimated from a small sample or small number of events, things do not work in the same way. The standard error estimate can go disastrously wrong. In a study of isolated intracardiac echogenic foci in foetuses, we found one trisomy-21 abnormality among 177 subjects (Prefumo *et al.*, 2001). The proportion was thus $1/177 = 0.00565$, or 5.65 per thousand. The usual 95% confidence interval using the normal approximation to the binomial distribution gives -5.4 to 16.6 per thousand, clearly impossible. The large sample assumption has broken down. Researchers will actually quote such impossible intervals and journals have been known to publish them! Sometimes, realising that the negative limit is impossible researchers will replace it by zero, but this, too, though better, is still wrong. The lower limit of the confidence interval cannot actually be zero in this example. Since we have found a case in the sample, it is not possible that there are no cases in the population.

There are a number of different methods to improve this interval (Newcombe, 1998). One of these uses a procedure based on the exact individual probabilities of the binomial distribution. The binomial distribution has two parameters, the number of independent observations n we make (e.g. number of patients) and the probability P that any given observation will be a 'yes'. This probability is what we are trying to estimate. We find the lower confidence limit as the value of P so that the probability of obtaining the observed number of 'yes's or more will be 0.025 and the upper limit as the value of P so that the probability of the observed number of 'yes's or fewer will be 0.025. These probabilities are obtained by summing the exact binomial probabilities for all the possible numbers of 'yes's equal to and beyond that observed. The calculations for such methods are extremely tedious, but not to a computer. For the echogenic foci data the 95% confidence interval by this method is 0.00014 to 0.03107, or 0.014 to 31 per thousand. This is an example of an exact method calculation, because it uses the exact probabilities of the distribution (see EXACT METHODS FOR CATEGORICAL DATA). There are several other computer-intensive methods that can be used, such as the BOOTSTRAP and those based on rank tests.

The confidence interval allows for what is called sampling variation. This means that it reflects the difference between estimate and population value likely in random samples from that population. However, it does not take into account other sources of variation, termed nonsampling variation. The sample that we have is from geographical space, in that it contains one hospital, as in the active management trial (Sadler, Davison and McCowan, 2000). Even the largest clinical trial will contain at most only a few hospitals and their patients. The hospitals are not chosen randomly, so the sample will differ from the population in an unknown and inestimable way. It is also a sample in time, in that we want the sample of patients seen in the past to tell us about patients whom we will see in the future. The sample may not be as good at estimating quantities in this wider population as the confidence interval suggests.

The interval quoted in the active management trial was a 95 % confidence interval and 95% of such intervals would contain the relative risk for the population. We could also calculate intervals for other percentages, e.g. a 99% interval, calculated so that 99% of possible intervals would contain the population estimate. For the Caesarean section relative risk the corresponding 99% confidence interval would be 0.52 to 1.81, wider than the 95% interval of 0.60 to 1.56 reported. In compensation, more of these intervals would contain the population value.

We could calculate a much narrower interval. A 50% confidence interval is calculated as estimate minus or plus 0.67 standard errors, compared to estimate minus or plus 1.96 standard errors for a 95% confidence interval. The 50% interval based on a large sample normal approximation is only 34% of the width of the 95% interval. This is not very useful as an estimate, as only 50% of such intervals contain the population value they are estimating. However, it shows that if we calculate 95% confidence intervals, we can say that for about 50% of samples the middle third of the 95% confidence interval will contain the population parameter. Thus, 95% is chosen as a standard confidence level as a reasonable compromise between width (or precision) and coverage probability (accuracy).

Significance tests and confidence intervals are closely related. Many null hypotheses are about the value of something we can also estimate, such as the difference in mean between two groups. It will usually be the case that if the NULL HYPOTHESIS value (difference or regression coefficient = 0, odds ratio or relative risk = 1.0) is contained within a 95% confidence interval then the P-value will be greater than 0.05. For example, in the Birthweight Study, we might want to test the null hypothesis that mean birth weight in the population is 3400 g. To test this, we subtract 3400 from the observed mean and divide by the standard error, 11.2. This ratio, $(3384-3400)/11.2=-1.43$, would be an observation from the standard normal distribution if the null hypothesis were true, giving $P=0.15$. Here the 95% confidence interval (3362 to 3406 g) includes the null hypothesis value for the mean, 3400 g, and $P>0.05$. Contrariwise, we might want to test the null hypothesis that the population mean birth weight was 3500 g. Now the test statistic is $(3384-3500)/11.2=-10.36$, giving $P<0.0001$. The null hypothesis value is not included in the confidence interval and the difference is significant. Thus the 95% confidence interval can be used to do a significance test at the 5% level.

For means and their differences there is an exact relationship between the usual confidence interval and the usual significance test, because the standard error is not related to the quantities being compared (means) and thus is not affected by the null hypothesis. It may not work for proportions, relative risks, odds ratios, etc. For example, let us test the null hypothesis that in the population the proportion of births at less than 37 weeks' gestation is 8%. Under the null hypothesis, the proportion is 0.08 and the standard error is $\sqrt{0.08(1-0.08)/1749}=0.006487$, not the same as the 0.06614 used for the confidence interval. The test statistic is $(0.08348-0.08)/0.006487=0.54$, $P=0.59$. The null hypothesis value of the proportion is within the confidence interval 0.071 to 0.096 and the difference is not significant. Now let us consider a null hypothesis value just outside the confidence interval 0.97. The standard error, if the null hypothesis were true, would be $\sqrt{0.097 \times (1-0.097)1749} = 0.007077$. The test statistic is $(0.08348 - 0.097)/0.007077 = -1.91$, $P = 0.056$, but not significant. This effect of the null hypothesis on the standard error is why we sometimes see odds ratios, relative risks and standardised mortality ratios where the 95% confidence interval includes 1.0, but the ratio is reported as significant.

Researchers are now encouraged to present results as confidence intervals instead of, or in addition to, *P*-values (Gardner and Altman, 1986). This approach is more informative than the practice of giving a *P*-value or stating 'significant' or 'not significant', as it provides an estimate of the size of the possible difference or ratio between the groups in the population. This is particularly useful when differences are not statistically significant, as it enables the reader to judge whether a potentially important difference could have been missed. *P*-values and confidence intervals both have their role and if possible both should be given. Most major medical journals now include in their recommendations to authors that the main results of studies be presented using confidence intervals (or their equivalent) and that authors should avoid relying solely on hypothesis testing.

Finally, some comments on the Bayesian perspective, there being two differing statistical philosophies, the Bayesian and the frequentist. At present few Bayesian analyses appear in the medical literature, although we may expect to see more of them in future (see BAYESIAN METHODS).

People often talk about a 95% confidence interval as including the unknown population value with probability 0.95, saying, for instance, there is a 95% chance that the true value lies within the computed 95% confidence interval. Now, it is true that if we set out to collect a new sample, the probability that its confidence interval will include the population value is 0.95. However, once the sample has been collected and the interval calculated, it either includes the population value or it does not, we just do not know which.

In strict frequentist terms, we cannot talk about the probability of the population parameter having any given value or range of values. It has a constant, albeit unknown, value with no probability distribution. A Bayesian is willing to think of the population value as a variable with a distribution, which represents the uncertainty in our estimate of it. Bayesians quote something called a CREDIBLE INTERVAL, which is a range of possible values that has a given probability of including the unknown population value. This probability is often set at 95 %. Thus a 95% credible interval is a set of values that is estimated to include the population value with probability 95 %, whereas a 95% confidence interval is a set of values chosen so that 95% of such sets would include the population value. For the proportion of births before 37 weeks, a Bayesian credible interval, assuming no prior knowledge, is 7.1% to 9.7%, virtually the same as the confidence interval (7.1% to 9.6%). The difference is academic, which is perhaps why academics have spent so much time arguing about it. *JMB*

Brooke, O. G., Anderson, H. R., Bland, J. M., Peacock, J. L. and Stewart, C. M. 1989: Effects on birth weight of smoking, alcohol, caffeine, socioeconomic factors, and psychosocial stress. *British Medical Journal* 298, 795–801. **Gardner, M. J. and Altman, D. G.** 1986: Confidence-intervals rather than p-values – estimation rather than hypothesis-testing. *British Medical Journal* 292, 746–50. **Newcombe, R. G.** 1998: Two-sided confidence intervals for the single proportion: comparison of seven methods. *Statistics in Medicine* 17, 857–72. **Prefumo, F., Presti, F., Mavrides, E., Sanusi, A. F., Bland, J. M., Campbell, S. and Carvalho, J. S.** 2001: Isolated echogenic foci in the fetal heart: do they increase the risk of trisomy 21 in a population previously screened by nuchal translucency? *Ultrasound in Obstetrics and Medicine* 18, 126–30. **Sadler, L. C., Davison, T. and McCowan, L. M.** 2000: A randomised controlled trial and meta-analysis of active management of labour. *British Journal of Obstetrics and Gynaecology* 107, 909–15.

confidence level See CONFIDENCE INTERVALS

confirmatory factor analysis This is a procedure for testing a hypothesised factor structure for a set of observed variables. The hypothesised structure will specify both the number of factors and which observed variables are related to which factors (Dunn, Everitt and Pickles, 1993). This contrasts with FACTOR ANALYSIS when used in its exploratory mode when the number of factors has to be determined in some ways from the data and no a priori constraints are placed on the factor structure. Confirmatory factor analysis is a theory-testing model as opposed to a theory-generating method like exploratory factor analysis. The first step in a confirmatory factor analysis involves the calculation of either a correlation or a COVARIANCE MATRIX for a set of observed variables. Then possibly a number of competing factor models are proposed, derived either from theory or previously performed exploratory factor analyses on other datasets. The models will differ in their specifications of 'free' and 'fixed' parameters. MAXIMUM LIKELIHOOD ESTIMATION is generally used to estimate the free parameters in a model. Confirmatory factor analysis models can be fitted using one of a number of available software packages (LISREL, EQS, MPLUS) and a variety of methods can be used to test the fit of a model and to compare the fit of two competing models.

As an example of where this approach might be applied, consider a psychiatrist who measures a number of variables on a sample of mentally ill patients. The psychiatrist believes that some of the observed variables are related to a patient's depression and others to anxiety, and he or she is particularly interested in estimating the correlation between these two, essentially, LATENT VARIABLES. To make things specific suppose there are six observed variables with the first three indicating depression and the remaining three, anxiety. The

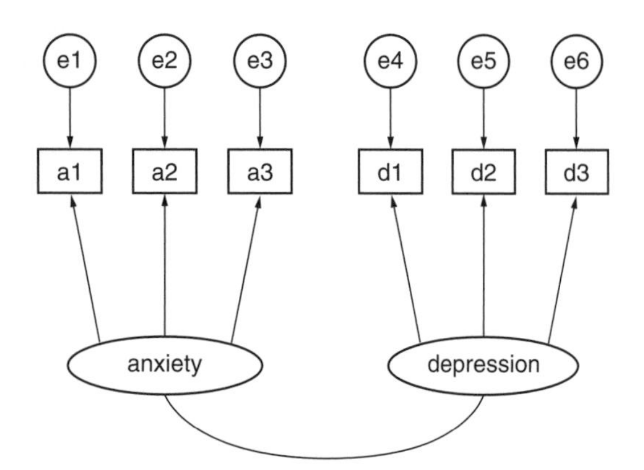

confirmatory factor analysis *Path diagram for depression and anxiety example*

correlated, two-factor model to be fitted is described graphically by the path diagram shown in the figure. Apart from the error variances, the parameters to be estimated are the loadings of the first three variables on factor one (depression) – variables four, five and six are constrained to have zero loadings on this variable – and the loadings of the last three variables on factor two (anxiety) – now the first three variables are constrained to have zero loadings. The estimated correlation between the latent variables, depression and anxiety will be a disattenuated correlation, i.e. one in which the effects of measurement errors in the observed variables have been effectively removed.

Detailed examples of the application of confirmatory factor analysis are given in Huba, Wingard and Bentler (1981) and Dunn, Everitt and Pickles (1993). *BSE*

Dunn, G., Everitt, B. S. and Pickles, A. 1993: *Modelling covariances and latent variables using EQS*. Boca Raton: CRC Press/Chapman & Hall. **Huba, G. J., Wingard, J. A. and Bentler, P. M.** 1981: A comparison of two latent variable causal models for adolescent drug use. *Journal of Personality and Serial Psychology* 40, 186–93.

confounding See BIAS IN OBSERVATIONAL STUDIES, CASE-CONTROL STUDIES

Consolidated Standards for Reporting Trials (CONSORT) statement This research tool was designed to improve the quality of reports of clinical trials (Begg *et al.*, 1996; Moher *et al.*, 2001). The core contribution of the CONSORT statement consists of a flow diagram (see the figure) and a checklist. The flow diagram enables reviewers and readers to grasp quickly how many eligible participants were randomly assigned to each arm of the trial and whether any imbalances are apparent regarding numbers of patients withdrawing from or failing to comply with their assigned treatment (see DROPOUTS). Large discrepancies or imbalances suggest the need for conducting not only INTENTION-TO-TREAT (ITT) analyses but also PER PROTOCOL analyses to seek corroboration. Such information is frequently difficult or impossible to ascertain from trial reports as they were reported in the past. The checklist identifies 21 items that should be incorporated in

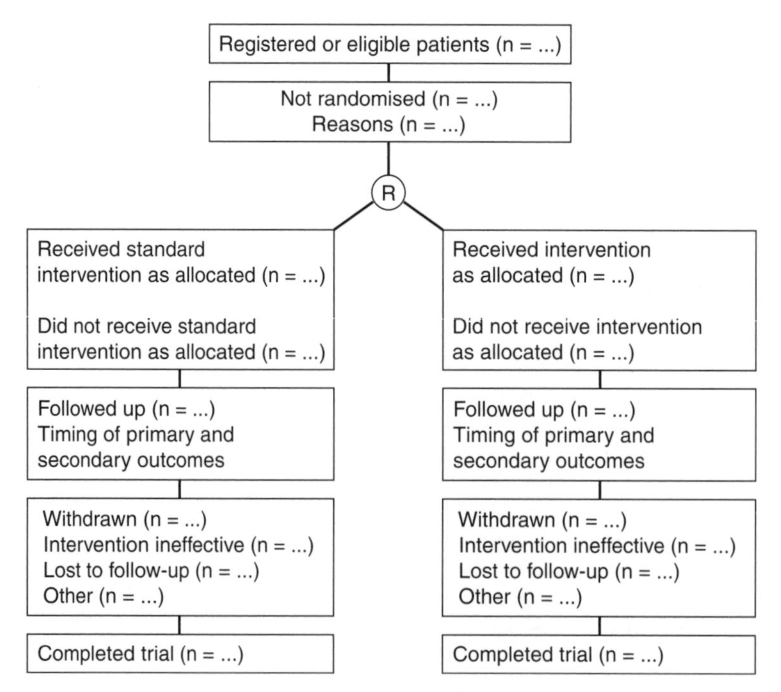

Consolidated Standards for Reporting Trials statement *Flow diagram of CONSORT statement*

the title, abstract, introduction, methods, results or conclusion of every randomized clinical trial. More details can be found at www.consort-statement.org. *BSE*

[See also CRITICAL APPRAISAL, STATISTICAL REFEREEING]

Begg, C., Cho, M., Eastwood, S., Morton, R., Moher, D., Ohlein, I. *et al*. 1996: Improving the quality of reporting of randomized clinical trials; the CONSORT statement. *Journal of the American Medical Association* 276, 637–9. **Moher, D., Shultz, K. F. and Altman, D. G.** 2001: The CONSORT statement: revised recommendations for improving the quality of reports of parallel-group randomized trials. *Annals of Internal Medicine* 134, 7–622.

consulting a statistician *'To consult the statistician after an experiment is finished is often merely to ask him to conduct a post-mortem examination. He can perhaps say what the experiment died of.'* So said R. A. Fisher, later Sir Ronald, widely considered the founding father of modern statistics, and of RANDOMISATION in particular, as long ago as 1938. His tongue-in-cheek message remains sage advice just as true today as a reminder of the single most important aspect of seeking statistical advice – to seek it early. Many novice researchers make the mistake of believing the statistician to be the numbers person only to be approached, and then with trepidation, once data have been collected. In actuality, a consultation with a statistician should be a positive experience and opportunity to assist planning all aspects of study design, meaning neither just the subsequent analysis nor the narrow matter of SAMPLE SIZE DETERMINATION.

Naturally, there are important differences in how statistical consulting takes place according to whether the setting is within a university, a hospital, a pharmaceutical company, a government agency and so on, due to the obvious differences between public and private sector employers, not to mention geographical differences from one continent to another. Statistical consulting can also take place in a variety of ways: telephone, email or face-to-face, or a mixture thereof.

This entry will focus on the most productive manner, namely face-to-face, since this maximises effective two-way communication. It also concentrates on those aspects of research projects that are reasonably consistent regardless of the particular environment, although the author's perspective is based on experience within academic settings. The remainder of this entry examines the sort of project-specific advice a statistician can give, notably including general guidance on preparing for a first meeting with a statistician and some observations on the interaction between the statistician and clinical researcher. What cannot be included, necessarily, is local advice on where to find a nearby consulting statistician in the first place. In the event none is available, one should consider using textbooks or WEB RESOURCES IN MEDICAL STATISTICS, or even travelling to attend a short course offering an introduction to the subject. For further details concerning technical content, in addition to the process, of statistical consultations, the reader is referred to the rest of this volume or else to one of several books, such as Hand and Everitt (1998), Derr (2000) or Cabera and McDougall (2001).

Broadly, research can be subdivided into a number of distinct stages as depicted in the figure. The worst time to first approach a statistician is at the post-refereeing stage of a submitted journal article. Consultant statisticians may be able to offer some remedial help at this late stage, but only on matters of analysis, interpretation or presentation. The most common reasons why statistical referees recommend rejection of submitted manuscripts to biomedical journals pertain to design issues, which is hardly surprising when one realises that fundamental flaws in study design simply cannot be retrieved by sophisticated analyses (see STATISTICAL REFEREEING). Thus, if the paper has not been rejected outright on statistical grounds, there may be hope for the manuscript after suitable revision. A statistician approached at such a late stage is likely to drop more than a subtle hint that it would be altogether more satisfactory essentially to heed Fisher's advice and request that the researcher come along sooner in a project's life cycle the next time!

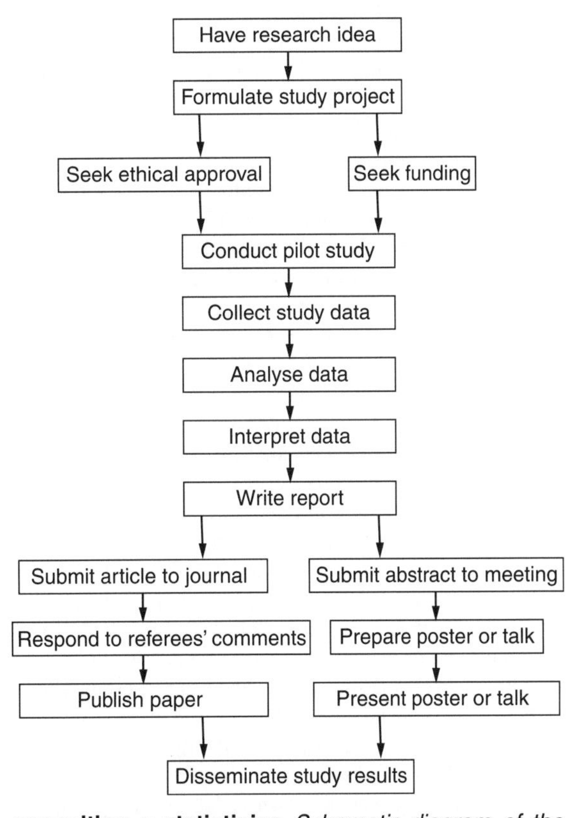

consulting a statistician *Schematic diagram of the research process, from initial thoughts through to dissemination of the study results. Statistical input should ideally be sought at the study formulation stage*

There is a temptation to think seeing a statistician is unnecessary if one has confidence in one's own statistical knowledge and ability (and access to relevant SOFTWARE). This can be a dangerous policy for the novice researcher, especially if the confidence turns out to be misplaced or handling the data more complex than envisaged. Even veterans of medical research with substantial statistical skills of their own can find consulting a statistician invaluable, despite the time and effort required in the midst of busy research agendas and clinical commitments. This is by no means just to delegate data-related tasks but to gain additional, independent input about the intended study from an altogether different perspective. Statisticians, after all, do not see the world of medicine and research in the same way as those on the front line dealing directly with patients, nor for that matter those working with test tubes in the laboratory.

What sort of help can a statistician offer? Clearly this depends on the nature of the project itself and the extent of the statistician's involvement. For example, if in an academic environment a student seeks advice on a research project forming a part of a degree, then involvement will be necessarily less than in a full collaboration. In the former case, the student needs to own the analyses and be able to defend them single-handedly, so that the role of the statistical consultant is to point in the right direction by recommending an appropriate choice of study design and method for data analysis. To help avoid becoming a surrogate supervisor by default it can be helpful to suggest that the student's project supervisor should also attend the consultation. It is important to clarify early on in the consultation process if it is expected to become a full collaboration, for then issues of payment, if indicated, and co-authorship (or acknowledgement for lesser statistical involvement) need to be discussed and agreed.

Payment for statistical advice remains a delicate matter and local rules would dictate. It is sensible, so as not to discourage those who most need statistical help, to have a policy whereby the first meeting (of say about an hour) is provided free of any direct charge to the consultee. Parker and Berman (1998) provide some helpful criteria for suggesting when authorship may or may not be appropriate. As a rule, if the finished piece of work could not have attained its statistical quality without the assistance of the consultant statistician, and more than just elementary descriptive or inferential statistics are involved, then the default ought to be co-authorship for the statistician. There is at least anecdotal evidence that statistical co-authorship enhances chances of publication in first-choice journals. Equally, there is a danger that statisticians' names can be used against their wishes to lend perhaps more credence than is due to some submitted papers or grant applications!

What should be brought to a first meeting? In order to make the most use of the time available it is best for the consultee to have made some specific preparations. A checklist can assist, perhaps in the format of a QUESTIONNAIRE to be completed in advance of the initial meeting. Useful questions to address both 'housekeeping' matters as well as more substantive issues concerning the project, include the following: *1. What is the single main aim of the project?* (A brief answer to this fundamental question at least ensures that the meeting can be focused.) *2. What stage is the project at right now?* (Options can be forming ideas/designing protocol/collecting data/analysis of data/writing up/referee's comments.) *3. What area(s) do you think you need help with?* (Some areas are formulating ideas/sample size calculation/designing protocol/making grant application/randomisation practicalities/carrying out the study/collecting data/managing data/analysis you are doing/checking your analysis/checking written report/responding to referee.) *4. What role would you like the statistician to play?* (These could be advisor/co-applicant/interpreter of results/co-author, although note that the statistician would reserve the right to decline the latter if authorship was felt to be inappropriate.) *5. Does this work form part of a dissertation or thesis?* (See comments above concerning student work.) *6. What is the source of patients or subjects and the criteria for selecting them?* (This allows an opportunity to discuss or review appropriate study design.) *7. How many subjects are required or available?* (If this is to be a topic for advice, there is a need to know clinically relevant differences in proportions involved and/or standard deviations for continuous outcomes.) *8. What is the main outcome measure?* (Again to focus attention on primary as opposed to secondary ENDPOINTS or, in the worst case, to ensure the project does pre-specify at least one endpoint.) *9. What is the main comparison or relationship of interest?* (To encourage being as specific as possible and to check for a suitable control group.) *10. What other quantities are being measured and when?* (For example, BASELINE MEASUREMENTS, covariates, secondary outcomes.) *11. What problems have been or are anticipated in data collection?* (To discuss, for example, accuracy, MISSING DATA, repeated measures, matching but essentially any potential BIAS.) *12. What are expected or hoped for results at the study's end?* (Again to focus on the real reason for performing the research.) *13. Are there any specific approaches to data analysis intended?* (For instance, the same method as in a previously published study, preferably with a hard copy to be handed over.) *14. Is there any further information you would like to give regarding the study?* (A suitable closing question to allow, one hopes, any pertinent facts to emerge.)

It is best if answers to the above catalogue of questions can be sent in advance of the meeting, along with a brief description of the project and copies of related documents to assist the statistician's understanding (e.g. protocol, grant application, ethics committee submission).

In terms of practicalities, the statistician may have some further expectations of the consultee to bring or transmit in advance of the first meeting at which data are to be analysed (recall, ideally, this is *not* at the first encounter!) Statisticians do not usually take on the more mundane data entry tasks, so would not be prepared to type in the numbers. They may express preferences for how data are presented electronically in terms of file type (e.g. Excel being a common choice) and media (floppy disks are fine, though somewhat old-fashioned; email attachments generally work better for small-to-moderate sized datasets, or USB pen drives more generally). In any event, it is always important to check for viruses to avoid spreading contaminated files. The layout of the data should ordinarily be as a spreadsheet, with well-labelled variable names, one column per variable and one row per subject. It is best to ask if there is a data entry preference when handling repeated measures data, but if in doubt the spreadsheet works well. In any case, data provided must be reasonably clean and free from data entry errors, although the odd OUTLIER is excusable.

Due to confidentiality issues (e.g. the UK's Data Protection Act 1998) there should not be any uncoded individual patient identifiers; i.e. names and addresses and other information that could be used to trace individuals must have been removed. Obviously, the anonymisation process must generate unique patient IDs in order to be fully reversible so that queries with data can be checked from original records that are stored elsewhere. While it is not a serious problem, it is better to code data numerically rather than alphanumerically. For example '1' and '2' for 'male' and 'female' respectively is better than use of 'M', 'm', 'male', 'Male', 'MALE', etc., especially as accidental leading or trailing blanks can add to potential confusion, possibly creating a needless missing data point on subsequent conversion to numeric format. In general, categorical variables should have a different number representing each group, with an accompanying description, or internal labelling, of how the categories are coded. Equally, missing data are better handled by inserting an obviously impossible value (e.g. '−99' when all other values are positive) rather than just leaving a spreadsheet cell blank. The statistician would rather be told about any such embedded code, however, to avoid unnecessary runs of software routines after noticing, for example, strange residuals in regression analyses.

An altogether less tangible item to bring along, but arguably the most important for a successful meeting, can be summarised as the *right attitude.* Statistical consultation involves a high degree of communication and mutual respect. Since areas of expertise are different, jargon is to be avoided – both by the statistician and the consultee. (Medics are not alone in having big words, or abbreviations and acronyms, to describe things that are obvious only to themselves!) Punctuality is important but it is understood that medical emergencies can and do occur that necessitate being late, in which case arranging for a telephone message is a simple courtesy or cutting short an ongoing meeting at a bleeper's notice. However, there is no such thing as a statistical emergency, so there is little excuse for the consultee who demands an immediate appointment with a statistician or expects results to be turned around within, say, 24 hours to meet his or her deadline for a grant, ethics or conference submission, particularly as such deadlines are typically known months in advance. Also bear in mind some consulting statisticians are new to their jobs. Just as some training of junior doctors occurs 'on the job', so too do junior statisticians have to learn, ideally under supervision from someone more experienced, by interacting with real clients in real consultations. The transition from a university degree course to a practising statistical consultant is never automatic. An attitude of patience is helpful in these circumstances, much as required by drivers stuck behind a learner struggling with hill starts (all were learner drivers once!).

To close, and in keeping with the spirit of Fisher's advice quoted earlier, it can be instructive to consider ways of having an *unhelpful* meeting between a medical researcher and statistician. So long as both parties can avoid making these mistakes, there is scope for real progress and genuine collaboration.

First, what are some of the ways a statistican upset medical colleagues? 1. Being too nit-picky, precise, detail-oriented and failing to see the big picture. 2. Being slow to respond to requests for appointments or to analyse data. 3. Being overly critical of genuine-but-flawed attempts to analyse data themselves. 4. Using unnecessary jargon. 5. Using unnecessarily complicated methods when simpler ones suffice. 6. Spending too much, or too little time, during the consultation. 7. Embarking on a mathematical lecture within a consultation. 8. Only expecting to meet on your home turf (despite owning a laptop). 9. Believing there is such a thing as an average patient. 10. Thinking EVIDENCE-BASED MEDICINE (EBM) means clinical experience counts for nothing compared to having a few well-honed CRITICAL APPRAISAL skills and a recently published META-ANALYSIS to hand.

Finally, how to upset your statistician? 1. Saying 'This will only take 5 minutes of your time', for it will not. 2. Arriving unnannounced, late or not at all (notwihstanding genuine emergencies). 3. Waiting until the grant or ethics application deadline is tomorrow and leaving no time for review of statistical input before sending the document off. 4. Dripfeeding data or hypotheses or telling half the story ('Oh, actually it's the same patient seen five times') or shifting between study aims. 5. Taking for granted – not considering acknowledgement or co-authorship or bothering to inform if that application or journal submission was ever successful or not. 6. Saying, in earshot, 'I just need the statto to crunch the numbers' and generally regarding the statistician as a technical service provider. 7. Expecting knowledge of specialist medical terminology. 8. Expecting poorly entered data to be cleaned or forgetting to run a virus check on your data. 9. Demanding 'What's the *P*-value?' or 'Can't you find one that is significant?' 10. Coming too late in research process and complaining about a statistical postmortem! *CRP*

Cabera, J. and McDougall, A. 2001: *Statistical consulting*. New York: Springer. **Derr, J.** 2000: *Statistical consulting: a guide to effective communication*. Pacific Grove, CA: Duxbury Press. **Hand, D. J. and Everitt, B. S.** (eds) 1987: *The statistical consultant in action*. Cambridge: Cambridge University Press. **Parker, R. A. and Berman, N. G.** 1998: Criteria for authorship for statisticians in medical papers. *Statistics in Medicine* 17, 2289–99.

contingency coefficient This is a measure of the strength of an association between two categorical variables. While the CHI-SQUARE TEST can detect an association between two variables, it is not a good measure of the strength of that association. This is because it is also dependent on the sample size and the number of categories into which the variables are classed. Typically, contingency coefficients are adjustments of the chi-square statistic, intended to remove the dependence on those factors. Because they are based on the chi-square statistic, any attempt to test the contingency coefficient for significance will merely resolve into repeating the chi-square test of independence.

The two most common contingency coefficients are Cramér's contingency coefficient (also known as Cramér's C, Cramér's V and occasionally Cramér's v) and Pearson's contingency coefficient (often just referred to as the contingency coefficient or as Pearson's coefficient of mean square contingency). For a table with r rows and c columns, with k being set as equal to the smaller of r and c, that produces a chi-square statistic of X^2 from n observations, the formulae for Cramér's and Pearson's coefficients are:

$$\text{Cramér's coefficient} = \sqrt{\frac{X^2}{n(k-1)}}$$

and

$$\text{Pearson's coefficient} = \sqrt{\frac{X^2}{X^2+n}}$$

While Cramér's coefficient can take values from zero to one, Pearson's coefficient can never reach one (the denominator is clearly always larger than the numerator). In fact, Pearson's coefficient has a known maximum of $\sqrt{(k-1)/k}$ so it is possible to rescale this coefficient to lie in the range 0 to 1.

While the use of these measures is popular in some fields, more so if we consider that the phi coefficient for a 2×2 table (see CORRELATION) is a special case of Cramér's coefficient, interpretation is not straightforward. Clearly, in some sense, the larger the coefficient is, the greater the association. However, the absolute value does not have any clear meaning and comparing correlation coefficients from two tables (especially tables of different dimensions) is not straightforward.

Contingency coefficients are widely used as a result of their convenience and in spite of their limitations. For 2×2 tables, odds ratios are possibly a better measure as it is easy to produce confidence intervals and they have a familiar interpretation. For larger tables with at least one ordered categorical variable a measure based on the Spearman rank correlation might be more appropriate. For further details see Goodman and Kruskal (1954), Fleiss (1981), Siegel and Castellan (1988) and Conover (1999). *AGL*

Conover, W. J. 1999: *Practical nonparametric statistics*. New York: John Wiley & Sons, Inc. **Fleiss, J. L.** 1981: *Statistical methods for rates and proportions*, 2nd edition. New York: John Wiley & Sons, Inc. **Goodman, L. A. and Kruskal, W. H.** 1954: Measures of association for cross-classifications. *Journal of the American Statistical Association* 49, 732–64. **Siegel, S. and Castellan Jr, N. J.** 1988: *Nonparametric statistics for the behavioural sciences*, 2nd edition. New York: McGraw-Hill.

contingency tables These are cross-tabulations that arise when a sample from some population is classified with respect to two or more qualitative variables. The first table shows a simple example involving two such variables each with three categories. A more complex contingency table that involves a classification with respect to three variables is shown in the second table.

contingency tables *Incidence of cerebral tumours*

		Type			
		A	B	C	Total
Site	I	23	9	6	38
	II	21	4	3	28
	III	34	24	17	75
Total		78	37	26	141

I, frontal lobe; II, temporal lobes; III, other cerebral areas. A, benign tumours; B, malignant tumours; C, other cerebral tumours.

contingency tables *Coronary heart disease*

	Blood pressure	Serum cholesterol 1	2	3	4
	1	2	3	3	4
CHD	2	3	2	1	3
(yes)	3	8	11	6	6
	4	7	12	11	11
	1	117	121	47	22
CHD	2	85	98	43	20
(no)	3	119	209	68	43
	4	67	99	46	33

Blood pressure: 1, <127 mmHg; 2, 127–146 mmHg; 3, 147–166 mmHg; 4, >167 mmHg.Serum cholesterol: 1, <200 mg/100 cm^2; 2, 200–219 mg/100 cm^2; 3, 220–259 mg/100 cm^2; 4, >260 mg/100 cm^2.

Contingency tables such as these two can be used to test various hypotheses about the variables from which they are formed. To begin with, we shall illustrate this using tables formed from just two variables (two-dimensional contingency tables), since these are the ones encountered most commonly in practice. The hypothesis of interest for two-dimensional tables is whether or not the two variables are independent.

This may be formulated more formally in terms of p_{ij}, the probability of an observation being in the *ij*th cell of the table, $p_{i\cdot}$, the probability of being in the *i*th row of the table, and $p_{\cdot j}$, the probability of being in the *j*th column of the table. The hypothesis of independence can now be written:

$$H_0 : p_{ij} = p_{i\cdot} \times p_{\cdot j}$$

Estimated values of $p_{i\cdot}$ and $p_{\cdot j}$ can be found from the relevant marginal totals ($n_{1\cdot}$, $n_{\cdot j}$) and overall sample size (n) as:

$$\hat{p}_{i\cdot} = \frac{n_{i\cdot}}{n}, \quad \hat{p}_{\cdot j} = \frac{n_{\cdot j}}{n}$$

These can then be combined to give the estimated probability of being in the *ij*th cell of the table under independence, $\hat{p}_{i\cdot} \times \hat{p}_{\cdot j}$. The frequencies to be expected under independence, E_{ij}, can then be obtained simply as:

$$E_{ij} = n \times \hat{p}_{i\cdot} \times \hat{p}_{\cdot j} = \frac{n_{i\cdot} \times n_{\cdot j}}{n}$$

The hypothesis of independence can now be assessed by comparing the observed (O_{ij}) and estimated (E_{ij}) expected frequencies using the familiar CHI-SQUARE TEST statistic:

$$X^2 = \sum_{i=1}^{r} \sum_{j=1}^{c} \frac{(O_{ij} - E_{ij})^2}{E_{ij}}$$

where r is the number of rows and c the number of columns in the table. If H_0 is true X^2 has, asymptotically, a CHI-SQUARE DISTRIBUTION with $(r - 1)(c - 1)$ degrees of freedom. For our first table, the estimated expected values under independence are given in the third table. Here $X^2 = 7.84$, which with 4 degrees of freedom gives an associated P-VALUE of 0.098. There is no evidence against the independence of site and type of tumour.

contingency tables Estimated expected values under the hypothesis of independence for data in the cerebral tumour table

		Type A	B	C	**Total**
	I	21.02	9.97	7.01	**38**
Site	II	15.49	7.35	5.16	**28**
	III	41.49	19.68	13.83	**75**
Total		**78**	**37**	**26**	**141**

Since the distribution of the test statistic is only a chi-square asymptotically there has been considerable work on trying to find when the sample size is sufficient for the P-values derived in this way to be valid and alternative procedures have been derived that do not rely on the asymptotic assumption (see EXACT METHODS FOR CATEGORICAL DATA).

When the contingency table is formed from more than two variables more than a single hypothesis may be of interest. We may, for example, wish to test the mutual independence of the variables forming the table or the conditional independence of two of the variables given a third and so on. For some hypotheses, estimated expected values can be found from particular marginal totals but for others an iterative scheme is needed (see Everitt, 1992, for details). In general the analysis of three-dimensional and higher contingency tables is best undertaken with the use of log-linear models. *BSE*

[See also CORRESPONDENCE ANALYSIS]

Everitt, B. S. 1992: *The analysis of contingency tables*, 2nd edition. Boca Raton: Chapman & Hall/CRC.

convenience sample A convenience sample is a nonrandom sample chosen due to its easy access. A convenience sample is unlikely to be representative of the population, the main disadvantage being that it is unclear how representative the sample is of the population of interest. One example is surveying people who walk by on the street. Another would be selecting patients who attend a clinic or doctors in a particular hospital. The main advantage is that the sample is simple to obtain and may save money.

A classic example is the use of medical students as study subjects when conducting medical research. If the study involves seeking opinions or, for that matter, measuring certain characteristics, such as height, one needs to bear in mind the fact that the sample is atypical of the population as a whole and, hence, limits conclusions to the population of all medical students, perhaps to ensure valid inference. For further details see Crawshaw and Chambers (1994). *SLV*

Crawshaw, J. and Chambers, J. A. 1994: *Concise course in A level statistics*, 3rd edition. Cheltenham: Stanley Thornes Publishers Ltd.

Cook's distance Cook's distance (Cook, 1977; Cook and Weisberg, 1999) is a measure of the influence of a case on the estimated parameters $\hat{\beta}$ of a linear regression. It measures the global impact of deleting the case on all the parameter estimates taken together and is the distance from $\hat{\beta}$ to $\hat{\beta}_{(i)}$, expressed in terms of confidence ellipsoids about $\hat{\beta}$, where $\hat{\beta}_{(i)}$ is the vector of parameters estimated with the ith case omitted.

For a dependent variable y_i, D_i is given by:

$$D_i = (y_i - \hat{y}_i)^2 \left[\frac{h_i}{(1-h_i)^2} \right]$$

where p is the number of independent variables, s^2 is the variance of the estimate and h_i is the leverage of the ith observation, given by the ith diagonal element of the so-called 'hat' matrix $\mathbf{H} = \mathbf{X}\,(\mathbf{X}^{\mathrm{T}}\mathbf{X})^{-1}\mathbf{X}^{\mathrm{T}}$, where $\mathbf{X}$ is the data matrix of independent values (see Cook and Weisberg, 1999).

For a point to be influential it must be both an outlier, i.e. have a high residual, and it must also have high leverage, i.e. be far from the centre of gravity of the points (see the figure). A general rule of thumb is to examine cases for which $D_i > 1$; alternatively, Hamilton (1992) has suggested examining cases for which $D_i > 4/n$. An informal approach is to sort the distances in order and examine the few cases with the highest distances. A large jump between these and the rest can suggest points worth investigating. Cases thus identified might be considered for removal or at least further investigation (subject to the caution that removal of OUTLIERS always requires). Analogous quantities for other models are available, e.g. for LOGISTIC REGRESSION (see Pregibon, 1981). If the interest is in one or more particular parameters in the regression, rather than the complete set taken as whole, 'dfbetas' can be computed: these estimate the changes in the individual parameters after deleting each case. *ML*

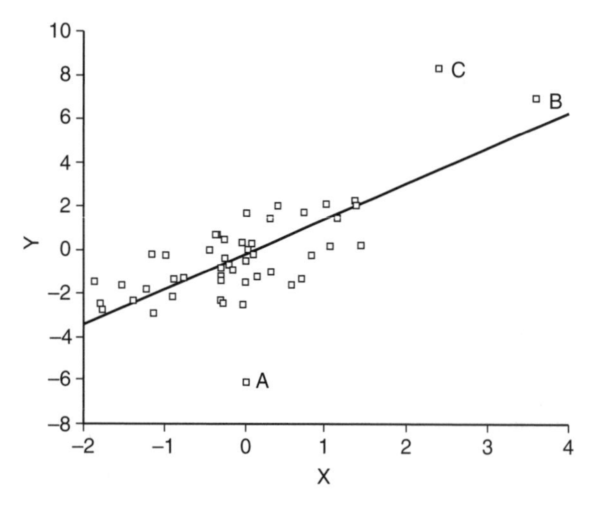

Cook's distance *Three points in a sample of 50, only one of which (C) has a high Cook's distance. Point A has a high residual but low leverage and B has a low residual but high leverage*

Cook, R. D. 1977: Detection of influential observations in linear regression. *Technometrics* 19, 15–18. **Cook, R. D. and Weisberg, S.** 1999: *Applied regression including computing and graphics*. New York: John Wiley & Sons, Inc. **Hamilton, L. C.** 1992: Regression with graphics. Belmont: Duxbury. **Pregibon, D.** 1981: Logistic regression diagnostics. *Annals of Statistics* 9, 705–24.

coplots See TRELLIS GRAPHS

COREC See ETHICAL REVIEW COMMITTEES

correlation Correlation is used to measure the strength of the linear relationship between two random variables. If we plot two variables on a SCATTERPLOT, their correlation is a measure of how closely the points lie to a straight line. We measure correlation by a correlation coefficient. The simplest of these is PEARSON'S CORRELATION COEFFICIENT, also known as the *product-moment correlation coefficient* or simply as the correlation coefficient. This is the ratio of the sum of products of differences from the MEAN divided by the square roots of the two sums of squares about the mean and is usually denoted by r:

$$r = \frac{\sum (y_i - \bar{y})(x_i - \bar{x})}{\sqrt{\sum (y_i - \bar{y})^2 \sum (x_i - \bar{x})^2}}$$

The confusing symbol 'r' (rather than 'c') is for historical reasons; it appears to have indicated 'regression' originally. It is now well established and if a medical paper uses '$r = \ldots$' without explanation, it usually means the correlation coefficient. When we want to distinguish between the correlation coefficient in a sample, r, and the correlation coefficient in the population from which the sample was drawn, we use 'ρ', the Greek letter 'rho', to denote the latter.

The figure (see page 103) shows some sample correlation coefficients. The coefficient is positive when large values of y are associated with large values of x, the variables being said to be positively correlated, as in (a), (b) and (c) in the figure.

The majority of observations will have either both observations greater than the mean or both less than the mean. In either case, observation minus mean will have the same sign, either positive or negative, for both variables and the product of these differences will be positive. Hence the sum of products will be positive and the correlation coefficient will be positive. The correlation coefficient is negative when small values of y are associated with large values of x, the variables being negatively correlated as in (g), (h) and (i) in the figure. The majority of observations will have one observation greater than the mean and the other less than the mean. Observation minus mean will have different signs for the two variables and the product of these differences will be negative. Hence the sum of products will be negative and the correlation coefficient will be negative. The correlation coefficient has a maximum value of +1 when the points all lie exactly on a straight line and the variables are positively correlated and a minimum of –1 when the points all lie exactly on a straight line and the variables are negatively correlated. When there is no linear relationship at all the coefficient is zero and the variables are said to be uncorrelated, as in (d) in the figure.

Correlation only measures the strength of the linear (i.e. straight line) relationship. Nonlinear relationships may be missed or underestimated by it. In the figure, (e) shows a strong relationship yet the correlation coefficient is zero and (f) shows an exact mathematical relationship, without any random variation, yet the correlation coefficient is less than one because the relationship is not a straight line.

We can test the NULL HYPOTHESIS that the population correlation is zero, i.e. that there is no linear relationship between the two variables, using a simple t-test. At least one of the two variables must follow a normal distribution and the observations must be independent. If we can assume this, we require only the value of r and the sample size n. Then, if the null hypothesis were true:

$$t = r\sqrt{\frac{n-2}{1-r^2}}$$

would follow a t-distribution with $n-2$ DEGREES OF FREEDOM. There are tables of this test in many books and almost all programs that calculate r also give the P-VALUE. As a result, correlation coefficients in medical papers are almost invariably followed by P-values, e.g. '$r=0.57$, $P<0.01$', whether the null hypothesis of zero correlation is plausible or not.

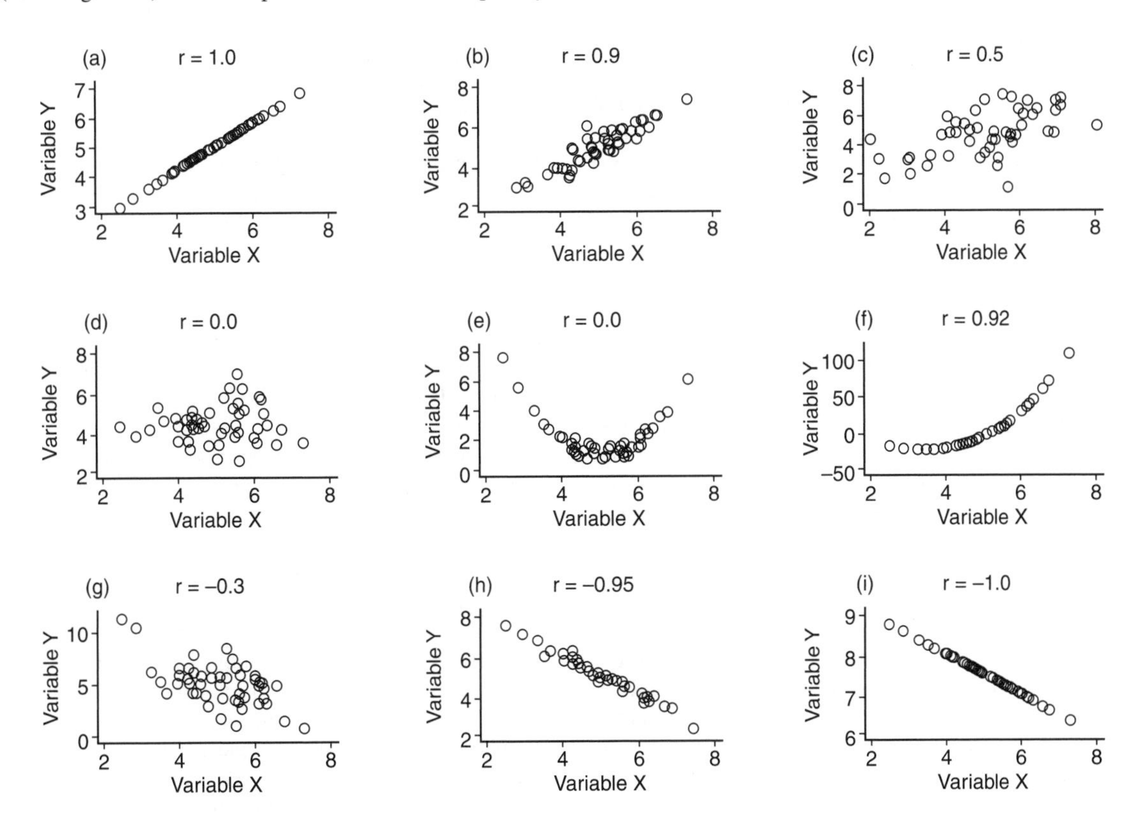

correlation *Nine correlation coefficients*

The distribution of the sample correlation coefficient is not a simple one, but it can be converted to the normal using a TRANSFORMATION known as Fisher's z-transformation:

$$z = \frac{1}{2}\log_e\left(\frac{1+r}{1-r}\right)$$

This works provided observations are independent and both variables follow normal distributions, a stronger assumption than that required for the test of significance. Provided it is met, the z-transformation can be used to find a confidence interval for r, the standard error on the transformed scale being $\sqrt{1/(n-3)}$. We can calculate the confidence interval for z and transform back. Curiously, many programs do not perform this. The standard error can also be used in a power calculation to estimate the sample size required to detect a relationship between two continuous variables. Bland (2000) and Machin *et al.* (1998) give formulae and tables.

We can still calculate correlation coefficients when normal assumptions are not met, but cannot use P-values or confidence intervals found by these methods. The assumption of independence is very important. It could be seriously misleading to take several observations from each subject and treat them as a simple sample for the calculation of correlation coefficients and their P-values and confidence intervals (Bland and Altman, 1994).

The correlation coefficient and regression equations between two variables are closely related. The proportion of variation explained by the regression is r^2, whether we have the regression of Y on X or of X on Y. There is only one correlation coefficient although there are two possible regression lines; correlation has no choice of dependent and independent (or outcome and predictor or explanatory) variables. The product of the two regression slopes will also be equal to r^2, which is sometimes called the *coefficient of determination.*

The tests of the null hypotheses of zero correlation and zero slope for the regression line give the same P-value. The two methods provide the same test for a linear relationship.

The product-moment correlation is only one of several correlation coefficients in use. There are two nonparametric rank correlation coefficients, SPEARMAN'S RHO and *Kendall's tau,* useful when the assumptions of normal distribution necessary for confidence intervals and significance tests are not tenable. The intraclass or INTRACLUSTER CORRELATION COEFFICIENT is used when, rather than two variables, we have two or more observations of the same variable on each subject. The *tetrachoric correlation coefficient,* seldom seen in practice in the modern literature, can be used when we have two underlying continuous variables but can only observe whether the subject is above or below some cut-off value for each, making both dichotomous. The *biserial correlation coefficient* can be used when one variable is continuous and the other dichotomous. These are not the same as the correlation coefficients found by simply making the dichotomous variable zero or one and calculating r, called the phi coefficient and point-biserial correlation coefficient respectively.

We can adjust the correlation between two variables for their mutual relationship with a third variable using a partial correlation coefficient. This is an estimate of the correlation between the two variables of interest for subjects who all have the same value of the third variable. Partial correlation is seldom seen now, multiple regression being preferred. There is also a partial rank correlation, using Kendall's approach. We can calculate a *multiple correlation coefficient,* usually denoted by R, which expresses the strength of the relationship between a chosen variable and several others. Time series (see TIME SERIES IN MEDICINE), where observations are measured successively over time, may show serial correlation or autocorrelation, where adjoining observations are correlated. A correlation matrix is the set of all the correlations between each pair of a set of variables and is the starting point for several multivariate techniques (see, for example, PRINCIPAL COMPONENT ANALYSIS). *JMB*

[See also SPEARMAN'S RHO (ρ)]

Bland, M. 2000: *An introduction to medical statistics,* 3rd edition. Oxford: Oxford University Press. **Bland, J. M. and Altman, D. G.** 1994: Correlation, regression and repeated data. *British Medical Journal* 308, 896. **Machin, D., Campbell, M. J., Fayers, P. and Pinol, A.** 1998: *Statistical tables for the design of clinical studies,* 2nd edition. Oxford: Blackwell.

correspondence analysis This is a technique for graphically displaying the associations among the categorical variables forming a CONTINGENCY TABLE in the form of a SCATTERPLOT. A correspondence analysis should ideally be seen as an extremely useful supplement to, rather than a replacement for, more formal inferential analysis such as a CHI-SQUARE TEST of the independence of the variables. Correspondence analysis provides a 'window' on to the data that may allow researchers easier access to the associated numerical results, facilitate discussion of the data and possibly generate interesting hypotheses about the data. The mathematics behind correspondence analysis (Greenacre, 1992; Everitt, 1997) leads to two sets of multidimensional coordinate values, one of which represents the categories of the row variable and the other the categories of the column variable. In general, the first two coordinate values representing each row and column category are used to provide a single scatterplot of the data. In the resulting diagram, the distance between a plotted row point and column point represents as accurately as possible how the corresponding cell of the contingency table departs from independence, as we shall illustrate using the data on age and boyfriends in the first table. The coordinates resulting from a correspondence analysis of these data are shown in the second table.

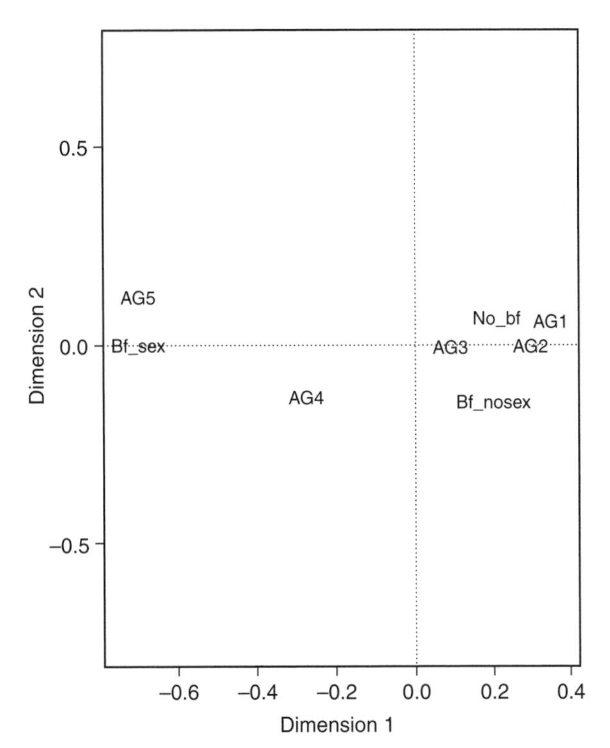

correspondence analysis *Two-dimensional correspondence analysis solution for age and boyfriends data*

correspondence analysis *Age and boyfriends contingency table*

	Age[a]				
	1	2	3	4	5
No boyfriend	21	21	14	13	8
Boyfriend/no sexual intercourse	8	9	6	8	2
Boyfriend/sexual intercourse	2	3	4	10	10
Totals	31	33	24	31	20

[a]Age groups: 1, <16 years; 2, 16–17 years; 3, 17–18 years; 4, 18–19 years; 5, 19–21 years. $\chi^2 = 20.6$, DF = 8, $p = 0.008$.

These coordinates can be plotted to give the scatterplot shown in the figure. Of most interest in correspondence analysis solutions is the joint interpretation of the points representing the row and column categories. It can be shown that row and column coordinates that are large and of the same sign correspond to a cell with considerably more observations than if independence held. Row and column coordinates that are large, but of opposite signs, imply a cell in the table with far fewer observations than required under the assumption of independence. Finally, small coordinate

correspondence analysis *Two-dimensional correspondence analysis coordinates for the row and column categories in the first table*

Category	x	y
No boyfriend	0.193	0.061
Boyfriend/no sexual intercourse	0.192	−0.143
Boyfriend/sexual intercourse	−0.732	0.000
Age group 1	0.355	0.055
Age group 2	0.290	0.000
Age group 3	0.103	0.000
Age group 4	−0.281	−0.134
Age group 5	−0.717	0.123

values close to the origin correspond to cells where the observed frequency is close to the expected frequency under independence. In the figure, for example, age group 5 and boyfriend/sexual intercourse both have large negative coordinate values on the first dimension; consequently the corresponding cell in the table contains more observations than would be the case under independence. Again, age group 5 and boyfriend/no sexual intercourse have coordinate values with opposite signs on both dimensions, implying that the corresponding cell in the table has fewer observations than expected under independence. *BSE*

Everitt, B. S. 1997: Annotation: correspondence analysis. *Journal of Child Psychology and Psychiatry* 38, 737–45. **Greenacre, M.** 1992: Correspondence analysis in medical research. *Statistical Methods in Medical Research* 1, 97–117.

cost–benefit analysis (CBA) See COST-EFFECTIVENESS ANALYSIS

cost-effectiveness analysis Cost-effectiveness analysis (CEA) is a tool for comparing costs and benefits in terms of patient outcomes (changes in health and welfare), so that the value for money of a proposed healthcare intervention can be judged. Cost–utility analysis (CUA) is a specific form of CEA, the main difference being that benefit is expressed in the form of patient preferences and converted to a measure such as the number of quality-of-life adjusted life-years (QALYs). CUA thus allows the comparison of competing health programmes that have very different sorts of outcome. CEA, and to a lesser extent CUA, now appear in many health services research studies, unlike cost–benefit analysis (CBA), in which benefits are expressed in purely monetary terms. Because of time limitations on clinical and health service trials, cost-effectiveness can often only be established for intermediate outcomes and it may be necessary to apply techniques such as SURVIVAL ANALYSIS to extrapolate values into the future (e.g. to predict mortality from risk factors). Drummond and McGuire (2001) discuss this point and other key methodological principles involved in CEA.

The incremental cost-effectiveness ratio (ICER) is a key measure in cost-effectiveness analysis and is defined (for a comparison of two treatments) as $\Delta C/\Delta E$, where ΔC is the mean difference in the costs of the treatments and ΔE is the mean difference in effectiveness. A treatment is considered cost-effective if $\Delta C/\Delta E < \lambda$, where λ is the maximum amount that the decision maker is willing to pay, or the 'ceiling ratio'. Note that the term 'incremental' stresses the comparison of the treatments with each other: the less useful 'average' cost-effectiveness ratio is often estimated separately for each treatment under study and compared without testing the differences statistically.

Many trials of new therapies collect data on the costs for individual patients, as well as the effectiveness in terms of changes in clinical status or quality of life, and can thus calculate ICERs as part of an economic evaluation. Indeed, regulatory bodies in most countries insist on such evaluation before considering drug licensing and provide guidelines as to acceptable methodology. Kobelt (2002) describes the workflow of an economic evaluation in relation to the typical drug development process, from pre-clinical studies to PHASE III TRIALS and marketing, and discusses the typical resource items that might be included and the economic perspectives from which such studies are performed.

In an economic evaluation, the term 'perspective' refers to the level at which the costs are to be considered. For example, the societal perspective would treat all costs as relevant, including loss to production due to illness, whereas a health service perspective might consider only direct treatment costs. As well as influencing the type of data collected, the perspective has a bearing on the summary statistics and type of analysis that might be considered appropriate. For example, where the perspective is that of a healthcare provider, it can be argued that it is the mean cost and the mean effectiveness that are relevant, rather than some other summary measure such as the MEDIAN. This is because the total cost aggregated over all patients (which is the important quantity for planning or budgeting purposes) is obtained by multiplying the mean cost by the number of patients.

Various types of uncertainty apply to economic data and all may need to be considered in CEA. For example, the amount of a service used by a patient will need to be multiplied by a unit cost (e.g. the cost per hour of employing a therapist) to obtain the total cost per patient for that service. The true unit costs may be available only as point estimates, e.g. in the form of approximate published values. Furthermore, costs and benefits may have to be discounted to take account of the preference for earlier benefits and/or later costs and this involves the application of discount rates, which are generally estimated. The impact of such deterministic sources of uncertainty is generally assessed through sensitivity analyses, in which ranges of plausible values are considered. Contrariwise, data that have been obtained from a random sample, e.g. on the service use of individual patients, are subject to stochastic rather than deterministic uncertainty, and this is generally expressed in the form of statistical quantities such as CONFIDENCE INTERVALS and P-VALUES.

Costs often have very skewed distributions and for this reason normal distribution theory may not be appropriate. Furthermore, if the emphasis is on the estimation of means as suggested here, simple TRANSFORMATIONS (e.g. log transformation) may not be appropriate since the quantity about which inferences are to be made would no longer be the MEAN (see Thompson and Barber, 2000). The use of generalised linear models is one solution since these can model mean values directly while also allowing for skewed distributions such as the GAMMA DISTRIBUTION. An alternative is to use nonparametric bootstrapping (see BOOTSTRAP), in which a large number of re-samples is generated, with individuals replicated by sampling with replacement. However, while widely used, there is a potential problem with bootstrapping applied to the ICER, as discussed later, and it has been criticised in more general terms by O'Hagan and Stevens (2002), who argue for a Bayesian framework (see BAYESIAN METHODS) for analysing cost-effectiveness data.

The ICER and some related quantities are illustrated and discussed by O'Brien and Briggs (2002), using an example relating to the costs of treatment and life-years gained in the Canadian Implantable Defibrillator Study (CIDS). The incremental cost-effectiveness plane for interpreting ICERs is shown in (a) in the first figure (see page 107). Each quadrant can be labelled according to the interpretation of an ICER falling within it and typically new treatments are more effective but also more costly and would appear in the NE quadrant. If the new treatment is both more effective and less costly than the control or comparison treatment (quadrant SE) then it is said to dominate. In this diagram the point denoted 'CIDS data' represents the observed (ΔC, ΔE) pair. The point estimate of the ICER is the slope of the line joining this point to zero (in this case the ICER is 49.115k/0.23, or \$213.5k per additional life-year gained).

The decision as to cost-effectiveness can be made for all ΔC, ΔE pairs consistent with the observed data, as represented by the bootstrapped values, a set of 1000 pairs for the CIDS data being shown in (b) in the first figure. Also shown are 95% confidence limits derived from the lines corresponding to the 26th and 975th replicates (ordered from the smallest to the largest ratio).

Ratios, even of normally distributed quantities, tend to be unstable and have discontinuities since both numerator and denominator could be 0. This can give rise to anomalies since the same ratio can arise from two opposing sets of results, one where the new treatment is less effective and more costly

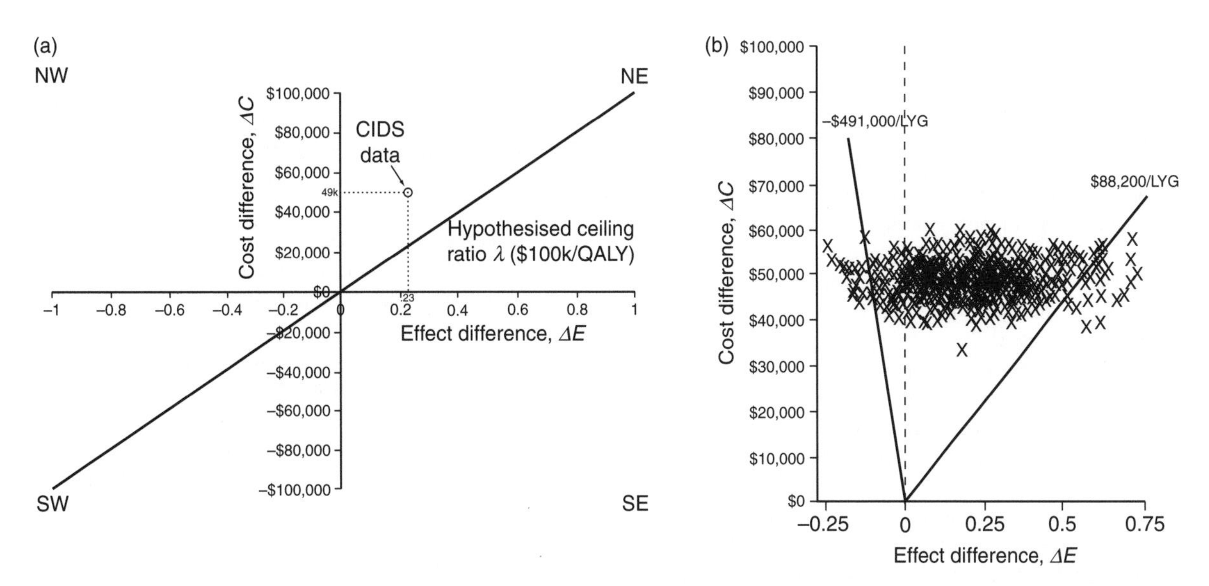

cost-effectiveness analysis *Incremental cost-effectiveness plane for Canadian Implantable Defibrillator Study data showing (a) the four quadrants, point estimates of ΔE and ΔC, from which the ICER is obtained, and a ceiling ratio line; (b) bootstrapped values for ΔE and ΔC and confidence limits for the ICER derived from these values*

(quadrant NW) and the other more effective and less costly (quadrant SE). When the ratios are ordered and used to estimate confidence limits there is no way to distinguish these cases. Thus it is usually recommended to plot the bootstrapped points in addition to calculating numerical confidence limits so that this situation can be detected.

One way around difficulties arising from using ratios is to cast the problem in a slightly different way, based on plotting the willingness-to-pay line on the cost-effectiveness plane. The region below this line (also called the *ceiling ratio line*) is the *acceptability* region. In this example the willingness-to-pay is $100k per QALY and the point estimate of the ICER is $214k per QALY, so the treatment would not be considered cost-effective. The proportion of the bootstrapped values falling in the acceptability region can then be plotted against a range of hypothetical values for the ceiling ratio to give a *cost-effectiveness acceptability curve*. This approach is useful where no fixed willingness-to-pay has

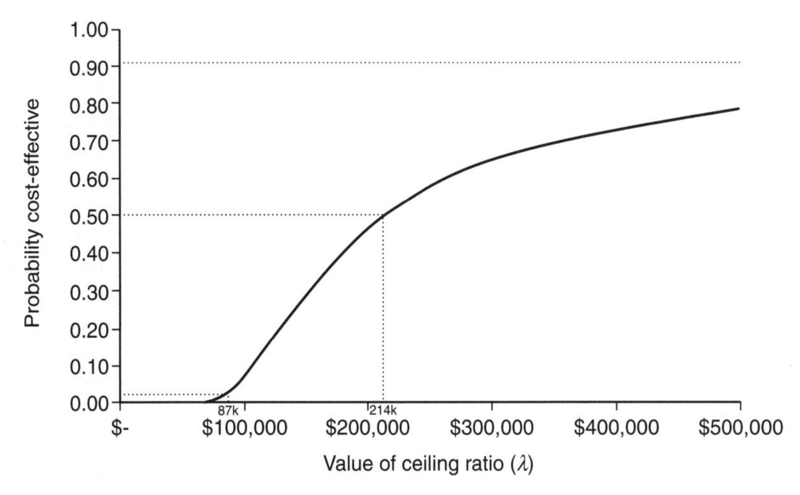

cost-effectiveness analysis *Cost-effectiveness acceptability curve for the Canadian Implantable Defibrillator Study data*

been established since it indicates how probable it is that a treatment will be cost-effective at any given hypothetical willingness-to-pay.

The second figure illustrates an acceptability curve for the CIDS data. For example, given that a healthcare provider considers \$214k as an acceptable upper limit, one might be able to claim that a new treatment is, say, 50% likely to be cost-effective. At \$87k it would only be about 2.5% likely to be cost-effective. This latter figure can be used as a lower 95% confidence limit, the upper 95% confidence limit being infinity in this case.

Another approach, mathematically equivalent to the cost acceptability curve technique, is to calculate the *net monetary benefit* (NMB), which re-expresses the ICER decision rule to give the quantity $\text{NMB} = \lambda\Delta E - \Delta C$. This is now expressed in monetary terms and because it is linear in ΔE and ΔC, parametric confidence limits are simple to calculate, given variances for ΔE and ΔC. The value of NMB can be plotted against λ, for a range of hypothetical values, and the value of λ for which the NMB is zero is a breakeven point, at which the likelihood of being cost-effective is 50%. Of course, the value \$214k is obtained as before. *ML*

Drummond, M. and McGuire, A. 2001: *Economic evaluation in health care – merging theory with practice*. Oxford: Oxford University Press. **Kobelt, G.** 2002: *Health economics: an introduction to economic evaluation*, 2nd edition. London: Office of Health Economics. **O'Brien, B. J. and Briggs, A. H.** 2002: Analysis of uncertainty in health care cost-effectiveness studies: an introduction to statistical issues and methods. *Statistical Methods in Medical Research* 11, 455–68. **O'Hagan, A. and Stevens, J. W.** 2002: Bayesian methods for design and analysis of cost-effectiveness trials in the evaluation of health care technologies. *Statistical Methods in Medical Research* 11, 469–90. **Thompson, S. G. and Barber, J. A.** 2000: How should cost data in pragmatic randomised trials be analysed? *British Medical Journal* 320, 1197–200.

cost–utility analysis (CUA) See COST-EFFECTIVENESS ANALYSIS

counterfactual model See CAUSAL MODELS

covariance See COVARIANCE MATRIX

covariance matrix This is a symmetric matrix in which the off-diagonal elements are the covariances of pairs of variables and the elements on the main diagonal are the variances of the variables. The sample covariance of two variables with sample values $(x_1, y_1), (x_2, y_2), \ldots, (x_n, y_n)$ is defined as:

$$\text{Cov}(x, y) = \frac{1}{n}\sum_{i=1}^{n}(x_i - \bar{x})(y_i - \bar{y})$$

where $\bar{x}$ is the arithmetic mean of the x variable and $\bar{y}$ the arithmetic mean of the y variable. The covariance matrix is often a better basis for the application of STRUCTURAL EQUATION MODELS than the correlation matrix. *BSE*

Cox's proportional hazards model See COX'S REGRESSION MODEL, PROPORTIONAL HAZARDS

Cox's regression model In 1972 David Cox developed the Cox (or proportional hazards) regression model. Since that time it has become probably the most widely used method of analysing time-to-event (survival) data. This model allows us to link three of the main components of such data for the first time: (a) an indicator variable reflecting whether the individual has experienced an event or not (i.e. has/has not been censored); (b) the length of time from entry in a study to the event or to the censoring time; (c) one or more explanatory variables, such as age, sex and treatment received, usually collected at the time of entry of an individual into a study. The popularity of the model is due to its relative computational simplicity, its interpretability, its ability/appearance to perform well in many situations and its incorporation into most major STATISTICAL PACKAGES.

The approach used was novel because it modelled the hazard function over time – made up from components (a) and (b) above – and related it to the explanatory variables – component (c). The hazard function can be thought of as the probability that someone now event-free will experience an event in the next small unit of time. The model makes no assumptions about the underlying distribution of hazards in the different groups and, indeed, this is left unestimated in the process of estimating the parameters in the model. The basic model relies on the assumption that the hazard functions are proportional across the groups being studied, i.e. the relative hazards experienced by any two groups of patients are constant over time.

The Cox model can be used to perform a number of different analyses for time-to-event data including: estimating a treatment effect in a study while adjusting for a number of explanatory or baseline variables, such as age, sex; assessing which of a number of explanatory variables are most important and consequently developing a prognostic index; performing stratified analyses; and assessing interactions between variables. It has also been extended to deal with situations where the relative hazard function changes over time, the so-called time-dependent Cox model, and for situations when there are deviations from proportional hazards for the hazard functions in different groups.

The Cox model can be written as:

$$h_1(t) = h_0(t)\exp(\beta_1 x_1 + \beta_2 x_2 + \cdots + \beta_k x_k)$$

where $h_1(t)$ is the hazard function in a given group, $h_0(t)$ is the hazard function in a baseline group (which remains unspecified), β_i are the regression coefficients and x_i are the explanatory variables (from $i = 1$ to k). Therefore, the hazard ratio (HR) between groups is:

$$h_1(t)/h_0(t) = \exp(\beta_1 x_1 + \beta_2 x_2 + \cdots + \beta_k x_k)$$

One should note the independence from time (t) of the hazard ratio on the right-hand side of this equation. Hazard ratios are relatively simple to interpret; they are the relative risk of one group experiencing the event to another group experiencing the event. Note that, as the baseline hazard is not estimated, only relative measures such as the hazard ratio are provided by the Cox model and thus no estimates are given in absolute terms using this method. Such estimates have to be calculated indirectly.

A variety of data types can be used in the model, including binary, categorical, ordered categorical and continuous variables. The number of variables one may include in a Cox model is theoretically limitless, but in practice it is limited by the number of events in the analysis. One guide is not to use more variables than the fourth root of the number of events. A more lenient guide is to have at least 15–20 events for each category combination.

For each variable being considered in a Cox model we test the NULL HYPOTHESIS that the variable is not important to the model, i.e. that the parameter value associated with the variable, β, is zero; this is equivalent to the hazard ratio (HR) for that variable, $\text{HR} = \exp(\beta) = e^0 = 1$. This can be tested with a z-statistic where $z = b/\text{SE}(b)$, where b is the estimate of the parameter and SE(b) is its standard error. Under the null hypothesis this should follow a normal distribution and thus P-VALUES can be calculated in the usual way.

We may assess models and the addition or removal of variables to models using a variety of different tests including the Wald test, LIKELIHOOD RATIO test and score test. The score test is the most complex and less commonly used test. The Wald test looks at the change in the overall χ^2 value between two models where the DEGREES OF FREEDOM is the number of different variables between the models. The likelihood ratio test compares the 'likelihoods' of the two models and takes a more general approach than the Wald test: it looks at how the included variables explain the variation in the model. This is, therefore, the preferred method for reasons of consistency and stability.

The time of each outcome event (failure time) is not actually relevant in a Cox model, but the ordering of these failures is. Therefore, consideration needs to be given to the order of failures in the event of failures with tied event times. These can be dealt with in a series of methods including marginal calculation, partial calculation, Efron approximation and Breslow approximation (see Kalbfleish and Prentice, 2002). The last of these is the simplest and is an adequate approximation if there are relatively few tied failures. Care should be taken when using the Cox model if there are too many tied event times.

As for normal linear regression, it is possible to assess the fit of the Cox model by calculating residuals. However, there are no unique residuals for the Cox model. Commonly used residuals are Schoenfield and Martingale residuals, although it can be difficult to interpret whichever are used. It is also possible to assess whether individual explanatory variables violate the proportional hazards assumption (see PROPORTIONAL HAZARDS) and therefore assess whether a variable should be included in the model. *MS/MP*

Cleves, M. A., Gould, W. W. and Gutierrez, R. G. 2003: *An introduction to survival analysis using Stata®*, revised edition. Texas: Stata Press. **Cox, D. R.** 1972: Regression models and life tables (with discussion). *Journal of the Royal Statistical Society* B34, 187–220. **Kalbfleisch, J. D. and Prentice, R. L.** 2002: *The statistical analysis of failure time data*, 2nd edition. New York: John Wiley & Sons, Inc. **Machin, D. and Parmar, M. K. B.** 1995: *Survival analysis: a practical approach*. London: John Wiley & Sons, Ltd.

Cramér's contingency coefficient See CONTINGENCY COEFFICIENT

credible interval When the aim of a Bayesian analysis (see BAYESIAN METHODS) is to provide a scientific inference about an unknown parameter all the required information about the uncertainties involved are contained in the POSTERIOR DISTRIBUTION. There is a sense in which the only true satisfactory inference summary is the complete 'picture' represented by the posterior distribution. Alternatively, a range of posterior distributions corresponding to a range of prior specifications allows a display of the sensitivity of 'conclusions' to 'assumptions'. Sometimes, however, providing a complete picture of the uncertainty in estimation by a posterior distribution is less convenient than providing a low-level summary of the message contained within it. Credible intervals are to Bayesian statistics as CONFIDENCE INTERVALS are to frequentist statistics; they provide a simple summary of the uncertainty associated with the estimate of an unknown parameter.

If we suppose that the posterior distribution for an unknown parameter δ is denoted by the posterior distribution $p(\delta)$ then an interval (δ_L, δ_U) is said to form a $100(1-\alpha)\%$ posterior credible for δ if

$$\int_{\delta_L}^{\delta_U} p(\delta)\,d\delta = 1-\alpha$$

There are infinitely many ways to determine a credible interval, some of which are illustrated in the first figure. Credible interval (1) is determined so that it excludes regions of equal posterior probability, each tail corresponding to a probability of $1-\alpha/2$. Credible interval (2) excludes a region of exactly α on the lower side extending to infinity on the right. In contrast, credible interval (3) excludes a region of exactly α content on the right beginning at zero. The final credible interval (4) is termed the highest posterior density (HPD) interval and is constructed so that every parameter value within the interval has a higher density than every value outside the interval and hence they are more likely values. It can be demonstrated that the HPD interval in any particular case determines the shortest credible interval. If the posterior distribution is symmetric then the equal-tailed and HPD intervals coincide. The concept of a credible interval generalises to more than a single parameter.

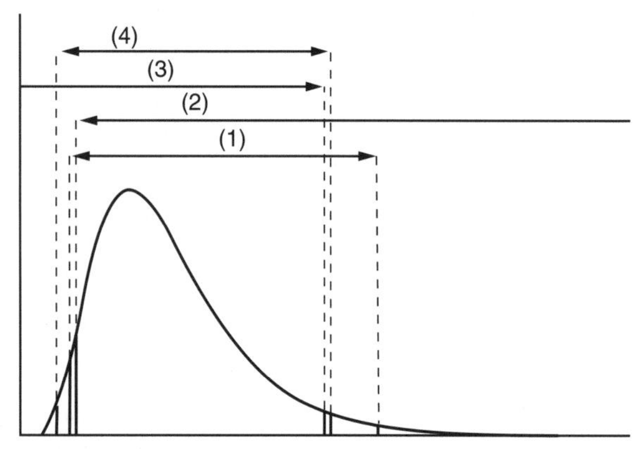

credible interval *Credible intervals for an unknown parameter*

The second figure illustrates a bivariate (two-parameter) HPD region constructed in exactly the same way as the univariate case but with every point within the region having a higher density than every point outside it. The specific example is taken from an extension of the Bradley–Terry paired-comparison model (see, for example, Imrey, 1998), accounting for ties. For a given probability content such a credible region has the smallest area. *AG*

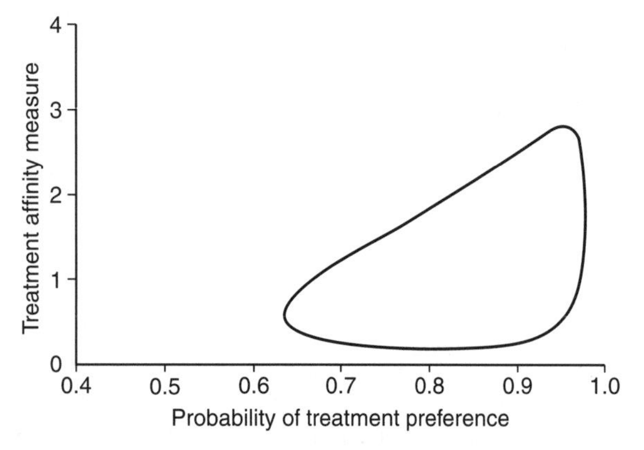

credible interval *Bivariate HPD region*

Imrey, P. B. 1998: Bradley–Terry model. In Armitage, P. and Colton, T. (eds) *Encyclopedia of biostatistics*. Chichester: John Wiley & Sons, Ltd.

critical appraisal This is a process that evaluates research reports and assesses their contribution to scientific knowledge and is typically applied to research papers in medical journals. A careful evaluation of the medical literature is important because the quality of research is variable, and often very poor. It is imprudent to assume that a paper is error free just because it has been published: even papers in well-respected journals contain faults that cast doubt on the conclusions. Altman has researched the extent and implications of errors in the medical literature, estimating that reviews have found statistical errors in about half of published papers (Altman, 1991a).

The problem of poor-quality research is set in the context of the increasing use of statistics in the medical literature. Altman (1991a) describes two surveys of research papers published in the *New England Journal of Medicine*, in 1978–1979 and 1990. In this time the proportion of papers containing nothing more than descriptive statistics fell from 27% to 11%, while the proportion using more complex statistical methods, such as SURVIVAL ANALYSIS, increased dramatically. A good understanding of statistical analysis, alongside an awareness of statistical issues surrounding research design and execution, therefore, is essential to effective appraisal of the medical literature.

Altman (1994) argued, in an editorial entitled 'The scandal of poor medical research', that research in the medical arena is often done with the aim of furthering a curriculum vitae, rather than promoting scientific knowledge. He suggests that: 'Much poor research arises because researchers feel compelled for career reasons to carry out research that they are ill equipped to perform, and nobody stops them.' The situation is compounded because the individual is 'expected to carry out some research with the aim of publishing several papers', the number of publications being 'a dubious indicator of ability to do good research; its relevance to the ability to be a good doctor is even more obscure'.

This culture, Altman argues, leads to poor-quality research. An additional difficulty arises because junior doctors typically move jobs frequently, but are nevertheless often expected to conduct research during their short tenures. This may lead to small sample sizes as well as inadequate time for training, planning, analysis and formulation of conclusions. Further problems may occur when investigators are expected to complete research initiated by their predecessors. Altman (1991b) suggests that easy access to computers and statistical packages unaccompanied by corresponding technical understanding, as well as inadequate statistical education, also contribute to the errors.

Elsewhere, Altman (1980) describes the grave ethical implications of poor-quality research. He argues that it is unethical to carry out bad scientific experiments since patients may be subject to unnecessary risk, discomfort and inconvenience, while other resources, including the researcher's time, are diverted from more valuable functions. The publication of erroneous results is also unethical since it may lead directly to patients receiving an inferior treatment. More subtle consequences are the encouragement to other researchers to replicate flawed methods or to do further research based on erroneous premises, as well as the difficulty in getting ethics committees to permit further research when it is thought that the 'correct' answer is known.

Many medical journals employ statistical reviewers in an attempt to improve the quality of research design, statistical analysis and presentation of results (see STATISTICAL REFEREEING). Goodman, Altman and George (1998) report that in a 1993–1995 study 37% of journals surveyed had a policy that guaranteed a statistical review before an acceptance decision.

Direct evidence of the effect of statistical reviewing is limited. However, Schor and Karten (1966) studied the implementation of a programme of statistical review at a leading medical journal. Of the 514 original contributions considered, 26% were judged statistically acceptable; this increased to 74% once these manuscripts had been published after statistical review. Gardner and Bond (1990) performed a similar study on 45 papers submitted to the *British Medical Journal*. They found that only 11% were initially considered suitable for publication, but after statistical review 84% were regarded to be of an acceptable statistical standard.

However, there is much research that has not undergone statistical review and reviewing itself is a subjective process. It is thus essential to read papers in the medical literature cautiously; the reputation of a journal is not a guarantee of the quality of research reported. Errors in research vary in their magnitude and impact and a major element of critical appraisal is therefore to evaluate their potential effects on conclusions.

The following describes some common errors in the design, analysis, presentation and interpretation of medical research. The comments are not comprehensive, but represent some of the more widespread and important errors made in the research process (see also pitfalls in medical research). Andersen's (1990) book, *Methodological errors in medical research*, contains a more complete description of errors illustrated by examples from the medical literature, although the author nevertheless describes it as an incomplete catalogue.

Medical research can be broadly divided into CLINICAL TRIALS, COHORT STUDIES, CASE-CONTROL STUDIES and CROSS-SECTIONAL STUDIES. Clinical trials are experimental studies where the investigator assigns participants to different interventions, preferably randomly. The well-conducted randomised controlled double-blind clinical trial comes the closest to establishing cause and effect between intervention and outcome in a single study.

Cohort studies, case-control studies and cross-sectional studies are all observational studies. Here the investigator observes participants without making any intervention. Conclusions are not considered as robust as those from experimental studies because factors controlling exposure may also be related to the outcome. However, observational studies are common because it is often impractical or unethical to assign participants to interventions; for example, it would not be possible to assign subjects to be smokers or nonsmokers. Each study design has advantages and disadvantages for specific research questions and an important initial consideration in critical appraisal is whether an appropriate experimental design has been employed. Several of the following criteria relate to clinical trials where rigorous design and conduct of studies is imperative if results are to be conclusive.

The vast majority of research studies cannot consider the whole population of interest and therefore a sample is selected. Results from this sample are then applied to the population of interest. If this inference is to be valid it is vital that the sample is representative of the population. A key concept is that of random selection; if the participants are randomly selected from the population then there is the best chance of the sample being truly representative. The research setting is often a pertinent consideration here; a study of childbirth in a maternity hospital receiving a high proportion of referrals for complications may not be representative of childbirths throughout the country. DROPOUT or refusal rates are also important issues because there is a strong possibility that those who do not take part in a study are systematically different from those who do. Although dropouts and refusals should be minimised, they are usually inevitable and a good research paper will describe the representative nature of a sample by reporting clearly the number originally selected, as well as the number completing the study. Reasons for dropout should be given, if possible, and any available characteristics of those who do not complete the study compared with those that do. All research papers must describe features of the study sample so that comparisons with the relevant population can be made.

A very common problem in experimental design is the lack of a pre-study sample size calculation. This indicates the number of participants required to be reasonably likely to detect a clinically significant effect. It is considered unethical to undertake a study with insufficient numbers to detect such an effect. Therefore it is important that a pre-study sample size calculation is performed and described in the research report, providing sufficient detail about the assumptions made, so that the calculation can be verified. Sample size considerations should be based on the primary outcome variable in a study, and should also allow for dropout or refusal rates.

In clinical trials the concepts of blinding and random allocation to intervention are essential aspects of experimental design. Blinding is necessary because BIAS may enter a study through a participant or observer knowing the intervention allocation. In a double-blind clinical trial neither the participant nor the observer is aware of the allocation. Blinding is clearly not always feasible, for example, in a trial comparing an intervention of physiotherapy to no physiotherapy among a sample of elderly patients who have had a fall. In some trials, it may only be possible to blind the participant and not the observer (known as single-blind trials), but it is important that the maximum level of blindness possible is used.

Random allocation to intervention is a further desirable feature of clinical trials. It is necessary that groups of participants receiving different interventions are as similar as possible so that any effects at the end of the trial are attributable only to differences in the intervention. RANDOMISATION optimises the chance that the groups will be as similar as possible. Unfortunately, many 'trials' are not planned, but instead are based on existing routinely collected data. Allocation to intervention in these instances is never truly random and is often particularly defective. For example, two surgeons in a hospital performed many operations to reduce snoring, each surgeon using a different technique. Analysis of several years of routine outcome data attempted to compare the two techniques. Here the surgeon was a confounder and it is not possible to deduce whether differences in outcome were due to the effects of the surgical technique or of the surgeons who operated.

The double-blind randomised controlled trial (see CLINICAL TRIALS) is considered the gold standard of medical research. If a research report states that blinding and random allocation have been used then it is important that the procedures employed in implementing each are described; it is insufficient to assume that authors understand the meaning of these terms.

Errors in research design would be reduced if statisticians were consulted more often in the early stages of the research process so that statistical issues could be considered throughout (see CONSULTING A STATISTICIAN). Unfortunately, errors in the design of experiments are nearly always impossible to correct and therefore the research may be fatally flawed.

Errors in statistical analysis are also widespread. Many statistical techniques make assumptions about the data to which they are applied, but a mistake often observed in research papers is that these assumptions have not been met. A common assumption is that of data conforming to a NORMAL DISTRIBUTION. It is, unfortunately, not always possible to tell whether a variable is normally distributed when the raw data cannot be inspected. However, summary statistics may be provided and for measurements that cannot be negative, which is often the case in medical research, it can be inferred that the data have a skewed distribution if the standard deviation is more than half the MEAN, although the converse is not necessarily true. When data do not conform to the assumption of normality they should either be transformed (see TRANSFORMATION) or NONPARAMETRIC METHODS used instead.

It may be clear from graphs or ranges that outliers are present in data. These can have a considerable effect on statistical analyses. Generally, however, values should not be altered or deleted if there is no evidence of a mistake. Instead if OUTLIERS are present a research paper should indicate that steps were taken to investigate their effects. Again, transformations, or nonparametric methods may be appropriate.

A common assumption of statistical tests is that all the observations are independent. However, multiple observations on one subject are not independent and should therefore not be analysed as such. For example, the results of hearing tests in the right and left ears of a group of study participants should not all be entered into an analysis where observations are assumed to be independent. Instead, the average of the right and left measurements could be taken, or the results from just the left or right ear might be chosen. It is also erroneous to analyse paired data ignoring the pairing. Paired data can arise when a one-to-one matched design has been used or when two measurements are made on the same subject, e.g. before and after treatment.

METHOD COMPARISON STUDIES are common in medical research and CORRELATION is often misused to assess agreement between the two. Correlation measures linear association, rather than agreement, so if one method always gives a value of exactly twice the other method a perfect correlation would be found although agreement is clearly lacking. Instead, agreement between two continuous variables should be assessed using the technique described by Bland and Altman (1986).

A major problem encountered in the analysis of medical research is that of multiple testing (see MULTIPLE COMPARISON PROCEDURES). Choosing the conventional significance level of 0.05 means that if 20 statistical tests were performed we would expect one to be significant purely by chance. Therefore the conclusions of a paper reporting one significant result among 20 tests performed should be tempered by this fact. It may be that an adjustment for multiple comparisons, such as a BONFERRONI CORRECTION, is appropriate. A distinction should always be made between prior hypotheses and those resulting from exploration of the data, so that the same data are not used for testing a hypothesis as for generating it (see POST HOC ANALYSIS).

Other errors in analysis abound, including the use of correlation to relate change to initial value, failure to take account of ordered categories or the evaluation of a diagnostic test only by means of SENSITIVITY and SPECIFICITY when the POSITIVE and NEGATIVE PREDICTIVE VALUES would be more informative. Whatever method of analysis is used, an important requirement of the research report is that all techniques employed are clearly specified for each analysis. Unusual or obscure methods should be referenced and methods that exist in more than one form, such as Pearson's or Spearman's correlation coefficient (see CORRELATION), must be identified unambiguously.

Errors can also be made in presentation, although these may have more trivial implications for the conclusions of a paper than the errors described earlier. Nevertheless, good presentation is important to ensure that the reader is not misled or confused. It is worth noting that a poor-quality paper may not necessarily describe poor-quality research, but if insufficient detail is provided in a report it is unsatisfactory to assume that the research has been performed acceptably.

P-values are often used in the medical literature to indicate statistical significance. However, it is preferable that CONFIDENCE INTERVALS are used in the presentation of results to give an immediate idea of the clinical significance of an effect. If a 95% confidence interval does not include zero (or, more generally, the value specified in the NULL HYPOTHESIS) then the *P*-value will be less than 0.05. Thus there is a close relation between confidence intervals and *P*-values (see TESTS AS CONFIDENCE INTERVALS), but confidence intervals additionally demonstrate the magnitude of the effect of interest.

MEASURES OF SPREAD should be quoted alongside MEASURES OF LOCATION to indicate variability around the average measurement. However, the $\pm$ notation is discouraged because its use to denote the STANDARD DEVIATION, the standard error and the half-width of a confidence interval has led to some ambiguity. Thus, rather than describing mothers in a study of pregnancy by saying 'the mean age of mothers was 28 ± 4.6 years', the data are better summarised as 'the mean age of mothers was 28 years (SD 4.6)'.

Since the vast majority of statistical analysis is now performed using computers, research papers should present exact *P*-values, which are far more informative, such as $P = 0.014$, rather than ranges, such as $P < 0.05$. The notation 'NS' for nonsignificant is even less revealing. However, there is no need to be specific below 0.0001. Authors must justify the appropriateness of a one-sided *P*-value quoted in a research paper. A one-sided *P*-value should only be used in the very rare situation where an observed difference could only have occurred in one direction. The decision to use a one-sided *P*-value should be made prior to the data analysis and hence not be dependent on the results.

Spurious precision is another common error in the medical literature that impairs the readability and credibility of a paper. When presenting results the precision of the original data must be borne in mind. Altman (1991b) suggests that means should not be quoted to more than one decimal place than the original data and standard deviations or standard errors to no more than two. Likewise, percentages need not be given to more than one decimal place and *P*-values need not have more than two significant figures.

Errors also arise in graphical presentation (see GRAPHICAL DECEPTION). Graphs that do not include a true zero on the vertical axis or that change scale in the middle of an axis can be misleading, as can the unnecessary use of three-dimensional effects. Other errors include the plotting of means without any indication of variability and the failure to show coincident points on a SCATTERPLOT.

Misinterpretation is common when *P*-values are presented. It must be remembered that the conventional cut-off of 0.05 is purely arbitrary. A frequent mistake is to interpret a value of, say, 0.045 as significant, but a value of 0.055 as not significant, when in reality there is very little difference between the two. There is also a prevailing belief that significant *P*-values are indicative of more successful research than nonsignificant *P*-values. This attitude is reflected in studies being described as 'positive' or 'negative', depending on the significance of the findings. Results should not be evaluated solely on the statistical significance of the findings, but also on their clinical significance (see CLINICAL VERSUS STATISTICAL SIGNIFICANCE). The use of confidence intervals is a helpful antidote to this problem.

A further serious error of interpretation is to interpret ASSOCIATION as causation. The only type of study where CAUSALITY can be inferred is a well-conducted randomised controlled trial. Otherwise, great care should be taken in the interpretation of results; in particular, the likely effect of confounders must be considered.

A final area where conclusions are often not treated with sufficient caution is that of inference from a sample to a population. Although a sample should theoretically be random, in practice this may not be realistic. Therefore a research paper should attempt to report any likely biases in the selection process and implications this may have for the findings reported.

When critically appraising a research paper it is helpful to have a checklist of issues to consider. A checklist is particularly useful because it is easier to spot errors than omissions and, as already noted, it is inappropriate to infer that a correct procedure was employed when the relevant information is not included. The *British Medical Journal* provides two checklists for use by its statistical reviewers that can be used when critically appraising a paper; these are published in Gardner, Machin and Campbell (1986) or can be found on the *British Medical Journal* website. One checklist is intended specifically for clinical trials and so includes questions relevant only to this study design; the other is for use with all other study types.

In conclusion, critical appraisal is an essential skill for users of the medical literature; it is important that readers have the confidence to question conclusions stated by the authors and the statistical knowledge to assess the methods used. The consequences of the range of errors described in this section can vary between reducing the readability of a paper to reversing the direction of the results. An important part of the critical appraisal process, therefore, is to make a judgement about the implications of any issues raised. A study should not be discarded because a single flaw is found, but, instead, a subjective assessment of the impact on the findings must be made. *SRC*

Altman, D. G. 1980: Statistics and ethics in medical research. *British Medical Journal* 281, 1182–4. **Altman, D. G.** 1991a: Statistics in medical journals: developments in the 1980s. *Statistics in Medicine* 10, 1897–913. **Altman, D. G.** 1991b: *Practical statistics for medical research*. London: Chapman & Hall. **Altman, D. G.** 1994: The scandal of poor medical research. *British Medical Journal* 308, 283–4. **Andersen, B.** 1990: *Methodological errors in medical research*. Oxford: Blackwell. **Bland, J. M. and Altman, D. G.** 1986: Statistical methods for assessing agreement between two methods of clinical measurement. *Lancet* 1, 307–10. **Gardner, M. J. and Bond, J.** 1990: An exploratory study of statistical assessment of papers published in the *British Medical Journal*. *Journal of the American Medical Association* 263, 1355–7. **Gardner, M. J., Machin, D. and Campbell, M. J.** 1986: Use of check lists in assessing the statistical content of medical studies. *British Medical Journal* 292, 810–12. **Goodman, S. N., Altman, D. G. and George, S. L.** 1998: Statistical reviewing policies of medical journals: caveat lector? *Journal of General Internal Medicine* 13, 753–6. **Schor, A. and Karten, I.** 1966: Statistical evaluation of medical journal manuscripts. *Journal of the American Medical Association* 195, 1123–8.

crossover trials These are trials in which patients are allocated to sequences of treatments with the object of studying differences between individual treatments or subsequences of treatments (Senn, 2002). That is to say, each patient is treated more than once and the responses under different treatments for the same patient can then be compared. This is best explained by considering some examples.

Suppose we are interested in general in comparing treatments A, B, C, etc., and that patients will be allocated to sequences of treatment of the form ABC, CBA, etc., where, for example, ABC means that the patient will receive A in a first period, B in a second and C in a third. When only two treatments are being compared, a very popular type of crossover design is one in which patients are allocated at random and usually in equal numbers to one of two sequences AB or BA. Such a trial was run by Graff-Lonnevig and Browaldh (1990) comparing the effects of single doses of inhaled formoterol (12 μg) and salbutamol (200 μg) in 14 moderately or severely asthmatic treatments. If we give the label A to formoterol and B to salbutamol, then children were allocated at random to one of the two sequences AB or BA. Where three treatments are being compared, patients may be allocated in equal numbers to one of three sequences forming a Latin square, either ABC, BCA and CAB or ACB, BAC and CBA, or it may be that both Latin squares would be employed, so that patients would be allocated in equal numbers to each of the six possible sequences involving each of the three treatments. For example, Dahlof and Bjorkman (1993) compared two doses of the potassium salt of diclofenac (50 mg or 100 mg) to placebo in the treatment of migraine in 72 patients. If A is placebo, B is the lower dose of diclofenac and C the higher one, then their design involved allocating patients to one of the six sequences ABC, ACB, etc.

More complex designs than this are possible. For example, it may sometimes be the case that the number of treatments that one wishes to study is greater than the number of periods in which it is considered realistic to treat patients. So-called *incomplete block designs*, in which patients receive suitable chosen subsets of the treatments to be investigated, are popular. At the other extreme, it may be that it is possible to treat patients in more periods than there are treatments, leading to so-called *replicate designs*. As we shall discuss, these are extremely useful for the purpose of studying an individual response.

Because crossover trials permit comparisons on a within-patient basis, they are efficient compared to parallel group trials and considerable savings in patient numbers are possible. However, crossover trials are clearly unsuitable for any condition in which death or cure is the outcome and their appropriate use is restricted to chronic diseases and treatments whose effects are reversible. Suitable conditions include asthma, rheumatism and migraine. However, it is not just the condition but the treatment and the ENDPOINT that determine the suitability of crossover trials. For example, they can be used to study blood pressure itself in short-term trials in hypertension but not the long-term sequelae of hypertension, such as, for example, stroke or kidney or eye damage. In asthma they are more suitable for studying the effects of beta-agonists, which are relatively short term and reversible, than those of steroids, which have longer-term effects.

In such conditions where crossover trials may be employed, it is nearly always the case that the sample size required to prove efficacy, even if a parallel group trial is used, is considerably less than that required to demonstrate safety of the drug. Hence, in Phase III, where safety considerations are extremely important, there is no point in reducing the sample size by employing crossover trials anyway (see PHASE III TRIALS, PHARMACOVIGILANCE). Consequently, some discussions of the comparative merits of crossover trials and parallel group trials that appear in the scientific literature are rather misleading. In practice, crossover trials are never an alternative for the major parallel group trials carried out in Phase III. They can, however, be extremely useful in Phases I and II for pharmacokinetic and pharmacodynamic modelling, for dose finding for tolerability in healthy volunteers and for efficacy using pharmacodynamic outcomes in patients (see PHASE I TRIALS, PHASE II TRIALS). They can also be useful elsewhere for answering certain specialist questions such as, for example, demonstrating equivalence (see EQUIVALENCE STUDIES: DESIGN) of generic and brand name products using so-called bioequivalence studies (Senn and Ezzet, 1999).

Unlike the parallel group trial, the basic unit of replication in a crossover design is not the patient but an episode of treatment. Since a general necessary assumption in standard analyses of experiments is that there is no interference between units, this is clearly potentially more problematic for crossover trials than for parallel group trials. It is inherently more plausible that the treatment given to a patient in an earlier period may affect the response for the same patient when being given a further treatment in a subsequent period, a phenomenon known as *carry-over*, than that the treatment given to one patient may affect another. (There are some cases where even parallel group trials may suffer from interference between units, in particular if infectious diseases are involved or if group therapy takes place. This may lead to cluster randomisation being necessary, but this is plausibly an infrequent problem.) In fact, carry-over is regarded as being the central (potential) difficulty of crossover trials and much of the considerable literature devoted to the design and analysis of these trials is concerned with matters to do with controlling for carry-over.

The phenomenon of carry-over means that it is prudent, indeed necessary, to employ a so-called *washout period*. This is a period between the measurement of the effect of one treatment and the next in which the effect of the previous treatment is allowed to dissipate. Washout treatments can be *passive*, if washout is allowed to occur without any treatment being given (Senn, 2002). This may seem the natural approach from the experimental point of view. It has, however, the disadvantage that the patient is expected to tolerate a period in which no therapy is offered. An alternative strategy is that of employing an *active* washout period (Senn, 2002). This might involve a near-immediate switch of the patient's therapy but a delay of measurement until a suitable period has taken place during which the effects of the previous treatment have disappeared. (For further details, see Senn, 2002.)

It seems plausible that crossover trials will be more vulnerable to dropouts than parallel group trials because of the greater demands on patient time that the former make, because dropouts in one period will also lead to loss of data from subsequent periods and because incomplete data will unbalance designs and lead to disproportionate losses in information.

It should be noted that in nearly all crossover trials, subjects are not recruited simultaneously. The exception is some designs in which healthy volunteers are used. For designs involving patients they must, of course, be treated when they present. Consequently, 'period' has a *relative* meaning in the context of crossover trials. For example, in an AB/BA crossover some patients will usually have completed both periods of treatment before others have even started in the trial.

A popular linear model for responses for a crossover trial with I patients in periods J with T treatments giving rise to L forms of carry-over may be expressed as follows (Jones and Kenward, 1989). We let the response in period j, $j = 1, \ldots, J$ on patient i, $i = 1, \ldots, I$ be Y_{ij}, the treatment given to that patient in that period be $t(i, j) = 1, \ldots, T$ and the form of carry-over be $l(i, j)$. Then we write:

$$Y_{ij} = \mu + \phi_i + \pi_j + \tau_{t(i,j)} + \lambda_{t(i,j)} + \varepsilon_{ij}.$$

Here μ is a grand mean, ϕ_i is an effect due to patient i, π_j is an effect due to period j, $\tau_{t(i,j)}$ is the effect of treatment $t(i,j)$, $\lambda_{l(i,j)}$ is the carryover effect of type $l(i,j)$ and the ε_{ij} are within-patient error terms usually assumed identically and independently distributed with variance σ^2 (say). The following points may be noted in connection with this model.

The model is severely overparameterised. However, interest centres on contrasts between the various τ terms and, given various restrictive assumptions about the carryover terms, these will usually be estimable.

Since there can be no carry-over in period 1, for each patient $\lambda_{l(i,1)} = 0$ for all i. In practice, to make progress in estimation, further restrictive assumptions are introduced about the carryover terms. There are two very popular choices. The first is to assume that any washout strategy has been successful and that all carryover terms are zero. The second is to assume that 'simple' carry-over applies and that carry-over depends only on the treatment given in the previous period, so that we may write $l(i, j) = t(i, j - 1), j \geq 2$.

This last assumption may seem more reasonable than the first but in practice there are very few imaginable circumstances under which the second assumption would apply if the first did not, as it seems plausible that if carry-over occurred the effect of the engendering treatment would be modified by the perturbed treatment (Senn, 2002). In practice, although designs can easily be found in which patient, period and treatment effects are orthogonal to each other, if carryover effects are included, the design matrix will usually be nonorthogonal and there will be a loss in efficiency.

For certain designs, for most purposes it makes no difference whether the patient effects ϕ_i are taken as FIXED EFFECTS or RANDOM EFFECTS. However, for incomplete block designs in which $T > J$, interpatient information will usually be recoverable by taking the patient effects as random, and this will also usually be the case if carryover effects are included in the model (Senn, 2002).

The following are a number of controversies and issues that are relevant to crossover designs. The most notorious controversy concerning carry-over has been in connection with the AB/BA design. For many years a popular approach to dealing with carry-over was the so-called two-stage procedure originally proposed by Grizzle (1965). He noted that in the presence of carry-over the treatment effect was not estimable and hence proposed that a preliminary test of carry-over be made. If carry-over were detected, the second period data should be ignored and a between-patient test using first-period data only should be employed. However, a subsequent paper by Freeman (1989) showed that this strategy was extremely biased as a whole and did not maintain nominal Type I error rates.

It is possible to adjust the two-stage procedure so that it maintains the overall Type I error rate (Senn, 1997; Wang and Hung, 1997) but it has less power than the strategy of simply ignoring carry-over and is not recommended (Senn, 1997).

Various extremely complicated designs and analysis strategies have been proposed for dealing with carry-over. They all make restrictive and unrealistic assumptions about the nature of carry-over, however, and they nearly all involve a penalty in terms of increased variances of estimators of the treatment effect (Senn, 2002).

This would seem to leave washout as the only reasonable strategy for dealing with carry-over. However, this approach is bound to leave some investigators unhappy because of the reliance it makes on judgement based on biology and pharmacology rather than strategies of design or analysis based on purely statistical principles.

More general error structures than those considered here are possible. In particular, one could allow for a true random effect, that is to say the possibility that different patients react differently to treatment. From one point of view we would then have ϕ_i as a random intercept parameter for each patient but then add a slope parameter for a given treatment for a given patient. For a given treatment these would then be assumed to be randomly distributed with unknown variance to be estimated. It is also possible, although this appears to be rarely attempted, to allow for an autocorrelation of the within-patient errors.

Baseline values at the beginning of each treatment period are sometimes collected in crossover trials but extreme care is required in their use. For many designs it is quite plausible that the outcome values may be unaffected by carry-over but that the baseline values would be. In that case, incorporating the baseline values in the analysis might introduce a BIAS due to carry-over that would not otherwise be present.

There have been various attempts to produce Bayesian analyses of crossover trials (Grieve, 1985). In theory this is attractive in that it permits compromise positions to be adopted between that of assuming that carry-over is absent to allowing that it may have any value at all. In practice, it is difficult to capture in the model the dependence that must inevitably exist between belief in the magnitude of the treatment effect and that of the carryover effect.

Ironically, despite the considerable potential of crossover trials to measure individual response to treatment, especially if replicate designs are employed whereby the number of periods exceeds the number of treatments so that $J > T$, and hence make a genuine contribution to the currently fashionable field of pharmacogenomics, this possibility has received most attention where it is least important, namely in investigating individual response to different formulations in the context of bioequivalence.

Despite some limitations of application and some difficulties in their use it would be wrong to conclude, however, that crossover trials have no place in drug development. They can be extremely efficient compared to parallel group trials and are far superior for the purpose of investigating true random effects. They are extremely valuable on occasion, in particular in pharmacokinetic studies and in dose finding in Phase II. *SS*

Dahlof, C. and Bjorkman, R. 1993: Diclofenac-K (50 and 100 mg) and placebo in the acute treatment of migraine. *Cephalalgia* 13, 2, 117–23. **Freeman, P.** 1989: The performance of the two-stage analysis of two-treatment, two-period cross-over trials. *Statistics in Medicine* 8, 1421–32. **Graff-Lonnevig, V. and Browaldh, L.** 1990: Twelve hours bronchodilating effect of inhaled formoterol in children with asthma: a double-blind cross-over study versus salbutamol. *Clinical and Experimental Allergy* 20, 429–32. **Grieve, A. P.** 1985: A Bayesian analysis of the two-period crossover design for clinical trials. *Biometrics* 41, 4, 979–90. **Grizzle, J. E.** 1965: The two-period change over design and its use in clinical trials. *Biometrics* 21, 467–80. **Jones, B. and Kenward, M. G.** 1989: *Design and analysis of cross-over trials.* London: Chapman & Hall. **Senn, S. J.** 1997: The case for cross-over trials in Phase III [letter; comment]. *Statistics in Medicine* 16, 17, 2021–2. **Senn, S. J.** 2002: *Cross-over trials in clinical research*, 2nd edition. Chichester: John Wiley & Sons, Ltd. **Senn, S. J. and Ezzet, F.** 1999: Clinical cross-over trials in Phase I. *Statistical Methods in Medical Research* 8, 3, 263–78. **Wang, S. J. and Hung, H. M.** 1997: Use of two-stage test statistic in the two-period crossover trials. *Biometrics* 53, 3, 1081–91.

cross-sectional studies The objective of a cross-sectional study is to determine the distribution of a variable or the joint distribution of more than one variable in a population. This may be accomplished by obtaining a representative sample of the population of interest through the use of a SIMPLE RANDOM SAMPLE, a *stratified random sample* or a complex survey design. Such a study is characterised by the fact that subjects are only observed at a single point in time even though the phenomena associated with the variables of interest may have evolved through a dynamic process that develops over time (Kleinbaum, Kupper and Morgenstern, 1982; Rothman and Greenland, 1998). However, because the study subjects are only observed at a single point in time, essential features in the temporal patterns will be missing, which renders it impossible to conduct a thorough longitudinal analysis of the phenomenon of interest. This approach is also sometimes used when conducting an epidemiologic study in which subjects are recruited without regard to their exposure or disease status, so that the information on each corresponds to the status of subjects at the time of the interview only.

In order to appreciate the inherent limitation that exists when observations are only observed at a single point in time, consider the hypothetical example illustrated by the SCATTERPLOT in the figure, part (a). Subjects are observed at different ages and the scatterplot suggests that the outcome tends to decrease as age increases. In contrast, a *longitudinal study* would observe subjects at multiple time points, thus enabling the investigator to track the development of the outcome over time. It is apparent that the tracking of individual subjects in our hypothetical example may have arisen either from an increase in the response as the subject ages (figure (b)) or a decrease (figure (c)). By only observing subjects at a single point in time, it is impossible to distinguish between age trends associated with enrolment into the study and those that evolve as each individual subject ages (Diggle, Liang and Zeger, 1994). This limitation is not resolved by conducting repeated cross-sectional studies carried out at different points in time, unless the studies are designed so as to obtain repeated assessments of the same individuals.

In epidemiology, a cross-sectional study contrasts with a COHORT STUDY and a CASE-CONTROL STUDY. In a cohort study, subjects are selected on the basis of their exposure status and then followed until the disease develops so that an investigator can directly assess the association with disease development. In a typical case-control study, incident cases are recruited near the time at which the disease is diagnosed and exposure is assessed by recall. In either case, an investigator would be studying incident cases while a cross-sectional study would be studying prevalent cases, i.e. cases that may have occurred at some time in the past. This would confound effects on disease incidence with effects on prognosis or survival. A cross-sectional study is especially prone to LENGTH-BIASED SAMPLING because the prevalent cases with a long period of illness before death would be more likely to enter the study than would a subject who died shortly after diagnosis.

Therefore, the design is primarily used in the study of diseases with relatively short-term effects. One example of such a study would be an attempt to discover causes of a food poisoning outbreak in a school, by identifying all students and assessing the specific food they had consumed and whether they had become ill.

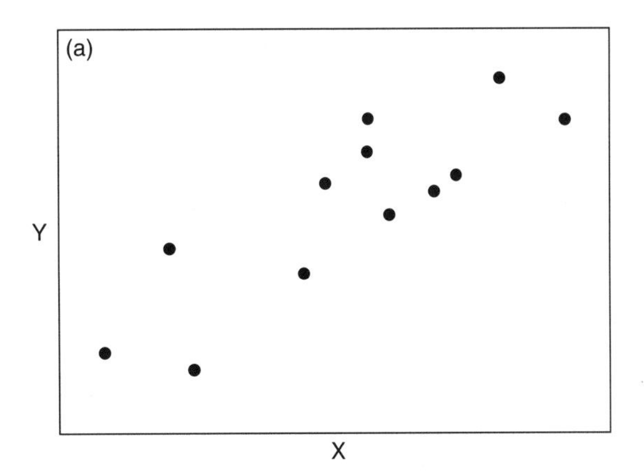

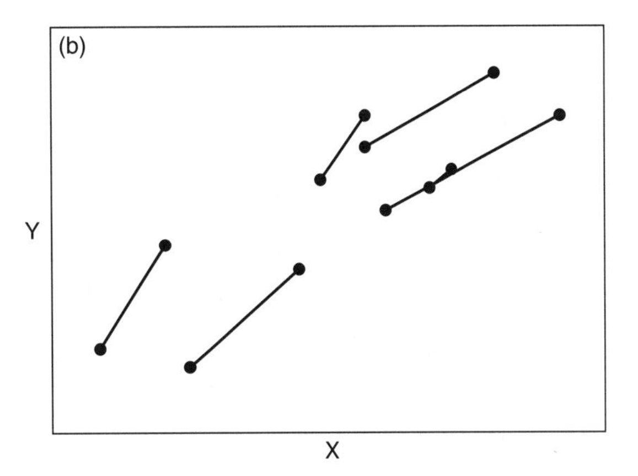

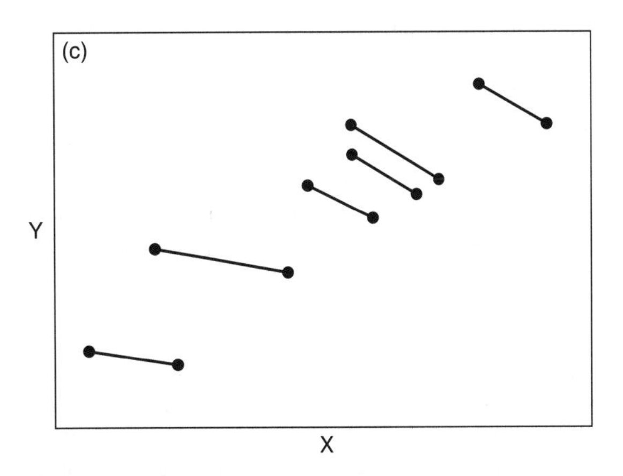

cross-sectional studies *Result from a hypothetical cross-sectional study (a) and corresponding longitudinal studies with increasing (b) and decreasing (c) time trends*

Another concern in a cross-sectional study of disease is whether there is a systematic BIAS or inaccuracy in the reporting of exposure by disease status. In some cases this can be avoided by using exposure measures that are not affected by disease, e.g. the determination of a particular genotype through the use of genomic analyses. However, if this cannot be avoided, then this potential source of bias may limit the strength or the association, as well as the strength of the evidence that the exposure of interest affects the aetiology of disease. *TRH*

Diggle, P. J., Liang, K.-Y. and Zeger, S. L. 1994: *Analysis of longitudinal data.* Oxford: Clarendon Press. **Kleinbaum, D. G., Kupper, L. L. and Morgenstern, H.** 1982: *Epidemiologic research: principles and quantitative methods.* Belmont: Lifetime Learning Publications. **Rothman, K. J. and Greenland, S.** 1998: *Modern epidemiology*. Philadelphia: Lippincott-Raven.

cross-validation See DISCRIMINANT FUNCTION ANALYSIS

crude birth/death rate See DEMOGRAPHY

cubic spline See SCATTERPLOT SMOOTHERS

cure models A cure model can be used in survival analysis when there are 'immunes' or 'long-term survivors' present in the data (Maller and Zhou, 1996). In such a setting, immune or cured subjects are censored since cure can never be observed, while susceptible subjects would eventually develop the event if followed for long enough. A typical example where a cure model might be appropriate would exhibit a Kaplan–Meier estimate of the marginal time-to-event distribution that levelled off at long follow-up times to a nonzero value (see KAPLAN–MEIER ESTIMATOR). An example is in studies of cancer for which a significant proportion of patients may be cured by the treatment.

A mixture model formulation is one approach to analysing such data (see FINITE MIXTURE DISTRIBUTIONS). Assume that a fraction p of the population are susceptibles and the remaining fraction are not; then the survival function $S(t)$ for the population is given by:

$$S(t) = pS_1(t) + (1-p)$$

where $S_1(t)$ is the survival function for the susceptible group and where covariates can affect both p and $S_1(t)$. Let (t_i, δ_i, Z_i) be the observations, where Z_i is a vector of covariates, t_i the observed or censored time and δ_i the censoring indicator. Let D_i indicate cure status for each subject denoted by $D_i = 1$ for a susceptible subject and $D_i = 2$ for cured. Thus each censored subject has either $D_i = 1$ and the event has not yet occurred or has $D_i = 2$. The incidence model is typically given by:

$$p(Z_i) = P(D_1 = 1|b, Z_1) = \frac{e^{b'Z_i}}{1 + e^{b'Z_i}}$$

Among susceptible individuals, the time to event has a distribution, such as a Weibull (Farewell, 1982):

$$S_1(t_i|D_i) = \exp\left[-\exp(\gamma' Z_i)t_i^{\alpha}\right]$$

An attractive feature of this model is the two separate components. The parameters b measure the effect of covariates on whether the event will occur and the parameters γ measure the effect of the covariates on when the event will occur given that the subject is susceptible. These two components are sometimes called incidence and latency and can have nice interpretations in given applications.

Different formulations can be used. Li and Taylor (2002) and Yamaguchi (1992) considered parametric and semi-parametric accelerated failure time models for the latency model. Kuk and Chen (1992), Peng and Dear (2000) and Sy and Taylor (2000) considered a semi-parametric proportional hazards model for the latency model.

One problem associated with the cure model is near-nonidentifiability (Farewell, 1986; Li, Taylor and Sy, 2001). This arises due to the lack of information at the end of the follow-up period, resulting in difficulties in distinguishing models with a high incidence of susceptibles and long tails of $S_1(t)$ from low incidence of susceptibles and short tails of $S_1(t)$. The incorporation of longitudinal data into the cure model is one way to reduce the problem (Law, Taylor and Sandler, 2002).

While the parameters in p and $S_1(t)$ have nice interpretations, in some applications the marginal survival distribution $S(t)$ and its dependence on Z may be of most interest. This distribution is easily obtained from the estimates of p and $S_1(t)$. Predicting the cure status of a censored subject may also be of interest. The formula to estimate the probability that a censored subject is in the susceptible group is given by:

$$P(D_i = 1|T_i > t_i) = \frac{pS_1(t_i)}{pS_1(t_i) + 1 - p}$$

The mixture cure model $S(t)$ does not in general have a proportional hazards structure. In order to keep this, however, nonmixture cure models have been proposed (Tsodikov, 1998; Chen, Ibrahim and Sinha, 1999). In these models, a bounded cumulative hazard is assumed: $\lim_{t\to\infty}\Lambda(t) = \theta$. One way to enforce this property is to write $\Lambda(t) = \theta F(t)$, where $F(t)$ is the distribution function of a nonnegative random variable. Then the survival distribution $S(t)$ for the population can be written as $S(t) = e^{-\theta F(t)}$, which has the cure rate $e^{-\theta}$. Covariates can be incorporated into the nonmixture cure model by assuming $\theta(Z_i) = \exp(\beta' Z_i)$.

Cure models are worthy of consideration for analysing data for which there is a strong scientific rationale for the existence of a cured group and empirical evidence of a nonzero limiting survival fraction, together with a substantial number of censored observations with long follow-up times.

An example where the cure model was applied arose from a study of 672 tonsil cancer patients treated with radiation therapy (Sy and Taylor, 2000). The radiation can eliminate all the cancer cells in the tonsil of some patients and thus the cancer will not reappear in the tonsil and the patient is regarded as being cured. If the radiation is not successful at eliminating all the cancer cells in the tonsil, those that remain will re-grow and become detectable, as a local recurrence, within about 3 years. This is a good situation where a cure model could be appropriate, because there is a scientific rationale for a cured group and because a Kaplan–Meier estimate of time of local recurrence will exhibit a long plateau region if there is sufficient follow-up in the data. For these data, there were 206 events of local recurrence and most patients had more than 3 years follow-up.

The main interest was in understanding the effect of the total dose of radiation and the overall treatment time between the start and the end of radiation on local recurrence. Other covariates, such as stage of the tumour and age of the patient were included in the analysis. A mixture cure model was used, with a logistic model for the incidence and a semi-parametric proportional hazards model for the latency. The results suggested that stage, dose and treatment time were strongly associated with whether the tumour recurred, as indicated by the parameters in the logistic model. Age, however, was not associated with the incidence. The estimates of the relative hazards parameters in the latency part of the model suggested that stage, dose and overall treatment time were not associated with when the recurrence would occur. The patient's age, however, was *strongly* associated with when recurrence would happen, given that the patient was not cured. The direction of the association was that younger patients would recur earlier. One possible interpretation of this is that young patients tend to have the same susceptibility to treatment as older patients but they tend to have faster growing cancers that will recur earlier if not cured. The initial size or stage of the tumour and how it is treated are important factors in determining whether a patient is cured, but are not important in determining how fast the tumour grows back after treatment if not cured. *JMGT*

Chen, M. H., Ibrahim, J. G. and Sinha, D. 1999: A new Bayesian model for survival data with a survival fraction. *Journal of the American Statistical Association* 94, 909–19. **Farewell, V. T.** 1982: The use of mixture models for the analysis of survival data with long-term survivors. *Biometrics* 83, 1041–6. **Farewell, V. T.** 1986: Mixture models in survival analysis: are they worth the risk? *The Canadian Journal of Statistics* 14, 257–62. **Kuk, A. Y. C. and Chen, C. H.** 1992: A mixture model combining logistic regression with proportional hazards regression. *Biometrika* 79, 531–41. **Law, N. J., Taylor, J. M. G. and Sandler, H.** 2002: The joint modelling of a longitudinal disease progression marker and the failure time process in the presence of cure. *Biostatics* 3, 547–63. **Li, C. S. and Taylor, J. M. G.** 2002: A semi-parametric accelerated failure time cure model. *Statistics in Medicine* 21, 3235–47. **Li, C. S., Taylor, J. M. G. and Sy, J. P.** 2001: Identifiability of cure models. *Statistics and Probability Letters* 54, 389–95. **Maller, R. A. and Zhou, X.** 1996: *Survival analysis with long-term survivors*. New York: John Wiley & Sons, Inc. **Peng, Y. and Dear, K. B. G.** 2000: A nonparametric mixture model for cure rate estimation. *Biometrics* 56, 237–43. **Sy, J. P. and Taylor, J. M. G.** 2000: Estimation in a Cox proportional hazards cure model. *Biometrics* 56, 227–36. **Tsodikov, A.** 1998: A proportional hazard model taking account of long-term survivors. *Biometrics* 54, 1508–15. **Yamaguchi, K.** 1992: Accelerated failure time regression models with a regression time model of surviving fraction: an application to the analysis of 'permanent employment'. *Journal of the American Statistical Association* 87, 284–92.

curtailment sampling See INTERIM ANALYSIS

D

data and safety monitoring boards These are committees of experts set up to monitor the safety of participants and validity and integrity of data in CLINICAL TRIALS. Some form of data and safety monitoring is called for in any trial to ensure minimal acceptable risks to trial participants and continually to reassess the risks versus benefits of trial interventions during the conduct to make sure that there is an *equipoise* in continuing the trial.

The International Conference on Harmonization defines good clinical practice (GCP) as 'an international ethical and scientific quality standard for designing, conducting, recording and reporting of trials that involve participation of human subjects'. Monitoring of trials for safety of participants, integrity of data leading to valid conclusions, adequate trial conduct and considerations for early termination to avoid unnecessary experimentation on human subjects is thus necessary to meet the stated GCP requirements.

The trial sponsor, the investigators and the institutional review board (IRB), also known as the ethics committee (EC), are at the frontline of safety monitoring for trial participants and they assume and share the responsibilities. However, the sponsor may elect to establish a data and safety monitoring board (DSMB), also known as an independent data monitoring committee (IDMC or DMC), and delegate part of its responsibilities to the DSMB. The establishment of a DSMB is recommended based on the recognition that monitoring of safety at regular intervals is essential to ensure safety of trial participants and that individuals directly involved in the conduct and management of a trial may not be suited for objective review of emerging interim data.

All clinical trials require monitoring of safety and efficacy data, but not all require monitoring by a DSMB independent of the sponsor and investigators. The degree and extent of such monitoring should depend on the potential risks associated with interventions, the severity of disease and ENDPOINTS of the trial and the method of monitoring on the size, scope and complexity. A DSMB is generally required for large, randomised controlled trials comparing mortality or major irreversible morbidity as a primary endpoint or pivotal trials for regulatory approval of marketing.

A DSMB is a body of experts who review accumulating data, both safety and efficacy, from an ongoing trial at regular intervals and advise the sponsor about the risks versus benefits and the scientific merit of continuing the trial. A typical DSMB is made up of people with pertinent expertise, including clinicians and scientists knowledgeable about the disease and interventions under investigation and a statistician knowledgeable about clinical trials methodology including methods for INTERIM ANALYSIS, to interpret the emerging data appropriately. A DSMB may also include a patient advocate or an ethicist. A DSMB is a separate entity from an institutional review board and its members should not be involved with the trial they monitor and have no conflict of interest, either scientific or financial.

A DSMB is primarily responsible for the appropriate oversight and monitoring of the conduct of trials for safety of participants and validity and integrity of the data. More specifically, the primary responsibilities of a DSMB include review of the study protocol and the plans for data and safety monitoring; evaluation of the progress of the study, including recruitment of trial participants, timeliness and completeness of follow-up, compliance with protocol procedures, performance of participating sites and other factors that may affect study outcome; assessments of risks versus benefits; making recommendations to the sponsor and the investigators concerning continuation, modifications to the protocol or termination of the trial; and communicating the findings from data and safety monitoring to the local IRBs.

A DSMB will allow difficult, mid-study decisions about the trials. DSMB members are provided unblinded data on the important outcome measurements at regular intervals or at intervals specified in the protocol. These unblinded data should be kept confidential from the sponsor and the investigators. A DSMB is responsible for making recommendations to the sponsor as to whether the trial should continue as originally planned or with modifications to the design, be temporarily suspended of enrolment or trial interventions until some uncertainty is adequately addressed or be terminated either because there is no longer equipoise among trial interventions or because it is highly unlikely that the trial can be successfully completed or meet its scientific goals.

The independence of the DSMB is intended to control the sharing of important comparative information and to protect the integrity of the trial from adverse impact resulting from premature knowledge about the emerging data. While small differences may be well accepted as nondefinitive, awareness of such differences may make investigators reluctant to enter patients on the trial, to limit entry to a certain subset of patients or to encourage patients to withdraw if they are assigned what they perceive as inferior intervention. Such tendencies will introduce biases and diminish the reliability

Encyclopaedic Companion to Medical Statistics: Second Edition Edited by Brian S. Everitt and Christopher R. Palmer

of the trial's eventual results or even preclude the completion of the trial. Limiting the access to unblinded interim data to a DSMB relieves the sponsor of the burden of deciding whether it is ethical to continue to randomise patients and helps protect the trial from biases in patient entry or evaluation.

A DSMB should have standard operating procedures and maintain records of all its meetings and deliberations, including interim results, and these should be available for review when the trial is completed. The DSMB standard operating procedures should specify meeting quorum, schedule and procedures, its decision-making rules and meeting follow-up. A DSMB should be consulted about the contents of interim reports that serve as a basis for the DSMB deliberations. A practical perspective on DSMBs and the recommendations for the operation and management of DSMBs can be found in Ellenberg, Fleming and DeMets (2002), reflecting a recent guidance for clinical trial sponsors on the establishment and operation of clinical trial data monitoring committee by the Food and Drug Administration of the US Department of Health and Human Services (www.fda.gov/cber/gdlns/clindatmon.htm).

Interim analyses of comparative trials are necessary to ensure that large differences between interventions do not go unnoticed, as well as to detect excess toxicity or unanticipated flaws in study designs. Routine reporting of toxicities or information about intervention administration helps ensure that interventions are being given safely and properly and improve trial quality. Routine reporting of outcome results, however, can harm study quality.

In general a DSMB would examine not only the trial data but also relevant external evidence from other sources. Its recommendation to the sponsor should be based on the interpretation of the results of the ongoing trial in the context of existing outside scientific data relevant to such interpretation. A final decision, as to whether or not to continue the trial, should not rely solely on a formal test of statistical significance.

The DSMB meetings provide a setting in which the clinical significance of early differences or lack thereof can be discussed openly with interim data and the complex statistical issues involved in sequential monitoring of a trial can be discussed at length. Focused discussions of the progress towards the scientific goals of a study are facilitated by a DSMB with access to unblinded data. Since intervention effects will be examined by a small group, the danger will be reduced that a promising trial will be informally stopped early with reduced accrual because of overinterpretation of interim results by the sponsor or the investigators. In addition, sequential monitoring rules are at best guidelines for complex decisions involving many aspects of a trial.

Deliberations and conclusions of the monitoring should be communicated to the sponsor and the IRBs without compromising the integrity of the trial. Recommendations resulting from monitoring activities should be reviewed by the study team and adequately addressed. Local IRBs should be provided feedback on a regular basis, including findings from adverse events and recommendations derived from data and safety monitoring. *KK*

Ellenberg, S., Fleming, T. R. and DeMets, D. L. 2002: *Data monitoring committees*. Boca Raton: Chapman & Hall/CRC.

data entry This process puts observations into electronic format for computer analysis. No successful statistical investigation takes place without the reliable and accurate collection of data and its conversion into a suitable electronic form for computerised analysis. While ostensibly a simple clerical process, it often suffers neglect in planning and execution that can jeopardise the smooth running of a research project.

The reliability of data collection is not specifically an issue for data entry, but we will see later how technological changes in data entry can encroach on the process. Most important for the majority of investigators is the accurate entry of data. In a formal research project such as a clinical trial there will be established and inviolable clerical procedures for data collection and entry that will help to ensure accuracy. But in many academic studies the researchers themselves will take responsibility for the complete collection and entry process.

With modern statistical packages and moderate information technology (IT) literacy on behalf of the user, this is a perfectly feasible and economic process for studies up to a few hundred variables and a few hundred cases. The spreadsheet data entry facilities of SPSS or Excel provide an easy way of entering data and, given that the researcher is entering the data, he or she can make checks during the procedure.

Two problems occur with this approach. One is simple clerical error or absent-mindedness in typing data, the other is a lack of an audit trail for changes in the spreadsheet. Dual entry is typically used to correct the first. Programs such as SPSS Data Entry or a program written in MS-Access or similar permit one user to enter data and then another to re-enter it from scratch. Any inconsistency is flagged and the appropriate variable checked. Such programs can also incorporate range checking. While taking extra effort, it is well worth the initial investment in design. Some argue against the administrative burden of double data entry, however, on the grounds that range checks, etc., will detect most clerical errors, yet it remains a sensible precaution, especially if temporary or external staff are to be used for data entry.

The audit of change is important, particularly if several researchers are reviewing the data. An individual correcting a variable may unwittingly invalidate another's previous analysis. It is good practice in these circumstances to set up a core dataset and then use a program to change individual data elements if they need revision. Thus, for example, a file with SPSS syntax language data transformation commands can be used to compute changes to an SPSS data file. It is then available for review by all in the team.

There is much interest in using personal digital assistants (PDAs), or internet browsers, for data entry, sometimes directly by the subject themselves. The superficial attractiveness of these procedures can be misleading. Transferring and merging data from a PDA is not necessarily simple and will usually require significant manual intervention. This can be a source of error and care needs to be taken in design to prevent this.

The use of a web page for data entry potentially gives access to many thousands of respondents. Setting up a reliable data entry page is not so simple. Browsing sessions often terminate for communications reasons mid-session and therefore program logic needs to identify successful completion. Care needs to be taken to identify unique data entry sessions by, for example, originating an IP address of the client browser. Data security is needed so a user cannot accidentally see other entries. A reasonable amount of programming effort is required to do this and will certainly require database, programming and HTML design skills to achieve it successfully. This does not mean it is not possible to design a simple web page to acquire data; it is just more complicated to acquire data both reliably and accurately.

Planning is the key to successful data entry. Before any data are collected the process flow for entering data into the computer package and its checking should be described and adhered to rigidly by the research team. Such discipline will pay off in smoothing the path to analysis. *CS*

[See also DATA MANAGEMENT]

data-dependent designs These are methods used for allocating treatments to patients in clinical trials that make constructive use of the emerging responses. Compared with traditional trial designs using predetermined sample sizes, data-dependent designs aim to impart some advantage to trial participants, reaching a conclusion sooner and/or exposing fewer to inferior therapy during the course of the trial. Such designs have been around in theory for at least as long as modern CLINICAL TRIALS, although their practical applications have hitherto been very limited. Other terms in use for similar methodological approaches for such trials include *flexible designs*, *dynamic designs*, ADAPTIVE DESIGNS and *learn-as-you-go designs*, although there is no apparent consensus on the nomenclature.

There are four broad categories of data-dependent designs, each of which shares the same spirit of learning from the accumulating data within the trial, as opposed to ignoring intermediate results until completion of the trial. These categories are: sequential, Bayesian, decision-theoretic and adaptive. Their descriptions given later in this entry deliberately avoid too much mathematical detail. Also, a distinction is drawn here from two other related types of clinical trial design not discussed further. First, there are designs that use MINIMISATION to incorporate knowledge of covariates of patients already entered into a trial (and hence self-modify according to treatment allocation though not to treatment response) and, second, there are trial designs intended to have an internal pilot study (see PILOT STUDIES).

Initially, however, it is worth considering briefly some historical background to help understand why modern clinical trials have emerged in the way they have. Typically they feature fixed sample sizes dictated by error probability considerations (see TYPE I ERROR and TYPE II ERROR rates), treatments being allocated equally (usually, to maximise POWERof a test) and results kept hidden (to all but a DATA AND SAFETY MONITORING BOARD (DSMB), if appointed) until the final analysis, which is conducted well after the final patient has been enrolled.

Interestingly, and tellingly, the roots of today's trials lie not in medicine but in agriculture. In the UK in the 1920s, R. A. Fisher began conducting crop field trials to try to determine which type of fertiliser produced higher yields of wheat. Realising there were more factors than could be listed (soil composition, aspect, slope, water and so on) that might influence total yield, Fisher (1926) pioneered RANDOMISATION to cope with the problem of balancing all the known and unknown variables as far as possible. It was a statistical masterstroke, for such use of an external chance mechanism alone could ensure that the comparison between fertilisers was fair and unbiased. Specifically, any difference observed in crop yield at harvest time could be attributed to the one factor that was known to be different between the groups, namely the fertiliser, all other factors being expected to be equal. Hence, inference from any observed differences between groups would link cause and effect (here, fertiliser and yield) as strongly as possible (see CAUSALITY).

The first medical application of randomisation came in the late 1940s, when A. B. Hill used Fisher's technique in a clinical trial testing streptomycin for the treatment of pulmonary tuberculosis. This was not without some controversy at the time, but Hill convinced sceptics by arguing that randomisation was also a fair way of allocating the scarce resource involved, given that the treatment was in strictly limited supply. This Medical Research Council sponsored trial became the first randomised controlled clinical trial to be published (MRC, 1948).

However, trials of today are fundamentally quite similar to those of 50 or more years ago, in that they typically involve equal allocation of treatments to patients, generally after performing a power calculation to determine a target number to be recruited. Thus, in a two-treatment comparative trial, half the patients customarily receive the standard and half the experimental treatment. As already mentioned, with the possible exception of DSMB committee members and a statistician conducting an INTERIM ANALYSIS, no one looks at the results until all the patients have been randomised and followed up. At the end of the trial it is possible that the experimental treatment is declared a statistically significant improvement and heralded as a clinical success. It is an ethical problem, however, if statistical 'failure' means the patient died and one can look back with some remorse wondering 'if only we had come to this conclusion sooner perhaps we could have saved some lives' (see ETHICS AND CLINICAL TRIALS). Even if the outcome is not as serious as death, the argument persists: Could fewer patients in the study have suffered on the way to reaching a valid conclusion?

This last question has motivated much research by ethically minded statisticians. Ironically, this work dates back at least as far as the first modern clinical trial, for the whole area of SEQUENTIAL ANALYSIS traces its history to the 1940s, World War II and US government-contracted statistician Abraham Wald (see Wald, 1943). His work, like Fisher's, was not in the medical area of application, but in ammunition testing, an altogether different example of seeking to cope with precious and limited resources. Medical application of sequential methods does seem entirely appropriate, after all, as patients arrive to be treated sequentially (they are not all waiting in line outside the doctor's office, or hospital clinic, at the start of a trial) and, similarly, results from some are available sooner than from others.

The rationale for sequential trials involves looking carefully at data *as they accrue* with a view to stopping just in time. Hence, the number of experimental units required is not fixed in advance but is a random variable. Theory shows that the expected numbers involved in a sequentially analysed randomised controlled trial is less than the corresponding fixed sample size trial, for any given power and level of significance. It is possible, when treatment groups fare broadly equally well, for a sequential trial to need slightly more patients overall compared with a trial using traditional design, but this would be quite unusual.

For better or worse, the clinical trial as conducted and analysed today is not in Wald's style of testing ammunition but rather in Fisher's application of fertiliser to fields of wheat. These two metaphors illustrate the fundamental difference between the statistics behind clinical trials that strive to learn-as-they-go and those that wait, literally, until harvest time before beginning to make scientific inferences. The reader may decide whether it is right that normative practice sees clinical trial volunteers afforded the same respect as the fertiliser rather than the ammunition.

Following Wald's pioneering research, sequential designs have evolved as sophisticated tools to assist those on DSMBs and hence can be considered mainstream, in contrast to the remaining design types discussed below. It should be said, however, that these methods are not routinely implemented as primary analytical tools for driving trials. Instead at best they are used as 'back seat drivers' to exert indirect influence on trial conduct. How do they do this? Essentially, as data accumulate, a test statistic can be plotted on a graph of treatment difference versus time, and trial recruitment can be recommended to terminate just as soon as a predetermined boundary is crossed.

This boundary may take on various shapes, the simplest being triangular with two possible options; either treatment A or B is declared better depending on which side of the triangle is crossed first. To allow for a third, nonconclusive, option with a predetermined maximum trial sample size, the boundary outline is modified to include a vertical line at a given point on the time (strictly 'information') axis. The idea is to stop the trial in favour of treatment A, say, if the upper line of the boundary is crossed first; B if the lower line; or else, conclude no clinically relevant difference between A and B if the vertical line is reached first.

There are variations on this theme with rules such as those derived by Pocock (1983) and by O'Brien and Fleming (1979) being popular examples. Thus it is not necessary to update the graph after every single observation. One can apply rules, called *group sequential methods*, that update after small batches of results become available. For more details refer to Jennison and Turnbull (2001). Statistical software for implementing these rules is readily available in several commercial packages (e.g. EaSt, PEST, S + SeqTrial).

One disadvantage with sequentially designed experiments is that their usefulness, namely their potential to learn while in progress, is self-limiting to trials having relatively rapid ENDPOINTS. Thus a sequential trial offers little benefit over a traditional, fixed sample size trial if the outcome remains unknown until years after randomisation. This may be so in breast cancer, for example, but is no limitation, for instance, in emergency medicine or in rapidly fatal diseases.

Turning to Bayesian designs, investigators start by eliciting a PRIOR DISTRIBUTION, either from a panel of clinical experts or from a reasonable selection of available theoretical distributions thought to mimic reality in terms of treatment success distributions (see BAYESIAN METHODS. For example, a BETA DISTRIBUTION with suitably chosen parameters can represent initial beliefs about a treatment's efficacy ranging from negatively skewed to uniformly distributed to positively skewed. In practice, there is virtue in choosing a prior that makes the experimental treatment appear initially a weak contender, so that positive results in favour of the treatment are not too dependent on initial choice of the prior. As the patients' results accumulate, the conditional distribution given the data thus far is evaluated – the so-called POSTERIOR DISTRIBUTION, amalgamating the prior and the LIKELIHOOD. Inference is based on the posterior, including the evaluation of CREDIBLE INTERVALS, analogous to CONFIDENCE INTERVALS in the frequentist context.

An advantage is the ease of interpretation of these intervals for they have more intuitive meaning to clinicians and patients. A disadvantage is the general lack of awareness of Bayesian methods since these are less often encountered than those from the frequentist school. This is reflected in the comparative lack of statistical textbooks, courses and software aligned to the Bayesian paradigm. Spiegelhalter, Freedman and Parmar (1994) provide an excellent overview of Bayesian methodology applied to clinical trials. Some see the subjective or arbitrary nature of the prior distribution involved as a weakness; others regard it as a positive opportunity to incorporate provisional information about the potential new treatment.

The third broad category of data-dependent designs involves the use of DECISION THEORY. Some experimental studies can be conducted with the resulting inference, in terms of how the information will be used to reach a practical decision concerning which treatment to recommend, as the driving force. For example, one can specify a criterion such as minimising expected successes lost, or maximising successes gained, over the course of a predetermined number of future patients, called the horizon, within and outside a comparative trial. Another criterion could be maximising the probability of correct selection of superior treatment. Either way, the focus is on the pragmatic need to make a decision and use one of the treatments or not once the trial is over in a direct attempt to balance the needs of current and future patients.

It is possible to discount future patients by putting more weight on present results, although this whole area can become mathematically quite intricate, especially when modelling with unconstrained 'multiarmed bandits' in the context of deciding among several treatments. Nevertheless, practical simplifications can be incorporated, such as limiting equal allocation among remaining treatments. In the case of just two treatments this amounts to allocating pairs of treatments until such time as it is optimal, by whatever criterion, to cease the comparative stage. After that one can switch all remaining patients within the horizon to the preferred treatment, or maybe enter them into a brand new randomised trial comparing this 'winner' with another novel treatment that is ready for a comparative trial. Thus, one is not constrained in actuality to put all remaining patients on to the indicated treatment, but one can act safely in the knowledge that the selection of the winner is working to the best available information, where 'best' is guaranteed until the original horizon is reached. (Note, in practice, the choice of horizon in absolute terms is not critical, for only an approximate size would need to be specified.)

Objections to the subjective nature of prior distributions involved in this type of decision-theoretic framework can be alleviated, for example, by appealing to minimax criteria. This means implementing a design that has good theoretical properties across a broad range of priors. Development of computer software to allow such designs to be implemented has been slower than for sequential methods, contributing to the current lack of use of decision-theoretic methods in practice.

The fourth category considered here, (response-)ADAPTIVE DESIGNS, is the most extreme type of data-dependent design. It incorporates the accruing information from the data to modify the treatment allocation probabilities away from 50:50 in the case of two treatments. Thus, for example, whereas the trial would start with equal allocation, as the data begin to favour one treatment even slightly, then it affects the odds of the next allocation being accordingly fractionally higher. In practice it works like this. Imagine a bag containing an equal number of red and blue balls. A red ball drawn indicates the next allocation is to treatment A; a blue ball, treatment B. If a success occurs a ball of the appropriate colour is added to the bag before the next drawing, and hence treatment allocation, takes place.

Adaptive designs were set back by a rather poor prototypical example of a mid-1980s trial (Bartlett, Roloff and Cornell, 1985) involving extracorporeal membrane oxygenation (ECMO) therapy, and which has received much attention in the statistical and medical literature. Ethicists, clinicians and statisticians have all contributed to the debate about this particular trial. It involved critically ill newborn babies and the relevant outcome in question really was a matter of life and death. In retrospect, it was clearly a mistake to begin this trial with precisely one ball of each colour in the bag instead of, say, ten of each. What ensued was a highly unbalanced distribution of treatment allocation (for ECMO babies generally lived, unlike many of those not on ECMO therapy) rendering sensible inference difficult, if not impossible. On the other hand, it can be said that since the ECMO trial, computing power and mobile technology, two prerequisites for successfully conducting an adaptive design, have taken huge leaps forward, making this design far more feasible to implement successfully than ever before.

These adaptive designs are the most controversial in the family of data-dependent designs. This is because they appear to react too quickly to early data, which may be subject to systematic bias, or time trends, and if not careful can begin to adapt too swiftly to chance results. There is also the criticism that if one treatment happens to be a PLACEBO, why should anything change after a success or a failure on such an inert substance? Nevertheless, with suitable cautions and awareness of the issues involved, adaptive designs can be a highly effective and ethically appealing design, despite once again the relative dearth of positive examples of their use so far.

A growing number of statisticians believe the 21st century will be characterised by more use of computer-intensive, data-dependent methods, so long as those responsible for conducting clinical trials are open to receiving suggestions on how to advance trial methodology. For further details, including when data-dependent methods are considered most suitable and a proposed strategy for their introduction, see Palmer (2002). In closing, it is worth remembering why one should consider using data-dependent designs. The primary reason is for their ethical advantage in terms of how patients within trials are regarded, without compromising the scientific rigour or usefulness of studies for the sake of future patients. There are secondary reasons besides, with benefits derived from the side effect of expecting fewer patients to be involved in learn-as-you-go trials compared with traditional trials. These benefits pertain to trial sponsors (notably the pharmaceutical industry), doctors, patients and ultimately the science of medicine itself. *CRP*

Bartlett, R. H., Roloff, D. W., Cornell, R. G. *et al*. 1985: Extracorporeal circulation in neonatal respiratory failure: a prospective randomised study. *Pediatrics* 76, 479–87. **Fisher, R. A.** 1926: The arrangement of field experiments. *Journal of the Ministry of Agriculture, Great Britain* 503–13. **Jennison, C. and Turnbull, B. W.** 2001: *Group sequential methods with applications to clinical trials*. London: Chapman & Hall/CRC. **MRC Streptomycin in Tuberculosis Trials Committee.** 1948: Streptomycin treatment for pulmonary tuberculosis. *British Medical Journal* 769–82. **O'Brien, P. C. and Fleming, T. R.** 1979: A multiple testing procedure for clinical trials. *Biometrics* 35, 549–56. **Palmer, C. R.** 2002: Ethics, data-dependent designs and the strategy of clinical trials: time to start learning-as-we-go? *Statistical Methods in Medical Research* 11, 381–402. **Pocock, S. J.** 1983: *Clinical trials: a practical approach*. Chichester: John Wiley & Sons, Ltd. **Spiegelhalter, D. J., Freedman, L. S. and Parmar, M. K. B.** 1994: Bayesian approaches to randomised trials. *Journal of the Royal Statistical Society, Series A* 157, 357–416. **Wald, A.** 1943: Sequential analysis of statistical data: report submitted to Applied Mathematics Panel National Defense Research Committee (declassified in Wald, A. 1947: *Sequential analysis*. New York, John Wiley & Sons, Inc.).

data dredging See POST HOC ANALYSES

data management This is the systematic management of a large structured collection of information. 'Data management' is always a component of data analysis, but is usually a more significant issue in large or multicentre studies where the data management features of software packages such as SPSS or Excel are inadequate. This will also be the case when the 'data model' of the study – the entities for which data are collected and the relationships between them – does not fit the standard rectangular data model or spreadsheet of the classical statistical package. Thus, for example, a study comparing treatment in hospitals may have three entities – hospital, ward and patient – that need data recording at each level. Longitudinal or repeated measurement studies also generate data that does not so easily fit the rectangular model.

Nonetheless, very many complicated studies are managed and analysed entirely in packages such as SPSS or SAS (see STATISTICAL PACKAGES). SAS in particular has very strong data management features. Several data files for each entity can be created, and the data merge features of these programs in order to prepare specific analyses. Because this merging of files is manual, more skill and experience on behalf of the researcher is needed and thorough documentation and understanding of merge procedures is essential to prevent error. Nonetheless, because only one programming language is used, the procedures are consistent and easier to learn. Although such an approach seems 'low-tech', there is much to recommend it for many studies.

The main weakness of this approach is the manual management of transaction updates and production of an audit trail. If, in the example above, more patients are recruited in a particular ward, then derivative files that include hospital or ward variables need to be recreated manually. In a very dynamic data environment this is tedious and also error prone. Equally the correction of values in one file similarly requires the recreation of all the derivative files.

An alternative is to consider using a formal data management tool, and this usually implies a database. With a fast modern PC, desktop database packages such as MS-Access are capable of managing datasets with some tens of thousands of cases and several hundred variables. Only the very largest studies will require a full SQL-compliant database, although there may be sound reasons for using the latter for security and access control.

Almost inevitably deployment of a database will require the production of data entry and update screens, a process that requires some programming ability. This is particularly the case if transaction control and an audit trail of changes are needed. Second, the database query statements needed to provide the appropriate rectangular matrix datasets for analysis can be complicated, and can require subtle understanding of SQL. Such in-depth expertise is not normally easily accessible in a research team and may be expensive to provide. Before deciding to use a database for data storage the research team should plan and budget for such skills to be available throughout the life of the project. Employing a programmer who then gets another job just before the end of the study can leave a research team without the support they need when wanting finally to analyse the data.

For this reason alone, it is often sensible to consider the acquisition of specialised clinical data management packages. These often include all the extra checks and forms necessary for formal CLINICAL TRIALS. Entering the appropriate terms in any web search engine will bring up several hundred companies offering products that are suitable – the difficulty will be in selecting one. Although there may be a seemingly significant initial cost (perhaps several thousand euros or dollars) the saving on development time, as well as the predictability and security of software operation, give a rapid payback.

At project conception it is usually possible to outline the extent of data management requirements dependent on the complexity of the problem. It is sensible for prior specification and budgeting of the software needed to take place, rather than awaiting project start and then developing ad hoc solutions. This will give the research team the security of control of the data over the project lifetime. *CS*

[See also DATA ENTRY]

data mining in medicine This is a branch of both computer science and statistics devoted to extracting useful knowledge from databases (also known as KDD, knowledge discovery in databases). In general, such knowledge is obtained by detecting various types of regularity and relation among the data, most often association rules, classification rules (see DISCRIMINANT FUNCTION ANALYSIS), linear and nonlinear dependencies and clusters (see CLUSTER ANALYSIS).

Depending on the context in which it is performed, data mining research may emphasise computational scalability of the algorithms or statistical significance of the results. The field benefits from a major injection of ideas and tools from general MACHINE LEARNING and pattern recognition and, as such, it is often considered also as part of ARTIFICIAL INTELLIGENCE.

Data mining (DM) is often described as an interactive process that involves both the computer and the human component. This is also why data visualisation is considered an essential part of the process. DM is more general than traditional statistical analysis, in the type of regularities that can be found (e.g. decision trees), in the size of the datasets (often in the range of millions of data items) and in the strong emphasis on visualisation of the data and automation of the analysis.

The application of DM to medicine has a long tradition. Automatic data collection in modern medicine is increasingly pushing towards the development and deployment of ools able to handle and analyse data in a computer-supported fashion.

Being able to detect sets of symptoms that are often simultaneously present (association analysis) can help predict which other symptoms may be observed (association rules). Observing many patient descriptions as well as their diagnoses may help find a rule to predict the diagnosis given a new patient (classification analysis). Spotting groups of similar patients can help customise the therapy (cluster analysis). Finally, being able to predict the expected cost of a patient based on his or her history can help insurance companies optimise their services (regression analysis – see MULTIPLE REGRESSION).

An early application of data mining in medicine is the decision tree learner ASSISTANT (Cestnik, Kononenko and Bratko, 1987; Witten and Frank, 1999). It was developed specifically to deal with the particular characteristics of medical datasets.

A whole new chapter in the application of DM techniques to biomedical data is being written with the introduction of genome-wide datasets. Genomic sequences for several organisms are now available online and the availability of high-throughput gene expression and proteomic data highlight the urgent need for efficient and flexible algorithms to extract the wealth of medical information contained in them. Datasets recording human genetic variability (SNPs) are soon expected and, with them, the possibility of correlating genotypic with phenotypic information (see GENETIC EPIDEMIOLOGY).

Classic examples of modern data mining methods are systems such as BLAST (Altschul *et al.*, 1990), which allows researchers to find related genetic sequences efficiently, together with a statistical assessment of the degree of similarity.

Significant biological discoveries are now routinely being made by combining DM methods with traditional laboratory techniques. For example, the discovery of novel regulatory regions for heat shock genes in *Caenorhabditis elegans* (Thakurta *et al.*, 2002) was made by mining vast amounts of gene expression and sequence data for significant patterns. *NC/TDB*

Altschul, S. F., Gish, W., Miller, W., Myers, E. W. and Lipman, D. J. 1990: Basic local alignment search tool. *Journal of Molecular Biology* 215, 403–10. **Bratko, I. and Kononenko, I.** 1987: Learning diagnostic rules from incomplete and noisy data. In Phelps, B. (ed.), *AI methods in statistics*. London: Gower Technical Press. **Cestnik, B., Kononenko, I. and Bratko, I.** 1987: ASSISTANT 86: a knowledge elicitation tool for sophisticated users. In Bratko, I. and Lavrac, N. (eds), *Progress in machine learning*. Wilmslow: Sigma Press. **Hand, D. J., Mannila, H. and Smyth, P.** 2001: *Principles of data mining (adaptive computation and machine learning)*. Cambridge, MA: MIT Press. **Lavrac, N.** 1999: Selected techniques for data mining in medicine. *Artificial Intelligence in Medicine* 16, 1, 3–23. **Thakurta, D. G., Palomar, L., Stormo, G. D., Tedesco, P., Johnson, T. E., Walker, D. W., Lithgow, G., Kim, S. and Link, C. D.** 2002: Identification of a novel *cis*-regulatory element involved in the heat shock response in *Caenorhabditis elegans* using microarray gene expression and computational methods. *Genome Research* 12, 5, 701–12. **Witten, I. H. and Frank, E.** 1999: *Data mining: practical machine learning tools and techniques with Java implementations*. London: Morgan Kauffmann.

data monitoring committee (DMC) See DATA AND SAFETY MONITORING BOARDS

decision theory This is an approach to the analysis of data that leads to choice between a number of alternative actions by consideration of the likely consequences. This is in contrast to the commonest form of analysis of data from a CLINICAL TRIAL that is based on hypothesis testing.

The decision-theoretic approach is most suitable for decisions in which the possible actions are entirely within the control of the decision maker. In a clinical trials setting, most suggestions for the use of decision theory have been in early-phase trials, the final outcome of which is usually a decision as to whether or not to continue with the clinical development programme.

In a decision-theoretic framework, the actions that will be taken as a result of the analysis are explicitly identified and a utility, or gain, assigned to each expressing the desirability of the action as a function of some unknown parameter. For example, in an early-phase drug trial, possible actions might be either conducting further clinical trials with the drug or abandoning the clinical development programme. The desirability of each of these actions depends on the true unknown efficacy. Information on the unknown parameter is summarised by a Bayesian posterior distribution (see BAYESIAN METHODS), indicating those values that are plausible given the observed data, and this can be used to obtain an expected value for the utility for each action. The action with the largest posterior expected utility may then be identified. This is the action that will be chosen by a rational decision maker whose preferences are accurately represented by the utility function.

A simple example based on that considered by Sylvester (1988) (see also the correction by Hilden, 1990) illustrates decision making in a PHASE II TRIAL. At the end of the trial a decision will be made as to whether or not to continue with Phase III development. The desirability of each of these two options is summarised by a utility function, which, if the observed data are binary (success/failure), depends on the true success rate, which will be denoted by p. Suppose that the success probability for the existing standard treatment is known; e.g. it may be known to be 0.5. Suppose also that if the Phase II trial is successful, some known number, denoted m, of patients will be treated with the new treatment in PHASE III TRIALS and that if it is found to be effective in the Phase III trial, a total of t additional patients will be treated with the new treatment. Patients treated with the standard treatment in the Phase III trial will receive the same treatment, regardless of the outcome of the Phase II trial, so need not be considered.

Suppose that the utility of each action can be measured by the number of future successes expected if that action is taken. If the Phase III trial is not conducted, the $m + t$ future patients will all receive the standard treatment. The success rate for these patients will then be 0.5, so that the expected number of future successes is $0.5\ (m+t)$. This does not depend on the success rate for the new drug since this will not be used for any further patients. If the Phase III trial is conducted, m patients will receive the new treatment in the trial, so that the expected number of successes for the patients in this trial is pm. If the Phase III trial is unsuccessful, the t further patients will receive the standard treatment and the expected number of successes for these patients will be $0.5t$. If the Phase III trial is successful, these patients will receive the new treatment so that the expected number of successes will be pt. We will assume that the Phase III trial will give the correct answer, so that it will be successful whenever $p > 0.5$, in which case the number of extra successes from treating the t future patients with the new rather than the standard treatment is the difference between pt and $0.5t$, which is $(p - 0.5)t$. The total expected number of successes if the Phase III trial is conducted will be $mp + 0.5t + (p - 0.5)tI(p > 0.5)$, where the indicator function $I(p > 0.5)$ is equal to 1 if $p > 0.5$ and 0 if $p \leq 0.5$. If p is large, this utility is large, reflecting the fact that continuation to Phase III is desirable, and if p is small, the utility is smaller as continuation to Phase III is undesirable.

If the value of p were known, we would take the action corresponding to the larger of the two utilities; i.e. we would continue to Phase III if the expected number of successes from doing so, $mp + 0.5t + (p - 0.5)tI(p > 0.5)$, was greater than the expected number from abandoning development of the experimental treatment, $0.5(m + t)$. In practice, of course, p is not known, but instead the information on p given by the observed data is summarised by its posterior distribution and the expected number of future successes from each action must be averaged over this distribution. The optimal action corresponding to the larger expected number of successes can then be selected.

In addition to making decisions at the end of a clinical trial, as illustrated here, decision theory can be used to make decisions before the study starts regarding the study design for clinical trials, as discussed by Sylvester (1988), and it is in this context that the approach is most often proposed. Design decisions considered might be those taken before the study starts regarding the sample size for a fixed sample size study or those taken during a sequential trial (see SEQUENTIAL ANALYSIS) about the future conduct of the trial. In the latter case, a method known as 'dynamic programming' may be used to obtain a sequence of optimal decisions by working backwards through the trial, considering each decision point in turn. Examples are given by Berry and Stangl (1996), who consider the problems of when to stop a sequential trial involving a single experimental treatment and of deciding which treatment to use for each patient in a sequential trial comparing an experimental treatment with a control.

Although the suggestion to use decision theory in clinical trials has a long history (see, for example, Colton, 1963), there has been little practical application (see DATA-DEPENDENT DESIGNS). The use of the approach has probably been limited by the difficulties associated with specification of appropriate utility functions. The detailed specification of the gain function also means that designs must be obtained with a particular type of trial in mind. One possible solution is to use what has been called a stylised Bayesian approach, as illustrated, for example, by Stallard, Thall and Whitehead (1999), in which parameters in the utility function are selected so as to lead to a design with attractive frequentist properties. *NS*

Berry, D. A. and Stangl, D. K. 1996: Bayesian methods in health-related research. In Berry, D. A. and Stangl, D. K. (eds), *Bayesian biostatistics*. New York: Marcel Dekker. **Colton, T.** 1963: A model for selecting one of two medical treatments. *Journal of the American Statistical Association* 58, 388–400. **Hilden, J.** 1990: Corrected loss calculation for Phase II trials. *Biometrics* 46, 535–8. **Stallard, N., Thall, P. F. and Whitehead, J.** 1999: Decision theoretic designs for Phase II clinical trials with multiple outcomes. *Biometrics* 55, 971–7. **Sylvester, R. J.** 1988: A Bayesian approach to the design of Phase II clinical trials. *Biometrics* 44, 823–36.

degrees of freedom This is an elusive concept that occurs throughout statistics. Essentially, the term degrees of freedom (DoF) means the number of independent units of information in a sample relevant to the estimation of a parameter or the calculation of a statistic. For example, in a 2 × 2 CONTINGENCY TABLE with a given set of marginal totals, only one of the four cell frequencies is free and the table is therefore said to have one degree of freedom. In many cases the term corresponds to the number of parameters in a model and in others to the number of parameters in a statistical distribution such as the *t*-DISTRIBUTION, the F-DISTRIBUTION and the CHI-SQUARE DISTRIBUTION. *BSE*

demography The study of population processes (Preston, Heuveline and Guillot, 2001). This entry provides a brief survey of the following topics: measures of fertility, measures of mortality, age standardisation, sources of data, historical demography and the demographic transition, population projection, population ageing and summary measures of population health.

Note that many demographic measures that are defined as 'rates' are not true rates in the sense that they are not measures of events per unit of person-time. These measures are identified by placing the 'rate' of their title in quotes.

We begin by discussing measures of fertility. Fertility refers to actual childbearing *performance*, not childbearing potential, which is called *fecundity*. The *crude birth rate* is the number of births per conventional unit of person-time. The person-time denominator is typically based on estimates of population size at mid-period multiplied by the length of the period. Where the period is a single calendar year (the usual circumstance) then this equals 1. For example, in England and Wales in 2001 there were 594 634 live births registered and the mid-year population was estimated at 52 084 500, giving a crude birth rate of 11.4 births per 1000 population per year.

The analogously calculated *crude death rate* may be subtracted from it to give the *rate of natural increase*. More specific measures of fertility are desirable because population age structure affects the childbearing potential of the population and because, at an individual level, births (unlike deaths) can be repeated and the likelihood of this happening depends on reproductive experience to that point. Thus, cumulative measures of individual fertility are also desirable.

The *general fertility 'rate'* (GFR) is calculated as the number of live births per conventional unit of female person-years in the age range 15 to 49, while the *total fertility 'rate'* (TFR) estimates the average number of babies that would be born per full reproductive lifetime – given current age-specific fertility rates. It equals the sum of the probabilities of giving birth in each of the years of life in which such a birth could occur, conventionally from 15 to 49.

The *gross reproduction 'rate'* (GRR) is the 'rate' at which mothers are reproducing themselves. It is an estimate of the average number of daughters that would be born to a woman during her lifetime if she passed through the childbearing ages experiencing the age-specific fertility rates of the population of interest. It can be estimated on the assumption that the proportion of female births is (approximately) 100/(100 + 105) = 0.488. The GRR is then 0.488 × TFR.

The *net reproduction 'rate'* (NRR) takes into account the prevailing mortality among women and thus estimates the extent to which each generation of mothers actually reproduce themselves, allowing for the proportion who die before reproducing. It can be calculated from a hypothetical birth cohort (e.g. of 1000 females) who are aged through the reproductive lifespan and exposed to the given death rates using lifetable methods. This yields an expected number of person-years lived by the cohort of potential mothers in each of the age intervals in which births could occur. Thus, formulaically (with Σ denoting 'sum of') the NRR = [0.488 × Σ(probabilities of giving birth in each of the years of life in which such a birth could occur × the person-years lived by the cohort at each age)]/number in the cohort.

By definition, a NRR of 1.0 equals 'replacement level fertility'. 'Zero population growth' will not typically be approached until several decades after the attainment of replacement level fertility because (previously) increasing populations typically have a higher proportion in the pre-reproductive ages than would obtain in the corresponding stationary population. This excess reproductive potential creates substantial momentum that is not slowed until the age structure approaches that of a stationary population. (See the discussion of stable population to follow.)

Turning to measures of mortality, the most basic is the *crude death rate*, analogous to the crude birth rate, giving the number of deaths per conventional unit of population time. Thus in 1999 in the US, 2 391 399 deaths were reported and the estimated mid-year population was 272 691 000, giving a crude death rate of 8.77 per 1000 population per year.

Death rates may be specific for sex, age group and cause: e. g. 8337 men aged 55–64 had their cause of death entered as heart attack (acute myocardial infarction) on their death certificates in England and Wales in 1990 – out of an estimated mean population at risk of 25.26×10^5, giving a rate specific for age, sex and cause of 330 per 10^5 per year.

Among other specific death 'rates' and ratios, an important one is the *infant mortality 'rate'*, conceptually, the probability of dying between birth and exact age 1 (${}_1q_0$ in lifetable notation). It is operationally defined, in relation, for example, to events occurring in a given calendar year, as:

$$\frac{\text{Number of deaths of liveborn infants who have not reached their first birthday}}{\text{Number of live births}} \times 1000$$

Note that this operational definition requires accurate counts of births and infant deaths and is therefore difficult to implement in the absence of a national vital statistics system with complete (e.g. greater than 95%) coverage. Only 75 of 191 member states of the World Health Organisation (WHO) met this criterion in 2000. Thus the infant mortality rate is most difficult to measure in those populations where infant mortality tends to be highest and of greatest public health importance. In such populations it is usually estimated using model lifetables starting from estimates of the *child mortality 'rate'*, which tend to be more robustly estimated.

Another important 'rate' is the *child (or under 5) mortality 'rate'*, conceptually, the probability of dying between birth and exact age 5 (${}_5q_0$ in lifetable notation), conventionally multiplied by 1000. It is the most robustly estimated measure of mortality in early life in low and middle income countries without comprehensive vital statistical systems. In such countries, it can be measured operationally (in demographic and health surveys and in censuses) by asking women of reproductive age about all the babies they have had and which of these have since died. There are standard demographic methods for using answers to these questions to estimate ${}_5q_0$. If details of dates of birth and death are available then mortality rates can be estimated directly. If only the numbers ever born and numbers alive (or dead) are known then the indirect method (also known as the 'Brass' method) is used. The WHO estimates that each year about 22% of global deaths occur in populations for which estimates of this type provide the only available evidence on mortality levels (at any age).

The *adult mortality 'rate'* is the probability of dying between exact age 15 and exact age 60 (${}_{45}q_{15}$). It is typically used by international agencies as a summary measure of adult mortality levels in low and middle income countries. High income countries with comprehensive vital statistical systems tend not to use this measure for their own purposes.

Maternal mortality is a topic of great policy interest globally but its measurement is fraught with difficulty. The WHO defines maternal death as the 'death of a woman while pregnant or within 42 days of termination of pregnancy, irrespective of the duration and the site of the pregnancy, from any cause related to or aggravated by the pregnancy or its management but not from accidental or incidental causes'. The *maternal mortality ratio* is the ratio of maternal deaths to live births × 100 000. Even in countries with the best vital statistical systems it is estimated that around one-third of maternal deaths are not identified as such by the ICD code assigned for the underlying cause of death. Elsewhere the ratio is subject to even greater uncertainty – making it unsuitable for comparisons between countries or over time. The global maternal mortality ratio for 1995 was estimated at 397/100 000 births with an uncertainty interval extending from 234 to 635 – emphasising the magnitude of the uncertainty associated with this measure.

In order to make fair comparisons, especially internationally, in demography it is necessary to standardise vital (birth and death) rates. Crude (unstandardised) death rates are poor guides for comparing the force of mortality in different populations: a retirement town, with many of its population in the older (dying) age ranges, will, purely as a function of its age structure, tend to have a very high crude death rate. The processes used to age-standardise can be conveniently described by distinguishing between the 'population of interest', i.e. the population whose vital rates are being characterised, and the *'reference' or 'standard' population*, i.e. either an artificial or a real population, used for standardisation.

Direct standardisation is one method requiring relatively precise estimates of the age-specific death rates in the population of interest. The standard population provides a standard age structure. The age-specific death rates of the population of interest are applied to the component age strata (of standardised size) of the standard population. In this way, each age stratum of the standard population is made to experience the same force of mortality as the corresponding age stratum in the population of interest. The resulting sum of deaths (in the standard population) is no longer influenced by the age structure of the population of interest and, when this sum is divided by the appropriate denominator (100 000 in this case), it yields a directly age-standardised death rate. A directly age-standardised death rate is, in effect, a weighted mean of the age-specific rates using a standard set of weights – with the weight for each age stratum being the standardised proportion it comprises of the total standard population.

The second method is called *indirect age standardisation.* If the population of interest is small or the deaths of interest are from an uncommon cause, the number of deaths occurring in some age strata may be too small to produce the stable estimates of the age-specific death rates that the direct method requires. In the indirect method, the age-specific death rates of a standard population are projected on to the age strata of the population of interest to give the number of deaths that would be expected in each age stratum on the basis of the standard rates. The ratio of the total observed deaths to the total 'expected deaths' is usually called the standardised mortality ratio (or SMR). Because SMRs are still influenced by the age structure of the populations of interest, each should, strictly, only be compared with the value for the standard population, i.e. with 1.00 (or 100 depending on which base is chosen).

Sources of data used for demographic measures can be illustrated for mortality. Around one-quarter of the 57 million deaths estimated to occur each year occur in countries where the vital statistical system has been judged to be at least 95% complete. Around 13% occur in populations whose vital statistical systems are less than 95% complete. For India, China and several smaller countries, vital rates are estimated using data from sample registration and surveillance systems. In these systems some 1% or so of the national population is covered by intensive surveillance for vital events. For populations in which around 22% of global deaths occur, child mortality can be estimated from survey and census returns on the numbers of children born and numbers still alive, even though there is little or no direct evidence on adult mortality levels. These are typically estimated using model lifetables to match plausible adult mortality levels to estimated child mortality. This leaves around 5% of deaths occurring in populations with no recent data on child or adult mortality.

Estimates of mortality in this last category of populations are entirely 'model based'; i.e. they are predicted from other known or estimated characteristics of the population. The calculation of death rates also requires estimates of populations at risk of dying. A minority of countries have regular censuses with coverage deemed complete. These countries estimate populations in intercensal years using adjacent censuses. At the other end of the data availability spectrum are populations with no recent censuses. For this group, bodies such as the Population Division of the UN have a long experience in preparing 'model-based' estimates of the size and age and sex distribution of the population, albeit with substantial levels of uncertainty. Thus, while mortality estimates are now prepared by international bodies such as the WHO for all components of the human population, many of these are subject to substantial uncertainty. The evolving philosophy of the WHO has been to make the best use of all available evidence and then to seek to quantify the level of uncertainty attaching to the resulting estimates. Life expectancy estimates published by WHO are now presented with uncertainty intervals. These intervals aim to quantify all sources of uncertainty, not just that associated with sampling error – hence their description as *uncertainty* rather than CONFIDENCE INTERVALS.

Historical demography is the branch of demography that studies how and why the force of mortality has changed through historical time, informing our understanding of the main determinants of human health, and is therefore of considerable interest. Historical demographers typically work their way backwards in time from more recent periods, with data that are readily available and of good reliability, to earlier periods where there are problems with either the availability or the quality of the available evidence. Mortality estimates based on a formalised system of data collection by parishes are available for Sweden from the mid-18th century. For England and Wales an official system of vital registration began in 1837. Before such systems were in place, parishes in England, for example, kept records of baptisms, burials and marriages. Historical demographers have used these records to 'reconstitute' families and from these genealogies have obtained both numerators (vital events) and denominators (estimates of person-time lived) for the estimation of vital rates. Family reconstitution has yielded estimates of fertility and mortality levels for England that now extend back to the 16th century. These constitute the longest such series for a North Atlantic society.

There have been two main findings from these data. First, it has been shown that the main means by which the English population adjusted to cyclic variation in economic fortunes, in the early modern period, was via the regulation of marriage (Wrigley *et al.*, 1997). When economic conditions became difficult, age at marriage increased and the proportion never marrying also increased. These departures from the pattern of universal early marriage as seen elsewhere have been characterised as the European marriage pattern. As nuptiality varied, so did fertility and with it the rate of population increase. Second, a high level of adult mortality in England in the early modern period has been observed. While somewhat more than 50% of those born survived to adulthood, among English males, for example, only around 30% of 15-year-olds could expect to survive to 65. It is of interest to note here that high levels of adult mortality were also typical of the poor agrarian society of India on the eve of its demographic transition. Around 1900, only 1 in 6 of 15-year-old Indian males could expect to survive to 65.

The overall transition from a 'pre-modern' to a 'late-modern' pattern of vital rates is described as the *demographic transition*. It begins with high mortality and fertility rates, followed by a period in which mortality declines in advance of the decline in fertility – a phase of the transition in which population growth accelerates. Fertility then declines – in an idealised form to reach replacement level (NRR = 1.0), with a new equilibrium being finally established with high survivorship. As has already been implied, the starting point for this demographic transition was more favourable in north-west Europe (in, say, the 17th century, when birth and death rates were 'submaximal') than in poor agrarian societies such as India (around, say, 1900, when birth and death rates were exceptionally high).

Turning from the past to the future, another important aspect of demography is making *population projections* and forecasts. Population projections, as the name implies, project existing populations forward in time under stated assumptions and in accord with established relationships between demographic parameters. Some projections may be known to be unrealistic but be carried out to explore 'what if' scenarios. Forecasts are those projections that are believed most likely to predict the future.

The standard method for projecting populations is known as the *cohort component* method. Typically, each 5-year age group in the population of interest is projected forward 5 calendar years at a time. It is depleted by expected losses to death and emigration and augmented by expected levels of immigration. At the beginning of life, births (to existing residents and to immigrants) are predicted. For these purposes, attention focuses on females to whom assumed fertility schedules are applied. A parallel exercise for males makes up the numbers. This exercise is repeated, starting again with the expected population in 5 years' time. The migration component usually introduces the largest levels of uncertainty into the calculations. Realistic assumptions entail nonlinear trends in fertility and perhaps also mortality so that the assumed rates need to be adjusted for each 5-year calendar period. Both the United Nations Population Division and the US Bureau of the Census prepare projections on 'high', 'medium' and 'low' assumptions for key inputs.

Thus, estimates for the size of the US population in 2050 vary by 102 million between low and high fertility assumptions, by 48 million between low and high mortality assumptions and 87 million between low and high migration assumptions. There is a general recognition that this scenario-based approach needs to be replaced by a more systematic approach to the quantification of uncertainty and its representation in probability distributions.

Understanding of the determination of age structure rests on the theory of *stable populations*. Stable populations emerge when the growth rate in the number of births is constant (or the schedule of age-specific fertility rates is constant), the schedule of age-specific death rates (i.e. the lifetable) is constant and there is no migration. In such populations, to which many historical populations approximate, various mathematical relationships hold between key parameters. The age distribution, the birth rate, the death rate and the growth rate are entirely determined by the fertility and mortality schedules. Populations that are not themselves currently approximating the stable model can nonetheless be said to have a 'stable equivalent', i.e. the population that would emerge if the birth and death schedules were allowed to act continuously. From this equivalence an 'intrinsic growth rate' may be determined.

One of the most striking and counterintuitive findings from stable population theory is that population age structure is very much more sensitive to changes in the fertility schedule than to changes in the mortality schedule (Coale, 1955). Thus with a gross reproductive rate of 2, increases in life expectancy from 40 to 60 years are associated with *reductions* in the mean age of the population. This is because increases in survival are proportionally greatest at each end of the lifespan. The increases in survival in the early years of life lead to larger cohorts of parents who in turn produce more children, keeping the base of the population pyramid extended. However, as fertility falls and life expectancy extends, proportions aged over 65 do increase.

Finally, as populations approach stationarity (sustained equality of birth and death rates), increases in survival are reflected in increased proportions of aged persons. On the way to such equilibrium, substantial perturbations may arise due to the passage of cohorts that are 'large' relative to those that immediately follow. These may have arisen from short periods of increased fertility, e.g. 'post-war baby boomers' in Western countries or from the last 'large' birth cohorts before subsequent substantial and rapid falls in fertility, e.g. in such countries as Japan, China and Italy. In the next half-century these presumptively transitional phenomena will result in periods of marked 'population ageing' when the relevant 'large' cohorts pass age 65. According to the UN Population Division's 'medium' variant projections, proportions aged over 65 will increase over the period 2010 to 2025 from 8.3% to 13.4% in China, from 20.4% to 24.4% in Italy and from 22.6% to an extraordinary 29.7% in Japan. By contrast, increases in the USA are expected to be more modest: from 16.1% to 18.1% (United Nations Population Division, 2009).

As populations approach stationarity, assumptions about limits to life expectancy become increasingly relevant. Oeppen and Vaupel (2002) have shown how demographers have repeatedly underestimated such limits. Mortality decline at high ages has continued in low mortality countries and has so far shown little evidence of slowing down at the highest ages.

Demography has played an important role in the development of methods for measuring the burden of disease and injury (Murray *et al.*, 2002). For example, the health-adjusted life expectancy (HALE) measure seeks to estimate the expectation of life in 'full health'. Time expected to be spent in less than full health is subtracted from total life expectancy, after weighting by the severity of the departure from full health. 'Health gap' measures, such as the disability-adjusted life-year (DALY) lost, estimate the hypothetical flows of 'lost healthy lifetime' arising from deaths and from onsets of disease and injury during the period of interest. For the 'years of life lost' component (and for long-term nonfatal health losses), gaps are estimated relative to a standard lifetable with a female life expectancy at birth of 82.5 years and a male life expectancy at birth of 80.0 years. Unlike health expectancy-type measures (such as HALE), health gap measures can be decomposed by allocating DALYs lost to the diseases and injuries responsible and also into the determinants of the diseases and injuries. *JP*

Coale, A. J. 1955: How the age distribution of a human population is determined. In *Proceedings of Cold Spring Harbour Symposia on Quantitative biology*, pp. 83–9. **Murray, C. J. L., Salomon, J. A., Mathers, C. D. and Lopez, A. D.** 2002: *Summary measures of population health: concepts, ethics, measurement and applications.* Geneva: World Health Organization (www.who.int/pub/smph/en/index.html). **Oeppen, J. and Vaupel, J. W.** 2002: Broken limits to life expectancy. Science 296, 1029–31. **Preston, S. H., Heuveline, P. and Guillot, M.** 2001: Demography: *measuring and modeling population processes.* Oxford: Blackwell. **United Nations Population Division** 2009: *World population prospects: the 2008 revision.* New York: UN Department of Economic and Social Affairs (http://www.un.org/esa/population/). **Wrigley, E. A., Davies, R. S., Oeppen, J. E. and Schofield, R. S.** 1997: *English population history from family reconstitution, 1580–1837.* Cambridge: Cambridge University Press.

dendrogram See CLUSTER ANALYSIS IN MEDICINE

density estimation This is the estimate of a probability distribution from a sample of observations. In many situations in medical research we may wish to use a sample of observation to estimate the frequency distribution or probability density of a variable of interest. Commonly this estimation problem is approached by simply constructing a HISTOGRAM of the data. However, the histogram may not be the most effective way of displaying the distribution of a variable, because of its dependence on the number of classes chosen. The problem becomes even more acute if two-dimensional histograms are used to estimate BIVARIATE DISTRIBUTIONS.

The density estimates provided by one- and two-dimensional histograms can be improved in a number of ways. If, of course, we were willing to assume a particular form for the distribution, e.g. normal, then density estimation would be reduced to estimating the parameters of the chosen density function. More commonly, however, we would like the data to 'speak for themselves' as it were, in which case we might choose one of a variety of the nonparametric density estimation procedures available. Perhaps the most common are the kernel density estimators, which are essentially smoothed estimates of the proportion of observations falling in intervals of some size. The essential components of such estimators are the kernel function and bandwidth or smoothing parameter. The kernel estimator is a sum of 'bumps' placed at the observations. The kernel function determines the shape of the bumps while the window width determines their width. Details of the mathematics involved are given in Silverman (1986) and Wand and Jones (1995), but the essence of the procedure can be gleaned from the first figure. Here the kernel

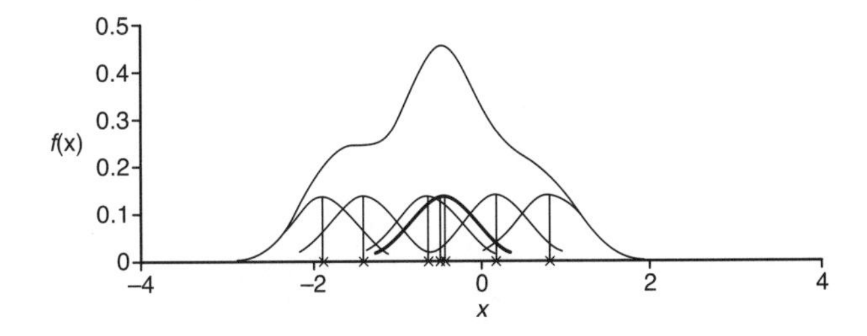

density estimation *Kernel estimate showing individual kernals (Silverman, 1986)*

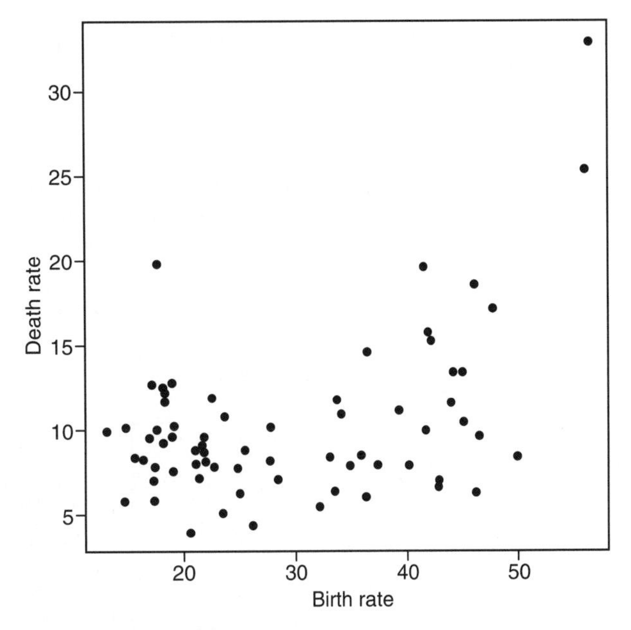

density estimation *Scatterplot of birth and death rates for 69 countries*

function is Gaussian, and the diagram shows the individual bumps at each observation as well as the density estimate obtained from adding them up.

The kernel density estimator considered as a series of bumps centred at the observations has a simple extension to two dimensions as described in, for example, Silverman (1986). Here we content ourselves with an example. The second figure shows a plot of birth and death rates for 69 countries and the third figure (in (a) and (b)) shows perspective plots of density estimates given by using different kernel functions.

Bivariate density estimates can also be useful when applied to the separate panels in a SCATTERPLOT MATRIX of data with more than two variables. As an example the fourth figure shows the scatterplot matrix of data consisting of three body

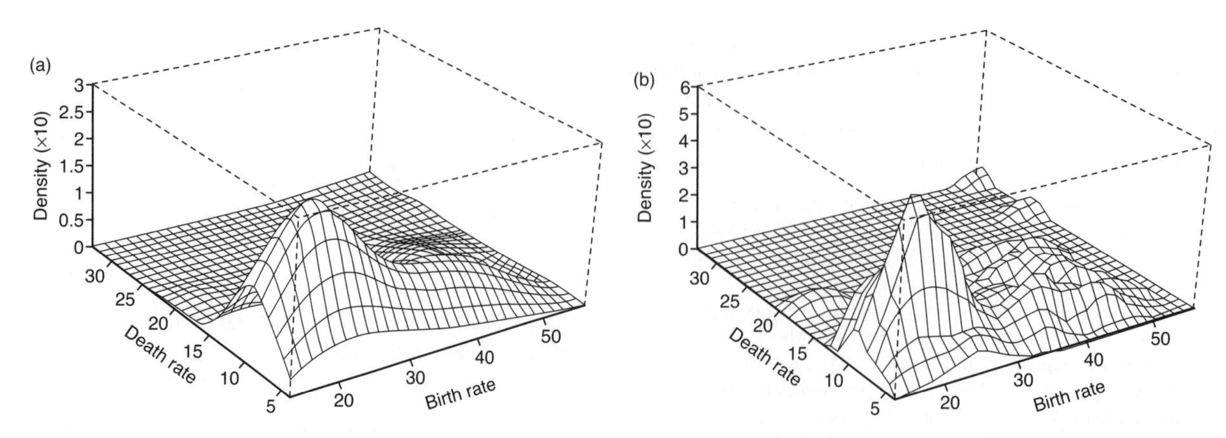

density estimation *Perspective plots of two density estimates for the birth and death rate date: (a) bivariate normal kernal; (b) Epanechnikov kernel*

measurements on 20 individuals with a contour plot of the appropriate estimated bivariate density function on each panel. There is clear evidence of two modes in the estimated densities, which is explained by the presence of men and women in the sample. *BSE*

Silverman, B. W. 1986: *Density estimates for statistics and data analysis*. London: CRC/Chapman & Hall. **Wand, M. P. and Jones, M. C.** 1995: *Kernel smoothing*. London: CRC/Chapman & Hall.

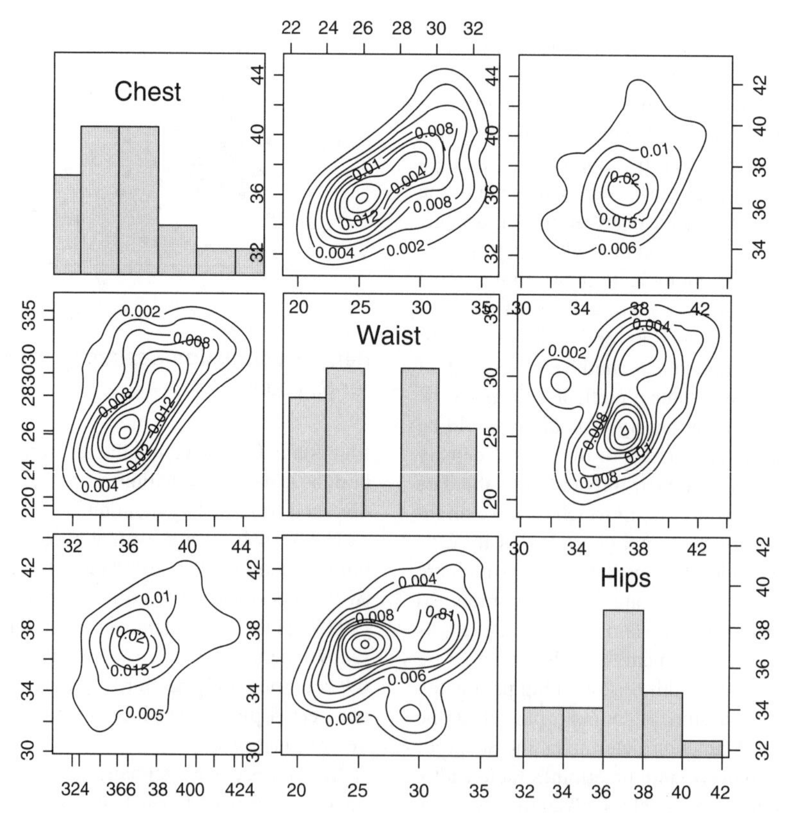

density estimation *Scatterplot matrix of body measurements data showing the estimated bivariate densities on each panel*

dental statistics Dentistry is concerned with the provision of care for the teeth, supporting tissues and the gums, and the treatment of diseases affecting these areas of the mouth. In the United Kingdom, the Social Survey Division of the Office for National Statistics carries out the Adult Dental Health Survey every 10 years (see, for example, the report on the 1998 survey by Kelly, Walker and Cooper, 2000). Participants are interviewed face to face and in addition those with natural teeth are asked to undergo a home dental examination. Using a random sample of several thousand individuals aged 16 years and over from England, Scotland, Wales and Northern Ireland, this survey yields information by constituent country on issues that include the number and condition of teeth, dental hygiene behaviour, patterns of dental visits and attitudes towards the provision of dental health care. In a similar manner the Child Dental Health Survey, involving schoolchildren aged between 5 and 15 years, has been conducted on a 10-yearly basis since 1973. Statistical analyses are provided in detailed official reports of these surveys.

For all dental specialties, current evidence is advanced through the conduct of suitably designed research studies that use statistical methods for data analysis. Developments in dental public health, especially aspects relevant to children, have attracted the most attention in the media. Worldwide, important public health themes have included the impact of the fluoridation of public water supplies on dental caries, the effect of the introduction of fluoride toothpaste on dental health, the decline in dental caries experienced by schoolchildren since the 1960s and the influence of socioeconomic factors and ethnicity on the provision and uptake of dental services.

In dental studies, possibly more so than for those of other types of health care, data from consecutive patients should not be treated as independent observations. In routine dental appointments, a significant number of patients are examined as part of a family group, with members being seen consecutively. Correlation between observations occurs because within a household, individuals tend to eat similarly and engage in the same type of routine dental care. For example, Metcalf *et al.* (2007) reported a within-household INTRACLASS CORRELATION COEFFICIENT (ICC) for the consumption of sugars of 0.578. Other types of cluster encountered include school classes in the study of adolescents and nursing homes in surveys of the elderly. Cluster randomisation (see CLUSTER RANDOMISED TRIALS) can be used to address this problem in randomised controlled CLINICAL TRIALS (Frenkel, Harvey and Newcombe, 2001).

Similarly, data from an individual's teeth cannot be treated as independent observations; an individual might, for example, have similar patterns of fillings on the left and right sides of the jaw. Consequently, in nearly all studies of teeth the unit of observation is taken as the individual rather than the tooth. Recently, with the development of more sophisticated statistical methods, studies that analyse individual teeth as correlated observations have started to appear, for instance that by Chuang *et al.* (2002) into possible factors influencing the survival time of dental implants. This paper also contains a useful review of SURVIVAL ANALYSIS techniques that have been applied in the modelling of dental implant failure.

Continuous and near-continuous dental data rarely follow a NORMAL DISTRIBUTION. For instance, salivary counts of bacteria generally show an extremely high positive skew between individuals. Consequently, NONPARAMETRIC METHODS have played an important role in data analysis; even at a basic level the use of the MANN-WHITNEY RANK SUM TEST is commonplace. Alternatively, taking logarithms of the observations can produce a much more symmetrical distribution and allow methods that require an assumption of normality to be considered.

In studies that involve clinicians performing a dental examination, a fraction of the participants might be reassessed in order to gauge examiner reproducibility. For instance, in a study of the dental heath of pre-school children, 10% of the participants were examined on a second occasion for this purpose (Godson and Williams, 1996). Generally, a form of the kappa statistic (Cohen, 1960, and see KAPPA AND WEIGHTED KAPPA) is used as the measure of agreement. SENSITIVITY, SPECIFICITY and the measurement of agreement are important in the comparison of results from screening with those from definitive findings. Bell *et al.* (2003) compared apparent characteristics of third molar ('wisdom') teeth from a radiological assessment with the actual features seen at surgery.

Populations may consist of two or more distinct groups representing different levels of dental care. One group of individuals might exercise a high level of care including the use of dental floss and disclosing tablets, another group might rely solely on the brushing of teeth, with a further group undertaking no regular dental care. For data consisting of counts, the POISSON DISTRIBUTION is therefore rarely appropriate. With the DMF score (sum of the numbers of decayed, missing and filled teeth) there is generally a higher proportion of individuals with a zero score (representing a perfect set of teeth – more than likely from the group exercising a high level of care) than this model would predict. Models that allow for OVERDISPERSION, such as the zero-inflated Poisson, have been found to provide a better fit for this type of data (Böhning *et al.*, 1999).

As with medical studies, data can consist of repeated measurements, for which appropriate methods of analysis are required. For instance, in a study of xerostomia (self-reported dryness of the mouth) and reduced production of saliva in individuals over a period of four years, Navazesh *et al.* (2003) applied repeated measures regression to data from 6-monthly assessments. META-ANALYSIS is increasingly being used as a tool in the review of specific issues in dental research. For example, Ismail and Hasson (2008) described a meta-analysis of the studies published between 1966 and 2006 into the association between fluoride supplements, dental caries and fluorosis, a dental condition that is characterised by staining of the teeth. Brief summaries of critical reviews can be found in the journal *Evidence-Based Dentistry*. *NCS*

Bell, G. W., Rodgers, J. A., Grime, R. J., Edwards, K. L., Hahn, M. R., Dorman, M. L., Keen, W. D., Stewart, D. J. C. and Hampton, N. 2003: The accuracy of dental panoramic tomographs in determining the root morphology of mandibular third molar teeth before surgery. *Oral Surgery, Oral Medicine, Oral Pathology, Oral Radiology and Endodontics* 95, 119–25. **Böhning, D., Dietz, E., Schlattmann, P., Mendonça, L. and Kirchner, U.** 1999: The zero-inflated Poisson model and the decayed, missing and filled teeth index in dental epidemiology. *Journal of the Royal Statistical Society A* 162, 195–209. **Chuang, S. K., Wei, L. J., Douglass, C. W. and Dodson, T. B.** 2002: Risk factors for dental implant failure: a strategy for the analysis of clustered failure-time observations. *Journal of Dental Research* 81, 572–7. **Cohen, J.** 1960: A coefficient of agreement for nominal scales. *Educational and Psychological Measurement* 20, 37–46. **Frenkel, H., Harvey, I. and Newcombe, R. G.** 2001: Improving oral health in institutionalised elderly people by educating caregivers: a randomised controlled trial. *Community Dentistry and Oral Epidemiology* 29, 289–97. **Godson, J. H. and Williams, S. A.** 1996: Oral health and health related behaviours among three-year-old children born to first and second generation Pakistani mothers in Bradford, UK. *Community Dental Health* 13, 27–33. **Ismail, A. I. and Hasson, H.** 2008: Fluoride supplements, dental caries and fluorosis: a systematic review. *Journal of the American Dental Association* 139, 1457–68. **Kelly, M., Walker, A. and Cooper, I.** 2000: *Adult Dental Health Survey: Oral Health in the United Kingdom 1998: a survey commissioned by the United Kingdom Health Departments carried out by the Social Survey Division of the Office for National Statistics in collaboration with the Dental Schools of Birmingham, Dundee, Newcastle and Wales, and the Central Survey Unit of the Northern Ireland Statistics and Research Agency*. London: Stationery Office. **Metcalf, P. A., Scragg, R. K. R., Stewart, A. W. and Scott, A. J.** 2007: Design effects associated with dietary nutrient intakes from a clustered design of 1 to 14-year-old children. *European Journal of Clinical Nutrition* 61, 1064–71. **Navazesh, M., Mulligan, R., Barron, Y., Redford, M., Greenspan, D., Alves, M. and Phelan, J.** 2003: A 4-year longitudinal evaluation of xerostomia and salivary gland hypofunction in the Women's Interagency HIV Study participants. *Oral Surgery, Oral Medicine, Oral Pathology, Oral Radiology and Endodontics*, 95, 693–8.

descriptive statistics These are summaries designed to encapsulate meaningful aspects of datasets. Here we focus on numerical descriptive statistics, GRAPHICAL DISPLAYS being considered separately. Individual observations are the basis of statistical analysis. However, when describing data it is rarely feasible to present all observations, and it is not always possible to illustrate the distribution using a graph. Therefore descriptive statistics are required to provide a numerical summary of the distribution.

MEASURES OF LOCATION are used to describe in a single figure the typical or representative level of all observations. The measures of location most often employed are the MEAN, MEDIAN, GEOMETRIC MEAN and MODE.

Variability around this central point is summarised by means of a MEASURE OF SPREAD. The RANGE, STANDARD DEVIATION and INTERQUARTILE RANGE are used as measures of spread.

Other aspects of a distribution are encapsulated by SKEWNESS, which measures how asymmetric the distribution is, and KURTOSIS, which quantifies its 'peakedness'.

It is important that descriptive statistics chosen are appropriate to the distribution of the data. Data that have an approximately symmetric distribution are usually summarised using the mean and standard deviation. On the other hand, the presence of skewness or OUTLIERS implies that the median and interquartile range are more appropriate. This is because they are based on ranks, and therefore make no assumptions about the distribution of the data. *SRC*

deviance Deviance is a measure of the extent to which a particular model differs from the saturated model for a dataset. Defined explicitly in terms of the difference in the LIKELIHOODS of the two models, deviance $= -2[\ln L_c - \ln L_s]$ where L_c and L_s are the likelihoods of the current model and the saturated model respectively. Large values of the deviance are encountered when L_c is small relative to L_s, indicating that the current model is poor. Small values of the deviance are obtained in the reverse case. Asymptotically, the deviance has a CHI-SQUARE DISTRIBUTION with DEGREES OF FREEDOM equal to the difference in the number of parameters in the two models. *BSE*

[See also GENERALISED LINEAR MODELS, LOG-LINEAR MODELS]

digit preference This is the personal and often subconscious BIAS that frequently occurs in the recording of observations. For example, a person may round the terminal digit of a number (i.e. the 5 in 624.75) systematically to a particular digit or set of digits that they prefer. Most frequently people will want to round to zero or five, although other numbers are of course possible. This is a problem when recording data from an analogue source, i.e. a clock or sphygmomanometer, and when recall or estimation is required, i.e. recalling the age at which menopause began or one's weight in kilos. It can also be a problem when using visual analogue scales or similar devices to garner information. In medicine the digit preference phenomenon is particularly troublesome in the recording of blood pressure, where it has been estimated that clinicians have up to a twelvefold bias in favour of the terminal digit zero (see the figure on page 137). This type of bias may have

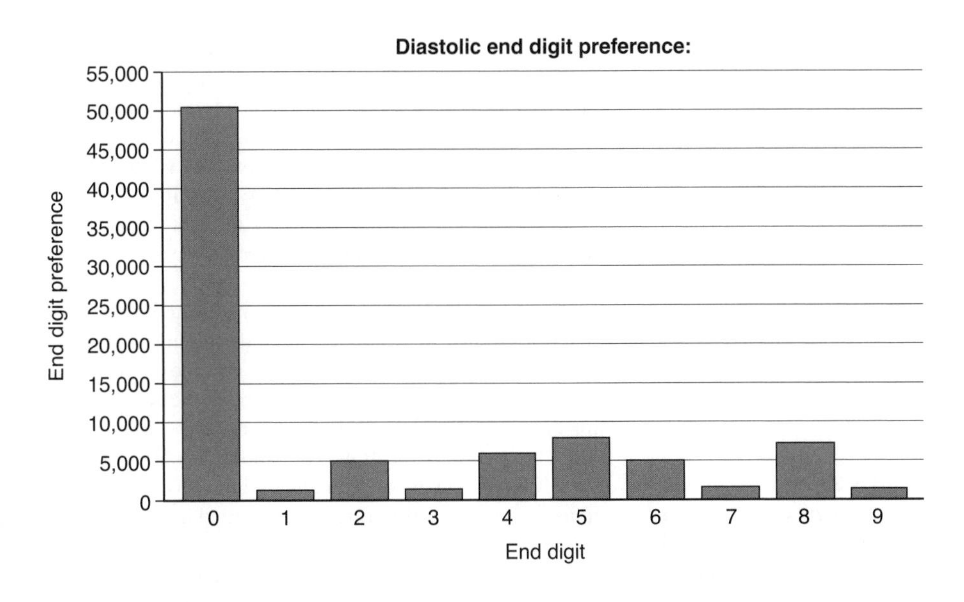

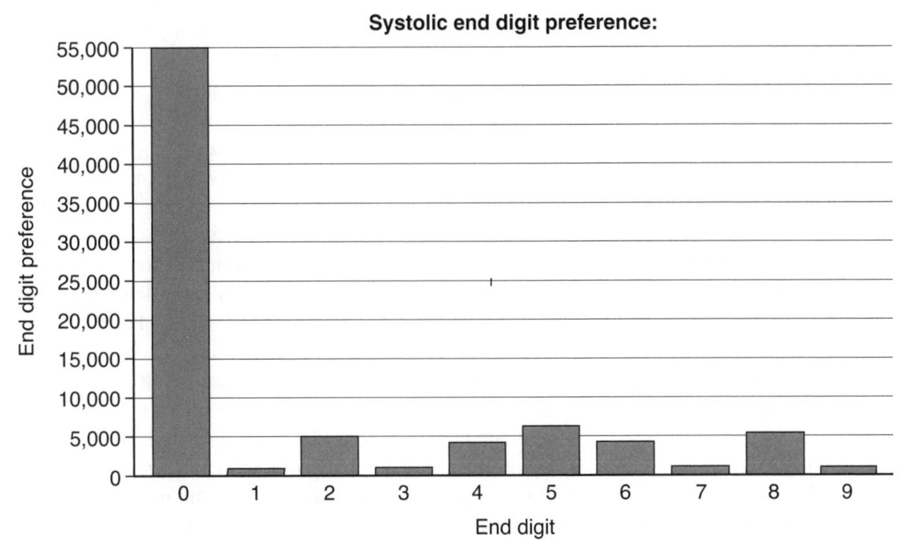

digit preference *Diastolic and systolic end digit preference*

grave implications for diagnosis and treatment, but its greatest effect is perhaps in epidemiological and other research studies where it can distort frequency distributions and reduce the power of statistical tests (see Hessel, 1986). Johnstone (2001) identifies a study in which digit preference may have led the authors to the wrong conclusion about the effect of a drug in the treatment of hypertension (Persson, Vitols and Yue, 2000).

There are times when it is appropriate to compare the distribution of terminal digits to a discrete uniform distribution in order to detect digit preference. However, one should not lose sight of the fact that, as demonstrated in Crawford, Johannes and Stellato (2002), it is possible for the distribution of terminal digits to be nonuniform, but still not the result of digit preference. *BSE/AGL*

[See also FRAUD DETECTION IN BIOMEDICAL RESEARCH]

Crawford, S. L., Johannes, C. B. and Stellato, R. K. 2002: Assessment of digit preference in self-reported year at menopause: choice of an appropriate reference distribution. *American Journal of*

Epidemiology 156, 676. **Hessel, P. A.** 1986: Terminal digit preference in blood pressure measurement: effects on epidemiological associations. *International Journal of Epidemiology* 15, 122–5. **Johnstone, G. D.** 2001: Letter. *British Medical Journal* 322, 110. **Persson, M., Vitols, S. and Yue, Q. Y.** 2000: Orlistat associated with hypertension. *British Medical Journal* 321, 87.

direct standardisation See DEMOGRAPHY

disability-adjusted life-year See DEMOGRAPHY

disattenuated correlation See CONFIRMATORY FACTOR ANALYSIS

discriminant function analysis This is a collection of methods aimed at optimally distinguishing between a priori groups of individuals, so that future unassigned individuals can be classified to one of the groups. To illustrate the problem in medical contexts, consider the following three situations.

First, patients entering hospital with jaundice could be suffering from one of a number of diseases. Some of these diseases require surgery, while others can be treated completely by medical means. Exact determination of disease may itself require surgery, which is to be avoided if at all possible, so it is hoped that diagnosis might be achieved via a battery of observations (signs, symptoms and laboratory measurements) taken on the patient. Such data are available on a sample of patients for whom either a biopsy or a *postmortem* examination has established the underlying disease, and hence the medical or surgical status, with certainty. Can we use these data to formulate a rule for predicting the status of a future patient from the battery of observations made on him or her?

Second, a retrospective study is conducted on patients undergoing surgery and the appearance or otherwise of postoperative pulmonary embolism is recorded for each patient alongside a range of other variables (e.g. age, obesity measure, number of cigarettes smoked per day, nature of disease, etc). Can the data be used to develop a screening index for predicting patients at high risk of appearance of postoperative pulmonary embolism?

Third, consider a prospective study being conducted on patients with thrombophlebitis. Each patient is monitored on a range of laboratory measurements; some patients develop embolic thrombosis, while others do not. Can those who develop it be predicted from the measurements monitored in the study?

These situations share some common characteristics and objectives. In each of them there are two distinct *groups* of individuals and observations have been recorded on a set of *features* (*variables* or *attributes*) for each individual. The hope is that these features are able to distinguish the two groups from each other well enough for the measurements taken on them to be capable of predicting the group membership of a future individual. The process of distinguishing between groups is known as *discrimination*, while the prediction of group membership of future individuals is termed *classification* or *allocation*. Typically, the best way of utilising the features is to combine them in some way, i.e. to form a *function* of them.

Discriminant function analysis then aims to find the best function for distinguishing between the groups, while formulating a *classification rule* will provide a means of predicting group membership. Frequently, the best function for discriminating between two groups also directly provides the best classification rule, but this is not always the case. There are many different potential discriminant functions and classification rules in any practical situation, so we need to be able to judge their performances in order to choose the best ones. The worth of a discriminant function can be assessed by any measure that estimates the separation between the groups using this function, while the worth of a classification rule is generally determined by estimating the probabilities of misclassifying future individuals with this rule.

The ideas and problems can be readily generalised to situations where there are more than two groups. For example, in the earlier jaundice case, we may be interested in discriminating between the actual *diseases* causing the condition rather than just between the group that requires surgery and the group that can be cured by medicine. The underlying principles remain the same, but the details become more complicated. For example, in discrimination we have to be able to assess the extent of separation among all groups (or all pairs of groups simultaneously), while in classification there are now many more potential mistakes that can be made in predicting an individual's group membership. For the purpose here, therefore, we will describe the methodology for the case of two groups and later merely indicate how the methods generalise to multiple group situations.

The first attempt at formulating and tackling the problem was made by Fisher (1936), who looked for the linear combination of features that maximally separates the two groups. He tackled this by maximising, over all linear combinations, the ratio of the squared difference between the group means to the pooled within-group VARIANCE of the observations, which is equivalent to maximising the squared t-statistic that tests for a difference between the two groups. The coefficients of the features in this linear combination are given by the elements of the vector formed on multiplying the vector of mean differences of the features by the inverse of their pooled COVARIANCE MATRIX. This very simple function has become known as *Fisher's linear discriminant function* (LDF).

Fisher derived his result in a purely practical setting, using actual data rather than statistical or probability models. Welch (1939) took a theoretical approach and showed that the classification rule maximising the a posteriori PROBABILITY of correct group membership is given by comparing the ratio of probability density functions of observations in the two populations against a given threshold value. This threshold, or cut-off, value depends on the prior probabilities of observations in each of the two populations; an individual is allocated to one group if the threshold is exceeded and otherwise to the other group. Welch also showed that if the two populations were characterised by multivariate normal distributions that had a common dispersion matrix then the resultant classification rule was a linear combination of the observed features, while if the dispersion matrices differed then a quadratic function was necessary. Subsequently, the theory was extended to encompass differential costs incurred in making classification errors and applied to practical datasets. It transpires that Fisher's LDF provides the best classification rule in the case of populations having MULTIVARIATE NORMAL DISTRIBUTIONS with a common dispersion matrix and that the cut-off value for classification is a simple ratio of costs of misclassification as well as prior probabilities in the two populations (see McLachlan, 1992, for details).

Fisher's derivation shows that his linear discriminant function will provide a good separation between groups for many practical situations, but the subsequent theoretical developments warn that this function may not provide good classification performance if populations are not multivariate normal with equal dispersion matrices. Alternative functions are necessary in such cases. Data arising in different contexts may have aspects that can readily be modelled by parametric distributions other than the normal, so various functions have been derived for specific types of data. Examples include functions based on the MULTINOMIAL DISTRIBUTION for discrete feature data, on the location model for mixed discrete and continuous feature data and on distributions such as the EXPONENTIAL DISTRIBUTION, the STUDENT'S t-DISTRIBUTION and the inverse normal for continuous nonnormal feature data (see McLachlan, 1992, for details).

An extreme assumption is that all features are independent, whence the class-conditional distributions can be simply estimated by products of marginal distributions. Although seemingly a totally inappropriate assumption in most practical situations, this method has had surprisingly good results on occasion (see Hand and Yu, 2001). However, approaches other than the parametric need to be sought if the resultant method is to be widely applicable.

Fix and Hodges (see Agrawala, 1977, pp. 261–322) therefore initiated a stream of research into nonparametric methods. Here the data alone determine the classification rule without imposition of any distributional assumptions, so that the methods can be applied in almost any context. Two of their ideas, which have undergone refinement but which retain their popularity, are nearest-neighbour and kernel methods. The *k-nearest-neighbour (kNN)* rule simply allocates an individual to the group that is in the majority among that individual's k nearest neighbours. Although simple in concept, this approach has several obvious questions that do not have easy answers: how do we measure the 'distance' between individuals in order to determine the nearest neighbours and how do we choose an optimal value of k in a given situation? These questions have been addressed in a number of contributions to the literature. The kernel method, by contrast, is a nonparametric method of estimating the probability densities in each group via the average of a so-called kernel function evaluated at each data point (see DENSITY ESTIMATION). Once the density functions have been estimated then computing their ratio at the point to be classified leads directly to an allocation rule. Fix and Hodges proposed a crude 'kernel' based on the empirical histograms of the data. Hand (1997) contains a good overview of both the kNN and the kernel methods, while McLachlan (1992) provides more technical detail.

A method intermediate between fully parametric models of the populations, in which the data are required to satisfy fairly strict assumptions, and the nonparametric approaches just described is the idea of *logistic discrimination* (LD). This was introduced in the early 1960s by Cox, Day, Kerridge and others, but was mainly developed in a series of papers well summarised by Anderson (1982). The idea is to estimate directly the posterior probability of group membership for an individual by a simple function of its observed features. Since this probability has to lie between zero and one, the simplest such function is logistic with argument given by a linear combination of the observed features. It can be shown that this form of posterior probability is either exhibited or approximately exhibited by a range of different parametric models including some of those already mentioned, so that the approach potentially has wide applicability. Moreover, it only requires estimation of a small number of parameters, as opposed to some of the parametric methods that involve many parameters, so it should be simpler to apply. Once some technical difficulties had been overcome the method proved both popular and useful and can now be found alongside Fisher's LDF in many software implementations.

Dramatic increases in computer power achieved towards the end of the 20th century led to an explosion of interest in computationally intensive methods. One of the first ideas in this vein was regularised discriminant analysis, where an optimal mixture of different types of discriminant function (such as linear and quadratic, for example) is sought by computationally optimising a criterion such as the cross-validated error rate (see later). This was followed by TREE-STRUCTURED METHODS such as classification and regression trees (CART), multivariate adaptive regression splines (MARS) and flexible discriminant analysis, while overshadowing all these approaches has been the development of NEURAL NETWORKS and SUPPORT VECTOR MACHINES among the computer science community. Unfortunately, the increasing reliance on computational power in these methods has turned each process into something of a 'black box', with results simply being produced at the end of a long series of computer operations and little chance being provided for either intervention in the process or interpretation of the underlying discriminant functions. More recently, therefore, attention has been given to computer-intensive enhancements of traditional methods. One general approach is *model averaging*, in which rather than seeking a single 'optimal' discriminant function of a given form many different such functions are derived and their average (in some sense) is used for future predictions. MARKOV CHAIN MONTE CARLO methods fall under this heading and much research is currently under way in their applications to discriminant functions. A good account of this work is given by Denison *et al.* (2002). The second general approach is in the use of *local models*, where instead of estimating parameters just once for all regions of the sample space, they are estimated separately for many subregions. Thus, for example, the optimal number k of nearest neighbours to use in kNN classification is estimated separately for all potential points to be classified.

Whichever method of discrimination or classification is chosen, a paramount consideration in practice is to obtain a reliable assessment of its performance. There are many possible measures that can be used for such assessment (see, for example, Hand, 1997, Chapter 6), but overwhelmingly the most prevalent in practice is the misclassification rate. Again, there are many possible ways of defining such a rate, but here we will just consider the one relevant to most practitioners: Given a particular classification rule formed from some sample data, what is the probability of misclassifying a future individual when using this rule?

It is possible to tackle this question theoretically, by postulating a probability model for the data and then following through with a sequence of probability calculations on implied SAMPLING DISTRIBUTIONS to arrive at a final value. Such calculations, however, frequently involve heavy simplifications to achieve tractability and stand or fall according to the appropriateness of the initial assumptions about the data. They have, therefore, long since been abandoned as genuine methods to use in practice and now usually serve only as benchmarks in simulation studies of properties of new methods. Attention instead has focused on purely data-based methods of assessment of performance, using the data from which the classification rule itself is constructed.

Given that the aim is to assess how often mistakes will be made in classifying *future* individuals and that in order to assess the accuracy of the classification we need to know the true group membership of each individual, an obvious method is to split the available sample data randomly into two portions. One portion, the *training* set, is used to form the classification rule, while the other, the *test* set, is used to assess its performance. Typically, the two sets are then combined and used to form the classification rule for actual use on future data. Such a process is known as *cross-validation*. Of course, it assumes that future samples are 'similar' in composition to the present ones and that the classification rule is stable over the different datasets. The former assumption is implicit in the classification procedure itself, but for the latter assumption to hold we really need large datasets. Problems arise when available samples are not large. Either the training set will be very small, so the training set classification rule may differ markedly from the final rule and the wrong rule will be assessed or the test set will be very small, so the assessment of the rule will be poor, or both drawbacks will occur.

An early attempt to solve this problem was by simply forming a classification rule from all available data and then reapplying it to the same data to assess its performance (resubstitution). However, it was soon realised that this method will provide a grossly overoptimistic assessment for small to medium samples. This is because most classification rules operate by optimising the group separation on the given data, so such a 'resubstitution' error rate represents the best achievable for the data and performance on genuinely future data will be much poorer. One possible solution is to conduct n distinct random training/test set divisions and to average the n resulting error rates as the final assessment of performance. This is known as *n-fold cross-validation*. Taking this process to the limit means removing each single observation in turn from the data, forming the classification rule using all the other observations and classifying the one that has been left out. The proportion of observations misclassified in this way then provides an estimate of the error rate of the rule. This is usually known as the *leave-one-out cross-validation* estimate.

The leave-one-out approach satisfactorily corrects for the known BIAS of the resubstitution error rate, but it has been shown to have the unsatisfactory property of a high variance. Thus, although it will give (approximately) the correct estimate *on average* over many replicates of equivalent datasets, any single application may yield an estimate far from the true value. An alternative line of attack was therefore developed in the 1980s using the idea of *bootstrapping*. Here the available data values are sampled with replacement to give a BOOTSTRAP sample, which is intended to mimic the drawing of a future sample from the populations under study, and relevant measures (such as the error rate) of the bootstrap sample are computed. This process is repeated for a large number of bootstrap samples and this enables distributions of the measures to be studied. In the context of classification error rates, many potential bootstrap corrections to the resubstitution error rate have been considered. The most popular appears to be the '632 bootstrap' estimate. A large number of bootstrap samples are generated, the classification rule is computed for each bootstrap sample and the observations *not* represented in that bootstrap sample are classified by the rule. If *ea* represents the error rate obtained in this way and *eb* represents the resubstitution error rate of the original data, then the 632 bootstrap error rate is given by $0.632ea + 0.368eb$. This appears satisfactorily to correct the optimistic bias of *eb*.

These data-based methods are applicable for assessment of any classification rule. However, the derivation of some rules itself requires assessment of error rates as part of the procedure. For example, estimation of the number k of nearest neighbours to use in kNN classification can be effected by trying all possible values of k in a given range and picking the one that produces the fewest misclassification errors. Simply quoting the resultant misclassification error rate again gives an overoptimistic assessment of performance of the method, because a parameter has been chosen to optimise such performance on the given data. The correct procedure here is to randomly divide the data into *three* portions: a training set, a validation set and a test set. The classification rule is formed from the training set and parameters are optimised by calculating error rates over the validation set. Having thus settled on all parameter values, final assessment of performance is conducted on the (truly independent) test set. The corresponding correction in the leave-one-out process is a *nested* leave-one-out: one observation is omitted and then the classification rule is formed *and optimised* using the remaining observations, nesting a second leave-one-out process within the first for the optimisation. The omitted observation is thus only classified once all parameters of the rule have been estimated.

Many of the above ideas were applied by Asparoukhov and Krzanowski (2001) in an empirical investigation of a range of different discriminant functions on binary data. Five datasets were used, of which the following four were medical: *pulmonary data* – 15 features to discriminate 144 patients who suffered postoperative pulmonary embolism and 246 who did not; *thrombosis data* – 15 features to discriminate 34 patients with embolic thrombosis from 68 patients without the condition; *epilepsy data* – 15 features to discriminate 81 children with craniocerebral trauma epilepsy from 48 without the condition; and *aneurysm data* – 17 features to discriminate 102 patients with dissecting aneurysm from 140 patients diagnosed with other similar diseases.

All features were already either binary in nature or were converted to binary form, the two categories in each case being scored 0 and 1. Each dataset was subjected to a number of different discriminant functions, but the table shows the error rates using those functions mentioned earlier. Each method was assessed exclusively using its leave-one-out error rates, so the best method on each dataset is the one with the lowest error rate.

discriminant function analysis *Error rates from seven methods applied to four datasets*

Discriminant procedure	*Pulmonary data*	*Thrombosis data*	*Epilepsy data*	*Aneurysm data*
Independence	0.146	0.265	0.209	0.021
Fisher LDF	0.159	0.294	0.163	0.041
Logistic	0.159	0.255	0.217	0.050
kNN	0.208	0.245	0.202	0.103
Kernel	0.172	0.265	0.217	0.054
Neural net	0.156	0.245	0.186	0.070
Vector support	0.213	0.245	0.209	0.058

These results demonstrate a typical empirical finding, namely that no single method is dominant in all cases and that each method performs well on at least some if not all datasets. The independence assumption works well on the pulmonary and aneurysm sets, but badly on the epilepsy data. Fisher's LDF is the best method on the epilepsy data, but the worst on the thrombosis data. NEURAL NETWORKS, SUPPORT VECTORS and kNN classifiers are the joint 'winners' on the thrombosis data, but have mixed results on the other datasets. Therefore, the message for practical applications is to try a range of potential methods before classifying individuals.

Extensions of all methods to the situation of multiple groups is straightforward in principle, although it may need careful computational implementation in practice. Fisher's LDF approach extends directly, by seeking the linear combination of features that maximises the ratio of between-group to within-group sums of squares. This is equivalent to maximising the F-ratio in standard ONE-WAY ANALYSIS OF VARIANCE, which is the multigroup extension of the two-group t-test for differences between the groups. The extra facet here is that more than one function results from this process; indeed, the number of functions will be the smaller of two values: the number of features present and one less than the number of groups. The resulting functions are known as *canonical variates* or *discriminant coordinates*, and plotting the original data against these functions as axes will highlight group differences pictorially (see CANONICAL CORRELATION ANALYSIS).

Welch's theoretical approach leads to classification being done via a series of pairwise comparisons, where ratios of each pair of population densities are in turn compared against cut-off thresholds until unique classification is achieved. This can be a somewhat protracted process, so recourse is usually made to one of the other methods. Logistic discrimination gives an estimated probability of group membership for each available group and allocation is made to the group with the highest probability. The *k*NN process gives allocation directly without any change in its definition. Kernel discrimination requires densities to be estimated separately for each group and allocation then follows directly from these densities, while all 'black box' routines deliver allocations for any number of groups. Likewise, all methods for estimating error rates naturally extend to the multigroup case. *WK*

[See also CANONICAL CORRELATION ANALYSIS]

Agrawala, A. K. (ed.) 1977: *Machine recognition of patterns.* New York: IEEE Press. **Anderson, J. A.** 1982: Logistic discrimination. In Krishnaiah, P. R. and Kanal, L. N. (eds), *Handbook of statistics*, Vol. 2. Amsterdam: North Holland, pp. 169–91. **Asparoukhov, O. K. and Krzanowski, W. J.** 2001: A comparison of discriminant procedures for binary variables. *Computational Statistics and Data Analysis* 38, 139–60. **Denison, D. G. T., Holmes, C. C., Mallick, B. K. and Smith, A. F. M.** 2002: *Bayesian methods for nonlinear classification and regression.* Chichester: John Wiley & Sons, Ltd. **Fisher, R. A.** 1936: The use of multiple measurements in taxonomic problems. *Annals of Eugenics* 7, 179–88. **Hand, D. J.** 1997: *Construction and assessment of classification rules.* Chichester: John Wiley & Sons, Ltd. **Hand, D. J. and Yu, K.** 2001: Idiot's Bayes – not so stupid after all? *International Statistical Review* 69, 385–98. **McLachlan, G. J.** 1992: *Discriminant analysis and statistical pattern recognition.* New York: John Wiley & Sons, Inc. **Welch, B. L.** 1939: Note on discriminant functions. *Biometrika* 31, 218–20.

disease clustering These are unusual aggregations of disease that appear on maps of disease incidence (Lawson, 2006, Chapter 6). They are areas of such elevated risk that they could not have arisen by chance alone. For example, concerns about the influence of industrial installations on the health of surrounding populations has given rise to the development of methods that seek to evaluate clusters of disease around such installations. These clusters are regarded as representing local adverse health risk conditions, possibly ascribable to environmental causes. However, it is also true that for many diseases the geographical incidence of disease will naturally display clustering at some spatial scale, even after the 'at-risk' population effects are taken into account. The reasons for such clustering of disease are various. First, it is possible that for some apparently noninfectious diseases there may be a viral agent, which could induce clustering. This has been hypothesised for childhood leukaemia. Second, other common but unobserved factors/variables could lead to observed clustering in maps. For example, localised pollution sources could produce elevated incidence of disease (e.g. road junctions could yield high carbon monoxide levels and hence elevated respiratory disease incidence) or a common treatment of diseases could lead to clustering of disease side-effects. The prescription of a drug by a particular medical practice could lead to elevated incidence of side-effects within that practice area.

Hence, there are many situations where diseases may be found to cluster, even when the aetiology does not suggest it should be observed. Because of this, it is important to be aware of the role of clustering methods, even when clustering per se is not the main focus of interest. In this case, it may be important to consider clustering as a background effect and to employ appropriate methods to detect such effects.

Two extreme forms of clustering can be defined. These two extremes represent the spectrum of modelling from nonparametric to parameteric forms and associated with these forms are appropriate statistical models and estimation procedures. First, as many researchers may not wish to specify a priori the exact form/extent of clusters to be studied, then a nonparametric definition is often the basis adopted. Without any assumptions about shape or form of the cluster, the most basic definition would be 'any area within the study region of significantly elevated risk' (Lawson, 2006, p. 104).

This definition is often referred to as *hot spot clustering*. In essence, any area of elevated risk, regardless of shape or extent, could qualify as a cluster, provided the area meets some statistical criteria. Note that it is not usual to regard areas of significantly low risk to be of interest, although these may have some importance in further studies of the aetiology of a particular disease.

Second, at the other extreme, we can define a parametric cluster form as 'the study region displays a prespecified cluster structure'. This definition describes a parameterised cluster form that would be thought to apply across the study region. Usually, this implies some stronger restriction on the cluster form and also some region-wide parameters that control the shape and size of clusters.

Nonspecific clustering is the analysis of the overall clustering tendency of the disease incidence in a study region. This is also know as *general clustering*. As such, the assessment of general clustering is closely akin to the assessment of spatial autocorrelation. Hence, any model or test relating to general clustering will assess some overall/global aspect of the clustering tendency of the disease of interest. This could be summarised by a model parameter (e.g. an autocorrelation parameter in an appropriate model) or by a test that assesses the aggregation of cases of disease. For example, the correlated prior distributions used in the Besag, York and Mollié (BYM) model (see SPATIAL EPIDEMIOLOGY). It should be noted at this point that the general clustering methods discussed above can be regarded as *nonspecific* in that they do not seek to estimate the spatial locations of clusters but simply to assess whether clustering is apparent in the study region. Any method that seeks to assess the locational structure of clusters is defined to be *specific*.

An alternative nonspecific effect has also been proposed in models for tract-count or case-event data. This effect is conventionally known as uncorrelated heterogeneity (or OVERDISPERSION, or extra-Poisson variation in the Poisson likelihood case).

Specific clustering concerns the analysis of the locations of clusters. This approach seeks to estimate the location, size and shape of clusters within a study region. For example, it is straightforward to formulate a nonspecific Bayesian model (see BAYESIAN METHODS) for case events or tract counts that includes heterogeneity. However, specific models or testing procedures are less often reported. Nevertheless, it is possible to formulate specific clustering models for the case-event and tract-count situation.

Another definition of clustering seeks to classify the methods based on whether the location or locations of clusters are known or not. Focused clustering is *specific* and usually seeks to analyse the clustering tendency of a disease or diseases around a known location. Often this location could be a *putative pollution source* or health hazard. Nonfocused clustering does not assume knowledge of a location of a cluster but seeks either to assess the locations of clustering within a map or to assess the overall clustering tendency. Hence, nonfocused clustering could be specific or nonspecific.

The literature of spatial epidemiology has developed considerably in the area of hypothesis testing and, more specifically, in the sphere of hypothesis testing for clusters (see, for example, Lawson and Kulldorff, 1999). Very early developments in this area arose from the application of statistical tests to spatiotemporal clustering, a particularly strong indicator of the importance of a spatial clustering phenomenon. As noted above, distinction should be made between tests for general (nonspecific) clustering, which assess the overall clustering pattern of the disease, and the *specific* clustering tests where cluster locations are estimated.

For case events, a few tests have been developed for nonspecific clustering. Specific nonfocused cluster tests address the issue of the location of putative clusters. These tests produce results in the form of locational probabilities or significances associated with specific groups of tract counts or cases. Openshaw and co-workers (Openshaw *et al.*, 1987) first developed a general method that allowed the assessment of the location of clusters of cases within large disease maps. The method was based on repeated testing of counts of disease within circular regions of different sizes. The statistical foundation of this method has been criticised and an improvement to the method was proposed by Besag and Newell (1991). An alternative statistic has been proposed by Kulldorff and Nagarwalla (1995) (the scan statistic). The test can be applied to both case events and tract counts. An evaluation of various tests for clustering has been made by Kulldorff, Tango and Park (2003). Focused tests have also developed and there is now a range of possible testing procedures (for a recent evaluation see Liu, Lawson and Ma, 2009). Cluster modelling has seen some development but has not developed as fully as testing procedures. Usually the successful models are Bayesian with prior distributions describing the clustering behaviour (see, for example, Lawson and Denison, 2002, and Lawson, 2009, Chapter 6, for recent examples). *AL*

Besag, J. and Newell, J. 1991: The detection of clusters in rare diseases. Journal of the Royal Statistical Society, Series A 154, 143. **Kulldorff, M. and Nagarwalla, N.** 1995: Spatial disease clusters: detection and inference. Statistics in Medicine 14, 799. **Kulldorff, M., Tango, T. and Park, P. J.** 2003: Power comparisons for disease clustering tests. Computational Statistics and Data Analysis 42, 665–84. **Lawson, A. B.** 2006: *Statistical methods in spatialepidemiology*, 2nd edition. Chichester: John Wiley & Sons, Ltd. **Lawson, A. B.** 2009: *Bayesian disease mapping: hierarchical modeling in spatial epidemiology*. New York: CRC Press. **Lawson, A. B. and Denison, D.** (eds) 2002: *Spatial cluster modelling*. London: Chapman & Hall. **Lawson, A. B. and Kulldorff, M.** 1999: A review of cluster detection methods. In *Disease mapping and risk assessment for public health*. Chichester: John Wiley & Sons, Ltd. **Liu, Y., Lawson, A. B. and Ma, B.** 2009: Evaluation of putative hazard tests under background risk heterogeneity. *Environmetrics* 20, 3, 260–74. **Openshaw, S. *et al*.** 1987: A mark 1 geographical analysis machine for the automated analysis of point data sets. *International Journal on Geographical Information Systems* 1, 335.

disease mapping See DISEASE CLUSTERING, SPATIAL EPIDEMIOLOGY

disease surveillance See SPATIOTEMPORAL DISEASE SURVEILLANCE

DMF score See DENTAL STATISTICS

dot plot This is a useful graphical display for data on some continuous variable recorded within the categories of a particular categorical variable. An example is shown in the figure on page 144. A dot plot is generally far more effective in communicating the pattern in the data than either a PIE CHART or a bar chart, particularly if the number of categories is reasonably large. *BSE*

dropout at random (DAR) See DROPOUTS

dropout completely at random (DCAR) See DROPOUTS

dropouts These are patients in a study, commonly a clinical trial, who fail to attend protocol-scheduled visits or assessments of a response variable taken after some particular time point in the study. Dropping out of a study implies that once an observation at a particular time point is missing so are all the remaining planned observations. Such missing

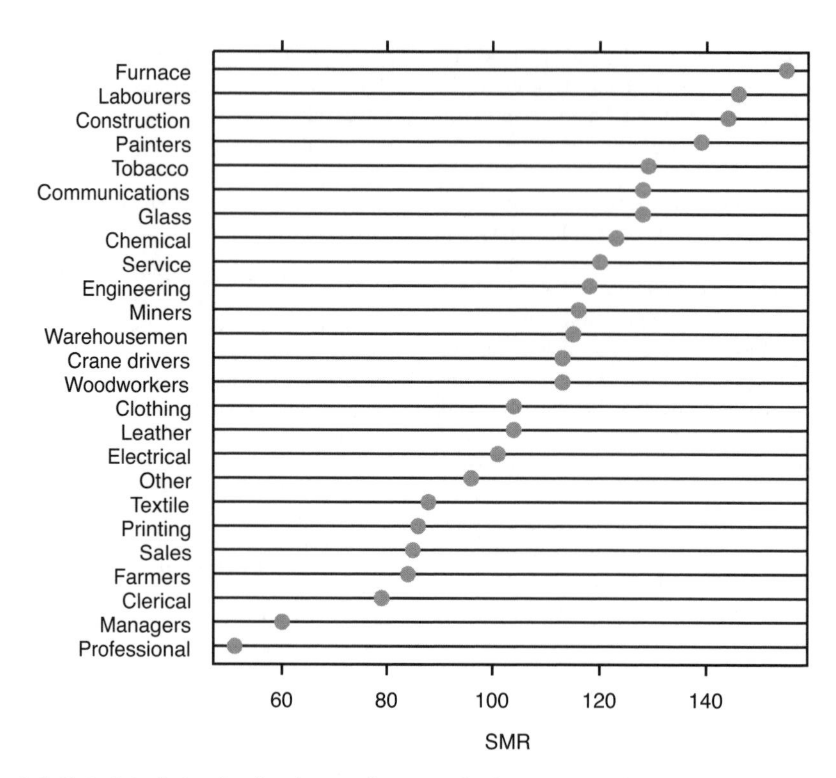

dot plot *Dot plot of standardised mortality rates for lung cancer in 25 occupational groups*

observations are a nuisance and the very best way to avoid problems with missing values is not to have any! If only a small proportion of the patients in the trial drop out it is unlikely that these will cause any major difficulties for analysis and so on. If, however, a substantial number of dropouts occur there is potential for making incorrect inferences and/or producing biased estimates if a valid analysis procedure is not used. In such cases consideration needs to be given to the reasons why individuals drop out and how the probability of dropping out depends on the response variable, since this has implications for which forms of analysis are suitable and which are not (Little, 1995). Three dropout mechanisms are usually differentiated based on the classification of missing values originally suggested by Rubin (1976).

Dropout at random (DAR). The dropout at random mechanism occurs when the dropout process depends on the outcome measures that have been observed in the past, but given this information is conditionally independent of all the future (unrecorded) values of the outcome variable following dropout. Here 'missingness' depends only on the observed data with the distribution of future values for a subject who drops out at time t being the same as the distribution of the future values of a subject who remains in at time t, if they have the same covariates and the same past history of outcome up to and including time t. Murray and Findlay (1988) provide an example of this type of missing value from a study of hypertensive drugs in which the outcome measure was diastolic blood pressure. The protocol of the study specified that the participant should be removed from the study when his or her blood pressure became too high. Here blood pressure at the time of dropout was observed before the participant dropped out, so although the missing value mechanism is not DCAR since it depends on the values of blood pressure, it is DAR, because dropout depends only on the observed part of the data.

Dropout completely at random (DCAR). Here the probability that a patient drops out does not depend on either the observed or missing values of the response. The observed (nonmissing) values effectively constitute a simple random sample of the values for all subjects. Possible examples include missing laboratory measurements because of a dropped test tube (if it was not dropped because of the knowledge of any measurement), the accidental death of a participant in a study or a participant moving to another area. Completely random dropout causes least problem for data analysis, but it is a strong assumption.

Nonignorable (sometimes referred to as informative) dropout. For this final type of dropout mechanism, missingness depends on the unrecorded missing values – observations are likely to be missing when the true values are systematically higher or lower than usual. A nonmedical example is when individuals with lower income levels or very high incomes are less likely to provide their personal income in an interview. In a medical setting, possible examples are a participant dropping out of a longitudinal study when his or her blood pressure became too high and this value was not observed or when their pain became intolerable and the associated pain value was not recorded. Dealing with data containing missing values of this type is not routine.

Simple methods of analysis for longitudinal data, e.g. COMPLETE CASE ANALYSIS or SUMMARY MEASURE ANALYSIS rely on the DCAR assumption; others, such as LINEAR MIXED-EFFECTS MODELS, involve only the weaker DAR restriction. Identifying the type of dropout mechanism for a particular dataset is rarely straightforward, although some useful informal procedures are described in Carpenter, Pocock and Lamm (2002). When informative dropouts are suspected the methods suggested by Rabe-Hesketh, Pickles and Skrondal (2001) may be useful. *BSE*

Carpenter, J., Pocock, S. and Lamm, C. J. 2002: Coping with missing data in clinical trials: a model based approach applied to asthma trials. *Statistics in Medicine* 21, 1043–66. **Little, R. J. A.** 1995: Modeling the dropout mechanism in repeated measure studies. *Journal of the American Statistical Association* 90, 1112–21. **Murray, G. D. and Findlay, J. G.** 1988: Correcting for the bias caused by dropouts in hypertension trials. *Statistics in Medicine* 7, 941–6. **Rabe-Hesketh, S., Pickles, A. and Skrondal, A.** 2001: GLAMM: a class of models and a STATA program. *Multilevel Modelling Newsletter* 13, 17–23. **Rubin, D. B.** 1976: Inference and missing data. *Biometrika* 63, 581–92.

dummy variables These comprise a set of variables, each with only two possible outcomes, that are used to represent a categorical explanatory variable in a statistical model that is designed to handle quantitative variables. Usually, the two outcomes allowed for each dummy variable are taken to be zero and unity. Each dummy variable takes the value zero when the parent categorical variable attains its predefined reference, or base, level. When the categorical variable attains any other level, one unique dummy variable takes the value unity; the rest take the value zero. A categorical variable with k outcomes will require $k - 1$ dummy variables to represent it.

dummy variables *Dummy variables (x_1 and x_2) used to represent smoking status for 14 people*

Smoking status	x_1	x_2
Current smoker	0	1
Ex-smoker	1	0
Ex-smoker	1	0
Current smoker	0	1
Never smoker	0	0
Never smoker	0	0
Never smoker	0	0
Current smoker	0	1
Never smoker	0	0
Current smoker	0	1
Ex-smoker	1	0
Ex-smoker	1	0
Never smoker	0	0
Ex-smoker	1	0

For example, consider a situation where the hypothesis is that blood pressure is different between those who currently smoke, used to smoke (but have since quit) and have never smoked. Assuming that suitable data have been collected, the hypothesis could be tested by fitting an ANALYSIS OF VARIANCE model with a smoking group as the explanatory variable and blood pressure as the dependent variable. This explanatory variable would have three levels (current/ex/never) and thus two DEGREES OF FREEDOM. An alternative approach would be to define two dummy variables to represent the smoking group, say:

$$x_1 = \begin{cases} 1 & \text{for ex-smokers} \\ 0 & \text{for never smokers} \end{cases}$$

$$x_2 = \begin{cases} 1 & \text{for current smokers} \\ 0 & \text{for never smokers} \end{cases}$$

(see the table) and then fit a MULTIPLE REGRESSION model with x_1 and x_2 as the explanatory variables. The regression sum of squares (with two degrees of freedom) in this regression model would be the same as that for the variable 'smoking group' in the earlier analysis of variance model. Both x_1 and x_2 are contrasts between, respectively, ex- and never smokers and between current and never smokers. In this sense, never smokers are the reference group for analyses of smoking status.

Dummy variables are useful in that they remove the necessity to develop separate statistical models for categorical and continuous explanatory variables. They also allow variables of mixed type to be handled within the same, single methodology.

Some computer software requires the user to define dummy variables for themselves, whereas others do the computation automatically once the particular variable for which dummy variables are required has been declared as a categorical variable. *MW*

dynamic designs See DATA-DEPENDENT DESIGNS

E

EaSt This software package allows the design and analysis of sequential trials. The name 'East' is derived from '**Ea**rly **st**opping'. As an alternative to standard, fixed-sample CLINICAL TRIALS, flexible clinical trials utilise group-sequential and adaptive methodologies to permit interim looks at accruing data with a view to making early stopping or sample size readjustment decisions, while preserving Type 1 error and POWER. Jennison and Turnbull (2001) provide a thorough treatment of sequential design and analysis. These methods have been incorporated into the East software package developed by Cytel Software Corporation (www.cytel.com).

East has three basic components: a design module, a simulation module and an interim monitoring module. The design module can be used to design two-arm superiority or noninferiority trials of normal, binomial or time-to-event ENDPOINTS. Extensions to more general endpoints are available through the use of an inflation factor that increases the sample size of a fixed-sample design by the appropriate amount so as to preserve power. A special design worksheet is provided for designing studies on the basis of maximum information rather than maximum sample size. Such designs provide the flexibility to adjust the sample size during the interim monitoring phase, to accommodate adjustments to important design parameters, such as patient-to-patient variability, that might have been misspecified at the design stage. Many families of stopping boundaries and spending functions are available, thus providing great flexibility for making early stopping decisions either for efficacy or futility or both.

Trials created by the design module can be simulated in the simulation module. Since the statistical theory underlying East utilises large-sample assumptions, the simulation module is a useful tool for verifying that the operating characteristics of the design are preserved for small or unbalanced studies. A special feature of the simulation module is the capability to simulate adaptive designs.

The interim monitoring module includes a worksheet that accepts the current value of the test statistic and the current sample size of information. It then re-computes the stopping boundaries based on the specified spending functions, determines if the boundary has been crossed and provides important interim results such as conditional power and repeated CONFIDENCE INTERVALS.

East additionally provides tables, graphs and reports that allow investigators to visualise and clearly demonstrate the features and results of the planned design. For example, one can plot stopping boundaries as a function of time or present power calculations in graphical or tabular form. *CM*

Jennison, C. and Turnbull, B. 2001: *Group sequential methods with applications to clinical trials*. New York: Chapman & Hall/CRC.

EBM See EVIDENCE-BASED MEDICINE

ecological fallacy See EPIDEMIOLOGY

ecological studies See EPIDEMIOLOGY

effectiveness See ADJUSTMENT FOR NONCOMPLIANCE IN CLINICAL TRIALS

efficacy See ADJUSTMENT FOR NONCOMPLIANCE IN CLINICAL TRIALS

eligibility criteria See INCLUSION AND EXCLUSION CRITERIA

EM algorithm This is a general computational method for calculating MAXIMUM LIKELIHOOD ESTIMATIONS with incomplete data, e.g. MISSING DATA or data containing CENSORED OBSERVATIONS. The algorithm is based on the notion that if we had the missing or censored observations we could estimate parameters of interest in the usual way and that if we knew the parameters we could impute the missing observations by setting them to their predicted values under the model. Consequently, given some initial values for the parameters we can proceed iteratively between computing predicted values, filling in the missing observations and then estimating the parameters using new 'complete data'. The algorithm is widely used in statistics and a detailed technical account is given in Laird (1998). *BSE*

Laird, N. 1998: EM algorithms. In Armitage, P. and Colton, T. (eds), *Encyclopedia of biostatistics*. Chichester: John Wiley & Sons, Ltd.

endpoints These describe a measurement or discrete event, related to the disease under investigation, which measures the effectiveness of an intervention. The definition of suitable endpoints depends on the disease under study. In serious diseases, such as coronary heart disease, endpoints such as 'mortality' provide a reliable measure of disease progression, while after curative surgery for cancer, 'recurrent cancer' would be an example of such a measure. These examples are 'binary' in nature (i.e. subjects have either had the event or not), but in some diseases it is more appropriate to measure disease progression in terms of a continuous outcome

Encyclopaedic Companion to Medical Statistics: Second Edition Edited by Brian S. Everitt and Christopher R. Palmer

measure (e.g. blood pressure) or an ordinal scale (e.g. a measure of the quality of life, such as the SF-36).

When designing a randomised trial (see CLINICAL TRIALS), it is usual to define primary and secondary endpoints on which judgements about the overall benefits and harms of treatment are to be made. The 'primary endpoint' (or 'target variable') is chosen as the chief measure of the effects of an intervention, on which analyses are to be conducted in order to assess the primary hypothesis originally stated in the study protocol (see PROTOCOLS IN CLINICAL TRIALS). Generally, the primary endpoint is a measure that is expected to be influenced favourably by the intervention, i.e. it is a measure of clinical efficacy. For example, based on long-term randomised trials of aspirin conducted among patients who had survived a heart attack, it was anticipated that aspirin would reduce the clinically important endpoint 'vascular mortality' in the first few weeks after a suspected acute heart attack, so this was the choice of the primary endpoint in the Second International Study of Infarct Survival (ISIS-2 Collaborative Group, 1988). Occasionally it may be appropriate for a primary endpoint to measure safety; e.g. if we were to compare a higher versus a lower dose of aspirin, the known pharmacology of aspirin might lead us to hypothesise that both regimens would have similar clinical efficacy, but that serious bleeding might be less frequent with the lower aspirin dose (Antithrombotic Trialists' Collaboration (ATC), 2002). In these circumstances, we might be advised to choose 'bleeding requiring transfusion' as a primary safety endpoint.

As well as defining a primary endpoint in the study protocol, together with the planned method of statistical analysis of that primary endpoint, it is usual to define a number of 'secondary endpoints'. These are measures that, when considered with the primary endpoint, provide helpful additional information about the clinical efficacy and safety of the intervention. In the ISIS-2 trial, for example, an assessment of the overall benefits and harms of aspirin (and of streptokinase) was facilitated by assessing the effects of these treatments on secondary endpoints such as re-infarction, haemorrhagic stroke and bleeds requiring transfusion (ISIS-2, 1988). In view of the fact that the aim of assessing effects on secondary endpoints is to make sound clinical judgements about the balance of benefit and harm, it is imperative to limit the number of the secondary endpoints. If too many are chosen, the likelihood that an ineffective treatment might appear to influence the risk of one or more such endpoints (i.e. a TYPE 1 ERROR) increases and it is difficult to interpret any data that result.

In circumstances where a number of clinically important endpoints are likely to be influenced favourably by a treatment, it may be useful to define a 'composite' (or 'combined') endpoint, where a subject is considered to have reached this endpoint if they experience one (or more) of the component endpoints. When a treatment has similar effects on each of the components of a composite endpoint, use of that composite endpoint increases the statistical power not only of the primary analysis but more especially of subgroup analyses aiming to assess whether the effects of treatment differ importantly among selected categories of patients of clinical interest. For example, to assess antiplatelet therapy for the prevention of occlusive vascular disease, which is a systemic disease affecting both cardiac and cerebral arteries, the composite endpoint 'myocardial infarction, stroke or vascular death' has been found to be useful, particularly when assessing effects in particular groups of patients (such as men and women, young and old, etc.) (ATC, 2002; CAPRIE Steering Committee, 1996). Several problems arise, however, when a composite endpoint is designed to assess a 'global effect' of treatment by grouping together the major anticipated benefits and harms. In these circumstances, an estimated effect on the global composite outcome that is not statistically significant could reflect a worthwhile benefit masked by a harm, and this would be entirely missed unless the benefits and harms are considered separately in a trial or META-ANALYSIS that is large enough to assess the effects on both.

In circumstances when it is not practical to assess the effects of treatment on clinical endpoints, it is possible to consider the use of a SURROGATE ENDPOINT for assessing treatment effects. In the past, however, many promising surrogate endpoints have proved unreliable. For this reason, it is inappropriate to rely on a nonvalidated surrogate endpoint to assess the effects of a drug that is to be used for a common and serious disease.

The choice of endpoints and the method by which they are to be analysed should be set out clearly in the study protocol. It is important that this choice is given careful thought, since it may often determine whether a trial succeeds in answering a clinically useful question, and it is difficult to change the choice of endpoints after the trial has closed without the risk of introducing serious bias. *CB*

Antithrombotic Trialists' Collaboration (ATC) 2002: Collaborative meta-analysis of randomised trials for prevention of death, myocardial infarction, and stroke in high-risk patients. *British Medial Journal* 324, 71–86. **CAPRIE Steering Committee** 1996: A randomised, blinded, trial of clopidogrel versus aspirin in patients at risk of ischaemic events. *The Lancet* 348, 1329–39. **ISIS-2 (Second International Study of Infarct Survival) Collaborative Group** 1988: Randomised trial of intravenous streptokinase, oral aspirin, both, or neither among 17,187 cases of suspected acute myocardial infarction: ISIS-2. *The Lancet* ii, 349–60.

entry criteria See INCLUSION AND EXCLUSION CRITERIA

EPIINFO See CASE-CONTROL STUDIES, STATISTICAL PACKAGES

epidemiology This is the study of disease and its risk factors (Clayton and Hills, 1993; Ashton, 1994; Stolley and Lasky, 1995; Woodward, 2005; Rothman, Greenland and Lash, 2008). Originally used to describe the study of epidemics, it now encompasses noncommunicable as well as infectious diseases. Simple examples would be an analysis of the trends over time in the number of cases of AIDS recorded in a certain country and a comparison of the age-specific death rates due to AIDS between countries. More complex examples would be an evaluation of the proximity of nuclear power stations to the homes of people diagnosed with leukaemia and an investigation into possible association between regular consumption of fast foods and subsequent development of cardiovascular disease.

Many of the major public health issues of the day have been the subject of scrutiny by epidemiologists, whose careful amassing of facts and figures, and subsequent statistical analyses, have demonstrated a strong case for causality. One of the first examples was that of John Snow, who used 19th century field data to demonstrate that cholera cases were clustered around the Broad Street water pump that was used by households around London's Golden Square (see HISTORY OF MEDICAL STATISTICS). His data fitted in with a theory, not generally accepted at the time, of the disease being carried by polluted water and with the practice of emptying sewerage into water that made its way to the well which the pump served (Chave, 1958). Whenever there is a new, unexplained, health problem, epidemiologists will usually be charged with plotting the course of the problem, sifting the available information in order to uncover the natural history and likely cause and helping to devise plans for avoidance of the problem in future. Examples would be the so-called 'Gulf War Syndrome' attributed to surviving soldiers from the Gulf War of 1990–1991 and the outbreak of SARS in 2002–2003, which was particularly severe in Asia and Canada (see the first figure). Epidemiologists tend to make considerable use of statistical methodology, but the profession requires a substantial level of medical knowledge also. In recent years, specialist areas of epidemiology have emerged, such as GENETIC EPIDEMIOLOGY, where epidemiological tools are applied to genetic data.

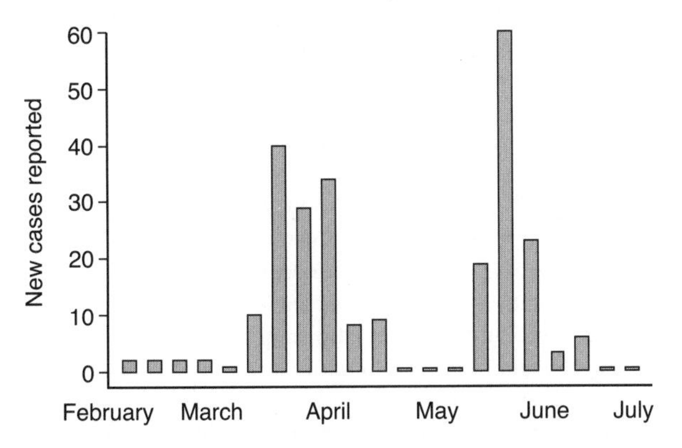

epidemiology *Number of new SARS cases, by week, in Canada in 2003 (WHO daily reports, with interpolation and extrapolation)*

Epidemiological studies form one of the two major types of study design in medical research – the other being CLINICAL TRIALS. The distinction between the two is that epidemiological investigations are observational whereas clinical trials involve *interventions*. For example, a large group of middle-aged women may be monitored for several years and the proportions of women who develop venous thrombosis compared between those women who did and those who did not use hormone replacement therapy (HRT) when the study began. This would be an epidemiological study. A corresponding clinical trial might involve taking a group of women who have never used HRT, but for whom there are no medical reasons why they should not take HRT. Some would be allocated to receive HRT and some to receive a placebo; both groups would be followed over time to make the same comparison as above. The great advantage of the epidemiological approach would be that results are obtained in real-life circumstances. Furthermore, there are ethical advantages compared with the approach of deciding who receives the factor of interest, even when the decision is made using some chance mechanism, as should be the case, whenever possible, in clinical trials. For example, it is unlikely that a clinical trial would be allowed that sought to study the effects of cigarette smoking, since it would be unethical to ask people to smoke. Yet some of the classic epidemiological studies have compared smokers and nonsmokers for their chance of disease. On the other hand, clinical trials (when feasible) generally offer more reliable results because they can be designed to minimise the effect of *confounding*, which is a common source of BIAS in epidemiological studies. For example, perhaps women who take HRT tend to be more educated than those who do not, and hence the relative effects seen cannot necessarily be attributed only to HRT in an epidemiological study. In the corresponding clinical trial, the two groups could be balanced in terms of the level of education achieved.

In making the distinction between epidemiological studies and clinical trials, it should be understood that the practising epidemiologist will make use of information from clinical trials wherever it contributes to their subject of interest. Indeed, many intervention studies are conducted by people who would consider themselves to be epidemiologists, who are, quite sensibly, using the best tools available for the job at hand. Furthermore, a randomized clinical trial will often be considered as the ultimate test of the epidemiological theory. For example, many epidemiological studies have found high consumption of food that is rich in certain vitamins, specifically those found in fruit and vegetables, to be protective against heart disease. This has given rise to the epidemiological hypothesis that these vitamins protect the heart, and thus to several clinical trials of vitamin supplementation. At the time of writing, the combined evidence from these trials is that such supplementation has no beneficial effect, leaving open the question of whether the findings in epidemiological studies are simply due to confounding.

Epidemiological investigations might simply be examinations of routinely collected data, such as registrations of death by cause, to search for seasonal patterns (as in the first figure) or differences in disease incidence by regions of the country, or examinations of cases of disease, to look for common factors or clusters of cases in time and space (see also DISEASE CLUSTERING). Such information identifies specific health problems and helps to formulate theories about the potential aetiology of the particular medical condition. The latter issue is sometimes addressed, using routine data, in a formal way through *ecological studies*. These are studies of data on average values of disease outcome and risk factor status from groups of people, typically those who live in specific regions. For example, St Leger, Cochrane and Moore (1979) plotted mortality from coronary heart disease (CHD) per thousand men aged 55–64 years in 18 countries against wine consumption (from industry sources) in the same countries (see the second figure). The data seemed to suggest an inverse relationship; e.g. France had the highest wine consumption and the lowest CHD rate.

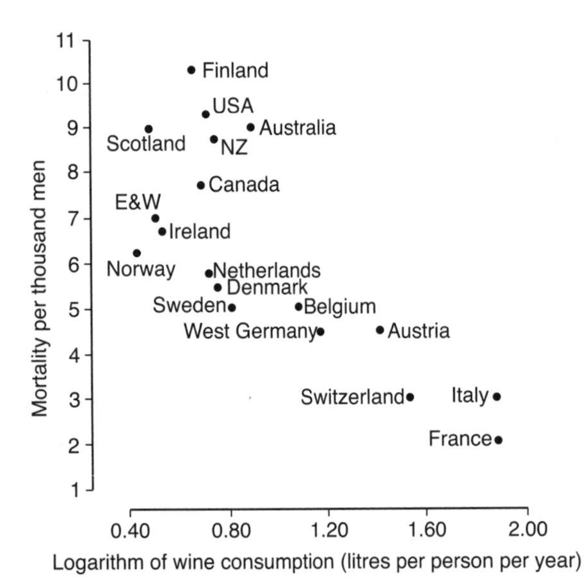

epidemiology *Death rates for men aged 55–64 years old in 1970 against the logarithm of wine consumption in 18 countries (St Leger, Cochrane and Moore, 1979)*

Routine, or other types of pre-existing (secondary) data are clearly relatively cheap and easy to collect, and will often be considered authoritative when derived from government sources or from international organisations, such as the World Health Organisation (WHO). However, often they are incomplete (such as when death registrations alone are used to examine morbidity), inadequate (such as when total numbers are recorded but not numbers within important demographic subgroups) and out of date. Ecological studies, used to investigate associations, often suffer from mismatching of the groups in the two data series, those for disease and risk factor. For instance, in the study of St Leger, Cochrane and Moore (1979) deaths were for a particular age/sex group whereas wine consumption was for the entire population. In most cases they offer little, or no, opportunity to control for confounding. Furthermore, there is no reason why relationships observed for groups should hold when individuals are observed – the so-called *ecological fallacy*. Thus, it could be that, while France does generally have a low rate of heart disease and a high consumption of wine, those Frenchmen who drink relatively little are the ones who suffer most CHD.

Ecological studies should be seen as hypothesis-generating tools rather than ways of deriving definitive information on relationships between risk factors and disease. Although many ecological studies may well give fallacious results, they may also be the first clue to an association that was not previously discussed. St Leger, Cochrane and Moore (1979) is a nice example of this; their demonstration that relatively low wine consumption might be protective against heart attack was initially met with scorn and was usually assumed to be the result of bias or confounding. However, subsequent research, using more reliable epidemiological study designs (see below), confirmed their hypothesis, which is now commonly accepted. Even so, the apparent relationship could have been spurious – for instance variations in the way heart disease is diagnosed across countries and confounding with other aspects of the diet besides wine might have explained away the inverse association between wine and heart disease. The ecological design generally cannot delve deeply enough to uncover such subtleties; collection of data from individuals is required.

There are three main types of epidemiological study that involve collection of new information from individual people: CROSS-SECTIONAL STUDIES (surveys), CASE-CONTROL STUDIES and COHORT STUDIES. Surveys are called cross-sectional because they occur at a single point in time (see the third figure on page 151). Generally, they involve the drawing of a representative sample of the entire population, although very occasionally the entire population is included, in which case a census has occurred. Surveys have the advantage, over using routine data, that the investigator can collect precisely the information required for the subsequent analysis, within practical constraints. They are particularly useful for descriptive purposes.

Epidemiological surveys typically include questions about disease states and levels of risk factors for these diseases. The answers can be used to estimate prevalence of disease and the distribution of the risk factors. For instance, a national population survey in Scotland (Tunstall-Pedoe *et al.*, 1997) included taking blood samples from all participants, from which serum cholesterol was measured. The results gave a picture of the distribution of cholesterol in Scotland at that time, allowing (for instance) an estimate of the number and percentage of Scots whose cholesterol level was above that considered 'safe' and thus were likely to benefit greatly from cholesterol-lowering treatment. Surveys can be made more accurate by using random sampling (to reduce bias), using sensible stratification (to increase precision) and taking a larger sample size (also to increase precision). The latter point needs some qualification: there may be no benefit from increasing sample size if this leads to increased bias error, for instance through taking less time, per subject, to ensure that accurate responses are solicited from questioning. Cluster sampling is often used for convenience, or simply to reduce costs, but does have the unfortunate effect of decreasing precision, measured, for example, by the width of the CONFIDENCE INTERVAL for the estimate obtained from the survey (e.g. the mean cholesterol).

Surveys have limited use in investigating CAUSALITY in associations because they are prone to the 'chicken and egg' effect – it is difficult, often impossible, to ascertain whether the observed value of the risk factor was a precursor, or a consequence, of the observed disease state. For instance, the Scottish survey described above included the question, 'Have you ever been told by a doctor that you suffered from a heart attack?' Comparing average cholesterol levels between those who did and did not report having had a heart attack gives a simple indication of whether cholesterol is associated with having had a heart attack. However, any such conclusion of causality may be spurious. A high cholesterol reading today could be the consequence of a recent heart attack, instead of the hypothesised effect of relatively high cholesterol increasing the risk of a heart attack. Another example would be a survey where a particular chronic disease is found to be less common among those who smoke. This may not mean that smoking tends to protect people against the disease; instead it may be that smokers, having developed the disease, give up the habit, thus leading to a predominance of the disease among those not smoking at the time of the survey. Although it may be possible to discover when the risk factor was first encountered (e.g. when smoking was taken up), the reliability of such information, often requiring long-term recall, may be poor. In any case, it is rare to be able to fix a time for the onset of the disease. Thus, surveys are not suited for investigations of causality, which are commonly the purpose of epidemiological investigations.

More reliable information on causality can be gleaned from a case-control study. A set of cases, people with the disease of interest, are identified – e.g. through hospital records – and the putative risk factor(s) of interest is(are) recorded for each case. In parallel, a contrasting set of controls, those without the disease, are selected and submitted to the same investigations as the cases. Comparison of the risk factor levels between cases and controls enables the risk factor–disease association to be assessed, usually measured by the ODDS RATIO. In principle, case-control studies compare disease status now with risk factor levels in the past, which is why they are often called retrospective studies (see the third figure). However, this is only really true if incident (new) cases are used, and even then certain potential risk factors (for instance markers of inflammation) might be elevated so soon after the disease hits that the 'chicken and egg' problem may occur. Even if this type of bias is not an issue, case-control studies are very susceptible to other kinds of bias, such as that caused by differential quality of information from cases compared with controls, the former group perhaps being of greater clinical interest to study investigators and hence observed or researched more thoroughly than controls. Berkson's bias may occur in certain case-control designs (see BERKSON'S FALLACY). Careful matching of controls to cases can increase precision, but still leaves the potential for several important sources of bias.

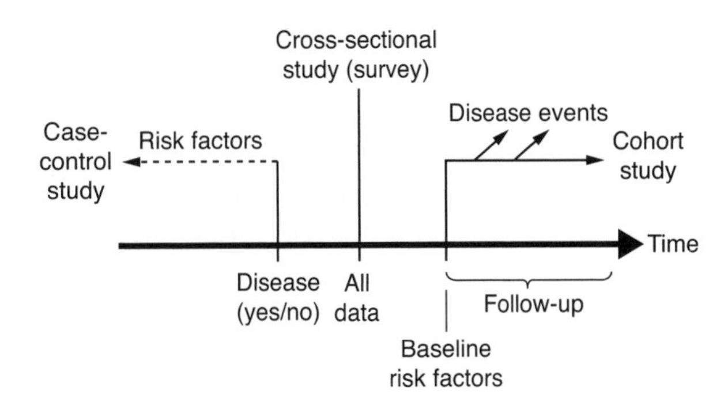

epidemiology *Schematic comparison of the three major study designs in epidemiology*

NESTED CASE-CONTROL STUDIES, on the other hand, are much less prone to bias error, as is also the case for the related case-cohort design. These are prospective studies based on the retrospective case-control concept. Although case-control studies are not the most reliable sources of information on causality for most epidemiological relationships, the case-control design is the design of choice in two situations. One is where the disease is so rare that any other kind of design is unlikely to produce enough cases of disease to obtain reliable estimates of association. The other is where the risk factor is transient, such as an outbreak of food poisoning. For some transient risk factors, a case-crossover study, where a case serves as his or her own control, might be advantageous. An example of this is a study of drivers involved in road traffic accidents where mobile (cell) telephone use just before the crash was compared with mobile phone use in equivalent periods when the subject had no crash (McEvoy *et al.*, 2005). Since drivers are their own controls, this design automatically controls for nontransient characteristics of the driver (e.g. age and sex) that may affect the risk of a crash.

In other situations the epidemiological design of choice is the cohort study. In the typical situation, a large group of people are surveyed at a point in time, or at least over a limited number of months, and several putative risk factors recorded. Over succeeding years (the follow-up) instances of disease and death may be recorded and related to the levels of a risk factor at initiation (baseline). Thus, cohort studies are said to be prospective (see the third figure). For example, after the initial cross-sectional study of the Scottish population mentioned above, the investigators arranged for any hospital admissions for coronary disease and deaths experienced by any of their sample (now called the study cohort) to be recorded. Tunstall-Pedoe *et al.* (1997) describe the relationship between coronary disease and 27 different risk factors for this study over a 7-year period. The advantage of the cohort study over the case-control study is that the time sequence of risk factor preceding disease can be established, strengthening the argument for causality. For instance, Tunstall-Pedoe and colleagues were able to conclude that high cholesterol levels tended to be followed by heart attacks, rather than the other way round.

Another advantage is that the cohort study can be used to investigate several risk factors and several diseases, whereas a case-control study is restricted to a single disease, that which defined the cases. One disadvantage is that they are time consuming, since it may take many years for enough cases of disease or death to occur to enable reliable estimation. Another is the likelihood of withdrawals, a special case of censoring, which is generally dealt with, at the analysis stage, through survival models. This leads to adoption of the hazard ratio as the measure of relative chance of disease or death in most cohort studies. Provided that pre-existing cases of disease at baseline are excluded, incident disease is measured in a cohort study. Cohort studies can be made more informative by re-measuring the cohort after baseline, in which case they are often called *longitudinal studies* (see LONGITUDINAL DATA). The repeat measurements may be used to obtain a more accurate picture of the true association between the risk factor (which will often change over time; e.g. smokers may quit during follow-up) and the disease outcome, perhaps through the use of mixed effects models. Sometimes only a subsample of the cohort is re-measured to enable correction of REGRESSION DILUTION BIAS.

A major question in many epidemiological investigations is whether it is reasonable to conclude that the hypothesised risk factor causes the disease in question (see also CAUSALITY). This issue was addressed by one of the pioneers of medical statistics and modern epidemiology, Sir Austin Bradford Hill. In 1965 he proposed a set of principles that describe ideal conditions for verifying a risk factor hypothesis (see also BRADFORD HILL'S CRITERIA). Some of these principles have been criticised as vague or impractical and often have been misinterpreted as rules, rather than guidelines. However, their use is widespread, if sometimes unconsciously. The principles include there being: a strong association between the putative risk factor and the disease (e.g. a large relative risk); consistency of the association in different settings (e.g. time and place); reversibility of the association (e.g. if the risk factor is removed the disease should have less likelihood of occurring); evidence of the risk factor preceding the disease (rather than vice versa); evidence of a biological gradient (a dose–response effect, meaning the more the risk factor, the more the disease); biological plausibility for the hypothesis (even if the mechanism is not yet understood); and lack of an alternative explanation for the observed association (e.g. confounding).

If all these principles hold, most epidemiologists would accept that there is truly good evidence to conclude causality. In real life, the situation is often less clear-cut and information on certain principles may be lacking or impossible to collect, so that judicious application of Bradford Hill's framework is required. In addition, their use has changed over time. For instance, relative risks of much lower levels than Bradford Hill seems to have envisaged are now routinely interpreted as supporting a causal hypothesis, for instance in debates over the effects of passive smoking. Conversely, the principle of consistency is now given more importance than in Bradford Hill's time due to the growth in popularity of META-ANALYSIS, as well as the developments in communication of research findings. Still the bottom line remains: anyone undertaking an epidemiological investigation would do well to judge their work against Bradford Hill's principles before attempting to ascribe causality.

Two important features of Bradford Hill's principles are the acknowledgment that data alone are insufficient (medical or other biological knowledge is crucial) and that final conclusions can only be drawn with comprehensive meta-analyses. Single epidemiological studies may fail to find a significant relationship due to small numbers, confounding effects or other biases. It is only by considering a range of studies that sensible conclusions may be drawn, not only because of the reliability of estimation afforded by large numbers, but also because variations in results might be explained by comparing results with study characteristics, such as by using META-REGRESSION. When epidemiological results are combined in meta-analysis it is usual to restrict use to case-control and cohort studies, because of the myriad biases inherent in ecological and cross-sectional designs. Even then, when there is a sufficient number of cohort studies, the tendency is to draw final 'best evidence' conclusions from the cohort studies (and, where available, nested case-control studies) alone. *MW*

Ashton, J. (ed.) 1994: *The epidemiological imagination*. Buckingham: Open University Press. **Bradford Hill, A.** 1965: The environment and disease: association or causation? *Proceedings of the Royal Society of Medicine* 58, 295. **Chave, S. P. W.** 1958: John Snow, the Broad Street pump and after. *The Medical Officer* 99, 347–9. **Clayton, D. and Hills, M.** 1993: *Statistical models in epidemiology*. Oxford: Oxford University Press. **McEvoy, S. P., Stevenson, M. R., McCartt, A. T., Woodward, M., Haworth, C., Palamara, P. and Cercarelli, R.** 2005: Role of mobile phones in motor vehicle crashes resulting in hospital attendance: a case-crossover study. *British Medical Journal* 331, 428. **Rothman, K. J., Greenland, S. and Lash, T. L.** 2008: *Modern epidemiology*, 3rd edition. Philadelphia, PA: Lippincott Williams and Wilkins. **St Leger, A. S., Cochrane, A. L. and Moore, F.** 1979: Factors associated with cardiac mortality in developed countries with particular reference to the consumption of wine. *Lancet* i, 1017–20. **Stolley, P. D. and Lasky, T.** 1995: *Investigating disease patterns. The science of epidemiology*. New York: W H Freeman. **Tunstall-Pedoe, H., Woodward, M., Tavendale, R., A'Brook, R. and McCluskey, M. K.** 1997: Comparison of the prediction by 27 different factors of coronary heart disease and death in men and women of the Scottish Heart Health Study: cohort study. *British Medical Journal* 315, 722–9. **Woodward, M.** 2005: *Epidemiology: study design and data analysis*, 2nd edition. Boca Raton, FL: Chapman and Hall/CRC Press.

EQS See STRUCTURAL EQUATION MODELLING SOFTWARE

equivalence studies See ACTIVE CONTROL EQUIVALENCE STUDIES

errors in hypothesis tests Neyman and Pearson (1933) proposed that the subjective view of the strength of evidence against the NULL HYPOTHESIS inherent in Fisher's significance tests be replaced with an objective, decision-theoretic approach to the results of experiments. In this approach, the investigator decides, in advance, a rule that states when the null hypothesis (e.g. that there is no association between a risk factor and a disease outcome, or no effect of a treatment) will be rejected or not rejected (accepted).

The null hypothesis may be rejected when it is in fact true, or alternatively we may fail to reject it when it is false. These are called TYPE I ERRORS and TYPE II ERRORS respectively (see the table). Neyman and Pearson suggested that by fixing, in advance, the Type I (α) and Type II (β) error rates, investigators would limit the number of mistakes made over many different experiments.

The figure shows the probabilities of occurrence of these two types of error, for a test at the 5% level, in the context of a normally distributed statistic such as the difference between two means. The *P*-VALUE (significance level) equals the probability of occurrence of a result as extreme as or more extreme than that observed if the null hypothesis were true. For example, part (a) shows that there is a 5 % probability that sampling variation alone will lead to a result significant at the 5 % level ($P < 0.05$).

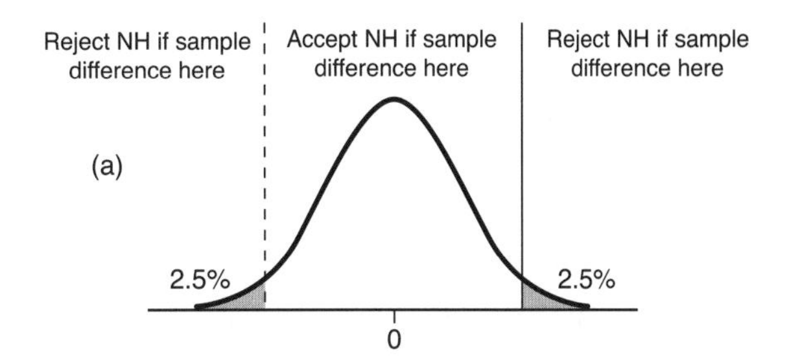

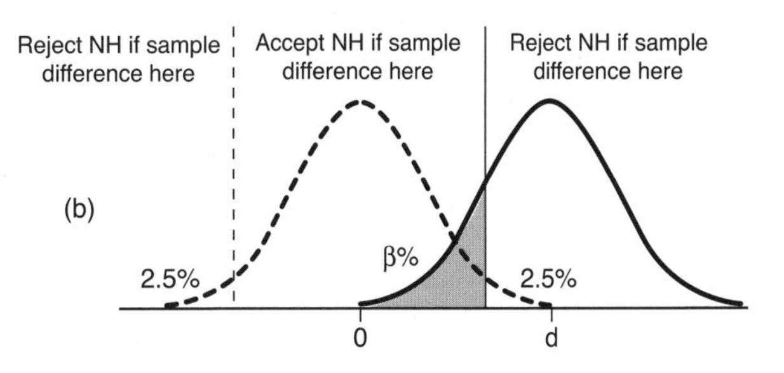

errors in hypothesis tests *Probabilities of occurrence of Type I and Type II errors, for a test at the 5% level*

errors in hypothesis tests *Types of error hypothesis tests*

	Reality	
Conclusion of significance test	*Null hypothesis is true*	*Null hypothesis is false*
Reject null hypothesis	*Type I error* (probability = significance level)	*Correct conclusion* (probability = power)
Do not reject null hypothesis	Correct conclusion (probability = 1 – significance level)	*Type II error* (probability = 1 –power)

The second type of error is that the null hypothesis is not rejected when it is false. This occurs because of overlap between the real sampling distribution of the sample difference about the population difference, d ($\neq$ 0), and the acceptance region for the null hypothesis based on the hypothesised sampling distribution about the incorrect difference, 0. This is illustrated in part (b). The shaded area shows the proportion (β%) of the real sampling distribution that would fall within the acceptance region for the null hypothesis, i.e. that would appear consistent with the null hypothesis at the 5 % level. The probability that we *do not* make a Type II error ($100 - \beta$%) equals the POWER of the test.

In the figure the following holds. Part (a) shows a Type I error: null hypothesis (NH) is *true* and the population difference = 0. The curve shows the sampling distribution of the sample difference. The shaded areas (total 5 %) give the probability that the null hypothesis is wrongly rejected. Part (b) shows a Type II error: null hypothesis is *false* and the population difference = $d \neq 0$. The continuous curve shows the real sampling distribution of the sample difference, while the dashed curve shows the sampling distribution under the null hypothesis. The shaded area is the probability (β%) that the null hypothesis fails to be rejected. *JS*

Neyman, J. and Pearson, E. 1933: On the problem of the most efficient tests of statistical hypotheses. *Philosophical Transactions of the Royal Society, Series A* 231, 289–337.

ethical review committees These committees are formally constituted and empowered groups charged with vetting and approving research protocols prior to study initiation to ensure sound ethics. The World Medical Association Declaration of Helsinki (2008) requires that experimental protocols be submitted to an 'ethical review committee, which must be independent of the investigator, the sponsor or any other kind of undue influence'. Such committees are variously known as institutional review boards (IRBs), research ethics boards (REBs) and human research ethics committees (HRECs). In the UK, research ethics committees (RECs) fall under the auspices of health authorities, ensuring independence from universities and teaching hospitals. The National Research Ethics Service (NRES) oversees the activities of about 100 authorised committees, a minority of which are 'recognised' to review applications for clinical trials of investigational medicinal products (CTIMPs). Historically, in the UK, a Central Office for Research Ethics Committees (COREC) fulfilled this role of overseeing regionally based Multicentre and (a much larger number of) Local Research Ethics Committees (abbreviated MRECs and LRECs respectively) (see the National Research Ethics Service at http://www.nres.npsa.nhs.uk/).

The perceived role of ethical review committees in practice varies both between and within countries. The Declaration of Helsinki requires that medical research 'conform to generally accepted scientific principles'. It is therefore the responsibility of the committee either to assess the scientific merit of each study or to satisfy itself that sufficient assessment has been undertaken.

Altman (1980) argued that poor use of statistics in medical research is unethical. First, if results cannot be trusted then the process is at best a waste of participants' time and may entail risk to participants without any possible benefit. Second, the process has also wasted scarce research resources. Finally, publication of incorrect conclusions may block or mislead future research, resulting in substandard patient care in the long term.

Statisticians on ethical review committees pay particular attention to the design of proposed studies: Is the design appropriate to the aims and are there reasonable safeguards to limit potential confounding and BIAS? Unlike many errors of analysis and interpretation, fundamental errors in design cannot be remedied at a later stage. Vail (1998) described several issues arising in practice for both experimental and observational studies.

In therapeutic intervention studies, the unbiased allocation of participants to groups is essential to avoid selection BIAS. The statistician should ensure the use of RANDOMISATION or MINIMISATION and the concealment of the randomization process so that the next allocation cannot be reliably guessed. BLINDING of clinicians, patients and outcome assessment, including the use of PLACEBO therapies, where appropriate, should be explained or the absence of blinding justified explicitly. It is also the statistician's role to ensure that, where intervention studies differ from the standard two-group parallel trial, justification is given and results will be valid. For example, misuse of CROSS-OVER TRIALS and failure to recognise clustering (see CLUSTER ANALYSIS IN MEDICINE) are common.

In observational studies the definition and selection of study groups is key and must be appropriate to the study aims. Use of individual-level matching should be explained. The statistician should also ensure that common sources of bias, such as REGRESSION TO THE MEAN and the HAWTHORNE EFFECT, have been adequately addressed.

In all quantitative studies the proposed sample size is a statistical issue. An excessively large (overpowered) study would waste resources and unnecessarily delay dissemination of potentially useful findings. In practice, such studies are rare. More commonly, a study that is too small (underpowered) risks patient involvement and resources with little chance of finding useful results. Newell (1978) argues that underpowered studies involving additional discomfort or risk to the patient are unethical. Whereas many statisticians consider all underpowered studies to be unethical, this view is not universally held (Edwards *et al.*, 1998), even for experimental studies. In general, statistical power calculations require several 'guesstimates' and there is no consensus on appropriate power, although 80% is often considered a minimum (see SAMPLE SIZE DETERMINATION entries).

The ethical review committee's statistical role in assessing analysis of studies is not easily defined. Occasionally it will be clear from a proposal that the planned analysis will be invalid or will be misinterpreted. For example, the investigators may propose to conclude equivalence of treatments from an underpowered, nonsignificant comparison, may confuse association with causation, may not recognise a need for case-mix adjustment or may simply propose a mathematically inappropriate analysis. In such cases, it may be considered unethical to allow the project to proceed without an undertaking to use appropriate statistical analyses. More usually, proposed analyses are sufficiently vague to cover valid as well as inappropriate analyses and individual committees vary in the extent to which they require detailed analysis plans.

Although there are good reasons for including a statistician in the membership of ethical review committees, Williamson *et al.* (2000) found low representation in a UK survey. Possible reasons for this include a shortage of statisticians who are qualified, available and willing to become involved. It may also reflect a lack of awareness of the benefits of statistical input: only 43 (29%) of 148 respondents without a statistician on their committee considered that they needed one. In the UK, official guidance since this survey requires each National Health Service REC to include expertise such that 'the rationale, aims, objectives and design of the research proposals ... can be effectively reconciled with the dignity, rights, safety and wellbeing' of participants, but falls short of requiring the input of a statistician. *AV*

[See also ETHICS AND CLINICAL TRIALS, SAMPLE SIZE DETERMINATION IN CLINICAL TRIALS, SAMPLE SIZE DETERMINATION IN CLUSTER RANDOMISED TRIALS, SAMPLE SIZE DETERMINATION IN OBSERVATIONAL STUDIES]

Altman, D. G. 1980: Statistics and ethics in medical research: misuse of statistics is unethical. *British Medical Journal* 281, 1182–4. **Edwards, S. J. L., Lilford, R. J., Braunholtz, D. A., Jackson, J. C., Hewison, J. and Thornton, J.** 1998: Ethical issues in the design and conduct of randomised clinical trials. *Health Technology Assessment* 2, 15. **Newell, D. J.** 1978: Type II errors and ethics. *British Medical Journal* 2, 1789. **Vail, A.** 1998: Experiences of a biostatistician on a UK research ethics committee. *Statistics in Medicine* 17, 2811–14. **Williamson, P., Hutton, J. L., Bliss, J., Blunt, J., Campbell, M. J. and Nicholson, R.** 2000: Statistical review by research ethics committees. *Journal of the Royal Statistical Society, Series A* 163, 5–13. **World Medical Association Declaration of Helsinki** 2008: *Ethical principles for medical research involving human subjects*. Seoul: 59th WMA General Assembly.

ethics and clinical trials Ethics and statistics meet one another head on, not exclusively but most acutely, in CLINICAL TRIALS. At first glance, one would think that the pair of disciplines, statistics and ethics, are poles apart, linked at best tenuously by their common, though misplaced, perception as being necessary but peripheral topics within a medical course curriculum. However, while their differences may be obvious, there are surprising similarities linking these diverse subjects. They are concerned, respectively, with the noble pursuits of what is true (at least numeric-based truth) and what is right, both amid uncertainty. For if there were no uncertainty, there would be nothing to pursue. One discipline appeals to PROBABILITY to describe what might, may or could happen; the other appeals to morality to describe what ought, must or should happen. Clinical trials are experiments incorporating a delicate, three-part mixture of theory, practice and ethics, with the utmost importance attached to matters of ethics, in view of the priceless nature of the experimental units involved. They raise ethical questions before they start, while still in progress, when they end and, often, long after they are finished too, whereas theoretical and practical aspects tend to be more limited in their scopes.

In an agricultural trial, it does not really matter if a field of wheat perishes under, say, fertiliser A. One may actually be quite happy to gain a clear-cut result that fertiliser B is superior to A, with happiness inversely proportional to the magnitude of the *P*-VALUE. However, when comparing drug A and drug B in a clinical setting, one must consider the people involved, not forgetting that statistical 'success' and 'failure' outcomes could be euphemisms for a patient's life and death. One may be happy to demonstrate a statistically significant difference between treatment groups and hence declare a positive result. Then again, one might think, 'Could not this conclusion have been reached sooner, with similar confidence, yet sparing the lives of some of those randomised to the inferior treatment?' This line of reasoning has motivated many statisticians to conduct research into DATA-DEPENDENT DESIGNS for clinical trials, including ADAPTIVE DESIGNS, sequential methods (see SEQUENTIAL ANALYSIS) and Bayesian and decision-theoretic approaches to the design and analysis of trials (see BAYESIAN METHODS).

Ethical concerns have, of course, been around for a longer time than statistics and the modern controlled clinical trial. A number of attempts to codify ethics for medical research in general (not just trials) have been made. The most famous such code is the World Medical Association (WMA) Declaration of Helsinki, first adopted in 1964 and updated periodically (1975, 1983, 1989, 1996, 2000, 2002, 2004 and 2008). An online version can be found at http://www.wma.net/en/30publications/10policies/b3/index.html. As a forerunner to this Declaration, in the aftermath of wartime atrocities in Nazi-occupied states, another set of international guidelines applicable to all types of medical research, the Nuremberg Code, was established. This has primary focus on the desire to protect defenceless subjects from unwilling participation in 'research'. See, for example, http://ohsr.od.nih.gov/guidelines/nuremberg.html for details.

Clearly, though, no set of regulations can be sufficient to guide researchers conducting any particular clinical trial and to protect its volunteers. Hence, recent decades have seen the emergence and growth of Institutional Review Boards or, more generally, ETHICAL REVIEW COMMITTEES. Their remit is to scrutinise study proposals, on a case-by-case basis, prior to granting study approval and to monitor ongoing research once underway.

The remainder of this entry will describe some general ethical principles, then outline important applications to clinical trials, emphasising statistical aspects, so that aspiring researchers may become familiar with these main areas for consideration when designing future clinical trials. Further details and discussion can be found in Edwards *et al.* (1998).

Bioethicists have developed various sets of principles by which to evaluate moral aspects of research. One such set, the 'four pillars of morality', are discussed at length by Gillon (1998) and by Beauchamp and Childress (2001). These pillars are *autonomy* (respecting a patient's right to self-govern), *beneficence* (doing good), *nonmalificence* (avoiding harm) and *justice* (being fair). An alternative set of three principles seeks to solve ethical decision-making problems by appealing to utilitarianism (a consequentialist approach asking 'What maximises total good minus total harm?') and duty-based and rights-based deontological principles. In brief, duty-based deontology argues for doing that which is intrinsically right (asking 'Is it right or wrong?') whereas a rights-based deontological approach bases decision-making on whether people are treated appropriately (asking 'Is he or she wronged?'). Thoughtful application of these sets of principles can help when faced with ethical quandaries, not limited to the conduct of clinical research. They serve useful purposes for those involved in trials, for they can be appealed to for justifying numerous research-related concepts. Among these are the role of RANDOMISATION, the need for obtaining informed consent, acceptability of blinding and use of PLACEBOS, and indeed the rationale behind trials in the first place.

Statisticians, by contrast, tend to believe a set of two ethical principles to be preferable to three or four, and make the simpler dichotomy into *collective ethics* and *individual ethics* (terms coined by Schwartz, Flamant and Lellouch, 1980). Applied to clinical trials, these concepts equate to doing what is right and best for future patients (those who stand to benefit from the results of a trial) and doing what is right and best for current patients (the volunteers in the trial) respectively. Indeed, a clinical trial can be thought of as a balance in delicate equilibrium between these two types of ethics (Pocock, 1983). Collective ethics, also known as research ethics, and individual ethics tend to be in direct competition with one another. If one adhered purely to collective ethics, there would be unacceptable human sacrifices, but equally, if one adhered purely to individual ethics, there would be little scope for making medical progress.

According to individual ethics, each patient in a trial should receive the best possible treatment, whereas according to collective ethics, each trial should yield the best scientific result possible. The tension is clear. A doctor rightly has to pay greatest attention to the needs of his or her patients. This is the essence of the Hippocratic Oath and fully supported by the above-mentioned Declaration of Helsinki. Among its precepts is 'In research on man, the interests of science and society should never take precedence over considerations related to the well-being of the subject', or, in other words, collective ethics can never be allowed to usurp individual ethics for the sake of scientific endeavour.

Turning to applications of ethics in clinical trials, it is convenient to categorise according to whether primarily affecting the period before, during or after the trial's recruitment phase. Chronologically, the first ethical consideration, then, is whether a proposed trial should be conducted at all. Quite often there is only a limited window of opportunity (meaning in calendar time) in which to conduct a randomised controlled trial, assuming of course the situation is one wherein randomisation is itself acceptable.

Why is there only a limited time window? Ethicists have coined the phrase, and notion, of *clinical equipoise*. It is a precondition for initiating a clinical trial and refers to the balance of clinical opinion among all doctors that needs to exist before it is ethical for a trial to begin. Thus, while it may not be possible for any one doctor to be perfectly balanced in their own mind concerning the relative merits of two or more treatments, perhaps including investigational new ones, it is quite possible that other doctors have preferences for the different treatments involved. Hence, a planned trial can be considered ethical on the basis of divergence of opinions, even if one has a slight personal preference, though not yet any firm evidence, in favour of one particular treatment. If, however, the weight of clinical opinion is too heavily in favour of a given treatment it can become too late to seize the opportunity to undertake a rigorous clinical trial. In turn, this means that the chance to secure best quality evidence to confirm, or overturn, such clinical (mere) opinion may be lost. EVIDENCE-BASED MEDICINE seeks to convert opinions (experience-based hunches, beliefs or gut feelings) into more objectively held evidence, based on sound data collection, gathered supremely through randomised controlled trials.

One certain point for debate at meetings of ethics committees relates to the process for obtaining each patient's informed consent to participate. This is not just because of the Nuremberg code, but a (happy) consequence of the composition of ethics committees being mandated to include a number of laypeople. Those nonmedically trained people may or may not understand the intricacies of treatments involved in a research proposal, but they will definitely be able to identify with prospective participants. Hence, committees will include discussion on the all- important patient information sheet, or its equivalent, that outlines in everyday language the risks and benefits involved in trial participation.

Informed consent from participants is one of the most important safeguards built into clinical trials. It means that subjects never take part in a trial against their will (and for this reason it can be helpful to refer to participants as 'volunteers'). Ideally consent should mean written, fully informed consent, although there are circumstances, usually emergencies, when witnessed verbal consent has to suffice. There are complications when the subjects of research include children, or those who are mentally ill, or comatose, or otherwise unable to understand the full implications of agreeing to enter a trial. In such cases, a proxy has to be appointed to serve as a spokesperson.

Ethical matters in general are heightened when dealing with special populations such as those just mentioned. In addition, one can add prisoners and medical students as special populations when undertaking research. Admittedly, it is unusual to put such groups together in the same breath, but in both cases, though for different reasons, there is a possible sense of coercion involved, meaning that one has to be extra careful when going about obtaining an individual's consent. Similarly, trials sponsored by developed nations that are to be conducted in developing nations are another major source of ethical conflict. This is especially true if wealthier nations stand to benefit more than the participating populations from the results of such foreign-based trials.

Another matter is exactly how much information is necessary to impart to the subjects at the time of trial recruitment. It surely includes relaying the uncertainty about which course of treatment is truly the best (for otherwise why perform a trial?) and that by volunteering they would be helping in the pursuit of medical progress to try and remove some of that uncertainty.

The act of giving informed consent is a statement from or on behalf of the trial participant that allows researchers to seek entry into the trial. Being randomised is not yet guaranteed, however, as there are strict eligibility requirements (see INCLUSION AND EXCLUSION CRITERIA) that may need to be checked after obtaining consent. Obviously, no subject should be forced into participating, but from the scientific viewpoint it is preferable to have as high a proportion as possible of those invited going on to participate. This is analogous to seeking a high response rate in a sample survey – it reduces selection bias and enhances the generalisability of the trial's results (see BIAS, QUESTIONNAIRES). Part of the informed consent process should include a brief justification of the need for randomisation within the trial as a general principle. In addition, patients must be informed of specific details of what will be expected of them, the likely risks involved, together with reassurance that they can withdraw from the trial at any time without compromising their future treatment or care.

So far informed consent has been discussed with regard to the prospective patient. It is their opportunity to decline to take part in the research. Next, consider the related decision that an investigator conducting research may have to make prior to involvement in a trial. Sometimes doctors or others are approached to become collaborators with someone else's research project. There is no formal equivalent to securing consent among investigators, but there are ethical considerations to be borne in mind. Chiefly, there is what is amusingly known as the *uncle test for randomisation*. It calls for the trial's investigators to answer affirmatively the question: 'Would you be willing to randomise either yourself or a close relative of yours (parent, spouse, sibling, child, uncle, etc.) into this trial?' That is, it seeks assurance that one's individual preferences are not so heavily biased towards one of the treatments that there would be a reluctance take the risk of receiving the least-favoured treatment for themselves or someone close to them. If investigators cannot honestly answer 'yes' to the uncle test then they simply should not enrol any patients into the trial (and certainly not succumb to financial inducements or other temptations to do so if the trial happens to be sponsored by deep-pocketed sources).

The use of PLACEBOS (inert substances made to mimic the appearance of active treatments) and of PLACEBO RUN-IN periods (to assess a patient's compliance with treatment schedules, possibly involving the withholding of their usual medication) within trials are controversial areas, attracting much attention from those with ethical viewpoints. Note that a footnote to the most recent version of the Declaration of Helsinki addresses the use of placebos in clinical trials, arguing in their favour in the right circumstances. The choice of an active treatment control or a placebo control group is an example of how trials are an amalgamation of theory, practice and ethics. Statistically, one needs fewer subjects to demonstrate a difference in efficacy between a new drug and a placebo than between two active drugs. However, the decision whether to use a placebo or not must not be dictated solely by sample size considerations, but primarily by whether it is ethically acceptable to put patients deliberately on to inactive treatment regimens. This further exemplifies the tension between individual and collective ethics. Similarly, the use of placebo run-ins prior to randomisation, while not necessarily completely unjustifiable, is harder to defend on ethical grounds (Senn, 1997).

The actual conduct of a clinical trial must be to highest scientific standards, for without such rigour the trial is compromised and the results unable to contribute meaningfully to medical progress. More positively, the well-conducted randomised controlled trial is rightly reckoned to be the most reliable source of evidence for or against any treatment. Note this includes adhering to the clinical trial's PROTOCOL, using a proper method for RANDOMISATION and, if blinded, employing suitable means to conceal treatment allocation effectively (see BLINDING). The protocol, among other things, must include the hypotheses being investigated along with the primary and secondary outcomes. It is unethical to look first at one's data and then decide to promote in importance a nonprimary outcome and demote the primary outcome, based on the actual findings. The excuse that unforeseen results may be more interesting does not justify departing from the protocol when publishing the study. See POST HOC ANALYSES for further discussion.

A matter not always given full consideration is what to do with samples that are collected during the course of the clinical trial. Ideally, the protocol would specify not only where and for how long sensitive information, or equally blood, or DNA samples, etc., are stored, but who has access to them in the future and under what conditions. For this ethical reason of protecting patient confidentiality, it is important that data and samples are stored securely and suitably coded, or even anonymised altogether. For the benefit of future researchers, however, it is preferable whenever possible to have a system that allows access to individual patient information. This is a matter that should be included within the patient informed consent process if longer term research use of patients' data, or samples, is envisaged.

Another issue is whether a trial should be allowed the chance to stop earlier than planned. That is, can patient recruitment cease before the originally anticipated recruitment levels have been reached? Scientifically, there is a penalty for stopping a trial unexpectedly once initiated, so only in extreme cases are trials interrupted. There can be pressure, e.g. from disease-specific patient support groups, to expedite a drug development process. This arises in part from the long timespan involved before a promising treatment can be safely marketed. The statistical implication of planning possibly to stop early is best described by analogy with making MULTIPLE COMPARISONS. Testing a dataset according to many different subsets will yield many false positive statistically significant findings. Similarly, multiple looks at the data at several INTERIM ANALYSES, each with the opportunity to stop the trial, make it more likely that an apparent treatment difference would emerge when in fact there is none. The penalty for unplanned early stopping is on the collective ethical side of the balance with individual ethics, for it is future patients who are denied the opportunity to learn more about the treatments under investigation. Then, again, one has also to consider the needs and rights of patients entered towards the end of the trial, so it is never a simple choice.

The accurate and timely reporting of clinical trials are yet further matters with ethical implications. Choosing not to report a trial because results show sponsors' products unfavourably is without excuse and a clear abuse of ethics. For the vast majority for which publication in journals is sought, then conforming to the CONSOLIDATED STANDARDS OF REPORTING TRIALS (CONSORT) STATEMENT guidelines is helpful, if not mandatory, as it is becoming for a growing number of biomedical journals. An online version of the latest statement can be accessed at http://www.consort-statement.org. Authors failing to declare relevant conflicts of interest, financial or otherwise, or bulk-ordering expensive reprints of an article from a journal's office prior to formal acceptance are further examples of inexcusable behaviour at the pre-publication stage.

In an editorial in the *British Medical Journal*, Altman (1994) suggested there was rather too much research happening, not all of the highest quality and not always undertaken for the right reasons. If a clinical trial can be conducted, then it should be, in preference to an observational study, in order to gain the best quality evidence. It must first pass ethical review and be conducted properly, with suitable sample size, randomisation, analysis plans, etc., detailed in the protocol, and it should also be reported according to highest standards. All this is a tall order, hence another reason for CONSULTING A STATISTICIAN early in the trial's life, for it is notoriously easy to fall into the PITFALLS OF MEDICAL RESEARCH. Finally, recalling the link between those not-so-disparate disciplines, forget not that bad statistics is bad ethics! *CRP*

Altman, D. G. 1994: The scandal of poor medical research (editorial). *British Medical Journal* 308, 283–4. **Beauchamp, T. and Childress, J.** 2001: *Principles of biomedical ethics*, 5th edition. New York: Oxford University Press. **Edwards, S. J. L., Lilford, R. J., Braunholz, D. A., Jackson, J. C., Hewison, J. and Thornton, J.** 1998: Ethical issues in the design and conduct of randomised controlled trials. *Health Technology Assessment* 2, 15. Online at http://www.hta.nhsweb.nhs.uk/execsumm/summ215.htm. **Gillon, R.** 1985: *Philosophical medical ethics*. Chichester: John Wiley & Sons, Ltd. **Schwartz, D., Flamant, R. and Lellouch, J.** 1980: *Clinical trials*, translated by Healy, M. J. R. London: Academic Press. **Senn, S. J.** 1997: Are placebo run-ins justified? *British Medical Journal* 314, 1191–3. **Pocock, S. J.** 1983: *Clinical trials: a practical approach*. Chichester: John Wiley & Sons, Ltd. **World Medical Association Declaration of Helsinki** 2008: *Ethical principles for medical research involving human subjects*. Seoul: 59th WMA General Assembly.

evidence-based medicine (EBM) Evidence-based medicine is the conscientious, explicit and judicious use of current best evidence in making diseases about the care of individual patients and is the definition of EBM given by one of its foremost proponents (Sackett *et al.*, 1996).

Alternative definitions that have appeared are very similar, stressing the aim of assessing and applying relevant evidence for better healthcare decision making and allowing clinicians to practise better medicine by being aware of the evidence in support of clinical practice and the strength of that evidence. The primary tools used to drive evidence-based medicine are RANDOMISED CLINICAL TRIALS and SYSTEMATIC REVIEWS AND META-ANALYSIS. However, as Sackett *et al.* (1996) make clear, evidence-based medicine is not restricted to these tools and, in particular circumstances, the best external evidence with which to answer a clinical question may involve CROSS-SECTIONAL STUDIES, genetic studies or immunological investigations. *BSE*

Sackett, D. L., Rosenberg, M. C., Gray, J. A., Haynes, R. B. and Richardson, W. 1996: Evidence-based medicine: what it is and what it isn't. *British Medical Journal* 312, 71–2.

exact methods for categorical data This is a collective term for analytical methods that require no distributional approximations in order to validate the resulting inference. The 'exact' label applies when the probability distribution of the appropriate test statistic is fully determined, requiring no assumptions about unknown population characteristics and no large-sample distributional justifications (e.g. using approximate normality). For expository reasons we illustrate the exact approach by focusing primarily on the *P*-VALUE yielded by a hypothesis test. It should be noted, however, that exact methods can also be used for estimation and computing exact CONFIDENCE INTERVALS.

A fundamental problem in statistical inference is summarising observed data in terms of a *P*-value. The *P*-value forms part of the theory of hypothesis testing and may be regarded as an index for judging whether to accept or reject the NULL HYPOTHESIS. A very small *P*-value is indicative of evidence against the null hypothesis, while a large *P*-value implies that the observed data are compatible with the null hypothesis. There is a long tradition of using the value 0.05 as the cut-off for rejection or acceptance of the null hypothesis. While this may appear arbitrary in some contexts, its almost universal adoption for testing scientific hypotheses has the merit of limiting the number of false–positive conclusions to at most 5 %. At any rate, no matter what cut-off one chooses, the *P*-value provides an important objective input for judging if the observed data are statistically significant. Therefore it is crucial that this number be computed accurately.

Since data may be gathered under diverse, often nonverifiable, conditions, it is desirable, for *P*-value calculations, to make as few assumptions as possible about the underlying data-generation process. In particular, one wishes to avoid making distributional assumptions, such as that the data came from a NORMAL DISTRIBUTION. This goal has spawned an entire field of statistics known as nonparametric statistics. In the preface to his book, *Nonparametrics: statistical methods based on ranks*, Lehmann (1975) traces the earliest development of a nonparametric test to Arbuthnot (1710), who came up with the remarkably simple yet popular sign test. In the 20th century, nonparametric methods received a major impetus from a seminal paper by Frank Wilcoxon (1945) in which he developed the now universally adopted WILCOXON SIGNED RANK TEST and the Wilcoxon rank sum test. The contributions of many researchers advanced this field – an excellent survey of these developments is given in Agresti (1992).

The research just mentioned, and the numerous papers, monographs and textbooks that followed in its wake, deal primarily with hypothesis tests involving continuous distributions. The data usually consisted of several independent samples of real numbers (possibly containing ties) drawn from different populations, with the objective of making distribution-free one, two or *K*-sample comparisons, performing goodness-of-fit tests and computing measures of association. Much earlier, Karl Pearson (1900) demonstrated that the large-sample distribution of a test statistic, based on the difference between the observed and expected counts of categorical data generated from multinomial, hypergeometric or POISSON DISTRIBUTIONS, is a CHI-SQUARE DISTRIBUTION. This work was found to be applicable to a whole class of discrete data problems. It was followed by many significant contributions and eventually evolved into the field of categorical data analysis. An excellent up-to-date textbook dealing with this continually growing field is Agresti (2002).

The techniques of nonparametric and categorical data inference are popular mainly because they make only minimal assumptions about how the data were generated; assumptions such as independent sampling or randomised treatment assignment. For continuous data one does not have to know the underlying distribution giving rise to the data. For categorical data mathematical models like the multinomial, Poisson or hypergeometric arise naturally from the independence assumptions of the sampled observations. Nevertheless, for both the continuous and categorical cases, these methods do require one assumption that is sometimes hard to verify.

They assume that the dataset is large enough for the test statistic to converge to an appropriate limiting normal or chi-square distribution. *P*-values are then obtained by evaluating the tail area of the limiting distribution, instead of actually deriving the true distribution of the test statistic and then evaluating its tail area. *P*-values based on the large-sample assumption are known as *asymptotic P-values*, while P-values based on deriving the true distribution of the test statistic are termed *exact P-values*. While one would prefer to use exact P-values for scientific inference they often pose formidable computational problems and so, as a practical matter, asymptotic P-values are used in their place. For large and well-balanced datasets this makes very little difference since the exact and asymptotic P-values are very similar. However, for small, sparse, unbalanced and heavily tied data, the exact and asymptotic P-values can be quite different and may lead to opposite conclusions concerning the hypothesis of interest. This was a major concern of Fisher, who stated in the preface to the first edition of *Statistical methods for research workers* (1925): 'The traditional machinery of statistical processes is wholly unsuited to the needs of practical research. Not only does it take a cannon to shoot a sparrow, but it misses the sparrow! The elaborate mechanism built on the theory of infinitely large samples is not accurate enough for simple laboratory data. Only by systematically tackling small problems on their merits does it seem possible to apply accurate tests to practical data.'

That Fisher's concern was justified is seen from the following example of a 3×9 sparse contingency table:

0	7	0	0	0	0	0	1	1
1	1	1	1	1	1	1	0	0
0	8	0	0	0	0	0	0	0

The Pearson CHI-SQUARE TEST is commonly used to test for row and column interaction. For our contingency table, the observed value of Pearson's statistic is 22.29 and the asymptotic P-value is the tail area to the right of 22.29 from a chi-square distribution with 16 DEGREES OF FREEDOM. This P-value is 0.1342, implying that there is no row and column interaction. However, we can also compute the tail area to the right of 22.29 from the exact distribution of Pearson's statistic. The exact P-value so obtained is 0.0013, implying that there is a strong row and column interaction.

We will conceptually describe further on how to compute an exact P-value for the chi-square statistic. However, even without knowing the technical details behind such a computation, an investigator comparing the asymptotic and exact P-values for this example might wonder at the disparity and not know which result is reliable. This example highlights the need to compute the exact P-value, rather than relying on asymptotic results, whenever the dataset is small, sparse, unbalanced or heavily tied. The trouble is that it is difficult to identify, a priori, that a given dataset suffers from these obstacles to asymptotic inference.

The concerns expressed by Fisher and others can be resolved if we directly compute exact P-values instead of replacing them by their asymptotic versions and hoping that these will be accurate. Fisher himself suggested the use of exact P-values for 2×2 tables (1925) as well as for data from randomised experiments (1956) (see FISHER'S EXACT TEST). For the 2×2 table, Fisher proposed permuting the observed data in all possible ways and comparing what was actually observed to what might have been observed. Thus exact P-values are also known as permutational P-values.

We demonstrate here how this approach can be used to obtain the exact distribution for the commonly used Pearson chi-square test. The table shows results from an entrance examination for fire fighters in a small US town. All five white applicants received a pass result, whereas the results for the other groups are mixed. Is this evidence that entrance exam results are related to race? Note that while there is some evidence of a pattern, the total number of observations is only 20. A statistically inclined researcher might proceed as follows:

Null hypothesis. Exam results and race of examinee are independent.

Alternative hypothesis. Exam results and race of examinee are not independent.

To test the hypothesis of independence, one would ordinarily use the Pearson chi-square test. The test statistic has the form:

$$\sum_{\text{Table cells}} \frac{(\text{Observed count} - \text{Expected count})^2}{\text{Expected count}}$$

exact methods for categorical data *Entrance examination results for fire fighters in a small US town*

	Race				
Test results	*White*	*Black*	*Asian*	*Hispanic*	*Row total*
Pass	5	2	2	0	9
No show	0	1	0	1	2
Fail	0	2	3	4	9
Column total	5	5	5	5	20

The distribution of this test statistic is *asymptotically* chi-square. Suppose we would like to conduct the test at the 0.05 level of significance for these data. Running this test, we obtain the following results:

Pearson chi-square	11.55556
Degrees of freedom	6
Significance	0.07265

Because the observed significance of 0.07265 is larger than 0.05, the researcher would conclude that exam results are independent of race of examinee. However, for this table the minimum expected cell frequency is 0.5, and all 12 cells in this table expected frequencies under the null hypothesis less than 5; i.e. since all the cells in the table have small expected counts, what does this mean? Does it matter? The term 'asymptotically' means 'given a sufficient sample size', although it is not easy to describe the sample size needed for the chi-square distribution to approximate well the exact distribution of the Pearson statistic. Two widely used rules of thumb are given by:

1. The minimum expected cell count for all cells should be at least 5.
2. For tables larger than 2×2, a minimum expected count of 1 is permissible as long as no more than about 20% of the cells have expected values below 5.

While these and other rules have been proposed and studied, in the end no simple rule covers all cases (see Agresti, 2002, for further discussion). In our case, in terms of sample size, number of cells relative to sample size or small expected counts, it appears that relying on an asymptotic result to compute a P-value might be problematic.

What if, instead of relying on the chi-square distribution to approximate the P-value, it were possible to use the true sampling distribution of the test statistic and thereby produce an exact P-value? Here we explain in an intuitive way how this P-value is computed and why it is exact. (For a more technical discussion, see Mehta, 1994.) The main idea is to evaluate our 3×4 cross-tabulation, relative to a 'reference set' of other 3×4 tables that are like it in every possible respect, except in terms of their reasonableness under the null hypothesis. It is generally accepted that this reference set consists of all 3×4 tables in the form of the observed table that have the same row and column margins as the observed table. This is a reasonable choice for a reference set, even when these margins are not naturally fixed in the original dataset, because they do not contain any information about the null hypothesis being tested. We refer to this as a *conditional exact approach.*

The exact P-value is then obtained by identifying all the tables in this reference set whose Pearson statistics equal or exceed 11.55556, the observed statistic, and summing their probabilities. This is an exact P-value because the probability of any table in the reference set of tables with fixed margins can be computed exactly under the null hypothesis.

For instance, the table:

5	2	2	0	9
0	0	0	2	2
0	3	3	3	9
5	5	5	5	20

is a member of the reference set. The Pearson statistic for this table yields a value of 14.67. Since this value is greater than the observed value 11.55556, we regard this member of the reference set as 'more extreme' than the observed table. Its exact probability is 0.000108 and will contribute to the exact P-value. In principle we can repeat this analysis for every single table in the reference set, identify all those that are at least as extreme as the original table and sum their exact probabilities. The exact P-value is this sum.

In fact, the exact P-value based on Pearson's statistic is 0.0398. At the 0.05 level of significance, a researcher would reject the null hypothesis and conclude that there is evidence that the exam results and race of examinee are related. This conclusion is the opposite of what would be concluded using the asymptotic approach, since the latter produced a P-value of 0.07625. The asymptotic P-value, however, is only an approximate estimate of the exact P-value. As the sample size goes to infinity the exact P-value converges to the chi-square-based P-value. Of course, the sample size for the current dataset is not infinite and we observe that this asymptotic result has fared rather poorly.

This conditional approach to exact inference is currently the most widely used method for exact inference. However, conditional methods have their drawbacks. They can be computationally intensive. Until the advent of modern computing, exact tests were generally infeasible for any dataset that was not relatively very small. Over the past two or three decades, progress in computing technology along with the development of efficient algorithms have made conditional exact methods available to more practitioners for solving an increasingly larger class of applied problems. Virtually all commercial statistical software packages now contain at least some exact options for analysing categorical data.

More fundamentally, conditional exact methods are sometimes criticised for their conservatism. By construction, exact conditional tests are guaranteed to control the Type 1 error rate at any desired level. (A Type 1 error occurs when we erroneously reject a null hypothesis.) This means, for example, that if you consistently use an exact P-value of 0.05 as your cut-off for deciding whether your results are statistically significant, this decision rule will limit your rate of declaring 'false positives' to at most 5%. However, since the exact distribution of the test statistic is discrete, you may not be able to achieve a Type 1 error rate of exactly 5%. Instead, the actual error rate of your decision rule will typically be smaller. This conservatism, which is entirely attributable to the discreteness of the test statistic, is the price you pay for exactness. The extent of the conservatism is not easy to determine. It does not manifest itself through exact P-values that are always larger than corresponding asymptotic P-values. On the contrary, there are plenty of examples (including the fire fighter data shown earlier) wherein the exact P-values are substantially smaller than the asymptotic P-values. Conservatism is a statement about a long-term error rate, rather than an individual P-value.

How serious is the conservatism? Not many empirical studies have been conducted to answer this question in general. For 2×2 tables, the extent of the conservatism has been investigated extensively and many alternatives have been proposed over the years to counter it. None of these other approaches has so far managed to dislodge conditional exact tests as the methods of choice, for reasons discussed in Yates (1984) and Barnard (1949, 1989). In fact, it has been demonstrated that the conservatism of both types of tests is negligible in this setting, indicating that conservatism is also likely to be a negligible factor in the more general $r \times c$ setting.

Conservatism is still an issue for the single 2×2 contingency table. One way of reducing conservatism that has recently received greater attention is to use an unconditional approach.

Much of the blame for conservatism is attributed to the reference set having fixed row and column margins. In the single 2×2 table especially, such conditioning makes the distribution of the test statistic rather discrete in small samples. On the other hand, eliminating nuisance parameters (e.g. the odds ratio for a 2×2 table) by conditioning on their sufficient statistics, i.e. the row and column margins, is at the heart of exact conditional inference. Unconditional tests usually rely on large sample theory to eliminate nuisance parameters from the distribution of the test statistic. This is acceptable for large datasets, but may not be accurate for small, sparse or unbalanced data. There does exist an alternative unconditional exact approach, however, that is valid in small samples.

Barnard (1947) proposed an unconditional exact test based on a minimax elimination of the nuisance parameter. The reference set was defined to be the set of all 2×2 tables with fixed row margins and all possible column margins. Since the reference set for Barnard's test does not fix the column margins, the distribution of the test statistic is less discrete than would be obtained by permuting the conditional reference set in which both margins are fixed. However, Barnard was not satisfied with his test and disavowed it two years later (Barnard, 1949).

Barnard was invoking Fisher's principle of *ancillarity*, whereby inference should be based on hypothetical repetitions of the original experiment, fixing those aspects of the experiment that are unrelated to the hypothesis under test. In more recent publications Barnard (1989) provides additional arguments against the test.

Some prominent statisticians have expressed regret at Barnard's disavowal. Others continue to favour inference based on the conditional reference set. At any rate most statisticians agree that the case for conditioning is especially persuasive with RANDOMISED CLINICAL TRIALS, for then one can argue that the sum of the responses from the two treatment groups is fixed in advance under the null hypothesis; i.e. subjects predisposed to respond will respond regardless of the treatment received, since the two treatments are identical under the null hypothesis. Permuting the reference set merely amounts to assigning the treatments to the patients in all possible ways.

Even if Barnard's method were accepted without reservation, it would be hard to implement for general $r \times c$ contingency tables. It calls for enriching the reference set by permitting the column margins to vary and then maximising over the unknown marginal probabilities. Such a process is difficult, computationally, for tables of higher dimension than 2×2.

Another way of addressing conservatism is by using 'flexible' significance levels. Conservatism really hinges on approaching data with a fixed significance level (say 5 %) in mind. Not all statisticians believe in fixed significance levels for decision making, however. Fisher (1973) stated: 'No scientific worker has a fixed level of significance at which from year to year, and in all circumstances, he rejects hypotheses; he rather gives his mind to each particular case in the light of his evidence and his ideas.'

Barnard (1989) has formalised this principle in terms of the 'flexible Fisher exact test' for the single 2×2 table for comparing two binomials. He proposes that we choose different significance levels for rejecting the null hypothesis depending on the observed sum of successes in the two binomial populations. The flexible Fisher exact test can be shown to be equivalent to the various alternatives that have been proposed over the years to counter the alleged conservatism of Fisher's exact test for 2×2 tables.

A third way of controlling conservatism is by using a continuity correction, such as the mid-P-value, which is obtained by subtracting half the probability of the observed statistic from the exact P-value. This modified P-value has been recommended by many statisticians (see, for example, Barnard, 1989) as a good compromise between reporting a possibly conservative exact P-value and relying on a randomised test to eliminate conservatism completely. However, the mid-P-value cannot guarantee, theoretically, that the Type 1 error rate will be limited to the desired level. By the same token, researchers have shown empirically that mid-P-values do in fact preserve the Type 1 error rate while reducing the conservatism of exact P-values for a single 2×2 contingency table and in k 2×2 contingency tables. The coverage of the mid-P confidence intervals was not compromised, but they were shorter on average than the corresponding exact intervals. *CCo/PSe/CM/NP*

Agresti, A. 1992: A survey of exact inference for contingency tables. *Statistical Science* 7, 1, 131–77. **Agresti, A.** 2002: *Categorical data analysis*, 2nd edition. New York: John Wiley & Sons, Inc. **Arbuthnot, J.** 1710: An argument for divine providence, taken from the constant regularity observed in the birth of both sexes. *Philosophical Transactions* 27, 186–90. **Barnard, G. A.** 1947: Significance tests for 2×2

tables. *Biometrika* 34, 123–38. **Barnard, G. A.** 1949: Statistical inference. *Journal of the Royal Statistical Society, Series B* 11, 115–39. **Barnard, G. A.** 1989: On alleged gains in power from lower *P*-values. *Statistics in Medicine* 8, 1469–77. **Fisher, R. A.** 1925: *Statistical methods for research workers*. Edinburgh: Oliver and Boyd. **Fisher, R. A.** 1956: *Statistical methods for scientific inference*. Edinburgh: Oliver and Boyd. **Fisher, R. A.** 1973: *Statistical methods and scientific Inference*, 3rd edition. London: Collier Macmillan. **Lehmann, E. L.** 1975: *Nonparametrics: statistical methods based on ranks*. San Francisco: Holden-Day. **Mehta, C. R.** 1994: The exact analysis of contingency tables in medical research. *Statistical Methods in Medical Research* 3, 135–56. **Pearson, K.** 1900: On the criterion that a given system of deviations from the probable in the case of a correlated system of variables is such that it can be reasonably supposed to have arisen from random sampling. *Philosophical Magazine Series 5* 50, 157–75. **Wilcoxon, F.** 1945: Individual comparisons by ranking methods *Biometrics* 1, 80–3. **Yates, F.** 1984: Test of significance for 2 × 2 contingency tables. *Journal of the Royal Statistical Society, Series A* 147, 426–63.

experimental design An experiment is a planned method of collecting data under controlled conditions, carried out so that the influence on the responses of one or more factors (which could, for example, represent different treatments in a CLINICAL TRIAL) may be assessed. The responses at the various levels of the factors are compared to see whether there are any differences that could indicate that the effect of one level of a factor is different from that of another.

Unlike industrial experimentation, the designs mainly used in medical research tend to be relatively simple in structure. This is partly because experiments involving patients are much more difficult to 'control' than experiments carried out under strict industrial conditions. Patients do not always comply with the medication regime and some may withdraw from treatment because they go on holiday, change GP, move house, change jobs, etc. It could even be the case that the benefits of a sophisticated design could be lost in the complexity of the final analysis. Although in the early phases of the development of a new drug, more involved experiments, such as dose-ranging experiments using animals or small numbers of healthy volunteers, might be conducted, it is usually regarded as safer when dealing with patients to employ designs that are simple and not so sensitive to MISSING DATA or incorrectly applied treatments. For this reason, the vast majority of clinical trials are carried out as *comparative randomised controlled* experiments using a *parallel group* structure (a completely randomised design) or as a simple *cross-over* experiment where each subject is exposed to each treatment using a predetermined specific treatment sequence.

In any design situation, it is important to consider the three basic principles of experimentation (Cox, 1958). These principles underlie all forms of good experimentation and are required if the conclusions are to possess the properties of *validity*, *precision* and *coverage*. To achieve validity, an experiment should be planned so that the conclusions are free from BIAS – either conscious or subconscious. It is not enough that the experimenter feels sure that they have not introduced personal bias or preferences into the experiment; it is a question of using a suitable experimental design and following the procedure laid down in the protocol. The first principle of experimentation is RANDOMISATION, which is used to avoid bias. There should be an allocation of treatment to individual subjects according to some randomisation procedure, which the experimenter cannot influence. Random number tables or a computer-generated randomisation should be used to allocate the treatments to the subjects. In a parallel group experiment, the subjects are randomly allocated to two or more separate arms of the study and receive one specific treatment throughout the treatment period. In a CROSSOVER TRIAL, patients are randomly allocated to a particular sequence of treatments. The simplest form of crossover experiment is the two-period, two-treatment, two-sequence crossover, also known as an AB/BA crossover. In clinical research, it has become common to use BLINDING as an additional feature to avoid bias. *Single-blinding* is where a treatment has been allocated to the patient at random, without the patient knowing which treatment he has received. *Double-blinding* is where neither the patient nor the investigator (or assessor) is aware of the specific treatment received.

When the object of an experiment is to compare the effects of different treatments, there should be a measure of the *precision* (standard error) of the estimates of the differences between the effects of the treatments. This can only be obtained if there is *replication*, the second principle of experimentation. To achieve this, the same treatment must be applied to different subjects. These repeated applications furnish a measure of the variation in the treatment effects that may be compared with the variation due to random error that would arise even if there were no differences between treatments. One of the main requirements of an experiment, particularly in a medical environment, where it could be unethical to carry out an experiment unless it can be shown that the estimates will have sufficient precision, is that the study must be of a sufficient size; i.e. there needs to be enough power in the experiment to detect a clinically important difference, if one exists. A power comparison, determining the sample sizes needed in a planned experiment, is an essential part of a clinical trial protocol (see PROTOCOLS FOR CLINICAL TRIALS).

The precision of an experiment depends not only on the number of replications used in the experiment but also on the inherent variability of the subjects studied. The variability will be smaller if the subjects are more homogeneous. However, in order to achieve wide *coverage* of the conclusions, the subjects used should be as varied as possible. For example, a trial to compare asthma treatments will tell us very little about the response for elderly patients if it is restricted to a narrow age range of young patients. If the results of the experiment are to apply to all patients, the experiment should include patients from a wide range of age groups. However, the desire to extend the coverage of an experiment may result in systematic errors due to the heterogeneity of the subjects. This would be particularly serious if the randomisation resulted in subjects exposed to one treatment being generally different from (e.g. younger than) those exposed to another treatment. One of the techniques for the control of this nonrandom systematic error is the technique of *stratifying* the subjects into homogeneous blocks and then randomising the treatments within blocks.

Stratification, the third principle of experimentation, leads to experimental designs in which the effects of different blocks may be taken into account in the analysis. Examples of stratified designs are randomised block designs, Latin square designs and incomplete block designs such as Youden squares, balanced incomplete blocks designs, group divisible designs and cyclic designs (see Cutler, 1998).

Other techniques for allowing for differences in the subjects, and therefore extending the coverage of the conclusions, involve the use of auxiliary information. For example, in a hypertension experiment, patients may be entered into the trial with different initial systolic blood pressures, so a simple comparison of the difference between the average systolic blood pressures for the patient groups after treatment will not give a true comparison of the treatments unless suitable adjustment for their initial (baseline) blood pressures is made. This adjustment can be made using the ANALYSIS OF COVARIANCE.

In addition to the parallel group and crossover designs already mentioned, factorial designs are being used more frequently in medical experimentation. In such experiments, more than one factor is involved, giving rise to treatments formed from different combinations of the factor levels. For example, one factor could be different drug treatments, while a second factor might consist of different levels of patient care. It is important in these designs not only to be able to assess the main effects of the different factors but also any interactions between them. One problem with such designs is that the number of treatment combinations increases rapidly so that it might be necessary to use fractional replications as well as confounding to reduce the size of the experiment and to accommodate stratification.

Response surface designs (Box and Draper, 1987) are used to model a surface representing the responses at different levels of a set of factors. The object of the experiment might be to determine the best levels at which to set a number of factors that affect the responses. Simple factorial designs may be used to fit first-order response surface models, but more complicated designs such as central composite designs are needed if higher order models are required.

Another type of experimental design, increasingly applied in clinical research, is the sequential design (see SEQUENTIAL ANALYSIS) or sequential group design (Whitehead, 1997), in which parallel groups of patients are studied. Such a trial continues until a clear benefit of one treatment is seen or until it is unlikely that any difference between treatments will emerge. The main advantage of these sequential trials is that they will often be shorter, and therefore involve fewer patients, when there is a large difference in the effectiveness of the two treatments. *PP*

[See also PHASE I TRIALS, PHASE II TRIALS, PHASE III TRIALS, PHASE IV TRIALS]

Box, G. E. P. and Draper, N. R. 1987: *Empirical model building and response surfaces*. New York: John Wiley & Sons, Inc. **Cox, D. R.** 1958: *Planning of experiments*. New York: John Wiley & Sons, Inc. **Cutler, D. R.** 1998: Incomplete block designs. In Armitage, P. and Colton, T. (eds) *Encyclopedia of biostatistics*. Chichester: John Wiley & Sons, Ltd. **Whitehead, J.** 1997: *The design and analysis of sequential clinical trials*. Chichester: John Wiley & Sons, Ltd.

expert systems Also known as knowledge-based systems, expert systems are computer programs that combine knowledge of some specific application domain with the general capability of drawing inferences from it, so as to be able to assist a decision-making process in specialist domains. Research in this field originated in the 1960s, flourishing in the 1970s and for some time dominating the mainstream of ARTIFICIAL INTELLIGENCE (AI). Many limitations to this paradigm are now known, along with its advantages.

One of the original motivations for the development of this line of research was the impossibility to build general-purpose problem solvers, mostly due to the surprisingly extensive use of commonsense knowledge that is required in such systems. In a very delimited area of expertise, however, this is much less of a problem. The area of human intellectual endeavour to be captured in an expert system is called the 'task domain' of the expert system. The primary goal of expert systems research is to make expert knowledge available to decision makers, by incorporating the expertise in an expert system, by means of a process known as 'knowledge extraction'.

Expert systems are often divided into two parts: a task-dependent 'knowledge base' and an 'expert system shell' that is independent of the task domain. In the classic approach the expertise in the knowledge base is represented in a symbolic way (e.g. by means of logic-type rules). The expert system shell, by way of contrast, contains an 'inference engine' that makes use of symbolic manipulations (e.g. logical inference) to reason with the information in the knowledge base. Some expert system shells also contain an explanation system (giving explanations about conclusions drawn and about questions asked) and a knowledge base editor (allowing modifications to be made in the knowledge base in an easy way). One last important part of expert system shells is the user interface, which allows user-friendly interaction with the expert system.

One of the first and most classical expert systems is PROSPECTOR (Duda *et al.*, 1977; Duda, 1980), which was used to evaluate the mineral potential of a geological site or region.

In medical applications expert systems first proved their usefulness by the development of MYCIN (Shortliffe *et al.*, 1975; Buchanan and Shortliffe, 1984). This is an interactive program developed at Stanford University that diagnoses certain blood infections, recommends treatment and can explain its reasoning in detail. In a controlled test, its performance equalled that of specialists. Unfortunately, legal and ethical issues related to the use of computers in medicine prevented the system being used in practice. Many other developments resulted from the MYCIN project, however – e.g. EMYCIN, the first expert shell developed from MYCIN. Another early example of a medical expert system is INTERNIST-I (Miller, Pople and Myers, 1982), which later evolved into Iliad. Today expert systems are used in different applications, including medical diagnosis, design (e.g. of large buildings), planning and scheduling (e.g. in logistics).

Recent research in expert systems focuses on the use of statistical reasoning methods such as belief networks, which in turn fall under the more general header of probabilistic GRAPHICAL MODELS. In modern AI, for example, much of the inferential procedures are based on statistical rather than logical inference. Another, related, trend is moving from a knowledge-intensive approach (with emphasis on producing a good knowledge base) to a data-intensive approach (letting the system improve by MACHINE LEARNING and data mining). *NC/TDB*

Buchanan, B. G. and Shortliffe, E. H. 1984: Uncertainty and evidential support. In *Rule-based expert systems: the MYCIN experiments of the Stanford heuristic programming project*. Reading, MA: Addison-Wesley. **Duda, R. O.** 1980: The PROSPECTOR system for mineral exploration. Menlo Park: Stanford Research Institute Final Report, Project 8172. **Duda, R. O., Hart, P. E., Nilsson, N. J., Reboh, R., Slocum, J. and Sutherland, G. L.** 1977: Development of a computer-based consultant for mineral exploration. Stanford Research Institute Annual Report, Project 5821 and 6415. **Miller, R. A., Pople, H. E. and Myers, J. D.** 1982: INTERNIST-I, an experimental computer-based diagnostic consultant for general internal medicine. *New England Journal of Medicine*, 19, August. **Shortliffe, E. H., Rhame, F. S., Axline, S. G., Cohen, S. N., Buchanan, B. G., Davis, R., Scott, A. C., Chavez-Pardo, R. and van Melle, W. J.** 1975: MYCIN: a computer program providing antimicrobial therapy recommendations. *Clinical Medicine* 34.

explanatory variables These are variables used as potential explanations of another variable in a statistical model. Thus, when using the simple linear regression model:

$$y = \alpha + \beta x + \varepsilon$$

we seek to explain the variation in the outcome variable, y, according to variation in the explanatory variable, x. Here, α and β are constants that specify how y is related to x; any residual, unexplained, variation is accounted for by the random error term, ε. Due to the widespread use of the symbols in this equation, an explanatory variable is often referred to as an 'x variable'. For example, Woodward and Walker (1994) used sugar consumption per head of population as an explanatory variable in a simple linear regression model to predict the average number of decayed, missing and filled teeth in 90 countries in an ecological study. In the analysis of association between these two variables, sugar consumption was taken as the explanatory variable, because the hypothesis was that consumption of sugar caused dental problems, rather than the other way round.

In general, several explanatory variables may be adopted as potential explanations of the outcome variable. For example, Bolton-Smith *et al.* (1991) used 15 explanatory variables in a MULTIPLE LINEAR REGRESSION model to predict the level of high-density lipoprotein (HDL)-cholesterol in 5236 women. Explanatory variables may be quantitative or categorical; the complete set of explanatory variables may be a mixture. For example, in the HDL-cholesterol study, the 15 variables included continuous measures, such as blood pressure, and categorical classifications, such as marital status.

Sometimes there may be a single explanatory variable that is the subject of the hypothesis of interest and the remaining explanatory variables in the statistical model are confounding, or prognostic, variables. For example, in a SURVIVAL ANALYSIS of the effects of cigarette smoking on lung cancer the explanatory variables in the fitted model might be age and smoking. Here, age is not of interest as a predictive variable in its own right, but is a potential confounder in the relationship between smoking and lung cancer. Including age as an explanatory variable enables adjustment for the effects of age. Sometimes, a model includes certain explanatory variables that represent interactions, for instance an age by smoking interaction might be included in the set of explanatory variables in order to see whether the effect of smoking differs by age. *MW*

Bolton-Smith, C., Woodward, M., Smith, W. C. S. and Tunstall-Pedoe, H. 1991: Dietary and non-dietary predictors of serum total and HDL-cholesterol in men and women: results from the Scottish Heart Health Study. *International Journal of Epidemiology* 20, 95–104. **Woodward, M. and Walker, A. R. P.** 1994: Sugar consumption and dental caries: evidence from 90 countries. *British Dental Journal* 176, 297–302.

exploratory factor analysis See FACTOR ANALYSIS

exponential distribution This is a single-parameter PROBABILITY DISTRIBUTION that often models the length of time to an event. If a random variable, X, takes an exponential distribution with parameter λ then this is sometimes written $X \sim \varepsilon(\lambda)$ for short. The exponential distribution has the density function:

$$f(x) = \lambda e^{-\lambda x}, \quad x \geq 0$$

which is monotonically decreasing from the MODE at $x = 0$. The distribution has a MEAN of $1/\lambda$ and VARIANCE of $1/\lambda^2$.

When $\lambda = 0.5$, the distribution is identical to a CHI-SQUARE DISTRIBUTION with two DEGREES OF FREEDOM. Changing λ rescales the density function on both axes, but leaves the shape unchanged. The function always intercepts the y axis at $y = \lambda$ and the 95th percentile (useful for calculating CONFIDENCE LIMITS) is always approximately $x = 3/\lambda$. (The shape can be seen illustrated in the figure accompanying the entry on the chi-square distribution. For further details on how the exponential distribution relates to other distributions, see Leemis (1986).)

The most interesting property of the exponential distribution is the lack of memory (LOM) property. This means that if the lifetime of a surgical instrument is distributed exponentially with parameter λ, then if after 2 years we note that it is still working, its remaining lifetime will be distributed exponentially with parameter λ; i.e. it is as if the process has been reset or the first 2 years have been forgotten.

To illustrate, Ainsworth *et al.* (2000) found that the length of time in days until discharge from high-dependency care for premature babies is exponentially distributed with a MEDIAN value of 6. The lack of memory property tells us that if a baby has already spent 2 days in such care then its median remaining time in high-dependency care is still 6 days; i.e. the median overall time for babies whose time exceeds 2 days is 8 days.

If one has n random variables that independently have exponential distributions with parameter λ, then their sum will be a GAMMA DISTRIBUTION with parameters λ and n. Coupled with the lack of memory property, this tells us that if the time to an event has an exponential distribution, the time to the second of 2 (or the third of 3, etc.) independent such events will have a gamma distribution. *AGL*

Ainsworth, S. B., Beresford, M. W., Milligan, D. W. A., Shaw, N. J., Matthews, J. N. S., Fenton, A. C. and Ward Platt, M. P. 2000: Pumactant and poractant alfa for treatment of respiratory distress syndrome in neonates born at 25–29 weeks' gestation: a randomised trial. *The Lancet* 355, 1387–92. **Leemis, L. M.** 1986: Relationships among common univariate distributions. *The American Statistician* 40, 2, 143–6.

exponential family This comprises distributions, including many of the common ones, whose distribution or density functions can be partitioned in a particular way. These distributions have some attractive features that enable easy computation and manipulation.

A distribution is defined by its distribution function (if a discrete distribution) or its density function (if a continuous distribution). In either case the defining function is a function of the data, x, and some parameters, θ. The distribution is in the exponential family if (1) the function can be written as a function of the parameters (e.g. $a(\theta)$) multiplied by a function of the data (e.g. $b(x)$) multiplied by the exponential of another function of the parameters (e.g. $c(\theta)$) multiplied by another function of the data (e.g. $d(x)$) and (2) the value of the parameters does not alter the range of possible values of the data.

To illustrate condition 1, the EXPONENTIAL DISTRIBUTION is unsurprisingly in the exponential family of distributions. The exponential distribution has the probability density function:

$$f(x) = \lambda e^{-\lambda x}, \; x \geq 0$$

which we need to show can be written in the form:

$$f(x) = a(\lambda)b(x)e^{c(\lambda)d(x)}$$

This can be achieved simply by setting $a(\lambda) = \lambda$, $b(x) = 1$, $c(\lambda) = -\lambda$, $d(x) = x$.

To illustrate condition 2, consider the continuous uniform distribution between 0 and θ that has the probability density function::

$$f(x) = \frac{1}{\theta}, \quad 0 \leq x \leq \theta$$

This can be parameterised in the correct manner, but different values of x become possible given different values of θ (e.g. 7 is possible if $\theta = 8$, but not if $\theta = 6$). Therefore this does not belong to the exponential family of distributions.

Other distributions that are in the exponential family of distributions include the exponential, the NORMAL, the BINOMIAL, the POISSON, the Bernoulli, the UNIFORM (if between fixed points), the BETA and the GAMMA – to name only some of the more common members.

In frequentist statistics, the distributions in the exponential family are useful because they possess properties that allow for both straightforward MAXIMUM LIKELIHOOD ESTIMATION and GENERALISED LINEAR MODELS, among others. For BAYESIAN METHODS, one attractive feature is that they are distributions that have natural conjugate PRIOR DISTRIBUTIONS (see Gelman *et al.*, 1995). For further details see Dobson (1990). *AGL*

Dobson, A. J. 1990: *An introduction to generalized linear models*. London: Chapman & Hall. **Gelman, A., Carlin, J. B., Stern, H. S. and Rubin, D. B.** 1995: *Bayesian data analysis*. Boca Raton: Chapman & Hall/CRC.

F

factor analysis This is a generic term for procedures that attempt to uncover whether the associations between a set of observed or manifest variables can be explained by the relationships of these variables to a small number of underlying LATENT VARIABLES (more usually referred to as *common factors* in this context). Factor analysis techniques attempt to discover the number and nature of the latent variables that explain the variation and more specifically covariation in the set of measured variables. The common factors are considered to contain the essential information in the larger set of observed variables.

The factor analysis model postulates that each observed variable can be expressed as a linear function of the common factors plus a residual term, i.e. a MULTIPLE LINEAR REGRESSION model for the observed variables on the common factors. The model implies that the covariances/correlations between the observed variables arise from their mutual relationships to the common factors. The COVARIANCE MATRIX of the observed variables predicted by the factor analysis model is a function of the regression coefficients of the observed variables on the common factors (the factor loadings) and the variances of the residual terms.

Estimation of both factor loadings and residual variances involves making the corresponding elements of the predicted and observed covariance matrices as close as possible in some sense. MAXIMUM LIKELIHOOD ESTIMATION is commonly used and has the advantage of providing a formal test of the number of factors needed adequately to represent the data. In general this test is to be preferred to informal tests such as KAISER'S RULE and the SCREE PLOT (see, for example, Preacher and MacCullum, 2003). After the initial estimation phase an attempt is made to simplify the often difficult task of interpreting the derived factors using a process known as FACTOR ROTATION. In general the aim is to produce a solution having what is known as *simple structure*, i.e. each common factor affects only a small number of specific observed variables. Rotated factors can be allowed to be independent or correlated, with the former often being chosen by default since it appears to provide a simpler solution. In particular circumstances, however, correlated factors might be considered to be a more realistic option.

A medical example of the application of factor analysis is provided in Whittick (1989). Attitudes to caregiving were examined in three groups of carers: mothers caring for a mentally handicapped child, mothers caring for a mentally handicapped adult and daughters caring for a parent with dementia. An attitude questionnaire containing 26 variables was developed and administered by post to 145 carers. The correlation matrix of the observed variables was subjected to factor analysis and a particular form of factor rotation giving independent factors. The three-factor solution could be interpreted as follows:

Factor 1. Negative aspects of caregiving with an emphasis on role conflict, family disruption and resentment about the caring role. Labelled as 'conflict'.

Factor 2. Positive aspects of caregiving with the emphasis on love for the dependant and satisfaction gained from the caregiving role. Labelled as 'love'.

Factor 3. Willingness to accept institutional care with an emphasis on its advantages. Labelled as 'institution'.

These three factors provide a concise and convenient description of a relatively complex dataset and were used as the basis of a further investigation of differences between the three groups of carers. Factor scores were calculated on each of these three factors for all 145 carers in the sample and a one-way ANALYSIS OF VARIANCE applied, giving results as shown in the table. The analysis of variance showed that there were significant differences between care groups on all three factors.

factor analysis *Calculation of factor scores on caregiving data*

Subscale	*Mother/ child*	*Mother/ adult*	*Daughter/ parent*	*F*
Conflict				
Mean	19.6	21.3	27.8	
SD	6.3	5.8	4.5	19.5
Love				
Mean	24.3	25.0	19.4	
SD	3.9	3.8	4.3	21.3
Institution				
Mean	7.5	8.3	12.9	
SD	3.5	2.9	2.7	34.5

Factor analysis as described in this section is more accurately called exploratory factor analysis, with the 'exploratory' implying that the investigator uses the analysis with no preconceived ideas about the factor structure to be

Encyclopaedic Companion to Medical Statistics: Second Edition Edited by Brian S. Everitt and Christopher R. Palmer

expected (except, of course, that it will be relatively simple and open to clear interpretation). In some situations, however, the researcher may have a theoretical factor structure in mind to be tested on a dataset. In such a case, confirmatory factor analysis may be used. *BSE*

[See also PRINCIPAL COMPONENT ANALYSIS]

Preacher, K. J. and MacCullum, R. C. 2003: Repairing Tom Swift's electric factor analysis machine. *Understanding Statistics* 2, 13–44. **Whittick, J. E.** 1989: Dementia and mental handicap: attitudes, emotional distress and caregiving. *British Journal of Medical Psychology* 62, 181–9.

factor rotation This is a procedure used in exploratory FACTOR ANALYSIS that aims to allow the factor analysis solution to be described as simply as possible. Such a process is possible because the exploratory factor analysis model does not possess a unique solution. Essentially, factor rotation tries to find an easily interpretable solution from among an infinitely large set of alternatives that each account for the variances and covariances of the observed variables equally well. Factor rotation is a way by which a solution is made more interpretable without changing its underlying mathematical properties.

The numerical techniques that are used for factor rotation aim for solutions in which each variable is highly loaded on at most one factor with factor loadings being either large and positive or near zero. In essence they try to alter the initial solution by making large loadings larger and small loadings smaller by optimising some suitable numerical criterion. Some methods of rotation give uncorrelated (orthogonal) factors, while others allow correlated (oblique) factors. As a general rule, if a researcher is primarily concerned with getting results that 'best fit' the data, then the factors should be rotated obliquely. If, however, there is more interest in the generalisability of results, then orthogonal rotation is probably to be preferred. For a full discussion of the pros and cons of the two forms of rotation see Preacher and MacCallum (2003).

We can illustrate factor rotation using the correlation matrix shown in the first table. The factor loadings in the initial two-factor solution for these correlations are shown in the second table, as are the rotated factor loadings (an orthogonal rotation was used). The factors in the rotated solution might be labelled as 'verbal' and 'mathematical'.

The lack of uniqueness of the factor loadings from an exploratory factor analysis once caused the technique to be viewed with a certain amount of suspicion (particularly by statisticians!), since, apparently, it allows investigators licence to consider a large number of solutions (each corresponding to a rotation of the factors) and to select the one closest to their a priori expectations (prejudices) about the factor structure of the data. However, such suspicion is largely misplaced because of the essential 'exploratory' nature of the factor analysis solution that is subjected to rotation. (See Everitt and Dunn, 2001, for further discussion.) *BSE*

factor rotation *Correlation coefficients of six school subjects*

Subject	*1*	*2*	*3*	*4*	*5*	*6*
1 French	1.00					
2 English	0.44	1.00				
3 History	0.41	0.35	1.00			
4 Arithmetic	0.29	0.35	0.16	1.00		
5 Algebra	0.33	0.32	0.19	0.59	1.00	
6 Geometry	0.25	0.33	0.18	0.47	0.46	1.00

factor rotation *Unrotated and rotated factor loadings*

	Unrotated loadings		*Rotated loadings*	
Variable	*1*	*2*	*1*	*2*
1 French	0.55	0.43	0.20	0.62
2 English	0.57	0.29	0.30	0.52
3 History	0.39	0.45	0.05	0.55
4 Arithmetic	0.74	−0.27	0.75	0.15
5 Algebra	0.72	−0.21	0.65	0.18
6 Geometry	0.59	−0.13	0.50	0.20

Everitt, B. S. and Dunn, G. 2001: *Applied multivariate data analysis*, 2nd edition. London: Arnold. **Preacher, K. J. and MacCallum, R. C.** 2003: Repairing Tom Swift's electric factor analysis machine. *Understanding Statistics* 2, 13–44.

factorial designs A term used in the context of a randomised trial (see CLINICAL TRIALS) to refer to a particular experimental design that allows two or more interventions to be evaluated in a statistically efficient way. In its simplest form, where treatments A and B are to be compared with their respective placebos (see the table), a 2×2 factorial design involves each patient being randomised twice, namely to either active A or placebo A and to either active B or placebo B. This design allows the separate effects of A and B to be assessed in the same sample of patients and, for a given sample size, is more powerful than a trial comparing A versus B versus no treatment.

The analysis of factorial design trials involves a number of steps, which may be illustrated by considering a hypothetical analysis of the trial depicted in the table. The first step is to assess the effects of treatment A among all patients allocated to A or matching placebo A. The most powerful means of assessing the effects of A is to compare all those allocated to

factorial designs *Schematic diagram of a 2 × 2 factorial trial of treatment A versus placebo A and of treatment B versus placebo B*

	Active A	*Placebo A*	
Active B	AB	0B	All B
Placebo B	A0	00	All non-B
	All A	All non-A	

Commentary on the table. The marginal analysis for the effectiveness of treatment A involves comparing the measure of the treatment effect among all those allocated treatment A (the total in the cell labelled 'All A') with the measure of the treatment effect among all those not allocated to A (the total in the cell labelled 'All non-A'). The marginal analysis for the effects of B is analogous ('All B' versus 'All non-B').

The test for interaction between the effects of A and B involves an appropriate measure of the difference between (a) the effects of A among subjects allocated to B and (b) the effects of A among patients not allocated to B. In the special situation where the outcome is binary (e.g. mortality) and treatment effects involve comparisons of proportions, then a ratio can be computed, so that the test in this table would be the ratio of the relative risks (AB/0B) and (A0/00).]

A (cells AB and A0 in the table) with all those allocated to placebo A (0B and 00). Analogously, the effects of B can then be estimated by comparing all those allocated to B (AB and 0B) with all those allocated Placebo B (A0 and 00). This 'marginal analysis' is the most statistically efficient analysis unless an 'interaction' exists such that the effects of A differ among patients allocated to B and among those allocated to placebo B (or vice versa, namely the effects of B differ among patients allocated to A or placebo A). It is necessary, therefore, to test for such an interaction in the routine analysis of factorial trials, because in situations where the effects of A are smaller among patients allocated to B than among those allocated to placebo B, the marginal analysis will underestimate the effects of B (and similarly the effects of B will be overestimated by the marginal analysis if allocation to B enhances the efficacy of A). In the example given, let us suppose that the primary outcome is binary (e.g. mortality), so that a test for interaction would test whether the primary comparison (e.g. relative risk) was statistically significantly different for the comparison of A versus placebo A among either patients allocated to B (i.e. AB versus 0B) or among those allocated to placebo B (i.e. A0 versus 00). In the rare situation where such an interaction is identified, and is clinically significant (i.e. its existence is relevant to drug selection), separate analyses of the effects of a treatment should be performed among all those allocated the interacting drug and all those allocated not to receive that drug.

In the past, some authors have expressed concerns about the potential for misleading estimates of effect arising from important interactions in factorial trials (Lubsen and Pocock, 1994). It should be noted, however, that such interactions appear to be quite rare. In a recent systematic review of factorial trials of treatments for myocardial ischaemia, for example, McAlister *et al.* (2003) found that only 2 of 31 (6%) comparisons demonstrated a statistically significant interaction between two treatments.

The factorial design is an especially versatile experimental design. For example, if there are good a priori reasons to suspect that two interventions might act synergistically (i.e. their effects in combination may be greater than strictly multiplicative), then a factorial design is the only design that can establish this reliably. Similarly, if the marginal analysis of a factorial trial suggests that two treatments, A and B, are effective, the absence of an interaction between A and B suggests that A will be similarly effective in the presence of B; i.e. the combination of A and B is more effective than B alone. The test for interaction between the effects of two treatments has low POWER and so would only be able to detect large differences in effectiveness when treatments are given alone or in combination. However, provided it can be established reliably that a treatment is effective, the existence of modest variation in the size of that effect (which is, after all, no more than would be expected biologically) may be of less immediate clinical relevance.

While factorial designs might well provide a useful tool for answering clinical questions more efficiently, there may occasionally be circumstances where such a design proves impractical. There are two particular situations to note. First, if two or more interventions are to be assessed in a factorial trial, subjects cannot be randomised if one of the treatments is considered to be definitely indicated (or is contraindicated). This may limit the proportion of a target population that is eligible for a trial. Second, trial participants who believe they have experienced an adverse drug reaction in a factorial trial may simply choose to discontinue both treatments, so if the price of assessing a speculative treatment is a sacrifice in compliance with a more promising treatment, then the price may not be worth paying. *CB*

Lubsen. J. and Pocock, S. J. 1994: Factorial trials in cardiology: pros and cons. *European Heart Journal* 15, 585–8. **McAlister, F. A., Straus S. E., Sackett, D. L. and Altman, D. G.** 2003: Analysis and reporting of factorial trials: a systematic review. *Journal of the American Medical Association* 289, 2545–53.

false negative rate A false negative test result in a diagnostic test study occurs when a person who has the disease when measured by a reference standard has a negative test result. The false negative rate is the proportion of individuals with false negative results out of all of those who have the disease. For example, when babies are screened for hearing loss using traditional 'distraction' tests, a number of babies with negative test results will be found later on to have significant hearing loss. These babies have had false negative results and the false negative rate is the number of babies with false negative results divided by the total number of babies with hearing loss. It can also be expressed as 1 – SENSITIVITY. *CLC*

[See also FALSE POSITIVE RATE]

false positive rate A false positive test result in a diagnostic test study occurs when a person who does not have the disease when measured by a reference standard has a positive test result. The false positive rate is the proportion of individuals with false positive results out of all of those who are disease free. For example, when newborn babies are screened for congenital hypothyroidism using blood spot tests, a number of babies with initial positive tests will be found to have normal values of thyroid hormone on repeat testing. These babies have had false positive results and the false positive rate is the number of babies with false positive results divided by the total number of babies who do not have congenital hypothyroidism. The false positive rate can also be expressed as 1 – SPECIFICITY.

While it is always desirable to avoid false positive results, this is particularly important in the context of population screening, where apparently healthy individuals are invited to undergo screening. Patients with false positive tests will be harmed, as they are likely to experience anxiety they otherwise would not have felt and, in order to clarify their disease status, will have to undergo further investigations that may carry some risk. Hence evaluations of screening should consider both the benefits to patients correctly identified as being at risk of the adverse consequences of a disease (as measured by the test SENSITIVITY) and the harm to patients with false positive diagnoses (as measured by the false positive rate). *CLC*

[See also FALSE NEGATIVE RATE]

FDA See FOOD AND DRUG ADMINISTRATION

F-distribution The F-distribution is the PROBABILITY DISTRIBUTION of the ratio of two variables, both of which have a CHI-SQUARE DISTRIBUTION. The F-distribution is defined by two parameters, often denoted m and n, known as the DEGREES OF FREEDOM of the distribution. If A is a random variable with a chi-square distribution with m degrees of freedom and B is a random variable independently distributed as chi-square with n degrees of freedom, then $nA/(mB)$ has an F-distribution with m and n (termed *numerator* and *denominator* respectively) degrees of freedom. For further details of the distribution see Grimmett and Stirzaker (1992).

The most common use of the F-distribution is in the ANALYSIS OF VARIANCE, where the ratio of two estimates of the variance, each of which independently has a chi-square distribution, is examined. For example, when looking at the effects of historical milk intake on hip bone density, Murphy *et al.* (1994) find that their variance ratio is approximately 3.8. Comparing this to an F-distribution with 2 and 245 degrees of freedom, the probability of such an extreme value is 0.0237.

Fortunately, in the medical literature one seldom has to deal with the probability density function (PDF) of the F-distribution. It is useful to know that the mean of the distribution, as defined here, is equal to $n/(n-2)$ as long as n is greater than 2, and that the distribution will be positively skewed but approaches symmetry as m and n become large.

The distribution was named 'F' by Snedecor (1934), a nomenclature later attributed to be in honour of R. A. Fisher, and so is occasionally referred to as Snedecor's F-distribution or some similar variant. *AGL*

Grimmett, G. R. and Stirzaker, D. R. 1992: *Probability and random processes*, 2nd edition. Oxford: Clarendon Press. **Murphy, S., Khaw, K.-T., May, H. and Compston, J. E.** 1994: Milk consumption and bone mineral density in middle aged and elderly women. *British Medical Journal* 308, 939–41. **Snedecor, G. W.** 1934: Calculation and interpretation of analysis of variance and covariance. *Ames Iowa Collegiate Press.*

finite mixture distributions These are PROBABILITY DISTRIBUTIONS that result from a weighted sum of a number of component distributions. Such distributions have a long history, apparently first being used by Karl Pearson in the 1890s to model a set of data on ratio of forehead to body length for 1000 crabs. These data were skewed and a possible reason suggested for this skewness was that the sample contained representatives of two types of crab but when the data were collected they had not been labelled as such. This led Pearson to propose that the distribution of the measurements on the crabs might be modelled by a weighted sum of two NORMAL DISTRIBUTIONS, with the two weights being the proportions of the crabs of each type (see Pearson, 1894). In mathematical terms, Pearson's suggested distribution for the measurements on the crabs was of the form:

$$f(x) = pN(x, \mu_1, \sigma_1) + (1-p)N(x, \mu_2, \sigma_2) \quad (1)$$

where p is the proportion of a type of crab for which the ratio of forehead to body length has mean μ_1 and standard deviation σ_1, and $(1-p)$ is the proportion of a type of crab for which the corresponding values are μ_2 and σ_2. In equation (1):

$$N(x, \mu_i, \sigma_i) = \frac{1}{\sqrt{2\pi}\sigma_i} \exp\left[-\frac{1}{2\sigma_i^2}(x-\mu_i)^2\right] \quad (2)$$

The distribution in equation (1) will be bimodal if the two component distributions are widely separated or will simply display a degree of skewness when the separation of the components is not so great (see BIMODIAL DISTRIBUTION).

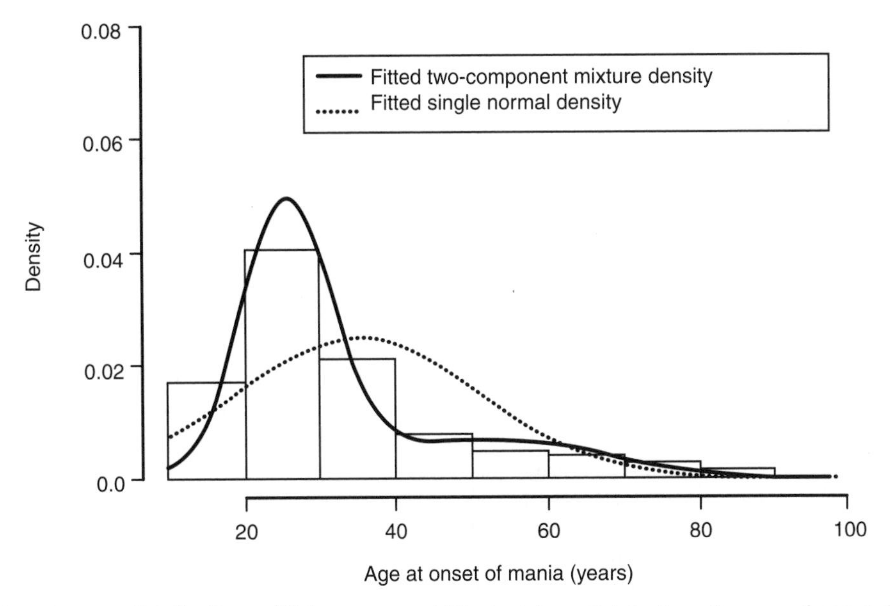

finite mixture distributions *Histograms and fitted mixture distributions for age of onset data*

Pearson's original estimation procedure for the five parameters in equation (1) was based on the method of moments (see Everitt and Hand, 1981), an approach that is now only really of historical interest. Nowadays, the parameters of a simple finite mixture model such as equation (1) or more complex examples with more than two components or other than univariate normal components would generally be quantified using the MAXIMUM LIKELIHOOD ESTIMATION, often involving the EM ALGORITHM. (Details are given in McLachlan and Peel, 2000.)

In some applications of finite mixture distributions the number of component distributions in the mixture is known a priori (this was the case for the crab data where two types of crab were known to exist in the region from which the data were collected). However, finite mixture distributions can also be used as the basis of a cluster analysis of data (see CLUSTER ANALYSIS IN MEDICINE), with each component of the mixture assumed to describe the distribution of the measurement (or measurements) in a particular cluster, and the maximum value of the estimated posterior probabilities of an observation being in a particular cluster being used to determine cluster membership. In such applications, the number of components of the mixture (i.e. the number of clusters in the data) will be unknown and therefore will also need to be estimated in some way. (This, too, is considered in McLachlan and Peel, 2000.)

As an example of the application of finite mixture distributions we will look at the age of onset of mania to investigate the possibility that there is an early onset group and a late onset group in the data. This subtype model implies that the age of onset distribution for mania will be a mixture with two components. To investigate this model, finite mixture distributions with normal components were fitted to the age of onset (determined as age on first admission) of 246 manic patients using the maximum likelihood estimation. Histograms of the data showing both the fitted two-component mixture distribution and a single normal fit are shown in the figure.

The LIKELIHOOD RATIO test for number of groups (see McLachlan and Peel, 2000) provides strong evidence that a two-component mixture provides a better fit than a single normal. *BSE*

Everitt, B. S. and Hand, D. J. 1981: *Finite mixture distributions*. London: Chapman & Hall/CRC. **McLachlan, G. J. and Peel, D.** 2000: *Finite mixture distributions*. New York: John Wiley & Sons, Inc. **Pearson, K.** 1894: Contributions to the mathematical theory of evolution. *Philosophical Transactions A* 185, 71–110.

Fisher's exact test This is a test of the association between the rows and columns of a two-way CONTINGENCY TABLE. The test is 'exact' in the sense that, under the hypothesis of no interaction between the rows and columns, the distribution of the associated test statistic is completely determined. An advantage of such exact methods is that they guarantee preservation of a researcher's pre-specified testing level (in this case, the probability of rejecting the hypothesis of no row and column interaction when the rows and columns are, in fact, not associated). The entry EXACT METHODS FOR CATEGORICAL DATA describes exact tests more generally, but one could say that Sir R. A. Fisher – with his method described here – was the father of exact tests. He developed what is popularly known as Fisher's exact test for a single 2×2 contingency table. He motivated his test through a British ritual, the drinking of tea. When drinking tea one afternoon during the 1920s with Sir Fisher and several other university associates, a British woman claimed to be able to distinguish whether milk or tea was added to the cup first. In order to test this claim, she was given eight cups of tea. In four of the cups, tea was added first and in the other four milk was added first. The order in which the cups were presented to her was randomised. She was told that there were three cups of each type, so that she should make four predictions of each order. This experiment is described by Fisher (1925) and more recently by Salsburg (2001). One possible result of the experiment is shown in the table.

Given this particular performance, could one conclude that she can distinguish whether milk or tea was added to the cup first? The experimental outcome displayed here shows that she guessed correctly more times than not, but, by the same token, the total number of trials is not very large and she might have guessed correctly by chance alone.

A statistically inclined researcher might proceed as follows:

Null hypothesis. The order in which milk or tea is poured in a cup and the taster's guess of the order are independent.

Alternative hypothesis. The taster can correctly guess the order in which milk or tea is poured in a cup.

Note that the alternative hypothesis is one-sided. Although there are two possibilities, that the woman guesses better than average or she guesses worse than average, we are only interested in detecting the alternative that she guesses better than average.

Suppose that the researcher decides to work at the 0.05 level of significance and decides to use the Pearson CHI-SQUARE TEST of independence. Results are as follows:

Pearson chi-square 2
Degrees of freedom 1
Significance 0.1573

Because the alternative hypothesis is one-sided (i.e. we are only interested in evidence in favour of the woman's ability to distinguish between milk first or tea first), one might have the reported significance, thereby obtaining 0.0786 as the observed P-VALUE. Because the observed P-value is greater than 0.05, the researcher might conclude that there is no evidence that the woman can correctly guess tea–milk order, although the observed level of 0.0786 is only marginally larger than the 0.05 level of significance used for the test.

It is easy to see from inspection of the table that the expected cell count under the NULL HYPOTHESIS of independence is 2 for every cell. Given the popular rules of thumb about expected cell counts (e.g. see EXACT METHODS FOR CATEGORICAL DATA), this raises concern about use of the one DEGREE OF FREEDOM chi-square distribution as an approximation to the distribution of the Pearson chi-square statistic for the table. Rather than rely on an approximation that has an asymptotic justification, suppose one instead uses an exact approach.

Fisher's exact test *Fisher's tea-tasting experiment*

Guess	*Pour*		*Row total*
	Milk	*Tea*	
Milk	3	1	4
Tea	1	3	4
Column total	4	4	4

We demonstrate here how this is accomplished. For the 2×2 table, Fisher noted that under the null hypothesis of independence, if one assumes fixed marginal frequencies for both the row and column counts, then a so-called hypergeometric distribution characterises the distribution of the four cell counts in the 2×2 table. A hypergeometric distribution can be thought of as the probability model for selecting a particular number of red balls (say x red balls) out of n total selections, without replacement, from a jar that contains r red balls and $m - r$ black balls. This distribution can be derived using basic probability rules, but it provides a useful tool for a variety of applications that include exact inference for contingency tables. (We suppress the details regarding the actual form of the hypergeometric distribution here, although one may find out more from any book containing a discussion of basic probability.)

In the case of our 2×2 table, if we fix the marginal counts we can see that, under the hypothesis of independence, the selection of milk first or tea first is like choosing red or black balls from a jar; i.e. the woman knows that four of the eight cups were prepared with milk first and the other four with tea first. If, in fact, she *cannot* tell the difference, then correctly choosing some number of milk-first cups from the eight-cup total is like randomly selecting red balls from a jar containing four red balls and four black. This fact enables one to calculate an exact P-value rather than rely on an asymptotic justification. In fact, the P-value for Fisher's exact test of independence in the 2×2 table is the sum of hypergeometric probabilities for outcomes *at least as favourable to the alternative hypothesis* as the observed outcome.

Let us apply this line of thought to the tea-drinking problem. In this example the experimental design itself fixes both marginal distributions, since the woman was asked to guess which four cups had the milk added first and therefore which four cups had the tea added first. Note that if we fix the marginal counts we can focus on the cell of the table that corresponds to the number of milk-first cups identified correctly – this value determines the other three cell values. In other words, assuming fixed row and column counts, one could observe the following table with the indicated probabilities:

Tea-tasting data

Number of milk-first cups correctly identified	*Table*			*Pr(Table)*	*P-value*
0	0	4	4	0.014	1.000
	4	0	4		
	4	4	8		
1	1	3	4	0.229	0.986
	3	1	4		
	4	4	8		
2	2	2	4	0.514	0.757
	2	2	4		
	4	4	8		
3	3	1	4	0.229	0.243
	1	3	4		
	4	4	8		
4	4	0	4	0.014	0.014
	0	4	4		
	4	4	8		

Note that the probability of each possible table in the reference set of 2×2 tables with the observed margins is obtained from the hypergeometric distribution just described, where the value x in this case represents the number of milk-first cups correctly identified by the woman; i.e. she expends four milk-first choices on the eight-cup total – four of which were *actually* prepared with the milk poured first. The P-values just displayed are the sums of probabilities for all outcomes at least as favourable (in terms of guessing correctly) as the one in question. For example, since the table actually observed has $x = 3$, the exact P-value is the sum of probabilities of all the tables for which x equals or exceeds 3. This works out to $0.229 + 0.014 = 0.243$.

Given such a relatively large P-value, one would conclude that the woman's performance does not furnish sufficient evidence that she can correctly guess milk–tea pouring order. Note that the approximate P-value for the Pearson chi-square test of independence was 0.0786, a dramatically different number. The exact test result leads to the same conclusion as the asymptotic test result, but the exact P-value is very different from 0.05 whereas the asymptotic P-value is only marginally bigger than 0.05.

In this example all four margins of the 2×2 table were fixed by design. In many cases, however, the margins are not fixed by design. Nevertheless, the reference set when computing Fisher's exact test is constructed using fixed row and column margins. We stress once again that whether or not the margins of the observed contingency table are naturally fixed is irrelevant to the method used to compute the exact test. In either case, one computes an exact P-value by examining the observed table in relation to all other tables in a reference set of contingency tables whose margins are the same as those of the actually observed table. We do not imply that other marginal outcomes are impossible under the conditions of the original experiment. However, since the inference is based on hypothetical repetitions of the original experiment, there is no logical problem with imagining that in these repetitions all outcomes whose row and column margins do not match the ones actually observed will be ignored. There are many compelling reasons for this conditioning, including the *ancillarity* principle, the sufficiency principle and the notion of eliminating unknown population characteristics (often called NUISANCE PARAMETERS) from the PROBABILITY DISTRIBUTION of the test statistic.

Beginning with Fisher, there is a rich body of literature justifying conditional inference along these lines. A more recent treatment of conditioning is provided by Yates (1984). The main advantage of conditioning is that the distribution of the observed table is known, thereby making exact inference possible.

Note further that while Fisher's exact test is traditionally associated with the single 2×2 contingency table, its extension to two-way tables with an arbitrary number of rows or columns was first proposed by Freeman and Halton (1951). Thus it is also known as the Freeman–Halton test. It is hence an alternative to the Pearson chi-square and the likelihood ratio tests for testing independence of row and column classifications in an unordered two-way contingency table.

The idea of conditional inference to eliminate nuisance parameters was first proposed by R. A. Fisher (1925). It is the driving force behind much of exact inference for categorical data. However, Barnard (1945) proposed an exact test that eliminates the nuisance parameter from a single 2×2 contingency table without conditioning on the marginal counts. This has been shown to be less conservative for 2×2 tables than the conditional test. The method is controversial, however (see Yates, 1984), and was subsequently disavowed by Barnard himself. Moreover, it does not readily extend to tables of higher dimension than 2×2. For these reasons Fisher's test remains the most widely used exact method for obtaining an exact test of association between two categorical variables. *CCo/PSe/CM/NP*

Barnard, G. A. 1945: A new test for 2×2 tables. *Nature* 156, 177. **Fisher, R. A.** 1925: *Statistical methods for research workers*. Edinburgh: Oliver and Boyd. **Freeman, G. H. and Halton, J. H.** 1951: Note on an exact treatment of contingency, goodness of fit and other problems of significance. *Biometrika* 38, 141–9. **Salsburg, D.** 2001: *The lady tasting tea*. New York: W.H. Freeman. **Yates, F.**

1984: Test of significance for 2 × 2 contingency tables. *Journal of the Royal Statistical Society, Series A* 147, 426–63.

Fisher's linear discriminant function (LDF)

See DISCRIMINANT FUNCTION ANALYSIS

Fisher's *z*-transformation

See CORRELATION

fishing

See POST HOC ANALYSES

five-number summary

See BOXPLOT

fixed effect

This is one of a set of effects on a response variable corresponding to a finite set of values taken by an EXPLANATORY VARIABLE. Fixed effects are included in a regression model to acknowledge that response tends to differ between the groups defined by the explanatory variable. By including fixed effects, the investigator can estimate the level of response for each separate group or estimate the effect of another variable of interest while controlling for the differences between groups.

Typical examples of explanatory variables include indicator variables for gender, drug treatment received in a clinical trial, ethnic background or centre in a multicentre study. If the variable defines k distinct groups in the dataset, e.g. k categories of ethnic background, the fixed group effects are incorporated by adding $k - 1$ regression parameters to the regression model. Fixed effects are appropriate when the investigator wishes to estimate or control for the effects of the k specific groups defined by the explanatory variable in the dataset of interest. An alternative approach to modelling the differences between groups is to assume RANDOM EFFECTS, by declaring the effects of the grouping variable to be drawn from a distribution of possible effects. This is appropriate when, for example, the 10 hospitals in a study are regarded as a sample drawn from all hospitals and the investigator is interested in the population of hospitals in general rather than the 10 particular hospitals recruited. *RT*

flexible designs

See DATA-DEPENDENT DESIGNS

Food and Drug Administration (FDA)

The food and drug regulatory body in the USA is the FDA. Its mission statement is: 'The FDA is responsible for protecting the public health by assuring the safety, efficacy, and security of human and veterinary drugs, biological products, medical devices, our nation's food supply, cosmetics and products that emit radiation. The FDA is also responsible for advancing the public health by helping to speed innovations that make medicines and foods more effective, safer and more affordable; and helping the public to get the accurate science-based information they need to use medicines and food to improve their health.'

For more details see www.fda.gov. BSE

[See also MEDICINES AND HEALTHCARE PRODUCTS REGULATORY AGENCY (MHRA), REGULATORY STATISTICAL MATTERS]

forest plot

This diagram is most commonly used in SYSTEMATIC REVIEWS AND META-ANALYSES of CLINICAL TRIAL data, but is also used for displaying the results from other types of studies. The plot consists of a diagram that shows both the estimated effect sizes from each study and the corresponding CONFIDENCE INTERVAL. An example from a meta-analysis of 28 CASE-CONTROL STUDIES concerned with the possible association between *Chlamydia trachomatis* and oral contraceptive use is shown in the first figure (see page 175). Here the effect size is the logarithm of the odds ratio. (Other examples are given in Sutton *et al.*, 2000.)

The point estimates are sometimes marked using square shapes of size proportional to the size of the study represented. This counteracts the viewer's eyes being drawn to the least significant studies, which have the widest confidence intervals and are therefore graphically more imposing. Sometimes, too, the individual lines are ordered by the date of study, by some index of quality study or by the point estimate of effect size.

The second figure (see page 175) contains both graphical and tabular elements. Data from each included study are summarised in horizontal rows on the diagram, with estimates of the treatment effect marked by a block and associated uncertainty depicted by lines extending between the upper and lower confidence limits. The size of the block varies between studies to reflect the weight given to each in the META-ANALYSIS, more influential studies having larger blocks. The overall estimate of effect is marked at the bottom of the plot as a diamond, the central points indicating the point estimate while the outer points mark the confidence limits. A vertical line is often drawn across the diagram at the meta-analytical point estimate. Forest plots of ratio effect measures (such as odds ratios, risk ratios – see RELATIVE RISK AND ODDS RATIO – and hazard ratios – see SURVIVAL ANALYSIS) are plotted on log scales so that the confidence intervals for individual studies and the overall estimate are symmetrical about their point estimates.

In addition to the graphical display, forest plots may numerically report the data for each trial from which the estimate is calculated, the estimate of effect and confidence interval and the percentage weight that the study contributes to the meta-analysis. The overall estimate may also be reported numerically stating the point estimate, a confidence interval, a test of statistical significance of the NULL HYPOTHESIS of no treatment effect and a test of homogeneity of no difference in effects between studies. From the plot it is often possible visually to assess the degree of heterogeneity in study results by noting the overlap of confidence intervals of individual studies with the meta-analytical point estimate.

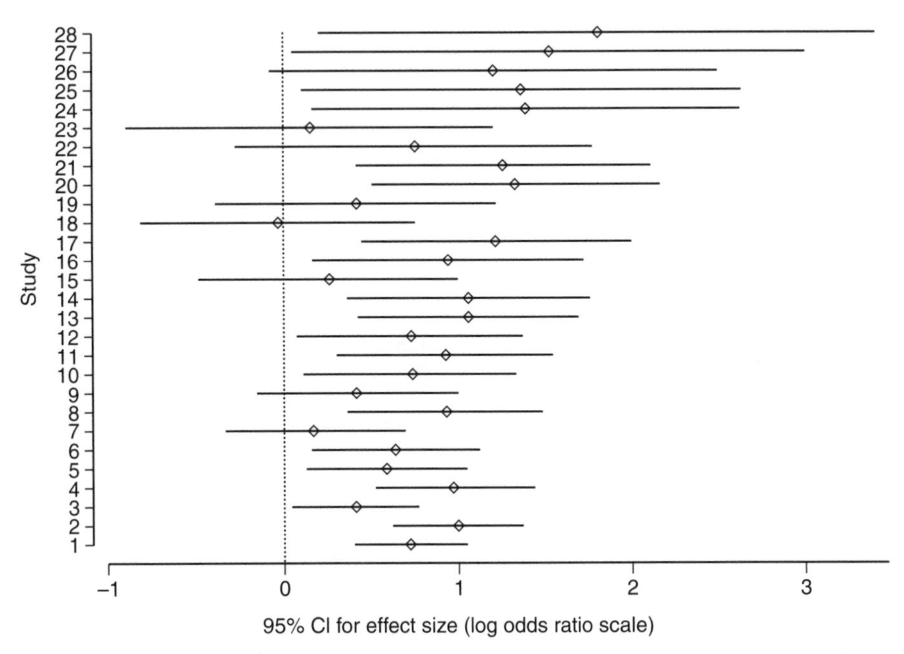

forest plot *Forest plot of log odd ratios for case-control studies of* Chlamydia trachomatis *and oral contraceptive use*

Forest plots may be supplemented in systematic reviews by *L'Abbé plots* (which plot event rates on treatment against event rates on control) and FUNNEL PLOTS. However, none of these diagrams can replace forest plots in their ability to report both the individual study data depicting effect estimates and uncertainty, as well as the meta-analytical summary estimates.

Lewis and Clarke (2001) reviewed the origins of the name 'forest' and identified that both the occasionally encountered spelling as *forrest* and capitalisation are inappropriate. *JD/BSE*

Lewis, S. and Clarke, M. 2001: Forest plot: trying to see the wood and the trees. *British Medical Journal* 322, 1479–80. **Moayyedi, P., Soo, S., Deeks, J., Forman, D., Mason, J., Innes, M. and Delaney, B.** 2000: Systematic review and economic evaluation of *Helicobacter pylori* eradication treatment for non-ulcer dyspepsia. *British Medical Journal* 321, 659–64. **Sutton, A. J.,**

Study	Treated still dyspeptic / n	Placebo still dyspeptic / n	Risk ratio (95% CI)	Weight
Blum	119/164	130/164	0.92 (0.81, 1.03)	14.9%
Koelz	67/89	73/92	0.95 (0.81, 1.11)	8.2%
McColl	121/154	143/154	0.85 (0.77, 0.93)	16.3%
Talley (ORCHID)	101/133	111/142	0.97 (0.85, 1.11)	12.3%
Talley (USA)	81/150	72/143	1.07 (0.86, 1.34)	8.4%
Miwa	33/48	28/37	0.91 (0.70, 1.18)	3.6%
Malfertheiner	269/460	143/214	0.88 (0.77, 0.99)	22.3%
Varannes	74/129	86/124	0.83 (0.68, 1.00)	10.0%
Froehlich	31/74	34/70	0.86 (0.60, 1.24)	4.0%
Overall fixed effect estimate (95% CI)			0.91 (0.86, 0.96)	P = 0.0002

Test for heterogeneity: Q=7.1, df=8, P=0.5

Risk ratio: 0.5 0.75 1.0 1.25 1.5

forest plot *Forest plot of the nine trials comparing* Helicobacter pylori *eradication therapy with placebo antibiotics (Moayyedi* et al.*, 2000)*

Abrams, K. R., Jones, D. R., Sheldon, T. A. and Song, F. 2000: *Methods in meta-analysis*. Chichester: John Wiley & Sons, Ltd.

forwards regression See LOGISTIC REGRESSION, MULTIPLE LINEAR REGRESSION

frailty A term generally used for unobserved individual heterogeneity, particularly in the analysis of SURVIVAL DATA. Such analysis is usually based on the assumption that the survival times of different individuals are independent and come from the same distribution, i.e. we assume a homogeneous population. In practice, populations are usually heterogeneous. We attempt to explain this heterogeneity by fitting models (such as COX'S REGRESSION MODELS) using explanatory variables. Frequently, even after adjusting for the explanatory variables, substantial heterogeneity remains, either due to lack of knowledge of or inability to measure all explanatory variables.

Frailty models were developed by Vaupel, Manton and Stallard (1979) to take into account this 'extra' variability. Individuals who have the same observed explanatory variable may differ in their underlying unobserved health status or frailty. As time progresses and frail individuals tend to die, the composition of the population with respect to frailty changes. Ignoring this can lead to biased estimation. An unmeasured random variable representing the unknown frailty is generally assumed to act multiplicatively on the baseline hazard function. These types of models are essentially MIXED or RANDOM EFFECT MODELS. The LIKELIHOOD functions for frailty models are quite complex and estimation is generally carried out using the EM ALGORITHM. Much research has been devoted to the choice of distributions for the frailty variable. A common choice is the GAMMA DISTRIBUTION. Klein and Moeschberger (1997) provide SAS macros that estimate the gamma frailty Cox regression model on their website. Other choices are discussed in Hougaard (2000). These frailty models have been widely applied. Hougaard (2000) provides a number of examples, including death due to malignant melanoma and times to catheter removal due to infection, for which the standard Cox regression approach is compared with the use of frailty models.

Frailty models are also used in the analysis of multivariate survival data to model dependence between times (Aalen, 1994). Such data can arise in several different ways. For example, time to recurrent events (such as epileptic seizures) on a subject will generally be dependent. Multivariate data also arises in connection with time to failure of similar organs (right eye, left eye) or lifetimes of related people (related genetically, by a common environment, etc.).

For simplicity, consider survival data on pairs of twins who are assumed to have a common frailty. This frailty is taken to be a random variable. Given the frailty for a specified pair, the individual hazard function is taken to be the multiple of frailty with a baseline hazard. The common frailty (of a pair) creates the dependence between the survival times within pairs. Assuming a distributional form for the frailty and averaging the conditional survival functions over the frailty distribution permits the derivation of a joint survival function. The resulting shared frailty models describe the dependence between the times. Details and more complicated multivariate frailty models are discussed by Hougaard (2000) along with various applications such as twin survival data. *DF/BR*

Aalen, O. O. 1994: Effects of frailty in survival analysis. *Statistical Methods in Medical Research* 3, 227–43. **Hougaard, P.** 2000: *Analysis of multivariate survival data*. New York: Springer Verlag. **Klein, J. P. and Moeschberger, M. L.** 1997: *Survival analysis*. New York: Springer Verlag. **Vaupel, J. W., Manton, K. G and Stallard, E.** 1979: The impact of heterogeneity in individual frailty on the dynamics of mortality. *Demography* 16, 439–54.

fraud detection in biomedical research Fraud in biomedical research comes in many guises (Lock and Wells, 1993). The boundary between fraud and simple carelessness is often fuzzy, although fraud is characterised by a *deliberate* attempt to deceive, which may be very hard to prove in the absence of positive external evidence or confession. Data discrepancies, such as transcription errors between the source documents and the data collection forms, may potentially be regarded as fraud if they occur in some systematic way or with abnormally high frequency, two circumstances that require a statistical assessment. In the USA, the term 'fraud' implies injury or damage to victims; hence the term 'misconduct' might be preferred. However, 'misconduct' also includes practices such as plagiarism, conflicts of interest, misuse of funds and other questionable research practices. In the UK, a Joint Consensus Conference on Misconduct in Biomedical Research held in October 1999 defined research misconduct as 'behaviour by a researcher, *intentional or not*, that falls short of good ethical and scientific standards'. Here the term 'fraud' refers specifically to *data fabrication* (making up data values) and *data falsification* (amending or eliminating data values), a use of the word that is at once more restrictive than is implied in normal conversation and less specific than in legal texts.

Scientific fraud (in the limited sense of data fabrication or falsification) is, in all likelihood, a rare phenomenon in biomedical research, although other misconduct may be more common. Some of the most widely publicised cases of fraud have taken place in randomised clinical trials and have created so much media attention that the uncritical observer may have been misled into thinking that the problem was far worse than it actually is. In all systematic investigations reported by cooperative groups and pharmaceutical companies, the proportion of investigators who were found to have committed fraud was less than 1% (Buyse *et al.*,

1999). In the audits performed by the United States FOOD AND DRUG ADMINISTRATION, the 'for-cause' investigations that followed revealed in most cases sloppiness or incompetence rather than fraud (Buyse *et al.*, 1999). However, there may be substantial bias in estimating the actual prevalence of fraud because of the natural tendency to suppress, conceal or minimise actual cases. In a recent cross-sectional survey of biostatisticians who were members of the International Society for Clinical Biostatistics in 1998, more than half of the respondents stated that they knew of a project in which fraud had occurred in the previous 10 years, while almost one-third of them had been engaged in a project in which fraud took place or was about to take place (Ranstam *et al.*, 2000). All in all, reliable data on the true prevalence of fraud are lacking.

The major difference between fraud and mere error lies in the 'intention to cheat' that defines fraud (Buyse *et al.*, 1999). This difference must, however, be qualified by the nature of the intent. Often investigators fabricate or falsify data to have complete records on all cases, or to eliminate OUTLIERS, not to modify the outcome of the experiment. Such data manipulations, if done independently of treatment assignment (e.g. in a blinded trial) introduce *noise* but no BIAS in the experiment. More serious cases of fraud involve fabricating complete patients or tampering the data in order to obtain a desirable result. These are cases where there is an expectation of gain in terms of prestige, advancement or money. These cases may also be the easiest to detect statistically, especially in MULTICENTRE TRIALS.

Some data items collected in clinical trials seem to be more prone to error and/or fraud than others. *Eligibility criteria* may be 'pushed' a little to make a patient eligible for the trial when in fact that patient does not strictly meet the criteria; many examples of fraud may have occurred because eligibility criteria were excessively restrictive and widening entry standards is often a good solution. *Repeated measurements* are requested repeatedly over time (such as, for instance, a battery of laboratory examinations), in which case data may be 'propagated' from the previous visit if the measurements are missing for a particular visit; such imputation of missing values is questionable at the time of the analysis, but certainly unacceptable when reporting the observations. *Adverse events* are likely to be underreported by some investigators (although such underreporting may reveal lack of interest or differences in interpretation rather than fraud). *Compliance data* are notoriously unreliable if they are based on the number of medications returned ('pill counts'); whenever compliance information is deemed important, it is advisable to use objective measurements based on blood or urine tests. Data fabrication has been detected in *patient diaries* through the colour and texture of the pen supposedly used on successive days by the patient, the patient's handwriting, etc.; the reliability of information collected in patient diaries can often be called into question, although electronic data capture can provide a more reliable alternative.

Some types of fraud are committed in trials conducted in a fastidious way, with lengthy case report forms, excessive requests for data clarification, etc. Often randomised clinical trials can be drastically simplified without loss of essential information. For instance, eligibility criteria can be simplified and left to the discretion of the investigator; the amount of data collected in the case report forms can be reduced, e.g. by eliminating much of the medical history and prior medications, concomitant medications, laboratory examinations, etc., that are not essential to the interpretation of the trial results; the follow-up of the patients in trials requiring prolonged observation can be as in routine clinical practice.

The traditional approach to fraud detection has involved monitoring visits to the centres or sites participating in the trial (Knatterud *et al.*, 1998). Some such onsite monitoring may be needed and useful, for many types of fraud would remain completely undiscovered were it not for the careful checks carried out during these visits. However, onsite monitoring is labour intensive and expensive and it, too, may fail to pick up fraudulent data. Moreover, the law of diminishing returns suggests that it is not cost effective to demand 100% verification of all source data. Monitoring activities can be limited to some random selection of the data, with the possible exception of data pertaining to the primary ENDPOINT of the trial. The random selection can be done at the level of the centres, the patients or the data items themselves. With such a random sampling scheme, one can estimate the overall data error rate with pre-specified precision and increase the amount of onsite monitoring, if the observed rate exceeds some upper limit. Another approach consists of visiting only the centres in which problems, errors or fraud are suspected.

A more innovative approach to detect fraud relies solely on statistics. The data of randomised CLINICAL TRIALS can be verified using statistical techniques that take advantage of their highly structured nature. Most DATA ENTRY and DATA MANAGEMENT software used for clinical trials perform basic checks, such as RANGE and consistency checks, but more extensive data checks typically occur at the end of the study along with other statistical analyses, far too late for corrective action. Batteries of checks using standard statistical techniques could be used early on in the course of a trial without large increases in costs and could save considerable time if problems were detected and corrected early.

The principles involved in uncovering fraud through statistical techniques rest on the difficulty of fabricating plausible data, particularly in high dimensions (Evans, 1998). Univariate observations can always be fabricated to fall close to the MEAN, although preserving their VARIANCE is more of a challenge to the inexperienced. Even the astute

fraud detection in biomedical research *Some statistical techniques that may be used to uncover fraud*

One variable at a time	Descriptive statistics Boxplot Frequency histogram Stem-and-leaf plots Tests for slippage
Several variables at a time	Cross-tabulation/scatterplot Correlation/regression Cook's distance Mahalanobis' distance Cluster analysis Discriminant analysis Chernoff faces Star (needle, spike) plots Hotelling's T^2 Tests for treatment contrasts
Repeated measurements	Autocorrelations Profiles Polynomial contrasts Runs tests
Calendar time	Residual plots Cusum Control charts

cheater who takes care in preserving both the mean and the variance may be tripped up by examination of the KURTOSIS of the distribution. Multivariate observations must in addition be consistent with the correlation structure between their individual components. In general, when data are fabricated to pass certain statistical tests, they are likely to fail on others (Haldane, 1948).

Another observation that can be used to check fabricated data is that humans are poor random number generators. Even informed people seem unable to generate long sequences of numbers that pass simple tests for randomness. Digit preference, especially terminal digit preference, or an excess of round numbers may easily reveal data fabrication. Benford's law may also be used to check the randomness of the first digit of all real numbers reported by a single individual (or a single centre). This law establishes that the probability of the first significant digit being equal to D ($D = 1, \ldots, 9$) is approximately given by a logarithmic distribution (Hill, 1998):

$$P(D) \approx \log(D+1) - \log(D)$$

Hence the frequency of 1 as a first digit should be as high as 30%, the frequency of 2 as a first digit should be close to 18%, while that of 9s should be lower than 5%, a result that runs against intuition. More sophisticated techniques are available to check the randomness of digits in a sequence of data values.

Statistical approaches may also take full advantage of the highly structured nature of clinical trials, which are prospective studies, entirely specified in a written protocol and data collection instrument (the 'case report form'), usually involving several centres and, when comparative, a randomly assigned treatment. Comparing each centre or treatment to the others in terms of the distribution of some variables, either taken in isolation (univariate approach) or jointly (multivariate approach), can detect unusual patterns in the data. Comparisons between centres are particularly informative if there are more than a few observations per centre (in which case fraud in any one centre may have a sizeable impact on the overall result). Such comparisons are useful with different types of fraud; for instance, the presence of outliers or the consistency in the effect of treatment may reveal fraud aimed at exaggerating the effect, while the presence of 'inliers' or underdispersion in the data may reveal invented cases.

Several univariate statistical techniques may be used to inspect the data (see the first table).

Statistical checks may reveal unusual data patterns that are often the mark of fraud (see the second table). Invented or manipulated data tend to have too little variance, no outliers or an abnormally flat distribution. Their distribution may be too close to a simple but implausible model, such as a NORMAL DISTRIBUTION with round numbers for the mean and STANDARD DEVIATION.

Since fraud usually occurs in a single centre (except in the unlikely situation of a coordinated fraud across several

fraud detection in biomedical research *Some patterns that may reveal fraud in clinical trial data*

One variable at a time	Digit preference Round number preference Too few or too many outliers Too little or too much variance Strange peaks Data too skewed
Several variables at a time	Multivariate inliers Multivariate outliers Leverage Too weak or too strong correlation
Repeated measurements	Interpolation Duplicates Invented patterns
Calendar time	Breach of randomisation Days of week (Sundays or holidays) Implausible accrual Time trends

centres), statistical checks must be performed within each centre as well as overall. A comparison of the results reported by different centres may reveal too little variability in one or more centres as compared to the overall variability. Perfect compliance with the protocol, for instance, may be the mark of fraud. Such a comparison may also reveal 'slippage' of one or more centres, the NULL HYPOTHESIS being that the means of the variable of interest are equal, but for random fluctuations, to the overall mean (Canner, Huang and Meinert, 1981).

Multivariate statistical techniques offer more checking possibilities, but they are seldom used in clinical trials, if at all. Multivariate statistical methods include correlations between several patient-related variables as well as comparisons between the randomised groups. Simple two-way cross-tabulations or SCATTERPLOTS for various pairs of variables can be compared across centres and any unusual patterns investigated further. Outlying observations, or outlying groups of observations coming from the same centre, can be detected more effectively in multidimensional space than in a single dimension. Moreover, in multidimensional space, inliers can be detected through the use of the Mahalanobis' distance just as well as outliers; inliers have an abnormally low Mahalanobis' distance (they fall too close to the multivariate mean), while outliers have an abnormally high Mahalanobis' distance (they fall too far from the multivariate mean). The detection of inliers may be more useful to detect fraud than the detection of outliers, because fabricated data will tend not to contain outliers that are at higher risk of being detected than are values close to the (multivariate) mean. Robust methods such as using ranks in place of the observations are advisable for the detection of outliers, because these can create severe departures from multivariate normality.

When, as is often the case, some variables are measured repeatedly over the course of the trial on the same patient, these measures lend themselves well to a variety of checks. Here again, an insufficient variability over time may reveal propagation of previous values rather than genuine observations. Sometimes the fraud involves a mechanism or computer algorithm for making up data.

In any trial with prolonged patient entry and follow-up, one can use calendar time to perform additional checks on the data. Simple checks can be performed on a specific day of the week, for instance, since certain events or examinations are unlikely to have taken place on a Sunday. Time intervals between successive visits and the number of visits per unit time provide further opportunities for checking the plausibility of a sequence of events. A comparison of treatment groups by week or month of RANDOMISATION can reveal suspect periods during which all treatments were not allocated with equal probability. Perfect compliance with the protocol in terms of dates may be a marker of fraud. More advanced checks can be useful, such as the stability of the variance of observations over time.

Randomised clinical trials constitute, by design, the most reliable type of medical experiment and their results are generally robust to occasional cases of data falsification and fabrication at some participating centres. The highly publicised case of fraud in the National Surgical Adjuvant Breast and Bowel Project (NSABP) provides a framework to examine the impact of such fraud on the results of clinical trials. Briefly, one of the investigators in breast cancer trials systematically altered some baseline patient data so that these patients became eligible for entry into the trials. The data subject to falsification were the date of surgery, the date of biopsy and estrogen receptor values. For example, in one study, the delay between the surgery and randomisation had been set to a maximum of 30 days by the trial protocol and dates were falsified for a few patients in whom this limit had been exceeded. The fraud was clearly not aimed at distorting the results of the trials one way or another and, indeed, a careful reanalysis of NSABP trial data with and without the fraudulent centre confirmed that the trial outcomes had not been materially affected by the fraud (Fisher *et al.*, 1995). In another large published trial in stroke, all data from one centre suspected of fraud were excluded from the analysis, again with negligible impact on the study results (ESPS2 Group, 1997). Yet this centre had contributed to the study of 452 of the 7054 patients involved overall!

Fraud is unlikely to affect the results of a trial if any of the following conditions hold: the fraud is limited to one or a few investigators (perhaps one centre in a multicentre setting) and/or to a few data items, provided that there are many investigators or centres; the fraud bears on secondary variables that have little or no effect on the primary endpoint of the trial; the fraud affects all treatment groups equally, and hence does not bias the results of the trial. Fraud committed without regard to the treatment assignments (e.g. prior to randomisation or in double-blind trials) generates noise but no bias. At least one of these conditions frequently holds and therefore fraud should not be expected to have a major impact on the results of multicentre clinical trials. One caveat is that where an increase in noise occurs, this can make dissimilar treatments appear similar. With a trend towards using equivalence or noninferiority trials for licensing purposes this is of concern and could result in ineffective medicines being licensed. *MB*

Buyse, M., George, S. L., Evans, S., Geller, N., Ranstam, J., Scherrer, B., Lesaffre, E., Murray, G., Edler, L., Hutton, J., Colton, T., Lachenbruch, P. and Verma, B. for the ISCB Subcommittee on Fraud 1999: The role of biostatistics in the prevention, detection and treatment of fraud in clinical trials. *Statistics in Medicine* 18, 3435–52. **Canner, P. L., Huang, Y. B. and Meinert, C. L.** 1981: On the detection of outlier clinics in medical and surgical trials: I. Practical considerations. *Controlled Clinical Trials* 2, 231–40. **ESPS2 Group** 1997: European Stroke Prevention Study 2. Efficacy and safety data. *Journal of Neurological Sciences* 151 (Suppl.), S1–S77. **Evans, S.** 1998: Fraud and misconduct in medical science. In Armitage, P. and Colton, T. (eds), *Encyclopaedia of biostatistics*. Chichester: John Wiley & Sons, Ltd. **Fisher, B., Anderson, S. and Redmond, C. K. *et al*.** 1995: Reanalysis and results after 12 years of follow-up in a randomised clinical trial comparing total mastectomy with lumpectomy with or without irradiation in the treatment of breast cancer. *New England Journal of Medicine* 333, 1456–61. **Haldane, J. B. S.** 1948: The faking of genetic results. *Eureka* 6, 21–8. **Hill, T. P.** 1998: The first-digit phenomenon. *American Scientist* 86, 358–63. **Knatterud, G. L., Rockhold, F. W., George, S. L., Barton, F. B., Davis, C. E., Fairweather, W. R., Honohan, T., Mowery, R. and O'Neill, R. T.** 1998: Guidelines for quality assurance procedures for multicenter trials: a position paper. *Controlled Clinical Trials* 19, 477–93. **Lock, S. and Wells, F.** (eds) 1993: *Fraud and misconduct in medical research*. London: BMJ Publishing Group. **Ranstam, J., Buyse, M., George, S. L., Evans, S., Geller, N., Scherrer, B., Lesaffre, E., Murray, G., Edler, L., Hutton, J., Colton, T. and Lachenbruch, P. for the ISCB Subcommittee on Fraud** 2000: The biostatistician's view of fraud in medical research. *Controlled Clinical Trials* 21, 415–27.

frequency distribution This describes the division of a sample of observations into a number of classes together with a count of the number of observations falling in each class. It acts as a useful tabular summary of the main features of a dataset, e.g. location, shape and spread. *BSE*

[See also HISTOGRAM, PROBABILITY DISTRIBUTION]

Friedman test This is a nonparametric equivalent to REPEATED MEASURES ANALYSIS OF VARIANCE, being an extension of the WILCOXON SIGNED RANK TEST to more than two groups or time points. It examines the ranks within a group and tests if the underlying continuous distribution is the same for each group. The NULL HYPOTHESIS is that there are no differences in MEDIANS between the groups. The alternative hypothesis is that there is at least one difference in medians. The data should be continuous in nature and should be a randomly selected sample measured at different time points or blocks of matched subjects randomly assigned to a group. Subjects or blocks of subjects should be independent. An extension to the test exists that allows repeated measures on each subject.

The test begins by constructing a two-way table with N (the number of subjects) rows and k (the number of time points) columns. Rank each row from lowest to highest, assigning the average rank to ties in the data. Find the sum of ranks in each of the columns. Calculate:

$$F_r = \frac{12}{Nk(k+1)}\left[\sum_{j=1}^{k} R_j^2\right] - [3N(k+1)]$$

Friedman test *FEV_1 (forced expiratory volume in 1 s) data from seven subjects recorded at three times a day*

Morning	*Rank*	*Afternoon*	*Rank*	*Evening*	*Rank*
0.25	1	0.4	2.5	0.4	2.5
0.56	1	0.87	2	1.06	3
0.63	2	1.45	3	0.25	1
0.65	2	3.02	3	0.45	1
0.74	2	1.07	3	0.28	1
0.97	1	1.29	2	1.98	3
1.91	3	0.15	1	0.27	2
Rank sum	**12**	—	**16.5**	—	**13.5**

where:

R_j = the sum of the ranks for column j
N = the number of subjects
k = the number of periods or conditions

Compare F_r to the critical value in Friedman tables and reject the NULL HYPOTHESIS if F_r is greater than or equal to the critical value. If N and k are sufficiently large then chi-square tables with $k-1$ DEGREES OF FREEDOM can be used instead of Friedman tables.

As an example a study measured FEV_1 at three different times of day to see if there was a difference. Data can be found in the table.

To compute the Friedman test statistic:

$$\begin{aligned} F_r &= \frac{12}{Nk(k+1)}\left[\sum_{j=1}^{k} R_j^2\right] - [3N(k+1)] \\ &= \frac{12}{7\times 3\times(3+1)} \times \left[12^2 + 16.5^2 + 13.5^2\right] \\ &\quad -[3\times 7\times(3+1)] \\ &= 1.5 \end{aligned}$$

From the tables ($N=7$, $k=3$, $\alpha=0.05$) the critical value is 7.714; as 1.5 is less than 7.714, there is insufficient evidence to reject the null hypothesis of no difference in medians between the three time points, so the medians can be considered unchanging across the time points.

When the Friedman test gives a significant result it is possible to do POST HOC TESTING to see where any differences lie. There are two ways of doing this.

Use the Wilcoxon signed rank test pairwise on the groups applying a correction for multiple testing such as a BONFERRONI CORRECTION. Alternatively, the average ranks in each of the groups can be compared. The null hypothesis is rejected if the absolute difference in mean ranks is greater than or equal to the critical value as shown:

$$\left|\bar{R}_i - \bar{R}_j\right| \geq Z_{\alpha/[k(k-1)]} \times \frac{\sqrt{k(k+1)}}{6N}$$

where:

$\bar{R}_i$ = the mean rank in period or condition i
$\bar{R}_j$ = the mean rank in period or condition j
$Z_{\alpha'}$ = the critical Z value for α'
$\alpha' = \alpha/[k(k-1)]$
k = the number of periods or conditions
N = the number of subjects

For further details see Pett (1997) and Conover (1999). *SLV*

Conover, W. J. 1999: *Practical nonparametric statistics*, 3rd edition. Chichester: John Wiley & Sons, Ltd. **Pett, M. A.** 1997: *Nonparametric statistics for health care research*. Thousand Oaks: Sage.

funnel plots Funnel plots are a graphical device, used to detect publication BIAS in SYSTEMATIC REVIEWS AND META-ANALYSES. For each study in a review the estimated treatment effect is plotted against a measure of trial precision such as the VARIANCE or STANDARD ERROR of the treatment effect or the study sample size (Light and Pillemar, 1984). In a departure from standard graphical practice, the plots conventionally depict precision on the vertical axis and treatment effect on the horizontal axis. The meta-analytical summary may be marked by a vertical line. When all study results are published it is expected that the studies will have a symmetrical distribution around the average effect line, the spread of studies with low precision being larger than that of studies with high precision, yielding a funnel-like shape. Some graphs mark the funnel with lines within which 95% of studies would fall were there no between-study heterogeneity. The choice of the measure of treatment effect (Tang and Liu, 2000) and the measure of precision (Sterne and Egger, 2001) makes a difference to the shape of the plot. Plots of treatment effects against standard errors are usually to be preferred, as the funnel will have straight rather than curvilinear sides.

Studies of the causes of publication bias have indicated that nonpublication is often linked to studies with nonsignificant P-VALUES, which tend to be the smaller studies reporting null or small effects (Dickersin, 1997). Suppression of these studies creates a visual hole in the plot. Unless the intervention has no effect, this will bias the META-ANALYSIS and induce asymmetry into the plot. Tests of publication bias, such as the Begg (Begg and Mazumdar, 1994) and Egger tests (Egger *et al.*, 1997), are testing for this asymmetry. The trim-and-fill method of investigating the impact of publication bias (Duval and Tweedie, 2000) imputes missing studies with results designed to remove asymmetry from the plot.

Interpretation of funnel plots is often visually difficult due to there being inadequate numbers of studies. Assessing the causes of funnel plot asymmetry is also difficult, as between-study heterogeneity, relationships betweens study quality and sample size, as well as publication bias can all cause similar patterns in funnel plots (Egger *et al.*, 1997). An example of a funnel plot is given in the SYSTEMATIC REVIEWS AND META-ANALYSIS entry. *JD*

[See also FOREST PLOT]

Begg, C. B. and Mazumdar, M. 1994: Operating characteristics of a rank correlation test for publication bias. *Biometrics* 50, 4, 1088–101. **Dickersin, K.** 1997: How important is publication bias? A synthesis of available data. *AIDS Education and Prevention* 9 (Suppl.), 15–21. **Duval, S. and Tweedie, R. L.** 2000: Trim-and-fill: a simple funnel plot based method of testing and adjusting for publication bias in meta-analysis. *Biometrics* 56, 2, 455–63. **Egger, M., Davey Smith, G., Schneider, M. and Minder, C.** 1997: Bias in meta-analysis detected by a simple, graphical test. *British Medical Journal* 315, 7109, 629–34. **Light, R. J. and Pillemar, D. B.** 1984: *Summing up: the science of reviewing research*. Cambridge, MA: Harvard University Press. **Sterne, J. A. C. and Egger, M.** 2001: Funnel plots for detecting bias in meta-analysis: guidelines on choice of axis. *Journal of Clinical Epidemiology* 54, 1046–55. **Tang, J.-L. and Liu, J. L. Y.** 2000: Misleading funnel plots for detection of bias in meta-analysis. *Journal of Clinical Epidemiology* 53, 477–84.

G

gamma distribution

This is a PROBABILITY DISTRIBUTION for nonnegative values defined by two parameters, here denoted λ and t (where t must be greater than 1), with the density function:

$$f(x) = \frac{1}{\Gamma(t)} \lambda^t x^{t-1} e^{-\lambda x}$$

Here t defines the shape of the distribution, while λ defines the scale on which the distribution is observed. The $\Gamma(t)$ expression is the gamma function that for integer values of t is equal to $(t-1)!$ (factorial). The mean of the distribution is t/λ and the VARIANCE is t/λ^2. For further details of the distribution see Grimmett and Stirzaker (1992).

If events take place as part of a Poisson process, i.e. the time between successive events can be modelled as taking the EXPONENTIAL DISTRIBUTION with parameter λ, then the time between any two events will take a gamma distribution. The gamma distribution can then arise as the distribution of the sum of exponentially distributed variables. It is also the case that the exponential distribution is a special case of the gamma distribution. The CHI-SQUARE DISTRIBUTION is also a special case, and given two variables A and B that have gamma distributions, it is possible to create a variable $A/(A+B)$ that has a BETA DISTRIBUTION. Gamma distributions with the same scale parameter will sum to another gamma distribution with that scale parameter.

The gamma distribution is always positively skewed (of little surprise since it is constrained at zero on the left-hand side, but the tail extends to infinity on the right-hand side), but the SKEWNESS diminishes as t tends to infinity, where the NORMAL DISTRIBUTION is the limiting distribution. For some illustration of the shapes that the gamma distribution can take see the illustration in the chi-square distribution entry. For further details of the relationships with other distributions see Leemis (1986).

The gamma is sometimes used to model times to event, as has been suggested, for example, in Phillips *et al.* (1994), where it is used to model the time to develop AIDS. Often, however, it is merely a convenient distribution for a quantity that cannot be negative. In Bayesian analyses (see BAYESIAN METHODS) (Gelman *et al.*, 1995), it is of use as the conjugate prior for the mean of a POISSON DISTRIBUTION, but most commonly as the conjugate prior for the inverse of the variance of a normal distribution. *AGL*

Gelman, A., Carlin, J. B., Stern, H. S. and Rubin, D. B. 1995: *Bayesian data analysis*. Boca Rotan: Chapman & Hall/CRC. **Grimmett, G. R. and Stirzaker, D. R.** 1992: *Probability and random processes*, 2nd edition. Oxford: Clarendon Press. **Leemis, L. M.** 1986: Relationships among common univariate distributions. *The American Statistician* 40, 2, 143–6. **Phillips, A. N., Sabin, C. A., Elford, J., Bofill, M., Janossy, G. and Lee, C. A.** 1994: Use of CD4 lymphocyte count to predict long-term survival free of AIDS after HIV infection. *British Medical Journal* 309, 309–13.

general fertility rate (GFR)

See DEMOGRAPHY

generalised additive models (GAMs)

These models allow possible nonlinear relationships between a response variable and one or more explanatory variables to be accounted for in a flexible manner. Generalised additive models are most useful in situations where the relationship between the variables is expected to be of a complex form, not easily fitted by standard methods or where there is no a priori reason for using a particular model. In generalised additive models, the $\beta_i x_i$ term of MULTIPLE LINEAR REGRESSION and LOGISTIC REGRESSION is replaced by a 'smooth' function of the explanatory variable x_i, as suggested by the observed data (Everitt, 2002).

The building blocks of generalised additive models are SCATTERPLOT SMOOTHERS such as locally weighted regression fits and spline functions. Generalised additive models work by replacing the regression coefficients found in other regression models by the fit from one or other of these 'smoothers'. In this way, the strong assumptions about the relationships of the response to each explanatory variable implicit in standard regression models are avoided. Details of how such models are fitted to data are given in Hastie and Tibshirani (1990).

Generalised additive models provide a useful addition to the tools available for exploring the relationship between a response variable and a set of explanatory variables. Such models allow possible nonlinear terms in the latter to be discovered and then perhaps to be modelled in terms of a suitable, more familiar, low-degree polynomial. Generalised additive models can deal with nonlinearity in covariates that are not the main interest in a study and adjust for such effects appropriately. An up-to-date technical account of GAMs is given in Wood (2006). *BSE*

[See also GENERALISED LINEAR MODELS]

Everitt, B. S. 2002: *Modern medical statistics*. London: Arnold. **Hastie, T. J. and Tibshirani, R. J.** 1990: *Generalized additive models*. Boca Raton: CRC Press/Chapman & Hall. **Wood, S. N.** 2006: *Generalized additive models*. London: Chapman & Hall/CRC.

Encyclopaedic Companion to Medical Statistics: Second Edition Edited by Brian S. Everitt and Christopher R. Palmer

generalised estimating equations (GEE) A popular method for analysing clustered data such as REPEATED MEASURES DATA.

Let there be $i=1,\ldots, n_j$ observations for each cluster j, $j=1,\ldots, N$. In a repeated measures setting, the clusters j would typically be subjects and the units i would be measurement occasions. The MEAN or expectation μ_{ij} of the responses y_{ij} given a vector of covariates $\mathbf{x}'_{ij}=(x_{1ij},\ldots, x_{pij})$ is modelled using a GENERALISED LINEAR MODEL:

$$g[\mathrm{E}(y_{ij}|\mathbf{x}_{ij})] \equiv g(\mu_{ij}) = \beta_0 + \beta_1 x_{1ij} + \cdots + \beta_p x_{pij}$$

where $g(\cdot)$ is a link function and β_0 to β_p are regression parameters. For example, an identity link gives a LINEAR REGRESSION MODEL and a logit link a LOGISTIC REGRESSION model.

The model is exactly the same as a generalised linear model for independent data. The interpretation of the regression parameters is therefore not affected by the nature of the within-cluster dependence between the responses. This is in contrast to RANDOM EFFECTS MODELS, where the regression parameters represent the *conditional* or *subject-specific* effects of covariates given the random effects. These effects will often differ from the *marginal* or population-averaged *effects* estimated using GEE. This difference is illustrated for a logit link (logistic regression) in the figure. Here the thin curves represent subject-specific relationships between the probability that the response equals 1 and a covariate x for a random intercept logistic regression model, where the horizontal shifts are due to different values of the random intercept. The thick curve represents the population-averaged relationship, formed by averaging the thin curves for each value of x. The slope of the population-averaged curve is flatter than the slopes of the subject-specific curves. Therefore the population-averaged regression parameters tend to be attenuated (closer to zero) relative to the subject-specific regression parameters. Note that the distinction between subject-specific and population-averaged effects disappears when an identity link is used and, for RANDOM INTERCEPT MODELS, when a log link is used.

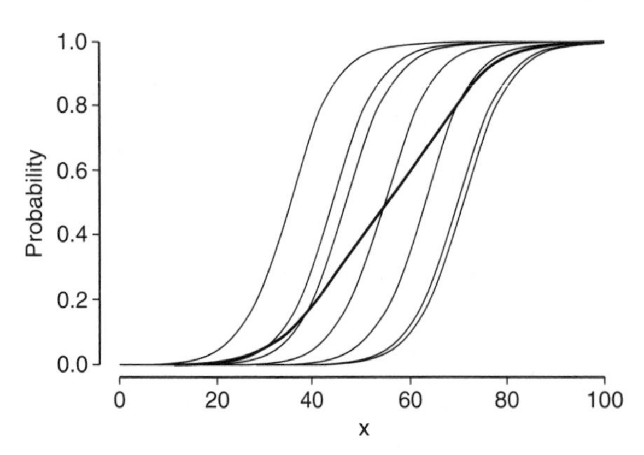

generalised estimating equations *Conditional and marginal logistic relationships*

As in conventional generalised linear models, the VARIANCES of the responses given the covariates are assumed to be $\mathrm{Var}(y_{ij}|\mathbf{x}_{ij})=\varphi V(\mu_{ij})$, where the variance function $V(\mu_{ij})$ is determined by the choice of distribution. For instance, for dichotomous responses, the Bernoulli distribution implies that $V(\mu_{ij})=\mu_{ij}(1-\mu_{ij})$, whereas for count data, the POISSON DISTRIBUTION implies that $V(\mu_{ij})=\mu_{ij}$. Since overdispersion is common in clustered data, the dispersion parameter φ is typically estimated even if the distribution requires $\varphi=1$.

The feature of GEE that differs from usual generalised linear models is that different responses y_{ij} and $y_{i'j}$ for a cluster j are allowed to be correlated given the covariates. These correlations are typically assumed to have a simple structure parameterised by a small number of α-parameters. One of the following correlation structures is commonly used.

Independence. The responses are independent given the covariates:

$$\mathrm{Cor}(y_{ij}, y_{i'j}|\mathbf{x}_{ij}, \mathbf{x}_{i'j}) = 0$$

Exchangeable. All pairs of responses for the same cluster have the same correlation given the covariates:

$$\mathrm{Cor}(y_{ij}, y_{i'j}|\mathbf{x}_{ij}, \mathbf{x}_{i'j}) = \alpha, i \neq i'$$

AR(1). The correlations between pairs of responses for the same cluster (given the covariates) fall off exponentially as the time lag between them increases (for longitudinal data only):

$$\mathrm{Cor}(y_{ij}, y_{i+t,j}|\mathbf{x}_{ij}, \mathbf{x}_{i+t,j}) = \alpha^t, |\alpha| < 1, t = 0, 1, \ldots, n_j - i$$

Unstructured. The correlation matrix of the responses given the covariates is estimated freely, without restrictions:

$$\mathrm{Cor}(y_{ij}, y_{i'j}|\mathbf{x}_{ij}, \mathbf{x}_{i'j}) = \alpha_{ii'}, i \neq i'$$

For given values of the regression parameters β_1 to β_p, the α-parameters of the *working correlation matrix* can be estimated along with the dispersion parameter φ (see Zeger and Liang, 1986, for details). These estimates can then be used in so-called *generalised estimating equations* to obtain estimates of the regression parameters. Since estimation of the correlation and dispersion parameters requires knowledge of the regression parameters and vice versa, the GEE algorithm proceeds by iterating between (a) estimation of the regression parameters using the correlation and dispersion parameters from the previous iteration and (b) estimation of the correlation and dispersion parameters using the regression parameters from the previous iteration. Eventually, the algorithm converges, producing the same estimates in successive iterations.

It has been demonstrated that the estimated *marginal* effects $\hat{\beta}_1$ to $\hat{\beta}_p$ are consistent; i.e. the estimates approach the true values as the number of clusters increases. Importantly, these estimates are 'robust' in the sense that they are consistent for misspecified correlation structures, assuming that the mean structure is correctly specified. Consistent estimates of the COVARIANCE MATRIX of the estimated marginal effects are next obtained by means of the so-called sandwich estimator.

The Madras Longitudinal Schizophrenia Study followed up 44 female patients monthly after their first hospitalisation for schizophrenia (see RANDOM EFFECT MODELS FOR DISCRETE LONGITUDINAL DATA for the full data). We will use GEE to investigate whether the course of illness differs between patients with early and late onset. The three variables considered are: [Month]: number of months since first hospitalisation; [Early]: early onset (1: before age 20, 0: at age 20 or later); [y]: repeated measures of thought disorder (1: present, 0: absent).

Here we consider a subset of the data, namely on whether thought disorder [y] was present or not at 0, 2, 4, 6, 8 and 10 months after hospitalisation. Letting y_{it} be the measurement of thought disorder at occasion i for patient j, we consider a dichotomous logistic regression model:

$$\ln\left(\frac{\Pr(y_{ij} = 1|\boldsymbol{x}_{ij})}{1-\Pr(y_{ij} = 1|\boldsymbol{x}_{ij})}\right) = \beta_0 + \beta_1 x_{1ij} + \cdots + \beta_p x_{pij}$$

We use GEE with independence and exchangeable working correlations to estimate a model with explanatory variables [Month], [Early] and the interaction [Early] × [Month], allowing us to investigate the linear trend of time (for the log odds) as well as differences between times of onset, not just in the overall odds of thought disorder but also in the trend over time (see the table).

The estimates assuming an exchangeable correlation structure suggest that there is a decline over time in the odds of thought disorder in the late-onset patients (odds ratio = exp(−0.28) = 0.76 per month). However, the early-onset patients do not appear to have an appreciably greater odds of thought disorder at baseline (odds ratio = exp(0.10) = 1.11) or a greater decline in the odds over time (odds ratio = exp(−0.04) = 0.96). The estimate of the correlation parameter α is 0.26. Estimates of the α-parameters are usually not reported because they are regarded as nuisance parameters.

A definite merit of GEE is that valid inferences are produced for population-averaged effects as long as the mean structure is correctly specified, even if the dependence structure is misspecified. However, there are also a number of limitations. The estimates are consistent only under the restrictive assumption that the probability that a response is missing does not depend on other responses for the same cluster, given the covariates (see MISSING DATA). Although GEE is often said to require that responses are missing completely at random (MCAR), missingness may in fact depend on covariates included in the model. Another limitation is that it is in general difficult to assess model adequacy in GEE; likelihood-based diagnostics are, for instance, not available. The use of GEE should furthermore be reserved to problems where marginal or population-averaged effects are of interest and it can be argued that it should be avoided in analyses of aetiology. This is because causal processes operate at the cluster or individual level, not at the population level. Unlike conditional effects, population-averaged effects also depend on the degree of heterogeneity in the population. Finally, the estimated regression parameters are no longer consistent if 'baseline' (initial) responses are included as covariates for longitudinal data (Crouchley and Davies, 1999).

See Diggle *et al.* (2002) for a thorough treatment of GEE and Lindsey and Lambert (1998) for a critical evaluation. *AS/SRH*

generalised estimating equations *Estimated regression parameters from GEE with independence and exchangeable correlation structures*

	GEE independent		*GEE exchangeable*	
	Est	*('Sandwich' SE)*	*Est*	*('Sandwich' SE)*
β_0 [Cons]	0.69	(0.35)	0.71	(0.35)
β_1 [Month]	−0.27	(0.07)	−0.28	(0.07)
β_2 [Early]	0.04	(0.67)	0.10	(0.69)
β_3 [Early] × [Month]	−0.04	(0.11)	−0.04	(0.11)

Crouchley, R. and Davies, R. B. 1999: A comparison of population average and random effects models for the analysis of longitudinal count data with base-line information. *Journal of the Royal Statistical Society, Series A* 162, 331–47. **Diggle, P. J., Heagerty, P., Liang, K.-Y. and Zeger, S. L.** 2002: *Analysis of longitudinal data*. Oxford: Oxford University Press. **Lindsey, J. K. and Lambert, P.** 1998: On the appropriateness of marginal models for repeated measurements in clinical trials. *Statistics in Medicine* 17, 447–69. **Zeger, S. L. and Liang, K.-Y.** 1986: Longitudinal data analysis for discrete and continuous outcomes. *Biometrics* 42, 121–30.

generalised linear model (GLM) This model forms a unified framework for regression models introduced in a landmark paper by Nelder and Wedderburn (1972) over 30 years ago. A wide range of statistical models including ANALYSIS OF VARIANCE, ANALYSIS OF COVARIANCE, MULTIPLE LINEAR REGRESSION and LOGISTIC REGRESSION are included in the GLM framework. A comprehensive technical account of the model is given in McCullagh and Nelder (1989) with a more concise description appears in Dobson (2001) and Cook (1998).

The term 'regression' was first introduced by Francis Galton in the 19th century to characterise a tendency to mediocrity, i.e. towards the average, observed in the offspring of parent seeds and used by Karl Pearson in a study of the heights of fathers and sons. The sons' heights tended, on average, to be less extreme than the fathers' (see REGRESSION TO THE MEAN). In essence, all forms of regression have as their aim the development and assessment of a mathematical model for the relationship between a response variable, y, and a set of q explanatory variables, $x_1, x_2, \ldots, x_q$. Multiple linear regression, for example, involves the following model for y:

$$y = \beta_0 + \beta_1 x_1 + \cdots + \beta_q x_q + \varepsilon$$

where $\beta_0, \beta_1, \ldots, \beta_q$ are regression coefficients that have to be estimated from sample data and ε is an error term assumed to be normally distributed with zero mean and a constant variance σ^2.

An equivalent way of writing the multiple regression model is:

$$y \sim N(\mu, \sigma^2)$$

where $\mu = \beta_0 + \beta_1 x_1 + \cdots + \beta_q x_q$. This makes it clear that this model is only suitable for continuous response variables with, conditional on the values of the explanatory variables, a NORMAL DISTRIBUTION with constant VARIANCE. Analysis of variance is essentially exactly the same model with x_1, x_2, …, x_q being dummy variable coding factor levels and interactions between factors; analysis of covariance is also the same model with a mixture of continuous and categorical explanatory variables.

The assumption of the conditional normality of a continuous response variable is one that is probably made more often than it is warranted. There are also many situations where such an assumption is clearly not justified. One example is where the response is a binary variable (e.g. improved/not improved) and another is where it is a count (e.g. number of correct answers in some testing situation). The question then arises as to how the multiple regression model can be modified to allow such responses to be related to the explanatory variables of interest. In the GLM approach, the generalisation of the multiple regression model consists of allowing the following three assumptions associated with this model to be modified: the response variable is normally distributed with a MEAN determined by the model; the mean can be modelled as a linear function of (possibly nonlinear transformations) of the explanatory variables, i.e. the effects of the explanatory variable on the mean are additive; and the variance of the response variable given the (predicted) mean is constant.

In a GLM, some transformations of the mean are modelled by a linear function of the explanatory variables and the distribution of the response around its mean (often referred to as the *error distribution*) is generalised usually in a way that fits naturally with a particular transformation. The result is a very wide class of regression models. The essential components of a GLM are: a linear predictor, η, formed from the explanatory variables:

$$\eta = \beta_0 + \beta_1 x_1 + \beta_2 x_2 + \cdots + \beta_q x_q$$

A transformation of the mean, μ, of the response variable is called the *link function*, $g(\mu)$. In a GLM it is $g(\mu)$ that is modelled by the linear predictor:

$$g(\mu) = \eta$$

In multiple linear regression and analysis of variance, the link function is the identity function. Other link functions include the log, logit, probit, inverse and power transformations, although the log and logit are those most commonly met in practice. The logit link, for example, is the basis of logistic regression.

The distribution of the response variable given its mean μ is assumed to be a distribution from the EXPONENTIAL FAMILY. Distributions in the exponential family include the normal distribution, the BINOMIAL DISTRIBUTION, POISSON DISTRIBUTION, GAMMA DISTRIBUTION and EXPONENTIAL DISTRIBUTION.

Particular link functions in GLMs are naturally associated with particular error distributions, e.g. the identity link with the Gaussian distribution, the logit with the binomial distribution and the log with the Poisson distribution. In these cases, the term *canonical link* is used.

The choice of PROBABILITY DISTRIBUTION determines the relationships between the variance of the response variable (conditional on the explanatory variables) and its mean.

This relationship is known as the *variance function*, denoted $\phi V(\mu)$. Both the Poisson and binomial distributions have variance functions that are completely determined by the mean. There is no free parameter for the variance since in applications of the generalised linear model with binomial or Poisson error distributions the dispersion parameter, ϕ, is defined to be one. However, in some applications this becomes too restrictive to account fully for the empirical variance in the data; in such cases it is common to describe the phenomenon as overdispersion. For example, if the response variable is the proportion of family members who have been ill in the past year, observed in a large number of families, then the individual binary observations that make up the observed proportions are likely to be correlated rather than independent. This nonindependence can lead to a variance that is greater (less) than that on the assumption of binomial variability. Observed counts often exhibit larger variance than would be expected from the Poisson assumption, a fact noted by Greenwood and Yule nearly a century ago (Greenwood and Yule, 1920). Greenwood and Yule's suggested solution to the problem was a model in which μ was a random variable with a gamma distribution leading to a NEGATIVE BINOMIAL DISTRIBUTION for the count.

Estimation of the parameters in a GLM is usually carried out through MAXIMUM LIKELIHOOD. Details are given in Dobson (2001). Having estimated the parameters, the question of the fit of the model for the sample data will need to be addressed. Clearly, a researcher needs to be satisfied that the chosen model describes the data adequately before drawing conclusions and making interpretations about the parameters themselves. In practice, most interest will lie in comparing the fit of competing models, particularly in the context of selecting subsets of explanatory variables so as to achieve a more parsimonious model. In GLMs a measure of fit is provided by a quantity known as the DEVIANCE. This is, essentially, a statistic that measures how closely the model-based fitted values of the response approximate the observed values; the deviance quoted in most examples of GLM fitting is actually -2 times the maximised log-likelihood for a model, so that differences in deviances of competing models give a LIKELIHOOD RATIO for comparing the models. A more detailed account of the assessment of fit for GLMs is given in Dobson (2002).

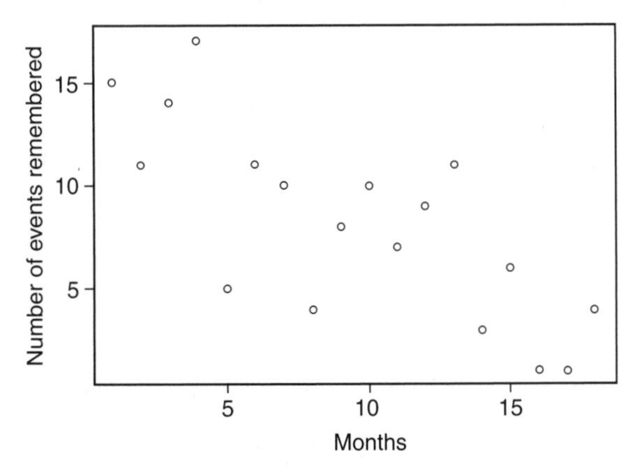

generalised linear model *Plot of recalled memories data*

We can illustrate an application of GLM using the data shown in the first table, which are given in Haberman (1978) and also in Seeber (1998). They arise from asking randomly chosen household members from a probability sample of a town in the USA which stressful events had occurred within the last 18 months and to report the month of occurrence of these events. A SCATTERPLOT of the data (see the first figure) indicates a decline in the number of events as these lay further in the past, the result perhaps of the fallibility of human memory.

generalised linear model *Distribution by months prior to interview of stressful events reported from subjects; 147 subjects reporting exactly one stressful event in the period from 1 to 18 months prior to interview*

Time	*y*
1	15
2	11
3	14
4	17
5	5
6	11
7	10
8	4
9	8
10	10
11	7
12	9
13	11
14	3
15	6
16	1
17	1
18	4

Since the response variable here is a count that can only take zero or positive values it would not be appropriate to use multiple linear regression here to investigate the relationship of recalls to time. Instead, we shall apply a GLM with a log link function so that fitted values are constrained to be positive and as an error distribution use the Poisson distribution, which is suitable for count data. These two assumptions lead to what is usually labelled *Poisson regression*. Explicitly, the model to be fitted to the mean number of recalls, μ, is:

$$\log(\mu) = \beta_0 + \beta_1 \text{time}$$

The results of the fitting procedure are shown in the second table. The estimated regression coefficient for time is −0.084, with an estimated STANDARD ERROR of 0.017. Exponentiating this last equation and inserting the estimated parameter values gives the model in terms of the fitted counts rather than their logs, i.e.:

$$\hat{\mu} = 16.5 \times 0.920^{\text{time}}$$

The scatterplot of the original data now also showing the fitted model is given in the second figure. The difference in deviance of the null model with no explanatory variables and one including time as an explanatory variable is large, which clearly indicates that the regression coefficient for time is not zero. *BSE*

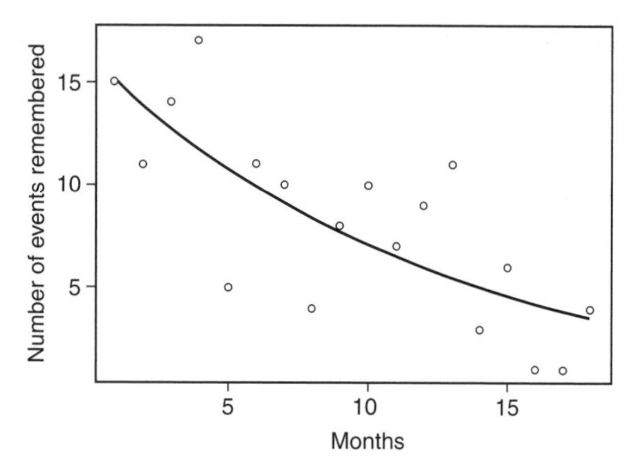

generalised linear model *Recalled memories data showing the fitted Poisson regression model*

generalised linear model *Results of a Poisson regression*

Covariates	*Estimated regression coefficient*	*Standard error*	*Estimate/ SE*
(Intercept)	2.803	0.148	18.920
Time	−0.084	0.017	−4.987

Dispersion parameter for Poisson family taken to be 1; null deviance: 50.84 on 17 degrees of freedom; residual deviance: 24.57 on 16 degrees of freedom.

[See also GENERALISED ESTIMATING EQUATIONS]

Cook, R. J. 1998: Generalised linear models. In Armitage, P. and Colton, T. (eds), *Encyclopedia of biostatistics*. Chichester: John Wiley & Sons, Ltd. **Dobson, A. J.** 2001: *An introduction to generalized linear models*, 2nd edition. Boca Raton: Chapman & Hall/CRC. **Greenwood, M. and Yule, G. U.** 1920: An inquiry into the nature of frequency distributions of multiple happenings. *Journal of the Royal Statistical Society* 83, 255. **Haberman, S.** 1978: *Analysis of qualitative data*, Vol. I. New York: Academic Press. **McCullagh, P. and Nelder, J. A.** 1989: *Generalised linear models*, 2nd edition. London: Chapman & Hall. **Nelder, J. A. and Wedderburn, R. W. M.** 1972: Generalised linear models. *Journal of the Royal Statistical Society, Series A* 135, 370–84. **Seeber, G. U. H.** 1998: Poisson regression. In Armitage, P. and Colton, T. (eds), *Encyclopedia of biostatistics*. Chichester: John Wiley & Sons, Ltd.

genetic epidemiology This is the study of the genetic aspects of the patterns of disease and other biological traits. Although, conceptually, genetic epidemiology is a branch of epidemiology, it has developed from distinct historical roots and employs different study designs, methods of statistical analysis and terminology.

Genetic epidemiology is usually distinguished from population genetics, which emphasises the genetics of populations over time and its relation to factors such as population structure and selection. The statistical component of these and related areas is referred to as *statistical genetics*.

The central themes in genetic epidemiology are the identification of genes related to disease or other traits and the evaluation of risks associated with different genetic variants. There is a major emphasis in genetic epidemiology on family studies (*pedigree analysis*). Indeed, some authors have considered that familial factors in disease, whether or not genetic, are an essential component of genetic epidemiology. Here we outline some of the main analytical approaches in genetic epidemiology.

A genetic epidemiological analysis often starts with descriptive studies of familial aggregation. *Twin studies* are important in this context, because comparison of MZ and DZ twins allows genetic effects to be separated from shared environmental influences on risk. For binary traits such as diseases, familial aggregation is often described in terms of *familial relative risks*. These are defined as the ratio of the risk of the disease in a relative of an affected individual to the risk of the disease in the general population. The size of the familial relative risk and its variation by type of relative can give clues to the genetic model underlying the disease. For quantitative traits, familial aggregation may be described in terms of correlations in trait values between relatives.

Another important statistical approach in genetic epidemiology is SEGREGATION ANALYSIS. The aim of segregation analysis is to fit different genetic models to diseases in pedigrees. These may include models involving a single genetic locus or multiple loci. The models are parameterised in terms of the frequencies of alleles and the risks of disease associated with each genotype. Measured or unmeasured environmental risk factors might also be included in the model. Models are usually fitted by computing likelihoods and using a MAXIMUM LIKELIHOOD ESTIMATION. Efficient algorithms, called peeling algorithms, are available for likelihood computation in large or complex pedigrees, although alternative approaches, such as a MARKOV CHAIN MONTE CARLO algorithm, are sometimes necessary.

An important concept in pedigree analysis is *ascertainment*. Analysis is usually conducted on a series of pedigrees that have been collected because of the presence of some trait. It is important to define the part of the data leading to the ascertainment of the pedigree and make appropriate adjustment for it in the analysis, by constructing the appropriate conditional likelihood.

There are two main approaches to the mapping of disease genes: GENETIC LINKAGE studies and ASSOCIATION studies. Genetic linkage studies are based on the inheritance of traits within families. They rely on the fact that loci that are close together on the same chromosome tend to be co-inherited (linked), whereas loci far apart will be inherited independently, due to process of *recombination* at meiosis. The PROBABILITY of a recombination between two genetic loci is called the *recombination fraction* (usually represented as θ). In a family with multiple cases of a disease, the diseased individuals will tend to share alleles at loci that are close to the disease gene. In this way, the entire genome can be examined for evidence of linkage using a limited number of genetic markers whose position in the genome is known. The statistical analysis of genetic linkage data aims to determine whether the pattern of co-inheritance of disease and marker genotype is different from what one would expect under the null hypothesis of no linkage.

Statistical analysis of genetic linkage studies can be of two types: parametric and nonparametric (or model free). In parametric linkage analysis, a particular disease model is specified and likelihoods are then constructed for different values of the recombination fraction. Linkage evidence is often summarised in terms of an *LOD score*, defined as $\log_{10}$ of the ratio of the likelihood of the data for a particular recombination fraction to the likelihood under no linkage.

Model-free linkage analysis is based on the sharing of marker alleles among affected individuals in pedigrees. The aim is to determine whether the number of marker alleles at a given time shared by affected individuals is greater than would be expected by chance. This approach is popular in the study of genetically complex disease where the disease model is unknown. It is often used in the study of affected sibling pairs, which is a common study disease for linkage in complex traits, but the approach has been generalised to more general pedigrees. Linkage analysis is implemented in several programs, including LINKAGE and GENEHUNTER.

Linkage analysis is a good approach for identifying genes that have a large effect on disease risk, but tend to lack power to identify loci that have a moderate effect on risk. Association studies evaluate directly the association between specific genetic variants and the trait of interest. They are the method of choice for identifying genes of weak effect. For diseases these are usually CASE-CONTROL STUDIES with unrelated cases and controls. Such studies can be analysed using standard case-control approaches. It is also possible to conduct association studies using within-family controls. This approach can eliminate problems of uncontrolled confounding. One commonly used design is to genotype a series of cases of the disease together with their two parents. The case genotypes are then compared with the alleles in the parents that are not transmitted to the affected case. This approach leads to the transmission distortion test (TDT).

Genetic association may arise due to a causal association with the polymorphism of interest or because of a true association with a neighbouring polymorphism. The latter can arise because genotypes at neighbouring polymorphisms tend to be correlated. Polymorphisms arise by mutation at one point in the history of a population, on a particular chromosome with a particular *haplotype* of marker alleles. A newly arising allele will therefore be in association with the alleles at neighbouring loci and this association will be maintained in the population if the loci are close together. This phenomenon is known as *linkage disequilibrium*. To elucidate fully the association at a particular locus, it may be necessary to extend the analysis to the joint effects of multiple markers. *DE*

[See also GENOTYPE, PHENOTYPE]

Balding, D. J., Bishop, M. and Cannings, C. (eds) 2001: *Handbook of statistical genetics*. Chichester: John Wiley & Sons, Ltd. **Khoury, M. J., Beaty, T. H. and Cohen, B. H.** 1993: *Fundamentals of genetic epidemiology*. Oxford: Oxford University Press. **Sham, P.** 1998: *Statistics in human genetics*. London: Arnold. **Terwilliger, J. D. and Ott, J.** 1994: *Analysis of human genetic linkage*, 3rd edition. Baltimore: Johns Hopkins University Press.

genetic linkage This is the nonindependent segregation of alleles at genetic loci close to one another on the same chromosome. Mendel's law of segregation states that an individual with the heterozygous genotype (Aa) has an equal probability of transmitting either allele (A or a) to an offspring. The same is true of any other locus with alleles B and b. Under Mendel's second law, that of independent assortment, the probabilities of transmitting the four possible combinations of alleles (AB, Ab, aB, ab) are all equal, namely one-quarter. This law is, however, only true for pairs of loci that are on separate chromosomes. For two loci that are on the same chromosome (known technically as syntenic), the probabilities of the four gametic classes (AB, Ab, aB, ab) are not equal, with an excess of the same allelic combinations as those that were transmitted to the individual from his or her parents. In other words, if the individual received the allelic combination AB from one parent and ab from the other, then he or she will transmit these same combinations with greater probability than the others (i.e. Ab and aB). The former allelic combinations are known as parental types and the latter recombinants. The strength of genetic linkage between two loci is measured by the recombination fraction, defined as the probability that a recombinant of the two loci is transmitted to an offspring. A recombination fraction ranges from 0 (complete linkage) to 0.5 (independent assortment, or the complete absence of linkage).

Recombinant gametes of two syntenic loci are generated by the crossing over of homologous chromosomes at certain semi-random locations during meiosis. The smaller the distance between two syntenic loci, the less likely that they will be separated by crossing over and therefore the smaller the recombination fraction. A recombination of 0.01 corresponds approximately to a genetic map distance of 1 centiMorgan (cM). The crossing-over rate varies between males and females and for different chromosomal regions, but on average a genetic distance of 1 cM corresponds approximately to a physical distance of one million DNA base pairs. The total genetic length of the human genome is approximately 3500 cM.

For many decades linkage analysis was restricted to Mendelian phenotypes such as the ABO blood groups and HLA antigens. Recent developments in molecular genetics have enabled a variety of naturally occurring polymorphisms (e.g. short sequence repeats, single nucleotide polymorphisms) to be detected and measured. Standard sets of such genetic markers, evenly spaced throughout the entire genome, have been developed for systematic linkage analysis to localise genetic variants that increase the risk of disease. This is a particularly attractive method of mapping the genes for diseases since no knowledge of the pathophysiology is required. For this reason, the use of linkage analysis to map disease genes is also called *positional cloning*.

Linkage analysis in humans presents interesting statistical challenges. For Mendelian diseases, the challenges are those of variable pedigree structure and size and the common occurrence of MISSING DATA. The standard method of analysis involves calculating the likelihood with respect to the recombination fraction between disease and marker loci or the map position of the disease locus in relation to a set of marker loci, while the disease model is assumed known (e.g. dominant or recessive). Traditionally, the strength of evidence for linkage is summarised as a lod score, defined as the common (i.e. base 10) logarithm of the ratio of the likelihood for a certain recombination fraction to that under no linkage. A lod score of 3 or more is conventionally regarded as significant evidence of linkage. For Mendelian disorders, 98 % of reports of linkage that meet this criterion have been subsequently confirmed. Linkage analysis has successfully localised and identified the genes for hundreds of Mendelian disorders.

Locus heterogeneity in linkage analysis refers to the situation where the mode of inheritance, but not the actual disease locus, is the same across different pedigrees. In other words, there are multiple disease loci that are indistinguishable from each other both in terms of manifestations at the individual level and in the pattern of familial transmission. Under these circumstances, the power to detect linkage is much diminished, even with lod scores modified to take account of locus heterogeneity, especially for samples consisting of small pedigrees.

For common diseases that do not show a simple Mendelian pattern of inheritance and are therefore likely to be the result of multiple genetic and environmental factors, linkage analysis is a more difficult task. For such diseases we typically would have an idea of the overall importance of genetic factors (i.e. HERITABILITY) but no detailed knowledge of genetic architecture in terms of the number of vulnerability genes or the magnitude of their effects. There are two major approaches to the linkage analysis of such complex diseases. The first is to adopt a lod score approach, but modified to allow for a number of more or less realistic models for the GENOTYPE–PHENOTYPE relationship and to adjust the largest lod score over these models for multiple testing. The second approach is 'model free' in the sense that a disease model does not have to be specified for the analysis. Instead, the analysis proceeds by defining some measure of allele sharing between individuals in a pedigree and relating the extent allele sharing to phenotypic similarity. One popular version of model-free linkage analysis is the affected sib-pair method, which is based on the detection of excessive allele sharing at a marker locus for a sample of sibling pairs where both members are affected by the disorder. The usual definition of allele sharing in model-free linkage analysis is *identity-by-descent*, which refers to alleles that are descended from (and are therefore replicates of) a single ancestral allele in a recent common ancestor. Algorithms for estimating the extent of local IBD from marker genotype data have been developed.

Methods of linkage analysis have been developed also for quantitative traits (e.g. blood pressure, body mass index). A particularly simple method is based on a regression of phenotypic similarity on allele sharing. A more sophisticated approach is based on a VARIANCE components model, in which a COMPONENT OF VARIANCE is specified to have COVARIANCE between relatives that is proportional to the extent of allele sharing between the relatives.

Regardless of the statistical method used for the linkage analysis of complex traits, there are two major inherent limitations of the approach. The first is that the sample sizes required to detect a locus with a small effect size are very large, potentially many thousands of families. The second is the low resolving power, in that the region that shows linkage is typically very broad, containing potentially hundreds of genes. For these reasons linkage is usually combined with an association strategy in the search for the genetic determinants of multigenic diseases.

PS

[See also ALLELIC ASSOCIATION, GENETIC EPIDEMIOLOGY, HAPLOTYPE ANALYSIS]

genome-wide association studies (GWAS)

See ALLELIC ASSOCIATION

genotype Genotype describes the genetic make-up of an individual or organism, usually referring to a particular gene or genetic locus. Different genotypes arise at loci in a genome where there are differences in DNA sequence – such loci are said to be *polymorphic* and the sequence differences are referred to as *alleles*. Alleles that are strongly associated with a particular disease or trait may be referred to as mutations. The genotype of an individual at a particular locus is then the combination of alleles at that location. In humans and other higher organisms, individuals inherit two copies of each gene, one from each parent (except those on the sex chromosomes in males). A genotype at a single locus therefore refers to a pair of alleles.

For example, the gene encoding the ABO blood group has three commonly recognised alleles, known as A, B and O. There are therefore six possible genotypes: (A,A), (A,B), (B,B), (A,O), (B,O) and (O,O).

One may also consider the genotypes at several loci together. For example, suppose there are two polymorphic loci in a particular gene or region with alleles A1 and A2 at the first locus and B1 and B2 at the second locus. The multilocus genotype for an individual may consist of A1–B1 on one chromosome and A2–B2 on the other. The combination of alleles at different loci on a given chromosome (e.g. A1–B1) is called a *haplotype*. *DE*

[See also GENETIC EPIDEMIOLOGY]

Balding, D. J., Bishop, M. and Cannings, C. (eds) 2001: *Handbook of statistical genetics*. Chichester: John Wiley & Sons, Ltd. **Sham, P.** 1998: *Statistics in human genetics*. London: Arnold.

geometric distribution This is the PROBABILITY DISTRIBUTION of the number of events required in order to observe a first 'success'. If we have a sequence of events, each of which independently has a probability of success p, then the probability mass function for the number of events required before observing a success is:

$$\Pr(X = x) = p(1-p)^x$$

and that for the number required to observe the first success is:

$$\Pr(X = x) = p(1-p)^{x-1}$$

Both are sometimes called the geometric distribution. Here we shall consider the second formula to be the definition. If X has this distribution, the mean of X is $1/p$ and the variance of X is $(1-p)/p^2$. The geometric distribution is a special case of the NEGATIVE BINOMIAL DISTRIBUTION. For further details of the distribution see Grimmett and Stirzaker (1992).

The interpretation of the probability mass function is straightforward. In order to observe the first success on the xth event, the xth event must be a success (with probability p) and the $x-1$ previous events must be failures (with probabilities $1-p$). Since the events are independent, the overall probability is obtained by simply multiplying together these individual probabilities (see PROBABILITY).

The event can be almost anything, from a generation of a family (perhaps for the purposes of modelling the number of generations until extinction of a species), to a nucleotide on a chromosome (for the purpose of modelling the length of chromosome before a mutation). Indeed, it is this last definition of event that Hilliker *et al.* (1994) use as they look to model the length of conversion tracts in terms of numbers of nucleotides, the length of a conversion tract being the number of nucleotides required before a termination ('success') of the tract is observed. Indeed, the majority of uses of this distribution in the medical sciences appear to be in the field of genetics.

The geometric distribution is a discrete analogue of the EXPONENTIAL DISTRIBUTION and a number of similarities should be apparent, e.g. the function of the distribution in modelling the time to an event. When the probability of a success, p, is small the continuous exponential distribution (with parameter p) can provide a good approximation to the discrete geometric distribution. The construction of the distribution in terms of independent events means that the 'lack of memory' property is also shared; i.e. observing two events to be failures does not alter the distribution of how many events need to be observed to have a success (independent events, by definition, cannot influence one another). *AGL*

Grimmett, G. R. and Stirzaker, D. R. 1992: *Probability and random processes*, 2nd edition. Oxford: Clarendon Press. **Hilliker, A. J., Harauz, G., Reaume, A. G., Gray, M., Clark, S. H. and Chovnick, A.** 1994: Meiotic gene conversion tract length distribution within the rosy locus of *Drosophila melanogaster*. *Genetics* 137, 1019–26.

geometric mean This is a MEASURE OF LOCATION used when data exhibit positive SKEWNESS, for example, such that the log-transformed data have an approximate NORMAL DISTRIBUTION. The (arithmetic) MEAN of the logged data is, like the individual values, on the log scale. The geometric mean is calculated by backtransforming (antilogging) this arithmetic mean.

The histogram in the figure illustrates the position of the arithmetic mean, geometric mean and MEDIAN of red cell folate measurements made on blood samples from a large random sample of women visiting a clinic. Since the data are skewed the median is a better measure of location than the arithmetic mean. However, the logged red cell folate data are normally distributed; in such circumstances the geometric mean is a good measure of location and is closer to the median than to the arithmetic mean.

Note that the geometric mean can only be used when all data points have values of above zero, since it is impossible without further TRANSFORMATION (e.g. by the addition of an initially large number to all observations) to take the log of zero or a negative number. Logs are typically taken to either base 10 or base e: both are equally valid. If the data are logged to base 10 (denoted *log* on most calculators), then the antilog to calculate the geometric mean must be to base 10 (denoted 10^x). If, however, the data are logged to base e (ln), known as the natural logarithm, then the antilog must be to base e (e^x). The red cell folate data were logged to base e and the arithmetic mean of these values was calculated as 5.777. The antilog to base e of 5.777 is $e^{5.777} = 323$.

In a RANDOMISED CONTROLLED TRIAL by Ng *et al.* (2002), the primary outcome was the length of hospital stay following computed tomography (CT) scan in patients with acute abdominal pain of uncertain aetiology. Since the length of hospital stay was expected to have a skewed distribution, the primary comparison was based on geometric mean length of stay. In their results section, the authors quoted the geometric mean for the two groups involved (5.3 and 6.4 days) but also, since the interpretation of geometric mean is more conducive to relative than to absolute differences, this was quoted as a 20 % increase in one group's length of stay, with a 95 % confidence interval from an 8 % shorter to a 56 % longer stay in hospital. *SRC*

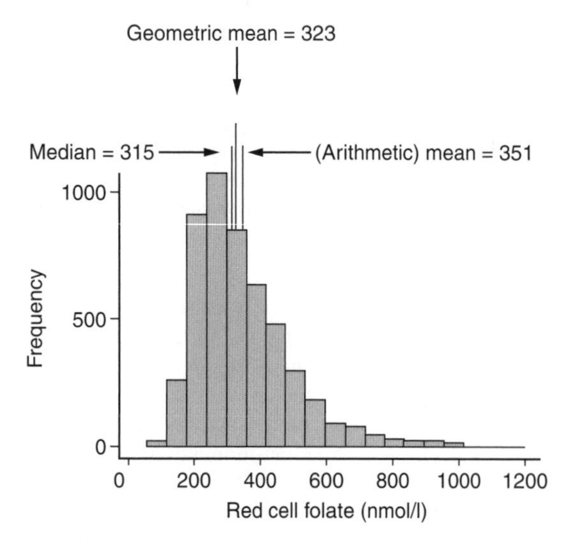

geometric mean *Median, geometric mean and arithmetic mean of red folate measurements on blood samples from 5052 women*

Ng, C. S., Watson, C. J. E., Palmer, C. R., See, T. C., Beharry, N. A., Howden, B. A., Bradley, J. A. and Dixon, A. K. 2002: Evaluation of early abdominopelvic computed tomography in patients with acute abdominal pain of unknown cause: prospective randomised study. *British Medical Journal* 325, 1387–91.

G-estimation This refers to an estimation technique designed to extract the causal effect of observed exposures in a randomised or observational setting (see CAUSALITY). It exploits an assumed independence (conditional on covariates) between the potential outcome under a reference treatment and the randomization indicator or the observed exposures.

The method typically starts from a structural nested model, which parameterises a causal contrast between the distribution of the observed outcome and a potential outcome under a reference treatment. When an INSTRUMENTAL VARIABLE R is available, the potential treatment free outcome will be (MEAN) independent of R. In an observational setting the 'no unmeasured confounders assumption' (NUCA) implies (mean) independence between a potential treatment-free outcome and the observed treatment, conditional on the confounders. In each case, G-estimation follows upon backtransforming the observed outcome using the causal parameters to an (expected) treatment-free outcome and then solving an independence equation.

The method has been especially valuable for estimation of the effect of time-varying exposures in the absence of unmeasured confounders when measured time-varying confounders may be intermediate variables on the causal path from exposure to outcome. This happens, for instance, when estimating the effect of AZT on T4-count, if an observed low T4-count leads to an increased dose of AZT (Robins, 1994). Standard regression does not work in this case since one must adjust for confounders to avoid bias, but cannot adjust for intermediate variables if one is to retain the full causal effect. Below, we explain the setup of G-estimation more formally for a structural nested mean model in that setting (see the figure).

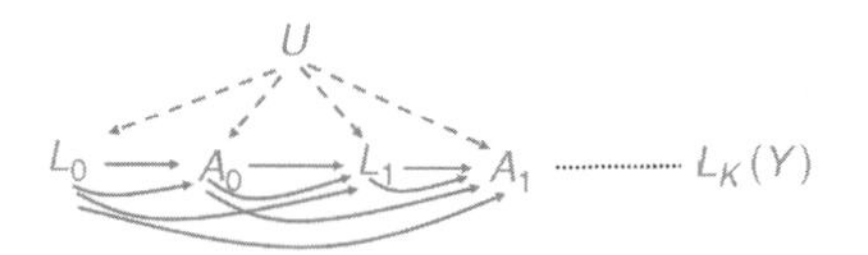

G-estimation *Structural nested mean model*

Let at time points $(t_0, t_1, t_2, \ldots, t_{K-1})$ a sequence of exposure levels $\underline{A}_{K-1} = (A_0, A_1, A_2, \ldots, A_{K-1})$ be observed, which could have been set to a different sequence $\underline{a}_{K-1} = (a_0, a_1, a_2, \ldots, a_{K-1})$ through some intervention or manipulation. One further observes covariates $\underline{L}_K = (L_0, L_1, \ldots, L_K)$ over those time points and at an additional final time point t_K, in response to the previously observed treatments. The ENDPOINT of interest, Y, is a well chosen function of the full sequence $\underline{L}_K$, which could be just L_K. A dynamic treatment strategy is then a rule G that assigns at each time point t_k a well-defined treatment $a_k(\underline{L}_k, \underline{A}_{k-1})$ in the function of the observed treatment and covariate history up to that point. With each such rule corresponds the potential outcome $Y(G)$ expressing what the outcome would have been had, possibly contrary to fact, treatment strategy G been followed.

To model the causal effect of (dynamic) treatments, the structural nested mean model (SNMM) considers at each time t_k how the expected outcome would change if a reference regime (e.g. placebo or treatment levels $a_j = 0$ for $j \geq k$) would be followed from then onwards. Conditional on observed treatment and covariate history $H_k = (\underline{L}_k, \underline{A}_{k-1})$, the expected mean difference between the observed and the potential outcome, is then parameterized. This can happen, for instance, through so-called blip functions, which express the additive effect of a final treatment blip a_k at time t_k on the mean potential outcome; i.e. with $(\underline{G}_{k-1}, a_k, 0_{k+})$ representing the treatment strategy G up to time t_{k-1}, followed by treatment a_k at time t_k and treatments 0 from then onwards, one could model $E(Y(\underline{G}_{k-1}, a_k, 0_{k+})| H_k) - E(Y(\underline{G}_{k-1}, 0_k, 0_{k+}) | H_k)$ as a linear function of the history $(\underline{L}_k, \underline{A}_{k-1})$.

The NUCA formally states that at each considered time, t_k, conditional on observed treatment and covariates history, H_k, the next level of observed treatment, A_k, is independent of the future response $Y(G)$ one would obtain under a pre-specified dynamic strategy G such as $Y(\underline{G}_{k-1}, 0_k, 0_{k+})$. Formally this is written as:

Assumption (1): $Y(G) \coprod A_k | \underline{L}_k, \underline{A}_k$

To justify Assumption (1), it is important to plan to measure the needed covariates L. As the figure indicates, this allows for direct effects between the measured variables, but not for unmeasured confounders: the dotted arrows should be absent, so we can discard the unmeasured U from the causal graph depicting the necessary direct effects.

An extension of the original approach yields a doubly robust estimator by involving an additional model for the distribution of the next treatment level given the history H_k. The corresponding estimator is then consistent when either this propensity model for the next treatment or the structural model is correct.

So far we have discussed parameter estimation for the SNMM. The above assumptions suggest also the G-null test of 'no causal effect of treatment A' on outcome. Indeed, under the strong null, the observed Y and counterfactual $Y(\underline{G})$'s coincide for any choice of treatment level A_k at any time point t_k. Hence assumption (1) then implies that the conditional distribution of the next treatment level A_k is independent of the future Y given H_k, the past treatment and covariate history. This can be tested by regressing Y on the past treatment and covariate history as well as A_k, and testing for independence of A_k in the context of that regression. Alternatively, one may test whether conditionally on the past H_k the distribution of the next treatment level A_k does not further depend on the observed outcome. An important advantage of this test is its robustness under the null to misspecification of the causal effect model.

To estimate $E(Y(G))$, the expected outcome under any (dynamic) treatment strategy G one wishes to evaluate, one needs the additional assumption of 'no current treatment interaction' (see, for instance, Section 4 in Robins, 1994). The assumption implies that the postulated mean causal effect of a_k is not restricted to the subset of individuals who happened to receive $A_k = a_k$, but holds for all concerned. In that case the so-called G-computation algorithm can be used with the estimated causal effect parameters to derive an estimate for $E(Y(G))$.

The G-estimation approach has generally proved to be quite rich. There is the doubly robust G-estimation which stays valid provided the structural nested causal model holds, when either a model for the conditional distribution of covariates is correct or the model for the conditional distribution of treatments is satisfied (Robins, 2000).

With correct censored survival outcomes (see SURVIVAL ANALYSIS – AN OVERVIEW), structural accelerated failure time models, allowing for time spans to be shrunk or expanded as a result of treatment received, have led to popular G-estimation approaches. The technique was further developed to estimate optimal dynamic treatment regimes (Robins, Orellana and Rotnitzky, 2008). Goetgeluk, Vansteelandt and Goetghebeur (2008) show how sequential G-estimation allows controlled direct effects quite simply to be estimated under conditional independence assumptions.

Several prominent applications of G-estimation of time-varying exposures have entered the literature. Most notably, Hernán and colleagues (2008) have applied it to analyse the effect of HRT on coronary heart disease, based on the Nurses' Health Study. They found results consistent with those from randomized trials hitherto thought to contradict observational findings. In general, implementation can be quite involved, not only because the assumption of no unmeasured confounders is hard to justify, but also due to the computational challenge. An implementation in Stata (see STATISTICAL PACKAGES) is proposed by Sterne and Tilling (2002). *EG*

Goetgeluk, S., Vansteelandt, S. and Goetghebeur, E. 2008: Estimation of controlled direct effects. *Journal of the Royal Statistical Society, Series B* 70, 1049–66. **Hernán, M. *et al.*** 2008: Observational studies analyzed like randomized experiments: an application to postmenopausal hormone therapy and coronary heart disease. *Epidemiology* 19, 766–79. **Robins, J.** 1994: Correcting for non-compliance in randomized trials using structural nested mean models. *Communications in Statistics* 23, 2379–412. **Robins, J.** 2000: Robust estimation in sequentially ignorable missing data and causal inference models. In *Proceedings of the American Statistical Association*, Section on *Bayesian statistical science*, 1999, pp. 6–10. **Robins, J., Orellana, L. and Rotnitzky, A.** 2008: Estimation and extrapolation of optimal treatment and testing strategies. *Statistics in Medicine* 27, 4678–721. **Sterne, J. and Tilling, K.** 2002: G-estimation of causal effects, allowing for time-varying confounding. *The Stata Journal* 2 164–82.

GLM See GENERALISED LINEAR MODEL

global scaling Measuring qualitative variables, such as quality of life, physical function, mental health and other complex variables, is a multiconceptual and multidisciplinary problem, and there are no standardised rules for recording. The theoretical definition is a conceptualisation of the variable, which means identification of the specific concepts and of the hierarchical structure to be studied. The variable might be considered as being unidimensional or composed of dimensions and subdimensions that in turn can be separated into different components. The operational definition defines each component in terms of items; each item comprises a single attribute to be recorded, often by an ordered categorical scale. The first figure shows a general scheme of a variable A that is operationally defined as being composed by the three subvariables, also called dimensions, labelled A_1, A_2 and A_3. The health-related quality of life questionnaire, Short Form(SF)-36 (see QUALITY OF LIFE MEASUREMENT), has this structure with nine dimensions and a total of 36 items (Svensson, 2001).

The variable	Subdimensions (subvariables)	Item scales
VARIABLE A	A_1	[ ]...[ ]
	A_2	[ ]...[ ]
		[ ]...[ ]
		[ ]...[ ]
		[ ]...[ ]
		[ ]...[ ]
		[ ]...[ ]
	A_3	[ ]...[ ]
		[ ]...[ ]
		[ ]...[ ]

global scaling *An illustration of the possible operational structure of a three-dimensional, multi-item (10 items) variable labelled A*

The purpose of using multi-items to measure a certain variable could be to increase the coverage over heterogeneous groups or to reflect various aspects of the same variable. Many items can also be used in order to identify the most significant sign of a certain status or dysfunction. The Postoperative Recovery Profile (PRP) Questionnaire includes 19 item variables of importance for recovery after surgery (Allvin *et al.*, 2009). Multidimensional multi-item instruments allow for interpretation at three distinct but related levels: the discrete item level, the dimensional, subvariable level and the global level. Data from each of these levels will provide an integrated picture of the individual. As the items cover different aspects of the same variable, a single global scale of the variable is required.

There are various approaches to aggregate multi-item assessments to a global scale.

Calculation of sum scores of item scale responses and transformation of this sum to a standardised score ranging from 0 to 100 is a very common approach to global scoring. However, the rank-invariant properties of data from scale assessments imply that adding scores is not appropriate and conclusions drawn from mathematical calculations on ordinal data may not be valid. Therefore, other approaches that take account of the nonmetric properties of ordinal data must be considered (see RANK INVARIANCE).

The rules for a global scaling of the variable should be based on the theoretical and operational frameworks of the variable. When the multi-items are constructed for identification of the most serious sign of a particular state, the maximum categorical level of the items could be an appropriate global score. Another appealing approach, especially for dichotomous data, is to use the number of indicators of the outcome of interest. In the PRP Questionnaire, each of the 19 items represent attributes, such as pain, gastrointestinal function and personal hygiene, which, when perceived as a problem, are a sign of not full recovery after surgery. Therefore the number of attributes of 'no problem' is the suggested global score of recovery, the global scale ranging from 0 to 19, where 19 is operationally defined as the score of being fully recovered (Allvin *et al.*, 2009).

A single score of multi-items can also be defined by the MEDIAN. When there are an odd number of items the median score is well defined by the ordered item response that comes halfway in the range of ordered item responses. In the case of an even number of items and the two central item responses differ the median cannot be defined as the average of these categorical values because of the nonnumerical properties of the data. Any of the two central categories and others in between the ordered set of item responses will serve as a median. For example, for an ordered set of six item responses, 'none, slight, slight, moderate, moderate, very severe', both 'slight' and 'moderate' will serve as a median. Then the category that reflects the greater degree of severity can be suitable as a global score, especially when the instrument is used to identify individuals at risk. In the example 'moderate' will be used as the global score.

A global scale with an optional number of ordered categories could also be constructed out of a theoretical

		Concept A			
		A1	A2	A3	A4
Concept B	B4	h Emotionally disturbed	f Dysfunctional	g Inadequate	h Emotionally disturbed
	B3	e Less functional	d Disengaged	d Disengaged	g Inadequate
	B2:	b Balanced	c Separated	d Disengaged	f Dysfunctional
	B1:	a Ideal	b Balanced	e Less functional	h Emotionally disturbed

global scaling *The main structure of the global scoring of family function conditional on the categorical levels of the subconcepts, here labelled A and B*

global scaling *Short descriptions of the criteria for the eight categories in the Swedish version of the Glasgow Outcomes Scale (S-GOS) as an example of a single global scale*

The S-GOS categories of outcome	*Short descriptions of criteria*
A. Death	Dead
B. Vegetative state	*Vegetative state*
	Conscious, but dependent
C. Severe disability, low	Minimal communication is possible by emotional response; dependent in all ADL activities
D. Severe disability, high	Partially independent in ADL; post-traumatic complaints/signs; resumption of previous life and work not possible
	Independent, but disabled
E. Moderate disability, low	Independence in ADL; unable to resume previous social and/or work activities; post-traumatic signs evident
F. Moderate disability, high	Post-traumatic signs are present, which, however, allow resumption of most former activities, either full-time or part-time
	May have mild residual effects
G. Good recovery, low	Capable of resuming normal occupational and social activities; there are minor physical or mental deficits or complaints
H. Good recovery, high	Full recovery without signs or symptoms

framework, professional knowledge and experience of the composite variable. FACES is a 20-item questionnaire of family function, which is a complex variable defined by two main concepts, each having three subdimensions and 10 item scales with four verbal descriptive scale categories, where one of the concepts could be illustrated by variable *A* in the first figure. The median approach was used for defining the global scale of each subdimension and of the two main concepts, here labelled A and B. In order to get more detailed information from the assessments the operational definition of an eight-point global scale of family function was suggested by the professional researcher based on the 4×4 combinations of median scores from the two main concepts, as shown in the second figure (Starke and Svensson, 2001). It is also possible to use numerical labels of categories, bearing in mind that the data will still be ordinal.

A similar approach to global scoring of the variable bodily pain assessed by two items in the SF-36 have been suggested as a rank-invariant alternative to the use of standardised sum scores (Svensson, 2001).

One way to avoid aggregation of multi-item scales is to construct a single hierarchical scale with multidimensional conditions for each ordered categorical level, such as in the Glasgow Outcomes Scale (GOS) (see the table) (Svensson and Starmark, 2002). *ES*

Allvin, R., Ehnfors, M., Rawal, N., Svensson, E. and Idvall, E. 2009: Development of a questionnaire to measure patient-reported postoperative recovery: content validity and intra-patient reliability. *Journal of Evaluation in Clinical Practice* 15, 411–19. **Starke, M. and Svensson, E.** 2001: Construction of a global assessment scale of family function, using a questionnaire. *Social Work in Health Care* 34, 131–42. **Svensson, E.** 2001: Construction of a single global scale for multi-item assessments of the same variable. *Statistics in Medicine* 20, 3831–46. **Svensson, E. and Starmark, J. E.** 2002: Evaluation of individual and group changes in social outcome after aneurysmal subarachnoid haemorrhage: a long-term follow-up study. *Journal of Rehabilitation Medicine* 34, 251–9.

graphical deception Graphical deception involves displays of data that may mislead the unwary either by design or by error. Consider, for example, the plot of the death rate per million from cancer of the breast, for several periods over the last three decades, shown in the first figure. The rate appears to undergo a rather alarming increase. However, when the data are re-plotted with the vertical scale starting at zero, as shown in the second figure, the increase is altogether less startling. The example illustrates that undue exaggeration or compression of the scales is best avoided when constructing graphs if you want to avoid the charge of graphical deception.

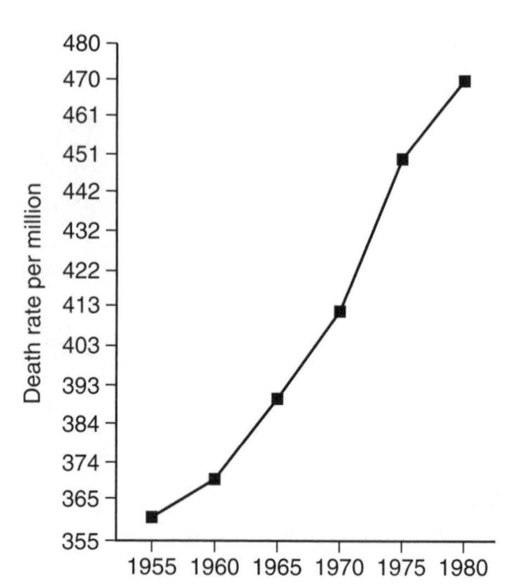

graphical deception *Death rates from cancer of the breast where the y-axis does not include the origin*

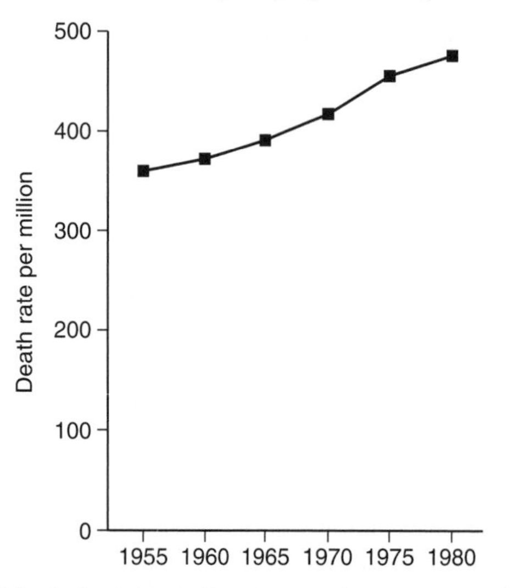

graphical deception *Death rates from cancer of the breast where the y-axis does include the origin*

A very common form of distortion introduced into the graphics, often popular in the media, is one where both dimensions of a two-dimensional figure or icon are varied simultaneously in response to changes in a single observed quantity. An example is shown in the third figure.

Another distortion made popular by graphics packages is the misuse of three-dimensionality, such as in PIE CHARTS, worsened by the ability to rotate the pie and detach slices at will. This can have the effect of inflating or masking a particular subcategory to suit the point being made. Leading journals prohibit the use of such unscientific devices. In the same way, bar charts and histograms should not have artificial three-dimensionality introduced as it confuses the reader when trying to read off axis values.

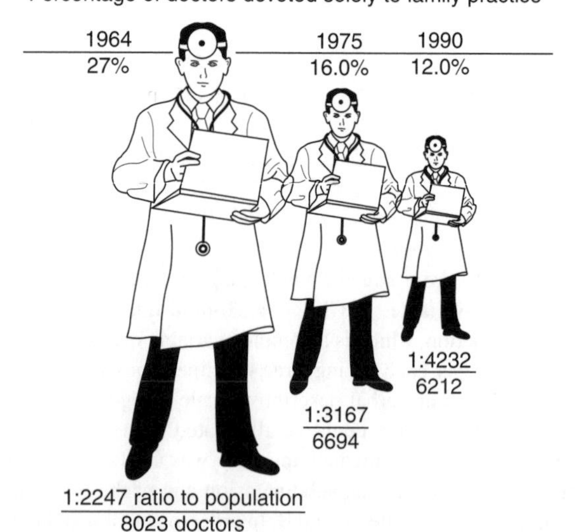

graphical deception *The shrinking family doctor (taken with permission from Tufte, 1983)*

Tufte (1983) quantifies the distortion in graphical displays with what he calls the lie factor of the display, defined as follows:

Lie factor = Size of effect in graph/size of effect in data

The lie factor for the shrinking doctors is 2.8.

Some suggested principles for avoiding graphical distortion leading to possible graphical deception taken from Tufte (1983) are: representation of numbers, as physically measured on the surface of the graphic itself, should be directly proportional to the numerical quantities represented; clear, detailed and thorough labelling should be used to defeat graphical distortion and ambiguity. Write out explanations of the data on the graphic itself. Label important events in the data; to be truthful and revealing, data graphics must bear on the heart of quantitative thinking, 'compared to what?' Graphics must not quote data out of context; above all else, show the data. *BSE*

[See also GRAPHICAL DISPLAYS]

Tufte, E. R. 1983: *The visual display of quantitative information*, Cheshire, CT: Graphics Press.

graphical displays These are procedures for visually displaying measured quantities by means of the combined use of points, lines, a coordinate system, numbers, symbols, words, shading and colour. It has been estimated that between 900 billion (9×10^{11}) and 2 trillion (2×10^{12}) images of statistical graphics are printed each year. Some of the advantages of graphical methods have been listed by Schmid (1954): in comparison with other types of presentation, well-designed charts are more effective in creating interest and in appealing to the attention of the reader; visual relationships as portrayed by charts and graphs are more easily grasped and more easily remembered; the use of charts and graphs saves time, because the essential meaning of large measures of statistical data can be visualised at a glance; charts and graphs provide a comprehensive picture of a problem that makes for a more complete and better balanced understanding than could be derived from tabular or textual forms of presentation; charts and graphs can bring out hidden facts and relationships and can stimulate, as well as aid, analytical thinking and investigation.

The last point in particular implies that perhaps the greatest value of a picture is when it forces us to notice what we never expected to see, although it should not be forgotten that humans are good at discerning subtle patterns that are really there (but equally good at imagining them when they are altogether absent!) – and graphs are sometimes constructed so as to mislead (see GRAPHICAL DECEPTION).

Many graphical displays used in medical research, e.g. the HISTOGRAM, PIE CHART and SCATTERPLOT, have been around for many years, but during the last two decades a wide variety of new methods have been developed with the aim of making this particular aspect of the examination of data as informative as possible. Graphical techniques have evolved that will provide an overview, hunt for special effects in data, indicate OUTLIERS, identify patterns, diagnose (and criticise) models and generally search for novel and unexpected phenomena. One example of these newer graphical techniques is TRELLIS GRAPHICS.

The current approach to statistical graphics largely arises from the 'visualisation' philosophy expounded by Cleveland (1985, 1993). There are two components to Cleveland's approach to displaying data: (1) *graphing*: visualisation implies a process in which information is encoded in visual displays and (2) *fitting*: fitting mathematical functions to data is needed as well as just a graphical display. Just graphing raw data, without fitting them and without graphing the fits and residuals, often leaves important aspects of the data undiscovered.

Visualisation is critical to data analysis. It provides a front line of attack, revealing intricate structure in data that cannot be absorbed in any other way and can lead to the discovery of unimagined effects as well as challenging imagined ones.

Good graphics will tell a convincing story about the data. In practice, large numbers of graphs may be needed and computers will generally be needed to draw them for the same reason that they are used for numerical analysis, speed and accuracy. *BSE*

[See also BOXPLOTS, GROWTH CHARTS, RESIDUAL PLOTS, SCATTERPLOT MATRICES]

Cleveland, W. S. 1985: *The elements of graphing data*. Summit, NJ: Hobart Press. **Cleveland, W. S.** 1993: *Visualizing data*. Summit, NJ: Hobart Press. **Schmid, C. F.** 1954: *Handbook of graphic presentation*. New York: Ronald.

graphical models Also known as *conditional independence graphs*, these models represent interrelationships in multivariate data pictorially. The graphs associated with these models depict the relationships between variables, with nodes representing random variables, and lines between them (*edges*) representing associations between them. The identification of independence between pairs of variables, conditional on the other variables in the model, enables graphs to be simplified by the omission of unnecessary edges. Models for multivariate normal data are called *graphical Gaussian models* or *covariance selection models*. The graphs associated with these models depict partial correlations between variables, conditional on all the other variables. Models for categorical data are based on the multinomial family of distributions and are called *graphical log-linear models*. Here the associations that are depicted are interaction terms in LOG-LINEAR MODELS. *Mixed models* that combine categorical and multivariate normal data (mixed models) can also be fitted.

The first figure (see page 198) shows some output from MIM, a software package dedicated to this type of analysis (see Edwards, 2002, for further details of the package and of this example, which relates to mathematics exam marks). The first matrix is a correlation matrix for five variables (*v*, *w*, *x*, *y*, *z*) and is followed by the partial correlation matrix corresponding to it. The full model with all pairwise partial correlations can be expressed by a model formula as *vwxzy* and the associated graph is shown on the left. A model specified as *vwx,xyz* retains those subsets of variables that are mutually (partially) correlated and sets to zero any insignificant partial correlation. Here, the choice of edges to be omitted (*vy*, *vz*, *wy*, *wz*) is fairly obvious. However, in a more complex situation the reduced model would typically be found by removing each pair of variables in turn on the basis of their statistical significance. The fitted partial correlation matrix corresponding to this reduced *vwx,xyz* model is shown as the third matrix, with the graph on the right. The position of the variables in the diagram is not important, only the links between them. In the reduced model, one can conclude, for example, that *x* is needed to predict any of the other variables and also that *y* and *z* are not needed to predict *v*, so long as *x* and *w* are available.

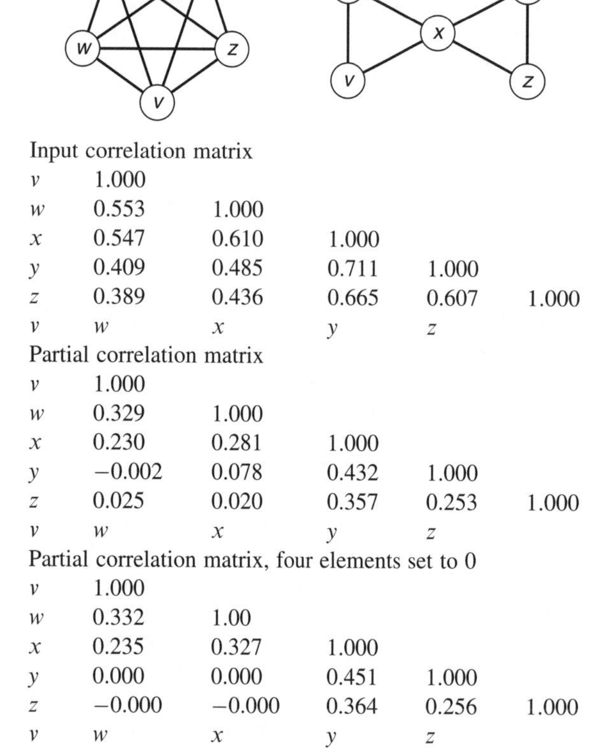

Input correlation matrix

v	1.000				
w	0.553	1.000			
x	0.547	0.610	1.000		
y	0.409	0.485	0.711	1.000	
z	0.389	0.436	0.665	0.607	1.000
v	*w*	*x*	*y*	*z*	

Partial correlation matrix

v	1.000				
w	0.329	1.000			
x	0.230	0.281	1.000		
y	−0.002	0.078	0.432	1.000	
z	0.025	0.020	0.357	0.253	1.000
v	*w*	*x*	*y*	*z*	

Partial correlation matrix, four elements set to 0

v	1.000				
w	0.332	1.00			
x	0.235	0.327	1.000		
y	0.000	0.000	0.451	1.000	
z	−0.000	−0.000	0.364	0.256	1.000
v	*w*	*x*	*y*	*z*	

graphical models *A full and a reduced graphical model with associated partial correlation matrices and the initial correlation matrix (mathematics marks data from Mardia, Kent and Bibby, 1979, analysed by Edwards, 2002)*

Graphical models are based not only on statistical distribution theory but also on concepts of mathematical graph theory, e.g. *cliques* and *acyclic graphs*. In this example, the subset of variables *vwx* is a *clique* since (a) all the vertices are joined and (b) any larger subset containing it does not have this property. *Acyclic* means that there are no paths from a node back to itself. Concepts such as these ensure that the illustrative graphs and their associated models do not contain redundant information and can be interpreted unambiguously. *Decomposability* is a criterion sometimes sought. This requires that models can be broken down into series of regressions; it can aid interpretation and allow certain exact significance tests to be applied. Other more familiar criteria may be applied to models. For example, only log-linear models that are *hierarchical* are usually considered, as is the case in standard log-linear analysis.

Graphical chain models, also known as *Bayesian networks*, represent a series of models that have a directional relationship to one another, i.e. one model precedes another in some sense, either through a natural ordering in time or through some assumed causal relationship. The variables for each component model are considered to be in a block and the blocks are ordered to form a chain. Associations within blocks are considered to be noncausal whereas those between blocks are considered potentially causal. From the second block onwards each model is conditional on the other variables in that block and those in all preceding blocks. The edges linking the components are termed *directed* and are denoted by arrows.

The second figure (see page 199) illustrates a relatively complex graphical chain model concerned with infant mortality in Malaysia (Mohamed, Diamond and Smith, 1998), using categorised data. Note that the convention is to show discrete variables (as they are in this example) as closed circles whereas continuous variables are depicted as open, as in the first figure. The components of chain models are shown as blocks (often enclosed in rectangles). In the second figure, parts (a) and (c) are simple graphical models showing associations between pairs of variables controlling for the others. The other parts show models that have a temporal or causal relationship with one another. A summary was produced from all the constituent chain models, from which it was concluded that neonatal mortality was directly associated with maternal education, ethnicity, state, year of birth, source of drinking water, birth interval, pre-maturity and sex. However, neonatal mortality was not associated with birth attendant, birthplace and antenatal care.

Estimation of models from data is generally performed by MAXIMUM LIKELIHOOD. Automatic selection of variables can be performed on the basis of the change in likelihood resulting from adding or removing edges in turn. This process typically starts either with the full or saturated model, containing all possible edges, followed by successive elimination of non-

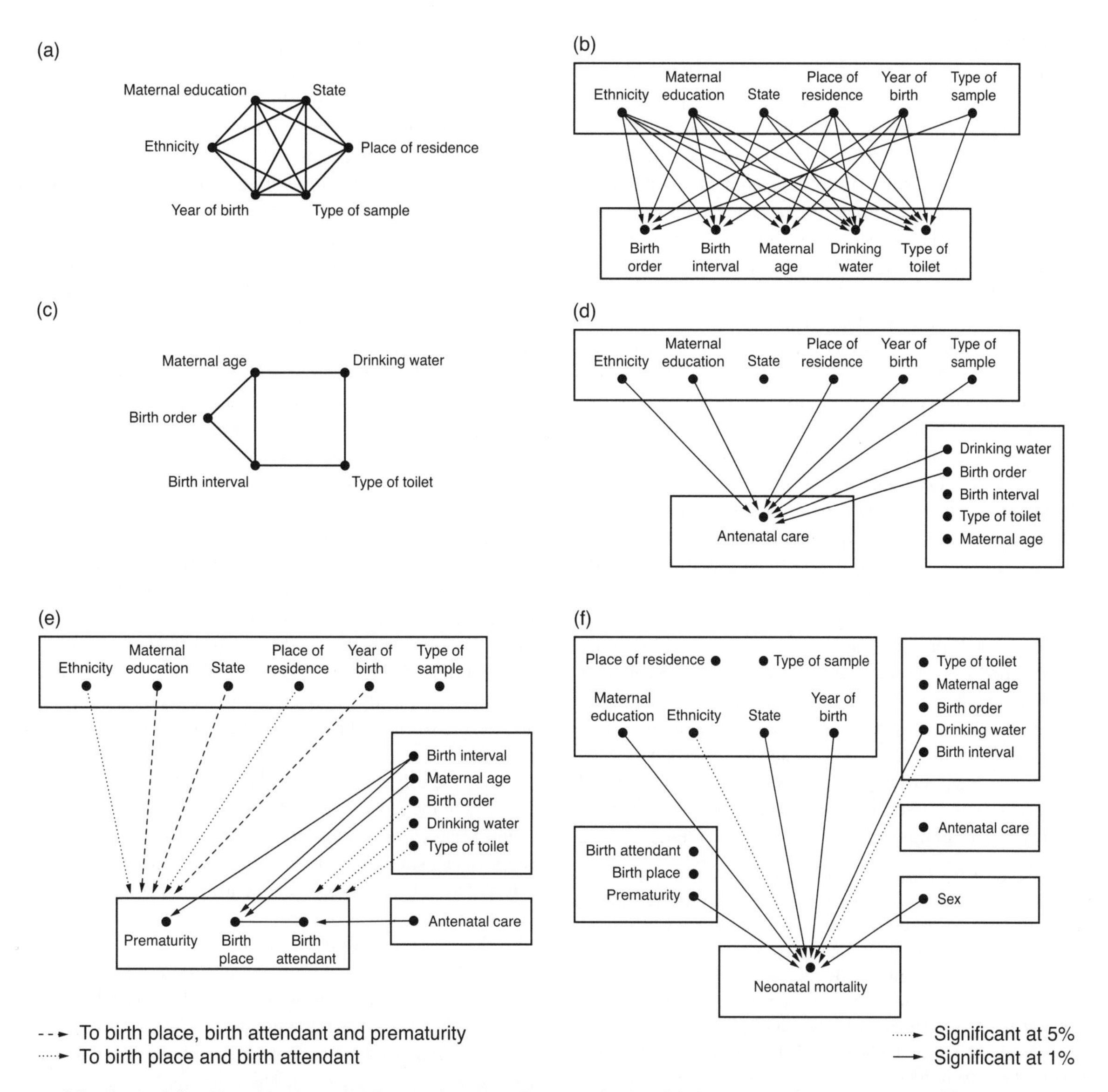

graphical models *Graphical model for each step of an analysis of infant mortality in Malaysia (from Mohamed, Diamond and Smith, 1998): (a) socioeconomic factors – intrablock associations; (b) socioeconomic factors and factors before pregnancy – interblock associations; (c) factors before pregnancy – intrablock associations; (d) direct associations between antenatal care and its potential determinants; (e) direct associations between factors at birth and their potential determinants; (f) direct associations between neonatal mortality and its determinants*

significant links, or with the null model followed by successive inclusion of links. In the case of multivariate normal data, summary statistics such as COVARIANCE MATRICES rather than raw data may be sufficient to fit models. However, diagnostic tests based on individual cases and data transformation are then not possible. Such tests include examination of residuals for OUTLIERS and assessment of normality if appropriate. *Box-Cox tests* can be used to decide on the appropriate transformation if this is indicated. Graphical Gaussian models assume that there are no interactions (partial correlations depending on the level of a third variable). In the case of continuous variables it may be advisable to dichotomise the data and refit the model as a log-linear model or a mixed model so that any interactions can be detected.

Many standard statistical techniques can be framed as special cases of graphical models, e.g. mixtures of multivariate normals (see FINITE MIXTURE DISTRIBUTIONS), or ANALYSIS OF VARIANCE. However, there are two situations in the analysis of medical data for which graphical models may be particularly useful. The first is when little is known about potential ASSOCIATIONS among a group of variables and where an exploratory approach is therefore called for; here a stepwise automatic selection process would probably be used to seek simple models consistent with the data. The other, as in the earlier infant mortality example, is where one has in mind a complex model of associations and causal links, the overall structure of which can be set out even though the details cannot be specified. The ability to depict the associations visually is particularly helpful in these two contexts. The numerical values of effect sizes, e.g. the partial correlations, are of course important, but it may be the qualitative information as to which pairs of variables are associated at any level (conditional on the rest) that may be of prime interest. This is most easily represented graphically rather than in tables.

A key paper is by Lauritzen and Wermuth (1989) and Edwards (2002) provides an application-oriented introduction making use of dedicated software MIM. Whittaker (1990) is another general text. Closely related techniques are path analysis and structural equation models. *Bayesian graphical modelling*, in which parameters and latent variables are included in the graph (in addition to the observed quantities), is discussed by Spiegelhalter (1998) and illustrated with an example relating to cancer incidence. *ML*

Edwards, D. 2002: *Introduction to graphical modelling*, 2nd edition. New York: Springer Verlag. **Lauritzen, S. L. and Wermuth, W. C.** 1989: Graphical models for associations between variables, some of which are qualitative and some quantitative. *Annals of Statistics* 17, 31–57. **Mardia, K. V., Kent, J. T. and Bibby, J. M.** 1979: *Multivariate analysis.* London: Academic Press. **Mohamed, W. N., Diamond, I. and Smith, P. W. F.** 1998: The determinants of infant mortality in Malaysia: a graphical chain modelling approach. *Journal of the Royal Statistical Society, Series A* 161, 349–66. **Spiegelhalter, D. J.** 1998: Bayesian graphical modelling: a case-study in monitoring health outcomes. *Applied Statistics* 47, 115–33. **Whittaker, J.** 1990: *Graphical models in applied multivariate statistics.* Chichester: John Wiley & Sons, Ltd.

gross reproduction rate (GRR) See DEMOGRAPHY

group sequential methods See DATA-DEPENDENT DESIGNS, INTERIM ANALYSIS, SEQUENTIAL ANALYSIS

growth charts These are GRAPHICAL DISPLAYS that present AGE-RELATED REFERENCE RANGES for anthropometry such as height and weight in childhood. Growth charts conventionally provide information on three or more QUANTILES of the age-related distribution, including the MEDIAN and other centiles (percentiles) placed symmetrically about the median. The first figure (see page 201) illustrates a growth chart of infant weight in British boys. There are nine centiles on the chart, equally spaced two-thirds of a unit apart on the Z-SCORE scale. There is also the growth curve of an infant followed over a 2-month period, showing marked growth faltering.

The growth chart is used in several distinct ways. First, it is a screening or diagnostic test, corresponding to the way age-related reference ranges are used. Measurements are plotted against age on the chart, and children whose measurement lies outside the reference range, i.e. above the top centile or below the bottom centile for the child's age, are considered to be at risk of a growth disorder and are referred for further investigation. The proportion of children screened depends on the centile used, e.g. 3 % below the 3rd centile or 10 % above the 90th centile. This assumes that the children are representative of the reference population on which the growth chart is based. The population PREVALENCE of a growth disorder is small, so the screening in rate corresponds closely to the FALSE POSITIVE RATE (100 % – SPECIFICITY) of the screening test, the vast majority of the selected children being free of the disorder. Note though that the growth chart provides no information of the true positive rate (SENSITIVITY) of the screening test, as this needs to be based on a representative sample of growth-disordered children, which is generally not available.

A second use of the growth chart is to quantify the centile position of the individual child, by seeing what centile curve their measurement is close to. For measurements in the body of the distribution the approximate centile can be obtained by interpolating between adjacent centile curves.

The third and most common use of the chart is an extension of this previous use to measurements on two or more occasions, which are plotted on the chart and the points joined together to make a growth curve. The child's growth velocity over time is assessed on the assumption of tracking, where the centile is expected to be constant and the growth curve parallel to the centile curves. This constant centile over time corresponds to a growth velocity that is close to the population MEDIAN, whereas growth-disordered children show 'centile crossing', i.e. they grow faster or slower than the implied reference median velocity, as seen, for example, in the first figure. It is this use that gives the name to growth charts, quantifying growth in individuals relative to the reference. It should be recognised, however, that growth charts used in this way lack statistical rigour – the reference data are cross-sectional rather than longitudinal, and contain no information about growth velocity. In addition the chart does not adjust for regression to the median.

Child centiles followed over time (assumed to be representative of the reference population) will be subject to regression towards the median, so that, for example, in a

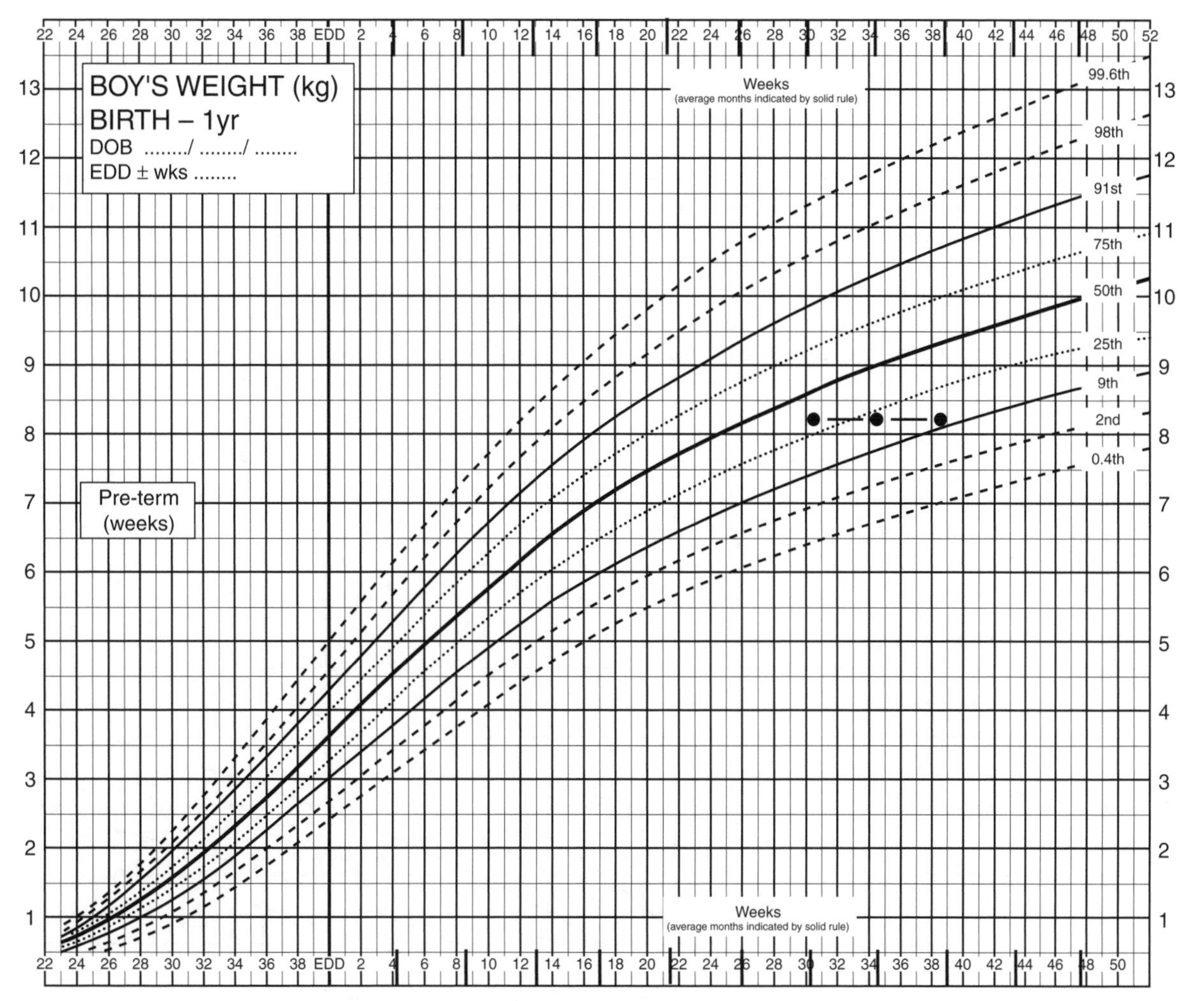

growth chart *Growth chart to assess weight in infancy, British boys 1990. The chart shows the 0.4th, 2nd, 9th, 25th, 50th, 75th, 91st, 98th and 99.6th centiles of the weight distribution by age. Also shown is the growth curve for a child measured at 7, 8 and 9 months (reproduced with permission of the Child Growth Foundation)*

group of children on the 9th centile on one occasion, their mean measurement centile will be closer to the median when followed up. The size of this centile crossing effect depends solely on the strength of the correlation between measurements on the two occasions. Height, for example, tracks very strongly in mid-childhood before puberty (the year-on-year correlation exceeds 0.97), so most children show very little height centile crossing over time and regression to the median is hard to detect. For more labile measurements like weight, body mass index or skinfold thickness, and particularly during periods of rapid growth such as early infancy or puberty, the age-on-age correlation may be as low as 0.5 and regression to the median is marked. This emphasises that charts to measure size are not ideal to assess growth, and vice versa.

A more useful tool to assess growth velocity is a dedicated velocity chart. Charts to assess growth velocity can be constructed in the same way as charts for size. Here each child from the reference population provides two measurements taken some pre-specified time interval apart (e.g. 1 year), which are converted to a velocity and then analysed to construct centile charts of velocity for age. The restricted time interval ensures that the contribution of measurement error to the total variability is fixed, as the amount of MEASUREMENT ERROR varies inversely as the time interval. Velocity charts are less satisfactory than size charts for three reasons: (a) the time interval requirement is restrictive as most children are not measured that regularly; (b) the chart does not adjust for regression to the median and (c) requiring two charts rather than one to assess each child's growth increases resource costs.

An alternative is to assess the child's growth velocity from the size chart. This exploits the principle that velocity is the rate of change of size, so that the slope of the line joining successive measurements is a measure of velocity. Velocity can be assessed in terms of the rate of change of either the measurement or the measurement centile (more accurately the z-score), and the latter corresponds directly to centile crossing. This is the principle behind 'thrive lines', a set of curves analogous to centile curves that are printed on a transparent plastic overlay and placed on the growth (size) chart to detect failure to thrive (i.e. poor growth). Centile curves represent the pattern (or direction) of growth in children who are tracking perfectly, i.e. whose centile remains constant over time. Such children are growing on the median velocity (ignoring regression to the median). In the same way, thrive lines can be drawn to reflect the pattern (or direction) of growth of children growing at some specified velocity centile other than the median. The second figure shows thrive lines for the 5th velocity centile, superimposed on the infant weight chart of the first figure. The child's growth curve in the second figure tracks along the

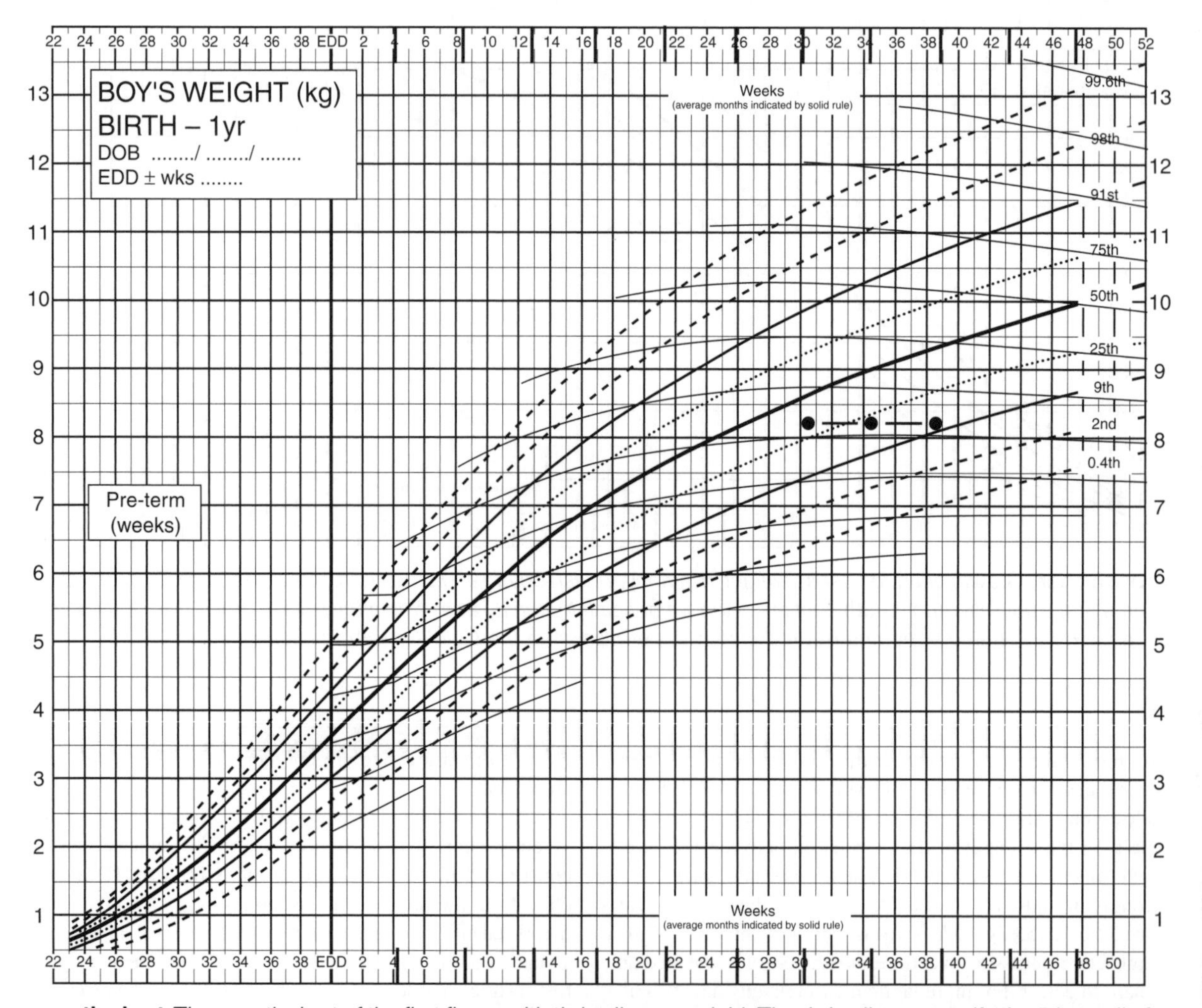

growth chart *The growth chart of the first figure with thrive lines overlaid. The thrive lines quantify the 5th centile for weight velocity over a 4-week period. The infant's growth curve tracks along the thrive lines, corresponding to the 5th weight velocity centile for the first month, and approximately the 1st centile for the whole 2-month period (reproduced with permission of the Child Growth Foundation)*

thrive lines for a period of 4 weeks, and this defines the growth rate as the 5th velocity centile. Only about 1 child in 20 grows more slowly than this. The time period is important, and the growth pattern becomes more extreme the longer it tracks the thrive lines. Therefore, as the child continues to track along the thrive lines for a further 4 weeks, i.e. 2 months altogether, this means that the velocity over the whole period is near the 1st centile, clearly of much greater concern. Note that the first figure highlights the centile crossing, but gives no clue as to how extreme the velocity centile is.

The main technical concern with growth charts is the representativeness of the underlying reference population. This depends on the nationality, ethnicity and timing of measurement of the child being assessed compared to the reference. For example, the British growth charts (e.g. the first figure) are based on ethnic Caucasian British children measured in 1990. They are therefore less appropriate for assessing Caucasian Dutch or ethnic minority British children, and will become progressively more out-of-date as time passes, due to the secular trend to increasing height and particularly weight. For further details see Tanner (1978), Cole (1998), Cole, Freeman and Preece (1998) and Ulijazsek, Johnston and Preece (1998). *TJC*

Cole, T. J. 1998: Presenting information on growth distance and conditional velocity in one chart: practical issues of chart design. *Statistics in Medicine* 17, 2697–707. **Cole, T. J., Freeman, J. V. and Preece, M. A.** 1998: British 1990 growth reference centiles for weight, height, body mass index and head circumference fitted by maximum penalized likelihood. *Statistics in Medicine* 17, 407–29. **Tanner, J. M.** 1978: *Foetus into man: physical growth from conception to maturity*. London: Open Books. **Ulijazsek, S. J., Johnston, F. E. and Preece, M. A.** 1998: *Cambridge encyclopedia of human growth and development*, pp. 440. Cambridge: Cambridge University Press.

Grubbs' test statistic

See OUTLIERS

H

haplotype analysis A haplotype refers to a combination of alleles transmitted from a parent to a child through a haploid nucleus in a gametic cell, although the term is often restricted to a combination of alleles that are in tight linkage on the same chromosome. Humans are diploid: an individual's genotype is derived from the union of a haplotype from the father and a haplotype from the mother. Haplotype analysis includes the estimation of population haplotype frequencies from sample genotype data, the inference of an individual's haplotype from genotype data and the investigation of possible associations between haplotypes and disease or other traits.

One important problem in haplotype analysis is that an individual's genotype may be consistent with multiple pairs of haplotypes. Thus, the genotype AaBb is consistent with haplotype pairs AB/ab and Ab/aB. In general, if there are m heterozygous loci in the genotype, then there are 2^{m-1} consistent haplotype pairs. The availability of genotype but not haplotype data can be regarded as a form of incomplete data, so that the estimation of haplotype frequencies from genotype data can be accomplished by an EM ALGORITHM. Other methods to haplotype frequency estimation have also been proposed, including Bayesian approaches that take account of the similarities between the haplotypes.

If the frequency of a haplotype deviates from the product of the frequencies of the constituent alleles, then the alleles are not independent but associated with each other, and are said to be in linkage disequilibrium. A number of measures of linkage disequilibrium are available for two diallelic loci, including D (the difference between haplotype frequency and the product of constituent allele frequencies), D' (D divided by the maximum possible D given the allele frequencies of the two loci) and R (the CORRELATION coefficient between numerically coded values for the alleles of the two loci). The strength of linkage disequilibrium between the markers in a region may reflect the recombination rate in that region (possibly determined by the local chromosome structure) and stochastic variation in the recombination and mutation history of the population.

A possible association between haplotype and disease is usually examined by estimating haplotype frequencies in cases and controls and testing whether these can be equated. Sometimes the association between a haplotype and a disease is stronger than the association between any of the constituent alleles and the disease. This happens when the true causal variant had originated in a mutation that occurred on a chromosome containing a particular combination of alleles, or when there is an interaction between the effects of alleles on the same chromosome (called *cis* interactions). Knowledge of haplotype structure is also important for the optimal choice of markers in association studies, leaving out any markers that are predictable by the others because of strong linkage disequilibrium. *PS*

[See also ALLELIC ASSOCIATION, GENETIC EPIDEMIOLOGY, GENETIC LINKAGE]

Hardy–Weinberg law This is a result concerning the frequency distribution of genotypes at a polymorphic genetic locus in a population under random mating. The Hardy–Weinberg law is an important result in population genetics that was derived independently by the English mathematician, G. H. Hardy, and the German physician, W. Weinberg, in 1908. For a genetic locus with two alternate sequence variants (alleles), the Hardy–Weinberg law states that half the frequency (expressed as a proportion) of the heterozygote genotype is equal to the square root of the product (i.e. the GEOMETRIC MEAN) of the frequencies of the two homozygous genotypes. An alternative way of stating the Hardy–Weinberg law is that the frequency of a homozygous genotype is equal to the square of the frequency of the constituent allele, while the frequency of a heterozygous genotype is equal to twice the product of the frequencies of the two constituent alleles. If the frequencies of alleles A and B are denoted by p and q, then the Hardy–Weinberg law states that the frequencies of the AA, AB and BB genotypes are given by p^2, $2pq$ and q^2 respectively. The Hardy–Weinberg law is therefore the result of the simple rule that the probability of two independent events is equal to the product of the probabilities of the two events.

The Hardy–Weinberg law can be violated in real populations or samples for many reasons. Populations that consist of noninterbreeding (i.e. stratified) subpopulations with different allele frequencies will tend to have an excess of individuals with homozygous genotypes. The characteristic Hardy–Weinberg ratios can be distorted by natural selection, where one or more genotypes confers a survival advantage over the others. The overall population ratios can be distorted at a locus that contains disease-predisposing variants in a sample of patients with the disease. Finally, the apparent distortions of the Hardy–Weinberg ratios for some loci in a set of genotype data can be the result of genotyping errors in the laboratory. Testing for Hardy–Weinberg proportions is therefore a routine part of data quality checks in genetic studies. For loci with two alleles a Pearson CHI-SQUARE TEST

Encyclopaedic Companion to Medical Statistics: Second Edition Edited by Brian S. Everitt and Christopher R. Palmer

with one degree comparing observed and predicted counts is standard, but for loci with more than two alleles a permutation-based test is preferable. *PS*

[See also ALLELIC ASSOCIATION, GENETIC EPIDEMIOLOGY, HERITABILITY]

Hawthorne effect This is a possible effect that might be produced in an experiment or study simply from subjects' awareness of participation in some form of scientific investigation. That individual behaviours might be altered because they know they are being studied was first said to have been demonstrated in a research project carried out at the Hawthorne Plant of the Western Electric Company in Cicero, Illinois, in the late 1920s. The major finding of the study was that, almost regardless of the experimental manipulation employed, the production of the workers seemed to improve. The implication of the effect is that people who are singled out for a study of any kind may improve their performance or behaviour, not because of any specific condition being tested but simply because of the attention they receive. A medical example suggested by Gail (1998) involves a study of methods to promote smoking cessation, in which it is necessary to contact study participants each year to determine smoking status. A further more recent medical example of the appearance of the Hawthorne effect is given in Fox, Brennan and Chasen (2008). The Hawthorne effect could distort study results if this repeated annual contact affected smoking behaviour or the reporting of smoking behaviour. *BSE*

Fox, N. S., Brennan, J. S. and Chasen, S. T. 2008: Clinical estimation of fetal weight and the Hawthorne effect. *European Journal of Obstetrics, Gynaecology and Reproductive Biology* 141, 111–14. **Gail, M. H.** 1998: Hawthorne effect. In Armitage, P. and Colton, T. (eds), *Encyclopedia of biostatistics*. Chichester: John Wiley & Sons, Ltd.

hazard function See PROPORTIONAL HAZARDS, SURVIVAL ANALYSIS – AN OVERVIEW

health-adjusted life expectancy (HALE) See DEMOGRAPHY

health services research Health services research, according to Bowling (2002), 'is concerned with the relationship between the provision, effectiveness and efficient use of health services and the health needs of the population. It is narrower than health research'. It thus entails measuring and evaluating the inputs, processes and outcomes of healthcare provision. Input and process information that is primarily aimed at assisting healthcare managers and providers, especially when collected on a routine basis, is probably more correctly considered as audit or quality assurance. Generally speaking, such routine data can rarely be used for research purposes, due to difficulties in maintaining standards in data collection. An exception would be a long-term case register containing data on all patients in a given area gathered in a strictly controlled and objective fashion. While most standard statistical methods are potentially applicable in health services research, some are more useful than others. This is because health services research is often relatively complex, involving as it does the analysis of different interventions, outcomes and levels of data simultaneously.

Because of this complexity, and also sometimes because of ethical issues, relatively unusual experimental or pseudo-experimental designs such as stepped wedge designs, preference trials and randomised consent designs are available in addition to the more standard experimental designs, such as individually or cluster randomised CLINICAL TRIALS. Observational studies, such as CROSS-SECTIONAL STUDIES, COHORT STUDIES and CASE-CONTROL STUDIES may be more appropriate or indeed the only feasible option for studying health services in naturalistic settings. For further information on various approaches, see the MRC guidance on complex interventions, which has been revised and updated from 2000 (Craig *et al.*, 2008).

A contrast between the typical healthcare trial and the typical pharmacological trial is that the latter usually focuses on the outcome for the individual patient and assesses some particular therapeutic intervention such as a drug, a surgical procedure or a psychological intervention. The remit of a typical healthcare trial, contrariwise, tends to be broader and more complex since it often involves the evaluation of one or more interventions, the environment in which they take place and the personnel administering them. The outcomes may be measured at the patient level but they may also be measured at other levels, such as the ward or the hospital, or indeed at several of these levels simultaneously. Three nested levels here might be patient, ward and hospital, and these would all need to be taken into account in a MULTILEVEL MODEL.

In a discussion aimed specifically at psychiatrists, but which is nevertheless generally applicable, Dunn (2001) draws attention to some of the problems inherent in health service trials. One of these is the HAWTHORNE EFFECT, in which there is a nonspecific or PLACEBO effect that is not directly associated with the specific content of the intervention but is rather due to the mere fact of participation in the study. Dunn points out that, in health service research, providers as well as patients may be subject to such an effect. Avoiding it is often more difficult in healthcare trials compared to clinical trials because blinding of participants may be impractical or unethical.

The definition of outcome is often quite problematic in health services research since interventions may potentially produce multiple and conflicting changes in several dimensions. Important statistical issues in this area are thus dealing with multiple significance testing and combining outcomes into summary statistics. Economic analysis, aimed at balancing the effectiveness of outcomes against the cost of providing interventions, is commonly performed in health services research (see COST-EFFECTIVENESS ANALYSIS). One issue that has to be addressed in this context is whether service use information, such as number of hospital admissions, should be regarded as an outcome in its own right or whether it should be considered purely on the cost side of the equation. The views of clinicians and health economists may differ on this point. Often outcomes are concerned with such concepts as patient satisfaction or QUALITY OF LIFE (Fayers and Machin, 2007), which may be difficult to define and capture.

Even once they have been defined conceptually, outcomes are not always straightforward to measure and often involve the use of QUESTIONNAIRES. The latter may be prone to test–retest imprecision, due to a subject's inconsistency or, indeed, genuine changes from one time point to the next, or disagreement between raters (in cases where the questionnaires are administered and interpreted by someone other than the subject). Methods for assessing MEASUREMENT ERROR are thus important in health services research. The analysis of the psychometric properties of instruments, such as their reliability and validity (see Streiner and Norman, 2008), may be necessary where instruments have been developed especially for a study. The treatment of MISSING DATA, and data quality in general, is also a relatively common issue arising in health services research. This is because there is generally less control over data collection in the community or a hospital, as opposed to an experimental laboratory or dedicated clinic. The standard CONSORT STATEMENT may need to be adapted (see Boutron *et al.*, 2008) for nonclinical outcomes.

Sometimes the focus in health services research is on aggregate data from high-level units such as hospitals or health authorities. For example, methods for comparing the performance of health providers in league tables may be required. Goldstein and Spiegelhalter (1996) discuss some of the issues arising from the comparison of institutional performance. Methods for analysing spatial statistics are used when the geographical location of the units is also important and such methods may be integrated with a geographical information system (GIS); this is a specialised form of database that holds complex geographical data so as to allow it to be visualised. Such methods may be aimed at identifying outlying disease clusters, examining the impact of area-wide interventions or measuring health inequalities and relating them to other area-wide data such as social deprivation. *ML*

[See also ECOLOGICAL STUDIES]

Boutron, I., Moher, D., Altman, D., Scultz, K. and Ravaud, P. 2008: Extending the CONSORT Statement to randomized trials of non-pharmacologic treatment: explanation and elaboration. *Annals of Internal Medicine* 148, 295–309. **Bowling, A.** 2002: *Research methods in health*, 2nd edition. Buckingham: Open University Press. **Craig, P., Dieppe, P., Macintyre, S., Michie, S., Nazareth, I. and Petticrew, M.** 2008: Developing and evaluating complex interventions: the new Medical Research Council guidance. *British Medical Journal*, **337,** a1655. **Dunn, G.** 2001: Statistical methods for measuring outcomes. In Thornicroft, G. and Tansella, M. (eds), *Mental health outcome measures*, 2nd edition. New York: Springer Verlag, pp. 5–18. **Fayers, P. M. and Machin, D.** 2007: *Quality of life: the assessment, analysis and interpretation of patient-reported outcomes*, 2nd edition. New York: John Wiley & Sons, Inc. **Goldstein, H. and Spiegelhalter, D. J.** 1996: League tables and their limitations: statistical issues in comparisons of institutional performance. *Journal of the Royal Statistical Society, Series A* 159, 385–443. **Streiner, D. L. and Norman, G. R.** 2008: *Health measurement scales: a practical guide to their development and use*, 4th edition. Oxford: Oxford University Press.

heritability In the broad sense, heritability is the proportion of the variance of a given trait that is explained by genetic differences in a population. In the narrow sense, genetic differences are restricted to those due to the additive effects of alleles. Heritability is a key concept in population genetics introduced by Sir R. A. Fisher, in close connection with his work on the ANALYSIS OF VARIANCE. Nonadditive genetic influences, which include interactions between alleles at the same locus (dominance) or at different loci (epistasis), are included in broad but not narrow heritability.

In humans, heritability is usually estimated by twin or adoption studies (see TWIN ANALYSIS). The classical twin design relies on the fact that monozygotic (MZ) twins are developed from the same fertilised egg and are therefore genetically identical, whereas dizygotic (DZ) twins are like ordinary brothers and sisters in being developed from two separate fertilised ova and therefore share on average 50 % of their genes. Given this fact, and under some additional assumptions (including the equality of the environmental similarity between MZ and DZ twins and the absence of dominance and epistasis), a simple estimate of heritability is given by twice the difference between the intraclass MZ and DZ correlations for the trait. This simple method of estimation for the heritability is known as Falconer's formula.

Adoption studies work under the assumption that any correlation between an adoptee and his or her biological family is entirely genetic in origin. In the absence of epistasis, twice the correlation between adoptee and biological parent provides an estimate of narrow heritability. Similarly, the intraclass correlation for MZ twins reared apart provides an estimate for the broad heritability.

A high heritability is sometimes misinterpreted as meaning that the trait is unlikely to respond to environmental changes. Heritability reflects on the genetic and environmental differences that exist in a particular population; it cannot be used to predict the consequences of environmental changes outside the normal range for the population. A familiar example is that the mental retardation that is invariably associated with the genetic condition phenylketonuria in a natural population can be prevented by the introduction of a low-phenylalanine diet in early infancy. *PS*

[See also GENETIC EPIDEMIOLOGY, GENETIC LINKAGE, QUANTITATIVE TRAIT LOCI]

hierarchical models See LOG-LINEAR MODELS

high-dimensional data This is a term used for datasets that are characterised by a very large number of variables and a much more modest number of observations. In the 21st century such datasets are collected in many areas, e.g. text/web data mining (see DATA MINING IN MEDICINE) and BIOINFORMATICS. The task of extracting meaningful statistical and biological information from such datasets presents many challenges for which a number of recent methodological developments may be helpful; for details see, for example, Francois (2008). *BSE*

Francois, D. 2008: *High-dimensional data analysis: from optimal metrics to feature selection.* VDM Verlag.

histogram This is a graphical representation of a frequency distribution in which each class interval is represented by a vertical bar whose base is the class interval and whose height is the number of observations in the class interval. When the class intervals are unequally spaced the histogram is drawn in such a way that the area of each bar is proportional to the frequency for that class interval. Scott (1979) considers how to choose the optimal number and width of classes in a histogram, for there are matters of choice. Two examples of histograms are shown in the figure.

The histogram is generally used for two purposes, counting and displaying the distribution of a variable, although it is relatively ineffective for both, with stem-and-leaf plots being better for counting and boxplots better for assessing distributional properties. *BSE*

Scott, D. W. 1979: On optimal and data-based histograms. *Biometrika* 66, 605–10.

historical controls This refers to the use of past data for the purpose of making comparisons with present data in a research context. Unfortunately, despite the appeal of desiring to make efficient use of previously collected resources, with information stored perhaps on a computer database, the use of historical controls is fraught with BIAS (Pocock, 1983). One cannot make reliable inferences in controlled CLINICAL TRIALS by comparing new data with old. The main reason why bias would be introduced is the lack of comparability at baseline between the two groups. Only

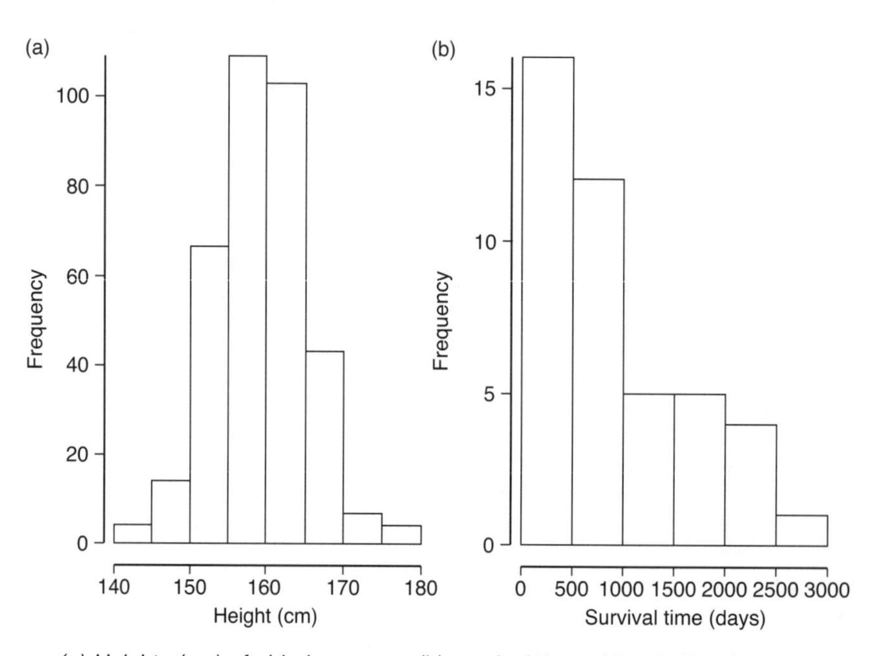

histogram *(a) Heights (cm) of elderly women; (b) survival times (days) of patients with leukaemia*

concurrent RANDOMISATION of eligible participants can bestow such between-group comparability, since randomising alone can seek to ensure treatment groups are balanced with respect to all the known and (innumerable) unknown risk factors. *CRP*

Pocock, S. J. 1983: *Clinical trials: a practical approach*. Chichester: John Wiley & Sons, Ltd.

historical demography

See DEMOGRAPHY

history of medical statistics

The first attempts at 'medical statistics' might perhaps be considered the early efforts to keep track of births and deaths through church records of weddings, christenings and burials. However, more ambitious statistical procedures than simple counting would have been largely unwelcome to physicians until well into the 17th century simply because they might have raised the unthinkable spectre of questioning the invulnerability most of them still claimed. Medical practices at the time were largely based on uncritical reliance on past experience, *post hoc, ergo propter hoc* reasoning, and veneration of the 'truth' as proclaimed by authoritative figures such as Galen (130–200), a Greek physician whose influence dominated medicine for many centuries. Such attitudes largely stifled any interest in experimentation or proper scientific investigation or explanation of medical phenomena. Even the few clinicians who did strive to increase their knowledge by close observation or simple experiment often interpreted their findings in the light of the currently accepted dogma.

Several authors have pointed out what must qualify as the world's earliest recorded comparative trial. Described in the biblical book of Daniel, hence circa 600 BC, Daniel and three colleagues expressed their preference not to be given food that had been prepared contrary to their beliefs. Their study involved a prior hypothesis and primary ENDPOINT, albeit rather subjective (facial appearance), and the trial duration was limited to just 10 days. The control group, which received the standard fare, was an unknown size, but, clearly, the treatment group, which received vegetables and water only, was small, at just four. The study turned out positively for Daniel. Despite modern-day criticism, notably lack of RANDOMISATION, no one could criticise Daniel for his influential choice of publication (see Daniel 1: 8–16, *Holy Bible*).

By the late 17th and early 18th centuries, medicine began its slow progress from a sort of mystical certainty to a scientifically more acceptable uncertainty about many of its procedures. The taking of systematic observations and carrying out of experiments became more widespread. John Graunt (1620–1674), son of a London draper, for example, published his *Natural and political observations made upon the bills of mortality* in 1662 and derived the first ever life table. Graunt was what might today be termed a vital statistician: he examined the risk inherent in the process of birth, marriage and death and used bills of mortality (weekly reports on the numbers and causes of death in an area) to compare one disease with another and one year with another by calculating mortality statistics. Graunt's work and ideas had considerable influence and bills of mortality were also introduced in Paris and other cities in Europe.

Early experimental work in medicine is illustrated by the example that is often quoted of James Lind's (1716–1794) study undertaken on board the ship the *Salisbury* in 1747. Lind assessed several different possible treatments for scurvy by giving each to a different pair of sailors with the disease. He observed that the two men given oranges and lemons made the most dramatic recovery, although it was to be another 40 years before the Admiralty was convinced enough by Lind's finding to issue lemon juice to members of the British Navy.

The 1700s also saw the first appearance of a procedure that looks remarkably similar to a modern-day SIGNIFICANCE TEST, specifically a SIGN TEST. This arose from John Arbuthnot's (1667–1735) endeavours to argue the case for Divine Providence in the stability of the ratio of number of men to women. Arbuthnot maintained that the guiding hand of a divine being was to be discerned in the nearly constant ratio of male to female christenings recorded annually in London over the years 1629–1710. The data presented by Arbuthnot (1710) showed that in each of the 82 years in this period, the annual number of male christenings had been consistently higher than the number of female christenings, but never very much higher. He then essentially tested a null hypothesis of 'chance' determination of sex at birth, against an alternative of Divine Providence, by calculating, under the assumption that the null hypothesis is true, a PROBABILITY defined by reference to the observed data. Arbuthnot's representation of chance in this context was the toss of a fair two-sided coin, in which case the distribution of births would be:

$$(1/2 + 1/2)^{82}$$

so that the observed excess of male christenings on each of 82 occasions had an extremely small probability, thus providing support for the Divine Providence hypothesis. Arbuthnot offered an explanation for the greater supply of males as a wise economy of nature, as the males are more subject to accidents and diseases, having to seek their food with danger. Therefore, provident nature to repair the loss brings forth more males. The near equality of the sexes is designed so that every male may have a female in the same country and of suitable age.

Other mathematical developments in the 18th century that were of special relevance for medical statistics included

Daniel Bernoulli's (1700–1782) development of the normal approximation to the BINOMIAL DISTRIBUTION, which was also used in studies of the stability of the sex ratio at birth.

The power of medical statistics in pursuing reform is illustrated by the work of Florence Nightingale (1820– 1907). In her efforts to improve the squalid hospital conditions in Turkey during the Crimean War, and in her subsequent campaigns to improve the health and living conditions of the British Army, the sanitary conditions and administration of hospitals and the nursing profession, Florence Nightingale was not unlike many other Victorian reformers. However, in one important respect she was very different, since she marshalled massive amounts of data, carefully arranged, tabulated and graphed, and presented this material to ministers, viceroys and others, to convince them of the justice of her case. No other major national cause had previously been championed through the presentation of sound statistical data and those who opposed Florence Nightingale's reforms went down to defeat because her data were unanswerable; their publication led to an outcry.

Another telling example of how careful arrangement of data was used in the 19th century to save lives is provided by the work of the epidemiologist John Snow (1813–1858). After an outbreak of cholera in central London in September 1854, Snow used data collected by the General Register Office and plotted the location of deaths on a map of the area and also showed the location of the area's 11 water pumps. The resulting map is shown in the figure. Examining the scatter over the surface of the map, Snow observed that nearly all the cholera deaths were among those who lived near the Broad Street pump. However, before claiming that he had discovered a possible causal connection, Snow made a more detailed investigation of the deaths that had occurred near some other pumps. He visited the families of 10 of the deceased and found that four of these, because they preferred its taste, regularly sent for water from the Broad Street pump. Three others were children who attended a school near the Broad Street pump. One other finding that initially confused Snow was that there were no deaths among workers in a brewery close to the Broad Street pump, a confusion that was quickly resolved when it became apparent that the workers drank only beer, never water. Snow's findings were sufficiently compelling to persuade the authorities to remove the handle of the Broad Street pump and, in days, the neighbourhood epidemic that had claimed more than 500 lives had ended.

Later in the 19th century and in the early 20th century, the work of people such as Sir Francis Galton (1822–1911), Wilhelm Lexis (1837–1914) and, in particular, Karl Pearson (1857–1936) began to change the emphasis in statistics from the descriptive to the mathematical. The concept of CORRELATION and its measurement by a correlation coefficient was introduced. Statistical inference began to develop and enter

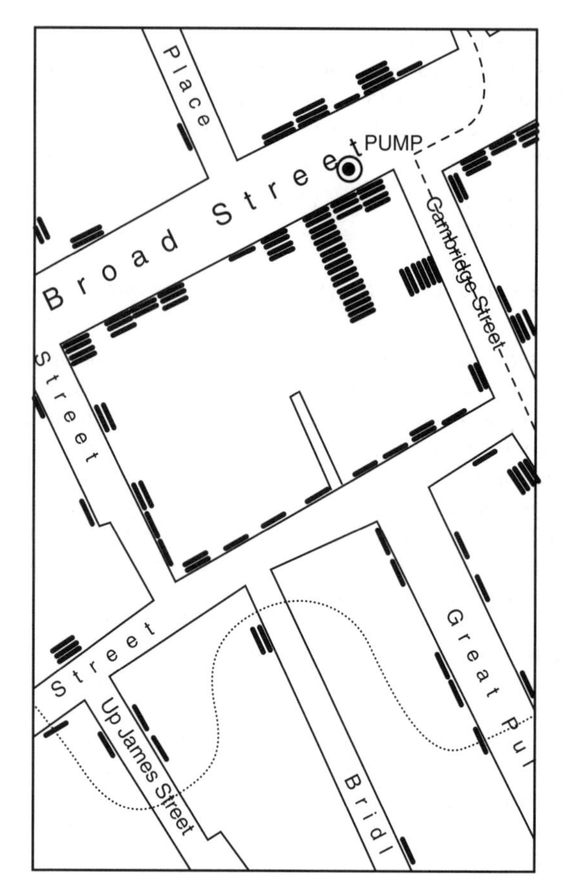

history of medical statistics *Snow's map of cholera deaths in the Broad Street area*

most areas of scientific investigation, including medical research. In 1909 Ronald Aylmer Fisher (later Sir Ronald) (1890–1962) entered Cambridge to study mathematics, the first step to becoming the most influential statistician of the 20th century. Fisher developed MAXIMUM LIKELIHOOD ESTIMATION, worked on evolutionary theory, made massive contributions to genetics and invented the ANALYSIS OF VARIANCE. However, Fisher's most important contribution to medical statistics was his introduction of randomisation as a principle in the design of certain experiments. In Fisher's case the experiments were in agriculture and were concerned with which fertilisers led to the greatest crop yields. Fisher divided agricultural areas into plots and randomly assigned the plots to different experimental fertilisers. The principle was soon adopted in medicine in studies to compare competing therapies for a particular condition, leading, of course, to the randomised CLINICAL TRIAL (RCT), described by eminent British statistician Sir David Cox as 'the most important contribution of 20th-century statistics'. The first properly performed randomised clinical trial is now generally acknowledged to be that published in 1948 by another giant of 20th century medical statistics, Sir Austin Bradford Hill (1897–1991), who investigated the use of streptomycin in the treatment of pulmonary tuberculosis. Nowadays, it is estimated that over 8000 RCTs are undertaken worldwide every year.

At about the time that Bradford Hill was busy with the first randomised clinical trial, another development was taking place, which, by revolutionising man's ability to calculate, was to have a dramatic effect on the science of statistics and the work of statisticians. The computer age was about to begin, although it would be some years before statisticians were entirely relieved of the burden of undertaking large amounts of laborious arithmetic on some pre-computer calculator. However, in the 1960s, the first statistical software packages began to appear, which made the application of many complex statistical procedures easy and routine.

The influence of increasing, inexpensive computing power on statistics continues to this day and over the last 20 years its almost universal availability has meant that research workers in statistics in general, and medical statistics in particular, no longer have to keep one eye on the computational difficulties when developing new methods of analysis. The result has been the introduction of many exciting and powerful new statistical methods many of which are of great importance in medical statistics. Notable examples to name but a few are BOOTSTRAP, COX'S REGRESSION, GENERALISED ESTIMATING EQUATIONS, LOGISTIC REGRESSION and MULTIPLE IMPUTATION.

In addition, BAYESIAN METHODS, at one time little more than an intellectual curiosity without practical implications because of their associated computational requirements, can now be applied relatively routinely. Many interesting examples are described in Congdon (2001). There seems little doubt that the remarkable success of medical statistics will continue into the 21st century. *BSE*

[See also DEMOGRAPHY, EPIDEMIOLOGY]

Arbuthnot, J. 1710: An argument for Divine Providence, taken from the constant regularity observ'd in the births of both sexes. *Philosophical Transactions of the Royal Society* 27, 186–90. **Congdon, P.** 2001: *Bayesian statistical modelling*. Chichester: John Wiley & Sons, Ltd.

hotspot clustering See DISEASE CLUSTERING

Huber-White estimate See CLUSTER RANDOMISED TRIALS

human research ethics board (HREB) See ETHICAL REVIEW COMMITTEES

hypothesis tests The testing of hypotheses is fundamental to statistics and arguments about appropriate ways to test hypotheses date back to disputes between the founders of statistical inference, during the first half of the 20th century. R. A. Fisher proposed SIGNIFICANCE TESTS as a means of examining the discrepancy between the data and a *null hypothesis* (e.g. the null hypothesis that there is no association between two variables). The *P*-VALUE (*significance level*) is the PROBABILITY that an association as large or larger than that observed in the data would occur if the null hypothesis were true. In Fisher's approach the null hypothesis is never proved or established, but is possibly *disproved*. Fisher advocated $P = 0.05$ (5 % significance) as a standard level for concluding that there is evidence against the hypothesis tested, although not as an absolute rule:

> If *P* is between .1 and .9 there is certainly no reason to suspect the hypothesis tested. If it is below .02 it is strongly indicated that the hypothesis fails to account for the whole of the facts. We shall not often be astray if we draw a conventional line at .05 (Fisher, 1950).
>
> In fact no scientific worker has a fixed level of significance at which from year to year, and in all circumstances, he rejects hypotheses; he rather gives his mind to each particular case in the light of his evidence and his ideas (Fisher, 1973).

For Fisher, interpretation of the *P*-value was ultimately for the experimenter; e.g. a *P*-value of around 0.05 might lead neither to belief nor disbelief in the null hypothesis, but to a decision to perform another experiment. To some extent, use of thresholds for significance resulted from the reduction in the size of statistical tables when only the quantiles of distributions (such as 0.1, 0.05 and 0.01) were tabulated.

Dislike of the subjective interpretation inherent in Fisher's approach led Neyman and Pearson (1933) to propose what they called *hypothesis tests*, which were designed to provide an objective, decision-theoretic approach to the results of experiments. Instead of focusing on evidence against a null hypothesis, Neyman and Pearson considered how to decide between two competing hypotheses, the null hypothesis and a specified *alternative hypothesis*. For example, the null hypothesis might state that the difference between the means of two normally distributed variables is zero, while the alternative hypothesis might state that this difference is 10.

Based on this paradigm, Neyman and Pearson argued that there were two types of error that could be made in interpreting the results of an experiment (see ERRORS IN HYPOTHESIS TESTS). We make a TYPE I ERROR if we reject the null hypothesis when it is, in fact, true, while we make a TYPE II ERROR if we accept the null hypothesis when it is, in fact, false. Neyman and Pearson then showed how to find optimal rules that would, in the long run, minimise the probabilities (the *Type I* and *Type II error rates*) of making these errors over a series of many experiments. The Type I error rate, usually denoted as α, is closely related to the *P*-value since if, for example, the Type I error rate is fixed at 5 % then we will reject the null hypothesis when $P < 0.05$. The Type II error rate is usually denoted as β and the power of the test (the probability that we do not make a Type II error if the alternative hypothesis is true) is $1 - \beta$. Based on these ideas, Neyman and Pearson were able to derive tests that were 'best' in the sense that they minimised the Type II error rate, given a particular Type I error rate.

It is important to realise that in this paradigm we do not attempt to infer whether the null hypothesis is true:

> No test based upon a theory of probability can by itself provide any valuable evidence of the truth or falsehood of a hypothesis. But we may look at the purpose of tests from another viewpoint. Without hoping to know whether each separate hypothesis is true or false, we may search for rules to govern our behaviour with regard to them, in following which we insure that, in the long run of experience, we shall not often be wrong (Neyman and Pearson, 1933).

To illustrate the differences between the two approaches, consider the hypothetical controlled trial of a new cholesterol-lowering drug, with results (mean post-treatment cholesterol) summarised in the table.

hypothesis tests *Results of a hypothetical controlled trial of a new cholesterol-lowering drug*

Group	*Number of participants*	*Mean cholesterol (mg/dl)*	*Standard deviation*
New drug	15	220	25
Placebo	15	205	25

Mean cholesterol has been reduced by 15 mg/dl; a reduction of this magnitude might lead to a substantial reduction in the risk of heart disease. An unpaired *t*-test gives $P = 0.11$. Based on Fisher's approach, the null hypothesis has not been disproved. However, a thoughtful investigator might, rather than discarding the drug, proceed to conduct a larger trial.

Application of the Neyman–Pearson approach requires the specification of both Type I and Type II error rates in advance, so we must specify a precise alternative hypothesis, e.g. that the mean reduction is 10 mg/dl. An investigator attempting to follow the Neyman–Pearson approach would need to report not only that the test was not significant at the 5 % level (Type I error rate 5 %) but also the pre-specified Type II error rate. However, the power of a study with 15 patients per group to detect a difference of 10 mg/dl is only 19.5 %. For a study that is too small, such as this one, there is no choice of Type I and Type II error rates that is satisfactory.

Had we done a POWER calculation on the basis that we wished to detect a difference of 10 mg/dl with 80 % power at 5 % significance, we would have found that we require a much larger study, with 99 patients in each group. The use of power calculations to ensure that studies are large enough to detect associations of interest is an enduring legacy of Neyman and Pearson's work.

Now that most statistical computer packages report precise *P*-values, there seems little justification in reporting the results of our drug trial as $P > 0.05$, $P > 0.1$ or 'NS' (nonsignificant) unless one is following a pre-specified choice of both Type I *and* Type II error rates. This is rarely the case: even in randomised trials we will usually investigate a number of hypotheses beyond the primary one for which the trial was designed. Therefore, in modern medical statistics, it is usual to report the precise *P*-value, together with the estimated difference and the CONFIDENCE INTERVAL for the difference. For example, for our hypothetical trial we could report that the MEAN reduction in cholesterol was 15 mg/dl (95 % CI –3.7 mg/dl to 33.7 mg/dl, $P = 0.11$). When we examine the confidence interval we see that the results are consistent either with a substantial and clinically important reduction in mean cholesterol or with a modest increase. Examining the confidence interval should help us avoid the common error of equating 'nonsignificance' with acceptance of the null hypothesis that the drug has no effect, regardless of the power of the study to detect differences of interest.

A number of books and articles discuss in more detail the testing of hypotheses, the arguments between the Fisher and Neyman–Pearson schools of inference and the case for Bayesian reasoning as an alternative (e.g. Cox, 1982; Oakes, 1986; Lehmann, 1993; Goodman, 1999a, 1999b; Sterne and Davey Smith, 2001). *JS*

Cox, D. R. 1982: Statistical significance tests. *British Journal of Clinical Pharmacology* 14, 325–31. **Fisher, R. A.** 1950: *Statistical methods for research workers*. London: Oliver and Boyd. **Fisher, R.A.** 1973: *Statistical methods and scientific inference*. London: Collins Macmillan. **Goodman, S. N.** 1999a: Toward evidence-based medical statistics. 1: the *P*-value fallacy. *Annals of International Medicine* 130, 995–1004. **Goodman, S. N.** 1999b: Toward evidence-based medical statistics. 2: the Bayes factor. *Annals of International Medicine* 130, 1005–13. **Lehmann, E. L.** 1993: The Fisher, Neyman–Pearson theories of testing hypotheses: one theory or two? *Journal of the American Statistical Association* 88, 1242–9. **Neyman, J. and Pearson, E.** 1933: On the problem of the most efficient tests of statistical hypotheses. *Philosophical Transactions of the Royal Society, Series A* 231, 289–337. **Oakes, M.** 1986: *Statistical inference*. Chichester: John Wiley & Sons, Ltd. **Sterne, J. A. and Davey Smith, G.** 2001: Sifting the evidence – what's wrong with significance tests? *British Medical Journal* 322, 226–31.

I

ICC Abbreviation for INTRACLUSTER CORRELATION COEFFICIENT

ICER Abbreviation for INCREMENTAL COST-EFFECTIVENESS RATIO. See COST-EFFECTIVENESS ANALYSIS

immune proportion This proportion indicates individuals who may not be subject to death, failure, relapse, etc., in a sample of censored survival times. The presence of such individuals may be indicated by a relatively high number of individuals with large censored survival times. Finite mixture distributions can be used to investigate such data. Specifically, the population is assumed to consist of two components. The first, which is present in proportion, p say, contains those individuals who are susceptible to some event of interest (death, relapse, etc.) and have, say, an exponential distribution for the time to the occurrence of the event. These individuals are subject to right censoring. The remaining proportion, $1-p$, of the population is assumed to be immune to, or cured of, the disease and for these individuals the event never happens. Consequently, observations on their survival times are always censored at the limit of follow-up. An important aspect of such analysis is to consider whether or not an immune proportion does in fact exist in the population (see, for example, Maller and Zhou, 1995). *BSE*

[See also CURE MODELS]

Maller, R. A. and Zhou, S. 1995: Testing for the presence of immune or cured individuals in censored survival data. *Biometrics* 51, 181–201.

imputation See MULTIPLE IMPUTATION

incidence The incidence of a disease is the number of new cases of the disease occurring within a specified period of time in a defined population. A time period of 1 year is most commonly used, but any appropriate length of time can be substituted. It is generally presented as a rate. Thus:

$$\text{Incidence rate} = \frac{\text{Number of new cases of the disease in one year}}{\text{Number in the population at risk}}$$

This assumes that the size of the study population remains constant over the time period for which the rate is calculated. Small increases or decreases in population size over a year, for example, can be dealt with by using the mid-year population as the denominator for the incidence rate.

This results in a number between 0 and 1, but for ease of presentation it is often expressed as a rate per 1000, per 100 000 or per 1 000 000 depending on the disease rarity. As an example, the incidence rate of colorectal cancer in males aged 60–64 in Scotland was 159 per 100 000 in the year 2006 compared to 206 per 100 000 in the year 2000 (NHS National Services Scotland: Information Services Division, www.isdscotland.org). Thus incidence rates can be used to measure risk and compare risks across time or between different populations.

This definition is rather simplistic because it ignores the fact that when new cases of the disease occur, the subject is no longer at risk and should ideally be removed from the denominator. It is also unsatisfactory for dealing with data from LONGITUDINALSTUDIES in which subjects may be followed up for varying lengths of time. For these studies the incidence rate can be defined as:

$$\text{Incidence rate} = \frac{\text{Number of new cases of the disease in the defined population}}{\text{Total length of time for which subjects have been followed up}}$$

The denominator gives the number of person-years of observation. Incidence rates defined in this way are often expressed as rates per 100 or per 1000 person-years of observation. (A more detailed discussion of incidence and incidence rates is given in Rothman, Greenland and Lash, 2008.)

Care should be taken to distinguish between incidence and PREVALENCE. Although the definitions appear similar at first sight, they are used for different purposes and it is essential to distinguish between them correctly. Further details can be found in Woodward (2004). *WHG*

Rothman, K. J., Greenland, S. and Lash, T. L. 2008: *Modern epidemiology*, 3rd edition. Philadelphia: Lippincott, Wilkins and Williams. **Woodward, M.** 2004: *Epidemiology: study design and data analysis*, 2nd edition. Boca Raton: Chapman & Hall.

inclusion and exclusion criteria These criteria operationalise the choice of study group, a choice that lies at the heart of the design of, and inference from, CLINICAL TRIALS. 'Inclusion' criteria define the population of interest; 'exclusion' criteria remove people for whom the study treatment is contraindicated or unlikely to be effective. Collectively, inclusion criteria and exclusion criteria comprise the *entry criteria* or *eligibility criteria*. Biological plausibility, the internal validity of the study, the epidemiological basis for generalisability and statistical power all play parts in selecting entry criteria and in making recommendations from the results of the trial. The selection of those to be enrolled in a trial often reflects a deliberate attempt to

Encyclopaedic Companion to Medical Statistics: Second Edition Edited by Brian S. Everitt and Christopher R. Palmer

select a study cohort homogeneous enough to allow a true treatment effect to become manifest, yet heterogeneous enough to permit reliable generalisation to a broader population. Clinical trials necessarily study people with more homogeneous characteristics than the patients to whom clinicians will apply the results.

Strict representativeness is relevant to the *generalisability* of clinical trials but is not essential to *inference* from them. In randomised studies, the logical basis for drawing conclusions lies in the act of RANDOMISATION. The process of concluding that the effect seen in a clinical trial will apply to another population is informal and subjective (Cowan and Wittes, 1994).

Homogeneity of the study population differs from homogeneity of the treatment effect. The former refers to a study group's sharing similar characteristics; the latter refers to an effect of treatment whose expected magnitude and direction would lead to the same recommendation for use or nonuse in identifiable subgroups. If a therapy affects a wide group of people quite similarly, then either a homogeneous or heterogeneous study group will provide similar answers regarding the magnitude of treatment effect.

An ideal study group would consist of a cohort for whom the treatment is effective and corresponding to whom is an identifiable population that will be treated. Defining such a study group before the trial is usually difficult. Available data are rarely sufficiently reliable to provide serious guidance about whom to include.

Early-phase studies typically define narrow entry criteria to establish preliminary safety or to demonstrate proof of concept (see PHASE I TRIALS, PHASE II TRIALS). Such trials often exclude children, pregnant and nursing women, the frail elderly and other vulnerable populations.

Later phase trials with narrow entry criteria specify the type of patient likely to benefit most and then test whether the treatment works for them (see PHASE III TRIALS, PHASE IV TRIALS). A study showing benefit in this narrow group of participants may lead to future trials with wider entry criteria. A treatment with important heterogeneity of effect requires a homogeneous study population.

Trials with wide entry criteria address whether the treatment under study works on average when applied to potential users. Wide entry criteria simplify screening and recruitment. Enrolling a wide range of people is consistent with assuming homogeneity of effect while affording the investigator a tentative glimpse at the likelihood of the truth of that assumption.

Biological plausibility should play a decisive role in selecting the range of people to enrol in a trial. Study entry criteria should aim to achieve heterogeneity when no convincing information at the start of the study suggests that sizeable differential effects are likely. As a heterogeneous study group leads to variation in the incidence of ENDPOINTS, increasing heterogeneity generally requires an increased sample size.

Defining entry criteria requires an operational definition of the disease in a treatment trial or a specification of who is at risk in a prevention trial. Allowing people with questionable diagnoses to enter a trial tends to attenuate the estimated treatment effect and hence decreases statistical power. Yet often the insistence on unequivocal documentation of diagnosis excludes many people who in fact would receive the treatment if the trial shows benefit (Yusuf, Held and Teo, 1994).

Trials must exclude people known to have contraindications to the treatments under study or those who are particularly vulnerable. Similarly, trials of therapies already known to be effective or ineffective in certain groups should exclude those groups of patients. Some randomised trials use an 'uncertainty principle' to guide entry (see MEGA-TRIAL). 'A patient can be entered if, and only if, the responsible clinician is substantially uncertain which of the trial treatments would be most appropriate for that particular patient' (Peto and Baigent, 1998).

Typical PROTOCOLS FOR CLINICAL TRIALS exclude people unlikely to finish a study or to adhere to the protocol. Many clinical trials have very few participants with some specific characteristics. A trial may exclude racial or ethnic groups, not because the entry criteria preclude their participation but because the clinics involved in the study do not have access to them.

In summary, trial designers should ensure that each entry criterion represents a defensible limitation on the study group; however, the fact of inclusion does not usually provide much information about the effect of treatment in specific groups of people. The argument that only by including, say, women and minorities, can one legitimately apply the results of the trial needs to be tempered with the fact that a trial rarely gives enough information about specific groups to learn much about the effect of treatment for them. When the trial is over, the results should usually be applied quite broadly, both to people whose demographic characteristics are similar and dissimilar to those in the trial; however, the medical community should maintain an intellectual stance open to suggestive data indicating differences.

The situation is more complicated for groups of people defined by such medical or physiologic variable as diagnosis, severity, prognostic features, prior history or concomitant medications, for often apparently biological cogent reasons justify exclusions. Here too a critical questioning of the reasons for exclusion is warranted; in many cases very few data are available to support even strongly held views.

Designers of clinical trials should construct entry criteria bearing in mind the purpose of the current trial, the available knowledge of the study treatments being tested, the likely studies that will follow the trial and how investigators, practising clinicians, patients and regulatory agencies will interpret the results in light of the entry criteria. *JW*

Cowan, C. and Wittes, J. 1994: Intercept studies, clinical trials, and cluster experiments: to whom can we extrapolate? *Controlled Clinical Trials* 15, 24–9. **Peto, R. and Baigent, C.** 1998: Trials: the next 50 years. *British Medical Journal* 317, 1170–1. **Yusuf, S., Held, P. and Teo, K. K.** 1994: Selection of patients for randomised controlled trials: implications of wide or narrow eligibility criteria. *Statistics in Medicine* 9, 73–86.

incomplete block designs See CROSSOVER TRIALS

incremental cost-effectiveness ratio (ICER) See COST-EFFECTIVENESS ANALYSIS

incubation period This is the time interval between the acquisition of infection and the appearance of symptomatic disease. Examples include the time between exposure to radiation or to a chemical carcinogen and the occurrence of cancer and the time from infection with HIV and the onset of AIDS.

The length of the incubation period depends on the disease, ranging from days, for instance, in the case of malaria to a number of years for HIV. The incubation period typically varies from individual to individual and may depend on the dose of the disease-causing agent received. Given this variability, it makes sense to talk about incubation period distribution. The incubation period distribution $F(t)$ represents the probability that the length of the incubation period is less than or equal to t time units.

Estimation and characterisation of $F(t)$ is important for a number of reasons. For diseases with short incubation periods, such as outbreaks, knowledge of the incubation period is essential to the investigation of the circumstances in which the disease has spread. In the case of diseases with long incubation periods, such as HIV or Creutzfeldt–Jakob disease, information on $F(t)$ is a necessary input to the estimation and projection of the evolution of the epidemic (see BACK-CALCULATION). Finally, it is very important to identify covariates that might affect the length of the incubation period for an effective clinical management of the patient.

The ideal setup to estimate the incubation period distribution is a COHORT STUDY where individuals are uninfected at enrolment and are followed up to observe both the occurrence of infection and the appearance of symptomatic disease. The resulting observations will be right censored as every individual will have either developed the disease or been censored by the end of the follow-up period (see CENSORED OBSERVATIONS). Classical survival analysis can be used to estimate $F(t)$ both nonparametrically, via KAPLAN–MEIER PLOTS, and parametrically, by fitting parametric models to the right-censored data. Usually, especially for diseases with a long incubation time, such cohort studies are difficult to set up. Estimation of the incubation period distribution is then carried out either using information on individuals who have already developed symptoms or following up cohorts of individuals who are already infected, but have not yet developed the disease. In either case, biased results can be obtained if estimation does not properly account for the sampling criteria by which individuals are included in the study. *DDA*

Brookmeyer, R. 1998: Incubation period of infectious diseases. In Armitage, P. and Colton, T. (eds), *Encyclopedia of biostatistics*, Vol. 1, pp. 2011–16. Chichester: John Wiley & Sons, Ltd. **Brookmeyer, R. and Gail, M. H.** 1994: *AIDS epidemiology: a quantitative approach*. New York: Oxford University Press.

indirect standardisation See DEMOGRAPHY

individual ethics See ETHICS AND CLINICAL TRIALS

infant mortality rate See DEMOGRAPHY

informative censoring/dropout See CENSORED OBSERVATIONS, DROPOUT, MISSING DATA

informative dropout Synonym for NONIGNORABLE DROPOUT

informed consent See ETHICS AND CLINICAL TRIALS

instantaneous death rate See SURVIVAL ANALYSIS – AN OVERVIEW

institutional review board (IRB) See ETHICAL REVIEW COMMITTEES

instrumental variables A variable that is highly correlated with an explanatory variable but has no direct influence on the response variable (i.e. its effect is mediated by the explanatory variable). Consider a situation in which we can assume that a response variable, Y, is linearly related to an explanatory variable, X, as follows:

$$Y = \alpha + \beta X + \varepsilon \tag{1}$$

where ε is a random deviation of a particular value of Y from that expected from its relationship with X. Typically, we wish to use a sample of (X, Y) pairs of measurements in order to estimate the unknown values of the parameters, α and β. The familiar ORDINARY LEAST SQUARES (OLS) estimator of β is equivalent to the ratio of the estimated COVARIANCE of X and Y to the estimated variance of X (this ratio is usually calculated by dividing the sum of the cross-products of the X and Y values from their respective mean by the sum of squares of the X values). It is possible to demonstrate that such an estimate is unbiased for β provided certain assumptions hold – the key one being that X and ε are uncorrelated.

Now, if we have an omitted variable, C, correlated with X, and such that the true model is, in fact:

$$Y = \alpha + \beta X + \gamma C + \delta \quad (2)$$

where δ is the random deviation of Y from that explained by the model. If we still proceed with our naïve OLS estimator as for equation (1) then we will obtain a biased estimate of β. This is a result of the fact that the correlation between X and ε is no longer zero. This is an example of what econometrists call endogeneity (see Wooldridge, 2003). In epidemiology, the variable C is known as a *confounder* (in this case a hidden confounder). In such circumstances, how might we obtain a valid estimate of β? The obvious answer is to measure C and fit equation (2). Another approach (much more common in economics than in medical applications) is to find a variable that is strongly correlated with X, but uncorrelated with the residual, ε. Such a variable is called an instrumental variable (IV) or instrument, for short.

Now let us consider a different circumstance. Suppose that the values of X are measured subject to error such that:

$$X = \tau + \nu \quad (3)$$

In which the ν values are random measurement errors with zero mean and assumed to be uncorrelated both with each other and with the true values, τ. The relationship we are really interested in is the following:

$$Y = \alpha + \beta\tau + \varepsilon \quad (4)$$

How do we estimate β? Again, using OLS in a regression of Y against X would produce a biased result (see ATTENUATION DUE TO MEASUREMENT ERROR). This is another example of the endogeneity problem. A similar situation holds when we attempt the comparative calibration of two measurement methods, both subject to measurement errors (see METHOD COMPARISON STUDIES). If we were in the fortuitous position of knowing the VARIANCE of the measurement errors in X, or the reliability of X, we would be able to make appropriate corrections. Another approach is again to find an instrumental variable – a variable that is strongly correlated with X but conditionally independent of Y given X.

Consider an instrumental variable, Z. The instrumental variable (IV) estimator of β in equation (2) or (4) is:

$$\widehat{\beta_{\text{IV}}} = \frac{\sum(Z-\bar{Z})(Y-\bar{Y})}{\sum(Z-\bar{Z})(X-\bar{X})} \quad (5)$$

which is equivalent to the ratio of the estimated covariance of Z and Y to the estimated covariance of Z and X. Typically this estimate is obtained through the use of a two-stage least squares (2SLS or TSLS) algorithm (see Wooldridge, 2003, for further details of the method, including the sampling distribution of the IV estimate). This algorithm is available in most large general-purpose software packages. Note that its validity is not dependent on any distributional assumptions concerning either Z or X. Both could be binary (yes/no) indicators, for example. For linear models, IV estimates can also be obtained with ease using structural equation modelling (see STRUCTURAL EQUATION MODELS and STRUCTURAL EQUATION MODELLING SOFTWARE).

As an example, an early medical application of instrumental variable methods was provided by Permutt and Hebel (1989). They describe a trial in which pregnant women were randomly allocated to receive encouragement to reduce or stop their cigarette smoking during pregnancy (the treatment group), or not (the control group) – indicated by the binary variable, Z. An intermediate outcome variable (X) was the amount of cigarette smoking recorded during pregnancy. The ultimate outcome (Y) was the birth weight of the newborn child. Readers will be familiar with evaluating the effect of RANDOMIZATION on the child's birth weight. However, what about the effect of smoking (X) on birth weight? Smoking is likely to have been reduced in the group subject to encouragement, but also in the control group (but, presumably, to a lesser extent). There are also likely to be hidden confounders (e.g. other health promoting behaviours) that are associated with both the mother's smoking during pregnancy and the child's birth weight. Smoking (X) is an endogenous treatment variable. The problem is solved by noting that randomization (Z) is an obvious candidate for the instrumental variable. If the intervention (i.e. encouragement to reduce smoking) works then randomization should be correlated with smoking during pregnancy. It is also a reasonable to assume that the effect of randomization is completely mediated by its effect on smoking (that there is no direct effect of randomization on outcome (the birth weight of the child).

Randomization (Z), in fact, is increasingly being used as an instrumental variable in the estimation of the effect of treatment received (X) on outcome (Y) in randomized controlled trials subject to nonadherence or noncompliance with the allocated treatment (see ADJUSTMENT FOR NONCOMPLIANCE IN RANDOMIZED CONTROLLED TRIALS). The potential for the use of instrumental variables in epidemiological investigations is illustrated by Greenland (2000) (see also MENDELIAN RANDOMISATION). Health economic applications are reviewed by Newhouse and McClellan (1998).

What about MEASUREMENT ERROR problems? Well, first note that for the example provided by Permutt and Hebel (1989) the IV estimate of the effect of mother's smoking on her child's birth weight is not attenuated by the inevitable measurement error in the number of cigarettes smoked by the mother. The IV estimator effectively copes with the simultaneous problems of confounding and measurement error. What about the problem solely due to measurement error? Here an obvious choice for an instrument is a measurement of the characteristic measured as X using a different procedure. Smoking (X) could be measured by self-report (in a diary, for example) and a suitable instrument (Z) might be a measurement of a biomarker of nicotine consumption (cotinine levels in the blood, for example). The key here is to be able to convince oneself of the conditional independence of Z (biomarker measurement) and outcome, Y (health status), given the fallible indicator of exposure, X (self-reported cigarette smoking). Dunn (2004) provides detailed descriptions of the use of instrumental variable methodology in the evaluation of measurement errors, mainly in the context of linear models, but also in latent class modelling of binary diagnostic test results. Nonlinear models are much more difficult to deal with and are well beyond the scope of this article (but see Stefanski and Buzas, 1995). *GD*

Dunn, G. 2004: *Statistical evaluation of measurement errors*. London: Arnold. **Greenland, S.** 2000: An introduction to instrumental variables for epidemiologists. *International Journal of Epidemiology* 29, 722–9 (Erratum, p.1102). **Newhouse, J. P. and McClellan, M.** 1998: Econometrics in outcomes research: the use of instrumental variables. *Annual Reviews of Public Health* 19, 17–34. **Permutt, T. and Hebel, J. R.** 1989: Simultaneous-equation estimation in a clinical trial of the effect of smoking and birth weight. *Biometrics* 45, 619–22. **Stefanski, L. and Buzas, J. S.** 1995: Instrumental variable estimation in binary regression measurement error models. *Journal of the American Statistical Association* 90, 541–9. **Wooldridge, J. M.** 2003: *Introductory econometrics: a modern approach*, 2nd edition. Mason, Ohio: South-Western.

integrated hazard function See SURVIVAL ANALYSIS – AN OVERVIEW

intention-to-treat (ITT) This is a principle used in the design, analysis and conduct of randomised CLINICAL TRIALS (Heritier, Gebski and Keech, 2003). It asserts that the effect of an intervention policy can best be assessed by comparing participants according to the intention to treat each participant (i.e. the planned intervention), rather than according to the actual intervention received. When the ITT principle is used, all participants allocated to an intervention group should be followed up, assessed and analysed as members of that group irrespective of their compliance to the planned intervention (ICH E9, 1999).

The purpose of RANDOMISATION in a controlled trial is to create groups that are similar, apart from chance variation, in all observed and unobserved characteristics that might affect the outcome. If analysis is not performed on the groups produced by the randomisation process, this key feature will be lost, which may lead to a biased comparison. For example, participants may not receive the allocated treatment due to a poor prognosis. In a trial comparing medical and surgical therapy in stable angina pectoris, a particularly high mortality rate was seen in participants allocated to surgery who did not receive it (see the first table). Participants who died before receiving surgery were either too sick for surgery or

intention-to-treat *Coronary artery bypass surgery in stable angina pectoris trial. Mortality at two years after randomisation by allocated and actual intervention (European Coronary Surgery Study Group, 1979)*

Allocated intervention	*Medical*	*Medical*	*Surgical*	*Surgical*
Actual intervention	Medical	Surgical	Surgical	Medical
Survivors	296	48	353	20
Deaths	27	2	15	6
Mortality	8.4%	4.0%	4.1%	23.1%

died before surgery could be done, and exclusion of such participants from one arm only introduces BIAS. The ITT analysis of these data would compare a mortality rate of 7.8 % (29/373) in those allocated to medical treatment with a rate of 5.3 % (21/394) in those allocated to surgery. If the six deaths that occurred in participants allocated to surgical intervention who died before receiving surgery (identified by 'Actual intervention = Medical' in the table) are not attributed to surgical intervention using an intention-to-treat analysis, surgery would appear to have a falsely low mortality rate.

Since protocol deviations and noncompliance are likely to occur in routine use of an intervention, ITT analysis can provide an estimate of the treatment effect, which reasonably reflects what might happen in clinical practice. It is therefore the most suitable approach for pragmatic trials that aim to measure the overall *effectiveness* of an intervention policy in

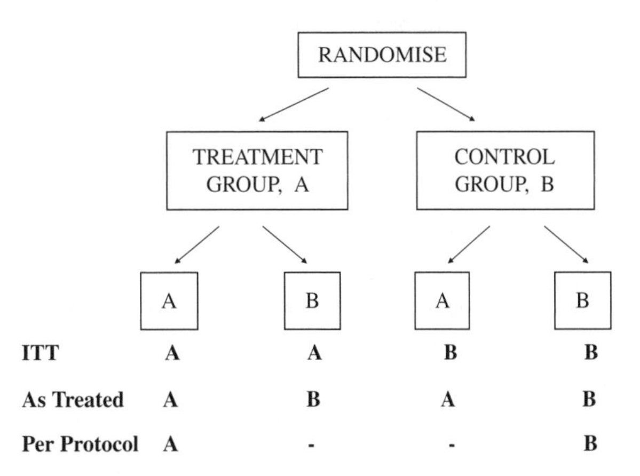

intention to treat *Graphical representation of group membership for how individuals following randomisation are considered for analysis purposes according to the principles of 'ITT', 'as treated' and 'per protocol'. Those, usually relatively few, individuals allocated to one group (A or B) but in actuality in receipt of alternative treatment (B or A, respectively) are handled differently, or if indicated "-" are dropped altogether, in analysis groups.*

routine practice. It is less suitable for explanatory trials, or explanatory analyses of pragmatic trials, which aim to measure the *efficacy* of an intervention under equalised conditions, but even here it may still be preferable to the alternatives.

Alternatives to ITT include PER PROTOCOL analysis, where only participants who comply with the allocated intervention are included, and AS TREATED analysis, where participants are analysed according to the intervention received rather than the randomised allocation (see the figure on page 217). Each of these analyses aims to estimate efficacy, rather than effectiveness as estimated in an ITT analysis. In the trial comparing medical and surgical therapy in stable angina pectoris, the intention to treat analysis gives an estimate of 2.5 % higher mortality with surgery (95 % confidence interval of −1.5 % to +5.5 %). Per protocol and as treated analyses are severely biased by their handling of the six deaths in patients randomised to surgery who were too sick or died too soon to receive surgery, giving statistically significant estimated increases in mortality with surgery of 4.3 % and 5.4 % respectively (see the second table). Ways to estimate efficacy while avoiding this bias are discussed in ADJUSTMENT FOR NONCOMPLIANCE IN RCTs.

If some participants lack outcome data, then a full ITT approach is not possible. *Last observation carried forward* is often used as a way to include all randomised participants in the analysis, but this introduces further assumptions that are rarely plausible. Instead, analysis should be based on plausible assumptions (see MISSING DATA and DROPOUTS), and sensitivity analysis should examine the potential impact of departures from these assumptions (Hollis and Campbell, 1999). All participants with post-randomisation data should be included in the analysis, even if they lack observations of the particular outcome variable of interest. This can be achieved by using MULTIPLE IMPUTATION or RANDOM EFFECTS MODELLING.

intention-to-treat *Different methods of analysis illustrated using mortality at two years after randomisation in the coronary artery bypass surgery in stable angina pectoris trial (European Coronary Surgery Study Group, 1979)*

	Medical % (n/N)	*Surgical (n/N)*	*Medical vs Surgical difference (95% CI)*
Intention-to-treat analysis	7.8% (29/373)	5.3% (21/394)	2.5% (−1.5%, 5.5%)
Per-protocol analysis	8.4% (27/323)	4.1% (15/368)	4.3% (0.7%, 8.2%)
As treated analysis	9.5% (33/349)	4.1% (17/418)	5.4% (1.9%, 9.3%)

It is often argued that ITT provides a conservative estimate of treatment effectiveness, which is a smaller effect than the true potential effectiveness of an intervention, since the estimated treatment effect is likely to be reduced by the inclusion of protocol deviations and noncompliance. This may be generally true for comparisons with PLACEBO, because any switching between groups will tend to dilute the estimated treatment effect. However, in comparisons between active treatments or when an effective rescue medication is available, an ITT analysis may not be conservative. For example, two equally good treatments will appear different on ITT analysis if clinicians are more likely to supplement one of the treatments with a more powerful agent. Particular care should be taken when using the ITT approach for adverse effects or safety data and for noninferiority or equivalence trials. In these situations the generally conservative answers provided by ITT may lead to inappropriate conclusions. Other analyses such as per protocol analysis are commonly carried out in these situations, although it may be preferable to avoid selection bias by using randomisation-based methods (see ADJUSTMENT FOR NONCOMPLIANCE IN RCTs).

Non-ITT analyses that exclude some randomised individuals are sometimes justified, provided that these exclusions are not associated with treatment allocation or outcome (Fergusson *et al.*, 2002). Ineligible participants who are randomised in error, or whose eligibility cannot be established before randomisation, could be excluded provided that the judgement of eligibility is based on information established before randomisation, and not influenced by the allocated intervention or outcome. Such exclusions should not be made if clinical practice requires treatment to be started before eligibility can be determined, since the most clinically relevant comparison is usually between all those randomised to treatment and all those randomised to control. Failure to start the allocated treatment in a double-blind trial is also sometimes a justified basis for exclusion. Whenever randomised participants are excluded from the analysis, it should be demonstrated that steps have been taken to avoid bias, such as the use of independent blinded assessment of eligibility. Even so, such analyses should not be described as ITT analyses. Finally, every effort should be made to avoid post-randomisation exclusions through appropriate design and execution of trials. *SH/IW*

[See also AVAILABLE CASE ANALSIS, COMPLETE CASE ANALYSIS]

European Coronary Surgery Study Group 1979: Coronary-artery bypass surgery in stable angina pectoris: survival at two years. *Lancet* i, 889–93. **Fergusson, D., Aaron, S., Guyatt, G. and Hebert, P.** 2002: Post-randomisation exclusions: the intention to treat principle and excluding patients from analysis. *British Medical Journal*, 325, 652–4. **Heritier, S. R., Gebski, V. J. and Keech, A. C.** 2003: Inclusion of patients in clinical trial analysis: the intention-to-treat principle. *Medical Journal of Australia* 179, 438–40. **Hollis, S. and**

Campbell, F. 1999: What is meant by 'intention to treat' analysis? *British Medical Journal* 319, 67–4. **International Conference on Harmonisation E9 Expert Working Group (ICH E9)** 1999: ICH harmonised tripartite guideline. Statistical principles for clinical trials. *Statistics in Medicine* 18, 15, 1905–42.

interim analysis This is performed at regular intervals for monitoring data and safety in clinical trials. An interim analysis refers to any analysis performed during the course of a trial and is often intended to compare intervention effects with respect to efficacy and safety prior to the formal completion of a trial. Because the number, methods and consequences of these comparisons affect the interpretation of the trial, all interim analyses should be carefully planned in advance and described in the protocol explicitly. When an interim analysis is planned with the intention of deciding whether or not to terminate a trial early, this is usually accomplished by one of three general methods known as group sequential methods, triangular tests and stochastic curtailment procedures.

The goal of such an interim analysis is to stop the trial early if the superiority of an intervention under study is clearly established, if the demonstration of a relevant difference in intervention effects becomes unlikely or if unacceptable adverse effects are apparent. Also, as a result of interim analyses, trial interventions may be modified or an experimental design, such as the enrolment inclusion and exclusion criteria or sample size requirement, changed. An ethical obligation to the study participants and even beyond the study demands that results be monitored during the study to protect study participants. If one intervention is substantially superior to the other, if there are unexpected adverse effects on either of the interventions or if the study is unlikely to give definitive answers to the study questions, continuing RANDOMISATION means that participants can be assigned to and subsequently treated with an inferior intervention or put to an unnecessary and unjustifiable experiment. The issues of early stopping due to unexpected adverse effects are less statistical in nature, unless safety is the primary outcome of interest to the investigators.

Suppose the response to intervention is normally distributed with means μ_A and μ_B for intervention arms A and B and known VARIANCE σ^2. We want to test the null hypothesis:

$$H_0 : \mu_A = \mu_B$$

against the alternative hypothesis H_1: $\mu_A \neq \mu_B$ or, equivalently, H_0: $\delta_\mu = 0$ against H_1: $\delta\mu \neq 0$, where $\delta_\mu = \mu_A - \mu_B$. Let $\bar{X}_A$ and $\bar{X}_B$ be the sample means respectively for interventions A and B and let n denote the number of participants per intervention per analysis. In a fixed sample study, one may use the test statistic:

$$Z = \frac{\bar{X}_A - \bar{X}_B}{\sqrt{(2\sigma^2/n)}}$$

to test the null hypothesis H_0. For a significance level α, one would reject H_0 if $|Z| \geq z_{1-\alpha/2}$, where $z_{1-\alpha/2}$ is the $1 - \alpha/2$ quantile of the standard NORMAL DISTRIBUTION.

GROUP SEQUENTIAL METHODS call for monitoring of the accumulating data periodically after groups of observations. One simple-minded approach is to reject the null hypothesis whenever the P-value is less than 0.05, say. The problem with this approach is that multiple looks at the 0.05 level lead to an overall level of significance greater than 0.05. More specifically, the actual TYPE I ERROR probability becomes 0.083 with two looks, 0.142 with five looks and becomes closer and closer to 1 with more and more looks. This phenomenon was aptly described as 'sampling to reach a foregone conclusion' by Anscombe (1954).

Suppose we plan to conduct interim analyses of the accumulating data up to K times after a pre-specified number of participants n on each intervention. The difference in intervention effects is measured at the kth interim analysis by:

$$\bar{X}_{Ak} - \bar{X}_{Bk} \sim N(\delta_\mu, 2\sigma^2/n)$$

where $\bar{X}_{Ak}$ and $\bar{X}_{Bk}$ are the sample means of n observations accumulated between the $(k-1)$th and the kth interim analyses on interventions A and B respectively. It could also be summarised by the standardised difference:

$$Y_k = \frac{\bar{X}_{Ak} - \bar{X}_{Bk}}{\sqrt{(2\sigma^2/n)}} \sim N(\delta^*, 1)$$

where $\delta^* = \delta_\mu/\sqrt{(2\sigma^2/n)}$. For the kth interim analysis, we consider the partial sum of independently and identically distributed normal random variables $Y_1, \ldots, Y_k$:

$$S_k = \sum_{j=1}^{k} Y_j \sim N(\delta^* k, k)$$

or equivalently the standardised test statistic:

$$Z_k = S_k/\sqrt{k} \sim N(\delta^* \sqrt{k}, 1)$$

and decide to reject H_0 or to continue to the next group, up to a maximum of K interim analyses.

The objective of a group sequential design is to derive a group sequential test that has desired operating characteristics, i.e. pre-specified Type I and Type II error probabilities. Thus a group sequential design for a trial requires choosing group sequential critical values, $c_1, \ldots, c_K$, such that one rejects H_0 after the kth interim analysis if the statistic $|Z_k|$ exceeds c_k for the first time. We do not reject the null hypotheses if $|Z_1| < c_1, \ldots, |Z_K| < c_K$. There are many different designs, i.e. many different choices, for the group sequential critical values. However, there are a few with known statistical justifications.

The group sequential test by Pocock (1997) uses the same critical value at each interim analysis. Specifically, the Pocock group sequential test rejects H_0 the first time when:

$$|Z_k| \geq c_k \equiv c_P \text{ or equivalently } |S_k| \geq b_k = c_P\sqrt{k}$$

Hence one has only to determine c_P as a function of the overall Type I error probability α and the maximum number of interim analyses K.

The group sequential test by O'Brien and Fleming (1979) uses larger critical values at earlier interim analyses so that it is difficult to reject H_0 early in the study and relaxes the criteria until, at the end, the critical value is close to the fixed sample critical value. Specifically, the O'Brien–Fleming group sequential test rejects H_0 the first time when:

$$|Z_k| \geq c_k = c_O\sqrt{(K/k)} \text{ or equivalently } |S_k| \geq b_k \equiv c_O\sqrt{K}$$

Again, one has only to determine c_O as a function of α and K.

The standard group sequential method has some limitations because of the requirements in the pre-specified maximum number of interim analyses and the equal increment in statistical information between interim analyses. There are, however, flexible group sequential procedures that make these requirements unnecessary based on the notion of an error spending function as proposed by Lan and DeMets (1983). Especially for trials with censored survival or repeated measures data, it is necessary to be flexible in the group sequential test since typical interim analyses take place after unequal increments in the information fraction. Also recent developments in group sequential methods allow early stopping in order to reject the alternative hypothesis just as the triangular tests do with the required flexibility, as proposed by Chang, Hwang and Shih (1998) and Pampallona, Tsiatis and Kim (2001).

Wald's sequential probability ratio test (1947) for H_0: $\theta = \theta_0$ versus H_1: $\theta = \theta_1$ is optimal in the sense that, among all tests with Type I and II error probabilities α and β, it minimises the expected sample sizes $E(N|\theta_0)$ and $E(N|\theta_1)$, where N is a random variable for the sequential sample size. However, $E(N|\theta)$ can be worse than the corresponding fixed sample size at or near $\theta = (\theta_0 + \theta_1)/2$. Moreover, the sample size is unbounded and, in particular, $\Pr(N \geq n) > 0$ for any given n. Thus the motivation in Anderson (1960) was to find a sequential test that would minimise $E(N|\theta)$ at $\theta = (\theta_0 + \theta_1)/2$, leading to the so-called TRIANGULAR TEST. Triangular tests have been further developed in Whitehead (1997), as described below.

In a general problem, the efficient score Z for the parameter of interest θ, which typically measures the difference, has the following asymptotic distribution according to LIKELIHOOD theory:

$$Z \overset{a}{\sim} N(\theta V, V)$$

where V denotes Fisher information. For a fixed sample test, the critical value c and Fisher information required for the study are determined to have Type I and II error probabilities α and β respectively, such that:

$$\Pr(|Z| \geq c;\ 0) = \alpha \quad \text{and} \quad \Pr(Z \geq c;\ \theta_1) = 1-\beta$$

where θ_1 is the hypothesised difference of interest. These two requirements lead to:

$$V = \left(\frac{z_{1-\alpha/2} + z_{1-\beta}}{\theta_1}\right)^2 \quad \text{and} \quad c = \frac{\left(z_{1-\alpha/2} + z_{1-\beta}\right)z_{1-\alpha/2}}{\theta_1}$$

According to Whitehead (1997), a triangular test is defined by the upper and lower boundaries of the form:

$$Z = a + cV \quad \text{and} \quad Z = -a + 3cV$$

respectively, with the apex of the triangle at $Z = 2a$ and $V = a/c$. In the special case where $\alpha/2 = \beta$:

$$a = -2\log\alpha/\theta_1 \quad \text{and} \quad c = \theta_1/4$$

A solution is possible for the general case as well when $\alpha/2 \ll \beta$, which is typically the case.

Since interim analyses are performed only a limited number of times, some adjustments need to be made in order to maintain the operating characteristics. This is accomplished by the so-called 'Christmas tree adjustment', which is described later. Suppose that (Z^*, V^*) denotes the value of sequential statistics at the time an upper boundary is crossed. The overshoot R is the vertical distance between the final point of the sample path and the continuous boundary defined as:

$$R = Z^* - (a + cV^*)$$

In order to account for the discreteness of the interim analyses, the continuous stopping criterion:

$$Z \geq a + cV$$

is replaced by:

$$Z \geq a + cV - A$$

where:

$$A = E(R;\theta)$$

In developing triangular tests, two different power requirements are specified. Traditional accounts of testing the null hypothesis H_0: $\theta = 0$ allow two outcomes in which 'H_0 is accepted' or 'H_0 is rejected'. However, three outcomes are possible in practice. The power function $C(\theta)$ is the probability of rejecting H_0 under the parameter value θ defined as:

$$C(\theta) = C^+(\theta) + C^-(\theta)$$

where $C^+(\theta)$ denotes the probability that H_0 is rejected and it is concluded that the experimental intervention is superior and $C^-(\theta)$ denotes the probability that H_0 is rejected and it is concluded that the experimental intervention is inferior. Obviously, for TWO-SIDED TESTS:

$$C(0) = \alpha \text{ with } C^{+}(0) = C^{-}(0) = \alpha/2$$

Two specific power requirements are either $C^{+}(\theta_1) = 1 - \beta$ or $C^{+}(\theta_1) = 1 - \beta = C^{-}(-\theta_1)$. These give rise to asymmetric or symmetric triangular tests for a two-sided test of the null hypothesis.

In curtailment sampling, one is interested in assessing the likelihood of a trend reversal. There are two possible ways: deterministic and stochastic. This notion has been found to be useful in consideration of early stopping for futility. In contrast, early acceptance of the null hypothesis is possible based on group sequential methods and triangular tests discussed earlier. An example of deterministic curtailment is curtailed sampling in sampling inspection in which trend reversal is impossible. Let S denote a test statistic that measures the difference in intervention effects and let the sample space Ω of S consist of disjoint regions, A and R, such that:

$$\Pr(S \in R|H_0) = \alpha \text{ and } \Pr(S \in A|H_1) = \beta$$

Let t denote the time of an interim analysis and let $D(t)$ denote the accumulated data up to time t. A deterministic curtailment test rejects or accepts the null hypothesis H_0 if:

$$\Pr(S \in R|D(t)) = 1 \text{ or } \Pr(S \in A|D(t)) = 1$$

respectively, regardless of whether H_0 or H_1 is true. Note that this procedure does not affect the Type I and II error probabilities.

As an example, consider testing the fairness of a coin, H_0: $\pi = 0.5$ versus H_1: $\pi \neq 0.5$. After tossing a coin 400 times, one will consider the total number of heads S and reject H_0 if $|Z| > 1.96$ at a significance level 0.05 where:

$$Z = \frac{S - 200}{\sqrt{(400 \times 0.5 \times 0.5)}}$$

or equivalently if $|S - 200| \geq 20$. After 350 tosses, we will reject H_0 for sure with 220 heads. With 210 heads, however, it depends on the future outcomes.

Consider a fixed sample test of H_0: $\theta = 0$ at a significance level α with power $1 - \beta$ to detect the difference $\theta = \theta_1$. The conditional probability of rejection of H_0, i.e. conditional power, at θ is defined as:

$$P_C(\theta) = \Pr(S \in R|D(t); \theta)$$

For some $\gamma_0, \gamma_1 > 1/2$, a stochastic curtailment test rejects the null hypothesis if:

$$P_C(\theta_1) \approx 1 \text{ and } P_C(0) > \gamma_0$$

or accepts the null hypothesis (rejects the alternative hypothesis) if:

$$P_C(0) \approx 0 \text{ and } P_C(\theta_1) < 1 - \gamma_1$$

According to Lan, Simon and Halperin (1982), the Type I and II error probabilities are inflated but remain bounded from above by:

$$\alpha' = \alpha/\gamma_0 \text{ and } \beta' = \beta/\gamma_1$$

Generally stochastic curtailment is very conservative and if $\gamma_0 = 1 = \gamma_1$, it becomes deterministic curtailment.

A formal significance test is only one factor in the complex decision process of whether to continue, modify or stop a trial. Interim analyses based on group sequential methods, triangular tests or stochastic curtailment procedures provide objective guidelines to the DATA AND SAFETY MONITORING BOARDS.

The choice for the method of interim analyses should depend on the desired operating characteristics of the study in terms of the early stopping property, the maximum sample size requirement and the expected sample size. For example, if the study continues through all K analyses, the group sequential design will accrue more participants than the fixed sample design, which is likely to occur if H_0 is true. However, if the study stops at an earlier interim analysis, the group sequential design will accrue fewer participants on average than the fixed sample design, which is likely to occur if H_1 is true.

A randomised Phase III trial should never be terminated in the early stages of recruitment merely because it is failing to reach the anticipated 'minimal' benefit envisaged at the design stage of the study. This is because early termination of a study in these circumstances will leave the associated CONFIDENCE INTERVAL unacceptably wide, thereby indicating the possibility of a plausible, and maybe worthwhile, advantage to one therapy even when there is no true difference in intervention effects. In such circumstances the level of uncertainty remains unacceptably high. *KK*

[See also DATA AND SAFETY MONITORING BOARDS]

Anderson, T. W. (1960). A modification of the sequential probability ratio test to reduce the sample size. *Annals of Mathematical Statistics* 31, 165–97. **Anscombe, F. J.** 1954: Fixed-sample-size analysis of sequential observations. *Biometrics* 10, 89–100. **Chang, M. N., Hwang, I. K. and Shih, W. J.** 1998: Group sequential designs using both Type I and Type II error probability spending functions. *Communications in Statistics, Part A – Theory and Methods* 27, 1323–39. **Lan, K. K. G. and DeMets, D. L.** 1983: Discrete sequential boundaries for clinical trials. *Biometrika* 70, 659–63. **Lan, K. K. G., Simon, R. and Halperin, M.** 1982: Stochastically curtailed testing in long-term clinical trials. *Sequential Analysis* 1, 207–19. **O'Brien, P. C. and Fleming, T. R.** 1979: A multiple testing procedure for clinical trials. *Biometrics* 35, 549–56. **Pampallona, S., Tsiatis, A. A. and Kim, K.** 2001: Spending functions for Type I and Type II error probabilities of group sequential trials. *Drug Information Journal* 72, 247–60. **Pocock, S. J.** 1977: Group sequential methods in the design and analysis of clinical trials. *Biometrika*

64, 191–9. **Wald, A.** 1947: *Sequential analysis*. New York: John Wiley & Sons, Inc. **Whitehead, J.** 1997: *The design and analysis of sequential clinical trials*, 2nd revised edition. Chichester: John Wiley & Sons, Ltd.

internal pilot study See PILOT STUDIES, SAMPLE SIZE DETERMINATION IN CLINICAL TRIALS

interquartile range This range is a MEASURE OF SPREAD defined as the interval between the values that are located one-quarter and three-quarters of the way through the sample when the observations are ordered. Thus, it encloses the middle 50 % of the data points.

For example, suppose the weights in kilograms of 11 elderly men from a community sample attending a clinic were:

Interquartile range

58 60 **61** 63 65 **66** 70 72 **77** 85 95

↑

Median

Then, the interquartile range is the interval between the third and ninth values, i.e. 61 kg to 77 kg, a difference of 16 kg.

The interquartile range is most informative if the upper and lower values are both quoted, rather than simply the interval between them. The lower value is known as the lower quartile or 25th percentile and the upper value as the upper quartile or 75th percentile.

If the number of observations + 1 is divisible by 4 then the interquartile range is simple to calculate. If this is not the case then the values for the interquartile range need to be interpolated. In general, the position of the lower quartile is calculated by multiplying the sample size plus one by 0.25, and by 0.75 in the case of the upper quartile. Therefore, if another man attends the clinic with a weight of 100 kg then the lower quartile is now at position $(12+1)/4 = 3^1/_4$:

58 60 61 63 65 66 70 72 77 85 95 100

↑ ↑

Lower quartile Upper quartile

Thus the lower quartile lies a quarter of the way between 61 and 63, interpolated as 61.5 kg. Similarly, the upper quartile lies three-quarters of the way between 77 and 85, interpolated as 83 kg.

The interquartile range is typically used as a measure of spread around the median. Like the median, it is useful when the data are not symmetrically distributed because it is not unduly affected by the presence of SKEWNESS or OUTLIERS. *SRC*

intraclass correlation coefficient See INTRACLUSTER CORRELATION COEFFICIENT

intracluster correlation coefficient (ICC) This is a measure that quantifies the extent of similarity among individual observations within clusters. For example, when a study collects data on patients from a number of different clinics, the intracluster correlation coefficient (ICC) represents the degree to which patients attending the same clinic are more similar than the patients attending different clinics. Also known as the intraclass correlation coefficient, the ICC takes values between 0 and 1, where the value 0 corresponds to the situation where individuals from the same cluster are no more alike than individuals from different clusters and higher values indicate greater similarity within clusters.

The ICC has been used extensively in several application areas. In health services research, it is used to measure the extent of similarity of patients within administrative units such as hospitals or geographical units such as towns. In family studies, it measures the degree of resemblance among members of the same family. In psychological research, it is used when examining reliability (see MEASUREMENT PRECISION AND RELIABILITY), where the same measurements are taken on subjects by different assessors. When individuals are sampled within clusters such as hospitals or families, the ICC representing within-cluster similarity is defined as the proportion of the variation between individuals that is explained by the variation between clusters.

Formally, this definition assumes a simple random effects model for the ENDPOINT of interest, which includes random cluster effects with VARIANCE σ_c^2 and individual residual effects with variance σ^2, and the ICC is defined as $\sigma_c^2/(\sigma_c^2+\sigma^2)$. The most common approach for estimating the ICC is to obtain estimates for σ_c^2 and σ^2 by fitting the RANDOM EFFECTS MODEL (Donner and Wells, 1986) to the data and to substitute these into the formula. CONFIDENCE INTERVALS for reporting alongside the ICC estimate are also obtained using the assumptions of the random effects model. In most settings, negative ICC values are regarded as implausible, so negative values obtained for ICC estimates are set to 0 and lower limits of confidence intervals are truncated at 0. When considering the ICC for a binary outcome, e.g. the presence or absence of a disease in family members, an ICC estimate can be obtained as described above, but the methods for constructing confidence intervals are based on different assumptions.

The simple model outlined here is not appropriate for all types of design. When measuring reliability, the definition of the ICC differs according to whether the focus is on ‘consistency’ or ‘absolute agreement’, as described by McGraw and Wong (1996). Some study designs require more complex models; e.g. when repeated measurements are available on patients within clusters or when wishing to estimate multiple ICC values simultaneously (Donner, 1986). *RT*

Donner, A. 1986: A review of inference procedures for the intraclass correlation coefficient in the one-way random effects model.

International Statistical Review 54, 67–82. **Donner, A. and Wells, G.** 1986: A comparison of confidence interval methods for the intraclass correlation coefficient. *Biometrics* 42, 401–12. **McGraw, K. O. and Wong, S. P.** 1996: Forming inferences about some intraclass correlation coefficients. *Psychological Methods* 1, 30–46.

inverse probability weighting (IPW) This refers to a general method of adjusting M-estimators (Everitt and Skrondal, 2010) for confounding or selection bias (e.g. due to CENSORED OBSERVATIONS, MISSING DATA or sample selection) when the unverifiable assumption of no unmeasured confounders, noninformative censoring or missingness at random is, respectively, met. The underlying idea is to filter spurious associations away from the data by weighting each subject's data inversely to the magnitude of those associations. In the process, one redresses imbalances so that issues of confounding or selection bias may subsequently be ignored in the analysis. We will illustrate IPW with two examples.

Consider first a setting where the relation between some exposure A and some outcome Y is of interest, but distorted by measured confounders L. Then weighting each subject's data by the reciprocal of the conditional density $f(A|L)$ of exposure A given confounders L, eliminates the association between A and L, and thereby eliminates confounding by L. This implies that standard measures of association between A and Y, when applied to the inversely weighted data, are no longer distorted by confounding due to L. In the special case where A is dichotomous, taking values 0 for the 'unexposed' and 1 for the 'exposed', the adjusted association between A and Y can thus be calculated as:

$$\sum_{i:A_i=1} w_{i1} Y_i \Big/ \sum_{i:A_i=1} w_{i1} - \sum_{i:A_i=0} w_{i0} Y_i \Big/ \sum_{i:A_i=0} w_{i0} \tag{6}$$

where the weights $w_{i1} = 1/\Pr(A_i = 1|L_i)$ and $w_{i0} = 1/\Pr(A_i = 0|L_i)$ can be estimated based on the fitted values of a LOGISTIC REGRESSION model for A, given L. More generally, adjustment for measured confounding can be accomplished via off-the-shelf software packages (see STATISTICAL PACKAGES) by fitting a regression model for the outcome, involving the exposure of interest only (e.g. $E(Y|A) = \alpha + \beta A$), while assigning each subject's data the given weight (i.e. w_{i1} for those with $A_i = 1$ and w_{i0} for those with $A_i = 0$).

The impact of inverse probability weighting is to standardize (see DEMOGRAPHY) the expected outcome in the exposed and unexposed respectively, with the total group as the reference population (Sato and Matsuyama, 2003). It follows that, when L includes all confounders of the relation between A and Y, the IPW estimate (6) can be interpreted as the change in average outcome that would be observed if the total group were exposed versus unexposed. As such, IPW provides an alternative to direct adjustment methods for measured confounders, where one involves models for the regression of exposure rather than outcome on the measured confounders.

Consider next a setting where the interest lies in a linear regression (see MULTIPLE LINEAR REGRESSION) of a completely observed outcome Y on to an incompletely observed covariate X. When missingness in X is influenced by the observed outcome and possibly also by extraneous, measured covariates Z, but has no residual association with the missing X, then a regression analysis of Y on X in the complete cases (i.e. those with complete data in X) may give biased results. When the association between missingness and its predictors is filtered away through inverse probability weighting of the complete cases, the missingness becomes completely at random so that an analysis of the reweighted complete cases becomes valid. This can be accomplished via off-the-shelf software packages by fitting a regression model for the outcome to the complete cases, while assigning each subject's data the weight $1/\Pr(R = 1|Y, Z)$, where R is a missingness indicator that assigns 1 to subjects with complete data and 0 to subjects with incomplete data in X. The missingness probability appearing in the weights can, for instance, be estimated via logistic regression analysis.

This idea that BIAS due to missing data can be corrected by weighting each of these subjects' observations by the inverse of the probability of observing complete data dates back to at least Horvitz and Thompson (1952). For many years, the IPW method gained little acceptance because of its imprecision relative to more popular missing data methods, such as MULTIPLE IMPUTATION. This has changed drastically over the past decade, since the seminal work of Robins, Rotnitzky and Zhao (1994), who demonstrated how the precision of IPW estimators could be greatly improved to the point where they become competitive with imputation estimators. The recent success of IPW is largely due to its ability to enable adjustment for time-varying confounding where standard regression adjustment fails (see MARGINAL STRUCTURAL MODELS). This ability results from its capacity to filter associations away from the data. IPW methods have the further advantage that they are generically applicable in a wide variety of settings, that they enable relatively simple sensitivity analyses to investigate the impact of violation of unverifiable (missing data) assumptions (Scharfstein, Rotnitzky and Robins, 1999) and that they are less prone to extrapolation than more standard adjustment methods (e.g. regression adjustment for confounders, multiple imputation) (Tan, 2008). The main limitation of IPW estimators is that they can be unstable and imprecise when some subjects receive large weights. This may happen when there is strong confounding, strongly informative censoring or missingness, or when a continuous exposure requires inverse weighting by a density. In that case, one must consider heuristic weight truncation procedures (Cole and Hernan, 2008) or recourse to more efficient (doubly robust) IPW estimation strategies (Robins *et al.*, 2008; Goetgeluk, Vansteelandt and Goetghebeur, 2008). The latter are typically also rewarding in terms of robustness against model misspecification, but are computationally more challenging.

Finally, the IPW method enjoys the unexpected property that plugging in estimated rather than known weights tends to increase efficiency. However, when standard statistical software packages (e.g. SAS, STATA, R; see statistical packages) are used for IPW, caution is needed in interpreting the standard errors provided by the software, because these ignore the imprecision of the estimated weights. By using routines that report so-called sandwich estimators (Everitt and Skrondal, 2010), conservative standard errors are obtained. *SV/EG*

Cole, S. R. and Hernan, M. A. 2008: Constructing inverse probability weights for marginal structural models. *American Journal of Epidemiology* 168, 656–64. **Everitt, B. S. and Skrondal, A.** 2010: *Cambridge dictionary of statistics,* 4th edition. Cambridge University Press, Cambridge. **Goetgeluk, S., Vansteelandt, S. and Goetghebeur, E.** 2008: Estimation of controlled direct effects. *Journal of the Royal Statistical Society, Series B* 70, 1049–66. **Horvitz, D. G. and Thompson, D. J.** 1952: A generalization of sampling without replacement from a finite universe. *Journal of the American Statistical Association* 47, 663–85. **Robins, J. M., Rotnitzky, A. and Zhao, L.-P.** 1994: Estimation of regression coefficients when some regressors are not always observed. *Journal of the American Statistical Association* 89, 846–66. **Robins, J. M., Sued, M., Lei-Gomez, Q. and Rotnitzky, A.** 2008: Performance of double-robust estimators when 'inverse probability' weights are highly variable. *Statistical Science* 22, 544–59. **Sato, T. and Matsuyama, Y.** 2003: Marginal structural models as a tool for standardization. *Epidemiology* 14, 680–6. **Scharfstein, D. O., Rotnitzky, A., and Robins, J. M.** 1999: Adjusting for non-ignorable drop-out using semiparametric non-response models. *Journal of the American Statistical Association* 94, 1096–120. **Tan, Z.** 2008: Understanding OR, PS, and DR. *Statistical Science* 22, 560–8.

ITT Abbreviation for INTENTION-TO-TREAT

J

Jaccard coefficient See CLUSTER ANALYSIS IN MEDICINE

jackknife method The jackknife was originally proposed as a method for estimating biases (Quenouille, 1949); soon afterwards it was applied to estimating standard errors. Subsequently it has been applied more widely. In particular, the 'jackknife after bootstrap', discussed later, is a useful tool for understanding the influence of individual observations on BOOTSTRAP analyses.

Here, we motivate the jackknife using a hypothetical example and illustrate it using a real dataset. We then briefly discuss its relationship to the bootstrap and 'jackknife after bootstrap' analyses.

Suppose we wish to estimate the average adult height in a population, which we denote by θ. To do this, we take a sample of seven adults from the population and calculate the average height in the sample. Let the first row of the figure represent the population, from which seven adults are sampled. Note that the numbers 1, ..., 7 index the sampled adults; they are not the actual heights. This sample is represented in the box in the second row. The average height of the seven adults in this sample, denoted by $\hat{\theta}$, is our estimate of θ. Suppose we now wish to calculate an estimate of the variability of $\hat{\theta}$ about θ. Hypothetically, we could do this by drawing a number of further samples, also of size 7, from the population and calculating the average adult heights in each of these.

Of course, this is impractical. However, it suggests an alternative approach. Suppose we draw a number of subsamples from the data and calculate the average height in each of these subsamples. The variation of the average height in these subsamples about the value in the observed data might be a reasonable estimate of the variation of $\hat{\theta}$ about the true adult height, θ.

One method of constructing these subsamples is the jackknife. The jackknife datasets are constructed by omitting one observation in turn from the observed dataset. Thus if the dataset has seven observations, there are seven jackknife datasets. These are shown in the third row in the figure.

Having obtained the jackknife datasets, we simply calculate the average height of the six adults in each jackknife dataset. By convention, these are denoted $\hat{\theta}_{(1)}, \ldots, \hat{\theta}_{(7)}$, where the subscript number indicates the *observation that is omitted*. Recall that we are interested in estimating the variability of $\hat{\theta}$. Let the average of the jackknife estimates be $\hat{\theta}_{(.)} = (\hat{\theta}_{(1)} + \hat{\theta}_{(2)} + \cdots + \hat{\theta}_{(7)})/7$ and let n equal the sample size (7 in this case). The jackknife estimate of VARIANCE is:

$$\hat{\sigma}^2_{\text{jack}} = \frac{n-1}{n}\sum_{i=1}^{n}\left(\hat{\theta}_{(i)} - \hat{\theta}_{(.)}\right)^2 \tag{1}$$

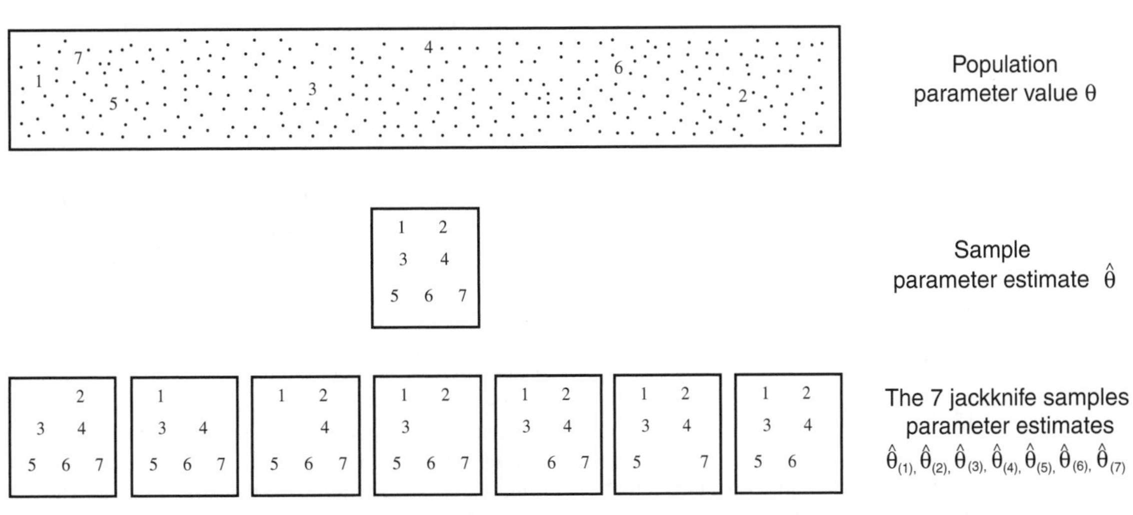

jackknife method *Schematic illustration of jackknife*

Encyclopaedic Companion to Medical Statistics: Second Edition Edited by Brian S. Everitt and Christopher R. Palmer

(Note, in passing, that by convention, the jackknife estimate of variance differs from the bootstrap estimate by a factor of $(n-1)/n$, but this is negligible unless n is small.) Thus the jackknife is motivated by an analogous version of the bootstrap principle, in which jackknife datasets replace bootstrap datasets. However, in comparison with bootstrap datasets, jackknife datasets are much less variable. Therefore, extra multiplying factors, like $n-1$ in the numerator of equation (1) are needed to make the jackknife work.

As an example, consider the data in the first table. We will use the jackknife to estimate the variance of the average change in the carbon monoxide transfer factor, which is $(33+2+24+27+4+1-6)/7 = 12.14$.

jackknife method *Data on the carbon monoxide transfer factor for seven smokers with chickenpox, measured on admission to hospital and after a stay of one week (Davison and Hinkley, 1997, p.67)*

Patient	*Entry*	*Week*	*Change = (Week−Entry)*
1	40	73	33
2	50	52	2
3	56	80	24
4	58	85	27
5	60	64	4
6	62	63	1
7	66	60	−6

The seven jackknife samples, and their corresponding means, are shown in the second table. As the statistic we are using is the MEAN, it turns out that $\hat{\theta}_{(.)} = (\hat{\theta}_{(1)} + \hat{\theta}_{(2)} + \cdots + \hat{\theta}_{(7)})/7 = \hat{\theta} = 12.14$. From equation (1), the jackknife estimate of variance is:

$$\hat{\sigma}^2_{\text{jack}} = \frac{6}{7}\Big\{(8.67-12.14)^2 + (13.83-12.14)^2 + (10.17-12.14)^2 + (9.67-12.14)^2 + (13.50-12.14)^2 + (14.00-12.14)^2 + (15.17-12.14)^2\Big\} = 5.81^2$$

Note that the jackknife estimate of variance agrees exactly with the result obtained using the usual formula for the variance of a mean. This is a feature of the statistic used in this example (the mean) and will not occur generally.

jackknife method *Jackknife samples for the 'change' data from the first table*

								Statistic (mean)
Data from first table	33	2	24	27	4	1	−6	$\hat{\theta} = 12.14$
1st jackknife sample	–	2	24	27	4	1	−6	$\hat{\theta}_{(1)} = 8.67$
2nd jackknife sample	33	–	24	27	4	1	−6	$\hat{\theta}_{(2)} = 13.83$
3rd jackknife sample	33	2	–	27	4	1	−6	$\hat{\theta}_{(3)} = 10.17$
4th jackknife sample	33	2	24	–	4	1	−6	$\hat{\theta}_{(4)} = 9.67$
5th jackknife sample	33	2	24	27	–	1	−6	$\hat{\theta}_{(5)} = 13.50$
6th jackknife sample	33	2	24	27	4	–	−6	$\hat{\theta}_{(6)} = 14.00$
7th jackknife sample	33	2	24	27	4	1	–	$\hat{\theta}_{(7)} = 15.17$

Apparent similarities of the jackknife and bootstrap are indicative of a deeper relationship. It turns out that, in situations where the jackknife works, it can be viewed as an approximation to the bootstrap (Efron and Tibshirani, 1993, p. 146). The attraction of the jackknife is that it is computationally easier (as there are only a finite number of jackknife samples). The disadvantage is that it only works in more restricted situations. In particular, the jackknife only works for statistics whose value changes smoothly as the data changes (Efron and Tibshirani, 1993, p. 148).

The mean is an example of a smooth statistic. The MEDIAN is not, however, because as the data values change it does not change smoothly. For example, the median of the change variable in the first table is 4. Suppose we now increase the second observation from its observed value, 2. While the new value for the second observation is less than 4, the median remains unchanged. As soon as it is greater than 4, the median changes. This lack of smoothness makes the jackknife fail for the median. By contrast, the bootstrap works for the median.

A useful application of the jackknife is known as the 'jackknife after bootstrap'. Following a bootstrap analysis, the bootstrap datasets are divided into groups: those that do not contain the first observation, those that do not contain the second and so on. In other words, we form jackknife groups from (i.e. after) generating the bootstrap datasets. The analysis of the bootstrap data can then be performed on each of these jackknife groups in turn. Marked differences in the results between the jackknife groups indicate that a particular observation, or group of observations, is strongly affecting the conclusions. For example, if jackknife group 1 (i.e. the group of bootstrap datasets that do not contain the first observation) results in a *P*-VALUE above 0.05, but all the other jackknife groups result in *P*-values below 0.05. This suggests that the results critically depend on the first observation and should be interpreted cautiously. In practice, 'jackknife after bootstrap' results are usually displayed graphically (e.g. Carpenter and Bithell, 2000; Davison and Hinkley, 1997, p. 117).

(For further reading, see Efron and Tibshirani, 1993, or Davison and Hinkley, 1997, both of which discuss the jackknife comprehensively.) *JRC*

[**Acknowledgement:** James R. Carpenter was supported by ESRC Research Methods Programme grant H333 250047, titled 'Missing data in multi-level models'.]

Carpenter, J. and Bithell, J. 2000: Bootstrap confidence intervals: when, which, what? A practical guide for medical statisticians. *Statistics in Medicine* 19, 1141–64. **Davison, A. C. and Hinkley, D. V.** 1997: *Bootstrap methods and their application*. Cambridge: Cambridge University Press. **Efron, B. and Tibshirani, R.** 1993: *An introduction to the bootstrap*. New York: Chapman & Hall. **Quenouille, M.** 1949: Approximate tests of correlation in time series. *Journal of the Royal Statistical Society, Series B*, 11, 18–44.

Jonckheere–Terpstra test This is a nonparametric test for ordered alternatives with the null hypothesis that there is no difference in group MEDIANS and the alternative hypothesis that the group medians increase in a specific predetermined sequence. It is used when the assumption that the independent variable is nominal in the KRUSKAL–WALLIS TEST is violated. As it allows the independent variable to have an order it is more powerful than the Kruskal–Wallis test when the groups are ordered.

The method begins by specifying the order of the groups, which need not be of equal size. Then cast the data into a two-way table with the groups in the prespecified order, with the group with the lowest median first and the data within the groups ordered from the smallest to the largest. Find the total number of times each value in the first group precedes a value in the subsequent groups. Add $^1/_2$ to each precedent count when a tie occurs. Repeat for the remaining groups and sum over the groups to give J, the test statistic. Compare this value to that as found in standard tables, for example, in Siegel and Castellan (1998).

To illustrate, Mcm-2 values were collected in a breast cancer study. The median Mcm-2 value was expected to increase with histological grade (see the data in the first table). This hypothesis was tested using the Jonckheere–Terpstra test, with intermediate calculations shown in the second table.

Jonckheere–Terpstra test *Mcm-2 values in a breast cancer study according to increasing histological grade*

Histological grade		
1	*2*	*3*
1.99	4.40	6.94
3.01	9.82	8.04
4.17	10.23	9.82
7.13	11.99	15.75
9.82	11.99	18.30
9.91	13.17	25.01
	13.20	26.40
		28.17

Jonckheere–Terpstra test *Derivation of the Jonckheere–Terpstra test statistic from data in first table*

Precedent counts		
Grades 1 and 2	*Grades 1 and 3*	*Grades 2 and 3*
7	8	8
7	8	5.5
7	8	5
6	7	5
5.5	5.5	5
5	5	5
		5
Total 37.5	**41.5**	**38.5**

Therefore, $J = 37.5 + 41.5 + 38.5 = 117.5$. From tables ($n_1 = 6$, $n_2 = 7$, $n_3 = 8$, $\alpha = 0.05$) the critical value is 99. As $117.5 > 99$, there is sufficient evidence to reject the null hypothesis and conclude that there is a significant increase in the median Mcm-2 value as the histological grade increases. For further details see Conover (1999). *SLV*

Conover, W. J. 1999: *Practical nonparametric statistics*, 3rd edition. Chichester: John Wiley & Sons, Ltd. **Siegel, S. and Castellan, N. J.** 1998: *Nonparametric statistics for the behavioral sciences*, 2nd edition. Maidenhead: McGraw-Hill.

journals in medical statistics There are thousands of published articles on medical statistics scattered very widely throughout the statistical and the biomedical literature. They reflect both the diversity and the complexity of this discipline and range from highly theoretical to the most mundane practical applications. This section provides a brief overview of this literature, together with an historical perspective of the rise of journals in medical statistics.

Papers on aspects of medical statistics have been published in the biomedical literature since the late 1920s. Their purpose is to explain and illustrate specific techniques, to

journals in medical statistics *Journals publishing papers in medical statistics*

Title	*1st vol.*	*Publisher*	*2003*			*2008*		
			Volume (issues)	*Pages*	*Papers*	*Volume (issues)*	*Pages*	*Papers*
Biometrika	1901	Biometrika Trust	90 (4)	994	81	95 (4)	1008	75
Psychological Bulletin	1904	APA	129 (6)	972	41	134 (6)	964	43
American Journal of Epidemiology(1)	1921	OUP	157/158(24)	2356	284	167/168(24)	2988	354
Psychometrika	1936	Psychometric Society	68 (4)	636	31	73 (4)	794	50
Biometrics(2)	1945	IBS	59 (4)	1198	128	64 (4)	1320	138
Journal of Epidemiology and Community Health(3)	1947	BMJ	57 (12)	996	204	62 (12)	1104	206
British Journal of Mathematical and Statistical Psychology(4)	1947	BPS	56 (2)	388	21	61 (2)	532	26
Journal of Clinical Epidemiology(5)	1955	Elsevier	56 (12)	1260	166	61 (12)	1300	172
Biometrical Journal	1959	Wiley-VCH	45 (8)	1042	72	50 (6)	1098	79
Clinical Pharmacology and Therapeutics	1960	Elsevier (Mosby)	73/74 (12)	1206	121	83/84 (13)	1810	256
Clinical Trials and Meta-Analysis(6)	1964	Elsevier	-	-	-	-	-	-
Drug Information Journal(7)	1967	DIA	37 (4)	450	46	42 (6)	640	64
International Journal of Clinical Pharmacology and Therapeutics(8)	1967	Dustri-Verlage	41 (12)	626	86	46 (12)	662	76
International Journal of Epidemiology	1972	OUP	32 (6)	1134	140	37 (7)	1508	196
Epidemiologic Reviews	1979	OUP	25 (1)	98	9	30 (1)	178	10
Contemporary Clinical Trials: Design, Methods, and Analysis(9)	1980	Elsevier	24 (8)	904	66	29 (6)	920	98
Statistics in Medicine	1982	Wiley	22 (24)	3932	247	27 (30)	6634	403
Annals of Epidemiology	1990	Elsevier	13 (11)	742	100	18 (12)	946	121
Epidemiology	1990	Kluwer	14 (7)	1018	125	19 (7)	1584	125
International Journal of Methods in Psychiatric Research	1991	Whurr (Wiley)	12 (4)	228	20	17 (6)	384	39
Journal of Biopharmaceutical Statistics	1991	Dekker (Taylor & Francis)	13 (4)	816	54	18 (6)	1236	81
Statistical Methods in Medical Research	1992	Arnold (Sage)	12 (6)	554	30	17 (6)	668	41
Lifetime Data Analysis	1995	Kluwer (Springer)	9 (4)	412	22	14 (4)	520	32
Journal of Epidemiology and Biostatistics(10)	1996	-	-	-	-	-	-	-
BioMed Central (BMC) Trials	2000	BioMed Central	- (2)	-	2	- (12)	-	75
Biostatistics	2000	OUP	4 (4)	650	44	9 (4)	786	56
BioMed Central (BMC) Medical Research Methodology	2001	BioMed Central	3 (1)	-	27	8 (1)	-	80
Statistical Modelling	2001	Arnold (Sage)	3 (4)	324	17	8 (4)	402	18
Pharmaceutical Statistics	2002	Wiley	2 (4)	308	24	7 (4)	308	29

Title	1st vol.	Publisher	2003			2008		
			Volume (issues)	Pages	Papers	Volume (issues)	Pages	Papers
Understanding Statistics	2002	Lawrence Erlbaum	2 (4)	280	16	-	-	-
Clinical Trials: Journal of the Society for Clinical Trials	2004	Arnold (Sage)	-	-	-	5 (6)	642	66
Epidemiologic Perspectives and Innovations	2004	BioMed Central	-	-	-	- (5)	-	8
Emerging Themes in Epidemiology	2004	BioMed Central	-	-	-	- (8)	-	25
International Journal of Biostatistics	2005	Berkeley Electronic Press	-	-	-	4 (1)	-	21

Paper counts for some journals are approximate as some Editorials and Commentaries have been excluded.
(1)Previously *American Journal of Hygiene; Journal of Hygiene*; (2)previously *Biometrics Bulletin*; (3)previously *British Journal of Social Medicine* then *British Journal of Preventive and Social Medicine; Journal of Epidemiology and Community Medicine; Epidemiology and Community Health*; (4)previously *British Journal of Psychology: Statistical Section* then *British Journal of Statistical Psychology*; (5)previously *Journal of Chronic Diseases*; (6)previously *Clinical Trials Journal* and from 1995 incorporated within *Controlled Clinical Trials*; (7)previously *Drug Information Bulletin*; (8)previously *International Journal of Clinical Pharmacology and Biopharmacy; International Journal of Clinical Pharmacology and Therapeutics; International Journal of Clinical Pharmacology, Therapy and Toxicology*; (9)previously *Controlled Clinical Trials: Design, Methods, and Analysis;* (10)later *Journal of Cancer Epidemiology and Prevention.*

point out incorrect applications and to make the readership generally aware of how, and why, statistics can lead to poor experimental designs, incorrect analysis and unjustified conclusions. Publication within the biomedical (instead of statistical) literature enables direct contact with researchers and greater impact through explanation within a specific medical context. These papers can be grouped broadly into seven areas (see Johnson and Altman, 2005, for more details): isolated papers on a particular statistical issue (e.g. the CHI-SQUARE TEST and INTENTION-TO-TREAT analysis); series of thematic papers dedicated to a narrow statistical area (e.g. systematic reviews and CONFIDENCE INTERVALS); series of papers covering broad areas of medical statistics (such as that by Bradford Hill published in *The Lancet* in 1937); guidelines (recent examples are CONSORT and its extensions for reporting clinical trials and STROBE for observational studies, all being available at www.equator-network.org); surveys of published papers reporting the frequency of usage of statistical techniques (and the statistical knowledge required to understand the research literature); reviews of published papers examining critically aspects of design, analysis, conduct, presentation and summary (all general medical journals, and many of the specialist ones, have been the subject of these); and SYSTEMATIC REVIEWS AND METAANALYSIS incorporating assessment of methodological quality.

Papers on the general theory of techniques employed within medical statistics, as well as specific applications, have been published for over a century in statistical journals such as *Journal of the American Statistical Association* and those produced by the Royal Statistical Society, London. However, the origins of medical statistics lie in the older disciplines of biometry, psychometry, epidemiology, demography and actuarial science, and it is the journals in these specialties that published many of the early papers in medical statistics; some of them continue to do so today (see the table for some examples).

Biometrika (from the Biometrika Trust) publishes papers 'containing original theoretical contributions of direct or potential value in applications', while *Biometrics* (the journal of the International Biometrics Society) 'promotes and extends the use of mathematical and statistical methods in pure and applied biological sciences'. The *Biometrical Journal*, also published over many years, aims for papers 'on the development of statistical and related methodology and its applications to problems arising in all areas of the life sciences, in particular medicine'. The three psychology journals, *Psychological Bulletin* (from the American Psychological Association), *Psychometrika* (from the Psychological Society) and the *British Journal of Mathematical and Statistical Psychology* (from the British Psychological Society), all publish articles on the development of quantitative methods in psychology.

The principal applications of medical statistics are within EPIDEMIOLOGY and CLINICAL TRIALS and it was the journals within these specific areas that next took up publication of papers on the broadest aspects of medical statistics, as it applied within each of them. The table provides a list of the principal (English language) journals in these areas together with some details of the size of the volumes published in 2003 (from the 1st edition of this volume) and 2008. All the epidemiological and clinical trials journals publish papers describing methodological developments (roughly 5 % of their content). Such papers are also found in the more specialised journals devoted to specific disease areas, e.g. the *British Journal of Cancer* and the *International Journal of Cancer*.

It was not until the 1980s that the discipline of medical statistics was finally recognised by publication of its own journals. *Statistics in Medicine*, which 'presents practical applications of statistics and other quantitative methods to medicine and its applied sciences, together with all aspects of the collection, analysis, presentation, and interpretation of medical data, and aims to enhance communication between statisticians, clinicians and medical researchers', started in 1981 and was soon followed by the review journal, *Statistical Methods in Medical Research*. The launch of these journals coincided with a period of huge increase in both the demand for medical statisticians and in the development of sophisticated modelling techniques, enabled particularly by increased computing power and fuelled by the need to forecast the requirements for future healthcare and the extent of the AIDS epidemic. By the turn of the century, the capacity of the current journals could not forestall further specialisation, represented by *Pharmaceutical Statistics*, 'concerned with the application and use of statistics in all stages of drug development', and *Statistical Modelling*, or the need for another journal of biostatistics and another for clinical trials. This expansion has continued since 2003 with the launch of four more journals as well as expansion in numbers of pages and papers in most of the journals listed in the table.

Medical statistics interacts with all quantitative areas of biomedicine and many qualitative areas as well. It is no surprise that in addition to the publications mentioned already, many papers are also found in established journals, such as *Medical Decision Making*, *Journal of Theoretical Biology* and *Multivariate Behavioural Research*, as well as those that lie at the interfaces with computing (*Computers in Biology and Medicine*, *Journal of Biomedical Computing* and *Statistics and Computing*) and mathematics (*Computational and Mathematical Methods in Medicine)*, artificial intelligence, quality of life and health economics, as well as the comparatively recent areas represented by *Evidence-Based Medicine* and *Journal of Bioinformatics*. Yet more journals will appear in the future, some as paper, others electronically, and with the welcome introduction of an important policy of open access. *TJ*

Johnson, T. and Altman, D. G. 2005: Medical journals, statistical articles. In Armitage, P. and Colton, T. (eds), *Encyclopedia of biostatistics*. Chichester: John Wiley & Sons, Ltd., 2nd edition, p. 3151–3162.

K

Kaiser's rule This is a rule for selecting a number of common factors in FACTOR ANALYSIS and the number of components in PRINCIPAL COMPONENTS ANALYSIS. The rule is to retain as many factors or components derived from the sample correlation matrix that have VARIANCES greater than one. The rationale for this rule is that factors (components) with variances greater than one account for at least as much variability as can be explained by a single observed variable. Those factors (components) with variances less than one account for less variability than does a single observed variable and so will usually be of little interest to the investigator. Further discussions of the rule are given in Floyd and Widaman (1995) and Preacher and MacCallum (2003). *BSE*

Floyd, F. J. and Widaman, K. F. 1995: Factor analysis in the development and refinement of clinical assessment instruments. *Psychological Assessment* 7, 286–99. **Preacher, K. J. and MacCullum, R. C.** 2003: Repairing Tom Swift's electric factor analysis machine. *Understanding Statistics* 2, 13–44.

Kaplan–Meier estimator Also known as the product limit estimator, it is one of the most commonly reported methods for describing (estimating) the survival distribution or survivor function (see SURVIVAL ANALYSIS – AN OVERVIEW), $S(t)$, of a homogeneous population in time-to-event (survival) studies with right-CENSORED OBSERVATIONS. Kaplan and Meier introduced it in 1958 as a nonparametric method for using the exact survival (or event) times of those subjects uncensored in the calculation of the survival (or event-free) probabilities. The method takes appropriate account of the information contained in censored observations through the definition of the number at risk. It requires, however, that the assumption of independent censoring hold to be applied validly.

The Kaplan–Meier estimator is based on the partitioning of the time during which subjects are observed in a sequence of nonoverlapping time bands (or time intervals), where each time band contains only one distinct (unique) uncensored event time and this event time is taken to occur at the start of the interval. If censored and uncensored times are tied, the convention is to consider the uncensored times to have occurred just before the censored times. The Kaplan–Meier estimate for the probability of surviving up to time t, $\widehat{S}(t)$, is then given by the product of CONDITIONAL PROBABILITIES for surviving each of these time intervals (given survival through the preceding time intervals) before or including t.

More precisely, suppose that there are n subjects drawn randomly from the population of interest, with observed survival times (both uncensored and censored) $t_1, \ldots, t_n$. Assume that there were d events occurring and that the r unique event times corresponding to these events can be arranged in ascending order as $t_{(1)} < t_{(2)} < \cdots < t_{(r)}$. Let the kth time band, I_k, start from $t_{(k)}$ and end at a time just before $t_{(k+1)}$, written $I_k = [t_{(k)}, t_{(k+1)})$. Also let n_k and d_k be the number of subjects at risk just before $t_{(k)}$ (i.e. at the beginning of the kth time band) and the number who experience the event of interest at $t_{(k)}$ (i.e. the number of events occurring in the kth time band) respectively. Then, the probability of surviving up to time t is estimated as:

$$\widehat{S}(t) = \prod_{j=1}^{k} \left(\frac{n_j - d_j}{n_j} \right)$$

for t lying within the kth time band, I_k. $\widehat{S}(t)$ is defined to be equal to 1 for any time less than the smallest uncensored time, $t_{(1)}$. If the largest observed survival time is an uncensored observation, then for all t greater than or equal to $t_{(r)}$, $\widehat{S}(t) = 0$. However, if the largest observed time is a censored observation, then the Kaplan–Meier estimate is not defined past this censored time. Observe also that the probability of 'surviving' remains constant throughout any given interval. Finally, note that standard errors based on a formula due to Greenwood can be attached to the Kaplan–Meier estimates of the survival probabilities so as to reflect the uncertainty inherent in them.

The information obtained from the Kaplan–Meier estimator can be displayed in tabular format or graphically as a KAPLAN–MEIER PLOT. The second table illustrates the use of the Kaplan–Meier estimator for the data in the first table describing a hypothetical study of 10 patients followed up for 365 days from diagnosis of small-cell lung cancer to death, if

Kaplan–Meier estimator *Survival times (days) of 10 small-cell lung cancer patients*

Survival times (days)	9	35	40	63	110	151	212	284	305	365
Status (1 = dead; 0 = censored)	1	0	0	1	1	1	1	0	1	0

Encyclopaedic Companion to Medical Statistics: Second Edition Edited by Brian S. Everitt and Christopher R. Palmer

Kaplan–Meier estimator *Kaplan–Meier estimate of the survivor function for the data given in the first table*

Interval k	*Time interval* I_k	*Number at risk* n_k	*Deaths* d_k	*Censored* c_k	*Conditional probability of surviving* $(n_k - d_k)/n_k$	*Cumulative probability* $\widehat{S}(t)$	*Standard error*
1	0 –	10	0	0	1	1	—
2	9 –	10	1	2	0.900	0.900	0.0949
3	63 –	7	1	0	0.857	0.771	0.1442
4	110 –	6	1	0	0.833	0.643	0.1679
5	151 –	5	1	0	0.800	0.514	0.1769
6	212 –	4	1	1	0.750	0.386	0.1732
7	305 –	2	1	1	0.500	0.193	0.1615

it has occurred. The notation c_k denotes the number of censored observations in the kth time band. Observe in the second table that the estimate $\widehat{S}(t)$ in the kth interval is obtained by multiplying the estimate of the conditional probability obtained in the kth interval to the estimate of $\widehat{S}(t)$ in the $(k-1)$th interval. For further details see Collett (2003). *BT*

Collett, D. 2003: *Modelling survival data in medical research*, 2nd edition. Boca Raton: Chapman & Hall/CRC. **Kaplan, E. L. and Meier, P.** 1958: Nonparametric estimation from incomplete observation. *Journal of the American Statistical Association* 53, 457–81.

Kaplan–Meier plot A popular way of displaying 'survival' information obtained from the KAPLAN–MEIER ESTIMATOR, the Kaplan–Meier plot or curve is a graph in which the horizontal axis represents the survival (or event) times and the vertical axis represents the survival (or event-free) PROBABILITIES. The plot starts from 1 (100% of subjects event free) at time 0 and declines towards 0 (all subjects have experienced the event of interest) with increasing time.

It is plotted as a step function, since the survival probability remains constant between successive (uncensored) event times and only drops instantaneously any time an event occurs. The graph only reaches 0 if the subject with the longest observed survival time experiences an event; otherwise the survival curve is undefined after this time. Note that CENSORED OBSERVATIONS occurring in the interval between two successive event times have no effect on the survival probability calculated for this interval, but will have an impact on the calculation of the survival probability at the start of the next interval through the number at risk. The figure shows an example of a Kaplan–Meier curve.

The Kaplan–Meier curve is the best description of the survival experience of a homogeneous group of subjects in a

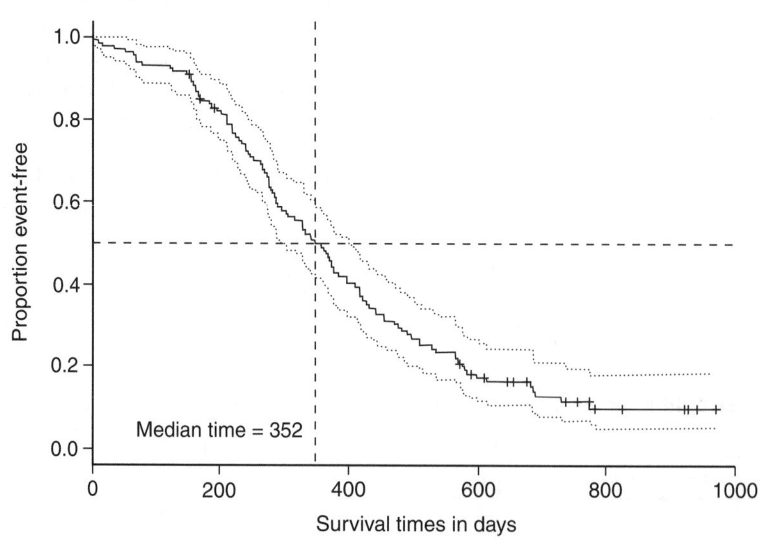

Kaplan–Meier plot *Kaplan–Meier curve with 95% confidence bands. The tick marks correspond to the occurrence of censoring*

study. The reading of the curve is straightforward. For example, if the estimated MEDIAN survival time (see SURVIVAL ANALYSIS – AN OVERVIEW) is to be reported (i.e. the time beyond which 50% of the subjects in the study are expected to survive or be event free), this can be found by extending a horizontal line from the survival probability of 0.5 on the vertical axis, to the point where the curve and this line intersect, and then dropping a vertical line to the time axis. The estimated median time is the point at which the vertical line cuts the time axis. This is illustrated in the figure. If, however, the survival curve is horizontal at the probability 0.5, then no unique value can be identified for the median survival time. In this situation, a reasonable estimate of the median time is the midpoint of the time interval over which the survival curve has a probability of 0.5.

Note that the survival curve shows the *pattern of mortality* (if death is the outcome of interest) *over time* and not the details. Thus any conclusions made based on the finer details of the curve are likely to be inaccurate. In particular, if the 'tail' of the survival curve is 'flat', then this does not mean that the risk to subjects still alive is nonexistent (i.e. evidence of a 'cured fraction' of patients). In fact, this may occur because the number of subjects under observation along the tail is small and therefore reliable estimates of the survival probabilities along it are not obtained. Also, drastic drops and large flat sections in other parts of the curve may be due to a large proportion of censored observations and may indicate inappropriate censoring.

Hence the survival curve is unreliable if based on small numbers at risk. There is also a natural tendency for the eye to be drawn to the right-hand end of the curve where it is least reliable. Therefore it is wise not to place too much confidence in the finer details of the curve, unless there is a valid reason (based possibly on prior knowledge) to do so. The overall picture is more reliable. Pocock, Clayton and Altman (2002) advocate placing confidence bands (or standard error bars) at a few regularly spaced time points on the curve in order to highlight the uncertainty in the estimated survival probabilities. In addition, they recommend that the number at risk at select time points should also be displayed. They further discuss whether 'survival curves' should go up or down (for it is sensible when events are rare to consider plotting the event rate instead of the event-free rate) and how far in time to extend the plot.

Finally, note that presenting more than one survival curve on a single diagram (e.g. curves for treated and untreated) is a useful way of informally comparing the survival experiences of different groups of homogeneous subjects and can be very informative. However, this diagram alone will not allow us to say with any confidence whether or not there are any real differences between these groups. The observed differences may be true differences, but equally could be due merely to chance variation. Assessing whether or not there are any real differences between groups can only be done, with any degree of confidence, by utilising formal statistical tests, such as the log rank test (Armitage, Berry and Matthews, 2002). For further details see Matthews and Farewell (1996) and Collett (2003). *BT*

Armitage, P., Berry, G. and Matthews, J. N. S. 2002: *Statistical methods in medical research*, 4th edition. Oxford: Blackwell Science. **Collett, D.** 2003: *Modelling survival data in medical research*, 2nd edition. London: Chapman & Hall/CRC. **Matthews, D. E. and Farewell, V. T.** 1996: *Using and understanding medical statistics*, 3rd edition. New York: Karger. **Pocock, S. J., Clayton, T. C. and Altman, D. G.** 2002: Survival plots of time-to-event outcomes in clinical trials: good practice and pitfalls. *The Lancet* 359, 1686–9.

kappa and weighted kappa Rating scales are commonly used for assessment of subjective variables, such as well-being, pain and satisfaction. Experts use scales in diagnostic judgements concerning, for example, the severity of disease and in the classification of physical, mental and social status. How reliable are judgements on scales? *Reliability* expresses the extent to which repeated assessments yield similar results. The Cohen's coefficient kappa (κ) is an agreement measure that takes into account the fact that two assessments could agree by chance (Cohen, 1960). Kappa is defined as the ratio between the percentage agreement (p_o) after excluding the chance expected agreement (p_e) and the chance-corrected maximum possible agreement:

$$\kappa = (p_o - p_e)/(1 - p_e)$$

Possible values range between -1 and 1, where the agreement that could occur by chance is zero kappa and a higher agreement than the chance expected will yield a positive value.

Kappa was developed for *categorical classifications*, where a disagreement to any other category is equally likely, but is commonly used for *ordered categorical (ordinal)* classifications, where disagreeing classifications concern adjacent rather than extreme categories. In the weighted kappa (κ_w) all observations are weighted and included in the calculation. Cicchetti (1976) proposed linear agreement weights: maximum weight ($=1$) for observations of total agreement and minimum weight ($=0$) for the most extreme disagreeing observations. For example, observations representing a disagreement of one, two and three categories between two four-point scale assessments are weighted 2/3, 1/3 and 0 respectively. A set of disagreement weights will have the opposite order. The κ_w with quadratic disagreement weights equals the INTRACLASS CORRELATION COEFFICIENT, provided there are equal marginal distributions (unbiased raters).

There are limitations with kappa. The maximum kappa, $\kappa = 1$, is obtainable only when unbiased raters agree completely. Kappa depends on the marginal distributions and on the number of categories; kappa increases when the

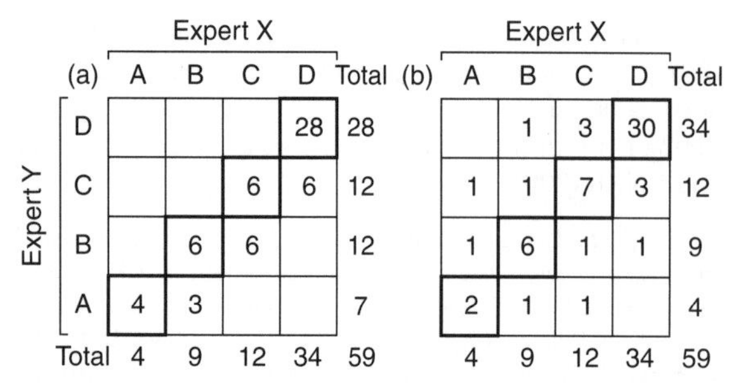

kappa and weighted kappa *Two examples of paired distributions of assessments made by experts X and Y when classifying 59 subjects in the ordered-scale categories A, B, C and D*

number of categories decreases. Therefore, kappa values from different studies are not comparable and using rules for interpretation saying that kappa larger than 0.6 represents good agreement is not meaningful. The κ_w depends on the choice of weights (Maclure and Willett, 1987; Altman, 2000). Two disagreement patterns inspired by data from a reliability study in neuroradiology illustrate the limitations of a summary measure of reliability.

Two experts, X and Y, independently judged 59 objects on a four-point scale, here labelled A, B, C and D. Two hypothetical frequency distributions of the pairs of judgements are given in the figures (parts (a) and (b)). The disagreement patterns differ, but the percentage agreements are similar: 75% (44 of 59) in (a) and 76% (45 of 59) in (b). The chance-expected agreement is determined by the marginal distributions and the number of chance-expected agreeing observations in (a) is $(4 \times 7 + 9 \times 12 + 12 \times 12 + 34 \times 28)/59 = 20.88$ and $\kappa = 0.61$. The kappa value of (b) is 0.60. Hence, the agreement measures are almost the same, but the observations are more dispersed in (b) than in (a). In (a) the experts agree completely in 44 and disagree just one category in 15 judgements, so, using Cicchetti weights, the weighted agreement is $(44 + (15 \times 2/3)) = 54$ and $\kappa_w = 0.84$, correspondingly $\kappa_w = 0.68$ in (b).

Besides the measures of kappa, the two patterns have different explanations. The marginal distributions in (a) differ, which indicates a systematic interrater disagreement concerning the use of the scale; X tends to use a higher categorical level than Y. Equal marginal distributions in (b) means a lack of such BIAS, so the disagreement between the experts can be regarded as occasional. For a detailed evaluation of interrater reliability the approach by Svensson *et al.* (1996), which identifies and measures the level of systematic disagreement separately from occasional disagreements, could be considered. *ES*

[See also MATCHED PAIRS ANALYSIS, ORDINAL DATA, RELIABILITY, RATING SCALES]

Altman, D. G. 2000: *Practical statistics for medical research.* Boca Raton: Chapman & Hall/CRC. **Cicchetti, D. V.** 1976: Assessing inter-rater reliability for rating scales: resolving some basic issues. *British Journal of Psychiatry* 129, 452–6. **Cohen, J.** 1960: A coefficient of agreement for nominal scales. *Educational and Psychological Measurement* 20, 37–46. **Maclure, M. and Willett, W. C.** 1987: Misinterpretation and misuse of the kappa statistic. *American Journal of Epidemiology* 126, 161–9. **Svensson, E., Starmark, J.-E., Ekholm, S., von Essen, C. and Johansson, A.** 1996: Analysis of inter-observer disagreement in the assessment of subarachnoid blood and acute hydrocephalus on CT scans. *Neurological Research* 18, 487–94.

Kendall's tau See CORRELATION

kernel density estimation See DENSITY ESTIMATION

k-nearest neighbour rule (kNN) See DISCRIMINANT FUNCTION ANALYSIS

knowledge discovery in databases (KDD) Synonym for data mining. See also DATA MINING IN MEDICINE

knowledge-based systems Synonym for expert systems. See EXPERT SYSTEMS

Kolmogorov–Smirnov test This is a nonparametric, goodness-of-fit test that takes two different forms, the one-sample and the two-sample form. The one-sample test compares the observed cumulative distribution function with a theoretical cumulative distribution function, e.g. the NORMAL DISTRIBUTION. The two-sample test compares two observed cumulative distribution functions with each other. The Kolmogorov–Smirnov test assumes that the theoretical distribution is completely specified, i.e. for each observed value, the value of the theoretical distribution can be calculated.

The parameters (e.g. the MEAN) of the theoretical distribution should be specified in advance and not derived from the data. For the one-sample case the NULL HYPOTHESIS is that there is no difference between the observed and the theoretical distribution. For the two-sample case, the null hypothesis is that there is no difference between the two distributions. The alternative hypothesis can be directional or nondirectional.

For the one-sample case, the method begins by ordering the observations from smallest to largest. Then calculate the empirical distribution function $S(x)$, which is the proportion of observations less than or equal to x. Calculate the value of the theoretical cumulative distribution function, $F(x)$. Calculate the test statistic:

$$T = \max|F(x) - S(x)|$$

Kolmogorov–Smirnov test *Raw data, expected cumulative distribution function, F(x), here normal with mean 3.5, standard deviation 1.75, empirical distribution function, S(x), their difference and absolute difference*

Mcm-2, x	*F(x)*	*S(x)*	*[F(x) – S(x)]*	*[S(x) – F(x)]*	*|S(x) – F(x)|*
0.54	0.045	0.05	0.045	0.005	0.005
0.97	0.074	0.1	0.024	0.026	0.026
1.27	0.101	0.15	0.001	0.049	0.049
1.99	0.194	0.2	0.044	0.006	0.006
2.02	0.199	0.25	−0.001	0.051	0.051
2.42	0.269	0.3	0.019	0.031	0.031
2.89	0.364	0.35	0.064	−0.014	0.014
3.01	0.39	0.4	0.04	0.01	0.01
3.13	0.416	0.45	0.016	0.034	0.034
3.26	0.445	0.5	−0.005	0.055	0.055
3.59	0.521	0.55	0.021	0.029	0.029
4.15	0.645	0.6	0.095	−0.045	0.045
4.17	0.649	0.65	0.049	0.001	0.001
4.27	0.67	0.7	0.02	0.03	0.03
4.3	0.676	0.75	−0.024	0.074	0.074
4.4	0.696	0.8	−0.054	0.104	0.104
4.68	0.75	0.85	−0.05	0.1	0.1
5.63	0.888	0.9	0.038	0.012	0.012
6.73	0.968	0.95	0.068	−0.018	0.018
6.94	0.975	1	0.025	0.025	0.025
Maximum	–	–	0.095	0.104	0.104

and compare this value to standard tables. This is a two-sided test with the alternative hypothesis:

$$F(x) \neq S(x)$$

There are two alternative one-sided tests and the hypotheses for these are $F(x) \leq S(x)$ and $S(x) \leq F(x)$, with test statistics

$$T_1 = \max\,[S(x) - F(x)] \text{ and } T_2 = \max\,[F(x) - S(x)]$$

respectively.

For the two-sample case, calculate $S(x)$ for each group and the test statistic is $T = \max\,|S_1(x) - S_2(x)|$; this can be compared to the tables and is a two-sided test. For a one-sided test, the test statistics are:

$$T_1 = \max\,[S_1(x) - S_2(x)] \text{ and } T_2 = \max\,[S_2(x) - S_1(x)],$$

depending on which distribution is expected to be greater.

As an example, Mcm-2 values are collected in a study. It is thought that they come from a normal distribution with mean 3.5 and STANDARD DEVIATION 1.75. The Kolmogorov–Smirnov one-sample test will be used to test this hypothesis. Therefore, $F(x)$ is calculated from a normal distribution with mean 3.5 and standard deviation 1.75 (see the first table).

The test statistic is shown in the final row of the first table as the maximal value of the final column for the two-sided case, with $T = \max\,|F(x) - S(x)| = 0.104$. Comparing $T = 0.104$ to

Kolmogorov–Smirnov test *Frequency distributions and empirical distribution functions, their differences and absolute difference*

Category	n_1	n_2	$S_1(x)$	$S_2(x)$	$[S_1(x) - S_2(x)]$	$[S_2(x) - S_1(x)]$	$\|S_1(x) - S_2(x)\|$
1	14	35	0.060	0.310	0.250	−0.250	0.250
2	43	45	0.245	0.708	0.463	−0.463	0.463
3	61	5	0.506	0.752	0.246	−0.246	0.246
4	29	10	0.631	0.841	0.210	−0.210	0.210
5	16	14	0.700	0.965	0.265	−0.265	0.265
6	70	4	1.000	1.000	0.000	0.000	0.000
Maximum	–	–	–	–	0.463	0.000	0.463

standard tables ($n = 20$, $\alpha = 0.05$) gives the value 0.294. As $0.104 < 0.294$, there is insufficient evidence to reject the null hypothesis. Therefore it can be concluded that it is plausible for these data to have come from a normal distribution with mean 3.5 and standard deviation 1.75.

Data for a two-sample example classified into six categories are shown in the second table, with:

$$T = \max|S_1(x) - S_2(x)| = 0.463$$

As the sample size is too large for the tables use the formula for $\alpha = 0.05$, which is:

$$1.36\sqrt{\frac{N_1 + N_2}{N_1 N_2}} = 1.36\sqrt{\frac{233 + 113}{233 \times 113}} = 0.156$$

As $0.463 > 0.156$ there is sufficient evidence to reject the null hypothesis that the distributions are the same; therefore it is concluded that the distributions are different.

Further information, including standard tables, is available in textbooks such as Pett (1997), Siegel and Castellan (1998) and Conover (1999) or in the software manual for StatXact (Mehta and Patel, 1998). *SLV*

Conover, W. J. 1999: *Practical nonparametric statistics*, 3rd edition. Chichester: John Wiley & Sons. **Mehta, C. and Patel, N.** 1998: *Stat-Xact 4 for Windows user manual.* Cytel Software Corporation. **Pett, M. A.** 1997: *Nonparametric statistics for healthcare research.* Thousand Oaks: Sage. **Siegel, S. and Castellan, N. J.** 1998: *Nonparametric statistics for the behavioural sciences*, 2nd edition. Maidenhead: McGraw-Hill.

Kruskal–Wallis test This nonparametric method is the extension of the MANN–WHITNEY RANK SUM TEST to more than two groups. It is more sensitive than the MEDIAN TEST as it uses the magnitude of the differences rather than the direction. It is less sensitive than the JONCKHEERE–TERPSTRA TEST if the groups have an inherent order, as the Kruskal–Wallis test looks for a difference between groups and the Jonckheere–Terpstra test looks for an increase over the groups. The Kruskal–Wallis test is derived from one-way ANALYSIS OF VARIANCE but uses ranks rather than the actual observations. It tests if k groups are drawn from populations with the same median and assumes that the data are a randomly selected set of observations and that the data are continuous in nature and in more than two groups. It also assumes an independence of the groups and observations within a group and that the groups have similarly shaped distributions.

To use the test, rank the whole sample from the smallest to the largest. Calculate the sum of the ranks in each group, the average rank in each group, $\bar{R}_i$, and the average rank for the whole sample, $\bar{R}$. Calculate the test statistic H:

$$H = \frac{12\sum_{i=1}^{k} n_i(\bar{R}_i - \bar{R})^2}{N(N+1)}$$

where n_i = the number of observations in group i and N = the total sample size. Compare H to the critical value of the CHI-SQUARE DISTRIBUTION with $k-1$ DEGREES OF FREEDOM. Reject the NULL HYPOTHESIS if H is bigger than the critical value.

To illustrate, suppose that, within a study, age is an important predictor and so whether there is a difference in age between three groups needs to be decided. The ages are shown in the table, as well as the ranks for each observation and the sum of the ranks and the average rank in each group.

Kruskal–Wallis test *Ages and their overall ranks within three groups*

Group 1	*Rank*	*Group 2*	*Rank*	*Group 3*	*Rank*
55	18	59	25.5	53	15.5
51	13	40	3.5	62	29
60	27.5	48	11	55	18
34	1.5	56	20	58	23
45	7.5	44	6	58	23
55	18	42	5	57	21
51	13	58	23	46	9.5
60	27.5	67	30	51	13
34	1.5	59	25.5	53	15.5
45	7.5	40	3.5	46	9.5
Rank sum	135		153		177
(average)	(13.5)		(15.3)		(17.7)

From standard tables, the critical value of the chi-square distribution with 2 degrees of freedom is 5.99, $1.15 < 5.99$. Therefore there is not sufficient evidence to reject the null hypothesis that the median age is the same in the three groups.

$$\bar{R} = \frac{N+1}{2} = \frac{30+1}{2} = 15.5$$

$$H = \frac{12\sum n_i(\bar{R}_i - \bar{R})^2}{N(N+1)}$$

$$= \frac{12 \times (10 \times (13.5-15.5)^2 + 10 \times (15.3-15.5)^2 + 10 \times (17.7-15.5)^2)}{30 \times (30+1)}$$

$$= 1.15$$

For further details, refer to texts by Pett (1997), Siegel and Castellan (1998) or Conover (1999). *SLV*

[See also MEDIAN TEST]

Conover, W. J. 1999: *Practical nonparametric statistics*, 3rd edition. Chichester: John Wiley & Sons, Ltd. **Pett, M. A.** 1997: *Non-parametrics for healthcare research.* Thousand Oaks: Sage. **Siegel, S. and Castellan, N. J.** 1998: *Nonparametric statistics for the behavioural sciences*, 2nd edition. Maidenhead: McGraw-Hill.

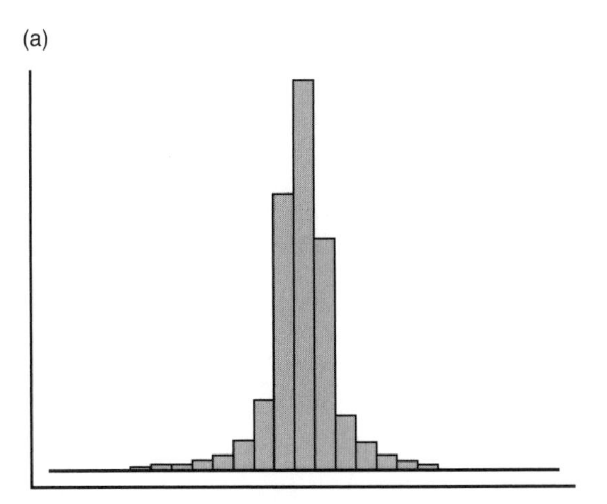

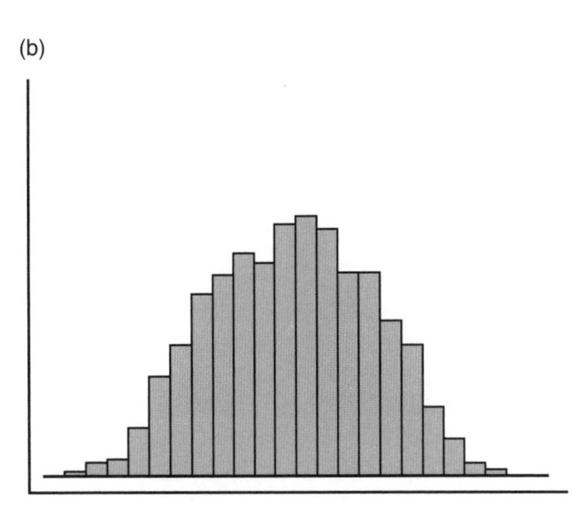

kurtosis *Histogram of distributions having (a) high and (b) low kurtosis*

kurtosis This is a term that describes 'peakedness' of a FREQUENCY DISTRIBUTION or a PROBABILITY DISTRIBUTION. Data that have high kurtosis have a sharp peak and decline rapidly at either side to leave long tails (part (a) in the figure). Data with low kurtosis are flatter (part (b) in the figure).

Mathematical measures of kurtosis can be used to describe distributions. The kurtosis of a NORMAL DISTRIBUTION is 3 and of a UNIFORM DISTRIBUTION is 1.8. However, some formulae for kurtosis subtract 3, so that the kurtosis of a normal distribution is 0 and that of a uniform distribution is -1.2. *SRC*

[See also DESCRIPTIVE STATISTICS]

L

L'Abbé plots See FOREST PLOT

large simple trial (LST) See MEGA-TRIAL

latent variables These are variables that cannot be measured directly but are assumed to be related to a number of observable or manifest variables. Consider, for example, a concept such as 'racial prejudice'. Clearly, direct measurement of this concept is not possible; however, one could, for example, observe whether a person approves or disapproves of a particular piece of government legislation designed to achieve a greater degree of racial equality, whether he or she numbers members of some ethnic minority among his or her friends and acquaintances, etc., and assume that these are, in some sense, indicators of the more fundamental variable, racial prejudice.

Although latent variables are the basis of FACTOR ANALYSIS and STRUCTURAL EQUATION MODELS, scepticism about methods based on such variables (particularly factor analysis) has not been uncommon among statisticians over the years. Latent variable modelling has often been viewed as a dubious exercise fraught with unverifiable assumptions and naive inferences regarding causality. For such critics the only thing in favour of latent variable models is that they occupy a rather obscure area of statistics, primarily confined to psychometrics. There are, however, a number of reasons that such criticisms are mistaken: ignoring latent variables often implies stronger assumptions than including them, latent variable modelling then being viewed as a sensitivity analysis of a simpler analysis excluding latent variables; many of the assumptions in latent variable modelling *can* be empirically assessed and some can be relaxed; latent variable modelling pervades modern mainstream statistics and are widely used in different disciplines – not only medicine but also economics, engineering, psychology, geography, marketing and biology (see Skrondal and Rabe-Hesketh, 2004).

This 'omnipresence' of latent variables is commonly not recognised, perhaps because latent variables are given different names in different areas of application, e.g. RANDOM EFFECTS, common factors and latent classes. In fact, latent variables can be used to represent a variety of phenomena, e.g. true variables measured with error, hypothetical constructs, unobserved heterogeneity, latent responses underlying categorical variables and missing data, to name but a few. *BSE*

[See also STRUCTURAL EQUATIONS MODELS]

Skrondal, A. and Rabe-Hesketh, S. 2004: *Generalized latent variable modeling*. Boca Raton: Chapman & Hall/CRC.

lead time bias See BIAS, SCREENING STUDIES

learn-as-you-go designs See DATA-DEPENDENT DESIGNS

least squares estimation This is a general method for estimating regression parameters in a model for expected outcomes (e.g. a linear regression model). Parameter values are chosen to minimise the sum of squared differences between the observed and expected outcomes. Least squares estimation is, for instance, the standard method of estimation in MULTIPLE LINEAR REGRESSION, where observations y_i are approximated by a linear function of explanatory variables x_i, e.g. $\beta_1 x_{i1} + \cdots + \beta_k x_{ik}$ with $\beta_1, \ldots, \beta_k$ unknown. When the model is linear and outcomes are independent and normally distributed with constant VARIANCE around their expectation, the method coincides with MAXIMUM LIKELIHOOD ESTIMATION and is efficient in large samples, i.e. yields precise estimators.

More generally, least squares estimators (LSEs) can be studied for any type of outcome distribution around the regression curve. They are also useful outside the context of statistical models. Least squares lines are, for instance, typically drawn as the best fitting line through a cloud of points. This is illustrated in the figure (see page 240), where we depict the LSE of a linear regression curve of body weight (in kg) as a function of body height (in cm) based on a random sample of 250 American men (Penrose, Nelson and Fisher, 1985). The curve is obtained by minimising the sum of squared distances e^2 over all observations. It contains the point where the sample-averaged body weight is predicted for men of average body height. Within the context of a regression model, predictions on the estimated regression curve are most precise at the centre of the data, but may be imprecise towards the tails. Extrapolations beyond the range of the data often cannot be trusted.

Least squares estimators are attractive due to their strong intuitive appeal, the stability and efficiency of the algorithm and their sound statistical properties in large samples. Under mild regularity conditions, large samples yield an LSE that is subject to normal variation around the true parameter. A 95% CONFIDENCE INTERVAL then becomes the LSE plus or minus 1.96 times the STANDARD ERROR.

When outcome variation is known to differ between observations, a more precise estimator is obtained by a weighted least squares estimator (WLSE), which minimises a sum of *weighted* squared differences between observed and

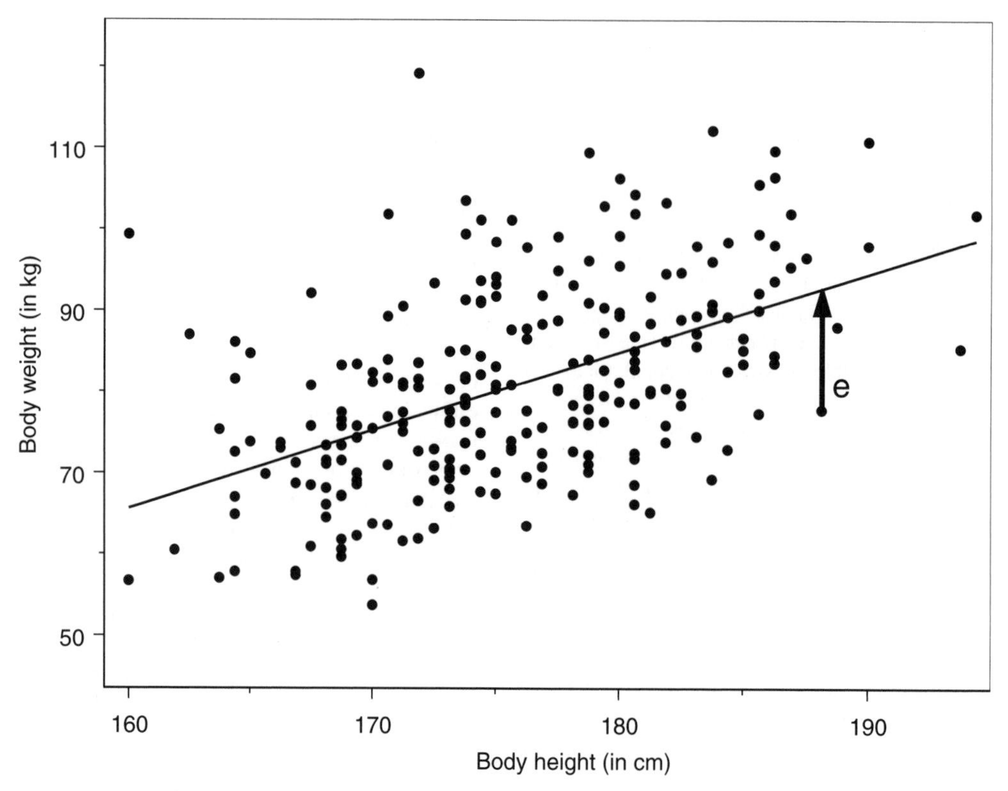

least squares estimation *Scatterplot and least squares regression line of body weight versus body height in a random sample of 250 American men*

expected outcomes. When the model is linear and the outcomes are independent, weights chosen as the inverse of the individual variances yield the most efficient estimator. When, furthermore, the outcomes are normally distributed, this WLSE then also coincides with the maximum likelihood estimate (MLE). In other instances, e.g. with logistic regression and many other generalised linear models, the optimal weights depend on the target parameter. Because the latter is unknown, a method of iteratively reweighted least squares is usually applied, whereby weights in each iteration depend on previous estimates for the unknown parameter.

Even though the LSE may be less precise than the MLE under certain models, least squares estimation is popular because it does not require a complete description of the sampling distribution of the observed data. For example, it is not necessary to assume that outcomes are normally distributed to derive least squares estimates of the unknown regression coefficients in the MULTIPLE LINEAR REGRESSION model. The model of interest for the means is all that is needed. The LSE is therefore immune to misspecification of the sampling distribution, unlike the MLE.

Further modifications of least squares estimation have been devised to achieve additional goals. For instance, to enhance robustness to outlying observations, L_1-regression is based on minimising the average absolute deviation between the observations and the regression function. *SV/EG*

Penrose, K. W., Nelson, A. G. and Fisher, A. G. 1985: Generalized body composition prediction for men using simple measurement techniques. *Medicine and Science in Sports and Medicine* 17, 189.

leave-one-out cross-validation See DISCRIMINANT FUNCTION ANALYSIS

length-biased sampling This form of sampling arises when items are sampled in proportion to their values on some variable of interest, e.g. a sampling scheme based on the number of patient visits. A BIAS may be introduced because some individuals are more likely to be selected than others simply because they make more frequent visits. The problem arises in SCREENING STUDIES, where the sample of cases detected is likely to contain an excess of slow-growing cancers compared to the sample diagnosed positive because of their symptoms. If length-biased sampling is ignored, the estimate of the true population MEAN can be greatly inflated. An example of length-biased sampling is described in Davidov and Zelen (2001) in the context of the assessment of familial risk of disease based on referent databases in which the larger the family, the greater the probability of finding the family in the databse. *BSE*

[See also BIAS, BIAS IN OBSERVATIONAL STUDIES, SAMPLING METHODS – AN OVERVIEW]

Davidov, O. and Zelen, M. 2001: Referent sampling, family history and relative risk: the role of length-biased sampling, *Biostatistics* 2, 173–81.

Levene's test This is used to test whether two or more groups have equal VARIANCE. The NULL HYPOTHESIS states that the variance of all groups is equal; the alternative hypothesis states that the variances are unequal for at least one pair. Equal variance of two or more groups is a frequent assumption for parametric tests, ANALYSIS OF VARIANCE for example, and so Levene's test can be used to verify this assumption. Levene's test is relatively simple and robust to departures from normality.

To perform Levene's test we begin by finding, for each group, the absolute differences between the observed values and the MEDIAN, MEAN or trimmed mean. The groups need not be of equal size. Whether to use the median, mean or trimmed mean depends on the underlying distribution of the data. If the data are symmetric and moderate tailed the mean provides the best power, using the trimmed mean performs best if the data are heavy tailed and the median performs best if the data are skewed. However, using the median provides good ROBUSTNESS for many types of nonnormal data while retaining good power (Wilcox, 1998). We complete Levene's test by performing an analysis of variance on these absolute differences (see the table).

Levene's test *Data resulting from applying Levene's test on three different treatment groups*

	Group 1	*Absolute differences*	*Group 2*	*Absolute differences*	*Group 3*	*Absolute differences*
	24	5.5	22	0.5	8	5
	23	4.5	21	0.5	12	1
	19	0.5	18	3.5	14	1
	18	0.5	25	3.5	15	2
	14	4.5	18	3.5	12	1
	6	12.5	22	0.5	13	0
	5	13.5	16	5.5	7	6
	21	2.5	17	4.5	14	1
	23	4.5	26	4.5	13	0
	10	8.5	23	1.5	15	2
Median	**18.5**		**21.5**		**13**	

For example, for treatment 1 a score of 23 differs by 4.5 points from the median, while a score of 14 also differs by 4.5 points from the median. The idea is that the larger the differences in some groups compared to others, the more the spread and, hence, the more likely it will be that the variance in the populations, from which they arose, is not the same. Therefore, a *one-way analysis of variance* on these differences will test this.

This results in an F-statistic of 4.21 with an associated P-value of 0.026 so we reject the null hypothesis that the variance of the groups is equal. More details can be found in Brown and Forsythe (1974). *MMB*

Brown, M. and Forsythe, A. 1974: Robust tests for the equality of variances. *Journal of the American Statistical Association* 69, 346, 364–7. **Wilcox, R.** 1998: Trimming and Winsorisation. In Armitage, P. and Colton, T. (eds), *Encyclopedia of biostatistics*. Chichester: John Wiley & Sons, Ltd.

Lexis diagram This is a descriptive tool used in epidemiology and demography, being a plot of individuals in a study on two timescales simultaneously. These timescales are most commonly calendar time and age. Each individual is then represented by a diagonal line at 45° to each axis, which begins at the calendar time and age at enrolment and ends at the calendar time and age at the event of interest (e.g. death) or censoring.

As an example, the table shows the year of birth, age at enrolment and age at death/censoring of four individuals enrolled in a study that began in 1975 and ended with follow-up in 2004. The corresponding Lexis diagram is shown in the figure (see page 242). A death is shown by a filled circle and a censoring event by an empty circle

One use of the Lexis diagram relates to estimating, or adjusting for, the effects of age and calendar time on the mortality (or morbidity) rate. In this application, age is divided into (e.g. 5-year) bands and calendar time is divided into (e.g. 5-year) periods. It is assumed that the mortality rate (or baseline mortality rate) is piecewise constant, i.e. is constant on each combination of age band and calendar period. To estimate the mortality rate in each of these band–period combinations, it is necessary to calculate the number of deaths and total time at risk in each band period.

Lexis diagram *Study data for four individuals*

Individual	*Birth year*	*Age at enrolment*	*Status*	*Age at death or censoring*
A	1940	35	Died	58
B	1951	24	Censored	53
C	1954	30	Died	48
D	1960	21	Censored	30

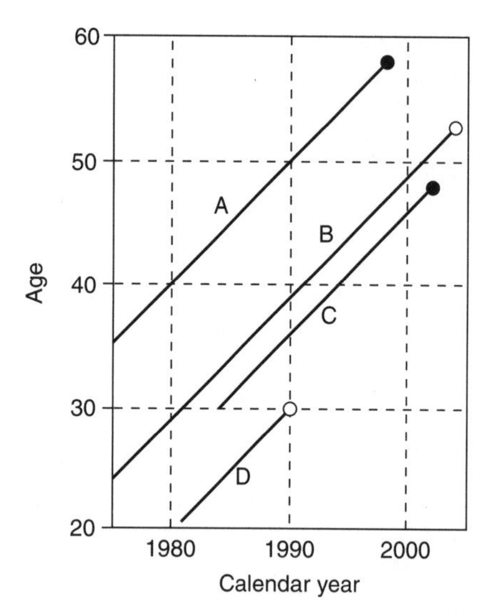

Lexis diagram *Lexis diagram for four individuals*

For example, if age bands and calendar periods of 10 years' duration are adopted, individual C joins the cohort in 1984 aged 30. He changes calendar period after 6 years, in 1990. Then he changes age band 4 years later, in 1994. Finally, he changes calendar period 6 years later, in 2000, and dies 2 years later, in 2002. The Lexis diagram representation makes it easy to see when he changes from one band-period to another: this happens whenever his diagonal line crosses a horizontal or vertical line. Individual C contributes 6, 4, 6 and 2 years at risk respectively to four band-periods and a death event to the last of these.

Variations on this simple Lexis diagram may be obtained by changing the timescales represented by the axes, by marking other symbols on the diagonal lines to represent other events of interest or by introducing colour to differentiate periods spent by an individual in different states. For further details see Keiding (1990), Goldman (1992) and Clayton and Hills (1993). *SRS*

Clayton, D. and Hills, M. 1993: *Statistical models in epidemiology*. Oxford: Blackwell Science Publications. **Goldman, A. I.** 1992: Events charts: visualizing survival and other timed-events data. *The American Statistician* 46, 13–18. **Keiding, N.** 1990: Statistical inference in the Lexis diagram. *Philosophical Transactions of the Royal Society of London, Series A* 332, 487–509.

life expectancy This popular summary measure is used in DEMOGRAPHY and EPIDEMIOLOGY for assessing the current health of a population or for health comparisons across populations. For a specific population, the standard

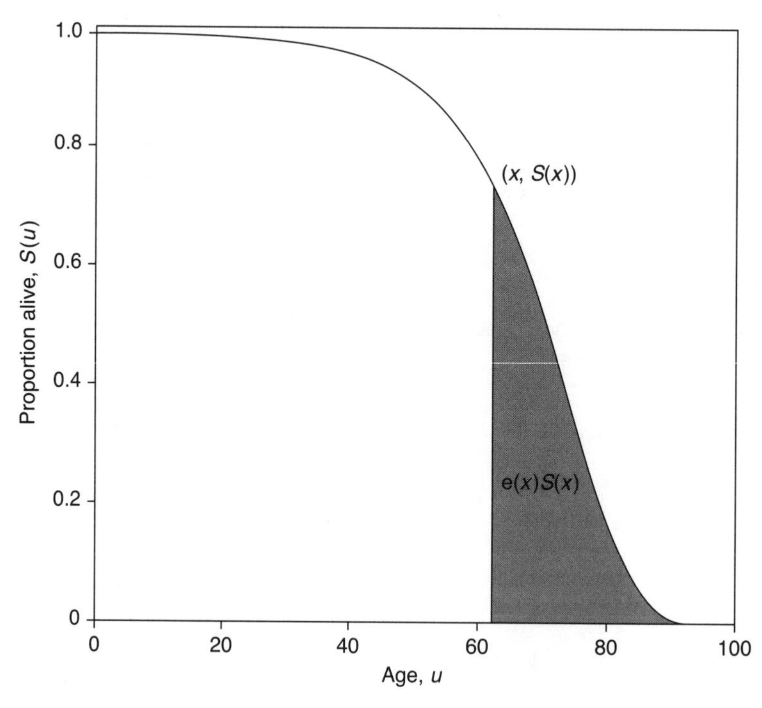

life expectancy *Survival curve over age. Shaded area corresponds to the area under the survival curve from age x onwards and is equal to e(x)S(x)*

definition of life expectancy at a given age is the average number of years an individual of that particular age has remaining if the age-specific mortality rates do not change in the future. These age-specific mortality rates can be obtained from the LIFE TABLES for that population, which may be stratified by variables known to be strongly associated with mortality, such as sex, calendar period and smoking status if available. The life table can be estimated either nonparametrically or parametrically (see, for example, Gaitatzis *et al.*, 2004, and Tom and Farewell, 2009).

In standard (actuarial) life table notation, the life expectancy at age x, e_x, is given by $e_x = T_x/l_x$, where T_x is the number of person-years lived between the exact age x and extinction, and l_x is the number of persons still alive at the exact age x. Thus, for example, the life expectancy at birth, e_0, in a given birth cohort or life table population is the average person-years lived from birth if the current age-specific mortality rates remain unchanged in the future.

The continuous time mathematical representation of life expectancy (also known as the mean residual lifetime) of an individual known to have survived to an age x in terms of the current force of mortality (i.e. hazard function over age; see SURVIVAL ANALYSIS – AN OVERVIEW), $\lambda(u)$, is given by

$$e(x) = \int_x^{\infty} \exp\left[-\int_x^{t} \lambda(u)\,\mathrm{d}u\right] \mathrm{d}t$$

and is equivalent to the area under the survival curve, $S(u)$, from age x onwards divided by the probability of surviving up to age x, $S(x)$ (see the figure on page 242).

Note that because the age-specific mortality rates are expected to change in the future, the life expectancy is not a measure of how long a specific individual of a given age in the population of interest is actually expected to live further. Although life expectancy is a long-standing and easily understood indicator of present population health, it has increasingly been seen as too crude for this purpose since it does not take into account the impact of chronic diseases and disability. Extensions of life expectancy to healthy life expectancy, disability-free life expectancy and, more generally, to life expectancies in various health states have been made and can be estimated through the fitting of MULTISTATE MODELS (see, for example, Butler *et al.*, 2008). *BT*

Butler, T. C., van den Hout, A., Matthews, F. E., Larsen, J. P., Brayne, C. and Aarsland, D. 2008: Dementia and survival in Parkinson disease – a 12-year population study. *Neurology* 70, 1017–22. **Gaitatzis, A., Johnson, A. L., Chadwick, D. W., Shorvon, S. D. and Sander, J. W.** 2004: Life expectancy in people with newly diagnosed epilepsy. *Brain* 127, 2427–32. **Tom, B. D. M. and Farewell, V. T.** 2009: Statistical methods for individual-level data in cohort mortality studies of rheumatic diseases. *Communications in Statistics – Theory and Methods* 38, 3472–87.

life tables Life tables are models that conveniently summarise the level of mortality in a population of interest. Their best-known function, life expectancy, has a ready intuitive meaning. Life table functions are independent of the age structure of the population whose mortality level they are used to summarise.

Period life tables are used to summarise the mortality experience during a given period, e.g. a calendar year. Cohort life tables summarise the experience of a defined cohort as it ages through calendar time. For the necessary mortality observations to be available to construct them, the relevant cohort has to be at least towards the end of its lifespan.

Full life tables have one row for each year of life, usually to age 110 (see the figure on page 244). Abridged life tables typically have one row for each 5 years of life except that there are usually separate rows for ages 0 to 1 and 1 to 5.

Constructing life tables. There are two main steps to building a life table.

First, there is the 'preliminary computation', in which the observed age and sex-specific death rates during the period of interest are converted into corresponding risks of death between two exact ages. Suppose, for example, that the observed central death rate in the population of interest for persons aged 40 to 44 last birthday is 0.003 404. (This is ${}_5\mathrm{M}_{40}$ in lifetable notation when referring to observations made in the population of interest, and is conventionally taken to be an unbiased estimator of the corresponding life table function ${}_5\mathrm{m}_{40}$.) The risk of death between exact age 40 and exact age 45 is given by

$${}_nq_x = \frac{n\,.\,{}_nm_x}{1+(1-{}_na_x)n\,.\,{}_nm_x} = \frac{5\,.\,{}_5m_{40}}{1+(1-{}_5a_{40})5\,.\,{}_5m_{40}}$$

$$= \frac{5\times 0.003404}{1\times(1-0.59)5\times 0.003404} = 0.01690$$

where ${}_5a_{40}$ is the fraction of the age interval lived, on average, by those who die during it. The risk of death across the age interval is close, but not equal to, the central death rate. (In this case, the central death rate times 5 – to take account of the age interval width – equals 0.017 02, slightly greater than the risk.)

Second, there is the computation of the life table proper (see the figure). In constructing the life table an initial hypothetical cohort of 100 000 (l_0, known as the radix) is subjected across each successive age interval to the calculated risks of death. Therefore, starting at birth ($x=0$), 100 000 are exposed to the risk of death before exact age 1, i.e. ${}_1q_0$, which in this example equals 0.020 06. This results in 2006 deaths in the interval (${}_1d_0$). The person-time lived in the interval (${}_1L_0$) is 1 year for all who survived it ($l_1 = 97\,994$) plus the time lived by those who died in the interval – which

Life table functions and notation

x is exact age x, i.e. the xth birthday.

n refers to the width of the age interval being considered. In a full life table where $n = 1$ it may be omitted.

e_x is life expectancy at exact age x.

${}_nm_x$ is the central death rate for persons aged between x and $x+n$ in the hypothetical life table population. It is estimated by ${}_nM_x$ (below).

${}_nM_x$ is the observed central death rate in the population of interest.

l_x is the number of persons still alive at exact age x.

${}_nq_x$ is the risk (probability) of death between exact ages x and $x+n$.

${}_np_x$ is the risk of surviving from exact age x to $x+n$ (equals $1 - {}_nq_x$).

${}_na_x$ is the average fraction of the interval lived by those who die between x and $x+n$.

${}_nL_x$ is the number of person-years lived between exact ages x and $x+n$.

T_x is the number of person-years lived between exact age x and the extinction of the hypothetical cohort.

life tables *Extract of first 6 and last 10 rows of a full life table for US white males in 1970*[a]

Age interval, period of life between 2 ages, x and $x+n$	Width of age interval in years	Proportion of persons alive at the beginning of age interval dying during interval[a]	Of 100 000 born alive: Number living at beginning of age interval	Of 100 000 born alive: Number dying during age interval	In stationary (life table) population with 100 000 born into it each year: Number of person-years of life lived in age interval	In stationary (life table) population with 100 000 born into it each year: Number of person-years of life lived in this and all subsequent intervals	Average number of years of life remaining at beginning of age interval (life expectancy)
x	n	${}_nq_x$	l_x	${}_nd_x$	${}_nL_x$	T_x	e_x
0	1	0.020 06	100 000	2006	98 252	6 793 828	67.94
1	1	0.001 16	97 994	114	97 037	6 695 576	68.33
2	1	0.000 83	97 880	81	97 840	6 597 639	67.41
3	1	0.000 72	97 799	71	97 763	6 499 799	66.46
4	1	0.000 59	97 728	57	97 700	6 402 036	65.51
5	1	0.000 54	97 671	52	97 645	6 304 336	64.55
	⋮	⋮	⋮	⋮	⋮	⋮	⋮
100	1	0.354 79	189	67	155	415	2.20
101	1	0.365 53	122	45	100	260	2.13
102	1	0.375 50	77	29	62	160	2.08
103	1	0.384 71	48	18	39	98	2.02
104	1	0.393 20	30	12	24	59	1.98
105	1	0.401 01	18	7	15	35	1.94
106	1	0.408 18	11	5	8	20	1.90
107	1	0.414 75	6	2	5	12	1.86
108	1	0.420 75	4	2	3	7	1.82
109	∞	1.000 00	2	2	4	4	1.79

[a] a_x is 0.129 for the first year of life and approximately 0.5 for all subsequent years.

equals ${}_1d_0 \times {}_1a_0$ (the fraction of the interval lived by those who died within it) or 2006×0.129, which equals 98 252, as shown, when added to l_1. (For economy, ${}_na_x$ is not shown in the table: it is 0.129 for the first year of life and approximately 0.5 thereafter.) The ${}_nL_x$ column is calculated in this way, one row at a time, to the end of the lifespan. A special rule is then needed for closing the last open-ended interval – representing, on the table shown, the person time lived beyond 109 – ${}_\infty L_{109}$. This is estimated by $l_{109}/{}_\infty M_{109}$. (The justification for this is that the remaining survival time is taken to be distributed exponentially with a mean of $1/{}_\infty M_{109}$.)

T_x is then summed back from the end of the lifespan, beginning, in this example, with T_{109}, which equals L_{109}.

Moving up one row, T_{108} then equals $T_{109} + {}_1L_{108}$ and so on back to T_0, where T_0 represents all the person-time lived by the 100 000 who set out, so the average person-time lived, or life expectancy e_0, is T_0/l_0 and more generally, for any age x, $e_x = T_x/l_x$.

Life table populations can be interpreted in two ways: (a) as fully hypothetical constructs or models in which 100 000 individuals are imagined, as it were, to be born in the same instant and then instantaneously subjected to the relevant risks of death throughout their hypothetical lifespans; or (b) as representing stationary populations experiencing constant, and equal, birth and death rates . In this latter interpretation, ${}_nL_x$ gives the expected number of individuals aged x to $x + n$ and T_0 gives the total population size. In such stationary populations there are l_0 births each year, so the birth rate $= l_0/T_0$ – the inverse of the life expectancy. Thus:

$$\text{Crude birth rate} = \text{Crude death rate} = 1/e_0$$

Uses of the life table. The l_x and q_x columns have many uses in summarising mortality risks in populations of interest. Thus infant mortality, conceptually defined as the risk of death before the first birthday, is ${}_1q_0$. The under 5 mortality 'rate' (actually the risk of death before age 5) is $1 - l_5/l_0$. The adult mortality 'rate' (${}_{45}q_{15}$, or the probability of death between 15 and 60) is $1 - l_{60}/l_{15}$. Similarly, the probability of survival between any two ages i and j is given by l_j/l_i.

Life tables have long been used to provide comparable summaries of mortality risks in populations. They are also serving as the basis of newer 'summary measures of average population health', which combine information on both the risks of premature death and of nonfatal illness and injury. Such summary measures may be either 'health expectancies' (such as 'health adjusted life expectancy') or 'health gaps' (such as DALYs (disability adjusted life-years) lost).

The d_x and l_x functions when plotted for a given population at successive time intervals show how the distribution of age at death changes as life expectancy has risen. One interpretation of recent trends in low mortality countries is that the rise in e_0 has been disproportionately due to a reduction in the more premature adult deaths. As this process continues, a larger and larger proportion of each generation survives until closer to the maximum lifespan. The distribution of deaths by age at death becomes concentrated at a high age – manifest as a reduced dispersion in the distribution of deaths by age (d_x) in the life table. The corresponding shift in pattern for the survivorship (l_x) function is for it to remain high until closer to the maximum lifespan and then fall sharply – a process described as the 'rectangularisation of the survival curve'. This 'optimistic' interpretation of recent trends is taken to imply that there has been no extension of the average duration of ill-health in the period immediately before death. For further details see Elandt-Johnson and Johnson (1980), Preston, Heuveline and Guillot (2001), Lopez *et al.* (2003) and Peeters *et al.* (2003). *JP*

Elandt-Johnson, R. C. and Johnson, N. L. 1980: *Survival models and data analysis*. New York: John Wiley & Sons, Inc. **Lopez, A. D., Ahmad, O. B., Guillot, M., Ferguson, B. D., Salomon, J. A., Murray, C. J. L. and Hill, K. H.** 2003: Life tables for 191 countries for 2000: data, methods, results. In Murray, C. J. L. and Evans, D. B. (eds), *Health systems performance assessment: debates, methods and empiricism*. Geneva: World Health Organization, pp. 335–53. **Peeters, A., Barendregt, J. J., Willekens, F., Mackenbach, J. P., Al Mamun, A. and Bonneux, L.** 2003: Obesity in adulthood and its consequences for life expectancy: a life-table analysis. *Annals of International Medicine* 138, 24–32. **Preston, S. H., Heuveline, P. and Guillot, M.** 2001: *Demography: measuring and modeling population processes*. Oxford: Blackwell.

likelihood The likelihood function plays two roles in STATISTICS. First, in its own right it provides a means for estimating unknown parameters by finding the value of the unknown parameter(s) that maximises it (maximum likelihood) as well as for comparing hypotheses (LIKELIHOOD RATIO). Second, it has a role in Bayesian statistics (see BAYESIAN METHODS).

Suppose interest lies in learning about the response of patients suffering from influenza symptoms to a new treatment. Data are collected from 10 patients, of whom six respond positively. What can be said about the unknown probability of a positive response π? By definition, the probability of a positive response is π and, of a negative response, $1 - \pi$. Suppose that we have observed six positive response and four negative responses in that order. The likelihood of this happening is $\pi^6 (1 - \pi)^4$. In practice, the order of observation of the responses is arbitrary and we could account for this by multiplying the likelihood we have calculated by the number of ways six positive responses and four negative responses could occur. If this is done the likelihood becomes:

$$\frac{6!}{4!2!}\pi^6(1-\pi)^4 = 15\pi^6(1-\pi)^4$$

which corresponds to a probability from a BINOMIAL DISTRIBUTION. For different values of π we can determine the likelihood and on this basis find the most likely value.

For example, if π has the value 0.1 the likelihood is $15 \times 0.2^6 \times 0.8^4 = 0.000\,009\,8$ and for the values of $\pi = 0.3$, 0.5, 0.7 and 0.9 the corresponding likelihood values are 0.002 63, 0.0146, 0.0143 and 0.000 80 respectively. Therefore, of these four values, 0.5 is the most likely. In fact, we can plot the likelihood values for all potential values of π and

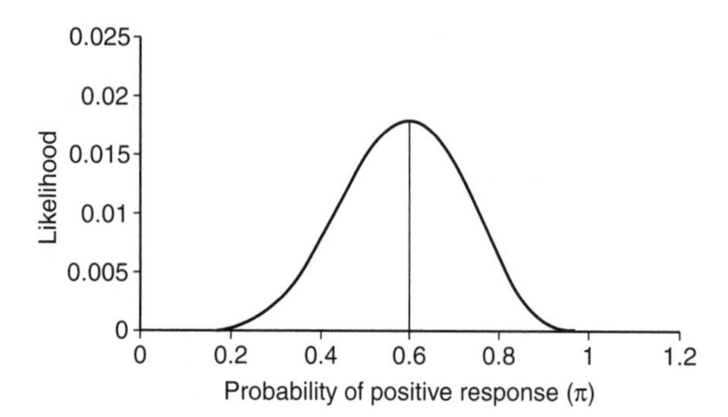

likelihood *Likelihood function based on six positive responses out of 10*

choose the value that gives the maximum, as in the first figure. From the figure, we can conclude that the most likely value for π is 0.6, as it gives the largest likelihood value. This value is the maximum likelihood estimator.

The same approach can be used for other types of data. For example, Altman (1991) gives the following data on the daily energy intake (kJ) of 11 healthy women: 5260, 5470, 5640, 6180, 6390, 6515, 6805, 7515, 7515, 8230, 8770. Assuming that these data arise from a NORMAL DISTRIBUTION with a common MEAN denoted μ and known STANDARD DEVIATION 1100 we can determine the likelihood as a function of the unknown μ and plot it as before. The second figure illustrates this, in which the maximum likelihood occurs at the value 6754.

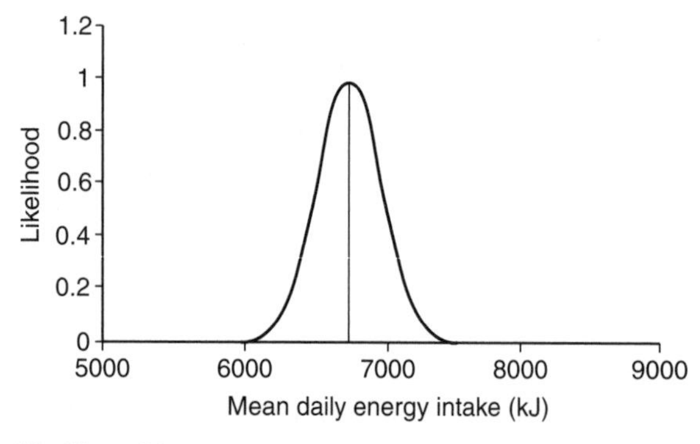

likelihood *Likelihood function for the mean daily energy intake based on a sample of 11 values*

In Bayesian statistics, the likelihood works to modify the PRIOR DISTRIBUTION to yield the POSTERIOR DISTRIBUTION and represents the information contained in the experiment about the parameter of interest. In a formal way, Bayesian analysis proceeds by calculating:

$$p(\theta|\text{Data}) \propto p(\text{Data}|\theta)p(\theta)$$

in which $p(\theta)$ is the prior distribution expressing initial beliefs in the parameter of interest, θ, $p(\theta|\text{Data})$ is the corresponding posterior distribution of beliefs and $p(\text{Data}|\theta)$ is the likelihood. If there is great prior uncertainty about the parameter of interest so that the prior distribution is essentially flat relative to the region in which the likelihood is peaked, then it has little impact on modifying prior beliefs. In such circumstances, the posterior distribution is essentially proportional to the likelihood so that posterior beliefs about the parameter are dictated by the location and shape of the likelihood. In particular, the posterior mode, the value believed to be the most likely after collecting data, is essentially equivalent to the value that maximises the likelihood, i.e. the MAXIMUM LIKELIHOOD ESTIMATION. *AG*

Altman, D. G. 1991: *Practical statistics for medical research.* London: Chapman & Hall.

likelihood ratio The likelihood ratio provides a method for comparing competing hypotheses based on the LIKELIHOOD calculated from experimental data. It also plays a role in Bayesian hypothesis testing (see BAYESIAN METHODS).

Suppose interest lies in learning about the response of patients suffering from influenza symptoms to a new treatment. Data are collected from 10 patients, of whom six respond positively. What can be said about the competing hypotheses H_1: $\pi = 0.3$ and H_2: $\pi = 0.7$? The likelihood of obtaining six positive results and four negative results is:

$$\frac{6!}{4!2!}\pi^6(1-\pi)^4 = 15\pi^6(1-\pi)^4$$

and this can be determined for the competing values of π under the pair of hypotheses. For hypothesis H_1 the value is 0.002 63, while that for H_2 is 0.0143. The ratio of these values is 5.44, the likelihood ratio of H_2 against H_1 indicating, in this instance, that hypothesis H_2 is almost $5^1/_2$ times as likely as H_1, which is strong evidence in favour of H_2 rather than H_1.

The Bayesian equivalent to this form of hypothesis testing is based on determining the ratio of the posterior probabilities of the hypotheses. Formally, we calculate:

$$p(H_i|\text{Data}) \propto p(\text{Data}|H_i)\,p(H_i), \quad i = 1, 2$$

in which $p(H_i)$ is the prior probability of hypothesis H_i, expressing initial beliefs in its veracity, $p(H_i|\text{Data})$ is the corresponding posterior probabilities and $p(\text{Data}|H_i)$ is the likelihood of the hypothesis giving rise to the data. By taking the ratio of the two expressions just given, the ratio of the posterior probabilities of the two hypotheses can be expressed as:

$$\frac{p(H_2|\text{Data})}{p(H_1|\text{Data})} = \frac{p(\text{Data}|H_2)}{p(\text{Data}|H_1)} \times \frac{p(H_2)}{p(H_1)}$$

The left-hand side of this expression is the posterior odds ratio, the first term on the right-side is the likelihood ratio and the second term is the prior odds ratio.

This form of Bayesian analysis is familiar in diagnostic testing. In that context the likelihood ratio is expressed as:

$$\text{Likelihood ratio} = \frac{\text{Probability (positive test result|disease)}}{\text{Probability (positive test result|no disease)}} = \frac{\text{sensitivity}}{1-\text{specificity}}$$

AG

Altman, D. G. 1991: *Practical statistics for medical research*. London: Chapman & Hall.

Likert scales These scales are used to measure the extent to which an individual agrees with a statement. A Likert scale typically has five levels, ranging from 'strongly disagree' to 'strongly agree'. One common alternative is to use an even number of options in order to avoid having a 'neutral' option. A typical Likert scale questionnaire item with five levels is the following:

In a proposed study of mild asthma, it is ethically acceptable to give some participants a placebo treatment.

- strongly disagree
- disagree
- neither agree nor disagree
- agree
- strongly agree

The data from a Likert scale are often coded as a number (e.g. 1 to 5) and it is typical for responses from multiple items to be summed or averaged to provide an overall score related to an underlying issue or LATENT VARIABLE.

When there are multiple items, it is recommended that the order of responses be reversed for some items, to help prevent subjects falling into a simple pattern of responses (e.g. always select 'strongly agree').

The data from one or more Likert scale items are often analysed as interval data. For this to be a justifiable approach, it is important to trial and develop the items properly, perhaps using a pilot study and test for validation and reliability (see MEASUREMENT PRECISION AND RELIABILITY). For further details see DeVellis (1991) and Streiner and Norman (1995). *PM*

[See also FACTOR ANALYSIS]

DeVellis, R. F. 1991: *Scale development: theory and applications*. London: Sage. **Streiner, D. L. and Norman, G. R.** 1995: *Health measurement scales: a practical guide to their development and use*, 2nd edition. Oxford: Oxford University Press.

limits of agreement This approach was developed by Bland and Altman (1986) to assess AGREEMENT in method comparison studies. Based on both graphical techniques and straightforward statistical calculations, it is simple to apply and interpret. It quantifies agreement between two methods through the mean difference (i.e. the estimate of the systematic BIAS of one method relative to the other) and the STANDARD DEVIATION of the differences between measurements taken by the methods on the same subjects (i.e. an indication of the variability of these differences across subjects).

The information provided by these summary statistics is commonly presented visually in a *Bland–Altman plot*, where the difference between measurements are plotted against the average of the measurements taken by the methods on the same subjects (see the figure). This summary information is displayed on the plot by three horizontal lines indicating the mean difference and the mean difference $\pm$ 1.96 standard deviation of the differences. The latter two lines represent the 95% limits of agreement. That is a range within which one would expect 95% of the differences to lie, under the assumptions of normally distributed differences and the uniformity of systematic bias and standard deviation across the whole range of measurements (as indicated by no evidence in the Bland and Altman plot of a relationship between the difference and the average in measurements taken by the two methods on each subject).

The assumption of normality of the differences is generally reasonable, as much of the between-subject variation is removed by the differencing of the measurements between methods; therefore what remains is the measurement error. However, even if this assumption is violated, there may not be any serious consequences resulting from the construction of the limits of agreement as already described. Thus 5% of the differences will still be expected to lie outside the range created by these limits, although most of these differences may be in the same direction.

If there is found to be a relationship between the difference and average, e.g. the plot shows a 'fanning out' effect of the differences as the average increases (i.e. the variability of the differences is increasing with the size of the measurement), then application of the limits of the agreement method, as described above, would produce limits that would be wider apart than needed for lower values of the average and narrower than expected at higher values. Thus it is better to try to either accommodate this relationship or remove it by suitably transforming the data (e.g. using a logarithmic transformation).

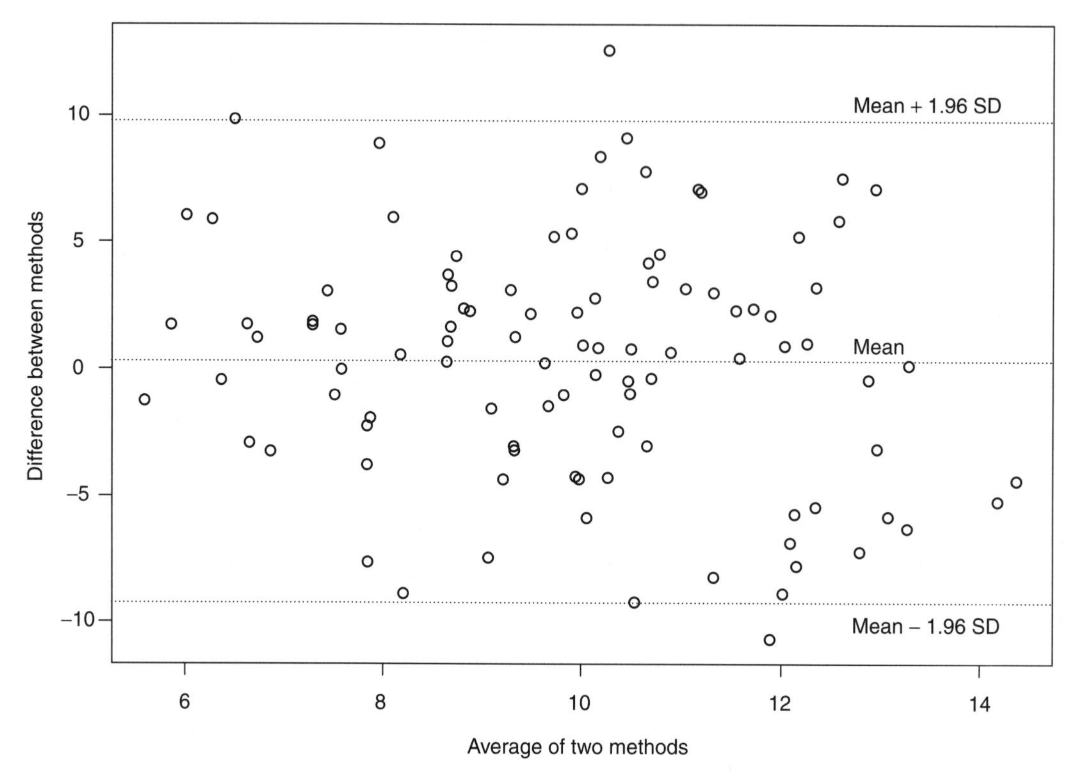

limits of agreement *Bland and Altman plot of the difference against the average for two methods*

Bland and Altman provide comprehensive expositions about the limits of the agreement approach, the issues that result when the assumptions of normality and uniformity of the bias and standard deviation are violated, and the ways to overcome these obstacles in their approach. They also discuss repeatability and replication, and further extensions to their approach. The reader is referred to their articles for more information (see the references below).

Before concluding this entry, further mention must be made of the purpose behind the limits of agreement approach. In medicine interest often lies in comparing two (or more) different methods or techniques for measuring some clinically important quantity, such as carotid stenosis or blood pressure. One of the methods may be already routinely established in clinical practice (e.g. intra-arterial digital subtraction angiography (DSA)), while the other may be a new technique (e.g. contrast-enhanced magnetic resonance angiography (CEMRA)) that needs to be evaluated. Both, however, measure the true quantity of interest with error. Thus neither is a true gold standard, and therefore the question of interest is not whether the new method, say CEMRA, accurately measures stenosis of the carotid artery, as assessed by the established method, DSA. Instead, it is 'Do the different methods of measurement agree sufficiently closely to allow either the new method to replace the old or both to be used interchangeably, with little or no differences arrived at in clinical conclusions or decisions?' For example, if CEMRA is shown to give sufficiently close measurements to DSA, then, as the latter is an invasive and expensive technique that carries with its use a small, but significant risk of stroke or death, justification for using CEMRA, which is a noninvasive technique, over DSA is obtained. Note finally that the setting of the level of acceptable agreement depends on the clinical context of the study, and should be made based on clinical judgement, not on statistical grounds. Further, this decision should be made, in general, a priori of the commencement of the study. *BT*

Altman, D. G. 1991: *Practical statistics for medical research*, London: Chapman & Hall. **Altman, D. G. and Bland, J. M.** 1983: Measurement in medicine: the analysis of method comparison studies. *The Statistician* 32, 307–17. **Bland, J. M. and Altman, D. G.** 1986: Statistical methods for assessing agreement between two methods of clinical measurement. *Lancet* i, 307–10. **Bland, J. M. and Altman, D. G.** 1999: Measuring agreement in method comparison studies. *Statistical Methods in Medical Research* 8, 135–60.

linear mixed-effects models Regression models that include both FIXED EFFECTS and RANDOM EFFECTS, which are also known as MULTILEVEL MODELS or hierarchical models. Mixed-effects models are fitted to data that have a hierarchical or clustered structure, in which individual observations are correlated within clusters. Examples of application areas include longitudinal data, where the measurements taken repeatedly on patients over time are correlated within patients, and multicentre studies, where the measurements taken on patients within centres are correlated. Mixed-effects models include random effects to allow for the correlation of observations within clusters. To illustrate the basic structure, we consider a simple example of a linear mixed-effects model. In a CLINICAL TRIAL comparing the effects of active treatment and control over time, suppose that y_{ij} is the response measured on subject j ($j=1, \ldots, J$) at time t_{ij} ($i=1, \ldots, I_j$), and let the treatment group allocated to each subject be indicated by $x_{ij}=0/1$. A simple mixed-effects model for the data y_{ij} includes random effects for patients to acknowledge that the response tends to differ between patients and that repeated measurements taken on the same patient are therefore alike. This model is written as:

$$y_{ij} = (\alpha + u_j) + \beta t_{ij} + \gamma x_{ij} + e_{ij}$$

where the u_j are random patient effects and the e_{ij} are random residual effects, which represent the variability between measurements within patients. The parameters α, β and γ are fixed effects that represent, respectively, the overall mean response in the control group (where $x_{ij}=0$), the trend in response over time and the treatment effect, which is constant over time. The two sets of random effects u_j and e_{ij} are independent and it is usual to assume these to be normally distributed, $u_j \sim N(0, \sigma_u^2)$ and $e_{ij} \sim N(0, \sigma_e^2)$. This basic mixed-effects model for the data can be extended in a number of interesting ways. For example, we could allow the trend in response over time to vary from one patient to another, we could include additional explanatory variables or we could allow the effect of treatment to vary over time. For a range of extended mixed-effects models and guidance on their interpretation, readers are referred to the full entry on this subject area, titled MULTILEVEL MODELS (see also Everitt and Pickles, 2004; Pinheiro and Bates, 2000). This entry provides details of methods and software for estimation of mixed-effects models and also covers topics such as handling of missing data and complex applications. *RT*

[See also GENERALISED ESTIMATING EQUATIONS, MULTILEVEL MODELS]

Everitt, B. S. and Pickles, A. 2004: *Statistical aspects of the design and analysis of clinical trials*. 2nd edition, London: Imperial College Press. **Pinheiro, J. C. and Bates, D. M.** 2000: *Mixed effects models in S and S-PLUS*. New York: Springer Verlag.

linear regression See MULTIPLE LINEAR REGRESSION

linkage disequilibrium See ALLELIC ASSOCIATION

LISREL See STRUCTURAL EQUATION MODELLING SOFTWARE

LMS method This is a method used for constructing AGE-RELATED REFERENCE RANGES, typically applied to GROWTH CHARTS. The underlying age-specific FREQUENCY DISTRIBUTION of the measurement (typically anthropometry such as height or weight, though it can be applied to any ratio scale measurement) is summarised by three age-varying parameters representing the first three moments of the distribution. The first is the MEDIAN as a MEASURE OF LOCATION, the second is the COEFFICIENT OF VARIATION or CV as a MEASURE OF SPREAD or scale and the third is the Box–Cox power transformation (see TRANSFORMATIONS) needed to adjust for SKEWNESS, as a measure of shape. KURTOSIS is assumed to match that of the NORMAL DISTRIBUTION, and is not estimated as such. Adjusting for skewness ensures a symmetric distribution, so that the MEAN on the transformed scale is also the median on the original scale. The CV is estimated rather than the STANDARD DEVIATION (SD), as the SD often increases with age in proportion to the mean, whereas the CV is relatively uncorrelated with age.

The original publication describing the LMS method used the notation λ for the Box–Cox POWER, μ for the median and σ for the CV – hence the LMS method. The three age-related curves are referred to as the L curve, M curve and S curve respectively, and together they allow any required QUANTILE of the distribution to be constructed as a smooth function of age. Equally they allow individual measurements to be expressed as a standardised residual or *z*-SCORE adjusted for skewness. See GROWTH CHARTS for an example of a centile chart constructed using the LMS method.

The LMS method is a semi-parametric regression model where the three parameters of the distribution are estimated as generalized additive cubic smoothing spline curves (see SCATTERPLOT SMOOTHERS) using penalized LIKELIHOOD. The only analytical decision to make when fitting the model is to specify the number of equivalent degrees of freedom (eDoF) required for each of the three smoothing spline curves, so that they are neither under- nor oversmoothed. Criteria such as AKAIKE'S INFORMATION CRITERION or the Bayesian information criterion (Everitt and Skrondal, 2010) are useful here to trade off improved fit against increased model complexity. For infant anthropometry the M curve is often steep at birth and progressively shallower with increasing age. Transforming the age scale can help here, e.g. with a square root transformation to stretch younger ages and shrink older ages, as this tends to linearise the M curve, simplify the curve shape and improve the fit.

The LMS method is a special case of the family of GENERALIZED ADDITIVE MODELS for location, scale and shape (GAMLSS). These are powerful models that can be applied to many different distributions, where up to four moments of the distribution are estimated as separate generalised additive regression models. For further details see Cole (1988), Cole and Green (1992) and Rigby and Stasinopoulos (2005). *TJC*

Cole, T. J. 1988: Fitting smoothed centile curves to reference data (with discussion). *Journal of the Royal Statistical Society, Series A* 151, 385–418. **Cole, T. J. and Green, P. J.** 1992: Smoothing reference centile curves: the LMS method and penalized likelihood. *Statistics in Medicine* 11, 1305–19. **Everitt, B. S. and Skrondal, A.** 2010: *Cambridge dictionary of statistics*, 4th edition. Cambridge: Cambridge University Press. **Rigby, R. A. and Stasinopoulos, D. M.** 2005: Generalized additive models for location, scale and shape (with discussion). *Applied Statistics* 54, 507–44.

local research ethics committee (LREC) See ETHICAL REVIEW COMMITTEES

locally weighted regression See SCATTERPLOT SMOOTHERS

loess See SCATTERPLOT SMOOTHERS

logistic discrimination See DISCRIMINANT FUNCTION ANALYSIS

logistic regression A form of regression analysis to be used when the response is a binary variable. Medical outcomes often have only two possibilities. Whether a patient is dead or alive is the most obvious, but the presence or absence of particular diagnoses, symptoms or signs are also examples of binary or dichotomous variables. Hypertension, obesity and airways obstruction are diagnoses that result from observing that a particular measurement is above or below a particular value, thus creating a binary outcome from a continuous measurement.

Methods for the analysis of binary data differ from those for continuous variables. First, the summary statistic to describe the results is a proportion or percentage of individuals who are dead (or alive), have the symptom or in general have the designated outcome. Data from a continuous variable are summarised by the MEAN and STANDARD DEVIATION or MEDIAN and INTERQUARTILE RANGE, as how variable the values are is required as well as a typical value, while for binary data the proportion or percentage tells us everything. Second, when we analyse a binary outcome in relation to EXPLANATORY VARIABLES we cannot use STUDENT'S *t*-TEST, ANALYSIS OF VARIANCE or MULTIPLE LINEAR REGRESSION, as the data are not normally distributed, do not have the same VARIANCE for groups with different outcome proportions and predictions of proportions must not fall outside the range zero to one, which can happen if multiple linear regression of a proportion is used.

Binary data can be analysed in relation to a single categorical explanatory variable using the CHI-SQUARE TEST, but very frequently it is necessary to analyse a binary outcome in relation to several explanatory variables, some or all of which may be continuous. For example, in a study that investigated whether reported wheeze is related to the use of gas for cooking it would be desirable to take age and gender into account, and also conditions in the home, such as an extractor fan, that might affect the concentration of the combustion product thought to be responsible for any increase in symptoms. Alternatively, we might want to analyse wheeze in relation to the use of gas for cooking and passive smoking simultaneously. To analyse binary data and adjust the relation to the factor of primary interest for confounding variables or to determine to which of several potential explanatory variables the outcome is related, an analogue of analysis of variance and multiple regression is required. Logistic regression meets these requirements.

An explanation of the method is easiest in relation to an example. Logistic regression has been used to describe the distribution of age at menarche in girls and the factors associated with early or delayed menarche. Roberts, Rozner and Swan (1971) carried out a cross-sectional survey of girls in South Shields, County Durham, in 1967. Data are shown in the first table.

logistic regression *Number of girls and number recorded as menstruating, by age group*

Age group	*No. of girls*	*No. menstruating*	*% menstruating*
11 – < 12	82	4	4.9
12 – < 13	304	76	25.0
13 – < 14	366	178	48.6
14 – < 15	351	285	81.2
15 – < 16	216	209	96.8

The percentage of girls who had reached menarche, of course, increased with age, being very low in the youngest age group and very high in the oldest. Had younger age groups been included the percentage menstruating would have been less than 4.9% and 0% if sufficiently low. Similarly, the percentage would have been close to 100% in older age groups. The relation of proportion or percentage menstruating to age can be described by an S-shaped (or 'sigmoid') curve. The data from the first table are plotted together with a fitted smooth sigmoid curve in the figure.

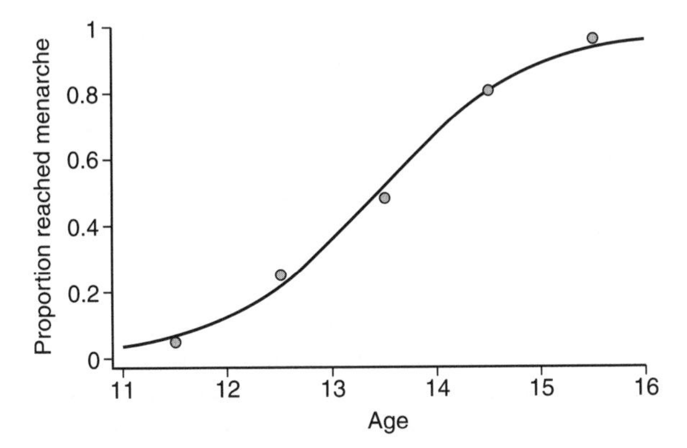

logistic regression *Cumulative logistic curve*

The curve shown is a cumulative logistic curve, selected from the family of such curves so that it best describes the data. Its formula is:

$$\pi = \frac{1}{1 + e^{-(\alpha+\beta x)}}$$

where π is the proportion menstruating by age x and α and β are the parameters that describe the best fitting curve. These parameters were estimated to be -18.40 and 1.37 respectively to fit the curve shown. Median age at menarche is estimated at $\pi = 0.5$, i.e. where $-(\alpha + \beta x) = 0$ or $x = -\alpha/\beta$, so was 13.4 years from these data. As the logistic distribution is symmetric this is an estimate of mean age at menarche.

The equation defining π can be rewritten:

$$\log_e \frac{\pi}{1-\pi} = \alpha + \beta x$$

The left-hand side of this equation is known as the *logistic transformation* of the proportion π. Its effect is to stretch the scale, so that the transformed variable can take values from minus infinity ($-\infty$), corresponding to $\pi = 0$, to plus infinity ($+\infty$), corresponding to $\pi = 1$, and also to linearise the relation with age. Fitting the logistic curve can therefore be achieved by least squares regression of the logistic transform of π on age (see LEAST SQUARES ESTIMATION), except that the transformation does not achieve homogeneity of variance and so iteratively weighted least squares regression is required. However, most modern computer programs use MAXIMUM LIKELIHOOD ESTIMATION, which also requires iteration, and individual rather than grouped data are usually analysed. Full specification of the binary outcome y for individuals requires that y is distributed as a binomial distribution with parameter 1 (here also known as a Bernoulli distribution) and success probability π.

The logistic curve is not the only curve that could be fitted to describe the data. A fitted cumulative NORMAL DISTRIBUTION would be almost indistinguishable from the logistic curve. Fitting a normal distribution before the days of electronic computing was known as *probit analysis* (see PROBIT MODELS), a probit being a normal equivalent deviate with 5 added to avoid negative numbers and was developed for use in pharmacology (Finney, 1971). The distribution of the dose of a toxic substance required to kill a given strain of animal is known as the *tolerance distribution*. It cannot usually be observed directly, but if groups of animals are given different doses of the drug and the proportions dying are recorded, then a sigmoid curve of proportion with dose is observed that describes the cumulative tolerance distribution.

Finney (1964) ascribed the logistic transformation to Berkson and showed the close agreement of the normal and logistic distributions, but favoured the normal distribution to describe the tolerance distribution of drug toxicity. Hence, in general, the normal or probit transformation was used when there was an underlying tolerance distribution. An exception was age at menarche; it became accepted that the logistic transformation should be used (Finney, 1971) as one study apparently found a better fit of the logistic transformation than of the normal distribution.

Just as linear regression can be extended to multiple regression and also incorporate categorical explanatory variables, so can logistic regression. A multiple logistic regression equation can be written:

$$\log_e \frac{\pi}{1-\pi} = \alpha + \beta_1 x_1 + \beta_2 x_2 + \beta_3 x_3 + \cdots$$

where the x_i can be continuous or dummy variables to indicate categories of groups. For example, Roberts, Danskin and Chinn (1975) analysed age at menarche in relation to family size, in categories of one, two, three, four and at least five children. In the model shown graphically in the paper, x_1 was age, and four dummy variables x_2 to x_5 were used to describe the differences in median age at menarche between the five family size groups, corresponding to fitting parallel sigmoid curves. The estimates b_i of the β_i are known as logistic regression coefficients.

To return to our first example: if the presence or absence of wheeze is the outcome and the presence or absence of a gas cooker the independent or explanatory variable, with no other factors considered for the moment, then if the dummy (indicator) variable x is 0 for absence and 1 for presence of a gas cooker, then we have:

$$\log_e \frac{\pi_{\text{nogas}}}{1-\pi_{\text{nogas}}} = \alpha$$

$$\log_e \frac{\pi_{\text{gas}}}{1-\pi_{\text{gas}}} = \alpha + \beta$$

Hence:

$$\log_e \left[\frac{\pi_{\text{gas}}}{1-\pi_{\text{gas}}}\right] - \log_e \left[\frac{\pi_{\text{nogas}}}{1-\pi_{\text{nogas}}}\right] = \beta$$

Each of the terms within brackets is an odds and a difference in log odds, when antilogged, is an ODDS RATIO. Antilogging both sides of the equation gives:

$$\frac{\pi_{\text{gas}}(1-\pi_{\text{nogas}})}{\pi_{\text{nogas}}(1-\pi_{\text{gas}})} = e^{\beta}$$

The odds ratio on the left-hand side of this equation is the odds of having wheeze in the presence of a gas cooker divided by the odds of having wheeze if no gas cooker is present. It will approximate the relative risk, or risk ratio, of wheeze in the presence of a gas cooker compared to a no gas cooker provided that the prevalence of wheeze is low. This relative risk is estimated by e^{β}. The differences of this example from the age at menarche example are that π for wheeze is unlikely to be greater than, say, 0.2, and no 'tolerance distribution' analogous to that of age at menarche is directly specified, although one can envisage this being the distribution of whatever product of combustion of gas is responsible for increased wheeze in people in homes with gas cookers. Even with such a distribution specified, it is unlikely that exposure would ever be high enough to cause 100% wheeze, so, in practice, only the lower portion of the sigmoid curve is relevant.

As with applications in general, it was the availability of software that led to an expansion in the use of logistic regression, in particular GLIM (generalised linear interactive modelling) in the early 1970s. Although now largely superseded, notably but not exclusively by Stata, GLIM enabled unbalanced analysis of variance, multiple linear regression and multiple logistic regression models to be fitted within the same framework. The application of logistic regression in epidemiology and public health journals showed a steep rise from around 1980 (Hosmer, Taber and Lemshow, 1991; Chinn, 2001). Odds ratios were used in epidemiology and, in particular, for the results of a CASE-CONTROL STUDY, before the widespread availability of computers and statistical software enabled easy fitting of multiple logistic regression models; therefore logistic regression was readily adopted by epidemiologists. It was also established as the appropriate method for the analysis of case-control studies with adjustment for confounding. When cases and controls are individually matched the method of analysis is conditional logistic regression. Most statistical software for logistic regression requires the binary outcome to be coded 0 and 1, with 1 for the 'positive' outcome.

Like all estimates from a sample, an odds ratio has an associated CONFIDENCE INTERVAL. The NULL HYPOTHESIS of no relation to an explanatory variable is an odds ratio of one or equivalently zero for the corresponding logistic regression coefficient. Older papers gave logistic regression coefficients with standard errors or 95% confidence intervals, but more recent papers give odds ratios with 95% confidence intervals.

For example, Somerville, Rona and Chinn (1988) gave logistic regression coefficients in a study of passive smoking by children in a survey of 5- to 11-year-old children in England and Scotland. One result is shown in the first line of the second table. The logistic regression coefficient of the symptom, reported by a parent in a self-administered questionnaire, 'chest EVER sounds wheezy or whistling', on passive smoking as measured by the total number of cigarettes per day reported to be smoked by the parents, was 0.011 with a standard error of 0.005. By calculating the coefficient $\pm 1.96 \times$ standard error, a 95% confidence for the logistic regression coefficient can be obtained. Antilogging (base e) the coefficient and each limit of its confidence interval gives the odds ratio and its 95% confidence interval in the second line. However, the odds ratio associated with exposure to just one cigarette a day is not very useful; 20 cigarettes a day represents a more common exposure of children who are exposed to passive smoking. To obtain the odds ratio associated with exposure to 20 cigarettes a day, multiply both the logistic regression coefficient and its standard error by 20 and repeat the confidence interval calculation and antilogging to obtain the third line of the table.

logistic regression *Alternative presentations of result of logistic regression analysis, illustrated by 'chest EVER sounds wheezy or whistling' in relation to passive smoking for children in the National Study of Health and Growth (Somerville, Rona and Chinn, 1988)*

Quantity presented	*Result*
Logistic regression coefficient ± standard error on total number of cigarettes smoked at home by father and mother	0.011 ± 0.005
Odds ratio per cigarette smoked (95 % confidence interval)	1.011 (1.001 to 1.021)
Odds ratio per 20 cigarettes smoked (95% confidence interval)	1.246 (1.024 to 1.516)

Although the evidence against the null hypothesis was not strong ($P \sim 0.028$) and the 95% confidence interval correspondingly wide, the results in the third line show that the size of the likely effect is not negligible, which could not be easily appreciated from either of the first two rows. Note that the confidence interval for the logistic regression coefficient is symmetric around the estimate, but that for an odds ratio it is not.

It is tempting to interpret the third line of the second table as meaning that exposure to 20 cigarettes smoked a day in the home results in an increased risk of wheeze of between 2.4% and 51.6%. This is interpreting an odds ratio as if it were a relative risk, which is only justified if the prevalence of wheeze is low, say less than 10% (Zhang and Yu, 1998). In this case it was 10.9%, so perhaps not too misleading, but it is easy to find examples of incorrect interpretation of odds ratios in the medical literature (Chinn, 2001). Although the fact that the odds ratio is biased away from the null value of 1 as an estimate of relative risk is well known to statisticians and epidemiologists, it is often conveniently ignored in the medical literature, especially in the reporting of results of CROSS-SECTIONAL STUDIES. In fact, it is possible to estimate relative risk directly, by binomial regression, but at the expense of the iterative model fitting sometimes failing to converge (Chinn, 2001).

Logistic regression is essential for the analysis of unmatched case-control studies and is likely to continue to be the most used method for the analysis of binary outcomes in cross-sectional studies. Statistically it cannot be faulted; it is in the reporting, and the fact that an odds ratio does not estimate relative risk directly, that the problem lies. Binary outcomes in COHORT STUDIES should be analysed by SURVIVAL ANALYSIS, unless the follow-up time is constant, which is rarely the case.

The *P*-VALUE associated with the odds ratio, to test a difference from 1, can be obtained by dividing the logistic regression coefficient by its STANDARD ERROR and comparing the result with the normal distribution, as the null hypothesis value for the logistic regression coefficient is zero. Note that the normal distribution is used rather than the *t*-DISTRIBUTION, as no residual standard deviation is estimated. This is because a binomial distribution is assumed for the observations, which is specified only by the expected proportion and does not involve a standard deviation. Alternatively, if the model were fitted by MAXIMUM LIKELIHOOD, the LIKELIHOOD RATIO test can be used and will usually give approximately the same answer for a single parameter. If model 1 is the model with the factor of interest included, with likelihood l_1, and model 2 that with it omitted, with likelihood l_2, then $-2\log(l_2/l_1)$ has a CHI-SQUARE DISTRIBUTION with DEGREES OF FREEDOM equal to the difference in the number of fitted parameters. This can be used to test the equivalence of several parameters, e.g. equal median age at menarche for girls from different sizes of family (Roberts, Danskin and Chinn, 1975).

Related to testing for association of outcome with risk factors is that of goodness of fit of the model. This is more difficult to assess than with a linear regression model, as individual values are each 0 or 1, so a plot of observed against fitted values, or of residuals, is uninformative. For associated reasons the overall likelihood ratio statistic cannot be used. Hosmer, Taber and Lemeshow (1991) give a number of plots that can be used and the necessary calculations are implemented in Stata.

Logistic regression as described here for a binary outcome is a special case of the more general multinomial, or polytomous, logistic regression for a categorical outcome with three or more possible values (see LOGIT MODELS FOR ORDINAL RESPONSES). It is also closely related to the LOG-LINEAR MODEL, which assumes a POISSON DISTRIBUTION for the count in each cell of a contingency table. Each is an example of a GENERALISED LINEAR MODEL.

Medical journals now frequently report results from multiple logistic regression, showing odds ratios, *P*-values and confidence intervals. These need to be read carefully, as seemingly similar tables may be used in different situations. The lines of the table may be for different binary outcomes or independent analyses of the same outcome with different explanatory variables. The odds ratios will often be adjusted, for a list of stated variables such as age and sex, although unadjusted odds ratios may also be shown. An example is shown in second table of Lawlor, Patel and Ebrahim (2003), in which odds ratios of falls in women aged 60 to 79 with drug use are given. Each row of the table gives results for one class of drug, while there are columns for 'crude', i.e. unadjusted, and fully adjusted odds ratios for each of three outcomes: any falls, two or more falls and falls where medical attention was given. The variables used to adjust the fully adjusted odds ratios are listed as a footnote to the table.

Other papers give odds ratios that are mutually adjusted for other factors shown in the same table of results, i.e. all the results come from a single multiple logistic regression and full information is given, while Lawlor, Patel and Ebrahim (2003) (described earlier) appear to have carried out 21 adjusted analyses (three outcomes by seven drug classes). (For an example of mutually adjusted odds ratios see Slap *et al.* (2003), first table in the abridged printed version, second table in the full electronic version.)

Particularly where all results shown are 'statistically significant' (SS), the reader needs to ascertain whether all factors in the model are shown and whether the final model was selected from a set of possible models. This is appropriate if either the question is 'What factors are associated with the outcome?' or a parsimonious model is required for prediction purposes and selected either by forwards or backwards stepwise elimination. However, as with a similar procedure with multiple regression, it must be understood that prediction on a further dataset will not be as good as on the one from which the prediction was derived, and exclusion or inclusion of factors with *P*-values close to the chosen critical value may not be reproducible.

By the same token, however, when there is a stated hypothesis, the odds ratio of interest should ideally be adjusted for all factors determined a priori to be of potential importance. Some of these may not be associated with outcome in the data at the conventional level of statistical significance, but adjustment can still affect the odds ratio of interest. It is useful when there may be controversy over the number of potentially confounding variables to be included to give both unadjusted, fully adjusted and, perhaps also, partially adjusted odds ratios. *SC*

Chinn, S. 2001: The rise and fall of logistic regression. *Australasian Epidemiologist* 4, 7–10. **Finney, D. J.** 1964: *Statistical method in biological assay*, 2nd edition. London: Griffin. **Finney, D. J.** 1971: *Probit analysis*, 3rd edition. Cambridge: Cambridge University Press. **Hosmer, D. W., Taber, S. and Lemeshow, S.** 1991: The importance of assessing the fit of logistic regression models: a case study. *American Journal of Public Health* 81, 1630–5. **Lawlor, D. A., Patel, R. and Ebrahim, S.** 2003: Association between falls in elderly women and chronic diseases and drug use: cross-sectional study. *British Medical Journal* 327, 712–15. **Roberts, D. F., Rozner, L. M. and Swan, A. V.** 1971: Age at menarche, physique and environment in industrial northeast England. *Acta Paediatrica Scandinavica* 60, 158–64. **Roberts, D. F., Danskin, M. J. and Chinn, S.** 1975: Menarcheal age in Northumberland. *Acta Paediatrica Scandinavica* 64, 845–52. **Slap, G. B., Lot, L., Huang, B., Daniyam, C. A., Zink, T. M. and Succop, P. A.** 2003: Sexual behaviour of adolescents in Nigeria: cross-sectional survey of secondary school students. *British Medical Journal* 326, 15–18. **Somerville, S., Rona, R. J. and Chinn, S.** 1988: Passive smoking and respiratory conditions in primary school children. *Journal of Epidemiology and Community Health* 42, 105–10. **Zhang, J. and Yu, K. F.** 1998: What's the relative risk? *Journal of the American Medical Association* 280, 1690–1.

logit models for ordinal responses A regression model is a statistical model for describing the relationship between one or more explanatory variables and the response (dependent) variable. The purpose of statistical modelling is to fit the best model from a medical and epidemiological point of view that describes this relationship. The statistical modelling of how the relationship between the explanatory variables and the response variable could be described depends on how the response variable is recorded. The *linear regression model* assumes continuous quantitative response values. When the response variable has only two possible values or is measured on a rating scale, a *logit transformation* of the response values will meet the assumption of continuity.

A simple *linear regression model* describes how much a continuous quantitative response variable (y) depends on the explanatory (x) variable by the expression $y = a + bx$, where a is a constant, the *intercept*, and b is the *regression coefficient*, which contains the important information about the dependence of y on x. According to the model, y will change b units when x increases 1 unit. In a *multiple regression model* a linear combination of several explanatory variables are included. The purpose could be to investigate how the response variable depends on all explanatory variables together. Some of the variables could also be included in the model as *confounding* factors, which means that they would disturb the relationship of interest if not being adjusted for.

In the case of only two responses, success/failure or diseased/nondiseased, the range of possible response values is between zero and one; e.g. when the probability of success is $p = 0.8$, then the probability of failure is $(1 - p) = 0.2$. As the modelling assumes unlimited possible continuous values, the explanatory variable will be linked to the response variable by a *logit transformation*. Then the *odds* of success is the ratio between the probability of success and the probability of failure: odds $= p/(1 - p)$. The logit of the proportion p is defined as the *log odds* $= \text{logit}(p) = \ln p/(1 - p)$, where ln denotes the logarithm to the base e. The regression model is called a (linear) LOGISTIC REGRESSION model, logit $(p) = a + bx$, and the multiple logistic regression model is $\text{logit}(p) = a + b_1x_1 + b_2x_2 + \cdots + b_kx_k$, when k explanatory variables are included. The interpretation of the relationship between an explanatory variable x and the probability p of success is that when x increases 1 unit, the odds will change e^b. For example, $\text{logit}(p) = 3.2 + 1.3x$ means that the logit(p), or the *log odds*, is predicted to change 1.3 for each unit of increase in x and hence the *odds* of success will change $e^{1.3} = 3.7$.

Logistic regression is commonly used to compare the odds of success between two groups of subjects with and without some prognostic property, such as smoking habits. For illustration, consider a model for having a specific disease, $\text{logit}(p) = 3.2 + 1.3$ age $+ 0.4$ smoking, where the prognostic variables are age (years) and smoking habits coded as smokers $= 1$ and nonsmokers $= 0$. Assuming the same age in the two groups, the logit for smokers is logit $(p_s) = 3.2 + 1.3$ age $+ 0.4$ and for nonsmokers $\text{logit}(p_{\text{nons}}) = 3.2 + 1.3$ age $+ 0$. The difference between these logits is $\text{logit}(p_s) - \text{logit}(p_{\text{nons}}) = 0.4$, which is a difference between the log odds of disease in smokers and nonsmokers.

This difference between logarithms is the same as a ratio, in this case the log odds ratio, lnOR. Thus, lnOR $= 0.4$, which was the regression coefficient associated with the variable smoking; then OR $= e^{0.4} = 1.5$. According to this example, we can predict that the odds of having the disease are related to smoking habits and are predicted to be 1.5 times larger in smokers than in nonsmokers, after adjustment for age.

The logit transformation makes it possible to model how a dichotomous response variable depends on the explanatory variables. The logit transformation is also suitable for *ordered categorical (ordinal) responses*, provided there is dichotomisation of the response categories. Consider a four-point scale with the categories 'none', 'slight', 'moderate' and 'severe'. Assume that the numbers of observations in the categories are n_1, n_2, n_3 and n_4 respectively and the total number of observations is n. The *cumulative, continuation-ratio* and *adjacent-categories logits* are three approaches to creating dichotomous datasets considering the ordered structure of the ordinal responses.

In the *cumulative logit*, also called the *proportional odds model*, the probability of being in the lower categories is compared with the probability of being in the higher. Empirically, the number of observations in categories representing lower levels is compared with the number of observations in the higher levels of the scale. There are $(m - 1)$ possible cut-off points between categories in a scale with m categories when creating cumulative logits. In the four-point scale there are three possible cumulative logits; when the cut-off point is the first category, 'none', the cumulative logit is $\ln [n_1/(n_2 + n_3 + n_4)]$; by moving the cut-off point one category at a time the cumulative logits will be $\ln [(n_1 + n_2)/(n_3 + n_4)]$ and $\ln [(n_1 + n_2 + n_3)/n_4]$. In cumulative logits, all data are used in each logit.

The first cumulative logit could be interpreted as the log odds of the response 'none', as compared with 'slight', 'moderate' and 'severe'. If the variable is pain this cut-off point seems reasonable. Absence of pain is compared with presence of pain, but the other cumulative logits could also be of interest in a logistic model.

In the *continuation-ratio approach*, the number of observations in one category is compared with the number of observations in all categories representing lower levels. In the four-point scale the continuation-ratio logits are $\ln[n_2/n_1]$, $\ln[n_3/(n_1 + n_2)]$ and $\ln[n_4/(n_1 + n_2 + n_3)]$.

In the *adjacent-categories logit*, adjacent categories are compared; in the four-point scale the logits are $\ln(n_2/n_1)$, $\ln(n_3/n_2)$ and $\ln(n_4/n_3)$ and this approach is also applicable to categorical/nominal data.

After dichotomisation, the logits for ordinal data can be used in the logistic regression model for dichotomous data and with corresponding interpretation of odds ratios, when evaluating possible relationships between dichotomised ordinal responses and some prognostic variable, when controlling for other prognostic or disturbing background variables. For further details see Agresti (1984), Altman (2000) and Campbell (2001). *ES*

[See also LINEAR REGRESSION, LOG-LINEAR MODELS, MULTIPLE REGRESSION MODELS]

Agresti, A. 1984: *Analysis of ordinal categorical data*. New York: John Wiley & Sons, Inc. **Altman, D. G.** 2000: *Practical statistics for medical research*. Boca Raton: Chapman & Hall/CRC. **Campbell, M. J.** 2001: *Statistics at square two*. Bristol: BMJ Books.

log-linear models These are models that serve to describe the relationships between frequencies (counts) and one or more variables that affect their size. In practice, log-linear models are most often used in connection with CONTINGENCY TABLES to describe the nature of associations between multiple nominal categorical variables. The analysis of contingency tables formed from three or more categorical variables will be the primary concern of this entry, since two-way contingency tables are dealt with in the entry mentioned earlier.

When a sample from some population is classified with respect to more than two qualitative variables, the resulting data can be displayed as a multiway contingency table. As an example, we consider the three-way contingency table resulting from classifying 1330 patients according to blood pressure, serum cholesterol and coronary heart disease (see the first table). In three-way tables 'layering' is

log-linear models *Cross-classification of patients with respect to three clinical variables discussed in Ku and Kullback (1974)*

Coronary		*Serum cholesterol*				*Total*
heart disease	*Blood pressure*	*< 200 mg/100cc*	*200–219*	*220–259*	*> 260*	
Yes	<127 mm Hg	2	3	3	4	12
	127–146	3	2	1	3	9
	147–166	8	11	6	6	31
	>167	7	12	11	11	41
Total		20	28	21	24	93
No	<127 mm Hg	117	121	47	22	307
	127–146	85	98	43	20	246
	147–166	119	209	68	43	439
	>167	67	99	46	33	245
Total		388	527	204	118	1237
Overall total		408	555	225	142	1330

used to accommodate the levels of the third categorical variable (heart disease).

Several independence hypotheses might be of interest in the three-way contingency table. These correspond to different combinations of *first-order relationships* between pairs of categorical variables:

(1) *mutual independence* of the three variables, i.e. none of the pairs of variables is associated;
(2) *partial independence*, i.e. an association exists between two of the variables, both of which are independent of the third;
(3) *conditional independence*, i.e. two of the variables are independent in each level of the third, but each may be associated with the third variable;
(4) *mutual association*, i.e. each pair of variables is associated within each level of the third variable.

In addition, the three variables in a three-way contingency table may display a more complex form of association, namely what is known as a *second-order relationship*. This means that the type and/or degree of association between two categorical variables is different in some or all levels of the remaining variable. In theory, in a k-dimensional table relationships up to $(k-1)$th order can be investigated but the interpretation of higher order relationships becomes increasingly more difficult.

For some of the hypotheses of interest in multiway tables, the corresponding expected values under the NULL HYPOTHESIS can be calculated directly from appropriate marginal totals, but for others some form of iterative fitting algorithm is needed (see Everitt, 1992, for details).

The basic idea of log-linear modelling is to translate the different hypotheses of interest in a multiway table into a sequence of statistical models so as to provide a systematic approach to the analysis of complex multidimensional tables and, in addition, to provide estimates of the magnitudes of effects of interest.

The analysis of three-dimensional tables poses entirely new conceptual problems as compared with those in two dimensions. However, the extension from tables of three dimensions to four or more, while becoming more complex in analysis and interpretation, poses no further new problems and here description of the analysis of higher order contingency tables will be in terms of those arising from three categorical variables.

The nomenclature used for dealing with the $r \times c$ table is easily extended to deal with a three-dimensional $r \times c \times l$ contingency table having r rows, c columns and l layers. The observed frequency in the ijkth cell is now represented by n_{ijk} for $i = 1, 2, \ldots, r$, $j = 1, 2, \ldots, c$, $k = 1, 2, \ldots, l$. The general model is:

$$\ln(\mathrm{F}_{ijk}) = \text{linear function of parameters}$$

where F_{ijk} are theoretical expected frequencies in a three-way table under a particular hypothesis. A *saturated model* for the F_{ijk}, i.e. a model that explains all the variation in the data, is given by:

$$\ln(F_{ijk}) = u + u_{1(i)} + u_{2(j)} + u_{3(k)} + u_{12(ij)} + u_{13(ik)} + u_{23(jk)} + u_{123(ijk)}$$

where u is an unknown parameter referred to as an 'overall mean effect' since all the other model terms are restricted to be deviation terms; $u_{1(i)}$ with $\sum_i u_{1(i)} = 0$ is an unknown deviation term that varies with the level of variable 1 and is called the 'main effect of variable 1'; $u_{2(j)}$ with $\sum_j u_{2(j)} = 0$ is an unknown deviation term that varies with the level of variable 2, the so-called 'main effect of variable 2'; $u_{3(k)}$ with $\sum_k u_{3(k)} = 0$ is an unknown deviation term that varies with the level of variable 3, the so-called 'main effect of variable 3'; $u_{12(ij)}$ with $\sum_i u_{12(ij)} = 0$ for all $j \in \{1, \ldots, c\}$ and $\sum_j u_{12(ij)} = 0$ for all $i \in \{1, \ldots, r\}$ is a further unknown deviation term for the ith category of variable 1 and the jth category of variable 2, the so-called 'interaction between variables 1 and 2'; $u_{13(ik)}$ with $\sum_i u_{13(ik)} = 0$ for all $k \in \{1, \ldots, l\}$ and $\sum_k u_{13(ik)} = 0$ for all $i \in \{1, \ldots, r\}$ is a further unknown deviation term for the ith category of variable 1 and the kth category of variable 3, the so-called 'interaction between variables 1 and 3'; $u_{23(jk)}$ with $\sum_j u_{23(jk)} = 0$ for all $k \in \{1, \ldots, l\}$ and $\sum_k u_{23(jk)} = 0$ for all $j \in \{1, \ldots, c\}$ is a further unknown deviation term for the jth category of variable 2 and the kth category of variable 3, the so-called 'interaction between variables 2 and 3'; $u_{123(ijk)}$ with $\sum_i u_{123(ijk)} = 0$ for all j and $\sum_j u_{123(ijk)} = 0$ for all i and k and $\sum_k u_{123(ijk)} = 0$ for all i and j is yet another unknown deviation term for the ith category of variable 1 within the jth category of variable 2 and the kth category of variable 3, the so-called 'three-way interaction'.

The main effect terms in the second to fourth of these terms serve to model the single variable marginal distributions. The two-way interaction terms in the fifth to seventh terms model the first-order relationships. Different combinations of absence/presence of the three two-way interactions correspond to the mutual, partial, conditional independence or mutual association hypotheses. The three-way interaction term in the eighth term models the two-way relationship. For example, for the data in our first table we might compare the following sequence of models:

(1) all cell frequencies are the same:

$$\ln(F_{ijk}) = u$$

(2) marginal totals for variable 2 (say cholesterol) and 3 (say heart disease) are equal:

$$\ln(F_{ijk}) = u + u_{1(i)}$$

log-linear models *Identification of an adequate log-linear model for the data in the first table*

	Model comparison			LR test		
	Model change	*Simpler model*	*More complex model*	*DF*	*Deviance change*	*P-value*
Step 1	Add interaction between blood pressure and cholesterol	Minimal model (4): mutual independence	Model (5): partial independence of heartdisease	9	24.45	0.0036
Step 2	Add interaction between blood pressure and heart disease	Model (5): partial independence of heart disease	Model (6): conditional independence of heart disease and cholesterol	3	30.45	< 0.0001
Step 3	Add interaction between cholesterol and heart disease	Model (6): conditional independence of heart disease and cholesterol	Model (7): mutual association between blood pressure, cholesterol and heart disease	3	19.28	0.0002
Step 4	Add three-way interaction	Model (7): mutual association between blood pressure, cholesterol and heart disease	Saturated model (8): all first-order and second-order relationships	9	4.77	0.85

(3) only marginal totals for variable 3 (heart disease) are equal:

$$\ln(F_{ijk}) = u + u_{1(i)} + u_{2(j)}$$

(4) the variables blood pressure, cholesterol and heart disease are mutually independent:

$$\ln(F_{ijk}) = u + u_{1(i)} + u_{2(j)} + u_{3(k)}$$

(5) variables 1 (blood pressure) and 2 (cholesterol) are associated and both are independent of variable 3 (heart disease):

$$\ln(F_{ijk}) = u + u_{1(i)} + u_{2(j)} + u_{3(k)} + u_{12(ij)}$$

(6) variables 2 (cholesterol) and 3 (heart disease) are conditionally independent given the level of variable 1 (blood pressure):

$$\ln(F_{ijk}) = u + u_{1(i)} + u_{2(j)} + u_{3(k)} + u_{12(ij)} + u_{13(ik)}$$

(7) all pairs of variables are associated:

$$\ln(F_{ijk}) = u + u_{1(i)} + u_{2(j)} + u_{3(k)} + u_{12(ij)} + u_{13(ik)} + u_{23(jk)}$$

(8) saturated model for the three-way table, including the second-order relationship:

$$\ln(F_{ijk}) = u + u_{1(i)} + u_{2(j)} + u_{3(k)} + u_{12(ij)} + u_{13(ik)} + u_{23(jk)} + u_{123(ijk)}$$

The model is analogous to a two-way ANALYSIS OF VARIANCE (ANOVA) – hence the use of the ANOVA terminology – but differs in a number of important respects: first, the data consist of counts rather than a score for each subject on some dependent variable; second, the model does not distinguish between independent and dependent variables. All categorical variables are treated alike as 'response' variables whose mutual associations are to be explored; third, whereas a linear combination of parameters is used in an ANOVA or regression model, in multiway tables the natural model is multiplicative and hence the counts are log-transformed to obtain a model in which parameters are combined additively; lastly, whereas the errors in an ANOVA or regression model are assumed to follow a normal distribution, appropriate distributions to model cell counts are the MULTINOMIAL DISTRIBUTION (for fixed sample size) or POISSON DISTRIBUTION (for random sample size).

The purpose of modelling a three-way table is to find the unsaturated model with fewest parameters that adequately predicts the observed cell frequencies. The LIKELIHOOD RATIO (LR) test principle can be employed formally to assess the improvement in model fit of a more complex model against a simpler model. The DEVIANCE of a model is defined as minus twice the log-likelihood ratio between the model fitted and a saturated model and represents a measure of model fit. For cell counts from a contingency table the deviance or log-likelihood ratio statistic for a particular model is calculated as:

$$X^2_{LR} = 2\sum_{i=1}^{r}\sum_{j=1}^{c} n_{ij} \times \ln(n_{ij}/E_{ij})$$

where the E_{ij} denote maximum likelihood estimates of the expected cell counts under the model. The likelihood ratio principle then states that an asymptotic test for a null hypothesis, which amounts to zero difference between two competing nested models, can be derived by comparing the difference in deviances with a CHI-SQUARE DISTRIBUTION with DEGREES OF FREEDOM equal to the number of extra parameters in the more complex model.

We carry out a series of LR tests to compare the sequence of models shown in the second table, model (4), which allows the marginal totals of all three variables to vary and is a good starting point since we are interested in the relationships between the variables rather than their marginal distributions. This model is usually referred to as the *minimal model* for a table. Adding the two-way interaction terms improves the model fit significantly compared to the simpler model in the previous step. Model (7), which includes all first-order relationships but no second-order relationship, provides an adequate fit for the data since the comparison with the saturated model does not indicate any lack of fit ($p = 0.85$).

It is important to note that attention must be restricted to HIERARCHICAL MODELS. These are such that, whenever a higher order effect is included in a model, the lower order effects composed from variables in the higher effects are also included. However, in practice, this restriction is of little consequence, since most tables can be described by a series of hierarchical models.

Identified associations are best understood by constructing tables of estimated cell counts under the final model. The final model for the data in the first table states that the association between blood pressure and cholesterol is the same for patients with or without heart disease, the association between blood pressure and heart disease is the same for each level of cholesterol and the association between cholesterol and heart disease is the same for each level of blood pressure. We can therefore assess three two-way tables of estimated cell counts (third table). For each of these two-way tables the levels of the third variable have been 'averaged out' (on the log scale) to provide a picture of the two-way interaction. To understand the nature of the interactions, odds of the categories of one variable can be calculated (e.g. of coronary heart disease) and compared between the categories of the second variable.

In essence the results indicate that there is a positive association between high blood pressure and the occurrence of coronary heart disease and, similarly, a positive association between high serum cholesterol level and coronary heart disease. The odds of coronary heart disease are estimated to more than triple when comparing the highest cholesterol (blood pressure) category with the lowest. The nature of the detected association between blood pressure and cholesterol is less clear. However, looking at the estimated odds of the second lowest blood pressure category to the largest it would appear that the odds of high

log-linear models *Cell counts estimated by the best log-linear model 'averaged' over the level of the third variable: (a) association between blood pressure and cholesterol; (b) association between heart disease and cholesterol; (c) association between heart disease and blood pressure*

(a)	*Serum cholesterol*			
Blood pressure	< 200 mg/100cc	200–219	220–259	> 260
< 127 mm Hg	20.12	20.59	11.08	7.51
127–146	14.25	15.90	9.36	6.40
147–166	27.83	47.37	20.90	17.53
> 167	22.60	33.38	21.55	19.74
Odds				
127–146 vs > 167	0.63	0.48	0.43	0.32

(b)	*Serum cholesterol*			
Heart disease	< 200 mg/100cc	200–219	220–259	> 260
Yes	4.5	5.75	4.31	4.53
No	94.29	125.13	50.15	28.46
Odds	0.048	0.046	0.086	0.159

(c)	*Blood pressure*			
Heart disease	< 127 mm Hg	127–146	147–166	> 167
Yes	2.95	2.24	7.57	10.10
No	62.82	52.09	91.77	56.09
Odds	0.047	0.043	0.082	0.18

blood pressure increase with increasing cholesterol level. When it has been decided how best to describe an interaction the relevant ODDS RATIOS and preferably CONFIDENCE INTERVALS should be reported. Associations between categorical variables can also be displayed graphically using CORRESPONDENCE ANALYSIS.

Log-linear modelling of cell counts is appropriate when a sample is classified with respect to several categorical variables and associations between their levels are of interest. In other words, all categorical variables are treated as dependent variables and none of the marginal totals is fixed by design. When one variable is viewed as the single dependent variable and the others as explanatory variables, either as a result of a study design that fixed some of the marginal totals (e.g. a COHORT STUDY) or simply because a directional relationship is of interest, models such as LOGISTIC REGRESSION for binary dependent variables and *multinomial logistic regression* for dependent variables with more than two categories are warranted. (For more details, see Agresti, 1996, and LOGIT MODELS FOR ORDINAL RESPONSES.)

While log-linear modelling of cell counts is most often used to analyse associations between categorical variables, the concept extends to any count data and also allows for effects of continuous variables. The total incidence is usually not fixed by design in such applications and the modelling is more generally referred to as POISSON REGRESSION. Even more generally, all the modelling approaches for counts mentioned so far can be considered special cases of GENERALISED LINEAR MODELS where a *link function* (e.g. the logarithm) is used to avoid predictions outside the possible range and the data modelled by a distribution from a class of distributions (e.g. Poisson or binomial). (For more details, see McCullagh and Nelder, 1989.) *SL*

Agresti, A. 1996: *An introduction to categorical data analysis*. New York: John Wiley & Sons, Inc. **Everitt, B. S.** 1992: *The analysis of contingency tables*, 2nd edition. Boca Raton: Chapman & Hall. **Ku, H. H. and Kullback, S.** 1974: Log-linear models in contingency table analysis. *American Statistician* 28, 115–22. **McCullagh, P. and Nelder, J.** 1989: *Generalized linear models*, 2nd edition. London: Chapman & Hall.

lognormal distribution This is A PROBABILITY DISTRIBUTION such that the natural logarithms of observations from the distribution are normally distributed. As a result, the distribution is always positively skewed (see SKEWNESS) and only produces positive observations. The distribution is usually defined by the standard parameters of the associated NORMAL DISTRIBUTION, so x is lognormally distributed with parameters μ and σ^2 if $\log(x)$ is normally distributed with parameters μ and σ^2. The density function of x is then:

$$f(x) = \frac{1}{x\sigma\sqrt{2\pi}}\exp\left(-\frac{[\log(x)-\mu]^2}{2\sigma^2}\right)$$

Although the need for an extra x in the leading denominator (compared to the density function of a normal distribution) may not be obvious, it becomes apparent that it is required when one considers the effective change of parameterisation that has taken place and its effect on the integral that defines the cumulative density function of the normal distribution. The distribution has MEAN $\exp(\mu + \sigma^2/2)$ and VARIANCE $\exp(2\mu + 2\sigma^2) - \exp(2\mu + \sigma^2)$.

If y $(=\log(x))$ has a normal distribution, this is often because y is the result of summing many independent but similarly distributed variables. The lognormal distribution then can arise from the multiplication of many independent but similarly distributed variables. Object sizes may often be lognormally distributed if they are the result of repeated (multiplicative) erosion processes or coagulation processes. Many sources of positively skewed data, e.g. survival times (see SURVIVAL ANALYSIS), may be adequately approximated by a lognormal distribution, although this is not a sufficient criterion for assuming that the lognormal distribution can model any positively skewed data. Marubini (1994), in fact, finds that breast cancer survival times can be modelled as coming from the lognormal distribution.

One particularly common use of the lognormal distribution is for modelling ratios. In particular, the CONFIDENCE INTERVALS for ODDS RATIOS and relative risks are often calculated by assuming that the ratio has come from a lognormal distribution.

It should be noted that lognormal data are often subjected to a log TRANSFORMATION and then treated as normal (Bland and Altman, 1996) rather than explicitly trying to use the density function given earlier. *AGL*

Bland, M. J. and Altman, D. G. 1996: Statistics notes: transforming data. *British Medical Journal* 312, 770. **Marubini, E.** 1994: When patients with breast cancer can be considered to be cured. *British Medical Journal* 309, 554–5.

LogXact LogXact is a companion product to STATXACT, featuring exact inference for binary data in the presence of covariates. An underlying LOGISTIC REGRESSION model is assumed. Both exact and asymptotic solutions are provided. LogXact additionally allows modelling of polychotomous responses (i.e. outcomes with more than two categories).

LogXact handles matched case-control data under general M:N matching using conditional LIKELIHOOD inference. Asymptotic inference is based on maximising the unconditional likelihood function for unstratified data and on maximising the conditional likelihood function for stratified data. Exact inference is based on generating the conditional distributions of the sufficient statistics for the regression coefficients of interest, NUISANCE PARAMETERS being eliminated by fixing their respective sufficient statistics at the observed values.

For a detailed discussion of the theory underlying exact logistic regression, references to numerical algorithms that perform the computations and several examples involving the analysis of biomedical data by LogXact, refer to Mehta and Patel (1995). LogXact also provides exact and asymptotic inference for POISSON REGRESSION. Reviews of LogXact are given by Lemeshow (1994) and Oster (2002).

The current version, LogXact 6, uses powerful Monte Carlo procedures that enable fast exact inference for a much larger class of datasets than those for which exact inference was previously thought feasible. For example, LogXact actually provides two variations of the Monte Carlo procedure. Neither dominates the other in terms of efficiency – for a given data analysis, the choice of model and the available computing memory will determine which computational method yields a solution the fastest. Moreover, LogXact gives the user flexibility in switching from one computational method to another without having to begin the analysis from the beginning. As exact conditional logistic regression can be time consuming, the incorporation of these refinements can allow an investigator to achieve significant time savings.

LogXact runs on Microsoft Windows NT/2000/XP as a standalone product. In addition, a special version, PROC-LogXact for SAS Users, is available as an external procedure that can be used with SAS for Microsoft Windows. *CCo/PSe/CM/NP*

Lemeshow, S. 1994: LogXact-Turbo: logistic regression software featuring exact methods. *Epidemiology* 5, 2, 259–60. **Mehta, C. R. and Patel, N. R.** 1995: Exact logistic regression: theory and applications. *Statistics in Medicine* 14, 2143–60. **Oster, R. A.** 2002: An examination of statistical software packages for categorical data analysis using exact methods. *The American Statistician* 56, 3, 235–46.

longitudinal data These data have the distinguishing feature that the response variable of interest and a set of explanatory variables (factors and/or covariates) are measured repeatedly over time. Such data arise frequently in medical studies, particularly, for example, from CLINICAL TRIALS. The main objective in collecting such data is to characterise change in the response variable over time and to determine the covariates most associated with any change. In many clinical trials, for example, primary interest will centre on the effect of the treatment group on changes in the response.

Because observations of the response variable are made on the same individual at different times, it is likely that these measurements will be correlated with each other rather than independently. This correlation must be accounted for adequately in order to draw valid and efficient inferences about how the covariates affect the response. Consequently models for longitudinal data (LINEAR MIXED-EFFECTS MODELS, GENERALISED ESTIMATING EQUATIONS) generally have two components: the first is essentially a regression model linking the average response to the covariates; the second is a model for the assumed covariance structure of the repeated measurements of the response. The estimated regression coefficients in the first part will be the parameters of most interest, with the parameters modelling the covariances being of less concern (they are essentially NUISANCE PARAMETERS). However, selecting an unsuitable model for the COVARIANCE structure of the repeated response values, i.e. one that does not match the observed structure, can adversely affect inferences on those parameters in which the investigator is most interested.

Several examples of the analysis of longitudinal data from clinical trials are given in Everitt and Pickles (2004). *BSE*

[See also DROPOUTS]

Everitt, B. S. and Pickles, A. 2004: *Statistical aspects of the design and analysis of clinical trials*. 2nd edition, London: ICP.

M

machine learning This is a branch of ARTIFICIAL INTELLIGENCE (AI) concerned with developing algorithms that can learn and generalise from examples. By 'learning' one means the acquisition of domain-specific knowledge resulting in increased predictive power.

The use of learning algorithms for data analysis can generally be divided into two stages. First, a training set of data is provided to the algorithm and used for selecting a 'hypothesis' (the learning phase). Then the selected hypothesis is validated on a set of known data to measure its predictive power (the validation phase) or used to make predictions on unseen data (the test phase).

A major problem in this setting is the risk that the selected hypothesis reflects specific features of the particular training set that are present due to chance instead of due to the underlying source generating it. This is called 'overfitting' or 'overtraining' and leads to reduced predictive power or generalisation. This risk is naturally higher with smaller training samples.

Motivated by the need to understand overfitting and generalisation, the last few years have seen significant advances in the mathematical theory of learning algorithms that have brought this field very close to certain parts of statistics, and modern machine learning methods tend to be more motivated by theoretical considerations (as is the case for SUPPORT VECTOR MACHINES and GRAPHICAL MODELS) and less by heuristics or analogies with biology (as was the case – at least originally – for NEURAL NETWORKS or genetic algorithms).

Modern machine learning is a very theoretical discipline, whose connections with AI are sometimes less obvious than its connections with multivariate statistics. Limited to the setting when the examples are all given together at the start, it is a valuable tool for data analysis. *NC/TDB*

[See also DATA MINING IN MEDICINE]

Mitchell, T. 1995: *Machine learning*. Maidenhead: McGraw-Hill. **Shawe-Taylor, J. and Cristianini, N.** 2004: *Kernel methods for pattern analysis*. Cambridge: Cambridge University Press (www .kernel-methods.net). **Witten, I. H. and Frank, E.** 1999: *Data mining: practical machine learning tools and techniques with Java implementations*. San Francisco: Morgan Kaufmann.

Mallows' C_p This criterion is used for the automated selection of variables in MULTIPLE LINEAR REGRESSION (Gorman and Toman, 1966; Mallows, 1973). Subsets of differing numbers of variables p are considered, from $p = 1$ to $p = k$, where k is the maximum number of variables. At each stage the criterion determines the optimal subset and a stopping rule is then used to decide the value of p.

Mallows' C_p criterion identifies the best subset of size p, i.e. the one that minimizes the following quantity:

$$C_p = \frac{SS_{res(p)}}{s^2} + 2(p-1) - n$$

where $SS_{res(p)}$ is the residual sum of squares based on a p-variable regression, s^2 is an estimate of σ^2 (the residual variance based on the full model with all k variables) and n is the sample size. The term $2(p-1) - n$ has the effect of penalising more complex models, the aim being to produce the simplest model that fits the data adequately. A plot of C_p versus p allows the various competitor subsets to be judged for different values of p and a formal stopping rule for p may be applied.

The rule suggested by Mallows is to stop when C_p is 'small' or close to $p - 1$. Gilmour (1996) discusses stopping rules and argues that the subset corresponding to the lowest C_p will tend to include at least one unimportant predictor variable; he proposes a modification of C_p to take account of this.

An example of the use of the criterion is given by Sutcliffe *et al.* (2001), who used multiple regression to determine cost predictors for patients with systemic lupus erythematosus. Mallows' C_p was used to determine how many variables were required in the model and the best-fitting model with this number of predictors. The next four best-fitting models with that number of predictors and the four best-fitting models with one more than this number were also found. This provided a set of candidate models for further analysis.

There are many alternative methods for subset selection. Hocking (1976) describes the advantages and disadvantages of several of them, including 'best subsets', which is computationally very intensive but globally optimal, and the more usual forward selection or backward elimination (or combination) stepwise methods based on F-tests for individual variables (see AUTOMATIC SELECTION METHODS). The latter are very widely used because of their inclusion in software packages, but they have been criticised by Hocking and many other statisticians as being potentially misleading. The Mallows' C_p criterion, while not globally optimal, is stepwise optimal and may be preferable to such methods. *ML*

[See also ALL SUBSETS REGRESSION]

Encyclopaedic Companion to Medical Statistics: Second Edition Edited by Brian S. Everitt and Christopher R. Palmer

Gilmour, S. G. 1996: The interpretation of Mallows' C_p statistic. *The Statistician* 45, 49–56. **Gorman, J. W. and Toman, R. J.** 1966: Selection of variables for fitting equations to data. *Technometrics* 8, 27–51. **Hocking, R. R.** 1976: The analysis and selection of variables in linear regression. *Biometrics* 32, 1–49. **Mallows, C. L.** 1973: Some comments on C_p. *Technometrics* 15, 661–75. **Sutcliffe, N., Clarke, A. E., Taylor, R., Frost, C. and Isenberg, D. A.** 2001: Total costs and predictors of costs in patients with systemic lupus erythematosus. *Rheumatology* 40, 37–47.

Mann–Whitney rank sum test This is the nonparametric version of the independent samples *t*-test (see STUDENT'S *t*-TEST), also known as the Wilcoxon rank sum test and the Mann–Whitney *U* test. Mann and Whitney, and Wilcoxon independently, derived the test, so the test statistic takes two different forms. The *U* statistic of Mann and Whitney is usually preferred as it has a useful interpretation. The Mann–Whitney test is applied to two independent samples, testing for a difference in shape and spread of the data between the two groups. With the addition of the assumption that the data from the two groups are similarly distributed, it also tests for a difference in MEDIANS or MEANS between the two groups. The other assumptions are that the data are randomly selected observations and that the data must be either continuous or ordinal in nature.

To carry out the test, first rank all the data from the smallest to the largest. Assign the average rank to any ties in the data. Calculate the sum of the ranks in each of the groups. Calculate U_1 and U_2:

$$U_1 = n_1 n_2 + \frac{n_1(n_1+1)}{2} - R_1$$

$$U_2 = n_1 n_2 + \frac{n_2(n_2+1)}{2} - R_2$$

where n_1 = the number of observations in group 1, n_2 = the number of observations in group 2, R_1 = the sum of the ranks assigned to group 1 and R_2 = the sum of the ranks assigned to group 2.

Calculate $U = \min(U_1, U_2)$. Compare U with the critical value of the Mann–Whitney *U* tables. The null hypothesis is rejected if the value of *U* is less than or equal to the critical value in the tables. The value of $U/(n_1 n_2)$ can be interpreted as the probability that a new observation from group 1 is less than a new observation from group 2.

As an example, data in the table show Mcm2 levels in two groups of people, seven with and eight without fibrosis of the liver. The groups do not have similar distributions. Note that the assigned rank for the lowest Mcm2 value observed, due to the tie, is midway between 1 and 2, exemplifying the convention mentioned earlier.

Mann–Whitney rank sum test *Mcm2 levels in two groups of people, one with and one without fibrosis of the liver*

	With fibrosis (group 1)							*Without fibrosis (group 2)*							
Mcm2 value	3	11	9	14	7	11	11	32	27	25	33	14	4	24	3
Rank	1.5	7	5	9.5	4	7	7	14	13	12	15	9.5	3	11	1.5
Sum of ranks	**41**							**79**							

$$U_1 = 7 \times 8 + \frac{7 \times (7+1)}{2} - 41 = 44$$

$$U_2 = 7 \times 8 + \frac{8 \times (8+1)}{2} - 79 = 13$$

Hence, $U = \min(44, 13) = 13$. From tables ($n_1 = 7$, $n_2 = 8$, $\alpha = 0.05$), the critical value is 10. As $13 > 10$, there is not sufficient evidence to reject the null hypothesis. Therefore, there is no evidence of a difference in spread or location between the two groups. There is a probability of 0.23 (=13/56) that a new observation from group 1 will be less than a new observation from group 2. For further details see Pett (1997), Hart (2001) and Swinscow and Campbell (2002). *SLV*

[See also MEDIAN TEST]

Hart, A. 2001: Mann–Whitney test is not just a test of medians: differences in spread can be important. *British Medical Journal* 323, 391–3. **Pett, M. A.** 1997: *Nonparametric statistics for healthcare research*. Thousand Oaks: Sage. **Swinscow, T. D. V. and Campbell, M. J.** 2002: *Statistics at square one*, 10th edition. London: BMJ Books.

Mann–Whitney U test Synonym for MANN–WHITNEY RANK SUM TEST

MANOVA See ANALYSIS OF VARIANCE

Mantel–Haenszel methods These are a collection of statistical methods for stratified, categorical data. When analysing data from an epidemiological study, one should be aware of the danger of confounding.

For example, in a CASE-CONTROL STUDY of the association between an industrial chemical and a particular cancer, 100 cases and 100 controls are recruited. When the data are analysed, the ODDS RATIO associated with the chemical is 0.91, suggesting no association or a possible protective effect. However, it is noticed that when the data are stratified by sex, the odds ratio in men is 1.29 and in women is 1.38, suggesting a possible harmful effect (data are shown in the table). The reason for this reversal in the odds ratio is that sex is a confounder in the association between exposure and disease. The disease is more common in women, but women are less likely to be exposed, and so exposure appears protective if one does not adjust for sex.

Mantel–Haenszel methods *Summary data from a case-control study, stratified by sex*

	Men		*Women*		*Total*	
	Case	*Control*	*Case*	*Control*	*Case*	*Control*
Exposed	7	11	4	1	11	12
Unexposed	34	69	55	19	89	88

One method for overcoming confounding is to stratify the data, as in this case where stratification was by sex. The statistic of interest (e.g. the odds ratio) is calculated for each stratum separately. It is then often desirable to combine these stratum-specific statistics into a single overall measure, to calculate a standard error for this and also to test a null hypothesis (e.g. that the odds ratio is one). If the number of subjects in each stratum is large, this may be done using MAXIMUM LIKELIHOOD METHODS (e.g. LOGISTIC REGRESSION), by introducing an additional parameter into the model for each stratum. However, when data are sparser, i.e. when the number of subjects in a stratum may be small, maximum likelihood may give biased estimates. In this situation, it is necessary to use either conditional maximum likelihood methods or Mantel–Haenszel methods. The latter have the advantage of being very straightforward to calculate and, for this reason, are popular. Mantel–Haenszel methods do not require that the numbers of individuals in each stratum be large, only that the total number of subjects be large enough. However, if even the total number of subjects is small, it is necessary to use 'exact' methods (see EXACT METHODS FOR CATEGORICAL DATA).

Mantel–Haenszel methods are available for estimating odds ratios from case-control data, rate ratios or rate differences from cohort data and odds ratios or risk ratios from case-cohort data. They may also be useful when analysing repeated-measures designs. When the exposure and the outcome (disease) are both binary, the analysis of a case-control study with stratification is an example of the analysis of multiple 2×2 tables: one table for each stratum and, in each table, two rows for exposure and two columns for outcome. Mantel–Haenszel methods also exist for the more general situation of multiple $I \times J$ tables, e.g. a case-control study with more than two possible exposure (or treatment) levels ($I > 2$) and/or more than two possible outcomes ($J > 2$). Both the exposure and outcome variables can be treated as either nominal or ordinal categorical variables.

Finally, when combining several stratum-specific estimates to form a single overall estimate, it is important to consider whether this is sensible. If the odds ratio (or other measure) appears to vary greatly from one stratum to another, possibly even being much greater than one in some strata and much less than one in other strata and this variation is more than would be expected by chance, a single summary measure may not be very meaningful. In this situation it is better to report the odds ratio estimate for each stratum separately. Thus, before calculating the overall odds ratio (or other measure), it is worth testing the null hypothesis of homogeneity, i.e. that the odds ratio does not vary from one stratum to another. The Breslow–Day test is one such test. For further details see Kuritz, Landis and Koch (1988), Clayton and Hills (1993) and Rothman and Greenland (1998). *SRS*

Clayton, D. and Hills, M. 1993: *Statistical models in epidemiology*. Oxford: Blackwell Science Publications. **Kuritz, S. J., Landis, J. R. and Koch, G. G.** 1988: A general overview of Mantel–Haenszel methods: applications and recent developments. *Annual Review of Public Health* 9, 123–60. **Rothman, K. J. and Greenland, S.** 1998: *Modern epidemiology*, 2nd edition. Philadelphia: Lippincott-Raven Publishers.

marginal structural models These are regression models for so-called counterfactual or potential outcomes Y_a, which express how the outcome of interest, Y, would have looked like if level a of the target exposure A had been received (Robins, Hernan and Brumback, 2000). The models are labelled 'marginal' because they are models for exposure effects at the population level, rather than within strata defined by covariate values. They are labelled 'structural' because, by construction, their parameters carry a causal interpretation. For instance, in the linear marginal structural model:

$$E(Y_a) = \alpha + \beta a,$$

the intercept α expresses what the expected outcome in the population would be if all subjects were unexposed (i.e., $a = 0$). The regression slope:

$$\beta = E(Y_{a+1}) - E(Y_a)$$

encodes the expected change in outcome that would result from a unit increase in the exposure. It compares potential outcomes for the 'same' subjects under different exposure levels and thereby can be interpreted as an average CAUSAL EFFECT of the exposure on the outcome. This contrasts with standard regression models:

$$E(Y|A = a) = \alpha' + \beta' a,$$

where the regression slope:

$$\beta' = E(Y|A = a + 1) - E(Y|A = a)$$

compares the expected outcome between differently exposed subgroups ($A = a + 1$ and $A = a$) of the population. When these subgroups are not inherently comparable, then β' (unlike β) cannot be interpreted as a causal exposure effect.

In standard regression models, comparability across treatment levels can be achieved by adjusting for measured confounders L of the exposure–outcome relationship. In marginal structural models, adjustment for confounding happens through a weighting procedure, called INVERSE PROBABILITY WEIGHTING (IPW), which works in two steps:

(1) A pseudo sample is constructed by weighting each subject with the probability of the observed exposure, given the observed confounders:

$$\frac{1}{\Pr(A|L)}.$$

For instance, a subject with confounder level $L = l$ who is exposed to level $A = 1$ is weighted with $1/\Pr(A = 1|L = l)$. Similarly, a subject with $L = l$ who is exposed to level $A = 0$ is weighted with $1/\Pr(A = 0|L = l)$. Here, $\Pr(A|L)$ can be estimated as the fitted value from a standard regression of A on L, e.g. a logistic regression if A is dichotomous.

(2) Because the weighting eliminates confounding, the marginal structural model can be fitted to the pseudo sample as if there were no confounding. This can be accomplished via off-the-shelf software packages by fitting the corresponding regression model $E(Y|A) = \alpha + \beta A$ while assigning each subject's data the given weight.

We emphasise that all methods of confounding adjustment – in particular standard regression adjustment and IPW – crucially rely on the assumption of no unmeasured confounding. This assumption holds if L contains all confounders of the exposure–outcome relationship, i.e. all factors that directly affect the exposure and are also associated with the outcome. This is visualised in the causal diagram of the first figure through the absence of an arrow from the unmeasured variables U to A. If not all confounders of the exposure–outcome relationship are included in the estimation of the weights $1/\Pr(A|L)$, then the assumption of no unmeasured confounding is violated, in which case the estimate of the causal exposure effect may be biased. The assumption of no unmeasured confounding is untestable and has to be defended by subject matter knowledge.

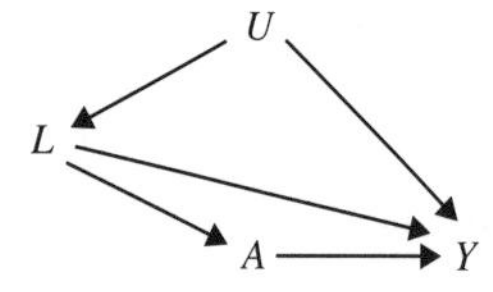

marginal structural models *Causal diagram representing the data generating mechanism, with A, exposure; L, measured confounders; Y, outcome; U, unmeasured variables that affect L and Y*

Because marginal structural models encode population-averaged effects, they can conveniently be used for standardisation (see DEMOGRAPHY) with the total group as the standard population (Sato and Matsuyama, 2003). Because the IPW estimation procedure involves the PROPENSITY SCORE, the resulting estimates inherit the properties of propensity score adjusted estimators. Marginal structural models are, however, most commonly adopted for assessing the effect of a time-varying exposure on an outcome in the presence of time-varying confounders. This is because these models avoid regression adjustment for time-varying confounders, which is fallible when the time-varying confounders are both affected by past exposure levels and affecting future exposure levels. For instance, CD4 count confounds the relationship between AZT treatment and survival in HIV infected subjects because it affects the physicians' assignment of particular AZT levels and is associated with survival. At the same time, it is affected by earlier AZT exposure levels. The reason why standard regression adjustment fails in this context is because, on the one hand, it eliminates indirect exposure effects that are mediated through these confounders (e.g. indirect effects of AZT on survival through its effect on the CD4 count) and, on the other hand, it may induce a so-called collider-stratification BIAS by which a spurious ASSOCIATION between exposure and outcome arises, even in the absence of an exposure effect. Inference for marginal structural models suffers neither of these two limitations because it involves no regression adjustment for confounding. The IPW procedure, which is used instead, is now slightly more involved as it must acknowledge the time-varying nature of the exposure. With A^t and L^t denoting the exposure (e.g. the AZT level) and confounders (e.g. the CD4 count) respectively, measured at study cycle t, the weights now take the form:

$$\frac{1}{\prod_{t=1}^{T} \Pr(A^t|\bar{A}^{t-1}, \bar{L}^t)}, \qquad (1)$$

where $\bar{A}^{t-1} = (A^1, A^2, \ldots, A^{t-1})$ and $\bar{L}^t = (L^1, L^2, \ldots, L^t)$ refer to exposure and confounder history respectively and where T is the end-of-study time. Next, a marginal structural model for time-varying exposures can be fitted. For instance, with Y denoting 1 for subjects who survive the end-of-study time and 0 otherwise, the marginal structural model:

$$\text{logit} \Pr(Y_{\bar{a}^T} = 1) = \alpha + \beta a^T + \gamma \sum_{s=1}^{T-1} a^s, \qquad (2)$$

is indexed by parameters:

$$\exp(\beta) = \frac{\text{odds}(Y_{(\bar{a}^{T-1},1)} = 1)}{\text{odds}(Y_{(\bar{a}^{T-1},0)} = 1)}$$

encoding the short-term effect of AZT on the odds of death, and $\exp(\gamma)$ capturing a long-term effect. These parameters can be estimated by fitting the corresponding regression model $\text{logit}\Pr(Y|\bar{A}^T = 1) = \alpha + \beta A^T + \gamma \sum_{s=1}^{T-1} A^s$ while assigning each subject's data the given weight.

Just as for point exposures, the IPW procedure for time-varying exposures crucially relies on the untestable assumption of no unmeasured confounding. For time-varying exposures, this assumption holds if all factors beside $(\bar{L}^{t-1}, \bar{A}^{t-1})$ that affect the exposure at time t and are associated with the outcome are contained in L^t, for each time t. This scenario is visualised in the causal diagram of the second figure, for $T = 2$, through the absence of an arrow from the unmeasured variables U to A^t. In this figure, the arrows from $(\bar{L}^{t-1}, \bar{A}^{t-1})$ to A^t indicate that the received AZT level at time t may be affected by the subjects' treatment and CD4 count history up to t. Similarly, the arrows from $(\bar{L}^{t-1}, \bar{A}^{t-1})$ to L^t indicate that the CD4 count at time t may be affected by the subjects' treatment and CD4 count history up to t. U represents all unmeasured variables (e.g. genetics, lifestyle factors) that affect a subject's health status over time.

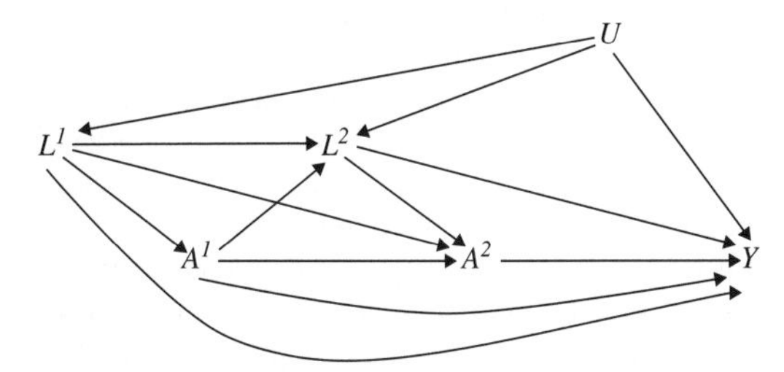

marginal structural models *Causal diagram, with A^t, exposure at time t; L^t, measured confounders at time t; Y, outcome; U, unmeasured variables that affect L^t and Y*

When standard statistical software packages (SAS, STATA, R-see STATISTICAL PACKAGES) are used for fitting marginal structural models through IPW, caution is needed in interpreting the STANDARD ERRORS provided by the software (see STATISTICAL PACKAGES), because these ignore the imprecision of the estimated weights. By using routines that report so-called sandwich estimators, conservative standard errors are obtained.

A major appeal to marginal structural models is that the underlying IPW estimation procedure straightforwardly generalises to more complex settings with time-varying outcomes (Hernan, Brumback and Robins, 2002) or survival ENDPOINTS (Hernan, Brumback and Robins, 2000). A drawback is that some subjects may have small probabilities $\Pr(A^t|\bar{A}^{t-1}, \bar{L}^t)$ at certain time points, so that they receive influential weights (1). This can make the IPW estimate unstable and imprecise. To some extent, this problem can be mitigated by using so-called stabilized weights, calculated as:

$$\frac{\prod_{t=1}^{T} \Pr(A^t|\bar{A}^{t-1})}{\prod_{t=1}^{T} \Pr(A^t|\bar{A}^{t-1}, \bar{L}^t)}.$$

When this is insufficient, progress can sometimes be made by including baseline covariates in the marginal structural model, i.e. covariates that precede A^1 in time (e.g. sex, ethnicity). For example, the model in (2) can be generalised to:

$$\text{logitPr}(Y_{\bar{a}^T} = 1|V = v) = \alpha + \beta a^T + \gamma \sum_{s=1}^{T-1} a^s + \delta v + \psi v a^T, \tag{3}$$

where V is a baseline covariate that is contained in the measured baseline confounder L^1, and ψ is a covariate–exposure interaction. When the model in (3) is fitted through IPW, the (stabilised) weights are modified as:

$$\frac{\prod_{t=1}^{T} \Pr(A^t|\bar{A}^{t-1}, V = v)}{\prod_{t=1}^{T} \Pr(A^t|\bar{A}^{t-1}, \bar{L}^t, V = v)}.$$

These weights are typically less influential. This adaptation may, however, be insufficient when the exposure has strong predictors or when it is measured on a continuous scale, in which case $\Pr(A^t|\bar{A}^{t-1}, \bar{L}^t)$ refers to the density of A^t, given $\bar{A}^{t-1}$ and $\bar{L}^t$. In that case, one must recourse to more efficient (doubly robust) estimation strategies or consider the related, but more complex class of structural nested models (Robins, 1997). The latter models have the additional advantage that, unlike marginal structural models, they can allow for modification of the exposure effect by time-varying covariates. *AJS/SV*

Hernan, M. A., Brumback, B. and Robins, J. M. 2000: Marginal structural models to estimate the causal effect of zidovudine on the survival of HIV-positive men. *Epidemiology* 11, 561–70. **Hernan, M. A., Brumback, B. and Robins, J. M.** 2002: Estimating the causal effect of zidovudine on CD4 count with a marginal structural model for repeated measures. *Statistics in Medicine*, 21, 1689–709. **Pearl, J.** 2000: *Causality: Models, Reasoning, and Inference*. Cambridge: Cambridge University Press. **Robins, J. M.** 1997: Causal inference from complex longitudinal data. In *Latent Variable Modeling and Applications to Causality*. New York: Springer Verlag. **Robins, J. M., Hernan, M. A. and Brumback, B.** 2000: Marginal structural models and causal inference in epidemiology. *Epidemiology* 11, 550–60. **Sato, T. and Matsuyama, Y.** 2003: Marginal structural models as a tool for standardization. *Epidemiology* 14, 680–6.

Markov chain Monte Carlo (MCMC) BAYES' THEOREM (1) provides a means for combining data, y, in the form of the LIKELIHOOD, $p(y|\theta)$, with external evidence in the form of a PRIOR DISTRIBUTION for θ, $p(\theta)$, to produce a POSTERIOR DISTRIBUTION, $p(\theta|y)$ (see BAYESIAN METHODS). However, in order to make inferences about either the posterior distribution itself or to obtain the posterior expectation of a function of the model parameters, θ, using Bayes' theorem (2), we have to evaluate often high-dimension integrals, which are only rarely analytically tractable. Consequently, much of Bayesian statistics over the last 30 years has been concerned with either parameterising models such that the integrals simplify or with the use of approximation methods (Bernardo and Smith, 1994). Such approximate methods fall into three broad categories: asymptotic approximations, e.g. Laplace approximations; numerical integration techniques, e.g. Gaussian quadrature; or simulation methods, e.g. Monte Carlo simulation (Bernardo and Smith, 1994):

$$p(\theta|y) = \frac{p(\theta)p(y|\theta)}{\int p(\theta)p(y|\theta)\mathrm{d}\theta} \tag{1}$$

$$E[f(\theta)|y] = \int f(\theta)p(\theta|y)\mathrm{d}\theta \tag{2}$$

Given a sample of values for θ from the joint posterior distribution, $p(\theta|y)$, $\{\theta^{(m)}, m = 1, \ldots, M\}$, then the posterior expectation of $f(\theta)$ can be approximated by:

$$E[f(\theta)|y] \approx \frac{1}{M}\sum_{m=1}^{M} f(\theta^{(m)}) \tag{3}$$

While the use of (3) appears appealing, in practice, generation of samples from often high-dimensional joint posterior distributions can be difficult. However, for (3) to hold, the samples generated need not be independent, but rather be from a Markov chain whose stationary distribution is, in fact, the posterior distribution. A Markov chain is a sequence of random variables $\{\theta^{(1)}, \theta^{(2)}, \ldots\}$ such that $\theta^{(i)}$ only depends on $\theta^{(i-1)}$ and not the rest of the random variables. Constructing such a chain then gives rise to Markov chain Monte Carlo simulation (Casella and George, 1992; Brooks, 1998).

The construction of a Markov chain with a stationary distribution that is the posterior distribution is relatively straightforward and was initially proposed by Metropolis *et al.* (1953) and later generalised by Hastings (1970) and is now referred to as the *Metropolis–Hastings algorithm*. At the ith of m iterations generate a candidate value for θ, θ^*, from a *proposal* distribution, $g(\theta|\theta^{(i-1)})$, and then with probability $\alpha(\theta^{(i-1)}, \theta^*)$ accept θ^*, i.e. $\theta^{(i)} = \theta^*$, or reject it, i.e. $\theta^{(i)} = \theta^{(i-1)}$, where $\alpha(\theta^{(i-1)}, \theta^*)$ is given by the following equation, which, in practice, is achieved by generating a value u from a uniform [0,1] distribution and, if $u \leq \alpha(\theta^{(i-1)}, \theta^*)$, accepting θ^*:

$$\alpha(\theta^{(i-1)}, \theta^*) = \min\left[1, \frac{p(\theta^*|y)g(\theta^{(i-1)}|\theta^*)}{p(\theta^{(i-1)}|y)g(\theta^*|\theta^{(i-1)})}\right] \tag{4}$$

Clearly, if $g(.)$ is still a multivariate distribution, the generation of samples may still be difficult. In practice, most applications of the Metropolis–Hastings algorithm use a single component proposal distribution (Gilks, Richardson and Spiegelhalter, 1996).

If the Metropolis–Hastings algorithm is *irreducible* then regardless of where it starts it will sample from the entire domain of $p(\theta|y)$ within a finite number of iterations and produce samples from the stationary distribution, i.e. $p(\theta|y)$. Thus, it should not be dependent on the starting values. Clearly, one way in which to verify irreducibility is to use the algorithm a number of times with different starting values and inspect the samples obtained. Even if the algorithm is irreducible it has to be run long enough so that it will 'forget' its starting values and, in practice, this is achieved by running the algorithm for a 'burn-in' period and discarding the first n samples and basing inferences on only the last $m - n$ samples. Of crucial importance, therefore, is the question of how large m and n should be. In practice, a combination of formal methods that have been advocated, together with knowledge of the statistical model and inspection of the samples obtained via sensitivity analyses to choices of m, n and the starting values, is the most pragmatic approach (Cowles and Carlin, 1996; Gilks, Richardson and Spiegelhalter, 1996).

Examination of the autocorrelation between the samples at various numbers of iterations apart can reveal algorithms that are *mixing* slowly, i.e. covering the whole of $p(\theta|y)$, and thus need to be run for considerable numbers of iterations. An alternative, often preferred, option is to consider the reparameterisation of the statistical model in order to increase the rate of mixing. In linear regression models, centring of covariates and, in the case of hierarchical models, *hierarchical centring* have been shown to have dramatic effects on the rate of mixing (Gelfand, Sahu and Carlin, 1995; Gilks, Richardson and Spiegelhalter, 1996).

A special case of the single component Metropolis–Hastings algorithm is the Gibbs sampler in which the proposal distributions are the set of full conditionals and the acceptance probability (4) is always equal to 1 (Geman and Geman, 1984; Gelfand and Smith, 1990). Thus, given a set of initial or starting values for the p parameters in a statistical model, $\{\theta_1^{(0)}, \ldots, \theta_p^{(0)}\}$, the Gibbs sampler at each iteration draws a sample from each of the conditional distributions in turn. Thus:

$$\begin{aligned}
\theta_1^{(1)} &\leftarrow p(\theta_1|\theta_2^{(0)}, \theta_3^{(0)}, \ldots, \theta_P^{(0)}, y) \\
\theta_2^{(1)} &\leftarrow p(\theta_2|\theta_1^{(1)}, \theta_3^{(0)}, \ldots, \theta_P^{(0)}, y) \\
&\vdots \\
\theta_P^{(1)} &\leftarrow p(\theta_P|\theta_1^{(1)}, \theta_2^{(1)}, \ldots, \theta_{P-1}^{(1)}, y)
\end{aligned} \tag{5}$$

Thus, the realisations $\{\theta_1^{(1)}, \ldots, \theta_1^{(m)}\}, \ldots, \{\theta_p^{(1)}, \ldots, \theta_p^{(m)}\}$ after m iterations provide samples from the marginal posterior distributions and on which inferences can be based. Sampling from the conditional distributions in (5) can be difficult unless they are univariate, although for many HIERARCHICAL MODELS they are, or they are log-concave, in which case *adaptive rejection sampling* may be used (Gilks, Richardson and Spiegelhalter, 1996). One particular appeal of the Gibbs sampler is that, in essence, it is simple to implement and while it can be programmed in a variety of computer languages and software packages, the development of user-friendly software such as BUGS AND W IN BUGS has promoted its widespread use in numerous applied settings in biomedical research (Gelman and Rubin, 1996). *KRA*

Bernardo, J. M. and Smith, A. F. M. 1994: *Bayesian theory*. Chichester: John Wiley & Sons, Ltd. **Brooks, S. P.** 1998: Markov chain Monte Carlo method and its applications. *Journal of the Royal Statistical Society, Series D* 47, 69–100. **Casella, G. and George, E.** 1992: Explaining the Gibbs sampler. *American Statistician* 46, 167–74. **Cowles, M. K. and Carlin, B. P.** 1996: Markov chain Monte Carlo convergence diagnostics: a comparative review. *Journal of the American Statistical Association* 91, 883–904. **Gelfand, A. E. and Smith, A. F. M.** 1990: Sampling-based approaches to calculating marginal densities. *Journal of the American Statistical Association* 85, 398–409. **Gelfand, A. E., Sahu, S. K. and Carlin, B. P.** 1995: Efficient parameterisations for normal linear mixed models. *Biometrika* 82, 479–88. **Gelman, A. and Rubin, D. B.** 1996: Markov chain Monte Carlo methods in biostatistics. *Statistical Methods in Medical Research* 5, 339–55. **Geman, S. and Geman, D.** 1984: Stochastic relaxation, Gibbs distributions and the Bayesian restoration of images. *IEEE Transactions on Pattern Analysis and Machine Intelligence* 6, 721–41. **Gilks, W. R., Richardson, S. and Spiegelhalter, D. J.** 1996: *Markov chain Monte Carlo methods in practice*. New York: Chapman & Hall. **Hastings, W. K.** 1970: Monte Carlo sampling methods using Markov chains and their applications. *Biometrika* 57, 97–109. **Metropolis, N., Rosenbluth, A. W., Teller, M. N. and Tellet, A. H.** 1953: Equations of state calculations by fast computing machine. *Journal of Chemical Physics* 21, 1087–91.

matched pairs analysis Different types of designs may lead to matched pairs analysis. Individually matched subjects in prospective studies, individually matched controls to cases in retrospective studies and pairs of data obtained when the same individual is measured twice are examples of matched pairs (see MATCHED SAMPLES). A sample of matched pairs consists of statistically dependent data and in statistical analysis the pair, not single subjects, should be the unit. Matched pairs analysis may concern concepts like change, difference and odds, but also AGREEMENT and ASSOCIATION.

Questions of change could include: Is there a difference in outcome due to different treatments between individually matched subjects? Is there a change in outcome within subjects before and after a treatment? Do patients prefer one treatment better than another? Statistical methods for matched pairs analysis of quantitative, ordered categorical and dichotomous data respectively will be presented.

The cholesterol level was measured in 20 students before and after a period of having a diet that was supposed to have a cholesterol-lowering effect. As each student was measured twice, the difference between the two values was the outcome variable. The changes in cholesterol ranged from −1.0 mmol/l (increase) to 0.8 mmol/l (decrease). The table shows three different statistical approaches to matched pairs analysis of quantitative data:

The mean approach. Provided that the dataset of differences is a sample from a NORMAL DISTRIBUTION, the paired STUDENT'S *t*-TEST of the null hypothesis of zero mean change can be used. The observed mean change was 0.23 mmol/l and according to the test (see the table) one can conclude that the diet will significantly decrease the mean cholesterol level in a representative population of about 0.04–0.42 mmol/l, which is the 95 % CONFIDENCE INTERVAL (CI).

The median approach. The WILCOXON SIGNED RANK TEST requires no assumptions about distribution of the differences in quantitative data. The median change was 0.2 mmol/l and according to the test the null hypothesis of no MEDIAN CHANGE can be rejected ($P = 0.01$).

The dichotomisation approach. The cholesterol level decreased in 16, increased in three and was unchanged in one student. If the null hypothesis of unchanged values were true one would expect about the same numbers of positive as negative differences; this comparison is performed by a *sign test*. Unchanged values provide no information about the direction of change and will be excluded. The BINOMIAL DISTRIBUTION is used for exact calculation of the PROBABILITY of getting the observed or even more extreme unbalance in negative and positive differences when the null hypothesis is true. The table shows that the probability of the observed or more extreme unbalance was 0.004, which is strong evidence that the diet will change the cholesterol level. The large sample approximation of the *one-sample sign test* (Altman, 2000) can be written as:

$$z_c = \frac{|r - np| - \frac{1}{2}}{\sqrt{np(1-p)}}$$

where r is the number of differences of one sign among n nonzero differences and p is the probability under the null hypothesis of having the actual sign $\left(p = \frac{1}{2}\right)$. In the example, $r = 16$, $n = 19$ and $z_c = 2.75$. The proportion of students with a decrease in cholesterol was 84 % and the 95 % CI (see Newcombe and Altman, 2000) deviates from that of the null hypothesis (50 %) (see the table).

Matched pairs analysis of ordered categorical data is applied to a dataset from a study in diagnostic radiology (Svensson *et al.*, 2002). The patient's perceived difficulty during each of two radiological examinations, here denoted CT and CO, was rated on a scale with the categories 'not at all', 'slightly', 'fairly' and 'very' difficult. Each of the 108 patients underwent both examinations, which means paired data (see the figure on page 268).

matched pairs analysis *Three different approaches to matched pairs analysis of change in S-cholesterol in a sample of 20 students*

Mean approach	*Median approach*	*Dichotomisation approach*
Mean change 0.23 mmol/l	Median change 0.2 mmol/l	Negative changes, 3 Positive changes, 16
Standard deviation 0.41 mmol/l	Quartiles (Q_1;Q_3) (0.1; 0.5) mmol/l	
Student's paired *t*: 2.527 significance level $P = 0.02$	Wilcoxon signed rank/ matched pairs/test $P = 0.01$	The exact sign test, $P = 0.004$ The approximate sign test: $z_c = 2.75$, $P = 0.006$
The 95% CI of mean change: (0.04 to 0.42) mmol/l	The 95% CI of median change: (0.1 to 0.5) mmol/l	The 95% CI for the proportion of students with decreased value: 62% to 94%

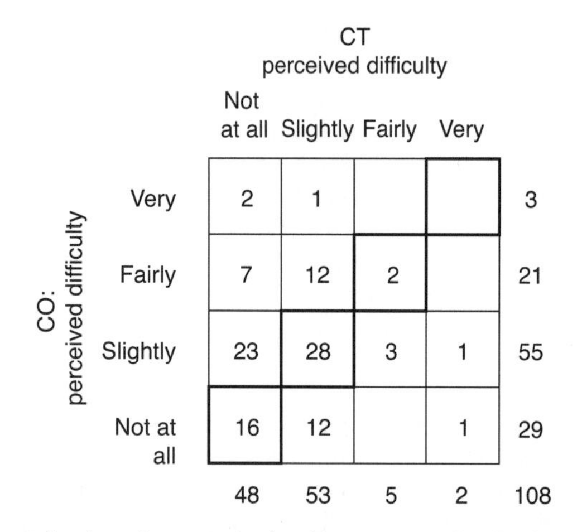

matched pairs analysis *Frequency distribution of paired data from evaluation concerning perceived difficulty concerning two radiological examinations (CT, CO)*

Data from scale assessments have an ordered structure only, which means that change is not defined by the difference. Therefore, the same statistical methods as for paired dichotomous data will be used. A common expression for *the sign test* is:

$$z_c = \frac{|b-c|-1}{\sqrt{b+c}}$$

where *b* and *c* denote the number of pairs with different categories. A McNEMAR'S TEST is an equivalent test (Bland, 1996; Altman, 2000). One approach to dichotomise the data is to compare the numbers of pairs below and above the diagonal of unchanged categories. For the data in the figure, the 17 patients, who rated the CT a higher level of difficulty than the CO, are compared with the 45 pairs above the diagonal. According to the sign test ($z_c = 3.43$, $P = 0.0006$), this observed unbalance in changes gives evidence enough to conclude that the CT is perceived as significantly less difficult than the CO examination.

An alternative way of dichotomising is by grouping the data in two categories, 'not at all difficult' (+) and 'difficult' (−), which contains three categories of difficulty. The table is simplified in two concordant and two discordant combinations of categories. The discordant pairs of 32 CT(+), CO(−) and 13 CT(−), CO(+) contain information about the difference in perceived difficulty between the examinations. The SIGN TEST ($z_c = 2.68$, $P = 0.007$) confirms that more patients will find the CT examination as 'not at all difficult' when compared with their rating of the CO examination.

As is evident from the figure, a larger proportion of the patients (17%) judged the CT as being 'not at all difficult', as 48 patients (44%) rated the CT and 29 (27%) rated the CO 'not at all difficult'. The 95% CI for the difference in the paired proportions, Δp, was from 5% to 29% according to the expression $\Delta p \pm 1.96 \times SE(\Delta p)$. The STANDARD ERROR (SE) is:

$$SE(\Delta p) = \frac{1}{n}\sqrt{b + c - \frac{(b-c)^2}{n}}$$

where n is the total number of patients ($n = 108$) and b and c are the numbers of discordant pairs (Altman, 2000; Newcombe and Altman, 2000).

In order to use a pair-matched CASE-CONTROL STUDY method we are interested in the exposure to the risk factor. Using retrospective case-control studies, individuals having a specific disease (e.g. lung cancer) are compared with individuals without the disease. Both the outcome variable (diseased, nondiseased) and the exposure to the risk factor (exposed, not exposed) are dichotomous. Within each pair, there are four possible combinations of disease status and exposure. Two sets of pairs are concordant, but information about the relationship between exposure and disease is given by the pairs with different exposure. Denote the number of pairs with only the case exposed n_{+-} and the number of pairs with the case unexposed n_{-+}. Providing nonzero numbers of discordant pairs, the odds ratio in matched pairs is calculated by $\mathrm{OR} = n_{+-}/n_{-+}$. An OR larger than unity indicates a relationship between exposure and disease, i.e. a higher odds of developing disease when exposed (McNeil, 1999). *ES*

[See also CORRELATION, KAPPA AND WEIGHTED KAPPA MATCHING, MATCHED SAMPLES]

Altman, D. G. 2000: *Practical statistics for medical research.* Boca Raton: Chapman & Hall/CRC. **Bland, M.** 1996: *An introduction to medical statistics*, 2nd edition. Oxford: Oxford Medical Press. **McNeil, D.** 1999: *Epidemiological research methods.* New York: John Wiley & Sons, Inc. **Newcombe, R. G. and Altman, D. G.** 2000: Proportions and their differences. In Altman, D. G., Machin, D., Bryant, T. N. and Gardner, M. J. (eds), *Statistics with confidence*, 2nd edition. Bristol: BMJ Books, pp. 45–57. **Svensson, M. H., Svensson, E., Lasson, A. and Hellström, M.** 2002: Patient acceptance of CT colonography and conventional colonoscopy: prospective comparative study in patients with or suspected of having colorectal disease. *Radiology* 222, 337–45.

matched samples This is a set of observations in which each observation in one sample is individually matched with one in every other sample. Paired samples consist of individually pair-matched observations. The individually matched observations are *statistically dependent* and should be regarded as one unit in statistical analysis, which means that the matched samples might be regarded as one group of dependent data. Hence, matched samples have an equal number of observations.

Different types of study, such as CROSS-SECTIONAL, CASE-CONTROL and *reliability* STUDIES, can be designed as matched samples. Self-pairing studies lead to matched samples, e.g. CROSSOVER TRIALS, *test-retest*, intraobserver studies and studies on paired organs. The purpose of matching is to create homogeneous pairs of observations with regard to important background properties, so the remaining difference between the observations within a pair could be ascribed the source of interest (effect of treatment, exposure, time, etc.). The choice and the number of matching variables should be carefully considered, as the matching variables cannot be used in the statistical evaluation of a possible relationship or explanation regarding the main variable.

Crossover design is ideal in randomised CLINICAL TRIALS for evaluation of the difference in effect between two treatments; one of the treatments could be a PLACEBO treatment. The variability between individuals in the sample is eliminated as each individual is its own control (*self-pairing*). Crossover studies are also called *changeover*, *within-subject* and *AB/BA crossover studies*. Each individual will get the two treatments, A and B, in random order, with a wash-out period in between in order to prevent the effect of the first treatment interacting with the second (carryover) effect (see the figure). The important assumption is that the patients will be in the same state when they receive each treatment. Therefore, this design is suitable for chronic states only. An alternative way of creating homogeneous pairs is by using individually matched pairs.

The aim of using matched pairs in cross-sectional studies is to evaluate the difference in effect between two or more treatments. Matching pairs is a preferable alternative method when the self-pairing crossover design is not appropriate to perform. The matched paired sample consists of individually matched pairs of subjects regarding some prognostic variables where each member of a pair is randomly given one of the treatments (see the figure). The comparison of the effects between the two treatments is made within each pair and could be defined as the difference in effect between the two treatments (additive effect) or as the ratio of the two treatment effects (multiplicative effect). Hence, each pair is treated as one unit in the statistical analysis.

Matched case-control studies. One aim of epidemiological studies is statistically to evaluate associations between risk factors and disease outcomes among individuals. Patients with a disease (cases) are compared with subjects without the disease (control) and the question of interest is their past experience of exposure to possible risk factors. Matched case-control studies consist of an individually matched

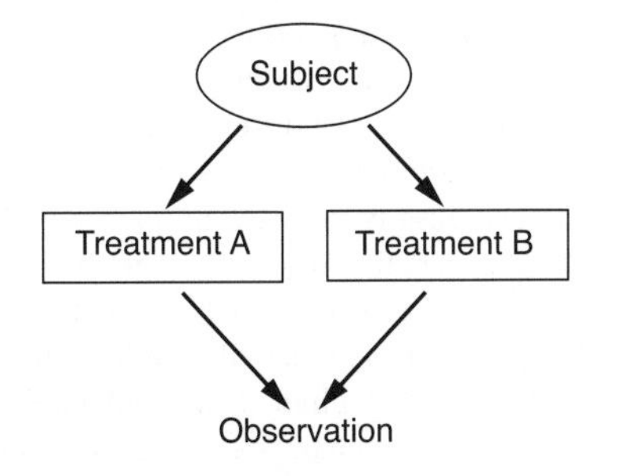

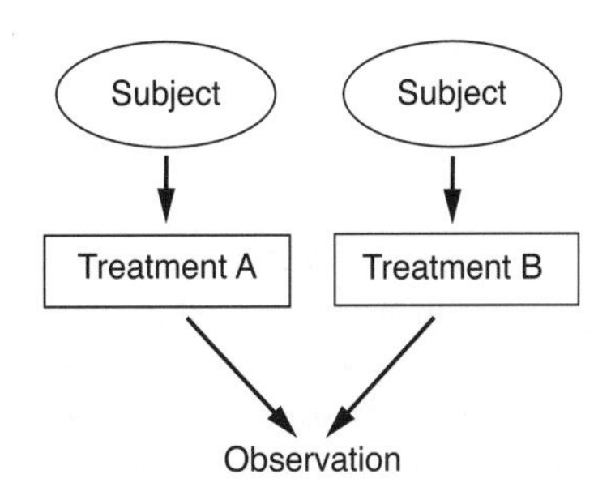

matched samples *Main differences between matched samples from crossover and matched pairs design (McNeil, 1999; Altman, 2000; Senn, 2000)*

control to each case. In the case of rare events, each case may be individually matched with more than one control. The first table shows the four different combinations of outcome and exposure in a matched pair case-control study, when the total number of pairs is n.

matched samples *Four different combinations of outcome and exposure in a matched case-control study*

		Case exposed	*Unexposed*	
Control	*Exposed*	n_{++}	n_{-+}	
	Unexposed	n_{+-}	n_{--}	
				n

Crossover study	*Matched pairs in cross-sectional study*
n subjects	$n+n$ subjects
Two treatments	One treatment
each	each
n observational	n observational
units in analysis	units in analysis

The distribution of data from unmatched case-control studies is also commonly presented in a 2×2 table. However, unmatched case-control data should be treated as two independent groups of data (one group of cases and one group of controls), which means that the number of observations equals the total numbers of cases and controls. The second table shows the frequency distribution of exposed and unexposed cases and controls respectively, when the matched pairs data of the first table are treated as unmatched (McNeil, 1999; Altman, 2000).

matched samples *2×2 frequency table of an unmatched case-control study, showing frequency distribution of the matched case-control data of the first table when treated as unmatched*

	Case	*Control*
Exposed	$n_{++}+n_{+-}$	$n_{++}+n_{-+}$
Unexposed	$n_{-+}+n_{--}$	$n_{+-}+n_{--}$
Total	n	n

Matched samples of ordered categorical data. Studies that involve *rating scales*, questionnaires and other types of categorical classifications often produce matched or paired samples of observations. *Quality* assessments of rating scales concern matched samples of ordered categorical data. In interobserver *reliability studies* the agreement between observers in their classifications of the same subjects is evaluated. The pairs of classifications of each subject are dependent. AGREEMENT studies could also concern an interscale comparison, which is a comparison between different scales for the same variable. Comparisons between self-rated ability and an expert-rated ability of the patient or between a child's opinion and the parents' judgement of it are also examples of the large variations of paired samples involving ordered categorical data. Matched samples are a natural consequence when multi-item instruments for qualitative variables, such as pain and quality of life, are used (Svensson, 2001).

In summary, a large number of different types of study create matched samples. The important feature they all have in common is that matched/paired samples of observations are dependent, a fact that should be taken into account in the statistical analysis. *ES*

[See also KAPPA AND WEIGHTED KAPPA, MATCHED PAIRS ANALYSIS, MATCHING]

Altman, D. G. 2000: *Practical statistics for medical research.* Boca Raton: Chapman & Hall/CRC. **McNeil, D.** 1999: *Epidemiological research methods.* New York: John Wiley & Sons, Inc. **Senn, S.** 2000: *Cross-over trials in clinical research.* Chichester: John Wiley & Sons, Ltd. **Svensson, E.** 2001: Construction of a single global scale for multi-item assessments of the same variable. *Statistics in Medicine* 20, 3831–46.

matching A study design technique of creating pairs of subjects that are homogeneous with respect to important background variables, which are not interesting for the actual study but could interfere with the variable of interest. Matching normally means that each subject is individually paired with another. In prospective clinical studies where the effects of two treatments are to be compared and the two treatments cannot be given to the same individual, matching means that two subjects with the same background properties (e.g. age, gender and some prognostic variables) are paired, one of whom is randomly given the treatment of interest and the other the PLACEBO or standard treatment. The main aim of the matching is to make the two treatment groups comparable by reducing the variability and possible systematic differences that could occur due to disturbing background variables (*confounding bias*). This means that the remaining difference of interest between the two members of a pair would be due to the different treatments.

In *retrospective* CASE-CONTROL STUDIES, matching means that each case is individually paired with a control subject with respect to the background variables and their exposure to the risk factor of interest is compared. Matching twins or siblings provides genetically similar individuals, which can be important in both clinical and epidemiological studies. Individually matched pairs can be regarded as 'artificial twins'.

A continuous matching variable, such as age, can be transformed into categories before pairing individuals, but the matching criterion for one subject to be matched with another can also be expressed in terms of a specified tolerance interval (*caliper matching*). The two individuals in a matched pair should be regarded as one unit and the data should be treated as dependent (paired) in the statistical analysis. For further details see McNeil (1999) and Altman (2000). *ES*

[See also MATCHED PAIRS ANALYSIS, MATCHED SAMPLES]

Altman, D. G. 2000: *Practical statistics for medical research*. Boca Raton: Chapman & Hall/CRC. **McNeil, D.** 1999: *Epidemiological research methods*. New York: John Wiley & Sons, Inc.

maximum likelihood estimation This refers to a general method of estimation for unknown parameters in a probabilistic/statistical model for observed data, for instance a LINEAR REGRESSION model. A well-chosen parameterisation of a discrete PROBABILITY or continuous density ensures that different parameter values determine different chances (densities) of observing a given dataset. The parameter value that maximises this chance (density) for an observed dataset is deemed most likely and is known as the maximum likelihood estimate (MLE).

The MLE is often derived in closed form by maximising the logarithm of the LIKELIHOOD of the data in a function of the unknown model parameters. The first figure depicts this log-likelihood function for the MEAN body weight based on a random sample of 250 American men (Penrose, Nelson and Fisher, 1985) with weights that are normally distributed with the known STANDARD DEVIATION equal to 12.3 kg. The solid vertical line indicates the MLE at 81.2 kg. With two parameters, the graph of the log-likelihood becomes a surface. In the second figure, we show contour lines of this surface for the intercept and slope in a linear model for the regression of body weight (in kg) on height (in cm). We assume that weights corresponding to fixed height measurements are normally distributed with the known standard deviation equal to 10.5 kg. The fitted regression curve is displayed in the entry on the LEAST SQUARES ESTIMATION (and seen in the figure there).

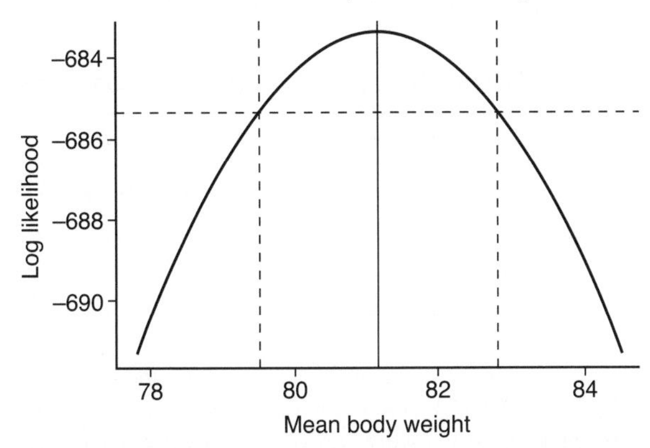

maximum likelihood estimation *Log-likelihood function for the mean body weight estimation based on a random sample of 250 American men*

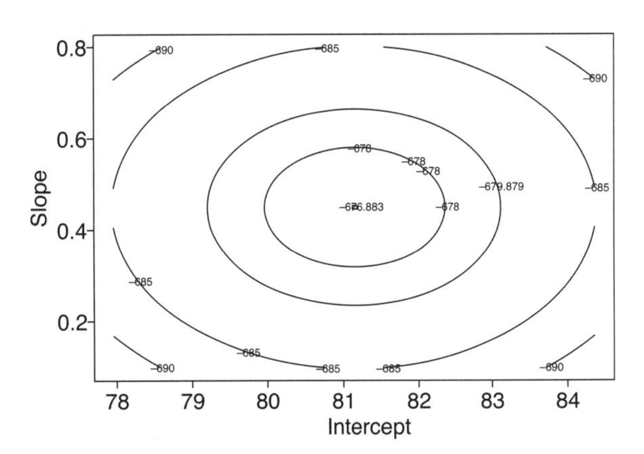

maximum likelihood estimation *Contour lines of the surface for the intercept and slope estimation in a linear model for the regression of body weight (in kg) on height (in cm)*

To maximise the log-likelihood in practice, one can set its derivative with respect to the target parameter, i.e. the score function, equal to zero and solve this for the unknown parameter. Alternatively, computational algorithms can search numerically for the maximum of the log-likelihood function or for a zero of the score function.

Maximum likelihood estimation plays a central role in statistics. It is the standard method of estimation, for instance, in linear, LOG-LINEAR MODELS and LOGISTIC REGRESSION. This is due to its strong intuitive appeal and nice statistical properties in large samples. Under quite general conditions, the MLE becomes normally distributed around the true parameter value as the sample size increases. Hence, 95% CONFIDENCE INTERVALS are obtained as the MLE plus or minus 1.96 times its STANDARD ERROR. For a p-dimensional parameter estimate beta with COVARIANCE V, the Wald confidence region is the ellipsoid consisting of beta_0-values whose distance from the estimator beta, i.e. $(\text{beta-beta}_0)^T V^{-1}$ (beta-beta_0), stays within the 95% percentile of the CHI-SQUARE DISTRIBUTION with p DEGREES OF FREEDOM.

More accurate p-dimensional 95% confidence regions are usually obtained as the set of values that are so likely that twice the log-likelihood is no further removed from the maximum achievable than the 95% percentile of the chi-square distribution with p degrees of freedom. In the first figure we depict this maximum deviation by the dashed horizontal line. The dashed vertical lines mark the 95% confidence interval 79.5–82.8 kg for mean body weight in American men. In the second figure, such a 95% confidence region for the intercept and slope contains all values enclosed by the contour at −679.879.

A flat log-likelihood in the neighbourhood of the MLE makes it hard to pinpoint the MLE and reveals limited information. Hence, a useful summary of the information about the target parameter is the curvature at the estimated value minus the second derivative of the log-likelihood function with respect to the unknown parameter, evaluated at the MLE. This is called the 'observed information'. Its expected value under the assumed model is known as the Fisher information. Its inverse approximates the VARIANCE of the MLE. The inverse of the observed information is often thought to yield the better variance estimate as it sits closer to the data.

The MLE is efficient when the assumed model holds and the sample size is large. This means that among all estimators expected to equal the true parameter value in large samples, no estimator will have smaller variance. The bad news is that the MLE relies on the correct specification of the probability model and may be biased otherwise. The MLE is usually also sensitive to outlying observations. To address these problems, alternative estimation techniques have been devised, such as least squares estimation, L1-regression or the generalised method of moments. These avoid specifying the entire sampling distribution of the observed data and trade robustness against outlying observations for efficiency. For further details see Cox and Hinkley (1979), Clayton and Hills (1993) and Jewell (2004). *SV/EG*

Clayton, D. and Hills, M. 1993: *Statistical models in epidemiology*. Oxford: Oxford University Press. **Cox, D. R. and Hinkley, D. V.** 1979: *Theoretical statistics*. Boca Raton: CRC Press. **Jewell, N. P.** 2004: *Statistics for epidemiology*. Boca Raton: CRC Press. **Penrose, K. W., Nelson, A. G. and Fisher, A. G.** 1985: Generalized body composition prediction for men using simple measurement techniques. *Medicine and Science in Sports and Medicine* 17, 189.

McNemar's test This is an approximation to the exact binomial test (see BINOMIAL DISTRIBUTION) used in situations where there are two paired outcomes of a binary nature (e.g. yes/no, dead/alive, infected/not infected). The test looks to see if the proportions of the response levels are the same in the two groups being considered.

This might occur when we have measurements taken on the same individuals at two different time points (i.e. the status of a recurrent condition), when two treatments are being tested on the same individuals (e.g. in CROSSOVER TRIALS), when paired individuals are entering a trial, when two tests are being performed on individuals (e.g. whether the individual has high blood pressure or is overweight) or when comparing the SENSITIVITY/SPECIFICITY of two classification methods. It should be clear that this is not an exclusive list.

Assuming that we have a set of paired yes/no outcomes, McNemar's test (McNemar, 1947) looks to answer the question as to whether the proportion of 'no's is different across the two groups/treatments/times. It does so by considering the discordant pairs, i.e. those pairs that return a 'yes' and a 'no' (as opposed to two 'no's or two 'yes's). If the proportion of 'no's is the same between two treatments, then for every 'yes–no' we would expect to see a 'no–yes'. Otherwise, there would be an imbalance in the number of 'no's between the two treatments and thus evidence that the proportions are not the same.

If we were to have 12 discordant pairs, 6 'yes–no's and 6 'no–yes's would be expected if the proportions were the same between the two groups. Deviation from this even split would be evidence of a difference in proportions and McNemar's test uses a normal approximation to evaluate the statistical significance of that evidence.

For example, Nicholson *et al.* (1998) perform a crossover trial to see if a vaccine for influenza can trigger asthma exacerbations. Individuals receive both the vaccine and a PLACEBO (in random order) and are observed to see in each case whether an exacerbation occurs. Of 256 people in the trial, 242 had the same reaction after both treatments and so provide no direct information about the difference between the treatments. Eleven have an exacerbation after the vaccine but not after the placebo and three after the placebo but not after the vaccine.

Under an assumption that there is no difference in proportions, we would expect these 14 discordant pairs to have a 7–7 split rather than the 11–3 split observed. However, the McNemar test is nonsignificant with a reported *P*-VALUE of 0.06.

The data for McNemar's test are typically displayed in a CONTINGENCY TABLE, as shown. There are various forms of the test statistic. Essentially, one is looking to compare $(b-c)/\sqrt{(b+c)}$ to the normal standard distribution (MEAN zero and VARIANCE one) or, equivalently, $(b-c)^2/(b+c)$ to the CHI-SQUARE DISTRIBUTION with one DEGREE OF FREEDOM (DoF). These statistics are usually modified to incorporate a continuity correction, the most commonly used of which is to subtract one from the absolute value of the numerator or the positive square root of the numerator.

McNemar's test *Tabulated paired binary outcome data*

		Response from treatment 1	
		No	*Yes*
Response from treatment 2	*No*	242 (a)	11 (b)
	Yes	3 (c)	0 (d)

For the example of Nicholson *et al.* (1998), the statistic is $(|11-3|-1)^2/(11+3) = 3.5$, which when compared to the chi-squared statistic with a DoF of 1, gives a *P*-value of 0.06, as previously stated. (Note that in order to have attained statistical significance at the usual 0.05 level, the statistic would have to exceed the value 3.84.)

As is common when data are well paired, it is the case here that the number of discordant pairs is small. This means that the normal approximation may not perform well. If the only resource is a table of the NORMAL DISTRIBUTION, then McNemar's test is clearly convenient, but today it is unlikely that researchers will not have the capability to perform the exact binomial test. Much literature is devoted to the problem of powering such trials and it is true that it is a simpler task to power for McNemar's test, but again, with modern computing resources, powering the exact binomial test is practicable. It is, therefore, difficult to advocate the use of this test.

McNemar's test is used for the study of paired dichotomous outcomes. For other types of paired data, other methods are available (see STUDENT'S *t*-TEST or the WILCOXON RANK SUM TEST, for example). If the data are matched triples rather than pairs, or in even greater numbers, then the COCHRAN Q-TEST is applicable. For further details see Zar (1999). *AGL*

McNemar, Q. 1947: Note on the sampling error of the differences between correlated proportions or percentages. *Psychometrika* 12, 153–7. **Nicholson, K. G., Nguyen-Van-Tam, J. S., Ahmed, A. H., Wiselka, M. J., Leese, J., Ayres, J., Campbell, J. H., Ebden, P., Eiser, N. M., Hutchcroft, B. J., Pearson, J. C. G., Willey, R. F., Wolstenholme, R. J. and Woodhead, M. A.** 1998: Randomised placebo-controlled crossover trial on effect of inactivated influenza vaccine on pulmonary function in asthma. *The Lancet* 351, 326–31. **Zar, J. H.** 1999: *Biostatistical analysis*, 4th edition. Englewood Cliffs, NJ: Prentice-Hall.

mean The mean is a MEASURE OF LOCATION giving a typical value of a set of observations and what is usually meant when the 'average' is referred to, although other DESCRIPTIVE STATISTICS can also be used as averages. Technically, the term 'mean' is shorthand for the 'arithmetic mean', to differentiate it from the GEOMETRIC MEAN. It is calculated by dividing the sum of all observations by the *number* of observations. For example, the ages in years of seven students in an undergraduate seminar are recorded as 18, 19, 19, 19, 20, 20 and 21. The mean age of the students is:

$$\frac{18 + 19 + 19 + 19 + 20 + 20 + 21}{7} = 19.4 \text{ years}$$

The mean is a suitable measure of location to use when the variable being summarised has an approximately symmetrical distribution. However, if SKEWNESS or OUTLIERS are present then the use of the mean is inappropriate, since it is unduly affected by a small number of values. For example, if a mature student aged 51 joins the undergraduate seminar just described then the mean age of the students becomes:

$$\frac{18 + 19 + 19 + 19 + 20 + 20 + 21 + 51}{8} = 23.4 \text{ years}$$

Here, a mean of 23.4 is not a suitable summary of the typical age of the students since it is strongly influenced by the age of the mature student. In such cases, measures of location such as the MEDIAN or geometric mean may be more appropriate. Again, the median may be preferred when summarising discrete data, such as family size, for no one has, say, 2.4 children!

The mean is usually denoted $\bar{x}$ or μ, although the latter technically refers to the mean of a population, rather than a sample (see SAMPLING DISTRIBUTION).

Altman (1991) suggests that a mean should not normally be quoted to more than one decimal place more than the raw data. The STANDARD DEVIATION is typically used as a MEASURE OF SPREAD around the mean. *SRC*

Altman, D. G. 1991: *Practical statistics for medical research*. London: Chapman & Hall.

measurement error This is a collective term for many different phenomena that arise when a measurement of a particular variable is made, but the measured value fails to match the true value for the subject. When someone's blood pressure is measured, for instance, the equipment for measuring it might be less than perfect or the observer may not be using it correctly. A sufficient indication of measurement error occurs when a repeat reading returns a different value from that obtained on the first occasion, even when the subject's *true* blood pressure has not changed.

The effect of measurement error is wide ranging and is relevant both for research studies and for clinical practice on individual patients. Measurement error might be systematic or random. Systematic error is where the measurement has a consistent tendency to overestimate (or consistently underestimate) the true value. A clock that is always five minutes fast would be an example of systematic overestimation of time of day. Systematic error may be identified by comparing measured values with known true values. In principle, it could then be removed by recalibrating the measuring instrument. This approach may be feasible if true values could be obtained from an alternative measuring instrument ('gold standard'). In practice, however, gold standards do not often exist and studying the AGREEMENT between two measuring instruments is the most realistic approach to quantifying systematic measurement error. The mean difference is a measure of systematic variation between the measuring instruments.

Random error is by its nature less predictable. In this scenario, the measured value neither consistently overestimates nor underestimates the true value, but may depart from the true value in an unpredictable manner. When two or more readings are taken on each of a number of individuals, random measurement error may be quantified in terms of a within-subject STANDARD DEVIATION (SD). This SD has the advantage of being expressed in the same units as the variable of interest, making it readily interpretable for clinicians with an intuitive grasp of the clinical meaning of the units; it is an absolute measure of measurement error. Its practical usefulness may be less, however, for a score derived from a subject's response to a questionnaire, since the score may not be used in routine clinical care. Few clinicians would hold an intuitive grasp of the meaning of values for such a score. One alternative approach, then, is to measure the amount of measurement error relative to the variability of subjects' true values, e.g. by use of the intraclass correlation coefficient (see INTRACLUSTER CORRELATION COEFFICIENT). This equals the ratio of between-subject variance to the sum of between- and within-subject variances.

Measurement error, of course, also occurs with categorical variables; the relative lack of agreement is popularly quantified using the kappa statistic (see KAPPA AND WEIGHTED KAPPA). Absolute measures of agreement for categorical variables are less well established. If quantifying systematic error in measurement of a categorical variable representing presence or absence of a condition, SENSITIVITY and SPECIFICITY are required. Again, they depend on a gold standard measure for the presence or absence of the condition.

In clinical medicine, knowledge of the likely size of measurement error for a particular variable will help interpretation of a single reading. An observer may know with 95 % confidence that the observed value for the subject is within two standard deviations of that subject's true value. Apparent changes in readings from one occasion to another may occur; if they are larger than twice the standard deviation of differences, they are likely to be true changes. Smaller apparent changes could be attributable to measurement error alone. It may, therefore, be more useful to carry out studies that assess intraobserver variation (within the observer) and interobserver variation (between observers) for this reason.

Many studies of measurement error will encounter both systematic and random error. Comparing two methods of measurement of the same variable may demonstrate a consistent tendency for method A to measure higher values than method B. If d represents the value given by method A minus that given by method B for a given subject, then $\bar{d}$ (the mean of d for all the subjects studied) can be calculated. If $\bar{d}$ is nonzero, systematic differences between methods A and B are suggested. However, the SD of d is also unlikely to equal zero; in other words, method A will not return exactly the same number of units higher than method B on every subject. This suggests that random differences, as well as systematic differences, exist between methods A and B. Bland and Altman (1986) have, therefore, recommended the calculation of approximate 95% limits of agreement, which are to be represented by $\bar{d} \pm 2 \times \mathrm{SD}(d)$.

Knowledge of size of measurement error is also required for interpretation of relationships between two or more variables, within research studies. Much of epidemiological research involves understanding the relationship between an 'exposure variable' and an outcome. Inappropriate use of a single measure of exposure may lead to REGRESSION DILUTION BIAS if the exposure is not measured precisely. In other words, the estimate of the effect of exposure will be too conservative. However, if the size of the measurement error can be quantified, the true magnitude of the relationship between the exposure and outcome can be properly estimated.

If the outcome variable in epidemiological or clinical studies is measured imprecisely, the estimate of the effect of exposure would not necessarily be biased, but the STANDARD ERRORS of estimates of exposure effects would be inflated. Thus CONFIDENCE INTERVALS would be too wide and P-VALUES too conservative.

Still further issues arise in epidemiological studies where a potential confounding variable is measured with error and included in an analysis of the relationship between an exposure and an outcome of interest. It is possible that the measurement error in the confounding variable fails to adjust for its effect properly, a phenomenon known as 'residual confounding'. *RM/JE*

[See also ATTENUATION TO MEASUREMENT ERROR, MEASUREMENT PRECISION AND RELIABILITY]

Bland, J. M. and Altman, D. G. 1986: Statistical methods for assessing agreement between two methods of clinical measurement. *The Lancet* 1, 8476, 307–10.

measurement precision and reliability If we repeatedly measure the concentration of glucose in a given blood sample, for example, or measure a patient's blood pressure on each of several successive days, then we are likely to obtain a set of measurements that are similar but not identical. The greater the similarity of the series of measurements then the more precise is the measurement procedure or method producing them. If we measure the variability of the replicated measurements by their variance (or STANDARD DEVIATION) then we can define the method's precision as the reciprocal of that variance (or standard deviation).

In the context of laboratory assays, say, if we hold the measurement conditions (laboratory, batch of reagents, equipment, equipment operator, temperature, etc.) as constant as possible then the measurements are being made under what are usually known as *repeatability conditions* (International Standards Institute, 1994a, 1994b). The precision of the measurements is then assessed from an estimate of the repeatability VARIANCE or repeatability standard deviation. The latter is often known for short as the *repeatability of the process*. The repeatability standard deviation is analogous to the psychometricians' standard error of measurement (Dunn, 2004). If the repeated measurements are taken in different laboratories, at different times, with different equipment, reagents and equipment operators, then the resulting variability of the measurements provides an assessment of reproducibility or generalizability of the measurement process (International Standards Institute, 1994a, 1994b). This is provided by an estimate of the reproducibility variance or reproducibility standard deviation (reproducibility, for short).

One key characteristic of a measuring instrument's or method's precision (the reciprocal of the repeatability variance or reproducibility variance, for example) is that it is scale dependent. If we measure weight in kilograms we will get a precision that is 1000 times as great as the precision if it were measured in grams (the standard deviation for measures using the former scale being one-thousandth of that for those using the latter). It is essential, therefore, that if we are interested in the relative precision of alternative methods of measurement then the scale of measurement is taken into account. Often the scale of measurement of one method relative to another is not defined a priori but needs to be established by experiment (see METHOD COMPARISON STUDIES).

Quite often investigators will be interested in the precision of their instrument or method compared to the variability of the characteristic in the population under clinical or scientific investigation. We postulate the following simple model for the observed measurement (X_i) on the ith individual within a population:

$$X_i = \tau_i + \varepsilon_i \quad (1)$$

Here τ_i is the ith individual's true value for the characteristic and ε_i is a random measurement error. We might be interested in a measure of precision relative to the variability of τ_i, i.e. the ratio $\sigma_\tau^2/\sigma_\varepsilon^2$ where the numerator is the VARIANCE of the true values and denominator the variance of the MEASUREMENT ERRORS. A more familiar index of relative precision is given by the RELIABILITY ratio or reliability coefficient, defined by:

$$\kappa_X = \sigma_\tau^2/(\sigma_\tau^2 + \sigma_\varepsilon^2) \quad (2)$$

If the measurement errors are uncorrelated with each other and with the true values then:

$$\sigma_X^2 = \sigma_\tau^2 + \sigma_\varepsilon^2 \quad (3)$$

The reliability ratio (or reliability, for short) is the proportion of the variation in the observed measurements that is explained by the variability in the underlying true values. It provides a measure of the attenuation that might be expected in the calculation of correlation between error-prone measurements or the attenuation in a regression coefficient when the independent variable is subject to measurement error (see ATTENUATION DUE TO MEASUREMENT ERROR). Note that reliability is not a fixed characteristic of a method or process (even if σ_ε^2 is assumed to be constant). The reliability ratio is a measure of how well the measurements distinguish between members of the relevant population, and as the heterogeneity of the population goes up (i.e. as σ_τ^2 increases) so does κ_X. As the homogeneity of the population increases (as σ_τ^2 decreases towards 0) the reliability tends towards zero. With a fixed σ_ε^2 it is quite straightforward to calculate the change in reliability as one arbitrarily changes the value of σ_τ^2. A relatively nontechnical introduction to reliability theory can be found in Carmines and Zeller (1979).

What if the variance of the measurement errors (σ_ε^2) is not independent of the characteristic (i.e. τ) being measured? It is, in fact, quite common to observe that the variance of the errors increases with the value of τ (i.e. the precision goes down with increasing τ). In this situation we need to be able to design a relatively sophisticated precision study (often an interlaboratory precision study) to provide data that can be used to model the relationship between the two. (This is beyond the scope of this entry, but see Dunn, 2004, for further information.)

Reliability ratios are usually estimated from data involving repeated measurements on each of an appropriate sample of subjects. The analysis usually involves the estimation of the relevant COMPONENTS OF VARIANCE (either using the traditional ONE-WAY ANALYSIS OF VARIANCE or RANDOM EFFECTS MODEL) and then using these to calculate the required reliability. The reliability as defined by equation (2) is equivalent to the correlation between repeated measurements, and this can be estimated by calculating an intraclass correlation coefficient (usually via a one-way analysis of variance). Generalizations of reliability and various versions of intraclass correlation (see INTRACLUSTER CORRELATION COEFFICIENT) can be found in Dunn (2004). In the case of binary measurements the intraclass correlation (intraclass kappa, see KAPPA AND WEIGHTED KAPPA) is equivalent to one of the chance-corrected agreement coefficients – Scott's π-statistic (Kraemer, 1979). *GD*

Carmines, E. G. and Zeller, R. A. 1979: *Reliability and validity assessment*. Thousand Oaks: Sage. **Dunn, G.** 2004: *Statistical evaluation of measurement errors*. London: Arnold. **International-Standards Institute** 1994a: *International Standard ISO 5725-1. Accuarcy (trueness and precision) of measurement methods and results. Part 1: General principles and definitions.* **International Standards Institute** 1994b: *International Standard ISO 5725-2. Accuarcy (trueness and precision) of measurement methods and results. Part 2: Basic method for the determination of the repeatability and reproducibility of a standard measurement method.* **Kraemer, H. C.** 1979: Ramifications of a population model for κ as a coefficient of reliability. *Psychometrika* 44, 461–71.

measures of central tendency See MEASURES OF LOCATION

measures of dispersion See MEASURES OF SPREAD

measures of fertility See DEMOGRAPHY

measures of location Also known collectively as 'averages', these are single-figure summaries intended to describe the typical or representative level of a set of observations. There are four measures of location commonly encountered.

As part of a study of childhood development, the IQ of 210 mothers was measured. The average IQ of the population is 100, but the average IQ in the sample of mothers was 106. This information indicates that on the whole these women had slightly higher IQs than would be expected from a truly representative sample.

The MEAN is what is usually being referred to when talking about the 'average'; the average mothers' IQ quoted above was technically the mean mothers' IQ. The other averages commonly used are the MEDIAN, GEOMETRIC MEAN and MODE.

The mean is calculated by dividing the sum of all the observations by the number of observations. However, the mean is not an appropriate summary statistic when SKEWNESS or OUTLIERS are present.

The median is an alternative measure of location in situations when the mean is not suitable. The median is the value that comes halfway when the observations are ordered. It is based on the ranks of the data and is therefore not dependent on the distribution of observations.

In the special case where the data are positively skewed such that the log-transformed data have a NORMAL DISTRIBUTION, the geometric mean is an alternative measure of location. The geometric mean is calculated by backtransforming (antilogging) the (arithmetic) mean of the logged values.

The fourth measure of location commonly used in statistics is the mode. This is simply the value that occurs most often and it is therefore useful in summarising categorical rather than continuous data. Its usefulness can, however, be limited. For instance, we could not say much in a study of smoking if the modal number of cigarettes smoked in a sample happened to be zero.

A measure of spread is often quoted alongside a measure of location to give information about the variability of observations around the average. *SRC*

measures of mortality See DEMOGRAPHY

measures of spread Once the average of a set of observations has been defined using a MEASURE OF LOCATION, it is helpful to know how widely the data are scattered around this typical value. Measures of spread are used to summarise this information numerically.

The most straightforward measure of spread is the RANGE: the interval between the minimum and maximum values in a set of observations. Although the range has the advantage of simplicity, it is only influenced by the most extreme observations in a dataset and is therefore not generally considered a good way of quantifying variability.

Instead, in the case where the data are approximately symmetrically distributed, the STANDARD DEVIATION is often quoted. This measure has the useful property that, especially when the data follow a NORMAL DISTRIBUTION, approximately 95% of the observations lie within two standard deviations of the MEAN.

The INTERQUARTILE RANGE is an alternative measure of spread used in situations when the standard deviation is not suitable, due to SKEWNESS or the presence of OUTLIERS. It is the interval between the values that are located a quarter and three-quarters of the way through the sample when the observations are ordered. Since it is based on the ranks of the data it is not dependent on the distribution of observations.

The VARIANCE is the square of the standard deviation. Although this quantity is frequently used in statistical analysis, it is not as useful as the standard deviation in describing the spread of observations because it is not in the same units as the original data. Whereas one can have an intuitive feel for, say, cm^2, it is less obvious how to cope with units such as $years^2$ or $mmHg^2$. *SRC*

median The median is a measure of location, being the central or 'halfway' value when the observations are ordered. Thus it is the middle value – in other words, half the data lie below it and half above. The median is also known as the *50th percentile*.

For example, in the following dataset, the heights of 11 women are recorded in centimetres:

154 157 157 158 159 **160** 161 162 162 163 169
↑
Median

The median is the 6th value when they are ordered, i.e. 160 cm.

If there is an even number of observations then the median is, by convention, the arithmetic mean of the *two* central values. Without such a convention, any value of an infinite number lying between the two central observations would be a median, but it is preferable to have a definition enforcing a unique value.

The MEAN is generally used as a MEASURE OF LOCATION when the data have a symmetrical distribution. If this is not the case then the median is often quoted because it is not unduly affected by the presence of SKEWNESS or OUTLIERS. For instance, a woman who is 185 cm tall is added to the heights dataset in our previous example:

154 157 157 158 159 160 161 162 162 163 169 185

Median = 160.5

This changes the median to 160.5 cm, a value that still gives a good indication of the average height of the women.

The INTERQUARTILE RANGE is typically used as a MEASURE OF SPREAD around the median. *SRC*

median regression See QUANTILE REGRESSION

median survival time See SURVIVAL ANALYSIS-AN OVERVIEW

median test The median test is a nonparametric test. It can be used when exact values of some observations are unknown, if they can be classified as above or below the overall MEDIAN. It is a useful alternative for testing differences in medians when the assumption of the similarity of distributions necessary for the MANN–WHITNEY RANK SUM TEST or KRUSKAL–WALLIS TEST is not met, although it is less powerful than these tests. The median test utilises the CHI-SQUARE TEST to determine whether two or more (k) independent groups are drawn from populations with the same median. The median test assumes that the data are randomly selected observations that are ordinal or continuous in nature and that there are at least two independent groups. In addition the assumptions of the chi-square test apply.

To perform the median test, first treat the data as a single sample and calculate the overall median. Classify each observation as below/equal or above the overall median. Calculate the total of each type in each group. Arrange these values into a $2 \times k$ CONTINGENCY TABLE, where the two rows are the classification against the overall median and the k columns are the groups. Carry out a chi-square test with $k-1$ DEGREES OF FREEDOM (DoF) on this table. Reject the null hypothesis of equal medians if the chi-square test proves significant, in which case post hoc pair-wise median tests can be carried out.

Example data are shown in the first table; they consist of mini mental state examination scores (MMSE) in three groups with different types of dementia. Group 1 has Alzheimer's disease, group 2 has frontotemporal dementia and group 3 has semantic dementia.

median test *MMSE scores for the three dementia groups*

Group number	*MMSE score*										
1	19	7	17	28	21	6	21	19	27	8	25
2	16	22	30	24	22	22	22	28	29	29	0
3	4	9	30	29	25	22	25	26	27	18	10

The overall median is 22, and the second table classifies the number of observations in each group against this overall median.

median test *Number below/equal or above the overall median in each group*

Group number	*≤ 22*	*>22*	*Total*
1	8	3	11
2	5	6	11
3	5	6	11
Total	18	15	33

To perform the chi-square test on this second table, first calculate the expected count in each cell by multiplying the row total by the column total and dividing by the overall total. This gives expected counts of 6, 6, 6 for the first column and 5, 5, 5 for the second. The chi-square test is then calculated as the sum over each cell, of the observed minus the expected counts squared, divided by the expected count, as in the following equation:

$$X^2 = \sum_{ij} \frac{(O_{ij}-E_{ij})^2}{E_{ij}}$$

$$= \frac{(8-6)^2}{6}+\frac{(5-6)^2}{6}+\frac{(5-6)^2}{6}+\frac{(3-5)^2}{5}+\frac{(6-5)^2}{5}+\frac{(6-5)^2}{5}$$

$$= 0.667+0.167+0.167+0.800+0.200+0.200=2.20$$

The critical value from the chi-square tables ($\alpha=0.05$, df $=2$) is 5.99. The value of 2.20 from the chi-square test is less than the critical value of 5.99; therefore there is insufficient evidence to reject the null hypothesis that the median MMSE scores are the same for all three dementia groups. Hence it is not possible to carry out post hoc pair-wise tests (see POST HOC ANALYSIS) between the groups. *SLV*

medical statistics – an overview Statistics may be defined as the science of collecting, analysing and interpreting numerical data. Medical statistics is the application of this science to medicine.

Statistics began as information concerning the state and this aspect is still a central part of medical statistics, as we collect and analyse information on national rates of birth, death and notifiable diseases. The term expanded to include many types of numerical data, collected for the purposes of administration (bed occupancy), research (CLINICAL TRIALS) and pleasure (batting averages) or any combination of the three. I was unable to think of a medical example for pleasure, so I will leave that as an exercise for the reader. To the statistician, statistics is all pleasure anyway.

Statistics is a skill as well as a science and a good statistician can take what seems to its owner to be a bewildering mass of data and find structure and meaning within it. What the data owner has to provide is the question that needs to be answered. The statistician can seldom provide that.

Indeed, when the statistician starts to wonder more about the substantive medical question than how to answer it, he or she ceases to be a statistician and becomes an epidemiologist, trialist or health service researcher. The true statistician is much more interested in the process of answering the question than the answer itself, in the journey rather than the destination.

Statistics is unusual among academic subjects in that its entire purpose is to solve problems in other disciplines. Medical statistics is a collaboration between statisticians and those whose main object is to increase understanding of disease, health and healthcare. In the history of statistics, many important innovations were made by those wishing to answer a question arising in their own discipline. Most of us cannot aspire to be such polymaths, however, and must content ourselves with providing one aspect of the collaboration. Most people from health fields who wish to use statistics really need to acquire two things: the ability to apply a few day-to-day statistical procedures correctly and a vocabulary with which to communicate with their statistician.

We can include in collecting data the design of studies: the decisions as to what data to collect and from where or whom and how they should be collected. A frequent complaint of consulting statisticians is that their clients do not come to them early enough (see CONSULTING A STATISTICIAN). The researcher arriving with the referee's comments on a rejected paper may be too late. If the design were fundamentally flawed, there is probably very little that can be done to rectify matters.

There is no substitute for sound design (see EXPERIMENTAL DESIGN) and statistics has a lot to offer. One of the greatest of all contributions to medicine is surely the invention and development of the randomised clinical trial, which, for the first time, enabled medical researchers to obtain reliable evidence on the relative efficacy of clinical treatments.

The basic principles of study design are fairly straightforward and easily understood, unlike other aspects of medical statistics. In experimental biological studies, whether clinical trials on human subjects or laboratory experiments on animals, tissue samples or cell cultures, RANDOMISATION with blind allocation is the key. We neglect it at our peril. Most medical experiments use very simple designs, such as fully randomised two-group comparisons or two-period crossovers (see CROSSOVER TRIALS). At most, we might have a simple factorial design, where two treatments are given in combinations of none, one or both. Much more complex designs are used in other areas of application, such as agriculture or the chemical industry.

Randomisation quickly gets complicated, however, as we often want to improve the comparability of groups by stratification or are forced by the nature of the treatment to have patients allocated in clusters rather than individually. We may want, within small blocks of patients, to allocate equal numbers to each of the treatments, to ensure that numbers in treatment groups are always similar. In a small trial on a variable subject group we may decide to improve comparability by MINIMISATION, ensuring that certain key variables will be balanced between the groups. These modifications in turn require changes to the planned analysis and hence to the sample size estimation. They must not be ignored.

In medical observational studies the usual statistical approach of random sampling is seldom possible, but this does not mean we should ignore sampling issues, rather that we should consider very carefully the representativeness of our sample. As in experimental design, the principles of epidemiological designs such as CASE-CONTROL, COHORT, CROSS-SECTIONAL and ecological STUDIES are easy to understand, but the details of their implementation using cluster or STRATIFIED SAMPLES, matched one to one or one to many, often require statistical insight and input. Clinical designs are often similar to those in epidemiology, but used for a different purpose: case series may be used to describe clinical experience, cohort studies may be used to describe the natural history of a disease, case-control designs arise in the evaluation of diagnostic tests, cross-sectional studies are used to investigate the properties of measurement methods.

Data collection is very important and here the principle is to ensure that what we collect is accurate, with a minimum of BIAS and error. Bias can be reduced by blind assessment, where the observer is unaware of the subject's status, and by BLINDING the subject to things that may influence response. In an experiment we may achieve this by concealing treatment allocations, e.g. with a PLACEBO, leading to the ideal of the double-blind randomised trial. In an observational study this is more difficult, but we may, for example, include the questions on our key outcome variable among many others to reduce the apparent emphasis. Statistics offers techniques to ensure that we cannot be certain what individual respondents have told us while still obtaining data for the sample as a whole, such as secret ballots and randomised response, but they are seldom used in healthcare studies. Perhaps the most important thing we must do is to convince our respondents that their answers are absolutely confidential.

Training of observers is also very important in maintaining data quality and we may wish to estimate the degree of observer variation and the effects of using different observers. To reduce measurement error we may need to consider the frequency of measurement and we can weigh the relative advantages of increasing the number of subjects and increasing the number of measurements made on each.

Sample size is one aspect of study design on which statisticians are asked to advise, but often far too late for them to have any real input. A question about the sample size the day before the deadline for a grant application is typical, when all that can be done is to provide some kind of justification for the sample size already chosen on feasibility grounds. To change it at that stage is usually quite impractical. To decide on sample size we need to think about the purpose of the study, what outcome variable we intend to use and the analysis that we propose to carry out. This is, in any case, an excellent discipline for any investigation. Too often people collect data without any idea of how they are to be analysed. It can be a shock when they discover that they have collected data for which the possible analysis is very complex, difficult to interpret, time consuming and expensive and cannot answer their most important questions.

Statistical analysis begins with simple graphical methods, such as HISTOGRAMS and scatter diagrams (see SCATTERPLOT), and tabulations, which begin to reveal the structure of data. To make this more manageable, we then use summary statistics such as MEANS, STANDARD DEVIATIONS, centiles and proportions. An analytical method of peculiarly medical application is the Kaplan–Meier survival method (see KAPLAN–MEIER ESTIMATOR), which enables us to estimate the cumulative survival of a group of subjects, some of whom are still surviving and who have been observed for varying lengths of time. Kaplan and Meier's paper (1958) has been reported to be the most highly cited statistical paper ever (Ryan and Woodall, 2005). Comparison of these summary statistics between different groups of subjects leads to the use of differences and ratios between groups. Investigating the strength of relationships between variables observed in tabulations can be done using RELATIVE RISKS AND ODDS RATIOS, investigating those observed in scatter diagrams by regression and correlation.

A frequent problem in medical data is that the variable of interest may be influenced by another, not of any interest in itself. National mortality data provide a good example, where the age structure of the population has a profound effect. Special age-standardisation methods have been developed for this, producing age-standardised mortality rates and standardised mortality ratios. Much more generally applicable to deal with such problems are MULTIPLE LINEAR REGRESSION (see LOGIT MODELS FOR ORDINAL RESPONSES) and its many offspring: logistic, ordered logistic, multinomial, Poisson (see GENERALISED LINEAR MODEL), COX'S REGRESSION MODEL, LOGISTIC REGRESSION, etc. Such techniques also allow us to analyse situations where we are interested in several predicting variables and want to look at their relative importance as predictors.

A key question in analysis of data is the correct unit of analysis. In a trial, for example, if we randomise individual patients to treatment, the patient will be the unit of analysis. However, if we allocate a group of patients together, forming a cluster, we must take this into account in the analysis. For example, we might allocate all the asthma patients in a primary care centre to receive an educational intervention or all to act as controls; the primary care centre becomes the unit of analysis. We might calculate the average of our outcome measurement for the cluster of patients in the practice and then compare two groups of clusters. If we want to include information collected at the level of the individual patient in such a study, we may have to use more complex, multilevel techniques. The same problem arises when we have multiple observations on a few patients or several tissue samples taken from each of a few organs. Ignoring the correct unit of analysis may be seriously misleading (see MULTILEVEL MODELS).

Another area when statistical analysis becomes essential is when we have several outcome variables and no clear primary one. This might happen if we have a battery of psychological tests or a QUESTIONNAIRE with many items. We may want to investigate the structure of these, summarise them into a single scale or analyse them all as a group. Methods developed in psychology, such as FACTOR ANALYSIS, enable us to do this.

In medical statistics, we usually have a sample of observations from some larger population, about which we want to draw some conclusion. For example, in a clinical trial the subjects are a sample of all the patients to whom we might wish to give the trial treatments, now and in the future. We use the trial to tell us about what would happen in this larger population. Even when we have data on the whole population, as in the case of national mortality rates, we often think of them as a sample of a hypothetical larger population so that we can investigate the reliability of differences and relationships found. This process of drawing conclusions about the larger population from the sample is called statistical inference. Users of statistics often find the concepts involved in inference quite difficult to master.

Most inference in the medical literature takes the form of CONFIDENCE INTERVALS and SIGNIFICANCE TESTS. These are methods from the frequentist formulation of statistics, one of two conceptual frameworks in which statistical inference is carried out. In this approach, we regard our sample as one of many we could have taken and then make deductions from the sample we do have about the many samples we do *not* have, and hence about the population from which they would all be drawn. The alternative is the Bayesian conceptual framework (see BAYESIAN METHODS). In this, we try to describe what we already know about the answer to our question in PROBABILITY terms and then see how these probabilities are modified by the data we have collected. Both approaches provide ways to take data from a sample and decide what they can tell us about the population. Both involve difficult concepts that can take years to master. Fortunately, it is possible to analyse data adequately even with a quite poor understanding of the underlying philosophy.

In the past, statisticians were divided into two warring camps, the Bayesians claiming that their methods had a securer philosophical foundation and frequentists claiming that Bayesian methods were impractical for real problems. In recent years the development of powerful computer-intensive Bayesian methods and computers fast enough to carry them out has led many more statisticians to make use of the Bayesian approach and the barriers are coming down.

Much statistical methodology consists of techniques to apply these fundamental concepts to inference for different types of design and data. The first methods developed were for large samples, estimating and comparing means, rates and proportions. These were followed by t-DISTRIBUTION-based methods for means of small samples, where the distribution of the observations themselves was important. This led to the use of transformations to allow data to be presented in a form that we knew how to analyse. It also led to the development of alternative methods, such as those based on rank order, which did not require such strong assumptions about the data. Small samples for proportions led to the development of exact probability methods such as FISHER'S EXACT TEST, methods that the advent of powerful and convenient computers has made feasible for large datasets, too. The list of statistical methods is long and growing. Almost all the analysis methods have inference attached to them, so that it is often not explicit which aspect of statistics we are doing.

When we have decided what our data can tell us about the wider world, we usually want to understand why the relationships that we have discovered have arisen. This brings us back in a circle to the design.

If we have a randomised, double-blind experiment we can usually conclude that the evidence supports the difference in treatment causing the difference in outcome. We do not, of course, conclude unequivocally that there is cause and effect; statistics is a discipline that instils caution and it is always possible, however unlikely, that we have an extreme sample producing a result atypical of its population. Statistics enables us to assess how cautious we need to be. If we have a study that is not blinded or randomised, we must consider very carefully the possible biases that may have been introduced.

If we have an observational study, we must always be very cautious in the interpretation. We must ask how good our sampling is and how comparable our groups are. We must remember that just because there is an ASSOCIATION between two variables we should not conclude that one causes the other. We must consider other factors that may be responsible for the relationship we have observed. If we are studying the aetiology of disease, in particular, we must beware of a rush to judgement. Guidelines for assessing the evidence for causality are available (such as the BRADFORD HILL CRITERIA), but it takes data from several different sources before we can apply them.

In the era of evidence-based healthcare, all healthcare professionals, whether doctors, nurses or therapists, must be able to understand and interpret the research evidence on which their practice should be based. Much of this evidence is quantitative and statistics is the key skill needed. Not only has the number of statistical analyses published and the number of statistical methods employed increased greatly, but also these analyses have achieved much greater prominence. They are no longer confined to the methods and results sections of papers, but now fill the abstracts, too. Familiarity with basic statistical ideas is inescapable for the healthcare professions.

Healthcare research, whether medical, nursing or therapy, is unusual in that it is mainly initiated and carried out by healthcare professionals. Medical research is done by doctors, nursing research by nurses. This does not happen in other fields. Educational research is not done by teachers, social research by social workers or agricultural research by farmers, but rather by professional researchers in academia and industry. It is quite an attractive idea that it is part of the role of doctors to add to medical knowledge, but it puts a great responsibility on them to do this to a high standard. Unfortunately, this is often not the case. Statistical analysis requires quite different habits of mind to those required for diagnosis, and the training of doctors, nurses and others is aimed at developing ways of thinking different to those learned by the mathematicians who specialise in statistics. Healthcare research requires many different skills and aptitudes and is much better done by collaboration between people from the disciplines possessing these. Rather than attempt to train a clinician in statistics to the level required for high-quality medical research, we should employ a statistician. Not only is this more effective, it is, regrettably, cheaper – they are not paid as much (or enough).

To work together, clinicians require familiarity with statistical ideas and vocabulary and close collaboration with statisticians should enhance this. Statisticians need to be familiar with the problems of carrying out research on particularly vulnerable human beings, whose needs must always be paramount, and the many special techniques that have been developed to do this. Close collaboration with clinical researchers should also further the education of the statisticians. Working together benefits clinicians, statisticians and the research itself, and hence is to the good of all of us. *JMB*

Kaplan, E. L. and Meier, P. 1958: Nonparametric estimation from incomplete observations. *Journal of the American Statistical Association* 53, 457–81. **Ryan, T. P. and Woodall, W. H.** 2005: The most-cited statistical papers. *Journal of Applied Statistics.*

Medicines and Healthcare products Regulatory Agency (MHRA) The Agency was formed on 1 April 2003 as the amalgamation of the former Medicines Control and Medical Devices Agencies, which were, respectively, the UK government agencies responsible for the assessment and licensing of pharmaceutical and biological human medicines and (human) medical devices. The Agency has a variety of responsibilities including assessment of new medicines/products, assessment of changes to existing medicines, post-marketing surveillance (see PHARMACOVIGILANCE), inspection of CLINICAL TRIAL conduct and manufacturing facilities, enforcement of regulations and so on.

The legal framework for the Agency's work is set out in the Medicines Act 1968: for medicines, an applicant is required to demonstrate adequate evidence of safety, quality and efficacy. To this end, the Agency employs a large number of quality, pre-clinical and clinical assessors. Most of the statistical work is done in conjunction with the clinical assessors, although statistical considerations often come into other areas such as assessment of pre-clinical safety studies, determination of product shelf-life and assessment of marketing/advertising claims. Companies apply for a 'marketing authorisation' (formerly called simply a 'licence'). Agency staff assess data and prepare assessment reports, which are considered by the Commission on Human Medicines (CHM) and some of its expert subcommittees. The CHM is a panel of independent experts (including practising doctors, pharmacologists, statisticians and lay members) that meets monthly and advises on the granting, or otherwise, of a marketing authorisation. The final decision on granting is made by the government minister responsible for health. In certain circumstances, companies may appeal against unfavourable decisions.

The MHRA works in close collaboration with other European national agencies and the European Medicines Agency (EMA), which is based in London. There are a variety of routes by which companies can apply for a marketing authorisation within Europe – including national licences in as many (or as few) EU member states as they wish or a centralised licence covering all member states. In the latter case, two member states will be allocated to complete a comprehensive assessment of the application but all other member states are given the opportunity to raise concerns. The European counterpart to the CHM is the Committee for Human Medicinal Products (CHMP), which meets monthly. The MHRA and EMA also work with other agencies across the world and, in particular, contribute to and follow guidelines jointly prepared by the International Conference on Harmonization (a collaborative effort between the major geographical regions of Europe, Japan and the USA).

The assessment of safety and efficacy from CLINICAL TRIALS is similar to that for refereeing a paper for a medical journal but explores much more detail (see REGULATORY STATISTICAL MATTERS). The law requires companies to submit details of all trials that have been conducted. Each of these trials will have detailed study reports running to hundreds of pages and further extensive appendices including: a copy of the protocol and any amendments; individual case reports for serious adverse events/reactions; line listings of individual patient data; possibly efficacy results presented separately for each participating centre; copies of investigators' curriculum vitae; documentation of quality and purity of product used in the trials; and so on. These appendices may run to hundreds of volumes and, hence, the need for a variety of disciplines within the assessment teams.

More information about the Agency, its work and regulations pertaining to medicinal products and medical devices is available at the MHRA website: www.mhra.gov.uk. *SD*

Medicines Control Agency (MCA) See MEDICINES AND HEALTHCARE PRODUCTS REGULATORY AGENCY

mega-trial This is a large-scale randomised trial (generally involving several thousand subjects) that is designed to detect the effects of one or more treatments on major ENDPOINTS, such as death or disability. The need for mega-trials arises because the vast majority of treatments have only moderate effects on such endpoints, typically producing relative reductions of, at most, a quarter. Any study aiming to detect such a moderate effect needs to be able to guarantee that any BIASES and random errors inherent in its design are substantially smaller than the expected treatment effect (Collins and MacMahon, 2001). This will ensure that the results of the study either confirm the presence of a moderate effect convincingly or, if the treatment is ineffective, provide clear evidence that this is so.

For a study to avoid moderate biases requires RANDOMISATION using a method that precludes knowledge of each successive allocation. Randomisation in CLINICAL TRIALS is intended to maximise the LIKELIHOOD that each type of patient will have been allocated in similar proportions to the different treatment strategies being investigated (Armitage, 1982). Randomisation requires that trial procedures are organised in a way that ensures that the decision to enter a patient is made irreversibly and without knowledge of which trial treatment a patient will be allocated. Even when studies are randomised, however, moderate biases can still be introduced by inappropriate analysis or interpretation.

The requirements for reliable assessment of moderate treatment effects are as follows: *negligible biases*, i.e. guaranteed avoidance of moderate biases involves proper randomisation (nonrandomised methods cannot guarantee the avoidance of moderate biases), involve: analysis by allocated treatments (i.e. an INTENTION-TO-TREAT analysis); chief emphasis on overall results (with no unduly data-derived subgroup analysis); and systematic META-ANALYSIS of all the relevant randomised trials (with no unduly data-dependent emphasis on the results from particular studies) and *small random errors*, i.e. guaranteed avoidance of moderate random errors, involve: use of large numbers (with minimal data collection since detailed statistical analyses of masses of data on prognostic features generally add little to the effective size of a trial); and systematic meta-analysis of all the relevant randomised trials.

One well-recognised circumstance is when patients are excluded after randomisation, particularly when the prognosis of the excluded patients in one treatment group differs from that in the other (such as might occur, for example, if noncompliers were excluded after randomisation).

While avoidance of moderate biases requires careful attention both to the randomisation process and to the analysis and interpretation of the available trial evidence, a study can only avoid moderate random errors if it accumulates a sufficiently large number of events. When major endpoints such as death affect only a small proportion of those randomised, very large numbers of patients need to be studied before estimates of treatment effect can be guaranteed to be statistically (and hence medically) convincing. In these circumstances, when a treatment has the potential to be used widely and hence confer large benefit (or harm), a mega-trial is the only type of study that is sufficiently reliable. For example, for an event that is expected to occur among 10% of subjects without active treatment, over 20 000 subjects are required to demonstrate a 20% relative risk reduction (i.e. from 10% to 8%) reliably (i.e. with 90% POWER at a TYPE I ERROR rate of 1%).

For a mega-trial to randomise large numbers of trial subjects, the main barriers to rapid recruitment need to be removed. To facilitate this, the information recorded at entry should be brief and should concentrate on those few clinical details that are of paramount importance (including at most only a few major prognostic factors and only a few variables that are thought likely to influence substantially the benefits or hazards of treatment). Similarly, the information recorded at follow-up should be limited to serious outcomes, adverse events and to approximate measures of compliance. (Other outcomes, such as surrogate endpoints, that are of interest but do not need to be studied on such a large scale, may best be assessed in separate smaller studies or in subsets of these large studies when this is practicable.) Keeping a trial as simple as possible increases the likelihood that it will be able to recruit large numbers of patients. For this reason, mega-trials are also known as 'large simple trials'.

For ethical reasons, randomisation is appropriate only if both the doctor and the patient feel substantially uncertain as to which trial treatment is best. The 'uncertainty principle' maximises the potential for recruitment within this ethical constraint. This says that a patient can be entered if, and only if, the responsible physician is substantially uncertain as to which of the trial treatments would be most appropriate for that particular patient. A patient should not be entered if the responsible physician or the patient are, for any medical or nonmedical reasons, reasonably certain that one of the treatments that might be allocated would be inappropriate for this particular individual (either in comparison with no treatment or in comparison with some other treatment that could be offered to the patient in or outside the trial).

If many hospitals are collaborating in a trial then wholehearted use of the uncertainty principle encourages heterogeneity in the resulting trial population and this, in mega-trials, may add substantially to the practical value of the results. Among the early trials of fibrinolytic therapy, for example, most of the studies had restrictive trial entry criteria that precluded the randomisation of elderly patients, so those trials contributed nothing of direct relevance to the important clinical question of whether treatment was useful among older patients. Other trials that did not impose an upper age limit, however, did include some elderly patients and were therefore able to show that age alone is not a contraindication to fibrinolytic therapy (Fibrinolytic Therapy Trialists' Collaborative Group, 1994). Mega-trials adopting the uncertainty principle to determine eligibility maximise heterogeneity of the study sample, which in turn ensures that their results are relevant to a very diverse range of future patients. *CB*

[See also ETHICS]

Armitage, P. 1982: The role of randomisation in clinical trials. *Statistics in Medicine* 1, 345–52. **Collins, R. and MacMahon, S.** 2001: Reliable assessment of the effects of treatment on mortality and major morbidity, I: clinical trials. *The Lancet* 357, 373–80. **Fibrinolytic Therapy Trialists' Collaborative Group** 1994: Indications for fibrinolytic therapy in suspected acute myocardial infarction: collaborative overview of early mortality and major morbidity results from all randomised trials of more than 1000 patients. *The Lancet* 343, 311–22.

Mendelian randomisation Mendelian randomisation refers to a method of leveraging improved causal inference (see CAUSALITY) from observational data through utilisation of the random assignment of an individual's genotype from their parental genotypes. It is justified by interpretations of the laws of Mendelian genetics. Assuming that the probability that a postmeiotic germ cell that has received any particular allele at segregation contributes to a viable conceptus is independent of environment (from Mendel's first law) and that genetic variants sort independently (from Mendel's second law), then at a population level these variants will not be associated with the confounding factors that generally distort conventional observational studies. Formally, random allocation of genotype occurs within family groups (from parents to offspring), and interpretation of such studies is closely analogous to that of randomised controlled trials (see CLINICAL TRIALS) (Davey Smith and Ebrahim, 2003). However, it has repeatedly been demonstrated that at a population level genetic variants are generally unrelated to potential confounding factors (Davey Smith *et al.*, 2008). Confounding by other genetic variants will only occur for variants located close together on the same chromosome, when they are said to be in linkage disequilibrium (LD) with each other.

The term 'Mendelian randomisation' was first applied in a study using the availability of genetically compatible siblings to evaluate the effectiveness of bone marrow transplant in haematopoietic cancer (Gray and Wheatley, 1991), an approach that is conceptually similar but distinct in terms of design (Davey Smith, 2006). The concept applied to population-based epidemiological studies was most clearly articulated by Katan (1986), who proposed that since polymorphic forms of the apolipoprotein ε (APOE) gene were related to different average levels of serum cholesterol, individuals with the genotype associated with lower average cholesterol should be expected to have a higher cancer risk if low cholesterol levels increased the risk of cancer. If, however, reverse causation or confounding generated the association between low cholesterol and cancer, then no association would be expected.

The conditions for, and assumptions underlying, a successful Mendelian randomisation study were elaborated in 2003 (Davey Smith and Ebrahim, 2003) when the general proposition that Mendelian randomization could be used to make causal inferences about the relationship between modifiable risk factors and disease outcomes was advanced. It was argued that if genetic variants are robustly related to different levels of exposure to a risk factor, then these genetic variants should be related to disease risk to the extent predicted by their influence on the risk factor. Mendelian randomisation implies that genotype–disease ASSOCIATIONS should not be affected either by confounding or by reverse causation, and many biases inherent in conventional observational studies may also be avoided (Davey Smith and Ebrahim, 2003). Such associations may therefore imply a causal effect of the risk factor on the disease outcome. However, these causal inferences from these studies may be undermined by issues including pleiotropy (the genotype has direct effects on more than one risk factor for the disease outcome), population stratification (population subgroups that experience both different disease rates and have different frequencies of genotypes of interest exist, resulting in confounded associations between genotype and disease) and canalization (buffering of the effect of genotype on disease by compensatory biological mechanisms) (Davey Smith and Ebrahim, 2003).

Thomas and Conti (2004) pointed out that the Mendelian randomization approach involves application of the method of INSTRUMENTAL VARIABLES, which is commonly used in econometrics. An instrumental variable satisfies the following assumptions: (a) it is associated with the exposure of interest, (b) it is independent of confounding factors and (c) it is independent of the outcome given the risk factors and confounding factors. Lawlor *et al.* (2008) reviewed instrumental variables methodology in the context of Mendelian randomisation studies. Because genetic variants often explain only a small proportion of the variance in the target risk factor, very large sample sizes may be needed to achieve precise estimates of the causal effect of the risk factor on disease outcomes. *JS/GDS*

Davey Smith, G. 2006: *Capitalising on Mendelian randomization to assess the effects of treatments.* James Lind Library; www.jameslindlibrary.org. **Davey Smith, G. and Ebrahim, S.** 2003: 'Mendelian randomisation': can genetic epidemiology contribute to understanding environmental determinants of disease? *International Journal of Epidemiology* 32, 1–22. **Davey Smith, G., Lawlor, D.A., Harbord, R., Timpson, N., Day, I. and Ebrahim, S.** 2008: Clustered environments and randomized genes: a fundamental distinction between conventional and genetic epidemiology. *PloS Medicine* 4, e352; DOI:10.1371/journal.pmed.0040352. **Gray, R. and Wheatley, K.** 1991: How to avoid bias when comparing bone marrow transplantation with chemotherapy. *Bone Marrow Transplantation*7(Suppl.), 3, 9–12. **Katan, M. B.** 1986: Apolipoprotein E isoforms, serum cholesterol, and cancer. *Lancet* 1, 8479, 507–8. **Lawlor, D. A., Harbord, R. M., Sterne, J. A. C., Timpson, N. and Davey Smith, G.** 2008: Mendelian randomization: using genes as instruments for making causal inferences in epidemiology. *Statistics in Medicine* 27, 8, 1133–63. **Thomas, D. C. and Conti, D. V.** 2004: Commentary: the concept of 'Mendelian randomization'. *International Journal of Epidemiology* 33, 21–5.

meta-analysis See SYSTEMATIC REVIEWS AND META-ANALYSIS

meta-regression This is an analysis of the relationship between study characteristics and study results in the context of a META-ANALYSIS.

Independent studies of the same problem, e.g. multiple CLINICAL TRIALS of a particular drug or multiple CASE-CONTROL STUDIES of the same exposure–disease ASSOCIATION, will inevitably differ in many ways. Some of the variation may cause the effects being evaluated to be different in different studies, a situation commonly known as heterogeneity. Meta-regression analyses are similar to traditional LINEAR REGRESSION analyses, a conceptual difference being that entire studies, rather than individuals, are the units of analysis. Characteristics of studies are used as explanatory (independent) variables and estimates of effect are used as outcome (dependent) variables. Regression coefficients describe how the effects across the studies increase per unit increase in the characteristic. Study characteristics might include numerical summaries of types of participants, variation in the implementation of an intervention or different methodological features. Estimates of effect may be, for example, ODDS RATIOS, hazard ratios or differences in mean responses, depending on the type of study and the nature of the outcome data. Ratio measures of effect are usually analysed on the (natural, or base e) log scale. Studies are weighted in the analysis to reflect imprecision in their results, the weights typically involving the inverse variances of the effect estimates.

A meta-regression may be a primary reason for assembling multiple studies, although meta-regressions are perhaps most commonly used as secondary analyses to investigate heterogeneity when a traditional meta-analysis was the primary objective of a review of studies. The study characteristics may be categorical or quantitative and several may be included in the same analysis. For categorical characteristics, meta-regression may be viewed as a generalisation of subgroup analyses, where the subgrouping is by studies rather than by participants. Meta-regression should ideally be conducted only as part of a thorough systematic review to ensure that the studies involved are reliably identified and appraised.

Notable examples of meta-regression analyses include an investigation of the dose–response relationship between aspirin and secondary prevention of stroke. Among clinical trials administering aspirin at different doses, no relationship was apparent between aspirin dose and the relative risk of recurrence (Johnson *et al.*, 1999). A second example is provided by Zeegers, Jellema and Ostrer (2003), who present a meta-analysis of observational studies comparing prostate cancer risk between people with and without family history of prostate cancer. They used meta-regression to perform several subgroup analyses to assess the robustness of their finding that a family history of the disease is associated with roughly a doubling of risk. Studies were broken down by study design, year of publication and ethnic group, among other characteristics. A third example that has inspired development of meta-regression methodology is an analysis describing a relationship between the geographical latitude of studies of the BCG vaccine and the relative risk of tuberculosis in those vaccinated versus those not (Berkey *et al.*, 1995).

A convenient illustration of a meta-regression analysis for a single characteristic is a simple SCATTERPLOT as in the figure, which shows the result of the BCG vaccine meta-regression. The circles represent studies, with the size of each circle proportional to the precision of the relative risk estimate from that study. The meta-regression line illustrates that the vaccine was observed to be more effective further away from the equator.

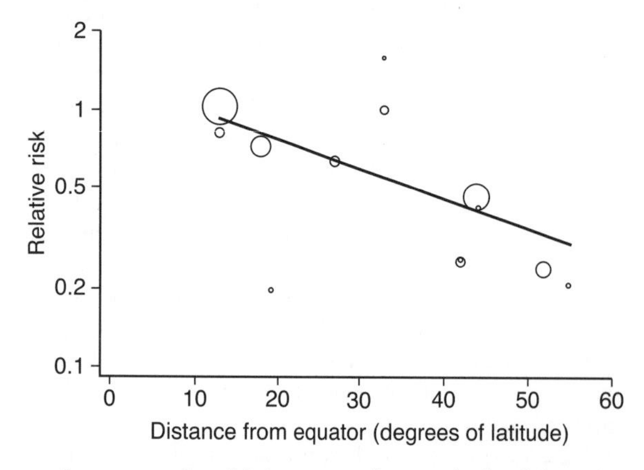

meta-regression *Meta-regression analysis of the relationship between effectiveness of BCG vaccine and latitude of study (data from Berkey et al., 1995)*

Meta-regression analyses involve observational comparisons across studies and may suffer from BIAS due to confounding, since studies similar in one characteristic may be similar in others. Causal relationships between characteristics and results can seldom be drawn with confidence. A particular problem is that in most situations the number of studies in a meta-regression is small while the number of potentially important characteristics is large. Thus any meta-regression analyses performed should be driven by a strong scientific rationale and ideally pre-specified and limited in number. It may be necessary to control for the possibility of false–positive findings since the risk of a TYPE I ERROR increases substantially when multiple meta-regression analyses are undertaken.

It is possible to summarise participant-level characteristics at the level of a study for use in a meta-regression. Thus the MEAN age of participants, the proportion of females or the average length of follow-up might be used as study-level characteristics. Such analyses should be interpreted carefully, as they may not adequately reflect true associations. For example, suppose the effect of an intervention truly depends on a patient's age. If several clinical trials each include a wide range of ages, but if the mean age is similar across trials, then a meta-regression relating mean age to size of effect will fail to detect the relationship that would be evident from within-trial analyses. When interest focuses on participant-level characteristics, a meta-analysis of individual participant-level data is the most reliable method of separating within-study from among-study relationships. Potential limitations of meta-regression, including those already mentioned, are discussed by Thompson and Higgins (2002).

In common with meta-analysis, meta-regression may be conducted assuming either a FIXED EFFECT model or a RANDOM EFFECTS MODEL. A fixed effect meta-regression assumes that the study characteristic(s) explain all of the interstudy variation in effects. It may be performed using weighted linear regression of the effect estimates on the study characteristics, weighting by the inverse VARIANCES of the effect estimates. However, the STANDARD ERRORS of the regression coefficients need to be corrected to account for the fact that the variances are known, by dividing them by the mean square error from the weighted regression.

A random effects meta-regression allows for variation in study effects that is not explained by the study characteristics. Such 'residual heterogeneity' of effect is commonly assumed to follow a NORMAL DISTRIBUTION analogous to a random effects meta-analysis. Random effects meta-regression requires more specialised software, although a convenient implementation is available for Stata (Sterne, Bradburn and Egger, 2001 and see also STATISTICAL PACKAGES). Since it is unlikely that heterogeneity can be fully explained by a finite selection of study characteristics, random effects meta-regression has been recommended as the default choice (Thompson and Higgins, 2002).

Meta-regression may be performed using alternative methods specific to the nature of the individual level outcome data when these are available. For example, when the outcome data from individuals in the studies are binary, then meta-regression may be undertaken using logistic regression.

Implementations of meta-regression that require special consideration include the relationship between effect estimates and underlying risk (due to correlation between effect estimates and risks), investigations of publication bias (again due to possible correlation between effect estimates and measures of precision) and nonindependent outcomes (e.g. entering subgroups from the same study). *JPTH*

Berkey, C. S., Hoaglin, D. C., Mosteller, F. and Colditz, G. A. 1995: A random effects regression model for meta-analysis. *Statistics in Medicine* 14, 395–411. **Johnson, E. S., Lanes, S. F., Wentworth III, C. E., Satterfield, M. H., Abebe, B. L. and Dicker, L. W.** 1999: A metaregression analysis of the dose–response effect of aspirin on stroke. *Archives of Internal Medicine* 159, 1248–53. **Sterne, J. A. C., Bradburn, M. J. and Egger, M.** 2001: Meta-analysis in Stata™. In Egger, M., Davey Smith, G. and Altman, D. G. (eds), *Systematic reviews in health care: meta-analysis in context*, 2nd edition. London: BMJ Publication Group. **Thompson, S. G. and Higgins, J. P. T.** 2002: How should meta-regression analyses be undertaken and interpreted? *Statistics in Medicine* 21, 1559–73. **Zeegers, M. P. A., Jellema, A. M. and Ostrer, H.** 2003: Empiric risk of prostate carcinoma for relatives of patients with prostate carcinoma: a meta-analysis. *Cancer* 97, 1894–903.

method comparison studies At its simplest, a method comparison study involves the measurement of a given characteristic on a sample of subjects or specimens by two different methods. To take a simple and familiar example, we could imagine a study in which the body temperature of each of, say, n patients was assessed once using an old mercury thermometer calibrated in degrees Fahrenheit (F) and again using a modern thermometer calibrated in degrees Celsius (C). If the true temperature of the ith individual (in degrees Celsius) is τ_i then the resulting set of measurements might be represented by the following two equations:

$$\begin{aligned} C_i &= \tau_i + \delta_i \\ F_i &= 32 + 1.8\tau_i + \varepsilon_i \end{aligned} \qquad (1)$$

The numbers 32 and 1.8 follow from the temperatures of freezing and boiling water (i.e. 0 °C ≡ 32 °F; 100 °C ≡ 212 °F). The key characteristic of the design is that the resulting data is a series of paired measurements (C_i, F_i). The δ and ε values in these two equations correspond to random measurement errors that are assumed to be uncorrelated both with each other and with the patient's true temperature. They are both assumed to have an average value of zero. They cannot be determined individually but statistical methods can be used to assess their variability (their variances, σ_δ^2 and σ_ε^2, respectively).

Now we will complicate matters by choosing to measure a characteristic, such as a tissue enzyme, using two different assay methods with indeterminate scales. Arbitrarily choosing one (X) to be the standard (or, indeed, it may be already recognised as the standard against which a new assay is being compared) and other (Y) to be the comparator, a realistic statistical model might have the form:

$$\begin{aligned} X_i &= \tau_i + \delta_i \\ Y_i &= \alpha + \beta\tau_i + \varepsilon_i \end{aligned} \qquad (2)$$

Here the values 32 and 1.8 have been replaced by unknown constants α and β respectively. Our task is now to take the n pairs of measurements (the X_i, Y_i) and use statistical methods to estimate α and β, together with the variances, σ_δ^2 and σ_ε^2. These are the parameters of the MEASUREMENT ERROR model (possibly with the addition of the variance of the true scores, σ_τ^2, depending on how the patients or specimens have been selected). Before describing how we might attempt to carry out this estimation, however, it will be useful to discuss briefly what we might wish to learn or decide from the results of a method comparison study.

We might wish to estimate the parameters of a relative calibration or measurement error model such as that described by the pair of equations in (2). It would obviously be of interest to know the values of α and β so that we might know how to convert the scale of X to that of Y, or vice versa. In particular, we might wish to establish whether $\alpha = 0$ or $\beta = 1$, or both (i.e. are the scales the same?). If the scales of measurement are the same (i.e. $\alpha = 0$ and $\beta = 1$) then the two measurement methods are the MEAN or average equivalent. In this case we might also wish to know whether the two methods are equally precise (or whether one is more precise than the other) by comparing the estimates of the VARIANCES of the measurement errors (i.e. comparing estimates of σ_δ^2 and σ_ε^2). If two methods are mean equivalent and their precisions are the same, then they are fully individually equivalent. In the theory of psychometric tests (applicable to the measurement of depression or anxiety, for example), tests that are mean equivalent or individual equivalent are referred to as being τ-equivalent or parallel respectively. Measurements using alternative methods that are individually equivalent or parallel are fully interchangeable without any loss of information.

Suppose, however, that we wish to evaluate and compare the precisions of two methods that are known not to be mean equivalent? How, for example, do we compare the performance (precision) of an old thermometer calibrated in degrees Fahrenheit with a new one in degrees Celsius? We would need first to convert the Fahrenheit measurements to degrees Celsius (or vice versa) and only then compare the variances of the measurement errors. For methods X and Y, the relevant ratio (i.e. relative precision) for this comparison is $\sigma_\varepsilon^2/(\beta^2\sigma_\delta^2)$. Here a direct comparison of σ_δ^2 and σ_ε^2 would provide the answer to the wrong question.

A less stringent question might involve asking whether the two measurements on a given patient are close enough. We do not ask whether two methods are exactly equivalent but whether, for all practical purposes, they are interchangeable. In this situation we may abandon the measurement model in the equations in (2) entirely and concentrate on the paired differences $(X_i - Y_i)$ as indicators of agreement between the two methods (Bland and Altman, 1986, 1999). If the agreement is good enough then we can for all practical purposes replace a measurement made using one of the methods by a corresponding measurement using the other one.

This is the rationale for the construction of LIMITS OF AGREEMENT (Bland and Altman, 1986, 1999). A very useful graphical summary to accompany these calculations is what is usually known as the Bland–Altman plot – a plot of the difference between the two measurements, $X_i - Y_i$, against their mean, $(X_i + Y_i)/2$ (Bland and Altman, 1986, 1999). In addition, one might wish to produce a simple Y versus X SCATTERPLOT, together with an estimate of their product–moment correlation and concordance CORRELATION (Lin, 1989).

Many investigators, however, will wish to go beyond testing for equivalence. They will wish to know, for example, whether Y is better than X. Is the new method an improvement on the old one? Or, contrariwise, is it worse? If this is the aim then we have no option but to collect the relevant data (the design might need to be more informative than those discussed so far), postulate realistic statistical models for the measurements and proceed to test whether the models are appropriate and, if so, to find what the estimates of the model's parameters tell us about the performance of the methods.

Returning to the statistical model described in (2), how do we estimate the parameters and test hypotheses concerning them, given a set of paired measurements (X_i, Y_i)? Well, the simple answer is that we cannot. There is insufficient information provided by the data to enable us to estimate these parameters. The technical phrase for this is the 'problem of model underidentification'. The only way to proceed is by making various assumptions concerning some of the parameters to produce a model that is identified and then to estimate the remaining parameters. Examples of these assumptions include (a) knowing the variance of the measurement errors of the standard (or its RELIABILITY), (b) assuming a common scale of measurement (i.e. that $\beta = 1$) or (c) knowing the relative sizes of the two measurement error variances (i.e. the ratio $\sigma_\varepsilon^2/\sigma_\delta^2$).

The trouble with each of these assumptions is that we are assuming something about the measurement methods that we would ideally have wished to study as part of the method comparison study. The other problem is that if the chosen assumption is not actually valid we are likely to finish up with the wrong conclusions. Consider, for example, the assumption that we know the ratio of the error variances. This leads to the use of a method known as orthogonal or *Deming's regression* (very popular in clinical chemistry). Typically the measurement error variances are estimated for each of the methods by repeatedly measuring the relevant characteristic on the same individual(s) or specimen(s). This enables us to estimate repeatability variances. These are only valid estimates of the measurement error variances (σ_ε^2 and σ_δ^2) if the repeated measurements do not have correlated measurement errors. Correlated measurement errors are almost universal and one should be very wary of the use of Deming's regression when they are known to be a possibility (Carroll and Ruppert, 1996).

The only really satisfactory way out is to use a more informative design involving one or more of the following features: replication using each of the methods, the use of instrumental variables (see INSTRUMENTAL VARIABLES) and the use of more than three different methods of measurement within the study. The other key feature of these studies should be an adequate sample size. Most method comparison studies are too small. Statistical analyses for the data arising from more of the informative designs, with more realistic measurement models, is beyond the scope of this entry, but the methods are described in considerable detail in Dunn (2004). The methods typically involve software developed for STRUCTURAL EQUATION MODELLING (see SOFTWARE FOR STRUCTURAL EQUATIONS MODELS). Methods for the comparison of binary measurements (diagnostic tests) can also be found in Dunn (2004). *GD*

Bland, J. M. and Altman, D. G. 1986: Statistical methods for assessing agreement between two methods of clinical measurement. *Lancet* i, 307–10. **Bland, J. M. and Altman, D. G.** 1999: Measuring agreement in method comparison studies. *Statistical Methods in Medical Research* 8, 135–60. **Carroll, R. J. and Ruppert, D.** 1996: The use and misuse of orthogonal regression in linear errors-in-variables models. *The American Statistician* 50, 1–6. **Dunn, G.** 2004: *Statistical evaluation of measurement errors*. London: Arnold. **Lin, L. I.-K.** 1989: A concordance correlation coefficient to evaluate reproducibility. *Biometrics* 45, 255–68 (see Corrections in *Biometrics* 56, 324–5).

MHRA See MEDICINES AND HEALTHCARE PRODUCTS REGULATORY AGENCY

microarray experiments These are studies in microbiology that are designed to measure the expression levels of genes in a particular organism, generally in response to some stimuli or conditions believed to stimulate the organism's genes, and hence can be used to assess probabilistically the risk of developing a disease, or assessing environmental sensitivity or adaptability, for which genetic expression has been identified. Presently, three technologies have been developed to measure these expression levels; all are based on the biological concept of hybridisation between matching nucleotides.

By comparing the expression levels of the full complement of genes from the organism with those from a 'normal' or 'control' individual, one can identify genes that are 'differentially expressed' (expressed at different levels by the two genomes). To the extent that a gene has been identified as either causative or highly associated with a particular disease (e.g. BRCA1 gene on chromosome 17 for suppression of an early-onset breast cancer tumour or the Rb gene on chromosome 13 for suppression of a retinoblastoma tumour), the risk of disease can be estimated and mechanisms of causation can be elucidated. In some cases, the analysis may permit early intervention to prevent the onset or progression of the disease (e.g. identification of an absent gene may permit early diagnosis and treatment). Similarly, adaptability or sensitivity to environmental contaminants (e.g. metals) can be compared among organisms that manifest differing levels of gene expression. (An example of such an ASSOCIATION has been identified in *Daphnia*, a freshwater crustacean that has a compact and well-characterised genome sequence, in their acclimation and adaptation to cadmium in lakes; cf. Shaw *et al.*, 2008.) The identification of genes responsible for an organism's health condition and response to the environment provides important clues on potential causes of disease.

The measurements in microarray experiments reflect levels of complementary deoxyribonucleic acid (cDNA), reversed-transcribed (as explained below) from cellular messenger ribonucleic acid (mRNA), not the levels of proteins that the organism manufactures in response to measured elevated levels of mRNA. (While not definitely proven, one assumes that an organism's mRNA levels would increase as a precursor to the manufacture of relevant protein products needed to respond to the stimulus.)

Presently, three technologies permit the measurement of gene expression levels on microarrays: treated glass slides with spotted and immobilized cDNA (*cDNA slides*), chips spotted with manufactured 25 base pair sequences of nucleotides found in genes (*oligonucleotide arrays*) and high-density chips with synthesised longer sequences of oligonucleotides (*high-density chips*). All three technologies are based on the biological concept of hybridisation between matching nucleotides, and can contain multiple copies of single-stranded genes or gene fragments, called probes, linked to a substrate or surface for binding with expressed transcripts from target tissues.

The genetic code for an organism is contained in organised strings of four nucleotides (A = adenine, C = cysteine, G = guanine, T = thymine), arranged in triplets such that each triplet codes for one of 20 amino acids. (Multiple triplets may code for the same amino acid.) Strings of amino acids are called peptides. Peptides can act independently in a cell or they can combine with other peptides to form complex proteins used by the organism for cell function. Genetic material known as deoxyribonucleic acid (DNA) is arranged in a double-stranded helical structure, with complementary base pairs on either side of the helix (AT/TA or CG/GC). In response to a stimulus to produce a protein, the coding genes in the DNA are transcribed into messenger RNA (mRNA) for translation into peptides. To test for gene expression, the investigator harvests cells from tissues of the types under study and the mRNA is reverse-transcribed into its more stable form (complementary DNA, or cDNA), split into smaller strands and denatured ('unzipped'), yielding single-stranded cDNA. The various tissue types are labelled with different fluorescing chemicals that can be detected by an instrument. Present instrumentation allows for the detection of two different chemicals that fluoresce at sufficiently different wavelengths that they can be readily distinguished, thus allowing the co-hybridisation of treatment and control samples on the same microarray for a direct comparison and minimizing technical variability, though the technology is not limited to only two scanning channels. For quantitative measurement, a single or a mixture of cDNA strands is placed on to the slide or chip containing the gene probes, and the strands of nucleotides in the target sample are allowed to bind (hybridise) to their matching partners. Spots on the slide or chip where hybridisation has occurred indicate gene products that are present in larger quantities and may have been expressed in response to the stimulus. With this technology, the reported intensity level at a particular location on the slide or chip is a summary of fluorescence measurements detected by an LCD (liquid crystal display) camera as a series of pixels that comprise the spot on the slide.

The three related technologies for measuring gene expression differ in the process that is used to manufacture the probes on the slide or chip. In cDNA slides, the probes typically are obtained from a cDNA library, which has thousands of bacterial colonies with cloned cDNA fragments. Once isolated in the bacterial hosts, the DNA fragments undergo a series of complex processes that amplify and then mechanically deposit them on the treated glass slide substrate. In oligonucleotide arrays, the gene fragments consist of specific DNA strings ('probes') of 25–70 manufactured nucleotides that are placed on the chip robotically. Commercial microarray manufacturers use either photolithography or digital mirror devices and photoreactant chemistries. In all cases, the DNA probes are placed in an array of rows and columns, hence the term 'microarray'.

The technologies also differ in the experimental protocols that yield quantitative gene expression data. For glass substrate microarrays, the hybridisation solution contains a mixture of two types of cells, control and experimental, whose mRNA is reverse-transcribed into the more stable cDNA and then labelled with two different fluorophores: control cells labelled with Cyanine 3, or Cy3 (green dye), and experimental cells,e.g. cells subjected to stress, heat, radiation or chemicals, or known to originate from disease tissue, labelled with Cyanine 5, or Cy5 (red dye). When the mRNA concentration is high in these samples, their cDNA will bind to their corresponding probes on the spotted cDNA slide; an optical detector in a laser scanner will measure the fluorescence at wavelengths corresponding to the green and red dyes (532 nm and 635 nm respectively). Good EXPERIMENTAL DESIGN will include technical replicates that interchange the dyes in a separate hybridisation experiment to account for imbalances in the signal intensities from the two types of fluorophores and expected BIASES from gene–dye interactions (e.g. possible degradation in the cDNA samples between the first scan at 532 nm and second scan at 635 nm). The ratio of the relative abundance of red and green dyes at these two wavelengths on a certain spot indicates relative mRNA concentration between the experimental and control samples at those genes. Thus, the gene expression levels in the genes under the experimental condition can be compared directly with those under the control condition. However, other technologies are limited to hybridising a single labelled sample of targets at one time, thus requiring the addition of control probes and alternate protocols for normalising the signal intensity data across replicates and across chips for reliable comparisons of gene expression between the experimental and control conditions.

Oligonucleotide arrays circumvent the possible inaccuracies that can arise in the preparation of cDNA probes and the control and experimental samples for spotted array slides, by using predefined and prefabricated sequences of 25–70 nucleotides to characterise each gene. Rather than mechanically depositing DNA, oligonucleotide probes can be synthesised directly on to the substrate. For arrays manufactured via the photolithography-like process, the probe cells measure 24×24 or 50×50 micrometres square and are divided in 8×8 pixels; arrays manufactured using photoreactant technology measure 13×13 micrometres and hence arrays can accommodate more probes. As with cDNA slides, cells from the target sample, labelled again with fluorophores, will hybridise to those squares on the chip that correspond to the complementary strands of the target sample's single-stranded cDNA. For these experiments, the target sample contains only one type of cell (e.g. treatment, or control); the assessment of expression is in comparison to the expression level on an adjacent probe, which is exactly the same as the gene probe except for certain nucleotides (e.g. 13th out of 25 nucleotides). This 'mismatch' (MM) for the 'perfect match' (PM) sequence is only a rough guide, since a target sample with elevated mRNA concentration for a certain gene may hybridise sufficiently to both the PM and MM probes. However, the results are believed to be less variable, since the probes on the chips are manufactured in more carefully controlled concentrations. Gene expression levels are measured again by a laser scanner that detects the optical energy in the pixels at the various probes (PM and MM) on the chip.

The analysis of the data (fluorescence levels at the various locations on the slide or chip) depends upon the technology. For cDNA experiments, the analysis usually involves the logarithm of the ratio of the expression levels between the target and control samples. For oligonucleotide experiments, the analysis involves a weighted linear combination of the logarithm of the PM expression level and the logarithm of the MM expression level (with some authors choosing zero for the weights of the MM values). Microarray analysis involves several considerations, including: the separation of 'spot' pixels from 'background' pixels and the determination of the expression level from the intensities recorded from the data 'spot' pixels; the adjustment of the calculated spot intensity for background ('background correction'); the normalisation of the range of fluorescence values from one experiment to another, particularly with oligonucleotide chips; experimental design of multiple slides or chips (Kerr and Churchill, 2001; Yang and Speed, 2002; Casella, 2008); data transformations (Yang *et al.*, 2002; Kafadar and Phang, 2003); statistical methods of inference and combining information from multiple cDNA experiments (Amaratunga, and Cabrera, 2001; Dudoit *et al.*, 2002) and from multiple oligonucleotide arrays (Efron *et al.*, 2001; Irizarry *et al.*, 2002) and adjustments for MULTIPLE COMPARISONS (Reiner, Yekutieli and Benjamini, 2002; Efron, 2004; Benjamini and Yekutieli, 2005). The 'low-level' analysis consists of the necessary 'pre-processing' steps, including data TRANSFORMATIONS (usually the logarithm) to address partially the nonnormality of the expression levels, and normalisation and background correction methods to adjust for different signal intensities across different microarray experiments and sources of variation arising from the chip manufacturing process and background intensity levels. The 'high-level' analysis usually involves clustering (see CLUSTER ANALYSIS IN MEDICINE) the gene expression levels into groups of genes that are believed to respond similarly, but no consensus has been achieved on the best methods for normalising, clustering and reducing the number of genes to consider as 'significantly differentially expressed' when searching for associations between disease and gene locations on chromosomes.

Microarray analyses have become a standard screening tool in the exploration and elucidation of mechanisms of disease and for studying the interface between the environment and the genome. They have also been used to understand the effect of certain exposures better, such as anthrax or anthrax-like organisms, or metals, on cells from human and animal populations, and hence to characterise better the risk of such agents to these populations (Human Genome Program, 2002). Examples of useful gene microarray-based investigations that potentially can improve public health practice include discoveries of candidates for biomarkers of disease; measurements of perturbations in the cell cycle under differing conditions; uncovering genetic underpinnings of numerous human, animal and plant diseases; and explorations of the impacts of environmental change on ecosystems and populations of humans, animals, plants and disease-causing organisms. An example of the potential utility of microarrays is in the tracking of antigenic drifts and shifts of influenza viruses and the changes in the virus–host relationship. Through these types of analyses, the fields of genomics, proteomics and statistics will contribute substantially to how we perceive, measure and address disease and environmental change. Genome- and proteome-scale microarrays will be used increasingly as a cost-effective means of quantifying risk and identifying precursors for diseases, for developing vaccines and for other public health and environmental initiatives. Through these types of analyses, the fields of genomics, proteomics and statistics will contribute substantially to how we perceive, measure and address disease and environmental change. *KK/KB*

[See also ALLELIC ASSOCIATION, GENETIC EPIDEMIOLOGY]

Amaratunga, D. and Cabrera, J. 2001: Analysis of data from viral DNA microchips. *Journal of the American Statistical Association* 96, 456, 1161–70. **Casella, G.** 2008: *Statistical design*. New York: Springer. **Dudoit, S., Yang, Y., Callow, M. and Speed, T.** 2002: Statistical methods for identifying differentially expressed genes in replicated cDNA microarray experiments. *Statistica Sinica* 12, 111–39. **Efron, B.** 2004: Large-scale simultaneous hypothesis testing: the choice of a null hypothesis. *Journal of the American Statistical Association* 99, 96–104. **Efron, B., Tibshirani, R., Storey, J., Tusher, V.** 2001: Empirical Bayes analysis of a microarray experiment. *Journal of the American Statistical Association* 96, 1151–60. **Human Genome Program** 2002: *US Dept of Energy Human Genome News* V12, N1-2, February 2002; http://www .ornl.gov/sci/techresources/Human_Genome/publicat/hg n/v12n1/ HGN121_2.pdf. **Irizarry, R. A., Hobbs, B., Collin, F., Beazer-Barclay, Y. C., Antonellis, K. J., Scherf, U. and Speed, T. P.** 2002: Exploration, normalization, and summaries of high density oligonucleotide array probe level data. *Biostatistics* 19, 185–93. **Kafadar, K. and Phang, T.** 2003: Transformations, background estimation, and process effects in the statistical analysis of microarrays. *Computational Statistics and Data Analysis* 44, 313–38. **Reiner, A., Yekutieli, D., Benjamini, Y.** 2002: Identifying differentially expressed genes using false discovery rate controlling procedures. *Bioinformatics* 19, 3, 368–75. **Shaw, J. R., Pfrender, B.D., Eads, R., Klaper, A., Callaghan, A., Colson, I., Jansen, B., Gilbert, D. and Colbourne, J. K.** 2008: *Daphnia* as an emerging model for toxicological genomics. In Hogstrand, C. and Kille, P. (eds), *Advances in experimental biology on toxicogenomics*. Elsevier, pp. 165–219. **Yang, Y. H. and Speed, T. P.** 2002: Design issues for cDNA microarray experiments. *Nature Reviews* 3, 579–88. **Yang, Y. H., Dudoit, S., Luu, P. and Speed, T. P.** 2002: Normalization for cDNA microarray data: a robust composite method addressing single and multiple slide systematic variation. *Nucleic Acids Research* 30, E15.

mid-*P*-value See EXACT METHODS FOR CATEGORICAL DATA

MIM See GRAPHICAL MODELS

minimisation This method is sometimes used to balance RANDOMISATION in a CLINICAL TRIAL when there are several factors on which it is considered necessary to try to force balance across the treatment groups. Simple randomisation will, in theory (or in 'the long run'), ensure that treatment groups are equally represented with respect to all known and unknown prognostic factors but, for any particular trial, this balance may not be as good as we would hope. When there are only a few factors for which balance is necessary (such as gender or stage of disease) then simple stratified randomisation may be sufficient. However, if there are more than two or three factors on which to try to balance, then the number of strata becomes excessive and the logistics of the trial become overwhelming. Minimisation was a method proposed by Taves (1974) and, more extensively, by Pocock and Simon (1975) as a way of balancing simultaneously for several factors (see also Pocock, 1983, pp. 84–6).

It is important to realise that in most trials patients arrive sequentially, rather than all being available as a 'pool' of patients at the beginning. Hence, when a patient of a certain demographic and/or disease state enrols, we do not know when (or even if) a similar patient will enrol subsequently. However, if two similar patients were to be available for a trial, it would be desirable to allocate one to each of the treatment groups (a method easily extendable to more than two treatment groups). If there were only one factor on which to balance the randomisation then for patients within each stratum we would (optimally) allocate them alternately between the treatments. If there is more than one factor, e.g. gender (male/female) and disease stage (early/progressive/advanced), we have to 'trade off' the benefits of allocating to one treatment in order to ensure an equal balance of males/females across the treatment groups – and simultaneously to ensure an equal balance of early/progressive/advanced patients across the treatment groups. Often to balance gender, we might be better off allocating the patient to one treatment but to balance for disease stage we might be better off allocating the patient to the other. Hence we use the term 'minimisation': to try to minimise the degree of imbalance across all the identified factors.

The following example is described by Day (1999) and concerns a trial randomising general practitioners to an intervention group or to control (see also Steptoe *et al.*, 1999). Three factors were identified on which to balance the groups: the Jarman score (a level of social status), the ratio of number of patients to hourly nurse practice hours ('low' or 'high') and the fundholding status of the practice (in three categories). Assume we are partway through the trial and the first 18 practices have been allocated as in the table. Balance looks reasonably good. Now assume that the next (the 19th) practice is of type: low Jarman score, high patient–practice nurse-hours and is a nonfundholder. We calculate 'scores' for these types of practice, which are $4+5+3=12$ for the intervention group and $3+4+3=10$ for the control group. Imbalance is 'in favour' of intervention, so by allocating this practice to control, we minimise the imbalance.

minimisation *Allocation of first 18 general practices and profile of 19th practice indicated*

	Prognostic factor	*Intervention group*	*Control group*	
	Jarman score			
→	Low	4	3	←
	Middle	3	5	
	High	2	1	
	Patient–practice hours per week			
	Low	4	5	
→	High	5	4	←
	Fundholding status			
→	Nonfundholder	3	3	←
	1st wave entry	4	3	
	2nd wave entry	2	3	

When Taves and then, the following year, Pocock and Simon published their early papers on this topic, they explained how simple the method is to use and, in particular, how, for a single institution, it is quite possible to 'minimise' on several factors with a simple card index system. In a MULTICENTRE TRIAL this would effectively be 'minimisation, stratified by centre'. With modern telephone and computer systems used for central randomisation it becomes even easier to use minimisation across centres (possibly using 'centre' as one of the minimisation factors).

It was mentioned earlier how an 'optimal' allocation could easily be determined in the case of a single factor but that would only be optimal in the sense of minimising the imbalance. Maintaining BLINDING is also important and most minimisation algorithms – particularly those run on computers – incorporate an element of randomisation within them so that, even with complete knowledge of all the patients in the study so far, it is not possible to guarantee correctly guessing the next patient assignment. Minimisation is not without controversy. Including such a random component satisfies most critics, but some (such as Rosenberger and Lachin, 2002) still consider that all the theoretical aspects of how the analysis should be done have not been fully worked out. *SD*

Day, S. 1999: Treatment allocation by the method of minimisation. *British Medical Journal* 319, 947–8. **Pocock, S. J.** 1983: *Clinical trials: a practical approach.* Chichester: John Wiley and Sons, Ltd. **Pocock, S. J. and Simon, R.** 1975: Sequential treatment assignment with balancing for prognostic factors in the controlled clinical trial. *Biometrics* 31, 103–15. **Rosenberger, W. F. and Lachin, J. M.** 2002: *Randomization in clinical trials.* New York: John Wiley and Sons, Inc. **Steptoe, A., Doherty, S., Rink, E., Kerry, S., Kendrick, T. and Hilton, S.** 1999: Behavioural counselling in general practice for the promotion of healthy behaviour among adults at increased risk of coronary heart disease: randomised trial. *British Medical Journal* 319, 943–7. **Taves, D. R.** 1974: Minimization: a new method of assigning patients to treatment and control groups. *Clinical Pharmacology and Therapeutics* 15, 443–53.

missing at random (MAR) See DROPOUTS, MISSING DATA

missing completely at random (MCAR) See DROPOUTS, MISSING DATA

missing data Well-designed statistical studies draw a representative sample from the study population by following a sampling plan and a detailed protocol. Often, some of the planned data are unavailable or otherwise absent from the database, hence the term 'missing data'. The data that would be observed if all intended measurements were obtained will be called 'potential data'. The potential data that are not missing, combined with an indicator of availability of each planned response, form the 'observed data'. The name 'missing data' may suggest that these data can simply be forgotten by the data analyst, but nothing could be further from the truth.

Missing data form one of the hardest challenges for data analysts. This is because the missed data can be intrinsically different from observed data in ways that are hard to predict, and thus leave a biased sample. For instance, when studying the evolution of CD4 counts over time, AIDS patients may fail to return for planned clinic visits, not only when they are sick as a result of low CD4 counts but also when they feel good and no longer in need of treatment. In view of this, three types of missing data are typically distinguished (Little and Rubin, 2002):

(1) The simplest situation occurs when the risk of missing a certain part of the data is the same for all subjects, regardless of their potential data values. This process is known as 'missing completely at random' (MCAR). It happens, for instance, in a study where very expensive

outcome measures are, by design, only gathered in a random subsample. In that case, missing observations can simply be deleted from the dataset and ignored in further analyses.

(2) A more realistic situation occurs when the risk of missing a certain part of the data is constant over the potential outcomes for that part among subjects for whom we observe the same outcomes on a well-chosen subset of variables. This condition is easily interpreted when dealing with baseline covariates that are always observed and a single outcome that can be missing. It becomes complex with nonmonotone missingness patterns. In either case, the data are then called 'missing at random' (MAR). This happens, for instance, in two-stage sampling designs where a subgroup of patients is invited to the second study cycle, depending on their first outcome. A naïve data analysis, which ignores missing data, may then be misleading, unless it conditions on the correct subset of observed data.

(3) When neither of these two constraints hold, missingness is called informative or nonignorable (NI). In a health survey, for instance, one may lose the uninterested or very busy respondents, who have their own disease profile.

Under each of the above scenarios the popular MAXIMUM LIKELIHOOD ESTIMATION method can be fruitfully employed for unbiased estimation of parameters in the study population. The challenge is then to propose a (parsimonious) model relating the distribution of observed and potential data. Typically, one chooses either a so-called 'selection model' or a 'pattern-mixture' model (Little and Rubin, 2002). The former adds to the usual model for the distribution of the potential data a model for the conditional distribution of being observed, given the potential data. The latter models the conditional distribution of the potential data for each level of the response indicator and adds to it a model for the distribution of the response indicator. In both cases one averages over all possible values of the missing data to find the observed data distribution that enters the maximum likelihood procedure. A very useful property of MAR is that maximum likelihood estimation can avoid the need to model the probability of being observed and still allow for inference on the potential data. This makes maximum likelihood estimation very popular in this setting.

Nonetheless, observed data likelihoods under MAR may have complex forms not covered by standard statistical packages. To help avoid lengthy computations in routine practice, the EM ALGORITHM and imputation techniques (see MULTIPLE IMPUTATION) have been devised. EM is an iterative algorithm that replaces the usual log-likelihood of the potential data by its conditional expectation, given the observed data. Maximum likelihood estimates are then obtained by maximizing it in the usual way and the expected log-likelihood is updated. Imputation methods 'fill in' the missing data by simulating from their distribution conditional on the available data. The resulting 'completed' dataset is then analysed using standard software as if no data were missing. The loss of information due to the missing data must, however, be recognised when STANDARD ERRORS are derived. Corrected standard errors have therefore been proposed based on the variation in estimates over different random imputations (Little and Rubin, 2002).

One drawback of the maximum likelihood approach is that estimates can be biased when the potential data model is misspecified. One may therefore choose to specify less features of the model and follow the Horvitz–Thompson principle, which helps achieve robustness against model misspecification (Preisser, Lohman and Rathouz, 2002). Here, the completely observed data are upweighted by the inverse conditional probability of being observed, given the potential data, to compensate for similar counterparts that are missing. This line of research has seen extensive developments in recent years and is entering statistical practice as software becomes more readily available.

Regardless of the adopted approach, observed data alone seldom contain information that distinguishes MAR from NI. One is statistically speaking blind and must rely on guidance from other sources to make progress. This is made abundantly clear by the pattern mixture approach. Indeed, the pattern with unobserved response completely lacks information on the data distribution, and unbiased inference needs unverifiable assumptions regarding the dependence of missingness on the potential data. Reassurance that 'missed' data are comparable to observed ones is found when data are 'missing by design', but is hard to obtain otherwise. It is hence very important at the design stage to plan to gather such information that helps determine the distribution of the missed data. In experimental studies one seeks to gather data over time that can help predict. Furthermore, a sensitivity analysis can be conducted by examining how estimates vary as different choices are postulated for the unknown outcome distribution in nonresponders. This practice of describing how conclusions vary over plausible but untestable assumptions is recommended (Kenward, Goetghebeur and Molenberghs, 2001; Scharfstein, Daniels and Robins, 2003).

While enormous progress has been made in statistical methodology for dealing with missing data, many problems remain in practice and in theory. The term 'missing data' is often misunderstood or the methods are abused. When answers to certain questions are intrinsically meaningless or undefined in certain categories of people (e.g. blood pressure of dead patients), it is very hard to justify missing data constructions that give nonresponders the outcome distribution of responders and base conclusions on outcomes averaged over both groups. Furthermore, the MAR assumption is frequently adopted for mathematical convenience but may be difficult to interpret or justify. The goal and relevance of any analysis, along with justified assumptions, must come first. In causal inference, for instance, one has fruitfully exploited missing data constructs (Vansteelandt and Goetghebeur, 2005). At other times one ignores valuable and simple MAR methods to fall for simplistic analyses that can be very misleading. Due caution is always necessary, additional thought is required in model selection and, with regard to missing data in general, the familiar adage holds: prevention is better than cure. *EG/SV*

[See also DROPOUTS]

Kenward, M. G., Goetghebeur, E. and Molenberghs, G. 2001: Sensitivity analysis for incomplete categorical data. *Statistical Modeling* 1, 31–48. **Little, R. J. A. and Rubin, D. B.** 2002: *Statistical analysis with missing data.* New York: John Wiley & Sons, Inc. **Preisser, J. S., Lohman, K. K., Rathouz, P. J.** 2002: Performance of weighted estimating equations for longitudinal binary data with drop-outs missing at random. *Statistics in Medicine*, 21, 3035–54. **Scharfstein, D. O., Daniels, M. J. and Robins, J. M.** 2003: Incorporating prior beliefs about selection bias into the analysis of randomized trials with missing outcomes. *Biostatistics* 4, 495–512. **Vansteelandt, S. and Goetghebeur, E.** 2005: Sense and sensitivity when correcting for observed exposures in randomized clinical trials *Statistics in Medicine* 24, 191–210.

MLWin See MULTILEVEL MODELS

mode The mode is a measure of location. It is simply the value that occurs most often. For example, the hair colour of 573 babies born at a UK maternity hospital is shown in the table. In this example, the mode is 'medium brown'.

Frequency distribution for hair colour

Hair colour	*Frequency*
Blond	41
Pale brown/blond	147
Medium brown	244
Dark brown	121
Black	14
Red	6

The mode, however, is of limited value in summarising continuous data. In contrast to the MEAN, the mode is not sensitive to OUTLIERS but need not be a unique value, as a distribution of data may be bimodal or multimodal (having two or more modes respectively). *SRC*

MPlus See STRUCTURAL EQUATION MODELLING

multicentre research ethics committee (MREC) See ETHICAL REVIEW COMMITTEES

multicentre trials These are studies that are carried out in several distinct centres, sites or units (hospitals, clinical departments, etc.). The first trials of new therapies in man (Phase I trials) require few subjects who must be monitored very tightly; therefore they are almost always carried out in a single centre. These trials are typically followed by medium sized multicentre efficacy trials (Phase II trials). If the results of these early trials are promising, then larger multicentre trials are carried out to confirm the efficacy and safety of the new therapies (Phase III trials).

Multicentre trials are often performed in several countries or even several continents. The conduct of multicentre trials can be overseen by a steering committee, headed by the study chair and consisting of persons designated to represent study centres, disciplines or activities.

Multicentre trials are needed primarily to accrue the number of subjects or patients required for the study over a reasonably short time period. For common diseases, multicentre trials may have several centres with large numbers of subjects per centre or, in the case of rare diseases, they may have a large number of centres with very few subjects per centre. Patients treated at different centres (let alone different countries or continents) may be expected to differ substantially in terms of their ethnicity, exposure to aetiologic or risk factors, living conditions and access to health resources, etc.

Such heterogeneity may be a drawback for two main reasons: first, the VARIANCE of the outcome of interest is increased because of the heterogeneity and, second, a treatment benefit in some patient subpopulations might be missed in a trial that accrued other patient subpopulations as well. On closer inspection, neither of these two reasons argues against multicentre trials. The increase in sample size that results from heterogeneity is usually negligible compared to the potential number of patients available at multiple centres. In many situations, no centre would be able to accrue the required number of patients; hence a single-centre trial would be infeasible. In addition, the results of a multicentre trial are more readily generalisable than those of a single-centre trial, because they are obtained in a patient sample more likely to reflect the population of interest. If a treatment is thought *a priori* to exert its benefit selectively in a patient subpopulation, a multicentre trial is still indicated with prior exclusion of patients unlikely to benefit.

An added advantage of multicentre trials is their ROBUSTNESS to fraud, delinquencies and other quality problems that may arise at a few centres. This was illustrated by a series of large multicentre trials in early breast cancer, in which exclusion of a fraudulent centre did not have any sizeable impact on the trial results (Peto *et al.*, 1997). In contrast, another highly publicised case of fraud in a trial of high-dose chemotherapy for advanced breast cancer had disastrous consequences, because its results were completely dominated by fraudulent data. Very few centres had taken part in this trial and fraud from a single investigator was sufficient to cause dramatic BIAS in the reported results (Weiss *et al.*,

2000). Published results of well-conducted multicentre trials often have a direct impact on clinical practice, while single-centre trials generally require to be reproduced on a larger scale before their results are accepted as valid.

The CLINICAL TRIAL PROTOCOL should encourage the participating centres to put the same procedures in place as regards patient management, measurement of treatment effects and other aspects of the study that may have a bearing on the therapeutic results. Some heterogeneity is unavoidable in multicentre trials, but such heterogeneity is unimportant so long as it does not directly affect the outcome of interest. In a randomised trial, in particular, differences between centres are ignorable if they do not impact the treatment effects of interest. Thus if one centre tended to recruit patients of poor prognosis and another centre tended to recruit patients of good prognosis, this difference would not compromise the trial results if the treatment effects were independent of prognosis. Such independence is generally unknown but postulated before the trial and it can, in fact, be studied within the trial itself. In the example just given, differences in prognostic mix between centres might be confounded by other centre-specific factors (such as concomitant medications, supportive care, etc.), and hence it is advisable to standardise trial procedures across all participating centres to the extent possible.

The logistics of a multicentre trial can be fairly complex and provisions must be made for drug shipment and storage, trial material distribution, ethical approval and compliance to local regulations, training of investigators and local staff to avoid variation in patient management, evaluation criteria, follow-up schemes, etc. All these issues can be discussed at investigator meetings and monitoring visits during the trial.

Statistical quality control checks can also be performed to identify discrepancies between centres that may call for more thorough investigations, especially in centres that are found to be clear OUTLIERS. Such checks are best performed while the trial is ongoing, so that remedial action can be taken early, and this can be facilitated by electronic data capture that feeds patient data to a central database in real time. There is no conclusive evidence that data quality is related to the number of patients enrolled in each centre (Sylvester *et al.*, 1981; Hawkins *et al.*, 1990).

In multicentre trials, RANDOMISATION of the patients is generally organised centrally, rather than performed in each centre. Centralised randomisation requires that all centres access a central resource, usually by internet, telephone or fax, to obtain the next treatment allocation. Such centralised control is useful to follow the accrual of patients into the trial and to check eligibility criteria in a uniform way prior to treatment allocation. Centralisation of the randomisation process also guarantees that it cannot be biased by foreknowledge of the next treatment assignment, which could happen in open-label trials with the use of randomisation lists. It is usually desirable to stratify the allocation by centre and more generally by important prognostic factors measured at baseline, such as severity of disease, patient status, age, etc. Such stratification can be implemented with permuted lists of treatment allocations or through dynamic allocation using MINIMISATION. Allocation through minimisation has the advantages of being able to take account of many prognostic factors and of being completely unpredictable at any given centre in the absence of information on patients already randomised in all other centres.

Multicentre trials are usually designed and their sample size calculated under the assumption that the treatment effect is the same in all centres. Whether this assumption is supported by the data can be tested formally. Assume that in the ith of I centres, the true treatment effect is given by τ_i and the estimate of τ_i, noted $\hat{\tau}_i$, is asymptotically normally distributed with variance v_i:

$$\hat{\tau}_i \sim N(\tau_i, v_i)$$

The measure of treatment effect is taken such that no treatment effect corresponds to $\tau_i = 0$. Interest focuses first and foremost on whether there is statistical evidence of an overall treatment effect, which can be tested through the test statistic:

$$X^2_{\text{treatment}} = \frac{\left(\sum_{i=1}^{I} \hat{\tau}_i w_i\right)^2}{\sum_{i=1}^{I} w_i} \sim \chi^2_1$$

where $w_i = v_i^{-1}$ denotes the inverse of the variance of the treatment effect in the ith centre. Under the null hypothesis of no treatment effect ($\tau_1 = \tau_2 = \cdots = \tau_I = 0$), this test statistic has an asymptotic χ^2 distribution (see CHI-SQUARE DISTRIBUTION) with one DEGREE OF FREEDOM.

Under the assumption of a common treatment effect in all trials ($\tau_1 = \tau_2 = \cdots = \tau_I = \tau$), the treatment effect is estimated by:

$$\hat{\tau} = \frac{\sum_{i=1}^{I} \hat{\tau}_i w_i}{\sum_{i=1}^{I} w_i}$$

In order words, this is a weighted average of the treatment effects in all centres. The presence of heterogeneity between centres can be tested through the test statistic:

$$X^2_{\text{heterogeneity}} = \sum_{i=1}^{I} (\hat{\tau}_i - \hat{\tau})^2 w_i \sim \chi^2_{I-1}$$

which has an asymptotic χ^2 distribution with $I - 1$ degrees of freedom. In practice, this test for heterogeneity in treatment effects between centres is not very informative, because it lacks POWER to detect true underlying differences, especially when there are many centres (I large) with few patients per centre.

Moreover, when statistical heterogeneity is found between centres, it may be difficult to ascribe it to a well-identified factor and the interpretation of the overall treatment effect may be controversial. The same test for heterogeneity is more useful when centres can be meaningfully combined according to a common characteristic (for instance all centres that have access to certain equipments or that use certain supportive treatments). When centres are thus combined for the purposes of statistical analysis, the grouping should be defined prospectively and blindly to treatment allocation and results in the various centres. A grouping of centres based solely on their sample sizes is unlikely to be informative. Even when the formal test of heterogeneity fails to reach statistical significance, heterogeneity can be explored through descriptive statistics or graphical displays of the treatment effects in individual centres or groups of centres. Large differences in treatment effects between centres would cause concern, especially if much of the overall effect was attributable to an unexpectedly large effect in a single centre or if treatment had a markedly negative effect in some centres – an overall positive treatment effect notwithstanding. Whenever substantial heterogeneity is found, attempts should be made to find an explanation in terms of identifiable features of trial management or subject characteristics. Such an explanation may suggest further analyses or appropriate interpretation. In the absence of an explanation, alternative estimates of the treatment effect may be required in order to substantiate the robustness of the trial results (International Conference on Harmonisation, 1998).

Regardless of the presence of statistical heterogeneity between centres, the statistical model adopted for the estimation and testing of treatment effects may account for centre through stratification or by inclusion of a fixed or random effect for centre in the model. If the number of subjects per centre is limited, centre effects are poorly estimated and the inclusion of centre effects in the model negatively affect the power of the treatment comparisons. In such cases, it is preferable to ignore the centre in the analysis. *MB*

Hawkins, B. S., Prior, M. J., Fisher, M. R. and Blackhurst, D. W. 1990: Relationship between rate of patient enrolment and quality of clinical center performance in two multicenter trials in ophthalmology. *Controlled Clinical Trials* 11, 374–94. **International Conference on Harmonisation** 1998: E-9 document: guidance on statistical principles for clinical trials. *Federal Register* 63, 179, 49583–98. **Peto, R., Collins, R., Sackett, D., Darbyshire, J., Babiker, A., Buyse, M., Stewart, H., Baum, M., Goldhirsch, A., Bonadonna, G., Valagussa, P., Rutqvist, L., Elbourne, D., Altman, D., Dalesio, O., Parmar, M., Hill, C., Clarke, M., Gray, R. and Doll, R.** 1997: The trials of Dr Bernard Fisher: a European perspective on an American episode. *Controlled Clinical Trials* 18, 1–13. **Sylvester, R., Pinedo, H., De Pauw, M., Staquet, M., Buyse, M., Renard, J. and Bonadonna, G.** 1981: Quality of institutional participation in multicenter clinical trials. *New England Journal of Medicine* 305, 852–5. **Weiss, R. B., Rifkin, R. M., Stewart, F. M., Theriault, R. L., Williams, L. A., Herman, A. A. and Beveridge, R. A.** 2000: High-dose chemotherapy for high-risk primary breast cancer: an on-site review of the Bezwoda study. *The Lancet* 355, 999–1003.

multicollinearity This term is used particularly in MULTIPLE LINEAR REGRESSION to indicate situations where the EXPLANATORY VARIABLES are linearly related, thus making the estimation of regression coefficients in the usual way essentially impossible. Including the sum or average of the explanatory variables as a variable would lead to this problem. For example, in a blood pressure study one cannot include among explanatory variables systolic blood pressure (SBP), diastolic blood pressure (DBP) and, additionally, a linear combination of the two, such as mean blood pressure, without causing the model to break down completely. Another example is using too many dummy variables to code a categorical explanatory variable.

In practice, of course, *approximate* multicollinearity, i.e. where one of the explanatory variables can be predicted with considerable accuracy from the other explanatory variables, will be of more cause for concern and can lead to inflated variances for the estimated regression coefficients. Some evidence for approximate multicollinearity can be found by looking at the multiple correlation coefficients (see CORRELATION) of each explanatory variable with the other explanatory variables; if any of these is close to one then multicollinearity should be suspected. There is no optimal way of dealing with multicollinearity but in many cases the simplest solution is to remove explanatory variables that are highly correlated or combine variables in some way. More details are given in Miles and Shevlin (2001). *BSE*

Miles, J. and Shevlin, M. 2001: *Applying regression and correlation*. London: Sage.

multidimensional scaling This technique is often used in psychology but less often in medicine. The basis of the method is a proximity matrix arising either directly from experiments in which subjects are asked to assess the similarity of pairs of stimuli or, indirectly, as a measure of the CORRELATION or COVARIANCE of a pair of stimuli derived from a number of measurements made on each. In some cases, high proximity values correspond to stimuli that are similar (similarities); in others, the reverse is the case (dissimilarities). As an example, the table shows judgements about various brands of cola made by a subject using a visual analogue scale with the anchor points 'same' (having a score of 0) and 'different' (having a score of 100). In this example, the resulting rating for a pair of colas is a dissimilarity-low value, indicating that the two colas are regarded as alike and vice versa. A similarity measure would have been obtained had the anchor points been reversed, although similarities are usually scaled to lie between zero and one.

multidimensional scaling *Dissimilarity data for all pairs of 10 colas for a subject*

	Subject 1									
Cola number	*1*	*2*	*3*	*4*	*5*	*6*	*7*	*8*	*9*	*10*
1	0									
2	16	0								
3	81	47	0							
4	56	32	71	0						
5	87	68	44	71	0					
6	60	35	21	98	34	0				
7	84	94	98	57	99	99	0			
8	50	87	79	73	19	92	45	0		
9	99	25	53	98	52	17	99	84	0	
10	16	92	90	83	79	44	24	18	98	0

Researchers with data in the form of proximity matrices are generally interested in uncovering any structure or pattern they may contain and multidimensional scaling aims to help by representing the observed proximities as a spatial or geometrical model in which the distances between the points (usually taken to be Euclidean) correspond in some way to the observed proximities. In general, this simply means that the larger the dissimilarity (or the smaller the similarity), the further apart should be the points representing them in the final geometrical model.

The required spatial model is defined by a set of d-dimensional points, each representing one of the stimuli of interest and a measure of the distance between these points. The objective of multidimensional scaling is to determine both the dimensionality of the model (i.e. the value of d) and the values of the coordinates. The coordinates of the points in the model that represent the proximities can be found in a variety of ways. One simple approach is to choose the coordinate values (for a given value of d) to minimise S, defined as:

$$S = \sum_{ij} (\delta_{ij} - d_{ij})^2$$

where δ_{ij} is the observed dissimilarity for stimuli i and j, and d_{ij} is the distance between the points representing stimuli i and j. Since the distances d_{ij} are a function of the coordinate values, so also is S. For various reasons, S is not generally a suitable function for comparing distances and dissimilarities and full details of more suitable criteria can be found in, for example, Everitt and Rabe-Hesketh (1997). This also includes a discussion of how many dimensions are needed to give an adequate fit of the geometrical model to the observed proximities.

An illustration how multidimensional scaling has been used in a medical setting is provided in Cliff *et al.* (1995). Here, a matrix of similarities is calculated in which each element is the number of months in which there were reported measles cases/deaths in both of a pair of areas. The greater the value of such a similarity, the greater the similarity of the time series of the occurrence of measles in the two areas. (In this study, the time series for each area considered consisted of monthly totals of measles cases for the 31-year period from January 1960 to December 1990.) The figure on page 296 shows the multidimensional scaling solutions corresponding to a one-, two- and three-dimensional solution. *BSE*

Cliff, A. D., Haggett, P., Smallman-Raynor, M. R., Stroup, D. F. and Williamson, G. D. 1995: The application of multidimensional scaling methods to epidemiological data. *Statistical Methods in Medical Research* 4, 102–23. **Everitt, B. S. and Rabe-Hesketh, S.** 1997: *The analysis of proximity data*. London: Arnold.

multilevel models Multilevel (also known as random effect, hierarchical and mixed) models are an extensive and flexible class of models for correlated data, which arise widely in medical statistics. For example, adult height or weight may be correlated with those of other family members and the chance of post-surgical infection may be correlated with that of other patients with the same surgical team. Further, many studies involve the repeated measurement of subjects' outcomes throughout follow-up (LONGITUDINAL DATA). Such observations are usually quite strongly correlated.

Multilevel models relax the assumption, required for ordinary least squares (OLS) regression, that each response is independent. They have their roots in agricultural experiments; indeed they embrace all the classical analysis of VARIANCE models. They have found ready application in social science, medical and economic research. A brief history is given by Kreft and de Leeuw (1998, p. 16).

The data structure is viewed as a series of levels (or hierarchies). For example, consider a multicentre trial where subjects' quantitative outcomes are recorded repeatedly over time. The first figure (on page 296) shows a possible data structure. Level 1 has the repeated observations that are nested within subjects at level 2. Subjects are in turn nested within centres at level 3.

A multilevel analysis enables us to allow correctly for, and model, the CORRELATION induced by this structure. If we have longitudinal data, we can investigate how subjects change with time, which could be quite different to the cross-sectional relationship (Diggle *et al.*, 2002, p. 16). In addition to the usual *fixed parameters* (whose interpretation is similar to their OLS counterparts), RANDOM EFFECTS are introduced to model the correlation structure, as described below. The mix of fixed and random effects gives rise to the term *mixed models*.

Once alerted, we see hierarchical structures everywhere: subjects within wards within counties, patients within hospitals within health authorities, and so on. Thus it is natural to ask what is gained by a multilevel model, and when they are unnecessary.

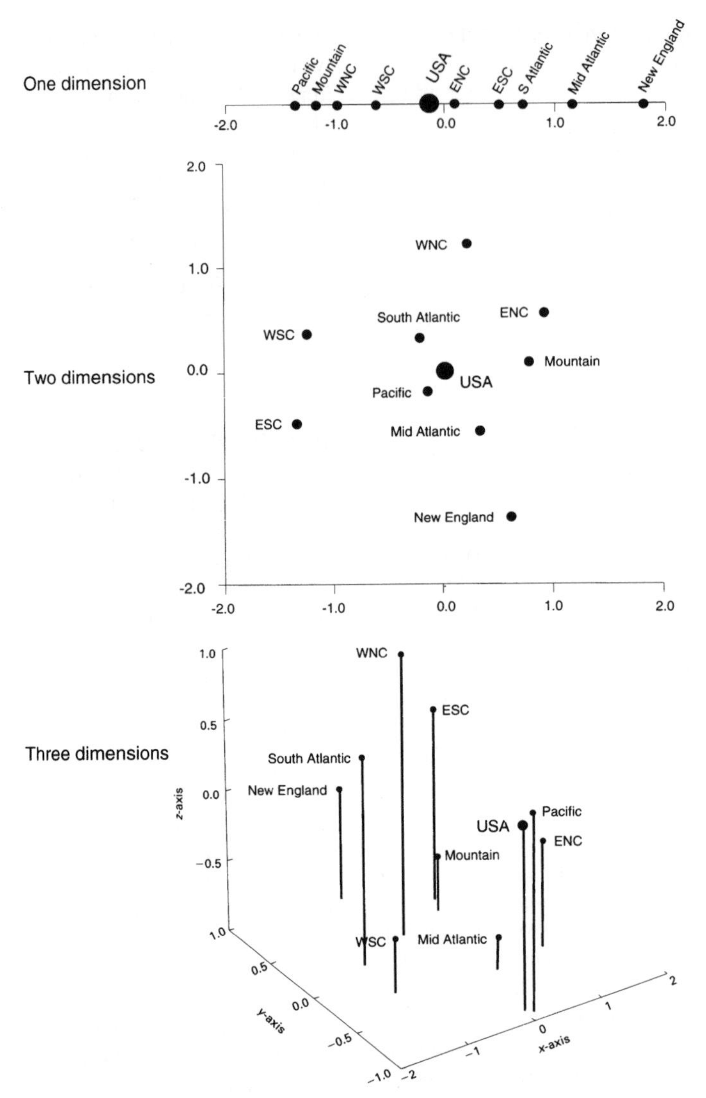

multidimensional scaling *MDS plots of the USA regions in one-, two- and three-dimensional space. Data are from monthly time series of reported measles cases 1960–1990. Taken from Cliff et al., 1995*

First, OLS STANDARD ERRORS are wrong when the data are multilevel. For example, subjects within a cluster are often similar to each other, i.e. not independent. They therefore convey less information about the value of a parameter than an independent (unclustered) sample of the same size (Goldstein, 2003, p. 23).

Level 3: Centres
Level 2: Subjects
Level 1: Observations

multilevel models *Data structure for a multicentre trial*

Second, OLS does not permit exploration of the variance structure. For example, we may wish to estimate the proportion of the total variance between subjects (the INTRACLUSTER CORRELATION COEFFICIENT (ICC), equation (1) below) or we may want to investigate how the variance changes as a function of covariates. Multilevel models will add little to an analysis when observations are effectively independent so that the ICC is close to zero. However, it is wise to be cautious as even a small ICC can have a nontrivial effect.

The plan of this article is as follows. First, the key ideas of multilevel models are outlined, followed by a discussion of commonly used algorithms for fitting multilevel models. Then extensions to discrete data are described and the relationship to GENERALISED ESTIMATING EQUATIONS (GEEs) is given. Medical applications and further extensions are discussed and then missing data, design and software. Some suggestions for further reading are given at the end of the entry.

Consider the multicentre trial of the first figure. Focusing on levels 1 and 2, we begin by describing the simplest model, which allows for correlation between the observations, before outlining how more flexible models can be built up. The idea is to generalise OLS regression. An OLS model would have a single regression line relating the average response to time. A multilevel model, however, can be thought of as extending this to include a regression line for each subject.

Thus, whereas in OLS regression the observations are distributed about a single regression line, in multilevel models we can view each subject's responses as distributed about their subject-specific regression line. The subject-specific regression lines are then distributed about the overall average regression line.

This is illustrated in the second figure. Here, the overall average relationship between the response and time is given by the bold line, $Y=\alpha+\beta t$. Five subject-specific regression lines are shown, which are parallel to this. Each subject's observations are distributed about their regression line. Five examples of this are given in the top half of the figure.

In this simple case, each subject's regression line is parallel to the overall average line. The distance between the *j*th subject specific line, $Y=(\alpha+u_j)+\beta t$ and the average line $Y=\alpha+\beta t$ is u_j (in the second figure, $j=1, 2, 3, 4$ or 5). These u_j are known as the subject-specific *random effects*, also known as the *level 2 residuals*. They are assumed to be normally distributed about zero.

The vertical distances between each subject's responses and their subject-specific regression line are known as the *level 1 residuals*. These are analogous to the residuals in OLS models and are likewise assumed to be normally distributed about zero.

The normal densities of the level 2 and some level 1 residuals are shown in the second figure. In the top half, we see five observations, marked '+'. The five vertical distances to their subject-specific regression are five level 1 residuals. The NORMAL DISTRIBUTION of these residuals is illustrated by the five normal densities about the subject-specific regression lines. Then, on the left-hand side, the level 2 residuals, $u_1, \ldots, u_5$ are shown. Their normal distribution is sketched on the left-hand side of the figure.

The parameters α and β in the figure are known as *fixed* parameters. They have a similar interpretation to their counterparts in the OLS models, so α is the average response at time zero and β is the average change in response per unit change in time. However, we have two new parameters, known as *random parameters*, which are the variance of the level 2 residuals, called σ_u^2, and the variance of the level 1 residuals, called σ_e^2. Thus, in the second figure, the density of the u_j is sketched on the left and has variance σ_u^2. The five densities of the level 1 residuals have common variance σ_e^2. Often, σ_u^2 is called the between-subject variance and σ_u^2 the within-subject variance.

The second figure represents the simplest multilevel model. As each u_j can be viewed as a random contribution to the intercept of the *j*th person's regression line, which is $(\alpha+u_j)$, this is often known as a RANDOM INTERCEPT MODEL. It is also

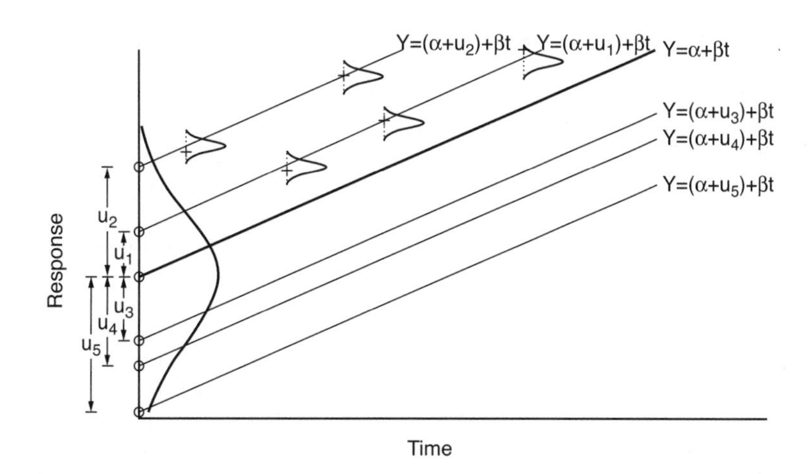

multilevel models *Schematic illustration of the random intercept model*

a simple example of a COMPONENTS OF VARIANCE model, as there is a single variance term corresponding to each level in the model (σ_u^2 for level 2, σ_e^2 for level 1).

The motivation for multilevel models was their ability to model the correlation structure of the data. We therefore consider the correlation structure implied by the random intercept model. To do this, we have to consider the variance of each observation and the COVARIANCE between observations.

First, consider the variance. In multilevel models, the random component is the RESIDUALS. Residuals from different levels are always assumed to be independent. Likewise, residuals corresponding to different units within a level (i.e. different observations within level 1 and different subjects within level 2) are assumed to be independent. The total variance of each observation is thus the sum of the variance of the residuals at each level. Thus, in the random intercept model, where each observation has a residual at level 1 and level 2, the variance of an observation is $\sigma_e^2 + \sigma_u^2$.

Second, consider the covariance. In the random intercept model of the second figure, different observations from the same subject, j, share a common random component, their level 2 residual u_j. Their covariance is therefore $\text{Cov}(u_j, u_j) = (u_j, u_j) = \sigma_u^2$. However, observations from different subjects share no common residuals. Their covariance is therefore zero.

Recalling that $\text{Cor}(A, B) = \text{Cov}(A, B)/\sqrt{\text{Var}(A)\,\text{Var}(B)}$, we see that the correlation structure implied by the random intercept model is:

$$\begin{cases} 1 & \text{same subject and time} \\ \rho = \sigma_u^2/\sqrt{(\sigma_u^2+\sigma_e^2)(\sigma_u^2+\sigma_e^2)} = \sigma_u^2/(\sigma_u^2+\sigma_e^2) & \text{same subject, different time} \\ 0 & \text{different subjects} \end{cases} \tag{1}$$

Thus the random intercept model of the second figure implies a fixed correlation, ρ, among a subject's responses, independent of how far apart in time they are. This is known as a *compound symmetry* or *exchangeable* correlation structure.

The correlation, ρ, in equation (1) is also known as the *intra level 2 unit*, or more commonly ICC; in random intercept models, ρ measures the proportion of total variance, which is between subjects. If $\rho = 0$, then observations are independent.

Consider how the random intercept model illustrated in the second figure compares to fitting an OLS line to each subject in turn. Such OLS lines would be unbiased estimates of each subject's true line. However, they might be imprecisely estimated, particularly if a subject has few observations. Conversely, the estimate of the overall average line ($\alpha + \beta t$) is a precise, but biased, estimate of each subject's true line. Both extremes are undesirable. By fitting a multilevel model, we compromise between the two extremes. The estimates of the u_j are known as 'best linear unbiased predictors' (BLUPs) and, as their name suggests, have certain optimality properties (Verbeke and Molenberghs, 2000, p. 80). The practical effect is that the subject-specific regression lines estimated by the multilevel model are drawn (or shrunk) closer to the mean line than the OLS estimates, and the fewer the observations on a subject, the more their line is drawn towards (borrows strength from) the MEAN regression line. This is often referred to as *shrinkage* in the literature.

Having fitted the random intercept model, we should examine the level 1 and level 2 residuals to check whether they are approximately normal, as the multilevel model assumes, and identify OUTLIERS. Level 2 residuals can also be used to distinguish outlying subjects; this has found wide application in medical settings.

For most longitudinal data, the correlation between observations declines as the time between them increases. Thus the fixed correlation structure of the random intercept model, ρ, is insufficient. A natural extension is to allow subjects to have their own slopes as well as their own intercepts, as illustrated in the third figure (on page 299). As before, the overall average regression line is $Y = \alpha + \beta t$. Now, however, the jth subject's regression line is $Y = (\alpha + u_j) + (\beta + v_j)t$. In the random intercept model the u_j were normally distributed with mean 0 and variance $\sigma_u{}^2$. In the RANDOM INTERCEPT AND SLOPE MODEL (u_j, v_j) have a BIVARIATE NORMAL DISTRIBUTION about $(0, 0)$.

As before, the level 1 residuals are the vertical distances between a subject's observations and their subject-specific regression line. The level 2 residuals are now (u_j, v_j), so we have two level 2 residuals per subject, representing the random intercept and slope respectively.

We can calculate the variance and covariance of observations in a similar way to the random intercept model, although the algebra is more involved. Then we can derive the correlation structure implied by this model. The variance of the responses is no longer constrained by the model to be constant; it can now increase with time. Further, the correlation between observations on the same subject can decline as the time between them increases. Hence this model is often more appropriate for longitudinal data.

The way the random intercept and slope model builds on the random intercept model suggests many further extensions. To begin with, if we have additional covariates, they too can have random effects. For example, if we include a treatment variable, subject-specific treatment effects can be estimated.

Levels can be added to the model to describe additional levels in the data. For example, the first figure shows that subjects are nested within centres. We can extend the random intercept model to include a random effect at the centre level. Such a model yields estimates of components of variance at each level (centre, subject and observation), so the proportion of the total variance between centres can be calculated. Further, the level 3 (centre) residuals can be examined to

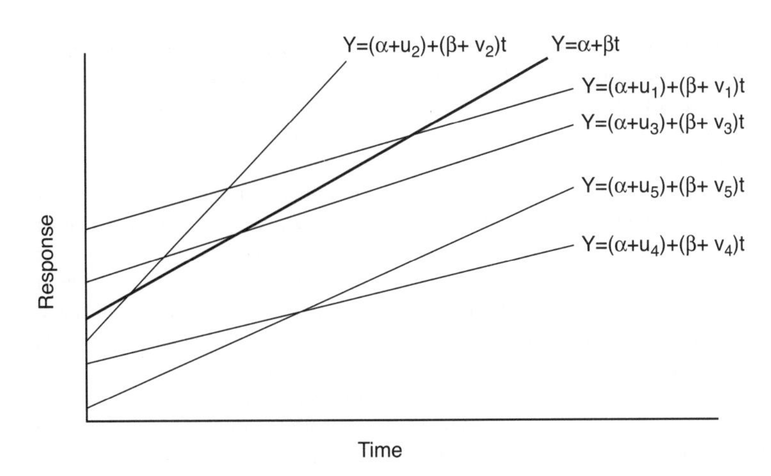

multilevel models *Schematic illustration of the random intercept and slope model*

indicate outliers and covariates can be given random centre-level terms as well as random subject-level terms.

The level 1 variance (which is analogous to the residual variance in OLS models) can also be modelled by covariates; e.g. male level 1 residuals may be more variable than those from females. This is known as modelling *complex variation.*

Sometimes the random intercept and slope model is not sufficiently flexible to model the correlation structure, particularly if observations are close together in time. Many options are possible; if subjects are observed at identical times then an attractive alternative is an *unstructured* COVARIANCE MATRIX, which imposes no parametric model on the covariance. Much has been written on this; see, for example, Verbeke and Molenberghs (2000, Chapter 16) and Diggle *et al.* (2002, Chapter 5).

Multilevel models for quantitative data are typically based on the multivariate normal distribution. Thus, the likelihood of the data can be written down and maximised using adaptations of Newton–Raphson techniques (for details, see Raudenbush and Bryk, 2002, Chapter 14). Alternatively, a Bayesian approach can be adopted (see the chapter by Clayton, D. G. in Gilks, Richardson and Spiegelhalter, 1996).

If likelihood methods are adopted, restricted maximum likelihood (REML) is usually used (Verbeke and Molenberghs, 2000, p. 43). This corrects the downward bias of maximum likelihood estimates of variance and requires negligible extra work computationally. However, changes in REML log-likelihoods cannot generally be used to compare nested models, so maximum likelihood may be preferred for model building (Goldstein, 2003, p. 36), although in uncommon situations with many fixed parameters the two can give quite different answers (Verbeke and Molenberghs, 2000, p. 198).

GENERALISED LINEAR MODELS (GLMs) extend OLS models to discrete responses. Analogously, *generalised linear mixed models* (GLMMs), sometimes called *nonlinear mixed models*, extend multilevel models to discrete responses.

As with GLMs, we model a function of the PROBABILITY that the response takes on a particular value. In GLMMs, however, for responses on the same subject, this probability shares a subject-specific term. For example, we can make the random intercept model, illustrated by the second figure, a GLMM by letting Y follow a binomial distribution and writing the overall regression line as $\text{logit}(\Pr\{Y=y\}) = \alpha + \beta t$. The subject specific regression line for subject j would be

$$\text{logit}(\Pr\{Y = y\}) = (\alpha + u_j) + \beta t \qquad (2)$$

and, as before, the level 2 residuals, u_j, would be normally distributed about zero with variance σ_u^2. Note that, as in GLMs, in GLMMs the level 1 variance is a fixed function of the mean. Therefore, there is no term corresponding to σ_e^2.

Also, as with GLMs, the function 'logit' in equation (2) is known as the link function. Alternative link functions (e.g. log, inverse normal) can be used together with other probability models such as the Poisson or negative binomial.

Unfortunately, fitting GLMMs is not nearly as straightforward as fitting multilevel models to quantitative data, because the LIKELIHOOD is much more difficult to compute. Three approaches, all discussed by Goldstein (2003), are commonly adopted.

The first approach is QUASI-LIKELIHOOD. There are two forms of this, penalised quasi-likelihood and marginalised quasi-likelihood. Both methods rely on approximations, which can be made to first or second order. The approximations involved mean that quasi-likelihood methods provide biased parameter estimates; in particular, estimates of variance components tend to be downwardly biased. This bias is most marked in data sets with few level 1 units per level 2 unit or probabilities close to boundaries (e.g. 1 or 0 for binary data). The bias is least for second-order penalised quasi-likelihood. Another drawback is that, with quasi-likelihood, no estimate of the log-likelihood is available for comparing models.

The second approach relies on numerical or Monte Carlo integration methods. This is computationally considerably more intensive if several random effects or levels are involved. Nevertheless, it is becoming increasingly feasible. An additional advantage is that these methods provide an estimate of the log-likelihood, which can be used for hypothesis testing and interval estimation.

The third approach is to adopt a Bayesian formulation with uninformative priors. Many common models are implemented in MLwiN (Rasbash *et al.*, 2000) and several models are described in the WinBUGS manual (Spiegelhalter, Thomas and Best, 1999).

Note that these methods can be extended to provide multilevel versions of more general multinomial models (see the chapter by Yang, M. in Leyland and Goldstein, 2001).

Multilevel models are likelihood based. An alternative class of methods, known as GENERALISED ESTIMATING EQUATIONS (GEEs), can also be used for multilevel data (see Diggle *et al.*, 2002, Chapter 11). GEEs model the mean and variance of the data only; unlike multilevel models, a PROBABILITY DISTRIBUTION for the data (e.g. normal) is not specified. Standard errors are often estimated robustly from the sampling variance of the residuals, using the Huber–White sandwich estimator (see HUBER–WHITE ESTIMATOR) (Diggle *et al.*, 2002, p. 80).

A theoretical advantage of GEEs is that the fixed parameter estimates are consistent (i.e. reliable if there is sufficient data) even if the covariance structure is wrongly specified; however, they may be inefficient if the covariance structure is substantially misspecified (Goldstein, 2003, p. 21). The drawback is that variance components are not explicitly modelled, but are treated as nuisance parameters, whereas from the multilevel modelling perspective the variance components contain useful insights.

This is directly related to an important, but subtle, difference between the two. Fixed parameter estimates from multilevel models estimate the effect of a covariate on a subject *conditional on the value of their subject-specific effects*. GEE parameter estimates are marginalised over subject-specific effects; they estimate the average effect of a covariate over the population the data are drawn from. For multilevel models for quantitative data, conditional and marginal estimates of fixed parameters coincide. For discrete data they do not; often marginal estimates are markedly smaller in magnitude (compare Tables 11.1 and 8.2 in Diggle *et al.*, 2002).

The appropriate approach adopted depends on the scientific question. If the primary aim is to model the average response as a function of covariates and time, and the correlation is a nuisance, GEEs may be preferred. The resulting parameter estimates are often known as *population averaged*. Conversely, if understanding of the variance structure is important, e.g. in investigating determinants of variation in growth rates, multilevel models are required. A complication with conditional models is that, because the interpretation of the fixed parameters is conditional on the variance model, if this is changed the interpretation is generally altered.

The literature on medical applications of multilevel models is vast and growing. A good starting point is the collection of papers in Leyland and Goldstein (2001), which includes models for growth data, spatial distribution of mortality and morbidity, and institutional comparisons. The latter is an important and widespread application of multilevel models.

Applications to META-ANALYSIS are discussed by Hardy and Thompson (1996) (quantitative data) and Turner *et al.* (2000) (binary data), CROSSOVER TRIALS by Jones and Kenward (2003, Chapters 5 and 6) and CLUSTER RANDOMISED TRIALS by Donner and Klar (2000).

So far, we have assumed that each subject at each time only has one response. However, the covariance model readily extends to allow multivariate responses at each time. For example, a subject's diastolic and systolic blood pressure can be modelled simultaneously (see the chapter by McLeod, A. in Leyland and Goldstein, 2001).

The multilevel framework can also be extended to handle time-to-event data, with subjects having repeated events and a common frailty (the commonly adopted term for a subject-specific random effect in survival analysis). Indeed, frailties at different levels of the hierarchy can be fitted (Singer and Willett, 2002).

Another extension is what is termed 'cross-classified' data. Here subjects are members of more than one hierarchy. For example, subjects may be nested within general practices and health authorities, but may also be nested within distinct neighbourhoods, served by a number of general practices. They therefore belong to more than one hierarchy. Parameter estimation is no longer always straightforward (Goldstein, 2003, Chapter 11).

Frequently in studies involving longitudinal follow-up, a proportion of the intended responses will be unobserved. An important advantage of multilevel models over classical techniques is that a complete set of observations on each subject included in the analysis is not required; subjects can still be included in the analysis with partially observed response data.

Further, if subjects are missing responses, or drop out, then provided that, given their observed data, the reason for the dropout does not depend on the unseen responses (the MISSING AT RANDOM, MAR, assumption), parameter estimates from multilevel models are still valid (Little and Rubin, 2002). Thus, if data are analysed using multilevel models and response data are MAR, ad hoc IMPUTATION techniques such as replacing a missing observation by the previous seen observation (LOCF) are not required; indeed they will generally introduce BIAS (Molenberghs *et al.*, 2004). Sensitivity to MAR can be assessed; see, for example, Carpenter, Pocock and Lamm (2002).

However, parameter estimates from GEEs are not valid under MAR; to guarantee their validity a stronger assumption, missing completely at random (MCAR), is required. This assumption states that the reason for a subject's unobserved data is independent of both their observed and unobserved data. Although GEEs can be modified to cope with MAR data, to do this efficiently requires a nontrivial multistage estimation process.

The above does not apply to missing covariate information; if a nontrivial degree of covariate information is missing, it usually needs to be recovered using appropriate data imputation methods.

The generality of multilevel models means that the distribution of many test statistics is only known under the null hypothesis, so simulation often has to be employed in sample size calculations. Simplifying assumptions enable progress in special cases (Diggle *et al.*, 2002, p. 24).

As multilevel models become more mainstream, software to fit the basic models for quantitative data is becoming increasingly available in standard packages. All the models described here can be fitted using MLwiN (Rasbash *et al.*, 2000); many can be fitted using PROCs MIXED, GLMMIX and NLMIXED in SAS version 8.x (SAS v. 8.1, 2002). For very large datasets, SAS is preferable. A comprehensive review of the capabilities of available packages is given on the MLwiN website, www.mlwin.com.

Bayesian model fitting can be performed with the WinBUGS package (Spiegelhalter *et al.*, 1999). This is very flexible, but the user is required to write the program to fit the model, and a degree of knowledge about MARKOV CHAIN MONTE CARLO methods is required.

Newcomers to multilevel modelling should start with one of the many excellent books now available. The least technical of these is Kreft and de Leeuw (1998), which gives a basic introduction to models for quantitative data, from a social science perspective. The software MLwiN (Rasbash *et al.*, 2000) also comes with an accessible manual and many examples. Raudenbush and Bryk (2002) give a much more extensive treatment, including discrete response models, with many detailed social science examples.

For the methodologically inclined, Verbeke and Molenberghs (2000) give a comprehensive overview for quantitative data from a longitudinal perspective. Examples are analysed in detail using mostly SAS (SAS v. 8.1, 2002) and some MLwiN (Rasbash *et al.*, 2000) and SPLUS (SPLUS v. 6, 2003). The latter half is given over to problems with missing data. Less detailed but more general is Diggle *et al.* (2002), who also come from a longitudinal standpoint, and discuss quantitative and discrete data, multilevel models, GEEs and transition models. Most recently, Goldstein (2003) gives a comprehensive account of the current state of multilevel modelling, including outlines of technical details and many illustrative examples. *JRC*

[**Acknowledgement:** James R. Carpenter was supported by ESRC Research Methods Programme grant H333250047, titled 'Missing data in multi-level models'.]

Carpenter, J., Pocock, S. and Lamm, C. J. 2002: Coping with missing data in clinical trials: a model based approach applied to asthma trials. *Statistics in Medicine* 21, 1043–66. **Diggle, P. J., Heagerty, P., Liang, K. Y. and Zeger, S. L.** 2002: *Analysis of longitudinal data*, 2nd edition. Oxford: Oxford University Press. **Donner, A. and Klar, N.** 2000: *Design and analysis of cluster randomization trials in health research.* London: Arnold. **Gilks, W. R., Richardson, S. and Spiegelhalter, D. J.** (eds) 1996: *Markov chain Monte-Carlo in practice.* London: Chapman & Hall. **Goldstein, H.** 2003: *Multilevel statistical models*, 2nd edition. London: Arnold. **Hardy, R. J. and Thompson, S. G.** 1996: A likelihood approach to meta-analysis with random effects. *Statistics in Medicine* 15, 619–29. **Jones, B. and Kenward, M. G.** 2003: *Design and analysis of cross-over trials*, 2nd edition. London: Chapman & Hall. **Kreft, I. and de Leeuw, J.** 1998: *Introducing multilevel modelling.* London: Sage. **Leyland, A. H. and Goldstein, H.** (eds) 2001: *Multilevel modelling of health statistics.* Chichester: John Wiley & Sons, Ltd. **Little, R. J. A. and Rubin, D. B.** 2002: *Statistical analysis with missing data*, 2nd edition. Chichester: John Wiley & Sons, Ltd. **Molenberghs, G., Thijs, H., Jansen, I., Beunkens, C., Kenward, M. G., Mallinkrodt, C. and Carroll, R. J.** 2004: Analyzing incomplete longitudinal clinical trial data. *Biostatistics* 5, 445–64. **Rasbash, J., Browne, W., Goldstein, H., Yang, M., Plewis, I., Healy, M., Woodhouse, G., Draper, D., Langford, I. and Lewis, T.** 2000: *A user's guide to MLwiN (version 2.1).* London: Institute of Education. **Raudenbush, S. W. and Bryk, A. S.** 2002: *Hierarchical linear models: applications and data analysis methods*, 2nd edition. London: Sage. **SAS v. 8.1** 2002: SAS Worldwide Headquarters, SAS Campus Drive, Cary, NC 27513-2414, USA, www.sas.com. **Singer, J. D. and Willett, J. B.** 2002: *Applied longitudinal data analysis: modelling change and event occurrence.* New York: Oxford University Press. **Spiegelhalter, D. J., Thomas, A. and Best, N. G.** 1999: *WinBUGS version 1.2 user manual.* Cambridge: MRC Biostatistics Unit. **SPLUS v. 6** 2003: Insightful Switzerland, Christoph Merian-Ring 11, 4153 Reinach, Switzerland. **Turner, R. M., Omar, R. Z., Yang, M., Goldstein, H. and Thompson, S. G.** 2000: A multilevel model framework for meta-analysis of clinical trials with binary outcomes. *Statistics in Medicine* 19, 3417–32. **Verbeke, G. and Molenberghs, G.** 2000: *Linear mixed models for longitudinal data.* New York: Springer Verlag.

multinomial distribution This is a generalisation of the BINOMIAL DISTRIBUTION to the case where more than two outcomes are possible for every 'trial'. Whereas the binomial distribution addresses the number of successes (and thus implicitly the number of failures also) in the case where every event or trial can only result in a success or a failure, the multinomial distribution models the numbers of each outcome in the case where each event or trial can have one of multiple outcomes. For example, Lossos *et al.* (2000) note that, when modelling genetic mutations in a situation with four rather than two distinct genotypes, it is necessary to extend the usual binomial model to a multinomial one.

In general, for n observations, each of which can independently take one of N mutually exclusive outcomes with probabilities p_1, p_2, ..., p_N (where $p_1+p_2+\cdots+p_N=1$), then the PROBABILITY of seeing x_1 observations achieving outcome 1, x_2 observations achieving outcome 2, etc., where $x_1+x_2+\cdots+x_N=n$, is given by:

$$\Pr[(X_1, X_2, \ldots, X_N) = (x_1, x_2, \ldots, x_N)]$$

$$= \frac{n!}{x_1!x_2!\cdots x_N!} p_1^{x_1} p_2^{x_2} \cdots p_N^{x_N}$$

where $n!$ (factorial n) is the product of all the integers up to and including n, namely, $n \times (n-1) \times (n-2) \times \cdots \times 3 \times 2 \times 1$, with 0! defined to be 1.

Note that since the data are multidimensional, there is no single mean value of the distribution as such, although (as for the binomial distribution) the expected number to be seen with outcome k is $p_k n_k$. *AGL*

Lossos, I. S., Tibshirani, R., Narasimhan, B. and Levy R. 2000: The inference of antigen selection on Ig genes. *Journal of Immunology* 165, 5122–6.

multiple comparisons Procedures for a detailed examination of where differences between a set of MEANS lie, usually applied after a significant F-test in an ANALYSIS OF VARIANCE has led to the rejection that all the means are equal. A large number of multiple comparison techniques has been proposed but no single technique is best in all situations. The major distinction between the techniques is how they control the inflation of the TYPE I ERROR that would occur if, for example, a simple STUDENT'S t-TEST was applied to test the equality of each pairs of means.

One very simple procedure for dealing with the inflation procedure is to judge the P-values from each t-test against a significance level of α/m rather than α, the nominal size of the Type I error, where m is the number of t-tests performed – this is known as the BONFERRONI CORRECTION. Many alternatives approaches are available, most of which are based on the usual t-statistic, but which differ in the choice of critical value against which the t-statistic is compared. A comprehensive account of multiple comparison procedures is given in Hsu (1996). *BSE*

Hsu, J. C. 1996: *Multiple comparisons*. London: Chapman & Hall.

multiple correlation coefficient See CORRELATION

multiple imputation This is a method by which missing values in a dataset are replaced by more than one, usually between 3 and 10, simulated versions. Each of the simulated complete datasets is then analysed by the method relevant to the investigation to hand and the results combined to produce estimates, STANDARD ERRORS and CONFIDENCE INTERVALS that incorporate missing data uncertainty. Introducing appropriate random error into the imputation process makes it possible to get approximately unbiased estimates of all parameters, although the data must be missing at random for this to be the case. The multiple imputations themselves are created by a Bayesian approach (see BAYESIAN METHODS), which requires specification of a parametric model for the complete data and, if necessary, a model for the mechanism by which data become missing. A comprehensive account of multiple imputation and details of associated software are given in Schafer (1997). *BSE*

[See also DROPOUTS]

Schafer, J. 1997: *The analysis of incomplete multivariate data*. Boca Raton: CRC/Chapman & Hall.

multiple linear regression This is a technique used to model, or characterise quantitatively, the relationship between a response variable, y, and a set of explanatory variables, x_1, x_2, ..., x_q. The explanatory variables are strictly assumed to be known or under the control of the investigator, i.e. they are not considered to be random variables. In practice, where this is rarely the case, the results from a multiple regression analysis are interpreted as being conditional on the observed values of the explanatory variables. The multiple regression model can be written as:

$$y = \beta_0 + \beta_1 x_1 + \cdots + \beta_q x_q + \varepsilon$$

where β_0 is an intercept and β_1, β_2, ..., β_q are regression coefficients that measure the change in the response variable associated with a unit change in the corresponding explanatory variable, conditional on the other explanatory variables remaining constant. If the explanatory variables are highly correlated such an interpretation is problematic. The residual, ε, is assumed to have a normal distribution with MEAN zero and VARIANCE σ^2. An alternative way of writing the multiple regression model is that y is distributed normally with mean μ and variance σ^2, where $\mu = \beta_0 + \beta_1 x_1 + \cdots + \beta_q x_q$. This formulation makes it clear that the model is only suitable for continuous response variables with, conditional on the values of the explanatory variables, a NORMAL DISTRIBUTION with constant variance. For a sample of n response values along with the corresponding values of the explanatory variables, the aim of multiple regression is to arrive at a set of values for the regression coefficients that make the values of the response variable predicted from the model as 'close' as possible to the observed response values. Estimation of the parameters of the model ($\beta_1, \beta_2, \ldots, \beta_q$) is usually by least squares (see Rawlings, Pantula and Dickey, 1998). The variation in the response variable can be partitioned into a part due to regression on the explanatory variables and a residual. This partition can be set out in an ANALYSIS OF VARIANCE-type table as shown.

multiple linear regression *An analysis of variance-type table*

Source	*DoF*	*SS*	*MS*	*MSR*
Regression	q	RGSS	RGSS/q	RGMS/ RSMS
Residual	$n-q-1$	RSS	RSS/($n-q-1$)	

Under the null hypothesis that all the regression coefficients, $\beta_1, \beta_2, \ldots, \beta_q$, are zero, the mean square ratio (MSR) in this table can be tested against an F-DISTRIBUTION with q and $n-q-1$ DEGREES OF FREEDOM. The residual mean square is an estimator of σ^2 and is used in calculating STANDARD ERRORS of the estimated regression coefficients (see Rawlings, Pantula and Dickey, 1998). The MULTIPLE CORRELATION COEFFICIENT is the correlation between the observed values of the response and the values predicted by the model. The square of the multiple correlation coefficient gives the proportion of the variance of the response that can be explained by the explanatory variables.

The overall test that all the regression coefficients in a multiple regression model are zero is seldom of great interest. The investigator is more likely to be concerned with assessing whether some subset of the explanatory variables might be almost as successful as the full set in explaining the variation in the response variable; i.e. a more parsimonious model is sought. Various procedures have been suggested to help in this search (see ALL SUBSETS REGRESSION, AUTOMATIC SELECTION PROCEDURES).

Once a final model has been settled on, the assumptions of the multiple linear regression approach for the data to hand need to be checked. One way to investigate the possible failings of a model is to examine what are known as *residuals*, defined as:

Residual =

Observed response value−Predicted response value

The n sample residuals can be plotted in a variety of ways to assess particular assumptions of the multiple regression model: a HISTOGRAM or STEM-AND-LEAF PLOT of the residuals can be useful in checking for normality of the error terms in the model; plots of the residuals against the corresponding values of each explanatory variables may help to uncover when the relationship between the response and an explanatory variable is more complex than that originally assumed – it may suggest that a quadratic term is needed to model a 'U-shape' or 'J-shape' apparent relationship; a plot of the residuals against the fitted values may identify that, for example, the presence of the multivariate OUTLIERS are worthy of further investigation and checking or perhaps that the variance of the response increases with the fitted values, suggesting that a transformation of the response should be considered. There are now many other regression diagnostics available (see, for example, Lovie, 1991).

To illustrate multiple regression we shall use the data shown in the second table. These data arise from a study of 20 patients with hypertension (Daniel, 1995). In practice, of course, there would be too few patients to allow a sensible analysis with seven explanatory variables. The response variable here is the mean arterial blood pressure (mmHg).

multiple linear regression *Data for 20 patients with hypertension*

	BP	*Age*	*Weight*	*BA*	*TimeHt*	*Pulse*	*Stress*
1	105	47	85.4	1.75	5.1	63	33
2	115	49	94.2	2.10	3.8	70	14
3	116	49	95.3	1.98	8.2	72	10
4	117	50	94.7	2.01	5.8	73	99
5	112	51	89.4	1.89	7.0	72	95
6	121	48	99.5	2.25	9.3	71	10
7	121	49	99.8	2.25	2.5	69	42
8	110	47	90.9	1.90	6.2	66	8
9	110	49	89.2	1.83	7.1	69	62
10	114	48	92.7	2.07	5.6	64	35
11	114	47	94.4	2.07	5.3	74	90
12	115	49	94.1	1.98	5.6	71	21
13	114	50	91.6	2.05	10.2	68	47
14	106	45	87.1	1.92	5.6	67	80
15	125	52	101.3	2.19	10.0	76	98
16	114	46	94.5	1.98	7.4	69	95
17	106	46	87.0	1.87	3.6	62	18
18	113	46	94.5	1.90	4.3	70	12
19	110	48	90.5	1.88	9.0	71	99
20	122	56	95.7	2.07	7.0	75	99

BP: Mean arterial blood pressure (mmHg); *Age:* Age in years; *Weight:* Weight in kg; *BA:* Body surface area (square metres); *TimeHt:* Duration of hypertension (years); *Pulse:* Basal pulse (beats/mim); *Stress:* Measure of stress.

multiple linear regression *Results for data in the second table*

Term	*Estimated regression coefficient*	*Standard error*	*T-value*	*P-value*
(Intercept)	−12.8705	2.5566	−5.0341	0.0002
Age	−0.7033	0.0496	−14.7710	0.0000
Weight	−0.9699	0.0631	−15.3691	0.0000
BA	−3.7765	1.5802	−2.3900	0.0327
TimeHT	−0.0684	0.0484	−1.4117	0.1815
Pulse	−0.0845	0.0516	−1.6370	0.1256
Stress	−0.0056	0.0034	−1.6328	0.1265

Residual standard error: 0.4072 on 13 degrees of freedom; Multiple R-squared: 0.9962.

The LEAST SQUARES ESTIMATIONS of the regression parameters are shown in the third table. The square of the multiple correlation coefficient is 0.99 and the mean squares ratio described above takes the value 560.6; tested against an F-distribution with 6 and 13 degrees of freedom the associated P-VALUE is extremely small. Clearly, the hypothesis that all the regression coefficients are zero can be safely rejected. For these data the sample size is too small for residual plots to be particularly informative. However, for interest, the figure shows a plot of residuals against fitted values. The plot gives no cause for concern in respect of the constant variance assumption. *BSE*

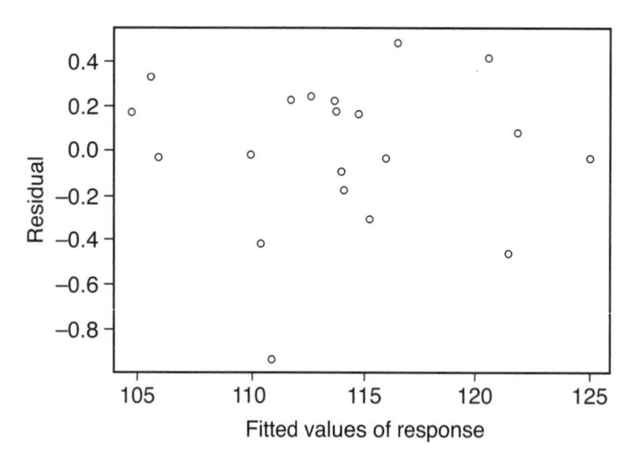

multiple linear regression *Residuals plotted against fitted values*

[See also GENERALISED LINEAR MODELS, LOGISTIC REGRESSION]

Daniel, W. 1995: *Biostatistics: a foundation for analysis in the health sciences*, 6th edition. New York: John Wiley & Sons, Inc. **Lovie, P.** 1991: Regression diagnostics. In Lovie, P. and Lovie, A. D. (eds), *New developments in statistics for psychology and the social sciences*. London: Routledge. **Rawlings, J. O., Pantula, S. G. and Dickey, D. A.** 1998: *Applied regression analysis: a research tool*. New York: Springer.

multiple record systems See CAPTURE–RECAPTURE METHODS

multiple testing This refers to carrying out multiple (more than one, but possibly very many) statistical SIGNIFICANCE TESTS. The problem is one of not controlling the overall TYPE I ERROR rate when we perform many significance tests. The Type I error is the probability of falsely rejecting the null hypothesis (H_0) when it is actually true. If we compare two treatments in a CLINICAL TRIAL, we generally state the null hypothesis to be that there is no difference in mean response (or in death rates, or cure rates, etc.) between the two treatments. This is not a statement about the data that we see in the trial (the sample means, $\bar{x}_i$) but rather one about the true (but unknown) population means, μ_i. The alternative (H_1) is simply the converse – i.e. that there *is* a difference between the treatments. Now, usually (although it is a very arbitrary yardstick), we reject H_0 and declare that a difference between treatments exists if the calculated P-VALUE is less than 5%. So for any single significance test, if the null hypothesis is true, implicitly we are accepting a risk of being wrong of 5% – and for many situations, many people consider that an adequately small risk. However, what happens if we perform more than one significance test to answer the same (or related) questions and we are prepared to reject the null hypothesis if either (or both) tests give $P < 0.05$? In the simplest case of two independent tests, the PROBABILITY of *either* test 1 *or* test 2 (or, indeed, both tests) giving us a small (say, <5%) P-value is $1 - 0.95^2$, which equals 0.0975 (or close to 10%). If we carried out three, four, five or even ten independent significance tests, the probabilities that at least one of them will give a small (<5%) P-value would be, respectively, 0.143, 0.186, 0.226 and 0.401. These are the FALSE POSITIVE ERROR RATES and it is apparent that, very quickly, the risk of *erroneously* declaring a statistically significant difference between the treatments (i.e. when all of the null hypotheses are true) becomes much (and unacceptably) higher than 5%. Therefore, we need methods to correct for this inflated Type I error rate.

The simplest method to use is the BONFERRONI CORRECTION. Using his very simplistic line of reasoning, if we are to carry out two significance tests but want to ensure that the overall chance of making a false positive error is kept at 5%, then we should test each of the two null hypotheses at the 2.5% level. Then the probability of *either* test 1 *or* test 2 (or both tests) giving us a small (now 'small' means less than 2.5%) P-value is $1 - 0.975^2$, which is close to 0.05. So if either or both tests meet this more stringent level of statistical significance, we can reject the null hypothesis and only run a 5% risk of making a false positive claim about a difference between the treatments. His idea extends very easily to more than two tests: for three we compare each calculated value of P with 0.05/3 (= 0.0167), for four tests we compare each calculated value of P with 0.05/4 (= 0.0125), and so on.

There are many reasons why multiple testing (particularly in CLINICAL TRIALS – but in other medical applications too) might occur, even in studies that just compare two treatments. Examples include: many ENDPOINTS, more than one time point (e.g. short-term response and long-term response), more than one analysis method (e.g. using a NONPARAMETRIC METHOD *and* its equivalent parametric test), different definitions of response to treatment (e.g. difference in means, or mean changes from baseline, or percentage change from baseline, or proportion of 'responders', etc.), differences within (and between) various subgroups (see SUBGROUP ANALYSIS), multiple INTERIM ANALYSES, and so on. In studies comparing more than two treatments, there are all these same possible examples of multiplicity *in addition to* those of many comparisons between all the different treatment arms. It is therefore quite clear that very quickly the simple scenario of one significance test of one primary endpoint can become unrealistic.

The Bonferroni method is simple but also very inefficient; it lacks statistical POWER to identify real differences between treatments (see, for example, Perneger, 1998). It can also lead to the very uncomfortable situation of two tests giving P-values of, say, 0.03 and 0.04. At first site, these *both* appear to demonstrate a statistically significant difference between the groups; they are both less than 0.05. However, because there are two tests, Bonferroni says we need to compare each of them against the more stringent criteria of 0.025, but now *neither* meets this level of stringency! Hence, various (in fact, numerous) other methods have been proposed – some are simple (albeit not as simple as Bonferroni) and some are very complex. The details are beyond the scope of this text although a good overview is given by, for example, Hochberg and Tamhane (1987).

We will illustrate ideas using two simple methods. One is by Holm (1979). He proposed ordering all (of k) calculated P-values P_1 (the largest) to P_k (the smallest). The smallest calculated P-value (P_k) should be compared with α/k; if $P_k < \alpha/k$ then we declare this difference to be statistically significant and move to the next smallest of the P-values, P_{k-1}. If $P_{k-1} < \alpha/(k-1)$ then we declare this difference also to be statistically significant. The procedure continues until all of the calculated P-values have been tested, or until one of the P-values fails to meet the criterion for statistical significance. In this case, the procedure stops and no more of the tests are considered significant.

Another method is called the 'closed testing' approach; it is best described by an example. If we wanted to compare two doses of an active drug with PLACEBO (or some other reference product) then it might seem that we need to carry out each test at a 2.5% significance level. However, assuming we believe that the dose–response relationship will be monotonically increasing then we can begin by just testing the highest dose against the reference at the standard 5% level. If the P-value is smaller than 5%, then the difference is declared statistically significant and the next dose (and its P-value) are considered. This P-value is *also* compared to 5% level and, if smaller than this, is declared as significant. The procedure could continue if there were several doses. Each test is carried out at the 5% significance level until one fails to meet it – then all testing stops and none of the other tests is considered significant. This is a much more powerful procedure than Bonferroni's although it has a major problem if a treatment turns out to have no effect on the first test (in this case the highest dose) but may have substantial effects at lower doses: none of the secondary tests can be considered 'significant' because the very first test (for the highest dose) failed. These two approaches (and there are many others) illustrate that there is no simple, single approach to solving multiplicity problems that is applicable in all situations. *SD*

Hochberg, Y. and Tamhane, A. C. 1987: *Multiple comparison procedures*. New York: John Wiley & Sons, Inc. **Holm, S.** 1979: A simple sequentially rejective multiple test procedure. *Scandinavian Journal of Statistics* 6, 65–70. **Perneger, T. V.** 1998: What's wrong with Bonferroni adjustments? *British Medical Journal* 316, 1236–8.

multistage cluster sample A nonprobabilistic method of sampling is used when members of a population are arranged in subgroups or clusters. In this method clusters are the sampling unit rather than individuals. Members within a cluster should be as different as possible whereas clusters, by way of contrast, should be as alike as possible. However, this condition is hard to satisfy and since two members of a cluster will be more alike than two from different clusters, it is better to have many small clusters than a few large clusters, as this reduces sampling error. Each cluster should be similar to the total population but on a smaller scale. Clusters must be distinct from each other and every member of the population should fall within a cluster. In some situations, it is necessary for all clusters to be of a similar size and this may require the pooling of some clusters. Otherwise, the PROBABILITY that a cluster is chosen can be made proportional to its size, so that bigger clusters are more likely to be chosen than smaller clusters and the probability of selecting an individual member of the cluster is inversely proportional to the size of the cluster.

For a single-stage cluster sample a list of the clusters is constructed. Then a random sample of the clusters is taken. This may be a simple random sample, with each cluster having an equal probability of being included in the sample, or it may be that the probability of being in the sample can be proportional to the size of the cluster. Once the clusters have been selected each member of the cluster is included in the sample.

For a two-stage cluster sample the method is the same as a single-stage cluster sample but once the clusters have been selected then a SIMPLE RANDOM SAMPLE is used to select the members of the cluster to be included in the sample. Clusters may also form larger clusters, in which case, multistage sampling would be used. First, the clusters would be sampled, followed by the subclusters. Depending on the makeup of the population there may be many stages to the multistage sampling.

Multistage cluster sampling was used in a study of violations of the international code of marketing of breast milk substitutes (Taylor, 1998). Here, the capital city of four chosen countries were the main clusters, districts were randomly selected subclusters, health facilities were randomly selected from the subclusters and mothers were systematically sampled from the health facilities.

The main advantage is that no sampling frame is required. The main disadvantage is that the sampling is nonrandom and sampling error increases by taking multiple samples, as there is sampling error at each stage. For further details see Crawshaw and Chambers (1994). *SLV*

Crawshaw, J. and Chambers, J. 1994: *A concise course in A level statistics*, 3rd edition. Cheltenham: Stanley Thornes Publishers Ltd.
Taylor, A. 1998: Violations of the international code of marketing of breast milk substitutes: prevalence in four countries. *British Medical Journal* 316, 1117–22.

multistate models These are often adopted for analysing event history (see SURVIVAL ANALYSIS-AN OVERVIEW) and LONGITUDINAL DATA. They are commonly applied in studies where subjects are followed up over time with respect to a (stochastic) process of interest that is observed to occupy exactly one of a finite number of discrete states at any point in time. Multistate models have been found to be extremely useful in medical areas such as psoriatic arthritis, where individuals may move between a number of disability states (e.g. mild, moderate and severe disability) over the longitudinal course of their disease (Husted *et al.*, 2005); in hepatitis C virus (HCV) disease progression studies where liver biopsy scores are used to determine the stage of HCV-related liver disease (Sweeting *et al.*, 2006); in describing the states, characterised by the occurrence of various events (acute graft versus host disease, chronic graft versus host disease, relapse and death in remission), where a leukaemia patient may enter following bone marrow transplantation (Keiding, Klein and Horowitz, 2001); and in other areas such as Alzheimer's disease, bronchiolitis obliterans syndrome, cancer, cognitive impairment, diabetic retinopathy, HIV/AIDS, studies of twins and in competing risks of death studies.

Multistate models are based on stochastic processes that move through a series of discrete states in continuous time. The movements between states are called transitions. States can be transient (movements out are allowed) or absorbing, if further transitions out are prohibited. The state structure describes the states and determines which transitions are allowed. Different state structures may characterise the same stochastic process and are thus not unique to the process. The state structure chosen depends on the questions of interest, the transparency of model assumptions and the ease with which to make inferences. This structure can be represented schematically with a multistate diagram in which boxes represent the various states and arrows between the boxes represent the possible transitions that can occur. Figures (a) to (e) present various multistate diagrams of commonly observed multistate process types.

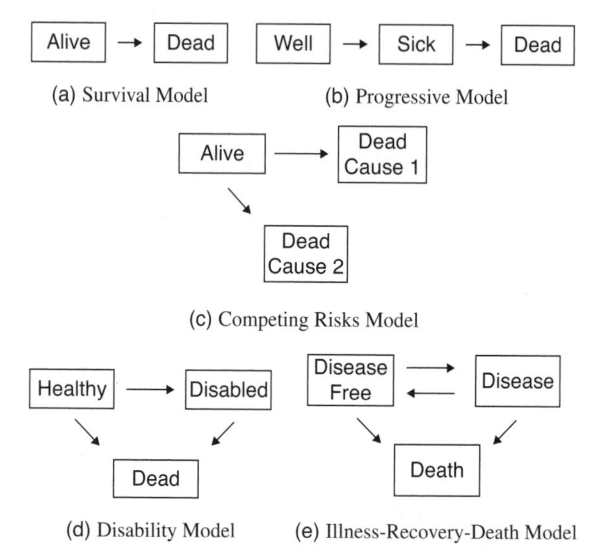

multistate models *Various diagrams of commonly observed multistate process types*

A multistate process can be specified fully either through its transition intensities (also known as hazard functions, see SURVIVAL ANALYSIS – AN OVERVIEW) or by its transition probabilities. The transition intensities are the instantaneous probabilities per time unit (i.e. the transition rates) of going from one state to another, given the history (development) of the process just prior to the times of the transitions. Where a transition from one state directly to another is impossible the corresponding transition intensity is zero. The transition probabilities represent the conditional probabilities of the process being in particular states at various times, given that the process was observed in specific states at earlier times and the histories of the process are up to these earlier times. For movements out of an absorbing state the transition probabilities will clearly be zero.

Mathematically, given a multistate process for a subject, $X(t)$, at time $t \geq 0$ with a finite discrete state space denoted by

$\Omega = \{1, 2, \ldots, w\}$ and history up to some time just prior to $s < t$, $H(s-)$, then the transition probability of moving from state i at time s to state j at time t is given by:

$$P_{ij}(s,t) = \Pr(X(t) = j | X(s) = i; H(s-))$$

and the transition intensity of making an i to j instantaneous transition (denoted by $i \rightarrow j$) at time t is given by:

$$\alpha_{ij}(t) = \lim_{\Delta t \rightarrow 0} \frac{\Pr(X(t + \Delta t) = j | X(t) = i; H(t-))}{\Delta t}$$

It is important to note that both the transition intensities and the transition probabilities are essential for the development, estimation and inference of multistate models. The transition intensities are important in the formulation of necessary models for the data, in terms of incorporating covariates that explain some of the heterogeneity across subjects and for representing the assumptions (e.g. time homogeneity, Markov or semi-Markov, piecewise constant or Weibull baseline intensities, proportional intensities (see PROPORTIONAL HAZARDS), etc.) being made. The transition probabilities, on the other hand, are important for constructing the likelihood function (see LIKELIHOOD) to be maximised and for making long-range predictions.

A simple and mathematically tractable example of a multistate model, which has been described often in the literature, is the time homogeneous Markov model. Here transition intensities are assumed constant over, or independent of, time (i.e. time homogeneous) and only depend on the history of the process through the current state (i.e. Markov assumption). In this special case, a model for the transition intensities, incorporating baseline explanatory variables, can be specified using a multiplicative structure and the proportional intensities (or hazards) assumption proposed by Sir David Cox in 1972, which is that the transition intensity, $\alpha_{ij}(t)$, of making an $i \rightarrow j$ transition at time t is given by:

$$\alpha_{ij}(t) = \alpha_{ij}^{0} \exp(\beta_{1ij} z_1 + \cdots + \beta_{pij} z_p)$$

where α_{ij}^{0} corresponds to a baseline intensity of making an $i \rightarrow j$ transition, which is being modified by the exponential of a linear combination of the baseline explanatory variables, $z_1, \ldots, z_p$. The p regression coefficients, $\beta_{1ij}, \ldots, \beta_{pij}$, associated with these baseline explanatory variables are assumed here to be transition intensity-specific, although constraints on them can lead to more parsimonious models. The exponential of the regression coefficients are interpreted as rate ratios.

Further extensions beyond simple time homogeneous Markov multistate models have been made. The readers are referred to review articles by Commenges (1999), Hougaard (1999), Andersen and Keiding (2002) and Meira-Machado *et al.* (2009) for further details.

Frequently in event history and longitudinal data studies in medical research, observation of the exact transition times may not occur. This may be because, by the end of follow-up, all individuals under study have not reached an absorbing state, which thus results in right censored observation times. Follow-up of some individuals may have only happened some time after the process began and various events may have occurred between the start of the process and the start of follow-up, which may result in left censoring if the times of these various events are unknown. Furthermore, there are many longitudinal studies where subjects are observed intermittently (i.e. discretely in time) and the times of transitions are interval censored (i.e. the exact times of transitions are unobserved), except possibly for an absorbing state such as death.

Finally, left truncation may occur when individuals come under observation only some known time after the 'natural'/ defined time origin of the process. For example, potential participants eligible for a study would only enter if they have not died by study commencement. The presence of these observation and selection schemes may pose special problems for valid inference, and assumptions on how these features may or may not be informative for the multistate process are required in order to construct the appropriate likelihood function. In most situations, at least initially, the 'sampling time process' is assumed to be noninformative (ignorable) for the multistate process. *BT*

Andersen, P. K. and Keiding, N. 2002: Multi-state models for event history analysis. *Statistical Methods in Medical Research* 11, 91–115. **Commenges, D.** 1999: Multi-state models in epidemiology *Lifetime Data Analysis* 5, 315–27. **Hougaard, P.** 1999: Multi-state models: a review. *Lifetime Data Analysis* 5, 239–64. **Husted, J. A., Tom, B. D., Farewell, V. T., Schentag, C. and Gladman, D. D.** 2005: Description and prediction of physical functional disability in psoriatic arthritis: a longitudinal analysis using a Markov model approach. *Arthritis Care and Research* 53, 404–9. **Keiding, N., Klein, J. P. and Horowitz, M. M.** 2001: Multistate models and outcome prediction in bone marrow transplantation. *Statistics in Medicine* 20, 1871–85. **Meira-Machado, L., de Uña-Álvarez, J., Cadarso-Suárez, C. and Andersen, P. K.** 2009: Multi-state models for the analysis of time-to-event data. *Statistical Methods in Medical Research* 18, 195–222. **Sweeting, M. J., De Angelis, D., Neal, K. R., Ramsay, M. E., Irving, W. L., Wright, M., Brant, L., Harris, H. E., Trent HCV Study Group, HCV National Register Steering Group** 2006: Estimated progression rates in three United Kingdom hepatitis C cohorts differed according to method of recruitment. *Journal of Clinical Epidemiology* 59, 144–52.

multivariate analysis of variance (MANOVA)

See ANALYSIS OF VARIANCE

multivariate normal distribution

This is a generalisation of the NORMAL DISTRIBUTION to more than one dimension and the PROBABILITY law that underlies many methods of multivariate analysis.

Any dataset comprises measurements made on a collection of individuals, and all measurements exhibit variation between individuals. A graphical presentation, such as a HISTOGRAM, of all the population values would show the range of variation as well as the relative preponderance of some values over others. Such a histogram would thus allow the calculation of probabilities of observing particular sample values, and this in turn would form a quantitative basis for methods of statistical analysis.

However, such histograms can never be obtained and must instead be modelled by theoretical PROBABILITY DISTRIBUTIONS. Many such distributions exist, but the one that is used most often in statistical analysis is the normal distribution. Many measurements have been empirically shown to follow normal distributions, while the CENTRAL LIMIT THEOREM provides a mathematical justification whenever the measurements represent either sums or means of quantities (see Nolan, 1998).

For a single variable, the normal distribution is characterised by a bell-shaped curve centred at its mean μ and with width governed by its variance σ^2; these are the parameters of the distribution. In the multivariate case, the curve becomes a surface in multidimensional space, while the parameters include not only the MEANS and VARIANCES of each variable but also the COVARIANCES between all pairs of variables.

The multivariate normal distribution is a relatively simple distribution to handle and it has a number of attractive properties. All marginal and conditional distributions of subsets of a multivariate normal set of measurements are themselves multivariate normal, as are all collections of linear combinations of these variables. In particular, the marginal or conditional distribution of any single such measurement, or the distribution of a single linear combination, is univariate normal. Since many techniques of multivariate analysis involve consideration of either linear combinations or subsets of measurements, these properties make the multivariate distribution a very attractive model to use as a basis for these techniques.

To delve any further into the multivariate normal distribution requires considerable technical mathematics. The interested reader is referred to any standard multivariate text, e.g. Morrison (1990). *WK*

Morrison, D. F. 1990: *Multivariate statistical methods*, 3rd edition. New York: McGraw-Hill. **Nolan, D.** 1998: Central limit theorem. In Armitage, P. and Colton, T. (eds), *Encyclopedia of biostatistics*. Chichester: John Wiley & Sons, Ltd.

N

negative binomial distribution This is the PROBABILITY DISTRIBUTION of the number of events required in order to observe k 'successes'. Contrast this with the BINOMIAL DISTRIBUTION, which models the number of successes that will occur given a fixed number of trials. Also note that, since the GEOMETRIC DISTRIBUTION models the number of events required to observe one success, it is a special case of the negative binomial.

If each event independently has a probability of success, p, then the probability mass function for the number of events, x, required before observing k successes is:

$$\Pr(X = x) = \frac{(x-1)!}{(k-1)!(x-k)!} p^k (1-p)^{x-k}$$

where $n!$ (factorial n) is given by the product of integers up to and including n, and $0!$ is defined to be 1. The MEAN of the distribution is k/p and the VARIANCE is $k(1-p)/p^2$.

The distribution can be generalised to the case where the k parameter is not an integer (by replacing the factorial terms with gamma functions as mentioned in GAMMA DISTRIBUTION), which then enables the following interpretation.

Suppose we have observations of count data from a population of size N, where each person's count will be independently distributed as Poisson with some parameter λ. In this case, we would expect the counts in the population to be distributed again as a POISSON DISTRIBUTION. Yet often the population exhibits more variance than can be explained by a Poisson distribution, an example of OVERDISPERSION. One reason for this might be that individuals do not share the same value for λ.

For example, Mwangi *et al.* (2008) show that counts of malaria episodes in 373 children show more variability than can be explained by a Poisson distribution, but show also that the negative binomial distribution provides a much better fit. This is attributed to variation in the susceptibility of the children, with some children being at increased risk of clinical malaria compared to others, and so the model assuming a common λ does not hold.

The negative binomial distribution is not the only distribution to allow for greater dispersion than the Poisson distribution, but, specifically, if values from individuals are Poisson distributed but the values of λ vary between individuals according to a gamma distribution, then the population frequencies will be distributed as a negative binomial distribution. Further discussion of this and other aspects of the distribution see Grimmett and Stirzaker (1992), Gelman *et al.* (1995) and Glynn and Buring (1996). *AGL*

Gelman, A., Carlin, J. B., Stern, H. S. and Rubin, D. B. 1995: Bayesian data analysis. Boca Raton: Chapman & Hall/CRC. **Glynn, R. J. and Buring, J. E.** 1996: Ways of measuring rates of recurrent events. *British Medical Journal* 312, 364–7. **Grimmett, G. R. and Stirzaker, D. R.** 1992: Probability and random processes, 2nd edition. Oxford: Clarendon Press. **Mwangi, T. W., Fegan, G., Williams, T. N., Kinyanjui, S. M., Snow, R. W., *et al.*** 2008 Evidence for over-dispersion in the distribution of clinical malaria episodes in children. *PLoS ONE*, 3(5), e2196.

negative predictive value (NPV) This is defined for a diagnostic test for a particular condition as the PROBABILITY that those who have a negative test do not actually have the condition under investigation as measured by a reference or 'gold' standard. (Contrast this with the POSITIVE PREDICTIVE VALUE.)

If the data are set out as in the table, then:

$$\text{NPV} = \frac{d}{c+d}$$

NPV can also be expressed as a percentage.

negative predictive value *General table of test results among a + b + c + d individuals sampled*

	Present	*Disease absent*	*Total*
Positive	*a*	*b*	***a + b***
Negative	*c*	*d*	***c + d***
Total	***a + c***	***b + d***	***a + b + c + d***

The NPV should be presented with CONFIDENCE INTERVALS (typically set at 95%) calculated using an appropriate method such as that of Wilson that will not produce impossible values (percentages greater than 100 or below 0) when NPV approaches extreme values. *CLC*

[See also FALSE NEGATIVE RATE, FALSE POSITIVE RATE, NEGATIVE PREDICTIVE VALUE SENSITIVITY, SPECIFICITY]

Altman, D. G., Machin, D., Bryant, T. N. and Gardner, M. J. 2000: *Statistics with confidence*, 2nd edition. London: BMJ Books.

nested case-control studies This is a form of CASE-CONTROL STUDY in which the cases and controls are drawn from within a larger study. In other words, they are nested within a parent study, which is usually a COHORT STUDY but sometimes a CROSS-SECTIONAL STUDY or a PREVALENCE study.

Encyclopaedic Companion to Medical Statistics: Second Edition Edited by Brian S. Everitt and Christopher R. Palmer

The nested nature of such studies provides the method's strength. One concern in the design of case-control studies is the appropriate choice of controls, all of whom should be eligible to be cases if they were to develop the disease. A case-control study nested within a cohort study overcomes this concern as a control within the cohort who developed the disease would be counted as a case. Usually, but not necessarily, the controls who are chosen are matched to the cases on various confounding factors such as age and sex.

The usual reason for conducting a nested case-control study, rather than analysing data on the entire cohort or survey, is economy. Usually more data are collected on the participants of the nested study than in the main study. Sometimes these data are derived from the analysis of stored samples, blood or urine, for example, or from further information being obtained from the participants.

In a study to examine the role of sex steroid hormones in relation to endometrial cancer, Lukanova *et al.* (2004) nested a case-control study within three large cohort studies from the Italy, Sweden and the United States. The cohorts comprised over 65 000 women from whom venous blood samples had been taken at enrolment in the cohorts. From within the cohorts, 124 cases of endometrial cancer were identified and two controls per case were chosen from within the same cohort as the case and matched on various factors including date of and age at blood donation. In this way, they were able to confine the processing of the samples to the 124 cases and corresponding controls, providing a great saving on processing samples from all participants in the three cohorts, yet without major loss of statistical POWER.

While the processing of blood samples is often a component of nested case-control studies, other samples can be the focus. In a study to assess the role of selenium in coronary heart disease in men, toenail clippings were obtained for selenium analysis (Yoshizawa *et al.*, 2003). The study was nested within the Health Professionals' Study in the United States, which is a cohort study of over 50 000 men. Within the cohort 470 participants developed coronary heart disease and a matched control was chosen for each one. Thus, fewer than 1000 toenail samples had to be analysed for selenium.

The cost of sample processing is not the only reason for conducting a nested case-control study. While information collected on the cohort is of interest, sometimes further data collection is required. For example, London *et al.* (2003) conducted a case-control study of breast cancer nested within a cohort study of more than 50 000 women. Their interest was in residential magnetic field exposure and for this they were able to focus on the 743 cases identified in the cohort as having breast cancer and a comparable number of controls. Detailed assessment of the magnetic field exposure was made in the homes of the selected participants. Data on other risk factors for breast cancer and possible confounding factors were already available in the data that had been collected for the entire cohort.

Case-control studies can also be nested within cross-sectional surveys. Baker *et al.* (2003) conducted a survey of approximately 3000 men in southern England to ascertain the prevalence of knee disorders in the general population. A nested case-control component considered the cases who had undergone knee surgery. The focus was on occupational and sporting activities. The activities undertaken by the cases at the birthday prior to their reported onset of symptoms were considered. For each case, five controls were selected matched to the case within 1 year of age. The activities undertaken by the controls at the same birthday as the case were then considered. The nested case-control study thus allowed the investigators to avoid bias due to the cases being likely to give up activities at an earlier age than the controls, because of knee pain. A matched analysis (see MATCHED PAIRS ANALYSIS) was required and thus the entire cross-sectional study was a weaker tool for this particular analysis than the matched nested case-control method.

A further variant on the method is to nest a case-control study within a large routine data collection system. Agerbo (2003) analysed the risk of suicide in relation to spouse's psychiatric illness or suicide. The data were obtained by linking the Danish population registers using the unique personal identification numbers assigned to all people in the country. All suicides were identified, as were 20 matched controls per case. All the spouses and children living with the cases and controls were identified from the registers, along with information on diagnoses from the Danish psychiatric register. The study showed that there was a greater risk of suicide among those whose spouse had been admitted to hospital with a psychiatric disorder or who had died, particularly if the death had been by suicide. Few countries are able to conduct such studies as the linkage between record systems is not possible or is not allowed, but where it is possible, such as in Scandinavia, the opportunities for such epidemiological studies are great.

In a cohort study in which cases are recruited prospectively it is possible that controls identified at one time point become cases later. This is particularly likely to happen if the disease is common. An example of this is a nested case-control study within a birth cohort in Sweden (Emenius *et al.*, 2003). Here the focus was on recurrent wheezing in children in relation to nitrogen dioxide exposure. Wheezing is common in childhood and cases were identified from assessment of the cohort at 1 and 2 years of age. Controls were chosen from within the cohort and matched to the cases on the day of birth. Three controls, selected to match cases identified at the age of 1 year, were found to be wheezing at the 2-year assessment and so were also included as cases at that time point. Such nested case-control studies in which controls can become cases are sometimes called case-cohort studies. *HI*

[See also BIRTH COHORT STUDIES, CASE-CONTROL STUDIES]

Agerbo, E. 2003: Risk of suicide and spouse's psychiatric illness or suicide: nested case-control study. *British Medical Journal* 327, 1025–6. **Baker, P., Reading, I., Cooper, C. and Coggon, D.** 2003: Knee disorders in the general population and their relation to occupation. *Occupational and Environmental Medicine* 60, 794–7. **Emenius, G., Pershagen, G., Berglind, N., Kwan, H.-J., Lewné, M., Nordvall, S. L. and Wickman, M.** 2003: NO_2 as a marker of air pollution, and recurrent wheezing in children: a nested case-control study within the BAMSE birth cohort. *Occupational and Environmental Medicine* 60, 876–81. **London, S. J., Pogoda, J. M., Hwang, K. L., Langholz, B., Monroe, R., Kolonel, L. N., Kaune, W. T., Peters, J. M. and Henderson, B. E.** 2003: Residential magnetic field exposure and breast cancer risk: a nested case-control study from a multiethnic cohort in Los Angeles County, California. *American Journal of Epidemiology* 158, 969–80. **Lukanova, A., Lundin, E., Micheli, A., Arslan, A., Ferrari, P., Rinaldi, S., Krogh, V., Lenner, P., Shore, R. E., Biessy, C., Muti, P., Riboli, E., Koenig, K. L., Levitz, M., Stattin, P., Berrino, F., Hallmans, G., Kaaks, R., Toniolo, P. and Zeleniuch-Jacquotte, A.** 2004: Circulating levels of sex steroid hormones and risk of endometrial cancer in postmenopausal women. *International Journal of Cancer* 108, 425–32. **Yoshizawa, K., Ascherio, A., Morris, J. S., Stampfer, M. J., Giovannucci, E., Baskett, C. K., Willett, W. C. and Rimm, E. B.** 2003: Prospective study of selenium levels in toenails and risk of coronary heart disease in men. *American Journal of Epidemiology* 158, 852–60.

net monetary benefit (NMB) See COST-EFFECTIVENESS ANALYSIS

net reproduction rate (NRR) See DEMOGRAPHY

neural networks This is a general class of algorithms for MACHINE LEARNING. A neural network can be described as a parameterised class of functions, specified by a weighted graph (the network's architecture). The weights associated with the edges of the graph are the parameters. Originally, neural networks were motivated by analogy with the structure of the brain. The nodes of the neural network correspond to the neurons and the edges to neuron interactions.

For directed graphs, we can distinguish recurrent architectures (containing cycles) and feedforward architectures (acyclic). A very important special case of feedforward networks is given by layered networks, in which the nodes of the graph are organised into layers such that connections are possible only between elements of two consecutive layers. The weight between the jth unit and the kth unit of successive layers $l-1$ and l in a network is indicated by w_{kj} and it is often assumed that all elements of a layer are connected to all elements of the successive layer (fully connected architecture). In this way, the connections between two layers $l-1$ and l can be represented by a weight matrix $\mathbf{W}_l$, whose entry at row k and column j corresponds to the weight w_{kj} of the edge from node j to node k in the successive layers (see the figure). It is customary to call the first layer the input layer and the last one the output layer. The remaining ones are called hidden layers.

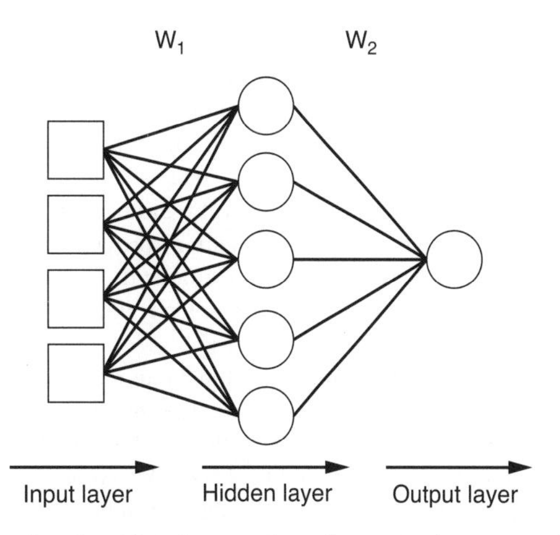

neural networks *Connections between layers on the weight matrix*

A 'perceptron' can be described as a network of this type with no hidden layers. It can also be seen as the building block of complex networks, in that each unit can be regarded as a perceptron (if instead of the transfer function one uses a threshold function, returning Boolean values). Therefore, layered feedforward neural networks as described above are also often referred to as 'multilayer perceptrons'.

In a layered network, the function is computed sequentially, assigning the value of the argument to the input layer, then calculating the activation level of the successive layers as described later, until the output layer is reached. The output of the function computed by the network is the activation value of the output unit.

All units in a layer are updated simultaneously and all the layers are updated sequentially, based on the output of the previous layer. The units of layer l calculate their output values y_l by a linear combination of the values at the previous layer y_{l-1}, followed by a nonlinear transformation $t{:}RØR$, as follows: $y_l = t(\mathbf{W}_l y_{l-1})$ where $\mathbf{W}_l$ is the edge weight matrix between layer $l-1$ and layer l, and where t is called the *transfer function*. A common choice for this transfer function is the logistic function:

$$f(z) = \frac{1}{1+e^{-z}}$$

Notice that each neural network thus represents a class of nonlinear functions parameterised by the weights whose values determine the input/output behaviour of the neural network. Training the network amounts to choosing the values of the weights automatically. For this, a (labelled) training dataset is needed and an error function for the performance of the network has to be fixed. Training a neural network can then be done by finding those weights that minimise the network's error on such samples (i.e. by fitting the network to the data).

More concretely, in the parameter space the error function evaluated on the training data translates to a cost function that associates each configuration of the edge weights with a given error on the training set. Such a function is typically nonconvex, so that it can be minimised only locally, which is often done by gradient descent. A technique known as 'backpropagation' provides a way to compute the necessary gradients efficiently, allowing the network to find a local minimum of the training error with respect to network weights.

The fact that the training algorithm is thus only guaranteed to converge to a local minimum implies that the solution is affected by the initial estimate for the weights. This is one of the major problems of neural networks. Also problematic is the design of the architecture (e.g. the size and the number of hidden layers), often chosen as the result of trial and error. Some such problems have been overcome by the introduction of the related method of support vector machines.

Other types of network arise from different design choices. For example, radial basis function networks use a different transfer function; Kohonen networks are used for clustering problems; Hopfield networks are used for combinatorial optimisation problems. Different training methods also exist. *NC/TDB*

[See also CLUSTER ANALYSIS IN MEDICINE]

Baldi, P. 1998: *Bioinformatics: a machine learning approach.* Cambridge, MA: MIT Press. **Bishop, C.** 1996: *Neural networks for pattern recognition.* Oxford: Oxford University Press. **Cristianini, N. and Shawe-Taylor, J.** 2000: *An introduction to support vector machines.* Cambridge: Cambridge University Press. **Mitchell, T.** 1995: *Machine learning.* Maidenhead: McGraw-Hill.

NMB Abbreviation for net monetary benefit. See COST-EFFECTIVENESS ANALYSIS

N-of-1 trials An N-of-1 (or single-patient) trial combines clinical practice with the well-established methodology of the RANDOMISED CONTROLLED TRIAL to compare the effectiveness of two or more treatment options within an individual. The N-of-1 trial offers a design that facilitates identification of responders and nonresponders to treatment and subsequent determination of optimum therapy for the individual. Indeed, within the context of the hierarchy of evidence-based study designs, it has recently been suggested that N-of-1 trials deliver the highest strength of evidence for making individual patient treatment decisions (Guyatt *et al.*, 2000).

In clinical practice, the clinician commonly performs a 'therapeutic trial' or 'trial of therapy', in which the individual patient receives a treatment and the subsequent clinical course determines whether treatment is judged effective and is continued. Such an approach has serious potential BIASES due to the PLACEBO effect, the natural history of the condition and the urge of the patient and clinician not to disappoint one another. The methodology of the N-of-1 trial at least partially overcomes some of these potential biases.

N-of-1 trials generally compare a single new therapy with a current standard therapy or a placebo. However, as with traditional randomised controlled trials, it is also possible to compare more than two treatment options. In an N-of-1 trial the individual serves as his or her own control, receiving all treatments under investigation. Ideally, such a trial is conducted as a double-blind (both the individual and outcome assessor blind to allocated treatment in any treatment period) multicrossover trial with three or more periods for each treatment. Repeated alternations between treatment periods with the new intervention and the control ensure several comparisons between the treatments. The trial design will, however, be tailored to the clinical entity and therapies involved.

The time commitment of such trials by both patients and health professionals is considerable. N-of-1 trials rely on cooperation between individual clinicians and patients. Hence, the patient's (and clinician's) commitment to the trial is essential for it to reach fruition. The duration of an N-of-1 trial will largely depend on the nature of the condition and the treatments under investigation, but is likely to continue for between several weeks and several months if not for longer. Hence, such trials are only effective for chronic and stable conditions where the natural history of the condition is unlikely to change dramatically over the course of the trial. Examples of their use in different clinical areas include osteoarthritis, gastroesophageal reflux disease, attention deficit hyperactivity disorder and chronic airflow limitation, among others (March *et al.*, 1994).

One problem encountered in N-of-1 trials, as in CROSSOVER TRIALS, is carry-over effects of treatment, which may reduce the estimated treatment effect. The therapies under investigation should therefore have a rapid onset and cessation of effect that will help to minimise any carryover effects. In addition, a washout period between treatments can be incorporated into the trial or a run-in period, where the first few days on each treatment are not evaluated. Because of the expense and time involved, it is important to determine at the outset whether an N-of-1 trial is really indicated for an individual; i.e. is the effectiveness of treatment in doubt for this specific individual? Full criteria (summarised earlier) that should be satisfied before an N-of-1 trial is commenced are provided by Guyatt *et al.* (1988).

When the individual patient and clinician are in agreement that an N-of-1 trial is justifiable, this design provides the additional opportunity to measure the symptoms that matter to the individual concerned. In addition to standardised and validated disease-specific and generic outcome measures, the individual is asked to identify their most troubling symptoms or problems associated with the illness that are important in their everyday lives. These then form the basis of a self-administered diary or questionnaire. It may be a daily diary or weekly summary depending on symptoms and treatment duration, but where possible several separate measurements should be taken within each treatment period. The opportunity to measure the symptoms that matter to the individual is a unique feature of N-of-1 trials.

In a classical randomised controlled trial the lowest experimental unit is the individual; in an N-of-1 trial it is the treatment period. Therefore the sample size in an N-of-1 trial is the number of treatment periods applied. Sample size or POWER calculations used with classical randomised controlled trials can also be used for N-of-1 trials. However, they make certain assumptions concerning the independence of data from each treatment period, which may not be reasonable. While a large number of treatment periods would increase the statistical power, the natural course of the clinical entity, therapy characteristics and patient compliance will generally put an upper limit on this number and thus statistical power will generally remain low.

Random assignment of subjects to treatments in classical randomised controlled trials is essential in order to obtain comparable groups with respect to explanatory and confounding variables. Correspondingly, random assignment of treatments to treatment periods is essential in N-of-1 trials. Once the number of treatment periods has been determined there are a number of ways of randomising the treatments to periods. The most recommended design when comparing two treatments (as is most commonly the case) is random allocation within pairs of treatment periods. For example, for the comparison of treatment A versus treatment B during eight treatment periods, the following RANDOMISATION schedule might be generated: AB AB BA AB. This approach avoids the possibility of several consecutive treatment periods with the same treatment.

In terms of the analyses of an N-of-1 trial an important first approach is to plot the data and examine the results visually. The more theoretical methods depend heavily on the type of randomisation used. When the paired design is employed, the simplest approach may be to perform a SIGN TEST, which examines the LIKELIHOOD of the individual preferring the same treatment within each pair of treatment periods. However, this does not assess the strength of the treatment effect, only the direction of it. A more powerful alternative is the STUDENT'S *t*-TEST (either paired or unpaired depending on randomisation). For such analyses the paired design is again preferable since it goes some way to reduce the impact of AUTOCORRELATION (i.e. the assumption of such a statistical test that observations from one treatment period to the next will be independent). Recording several measurements within each treatment period and comparing averages across the periods can reduce this problem further. Parametric tests also make the assumption of normality and nonparametric tests may alternatively be used. In addition, BAYESIAN METHODS are available (Zucker *et al.*, 1977) for combining information from a series of N-of-1 trials. When an individual's N-of-1 trial has been completed the results will be summarised and disseminated during a feedback session between the clinician and patient to inform future treatment. *SB*

Guyatt, G., Sackett, D., Adachi, J., Roberts, R., Chong, J., Rosenbloom, D. *et al.* 1988: A clinician's guide for conducting randomized trials in individual patients. *Canadian Medical Association Journal* 139, 497–503. **Guyatt, G. H., Haynes, R. B., Jaeschke, R. Z., Cook, D. J., Green, L., Naylor, D. *et al.* for the Evidence-Based Medicine Working Group** 2000: User's guides to the medical literature XXV. Evidence-based medicine: principles for applying the users' guides to patient care. *Journal of the American Medical Association* 284, 1290–6. **March, L., Irwig, L., Schwarz, J., Simpson, J., Chock, C. and Brooks, P.** 1994: N of 1 trials comparing a non-steroidal anti-inflammatory drug with paracetamol in osteoarthritis. *British Medical Journal* 309, 1041–5. **Zucker, D. R., Schmid, C. H., McIntosh, M. W., D'Agostino, R. B., Selker, H. P. and Lau, J.** 1997: Combining single patient (N-of-1) trials to estimate population treatment effects and to evaluate individual patient responses to treatment. *Journal of Clinical Epidemiology* 50, 401–10.

noncompliance See ADJUSTMENT FOR NONCOMPLIANCE IN RCTs

nonignorable dropout See DROPOUTS, MISSING DATA

noninferiority See ACTIVE CONTROL EQUIVALENCE STUDIES

noninformative censoring See CENSORED OBSERVATIONS, SURVIVAL ANALYSIS

noninformative censoring/dropout See CENSORED OBSERVATIONS, DROPOUTS, MISSING DATA

nonparametric methods – an overview These are inferential methods used when the assumptions of parametric methods are violated or the sample size is small, i.e. fewer than 25–30 in each group. Nonparametric methods do not assume that data are normally distributed as parametric methods do, although they usually have their own assumptions. Several situations suggest the use of nonparametric methods, including: when the independent and/or dependent variables are nominal in measurement; when the data are ordered with many ties; when the data are rank ordered; when there is a small sample size or unequal groups; when the dependent variable has a distribution other than a NORMAL DISTRIBUTION; when the groups have unequal variances; when there are unequal pairwise CORRELATIONS across repeated measurements; when the data has notable OUTLIERS.

Nonparametric methods have common characteristics, including: independence of observations; few assumptions;

dependent variable may be categorical; focus on rank ordering or frequencies; hypotheses in terms of rank, MEDIANS or frequencies; sample sizes are less stringent.

Most parametric methods have at least one nonparametric alternative. Some may have several and which to choose depends on which assumptions are met and what is to be shown with the data.

There are two nonparametric correlation coefficients; they are nonparametric versions of PEARSON'S CORRELATION COEFFICIENT. These are SPEARMAN'S RANK CORRELATION COEFFICIENT, also known as SPEARMAN'S RHO (ϱ) coefficient, and KENDALL'S TAU (τ) coefficient. There are several different versions of Kendall's τ including a, b and c. Spearman's rank is the Pearson correlation calculated on the ranks of the data rather than the raw data. It is therefore often preferred due to its similarity to Pearson; in fact, if the data are normally distributed both coefficients give numerically similar answers. Spearman's rank is difficult to interpret if there are many ties in the data, so in this situation Kendall's τ is often preferred. Kendall's τ can be extended to give a PARTIAL CORRELATION COEFFICIENT; this finds the correlation between variables while controlling for the effects of a third variable.

Nonparametric methods can also be used to analyse CONTINGENCY TABLES. The most common of these methods is the CHI-SQUARE TEST of independence, which is used if both variables in the contingency table are nominal. If the assumptions of the chi-square test are not met then FISHER'S EXACT TEST or its extension, the Fisher–Freeman–Houlton test, can be used instead. Both the chi-square and Fisher's exact tests can only be used if the groups are independent. If there is an ordering in the data within a contingency table then a test for trend may be more powerful than a test of association. In a $2 \times c$ table, the chi-square test for trend can be used to look for a trend in proportions between the two groups. In an $r \times c$ table where both variables are ordered, then linear-by-linear ASSOCIATION, also known as the Mantel–Haenszel test for trend, can be used (see MANTEL–HAENSZEL METHODS).

MCNEMAR'S TEST is used to analyse binary data in two groups where the groups are paired. These can be data before and after some event or matched pairs. McNemar's test takes two different forms depending on the sample size. There are two extensions to McNemar's test; these are the COCHRAN Q-TEST, which is for more than two time points or more than two groups with a binary outcome, and the Stuart–Maxwell test, which is used when there are two paired groups with more than two outcomes. If agreement rather than association is of interest then the kappa coefficient can be used. Kappa measures agreement while adjusting for chance agreement. There are three forms of kappa, simple or Cohen's kappa for agreement between two raters rating on the same scale, weighted kappa, which takes into account the degree of disagreement and multirater kappa, which allows for more than two raters (see KAPPA AND WEIGHTED KAPPA).

Contingency tables can also be analysed using contingency coefficients. The phi coefficient is used to give strength to the association found in a significant chi-square test of association in a 2×2 table. Cramer's V, also called Cramer's C, is the extension of the phi coefficient to an $r \times c$ table and should only be used when the chi-square test has already proved to be significant. If both variables in the contingency table are ordered, the gamma statistic, G, can be used to measure the strength of association. If there is a special distinction between the ordered variables, e.g. one is the dependent and one is the independent, Somer's D can be used: this is asymmetric and gives a different answer if the variables are interchanged.

Goodness-of-fit tests are usually nonparametric. These tests are used to see if a sample distribution is similar to a pre-specified distribution or not. The binomial test is used for dichotomous data, i.e. data that can only take two outcomes. It sees whether such an extreme split into the two groups is likely to have occurred by chance or not. If there are more than two outcomes then the chi-square goodness-of-fit test could be used instead. The KOLMOGOROV–SMIRNOV TEST takes two different forms; the first is the two-sample test. This test compares the distribution of two different groups to see if they are similar or not. This is done by seeing if the largest difference between the distributions of the two groups could have occurred by *chance alone*. The one-sample test compares the observed data to a theoretical distribution to see if the largest differences between the two distributions could have occurred by chance. It is often used to test if data are normally distributed enough to use parametric tests. However, it should only be used if the parameters of the distributions can be specified in advance. If this is not the case then the Lilliefors test, which is similar to the Kolmogorov–Smirnov test but allows the MEAN and STANDARD DEVIATION to be estimated from the data rather than being specified in advance, should be used. The Shapiro–Wilks test is also similar to the Kolmogorov–Smirnov test and is usually used to compare to a normal or EXPONENTIAL DISTRIBUTION. The runs test sees whether the order of occurrence of two values of a variable is random. A run is a sequence of like observations. If a sample contains too many or too few runs then that sample may not be random.

Nonparametric tests for two samples are nonparametric versions of the t-test. One of these tests is the MANN–WHITNEY RANK SUM TEST, which tests for a difference in spread and location or medians between two independent groups. The hypothesis tested depends on whether the assumption of similarity of distributions is met or not. The SIGN TEST tests for a difference in medians between two paired groups but is less powerful than the WILCOXON SIGNED RANK TEST for doing the same. The Wilcoxon signed rank test can only be used if there is a similarity of difference scores about the true median difference.

There are many nonparametric versions of ANALYSIS OF VARIANCE; the most commonly used is the KRUSKAL–WALLIS TEST, which is the extension of the Mann–Whitney rank sum test to more than two groups. The MEDIAN TEST can be used when the assumptions of the Kruskal–Wallis test are violated; it compares medians between k groups. The JONCKHEERE–TERPSTRA TEST is used when the independent variable is ordinal and looks for an increase in medians rather than a difference in medians, the order of the groups having been specified a priori. If there are repeated measures in the independent groups then the Scheirer–Ray–Hare test can be used; this is an extension of the Kruskal–Wallis test to allow for repeated measures within independent groups. The FRIEDMAN TEST is the nonparametric version of repeated measures analysis of variance. It is used for multiple paired samples. There is a multivariate extension of the Friedman test, the Quade test, which is similar to the Friedman test but takes account of the range of the data within a block. If the independent variable is ordinal then the Page test for ordered alternatives is more powerful than the Friedman test; it tests for an increase in medians rather than a difference in medians but the order of the groups must be specified a priori.

As the size of the samples used in nonparametric methods increase, the test statistics tend towards a normal or CHI-SQUARE DISTRIBUTION. Therefore, when the sample size is large enough a normal or chi-square approximation to the test statistic is used to calculate the P-VALUE. It is important to report this asymptotic P-value only when the sample size is large enough. If the sample size is small, the exact P-value should be quoted instead since it is more appropriate and more accurate than the asymptotic one. For further details see Pett (1997), Siegel and Castellan (1998) and Conover (1999). *SLV*

Conover, W. J. 1999: *Practical nonparametric statistics*, 3rd edition. Chichester: John Wiley & Sons, Ltd. **Pett, M. A.** 1997: *Nonparametric statistics for health care research*. Thousand Oaks: Sage. **Siegel, S. and Castellan, N. J.** 1998: *Nonparametric statistics for the behavioral sciences*, 2nd edition. New York: McGraw-Hill.

nonresponse bias This occurs in all types of study when there is a systematic difference between the characteristics of those who choose to participate and those who do not.

In surveys, it is common practice to select a representative sample from the target population and collect data by means of a QUESTIONNAIRE. Some of the sample will not respond, either because they cannot be contacted or because they do not wish to participate. If the nonresponders differ in a systematic way in terms of characteristics (such as age, sex or deprivation) that are related to the response variable(s), biased estimates will result. If basic demographic information is available for the entire sample, including the nonrespondents, this information may be of use in adjusting estimates to account for the BIAS. However, the best approach is to maximise the response rate by using techniques such as incentives, reminders or enclosing stamped addressed envelopes. Edwards *et al.* (2002) summarise the evidence for a range of these techniques.

Nonresponse bias can also be a problem in comparative studies. For example, the classic Salk Polio Vaccine Trial of 1954 was conducted as a randomised, double-blind, placebo-controlled trial (see CLINICAL TRIALS) in some health departments in the USA (Tanur *et al.*, 1989; Meldrum, 1998). Parents of almost 750 000 children aged 6–7 years were asked for permission to include their child in the randomised trial but 45% refused. Those who consented were randomly allocated to Salk vaccine or PLACEBO. All children, including those who refused inoculation, were followed up for one year. The INCIDENCE rates of polio in the following year were 28 per 100 000 in the vaccinated group, 71 per 100 000 in the placebo group and 46 per 100 000 in those who refused. Although other more subtle biases may be present, the large statistically significant difference in polio incidence between those randomised to placebo and those who refused is largely due to nonresponse bias. Those who refused were more likely to be from deprived households who were known to have a *lower* incidence of polio in children aged 6–7 years. In this randomised trial, a valid comparison between vaccine and placebo among volunteers is obtained, giving convincing evidence of the effect of the vaccine. The nonresponse bias here is an interesting side issue, but a nonresponse bias of this magnitude in an observational study would be a serious problem.

Nonresponse can also occur among individuals lost from observation in LONGITUDINAL STUDIES. Any resulting bias is known as withdrawal bias or loss to follow-up bias (see DROPOUTS). *WHG*

[See also BIAS, BIAS IN OBSERVATIONAL STUDIES]

Edwards, P., Roberts, I., Clarke, M. *et al*. 2002: Increasing response rates to postal questionnaires: systematic review. *British Medical Journal* 324, 1183–91. **Meldrum, M.** 1998: 'A calculated risk': the Salk Polio Vaccine Trial of 1954. *British Medical Journal* 317, 1233–6. **Tanur, J. M. *et al*.** 1989: *Statistics: a guide to the unknown*, 3rd edition. Pacific Grove, CA: Wadsworth and Brooks/Cole.

normal distribution Probably the most important of the PROBABILITY DISTRIBUTIONS, the normal distribution takes the form of the familiar, symmetric, unimodal, bell-shaped curve, as illustrated in the figure on page 316. Indeed, it is often referred to as the *bell-shaped distribution* or *Gaussian distribution*. The normal distribution has a number of properties that are appealing.

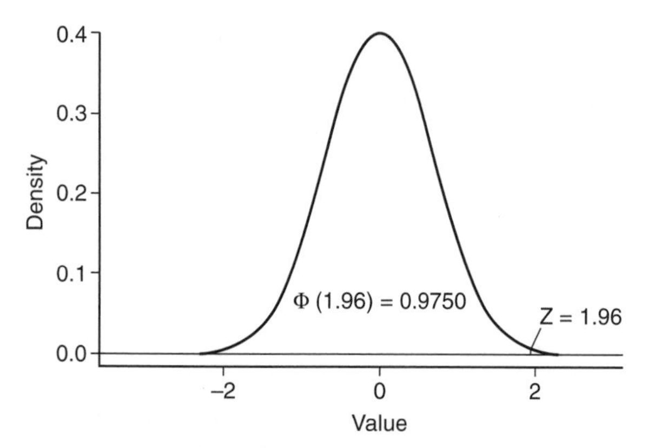

normal distribution *Illustrating the form of the normal (or Gaussian) distribution. Illustrated here is the standard normal distribution with mean zero and variance of one. Marked is the line corresponding to a value for the z-score of 1.96. As can be seen, 97.5% of the distribution lies to the left of 1.96 and 2.5% of the distribution is associated with larger values. The probability of seeing a value greater in magnitude than 1.96 is 2×2.5% as the distribution is symmetric about zero*

First and foremost, there is a mathematical result called the CENTRAL LIMIT THEOREM, which, broadly speaking, tells us that if a sample is taken from a single population and all observations in that sample are independent then the sample MEAN will be approximately normally distributed, with the approximation improving with the size of the sample. It is also the distribution that leads to a statistical model for a regression resulting in the same parameter estimates as a least squares regression, a property famously exploited by Gauss (hence Gaussian).

The normal distribution is related to many others. The F-DISTRIBUTION, CHI-SQUARE DISTRIBUTION, t-DISTRIBUTION and LOGNORMAL DISTRIBUTION can all be derived from it. It is what is known as a limiting distribution for, and thus can be a good approximation to, the BETA and GAMMA DISTRIBUTIONS and even for discrete distributions such as the BINOMIAL and POISSON DISTRIBUTIONS. For further details on how the normal distribution relates to other distributions see Leemis (1986).

The normal distribution is defined by two parameters: the mean, μ, and the STANDARD DEVIATION, σ. The density function for a variable X following a normal distribution with parameters μ and σ is:

$$f(x) = \frac{1}{\sigma\sqrt{2\pi}}\exp\left[-\frac{(x-\mu)^2}{2\sigma^2}\right]$$

By transforming the data X to create $Z = (X-\mu)/\sigma$ one can reduce the data to that with the standard normal density function with mean zero and standard deviation of one:

$$f(z) = \frac{1}{\sqrt{2\pi}}\exp\left[-\frac{z^2}{2}\right]$$

Performing such a transformation is sometimes referred to as calculating the z-SCORE of the data.

Since it is symmetric and unimodal, the median and the mode of the distribution equal the mean (μ). The distribution can be recentred by changing μ or rescaled by changing σ, but the shape of the distribution remains unchanged. Because of this, properties of the shape such as SKEWNESS and KURTOSIS are constant for all normal distributions. There is an added benefit that 95% of the density will lie within 1.96 (or approximately two) standard deviations of the mean. This result is invaluable in calculating confidence intervals or performing tests on the mean.

Despite the name, it is not always normal to see a normal distribution! Indeed, there are times when it would be a positively *ab*normal distribution. However, a number of statistical procedures such as ANALYSIS OF VARIANCE (and its variants) and the STUDENT'S t-TEST do rely on properties of the normal distribution. Other techniques and procedures that might make assumptions of normality include PEARSON'S CORRELATION COEFFICIENT, LINEAR REGRESSION, PRINCIPAL COMPONENT ANALYSIS and FACTOR ANALYSIS. This leaves the problem of testing to see if data or residuals are normally distributed.

The normal distribution always stretches from minus infinity to infinity and, as such, while it can often provide an adequate approximation to a bounded distribution, if the data have to be positive, or between 0 and 1, or are constrained by the inclusion/exclusion criteria of a trial, then this might be cause for alarm. Essentially, if there are no observations near the boundary, i.e. the density of the distribution at the boundary is negligible, the approximation may well be fine. For further information on the normal distribution see Altman and Bland (1995) and Armitage and Colton (1998).

There are a number of tests available to seek evidence of nonnormality. One can test the skewness and kurtosis of the sample, but this is not generally advisable. One can conduct the KOLMOGOROV–SMIRNOV TEST, but this can be oversensitive, as can the Shapiro–Wilk W test, another popular alternative. A graphical assessment of the distribution, either through HISTOGRAMS or QUANTILE–QUANTILE (Q–Q) PLOTS, will often suffice.

If one decides that a sample of data is not normally distributed, one has the choice of using a method that makes no assumption of normality (e.g. a NONPARAMETRIC METHOD) or performing a TRANSFORMATION of the data so that they approximate normality. For example, taking the logarithm of ratios often leaves them acceptably approximated by the normal distribution. This is the most common way of estimating confidence intervals for ODDS RATIOS and relative risks (see RELATIVE RISK AND ODDS RATIO).

It is not often that one is required to perform a normal test (as opposed to a test for normality), as this requires knowledge of the standard deviation of the distribution. If the standard deviation is to be estimated from the sample, then a t-test should be performed. There are, however, tables of probabilities associated with z-scores in many texts Lindley and Scott, 1984). These are normally the probabilities of a standard normal variable taking a value less than the z-score, a function (the distribution function) usually denoted by the upper case Greek letter phi Φ. If the z-score is positive, then the PROBABILITY of observing a score as large in magnitude is $2(1-\Phi(z))$. Alternatively, if the z-score is negative, the probability is $2\Phi(z)$. When the z-score is zero, it is taking the mean value and so $\Phi(0)=0.5$ as the distribution is symmetric about zero.

As an example of its use, Kanis (2002) models the density of bone minerals as a normal distribution and by so doing is able to calculate the effect of several variables on fracture risks. *AGL*

[See also MULTIVARIATE NORMAL DISTRIBUTION]

Altman, D. G. and Bland, M. J. 1995: The normal distribution. *British Medical Journal* 310, 298. **Armitage, P. and Colton, T.** (eds) 1998: *Encyclopaedia of biostatistics*. Chichester: John Wiley & Sons, Ltd. **Kanis, J. A.** 2002: Diagnosis of osteoporosis and assessment of fracture risk. *The Lancet* 359, 1929–36. **Leemis, L. M.** 1986: Relationships among common univariate distributions. *The American Statistician* 40, 2, 143–6. **Lindley, D. V. and Scott, W. F.** 1984: *New Cambridge elementary statistical tables*. Cambridge: Cambridge University Press.

normal probability plot See PROBABILITY PLOT

nQuery Advisor This is a software package useful for determining sample sizes when planning research studies. Details are available from Statistical Solutions Ltd, 8 South Bank, Crosse's Green, Cork, Ireland, www.statsol.ie/nquery/nquery.htm. *BSE*

nuisance parameter This is a parameter of a model in which there is little or no scientific interest but whose presence is needed to make valid inferences and estimates of the parameters that are of real interest. An example of a nuisance parameter is the VARIANCE of the random effect terms in a RANDOM INTERCEPT MODEL (see MULTILEVEL MODELS). *BSE*

null hypothesis See HYPOTHESIS TESTS

number needed to treat (NNT) Often a useful way to report the results of a randomised clinical trial, the NNT is the estimated number of patients who need to be treated with the new treatment rather than the standard treatment for one additional patient to benefit. The NNT is calculated as one divided by the absolute risk reduction (ARR), where the latter is simply the absolute value of the difference between the control group event rate and the experimental group event rate. The concept of NNT can equally well be applied to harmful outcomes as well as benefical ones, when instead it becomes the number needed to harm (NNH).

For example, in a study into the effective use of intensive diabetes therapy on the development and progression of neuropathy, 9.6% of patients randomised to usual care and 2.8% of patients randomised to intensive therapy suffered from neuropathy. Consequently,

$$\text{ARR} = |9.6\% - 2.8\%| = 6.8\%$$

leading to

$$\text{NNT} = 1/\text{ARR} = 1/6.8\% = 14.7$$

which is rounded up to 15. This means 15 diabetic patients need to be treated with intensive therapy to prevent one from developing neuropathy. (This example is given on the website of the Centre for Evidence Based Medicine.)

Altman (1998) shows how to calculate a confidence interval for NNT, although this is not considered helpful if the 95% confidence interval for ARR includes the value zero, as this gives rise to a nonfinite CONFIDENCE INTERVAL for NNT. Walter (2001) illustrates some statistical properties of NNT and similar measures, with examples drawn from different types of study design. While there have been a few critics of NNT as an index, most medical statisticians, including Altman and Deeks (2000) defend the concept as a useful communication tool when presenting results from clinical studies. Bandolier, an Oxford-based, independent research group promoting evidence-based medicine, maintain a useful website with further information on NNTs and their applications, www.jr2.ox.ac.uk/bandolier/booth/booths/NNTs.html. *BSE/CRP*

Altman, D.G. 1998: Confidence intervals for the number needed to treat. *British Medical Journal* 317, 1309–12. **Altman, D. G. and Deeks, J. J.** 2000: Comment on the paper by Hutton. *Journal of the Royal Statistical Society, Series A* 163, 415–16. **Walter, S. D.** 2001: Number needed to treat (NNT): estimation of a measure of clinical benefit. *Statistics in Medicine* 20, 3947–62.

number needed to harm (NNH) See NUMBER NEEDED TO TREAT (NNT)

O

observational studies There are situations where medical research has to be conducted using study designs that do not involve RANDOMISATION. These observational studies include a range of different study types, four main types of which are described and illustrated briefly below: CASE-CONTROL, COHORT, CROSS-SECTIONAL and ECOLOGICAL STUDIES (see EPIDEMIOLOGY). Case reports (or case series), which involve only studying patients with specific diagnoses, are sometimes included under this heading. While providing valuable information about characteristics of patients, from the scientific perspective they are limited and, therefore, are considered no further.

Interest in most observational studies typically focuses on studying the relationship between disease and exposure. Exposure data are collected for those diagnosed with a disease and also for those who are disease free. In the absence of randomisation, observational studies are particularly prone to problems associated with confounding, and this needs to be considered at the design and analysis stage of such studies. SAMPLE SIZE DETERMINATION IN OBSERVATIONAL STUDIES are described elsewhere.

In *ecological studies*, the unit of analysis is a group of individuals, or 'community', where each community provides its measure of disease outcome and exposure. Examining the strength of ASSOCIATION between disease and exposure in these studies is usefully examined graphically using a scatter diagram (see SCATTERPLOT). A study examining the association between incidence of squamous-cell carcinoma of the eye and ambient levels of solar ultraviolet is one such example (Newton *et al.*, 1996). When this suggests a linear relationship between the disease and exposure a regression analysis can be informative. In this example, the incidence of squamous-cell carcinoma was found to decrease by 29 % per unit reduction in ultraviolet exposure, a finding that was highly statistically significant ($p < 0.0001$).

Associations observed between disease and exposures in ecological studies may not necessarily reflect the pattern seen when individuals are the unit of analysis. *Cross-sectional studies*, in contrast, provide the opportunity to study data on disease and exposure on individuals at a particular point in time. In a cross-sectional study of elderly women, for example, interest focused on the prevalence of falls (Lawlor, Patel and Ebrahim, 2003). Such disease measures can be examined in relation to one or more exposures also recorded at that time. The effects of chronic diseases and drug use on the prevalence of falls were of interest and data on these factors were also recorded at that time. The findings from this study illustrate the importance of considering confounding factors. While the prevalence of falls increased with increasing numbers of simultaneously occurring chronic diseases, no such relation was found with the number of drugs used after adjusting for such factors (Lawlor, Patel and Ebrahim, 2003).

Case-control and *cohort studies* both offer the important advantage of measuring disease and exposure at different points in time. Case-control studies proceed by identifying a representative group of individuals diagnosed with a certain condition ('cases') and a representative group of individuals who are disease free ('controls') (see the first figure on page 320). Information regarding specific exposures of interest is collected and compared in cases and controls using odds ratios (see RELATIVE RISK AND ODDS RATIO), which are estimated by dividing the odds of exposure in cases by the odds of exposure in controls. Case-control studies have particular appeal for studying rare diseases and it is noteworthy that this was the study design first used to demonstrate the link between smoking and lung cancer (Doll and Hill, 1950). They do, however, rely on individuals being able to recall their past exposure histories accurately: RECALL BIAS is a particular problem in case-control studies.

Confounding in case-control studies can be taken into account by matching cases and controls at the study design, but allowance must be made for this in subsequent data analyses. The United Kingdom Childhood Cancer Study provides such an example, where interest focused on the relationship between neonatal exposure to vitamin K and childhood cancer (Fear *et al.*, 2003). Here, cases were children diagnosed with cancer and controls were those without cancer. One measure of 'exposure' was whether neonatal vitamin K had been received orally or by the intramuscular (im) route. Cases and controls were matched on sex, month and year of birth, and region of residence at diagnosis. The odds ratio for cancer in children who had received vitamin K orally compared to those who received it by the im route was 1.09 (95 %, CI 0.94 to 1.61) after adjusting for the matching and other confounding factors. This study did not, therefore, provide support for an association between exposure of neonates to vitamin K and subsequent risk of childhood cancer.

When ordered according to the strength of scientific evidence that observational studies have the potential to provide, cohort studies hold the top position. In its simplest form, this involves identifying disease-free individuals, who are classified according to an exposure at a particular point in

Encyclopaedic Companion to Medical Statistics: Second Edition Edited by Brian S. Everitt and Christopher R. Palmer

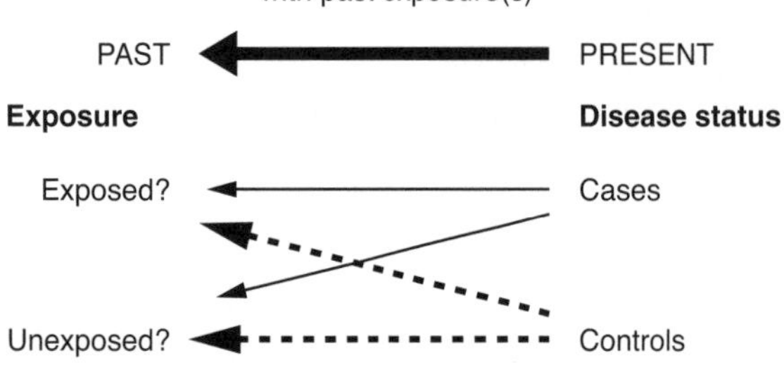

observational studies *Principles of a case-control study design*

time and followed up to determine which of them develops the disease of interest (see the second figure). In contrast to case-control studies that rely solely on ODDS RATIOS, cohort studies provide the opportunity to measure disease in many different ways using absolute (e.g. risks, rates or odds) or relative (risk ratios, rate ratios or odds ratios) measures. Cohort studies also have the important advantage of collecting exposure data prior to disease occurrence, i.e. avoiding recall bias, and studying a range of different health outcomes. Such studies, however, have the disadvantages of requiring larger study size and taking much longer to conduct, particularly for rare diseases.

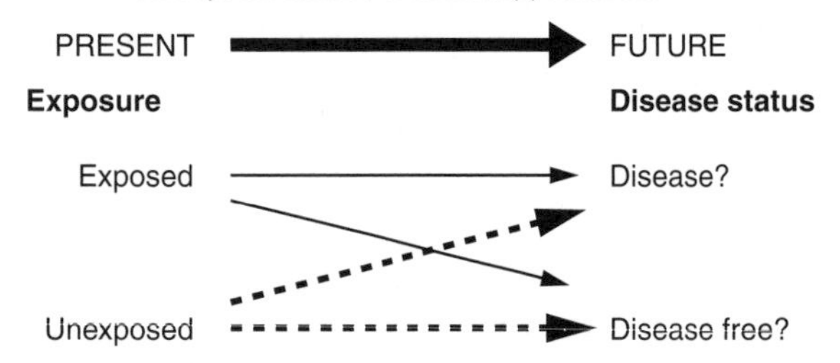

observational studies *Principles of a cohort study design*

Studying disease rates in a specific cohort often requires identifying a separate comparison cohort. A cohort study of cancer incidence in 51 721 UK service personnel deployed in the 1991 Gulf War, for example, also involved assembling a cohort of 50 755 active service personnel who were not deployed in that war (era cohort) (Macfarlane *et al.*, 2003). In order to take account of confounding, members of the Gulf War veterans, cohort members were matched to members of the era cohort according to age, sex, rank, service and level of fitness. The main outcome measure here was the INCIDENCE rate ratio (IRR). After adjusting for confounding factors, this study suggested no difference in cancer rates between these two cohorts (IRR = 0.99, 95%, CI 0.83 to 1.17). When the outcome studied was other symptoms of ill-health, however, Gulf War veterans were found to have an excess of illness at follow-up (Hotopf *et al.*, 2003). *LMC*

Doll, R. and Hill, A. B. 1950: Smoking and carcinoma of the lung. Preliminary report. *British Medical Journal* ii, 739–48. **Fear, N. T., Roman, E., Ansell, P., Simpson, J., Day, N. and Eden, O. B.** 2003: Vitamin K and childhood cancer: a report from the United Kingdom Childhood Cancer Study. *British Journal of Cancer* 89, 1228–31. **Hotopf, M., David, A. S., Hull, L., Nikalaou, V., Unwin, C. and Wessely, S.** 2003: Gulf war illness – better, worse, or just the same? A cohort study. *British Medical Journal* 327, 1370–2. **Lawlor, D. A., Patel, R. and Ebrahim, S.** 2003: Association between falls in elderly women and chronic diseases and drug use: cross-sectional survey. *British Medical Journal* 327, 712–17. **Macfarlane, G. J., Biggs, A.-M., Maconochie, N., Hotopf, M., Doyle, P. and Lunt, M.** 2003: Incidence of cancer among UK Gulf war veterans: cohort study. *British Medical Journal* 327, 1373–7. **Newton, R., Ferlay, J., Reeves, G., Beral, V. and Parkin, D. M.** 1996: Effect of ambient solar ultraviolet radiation on incidence of squamous-cell carcinoma of the eye. *The Lancet* 347, 1450–1.

odds ratio See RELATIVE RISK AND ODDS RATIO

one-sample *t*-test See STUDENT'S *t*-TEST

one-sided tests In hypothesis tests, we try to distinguish between chance variation in a dataset and a genuine effect. We do this by comparing the NULL HYPOTHESIS, which states that there is no difference between the populations in which the data arose, to the alternative hypothesis, which states that there *is* a difference. For a one-sided test, this alternative hypothesis specifies the direction of the difference, i.e. we wish to distinguish chance variation from a decrease or increase in comparison to the null hypothesis. The *P*-VALUE for a one-sided test is calculated by considering only one side of the test statistic's distribution.

One-sided tests are used in situations where a genuine difference can only occur in one pre-specified direction and any differences seen in the opposite direction are a result of mere chance. For example, in Völzke *et al.* (2002) only increases in cardiovascular risk were looked at, as a lower cardiovascular risk was not biologically plausible for the study group. One-sided tests are also useful in situations where we are only interested in differences in one direction. For example, when introducing a new cheaper and more convenient diagnostic test we might only be interested in whether it is less accurate than the current test.

Tests should never be one-sided unless there is strong evidence present prior to data collection, suggesting that any change from the null hypothesis must be in one (specified) direction only. Using one-sided tests makes it easier to reject the null hypothesis when the alternative is true; thus one-sided tests are attractive to those who define success as having a P-value less than the significance level. A one-sided test should not be used just because a difference in a particular direction is expected, as things do not always turn out as planned. More details can be found in Bland and Altman (1994). *MMB*

Bland, J. M. and Altman, D.G. 1994: Statistics notes: one- and two-sided tests of significance. *British Medical Journal* 309, 248. **Völzke, H., Engel, J., Kleine, V., Schwahn, C., Dahm, J. B., Eckel, L. and Rettig, R.** 2002: Angiotensin I-converting enzyme insertion/deletion polymorphism and cardiac mortality and morbidity after coronary artery bypass graft surgery. *Chest* 122, 31–6.

one-way analysis of variance See ANALYSIS OF VARIANCE (ANOVA)

ordered categorical data See ORDINAL DATA

ordinal data Data that have been collected from research studies generally fall into three main categories: (1) nominal, (2) interval and (3) ordinal. In the case of ordinal data, the most appropriate methods for analysis are those that take advantage of the ordering of the response categories. These methods are collectively termed ordinal regression models. In the literature, there are six main types of ordinal regression model, including the following: polytomous model, proportional odds model, unconstrained/constrained partial proportional odds model, adjacent category model, continuation ratio model and stereotype model.

Ordinal regression models have been included in the broader category of GENERALISED LINEAR MODELS and therefore consist of the usual components: a *random component*, which identifies the probability distribution of the response variable; a *systematic component*, which specifies a linear function of explanatory variables; and a *link*, which describes the functional relationship between the systematic component and the expected value of the random component. All ordinal regression models can be expressed as:

$$F(\pi) = \alpha_j + X_1\beta_{j1} + X_2\beta_{j2} + \cdots + X_p\beta_{jp} \qquad (1)$$

where $F(\pi)$ denotes the link function, also known as the *logit* or *log odds* and this function includes the 'cumulative' logits, the 'continuation ratio' logits, the 'adjacent category' logits and 'generalised' logits. The α_j and $\beta_{j1}, \ldots, \beta_{jp}$ are the parameters to be estimated based on the *j*th cut-point (this is the point at which the scale is dichotomised) and the $X_1, \ldots, X_p$ are the covariates measured on the subjects in the study. If we let $\Pr(Y = y_j)$ denote the PROBABILITY that a subject falls into the y_j category, then using equation (1) we can express the logit functions for various ordinal regression models. For the purpose of illustration, the five-point ordinal scale from the health status question on the SF-36 Health Survey (Ware and Gandek, 1998) is used. This question asks: 'In general would you say your health is "*excellent*", "*very good*", "*good*", "*fair*" or "*poor*"?'

The generalised logit or polytomous model is a straightforward extension of the LOGISTIC REGRESSION model for binary response and accommodates for multinomial responses (Agresti, 1984). It does not, however, take account of the ordering of the categories.

To indicate the form of the polytomous model, let X_1 and X_2 denote covariates of interest and y be the response measured on the ordinal scale of the health status question. Then taking the last category (i.e. '*poor*') as referent, the polytomous model is expressed as:

$$\log\left[\frac{\Pr(Y = y_j)}{\Pr(Y = y_4)}\right] = \alpha_j + X_1\beta j_1 + X_2\beta j_2, \quad j = 1, 2, 3, 4 \qquad (2)$$

There are four logits functions based on the cut-points '*excellent*' versus '*poor*', '*very good*' versus '*poor*'; '*good*' versus '*poor*' and '*fair*' versus '*poor*'. The logit functions are expressed in terms of the four cut-point-specific intercept parameters (α_j) and for each of the covariates X_1 and X_2, the four cut-point-specific regression coefficients are β_{j1} and β_{j2} respectively.

For a given covariate, say X_1, the parameters β_{j1} corresponds to the four log-odds of $(Y = y_j)$, relative to the referent category $(Y = y_4)$. Exponentiating the regression coefficients β_{j1} results in the cut-point-specific ODDS RATIOS comparing $(Y = y_j)$ versus $(Y = y_c)$ for a unit increase in the levels of X_1 having adjusted X_2.

The prime feature of the proportional odds model is that a single summary measure (in terms of an odds ratio) is used to summarise the relationship of the ordinal response and the covariates. The proportional odds model (sometimes known as the *cumulative logit model*) allows for the ordering of the response categories through the use of cumulative probabilities.

The proportional odds model was first introduced by Walker and Duncan (1967) and their model was based on cumulative probabilities. McCullagh (1980) considered their model in great detail and derived from it the *proportional odds model*. For this latter model, it is assumed that one can combine the cut-points of the response into a single model, in which the same slope parameter β is used for each logit. The proportional odds model fitted for the ordinal scale of the health status question would take on the form:

$$\log\left[\frac{\Pr(Y \leq y_j)}{\Pr(Y > y_j)}\right] = \alpha_j + X_1\beta_1 + X_2\beta_2, \quad j = 1, 2, 3, 4 \qquad (3)$$

Here the logits are based on the four cut-points: '*excellent*' versus ('*very good*', '*good*', '*fair*', '*poor*'); ('*excellent*', '*very good*') versus ('*good*', '*fair*', '*poor*'); ('*excellent*', '*very good*', '*good*') versus ('*fair*', '*poor*'); ('*excellent*', '*very good*', '*good*', '*fair*') versus '*poor*'. There are four intercept parameters $\{\alpha_j\}$ in this model. As j increases, these parameters increase, reflecting an increase in the logits, as additional probabilities are added into the numerator (i.e. $\alpha_1 \leq \alpha_2 \cdots \leq \alpha_4$). Also the β_1 and β_2 are the common slope parameters over the four cut-points for each covariate respectively.

There are two assumptions of the proportional odds model: (1) the existence of an underlying continuous variable – not all ordinal scales will have an underlying continuum (e.g. the total score); the proportional odds model can still be used in such circumstances, the only drawback being that the interpretation of the parameters becomes difficult; (2) homogeneity in the cut-point-specific regression parameters (known as the *proportional odds assumption*). Prior to fitting the proportional odds model, it is important that the assumption of proportionality be checked, either graphically or formally, for instance using the score test (Peterson and Harrell, 1990).

An appealing requirement for ordinal data is that the model should in some sense be invariant under a reversal of category order. This implies that the magnitude of the summary estimates does not depend on the direction employed in modelling the outcome, i.e. whether the cut-points are formed using increasing or decreasing levels of severity. However, the sign of the β parameter is changed and the $\{\alpha_j\}$ reverse sign and order.

The proportional odds model is also invariant under the collapsability of the response categories. Hence, if two adjacent response categories are pooled together and the cut-point removed, the estimates of β should remain essentially unchanged, although the $\{\alpha_j\}$ are affected.

For the covariate, say X_1, the parameter β_1 corresponds to the global log-odds over all the four cut-points. The exponential of the regression coefficients β_1, results in a single estimate of the odds ratios for a unit increase in the levels of X_1 having adjusted X_2.

The proportional odds model and the partial proportional odds models are collectively termed *cumulative logit models*.

In practice, it is often difficult to find data for which a proportional odds model is a plausible description. There is, therefore, a need for a model that permits partial proportional odds where some explanatory variables may meet the proportional odds assumption and others may not. Thus, the primary reason for the formulation of the 'partial proportional odds models' was to relax the stringent assumption of a constant odds ratio presented by the proportional odds model. The assumption that a constant slopes model holds, when in fact, for a given variable, a constant log-odds ratio is not representative of all the log-odds ratios over the cut-points, can lead to the formulation of an incorrect model.

The partial proportional odds models were initiated by the work of Peterson and Harrell (1990) and in general there are two types of partial proportional odds model: *the unconstrained partial proportional odds model*, for which no constraints are placed in the estimation of the parameters, and the *constrained partial proportional odds model*, for which a certain relationship may have been observed between the log-odds ratios and j, the point of dichotomisation. Such a relationship may be linear, for example, in which case a linear constraint is placed on the parameters of the model.

The cut-points that are used for the partial proportional odds models are the same as for the proportional odds model.

Assuming X_1 has proportional odds and X_2 does not have proportional odds, then the unconstrained partial proportional odds model for the ordinal scale of the health status question takes the form:

$$\log\left[\frac{\Pr(Y \leq y_j)}{\Pr(Y > y_j)}\right] = \alpha_j + X_1\beta_1 + X_2\beta_2 + T_2\gamma_{j2} \qquad (4)$$

$$j = 1,\ 2,\ 3, 4$$

Here the β_1 and β_2 are the regression coefficients associated with the two covariates of interest. The T_2 is the covariate which is a subset of the X_2 for which the proportional odds assumption either is not assumed or is to be tested and γ_{j2} are the regression coefficients associated with T_2, so that $T_2\gamma_{j2}$ is an increment associated only with the jth cumulative logit and $\gamma_{12} = 0$. If $\gamma_{j2} = 0$ for all j, then this model reduces to the proportional odds model. Thus a simultaneous test of the proportional odds model assumption is a test of the NULL HYPOTHESIS that $\gamma_{j2} = 0$ for all $j = 2, 3, 4$. Since $\gamma_{12} = 0$, the model uses only $\alpha_1 + X_2\beta j_2$ to estimate the odds ratio associated with the dichotomisation of the y-response categories into the first category versus the rest of the categories, where the estimation of the odds ratios associated with the remaining cumulative probabilities involve incrementing $\alpha_j + X_2\beta_2$ by $T_2\gamma_{j2}$.

Given that the relationship of a covariate and the response is represented with nonproportional odds, then for the individual cut-point-specific odds ratios, often a certain type of trend may be anticipated; e.g. a linear trend may be expected. In such a case, a constraint can be placed on the parameters in the model, so that the trend is taken into account. When the constraints are incorporated into the unconstrained partial proportional odds model, for the scale of the health status question, this model takes the form:

$$\log\left[\frac{\Pr(Y \leq y_j)}{\Pr(Y > y_j)}\right] = \alpha_j + X_1\beta_1 + X_2\beta_2 + T_2\gamma_2\Gamma_j \qquad (5)$$

$$j = 1,\ 2,\ 3,\ 4$$

Here the Γ_j are fixed pre-specified scalars and $\Gamma_1 = 0$. The new parameter γ_2 is not subscripted by j. Although γ_2 depends on j, it is multiplied by the fixed constant scalar Γ_j in the calculation of the jth cumulative logit.

The underlying assumptions of the partial proportional odds models are as for the proportional odds model.

Given covariate X_2, the test of whether a single γ_2 parameter fits the data as well as 3 (number of categories − 2) γ_{j2} parameters can be obtained by using the LIKELIHOOD RATIO test. Here we compare the log-likelihood of unconstrained and constrained models. This gives an approximate chi-square with 2 (number of categories − 2) − 1 DEGREES OF FREEDOM (DoF).

The adjacent category model utilises single-category probabilities rather than cumulative probabilities. Agresti (1984) states that when the response categories have a natural ordering, logit models should utilise that ordering. One can incorporate the ordering directly in the way we construct the logits. Like the proportional odds model the adjacent categories logit model implies stochastic orderings of the response distributions for different predictor values.

Agresti (1989) describes the adjacent category logistic model as modelling the ratio of the two probabilities $\Pr(Y=y_j)$ and $\Pr(Y=y_{j+1})$, $j=1, 2, 3, 4$. The cut-points that are used for the adjacent categories, given the scale of the health status question, would be: '*excellent*' versus '*very good*'; '*very good*' versus '*good*'; '*good*' versus '*fair*'; '*fair*' versus '*poor*'.

There are two types of adjacent category logit model. The constant slope adjacent category model has the following representation:

$$\log\left[\frac{\Pr(Y=y_j)}{\Pr(Y=y_{j+1})}\right] = \alpha_j + X_1\beta_1 + X_2\beta_2 \quad j=1, 2, 3, 4 \tag{6}$$

Manor, Matthews and Power (2000) described the adjacent category model in a slightly different way. Their version of the model is:

$$\log\left[\frac{\Pr(Y=y_j)}{\Pr(Y=y_{j+1})}\right] = \alpha_j + X_1\beta_{j1} + X_2\beta_{j2} \quad j=1, 2, 3, 4 \tag{7}$$

In model (6), for a given covariate, say X_1, the parameter β_1 corresponds to the log-odds of falling in categories '*excellent*' versus '*very good*'; '*very good*' versus '*good*'; '*good*' versus '*fair*'; '*fair*' versus '*poor*'. If the exponential is taken, this results in the global odds ratios for the adjacent categories for each unit increase in the levels of X_1. On a similar note, model (7) provides the cut-point-specific adjacent category odds ratios, and for the health status question scale there are four of these, for each unit increase in the levels of a given covariate.

Given an ordinal scale, where one is particularly interested in assessing the relative chance of a given rating, against all more favourable ones, then one would normally consider employing the *continuation ratio logits*. The continuation ratio model is best suited to circumstances in which the individual categories are of particular interest. It is well suited for failure time data and outcomes that measure threshold points, where individuals at a given level of an outcome must have passed through all previous levels of an outcome.

The logits for the continuation ratio model are based on the cut-points: '*excellent*' versus ('*very good*', '*good*', '*fair*', '*poor*'); '*very good*' versus ('*good*', '*fair*', '*poor*'); '*good*' versus ('*fair*', '*poor*'); '*fair*' versus '*poor*'. As for the adjacent category models, there are two versions of the continuation ratio model.

The form of the continuation ratio model was initially formulated by Feinberg (1980) and originated from survival time data. Various forms of the model exist; the most common is the forward formulation model and is written as:

$$\log\left[\frac{\Pr(Y=y_j)}{\Pr(Y>y_j)}\right] = \alpha_j + X_1\beta_1 + X_2\beta_2 \quad j=1, 2, 3, 4 \tag{8}$$

Model (8) is described by Cole and Ananth (2001) as a *fully constrained continuation ratio* model. It allows the cut-point-specific continuation ratios to be described by a single regression parameter (in a similar way to the proportional odds model). This model represents the probability of being in category j, conditional on being in a category greater than j. The intercept parameters are denoted by $\{\alpha_j\}$ and are the same as the cumulative logit model, but are not necessarily ordered for the continuation ratio model. Essentially, this model can be viewed as the ratio of the two conditional probabilities, $\Pr(Y=y_j/Y\,\varepsilon\,y_j)$ and $\Pr(Y>y_j/Y\,\varepsilon\,y_j)$, i.e. one models the odds of falling in category j as opposed to higher than category j, given that one has been in category j or higher.

By viewing the outcome as going from more to less severe, this model can be applied in reverse and forms the backward continuation ratios $\Pr(Y=y_i)/\Pr(Y<y_j)$. Because of the conditioning on adjacent cut-points, the continuation ratio, unlike the proportional odds, is affected by the direction chosen for the response variable, and the forward and backward ratios are not equivalent and yield different results. Thus, the continuation ratio model is not invariant under the reversal of categories unless Y is binary, in which case one has to be careful which continuation ratio model one uses.

Another form is the different slopes continuation ratio model and for this model the regression parameters are allowed to vary by the cut-point. This model is written as:

$$\log\left[\frac{\Pr(Y=y_j)}{\Pr(Y>y_j)}\right] = \alpha_j + X_1\beta_{j1} + X_2\beta_{j2} \quad j=1, 2, 3, 4 \tag{9}$$

In this case, the multinomial likelihood factors into a product of the binomial likelihoods for the separate logits. The continuation ratio model has the advantage that the $c-1$ logits produced when fitting the ratios in relation to the cut-points are asymptotically independent of each other. Thus, the estimation of the parameters in each of the $c-1$ logits can be carried out separately, using the method of maximum likelihood, and the summation of the individual chi-square statistics gives the overall goodness-of-fit statistics for the set of the logit models. In practice, the continuation ratio model can be fitted in any statistical package that includes binary logistic regression, after suitable restructuring of the data. As the fully constrained model is nested within the different slopes continuation ratio model, the difference in -2 log-likelihood (deviance) provides a test of the validity of the assumption that the threshold-specific continuation ratios are equal and is distributed as a χ^2 variate under the null with DoFs equal to the difference in the number of parameters between the nested models.

The odds ratios for the different slopes and fully constrained continuation ratio models can be described in a similar way to those of the adjacent category models, as described earlier.

The stereotype ordinal regression model was introduced by Anderson (1984) as part of a general model for discrete multivariate outcomes and also arises naturally in the context of truly discrete outcomes.

The stereotype model is a derivative of the polytomous logistic model (2). The polytomous model provides the best possible fit to the data, at the cost of a large number of parameters that can be difficult to interpret. The stereotype model aims to reduce the number of parameters by imposing constraints without reducing the adequacy of the model.

For the health status scale, the starting point for the stereotype model is to impose a structure on the β_{j1} and β_{j2} such that:

$$\beta_{j1} = -\phi_j\beta_1 \text{ and } \beta_{j2} = -\phi_j\beta_2, j = 1, 2, 3, 4 \quad (10)$$

Then model (2) becomes:

$$\log\left[\frac{\Pr(Y = y_j)}{\Pr(Y = y_4)}\right] = \alpha_j - \phi_j(X_1\beta_1 + X_2\beta_2) \quad (11)$$

$$j = 1, 2, 3, 4$$

There are certain features that are specifically relevant to the stereotype ordinal regression model and these include the *dimensionality* of the model, *distinguishable* y-response categories and the *ordering* of the y-response categories with respect to the covariates.

The dimensionality of the model for the y-response and the covariates is determined by the number of linear functions required to describe the relationship. If only one linear function is used to describe the relationship between the ordinal response and a set of predictors, model (11) is a *one-dimensional stereotype model*. One-dimensional relationships are much more common in the literature compared to the higher dimensions.

Having decided on the dimensionality of the model, there are questions about ordering and model simplification, perhaps using distinguishability as a criterion. The concept of *indistinguishability* is described when a given covariate, X_1, affects two response categories y_j and y_{j+1} in an identical manner (thus X_1 is not predictive between the two categories): we then say that these two categories are indistinguishable with respect to X_1. The hypothesis that $y = y_s$ is indistinguishable from $y = y_t$, with respect to the covariates, takes the form H_0: $\beta_s = \beta_t$. In the one dimension stereotype model, this is equivalent to asking whether there are differences among the ϕ_j (H_0: $\phi_s = \phi_t$).

In the regression models discussed so far (with the exception of the polytomous model), the regression parameters and consequently the logits are based on the ordering of the y-response categories. Therefore the proportional odds, continuation ratio and adjacent category models assess the ASSOCIATION of y-response and the covariates conditional of the order that the categories occur. In this case the ordering is 'inbuilt' and assumed a priori.

In many cases, one cannot be too certain about the relevance of the ordering of the response categories. The stereotype model is based on the polytomous model and therefore uses generalised logits. The polytomous model does not have the mechanism to account for the ordering of the y-response categories. Anderson (1984) took the latter model and assessed the relationship of the y-response and a given covariate. If the individual cut-point-specific regression parameters were ordered (leading to the stereotype model), then one could assume that an ordered nature existed in the response categories. This is quite different from the 'ordinality' aspect of the proportional odds, continuation ratio and adjacent category models. For these latter models, the ordered categories are accounted for through the formation of the logits. Therefore they are not necessarily ordered with respect to the covariates and in a sense it is not necessary to have any regressor variables. By contrast, in the stereotype model, the 'ordinality' only reveals itself through assessing the relationship of the y-response and covariates.

If ordering is appropriate, the model orders the β_j (in the polytomous model) instead of ordering the odds or the link function. The 'ordering' is more directly tied to the effects of the explanatory variables and becomes a testable statement. If the dimensionality is one, ordering of the odds ratios is easily verified. If $\beta_j > 0$ and the odds ratio form a decreasing sequence $e^{\phi 1\beta_j} \geq e^{\phi 2\beta_j} \geq e^{\phi 3\beta_j} \geq \ldots \geq e^{\phi 5\beta_j} \geq 1$, then:

$$\phi_1 = 1 \geq \phi_2 \geq \cdots \geq \phi_5 = 0 \quad (12)$$

Note that (12) is not strictly ordinal, as adjoining categories may be indistinguishable. If (12) is satisfied, then the effect of the covariates upon the first odds ratio is greater than its effect on the second and so on.

For the health status scale, model (11) has a standard multinomial intercept with four parameters for a response variable. It estimates three independent scale values of $\{\phi_j\}$ for the response factor and a single beta parameter for each independent variable. The larger the difference between any two ϕ_j values, the more the log odds between the outcomes is affected by the independent variables. The ϕ_1 parameters show how the independent variable, X_1, affects the log odds of higher versus lower scores, where 'higher' and 'lower' is defined by the ϕ_j scale.

Most of the ordinal regression models detailed here can be fitted in well-known statistical software packages. However, there are some models (e.g. the partial proportional odds models) that are not very well accommodated for, and in the literature these model cannot be easily fitted and are more computational intensive.

The goodness-of-fit for a c-category model is a natural extension of that for the two categories. However, methods to assess the residual, where there is lack-of-fit are underdeveloped and require further research.

Indices such as AKAIKE'S INFORMATION CRITERIA (AIC) are often used to compare different models from the same data. However, the use of such an index has been rarely cited and it would be worthwhile to exploit and evaluate such an index in the context of ordinal regression models. *RL*

Agresti, A. 1984: *Analysis of ordinal categorical data.* New York: John Wiley & Sons, Inc. **Agresti, A.** 1989: Tutorial on modelling ordered categorical response data. *Psychological Bulletin* 105, 290–301. **Anderson, J. A.** 1984: Regression and ordered categorical variables (with discussion). *Journal of the Royal Statistical Society, Series B* 46, 1–30. **Cole, S. and Ananth, C.** 2001: Regression models for unconstrained, partially or fully constrained continuation odds ratios. *International Epidemiological Association* 30, 1379–82. **Feinberg, B.** 1980: *Analysis of cross-classified data,* 2nd edition. Cambridge, MA: MIT Press. **Manor, O., Matthews, S. and Power, C.** 2000: Dichotomous or categorical response? Analysing self-rated health and lifetime social class. *International Journal of Epidemiology* 29, 149–57. **McCullagh, P.** 1980: Regression models for ordinal data (with discussion). *Journal of the Royal Statistical Society, Series B* 42, 109–42. **Peterson, B. and Harrell, F.** 1990: Partial proportional odds models for ordinal response variables. *Applied Statistics* 39, 205–17. **Walker, S. and Duncan, D.** 1967: Estimation of the probability of an event as a function of several independent variables. *Biometrika* 54, 167–79. **Ware, J. and Gandek, B.** 1998: Overview of the SF-36 health survey and the international quality of life assessment (IQOLA) project. *Journal of Clinical Epidemiology* 51, 11, 903–12.

ordinary least squares (OLS) See LEAST SQUARES ESTIMATION

outliers These are observations judged to be too far from their group average. Outliers may genuinely come from long-tailed distributions, but they may also be irrelevant or erroneous observations that need to be expunged from the datasets before analysis, with due precautions.

Outliers in the distribution of a single variable (*univariate* outliers) are defined in terms of the data spread (Ramsey and Schafer, 2002). The figure shows the BOXPLOT for some data with a MEDIAN value equal to 60, first quartile equal to 40 and third quartile equal to 80.

The INTERQUARTILE RANGE of these data (shown as the central box) is thus equal to 40. Outliers are defined as all observations that are more than 1.5 interquartile ranges away from the box. Some STATISTICAL PACKAGES use another convention of showing extreme values more than 3 interquartile ranges away from the box. In the figure, all values exceeding 140 ($= 80 + 1.5 \times 40$) are outliers (shown as circles).

When the data come from a NORMAL DISTRIBUTION, a test can be used to detect univariate outliers. *Grubbs' test statistic* is defined as the largest absolute deviation from the sample MEAN in units of the sample STANDARD DEVIATION: max $| y_i - m|/s$, where y_i is the ith observation, m is the sample mean and s the sample standard deviation. Critical values for this test statistic can be computed from the t-DISTRIBUTION with $N - 2$ DEGREES OF FREEDOM, where N is the sample size (Grubbs, 1969).

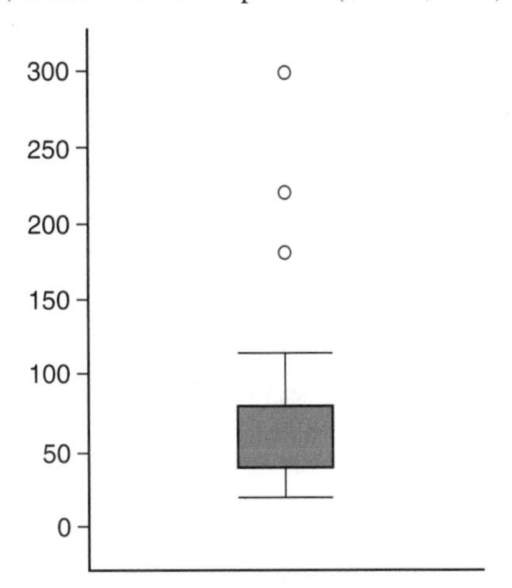

outliers *A boxplot showing three outliers*

Some statistical tests are resistant to outliers, but many are not. The simplest example arises from the calculation of a sample mean: the mean is very sensitive to outliers; in contrast, the median is resistant to outliers. Likewise, the comparison of two sample means through a t-TEST is sensitive to outliers; in contrast, a test that uses the ranks of the observations, rather than the observations themselves, is resistant to outliers (such a rank test is nonparametric and is also robust to deviations from normality).

Outliers in the distribution of several variables (multivariate outliers) are defined in terms of the distance of each observation to the multivariate mean. The Mahalanobis' distance is computed by standardising the variables of interest (subtracting the mean and dividing by the standard deviation) and summing the squares of these standardised variables. The sum approximately follows a CHI-SQUARE DISTRIBUTION with m degrees of freedom, if m variables are considered.

Multivariate outliers can exert undue influence on the analysis, particularly if multivariate regression is used. Case-influence statistics such as the leverage, Cook's distance or studentised residual, are useful to detect influential observations and to assess their impact on the results of the analyses (Ramsey and Schafer, 2002). *MB*

[See also STUDENT'S t-TEST]

Grubbs, F. 1969: Procedures for detecting outlying observations in samples. *Technometrics* 11, 1–21. **Ramsey, F. L. and Schafer, D. W.** 2002: *The statistical sleuth. A course in methods in data analysis*. Pacific Grove, CA: Duxbury.

overdispersion Overdispersion occurs whenever the outcome (or the response variable in a regression model) has a larger VARIANCE than that predicted by whatever model is being used. It is not usually a problem in regression models with a continuous outcome and a normally distributed error term, as the NORMAL DISTRIBUTION has separate parameters for the variance and MEAN. Overdispersion arises more commonly in the case of discrete variables – usually either count or binary variables. For the analysis of count variables, a usual assumption is that they follow the POISSON DISTRIBUTION. For binary variables, the BINOMIAL DISTRIBUTION is often assumed. In each case, the distribution has only one parameter, so the variance is determined by the mean. The Poisson distribution assumes that the mean and variance of the distribution are equal. The binomial distribution assumes that the variance is the mean multiplied by (1 – the probability of success). Overdispersion occurs when the actual variance seen is greater than that predicted by the Poisson or binomial distributions.

An example in ageing research is the analysis of activities of daily living (ADL) scores in a trial of a prehabilitation programme (Byers *et al.*, 2003). ADL were scored on a 16-point scale and had a positively skewed distribution with a mode of 0 and mean of 2.8. The variance was 16.4, indicating considerable overdispersion.

There are two common causes of overdispersion (Agresti, 1990). These are:

(a) positive correlation (rather than independence) between observations;
(b) the true sampling distribution, being a mixture of Poisson distributions.

The latter could be caused by heterogeneity among subjects. For example, suppose that the distribution of ADL for women was Poisson with a mean of 9 and for men was Poisson with a mean of 4. For a group of equal numbers of men and women, ADL would have a distribution with a mean of approximately 6.5 but a variance of approximately 13, thus showing overdispersion.

Overdispersion can be examined by comparing the variance of a set of observations to the predicted variance under an assumed distribution (as above). However, overdispersion can also be examined using regression models. For example, using a Poisson regression model, if there is no overdispersion and the model is correctly specified, the DEVIANCE of the model would be expected to equal the number of DEGREES OF FREEDOM (Lindsay *et al.*, 2002). If the deviance is larger than the number of degrees of freedom, this is traditionally taken to indicate overdispersion. For example, in the ADL example (above), the ratio of deviance to number of degrees of freedom was 4. Lindsay *et al.* recommend that a deviance greater than twice the number of degrees of freedom should be taken as an indication that overdispersion ought to be examined.

Overdispersion can result in underestimation of STANDARD ERRORS, if the overdispersion is not taken into account. Traditionally, variances have been adjusted by multiplying them by an inflation factor. This inflation factor (or 'heterogeneity factor') is equal to the deviance divided by the number of degrees of freedom. It will thus be greater than 1 if there is overdispersion. Inference is then based on the MAXIMUM LIKELIHOOD ESTIMATIONS obtained by the fitting of the Poisson model to the overdispersed data, but with multiplication of the standard errors by the square root of the inflation factor.

However, in the era of fast statistical programming, overdispersion can be investigated and taken into account using statistical models, rather than merely correcting the standard errors (Lindsay *et al.*, 2002). For count data, the standard alternative to the Poisson model is the negative binomial model. This model includes a disturbance or error term, i.e. it assumes that the mean varies randomly in the population. For binary data, the standard alternative to the binomial model is the beta-binomial, which assumes that the PROBABILITY has a BETA DISTRIBUTION. Inclusion of appropriate covariates in a simple Poisson or binomial model may also be a way of accounting for overdispersion.

Lindsay *et al.* (2002) examined the presence and effect of overdispersion in data on the annual number of blackgrouse and the effect of climate on this number. Using the Poisson distribution, there was evidence of overdispersion, with the model having a deviance of 29 on 18 degrees of freedom (inflation factor of 1.6). However, allowing for overdispersion using the negative binomial model gave parameter estimates and standard errors that were very close to those from the Poisson model. Another example has an inflation factor of 1.95, and yet the negative binomial model fitted no better than the Poisson model. In this case, from the model-fitting there was no evidence of overdispersion – and yet using the inflation factor would have multiplied the standard errors by 1.4. The recommendation from this paper is that the deviance can be used to indicate possible overdispersion, but this should then be investigated using model-based techniques, rather than merely inflating the standard errors. *KT*

Agresti, A. 1990: *Categorical data analysis*. New York: John Wiley and Sons, Inc. **Byers, A. L., Allore, H., Gill, T. M. and Peduzzi, P. N.** 2003: Application of negative binomial modeling for discrete outcomes: a case study in aging research. *Journal of Clinical Epidemiology* 56, 6, 559–64. **Lindsay, J., Laurin, D., Verreault, R., Hebert, R., Helliwell, B., Hill, G. B. *et al*.** 2002: Risk factors for Alzheimer's disease: a prospective analysis from the Canadian Study of Health and Aging. *American Journal of Epidemiology* 156,5, 445–53.

[See also QUASI-LIKELIHOOD]

overmatching See CASE-CONTROL STUDIES

P

paired *t*-test See STUDENT'S *t*-TEST

partial correlation coefficient See CORRELATION

partial likelihood This is a function, consisting of a product of conditional LIKELIHOODS, used in certain situations for estimation and hypothesis testing.

The most commonly used partial likelihood is that used in COX'S REGRESSION MODEL. In this model, it is assumed that the hazard of failure at time t for an individual with covariates $X = (X_1, \ldots, X_K)$ is $\lambda_0(t) f(X, \beta)$, where $f(X, \beta) = \exp\left(\sum_{k=1}^{K} \beta_k X_k\right)$, $\beta = (\beta_1, \ldots, \beta_K)$ are the hazard ratio parameters of interest and $\lambda_0(t)$ is the unknown baseline hazard function. No assumptions are made about the form of $\lambda_0(t)$.

In the full likelihood function for Cox's model, information about β is tied up with information about the nuisance function $\lambda_0(t)$. Cox's partial likelihood sacrifices some of the information about β contained in the data, in order to eliminate this dependence on $\lambda_0(t)$.

Let t_j denote the time of the jth failure and let R_j denote the risk set at time t_j, i.e. the set of individuals who, just before time t_j, remained not failed and not censored. Then, conditional on the risk set R_j and the fact that a failure took place at time t_j, the PROBABILITY that it was person i who failed is:

$$\frac{\lambda_0(t_j) f(X_i, \beta)}{\sum_{l \in R_j} \lambda_0(t_j) f(X_l, \beta)} = \frac{f(X_i, \beta)}{\sum_{l \in R_j} f(X_l, \beta)}$$

The partial likelihood is the product, over all the failure times, of these conditional probabilities. It uses not the actual failure times, but the ranks of the failure times. Note that it does not depend on $\lambda_0(t)$.

Intuitively, since no assumptions are being made about the form of the baseline hazard and since censoring (see CENSORED OBSERVATIONS) is assumed to be noninformative, information about the actual times of failure and the censoring events would not be expected to reveal much about the parameters of interest, β, and indeed, this turns out to be the case. Thus, conditioning on these leads to very little loss of information about β.

Less obviously, it also turns out that this partial likelihood function, although not a proper likelihood function, has similar statistical properties. Thus, it may be used to estimate β, to calculate a COVARIANCE MATRIX for β and for LIKELIHOOD RATIO hypothesis testing.

A more general definition of the partial likelihood can be found in, for example, Kalbfleisch and Prentice (2002). For further details see Clayton and Hills (1993) and Collett (2003). *SRS*

Clayton, D. and Hills, M. 1993: *Statistical models in epidemiology*. Oxford: Oxford Science Publications. **Collett, D.** 2003: *Modelling survival data in medical research*, 2nd edition. London: Chapman & Hall. **Kalbfleisch, J. D. and Prentice, R. L.** 2002: *The statistical analysis of failure time data*, 2nd edition. Chichester: John Wiley & Sons, Ltd.

path analysis This is a tool for evaluating the interrelationships among a set of observed (or latent) variables based on their correlational structure. The postulated relationships between the variables are often illustrated graphically by means of a path diagram, in which single-headed arrows indicate the direct influence of one variable on another and curved double-headed arrows indicate correlated variables. (For an example of such a diagram see CONFIRMATORY FACTOR ANALYSIS.) Originally introduced for simple regression models for observed variables, the method has now become the basis for more sophisticated procedures such as confirmatory factor analysis and use of structural equation models, involving both manifest and latent variables. *BSE*

path diagram See PATH ANALYSIS

pattern recognition See CLUSTER ANALYSIS IN MEDICINE, SUPPORT VECTOR MACHINES

Pearson's contingency coefficient See CONTINGENCY COEFFICIENT

Pearson's correlation coefficient See CORRELATION

per protocol This term is used to describe a subset of participants in a randomised clinical trial who complied with the protocol (see PROTOCOLS FOR CLINICAL TRIALS). It is also used to describe an analysis based only on these participants.

The per protocol dataset consists of those participants who complied with the protocol sufficiently to ensure that their data would be likely to show the effects of treatment. Aspects of compliance that could be considered include exposure to treatment and violations of the entry criteria. For example,

Encyclopaedic Companion to Medical Statistics: Second Edition Edited by Brian S. Everitt and Christopher R. Palmer

participants could be excluded if they took less than 80% of the prescribed treatment, if they used an additional treatment that could also affect outcome or if retrospective evaluation shows that they did not meet eligibility criteria. The rules for inclusion and exclusion of participants from the per protocol dataset should be carefully specified before treatment allocations are unblinded. Any decisions made after the data are unblinded should be regarded with suspicion.

The rationale for per protocol analysis is to estimate the efficacy of an intervention, undiluted by factors such as noncompliance, lack of eligibility and additional treatments. It is therefore most often used in explanatory trials that aim to measure the efficacy of an intervention under equalised conditions, rather than in pragmatic trials that aim to measure the effectiveness of an intervention policy in routine practice (Roland and Torgerson, 1998).

However, per protocol analysis is subject to SELECTION BIAS because some participants are excluded after RANDOMISATION. Selection bias may be less when the nature and number of exclusions is similar in different randomised groups, but BIAS can still occur in this situation. Therefore per protocol analysis should only be used when estimating efficacy is more important than avoiding selection bias. When a per protocol analysis is done, other analyses, such as INTENTION-TO-TREAT (ITT), should also be reported.

One specific situation where both per protocol and intention-to-treat analyses are routinely reported is in EQUIVALENCE STUDIES (ICH E9, 1999). ITT is anticonservative for equivalence trials because the inclusion of participants who do not comply or switch treatments causes dilution of differences. However, since per protocol analysis is always potentially biased, it is best to do both and carefully characterise exclusions from the per protocol analysis (Jones *et al.*, 1996).

Participants without outcome measurements may also sometimes be excluded from a per protocol analysis. However, this may lead to biased results and the potential impact of missing data should be considered (Shih, 2002). *SH/IW*

International Conference on Harmonisation E9 Expert Working Group (ICH E9) 1999: ICH harmonised tripartite guideline. Statistical principles for clinical trials. *Statistics in Medicine* 18, 15, 1905–42. **Jones, B., Jarvis, P., Lewis, J. A. and Ebbutt, A. F.** 1996: Trials to assess equivalence: the importance of rigorous methods. *British Medical Journal* 313, 36–9. **Roland, M. and Torgerson, D. J.** 1998: Understanding controlled trials: what are pragmatic trials? *British Medical Journal* 316, 285. **Shih, W. J.** 2002: Problems in dealing with missing data and informative censoring in clinical trials. *Current Controlled Trials in Cardiovascular Medicine* 3, 4.

person-years at risk

This is time in years summed over several individuals. In any COHORT STUDY of incidence rate or rate ratios, different members of the cohort will be at risk for different amounts of time.

Some members may enter the risk set later, because they joined the cohort later, and some will leave earlier, because they were censored or experienced the event of interest. The length of time at risk (measured in years) experienced by each of a set of individuals, summed over those individuals, is the total person-years at risk for that set.

Incidence rates (measured in units of per person-year) are calculated by dividing the number of events occurring in a set of individuals while they are at risk by their total time at risk (measured in person-years).

Notice the implicit assumption being made here: one person at risk for 10 years is equivalent to a group of 10 people at risk for 1 year; both yield 10 person-years at risk. For further details see Rothman and Greenland (1998). *SRS*

Rothman, K. J. and Greenland, S. 1998: *Modern epidemiology*, 2nd edition. Philadelphia: Lippincott-Raven Publishers.

PEST

See SEQUENTIAL ANALYSIS

pharmacokinetics/pharmacodynamics (PK/PD)

See PHASE I TRIALS

pharmacovigilance

The terms 'post-marketing surveillance' and 'pharmacovigilance' are often used synonymously to refer to monitoring (both actively and prospectively or reacting to spontaneously occurring safety concerns) licensed pharmaceutical and biological medicinal products. At the time a marketing authorisation is granted (see MEDICINES AND HEALTHCARE PRODUCTS REGULATORY AGENCY) there may have been between a few hundred and a few thousand patients studied. Typically, each of the PHASE III CLINICAL TRIALS may have only randomised a few hundred patients to the new, experimental treatment. It is quite likely, therefore, that rare reactions (perhaps occurring in one in a thousand patients or fewer) may never have been seen in such trials. No medicine is ever completely safe and it is therefore important that once widespread use of a new one has begun, safety signals are monitored and followed up. Rare events may manifest themselves, as may adverse interactions with other drugs commonly used but in whom interaction studies were never carried out.

Pharmacovigilance is a relatively new science, its grounds perhaps being set out by Finney (1971), who suggested that 'the primary duty of a drug monitoring system is less to demonstrate danger or to estimate incidence than to initiate suspicion'. It might be argued that once 'danger' has been determined, the monitoring process is too late and patients have already been injured (or may have died). What we need is a system that can 'initiate suspicion' so that preventive measures can be taken. Evans (2000) discusses a similar theme.

The great challenge for phamacovigilance is the uncontrolled nature of the data. Marketing authorisations are generally based on randomised, double-blind, PLACEBO (or active) controlled studies. Assessing post-marketing data is much closer to working in EPIDEMIOLOGY. Various national systems for reporting unexpected adverse reactions exist, including the 'Yellow Card' system in the United Kingdom and the 'MedWatch' system in the USA. These (and other) systems rely on reports from a variety of sources including doctors, pharmacists, pharmaceutical companies and sometimes from patients themselves. However, they rely on the judgement of whether an adverse reaction needs to be considered. By the very nature of the rarity of the reactions that pharmacovigilance systems are trying to identify, doctors may not realise that an adverse event in a particular patient has anything to do with the medication(s) he or she is receiving. Conversely, reporting any and all adverse events experienced by patients and highlighting a *possible* relationship to any or all of the medications the patient is receiving would overburden any doctor and monitoring system. A middle ground needs to be found. A further problem arises when a 'new' adverse reaction is suspected and reported (in the scientific or lay press) and then the incidence of spontaneous reports often suddenly increases. The system swings from (almost always) underreporting to (occasionally) overreporting.

Various methods have been proposed to try to overcome problems of over- and underreporting. Systems exist to record all medical events (prescriptions, illnesses, etc.) for samples of patients (say, all patients of selected general practitioners). These are better at overcoming selective reporting and, although they are based on samples, they are typically much larger samples than can be recruited into CLINICAL TRIALS. They also represent samples from practical experience in the community, rather than the highly controlled environment of a clinical trial.

For further reading, van der Heijden *et al.* (2002) proposed statistical methods that use 'all reports of reactions' as a means of adjusting for a general level of underreporting. Grigg, Farewell and Spiegelhalter (2003) have reviewed statistical methods taken from ideas in quality control to monitor streams of reports (adverse reactions and others) in 'real time', while Strom (2000) gives an excellent and detailed coverage of issues and methods used in pharmacovigilance. *SD*

Evans, S. J. W. 2000: Pharmacovigilance: a science or fielding emergencies? *Statistics in Medicine* 19, 3199–209. **Finney, D. J.** 1971: Statistical aspects of monitoring for dangers in drug therapy. *Methods of Information in Medicine* 10, 1–8. **Grigg, O. A., Farewell, V. T. and Spiegelhalter, D. J.** 2003: Use of risk-adjusted CUSUM and RSPRT charts for monitoring in medical contexts. *Statistical Methods in Medical Research* 12, 147–70. **Strom, B.** (ed.) 2000: *Pharmacovigilance*, 3rd edition. Chichester: John Wiley & Sons, Ltd. **van de Heijden, P. G. M., van Puijenbroek, E. P., van Buuren, S. and van der Hofstede, J. W.** 2002: On the assessment of adverse drug reactions from spontaneous reporting systems: the influence of under-reporting on odds ratios. *Statistics in Medicine* 21, 2027–44.

Phase I trials These are CLINICAL TRIALS carried out in the early development of a drug, after animal toxicology studies have been completed, when the drug will first go into man. They involve human volunteers, healthy or patient, and as such the subject of such trials expects no therapeutic benefit. The focus of Phase I trials is on the safety and tolerability of drugs and pharmacokinetics (PK), which can be considered as what the body does to the drug, and pharmacodynamics (PD), what the drug does to the body. Pharmacokinetics involves measurement of drug concentrations in the body, determined by taking blood samples at specific times throughout the study. These are then summarised using such measures as AREA UNDER THE CURVE (AUC), representing total exposure, and maximum drug concentration ($C_{\max}$). Other measures that might be used are time to maximum drug concentration ($t_{\max}$) and elimination half-life. It is not possible to cover these topics in detail here, but Roland and Tozer (1995) give more details and applications for these measurements as well as how to look at the relationships between them, using PK/PD modelling.

Phase I trials can also be carried out prior to drug submission or when a new formulation is being developed. Examples of such studies are bioequivalence, drug interactions (both pharmacokinetic and pharmacodynamic), the effect of food and certain special populations, such as renal or hepatic impairment, elderly or males versus females.

These trials can also be carried out in early development to give a company a general idea of the effect studied, but a confirmatory study will inevitably be required during submission to a regulatory body. They are highly regulated studies, with design and analysis being very standard. Further details on the studies can be found at the FDA website: http://www.fda.gov/cder/guidance.

First-into-man studies take place after completion of toxicology tests. The primary objective of Phase I trials should always be the examination of the safety and tolerability; however, PK and PD data are usually collected and analysed.

Since the drug has not been previously administered to humans the doses have to be escalated, meaning that each subject starts on a low dose and proceeds through the doses to progressively higher ones. Obviously, data have to be reviewed prior to any escalation and, if there is any cause for concern over safety, escalation will not occur, remembering that the safety of subjects is most important.

The table gives examples of dosing escalation schemes showing just single cohorts. In reality, there is likely to be up to 24 subjects in six or eight cohorts, although this depends on how many subjects can be recruited and how many safety data need to be collected.

Phase I trials *Examples of a rising dose scheme*

Example 1

Subject	*Period 1*	*Period 2*	*Period 3*	*Period 4*
1	10 mg	20 mg	30 mg	Placebo
2	10 mg	Placebo	20 mg	30 mg
3	10 mg	20 mg	Placebo	30 mg
4	Placebo	10 mg	20 mg	30 mg

Example 2

Subject	*Period 1*	*Period 2*	*Period 3*
1	10 mg	Placebo	30 mg
2	10 mg	20 mg	Placebo
3	Placebo	20 mg	30 mg

Notice that both designs incorporate a PLACEBO period for each subject. This is so that the measurements from an active period can be put into some context and it can be determined, for example, whether headache is a real drug effect or merely a result of being in a clinical trial.

Much of the analysis for these types of study is likely to be data driven, since they are mainly exploratory. Such things as dose or exposure response curves, with safety and pharmacokinetics, can be examined, as can the determination of a maximum tolerated dose. In addition, Bayesian techniques (see BAYESIAN METHODS) can be used optimally to design and analyse these studies (Whitehead *et al.*, 2001).

The TGN1412 Phase I trial (Senn *et al.*, 2007) raised important issues regarding first-time-in-human trials of new treatments. In this trial all six healthy volunteers exposed to the active drug suffered immune reactions with severe and in some cases long-term sequelae. As a consequence of this trial the UK Secretary of State for Health set up an expert panel to investigate Phase I trials. Their report includes 22 recommendations and was published in November 2006 (Expert Scientific Group on Phase One Clinical Trials, 2006). The ABPI/BIA also published a report on the TGN1412 study (Early Stage Clinical Trial Taskforce, 2006). Senn *et al.* (2007) discuss statistical aspects of the trial, provide identification of shortcomings and make recommendations for future trials. *AB*

Early Stage Clinical Trial Taskforce 2006: Joint ABPI/BIA Report. London: Association of the British Pharmaceutical Industry/BioIndustry Association. **Expert Scientific Group on Phase One Clinical Trials** 2006: Final Report. London: The Stationery Office. **Rowland, M. and Tozer, T. N.** 1995: *Clinical pharmacokinetics – concepts and applications*, 3rd edition. Baltimore: Williams and Wilkins. **Whitehead, J., Zhou, Y., Patterson, S., Webber, D. and Francis, S.** 2001: Easy-to-implement Bayesian methods for dose-escalation studies in healthy volunteers. *Biostatistics* 2, 47–61. **Senn, S., Amin, D., Bailey, R. A., Bird, S. M., Bogacka, B., Colman, P., Garrett, A., Grieve, A. and Lachmann, P.** 2007: Statistical issues in first-in-man studies. *Journal of the Royal Statistical Society, Series A* 170, 517–79.

Phase II trials This is a CLINICAL TRIAL of a new agent or procedure in which the primary objective is typically to determine whether it has sufficient therapeutic efficacy with an acceptable safety profile in patients to warrant further testing and development in additional Phase II trials or large Phase III trials. Before starting a Phase II trial, however, a safe dose and schedule of the new drug or procedure has to be established in earlier dose-finding Phase I trials. As such, Phase II trials are similar to the therapeutic exploratory studies according to the ICH Harmonised Tripartite Guideline E8, and start with the studies in which the primary objective is exploration of and screening for therapeutic efficacy.

Phase II trials often employ study designs such as historical controlled studies in which comparison is made with the efficacy from the historical controls or self-controlled studies in which comparison is made with the baseline status. Phase II trials are often conducted in patients who meet narrowly defined eligibility criteria in order to ensure homogeneity in patient baseline characteristics. ENDPOINTS of Phase II trials can be biological or a surrogate of clinical outcome and sometimes Phase II trials may be further classified as IIa or IIb accordingly. Also, in Phase II trials additional objectives such as exploration of other study endpoints, therapeutic regiments including concomitant medications or target patient populations are evaluated and analyses for these objectives are, by necessity, exploratory and involve many subset analyses.

In the early development of a new therapeutic agent or procedure, the dose and schedule to be used in subsequent trials are determined in Phase I trials. For example, with traditional cytotoxic agents for cancer treatment, this dose is generally known as the maximum tolerated dose (MTD), although other choices, e.g. dose level before the MTD, may be used. In subsequent Phase II trials, patients are treated at the dose level established as safe and acceptable in Phase I trials to screen if the treatment has sufficiently promising clinical activity or therapeutic efficacy, usually evaluated by the PROBABILITY of treatment success, e.g. objective response in cancer, for further investigation and development.

Since Phase II trials serve as initial screening, it is desirable to achieve the goals of the study with a minimal number of patients so that as few patients are given inactive treatment as possible. Also it is important to minimise the Type II error probability of reaching a false negative conclusion (see FALSE NEGATIVE RATE). To this end, sequential designs have been proposed in which a fixed number of patients are accrued in each stage and the study is stopped early if the observed number of treatment successes is too small.

Gehan (1961) was the first to propose a two-stage design for screening in Phase II trials in which 14 patients are initially accrued during stage 1. The study stops if no treatment success is observed and otherwise another cohort of patients are accrued in stage 2. The number of patients during stage 1 is chosen to keep the probability of early termination very small, say <0.05, when the treatment is active with the probability of treatment success $p \geq 0.20$. The sample size for stage 2 is determined to achieve a specified level of confidence in the estimation of the probability of treatment success with the maximum sample size. Later this notion of two-stage designs was formalised as a test of statistical hypotheses regarding the probability of treatment success, as in Simon (1989). Multistage designs have also been formalised by Fleming (1982), among others.

In general, a two-stage design is defined by the numbers of patients to be accrued, n_1 and n_2, and the boundary values, r_1 and r, during stages 1 and 2 respectively, and is denoted as $(r_1/n_1, r/n)$ where $n = n_1 + n_2$ is the maximum sample size. During stage 1, n_1 patients are initially enrolled and treated. If the number of treatment successes during stage 1 is less than or equal to r_1, the trial is terminated for lack of therapeutic efficacy and it is concluded that the treatment does not warrant further investigation. Otherwise, the study is continued to stage 2, during which an additional n_2 patients are enrolled and treated. If the total number of treatment successes after stage 2 exceeds r, it is concluded that the treatment has sufficient therapeutic efficacy and it is, therefore, considered for further investigation. Early acceptance of the treatment is not considered here, as there are no ethical reasons to do so when there is evidence of therapeutic efficacy. The ethical imperative for early termination occurs when the treatment has unacceptable therapeutic efficacy. Furthermore, when there is sufficient therapeutic efficacy, there is often interest and desire in studying additional patients to assess the safety of treatment more extensively.

The sample sizes and the boundary values for such two-stage designs can be determined based on a test of hypothesis. Consider testing $H_0: p \leq p_0$ against $H_1: p \geq p_1$ with Type I and Type II error probabilities α and β, where p denotes the probability of treatment success. The value of p_0 is chosen to represent the maximally unacceptable level of therapeutic efficacy and the value of p_1 is chosen to represent the minimally acceptable level of therapeutic efficacy. These parameters $(p_0, p_1, \alpha, \beta)$ are the design parameters. Then the probability of early termination because of insufficient therapeutic efficacy indicated by no more than r_1 treatment successes is given by $P_{ET} = B(r_1; n_1, p)$, where B denotes the BINOMIAL DISTRIBUTION function and the expected sample size for the true value of p is determined by:

$$E(N|p) = n_1 + (1 - P_{ET})n_2$$

The probability of rejecting a treatment with a success probability P is thus given by:

$$B(r_1; n_1, p) + \sum_{x=r_1+1}^{\min|n_1, r|} b(x; n_1, P)B(r-x; n_2, p)$$

where b denotes the binomial probability mass function.

Simon (1989) proposed two-stage designs that are optimal in the sense that the expected sample size $E(N|p)$ is minimised when $p = p_0$ with insufficient therapeutic efficacy subject to the Type I and II error probabilities. Despite the minimum expected sample size, the maximum sample size $n = n_1 + n_2$ for the optimal design can be much larger than other designs, which is undesirable. Therefore, Simon (1989) also suggested the minimax design, which minimises the maximum sample size, again subject to the same Type I and II error probabilities. However, Simon's minimax and optimal designs sometimes result in quite different sample size requirements. For example, the minimax design can have a much larger expected sample size $E(N|p_0)$ than the optimal design and the optimal design can have a much larger maximum sample size n than the minimax design. For example, with the design parameters $(p_0, p_1, \alpha, \beta) = (0.3, 0.5, 0.05, 0.15)$, the minimax design is $(r_1/n_1, r/n) = (14/37, 17/42)$ and the optimal design is (7/21, 19/48), whereas the expected sample size is 37.6 and 28.5 for the minimax and the optimal design respectively.

Jung *et al.* (2004) used a Bayesian decision-theoretic criterion of admissibility to define a class of designs based on a loss function, which is a weighted average of the maximum sample size and the expected sample size under the NULL HYPOTHESIS, the two criteria used by Simon (1989). The admissible designs include Simon's minimax and optimal designs as special cases and compromises between Simon's minimax and optimal designs. For our earlier example, one can find an admissible design (4/15, 18/45) with a maximum sample size of 45, which is a compromise between the minimax and the optimal designs and an expected sample size of 29.5, which is close to the expected sample for the optimal design. These compromise admissible designs can be found on a convex hull formed by Simon's minimax and optimal designs. The admissible designs of Jung *et al.* (2004) can be easily generalised to any number of stages. *KK*

Fleming, T. R. 1982: One sample multiple testing procedures for Phase II clinical trials. *Biometrics* 38, 143–51. **Gehan, E. A.** 1961: The determination of the number of patients required in a follow-up trial of a new chemotherapeutic agent. *Journal of Chronic Diseases* 13, 346–53. **Jung, S.-H., Lee, T., Kim, K. and Geroge, S. L.** 2004: Admissible two-state designs for Phase II cancer clinical trials. *Statistics in Medicine* 23, 561–9. **Simon, R.** 1989: Optimal two-stage designs for Phase II clinical trials. *Controlled Clinical Trials* 10, 1–10.

Phase III trials New medicines, in particular new drugs, are generally developed in a phased process. Phase I trials and Phase II trials are small trials, the former generally in volunteers, the latter in patients, and tend to be learning or exploratory trials. In contrast, Phase III trials are larger and are confirmatory. Phase III trials provide confirmation of the properties of the new medicine that have been discovered in early phases of the development programme.

In order to be approved for marketing, a new drug must be shown to constitute a worthwhile contribution to medical treatment. Phase III trials are designed to identify an appropriate population of patients who are better treated with the new drug than with other treatments. Additionally, they should identify those patients who do not benefit from treatment with the new drug and those who may be harmed by its use. The prime goal of such trials is to recommend treatment strategies, including the appropriate dose, or dose titration, for the prescribing physician. This goal is achieved by the detailed label whose content is informed by the results of all three phases of drug development, although primarily Phase III trials.

Phase III clinical trials will generally exhibit all the characteristics that have come to be associated with the 'mythical' gold standard trial. They will generally be randomised, be run double-blind, PLACEBO controlled and conducted in a well-defined population of patients. They will most often be parallel group, fixed sample trials. However, there is a growing appreciation that group sequential, adaptive or flexible designs provide drug sponsors with the opportunity to manage their drug development programmes in a more efficient way. This class of trials is characterised by the ability to make changes to the initial setting of the trial, in such a way as to protect the TYPE I ERROR. The changes include the dropping of treatment arms for inefficacy (futility), including the stopping of the trial, resample sizing based on learning about either the variability in the trial and/or the estimated effect and early stopping because of mounting evidence of a large beneficial effect. The ability of drug sponsors to stop individual trials early because of evidence of a large beneficial effect may be restricted by the requirement for a prescribed level of exposure of patients to the exploratory medicine. For example, it may be that a drug sponsor is required to have 500 patients exposed to the exploratory medicine for at least 12 months. Phase III trials are normally conducted in parallel groups and while it is not unknown for a trial to include more than one dose of the experimental medicine, such trials are not the norm since the appropriate dose will normally have been chosen in Phase II.

Generally, two Phase III trials are required for the registration of a new medicine. One reason for this is that to require two trials each to be significant at the one-sided 0.025 level corresponds to an overall Type I error of 0.000625 corresponding to a FALSE POSITIVE RATE of 1/1600. The use of a one-sided SIGNIFICANCE LEVEL is necessary since we clearly require both tests to be positive, that is the treatment effects are in the same direction. It can be shown that if this is the only requirement, then there are more efficient ways to ensure its achievement other than requiring the individual trials to be separately significant. More recently, the Federal Drugs Administration (FDA) in the USA has set conditions under which it is possible to register a new medicine on the basis of a single, large CLINICAL TRIAL with a smaller than usual Type I error. The circumstances tend to be when there is a critical need among seriously ill patients. In such circumstances it is not unusual for standard practices to be relaxed. One well-known example is in the development of treatment for HIV and AIDS in which regulators, under pressure from patients, relaxed standard clinical practice.

The choice of control in Phase III trials is a matter of considerable debate. It was noted earlier that Phase III trials are generally placebo controlled. There are, however, circumstances in which it is not possible ethically to require patients to be treated with a placebo. Such circumstances typically, although not exclusively, arise when the disease under investigation is associated with a mortality outcome. The recent International Conference on Harmonization (ICH) guideline (E10), entitled 'Choice of control group and related issues in clinical trials', addresses a number of important issues when an active control group is chosen and the primary aim of the study is to demonstrate noninferiority. Of prime importance in this context is what E10 terms *assay sensitivity*. Assay sensitivity is the ability of a clinical trial to differentiate between effective, minimally effective and ineffective treatments. Assay sensitivity is important in all trials, but has particular implications in trials whose prime purpose is to demonstrate noninferiority. If a drug sponsor intends to claim effectiveness of a new medicine by showing it to be noninferior to active control, but the trial lacks assay sensitivity, it is possible that an ineffective treatment will be found to be noninferior and thereby lead to an erroneous conclusion of efficacy.

The E10 guideline points to a number of issues that need to be addressed: Was the appropriate patient population chosen? Was the appropriate dose of the active comparator used? Was treatment with the active comparator of the appropriate duration? How was the noninferiority margin, which defines noninferiority, chosen? Is there historical evidence that the design has in the past shown itself to be capable of distinguishing effective from ineffective treatments?

After the conclusion of the experiment, the guideline requires a demonstration that the results obtained on the active comparator are similar to those obtained in previous trials. *AG*

Phase IV trials No totally adequate definition has yet been arrived at. One working definition, however, is that these types of trial encompass all the studies undertaken after obtaining a marketing licence. While studies conducted in the run-up to the registration process are regarded as adequate for the purpose of determining whether a new drug is efficacious and in large part safe, they do, by their very nature, leave important questions unanswered. As an example, during the pre-registration phase, the toxicity of a drug could not have been accurately assessed in a Phase III clinical trial if the incidence of agranulocytosis was 1 in 20 000 or less. In general, Phase II and Phase III studies are restricted because they are based on limited numbers of patients, a limited duration of patient exposure and a restricted population of patients.

Phase IV trials are designed to answer detailed questions about the practical use of a drug. There are many different aspects of this, not all of which will be studied at the same time. Such trials may be conducted to investigate other doses or the scheduling of a new treatment; they may be used to look at side-effects in more detail, particularly in long-term chronic usage; they can be used to investigate drug efficacy in long-term usage where, for example, the course of a disease may be modified over a period of months or years; they can be used to collect comparative data of a long-term usage if the original studies were restricted to a comparison against PLACEBO; they can be used to investigate new uses and indications leading, potentially, to a new submission; or they can be used to investigate alternative populations.

Phase IV trials are usually conducted after a drug has already been through the full trial process and has already got a licence from the authorities for general use, although Phase IV trials can also be run during the registration process itself. In many aspects, Phase IV trials are similar to Phase III trials. They will, in the main, be randomised and controlled, although not generally placebo controlled; furthermore, some of their objectives may be investigated by *uncontrolled* studies.

It is important to distinguish between Phase IV trials and post-marketing surveillance (PMS) or monitored release studies (MRS). A PMS trial will tend to be observational and noninterventional and will be conducted primarily to monitor safety in a new medicine shortly after it has begun to be prescribed in daily practice. While there may be simple measures of efficacy included so that additional risk/benefit judgements may be made, this is not the prime purpose of the trial. A further type of study that may be conducted post-registration is a so-called 'seeding' study. A 'seeding study' involves the drug sponsor providing a new medicine to physicians in order to familiarise them with its use and to encourage them to prescribe it. In such studies there is neither intention nor attempt to gather data that could provide useful scientific or medical information.

More recently Phase IV trials have been used in order to collect health economics, also known as outcome research, data concerning resource utilisation. Such data are required in order to provide information on the cost-effectiveness of new medicines for reimbursement agencies. Although this information may be required at, or shortly after, registration of a new medicine, forcing its collection during Phase II and Phase III trials – so-called piggybacking – may not be appropriate. Phase II and Phase III trials are highly controlled scientific experiments designed to minimise variation and to make it possible to detect small, but clinically relevant, treatment effects. In one sense, therefore, they are artificial and do not represent the environment in which new medicines will ultimately be prescribed.

This has a number of consequences. First, protocols in Phase III trials may include investigations that are nonstandard in the clinical treatment of patients but are entirely appropriate in a CLINICAL TRIAL to monitor the safety of patients. If these 'extra' investigations are allocated against the new medicine their cost may BIAS the overall evaluation of its cost and hence its cost-effectiveness. Second, health outcomes may be more variable than the clinical ENDPOINTS studied in Phase III trials. If, therefore, health economic outcomes are piggybacked on a Phase II trial this is likely to increase the sample size and lead to an even greater cost of drug development than hitherto. Ideally, health economic studies should be conducted in a more naturalistic setting. They will tend to be large trials, using endpoints that are patient centric so that quality of life outcomes (see QUALITY OF LIFE MEASUREMENT) may predominate over clinical outcomes and they will tend to be simple with the minimum of information being collected. Such trials may more appropriately be carried out in a Phase IV programme. *AG*

phenotype This term refers to the observable characteristics of an individual or organism, as distinct from its GENOTYPES, which, before modern molecular genetics, were not observable. Although many genotypes are also now measurable, the distinction is still useful.

Phenotype covers a wide range of possibilities, including diseases (affected or not affected), quantitative traits (e.g. blood pressure or height) and biological measurements (e.g. blood glucose levels). In some cases, there is a close relationship between a particular gene and a particular phenotype. For example, in the ABO example, there are four main blood groups, as measured by serology, determined by the ABO gene: A, B, AB and O. These phenotypes are determined by the genotypes: individuals with genotypes (A,O) and (A,A) are group A; genotypes (B,O) and (B,B) give rise to group B; (A,B) gives rise to group AB; and (O,O) to group O.

Most phenotypes are, however, related to genotypes in a more complex fashion – most diseases and other common phenotypes are related to genotypes at many genetic loci and are also influenced by environmental or lifestyle factors. The PROBABILITIES with which particular genotypes are associated with particular phenotypes are referred to as *penetrances*. *DE*

[See also GENETIC EPIDEMIOLOGY, TWIN ANALYSIS]

Balding, D. J., Bishop, M. and Cannings, C. (eds) 2001: *Handbook of statistical genetics*. Chichester: John Wiley & Sons, Ltd. **Sham, P.** 1998: *Statistics in human genetics*. London: Arnold.

pie chart This is a graphical display in which a series of frequencies or percentages are represented by sections of a circle having areas proportional to the observed values. An example is given in the figure.

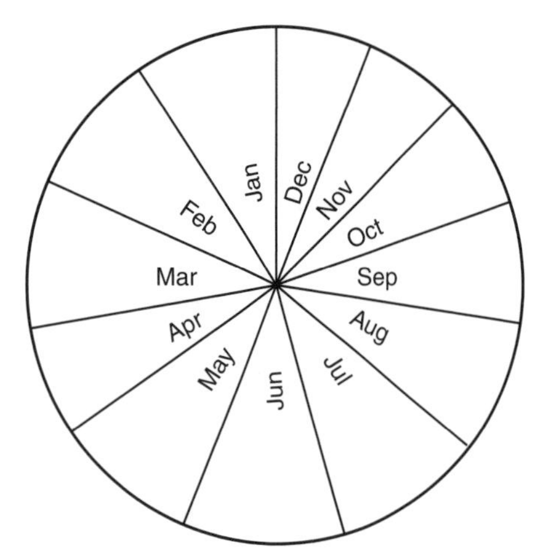

pie chart *Frequencies of first births in each month in a Swiss town* (sample size is 500)

Although very popular in the media, both the general and scientific use of pie charts has been severely criticised (see Tufte, 1983; Wainer, 1997) and tables are preferable for most small datasets. Among particular dangers to be aware of concerning pie charts are the following: distortion of the basic shape from a circle, misleading use of three-dimensionality, detaching of slices from the pie, rotation to promote or hide a given slice and failure to show sample size on which the proportions are based. *BSE*

[See also BAR CHART, DOTPLOT, HISTOGRAM, STEM-AND-LEAF PLOT]

Tufte, E. R. 1983: *The visual display of quantitative information*. Cheshire, CT: Graphics Press. **Wainer, H.** 1997: *Visual revelations*. New York: Springer.

pilot studies These are small-scale research experiments primarily undertaken to inform or improve the conduct of related future research. In general, pilot studies do not in themselves aim to generate scientifically useful evidence. Instead, pilot studies are conducted for a variety of reasons, including: testing feasibility and appropriateness of data collection (e.g. are sufficient manpower and time resources planned to gather required information?); identifying problems with QUESTIONNAIRE wording, if extracting information in this manner (e.g. do patients' responses indicate that questions have been properly understood and answered?); identifying problems with data processing (e.g. does a null, or a blank, response mean an answer is missing, or not asked, or no response, or an unknown answer?); training observers or interviewers to equally high standards to help ensure uniformity of data quality, whenever multiple individuals are involved in data collection and processing, as in most large or multicentre studies (e.g. is there a learning curve in how data are extracted and do all attain a sufficient level of expertise?); assessing, at least initially, anticipated variations of responses, for this can help sample size estimation for the main study (e.g. when such information is either not available or not applicable from the published literature, being better than relying on 'educated guesswork' alone); estimating main study duration and cost (e.g. if uncertain, both time and money estimates may need to be firmed up before submitting a realistic budget to potential sponsors).

Although often on a hidden agenda, a well-conducted pilot study can also help serve to convince potential sponsors of the main study to follow and that the research team is indeed capable of performing their intended study, once awarded the necessary funding. Pilot studies do not need to be large to serve their purposes. There may be the need for several drafts before the final version of a questionnaire is deemed most suitable, although one pilot is generally sufficient and its cost and effort are usually well rewarded.

On the negative side, pilot studies can cause results from an investigation to be delayed, a classic case of the compromise between speed and quality. (This is reminiscent of the service salesperson's banter: 'You can have it fast, cheap or reliable – pick any one!') However, various attempts to circumvent this problem have been proposed that involve internal pilot studies. These are study designs that seek to incorporate a seamless transition from the pilot to the main study, in which the initial information arising helps to dictate the choice of the overall study sample size, a similar motivation to some DATA-DEPENDENT DESIGNS for CLINICAL TRIALS. See, for instance, contributions by Burkett and Day (1994), Wittes *et al.* (1999) and Zucker *et al.* (1999) for a flavour of these developments.

One warning about running a pilot study is the potential for BIAS brought about by inappropriately combining results from the pilot and the main studies, especially if changes occur after the pilot phase. This is so even if these changes are as innocuous looking as rephrasing questions from positive to negative or from open to closed (see QUESTIONNAIRES). Pilot patients will be in a different timeframe, so beware of any temporal or seasonal variations in responses. Furthermore, eligibility criteria for inclusion in the main study and in the pilot study may differ in important ways, meaning it would not be obvious to which population (if any) inferences from such a hybrid sample might apply.

In conclusion, pilot studies are well worth the initial effort involved. Too few are undertaken in advance of major studies. This may be attributable in part to too much research occurring on a tight timescale to accommodate individuals' medical career steps. Also, it is more than a pity that much published research is labelled as a 'pilot' as if this were somehow a post-experimental justification for inadequate, or, worse still, absent, design planning. For one's own research, always run a pilot study unless there is a sound reason not to do so. Expect to discard its results from the main analysis, unless it causes nothing to change and it was pre-planned as an internal pilot study. *CRP*

Burkett, M. A. and Day, S. J. 1994: Internal pilot studies for estimating sample size. *Statistics in Medicine* 13, 2455–64. **Wittes, J. T., Schabenberger, O., Zucker, D. M., Brittain, E. and Proschan, M.** 1999: Internal pilot studies I: Type I error rate of the naïve *t*-test. *Statistics in Medicine* 18, 3481–91. **Zucker, D. M., Wittes, J. T., Schabenberger, O. and Brittain, E.** 1999: Internal pilot studies II: comparison of various procedures. *Statistics in Medicine* 18, 3493–509.

pitfalls in medical research Many people working in healthcare become involved in the design, execution, analysis and dissemination of medical research at some point in their careers. Doctors are taught about medical research and in the course of their professional training may well be expected to be actively involved in research. Other health professionals and even managers are also likely to be required to be active researchers. Consumers (that band of once passive recipients of healthcare formerly known as patients) are also increasingly contributing to the commissioning and design of research. There will always be those for whom research is at the centre of their professional lives and those for whom it is at the periphery, but many individuals are expected at some point to be researchers. Research, however, presents many pitfalls for the inexperienced, some of which are described here in the hope that this may assist future researchers to avoid them.

Research and clinical audit are sometimes confused and the novice researcher may be uncertain what kind of activity it is they are undertaking. The UK NHS working definition of research is 'the attempt to derive generalisable new knowledge by addressing clearly defined questions with systematic and rigorous methods' (Department of Health, 2001), while clinical audit can be defined as 'a quality improvement process that seeks to improve patient care and outcomes through systematic review of care against explicit criteria and the implementation of change' (National Institute for Clinical Excellence (NICE), 2002).

The distinction is not an academic one, as research has wider ethical and governance implications than audit, as patients must be protected and the scientific quality of the research guaranteed (World Medical Association (WMA), 2002). For example, access to confidential patient information for research purposes requires independent ethical review and generally the informed consent of the patient, but access for an audit activity aimed at improving patient care does not. Appropriate review for research should be obtained, including scientific review, ethical review from a research ethics committee or institutional review board and other approvals. For example, in England and Wales, approval must be sought from each NHS trust hosting the research (see ETHICAL REVIEW COMMITTEES). Hence the first pitfall may sometimes be not to recognise research activity, thus denying participants the protection offered by ethical and scientific review.

Many individuals are involved in and contribute to research, but they do not all have the same experience or expertise. It follows that researchers who do not seek out the range of skills needed for their projects are likely to run into problems. Successful medical research no longer involves a single scientist working in isolation, if, indeed, it ever did. Major grant-awarding bodies look not only for individuals with research track records but also for a research team with appropriate clinical and other collaborators, including explicitly identified statistical expertise. Even a relatively junior and isolated researcher should seek to identify a research team that will support his or her project. This should ideally include senior clinicians and investigators with deeper knowledge of the research area, statistical support and advice (often available in research active institutions). Peer support enables researchers to benefit from shared experience.

Essential though it may be, collaboration, or, more particularly, *academic supervision*, can cause difficulties for researchers. Collaborators and supervisors may have overambitious views of what their researchers can achieve and may underestimate the time required. Supervisors may have their own special interests that are not shared by students or that do not constitute appropriate research for the students' objectives, particularly academic ones. Indeed, more than one supervisor may have conflicting special interests. There are no easy answers, but early development of an agreed protocol for the research and a project plan for student's time offers some protection against conflicting interests and changes of direction.

Why might failing to seek statistical collaboration be a pitfall? This can be answered by considering what a statistician can contribute to a research project. Researchers whose memory of learning basic statistics consists of a succession of hypothesis tests might be surprised to find that a consultation with a statistician is unlikely to focus on the statistical tests required in the proposed research. Discussions will focus on, first, the question that the research will address, moving on to the appropriate research design to answer that question and the data to be collected. Only then can the proposed analysis and the required sample size be considered. The statistician contributes to the whole research design, not merely the statistical analysis (see CONSULTING A STATISTICIAN).

If the role of statistics in research is about design as much as it is about analysis, what problems are encountered where the research design is inadequate? The pitfalls involved in proceeding with research without a clearly thought-out research design are several: the finished research may answer no clear question; although the researcher had a clear question, an inappropriate design that cannot give a clear answer has been used; some of the information needed properly to answer the question was not collected; the study size may be too small to answer the researcher's question.

Why does it matter that a research proposal should address a clear question? Medical research progresses by framing hypotheses and addressing answerable questions, and patient involvement in research cannot be justified when this does not apply. Without a research question, a statistician cannot develop a research design, just as an architect cannot develop a drawing unless he knows whether a house or a railway station is required. Research questions and research designs are so closely linked that once a researcher has framed a clear question, the appropriate design often becomes apparent. Without an a priori answerable question, the data collected may well be inadequate to provide the answer to any useful question. Some researchers might find an evidence-based approach to medicine (see EVIDENCE-BASED MEDICINE) helpful in thinking through clinical research questions. Different types of clinical question, e.g. concerning treatment, prognosis or diagnosis, can be framed, and then a research protocol with the appropriate design to answer the question can be developed (Sackett *et al.*, 2000). For example, a question concerning a medical intervention needs to define the population to be studied, an intervention and a comparator and an outcome.

The design most likely to control for confounding in an intervention study and thus provide high-quality evidence is a randomised controlled trial (see CLINICAL TRIALS). In specifying the question, the preferred research design has also been determined. Not all research has a clinical focus – it might be epidemiological or laboratory research – but the researcher will still need to have clearly formulated, refutable hypotheses, if a protocol designed to test them is to be developed.

A researcher may have developed a question, but is it the right question? A further pitfall is that the research question may already have been adequately answered. Would the proposed research add anything new to the existing literature? Once the component parts of an initial question, e.g. the patient population, the precise intervention, the most valid comparators and outcomes, are clearly defined, they can be turned into search terms to be used in reviewing the medical literature, perhaps to be used in a formal systematic review (see SYSTEMATIC REVIEWS AND META-ANALYSIS) (Chalmers and Altman, 2001). If the question has been answered or there is a good-quality study in progress likely to give a definitive answer, then continuing to plan a new study is neither ethical nor cost-effective. Even where a thorough literature search confirms that the research question has not previously been adequately answered, it may be that the literature or contact with experts, clinical colleagues and patients suggests that the initial question was not clinically relevant or has already been partially answered, and thus the question may need to be modified before developing a research protocol.

Once the research question has been framed, inappropriate research design presents the next pitfall. It is possible for a researcher to have a clear research question, but to have chosen a research design that cannot answer it. An epidemiologist evaluating the results of a study will consider whether they are a chance finding, are explained by confounding (when more than one factor are associated both with each other and the outcome so that it is impossible to say what the true effect of each is on the outcome) or by BIAS (Rothman, 2002). The research design must answer the research question in a way that minimises the influence of chance, bias and confounding.

Bias is any process that causes the study results systematically to depart from the true result. The selection of participants can be biased, perhaps because the sample is drawn from a highly selected hospital population but the intention is to generalise the results to the whole population (see SELECTION BIAS). The measurement of outcomes can be biased by poor ascertainment or poor response, by bias in the recording of information or by inadequate follow-up. Sometimes proposed research has a biased design because the researcher starts with the data that are most easily available, not with the data that can answer the question. Suppose the accuracy of a blood test is under consideration and a researcher has access to test results and associated patient records. Suppose that diagnosis requires expensive tests. It is likely that only the most severe and quite probably only those with positive test results will have the expensive investigations and the results will reflect this bias, inflating the apparent accuracy of the test. If the researcher had started by considering an optimal design, then a COHORT STUDY including a representative sample of patients, all of whom receive both tests, would be the preferred design. Underpinning clearly framed research questions there should be an understanding of how research design can as far as possible avoid bias and confounding, whether that understanding is mediated through a traditional approach to epidemiology or to the design of experiments or by evidence-based medicine.

A further pitfall in the design of a research study is that the data collected may be inadequate for the purpose. The chosen outcome measures may not adequately measure the chosen outcome. Standardised instruments (including QUESTIONNAIRES and psychological tests) with known validity and reliability should be used where possible (see MEASUREMENT ERROR AND RELIABILITY). If researchers are developing their own outcome measures, whether laboratory tests or questionnaires, they should investigate the proposed outcome's properties. All questionnaires should be piloted, preferably using the target population (see PILOT STUDIES). Investigators should be wary of confusing process with outcomes. Some potential outcome measures will not be available for all patients; e.g. the gold standard for clinical diagnosis may depend on histological confirmation of disease that can only be obtained in the most ill patients. Hence, protocol outcome definitions may need to be pragmatically adapted to the realities of the clinical setting.

Important data items may be omitted. For example, it is not unknown for inexperienced researchers interested in the survival of their patients carefully to collect information about deaths among their patients, but not explicitly to record the last date living patients were seen, without which SURVIVAL ANALYSIS is impossible. Important confounding factors (age and sex should almost always be recorded) may be omitted. Consultation with clinical and statistical collaborators and careful reading of previous relevant studies will prevent some of these mistakes.

It is equally possible to collect far more data than is necessary, particularly when many tests are repeated as part of clinical care or research is designed by large, multidisciplinary committees comprised of specialists, where each wants to promote his or her own area of clinical interest. Decisions should be made at the outset on which time points data will be collected and what will be measured. A researcher might, for example, want to know the lowest white cell count following a bone marrow transplant or the time point at which the white cell count recovered, but will certainly not need the results of daily blood tests. All of this should be specified in the research protocol, not after the end of data collection (see PROTOCOLS FOR CLINICAL TRIALS).

Researchers are often disappointed when their apparently interesting data do not show statistically significant results and thus may have occurred simply by chance. To avoid this disappointment, a researcher must before the research starts calculate what sample size is required in order to be reasonably certain that, if the desired results are obtained, they are precise enough to be convincing, as demonstrated by a narrow CONFIDENCE INTERVAL or by the result of a hypothesis test that is statistically significant at a pre-specified level. To estimate how many patients are needed (a sample size calculation) the researcher must typically specify what the researcher expects to find, e.g. the outcome in the control group, and what the researcher hopes to detect, e.g. a clinically important difference that might be achieved with a new treatment. The researcher also needs to specify acceptable statistical power so that there is a reasonable chance of detecting such a difference if it does exist (the minimum acceptable is 80%, which will detect a true result in 80% of samples with a given level of statistical significance) and the significance level (typically $\alpha = 0.05$) or precision (i.e. the confidence interval, typically 95%).

Statistical aspects of sample size calculations are straightforward, given appropriate reference books, tables (e.g. Machin *et al.*, 1997) or software (e.g. NQUERY ADVISOR), but deciding what is a worthwhile outcome is not (see various SAMPLE SIZE DETERMINATION entries). It is easy to manipulate the inputs to the calculation by adjusting the researcher's expectations and the statistical power required. All too often a researcher approaches a statistician for a sample size calculation in the expectation the answer will be precisely the number of patients available. It is important, however, to explore what the sample size requirements would be if the initial assumptions were wrong. Researchers should aim for samples larger than the minimum necessary to identify a substantial improvement over standard treatment.

There may be few data to inform calculations and the temptation is to underestimate sample size (e.g. by assuming implausibly small STANDARD DEVIATIONS). As an illustration of what can go wrong, consider a placebo-controlled trial where there is good reason to expect a 60% response with the intervention and where $\alpha = 0.05$ and POWER is 90%. PLACEBO response rates are notoriously variable. If an overoptimistic researcher assumes a 30% placebo response and the actual rate is only 5% higher, then 50% more patients will be required (see the figure on page 339).

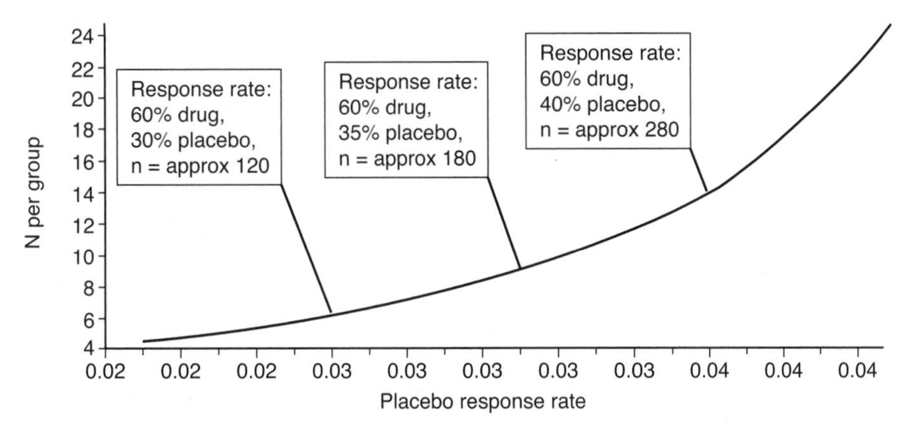

pitfalls in medical research *Impact of different levels of placebo response on sample size*

Determination of sample size is not simply a statistical calculation. Researchers often overestimate the availability of patients. In practice, patients do not always consent to take part, default from clinics, drop out, move house and even die at inconvenient times (see DROPOUTS). Medical records are often indefinitely unavailable if not explicitly lost. A generous allowance should be added to the sample size to allow for such contingencies.

What can be done if the researcher cannot recruit enough patients to meet a realistically estimated sample size? If a proposed study cannot answer the proposed question, it would be unethical to ask patients to take part and the study should not go ahead. If a single-centre study had been proposed, it might be possible to find collaborators in other centres in order to obtain sufficient numbers, although MULTI-CENTRE TRIALS have further potential pitfalls compared to single-centre studies.

After careful consideration of the research question and the design, the researcher should have developed a research protocol (see PROTOCOLS FOR CLINICAL TRIALS). This should be sufficiently detailed to provide a clear picture of what the study will involve. Studies based on inadequate or even nonexistent protocols are likely to fail to provide data that match the researcher's objectives. If the study aims are unclear in the protocol, if it does not clearly specify which patients are to be recruited, what is to happen to them and what data are to be collected, then everyone involved in carrying out the study will share the same uncertainties. While not all studies need the detailed protocols of commercial drug trials, the protocol should describe the research aims and lay out the activities necessary to achieve them. The researcher might ask the question: 'Is there enough information in the protocol for someone from outside the research team to understand why the study is being carried out and what would be needed to replicate it?' If the answer is 'no', there is the danger that the research project will not follow the research design and will provide meaningless or biased results. The research protocol is at the heart of good clinical practice (GCP) in research, the code of conduct that aims to protect the rights and safety of research subjects and to ensure the scientific quality of the research. While some aspects of GCP are specific to drug trials (International Conference on Harmonization (ICH), 1996), most of it applies to all clinical studies (Medical Research Council (MRC), 1998, 2000). Scientific and ethical review of the protocol ensures that the research design is valid and ethical. Adherence to the protocol while the research is in progress should ensure that the study aims are met.

Many novice researchers have little awareness of GCP, but adhering to its principles would avoid many pitfalls for the research project in progress. The principal investigator and other suitably qualified personnel involved in the research should be identified and, particularly in drug trials, responsibilities should be explicitly delegated in writing. In commonsense terms, this means that all members of the research team should understand their role in the project and should receive adequate training. The scientific quality of the study is assured by adherence to the protocol, but this is only meaningful if it is possible to monitor that adherence. This is done (extensively in commercially sponsored trials) by comparing the study documentation, the 'case report forms', against the source documentation, e.g. patient notes and laboratory records. Thus, not only should the study be meticulously documented but also patients' records need to be maintained to high standards.

Even where a study is not likely to be audited in any detail, adherence to the protocol, careful study documentation and maintenance of high-quality patient records will help ensure that all subjects successfully complete the study and will contribute to the quality of the data collected. Conversely, in studies where standards of record keeping are low, even the investigators will find it difficult to find out what was actually done, should they need to go back to the study records and raw data. Study records must be kept securely and, in clinical research, separately from the patient record, which should contain all important information, including that the patient has given informed consent. Study records should be archived after completion of the study, so that there is an audit trail should there be any queries concerning the way the study was carried out or the accuracy of the data.

A researcher will not be able to start a project as soon as the protocol is finalised. Enough time must be allowed at the start of the project to obtain research ethics approval, any institutional approvals required and any regulatory approvals required, e.g. a clinical trial authorisation for a trial of a medicinal product in the UK (see MEDICINES AND HEALTHCARE PRODUCTS REGULATORY AGENCY (MHRA)). The time required should not be underestimated, as research ethics committees generally have queries that must be answered before final approval is given. Often these queries concern incomplete application forms or badly drafted patient information and might have been avoided if guidance on drafting forms and leaflets had been followed. Never allow only the minimum possible time for approvals to be obtained.

Even after all the data have been collected, some further pitfalls remain. Before preparing the data for analysis, the coding of data items should be considered, as tidying up a messy dataset can add considerably to the time needed for data analysis. Codes should be allocated for missing data, taking care to distinguish genuine missing data from 'unknown' and zero entries. Each patient should be allocated a study number, so that individuals can be easily identified in data analysis if necessary. The use of a database with an appropriate form for DATA ENTRY incorporating checks on the data entered can facilitate quick and accurate data entry. Databases are preferred to spreadsheets as it is easier to ensure the correct data are entered for a particular case. If, however, a spreadsheet is used, one row should be used for all the information on a single patient and care should be taken when sorting the data to ensure that the correct data remains attached to the correct patient. At least a sample of data should be checked by double data entry or proofreading (see DATA MANAGEMENT).

Data should be kept securely and the password protected to protect patient confidentiality, should always be anonymised and should be backed up regularly so that at worst only the last few cases entered are lost. Data protection is a legal requirement as well as good practice: losing precious data when a portable computer is stolen is bad enough; it is worse if the backup copies were in the computer bag; it is far more serious if confidential personal information is lost or revealed.

Where available, STATISTICAL PACKAGES offer more flexibility and speed in statistical analysis than spreadsheets and databases and, an important factor, the statistical calculations made have been checked. Data can be imported from databases, spreadsheets and other formats. Menu-driven statistical packages are easy to use and offer extensive help files, but that can in itself present a hazard. A statistical method should only be used if the assumptions underlying it are met by the data to which it is applied. Advice should be sought if a researcher does not understand the application of assumptions behind the statistical methods they are using else the researcher risks presenting superficially sophisticated but meaningless results.

Poor or nonexistent planning often means that a research project is never successfully completed. Good project management from the outset, however, increases a project's chance of successful completion. Research projects that have received substantial funding will be expected to have project plans monitored by a management committee to help ensure that the project achieves its aims on time and within budget. This approach can be usefully applied to smaller scale projects. The quality of the desired outcome should be specified and the time and other resources available (or else which must be obtained) must be identified. For example, a doctor in a training post planning some research might have a project plan that specifies: the quality of the research must be at least adequate for a conference presentation; the resources available are the researcher's time, some input into the research design from others, a little local funding and some secretarial time; and the time constraints are that sufficient results must be available by the conference abstract deadline and the project must be completed by the conference.

It is important to be realistic when considering both the resources and the time available. A project completed over a year should allow a realistic amount of time for Christmas and other public holidays, vacations and illness (the researcher's, their supervisor's and the patients'), as well as building in some time for unexpected contingencies. Once these constraints have been identified, the researcher should plan backwards from the ENDPOINT. He or she should ask the questions: Is there enough time and resources to achieve quality? If not, can the timescale be adjusted or further resources sought (which will itself take time)? Would a less ambitious project meet quality standards?

The timelines can be plotted and an analysis made of the critical pathways where delay might occur. What steps will hold everything up (e.g. obtaining research ethics approval, retrieving records)? Once this has been identified the researcher should focus on how to achieve those steps in a timely fashion. Milestones critical to the success of the project should be identified and targets set for achieving them. A management committee can be set up to monitor progress against targets.

The researcher who wishes to avoid pitfalls will have clearly specified research questions, appropriate research designs, adequately detailed protocols and well-planned projects carried out in line with good practice. Statisticians contribute to all of these objectives. Statisticians, like doctors, are most effectively consulted in the earliest stages and the consequences of not taking this advice are well known. Perhaps, then, the biggest pitfall in medical research is to forget that statistical collaboration should start at the beginning and last throughout the life of a research project. *CLC*

Chalmers, I. and Altman, D. G. (eds) 2001: *Systematic reviews in health care: meta-analysis in context*. London: BMJ Books. **Department of Health** 2001: *Research governance framework for health and social care*. London: Department of Health. **International Conference on Harmonization (ICH)** 1996: E6 document: guideline for good clinical practice. ICH harmonised tripartite guideline. *Federal Register*. **Machin, D., Campbell, M., Fayers, P. and Pinol, A.** 1997: *Sample size tables for clinical studies*, 2nd edition. Oxford: Blackwell Sciences Ltd. **Medical Research Council (MRC)** 1998: *MRC guidelines for good practice in clinical trials*. London: Medical Research Council. **Medical Research Council (MRC)** 2000: *MRC ethics series. Good research practice*. London: Medical Research Council. **National Institute for Clinical Excellence (NICE)** 2002: *Principles for best practice in clinical audit*. Oxford: Radcliffe Medical Press. **Rothman, K. J.** 2002: *Epidemiology*. Oxford: Oxford University Press. **Sackett, D. L., Straus, S. E., Richardson, W. S., Rosenberg, W. and Haynes, R. B.** 2000: *Evidence-based medicine: how to practice and teach EBM*, 2nd edition. London: Churchill Livingstone. **World Medical Association (WMA)** 2002: Declaration of Helsinki: ethical principles for medical research involving human subjects (amendment). In 52nd WMA General Assembly, Edinburgh.

placebo This consists of a treatment that mimics a potentially active treatment in every respect except the ingredient or other feature through which the treatment is assumed to exert its effects. A placebo can be pharmacological (e.g. a tablet or an injection), physical (e.g. a manipulation) or psychological (e.g. a conversation). The ideal placebo is indistinguishable from its active counterpart in all respects other than its effects (appearance, taste, etc.). Placebos should avoid any risk inherent in the active intervention. For instance, in ophthalmology, a trial of intraocular injections of a new drug could use 'sham' subconjunctival injections of saline as the appropriate placebo.

Randomised clinical trials in which a placebo is used can be conducted with BLINDING or masking of the treatments. In a single-blind trial, the investigator is aware of the treatment but the subject is not, or vice versa. In a double-blind trial neither the subject nor the investigator is aware of the treatment received. Blinding limits the occurrence of conscious and unconscious BIAS arising from the influence that the knowledge of treatment may have on the recruitment and allocation of subjects, their subsequent care, the response of subjects to treatment, the assessment of ENDPOINTS, the handling of withdrawals and so on (International Conference on Harmonization (ICH), 1998).

Placebos can sometimes be useful, even in so-called active control trials, i.e. trials comparing two or more active treatments, say A and B. A first situation requiring the use of placebos is when blinding of treatment is deemed essential but A and B cannot be made identical. In this case, each subject can be allocated randomly to two sets of treatment ('double-dummy'): either A and placebo for B or B and placebo for A. Another use of placebos in active control trials is when two active treatments A and B must be compared, but a placebo arm can also reasonably be contemplated. In that case, the treatment contrast of interest is A versus B and the placebo group is of little use if A differs from B. However, if A does not differ from B, the contrasts of A or B versus placebo may suggest whether A and B are equally active or equally inactive.

The Declaration of Helsinki states that: 'The benefits, risks, burdens and effectiveness of a new method should be tested against those of the best current prophylactic, diagnostic and therapeutic methods. This does not exclude the use of placebo, or no treatment in studies where no proven prophylactic, diagnostic or therapeutic method exists' (World Medical Association (WMA), 2002). The appropriateness of placebo control versus active control should be considered on a trial-by-trial basis. For serious illnesses, when a therapeutic treatment has been shown to be efficacious, a placebo-controlled trial is unethical. A placebo may also be less necessary in these cases because the assessment of 'hard' endpoints such as death or a major clinical event are unaffected by knowledge of the treatment.

In a placebo-controlled trial, the estimated treatment effect represents any effect of the active treatment A over and above that of the placebo (δ_A in the figure). This is generally the effect of interest. In contrast, if the active treatment is compared to an untreated control group, the estimated treatment effect includes any placebo effect ($\delta_A + \delta_P$ in the figure). If there is interest in estimating the placebo effect on its own (δ_P in the figure), the trial should randomise patients between an untreated control group and a placebo group, as shown in the figure, in order to control for the natural evolution of the disease (e.g. spontaneous regressions).

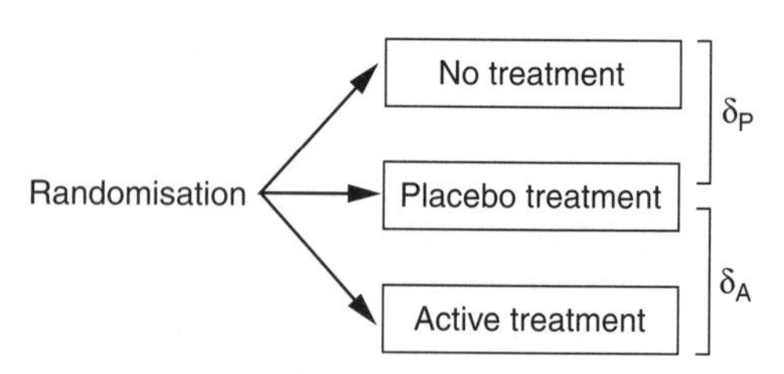

placebo *A trial design randomising patients to a no-treatment control group, a placebo treatment group or an active treatment group*

Is there such a thing as a placebo effect? A seminal paper on placebos (Beecher, 1955), largely responsible for the general adoption of the double-blind study design, reported an average placebo response rate of about one-third in 26 studies. Another paper (Roberts *et al.*, 1993) suggested that the effects of placebos could be much greater 'under conditions of heightened expectations'. This claim is not supported by a recent systematic review of 130 CLINICAL TRIALS in which patients were randomly assigned to either placebo or no treatment (Hrobjartsson and Gotzsche, 2001). Outcomes were binary in 32 trials and continuous in 82 trials. As compared with no treatment, placebo had no significant effect on binary outcomes, regardless of whether these outcomes were subjective or objective. For the trials with continuous outcomes, placebo showed a statistically significant beneficial effect on subjective outcomes, but the effect decreased with increasing sample size, indicating a possible bias in small trials. *MB*

Beecher, H. K. 1955: The powerful placebo. *Journal of the American Medical Association* 159, 1602–6. **Hrobjartsson, A. and Gotzsche, P. C.** 2001: Is the placebo powerless? An analysis of clinical trials comparing placebo with no treatment. *New England Journal of Medicine* 344, 1594–60. **International Conference on Harmonization (ICH)** 1998: E9 document: guidance on statistical principles for clinical trials. *Federal Register* 63, 179, 49583–98. **Roberts, A. H., Kewman, D. G., Mercier, L. and Hovell, M.** 1993: The power of nonspecific effects in healing: implications for psychosocial and biological treatments. *Clinical Psychology Review* 13, 375–91. **World Medical Association (WMA)** 2002: Declaration of Helsinki: ethical principles for medical research involving human subjects (amendment). In 52nd WMA General Assembly, Edinburgh.

placebo run-in In some CLINICAL TRIALS, especially trials of preventive (rather than therapeutic) interventions, the enrolled subjects go through a run-in placebo period prior to RANDOMISATION. The run-in period is useful to screen the subjects who are likely to comply with the intervention and to avoid randomising subjects who are unlikely to do so, thereby diluting any benefit of the intervention being tested. Run-in periods allow investigators to document why eligible subjects refuse trial enrolment or fail to be randomised at the end of the run-in period. They may also be used to evaluate the feasibility of strategies designed to promote trial enrolment and adherence. The disadvantage of selecting good compliers in a run-in period is that these subjects are not representative of the targeted population, thereby compromising the external validity of the trial.

For instance, a chemoprevention trial in patients at high risk of a recurrence of a head and neck cancer included an 8-week placebo run-in period (Hudmon, Chamberlain and Frankowski, 1997). Of 391 former cancer patients who entered the run-in period, 356 were randomised; the others were no longer interested in trial participation ($n = 20$), did not return within 10 weeks of enrolment date ($n = 3$), did not achieve a drug adherence level of at least 75% ($n = 9$) or were not randomised for another reason ($n = 3$). The most significant predictors of run-in outcome (randomised or not randomised) were education level and Karnofsky performance score. The odds of randomisation were more than twice higher in subjects with a good Karnofsky performance score and those with more than a high school education. *MB*

Hudmon, K. S., Chamberlain, R. M. and Frankowski, R. F. 1997: Outcomes of a placebo run-in period in a head and neck cancer chemoprevention trial. *Controlled Clinical Trials* 18, 228–40.

point-biserial correlation coefficient See CORRELATION

Planning and Evaluation of Sequential Trials (PEST) See SEQUENTIAL ANALYSIS

Poisson distribution This is a PROBABILITY DISTRIBUTION of the number of (rare) events occurring in a fixed time or area. Whereas the BINOMIAL DISTRIBUTION is used to model the number of 'successes' observed given the number of 'trials' taking place (and the independent PROBABILITY of success in any one trial), the Poisson distribution has the advantage that the number of trials need not be known (although it still needs to be large).

For example, Armitage *et al.* (1999) model the number of cases of juvenile-onset Crohn's disease in Scotland as being Poisson random variables. This is a typical example of its use, as the risk for an individual is small, but the population of at-risk individuals, while unknown, can be presumed to be large.

Although the distribution was identified before he worked on it, it is named after the French mathematician Siméon Denis Poisson (1781–1840) and hence the 'P' in 'Poisson' is always capitalised. The distribution is defined by a single parameter (here we use λ) that represents both the MEAN and the VARIANCE and so must be positive. If X is a random variable taking a Poisson distribution with parameter λ, then the probability that we will observe X taking the value x, the probability mass function of X, is:

$$\Pr(X = x) = \frac{\lambda^x e^{-\lambda}}{x!}$$

for nonnegative integer values of x.

The characteristic that the mean and variance are the same is one of the signatures of the Poisson distribution and of use in identifying the correct model when we have several observations from the distribution. Note that:

$$\Pr(X = x) = \frac{\lambda}{x}\Pr(X = x-1)$$

and so the mode of the distribution is the largest integer value smaller than λ (since this is the last value for which λ/x is greater than one and thus the last value for which the probability is increasing) or, if λ is an integer, λ and $\lambda - 1$ are both modal values.

The Poisson distribution is the limiting distribution for a binomial distribution, i.e. as n increases in the binomial distribution and p decreases to keep np constant, then the Poisson distribution provides an increasingly better approximation. As λ increases in the Poisson distribution, then the NORMAL DISTRIBUTION provides a better and better approximation. It is of no surprise then to learn that the Poisson distribution is skewed, but that this SKEWNESS decreases as λ increases. For further details on how the Poisson distribution relates to other distributions see Leemis (1986).

While the normal distribution may provide a good approximation to the Poisson data, techniques such as the t-test may often not be valid for comparing two groups because the assumption of equal variances only holds under the NULL HYPOTHESIS. In this case, a square-root TRANSFORMATION of the data can stabilise the variances while leaving a distribution that is still reasonably approximated by the normal (Altman and Bland, 1995). In practice, the data may suffer from OVERDISPERSION; i.e. there is more variation in the data than would be anticipated from a Poisson model. This may be a result of heterogeneity in the population; the probability of a 'success' may not be constant. Such variation may be accounted for by using a NEGATIVE BINOMIAL DISTRIBUTION or if the sources of variation are known it may be modelled via Poisson regression (see GENERALISED LINEAR MODEL). For further reading see Armitage and Colton (1998). *AGL*

Altman, D. G. and Bland, M. J. 1995: Transforming data. *British Medical Journal* 312, 770. **Armitage, P. and Colton, T.** (eds) 1998: *Encyclopaedia of biostatistics*. Chichester: John Wiley & Sons, Ltd. **Armitage, E., Drummond, E. H., Ghosh, S. and Ferguson, A.** 1999: Incidence of juvenile-onset Crohn's disease in Scotland. *The Lancet* 353, 1496–7. **Leemis, L. M.** 1986: Relationships among common univariate distributions. *The American Statistician* 40, 2, 143–6.

Poisson regression See GENERALISED LINEAR MODEL (GLM)

population projection See DEMOGRAPHY

positive predictive value (PPV) PPV is defined for a diagnostic test for a particular condition as the PROBABILITY that those who have a positive test actually have the condition as measured by a reference or 'gold' standard (contrast with the NEGATIVE PREDICTIVE VALUE).

If the data are set out as in the table, then PPV $= a/(a + b)$. PPV can also be expressed as a percentage. PPV depends on the prevalence of disease in the target population and this has important consequences in the context of population screening. In contrast, test SENSITIVITY and SPECIFICITY often remain the same when a test is applied in different populations. For example, suppose that test sensitivity and test specificity both equal 95% and 1 000 000 people are tested in a screening programme. If the prevalence of the disease in the population is 10%, i.e. 1 in 100, then 95 000 out of the 140 000 with positive tests would have the disease and PPV = 68%. If the prevalence of the disease in the population, however, is only 0.1%, i.e. 1 in 1000, then 950 out of the 50 900 with positive tests would have the disease and PPV = 1.86%.

positive predictive value *General table of test results among a+b+c+d individual samples*

	Disease		
	Present	*Absent*	*Total*
Positive	a	b	$a+b$
Negative	c	d	$c+d$
Total	$a+c$	$b+d$	$a+b+c+d$

Thus even with a relatively high prevalence of the condition in the population, many false positive test results may be generated (see FALSE POSITIVE RATE, with the attendant harm to the patient from anxiety and further investigations. Hence the usefulness of a diagnostic test depends not only on the test characteristics, i.e. the test's sensitivity and specificity, but also on the prevalence of the condition in the population in which the test is used. As disease prevalence is likely to be lower in community populations than in hospital populations, a diagnostic test may be useful in a hospital setting where there is a high prior probability that a patient has the disease, but may not be useful as a population screening test.

The PPV should be presented with CONFIDENCE INTERVALS (typically set at 95%) calculated using an appropriate method such as that of Wilson, which will not produce impossible values (percentages greater than 100 or below 0) when PPV approaches extreme values. *CLC*

[See also FALSE POSITIVE RATE, POPULATION SCREENING, PREVALENCE, TRUE POSITIVE RATE]

Altman, D. G., Machin, D., Bryant, T. N. and Gardner, M. J. 2000: *Statistics with confidence*, 2nd edition. London: BMJ Books.

posterior distribution This is a PROBABILITY DISTRIBUTION that represents the information, or beliefs, associated with a parameter of interest after data are collected or an experiment conducted. The posterior distribution is produced by combining the PRIOR DISTRIBUTION, which represents information about the parameter before the collection of data, with the LIKELIHOOD function, which represents the information about the parameter contained in the experimental data, by the use of BAYES' THEOREM.

Formally the posterior distribution is derived as follows:

$$p(\theta|\text{Data}) \propto p(\text{Data}|\theta)p(\theta)$$

in which $p(\theta)$ is the prior distribution expressing initial beliefs in the parameter of interest, θ, $p(\theta|\text{Data})$ is the corresponding posterior distribution of beliefs and $p(\text{Data}|\theta)$ is the likelihood. The posterior distribution provides the full information concerning the parameter of interest. It is often useful to display the posterior on the same plot as the prior distribution and likelihood, in a so-called triplot. This allows the reader to understand to what extent the prior distribution is contributing to the overall inference. Examples of these plots are shown in the figure. In figure (a) an example is shown in which the prior distribution dominates the likelihood and almost fully determines the posterior. Figure (b), in contrast,

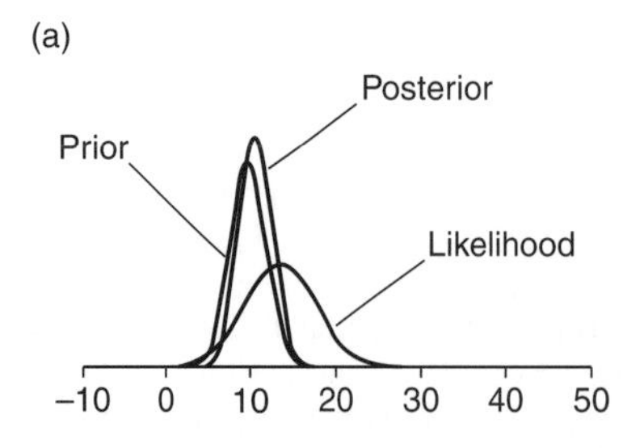

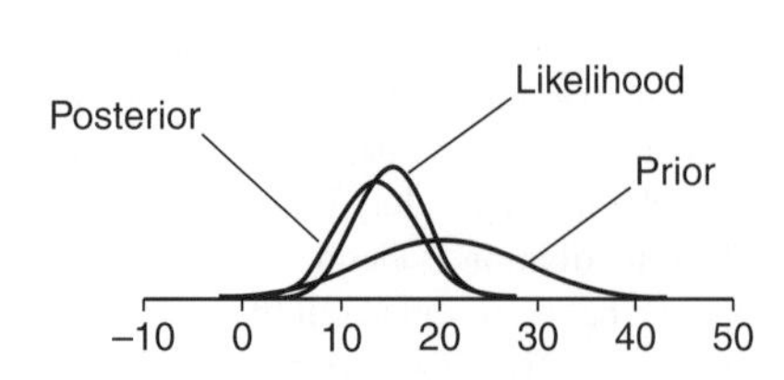

posterior distribution *Examples of triplots of prior distribution, likelihood and posterior distribution*

shows an example in which the prior distribution is weak and the posterior is essentially determined by the likelihood.

It is sometimes convenient to represent this posterior information by summaries. As posterior estimates, the posterior MEAN and posterior MODE are most often used and the uncertainty in the estimate is often reported by a posterior CREDIBLE INTERVAL.

An important consideration following a Bayesian analysis is the reporting of the results. The statistician's job is not over when the analysis is completed and a report written, because it is at this stage that thought needs to be given to the transmission of information to diverse groups of remote customers. For example, in the context of a pharmaceutical drug trial, there are at least three groups of individuals who interact with each other during the drug development process. These groups are the experimenters, the reviewers and the consumers. The aim of the experimenters, among whom are individual pharmaceutical companies, research organisations and clinicians, is to influence the customers, who are the doctors treating patients. They do this by providing them with information that has, in a sense, been 'sanitised' to ensure objectivity by the reviewers, who are the editors of journals and regulatory authorities. Because they each have different motivations it is not at all clear that there will be a single 'parcel of information' appropriate for each customer group.

One approach is to provide a range of posterior distributions based on a 'community of priors'. This approach works well if the community is broad enough to cover many different prior beliefs about treatment effects. Alternatively, the likelihood function can be reported allowing remote customers to input their own prior distributions to derive the appropriate posterior distribution. This approach will only work if the remote customers are able to carry out the calculations, which may limit its use to only the simplest cases. *AG*

post hoc analyses These are analyses that were not specified in advance of data collection. Such analyses tend to be regarded with suspicion, because of the possibility that an observed ASSOCIATION may have been selected from among a large number of potential associations that were examined but not reported ('data dredging'). Such multiple comparisons can lead to inflation of TYPE I ERROR rates. If the number of associations tested is known then formal procedures such as the BONFERRONI CORRECTION may be used to correct *P*-VALUES (SIGNIFICANCE LEVELS). However, these methods have substantial disadvantages, one in particular being that they are highly conservative.

Unfortunately, the notion that the formulation of prior hypotheses is a guarantor against being misled is itself misleading. If we do 100 randomised trials of useless therapies, each testing only one hypothesis and only performing one statistical test, all statistically significant results will be spurious. Furthermore, it is impossible to police claims that reported associations were examined because of existing hypotheses. This notion has been satirised by Cole (1993), who announced that, using a computer algorithm, he had generated every possible hypothesis in EPIDEMIOLOGY so that all statistical tests are now of a priori hypotheses. In practice, the best approach is to report accurately the context in which an association was examined and to regard findings selected from among many comparisons as requiring confirmation by further research. *JS/GDS*

Cole, P. 1993: The hypothesis generating machine. *Epidemiology* 4, 3, 271–3.

power See SAMPLE SIZE DETERMINATION IN CLINICAL TRIALS

power transformation See TRANSFORMATIONS

P-P plots See PROBABILITY PLOTS

prevalence The prevalence, or point prevalence, of a disease is the number of cases of a disease that exist at a specified point in time in a defined population. It is generally presented as a rate. Thus:

$$\text{Prevalence} = \frac{\text{Number of cases of a disease at a particular point in time}}{\text{Number in the population at that point in time}}$$

This results in a number between 0 and 1, but for ease of presentation it is often expressed as a rate per 1000, per 100 000 or per 1 000 000, depending on the disease rarity. As an example, the number of males living with colorectal cancer in Scotland on 31 December 2005 was reported as 9880, giving a prevalence rate of 401 per 100 000 Scottish males (NHS National Services Scotland, Information Services Division; www.isdscotland.org).

Care should be taken to distinguish between prevalence and INCIDENCE. Although the definitions appear similar at first sight, they are used for different purposes and it is essential to distinguish between them correctly.

The prevalence of a disease clearly depends on the incidence and also on the duration of the disease. A disease with a high incidence rate, from which most sufferers die very quickly, will have a low prevalence at any point in time. Conversely, a chronic disease with a low incidence rate may have a high prevalence if the duration is long. If the incidence and average duration of the disease remain approximately constant over time the prevalence and incidence will be related by:

$$\text{Prevalence} = \text{Incidence} \times \text{Duration}$$

Prevalence is of most use in determining the burden of chronic disease in the population and therefore is useful in allocating resources and planning healthcare services. It is of limited use in epidemiological studies of disease aetiology because it does not measure RISK (Woodward, 2004). Using prevalence to assess disease burden requires care when dealing with curable diseases. The prevalence rate quoted previously for colorectal cancer in Scottish males in 2005 (401 per 100 000) includes cases first diagnosed up to 20 years previously. A more realistic assessment of current burden for males is obtained from the number of prevalent cases diagnosed within the previous year, which was 1463 (59 per 100 000 Scottish males) on 31 December 2005 (NHS National Services Scotland, Information Services Division; www.isdscotland.org).

Paradoxically, some important advances in treatment can lead to an increase in prevalence of a disease. For example, the introduction of insulin as a treatment for diabetes resulted in a reduction in the number of deaths and an increase in the prevalence of diabetes in the population.

A more detailed discussion of prevalence and the relationship between prevalence, incidence and disease duration is given in Rothman, Greenland and Lash (2008). *WHG*

[See also INCIDENCE]

Rothman, K. J., Greenland, S. and Lash, T. L. 2008: *Modern epidemiology*, 3rd edition. Philadelphia: Lippincott Wilkins and Williams. **Woodward, M.** 2004: *Epidemiology: study design and data analysis*, 2nd edition. Boca Raton: Chapman & Hall.

primary endpoint See ENDPOINT

principal component analysis This is an exploratory technique, mainly used in partitioning the variation present in a quantitative multivariate dataset and in examining the data to highlight their important patterns or features.

Suppose that p observations have been taken on each of n individuals and the values have been collected into a data matrix having n rows and p columns. For example, Jackson (1991, pp. 107–9) provides a dataset on hearing loss as assessed by audiometry for 100 males aged 39. Each subject had decibel loss measured at four frequencies for each of two ears, yielding a 100×8 data matrix with values between -10 and 99 in each cell. Other instances of multivariate medical datasets might arise when screening patients, e.g. when a variety of measurements are routinely taken on all individuals signing up for a health centre, or in disease characterisation, where measurements are taken on variables that should distinguish sufferers from healthy individuals.

One of the main stumbling blocks in trying to assimilate a multivariate data matrix is the fact that ASSOCIATIONS exist between the columns. For quantitative data, VARIANCE measures scatter while COVARIANCE or CORRELATION measures association. The table illustrates such associations for the audiometry data. The variables are denoted by ear (L/R) plus frequency (500, 1000, 2000, 4000 Hz); the values down the left-to-right diagonal give the variances of the variables; the entries below the diagonal give the covariances between pairs of variables; and the entries above the diagonal give the corresponding correlations.

In essence, principal component analysis simply transforms the p measured variables $x_1, x_2, \ldots, x_p$ into a set of linear combinations $y_1, y_2, \ldots, y_p$ (i.e. $y_i = a_{i1}x_1 + a_{i2}x_2 + \cdots + a_{ip}x_p$ for all i, where the coefficients a_{ij} are suitable constants), which are mutually uncorrelated and which successively maximise the variance of such linear combinations. In other words, y_1 is the linear combination of the x that has the greatest possible sample variance among all linear combinations for the given dataset, y_2 is the linear combination of the x that has the next largest variance and so on. Technically, the variances of the y are given by the eigenvalues of the COVARIANCE MATRIX of the original x and the coefficients a_{ij} are given by the elements of the corresponding eigenvectors. These quantities are obtainable from the raw data in all standard statistical software packages; the y are known as the *principal components* of the data.

Principal components are useful for multivariate data exploration in various ways. It is often the case that each y can be interpreted by relating the size of the coefficients a_{ij} to the variables that they multiply and hence a substantive meaning can be attached to the component. Since the components are arranged in decreasing order of variance, such interpretation will therefore identify the main sources of variation among the sample members. Second, by considering the variance of each component in relation to the total

principal component analysis *Variances, covariances and correlations for audiometry data*

	L500	*L1000*	*L2000*	*L4000*	*R500*	*R1000*	*R2000*	*R4000*
L500	41.07	0.78	0.40	0.26	0.70	0.64	0.24	0.20
L1000	37.73	57.32	0.54	0.27	0.55	0.71	0.36	0.22
L2000	28.13	44.44	119.70	0.42	0.24	0.45	0.70	0.33
L4000	32.10	40.83	91.21	384.78	0.18	0.26	0.32	0.71
R500	31.79	29.75	18.64	25.01	50.75	0.66	0.16	0.13
R1000	26.30	34.24	31.21	33.03	30.23	40.92	0.41	0.22
R2000	14.12	25.30	71.26	57.67	10.52	24.62	86.30	0.37
R4000	25.28	31.74	68.99	269.12	18.19	27.22	67.26	373.66

variance of all components, it is often apparent that just a few components account for most of the variability in the sample and the remaining components can be ignored as essentially representing the 'noise' in the system. A typical rule-of-thumb employed by practitioners is to retain only as many components as account for around 75% to 80% of the total variance. Each sample individual's value (or *score*) is easily obtained on each principal component by applying the definition of the component to the original sample data. Taking the scores on the first two principal components and plotting the individuals in a SCATTERPLOT from these two sets of values then gives the best two-dimensional view of the data. This enables any OUTLIERS or groupings of individuals to be identified and other patterns in the data can also be readily discerned. Finally, since the principal components are uncorrelated, the scores on the y can form useful input to further statistical analysis.

However, forming a linear combination of the x only really makes sense if they are all similar entities. Thus, if they are all different (e.g. a height, a weight and a count, say) or if they are similar but of very varying magnitudes (e.g. height of individual, length of leg, length of arm and head circumference) then it is preferable to *standardise* the variables before analysis. In this case, the *correlation* rather than the covariance matrix should form the basis of the calculations. It is important to be aware that analysis of the standardised data gives different results from analysis of the unstandardised data, so careful consideration is needed at the outset to decide which analysis is the more appropriate.

To illustrate these ideas, consider a principal component analysis of the audiometry data. First, although all units of the eight variables are the same (Hz), it is evident from the diagonal elements of the table that the variances of the two highest frequencies (384.78, 373.66) are nearly 10 times the size of those of some lower frequencies (e.g. 40.92, 41.07). Hence standardisation is warranted and the analysis should be conducted on the correlation matrix. The eigenvalues of this matrix, i.e. the variances of the principal components, are found to be 3.93, 1.62, 0.98, 0.47, 0.34, 0.31, 0.20 and 0.15. The first four components have a combined variance of 6.0, which is 75% of the overall variance of 8.0, so we can effectively replace the eight original variables by the first four principal components.

How might we interpret these components? The first one is:

$$y_1 = 0.40x_1 + 0.42x_2 + 0.37x_3 + 0.28x_4 + 0.34x_5 \\ +0.41x_6 + 0.31x_7 + 0.25x_8$$

where the numbering is from L500 for x_1 through to R4000 for x_8. The coefficients are all positive and of approximately the same size, so the component is approximately proportional to the sum of the x. This represents the overall hearing level of an individual and implies that individuals who suffer loss at some frequencies are likely to suffer loss at the other frequencies also. The main source of variation among the sample is thus in terms of the individual's overall hearing level. The second component is:

$$y_2 = -0.32x_1 - 0.23x_2 + 0.24x_3 + 0.47x_4 - 0.39x_5 \\ -0.23x_6 + 0.32x_7 + 0.51x_8$$

The coefficients are similar between the left and right ear for each frequency but now show a 'contrast' – negative coefficients attached to low frequencies (500 and 1000 Hz), positive coefficients attached to high frequencies (2000 and 4000 Hz). It is known that hearing loss as individuals age is first noticeable at high frequencies, so the second most important source of variation among sample members is in terms of this form of hearing loss. The third component is:

$$y_3 = 0.16x_1 - 0.05x_2 - 0.47x_3 + 0.43x_4 + 0.26x_5 \\ -0.03x_6 - 0.56x_7 + 0.43x_8$$

'Small' coefficients correspond to 'unimportant' variables, so we can conclude that the third most important source of variation among sample members is a contrast between just the two higher frequencies. Finally, the fourth component has negative coefficients for the first four variables and positive coefficients for the other four variables and is therefore a contrast between left and right ears. Thus we have been able to characterise the main sources of variability among sample members. A scatterplot of the scores on the first two components reveals three potential outliers in the sample, which should be checked for aberrant values and possibly removed from further analysis, but no other structure of interest.

In addition to its role in multivariate exploration and description, principal component analysis is often also used either as a prelude to, or in conjunction with, a range of other techniques such as variable selection or orthogonal regression. (See Jackson, 1991, and Jolliffe, 2002.) *WK*

[See also CORRESPONDENCE ANALYSIS, FACTOR ANALYSIS]

Jackson, J. E. 1991: *A user's guide to principal components*. New York: John Wiley & Sons, Inc. **Jolliffe, I. T.** 2002: *Principal component analysis*, 2nd edition. New York: Springer.

principal stratification This is a method for the estimation of treatment effects in randomised CLINICAL TRIALS adjusting for an intermediate outcome that is not the primary ENDPOINT. It involves classifying subjects into classes that are defined by their joint potential responses of the intermediate variable to all possible random allocations. These classes are known as principal strata, which have the property that they are independent of treatment allocation and can be handled in the analysis in an analogous way to pre-randomisation variables. Frangakis and Rubin (2002) introduced the concept of principal effects, which compare treatments within principal strata (within-class or stratum-specific INTENTION-TO-TREAT (ITT) effects). The principal strati-fication method allows for hidden confounding (i.e. selection effects) between intermediate and final outcomes in the evaluation of treatment–effect mediation and post-randomisation sources of treatment–effect heterogeneity (see ADJUSTMENT FOR NONCOMPLIANCE IN RANDOMISED CONTROLLED TRIALS and DIRECT AND INDIRECT EFFECTS).

We now illustrate principal stratification in the context of assessing direct and indirect effects of RANDOMISATION on an outcome. Consider a binary randomisation variable ($Z_i = 0, 1$) for subject i and a binary intermediate variable ($M_i = 0, 1$). We define $M_i(z)$ to be the potential values of the intermediate variable after random allocation, such that $M_i(0)$ is the value of the intermediate outcome if randomised to the control condition ($Z_i = 0$) and $M_i(1)$ is the value if randomised to treatment ($Z_i = 1$). These are referred to as potential values as only one of these values can be observed in each subject, depending on the value of Z_i.

The table shows there are four distinct classes for the joint combination of $M_i(1)$ and $M_i(0)$. These classes are the principal strata. Note that a subject's stratum membership is not usually known: e.g. a subject with $Z_i = 1$ and $M_i = 1$ is only known to belong to class 2 or 4. In general, the stratum-specific average treatment effects τ_1, τ_2, τ_3 and τ_4 may differ. The direct effects of treatment are measured by τ_1 and τ_2, since in these classes there is no change in the value of the intermediate variable between random allocations. Since class membership is independent of treatment allocation, the intention-to-treat effect for all subjects τ is the weighted average of the effects within strata:

$$\tau = \sum_j \pi_j \tau_j$$

principal strata *The four possible principal strata with a binary mediator (M) and binary randomisation (Z), the proportion of the sample in each strata and the stratum-specific treatment effect*

Class (stratum)	$M_i(0)$	$M_i(1)$	$(M_i(0), M_i(1))$	*Proportion of subjects*	*Treatment effect*
1	0	0	(0,0)	π_1	τ_1
2	1	1	(1,1)	π_2	τ_2
3	1	0	(1,0)	π_3	τ_3
4	0	1	(0,1)	π_4	τ_4

We now consider how the treatment effects τ_1, τ_2, τ_3 and τ_4 may be estimated. Within each principal stratum C ($C = 1$ to 4), and given a set of k measured covariates, $\boldsymbol{X}$, we may have a model for the outcome

$$Y = \sum_k \varphi_k X_k + \tau_c z + \xi$$

Superficially, it looks straightforward to estimate τ_c since it is an effect of randomisation, but, of course, we typically do not know to which stratum each subject belongs since the class, C, is frequently not identified. In order to identify this, we require baseline covariates that are predictors of stratum membership C.

For estimation using principal stratification, we would construct a regression (latent class) model to predict stratum membership using baseline covariates, $\boldsymbol{X}$. We would then simultaneously model the ITT effects on outcome within the principal strata. The estimation proceeds by specifying a full probability model using MAXIMUM LIKELIHOOD ESTIMATION. Alternatively, a fully Bayesian approach (see BAYESIAN METHODS) to estimation could be used. If we have missing outcome data (see MISSING DATA) we can also simultaneously fit a third model predicting missing outcomes, based on the assumptions of latent ignorability (see Frangakis and Rubin, 1999).

We now illustrate these concepts with some more examples where principal stratification is commonly applied. Consider a simple example where we have random allocation of subjects to receive either treatment or no treatment (control). Those allocated to the control group cannot get access to the treatment (there is no contamination), but those subjects in the treatment group can fail to receive the treatment, and so end up in the control group. In this case we can define two principal strata: compliers and noncompliers. The first principal stratum of compliers are those subjects who are treated if they are allocated to the treatment group, and not treated if allocated to the control group. The second principal stratum is the noncompliers who will never receive the treatment, regardless of their individual random allocation. It is possible to identify these two classes in those subjects who are allocated to treatment, but they remain hidden (latent) in the control group. The ITT effect within the principal strata of compliers is known as the COMPLIER AVERAGE CAUSAL EFFECT, which is a special case of a principal effect (see ADJUSTMENT FOR NONCOMPLIANCE IN RANDOMISED CONTROLLED TRIALS). It can be estimated as described above or using INSTRUMENTAL VARIABLES methods.

Frangakis and Rubin (2002) introduced the concept of principal surrogacy to apply principal stratification to analysing surrogate or biomarker outcomes (see SURROGATE ENDPOINTS). The authors provide an example with random allocation to standard treatment ($z = 0$) or a new treatment ($z = 1$) where the outcome was survival time and the putative surrogate M is a measure of CD4 count at two months (1 = high and 0 = low). The table again illustrates the four basic principal strata: class 1 is those subjects who will have a low CD4 count unaffected by treatment; class 2 is those subjects who will have a high CD4 count regardless of treatment; class 3 is those subjects who have a higher CD4 count under the standard treatment than the new treatment; and class 4 is those subjects who have a lower CD4 count under the standard treatment than the new treatment. CD4 count is then defined as a principal surrogate for survival time if the group of subjects with no ITT effect on CD4 count also have no ITT effect on survival time. Mathematically, these are the groups in which $M_i(1) = M_i(0)$, which would be classes 1 and 2 in the table, and we have then identified a principal surrogate if $\tau_1 = \tau_2 = 0$. Principal surrogates allow estimation of true causal effects, as opposed to statistical surrogates (Prentice, 1989) because they avoid post-randomisation SELECTION BIAS (unmeasured confounding), which may be present between the surrogate and true outcomes.

Although the basic idea of principal stratification is the estimation of ITT effects within principal strata, it is possible to fit alternative explanatory models nested within principal strata. For example, typically we are interested in a univariate response, but we could investigate the advantages of simultaneously estimating effects for two or more different outcomes (i.e. multivariate responses). In the context of treatment compliance, Jo and Muthén (2002) have investigated the use of the latent growth curve or trajectory models for longitudinal outcome data.

Recently there has been an extended use of principal stratification in assessing mediation of treatment effects by intermediate variables (see Jo, 2008; Gallop *et al.*, 2009; Emsley, Dunn and White, 2010). The latter authors define principal strata according to measures of therapeutic alliance in a trial with random allocation to either treatment or control (no treatment), and where those allocated to control group cannot get access to treatment. Therapeutic alliance is only measured in the treatment group. The strata are defined as being: principal stratum 1 is the low alliance strata, where subjects would have a low therapeutic alliance when allocated to the treatment group, and are not treated when allocated to control; and principal stratum 2 is the high alliance strata, where subjects would have a high therapeutic alliance when allocated to the treatment group, and are not treated when allocated to the nontreatment group. As previously, it is possible to identify these two classes in those allocated to the treatment group but they remain hidden in the control group. The authors then examine ITT effects and explanatory models for a dose–response relationship within the principal strata.

Principal stratification is most commonly used with binary treatment and binary intermediate variables. However, the framework can be extended to settings with multivariate, time-dependent and continuous intermediate variables, and to multiple random allocations (Frangakis and Rubin, 2002). *RE/GD*

Emsley, R. A., Dunn, G. and White, I. R. 2010: Mediation and moderation of treatment effects in randomised controlled trials of complex interventions. *Statistical methods in medical research* (in press). **Frangakis, C. E. and Rubin, D. B.** 1999: Addressing complications of intention-to-treat analysis in the combined presence of all-or-none treatment–noncompliance and subsequent missing outcomes, *Biometrika* 86, 2, 365–79. **Frangakis, C. E. and Rubin, D. B.** 2002: Principal stratification in causal inference. *Biometrics* 58, 1, 21–9. **Gallop, R., Small, D. S., Lin, J. Y., Elliott, M. R., Joffe, M. and Ten Have, T. R.** 2009: Mediation analysis with principal stratification. *Statistics in Medicine* 28, 7, 1108–30. **Jo, B.** 2008: Causal inference in randomized experiments with mediational processes. Psychological Methods 13, 4, 314–36. **Jo, B. and Muthén, B. O.** 2002: Longitudinal studies with intervention and noncompliance: estimation of causal effects in growth mixture modeling. In Duan, N. and Reise, S. (eds), *Multilevel modeling: methodological advances, issues, and applications*. Lawrence Erlbaum Associates, pp. 112–39. **Prentice, R. L.** 1989: Surrogate endpoints in clinical-trials – definition and operational criteria. *Statistics in Medicine* 8, 4, 431–40.

prior distribution This is a PROBABILITY DISTRIBUTION that represents the information, or beliefs, associated with a parameter of interest before data are collected or an experiment conducted. The prior distribution is an essential component of a Bayesian analysis (see BAYESIAN METHODS) and differentiates it from traditional, so-called frequentist, analysis. The prior distribution is used together with the LIKELIHOOD function, which represents the information about the parameter contained in the experimental data, to produce the posterior distribution from BAYES' THEOREM. Formally, a Bayesian analysis proceeds from the following formula:

$$p(\theta|\text{Data}) \propto p(\text{Data}|\theta)p(\theta)$$

in which $p(\theta)$ is the prior distribution expressing initial beliefs in the parameter of interest, θ, $p(\theta|\text{Data})$ is the corresponding POSTERIOR DISTRIBUTION of beliefs and $p(\text{Data}|\theta)$ is the likelihood.

There are many types of prior distribution and many ways to derive them. Prior distributions can be based on information available in the literature – historical data – and quantified through a formal process of meta-analysis (see SYSTEMATIC REVIEWS AND META-ANALYSIS). In the pharmaceutical industry, prior distributions about potential treatment effects can be determined from existing clinical databases of similar drugs from the same class of compounds. Where existing data are not available, experts can be used to elicit prior distributions by a process of questioning and refinement. Formalistic prior distributions can be determined to represent cases in which there is little or no relevant prior information. Such prior distributions are normally called noninformative and a Bayesian analysis with a noninformative prior will usually give very similar conclusions to those obtained using a frequentist analysis. However, the output and presentation of the analysis will be very different from standard approaches and is often found to be more intuitive and helpful by nonstatisticians. Finally, a set of prior distributions may also be used in order that a range of assumptions may be tested against their subsequent conclusions. For example, a community of priors could consist, in addition to a neutral uninformative prior, of priors that represent both 'pessimistic' and 'optimistic' scenarios. A 'pessimistic' prior would be one in which, for example, there would be considerable doubt about the effectiveness of a treatment; an 'optimistic' prior, in contrast, would represent strong prior belief that the treatment is effective.

The use of prior distributions in a Bayesian analysis is not without controversy. From one perspective, their use is a strength in that it allows the scientist to access more information and thereby produce stronger inferences. This strength, however, can also be regarded as a weakness and gives rise to any number of questions. First, if the prior is based on historical data, is it relevant to the current investigation? If it is elicited from experts to what extent are the resulting inferences subjective and hence of less credibility than inferences based on data alone? In practice, the influence of the prior distribution may be minimal if the information contained in experimental data outweighs that contained in the prior, which will generally be the case if the experimental study has a large sample size *AG*

probability The notion of probability has two connotations: (1) as a mathematical discipline concerning the study of uncertainty and (2) as a numerical scale from zero to one to describe the frequency of occurrence, or degree-of-belief in, a given event.

The first definition is akin to statistics as a discipline, with broad distinction between the two being that whereas probability seeks to learn about the sample given characteristics of the population, statistics seeks to learn about the population from given characteristics of the sample. Their interrelationship is such that the theory of probability is the backbone of statistics. However, the second connotation of probability will be our focus here.

Any chance event happens according to some numerical measure between zero and one, with these extremes representing the probability of impossible and certain events respectively. In practice, most probabilities lie between these extremes (see the first figure). Sometimes, probability is expressed on an entirely equivalent 0–100% scale, conventionally known as *chance*, if so.

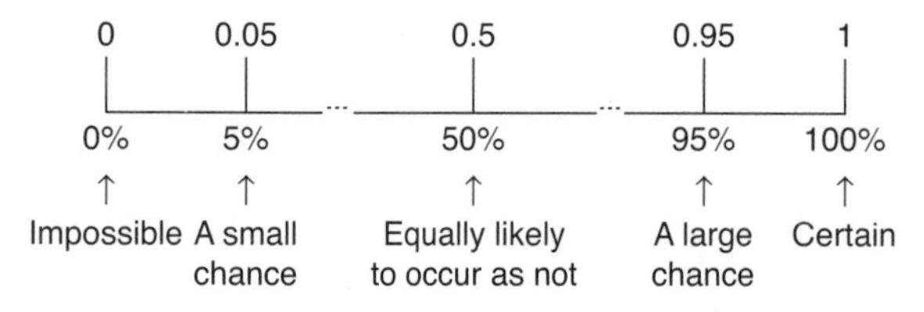

probability *Probability quantifies the scale of uncertainty. Note that probabilities of 0.05 and 0.95 are simply convenient round numbers towards the ends of the scale, with 'small' and 'large' being just (nonstandard) descriptive labels*

Usual notation to describe the probability of an event is to write the event in brackets or parentheses, namely {...}, [...] or (...), immediately after an abbreviation of probability to P, Pr or Prob, with no consensus about capitalising the initial letter. Thus:

$$\Pr\{\text{a male is pregnant}\} = 0$$

$$\text{P}[\text{a coin lands heads uppermost}] = 0.5$$

$$\text{prob}(\text{baby is either male or female}) = 1$$

are just some of the many ways that simple probability statements can be written. For the sake of brevity, the event being considered is often reduced to a single letter, so, for example, when considering the coin-tossing scenario, one might say $\Pr\{H\} = \Pr\{T\} = 0.5$, where H and T are shorthand for landing heads and tails respectively, it being understood from the context that only one toss of a fair coin at a time is being considered. (Note that a generalisation to describe chances of sequences of events when tossing a coin three times, say, could be Pr{HHH} or Pr{2H, 1T}, although one has to be mindful to distinguish whether one means in the latter case the ordered sequence H, H, then T or *any* sequence with two Hs and one T.)

It is important to know there are two ways of interpreting probability, as follows: (1) long-run average proportion or 'frequentist' view and (2) degree-of-belief or 'subjective' view.

The first type applies strictly only to those events that are repeatable, under assumed identical conditions, such as idealised coin tossing or die rolling, whereas the second applies universally, even to one-off events. Arguably, in medicine, all events are one-offs as they refer to individual patients with their own unique set of symptoms and treatments, genetic and environmental backgrounds and occur in particular places and times. Incidentally, statisticians need reminding from time to time there is no such thing as an average patient! (See CONSULTING A STATISTICIAN.) Largely due to mathematical convenience and history (see HISTORY OF MEDICAL STATISTICS), it is the frequentist definition that is by far the more commonly used in statistics, dominating almost all courses, textbooks and software.

However, there is a growing tendency among biostatisticians nowadays to favour the subjective formulation of probability, being the cornerstone of the Bayesian approach (see BAYESIAN METHODS). While some may criticise this approach for being too subjective and somehow less scientific than the frequentist approach, others argue that the ability to tailor results to one's own beliefs is actually advantageous and especially beneficial for applications in medical research.

At least the two philosophical approaches do not lead to conflicting results, even though the subjective approach is undoubtedly more intuitively appealing and more obviously applicable. It is helpful to bear in mind both definitions when thinking about probabilities and their interpretation. To illustrate, one can say Pr{H} = 0.5 *either* because the long-run ratio of number of heads to number of tosses approaches the value one-half as the number of tosses increases indefinitely *or* because in a single toss you reckon it is 50:50, being equally likely to fall tails as heads. Thus, a statement such as Pr{a person's blood group is type O} = 0.47 can, and does, mean both that in a large random sample of the population we can expect 47 in every 100 people to have blood group O and that, in the absence of further information, one's degree-of-belief about an individual patient being type O is 0.47.

How are probabilities assigned numerically? In simple situations, they are enumerated by appealing to the somewhat circular notion of 'equally likely' outcomes. Thus, for example, a standard die of six sides is equally likely to land any side facing upwards – hence the deduction Pr{getting a 6 on a single roll} = 1/6. This line of reasoning is unrealistic in medical applications, however. In practice, we assign probabilities by gathering (preferably large amounts of) data from random samples, in order to compute the average frequencies. For instance, it may be that based on a large, national, randomly selected cohort that the proportions of blood types O, A, B and AB are, respectively, 0.47, 0.42, 0.08 and 0.03, leading to the earlier pair of statements about type O blood group.

Notice in this simple example that the proportions representing the probabilities sum to one. This is no surprise given that the list of ABO types was exhaustive. A PROBABILITY DISTRIBUTION more generally describes the theoretical distribution of the total probability of one. These can be either for discrete values (some classic examples being BINOMIAL, POISSON and GEOMETRIC DISTRIBUTIONS) for which respective formulae for the probability density function sum to one or else continuous distributions (e.g. NORMAL, UNIFORM and CHI-SQUARE DISTRIBUTIONS) for which the area under the probability distribution curve integrates to one. A HISTOGRAM can be thought of as a pictorial representation of a probability distribution. The larger the sample on which it is based, the closer the histogram's shape approximates the underlying population probability distribution.

Next consider the combining of probabilities to quantify chances of more complex interactions of events. There are two rules that apply for special types of events that are said to be 'mutually exclusive' or 'independent' respectively, as follows:

1. *Addition rule*. The probability that any of two or more mutually exclusive events occurs is given by the sum of their individual probabilities. For example:

   $$\begin{aligned}&\Pr\{\text{person is type B } \textit{or} \text{ type AB}\}\\ &= \Pr\{\text{person is type B}\} + \Pr\{\text{person is type AB}\}\\ &= 0.08 + 0.03 = 0.11\end{aligned}$$

   Mutual exclusivity, or incompatibility, of events simply means that if one event occurs, then other events are precluded (e.g. if someone is of blood type B, they cannot also be simultaneously of type AB).
2. *Multiplication rule*. The probability that two (or more) independent events both (or all) occur together is given by the product of their individual probabilities. For example:

   $$\begin{aligned}&\Pr\{\text{two unrelated people are both type A}\}\\ &= \Pr\{\text{first is type A}\} \times \Pr\{\text{second is type A}\}\\ &= 0.42 \times 0.42 = 0.176\end{aligned}$$

   Independence of events means the outcome of one has no bearing on the other(s), such as here the blood types of unrelated individuals.

Note that the concept of independence or dependence of events plays a fundamental role in EPIDEMIOLOGY. In CASE-CONTROL STUDIES, patients with a disease (the 'cases') are compared with disease-free individuals (the 'controls') with respect to some possible risk factor. For instance, men with testicular cancer might be compared with controls regarding, say, milk consumption during adolescence. If more cases were 'exposed to the risk' of high milk consumption than controls, then the probabilities of exposure being different might suggest a possible causative link. In other words, the events {having the disease} and {having had the exposure} would not be independent events.

Thus, an observational study can be thought of in probabilistic terms as seeking to demonstrate whether these two events are dependent or independent of one another. If deemed to be dependent, there is an ASSOCIATION (note, not the same as *causation*; see CAUSALITY) linking the exposure to the disease, a matter of central importance.

To explore this further, one needs to introduce the concept of CONDITIONAL PROBABILITY. In brief, probabilities of events can change in the light of unfolding information, in particular whether or not another event is known to have occurred. To illustrate this, consider that the probability someone has tuberculosis is no longer the same if it is known that their skin test for tuberculosis was positive. Letting D denote the event {person has disease} and T+ denote the event {skin test is positive}, then interest focuses on the probability of {D given T+}, which is written Pr{D|T+} for short. The '|' sign is read as 'given', meaning events written after it are already known to have occurred. Now Pr{D|T+} is not equal to Pr{D}, since the skin test result, obviously, is not independent of the presence of the disease. Equally, no test is 100% accurate, for if it were in this example, knowledge of the skin test result would be equivalent to knowing the tuberculosis status, but in reality no test is this reliable (see POSITIVE PREDICTIVE VALUE).

In general, the conditional probability of any one event A given another (nonimpossible) event B is defined as Pr{A|B} = Pr{A and B}/Pr{B}, where '/' represents division. If it is the case that A and B are independent, then the multiplication rule says that the numerator Pr{A and B} = Pr{A} × Pr{B} and so it follows that Pr{A|B} = Pr{A}. Indeed, this is an alternative definition of independence of events, namely that the conditional probability of an event, A, on another, B, is the same as the unconditional probability of event A. (See the related and important result known as BAYES' THEOREM.)

An application of probability is in the use of the summary measure odds ratio (OR) or relative risk (RR) (see RELATIVE RISK AND ODDS RATIOS). Odds is simply an alternative way of expressing probability, adopted and favoured by bookmakers perhaps to make it less obvious their total probabilities within a given horse race, say, sum to more than one in order to guarantee themselves a long-run profit margin. If a probability of an event is denoted by p, then the odds for the same event equal $p/(1-p)$.

The odds of an event having probability 1/2 is quoted as evens. Probabilities below one-half are quoted as the number of chances of failing to the number of chances of occurring, for a suitable total number of chances to avoid fractions. For example, a probability 1/8 is the same an odds 7 to 1; probability 2/5 is odds 3 to 2; probability 1/21 is odds 20 to 1. Probabilities of events greater than one-half are expressed with larger number first and then 'on' (instead of 'against', the default assumption); e.g. odds of '2 to 1 on' means a probability of 2/3, whereas 2 to 1 against is the complementary probability of 1/3. Odds are virtually never used in health applications by themselves, but do occur in the context of odds ratios when it is convenient, especially when analysing case-control studies to have a comparative measure of relative probabilities across two groups, typically cases versus controls. When using LOGISTIC REGRESSION models also, computer output usually displays odds ratios and their CONFIDENCE INTERVALS in association with categorical variables.

The RR is defined as the ratio of two conditional probabilities:

$$\text{RR} = \Pr\{\text{disease}|\text{exposed}\}/\Pr\{\text{disease}|\text{not exposed}\}$$

If the disease status and exposure to a risk factor are independent, as when there is no causal link between them, then RR = Pr{disease}/Pr{disease} = 1, whereas if exposure and disease are dependent the RR will not equal 1.

Finally, there are two further rules for handling probability one can consider. The first of these, referring to any events A and B, regardless of whether independent or not, mutually exclusive or not, says:

$$\Pr\{\text{A or B}\} = \Pr\{\text{A}\} + \Pr\{\text{B}\} - \Pr\{\text{A and B}\}$$

This rule is best visualised with a Venn diagram showing overlapping events, A and B (see the second figure). Notice that it reduces to the addition law just given if A and B happen to be mutually exclusive, for then Pr{A and B} = 0, being impossible to occur together, so that a Venn diagram representation would show no overlap between sets representing A and B.

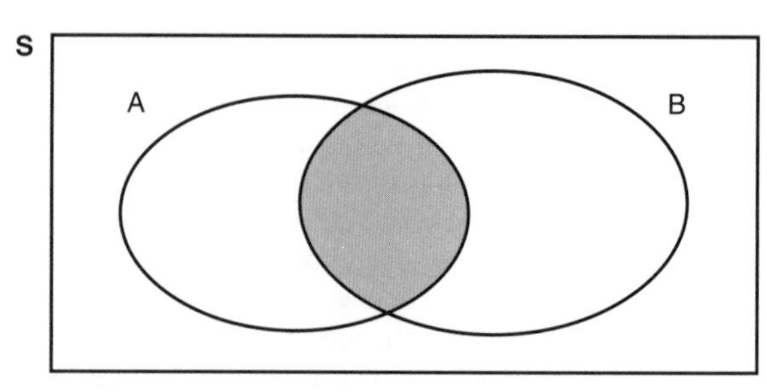

probability *Venn diagram depicting two general events A and B among all those possible within 'state space' S. The general rule says that the probability of lying within the region covered by A or B is that of the two individual regions minus the double-counted shaded region where A and B overlap*

The second general rule for combining probabilities is as follows:

$$\Pr\{\text{A and B}\} = \Pr\{\text{A}\} \times \Pr\{\text{B}|\text{A}\}$$

Note, by symmetry, that this can equivalently be stated as:

$$\Pr\{\text{A and B}\} = \Pr\{\text{B}\} \times \Pr\{\text{A}|\text{B}\}$$

Examples of the application of this rule are found in SCREENING STUDIES, where it is useful to consider SENSITIVITY and SPECIFICITY as well as FALSE POSITIVE and FALSE NEGATIVE RATES.

Other common uses of probability abound, for instance in the summary measures of incidence and prevalence, survival rates within life tables and applications in GENETIC EPIDEMIOLOGY to mention but a few. The most common sighting of probability in medical journals, for better or worse, is due to the preference to summarise evidence from studies in terms of P-VALUES, measures of probability used to assess plausibility of the NULL HYPOTHESIS of chance variation.

For further reading at a recreational level, Everitt (2008) provides an accessible account of the role of probability including health and research applications, while more formal accounts can be found in just about any textbook of medical statistics. *CRP*

Everitt, B. S. 2008: *Chance rules: an informal guide to probability, risk and statistics*. 2nd edition, New York: Springer.

probability density function (PDF) See PROBABILITY DISTRIBUTION

probability distribution This is a statement of all the outcome values that a variable can take, and the individual probabilities of obtaining those values. If outcomes are discrete (as in categorical variables, or variables that can only take integer numbers), we have two paired lists, one of outcomes and one of PROBABILITIES. For example, when interested in the number of heads achieved when tossing four fair coins, we have outcomes and probabilities:

Outcome (x)	Probability $\Pr(X=x)$
0	0.0625
1	0.2500
2	0.3750
3	0.2500
4	0.0625

This list of probabilities defines our probability distribution. Note that the list of probabilities must sum to one. However, even with five possible outcomes such an approach is becoming cumbersome, and we prefer to relate the outcome and probability mathematically. For our example, we note that the outcome, x, and probability, $\Pr(X=x)$, are related as:

$$\Pr(X = x) = \frac{3}{2x!(4-x)!}$$

where $x!$ (x factorial) is the product of all the integers up to and including x, and $0!$ is defined to be one. We refer to $\Pr(X=x)$ as the probability mass function. To define our probability distribution, we can just provide the mass function (if one exists). Note that in our example the distribution is an example of a BINOMIAL DISTRIBUTION. Other discrete examples are the POISSON, GEOMETRIC and NEGATIVE BINOMIAL DISTRIBUTIONS.

If the outcomes are on a continuous scale, e.g. the length of a tumour, then there are an infinite number of possible outcomes. Since the total probability must be equal to one, then generally the probability of the outcome being exactly a particular value is 0. In this case, rather than defining a distribution function, we define a probability density function (PDF). This is a curve, the area under which represents probabilities. The total area under the curve represents the total probability and therefore must be equal to one. In general, the area under the curve between any two values gives the probability that the outcome will lie between those two values.

It is often the case that we will work with the distribution function (or cumulative distribution function), which gives the probability of being less than a particular value and so equates to the area under the curve to the left of a particular value. As an example consider the success rate of a new type of operation. Suppose that it is equally likely to be anywhere from 0 (never successful) to 1 (always successful). Our probability density function will be a horizontal line (at $y=1$ since the area under the curve must be equal to one) (see the figure). The probability that the operation is successful more than 80% of the time is the area under the curve between 0.8 and 1, which is equal to 0.2. Note that our example is an example of the BETA DISTRIBUTION. Other continuous probability distributions include the CHI-SQUARE, EXPONENTIAL, F, GAMMA, LOGNORMAL, t and (most importantly) NORMAL DISTRIBUTIONS. *AGL*

probability plots These are plots for comparing two PROBABILITY DISTRIBUTIONS or for assessing assumptions about the probability distribution of a sample of observations.

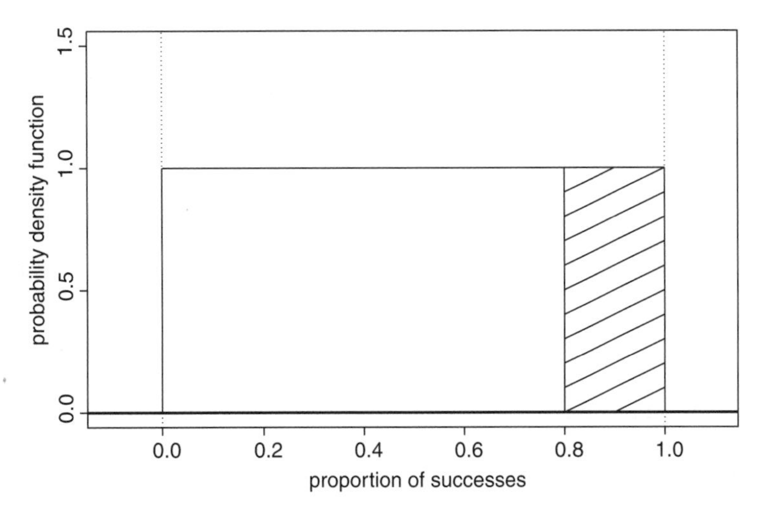

probability distribution *The probability density function for the success rate of an operation, when all rates are equally likely. Note that the area under the curve is equal to one and the area under the curve between 0.8 and 1 is 0.2*

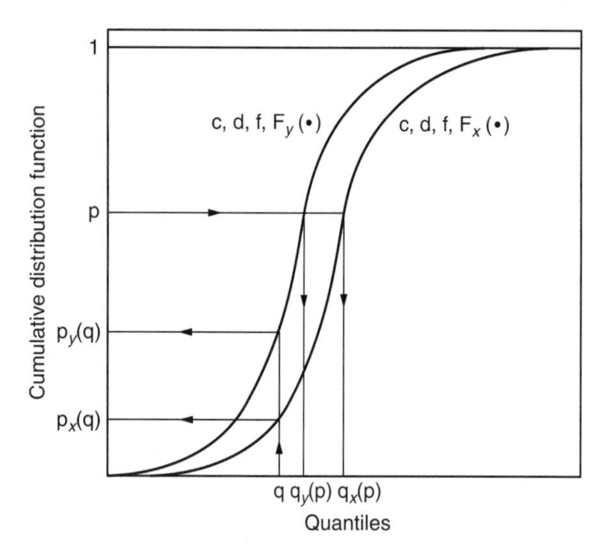

probability plots *Diagram illustrating P-P and Q-Q plots*

There are two basic types, the *probability–probability plot* (P-P plot) and the *quantile–quantile plot* (Q-Q plot). Each type is illustrated in the first figure.

A plot of points whose coordinates are the cumulative probabilities $\{p_x(q), p_y(q)\}$ for different values of q is a probability–probability plot, whereas a plot of the points whose coordinates are the *quantiles* $\{q_x(p), q_y(p)\}$ for different values of p is a quantile–quantile plot. As an example, a quantile–quantile plot for investigating the assumption that a set of data is from a NORMAL DISTRIBUTION would involve plotting the ordered sample values $y_{(1)}, y_{(2)}, \ldots, y_{(n)}$ against the quantiles of a standard normal distribution, i.e.:

$$\Phi^{-1}[p_i]$$

where usually:

$$p_i = \frac{i-1/3}{n+1/3} \text{ and } \Phi(x) = \int_{-\infty}^{x} \frac{1}{\sqrt{2\pi}} e^{-\frac{1}{2}u^2} du$$

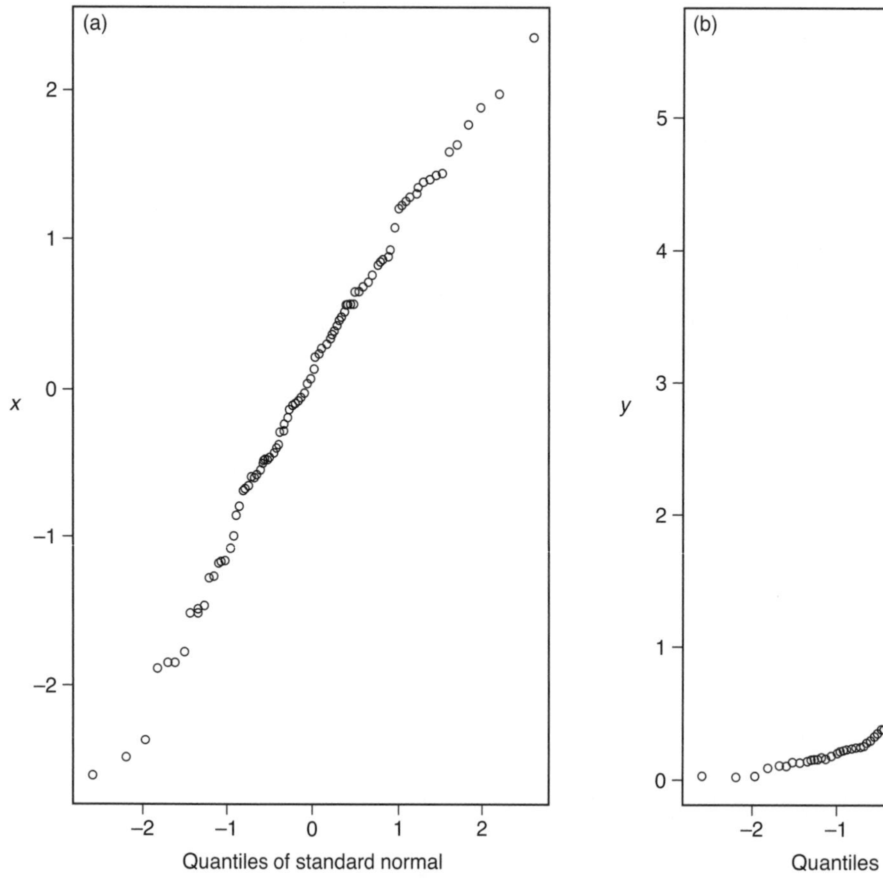

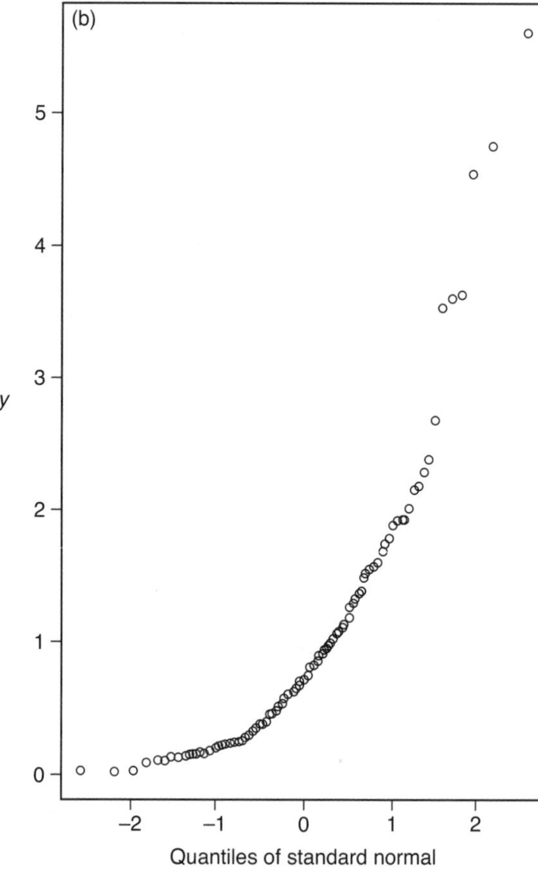

probability plots *Two normal probability plots: (a) 100 observations from a normal distribution and (b) 100 observations from an exponential distribution*

This is usually known as a *normal probability plot*. Two such plots are shown in the second figure on page 354: the first (a) is the probability plot for 100 points generated from a normal distribution and the second (b) is the corresponding plot for 100 points generated from an EXPONENTIAL DISTRIBUTION. Plot (a) is essentially linear, confirming that the sample is close to normally distributed. Plot (b) by contrast is clearly concave, indicating the presence of right SKEWNESS. KURTOSIS if present would show itself as an S-shaped plot. *BSE*

probit model A model for dichotomous or ordinal responses that can be defined either as a GENERALISED LINEAR MODEL or by using a latent response formulation. In a generalised linear model, the expectation μ_i of the response for unit i given covariates $\mathbf{x}_i = (x_{1i}, \ldots, x_{ki})'$ is modelled as $g(\mu_i) = \beta_0 + \beta_1 x_{1i} + \cdots + \beta_k x_{ki} = \mathbf{x}_i'\boldsymbol{\beta}$, where $g(\cdot)$ is a link function and $\boldsymbol{\beta}$ are regression parameters. In a binary probit model (for dichotomous responses), a probit link $\Phi^{-1}(\cdot)$ is specified as:

$$\Phi^{-1}(\mu_i) = \mathbf{x}_i'\boldsymbol{\beta} \text{ or } \mu_i = \Phi(\mathbf{x}_i'\boldsymbol{\beta}) \qquad (1)$$

where $\Phi(\cdot)$ is the standard normal cumulative distribution function. The distribution of y_i is then specified as a Bernoulli with MEAN μ_i. Proportions out of a total of n_i trials can be modelled using the same link function by specifying a BINOMIAL DISTRIBUTION with denominator n_i.

The same model can be formulated by specifying a linear regression model for an underlying latent (unobserved) continuous response y_i^*:

$$y_i^* = \mathbf{x}_i'\boldsymbol{\beta} + \varepsilon_i$$

with ε_i having standard NORMAL DISTRIBUTION.

The observed dichotomous response y_i then represents an indicator for y_i^* exceeding zero:

$$y_i = \begin{cases} 0 \text{ if } y_i^* \le 0 \\ 1 \text{ if } y_i^* > 0 \end{cases} \qquad (2)$$

This latent response formulation is equivalent to a generalised linear model with a probit link and Bernoulli distribution:

$$\begin{aligned} \mu_i &\equiv \mathrm{E}(y_i|\boldsymbol{x}_i'\boldsymbol{\beta}) = \Pr(y_i = 1|\boldsymbol{x}_i'\boldsymbol{\beta}) = \Pr(y_i^* > 0|\boldsymbol{x}_i'\boldsymbol{\beta}) \\ &= \Pr(\boldsymbol{x}_i'\boldsymbol{\beta} + \varepsilon_i > 0) = \Pr(\varepsilon_i > -\boldsymbol{x}_i'\boldsymbol{\beta}) \\ &= Pr(\varepsilon_i \le \boldsymbol{x}_i'\boldsymbol{\beta}) = \Phi(\boldsymbol{x}_i'\boldsymbol{\beta}) \end{aligned}$$

where the penultimate equality hinges on the symmetry of the normal density of ε_i.

Consider an ordinal response variable with $S+1$ categories 0, . . ., S. An ordinal probit model can be specified using the latent response formulation by extending the threshold model in (2) as follows:

$$y_i = \begin{cases} 0 & \text{if} & -\infty < & y_i^* & \le \kappa_1 \\ 1 & \text{if} & \kappa_1 < & y_i^* & \le \kappa_2 \\ \vdots & \vdots & \vdots & \vdots & \vdots \\ S & \text{if} & \kappa_S < & y_i^* & \le \infty \end{cases}$$

where κ_s, $s = 1, \ldots, S$ are threshold parameters.

Binary and ordinal logit models can be defined in the same way as their probit counterparts by replacing the link in equation (1) by a logit link and the distribution in (2) by a logistic distribution. LOGIT MODELS are more popular than probit models because the regression parameters can be interpreted as log odds ratios and because the models can be used for CASE-CONTROL STUDIES. One advantage of probit models is that they can easily be extended to the multivariate case by using several latent responses having a MULTIVARIATE NORMAL DISTRIBUTION. For further details see Finney (1971). *SRH/AS*

Finney, D. J. 1971: *Probit analysis*. Cambridge: Cambridge University Press.

product-limit estimator See KAPLAN–MEIER ESTIMATOR

product-moment correlation coefficient A synonym for Pearson's correlation coefficient. See CORRELATION

propensity scores In OBSERVATIONAL STUDIES aimed at comparing treatment arms investigators have no control over the treatment assignment. Therefore in contrast to randomised controlled trials (RCTs) (see CLINICAL TRIALS), estimates of group differences in outcome derived from such studies may be subject to confounding by observed pre-treatment covariates. The propensity score for an individual is defined as the CONDITIONAL PROBABILITY of being treated (or more generally of being in the group of interest), given the individual's covariate values. A sample's propensity scores can be used to balance the covariates in the two groups, thus creating a 'quasi-randomised' experiment and avoid bias. The approach has been applied in many fields including medicine, epidemiology and health services research in attempts to derive causal effect estimates from observational studies.

As a motivating example consider the prospective cohort study (see COHORT STUDIES) recently utilised by Ye and Kaskutas (2009) to investigate the effect of Alcoholics Anonymous (AA) meeting attendance on alcohol abstinence. Observational studies had consistently found strong dose–response relationships between AA meeting attendance and abstinence but to date there was little evidence of such a relationship from experimental studies. The relationship between the treatment 'AA attendance' and the outcome 'abstinence' in observational studies is potentially subject to confounding, in that there may be a number of observable pre-treatment variables such as alcohol problem severity, self-motivation and coercion by others that affect study participants' decisions to go to AA meetings and independently contribute to them becoming abstinent. The authors employ propensity score methods to adjust the AA effect estimate for SELECTION BIAS due to observed confounders.

The ability of the propensity score to balance groups with respect to a large set of covariates was confirmed in a seminal paper by Rosenbaum and Rubin (1983). The propensity score is the coarsest balancing score, which is a function of the observed covariates such that the conditional distribution of the covariates given the balancing score is the same for treated and untreated individuals. Under a *strongly ignorable treatment assignment*, i.e. given the observed covariates treatment assignment is not determined by the potential outcomes, at any value of the balancing score, the difference between the treatment and control group means is an unbiased estimate of the average treatment effect at that value of the balancing score. Consequently, with a strongly ignorable treatment assignment:

- pair matching on propensity scores;
- subclassification on propensity scores (also referred to as stratification);
- and covariance adjustment (also referred to as regression adjustment)

can produce unbiased estimates of causal treatment effects.

Propensity score matching is part of the design of an observational study and refers to the procedure whereby individuals in one group are matched to individuals from the other group on the basis of similar propensity scores. Typically, in observational studies the control group is much larger than the treated group, thus making it feasible to select a subset of controls that match the treated individuals. Matching can be one-to-one or one-to-*k* individuals and at the end of the matching process the sample is reduced to a smaller analysis sample. When controls are matched to treated individuals the analysis estimates the *average treatment effect on the treated* (ATT*)*; conversely, if treated individuals are matched to controls the *average treatment effect on the untreated* (ATU) is estimated. There has been some discussion as to the mechanics of the matching process (random sampling of treated or untreated individuals, matching with or without replacement, definition of 'similar' propensity scores, etc.; for a review see, for example, D'Agostino, 1998) and a number of propensity score matching algorithms have been put forward (e.g. in general purpose packages such as SAS and Stata). It is important to realise that when propensity score matching is carried out as part of the design of the study this needs to be acknowledged by the statistical analyses method (as in matched CASE-CONTROL STUDIES), e.g. by including random effects for matched pairs or by allowing for within matched pair correlations by other approaches such as GENERALIZED ESTIMATING EQUATIONS (GEE). This is clearly a point not always appreciated by the practitioner (see Austin, 2008).

Stratification by propensity scores refers to the technique whereby individuals are grouped into strata on the basis of their propensity scores, treated and control subjects are contrasted within each stratum and a weighted average of these differences constructed to estimate the average treatment effect (ATE, weight = stratum size), the ATT (weight = number of treated in stratum) or the ATU (weight = number of untreated in stratum) respectively. Rosenbaum and Rubin (1984) showed that Cochran's (1968) result, which states that five strata each containing 20% of the subjects remove 90% of the selection bias, holds for propensity score stratification. Thus stratification according to propensity score quintiles from the combined group is commonly employed by practitioners.

A third technique used to achieve propensity score conditioning is the inclusion of the propensity score as a covariate in the analysis model, the so-called 'regression adjustment'. However, while very simple, this technique further assumes that average treatment effects do not vary with the value of the propensity score and requires knowledge of the functional relationship between propensity scores and outcomes; thus propensity score matching or stratification are the preferred techniques.

A big issue for the practitioner is the construction of the propensity scores. These are not typically known and have to be estimated from the sample data at hand. A standard approach is to use DISCRIMINANT FUNCTION ANALYSIS or modelling techniques for the binary outcome 'assignment to the group of interest/assignment to the comparison group', such as LOGISTIC REGRESSION or the PROBIT MODEL, to establish a rule for predicting the probability of being assigned to the group of interest, and then to estimate this probability from the relevant covariate values for each member of the original sample. Note that the underlying theory that supports the use of estimated propensity scores is based on the estimated propensity scores producing *sample* balance (see Rosenbaum and Rubin, 1983). Therefore, one tends to be overinclusive rather than underinclusive in the modelling of treatment assignment in this context, including nonlinear and interaction effects where possible. For example, Brookhart and colleagues (2006) show that standard model-building tools for predictive modelling will not always lead to good propensity score models, in particular in smaller studies. These authors further suggest that covariates related to the outcome should always be included since they add precision without adding bias. If the covariate is a confounder, i.e. also related to treatment assignment, then its inclusion will also decrease BIAS. In contrast, covariates related to treatment assignment but not to the outcome will decrease precision without decreasing bias.

As the use of (estimated) propensity scores is motivated by their ability to balance covariates across groups after matching or stratification, such sample balance should be checked before proceeding to interpret treatment effects obtained using propensity scores as adjusted for selection bias. However, as Austin (2008) points out, in contrast to other applications, the statistic for assessing covariate balance should measure a property of the sample (and not of an underlying population) and the sample size should not affect its value. This requirement rules out the use of statistical hypothesis testing. A number of appropriate statistics have been proposed, with perhaps the most commonly used one being the standardised difference, defined as the absolute group difference in sample means divided by an estimate of the pooled STANDARD DEVIATION (not STANDARD ERROR). In addition, when using propensity score matching the overlap between the distributions of the propensity scores between the treated and the control individuals should be assessed in order to be able to comment on the representativeness of the matched sample.

As an example consider the first table, which summarizes the distribution of potential confounders in the AA and abstinence study before and after propensity score matching. Propensity scores were estimated using logistic regression modelling. Then 102 nonattendees were matched to 282 attendees using the Stata-user written command PSMATCH2. Propensity score matching performed well in terms of balancing the potential confounders, with most standardised differences between the AA attendee and nonattendee groups reduced to below 10% of

propensity scores *Pre-treatment covariate distributions before and after propensity score matching (extract from Table 1 in Ye and Kaskutas, 2009)*

	AA attender (n = 336) mean (SD)	*AA nonattender (n = 233) mean (SD)*	*Standardized difference in %*	
			Before matching (whole sample)	*After propensity score matching*
Demographics				
Male	0.59 (0.49)	0.56 (0.50)	6.93	0.00
Mean age	38.8 (10.1)	36.8 (11.5)	18.4	−11.1
Ethnicity				
White	0.63 (0.48)	0.58 (0.49)	9.67	4.32
Black	0.26 (0.44)	0.26 (0.44)	−0.36	−4.02
Others	0.11 (0.32)	0.16 (0.37)	−13.3	−1.03
Marital				
Married	0.34 (0.48)	0.45 (0.50)	−21.4	−0.72
Sep/Div/Widow	0.36 (0.48)	0.28 (0.45)	16.8	−9.09
Single	0.30 (0.46)	0.27 (0.45)	5.72	−10.1
Level of education	3.34 (1.02)	3.18 (0.98)	15.7	1.40
Motivation				
Readiness to change index	50.0 (6.7)	46.6 (7.5)	48.3	−11.5
Coercion				
Number who pressure you to get treatment	1.85 (1.31)	1.58 (1.16)	22.6	16.8
Number who give you ultimatum	0.58 (0.80)	0.53 (0.72)	5.96	0.93
Problem severity				
ASI composite alcohol score	0.43 (0.32)	0.34 (0.30)	30.8	6.03
Number of dependence symptoms	5.20 (2.77)	3.69 (2.67)	55.8	9.84
Number of alcohol-related consequences	1.42 (1.42)	0.99 (1.15)	33.4	−3.81
...				

the respective standard deviation. For example, the standardised difference in dependence symptoms was reduced from 56% of a standard deviation to 9.8%. However, the matched sample had a higher average propensity score than the original sample due to the target group being matched to (AA attendees) having higher propensities than the sample as a whole.

The second table shows the estimated AA effects on abstinence before and after propensity score matching. As expected, the ODDS RATIO associated with AA attendance is reduced considerably (from 3.6 to 2.2) after adjusting for observed confounders by propensity score matching. The bias adjustment was repeated using propensity score stratification. Study participants were classed into one of five strata using propensity score quintiles. This allowed the use of the whole sample of 334 AA attendees and 228 nonattendees. A stratified analysis then estimated the AA effect on abstinence within each stratum (see the middle section in the second table). This gave varying odds ratio estimates with larger AA effects observed in the strata with lower AA attendance propensity. The combined odds ratio was 3, again suggesting that part of the unadjusted AA effect was due to observed baseline confounders.

One might ask: 'Why use propensity score matching or stratification instead of conventional matching or stratification by covariates?' The answer is that the propensity score approach has the same aim, i.e. bias reduction, but the use of propensity scores has practical advantages, especially when there are a large number of covariates to consider. Even when there are only a few covariates, it can be difficult to find controls that match the treated individuals on all the covariate values. Propensity score matching overcomes this problem by allowing the investigator to control for multiple covariates by using only a single scalar matching variable. Similarly, in the context of stratification the number of possible strata grows exponentially with the number of binary background characteristics, and this will eventually lead to only one of the groups being present in some of the strata. Again the propensity score as a scalar summary of the covariates is useful as it can balance the distribution of the covariates across the groups within the strata without requiring a large number of strata.

propensity scores *Effect of AA attendance on abstinence estimated before and after propensity score adjustment (extract from Table 2 in Ye and Kaskutas, 2009)*

	AA attendee at follow-up	*n*	*Rate of abstinence at follow-up mean (SE)*	*Differences in rate of abstinence mean (SE)*	*Odds ratio associated with AA attendance*
Before propensity score adjustment	No	228	0.377 (0.032)		
	Yes	334	0.686 (0.025)	−0.308 (0.041)[c]	**3.60**
Propensity score stratification Subclass (from lowest AA propensity to highest AA propensity)					
1	No	93	0.323 (0.049)		
	Yes	20	0.700 (0.105)	−0.377 (0.116)[b]	4.90
2	No	68	0.368 (0.059)		
	Yes	43	0.744 (0.067)	−0.377 (0.091)[c]	5.00
3	No	37	0.432 (0.083)		
	Yes	75	0.707 (0.053)	−0.274 (0.095)[b]	3.16
4	No	21	0.476 (0.112)		
	Yes	92	0.652 (0.050)	−0.176 (0.117)	2.06
5	No	9	0.556 (0.176)		
	Yes	104	0.673 (0.046)	−0.118 (0.165)	1.65
Adjusted	No	228	0.431 (0.106)		
	Yes	334	0.696 (0.068)	−0.265 (0.126)[a]	**3.02**
Propensity score matching (nearest distance with replacement)					
Matching AA attendee group	No	102	0.479 (0.067)		
	Yes	282	0.670 (0.028)	−0.191 (0.073)[b]	**2.21**

[a] $p < 0.05$, [b] $p < 0.01$, [c] $p < 0.001$.

Finally, ATE estimates are sometimes obtained by weighting observations inversely to the probability of selecting their treatment group (which is the propensity score for the treated individuals and 1 – propensity score for the controls). This is a particular application of INVERSE PROBABILITY WEIGHTING (IPW) estimators and here the motivation is somewhat different compared to propensity score matching or stratification. Inverse weighting methods are used in survey sampling and for MISSING DATA problems when the sample used to draw inferences is not a simple random sample (SRS). Rather, members of certain subpopulations are under- or over-sampled and thus in the analysis each individual is weighted relative to the inverse probability of being sampled. The concept can be applied to deal with selection bias by viewing the (hypothetical) sample of potential outcomes under treatment and no treatment as an SRS from the population of interest. The observed outcomes are then realised nonrandomly with the probability of observing a potential outcome being the probability of selecting its treatment group.

In summary, propensity scores are a useful tool to quasi-randomise observational studies. The approach can have benefits in terms of generalisability to wider populations than those typically studied in clinical trials, reduced cost and time in obtaining the data and ability to investigate smaller effect sizes due to larger sample sizes. However, it needs to be emphasized that the approach can only reduce selection bias due to observed confounders. Thus there remains a need for experimental approaches (e.g. RCTs) or analysis approaches (e.g. the use of INSTRUMENTAL VARIABLES) that can address confounding by unobserved variables. (See websites http://www2.sas.com/proceedings/sugi26/p214-26.pdf and http://ideas.repec.org/c/boc/bocode/s432001.html). *SL*

Austin, P. C. 2008: A critical appraisal of propensity-score matching in the medical literature between 1996 and 2003. *Statistics in Medicine* 27, 2037–49. **Brookhart, M. A., Schneeweiss, S., Rothman, K. J., Glynn, R. J., Avorn, J. and Stürmer, T.** 2006: Variable selection for propensity score models. *American Journal of Epidemiology* 163, 1149–56. **Cochran, W. G.** 1968: The effectiveness of adjustment by subclassification in removing bias in observational studies. *Biometrics* 24, 205–13. **D'Agostino, R. B.** 1998: Tutorial in biostatistics: propensity score methods for bias reduction in the comparison of a treatment to non-randomised control group. *Statistics in Medicine* 17, 2265–81. **Rosenbaum, P. R. and Rubin, D. B.** 1983: The central role of the propensity score in observational studies for causal effects. *Biometrika* 70, 41–55. **Rosenbaum, P. R. and Rubin, D. B.** 1984 Reducing bias in observational studies using subclassification on the propensity score. *Journal of the American Statistical Association* 79, 516–24. **Ye, Y. and Kaskutas, L. A.** 2009: Using propensity scores to adjust for selection bias when assessing the effectiveness of Alcoholics Anonymous in observational studies. *Drug and Alcohol Dependence* 104, 56–64.

proportional hazards When comparing two groups with time-to-event data the summary statistic that is usually employed is the hazard rate in one group compared to the hazard rate in the other group, which is usually called the *hazard ratio*. If the hazard ratio is constant over time, i.e. it is *independent* of time, then the hazards in the two groups are said to be proportional – there is a constant multiplicative relationship between the two hazard rates.

Mathematically, this relationship takes the form:

$$h_1(t)/h_0(t) = \exp(\beta_1 X_1 + \beta_2 X_2 + \cdots)$$

where $h_0(t)$ and $h_1(t)$ are the hazard functions in the two groups (which may vary over time), X_1, X_2, ... are the explanatory variables and β_1, β_2, ... are the coefficients estimated from the data.

The assumption of proportional hazards underlies the inclusion of any variable in COX'S REGRESSION MODEL. Therefore, it is usually important to assess that the hazards are approximately proportional across different groups before including a variable in a Cox model. There is, however, no unique or completely satisfactory way in which to test this assumption. Many approaches have been suggested and here we present one numerical and one graphical approach that seem to perform as well as most others (Persson, 1991).

A numerical approach was introduced by Cox himself in his original paper introducing the model. It involves including a time-dependent covariate in the model. Thus for one variable, the model would look like $h_1(t)/h_0(t) = \exp(\beta_1 X_1 + \gamma X_1(t))$, with a simple extension for more variables. The NULL HYPOTHESIS of proportional hazards, i.e. $\gamma = 0$, is tested by fitting the model and assessing this null hypothesis in the usual way.

One simple approach to assess this assumption graphically is to use a complementary log plot. This is a plot of log t against $\log\{-\log[S(t)]\}$, where t is time from baseline in whatever unit is being used and $S(t)$ is the surviving proportion at time t (see the figure on page 360). If the assumption of proportional hazards holds true then the curves for each group should be approximately parallel (i.e. the curves will be identical but vertically shifted by a constant). One may formally test that the 'slopes' of these curves do not differ, but this is not usually very helpful or informative (e.g. assuming that the curves are approximately linear) and thus usually we are left with visual inspection alone, to assess whether the curves are indeed parallel. This is true of all graphical methods. Except for gross departures from proportional hazards or as a supplement to numerical methods, these methods cannot be widely recommended. *MS/MP*

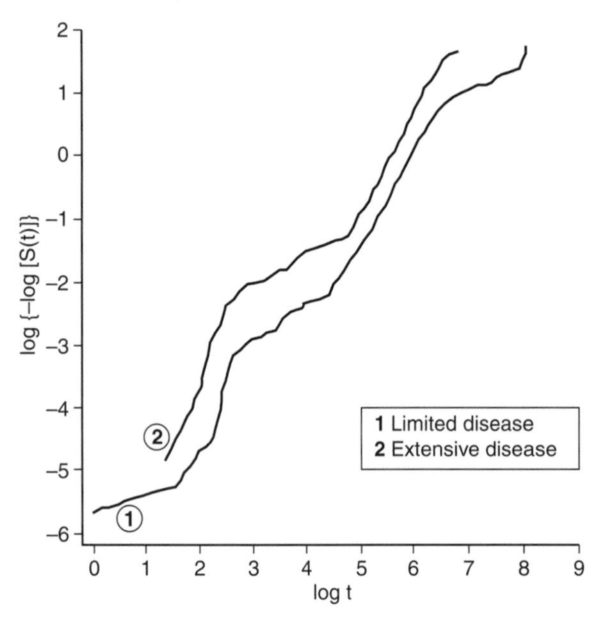

proportional hazards *Graph of log {– log [S(t)]} against log t in patients treated with ECMV chemotherapy with limited and extensive small cell lung cancer (from Machin and Parmar, 1995)*

Cleves, M. A., Gould, W. W. and Gutierrez, R. G. 2003: *An introduction to survival analysis using Stata®*, revised edition. Texas: Stata Press. **Machin, D. and Parmar, M. K. B.** 1995: *Survival analysis: a practical approach*. Chichester: John Wiley & Sons, Ltd. **Persson, I.** 1991: Essays on the assumption of proportional hazards in Cox's regression. *Acta Universitatis Upsalaiensis*. Uppsala: Comprehensive Summaries of Uppsala Dissertations from the Faculty of Social Sciences. **Piantodosi, S.** 1997: *Clinical trials*. New York: Wiley Interscience.

protocols for clinical trials These are formal documents outlining proposed procedures for carrying out a CLINICAL TRIAL. Protocols for clinical trials serve a variety of masters; consequently, a well-constructed protocol discusses a range of issues relevant to the trial it describes. The principal investigator uses the protocol as a tool to describe the scientific rationale behind the study and the goals of the trial. The study staff depends on the protocol for operational guidance on whom to recruit, what study procedures to perform and how and when to perform them. The protocol must include information about safeguards planned for the participants so that institutional boards and ETHICAL REVIEW COMMITTEES can assess the appropriateness of the plans to protect the participants in the research. If the trial is to be used as part of a regulatory submission, the protocol should include enough information so that when the trial is complete and the final results presented, the regulators will be able to assess the quality of the study, the results and the statistical interpretation. Finally, in anticipation of eventual publication of the results, the investigators, in writing the protocol, should consider the types of data they expect to include in publications.

An ideal protocol addresses these multiple masters in a way that is well organised, easily readable, unambiguous and internally consistent. Several guidelines are available to help the writer construct a protocol that covers all important aspects, notably those produced by the International Committee of Harmonization (in particular E3, E9 and E10) (www.ich.org), the various guidances and the 'points to consider' documents provided by the United States Food and Drug Administration on disease-specific protocols (www.fda.org), as well as the CONSORT statement on recommendations for reporting parallel-group randomised trials (Moher, Schultz and Altman, 2001). All of these provide useful advice to the constructors of clinical protocols. These documents are guidelines; a specific protocol should incorporate the relevant features described in these and related documents, individually tailored to the study at hand. A protocol that reads as if it were constructed of boilerplate sections from various previous protocols, or uncritically edited versions of standard templates, is not only boring to read but, worse, may lead to faulty compliance to its procedures.

Most well-constructed protocols share a similar structure. They begin with a section discussing the disease under study. Next comes a description of the context of the particular trial including information on the intervention under study. Depending on the nature of the study, this section might describe the pharmacology of the product and the justification for the particular dose or doses under study, the mechanism and structure of the device or the justification for the particular behavioural intervention. Another section describes the aims and goals of the study. The remainder of this entry assumes that the interventions under study will be drugs or biologics; however, the general considerations apply also to trials of devices and behavioural interventions.

The protocol will carefully define the population of interest, specify the entry criteria and delineate the important facets of the study design. The heart of the protocol will contain a description of the procedures for screening, enrolment and subsequent visits. The actual content of these sections will depend on the design and purpose of the study. PHASE I TRIALS or dose escalation studies, which are typically not hypothesis-driven, aim to provide evidence that the product is sufficiently safe to allow administration to more people. Consequently, the section describing the design of this type of study should address the considerations, statistical and otherwise, that led to the particular sample size and to the choice of criteria for dose escalation. The protocol will describe the criteria by which the various doses under consideration will be evaluated and the methods by which dose escalation will proceed.

Typical PHASE II TRIALS studying nonclinical ENDPOINTS as well as both PHASE III and later confirmatory PHASE IV TRIALS studying clinical endpoints aim to test hypotheses. A protocol for such a trial should unambiguously describe the primary endpoint, the primary hypothesis and the statistical methods planned to test that hypothesis. A section on statistical POWER should justify the sample size. If the study has a control group, the protocol should describe its nature and should defend its choice. If the study is testing equivalence or noninferiority, the protocol should specify the appropriate equivalence or noninferiority margin (see ACTIVE CONTROL EQUIVALENCE STUDIES). A section on secondary endpoints should present a cogent rationale for each with a discussion of how each one will yield information that augments the data provided by the primary endpoint.

All protocols should contain a statistical section that specifies which participants the primary analysis will include, how missing data will be handled and the planned statistical methods. Early phase studies typically use descriptive and exploratory statistical methods, for such studies aim to produce data relevant to the design of subsequent trials. In Phase III or later trials, however, where the purpose is to test hypotheses, the statistical section should describe a rigorous approach to analysis that preserves the TYPE I ERROR rate. Failure to define the primary endpoint clearly, to select a rigorous statistical approach and to specify statistical analyses unambiguously jeopardises the ultimate validity of the inference from the trial.

Protocols for randomised clinical trials should describe the methods of RANDOMISATION and, if relevant, the nature of any stratification. If randomisation is blocked, the protocol should not include the block size because making that information available can potentially lead the investigator to deduce the treatment given to some participants. For studies involving BLINDING, a section should describe the methods used to conceal treatment allocation. Generally, protocols should discuss the methods planned for ensuring unbiased assessment of outcome.

The final sections of the protocol provide rules for drug disposition, handling unexpected adverse events, monitoring safety, adhering to regulatory guidelines and administrative matters essential to the conduct of the study. (These sections, unfortunately, sometimes read as if the writers had become tired by the time they reached them. If the trial is studying ovarian cancer, for instance, the use of the pronoun 'he', or even the more politically correct 'he or she', clearly indicates that the writer simply pasted the material from another protocol!)

The protocol must describe the plans for monitoring the safety of the participants during the study, the methods of follow-up and the way in which the investigators plan to protect the participant's rights and confidentiality.

The protocol should be clear enough that the designers of the case report forms can use it to construct the forms.

The inclusion of all this necessary material makes many protocols very long. To ensure that the clinic staff implementing the protocol understands the purpose of the trial and the procedures, the document should contain two summaries. One, which generally comes at the beginning of the document, is a two- to (approximately) four-page synopsis briefly describing the product or other intervention, the objectives, the study design, the study population, the dosing and dosing regimen, the primary and secondary endpoints and the statistical plans for these endpoints along with a discussion of the power. The second summary is a one- or two-page flowchart listing the procedures to be performed at each visit. A helpful aid to accurate implementation of the protocol is a laminated pocket-sized card containing a miniature, but legible, version of this flowchart.

In summary, the protocol for a clinical trial should justify the study and describe its procedures, hypotheses, statistical plans and administrative guidelines. A clearly written protocol with close connection between the trial's goals, design and analysis plays a crucial role in implementing a study that is likely to result in a correct inference. *JW*

Moher, D., Shultz, K. F. and Altman, D. G. 2001: The CONSORT statement: revised recommendations for improving the quality of reports of parallel-group randomized trials. *Annals of Internal Medicine* 134, 7–622.

publication bias Publication bias refers to the publication or nonpublication of research findings, depending on the nature and direction of the results. Its potential to undermine the validity of medical research was noted by Begg and Berlin (1988). There is extensive empirical evidence for the existence of publication bias. Scherer *et al.* (2007) reviewed 79 studies describing subsequent full publication of research initially presented in abstract or short report form. Only about half of abstracts presented at conferences were later published in full, and subsequent publication was associated with factors such as 'positive' (usually equated with statistically significant) results, acceptance for oral presentation (versus poster presentation), clinical research (versus basic research) and randomized trial design (versus other study designs). Hopewell *et al.* (2009) reviewed COHORT STUDIES of registered CLINICAL TRIALS, in which investigators were subsequently contacted to determine the publication status of each completed study. Publication was more likely if results were positive (defined as results classified by the investigators as statistically significant ($P < 0.05$), or perceived as striking or important, or showing a positive direction of effect) rather than negative. Other factors such as the study size, funding source and academic rank and the sex of the primary investigator were not consistently associated with the PROBABILITY of publication.

Publication bias should be seen as one of a number of reporting BIASES, which also include time lag bias (the rapid or delayed publication of research findings), multiple publication bias (the multiple or singular publication of research findings), location bias (the publication of research findings in journals with different ease of access or levels of indexing in standard databases), citation bias (the citation or noncitation of research findings), language bias (the publication of research findings in a particular language) and outcome reporting bias (the selective reporting of some outcomes but not others), depending on the nature and direction of the results (Sterne, Egger and Moher, 2008). Empirical evidence that selective reporting of outcomes *within* studies is an important threat to the validity of research findings has emerged from a series of cohort studies conducted by Chan *et al.* (2004).

Begg and Berlin (1988) noted that an asymmetric appearance of a FUNNEL PLOT might suggest that a meta-analysis (see SYSTEMATIC REVIEWS AND META ANALYSIS) has been affected by publication bias. The subjectivity inherent in interpretation of graphical displays (see GRAPHICAL DECEPTION) has led a number of authors to propose that statistical tests for funnel plot asymmetry might be used to diagnose publication bias. The most widely used such test was proposed by Egger *et al.* (1997). These authors also noted that publication bias is only one of a number of possible causes of funnel plot asymmetry, which may also result from other reporting biases, methodological flaws that lead to spuriously inflated effects in smaller studies, true heterogeneity or chance. Therefore tests for funnel plot asymmetry should be seen as examining 'small-study effects' – a tendency for effects estimated in smaller studies to differ from those estimated in larger studies (Sterne, Gavaghan and Egger, 2000). Publication bias is one of a number of possible explanations for small-study effects (Egger *et al.*, 1997; Sterne, Gavaghan and Egger, 2000; Sterne, Egger and Moher, 2008). Statistical problems that can affect the Egger test have led a number of authors to propose alternative tests for funnel asymmetry. Sterne, Egger and Moher (2008) reviewed a number of such tests, and provided recommendations on testing for funnel plot asymmetry. *JS*

Begg, C, B. and Berlin, J. A. 1988: Publication bias: a problem in interpreting medical data. *Journal of the Royal Statistical Society, Series A* 151, 419–63. **Chan, A. W., Hróbjartsson, A., Haahr, M. T., Gøtzsche, P. C. and Altman, D. G.** 2004: Empirical evidence for selective reporting of outcomes in randomized trials: comparison of protocols to published articles. *Journal of the American Medical Association* 291, 2457–65. **Egger, M., Smith, G. D., Schneider, M. and Minder, C.** 1997: Bias in meta-analysis detected by a simple, graphical test. *British Medical Journal* 315, 629–34. **Hopewell, S., Loudon, K., Clarke, M. J., Oxman, A. D. and Dickersin, K.** 2009: Publication bias in clinical trials due to statistical significance or direction of trial results. *Cochrane Database of Systematic Reviews* Issue 1, Art. No.: MR000006. DOI: 10.1002/14651858.MR000006. pub3. **Scherer, R. W., Langenberg, P. and von Elm, E.** 2007: Full publication of results initially presented in abstracts. *Cochrane Database of Systematic Reviews* Issue 2, Art. No.: MR000005. DOI: 10.1002/14651858.MR000005.pub3. **Sterne, J. A. C., Egger, M. and Moher, D. on behalf of the Cochrane Bias Methods Group** 2008: Chapter 10: Addressing reporting biases. In Higgins, J. P. T. and Green, S. (eds), *Cochrane Handbook for systematic reviews of interventions*. Chichester: John Wiley & Sons, Ltd. **Sterne, J. A. C., Gavaghan, D. and Egger, M.** 2000: Publication and related bias in meta-analysis: power of statistical tests and prevalence in the literature. *Journal of Clinical Epidemiology* 53, 1119–29.

P-values These were introduced by R. A. Fisher (1925) as a means of assessing the evidence against a NULL HYPOTHESIS. Often, such a null hypothesis states that there is no association between two variables, e.g. between hypertension and subsequent heart disease. The *P*-value is defined as the PROBABILITY, if the null hypothesis were true, that we would have observed an ASSOCIATION as large as we did by chance. If this probability is small we have evidence *against* the null hypothesis; in other words, the smaller the *P*-value, the stronger the evidence against the null hypothesis.

To illustrate the calculation of a *P*-value consider the results, displayed in the table, of a study to investigate whether smoking reduces lung function. Forced vital capacity (FVC) (a test of lung function) was measured in 100 men aged 25–29, of whom 36 were smokers and 64 nonsmokers. (For simplicity we will use large-sample formulae.)

The mean FVC in smokers was 4.7 litres compared with 5.0 litres in nonsmokers. The difference in mean FVC, $\bar{x}_1 - \bar{x}_0$, is therefore $4.7 - 5.0 = -0.3$ litres. The STANDARD DEVIATION (SD) in both groups was 0.6 litres.

If the null hypothesis is true, then the MEAN of the SAMPLING DISTRIBUTION of $(\bar{x}_1 - \bar{x}_0)$ is zero. The large-sample formulae state that the sampling distribution of $(\bar{x}_1 - \bar{x}_0)$ is normal: its STANDARD ERROR (SE) is derived from the standard errors of $\bar{x}_1$ and $\bar{x}_0$) as:

P-values *Results of a study to investigate whether smoking reduces lung function*

Group	*Number of men*	*Mean FVC*	*Standard deviation*	*Standard error of mean FVC*
Nonsmokers (0)	$n_0 = 64$	$\bar{x}_0 = 5.0$	$s_0 = 0.6$	$SE_0 = 0.6/\sqrt{64} = 0.075$
Smokers (1)	$n_1 = 36$	$\bar{x}_1 = 4.7$	$s_1 = 0.6$	$SE_1 = 0.6/\sqrt{36} = 0.100$

$$\text{SE} = \sqrt{(\text{SE}_0^2 + \text{SE}_1^2)} = \sqrt{0.1^2 + 0.075^2} = 0.125 \text{ litres}$$

The *test statistic* $z = (\bar{x}_1 - \bar{x}_0)/\text{SE}$ measures by how many standard errors the mean difference $(\bar{x}_1 - \bar{x}_0)$ differs from the null value of 0. In this example:

$$z = (-0.3)/0.125 = -2.4$$

The difference between the means is therefore 2.4 standard errors below 0. The figure shows that the probability of getting a difference of −2.4 standard errors or fewer (the area under the curve to the left of −2.4) is 0.0082 (this can be found using a computer or statistical tables). This probability is known as the *one-sided P-value*.

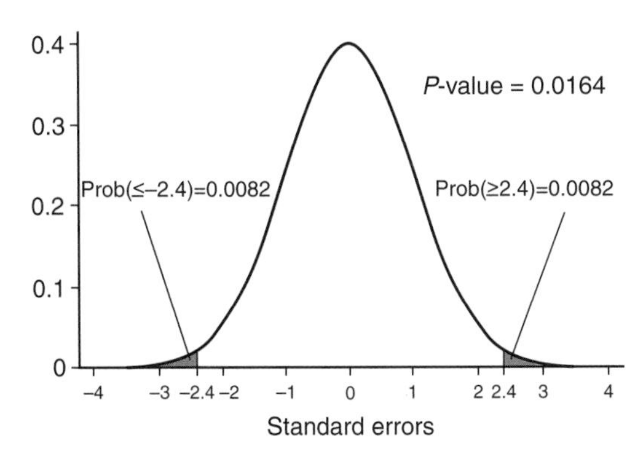

***P*-values** *Using test statistic z to determine the probability of getting a difference of −2.4 standard error or fewer*

By convention, we usually use *two-sided P-values*. The justification for this is that our assessment of the probability that the result is due to chance should be based on how extreme is the *size* of the departure from the null hypothesis and not its direction. We therefore include the probability that the difference might (by chance) have been in the opposite direction: the mean FVC might have been greater in smokers than nonsmokers. Because the NORMAL DISTRIBUTION is symmetrical, this probability is also 0.0082. The 'two-sided' *P*-value – the probability of observing a difference at least as extreme as 2.4, if the null hypothesis of no difference is correct – is thus found to be 0.0164 (= 0.0082 + 0.0082). Such a *P*-value provides evidence *against* the null hypothesis and suggests that smoking does, indeed, affect FVC.

While most standard statistical computer packages include *P*-values as part of their standard output, the interpretation of *P*-values causes confusion. For the reasons discussed in the entry on HYPOTHESIS TESTS, there is little justification for dividing results into 'significant' and 'nonsignificant' according only to whether the *P*-value is less than 0.05 (or on the basis of any other threshold). Three errors common in the interpretation of *P*-values are as follows.

First, potentially clinically important associations observed in small studies, for which the *P*-value is more than 0.05, are denoted as nonsignificant and ignored. To protect ourselves against this error, we should always consider the range of possible values for the association shown by the confidence interval as well as the *P*-value.

Second, statistically significant ($P < 0.05$) findings are assumed to result from real associations, whereas by definition an average of 1 in 20 comparisons in which the null hypothesis is true will result in $P < 0.05$.

Third, statistically significant ($P < 0.05$) findings are assumed to be of clinical importance, whereas given a sufficiently large sample size, even an extremely small association in the population will be detected as different from the null hypothesis value of zero.

Based on considerations of the power of studies and the proportion of null hypotheses that are, in fact, false, Sterne and Davey Smith (2001) adapted the work of Oakes (1986) to suggest that in situations typical of medical statistics *P*-values less than 0.001 could be considered to provide strong evidence against the null hypothesis. However, the interpretation of *P*-values will always depend on the context in which they were generated. For example, Wacholder and Chanock (2004) suggested that in the context of molecular epidemiological studies, in which many thousands or even millions of single-nucleotide polymorphisms (SNPs) may be tested for associations with a disease outcome, it may be appropriate to consider only *P*-values less than 10^{-4}, or even 10^{-6}, as providing evidence of a real association. *JS*

Fisher, R. A. 1925: *Statistical methods for research workers*. Edinburgh: Oliver and Boyd. **Oakes, M.** 1986: *Statistical inference*. Chichester: John Wiley & Sons, Ltd. **Sterne, J. A. and Davey Smith, G.** 2001: Sifting the evidence – what's wrong with significance tests? *British Medical Journal* 322, 226–31. **Wacholder, S. and Chanock, R. M.** 2004: Assessing the probability that a positive report is false: an approach for molecular epidemiology studies. *Journal of the National Cancer Institute* 96, 434–42.

Q

Q-Q plots See PROBABILITY PLOTS

quality of life (QoL) measurement This is a standardised subjective approach to measuring a person's perception of their own health by using numerical scoring systems and may include one or several dimensions of quality of life (QoL). Quality of life is a complex concept with multiple aspects. These aspects (usually referred to as domains or dimensions) can include: cognitive functioning; emotional functioning; psychological well-being; general health; physical functioning; physical symptoms and toxicity; role functioning; sexual functioning; social well-being and functioning; spiritual/existential issues and many more. This broad definition of QoL includes scales or instruments that ask general questions, such as 'In general, how would you rate your health now?' and more specific questions on particular symptoms, and side-effects, such as 'During the past week have you felt nauseated?' These measurement scales all have the common feature of using a standardised approach to assessing a person's perception of their own health by using numerical scoring systems and may include one or several dimensions of quality of life.

Researchers have used a variety of names to describe quality of life measurement scales. Some prefer to use the term *health-related quality of life* (HRQoL or HRQL) to stress that we are only concerned with health aspects. Others have used the terms *health status* or *self-reported health*. The United States FOOD AND DRUG ADMINISTRATION has adopted the term *patient reported outcome* (PRO) in its guidance to the pharmaceutical industry for supporting labelling claims for medical product development (Food and Drug Administration, 2006). The UK Medical Research Council (2009) used a similar term *patient-reported outcome measures* (PROMs). However, not all people who complete such outcomes are ill and patients and hence PRO could legitimately stand for *person-reported outcome*. Mostly, we shall assume that the quality of life instrument or outcome is self-reported by the person whose experience we are interested in, but it could be completed by another person or proxy. The term *health outcome assessment* has been put forward as an alternative that avoids specifying the respondent. This article will follow convention and use the now well-established term *quality of life* (QoL).

There no formally agreed definition of QoL, so most investigators get around this problem by describing what they mean by QoL, and then letting the items (questions) in their QUESTIONNAIRE speak for themselves. Some QoL instruments focus upon a single concept (or dimension), such as physical functioning. Other QoL instruments have several dimensions, such as physical, emotional and social functioning. Since there are many potential dimensions of QoL it is impractical to assess all of these concepts simultaneously in one instrument. Furthermore, QoL is a subjective concept, because symptoms, such as pain or depression and even physical functioning, are experienced by the individual patient and therefore they cannot entirely be assessed by 'objective' measures. So how do we actually measure QoL?

Simplistically, QoL measures represent a standardised approach to assessing a patient's perception of their own health, using numerical scoring, and can include symptoms, function and well-being. The concepts forming the various QoL dimensions are subjective measures and should best be evaluated by asking the patient.

The Medical Outcomes Study (MOS) *Short Form* (SF)-36 is the most commonly used QoL measure in the world today. It originated in the USA (Ware and Sherbourne, 1992), but has been validated for use in the United Kingdom (Brazier *et al.*, 1992). It contains 36 questions measuring health across eight dimensions: physical functioning (PF) 10 items; role limitation because of physical health (RP) 4 items; social functioning (SF) 2 items; vitality (VT) 4 items; bodily pain (BP) 2 items; mental health (MH) 5 items; role limitation because of emotional problems (RE) 3 items; and general health (GH) 5 items. (The first figure on page 366 shows the 10 questions that make up the physical function dimension of the SF-36.)

The responses to the 36 individual questions are classified into a mixture of binary (yes/no) and three-, five- and six-point ordered response categories. In planning and analysis, the question responses are often analysed by assigning equally spaced numerical scores to the ordinal categories (e.g. 1 = 'yes, limited a lot', 2 = 'yes, limited a little' and 3 = 'no, not limited at all', for the 10 items in the figure). The raw scores across similar questions (e.g. the 10 physical functioning items shown in the figure) are summed to generate a raw dimension score. Thus the 10-item physical function scale of the SF-36, with items scored 1 to 3, would yield a raw score ranging from 10 to 30. Finally, these raw dimension scores are then transformed to generate a QoL score from 0 to 100, where 100 indicates 'good health'.

Encyclopaedic Companion to Medical Statistics: Second Edition Edited by Brian S. Everitt and Christopher R. Palmer

HEALTH AND DAILY ACTIVITIES

The following questions are about activities that you might do during a typical day. Does your health limit you in these activities? If so, how much?

ACTIVITIES	Yes, limited a lot	Yes, limited a little	No, not limited at all
a. **Vigorous activities**, such as running, lifting heavy objects, participating in strenuous sports	1	2	3
b. **Moderate activities**, such as moving a table, pushing a vacuum cleaner, bowling or playing golf	1	2	3
c. Lifting or carrying groceries	1	2	3
d. Climbing **several** flights of stairs	1	2	3
e. Climbing **one** flight of stairs	1	2	3
f. Bending, kneeling or stooping	1	2	3
g. Walking **more than a mile**	1	2	3
h. Walking **half a mile**	1	2	3
i. Walking **100 yards**	1	2	3
j. Bathing and dressing yourself	1	2	3

quality of life (QoL) measurement *The 10 questions that make up the SF-36 physical function dimension (Brazier* et al., *1992)*

In our PF dimension example this transformation is achieved by:

$$\left[\frac{\text{Raw score}-\text{Lowest possible score}}{\text{Range of possible scores}}\right]\times 100$$

$$\text{i.e.}\left[\frac{\text{Raw score}-10}{20}\right]\times 100$$

Fayers and Machin (2007) term this procedure the 'standard scoring method' and this basic procedure is used to score many QoL instruments besides the SF-36.

The SF-36 is an example of a QoL instrument that is intended for general use, irrespective of the illness or condition of the patient. Such instruments are often termed *generic* measures and may often be applicable to healthy people too, and hence used in population surveys. The second figure on page 367 shows the distribution of the eight main dimensions of the SF-36 from a general population survey of United Kingdom residents (Brazier *et al.*, 1992). The third figure on page 368 shows how physical functioning in the general population (Walters, Munro and Brazier, 2001) declines rapidly with increasing age.

The SF-36 is also an example of a *profile* QoL measure since it generates eight separate scores for each dimension of health (fourth figure on page 368). Other generic profile instruments such as the Sickness Impact Profile (SIP) and Nottingham Health Profile (NHP) are described in Bowling (2004) and Fayers and Machin (2007). Conversely, some other QoL measures generate a single summary score or *single index*, which combines the different dimensions of health into a single number. An example of a single index QoL outcome is the EuroQol, or EQ-5D as it is now named (Fayers and Machin, 2007).

Generic instruments are intended to cover a wide range of conditions and have the advantage that the scores from patients with various diseases may be compared against each other and against the general population. For example, the fourth figure compares the mean SF-36 dimension scores of a group of young male cancer survivors aged 25–44 with an age and sex matched control sample (Greenfield *et al.*, 2007). The cancer survivors sample has a lower QoL on all eight dimensions of the SF-36 than the control sample. On the other hand, generic instruments may fail to focus on the issues of particular concern to patients with disease, and may often lack the SENSITIVITY to detect differences that arise as a consequence of treatments that are compared in CLINICAL TRIALS. This has led to the development of *condition-* or *disease-specific* QUESTIONNAIRES. Disease-specific QoL measurement scales are comprehensively reviewed by Bowling (2001). Examples of disease-specific QoL questionnaires described in Fayers and Machin (2007) include the cancer-specific 30-item European Organisation for Research and Treatment of Cancer (EORTC) QLC-30 questionnaire and the cancer-specific 30-item Rotterdam Symptom Checklist (RSCL).

The instruments described above claim to measure general QoL and usually include at least one question about overall QoL or health. Sometimes investigators may wish to explore particular aspects or concepts in greater depth. There are also instruments for specific aspects of QoL. These specific aspects may include anxiety and depression, physical functioning, pain and fatigue. Examples of instruments that evaluate specific aspects of QoL are again described in Fayers and Machin (2007) and include: the Hospital Anxiety and Depression Scale (HADS) and the Beck Depression Inventory (BDI) instruments for measuring anxiety and depression; the McGill Pain Questionnaire (MPQ) for the measurement of pain; the Multidimensional Fatigue Inventory (MFI) for assessing fatigue and the Barthel Index of Disability (BID) for assessing disability and functioning.

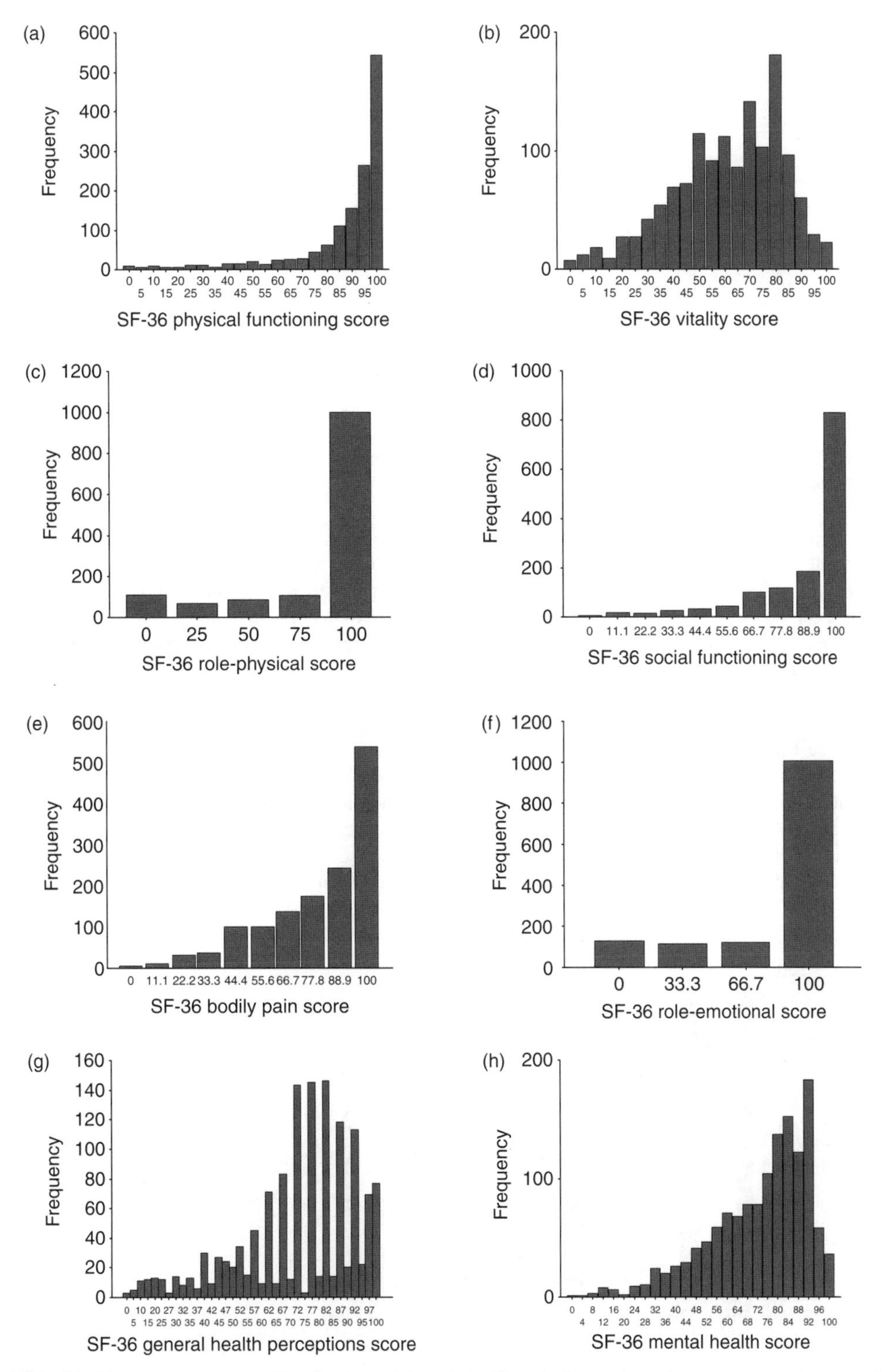

quality of life (QoL) measurement *Distribution of the eight SF-36 dimensions from a general population survey (n = 1372), where a score of 100 indicates 'good health' (data from Brazier et al., 1992)*

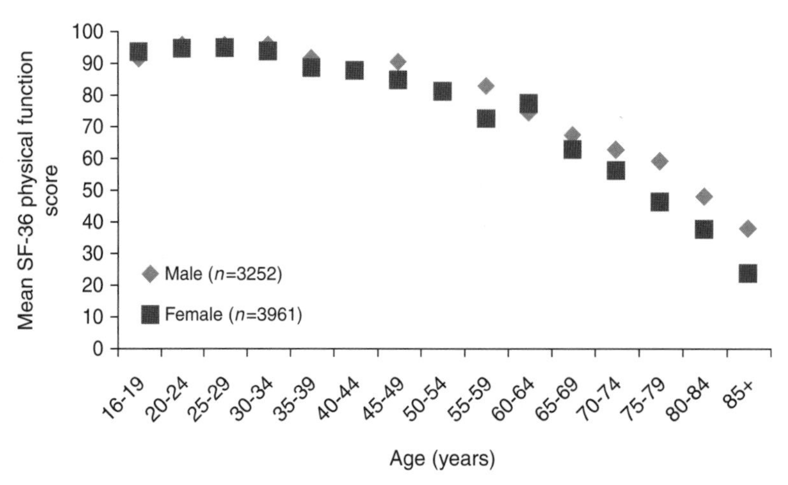

quality of life (QoL) measurement *Mean SF-36 physical function scores by age and sex (data from Walters, Munro and Brazier, 2001)*

The historical development of QoL assessment is briefly discussed in Fayers and Machin (2007). One of the first instruments that broadened the assessment of patients beyond physiological and clinical examination was the Karnofsky Performance Scale proposed in 1947 for the use in clinical settings. Over the following years a number of other scales were developed to assess functionally ability, such as the Barthel Index. The next generation of questionnaires from 1980 onwards, such as the SIP and NHP, attempted to quantify general health status and not just functional ability.

The functional ability dimensions of the generic instruments may not be responsive enough to detect the 'small' changes in physical functioning experienced by patients undergoing treatment. For example, the least 'difficult' item of the SF-36 physical function dimension (see the first figure) is question 3j, 'Bathing or dressing yourself'. Older adults with functioning problems may even have difficulty completing this item. For example, a general population survey of older adults aged 65 or more (Walters, Munro and Brazier, 2001) found that 6.5% of respondents were at the 'floor' (i.e. scored 0) on the original 10-item PF dimension.

There are several solutions to this problem of unresponsive QoL measures, including extending the scales (by adding extra questions), computer adaptive testing (CAT) and patient generated measures (PGM). With the increasing use

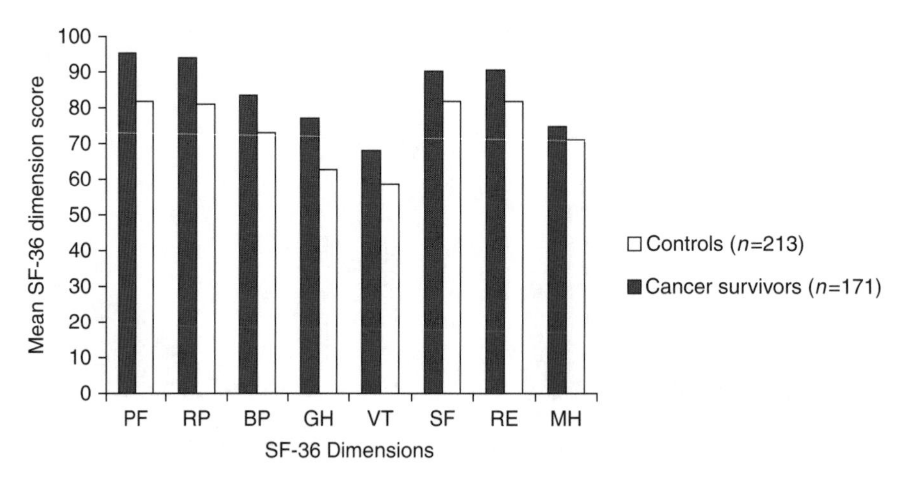

quality of life (QoL) measurement *Profile of mean SF-36 dimension scores for a sample of young male cancer survivors (aged 25–44) compared with n age and sex matched control group (data from Greenfield et al., 2007)*

of information technology we could use CAT to assess QoL. This would involve the use of 'tailored' or adapted tests. For example, to assess physical functioning (using the SF-36 say), if a patient cannot manage a short walk, why ask questions about long walks or running? With the use of CAT, the computer software can select questions of appropriate difficulty on the basis of earlier responses. This can result in benefits of more precise grading of ability and fewer questions put to each person.

A simpler solution to the problem of nonresponsive generic QoL instruments is to use patient generated measures (PGMs). These are a set of QoL instruments that ask patients to select their own dimensions and/or items. PGMs have the advantages that they are more relevant to the patients and likely to be more sensitive to change. The disadvantages of PGMs are the reduced comparability between patients, the dimensions may change before and after treatment and they are more complex to administer. Fayers and Machin (2007) give two examples of PGMs, the Patient Generated Index (PGI) and the Schedule for the Evaluation of Individual Quality of Life (SEIQOL).

There is a continuing philosophical debate about the meaning of QoL and about what should be measured. Despite this, it is still important to measure health-related quality of life (QoL) as well as clinical and process-based outcomes. This is because 'all of the these (QoL) concepts reflect issues that are of fundamental importance to patients' well being. They are all worth investigating and quantifying' (Fayers and Machin, 2007).

Most investigators treat the 'summated scores' from the QoL instruments as if they are from a continuous distribution. This is probably not an unreasonable assumption, particularly if we believe that there exists an underlying continuous latent variable that measures QoL, and that the actual measured outcomes are ordered categories that reflect contiguous intervals along this continuum.

Most QoL outcome measures, such as the SF-36, which use the standard scoring method described previously, generate data with discrete, bounded and nonstandard distributions (see the second figure). The scaling of QoL measures such as the SF-36 may lead to several problems in determining sample size and analysing the data (Walters, Campbell and Lall, 2001; Walters, Campbell and Paisley, 2001). The apparent continuum hides the fact that only a few discrete values are possible. For example, the role physical (RP) dimension of the SF-36 is scored on a 0 to 100 scale but there are only five possible categories/scores, e.g. 0, 25, 50, 75 and 100 (see the second figure, part (c)). Also, the scale may not be linear. For example, using the SF-36 RP dimension, is a change of score from 0 to 25 the same as a change from 75 to 100?

Another common concern is a floor or ceiling effect. Patients cannot be worse than the worst category or better than the best category. (In the case of the SF-36 score either 0 or 100). For some populations the level is wrong and most people score on either the best category or the worst category. Floor and ceiling effects are more likely to be a problem in longitudinal studies because they limit the ability of the instrument to detect an improvement or deterioration in a patient's QoL over time. Part (c) of the second figure shows that for the RP dimension of the SF-36 over 72 % (1000/1372) of the general population sample had scored 100 and were at the ceiling of the distribution.

Furthermore, methods based on the NORMAL DISTRIBUTION (such as MULTIPLE LINEAR REGRESSION) assume that the outcome variable has a constant VARIANCE. The variances of changes may depend on initial values. This is a common problem with range-limited values. Patients may enter the study with a wide variety of scores, but tend always to increase their scores. Thus patients who score lower at the start of the study have more range to improve than those who are already close to the maximum.

Another issue is that normal approximations may not apply. Since the data are in fact categorical, they may require different techniques of analysis. By definition, no ordinal variable can be normally distributed, although in some cases a normal approximation will suffice. Also, it is difficult to quantify an effect size (e.g. a desirable difference in MEAN score between groups) in advance and another concern in QoL measurement is that MISSING DATA are likely, e.g. in questionnaires that ask 'How far can you walk?' when the patient is in a wheelchair.

The advantages in being able to treat QoL scales as continuous and normally distributed are simplicity in sample size estimation and statistical analysis. Therefore, it is important to examine such simplifying assumptions for different instruments and their scales. Since QoL outcome measures may not meet the distributional requirements (usually that the data have a normal distribution) of parametric methods of sample size estimation and analysis, NONPARAMETRIC METHODS are often used to analyse QoL data.

Conventional methods of analysis of QoL outcomes are extensively described in Fairclough (2002), Fayers and Machin (2007) and Walters (2009). The papers by Walters, Campbell and Lall (2001), Walters, Campbell and Paisley (2001) and Walters and Campbell (2005) discuss alternative ways of determining sample size and analysing QoL outcomes, including the use of the proportional odds model for ordinal data and the nonparametric BOOTSTRAP computer simulation method.

There are numerous QoL instruments now available (and these are extensively described in Bowling, 2001, 2004). By far the easiest way to assess QoL is to use an *off-the-shelf* instrument rather than designing your own. So how do you choose between the various QoL instruments? This fundamentally depends on the purpose of the study and the reliability, validity, responsiveness and practicality of the instrument. A *belt and braces* approach is recommended for the assessment of QoL, meaning both a generic and condition-specific instrument should be used.

The interested reader is referred to Fayers and Machin (2007) for comprehensive guidelines on assessing, analysing and interpreting QoL data. Alternatively, Fairclough (2002) concentrates more deeply on the design and analysis of QoL studies in clinical trials. Walters (2009) presents practical and pragmatic guidelines for the design (i.e. sample size estimation) and analysis of trials involving QoL measures. Finally, Bowling (2001, 2004) provides an extensive review of generic and disease-specific QoL measurement scales. *SJW*

Bowling, A. 2001: *Measuring disease: a review of disease-specific quality of life measurement scales*, 2nd edition. Buckingham: Open University Press. **Bowling, A.** 2004: *Measuring health: a review of quality of life measurement scales*, 2nd edition. Buckingham: Open University Press. **Brazier, J. E, Harper, R., Jones, N. M. B., O'Cathain, A., Thomas, K. J., Usherwood,T. and Westlake, L.** 1992: Validating the SF-36 health survey questionnaire: new outcome measure for primary care. *British Medical Journal* 305, 160–4. **Fairclough, D. L.** 2002: *Design and analysis of quality of life studies in clinical trials*. New York: Chapman & Hall. **Fayers, P. M. and Machin, D.** 2007: *Quality of life: the assessment, analysis and interpretation of patient-reported outcomes*, 2nd edition. Chichester: John Wiley & Sons, Ltd. **Food and Drug Administration** 2006: *Guidance for industry: patient-reported outcome measures: use in medical product development to support labelling claims* (draft). New York: Food and Drug Administration. **Greenfield, D. M., Walters, S. J., Coleman, R. E., Hancock, B. W., Eastell, R., Davies, H. A., Snowden, J. A., Derogatis, L., Shalet, S. M. and Ross, R. J.** 2007: Prevalence and consequences of androgen deficiency in young male cancer survivors in a controlled cross-sectional study. *Journal of Clinical Endocrinology and Metabolism* 92, 9, 3476–82. **Medical Research Council** 2009: *Patient reported outcome measures (PROMs): identifying UK research priorities*. Report of an MRC Workshop, 12 January 2009, Royal College of Physicians, London. London: Medical Research Council. **Walters, S. J.** 2009: *Quality of life outcomes in clinical trials and health care evaluation: a practical guide to analysis and interpretation*. Chichester: John Wiley & Sons, Ltd. **Walters, S. J. and Campbell, M. J.** 2005: The use of bootstrap simulation methods for determining sample sizes for studies involving health-related quality of life measures. *Statistics in Medicine* 24, 1075–102. **Walters, S. J., Campbell, M. J. and Lall, R.** 2001: Design and analysis of trials with quality of life as an outcome: a practical guide. *Journal of Biopharmaceutical Statistics* 11, 3, 155–76. **Walters, S. J., Campbell, M. J. and Paisley, S.** 2001: Methods for determining sample sizes for studies involving health-related quality of life measures: a tutorial. *Health Services and Outcomes Research Methodology* 2, 83–99. **Walters, S. J., Munro, J. F. and Brazier, J. E.** 2001: Using the SF-36 with older adults: cross-sectional community based survey. *Age and Ageing* 30, 337–43. **Ware Jr, J. E. and Sherbourne, C. D.** 1992: The MOS 36-item short-form health survey (SF-36). I. Conceptual framework and item selection. *Medical Care* 30, 473–83.

quantile–quantile (Q-Q) plots See PROBABILITY PLOTS

quantile regression This is a statistical regression method that models any specified QUANTILE (e.g. MEDIAN, first quartile, 90th percentile) of a continuous dependent variable given a set of EXPLANATORY VARIABLES. It is analogous to linear regression (see MULTIPLE LINEAR REGRESSION), which models the MEAN of the dependent variable instead. When applied to the median, quantile regression is known as *median regression*.

Quantile regression has several appealing features, some of which are illustrated in the following four examples. Although inspired by real-life applications, all the examples presented are fictitious. In them, descriptions and interpretations are kept as concise as possible and may occasionally be simplistic. It is hoped that they may nevertheless facilitate understanding of the prominent features of quantile regression.

Example 1. Quantiles are of substantive interest. Forced vital capacity (FVC) measures the total volume of air one can exhale after a deep inhalation and is commonly used along with other indexes to evaluate lung function. Lower values of FVC may be indicative of some pulmonary disorder. In clinical practice it is of great interest to compare an individual's observed measure with reference values of normal. FVC is known to change physiologically along with age, sex and height. Reference values should therefore be age-, sex- and height-specific. The first figure (on page 371) shows a SCATTERPLOT of FVC against age measured on healthy, nonsmoking men. The lines depict the 5th percentile estimated by quantile regression (solid) and mean FVC by LINEAR REGRESSION (dashed). The lines are estimated for 1.8-m tall men. The 5th percentile line can be interpreted as follows: at any given age, 95 % of healthy 1.8-m tall men are expected to measure above the line. Observed FVC measures that fall below are typically considered subnormal. Note that mean FVC estimated by linear regression is hardly of any interest in this context. FVC measures of perfectly healthy individuals are expected to fall above and below the mean line. Insofar as they are not too low, they should raise no suspicion. Quantile regression may estimate other percentiles of normal (e.g. the 1st, the 10th). Quantiles are of research interest in many other settings, which include, for instance, a median lethal dose in toxicology, percentiles of seawater concentration of chemicals in environmental studies, median survival time in CLINICAL TRIALS and 90th

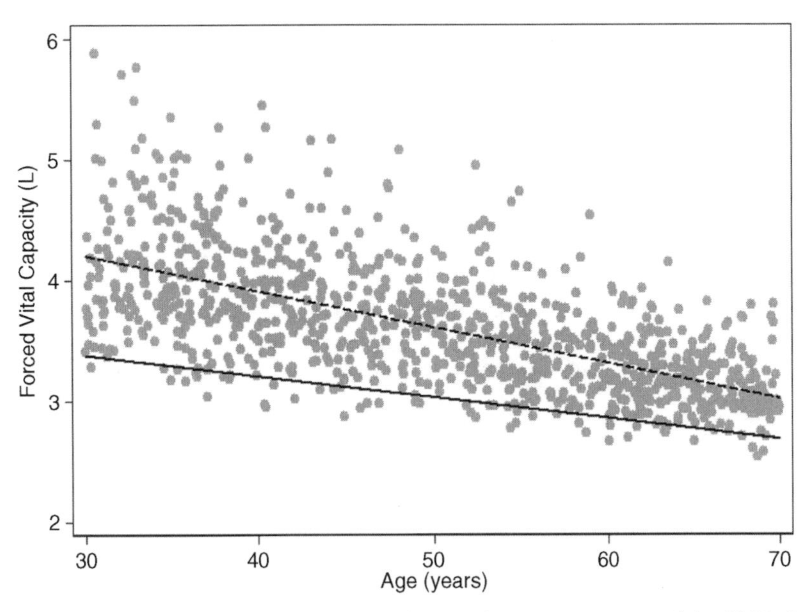

quantile regression *Scatterplot of forced vital capacity against age measured in 1000 fictitious individuals with the 5th percentile estimated by quantile regression (solid line) and the mean estimated by linear regression (dashed line)*

percentile of the time from an emergency call to admission in a hospital in emergency medicine.

Example 2. Quantiles provide insight. Body mass index (BMI = weight/height-squared, in kg/m^2) is often used when studying obesity. The second figure on page 372 shows a scatterplot of BMI against age in sedentary children (left-hand panel) and in children on a physical activity programme (right-hand panel). The solid lines in each panel represent from bottom to top the estimated 5th, 25th, 50th, 75th and 95th percentiles. At 10 years of age the distribution of BMI values in the two groups look similar. With ageing, however, the two distributions separate. The larger BMI values are estimated to grow higher in the sedentary population than in the active population. However, the lower 50 % of the BMI values in the two populations seem not to be conspicuously impacted by a sedentary lifestyle. Indeed, the slopes estimated by quantile regression do not differ significantly between the two groups for any percentile below the median. Linear regression (not shown) would provide estimates for the slopes of mean BMI. They would show a diluted, average effect, which could allow but a partial understanding of the complex impact of the physical activity programme on BMI.

Example 3. Quantiles allow variable transformation. Uranium is a naturally occurring alpha-emitting radionuclide and a toxic heavy metallic element with carcinogenic potential. Groundwater concentrations of uranium are measured in the vicinity of a polluting source. The third figure shows the scatterplot of uranium concentrations (left-hand panel) and the logarithm transform of uranium (right-hand panel) against distance (miles) from the source. The solid lines represent the 5th, 50th and 95th percentiles of uranium and the dashed line its mean. Modelling the relationship between uranium and distance is simpler on the logarithm scale, where it is approximately linear, than on the untransformed scale. Quantile regression allows transformation of the dependent variable. The quantiles of uranium are estimated on the logarithmic scale and then transformed back to the untransformed scale. In the untransformed scale the estimated quantile curves are thus constrained to be positive, which is clearly desirable. In general, in linear regression the dependent variable should not be transformed, despite this being common practice. Inference on transformed outcome would carry no information about untransformed outcome, unless strong distributional assumptions were made. A direct application of linear regression to untransformed uranium, however, produces nonsensical negative estimates of mean uranium. The nonlinear relationship between mean uranium and distance should instead be modelled with some other appropriate method (e.g. splines – see SCATTERPLOT SMOOTHERS – and nonlinear regression methods). Even then, however, inference about the mean only may be unsatisfactory. In the presence of skewed distributions the mean may be highly affected by few

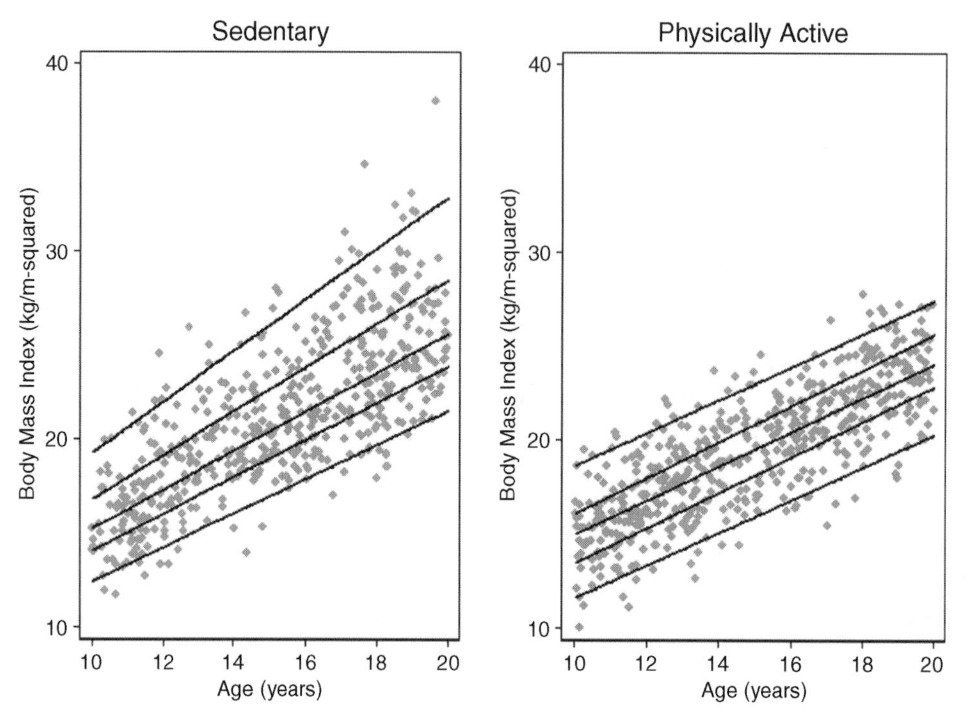

quantile regression *Scatterplot of body mass index against age in 500 fictitious sedentary children (left-hand panel) and 500 fictitious children on a physical activity program (right-hand panel) with the 5th, 25th, 50th, 75th and 95th percentiles estimated by quantile regression (lines bottom to top)*

unusually large values. Inference about a set of quantiles with quantile regression generally permits more complete inference.

Example 4. Quantiles are robust to outliers and measurement error. Sample data may sometimes contain unusually large or unusually small values, often referred to as OUTLIERS. Outliers may occur because they are present in the population from where the sample is drawn or because of measurement errors. Both cases are extremely frequent in real applications. When outliers are present, the median may be a better summary statistic than the mean to assess the location of a distribution because, unlike the mean, it is largely unaffected by them. When the distribution of some variable given the independent variables has unusually large or small values, the median may be more efficient than the mean, in that it has more POWER and gives narrower CONFIDENCE INTERVALS. If, for example, the distribution is normal, then the median is less efficient than the mean; if it is exponential they are equally efficient; if it is a STUDENT'S t-DISTRIBUTION with 3 DEGREES OF FREEDOM the median is more efficient. The robustness to outliers and measurement error applies to quantile regression as well and makes it preferable to linear regression when these issues may be relevant. The fourth figure shows a scatterplot of weight against height. The solid line represents median regression and the dashed line linear regression. The two outliers affect linear regression but not median regression. The slope of the estimated median weight is statistically significantly different from zero, while the slope of mean weight is not. Further, the confidence interval for the estimated slope of the mean is about 50 % larger than that of the median.

Quantile regression models can be easily estimated with most of the widely available STATISTICAL SOFTWARE (e.g. SAS, Stata, R/Splus). For instance, suppose it is of interest to infer median blood pressure in two populations: females and males (or treatment and placebo, exposed to some risk factor and unexposed). Given sample data from each population, median regression provides confidence intervals and P-VALUES for the medians in the two populations and their difference. In this case, the median regression model would have blood pressure as the dependent variable and sex as a binary independent variable, just as linear regression would have the same variables to infer about mean blood pressure in the two populations.

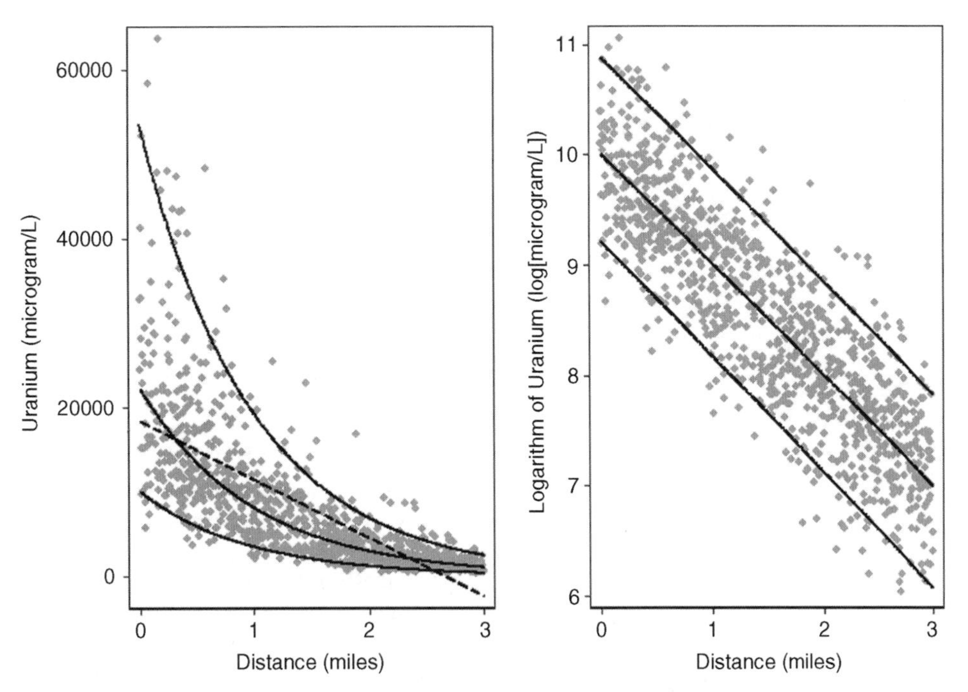

quantile regression *Scatterplot of 1000 fictitious measures of groundwater uranium concentration against distance from a polluting source in the natural scale (left-hand panel) and in the logarithmic scale (right-hand panel) with the 5th, 50th and 95th percentiles estimated by quantile regression (solid lines bottom to top) and the mean estimated by linear regression (dashed line)*

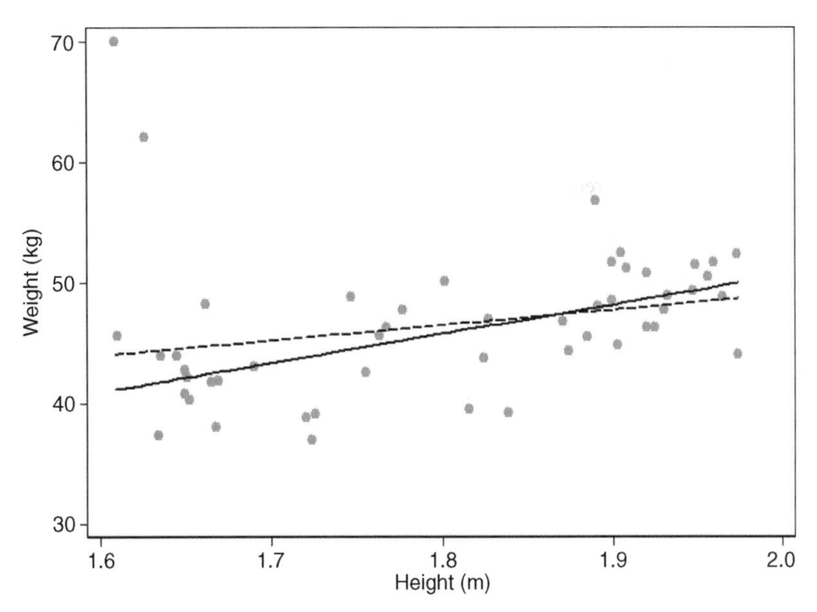

quantile regression *Scatterplot of weight against height in 50 fictitious individuals with median weight estimated by quantile regression (solid line) and mean weight estimated by linear regression (dashed line)*

On a technical note, quantile regression models are estimated by a simple and fast iterative algorithm, an adaptation of the simplex algorithm from mathematical linear programming. Confidence intervals and *P*-values may be obtained from large-sample approximations, rank-score test inversions or the BOOTSTRAP. The latter has been shown to perform better than the others and is generally recommended.

Quantile regression is a flexible, insightful and efficient statistical regression method. Its extensions are an active area of research and include, for example, methods for estimation of conditional quantiles with censored data, clustered data (see CLUSTER RANDOMISED TRIALS and CLUSTERED BINARY DATA), LONGITUDINAL DATA and count data. Given its appealing features, quantile regression is becoming increasingly popular in biomedical research, and its use has been recommended in editorials, commentaries and research articles of numerous reputed journals. For further details see, among others, Koenker and Hallock (2001), Cade and Noon (2003), Austin *et al.* (2005), Koenker (2005), Gillman and Kleinman (2007) and Bottai, Cai and McKeown (2010). *MtB*

Austin, P. C., Tu, J. V., Daly, P. A. and Alter, D. A. 2005: Tutorial in biostatistics: the use of quantile regression in health care research: a case study examining gender differences in the timeliness of thrombolytic therapy. *Statistics in Medicine* 24, 791–816. **Bottai, M., Cai, B. and McKeown, E. R.** 2010: Tutorials in biostatistics: logistic quantile regression for bounded outcomes. *Statistics in Medicine* 29, 309–17. **Cade, B. S. and Noon, B. R.** 2003: A gentle introduction to quantile regression for ecologists. *Frontiers in Ecology and the Environment* 1, 412–20. **Gillman, M. W. and Kleinman, K.** 2007: Invited commentary: antecedents of obesity – analysis, interpretation, and use of longitudinal data. *American Journal of Epidemiology* 166, 14–16. **Koenker, R.** 2005: *Quantile regression.* New York: Cambridge University Press. **Koenker, R. and Hallock, K. F.** 2001: Quantile regression. *Journal of Economic Perspectives* 15, 143–56.

quantiles Cut-points that split either a sample of ordered data, or a PROBABILITY DISTRIBUTION, into regions of pre-specified and often equal size. The MEDIAN is the simplest quantile, which splits the sample into two equal halves. Similarly tertiles, quartiles, quintiles, deciles and centiles (or percentiles) split the sample into respectively thirds, quarters, fifths, tenths and hundredths. There are one fewer cut-points than regions, numbered from the lower to the upper tail of the distribution. For example, the three quartiles split the sample into four equal regions: a quarter below the lower quartile, a quarter above the upper quartile and a quarter either side of the mid-quartile (or median).

Sample quantiles are required in a PROBABILITY PLOT. There are several ways to calculate them; Hyndman and Fan (1996) recommend the median-unbiased estimator:

$$p_i = \frac{i-\frac{1}{3}}{n+\frac{1}{3}}$$

where p_i is the cumulative probability corresponding to the ith quantile, i.e. the ith ordered value in a sample of size n.

The alternative epidemiological definition of quantiles is the regions themselves, so that with this definition quintiles, for example, correspond to the five regions of the distribution rather than the four cut-points. The two definitions are often confused. *TJC*

[See also INTERQUARTILE RANGE and GROWTH CHARTS]

Hyndman, R. J. and Fan, Y. 1996: Sample quantiles in statistical packages. *American Statistician* 50, 361–5.

quantitative trait loci (QTL) These are chromosomal locations of functional variants that affect continuous characteristics (e.g. height, blood pressure) or common diseases that are thought to have an underlying continuous liability (e.g. hypertension). The mode of inheritance of such traits is usually consistent with a polygenic model that assumes multiple small genetic and environmental effects. The term QTL is sometimes used to describe all the constituent loci in the polygenic model, but is more often restricted to the loci that have relatively major and therefore potentially detectable effects, while the rest are known collectively as the *residual polygenic background*. Recent developments in molecular genetics have made available multiple genetic markers throughout the genome and enabled the localisation and detection of individual QTL by linkage and association strategies.

The most popular method of QTL linkage analysis is based on an extension of the variance components model for partitioning phenotypic VARIANCE into genetic and environmental components. Traditional variance components models in genetics rely on different genetic relationships having different extents of genetic sharing and therefore different magnitudes of CORRELATION for the genetic component of the trait. For example, monozygotic twins share all their genes and have a genetic correlation of 1, whereas dizygotic twins share half their genes and have a genetic correlation of 0.5, so that a greater trait similarity between monozygotic twins than between dizygotic twins would suggest the presence of a genetic component.

The extension involves introducing a COMPONENT OF VARIANCE that is correlated between relatives to the same extent as the proportion of alleles they share at the QTL (in the identity-by-descent sense). In other words, relatives who share both alleles at the QTL will be completely correlated for the effects of the QTL and similarly those sharing one or none of the alleles will have correlations of 0.5 and 0 respectively for the effects of the QTL. A genome scan for QTL linkage would involve estimating the extent of allele sharing between family members in a sample from marker genotype data and systematically testing each chromosomal location for a significant QTL component.

Another approach to QTL linkage analysis is based on linear regression of some measure of trait similarity on the proportion of allele sharing for the relative pairs in a sample of families. The original method, proposed by Haseman and Elston, uses the square of the trait difference between relatives as the measure of trait similarity, but more recent work has shown that another definition based on a weighted sum of the squared difference and squared sum (of the mean-centred variables) is more powerful. The original regression approach was restricted to sibling pairs but it has been extended to general pedigrees.

A major problem of QTL linkage analysis is that the POWER to detect a QTL decreases rapidly with decreasing QTL effect size. If random population samples are studied, a sample size of tens of thousands is required for adequate power to detect a QTL that accounts for as much as 10 % of the trait variance. Selective genotyping on the basis of informativeness has been proposed as a method for reducing the cost of linkage studies; typically families are selected for genotyping only if they contain individuals at the extreme of the trait distribution. Both variance components and regression methodologies can be modified to deal with samples with selective genotyping.

ASSOCIATION analysis is complementary to linkage analysis for the localisation and identification of QTL. Both family and unrelated designs are used for QTL association studies, the former providing for the possibility of a within-family test that is robust to population stratification. Typically, QTL association data are analysed by LINEAR REGRESSION or an extension of linear regression to family data that makes allowance for correlated data. A popular method assumes that the trait has a MULTIVARIATE NORMAL DISTRIBUTION within a family, where the means are determined by a linear function of allelic effects and the COVARIANCE structure is determined by degree of genetic relationship, and possibly local allele sharing, between relative pairs. *PS*

[See also GENETIC EPIDEMIOLOGY, TWIN ANALYSIS]

Sham, P. 1997: *Statistics in human genetics*. London: Arnold.

quasi-likelihood This is a generalisation of the GENERALISED LINEAR MODELS approach. In generalised linear model (GLM) methods, such as Gaussian, Poisson and LOGISTIC REGRESSION, the distribution of the response variable is assumed to be one of the EXPONENTIAL FAMILY of distributions. The unknown parameters of the model are estimated by maximising the LIKELIHOOD function. This likelihood function is based on a specification of the whole distribution of the response variable conditional on the covariates.

However, it turns out that the MAXIMUM LIKELIHOOD ESTIMATES and the estimated variance–COVARIANCE MATRIX (and so STANDARD ERRORS) depend only on the first two moments (the MEAN and VARIANCE) of the distribution of the response conditional on the covariates. The quasi-likelihood approach makes use of this property of GLMs. Models are fitted in which only the link and variance functions (the functions that determine how the mean and variance of the response depend on the covariates) are specified, rather than the whole distribution of the response. This may be done even if the link and variance functions do not correspond to a member of the exponential family.

An example may make this clearer. Suppose that $Y_1, \ldots, Y_N$ are the numbers of tumours induced in N mice. We might model the association between the expected number of tumours, $E(Y)$, and some covariate of interest, X, using Poisson regression with the log-link function. Therefore, $\log E(Y) = \alpha + \beta X$, where α and β are unknown parameters. This model includes the assumption that $\text{Var}(Y) = E(Y)$. In many experiments, however, there is overdispersion, i.e. the variance of the response is greater than its expected value. This will lead to the standard errors of α and β being underestimated and the TYPE I ERROR rate of any hypothesis test being inflated.

Instead, one might adopt a quasi-likelihood approach. One possibility would be to assume again that $\log E(Y) = \alpha + \beta X$, but that $\text{Var}(Y) = \phi E(Y)$, where ϕ is some unknown parameter. If $\phi = 1$, we have the POISSON DISTRIBUTION; if $\phi > 1$, there is OVERDISPERSION. Also, note that if $\phi > 1$, the response distribution does not belong to the exponential family.

A generalisation of quasi-likelihood consists of GENERALISED ESTIMATING EQUATIONS (GEEs). Whereas quasi-likelihood is for independent responses, GEEs allow responses to be correlated. This might be of use, for example, for repeated-measures data (see REPEATED MEASURES ANALYSIS OF VARIANCE). For further details see McCullogh and Nelder (1983) and Diggle *et al.* (2002). *SRS*

Diggle, P., Heagerty, P., Liang, K.-Y. and Zeger, S. 2002: *Analysis of longitudinal data*, 2nd edition. Oxford: Oxford University Press.
McCullogh, P. and Nelder, J. A. 1983: *Generalised linear models*. London: Chapman & Hall.

questionnaires These are a means of collecting information from participants in a study. They are useful in many research settings and good design is paramount to ensure that results are informative.

Questionnaires can be self-administered or interviewer-administered, in which case this might be done face to face or over the telephone. Information gained from interviewer-administered questionnaires is more complete and is usually thought to be more accurate, because the interviewer is able to provide additional guidance to the respondent. Interviewer-administered questionnaires must be used with respondents who are illiterate or semi-literate in the language of the questionnaire. Self-administered questionnaires, by way of contrast, are cheaper to use and are generally quicker for the respondent to complete. Self-administered questionnaires are often considered a more appropriate technique when questions are of a very sensitive nature, such as enquiries about illegal drug use.

When developing a questionnaire the precise issues of interest should be considered carefully and numbers of questions apportioned correspondingly. Time spent conducting a PILOT STUDY on a small sample from the target population is rarely wasted; analysis of the process and responses will highlight problems with timing, omission of questions or misunderstanding of instructions.

There are two major types of question that can be included in any questionnaire: open and closed. Open questions ask respondents to reply in their own words. For example, the following open question could be included in a survey about children's attitudes to smoking: 'How did you feel when you had your first cigarette?'

Such questions have the advantage that the respondent is not influenced by the researcher's suggestions and is able to provide a more detailed reply. However, supplying such answers takes more time and effort on the part of the respondent and the process of coding answers is time consuming and can be complex.

Closed questions provide a set of responses from which the individual chooses their answer(s). For example, the open question from earlier could be made into a closed question by supplying the following list of possible responses.

☐ I felt grown up
☐ I enjoyed it
☐ I was disappointed
☐ I felt ill
☐ I felt guilty
☐ Other ....

When using a closed question in a face-to-face interviewer-administered questionnaire, it is helpful to list possible responses on a flashcard so that the participant can easily see all the options. Closed questions must provide all possible responses or include a category entitled 'other', as shown here. Note that a question that includes the 'other' category (as in this example) is sometimes deemed 'semi-open'. A pilot study can be useful to determine popular responses in the 'other' category, which can then be included as defined options on the final questionnaire.

To standardise responses, categories should be qualified as far as possible. For example, use the descriptions on the right of the following frequencies, rather than those on the left, in response to the question 'How often do you eat chocolate?'

☐ Not often	☐ Less than once a week
☐ Fairly often	☐ Between one and seven times a week
☐ Very often	☐ More than once a day

A further possible mistake to avoid is to have categories that are not mutually exclusive, e.g. by asking a participant to indicate their age by ticking a box below:

☐ Under 18
☐ 18–25
☐ 25–30
☐ 30–40
☐ 40–55
☐ 55 or older

When deciding on categories it can be helpful to ensure that they will be comparable with external data, such as ethnic groupings used by government bodies.

Scales are a specific type of closed question. Two commonly used scales are the LIKERT SCALE and the VISUAL ANALOGUE SCALE. The Likert scale requires a participant to choose a response indicating their level of agreement with a statement. For example, the participant might be asked whether they agree with the statement 'I am restricted in my activities because of pain'. He or she would have to choose one of the responses that follow.

☐ Strongly agree
☐ Agree
☐ Neither agree nor disagree
☐ Disagree
☐ Strongly disagree

The visual analogue scale requires participants to indicate their response on a continuous scale, marked at either end. For example, the respondent might be asked to indicate their level of pain following an operation with a cross on the following scale.

Worst pain imaginable ⎯⎯⎯⎯⎯⎯⎯⎯⎯⎯ No pain

Once the data have been collected the researcher must measure the distance from the lower end of the scale to convert the response into a score. For convenience, the scale is often 10 cm long and measurements are taken to the nearest millimetre.

It is helpful if questions are as specific as possible, so that instead of asking 'Do you have a car?' ask 'Is there a car or van available for private use by you or a member of your household?' It is also important that the wording of questions is not ambiguous. For example, the question 'Where do you live?' might elicit responses about geographical locations or types of accommodation.

Hedges (1979) describes other ways in which the wording of questions affects responses. In answering question (a) in the following questionnaire, 82% of respondents replied that they take enough care of their health, whereas only 68% gave this response to question (b):

(a) Do you feel you take enough care of your health or not?
(b) Do you feel you take enough care of your health or do you think you could take more care of your health?

The description of the alternative to taking enough care of your health may have influenced respondents. However, it is also notable that the two options in question (b) are not necessarily mutually exclusive: a respondent may be aware that he or she could take *more* care of his or her health, but at the same time considers that he or she takes *enough* care.

A respondent must only be asked one question at a time. The enquiry 'Were you satisfied with the treatment you received in hospital and at home?' would be better split into two separate questions. Difficulties also arise when respondents are asked questions that are irrelevant to them. If a questionnaire distributed to an elderly population asks whether they get breathless when doing housework, an individual who never does any housework might answer 'no' for this reason.

For some measurements, such as birth weight, both imperial and metric units are in common use. More accurate responses will result if the respondent is allowed to report on either scale and the conversion done at the analysis stage.

The layout of questionnaires is also important. The form should be easy to read, particularly for self-administered questionnaires; it helps if there is plenty of white space on each page. Questions on the same topic are best grouped together and 'transitions' such as 'We would now like to find out about the health of your family' are useful. If sections on a questionnaire are to be skipped by some respondents, then this should be made as clear as possible, perhaps by the use of arrows.

Sampling considerations are as important in questionnaire-based studies as in any others and as such it is vital to use a representative sample from the population to whom the results are to be applied. Nonresponse often introduces BIAS (see NONRESPONSE BIAS) and this can be a considerable problem when using self-administered questionnaires, particularly postal questionnaires. To minimise nonresponse, questionnaires should be kept concise, easy to read and should not begin with personal or difficult questions that discourage the participant from starting. A covering letter explaining the reason for the research and the investigator's credentials may also help motivate the respondent. Pre-paid return envelopes should be included with postal questionnaires and one or preferably two reminders sent to those who do not respond, enclosing further copies of the questionnaire and pre-paid return envelopes. The investigator must make every effort to ensure that names and addresses used are current.

Before a questionnaire is used the issues of validity and reliability should be addressed (see MEASUREMENT PRECISION AND RELIABILITY). Validity assesses whether a questionnaire measures what it intends to measure, while reliability evaluates the consistency of the questionnaire when it is administered repeatedly to the same individual. It is therefore important that the questionnaire is known to be valid and reliable before time and resources are invested in the study. McDowell and Newell (1996) give information on how validity and reliability can be assessed.

An important consideration before writing a new questionnaire is whether a suitable instrument already exists. The use of an existing questionnaire saves time in writing and piloting. Also, information may be available regarding validity and reliability, and results could be more comparable with those from other studies. Again, McDowell and Newell (1996) provide detailed descriptions of existing questionnaires for measuring aspects of health such as depression, pain and quality of life. *SRC*

Hedges, B. M. 1979: Question wording effects: presenting one or both sides of a case. *The Statistician* 28, 83–99. **McDowell, I. and Newell, C.** 1996: *Measuring health: a guide to rating scales and questionnaires*, 2nd edition. Oxford: Oxford University Press.

quota sample Quota sampling is a nonrandom SAMPLING METHOD. Before the sample is chosen, the population is divided into groups according to certain characteristics, e.g. age, sex or smoking status. The interviewer is then told to interview a specified number of people within each group, but is given no instructions on how to find the people to interview. Quota sampling is often used in opinion polls and in market research surveys.

Quota sampling has the advantage that it is quick and easy to do. Any member of the sample can be replaced with another member with the same characteristics, which is not the case in random sampling.

A major disadvantage of quota sampling is that, as it is completely nonrandom, there is likely to be a great deal of BIAS in the selection process. The interviewer is more likely to approach people who are easy to question or who appear cooperative. It is also difficult to find out about those who do not cooperate, since they are replaced in the sample. However, if no sampling frame exists then quota sampling may be the only practical method of obtaining a sample.

As an example, in a paper assessing the priorities for allocation of donor liver grafts (Neuberger *et al.*, 1998), quota sampling was used to choose members of the general public to be included. The quota was designed so that the sample would be 'nationally representative' and included 1000 people aged 15 or above. It was based on 10-cell quota for sex, household tenure, age and work status. Quota sampling was also used to choose the regions from which the family doctors came, quotas being based on region, with one practitioner per practice. Within regions the selection of practices was random. For further details see Crawshaw and Chambers (1994). *SLV*

Crawshaw, J. and Chambers, J. 1994: *A concise course in A level statistics*, 3rd edition. Cheltenham: Stanley Thornes Publishers. **Neuberger, J., Adams, D., MacMaster, P., Maidment, A. and Speed, M**. 1998: Assessing priorities for allocation of donor liver grafts: survey of public and clinicians. *British Medical Journal* 317, 172–5.

R

R This free statistical software offers extensive data analysis and graphics facilities. R runs on many different computer operating systems (including Windows, MacOS and various forms of Linux), and provides a wide range of statistical analyses and has very powerful facilities for producing publication-quality graphics. R is used worldwide by researchers in both universities and industry (including the pharmaceutical industry).

In addition to the predefined statistical analyses and graphics capabilities, R provides a fully featured programming language for manipulating data and for creating new analysis and graphics functions. This programming language is similar to the S language, so code that is written for S-Plus will often run in R without modification. The basic functionality of R can be extended by loading add-on packages, of which there are now several thousand available (see the Comprehensive R Archive Network, http://cran.r-project.org/).

The default user interface for R is a command line, but a number of GUI interfaces are available via related software projects and add-on packages.

For more information on R, see the R homepage (http://www.r-project.org/), which has links to download sites, mailing lists, documentation, add-on packages and related software projects. The documentation on this website includes several book-length introductions to using R that can be downloaded for free. There are also many published books: Dalgaard's *Introductory statistics with R* (2008) provides an entry-level statistical context while Murrell's *R graphics* (2005) focuses on producing a variety of charts and figures. Venables and Ripley's *Modern applied statistics with S* (2002) provides a more sophisticated treatment. Some of Everitt and Rabe-Hesketh's *Analyzing medical data using S-Plus* (2001) and all of Everitt and Hothorn's *A handbook of statistical analyses using R* (2006) are also applicable. *PM*

Dalgaard, P. 2008: *Introductory statistics with R*, 2nd edition. New York: Springer. **Everitt, B. and Hothorn, T.** 2006: *A handbook of statistical analyses using R*. Boca Raton, FL: Chapman & Hall/CRC. **Everitt, B. and Rabe-Hesketh, S.** 2001: *Analyzing medical data using S-Plus*. New York: Springer. **Murrell, P.** 2005: *R graphics*. Boca Raton, FL: Chapman & Hall/CRC. **Venables, W. N. and Ripley, B. D.** 2002: *Modern applied statistics with S*, 4th edition. New York: Springer.

random effect This is one of a set of effects on a response variable corresponding to a set of values taken by an explanatory variable. Random effects are included in a regression model to acknowledge that response tends to differ between the groups defined by the explanatory variable. By including random effects, the investigator can estimate the MEAN level of response across groups and the extent to which response varies between groups or estimate the effect of another variable of interest while controlling for the differences between groups.

Typical examples of explanatory variables include indicator variables for hospitals in a national survey, centres in a multicentre study, patients in a longitudinal dataset or studies in a META-ANALYSIS. If the variable defines k distinct groups in the dataset, e.g. if k hospitals are recruited for a survey, the k random hospital effects in the present survey are assumed to be drawn from a distribution of effects associated with the population of hospitals in general. It is common to assume random effects to be drawn from a NORMAL DISTRIBUTION. The variance of this distribution is estimated in the analysis and represents the extent of variation between the groups. For example, researchers may be interested in the variability between hospitals in admission rates or the variability between family doctors in prescribing lipid-lowering drugs.

Random effects are appropriate when the investigator wishes to estimate or control for the distribution of the group effects defined by the explanatory variable over the population of possible groups. When fitting random effects, the groups are assumed to be exchangeable, which means that the investigator has no reason to distinguish one group from another and would (in principle) be prepared to mix up the names of the groups in the dataset before carrying out the analysis. An alternative approach to modelling the differences between groups is to model these as FIXED EFFECTS. This is appropriate when the focus is on the group effects for the k specific groups included in the dataset, e.g. when estimating the response of patients to each of three treatments compared in a CLINICAL TRIAL. *RT*

random effects models This is essentially a synonym for MULTILEVEL and LINEAR MIXED EFFECTS MODELS. This term is commonly used to represent regression models that include both FIXED EFFECTS and RANDOM EFFECTS. *RT/BSE*

random effects models for discrete longitudinal data Discrete responses include categorical responses (e.g. dichotomous or ordinal) and integers (e.g. counts). We can use well-known models such as LOGISTIC REGRESSION and POISSON REGRESSION to model the effects of covariates on the responses. However, an extra complication with LONGITUDINAL DATA is that the responses on the same subject tend to be dependent over time. For instance, in the example considered later, some subjects are consistently more prone to thought disorder than others. While some of this dependence is due to subject-specific covariates that we can include in the model, some extra dependence usually remains even after controlling or adjusting for the covariates. This extra dependence, which may be due to omitted covariates, can be modelled using RANDOM EFFECTS. Our focus here is on random effects models for dichotomous responses but we briefly consider RANDOM EFFECTS MODELS for counts or incidence rates later.

For dichotomous and ordinal responses the most popular models are logistic regression and probit regression (see PROBIT MODEL). Let y_{ij} be the measurement at occasion $i = 1, \ldots, n_j$ for a subject $j = 1, \ldots, N$. The simplest random effects logistic regression model includes a subject-specific random intercept ζ_{0j} to model the dependence among the repeated measurements. Here the log of the odds of a '1' response versus a '0' response is modelled as:

$$\ln\left(\frac{\Pr(y_{ij}=1|\mathbf{x}_{ij},\zeta_{0j})}{\Pr(y_{ij}=0|\mathbf{x}_{ij},\zeta_{0j})}\right) = (\beta_0+\zeta_{0j}) + \beta_1 x_{1ij} + \cdots + \beta_p x_{pij},$$

where $\mathbf{x}_{ij} = (x_{1ij}, \ldots, x_{pij})$ are covariates with regression coefficients β_1 to β_p, β_0 is the mean intercept and ζ_{0j} is the deviation of subject j's intercept from the MEAN. The covariates typically include time or functions of time to model changes in the log odds over time. The random intercept ζ_{0j} is assumed to be normally distributed with zero mean. Inclusion of ζ_{0j} allows the overall log odds of the response to vary between subjects, even after controlling for the covariates. Since ζ_{0j} remains constant over time, the log odds and therefore the PROBABILITY of a '1' response for a given subject is either greater than expected given the covariates at all occasions (if $\zeta_{0j} > 0$) or smaller than expected (if $\zeta_{0j} < 0$), producing the required within-subject dependence. The random intercept can be interpreted as the component of the effects of all omitted covariates on the log odds that is constant over time and uncorrelated with the included covariates.The random intercept logistic regression model can equivalently be expressed in terms of a latent (unobserved) response y^*_{ij} (see LATENT VARIABLES) underlying the observed response y_{ij}, where $y_{ij} = 1$ if $y^*_{ij} > 0$ and $y_{ij} = 0$ otherwise. The logistic regression model becomes a linear regression model for the latent response:

$$y^*_{ij} = \beta_0 + \beta_1 x_{1ij} + \cdots + \beta_p x_{pij} + \zeta_{0j} + \varepsilon_{ij}$$

where ε_{ij} is independent of ζ_{0j} and has a logistic distribution with mean zero and VARIANCE $\pi^2/3$.

The strength of the residual within-subject dependence can be expressed by the intraclass CORRELATION for repeated latent responses y^*_{ij} and $y^*_{i'j}$:

$$\rho = \mathrm{Cor}\left(y^*_{ij}, y^*_{i'j}|x_{ij}, x_{i'j}\right) = \mathrm{Cor}(\zeta_{0j}+\varepsilon_{ij}, \zeta_{0j}+\varepsilon_{i'j}) = \frac{\mathrm{Var}(\zeta_{0j})}{\mathrm{Var}(\zeta_{0j}) + \pi^2/3}.$$

The random intercept model assumes that the log odds change in the same way over time for all subjects with the same covariate values. Since this may be unrealistic, we can allow the linear growth or rate of change of the log odds to vary randomly between subjects by including a random slope ζ_{1j} of time in the model, giving the random coefficient model:

$$\ln\left(\frac{\Pr(y_{ij}=1|\mathbf{x}_{ij},\zeta_{0j},\zeta_{1j})}{\Pr(y_{ij}=0|\mathbf{x}_{ij},\zeta_{0j},\zeta_{1j})}\right) = (\beta_0+\zeta_{0j}) + (\beta_1+\zeta_{1j})x_{1ij} + \beta_2 x_{2ij} + \cdots + \beta_p x_{pij},$$

where x_{1ij} represents the time at measurement occasion i for subject j, β_1 is the mean slope and ζ_{1j} is the deviation of subject j's slope from the mean slope. The random intercept ζ_{0j} and slope ζ_{1j} are typically assumed to have a BIVARIATE NORMAL DISTRIBUTION with zero means. The model can be extended by including further random effects for other variables, for instance polynomials in time to model variability in nonlinear growth.

MAXIMUM LIKELIHOOD ESTIMATION is the state-of-the-art method for estimating RANDOM EFFECTS MODELS for discrete data. The marginal LIKELIHOOD is obtained by 'integrating out' the random effects. When the random effects are multinormal, the integration is typically performed using Gaussian quadrature or the superior adaptive quadrature approach (e.g. Rabe-Hesketh, Skrondal and Pickles, 2005). Computationally efficient but rather crude approximations such as penalised QUASI-LIKELIHOOD or marginal quasi-likelihood are commonly used. Sometimes the distribution of the random effects is left unspecified and nonparametric maximum likelihood is used (e.g. Aitkin, 1999).

The Madras Longitudinal Schizophrenia Study followed up patients monthly after their first hospitalisation for schizophrenia. We will use random effects logistic regression to investigate whether the course of illness differs between patients with early and late onset. The variables considered are: [Month]: number of months since first hospitalisation; [Early]: early onset (1: before age 20, 0: at age 20 or later); [y]: repeated measures of thought disorder (1: present, 0: absent)

The first table contains a subset of the data, namely on whether thought disorder [y] was present or not for 44 female patients at 0, 2, 4, 6, 8 and 10 months after hospitalisation.

random effects models for discrete longitidunal data *Data on thought disorder*

j	*early*	y_0	y_2	y_4	y_6	y_8	y_{10}
1	0	1	1	1	0	0	0
6	1	0	0	0	0	0	0
10	0	1	1	0	0	0	0
13	0	0	0	0	0	0	0
14	0	1	1	1	1	1	1
15	0	1	1	0	0	0	0
16	0	1	0	0	1	0	0
22	0	1	1	1	0	0	0
23	0	0	0	0	1	0	0
25	1	0	0	0	0	0	0
27	1	1	1	1	1	0	1
28	0	0	0				
31	1	1	1	1	0	0	0
34	0	1	1	0	0	0	0
36	1	1	1	0	0	0	0
43	0	1	1	0	1	0	0
44	0	0	1	0	0	0	0
45	0	1	1	1	0	1	0
46	0	1	1	1	1	0	0
48	0	0	0	0	0	0	0
50	0	0	0	0	1	1	1
51	1	0	1	1	1	0	0
52	0	1	1	1	0	1	0
53	0	1	0	0	0	0	0
56	1	1	0	1	0	0	0
57	0	0	0	1	0	0	0
59	0	1	1	0	0	0	0
61	0	0	0	0	0	0	0
62	1	1	1	1	0	0	0
65	1	0	0	0	0	0	0
66	0	0	0	0	0	0	0
67	0	0	1	0	0	0	0
68	0	1	1	1	1	1	1
71	0	1	1	0	1	0	0
72	0	1	0	0	0	0	0
75	1	1	0	0	0		
76	0	0	1				
77	1	0	0	1	0	0	0
79	0	1					
80	0	1	1	1	0	0	0
85	1	1	1	1	1	0	0
86	0	0	1				
87	0	1	0	0	0	0	
90	0	1	1	1	0	0	0

Maximum likelihood estimates based on adaptive quadrature are given in the second table (estimates were obtained using gllamm: http://www.gllamm.org). For the random intercept model the odds of thought disorder decrease over time in late-onset women with an estimated subject-specific ODDS RATIO of $\exp(-0.40) = 0.67$ per month. Early onset patients do not seem to have a higher odds of thought disorder at first hospitalisation (estimated subject-specific odds ratio $\exp(0.05) = 1.05$). Early-onset patients appear to have a greater decline in their odds of thought disorder over time. The odds ratios should be interpreted as conditional on the random intercept, sometimes referred to as subject-specific. For time, the odds ratio can be viewed as a within-subject comparison and for subject-specific covariates such as early onset, the comparison is between two patients having the same value of the random intercept. The intraclass correlation of the latent responses is estimated as $\hat{\rho} = 0.46$.

random effects models for discrete longitudinal data *Repeated measurements of thought disorder: maximum likelihood estimates for dichotomous logistic regressions with random intercept and with random intercept and random slope for [Month]*

	Random intercept model	*Random coefficient model*
	Est (SE)	*Est (SE)*
Fixed part		
β_0 [Cons]	1.01 (0.46)	1.36 (0.66)
β_1 [Month]	−0.40 (0.08)	−0.51 (0.13)
β_4 [Early]	0.05 (0.88)	0.04 (1.21)
β_5 [Early] × [Month]	−0.07 (0.14)	−0.06 (0.21)
Random part		
Var(ζ_{0j})	2.76 (1.24)	7.17 (4.03)
Var(ζ_{1j})		0.14 (0.09)
Cov(ζ_{1j},ζ_{0j})		−0.71 (0.53)
Log-likelihood	−124.75	−121.20

Estimates for the random coefficient model are also reported in the second table. The random slope variance is estimated as 0.14 and the COVARIANCE between the intercept and slope as −0.71, corresponding to a correlation of −0.70. Therefore those at higher risk of thought disorder at the time of hospitalisation experience a greater reduction in their risk over time than those at lower risk. It is important to note that the random intercept variance and the correlation between the random intercept and coefficient are interpreted at [Month] = 0. (Subtracting 5 months from [Month] yields an estimated correlation close to zero.)

To gain more insight into the model, we have plotted the conditional or *subject-specific* probabilities of thought disorder given various values of the random intercept (±3) and slope (±0.4) for women with early onset. These are shown as dashed curves in the figure on page 382, where the dotted

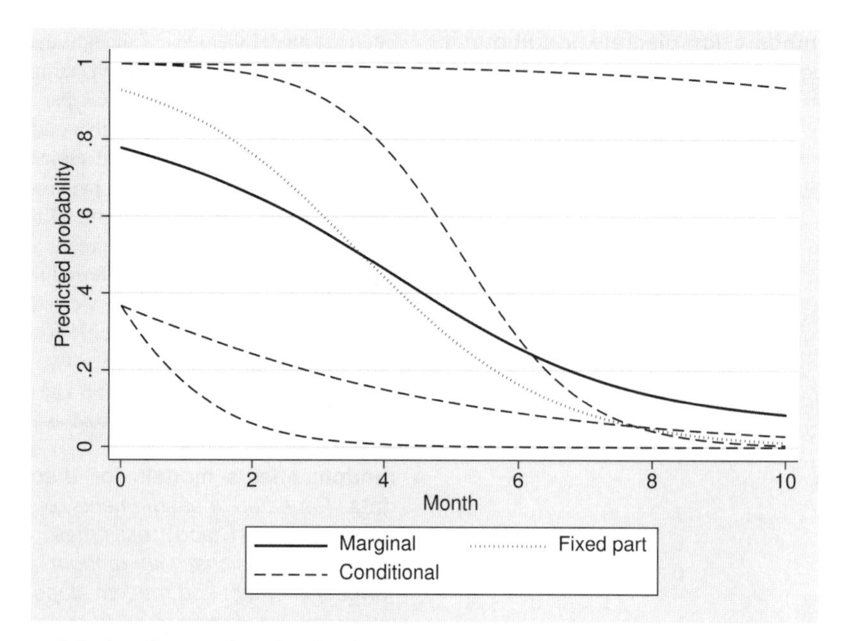

random effects models for discrete longitudinal data *Conditional and marginal predicted probabilities of thought disorder for women with early onset. Dotted curve is for conditional probability from the random coefficient model when both the random intercept and slope are zero*

curve is the CONDITIONAL PROBABILITY for random intercept and slope both equal to their population means of zero, thus representing a 'typical' individual. Also shown as a solid bold curve is the *population average* or marginal probability of thought disorder obtained by integrating the conditional probability over the random effects distribution.

Note that the population average curve is considerably flatter than that of a typical patient. Such attenuation of the effects of covariates in marginal models compared with conditional models is a well-known phenomenon for dichotomous responses (see GENERALISED ESTIMATING EQUATIONS).

For counts or incidence rates, the most common model is Poisson regression. As for the random effects logistic model, RANDOM INTERCEPT MODELS and different kinds of random coefficient models can be specified. Note that the marginal or population-averaged effects equal the conditional or subject-specific effects for the random intercept Poisson model.

Random effects models for discrete longitudinal data can also be used for multilevel designs with more than two levels, for instance where repeated measures are nested in patients who are nested in hospitals. Rabe-Hesketh and Skrondal (2008) give a general treatment of random effects models for discrete data. *SRH/AS*

Aitkin, M. 1999: A general maximum likelihood analysis of variance components in generalised linear models. *Biometrics* 55, 117–28. **Rabe-Hesketh, S. and Skrondal, A.** 2008: *Multilevel and longitudinal modeling using Stata*, 2nd edition. College Station, TX: Stata Press. **Rabe-Hesketh, S., Skrondal, A. and Pickles, A.** 2005: Maximum likelihood estimation of limited and discrete dependent variable models with nested random effects. *Journal of Econometrics* 128, 301–23.

random forest This is a nonparametric method for regression, classification and survival analysis (see SURVIVAL ANALYSIS – AN OVERVIEW), which works especially well for exploratory variables with nonlinear influence on the response, in the presence of higher order interactions and for HIGH-DIMENSIONAL DATA (Breiman, 2001). The method is motivated by the idea of growing an ensemble of trees on BOOTSTRAP samples of the original data. Predictions are computed by averaging (regression) the predictions of multiple trees or by a majority vote of class predictions (classification). Using the out-of-bootstrap observations, random forests compute an honest estimate of the generalisation error, which can be used to assess the model fit. Empirically, the algorithm has been found to be rather insensitive to the choice of hyperparameters. A drawback is that the fitted models are hard to interpret. Variable importance measures are commonly used to derive a ranking of the exploratory variables with respect to their influence on the response. However, these measures have been questioned by various authors and remain a matter of debate. Applications of random forests in medicine and biology include large-scale

ASSOCIATION studies for complex genetic diseases, e.g. to detect SNP–SNP (single-nucleotide polymorphism) interactions in the case-control context by means of computing a random forest variable importance measure for each polymorphism. Prediction of PHENOTYPES based on amino acid or DNA sequence is another important area to which random forests have been applied. *TH*

[See also BOOSTING]

Breiman, L. 2001: Random forests. *Machine Learning* 45, 1, 5–32.

random intercept model See LINEAR MIXED EFFECTS MODEL

random intercept and slope model See LINEAR MIXED EFFECTS MODEL

randomisation This is the process by which patients should be assigned to treatments in a CLINICAL TRIAL. At the outset we should contrast random assignment (randomisation) with random sampling: the latter is the process by which we select individuals to take part in our experiment and forms the basis of our concluding that the results apply to a broad population (see SAMPLING METHODS – AN OVERVIEW). If we were to recruit only young healthy males to a clinical trial or any other form of research, it would be unreasonable to expect our results to apply to a broad population of men, women, adults and children.

Unfortunately, random sampling is almost never possible in the context of medical research although we may still be able to make reasonable reference to a future population to which our results may apply. Little more will be said here about random sampling.

The key reason for using randomisation to decide which patient receives which treatment is to eliminate BIAS (Altman and Bland, 1999). Bias may obviously be introduced deliberately but it can also be introduced inadvertently. Selecting healthier patients or those with fewer – or less severe – symptoms to receive one treatment rather than another would probably influence the outcome from a clinical trial such that we might incorrectly infer that one treatment was better at relieving symptoms than another.

From this it should be clear that not only must the sequence be random but those responsible for recruiting patients must not know what the sequence is – otherwise it would still be possible to enrol the 'milder' patients on to one treatment arm and the 'more severe' patients on to the other. This is also a reason against 'alternate' allocation of one treatment followed by the other (Chalmers, 1999). The strength of a clinical trial should be that causality can be inferred because the two (or more) treatment groups are balanced – at least in the probabilistic sense – for all factors except the treatment received.

The table contains an extract of the table of random digits published by Altman (1991, p. 540). There are many ways of using such a table to assign patients randomly to two treatment groups. The simplest approach might be to assign even numbers to receive treatment and odd numbers to receive placebo, or control. Reading across the first row the 1st patient is assigned number 47 and so receives placebo; the 2nd patient assigned number 44 would receive active treatment; the 3rd, 4th, 5th and 6th (assigned 76, 60, 72 and 56 respectively) all receive active treatment; the 7th patient (99) receives placebo and so on. We could split the numbers up so that the 1st patient is assigned number 4 and the 2nd number 7 (receiving active and placebo treatments respectively), and the next patients are assigned 4, 4, 7, 6, 6, etc.

randomisation *Random numbers*

47	44	76	60	72	56	99	20	20	52
31	60	26	13	69	74	80	71	48	73
72	89	83	91	86	62	78	86	95	07
31	40	99	54	61	99	32	30	43	80
03	49	79	75	46	76	56	99	54	46
36	61	26	31	49	40	74	86	32	36
91	72	12	92	31	66	91	99	48	42
42	73	76	68	86	75	21	91	72	38
32	95	21	17	27	63	06	14	24	05
57	24	32	29	46	60	82	90	81	31

Alternatively, we could choose to read down the columns instead of across the rows. Any one of these procedures is perfectly valid and will result in an unbiased and (in the long run) balanced assignment between the two treatment groups, provided that the rule for using the random number digits is set out in advance. If we had three treatment groups then we could use the same tables but use the numbers 1, 2, 3, for assignment to treatment one; 4, 5, 6 for assignment to treatment two; and 7, 8, 9 for assignment to treatment three or placebo. An occurrence of a zero is ignored and the next digit used instead. Other rules can be established for assignment to any number of treatments. This is referred to as 'simple randomisation'. We shall make reference to this table of random digits later to illustrate different forms of randomisation.

Reference was made earlier to 'balanced' treatment groups. Balance is one of the most important aspects of a clinical trial because it ensures that we are making a fair comparison between the treatments and neither one nor the other is predisposed to showing a better or worse response.

There are three different aspects to balance that are useful to discuss: ensuring the same number of patients receive each of the treatments; ensuring that demographic data and disease severity data are similar between the two treatment groups; and ensuring that factors unknown to the experimenter but that may nevertheless influence outcome are also balanced between the treatment groups. Randomisation is quite a remarkable tool – not only does it balance the treatment groups for all the known important prognostic factors but it also balances for any factors that may be important but we may not know about.

In the example given using line 1 of the table, among the first 10 patient assignments only two of them are odd numbers and so would receive the placebo (patient numbers 1 and 7). Superficially, this seems to be a concern although, in fact, if we randomise a sufficiently large number of patients using the full random number table then we should – on average – assign an equal number of patients to each of the treatment arms. Trials of only 10 patients are extremely rare and would probably not be very convincing whatever the results. However, in a relatively small study, perhaps of some complex surgical technique requiring a lot of skill on the part of the surgeon, we might be concerned if so few of the early patients were assigned to PLACEBO (or *sham* treatment).

It is quite possible that the skills of the surgeon might improve over time and so the overall outcome (regardless of treatment group) might tend to improve as the trial progresses. Therefore having an imbalance in the number of patients assigned to one or other treatment very early on could introduce a bias between the treatment groups. For this reason, we often use blocking to ensure that at regular intervals the number of patients assigned to each treatment group is the same.

In the simplest case of two treatment groups, we may use block sizes of 4. This would mean that within every group of four patients, two are assigned to treatment and two are assigned to placebo. We would not know which two patients receive either treatment or placebo, neither would it be the same two in every block of four treatment assignments. When we have two treatments (call them A and B) and assignment is in blocks of size four, there are six possible configurations of each block: AABB, ABAB, ABBA, BBAA, BABA and BAAB. Each of these blocks should be equally likely to occur. With two treatments, the block size does not have to be four but it could be any multiple of the number of treatments. Similarly, if there are four or more treatments then the block size might be anything from twice the number of treatments upwards.

The advantage of using blocks should be quite clear – not only will the total number of patients assigned to each treatment be the same at the end of the study but also, on a regular basis (perhaps once every four patients), the number assigned to each treatment will also be the same. These types of processes are referred to as 'blocked randomisation' or 'restricted randomisation'.

There can be disadvantages to blocking since it can compromise blinding and so allow bias to be introduced. Consider an extreme situation where there are two treatments and the block size is two. If, for some reason, the assignment of the first patient were to be known (possibly because of typical adverse reactions or even some inadvertent unblinding of the treatment), then the identity of the next treatment is necessarily known. This is typically why a block size equal to the number of treatments would not be used. Even if the block size is twice the number of treatments, if one of the early patients in the block is unblinded, then still the probability of future assignments will not be 0.5. Such potential unblindings are arguments against blocking – or at least in favour of longer blocks, but the longer the block the less the balance on a regular basis and so a compromise has to be found. One strategy that is sometimes used is to vary the block length, perhaps between blocks of four treatments and blocks of six treatments. If the fifth patient were to be unblinded then the investigator would not know if this is the penultimate patient in a block of size six or if it were the first patient following a block of size four. This strategy can greatly help to eliminate possible biases due to unblinding but does increase complication in packing treatment.

This strategy of blocking ensures that there is balance on a regular basis through the duration of the trial, but a further feature upon which balance may be desired is that of demographic or disease severity factors. In the simplest case, consider that gender is a factor highly prognostic of treatment outcome so it is important to ensure that there is not an imbalance of men or women assigned to one or other treatment. Simple randomisation as just described should ensure that this is the case but only in a long-term or probabilistic sense and in any particular trial, if there were some imbalance, then this might bring into question the validity or reliability of the results.

Stratification is a very simple mechanism that uses different randomisation sequences for the different strata on which treatment balance is necessary. Using the random digits in the table, we may decide that the first five rows should be used for assigning men to either treatment (even numbers) or control (odd numbers) and that the second five rows should be used for similarly assigning women to either treatment or control. A common misconception is that stratification ensures equal numbers of men and women on each treatment, but considering random sampling, described at the beginning of this section, this is not necessarily the case. The proportion of men and women in the target population with the disease that is being studied may not be equal. The proportion of men and women in the target population who are prepared to take part in the clinical trial may not be equal. Stratification ensures that *of the men* who take part in the trial, half of them will be assigned to treatment and half to control and that *of the women* who take part in the trial, similarly, half will receive treatment and half will receive control. Considering our earlier discussion concerning blocking, it will be evident that the equal allocation to treatment and placebo of the men (and of the women) will only be the case in the long term and it is quite common within stratified randomisation to also include blocking. This introduces very little extra complexity.

One practicality of preparing stratified randomisation sequences is that, instead of one sequence being prepared for the entire trial, two or more sequences are prepared (one for each stratum). The methods used to introduce blocking into the single sequence used for a trial without stratification is simply replicated in each of the stratum-specific randomisation sequences. In MULTICENTRE TRIALS, the most common feature of the randomisation scheme is stratification by investigator or centre. This ensures that each centre uses all of the treatments and there is no risk that, within any particular centre, all patients might receive only one of the treatments.

It is quite possible to stratify on more than one factor. The simplest approach where this applies is in a multicentre study where there is one (noncentre) factor on which stratification is needed. Different randomisation sequences are prepared for each centre so that we have stratification by centre. In fact, each centre would be given two randomisation sequences, e.g. one to be used for males and one to be used for females. Such a study then has two stratification factors (centre and gender), although, in practice, each investigator would only see and use one stratification factor, that of gender. When there are two or more important prognostic factors but both needs to be balanced for, then stratification begins to become rather more complex.

Consider the case where we wish to stratify by gender (male and female) and also stage of disease (early or advanced). We now need four randomisation sequences: the first can be used for females with early stage disease, the second for females with advanced stage disease, the third for males with early stage disease and the fourth for males with advanced stage disease. The potential for using the wrong randomisation sequence obviously becomes higher in this situation and if any of the factors has more than two levels or if there are more than two or three factors then the number of sequences usually becomes prohibitively large.

The logistics of randomly assigning patients to treatment can be eased by central (often telephone or web-based) randomisation schemes where the investigator does not have to concern himself or herself with using different randomisation sequences for different types of patients. If medication is supplied to the investigational site in sequentially numbered packages then a central randomisation system, if given details of the levels of the stratum for any particular patient, can simply inform the investigator which treatment pack to assign to that patient. The details of multiple randomisation sequences need not then concern the investigator.

Despite the fact that the investigator does not need to concern himself or herself with the multiple randomisation schemes, somebody does! The additional complexities of treatment packaging and assignment should not be underestimated and the potential for errors in randomising patients to treatments must be considered. The risk of introducing errors can become quite high. Also, it is quite possible, with many strata and relatively few patients, that some of the combinations of strata values may occur only very infrequently or not at all. Because of the desire to try to balance within these stratification factors there can sometimes be a problem that the overall balance between the treatment groups (i.e. the *total* number of patients on each treatment) may now become unbalanced. Methods exist to try to find a compromise between balancing individual factors and balancing the total treatment assignment.

Most trials are arranged such that the sequence of randomisation codes is established before patients are recruited to the trial and medication is then pre-packed and despatched to investigators. In this case, although the sequence is unknown to the investigators and the patients it is fixed and potentially known to the study statistician or those responsible for packing the medication. Provided patients are assigned sequentially, balance (on average) will be maintained. Adaptive designs change the randomisation sequence as the trial progresses. One method is called MINIMISATION, which helps to solve the problem discussed earlier of multiple factors on which balance is required when stratified randomisation becomes too complex.

Another type of ADAPTIVE RANDOMISATION is that called 'response-adaptive randomisation' (see Wei and Durham, 1978). These methods are rarely used but they have an intriguing and appealing ethical basis. They can be used when the response to treatment for each patient is known relatively quickly and the recruitment of patients is relatively slow. Such a design would start with a traditional, equally balanced, randomisation scheme and the first 20 or 30 patients may be assigned randomly to each of the treatment groups. Thereafter if one treatment is appearing to be superior to the other then the randomisation probabilities begin to change in favour of the most advantageous treatment. Early results from trials can be very unreliable and it is quite possible that the treatment appearing to be beneficial early on may subsequently appear less good than the alternative treatment. In this case, the allocation probabilities would then change back through 0.5 to again favour the treatment that is emerging as 'best'. (See DATA-DEPENDENT DESIGNS for further discussion.)

A particular application of randomisation in the design of clinical trials is the randomised consent design proposed by Zelen (1979, 1990). Such designs are used in highly pragmatic trials – i.e. those that are intended to mimic, as closely as possible, true clinical practice. They encompass the constraint that some patients may not wish to receive certain treatments so that, even if the physician might prescribe a particular treatment, the patient may wish not to take it.

Subjects are randomised to one of two treatment groups and those who are randomised to receive standard therapy then receive that therapy. Those who are randomised to receive the new, experimental treatment are asked to consent to receive that treatment or, if they prefer, they can receive the standard treatment. Their preference is respected and they receive the treatment of their choice.

It is absolutely critical that such trials are analysed by the intention-to-treat principle, so that patients are analysed according to which treatment they were randomised and not which treatment they actually received. Such a trial is then intended to answer the question about what would happen if patients are *prescribed* a particular treatment, considering the inevitable problem that some patients prescribed a particular treatment will choose not to take it.

A related type of trial is called *Wennberg's design*, where patients are randomised to receive either the treatment of their own choice or a standard, specified experimental treatment. Such a design is, again, highly pragmatic, but it is very difficult to judge the benefit of one treatment strategy over another for future patients since the results of the study are highly dependent on the preferences of the patients who take part.

Another use of randomisation is in testing treatment policies following an existing treatment regime. Examples typically include answering the question: 'If patients do not respond to a low dose, are they likely to respond to a higher dose?' Very commonly, this question is addressed as follows: patients are randomised to treatment or control and efficacy assessments made at an appropriate follow-up time after randomisation. Treatment may have been shown to be, on average, superior to control, yet not all patients will have responded to the new active treatment. Those who have not responded are now treated with a higher dose (or some other modification of the treatment regime) and subsequent efficacy or response rates are recorded.

The criterion for being included in such a follow-up is that the patient has not responded to the initial treatment and so any response at all is considered indicative of an additional benefit of the additional treatment regime. The question, of course, remains as to what would have happened to these patients had they either continued with the existing medication or followed some other course of treatment instead. The appropriate means of evaluating such a question is illustrated in the figure.

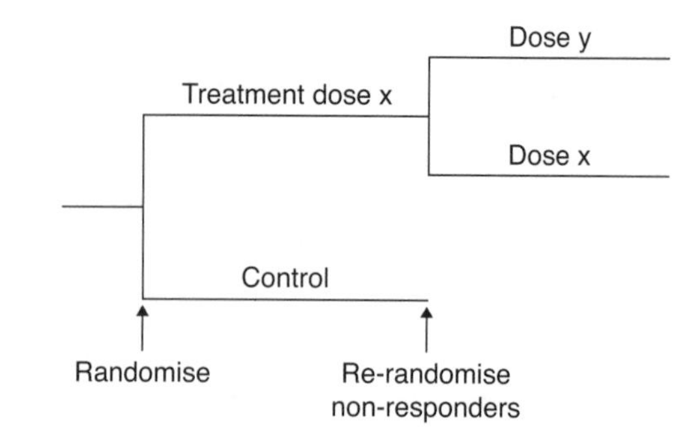

randomisation *Example of re-randomising 'non-responders'*

Patients who are to be included in this follow-up regime should be randomised to either receive the new modified treatment (perhaps a dose increase) or to continue on existing treatment (usually meaning at the same dose). Again, note the criterion for being included in this follow-up study is that the patient has not responded to the initial treatment, and in such a design it is not unusual to see some of the patients who continue on the identical treatment to respond now. This design, however, allows us to see possibly differential response rates in the patients who have continued on exactly the same treatment as those who have been randomised to receive the modified treatment. This then allows a proper assessment of the beneficial effects of changing the treatment regime as opposed to continuing with the same regime.

An ethical debate often arises around randomisation (see ETHICS AND CLINICAL TRIALS). How can it be right to choose treatment for a patient based on chance alone (or the 'throw of a die', to use a more emotional phrase)? This is an important consideration but many counter with the argument that it can be unethical *not* to randomise patients into clinical trials.

The ethics of clinical trials is a very broad subject and cannot be fully covered here. Where genuine doubt exists, however, as to the relative benefits of one treatment over another (a state often called 'equipoise'), then not randomising patients into trials can lead to misleading or even false judgements about the relative efficacy of different therapies.

Even where no alternative treatment exists for life-threatening conditions and a new potential therapy might offer the 'only hope' for a patient, without randomising patients between this new (possible) treatment and placebo, we can never gain a true understanding of the risks and benefits of the new treatment.

Folklore often then suggests that the new treatment is, in fact, better than placebo when it has never been properly tested, and it then becomes impossible (even if still not unethical) to carry out a randomised trial. Randomised consent designs and some of the adaptive designs discussed in this section can help balance the ethical arguments. Using 'unequal' randomisation (assigning more patients to some treatment arms than to others – or to placebo) can also help.

Randomisation is one of the most fundamental and one of the most important considerations in the design of a clinical trial. In contrast to observational research and epidemiology, randomisation provides the basis for assigning causality of response to the assigned treatment. For a much fuller discussion of randomisation, readers are referred to Rosenberger and Lachin (2002). *SD*

Altman, D.G. 1991: *Practical statistics for medical research.* London: Chapman & Hall. **Altman, D. G. and Bland, M.** 1999: Treatment allocation in controlled trials: why randomise? *British Medical Journal* 318, 1209. **Chalmers, I.** 1999: Why transition from alternation to randomisation in clinical trials was made (letter to the editor). *British Medical Journal* 319, 1372. **Rosenberger, W. F. and Lachin, J. M.** 2002: *Randomization in clinical trials.* New York: John Wiley & Sons, Inc. **Wei, L. J. and Durham, S. D.** 1978: The randomized play-the-winner rule in medical trials. *Journal of the American Medical Association* 73, 840–3. **Zelen, M.** 1979: A new design for randomized clinical trials. *New England Journal of Medicine* 300, 1242–5. **Zelen, M.** 1990: Randomized consent designs for clinical trials: an update. *Statistics in Medicine* 9, 645–6.

randomised controlled trials (RCT)

See CLINICAL TRIALS

randomised-response technique

This is a technique that aims to get accurate answers to a sensitive question that respondents might be reluctant to answer truthfully, for example, 'Have you ever had an abortion?' The randomised response technique protects the respondent's anonymity by offering both the question of interest and an innocuous question, which has a known probability (α), of yielding a 'yes' response, for example:

1. [Flip a coin.] Have you ever had an abortion?
2. [Flip a coin.] Did you get a head?

A random device is then used by the respondent to determine which question to answer. The outcome of the randomising device is seen only by the respondent, not by the interviewer. Consequently, when the interviewer records a 'yes' response, it will not be known whether this was a 'yes' to the first or second question (Warner, 1965). If the PROBABILITY of the random device posing question one (p) is known, it is possible to estimate the proportion of 'yes' responses to question one (π) from the overall proportion of 'yes' responses ($P = n_1/n$), where n_1 is the total number of 'yes' responses in the sample size n:

$$\hat{\pi} = P - (1-p)\alpha$$

So, for example, if $p = 0.60$ (360/600), $p = 0.80$ and $\alpha = 0.5$ then $\hat{\pi} = 0.125$. The estimated variance of $\hat{\pi}$ is:

$$\mathrm{Var}(\hat{\pi}) = \hat{\pi}(1-\hat{\pi}) + (1-p)^2\alpha(1-\alpha) + p(1-p)[p(1-\alpha) + \alpha(1-p)]$$

For the example here, this gives $\hat{\pi} = 0.0004$.

Further examples of the application of the technique are given in Chaudhuri and Mukerjee (1988). *BSE*

Chaudhuri, A. and Mukerjee, R. 1988: *Randomized response: theory and techniques.* New York: Marcel Dekker. **Warner, S. L.** 1965: Randomized response: a survey technique for eliminating evasive answer bias. *Journal of the American Statistical Association* 60, 63–9.

range

The range is simply the interval between the minimum and maximum values in a set of observations. The range is a MEASURE OF SPREAD, although it is of limited use because it is dependent only on the extreme (and possibly unusual) observations, and not on the majority of the values. For this reason, the INTERQUARTILE RANGE is often a preferred measure. However, if the sample size is very small the range may be considered to be a useful summary statistic. It is more informative if the minimum and maximum are both quoted (e.g. 21 to 54), rather than merely the difference between the two (e.g. 33). *SRC*

rank invariance

Qualitative variables are commonly measured by different types of rating scales, with a discrete number of ordered categories or on a VISUAL ANALOGUE SCALE (VAS) having a continuous range of possible values between the ENDPOINTS of a straight line. The measurement level of data from scale assessment is called ordinal having rank-invariant properties only. Such data remain invariant in all order-preserving transformations, which means that the category labels do not represent any mathematical value except the categorical order of the scale. Furthermore, ORDINAL DATA, irrespective of the type of labelling, only contain information about ordering and not about magnitude or distance between the categories. Thus, one succession of labels can be replaced by another, e.g. numerals by letters or by a set of increasing numbers of symbols or by a pictogram. For example, the five categories of a five-point scale are often labelled with the figures 1 to 5, but the categorical assignment could be any other set of ordered figures, such as 0, 25, 50, 75, 100. The lack of mathematical meaning of the categorical labels implies, for example, that the intermediate numerals between the labels 50 and 75 do not exist. Statistical methods must therefore be unaffected by any kind of re-labelling of categorical scores. This means that rank-based methods should be used, for example MEDIAN, QUARTILES or other centiles for description. (For further details see Stevens, 1946, 1996, Hand, 1996, and Svensson, 2001.) *ES*

See also [RANKING PROCEDURES, AGREEMENT, GLOBAL SCALING, VALIDITY]

Hand, D. J. 1996: Statistics and the theory of measurement. *Journal of the Royal Statistical Society, Series A* 159, 445–92. **Stevens, S. S.** 1946: On the theory of scales of measurement. *Science* 103, 677–80. **Stevens, S. S.** 1955: On the averaging of data. *Science* 121,113–16. **Svensson, E.** 2001: Guidelines to statistical evaluation of data from ratings scales and questionnaires. *Journal of Rehabilitation Medicine* 33, 47–8.

ranks, ranking procedures Nonparametric statistical methods (see NONPARAMETRIC METHODS – AN OVERVIEW) are useful for all types of data; while parametric statistical methods are applicable to quantitative data that meet the criteria of being normally distributed (see NORMAL DISTRIBUTION) or other known PROBABILITY DISTRIBUTIONS. A common approach of nonparametric statistical methods is to transform data to ranks. A ranking of n ordered observations is a set of numerical ranks $\{1, 2, \ldots, n\}$ that will represent the observations in statistical analyses. The rank sum is $\frac{1}{2}n(n+1)$ and the mean rank is $\frac{1}{2}(n+1)$, where n is the number of observations. Assessments on rating scales with a limited number of possible categories imply that groups of observations will share the same category, and these observations will share the same rank value, which often is the MEAN of the ranks that belong to the group of observations, so-called tied ranks.

The calculations of the MANN–WHITNEY RANK SUM TEST of the difference between two independent groups of data and of SPEARMAN'S RANK CORRELATION COEFFICIENT are based on this type of rank transformation (Siegel and Castellan, 1988; Gibbons and Chakraborty, 2003).

The first figure shows the frequency distribution of pairs of data from psychiatric assessments of the severity of fatigue and the level of concentration difficulties in 43 patients. The assessments of the two variables are made on rating scales having five ordered categories denoted F1, ..., F5 and C1, ..., C5, where F5 and C5 represent lack of symptom or difficulty. According to the frequency distribution of assessments on the fatigue scale 16 patients were judged to the category F1, which represents the most severe level of fatigue, and these will share the ranks from 1 to 16, the mean rank being 8.5. The seven patients in the category F2 will share the ranks 17 to 23, the mean rank being $16 + \frac{1}{2}(7 + 1) = 20$. The mean ranks of the two sets of distributions are shown in this figure. The pairs of cell frequencies are replaced by pairs of ranks when the relationship

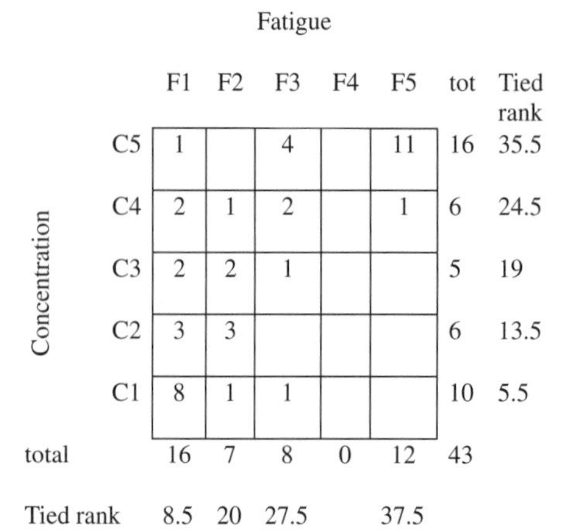

Concentration \ Fatigue	F1	F2	F3	F4	F5	tot	Tied rank
C5	1		4		11	16	35.5
C4	2	1	2		1	6	24.5
C3	2	2	1			5	19
C2	3	3				6	13.5
C1	8	1	1			10	5.5
total	16	7	8	0	12	43	
Tied rank	8.5	20	27.5		37.5		

ranks, ranking procedures *The frequency distribution of pairs of data from psychiatric assessments of the severity of fatigue and the level of concentration difficulties in 43 patients. The rating scales have five ordered categories denoted F1, ..., F5 and C1, ..., C5, where F5 and C5 represent a lack of symptom or difficulty. The two sets of marginal distributions and the tied rank values of the marginal frequencies are given*

between the severity of fatigue and concentration difficulty is calculated by Spearman's rank correlation coefficient. This means that the observation (F2, C1) will get the pair of tied ranks (20; 5.5) and the three observations (F2, C2) (20; 13.5), and so on. Spearman's rank correlation coefficient, when adjusted for tied observations, is 0.75.

A bivariate ranking approach developed for analysis of paired ordinal data regarding agreement and disagreement that takes account of the information given by the pairs of data is suggested by Svensson (1997). In this augmented ranking approach (aug-rank), the ranks are tied to the pairs of data, which means to the observations in the cells of a square CONTINGENCY TABLE or to the points of a SCATTERPLOT of data from VISUAL ANALOGUE SCALE (VAS) assessments. This means that the augmented rank of the assessments X depends on the pairing with Y. The second figure (on p. 389) part (a) shows the paired distribution of 50 assessments made by two raters labelled X and Y. The three individuals categorised A by rater X are found in the cells (A;A), (A;A) and (A;B), which means that rater Y has assessed one of these individuals to a higher category than has rater X, and this individual will therefore be given a higher aug-rank X-value. The aug-rank X-values of these three pairs are therefore 1.5, 1.5 and 3 respectively (see part (b)). According to rater Y, 14 individuals are categorised A, but (2, 8, 3, 1) of them are categorised A, B, C, D respectively by X, and therefore the aug-rank Y-values of these four groups of individuals will differ (see part (b)).

(a)

RATER Y \ RATER X	A	B	C	D
D			1	1
C		2	2	14
B	1	1	11	3
A	2	8	3	1

ranks, ranking procedures *(a) The paired distribution of interrater assessments of 50 individuals by a four-point scale (A, B, C, D). The agreement diagonal is marked*

(b) (Aug-rank-X; Aug-rank-Y)

	A	B	C	D
D			(31;49)	(50;50)
C		(13.5;31.5)	(29.5;33.5)	(42.5;41.5)
B	(3;15)	(12;16)	(23; 22)	(34;29)
A	(1.5;1.5)	(7.5;6.5)	(16;12)	(32;14)

ranks, ranking procedures *(b) The paired distribution of aug-ranks of the frequency distribution of paired assessments X and Y of (a)*

This aug-rank approach to taking account of information from the pairs of ordered categorical assessments when replacing paired ordinal data with pairs of aug-ranks makes it possible to identify and separately analyse a possible systematic component of observed disagreement from the occasional, noise, variability (see AGREEMENT).

A complete agreement in all pairs of aug-ranks defines the rank-transformable pattern of agreement (RTPA), which is uniquely related to the two marginal distributions. The RTPA is the distribution of pairs that is expected when the observed disagreement is completely explained by a systematic disagreement and in the case of complete agreement. *ES*

[See also VALIDITY OF SCALES]

Gibbons, J. D. and Chakraborty, S. 2003: *Nonparametric statistical inference*, 4th edition, revised and expanded. New York: Marcel Dekker. **Siegel, S. and Castellan, N. J.** 1988: *Nonparametric statistics for the behavioral sciences*, 2nd edition. New York: McGraw-Hill. **Svensson, E.** 1997: A coefficient of agreement adjusted for bias in paired ordered categorical data. *Biometrical Journal* 39, 643–57.

reading the medical literature See CRITICAL APPRAISAL

recall bias This is a BIAS that can occur in CASE-CONTROL STUDIES due to cases being more likely to recall having been exposed than are controls. In some case-control studies retrospective data on exposure are obtained from historical records.

However, in most situations such data are not available and data are instead obtained by interviewing cases and controls (or their relatives). When this is done there is a chance that cases may be more likely to remember having been exposed than are controls. Even where there is no genuine difference in frequency of exposure between cases and controls, this differential recall may cause an apparent difference, so that the exposure appears to be associated with disease.

For example, in a case-control study of congenital malformations, mothers are asked about prior exposures to infectious diseases, drugs, environmental pollutants, etc. It is quite plausible that a mother who has given birth to a malformed child will be more interested in the study and make more effort to remember instances of past exposure. It is also possible for recall bias to operate in the opposite direction: e.g. if, through shame, cases were less likely than controls to admit exposure. (For further details see Hennekens and Buring, 1987, and Rothman and Greenland, 1998.) *SRS*

[See also BIAS IN OBSERVATIONAL STUDIES]

Hennekens, C. H. and Buring, J. E. 1987: *Epidemiology in medicine*. New York: Little, Brown and Company. **Rothman, K. J. and Greenland, S.** 1998: *Modern epidemiology*, 2nd edition. Philadelphia: Lippincott-Raven Publishers.

receiver operating characteristic (ROC) curve Diagnostic testing plays an increasingly important role in modern medicine and the ROC curve is a common graphical tool for displaying the discriminatory ability of a diagnostic marker (test) in distinguishing between diseased and healthy subjects. The outcome of a diagnostic test can be dichotomous (positive, negative), ordinal (e.g. normal, questionable, abnormal) or continuous (e.g. PSA measurements). The ROC curve arises only for ordinal and continuous outcomes.

A diagnostic marker is generally evaluated by comparison to a definite gold standard procedure/test. Such gold standards are often complicated to conduct, intrusive, not sufficiently timely or expensive. This motivates the search for inexpensive, easily measurable and reliable alternatives.

A subject is assessed as diseased or healthy depending on whether the corresponding marker value is above or below a given threshold. Associated with any threshold value are the PROBABILITY of a true positive (SENSITIVITY) and the probability of a true negative (SPECIFICITY). The ROC curve presents graphically the trade-off between sensitivity and specificity for every possible threshold value. By convention, the plot displays the specificity on the y axis and 1 – sensitivity on the x axis.

Consider, for example, ORDINAL DATA arising in the context of medical imaging. A reader is presented with images from diseased and healthy subjects and is required to rate each image on a discrete ordinal scale (1–5): (1) definitely normal, (2) probably normal, (3) questionable, (4) probably abnormal, (5) definitely abnormal. Suppose the results are as shown in the first table or equivalently expressed as cumulative percentages, as in the second table.

receiver operating characteristic (ROC) curve *Ordinal data results from a medical imaging study*

	Rating					
True disease status	(1)	(2)	(3)	(4)	(5)	**Total**
Healthy	28	14	5	2	1	**50**
Diseased	2	4	10	14	20	**50**
Total	**30**	**18**	**15**	**16**	**21**	**100**

receiver operating characteristic (ROC) curve *Cumulative percentage results from a medical imaging study*

	Rating				
True disease status	(1)	(2)	(3)	(4)	(5)
Healthy	56%	84%	94%	98%	100%
Diseased	4%	12%	32%	60%	100%

receiver operating characteristic (ROC) curve *Specificity and sensitivity pairs*

True disease status		(1)	(2)	(3)	(4)	(5)
Specificity	0%	56%	84%	94%	98%	100%
Sensitivity	100%	96%	88%	68%	40%	0%

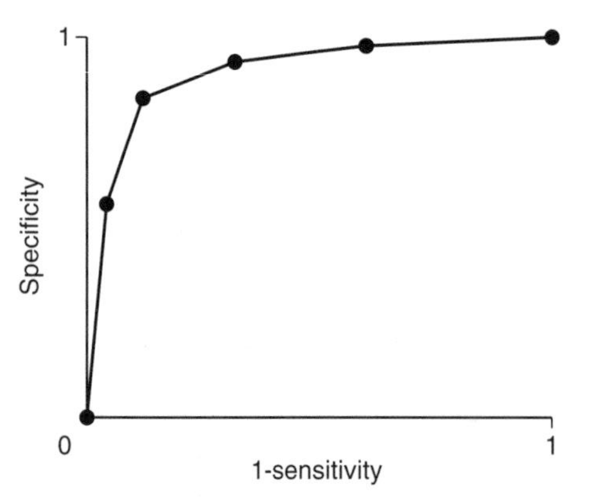

receiver operating characteristic (ROC) curve *A typical ROC curve*

If we use a threshold of (1), we would have a specificity of 56% and a sensitivity of 96% (100% – 4%). For a threshold of (2), the specificity is 84% while the sensitivity is 88% (100% – 12%). If we ignored the test and called everyone diseased then the specificity is 0% with a corresponding sensitivity of 100%. This gives rise to the specificity/sensitivity pairs in the third table. The figure shows the plot of specificity versus 1 – sensitivity, the ROC curve or plot.

A diagnostic marker shows good discriminatory ability if both sensitivity and specificity are high for a reasonable range of threshold values. In terms of the ROC plot, this means that the closer the curve comes to the left-hand border and then the top border the better the marker. The closer the curve is to the diagonal (45°) line the worse the discriminatory accuracy of the marker.

For continuous data, the calculations are carried out similarly. However, for this situation the number of outcome values will be much larger and usually only one of sensitivity or specificity will change when the threshold value is increased to its next observed value. Consequently, for data arising from a continuous scenario, the resulting ROC plot will tend to look like a step function. Both for the continuous case and the ordinal situation that has arisen from an underlying continuous mechanism, the *true* underlying ROC curve is a smooth function. Many methods, nonparametric, semiparametric and parametric, have been developed to provide smooth estimation procedures for the ROC curve. In addition, methodology for adjusting the ROC curve for covariate information and selection bias has been proposed. In certain situations the diagnostic markers are not measured directly on each subject but are taken on pooled groups of subjects. Faraggi, Reiser and Schisterman (2003) describe estimating the ROC curve for such pooled data.

The discriminatory power of the diagnostic marker that is indicated graphically by the ROC curve is often summarised by a one-number index. The AREA UNDER THE ROC CURVE (AUC) is the most commonly used index. Sometimes, only a particular range of sensitivity values is of interest and a partial area is computed as the area under the curve over the range of sensitivities considered important. An alternative index due to Youden (see Greiner, Pfeiffer and Smith, 2000) is to compute max{sensitivity + specificity} – 1, where the maximisation is carried out over all pairs of sensitivity and specificity values.

The ROC curve is also useful in assessing the discriminatory power of statistical models and classifiers for binary outcomes. For further details see Shapiro (1999), Zhou, Obuchowski and McClish (2002) and Pepe (2003). *DF/BR*

Faraggi, D., Reiser, B. and Schisterman, E. F. 2003: ROC curve analysis for biomarkers based on pooled assessments. *Statistics in Medicine* 22, 2515–27. **Greiner, M., Pfeiffer, D. and Smith, R. M.**

2000: Principles and practical application of the receiver-operating characteristic analysis for diagnostic tests. *Preventive Veterinary Medicine* 45, 23–41. **Pepe, M. S.** 2003: *The statistical evaluation of medical tests for classification and prediction*. Oxford: Oxford University Press. **Shapiro, D. E.** 1999: The interpretation of diagnostic tests. *Statistical Methods in Medical Research* 8, 113–34. **Zhou, X. H., Obuchowski, N. A. and McClish, D. K.** 2002: *Statistical methods in diagnostic medicine*. New York: John Wiley & Sons, Inc.

regression dilution bias The term applies to any setting where one wishes to assess the relationship between a variable X that is measured with error and an outcome variable Y. When observed values of X are used as estimates of true values, the true relationship between X and Y is underestimated; this is regression dilution bias. In epidemiology, the term is often used to describe the situation where relationships between risk exposures of interest (such as blood pressure) and the risk of a particular disease occurring (such as a heart attack) are underestimated because of the use of single 'baseline' measurements of the risk exposure as estimates of the true underlying level.

The situation arises because studies that aim to identify risk factors for a particular disease usually take 'baseline' assessments of individuals and relate these measurements to disease events observed over a particular follow-up period. However, BASELINE MEASUREMENTS often do not reflect the patient's true usual level during the period of follow-up because of MEASUREMENT ERRORS, short-term 'random' fluctuations from the individual's average level and longer term true changes. Although these effects are random, meaning that the baseline measurement is just as likely to over- as underestimate the patient's true level, the differences between patients estimated from a baseline sample (the between-person variation) exaggerate the true differences that really exist between those patients over a period of time (because the estimated VARIANCE consists of both within- and between-person variations). In other words, the differences in the level of the risk exposure between the study participants are not as large as one would estimate from the baseline sample alone.

The effect of this on the estimated relationship between the risk exposure and the risk of disease is shown in the figure where, for the sake of illustration, it assumed that the risk exposure is positively associated with the risk of disease and that a unit increase in the exposure leads to a proportional increase in the risk of disease (a log-linear relationship).

The solid line in the figure shows the 'apparent' relationship between the risk exposure and the risk of disease obtained when using the baseline measurement levels as estimates of true usual levels. However, as already described, the true differences in exposure levels between the individuals are not likely to be as great as it would seem, in which case the difference in disease risk between those at the top and those at the bottom of the risk exposure distribution should correspond to a narrower range of values on the horizontal axis than suggested.

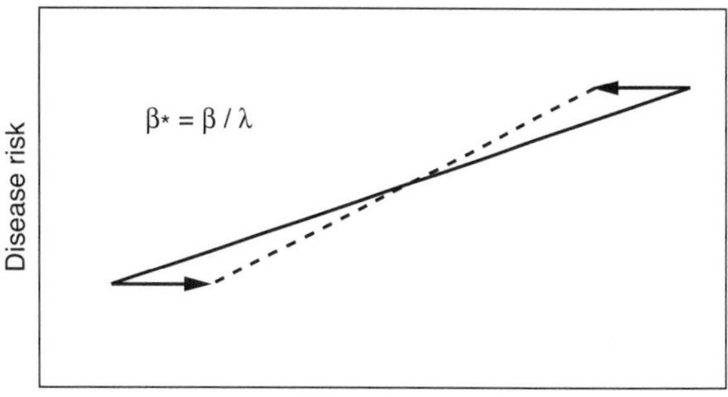

regression dilution bias *Effects of regression dilution bias*

The true relationship between the usual risk exposure level and disease risk may therefore be obtained by 'shrinking' the line towards the middle by some predetermined amount (as indicated), so that the true relationship (shown by the dashed line) may be obtained. This line indicates the relationship between usual levels of the risk exposure and the risk of disease and, as can be seen, its slope (the strength of association) is greater than would otherwise be estimated. The extent of regression dilution bias (the difference between the two slopes) can usually be estimated by taking repeated measurements on individuals at various periods throughout the follow-up period, thus enabling estimation of the amount of within-person variation likely to be present. Thus, the apparent regression slope β may be appropriately corrected by a factor λ (the regression dilution ratio that may be estimated by the intraclass correlation) in order to obtain the true slope $\beta^* = \beta/\lambda$.

The concept of regression dilution bias is a very general one and is likely to apply in any setting where interest lies in the association between usual or average exposure levels over an exposure period and disease risk over that period. In situations where the interest is not in assessing relations with usual exposure levels, however, or where the outcome of interest is not determined by usual levels, correction for regression dilution bias would not be appropriate.

For a practical example, see the study into blood pressure differences conducted by MacMahon *et al.* (1990), which adjusted for the effects of regression dilution bias. *RM/JE*

[See also REGRESSION TO THE MEAN]

MacMahon, S., Peto, R., Cutler, J., Collins, R., Sorlie, P., Neaton, J. ***et al.*** 1990: Blood pressure, stroke and coronary heart disease. Part 1. Prolonged differences in blood pressure: prospective observational studies corrected for the regression dilution bias. *Lancet* 335, 765–74.

regression to the mean This is a phenomenon where an unusual or extreme value for a given variable is followed by a much less extreme value when re-measurement takes place. The phenomenon is an inevitable consequence of MEASUREMENT ERROR. A particularly well-known scenario concerns apparent hypertension in a subject who records a very high blood pressure reading. Re-measurement is very likely to result in a lower blood pressure reading, even if the subject has not been treated. The blood pressure has 'regressed' (gone back) towards the true underlying MEAN for that subject.

The phenomenon was first noted by Francis Galton in the 19th century when he compared the heights of parents with those of their children. Galton demonstrated that the children with tall parents tended to be *smaller* than the parents, while children with small parents tended to be *taller* than the parents. Yet the phenomenon also worked the other way round: parents of tall children tended to be smaller than the children and parents of small children tended to be taller. Consequently, when Galton plotted the heights of children against the heights of parents, he observed that the line that best fitted the data had a slope less than one. Indeed, that is how the 'regression line', nowadays used more widely for relating any pair of continuous variables, obtained its name.

The extent of regression to the mean may be estimated by measuring a group of individuals twice. PEARSON'S CORRELATION COEFFICIENT will estimate the extent to which single individual values would be expected to regress to the mean. Another method involves dividing the distribution of the first measurements on the individuals into fifths (MacMahon *et al.*, 1990). We calculate the means for those in the top fifth and those in the bottom fifth. For these two extreme groups of individuals, we then calculate the mean of their measurements on the *second* occasion. The means for the two groups on the second occasion will be more similar than on the first occasion. Indeed, the ratio of the difference in means on the second occasion to the difference in means obtained for the first occasion has been recommended as a good method of correcting for REGRESSION DILUTION BIAS.

Many examples have been provided about regression to the mean in various branches of medicine (see, for example, Bland and Altman, 1994a; 1994b; Morton and Torgerson, 2003). Individual patients, when treated for unusually high blood pressure, will be likely to improve; the mistake is to attribute such improvement to the treatment rather than regression to the mean. Similarly, public health action to prevent a given disease may be targeted at a geographical area where incidence of a given disease has been observed as unusually high. Incidence of the disease will be likely to decline subsequently, but that may have happened even in the absence of the public health intervention.

In both these examples, the true test of the intervention involves an experiment that compares the change following intervention with that seen in a control group who, despite having unusually high initial values observed, are not subjected to the intervention. If the improvement observed in the intervention group was greater than that seen in the control group, then regression to the mean was not the solely responsible factor.

Because of regression to the mean, the British Hypertension Society recommends at least two readings on each of several occasions (Ramsay *et al.*, 1999). This is partly to address the possibility of a poor measurement technique for single readings. However, the blood pressure at the time of the reading may have been genuinely higher than that normally experienced by the individual. The actual clinical objective is to ascertain the individual's true underlying mean value, and repeating the measurement a week later may help to address this. *RM/JE*

Bland, J. M. and Altman, D. G. 1994a: Regression towards the mean. *British Medical Journal* 308, 1499. **Bland, J. M. and Altman, D. G.** 1994b: Some examples of regression towards the mean. *British Medical Journal* 309, 780. **MacMahon, S., Peto, R., Cutler, J., Collins, R., Sorlie, P., Neaton, J. *et al.*** 1990: Blood pressure, stroke, and coronary heart disease. Part 1. Prolonged differences in blood pressure: prospective observational studies corrected for the regression dilution bias. *Lancet* 335, 765–74. **Morton, V. and Torgerson, D. J.** 2003: Effect of regression to the mean on decision making in health care. *British Medical Journal* 326, 1083–4. **Ramsay, L. E., Williams, B., Johnston, G. D., MacGregor, G. A., Poston, L., Potter, J. F. *et al.*** 1999: British Hypertension Society guidelines for hypertension management: summary. *British Medical Journal* 319, 630–5.

regulatory statistical matters National and international guidances have been written to set out (in greater or lesser levels of detail) how CLINICAL TRIALS should be planned, executed, analysed and reported for regulatory submissions. The earliest documents were national or regional: FOOD AND DRUG ADMINISTRATION (1988), Ministry of Health and Welfare (1992) and the CHMP Working Party on Efficacy of Medicinal Products (1995). Later, the International Conference on Harmonization (ICH) began to coordinate guidance in all areas of the regulated pharmaceutical industry (not just statistics and not just clinical trials). They produced the guidance known as 'ICH E9' (see ICH E9 Expert Working Group, 1999), which was adopted by the ICH Steering Committee in February 1998.

The guidance is focused on statistical principles and not on specific methods. It is also aimed at a target audience of nonstatisticians as well as statisticians, and so should be mostly understandable to the target audience of the current volume. It is the responsibility of a sponsor company to produce convincing evidence of quality, safety and efficacy of a new medicinal product. While the convincing nature of such evidence may be influenced by the specific method of analysis used in a clinical trial, it is more likely to be influenced by the more fundamental issues of design and conduct of the studies. Trials submitted for regulatory submission have to be described and reported in great detail (see MEDICINES AND HEALTHCARE PRODUCTS REGULATORY AGENCY). ICH E9 explains that it is important that all the key features relating to the design, monitoring and analysis of a trial (and sets of trials) should be set out in detailed study protocols (see PROTOCOLS FOR CLINICAL TRIALS). There should then be a traceable record of any changes made to the original protocol, including when the decisions were made and when the changes were implemented. In this way, the protocol, its documented changes and the final clinical study report should all link together.

The ICH E9 document is mostly aimed at confirmatory Phase III trials, although much of its content may be thought of as good statistical practice and applicable more widely. However, within the context of Phase III trials, the INTENTION-TO-TREAT principle (and, therefore, 'pragmatic' trials) is stressed in preference to 'explanatory' trials, where a PER PROTOCOL design and analysis may have a greater part to play. The document introduced a new term, 'full analysis set', referring to the data that are used to try to address the intention-to-treat question. This was a compromise situation, recognising that it is often not possible to include every patient in the analysis for a true intention-to-treat answer. This term 'full analysis set' has not really caught on as a concept outside of the regulatory field and perhaps not within it either; 'intention-to treat', however, is firmly fixed in the language.

The definition (and specification) of primary and secondary ENDPOINTS is clearly important and covered in detail, but further consideration is also given to 'composite variables' and 'multiple primary' (now more usually called 'co-primary') endpoints. Many trials use composite variables as primary (or secondary) endpoints – examples include psychiatry, where rating scales are used, dermatology (lesion counts), arthritis (total joint scores), dentistry (number of eroded surfaces) and so on. 'Global assessment scores' are widely used in many therapeutic areas and are a clear example of a composite endpoint (even though it may not be clear what the components are). It is important that endpoints (particularly primary endpoints and, more especially, composite endpoints) are well validated both for statistical properties but also for face validity. Co-primary endpoints present further difficulties but obvious examples exist – particularly in cardiology studies. Reduction in 'all cause mortality, recurrent myocardial infarction, and stroke' is an endpoint often used in myocardial infarction treatment trials. While there may be a hierarchy to such endpoints (death being worse than MI and stroke), there are also elements that are not hierarchical (recurrent MI is not necessarily 'better' or 'worse' than having a stroke).

The fundamental techniques to avoid BIAS are considered to be BLINDING and RANDOMISATION. Each of these is covered in some detail. Fair consideration is given to the difficulties (in some circumstances) of designing trials to be double-blind or of maintaining the blind through the course of the study – but limitations of studies that are not blinded are outlined. Randomisation, including stratified and adaptive (such as MINIMISATION), methods are discussed. Their benefits are explained although some concern is raised about the use of adaptive designs. If such methods are used, it remains for the sponsor to produce convincing evidence that the analysis is adequate to account for the assignment method. SENSITIVITY analyses (used here and in other contexts) may help to address this.

Basic design issues (parallel groups, cross-over and factorial designs) are described and the relative advantages and disadvantages discussed. CROSSOVER TRIALS, in particular, can be a very efficient method of experimentation but are highly prone to difficulties in the face of carryover effects between periods. This issue was particularly highlighted by Freeman (1989) but seemed to struggle to become recognised. The ICH E9 document clearly states in relation to carryover and the necessary chronic and stable nature of the disease: 'The fact that these conditions are likely to be met should be established in advance of the trial by means of prior information and data.' This was a significant step forward in the use of (or restriction of) crossover trials in regulatory work.

Another substantive step was in the explicit recognition and contrast between trials to show superiority (which are what most people think of when discussing clinical trials) and trials to show noninferiority or equivalence. Much has been written on this subject recently; although some was written before ICH E9 was published, until then it was still rather a new and unclear concept. Within the same section is comment on trials to show a dose–response relationship, another area previously suffering from lack of clear consideration.

A whole raft of other issues is also discussed, such as handling MISSING VALUES and OUTLIERS, data transformations, estimation versus hypothesis testing, multiplicity, subgroup analyses and interactions, use of baseline covariates and so on. The document is wide in its scope, covering safety as well as efficacy.

Overall, the ICH E9 guidance has been highly influential both within and outside the pharmaceutical industry. However, as some problems begin to be solved, others come to the forefront and it is now recognised that further guidance on topics covered, perhaps rather sparingly, is necessary. To this end, the European Committee for Human Medicinal

Products (CHMP) has set up various subcommittees, notably its Efficacy Working Party, which has identified various areas that need further explanation or clarity. This group has developed a number of 'guidelines', 'points to consider' documents and 'reflection papers' (not just in statistical areas). Currently, there are eight agreed documents relating specifically to statistical/methodological issues: switching between superiority and non-inferiority (August 2000); applications with meta-analyses and one pivotal study (May 2001); missing data (November 2001); multiplicity issues in clinical trials (September 2002); adjustment for baseline covariates (May 2003); data monitoring committees (January 2006); choice of non-inferiority margin (January 2006); and confirmatory trials with adaptive designs (October 2007). The missing data guidance is being updated and new guidance is being prepared on sub-group analyses. Numerous other guidance documents also cover issues relating to statistical points. All of these documents have been written to help clinical assessors to evaluate applications for Marketing Authorisations; they have not been written to guide statisticians; they have not been written to guide pharmaceutical companies. However, it is clear that they receive a lot of attention from these latter groups, and worthily so. They are all freely available on the EMA website at http://www.ema.europa.eu/ema. These are European documents and have no formal status outside of the EU but they are widely recognised as valuable and important guidances. In parallel, the US FDA has issued guidance on the use of data monitoring committees (March 2006), Bayesian statistics in medical device trials (February 2010), adaptive designs (February 2010) and non-inferiority studies (March 2010). *SD*

CHMP Working Party on Efficacy of Medicinal Products 1995: Biostatistical methodology in clinical trials in marketing authorizations for medicinal purposes. *Statistics in Medicine* 14, 1659–82. **Food and Drug Administration** 1988: *Guideline for the format and content of the clinical and statistical sections of new drug applications.* Rockville, Maryland: FDA US Department of Health and Human Services. **Freeman, P. R.** 1989: The performance of the two-stage analysis of two-treatment, two-period crossover trials. *Statistics in Medicine* 8, 1421–32. **ICH E9 Expert Working Group** 1999: Statistical principles for clinical trials: ICH harmonised tripartite guideline. *Statistics in Medicine* 18, 1905–42. **Ministry of Health and Welfare** 1992: *Guideline for the statistical analysis of clinical trials* (in Japanese). Tokyo: MHW Pharmaceutical Affairs Bureau.

relative risk and odds ratio These two approaches measure the effect of a risk factor. The way in which risk is presented can have an influence on how the associated risks are perceived. For example, one might be worried to hear that occupational exposure at one's place of work doubled the risk of a serious disease compared to some other occupation. However, the statement that the risk had increased from one in a million to two in a million might be less worrisome. In the first case, it is a *relative* risk that is presented, and, in the second, an *absolute* risk.

Both the relative risk and the ODDS RATIO are measures of relative risk/chance. They measure how the chance of an event, typically of getting a disease, varies between two categories, typically a group that is exposed to a risk factor and one that is not. Explicit definitions are most easily understood in terms of a 2×2 table of exposure by disease, as shown in the table. Therein, N denotes the population size and a, b, c and d the absolute frequencies of the respective combinations of the levels of the risk factor (exposed or not exposed) and the disease factor (present or absent).

The (absolute) risk of the disease within a subpopulation is defined as the proportion of subjects within the group that have the disease, i.e. r(Disease present|Exposed group) = $a/(a + b)$ and r(Disease present|Nonexposed group) = $c/(c + d)$. As a result the risk ratio or relative risk, RR, of disease comparing the exposed with the nonexposed subpopulation is given by the ratio RR = $a/(a + b)$. For example, a relative risk of lung cancer comparing smokers with nonsmokers of 2 would be interpreted as doubling the risk of lung cancer when smoking.

relative risk and odds ratio *Two-way classification of exposure by disease*

		Disease		
		Present	*Absent*	***Total***
Risk factor	Exposed	a	b	$a + b$
	Not exposed	c	d	$c + d$
	Total	$a + c$	$b + d$	$N = a + b + c + d$

In contrast, the odds of the disease within a subpopulation is defined as the ratio of subjects within the group that have the disease to those that do not, i.e. o(Disease present|Exposed group) = a/b and o(Disease present |Nonexposed group) = c/d. As a result the odds ratio (OR) of disease comparing the exposed with the nonexposed subpopulation is given by the ratio OR = (a/b)/(c/d). For example, an odds ratio of high blood pressure comparing treatment A with treatment B of 0.8 would be interpreted as a 20% reduction in the odds of high blood pressure under treatment A.

One can evaluate, say, 95% CONFIDENCE INTERVALS for the population OR or RR by first transforming into the natural (base e) log scale and then exponentiating the lower and upper limits. The formula for the STANDARD ERROR of the log (OR) is memorable as the square root of the sum of the reciprocals of the entries in the 2×2 table of disease and risk factor. (The formula is given explicitly in the entry on CASE-CONTROL STUDIES.)

Usually, the preferred choice of expressing the effect of exposure on disease is the RR. However, not all study designs allow the estimation of this parameter. A CROSS-SECTIONAL STUDY might fix the sample size and in a COHORT STUDY the sizes of the cohorts of exposed and nonexposed subjects (the row totals in the table) are under the control of the investigator. Both these restrictions allow the risk of disease within the exposure groups and hence the RR to be estimated.

In contrast, in a case-control study the number of subjects with the disease and the number of subjects without the disease (the column totals in the table) are chosen by the investigator, making it impossible to estimate any risk of disease from the sample data. However, since the odds ratio of disease comparing an exposed with a nonexposed population is the same as the odds ratio of having been exposed to a risk factor comparing cases with controls, the odds ratio can be estimated from all the designs mentioned above. Case-control studies are frequently carried out in practice and this explains the widespread use of the OR as a measure of effect size. In addition, if a disease is relatively rare in the population (say below 5%) the OR can be used as an approximation to the RR. For further details see Dunn and Everitt (1995). *SL*

[See also CONTINGENCY TABLES, EPIDEMIOLOGY, LOGISTIC REGRESSION, MANTEL–HAENSZEL TEST]

Dunn, G. and Everitt, B. S. 1995: *Clinical biostatistics: an introduction to evidence-based medicine*. London: Arnold.

relative survival This is the ratio of the observed survival of a given cohort of patients with the disease of interest to the survival that this group would be expected to experience based on the LIFE TABLE of a comparable group without the disease (suitably matched for age, sex, calendar time and possibly other covariates) from the background or reference population from which they were sampled. It is routinely applied to population-based cancer studies of survival utilising cancer registry data. However, this measure can equally be applied to population-based studies of other chronic diseases, such as cardiovascular disease.

Relative survival (Ederer, Axtell and Cutler, 1961) was developed as a technique for adjusting the survival curve (see SURVIVAL ANALYSIS – AN OVERVIEW) obtained from the study cohort for 'the deaths that would have occurred anyway'. In the case of cancer, the relative survival separates the risk of death attributable directly to the cancer from the background risk of death from all other causes, and thus attempts to get at the 'net effect' of cancer on survival. It reflects the impact of the disease on mortality.

If the cause-specific death information in the study is available and, importantly, accurate then a measure of the net effect of disease on survival could easily be obtained from a cause-specific analysis where only those deaths due directly to the disease of interest are considered to be events, while observation times of individuals who have not yet died or have died from other causes are considered to be right-censored. A straightforward application of survival analysis methods (see SURVIVAL ANALYSIS – AN OVERVIEW) will then provide an estimate of the disease-specific survival curve when deaths from other causes are removed. This estimate of the disease-specific survival curve is known as the 'net survival curve' (Estève *et al*., 1990).

In practice, information on cause of death, if available, tends to be unreliable in population-based research. Here cause of death is usually obtained via death certificates and this information has been shown to be inaccurate for both cancer and heart disease in terms of coding correctly the primary cause of death (Lauer *et al*., 1999; Welch and Black, 2002; Mant *et al*., 2006). In this situation a cause-specific (or competing risk) analysis approach should not be used, as the estimation of the net survival would be biased due to misclassification of the cause of death. However, relative survival can still be used here as it does not require information on cause of death (whether attributable to the disease or otherwise) to be available. All that is required are the occurrences of the deaths in the cohort during the study period and the external life table information from the comparison group, which is commonly obtained from national mortality data.

It should be noted that under certain conditions the net survival curve and the relative survival function may be estimating the same 'net effect' of disease on survival. This can be clearly seen if the definition for relative survival is more precisely (i.e. mathematically) written out as:

$$S_{\text{rel}}(t) = \frac{S_{\text{obs}}(t)}{S_{\text{exp}}(t)} \quad (1)$$

where $S_{\text{rel}}(.)$, $S_{\text{obs}}(.)$ and $S_{\text{exp}}(.)$ are the relative survival function and the observed and expected survival curves respectively. If, additionally, it is assumed that the relative survival function, $S_{\text{rel}}(.)$, corresponds to a 'proper' survival curve (in the mathematical sense), then equation (1) is equivalent to:

$$\lambda_{\text{obs}}(t) = \lambda_{\text{rel}}(t) + \lambda_{\text{exp}}(t) \quad (2)$$

where $\lambda_{\text{rel}}(.)$, $\lambda_{\text{obs}}(.)$ and $\lambda_{\text{exp}}(.)$ are the hazard functions (see SURVIVAL ANALYSIS– AN OVERVIEW) associated with $S_{\text{rel}}(.)$, $S_{\text{obs}}(.)$ and $S_{\text{exp}}(.)$ respectively. Thus equation (2) corresponds to the cause-specific/competing risks (or additive hazards) model, where λ_{rel} represents the disease-specific hazard and λ_{exp} the 'other causes' – specific hazard. In the relative survival literature, λ_{rel} is referred to as the excess mortality or excess hazard.

A number of approaches exist for estimating relative survival. The majority are based on regression models developed on the 'hazards scale', where covariates that affect mortality (either mortality from the disease or from other causes) can be easily incorporated. The most commonly used relative regression approaches are those based on additive hazards (2), although multiplicative models (see PROPORTIONAL HAZARDS) exist as well as those based on other approaches (e.g. FINITE MIXTURE DISTRIBUTIONS, TRANSFORMATIONS, etc.). Estimation of these models can be based on the full LIKELIHOOD or other approaches (e.g. EM ALGORITHM, grouped data, etc.). The reader is referred to articles by Hakulinen and Tenkanen (1987), Estève *et al.* (1990), Andersen *et al.* (1999), Dickman *et al.* (2004), Pohar and Stare (2006), Nelson *et al.* (2007) and Perme, Henderson and Stare (2009) for further details. *BT*

Andersen, P. K., Horowitz, M. M., Klein, J. P., Socie, G., Stone, J. V. and Zhang,M.-J. 1999: Modelling covariate adjusted mortality relative to a standard population. *Statistics in Medicine* 18, 1529–40. **Dickman, P. W., Sloggett, A., Hills, M. and Hakulinen, T.** 2004: Regression models for relative survival. *Statistics in Medicine* 23, 51–64 **Ederer, F., Axtell, L. M. and Cutler, S. J.** 1961: The relative survival rate: a statistical methodology. *National Cancer Institute Monograph* 6, 101–21. **Estève, J., Benhamou, E., Crosdale, M. and Raymond, L.** 1990: Relative survival and the estimation of net survival: elements for further discussion. *Statistics in Medicine* 9, 529–38. **Hakulinen, T. and Tenkanen, L.** 1987: Regression analysis of relative survival rates. *Journal of the Royal Statistical Society, Series C (Applied Statistics)* 36, 309–17. **Lauer, M. S., Blackstone, E. H., Young, D. B. and Topol, E. J.** 1999: Cause of death in clinical research: time for a reassessment? *Journal of the American College of Cardiology* 34, 618–20. **Mant, J., Wilson, S., Parry, J., Bridge, P., Wilson, R., Murdoch, W., Quirke, T., Davies, M., Gammage, M., Harrison, R. and Warfield, A.** 2006 Clinicians didn't reliably distinguish between different causes of cardiac death using case histories. *Journal of Clinical Epidemiology* 59, 862–7. **Nelson, C. P., Lambert, P. C., Squire, I. B. and Jones, D. R.** 2007: Flexible parametric models for relative survival, with application in coronary heart disease. *Statistics in Medicine* 26, 5486–98. **Perme, M. P., Henderson, R. and Stare, J.** 2009 An approach to estimation in relative survival regression. *Biostatistics* 10, 139–46. **Pohar, M. and Stare, J.** 2006: Relative survival analysis in R. *Computer Methods and Programs in Biomedicine* 81, 272–8. **Welch, H. G. and Black, W. C.** 2002: Are deaths within 1 month of cancer-directed surgery attributed to cancer? *Journal of the National Cancer Institute* 94, 1066–70.

reliability See MEASUREMENT PRECISION AND RELIABILITY

repeatability See MEASUREMENT PRECISION AND RELIABILITY

repeated measures analysis of variance This is a test to see if the MEAN varies with either (or both of) a categorical factor and time. The TWO-WAY ANALYSIS OF VARIANCE seeks to partition the variation in a sample into that due to one factor, that due to a second factor and the residual variation that cannot be explained by either factor. The repeated measures ANALYSIS OF VARIANCE (ANOVA) is a special case of the two-way analysis of variance, where one of the categorical factors is the time at which the measurement is taken.

As with other versions of the analysis of variance, the repeated measures ANOVA is usually employed following a designed experiment. If an experimental design decrees that measurements will be taken at baseline, after 3 months, after 7 months and after a year, then it is sensible to view time as a factor of four levels. Contrariwise, if we have naturally arising times (e.g. times at which patients choose to visit their GP) then viewing them as categorical will be difficult and repeated measures ANOVA will probably not be appropriate.

Cadogan *et al.* (1997) investigate the effect of increasing milk consumption on the mean bone density in adolescent girls. Measurements were taken at 0, 6, 12 and 18 months and the treatment was a two-level factor. For these reasons, the repeated measures ANOVA was the choice of analysis technique.

As for the other analysis of variance techniques, repeated measures ANOVA can be extended to more complicated situations and for greatest flexibility can be viewed from within a regression framework. For further details see Altman (1991). *AGL*

[See also LINEAR MIXED-EFFECTS MODELS]

Altman, D. G. 1991: *Practical statistics for medical research.* London: Chapman & Hall. **Cadogan, J., Eastell, R., Jones, N. and Barker, M. E.** 1997: Milk intake and bone mineral acquisition in adolescent girls: randomised, controlled intervention trial. *British Medical Journal* 315, 1255–60.

replicate designs See CROSSOVER TRIALS

reproducibility See MEASUREMENT PRECISION AND RELIABILITY

re-randomisation designs See RANDOMISATION

resampling See BOOTSTRAP

research ethics board (REB) See ETHICAL REVIEW COMMITTEES

research ethics committee (REC) See ETHICAL REVIEW COMMITTEES

residual confounding See MEASUREMENT ERROR

residuals See MULTIPLE LINEAR REGRESSION

response feature analysis See SUMMARY MEASURE ANALYSIS

response variable See ENDPOINT

restricted maximum likelihood (REML) See LIKELIHOOD, MULTILEVEL MODELS

robustness This is a property of statistical procedures that implies that they continue to work well even when there are departures from the assumptions on which they were based.

Many standard statistical procedures require that certain assumptions hold for the underlying theory to be applicable. For example, the TWO-SAMPLE *t*-TEST requires that the data are normally distributed with the same VARIANCE. In general, statistical procedures are considered to be robust if they are insensitive to small deviations from the underlying assumptions. If the optimal procedure requires the assumption of normality, then the corresponding robust procedures would not be influenced by departures from normality arising from slightly longer or shorter tails or slight SKEWNESS in the underlying distribution, which could result from the presence of a small proportion of OUTLIERS or spurious values. Robust procedures are ones such that these outliers, if they occurred, would have little effect on the analysis of the data.

Perhaps the most common estimator of the population MEAN is the sample mean, but this is not robust against quite small departures from the assumption of normality, being particularly sensitive to outliers. The MEDIAN, by way of contrast, although less efficient than the sample mean when the assumptions hold, is much more robust since it is hardly affected by the presence of outliers. A 'good' estimator is one that combines the properties of high efficiency and robustness.

One of the major problems in the development of robust procedures concerns the appropriate set of criteria that the procedure should satisfy; different criteria have led to different robust methods. Procedures that maintain the Type I error (declared significance level) or the CONFIDENCE LEVEL are known as *validity robust*, while those that maintain high power or size of CONFIDENCE INTERVAL are known as *efficiency robust*. Even within these broad categories there might be different kinds of departure from the assumptions that could be considered. These competing influences have resulted in a wide range of robust estimation methods being developed; these include M-estimators (based on MAXIMUM LIKELIHOOD ESTIMATION), L-estimators (based on linear functions of order statistics) and R-estimators (based on ranking methods).

The M-estimators (which include the sample mean and median as special cases) are based on bounded or re-descending weight functions, which give lower (or even zero) weight to extreme observations and usually involve an iterative solution of the resulting likelihood equations. The L-estimators include trimmed means and Winsorised means (and, again, the sample mean and median as special cases), while the R-estimators lead to Wilcoxon and Mann–Whitney procedures based on signed ranks see WILCOXON SIGNED RANK TEST and MANN–WHITNEY RANK SUM TEST. Generally all these alternative robust methods lead to reasonably high efficiency and wider applicability than the 'optimal' procedures. These ideas have been extended to multivariate data and to regression problems. There is now an extensive literature on robust regression, robustness in scientific modelling and in experimental design. *PP*

Davies, P. L. 1993: Aspects of robust linear regression. *Annals of Statistics* 21, 1843–99. **Rousseeuw, P. J. and Leroy, A. M.** 1987: *Robust regression and outlier detection.* New York: John Wiley & Sons, Inc. **Staudte, R. G. and Sheather, S. J.** 1990: *Robust estimation and testing*. New York: John Wiley & Sons, Inc. **Wilcox, R.** 1998: Trimming and Windorisation. In Armitage, P. and Colton T. (eds), *Encyclopedia of biostatistics*. Chichester: John Wiley & Sons, Ltd.

ROC curves See RECEIVER OPERATING CHARACTERISTIC (ROC) CURVE

S

sample size determination in clinical trials

One hallmark of a well-designed study is to have a formally estimated required sample size before the study commences. Awareness of the importance of this has led to increasing numbers of medical journals demanding that full justification of the sample size chosen is published with reports of trials. The *British Medical Journal*, the *Journal of the American Medical Association* and numerous other journals issue checklists for authors of papers on CLINICAL TRIALS, in which there is a question relating to sample size justification. Investigators, grant-awarding bodies and biotechnology companies all wish to know how much a study is likely to cost them. They would also like to be reassured that their effort (and money) is well spent, by assessing the LIKELIHOOD that the study will give unequivocal results.

Providing a sample size is not simply a matter of providing a single number from a set of tables but is a two-stage process. At the preliminary stages, 'ballpark' figures are required that enable the investigator to judge whether or not to start the detailed planning of the study. If a decision is made to proceed, then a subsequent stage is to refine the calculations for the formal study protocol itself (see PROTOCOLS FOR CLINICAL TRIALS).

When a clinical trial is designed, the investigator must make a realistic assessment of the potential benefit (the anticipated effect size) of the proposed test therapy. The history of clinical trials research suggests that, in certain circumstances, rather ambitious or overoptimistic views of potential benefit have been claimed at the design stage. This has led to trials of inadequate size for the questions posed.

If too few subjects are involved, the trial may be a waste of time because realistic medical improvements are unlikely to be distinguished from chance variation. A small trial with no chance of detecting a clinically meaningful difference between treatments is unfair to all the subjects put to the risk and discomfort of the clinical trial. Too many subjects is a waste of resource and may be unfair as a larger than necessary number of subjects receive inferior treatment if one treatment could have been shown to be more effective with fewer patients.

The traditional approach to sample size determination is by consideration of significance or HYPOTHESIS TESTS. Suppose we wish to compare two groups with a continuous outcome variable. We set up a NULL HYPOTHESIS that the two population MEANS, μ_0 and μ_1, are equal. We carry out a SIGNIFICANCE TEST to test this hypothesis. We calculate the observed difference in means $\bar{d}$. This significance test results in a P-VALUE, which is the PROBABILITY of getting the observed result, $\bar{d}$, or one more extreme, if the null hypothesis is true, by chance. If the P-value obtained from a trial is less than or equal to α, then one rejects the null hypothesis and concludes that there is a statistically significant difference between treatments. The value we take for α is arbitrary, but conventionally either 0.05 or 0.01. Contrariwise, if the P-value is greater than α, we do not reject the null hypothesis.

Even when the null hypothesis is, in fact, true there is still a risk of rejecting it. To reject the null hypothesis when it is true is to make a TYPE I ERROR. Plainly the associated probability of rejecting the null hypothesis when it is true equals α. The quantity α is interchangeably termed the test size,

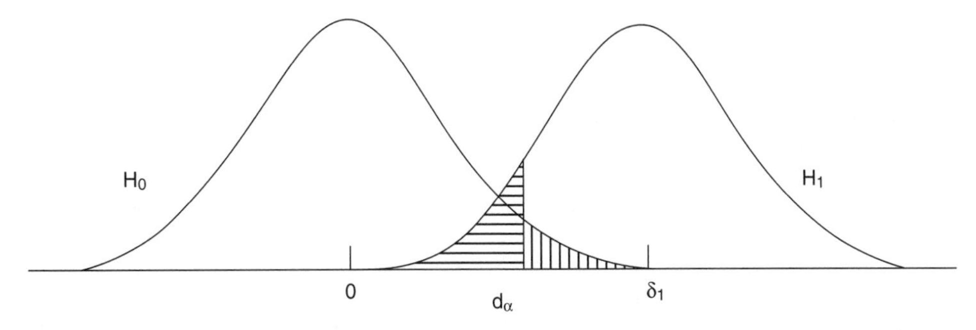

sample size determination in clinical trials *The left-hand curve shows the distribution of $\bar{d}$ under the null hypothesis. If $\bar{d} > d_\alpha$ then H_0 is rejected. The vertically hatched area represents the Type I error α. The right-hand curve shows the distribution of $\bar{d}$ under the alternative hypothesis that the difference in means is δ_1 and the horizontal hatched area represents the Type II error, β*

Encyclopaedic Companion to Medical Statistics: Second Edition Edited by Brian S. Everitt and Christopher R. Palmer

SIGNIFICANCE LEVEL or probability of a Type I (or false positive) error.

The left-hand curve in the figure on page 399 shows the expected distribution of the observed difference $\bar{d}$ under the null hypothesis, centred at zero. If $\bar{d}$ is greater than some value d_α, which is determined so that the shaded area to the left is equal to α, then H_0 is rejected.

The clinical trial could yield an observed difference $\bar{d}$ that would lead to a P-value above α, even though the null hypothesis is not true, i.e. μ_0 is indeed not equal to μ_1. In such a situation, we then accept (more correctly phrased as 'fail to reject') the null hypothesis although it is truly false. This is called a TYPE II (false negative) ERROR and the probability of this is denoted by β. The probability of a Type II error is based on the assumption that the null hypothesis is not true, i.e. $\delta = \mu_0 - \mu_1 \neq 0$. There are clearly many possible values of δ in this instance and each would imply a different alternative hypothesis, H_1, and a different value for the probability β.

The POWER is defined as one minus the probability of a Type II error and thus the power equals $1 - \beta$; i.e. the *power* is the probability of obtaining a 'statistically significant' P-value if the null hypothesis is truly false.

The right-hand curve of the figure illustrates the distribution of $\bar{d}$ under the alternative hypothesis H_1, centred on the expected difference in means $\delta_1 = \mu_0 - \mu_1$. If δ, α and β are all fixed, it would appear there is nothing left to vary. However, the distribution of $\bar{d}$ depends on the number of subjects in the two groups. With more subjects the STANDARD ERROR decreases, so the curves become narrower and so for a fixed α the value of β decreases. The sample size calculation is a compromise between the power $(1 - \beta)$, the effect size δ and the sample size n.

A key element in the design is the 'effect size' that it is reasonable to plan to observe – should it exist. Sometimes there is prior knowledge, which then enables an investigator to anticipate what effect size between groups is likely to be observed, and the role of the study or trial is to confirm that expectation. In some situations, it may be possible to state that, for example, only a doubling of MEDIAN survival would be worthwhile to demonstrate in a planned trial. This might be because the new treatment, as compared to standard, is expected to be so toxic that only if substantial benefit could be shown would it ever be used. In such cases the investigator may have definite opinions about the difference that it is pertinent to detect.

In practice, a range of plausible effect size options are considered before the final effect size is agreed. For example, an investigator might specify a scientific or clinically useful difference that it is hoped could be detected and would then estimate the required sample size on this basis. The calculations might then indicate that an extremely large number of subjects is required. As a consequence, the investigator may next define a revised aim of detecting a rather larger difference than that originally specified. The calculations are repeated and perhaps the sample size now becomes realistic in that new context.

One additional problem when planning comparative clinical trials is that investigators are often optimistic about the magnitude of the improvement of new treatments over the standard. This optimism is understandable, since it can take considerable effort to initiate a trial and, in many cases, the trial would only be launched if the investigator is enthusiastic about the new treatment and is sufficiently convinced about its potential efficacy. However, experience suggests that as trials progress there is often a growing realism that, even at best, the initial expectations were optimistic. There is ample historical evidence to suggest that trials that set out to detect large treatment differences nearly always result in 'no significant difference was detected'. In such cases, there may have been a true and worthwhile treatment benefit that has been missed, since the level of detectable differences set by the design was unrealistically high and hence the sample size too small to establish the true (but less optimistic) size of the benefit.

The way in which possible effect sizes are determined will depend on the specific situation under consideration. For example, if a study is repeating one already conducted then very detailed information may be available on the options for the effect size suitable for planning the new study. Estimates of the anticipated effect size may be obtained from the available literature or formal META-ANALYSES of related studies or may be elicited from expert opinion. For clinical trials, in circumstances where there is little prior information available, Cohen (1988) has proposed a standardised effect size, Δ. In the case when the difference between two treatments 1 and 2 is expressed by the difference between their means $(\mu_1 - \mu_2)$ and σ is the standard deviation (SD) of the ENDPOINT variable, which is assumed to be a continuous measure, then $\Delta = (\mu_1 - \mu_2)/\sigma = \delta/\sigma$. A value of $\Delta \leq 0.2$ is considered a small standardised effect, $\Delta \approx 0.5$ as moderate and $\Delta \geq 1$ as large. Experience has suggested that, in many clinical areas, these can be taken as a good practical guide for design purposes.

In intermediate situations for clinical trials at least, Bayesian approaches to obtaining a distribution of effect size have been suggested by Spiegelhalter, Freedman and Parmar (1994) (see BAYESIAN METHODS). These involve obtaining views on likely effect size from a survey of relevant experts and combining their responses into a PRIOR DISTRIBUTION of plausible effect sizes from their responses. Subsequently, this prior distribution is then combined with the data obtained from the trial once conducted to give a POSTERIOR DISTRIBUTION concerning the true effect size from which conclusions with regard to efficacy are then drawn. This approach has also been advocated by Tan *et al.* (2003) who suggest how information, from whatever source, may be synthesised into a prior distribution for the anticipated effect size that is then utilised for planning purposes.

Next, some theory and formulae are presented to show their derivation. In practice, one can refer in simple situations to a graphical approach that yields a suitably approximate sample size (see Altman's nomogram) (Altman, 1991). In a trial comparing two groups, with n subjects per group, if we assume that the outcome variable is continuous, $\bar{x}_1$ and $\bar{x}_2$ summarise the respective means of the observations taken. Further, if the data are normally distributed with equal (population) SDs, σ, then the standard errors (SE) are SE $(\bar{x}_1) = \text{SE}(\bar{x}_2) = \sigma/\sqrt{n}$. The two groups are compared using $\bar{d} = \bar{x}_1 - \bar{x}_2$ with $\text{SE}(\bar{d}) = \sigma\sqrt{2/n}$. Here we assume that $\text{SE}(\bar{d})$ is the same when the null hypothesis, H_0, of no difference is true and when the alternative hypothesis, H_A, that there is a difference of size δ_1 is true.

Under the null hypothesis, H_0, the critical value d_α is determined by:

$$\frac{d_\alpha - 0}{\sigma\sqrt{\frac{2}{n}}} = z_{1-\alpha} \quad (1)$$

In contrast, under the assumption that the alternative hypothesis, H_1, is true, $\bar{d}$ now has mean δ_1 but the same $\text{SE}(\bar{d}) = \sigma\sqrt{2/m}$. In this, case the probability that $\bar{d}$ exceeds d_α must be $1-\beta$, and this implies that:

$$\frac{d_\alpha - \delta}{\sigma\sqrt{\frac{2}{n}}} = -z_{1-\beta} \quad (2)$$

Solving the two expressions (1) and (2) for d_α and rearranging, we obtain the sample size for each group in the trial as:

$$n = 2\frac{\sigma^2}{\delta^2}(z_{1-\alpha} + z_{1-\beta})^2 = \frac{2(z_{1-\alpha} + z_{1-\beta})^2}{\Delta^2} \quad (3)$$

This is termed the *fundamental equation* as it arises, in one form or another, in many situations for which sample sizes are calculated.

The use of equation (3) for the case of a two-tailed test, rather than the one-tailed test, involves a slight approximation since $\bar{d}$ is also statistically significant if it is less than $-d_\alpha$. However, with $\bar{d}$ positive the associated probability of observing a result smaller than $-d_\alpha$ is negligible. Thus, for the case of a two-sided test, we simply replace $z_{1-\alpha}$ in equation (3) by $z_{1-\alpha/2}$.

For the commonly occurring situation of $\alpha = 0.05$ (two-sided) and $\beta = 0.2$, we find that equation (3) simplifies to:

$$n = \frac{16}{\Delta^2} \quad (4)$$

This basic equation has to be modified to adapt to the specific experimental design, the allocation ratio (i.e. the possibility of the design stipulating unequal subject numbers in each group), the particular type of endpoint under consideration as well as, for CLINICAL TRIALS, the type of RANDOMISATION involved.

Viljanen *et al.* (2003) specified $\alpha = 0.05$, $\beta = 0.2$ and the anticipated effect size, $\delta = 1$ unit with standard deviation 2 units. From this, the standardised effect size is $\Delta = 0.5$ and from equation (4) we find $n = 64$ patients per group are required.

When dealing with binary outcome variables the sample size derivations for binary data are similar to that for continuous data, but one has to specify two proportions π_1 and π_2 and the effect size is the difference $\delta = \pi_1 - \pi_2$. Campbell, Julious and Altman (1995) give tables for the sample sizes required for the comparison of two binomial proportions, and this is shown in the first table.

sample size determination in clinical trials *Components necessary to estimate the size of a study*

Effect size, δ	Anticipated (planning) size of the difference between the two groups
Type I error, α	Equivalently, the test size or significance level of the statistical test to be used in the analysis
Type II error, β	Equivalently, the power, $1-\beta$ (usually expressed as a percentage)

As an example of use of the table, we consider the trial by Viljanen *et al.* (2003). They showed that the proportion of people in their control group with neck pain who had been on sick leave over 12 months was 15%. Suppose we wished to design a trial that proposed to reduce this to 10%. Then from the second table we would require 686 (say 700) people per group with 80% power at 5% significance.

The point to note here is that although this is a 33% reduction in the rate, it still requires a large trial. In general, the binary outcomes will often require large trials because they contain much less information than a continuous outcome. For example, suppose the body mass index (BMI) in a population was about 28 kg/m^2, with a standard deviation of 2 kg/m^2. This means that about 16% of the group are defined as obese (BMI > 30 k/m^2). Suppose a trial tried to reduce this absolute proportion by 5%, to about 11%. Then from the second table we would need about 686 people to detect this with 5% significance and 80% power.

However, to obtain about 11% obese persons in the population we would have to reduce the mean BMI to 27.54. Thus the standardised effect size is 0.23 = (28 − 27.54)/2 and from equation (4) we would require about 300 patients per group for 80% power at 5% significance level, or less than half the equivalent sample

sample size determination in clinical trials *Sample sizes to detect a difference in two proportions, π_1 and π_2, at a 5% significance level with 80% power*

	π_2																			
π_1	*0.05*	*0.10*	*0.15*	*0.20*	*0.25*	*0.30*	*0.35*	*0.40*	*0.45*	*0.50*	*0.55*	*0.60*	*0.65*	*0.70*	*0.75*	*0.80*	*0.85*	*0.90*	*0.95*	*1.00*
0.00	152	74	48	35	27	22	18	15	13	11	10	8	7	6	6	5	4	4	3	2
0.05		435	141	76	49	36	27	22	18	15	12	11	9	8	7	6	5	4	4	3
0.10			686	199	100	62	43	32	25	20	16	14	11	10	8	7	6	5	4	4
0.15				906	250	121	73	49	36	27	22	17	14	12	10	8	7	6	5	4
0.20					1094	294	138	82	54	39	29	23	18	15	12	10	8	7	6	5
0.25						1251	329	152	89	58	41	31	24	19	15	12	10	8	7	6
0.30							1377	356	163	93	61	42	31	24	19	15	12	10	8	6
0.35								1471	376	170	96	62	43	31	24	18	14	11	9	7
0.40									1534	388	173	97	62	42	31	23	17	14	11	8
0.45										1565	392	173	96	61	41	29	22	16	12	10

size required for the binary outcome. The moral of the story is to try and have continuous outcome variables if possible.

Commonly, the number of patients that can be included in a study is governed by nonscientific forces such as time, money and human resources. Thus with a predetermined sample size, the researcher may then wish to know the probability of detecting a certain effect size with a study confined to this size. If the resulting power is small, say less than 50%, then the investigator may decide that the study should not go ahead. A similar situation arises if the type of subject under consideration is uncommon, as would be the case with a clinical trial in rare disease groups. In either case, the sample size is constrained and the researcher is interested in finding the size of effects that could be established for a reasonable power of, say, 80%. Thus the output from a sample size calculation should be a range of possible sample sizes against the effect sizes detectable with a number of levels of power.

In order to calculate the sample size of a study one must first have suitable background information together with some idea as to what is a realistic difference to seek. Sometimes such information is available as prior knowledge from the literature or other sources; at other times, a PILOT STUDY may be conducted.

Traditionally, a pilot study is a distinct preliminary investigation, conducted before embarking on the main trial. However, Wittes and Brittain (1990) have explored the use of an internal pilot study. The idea here is to plan the clinical trial on the basis of best available information, but to regard the first patients entered as the internal pilot. When data from these patients have been collected, the sample size can be reestimated with the revised knowledge so generated.

Two vital features accompany this approach: first, the final sample size should only ever be adjusted *upwards*, never down, and, second, one should only use the internal pilot in order to improve the estimation factors that are independent of the treatment variable. This second point is crucial. It means that when comparing the means of two groups, it is valid to reestimate the planning SD, σ_{Plan}, but not δ_{Plan}. Both these points should be carefully observed to avoid distortion of the subsequent significance test and a possible misleading interpretation of the final study results.

The advantage of an internal pilot is that it can be relatively large – perhaps half of the anticipated patients. It provides an insurance against misjudgement regarding the baseline planning assumptions. It is, nevertheless, important that the intention to conduct an internal pilot study is recorded at the outset and that full details are given in the study protocol. An internal pilot is an example of an ADAPTIVE DESIGN (Bauer, 2008), which is a more recent, flexible approach to clinical trial design.

In studies involving a single group, sample size calculations are couched in terms of the CONFIDENCE INTERVAL. Thus for a given study endpoint, for example, the mean systolic blood pressure (SBP), the hypotensive proportion or the median duration of fever, calculated from the subjects in a case series or a cross-sectional survey, it is usual also to quote the corresponding confidence interval (CI). Thus, when planning a case series survey, it would be appropriate to define ω, the width of the desired CI. This width will depend on the variability from subject to subject (which we cannot control) and the number of subjects in the case series. We assume the object of the study is to estimate a population mean μ, and this is thought to be close to μ_{Plan}. Further, if the data can be assumed to follow a NORMAL DISTRIBUTION, then provided we choose a relatively large sample size n, the 100 $(1-\alpha)\%$ CI for the population mean μ is likely to be close to:

$$\mu_{\text{Plan}} - z_{1-\alpha/2}\sqrt{\frac{\sigma^2}{n}} \text{ to } \mu_{\text{Plan}} + z_{1-\alpha/2}\sqrt{\frac{\sigma^2}{n}} \quad (5)$$

Here σ is the standard deviation, which summarises the subject-to-subject variation.

The width, ω, of this CI is obtained from the difference between the upper and lower limits of equation (5) as:

$$\omega = 2 \times z_{1-\alpha/2} \times \sqrt{\frac{\sigma^2}{n}} \quad (6)$$

Thus, for a planning value ω, the number of subjects n required is obtained by reorganising equation (6) to give the required study size as:

$$n = 4\left[\frac{\sigma^2}{\omega^2}\right] z^2_{1-\alpha/2} \quad (7)$$

In practice, to calculate n_{Plan}, a value of σ_{Plan} as well as ω_{Plan} or a value for their ratio has to be provided. The actual value of μ_{Plan} does not feature in this calculation. Once the study is completed, the sample mean $\bar{x}$ replaces μ_{Plan} and the sample standard deviation, s, replaces σ_{Plan} in the calculation of the CI of equation (5).

For example, Weir, Fiaschi and Machin (1998) give the mean latency of the auditory P300 measured in 19 right-handed patients with schizophrenia as 346 ms with SD = 27 ms. Using equation (5), the corresponding 95% CI is from 334 to 358 ms. The width of this CI is $\omega = 358 - 334 = 24$ ms.

If the study were to be repeated but in (say) left-handed patients, how many would be required to obtain a narrower width of the CI set at 20 ms? In this case, $\omega_{\text{Plan}} = 20$ ms, and, assuming the same SD of 27 ms, equation (7) suggests $4 \times (27/20) \times (1.96)^2 = 28.1$ or approximately 30 patients.

Many clinical trials are designed to show that treatments are effectively equivalent, rather than different (see EQUIVALENCE STUDIES). In Phase II trials one might like to show that a generic drug is equivalent to a standard one in terms of its pharmacokinetics. This *bioequivalence* is often phrased in terms of the AREA UNDER THE CURVE (AUC) of the serum levels of the drug after consumption. Since it is impossible to prove equivalence, one has to specify in advance a difference, δ, within which one is willing to concede that there is, in fact, no difference. For bioequivalence, the convention is to accept that two drugs are equivalent if their AUCs are within 20% of each other, or the ratio is between 0.8 and 1.25. Further details are given in Diletti, Hauschke and Steinijans (1991).

Cohen (1988) is the classical reference for sample size calculations. The book by Machin *et al.* (2008) gives details of sample size calculations for a large number of other designs, such as studies with more than two groups, with ordinal or survival outcomes and for paired data. Hints for sample size calculations are given by Lenth (2001). Nowadays there is much software, both commercial and freely available on the web, for performing sample size calculations; see, for example, http://www.stat.uiowa.edu/~rlenth/Power/.

Sample size calculations have been criticised by numerous authors on a number of grounds, such as that they depend on only one endpoint and yet trials will have several, that any size study can be justified by judicious choice of endpoint and power, that often an investigator has no idea of what a meaningful effect size is and that they concentrate on significance tests when in fact the purpose of most experiments is estimation. However, Williamson *et al.* (2000) gave a vigorous defence of the practice that forces an investigator a priori to name the main outcome variable, which can then be checked in the analysis, to protect against data dredging and to prevent investigators embarking on studies that scientifically have little chance of getting meaningful results. *MJC*

Altman, D. G. 1991: *Practical statistics for medical research.* London: Chapman & Hall/CRC. **Bauer, P.** 2008: Adaptive designs: looking for a needle in a haystack – a new challenge in medical research. *Statistics in Medicine* 27, 1565–80 **Campbell, M. J., Julious, S. A. and Altman, D. G.** 1995: Sample sizes for binary, ordered categorical and continuous outcomes in two group comparisons. *British Medical Journal* 311, 1145–8. **Cohen, J.** 1988: *Statistical power analysis for the behavioral sciences*, 2nd edition. Mahwah: Lawrence Erlbaum. **Diletti, E., Hauschke, D. and Steinijans, V. W.** 1991: Sample size determination for bioequivalence assessment by means of confidence intervals. *International Journal Clinical Pharmacology, Therapy and Toxicology* 29, 1–8. **Lenth, R. V.** 2001: Some practical guidelines for effective sample size determination. *The American Statistician* 55, 187–93. **Machin, D., Campbell, M. J., Tan, S. B. and Tan, S. H.** 2008: *Sample size tables for clinical studies*, 3rd edition. Chichester: Wiley-Blackwell. **Spiegelhalter, D. J., Freedman, L. S. and Parmar, M. K. B.** 1994: Bayesian approaches to randomized trials (with discussion). *Journal of the Royal Statistical Society, Series A* 157, 357–416. **Tan, S.-B., Dear, K. B. G., Bruzzi, P. and Machin, D.** 2003: Towards a strategy for randomised clinical trials in rare cancers. *British Medical Journal* 327, 47–9. **Viljanen, M., Malmivaara, A., Uitte, J., Rinne, M., Palmroos, P. and Laippala, P.** 2003: Effectiveness of dynamic muscle training, relaxation training or ordinary activity for chronic neck pain: randomised controlled trial. *British Medical Journal* 327, 475–7. **Williamson, P., Hutton, J. L., Bliss, J., Blunt, J., Campbell, M. J., Nicholson, R.** 2000: Statistical review by research ethics committees. *Journal of the Royal Statistical Society, Series A* 163, 5–13. **Weir, N. H., Fiaschi, K. and Machin, D.** 1998: The distribution of latency of the auditory P300 in schizophrenia and depression. *Schizophrenia Research* 31, 151–8. **Wittes, J. and Brittain, E.** 1990: The role of internal pilot studies in increasing the efficiency of clinical trials. *Statistics in Medicine* 9, 65–72.

sample size determination in cluster randomised trials When designing CLUSTER RANDOMISED TRIALS, the sample size should be carefully chosen, as when designing conventional CLINICAL TRIALS that randomise individual patients. The use of cluster randomisation must be taken into account at the design stage of the trial; the total number of patients required is larger under cluster randomisation than under individual randomisation, so a trial that randomises clusters without increasing the sample size will lack POWER. In a cluster randomised trial, the responses of patients from the same cluster cannot be assumed to be independent, because patients within a cluster are more similar than patients from different clusters.

Standard formulae for sample size determination assume the outcomes for patients in the planned study to be independent and this assumption is invalid in a cluster randomised trial. The size of a cluster trial has two components: the number of clusters recruited and the number of patients recruited from each cluster, which is referred to as the cluster size (not usually equal to the population size for each cluster, e.g. the number of patients in a hospital catchment area). To calculate how many patients are required in a cluster randomised trial, sample sizes given by standard formulae should be inflated by a factor known as the design effect. The design effect is equal to $[1 + (\bar{n}-1)\rho]$, where $\bar{n}$ is the average cluster size and ρ is the INTRACLUSTER CORRELATION COEFFICIENT (ICC), which represents the anticipated extent of similarity within clusters. For a cluster trial employing paired or STRATIFIED RANDOMISATION rather than SIMPLE RANDOMISATION, more complicated formulae are needed to calculate the sample size (Donner and Klar, 2000).

Deciding to randomise clusters rather than individual patients can have a substantial impact on the sample size required. Consider, for example, designing a trial to detect a difference of 0.25 STANDARD DEVIATIONS in total cholesterol at a (two-sided) 5% significance level. If the trial were to randomise individual patients, a total of 504 patients would provide 80% power to detect this difference. If choosing to randomise general practices (for example), the sample size required to provide the same level of power depends heavily on the anticipated value for the ICC, as shown in the table.

sample size determination in cluster randomised trials *Sample sizes that provide 80% power to detect the specified difference in a cluster randomised trial, at different levels of ICC*

Total sample size	*Average cluster size*	*ICC*
752	50	0.01
1004	100	0.01
998	50	0.02
1502	100	0.02
1740	50	0.05
2974	50	0.10

Even when the ICC is expected to be small, cluster randomised trials can require considerably increased numbers of patients in comparison with individually randomised trials, especially when cluster sizes are large. This demonstrates the flaw of an approach occasionally used in the past, in which researchers who anticipated a small ICC neglected to allow at all for the use of cluster randomisation in their design. The desired level of power for a cluster trial is more easily achieved through raising the planned number of clusters than through raising the planned average cluster size. In some settings, however, the number of clusters available for recruitment may be limited or fixed in advance because of administrative or financial constraints. The occurrence of DROPOUTS in clinical trials is always undesirable, but in a cluster randomised trial there is the possibility that entire clusters will drop out. Because loss of clusters can seriously reduce the power of a trial, every attempt must be made to retain all clusters recruited and some authors recommend identifying a reserve of potential substitutes before starting the trial (Donner and Klar, 2000).

The value assumed for the ICC when calculating the sample size for a cluster randomised trial is usually based on available estimates for ICC values in similar settings. For example, in designing the hypothetical trial discussed above, the ICC value ρ used in the formula for the design effect would ideally be based on ICC estimates representing similarity of total cholesterol measurements within general practices. In order that researchers planning trials have a good chance of locating relevant information on likely values for the ICC in their trial, it has been recommended that completed cluster trials publish the ICC estimates for all outcomes collected. In addition, some research groups are collating published and unpublished estimates into ICC databases, such as that of Ukoumunne *et al.* (1999). Even when completely relevant information exists, available ICC estimates tend to be imprecise; i.e. the CONFIDENCE INTERVAL associated with the estimate tends to include a wide range of ICC values. The table demonstrates that if the ICC value in the future trial is higher than the value allowed for at the design stage, there could be a serious loss of trial power. For this reason, several researchers have recommended using a conservative value for the ICC or taking into account the uncertainty in the ICC estimates used (Feng and Grizzle, 1992; Turner, Prevost and Thompson, 2004). Alternatively, an internal pilot study design could be used, in which the sample size is recalculated some way into the trial on the basis of the current ICC estimate (Lake *et al.*, 2002). *RT*

Donner, A. and Klar, N. 2000: *Design and analysis of cluster randomisation trials in health research.* London: Arnold. **Feng, Z. and Grizzle, J. E.** 1992: Correlated binomial variates: properties of estimator of intraclass correlation and its effect on sample size calculation. *Statistics in Medicine* 11, 1607–14. **Lake, S., Kammann, E., Klar, N. and Betensky, R.** 2002: Sample size re-estimation in cluster randomization trials. *Statistics in Medicine* 21, 1337–50. **Turner, R. M., Prevost, A. T. and Thompson, S. G.** 2004: Allowing for imprecision of the intracluster correlation coefficient in the design of cluster randomized trials. *Statistics in Medicine* 23, 1195–214. **Ukoumunne, O. C., Gulliford, M. C., Chinn, S., Sterne, J. A. C. and Burney, P. G. J.** 1999: Methods for evaluating area-wide and organisation-based interventions in health and health care: a systematic review. *Health Technology Assessment* 3, 5.

sample size determination in observational studies Consideration of sample size is as important for OBSERVATIONAL STUDIES as it is for randomised controlled trials (see CLINICAL TRIALS), the crucial issue being whether the study will be large enough to answer the research question (or questions) with sufficient statistical precision or POWER. As with randomised controlled trials, the degree of statistical uncertainty decreases with increasing sample size and, in general, sample size requirements are substantially greater when studying rarer diseases or less common exposures. Observational studies that are too small may fail to detect important effects or produce estimates too imprecise to be useful, while those that are too large can waste resources and take too long to produce results. At the same time, it should be noted that study size guidelines are not intended as rigid rules. It may be, for example, that the research question is of sufficient importance to outweigh an inadequate, yet unavoidable, sample size (e.g. research into a rare cancer). Although a single study may be too small to provide definitive results, if well designed it will still have the potential to make an important contribution to existing evidence. In contrast, a poorly designed study will be at high risk of producing biased results. The problem of BIAS will not diminish by increasing the sample size.

Sample size requirements for observational studies should be critically assessed at the study design stage. Regardless of the type of study being planned, the first step involves identifying the primary health outcome of interest and the primary exposure. Sample size calculations should then be carried out for a range of different scenarios. Particular consideration here should be given to the extent of missing data to be expected. Individuals selected for interview in a CASE-CONTROL STUDY using hospital records, for example, may not all be traced and some of those successfully traced may refuse to take part. In COHORT STUDIES, individuals may be lost to follow-up, particularly when the study period extends over several years. Sample size calculations should be adjusted to take account of the likely effect of these possibilities. If there is more than one outcome measure or more than one exposure being studied, these should be ordered in terms of priority. Once the sample size required to achieve the needs for the primary research question has been determined, this can subsequently be used to examine its adequacy for additional research questions.

The specific type of outcome measure(s) to be studied is an important determinant of sample size requirements for observational studies. Outcome measures in CROSS-SECTIONAL STUDIES, for example, can range from binary proportion or PREVALENCE (e.g. obesity) to those that are continuous where interest may focus on means (e.g. mean systolic blood pressure). In general, sample size requirements will be much larger when the outcome measure is binary.

Sample size requirements can be addressed from two different perspectives: the statistical power to be achieved from a test of statistical significance or the level of precision to be attained from estimation. Standard formulae required when estimating a single prevalence rate or MEAN, or comparing prevalence (or means) in two groups, are provided elsewhere (Kirkwood and Sterne, 2003). Computer programs required to perform these calculations are also available (Statcorp, 2003).

For brevity, the remainder of this entry considers certain issues associated with sample size requirements for case-control studies and cohort studies only where interest focuses on a binary exposure. Consideration of more advanced issues for these studies, such as tests for trend or interaction, is addressed elsewhere (Smith and Day, 1984; Breslow and Day, 1987).

Sample size requirements in case-control studies depend on the expected magnitude of the ODDS RATIO and the prevalence of exposure in the controls. In order to determine the study size required to achieve adequate statistical power, the number required in each group (cases and controls) can be determined from the formula used for comparing two proportions (Kirkwood and Sterne, 2003; Statcorp, 2003). In general, the closer the odds ratio to be detected is to the null value (1.0) the larger the sample size required, while the closer the prevalence is to 0.5 (or 50%) the smaller is the required sample size.

When the number of cases is fixed, as often occurs in very rare diseases, statistical precision (and power) can be improved by increasing the number of controls per case. The study size required to achieve a certain level of precision can be assessed by examining the width of 95% confidence intervals (95% CIs) for the *odds ratio*. The first figure on page 406 shows the effect of increasing the number of controls per case on precision where the expected odds ratio is 2.0, the prevalence of exposure in controls is 25% and the number of cases is fixed at 100. As may be seen, the precision of the estimate increases with the number of controls per case, but little gain is observed beyond four controls per case. In general, unless the odds ratio is substantially different from unity there is little advantage in having more than four controls per case (Breslow and Day, 1987).

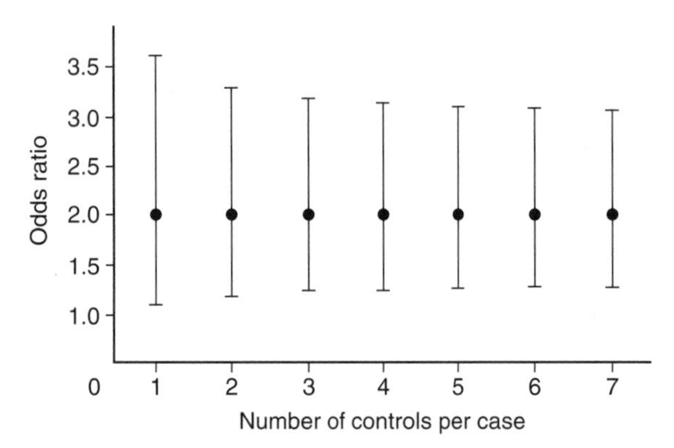

sample size determination in observational studies *Effect of increasing the number of controls per case on 95% confidence intervals for an odds ratio of 2.0 with 25% of controls exposed*

One potential route for improving the efficiency of a case-control study can be achieved at the design stage by matching cases and controls in relation to one or more specific confounders. Matching cases and controls on the basis of strong confounding factors can increase the precision (and power) of a study and also offers the potential of a smaller study size requirement. However, it should be noted that matching does not always yield such gains. In particular, unless the confounding factor is strongly related to the disease there may be little benefit from matching (Smith and Day, 1984).

Unlike case-control studies, cohort studies provide the opportunity to estimate the absolute magnitude of disease risks, rates or odds as well as the corresponding measures of effect (risk ratios, rate ratios or odds ratios). As a result, when considering the study size requirements in cohort studies it is particularly important to decide on the primary outcome measure in advance. When planning to compare disease occurrence in two groups (e.g. exposed and unexposed individuals) using a test of statistical significance, sample size requirements depend on: the magnitude of the ratio of (or difference in) disease outcomes to be detected, the level of disease occurrence expected in the unexposed group, the level of statistical significance and the statistical power required. The required formulae for these calculations are provided elsewhere (Kirkwood and Sterne, 2003).

Alternatively, the level of statistical power can be determined for a variety of different study sizes. The second figure shows power curves obtained for a cohort study where the rate in the unexposed is 10 per 1000 PERSON-YEARS AT RISK and the two-sided level of statistical significance is 5%. Two different scenarios are shown. The lower curve shows the power to detect a rate ratio of 1.5 and the upper curve a rate ratio of 2.0. If the true rate ratio is 2.0 and the aim is to achieve a minimum of 90% power, this suggests that a study size of 3000 will be required in each group, i.e. a total study size of 6000. This could be achieved by studying 3000 individuals in each group for 1 year or 1500 in each group for 2 years and so on. Similar study sizes will only achieve 40% power, however, if the true rate ratio is only 1.5.

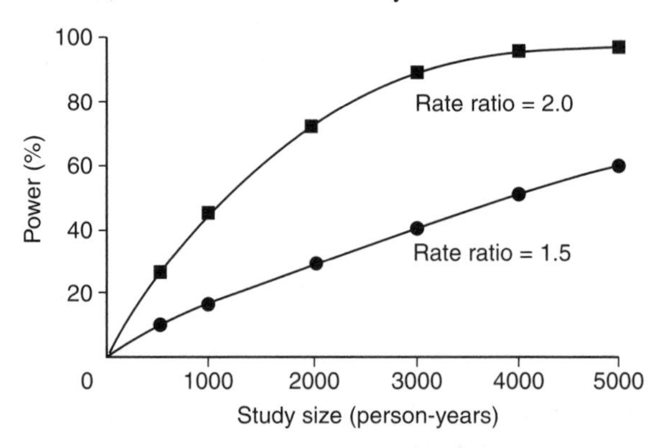

sample size determination in observational studies *Power to detect the rate ratio in cohort studies with the rate in unexposed= 10 per 1000, with 5% level of statistical significance (two-sided)*

Sample size requirements for cohort studies can also be considered from the perspective of precision of estimates, as addressed in case-control studies above. Again, consideration should also be given to the need to adjust for confounding factors, as this will tend to increase study size requirements (Breslow and Day, 1987). Assessing the effect of exposure on disease experienced by a cohort of individuals exposed to a particular substance may involve comparing the events observed in the cohort with those expected on the basis of rates in a 'standard' population. Sample size requirements for this scenario are provided elsewhere (Breslow and Day, 1987). *LMC*

Breslow, N. E. and Day, N. E. 1987: *Statistical methods in cancer research*, Vol. II, *The design and analysis of cohort studies*. Lyon: International Agency for Research on Cancer. **Kirkwood, B. R. and Sterne, J. A. C.** 2003: *Essential medical statistics*, 2nd edition. Oxford: Blackwell. **Smith, P. G. and Day, N. E.** 1984: The design of case-control studies: the influence of confounding and interaction effects. *International Journal of Epidemiology* 13, 356–65. **Statcorp** 2003: *Statistical software: release 8*, Vol. 4 (Sampsi program). College Station, TX: Stata Corporation.

sample size reestimation At the design stage of a CLINICAL TRIAL there are some uncertain parameters. For example, the sample size of a study will be based on an estimation of variability (see SAMPLE SIZE DETERMINATION IN CLINICAL TRIALS). Other things that are estimated, but have some uncertainty, include hazard rates or a group event rate (see PROPORTIONAL HAZARDS). There are also situations where the primary objective of the study relies on the exposure to a drug. This might be needed to assess safety and a minimum amount of exposure may be needed. With this uncertainty surrounding initial assumptions it may be wise to consider modifying the total sample size based on an interim assessment of the data. If the assumptions appear to be incorrect it might be possible to make mid-course adjustments based on the data collected. Chuang-Stein *et al.* (2006) provide a thorough review and recommendations for sample size reestimation for PHASE III TRIALS and PHASE IV TRIALS, but the principles can also be applied in any study where the adjustment of the sample size is needed.

Methodology to address sample size reestimation is wide and varied, and each comes with an accompanying decision rule. This decision rule is important in order to understand the operating characteristics of the resulting design. For confirmatory trials the control of a TYPE I ERROR at the designated level (usually 5%) is paramount and any adjustment to sample size must ensure this. With the use of simulation one can investigate whether the TYPE I ERROR will be controlled.

It should be noted that traditional group sequential methodology by performing interim analyses is based on the statistical information accrued, e.g. in an event-driven trial, and then adjustment based on the number of events required at the end of trial.

Sample size reestimation can be conducted in a blinded or unblinded fashion and the considerations for both will be discussed. Blinded sample size reestimation involves a review of the sample size based on a nuisance parameter such as the variance of a continuous ENDPOINT or the underlying rate for a binary event. Because it is blinded there will be no indication as to the observed treatment effect, and the sample size reestimation will be based on the original assumptions. Gould (2001) reviewed methods of this kind and showed that they were comparable in performance to those using unblinded methodology. Because there is potentially little BIAS introduced owing to the review being blinded, these types of reviews are acceptable to regulatory agencies.

If there is greater uncertainty about the treatment difference as well as the NUISANCE PARAMETER, then one might want to choose to do an unblinded sample size reestimation. However, there would need to be procedural controls to minimise the effect of any potential bias that could be introduced by unblinding the data. Unblinded sample size reestimation presents more of an issue because of this potential inherent bias they could introduce, but with more knowledge they should perform better than in the blinded case. However, there is the potential to increase the Type I error rate. In order to control the overall Type I error the employment of combination tests is undertaken. These are methods where the *P*-VALUES before and after the adaptation are combined. The Fisher combination test is commonly used and Lehmacher and Wassmer (1999) suggest an inverse normal method for combining the stages.

Any unblinded sample size reestimation should be undertaken with caution. Access to the data can potentially lead to operational bias and so put the integrity of the study in doubt. The analysis and results need to be properly managed as awareness of the methods and resulting sample size modification could lead to inference about the treatment difference. In addition, unblinded sample size reestimation is likely to be less acceptable to regulatory agencies.

Chuang-Stein *et al.* (2006) warn against using the interim treatment effect as a parameter of interest for the following reasons. First, they can be inefficient with interim treatment effects being highly variable, leading to unreliable estimates of the sample size required. Second, the actual effect size used in the original sample size calculation is not an expectation of the magnitude of the effect, but is more like the clinically relevant difference. Using a variable point estimate to determine the future sample size might actually not be achieving the desired objective. They make the following recommendation: 'Before implementation methods that modify sample size based on the interim treatment effect estimates, it should be strongly considered whether the sample size determination objectives can be achieved, statistically and procedurally, using either an appropriate group sequential scheme or an adaptive scheme that does not utilize the interim observed effect for reestimating sample size.'

It is also recommended that the number of reestimations are kept to a minimum and usually the objective can be achieved with one. It is also recommended that following a sample size reestimation the total number of subjects should be either increased or stay the same. Reducing patient numbers can have inherent problems, particularly if a required amount of safety data is required. *AB*

Chuang-Stein, C., Anderson, K., Gallo, P. and Collins, S. 2006: Sample size reestimation: a review and recommendations. *Drug Information Journal* 40, 475–84. **Gould, A. L.** 2001: Sample size re-estimation: recent developments and practical considerations. *Statistics in Medicine* 20, 2625–43. **Lehmacher, W. and Wassmer, G.** 1999: Adaptive sample size calculation in group sequential trials. *Biometrics* 55, 1286–90.

sampling distributions These are PROBABILITY DISTRIBUTIONS of statistics calculated from random samples of a particular size. When we draw a sample from a population, it is just one of the many samples we could take. If we calculate a statistic from the sample, such as a MEAN or proportion, this will vary from sample to sample. The means or proportions from all the possible samples form the sampling distribution. To illustrate this with a simple example, we could put lots numbered 1 to 9 into a hat and sample by drawing one out, replacing it, drawing another out, and so on. Each number would have the same chance of being chosen

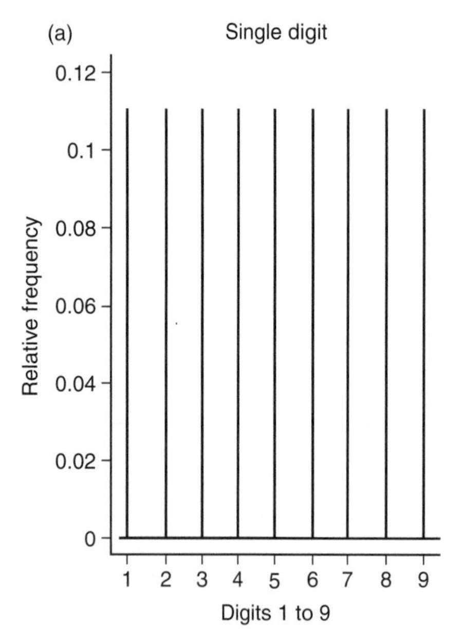

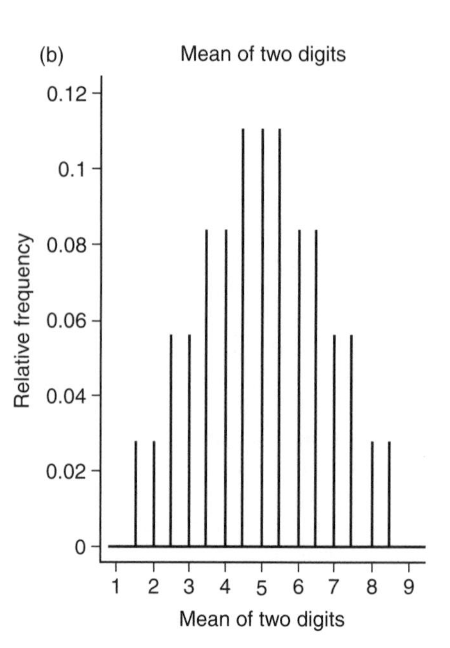

sampling distributions *Sampling distribution for a single digit drawn at random and for the mean of two digits drawn together*

each time and the sampling distribution would be as part (a) in the figure. Now we change the procedure, draw out two lots at a time and calculate the average. There are 36 possible pairs, and some pairs will have the same average (e.g. 1 and 9, 4 and 6, both having the average 5.0). The sampling distribution of this average is shown in part (b) in the figure. Notice that it has a different shape to (a). The sampling distribution of a statistic does not necessarily have the same shape as the distribution of the observations themselves, which we call the parent distribution.

If we know the sampling distribution it can help us draw conclusions about the population from the sample, using CONFIDENCE INTERVALS and SIGNIFICANCE TESTS. We often use our sample statistic as an estimate of the corresponding value in the population, e.g. using the sample mean to estimate the population mean. The sampling distribution tells us how far from the population value the sample statistic is likely to be.

In most circumstances, we do not know what the sampling distribution is. However, we do not need to take many samples to estimate it. We can do this from a single sample only. Theory tells us into what general family of distributions the sampling distribution will fall and we can estimate which member of this family the sampling distribution is. For example, if we follow a case series of 100 patients and find that 89 of them have a satisfactory outcome, we would expect the sampling distribution from which the statistic 89 comes to be a member of the binomial family. The particular member of that family is estimated to be the BINOMIAL DISTRIBUTION with parameters $n = 100$ and $p = 89/100 = 0.89$. For the mean of a large sample, we would expect a normal sampling distribution and we estimate the mean and VARIANCE of this NORMAL DISTRIBUTION from the mean and variance of the sample.

If the sample statistic is used as an estimate, we call the STANDARD DEVIATION of the sampling distribution the STANDARD ERROR. Rather confusingly, we use this term both for the unknown standard deviation of the sampling distribution and for the estimate of this standard deviation found from the data. *JMB*

sampling methods – an overview Sampling is a way of choosing a subset of the population of interest, within which it is easier to study the properties that are of interest within the main population. The population is the group of people or items under investigation. The population may be small, large or infinite. The population is the entire set of units to which the results will be extrapolated. Once the precise purpose of a study has been defined then the target population should be decided. The aim of the process of data collection is to draw conclusions about this population. A sample from the population is the set of all units about which information is collected during the investigation.

The sampling units are the individual members that make up the population. It is important that the sampling units are clearly defined and care must be taken to choose the correct sampling unit to answer the question under consideration. A sampling frame is a list of all sampling units in the target population. It may not be possible, or indeed be too expensive or time consuming, to obtain a complete sampling frame for the population. It is important that if one is used it is as accurate and up to date as possible. It should be as free as possible from omissions and duplications. If MULTISTAGE CLUSTER SAMPLING is being used, then more than one sampling frame will be needed. In this case one sampling frame will be needed for each stage of sampling. The sampling units for the first stage of sampling are called the primary sampling units. Those for the final stage of sampling are called the listing units.

Information is collected using a survey. There are two different forms of survey: a census and a sample survey. A census is a survey that includes every member of a population. A census is often carried out if the population is small enough. In many countries, a population census is carried out every 10 years. When populations are large then a census may be very expensive and time consuming. If the population is very large it might not be possible to survey every member. In some circumstances it may not be sensible to carry out a census.

When a census is not possible or is too difficult then a sample survey can be used instead. This is where less than the entire population is included in the survey. If a representative sample is taken then an accurate picture of the overall population can be obtained. Valid inferences can only be made from randomly selected subsamples. BIAS contaminates the study if the samples are chosen nonrandomly. SELECTION BIAS is the most common form of bias in samples.

The statistical objectives of an investigation are: to make inferences about a population by analysing sample data, to make assessments of the extent of uncertainty in these interferences and to design the process and extent of sampling to form a basis for valid and accurate inferences.

Samples can be chosen in two ways, by PROBABILITY/random sampling or by nonprobability/nonrandom sampling. Random sampling is where the probability of getting any particular sample can be calculated from a probability model. Usually each unit has a known, possibly equal, probability of being chosen to be included in the sample. In nonprobability sampling there is an uneven and unknown chance of being included in the sample and should only be used with caution when making inference to a general population. Nonprobability sampling is often used, however, as it can be less expensive and time consuming than probabilistic sampling.

There are several examples of nonprobability sampling. CONVENIENCE SAMPLES are chosen for ease of access. They might be patients within a clinic or doctors who work in the same hospital. Snowball sampling is where the first respondent recommends a personal contact or a friend, and so on until no new members of the sample are found. Purposive or judgement sampling is where the investigator decides who should be included in the sample; they are usually selected to be representative of the population. For example, to estimate the number of blood samples drawn in a year in a clinic, a few typical days could be chosen and the records reviewed. The main problem with this sort of sampling is that there is no insight into the reliability of the estimates. If only a few days could be looked at then nonrandom sampling might include some atypical days that would make the estimate inaccurate. A case study is limited to one group or, in the case of N of 1, to one individual. In a QUOTA SAMPLE, the sample is chosen so that there are a certain number of units or individuals in each category. This is often used with convenience samples; the person carrying out the convenience sample may be told to interview a certain number of, say, males under 25, usually with no instruction on how to select those to be included.

The simplest type of random sampling is the SIMPLE RANDOM SAMPLE. This can be carried out in two ways, with or without replacement. Simple random sampling is equivalent to putting all the units in a hat and drawing one out. When the second selection is made there is a choice of replacing the first unit first or not. In the former case each item can be drawn more than once; this is sampling with replacement. In the latter case each unit can be chosen once at the most; this situation is sampling without replacement. Sampling without replacement is more precise than sample with replacement.

SYSTEMATIC SAMPLING is similar to simple random sampling, with each unit having an equal and known probability of being chosen to be included in the sample. In systematic sampling a random starting point is chosen and then every kth unit is chosen to be included. It should be ensured that this does not hide a pattern in the data; e.g. if every $(k-1)$th element has a fault then it is possible that this could be hidden by choosing every kth element.

If there are obvious subgroups or 'strata' within the population then STRATIFIED SAMPLING may be more efficient than simple random sampling. In this case the population is separated into the strata and simple random samples taken in each of the strata. Each stratum is then represented proportionately in the sample. If the population forms distinct strata then stratified sampling may give more precise information than a simple random sample and therefore maybe more efficient.

If the population is arranged in a hierarchical structure then multistage cluster sampling could be used. In a single-stage cluster sample, a sample of the clusters is chosen at random and then a random sample of units is chosen from within this selection of clusters. In a multistage sample a random sample of clusters is chosen and then a random sample of clusters within these clusters. This is repeated and eventually a random selection of units is chosen within a cluster.

If a probability sample can be obtained, it is better to do this than to obtain a nonprobability sample, as a probability sample should be less biased. A probability sample is representative of the population and can therefore be extrapolated to the population from which it was drawn. This is not the case with a nonprobability sample. There is no way of knowing how representative a nonprobability sample is of the population.

The main sources of bias in sampling are lack of a good sampling frame, the wrong choice of sampling unit, nonresponse by chosen units, those that are introduced by the person gathering the data and self-selection bias. It is important if a sampling frame is to be used that it is a good one and is up to date and free from duplications and omissions. If this is not the case, then the probability that each unit be included in the sample will not be equal, as some units will have a probability of 0 of being included in the sample, as they are not in the frame. If, for example, the telephone directory is used as the sampling frame then all those who are ex-directory or do not have a phone will not be included. The electoral register also misses people. Some units may have more than double the chance of another of being included in the sample, as they are included more than once in the sampling frame.

If the wrong sampling unit is used then the correct inferences might not be drawn from the results as a slightly different question might be being answered; e.g. if individuals are sampled instead of households, then the same event or experience might be referred to by two individuals and counted twice instead of once. Nonresponse by particular units may be due to being unable to locate the particular unit chosen, the person may refuse to respond or the question may be misunderstood. QUESTIONNAIRES should be worded clearly, be unambiguous and easy to understand; they should also be neutrally worded to avoid pointing towards a particular response. The interviewer can introduce bias by not interviewing people who look uncooperative or the way in which a question is asked may influence the answer given to the question. Self-selection bias is due to those volunteering to be selected to be in the sample being systematically different to those who have not volunteered. Self-selected samples are unlikely to be representative of the population.

The size of a sample to be taken is a very important consideration (see SAMPLE SIZE DETERMINATION). There are many formulae in existence to calculate the correct size for a sample. It is important to make sure that the sample is large enough to make the inferences required from the sample. However, being unbiased is more important than the size of the sample and any sample is only representative of the population from which it was drawn. Only with caution should any extrapolations be made beyond that population. For further details see Crawshaw and Chambers (1994) and Levy and Lemeshow (1999). *SLV*

Crawshaw, J. and Chambers, J. 1994: *A concise course in A level statistics*, 3rd edition. Cheltenham: Stanley Thornes Publishers Ltd. **Levy, P. S. and Lemeshow, S.** 1999: *Sampling of populations: methods and applications*, 3rd edition. Chichester: John Wiley & Sons, Ltd.

SAS
See STATISTICAL SOFTWARE

saturated model
See LOG-LINEAR MODELS

scatterplot
An *xy* plot of the values of two, usually continuous, variables that have been recorded on a sample of individuals. Such plots have been in use since at least the 18th century and they have many advantages for an intitial examination of bivariate data. Indeed, according to Tufte (1983): 'The relational graphic – in its barest form the scatterplot and its variants – is the greatest of all graphical designs. It links at least two variables, encouraging and even imploring the viewer to assess the possible causal relationship between the plotted variables. It confronts causal theories that x causes y with empirical evidence as to the actual relationship between x and y.'

Such a plot links the two variables, allows any relationship between them to be visually assessed and may help in identifying OUTLIERS or distinct groups of observations ('clusters'). The appropriate scatterplot should *always* be used when interpreting the numerical value of an estimated correlation between two variables. An example of a scatterplot in which the mortality rate from malignant melanoma of the skin for white males is plotted against latitude of the centre of the state for each state on the US mainland is shown in part (a) of the figure (page 411). The plot clearly demonstrates that mortality is strongly related to latitude. In many cases, scatterplots can be made more useful by adding the estimated regression line of the two variables or a locally weighted regression fit (see SCATTERPLOT SMOOTHERS). Both possibilities are illustrated in part (b) in the figure.

Many other examples of interesting scatterplots are given in Tufte (1983). *BSE*

[See also SCATTERPLOT MATRICES]

Tufte, E. R. 1983: *The visual display of quantitative information.* Cheshire, CT: Graphics Press.

scatterplot matrices
This is a convenient arrangement of all the pairwise SCATTERPLOTS of the variables in a set of multivariate data that aids both the understanding of the relationships between the variables and in uncovering any unusual features of the data, e.g. possible OUTLIERS (Cleveland, 1994). In a scatterplot matrix the separate scatterplots are arranged in the form of a square grid with the same number of rows and columns as the number of variables. Each panel of the grid contains a scatterplot of one pair of variables. The upper left-hand triangle of the grid contains all

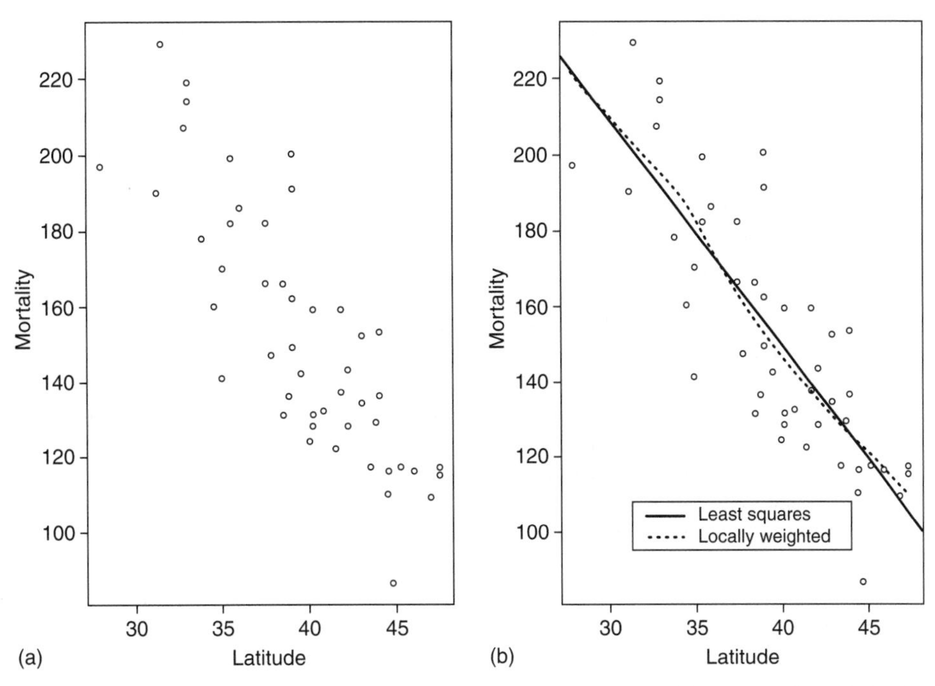

scatterplot *(a) Scatterplot of mortality against latitude, (b) scatterplot as in (a) with locally weighted regression fit added*

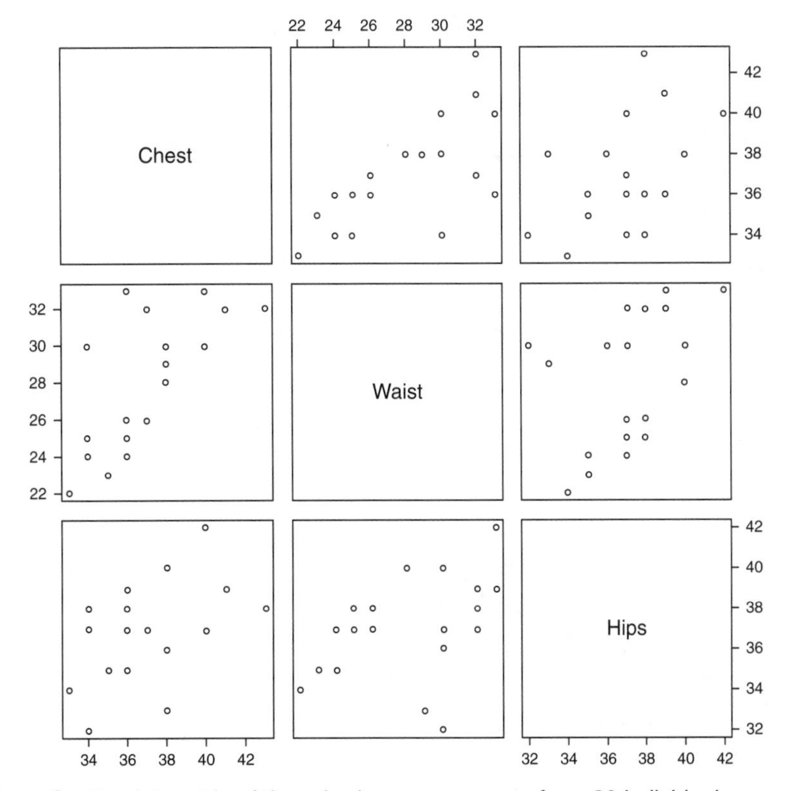

scatterplot matrices *Scatterplot matrix of three body measurements from 20 individuals*

pairs of scatterplots, as does the lower right-hand triangle. The reason for including both the upper and lower triangles in the diagram, despite the seeming redundancy, is that it enables a row and columns to be visually scanned to see one variable against all others, with the scale for the one variable lined up along the horizontal or the vertical. An example of the basic scatterplot matrix for three body measurements taken on 20 individuals is shown in the figure (page 411). The diagram illustrates the varying strengths of the positive relationships between each pair of variables but also points to the possibility that there are relatively distinct groups of observations in the data in the waist/hips scatterplot. Here the explanation of these possible 'groups' is straightforward – they simply correspond to the 10 men and 10 women in the data. The panels in a scatterplot matrix can often be enhanced in some way in an attempt to make the diagram more useful. An example is shown in the DENSITY ESTIMATION entry. *BSE*

[See also TRELLISGRAPHICS]

Cleveland, W. S. 1994: *Visualizing data*. Summit, NJ: Hobart Press.

scatterplot smoothers These are smooth, generally nonparametric curves added to a SCATTERPLOT to aid in understanding the relationships between the two variables forming the plot. They are often a useful alternative to the more familiar parametric curves such as simple linear or polynomial regression fits when the bivariate data plotted is too complex to be described by a simple parametric family.

The simplest scatterplot smoother is a locally weighted regression or *loess fit*, first suggested by Cleveland (1979). In essence, this approach assumes that the variables x and y are related by the equation:

$$y_i = g(x_i) + \varepsilon_i$$

where g is a 'smooth' function and the ε_i are random variables with mean zero and constant scale. Values $\hat{y}_i$, used to 'estimate' the y_i at each x_i, are found by fitting polynomials using weighted least squares with large weights for points near to x_i and small weights otherwise. Therefore, smoothing takes place essentially by local averaging of the y values of observations having predictor values close to a target value.

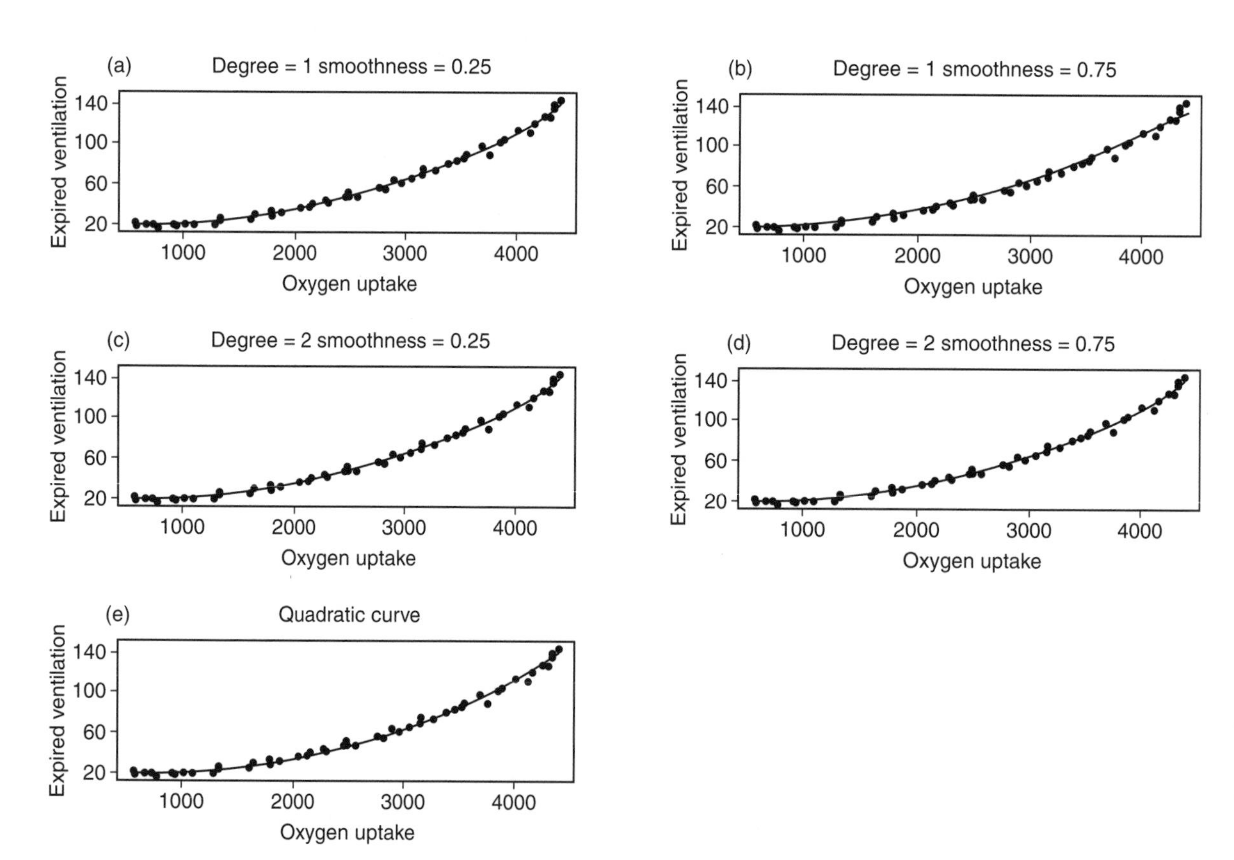

scatterplot smoothers *Locally weighted regression fits for oxygen uptake data*

Two parameters control the shape of a loess curve: the first is a smoothing parameter, α, with larger values leading to smoother curves – typical values are $^1/_4$ to 1. The second parameter, λ, is the degree of certain polynomials that are fitted by the method; λ can take values 1 or 2. In any specific application, the choice of the two parameters must be based on a combination of judgement and of trial and error. Residual plots may, however, be helpful in judging a particular combination of values.

The use of locally weighted regression is demonstrated in the first figure (page 412) for data collected on the oxygen uptake and the expired ventilation of a number of subjects performing a standard exercise task. In this figure, parts (a), (b), (c) and (d) show plots of the data with added locally weighted regression fits with different values of λ and α. Here the four fitted curves are very similar and, in this relatively simple case, each of them is almost identical to a fitted polynomial containing a quadratic term in oxygen uptake – see (e) in the figure.

An alternative smoother that can often usefully be applied to bivariate data is some form of *spline function*. (In its nontechnical use, a spline is a term for a flexible strip of metal or rubber used by a draftsman to draw curves.) Spline functions are polynomials within intervals of the x variable that are connected across different values of x. The second figure, for example, shows a linear spline function, i.e. a piecewise linear function, of the form:

$$f(x) = \beta_0 + \beta_1 X + \beta_2 (X-a)_+ + \beta_3 (X-b)_+ + \beta_4 (X-c)_+$$

where

$$\begin{aligned}(u)_+ &= 0, \quad u > 0\\ &= 0, \quad u \leq 0\end{aligned}$$

The interval ENDPOINTS, a, b and c, are called *knots*. The number of knots can vary according to the amount of data available for fitting the function (see Harrell, 2001).

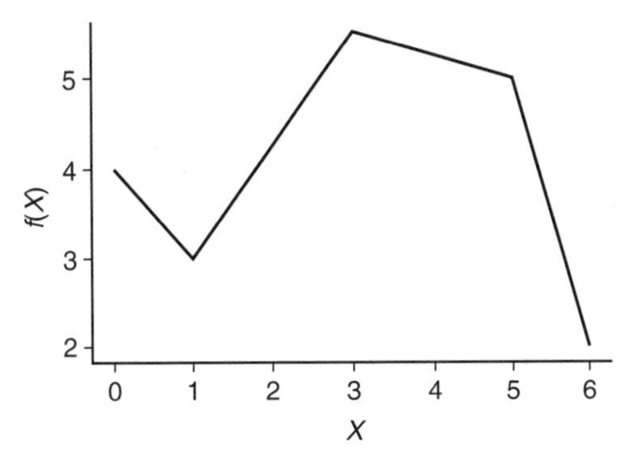

scatterplot smoothers *A linear spline function with knots at a = 1, b = 3, c = 5*

The linear spline is simple and can approximate some relationships, but it is not smooth and so will not fit highly curved functions well. The problem is overcome by using piecewise polynomials, in particular cubics, which have been found to have nice properties with good ability to fit a variety of complex relationships. The result is a *cubic spline*, which arises formally by seeking a smooth curve $g(x)$ to summarise the dependence of y on x, which minimises the expression:

$$\sum [y_i - g(x_i)]^2 + \gamma \int g''(x)^2 dx$$

where $g''(x)$ represents the second derivative of $g(x)$ with respect to x. Although when written formally this criterion looks a little formidable, it is really nothing more than an effort to govern the trade-off between the goodness-of-fit of the data (as measured by $\sum [y_i - g(x_i)]^2$) and the 'wiggliness' or departure of linearity of g as measured by $\int g''(x)^2 dx$; for a linear function, this latter part would be zero. The parameter λ governs the smoothness of g, with larger values resulting in a smoother curve. The solution is a cubic spline, i.e. a series of cubic polynomials joined at the unique observed values of the explanatory variable, x_i. (For more details, see Friedman, 1991.)

The 'effective number of parameters' (analogous to the number of parameters in a parametric fit) or DEGREES OF FREEDOM of a cubic spline smoother is generally used to specify its smoothness rather than λ directly. A numerical search is then used to determine the value of λ corresponding to the required degrees of freedom. The complexity of a cubic spline is approximately the same as a polynomial of degree one less than the degrees of freedom. However, the cubic spline smoother 'spreads out' its parameters in a more even way and hence is much more flexible than polynomial regression.

We shall illustrate the use of cubic splines by fitting such a curve to the monthly deaths from bronchitis, emphysema and asthma in the UK from 1974 to 1979 for men and women. A scatterplot of the data and the fitted cubic spline is shown in the third figure (page 414).

For these data, locally weighted regression is not so successful in representing the data. The fourth figure (page 414) shows a number of plots of the data with added locally weighted regression fits, again with different values of λ and α. Here the characteristic cyclical nature of the data is only picked up with $\lambda = 2$ and $\alpha = 0.25$. In the other three diagrams the amount of smoothing is too great to reveal the structure in the data. *BSE*

Cleveland, W. S. 1979: Robust locally weighted regression and smoothing scatterplots. *Journal of the American Statistical Association* 74, 829–36. **Friedman, J. H.** 1991: Multiple adaptive regression splines. *Annals of Statistics* 19, 1–67. **Harrell, F. E.** 2001: *Regression modelling strategies with applications to linear models, logistic regression and survival analysis*. New York: Springer.

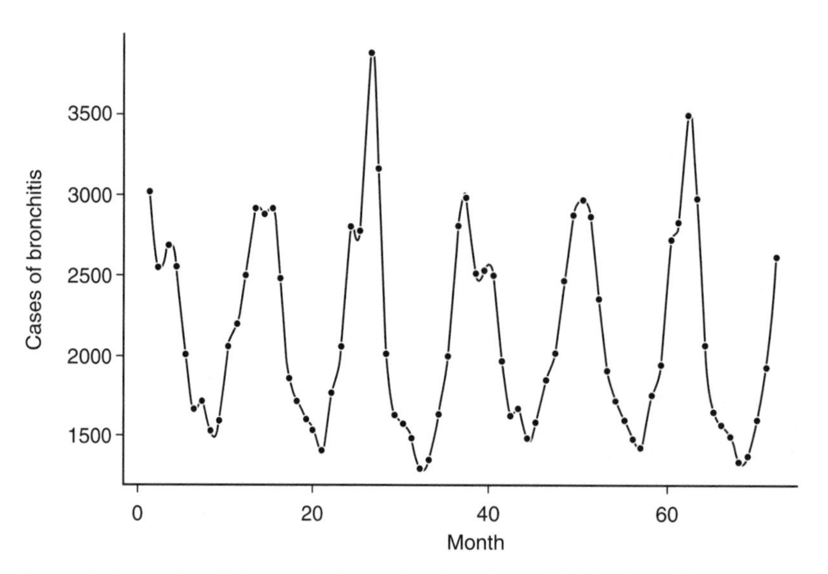

scatterplot smoothers *Cubic spline fit for monthly deaths from bronchitis in the UK, 1974–1979*

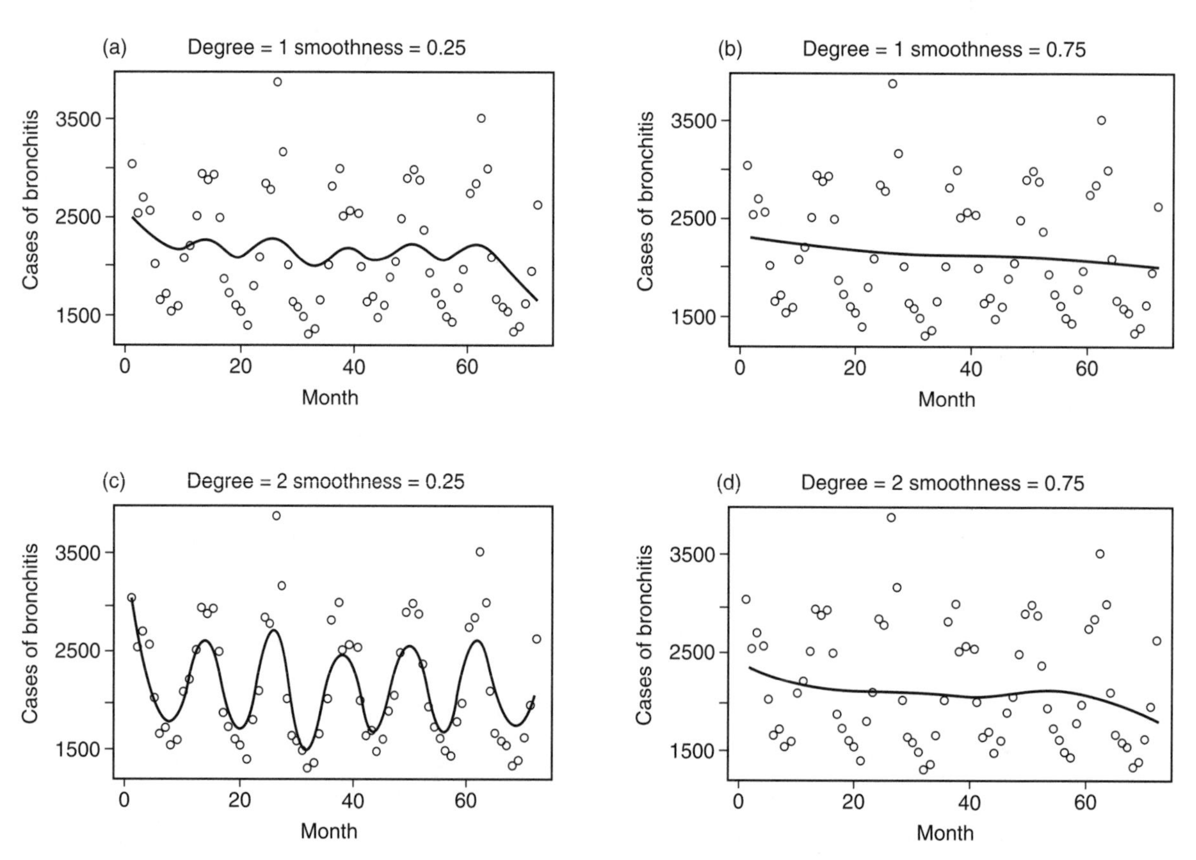

scatterplot smoothers *Locally weighted regression fits for monthly deaths from bronchitis in the UK, 1974–1979*

scree plot This is a SCATTERPLOT of the VARIANCES of the factors in a factor analysis or the components in a PRINCIPAL COMPONENT ANALYSIS against their RANKS in terms of magnitude. The plot can be used to provide an informal estimate of the number of factors (components) by retaining as many factors (components) as there are variances that fall before the last large drop on the plot. An example of such a plot that suggests three factors is shown in the figure. Other examples are given in Preacher and MacCallum (2003). *BSE*

[See KAISER'S RULE]

Preacher, K. J. and MacCullum, R. C. 2003: Repairing Tom Swift's electric factor analysis machine. *Understanding Statistics* 2, 13–44.

screening studies These are planned investigations to determine the effect of administering a diagnostic test to detect the presence or absence of preclinical disease in asymptomatic individuals (screening). The encounter is initiated by the health professional, rather than by the patient, since no clinical symptoms are apparent that otherwise would drive the patient to seek medical diagnosis. The goal of screening is to separate the population into two groups: those with a high versus low probability of the given disorder, usually one that is perceived to be a serious public health condition. Implicit to screening is the assumption of a clearly recognisable outcome that is indicative of preclinical disease and the assumption that early diagnosis is beneficial in some way, such as better prognosis, safer treatment, less invasive medical procedures, higher 'quality of life' or reduced chances of mortality. Examples of diagnostic tests used in screening are: 1. mammography, to detect preclinical breast cancer disease in women; 2. a blood test to detect prostate specific antigen (PSA), as high levels in men are thought to be associated with preclinical disease of prostate cancer; 3. blood pressure and cholesterol levels, as high levels of both are associated with cardiac disease.

Screening tests are not without costs (cost of examination; costs of false positive results arising from follow-up laboratory procedures; costs of false negative results arising from false hope of disease-free status).

Screening studies are designed to quantify: the nature of the 'benefit' (e.g. reduction in mortality, extended survival time, measures of quality of life); the target population that is expected to benefit from screening (in terms of age/gender/ethnic groups); and the error rates (false positives and false negatives). The FALSE POSITIVE RATE is the PROBABILITY that the test asserts 'disease' when, in fact, no disease is present (conversely, SPECIFICITY, or the probability of obtaining a negative result when disease is absent); the FALSE NEGATIVE

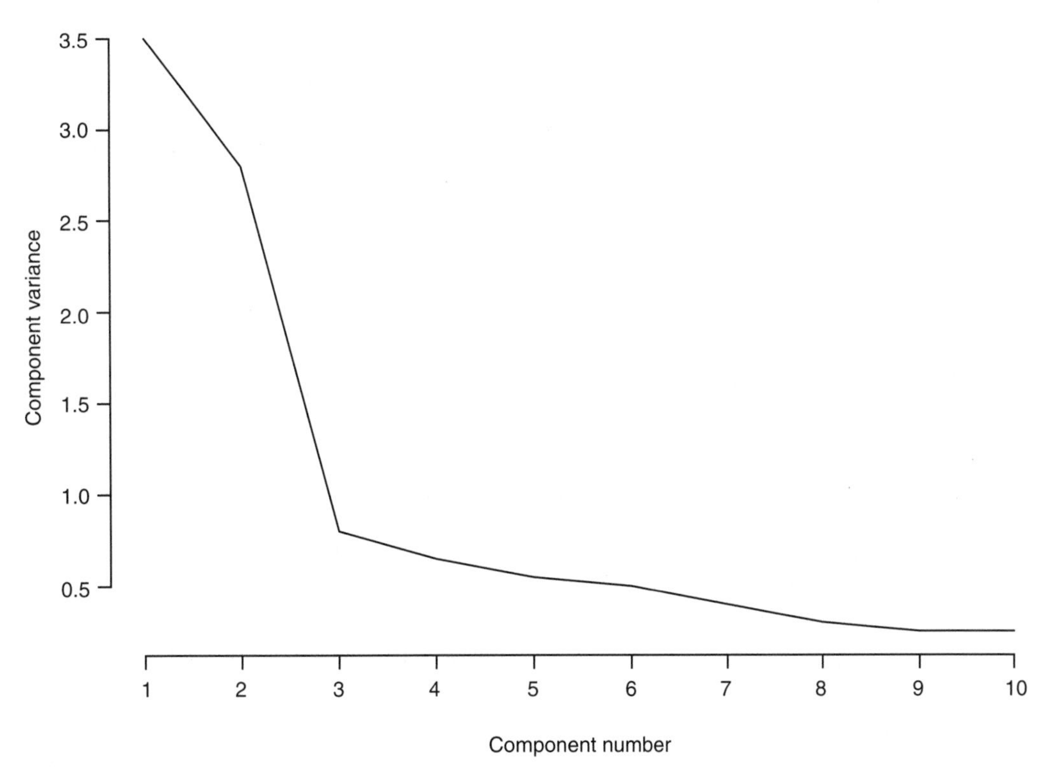

scree plot *A scree plot for a principal component analysis of a correlation matrix of 10 observed variables*

RATE is the probability that the test asserts 'no disease' when in fact disease is present (conversely, SENSITIVITY, or the probability of obtaining a positive result when disease is present). For diagnostic purposes, sensitivity should be high, while, for screening purposes, specificity should be high, to avoid unnecessary follow-up testing of disease-free individuals. Screening tests are indicated when the benefits are judged to outweigh the potential drawbacks (costs, risks of false positives and false negatives, etc.).

Because nonrandomised trials are subject to SELF-SELECTION BIAS, randomised screening trials offer the best and most reliable mechanism for evaluating the potential benefit from screening. In randomised screening trials, study arm participants are offered screening at regular intervals and the control arm participants follow their 'usual medical care'. Due to the cost of screening, such trials are usually conducted using a 'stop-screen design', in which screening is offered for a limited time only (e.g. annual screens for 3 to 5 years).

Several important differences between screening studies, used to evaluate the potential benefit of a screening intervention, and CLINICAL TRIALS, for the evaluation of a specific therapeutic intervention, make the design and analysis of screening studies very challenging. In clinical treatment trials, the cases are specified in the protocol to be comparable in both the study and control arms of the trial; in screening trials, participants are initially asymptomatic and are randomised to the study ('offered screening') or control ('follow usual medical care') arms of the trial and cases evolve as the study progresses. If the screening test is successful, then cases will arise sooner in the study arm than in the control arm, so survival times (time of ENDPOINT minus time of diagnosis) will be longer in the study arm than in the control arm, even in the absence of a screening benefit. This BIAS in the evaluation of screening is known as 'lead time bias'. Also, because cases with longer pre-clinical disease durations are more likely to be detected by screening than cases with shorter pre-clinical durations, the cases that arise in the study arm of a screening trial are more likely to be less aggressive and hence have a more favourable prognosis, even in the absence of a screening benefit. This phenomenon is known as LENGTH-BIASED SAMPLING. Study arm participants also experience 'overdiagnosis bias', or the tendency of the screening test to suggest apparent but truly nonthreatening disease. (In an ideal world, this bias would not affect the results if further diagnostic tests later eliminate these individuals as cases of disease.) Finally, noncompliance in both arms is inevitable: some participants in the study arm may refuse screening, while some in the control arm may seek screening. Thus, the cases that arise in the two arms of a screening trial may not be comparable as they are in a treatment trial. RANDOMISATION ensures that the participant characteristics are the same in both arms, including those that lead to noncompliance of either type in either arm, arguing for an INTENTION-TO-TREAT analysis (Byar *et al.*, 1976).

The most common measures used to evaluate screening are *reduction in mortality* (comparison of death rates) and *mean benefit time* (difference in the MEAN survival time between the time of entry into the trial and the case endpoint). Randomisation ensures that: 1. the participant characteristics are the same in the two trial arms, including those that lead to noncompliance of either type in either arm, and 2. the elimination of bias due to lead time, when survival is measured from the time of entry into trial.

Statistical methods to estimate the benefit (reduction in mortality or extended survival time), lead time and the effect of length-biased sampling have been proposed. For overviews of the issues related to screening and for statistical methodology of design and analysis of screening studies, see Zelen and Feinleib (1969), Zelen (1976), Goldberg and Wittes (1981), Prorok and Connor (1986), Gastwirth (1987), Shapiro *et al.* (1988), Prorok, Connor and Baker (1990), Connor and Prorok (1994), Kafadar and Prorok (1994, 1996, 2003, 2005) and Baker, Kramer and Prorok (2002).

Screening studies are also used to evaluate the outcomes of designed trials to screen drug compounds for their potential to be biologically active. A typical drug screening protocol may involve several stages based on the response of the compound to various reactions: e.g. 'Conduct experiment 1; if the energy from the reaction is less than a specified level, reject the compound; otherwise, conduct experiment 2; if the second reaction is less than a second specified level, reject; otherwise, submit the compound for further testing.' The evaluation of such drug screening designs involves the same kind of consideration as the evaluation of randomised screening trials used on human subjects, described earlier. See Roseberry and Gehan (1964) and Schultz *et al.* (1973) for designs and analysis of drug screening trials, as well as related articles in the literature on screening designs to detect unacceptable products in manufacturing. *KKa*

Baker, S. G., Kramer, B. S. and Prorok, P. C. 2002: Statistical issues in randomised trials of cancer screening. *British Medical Council Medical Research Methodology* 2, 11; www.biomedcentral.com/1471-2288/2/11. **Byar, D. P., Simon, R. M., Friedewald, W. T. *et al.*** 1976: Randomized clinical trials: perspectives on some recent ideas. *New England Journal of Medicine* 295, 74–80. **Connor, R. J. and Prorok, P. C.** 1994: Issues in the mortality analyses of randomized controlled trials of cancer screening. *Controlled Clinical Trials* 15, 81–99. **Gastwirth, J. L.** 1987: The statistical precision of medical screening procedures. *Statistical Science* 2, 213–38. **Goldberg, J. D. and Wittes, J. T.** 1981: The evaluation of medical screening procedures. *The American Statistician* 35, 4–11. **Kafadar, K. and Prorok, P. C.** 1994: A data-analytic approach for estimating lead time and screening benefit based on survival curves in randomised trials. *Statistics in Medicine* 13, 569–86. **Kafadar, K. and Prorok, P. C.** 1996: Computer simulation experiments of randomized screening trials. *Computational Statistics and Data Analysis* 23, 263–91. **Kafadar, K. and Prorok, P. C.** 2003: Alternative definitions of comparable case groups and estimates of lead time and benefit time in randomized cancer screening trials. *Statistics in Medicine* 21, 83–111. **Kafadar, K. and Prorok, P. C.** 2005: Computational methods in medical decision making: to screen or not to screen? *Statistics in Medicine* 24, 569–81. **Prorok, P. C. and Connor, R. J.** 1986: Screening for the early detection of cancer. *Cancer Investigations* 4, 225–38. **Prorok, P. C., Connor, R. J. and Baker, S. G.** 1990: Statistical considerations in cancer screening programs. *Urologic Clinics of North America* 17, 699–708. **Roseberry, T. D. and Gehan, E. A.** 1964: *Biometrics* 20, 73–84. **Schultz, J. R., Nichol, F. R., Elfring, G. L. and Weed, S. D.** 1973: Multiple-stage procedures for drug screening. *Biometrics* 29, 293–300. **Shapiro, S., Venet, W., Strax, P. and Venet, L.** 1988: *Periodic screening for breast cancer: the health insurance plan project and its sequelae, 1963–1986*. Baltimore: Johns Hopkins University Press. **Zelen, M.** 1976: Theory of early detection of breast cancer in the general population. In Heuson, J. C., Mattheiem, W. H. and Rozenweig, M. (eds), *Breast cancer: trends in research and treatment*. New York: Raven Press, pp. 287–301. **Zelen, M. and Feinleib, M.** 1969: On the theory of screening for chronic diseases. *Biometrika* 56, 601–13.

seamless Phase II/III trials Traditional drug development follows several distinct phases of development through to registration. PHASE I TRIALS are usually followed by PHASE II TRIALS in order to choose the optimal dose, and then after some planning time PHASE III TRIALS are initiated. Although it is highly desirable to reduce the time between Phase II and Phase III there is usually a minimum time that teams want to spend planning Phase III. It should be an objective to reduce this time.

One possible way of reducing this time is to carry out a seamless Phase II/III design. In this type of design the time between Phase II and Phase III is reduced and the two are combined into a single trial. An adaptive seamless trial is one in which the final analysis will use data from patients enrolled before and after the adaptation. The figure gives an illustration of the difference between a traditional Phase II/Phase III approach and a seamless learn/confirm approach.

There are certain considerations that need to be taken into account for a seamless design to be feasible. The most important is the time a patient needs to be followed to reach the ENDPOINT, which is to be used to make the dose selection. If the time to reach the endpoint is short in relation to patient recruitment then enrolment can continue while the decision is made and the number of overrunning patients, i.e. those randomised to dose not taken forward, will be minimised. However, if this time is long then the number of overrunning patients will be more significant and a seamless study less applicable.

It is also advisable to use a well-established endpoint or surrogate marker (see SURROGATE ENDPOINTS) when

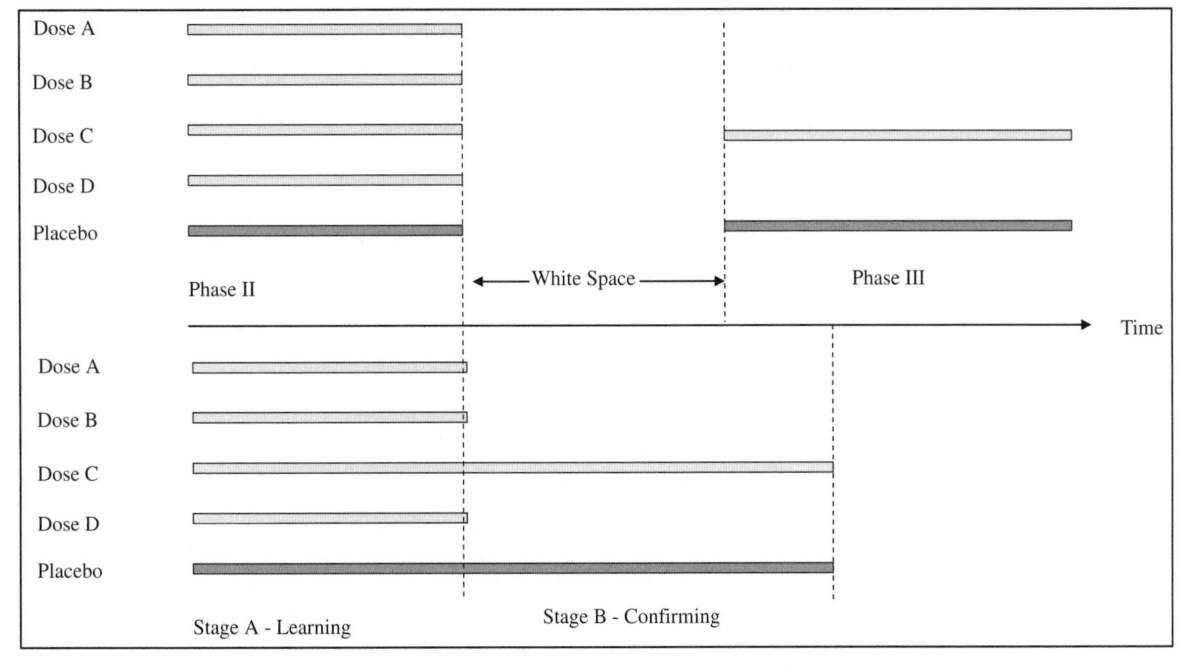

seamless Phase II/III trials *Comparison of the traditional Phase II/Phase III approach (top panel) and seamless learning/confirming (bottom panel)*

implementing a seamless design. Where the goal of Phase II is to establish an endpoint for Phase III, it is likely that a seamless trial will be accepted.

There are also some logistical considerations that need to be taken into account, particularly around the drug supply and drug packaging. To this extent drug development programmes that do not have complicated or expensive regimens are more suited to seamless designs.

Regulatory agencies are also likely to have many questions around the use of a seamless study and are likely to require a second confirmatory study. It is unlikely that a seamless Phase II/III study will be accepted as a single confirmatory study.

One other consideration is that of maintaining the blind until all data are frozen at the end of the Phase III part. This then requires the use of an independent data monitoring committee to make the decision to switch to Phase III.

There are many methods of analysis that can be used in a seamless study. All methods must be seen to control the overall TYPE I ERROR, as this is a paramount requirement for the regulatory agencies and is not negotiable for a confirmatory study.

Todd and Stallard (2005) consider a group sequential method (see INTERIM ANALYSIS) that incorporates a treatment selection based on a short-term endpoint, followed by a confirmatory phase that uses a longer term endpoint. Bauer and Kieser (1999) consider the use of *P*-VALUE combination tests in a seamless trial by combining the information before and after the adaptation. Inoue, Thall and Berry (2002) and Schmidli, Bretz and Racine-Poon (2007) consider Bayesian decision rules to decide which dose to take forward to the confirmatory phase. However, while Inoue, Thall and Berry (2002) use BAYESIAN METHODS in both the Phase II and Phase III parts of the study, Schmidli, Bretz and Racine-Poon (2007) use Bayesian methods for dose selection but traditional frequentist methods in Phase III, using a combination test to control the Type I error. Whichever method is used, simulations need to be carried out to understand the operating characteristics of the design. Submission of these designs to regulatory agencies would require these simulations. A more thorough coverage of the considerations, operational aspects and examples are given in Maca *et al.* (2006).

Sponsor representation in seamless designs is controversial, but in a seamless adaptive design there can be more motivation for sponsor participation in the dose decision process. *AB*

Bauer, P. and Kieser, M. 1999: Combining different phases in the development of medical treatments within a single trial. *Statistics in Medicine* 18, 1833–48. **Inoue, L. Y. T., Thall, P. F. and Berry, D. A.** 2002: Seamlessly expanding a randomized Phase II trial to Phase III. *Biometrics* 58, 823–31. **Maca, J., Bhattacharya, S., Dragalin, V., Gallo, P. and Krams, M.** 2006: Adaptive seamless Phase II/III designs – background, operational aspects and examples. *Drug Information Journal* 40, 463–73. **Schmidli, H., Bretz, F. and Racine-Poon, A.** 2007: Bayesian predictive power for interim adaptation in seamless Phase II/III trials where the endpoint is survival up to some specified timepoint. *Statistics in Medicine* 26, 4925–38. **Todd, S. and Stallard, N.** 2005: A new clinical trial design combining Phases II and III: sequential designs with treatment selection and a change of endpoint. *Drug Information Journal* 39, 109–18.

secondary endpoints See ENDPOINTS

segregation analysis The observation of characteristic segregation ratios among the offspring of particular parental crosses was first made by the Austrian monk Gregor Mendel (1822–1884) in his experiments on the garden pea. These observations enabled him to formulate a theory of genetic transmission from parent to offspring.

Mendel studied discrete traits (called PHENOTYPES) in the garden pea (e.g. smooth versus wrinkled seed) and, after many generations of inbreeding, obtained pure lines (with uniform phenotype, e.g. all having smooth seeds, over many generations) for each trait. When two pure lines with different phenotypes (e.g. smooth and wrinkled seeds) are crossed, all the offspring (called the *F1 generation*) were of the same phenotype (e.g. all smooth). The trait that is uniformly present in the F1 generation is said to be *dominant*, while the absent alternative is said to be *recessive*. When F1 individuals are crossed with the recessive pure line (which is called a 'back-cross'), half the offspring had the dominant phenotype and the other half the recessive phenotype. When two F1 individuals were crossed (which is called an 'inter-cross'), three-quarters of the offspring had the dominant phenotype and one-quarter the recessive phenotype. These characteristic 1:1 and 3:1 ratios are called *segregation ratios*.

Segregation ratios are explained by the fact that each individual receives a complete set of genes from both parents, so that each gene is present in duplicate. When there are different forms of the same gene, each form (or *allele*) may correspond to a different phenotype, but when an individual has two different alleles (i.e. is *heterozygous*), the phenotype of one of the alleles (the recessive allele) is completely masked by the phenotype of the other allele (the dominant allele). Thus the F1 generation from two different pure (i.e. *homozygous*) lines will be all heterozygous and therefore display the dominant phenotype. A back-cross will result in half the offspring having the heterozygous genotype and the other half having the homozygous recessive genotype. An inter-cross will result in half the offspring being heterozygote, one-quarter being homozygous dominant and one-quarter homozygous recessive.

Classical segregation analysis is the examination of the offspring of different mating types to see if Mendelian segregation ratios are present. When such ratios are observed, the inference is made that the phenotype in question is determined by a single underlying genetic locus. Complex segregation analysis is a further development of this method for traits in which Mendelian segregation ratios may be masked by complexities such as the involvement of background genetic or environmental factors in addition to a locus of major effect. *PS*

[See also ALLELICASSOCIATION, GENETIC EPIDEMIOLOGY, GENETIC LINKAGE, GENOTYPE, PHENOTYPE]

selection bias Selection bias occurs when there are systematic differences between those who are selected for study and those who are not selected, so that the selected sample is not representative of the target population. For example, in a survey of the smoking habits of 14 year olds, a convenient sampling frame would be children attending schools in a defined geographical area. However, not all 14 year olds will be included in this sampling frame and if the reasons for exclusion are associated with the smoking habit a biased estimate of the prevalence of smoking will be obtained. Another area where the choice of sampling frame might lead to BIAS is in telephone sampling, where households without telephones would be systematically excluded.

Even when an appropriate sampling frame is used for a survey, nonrandom sampling can lead to biased estimates. For example, in a study of overcrowding, an appropriate sampling frame might be all households in an electoral ward or postcode sector, listed in order of postal address. However, a systematic sample of every eighth household might over-represent certain types of accommodation, such as flats on a particular floor (e.g. ground floor or top floor) in tenement blocks of eight. If the average number of people per household differs systematically between floors, this is likely to lead to a biased estimate of overcrowding. Ideally, probability sampling methods should be used to avoid selection bias in surveys (see SAMPLING METHODS – AN OVERVIEW).

One type of study that is almost never carried out on a random sample of the target population is a randomised CLINICAL TRIAL. Trials rely on random allocation to treatment groups for their internal validity but, because of tight eligibility criteria for patient selection, those in the trial may not be representative of all patients with the condition being treated.

Epidemiological studies, especially CASE-CONTROL STUDIES, are susceptible to selection bias. In case-control studies it can be extremely difficult to obtain a control group that is representative of all noncases in the same target population that the cases arise from. This can result in biased estimates of the ODDS RATIO in either direction, depending on the form of selection bias. These issues are discussed in detail in Sackett (1979) and Ellenberg (1994). Even in carefully designed OBSERVATIONAL STUDIES it can be difficult or impossible to rule out selection bias as a possible explanation for an observed association (Boydell *et al.*, 2001).

Selection bias can occur in many other contexts. For example, Kho *et al.* (2009) describe how it can result from the requirement for written informed consent in studies of medical records. W*HG*

Boydell, J., van Os, J., McKenzie, K. *et al.* 2001: Incidence of schizophrenia in ethnic minorities in London: ecological study into interactions with environment. *British Medical Journal* 323, 1336–8. **Ellenberg, J. H.** 1994: Selection bias in observational and experimental studies. *Statistics in Medicine* 13, 557–67. **Kho, M. E., Duffett, M., Willison, D. J. *et al.*** 2009: Written informed consent and selection bias in observational studies using medical records: systematic review. *British Medical Journal* 338, b866, DOI: 10.1136/bmj.b866. **Sackett, D. L.** 1979: Bias in analytic research. *Journal of Chronic Diseases* 32, 51–63.

sensitivity This is a measure of how well an alternative test performs when it is compared with the reference or 'gold' standard test for the diagnosis of a condition. Sensitivity is the proportion of patients who are correctly identified by the test as having the condition out of all patients who have the condition. Sensitivity may also be expressed as a percentage and is the counterpart to SPECIFICITY.

The reference standard may be the best available diagnostic test or may be a combination of diagnostic methods, including following up patients until all with the disease have presented with clinical symptoms. For example, in a study of mammography, the reference standard for breast cancer would include all women who went on to develop breast cancer, whether they were first diagnosed radiologically, histologically or symptomatically. Thus, the best design when a diagnostic test is evaluated against a reference standard is a COHORT STUDY with complete follow-up.

When the data are set out as in the table:

$$\text{Sensitivity} = \frac{a}{a + c}$$

sensitivity *General table of test results among $a + b + c + d$ individuals sampled*

		Disease		
		Present	*Absent*	*Total*
Test	Positive	a	b	$a + b$
	Negative	c	d	$c + d$
	Total	$a + c$	$b + d$	$a + b + c + d$

Sensitivity should be presented with CONFIDENCE INTERVALS, typically set at 95%, calculated using an appropriate method such as that of Wilson (described in Altman *et al.*, 2000), which will produce asymmetric confidence intervals without impossible values, i.e. that will not give values for the upper confidence interval >1 when sensitivity approaches 1 and the sample size is small.

Where a test result is a continuous measurement, e.g. liver enzymes in serum, a cut-off point for abnormal values is chosen. If a lower value is chosen, then sensitivity will be relatively high, but specificity relatively low.

The impact of all possible cut-off points can be displayed graphically in a RECEIVER OPERATING CHARACTERISTIC (ROC) CURVE by plotting sensitivity at each cut-off point on the y axis against 1 – specificity at each cut-off point on the x axis. The choice of cut-off point is not, however, solely a statistical decision, as the balance between the FALSE POSITIVE RATE and the FALSE NEGATIVE RATE should be related to the clinical context and consequences of wrong diagnosis for the patient and healthcare system.

A sample size calculation for sensitivity can be made by specifying a confidence interval (e.g. 95%) and an acceptable width for the lower bound of the confidence interval. Where the anticipated sensitivity is high and the sample size small, a 'small sample' method should be used: a sample size table can be found in Machin *et al.* (1997). *CLC*

[See also LIKELIHOOD RATIO, NEGATIVE PREDICTIVE VALUE, POSITIVE PREDICTIVE VALUE, TRUE POSITIVE RATE]

Altman, D. G., Machin, D., Bryant, T. N. and Gardner, M. J. 2000: *Statistics with confidence*, 2nd edition. London: BMJ Books.
Machin, D., Campbell, M., Fayers, P. and Pinol, A. 1997: *Sample size tables for clinical studies*, 2nd edition. Oxford: Blackwell Sciences Ltd.

sequential analysis A method allowing hypothesis tests to be conducted on a number of occasions as the data accumulate through the course of a CLINICAL TRIAL. A trial monitored in this way is usually called a sequential trial. This approach is in contrast to the use of a standard fixed sample size trial design, in which a single hypothesis test is conducted at the end of a trial, usually when some specified sample size has been attained, with no allowance to collect further data and repeat the test. Sequential analysis methods are attractive in clinical trials since, for ethical reasons, it is often important to analyse the data as they accumulate and to stop the study as soon as the presence or absence of a treatment effect is indicated sufficiently clearly. Although the total sample size for a sequential trial is not fixed in advance – it depends on the observed data – an additional advantage of sequential methodology is that trials may be constructed so that the expected sample size is smaller than that for a fixed sample size trial with the same Type I error rate and POWER.

Suppose that in a clinical trial we wish to compare two groups of patients, with one receiving the experimental treatment and the other the control treatment. Formally, we define some measure of the treatment difference between the experimental and control groups, which we will denote by θ. This treatment difference may, for example, be measured by the difference between the MEAN response for a normally distributed ENDPOINT, the log-odds ratio for a binary endpoint or the log-hazard ratio for a survival time endpoint. We generally wish to test the NULL HYPOTHESIS that there is no difference between the treatment groups, i.e. that $\theta = 0$.

In a standard fixed sample size test, some test statistic is obtained and compared with a critical value. The critical value is chosen so as to give a specified Type I error rate, i.e. to ensure that the risk of concluding that there is a treatment difference when, in fact, the treatments are identical is controlled, usually to be no more than 5%. If this standard hypothesis test is repeated at a number of INTERIM ANALYSES, there are a number of opportunities to conclude that the treatments are different. The risk of doing so on at least one occasion therefore increases above 5%, so that the overall Type I error rate thus exceeds 5% and a valid test is no longer provided. This problem is addressed by sequential analysis, in which the repeated hypothesis tests are conducted in such a way as to maintain an overall Type I error rate for the sequential trial as a whole.

Although sequential monitoring methods have been proposed based on a range of possible test statistics (see, for example, Jennison and Turnbull, 2000, for a discussion of possible methods) a general sequential approach is based on the use of the efficient score statistic (see Whitehead, 1997), as a measure of the treatment difference. Large positive values correspond to an indication of superiority of the experimental treatment, large negative values to an indication of superiority of the control treatment, while values close to zero indicate little difference between the treatments. The exact form of the score statistic depends on the type of data used and the way in which the treatment difference is measured. As an example, for binary data, with the treatment difference measured by the log-odds ratio, if equal numbers of patients have received the experimental and control treatments, the score statistic is half of the difference in observed numbers of successes on the experimental and control arms. For survival data, with the treatment difference measured by the log-hazard ratio, the score statistic is the log-rank statistic (see SURVIVAL ANALYSIS).

In a sequential trial, a number of interim analyses are conducted. The value of the score statistic is calculated at each interim analysis together with the observed Fisher's information, a quantity related to the sample size summarising the amount of information available. If, at any interim analysis, the value of the score statistic is sufficiently large, the trial is stopped and it is concluded that the experimental treatment is superior to the control treatment. If the score statistic is too small, the trial is stopped and, depending on the way in which the test is constructed, it may either be concluded that the experimental treatment is inferior to the control or that there is insufficient evidence to distinguish between the two treatments. If neither criterion is met, that is for intermediate values of the score statistics, the trial continues to the next interim analysis. Graphically, the observed values of the score statistic may be plotted against the values of the information. As the information available increases throughout the trial the plotted points form what is called a *sample path*. At each interim analysis, the sample path is compared with upper and lower critical values, with the trial stopped as soon as the score statistic lies either above the upper critical value or below the lower critical value. The critical values, which, in general, take different values at the different interim analyses, thus define a continuation region. As already explained, the problem of sequential analysis is the calculation of the critical values so as to give a specified Type I error rate, for example, of 5%.

As the choice of critical values to achieve this aim is not unique, problems of appropriate choices for use in a sequential clinical trial setting are also of interest. In particular, in contrast to fixed sample size hypothesis tests, asymmetric sequential methods are possible. A fixed sample size test that is designed to have specified power, say 90%, to detect a treatment effect of given size, say $\theta = \theta_1$, has equal power to detect the opposite treatment effect of the same magnitude, i. e. $\theta = -\theta_1$. A sequential test may be constructed to have power 0.9 to detect $\theta = \theta_1$, but lower power to detect $\theta = -\theta_1$. Such a sequential test may have a smaller expected sample size when $\theta = -\theta_1$ than when $\theta = \theta_1$. This is sometimes desirable in clinical trials, when it is advantageous to stop a trial as soon as possible if the experimental treatment appears to be inferior to the control and there is no desire to continue recruiting patients to test whether or not this inferiority is statistically significant.

The method based on the score statistic is a very flexible one, since, as shown by Scharfstein, Tsiatis and Robins (1997) for a wide range of problems, conditional on the observed information values, the score statistics at the interim analyses are approximately normally distributed. This means that critical values can be obtained based on this normality to provide sequential tests that can be used for many different types of data and choices of measure for the treatment difference. Two distinct approaches to the calculation of the critical values with which the efficient score statistics are compared have been developed. The first, which is sometimes called the *boundaries approach*, is based on modelling a continuous sample path. The second uses the assumed normality of the score statistics directly, evaluating the critical values via a recursive numerical integration technique, with the form of the sequential test often specified by what is called a *spending function*. The two approaches are described in detail later. A more general approach, the adaptive design method, which is not based on the asymptotic normality of the score statistics, is also briefly described. After a brief example, the problem of analysis at the end of a sequential trial is then discussed. We then continue with a description of the related area of response-driven designs and end with some comments on the role of a DATA AND SAFETY MONITORING COMMITTEE in a sequential clinical trial.

In the boundaries approach, the approximate NORMAL DISTRIBUTION of the score statistics evaluated at the interim analyses means that the observed values can be considered as points on a Brownian motion with drift equal to the treatment difference, observed at times given by the observed information. This has led to the consideration of the abstract concept of continuous monitoring, in which the value of the test statistic is taken to be observed at all times rather than at the discrete times given by the interim analyses. The plotted sample path thus forms a continuous line, which is compared with continuous boundaries, which may be expressed as functions of the information level. Many of the theoretical developments in sequential analysis have been based on consideration of this problem. A consequence of this formulation is that, since the sample path is considered to be continuous, the trial stops exactly on a boundary, whereas for a discretely monitored trial, there is some overshoot of the critical value when the trial stops.

The boundaries approach stems from the work of Wald (1947) who developed the sequential probability ratio test (SPRT) for the testing of armaments during the Second World War. In Wald's SPRT after each observation, the LIKELIHOOD RATIO for the simple alternative hypothesis relative to the null hypothesis is calculated and the test continues so long as this likelihood ratio falls within some fixed range, equivalent to the plotted values of the score statistic lying between two parallel straight boundaries. Wald derived stopping limits so as to give a test with a specified Type I error rate and power under the assumption of continuous monitoring. Among all tests with the same properties, the SPRT minimises the expected sample size when either the null or alternative hypothesis holds. However, the parallel boundaries give a test that, although it terminates with probability 1, has no finite maximum sample size. This feature makes it unsuitable for many clinical trials.

Following the work of Wald, a number of alternative forms for boundaries that maintain the overall Type I error rate have been proposed. Whitehead (1997) describes a wide range of such tests. One form that is particularly commonly used in sequential clinical trials is the *triangular test*. This test has straight boundaries that form a triangular-shaped continuation region. The test approximately minimises the maximum expected sample size among all tests with the same error rates and has a high probability of stopping with a sample size below that of the equivalent fixed sample size test.

The critical values obtained using the boundaries approach maintain the overall Type I error rate for a continuously monitored test. In practice, monitoring is necessarily discrete, since even if an interim analysis is conducted after observation of each patient, the information will increase in small steps. This means that if the critical values from the boundaries approach are used, the Type I error rate will be less than the planned level of, for example, 5%. Whitehead (1997) has proposed a correction to modify the continuous boundaries to allow for the discretely monitored sample path. This correction brings in the critical values by an amount equal to the expected overshoot of the discrete sample path. The correction is particularly accurate for the triangular test.

In general, specialist software is needed for the construction of critical values using the boundaries approach. A commercially available software package, Planning and Evaluation of Sequential Trials (PEST), is available from Medical and Pharmaceutical Statistics Research Unit at Lancaster University for the calculation of the boundaries.

An alternative approach was a recursive numerical integration method for calculation of the overall Type I error rate for a sequential trial with specified critical values under the assumption that the score statistics observed at the interim analyses are normally distributed (Armitage, McPherson and Rowe, 1969). As well as demonstrating the effect of conducting interim analyses without adjusting for MULTIPLE TESTING, this method allows the construction of critical values to maintain an overall Type I error rate of, say, 5%. Using this approach, Pocock (1977) and O'Brien and Fleming (1979) calculated critical values for sequential tests that preserve the overall Type I error rate to be 5% when, for O'Brien and Fleming's design, the critical values with which the score statistics are compared are the same at each interim analysis, and, for Pocock's design, the critical values correspond to the same *P*-VALUE for a conventional analysis performed at each interim analysis. The critical values obtained were tabulated to allow easy implementation without the need for additional computation. Although these methods, particularly that proposed by O'Brien and Fleming, remain in use, they are not always the most appropriate designs in the clinical trial setting. Pocock's design has been criticised because it has a relatively high chance of leading to rejection of the null hypothesis very early in the trial. O'Brien and Fleming's design, in contrast, is unlikely to stop early in the trial unless there is very strong evidence of a treatment difference. If the two treatments are very similar, both designs are likely to lead to a trial requiring more patients than the equivalent fixed sample size trial.

A more flexible design approach is provided by the spending function method proposed by Lan and DeMets (1983). In this approach, the total overall Type I error rate of, say, 5% is considered to be spent through the course of the trial, with the rate at which it is spent controlled by the specified spending function. Not only does this introduce flexibility in the choice of the shape of the stopping boundaries but it also, in contrast to the tests of Pocock and O'Brien and Fleming, allows construction of a test that maintains the Type I error rate if interim analyses are not taken at the planned times. Many forms can be used for the spending function, but families of functions to give tests with certain properties have been proposed. A thorough review of the approach is given by Jennison and Turnbull (2000). As with the boundaries approach, specialist software is required to calculate the critical values. The software package EAST produced by Cytel Software Corporation and the S-PLUS module SeqTrial produced by MathSoft perform the necessary calculations.

An alternative to the sequential design approaches based on the assumption of normality for score statistics just described is the adaptive design approach described by Bauer and Köhne (1994). Although the ideas can be extended to trials with greater numbers of stages, Bauer and Köhne focus on a two-stage design and assume that the data from each stage are independent of those from the other stage. Suppose that a standard hypothesis test of the null hypothesis that there is no treatment difference is conducted based on the data obtained from each stage, leading to two *P*-values, p_1 and p_2. A result of Fisher cited by Bauer and Köhne shows that, if there is no treatment difference, $-2\log(p_1p_2)$ follows a CHI-SQUARE DISTRIBUTION on 4 DEGREES OF FREEDOM, allowing the data from the two stages to be combined in a single test.

The fact that the only assumption made is the independence of data from the two stages means that this approach has great flexibility, enabling changes to many features of the trial design without invalidating the final test. The most common change discussed is modification of the sample size of the second stage based on the predicted power of the trial at the end of the first stage, but possible changes go far beyond this to include changes of the endpoint being measured and the null hypothesis being tested. The adaptive design approach has been criticised, however, for the fact that the test statistic, $-2\log(p_1p_2)$, is not a sufficient statistic for the treatment difference. This leads to a lack of power for the test, so that, if the flexibility of the adaptive design is not utilised, a sequential test based on the boundaries approach or the spending function method can be found that is as powerful and has smaller expected sample size.

As an example of a sequential analysis, the figure (page 423) shows results from the analysis of a small trial to assess the efficacy of Viagra in men suffering erectile dysfunction as a result of spinal cord injury. Eligible men with a regular female partner, who were attending clinics in Southport, Belfast and Stoke Mandeville, were randomised between Viagra and a matching PLACEBO pill. After four weeks they were asked whether the treatment received had improved their erections. The trial was designed using the boundaries approach with the triangular test being chosen as an appropriate design. The solid lines on the figure illustrate the continuation region for this test when the efficient score statistic, Z, is plotted against the observed Fisher's information, V. The trial continues until the values of Z and V lead to a point outside this triangular region.

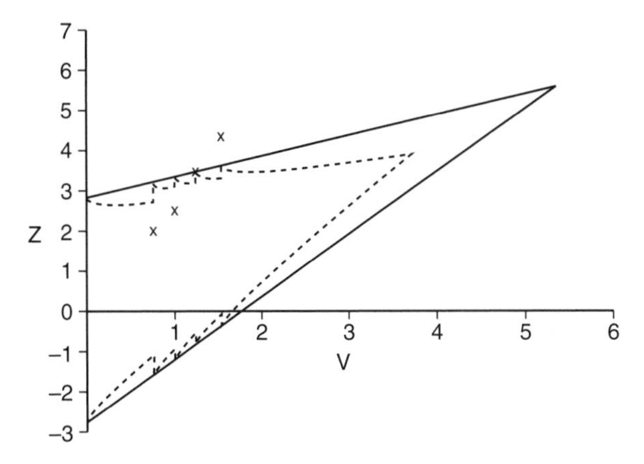

sequential analysis *Continuation region and sample path for a clinical trial of Viagra in men with spinal cord injury*

At the first interim analysis, 12 men had completed four weeks' treatment with 5/6 on Viagra and 1/6 on placebo reporting improvement. The first plotted point on the figure represents these data. To allow for the fact that the trial is not monitored continuously, the boundaries are adjusted using the so-called Christmas tree correction, so that the plotted point is compared with the inner dotted boundaries shown. As the point is between these boundaries, recruitment to the trial continued. At the third interim analysis, the observed improvement rates were 8/10 on Viagra and 1/10 on placebo. On the basis of these data, the upper boundary was reached and recruitment closed. When the results on the 6 men under treatment were added, the improvement rates became 9/12 and 1/14 respectively, leading to the fourth plotted point. The design allowed a strong positive conclusion to be drawn after only 26 men had been treated. This is in comparison with a conventional fixed sample size trial, for which 57 subjects, more than twice as many, would have been required for a design of the same power.

The methods described earlier mainly lead to tests conducted at a number of interim analyses of the data obtained in a clinical trial, with the possibility of stopping the trial as soon as sufficient evidence of a treatment effect, or the lack or such an effect, is obtained. Much more general methods can be envisaged in which many aspects of a clinical trial design may be reconsidered following an interim analysis. Such methods are sometimes referred to as DATA-DEPENDENT DESIGNS, or response-driven designs, or, rather confusingly given that the same terms are used for the different approaches described here, as either sequential designs or adaptive designs. A review of response-driven design methods is given by Rosenberger (1996).

A simple response-driven design is the play-the-winner design for a clinical trial comparing two treatments on the basis of a success/failure endpoint. The purpose of this design is to replace the random allocation of treatment to patients with a method that leads to more patients receiving the superior treatment. Since, of course, it is not known at the beginning of the trial which treatment is superior, the first patient may be assigned to a treatment at random. If a success is observed from the treatment of this patient, the next patient receives the same treatment. If a failure is observed, the next patient receives the other treatment. Each subsequent patient receives the same treatment as the previous one if a success was observed and the other treatment if a failure was observed. In practice, the simple play-the-winner rule is generally modified to include some random element. Several other rules with different properties, but with the common aim of assigning more patients to the most successful treatment arm, have also been suggested.

Response-driven designs have found most use in early-phase clinical trials for dose finding. Here, the dose of the experimental treatment that is to be given to each patient in the trial is determined depending on the responses from patients treated earlier. Often the aim is to determine a dose that leads to a certain proportion of patients experiencing some event; in trials in oncology, for example, toxicity rates of 20% are often considered optimal. The use of response-driven designs in such trials means that the optimal dose can be efficiently estimated without exposing patients to large doses that may be highly toxic.

The use of a sequential stopping rule in a clinical trial means that many of the standard analysis methods are no longer appropriate. Suppose that a sequential trial has stopped at some interim analysis with the test statistic exceeding the upper critical value, i.e. with the conclusion that the experimental treatment is superior to the control treatment. The trial has stopped precisely because of the large observed value of the random score statistic. This means that a standard unbiased estimate of the treatment difference based on the observed value of the test statistic, e.g. the common MAXIMUM LIKELIHOOD ESTIMATION, will, on average, overestimate the true value of the treatment difference. The P-value from a standard analysis will, in a similar way, on average be too small; i.e. it will overstate the evidence against the null hypothesis. Special methods of analysis allowing for the sequential monitoring have been developed. These are described in detail by Whitehead (1997) and Jennison and Turnbull (2000) and are implemented in the software packages PEST, EaSt and SeqTrial.

In large-scale clinical trials, monitoring of accumulating data is commonly undertaken by an independent data and safety monitoring committee (DMC). The primary role of such a committee is to ensure the safety of patients recruited to the trial. It is therefore natural, in a sequential clinical trial, that the DMC should be involved in the interim analyses conducted to assess the treatment difference and in decisions of whether or not the study should be stopped. The involvement of a DMC, the use of a carefully chosen sequential stopping rule, approved by the DMC before the start of the study, and a final analysis that allows for the sequential monitoring provide a clinical trial that can be stopped when appropriate without compromising the statistical integrity of the results obtained. *NS*

Armitage, P., McPherson, C. K. and Rowe, B. C. 1969: Repeated significance tests on accumulating data. *Journal of the Royal Statistical Society, Series A* 132, 235–44. **Bauer, P. and Köhne, K.** 1994: Evaluation of experiments with adaptive interim analyses. *Biometrics* 50, 1029–41. **Jennison, C. and Turnbull, B. W.** 2000: *Group sequential methods with applications to clinical trials*. Boca Raton: Chapman & Hall/CRC. **Lan, K. K. G. and DeMets, D. L.** 1983: Discrete sequential boundaries for clinical trials. *Biometrika* 70, 659–63. **O'Brien, P. C. and Fleming, T. R.** 1979: A multiple testing procedure for clinical trials. *Biometrics* 35, 549–56. **Pocock, S. J.** 1977: Group sequential methods in the design and analysis of clinical trials. *Biometrika* 64, 191–9. **Rosenberger, W. F.** 1996: New directions in adaptive designs. *Statistical Science* 11, 137–49. **Scharfstein, D. O., Tsiatis, A. A. and Robins, J. M.** 1997: Semiparametric efficiency and its implications on the design and analysis of group-sequential studies. *Journal of the American Statistical Association* 92, 1342–50. **Wald, A.** 1947: *Sequential analysis*. New York: John Wiley & Sons. **Whitehead, J.** 1997: *The design and analysis of sequential clinical trials*. Chichester: John Wiley & Sons, Ltd.

SeqTrial See SEQUENTIAL ANALYSIS

sequential probability ratio test (SPRT) See SEQUENTIAL ANALYSIS

shrinkage See MULTILEVEL MODELS

sign test This is one of the oldest nonparametric methods and one of the most simple. It is so named because it uses the sign of the differences rather than their magnitude and is therefore less sensitive than the WILCOXON SIGNED RANK TEST. It can be used for two samples that are matched or paired with a NULL HYPOTHESIS that the MEDIANS are not different between the two groups. Alternatively, it can be used in the one sample case to compare to a particular value, e.g. the median, where the null hypothesis is that the group median is not different to the proposed median. It is a nonparametric version of both the paired and the ONE-SAMPLE *t*-TEST.

For the two-sample case, find the sign of the difference between the two values in the pair. Calculate N, the number of differences showing a sign. For the one-sample case, find the sign of the difference between each subject's value and the value of interest. Calculate N, the number of observations that are different to the value of interest. Then for both cases let x be the number of fewer signs, $x = \min(+s, -s)$, and compare x to the critical region of the BINOMIAL DISTRIBUTION, N, 1/2. Reject the null hypothesis if x is less than or equal to the critical value.

As part of a study, the general health section of the SF-36 was collected. The subject's values (shown in the first table) are to be compared to the expected value of 72 within the population. There are 5 plus signs, 9 minus signs and 1 tie; therefore $x = 5$ and $N = 14$. From the tables of the binomial distribution ($N = 14$, $p = 1/2$) the critical value is 3. As 5 is greater than 3 there is insufficient evidence to reject the null hypothesis, so it is concluded that this group's general health scores are not different from those expected in the population.

General health scores were collected on this group of subjects at a second time point; the scores at this time point are shown in the second table. This time, there are 7 plus signs, 8 minuses and *no* ties. Therefore $x = 7$ and $N = 15$. Compare $x = 7$ to the critical value of the binomial distribution 15, $1/2$. This value is 3; as 7 is greater than 3 there is insufficient evidence to reject the null hypothesis. Therefore the general health scores are not different at the two time points. For further details see Pett (1997) and Siegel and Castellan (1998). *SLV*

Pett, M. A. 1997: *Nonparametric statistics for health care research*. Thousand Oaks: Sage. **Siegel, S. and Castellan, N. J.** 1998: *Nonparametric statistics for the behavioral sciences*, 2nd edition. New York: McGraw-Hill.

sign test *Subject's values in the general health section of the SF-36, using signs from the sign test*

GH value	60	55	75	100	55	60	50	60	72	40	90	75	70	75	55
Sign	−	−	+	+	−	−	−	−	=	−	+	+	−	+	−

sign test *Second recording of subject's values in the general health section of the SF-36*

Time 1	60	55	75	100	55	60	50	60	72	40	90	75	70	75	55
Time 2	40	45	100	50	70	95	95	65	85	55	70	45	75	65	50
Sign	+	+	−	+	−	−	−	−	−	−	+	+	−	+	+

significance tests and significance levels

Significance tests were introduced by R. A. Fisher (1925) as a means of assessing the evidence against a NULL HYPOTHESIS. Often, such a null hypothesis states that there is no association between two variables: e.g. between hypertension and subsequent heart disease. Significance tests are conducted by calculating the *P*-VALUE, defined as the PROBABILITY, if the null hypothesis were true, that we would have observed an association as large as we did by chance. The term significance level is sometimes used as a synonym for the *P*-value. If the *P*-value is small we have evidence *against* the null hypothesis: the entry on *P*-values describes their calculation and interpretation in more detail. Fisher suggested that if the *P*-value is sufficiently small then the result of the test should be regarded as providing evidence against the null hypothesis. He advocated that a conventional line be drawn at 5% significance (although he rejected fixed rules) and described results of experiments in which the *P*-value was sufficiently small as *statistically significant*. Sterne and Davey Smith (2001) have argued that in situations typical of modern medical research, *P*-values of around 0.05 provide only modest evidence against the null hypothesis.

A different use of the phrase significance level arises from the hypothesis testing approach to the interpretation of experiments advocated by Neyman and Pearson (1933), who showed how to find optimal rules that would minimise the TYPE I and TYPE II ERROR rates over a series of many experiments. We make a Type I error if we reject the null hypothesis when it is in fact true, while we make a Type II error if we accept the null hypothesis when it is, in fact, false (see HYPOTHESIS TESTS).

The Type I error rate, usually denoted as α, is closely related to the *P*-value since if, for example, the Type I error rate is fixed at 5%, then we will reject the null hypothesis when $P < 0.05$. Therefore, researchers using the Neyman–Pearson approach often report simply that the *P*-value for their test was less than their chosen *significance level*. There is, however, an important distinction between the use of the term significance level to refer to the evidence against the null hypothesis provided by a particular experiment (Fisher's approach) and the choice of a fixed significance level that, together with the Type II error rate, will be used to determine our behaviour with regard to the results. Goodman (1999) discusses the confusion caused by the failure to appreciate this distinction in more detail. *JS*

Fisher, R. A. 1925: *Statistical methods for research workers*. Edinburgh: Oliver & Boyd. **Goodman, S. N.** 1999: Toward evidence-based medical statistics. 1: The *P*-value fallacy. *Annals of Internal Medicine* 130, 995–1004. **Neyman, J. and Pearson, E.** 1933: On the problem of the most efficient tests of statistical hypotheses. *Philosophical Transactions of the Royal Society, Series A* 231, 289–337. **Sterne, J. A. and Davey Smith, G.** 2001: Sifting the evidence – what's wrong with significance tests? *British Medical Journal* 322, 226–31.

simple random sample

This is the most basic sampling technique. It is where a smaller group, a sample, is chosen by chance from a population. Each member of the population has an equal and known probability of being chosen to be in the sample. Each sample of a given size also has an equal probability of being chosen from the population. Sampling is usually done without replacement, so that each member of the population can only be selected for inclusion in the sample once.

To choose a random sample, first a list is needed of every member of the population to be sampled: this is the sampling *frame*. Each member of this list is then assigned a number from 1 to *N* (where *N* is the total size of the population) in any order. Each member of the sample then has a probability of 1/*N* of being in the sample. A random number generator, or table, is then used to select a random number. The member of the population assigned that number is then selected to be included in the sample. This process is repeated until a sample of the required size is obtained.

For example, suppose that a survey of doctors' opinions is to be carried out. There are 500 doctors in a hospital and a 10% sample is to be collected. First, the sampling frame needs to be obtained – a list of all doctors in the hospital. Next, each doctor is assigned a number from 1 to 500, e.g. in alphabetical order or the order on the list. Now look at a random number table, which gives the following numbers, say:

28049	11632	68254	14217	44612	05049
16831	13213	76103	07222	31852	43501

Therefore, the sample would include doctors numbered 280, 491, 163, 268, 254, 142, 174, 461, 205, 49, 168, 311, 321, 376, 103, 72, 223, 185, 243, 501. As 501 is outside the range of the numbers assigned it is ignored. Care is needed so as not to ignore leading zeros or else some numbers might be inadvertently overlooked.

The main advantage of this method of sampling is the lack of classification error, as no information needs to be known about items except that they are in the population. It is useful when little is known about the population, only that it is likely to be homogeneous. The main disadvantage is it might not be possible to find the sampling frame. In the example given earlier, there might not be a list of all the doctors in the hospital, meaning that a different method would need to be used. For further details see Crawshaw and Chambers (1994). *SLV*

Crawshaw, J. and Chambers, J. 1994: *A concise course in A level statistics*, 3rd edition. Cheltenham: Stanley Thornes Publishers Ltd.

simple randomisation

See RANDOMISATION

Simpson's paradox

This is the observation that a measure of association between two categorical variables may be identical within the levels of a third categorical variable, but can take on an entirely different value when the variable is disregarded and the association measure calculated from the pooled data.

As an example consider the three-way contingency table shown in the table. Infants born in two clinics during a certain time period were categorised according to survival and amount of pre-natal care received.

Simpson's paradox Three-way classification of infant survival and amount of pre-natal care in two clinics, taken from Everitt (1992)

		Infant survival	
Clinic	*Amount of pre-natal care*	*Died*	*Survived*
A	Less	3	176
	More	4	293
B	Less	17	197
	More	2	23

Calculated within clinics, the odds of survival vary little between the two pre-natal care groups (ODDS RATIO (OR) for survival, comparing higher with lower amount of care, clinic A: OR = 1.25, clinic B: OR = 0.99) and the corresponding CHI-SQUARED TESTS of independence of survival and amount of care do not reach significance. If, however, the data are collapsed over clinics, the odds ratio becomes OR = 2.82 and is statistically significant according to a chi-squared test, and the conclusion would be that the amount of care and survival are related.

Such a situation occurs when the third variable is associated with both the other variables and, therefore, confounds the association between the variables of interest. Here, relatively more pre-natal care is given in clinic A and the survival percentage is also higher in clinic A than B. Therefore, to some extent the pooled measure of ASSOCIATION between survival and pre-natal care measures both the association with pre-natal care as well as that with clinics.

To take account of the levels of a confounding variable, such as a clinic, a pooled within-level measure of association can be constructed (see MANTEL–HAENSZEL METHODS) or a statistical model can be used to adjust the association of interest for the confounder (see LOGISTIC REGRESSION, LOG-LINEAR MODELS). *SL*

Everitt, B. S. 1992: *The analysis of contingency tables*, 2nd edition. Boca Raton: Chapman & Hall/CRC.

skewness Data are described as skewed if they have an asymmetric distribution. When the tail of the distribution is on the right-hand side (see part (a) in the first figure) the data have positive or right-hand skew. When the tail of the distribution is on the left-hand side (see part (b) in the first figure) the data have negative or left-hand skew.

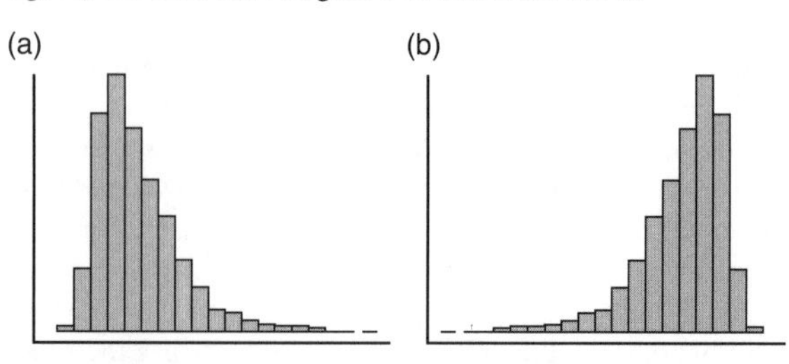

skewness *Example of right-hand and left-hand skew*

Many distributions encountered in analyses of medical data are positively skewed. For example, leptin, a fat-related growth hormone, was measured in umbilical cord blood samples taken from 407 babies born at 37 weeks' gestation or later. The distribution of the results is given in the second figure; the data are positively skewed since relatively few babies have cord leptin levels above 20 ng/ml.

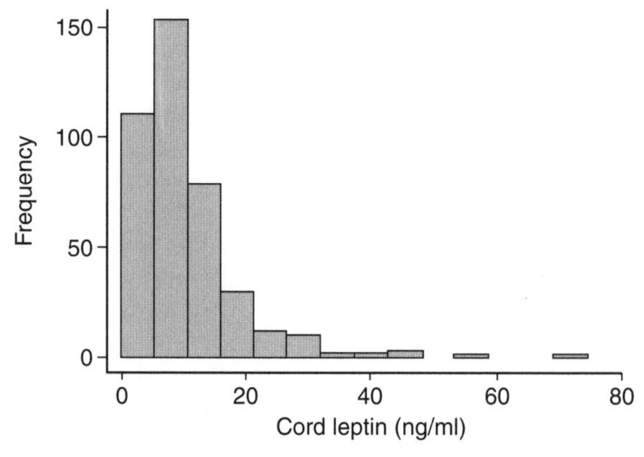

skewness *Recordings of leptin in umbilical cord blood samples*

Analysis of skewed data can proceed either using the RANKS of the data or using transformed values. Analyses using ranks are known as nonparametric or distribution-free methods, because they make no assumptions about the distribution of the data. When describing skewed data using nonparametric methods the MEDIAN is a suitable MEASURE OF LOCATION.

Alternative analysis techniques are based on transformed values. These use parametric methods, which rest on the assumption that the data have a particular distribution, usually a NORMAL DISTRIBUTION. Although skewed data do not conform to this assumption, it may be possible to apply a mathematical TRANSFORMATION to the data so that they do. When the data are positively skewed it is often found that the logarithmic (log) transformation is appropriate. If the leptin data are logged they have an approximately normal distribution, as shown in the third figure. When describing skewed data in this situation then the GEOMETRIC MEAN is an appropriate parametric measure of location.

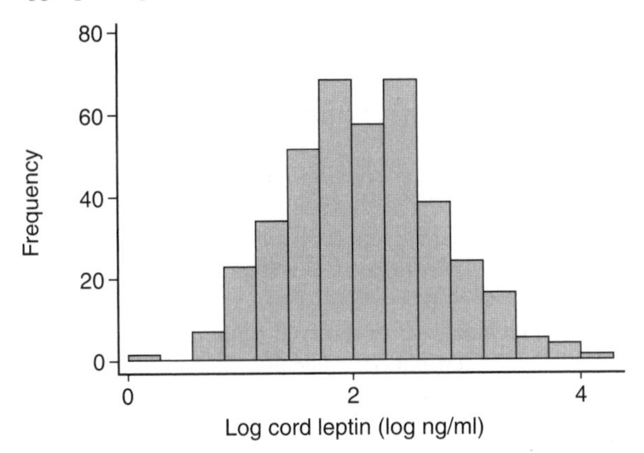

skewness *Logarithmic transformation of leptin readings*

Mathematical measures of skewness can be used to describe distributions. Data with a symmetric distribution, such as the normal distribution, have a skewness of zero. Positive values for skewness indicate a positively skewed distribution whereas negative values for skewness indicate a negative skew. The skewness of the raw cord leptin measurements is 2.7, whereas that of the log-transformed measurements is 0.2, which is considerably closer to zero. *SRC*

software See STATISTICAL PACKAGES

spatial epidemiology This is the analysis of epidemiological or public health data that are geographically referenced. Typically the data arises in two forms: either (a) the residential address of cases of disease are known or (b) arbitrary small areas such as census tracts, zip codes or postcodes have counts of disease observed within them. The locational information is used in the analysis, usually to make inferences about spatial health effects.

Often hypotheses of interest in spatial EPIDEMIOLOGY focus on whether the residential address of cases of disease yields insight into etiology of the disease or, in a public health application, whether adverse environmental health hazards exist locally within a region (as exemplified by local increases in disease risk). For example, in a study of the relationship between malaria endemicity and diabetes in Sardinia a strong negative relationship has been found. This relation had a spatial expression and the geographical distribution of malaria was important in generating explanatory models for the relation (Bernardinelli *et al.*, 1999).

In public health practice, it is of considerable importance to be able to assess whether localised areas that have larger than expected numbers of cases of disease are related to any underlying environmental cause. Here spatial evidence of a link between cases and a source is fundamental in the analysis. Evidence such as a decline in risk with distance from the putative source of hazard or elevation of risk in a preferred direction is important in this regard (see, for example, Lawson, 2001, 2006, Chapter 7; Lawson *et al.*, 1999).

There are four main areas where statistical methods have seen development in spatial epidemiology: DISEASE MAPPING, DISEASE CLUSTERING, ecological analysis and disease map surveillance. Before looking in detail at each of these areas, it is appropriate to consider some common themes or issues that arise in all areas of the subject.

A fundamental feature of data available for analysis in spatial epidemiology is that it is usually discrete (either in the form of a point process or counting process), and the cases of concern arise from within a local human population that varies in spatial density and in susceptibility to the disease of interest. Hence any model or test procedure must make allowance for this background (nuisance) population effect. The background population effect can be allowed for in a variety of ways.

For count data it is commonplace to obtain expected rates for the disease of interest based on the age–sex structure of the local population, and some crude estimates of local relative risk are often computed from the ratio of observed to expected counts (e.g. STANDARDISED MORTALITY/incidence RATIOS, or SMRs). For case event data, expected rates are not available at the resolution of the case locations and the use of the spatial distribution of a control disease has been advocated. In that case the spatial variation in the case disease is compared to the spatial variation in the control disease. A major issue in this approach is the correct choice of control disease. It is important to choose a control that is matched to the age–sex structure of the case disease but is unaffected by the feature of interest. For example,

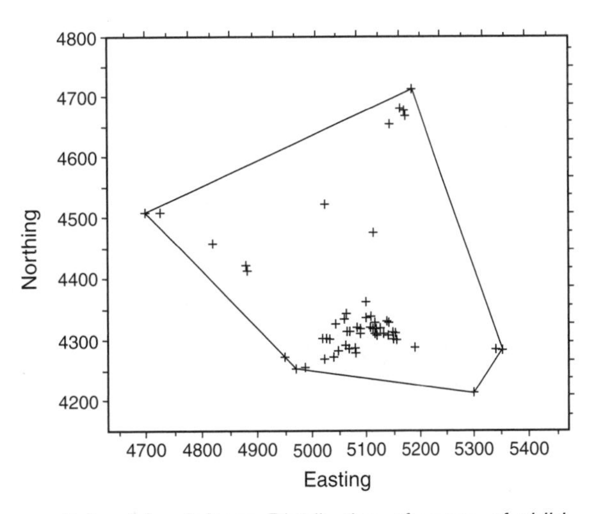

spatial epidemiology *Distribution of cases of childhood lymphoma and leukaemia in Humberside, UK, 1974–1986*

in the analysis of cases around a putative health hazard, a control disease should not be affected by the health hazard. Counts of control disease cases could also be used instead of expected rates when analysing count data. The first (see page 427) and second figures display case event and control data maps for a region of the UK for a fixed time period. The third figure displays a typical count data example.

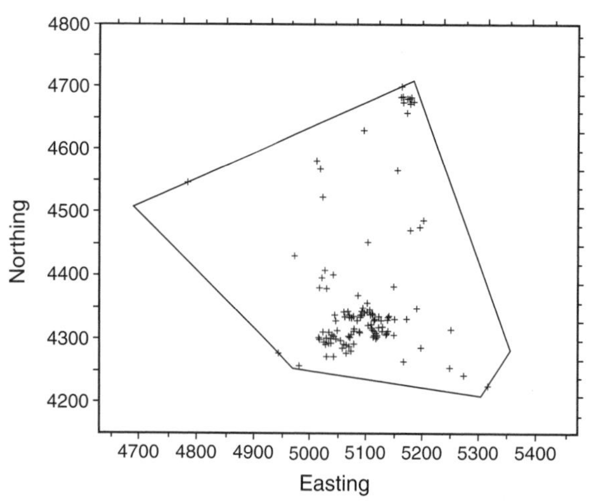

spatial epidemiology *Control distribution: distribution of a sample of live births from the birth register in Humberside, UK, 1974–1986*

For case event data, locations often represent residential addresses of cases and the cases arise from a heterogeneous population that varies both in spatial density and in susceptibility to disease. A heterogeneous Poisson process model is often assumed as a starting point for further analysis. The focus of interest for making inference regarding parameters describing excess risk lies in a relative risk function, which is included in the first-order intensity of the Poisson process.

It is possible that population or environmental heterogeneity may be unobserved in the data set. This could be because either the population background hazard is not directly available or the disease displays a tendency to cluster (perhaps due to unmeasured covariates). The heterogeneity could be spatially correlated, or it could lack CORRELATION, in which case it could be regarded as a type of OVERDISPERSION. One can include such unobserved heterogeneity within the framework of conventional models as a RANDOM EFFECT.

A considerable literature has developed concerning the analysis of count data in spatial epidemiology (e.g. see reviews in Elliott *et al.*, 2000; Lawson, 2001, 2006; Lawson and Williams, 2001; Lawson *et al.*, 2003).

The usual model adopted for the analysis of region counts assumes that the counts are independent Poisson random variables with parameter λ_i in the *i*th region. This model may be extended to include unobserved heterogeneity between regions by introducing a prior distribution for the log relative risks (log λ_i). Incorporation of such heterogeneity has become a common approach and the Besag, York and Mollié (BYM) model is now a standard model. A full Bayesian analysis using this model (see BAYESIAN METHODS) is available on WINBUGS and recently many extensions of this model have been proposed (for various examples see Lawson, 2009). *AL*

spatial epidemiology *Distribution of counts of sudden infant death (SID) within the counties of North Carolina, USA, 1974–1978*

Bernardinelli, L., Pascutto, C. *et al.* 1999: Ecological regression with rrrors in covariates: an application. In Lawson, A. B., Biggeri, A., Boehning, D., Lesaffre, E. *et al.* (eds), *Disease mapping and risk assessment for public health*. New York: John Wiley **& Sons, Inc., p.** 329. **Elliott, P., Wakefield, J. *et al.*** (eds) 2000: *Spatial epidemiology: methods andapplications*. London: Oxford University Press. **Lawson, A. B.** 2001: *Statistical methods in spatial epidemiology*. New York: John Wiley & Sons, Inc. **Lawson, A. B.** 2006: *Statistical methods in spatial epidemiology*, 2nd edition. New York: John Wiley & Sons, Inc. **Lawson, A. B.** 2009: *Bayesian disease mapping: hierarchical modeling in spatial epidemiology*. New York: CRC Press. **Lawson, A. B. and Williams, F. L. R.** 2001 *An introductory guide to disease mapping*. New York: John Wiley & Sons, Inc. **Lawson, A., Biggeri, A. *et al.*** 1999: A review of modelling approaches in health risk assessment around putative sources. In Lawson, A. B., Biggeri, A., Boehning, D., Lesaffre, E. *et al.* (eds), *Disease mapping and risk assessment for public health*. New York: John Wiley **&** Sons, Inc., pp. 231–45. **Lawson, A. B., Browne, W. *et al.*** 2003: *Disease mappingin WinBUGS andMLwiN*. New York: John Wiley & Sons, Inc.

spatio-temporal disease surveillance This is the detection of aberrations in health or disease, usually as they arise. This definition stresses both the unusual nature of the disease event and also the importance of temporal change in surveillance. How 'unusual' an event or sequence of events is becomes, of course, an issue in the design of any surveillance system. The Centers for Disease Control (CDC) define surveillance as: '... the ongoing, systematic collection, analysis, and interpretation of health data essential to the planning, implementation, and evaluation of public health practice, closely integrated with the timely dissemination of these data to those who need to know. The final link of the surveillance chain is the application of these data to prevention and control. A surveillance system includes a functional capacity for data collection, analysis, and dissemination linked to public health programs' (Thacker, 1994).

This definition stresses the collection, analysis and dissemination of data in a timely manner, and hence it is very broad and stresses the focus on public health needs. However, it is possible to distinguish two basic types of surveillance that play different roles in public health activities.

First, *retrospective* surveillance concerns the collection of historical data on disease occurrence and its examination. The purpose of such analysis may be to inform decision makers as to temporal or spatial trends and other features of disease behaviour. This form of surveillance is closely associated with classical epidemiological analysis (see EPIDEMIOLOGY), and differs mainly in its focus on public health needs.

Second, *prospective* surveillance is the online or active examination of disease data to discover changes in disease at the time or close to the time of occurrence. In this case, monitoring of disease occurrence is done 'as data arrive' so that decisions can be made concerning outbreaks of disease. The importance of the this form of surveillance has been heightened with the recent rise in terrorism and the potential threat to health from bioterrorism.

The release of toxic or highly infectious agents into a population would be of grave concern in this context and so it is now important that fast and accurate surveillance of disease be undertaken to detect changes as early as possible.

The design of disease surveillance systems must consider some of the following issues:

1. *Early detection.* It is important to detect changes to disease incidence as early as possible. For example, for certain infectious diseases it may take up to 7 days to receive laboratory confirmation of cases. However, such a delay may be unacceptable if a serious event had occurred. Hence ways to speed up detection could be important.
2. *Syndromic methods.* The use of ancillary information is required in order to speed up detection of population health aberrations. A formal definition is given by (Sosin, 2003): '...the public health term Syndromic Surveillance has been applied to systematic and ongoing collection, analysis and interpretation of data that precede diagnosis (e.g. laboratory test requests, emergency department chief complaint, ambulance response logs, prescription drug purchases, school or work absenteeism, as well as signs and symptoms recorded during acute care visits) and that can signal a sufficient probability of an outbreak to warrant public health investigation.' Often covariate information could be useful in helping to establish the character of an outbreak. The first figure (page 430) displays the time series of pharmaceutical sales and gastrointestinal disease reporting for a Canadian example (see for further examples Lawson and Kleinman, 2005).
3. *Sensitivity and specificity of detection methods.* The calibration of a surveillance system is very important in that false aberration alarms (false positives) could lead to unnecessary public alarm, while false aberration negatives could lead to health disasters. Farrington and Andrews, 2004, and Le Strat, 2005, provide more detail on these issues.
4. *Which disease and what to look for?* In retrospective surveillance the disease is usually known and the features of interest are also known (e.g. trends). In prospective surveillance, particularly where bioterrorism may play a role, there could be little a priori knowledge of the disease and the aberration to look for. Hence, detection systems must have the capability to deal with multiple diseases and possibly multiple forms of aberration. This is known as *multivariate–multifocus* surveillance. Because of the potentially huge database searching problem that results from this, DATA MINING approaches have been adopted (see, for example, Wong *et al.*, 2002, and Madigan, 2005).

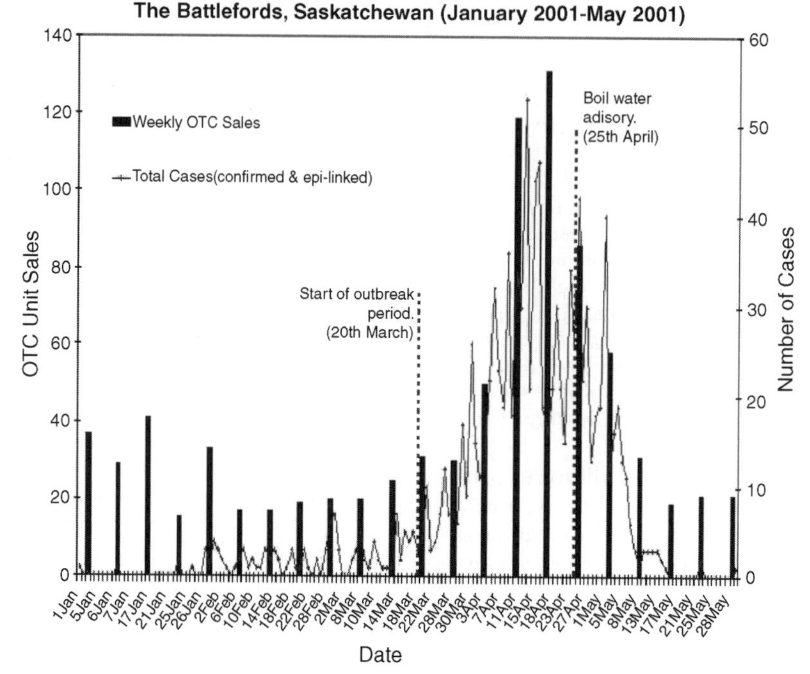

spatio-temporal disease surveillance *Battlefords Saskatchewan: epidemic curve and time series of over-the-counter (OTC) phamaceutical sales January to May 2001 (from Edge et al., 2004, Canadian Journal of Public Health)*

Aberration types

1. *Temporal.* In temporal surveillance a time series of events is available (usually either as counts of disease in intervals or as a point process of reporting or notification times). With time series of discrete counts, the aberrations that might be of interest (suggesting 'important' disease changes) are highlighted in the second figure, which shows that there are three main changes (A,B,C). The first aberration (A) is a sharp rise in risk (jump) or changepoint in level. The second aberration (B) that might be of interest is a cluster of risk (an increase and then decrease in risk). This can only be detected retrospectively of course. The third aberration (C) is an overall process change where the level of the process is changed but also the variability is increased or decreased. When a point process of event times is observed, action may need to be taken whenever a new event arrives. Aberrations that are found in point processes can take the form of unusual aggregations of points (temporal clusters), sharp changes in the rate of occurrence (jumps), and overall change (as above).

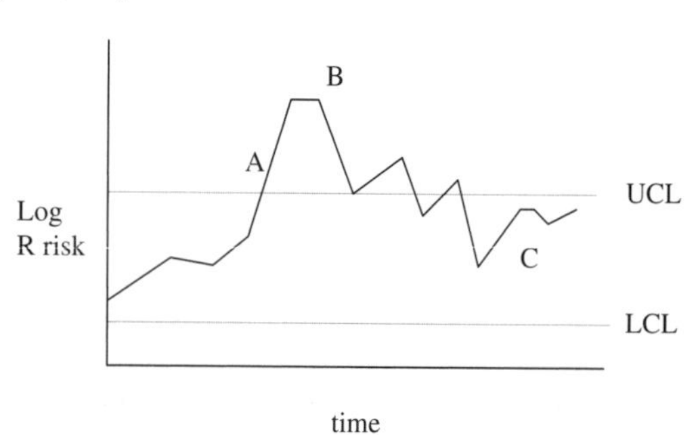

spatio-temporal disease surveillance *Schematic features of changes/aberrations found in temporal surveillance*

2. *spatio-temporal.* If a spatial domain (disease map) is to be monitored then spatial and spatio-temporal aberrations must be considered. Spatial aberrations could consist of discontinuities in risk between regions (jumps in

risk across region boundaries), spatial clusters of risk (localised aggregations of cases of disease) and a long-range trend in risk. When disease maps are monitored in time then *changes* in the above features might be of interest. The development of new clusters of (say) infectious disease in time might signal localised outbreaks, for example.

Multivariate issues

In health surveillance systems designed for prospective surveillance many of the above features would have to be detectable across a wide range of diseases. Not only would whole disease streams be monitored, but subsets of disease streams may be required to be examined. For example, for early detection of effects it may be important to monitor frail subsets (such as old or young age groups). Also, it may be that CORRELATIONS between subsets may be important when making detection decisions. In addition, if disease maps are to be monitored over time as well as time series, then the problem grows considerably. The *Biosense* system developed by CDC (www.cdc.gov/phin/component-initiatives/biosense/index.html) has a multivariate and mixed stream (spatial and temporal) capacity.

Models

One of the current concerns about large-scale surveillance systems is the ability to model correctly the variation in disease and to calibrate properly SENSITIVITY and specificity with such a multivariate and multifocus task. The application of Bayesian hierarchical modelling has been advocated for surveillance purposes and this may prove to be useful in its ability to deal with large-scale systems evolving in time in a natural way (Zhou and Lawson, 2008). For example, recursive Bayesian learning could be important. Clearly computational issues could be paramount here, given the potentially multivariate nature of the problem, and the need to optimise computation is paramount. Sequential Monte Carlo methods have been proposed to deal with computational speedups (Kong, Lai and Wong, 1994; Doucet, De Freita and Gordon, 2001; Doucet, Godsill and Andrieu, 2005; Vidal Rodeiro and Lawson, 2006) as have novel algorithmic speedups (Neill and Moore, 2005). *AL*

[See also TIME SERIES IN MEDICINE]

Doucet, A., De Freita, N. and Gordon, N. 2001: *Sequential Monte Carlo methods in practice*. New York: Springer. **Doucet, A., Godsill, S. and Andrieu, C.** 2005: On sequential Monte Carlo sampling methods for Bayesian filtering. *Statistics and Computing* 10 197–208. **Farrington, P. and Andrews, N.** 2004: Outbreak detection: application to infectious disease surveillance. In Brookmeyer, R. and Stroup, D. (eds), *Monitoring the health of populations: statistical principles and methods for public health surveillance*. New York: Oxford University Press. **Kong, A., Lai, J. and Wong, W.** 1994: Sequential imputations and {B}ayesian missing data problems. *Journal of the American Statistical Association* 89, 278. **Lawson, A. B. and Kleinman, K.** 2005: *Spatial and syndromic surveillance for public health*. New York: John Wiley & Sons, Inc. **Le Strat, Y.** 2005: Overview of temporal surveillance. In Lawson, A. B. and Kleinman, K. (eds), *Spatial and syndromic surveillance for public health*. New York: John Wiley & Sons, Inc. **Madigan, D.** 2005: Bayesian data mining for health surveillance. In Lawson, A. B. and Kleinman, K. (eds), *Spatial and syndromic surveillance for public health*. New York: John Wiley & Sons, Inc. **Neill, D. and Moore, A. W.** 2005: Efficient scan statistic computations. In Lawson, A. B. and Kleinman, K. (eds), *Spatial and syndromic surveillance for public health*. New York: John Wiley & Sons, Inc. **Sosin, D.** 2003: Draft framework for evaluating syndromic surveillance systems. *Journal of Urban Health* 80, i8–i13. **Thacker, S.** 1994: Historical development. In Teusch, S. and Churchill, R. (eds), *Principles and practice of public health surveillance*. New York: Oxford University Press. **Vidal Rodeiro, C. and Lawson, A. B.** 2006: Online updating of space-time disease – surveillance models via particle filters. *Statistical Methods in Medical Research* 15, 1–22. **Wong, W., Moore, A., Cooper, G. and Wagner, M.** 2002: Rule-based anomaly pattern detection for detecting disease outbreaks. In *18th National Conference on Artificial intelligence*. Cambridge, MA: MIT Press. **Zhou, H. and Lawson, A. B.** 2008: EWMA smoothing and Bayesian spatial modeling for health surveillance. *Statistics in Medicine* 27, 5907–28.

Spearman's rank correlation coefficient

See CORRELATION

Spearman's rho (ρ)

Also known as *Spearman's rank correlation coefficient*, this is a measure of the relationship between two variables that uses only the rankings of the observations. If the ranked values of the two variables for a set of n individuals are a_i and b_i, with $d_i = a_i - b_i$, then the coefficient is defined explicitly as:

$$\rho = 1 - \frac{6\sum_{i=1}^{n} d_i^2}{n^3 - n}$$

In essence, ρ is simply Pearson's product moment correlation coefficient (see CORRELATION) between the rankings a and b. We can illustrate the coefficient on the data shown in the table, which were collected to investigate the relationship between MEAN annual temperature and the mortality rate for a type of breast cancer in women. The data relate to certain regions of Great Britain, Norway and Sweden (see Lea, 1965). Here, the Spearman correlation is 0.90 and Pearson's product moment correlation 0.87. In general, the Spearman coefficient is more robust against the presence of outliers. *BSE*

Spearman's rho (ρ) *Breast cancer mortality and temperature*

Mean annual temperature (°F)	*Mortality index*
51.3	102.5
49.9	104.5
50.0	100.4
49.2	95.9
48.5	87.0
47.8	95.0
47.3	88.6
45.1	89.2
46.3	78.9
42.1	84.6
44.2	81.7
43.5	72.2
42.3	65.1
40.2	68.1
31.8	67.3
34.0	52.5

[See also CORRELATION, NONPARAMETRIC METHODS – AN OVERVIEW]

Lea, A. J. 1965: New observations on distribution of neoplasms of female breast cancer in certain European countries. *British Medical Journal* 1, 488–90.

specificity This is a measure of how well an alternative test performs when it is compared with the reference of 'gold' standard test for diagnosis of a condition. Specificity is the proportion of patients correctly identified as free from the condition by the diagnostic test out of all patients who do not have the condition. Specificity may also be expressed as a percentage and is the counterpart to SENSITIVITY.

The reference standard may be the best available diagnostic test or may be a combination of diagnostic methods, including following up patients until all patients with the disease have presented with clinical symptoms. It follows that the best design when a diagnostic test is evaluated against a reference standard is a COHORT STUDY. Investigators should consider whether verification BIAS is present: this occurs when obtaining a negative result on one diagnostic test influences the chances of a patient going on to have further tests, so that some patients with the condition never receive the correct diagnosis.

When the data are set out as in the table:

$$\text{Specificity} = \frac{d}{b+d}$$

specificity *General table of test results among $a + b + c + d$ individuals sampled*

	Disease		
	Present	*Absent*	*Total*
Positive	a	b	$a+b$
Negative	c	d	$c+d$
Total	$a+c$	$b+d$	$a+b+c+d$

Specificity should be presented with CONFIDENCE INTERVALS, typically set at 95%, calculated using an appropriate method such as that of Wilson (described in Altman *et al.*, 2000) that will not produce impossible values, i.e. that will not give values for the upper confidence interval > 1 when specificity approaches 1 and the sample size is small.

Where a test result is a continuous measurement, for example, HDL cholesterol, a cut-off point for abnormal values is chosen. If a higher value is chosen, then specificity will be relatively high, but sensitivity relatively low. The impact of all possible cut-off points can be displayed graphically in a RECEIVER OPERATING CHARACTERISTIC (ROC) curve. The choice of cut-off point is not, however, solely a statistical decision, as the balance between the FALSE POSITIVE RATE and the FALSE NEGATIVE RATE should be related to the clinical context and consequences of wrong diagnosis both for the patient and the healthcare system.

A sample size calculation for specificity can be made by stipulating a confidence interval (e.g. 95%) and an acceptable width for the lower bound of the confidence interval. Where the anticipated specificity is high and the sample size is small, a 'small sample' method should be used: a sample size table is included in Machin *et al.* (1997). *CLC*

[See also NEGATIVE PREDICTIVE VALUE, POSITIVE PREDICTIVE VALUE, TRUE POSITIVE RATE]

Altman, D. G., Machin, D., Bryant, T. N. and Gardner, M. J. 2000: *Statistics with confidence*, 2nd edition. London: BMJ Books.
Machin, D., Campbell, M., Fayers, P. and Pinol, A. 1997: *Sample size tables for clinical studies*, 2nd edition. Oxford: Blackwell Sciences Ltd.

spending function See SEQUENTIAL ANALYSIS

spline function See SCATTERPLOT SMOOTHERS

S-Plus See STATISTICAL PACKAGES

SPSS See STATISTICAL PACKAGES

stable population See DEMOGRAPHY

stacked bar chart See BAR CHART

standard deviation This is a measure of spread intended to give an indication of the spread of a series of values ($x_1, x_2, \ldots, x_n$) about their MEAN($\bar{x}$).

Taking the average of the differences from the mean may initially seem a good measure of their spread, but in fact this is always zero. Therefore, the standard deviation is based on the average of the squared differences from the mean, since these are all positive. Taking the square root of this result gives a measure that is in the same units as the original values. Thus, the standard deviation (s) is calculated using the following formula. Here n is the number of observations, i takes values from 1 to n and the $\sum$ notation denotes the sum, i.e. $(x_1-\bar{x})^2 + (x_2-\bar{x})^2 + \cdots + (x_n-\bar{x})^2$:

$$s = \sqrt{\frac{\sum (x_i-\bar{x})^2}{n-1}}$$

Note that the formula indicates division by $n-1$, rather than n, when taking the average of the squared differences. This gives a result that is a better estimate of the standard deviation in the whole population, which is being estimated from the sample available.

The standard deviation can be denoted SD, *sd*, s or σ, although the last technically refers to the standard deviation of a population, rather than a sample. To calculate the standard deviation by hand there is a more convenient and mathematically equivalent formula:

$$s = \sqrt{\frac{\sum x_i^2 - \left[(\sum x_i)^2/n\right]}{n-1}}$$

As an example, the bone mineral content (BMC) of 10 babies was measured using dual energy X-ray absorptiometry (DXA). The measurements in grams were: 46.6, 46.9, 49.2, 49.8, 53.2, 61.1, 68.1, 73.1, 77.1 and 78.6.

It is simple to calculate that the sum of the observations $\sum x_i = 603.7$ and the sum of the squares of the observations $\sum x_i^2 = 37\,938.89$. Thus:

$$s = \sqrt{\frac{37\,938.89 - [(603.7)^2/10]}{9}} = 12.88 \text{ g}$$

The VARIANCE of a set of measurements is the square of their standard deviation. Although the variance has many uses, the standard deviation is a more meaningful descriptive statistic because it is in the same units as the raw data. Whereas square millimetres, mm^2, may have an obvious interpretation, square millimetres of mercury, $mmHg^2$, does not. Altman (1991) suggests that standard deviations may be quoted with one or two more decimal places than the original values.

The standard deviation is typically used as a measure of spread alongside the mean and is most appropriate when the data are approximately symmetrically distributed. It has the useful property that when the data follow a NORMAL DISTRIBUTION then approximately 95% of the observations will be within two standard deviations of the mean. The figure shows the case of a standard normal distribution, which has a mean of 0 and a standard deviation of 1. *SRC*

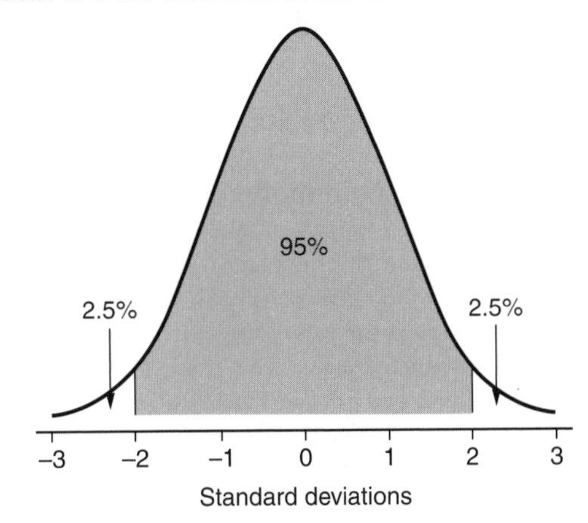

standard deviation *Standard normal distribution, with mean of 0 and SD of 1*

Altman, D. G. 1991: *Practical statistics for medical research*. London: Chapman & Hall.

standard error This is the STANDARD DEVIATION of the SAMPLING DISTRIBUTION of a statistic. For example, the standard error of the sample MEAN of n observations is $\sigma/\sqrt{n}$, where σ^2 is the VARIANCE of the original observations.

A useful aide-memoire to distinguish when to use standard deviation (SD) and when to use standard error (SE) is to recall: 'SD for description, SE for estimation.' In particular, when describing patient characteristics in a sample, as in a research paper's typical Table 1, means and SDs should be reported, whereas when seeking to learn from the sample and apply results to the relevant population, i.e. performing statistical inference either by HYPOTHESIS TESTS or estimation by CONFIDENCE INTERVALS, then the standard error is used. The SE is necessarily smaller than the SD and it is wrong to use SE as a MEASURE OF SPREAD when describing samples.

More generally, standard errors can be attached to any sample-based quantity, not just the mean of a single sample of continuously distributed data, as just discussed.

The general form of a large-sample 95% confidence interval for a population parameter (numerical characteristic) is the sample-based point estimate ± 1.96 (standard errors), where 1.96 arises from the standard NORMAL DISTRIBUTION and the standard error is that of the point estimate, itself the best sample-based guess for the value of the parameter. For two-sample inference, this is usually a quantity such as the difference in population means, for continuous data, or the difference in population proportions, for categorical data. *BSE*

standard population See DEMOGRAPHY

statistical consulting See CONSULTING A STATISTICIAN

standardised mortality ratio (SMR) See DEMOGRAPHY

STATA See STATISTICAL PACKAGES

statistical methods in molecular biology Molecular biology is the branch of biology that studies the structure and function of biological macromolecules of a cell, and especially their genetic role. Three types of macromolecules are the main subjects of interest: deoxyribonucleic acids (DNA), ribonucleic acids (RNA) and proteins. Genetic information is encoded in the DNA and inherited from parents to children and when expressed, a gene, the basic unit of inheritance, is first transcribed to messenger RNA, which then carries the information to a cellular machinery (ribosome) for protein production. This basic principle of the information flow in biology is often referred to as the 'central dogma', put forward by Francis Crick in 1958. A central goal of molecular biology is to decipher the genetic information and understand the regulation of protein synthesis and interaction in cellular processes.

The rapid advance of biotechnology in the past few decades has facilitated manipulation of these important biopolymers and allowed scientists to clone, sequence and amplify DNA. As a result, a large amount of biological sequence and structural information has been generated and deposited into public accessible databases. The phenomenal growth of biological data is underpinned by the developments of high-throughput DNA sequencing and microarray technologies and the recent progresses in giant research projects such as the human genome project that produced the sequence of the human genome.

The word 'genome' refers to the entire collection of genetic material of an organism. These advances result in many complex and massive datasets, sometimes decoupled from specific biological questions under investigation. The need to extract scientific insights from these rich data by computational and analytic means has spawned the new field of bioinformatics and computational molecular biology, which deals with storage, retrieval and analysis of biological data. These can consist of information stored in the genetic code, but also experimental results from various sources, patient statistics and scientific literature. Bioinformatics is highly interdisciplinary, using techniques and concepts from informatics, statistics, mathematics, physics, chemistry, biochemistry and linguistics. Nowadays, various biological databases and practical applications of bioinformatics are readily available through the internet and are widely used in biological and medical research.

A wide spectrum of statistical methods has been successfully applied in bioinformatics, ranging from the basic summary statistics and exploratory data analysis tools, to sophisticated hidden Markov models and Bayesian resampling methods (see BAYESIAN METHODS, MARKOV CHAIN MONTE CARLO). Analyses in bioinformatics focus on three types of datasets: genome sequences, macromolecule structures and large-scale functional genomics experiments. Various other data types are also involved, such as taxonomy trees, sequence polymorphisms, relationship data from metabolic pathways, patient statistics, text from scientific literature and so on.

DNA sequences are the primary data from the sequencing projects and they only become really valuable through multiple layers of annotation and organisation. Several areas of bioinformatics analysis are relevant when dealing with DNA and protein sequences: sequence assembly, to establish the correct order of sequence contigs for a contiguous sequence; prediction of functional units, to identify subsets of sequences that code for various functional signals such as protein coding genes, promoters, splice sites, regulatory elements; and sequence comparison and database search, to retrieve data efficiently from organised databases.

Most of these analyses involved *sequence alignment*, one of the classic problems in the early development of bioinformatics. Sequence alignment is the basic tool that allows us to determine the similarity of two or more sequences and infer components that might be conserved through evolution and natural selection. To align two protein sequences, similarity scores are assigned to all possible pairs of residues and the sequences are aligned to each other so as to maximise the sum total of scores in the sequence pairings induced by the alignment. *Dynamic programming*-based algorithms were developed to overcome the large search space for the solution of optimal global and local alignment problems (Needleman and Wunsch, 1970; Smith and Waterman, 1981).

Dynamic programming is a general algorithmic technique that solves an optimisation problem by recursively using 'divide and conquer' for its subproblems. Faster heuristic word-based alignment algorithms were later introduced for large database similarity searches (BLAST by Altschul *et al.*, 1990; FASTA by Pearson and Lipman, 1988). These algorithms build alignments by extending or joining common short patterns ('words') that are computationally efficient, but often yield suboptimal solutions. The interpretation of alignment scores and database search results was aided by statistical significance derived from simulations and PROBABILITY theory of extreme value distributions under the framework of standard statistical hypothesis testing (Karlin and Altschul, 1990). These classic results have become indispensable tools for biomedical researchers and computational biologists to analyse molecular sequence data.

Statistical models are also routinely used to construct probabilistic profiles to characterise the regularity of biological signals based on collections of pre-aligned sequences and to increase SENSITIVITY of searches. For example, a block-based product multinomial model can be used to describe the position-specific base distributions of the 5′ splice site (exon–intron junction) signal in humans (see the figure), which gives a richer representation of the sequence motif than the consensus CAG|GTGAG ('|' indicates the exon–intron junction). A *position-specific scoring matrix* can be derived subsequently using logarithms of the ODDS RATIO of the signal to background base to evaluate matches of new query sequences to the sequence motif and to quantify the *information content* of the signal sequence pattern. The information content of a signal is defined as the average score of random sequence matches, measured in 'bits' using the log (base two) odds ratio scores that represent the number of 0–1s necessary to code for this signal in a binary coding system. For instances, the human 5′ splice site depicted in the figure contains 8 bits of information, meaning that 'decoy' splice sites will be observed roughly every $2^8 = 256$ bases in random sequence. Note that the information content can also be formulated as the *relative entropy* (or *Kullback–Leibler distance*) of the signal to background nucleotide frequency distributions in the context of information theory. More sophisticated models and scoring matrices are also available to capture dependencies among neighbouring positions using *Markov models* and others.

Another area of biological sequence analysis that relies heavily on statistical reasoning is gene finding or, more generally, predicting complex features from a sequence. The goal of protein-coding gene finding is to locate gene features such as exons and introns in a DNA genomic sequence, which is the essential first-pass annotation of the genome project products. In addition to inferring *homologous* (evolutionarily related) gene structures from database similarity searches, statistical ab initio gene-finding programmes have been developed to integrate all known features and 'grammars' of protein-coding genes in a probabilistic model.

Hidden Markov models (HMMs) are at the heart of the most popular gene finders (Genscan by Burge and Karlin, 1997, and reviewed in Durbin *et al.*, 1998). HMMs were originally developed in the early 1970s by electrical engineers for the problem of speech recognition – to identify what sequence of phonemes (or words) was spoken from a long sequence of category labels representing the speech signal. The resemblance of the gene-finding problem to speech recognition and the way HMMs are formulated make them especially suited in this context. In addition, HMMs are theoretically well-founded models, combining probabilistic modelling and formal language theory that guarantees 'sensible' predictions that obey specified grammatical rules even though they might not be the correct genes.

There are also well-documented and computationally efficient methods for parameter estimation (e.g. expectation–maximisation) and optimisation (Viterbi algorithm). A Markov chain is a series of random events occurring with probabilities conditionally dependent on the state of the preceding event(s). A hidden Markov model is a Markov chain in which each state generates an observation according to some rule (usually stochastic). The objective is to infer the hidden state sequence that maximises the posterior probability of the observed event sequence given the model. For example, the hidden states may represent words or phonemes and the observations are the acoustic signal.

AAGGTGCTGTG
CAGGTGAGTGG
AATGTACGTGT
CAGGTGAGCGG
CAGGTATGCGG
AAGGTAAAGTT
CAGGTGAGCCC
GCGGTAAGAGG
GGGGTGAGTCA
GAGGTGTGTGC
CAGGTAATCAA
ACGGTAAGCCC
GTGGTGAGCGG
AAGGTGGGTGC
GAGGTGAGAGG
AAGGTGAGGGC
CAGGTAAGGCA
CAGGTGAGCCT
. . .

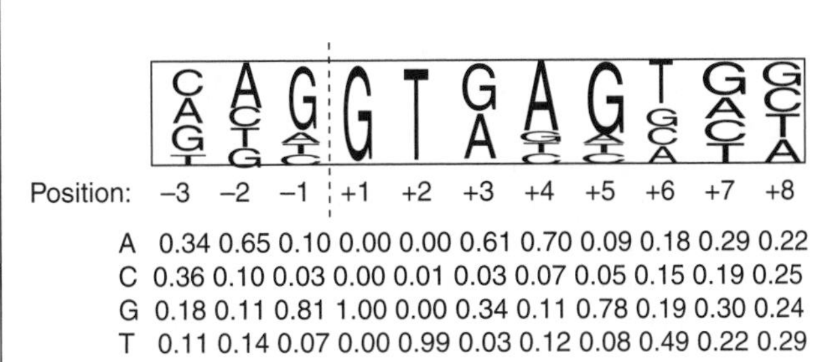

	−3	−2	−1	+1	+2	+3	+4	+5	+6	+7	+8
A	0.34	0.65	0.10	0.00	0.00	0.61	0.70	0.09	0.18	0.29	0.22
C	0.36	0.10	0.03	0.00	0.01	0.03	0.07	0.05	0.15	0.19	0.25
G	0.18	0.11	0.81	1.00	0.00	0.34	0.11	0.78	0.19	0.30	0.24
T	0.11	0.14	0.07	0.00	0.99	0.03	0.12	0.08	0.49	0.22	0.29

statistical methods in molecular biology *The human 5′ splice site (exon–intron junction signal)*

Motif discovery is an area under active research and has benefited from sophisticated modern statistical techniques. In a typical setting, a collection of sequences derived from MICROARRAY EXPERIMENTS or various sources are believed to share common sequence motifs that often represent functional domains or regulatory elements, and the challenge is to find the unknown signals and locate them in individual sequences. One approach is to formulate the multiple alignment information as MISSING DATA and infer them together with other parameters of the statistical model, given only the sequences as observables. Advanced statistical modelling and iterative computation techniques such as the EM ALGORITHM and Markov chain Monte Carlo are typically used for simultaneous model estimation (Liu, Neuwald and Lawrence, 1999).

The function of a protein is determined by its three-dimensional structure. The problem of predicting the three-dimensional structure of a protein from its amino acid sequence (or the protein-folding problem, because proteins are capable of quickly folding into their stable, unique three-dimensional structure, starting from a random coil conformation without additional genetic mechanisms) is one of biggest challenges in bioinformatics. There are three major lines of approaches for protein structure prediction: comparative modelling, fold recognition and ab initio prediction.

Comparative modelling makes use of sequence alignment and database searches and builds on the fact that evolutionarily related proteins with similar sequences have a similar structure. For proteins without a homologous sequence of known structure, the approach of 'threading' has been developed. It is assumed that a small collection of 'folds', perhaps several hundreds in number, can be used to model the majority of protein domains in all organisms. The protein-folding problem is thus reduced to the tasks of classifying the query protein based on its primary sequence into one of the folding classes in a database of known three-dimensional structures. This classification is often accomplished using complicated statistical models such as Gibbs sampling and HMMs to parameterise the fit of a sequence to a given fold and solve the optimisation problem accordingly.

Analogous to the gene-finding problem, one may attempt to compute a protein's structure directly from its sequence, based on biophysical understanding of how the three-dimensional structure of proteins is attained. The challenge can be broken down into two components: devising a scoring function that can distinguish between correct and incorrect structures and a search method to explore the conformational space efficiently. If successful, direct folding certainly would give a deeper insight than the 'top-down' threading or homology modelling approaches. However, currently no reliable method has yet emerged in this category.

During the past few years, the development of DNA array technology has scaled up the traditionally one-gene-at-a-time functional studies to allow the monitoring of hundreds of thousands of genes simultaneously. A large number of statistical issues arise in connection with these studies and these have fostered unprecedented conversation and collaborations between biologists and statisticians to establish means to plan, process and analyse these massive datasets. Many branches of statistics have been revived and/or extended by their recent applications in the analysis of functional genomics and molecular data, including DATA MINING methods to discover and classify patterns, MULTIPLE TESTING procedures to adjust *P*-VALUES to control false discovery rates and meta-analysis (see SYSTEMATIC REVIEWS AND META-ANALYSIS) to combine experimental results from various sources. New statistical methods will soon be needed when combining information from multiple distinct data types (sequence, gene expression, protein structures, sequence variation and phenotypes) for the same subjects. *RFY*

Altschul, S. F., Gish, W., Miller, W., Myers, E. W. and Lipman, D. 1990: Basic local alignment search tool. *Journal of Molecular Biology* 215, 403–10. **Burge, C. B. and Karlin, S.** 1997: Prediction of complete gene structures in human genomic DNA. *Journal of Molecular Biology* 268, 78–94. **Durbin, R., Eddy, S., Krogh, A. and Mitchison, G.** 1998: *Biological sequence analysis: probabilistic models of proteins and nucleic acids*. Cambridge: Cambridge University Press. **Karlin, S. and Altschul, S. F.** 1990: Methods for assessing the statistical significance of molecular sequence features by using general scoring schemes. *Proceedings of the National Academy of Sciences of the United States of America* 87, 2264–8. **Liu, J. S., Neuwald, A. and Lawrence, C.** 1999: Markovian structures in biological sequence alignments. *Journal of the American Statistical Association* 94, 1–15. **Needleman, S. B. and Wunsch, C. D.** 1970: A general method applicable to the search for similarities in the amino acid sequence of two proteins. *Journal of Molecular Biology* 48, 443–53. **Pearson, W. R. and Lipman, D. J.** 1988: Improved tools for biological sequence comparison. *Proceedings of the National Academy of Sciences of the United States of America* 85, 2444–8. **Smith, T. H. and Waterman, M. S.** 1981: Identification of common subsequences. *Journal of Molecular Biology* 147, 195–7.

statistical packages In 2010 the Association for Survey Computing (ASC) website (www.asc.org.uk) listed some around 200 statistical packages. Many of these have been under development for nearly 40 years and therefore it is both a very mature and diverse software market.

While many of these around 200 packages are developed for niche markets, there are still several generic software suites. It seems almost invidious to try to select and discuss individual packages. However, there are clearly some well-known and long-established packages, and to many the term 'statistical package' is almost synonymous with SPSS™ or possibly SAS™. Given the variety of analyses that these packages offer, they can meet most user needs. It would seem likely that a virtual monopoly should exist, but in fact there have been new entrants gaining popularity. Comparing these is instructive about trends in the development of statistical software. The packages in the first table are the ones on which we will concentrate here.

statistical packages *Major statistical packages*

Major statistical packages	
SPSS	www.spss.com
SAS	www.sas.com
STATA	www.stata.com
S-Plus	www.insightful.com

The prevalence of these major packages notwithstanding, there are other packages, as listed in the second table, although these will not be further discussed. Competition has been good for the development of programs and potential purchasers should always be aware of options outside the norm that may well fit their requirements. Together with the ASC website (given earlier), it will always be profitable to make comparisons when purchasing.

statistical packages *Other major statistical packages*

Other major statistical packages	
Genstat	www.vsn-intl.com
STATISTICA	www.statsoft.com
NCSS	www.ncss.com
SYSTAT	www.systat.com

Naturally enough, one wants a statistical package to do statistics and the leading packages cover a wide range. These include basic descriptive statistics, including EDA-style charting, comprehensive cross-tabulation analysis, means testing, the general linear model, multivariate methods, data reduction and clustering, nonparametrics, log-linear modelling, time series – and more.

The conversion in the late 1980s–early 1990s of the packages SPSS and SAS to run on desktop PCs seemed to cause a hiatus in the development of statistical methodology within these suites. Quite possibly, one of the main reasons for this was the need to develop new user interfaces, as an alternative the command-line format previously used on mainframe and minicomputers. With the DOS interface model being rapidly succeeded by that of Windows™, major consecutive design changes were needed. This did seem to leave a window of opportunity for new entrants to the market, which could write directly using modern programming architectures.

S-Plus is perhaps the earliest example of this, initially written for the UNIX system and then subsequently ported to PCs. The design was conceptually novel, based on the notion of an extensible statistical calculator. It provides advanced graphics facilities and has become popular with professional statisticians for its ability to develop analysis methodologies, rather than being tied to a rigid framework. Over time S-Plus has developed to add extensive user interface enhancements as well as larger statistical libraries. The public domain 'R' (www.r-project.org) is based on a similar philosophy to S-Plus (see R).

STATA has become a very popular alternative for similar reasons. Starting out as a command-line-driven program, it has matured over the years to offer a windowing interface in addition. Its attractiveness to researchers has been a modern approach to statistical testing, as well as its ability to incorporate new methodologies quickly. Not only do the developers have an architecture that permits easy incremental expansion, users themselves can program their own procedures. This has gained the support of the professional statistical community, who through their educative role have promoted the package's popularity.

Partly as a result of competition, packages have also begun to differentiate themselves in terms of extending extra support to the whole data analysis process. While the actual test result remains the core of any analysis, data management is far more demanding in terms of time. The resources needed to support DATA MANAGEMENT in a MULTICENTRE clinical TRIAL are significantly larger than those for a classical experiment. In these scenarios, managing and manipulating data prior to analysis becomes very important.

SAS has long specialised in data management support, with flexible procedures for merging and manipulating datasets, as well as links to database packages. In the pharmaceutical industry SAS is almost a de facto standard for major analyses, reflecting its ability to support the strong audit requirements in the industry. To a certain extent other packages have been restricted to the rectangular data model (or spreadsheet) view of data, although all are now improving these features.

One direct effect of the development of statistical packages has been to introduce the possibility of statistical data analysis to a wider audience than just statisticians. Since these users are often in finance and commerce, they represent a significant revenue stream to package producers and making the program user friendly for nonspecialist audiences has become a priority for some. SPSS's menu-driven 'point-and-click' interface, for example, epitomises this model. In contrast, the command-line models of SAS, STATA or SPlus require more dedicated training, although as noted earlier all have developed similar facilities. (STATA 8 introduced a menu-driven interface in 2003 to complement its traditional command-line orientation.)

Integrating advanced data-entry features with a statistical analysis package is common. The predominant spreadsheet data entry model can be enhanced to include data entry forms, data checking and audit. The large packages such as SPSS and SAS provide 'add-on' programs for this. Other programs provide direct database links so that data entry can be provided in a normal programming package such as Microsoft Access and then directly imported for analysis.

While traditionally the results of an analysis are interpreted and then incorporated into a final report, packages have begun to differentiate themselves on their ability to produce tables and results that can be directly pasted into a presentation quality report. Packages vary widely on their ability to do this and support can be patchy. SPSS provides a very good ability to move results tables, but the exported graphics are not of such a good quality. STATA, by contrast, does not offer sophisticated export of results, but has in its latest versions excellent graphical output. SAS offers full programmable reporting features that are very flexible, but challenging for the naive user.

While the main focus of any statistical user is on the large packages, dedicated packages still have a role. As an example, programs such as NQUERY (www.statsol.ie), dedicated to sample size estimation, do one particular job very well and are popular as a result. The lone, innovative researcher (an example perhaps being MX found at www.vcu.edu/mx/) is also a likely producer of innovative software.

An important dimension for the individual consumer can be price. Some of the major packages have prices that match their capabilities: the single researcher, particularly in the commercial sector, may find this an important factor in choice. All the relevant websites can give guidance on obtaining price quotations.

Rather than ossifying, the marketplace for statistical software is healthy and researchers can find themselves well supported with a choice of diverse packages. *CS*

statistical parametric map See STATISTICS IN IMAGING

statistical refereeing There have been hundreds of review articles published in the biomedical literature that point out statistical errors in the design, conduct, analysis, summary and presentation of research studies. The contents of every general medical journal (most notably *Annals of Internal Medicine*, *British Medical Journal*, *Journal of the American Medical Association*, *Lancet* and *New England Journal of Medicine*), as well as of many specialist ones, have been subjected to this intense scrutiny sometimes frequently. These review articles have focused on particular statistical tests, frequency of usage and correct application of techniques of statistical analysis, design of CLINICAL TRIALS and epidemiological studies, use of POWER calculations and CONFIDENCEINTERVALS and many other aspects.

Their almost universal conclusion is that a substantial percentage of research studies, perhaps as many as 50%, published in the biomedical literature contains errors of sufficient magnitude to cast some doubt on the validity of the conclusions that have been drawn. This does not mean that the conclusions are wrong, but it does imply that they may not be right, and this inevitably leads to serious concern about the consequences both for understanding of disease and for the treatment of patients.

One solution to this problem has been the introduction of medical statisticians into the peer review process. Some have advocated that all submitted papers should be scrutinised in this way, arguing that statistical review of those that are not published, no matter how poor, will at least lead to higher standards in research and improvement in future papers. In view of the very large number of biomedical journals and the huge numbers of papers submitted for publication every year, such a remedy is impracticable. An alternative, now used by several journals, is to divide the peer review process into two stages, whereby papers considered by the editors as candidates for publication are sent first to subject matter referees (physicians, surgeons, epidemiologists, etc.) and those recommended for publication by them are then sent to statisticians for further specialist review.

The process of statistical review is complex, requires sophisticated judgement and varies considerably in its application to every section of a paper (abstract, introduction, methods, results and discussion). Altman (1998) reviews some of the difficulties and provides practical examples of both definite errors and matters of judgement, within study design, analysis, presentation and interpretation. There are 12 broad aims of statistical review that can be summarised as follows: to prevent publication of studies that have a fundamental flaw in design; to prevent publication of papers that have a fundamental flaw in *interpretation*; to ensure that key aspects of background, design and methods of analysis are reported clearly; to ensure that key features of the design are reflected in the analysis; to ensure that the best methods of analysis, appropriate to the data, are used; to ensure that the presentation of results is adequate and employs summary statistics that are justified by the design, the data and the analysis; to ensure that tables are accurate and are consistent both with the text and with each other; to ensure that the style of figures is appropriate, that they are consistent with text and tables and not unduly repetitious of other content; to guard against excessive analysis and spurious accuracy; to ensure that conclusions are justified by the results; to ensure that content of the discussion is justified by the results and, in particular, that it avoids generalisation far beyond the confines of the paper; and, finally, to ensure that the abstract accords with the paper.

The statistical reviewer may also comment on subject matter when an expert within the medical specialty of the paper, but will not indicate typos, except when these are critical for accuracy within formulae or text. Indeed, pointing out inconsequential typos is not part of any aspect of any review process; they should be disregarded by expert reviewers and left entirely to the journal's copyeditor!

Since statistical review is complicated and, for the reviewer, sometimes excessively tedious, with the necessity of making very similar, sometimes the same, comments about manuscript after manuscript, detailed statistical guidelines and checklists have been written with the specific intention of helping authors (and reviewers). These have been supported by the editors of many biomedical journals and referred to in the journal's guidelines to authors. Examples can be found in Altman *et al.* (2000) and Gardner *et al.* (2000). Those most widely used for clinical trials are the CONSORT guidelines (Moher, Schulz and Altman, 2001, updated 2010), for which there is accompanying explanation (Altman *et al.*, 2001, also updated 2010), and extension to cluster trials, noninferiority and equivalence randomised trials, herbal medicine interventions, nonpharmacological interventions, harms, abstracts and pragmatic trials (see www.consort-statement.org). The checklist that forms part of the CONSORT statement is intended to accompany a submitted paper and to indicate where in the manuscript each item in the checklist has been addressed, thus serving as a useful reminder to authors and an aide to referees. There are also recent guidelines for reporting META-ANALYSIS (PRISMA, which supercedes QUORUM), for observational studies (STROBE) and for genetic ASSOCIATION studies (STREGA); details can be found through the EQUATOR network (www.equator-network.org).

Statistical review is intended to be helpful and constructive; it should also reassure authors and readers that published papers are sound. However, it is not always seen from this perspective and editors of journals need to be vigilant in ensuring that it does not become a focus for controversy and dispute, as can happen, for example, when authors parade the views of 'their own statistician' to counter comments from a referee. There is at present little incentive for statisticians to engage in such review – it does not enhance their careers, there is no specific training for it, small (if any) remuneration, it is time consuming and 'the only likely concrete consequence of good reviewing is future requests for more reviews' (Bacchetti, 2002). Bacchetti also points out that statistics is a rich area for finding mistakes and, when coupled with 'the notion that finding flaws is the key to high quality peer review', can lead to 'finding flaws that are not really there'. This reinforces the need for sound statistical judgement. Statisticians may also have to counter mistaken criticisms from subject matter reviewers with limited statistical knowledge (Bacchetti, 2002).

The final part of statistical review is usually a recommendation to the journal's editor either to accept, accept with revision, revise and resubmit, or reject the paper. The distinction between the second and third is sometimes difficult and can only be made by balancing the extent and nature of the revisions against the capabilities of the authors as evinced from the submitted paper. Rejection by the statistician can also lead to provocation, especially as authors will be aware that their 'subject matter' peers have already judged it sound. In 1937 the *Lancet*'s leading article that heralded the series of classic papers by Bradford Hill on *The Principles of Medical Statistics* forewarned: 'It is exasperating, when we studied a problem by methods that we have spent laborious years in mastering, to find our conclusions questioned, and perhaps refuted, by someone who could not have made the observations himself. It requires more equanimity than most of us possess to acknowledge that the fault is in ourselves.'

Authors of papers are advised to read statistical reviews carefully, put them aside for 48 hours and only then start to think about how to respond. For further information and discussion see Rubinstein (2005), Smith (2005), Ware (2005). *TJ*

Altman, D. G. 1998: Statistical reviewing for medical journals. *Statistics in Medicine* 17, 2610–74. **Altman, D. G., Gore, S. M., Gardner, M. J. and Pocock, S. J.** 2000: *Statistical guidelines for contributors to medical journals.* In Altman, D. G., Machin, D., Bryant, T. N. and Gardner, M. J. (eds), *Statistics with confidence*, 2nd edition. London: BMJ Books, 171–90. **Altman, D. G., Schulz, K. F., Moher, D., Egger, M., Davidoff, F., Elbourne, D., Gotzsche, P. C. and Lang, T. for the CONSORT Group** 2001: The revised CONSORT statement for reporting randomised trials: explanation and elaboration. *Annals of Internal Medicine* 134, 663–94. **Bacchetti, P.** 2002: Peer review of statistics in medical research: the other problem. *British Medical Journal* 324, 1271–73. **Gardner, M. J., Machin, D., Campbell, M. J. and Altman, D. G.** 2000: *Statistical checklists.* In Altman, D. G., Machin, D., Bryant, T. N. and Gardner, M. J. (eds), *Statistics with confidence*, 2nd edition. London: BMJ Books, 191–201. **Moher, D., Schulz, K. F. and Altman, D. G. for the CONSORT Group** 2001: The CONSORT statement: revised recommendations for improving the quality of reports of parallel-group randomised trials. *Annals of Internal Medicine* 134, 657–62. **Rubinstein, L. V.** 2005: *Statistical review for medical journals, guidelines for authors.* In Colton, T. and Armitage, P. (eds). *Encyclopedia of Biostatistics*, 2nd edition. 2005, Chichester: John Wiley & Sons Ltd, pages 5190–5192. **Smith, R.** 2005: *Statistical review for medical journals, journal's perspective;* In Colton, T. and Armitage, P. (eds). *Encyclopedia of Biostatistics*, 2nd edition. 2005, Chichester: John Wiley & Sons Ltd, pages 5193–5196. **Ware, J. H.** 2005: *Statistical review for medical journals;* In Colton, T. and Armitage, P. (eds). *Encyclopedia of Biostatistics*, 2nd edition. 2005, Chichester: John Wiley & Sons Ltd, pages 5186–5190.

statistics in imaging This is the use of statistical techniques to analyse and quantify information contained in digital image format. Imaging is widely used in medicine to visualise objects, structures and even physical processes *in vivo* and *in vitro*. A significant advantage in medical imaging is the ability to visualise structures or processes without relying on surgical operations. Thus, animals may be recycled in drug discovery and development or patients may not suffer from intrusive procedures. The ability to acquire information without intrusive procedures is also a disadvantage to medical imaging. This raises the issue of surrogate imaging ENDPOINTS (or *surrogacy*); i.e. how well do the conclusions from an imaging experiment correspond to physical properties obtained from an intrusive procedure?

Although the human visual system is very good at extracting information from images, the sheer amount of data being produced creates the common problem of 'not enough time to look at everything'. Statistical techniques using computers enable researchers and clinicians to summarise large numbers of images rapidly so that patterns, trends, regions of activation, etc., may be identified and quantified. Besides the amount of information, medical imaging systems also see beyond the visible light spectrum and are able to process information from a wide range of the electromagnetic spectrum.

Examples of medical imaging systems include conventional radiology (X-rays), angiography (imaging of a system of blood vessels using X-rays), positron emission tomography (PET), X-ray transmission computed tomography (CT), magnetic resonance imaging (MRI), microscopy, single photon emission (computed) tomography (SPET or SPECT), spectroscopy and ultrasound imaging. Even electroencephalograms (EEGs) or magnetoencephalograms (MEGs) are examples of imaging systems, albeit with very poor spatial resolution when compared to MRI or PET.

An image is a two-dimensional function that depends on spatial coordinates, where the amplitude of the function represents the brightness or grey level of the image at a particular point. Individual elements of the image are known as picture elements, pixels for short. Images may be collected to form a three-dimensional data structure, or volume, where the individual elements are called voxels. This is common in, for example, MRI and PET, where an experiment on a single subject will involve acquiring information in three spatial dimensions and in time.

Traditional statistical techniques in image analysis include areas such as signal and morphological processing. Signal processing applications include image enhancement, image restoration, colour image processing, wavelets and compression. Morphological processing assumes that set theory may be applied to manipulate structures present in an image.

A relatively new area of research in imaging is the use of MRI in functional or pharmacological studies of the brain. Functional MRI (fMRI) is now well developed and seeks to associate brain functions (human or animal) with specific regions of the brain. Pharmacological MRI (phMRI) is relatively new and seeks to associate pharmacokinetics with specific regions of the (animal) brain. Although group studies are widespread, consider a single-subject analysis from a typical fMRI experiment. After data acquisition, a set of images associated with distinct slices of the brain is available for analysis. Each slice will have a time sequence associated with it; i.e. the imaging experiment contains both spatial and temporal information. Given knowledge of the study design, the goal is to identify regions of the brain where significant activation was observed, where activation is measured by the intensity of the signal observed in the fMRI experiment. Signal intensity is related to the ratio of oxygenated and deoxygenated blood locally in the brain.

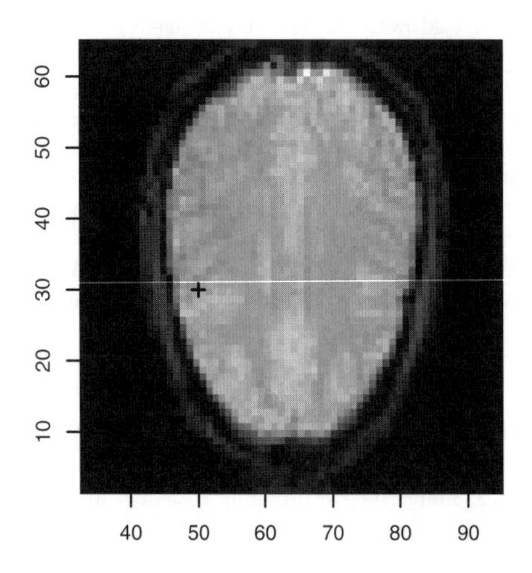

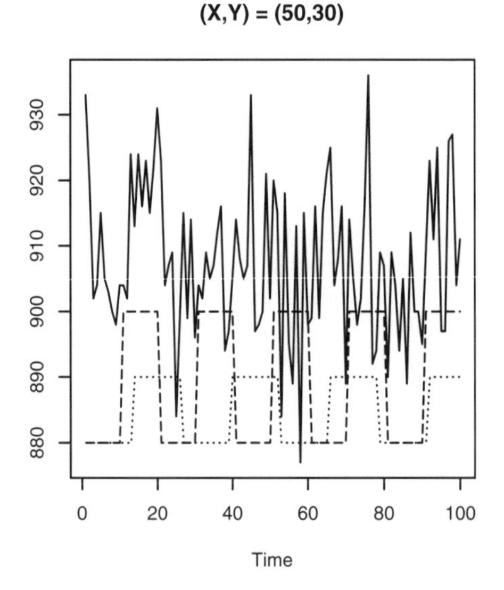

statistics in imaging *Example of an MRI slice (left) and voxel time course (right). The experimental design has been superimposed on the time course plot where the visual stimulation is shown by a dashed line and the audio stimulation is shown by a dotted line (data provided by the Brain Mapping Unit, Department of Psychiatry, University of Cambridge)*

The time course in the figure (page 440) shows a typical slice from an MRI experiment and the study design of on/off sequences for visual (dashed line) and auditory (dotted line) stimuli. Each voxel in the image has an associated time course; a mask that eliminates nonbrain voxels is typically used to focus the data analysis. LINEAR REGRESSION, or, more fully, fitting the GENERALISED LINEAR MODEL (GLM), is performed on each voxel using the experimental design, convolved with a function to model the haemodynamic response of the patient, as the independent variable. Trend removal is an important step and may be applied as a pre-processing step or by incorporating low-frequency terms explicitly in the GLM. The typical assumption of independence between observations is not true in fMRI data; methods such as pre-whitening, autoregressive modelling and least squares with adjustment for correlated errors are attempts to overcome the limitations of ordinary least squares.

Fitting the GLM to fMRI data may be performed on an individual voxel, on a cluster of voxels known as a region of interest (ROI), where the data are averaged in space to produce a single time course, or on every brain voxel in the image. For the first two cases, standard theory for statistical inference on regression models may be applied. For the third case, techniques such as Gaussian random field theory, resampling (see BOOTSTRAP) and adjustments by multiple comparison procedures have been used. Regardless of which method is applied, a set of voxels is obtained where significant activation during the experiment was detected. Researchers then relate the images to the anatomical regions identified in the activation image, also known as a statistical parametric map (SPM).

Information from a group of patients may be combined or compared by first registering all images with a standard brain. The most common brain atlas used is the Talairach atlas. Then, a random effects or fixed effects model (see LINEAR MIXED EFFECTS MODELS) may be used to apply a statistical hypothesis test between groups of subjects in the experiment. For more details see Serra (1982), Glasbey and Horgan (1995), Moonen and Bandettini (1999), Gonzalez and Woods (2002) and Worsley *et al.* (2002). *BW*

Glasbey, C. A. and Horgan,G. W. 1995: *Image analysis for the biological sciences*. Chichester: John Wiley & Sons, Ltd. **Gonzalez, R. C. and Woods, R. E.** 2002: *Digital image processing*, 2nd edition. Englewood Cliffs, NJ: Prentice Hall. **Moonen, C. T. W. and Bandettini,P. A.** (eds) 1999: *Functional MRI*. Berlin: Springer-Verlag. **Serra, J.** 1982: *Image analysis and mathematical morphology*. London: Academic Press. **Worsley, K. J., Liao, C. H., Aston, J., Petre, V., Duncan, G. H., Morales, F. and Evans, A. C.** 2002: A general statistical approach for fMRI data. *NeuroImage* 15, 1, 1–15.

StatXact This is a specialised software package for the exact analysis of small-sample categorical and nonparametric data with special emphasis on data in the form of contingency tables. The term 'small-sample' applies equally to datasets with only a few observations, to large but unbalanced datasets or to CONTINGENCY TABLES with zeros and small cell counts in some of the cells but large cell counts in other cells.

In these settings, StatXact produces exact P-VALUES and exact CONFIDENCE INTERVALS instead of relying on possibly unreliable large-sample theory for its inferences. The inference is based on generating permutation distributions of the appropriate test statistics in a conditional reference set.

Different reviews of StatXact are given by Lynch, Landis and Localio (1991), Wass (2000) and Oster (2002). The current version, StatXact 6, offers exact P-values for one-, two- and K-sample problems, 2×2, $2 \times c$ and $r \times c$ contingency tables and measures of ASSOCIATION. The data may be either unstratified or stratified. Both independent and blocked samples are accommodated. This version computes the exact confidence interval for ODDS RATIOS that arise from 2×2 and $2 \times c$ contingency tables, as well as an exact confidence interval for the MEDIAN shift parameter in an ordered $2 \times c$ contingency table. StatXact offers procedures that cater explicitly to binomial data, Poisson data, nominal categorical data, ordered categorical data, ordered correlated categorical data, continuous complete data and continuous right-censored data. For comparing two proportions (either from dependent or independent samples), StatXact provides the exact unconditional confidence interval for a difference in proportions or the ratio of two proportions and computes exact P-values for tests of equivalence and noninferiority.

In addition to tools for exact inference, StatXact also provides exact power and sample-size calculations for study designs involving one, two or several binomial populations. In the two-binomial case, these features include exact power and sample-size calculations for designing noninferiority and equivalence studies.

In case the computation of an exact P-value becomes infeasible due to the lack of either time or computing memory, StatXact produces an unbiased estimate of the exact P-value to at least two decimal digits of accuracy using efficient Monte Carlo simulation strategies (see MARKOV CHAIN MONTE CARLO). The user can arbitrarily increase the number of Monte Carlo simulations in order to increase the accuracy.

StatXact 6 runs on Microsoft Windows NT/2000/XP as a standalone product. In addition, a special version, StatXact PROCs for SAS Users, is available as external SAS procedures for both the Microsoft Windows and Unix operating systems. *CCo/PSe/CM/NP*

Lynch, J. C., Landis, J. R. and Localio, A. R. 1991: StatXact. *The American Statistician* 45, 2, 151–4. **Oster, R. A.** 2002: An examination of statistical software packages for categorical data analysis using exact methods. *The American Statistician* 56, 3, 235–46. **Wass, J. A.** 2000: StatXact 4 for Windows. *Biotech Software and Internet Report* 1, 1, 17–23.

stem-and-leaf plot Essentially, this is an enhanced HISTOGRAM in which the actual data values are retained for inspection. Observed values are each divided into a suitable 'stem' and 'leaf', e.g. the tens figure and the units figure in many examples, and then all the leaves corresponding to a particular stem are listed (usually horizontally) next to the value of the stem. An example is shown in the figure.

```
14 : 2
14 : 555
14 : 67777
14 : 889
15 : 000000111111
15 : 222222222222333333333333333333
15 : 444444444445555555555555555555555
15 : 66666666666666666666677777777777777777777
15 : 88888888888888888888888888888888899999999999999999
16 : 00000000000000000000011111111111111111111
16 : 22222222222222222233333333333333333333333333333333
16 : 4444444444444444445555555555555555555
16 : 666666666667777777
16 : 88888899999999
17 : 00000000000111
17 : 333
17 : 4
17 : 67
17 : 88
```

stem-and-leaf plot *A stem-and-leaf plot for the heights in centimetres of 351 elderly women*

The plot combines the visual picture of the data provided by the histogram with a display of the ordered data values. The design of stem-and-leaf plots is discussed in Velleman and Hoaglin (1981). It is important to use a typeface for which each digit occupies equivalent space, otherwise a key feature of being 'a histogram on its side' is lost. *BSE*

Velleman, P. F. and Hoaglin, D. C. 1981: Applications, basics, and computing of exploratory data analysis. Boston: Duxbury.

stepwise regression See LOGISTIC REGRESSION, MULTIPLE LINEAR REGRESSION

stochastic process This is any system that develops in accordance with probabilistic laws, usually in time but sometimes in space and possibly even in both time and space. For example, the spread of an epidemic is a stochastic process and its development can be tracked in time, across some terrain or at the conjunction of both time and position.

The constituents of a stochastic process are its *state*, X say, and its *indexing variable(s)*, s or t. The state is the primary measure of interest, such as number of individuals ill, while the indexing variable denotes either the time (t) or the position (s) at which the state is measured. A discrete indexing variable is usually shown as a subscript, but a continuous index appears within traditional function notation. For example, suppose that the state of the epidemic is the number of individuals who are ill. Then X_t would denote the number of individuals ill at time t if observations were taken at the start of each day, while $X(s)$ would denote the number of individuals ill at position s measured continuously in space. Of course, the state of the process can also be either discrete (e.g. number of individuals ill) or continuous (e.g. ECG reading of a cardiac patient).

An essential ingredient in a stochastic process is the *dependence* of either successive or neighbouring observations. Different assumptions about the dependence structure lead to different types of stochastic process, which can be used as models for many observations collected in practice. The objective is usually to derive theoretical PROBABILITIES for the various states of the system and thus to use these probabilities either for predicting the future behaviour of the system or for gaining some understanding of its mechanism. Many practical systems can be modelled adequately by assuming a Markovian dependence structure, in which the PROBABILITY DISTRIBUTION of X depends only on the most recent or neighbourly value. Standard stochastic processes that accord with such an assumption include random walks, Markov chains, branching processes, birth-and-death processes, queues and Poisson processes. Jones and Smith (2001) provide an accessible introduction to the mathematics of such processes. Some classical applications of stochastic models to medicine are described in Gurland (1964).

Successful uses of Markov models in medical contexts range in time and application from the planning of patient care (Davies, Johnson and Farrow, 1975) to resource provision (Davies and Davies, 1994) and the cost-effectiveness of vaccines (Byrnes, 2002). Many more examples can be found in journals such as *Health Care Management Science*. *WK*

Byrnes, G. B. 2002: A Markov model for sample size calculation and inference in vaccine cost-effectiveness studies. *Statistics in Medicine* 21, 3249–60. **Davies, R. and Davies, H. T. O.** 1994: Modelling patient flows and resource provision in health systems. *Omega, International Journal of Management Science* 22, 123–31. **Davies, R., Johnson, D. and Farrow, S.** 1975: Planning patient care with a Markov model. *Operational Research Quarterly* 26, 599–607. **Gurland, J.** (ed.) 1964: *Stochastic models in medicine and biology*. Madison, WI: University of Wisconsin Press. **Jones, P. W. and Smith, P.** 2001: *Stochastic processes, an introduction*. London: Arnold.

stratified randomisation See RANDOMISATION

stratified sampling Stratified sampling occurs within defined strata of some population. This should be carried out when the population contains easily identifiable subpopulations. If the sizes of the strata are different then proportional allocation should be used. If the STANDARD DEVIATIONS are known in advance then optimal or Neyman allocation can be used to minimise the VARIANCE of the estimate of the population MEAN. If they are unknown it is possible to use a pilot study to estimate the standard deviations.

The method is as follows. Define the strata that the population falls into. Decide if the strata are of a similar size and if the standard deviations are known. For similar sized strata use simple random sampling to select members of each stratum. If the sizes are different then the number in each stratum is proportional to stratum size. Then simple random sampling is used to obtain the correct number in each stratum. If the standard deviation is known in advance then for a fixed population size, n is obtained by choosing n_j so that:

$$\frac{n_j}{n} \approx \frac{N_j S_j}{\sum_{m=1}^{s} N_m S_m}$$

where N_j is the number in the stratum, S_j is the standard deviation of values of items within the strata, n is the fixed population size, n_j is the number to be chosen by simple random sampling from the stratum and s is the number of strata.

Thornhill *et al.* (2000) used stratified sampling in a study of disability following head injury. The patients were stratified according to the Glasgow coma score. The mild and unclassified patients were further stratified by the presenting hospital and a simple random sample was taken.

In general, if the population can be separated into distinguishable strata then the estimates from stratified sampling will be more precise than from a SIMPLE RANDOM SAMPLE and therefore it can be efficient. The disadvantages are that it can be difficult to choose the strata, it is not useful without homogeneous subgroups, it can require accurate information about the population and it can be expensive. For more details see Crawshaw and Chambers (1994) and Upton and Cook (2002). *SLV*

Crawshaw, J. and Chambers, J. 1994: *A concise course in A level statistics*, 3rd edition. Cheltenham: Stanley Thornes Publishers Ltd. **Thornhill, S., Teasdale, G. M., Murray, G. D., McEwen, J., Roy, C. W. and Penny, K. I.** 2000: Disability in young people and adults one year after head injury: prospective cohort study. *British Medical Journal* 320, 1631–5. **Upton, G. and Cook, I.** 2002: *Dictionary of statistics*. Oxford: Oxford University Press.

structural equation modelling software The four most commonly used packages for fitting structural equation models are:

EQS (http://www.mvsoft.com/)
LISREL(http://www.ssicentral.com/)
MPlus(http://www.statmodel.com/order.html)
AMOS(http://www.amosdevelopment.com/)

All four allow the fitting of complex models relatively easily, although MPlus is possibly the most flexible. Each package's website provides specific information on their capabilities, as well as availability and cost. *BSE*

structural equation models The operational definition provided by Pearl (2000, p. 160) states: 'An equation $y=\beta x + \varepsilon$ is said to be *structural* if it is to be interpreted as follows: In an ideal experiment where we control X to x and any other set Z of variables (not containing X or Y) to z, the value of Y is given by $\beta x + \varepsilon$, where ε is not a function of the settings x and z.'

The key word here is 'control'. We are observing values of Y after manipulating or fixing the values of X. The model implies that the values of Y, in fact, are determined by the values of X. A structural equation model is a description of the causal effect of X on Y. It is a CAUSAL MODEL, and the parameter β is a measure of the causal effect of X on Y. It should be clearly distinguished from a linear regression equation that simply describes the ASSOCIATION between two random variables, X and Y. If we are able, in practice, to intervene and control the values of X (by random allocation, for example) then it is straightforward to use the resulting data to obtain a valid estimate of β. If, however, we do not have control of X, but can only observe the values of X and Y (and Z), as in an epidemiological or other type of OBSERVATIONAL STUDY, for example, this does not invalidate the above operational definition, but the challenge for the data analyst is to find a valid (i.e. unbiased) estimate of the causal parameter β under these circumstances.

The equation $y=\beta x + \varepsilon$ is, of course, a description of a very simple structural model. It is common to collect data on several response variables (Ys) and several explanatory variables (Xs) and to construct a series of structural equations of the following form:

$$y_j = \sum_i \beta_i x_i + \sum_{k \neq j} \beta_k y_k + \varepsilon_j \qquad (i = 1 \text{ to } I; j,k = 1 \text{ to } J) \tag{8}$$

in which several of the β values will be fixed to be zero, a priori. The others are to be estimated from the data. The form of the equations defined in (1) – i.e. the structural theory that determines the pattern of β values to be estimated and those fixed at zero - is determined by the investigator's prior knowledge or hypotheses concerning the causal processes generating the data. Quoting Byrne (1994, p. 3): 'Structural modeling (SEM) is a statistical methodology that takes a hypothesis-testing (i.e. confirmatory) approach to the multivariate analysis of a structural theory bearing on some phenomenon.'

Typically, SEM involves (a) the specification of a set of structural equations, (b) representation of these structural equations using a graphical model (a path diagram – see later), (c) simultaneously fitting the set of structural equations to a given set of data in order to estimate the β values and to test the adequacy of the model. If the model fails to fit then the investigator may revise the model and try again. The success of the exercise is likely to be highly dependent upon the quality of the investigator's prior knowledge of the likely causal mechanisms under test and how much thought he or she has given to the design of the study in the first place. Good design and subsequent statistical analyses require technical knowledge, skill and experience. For technical knowledge, readers are referred to introductory texts by Dunn, Everitt and Pickles (1993), Byrne (1994) and Shipley (2000), and to the advanced monograph by Bollen (1989). Discussion of SEM in the context of recent work on causal inference can be found in Pearl (2000, 2009) and, again, in Shipley (2000). Traditionally, SEM has concentrated on structural models for quantitative data, which are usually assumed to be multivariate normal. Extensions from the traditional linear structural equations (i.e. LINEAR REGRESSION) to generalized linear structural equations are discussed by Skrondal and Rabe-Hesketh (2004).

It is frequently the case that we cannot measure constructs directly, or at least not without considerable MEASUREMENT ERROR. This gives rise to the idea of LATENT VARIABLES. These are characteristics that are not directly observable. They may be straightforward concepts such as height, weight, amount of exposure to a known toxin, or concentration of a given metabolite in blood or urine, but we explicitly acknowledge that they cannot be measured without error. The observed measurement is a manifest or indicator variable, while the corresponding unknown, but true, value is a latent variable. However, latent variables may be more abstract theoretical constructs that are introduced to explain COVARIANCE between manifest or indicator variables.

An example of this last type is the set of scores on a battery of cognitive tests that are assumed in some way to reflect a subject's cognitive ability or general intelligence. Another example could be a set of symptom severity scores (the manifest variables), which are assumed to be indicators of a patient's overall degree of depression (the latent variable). Typically, a data analyst will propose a formal measurement model (usually equivalent to some form of factor analysis representation) to relate the observed measurements with the underlying latent variables. We can then proceed to propose structural or causal hypotheses involving the latent variables instead of the fallible (error-prone) indicators. We start, for example, with a COVARIANCE MATRIX for the observed variables. We fit a general structural equation model to this covariance or moments matrix. This procedure will involve the simultaneous fitting of the measurement equations for the relevant latent variables and their corresponding indicators and of the structural equations thought to reflect the assumed causal relationships between the latent variables. Specialist software packages are now widely available for such analyses (see STRUCTURAL EQUATION MODELLING SOFTWARE).

Structural equation models are very often represented by a graphical structure known as a path diagram (see PATH ANALYSIS). In a path diagram the proposed relationships between variables (whether manifest or observed) are represented either by a single-headed arrow (indicating the direction of a causal effect) or a double-headed one (indicating CORRELATION). The observed or manifest variables are usually placed within a rectangular square box, while latent variables are placed within an oval or a circle. Random measurement errors and residuals from structural equations, although they are strictly speaking latent variables, are not traditionally placed within a circle or oval. Path diagrams are very closely related to the graphical representations (directed acyclic graphs, or DAGs, for example; see GRAPHICAL MODELS) that have relatively recently been developed elsewhere (see Pearl, 2000, 2009, for example). Two simple examples of path diagrams are shown in the two figures. A detailed explanation will be given in the following section.

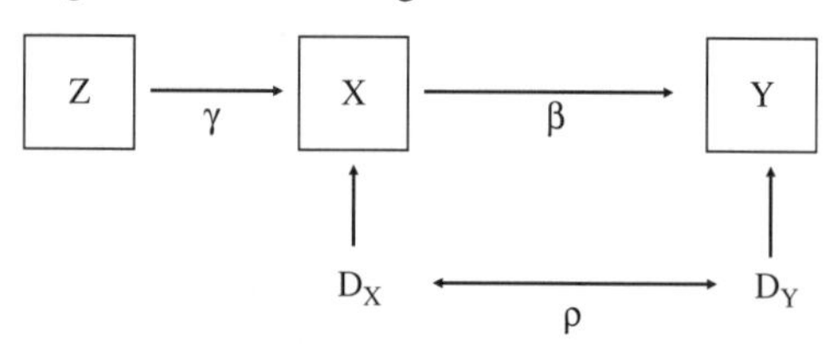

structural equation models *Path diagram to represent the structural equations linking encouragement to stop smoking during pregnancy (Z), the amount smoked during pregnancy (X) and the birth weight of the child (Y). D_X and D_Y are randomly distributed residuals*

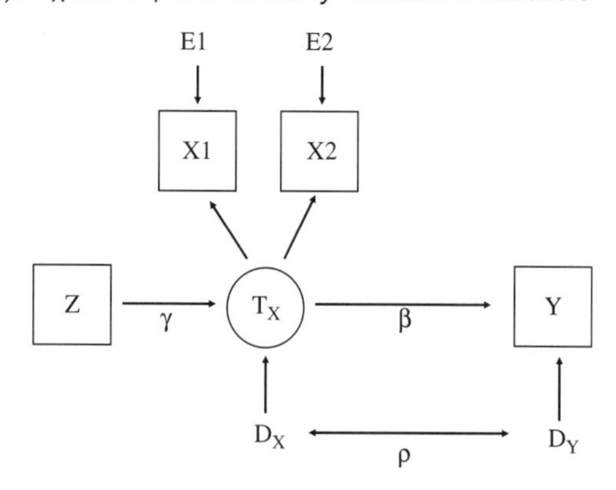

structural equation models *Path diagram to represent the structural equations linking encouragement to stop smoking during pregnancy (Z), the true amount smoked during pregnancy (T_X) and the birth weight of the child (Y). D_X and D_Y are randomly distributed residuals. X1 and X2 are error-prone indicators of smoking, with uncorrelated measurement errors E1 and E2 respectively*

For an example, Permutt and Hebel (1989) describe a trial in which pregnant women were randomly allocated to receive encouragement to reduce or stop their cigarette smoking during pregnancy (the treatment group) or not (the control group) – indicated by the binary variable, Z. An intermediate outcome variable (X) was the amount of cigarette smoking recorded during pregnancy. The ultimate outcome (Y) was the birth weight of the newborn child. Smoking is likely to have been reduced in the group subject to encouragement, but also in the control group (although, presumably, to a lesser extent). There are also likely to be hidden confounders (e.g. other health promoting behaviours) that are associated with both the mother's smoking during pregnancy and the child's birth weight. Smoking (X) is an endogenous treatment variable – the above confounding will result in the residual from a structural equation model to explain the level of smoking by RANDOMISATION to receive encouragement being correlated with the residual from the structural equation linking observed levels of smoking to the birth weight of the child. We assume that there is no direct effect of randomization (Z) on outcome (Y); the effect of Z on Y is an indirect one through smoking (X); i.e. Z is an INSTRUMENTAL VARIABLE. Ignoring the intercept terms, the two structural equations are the following:

$$X = \gamma Z + D_X \text{ and } Y = \beta X + D_Y$$

In fitting these two models to the appropriate data we acknowledge the correlation (ρ) between the residuals, D_X and D_Y (those components of X and Y not explained by Z and X respectively). The overall model is illustrated by the first figure (page 444). Now, what if we acknowledge that smoking levels cannot be measured accurately and we decide to obtain two different measurements on each person in the trial ($X1$ and $X2$, say, being self-reported numbers of packs per day, obtained at 6 months and 8 months into the pregnancy)? The true level of smoking is now represented by the variable T_X. Our measurement model is represented by the two equations:

$$X1 = T_X + E1 \text{ and } X2 = T_X + E2$$

We assume that the $E1$ and $E2$ measurement errors are uncorrelated and that there is no change in the true level of smoking between the two times. The revised structural equations now use T_X rather than X, as follows:

$$T_X = \gamma Z + D_X \text{ and } Y = \beta T_X + D_Y$$

The corresponding path diagram is shown in the second figure (page 444). Note that not all of the model parameters implied by the model in the second figure can be estimated. The model is too complex for the data at hand. The model as a whole is said to be underidentified, but the good news is that we can still estimate β, the parameter most likely to be of interest to the investigator. Problems of underidentification are beyond the scope of this entry, but are covered by the standard textbooks on structural equations modelling referenced below. *GD*

Bollen, K. A. 1989: *Structural equations with latent variables*. New York: John Wiley & Sons, Inc. **Byrne, B. M.** 1994: *Structural equation m odeling with EQS and EQS/Windows*. Thousand Oaks, CA: Sage Publications. **Dunn, G., Everitt, B. S. and Pickles, A.** 1993: *Modelling covariances and latent variables using EQS*. London: Chapman & Hall. **Pearl, J.** 2000; 2nd edition, 2009: *Causality*. Cambridge: Cambridge University Press. **Permutt, T. and Hebel, J. R.** 1989: Simultaneous-equation estimation in a clinical trial of the effect of smoking and birth weight. *Biometrics* 45, 619–22. **Shipley, B.** 2000: *Causes and correlation in biology*. Cambridge: Cambridge University Press. **Skrondal, A. and Rabe-Hesketh, S.** 2004: *Generalized latent variable modeling: multilevel, longitudinal and structural equations models*. Boca Raton, FL: Chapman & Hall/CRC.

Student's *t*-distribution See *t*-DISTRIBUTION

Student's *t*-test William Sealy Gosset, who worked under the pseudonym of 'Student', developed the Student's *t*-test. The Student's *t*-test is commonly referred to merely as the *t*-test. The simplest use of the *t*-test is in comparing the MEAN of a sample to some specified population mean this is usually called the one-sample *t*-test. The *t*-test can be modified to compare the means of two independent samples (the two-sample *t*-test) and for paired data to compare the differences between the pairs (the paired *t*-test).

Student's *t*-test is a parametric test and certain assumptions are made about the data. These are that the observations within each group (with independent samples) or the differences (with paired samples) are approximately normally distributed and for the two-sample case we also require the two groups to have similar VARIANCES. If the sample data does not meet these assumptions then the analysis is seriously flawed. However, the *t*-test is 'robust' and is not greatly affected by a moderate failure to meet the assumptions.

The one-sample *t*-test can be used to compare the mean of a sample to a certain specified value. This value is usually the population mean. The NULL HYPOTHESIS states that there is no significant difference between the sample mean and the population mean and the alternative hypothesis states that there is a significant difference between the sample mean and the population mean. The assumption we make is that the data are a random sample of independent observations from an underlying normal distribution. The test statistic t is given by:

$$t = \frac{\text{Sample mean} - \text{Hypothesised mean}}{\text{Standard error of sample mean}}$$

This is compared against the *t*-DISTRIBUTION with $n - 1$ DEGREES OF FREEDOM, where n is the sample size. So t is the deviation of a normal variable from its hypothesised mean measured in STANDARD ERROR units. The standard error of the sample mean is estimated by $\sqrt{(s/n)}$, where s is the sample STANDARD DEVIATION.

For example, suppose BMI values for a sample of 25 people were measured and a mean value of 24.5 was found with a sample standard deviation of 2.5. To test if this sample mean BMI is significantly different from a population mean

BMI of 26 we can use the one-sample t-test, where our null hypothesis is that there is no difference between the sample mean of 24.5 and the population mean of 26. This allows us to calculate the test statistic as follows:

$$t = \frac{24.5 - 26}{\sqrt{\frac{6.25}{25}}} = -3.0$$

Using tables for t-distribution with $(n-1) = 24$ degrees of freedom. We find a P-VALUE of 0.0062. The result is statistically significant and we therefore accept the alternative hypothesis that the mean BMI of the sample is significantly different from 26.

We can use the two-sample t-test to determine the statistical significance of an observed difference between the mean values of some variable between two subgroups or between separate populations. For example, we could look at the differences in heights between males and females. The test statistic for the two-sample t-test is given by:

$$t = \frac{\text{Difference in sample means} - \text{Difference in hypothesised means}}{\text{Standard error of the difference in the two sample means}}$$

Frequently the null hypothesis of interest is whether the two groups have equal means and the corresponding two-sided alternative hypothesis is that the means are in fact different. For example, when comparing the mean outcome for two different treatments is the difference in means observed a statistically significant one? In this case the test statistic reduces to:

$$t = \frac{\text{Difference in the two sample means}}{\text{Standard error of the difference in the two sample means}}$$

This is then compared to the t-distribution with $n_1 + n_2 - 2$ degrees of freedom, where n_1 is the sample size for the first group and n_2 is the sample size for the second group. The standard error of the difference in the two-sample means is given by:

$$\text{SE}(\bar{x}_1 - \bar{x}_2) = \sqrt{\frac{s_p^2}{n_1} + \frac{s_p^2}{n_2}}$$

where

$$s_p^2 = \frac{(n_1 - 1)s_1^2 + (n_2 - 1)s_2^2}{n_1 + n_2 - 2}$$

and s_1 and s_2 are the standard deviations for groups one and two respectively.

For the paired t-test, the data are dependent, i.e. there is a one-to-one correspondence between the values in the two samples. Paired data can occur from two measurements on the same person, e.g. before and after treatment or the same subject measured at different times. It is incorrect to analyse paired data ignoring the pairing in such circumstances, as important information is lost. Some factors you do not control in the experiment will affect the before and the after measurements equally, so they will not affect the difference between before and after. By looking only at the differences, a paired t-test corrects for these factors.

The two-sample paired t-test usually tests the null hypothesis that the population mean of the paired differences of the two samples is zero. We assume that the paired differences are independent. To perform the paired t-test we calculate the difference between each set of pairs and then perform a one-sample t-test on the differences with the null hypothesis that the population mean of the differences is equal to zero. More details can be found in Altman (1991). *MMB*

Altman, D. G. 1991: *Practical statistics for medical research*. London: Chapman & Hall.

subgroup analysis This form of analysis is often employed in CLINICAL TRIALS in an attempt to identify particular subgroups of patients for whom a treatment works better (or worse) than for the overall patient population. For example, does a treatment work better for men than for women? Such a question is a natural one for clinicians to ask since they do not treat 'average' patients and, when confronted with a female patient with a certain condition, would like to know whether the accepted treatment for the condition works, say, less well for women.

Assessing whether the effect of treatment varies according to the value of one or more patient characteristics is relatively straightforward from a statistical viewpoint, involving nothing more than testing a treatment by covariate interaction. However, many statisticians would caution against such analyses and, if undertaken at all, suggest that they are interpreted extremely cautiously in the spirit of 'exploration' rather than anything more formal. The reasons for such caution are not difficult to identify.

First, trials can rarely provide sufficient POWER to detect such subgroup/interaction effects; clinical trials accrue sufficient participants to provide adequate precision for estimating quantities of primary interest, usually overall treatment effects. Confining attention to subgroups almost always results in estimates of inadequate precision. A trial just large enough to evaluate an overall treatment effect reliably will almost inevitably lack precision for evaluating differential treatment effects between different population subgroups.

Second, RANDOMISATION ensures that the overall treatment groups in a clinical trial are likely to be comparable. Subgroups may not enjoy the same degree of balance in patient characteristics.

Finally, there are often many possible prognostic factors in the baseline data, e.g. age, gender, race, type or stage of disease, from which to form subgroups, so that analyses may quickly degenerate into 'data dredging', from which arises the potential for post hoc emphasis on the subgroup analysis giving results of most interest to the investigator, with undue emphasis given to results deemed 'statistically significant' contributing, in turn, to a preponderance of '$p < 0.05$' results published in the medical literature (an excess of false positive findings, therefore).

Other potential dangers of subgroup analysis can be found in detail in Pocock *et al.* (2002). *BSE*

Pocock, S. J., Assmann, S. E., Enos, L. E. and Kasten, L. E. 2002: Subgroup analysis, covariate adjustment and baseline comparisons in clinical trial reporting: current practice and problems. *Statistics in Medicine* 21, 2917–30.

sufficient-component cause model See CAUSAL MODELS

summary measure analysis This is a relatively straightforward approach to the analysis of LONGITUDINAL DATA, in which the repeated measurements of a response variable made on each individual in the study are reduced in some way to a single number that is considered to capture an essential feature of the response over time. In this way, the multivariate nature of the repeated observations is transformed to a univariate measure. The approach has been in use for many years – see, for example, Oldham (1962) and Matthews *et al.* (1989). The most important consideration when applying a summary measure analysis is the choice of a suitable summary measure, a choice that needs to be made before any data are collected. The measure chosen needs to be relevant to the particular questions of interest in the study and in the broader scientific context in which the study takes place. A wide range of summary measures has been proposed, as shown in the first table. According to Frison and Pocock (1992), the average response over time is often likely to be the most relevant, particularly in CLINICAL TRIALS.

Having chosen a suitable summary measure, analysis will involve nothing more complicated than the application of Student's t-test or calculation of a CONFIDENCE INTERVAL for the group difference when two groups are being compared or a one-way ANALYSIS OF VARIANCE when there are more than two groups. If considered more appropriate because of the distributional properties of the selected summary measure, then analogous NONPARAMETRIC METHODS might be used.

The summary measure approach can be illustrated using the data shown in the second table, which come from a study of alcohol dependence. Two groups of subjects, one with severe dependence and one with moderate dependence on alcohol, had their salsolinol excretion levels (in millimoles) recorded on four consecutive days.

summary measure analysis *Salsolinol excretion data*

	Day			
Subject	*1*	*2*	*3*	*4*
Group 1 (moderate dependence)				
1	0.33	0.70	2.33	3.20
2	5.30	0.90	1.80	0.70
3	2.50	2.10	1.12	1.01
4	0.98	0.32	3.91	0.66
5	0.39	0.69	0.73	3.86
6	0.31	6.34	0.63	3.86
Group 2 (severe dependence)				
7	0.64	0.70	1.00	1.40
8	0.73	1.85	3.60	2.60
9	0.70	4.20	7.30	5.40
10	0.40	1.60	1.40	7.10
11	2.50	1.30	0.70	0.70
12	7.80	1.20	2.60	1.80
13	1.90	1.30	4.40	2.80
14	0.50	0.40	1.10	8.10

summary measure analysis *Possible summary measures (from Matthews et al., 1989)*

Type of data	*Question of interest*	*Summary measure*
Peaked	Is overall value of outcome variable the same in different groups?	Overall mean (equal time intervals) or area under curve (unequal intervals)
Peaked	Is maximum (minimum) response different between groups?	Maximum (minimum) value
Peaked	Is time to maximum (minimum) response different groups?	Time to maximum (minimum) respons
Growth	Is rate of change of outcome different between groups?	Regression coefficient
Growth	Is eventual value of outcome different between groups?	Final value of outcome or difference between last and first values or percentage change between first and last values
Growth	Is response in one group delayed relative to the other?	Time to reach a particular value (e.g. a fixed percentage of baseline)

Using the mean of the four measurements available for each subject as the summary measure leads to the results shown in the third table. There is no evidence of a group difference in salsolinol excretion levels.

summary measure analysis *Results from using the mean as a summary measure for the data in the second table*

	Moderate	*Severe*
Mean	1.80	2.49
sd	0.60	1.09
n	6	8
$t = -1.40$, df $= 12$, $P = 0.19$		
95% CI: $[-1.77, 0.39]$		

A possible alternative to the use of the MEAN as a summary measure is to use the maximum excretion rate recorded over the four days. Applying the WILCOXON RANK SUM TEST to this summary measure results in a test statistic of 36 and associated P-VALUE of 0.28.

The summary measure approach to the analysis of longitudinal data can accommodate missing data but the implicit assumption is that these are missing completely at random (see DROPOUTS). *BSE*

[See also AREA UNDER CURVE]

Frison, L. and Pocock, S. J. 1992: Repeated measures in clinical trials: analysis using mean summary statistics and its implications for design. *Statistics in Medicine* 11, 1685–704. **Matthews, J. N. S., Altman, D. G., Campbell, M. J. and Royston, P.** 1989: Analysis of serial measurements in medical research. *British Medical Journal* 300, 23–35. **Oldham, P. D.** 1962: A note on the analysis of repeated measurements of the same subjects. *Journal of Chronic Disorders* 15, 969–77.

support vector machines These are algorithms for learning complex classification and regression functions, belonging to the general family of 'kernel methods' discussed later. Their computational and statistical efficiency recently made them one of the tools of choice in certain biological DATA MINING applications.

Support vector machines (SVMs) work by embedding the data into a feature space by means of kernel functions (the so-called 'kernel trick'). In the binary classification case, a separating hyperplane that separates the two classes is sought in this feature space. New data points will be classified into one of both classes according to their position with respect to this hyperplane. SVMs owe their name to their property of isolating a (often small) subset of data points called 'support vectors', which have interesting theoretical properties.

The SVM approach has several important virtues when compared with earlier approaches: the choice of the hyperplane is founded on statistical arguments; the hyperplane can be found by solving a convex (quadratic) optimisation problem, which means that training an SVM is not subject to local minima; when a nonlinear kernel function is used, the hyperplane in the feature space can correspond to a complex (nonlinear) decision boundary in the original data domain. Even more interestingly, kernel functions can be defined not only on vectorial data but on virtually any kind of data, making it possible to classify strings, images, trees or nodes in a graph; the classification of unseen data points is generally computationally cheap and depends on the number of support vectors.

First introduced in 1992, support vector machines are now one of the standard tools in PATTERN RECOGNITION applications, mostly due to their computational efficiency and statistical stability. In recent years, extensions of this algorithm to deal with a number of important data analysis tasks have been proposed, resulting in the general family of 'kernel methods' (Shawe-Taylor and Cristianini, 2004) (see DENSITY ESTIMATIONS).

The kinds of relation detected by kernel methods include classifications, regressions, clustering (see CLUSTER ANALYSIS IN MEDICINE), principal components (see PRINCIPAL COMPONENT ANALYSIS), canonical correlations (see CANONICAL CORRELATION ANALYSIS) and many others. In the same way as with SVMs, the kernel trick allows these methods to be applied in a feature space that is induced by this kernel, making kernel methods applicable to virtually any kind of data.

Elegantly, the development of kernel methods can always be decomposed into two modular steps: the kernel design, on the one hand, and the choice of the algorithm, on the other hand. The kernel design part implicitly defines the feature space, which should contain all available information that is relevant for the problem at hand. The choice of the algorithm (which needs to be written in terms of kernels) can be done independently from the kernel design.

As with SVMs, most kernel methods reduce their training phase to optimising a convex cost function or to solving a simple eigenvalue problem, hence avoiding one of the main computational pitfalls of NEURAL NETWORKS. However, since they often implicitly make use of very high dimensional spaces, kernel methods run the risk of overfitting. For this reason, their design needs to incorporate principles of statistical learning theory, which help to identify the crucial parameters that need to be controlled in order to avoid this risk (see Vapnik, 1995). For further reference on SVMs, see Cristianini and Shawe-Taylor (2000). *NC/TDB*

Cristianini, N. and Shawe-Taylor, J. 2000: *An introduction to support vector machines*. Cambridge: Cambridge University Press (www.support-vector.net). **Shawe-Taylor, J. and Cristianini, N.**

2004: *Kernel methods for pattern analysis*. Cambridge: Cambridge University Press (www.kernel-methods.net). **Vapnik, V.** 1995: *The nature of statistical learning theory*. New York: Springer.

surrogate endpoints These are ENDPOINTS that can replace a clinical endpoint for the purpose of assessing the effects of new treatments earlier, at lower cost, or with greater statistical SENSITIVITY. Surrogate endpoints can include measurements of a biomarker, defined as 'a characteristic that is objectively measured and evaluated as an indicator of normal biological processes, pathogenic processes, or pharmacologic responses to a therapeutic intervention' (Biomarkers Definitions Working Group, 2001). Use of a biomarker as a surrogate endpoint can also be useful if the final endpoint measurement is unduly invasive or uncomfortable. For an endpoint to be a surrogate for a clinical endpoint it must be a measure of disease such that: (a) the size (or frequency) correlates strongly with that clinical endpoint (e.g. blood pressure is positively correlated with the risk of stroke) and (b) treatments producing a change in the surrogate endpoint also modify the risk of that particular clinical endpoint (e.g. reducing blood pressure reduces the risk of stroke).

Surrogate endpoints are routinely used in early drug development, where interest focuses on showing that new treatments have enough activity to warrant further research. In confirmatory PHASE III TRIALS, however, interest focuses on showing that new treatments have the anticipated clinical benefits, and in such situations surrogate endpoints can only be used if they have undergone rigorous statistical evaluation (or 'validation') (Burzykowski, Molenberghs and Buyse, 2005). Indeed, some promising surrogate endpoints have proven to be unreliable predictors of clinical benefits. For example, cardiac arrhythmia was believed to be a good surrogate endpoint for mortality after an acute heart attack, since in these circumstances patients with a higher risk of such an arrhythmia have a greater risk of death. However, several drugs (e.g. lignocaine, flecainide) that prevent arrhythmias after a heart attack actually increase mortality (Echt *et al.*, 1991). Similarly, some blood pressure-lowering drugs (such as angiotensin-converting enzyme inhibitors) have much larger effects on vascular mortality than might be predicted from their effects on blood pressure (Heart Outcomes Prevention Evaluation Study Investigators, 2000). In contrast, disease-free survival has recently been validated as an acceptable surrogate for overall survival in patients with colorectal cancer treated with fluoropyrimidines (Sargent *et al.*, 2005).

Prentice (1989) proposed a definition and operational criteria for the validation of surrogate endpoints. Although the strict criteria proposed by Prentice seem too stringent to ever be met in practice, his landmark paper sparked interest in developing statistical methods that could be used to show that a surrogate is acceptable (or 'validated') for the purposes of assessing a specific class of treatments in a specific disease setting. One approach consists of using a MULTILEVEL MODEL to show that the surrogate endpoint predicts the true endpoint ('individual-level' surrogacy), and that the effects of a treatment on the surrogate endpoint predict the effects of the treatment on the true endpoint ('trial-level' surrogacy) (Buyse *et al.*, 2000). The latter condition requires data to be available from several units, usually from a META-ANALYSIS of several trials. Another approach consists of using a CAUSAL MODEL to compare the causal effect of treatment on the true endpoint in patients for whom treatment does, and does not, affect the surrogate. See Weir and Walley (2006) for a review of the terminology and surrogate validation models. *CB/MB*

Biomarkers Definitions Working Group 2001: Biomarkers and surrogate endpoints: preferred definitions and conceptual framework. *Clinical Pharmacology and Therapeutics* 69, 89–95. **Burzykowski, T., Molenberghs, G. and Buyse, M.** (eds) 2005: *Evaluation of surrogate endpoints*. Springer Verlag. **Buyse, M., Molenberghs, G., Burzykowski, T., Renard, D. and Geys, H.** 2001: The validation of surrogate endpoints in meta-analyses of randomized experiments. *Biostatistics* 1, 49–67. **Echt, D. S., Liabson, P. R., Mitchell, L. B. *et al.* and the CardiacArrhythmia Suppression Trial (CAST) Investigators** 1991: Mortality and morbidity in patients receiving encainide, flecainide, or placebo: the cardiac arrhythmia suppression trial. *New England Journal of Medicine* 324, 781–8. **Heart Outcomes Prevention Evaluation Study Investigators** 2000: Effects of an angiotensin-converting enzyme inhibitor, ramipril, on death from cardiovascular causes, myocardial infarction, and stroke in high-risk patients. *New England Journal of Medicine* 342, 145–53. **Prentice, R. L.** 1989: Surrogate endpoints in clinical trials: definition and operational criteria. *Statistics in Medicine* 8, 431–40. **Sargent, D., Wieand, S., Haller, D. G. *et al.*** 2005: Disease-free survival (DFS) vs overall survival (OS) as a primary endpoint for adjuvant colon cancer studies: individual patient data from 20,898 patients on 18 randomized trials. *Journal of Clinical Oncology* 23, 8664–70. **Weir, C. J. and Walley, R. J.** 2006: Statistical evaluation of biomarkers as surrogate endpoints: a literature review. *Statistics in Medicine* 25, 183–203.

survival analysis – an overview This covers methods for the analysis of time-to-event data, e.g. survival times. Survival data occur when the outcome of interest is the time from a well-defined time origin to the occurrence of a particular event or ENDPOINT. If the endpoint is the death of a patient the resulting data are, literally, survival times. However, other endpoints are possible, e.g. the time to relief or recurrence of symptoms. Such observations are often referred to as time-to-event data although survival data is commonly used as a generic term. Standard statistical methodology is not usually appropriate for such data, for two main reasons.

First, the distribution of survival time in general is likely to display positive SKEWNESS and so assuming normality for an analysis (as done, for example, by a *t*-TEST or a regression) is probably not reasonable.

Second, more critical than doubts about normality, however, is the presence of censored observations, where the survival time of an individual is referred to as censored when the endpoint of interest has not yet been reached (more precisely, right censored). For true survival times this might be because the data from a study are analysed at a time point when some participants are still alive. Another reason for censored event times is that an individual might have been lost to follow-up for reasons unrelated to the event of interest, e.g. due to moving to a location that cannot be traced or due to accidental death (see DROPOUTS). When censoring occurs all that is known is that the actual, but unknown, survival time is larger than the censored survival time.

Specialised statistical techniques developed to analyse such censored and possibly skewed outcomes are known as survival analysis. An important assumption made in standard survival analysis is that the censoring is noninformative, i.e. that the actual survival time of an individual is independent of any mechanism that causes that individual's survival time to be censored. For simplicity, this description also concentrates on techniques for continuous survival times – the analysis of discrete survival times is described in Collett (2003).

As an example, consider data that arise from a double-blind, randomised controlled clinical trial (RCT) (see CLINICAL TRIALS) to compare treatments for prostate cancer (placebo versus 1.0 mg of diethylstilbestrol (DES) administered daily by mouth). The full dataset is given in Andrews and Herzberg (1985) and the first table shows the first seven of a subset of 38 patients used here and discussed in Collett (2003).

In this study, the time of origin was the date on which a cancer sufferer was randomised to a treatment and the endpoint is the death of a patient from prostate cancer. The survival times of patients who died from other causes or were lost during the follow-up process are regarded as right censored. The 'status' variable in the first table takes the value unity if the patient has died from prostate cancer and zero if the survival time is censored. In addition to survival times, a number of prognostic factors were recorded, namely the age of the patient at trial entry, their serum haemoglobin level in gm/100 ml, the size of their primary tumour in cm^2 and the value of a combined index of tumour stage and grade (the Gleason index with larger values indicating more advanced tumours). The main aim of this study was to compare the survival experience between the two treatment groups.

In general, to describe survival two functions of time are of central interest – the *survival function* and the *hazard function*. These are described in some detail next. The survival function $S(t)$ is defined as the probability that an individual's survival time, T, is greater than or equal to time t, i.e.:

$$S(t) = \text{Prob}(T \geq t)$$

The graph of $S(t)$ against t is known as the survival curve. The survival curve can be thought of as a particular way of displaying the frequency distribution of the event times, rather than by, say, a HISTOGRAM. When there are no censored observations in the sample of survival times, the survival function can be estimated by the empirical survivor function:

$$\hat{S}(t) = \frac{\text{Number of individual s with survival times} \geq t}{\text{Number of individual s in the data set}}$$

Since every subject is 'alive' at the beginning of the study and no one is observed to survive longer than the largest of the observed survival times then:

$$\hat{S}(0) = 1 \text{ and } \hat{S}(t_{\max}) = 1$$

Furthermore, the estimated survivor function is assumed constant between two adjacent death times, so that a plot of $\hat{S}(t)$ against t is a step function that decreases immediately after each 'death'.

This simple method cannot be used when there are censored observations since the method does not allow for information provided by an individual whose survival time is censored before time t to be used in the computing of the

survival analyisis *Survival times of prostate cancer patients*

Patient number	*Treatment (1 = placebo, 2 = DES)*	*Survival time (months)*	*Status (1 = died, 0 = censored)*	*Age (years)*	*Serum haem. (gm/100 ml)*	*Size of tumour (cm²)*	*Gleason index*
1	1	65	0	67	13.4	34	8
2	2	61	0	60	14.6	4	10
3	2	60	0	77	15.6	3	8
4	1	58	0	64	16.2	6	9
5	2	51	0	65	14.1	21	9
6	1	51	0	61	13.5	8	8
7	1	14	1	73	12.4	18	11
...	...	...	...	...	...	...	...

estimate at t. The most commonly used method for estimating the survival function for survival data containing censored observations is the product-limit or KAPLAN–MEIER ESTIMATOR. The essence of this approach is the use of a product of a series of conditional probabilities. One alternative estimator for censored survival times, derived differently but in practice often similar, is the Nelson–Aalen estimator. Approximate STANDARD ERRORS and pointwise symmetric or asymmetric CONFIDENCE INTERVALS for the survival function at a given time can be derived to determine the precision of the estimator – details are given in Collett (2003).

The Kaplan–Meier estimators of the survivor curves for the two prostate cancer treatments are shown graphically in the figure. The survivor curves are step functions that decrease at the time points when participants died of the cancer. The censored observations in the data are indicated by the 'cross' marks on the curves. In our patient sample there is approximately a difference of 20% in the proportion surviving for at least 50 to 60 months between the treatment groups.

Since the distribution of survival times tends to be positively skewed the MEDIAN is the preferred summary measure of location. The median survival time is the time beyond which 50% of the individuals in the population under study are expected to survive and, once the survivor function has been estimated by $\hat{S}(t)$, can be estimated by the smallest observed survival time, t_{50}, for which the value of the estimated survivor function is less than 0.5. The estimated median survival time can be read from the survival curve by finding the smallest value on the x axis for which the survival proportion reaches less than 0.5. The figure shows that the median survival in the placebo group can be estimated as 69 years while an estimate for the DES group is not available since survival exceeds 50% throughout the study period. A similar procedure can be used to estimate other percentiles of the distribution of the survival times and approximate confidence intervals can be found once the variance of the estimated percentile has been derived from the VARIANCE of the estimator of the survivor function.

In the analysis of survival data, it is often of some interest to assess which periods have the highest and which the lowest chance of death (or whatever the event of interest happens to be) among those people alive at the time. The appropriate quantity for such risks is the hazard function, $h(t)$, defined as the (scaled) PROBABILITY that an individual experiences an event in a small time interval δt, given that the individual has survived up to the beginning of the interval. The hazard function therefore represents the instantaneous death rate for an individual surviving to time t. It is a measure of how likely an individual is to experience an event as a function of the age of the individual. The hazard function may remain constant, increase or decrease with time or take some more complex form. The hazard function of death in human beings, for example, has a 'bathtub' shape. It is relatively high immediately after birth, declines rapidly in the early years and then remains relatively constant until beginning to rise during late middle age.

A Kaplan–Meier type estimator of the hazard function is given by the proportion of individuals experiencing an event in an interval per unit time, given that they have survived to the beginning of the interval. However, the estimated hazard function is generally considered 'too noisy' for practical use. Instead, the cumulative or integrated hazard function, which is derived from the hazard function by summation, is usually displayed to describe the change in hazard over time.

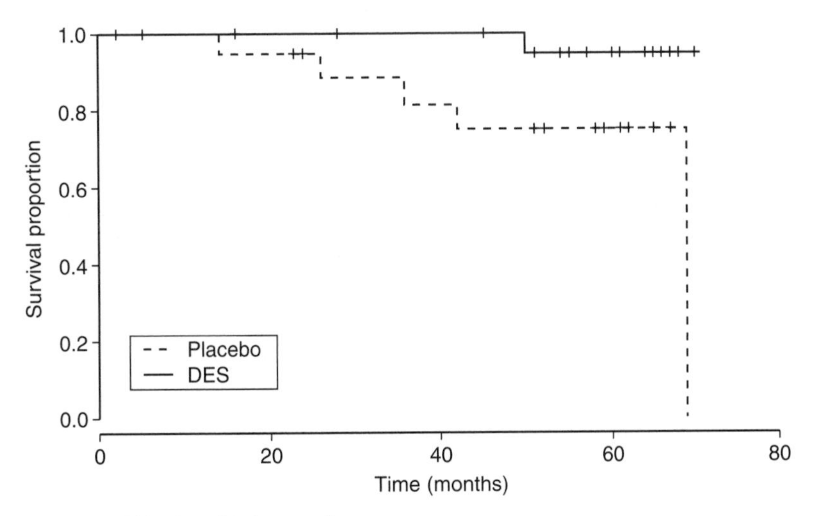

survival analysis *Display of Kaplan–Meier survivor curves*

In addition to comparing survivor functions graphically, a more formal statistical test for a group difference is often required in order to compare survival times analytically. In the absence of censoring, a nonparametric test such as the Mann–Whitney test could be used (see MANN–WHITNEY RANK SUM TEST). In the presence of censoring, the log-rank or Mantel–Haenszel test is the most commonly used nonparametric test (see MANTEL–HAENSZEL METHODS). It tests the NULL HYPOTHESIS that the population survival functions $S_1(t)$, $S_2(t)$, ..., $S_k(t)$ are the same in k groups.

Briefly, the test is based on computing the expected number of deaths for each observed 'death' time in the dataset, assuming that the chances of dying, given that subjects are at risk, are the same in the groups. The total number of expected deaths is then computed for each group by adding the expected number of deaths for each failure time. The test finally compares the observed number of deaths in each group with the expected number of deaths using a CHI-SQUARE TEST with $k-1$ DEGREES OF FREEDOM (see Hosmer and Lemeshow, 1999).

The log-rank test statistic, X^2, weights contributions from all failure times equally. Several alternative test statistics have been proposed that give differential weights to the failure times. For example, the generalised Wilcoxon test (or Breslow test) uses weights equal to the number at risk. For the prostate cancer data in the first table the log-rank test ($X^2 = 4.4$ on 1 degree of freedom, $P = 0.036$) detects a significant group difference in favour of longer survival on DES treatment while the Wilcoxon test, which puts relatively more weight on differences between the survival curves at earlier times, fails to reach significance at the 5% test level ($X^2 = 3.4$ on 1 degree of freedom, $P = 0.065$).

Modelling survival times is useful especially when there are several explanatory variables to consider. For example, in the prostate cancer trial patients were randomised to treatment groups so that the theoretical distributions of the diagnostic factors were the same in the two groups. However, empirical distributions in the patient sample might still vary and if the prognostic variables are related to survival they might confound the group difference. A survival analysis that 'adjusts' the group difference for the prognostic factor(s) is needed. The main approaches used for modelling the effects of covariates on survival can be divided roughly into two classes – models based on assuming proportional hazards and models for direct effects on the survival times.

The main technique used for modelling survival times is due to Cox (1972) and is known as the PROPORTIONAL HAZARDS model or, more simply, Cox's regression (see COX'S REGRESSION MODEL). In essence, the technique acts as the analogue of multiple regression for survival times containing censored observations, for which multiple regression itself is clearly not suitable. Briefly, the procedure models the hazard function and central to it is the assumption that the hazard functions for two individuals at any point in time are proportional, the so-called proportional hazards assumption. In other words, if an individual has a risk of 'death' at some initial time point that is twice as high as another individual, then at all later times the risk of death remains twice as high. Cox's model is made up of an unspecified baseline hazard function, $h_0(t)$, which is then multiplied by a suitable function of an individual's explanatory variable values, to give the individual's hazard function. The interpretation of the regression parameter of the ith covariate, β_i, is that $\exp(\beta_i)$ gives the hazard or INCIDENCE rate change associated with an increase of one unit in the ith covariate, all other explanatory variables remaining constant.

Cox's regression is considered a semi-parametric procedure because the baseline hazard function, $h_0(t)$, and by implication the PROBABILITY DISTRIBUTION of the survival times, does not have to be specified. The baseline hazard is left unspecified; a different parameter is essentially included for each unique survival time. These parameters can be thought of as NUISANCE PARAMETERS whose purpose is merely to control the parameters of interest for any changes in the hazard over time.

Cox's regression can be used to model the prostate cancer survival data. To start with, a model containing only the single treatment factor is fitted. The estimated regression coefficient of a DES indicator variable is -1.98 with a standard error of 1.1. This translates into an (unadjusted) hazard ratio of $\exp(-1.98) = 0.138$. In other words, DES treatment is estimated to reduce the hazard of immediate death by 86.2% relative to PLACEBO treatment. According to a LIKELIHOOD RATIO (LR) test, the unadjusted effect of DES is statistically significant at the 5% level ($X^2 = 4.55$ on 1 degree of freedom, $P = 0.033$).

For the prostate cancer data, it is of interest to determine the effect of DES after controlling for the other prognostic variables. Likelihood ratio tests showed that dropping age and serum haemoglobin from a model that contains the treatment indicator variable and all four prognostic variables did not significantly worsen the model fit (at the 10% level); the fit of the final model is shown in the second table. After adjusting for the effects of tumour size and stage the hazard reduction for DES relative to placebo treatment is reduced to 67.1% and is no longer statistically significant (LR test: $X^2 = 0.48$ on 1 degree of freedom, $P = 0.49$). Both tumour size and Gleason index have a hazard ratio above unity, indicating that increases in tumour size and advanced stages are estimated to increase the chance of death.

Cox's model does not require specification of the probability distribution of the survival times. The hazard function is not restricted to a specific form and as a result the

survival analysis *Parameter estimates from Cox's regression of survival on treatment group, tumour size and Gleason index*

Predictor variable	*Effect estimate*			*95% CI for exp(β)*	
	Regression coefficient ($\hat{\beta}$)	*Standard error* $\left(\sqrt{\text{var}(\hat{\beta})}\right)$	*Hazard ratio* $\left(\exp(\hat{\beta})\right)$	*Lower limit*	*Upper limit*
DES	−1.113	1.203	0.329	0.031	3.47
Tumour size	0.0826	0.048	1.086	0.990	1.19
Gleason index	0.7102	0.338	2.034	1.049	3.95

survival analysis *Parameter estimates from log-logistic accelerated failure time model of survival on treatment group, tumour size and Gleason index*

Predictor variable	*Effect estimate*			*95% CI for exp(−α)*	
	Regression coefficient($\hat{\alpha}$)	*Standard error*$(\sqrt{\text{var}(\hat{\alpha})})$	*Acceleration factor*$(\exp(-\hat{\alpha}))$	*Lower limit*	*Upper limit*
DES	0.628	0.550	0.534	0.182	1.568
Tumour size	−0.031	0.022	1.031	0.988	1.077
Gleason index	−0.335	0.203	1.393	0.939	2.080

semi-parametric model has considerable flexibility and is widely used. However, if the assumption of a particular probability distribution for the data is valid, inferences based on such an assumption are more precise. For example, estimates of hazard ratios or median survival times will have smaller standard errors.

A fully parametric proportional hazards model makes the same assumptions as Cox's regression but in addition also assumes that the baseline hazard function, $h_0(t)$, can be parameterised according to a specific model for the distribution of the survival times. Survival time distributions that can be used for this purpose, i.e. that have the proportional hazards property, are principally the EXPONENTIAL, Weibull and Gompertz DISTRIBUTIONS. Different distributions imply different shapes of the hazard function, and in practice the distribution that best describes the functional form of the observed hazard function is chosen – for details see Collett (2003).

A family of fully parametric models that accommodate direct multiplicative effects of covariates on survival times and hence do not have to rely on proportional hazards are *accelerated failure time models*. A wider range of survival time distributions possesses the accelerated failure time property, principally the exponential, Weibull, log-logistic, generalised GAMMA or LOGNORMAL DISTRIBUTIONS. In addition, this family of parametric models includes distributions (e.g. the log-logistic distribution) that model unimodal hazard functions while all distributions suitable for the proportional hazards model imply hazard functions that increase or decrease monotonically. The latter property might be limiting, for example, for modelling the hazard of dying after a complicated operation that peaks in the post-operative period.

The general accelerated failure time model for the effects of p explanatory variables, $x_1, x_2, \ldots, x_p$, can be represented as a log-linear model for survival time, T, namely:

$$\ln(T) = \alpha_0 + \sum_{i=1}^{p} \alpha_i x_i + \text{error}$$

where $\alpha_1, \ldots, \alpha_p$ are the unknown coefficients of the explanatory variables and α_0 an intercept parameter. The parameter α_i reflects the effect that the ith covariate has on log-survival time with positive values indicating that the survival time increases with increasing values of the covariate and vice versa. In terms of the original timescale, the model implies that the explanatory variables measured on an individual act multiplicatively and so affect the speed of progression to the event of interest.

The interpretation of the parameter α_i is therefore that exp (α_i) gives the factor by which any survival time percentile (e.g. the median survival time) changes per unit increase in x_i, all other explanatory variables remaining constant. Expressed differently, the probability that an individual with covariate value $x_i + 1$ survives beyond t is equal to the probability that an individual with value x_i survives beyond $\exp(-\alpha_i)t$. Hence $\exp(-\alpha_i)$ determines the change in the speed with which individuals proceed along the timescale, and the coefficient is known as the acceleration factor of the ith covariate.

Software packages typically use the log-linear formulation. The regression coefficients from fitting a log-logistic accelerated failure time model to the prostate cancer survival times using treatment, size of tumour and Gleason index as predictor variables are shown in the third table. The negative regression coefficients suggest that the survival times tend to be shorter for larger value of tumour size and Gleason index. The positive regression coefficient for the DES treatment indicator suggests that survival times tend to be longer for individuals assigned to the active treatment after adjusting for the effects of tumour size and stage. The estimated acceleration factor for an individual in the DES group compared with the placebo group is $\exp(-0.628) = 0.534$; i.e. DES is estimated to slow down the progression of the cancer by a factor of about 2. While possibly clinically relevant, this effect is, however, not statistically significant (LR test: $X^2 = 1.57$ on 1 degree of freedom, $P = 0.211$).

In summary, survival analysis is a powerful tool for analysing time-to-event data. The classical techniques, Kaplan–Meier estimation, Cox's regression and accelerated failure time modelling, are implemented in most general purpose STATISTICAL PACKAGES, with the S-Plus package having particularly extensive facilities for fitting and assessing nonstandard Cox models.

The area is complex and one of active current research. For more recent advances, such as frailty models to include RANDOM EFFECTS, MULTISTATE MODELS to model different transition rates and models for competing risks, the reader is referred to Andersen (2002), Crowder (2001) and Hougaard (2000). *SL*

Andersen, P. K. (ed.) 2002: *Multistate models, statistical methods in medical research 11*. London: Arnold. **Andrews, D. F. and Herzberg, A. M.** 1985: *Data*. New York: Springer. **Collett, D.** 2003: *Modelling survival data in medical research*, 2nd edition. London: Chapman & Hall/CRC. **Cox, D. R.** 1972: Regression models and life tables (with discussion). *Journal of the Royal Statistical Society, Series B* 74, 187–220. **Crowder, K. J.** 2001: *Classical competing risks*. Boca Raton, FL: Chapman & Hall/CRC. **Hosmer, D. W. and Lemeshow, S.** 1999: *Applied survival analysis*. New York: John Wiley & Sons, Inc. **Hougaard, P.** 2000: *Analysis of multivariate survival data*. New York: Springer.

survival curve See KAPLAN–MEIER ESTIMATION, SURVIVAL ANALYSIS-AN OVERVIEW

survival function See SURVIVAL ANALYSIS

systematic reviews and meta-analysis This is an approach to the combining of results from the many individual CLINICAL TRIALS of a particular treatment or therapy that may have been carried out over the course of time. Such a procedure is needed because individual trials are rarely large enough to answer the questions we want to answer as reliably as we would like. In practice, most trials are too small for adequate conclusions to be drawn about potentially small advantages of particular therapies. Advocacy of large trials is a natural response to this situation, but it is not always possible to launch very large trials before therapies become widely accepted or rejected prematurely. An alternative possibility is to examine the results from all relevant trials, a process that involves two components, one *qualitative*, i.e. the extraction of the relevant literature and description of the available trials, in terms of their relevance and methodological strengths and weaknesses (the *systematic review*), and the other *quantitative*, i.e. mathematically combining results from different studies, even on occasions when these studies have used different measures to assess outcome. This component is known as a *meta-analysis* (Normand, 1999).

Informal synthesis of evidence from different studies is, of course, nothing new, but it is now generally accepted that meta-analysis gives the systematic review an objectivity that is inevitably lacking in the classical review article and can also help the process to achieve greater precision and generalisability of findings than any single study. There remain sceptics who feel that the conclusions from a meta-analysis often go far beyond what the technique and the data justify, but despite such concerns, the demand for systematic reviews of healthcare interventions has developed rapidly during the last decade, initiated by the widespread adoption of the principles of EVIDENCE-BASED MEDICINE both among healthcare practitioners and policy-makers. Such reviews are now increasingly used as a basis for both individual treatment decisions and the funding of healthcare and healthcare research worldwide. This growth in systematic reviews is reflected in the current state of the COCHRANE COLLABORATION database containing as it does more than 1200 complete systematic reviews, with a further 1000 due to be added soon.

Systematic reviews and the subsequent meta-analysis have a number of aims: to review systematically the available evidence from a particular research area; to provide quantitative summaries of the results from each study; to combine the results across studies if appropriate – such combination of results leads to greater statistical power in estimating treatment effects; to assess the amount of variability between studies; to estimate the degree of benefit associated with a particular study treatment; to identify study characteristics associated with particularly effective treatments.

Ideally, the trials included in a systematic review should be clinically homogeneous. For example, they might all study a similar type of patient for a similar duration with the same treatment in the two arms of each trial. In practice, of course, the trials included are far more likely to differ in some aspects, such as eligibility criteria, duration of treatment, length of follow-up and how ancillary care is used. On occasions, even treatment itself may not be identical in all the trials. This implies that, in most circumstances, the objective of a systematic review *cannot* be equated with that of a single large trial, even if that trial has wide eligibility. While a single trial focuses on the effect of a specific treatment in specific situations, a meta-analysis aims for a more generalisable conclusion about the effect of a generic treatment policy in a wider range of areas.

When the trials included in a systematic review do differ in some of their components, therapeutic effects may very well be different, but these differences are likely to be in the *size* of the effects rather than their direction. It would, after all, be extraordinary if treatment effects were exactly the same when estimated from trials in different countries, in different populations, in different age groups or under different treatment regimens. If the studies were big enough it would be possible to measure these differences reliably, but in most cases this will not be possible. However, meta-analysis allows the investigation of sources of possible heterogeneity in the results from different trials, as we shall see later, and discourages the common, simplistic and often misleading interpretation that the results of individual clinical trials are in conflict because some are labelled 'positive' (i.e. statistically significant) and others 'negative' (i.e. statistically nonsignificant). A systematic approach to synthesising information can often both estimate the degree of benefit from a particular therapy and whether the benefit depends on specific characteristics of the studies.

The selection of studies is the greatest single concern in applying meta-analysis and there are at least three important components of the selection process, namely breadth, quality and representativeness (Pocock, 1996). Breadth relates to the decision as to whether to study a very specific narrow question (e.g. the same drug, disease and setting for studies following a common protocol) or a more generic problem (e.g. a broad class of treatments for a range of conditions in a variety of settings). The broader the meta-analysis, the more difficulty there is in interpreting the combined evidence as regards future policy. Consequently, the broader the meta-analysis, the more it needs to be interpreted qualitatively rather than quantitatively.

Quality and reliability of a systematic review is dependent on the quality of the data in the included studies, although criticisms of meta-analyses for including original studies of questionable quality are typical examples of shooting the messenger who bears bad news. Aspects of quality of the original articles that are pertinent to the reliability of the meta-analysis include a valid RANDOMISATION process (we are assuming that in meta-analysis of clinical trials, only randomised trials will be selected), MINIMISATION of potential BIASES introduced by DROPOUTS, acceptable methods of analysis, level of BLINDING and recording of adequate clinical details. Several attempts have been made to make this aspect of meta-analysis more rigorous by using the results given by applying specially constructed quality assessments scales to assess the candidate trials for inclusion in the analysis. Determining quality would be helped if the results from so many trials were not so poorly reported. In the future, this may be improved by the CONSORT statement (CONSOLIDATED STANDARDS FOR REPORTING TRIALS).

The representativeness of the studies in a systematic review depends largely on having an acceptable search strategy. Once the researcher has established the goals of the systematic review, an ambitious literature search needs to be undertaken, the literature obtained and then summarised. Possible sources of material include the published literature, unpublished literature, uncompleted research reports, work in progress, conference/symposia proceedings, dissertations, expert informants, granting agencies, trial registries, industry and journal hand searching. The search will probably begin by using computerised bibliographic databases of published and unpublished research review articles, for example, MEDLINE. This is clearly a sensible strategy, although there is some evidence of deficiencies in MEDLINE when searching for RANDOMISED CONTROLLED TRIALS.

Ensuring that a meta-analysis is truly representative can be problematic. It has long been known that journal articles are not a representative sample of work addressed to any particular area of research. Research with statistically significant results is potentially more likely to be submitted and published than work with null or nonsignificant results, particularly if the studies are small. The problem is made worse by the fact that many medical studies look at multiple outcomes and there is a tendency for only those outcomes suggesting a significant effect to be mentioned when the study is written up. Outcomes that show no clear treatment effect are often ignored and so will not be included in any later review of studies looking at those particular outcomes. Publication bias is likely to lead to an overrepresentation of positive results.

Clearly it becomes of some importance to assess the likelihood of publication bias in any meta-analysis reported in the literature. A well-known informal method of investigating this potential problem is the so-called FUNNEL PLOT, usually a plot of a measure of a study's precision (e.g. one over the STANDARD ERROR) against effect size. The most precise estimates (e.g. those from the largest studies) will be at the top of the plot and those from less precise or smaller studies at the bottom. The expectation of a 'funnel' shape in the plot relies on two empirical observations. First, the variances of studies in a meta-analysis are not identical, but are distributed in such a way that there are fewer precise studies and rather more imprecise ones and, second, at any fixed level of VARIANCES, studies are symmetrically distributed about the MEAN.

Evidence of publication bias is provided by an absence of studies on the left-hand side of the base of the funnel. The assumption is that, whether because of editorial policy or author inaction or some other reason, these studies (which are not statistically significant) are the ones that might not be published. An example of a funnel plot suggesting the possible presence of publication bias is given in the figure (taken from Duval and Tweedie, 2000).

Various proposals have been made as to how to test for publication bias in a systematic review although none of these is wholly satisfactory. The danger of the testing approach is the temptation to assume that, if the test is not significant, there is no problem and the possibility of publication bias can be conveniently ignored. In practice, however, publication bias is very likely endemic to all empirical research and so should be assumed present, whatever the result of some testing procedures with possibly low POWER.

Once the studies for systematic review have been selected and the possible problems of publication bias addressed, effect sizes and variance estimates are extracted from the selected papers, reports, etc., and subjected to a meta-analysis, in which the aim is to provide a global test of significance for the overall NULL HYPOTHESIS of no effect in all studies and to calculate an estimate and a CONFIDENCE INTERVAL of the overall effect size.

Two models are usually considered, one involving FIXED EFFECTS and the other RANDOM EFFECTS (Fleiss, 1993; Sutton *et al.*, 2000). The former assumes that the true effect is the same for all studies whereas the latter assumes that individual studies have different effect sizes that vary randomly around the overall mean effect size. Thus the random effects model specifically allows for the existence of both between study heterogeneity and within-study variability. When the research question concerns whether treatment has produced an effect, on the average, in the set of studies being analysed, then the fixed effects model for the studies may be the more appropriate; here there is no interest in generalising the results to other studies. Many statisticians believe, however, that the random effects model is more appropriate than a fixed effects model for meta-analysis, because between-study variation is an important source of uncertainty that should not be ignored when assigning uncertainty into pooled results. Tests of homogeneity are available, i.e. a test that the between-study variance component is zero – if it is, a fixed effects model is considered justified. Such a test is, however, likely to be of low power for detecting departures from homogeneity and so its practical consequences are probably quite limited.

The essential feature of both the fixed and random effects models for meta-analysis is the use of a weighted mean of treatment effect sizes from the individual studies, with the weights usually being the reciprocals of the associated

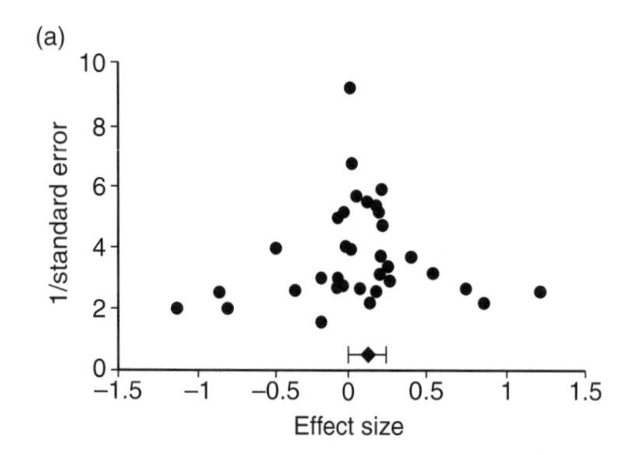

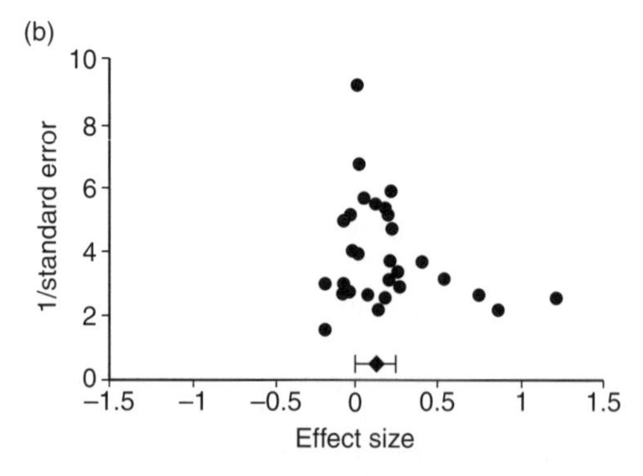

systematic reviews and meta-analysis *(a) Funnel plot of 35 simulated studies and meta-analysis with true effect size of zero: estimated effect size is 0.080 with a 95% confidence interval of [−0.018,0.178]; (b) funnel plot as in (a) with five 'leftmost' studies suppressed; overall effect size is now estimated as 0.124 with a 95% confidence interval of [0.037,0.210]. Reprinted from Duval and Tweedie, 2000, with permission from* The Journal of the American Statistical Association. *Copyright 2000 by the American Statistical Association. All rights reserved*

variances. Effect sizes might be standardised mean differences for continuous RESPONSE VARIABLES or RELATIVE RISKS AND ODDS RATIOS for binary outcomes. Both fixed effects and random effects models result in a test of zero effect size and a confidence interval for effect size. However, it should be remembered that, in general, a more important aspect of meta-analysis is often the exploration of the likely heterogeneity of effect sizes from the different studies. Random effect models, for example, allow for such heterogeneity but they do not offer any way of exploring and potentially explaining the reasons study results vary. In other words, random effects models do not 'control for', 'adjust for' or 'explain away' heterogeneity. Understanding heterogeneity should perhaps be the primary focus of the majority of meta-analyses carried out in medicine.

The examination of heterogeneity may begin with formal statistical tests for its presence, but even in the absence of statistical evidence of heterogeneity, exploration of the relationship of effect size to study characteristics may still be valuable. The question of importance is, what causes heterogeneity in systematic reviews of clinical trials? Study of the causes of heterogeneity of treatment effects in a meta-analysis often involves the technique generally known as META-REGRESSION. Essentially, this is nothing more than a weighted regression analysis with effect size as the dependent variable, a number of study characteristics as explanatory variables and weights usually being the reciprocal of the sum of the estimated variance of a study and the estimated between-study variance, although other more complex approaches have been described. Meta-regression can, like subgroup analysis within a single clinical trial, quickly become little more than DATA DREDGING. This danger can be partially dealt with at least by pre-specification of the covariates, which will be investigated as potential sources of heterogeneity.

As an example of the systematic review and associated meta-analysis we shall consider transcranial magnetic stimulation (TMS) for the treatment of depression. Such treatment involves placing a high-intensity magnetic field of brief duration at the scalp surface to induce an electrical field at the cortical surface that can alter neuronal function. Repetitive TMS (rTMS) involves applying trains of these magnetic pulses. In humans rTMS has been shown to produce changes in frontal lobe blood flow and to normalise the response to dexmethasone in depression. Since trials in the late 1990s, rTMS has been proposed as a treatment for drug-resistant depression, schizophrenia and mania. McNamara *et al.* (2001) report a systematic review of the published data, in which RANDOMISED CONTROLLED TRIALS were searched for using a variety of databases, including Medline and Embase. Sixteen published clinical trials of rTMS for depression were identified, but eight were excluded because there was no randomised control group and a further three excluded for reasons given in the original paper. The results from the five trials accepted for the meta-analysis are shown in the table.

systematic reviews and meta-analysis *Data for five RCTs of rTMS*

		rTMS	*Placebo*
Trial 1	Improved	11	6
	Not improved	6	11
Trial 2	Improved	7	1
	Not improved	1	4
Trial 3	Improved	8	2
	Not improved	4	4
Trial 4	Improved	4	1
	Not improved	6	10
Trial 5	Improved	17	8
	Not improved	18	24

The results from both the fixed effects and random effects models are, for these data, exactly the same. The overall effect size (log odds ratio) is estimated to be 1.33 with a standard error of 0.37, leading to an estimated odds ratio of 3.78 with 95% confidence interval (1.83, 7.81). *BSE*

[See also FOREST PLOT]

Duval, S. and Tweedie, R. L. 2000: Nonparametric 'trim and fill' method of accounting for publication bias in meta-analysis. *Journal of the American Statistical Association* 95, 89–98. **Fleiss, J. L.** 1993: The statistical basis of meta-analysis. *Statistical Methods in Medical Research* 2, 121–45. **MacNamara, B., Ray, J. L., Arthurs, O. J., and Boniface, S.** 2001: Transcrannial magnetic stimulation for depression and other psychiatric disorders. *Psychological Medicine* 31, 1141–6. **Normand, S. T.** 1999: Meta-analysis: formulating, evaluating, combining and reporting. *Statistics in Medicine* 18, 321–59. **Pocock, S. J.** 1996: Clinical trials: a statistician's perspective. In Armitage, P. and David, H. A. (eds), *Advances in biometry*. Chichester: John Wiley & Sons, Ltd. **Sutton, A. J., Abrams, K. R., Jones, D. R. and Sheldon, T. A.** 2000: *Methods for meta-analysis in medical research*. Chichester: John Wiley & Sons, Ltd.

systematic sample Every kth element in a list is included in the sample. To obtain such a sample, begin with the sampling frame and arrange in an order, which may be alphabetical or some other order. Then select the number of samples to be taken. Select a random starting point in the list. Divide the size of the population by the number of samples to be taken. This is the length of interval, k. Then every kth unit, depending on the starting point, is included in the sample until the number of samples to be taken is reached. This may mean starting again from the beginning of the list.

For example, Little, Keefe and White (1995) used a systematic sample when studying melanoma patients in a general practice. Every 125th patient on the general practice register was selected to be included; this was to yield a minimum of 60 individuals.

The main advantage of systematic sampling is that it is a quick and easy-to-use sampling method, particularly when dealing with large samples, where it is often used in preference to SIMPLE RANDOM SAMPLES. However, if there is a periodic cycle within the sampling frame then estimates obtained from systematic sampling may be incorrect. If the periodic cycle is recognised then the starting point and the length of interval between chosen items can be varied. Systematic sampling can often only be used when there is a sampling frame available that can be ordered in some way. For further details see Crawshaw and Chambers (1994) and Upton and Cook (2002). *SLV*

Crawshaw, J. and Chambers, J. 1994: *A concise course in A level statistics*, 3rd edition. Cheltenham: Stanley Thornes Publishers Ltd. **Little, P., Keefe, M. and White, J.** 1995: Self-screening for risk of melanoma: validity of self-mole counting by patients in a single general practice. *British Medical Journal* 310, 912–16. **Upton, G. and Cook, I.** 2002: *Dictionary of statistics*. Oxford: Oxford University Press.

T

t-distribution Also known as Student's t-distribution, this is the distribution of the estimate of the MEAN of a NORMAL DISTRIBUTION when the STANDARD ERROR has also been estimated. The distribution is used when performing STUDENT'S t-TEST. If we have a normal distribution of known VARIANCE σ^2, then we can test to see if the mean of a set of observations $x_1, x_2, \ldots, x_n$, denoted m, is consistent with a hypothesised mean μ by calculating a CONFIDENCE INTERVAL for μ. This is done by considering that:

$$N = \frac{m-\mu}{\sqrt{\sigma^2/n}}$$

will have an approximately standard normal distribution mean (mean 0, standard deviation 1), denoted $N(0,1)$ and calculating a 95 % confidence interval for μ as:

$$(m-1.96\sqrt{\sigma^2/n}) < \mu < (m+1.96\sqrt{\sigma^2/n})$$

where 1.96 is the critical value from the $N(0,1)$ distribution.

If the variance is also being estimated, instead of knowing the population variance σ^2, we must use the sample estimate s^2. Now if we construct the statistic:

$$T = \frac{m-\mu}{\sqrt{s^2/n}}$$

will have a t-distribution with $n-1$ DEGREES OF FREEDOM, written $t(n-1)$. A 95 % confidence interval for μ can then be expressed as:

$$(m-t_{0.025}\sqrt{s^2/n}) < \mu < (m+t_{0.975}\sqrt{s^2/n})$$

where $t_{0.025}$ and $t_{0.975}$ are the critical values from the $t(n-1)$ distribution. These values, which are chosen to ensure 2.5 % of the probability density lies in each tail, can be found from tables (Lindley and Scott, 1984) or computer packages. As long as n is at least three, like the $N(0,1)$ distribution, the t-distribution has a zero mean and is symmetric, but the variance is $(n-1)/(n-3)$. As the sample size, n, increases, the variance approaches 1, but for small sample sizes the variance will be greater, which reflects the uncertainty in the estimation of σ^2.

The t-distribution is related to other common distributions. If we compare the statistics N and T, we will see that T is merely N divided through by $\sqrt{s^2/\sigma^2}$. Now, s^2/σ^2 is known to have a CHI-SQUARED DISTRIBUTION with $n-1$ degrees of freedom and so we see that, more generally, the t-distribution with $n-1$ degrees of freedom arises when an $N(0,1)$ variable is multiplied by the square root of $(n-1)$ and divided by the square root of a $\chi^2(n-1)$ variable.

Having observed this, if we now square T, we can see that it is $(n-1)$ times the square of an $N(0,1)$ variable divided by a $\chi^2(n-1)$ variable. Now the square of an $N(0,1)$ variable is a $\chi^2(1)$ variable, so the square of T is $(n-1)$ times a $\chi^2(1)$ variable divided by 1 times a $\chi^2(n-1)$ variable. The division of one chi-square variable by another independent one with the correct multipliers is known to generate an F-distribution, and so we can see that the square of a t-distributed variable (with $n-1$ degrees of freedom) will have an F-DISTRIBUTION (with 1 and $n-1$ degrees of freedom). For further reading see Leemis (1986), Altman (1991) and Jones (2008). *AGL*

Altman, D. G. 1991: *Practical statistics for medical research.* London: Chapman & Hall. **Jones, M. C.** 2008: The t family and their close and distant relations *Journal of the Korean Statistical Society* 37, 293–302. **Leemis, L. M.** 1986: Relationships among common univariate distributions. *The American Statistician* 40, 2, 143–6. **Lindley, D. V. and Scott, W. F.** 1984: *New Cambridge elementary statistical tables.* Cambridge: Cambridge University Press.

teaching medical statistics Four main groups of people are expected to learn medical statistics: undergraduate students of medicine and other healthcare professional subjects, healthcare practitioners, researchers in healthcare and would-be medical statisticians. For all these groups we must select the most appropriate material from the huge amount available (even to master the contents of the journal *Statistics in Medicine* would take me several lifetimes). This material will, in turn, partly determine the teaching method.

Students rarely have much time in their crowded curriculum for statistics and rarely have much natural sympathy for the subject. They see their futures as practical people busy saving lives and caring for the sick, not analysing data or reading journals. They are also usually at the age of maximum confidence in their own infallibility and hence difficult to persuade that they might be mistaken in their image of their future roles. It is easier to persuade them of the relevance of reading evidence than number crunching, and such courses do better if they concentrate on the understanding of research publications. I have found that my conventional lectures are of little value to this group and seminars where they discuss papers, backed up by printed notes or web pages on the statistical principles, are more effective.

Encyclopaedic Companion to Medical Statistics: Second Edition Edited by Brian S. Everitt and Christopher R. Palmer

Lectures work better when students who have been challenged with this material are then able to ask a statistician to explain things that puzzle them. The concepts acquired in this way are more likely to be backed up by other parts of their course than the calculation of CHI-SQUARED TESTS. If students can be equipped with the basic ideas of variability, measurement, RANDOMISATION, estimation and significance, we have done well. The machines can do the sums.

Increasing numbers of healthcare students are taught by problem-based learning (PBL), a system intended to prepare students for a life of EVIDENCE-BASED MEDICINE. Statistics is rarely taught as part of the core PBL programme, but is instead taught as a separate addition to PBL cases, or in a separate, parallel lecture- or seminar-based course, or not at all. This is bad news not only for medical statistics (and those who teach it) but also for medicine. Surely the skills needed for the interpretation of evidence should be central in a course preparing students for evidence-based practice. It happens because tutors, mainly laboratory scientists or clinicians, feel insecure about teaching statistics and because the 'problems' of PBL are usually descriptions of a patient. Tutors need to be convinced that they do not need to know the subject to facilitate students' mutual education and course organisers need to be convinced that problems can be a publication or a community problem, that the patient case is not the only way.

Healthcare practitioners are usually taught statistics as part of study for a higher professional qualification. The key application is still the interpretation of numerical evidence, mainly in the context of published research. However, they often have the more immediate goal of passing a demanding examination with a high failure rate. Some of these examinations include some quite advanced statistics, such as those in radiotherapy or public health. The teacher can make use of this by collaborating with the students to defeat the examiner and concentrating on past questions. I find that starting with a few multiple-choice questions to identify areas of difficulty and then explaining the answers the students get wrong works very well. Once the basics have been covered in this way, past examination questions form the ideal motivator. It is for the examiners to design their tests so that in order to pass them the students must learn what the examiners think they need to know.

For those who do not have to satisfy an examiner but simply wish to understand their own subject's literature better, indirect teaching is frequently used. Many journals have carried long series of articles on statistics intended to help their readers understand what is published, a practice that began with the early ground-breaking *Lancet* articles by Bradford Hill in the 1930s and continues still.

Researchers have very different needs from practitioners. They must acquire the skills to design studies and analyse data. Understanding of concepts, while still central, is not enough. Practical skills are usually developed in hands-on computing practical classes, preferably using software of the type that they will use in their own research. Lectures have a more natural place in this teaching, as methods and their applications and limitations can be described. We can even risk a few mathematical formulae without too much discouragement of well-motivated students. The opportunity to discuss their own projects is very attractive to these students.

Textbooks are particularly important to this group. At one time the market was flooded with poor books on medical statistics (Bland and Altman, 1987), but there are now many good ones. Another source of statistical education for researchers comes from individual discussions of their projects with a statistician. This is a two-way street, as they educate the statistician about the research topic and medicine in general. I have learned so much from the people who have come to me for help.

For new statisticians, statistics is usually a Master's course taken by graduates in mathematics or other quantitative subjects. It is possible to study statistics as a branch of mathematics without real data making much of an appearance, but if students have chosen to study medical statistics specifically we would expect them to want a practical course with the focus on application to real problems. Clearly, they must become familiar with the common techniques of design and analysis and should be able to analyse data within both the frequentist and the Bayesian frameworks (see BAYESIAN METHODS). As nearly all statistical analysis is now done using general-purpose statistical software, they should learn the basics of the software they are likely to meet. At the time of writing, SAS, Stata and BUGS would be contenders for the programs of choice, but familiarity with other widely used or specialist software could be included. Statisticians need not only technical skills but also the ability to collaborate with and give advice to members of other disciplines. Experience is the best teacher, but experience is what you get just after you needed it. We would like to give our students a bit of experience before they are plunged into real-life problems. Medicine and other healthcare professions have much to teach us here. I used to run a session for MSc students in medical statistics where I invited clinical researchers who sought my advice to come and get it in front of a live audience. I pointed out to them that if I went to consult them, they would do it with an audience of medical students. Perhaps we could incorporate this type of advisory clinic into our teaching.

We want to enable our students not only to use the current set of statistical methods but also to develop new ones where these are needed. To this end they need some theory as well as the practice of statistics. I think that statisticians should also have a secure basis for thinking that the statistical methods that we routinely use are in some way the best methods we could use and for this reason a theoretical course will provide valuable grounding, even though they may never use it again.

We should not think that students will learn and retain all we teach them or that if we do not teach them something they will never know it. Being taught is only a part of learning and good students will continue to learn throughout their careers. What we must try to do is to give them the desire to retain what they have learned already and the ability to add to their knowledge whenever they need to. *JMB*

Bland, J. M. and Altman, D. G. 1987: Caveat doctor: a grim tale of medical statistics textbooks. *British Medical Journal* 295, 979.

tetrachoric correlation coefficient See CORRELATION

thrive lines See GROWTH CHARTS

time-dependent variables Time-dependent covariates, also known as time-varying covariates or updating covariates, are variables that can change their value over time. They are particularly important for prognostic models, such as COX'S REGRESSION MODEL. They should be distinguished from fixed covariates, which are measurable at baseline and do not change with time. Examples of fixed covariates are race and sex. Age varies with time, but is completely predictable from baseline data and so is not included among time-dependent covariates. Time-dependent covariates may be classed as being internal and external (Altman and de Stavola, 1994). External factors impact on outcome but do not explicitly reference time, e.g. the half-life of a drug treatment; whereas internal factors are measurements taken at set times relating to the individual or their condition, e.g. blood pressure or blood markers.

The reason for considering the inclusion of time-dependent covariates is that including only baseline variables may ignore a great deal of potential prognostic information. Therefore, the inclusion of time-dependent covariates may substantially increase the potential detail and accuracy in a model. For example, increases (or decreases) over time in patients' blood pressure may be a better predictor of future prognosis than a single baseline value of blood pressure.

The Cox model can be extended to include time-dependent covariates instead of, or in addition to, fixed covariates. In simplest terms, the hazard for a time-dependent covariate takes the form: $h(t) = \exp[\gamma z(t)]$, where $h(t)$ is the hazard at time t and z is a time-dependent covariate and γ is its coefficient value. As for fixed covariates, all data types can be entered as time-dependent covariates into a Cox regression model. It is important to assess the assumption of PROPORTIONAL HAZARDS, once any time-dependent covariate has been taken into account.

These variables do add additional complications to any model. First, they require the dataset being analysed to contain additional variables or additional observations (depending on the dataset's structure). Second, it can be difficult to obtain complete data on these variables, especially with increasing time. MISSING DATA can be problematic. Third, these variables effectively increase the choice of Cox models available for consideration. One must ensure that issues of multiplicity of testing are addressed. Finally, there are issues of interpretation. Including time-dependent covariates in a model may be practically simple, but the greater difficulty lies in interpreting the data: one must be sure how any variable would be interpreted before including it in a model. Simply, the hazard ratio for a time-dependent covariate represents an additional change in risk associated with a change in this variable over time. For example, when considering bone pain as an outcome after treatment for prostate cancer, one may wish to record the development of osteoarthritis over time as the second condition may increase the risk of bone pain. When interpreting output it can be complicated trying to tease cause from effect with such variables. *MS/MP*

Altman, D. G. and de Stavola, B. L. 1994: Practical problems in fitting a proportional hazards model to data with updated measurements of the covariates. *Statistics in Medicine* 13, 4, 301–41. **Cleves, M. A., Gould, W. W. and Gutierrez, R. G.** 2003: *An introduction to survival analysis using Stata®*, revised edition. Texas: Stata Press. **Machin, D. and Parmar, M. K. B.** 1995: *Survival analysis: a practical approach*. Chichester: John Wiley & Sons, Ltd. **Piantodosi, S.** 1997: *Clinical trials*. New York: Wiley Interscience.

time series in medicine Chatfield (1989) has defined a time series as 'a sequence of observations ordered in time'. In medicine and medical research, observations are often ordered in time and special techniques have evolved to deal with them. It is helpful to think of three types of time series: (1) single series, often long; (2) more than one series, each of moderate length; and (3) many shorter series.

There are at least three reasons for collecting a single time series:

(a) To predict some future event. An example might be measuring creatinine clearance from kidney failure patients where the main aim is to predict complete kidney failure.
(b) To test whether some event in time has an effect on subsequent outcomes. These are sometimes called before-and-after studies or interrupted times series (Glass *et al.*, 1975). Examples include the effect of seat belt legislation on deaths due to car accidents, effect of NHS Direct on consultation to a general practitioner and behavioural experiments in psychology.
(c) To look for trends and rhythms in the series. An example would be a spectral analysis of the electroencephalogram (EEG) signal to measure the strength of alpha waves.

For series of more moderate length a common theme is to examine whether there is an ASSOCIATION between two series. Examples include studies to examine relationships between cot deaths and environmental temperature and daily deaths from heart disease and air pollution.

Shorter series are often dealt with under the term 'repeated measures'. The reason for measuring observations over time is that a series would more accurately reflect the action of treatment than a single measure at one point in time. A typical example would be repeated measures in a CLINICAL TRIAL, such as blood pressure measured monthly for a year. Summary measures (Matthews *et al.*, 1990) such as the AREA UNDER THE CURVE or the slope of the response over time are often the outcomes of interest. Repeating observations can improve the accuracy of the estimates of treatment effects.

The most basic time series model is the autoregressive (AR (1)) model (Diggle, 1990). Given a time series x_t, from which the MEAN value has been subtracted an AR(1) is given by:

$$x_t = ax_{t-1} + \varepsilon_t, \text{ where } -1 < \alpha < 1$$

This model is often called a *Markov* model because the value at one point in time only depends on the value at the point immediately preceding it. The model is easily extended to AR(p), where $p > 1$. For $p = 2$, certain values of the coefficients can give models in which cycles appear. Some forecasting models use autoregressive models, but in medicine these are rarely used because usually one is more interested in estimating trends, which are more easily fitted using convention models.

The complementary model to the AR(1) is the moving average model MA(1):

$$x_t = \varepsilon_t + \beta\varepsilon_{t-1}, \text{ where } -1 < \beta < 1$$

This model is less commonly used than an AR(1) but again can be easily extended to MA(q), $q > 1$.

These models can be combined to produce an autoregressive moving average ARMA model. This has been found to model many time series and requires only low values of p and q. The procedure of fitting these models is often known as *Box–Jenkins* modelling (Box and Jenkins, 1976). In general, this type of modelling is more common in the forecasting and control of industrial processes, although on occasion it has been applied in medicine.

Perhaps the most common feature of the analysis of clinical signals is to look for regularly occurring features or rhythms (Campbell, 1996). For humans to maintain stable bodily functions, clinical signals must be constrained to lie within certain limits. This is done using nonlinear feedback loops, which tends to make patterns within signals recur regularly. A simple example will illustrate the point. To remain healthy, humans must maintain blood pressure to within certain narrow limits. Blood pressure is mediated through the baroreceptors, located in the wall of the aortic arch and in the wall of the carotid sinus. If blood pressure is too high, then signals from the baroreceptors result in vasodilation, which drops the blood pressure. If pressure is too low, then vasoconstriction occurs to increase the blood pressure. The feedback mechanism is thought to be nonlinear and incorporates a delay and, for these reasons, at-rest rhythms can occur spontaneously.

Periodogram analysis involves decomposing a signal into individual frequency components where the amplitude of these components is proportional to the 'energy' of the signal at that frequency. It is a convenient method for summarising a long time series and is a natural procedure if we believe there are rhythms in the data. Periodogram analysis is the method of choice for the analysis of clinical signals. The problem with the periodogram is that it is an inconsistent estimator, in that its VARIANCE does not reduce as the sample size increases. To achieve a consistent estimator various smoothing techniques (known as 'windows') are applied to the periodogram, so that it estimates what is known as the *spectrum*.

There are three major components to be found in a typical heart rate spectrum and these are also present in the blood pressure spectrum. A region of activity occurs at around 0.25 Hz, which is attributable to respiration (respiratory sinus arrhythmia) and this is thought to be a marker of vagal (parasympathetic) activity. A second component at around 0.1 Hz arises from spontaneous vasomotor activity within the blood pressure control system and is mediated by vagal and sympathetic activity. A third, low-frequency component at around 0.04 Hz is thought to arise from thermoregulatory activity. An example is given in the first figure on page 463 (Bernardi *et al.*, 2001), which shows the effect of recitation of mantras or prayers on the spectrum of respiration, heart rate (RR interval) blood pressure and mid-cerebral blood flow. It can be seen that recitation concentrates the power of the signal at a cycle with a frequency of about 6 cycles per minute (0.1 Hz).

Some signals are essentially continuous, whereas others are discrete. For example, the heart rate is measured from surface electrodes on the chest from the electrocardiogram (ECG). Although the ECG is continuous, the heart rate is usually derived from the 'R' wave in the ECG, which is a sudden spike just preceding the ventricular contraction. Thus, the heart beat signal is essentially a point process. Some authors have analysed the interbeat intervals, thus arriving at a spectrum, which estimates frequencies per beat, rather than per unit time. Others sample the heart rate (or RR interval) signal at regular intervals or filter the point process to produce a continuous signal that can be sampled.

The electroencephlogram (EEG) is electrical activity of the brain measured by electrodes at the surface of the skull. There is an immense amount of literature devoted to the spectral analysis of EEGs. In particular, six spectral peaks can be identified. These peaks, with a typical range of frequencies

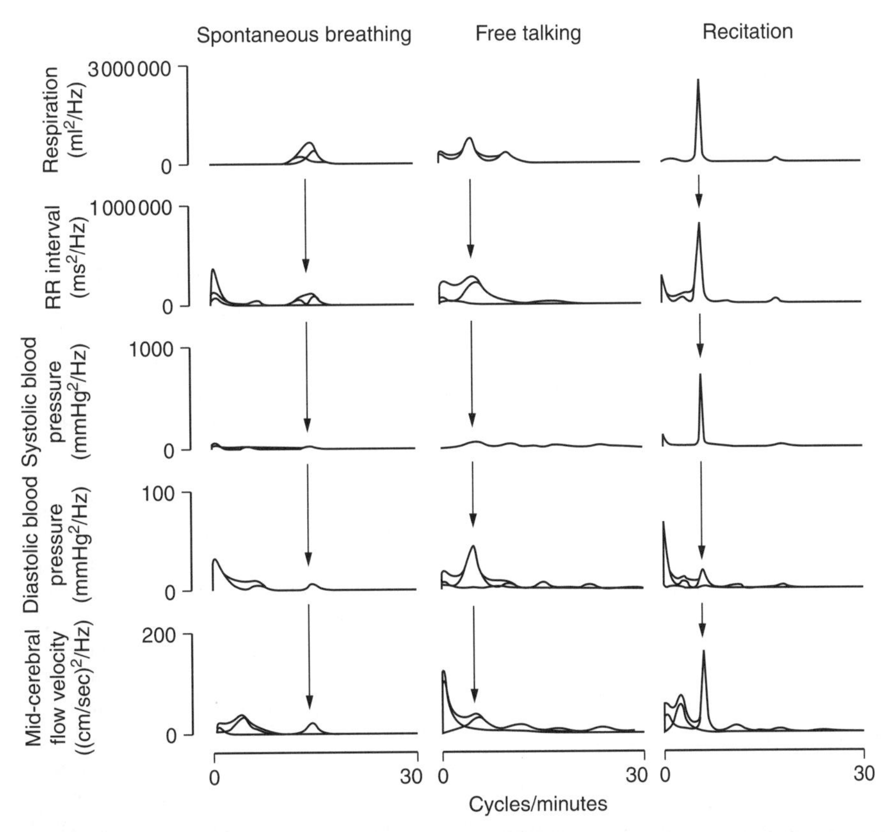

time series in medicine *Spectrum showing effects (in one subject) of rhythmic rituals compared with spontaneous breathing, on respiratory and cardiovascular rhythms. Note slow rhythmic oscillations (approximately 6/min) in all signals during recitation (Bernardi et al., 2001,* British Medical Journal *323, 1446–9, with permission from the BMJ Publishing Group)*

are: delta 1 (0.5–2.0 Hz), delta 2 (2.0–4.0 Hz), theta (4.0–8.0 Hz), alpha (8.0–12.0 Hz), sigma (12.0–14.0 Hz) and beta (14.0–20.0 Hz). The peaks can be used, for example, to classify different levels of sleep. Recently there has been interest in describing neural processes in the context of nonlinear dynamics and, in particular, in the rapidly evolving field of *deterministic chaos*.

An assumption before applying spectral analysis is that the signal is stationary (i.e. the mean and variance do not change). However, medical signals are not stationary in the usual sense. They contain rhythms that may come and go in the time interval, the frequencies may vary or amplitudes of cycles at certain frequencies increase or decrease. Spectral analysis considers the entire time interval and so cycles that only occur in part of the interval will have their spectral peaks attenuated by the low power in other parts of the interval. One solution is to divide the series up into sections and compute the spectrum for each section. The difficulty here is that it is not realistic to think of a signal being stationary in sections. A better intuitive model is one in which the signal 'evolves' slowly so that the nonstationary component is slow in comparison with the signal in which we are interested. The newly developed field of WAVELET ANALYSIS is used to analyse signals of this kind.

The main problem with time series is that serial correlation invalidates one of the main assumptions of conventional regression, namely that the errors in the model are independent of each other. The second problem is that if we are interested in the relationship between two time dependent variables y_t and x_t, any other variable associated with time will be a confounder between them. For example, if both series increase with time, then they will appear correlated. This has been the source of much amusement, with positive correlations such as those between the annual population of Holland and the number of storks' nests and sales of ice cream and deaths by drowning being quoted as evidence of causation!

To make progress one has to try and fit a model that removes the confounder variables. For a continuous outcome suppose the model is $y_t = \boldsymbol{\beta}'\boldsymbol{x}_t + \upsilon_t$, $t = 1, \ldots, n$, where y_t is the dependent value measured at time t, $\boldsymbol{x}_t$ a vector of confounder and predictor variables and $\boldsymbol{\beta}$ a vector of regression coefficients.

If the residuals are serially correlated, then ordinary least squares does not provide valid estimates of the STANDARD ERRORS of the parameters. If we assume that $\boldsymbol{x}_t$ and υ_t are generated by AR(1) processes with parameters α and γ, then, using ordinary least squares to estimate β, the ratio of the estimated variance to the true variance is approximately $(1 - \alpha\gamma)/(1 + \alpha\gamma)$. In general, $\boldsymbol{x}_t$ and υ_t are likely to be positively correlated. Thus the effect of ignoring serial CORRELATION is to give artificially low estimates of the standard error (SE) of the regression coefficients. This means declaring significance more often than the significance level would suggest, under the NULL HYPOTHESIS of no ASSOCIATION.

Assuming α is known, a method of generalised least squares known as the Cochrane–Orcutt procedure (Cochrane and Orcutt, 1949) can be employed. Write $y^*_t = y_t - \alpha y_{t-1}$ and $x^*_t = x_t - \alpha x_{t-1}$. Obtain an estimate of β using ordinary least squares on y^*_t and x^*_t. However, since α will not usually be known it can be estimated from the ordinary least squares residuals e_t by:

$$a = \sum_{t=2}^{n} e_t e_{t-1} \Big/ \sum_{t=2}^{n} e^2_{t-1}$$

This leads to an iterative procedure in which we can construct a new set of transformed variables and thus a new set of regression estimates and so on until convergence.

The iterative Cochrane–Orcutt procedure can be interpreted as a stepwise algorithm for computing MAXIMUM LIKELIHOOD ESTIMATIONS of α and β where the initial observation y_1 is regarded as fixed. If the residuals can be assumed to be normally distributed then full maximum likelihood methods are available, which estimate α and β simultaneously, and this can be generalised to higher order autoregressive models. These models can be fitted using (say) PROC AUTOREG or AUTOREGRESSION in the computer packages SAS and SPSS respectively (see STATISTICAL PACKAGES). However, AUTOCORRELATION of residuals can appear because the wrong model is being fitted. For example, if the true response was quadratic and a linear model was fitted, the errors would appear as a group of negative errors, a group of positive errors and then a group of negative errors. It is a better strategy to obtain a good model than using an autoregressive error model as a panacea for models that simply do not fit.

Interrupted time series are often either before-and-after treatment for single subjects or before-and-after intervention for populations. An important question for the analysis is whether the data are correlated. The main reason for correlation might be because the same subject is measured before and after. However, if we removed the subject effect, the data may be independent and so, for example, a two-sample STUDENT'S t-TEST would be valid.

One could look for three different sorts of effect on an outcome of an intervention at one point in time: (a) a change in slope, (b) a change in level or (c) a combination of a change in slope and a change in level.

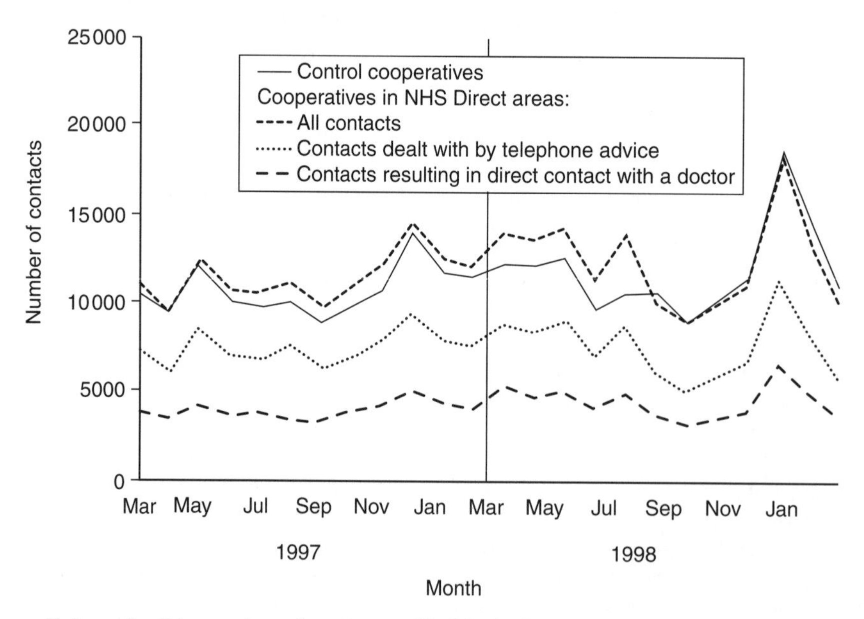

time series in medicine *Monthly number of contacts with GPs before and after introduction of NHS Direct (Munro et al., 2000, British Medical Journal 321, 150–3, with permission from the BMJ Publishing Group)*

NHS Direct is a telephone system designed to relieve pressure on general practitioners (GPs). It was introduced into the UK in 2001 (Munro *et al.*, 2000). We are interested whether it has an effect on the number of telephone calls to GPs. The most likely model is a change in slope (see the second figure on page 464).

One simple method is the following. We make the origin for time the point at which the intervention occurred. The model is $y_t = \alpha + \beta_1 t + \beta_2 t' + \varepsilon_t$, where y_t is the monthly number of calls to selected practices in month t and $t' = 0$ if $t < 0$, $t' = t$ if $t > 0$. Thus a test of the effect is to test whether $\beta_2 = 0$. As stated earlier, conventional regression will give invalid results if the errors are serially correlated and so we need to check the serial correlation of ε_t. Again, we can use certain statistical packages to fit the model assuming the errors are generated by an autoregressive process.

Many epidemiological series consist of counts and require Poisson regression rather than ordinary linear regression. We can also employ a method similar to the Cochrane–Orcutt method to allow for serial correlation and use GENERALISED ESTIMATING EQUATIONS to estimate the parameters.

Campbell (1994) analysed the dependence of daily deaths from sudden infant death syndrome (SIDS) in England and Wales from 1979 to 1983 on the mean daily environmental temperature measured in London. The input was mean daily temperature and the output daily deaths due to SIDS. There is clear seasonality in the mortality series but this does not mean that there is a causal relationship between temperature and cot deaths since many factors behave seasonally, such as length of day and rainfall. It is only when these effects are removed can we deduce a possible relationship. A model was fitted that removed seasonality and then included a linear temperature effect. The coefficient associated with mean temperature 3–5 days before the death was –0.041 (SE 0.005). We interpret this as saying that a 1 °C drop in temperature is associated with a rise in SIDS by about 4 %. Further investigations demonstrated that the relationship was approximately linear.

We can test the residuals for autocorrelation, using tests such as the Durbin–Watson test (first-order AR) and the Ljung–Box (general order). However, one should ask if it is sensible to test for serial correlation and only include serial correlation in the model if the test is significant. One should also ask why the data are serially correlated.

Serial correlation could be split into *intrinsic* correlation (endogenous) and *extrinsic* correlation (exogenous). Intrinsic correlation means that the value at a particular time depends directly on the value at an earlier time. Examples include: serum cholesterol at different times, population in age groups in successive years, epidemics of measles. Extrinsic correlation occurs because both variables depend on some third (time-dependent) variable. Examples include daily SIDS, where the deaths are not caused by epidemics and are really unrelated to each other except through (say) the weather.

We will not cover repeated measures in detail here. Commonly they arise when individuals have measurements taken repeatedly over time (see REPEATED MEASURES ANALYSIS OF VARIANCE). Often the serial correlation aspect of the data can be removed by the simple expedient of using summary measures (Matthews *et al.*, 1990). If not, usually either a simple AR(1) model is assumed, or what is known as an exchangeable correlation model or compound symmetry. This is generated by a model of the form:

$$y_{it} = \mu + \alpha_i + \varepsilon_{it}$$

where y_{it} is an outcome at time t on subject i, α_i is the effect of subject i, which is assumed normally distributed with variance σ_α^2, $E(\alpha_i\,\alpha_j) = 0$ when $i \neq j$ and ε_{ij} has variance σ^2.

The effect of this is to generate a covariance matrix with σ_α^2 on the off-diagonal and $\sigma^2 + \sigma_\alpha^2$ on the diagonal terms. Although one would expect measurements made further away to be less correlated (i.e. perhaps an AR(1)), in practice compound symmetry has been found to be a reasonable assumption in many cases.

We need to distinguish between methods where serial correlation is an important part of the model, such as for prediction, and where it is simply a nuisance. If it is a nuisance, then we need to examine intrinsic and extrinsic correlation. We should allow for serial correlation in regression modelling. Often serial correlation can be 'made to go away' and so the time series aspect is not a major concern. Compound symmetry is a useful assumption for repeated measures in RANDOMISED CONTROLLED TRIALS. *MJC*

Bernardi, L., Sleight, P., Bandinelli, G., Cencetti, S., Fattorini, L., Wdowczyc-Szulc, J. and Lagi, A. 2001: Effect of rosary prayer and yoga mantras on autonomic cardiovascular rhythms: comparative study. *British Medical Journal* 323, 1446–9. **Box, G. E. P. and Jenkins, G.** 1976: *Time series analysis: forecasting and control.* San Francisco: Holden Day. **Campbell, M. J.** 1994: Time series regression for counts: an investigation into the relationship between sudden infant death syndrome and environmental temperature. *Journal of the Royal Statistical Society, Series A* 157, 191–208. **Campbell, M. J.** 1996: Spectral analysis of clinical signals: an interface between medical statisticians and medical engineers. *Statistical Methods in Medical Research* 5, 51–66. **Chatfield, C.** 1989: *The analysis of time series: an introduction*, 4th edition. London: Chapman & Hall. **Cochrane, D. and Orcutt, G. H.** 1949: Application of least squares regression to relationships containing autocorrelated error terms. *Journal of the American Statistical Association* 44, 32–61. **Diggle, P. J.** 1990: *Time series. A biostatistical introduction.* Oxford: Oxford Science Publications. **Glass, G. V., Wilson, V. L. and Gottman, J. M. *et al.*** 1975: *Design and analysis of time series experiments.* Colorado: Colorado Associated Press. **Matthews, J. N. S., Altman, D. G., Campbell, M. J. and Royston, J. P.** 1990: Analysis of serial measurements in medical research. *British Medical Journal* 300,

230–5. **Munro, J., Nicholl, J., O'Cathain, A. and Knowles, E.** 2000: Impact of NHS Direct on demand for immediate care: observational study. *British Medical Journal* 321, 150–3.

time trade-off technique See VON NEUMAN–MORGENSTERN STANDARD GAMBLE

total fertility rate (TRF) See DEMOGRAPHY

transformations The use of transformations in statistics has a long history. For example, the Wilson–Hilferty cube root transformation for CHI-SQUARE DISTRIBUTIONS, the Fisher z-TRANSFORMATION for CORRELATIONS, the use of logarithms for biological data and the arc-sine root transformations for proportions are well-known procedures. In most cases the use of transformations is not an end in itself, but rather a means to an end. The ultimate benefit is usually not what the transformation directly achieves, but rather that it allows subsequent analysis to be simpler, more revealing or more accurate. What is most important is how the transformation aids in the interpretation and description of the data.

The transformations may be applied to observations, either response or explanatory variables, or to parameters or statistics, or they might be an explicit part of a statistical model. The purposes of using transformations include: (a) to reduce nonadditivity or nonnormality in ANALYSIS OF VARIANCE (ANOVA), or more generally to improve the agreement between the observations and the assumptions in a model, (b) to reduce SKEWNESS or achieve approximate symmetry, (c) to describe the structure of observations and (d) to simplify the relationship between variables.

The power transformation x^λ or Box–Cox transformation $x^{(\lambda)} = (x^\lambda - 1)/\lambda$ are simple monotonic transformations that are frequently used; however, they can be applied only to nonnegative x. The Box–Cox transformation has the advantage that the limiting transformation as $\lambda \to 0$ is $\log(x)$. To ease interpretation, it is sometimes preferable to limit λ to a finite set, such as $(-2, -1, -1/2, 0, 1/4, 1/3, 1/2, 2/3, 1, 2)$. A general discussion of a variety of aspects in the use of power transformations can be found in Box and Cox (1964), the review article by Sakia (1992) and the book by Carroll and Ruppert (1988).

A common use of transformations is in a regression or ANOVA setting. It is common for the VARIANCE of the response to increase as the expected value increases. In such cases a POWER transformation of the response variable can sometimes achieve approximate homogeneity of variance. Also when the variance does increase with the MEAN it is fairly common for the observations, or RESIDUALS from the mean, to be positively skewed, for which the same transformation may have the added benefit of substantially reducing the skewness.

Overall in regression settings, transformations of the response variable are used in the hope that they will give a correct structure in the systematic part of the model and also achieve homogeneous and Gaussian error distributions. It is unlikely that a single transformation will achieve all of these exactly. It is the correct systematic that is the most important of the three aspects to achieve.

For a set of observations $\mathbf{Y}^{\mathrm{T}} = (Y_1, \ldots, Y_n)$, Box and Cox proposed the model:

$$Y_i^{(\lambda)} = \mathbf{X}_i\beta + e_i \qquad (1)$$

where $\mathbf{X}_i$ is the ith row of the design matrix $\mathbf{X}$, e_i are independent, $e_i \sim N(0, \sigma^2)$. In this model a single λ is assumed to achieve the three objectives of a simple systematic structure, homogeneity of variance and normal errors.

A problematic aspect of equation (1) is that the interpretation of the parameter β depends on λ. However, various aspects of β, in particular, the direction of β (represented by $\beta/(\text{length}(\beta))$, say) or the ratio of two regression coefficients, β_1/β_2, which measures the relative importance of one explanatory variable to another, both have interpretations that are not dependent on λ.

For models involving transformations some aspects of inference for the regression coefficient, β, have been controversial. In an unconditional approach λ is treated as a parameter on an equal footing with all the other parameters; however, the interpretation of β depends on the estimated λ, and the variance of β is very large. In the conditional approach inference about β is made on the estimated transformed scale, ignoring the fact that λ has been estimated from the observations, which is not entirely satisfactory because it ignores the uncertainty associated with the estimation of λ. Aspects of inference that are less controversial are tests of $\beta = 0$, hypotheses that have an interpretation irrespective of λ.

In some applications transformations back to the original Y scale are desirable and can be particularly useful for graphical display. Caution is necessary in transforming interpretations from one scale to the other; e.g. a lack of interaction on the transformed scale should not be interpreted as a lack of interaction on the original scale. Power transformations allow predictions back to the original scale because they are monotonic. For model (1), the quantity $1 + \lambda(\mathbf{X}_0\beta)^{1/\lambda}$, when $\lambda \neq 0$, or $\exp(\mathbf{X}_0\beta)$, when $\lambda = 0$, is the predicted median of the distribution of Y given $\mathbf{X}_0$.

In multiple regression modelling it is common to consider transformations of explanatory variables. In the model:

$$Y_i = \alpha + \beta X_i^{(\lambda)} + e_i \qquad (2)$$

for scalar X, where $e_i \sim N(0, \sigma^2)$, the purpose of adding the extra parameter λ is to fit the systematic structure of the model better, with the requirement of homogeneity of variance and normality of e_i playing a lesser role. Whether the transformation achieves symmetry or normality of the marginal distribution of $X^{(\lambda)}$ is usually of less importance, other than to reduce the SENSITIVITY of OUTLIERS in X or unless one needs to model the X distribution.

Carroll and Ruppert (1988) have developed an approach for nonlinear regression models in which the systematic part of the model, $f(\mathbf{X}, \beta)$, is known through subject matter considerations. For example, the Michaelis–Menten equation for enzyme reactions is $Y=\beta_0 X/(\beta_1+X)$. The same transformation is applied to both sides of this equation by assuming the model $Y_i^{(\lambda)}=(f(\mathbf{X}_i,\beta))^{(\lambda)}+e_i$, where $e_i \sim N(0, \sigma^2)$. The model assumes that the untransformed relationship already fits the MEDIAN of the data adequately, but that the residuals exhibit heteroscedasticity and/or nonnormality. The main aim of the transform-both-sides approach is to make the residuals normal with constant variance, hence improving properties and inference associated with estimates of β. An important aspect of the approach is that the interpretation of β does not depend on λ.

The accelerated failure time model for censored survival data, $\log(T_i)=\mathbf{X}_i\beta+e_i$, where $e_i \sim N(0, \sigma^2)$, can be viewed as a special case of the power transformation model. In the extension of the Box–Cox procedure to multivariate data a separate power transformation parameter is assumed for each component of the multivariate vector. Solomon and Taylor (1999) considered Box–Cox transformations for components of variance models.

Power transformations have been used to assist in estimating regression centiles, which have application in establishing reference ranges. In the LMS METHOD at each fixed value of a scalar variable x the distribution of Y is assumed to be normal following a power transformation; from this the percentiles can be easily calculated. The model assumes that the median of Y, the power transformation parameter and the scale parameter of the NORMAL DISTRIBUTION all vary smoothly as a function of x.

In AIDS studies, viral load and CD4 count, a measure of the immune system, are frequently measured and are important indicators of disease progression. Both of these variables are highly skewed. Relatively complicated longitudinal and joint longitudinal–survival models have been applied to serial measurements of these markers. Nearly everyone uses the logarithm of viral load, either $\log_e$, $\log_{10}$ or $\log_2$. For CD4 counts some authors use the log for ease of interpretation, whereas others use $CD4^{1/4}$ or $CD4^{1/2}$ to ensure that the assumptions in their models, such as homogeneity of variance and symmetry for the MEASUREMENT ERROR, were satisfied for their data.

PSA is a common blood test used in prostate cancer studies both for screening and to monitor disease progression. PSA is quite skew and it is natural to consider a logarithm transformation because the PSA value is thought to be roughly proportional to the volume of the tumour, which grows approximately exponentially. In practice $\log(\text{PSA}+1)$ has been used, because PSA can be close to or even equal to zero, causing a standard log transformation to produce too many large negative values or not be calculable. For further details see Cole and Green (1992), Tsiatis, DeGruttola and Wulfsohn (1995), Slate and Cronin (1997) and Wang and Taylor (2001). *JMGT*

Box, G. E. P. and Cox, D. R. 1964: An analysis of transformations (with discussion). *Journal of the Royal Statistical Society, Series B*, 26, 211–52. **Carroll, R. J. and Ruppert, D.** 1988: *Transformation and weighting in regression*. London: Chapman & Hall. **Cole, T. J. and Green, P. J.** 1992: Smoothing reference centile curves: the LMS method and penalized likelihood. *Statistics in Medicine* 11, 1305–19. **Sakia, R. M.** 1992: The Box–Cox transformation technique: a review. *The Statistician*, 41, 169–78. **Slate, E. H. and Cronin, K. A.** 1997: Changepoint modeling of longitudinal PSA as a biomarker for prostate cancer. *Case Studies in Bayesian Statistics* III, Springer-Verlag, pp. 435–56. **Solomon, P. J. and Taylor, J. M. G.** 1999: Orthogonality and transformations in variance components model. *Biometrika* 86, 289–300. **Tsiatis, A. A., DeGruttola, V. and Wulfsohn, M. S.** 1995: Modeling the relationship of survival to longitudinal data measured with error. Applications to survival and CD4 counts in patients with AIDS. *Journal of the American Statistical Association* 90, 27–37. **Wang, Y. and Taylor, J. M. G.** 2001: Jointly modeling longitudinal and event time data: application in AIDS studies. *Journal of the American Statistical Association* 96, 895–905.

tree-structured methods (TSM) These are methods designed to produce interpretable prediction rules by subdividing data into subgroups that are homogeneous with respect to both covariates and outcome. Predictions flow from this outcome constancy, with simple subgroup summaries sufficing. The interpretability of the attendant prediction rules derives from the simple, recursive fashion by which the covariates are employed in eliciting the subgroups. As a consequence of this simplicity tree-structured methods have enjoyed widespread popularity, particularly in biomedical settings. However, of course, all this simplicity belies a number of issues, especially pertaining to prediction performance, that have spawned considerable recent research activity.

TSM prediction rules can be developed for both categorical and continuous outcomes, reflecting classification and regression problems respectively. It was the correspondingly named monograph *Classification and regression trees* by Breiman *et al.* (1984) that, by way of establishing a methodological framework and providing several compelling applications, fuelled the subsequent popularity of TSM. Indeed, tree-structured methods are frequently referred to by the monograph's title-based acronym, CART. The terminology 'recursive partitioning' is also commonplace, with another relevant monograph being that by Zhang and Singer (1999). The 'tree' terminology itself derives from the companion graphical depictions of the fitted models (see the first figure). As an historical note, the forerunners of TSM date to the 1960s.

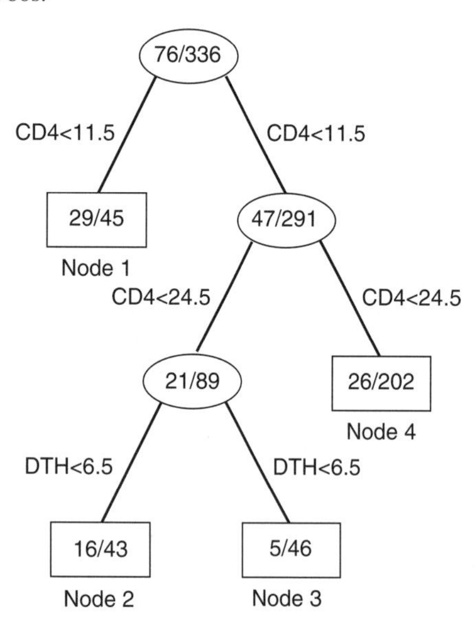

tree-structured methods *Log-rank survival tree*

The basic TSM paradigm, as developed by Breiman *et al.* (1984) and outlined later, has been extended in many directions. Of particular importance from a biomedical standpoint are extensions to survival outcomes and LONGITUDINAL DATA. The resultant methods are described in an overview article (Segal, 1995). Here, illustration of TSM will make recourse to the SURVIVAL ANALYSIS applications since this allows exposition of TSM fundamentals, as well as showcasing a setting for which tree concepts seem particularly well suited.

The central thrust of tree techniques is the elicitation of subgroups. Within these subgroups covariates are homogeneous and between subgroups outcomes are distinct. Therefore, in clinical settings with survival outcomes, interpretation in terms of prognostic group identification is frequently possible. Creation of the subgroups according to a tree structure (binary recursion) mimics, at least simplistically, medical decision making: if the patient is female, has a family history of breast cancer and is over 40, then annual mammograms are recommended. Similarly, given a survival tree, it is straightforward to classify a new patient to a prognostic group by simply answering the sequence of yes/no (binary) questions or splits that give rise to each subgroup or node.

It is reasonable to assess whether this goal of subgroup extraction requires new methodology. Could not, for example, the COX REGRESSION MODEL (Cox, 1972) be employed for this purpose? Suppose, without loss of generality, that in fitting a PROPORTIONAL HAZARDS model with three continuous covariates we obtain positive coefficients for each; i.e. each variable is adverse: increased values of each are associated with elevated risk. Thus, we might try to create a high-risk group by combining individuals who have high values for all three covariates. However, this approach may fail due to no patients possessing such a covariate profile.

Alternatively, we could compute a risk score for each member of the sample based on substitution of the actual covariate profiles into the LOG-LINEAR MODEL using the fitted coefficients. Then a high-risk stratum could be obtained by selecting the desired percentile of the sample risk scores. The difficulty here is that individuals with potentially disparate covariate values are combined and hence the resultant risk group is hard to label or interpret.

In addition to identifying important prognostic groups, which can be thought of as local interactions, survival tree techniques can also be informative about individual covariates. This derives from single splitting (subdivision) being revealing about threshold effects for time-independent covariates or change points in the case of time-dependent covariates. Also, repeated splitting on a given covariate can be revealing about more complex nonlinearities. However, use of (smoothed) martingale residual plots (Therneau, Grambsch and Fleming, 1990) is arguably a more direct way for determining appropriate functional form. Further, tree methods in general are not geared towards making global assessments of a covariate's importance. This is for a variety of reasons.

First, if a covariate is used (to define a split) in just one branch of the tree, then it is problematic trying to gauge its overall importance. Second, masking whereby a covariate selected as (the best) split variable precludes another, almost as good, covariate from emerging complicates covariate evaluation. (Splitting criteria are discussed later; these allow determination as to which covariate constitutes the best split variable. Most software implementations of TSM provide output detailing several of the top competing splits (not just the best), as well as measures of overall covariate importance. The related issue of instability is further discussed in the context of improving TSM predictive performance.) Finally, covariate splits are selected by optimising a split criterion and are therefore highly adaptive. While some corresponding distributional results have been obtained, difficulties remain in assigning significances to a sequence of splits, and hence to formally appraising covariate importance.

The prescription for tree construction advanced by Breiman *et al.* (1984) has served as the foundation for many extensions and refinements and therefore is worth detailing. Their approach features four constituent components: a set of questions, or splits, phrased in terms of the covariates that serve to partition the covariate space. A tree structure derives from the recursive application of these questions and a binary tree results if the questions are binary (yes/no). The subgroups created by assigning cases according to these splits are termed *nodes*; a split function (or split criterion) $\phi(s, t)$ can be evaluated for any split s of any node t. The split function is used to assess the worth of the competing splits and a means for determining appropriate tree size and statistical summaries for the nodes of the tree.

The first item defines what sort of subdivisions are permitted – these are the allowable splits. Binary splits are generally used, mostly for computational reasons. These have the flavour: 'Is age less than 45?' or 'Is ethnicity Asian, black or Hispanic?'. The answers to such questions induce a partition, or split, of the covariate space: cases for which the answer is 'yes' belong to the corresponding region while those for which the answer is 'no' belong to the complementary region. The allowable splits satisfy the following constraints: each split depends on the value of only a single covariate; for ordered (continuous or categorical) covariates, X_j, only splits resulting from questions of the form 'Is $X_j \leq c$?' for $c \in$ domain (X_j) are considered. Thus ordering is preserved (see the first question); for unordered categorical predictors all possible splits into disjoint subsets of the categories are allowed (see the second question).

The allowable splits are formulated in this fashion in order to balance flexibility and interpretability of the fitted models with computational feasibility. While variants and extensions have been subsequently promoted, this formulation underlies most implementations.

Given a set of allowable splits a tree is grown as follows: for each subgroup or node: (a) examine every allowable split on each predictor variable and (b) select and execute (create left and right daughter nodes) the best of these splits. The initial or root node comprises the entire sample. Steps (a) and (b) are then reapplied to each of the daughter nodes and so on. It is this reapplication that gives rise to the *recursive partitioning* terminology. The determination of tree size (how many splits), the third component of the paradigm, is important yet complicated – details are deferred to Breiman *et al.* (1984) and Segal (1995). Thus, it remains to define what constitutes a best split; this is the province of the second component.

Best splits are decided by optimising a split function $\phi(s, g)$ that can be evaluated for any splits of any node g. For regression (i.e. continuous outcomes) Breiman *et al.* (1984) describe two possibilities: least squares (LS), detailed later, and least absolute deviations. Let g designate a node of the tree; i.e. g contains a subsample of cases $\{(\boldsymbol{x}'_i, y_i)g$ where $\boldsymbol{x}'_i = (x_{i1}, x_{i2}, \ldots, x_{ip})$ is the vector of observed covariate values and y_i is the observed outcome for the ith case. Let N_g be the total number of cases in g and let $\bar{y}(g) = (1/N)\sum_{i \in g} y_i$ be the outcome average for node g. Then the within-node sum of squares is given by $\mathrm{SS}(g) = \sum_{i \in g}(y_i - \bar{y}(g))^2$. Now suppose a split s partitions g into left and right daughter nodes g_{L} and g_{R}. The LS split function is $\phi(s, g) = \mathrm{SS}(g) - \mathrm{SS}(g_{\mathrm{L}}) - \mathrm{SS}(g_{\mathrm{R}})$ and the best split s^* of g is the split such that $\phi(s^*, g) = \max_{s \in \Omega}(s, g)$, where Ω is the set of all allowable splits s of g. An LS regression tree is constructed by recursively splitting nodes so as to maximise the above ϕ function. The function is such that we create smaller and smaller nodes of progressively increased homogeneity on account of the nonnegativity of ϕ: $\phi \geq 0$, since $\mathrm{SS}(g) \geq \mathrm{SS}(g_{\mathrm{L}}) + \mathrm{SS}(g_{\mathrm{R}})\, \forall s$.

It is worth noting that a tree grown in accordance with this LS split function will coincide with a tree grown using a two-sample t-statistic as the split function if the latter uses a pooled estimate of VARIANCE. Selecting the split that makes the resultant t-statistic maximal can be viewed as optimising node separation as measured by the difference in the respective node averages.

Modifications to the split function are a primary means for expanding the scope of TSM. Several such modifications have been proposed to enable handling of (censored) survival outcomes. One suite of such split functions is based on notions of between-node separation, analogous to use of the t-statistic. The log-rank statistic provides a familiar and readily implemented example. The resultant rewarding of subgroups that are internally homogeneous with regard to covariates (as imparted by allowable splits), yet externally different, dovetails with the objective of identifying distinct prognostic groups. Further, use of the log-rank statistic as a split function allows additional accommodation of left-truncated survival times as well as time-dependent covariates (see TIME-DEPENDENT VARIABLES).

We present an illustrative example of TSM with survival ENDPOINTS that pertains to HIV disease progression. The latency or incubation period for AIDS (i.e. the time from HIV infection to an AIDS diagnosis) is both long and variable. In order to try to explain this variability in terms of immune function decay, markers of immune function are regularly measured on longitudinally followed cohorts of HIV seropositive and seroconverting individuals. In particular, counts of CD4+T lymphocytes have been widely used both to follow the course of immune function loss and to predict time to AIDS or death. Here we consider an additional marker, delayed-type hypersensitivity (DTH) skin tests, as a putative supplement to CD4.

Use of DTH is motivated in part by deficiencies of CD4: quantification of peripheral blood CD4 depletion underestimates the severity of the HIV-induced loss of antigen-specific cellular immunity and provides no guide as to which antigen-specific responses have been lost. More sensitive measures of antigen-specific cellular immunity are therefore required as an adjunct to the monitoring of CD4 counts. Testing of cutaneous DTH responses to recall antigens provides a direct measure of cell mediated antigen-specific responses *in vivo*.

Assessment of these markers made recourse to the Western Australian HIV database. Patients in Western Australia with HIV infection have been managed at a single specialist referral centre since the first AIDS case was confirmed in

August 1983. Both HIV-infected and at-risk individuals were followed regularly. The closely scheduled (bimonthly) visits and early inception of the cohort provides a good opportunity for marker evaluation. Further details on the cohort, markers, handling seroprevalent (HIV positive at enrolment) subjects in the context of survival analysis, treating markers as time-dependent covariates and complementary Cox proportional hazards results are given in Segal *et al.* (1995).

Results from a TSM analysis of the survival endpoint time-to-AIDS are presented in the two figures for tree-structured methods. The log-rank statistic was used as a split function, although results were insensitive to this choice. The covariates used were age, CD4 and DTH, all measured at enrolment. CD4 is expressed as the percentage of T-lymphocytes that are CD4 positive. This is in accord with studies that, while noting strong correlations between this and other measures (CD4 count, CD4/CD8 ratio), found CD4 % to be the most prognostic for time-to-AIDS and exhibiting the smallest variability on repeated determinations.

The initial sample features 336 individuals, 76 of whom progressed to AIDS, as depicted in the uppermost (root) node. This sample is subdivided on the basis of CD4 %, the optimal cut-off point being 11.5 %, as shown on the emanating branches. In determining this split all possible cut-off points on all three (continuous) covariates were evaluated; i.e. corresponding log-rank statistics were computed. The selected split was maximal among all these statistics. The 291 individuals with CD4 % exceeding 11.5 % are also subdivided on the basis of CD4 % – covariates can be used repeatedly. Examination of these CD4-based splits affirms the anticipated: the subgroups with higher CD4 % have superior survival.

Indeed, the Kaplan–Meier curves (seen in the second figure) for the respective terminal nodes (rectangles in the first figure) showcase dramatic differences between the CD4 extremes, with survival prospects for the low CD4 group (node 1) being dismal in comparison with the high group (node 4). The 89 individuals with intermediary CD4 % ($11.5 < \text{CD4} < 24.5$) are further partitioned, this time on the basis of DTH. The optimal cut-off point is at a DTH value of 6.5 mm. The second figure again reveals survival differences for those in the two DTH-defined subgroups (node 2 versus node 3). Thus, it is possible that DTH can serve to augment CD4 % as a marker for HIV progression: for individuals whose CD4 % values are intermediary, rather than extreme, additional prognostic information may be obtained from their DTH values. It is interesting to contrast the splits with smoothed martingale residual plots obtained from a null Cox proportional hazards model. Such graphics are useful for informing an appropriate covariate functional form (Therneau, Grambsch and Fleming, 1990). Each split conforms to a step function approximation of the smoothed martingale residuals, which in turn are suggestive of threshold effects.

Of course, TSM are not without shortcomings. The primary deficiencies pertain to the interrelated concerns of instability, modest prediction performance and inefficiency in capturing underlying smooth response surfaces. Instability refers to the frequently large impact (on tree topology – selected covariates and cut-off points) that can result from small changes in data and/or inputs. This instability in part leads to relatively modest prediction performance by way of the associated large prediction variances that are exhibited when applying a TSM to independent test data.

A number of so-called committee or ensemble prediction methods have recently emerged that improve performance by

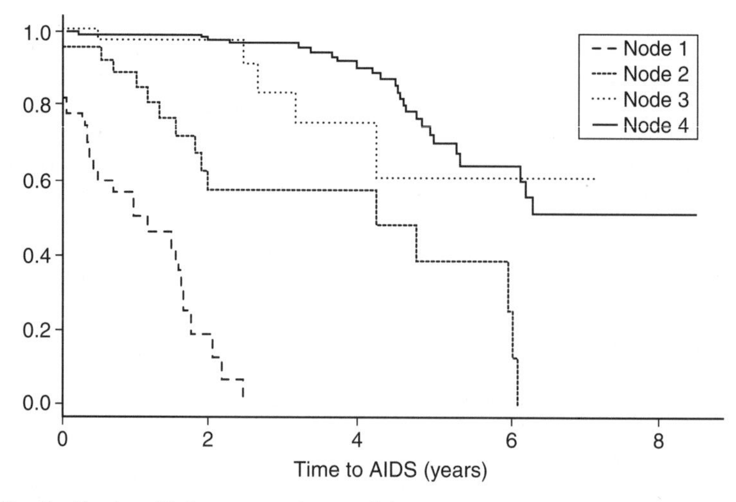

tree-structured methods *Kaplan–Meier curves: log-rank tree*

reducing this variability by way of strategic creation and combination of (many) individual TSMs. Bagging, BOOSTING and RANDOM FORESTS are examples; see Breiman (2001). Nonetheless, by virtue of their ready interpretability, TSMs remain a valuable tool for a wide range of biomedical problems. *MRS*

Breiman, L. 2001: Statistical modelling: the two cultures. *Statistical Science* 16, 199–215. **Breiman, L., Friedman, J. H., Olshen, R. A. and Stone, C. J.** 1984: *Classification and regression trees*. Belmont: Wadsworth. **Cox, D. R.** 1972: Regression models and life-tables (with discussion). *Journal of the Royal Statistical Society, Series B* 34, 187–220. **Segal, M. R.** 1995: Extending the elements of tree-structured regression. *Statistical Methods in Medical Research* 4, 219–36. **Segal, M. R., James, I. R., French, M. A. H. and Mallal, S.** 1995: Statistical issues in the evaluation of markers of HIV progression. *International Statistical Review* 63, 179–97. **Therneau, T. M., Grambsch, P. M. and Fleming, T. R.** 1990: Martingale-based residuals for survival models. *Biometrika* 77, 147–60. **Zhang, H. and Singer, B.** 1999: *Recursive partitioning in the health sciences*. New York: Springer.

trellis graphics This is an approach used to examining high-dimensional structure in data by means of one-, two- and three-dimensional graphs. The problem addressed is how observations on one or more variables depend on the values of other variables. The essential feature of the approach is the multiple conditioning that allows some types of graphic involving one or more variables to be displayed several times, each time as it appears when one or more other variables take particular values. The simplest example of a trellis graphic is the coplot, which is a SCATTERPLOT of two variables conditioned on the values taken by a third variable.

For example, the first figure shows a coplot of mortality versus latitude conditioned on population size. In this diagram, the panel at the top of the figure is known as the 'given' panel; those below are 'dependence' panels. Each rectangle in the given panel specifies a range of values of population size. On a corresponding dependence panel, mortality is plotted against latitude for those countries whose population sizes lie in the particular interval. To match population size intervals to a dependence panel, the latter are examined in order from left to right in the bottom row and then again from left to right in subsequent rows. The association between higher values of mortality and lower values of latitude (and vice versa) is seen to hold for all levels of population size.

A more complex example of a trellis graphic is shown in the second figure on page 472. Here a three-dimensional plot of mortality, latitude and longitude is given for four ranges of population size.

Several other examples of the use of trellis graphics are given in Verbyla *et al.* (1999). *BSE*

[See also SCATTERPLOT MATRICES]

Verbyla, A. P., Cullis, B. R., Kenward, M. G. and Welham, S. J. 1999: The analysis of designed experiments and longitudinal data using smoothing splines (with discussion). *Applied Statistics* 48, 269–312.

triangular test See SEQUENTIAL ANALYSIS

***t*-test** See STUDENT'S *t*-TEST

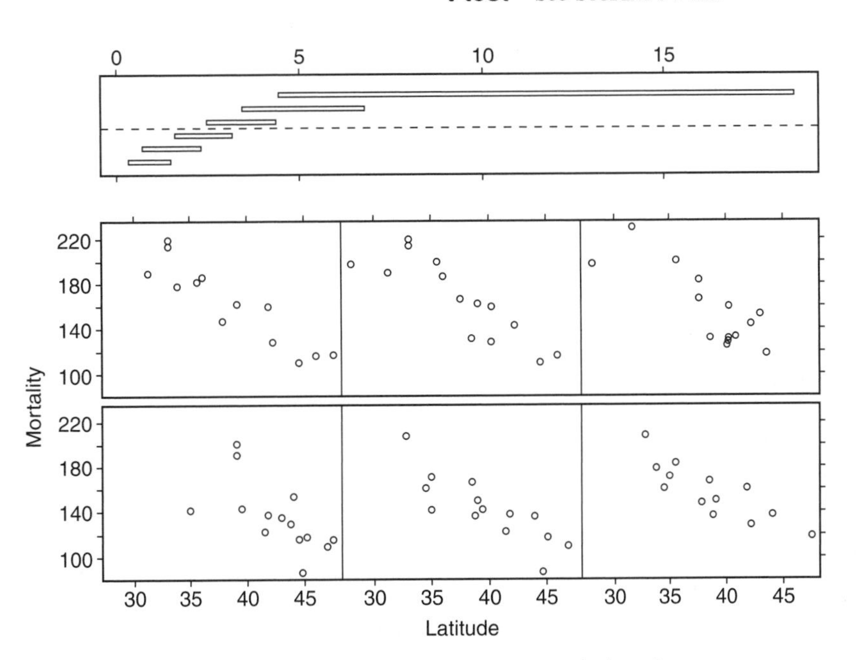

trellis graphics *Coplot of mortality versus latitude conditioned on population size*

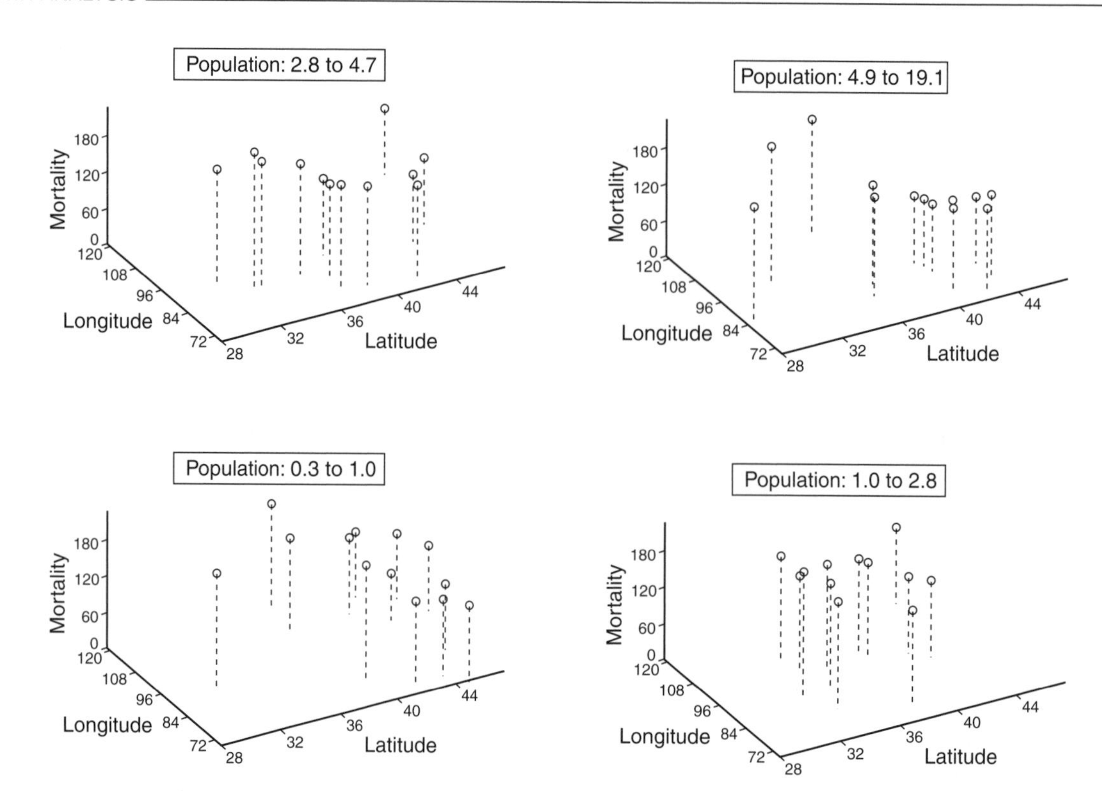

trellis graphics *Trellis graphic of mortality, latitude and longitude conditioned on population size*

twin analysis The analysis of diseases or other phenotypes in twins is an important tool in GENETIC EPIDEMIOLOGY. Twins may be either monozygotic (MZ), i.e. arising from the same fertilised egg, or dizygotic (DZ). MZ twins are genetically identical to one another. DZ twins, contrariwise, are genetically equivalent to siblings and share, on average, half their genetic material. Since both MZ and DZ twin pairs share the same environment (at least *in utero* and early life) they provide a well-controlled study design to evaluate genetic influences. For any phenotype that has a genetic component, MZ twins will tend to be more similar than DZ twins. MZ and DZ may be distinguished in studies by QUESTIONNAIRE (e.g. 'Are you like two peas in a pod?') or more objectively by DNA analysis.

For quantitative traits, the extent of concordance between MZ and DZ twins can be expressed in terms of *intraclass correlations*. Simple comparisons can then be made using F-tests. Since these tests assume normality, some transformation of the trait may be required.

More general analysis of twin data is usually conducted in the framework of a VARIANCE COMPONENTS analysis. Thus, we think of the trait X as being decomposed into a sum of components:

$$X = G + E + R$$

where G is the genetic component, E is a component due to shared environmental factors and R is a residual component. The proportion of the trait variance due to genetic component G is called the *heritability* of the trait (often written H). The genetic component can be further decomposed into an *additive* genetic component A and a nonadditive (or *dominance*) component D. An additive component in this sense is such that the value for any individual is the MEAN of the values in their two parents. If the proportions of the trait variance due to additive and nonadditive genetic factors and shared environment are ρ_A, ρ_D and ρ_E, then the correlations between MZ and DZ twin pairs will be given by:

$$\rho_{MZ} = \rho_A + \rho_D + \rho_E$$

$$\rho_{MZ} = 1/2\rho_A + 1/4\rho_D + \rho_E$$

Since twin studies only allow two CORRELATIONS to be estimated, it is not possible to estimate the additive, dominance and shared environmental components simultaneously. A common assumption is that the dominance component is zero, so that the genetic component is purely additive. Data on other relatives may allow all three components to be estimated. Where necessary, other covariates may be included in this framework. Age may be a particularly important covariate to include since twin pairs are identical in age.

An important assumption in these analyses is that the shared environment component E is identical in MZ and DZ twins. This assumption may sometimes be violated if certain *in utero* effects are important.

Twin data for binary traits are often expressed in terms of the concordance rate, defined as the proportion of twins of affected individuals who are themselves affected. A higher concordance rate in MZ twins is expected for traits with a genetic component. Concordance rates may be calculated in two ways. If the starting point is a series of affected individuals (e.g. from hospital records), the concordance rate can be estimated straightforwardly by identifying those individuals with twins and calculating the rates in the co-twins. If, however, the starting point is a register of twins, the concordance rate can be calculated from the number of pairs where both twins are affected (concordant) and where only one twin is affected (discordant), thus:

$$\frac{2 \times \#\text{ concordant pairs}}{2 \times \#\text{ concordant pairs} + \#\text{ discordant pairs}}$$

The factor of two is necessary to allow for the fact that concordant twins may be ascertained through either twin. Studies based on twin registers can be seriously biased if concordant twin pairs are more likely to be identified. For this reason, the best twin data on disease come from population-based registers with record linkage to medical records, as are available, for example, in Scandinavia.

More general methods of analysing binary twin data have been developed. A common approach is to extend the ideas of LOGISTIC REGRESSION or, for a chronic disease, POISSON or COX REGRESSION. The genetic and shared environmental effects are modelled by including RANDOM EFFECTS terms in the model in addition to any fixed covariates. *DE*

[See also GENETIC EPIDEMIOLOGY, GENOTYPE, PHENOTYPE]

Balding, D. J., Bishop, M. and Cannings, C. (eds) 2001: *Handbook of statistical genetics*. Chichester: John Wiley & Sons, Ltd. **Sham, P.** 1998: *Statistics in human genetics*. London: Arnold.

two-dimensional contingency table See CONTINGENCY TABLES

two-sample t-test See STUDENT'S *t*-TEST

two-sided tests In hypothesis tests we try to distinguish between chance variation in a dataset and a genuine effect. We do this by comparing the NULL HYPOTHESIS, which states that there is no difference between the populations in which the data arose, and the alternative hypothesis, which states that there is, in fact, a difference. If no direction for the difference is specified by either the null hypothesis or the alternative hypothesis, we have a two-sided test, sometimes referred to as a two-tailed test. We are therefore looking for a difference but are equally interested in differences in either direction.

For instance, when comparing a new treatment to an existing one we would be interested in detecting differences both in favour of or against the new treatment. In the majority of cases, the two-sided alternative is the appropriate one as it allows for the uncertainty about the direction of an effect that is often present. Whether or not to use a two-sided test should be decided based on the design of the study. Unless the study specifically seeks to detect an upward or downward change determined in advance, two-sided tests should be used. It is usually assumed that the *P*-VALUE reported from a specific statistical test is two-sided unless stated otherwise. More details can be found in Bland and Altman (1994). *MMB*

[See also ONE-SIDED TESTS]

Bland, J. M. and Altman, D. G. 1994: Statistics notes: one- and two-sided tests of significance. *British Medical Journal* 309, 248.

two-way analysis of variance This is a test to see if the mean varies with either (or both) of two categorical factors. The one-way ANALYSIS OF VARIANCE seeks to partition the variation in a sample into that due to the group factor (the between-groups sum of squares) and the residual variation that cannot be explained by a factor (the within-groups sum of squares).

In a two-way analysis of variance, there are two factors that define the groups and each factor explains some of the variation. It is therefore necessary to partition the variation in the sample into that due to factor A, that due to factor B and the residual variation.

The total variation/sum of squares (total SS) can be calculated in the same way as for one-way analysis of variance. The sum of squares due to factor A is the sum over all individuals of the squared differences between the overall MEAN and the mean value associated with the level of the factor appropriate to the individual. The sum of squares due to factor B can be calculated in a similar manner. The residual sum of squares is then calculated as the total sum of squares minus the sums of squares due to factors A and B.

If we wish to calculate the sum of squares due to the interaction between the factors, then this is calculated as the sum over all individuals of the squared difference between the mean for the appropriate combination of the factors and the sum of the means associated with the relevant levels of the individual factors less the overall mean.

Like the one-way flavour, the two-way ANOVA can be presented as a table. If factor A has k levels and factor B has j levels, then the statistics to test for factor effects are presented in the first table. The F-statistic associated with factor A will be compared to an F-DISTRIBUTION with $k - 1$ and

two-way analysis of variance *Two-way ANOVA table for main factor effects only. Entries in bold must be calculated directly from the data; the other entries follow in the manner indicated*

Source of variance	*Degrees of freedom*	*Sums of squares*	*Mean squares*	*F*	*P-value*
Factor A	**k–1**	**A SS**	A MS = A SS/(k–1)	$\frac{\text{A MS}}{\text{Res MS}}$	p
Factor B	**j–1**	**B SS**	B MS = B SS/(j–1)	$\frac{\text{B MS}}{\text{Res MS}}$	p
Residual	**N–k–j+1**	Res SS = Total SS – (A SS + B SS)	Res MS = Res SS/(N–k–j+1)		
Total	**N–1**	**Total SS**			

two-way analysis of variance *Two-way ANOVA table when an interaction effect is included. Entries in bold must be calculated directly from the data; the other entries follow in the manner indicated*

Source of variance	*Degrees of freedom*	*Sums of squares*	*Mean squares*	*F*	*P-value*
Factor A	**k–1**	**A SS**	A MS = A SS/(k–1)	$\frac{\text{A MS}}{\text{Res MS}}$	p
Factor B	**j–1**	**B SS**	B MS = B SS/(j–1)	$\frac{\text{B MS}}{\text{Res MS}}$	p
Interaction	**(k–1)(j–1)**	**Int SS**	Int MS = Int SS/((k–1)(j–1))	$\frac{\text{Int MS}}{\text{Res MS}}$	p
Residual	**N–kj**	Res SS = Total SS – (Int SS + A SS + B SS)	Res MS = Res SS/(N–kj)		
Total	**N–1**	**Total SS**			

$N-k-j+1$ DEGREES OF FREEDOM. That associated with factor B will be compared to an F-distribution with $j-1$ and $N-k-j+1$ degrees of freedom. If the interaction term is required, then the analysis is as shown in the second table.

Reneman *et al.* (2001) use two-way ANOVA to compare alcohol use in eight groups defined by gender ($k=2$) and four levels of ecstasy use ($j=4$). With 69 people in the study and no interaction being considered, the test for a gender effect is conducted by comparing the F-statistic to an F-distribution with 1 and 64 degrees of freedom. For further reading see Armitage and Berry (1987). *AGL*

[See also GENERALISED LINEAR MODEL]

Armitage, P. and Berry, G. 1987: *Statistical methods in medical research*. Oxford: Blackwell. **Reneman, L. *et al*.** 2001: Effects of dose, sex, and long-term abstention from use on toxic effects of MDMA (ecstasy) on brain serotonin neurons. *Lancet* 358, 1864–9.

Type I and Type II errors After every HYPOTHESIS TEST the decision to accept or reject the NULL HYPOTHESIS is made. This decision can, however, lead to two possible errors. First, we can obtain a significant result, and thus reject the null hypothesis, when the null hypothesis is in reality true. This is termed a Type I error, and can be considered as a false positive result. Secondly, we may obtain a nonsignificant result when the null hypothesis is not true, in which case the error is called a Type II error, which may be considered a false negative result. A Type II error thus refers to a mistaken failure to reject the null hypothesis when the alternative hypothesis is true and there is a real difference between the study groups.

The Type I error rate is no more than the so-called SIGNIFICANCE LEVEL, a PROBABILITY frequently denoted by alpha, α. Thus the significance level represents the chance that the null hypothesis is rejected when it is actually true. For every hypothesis test the significance level should be decided upon beforehand; the typical value chosen for this is 0.05. Consequently, over many trials, 5% or 1 in 20 are expected to yield false positive results. Sometimes, however, smaller values for alpha are used in particular to help deal with the problem of MULTIPLE TESTING (see MULTIPLE COMPARISONS).

The probability of making a Type II error is represented by the Greek letter beta (β). The Type II error is closely related to the POWER of a test. The statistical power of a test can be thought of as the chance or probability that a study of a given size would detect as statistically significant a real difference of a given magnitude, defined as $1-\beta$ or, more usually, power = $100(1-\beta)$ %. For example, when $\beta=0.2$ there is only an 80 % chance of the test detecting the particular alternative hypothesis when actually true.

In designing a study, both types of error should ideally be minimized. It is common to fix beta in advance by choosing an appropriate sample size. We do this by calculating the necessary sample size for a study to have a high probability of finding a true effect of a given magnitude (see SAMPLE SIZE DETERMINATION entries). More details can be found in Altman (1991). *MMB*

[See also ERRORS IN HYPOTHESIS TESTS]

Altman, D. G. 1991: *Practical statistics for medical research*. London: Chapman & Hall.

U

uncle test for randomisation See ETHICS AND CLINICAL TRIALS

uniform distribution This is the PROBABILITY DISTRIBUTION whereby all outcomes (within a RANGE) are equally likely. There is a discrete uniform distribution and a continuous uniform distribution (also known as the rectangular distribution), and we consider the discrete version first. If a random variable has a discrete uniform distribution on the integers from a to $a + b$, then we can write the probability mass function as:

$$P(X = x) = \frac{1}{b + 1}, \quad a \leq x \leq b$$

Expressed in this way, the distribution has a MEAN of $a + b/2$ and a VARIANCE of $(b^2 + 2b)/12$. As all PROBABILITIES are the same, it is of course symmetric about the mean.

The discrete uniform distribution is used in assessing DIGIT PREFERENCE. If durations of operations are being recorded to the nearest minute, and the operations typically take between two and four hours, then it might be presumed that the distribution of the terminal digit of the minutes would be approximately uniform from 0 to 9. The observed digits can be compared to the uniform distribution to detect any BIAS. The most common use of the uniform distribution in the medical sciences is perhaps in generating random sequences for CLINICAL TRIALS.

The probability density function of the continuous uniform distribution appears as a rectangle on a graph (hence its alternative name). Since the area of the rectangle must be equal to 1, the value of the density function is defined by the range of plausible observations (to keep the area of the rectangle constant, the height is defined by the width). The probability density function for the continuous uniform distribution is:

$$f(x) = \frac{1}{b - a}, \quad a \leq x \leq b$$

In this formulation, the mean is $(a + b)/2$ and the variance is $(b - a)^2/12$. As with the discrete uniform distribution, the distribution is symmetric about the mean. Note that when $a = 0$ and $b = 1$, this is a special case of the BETA DISTRIBUTION.

The cumulative distribution function maps observations from a distribution to probabilities that should be uniformly distributed between 0 and 1. Conversely, the inverse of the cumulative distribution function maps uniform observations between 0 and 1 to observations from the distribution in question. The first property provides a useful check of model fit. The second allows us to generate random observations from many distributions, if we can generate random uniformly distributed data and write out the inverse of the cumulative distribution function. For further reading see Altman (1991). *AGL*

Altman, D. G. 1991: *Practical statistics for medical research*. London: Chapman & Hall.

Encyclopaedic Companion to Medical Statistics: Second Edition Edited by Brian S. Everitt and Christopher R. Palmer

V

validity of scales The concepts of quality of assessments by rating scales and multiscale QUESTIONNAIRES are validity and reliability (see MEASUREMENT PRECISION AND RELIABILITY). A rating scale is valid if it measures what it is intended to measure in the specific study. The validity of subjective assessments is relative, study specific and cannot be determined absolutely. Therefore there are various concepts of validity, each addressing a specific type of quality assessment. The main concepts are criterion, construct and content validity, but a large number of subconcepts are used. The meaning of these concepts is not univocal and depends on applications and research paradigms. *Criterion validity* refers to the conformity of a scale to a true state or a gold standard, and depending on the purpose of the study subconcepts like clinical, predictive and concurrent validity will be used. In the absence of a true state or a gold standard, *construct validity*, referring to the consistency between scales having the same theoretical definition, and subconcepts like convergent, descriptive, divergent, factorial, translation validity and parallel reliability have been used in studies. Discriminative rating scales are used to distinguish between individuals or groups; when no external criterion is available, *discriminant validity* is to be assessed. Parallel reliability refers to the interchangeability of scales.

The concept *content validity* refers to the completeness of the scale or multiscale questionnaire in the coverage of important areas. Subconcepts like face, ecological, decision, consensual, sampling validity, comprehensiveness and feasibility have been used in studies (Svensson, 1993).

Assessments on RATING SCALES generate ordinal data having rank-invariant properties only, which means that the responses indicate a rank order and not a mathematical value. The results of statistical treatments of data must not be changed when relabelling the ordered responses (see RANK INVARIANCE). Appropriate statistical methods for evaluation of criterion and construct validity often refer to the order consistency or to the relationship between the scales of comparison.

The SCATTERPLOT of low back pain perceived by 48 patients with at least 4 years of pain history made on a VISUAL ANALOGUE SCALE (VAS) and a verbal descriptive scale (VDS-5) having the five categories of perceived pain, 'none', 'negligible', 'moderate', 'rather severe', 'very severe', is shown in the first figure (Svensson *et al.*, 2009). As evident from the plot there is large overlapping; e.g. patients having 'moderate pain' on the VDS used VAS positions from 28 to 73, and the same VAS positions were used by patients with 'rather severe pain'. The proportion of overlapping pairs among all possible different pairs of data defines the measure of disorder D. In this case D equals 0.07, which means that 7 % of all possible combinations of different pairs are disordered.

The expected pattern of complete order consistency, the rank-transformable pattern of agreement (RTPA), is constructed by pairing off the two sets of distributions of data

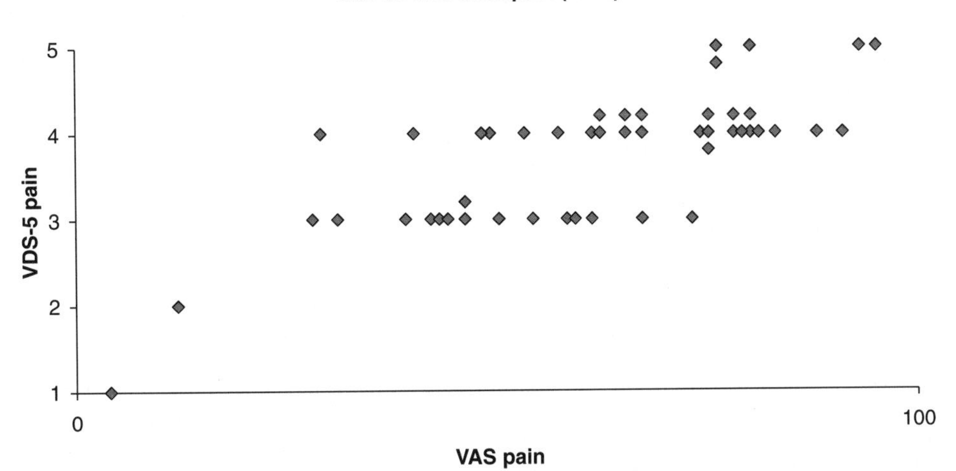

validity of scales *The distribution of paired assessments of back pain on a visual analogue scale for pain (VAS pain) and a five-point verbal descriptive scale for pain (VDS-5 pain)*

Encyclopaedic Companion to Medical Statistics: Second Edition Edited by Brian S. Everitt and Christopher R. Palmer

against each other. The measure of disorder expresses the observed dispersion of pairs from this order consistent distribution of interchangeability between the scales. The cut-off response values for interscale calibration are also provided by the RTPA, and it is obvious that there is no linear correspondence between VAS and discrete scale assessments (see the second figure) (see RANKS, RANKING PROCEDURES) (Svensson, 1993, 2000a, 2000b).

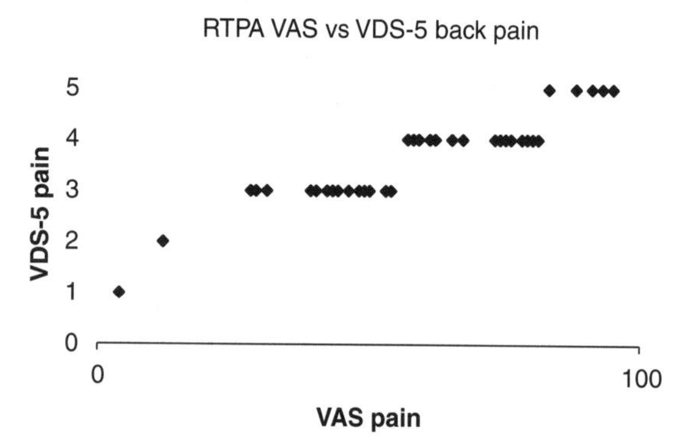

validity of scales *The rank-transformable pattern of agreement (RTPA), uniquely defined by the two sets of frequency distributions of data in the first figure*

There are other measures that could be applied to evaluation of various kinds of validity of scales. Depending on the purpose, SPEARMAN'S RANK CORRELATION COEFFICIENT, Goodman–Kruskal's gamma (see Everitt, 1992) and Kendall's tau (see CORRELATION) could be suitable. Spearman's rank-order correlation coefficient is a commonly used nonparametric measure of ASSOCIATION. However, a strong association does not necessary mean a high level of order consistency, and does not indicate that two scales are interchangeable. The Pearson correlation coefficient (see CORRELATION), STUDENT'S *t*-TEST and the ANALYSIS OF VARIANCE are also common in validity studies. A serious drawback is that these methods assume normally distributed quantitative data (see NORMAL DISTRIBUTION), and such requirements are not met by data from rating scales (Svensson, 2000b). *ES*

Everitt, B. S. 1992: *The analysis of contingency tables*, 2nd edition. London: Chapman & Hall. **Svensson, E.** 1993: *Analysis of systematic and random differences between paired ordinal categorical data* (dissertation). Göteborg: Göteborg University. **Svensson, E.** 2000a: Comparison of the quality of assessments using continuous and discrete ordinal rating scales. *Biometrical Journal* 42, 417–34. **Svensson, E.** 2000b: Concordance between ratings using different scales for the same variable. *Statistics in Medicine* 19, 3483–96. **Svensson, E., Schillberg, B., Kling, A-M. and Nyström, B.** 2009: The balanced inventory for spinal disorders. The validity of a disease specific questionnaire for evaluation of outcomes in patients with various spinal disorders. *Spine* 34, 1976–83.

variance The variance is the square of the STANDARD DEVIATION. It is calculated using the following formula, in which n is the number of observations, i takes values from 1 to n and the Σ notation denotes the sum, i.e. $(x_1-\bar{x})^2 + (x_2-\bar{x})^2 + \cdots + (x_n-\bar{x})$:

$$s^2 = \frac{\sum (x_i-\bar{x})^2}{n-1}$$

Both s^2 and σ^2 are used to indicate the variance. Technically, the former refers to the variance of the sample and the latter to the variance of the population, which is being estimated by the sample and is marginally smaller, since the divisor is n instead of $n-1$ in the formula. When quoting a mean to summarise data, it is also customary to quote a sample standard deviation. This is the square root of the sample variance, and is in the same units as the raw data. *SRC*

[See also COVARIANCE MATRIX]

variance components See COMPONENTS OF VARIANCE

variogram This is a procedure that provides a description of the autocorrelation (see CORRELATION in line series or spatial clusters. It is the latter that forms the focus for the following account. It is important to describe and model this autocorrelation so as to incorporate it into estimation and prediction procedures. For example, consider disease incidences measured at spatial locations. To construct a map, one would need to interpolate the incidence value for the locations at which it was not observed and in the absence of large-scale spatial trend such predictions should give larger weights to nearby locations if the autocorrelation were increasing with decreasing distances.

The variogram is based on the semi-variance $\gamma(x, y; h_x, h_y)$, which measures half the variance of the difference between two values of an outcome, Z, observed at two spatial locations referenced by the spatial coordinates (x, y) and $(x+h_x, y+h_y)$. Strictly speaking, the theoretical variogram is defined as twice the semi-variance, i.e.:

$$2\gamma(x, y; h_x, h_y) = E\left\{\left[Z(x, y)-Z(x+h_x, y+h_y)\right]^2\right\}$$

but the semi-variance itself is usually referred to as the variogram. It represents an (inverse) measure of the statistical dependency of the variables at locations (x, y) and $(x+h_x, y+h_y)$.

In all generality, the variogram is a function of both the location (x, y) and the distance and direction (h_x, h_y). Hence, to estimate it, replicate observations at each location would be needed. In practice, only one such realisation is available. To overcome this, the intrinsic hypothesis is introduced, which makes assumptions about the difference $Z(x, y) - Z(x+h_x, y+h_y)$. It states that for the spatial area under investigation (1) the expectation of the difference is zero, i.e. that there is no spatial trend, and (2) the variance of the difference depends only on the distance vector (h_x, h_y) and not the location. For variograms that reach an asymptote (so-called *bounded* variograms) this hypothesis is equivalent to assuming second-order stationarity of the measures themselves.

Under this assumption it is possible to estimate the variogram from the data. For simplicity, further assuming that the variogram is isotropic, i.e. that it only depends on the distance, h, between two locations and not the direction of (h_x, h_y), the variogram can be estimated by the empirical variogram:

$$\hat{\gamma}(h) = \frac{1}{2|N(h)|}\sum_{N(h)} (z_i - z_j)^2$$

where $N(h)$ is the set of distinct location pairs (i, j) that are distance h apart, $|N(h)|$ the number of such pairs and z_i and z_j the observed values at these locations. To achieve reasonable numbers an estimation is carried out at discrete lags and distance bins are allowed for. Care has to be taken when choosing the number of lags, lag increments and bin widths (see Cressie, 1993).

Since the main goal of variogram analysis is to find a parsimonious description of the spatial autocorrelation structure of a variable, a variogram of a particular functional form is usually fitted to the empirical variogram. Suitable monotonically increasing functions for bounded variograms are defined through three parameters: the *nugget effect* represents microscale variation or measurement error, the *sill* represents the variance of the outcome measure and the *effective range* is the distance at which autocorrelation becomes negligible.

For an example, the open symbols in the figure show an empirical variogram to which a spherical variogram function (a particular choice of functional form) was fitted. The curve is fully described by the parameter estimates (effective range = 0.8, sill = 1.75, nugget = 0.7) and indicates small-scale spatial autocorrelation up to a distance of 0.8.

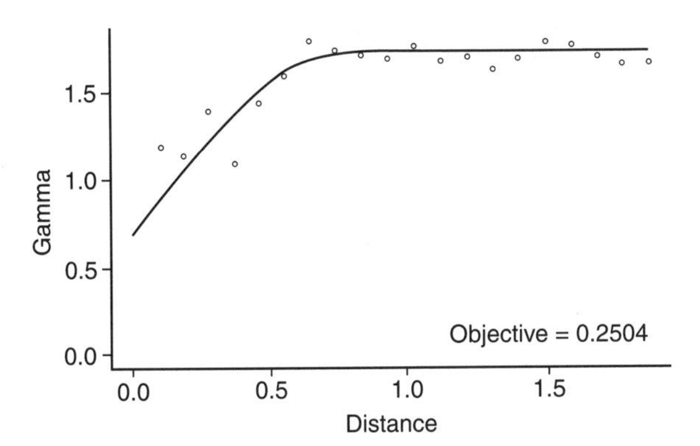

variogram *Spherical variogram model (curve) fitted to an empirical variogram (open symbols) by optimising an objective function*

Once a variogram function has been identified and fitted, this function is usually considered as known and fed into a prediction routine to specify the interpolation weights (see Lawson, 1998), although simultaneous maximum likelihood estimation of the variogram parameters and possible trend parameters is considered preferable (see Cressie, 1993). *SL*

Cressie, N. A. C. 1993: *Statistics for spatial data*. New York: John Wiley & Sons, Inc. **Lawson, A. B.** 1998: Statistical map. In Armitage, P. and Colton, T. (eds), *Encyclopedia of biostatistics*. Chichester: John Wiley & Sons, Ltd.

velocity charts See GROWTH CHARTS

visual analogue scale These are scales used to measure a subjective assessment, such as 'amount of pain' or 'level of anxiety', particularly when the assessment is believed to lie along a continuum rather than only taking a discrete set of values. The item consists of a line, typically 10 cm in length, with lowest and highest values indicated by labels at each end. The subject is expected to place a mark on the line to represent his or her assessment. An example follows (with a cross indicating where a subject has placed a mark):

How much pain do you feel?

No pain |——X————| Unbearable pain

The data from a visual analogue scale are recorded by measuring how far along the line from the left end the subject has placed a mark.

It is important to remember that, although it is possible to record the data from a visual analogue scale with great accuracy, the value is very subjective. In the example, the subject's response may be recorded as 1.1 (because it is 1.1 cm from the left-hand end of the line), but there is no objective unit on this value. If another subject records a value of 2.1, it is not necessarily the case that this subject experiences more pain than the first subject. However, if the first subject is measured again (say, after a month) and gives a score of 2.1, it is possible to interpret this to mean that the first subject is now experiencing more pain than previously.

It is also important to remember that a visual analogue scale is unlikely to be linear. For example, a distance of 1 cm at one end of the scale does not necessarily represent the same difference as a distance of 1 cm at the other end of the scale. This cautions against the use of standard methods for continuous data; a common recommendation is to analyse ranks of the scores rather than the raw scores. For further details see Altman (1991) and Streiner and Norman (1995). *PM*

Altman, D. G. 1991: *Practical statistics for medical research*. London: Chapman & Hall. **Streiner, D. L. and Norman, G. R.**

1995: *Health measurement scales: a practical guide to their development and use*, 2nd edition. Oxford: Oxford University Press.

Von Neumann–Morgenstern standard gamble

This classic method of measuring preferences in health economics was first presented in von Neumann and Morgenstern (1953). The method uses hypothetical lotteries as a means of measuring people's preferences when faced with a choice between treatment that offers potential benefit in quality of life (see QUALITY OF LIFE MEASUREMENT), but with the trade-off that there is a finite possibility that the patient will not survive treatment. An individual might be asked to choose between the certainty of surviving for a fixed period in a particular state of ill health and a gamble between surviving for the same period without disability, on the one hand, and immediate death, on the other. The PROBABILITY of surviving without disability, as opposed to dying, is then varied until the person shows no preference between the certain option and the gamble. This probability then defines the utility of an individual for the disabled state between 0 and 1, whose ENDPOINTS are death and perfect health.

Because few patients are accustomed to dealing in probabilities, an alternative procedure called the time trade-off technique is often suggested. This begins by estimating the likely remaining years of life for a healthy subject, using actuarial tables, and then the following question is asked: 'Imagine living the remainder of your natural span (an estimated number of years would be given) in your present state. Contrast this with the alternative that you remain in perfect health for fewer years. How many years would you sacrifice if you could have perfect health?'

An example of the use of the standard gamble approach is given in Petrou and Campbell (1997) who use it to estimate utilities for a range of health states in colorectal carcinoma. They were able to demonstrate that the quality of life benefits of stabilisation in the treatment of advanced metastatic colorectal cancer were rated almost as highly as those of a partial response. They also showed that the benefits of irinotecan, a drug licensed for the treatment of metastatic colorectal cancer in patients who had failed an established 5-FU-containing regimen, outweighed the short-term impact of toxicity in those patients who achieved at least stabilisation of their disease. *BSE*

Petrou, S. and Campbell, N. 1997: Stabilisation in colorectal cancer. *International Journal of Palliative Nursing* 3, 275–80. **Von Neumann, J. and Morgenstern, O.** 1953: *Theory of games and economic behavior*. New York: John Wiley & Sons, Inc.

W

washout period See CROSSOVER TRIALS

wavelet analysis This is a method of representing a function by projecting it on to a collection of basis functions derived from a single wavelet (often referred to as the mother wavelet $\psi(t)$). All basis functions (wavelets) required in the analysis are translated (shifted) and dilated (stretched) versions of the mother wavelet. Unlike the Fourier transform (see Subba Rao, 1998), whose basis functions are derived from sine and cosine waves with persistent oscillations, the basis functions for the wavelet transform are nonzero and oscillate for a short interval. As a result, the wavelet transform simultaneously localises information from a function in both time and frequency. For functions with time-varying characteristics or sudden changes, the wavelet transform has proved quite useful.

Two main flavours of the wavelet transform are the continuous wavelet transform (CWT) and the discrete wavelet transform (DWT). They differ by the fact that the transform works with continuous or discrete translations and dilations, respectively, of the wavelet function. The CWT is a highly redundant transform with the family of wavelets being computed via $\psi(at + b)$, where a and b are real numbers. In general, the number of wavelet coefficients is much greater than the number of observations. Popular continuous wavelet functions include the Morlet, Mexican hat and first derivative of a Gaussian density function (top row of the figure). Notice that all the wavelet functions oscillate – i.e. they have both positive and negative values – and the Morlet wavelet is complex valued (the real and imaginary portions are plotted using different line types).

The DWT uses a wavelet that is translated and dilated by discrete values, i.e. of the form $\psi(2^j t + k2^j)$ where j and k are integers. The parameter j is commonly referred to as the *scale*. The DWT may be an orthogonal or biorthogonal transform depending on the wavelet function. Popular orthogonal discrete wavelet functions include the Haar and Daubechies families of wavelets (bottom row of the figure). Although the discrete wavelet functions displayed look continuous they are derived from two, four and eight unique values – from left to right. Notice, the discrete wavelet functions are not symmetric, except for the Haar wavelet, and much less smooth when compared to continuous wavelet functions.

A compromise between the CWT and DWT is partially achieved by using the translation invariant DWT where a discrete wavelet function is applied to all possible integer shifts of the data in time via $\psi(2^j t + k)$. This results in a redundant (*not* orthogonal) transform in time with the same number of scales as the DWT; each scale is a distinct range of frequencies.

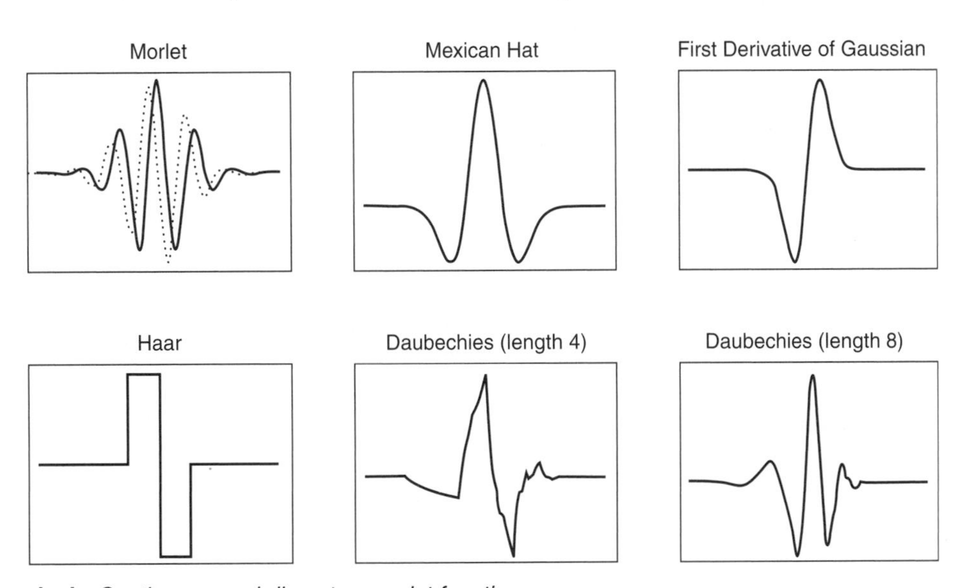

wavelet analysis *Continuous and discrete wavelet functions*

Encyclopaedic Companion to Medical Statistics: Second Edition Edited by Brian S. Everitt and Christopher R. Palmer

The DWT is most commonly used in nonparametric regression (see Hazelton, 1998) (using a technique known as *wavelet denoising*). The key contribution of the wavelet transform is that both smooth and abrupt changes in the signal will result in a few large wavelet coefficients while the noise will be dispersed throughout the entire vector of wavelet coefficients. Thus, thresholding all wavelet coefficients will eliminate the noise and preserve the features of interest. Wavelet denoising has been adapted to cases where the noise exhibits autocorrelation and may be non-Gaussian (e.g. Poisson distributed). It is very important when performing a wavelet analysis to select the wavelet function (discrete or continuous) that best matches the features inherent in the function of interest. This will concentrate the function into a small number of wavelet coefficients, simplifying the analysis.

The wavelet transform may be applied to data of several dimensions, e.g. in time series analysis (one dimension), image analysis (two dimensions) or spatiotemporal analysis (three or more dimensions). For further details behind the theory and applications to medicine and time series, see Aldroubi and Unser (1996), Chui (1997) and Percival and Walden (2000). *BW*

Alroubi, A. and Unser, M. (eds) 1996: *Wavelets in medicine and biology*. Boca Raton: Chapman & Hall/CRC. **Chui, C. K.** 1997: *Wavelets: a mathematical tool for signal analysis*. SIAM monographs on mathematical modeling and computation. Philadelphia: Society for Industrial and Applied Mathematics. **Hazelton, M. L.** 1998: Nonparametric regression. In Armitage, P. and Colton, T. (eds), *Encyclopedia of biostatistics*. Chichester: John Wiley & Sons, Ltd. **Percival, D. B. and Walden, A. T.** 2000: *Wavelet methods for time series analysis*. Cambridge: Cambridge University Press. **Subba Rao, T.** 1998: Fast Fourier transform. In Armitage, P. and Colton, T. (eds) *Encyclopedia of biostatistics*. Chichester: John Wiley & Sons, Ltd.

web resources in medical statistics The growth of the internet has revolutionised access to information for everyone from academics to the general public. The potential uses of the internet in the area of medical statistics are vast and here we can only provide an overview of these uses as they currently stand. As the internet continues to develop, its potential for use in this area can only increase.

Internet resources in medical statistics can be grouped loosely under the following headings: sources of routinely collected data; reference (online encyclopaedias, dictionaries, lecture notes); email discussion lists; statistical software, reviews and downloads; e-journals; and datasets for use in teaching.

For routinely collected data, the World Health Organization (WHO) website (www.who.int/en) is a good place to start. A number of databases are available to browse, including the WHO Statistical Information System (WHOSIS), containing national statistics on mortality, morbidity, risk factors, service coverage and health systems, and a Global Health Atlas, containing statistics for infectious diseases at country, regional and global levels.

Many countries have their own websites for routinely collected data, e.g. National Statistics Online for the UK as a whole, which incorporates health statistics (www.statistics.gov.uk), Scottish Health Statistics (www.isdscotland.org) and the CDC National Centre for Health Statistics in the United States (www.cdc.gov/nchs). Data are routinely available from all of these sites in summary tables and charts and some sites allow access to some of the data in the form of Excel spreadsheets, which can be customised by the user. Most national sites provide information from censuses, mortality data, morbidity data and information on usage and performance of health services. If any of these links become redundant in the future, many university libraries will continue to maintain up-to-date, accessible links to the latest information on their web pages. For example, the Glasgow University Library website (www.lib.gla.ac.uk and select the link to *Maps, Official Publications and Statistics*), which can be accessed by anyone, has an excellent page of links to local, national and international data sources and is regularly updated.

The internet is increasingly a source of good reference material in medical statistics. There are many online dictionaries, glossaries, encyclopaedias, sets of lecture notes, lists of statisticians and statistical bodies (e.g. The Royal Statistical Society: www.rss.org.uk), interactive training websites (e.g. Computer-Assisted Statistics Teaching: http://cast.massey.ac.nz), Java applets for use in teaching (e.g. www.stat.sc.edu/~west/javahtml; Rice Virtual Lab in Statistics: http://onlinestatbook.com), free statistical software (e.g. Epi Info: www.cdc.gov/EpiInfo; StatCrunch: www.statcrunch.com; and R: http://cran.r-project.org) and e-journals.

It is likely that links to some of these materials, especially lecture notes and teaching materials, will be volatile, but a reasonably up-to-date list (including all of the links listed in this article) should be obtained by a carefully worded search in *Google*. There are many other widely available search engines including some that are dedicated to resources in the area of health and medicine, such as *Intute* (www.intute.ac.uk/medicine) and *Medscape* (www.medscape.com). Also, *Wikipedia* (www.wikipedia.org) has many entries on topics related to medical statistics of varying quality and depth. It is probably safer at present, however, to rely on articles that are available from a reputable university website, rather than those in *Wikipedia*.

Most leading medical journals have their own websites with useful links. They also generally allow free access to abstracts of papers and sometimes also to the full text of papers. For example, the *British Medical Journal* (www.bmj.com), *The Lancet* (www.thelancet.com) and the *New England Journal of Medicine* (http://content.nejm.org) have published many articles on statistical methods, which are written at a level accessible to nonstatisticians, as well as reports on CLINICAL TRIALS and SYSTEMATIC REVIEWS, which (from a teaching point of view) illustrate the use of these methods.

The development and use of the internet has taken place at such a phenomenal rate that it is difficult to predict what further changes will take place in the future. The one thing that is certain is that its importance as a resource for all those with an interest in medical statistics will increase. *WHG*

weighted kappa See KAPPA AND WEIGHTED KAPPA

weighted least squares estimator (WLSE) See LEAST SQUARES ESTIMATION

Wennberg's design See RANDOMISATION

when to use which test? This is a question raised by many clinical researchers wanting to grasp basic statistics and hence feel equipped to analyse their own data. However, there are several reasons why it may not be the most appropriate question to ask and it is instructive to begin by considering why not.

First, many medical statisticians, and informed medical journal editors, nowadays prefer analyses by CONFIDENCE INTERVALS instead of HYPOTHESIS TESTS. There are numerous reasons for this, some of which are intrinsic problems with the inferential procedure of hypothesis testing, while others are due to this procedure's historical misuse, overuse or abuse. Thus, the first advice is to counter the question with: 'Are you sure you want to do a hypothesis test?'

Second, the question betrays a false view of statistics as if there were a formulaic approach with one size fitting all. Unfortunately for those seeking quick answers to apparently simple questions, this is not the case, as doctors may know from their own experiences of calling differential diagnoses from outwardly similar signs and symptoms. Equally, medical statistics is a diverse discipline with no two research studies being quite the same and hence different analytical strategies may apply in similar-looking circumstances.

Third, it may be inappropriate to perform a hypothesis test if the topic of investigation is only raised by the data themselves and not by a pre-existing research question (see PITFALLS IN MEDICAL STATISTICS).

Fourth, research must never be seen as a chase to get a P-VALUE and preferably one less than 0.05, for among other things this confuses CLINICAL VERSUS STATISTICAL SIGNIFICANCE.

Consequently, we may here not fully answer the question as posed, but with suitable precautions and having decided a test is appropriate, will go as far as identifying the proper test to use in commonly encountered univariate situations. Fuller details concerning methods and applications of each test procedure mentioned can be found under their specific entry. Note also that one can easily find guides about choice of statistical procedures on the internet (see WEB RESOURCES IN

when to use which test? *When to use which test, according to number and type of samples and outcome (or response) variable measured. Some entries contain alternative tests. Last two rows indicate circumstances to use various approaches for assessing association or agreement. The table is not a complete categorisation of all possible tests and data types but includes those referred to elsewhere under individually named entries*

		Nominal (categorical)	*Ordered categorical or continuous and non normal*	*Continuous and normally distributed*
One sample		χ^2-test	Kolmogorov–Smirnov test, Sign test	Student's test
Two samples	Independent	χ^2-test ($2 \times k$), Fisher's exact test	Mann–Whitney rank sum test	Unpaired t-test
	Paired	McNemar's test	Wilcoxon signed rank test, Sign test	Paired t-test
Multiple samples ($k > 2$)	Independent	χ^2-test ($r \times k$)	Kruskal–Wallis test, Jonckheere-Terpstra test	Analysis of variance (ANOVA)
	Related	Cochran Q-test	Friedman test	Repeated measures ANOVA
Association between two variables		Contingency coefficient	Spearman's rank correlation, Kendall's tau correlation	Pearson product-moment correlation
Agreement between two variables		Kappa coefficient	Weighted kappa coefficient	Limits of agreement

MEDICAL STATISTICS); e.g. www.whichtest.info/index.htm may prove helpful.

Three factors influence choice of statistical test: the nature of the response (type of data being analysed); the number of groups sampled (one, two or many); and, if more than one, the nature of the sampling (matched or independent).

The response or outcome variable can be continuous and approximately normally distributed or dichotomous (a binary 'yes/no' outcome) or intermediate to these in a variety of ways. For example, the response variable could be in ordered categories (see ANALYSIS OF ORDINAL DATA). Otherwise, the response variable could be continuous and nonnormally distributed, being skewed or containing OUTLIERS, perhaps. In either of these latter cases it is appropriate to apply one of the many NONPARAMETRIC METHODS.

In the special case of the response variable being the time until an event, which may or may not have occurred by the time of analysis (strictly, database closure), then survival methods would be used to handle the CENSORED OBSERVATIONS. Notably, this entails a version of the log-rank test or one of its alternatives and can be stratified or not depending on the structure of the data (see SURVIVAL ANALYSIS).

The number of groups being sampled is generally obvious, although sometimes care must be taken about analysing the correct statistical unit. In a cluster randomised study, for instance, it is the clusters that need to be analysed, not the individuals forming the clusters. When repeated measurements are taken, while more sophisticated approaches can be adopted, the simplest is to convert each individual's data into a suitable summary statistic prior to analysis. For example, this statistic might be the AREA UNDER THE CURVE, slope of the regression line or MEAN observation, etc., depending on whatever was previously decided to be the most clinically meaningful. In practice, note that statistical convenience should not be the criteria for choosing among possible summary statistics (see SUMMARY MEASURE ANALYSIS).

Lastly, the relationship among groups is crucial for deciding on the correct testing procedure. In the simplest case involving two groups, one needs to know whether sample data were *paired* (also known as *matched, related* or *dependent*) or unpaired (*unmatched, unrelated* or *independent*). This is usually straightforward, e.g. whenever data are collected on the same patients before and after an intervention or when a pair of organs (ear, eye, hand, kidney, etc.) is measured within the same person or when twins are studied within a controlled experiment. It can be less clear how best to analyse data in certain CASE-CONTROL STUDIES, matched by sex and age to within a fixed number of years, however. This is because, here, the purpose of matching is to create broadly comparable groups according to basic demographic status, rather than attempting to achieve precisely well-matched pairs (see MATCHING).

The table shows, according to the three basic criteria, when to use which test method in the simplest cases. For completion, it also indicates which procedure applies when assessing ASSOCIATION or AGREEMENT. Again, further details can be found under individually named entries.

However, as emphasised throughout, confidence intervals are preferred to tests and for more informative analyses still, modelling or regression techniques can be better still. These provide mutually adjusted results for important confounders, an altogether more satisfactory approach to handling data and superior to expecting it to be adequately described by a P-value, as if relationships within the data could possibly be encapsulated by a single number, a hopelessly false ambition. Nevertheless, viewed positively and correctly, the right hypothesis test can serve to rule out chance as an explanation for discrepant data apparent in one or more random samples, and lead the investigator on towards a fuller analysis of the data collected and, in turn, a deeper clinical understanding. *CRP*

[See also EXACT METHODS FOR CATEGORICAL DATA, HYPOTHESIS TESTS]

Wilcoxon rank sum test See MANN–WHITNEY RANK SUM TEST

Wilcoxon signed rank test This is a nonparametric version of the paired t-test (see STUDENT'S t-TEST) used for two groups that are either matched or paired. It is more sensitive than the SIGN TEST as it uses the magnitude of the differences between the pairs not simply the sign of the difference. It gives more weight to pairs that show large differences than those that show small differences. The Wilcoxon signed rank test tests the assumption that the sum of the positive RANKS equals the sum of the negative ranks. The test statistic is the smaller of the sum of the positive and the sum of the negative ranks. The data should be continuous or ordinal in

Wilcoxon signed rank test *Mcm2 and Ki67 values, data from a study of patients with cancer*

Patient	*Mcm2*	*Ki67*	*Difference*	*Rank of difference*	*Signed rank of difference*
1	14.78	14.78	0	–	–
2	7.96	8.68	−0.72	1	−1
3	10.89	1.57	9.32	2	+2
4	12.10	1.85	10.25	3	+3
5	18.23	5.84	12.39	4	+4
6	16.40	3.04	13.36	5	+5
7	18.02	3.96	14.06	6	+6
8	23.35	8.16	15.19	7	+7
9	26.70	8.40	18.30	8	+8

nature. The paired differences should be independent and symmetrical about the true MEDIAN difference.

First, find the difference in values for each pair (variable 1 – variable 2). Then rank the magnitude of the differences, smallest to largest, assigning the average rank to ties in the differences and no rank to zero differences. Find the sum of the ranks for the positive differences, W^+, and the sum of the ranks for the negative differences, W^-. Find N, the total number of differences not including ties. Find the critical value from standard tables and compare $W = \min(W^+, W^-)$; reject the NULL HYPOTHESIS if W is less than or equal to the critical value. If $W^+ > W^-$, then variable 1 tends to be greater than variable 2 and vice versa.

As part of a study Mcm2 and Ki67 values were compared to see if there was a difference between the values in patients with cancer. Data are shown in the table. A plot of the differences shows that they are plausibly symmetric so the assumption of symmetry holds.

Take $W^+ = 35$ and $W^- = 1$, with $W = \min(W^+, W^-) = 1$ and $N = 8$. From standard tables ($N = 8$, $\alpha = 0.05$) and the critical value is 3. As 1 is less than 3, there is sufficient evidence to reject the null hypothesis. Therefore there is a difference between Mcm2 and Ki67 values. Mcm2 values tend to be higher than Ki67 values. For further details see Pett (1997), Siegel and Castellan (1998) and Swinscow and Campbell (2002). *SLV*

Pett, M. A. 1997: *Non-parametrics for health care research*. Thousand Oaks: Sage. **Siegel, S. and Castellan, N. J.** 1998: *Non-parametric statistics for the behavioral sciences*, 2nd edition. New York: McGraw-Hill. **Swinscow, T. D. V. and Campbell, M. J.** 2002: *Statistics at square one*, 10th edition. London: BMJ Books.

WinBUGS See BUGS AND WINBUGS

WLSE Abbreviation for weighted least squares estimator. See LEAST SQUARES ESTIMATION

Yates' correction This is an adjustment made to a CHI-SQUARE TEST when the number of observations is small. Yates' correction is an example of a continuity correction and is designed specifically for 2×2 frequency tables.

The chi-square test uses the CHI-SQUARE DISTRIBUTION to determine whether a set of observed counts from a study (the number of observations in each cell of a frequency table) differ significantly from the expected counts predicted by a hypothesis. This use of the chi-square distribution is an approximation based on the use of the NORMAL DISTRIBUTION to approximate the distribution of the number of observations in each cell of the frequency table.

The use of the normal distribution is only an approximation because the number of observations in a cell of a frequency table is a discrete value (it can only take non-negative integer values: 0, 1, 2, ...) whereas the normal distribution is continuous (it can take any value). When the number of observations is large, this difference becomes irrelevant (the approximation becomes very good), but when the number of observations is small the approximation can mean that the standard chi-square test result is invalid (the P-VALUE reported by the test may be too low).

Yates' correction is an attempt to take account of the approximation in the calculation of the chi-square test statistic, χ^2. The effect of the correction is to decrease the size of χ^2, which increases the P-value from the test. This means that Yates' correction will always give a more conservative test; it will be less likely to report a significant result.

There is no universal agreement on when Yates' correction should be applied. With modern computer processing speeds and memory capacity it is now usually feasible to use exact methods instead, which avoid approximations altogether (e.g. FISHER'S EXACT TEST). For further details see Altman (1991). *PM*

Altman, D. G. 1991: *Practical statistics for medical research*. London: Chapman & Hall.

Encyclopaedic Companion to Medical Statistics: Second Edition Edited by Brian S. Everitt and Christopher R. Palmer

Z

z-score This is a standardised score of a random variable used to simplify comparisons across samples measured on different scales. Sample values X are converted to z-scores Z using the formula:

$$Z = \frac{X - M}{\text{SD}}$$

where the MEAN M and STANDARD DEVIATION (SD) for X are obtained from either the sample or another population. The mean and SD of Z are generally close to zero and one respectively. In the context of LINEAR REGRESSION the z-score is equivalent to the standardized residual. If the underlying sample distribution is known, e.g. a NORMAL DISTRIBUTION, the z-score can be expressed as a cumulative probability or QUANTILE. For example, a z-score of -0.67 corresponds to the lower quartile or 25th centile, and positive z-scores relate to quantiles in the upper half of the distribution.

The LMS METHOD extends the definition of the z-score to adjust for SKEWNESS:

$$Z = \frac{\left({}^{X}\!/_{M}\right)^{L} - 1}{L \times S}$$

where M is the median of X, S the coefficient of variation (i.e. the SD divided by the mean) and L the Box–Cox power TRANSFORMATION to remove the skewness, as estimated by regression methods. When $L = 1$ (i.e. a normal distribution) this simplifies to the first definition. *TJC*

Encyclopaedic Companion to Medical Statistics: Second Edition Edited by Brian S. Everitt and Christopher R. Palmer